The Wiley/NBS
Registry of
Mass Spectral Data

VOLUME 6

The Wiley/NBS Registry of Mass Spectral Data

VOLUME 6

Fred W. McLafferty

Cornell University
Ithaca, New York

Douglas B. Stauffer

Palisade Corporation
Newfield, New York

This work is a combination of the revisions of the
following two books and their database versions:

REGISTRY OF MASS SPECTRAL DATA
by Einar Stenhagen, Sixten Abrahamsson, and Fred W. McLafferty
and
EPA/NIH MASS SPECTRAL DATA BASE
by S.R. Heller and G.W.A. Milne
and its two supplements

WILEY

A Wiley-Interscience Publication

JOHN WILEY & SONS

NEW YORK · CHICHESTER · BRISBANE · TORONTO · SINGAPORE

Library of Congress Cataloging in Publication Data:

McLafferty, Fred W.
   The Wiley/NBS registry of mass spectral data/Fred W. McLafferty,
Douglas B. Stauffer.
      p.   cm.
   "A combination of the revisions of the following two books and
their data base versions: Registry of mass spectral data, by Einar
Stenhagen, Sixten Abrahamsson, and Fred W. McLafferty; and EPA/NIH
mass spectral data and its two supplements."
   "A Wiley-Interscience publication."
   Includes bibliographies and index.
   ISBN 0-471-62886-7 (set)
   1. Mass spectrometry—Tables.   I. Stauffer, Douglas B.
II. Stenhagen, Einar.   Registry of mass spectral data.   III. Heller,
Stephen R., 1943–   EPA/NIH mass spectral data base.   IV. John Wiley
& Sons.   V. United States.   National Bureau of Standards.
VI. Title.   VII. Title: Registry of mass spectral data.
QC454.M3M395 1988
539'.6'028—dc19   87-31645
                                                              CIP

Printed in the United States of America

10 9 8 7 6 5 4 3 2 1

The Wiley/NBS
Registry of
Mass Spectral Data

VOLUME 6

466   C20 H18 N8 O2 S2   IC-1202-0-0   Bis(2-anilino-4-methoxy-s-triazin-6-yl) disulphide;   W   124651

466   C20 H26 Fe O9   86813-59-4   KC-1983-621-0   Tricarbonyl {2-η5-[dimethyl-4-isopropoxy-1-(2-methoxyethyl) cyclohexa-2,4-dienylmalonate]}iron;   W   124652

466   C20 H26 Fe O9   86813-60-7   KC-1983-622-0   Tricarbonyl{2-η5-[dimethyl-2-isopropoxy-5-(2-methoxyethyl)cyclohexa-2,4-dienylmalonate]}iron;   W   124653

466   C21 H12 F9 P   13406-29-6   HE-1982-0-0   Phosphine, tris[4-(trifluoromethyl)phenyl]-;; PHOSPHINE, TRI-(P-TRIFLUOROPHENYL)-;   W/NBS   61574

466   C21 H22 N8 O5   2522-28-3   F-34-2391-0   N-METHYL-4-(N-METHYL-4(N-METHYL-H-NITROPYMOLE-2-CARBOXYAMIDO)-PYRROLE-2-CARBOXYAMIDE)-PYRROLE-2-CARBOXYAMIDO-β-PROPIONITRILE-;   W   61575

466   C22 H19 F5 Fe O2   41684-55-3   EP-6711-0-0   Iron, dicarbonyl[(4a,4b,9a,10,10aη)-1,3,4,5,6,7,8,9-octahydrobenz[a]azulen-4a(2H)-yl](pentafluorophenyl)-;; 1,2-CYCLOHEPTENO-3,4-CYCLOHEXENO-CYCLOPENTADIENYL-PENTAFLUOROPHENYL-DICARBONYL-IRON;   W/NBS   61576

466   C22 H20 Co2 S2   HE-1986-2037-0   DICYCLOPENTADIENYL-2,4-DIPHENYL-1,3-DICOBALTA-2,4-DITHIACYCLOBUTANE;   W   124654

466   C22 H21 F7 O3   2192-60-1   AD-0-2743-0   Estrone, heptafluorobutyrate;; 17-OXO-1,3,5,(10)-ESTRATRIEN-3-YL HEPTAFLUOROBUTYRATE;   W/NBS   61577

466   C22 H26 O3 S4   26416-50-2   NS--29551-0-0   Carbonodithioic acid, S,S'-[(4-methoxyphenyl)phenylmethylene] O,O'-bis(1-methylethyl) ester;;   NBS   61578

466   C22 H34 Fe O3 S2   AH-113-970-0   meso-1,15-Dimethyl-2,14-dithia-5,8,11-trioxa[15](1,1')ferrocenophane;   W   124655

466   C22 H34 Fe O3 S2   AH-113-970-0   rac-1,15-Dimethyl-2,14-dithia-5,8,11-trioxa[15](1,1')ferrocenophane;   W   124656

466   C22 H40 Cl2 N2 O4   86296-67-5   O-18-38-2   3-Chloro-1-cyclohexyloxy-3-methyl-2-nitrosobutane;   W 124657

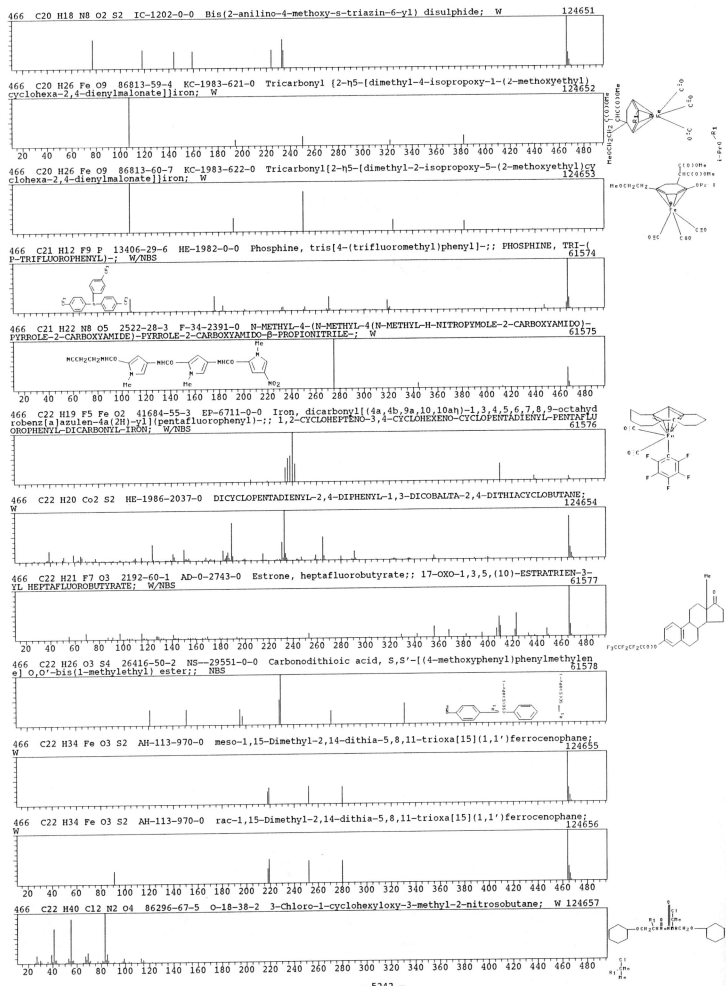

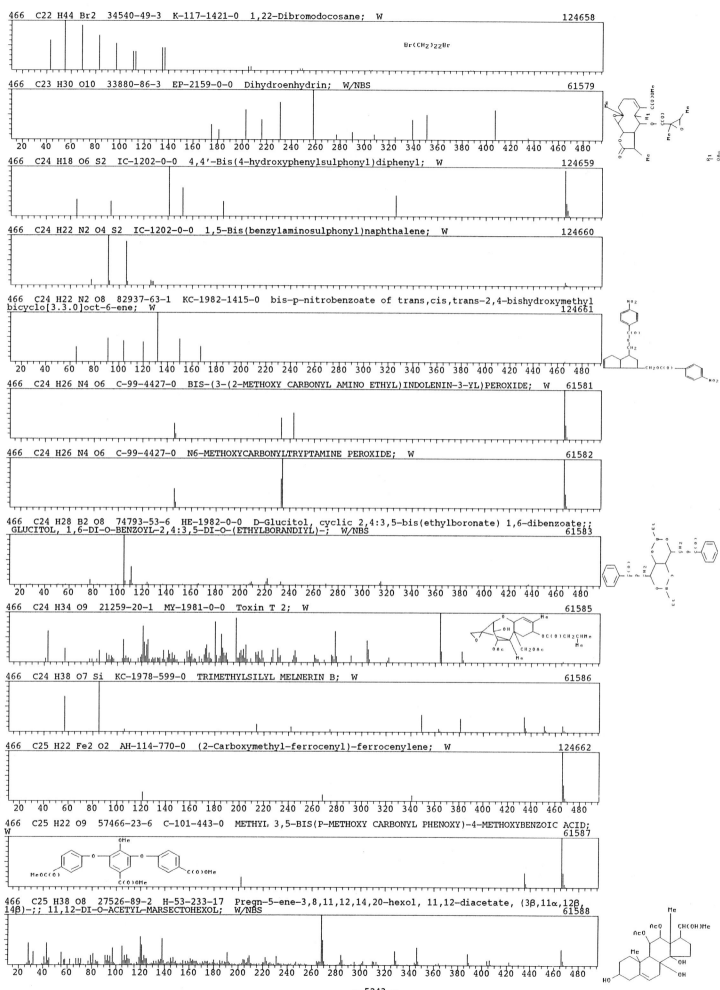

466   C22 H44 Br2   34540-49-3   K-117-1421-0   1,22-Dibromodocosane;   W   124658

Br(CH2)22Br

466   C23 H30 O10   33880-86-3   EP-2159-0-0   Dihydroenhydrin;   W/NBS   61579

20 40 60 80 100 120 140 160 180 200 220 240 260 280 300 320 340 360 380 400 420 440 460 480

466   C24 H18 O6 S2   IC-1202-0-0   4,4'-Bis(4-hydroxyphenylsulphonyl)diphenyl;   W   124659

466   C24 H22 N2 O4 S2   IC-1202-0-0   1,5-Bis(benzylaminosulphonyl)naphthalene;   W   124660

466   C24 H22 N2 O8   82937-63-1   KC-1982-1415-0   bis-p-nitrobenzoate of trans,cis,trans-2,4-bishydroxymethyl
bicyclo[3.3.0]oct-6-ene;   W   124661

20 40 60 80 100 120 140 160 180 200 220 240 260 280 300 320 340 360 380 400 420 440 460 480

466   C24 H26 N4 O6   C-99-4427-0   BIS-(3-(2-METHOXY CARBONYL AMINO ETHYL)INDOLENIN-3-YL)PEROXIDE;   W   61581

466   C24 H26 N4 O6   C-99-4427-0   N6-METHOXYCARBONYLTRYPTAMINE PEROXIDE;   W   61582

466   C24 H28 B2 O8   74793-53-6   HE-1982-0-0   D-Glucitol, cyclic 2,4:3,5-bis(ethylboronate) 1,6-dibenzoate;;
GLUCITOL, 1,6-DI-O-BENZOYL-2,4:3,5-DI-O-(ETHYLBORANDIYL)-;   W/NBS   61583

20 40 60 80 100 120 140 160 180 200 220 240 260 280 300 320 340 360 380 400 420 440 460 480

466   C24 H34 O9   21259-20-1   MY-1981-0-0   Toxin T 2;   W   61585

466   C24 H38 O7 Si   KC-1978-599-0   TRIMETHYLSILYL MELNERIN B;   W   61586

466   C25 H22 Fe2 O2   AH-114-770-0   (2-Carboxymethyl-ferrocenyl)-ferrocenylene;   W   124662

20 40 60 80 100 120 140 160 180 200 220 240 260 280 300 320 340 360 380 400 420 440 460 480

466   C25 H22 O9   57466-23-6   C-101-443-0   METHYL, 3,5-BIS(P-METHOXY CARBONYL PHENOXY)-4-METHOXYBENZOIC ACID;
W   61587

466   C25 H38 O8   27526-89-2   H-53-233-17   Pregn-5-ene-3,8,11,12,14,20-hexol, 11,12-diacetate, (3β,11α,12β,
14β)-;;   11,12-DI-O-ACETYL-MARSECTOHEXOL;   W/NBS   61588

20 40 60 80 100 120 140 160 180 200 220 240 260 280 300 320 340 360 380 400 420 440 460 480

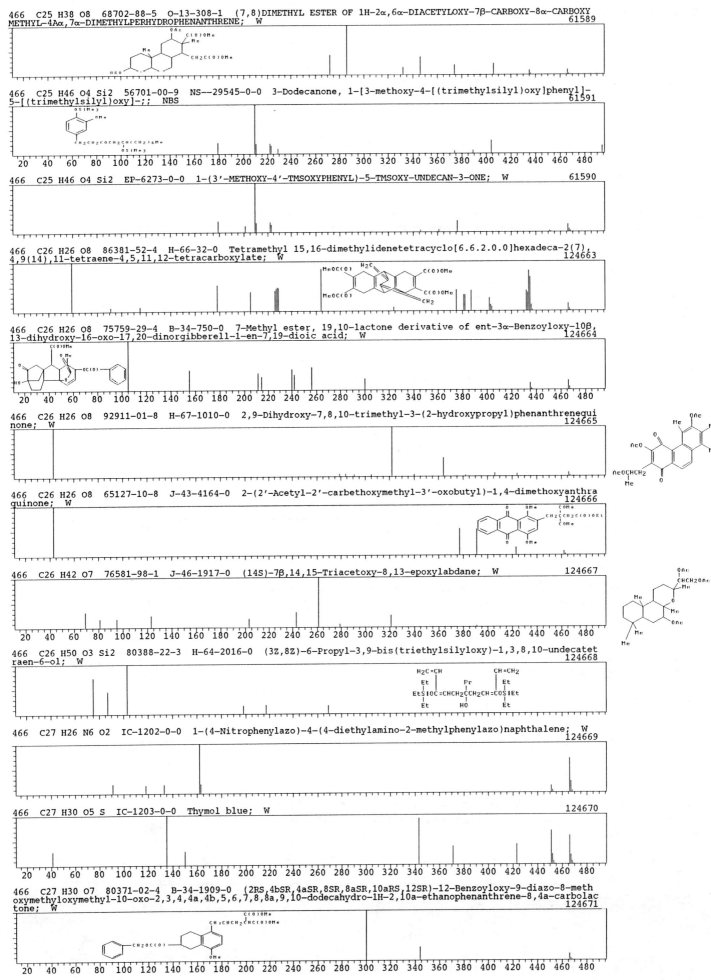

466   C25 H38 O8   68702-88-5   O-13-308-1   (7,8)DIMETHYL ESTER OF 1H-2α,6α-DIACETYLOXY-7β-CARBOXY-8α-CARBOXY
METHYL-4Aα,7α-DIMETHYLPERHYDROPHENANTHRENE;   W   61589

466   C25 H46 O4 Si2   56701-00-9   NS--29545-0-0   3-Dodecanone, 1-[3-methoxy-4-[(trimethylsilyl)oxy]phenyl]-
5-[(trimethylsilyl)oxy]-;;   NBS   61591

466   C25 H46 O4 Si2   EP-6273-0-0   1-(3'-METHOXY-4'-TMSOXYPHENYL)-5-TMSOXY-UNDECAN-3-ONE;   W   61590

466   C26 H26 O8   86381-52-4   H-66-32-0   Tetramethyl 15,16-dimethylidenetetracyclo[6.6.2.0.0]hexadeca-2(7),
4,9(14),11-tetraene-4,5,11,12-tetracarboxylate;   W   124663

466   C26 H26 O8   75759-29-4   B-34-750-0   7-Methyl ester, 19,10-lactone derivative of ent-3α-Benzoyloxy-10β,
13-dihydroxy-16-oxo-17,20-dinorgibberell-1-en-7,19-dioic acid;   W   124664

466   C26 H26 O8   92911-01-8   H-67-1010-0   2,9-Dihydroxy-7,8,10-trimethyl-3-(2-hydroxypropyl)phenanthrenequi
none;   W   124665

466   C26 H26 O8   65127-10-8   J-43-4164-0   2-(2'-Acetyl-2'-carbethoxymethyl-3'-oxobutyl)-1,4-dimethoxyanthra
quinone;   W   124666

466   C26 H42 O7   76581-98-1   J-46-1917-0   (14S)-7β,14,15-Triacetoxy-8,13-epoxylabdane;   W   124667

466   C26 H50 O3 Si2   80388-22-3   H-64-2016-0   (3Z,8Z)-6-Propyl-3,9-bis(triethylsilyloxy)-1,3,8,10-undecatet
raen-6-ol;   W   124668

466   C27 H26 N6 O2   IC-1202-0-0   1-(4-Nitrophenylazo)-4-(4-diethylamino-2-methylphenylazo)naphthalene;   W
                                                                                                    124669

466   C27 H30 O5 S   IC-1203-0-0   Thymol blue;   W   124670

466   C27 H30 O7   80371-02-4   B-34-1909-0   (2RS,4bSR,4aSR,8SR,8aSR,10aRS,12SR)-12-Benzoyloxy-9-diazo-8-meth
oxymethyloxymethyl-10-oxo-2,3,4,4a,4b,5,6,7,8,8a,9,10-dodecahydro-1H-2,10a-ethanophenanthrene-8,4a-carbolac
tone;   W   124671

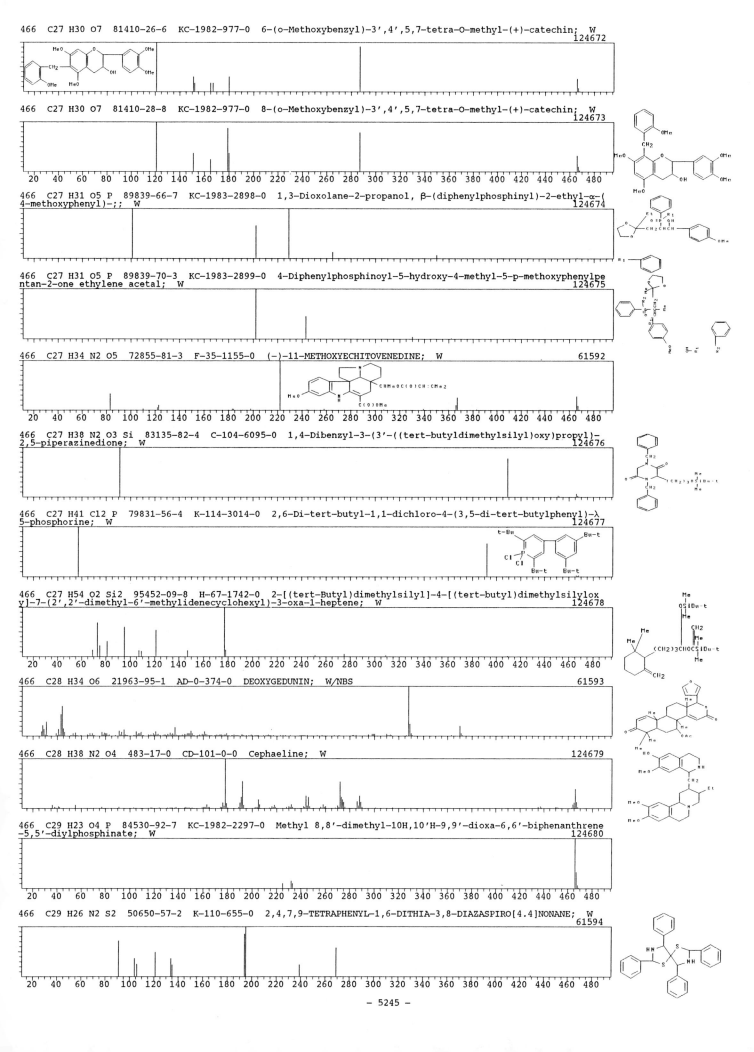

466   C27 H30 O7   81410-26-6   KC-1982-977-0   6-(o-Methoxybenzyl)-3',4',5,7-tetra-O-methyl-(+)-catechin;   W
124672

466   C27 H30 O7   81410-28-8   KC-1982-977-0   8-(o-Methoxybenzyl)-3',4',5,7-tetra-O-methyl-(+)-catechin;   W
124673

466   C27 H31 O5 P   89839-66-7   KC-1983-2898-0   1,3-Dioxolane-2-propanol, β-(diphenylphosphinyl)-2-ethyl-α-(4-methoxyphenyl)-;;   W
124674

466   C27 H31 O5 P   89839-70-3   KC-1983-2899-0   4-Diphenylphosphinoyl-5-hydroxy-4-methyl-5-p-methoxyphenylpentan-2-one ethylene acetal;   W
124675

466   C27 H34 N2 O5   72855-81-3   F-35-1155-0   (−)-11-METHOXYECHITOVENEDINE;   W
61592

466   C27 H38 N2 O3 Si   83135-82-4   C-104-6095-0   1,4-Dibenzyl-3-(3'-((tert-butyldimethylsilyl)oxy)propyl)-2,5-piperazinedione;   W
124676

466   C27 H41 Cl2 P   79831-56-4   K-114-3014-0   2,6-Di-tert-butyl-1,1-dichloro-4-(3,5-di-tert-butylphenyl)-λ5-phosphorine;   W
124677

466   C27 H54 O2 Si2   95452-09-8   H-67-1742-0   2-[(tert-Butyl)dimethylsilyl]-4-[(tert-butyl)dimethylsilyloxy]-7-(2',2'-dimethyl-6'-methylidenecyclohexyl)-3-oxa-1-heptene;   W
124678

466   C28 H34 O6   21963-95-1   AD-0-374-0   DEOXYGEDUNIN;   W/NBS
61593

466   C28 H38 N2 O4   483-17-0   CD-101-0-0   Cephaeline;   W
124679

466   C29 H23 O4 P   84530-92-7   KC-1982-2297-0   Methyl 8,8'-dimethyl-10H,10'H-9,9'-dioxa-6,6'-biphenanthrene-5,5'-diylphosphinate;   W
124680

466   C29 H26 N2 S2   50650-57-2   K-110-655-0   2,4,7,9-TETRAPHENYL-1,6-DITHIA-3,8-DIAZASPIRO[4.4]NONANE;   W
61594

− 5245 −

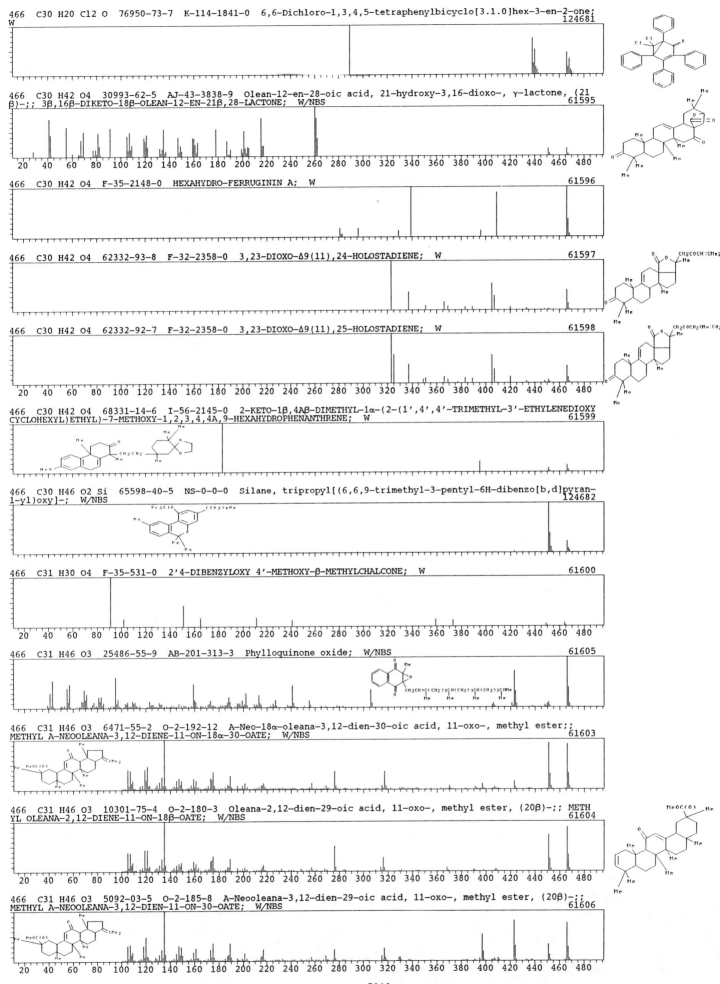

466   C30 H20 Cl2 O   76950-73-7   K-114-1841-0   6,6-Dichloro-1,3,4,5-tetraphenylbicyclo[3.1.0]hex-3-en-2-one;
W   124681

466   C30 H42 O4   30993-62-5   AJ-43-3838-9   Olean-12-en-28-oic acid, 21-hydroxy-3,16-dioxo-, γ-lactone, (21
β)-;;   3β,16β-DIKETO-18β-OLEAN-12-EN-21β,28-LACTONE;   W/NBS   61595

466   C30 H42 O4   F-35-2148-0   HEXAHYDRO-FERRUGININ A;   W   61596

466   C30 H42 O4   62332-93-8   F-32-2358-0   3,23-DIOXO-Δ9(11),24-HOLOSTADIENE;   W   61597

466   C30 H42 O4   62332-92-7   F-32-2358-0   3,23-DIOXO-Δ9(11),25-HOLOSTADIENE;   W   61598

466   C30 H42 O4   68331-14-6   I-56-2145-0   2-KETO-1β,4Aβ-DIMETHYL-1α-(2-(1',4',4'-TRIMETHYL-3'-ETHYLENEDIOXY
CYCLOHEXYL)ETHYL)-7-METHOXY-1,2,3,4,4A,9-HEXAHYDROPHENANTHRENE;   W   61599

466   C30 H46 O2 Si   65598-40-5   NS-0-0-0   Silane, tripropyl[(6,6,9-trimethyl-3-pentyl-6H-dibenzo[b,d]pyran-
1-yl)oxy]-;   W/NBS   124682

466   C31 H30 O4   F-35-531-0   2'4-DIBENZYLOXY 4'-METHOXY-β-METHYLCHALCONE;   W   61600

466   C31 H46 O3   25486-55-9   AB-201-313-3   Phylloquinone oxide;   W/NBS   61605

466   C31 H46 O3   6471-55-2   O-2-192-12   A-Neo-18α-oleana-3,12-dien-30-oic acid, 11-oxo-, methyl ester;;
METHYL A-NEOOLEANA-3,12-DIENE-11-ON-18α-30-OATE;   W/NBS   61603

466   C31 H46 O3   10301-75-4   O-2-180-3   Oleana-2,12-dien-29-oic acid, 11-oxo-, methyl ester, (20β)-;;   METH
YL OLEANA-2,12-DIENE-11-ON-18β-OATE;   W/NBS   61604

466   C31 H46 O3   5092-03-5   O-2-185-8   A-Neooleana-3,12-dien-29-oic acid, 11-oxo-, methyl ester, (20β)-;;
METHYL A-NEOOLEANA-3,12-DIEN-11-ON-30-OATE;   W/NBS   61606

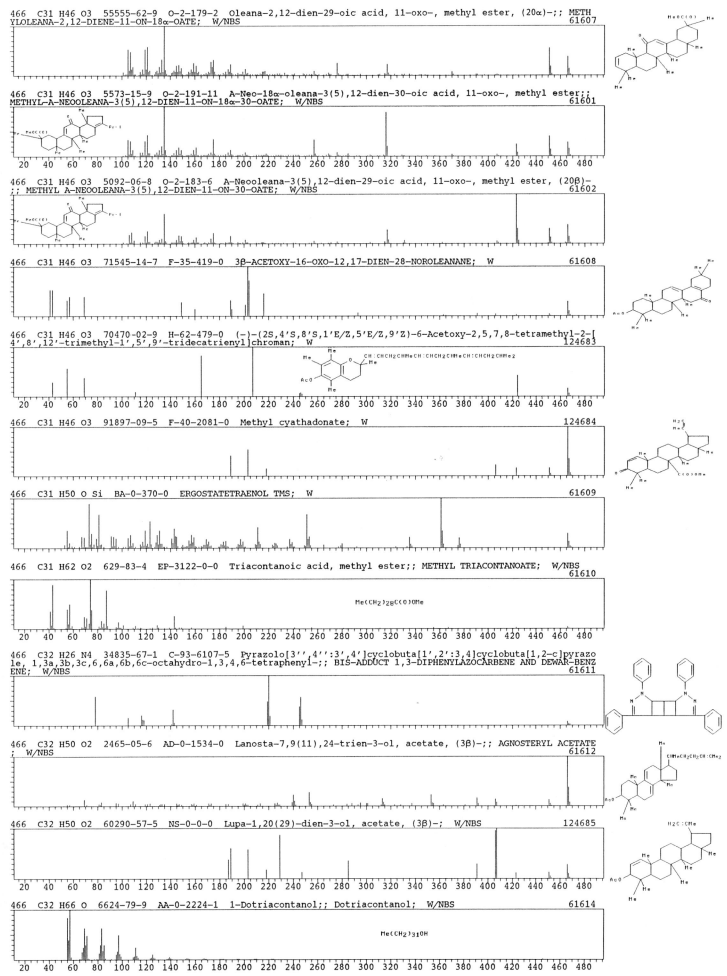

466 C31 H46 O3 55555-62-9 O-2-179-2 Oleana-2,12-dien-29-oic acid, 11-oxo-, methyl ester, (20α)-;; METH
YLOLEANA-2,12-DIENE-11-ON-18α-OATE; W/NBS
61607

466 C31 H46 O3 5573-15-9 O-2-191-11 A-Neo-18α-oleana-3(5),12-dien-30-oic acid, 11-oxo-, methyl ester;;
METHYL-A-NEOOLEANA-3(5),12-DIEN-11-ON-18α-30-OATE; W/NBS
61601

466 C31 H46 O3 5092-06-8 O-2-183-6 A-Neooleana-3(5),12-dien-29-oic acid, 11-oxo-, methyl ester, (20β)-
;; METHYL A-NEOOLEANA-3(5),12-DIEN-11-ON-30-OATE; W/NBS
61602

466 C31 H46 O3 71545-14-7 F-35-419-0 3β-ACETOXY-16-OXO-12,17-DIEN-28-NOROLEANANE; W
61608

466 C31 H46 O3 70470-02-9 H-62-479-0 (-)-(2S,4'S,8'S,1'E/Z,5'E/Z,9'Z)-6-Acetoxy-2,5,7,8-tetramethyl-2-[
4',8',12'-trimethyl-1',5',9'-tridecatrienyl]chroman; W
124683

466 C31 H46 O3 91897-09-5 F-40-2081-0 Methyl cyathadonate; W
124684

466 C31 H50 O Si BA-0-370-0 ERGOSTATETRAENOL TMS; W
61609

466 C31 H62 O2 629-83-4 EP-3122-0-0 Triacontanoic acid, methyl ester;; METHYL TRIACONTANOATE; W/NBS
61610

Me(CH2)28C(O)OMe

466 C32 H26 N4 34835-67-1 C-93-6107-5 Pyrazolo[3'',4'':3',4']cyclobuta[1',2':3,4]cyclobuta[1,2-c]pyrazo
le, 1,3a,3b,3c,6,6a,6b,6c-octahydro-1,3,4,6-tetraphenyl-;; BIS-ADDUCT 1,3-DIPHENYLAZOCARBENE AND DEWAR-BENZ
ENE; W/NBS
61611

466 C32 H50 O2 2465-05-6 AD-0-1534-0 Lanosta-7,9(11),24-trien-3-ol, acetate, (3β)-;; AGNOSTERYL ACETATE
; W/NBS
61612

466 C32 H50 O2 60290-57-5 NS-0-0-0 Lupa-1,20(29)-dien-3-ol, acetate, (3β)-; W/NBS
124685

466 C32 H66 O 6624-79-9 AA-0-2224-1 1-Dotriacontanol;; Dotriacontanol; W/NBS
61614

Me(CH2)31OH

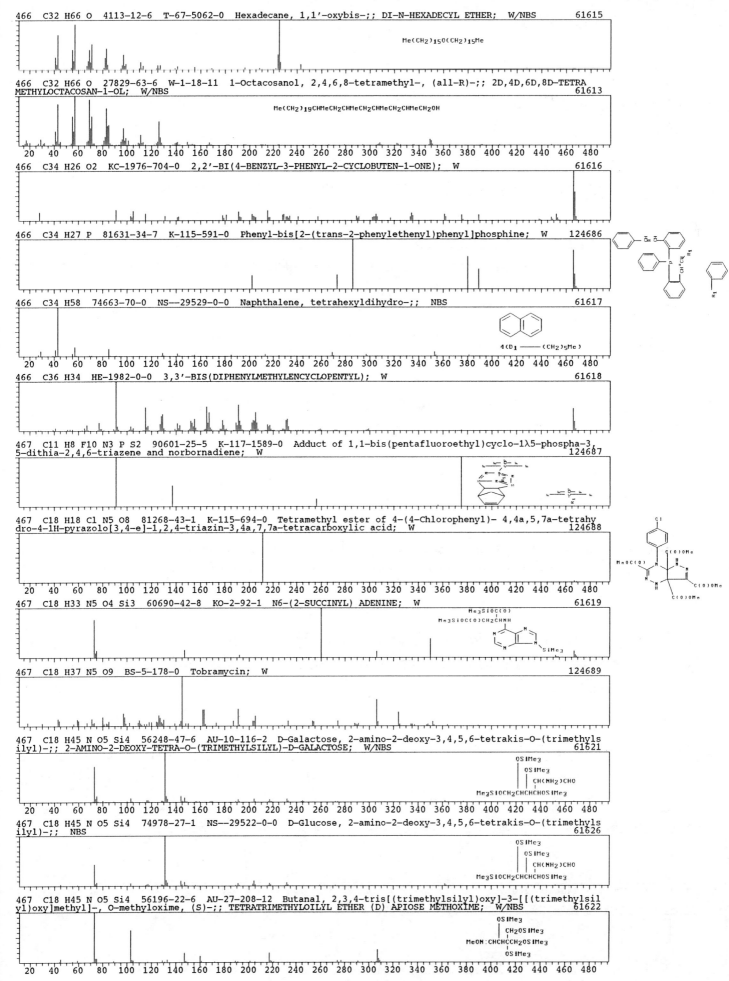

466   C32 H66 O   4113-12-6   T-67-5062-0   Hexadecane, 1,1'-oxybis-;; DI-N-HEXADECYL ETHER;   W/NBS   61615

Me(CH₂)₁₅O(CH₂)₁₅Me

466   C32 H66 O   27829-63-6   W-1-18-11   1-Octacosanol, 2,4,6,8-tetramethyl-, (all-R)-;;   2D,4D,6D,8D-TETRA
METHYLOCTACOSAN-1-OL;   W/NBS   61613

Me(CH₂)₁₉CHMeCH₂CHMeCH₂CHMeCH₂CHMeCH₂OH

466   C34 H26 O2   KC-1976-704-0   2,2'-BI(4-BENZYL-3-PHENYL-2-CYCLOBUTEN-1-ONE);   W   61616

466   C34 H27 P   81631-34-7   K-115-591-0   Phenyl-bis[2-(trans-2-phenylethenyl)phenyl]phosphine;   W   124686

466   C34 H58   74663-70-0   NS--29529-0-0   Naphthalene, tetrahexyldihydro-;;   NBS   61617

466   C36 H34   HE-1982-0-0   3,3'-BIS(DIPHENYLMETHYLENCYCLOPENTYL);   W   61618

467   C11 H8 F10 N3 P S2   90601-25-5   K-117-1589-0   Adduct of 1,1-bis(pentafluoroethyl)cyclo-1λ5-phospha-3,
5-dithia-2,4,6-triazene and norbornadiene;   W   124687

467   C18 H18 Cl N5 O8   81268-43-1   K-115-694-0   Tetramethyl ester of 4-(4-Chlorophenyl)- 4,4a,5,7a-tetrahy
dro-4-1H-pyrazolo[3,4-e]-1,2,4-triazin-3,4a,7,7a-tetracarboxylic acid;   W   124688

467   C18 H33 N5 O4 Si3   60690-42-8   KO-2-92-1   N6-(2-SUCCINYL) ADENINE;   W   61619

467   C18 H37 N5 O9   BS-5-178-0   Tobramycin;   W   124689

467   C18 H45 N O5 Si4   56248-47-6   AU-10-116-2   D-Galactose, 2-amino-2-deoxy-3,4,5,6-tetrakis-O-(trimethyls
ilyl)-;;   2-AMINO-2-DEOXY-TETRA-O-(TRIMETHYLSILYL)-D-GALACTOSE;   W/NBS   61621

467   C18 H45 N O5 Si4   74978-27-1   NS--29522-0-0   D-Glucose, 2-amino-2-deoxy-3,4,5,6-tetrakis-O-(trimethyls
ilyl)-;;   NBS   61626

467   C18 H45 N O5 Si4   56196-22-6   AU-27-208-12   Butanal, 2,3,4-tris[(trimethylsilyl)oxy]-3-[[(trimethylsil
yl)oxy]methyl]-, O-methyloxime, (S)-;;   TETRATRIMETHYLOILYL ETHER (D) APIOSE METHOXIME;   W/NBS   61622

467 C18 H45 N O5 Si4 56196-07-7 AU-27-202-3 D-Xylose, 2,3,4,5-tetrakis-O-(trimethylsilyl)-, O-methyloxi
me;; TETRAKISTRIMETHYLSILYL ETHER XYLOSE (D) METHOXIME; W/NBS   61623

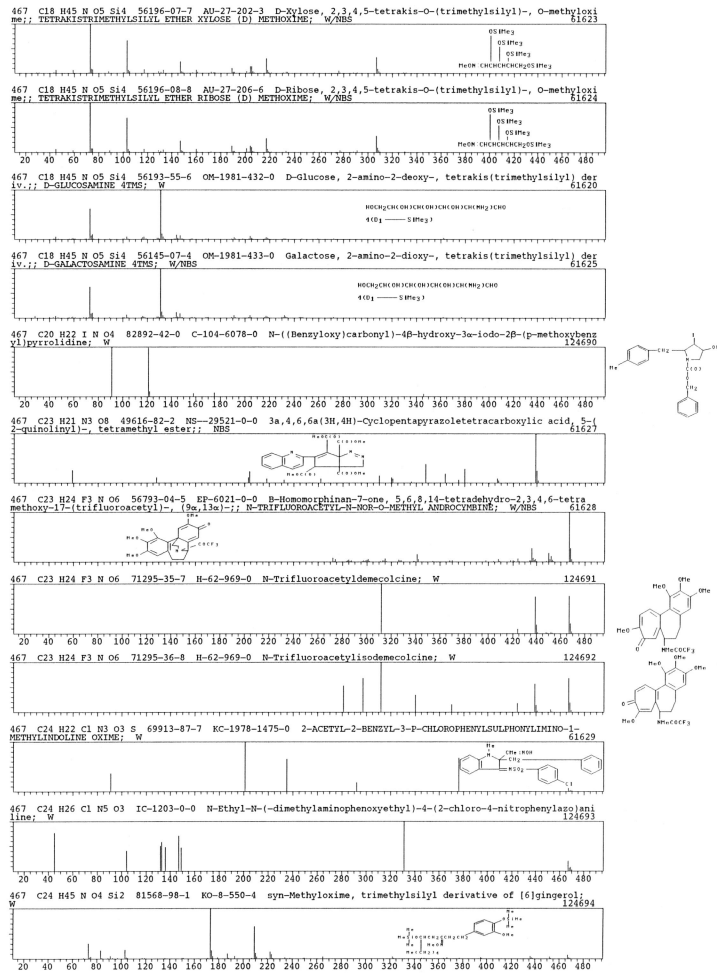

OSIMe3
OSIMe3
OSIMe3
MeON:CHCHCHCHCH2OSIMe3

467 C18 H45 N O5 Si4 56196-08-8 AU-27-206-6 D-Ribose, 2,3,4,5-tetrakis-O-(trimethylsilyl)-, O-methyloxi
me;; TETRAKISTRIMETHYLSILYL ETHER RIBOSE (D) METHOXIME; W/NBS   61624

OSIMe3
OSIMe3
OSIMe3
MeON:CHCHCHCHCH2OSIMe3

467 C18 H45 N O5 Si4 56193-55-6 OM-1981-432-0 D-Glucose, 2-amino-2-deoxy-, tetrakis(trimethylsilyl) der
iv.;; D-GLUCOSAMINE 4TMS; W   61620

HOCH2CH(OH)CH(OH)CH(OH)CH(NH2)CHO
4(D1 ——— SIMe3)

467 C18 H45 N O5 Si4 56145-07-4 OM-1981-433-0 Galactose, 2-amino-2-dioxy-, tetrakis(trimethylsilyl) der
iv.;; D-GALACTOSAMINE 4TMS; W/NBS   61625

HOCH2CH(OH)CH(OH)CH(OH)CH(NH2)CHO
4(D1 ——— SIMe3)

467 C20 H22 I N O4 82892-42-0 C-104-6078-0 N-((Benzyloxy)carbonyl)-4β-hydroxy-3α-iodo-2β-(p-methoxybenz
yl)pyrrolidine; W   124690

467 C23 H21 N3 O8 49616-82-2 NS--29521-0-0 3a,4,6,6a(3H,4H)-Cyclopentapyrazoletetracarboxylic acid, 5-(
2-quinolinyl)-, tetramethyl ester;; NBS   61627

467 C23 H24 F3 N O6 56793-04-5 EP-6021-0-0 B-Homomorphinan-7-one, 5,6,8,14-tetradehydro-2,3,4,6-tetra
methoxy-17-(trifluoroacetyl)-, (9α,13α)-;; N-TRIFLUOROACETYL-N-NOR-O-METHYL ANDROCYMBINE; W/NBS   61628

467 C23 H24 F3 N O6 71295-35-7 H-62-969-0 N-Trifluoroacetyldemecolcine; W   124691

467 C23 H24 F3 N O6 71295-36-8 H-62-969-0 N-Trifluoroacetylisodemecolcine; W   124692

467 C24 H22 Cl N3 O3 S 69913-87-7 KC-1978-1475-0 2-ACETYL-2-BENZYL-3-P-CHLOROPHENYLSULPHONYLIMINO-1-
METHYLINDOLINE OXIME; W   61629

467 C24 H26 Cl N5 O3 IC-1203-0-0 N-Ethyl-N-(-dimethylaminophenoxyethyl)-4-(2-chloro-4-nitrophenylazo)ani
line; W   124693

467 C24 H45 N O4 Si2 81568-98-1 KO-8-550-4 syn-Methyloxime, trimethylsilyl derivative of [6]gingerol;
W   124694

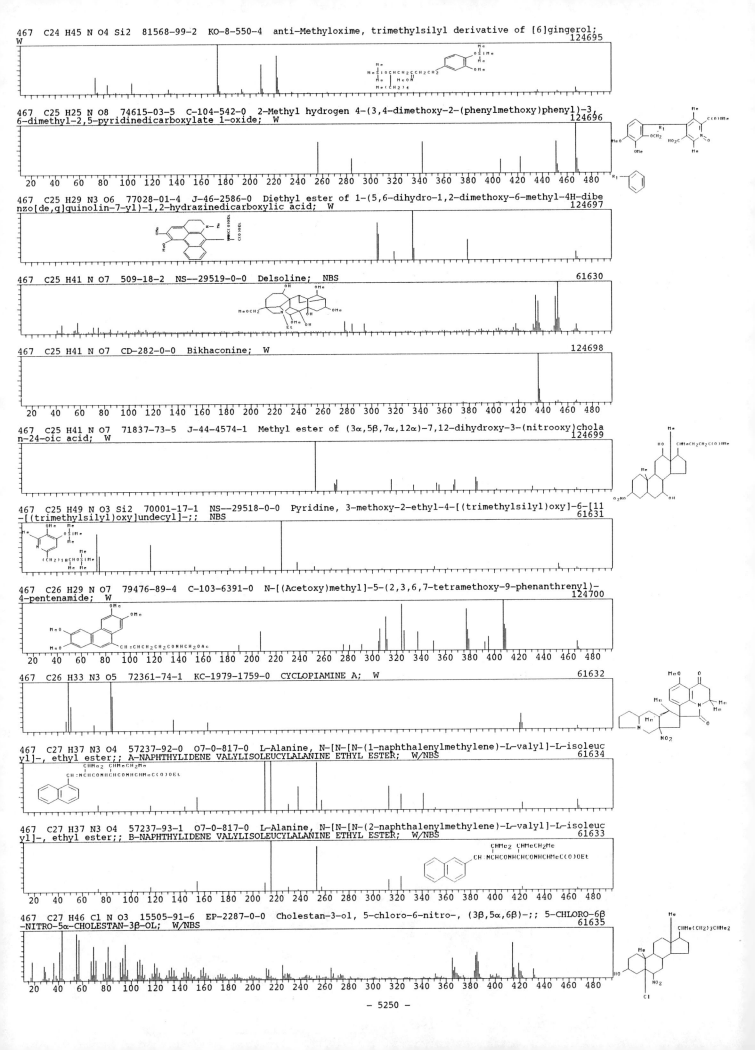

467   C24 H45 N O4 Si2   81568-99-2   KO-8-550-4   anti-Methyloxime, trimethylsilyl derivative of [6]gingerol;
W   124695

467   C25 H25 N O8   74615-03-5   C-104-542-0   2-Methyl hydrogen 4-(3,4-dimethoxy-2-(phenylmethoxy)phenyl)-3,
6-dimethyl-2,5-pyridinedicarboxylate 1-oxide;   W   124696

467   C25 H29 N3 O6   77028-01-4   J-46-2586-0   Diethyl ester of 1-(5,6-dihydro-1,2-dimethoxy-6-methyl-4H-dibe
nzo[de,q]quinolin-7-yl)-1,2-hydrazinedicarboxylic acid;   W   124697

467   C25 H41 N O7   509-18-2   NS--29519-0-0   Delsoline;   NBS   61630

467   C25 H41 N O7   CD-282-0-0   Bikhaconine;   W   124698

467   C25 H41 N O7   71837-73-5   J-44-4574-1   Methyl ester of (3α,5β,7α,12α)-7,12-dihydroxy-3-(nitrooxy)chola
n-24-oic acid;   W   124699

467   C25 H49 N O3 Si2   70001-17-1   NS--29518-0-0   Pyridine, 3-methoxy-2-ethyl-4-[(trimethylsilyl)oxy]-6-[11
-[(trimethylsilyl)oxy]undecyl]-;;   NBS   61631

467   C26 H29 N O7   79476-89-4   C-103-6391-0   N-[(Acetoxy)methyl]-5-(2,3,6,7-tetramethoxy-9-phenanthrenyl)-
4-pentenamide;   W   124700

467   C26 H33 N3 O5   72361-74-1   KC-1979-1759-0   CYCLOPIAMINE A;   W   61632

467   C27 H37 N3 O4   57237-92-0   O7-0-817-0   L-Alanine, N-[N-[N-(1-naphthalenylmethylene)-L-valyl]-L-isoleuc
yl]-, ethyl ester;;   A-NAPHTHYLIDENE VALYLISOLEUCYLALANINE ETHYL ESTER;   W/NBS   61634

467   C27 H37 N3 O4   57237-93-1   O7-0-817-0   L-Alanine, N-[N-[N-(2-naphthalenylmethylene)-L-valyl]-L-isoleuc
yl]-, ethyl ester;;   B-NAPHTHYLIDENE VALYLISOLEUCYLALANINE ETHYL ESTER;   W/NBS   61633

467   C27 H46 Cl N O3   15505-91-6   EP-2287-0-0   Cholestan-3-ol, 5-chloro-6-nitro-, (3β,5α,6β)-;;   5-CHLORO-6β
-NITRO-5α-CHOLESTAN-3β-OL;   W/NBS   61635

- 5250 -

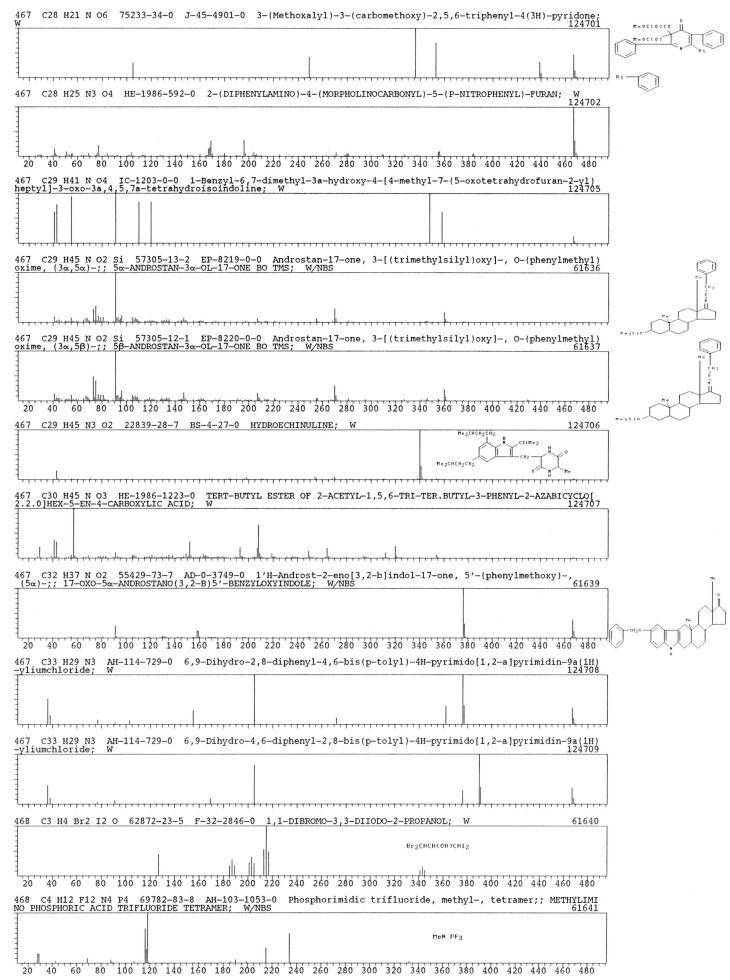

467 C28 H21 N O6   75233-34-0   J-45-4901-0   3-(Methoxalyl)-3-(carbomethoxy)-2,5,6-triphenyl-4(3H)-pyridone;
W   124701

467 C28 H25 N3 O4   HE-1986-592-0   2-(DIPHENYLAMINO)-4-(MORPHOLINOCARBONYL)-5-(P-NITROPHENYL)-FURAN; W
                                                                                                              124702

467 C29 H41 N O4   IC-1203-0-0   1-Benzyl-6,7-dimethyl-3a-hydroxy-4-[4-methyl-7-(5-oxotetrahydrofuran-2-yl)
heptyl]-3-oxo-3a,4,5,7a-tetrahydroisoindoline; W   124705

467 C29 H45 N O2 Si   57305-13-2   EP-8219-0-0   Androstan-17-one, 3-[(trimethylsilyl)oxy]-, O-(phenylmethyl)
oxime, (3α,5α)-;; 5α-ANDROSTAN-3α-OL-17-ONE BO TMS; W/NBS   61636

467 C29 H45 N O2 Si   57305-12-1   EP-8220-0-0   Androstan-17-one, 3-[(trimethylsilyl)oxy]-, O-(phenylmethyl)
oxime, (3α,5β)-;; 5β-ANDROSTAN-3α-OL-17-ONE BO TMS; W/NBS   61637

467 C29 H45 N3 O2   22839-28-7   BS-4-27-0   HYDROECHINULINE; W   124706

467 C30 H45 N O3   HE-1986-1223-0   TERT-BUTYL ESTER OF 2-ACETYL-1,5,6-TRI-TER.BUTYL-3-PHENYL-2-AZABICYCLO[
2.2.0]HEX-5-EN-4-CARBOXYLIC ACID; W   124707

467 C32 H37 N O2   55429-73-7   AD-0-3749-0   1'H-Androst-2-eno[3,2-b]indol-17-one, 5'-(phenylmethoxy)-,
(5α)-;; 17-OXO-5α-ANDROSTANO(3,2-B)5'-BENZYLOXYINDOLE; W/NBS   61639

467 C33 H29 N3   AH-114-729-0   6,9-Dihydro-2,8-diphenyl-4,6-bis(p-tolyl)-4H-pyrimido[1,2-a]pyrimidin-9a(1H)
-yliumchloride; W   124708

467 C33 H29 N3   AH-114-729-0   6,9-Dihydro-4,6-diphenyl-2,8-bis(p-tolyl)-4H-pyrimido[1,2-a]pyrimidin-9a(1H)
-yliumchloride; W   124709

468 C3 H4 Br2 I2 O   62872-23-5   F-32-2846-0   1,1-DIBROMO-3,3-DIIODO-2-PROPANOL; W   61640

468 C4 H12 F12 N4 P4   69782-83-8   AH-103-1053-0   Phosphorimidic trifluoride, methyl-, tetramer;; METHYLIMI
NO PHOSPHORIC ACID TRIFLUORIDE TETRAMER; W/NBS   61641

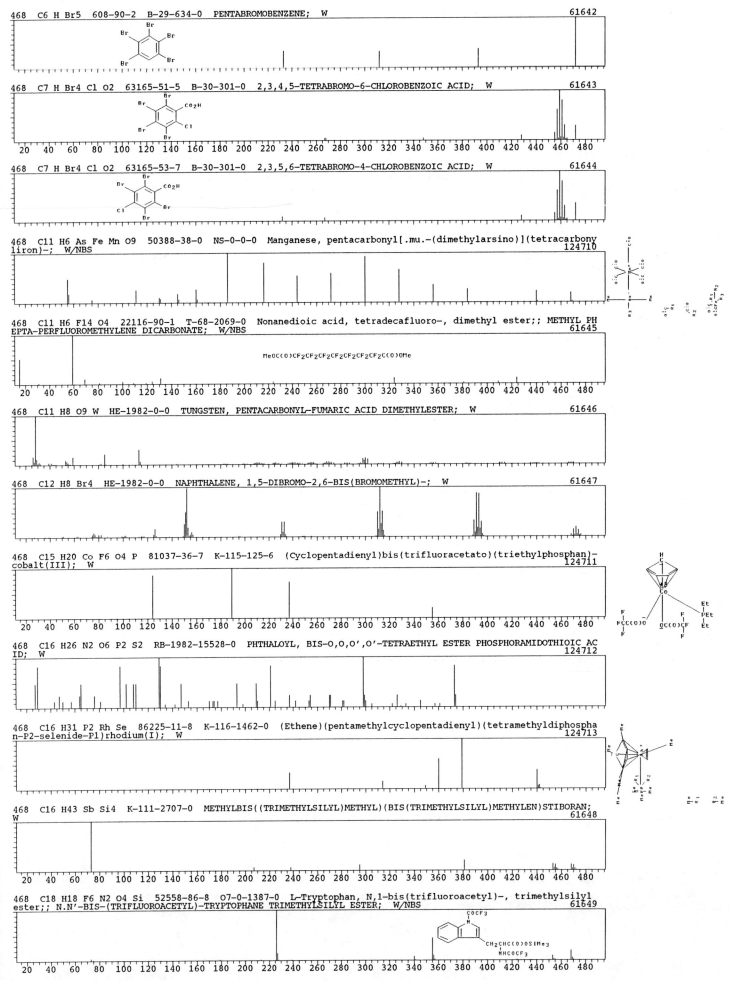

468   C6 H Br5   608-90-2   B-29-634-0   PENTABROMOBENZENE;   W   61642

468   C7 H Br4 Cl O2   63165-51-5   B-30-301-0   2,3,4,5-TETRABROMO-6-CHLOROBENZOIC ACID;   W   61643

468   C7 H Br4 Cl O2   63165-53-7   B-30-301-0   2,3,5,6-TETRABROMO-4-CHLOROBENZOIC ACID;   W   61644

468   C11 H6 As Fe Mn O9   50388-38-0   NS-0-0-0   Manganese, pentacarbonyl[.mu.-(dimethylarsino)](tetracarbonyliron)-;   W/NBS   124710

468   C11 H6 F14 O4   22116-90-1   T-68-2069-0   Nonanedioic acid, tetradecafluoro-, dimethyl ester;;   METHYL PHEPTA-PERFLUOROMETHYLENE DICARBONATE;   W/NBS   61645

MeOC(O)CF2CF2CF2CF2CF2CF2CF2C(O)OMe

468   C11 H8 O9 W   HE-1982-0-0   TUNGSTEN, PENTACARBONYL-FUMARIC ACID DIMETHYLESTER;   W   61646

468   C12 H8 Br4   HE-1982-0-0   NAPHTHALENE, 1,5-DIBROMO-2,6-BIS(BROMOMETHYL)-;   W   61647

468   C15 H20 Co F6 O4 P   81037-36-7   K-115-125-6   (Cyclopentadienyl)bis(trifluoracetato)(triethylphosphan)-cobalt(III);   W   124711

468   C16 H26 N2 O6 P2 S2   RB-1982-15528-0   PHTHALOYL, BIS-O,O,O',O'-TETRAETHYL ESTER PHOSPHORAMIDOTHIOIC ACID;   W   124712

468   C16 H31 P2 Rh Se   86225-11-8   K-116-1462-0   (Ethene)(pentamethylcyclopentadienyl)(tetramethyldiphosphan-P2-selenide-P1)rhodium(I);   W   124713

468   C16 H43 Sb Si4   K-111-2707-0   METHYLBIS((TRIMETHYLSILYL)METHYL)(BIS(TRIMETHYLSILYL)METHYLEN)STIBORAN;   W   61648

468   C18 H18 F6 N2 O4 Si   52558-86-8   O7-0-1387-0   L-Tryptophan, N,1-bis(trifluoroacetyl)-, trimethylsilyl ester;;   N,N'-BIS-(TRIFLUOROACETYL)-TRYPTOPHANE TRIMETHYLSILYL ESTER;   W/NBS   61649

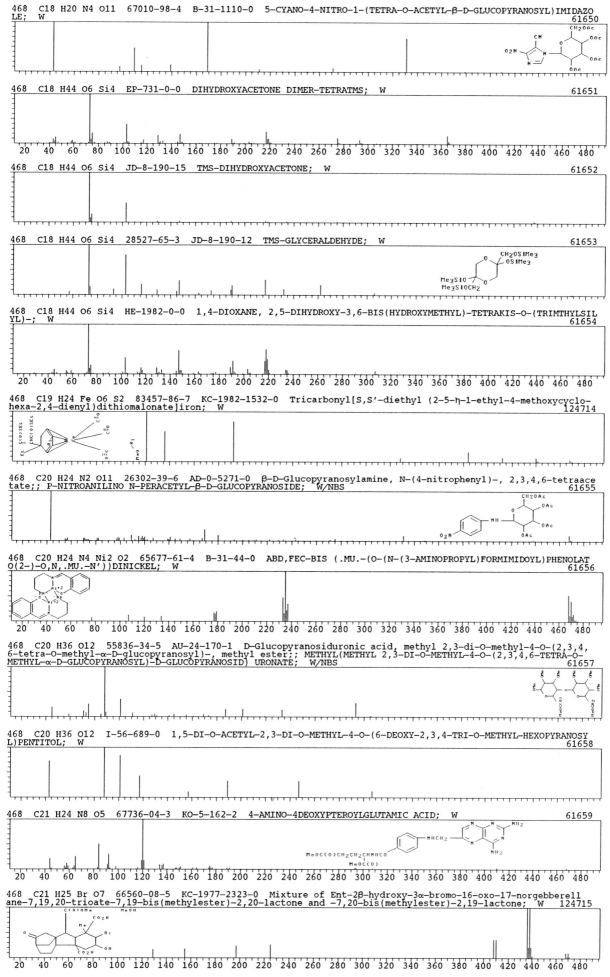

468 C18 H20 N4 O11 67010-98-4 B-31-1110-0 5-CYANO-4-NITRO-1-(TETRA-O-ACETYL-β-D-GLUCOPYRANOSYL)IMIDAZOLE; W
61650

468 C18 H44 O6 Si4 EP-731-0-0 DIHYDROXYACETONE DIMER-TETRATMS; W
61651

468 C18 H44 O6 Si4 JD-8-190-15 TMS-DIHYDROXYACETONE; W
61652

468 C18 H44 O6 Si4 28527-65-3 JD-8-190-12 TMS-GLYCERALDEHYDE; W
61653

468 C18 H44 O6 Si4 HE-1982-0-0 1,4-DIOXANE, 2,5-DIHYDROXY-3,6-BIS(HYDROXYMETHYL)-TETRAKIS-O-(TRIMTHYLSILYL)-; W
61654

468 C19 H24 Fe O6 S2 83457-86-7 KC-1982-1532-0 Tricarbonyl[S,S'-diethyl (2-5-η-1-ethyl-4-methoxycyclohexa-2,4-dienyl)dithiomalonate]iron; W
124714

468 C20 H24 N2 O11 26302-39-6 AD-0-5271-0 β-D-Glucopyranosylamine, N-(4-nitrophenyl)-, 2,3,4,6-tetraacetate;; P-NITROANILINO N-PERACETYL-β-D-GLUCOPYRANOSIDE; W/NBS
61655

468 C20 H24 N4 Ni2 O2 65677-61-4 B-31-44-0 ABD,FEC-BIS (.MU.-(O-(N-(3-AMINOPROPYL)FORMIMIDOYL)PHENOLATO(2-)-O,N,.MU.-N'))DINICKEL; W
61656

468 C20 H36 O12 55836-34-5 AU-24-170-1 D-Glucopyranosiduronic acid, methyl 2,3-di-O-methyl-4-O-(2,3,4,6-tetra-O-methyl-α-D-glucopyranosyl)-, methyl ester;; METHYL(METHYL 2,3-DI-O-METHYL-4-O-(2,3,4,6-TETRA-O-METHYL-α-D-GLUCOPYRANOSYL)-D-GLUCOPYRANOSID) URONATE; W/NBS
61657

468 C20 H36 O12 I-56-689-0 1,5-DI-O-ACETYL-2,3-DI-O-METHYL-4-O-(6-DEOXY-2,3,4-TRI-O-METHYL-HEXOPYRANOSYL)PENTITOL; W
61658

468 C21 H24 N8 O5 67736-04-3 KO-5-162-2 4-AMINO-4DEOXYPTEROYLGLUTAMIC ACID; W
61659

468 C21 H25 Br O7 66560-08-5 KC-1977-2323-0 Mixture of Ent-2β-hydroxy-3α-bromo-16-oxo-17-norgebberellane-7,19,20-trioate-7,19-bis(methylester)-2,20-lactone and -7,20-bis(methylester)-2,19-lactone; W 124715

468 C21 H25 Br O7 81826-63-3 KC-1982-703-0 ent-13-Acetoxy-3α-bromo-10-hydroxy-16-oxo-17,20-dinor-gibber
ellane-7,19-dioic acid; W   124716

468 C21 H29 I N2 O2 J-49-4143-0 16-Hydroxytetrahydro-1,21-dimethylmeloscine iodide; W   124717

468 C21 H48 O7 Si2 72347-77-4 EP-2910-0-0 Silicic acid (H6Si2O7), tributyl tripropyl ester;; TRIPROPOXY
TRIBUTOXYSILOXANE; W/NBS   61660

468 C23 H16 O11 16110-51-3 CD-137-0-0 Cromoglycic acid; W   124719

468 C23 H16 O11 80234-48-6 SB-35-517-0 8,9-Diacetoxy-2,3-dihydro-2-methoxyspiro[4H-benzofuro[2,3-g]-1-
benzopyran-4,2'(5H)-furan]-5,5',11-trione; W   124718

468 C23 H24 N4 O7 69924-48-7 F-34-2935-0 ETHYL 2-METHYL-3-METHOXY CARBONYL-1-(3-METHYL-3-(2,4-DINITROBE
NZOYL AMINO)PROPYL INDOLE; W   61661

468 C23 H24 N4 O7 SS-3-165-0 3-Carboethoxy-1-[3-(3,5-dinitrobenzoylamino)-3-methylpropyl]-2-methylindole
; W   124720

468 C23 H27 F7 O2 60881-91-6 AN-85-343-0 3-HEPTAFLUOROBUTANOYLOXY-3,5-ANDROSTADIENE; W   61662

468 C23 H27 F7 O2 60881-92-7 AN-85-343-0 3-HEPTAFLUOROBUTANOYLOXY-2,4-ANDROSTADIENE; W   61663

468 C23 H33 Br O5 20918-48-3 AD-0-3117-0 5α-Androstan-6-one, 2α-bromo-3β,17β-dihydroxy-, diacetate;; 2α
-BROMO-3β,17β-DIACETOXY-5α-ANDROST-6-ONE; W/NBS   61664

468 C23 H36 N2 O6 S 28417-16-5 O-3-799-9 Glutamic acid, N-[N-(1-adamantylcarbonyl)-L-methionyl]-, di
methyl ester, L-;; ADCO-MET-GLU-(OME)2; W/NBS   61665

468 C24 H17 Cr O5 P 52615-02-8 AG-131-52-3 (Pentacarbonyl)(triphenylphosphoniummethylide)chromium; W
                                                                                               124721

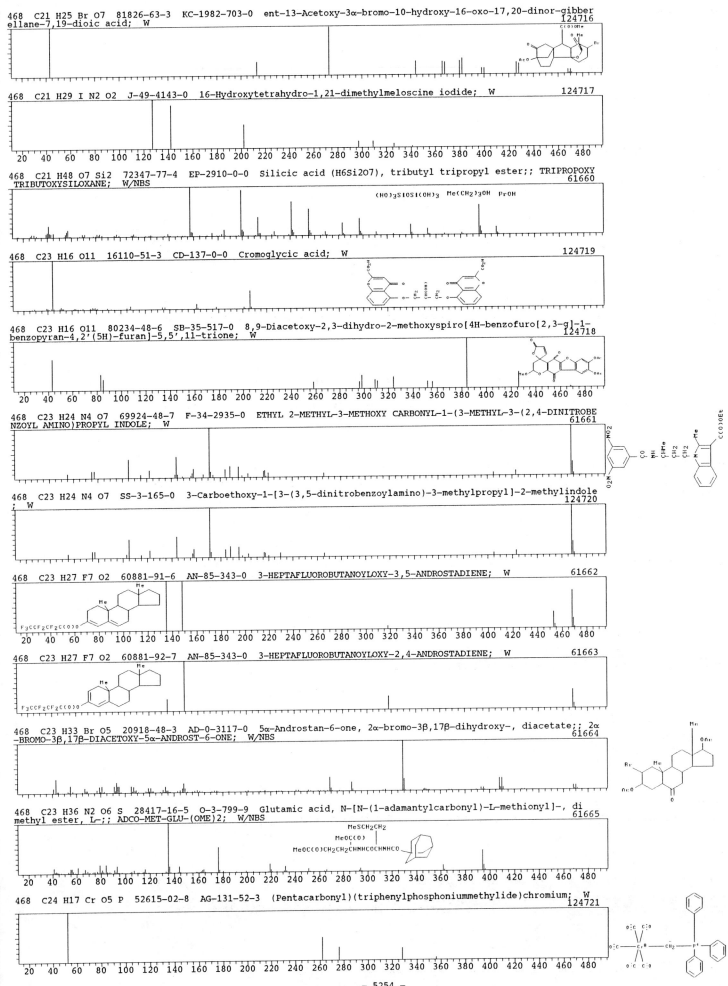

- 5254 -

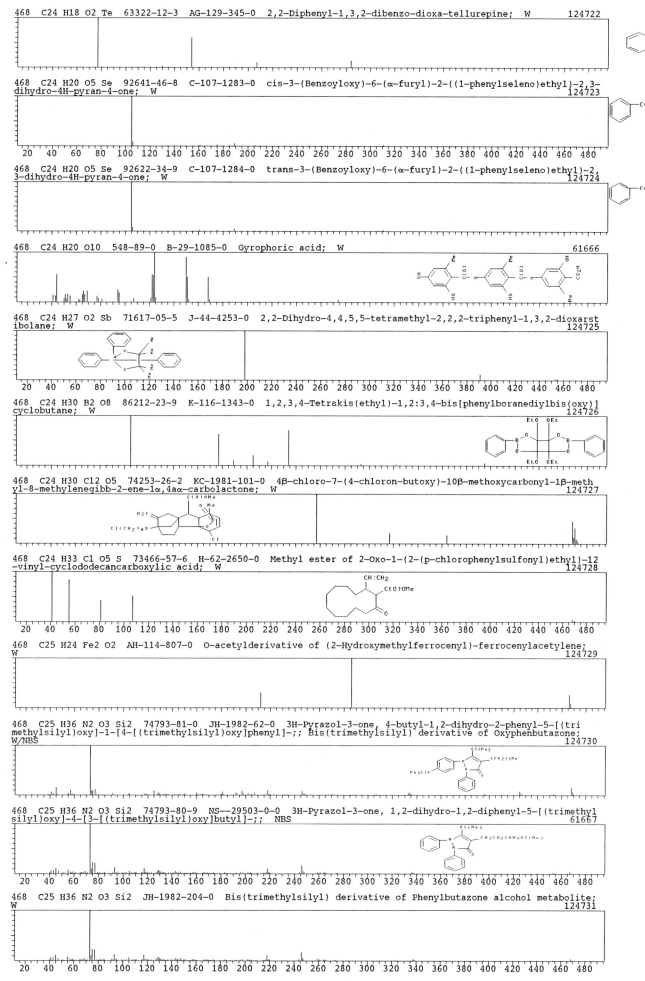

468 C24 H18 O2 Te 63322-12-3 AG-129-345-0 2,2-Diphenyl-1,3,2-dibenzo-dioxa-tellurepine; W 124722

468 C24 H20 O5 Se 92641-46-8 C-107-1283-0 cis-3-(Benzoyloxy)-6-(α-furyl)-2-((1-phenylseleno)ethyl)-2,3-dihydro-4H-pyran-4-one; W 124723

468 C24 H20 O5 Se 92622-34-9 C-107-1284-0 trans-3-(Benzoyloxy)-6-(α-furyl)-2-((1-phenylseleno)ethyl)-2,3-dihydro-4H-pyran-4-one; W 124724

468 C24 H20 O10 548-89-0 B-29-1085-0 Gyrophoric acid; W 61666

468 C24 H27 O2 Sb 71617-05-5 J-44-4253-0 2,2-Dihydro-4,4,5,5-tetramethyl-2,2,2-triphenyl-1,3,2-dioxarstibolane; W 124725

468 C24 H30 B2 O8 86212-23-9 K-116-1343-0 1,2,3,4-Tetrakis(ethyl)-1,2:3,4-bis[phenylboranediylbis(oxy)]cyclobutane; W 124726

468 C24 H30 Cl2 O5 74253-26-2 KC-1981-101-0 4β-chloro-7-(4-chloron-butoxy)-10β-methoxycarbonyl-1β-methyl-8-methylenegibb-2-ene-1α,4aα-carbolactone; W 124727

468 C24 H33 Cl O5 S 73466-57-6 H-62-2650-0 Methyl ester of 2-Oxo-1-(2-(p-chlorophenylsulfonyl)ethyl)-12-vinyl-cyclododecancarboxylic acid; W 124728

468 C25 H24 Fe2 O2 AH-114-807-0 O-acetylderivative of (2-Hydroxymethylferrocenyl)-ferrocenylacetylene; W 124729

468 C25 H36 N2 O3 Si2 74793-81-0 JH-1982-62-0 3H-Pyrazol-3-one, 4-butyl-1,2-dihydro-2-phenyl-5-[(trimethylsilyl)oxy]-1-[4-[(trimethylsilyl)oxy]phenyl]-;; Bis(trimethylsilyl) derivative of Oxyphenbutazone; W/NBS 124730

468 C25 H36 N2 O3 Si2 74793-80-9 NS--29503-0-0 3H-Pyrazol-3-one, 1,2-dihydro-1,2-diphenyl-5-[(trimethylsilyl)oxy]-4-[3-[(trimethylsilyl)oxy]butyl]-;; NBS 61667

468 C25 H36 N2 O3 Si2 JH-1982-204-0 Bis(trimethylsilyl) derivative of Phenylbutazone alcohol metabolite; W 124731

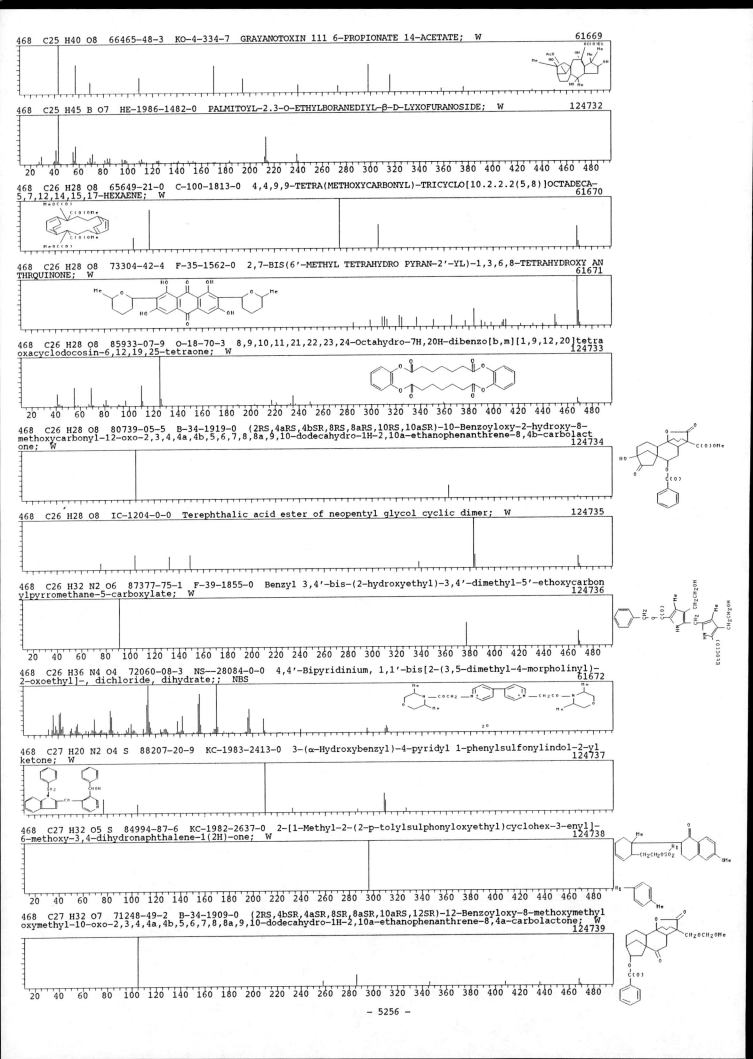

468   C25 H40 O8   66465-48-3   KO-4-334-7   GRAYANOTOXIN 111 6-PROPIONATE 14-ACETATE;   W   61669

468   C25 H45 B O7   HE-1986-1482-0   PALMITOYL-2.3-O-ETHYLBORANEDIYL-β-D-LYXOFURANOSIDE;   W   124732

468   C26 H28 O8   65649-21-0   C-100-1813-0   4,4,9,9-TETRA(METHOXYCARBONYL)-TRICYCLO[10.2.2.2(5,8)]OCTADECA-
5,7,12,14,15,17-HEXAENE;   W   61670

468   C26 H28 O8   73304-42-4   F-35-1562-0   2,7-BIS(6'-METHYL TETRAHYDRO PYRAN-2'-YL)-1,3,6,8-TETRAHYDROXY AN
THRQUINONE;   W   61671

468   C26 H28 O8   85933-07-9   O-18-70-3   8,9,10,11,21,22,23,24-Octahydro-7H,20H-dibenzo[b,m][1,9,12,20]tetra
oxacyclodocosin-6,12,19,25-tetraone;   W   124733

468   C26 H28 O8   80739-05-5   B-34-1919-0   (2RS,4aRS,4bSR,8RS,8aRS,10RS,10aSR)-10-Benzoyloxy-2-hydroxy-8-
methoxycarbonyl-12-oxo-2,3,4,4a,4b,5,6,7,8,8a,9,10-dodecahydro-1H-2,10a-ethanophenanthrene-8,4b-carbolact
one;   W   124734

468   C26 H28 O8   IC-1204-0-0   Terephthalic acid ester of neopentyl glycol cyclic dimer;   W   124735

468   C26 H32 N2 O6   87377-75-1   F-39-1855-0   Benzyl 3,4'-bis-(2-hydroxyethyl)-3,4'-dimethyl-5'-ethoxycarbon
ylpyrromethane-5-carboxylate;   W   124736

468   C26 H36 N4 O4   72060-08-3   NS--28084-0-0   4,4'-Bipyridinium, 1,1'-bis[2-(3,5-dimethyl-4-morpholinyl)-
2-oxoethyl]-, dichloride, dihydrate;;   NBS   61672

468   C27 H20 N2 O4 S   88207-20-9   KC-1983-2413-0   3-(α-Hydroxybenzyl)-4-pyridyl 1-phenylsulfonylindol-2-yl
ketone;   W   124737

468   C27 H32 O5 S   84994-87-6   KC-1982-2637-0   2-[1-Methyl-2-(2-p-tolylsulphonyloxyethyl)cyclohex-3-enyl]-
6-methoxy-3,4-dihydronaphthalene-1(2H)-one;   W   124738

468   C27 H32 O7   71248-49-2   B-34-1909-0   (2RS,4bSR,4aSR,8SR,8aSR,10aRS,12SR)-12-Benzoyloxy-8-methoxymethyl
oxymethyl-10-oxo-2,3,4,4a,4b,5,6,7,8,8a,9,10-dodecahydro-1H-2,10a-ethanophenanthrene-8,4a-carbolactone;   W
                                                                                                           124739

468   C27 H33 Br O2   93297-57-5   K-117-3090-0   (1RS,3aRS,6aSR,6bSR,9aRS,9bSR,9cSR)-1-(4-Bromobenzoyloxy)-1,
2,3,5,6,6b,7,8,99b-decahydro-6b-methyl-4H-3a,6a:9a,9c-diethanocyclopent[a]acenaphthylene;   W   124740

468   C27 H36 N2 O3 S   79135-45-8   J-46-4642-0   (3α,7aα,7bβ,12β,14aβ,14bα)-4-Methyl-(tetradecahydro-14-oxo-
3,7a:7b,12-dimethanocyclobuta[!,2:3,4]dicyclononen-1(8H)-ylidene)hydrazide benzenesulfonic acid;   W   124741

20   40   60   80   100   120   140   160   180   200   220   240   260   280   300   320   340   360   380   400   420   440   460   480

468   C27 H40 N2 O3 Si   KO-9-198-9   Trimethylsilyloxime derivative of N-(cyclopropylmethyl)-8β-n-propyldihyd
ronorcodeinone;   W   124742

468   C27 H50 O2 P2   HE-1986-2024-0   1,3-BIS(DICYCLOHEXYLPHOSPHINE OXIDE-P-YL)-PROPANE;   W   124743

468   C28 H20 O2 Se   89936-27-6   J-49-2058-0   2,4,5-Triphenyl-2-benzoyl-2H-1,3-oxaselenole;   W   124744

20   40   60   80   100   120   140   160   180   200   220   240   260   280   300   320   340   360   380   400   420   440   460   480

468   C28 H24 N2 O5   3249-94-3   O7-0-317-0   5'-O-TRITYL-2,2'-ANHYDROURIDINE;   W/NBS   61673

468   C28 H28 N4 O3   75841-16-6   Y-17-702-0   (1,2-Bisanilino-2-(4-acetyl-5-methyl-2-yl)-ethanone;   W   124745

468   C28 H36 O6   62501-55-7   KC-1977-13-0   HEXAHYDROPENTADESMAX ANTHENE;   W   61674

20   40   60   80   100   120   140   160   180   200   220   240   260   280   300   320   340   360   380   400   420   440   460   480

468   C28 H36 O6   KC-1982-2838-0   (22R)-14α,17β,20α(F)-trihydroxy-1-oxo-witha-2,4,6,24-tetraenolide;   W
                                                                                                       124746

468   C28 H36 O6   KC-1982-2838-0   (22R)-(8α,H)-14α,17β,20α(F)-trihydroxy-1-oxowitha-2,4,6,24-tetraenolide;
W   124747

468   C28 H56 O3 Si   56847-14-4   EP-6965-0-0   Tetracosenoic acid, 2-[(trimethylsilyl)oxy]-, methyl ester;;
METHYL 2-TMSOXYTETRACOS-X-ENOATE;   W/NBS   61675

OSiMe3
|
MeOC(O)CH(CH2)21Me

20   40   60   80   100   120   140   160   180   200   220   240   260   280   300   320   340   360   380   400   420   440   460   480

468   C29 H24 O6   IC-1204-0-0   2,2'-Dimethoxy-4,4'-dibenzoyl-5,5'-dihydroxydiphenylmethane;   W   124748

468   C29 H26 O2 P2   89243-76-5   K-117-362-0   1,3-Bis(diphenylphosphino)-2,2-dimethyl-1,3-propandion;   W
                                                                                                       124749

20   40   60   80   100   120   140   160   180   200   220   240   260   280   300   320   340   360   380   400   420   440   460   480

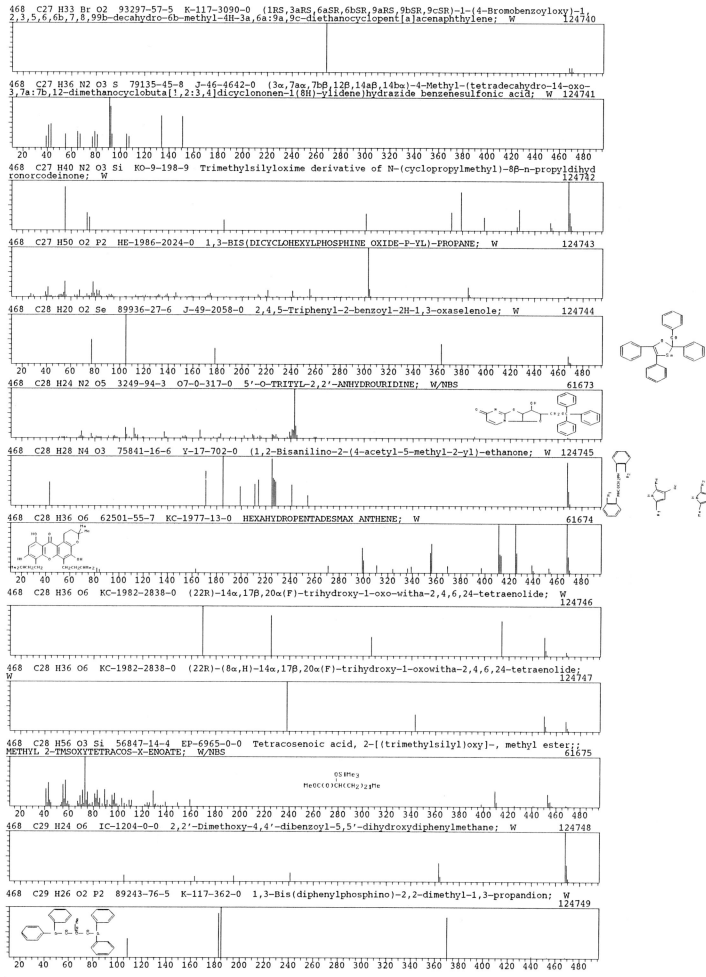

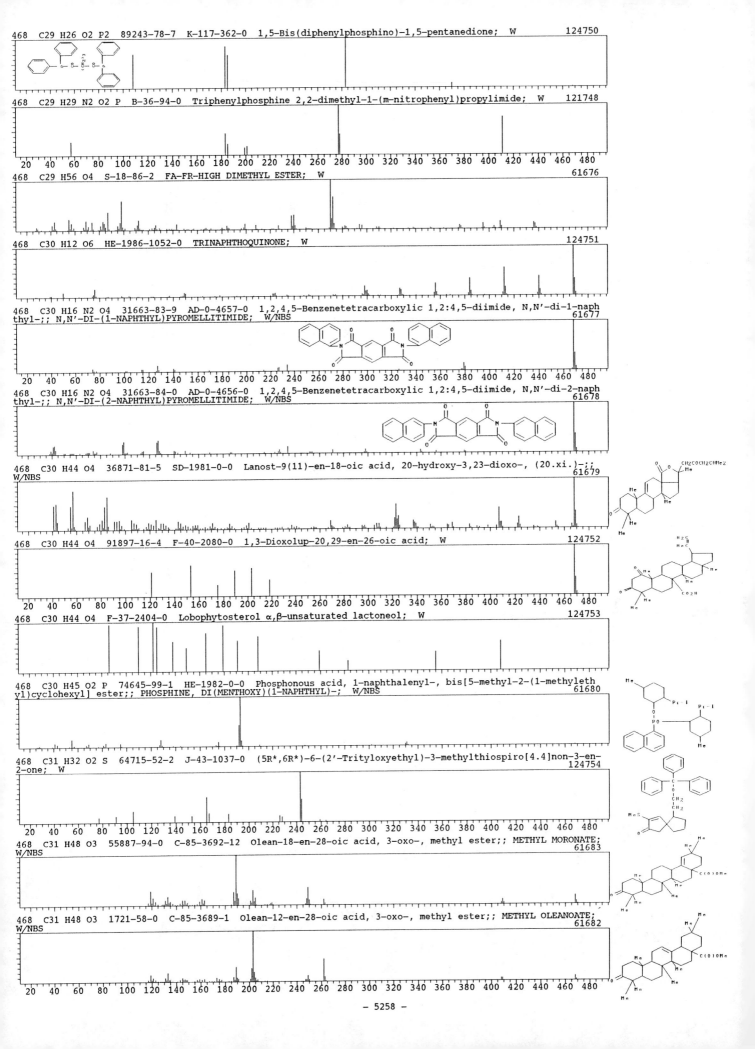

468 C29 H26 O2 P2 89243-78-7 K-117-362-0 1,5-Bis(diphenylphosphino)-1,5-pentanedione; W   124750

468 C29 H29 N2 O2 P B-36-94-0 Triphenylphosphine 2,2-dimethyl-1-(m-nitrophenyl)propylimide; W   121748

468 C29 H56 O4 S-18-86-2 FA-FR-HIGH DIMETHYL ESTER; W   61676

468 C30 H12 O6 HE-1986-1052-0 TRINAPHTHOQUINONE; W   124751

468 C30 H16 N2 O4 31663-83-9 AD-0-4657-0 1,2,4,5-Benzenetetracarboxylic 1,2:4,5-diimide, N,N'-di-1-naph
thyl-;; N,N'-DI-(1-NAPHTHYL)PYROMELLITIMIDE; W/NBS   61677

468 C30 H16 N2 O4 31663-84-0 AD-0-4656-0 1,2,4,5-Benzenetetracarboxylic 1,2:4,5-diimide, N,N'-di-2-naph
thyl-;; N,N'-DI-(2-NAPHTHYL)PYROMELLITIMIDE; W/NBS   61678

468 C30 H44 O4 36871-81-5 SD-1981-0-0 Lanost-9(11)-en-18-oic acid, 20-hydroxy-3,23-dioxo-, (20.xi.)-;;
W/NBS   61679

468 C30 H44 O4 91897-16-4 F-40-2080-0 1,3-Dioxolup-20,29-en-26-oic acid; W   124752

468 C30 H44 O4 F-37-2404-0 Lobophytosterol α,β-unsaturated lactoneol; W   124753

468 C30 H45 O2 P 74645-99-1 HE-1982-0-0 Phosphonous acid, 1-naphthalenyl-, bis[5-methyl-2-(1-methyleth
yl)cyclohexyl] ester;; PHOSPHINE, DI(MENTHOXY)(1-NAPHTHYL)-; W/NBS   61680

468 C31 H32 O2 S 64715-52-2 J-43-1037-0 (5R*,6R*)-6-(2'-Trityloxyethyl)-3-methylthiospiro[4.4]non-3-en-
2-one; W   124754

468 C31 H48 O3 55887-94-0 C-85-3692-12 Olean-18-en-28-oic acid, 3-oxo-, methyl ester;; METHYL MORONATE;
W/NBS   61683

468 C31 H48 O3 1721-58-0 C-85-3689-1 Olean-12-en-28-oic acid, 3-oxo-, methyl ester;; METHYL OLEANOATE;
W/NBS   61682

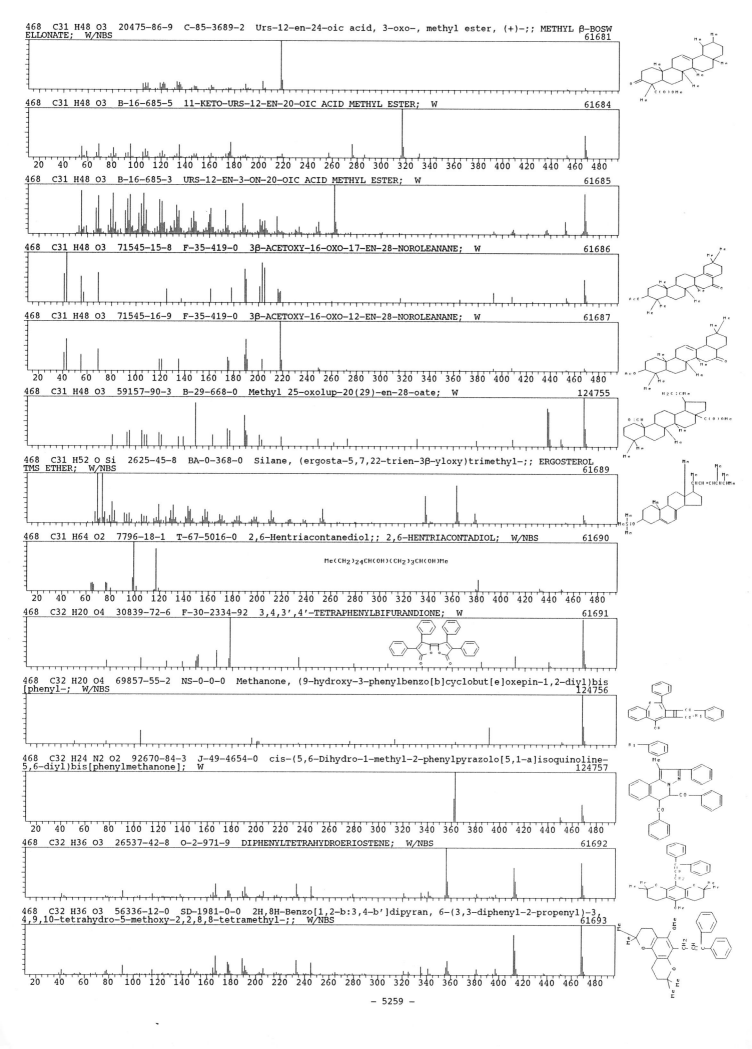

468 C31 H48 O3 20475-86-9 C-85-3689-2 Urs-12-en-24-oic acid, 3-oxo-, methyl ester, (+)-;; METHYL β-BOSW
ELLONATE; W/NBS   61681

468 C31 H48 O3 B-16-685-5 11-KETO-URS-12-EN-20-OIC ACID METHYL ESTER; W   61684

468 C31 H48 O3 B-16-685-3 URS-12-EN-3-ON-20-OIC ACID METHYL ESTER; W   61685

468 C31 H48 O3 71545-15-8 F-35-419-0 3β-ACETOXY-16-OXO-17-EN-28-NOROLEANANE; W   61686

468 C31 H48 O3 71545-16-9 F-35-419-0 3β-ACETOXY-16-OXO-12-EN-28-NOROLEANANE; W   61687

468 C31 H48 O3 59157-90-3 B-29-668-0 Methyl 25-oxolup-20(29)-en-28-oate; W   124755

468 C31 H52 O Si 2625-45-8 BA-0-368-0 Silane, (ergosta-5,7,22-trien-3β-yloxy)trimethyl-;; ERGOSTEROL
TMS ETHER; W/NBS   61689

468 C31 H64 O2 7796-18-1 T-67-5016-0 2,6-Hentriacontanediol;; 2,6-HENTRIACONTADIOL; W/NBS   61690

Me(CH2)24CH(OH)(CH2)3CH(OH)Me

468 C32 H20 O4 30839-72-6 F-30-2334-92 3,4,3',4'-TETRAPHENYLBIFURANDIONE; W   61691

468 C32 H20 O4 69857-55-2 NS-0-0-0 Methanone, (9-hydroxy-3-phenylbenzo[b]cyclobut[e]oxepin-1,2-diyl)bis
[phenyl-; W/NBS   124756

468 C32 H24 N2 O2 92670-84-3 J-49-4654-0 cis-(5,6-Dihydro-1-methyl-2-phenylpyrazolo[5,1-a]isoquinoline-
5,6-diyl)bis[phenylmethanone]; W   124757

468 C32 H36 O3 26537-42-8 O-2-971-9 DIPHENYLTETRAHYDROERIOSTENE; W/NBS   61692

468 C32 H36 O3 56336-12-0 SD-1981-0-0 2H,8H-Benzo[1,2-b:3,4-b']dipyran, 6-(3,3-diphenyl-2-propenyl)-3,
4,9,10-tetrahydro-5-methoxy-2,2,8,8-tetramethyl-;; W/NBS   61693

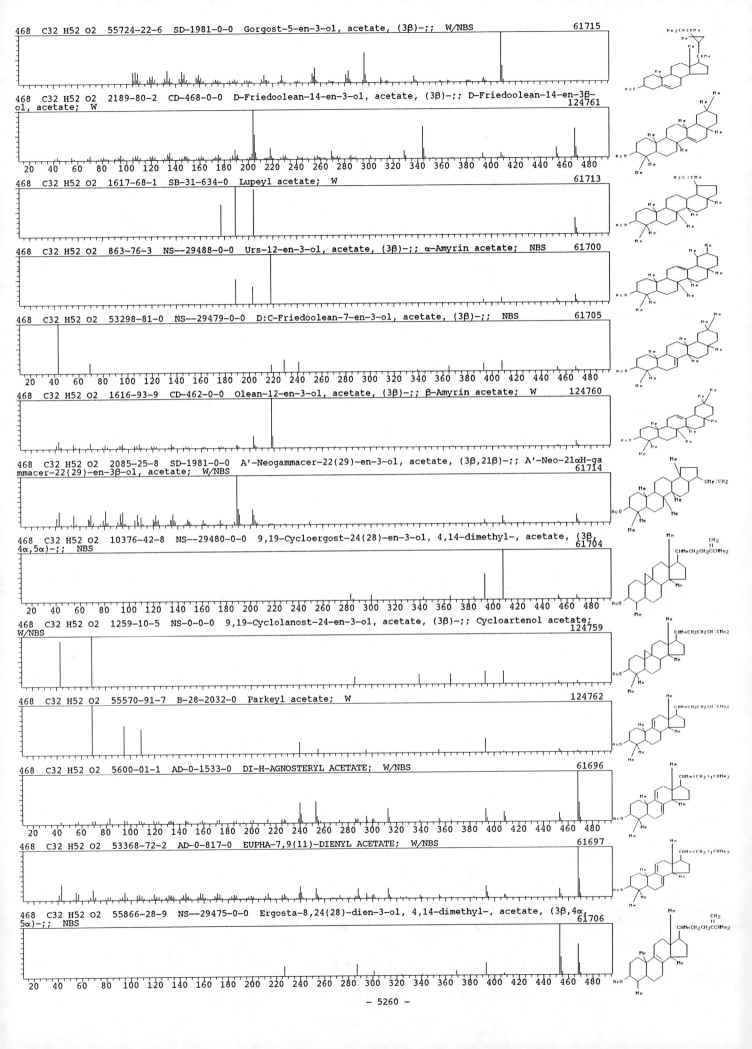

468　C32 H52 O2　55724-22-6　SD-1981-0-0　Gorgost-5-en-3-ol, acetate, (3β)-;;　W/NBS　　61715

468　C32 H52 O2　2189-80-2　CD-468-0-0　D-Friedoolean-14-en-3-ol, acetate, (3β)-;; D-Friedoolean-14-en-3β-ol, acetate;　W　　124761

468　C32 H52 O2　1617-68-1　SB-31-634-0　Lupeyl acetate;　W　　61713

468　C32 H52 O2　863-76-3　NS--29488-0-0　Urs-12-en-3-ol, acetate, (3β)-;; α-Amyrin acetate;　NBS　　61700

468　C32 H52 O2　53298-81-0　NS--29479-0-0　D:C-Friedoolean-7-en-3-ol, acetate, (3β)-;;　NBS　　61705

468　C32 H52 O2　1616-93-9　CD-462-0-0　Olean-12-en-3-ol, acetate, (3β)-;; β-Amyrin acetate;　W　　124760

468　C32 H52 O2　2085-25-8　SD-1981-0-0　A'-Neogammacer-22(29)-en-3-ol, acetate, (3β,21β)-;; A'-Neo-21αH-gammacer-22(29)-en-3β-ol, acetate;　W/NBS　　61714

468　C32 H52 O2　10376-42-8　NS--29480-0-0　9,19-Cycloergost-24(28)-en-3-ol, 4,14-dimethyl-, acetate, (3β,4α,5α)-;;　NBS　　61704

468　C32 H52 O2　1259-10-5　NS-0-0-0　9,19-Cyclolanost-24-en-3-ol, acetate, (3β)-;; Cycloartenol acetate; W/NBS　　124759

468　C32 H52 O2　55570-91-7　B-28-2032-0　Parkeyl acetate;　W　　124762

468　C32 H52 O2　5600-01-1　AD-0-1533-0　DI-H-AGNOSTERYL ACETATE;　W/NBS　　61696

468　C32 H52 O2　53368-72-2　AD-0-817-0　EUPHA-7,9(11)-DIENYL ACETATE;　W/NBS　　61697

468　C32 H52 O2　55866-28-9　NS--29475-0-0　Ergosta-8,24(28)-dien-3-ol, 4,14-dimethyl-, acetate, (3β,4α,5α)-;;　NBS　　61706

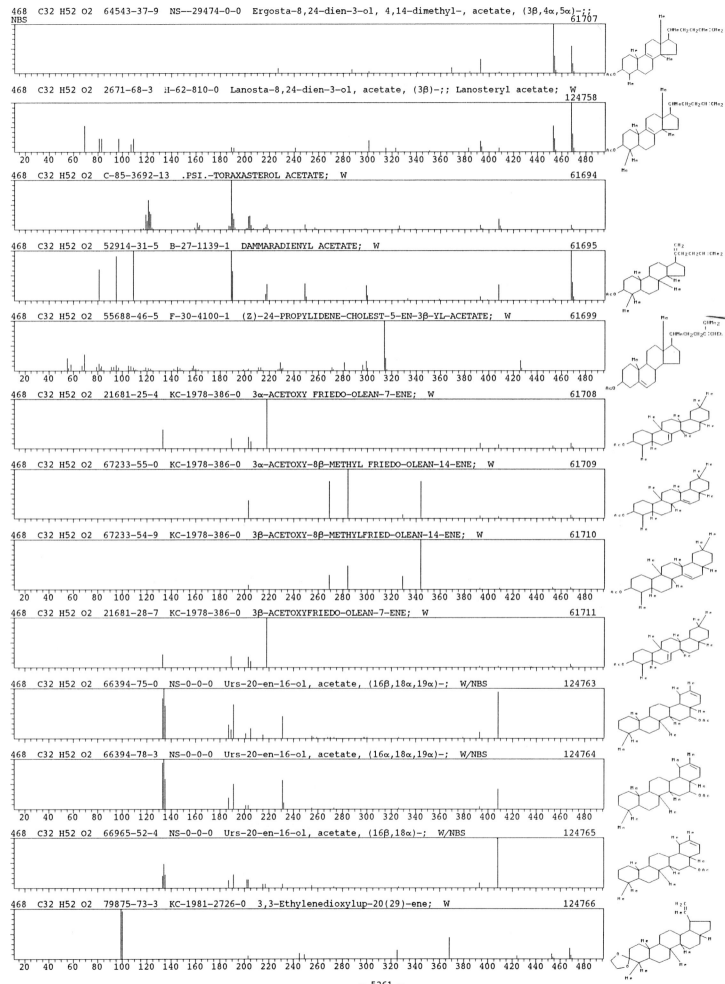

468   C32 H52 O2   64543-37-9   NS--29474-0-0   Ergosta-8,24-dien-3-ol, 4,14-dimethyl-, acetate, (3β,4α,5α)-;;
NBS   61707

468   C32 H52 O2   2671-68-3   N-62-810-0   Lanosta-8,24-dien-3-ol, acetate, (3β)-;; Lanosteryl acetate;   W
                                                                                                               124758

468   C32 H52 O2   C-85-3692-13   .PSI.-TORAXASTEROL ACETATE;   W   61694

468   C32 H52 O2   52914-31-5   B-27-1139-1   DAMMARADIENYL ACETATE;   W   61695

468   C32 H52 O2   55688-46-5   F-30-4100-1   (Z)-24-PROPYLIDENE-CHOLEST-5-EN-3β-YL-ACETATE;   W   61699

468   C32 H52 O2   21681-25-4   KC-1978-386-0   3α-ACETOXY FRIEDO-OLEAN-7-ENE;   W   61708

468   C32 H52 O2   67233-55-0   KC-1978-386-0   3α-ACETOXY-8β-METHYL FRIEDO-OLEAN-14-ENE;   W   61709

468   C32 H52 O2   67233-54-9   KC-1978-386-0   3β-ACETOXY-8β-METHYLFRIED-OLEAN-14-ENE;   W   61710

468   C32 H52 O2   21681-28-7   KC-1978-386-0   3β-ACETOXYFRIEDO-OLEAN-7-ENE;   W   61711

468   C32 H52 O2   66394-75-0   NS-0-0-0   Urs-20-en-16-ol, acetate, (16β,18α,19α)-;   W/NBS   124763

468   C32 H52 O2   66394-78-3   NS-0-0-0   Urs-20-en-16-ol, acetate, (16α,18α,19α)-;   W/NBS   124764

468   C32 H52 O2   66965-52-4   NS-0-0-0   Urs-20-en-16-ol, acetate, (16β,18α)-;   W/NBS   124765

468   C32 H52 O2   79875-73-3   KC-1981-2726-0   3,3-Ethylenedioxylup-20(29)-ene;   W   124766

- 5261 -

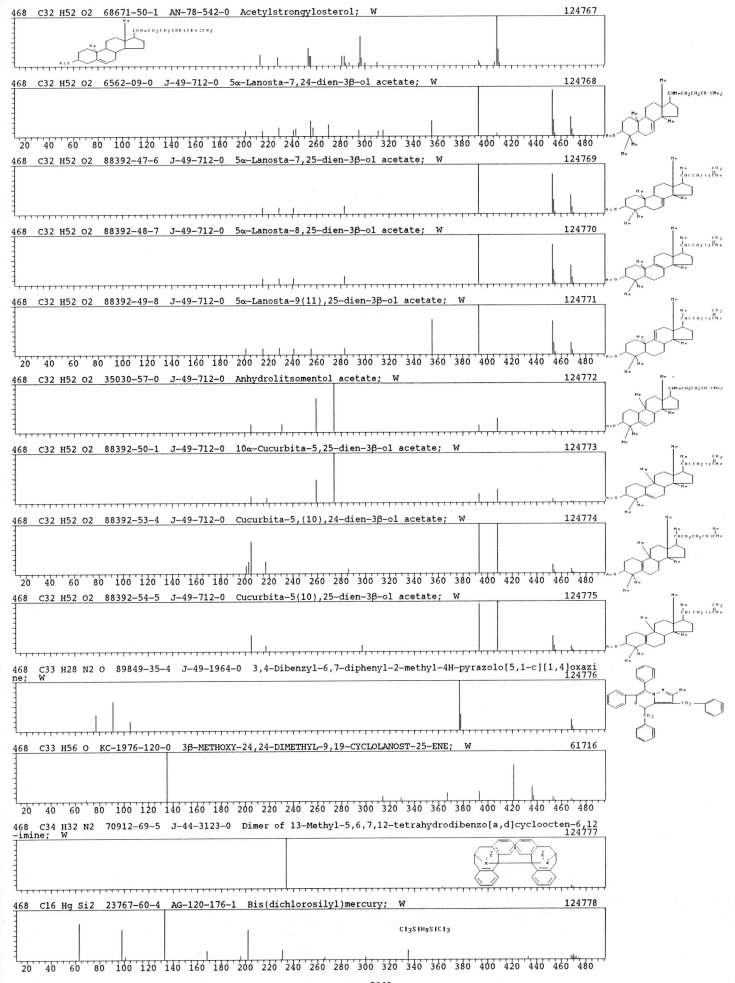

468   C32 H52 O2   68671-50-1   AN-78-542-0   Acetylstrongylosterol;   W   124767

468   C32 H52 O2   6562-09-0   J-49-712-0   5α-Lanosta-7,24-dien-3β-ol acetate;   W   124768

468   C32 H52 O2   88392-47-6   J-49-712-0   5α-Lanosta-7,25-dien-3β-ol acetate;   W   124769

468   C32 H52 O2   88392-48-7   J-49-712-0   5α-Lanosta-8,25-dien-3β-ol acetate;   W   124770

468   C32 H52 O2   88392-49-8   J-49-712-0   5α-Lanosta-9(11),25-dien-3β-ol acetate;   W   124771

468   C32 H52 O2   35030-57-0   J-49-712-0   Anhydrolitsomentol acetate;   W   124772

468   C32 H52 O2   88392-50-1   J-49-712-0   10α-Cucurbita-5,25-dien-3β-ol acetate;   W   124773

468   C32 H52 O2   88392-53-4   J-49-712-0   Cucurbita-5,(10),24-dien-3β-ol acetate;   W   124774

468   C32 H52 O2   88392-54-5   J-49-712-0   Cucurbita-5(10),25-dien-3β-ol acetate;   W   124775

468   C33 H28 N2 O   89849-35-4   J-49-1964-0   3,4-Dibenzyl-6,7-diphenyl-2-methyl-4H-pyrazolo[5,1-c][1,4]oxazine;   W   124776

468   C33 H56 O   KC-1976-120-0   3β-METHOXY-24,24-DIMETHYL-9,19-CYCLOLANOST-25-ENE;   W   61716

468   C34 H32 N2   70912-69-5   J-44-3123-0   Dimer of 13-Methyl-5,6,7,12-tetrahydrodibenzo[a,d]cycloocten-6,12-imine;   W   124777

468   C16 Hg Si2   23767-60-4   AG-120-176-1   Bis(dichlorosilyl)mercury;   W   124778

Cl3SiHgSiCl3

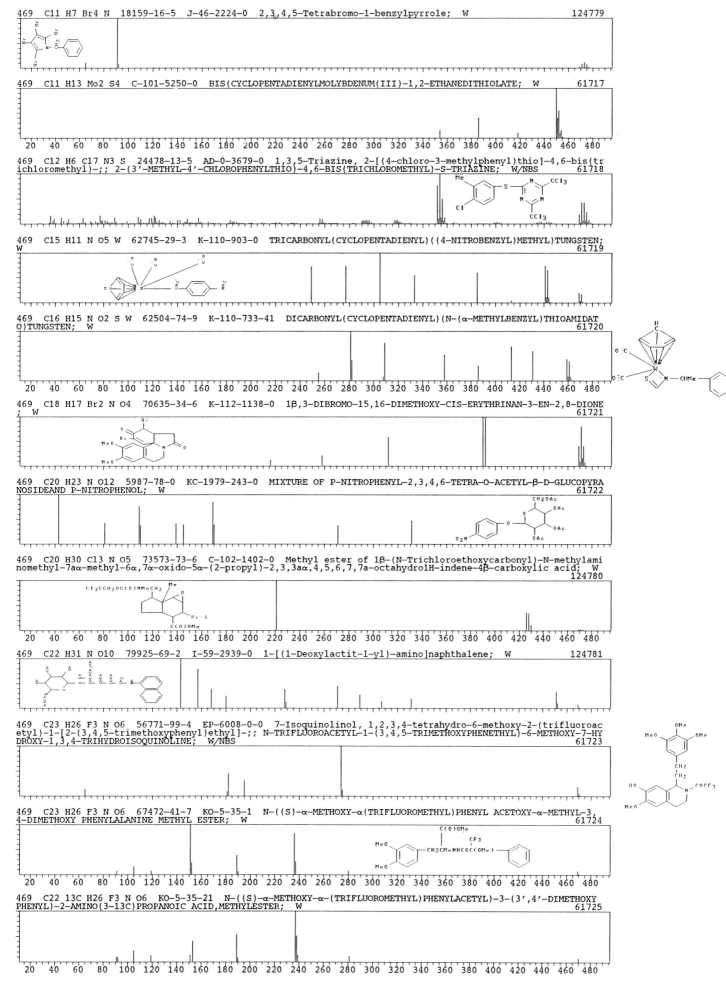

469   C11 H7 Br4 N   18159-16-5   J-46-2224-0   2,3,4,5-Tetrabromo-1-benzylpyrrole;   W   124779

469   C11 H13 Mo2 S4   C-101-5250-0   BIS(CYCLOPENTADIENYLMOLYBDENUM(III)-1,2-ETHANEDITHIOLATE;   W   61717

469   C12 H6 Cl7 N3 S   24478-13-5   AD-0-3679-0   1,3,5-Triazine, 2-[(4-chloro-3-methylphenyl)thio]-4,6-bis(trichloromethyl)-;;   2-(3'-METHYL-4'-CHLOROPHENYLTHIO)-4,6-BIS(TRICHLOROMETHYL)-S-TRIAZINE;   W/NBS   61718

469   C15 H11 N O5 W   62745-29-3   K-110-903-0   TRICARBONYL(CYCLOPENTADIENYL)((4-NITROBENZYL)METHYL)TUNGSTEN;   W   61719

469   C16 H15 N O2 S W   62504-74-9   K-110-733-41   DICARBONYL(CYCLOPENTADIENYL)(N-(α-METHYLBENZYL)THIOAMIDATO)TUNGSTEN;   W   61720

469   C18 H17 Br2 N O4   70635-34-6   K-112-1138-0   1β,3-DIBROMO-15,16-DIMETHOXY-CIS-ERYTHRINAN-3-EN-2,8-DIONE;   W   61721

469   C20 H23 N O12   5987-78-0   KC-1979-243-0   MIXTURE OF P-NITROPHENYL-2,3,4,6-TETRA-O-ACETYL-β-D-GLUCOPYRANOSIDEAND P-NITROPHENOL;   W   61722

469   C20 H30 Cl3 N O5   73573-73-6   C-102-1402-0   Methyl ester of 1β-(N-Trichloroethoxycarbonyl)-N-methylaminomethyl-7aα-methyl-6α,7α-oxido-5α-(2-propyl)-2,3,3aα,4,5,6,7,7a-octahydro1H-indene-4β-carboxylic acid;   W   124780

469   C22 H31 N O10   79925-69-2   I-59-2939-0   1-[(1-Deoxylactit-1-yl)-amino]naphthalene;   W   124781

469   C23 H26 F3 N O6   56771-99-4   EP-6008-0-0   7-Isoquinolinol, 1,2,3,4-tetrahydro-6-methoxy-2-(trifluoroacetyl)-1-[2-(3,4,5-trimethoxyphenyl)ethyl]-;;   N-TRIFLUOROACETYL-1-(3,4,5-TRIMETHOXYPHENETHYL)-6-METHOXY-7-HYDROXY-1,3,4-TRIHYDROISOQUINOLINE;   W/NBS   61723

469   C23 H26 F3 N O6   67472-41-7   KO-5-35-1   N-((S)-α-METHOXY-α(TRIFLUOROMETHYL)PHENYL ACETOXY-α-METHYL-3,4-DIMETHOXY PHENYLALANINE METHYL ESTER;   W   61724

469   C22 13C H26 F3 N O6   KO-5-35-21   N-((S)-α-METHOXY-α-(TRIFLUOROMETHYL)PHENYLACETYL)-3-(3',4'-DIMETHOXYPHENYL)-2-AMINO(3-13C)PROPANOIC ACID,METHYLESTER;   W   61725

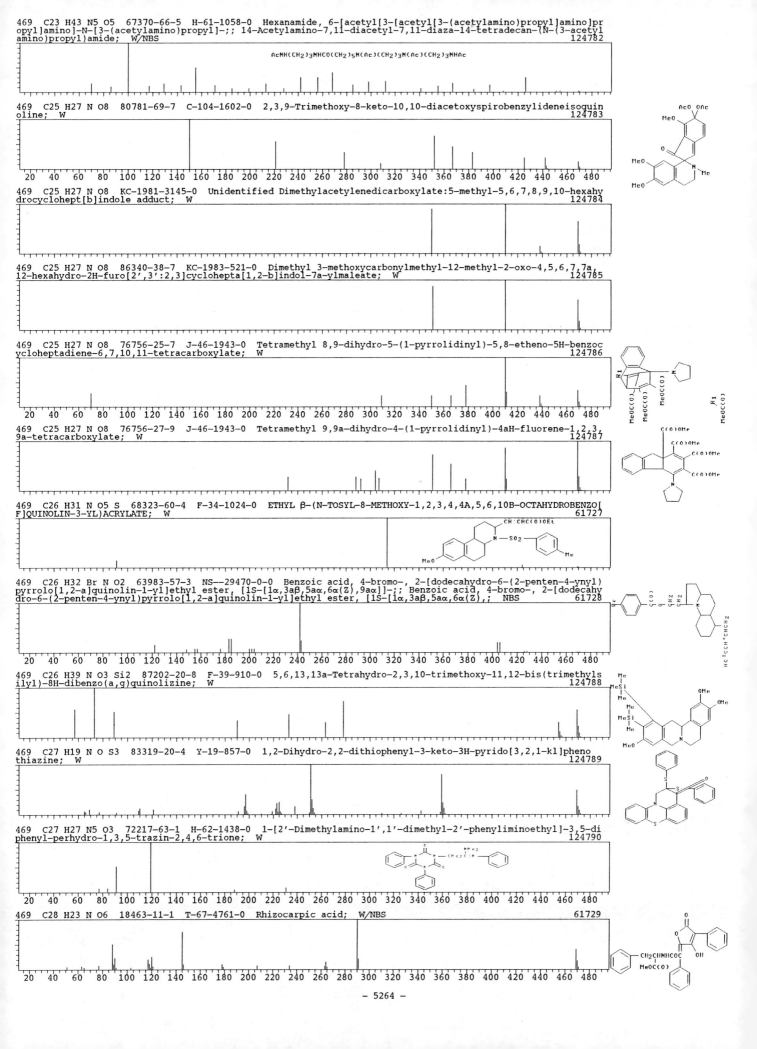

469 C23 H43 N5 O5 67370-66-5 H-61-1058-0 Hexanamide, 6-[acetyl[3-[acetyl[3-(acetylamino)propyl]amino]propyl]amino]-N-[3-(acetylamino)propyl]-;; 14-Acetylamino-7,11-diacetyl-7,11-diaza-14-tetradecan-(N-(3-acetylamino)propyl)amide; W/NBS
124782

AcNH(CH₂)₃NHCO(CH₂)₅N(Ac)(CH₂)₃N(Ac)(CH₂)₃NHAc

469 C25 H27 N O8 80781-69-7 C-104-1602-0 2,3,9-Trimethoxy-8-keto-10,10-diacetoxyspirobenzylideneisoquinoline; W
124783

469 C25 H27 N O8 KC-1981-3145-0 Unidentified Dimethylacetylenedicarboxylate:5-methyl-5,6,7,8,9,10-hexahydrocyclohept[b]indole adduct; W
124784

469 C25 H27 N O8 86340-38-7 KC-1983-521-0 Dimethyl 3-methoxycarbonylmethyl-12-methyl-2-oxo-4,5,6,7,7a,12-hexahydro-2H-furo[2',3':2,3]cyclohepta[1,2-b]indol-7a-ylmaleate; W
124785

469 C25 H27 N O8 76756-25-7 J-46-1943-0 Tetramethyl 8,9-dihydro-5-(1-pyrrolidinyl)-5,8-etheno-5H-benzocycloheptadiene-6,7,10,11-tetracarboxylate; W
124786

469 C25 H27 N O8 76756-27-9 J-46-1943-0 Tetramethyl 9,9a-dihydro-4-(1-pyrrolidinyl)-4aH-fluorene-1,2,3,9a-tetracarboxylate; W
124787

469 C26 H31 N O5 S 68323-60-4 F-34-1024-0 ETHYL β-(N-TOSYL-8-METHOXY-1,2,3,4,4A,5,6,10B-OCTAHYDROBENZO[F]QUINOLIN-3-YL)ACRYLATE; W
61727

469 C26 H32 Br N O2 63983-57-3 NS--29470-0-0 Benzoic acid, 4-bromo-, 2-[dodecahydro-6-(2-penten-4-ynyl)pyrrolo[1,2-a]quinolin-1-yl]ethyl ester, [1S-[1α,3aβ,5aα,6α(Z),9aα]]-;; Benzoic acid, 4-bromo-, 2-[dodecahydro-6-(2-penten-4-ynyl)pyrrolo[1,2-a]quinolin-1-yl]ethyl ester, [1S-[1α,3aβ,5aα,6α(Z),; NBS
61728

469 C26 H39 N O3 Si2 87202-20-8 F-39-910-0 5,6,13,13a-Tetrahydro-2,3,10-trimethoxy-11,12-bis(trimethylsilyl)-8H-dibenzo(a,g)quinolizine; W
124788

469 C27 H19 N O S3 83319-20-4 Y-19-857-0 1,2-Dihydro-2,2-dithiophenyl-3-keto-3H-pyrido[3,2,1-kl]phenothiazine; W
124789

469 C27 H27 N5 O3 72217-63-1 H-62-1438-0 1-[2'-Dimethylamino-1',1'-dimethyl-2'-phenyliminoethyl]-3,5-diphenyl-perhydro-1,3,5-trazin-2,4,6-trione; W
124790

469 C28 H23 N O6 18463-11-1 T-67-4761-0 Rhizocarpic acid; W/NBS
61729

- 5264 -

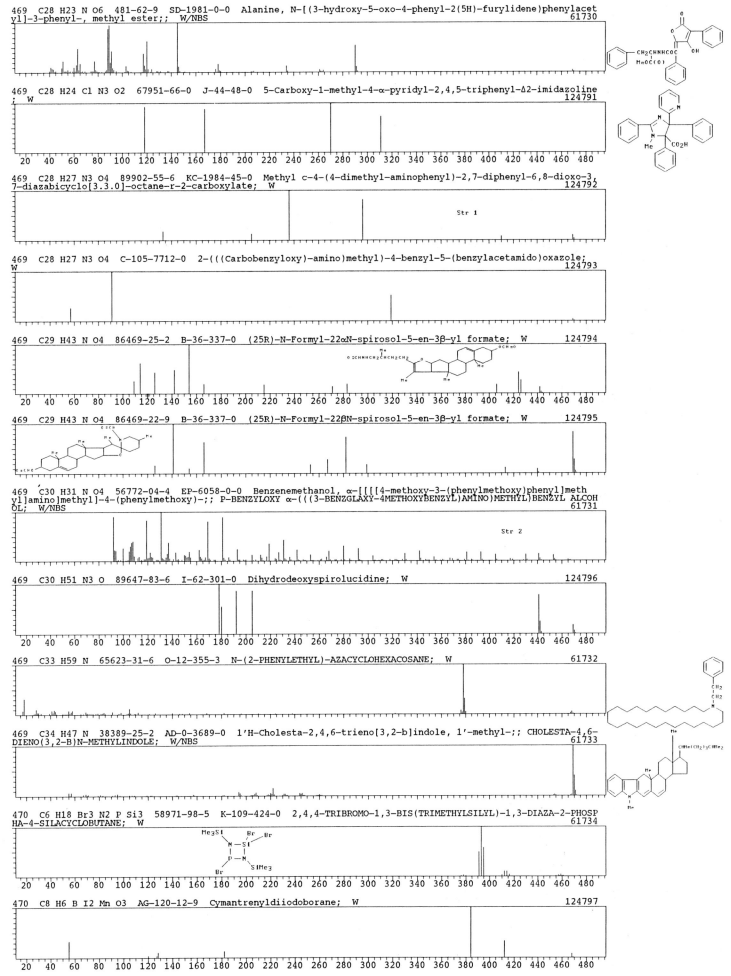

469 C28 H23 N O6 481-62-9 SD-1981-0-0 Alanine, N-[(3-hydroxy-5-oxo-4-phenyl-2(5H)-furylidene)phenylacetyl]-3-phenyl-, methyl ester;; W/NBS
61730

469 C28 H24 Cl N3 O2 67951-66-0 J-44-48-0 5-Carboxy-1-methyl-4-α-pyridyl-2,4,5-triphenyl-Δ2-imidazoline; W
124791

469 C28 H27 N3 O4 89902-55-6 KC-1984-45-0 Methyl c-4-(4-dimethyl-aminophenyl)-2,7-diphenyl-6,8-dioxo-3,7-diazabicyclo[3.3.0]-octane-r-2-carboxylate; W
124792
Str 1

469 C28 H27 N3 O4 C-105-7712-0 2-(((Carbobenzyloxy)-amino)methyl)-4-benzyl-5-(benzylacetamido)oxazole; W
124793

469 C29 H43 N O4 86469-25-2 B-36-337-0 (25R)-N-Formyl-22αN-spirosol-5-en-3β-yl formate; W
124794

469 C29 H43 N O4 86469-22-9 B-36-337-0 (25R)-N-Formyl-22βN-spirosol-5-en-3β-yl formate; W
124795

469 C30 H31 N O4 56772-04-4 EP-6058-0-0 Benzenemethanol, α-[[[[4-methoxy-3-(phenylmethoxy)phenyl]methyl]amino]methyl]-4-(phenylmethoxy)-;; P-BENZYLOXY α-(((3-BENZGLAXY-4METHOXYBENZYL)AMINO)METHYL)BENZYL ALCOHOL; W/NBS
61731
Str 2

469 C30 H51 N3 O 89647-83-6 I-62-301-0 Dihydrodeoxyspirolucidine; W
124796

469 C33 H59 N 65623-31-6 O-12-355-3 N-(2-PHENYLETHYL)-AZACYCLOHEXACOSANE; W
61732

469 C34 H47 N 38389-25-2 AD-0-3689-0 1'H-Cholesta-2,4,6-trieno[3,2-b]indole, 1'-methyl-;; CHOLESTA-4,6-DIENO(3,2-B)N-METHYLINDOLE; W/NBS
61733

470 C6 H18 Br3 N2 P Si3 58971-98-5 K-109-424-0 2,4,4-TRIBROMO-1,3-BIS(TRIMETHYLSILYL)-1,3-DIAZA-2-PHOSPHA-4-SILACYCLOBUTANE; W
61734

470 C8 H6 B I2 Mn O3 AG-120-12-9 Cymantrenyldiiodoborane; W
124797

- 5265 -

470   C8 H18 Br2 F4 N2 O2 P2   94445-88-2   AH-115-892-0   2,2,4,4-Tetrafluoro-1,3-dimethyl-2,4-bis(3-bromoprop
oxy)-1,3,2λ5,4λ5-diazadiphosphetidine;   W   124798

470   C9 F5 O5 Re   15038-33-2   T-68-1338-0   Rhenium, pentacarbonyl(2,3,3,4,4-pentafluoro-1-cyclobuten-1-yl)-
-;; PENTAFLUORO-CYCLOBUT-1-ENYL PENTABARBONYLRHENIUM COMPLEX;   W/NBS   61735

470   C10 Cl10   2227-17-0   PG-1982-892-0   Decachlor;   W   79214

470   C12 Cl3 F10 P   6779-72-2   AD-0-2733-0   Phosphorane, trichlorobis(pentafluorophenyl)-;; BIS(PENTAFLUORO
PHENYL) TRICHLOROPHOSPHORANE;   W/NBS   61737

470   C15 H16 F6 O10   KO-5-672-1   PYRUVATED METHYL GALACTOSIDES BIS-TRIFLUOROACETATE OR PYRUVATED DERIVATIVE
OF TRIFLUOROACETIC ACID-METHYL GALACTOSIDES;   W   57947

470   C16 H15 F5 N4 O5 S   JH-1982-86-0   N'-Methyl derivative of Sulfadimethoxineperfluoropropionylamide;   W
                                                                                                        124799

470   C16 H42 O6 Si5   KS-71-737-9   DIMETHYLSILYL 2,3,4,6-O-DIMETHYLSILYL-GLUCOPYRANOSE;   W   61738

470   C17 H16 Br2 N2 O4   59400-14-5   B-30-2504-0   5,5'-DIBROMO-2,2'-(PROPANE-1,3-DIYLDIIMINO)BIS(BENZOIC AC
ID);   W   61739

470   C20 H48 Cl2 Si4   AG-131-190-0   Dichlorotetra-t-butyltetramethyltetrasilane;   W   124197

470   C21 H14 Cl4 O4   21505-32-8   NS--29463-0-0   1,3-Benzodioxole, 4,5,6,7-tetrachloro-2,2-bis(4-methoxyphen
yl)-;; Benzene, 1,2-[[bis(p-methoxyphenyl)methylene]dioxy]-3,4,5,6-tetrachloro-;   NBS   61740

470   C21 H26 O12   511-89-7   O-16-85-1   PLUMIERIDE;   W   61741

470   C21 H31 Br N2 O5   67705-71-9   KC-1978-564-0   O-BROMOPHENYL HYDRAZONE OF 3R, 4R-DIHYDROXY-5S-(2S,3S-EPO
XY-5S-HYDROXY-4S-METHYLHEXL)TETRAHYDROPYRAN-2S-YLACETONE;   W   61742

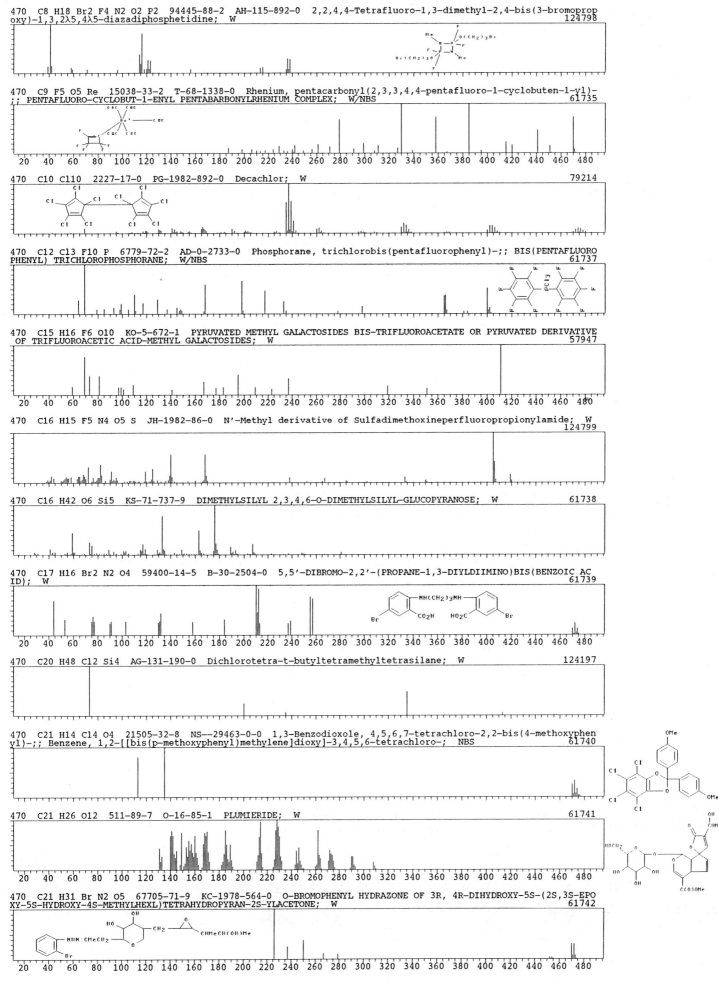

470 C22 H21 O4 Sb 1538-62-1 AG-116-205-3 Triphenylantimony diacetate; W   124801

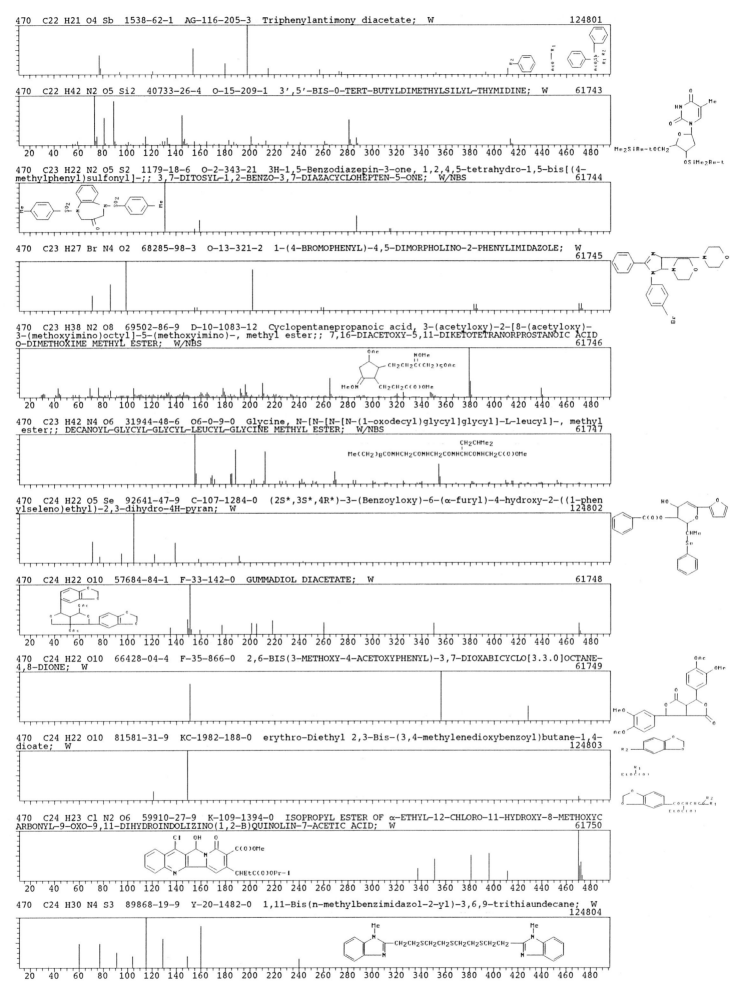

470 C22 H42 N2 O5 Si2 40733-26-4 O-15-209-1 3',5'-BIS-O-TERT-BUTYLDIMETHYLSILYL-THYMIDINE; W   61743

470 C23 H22 N2 O5 S2 1179-18-6 O-2-343-21 3H-1,5-Benzodiazepin-3-one, 1,2,4,5-tetrahydro-1,5-bis[(4-
methylphenyl)sulfonyl]-;; 3,7-DITOSYL-1,2-BENZO-3,7-DIAZACYCLOHEPTEN-5-ONE; W/NBS   61744

470 C23 H27 Br N4 O2 68285-98-3 O-13-321-2 1-(4-BROMOPHENYL)-4,5-DIMORPHOLINO-2-PHENYLIMIDAZOLE; W
                                                                                                            61745

470 C23 H38 N2 O8 69502-86-9 D-10-1083-12 Cyclopentanepropanoic acid, 3-(acetyloxy)-2-[8-(acetyloxy)-
3-(methoxyimino)octyl]-5-(methoxyimino)-, methyl ester;; 7,16-DIACETOXY-5,11-DIKETOTETRANORPROSTANOIC ACID
O-DIMETHOXIME METHYL ESTER; W/NBS   61746

470 C23 H42 N4 O6 31944-48-6 O6-0-9-0 Glycine, N-[N-[N-[N-(1-oxodecyl)glycyl]glycyl]-L-leucyl]-, methyl
ester;; DECANOYL-GLYCYL-GLYCYL-LEUCYL-GLYCINE METHYL ESTER; W/NBS   61747

470 C24 H22 O5 Se 92641-47-9 C-107-1284-0 (2S*,3S*,4R*)-3-(Benzoyloxy)-6-(α-furyl)-4-hydroxy-2-((1-phen
ylseleno)ethyl)-2,3-dihydro-4H-pyran; W   124802

470 C24 H22 O10 57684-84-1 F-33-142-0 GUMMADIOL DIACETATE; W   61748

470 C24 H22 O10 66428-04-4 F-35-866-0 2,6-BIS(3-METHOXY-4-ACETOXYPHENYL)-3,7-DIOXABICYCLO[3.3.0]OCTANE-
4,8-DIONE; W   61749

470 C24 H22 O10 81581-31-9 KC-1982-188-0 erythro-Diethyl 2,3-Bis-(3,4-methylenedioxybenzoyl)butane-1,4-
dioate; W   124803

470 C24 H23 Cl N2 O6 59910-27-9 K-109-1394-0 ISOPROPYL ESTER OF α-ETHYL-12-CHLORO-11-HYDROXY-8-METHOXYC
ARBONYL-9-OXO-9,11-DIHYDROINDOLIZINO(1,2-B)QUINOLIN-7-ACETIC ACID; W   61750

470 C24 H30 N4 S3 89868-19-9 Y-20-1482-0 1,11-Bis(n-methylbenzimidazol-2-yl)-3,6,9-trithiaundecane; W
                                                                                                          124804

470   C24 H39 I O   69101-83-3   C-100-7683-0   3α,5-CYCLO-6β-METHOXY-23-IODONORCHOLANE;   W   61751

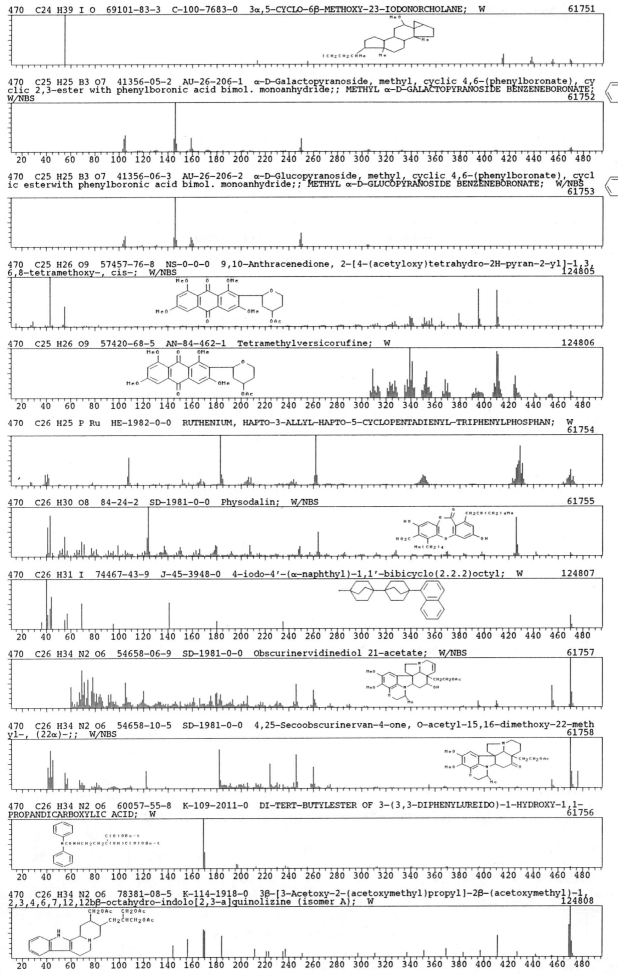

470   C25 H25 B3 O7   41356-05-2   AU-26-206-1   α-D-Galactopyranoside, methyl, cyclic 4,6-(phenylboronate), cy
clic 2,3-ester with phenylboronic acid bimol. monoanhydride;; METHYL α-D-GALACTOPYRANOSIDE BENZENEBORONATE;
W/NBS   61752

470   C25 H25 B3 O7   41356-06-3   AU-26-206-2   α-D-Glucopyranoside, methyl, cyclic 4,6-(phenylboronate), cycl
ic esterwith phenylboronic acid bimol. monoanhydride;; METHYL α-D-GLUCOPYRANOSIDE BENZENEBORONATE;   W/NBS
                                                                                                               61753

470   C25 H26 O9   57457-76-8   NS-0-0-0   9,10-Anthracenedione, 2-[4-(acetyloxy)tetrahydro-2H-pyran-2-yl]-1,3
6,8-tetramethoxy-, cis-;   W/NBS   124805

470   C25 H26 O9   57420-68-5   AN-84-462-1   Tetramethylversicorufine;   W   124806

470   C26 H25 P Ru   HE-1982-0-0   RUTHENIUM, HAPTO-3-ALLYL-HAPTO-5-CYCLOPENTADIENYL-TRIPHENYLPHOSPHAN;   W
                                                                                                               61754

470   C26 H30 O8   84-24-2   SD-1981-0-0   Physodalin;   W/NBS   61755

470   C26 H31 I   74467-43-9   J-45-3948-0   4-iodo-4'-(α-naphthyl)-1,1'-bibicyclo(2.2.2)octyl;   W   124807

470   C26 H34 N2 O6   54658-06-9   SD-1981-0-0   Obscurinervidinediol 21-acetate;   W/NBS   61757

470   C26 H34 N2 O6   54658-10-5   SD-1981-0-0   4,25-Secoobscurinervan-4-one, O-acetyl-15,16-dimethoxy-22-meth
yl-, (22α)-;;   W/NBS   61758

470   C26 H34 N2 O6   60057-55-8   K-109-2011-0   DI-TERT-BUTYLESTER OF 3-(3,3-DIPHENYLUREIDO)-1-HYDROXY-1,1-
PROPANDICARBOXYLIC ACID;   W   61756

470   C26 H34 N2 O6   78381-08-5   K-114-1918-0   3β-[3-Acetoxy-2-(acetoxymethyl)propyl]-2β-(acetoxymethyl)-1,
2,3,4,6,7,12,12bβ-octahydro-indolo[2,3-a]quinolizine (isomer A);   W   124808

470 C26 H34 N2 O6 78419-72-4 K-114-1918-0 3β-[3-Acetoxy-2-(acetoxymethyl)propyl]-2β-(acetoxymethyl)-1,
2,3,4,6,7,12,12bβ-octahydro-indolo[2,3-a]quinolizine (isomer B); W
124809

470 C26 H28 D6 N2 O6 78381-09-6 K-114-1918-0 3β-[3-Acetoxy-2(acetoxy dideuteromethyl)-3,3-dideutero pro
pyl]-2β-(acetoxymethyl)-1,2,3,4,6,7,12,12bβ-octahydroindolo[2,3-a]quinolizine; W
124810

470 C26 H38 N4 O4 18658-42-9 C-91-5621-1 Ceanothine C;; CEANOTHENE-C; W/NBS
61759

470 C27 H18 O8 39657-83-5 NS--29453-0-0 Difuro[2',3':5,6:3'',2'':7,8]perylo[1,12-def][1,3]dioxepin-8,
15-dione,10,11,12,13-tetrahydro-1,7-dihydroxy-10,13-dimethyl-;; NBS
61760

470 C27 H20 Cl2 N4 55816-21-2 SB-29-283-0 1,7-DI-(4-CHLOROPHENYL)-1,7-DIPHENYL-2,3,5,6-TETRAZOHEPTA-1,
3,6-TRIENEN*2,N*4-DI(4-CHLOROPHENYLPHENYLMETHYLENE)-FORMOHYDRAZIDE HYDRAZONE; W
61761

470 C27 H20 Cl2 N4 55816-20-1 SB-29-283-0 1,7-DI-(2-CHLOROPHENYL-2,3,5,6-TETRAZOHEPTA-1,3,6-TRIENEN*2,
N*4-DI(2-CHLOROPHENYLPHENYLMETHYLENE)-FORMOHYDRAZIDE HYDRAZONE; W
61762

470 C27 H26 Sn 79523-79-8 O-18-224-1 (2-Phenylpropyl)triphenyltin; W
124811

470 C27 H34 O7 B-34-1909-0 (2RS,4aSR,4bSR,8SR,8aSR,10SR,10aRS,12SR)-12-Benzoyloxy-10-hydroxy-8-methoxy
methyloxymethyl-2,3,4,4a,4b,5,6,7,8,8a,9,10-dodecahydro-1H-2,10a-ethanophenanthrene-8,4b-carbolactone; W
125281

470 C27 H42 O3 Si2 KO-2-268-4 4'-HYDROXYCANNABINOL, DI-TMS ETHER; W
61763

470 C27 H50 O6 55429-67-9 AA-0-2225-1 Eicosanoic acid, 2,3-bis(acetyloxy)propyl ester;; 1-ARACHIDYL-2,
3-DIACETIN; W/NBS
61765

470 C27 H50 O6 55429-68-0 AA-0-2224-2 Eicosanoic acid, 2-(acetyloxy)-1-[(acetyloxy)methyl]ethyl ester;;
2-ARACHIDYL-1,3-DIACETIN; W/NBS
61764

470 C27 H50 O6 538-23-8 EP-5343-0-0 Octanoic acid, 1,2,3-propanetriyl ester;; Glyceryl trioctanoate;
W/NBS
61766

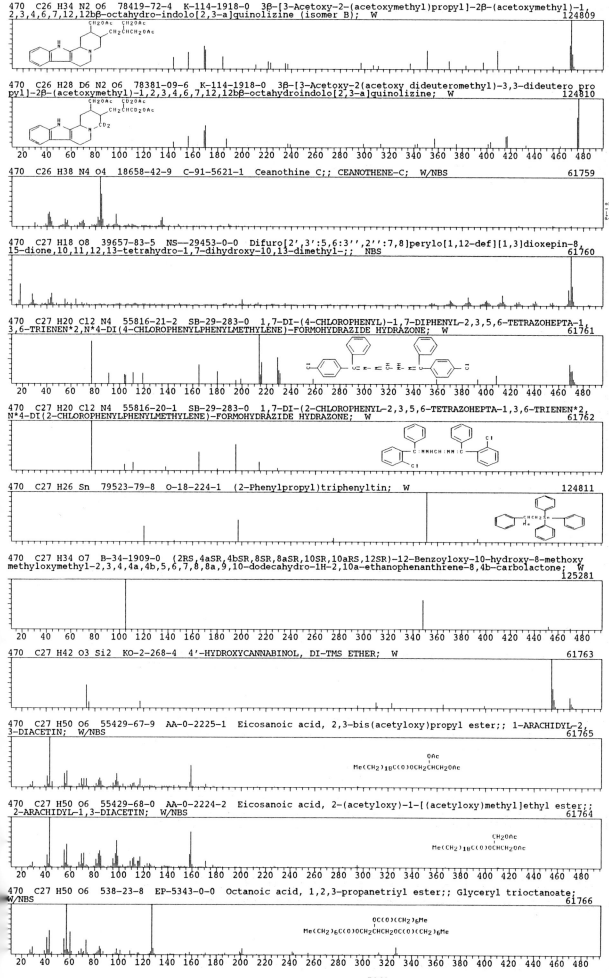

470 C28 H14 N4 S2 85693-93-2 K-116-989-0 [1]Benzothieno[2',3'-3,4]thieno[3'',2''-7,8]cycloocta[1,2-b:5,6-b']diquinoxaline; W   124812

470 C28 H22 O5 S 87712-59-2 J-49-84-0 α-(Tosyloxy)-9-(p-methoxybenzylidene)xanthene; W   124813

470 C28 H22 O7 69122-01-6 KC-1978-1035-0 DIMETHYL SPIRO 26-OXABICYCLO(2.2.1)HEXANE-5,3'-PHTHALIDE)-2,3-CIS-DICARBOXYLATE; W   61767

470 C28 H22 O7 69122-01-6 KC-1978-1035-0 DIMETHYL SPIRO(6-OXABICYCLO(2.1.1)-HEXANE-5,3'-PHTHALIDE)-2,3-TRANS-DICARBOXYLATE; W   61768

470 C28 H38 O4 S 59385-10-3 H-59-403-12 ALL-E-3,7-DIMETHYL-7-(PHENYLSULFONYL)-9-(2,6,6-TRIMETHYL-1-CYCLOHEXEN-1-YL)-2,4,8-NONATRIENYL-ACETATE; W   61769

470 C28 H38 O4 S 67689-07-0 J-43-4772-0 1-Acetoxy-3,7-dimethyl-5-(phenylsulfonyl)-9-(2,6,6-trimethylcyclohexen-1-yl)nona-2,6,8-triene; W   124814

470 C28 H38 O6 5119-48-2 J-4-8363-1 Withaferin A; W/NBS   61770

470 C28 H38 O6 63646-82-2 H-60-969-0 Furan, 2,2'-[1,2-ethanediylbis(oxy)]bis[tetrahydro-5-(2-methoxy-4-methylphenyl)-5-methyl-;; ETHYLENE GLYCOL DI(5-(2'-METHOXY-4'-METHYL-PHENYL)-5-METHYL-TETRAHYDROFURAN-2-YL) ETHER; W/NBS   61771

470 C28 H38 O6 61621-86-1 B-29-1986-0 6,6'-(TETRADECANE-1,14-DIYL)DI(2-METHOXY-1,4-BENZOQUINONE); W   134025

470 C28 H58 O3 Si 56784-04-4 OM-1981-284-0 Tetracosanoic acid, 2-[(trimethylsilyl)oxy]-, methyl ester;; METHYL 2-TMSOXYTETRACOSANOATE; W   61773

470 C30 H18 N2 S2 35461-38-2 AD-0-5854-0 3H,8H-Benzo[c]benzo[6,7]phenothiazino[4,3-h]phenothiazine;; 7,12-DIHYDRO-BENZO[4]PHENOTHIAZINO[3,4-C]BENZO[D]PHENOTHIAZINE; W/NBS   61774

470 C30 H21 F3 S 63072-27-5 Y-14-200-0 4-(P-TRIFLUOROMETHYLPHENYL)-2,4,6-TRIPHENYLTHIOPYRAN; W   61775

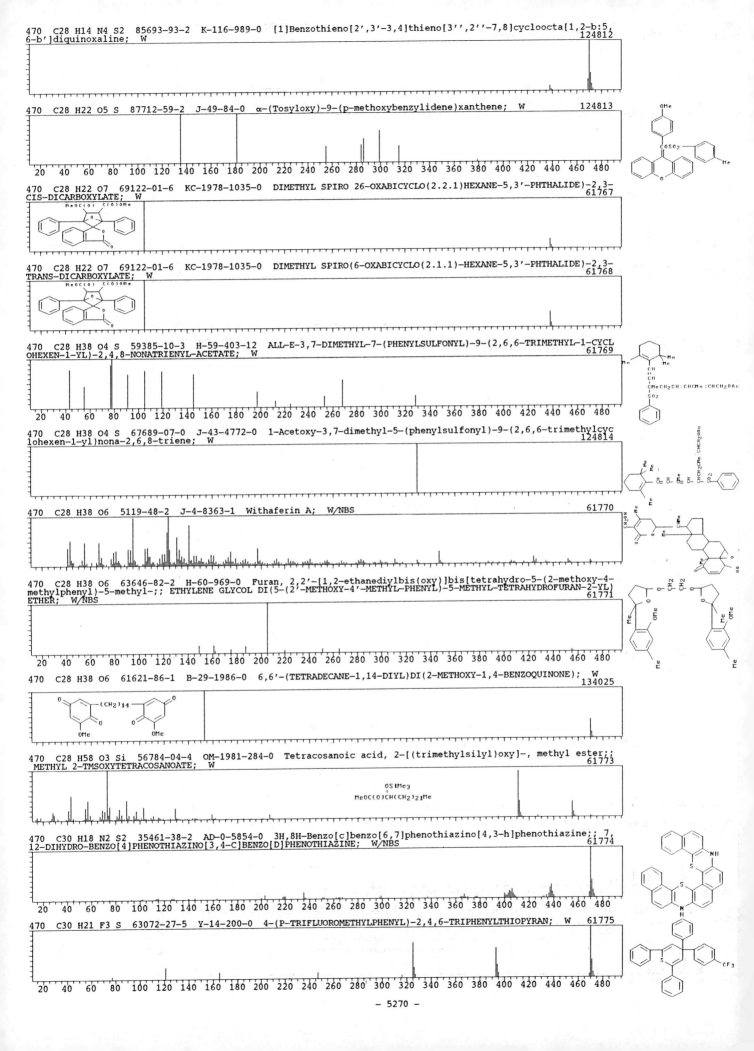

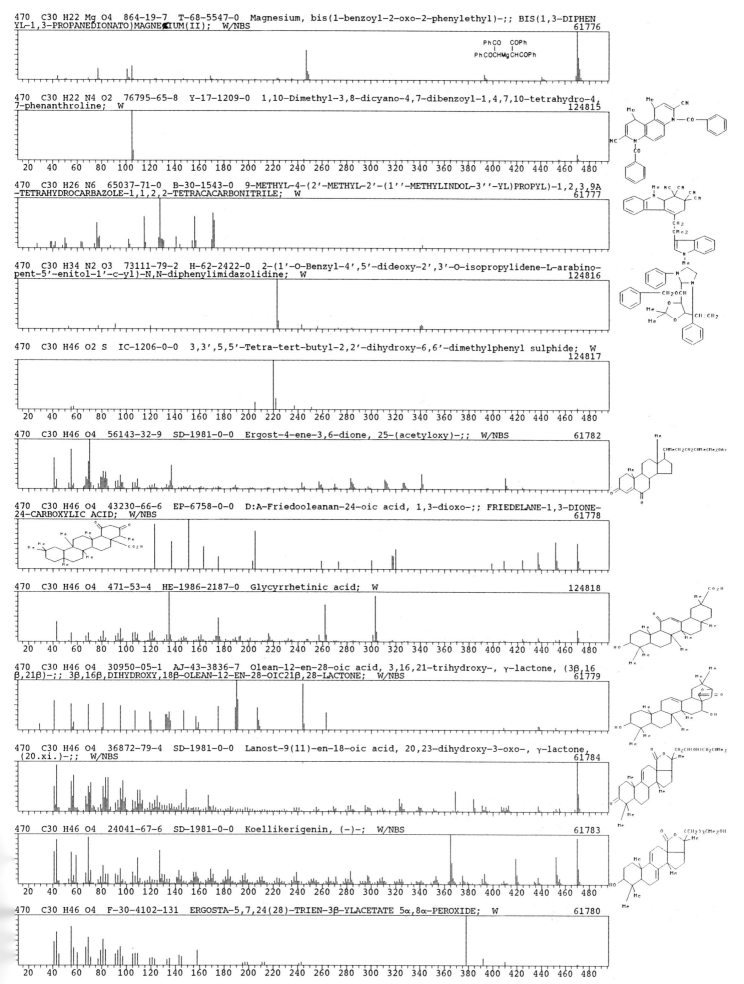

470 C30 H22 Mg O4 864-19-7 T-68-5547-0 Magnesium, bis(1-benzoyl-2-oxo-2-phenylethyl)-;; BIS(1,3-DIPHEN
YL-1,3-PROPANEDIONATO)MAGNESIUM(II); W/NBS   61776

PhCO COPh
PhCOCHMgCHCOPh

470 C30 H22 N4 O2 76795-65-8 Y-17-1209-0 1,10-Dimethyl-3,8-dicyano-4,7-dibenzoyl-1,4,7,10-tetrahydro-4,
7-phenanthroline; W   124815

470 C30 H26 N6 65037-71-0 B-30-1543-0 9-METHYL-4-(2'-METHYL-2'-(1''-METHYLINDOL-3''-YL)PROPYL)-1,2,3,9A
-TETRAHYDROCARBAZOLE-1,1,2,2-TETRACACARBONITRILE; W   61777

470 C30 H34 N2 O3 73111-79-2 H-62-2422-0 2-(1'-O-Benzyl-4',5'-dideoxy-2',3'-O-isopropylidene-L-arabino-
pent-5'-enitol-1'-c-yl)-N,N-diphenylimidazolidine; W   124816

470 C30 H46 O2 S IC-1206-0-0 3,3',5,5'-Tetra-tert-butyl-2,2'-dihydroxy-6,6'-dimethylphenyl sulphide; W
                                                                          124817

470 C30 H46 O4 56143-32-9 SD-1981-0-0 Ergost-4-ene-3,6-dione, 25-(acetyloxy)-;; W/NBS   61782

470 C30 H46 O4 43230-66-6 EP-6758-0-0 D:A-Friedooleanan-24-oic acid, 1,3-dioxo-;; FRIEDELANE-1,3-DIONE-
24-CARBOXYLIC ACID; W/NBS   61778

470 C30 H46 O4 471-53-4 HE-1986-2187-0 Glycyrrhetinic acid; W   124818

470 C30 H46 O4 30950-05-1 AJ-43-3836-7 Olean-12-en-28-oic acid, 3,16,21-trihydroxy-, γ-lactone, (3β,16
β,21β)-;; 3β,16β,DIHYDROXY,18β-OLEAN-12-EN-OIC21β,28-LACTONE; W/NBS   61779

470 C30 H46 O4 36872-79-4 SD-1981-0-0 Lanost-9(11)-en-18-oic acid, 20,23-dihydroxy-3-oxo-, γ-lactone,
(20.xi.)-;; W/NBS   61784

470 C30 H46 O4 24041-67-6 SD-1981-0-0 Koellikerigenin, (-)-; W/NBS   61783

470 C30 H46 O4 F-30-4102-131 ERGOSTA-5,7,24(28)-TRIEN-3β-YLACETATE 5α,8α-PEROXIDE; W   61780

- 5271 -

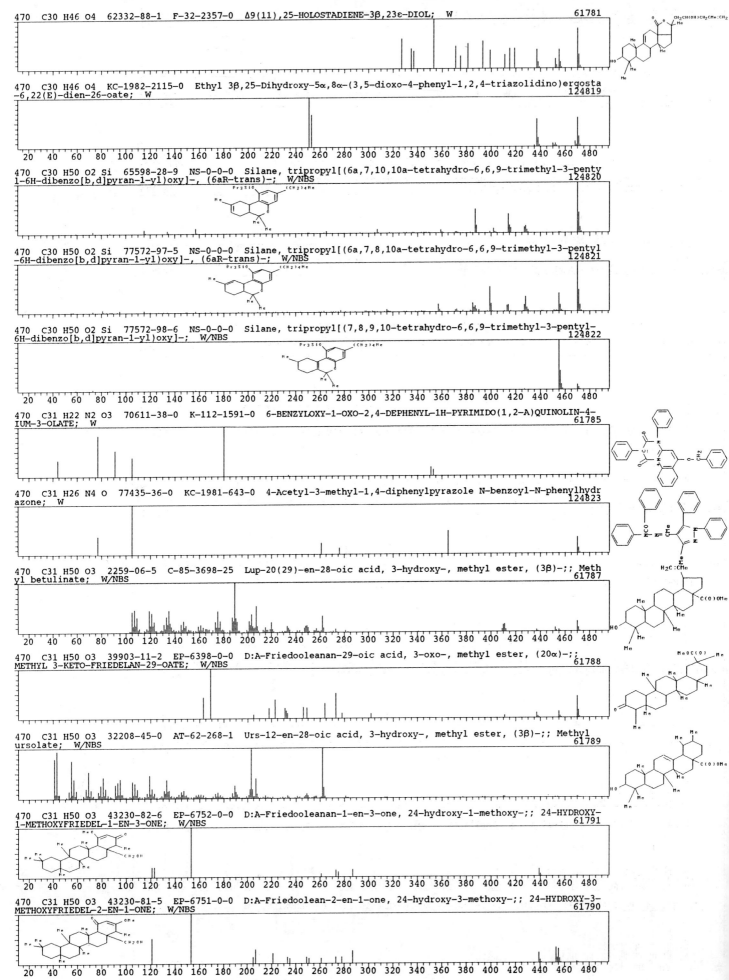

470   C30 H46 O4   62332-88-1   F-32-2357-0   Δ9(11),25-HOLOSTADIENE-3β,23ε-DIOL;   W   61781

470   C30 H46 O4   KC-1982-2115-0   Ethyl 3β,25-Dihydroxy-5α,8α-(3,5-dioxo-4-phenyl-1,2,4-triazolidino)ergosta
-6,22(E)-dien-26-oate;   W   124819

470   C30 H50 O2 Si   65598-28-9   NS-0-0-0   Silane, tripropyl[(6a,7,10,10a-tetrahydro-6,6,9-trimethyl-3-penty
l-6H-dibenzo[b,d]pyran-1-yl)oxy]-, (6aR-trans)-;   W/NBS   124820

470   C30 H50 O2 Si   77572-97-5   NS-0-0-0   Silane, tripropyl[(6a,7,8,10a-tetrahydro-6,6,9-trimethyl-3-pentyl
-6H-dibenzo[b,d]pyran-1-yl)oxy]-, (6aR-trans)-;   W/NBS   124821

470   C30 H50 O2 Si   77572-98-6   NS-0-0-0   Silane, tripropyl[(7,8,9,10-tetrahydro-6,6,9-trimethyl-3-pentyl-
6H-dibenzo[b,d]pyran-1-yl)oxy]-;   W/NBS   124822

470   C31 H22 N2 O3   70611-38-0   K-112-1591-0   6-BENZYLOXY-1-OXO-2,4-DEPHENYL-1H-PYRIMIDO(1,2-A)QUINOLIN-4-
IUM-3-OLATE;   W   61785

470   C31 H26 N4 O   77435-36-0   KC-1981-643-0   4-Acetyl-3-methyl-1,4-diphenylpyrazole N-benzoyl-N-phenylhydr
azone;   W   124823

470   C31 H50 O3   2259-06-5   C-85-3698-25   Lup-20(29)-en-28-oic acid, 3-hydroxy-, methyl ester, (3β)-;;   Meth
yl betulinate;   W/NBS   61787

470   C31 H50 O3   39903-11-2   EP-6398-0-0   D:A-Friedooleanan-29-oic acid, 3-oxo-, methyl ester, (20α)-;;
METHYL 3-KETO-FRIEDELAN-29-OATE;   W/NBS   61788

470   C31 H50 O3   32208-45-0   AT-62-268-1   Urs-12-en-28-oic acid, 3-hydroxy-, methyl ester, (3β)-;;   Methyl
ursolate;   W/NBS   61789

470   C31 H50 O3   43230-82-6   EP-6752-0-0   D:A-Friedooleanan-1-en-3-one, 24-hydroxy-1-methoxy-;;   24-HYDROXY-
1-METHOXYFRIEDEL-1-EN-3-ONE;   W/NBS   61791

470   C31 H50 O3   43230-81-5   EP-6751-0-0   D:A-Friedooolean-2-en-1-one, 24-hydroxy-3-methoxy-;;   24-HYDROXY-3-
METHOXYFRIEDEL-2-EN-1-ONE;   W/NBS   61790

470   C31 H50 O3   AT-64-3181-2   METHYL COMMATE B;   W   61786

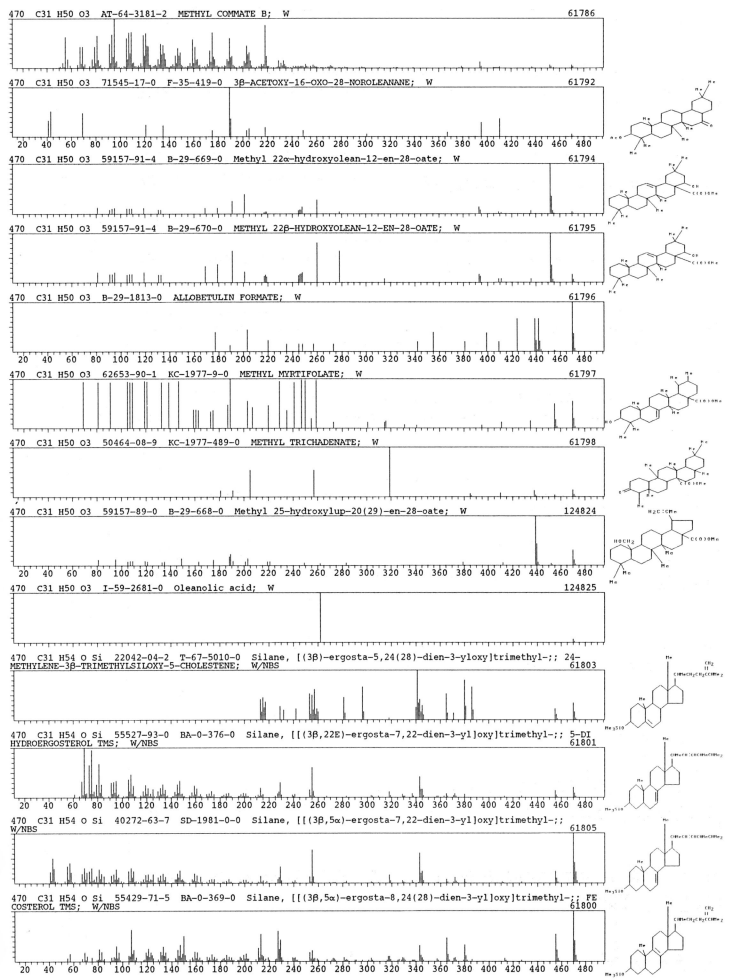

470   C31 H50 O3   71545-17-0   F-35-419-0   3β-ACETOXY-16-OXO-28-NOROLEANANE;   W   61792

470   C31 H50 O3   59157-91-4   B-29-669-0   Methyl 22α-hydroxyolean-12-en-28-oate;   W   61794

470   C31 H50 O3   59157-91-4   B-29-670-0   METHYL 22β-HYDROXYOLEAN-12-EN-28-OATE;   W   61795

470   C31 H50 O3   B-29-1813-0   ALLOBETULIN FORMATE;   W   61796

470   C31 H50 O3   62653-90-1   KC-1977-9-0   METHYL MYRTIFOLATE;   W   61797

470   C31 H50 O3   50464-08-9   KC-1977-489-0   METHYL TRICHADENATE;   W   61798

470   C31 H50 O3   59157-89-0   B-29-668-0   Methyl 25-hydroxylup-20(29)-en-28-oate;   W   124824

470   C31 H50 O3   I-59-2681-0   Oleanolic acid;   W   124825

470   C31 H54 O Si   22042-04-2   T-67-5010-0   Silane, [(3β)-ergosta-5,24(28)-dien-3-yloxy]trimethyl-;; 24-
METHYLENE-3β-TRIMETHYLSILOXY-5-CHOLESTENE;   W/NBS   61803

470   C31 H54 O Si   55527-93-0   BA-0-376-0   Silane, [[(3β,22E)-ergosta-7,22-dien-3-yl]oxy]trimethyl-;; 5-DI
HYDROERGOSTEROL TMS;   W/NBS   61801

470   C31 H54 O Si   40272-63-7   SD-1981-0-0   Silane, [[(3β,5α)-ergosta-7,22-dien-3-yl]oxy]trimethyl-;;
W/NBS   61805

470   C31 H54 O Si   55429-71-5   BA-0-369-0   Silane, [[(3β,5α)-ergosta-8,24(28)-dien-3-yl]oxy]trimethyl-;; FE
COSTEROL TMS;   W/NBS   61800

470 C31 H54 O Si 55429-69-1 BA-0-381-0 Silane, trimethyl[[(3β,4α,5α)-4-methylcholesta-8,24-dien-3-yl]oxy]-;; 4α-METHYLZYMOSTEROL TMS; W/NBS
61802

470 C31 H54 O Si 69688-07-9 NS--29428-0-0 Silane, trimethyl[[(3β,5E,7E,22E)-9,10-secoergosta-5,7,22-trien-3-yl]oxy]-;; NBS
61804

470 C31 H54 O Si 55429-70-4 BA-0-366-0 Silane, trimethyl[[(3β,4α)-4-methylcholesta-8(14),24-dien-3-yl]oxy]-;; 4α-METHYLCHOLESTA-8(14),24-DIEN-3β-OL TMS; W/NBS
61799

470 C32 H22 O4 61549-17-5 NS--29427-0-0 2-Anthracenecarboxylic acid, 1,2-ethanediyl ester;; NBS 61806

470 C32 H22 O4 72193-17-0 SB-33-406-0 6,12-BIS(2-HYDROXYBENYYL)-BENZO(1,2-B:4,5-B')BISBENZOFURAN; W
124826

470 C32 H22 O4 92012-85-6 J-49-4171-0 4-(Acetyloxy)-3-(9-anthracenyl)-1,4-diphenyl-3-butene-1,2-dione; W
124827

470 C32 H22 S2 84264-12-0 H-65-1340-0 x-Methyl-2-[5'(9''-fluorenyliden)bicyclo[4.4.1]undeca-3,6,8,10'-tetraen-2'-yliden]-1,3-benzodithiol; W
124828

470 C32 H26 N2 O2 68803-52-1 B-31-2036-0 4-(2-OXO-1,5-DIPHENYL-2,3-DIHYDROPYRROL-3-YLIDENE)-N,4-DIPHENYLBUTANAMIDE; W
61808

470 C32 H26 N2 O2 IC-1207-0-0 1,5-Diphenyl-3-(α,N-diphenyl-carbamoyl-propylidene)pyrrol-2-one; W 124829

470 C32 H54 O2 55724-21-5 SD-1981-0-0 Ergost-5-en-3-ol, 22,23-dimethyl-, acetate, (3β)-;; W/NBS 61824

470 C32 H54 O2 6252-45-5 SD-1981-0-0 2H-Pyran, 2-[[(3β)-cholest-5-en-3-yl]oxy]tetrahydro-;; Cholesterol tetrahydropyran-2-yl ether; W/NBS
61825

470 C32 H54 O2 54498-61-2 SD-1981-0-0 Cholestan-3-one, 2-(1-methylethylidene)-, cyclic 1,2-ethanediyl acetal, (5α)-;; W/NBS
61821

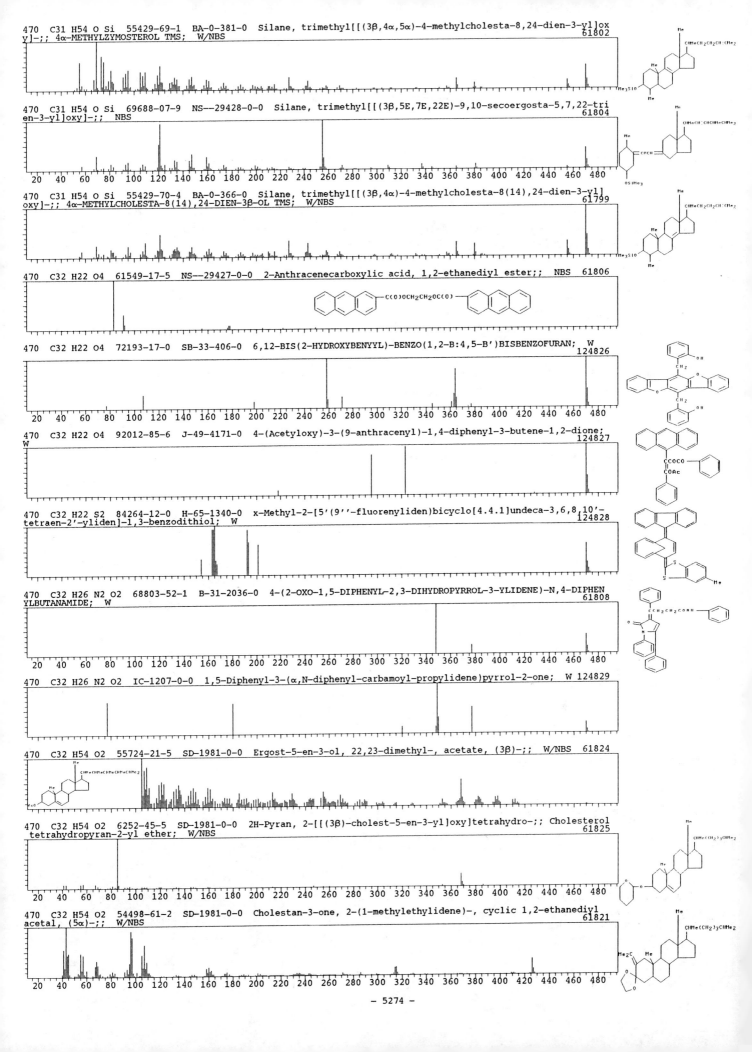

470 C32 H54 O2 54498-62-3 SD-1981-0-0 Cholestan-3-one, 2-(1-methylethenyl)-, cyclic 1,2-ethanediyl acetal, (5α)-;; W/NBS

61822

470 C32 H54 O2 55637-47-3 AD-0-813-0 Lanost-24-en-3-ol, acetate, (3β,13α,14β,17α)-;; DIHYDROBUTYROSPERMYL ACETATE; W/NBS

61815

20 40 60 80 100 120 140 160 180 200 220 240 260 280 300 320 340 360 380 400 420 440 460 480

470 C32 H54 O2 56588-24-0 SD-1981-0-0 D:A-Friedooleanan-7-ol, acetate, (7α)-;; W/NBS

61820

470 C32 H54 O2 54498-65-6 SD-1981-0-0 Lupan-3-one, cyclic 1,2-ethanediyl acetal;; W/NBS

61823

470 C32 H54 O2 4488-99-7 AD-0-812-0 Lanost-7-en-3-ol, acetate, (3β)-;; LANOST-7-ENYL ACETATE; W/NBS

61813

20 40 60 80 100 120 140 160 180 200 220 240 260 280 300 320 340 360 380 400 420 440 460 480

470 C32 H54 O2 55515-27-0 AD-0-814-0 Lanost-7-en-3-ol, acetate, (3β,9β,13α,14β,17α)-;; 9β-EUPH-7-ENYL ACETATE; W/NBS

61814

470 C32 H54 O2 38602-32-3 AD-0-811-0 Dammar-13(17)-en-3-ol, acetate, (3β)-;; Isoeuphenyl acetate; W/NBS

61816

470 C32 H54 O2 55515-26-9 AD-0-816-0 Lanost-9(11)-en-3-ol, acetate, (3β,8α,13α,14β,17α)-;; 8α-EUPH-9(11)-ENYL ACETATE; W/NBS

61812

20 40 60 80 100 120 140 160 180 200 220 240 260 280 300 320 340 360 380 400 420 440 460 480

470 C32 H54 O2 1180-88-7 SD-1981-0-0 Lanost-9(11)-en-3-ol, acetate, (3β)-;; Lanost-9(11)-en-3β-yl acetate; W/NBS

61818

470 C32 H54 O2 4575-74-0 SD-1981-0-0 9,19-Cyclolanostan-3-ol, acetate, (3β)-;; Cycloartanyl acetate; W/NBS

61819

470 C32 H54 O2 1724-19-2 AD-0-809-0 Lanost-8-en-3-ol, acetate, (3β)-;; LANOST-8-ENYL ACETATE; W 61809

20 40 60 80 100 120 140 160 180 200 220 240 260 280 300 320 340 360 380 400 420 440 460 480

470 C32 H54 O2 38602-31-2 AD-0-810-0 Lanost-8-en-3-ol, acetate, (3β,13α,14β,17α)-;; EUPH-8-ENYL ACETATE; W/NBS

61810

20 40 60 80 100 120 140 160 180 200 220 240 260 280 300 320 340 360 380 400 420 440 460 480

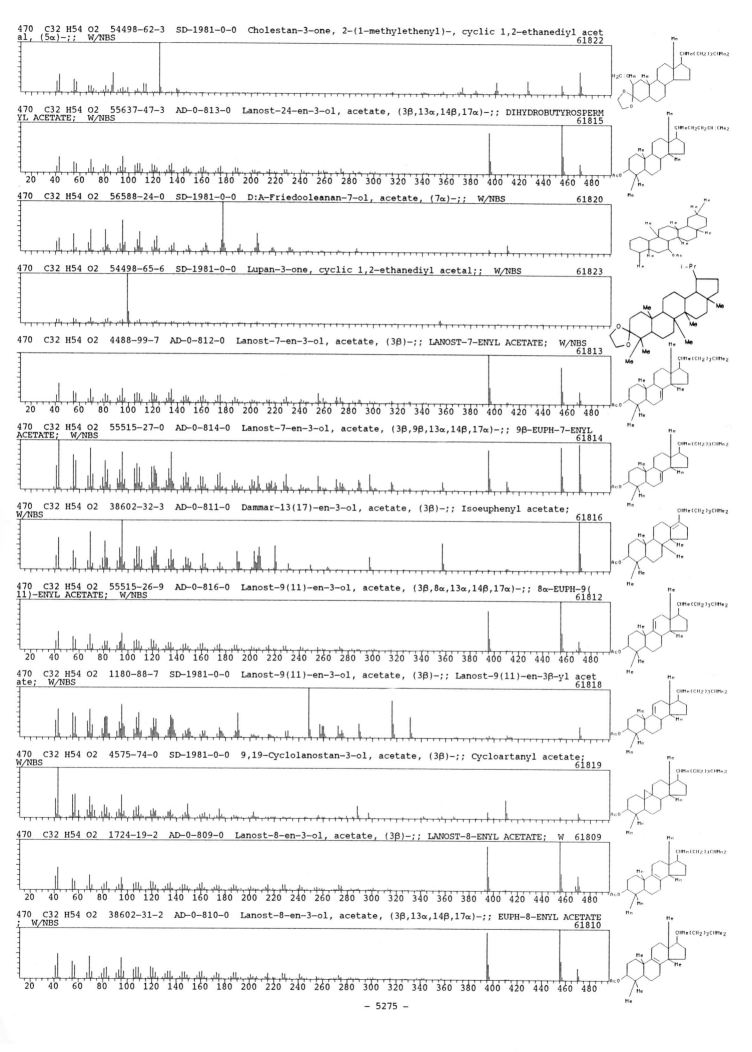

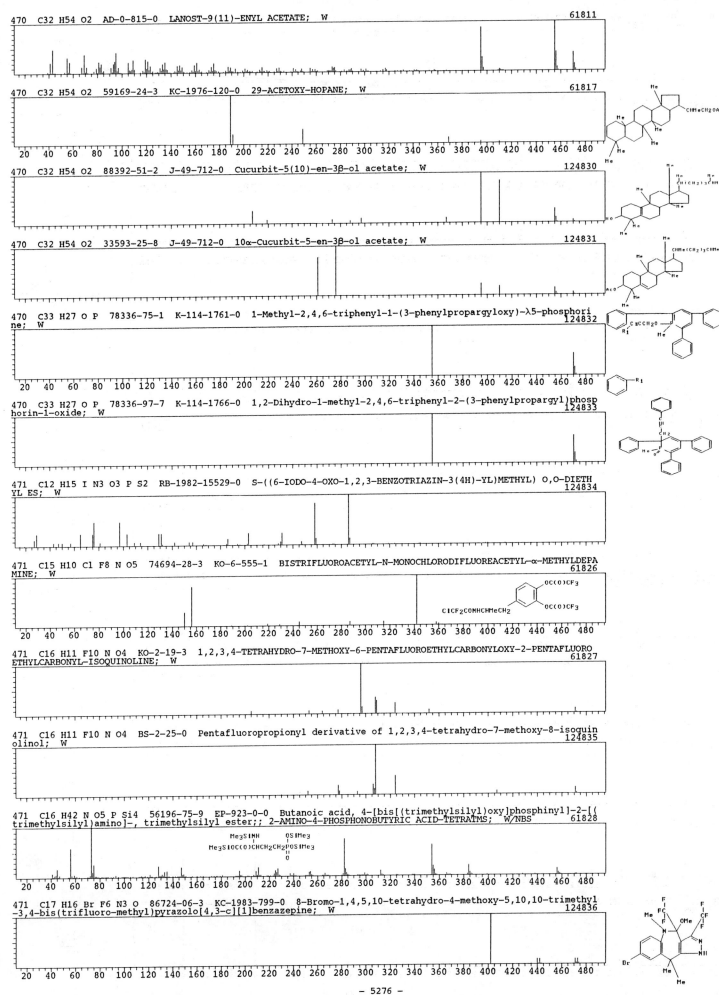

470   C32 H54 O2   AD-0-815-0   LANOST-9(11)-ENYL ACETATE; W   61811

470   C32 H54 O2   59169-24-3   KC-1976-120-0   29-ACETOXY-HOPANE; W   61817

20 40 60 80 100 120 140 160 180 200 220 240 260 280 300 320 340 360 380 400 420 440 460 480

470   C32 H54 O2   88392-51-2   J-49-712-0   Cucurbit-5(10)-en-3β-ol acetate; W   124830

470   C32 H54 O2   33593-25-8   J-49-712-0   10α-Cucurbit-5-en-3β-ol acetate; W   124831

470   C33 H27 O P   78336-75-1   K-114-1761-0   1-Methyl-2,4,6-triphenyl-1-(3-phenylpropargyloxy)-λ5-phosphori
ne; W   124832

470   C33 H27 O P   78336-97-7   K-114-1766-0   1,2-Dihydro-1-methyl-2,4,6-triphenyl-2-(3-phenylpropargyl)phosp
horin-1-oxide; W   124833

471   C12 H15 I N3 O3 P S2   RB-1982-15529-0   S-((6-IODO-4-OXO-1,2,3-BENZOTRIAZIN-3(4H)-YL)METHYL) O,O-DIETH
YL ES; W   124834

471   C15 H10 Cl F8 N O5   74694-28-3   KO-6-555-1   BISTRIFLUOROACETYL-N-MONOCHLORODIFLUOREACETYL-α-METHYLDEPA
MINE; W   61826

20 40 60 80 100 120 140 160 180 200 220 240 260 280 300 320 340 360 380 400 420 440 460 480

471   C16 H11 F10 N O4   KO-2-19-3   1,2,3,4-TETRAHYDRO-7-METHOXY-6-PENTAFLUOROETHYLCARBONYLOXY-2-PENTAFLUORO
ETHYLCARBONYL-ISOQUINOLINE; W   61827

471   C16 H11 F10 N O4   BS-2-25-0   Pentafluoropropionyl derivative of 1,2,3,4-tetrahydro-7-methoxy-8-isoquin
olinol; W   124835

471   C16 H42 N O5 P Si4   56196-75-9   EP-923-0-0   Butanoic acid, 4-[bis[(trimethylsilyl)oxy]phosphinyl]-2-[(
trimethylsilyl)amino]-, trimethylsilyl ester;; 2-AMINO-4-PHOSPHONOBUTYRIC ACID-TETRATMS; W/NBS   61828

20 40 60 80 100 120 140 160 180 200 220 240 260 280 300 320 340 360 380 400 420 440 460 480

471   C17 H16 Br F6 N3 O   86724-06-3   KC-1983-799-0   8-Bromo-1,4,5,10-tetrahydro-4-methoxy-5,10,10-trimethyl
-3,4-bis(trifluoro-methyl)pyrazolo[4,3-c][1]benzazepine; W   124836

20 40 60 80 100 120 140 160 180 200 220 240 260 280 300 320 340 360 380 400 420 440 460 480

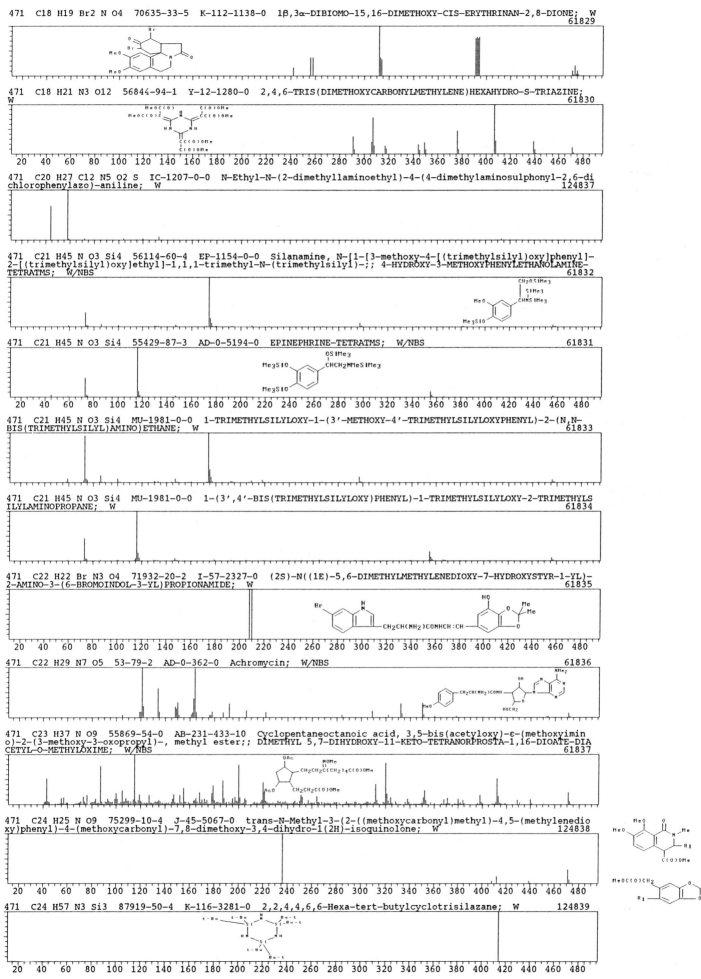

471   C18 H19 Br2 N O4   70635-33-5   K-112-1138-0   1β,3α-DIBIOMO-15,16-DIMETHOXY-CIS-ERYTHRINAN-2,8-DIONE;   W
61829

471   C18 H21 N3 O12   56844-94-1   Y-12-1280-0   2,4,6-TRIS(DIMETHOXYCARBONYLMETHYLENE)HEXAHYDRO-S-TRIAZINE;
W
61830

20   40   60   80   100   120   140   160   180   200   220   240   260   280   300   320   340   360   380   400   420   440   460   480

471   C20 H27 Cl2 N5 O2 S   IC-1207-0-0   N-Ethyl-N-(2-dimethyllaminoethyl)-4-(4-dimethylaminosulphonyl-2,6-di
chlorophenylazo)-aniline;   W   124837

471   C21 H45 N O3 Si4   56114-60-4   EP-1154-0-0   Silanamine, N-[1-[methoxy-4-[(trimethylsilyl)oxy]phenyl]-
2-[(trimethylsilyl)oxy]ethyl]-1,1,1-trimethyl-N-(trimethylsilyl)-;;   4-HYDROXY-3-METHOXYPHENYLETHANOLAMINE-
TETRATMS;   W/NBS   61832

471   C21 H45 N O3 Si4   55429-87-3   AD-0-5194-0   EPINEPHRINE-TETRATMS;   W/NBS   61831

20   40   60   80   100   120   140   160   180   200   220   240   260   280   300   320   340   360   380   400   420   440   460   480

471   C21 H45 N O3 Si4   MU-1981-0-0   1-TRIMETHYLSILYLOXY-1-(3'-METHOXY-4'-TRIMETHYLSILYLOXYPHENYL)-2-(N,N-
BIS(TRIMETHYLSILYL)AMINO)ETHANE;   W   61833

471   C21 H45 N O3 Si4   MU-1981-0-0   1-(3',4'-BIS(TRIMETHYLSILYLOXY)PHENYL)-1-TRIMETHYLSILYLOXY-2-TRIMETHYLS
ILYLAMINOPROPANE;   W   61834

471   C22 H22 Br N3 O4   71932-20-2   I-57-2327-0   (2S)-N((1E)-5,6-DIMETHYLMETHYLENEDIOXY-7-HYDROXYSTYR-1-YL)-
2-AMINO-3-(6-BROMOINDOL-3-YL)PROPIONAMIDE;   W   61835

20   40   60   80   100   120   140   160   180   200   220   240   260   280   300   320   340   360   380   400   420   440   460   480

471   C22 H29 N7 O5   53-79-2   AD-0-362-0   Achromycin;   W/NBS   61836

471   C23 H37 N O9   55869-54-0   AB-231-433-10   Cyclopentaneoctanoic acid, 3,5-bis(acetyloxy)-ε-(methoxyimin
o)-2-(3-methoxy-3-oxopropyl)-, methyl ester;;   DIMETHYL 5,7-DIHYDROXY-11-KETO-TETRANORPROSTA-1,16-DIOATE-DIA
CETYL-O-METHYLOXIME;   W/NBS   61837

471   C24 H25 N O9   75299-10-4   J-45-5067-0   trans-N-Methyl-3-(2-((methoxycarbonyl)methyl)-4,5-(methylenedio
xy)phenyl)-4-(methoxycarbonyl)-7,8-dimethoxy-3,4-dihydro-1(2H)-isoquinolone;   W   124838

20   40   60   80   100   120   140   160   180   200   220   240   260   280   300   320   340   360   380   400   420   440   460   480

471   C24 H57 N3 Si3   87919-50-4   K-116-3281-0   2,2,4,4,6,6-Hexa-tert-butylcyclotrisilazane;   W   124839

20   40   60   80   100   120   140   160   180   200   220   240   260   280   300   320   340   360   380   400   420   440   460   480

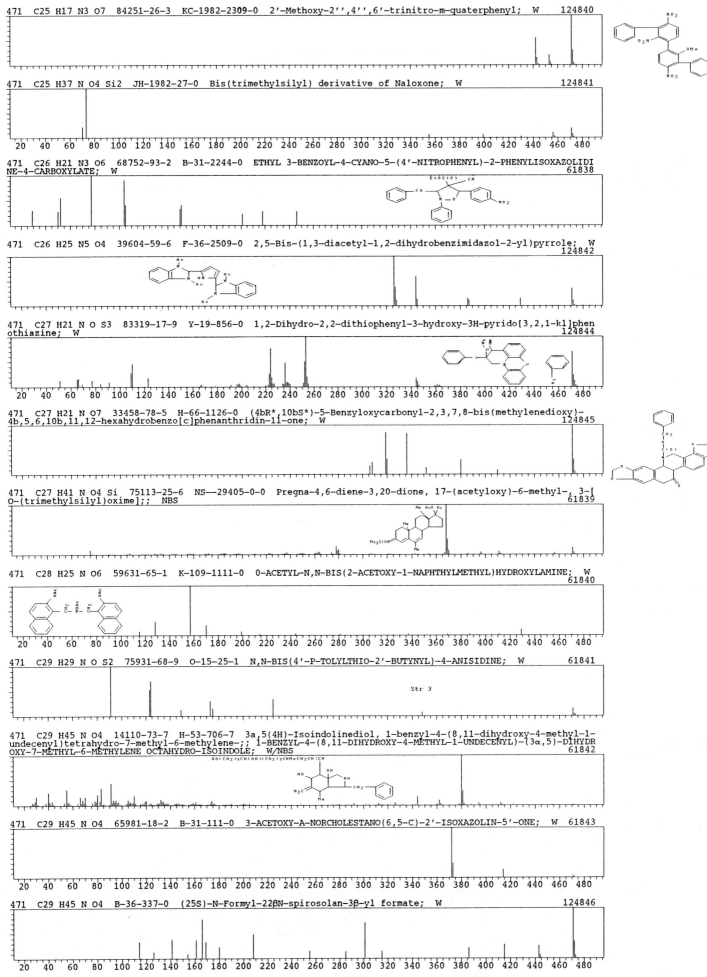

471   C25 H17 N3 O7   84251-26-3   KC-1982-2309-0   2'-Methoxy-2'',4'',6'-trinitro-m-quaterphenyl;   W   124840

471   C25 H37 N O4 Si2   JH-1982-27-0   Bis(trimethylsilyl) derivative of Naloxone;   W   124841

471   C26 H21 N3 O6   68752-93-2   B-31-2244-0   ETHYL 3-BENZOYL-4-CYANO-5-(4'-NITROPHENYL)-2-PHENYLISOXAZOLIDINE-4-CARBOXYLATE;   W   61838

471   C26 H25 N5 O4   39604-59-6   F-36-2509-0   2,5-Bis-(1,3-diacetyl-1,2-dihydrobenzimidazol-2-yl)pyrrole;   W   124842

471   C27 H21 N O S3   83319-17-9   Y-19-856-0   1,2-Dihydro-2,2-dithiophenyl-3-hydroxy-3H-pyrido[3,2,1-kl]phenothiazine;   W   124844

471   C27 H21 N O7   33458-78-5   H-66-1126-0   (4bR*,10bS*)-5-Benzyloxycarbonyl-2,3,7,8-bis(methylenedioxy)-4b,5,6,10b,11,12-hexahydrobenzo[c]phenanthridin-11-one;   W   124845

471   C27 H41 N O4 Si   75113-25-6   NS--29405-0-0   Pregna-4,6-diene-3,20-dione, 17-(acetyloxy)-6-methyl-, 3-[O-(trimethylsilyl)oxime];;   NBS   61839

471   C28 H25 N O6   59631-65-1   K-109-1111-0   O-ACETYL-N,N-BIS(2-ACETOXY-1-NAPHTHYLMETHYL)HYDROXYLAMINE;   W   61840

471   C29 H29 N O S2   75931-68-9   O-15-25-1   N,N-BIS(4'-P-TOLYLTHIO-2'-BUTYNYL)-4-ANISIDINE;   W   61841

Str 3

471   C29 H45 N O4   14110-73-7   H-53-706-7   3a,5(4H)-Isoindolinediol, 1-benzyl-4-(8,11-dihydroxy-4-methyl-1-undecenyl)tetrahydro-7-methyl-6-methylene-;; 1-BENZYL-4-(8,11-DIHYDROXY-4-METHYL-1-UNDECENYL)-(3α,5)-DIHYDROXY-7-METHYL-6-METHYLENE OCTAHYDRO-ISOINDOLE;   W/NBS   61842

471   C29 H45 N O4   65981-18-2   B-31-111-0   3-ACETOXY-A-NORCHOLESTANO(6,5-C)-2'-ISOXAZOLIN-5'-ONE;   W   61843

471   C29 H45 N O4   B-36-337-0   (25S)-N-Formyl-22βN-spirosolan-3β-yl formate;   W   124846

471   C30 H37 N3 O2   72363-45-2   H-62-1517-0   Quinoxaline derivative of Aspochalasin A;   W   124847

471   C32 H25 N O3   75155-53-2   KC-1983-463-0   1,3,-diphenylbenz[f]inden-2-one-trimethylmaleimide adduct;   W   124848

471   C32 H26 N O P   67464-45-3   J-43-4272-0   1a,9b-Dihydrophenanthr[9,10-b]azirin-1-yl)triphenylphosphonium
hydroxide;   W   124849

471   C33 H33 N3   88129-03-7   AH-114-731-0   3,4,6,7,8,9-Hexahydro-2,8-diphenyl-4,6-bis(p-tolyl)-2H-pyrimido[
1,2-a]pyrimidin-9a(1H)-yliumchloride;   W   124850

471   C33 H45 N O   38389-30-9   AD-0-3691-0   1'H-Cholesta-2,4-dieno[3,2-b]indol-6-one;;   6-OXO-CHOLEST-4-ENO(
3,2-B)INDOLE;   W/NBS   61844

471   C34 H49 N   55429-64-6   AD-0-3732-0   1'H-Cholesta-2,5-dieno[3,2-b]indole, 1'-methyl-;;   CHOLEST-5-ENO(3,
2-B)N-METHYLINDOLE;   W/NBS   61845

472   C6 H3 I3 O   609-23-4   NS--29358-0-0   Phenol, 2,4,6-triiodo-;;   2,4,6-Triiodophenol;   NBS   61846

472   C6 H3 I3 O   RB-1982-14867-0   TRIIODOPHENOL;   W   124851

472   C10 H2 Cl10   55570-84-8   DT-0-84-1   1,3,4-Metheno-1H-cyclobuta[cd]pentalene, 1a,2,2,3,3a,4,5,5,5a,6-de
cachlorooctahydro-;;   1,3,4-Metheno-1H-cyclobuta[cd]pentalene, 1,1a,2,2,3,4,5,5,5a,5b-decachlorooctahydro-;;
W/NBS   61847

472   C10 H2 Cl10   RB-1982-14204-0   DECACHLORO-4,7-METHANOINDENE;   W   124852

472   C10 H2 Cl10   RB-1982-14227-0   PENTACYCLODECACHLORODECANE;   W   124853

472   C10 H12 Cl2 Fe2 O6 P2   20274-11-7   AD-0-1594-0   Iron, hexacarbonyldichlorobis[.mu.-(dimethylphosphin
o)]di-;;   .mu.-TETRAMETHYLDIPHOSPHINE-BIS(CHLOROTRICARBONYLIRON);   W/NBS   61848

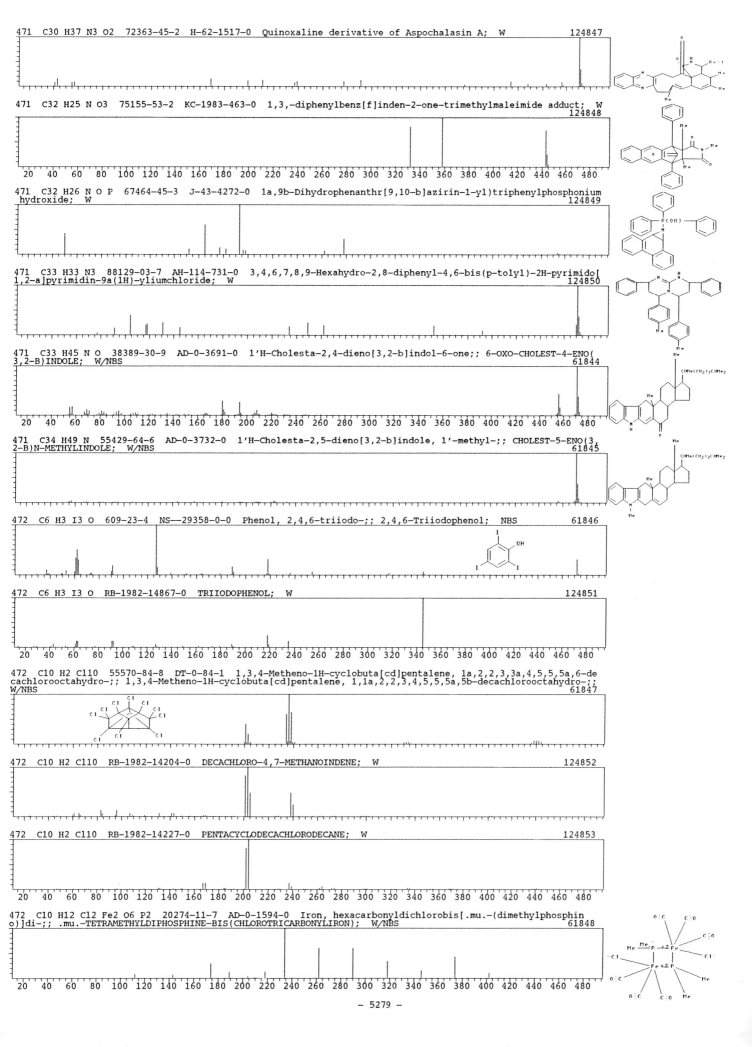

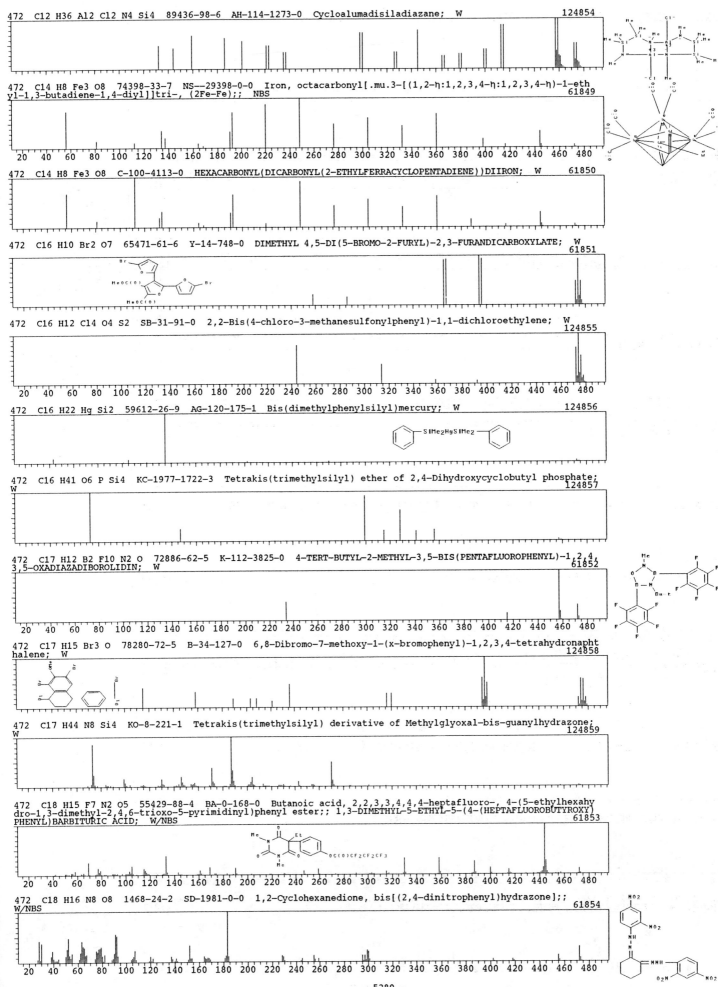

472 C12 H36 Al2 Cl2 N4 Si4 89436-98-6 AH-114-1273-0 Cycloalumadisiladiazane; W   124854

472 C14 H8 Fe3 O8 74398-33-7 NS--29398-0-0 Iron, octacarbonyl[.mu.3-[(1,2-η:1,2,3,4-η:1,2,3,4-η)-1-eth
yl-1,3-butadiene-1,4-diyl]]tri-, (2Fe-Fe);; NBS   61849

472 C14 H8 Fe3 O8 C-100-4113-0 HEXACARBONYL(DICARBONYL(2-ETHYLFERRACYCLOPENTADIENE))DIIRON; W   61850

472 C16 H10 Br2 O7 65471-61-6 Y-14-748-0 DIMETHYL 4,5-DI(5-BROMO-2-FURYL)-2,3-FURANDICARBOXYLATE; W   61851

472 C16 H12 Cl4 O4 S2 SB-31-91-0 2,2-Bis(4-chloro-3-methanesulfonylphenyl)-1,1-dichloroethylene; W   124855

472 C16 H22 Hg Si2 59612-26-9 AG-120-175-1 Bis(dimethylphenylsilyl)mercury; W   124856

472 C16 H41 O6 P Si4 KC-1977-1722-3 Tetrakis(trimethylsilyl) ether of 2,4-Dihydroxycyclobutyl phosphate;
W   124857

472 C17 H12 B2 F10 N2 O 72886-62-5 K-112-3825-0 4-TERT-BUTYL-2-METHYL-3,5-BIS(PENTAFLUOROPHENYL)-1,2,4,
3,5-OXADIAZADIBOROLIDIN; W   61852

472 C17 H15 Br3 O 78280-72-5 B-34-127-0 6,8-Dibromo-7-methoxy-1-(x-bromophenyl)-1,2,3,4-tetrahydronapht
halene; W   124858

472 C17 H44 N8 Si4 KO-8-221-1 Tetrakis(trimethylsilyl) derivative of Methylglyoxal-bis-guanylhydrazone;
W   124859

472 C18 H15 F7 N2 O5 55429-88-4 BA-0-168-0 Butanoic acid, 2,2,3,3,4,4,4-heptafluoro-, 4-(5-ethylhexahy
dro-1,3-dimethyl-2,4,6-trioxo-5-pyrimidinyl)phenyl ester;; 1,3-DIMETHYL-5-ETHYL-5-(4-(HEPTAFLUOROBUTYROXY)
PHENYL)BARBITURIC ACID; W/NBS   61853

472 C18 H16 N8 O8 1468-24-2 SD-1981-0-0 1,2-Cyclohexanedione, bis[(2,4-dinitrophenyl)hydrazone];;
W/NBS   61854

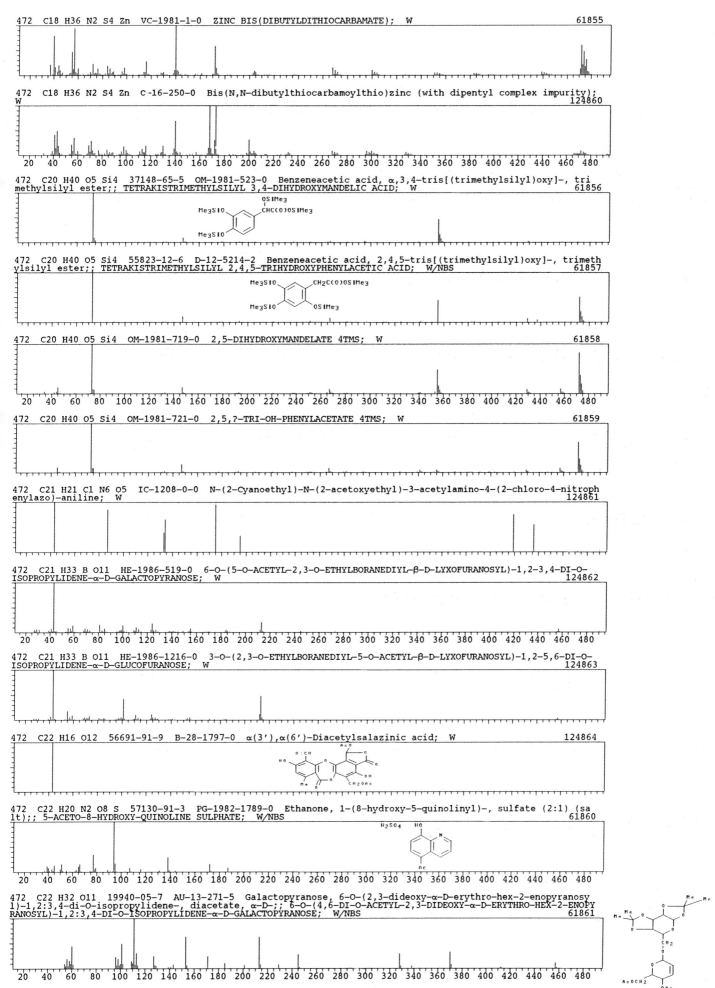

472   C18 H36 N2 S4 Zn   VC-1981-1-0   ZINC BIS(DIBUTYLDITHIOCARBAMATE);   W   61855

472   C18 H36 N2 S4 Zn   C-16-250-0   Bis(N,N-dibutylthiocarbamoylthio)zinc (with dipentyl complex impurity);
W   124860

472   C20 H40 O5 Si4   37148-65-5   OM-1981-523-0   Benzeneacetic acid, α,3,4-tris[(trimethylsilyl)oxy]-, tri
methylsilyl ester;;   TETRAKISTRIMETHYLSILYL 3,4-DIHYDROXYMANDELIC ACID;   W   61856

472   C20 H40 O5 Si4   55823-12-6   D-12-5214-2   Benzeneacetic acid, 2,4,5-tris[(trimethylsilyl)oxy]-, trimeth
ylsilyl ester;;   TETRAKISTRIMETHYLSILYL 2,4,5-TRIHYDROXYPHENYLACETIC ACID;   W/NBS   61857

472   C20 H40 O5 Si4   OM-1981-719-0   2,5-DIHYDROXYMANDELATE 4TMS;   W   61858

472   C20 H40 O5 Si4   OM-1981-721-0   2,5,?-TRI-OH-PHENYLACETATE 4TMS;   W   61859

472   C21 H21 Cl N6 O5   IC-1208-0-0   N-(2-Cyanoethyl)-N-(2-acetoxyethyl)-3-acetylamino-4-(2-chloro-4-nitroph
enylazo)-aniline;   W   124861

472   C21 H33 B O11   HE-1986-519-0   6-O-(5-O-ACETYL-2,3-O-ETHYLBORANEDIYL-β-D-LYXOFURANOSYL)-1,2-3,4-DI-O-
ISOPROPYLIDENE-α-D-GALACTOPYRANOSE;   W   124862

472   C21 H33 B O11   HE-1986-1216-0   3-O-(2,3-O-ETHYLBORANEDIYL-5-O-ACETYL-β-D-LYXOFURANOSYL)-1,2-5,6-DI-O-
ISOPROPYLIDENE-α-D-GLUCOFURANOSE;   W   124863

472   C22 H16 O12   56691-91-9   B-28-1797-0   α(3'),α(6')-Diacetylsalazinic acid;   W   124864

472   C22 H20 N2 O8 S   57130-91-3   PG-1982-1789-0   Ethanone, 1-(8-hydroxy-5-quinolinyl)-, sulfate (2:1) (sa
lt);;   5-ACETO-8-HYDROXY-QUINOLINE SULPHATE;   W/NBS   61860

472   C22 H32 O11   19940-05-7   AU-13-271-5   Galactopyranose, 6-O-(2,3-dideoxy-α-D-erythro-hex-2-enopyranosy
l)-1,2:3,4-di-O-isopropylidene-, diacetate, α-D-;;   6-O-(4,6-DI-O-ACETYL-2,3-DIDEOXY-α-D-ERYTHRO-HEX-2-ENOPY
RANOSYL)-1,2:3,4-DI-O-ISOPROPYLIDENE-α-D-GALACTOPYRANOSE;   W/NBS   61861

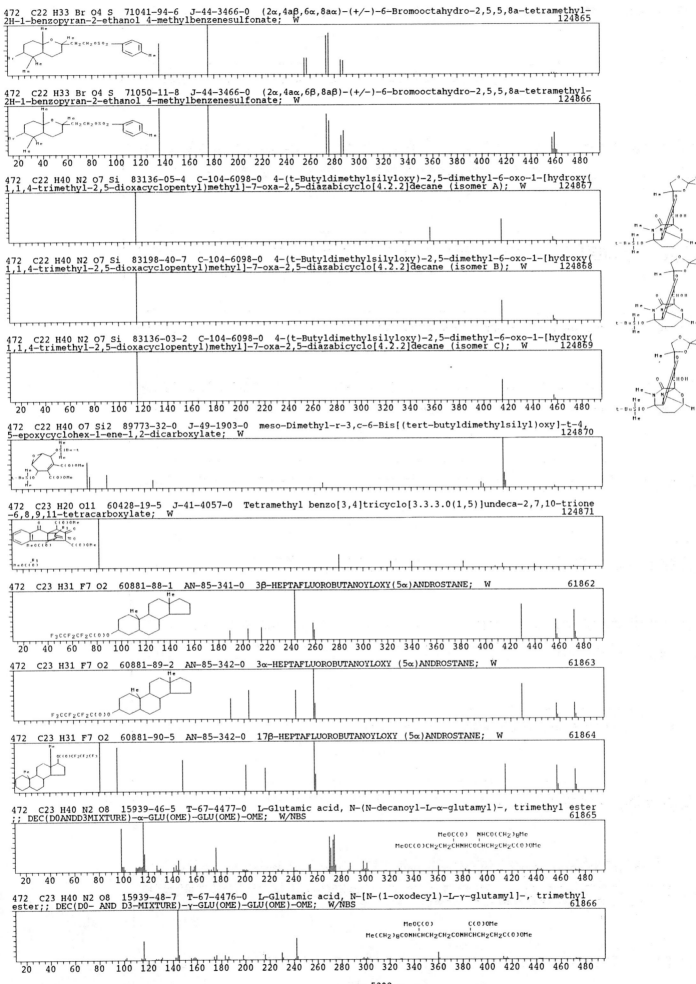

472 C22 H33 Br O4 S 71041-94-6 J-44-3466-0 (2α,4aβ,6α,8aα)-(+/-)-6-Bromooctahydro-2,5,5,8a-tetramethyl-
2H-1-benzopyran-2-ethanol 4-methylbenzenesulfonate; W   124865

472 C22 H33 Br O4 S 71050-11-8 J-44-3466-0 (2α,4aα,6β,8aβ)-(+/-)-6-bromooctahydro-2,5,5,8a-tetramethyl-
2H-1-benzopyran-2-ethanol 4-methylbenzenesulfonate; W   124866

472 C22 H40 N2 O7 Si 83136-05-4 C-104-6098-0 4-(t-Butyldimethylsilyloxy)-2,5-dimethyl-6-oxo-1-[hydroxy(
1,1,4-trimethyl-2,5-dioxacyclopentyl)methyl]-7-oxa-2,5-diazabicyclo[4.2.2]decane (isomer A); W   124867

472 C22 H40 N2 O7 Si 83198-40-7 C-104-6098-0 4-(t-Butyldimethylsilyloxy)-2,5-dimethyl-6-oxo-1-[hydroxy(
1,1,4-trimethyl-2,5-dioxacyclopentyl)methyl]-7-oxa-2,5-diazabicyclo[4.2.2]decane (isomer B); W   124868

472 C22 H40 N2 O7 Si 83136-03-2 C-104-6098-0 4-(t-Butyldimethylsilyloxy)-2,5-dimethyl-6-oxo-1-[hydroxy(
1,1,4-trimethyl-2,5-dioxacyclopentyl)methyl]-7-oxa-2,5-diazabicyclo[4.2.2]decane (isomer C); W   124869

472 C22 H40 O7 Si2 89773-32-0 J-49-1903-0 meso-Dimethyl-r-3,c-6-Bis[(tert-butyldimethylsilyl)oxy]-t-4,
5-epoxycyclohex-1-ene-1,2-dicarboxylate; W   124870

472 C23 H20 O11 60428-19-5 J-41-4057-0 Tetramethyl benzo[3,4]tricyclo[3.3.3.0(1,5)]undeca-2,7,10-trione
-6,8,9,11-tetracarboxylate; W   124871

472 C23 H31 F7 O2 60881-88-1 AN-85-341-0 3β-HEPTAFLUOROBUTANOYLOXY(5α)ANDROSTANE; W   61862

472 C23 H31 F7 O2 60881-89-2 AN-85-342-0 3α-HEPTAFLUOROBUTANOYLOXY (5α)ANDROSTANE; W   61863

472 C23 H31 F7 O2 60881-90-5 AN-85-342-0 17β-HEPTAFLUOROBUTANOYLOXY (5α)ANDROSTANE; W   61864

472 C23 H40 N2 O8 15939-46-5 T-67-4477-0 L-Glutamic acid, N-(N-decanoyl-L-α-glutamyl)-, trimethyl ester
;; DEC(D0ANDD3MIXTURE)-α-GLU(OME)-GLU(OME)-OME; W/NBS   61865

472 C23 H40 N2 O8 15939-48-7 T-67-4476-0 L-Glutamic acid, N-[N-(1-oxodecyl)-L-γ-glutamyl]-, trimethyl
ester;; DEC(D0- AND D3-MIXTURE)-γ-GLU(OME)-GLU(OME)-OME; W/NBS   61866

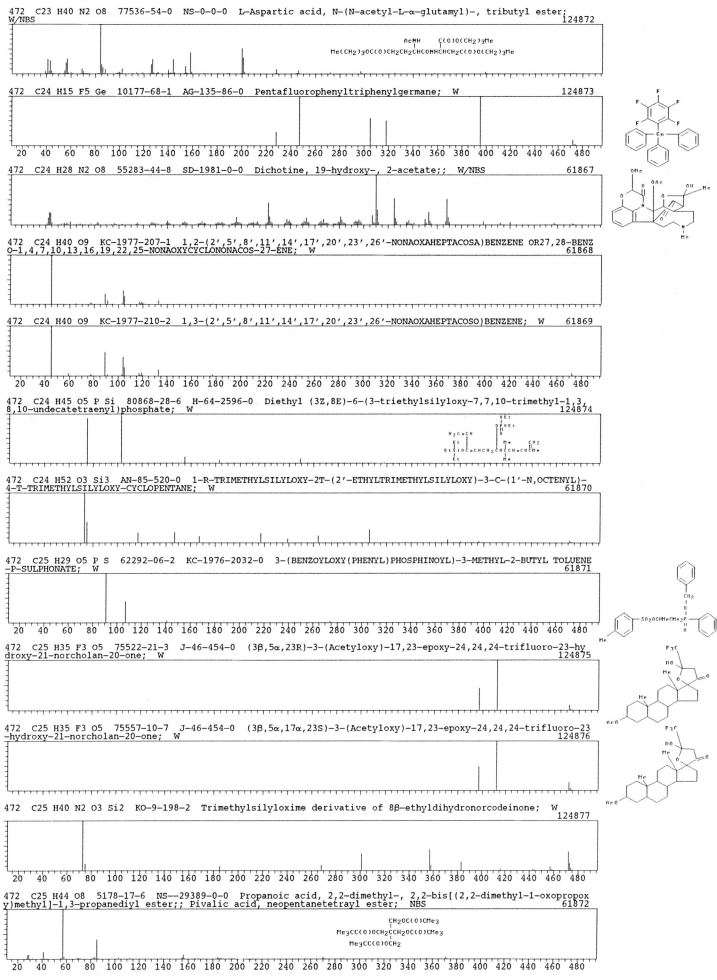

472　C23 H40 N2 O8　77536-54-0　NS-0-0-0　L-Aspartic acid, N-(N-acetyl-L-α-glutamyl)-, tributyl ester;
W/NBS　124872

```
                              AcNH       C(O)O(CH2)3Me
             Me(CH2)3OC(O)CH2CH2CHCONHCHCH2C(O)O(CH2)3Me
```

472　C24 H15 F5 Ge　10177-68-1　AG-135-86-0　Pentafluorophenyltriphenylgermane; W　124873

20　40　60　80　100　120　140　160　180　200　220　240　260　280　300　320　340　360　380　400　420　440　460　480

472　C24 H28 N2 O8　55283-44-8　SD-1981-0-0　Dichotine, 19-hydroxy-, 2-acetate;; W/NBS　61867

472　C24 H40 O9　KC-1977-207-1　1,2-(2',5',8',11',14',17',20',23',26'-NONAOXAHEPTACOSA)BENZENE OR27,28-BENZ
O-1,4,7,10,13,16,19,22,25-NONAOXYCYCLONONACOS-27-ENE; W　61868

472　C24 H40 O9　KC-1977-210-2　1,3-(2',5',8',11',14',17',20',23',26'-NONAOXAHEPTACOSO)BENZENE; W　61869

20　40　60　80　100　120　140　160　180　200　220　240　260　280　300　320　340　360　380　400　420　440　460　480

472　C24 H45 O5 P Si　80868-28-6　H-64-2596-0　Diethyl (3Z,8E)-6-(3-triethylsilyloxy-7,7,10-trimethyl-1,3,
8,10-undecatetraenyl)phosphate; W　124874

```
                                         OEt
                                         O POEt
                                       H2C=CH ‖
                                             O
                     Et        Me      CH2
                  EtSiOC=CHCH2CHCCH=CHCMe
                     Et        Me
```

472　C24 H52 O3 Si3　AN-85-520-0　1-R-TRIMETHYLSILYLOXY-2T-(2'-ETHYLTRIMETHYLSILYLOXY)-3-C-(1'-N,OCTENYL)-
4-T-TRIMETHYLSILYLOXY-CYCLOPENTANE; W　61870

472　C25 H29 O5 P S　62292-06-2　KC-1976-2032-0　3-(BENZOYLOXY(PHENYL)PHOSPHINOYL)-3-METHYL-2-BUTYL TOLUENE
-P-SULPHONATE; W　61871

20　40　60　80　100　120　140　160　180　200　220　240　260　280　300　320　340　360　380　400　420　440　460　480

472　C25 H35 F3 O5　75522-21-3　J-46-454-0　(3β,5α,23R)-3-(Acetyloxy)-17,23-epoxy-24,24,24-trifluoro-23-hy
droxy-21-norcholan-20-one; W　124875

472　C25 H35 F3 O5　75557-10-7　J-46-454-0　(3β,5α,17α,23S)-3-(Acetyloxy)-17,23-epoxy-24,24,24-trifluoro-23
-hydroxy-21-norcholan-20-one; W　124876

472　C25 H40 N2 O3 Si2　KO-9-198-2　Trimethylsilyloxime derivative of 8β-ethyldihydronorcodeinone; W　124877

20　40　60　80　100　120　140　160　180　200　220　240　260　280　300　320　340　360　380　400　420　440　460　480

472　C25 H44 O8　5178-17-6　NS--29389-0-0　Propanoic acid, 2,2-dimethyl-, 2,2-bis[(2,2-dimethyl-1-oxopropox
y)methyl]-1,3-propanediyl ester;; Pivalic acid, neopentanetetrayl ester; NBS　61872

```
                            CH2OC(O)CMe3
              Me3CC(O)OCH2CH2OC(O)CMe3
                  Me3CC(O)OCH2
```

20　40　60　80　100　120　140　160　180　200　220　240　260　280　300　320　340　360　380　400　420　440　460　480

472   C25 H52 O4 Si2   75299-46-6   NS--29388-0-0   9-Octadecenoic acid, 12,13-bis[(trimethylsilyl)oxy]-, meth
yl ester;;   NBS   61873

OSiMe3
OSiMe3
MeOC(O)(CH2)7CH:CHCH2CHCH(CH2)4Me

472   C26 H16 O9   KC-1983-1812-0   1,5-Bis(psoralen-8-yloxy)-3-oxapentene;   W   124878

472   C26 H20 Cl4   72188-20-6   C-102-675-0   endo,exo,exo-1,2,3,4-Tetrachloro-4a,5,5a,6,11,11a,12,12a-octahy
dro-6,11(1'2')-benzeno-1,4:5,12-dimethanonaphthacene;   W   124879

472   C26 H21 Br N2 O2   Y-20-381-0   3-(p-Methoxy phenyl)-5,7-diphenyl-2H-pyrido[2,1-f]-1,3,4-oxadiazinium br
omides;   W   124880

472   C26 H23 O Sb   70080-24-9   K-116-476-0   2-(Diphenylstibino)-1,1-diphenylethanol;   W   124881

C(OH)CH2Sb

472   C26 H26 Cl2 O4   74253-29-5   KC-1981-102-0   10β-benzyloxycarbonyl-4β,7-dichloro-1β-methyl-8-methylenegi
bb-2-ene-1α,4aα-carbolactone;   W   124882

472   C26 H32 O8   7323-92-4   NS--29386-0-0   Estra-1,3,5(10)-triene-3,6α,7β,17β-tetrol, tetraacetate;;   NBS
                                                                                                    61875

472   C26 H32 O8   69688-24-0   NS--29384-0-0   Estra-1,3,5(10)-triene-3,6,7,17-tetrol, tetraacetate, (6β,7α,17
β)-;;   NBS   61876

472   C26 H32 O8   491-47-4   SD-1981-0-0   Olivetoric acid;   W/NBS   61879

CH2CO(CH2)4Me   (CH2)4Me
                       CO2H
            C(O)O
HO   OH   OH

472   C26 H32 O8   66610-36-4   NS--29385-0-0   Spiro[cyclopropane-1,2'(1'H)-phenanthrene]-1',4'(3'H)-dione,
3',9',10'-tris(acetyloxy)-4'b,5',6',7',8',8'a,9',10'-octahydro-2,4'b,7'-trimethyl-8'-methylene-, [2'S-[2'α(
R*),3'α,4'bβ,7'β,8'α,9'β,10'α]]-;; Spiro[cyclopropane-1,2'(1'H)-phenanthrene]-1',4'(3'H)-dione, 3',9',10'-
tris(acetyloxy)-4'b,5',6',7',8',8'a,9',10'-octahydro-2,4'b,7'-trimethyl-8;   NBS   61874

472   C26 H32 O8   63529-36-2   B-30-604-0   8-HYDROXY-3,4-DIMETHOXY-11-OXO-1,6-DEPENTYL-11H-DIBENZO[B,E][1,4]
DIOXEPIN-7-CARBOXYLIC ACID;   W   61877

HO   (CH2)4Me
            O
HO2C   OMe
Me(CH2)4   O   OMe

472   C26 H32 O8   65714-82-1   H-60-2777-0   6,11,12-TRIACETOXY-14-HYDROXY-ABIETA-5,8,11,13-TETRAENE-7-ONE;   W
                                                                                                    61878

AcO   OAc
          Pr-i
Me   OH

Me   OAc
O
Me
Me

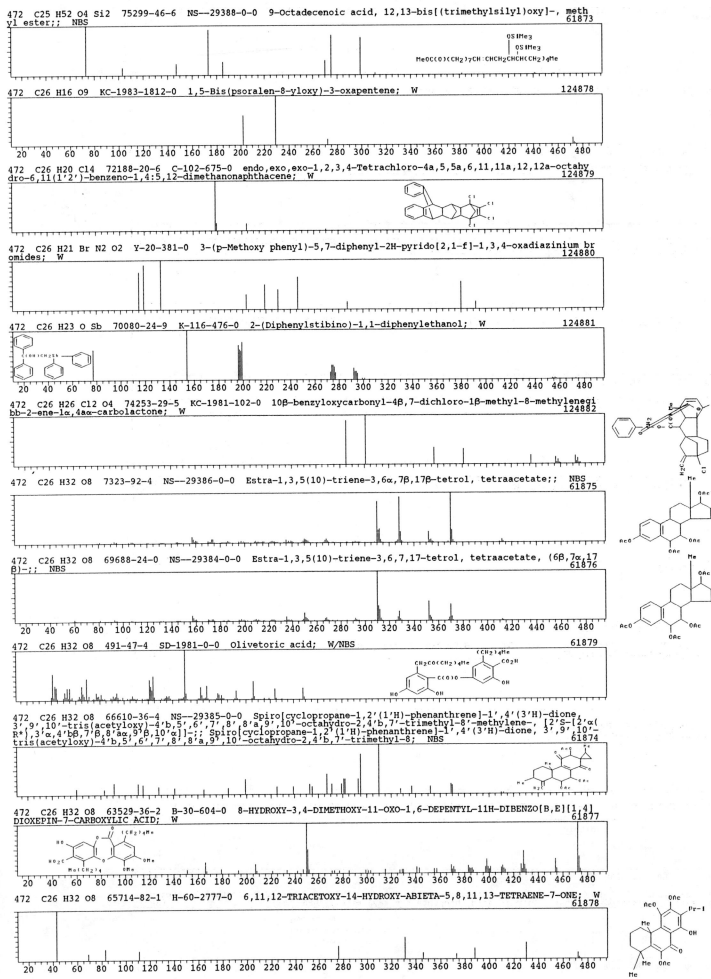

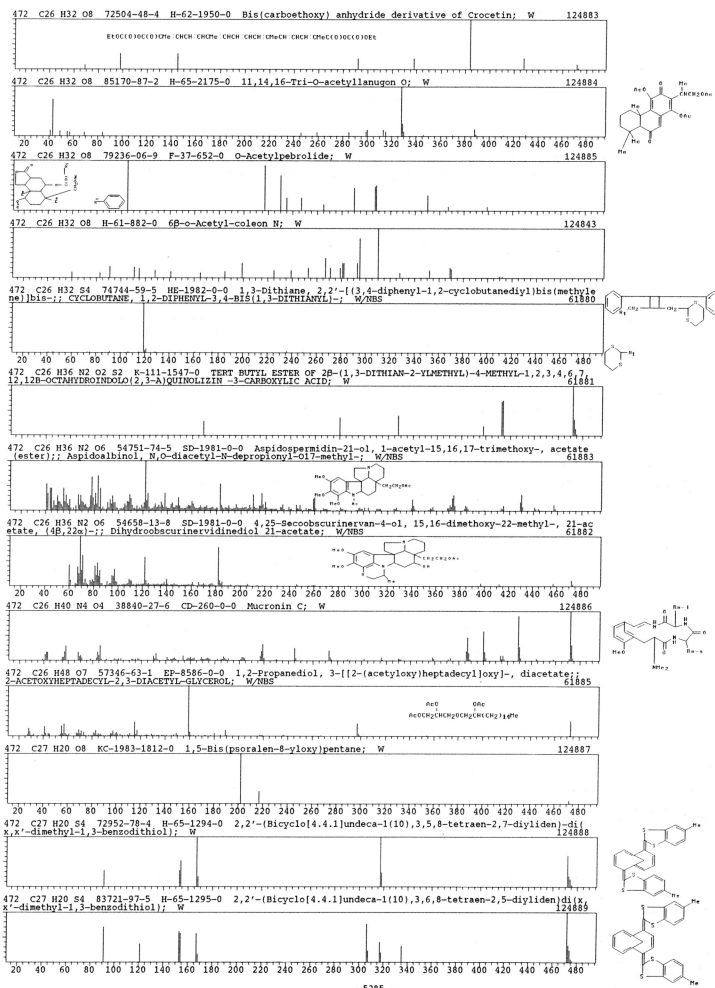

472   C26 H32 O8   72504-48-4   H-62-1950-0   Bis(carboethoxy) anhydride derivative of Crocetin;   W   124883

EtOC(O)OC(O)CMe:CHCH:CHCHMe:CHCH:CHCH:CMeCH:CHCH:CMeC(O)OC(O)OEt

472   C26 H32 O8   85170-87-2   H-65-2175-0   11,14,16-Tri-O-acetyllanugon O;   W   124884

472   C26 H32 O8   79236-06-9   F-37-652-0   O-Acetylpebrolide;   W   124885

472   C26 H32 O8   H-61-882-0   6β-o-Acetyl-coleon N;   W   124843

472   C26 H32 S4   74744-59-5   HE-1982-0-0   1,3-Dithiane, 2,2'-[(3,4-diphenyl-1,2-cyclobutanediyl)bis(methylene)]bis-;; CYCLOBUTANE, 1,2-DIPHENYL-3,4-BIS(1,3-DITHIANYL)-;   W/NBS   61880

472   C26 H36 N2 O2 S2   K-111-1547-0   TERT BUTYL ESTER OF 2β-(1,3-DITHIAN-2-YLMETHYL)-4-METHYL-1,2,3,4,6,7,12,12B-OCTAHYDROINDOLO(2,3-A)QUINOLIZIN -3-CARBOXYLIC ACID;   W   61881

472   C26 H36 N2 O6   54751-74-5   SD-1981-0-0   Aspidospermidin-21-ol, 1-acetyl-15,16,17-trimethoxy-, acetate (ester);; Aspidoalbinol, N,O-diacetyl-N-depropionyl-O17-methyl-;   W/NBS   61883

472   C26 H36 N2 O6   54658-13-8   SD-1981-0-0   4,25-Secoobscurinervan-4-ol, 15,16-dimethoxy-22-methyl-, 21-acetate, (4β,22α)-;; Dihydroobscurinervidinediol 21-acetate;   W/NBS   61882

472   C26 H40 N4 O4   38840-27-6   CD-260-0-0   Mucronin C;   W   124886

472   C26 H48 O7   57346-63-1   EP-8586-0-0   1,2-Propanediol, 3-[[2-(acetyloxy)heptadecyl]oxy]-, diacetate;; 2-ACETOXYHEPTADECYL-2,3-DIACETYL-GLYCEROL;   W/NBS   61885

472   C27 H20 O8   KC-1983-1812-0   1,5-Bis(psoralen-8-yloxy)pentane;   W   124887

472   C27 H20 S4   72952-78-4   H-65-1294-0   2,2'-(Bicyclo[4.4.1]undeca-1(10),3,5,8-tetraen-2,7-diyliden)-di(x,x'-dimethyl-1,3-benzodithiol);   W   124888

472   C27 H20 S4   83721-97-5   H-65-1295-0   2,2'-(Bicyclo[4.4.1]undeca-1(10),3,6,8-tetraen-2,5-diyliden)di(x,x'-dimethyl-1,3-benzodithiol);   W   124889

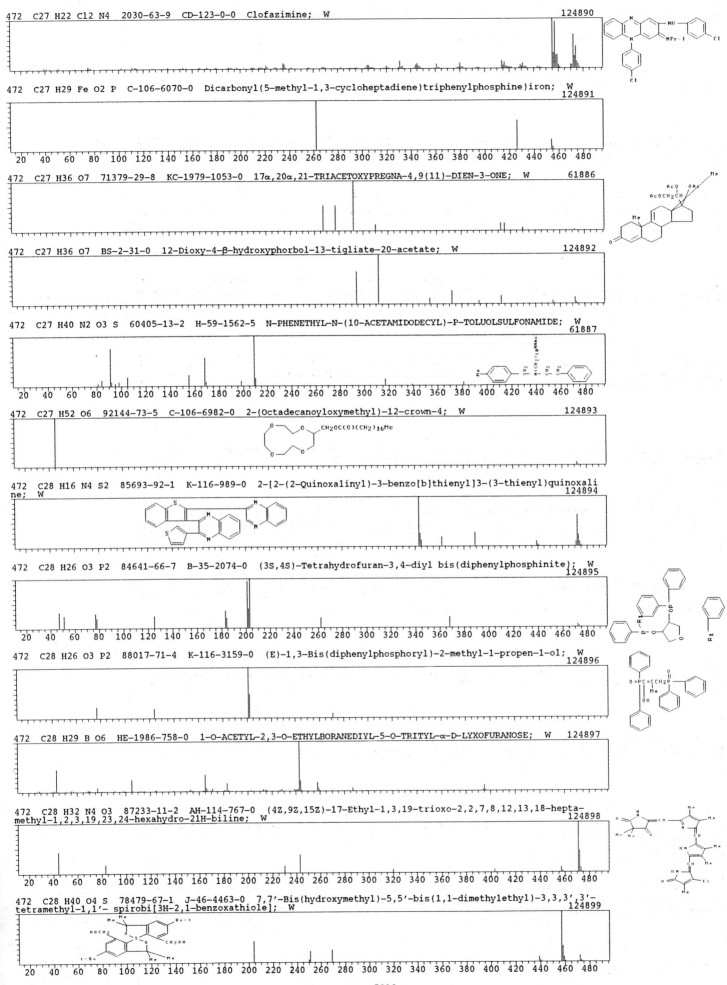

472 C27 H22 Cl2 N4 2030-63-9 CD-123-0-0 Clofazimine; W   124890

472 C27 H29 Fe O2 P C-106-6070-0 Dicarbonyl(5-methyl-1,3-cycloheptadiene)triphenylphosphine)iron; W
124891

472 C27 H36 O7 71379-29-8 KC-1979-1053-0 17α,20α,21-TRIACETOXYPREGNA-4,9(11)-DIEN-3-ONE; W   61886

472 C27 H36 O7 BS-2-31-0 12-Dioxy-4-β-hydroxyphorbol-13-tigliate-20-acetate; W   124892

472 C27 H40 N2 O3 S 60405-13-2 H-59-1562-5 N-PHENETHYL-N-(10-ACETAMIDODECYL)-P-TOLUOLSULFONAMIDE; W
61887

472 C27 H52 O6 92144-73-5 C-106-6982-0 2-(Octadecanoyloxymethyl)-12-crown-4; W   124893

472 C28 H16 N4 S2 85693-92-1 K-116-989-0 2-[2-(2-Quinoxalinyl)-3-benzo[b]thienyl]3-(3-thienyl)quinoxali
ne; W   124894

472 C28 H26 O3 P2 84641-66-7 B-35-2074-0 (3S,4S)-Tetrahydrofuran-3,4-diyl bis(diphenylphosphinite); W
124895

472 C28 H26 O3 P2 88017-71-4 K-116-3159-0 (E)-1,3-Bis(diphenylphosphoryl)-2-methyl-1-propen-1-ol; W
124896

472 C28 H29 B O6 HE-1986-758-0 1-O-ACETYL-2,3-O-ETHYLBORANEDIYL-5-O-TRITYL-α-D-LYXOFURANOSE; W 124897

472 C28 H32 N4 O3 87233-11-2 AH-114-767-0 (4Z,9Z,15Z)-17-Ethyl-1,3,19-trioxo-2,2,7,8,12,13,18-hepta-
methyl-1,2,3,19,23,24-hexahydro-21H-biline; W   124898

472 C28 H40 O4 S 78479-67-1 J-46-4463-0 7,7'-Bis(hydroxymethyl)-5,5'-bis(1,1-dimethylethyl)-3,3,3',3'-
tetramethyl-1,1'- spirobi[3H-2,1-benzoxathiole]; W   124899

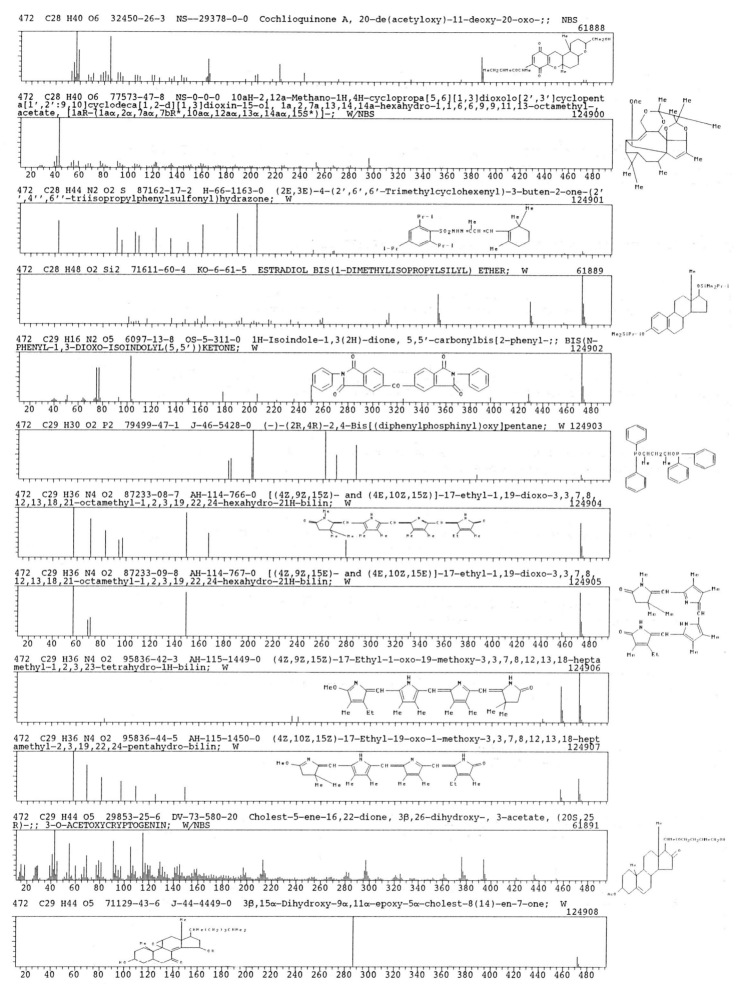

472   C28 H40 O6   32450-26-3   NS--29378-0-0   Cochlioquinone A, 20-de(acetyloxy)-11-deoxy-20-oxo-;;   NBS
61888

472   C28 H40 O6   77573-47-8   NS-0-0-0   10aH-2,12a-Methano-1H,4H-cyclopropa[5,6][1,3]dioxolo[2',3']cyclopent
a[1',2':9,10]cyclodeca[1,2-d][1,3]dioxin-15-ol, 1a,2,7a,13,14,14a-hexahydro-1,1,6,6,9,9,11,13-octamethyl-
acetate, [1aR-(1aα,2α,7aα,7bR*,10aα,12aα,13α,14aα,15S*)]-;   W/NBS
124900

472   C28 H44 N2 O2 S   87162-17-2   H-66-1163-0   (2E,3E)-4-(2',6',6'-Trimethylcyclohexenyl)-3-buten-2-one-(2'
',4'',6''-triisopropylphenylsulfonyl)hydrazone;   W
124901

472   C28 H48 O2 Si2   71611-60-4   KO-6-61-5   ESTRADIOL BIS(1-DIMETHYLISOPROPYLSILYL) ETHER;   W
61889

472   C29 H16 N2 O5   6097-13-8   OS-5-311-0   1H-Isoindole-1,3(2H)-dione, 5,5'-carbonylbis[2-phenyl-;;   BIS(N-
PHENYL-1,3-DIOXO-ISOINDOLYL(5,5'))KETONE;   W
124902

472   C29 H30 O2 P2   79499-47-1   J-46-5428-0   (-)-(2R,4R)-2,4-Bis[(diphenylphosphinyl)oxy]pentane;   W 124903

472   C29 H36 N4 O2   87233-08-7   AH-114-766-0   [(4Z,9Z,15Z)- and (4E,10Z,15Z)]-17-ethyl-1,19-dioxo-3,3,7,8,
12,13,18,21-octamethyl-1,2,3,19,22,24-hexahydro-21H-bilin;   W
124904

472   C29 H36 N4 O2   87233-09-8   AH-114-767-0   [(4Z,9Z,15E)- and (4E,10Z,15E)]-17-ethyl-1,19-dioxo-3,3,7,8,
12,13,18,21-octamethyl-1,2,3,19,22,24-hexahydro-21H-bilin;   W
124905

472   C29 H36 N4 O2   95836-42-3   AH-115-1449-0   (4Z,9Z,15Z)-17-Ethyl-1-oxo-19-methoxy-3,3,7,8,12,13,18-hepta
methyl-1,2,3,23-tetrahydro-1H-bilin;   W
124906

472   C29 H36 N4 O2   95836-44-5   AH-115-1450-0   (4Z,10Z,15Z)-17-Ethyl-19-oxo-1-methoxy-3,3,7,8,12,13,18-hept
amethyl-2,3,19,22,24-pentahydro-bilin;   W
124907

472   C29 H44 O5   29853-25-6   DV-73-580-20   Cholest-5-ene-16,22-dione, 3β,26-dihydroxy-, 3-acetate, (20S,25
R)-;; 3-O-ACETOXYCRYPTOGENIN;   W/NBS
61891

472   C29 H44 O5   71129-43-6   J-44-4449-0   3β,15α-Dihydroxy-9α,11α-epoxy-5α-cholest-8(14)-en-7-one;   W
124908

472   C29 H44 O5   IC-1209-0-0   (3β,5α)-3-Acetyloxy-spirostan-12-one;   W   124909

472   C30 H20 N2 O4   67390-08-3   K-111-2178-0   1,1',3,3'-TETRAPHENYL-3,3'-BIAZETIDIN-2,2',4,4'-TETRAONE;   W
61892

20 40 60 80 100 120 140 160 180 200 220 240 260 280 300 320 340 360 380 400 420 440 460 480

472   C30 H20 N2 O4   75167-73-6   Y-17-765-0   4-(3',4'-Diphenylisoxazolin-5'-on-2'-yl)-3,4-diphenylisoxazolin
-5-one;   W   124910

472   C30 H32 O5   65690-36-0   H-60-3043-0   (+,-)-CURVULARIN DIBENZYL ETHER;   W   61893

472   C30 H48 O4   64543-32-4   NS--29374-0-0   D:A-Friedooleanan-28-oic acid, 21-hydroxy-3-oxo-, (21α)-;;   NBS
61894

472   C30 H48 O4   36872-80-7   SD-1981-0-0   Lanost-9(11)-en-18-oic acid, 3,20,23-trihydroxy-, (3β,20.xi.)-;;
W/NBS   61897

472   C30 H48 O4   F-32-2357-0   ΔA(11)OLOSTENE-3β,23ε-DIOL;   W   61895

472   C30 H48 O4   57706-70-4   B-28-2031-0   3-ACETOXY-4,4,14-TRIMETHYLCHOL-11-EN-24-OIC ACID METHYL ESTER;   W
61896

20 40 60 80 100 120 140 160 180 200 220 240 260 280 300 320 340 360 380 400 420 440 460 480

472   C30 H48 O4   70117-01-0   F-40-2080-0   1β,3β-Dihydroxylup-20,29-en-26-oic acid;   W   124911

472   C30 H48 O4   54145-67-4   KC-1983-2762-0   3β-Hydroxy-24-oxo-16,17-seco-5α-dammar-17(20)Z-eno-16,30-lact
one;   W   124912

472   C30 H48 O4   CD-462-0-0   Cyclamiretin D;   W   124913

472   C30 H48 O4   78835-09-3   KC-1983-2465-0   Kokoona triterpene A;   W   124914

Str 4

472   C30 H48 O4   54878-50-1   KC-1983-1124-0   (20R,24R)-3β,25-Dihydroxy-15α,30-cyclo-20,24-epoxydammaran-16-
one;   W   124915

20 40 60 80 100 120 140 160 180 200 220 240 260 280 300 320 340 360 380 400 420 440 460 480

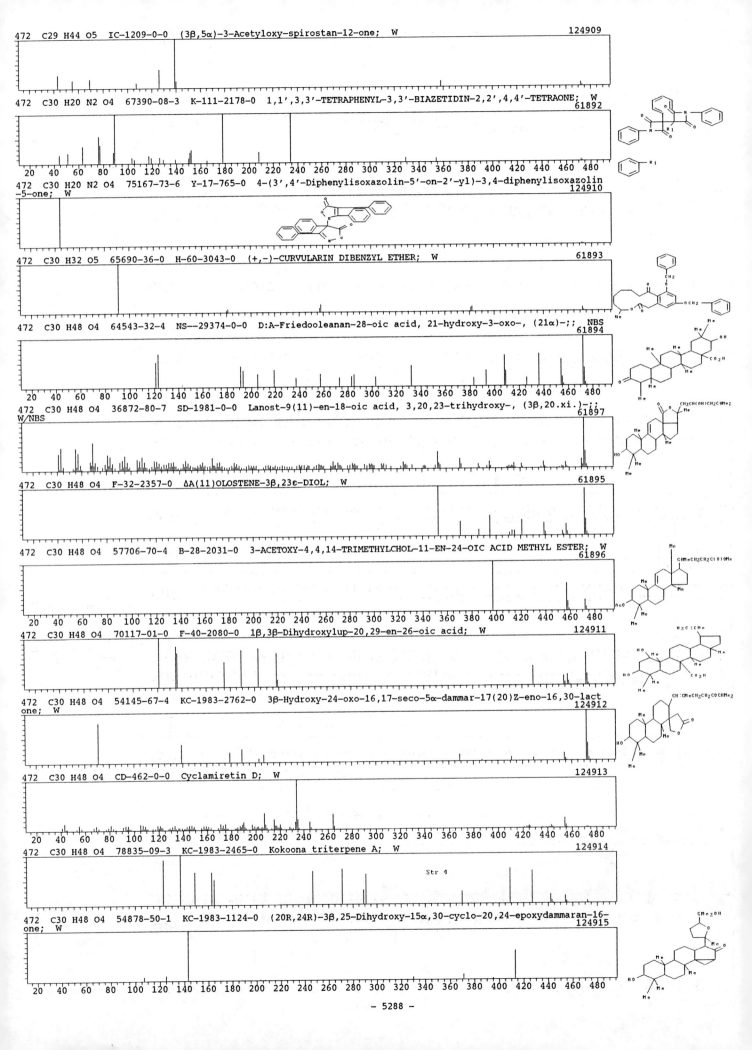

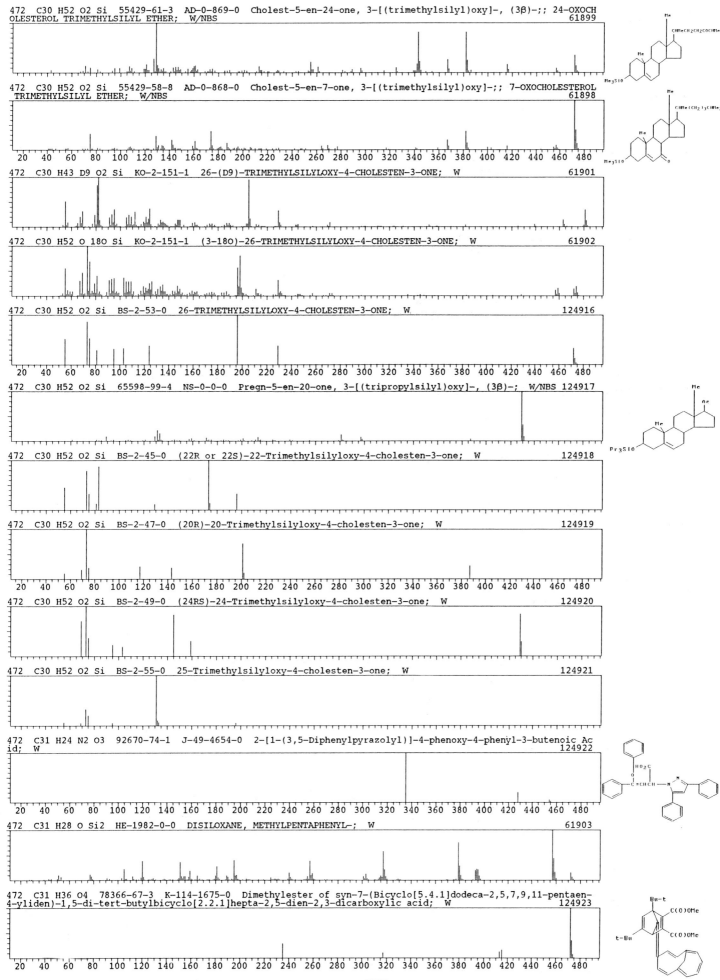

472 C30 H52 O2 Si 55429-61-3 AD-0-869-0 Cholest-5-en-24-one, 3-[(trimethylsilyl)oxy]-, (3β)-;; 24-OXOCH
OLESTEROL TRIMETHYLSILYL ETHER; W/NBS
61899

472 C30 H52 O2 Si 55429-58-8 AD-0-868-0 Cholest-5-en-7-one, 3-[(trimethylsilyl)oxy]-;; 7-OXOCHOLESTEROL
TRIMETHYLSILYL ETHER; W/NBS
61898

472 C30 H43 D9 O2 Si KO-2-151-1 26-(D9)-TRIMETHYLSILYLOXY-4-CHOLESTEN-3-ONE; W
61901

472 C30 H52 O 18O Si KO-2-151-1 (3-18O)-26-TRIMETHYLSILYLOXY-4-CHOLESTEN-3-ONE; W
61902

472 C30 H52 O2 Si BS-2-53-0 26-TRIMETHYLSILYLOXY-4-CHOLESTEN-3-ONE; W
124916

472 C30 H52 O2 Si 65598-99-4 NS-0-0-0 Pregn-5-en-20-one, 3-[(tripropylsilyl)oxy]-, (3β)-; W/NBS 124917

472 C30 H52 O2 Si BS-2-45-0 (22R or 22S)-22-Trimethylsilyloxy-4-cholesten-3-one; W
124918

472 C30 H52 O2 Si BS-2-47-0 (20R)-20-Trimethylsilyloxy-4-cholesten-3-one; W
124919

472 C30 H52 O2 Si BS-2-49-0 (24RS)-24-Trimethylsilyloxy-4-cholesten-3-one; W
124920

472 C30 H52 O2 Si BS-2-55-0 25-Trimethylsilyloxy-4-cholesten-3-one; W
124921

472 C31 H24 N2 O3 92670-74-1 J-49-4654-0 2-[1-(3,5-Diphenylpyrazolyl)]-4-phenoxy-4-phenyl-3-butenoic Ac
id; W
124922

472 C31 H28 O Si2 HE-1982-0-0 DISILOXANE, METHYLPENTAPHENYL-; W
61903

472 C31 H36 O4 78366-67-3 K-114-1675-0 Dimethylester of syn-7-(Bicyclo[5.4.1]dodeca-2,5,7,9,11-pentaen-
4-yliden)-1,5-di-tert-butylbicyclo[2.2.1]hepta-2,5-dien-2,3-dicarboxylic acid; W
124923

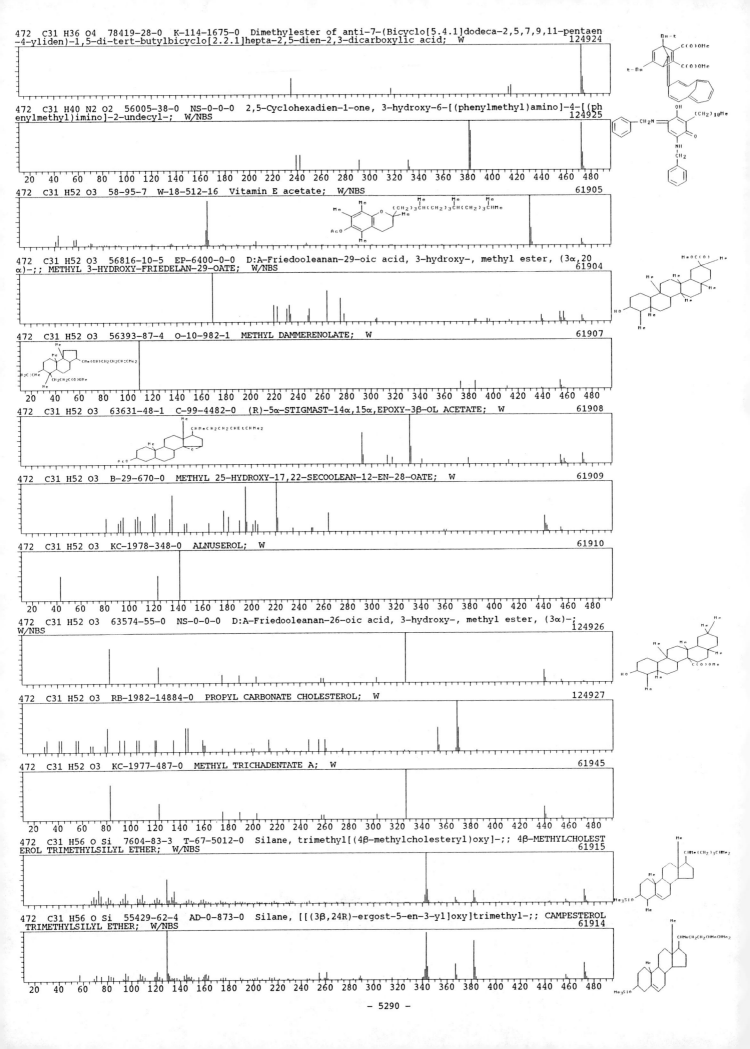

472 C31 H36 O4 78419-28-0 K-114-1675-0 Dimethylester of anti-7-(Bicyclo[5.4.1]dodeca-2,5,7,9,11-pentaen
-4-yliden)-1,5-di-tert-butylbicyclo[2.2.1]hepta-2,5-dien-2,3-dicarboxylic acid; W   124924

472 C31 H40 N2 O2 56005-38-0 NS-0-0-0 2,5-Cyclohexadien-1-one, 3-hydroxy-6-[(phenylmethyl)amino]-4-[(ph
enylmethyl)imino]-2-undecyl-; W/NBS   124925

472 C31 H52 O3 58-95-7 W-18-512-16 Vitamin E acetate; W/NBS   61905

472 C31 H52 O3 56816-10-5 EP-6400-0-0 D:A-Friedooleanan-29-oic acid, 3-hydroxy-, methyl ester, (3α,20
α)-;; METHYL 3-HYDROXY-FRIEDELAN-29-OATE; W/NBS   61904

472 C31 H52 O3 56393-87-4 O-10-982-1 METHYL DAMMERENOLATE; W   61907

472 C31 H52 O3 63631-48-1 C-99-4482-0 (R)-5α-STIGMAST-14α,15α,EPOXY-3β-OL ACETATE; W   61908

472 C31 H52 O3 B-29-670-0 METHYL 25-HYDROXY-17,22-SECOOLEAN-12-EN-28-OATE; W   61909

472 C31 H52 O3 KC-1978-348-0 ALNUSEROL; W   61910

472 C31 H52 O3 63574-55-0 NS-0-0-0 D:A-Friedooleanan-26-oic acid, 3-hydroxy-, methyl ester, (3α)-;
W/NBS   124926

472 C31 H52 O3 RB-1982-14884-0 PROPYL CARBONATE CHOLESTEROL; W   124927

472 C31 H52 O3 KC-1977-487-0 METHYL TRICHADENTATE A; W   61945

472 C31 H56 O Si 7604-83-3 T-67-5012-0 Silane, trimethyl[(4β-methylcholesteryl)oxy]-;; 4β-METHYLCHOLEST
EROL TRIMETHYLSILYL ETHER; W/NBS   61915

472 C31 H56 O Si 55429-62-4 AD-0-873-0 Silane, [[(3β,24R)-ergost-5-en-3-yl]oxy]trimethyl-;; CAMPESTEROL
TRIMETHYLSILYL ETHER; W/NBS   61914

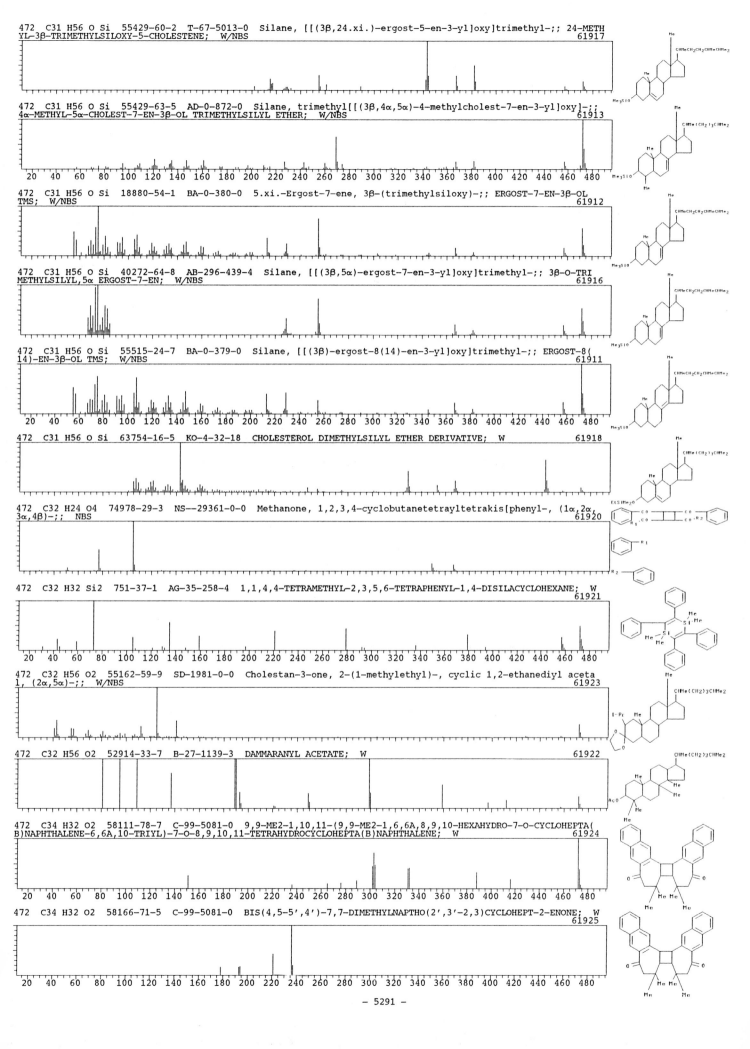

472   C31 H56 O Si   55429-60-2   T-67-5013-0   Silane, [[(3β,24.xi.)-ergost-5-en-3-yl]oxy]trimethyl-;; 24-METH
YL-3β-TRIMETHYLSILOXY-5-CHOLESTENE;   W/NBS   61917

472   C31 H56 O Si   55429-63-5   AD-0-872-0   Silane, trimethyl[[(3β,4α,5α)-4-methylcholest-7-en-3-yl]oxy]-;;
4α-METHYL-5α-CHOLEST-7-EN-3β-OL TRIMETHYLSILYL ETHER;   W/NBS   61913

472   C31 H56 O Si   18880-54-1   BA-0-380-0   5.xi.-Ergost-7-ene, 3β-(trimethylsiloxy)-;; ERGOST-7-EN-3β-OL
TMS;   W/NBS   61912

472   C31 H56 O Si   40272-64-8   AB-296-439-4   Silane, [[(3β,5α)-ergost-7-en-3-yl]oxy]trimethyl-;; 3β-O-TRI
METHYLSILYL,5α ERGOST-7-EN;   W/NBS   61916

472   C31 H56 O Si   55515-24-7   BA-0-379-0   Silane, [[(3β)-ergost-8(14)-en-3-yl]oxy]trimethyl-;; ERGOST-8(
14)-EN-3β-OL TMS;   W/NBS   61911

472   C31 H56 O Si   63754-16-5   KO-4-32-18   CHOLESTEROL DIMETHYLSILYL ETHER DERIVATIVE;   W   61918

472   C32 H24 O4   74978-29-3   NS--29361-0-0   Methanone, 1,2,3,4-cyclobutanetetrayltetrakis[phenyl-, (1α,2α,
3α,4β)-;;   NBS   61920

472   C32 H32 Si2   751-37-1   AG-35-258-4   1,1,4,4-TETRAMETHYL-2,3,5,6-TETRAPHENYL-1,4-DISILACYCLOHEXANE;   W
61921

472   C32 H56 O2   55162-59-9   SD-1981-0-0   Cholestan-3-one, 2-(1-methylethyl)-, cyclic 1,2-ethanediyl aceta
1, (2α,5α)-;;   W/NBS   61923

472   C32 H56 O2   52914-33-7   B-27-1139-3   DAMMARANYL ACETATE;   W   61922

472   C34 H32 O2   58111-78-7   C-99-5081-0   9,9-ME2-1,10,11-(9,9-ME2-1,6,6A,8,9,10-HEXAHYDRO-7-O-CYCLOHEPTA(
B)NAPHTHALENE-6,6A,10-TRIYL)-7-O-8,9,10,11-TETRAHYDROCYCLOHEPTA(B)NAPHTHALENE;   W   61924

472   C34 H32 O2   58166-71-5   C-99-5081-0   BIS(4,5-5',4')-7,7-DIMETHYLNAPTHO(2',3'-2,3)CYCLOHEPT-2-ENONE;   W
61925

472   C34 H32 O2   C-99-5081-0   BIS(4,5-5',4')-6,6-DIMETHYLNAPTHO(2',3'-2,3)CYCLOHEPT-2-ENONE;   W   61926

472   C34 H32 O2   90388-42-4   J-49-2540-0   9,24-Dihydroxy-8,15,23,30-tetramethyl[2.1.2.1]metacyclophane-1,6-
dienes;   W   124928

472   C34 H36 N2   92345-73-8   C-106-7173-0   α-Mesitylacetophenone ketazine;   W   124929

472   C34 H48 O   74421-21-9   HE-1982-0-0   Cholest-4-en-3-one, 2-(phenylmethylene)-;; 4-CHOLESTEN-3-ONE, 2-BE
NZYLIDEN-;   W/NBS   61927

472   C36 H24 O   78823-54-8   J-46-4432-0   5,6,11,12-Tetrahydro-5,12-diphenyl-6,11[1',2']benzeno-5,12-epoxyna
phthacene;   W   124930

473   C3 H9 Cl4 N3 O2 P2 S Sn   41083-44-7   EP-6676-0-0   4H-1,2,4,6,3,5-Thiatriazadiphosphorine, 3,3,5,5-tet
rachloro-3,3,5,5-tetrahydro-4-(trimethylstannyl)-, 1,1-dioxide;; N-(TRIMETHYLSTANNYL)-3,3,5,5-TETRACHLORO-
1,2,4,6,3,5-4H-THIATRIAZAPHOSPHORO(V)IN-S,S-DIOXIDE;   W/NBS   61928

473   C10 H10 Br2 N3 O3 P S2   RB-1982-15530-0   PHOSPHORODITHIOIC ACID, S-((6,8-DIBROMO-4-OXO-1,2,3-BENZOTRIA
ZIN-3;   W   124931

473   C15 H40 N O6 P Si4   55429-50-0   BA-0-294-0   2-Propanone, 1,3-bis[(trimethylsilyl)oxy]-, O-[bis[(tri
methylsilyl)oxy]phosphinyl]oxime;; O-TETRAKIS(TRIMETHYLSILYL)-DIHYDROXYACETONE-PHOSPHATE OXIME;   W/NBS   61931

473   C15 H4 D36 N O6 P Si4   55429-48-6   BA-0-295-0   O-TETRAKIS(TRIMETHYLSILYL-D9)-DIHYDROXYACETONE-PHOSPH
ATE OXIME;   W   61932

473   C15 H40 N O6 P Si4   55429-92-0   NS--29356-0-0   Phosphoric acid, bis(trimethylsilyl) 2-[(trimethylsily
l)oxy]-3-[[(trimethylsilyl)oxy]imino]propyl ester, (.+-.)-;;   NBS   61935

473   C15 H40 N O6 P Si4   BA-0-281-0   O-TETRAKIS(TRIMETHYLSILYL)-DL-GLYCERALDEHYDE-3-PHOSPHATE OXIME;   W
                                                                                                           61929

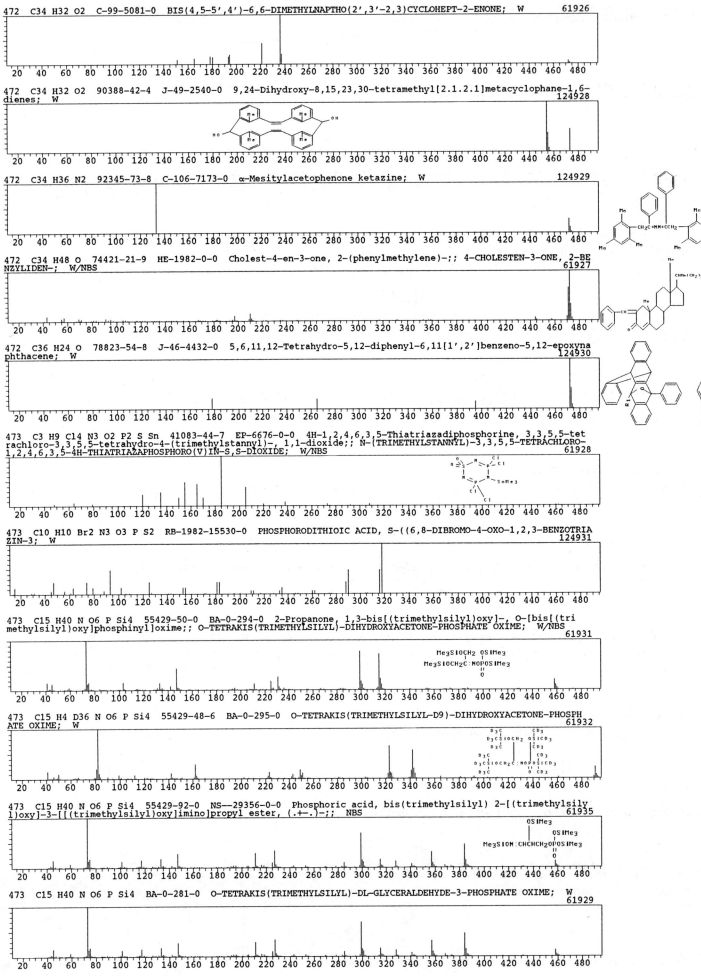

473　C15 H4 D36 N O6 P Si4　BA-0-284-0　O-TETRAKIS(TRIMETHYLSILYL-D9)-DL-GLYCERALDEHYDE-3-PHOSPHATE OXIME;
W
61930

473　C17 H17 Br2 N O5　69640-20-6　KC-1978-1593-0　DIETHYL 2-(5-BROMO-2-METHOXYPHENYL)PYRROLE-5-BROMO-3,4-
DICARBOXYLATE; W
61936

473　C17 H17 Br2 N O5　40808-65-9　KC-1978-1592-0　DIETHYL 2-(3,5-DIBROMO-2-METHOXYPHENYL)PYRROLE-3,4-DICAR
BOXYLATE; W
61937

473　C20 H13 Br2 N O3　77495-82-0　NS-0-0-0　Benzoic acid, 2-(benzoylphenylamino)-3,5-dibromo-; W/NBS
124932

473　C20 H40 N O4 P Si3　56227-26-0　EP-958-0-0　Phosphonic acid, [1-[(1-methylethylidene)amino]-2-[4-[(tri
methylsilyl)oxy]phenyl]ethyl]-, bis(trimethylsilyl) ester;; 1-AMINO-2(4-HYDROXYPHENYL)ETHYLPHOSPHONIC-ACETO
NE SCHIFF-DITMS; W/NBS
61938

473　C22 H36 I N O2　41080-94-8　EP-6520-0-0　Benzene, 1,3,5-tris(2,2-dimethylpropyl)-2-iodo-4-methyl-6-nit
ro-;; 1-IODO-3-METHYL-2,4,6-TRINEOPENTYL-5-NITROBENZENE; W/NBS
61939

473　C25 H23 N5 O5　89965-37-7　AH-115-195-0　1-(4'-N,N-Dimethylaminophenyl)-3-(4'-nitrophenyl)-5-methyl-5-
phenylamido-imidazolidin-2,4-dione; W
124933

473　C25 H31 N O8　62787-00-2　K-110-486-0　SENAMPELIN (A); W
61940

473　C25 H31 N O8　62860-52-0　K-110-486-0　MIXTURE OF SENAMPELIN B AND SENAMPELIN A; W
61941

473　C27 H23 N O7　33458-79-6　H-66-1127-0　(4bR*,10bS*,11R*)-5-Benzyloxycarbonyl-2,3,7,8-bis(methylenediox
y)-4b,5,6,10b,11,12-hexahydrobenzo[c]phenanthridin-11-ol; W
124934

473　C27 H27 N3 O3 S　78381-21-2　K-114-1923-0　Methylester of 2β-(Cyanmethyl)-1,2,3,4,6,7,12,12bβ-octahy
dro-4-oxo-α-(phenylthio)indolo[2,3-a]quinolizin-3α-propanoic acid; W
124935

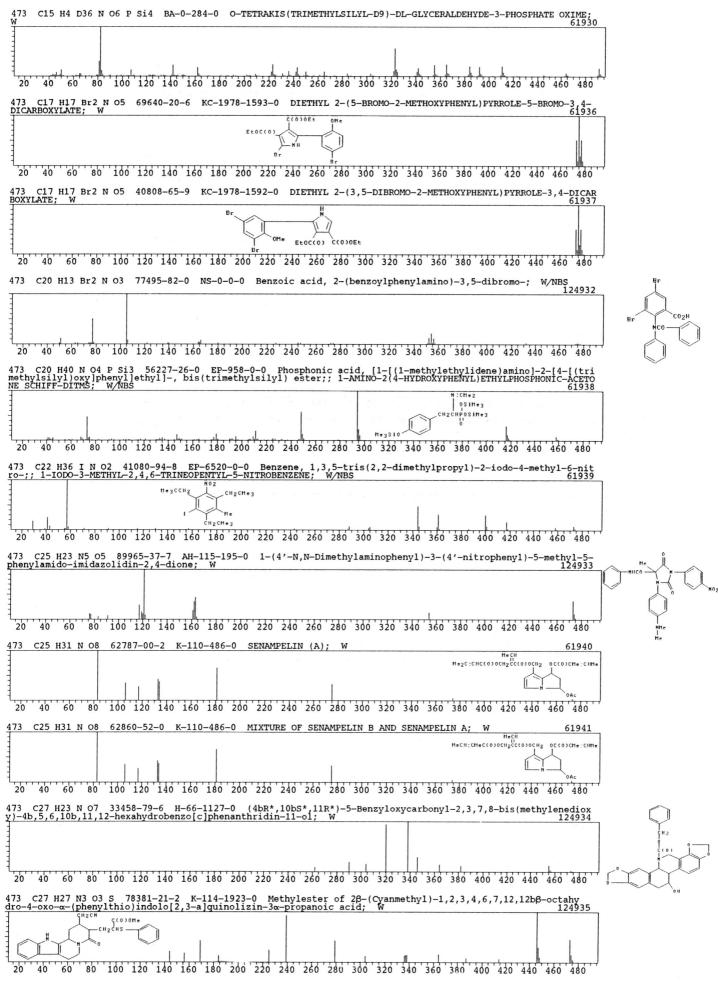

473 C27 H43 N O4 Si 74299-45-9 NS--29352-0-0 Pregn-4-ene-3,20-dione, 17-(acetyloxy)-6-methyl-, 3-[O-(tr
imethylsilyl)oxime], (6α)-;; NBS
61942

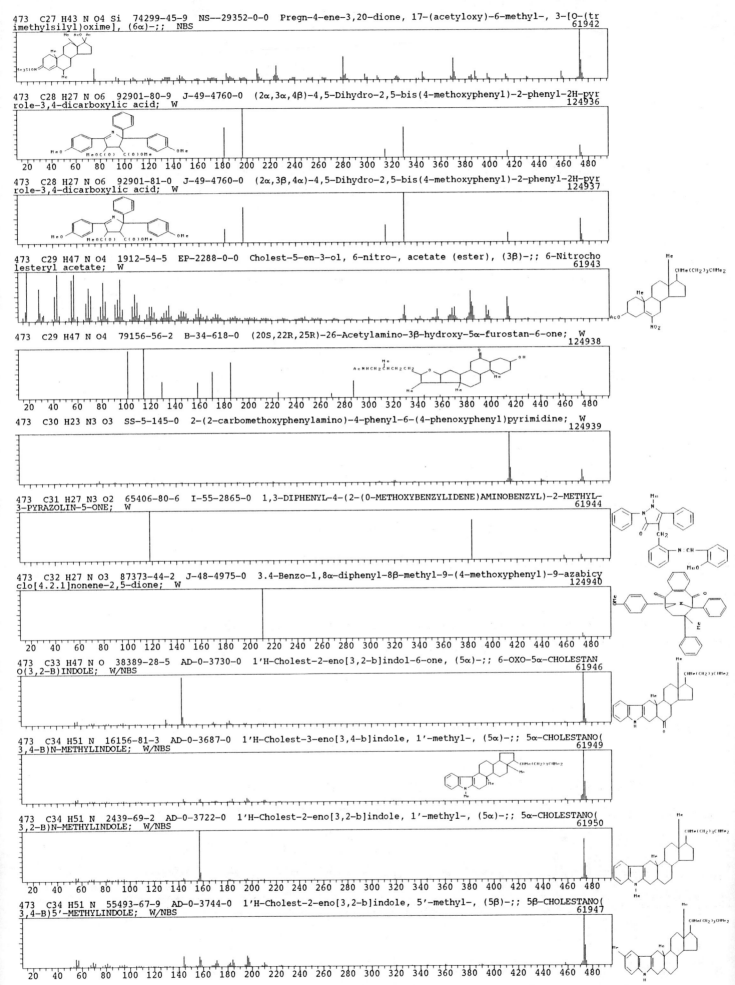

473 C28 H27 N O6 92901-80-9 J-49-4760-0 (2α,3α,4β)-4,5-Dihydro-2,5-bis(4-methoxyphenyl)-2-phenyl-2H-pyr
role-3,4-dicarboxylic acid; W
124936

473 C28 H27 N O6 92901-81-0 J-49-4760-0 (2α,3β,4α)-4,5-Dihydro-2,5-bis(4-methoxyphenyl)-2-phenyl-2H-pyr
role-3,4-dicarboxylic acid; W
124937

473 C29 H47 N O4 1912-54-5 EP-2288-0-0 Cholest-5-en-3-ol, 6-nitro-, acetate (ester), (3β)-;; 6-Nitrocho
lesteryl acetate; W
61943

473 C29 H47 N O4 79156-56-2 B-34-618-0 (20S,22R,25R)-26-Acetylamino-3β-hydroxy-5α-furostan-6-one; W
124938

473 C30 H23 N3 O3 SS-5-145-0 2-(2-carbomethoxyphenylamino)-4-phenyl-6-(4-phenoxyphenyl)pyrimidine; W
124939

473 C31 H27 N3 O2 65406-80-6 I-55-2865-0 1,3-DIPHENYL-4-(2-(O-METHOXYBENZYLIDENE)AMINOBENZYL)-2-METHYL-
3-PYRAZOLIN-5-ONE; W
61944

473 C32 H27 N O3 87373-44-2 J-48-4975-0 3.4-Benzo-1,8α-diphenyl-8β-methyl-9-(4-methoxyphenyl)-9-azabicy
clo[4.2.1]nonene-2,5-dione; W
124940

473 C33 H47 N O 38389-28-5 AD-0-3730-0 1'H-Cholest-2-eno[3,2-b]indol-6-one, (5α)-;; 6-OXO-5α-CHOLESTAN
O(3,2-B)INDOLE; W/NBS
61946

473 C34 H51 N 16156-81-3 AD-0-3687-0 1'H-Cholest-3-eno[3,4-b]indole, 1'-methyl-, (5α)-;; 5α-CHOLESTANO(
3,4-B)N-METHYLINDOLE; W/NBS
61949

473 C34 H51 N 2439-69-2 AD-0-3722-0 1'H-Cholest-2-eno[3,2-b]indole, 1'-methyl-, (5α)-;; 5α-CHOLESTANO(
3,2-B)N-METHYLINDOLE; W/NBS
61950

473 C34 H51 N 55493-67-9 AD-0-3744-0 1'H-Cholest-2-eno[3,2-b]indole, 5'-methyl-, (5β)-;; 5β-CHOLESTANO(
3,4-B)5'-METHYLINDOLE; W/NBS
61947

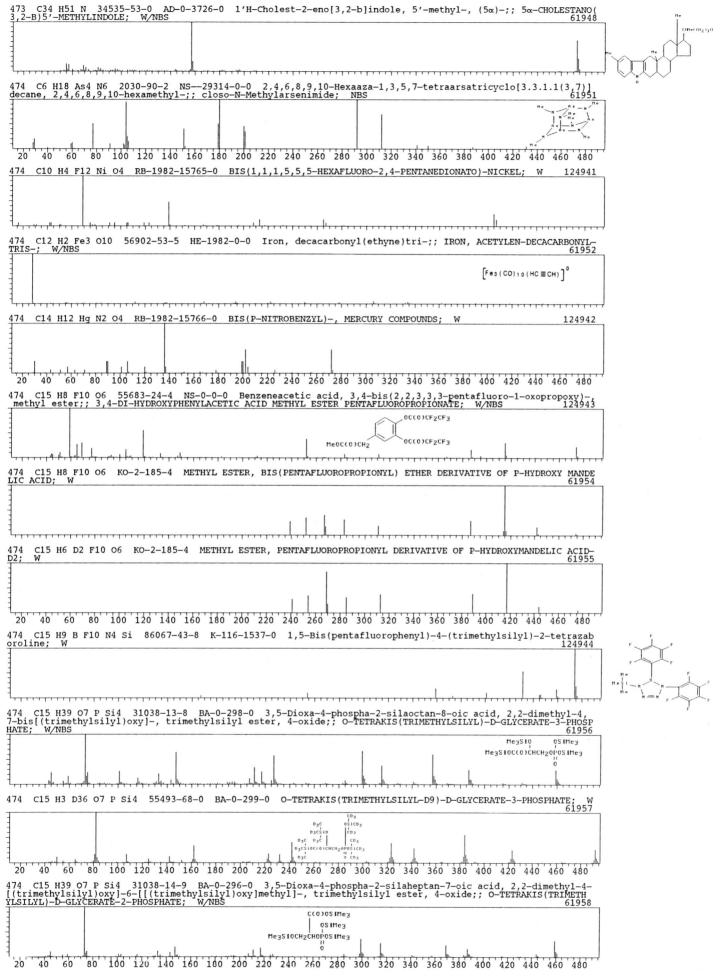

473 C34 H51 N 34535-53-0 AD-0-3726-0 1'H-Cholest-2-eno[3,2-b]indole, 5'-methyl-, (5α)-;; 5α-CHOLESTANO(3,2-B)5'-METHYLINDOLE; W/NBS
61948

474 C6 H18 As4 N6 2030-90-2 NS--29314-0-0 2,4,6,8,9,10-Hexaaza-1,3,5,7-tetraarsatricyclo[3.3.1.1(3,7)]decane, 2,4,6,8,9,10-hexamethyl-;; closo-N-Methylarsenimide; NBS
61951

474 C10 H4 F12 Ni O4 RB-1982-15765-0 BIS(1,1,1,5,5,5-HEXAFLUORO-2,4-PENTANEDIONATO)-NICKEL; W 124941

474 C12 H2 Fe3 O10 56902-53-5 HE-1982-0-0 Iron, decacarbonyl(ethyne)tri-;; IRON, ACETYLEN-DECACARBONYL-TRIS-; W/NBS
61952

$$[Fe_3(CO)_{10}(HC≡CH)]^0$$

474 C14 H12 Hg N2 O4 RB-1982-15766-0 BIS(P-NITROBENZYL)-, MERCURY COMPOUNDS; W 124942

474 C15 H8 F10 O6 55683-24-4 NS-0-0-0 Benzeneacetic acid, 3,4-bis(2,2,3,3,3-pentafluoro-1-oxopropoxy)-methyl ester;; 3,4-DI-HYDROXYPHENYLACETIC ACID METHYL ESTER PENTAFLUOROPROPIONATE; W/NBS 124943

474 C15 H8 F10 O6 KO-2-185-4 METHYL ESTER, BIS(PENTAFLUOROPROPIONYL) ETHER DERIVATIVE OF P-HYDROXY MANDELIC ACID; W
61954

474 C15 H6 D2 F10 O6 KO-2-185-4 METHYL ESTER, PENTAFLUOROPROPIONYL DERIVATIVE OF P-HYDROXYMANDELIC ACID-D2; W
61955

474 C15 H9 B F10 N4 Si 86067-43-8 K-116-1537-0 1,5-Bis(pentafluorophenyl)-4-(trimethylsilyl)-2-tetrazaboroline; W 124944

474 C15 H39 O7 P Si4 31038-13-8 BA-0-298-0 3,5-Dioxa-4-phospha-2-silaoctan-8-oic acid, 2,2-dimethyl-4,7-bis[(trimethylsilyl)oxy]-, trimethylsilyl ester, 4-oxide;; O-TETRAKIS(TRIMETHYLSILYL)-D-GLYCERATE-3-PHOSPHATE; W/NBS
61956

474 C15 H3 D36 O7 P Si4 55493-68-0 BA-0-299-0 O-TETRAKIS(TRIMETHYLSILYL-D9)-D-GLYCERATE-3-PHOSPHATE; W
61957

474 C15 H39 O7 P Si4 31038-14-9 BA-0-296-0 3,5-Dioxa-4-phospha-2-silaheptan-7-oic acid, 2,2-dimethyl-4-[(trimethylsilyl)oxy]-6-[[(trimethylsilyl)oxy]methyl]-, trimethylsilyl ester, 4-oxide;; O-TETRAKIS(TRIMETHYLSILYL)-D-GLYCERATE-2-PHOSPHATE; W/NBS
61958

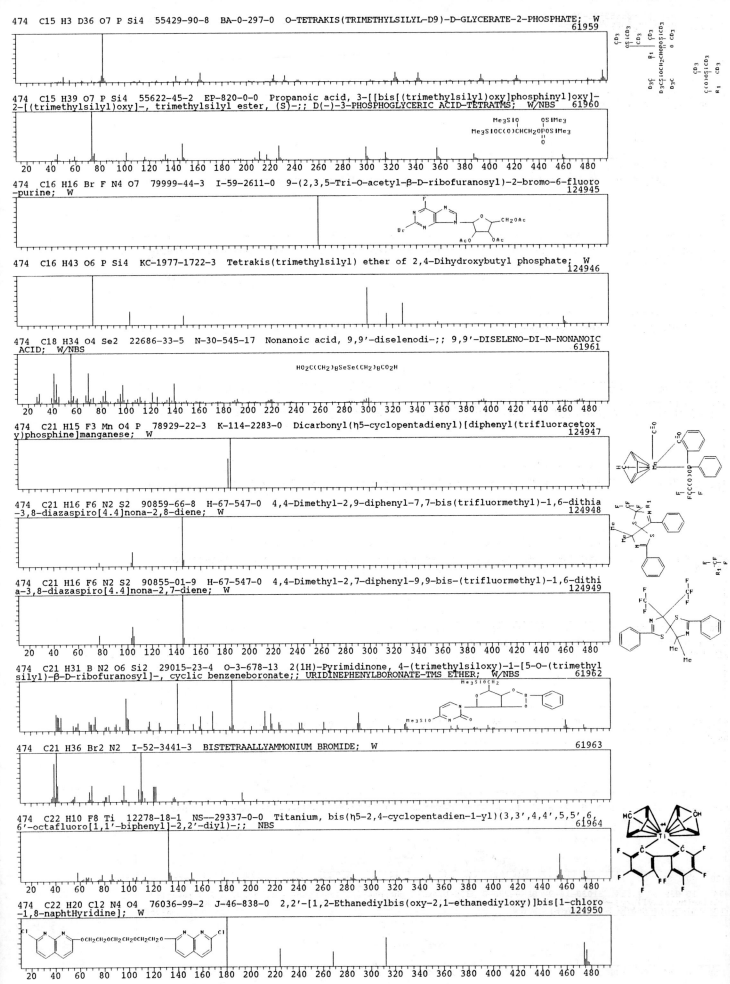

474   C15 H3 D36 O7 P Si4   55429-90-8   BA-0-297-0   O-TETRAKIS(TRIMETHYLSILYL-D9)-D-GLYCERATE-2-PHOSPHATE;   W
61959

474   C15 H39 O7 P Si4   55622-45-2   EP-820-0-0   Propanoic acid, 3-[[bis[(trimethylsilyl)oxy]phosphinyl]oxy]-
2-[(trimethylsilyl)oxy]-, trimethylsilyl ester, (S)-;; D(-)-3-PHOSPHOGLYCERIC ACID-TETRATMS;   W/NBS   61960

Me3SiO   OSIMe3
Me3SiOC(O)CHCH2OPOSIMe3
                  ||
                  O

474   C16 H16 Br F N4 O7   79999-44-3   I-59-2611-0   9-(2,3,5-Tri-O-acetyl-β-D-ribofuranosyl)-2-bromo-6-fluoro
_purine;   W
124945

474   C16 H43 O6 P Si4   KC-1977-1722-3   Tetrakis(trimethylsilyl) ether of 2,4-Dihydroxybutyl phosphate;   W
124946

474   C18 H34 O4 Se2   22686-33-5   N-30-545-17   Nonanoic acid, 9,9'-diselenodi-;; 9,9'-DISELENO-DI-N-NONANOIC
ACID;   W/NBS   61961

HO2C((CH2)8SeSe((CH2)8CO2H

474   C21 H15 F3 Mn O4 P   78929-22-3   K-114-2283-0   Dicarbonyl(η5-cyclopentadienyl)[diphenyl(trifluoracetox
y)phosphine]manganese;   W   124947

474   C21 H16 F6 N2 S2   90859-66-8   H-67-547-0   4,4-Dimethyl-2,9-diphenyl-7,7-bis(trifluormethyl)-1,6-dithia
-3,8-diazaspiro[4.4]nona-2,8-diene;   W   124948

474   C21 H16 F6 N2 S2   90855-01-9   H-67-547-0   4,4-Dimethyl-2,7-diphenyl-9,9-bis-(trifluormethyl)-1,6-dithi
a-3,8-diazaspiro[4.4]nona-2,7-diene;   W   124949

474   C21 H31 B N2 O6 Si2   29015-23-4   O-3-678-13   2(1H)-Pyrimidinone, 4-(trimethylsiloxy)-1-[5-O-(trimethyl
silyl)-β-D-ribofuranosyl]-, cyclic benzeneboronate;;   URIDINEPHENYLBORONATE-TMS ETHER;   W/NBS   61962

474   C21 H36 Br2 N2   I-52-3441-3   BISTETRAALLYAMMONIUM BROMIDE;   W   61963

474   C22 H10 F8 Ti   12278-18-1   NS--29337-0-0   Titanium, bis(η5-2,4-cyclopentadien-1-yl)(3,3',4,4',5,5'6,
6'-octafluoro[1,1'-biphenyl]-2,2'-diyl)-;;   NBS   61964

474   C22 H20 Cl2 N4 O4   76036-99-2   J-46-838-0   2,2'-[1,2-Ethanediylbis(oxy-2,1-ethanediyloxy)]bis[1-chloro
-1,8-naphtHyridine];   W   124950

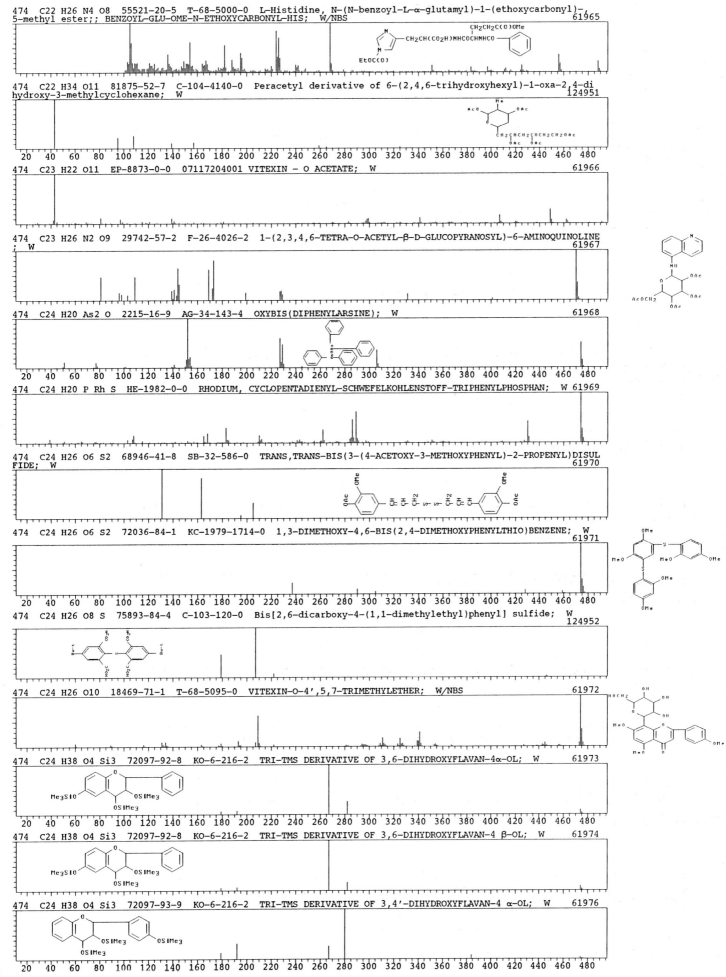

474　C22 H26 N4 O8　55521-20-5　T-68-5000-0　L-Histidine, N-(N-benzoyl-L-α-glutamyl)-1-(ethoxycarbonyl)-
5-methyl ester;; BENZOYL-GLU-OME-N-ETHOXYCARBONYL-HIS;　W/NBS
61965

474　C22 H34 O11　81875-52-7　C-104-4140-0　Peracetyl derivative of 6-(2,4,6-trihydroxyhexyl)-1-oxa-2,4-di
hydroxy-3-methylcyclohexane;　W
124951

474　C23 H22 O11　EP-8873-0-0　07117204001　VITEXIN - O ACETATE;　W
61966

474　C23 H26 N2 O9　29742-57-2　F-26-4026-2　1-(2,3,4,6-TETRA-O-ACETYL-β-D-GLUCOPYRANOSYL)-6-AMINOQUINOLINE
; 　W
61967

474　C24 H20 As2 O　2215-16-9　AG-34-143-4　OXYBIS(DIPHENYLARSINE);　W
61968

474　C24 H20 P Rh S　HE-1982-0-0　RHODIUM, CYCLOPENTADIENYL-SCHWEFELKOHLENSTOFF-TRIPHENYLPHOSPHAN;　W 61969

474　C24 H26 O6 S2　68946-41-8　SB-32-586-0　TRANS,TRANS-BIS(3-(4-ACETOXY-3-METHOXYPHENYL)-2-PROPENYL)DISUL
FIDE;　W
61970

474　C24 H26 O6 S2　72036-84-1　KC-1979-1714-0　1,3-DIMETHOXY-4,6-BIS(2,4-DIMETHOXYPHENYLTHIO)BENZENE;　W
61971

474　C24 H26 O8 S　75893-84-4　C-103-120-0　Bis[2,6-dicarboxy-4-(1,1-dimethylethyl)phenyl] sulfide;　W
124952

474　C24 H26 O10　18469-71-1　T-68-5095-0　VITEXIN-O-4',5,7-TRIMETHYLETHER;　W/NBS
61972

474　C24 H38 O4 Si3　72097-92-8　KO-6-216-2　TRI-TMS DERIVATIVE OF 3,6-DIHYDROXYFLAVAN-4α-OL;　W
61973

474　C24 H38 O4 Si3　72097-92-8　KO-6-216-2　TRI-TMS DERIVATIVE OF 3,6-DIHYDROXYFLAVAN-4 β-OL;　W
61974

474　C24 H38 O4 Si3　72097-93-9　KO-6-216-2　TRI-TMS DERIVATIVE OF 3,4'-DIHYDROXYFLAVAN-4 α-OL;　W
61976

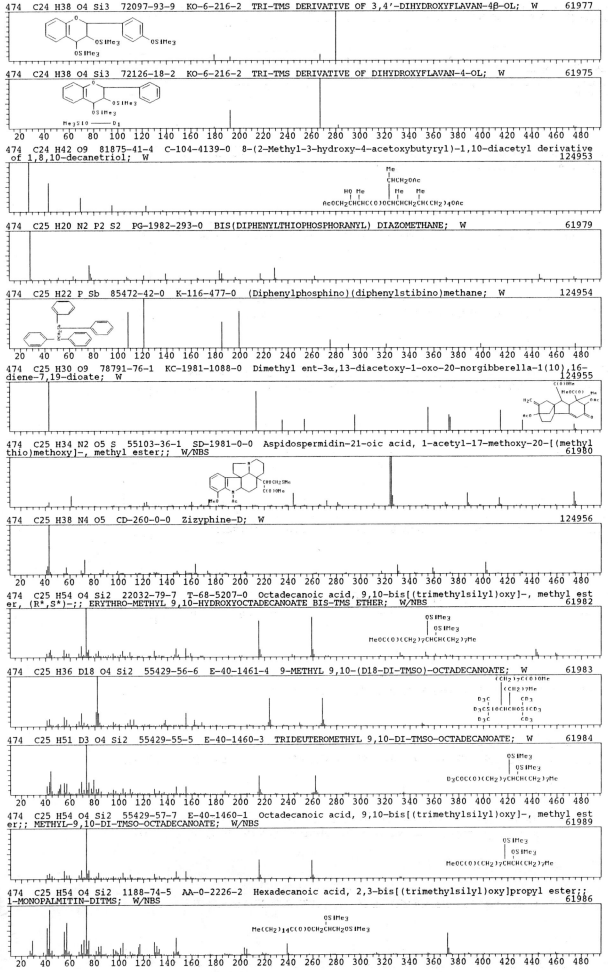

474 C24 H38 O4 Si3 72097-93-9 KO-6-216-2 TRI-TMS DERIVATIVE OF 3,4'-DIHYDROXYFLAVAN-4β-OL; W   61977

474 C24 H38 O4 Si3 72126-18-2 KO-6-216-2 TRI-TMS DERIVATIVE OF DIHYDROXYFLAVAN-4-OL; W   61975

474 C24 H42 O9 81875-41-4 C-104-4139-0 8-(2-Methyl-3-hydroxy-4-acetoxybutyryl)-1,10-diacetyl derivative
of 1,8,10-decanetriol; W   124953

474 C25 H20 N2 P2 S2 PG-1982-293-0 BIS(DIPHENYLTHIOPHOSPHORANYL) DIAZOMETHANE; W   61979

474 C25 H22 P Sb 85472-42-0 K-116-477-0 (Diphenylphosphino)(diphenylstibino)methane; W   124954

474 C25 H30 O9 78791-76-1 KC-1981-1088-0 Dimethyl ent-3α,13-diacetoxy-1-oxo-20-norgibberella-1(10),16-
diene-7,19-dioate; W   124955

474 C25 H34 N2 O5 S 55103-36-1 SD-1981-0-0 Aspidospermidin-21-oic acid, 1-acetyl-17-methoxy-20-[(methyl
thio)methoxy]-, methyl ester;; W/NBS   61980

474 C25 H38 N4 O5 CD-260-0-0 Zizyphine-D; W   124956

474 C25 H54 O4 Si2 22032-79-7 T-68-5207-0 Octadecanoic acid, 9,10-bis[(trimethylsilyl)oxy]-, methyl est
er, (R*,S*)-;; ERYTHRO-METHYL 9,10-HYDROXYOCTADECANOATE BIS-TMS ETHER; W/NBS   61982

474 C25 H36 D18 O4 Si2 55429-56-6 E-40-1461-4 9-METHYL 9,10-(D18-DI-TMSO)-OCTADECANOATE; W   61983

474 C25 H51 D3 O4 Si2 55429-55-5 E-40-1460-3 TRIDEUTEROMETHYL 9,10-DI-TMSO-OCTADECANOATE; W   61984

474 C25 H54 O4 Si2 55429-57-7 E-40-1460-1 Octadecanoic acid, 9,10-bis[(trimethylsilyl)oxy]-, methyl est
er;; METHYL-9,10-DI-TMSO-OCTADECANOATE; W/NBS   61989

474 C25 H54 O4 Si2 1188-74-5 AA-0-2226-2 Hexadecanoic acid, 2,3-bis[(trimethylsilyl)oxy]propyl ester;;
1-MONOPALMITIN-DITMS; W/NBS   61986

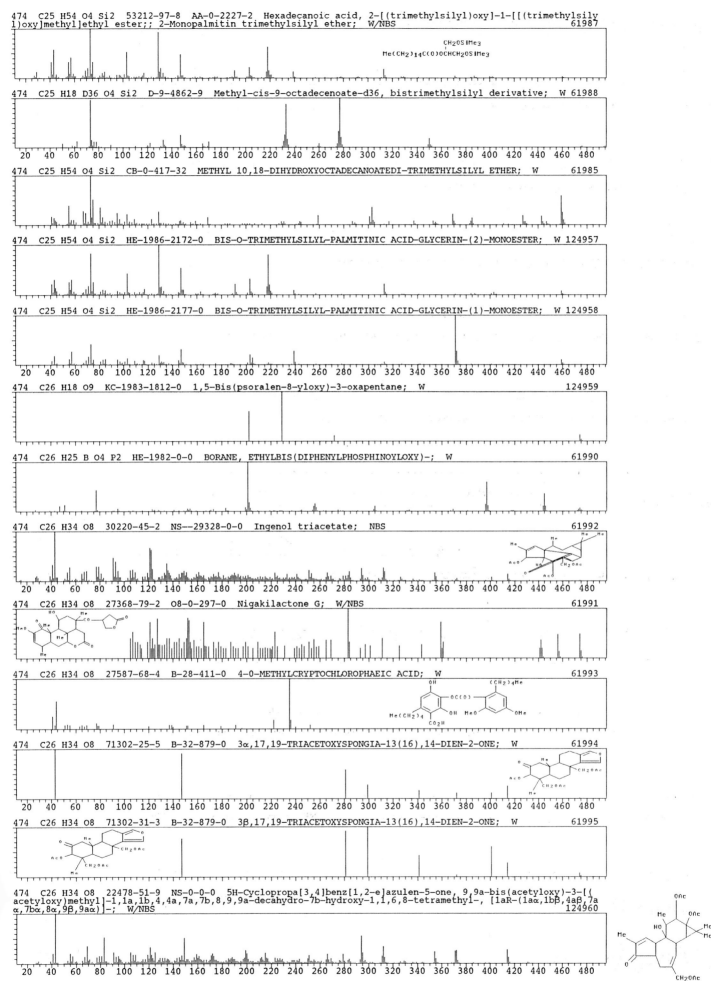

474   C25 H54 O4 Si2   53212-97-8   AA-0-2227-2   Hexadecanoic acid, 2-[(trimethylsilyl)oxy]-1-[[(trimethylsilyl)oxy]methyl]ethyl ester;; 2-Monopalmitin trimethylsilyl ether;   W/NBS   61987

CH₂OSiMe₃
Me(CH₂)₁₄C(O)OCHCH₂OSiMe₃

474   C25 H18 D36 O4 Si2   D-9-4862-9   Methyl-cis-9-octadecenoate-d36, bistrimethylsilyl derivative;   W 61988

474   C25 H54 O4 Si2   CB-0-417-32   METHYL 10,18-DIHYDROXYOCTADECANOATEDI-TRIMETHYLSILYL ETHER;   W   61985

474   C25 H54 O4 Si2   HE-1986-2172-0   BIS-O-TRIMETHYLSILYL-PALMITINIC ACID-GLYCERIN-(2)-MONOESTER;   W 124957

474   C25 H54 O4 Si2   HE-1986-2177-0   BIS-O-TRIMETHYLSILYL-PALMITINIC ACID-GLYCERIN-(1)-MONOESTER;   W 124958

474   C26 H18 O9   KC-1983-1812-0   1,5-Bis(psoralen-8-yloxy)-3-oxapentane;   W   124959

474   C26 H25 B O4 P2   HE-1982-0-0   BORANE, ETHYLBIS(DIPHENYLPHOSPHINOYLOXY)-;   W   61990

474   C26 H34 O8   30220-45-2   NS--29328-0-0   Ingenol triacetate;   NBS   61992

474   C26 H34 O8   27368-79-2   O8-0-297-0   Nigakilactone G;   W/NBS   61991

474   C26 H34 O8   27587-68-4   B-28-411-0   4-0-METHYLCRYPTOCHLOROPHAEIC ACID;   W   61993

474   C26 H34 O8   71302-25-5   B-32-879-0   3α,17,19-TRIACETOXYSPONGIA-13(16),14-DIEN-2-ONE;   W   61994

474   C26 H34 O8   71302-31-3   B-32-879-0   3β,17,19-TRIACETOXYSPONGIA-13(16),14-DIEN-2-ONE;   W   61995

474   C26 H34 O8   22478-51-9   NS-0-0-0   5H-Cyclopropa[3,4]benz[1,2-e]azulen-5-one, 9,9a-bis(acetyloxy)-3-[(acetyloxy)methyl]-1,1a,1b,4,4a,7a,7b,8,9,9a-decahydro-7b-hydroxy-1,1,6,8-tetramethyl-, [1aR-(1aα,1bβ,4aβ,7aα,7bα,8α,9β,9aα)]-;   W/NBS   124960

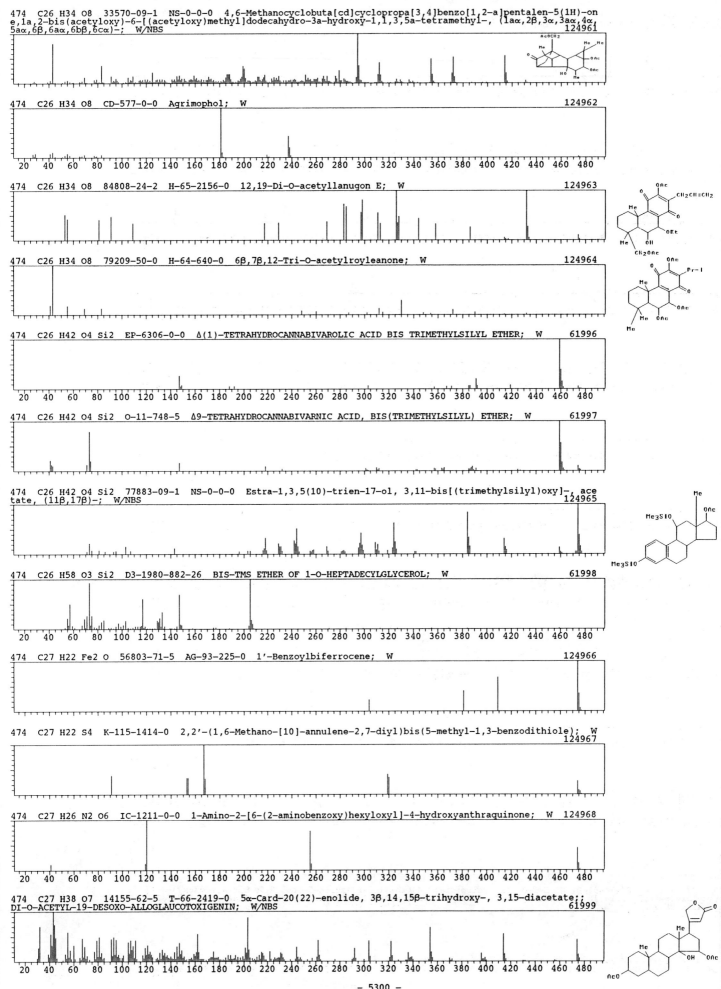

474   C26 H34 O8   33570-09-1   NS-0-0-0   4,6-Methanocyclobuta[cd]cyclopropa[3,4]benzo[1,2-a]pentalen-5(1H)-one,1a,2-bis(acetyloxy)-6-[(acetyloxy)methyl]dodecahydro-3a-hydroxy-1,1,3,5a-tetramethyl-, (1aα,2β,3α,3aα,4α,5aα,6β,6aα,6bβ,6cα)-;   W/NBS   124961

474   C26 H34 O8   CD-577-0-0   Agrimophol;   W   124962

474   C26 H34 O8   84808-24-2   H-65-2156-0   12,19-Di-O-acetyllanugon E;   W   124963

474   C26 H34 O8   79209-50-0   H-64-640-0   6β,7β,12-Tri-O-acetylroyleanone;   W   124964

474   C26 H42 O4 Si2   EP-6306-0-0   Δ(1)-TETRAHYDROCANNABIVAROLIC ACID BIS TRIMETHYLSILYL ETHER;   W   61996

474   C26 H42 O4 Si2   O-11-748-5   Δ9-TETRAHYDROCANNABIVARNIC ACID, BIS(TRIMETHYLSILYL) ETHER;   W   61997

474   C26 H42 O4 Si2   77883-09-1   NS-0-0-0   Estra-1,3,5(10)-trien-17-ol, 3,11-bis[(trimethylsilyl)oxy]-, acetate, (11β,17β)-;   W/NBS   124965

474   C26 H58 O3 Si2   D3-1980-882-26   BIS-TMS ETHER OF 1-O-HEPTADECYLGLYCEROL;   W   61998

474   C27 H22 Fe2 O   56803-71-5   AG-93-225-0   1'-Benzoylbiferrocene;   W   124966

474   C27 H22 S4   K-115-1414-0   2,2'-(1,6-Methano-[10]-annulene-2,7-diyl)bis(5-methyl-1,3-benzodithiole);   W   124967

474   C27 H26 N2 O6   IC-1211-0-0   1-Amino-2-[6-(2-aminobenzoxy)hexyloxyl]-4-hydroxyanthraquinone;   W   124968

474   C27 H38 O7   14155-62-5   T-66-2419-0   5α-Card-20(22)-enolide, 3β,14,15β-trihydroxy-, 3,15-diacetate;; DI-O-ACETYL-19-DESOXO-ALLOGLAUCOTOXIGENIN;   W/NBS   61999

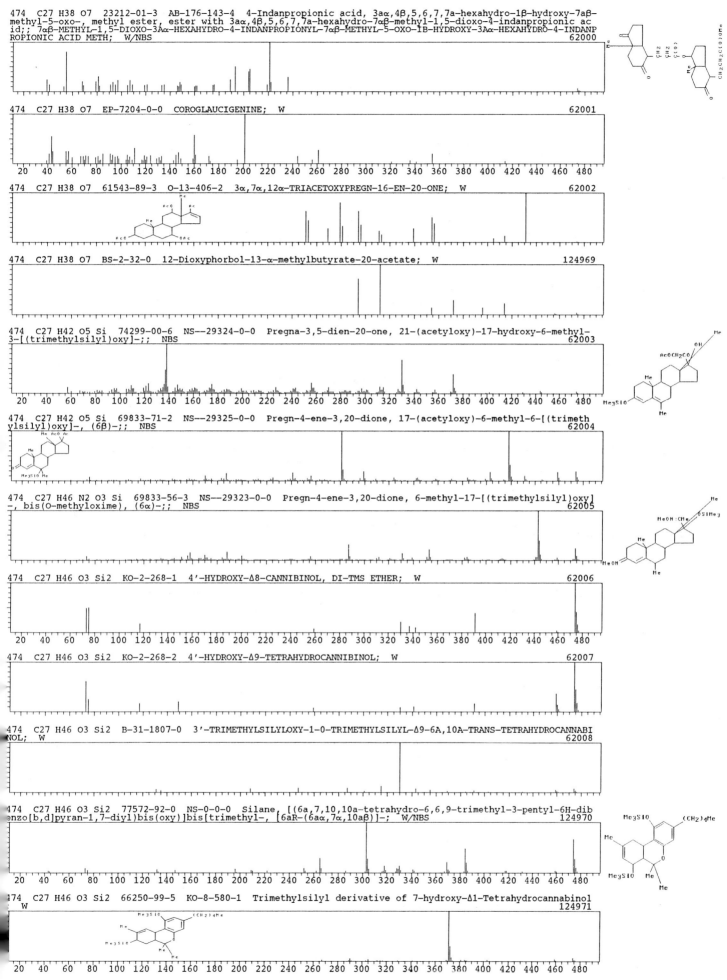

474 C27 H38 O7 23212-01-3 AB-176-143-4 4-Indanpropionic acid, 3aα,4β,5,6,7,7a-hexahydro-1β-hydroxy-7aβ-methyl-5-oxo-, methyl ester, ester with 3aα,4β,5,6,7,7a-hexahydro-7αβ-methyl-1,5-dioxo-4-indanpropionic acid;; 7αβ-METHYL-1,5-DIOXO-3Aα-HEXAHYDRO-4-INDANPROPIONYL-7αβ-METHYL-5-OXO-1B-HYDROXY-3Aα-HEXAHYDRO-4-INDANPROPIONIC ACID METH; W/NBS   62000

474 C27 H38 O7 EP-7204-0-0 COROGLAUCIGENINE; W   62001

474 C27 H38 O7 61543-89-3 O-13-406-2 3α,7α,12α-TRIACETOXYPREGN-16-EN-20-ONE; W   62002

474 C27 H38 O7 BS-2-32-0 12-Dioxyphorbol-13-α-methylbutyrate-20-acetate; W   124969

474 C27 H42 O5 Si 74299-00-6 NS--29324-0-0 Pregna-3,5-dien-20-one, 21-(acetyloxy)-17-hydroxy-6-methyl-3-[(trimethylsilyl)oxy]-;; NBS   62003

474 C27 H42 O5 Si 69833-71-2 NS--29325-0-0 Pregn-4-ene-3,20-dione, 17-(acetyloxy)-6-methyl-6-[(trimethylsilyl)oxy]-, (6β)-;; NBS   62004

474 C27 H46 N2 O3 Si 69833-56-3 NS--29323-0-0 Pregn-4-ene-3,20-dione, 6-methyl-17-[(trimethylsilyl)oxy]-, bis(O-methyloxime), (6α)-;; NBS   62005

474 C27 H46 O3 Si2 KO-2-268-1 4'-HYDROXY-Δ8-CANNIBINOL, DI-TMS ETHER; W   62006

474 C27 H46 O3 Si2 KO-2-268-2 4'-HYDROXY-Δ9-TETRAHYDROCANNIBINOL; W   62007

474 C27 H46 O3 Si2 B-31-1807-0 3'-TRIMETHYLSILYLYLOXY-1-0-TRIMETHYLSILYL-Δ9-6A,10A-TRANS-TETRAHYDROCANNABINOL; W   62008

474 C27 H46 O3 Si2 77572-92-0 NS-0-0-0 Silane, [(6a,7,10,10a-tetrahydro-6,6,9-trimethyl-3-pentyl-6H-dibenzo[b,d]pyran-1,7-diyl)bis(oxy)]bis[trimethyl-, [6aR-(6aα,7α,10aβ)]-; W/NBS   124970

174 C27 H46 O3 Si2 66250-99-5 KO-8-580-1 Trimethylsilyl derivative of 7-hydroxy-Δ1-Tetrahydrocannabinol; W   124971

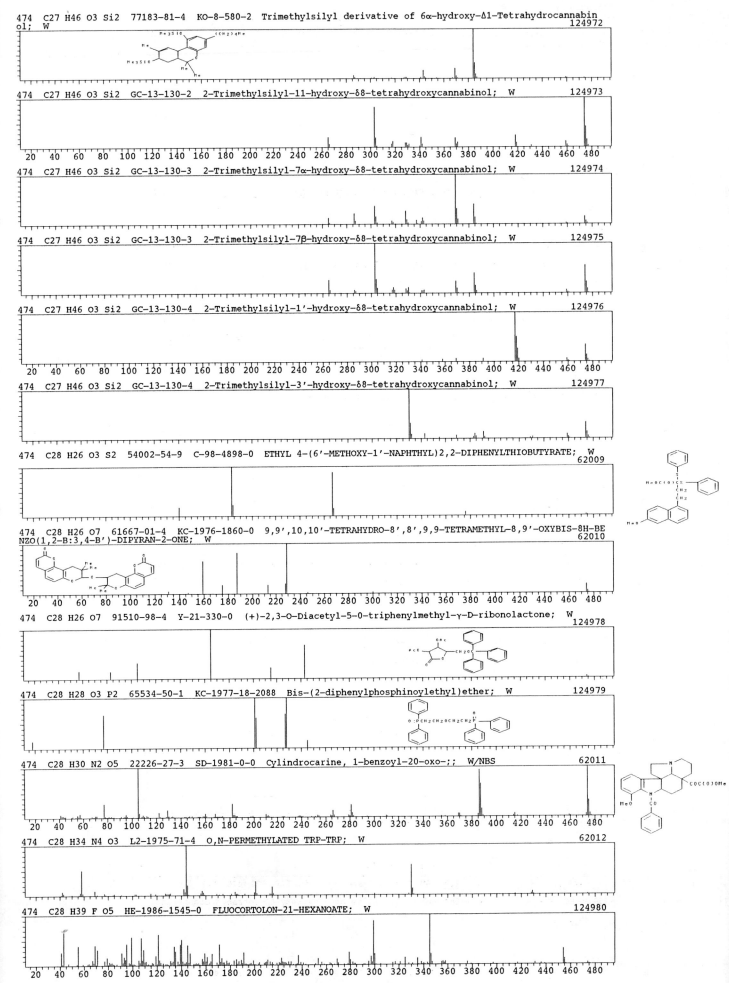

474 C27 H46 O3 Si2 77183-81-4 KO-8-580-2 Trimethylsilyl derivative of 6α-hydroxy-Δ1-Tetrahydrocannabinol; W
124972

474 C27 H46 O3 Si2 GC-13-130-2 2-Trimethylsilyl-11-hydroxy-δ8-tetrahydroxycannabinol; W
124973

474 C27 H46 O3 Si2 GC-13-130-3 2-Trimethylsilyl-7α-hydroxy-δ8-tetrahydroxycannabinol; W
124974

474 C27 H46 O3 Si2 GC-13-130-3 2-Trimethylsilyl-7β-hydroxy-δ8-tetrahydroxycannabinol; W
124975

474 C27 H46 O3 Si2 GC-13-130-4 2-Trimethylsilyl-1'-hydroxy-δ8-tetrahydroxycannabinol; W
124976

474 C27 H46 O3 Si2 GC-13-130-4 2-Trimethylsilyl-3'-hydroxy-δ8-tetrahydroxycannabinol; W
124977

474 C28 H26 O3 S2 54002-54-9 C-98-4898-0 ETHYL 4-(6'-METHOXY-1'-NAPHTHYL)2,2-DIPHENYLTHIOBUTYRATE; W
62009

474 C28 H26 O7 61667-01-4 KC-1976-1860-0 9,9',10,10'-TETRAHYDRO-8',8',9,9-TETRAMETHYL-8,9'-OXYBIS-8H-BENZO(1,2-B:3,4-B')-DIPYRAN-2-ONE; W
62010

474 C28 H26 O7 91510-98-4 Y-21-330-0 (+)-2,3-O-Diacetyl-5-0-triphenylmethyl-γ-D-ribonolactone; W
124978

474 C28 H28 O3 P2 65534-50-1 KC-1977-18-2088 Bis-(2-diphenylphosphinoylethyl)ether; W
124979

474 C28 H30 N2 O5 22226-27-3 SD-1981-0-0 Cylindrocarine, 1-benzoyl-20-oxo-;; W/NBS
62011

474 C28 H34 N4 O3 L2-1975-71-4 O,N-PERMETHYLATED TRP-TRP; W
62012

474 C28 H39 F O5 HE-1986-1545-0 FLUOCORTOLON-21-HEXANOATE; W
124980

474 C28 H42 O4 S 75893-97-9 C-103-122-0 Bis[2-(1-hydroxy-1-methyethyl)-4-(1,1-dimethyethyl-6-(hydroxy
methyl)phenyl] sulfide; W   124981

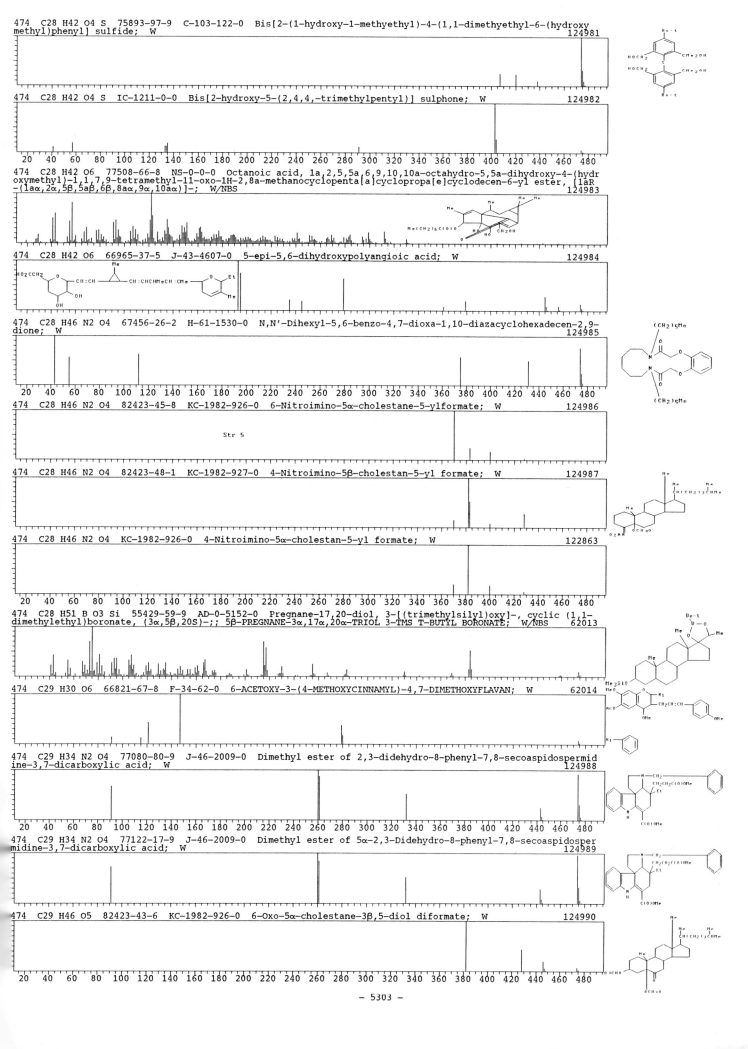

474 C28 H42 O4 S IC-1211-0-0 Bis[2-hydroxy-5-(2,4,4,-trimethylpentyl)] sulphone; W   124982

474 C28 H42 O6 77508-66-8 NS-0-0-0 Octanoic acid, 1a,2,5,5a,6,9,10,10a-octahydro-5,5a-dihydroxy-4-(hydr
oxymethyl)-1,1,7,9-tetramethyl-11-oxo-1H-2,8a-methanocyclopenta[a]cyclopropa[e]cyclodecen-6-yl ester, [1aR
-(1aα,2α,5β,5aβ,6β,8aα,9α,10aα)]-; W/NBS   124983

474 C28 H42 O6 66965-37-5 J-43-4607-0 5-epi-5,6-dihydroxypolyangioic acid; W   124984

474 C28 H46 N2 O4 67456-26-2 H-61-1530-0 N,N'-Dihexyl-5,6-benzo-4,7-dioxa-1,10-diazacyclohexadecen-2,9-
dione; W   124985

474 C28 H46 N2 O4 82423-45-8 KC-1982-926-0 6-Nitroimino-5α-cholestane-5-ylformate; W   124986

Str 5

474 C28 H46 N2 O4 82423-48-1 KC-1982-927-0 4-Nitroimino-5β-cholestan-5-yl formate; W   124987

474 C28 H46 N2 O4 KC-1982-926-0 4-Nitroimino-5α-cholestan-5-yl formate; W   122863

474 C28 H51 B O3 Si 55429-59-9 AD-0-5152-0 Pregnane-17,20-diol, 3-[(trimethylsilyl)oxy]-, cyclic (1,1-
dimethylethyl)boronate, (3α,5β,20S)-;; 5β-PREGNANE-3α,17α,20α-TRIOL 3-TMS T-BUTYL BORONATE; W/NBS   62013

474 C29 H30 O6 66821-67-8 F-34-62-0 6-ACETOXY-3-(4-METHOXYCINNAMYL)-4,7-DIMETHOXYFLAVAN; W   62014

474 C29 H34 N2 O4 77080-80-9 J-46-2009-0 Dimethyl ester of 2,3-didehydro-8-phenyl-7,8-secoaspidospermid
ine-3,7-dicarboxylic acid; W   124988

474 C29 H34 N2 O4 77122-17-9 J-46-2009-0 Dimethyl ester of 5α-2,3-Didehydro-8-phenyl-7,8-secoaspidosper
midine-3,7-dicarboxylic acid; W   124989

474 C29 H46 O5 82423-43-6 KC-1982-926-0 6-Oxo-5α-cholestane-3β,5-diol diformate; W   124990

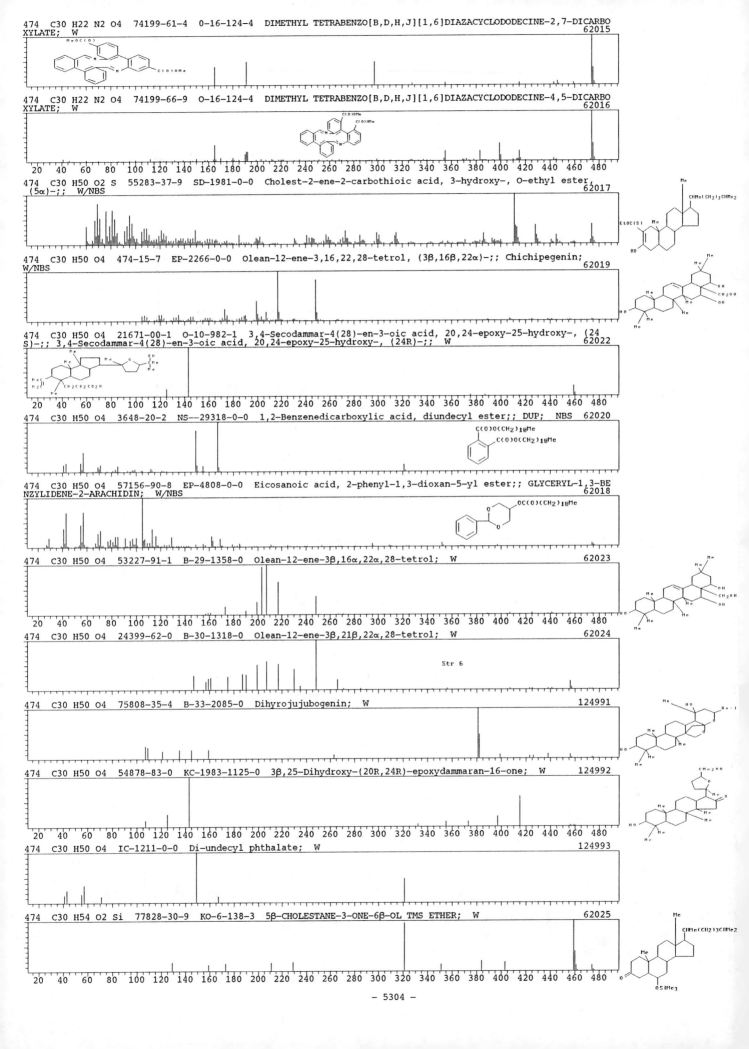

474 C30 H22 N2 O4 74199-61-4 0-16-124-4 DIMETHYL TETRABENZO[B,D,H,J][1,6]DIAZACYCLODODECINE-2,7-DICARBO
XYLATE; W
62015

474 C30 H22 N2 O4 74199-66-9 0-16-124-4 DIMETHYL TETRABENZO[B,D,H,J][1,6]DIAZACYCLODODECINE-4,5-DICARBO
XYLATE; W
62016

474 C30 H50 O2 S 55283-37-9 SD-1981-0-0 Cholest-2-ene-2-carbothioic acid, 3-hydroxy-, O-ethyl ester,
(5α)-;; W/NBS
62017

474 C30 H50 O4 474-15-7 EP-2266-0-0 Olean-12-ene-3,16,22,28-tetrol, (3β,16β,22α)-;; Chichipegenin;
W/NBS
62019

474 C30 H50 O4 21671-00-1 O-10-982-1 3,4-Secodammar-4(28)-en-3-oic acid, 20,24-epoxy-25-hydroxy-, (24
S)-;; 3,4-Secodammar-4(28)-en-3-oic acid, 20,24-epoxy-25-hydroxy-, (24R)-;; W
62022

474 C30 H50 O4 3648-20-2 NS--29318-0-0 1,2-Benzenedicarboxylic acid, diundecyl ester;; DUP; NBS
62020

474 C30 H50 O4 57156-90-8 EP-4808-0-0 Eicosanoic acid, 2-phenyl-1,3-dioxan-5-yl ester;; GLYCERYL-1,3-BE
NZYLIDENE-2-ARACHIDIN; W/NBS
62018

474 C30 H50 O4 53227-91-1 B-29-1358-0 Olean-12-ene-3β,16α,22α,28-tetrol; W
62023

474 C30 H50 O4 24399-62-0 B-30-1318-0 Olean-12-ene-3β,21β,22α,28-tetrol; W
62024

Str 6

474 C30 H50 O4 75808-35-4 B-33-2085-0 Dihyrojujubogenin; W
124991

474 C30 H50 O4 54878-83-0 KC-1983-1125-0 3β,25-Dihydroxy-(20R,24R)-epoxydammaran-16-one; W
124992

474 C30 H50 O4 IC-1211-0-0 Di-undecyl phthalate; W
124993

474 C30 H54 O2 Si 77828-30-9 KO-6-138-3 5β-CHOLESTANE-3-ONE-6β-OL TMS ETHER; W
62025

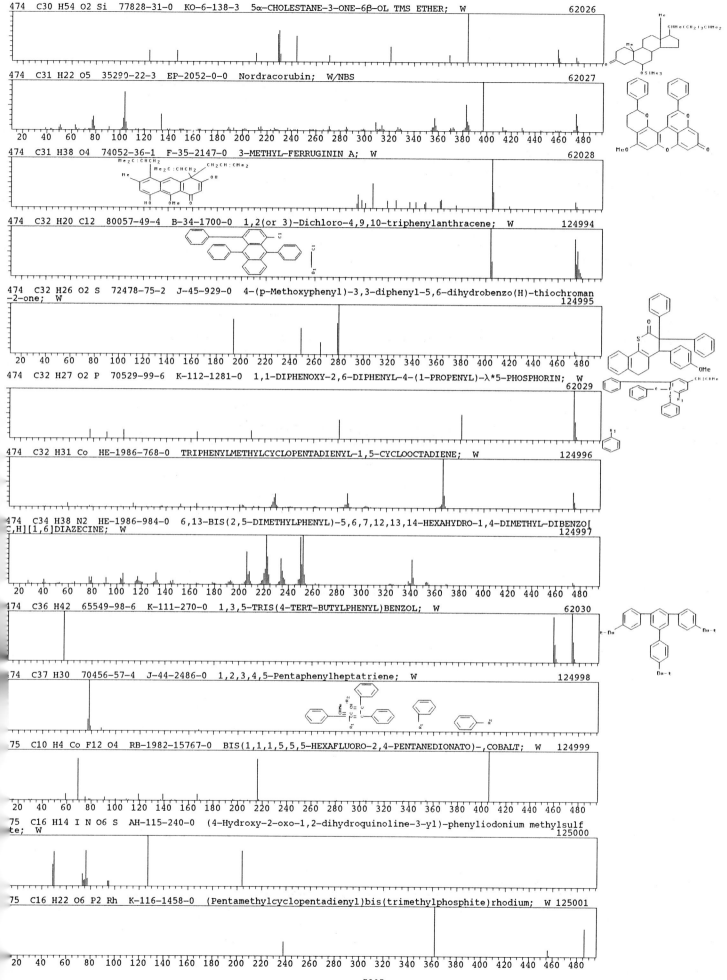

474   C30 H54 O2 Si   77828-31-0   KO-6-138-3   5α-CHOLESTANE-3-ONE-6β-OL TMS ETHER;   W   62026

474   C31 H22 O5   35290-22-3   EP-2052-0-0   Nordracorubin;   W/NBS   62027

474   C31 H38 O4   74052-36-1   F-35-2147-0   3-METHYL-FERRUGININ A;   W   62028

474   C32 H20 Cl2   80057-49-4   B-34-1700-0   1,2(or 3)-Dichloro-4,9,10-triphenylanthracene;   W   124994

474   C32 H26 O2 S   72478-75-2   J-45-929-0   4-(p-Methoxyphenyl)-3,3-diphenyl-5,6-dihydrobenzo(H)-thiochroman
-2-one;   W   124995

474   C32 H27 O2 P   70529-99-6   K-112-1281-0   1,1-DIPHENOXY-2,6-DIPHENYL-4-(1-PROPENYL)-λ*5-PHOSPHORIN;   W
          62029

474   C32 H31 Co   HE-1986-768-0   TRIPHENYLMETHYLCYCLOPENTADIENYL-1,5-CYCLOOCTADIENE;   W   124996

474   C34 H38 N2   HE-1986-984-0   6,13-BIS(2,5-DIMETHYLPHENYL)-5,6,7,12,13,14-HEXAHYDRO-1,4-DIMETHYL-DIBENZO[
C,H][1,6]DIAZECINE;   W   124997

474   C36 H42   65549-98-6   K-111-270-0   1,3,5-TRIS(4-TERT-BUTYLPHENYL)BENZOL;   W   62030

474   C37 H30   70456-57-4   J-44-2486-0   1,2,3,4,5-Pentaphenylheptatriene;   W   124998

75   C10 H4 Co F12 O4   RB-1982-15767-0   BIS(1,1,1,5,5,5-HEXAFLUORO-2,4-PENTANEDIONATO)-,COBALT;   W   124999

75   C16 H14 I N O6 S   AH-115-240-0   (4-Hydroxy-2-oxo-1,2-dihydroquinoline-3-yl)-phenyliodonium methylsulf
te;   W   125000

75   C16 H22 O6 P2 Rh   K-116-1458-0   (Pentamethylcyclopentadienyl)bis(trimethylphosphite)rhodium;   W 125001

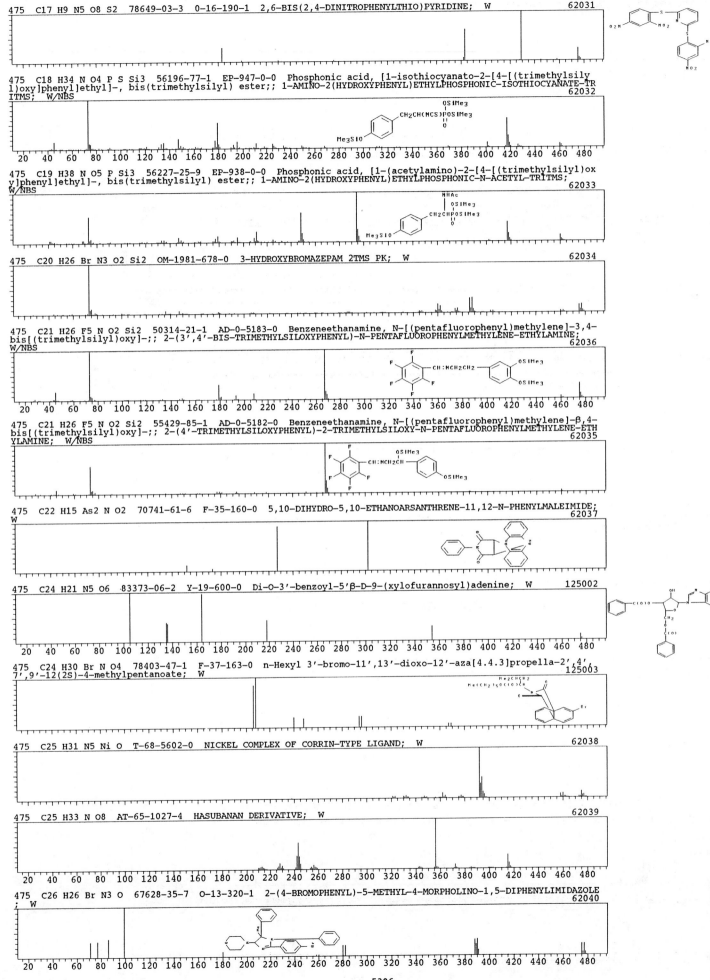

475   C17 H9 N5 O8 S2   78649-03-3   0-16-190-1   2,6-BIS(2,4-DINITROPHENYLTHIO)PYRIDINE;   W
62031

475   C18 H34 N O4 P S Si3   56196-77-1   EP-947-0-0   Phosphonic acid, [1-isothiocyanato-2-[4-[(trimethylsilyl)oxy]phenyl]ethyl]-, bis(trimethylsilyl) ester;; 1-AMINO-2(HYDROXYPHENYL)ETHYLPHOSPHONIC-ISOTHIOCYANATE-TRITMS;   W/NBS
62032

475   C19 H38 N O5 P Si3   56227-25-9   EP-938-0-0   Phosphonic acid, [1-(acetylamino)-2-[4-[(trimethylsilyl)oxy]phenyl]ethyl]-, bis(trimethylsilyl) ester;; 1-AMINO-2(HYDROXYPHENYL)ETHYLPHOSPHONIC-N-ACETYL-TRITMS; W/NBS
62033

475   C20 H26 Br N3 O2 Si2   OM-1981-678-0   3-HYDROXYBROMAZEPAM 2TMS PK;   W
62034

475   C21 H26 F5 N O2 Si2   50314-21-1   AD-0-5183-0   Benzeneethanamine, N-[(pentafluorophenyl)methylene]-3,4-bis[(trimethylsilyl)oxy]-;; 2-(3',4'-BIS-TRIMETHYLSILOXYPHENYL)-N-PENTAFLUOROPHENYLMETHYLENE-ETHYLAMINE;   W/NBS
62036

475   C21 H26 F5 N O2 Si2   55429-85-1   AD-0-5182-0   Benzeneethanamine, N-[(pentafluorophenyl)methylene]-β,4-bis[(trimethylsilyl)oxy]-;; 2-(4'-TRIMETHYLSILOXYPHENYL)-2-TRIMETHYLSILOXY-N-PENTAFLUOROPHENYLMETHYLENE-ETHYLAMINE;   W/NBS
62035

475   C22 H15 As2 N O2   70741-61-6   F-35-160-0   5,10-DIHYDRO-5,10-ETHANOARSANTHRENE-11,12-N-PHENYLMALEIMIDE; W
62037

475   C24 H21 N5 O6   83373-06-2   Y-19-600-0   Di-O-3'-benzoyl-5'β-D-9-(xylofurannosyl)adenine;   W   125002

475   C24 H30 Br N O4   78403-47-1   F-37-163-0   n-Hexyl 3'-bromo-11',13'-dioxo-12'-aza[4.4.3]propella-2',4',7',9'-12(2S)-4-methylpentanoate;   W   125003

475   C25 H31 N5 Ni O   T-68-5602-0   NICKEL COMPLEX OF CORRIN-TYPE LIGAND;   W
62038

475   C25 H33 N O8   AT-65-1027-4   HASUBANAN DERIVATIVE;   W
62039

475   C26 H26 Br N3 O   67628-35-7   O-13-320-1   2-(4-BROMOPHENYL)-5-METHYL-4-MORPHOLINO-1,5-DIPHENYLIMIDAZOLE; W
62040

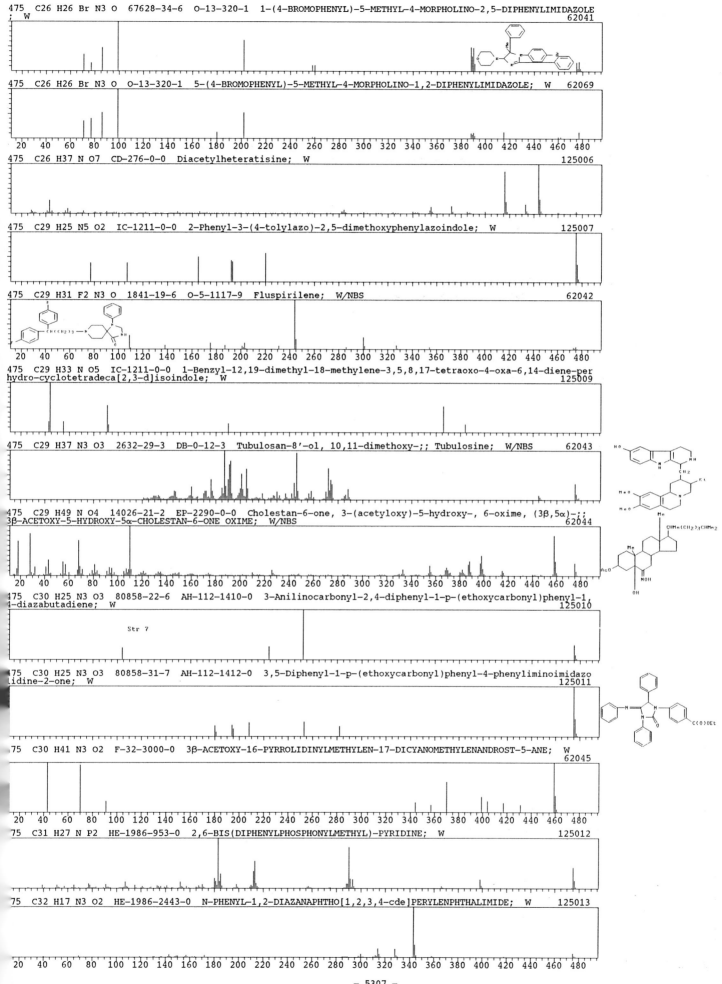

475 C26 H26 Br N3 O 67628-34-6 O-13-320-1 1-(4-BROMOPHENYL)-5-METHYL-4-MORPHOLINO-2,5-DIPHENYLIMIDAZOLE
; W   62041

475 C26 H26 Br N3 O O-13-320-1 5-(4-BROMOPHENYL)-5-METHYL-4-MORPHOLINO-1,2-DIPHENYLIMIDAZOLE; W   62069

20 40 60 80 100 120 140 160 180 200 220 240 260 280 300 320 340 360 380 400 420 440 460 480

475 C26 H37 N O7 CD-276-0-0 Diacetylheteratisine; W   125006

475 C29 H25 N5 O2 IC-1211-0-0 2-Phenyl-3-(4-tolylazo)-2,5-dimethoxyphenylazoindole; W   125007

475 C29 H31 F2 N3 O 1841-19-6 O-5-1117-9 Fluspirilene; W/NBS   62042

20 40 60 80 100 120 140 160 180 200 220 240 260 280 300 320 340 360 380 400 420 440 460 480

475 C29 H33 N O5 IC-1211-0-0 1-Benzyl-12,19-dimethyl-18-methylene-3,5,8,17-tetraoxo-4-oxa-6,14-diene-per
hydro-cyclotetradeca[2,3-d]isoindole; W   125009

475 C29 H37 N3 O3 2632-29-3 DB-0-12-3 Tubulosan-8'-ol, 10,11-dimethoxy-;; Tubulosine; W/NBS   62043

475 C29 H49 N O4 14026-21-2 EP-2290-0-0 Cholestan-6-one, 3-(acetyloxy)-5-hydroxy-, 6-oxime, (3β,5α)-;;
3β-ACETOXY-5-HYDROXY-5α-CHOLESTAN-6-ONE OXIME; W/NBS   62044

20 40 60 80 100 120 140 160 180 200 220 240 260 280 300 320 340 360 380 400 420 440 460 480

475 C30 H25 N3 O3 80858-22-6 AH-112-1410-0 3-Anilinocarbonyl-2,4-diphenyl-1-p-(ethoxycarbonyl)phenyl-1,
4-diazabutadiene; W   125010

Str 7

475 C30 H25 N3 O3 80858-31-7 AH-112-1412-0 3,5-Diphenyl-1-p-(ethoxycarbonyl)phenyl-4-phenyliminoimidazo
lidine-2-one; W   125011

75 C30 H41 N3 O2 F-32-3000-0 3β-ACETOXY-16-PYRROLIDINYLMETHYLEN-17-DICYANOMETHYLENANDROST-5-ANE; W
                                                                                                      62045

20 40 60 80 100 120 140 160 180 200 220 240 260 280 300 320 340 360 380 400 420 440 460 480

75 C31 H27 N P2 HE-1986-953-0 2,6-BIS(DIPHENYLPHOSPHONYLMETHYL)-PYRIDINE; W   125012

75 C32 H17 N3 O2 HE-1986-2443-0 N-PHENYL-1,2-DIAZANAPHTHO[1,2,3,4-cde]PERYLENPHTHALIMIDE; W   125013

20 40 60 80 100 120 140 160 180 200 220 240 260 280 300 320 340 360 380 400 420 440 460 480

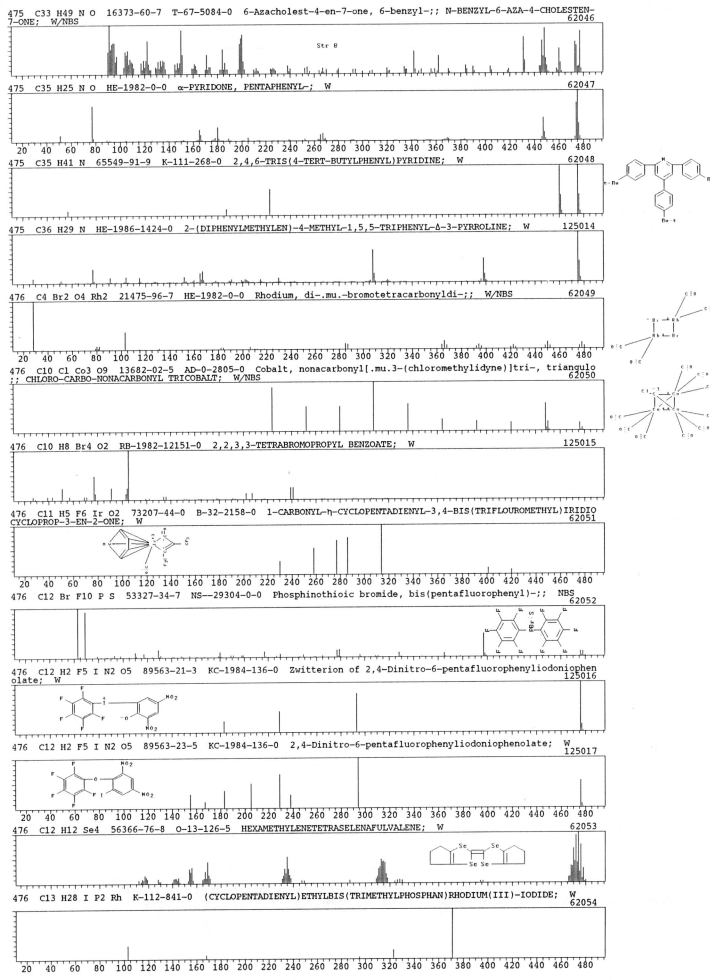

475 C33 H49 N O 16373-60-7 T-67-5084-0 6-Azacholest-4-en-7-one, 6-benzyl-;; N-BENZYL-6-AZA-4-CHOLESTEN-7-ONE; W/NBS
62046

Str 8

475 C35 H25 N O HE-1982-0-0 α-PYRIDONE, PENTAPHENYL-; W
62047

475 C35 H41 N 65549-91-9 K-111-268-0 2,4,6-TRIS(4-TERT-BUTYLPHENYL)PYRIDINE; W
62048

475 C36 H29 N HE-1986-1424-0 2-(DIPHENYLMETHYLEN)-4-METHYL-1,5,5-TRIPHENYL-Δ-3-PYRROLINE; W 125014

476 C4 Br2 O4 Rh2 21475-96-7 HE-1982-0-0 Rhodium, di-.mu.-bromotetracarbonyldi-;; W/NBS
62049

476 C10 Cl Co3 O9 13682-02-5 AD-0-2805-0 Cobalt, nonacarbonyl[.mu.3-(chloromethylidyne)]tri-, triangulo ;; CHLORO-CARBO-NONACARBONYL TRICOBALT; W/NBS
62050

476 C10 H8 Br4 O2 RB-1982-12151-0 2,2,3,3-TETRABROMOPROPYL BENZOATE; W 125015

476 C11 H5 F6 Ir O2 73207-44-0 B-32-2158-0 1-CARBONYL-η-CYCLOPENTADIENYL-3,4-BIS(TRIFLOUROMETHYL)IRIDIO CYCLOPROP-3-EN-2-ONE; W
62051

476 C12 Br F10 P S 53327-34-7 NS--29304-0-0 Phosphinothioic bromide, bis(pentafluorophenyl)-;; NBS
62052

476 C12 H2 F5 I N2 O5 89563-21-3 KC-1984-136-0 Zwitterion of 2,4-Dinitro-6-pentafluorophenyliodoniophenolate; W
125016

476 C12 H2 F5 I N2 O5 89563-23-5 KC-1984-136-0 2,4-Dinitro-6-pentafluorophenyliodoniophenolate; W
125017

476 C12 H12 Se4 56366-76-8 O-13-126-5 HEXAMETHYLENETETRASELENAFULVALENE; W
62053

476 C13 H28 I P2 Rh K-112-841-0 (CYCLOPENTADIENYL)ETHYLBIS(TRIMETHYLPHOSPHAN)RHODIUM(III)-IODIDE; W
62054

476   C14 H23 O5 Tl   54156-79-5   C-104-3985-0   2-((Diacetato)thallio)-3-methoxycyclononene;   W   125018

476   C14 H32 Cl Co P2 Sn   K-112-831-0   (CHLOROTRIMETHYLTIN)(CYCLOPENTADIENYL)BIS(TRIMETHYLPHOSPHAN)COBALT;
W   62055

476   C15 H9 Co F12   HE-1986-2015-0   CYCLOPENATDIENYL-(HAPTO-4-1,2,3,4-TETRAKIS(TRIFLUOROMETHYL)-CYCLOHEXA-
1,3-DIENE)COBALT;   W   125019

476   C16 H10 F6 O6 S2   89849-96-7   J-49-2277-0   (Z)-1,2-Diphenylvinylene-1,2-bis(trifluoromethanesulfonate)
;   W   125020

476   C16 H12 Fe2 O10   55450-73-2   AD-0-2442-0   Iron, [.mu.-[(1,2,3,4-η:1,2-η)-1,4-bis(ethoxycarbonyl)-1,3-
butadiene-1,4-diyl]]di-, (Fe-Fe);; 1,1,1-TRICARBONYL-FERROL-2,5,-DIETHOXYCARBONYL-TRICARBONYLIRON-II;
W/NBS   62056

476   C18 H10 Br2 N2 O4   59400-17-8   B-30-2504-0   5,5'-Dibromo-1,1-(ethane-1,2-diyl)bis(2,3-dihyroindole-2,
3-dione);   W   125021

476   C19 H44 N2 O4 Si4   OM-1981-388-0   N-ACETYLGLUTAMINE 4TMS;   W   62057

476   C20 H14 Br2 O4   67601-97-2   B-31-1617-0   DIMETHYL 1,2-DIBROMO-1,2-DIHYDROCYCLOBUTA(1)PHENANTHRENE-1,2-
DICARBOXYLATE;   W   62058

476   C20 H18 Cl2 N6 O4   63376-63-6   K-110-1728-0   1-(1,2-DICHLORO-2-(5-METHOXY-4-PHENYL-1,2,4-TRIAZOL-3-YLO
XY)ETHYL)-2-METHYL)-2-METHYL-4-PHENYL-1,2,4-TRIAZOLIDIN-3,5-DIONE;   W   62059

476   C20 H18 Cl2 N6 O4   63376-62-5   K-110-1728-0   1-(1,2-DICHLORO-2-(1-METHYL-5-OXO-4-PHENYL-Δ*2-1,2,4-TRIA
ZALIN-3-YLOXY)ETHYL)-2-METHYL-4-PHENYL-1,2,4-TRIAZOLIDIN-3,5-DIONE;   W   62060

476   C21 H24 Fe O7 S   KC-1982-2475-0   Toluene-p-sulphonate of tricarbonyl{1--4-η[5-(2-cyanoethyl)-5-ethyl-
4-methoxycyclohexa-1,3-diene]}iron;   W   125022

476   C21 H32 O12   56246-40-3   K-102-1259-1   Pentanoic acid, 4-methyl-5-[(2,3,4,6-tetra-O-acetyl-D-glucopyra
nosyl)oxy]-, methyl ester;; Δ-HYDROXY-γ-METHYL-VALERIAN ACID METHYLESTER GLUCOSIDE TETRAACETATE;   W/NBS
                                                                                                         62061

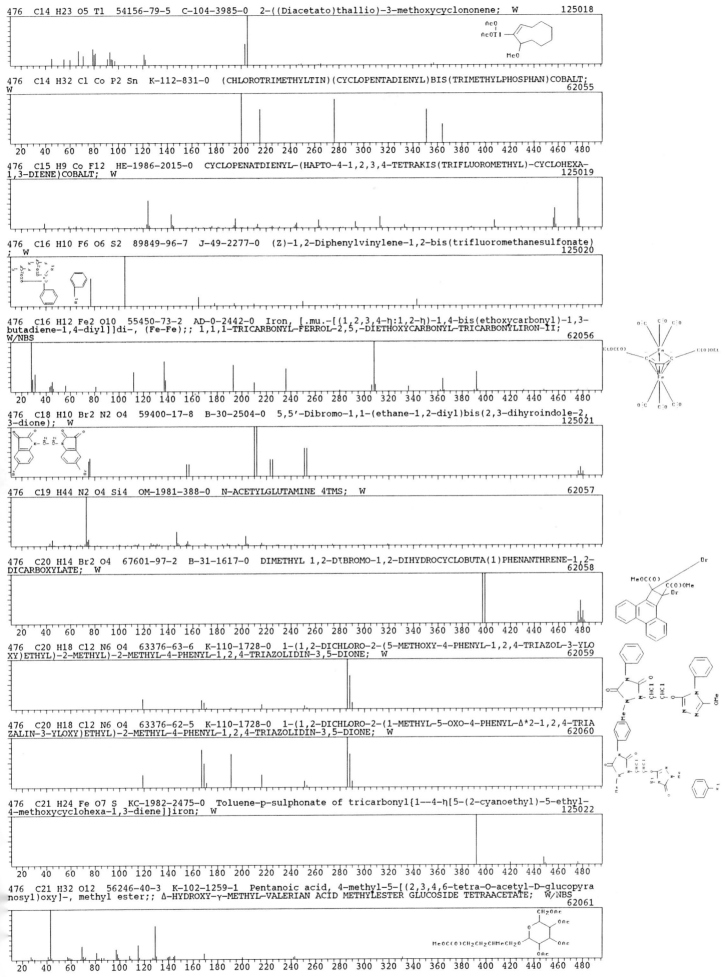

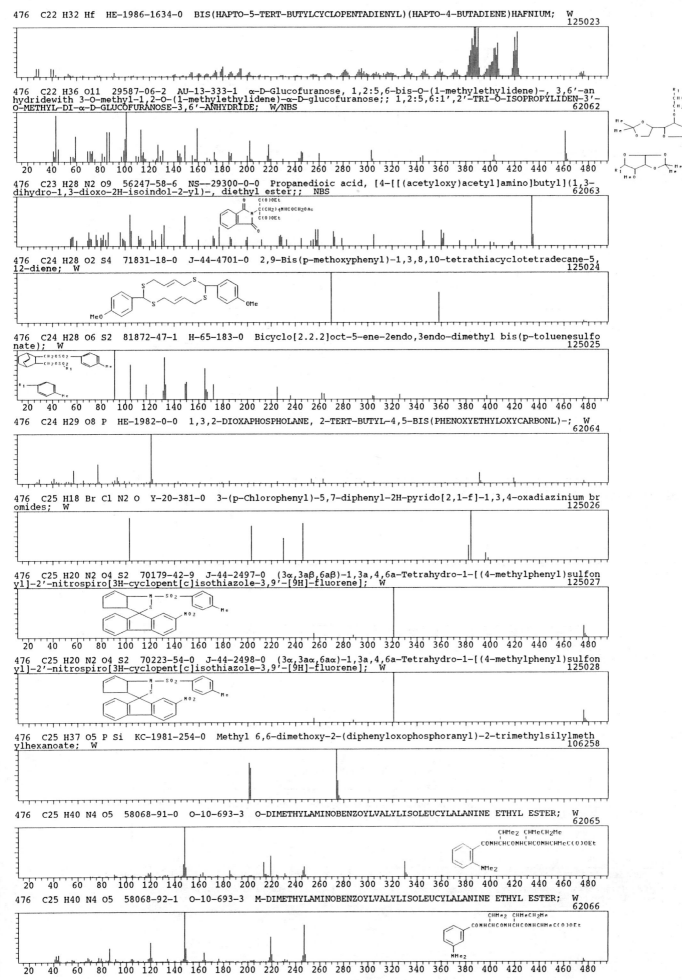

476 C22 H32 Hf HE-1986-1634-0 BIS(HAPTO-5-TERT-BUTYLCYCLOPENTADIENYL)(HAPTO-4-BUTADIENE)HAFNIUM; W
125023

476 C22 H36 O11 29587-06-2 AU-13-333-1 α-D-Glucofuranose, 1,2:5,6-bis-O-(1-methylethylidene)-, 3,6'-an
hydridewith 3-O-methyl-1,2-O-(1-methylethylidene)-α-D-glucofuranose;; 1,2:5,6:1',2'-TRI-O-ISOPROPYLIDEN-3'-
O-METHYL-DI-α-D-GLUCOFURANOSE-3,6'-ANHYDRIDE; W/NBS
62062

476 C23 H28 N2 O9 56247-58-6 NS--29300-0-0 Propanedioic acid, [4-[[(acetyloxy)acetyl]amino]butyl](1,3-
dihydro-1,3-dioxo-2H-isoindol-2-yl)-, diethyl ester;; NBS
62063

476 C24 H28 O2 S4 71831-18-0 J-44-4701-0 2,9-Bis(p-methoxyphenyl)-1,3,8,10-tetrathiacyclotetradecane-5,
12-diene; W
125024

476 C24 H28 O6 S2 81872-47-1 H-65-183-0 Bicyclo[2.2.2]oct-5-ene-2endo,3endo-dimethyl bis(p-toluenesulfo
nate); W
125025

476 C24 H29 O8 P HE-1982-0-0 1,3,2-DIOXAPHOSPHOLANE, 2-TERT-BUTYL-4,5-BIS(PHENOXYETHYLOXYCARBONL)-; W
62064

476 C25 H18 Br Cl N2 O Y-20-381-0 3-(p-Chlorophenyl)-5,7-diphenyl-2H-pyrido[2,1-f]-1,3,4-oxadiazinium br
omides; W
125026

476 C25 H20 N2 O4 S2 70179-42-9 J-44-2497-0 (3α,3aβ,6aβ)-1,3a,4,6a-Tetrahydro-1-[(4-methylphenyl)sulfon
yl]-2'-nitrospiro[3H-cyclopent[c]isothiazole-3,9'-[9H]-fluorene]; W
125027

476 C25 H20 N2 O4 S2 70223-54-0 J-44-2498-0 (3α,3aα,6aα)-1,3a,4,6a-Tetrahydro-1-[(4-methylphenyl)sulfon
yl]-2'-nitrospiro[3H-cyclopent[c]isothiazole-3,9'-[9H]-fluorene]; W
125028

476 C25 H37 O5 P Si KC-1981-254-0 Methyl 6,6-dimethoxy-2-(diphenyloxophosphoranyl)-2-trimethylsilylmeth
ylhexanoate; W
106258

476 C25 H40 N4 O5 58068-91-0 O-10-693-3 O-DIMETHYLAMINOBENZOYLVALYLISOLEUCYLALANINE ETHYL ESTER; W
62065

476 C25 H40 N4 O5 58068-92-1 O-10-693-3 M-DIMETHYLAMINOBENZOYLVALYLISOLEUCYLALANINE ETHYL ESTER; W
62066

- 5310 -

476   C25 H40 N4 O5   58068-93-2   O-10-693-3   P-DIMETHYLAMINOBENZOYLVALYLISOLEACYLALANINE ETHYL ESTER;   W
62067

476   C25 H40 O5 Si2   HE-1986-2216-0   BIS-O-TRIMETHYLSILYL-GIBBERELLIN A4;   W
125029

476   C26 H16 N6 O4   21589-16-2   AD-0-5478-0   Phenazine, 1,1'-ethylenebis[2-nitro-;; 1,2-BIS(2-NITROPHENAZIN
-1-YL)ETHANE;   W/NBS
62068

476   C26 H20 O9   74695-01-5   B-33-2539-0   4,5,9,10-Tetramethoxydinaphtho[2,1-b:1',2'-d]furan-2,12-dicarboxy
lic acid;   W
125031

476   C26 H28 N4 O5   KO-3-158-2   2-(O-ETHOXYCARBONYLPHENYLAZO)-3-(2-METHOXYCARBONYLETHYL)-4-METHYL-5-((2,5-
DIHYDRO-3-VINYL-4-METHYL-5-OXO-2-PYRROLYLIDENE)METHYL)PYRROLE;   W
62070

476   C26 H28 N4 O5   KO-3-158-2   2-(O-ETHOXYCARBONYLPHENYLAZO)-3-(2-METHOXYCARBONYLETHYL)-4-METH
YL-5-((2,5-DIHYDRO-3-METHYL-4-VINYL-5-OXO-2-PYRROLYLIDENE)METHYL)PYRROLE;   W
62071

476   C26 H28 N4 O5   KO-3-158-2   2-(O-ETHOXYCARBONYLPHENYLAZO)-3-VINYL-4-METHYL-5-((2,5-DIHYDRO-4-(2-METHOXY
CARBONYLETHYL)-5-OXO-2-PYRROLYLIDENE)METHYL)PYRROLE;   W
62072

476   C26 H28 N4 O5   62290-65-7   KO-3-158-2   2-(O-ETHOXYCARBONYLPHENYLAZO)-3-METHYL-4-VINYL-5-((2,5-DIHYDRO-
3-METHYL-4-(2'-METHOXYCARBONYLETHYL)-5-OXO-2-PYRROLYLIDENE)METHYL)PYRROLE;   W
62073

476   C26 H28 N4 O5   73108-51-7   H-62-2790-0   5'-Deoxy-2',3'-O-isopropylidene-5',5'-(N,N'-diphenylethylenedi
amino)-uridine;   W
125032

476   C26 H33 Cl O6   80902-83-6   KC-1981-3210-0   3-(3-chloro-3-methylbutyl)-4,6-dimethoxy-2-hydroxyphenyl 3,
4-dihydro-2,2-dimethyl-8-methoxy-2H-1-benzopyran-5-yl ketone;   W
125033

476   C26 H36 O4 S2   K-116-3843-0   8,11,20,23-Tetramethoxy-2,17-dithia[6.6]paracyclophane;   W
125034

476   C26 H36 O8   77508-64-6   NS-0-0-0   1H-Cyclopropa[3,4]benz[1,2-e]azulene-5,7b,9,9a-tetrol, 1a,1b,4,4a,5,
7a,8,9-octahydro-3-(hydroxymethyl)-1,1,6,8-tetramethyl-, 5,9,9a-triacetate, [1aR-(1aα,1bβ,4aβ,5β,7aα,7bα,
8α,9β,9aα)]-;   W/NBS
125035

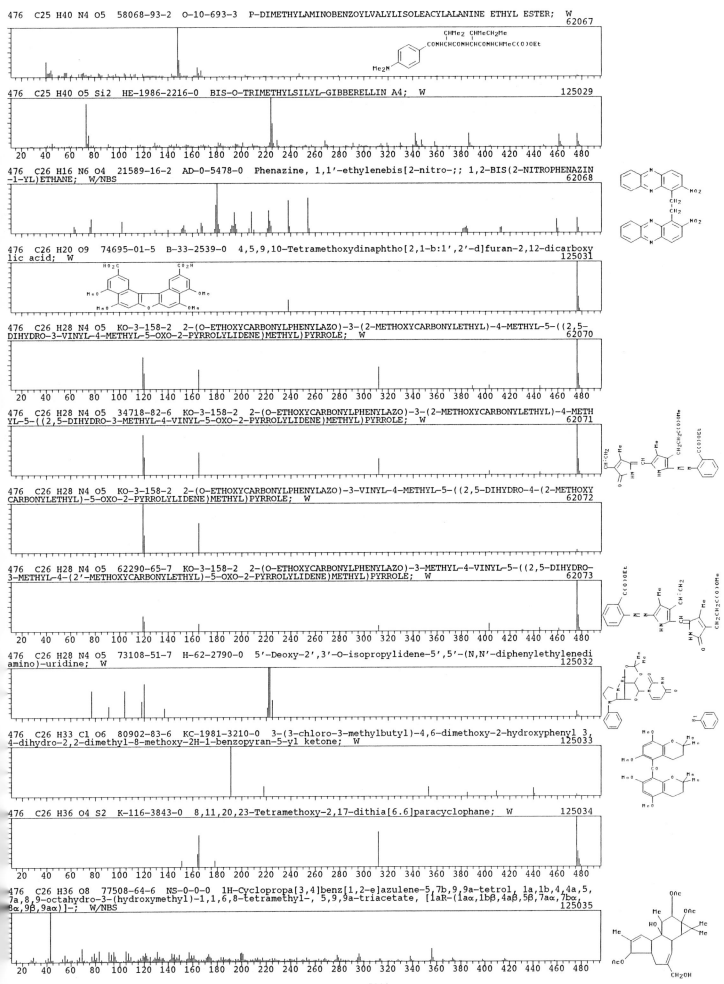

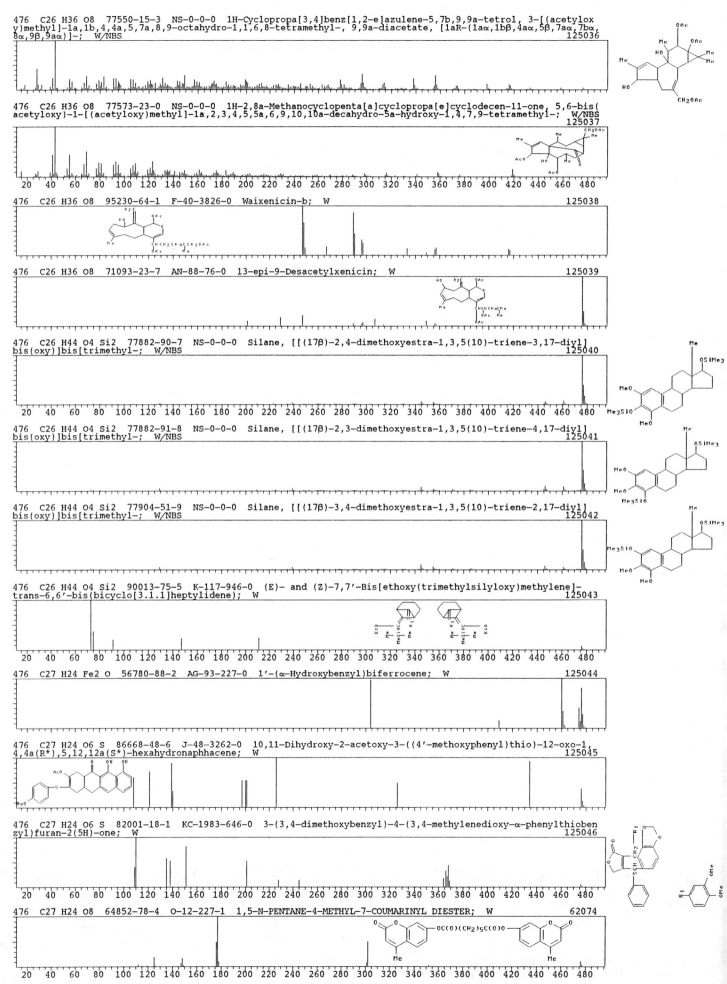

476 C26 H36 O8 77550-15-3 NS-0-0-0 1H-Cyclopropa[3,4]benz[1,2-e]azulene-5,7b,9,9a-tetrol, 3-[(acetylox
y)methyl]-1a,1b,4,4a,5,7a,8,9-octahydro-1,1,6,8-tetramethyl-, 9,9a-diacetate, [1aR-(1aα,1bβ,4aα,5β,7aα,7bα,
8α,9β,9aα)]-; W/NBS
125036

476 C26 H36 O8 77573-23-0 NS-0-0-0 1H-2,8a-Methanocyclopenta[a]cyclopropa[e]cyclodecen-11-one, 5,6-bis(
acetyloxy)-1-[(acetyloxy)methyl]-1a,2,3,4,5,5a,6,9,10,10a-decahydro-5a-hydroxy-1,4,7,9-tetramethyl-; W/NBS
125037

476 C26 H36 O8 95230-64-1 F-40-3826-0 Waixenicin-b; W
125038

476 C26 H36 O8 71093-23-7 AN-88-76-0 13-epi-9-Desacetylxenicin; W
125039

476 C26 H44 O4 Si2 77882-90-7 NS-0-0-0 Silane, [[(17β)-2,4-dimethoxyestra-1,3,5(10)-triene-3,17-diyl]
bis(oxy)]bis[trimethyl-; W/NBS
125040

476 C26 H44 O4 Si2 77882-91-8 NS-0-0-0 Silane, [[(17β)-2,3-dimethoxyestra-1,3,5(10)-triene-4,17-diyl]
bis(oxy)]bis[trimethyl-; W/NBS
125041

476 C26 H44 O4 Si2 77904-51-9 NS-0-0-0 Silane, [[(17β)-3,4-dimethoxyestra-1,3,5(10)-triene-2,17-diyl]
bis(oxy)]bis[trimethyl-; W/NBS
125042

476 C26 H44 O4 Si2 90013-75-5 K-117-946-0 (E)- and (Z)-7,7'-Bis[ethoxy(trimethylsilyloxy)methylene]-
trans-6,6'-bis(bicyclo[3.1.1]heptylidene); W
125043

476 C27 H24 Fe2 O 56780-88-2 AG-93-227-0 1'-(α-Hydroxybenzyl)biferrocene; W
125044

476 C27 H24 O6 S 86668-48-6 J-48-3262-0 10,11-Dihydroxy-2-acetoxy-3-((4'-methoxyphenyl)thio)-12-oxo-1,
4,4a(R*),5,12,12a(S*)-hexahydronaphhacene; W
125045

476 C27 H24 O6 S 82001-18-1 KC-1983-646-0 3-(3,4-dimethoxybenzyl)-4-(3,4-methylenedioxy-α-phenylthioben
zyl)furan-2(5H)-one; W
125046

476 C27 H24 O8 64852-78-4 O-12-227-1 1,5-N-PENTANE-4-METHYL-7-COUMARINYL DIESTER; W
62074

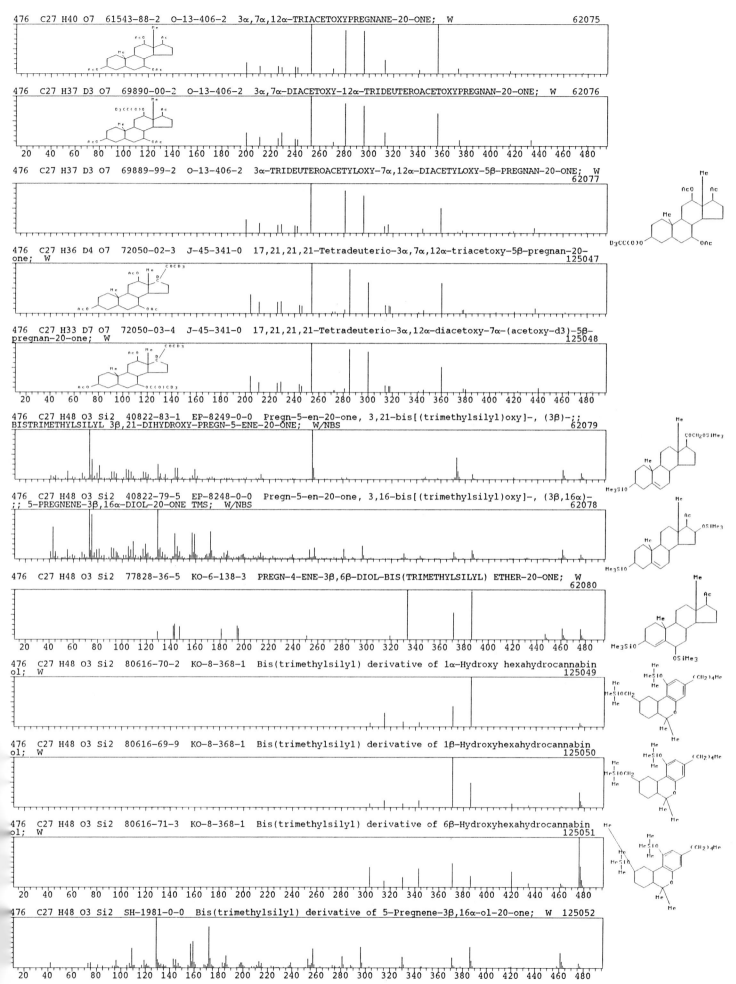

476   C27 H40 O7   61543-88-2   O-13-406-2   3α,7α,12α-TRIACETOXYPREGNANE-20-ONE;   W   62075

476   C27 H37 D3 O7   69890-00-2   O-13-406-2   3α,7α-DIACETOXY-12α-TRIDEUTEROACETOXYPREGNAN-20-ONE;   W   62076

476   C27 H37 D3 O7   69889-99-2   O-13-406-2   3α-TRIDEUTEROACETYLOXY-7α,12α-DIACETYLOXY-5β-PREGNAN-20-ONE;   W
62077

476   C27 H36 D4 O7   72050-02-3   J-45-341-0   17,21,21,21-Tetradeuterio-3α,7α,12α-triacetoxy-5β-pregnan-20-one;   W   125047

476   C27 H33 D7 O7   72050-03-4   J-45-341-0   17,21,21,21-Tetradeuterio-3α,12α-diacetoxy-7α-(acetoxy-d3)-5β-pregnan-20-one;   W   125048

476   C27 H48 O3 Si2   40822-83-1   EP-8249-0-0   Pregn-5-en-20-one, 3,21-bis[(trimethylsilyl)oxy]-, (3β)-;;
BISTRIMETHYLSILYL 3β,21-DIHYDROXY-PREGN-5-ENE-20-ONE;   W/NBS   62079

476   C27 H48 O3 Si2   40822-79-5   EP-8248-0-0   Pregn-5-en-20-one, 3,16-bis[(trimethylsilyl)oxy]-, (3β,16α)-
;; 5-PREGNENE-3β,16α-DIOL-20-ONE TMS;   W/NBS   62078

476   C27 H48 O3 Si2   77828-36-5   KO-6-138-3   PREGN-4-ENE-3β,6β-DIOL-BIS(TRIMETHYLSILYL) ETHER-20-ONE;   W
62080

476   C27 H48 O3 Si2   80616-70-2   KO-8-368-1   Bis(trimethylsilyl) derivative of 1α-Hydroxy hexahydrocannabinol;   W   125049

476   C27 H48 O3 Si2   80616-69-9   KO-8-368-1   Bis(trimethylsilyl) derivative of 1β-Hydroxyhexahydrocannabinol;   W   125050

476   C27 H48 O3 Si2   80616-71-3   KO-8-368-1   Bis(trimethylsilyl) derivative of 6β-Hydroxyhexahydrocannabinol;   W   125051

476   C27 H48 O3 Si2   SH-1981-0-0   Bis(trimethylsilyl) derivative of 5-Pregnene-3β,16α-ol-20-one;   W   125052

476   C27 H48 O3 Si2   KO-9-413-3   Bis(trimethylsilyl)derivative of 3β,11α-dihydroxypregn-5-en-20-one;   W
125053

476   C27 H48 O3 Si2   KO-9-413-3   Bis(trimethylsilyl)derivative of 3β,20β-dihydroxypregn-5,17-diene;   W
125054

476   C27 H48 O3 Si2   KO-9-413-3   Bis(trimethylsilyl)derivative of 3β,21-dihydroxypregn-5-en-20-one;   W
125055

476   C28 H16 N2 O6   78588-55-3   0-16-145-1   3,3-BIS(P-MALEIMIDOPHENYL)PHTHALIDE;   W
62082

476   C28 H16 N2 O6   87618-38-0   KC-1983-1273-0   4-(2-Nitrobenzyl)-2-(2-nitrophenyl)-6H-phenaleno[1,9-bc]pyr
an-6-one;   W
125056

476   C28 H23 F3 N2 O2   88341-23-5   KC-1983-2428-0   4,4a-Dihydro-9-methyl-4a-[2,2,2-trifluoro-1-(2-hydroxy-
9-methylcarbazol-3-yl)ethyl]carbazol-2(3H)one;   W
125057

476   C28 H28 O7   73020-05-0   B-34-1131-0   (5S,7R,8S,9R)-9-Benzoyloxy-3',4-dimethoxy-7,8-dimethyl-2,3-methyl
enedioxy-6,7,8,9-tetrahydrospiro[5H-benzocycloheptene-5,1'-cyclohexa-2',5'-dien]-4'-one;   W
125058

476   C28 H32 N2 O5   56143-44-3   SD-1981-0-0   Aspidospermidin-21-oic acid, 1-benzoyl-20-hydroxy-17-methoxy-,
methyl ester;;   W/NBS
62084

476   C28 H32 N2 O5   71522-13-9   I-57-1701-0   N-(2-(1'-BENZYL-3'-(2''-KETO-3''-CARBOXYMETHYLCYCLOHEXYL)OXIND
OL-3'-YL)-ETHYL(-N-METHYLACETAMIDE;   W
62083

476   C28 H32 N2 O5   CD-14-0-0   N-Benzoyl-20-hydroxycylindrocarine;   W
125059

476   C28 H44 O6   33628-48-7   SD-1981-0-0   Deoxycholic acid diacetate;   W/NBS
62087

476   C28 H44 O6   60354-47-4   NS--29294-0-0   24-Norcholan-23-oic acid, 3,12-bis(acetyloxy)-, methyl ester,
(3α,5β,12α)-;;   NBS
62085

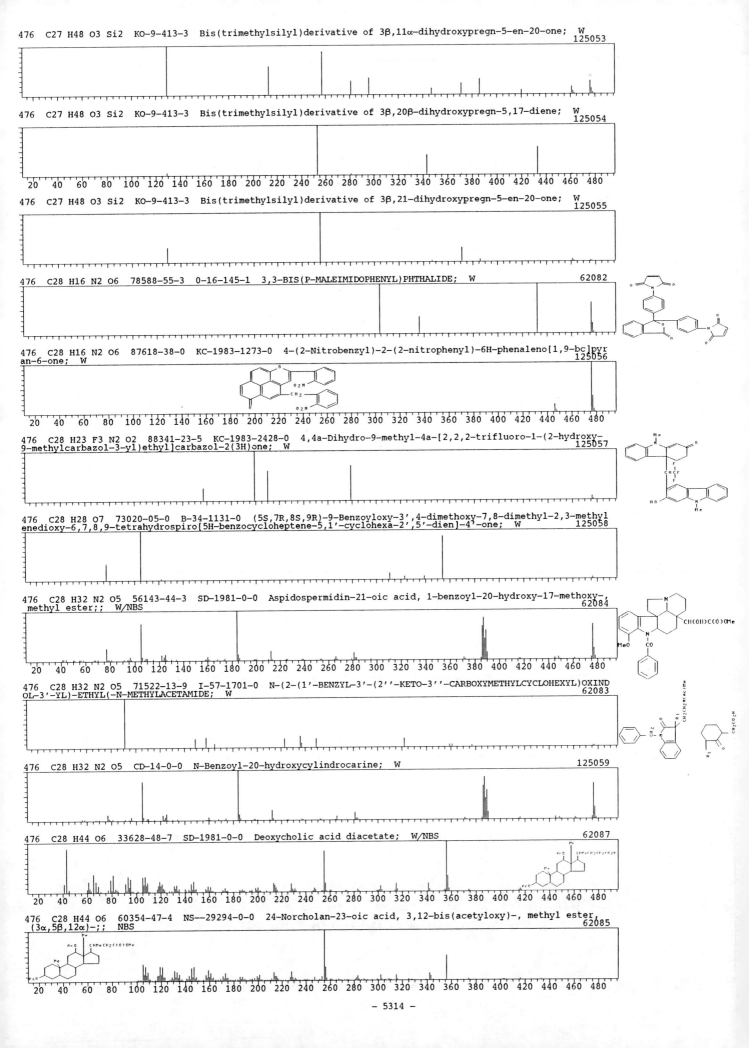

476 C28 H44 O6 60354-48-5 NS--29293-0-0 24-Norcholan-23-oic acid, 3,11-bis(acetyloxy)-, methyl ester, (3α,5β,11α)-;; NBS
62086

476 C28 H41 D3 O6 72049-93-5 J-45-339-0 Methyl 3α-(acetoxy-d3)-7α-acetoxy-24-nor-5β-cholanoate; W
134032

476 C28 H45 Br O 83511-90-4 C-104-6732-0 (23S,24R,28S)-23,24-(Bromomethylene)-cholest-5-en-3β-ol; W
125060

476 C28 H52 O2 Si2 BS-2-50-0 3β,22-Di(trimethylsilyloxy)-23,24-dinor-5-cholene; W
125061

476 C29 H32 O6 61110-24-5 KC-1976-1575-0 (2S)-8-TRANS-(2-(2',4',6'-TRIMETHOXY-3'-METHYLPHENYL)ETHENYL)-5,7-DIMETHOXY FLAVAN; W
62088

476 C29 H33 Cl N2 O2 34552-83-5 KO-6-255-2 LOPERAMIDE; W
62089

476 C29 H48 O S2 1107-85-3 T-68-5502-0 Cholest-5-en-19-al, 3β-hydroxy-, cyclic ethylene mercaptal;; C19 -ETHYLENE THIOKETAL CHOLESTEROL; W/NBS
62090

476 C29 H48 O5 10473-42-4 O-8-5-5 5,6-Secocholestan-6-oic acid, 3-(acetyloxy)-5-oxo-, (3β)-;; 3β-ACETOXY-5-KETO-5,6-SECOCHOLESTAN-6-OIC ACID; W/NBS
62091

476 C29 H48 O5 84681-05-0 C-105-1978-0 3β-Acetoxy-10-hydroxy-6-oxo-5,6,5,10-disecocholestan-5-oic acid; W
125064

476 C30 H20 O2 S2 19018-13-4 T-68-5539-0 Ethanone, 2,2'-(1,3-dithietane-2,4-diylidene)bis[1,2-diphenyl-;; 2,4-BIS(2-OXO-1,2-DIPHENYLETHYLIDENE)-1,3-DITHIETANE; W/NBS
62092

476 C30 H28 N4 O2 30563-73-6 NS--29289-0-0 Tetrabenzo[e,i,o,s][1,4,7,11,14,18]dioxatetraazacycloeicosine, 11,12,13,14,26,27-hexahydro-;; NBS
62093

476 C30 H32 N6 80430-70-2 K-114-3681-0 1,1',1'',1''',1'''',1'''''-Hexamethyl-2,2':5',2'':5'',2''':5''',2'''':5'''',2'''''-sexipyrrole; W
125065

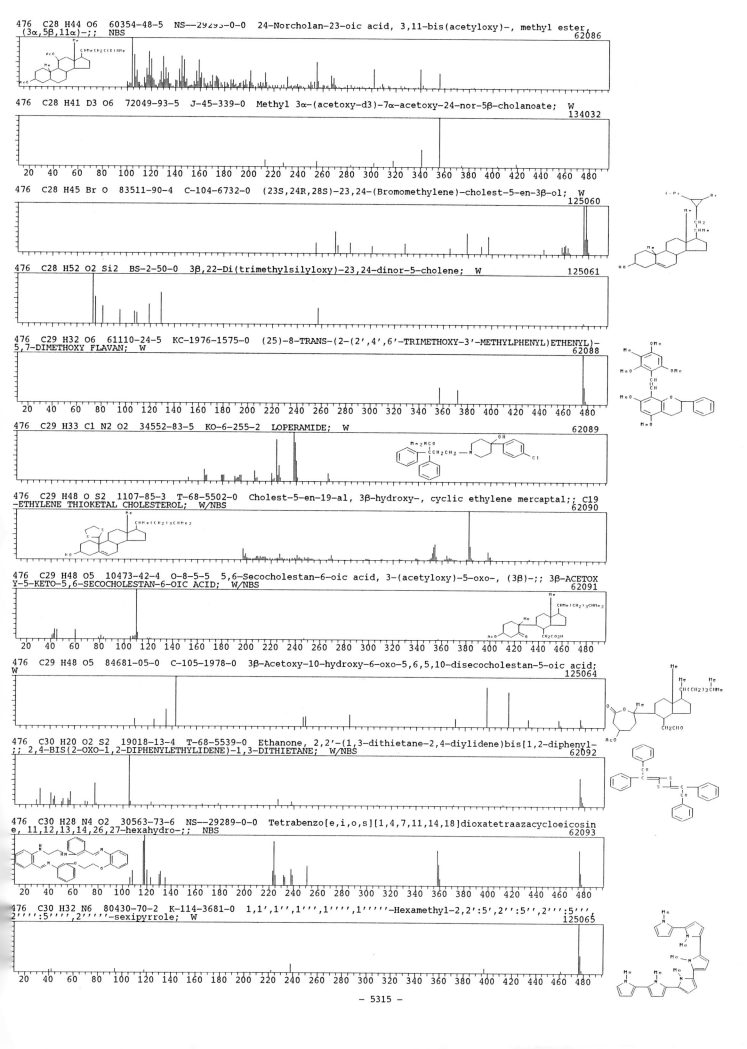

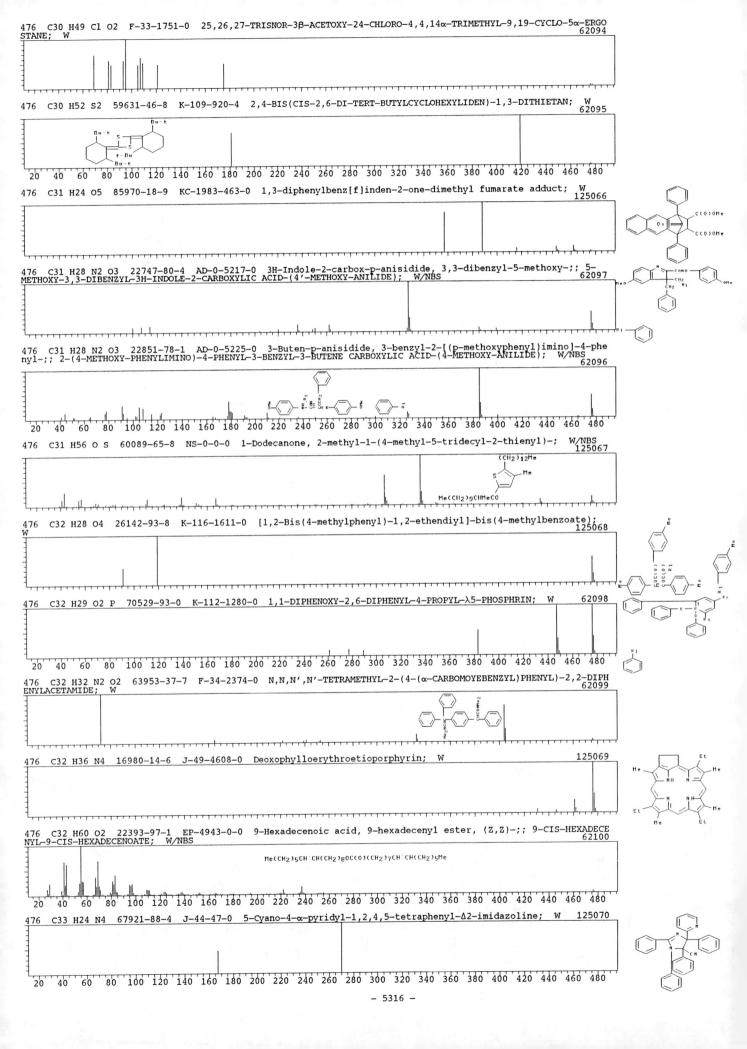

476   C30 H49 Cl O2   F-33-1751-0   25,26,27-TRISNOR-3β-ACETOXY-24-CHLORO-4,4,14α-TRIMETHYL-9,19-CYCLO-5α-ERGO
STANE;   W
62094

476   C30 H52 S2   59631-46-8   K-109-920-4   2,4-BIS(CIS-2,6-DI-TERT-BUTYLCYCLOHEXYLIDEN)-1,3-DITHIETAN;   W
62095

476   C31 H24 O5   85970-18-9   KC-1983-463-0   1,3-diphenylbenz[f]inden-2-one-dimethyl fumarate adduct;   W
125066

476   C31 H28 N2 O3   22747-80-4   AD-0-5217-0   3H-Indole-2-carbox-p-anisidide, 3,3-dibenzyl-5-methoxy-;; 5-
METHOXY-3,3-DIBENZYL-3H-INDOLE-2-CARBOXYLIC ACID-(4'-METHOXY-ANILIDE);   W/NBS
62097

476   C31 H28 N2 O3   22851-78-1   AD-0-5225-0   3-Buten-p-anisidide, 3-benzyl-2-[(p-methoxyphenyl)imino]-4-phe
nyl-;; 2-(4-METHOXY-PHENYLIMINO)-4-PHENYL-3-BENZYL-3-BUTENE CARBOXYLIC ACID-(4-METHOXY-ANILIDE);   W/NBS
62096

476   C31 H56 O S   60089-65-8   NS-0-0-0   1-Dodecanone, 2-methyl-1-(4-methyl-5-tridecyl-2-thienyl)-;   W/NBS
125067

476   C32 H28 O4   26142-93-8   K-116-1611-0   [1,2-Bis(4-methylphenyl)-1,2-ethendiyl]-bis(4-methylbenzoate);
W
125068

476   C32 H29 O2 P   70529-93-0   K-112-1280-0   1,1-DIPHENOXY-2,6-DIPHENYL-4-PROPYL-λ5-PHOSPHRIN;   W
62098

476   C32 H32 N2 O2   63953-37-7   F-34-2374-0   N,N,N',N'-TETRAMETHYL-2-(4-(α-CARBOMOYEBENZYL)PHENYL)-2,2-DIPH
ENYLACETAMIDE;   W
62099

476   C32 H36 N4   16980-14-6   J-49-4608-0   Deoxophylloerythroetioporphyrin;   W
125069

476   C32 H60 O2   22393-97-1   EP-4943-0-0   9-Hexadecenoic acid, 9-hexadecenyl ester, (Z,Z)-;; 9-CIS-HEXADECE
NYL-9-CIS-HEXADECENOATE;   W/NBS
62100

476   C33 H24 N4   67921-88-4   J-44-47-0   5-Cyano-4-α-pyridyl-1,2,4,5-tetraphenyl-Δ2-imidazoline;   W   125070

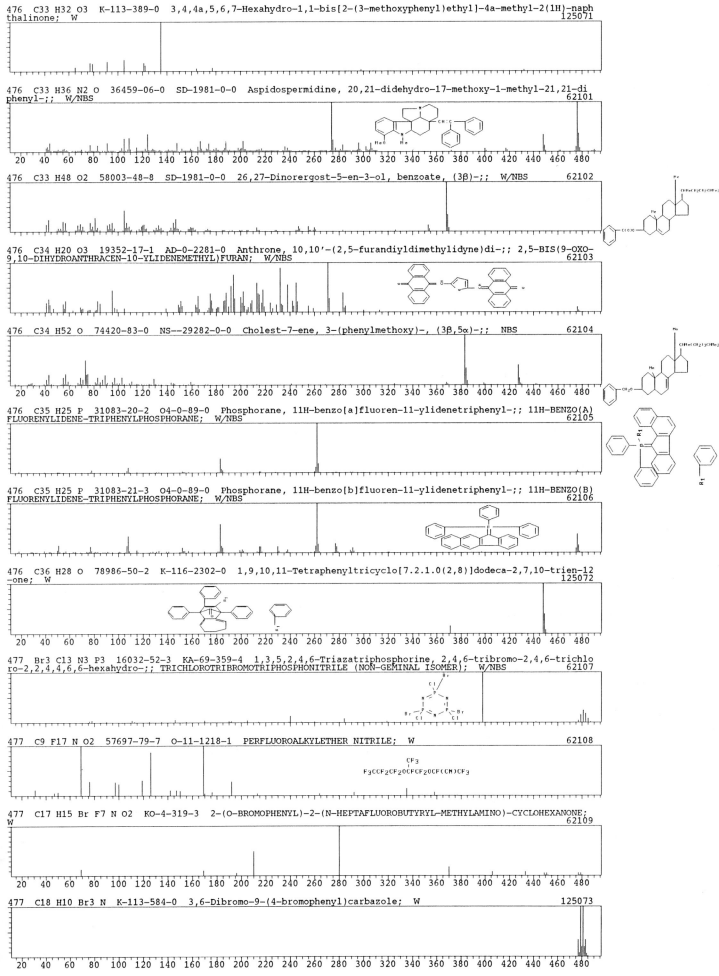

476 C33 H32 O3   K-113-389-0   3,4,4a,5,6,7-Hexahydro-1,1-bis[2-(3-methoxyphenyl)ethyl]-4a-methyl-2(1H)-naphthalinone; W
125071

476 C33 H36 N2 O   36459-06-0   SD-1981-0-0   Aspidospermidine, 20,21-didehydro-17-methoxy-1-methyl-21,21-diphenyl-;; W/NBS
62101

476 C33 H48 O2   58003-48-8   SD-1981-0-0   26,27-Dinorergost-5-en-3-ol, benzoate, (3β)-;; W/NBS
62102

476 C34 H20 O3   19352-17-1   AD-0-2281-0   Anthrone, 10,10'-(2,5-furandiyldimethylidyne)di-;; 2,5-BIS(9-OXO-9,10-DIHYDROANTHRACEN-10-YLIDENEMETHYL)FURAN; W/NBS
62103

476 C34 H52 O   74420-83-0   NS--29282-0-0   Cholest-7-ene, 3-(phenylmethoxy)-, (3β,5α)-;;   NBS
62104

476 C35 H25 P   31083-20-2   O4-0-89-0   Phosphorane, 11H-benzo[a]fluoren-11-ylidenetriphenyl-;; 11H-BENZO(A)FLUORENYLIDENE-TRIPHENYLPHOSPHORANE; W/NBS
62105

476 C35 H25 P   31083-21-3   O4-0-89-0   Phosphorane, 11H-benzo[b]fluoren-11-ylidenetriphenyl-;; 11H-BENZO(B)FLUORENYLIDENE-TRIPHENYLPHOSPHORANE; W/NBS
62106

476 C36 H28 O   78986-50-2   K-116-2302-0   1,9,10,11-Tetraphenyltricyclo[7.2.1.0(2,8)]dodeca-2,7,10-trien-12-one; W
125072

477 Br3 Cl3 N3 P3   16032-52-3   KA-69-359-4   1,3,5,2,4,6-Triazatriphosphorine, 2,4,6-tribromo-2,4,6-trichloro-2,2,4,4,6,6-hexahydro-;; TRICHLOROTRIBROMOTRIPHOSPHONITRILE (NON-GEMINAL ISOMER); W/NBS
62107

477 C9 F17 N O2   57697-79-7   O-11-1218-1   PERFLUOROALKYLETHER NITRILE; W
62108

477 C17 H15 Br F7 N O2   KO-4-319-3   2-(O-BROMOPHENYL)-2-(N-HEPTAFLUOROBUTYRYL-METHYLAMINO)-CYCLOHEXANONE; W
62109

477 C18 H10 Br3 N   K-113-584-0   3,6-Dibromo-9-(4-bromophenyl)carbazole; W
125073

477   C18 H39 N3 O6 Si3   RB-1982-15769-0   TRIS(2-TRIMETHYLSILOXYETHYL)ISOCYANURIC ACID;   W   125074

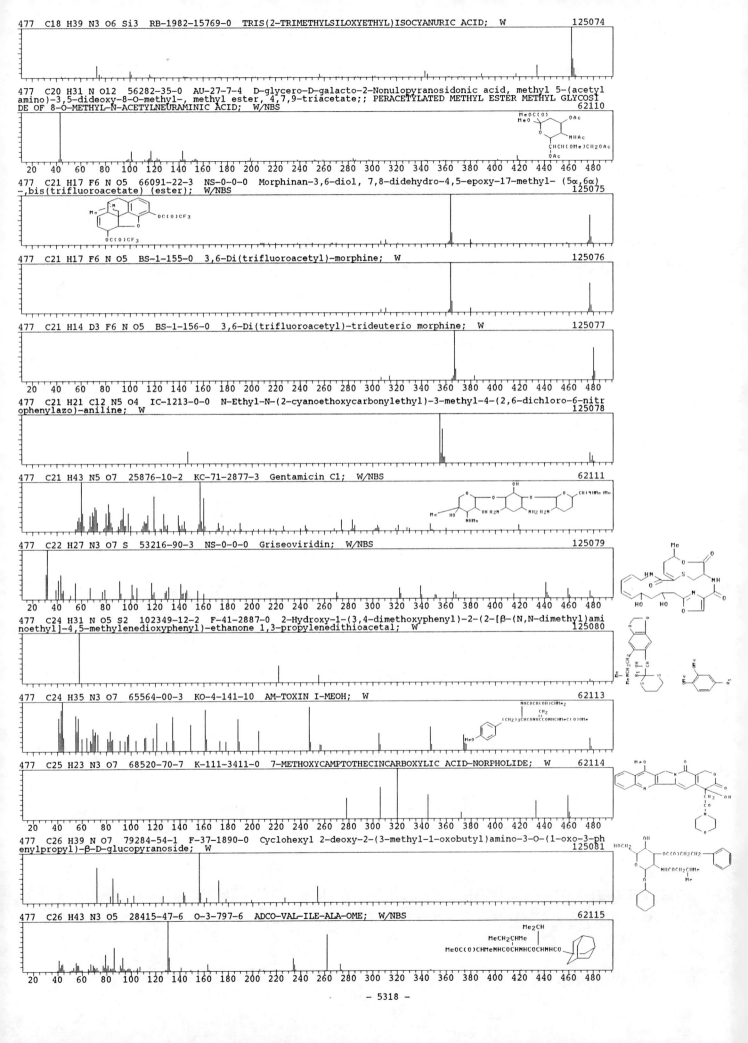

477   C20 H31 N O12   56282-35-0   AU-27-7-4   D-glycero-D-galacto-2-Nonulopyranosidonic acid, methyl 5-(acetyl
amino)-3,5-dideoxy-8-O-methyl-, methyl ester, 4,7,9-triacetate;;   PERACETYLATED METHYL ESTER METHYL GLYCOSI
DE OF 8-O-METHYL-N-ACETYLNEURAMINIC ACID;   W/NBS   62110

477   C21 H17 F6 N O5   66091-22-3   NS-0-0-0   Morphinan-3,6-diol, 7,8-didehydro-4,5-epoxy-17-methyl- (5α,6α)
-,bis(trifluoroacetate) (ester);   W/NBS   125075

477   C21 H17 F6 N O5   BS-1-155-0   3,6-Di(trifluoroacetyl)-morphine;   W   125076

477   C21 H14 D3 F6 N O5   BS-1-156-0   3,6-Di(trifluoroacetyl)-trideuterio morphine;   W   125077

477   C21 H21 Cl2 N5 O4   IC-1213-0-0   N-Ethyl-N-(2-cyanoethoxycarbonylethyl)-3-methyl-4-(2,6-dichloro-6-nitr
ophenylazo)-aniline;   W   125078

477   C21 H43 N5 O7   25876-10-2   KC-71-2877-3   Gentamicin C1;   W/NBS   62111

477   C22 H27 N3 O7 S   53216-90-3   NS-0-0-0   Griseoviridin;   W/NBS   125079

477   C24 H31 N O5 S2   102349-12-2   F-41-2887-0   2-Hydroxy-1-(3,4-dimethoxyphenyl)-2-(2-[β-(N,N-dimethyl)ami
noethyl]-4,5-methylenedioxyphenyl)-ethanone 1,3-propylenedithioacetal;   W   125080

477   C24 H35 N3 O7   65564-00-3   KO-4-141-10   AM-TOXIN I-MEOH;   W   62113

477   C25 H23 N3 O7   68520-70-7   K-111-3411-0   7-METHOXYCAMPTOTHECINCARBOXYLIC ACID-NORPHOLIDE;   W   62114

477   C26 H39 N O7   79284-54-1   F-37-1890-0   Cyclohexyl 2-deoxy-2-(3-methyl-1-oxobutyl)amino-3-O-(1-oxo-3-ph
enylpropyl)-β-D-glucopyranoside;   W   125081

477   C26 H43 N3 O5   28415-47-6   O-3-797-6   ADCO-VAL-ILE-ALA-OME;   W/NBS   62115

477 C26 H47 N O3 Si2 56210-92-5 SW-0-34-0 Androst-5-en-16-one, 3,17-bis[(trimethylsilyl)oxy]-, O-methyl oxime, (3β,17β)-;; BISTRIMETHYLSILYL 3β,17β-DIHYDROXY-ANDROST-5-ENE-16-ONEMETHOXIME; W   79244

477 C26 H47 N O3 Si2 55557-05-6 EP-8221-0-0 Androst-5-en-17-one, 3,16-bis[(trimethylsilyl)oxy]-, O-methyloxime, (3β,16α)-;; 5-ANDROSTENE-3β,16α-DIOL-17-ONE MO TMS; W/NBS   62118

477 C26 H47 N O3 Si2 69597-48-4 NS--29271-0-0 Androst-5-en-17-one, 3,16-bis[(trimethylsilyl)oxy]-, O-methyloxime, (3β,16β)-;; NBS   62122

477 C26 H47 N O3 Si2 55836-47-0 AV-53-75-4 Androst-4-en-3-one, 17,19-bis[(trimethylsilyl)oxy]-, O-methyloxime, (17β)-;; 19-HYDROXYTESTOSTERONE METHOXIME-BIS-TRIMETHYLSILYL ETHER; W/NBS   62119

477 C26 H47 N O3 Si2 69688-34-2 SH-1981-0-0 Androst-4-en-3-one, 16,17-bis[(trimethylsilyl)oxy]-, O-methyloxime, (16α,17β)-;; ANDROST-4-EN-16α,17β-OL-3-ONE MO TMS; W/NBS   62123

477 C26 H47 N O3 Si2 SW-0-176-0 BISTRIMETHYLSILYL 3β,7α-DIHYDROXY-ANDROST-5-ENE-17-ONE METHOXIME; W   62116

477 C26 H47 N O3 Si2 SW-0-177-0 BISTRIMETHYLSILYL 3β,11β-DIHYDROXY-ANDROST-5-ENE-17-ONE METHOXIME; W   62120

477 C26 H47 N O3 Si2 SW-0-174-0 BISTRIMETHYLSILYL 3β,17β-DIHYDROXY-ANDROST-5-ENE-7-ONE METHOXIME; W   62121

477 C26 H47 N O3 Si2 SH-1981-0-0 ANDROST-4-EN 7α,17β-OL-3-ONE MO-TMS; W   62124

477 C26 H47 N O3 Si2 SH-1981-0-0 ANDROST-4-EN-2β,17β-OL-3-ONE MO TMS; W   62125

477 C26 H47 N O3 Si2 SH-1981-0-0 ANDROST-4-EN-1β,17β-DIOL-3-ONE METHYL OXIME DI-TMS DERIVATIVE; W 62126

477 C26 H47 N O3 Si2 SH-1981-0-0 Monomethyloxime, bis(trimethylsilyl)- 6β-Hydroxytestosterone; W 125082

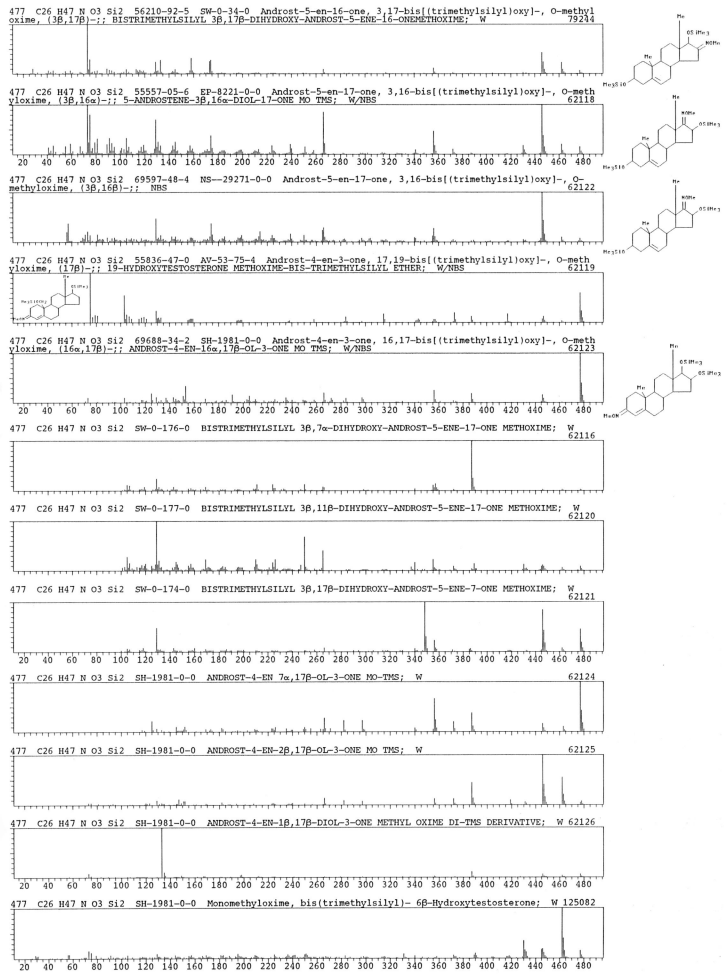

477   C26 H47 N O3 Si2   SH-1981-0-0   Methyloxime, trimethylsilyl- 16-α-Hydroxydehydroepiandrosterone or Methyloxime, trimethylsilyl- 3α,16α-Dihydroxy-androstene-17-one;   W   125083

477   C26 H47 N O3 Si2   SH-1981-0-0   Monomethyloxime, trimethylsilyl-6α-Hydroxytestosterone;   W   125084

20 40 60 80 100 120 140 160 180 200 220 240 260 280 300 320 340 360 380 400 420 440 460 480

477   C27 H18 N3 O3 Sc   OS-5-433-0   Triquinolin-8-olatoscadium(3);   W   125085

477   C27 H27 N O7   86340-47-8   KC-1983-523-0   Trimethyl 9-oxo-5,6,7a,8,9,10-hexahydro-7a,10-etheno-4H-pyrido[3,2,1-jk]carbazole-8-spirocyclopentane-11,12,13-tricarboxylate;   W   125086

477   C28 H31 N O4 S   74974-73-5   C-104-5725-0   Ethyl (4S)-(E)-4-(N-benzyl-p-toluenesulfonamido)-2-methyl-5-phenylpent-2-enoate;   W   125087

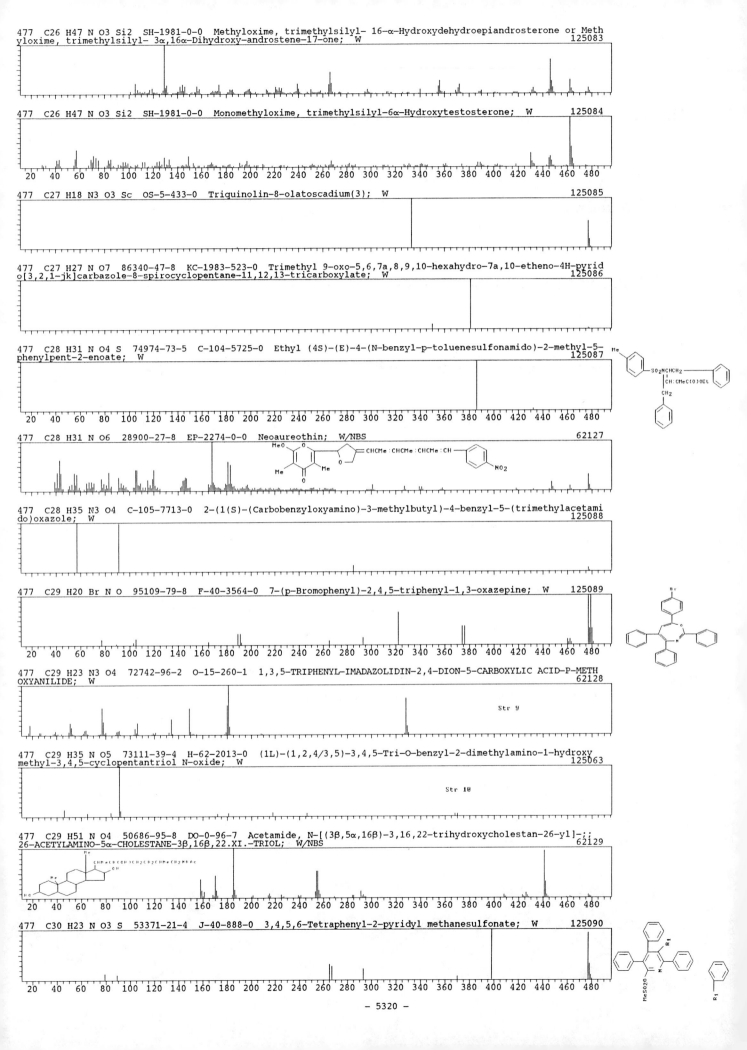

20 40 60 80 100 120 140 160 180 200 220 240 260 280 300 320 340 360 380 400 420 440 460 480

477   C28 H31 N O6   28900-27-8   EP-2274-0-0   Neoaureothin;   W/NBS   62127

477   C28 H35 N3 O4   C-105-7713-0   2-(1(S)-(Carbobenzyloxyamino)-3-methylbutyl)-4-benzyl-5-(trimethylacetamido)oxazole;   W   125088

477   C29 H20 Br N O   95109-79-8   F-40-3564-0   7-(p-Bromophenyl)-2,4,5-triphenyl-1,3-oxazepine;   W   125089

20 40 60 80 100 120 140 160 180 200 220 240 260 280 300 320 340 360 380 400 420 440 460 480

477   C29 H23 N3 O4   72742-96-2   O-15-260-1   1,3,5-TRIPHENYL-IMADAZOLIDIN-2,4-DION-5-CARBOXYLIC ACID-P-METHOXYANILIDE;   W   62128

Str 9

477   C29 H35 N O5   73111-39-4   H-62-2013-0   (1L)-(1,2,4/3,5)-3,4,5-Tri-O-benzyl-2-dimethylamino-1-hydroxymethyl-3,4,5-cyclopentantriol N-oxide;   W   125063

Str 10

477   C29 H51 N O4   50686-95-8   DO-0-96-7   Acetamide, N-[(3β,5α,16β)-3,16,22-trihydroxycholestan-26-yl]-;; 26-ACETYLAMINO-5α-CHOLESTANE-3β,16β,22.XI.-TRIOL;   W/NBS   62129

20 40 60 80 100 120 140 160 180 200 220 240 260 280 300 320 340 360 380 400 420 440 460 480

477   C30 H23 N O3 S   53371-21-4   J-40-888-0   3,4,5,6-Tetraphenyl-2-pyridyl methanesulfonate;   W   125090

20 40 60 80 100 120 140 160 180 200 220 240 260 280 300 320 340 360 380 400 420 440 460 480

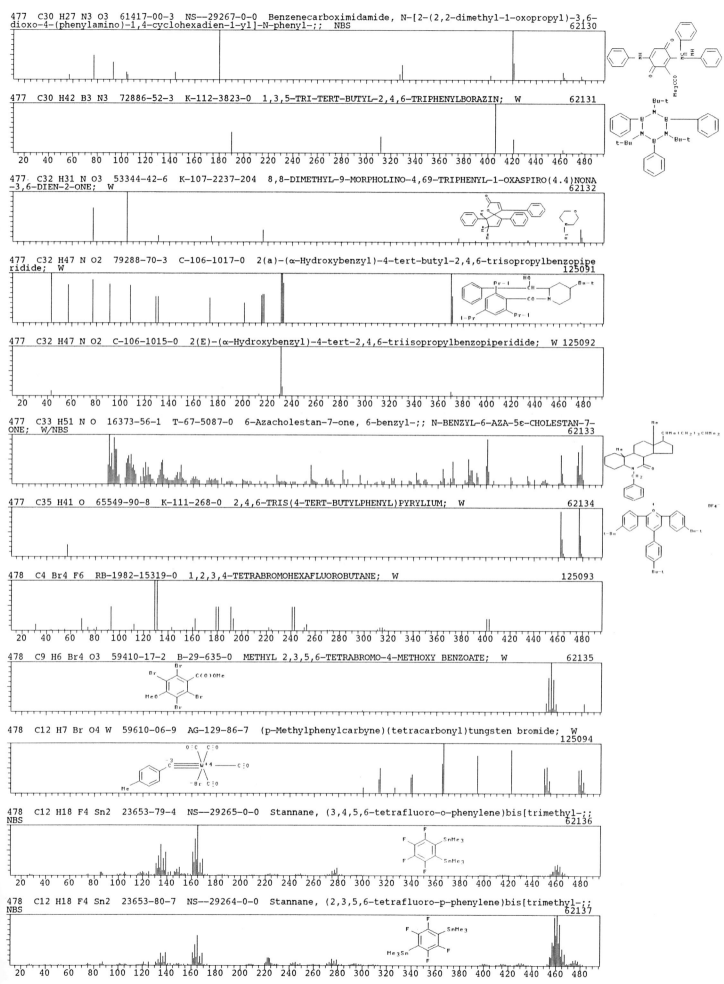

477 C30 H27 N3 O3 61417-00-3 NS--29267-0-0 Benzenecarboximidamide, N-[2-(2,2-dimethyl-1-oxopropyl)-3,6-dioxo-4-(phenylamino)-1,4-cyclohexadien-1-yl]-N-phenyl-;; NBS
62130

477 C30 H42 B3 N3 72886-52-3 K-112-3823-0 1,3,5-TRI-TERT-BUTYL-2,4,6-TRIPHENYLBORAZIN; W
62131

477 C32 H31 N O3 53344-42-6 K-107-2237-204 8,8-DIMETHYL-9-MORPHOLINO-4,69-TRIPHENYL-1-OXASPIRO(4.4)NONA-3,6-DIEN-2-ONE; W
62132

477 C32 H47 N O2 79288-70-3 C-106-1017-0 2(a)-(α-Hydroxybenzyl)-4-tert-butyl-2,4,6-trisopropylbenzopiperidide; W
125091

477 C32 H47 N O2 C-106-1015-0 2(E)-(α-Hydroxybenzyl)-4-tert-2,4,6-triisopropylbenzopiperidide; W 125092

477 C33 H51 N O 16373-56-1 T-67-5087-0 6-Azacholestan-7-one, 6-benzyl-;; N-BENZYL-6-AZA-5ε-CHOLESTAN-7-ONE; W/NBS
62133

477 C35 H41 O 65549-90-8 K-111-268-0 2,4,6-TRIS(4-TERT-BUTYLPHENYL)PYRYLIUM; W
62134

478 C4 Br4 F6 RB-1982-15319-0 1,2,3,4-TETRABROMOHEXAFLUOROBUTANE; W
125093

478 C9 H6 Br4 O3 59410-17-2 B-29-635-0 METHYL 2,3,5,6-TETRABROMO-4-METHOXY BENZOATE; W
62135

478 C12 H7 Br O4 W 59610-06-9 AG-129-86-7 (p-Methylphenylcarbyne)(tetracarbonyl)tungsten bromide; W
125094

478 C12 H18 F4 Sn2 23653-79-4 NS--29265-0-0 Stannane, (3,4,5,6-tetrafluoro-o-phenylene)bis[trimethyl-;; NBS
62136

478 C12 H18 F4 Sn2 23653-80-7 NS--29264-0-0 Stannane, (2,3,5,6-tetrafluoro-p-phenylene)bis[trimethyl-;; NBS
62137

478 C15 H9 F11 O5 KO-9-303-1 Pentafluoropropionyl,hexafluoroisopropyl derivative of homovanillac acid;
W
125095

478 C15 H21 Cl3 N2 O3 S3 RB-1982-11224-0 CARBAMIC ACID, N,N-DIETHTHIO-S(2,3,4-TRICHLORO-6-(N,N-DIETHYLSU
LFAMOYL)PHENYL)ESTER; W
125096

478 C18 H5 B F10 N4 86067-38-1 K-116-1537-0 1,5-Bis(pentafluorophenyl)-4-phenyl-2-tetrazaboroline; W
125097

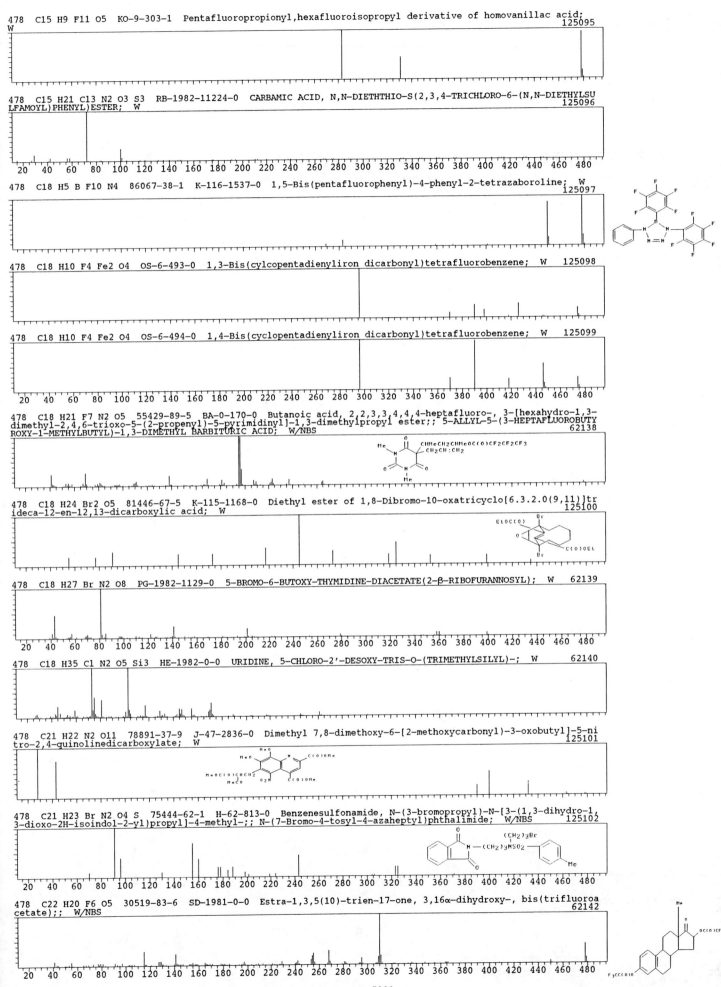

478 C18 H10 F4 Fe2 O4 OS-6-493-0 1,3-Bis(cylcopentadienyliron dicarbonyl)tetrafluorobenzene; W 125098

478 C18 H10 F4 Fe2 O4 OS-6-494-0 1,4-Bis(cyclopentadienyliron dicarbonyl)tetrafluorobenzene; W 125099

478 C18 H21 F7 N2 O5 55429-89-5 BA-0-170-0 Butanoic acid, 2,2,3,3,4,4,4-heptafluoro-, 3-[hexahydro-1,3-
dimethyl-2,4,6-trioxo-5-(2-propenyl)-5-pyrimidinyl]-1,3-dimethylpropyl ester;; 5-ALLYL-5'-(3-HEPTAFLUOROBUTY
ROXY-1-METHYLBUTYL)-1,3-DIMETHYL BARBITURIC ACID; W/NBS
62138

478 C18 H24 Br2 O5 81446-67-5 K-115-1168-0 Diethyl ester of 1,8-Dibromo-10-oxatricyclo[6.3.2.0(9,11)]tr
ideca-12-en-12,13-dicarboxylic acid; W
125100

478 C18 H27 Br N2 O8 PG-1982-1129-0 5-BROMO-6-BUTOXY-THYMIDINE-DIACETATE(2-β-RIBOFURANNOSYL); W 62139

478 C18 H35 Cl N2 O5 Si3 HE-1982-0-0 URIDINE, 5-CHLORO-2'-DESOXY-TRIS-O-(TRIMETHYLSILYL)-; W 62140

478 C21 H22 N2 O11 78891-37-9 J-47-2836-0 Dimethyl 7,8-dimethoxy-6-[2-methoxycarbonyl)-3-oxobutyl]-5-ni
tro-2,4-quinolinedicarboxylate; W
125101

478 C21 H23 Br N2 O4 S 75444-62-1 H-62-813-0 Benzenesulfonamide, N-(3-bromopropyl)-N-[3-(1,3-dihydro-1,
3-dioxo-2H-isoindol-2-yl)propyl]-4-methyl-;; N-(7-Bromo-4-tosyl-4-azaheptyl)phthalimide; W/NBS 125102

478 C22 H20 F6 O5 30519-83-6 SD-1981-0-0 Estra-1,3,5(10)-trien-17-one, 3,16α-dihydroxy-, bis(trifluoroa
cetate);; W/NBS
62142

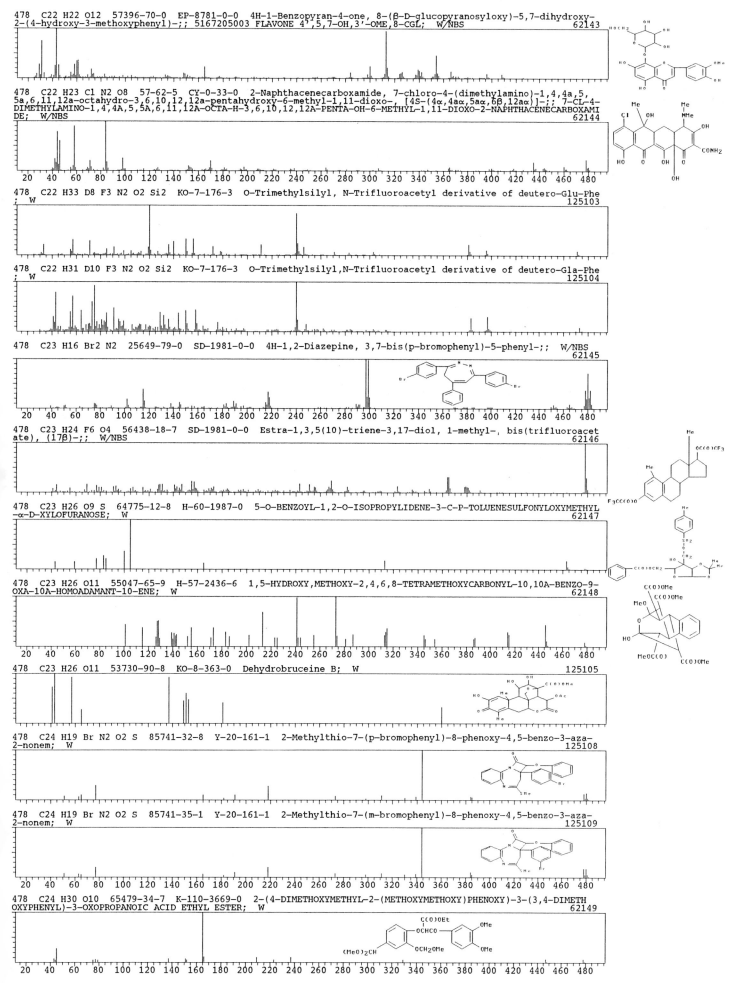

478   C22 H22 O12   57396-70-0   EP-8781-0-0   4H-1-Benzopyran-4-one, 8-(β-D-glucopyranosyloxy)-5,7-dihydroxy-2-(4-hydroxy-3-methoxyphenyl)-;;   5167205003 FLAVONE 4',5,7-OH,3'-OME,8-CGL;   W/NBS
62143

478   C22 H23 Cl N2 O8   57-62-5   CY-0-33-0   2-Naphthacenecarboxamide, 7-chloro-4-(dimethylamino)-1,4,4a,5,5a,6,11,12a-octahydro-3,6,10,12,12a-pentahydroxy-6-methyl-1,11-dioxo-, [4S-(4α,4aα,5aα,6β,12aα)]-;; 7-CL-4-DIMETHYLAMINO-1,4,4A,5,5A,6,11,12A-OCTA-H-3,6,10,12,12A-PENTA-OH-6-METHYL-1,11-DIOXO-2-NAPHTHACENECARBOXAMIDE; W/NBS
62144

478   C22 H33 D8 F3 N2 O2 Si2   KO-7-176-3   O-Trimethylsilyl, N-Trifluoroacetyl derivative of deutero-Glu-Phe ; W
125103

478   C22 H31 D10 F3 N2 O2 Si2   KO-7-176-3   O-Trimethylsilyl,N-Trifluoroacetyl derivative of deutero-Gla-Phe ; W
125104

478   C23 H16 Br2 N2   25649-79-0   SD-1981-0-0   4H-1,2-Diazepine, 3,7-bis(p-bromophenyl)-5-phenyl-;;   W/NBS
62145

478   C23 H24 F6 O4   56438-18-7   SD-1981-0-0   Estra-1,3,5(10)-triene-3,17-diol, 1-methyl-, bis(trifluoroacetate), (17β)-;; W/NBS
62146

478   C23 H26 O9 S   64775-12-8   H-60-1987-0   5-O-BENZOYL-1,2-O-ISOPROPYLIDENE-3-C-P-TOLUENESULFONYLOXYMETHYL-α-D-XYLOFURANOSE; W
62147

478   C23 H26 O11   55047-65-9   H-57-2436-6   1,5-HYDROXY,METHOXY-2,4,6,8-TETRAMETHOXYCARBONYL-10,10A-BENZO-9-OXA-10A-HOMOADAMANT-10-ENE; W
62148

478   C23 H26 O11   53730-90-8   KO-8-363-0   Dehydrobruceine B; W
125105

478   C24 H19 Br N2 O2 S   85741-32-8   Y-20-161-1   2-Methylthio-7-(p-bromophenyl)-8-phenoxy-4,5-benzo-3-aza-2-nonem; W
125108

478   C24 H19 Br N2 O2 S   85741-35-1   Y-20-161-1   2-Methylthio-7-(m-bromophenyl)-8-phenoxy-4,5-benzo-3-aza-2-nonem; W
125109

478   C24 H30 O10   65479-34-7   K-110-3669-0   2-(4-DIMETHOXYMETHYL-2-(METHOXYMETHOXY)PHENOXY)-3-(3,4-DIMETHOXYPHENYL)-3-OXOPROPANOIC ACID ETHYL ESTER; W
62149

478   C24 H34 N2 O4 S2   K-117-1668-0   3,3,7,7-Tetramethyl-1,5-diazacyclooctane;   W   125110

478   C24 H34 N2 O8   77236-73-8   K-114-142-0   (3RS,3'RS,4RS0-5'-T-Butoxycarbonyl-3,4-dihydro-4'-[2-(methoxyc
arbonyl)ethyl]-3-(1-methoxyethyl)-3',4-dimethyl-5-oxo-2,2'-pyrromethene-meso-carboxylic acid;   W   125111

478   C24 H34 N2 O8   77286-25-0   K-114-142-0   (3RS,3'SR,4RS)-5'-T-Butoxycarbonyl-3,4-dihydro-4'-[2-(methoxyc
arbonyl)ethyl]-3-(1-methoxyethyl)-3',4-dimethyl-5-oxo-2,2'-pyrromethene-meso-carboxylic acid;   W   125112

478   C25 H18 O10   56691-90-8   B-28-1797-0   Benzyl salazinate;   W   125113

478   C25 H23 Br N2 O3   77143-53-4   O-18-364-1   2-(α-Bromo-β-ethoxy-p-methoxyphenethyl)-3-phenyl-4(3H)-quina
zolinone;   W   125114

478   C25 H34 O7 S   82039-61-0   B-35-502-0   p-Toluenesulfinate of 1,2:5,6-di-O-cyclohexylidene-α-D-glucofura
nose;   W   125115

478   C25 H34 O7 S   82769-93-5   C-104-5737-0   (1SR,2RS,3RS,4RS,6SR)-2,6-Dimethyl-4-(phenylsulfonyl)-4-(2-pro
penyl)-3-(3-(methoxycarbonyl)-2-oxopropyl)cyclohexane-1,2-diol 1,2-acetonide;   W   125116

478   C25 H34 O9   68702-89-6   O-13-308-1   (7,8)DIMETHYL ESTER OF 1H-6β,9β-DIACETYLOXY-7β-CARBOXY-8α-CARBOXY
METHYL-4Aα,7α-DIMETHYL-2-OXOPERHYDROPHENANTHRENE;   W   62150

478   C25 H34 O9   74269-90-2   F-35-2395-0   8Aα-BENZYLOXYMETHYL-5α,7α-DIACETOXY-2β-METHOXY-6β-METHOXYCARBONYL
METHYL-4Aα-PERHYDROISOCHROMANE;   W   62151

478   C25 H34 O9   91156-85-3   J-49-3258-0   (+-)-(1α,4bβ,10α)-Dimethyl 8,8-(Ethylenedioxy)-2-methoxy-7-[(meth
oxymethyl)oxy]-1-methylgibba-2,4a(10a)-diene-1,10-dicarboxylate;   W   125117

478   C25 H34 O9   91127-27-4   J-49-3258-0   (+-)-(1α,4bβ,10β)-Dimethyl 8,8-(ethylenedioxy)-2-methoxy-7-[(meth
oxymethyl)oxy]-1-methylgibba-2,4a(10a)-diene-1,10-dicarboxylate;   W   125118

478   C25 H33 D O9   J-49-3258-0   (+/-)-(1α,4bβ,10α)-Dimethyl 8,8-(ethylenedioxy)-2-methoxy-7-[(methoxymeth
yl)oxy]-10-deutero-1-methylgibba-2,4a(10a)-diene-1,10-dicarboxylate;   W   125119

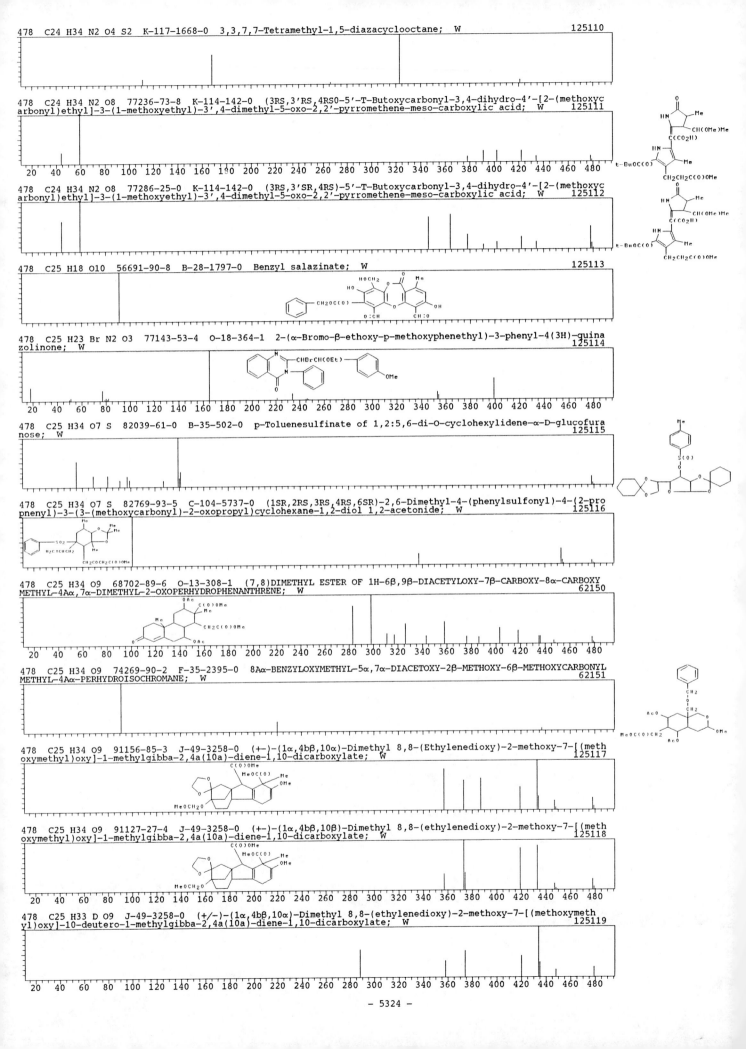

478 C25 H33 D O9 J-49-3258-0 (+/-)-(1β,4bβ,10β)-Dimethyl 8,8-(ethylenedioxy)-2-methoxy-7-[(methoxymeth yl)oxy]-10-deutero-1-methylgibba-2,4a(10a)-diene-1,10-dicarboxylate; W   125005

478 C25 H50 O8 20207-67-4 T-67-4643-0 Octadecanoic acid, 9,10,12,13,15,16-hexamethoxy-, methyl ester;; 9,10,12,13,15,16-HEXAMETHOXY-METHYL OCTADECANOATE; W/NBS   62152

MeOC(O)(CH2)7CH(OMe)CH(OMe)CH2CH(OMe)CH(OMe)CH2CH(OMe)CH(OMe)Et

478 C25 H50 O8 20207-68-5 T-67-4642-0 Octadecanoic acid, 6,7,9,10,12,13-hexamethoxy-, methyl ester;; 6, 7,9,10,12,13-HEXAMETHOXY-METHYL OCTADECANOATE; W/NBS   62153

MeOC(O)(CH2)4CH(OMe)CH(OMe)CH2CH(OMe)CH(OMe)CH2CH(OMe)CH(OMe)(CH2)4Me

478 C26 H22 O9 69813-88-3 C-105-1612-0 Epoxy triacetate derivative of 4-demethoxy-7,9-dideoxydaunomycin one; W   125120

478 C26 H22 O9 58124-25-7 AC-1975-1808-0 5,7-Dihydroxy-2',4'-dimethoxy-3-(2,4-dimethoxy-benzoyl)-flavo ne; W   125121

478 C26 H24 N6 Ni 74834-20-1 Y-17-443-0 6,13-Di-(2-acetonitrile)-tetramethyldibenzo[b,i][1,4,8,11]tetra azacyclotetradeca-4,6,11,13-tetraene; W   125122

478 C26 H30 N4 O5 62290-61-3 KO-3-158-2 2-(O-ETHOXYCARBONYLPHENYLAZO)-3-(2-METHOXYCARBONYLETHYL)-4-METH YL-5-((2,5-DIHYDRO-3-ETHYL-4-METHYL-5-OXO-2-PYRROLYLIDENE)METHYL)PYRROLE; W   62154

478 C26 H30 N4 O5 62290-62-4 KO-3-158-2 2-(O-ETHOXYCARBONYLPHENYLAZO)-3-(2-METHOXYCARBONYLETHYL)-4-METH YL-5-((2,5-DIHYDRO-3-METHYL-4-ETHYL-5-OXO-2-PYRROLYLIDENE)METHYL)PYRROLE; W   62155

478 C26 H30 N4 O5 62290-66-8 KO-3-158-2 2-(O-ETHOXYCARBONYLPHENYLAZO)-3-ETHYL-4-METHYL-5-((2,5-DIHYDRO- 4-(2-METHOXYCARBONYLETHYL)-5-OXO-2-PYRROLYLIDENE)METHYL)PYRROLE; W   62156

478 C26 H30 N4 O5 62290-64-6 KO-3-158-2 2-(O-ETHOXYCARBONYLPHENYLAZO)-3-METHYL-4-ETHYL-5-((2,5-DIHYDRO- 3-METHYL-4-(2-METHOXYCARBONYLETHYL)-5-OXO-2-PYRROLYLIDENE)METHYL)PYRROLE; W   62157

478 C26 H38 O8 73480-56-5 H-62-2752-0 7α,17β-Diacetoxy-3β,19-epoxy-3α-methoxy-4,4-dimethyl-8,14-seco- 5α,13α-androstan-8,14-dione; W   125123

478 C26 H46 O4 Si2 41577-91-7 O-9-45-3 Prosta-5,10,13-trien-1-oic acid, 9-oxo-15-[(trimethylsilyl)oxy] -, trimethylsilyl ester, (5Z,13E,15S)-;; BISTRIMETHYLSILYL-PGA2; W/NBS   62159

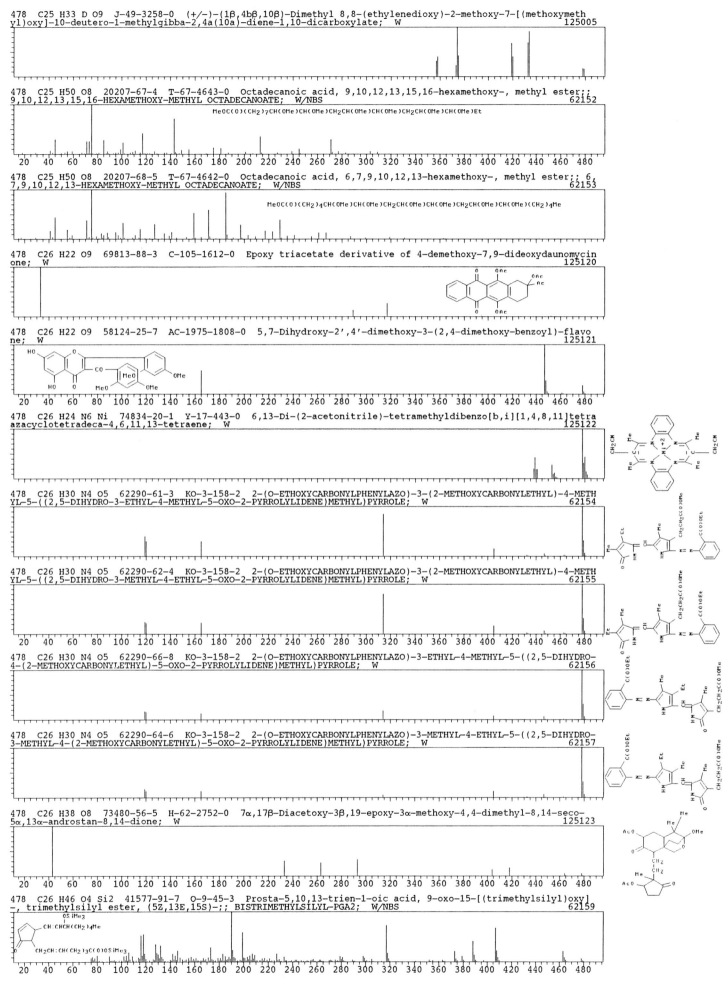

478 C26 H46 O4 Si2 53044-54-5 O-9-45-3 Prosta-5,8(12),13-trien-1-oic acid, 9-oxo-15-[(trimethylsilyl)ox
yl]-, trimethylsilyl ester, (5Z,13E,15S)-;; BISTRIMETHYLSILYL-PBG2; W/NBS   62158

Me3SiO
Me(CH2)4CHCH:CH
Me3SiOC(O)(CH2)3CH:CHCH2

478 C27 H34 N4 O4 21761-50-2 K-102-53-3 Aralionine, debenzoyl-;; DESBENZOYL-ARALIONINE; W/NBS   62160

478 C27 H39 Co Si2 63516-98-3 C-99-4061-0 1-(BENZOCYCLOPENTEN-5-YL)-3-(CYCLOPENTADIENYL COBALT-2,3-DI-(
TRIMETHYL SILYL)CYCLOBUTACLIENYL)PROPANE; W   62161

478 C27 H39 P Si3 18848-96-9 HE-1982-0-0 Phosphine, tris[4-(trimethylsilyl)phenyl]-;; Phosphine, tris[
p-(trimethylsilyl)phenyl]-; NBS   58740

478 C27 H40 Cl2 N2 O I-55-2675-0 2-METHYL-5-ETHYL-2-(5'-ETHYL-2'-PICOLYL)-3-(1''-HYDROXY-3''-METHYLBUTY
L)-6-PHENYL-4,5-DIDEHYDROPIPERIDINE DIHYDROCHLORIDE; W   62163

478 C27 H42 O7 72049-94-6 J-45-339-0 Methyl ester of (3α,5β,7α,12β)-7-(acetyloxy)-3,12-dihydroxy-11-oxo
cholan-24-oic acid; W   125124

478 C27 H43 Br O2 4988-84-5 AD-0-3874-0 5α-Spirostan, 23-bromo-, (22S,23R,25R)-;; 23-BROMODEOXYTIGOGEN
IN; W/NBS   62164

478 C27 H43 Br O2 4947-69-7 AD-0-3876-0 5α-Spirostan, 23-bromo-, (22S,23S,25R)-;; 23S-BROMO-(25R)-5α-SP
IROSTAN; W/NBS   62165

478 C27 H50 O3 Si2 41164-18-5 J-8-3557-2 Pregn-5-en-20-ol, 3,17-bis[(trimethylsilyl)oxy]-, (3β,20S)-;;
3β,17α-BIS(TRIMETHYLSILYLOXY)-5-PREGNEN-20α-OL; W/NBS   62173

478 C27 H50 O3 Si2 41259-41-0 J-8-3557-1 Pregn-4-en-17-ol, 3,20-bis[(trimethylsilyl)oxy]-, (3β,20S)-;;
3β,20α-BIS(TRIMETHYLSILYLOXY)-5-PREGNEN-17α-OL; W/NBS   62172

478 C27 H50 O3 Si2 33287-45-5 EP-8439-0-0 5β-Pregnan-20-one, 3α,21-bis(trimethylsiloxy)-;; BISTRIMETHYL
SILYL 3α,21-DIHYDROXY-5β-PREGNANE-20-ONE; W/NBS   62166

478 C27 H50 O3 Si2 33287-43-3 EP-8440-0-0 5β-Pregnan-20-one, 3α,6α-bis(trimethylsiloxy)-;; BISTRIMETHYL
SILYL 3α,6α-DIHYDROXY-5β-PREGNANE-20-ONE; W/NBS   62169

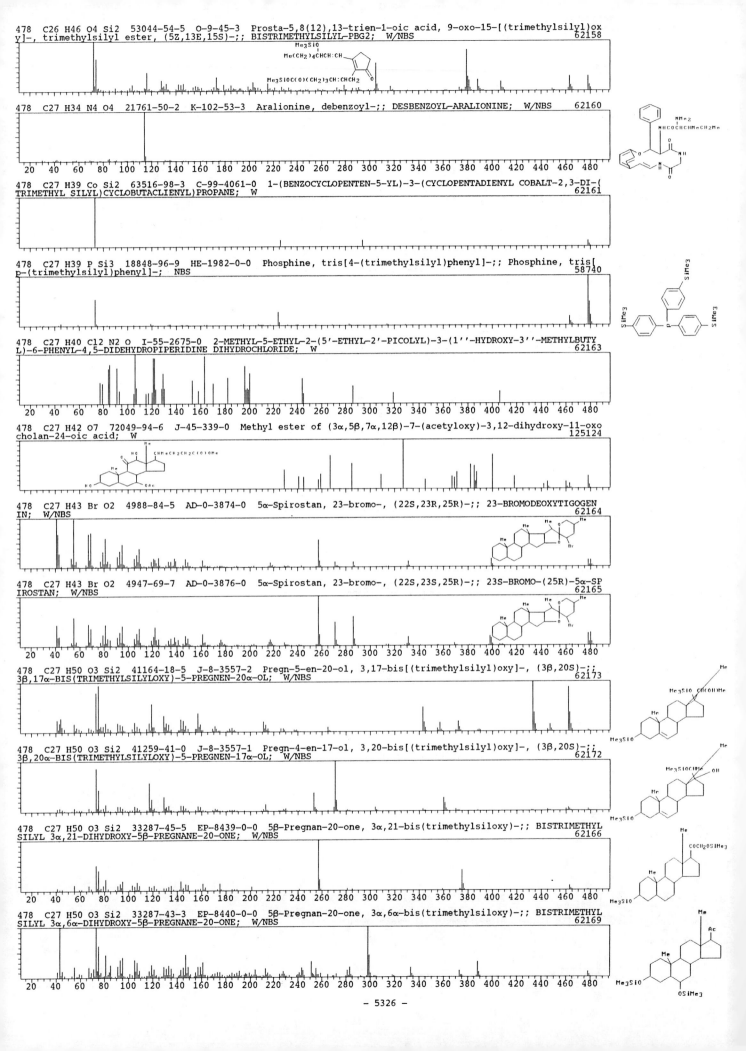

478 C27 H50 O3 Si2 56247-72-4 N-29-538-3 Pregnane-20-one, 3,19-bis[(trimethylsilyl)oxy]-, (3α,5α)-;;
3α,19-TRIMETHYLSILYLOXY-5α-PREGNAN-20-ONE; W/NBS   62167

478 C27 H50 O3 Si2 56247-71-3 N-29-539-4 Pregnan-20-one, 3,16-bis[(trimethylsilyl)oxy]-, (3α,5α,16α)-;;
3α,16α-BIS(TRIMETHYLSILYLOXY)-5α-PREGNANE-20-ONE; W/NBS   62168

478 C27 H50 O3 Si2 SW-0-192-0 BISTRIMETHYLSILYL 3β,21-5β-DIHYDROXY-5β-PREGNANE-20-ONE; W   62170

478 C27 H50 O3 Si2 SW-0-195-0 BISTRIMETHYLSILYL 3α,15α-DIHYDROXY-5α-PREGNANE-20-ONE;   W   62171

478 C27 H50 O3 Si2 SW-0-193-0 BISTRIMETHYLSILYL 3β,21-5α-DIHYDROXY-5β-PREGNANE-20-ONE; W   62174

478 C27 H50 O3 Si2 AH-106-1421-17 3α,16α-DIHYDROXY-5β-PREGNAN-20-ONE, DI-TMS ETHER; W   62175

478 C27 H50 O3 Si2 SH-1981-0-0 Bis(trimethylsilyl) derivative of 3β,15α-Dihydroxy-5β-pregnane-20-one; W
                                                                                          125125

478 C27 H50 O3 Si2 SH-1981-0-0 Bis(trimethylsilyl) derivative of 3β,15β-Dihydroxy-5α-pregnane-20-one; W
                                                                                          125126

478 C27 H50 O3 Si2 SH-1981-0-0 Bis(trimethylsilyl) derivative of 3β,11α-Dihydroxy-5α-pregnan-20-one; W
                                                                                          125127

478 C27 H50 O3 Si2 SH-1981-0-0 Bis(trimethylsilyl) derivative of 5α-Pregnan-3α,16α-diol-20-one; W
                                                                                          125128

478 C27 H50 O3 Si2 SH-1981-0-0 Bis(trimethylsilyl) derivative of 5α-Pregnan-3β,19-diol-20-one; W 125129

478 C27 H50 O3 Si2 SH-1981-0-0 Bis(trimethylsilyl) derivative of 5α-Pregnan-3α,15α-diol-20-one; W
                                                                                          125130

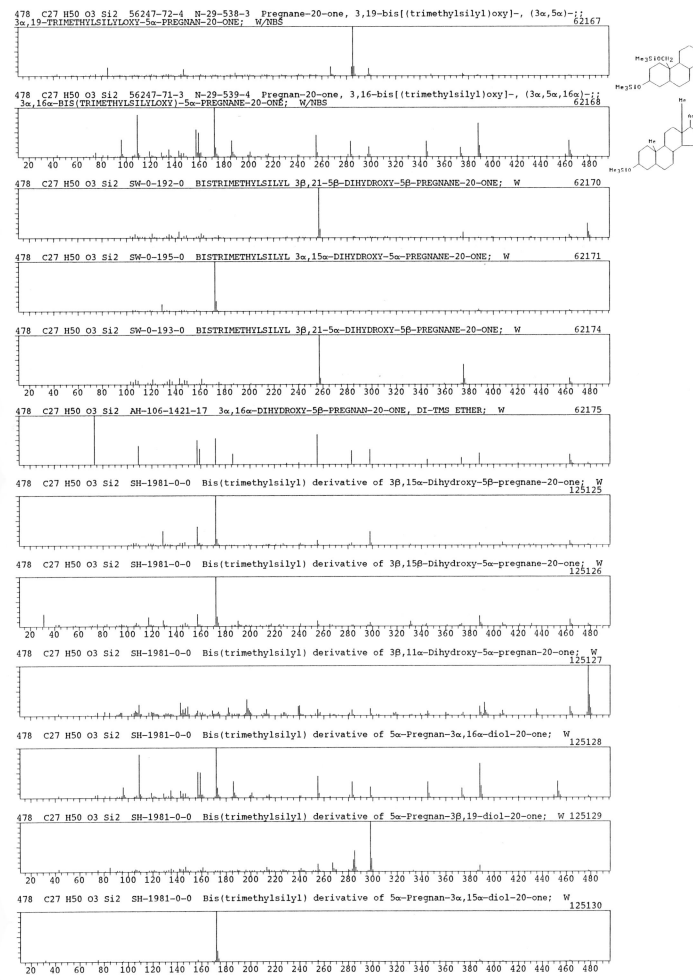

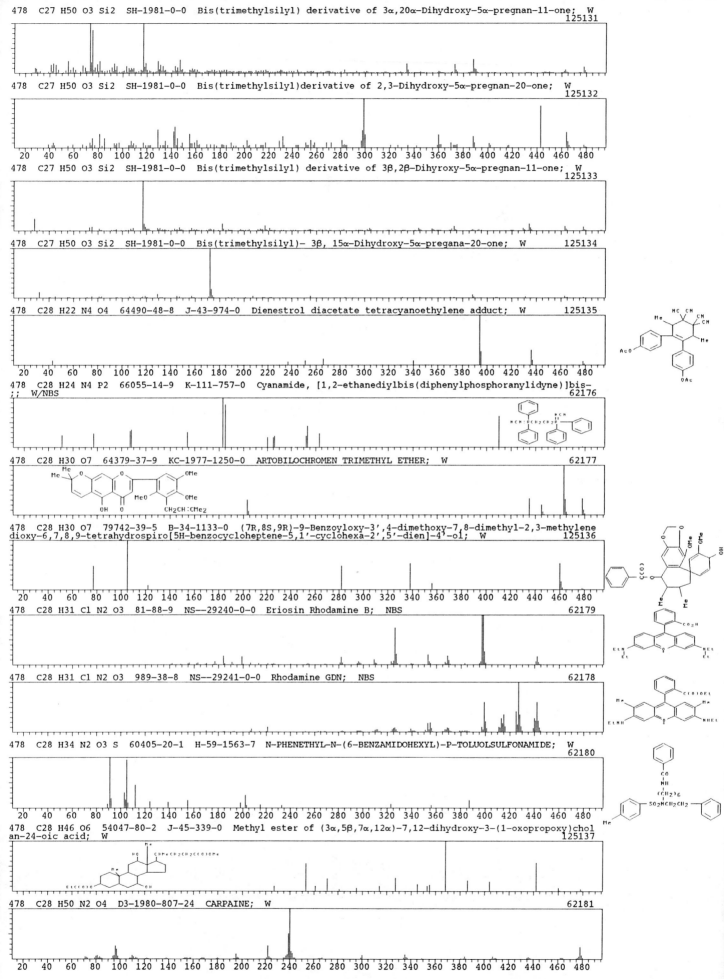

478 C27 H50 O3 Si2 SH-1981-0-0 Bis(trimethylsilyl) derivative of 3α,20α-Dihydroxy-5α-pregnan-11-one; W
125131

478 C27 H50 O3 Si2 SH-1981-0-0 Bis(trimethylsilyl)derivative of 2,3-Dihydroxy-5α-pregnan-20-one; W
125132

478 C27 H50 O3 Si2 SH-1981-0-0 Bis(trimethylsilyl) derivative of 3β,2β-Dihyroxy-5α-pregnan-11-one; W
125133

478 C27 H50 O3 Si2 SH-1981-0-0 Bis(trimethylsilyl)- 3β, 15α-Dihydroxy-5α-pregana-20-one; W   125134

478 C28 H22 N4 O4 64490-48-8 J-43-974-0 Dienestrol diacetate tetracyanoethylene adduct; W   125135

478 C28 H24 N4 P2 66055-14-9 K-111-757-0 Cyanamide, [1,2-ethanediylbis(diphenylphosphoranylidyne)]bis-
;; W/NBS   62176

478 C28 H30 O7 64379-37-9 KC-1977-1250-0 ARTOBILOCHROMEN TRIMETHYL ETHER; W   62177

478 C28 H30 O7 79742-39-5 B-34-1133-0 (7R,8S,9R)-9-Benzoyloxy-3',4-dimethoxy-7,8-dimethyl-2,3-methylene
dioxy-6,7,8,9-tetrahydrospiro[5H-benzocycloheptene-5,1'-cyclohexa-2',5'-dien]-4'-ol; W   125136

478 C28 H31 Cl N2 O3 81-88-9 NS--29240-0-0 Eriosin Rhodamine B; NBS   62179

478 C28 H31 Cl N2 O3 989-38-8 NS--29241-0-0 Rhodamine GDN; NBS   62178

478 C28 H34 N2 O3 S 60405-20-1 H-59-1563-7 N-PHENETHYL-N-(6-BENZAMIDOHEXYL)-P-TOLUOLSULFONAMIDE; W
62180

478 C28 H46 O6 54047-80-2 J-45-339-0 Methyl ester of (3α,5β,7α,12α)-7,12-dihydroxy-3-(1-oxopropoxy)chol
an-24-oic acid; W   125137

478 C28 H50 N2 O4 D3-1980-807-24 CARPAINE; W   62181

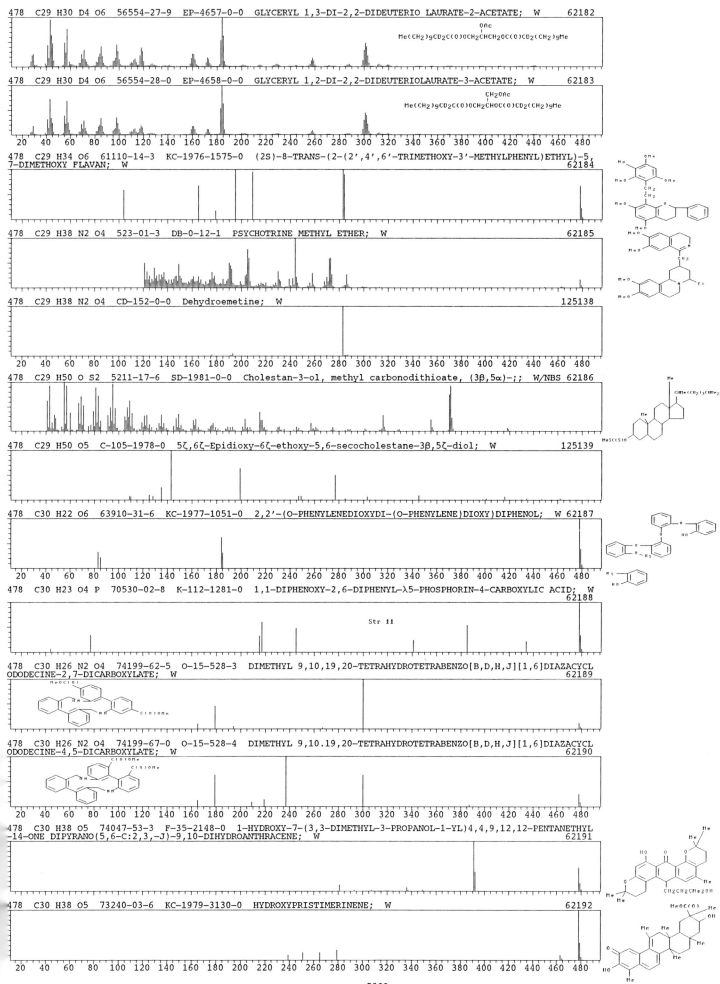

478   C29 H30 D4 O6   56554-27-9   EP-4657-0-0   GLYCERYL 1,3-DI-2,2-DIDEUTERIO LAURATE-2-ACETATE; W   62182

OAc
|
Me(CH2)9CD2C(O)OCH2CHCH2OC(O)CD2(CH2)9Me

478   C29 H30 D4 O6   56554-28-0   EP-4658-0-0   GLYCERYL 1,2-DI-2,2-DIDEUTERIOLAURATE-3-ACETATE; W   62183

CH2OAc
|
Me(CH2)9CD2C(O)OCH2CHOC(O)CD2(CH2)9Me

478   C29 H34 O6   61110-14-3   KC-1976-1575-0   (2S)-8-TRANS-(2-(2',4',6'-TRIMETHOXY-3'-METHYLPHENYL)ETHYL)-5,
7-DIMETHOXY FLAVAN; W   62184

478   C29 H38 N2 O4   523-01-3   DB-0-12-1   PSYCHOTRINE METHYL ETHER; W   62185

478   C29 H38 N2 O4   CD-152-0-0   Dehydroemetine; W   125138

478   C29 H50 O S2   5211-17-6   SD-1981-0-0   Cholestan-3-ol, methyl carbonodithioate, (3β,5α)-;; W/NBS 62186

478   C29 H50 O5   C-105-1978-0   5ζ,6ζ-Epidioxy-6ζ-ethoxy-5,6-secocholestane-3β,5ζ-diol; W   125139

478   C30 H22 O6   63910-31-6   KC-1977-1051-0   2,2'-(O-PHENYLENEDIOXYDI-(O-PHENYLENE)DIOXY)DIPHENOL; W 62187

478   C30 H23 O4 P   70530-02-8   K-112-1281-0   1,1-DIPHENOXY-2,6-DIPHENYL-λ5-PHOSPHORIN-4-CARBOXYLIC ACID; W
62188

Str 11

478   C30 H26 N2 O4   74199-62-5   O-15-528-3   DIMETHYL 9,10,19,20-TETRAHYDROTETRABENZO[B,D,H,J][1,6]DIAZACYCL
ODODECINE-2,7-DICARBOXYLATE; W   62189

478   C30 H26 N2 O4   74199-67-0   O-15-528-4   DIMETHYL 9,10.19,20-TETRAHYDROTETRABENZO[B,D,H,J][1,6]DIAZACYCL
ODODECINE-4,5-DICARBOXYLATE; W   62190

478   C30 H38 O5   74047-53-3   F-35-2148-0   1-HYDROXY-7-(3,3-DIMETHYL-3-PROPANOL-1-YL)4,4,9,12,12-PENTANETHYL
-14-ONE DIPYRANO(5,6-C:2,3,-J)-9,10-DIHYDROANTHRACENE; W   62191

478   C30 H38 O5   73240-03-6   KC-1979-3130-0   HYDROXYPRISTIMERINENE; W   62192

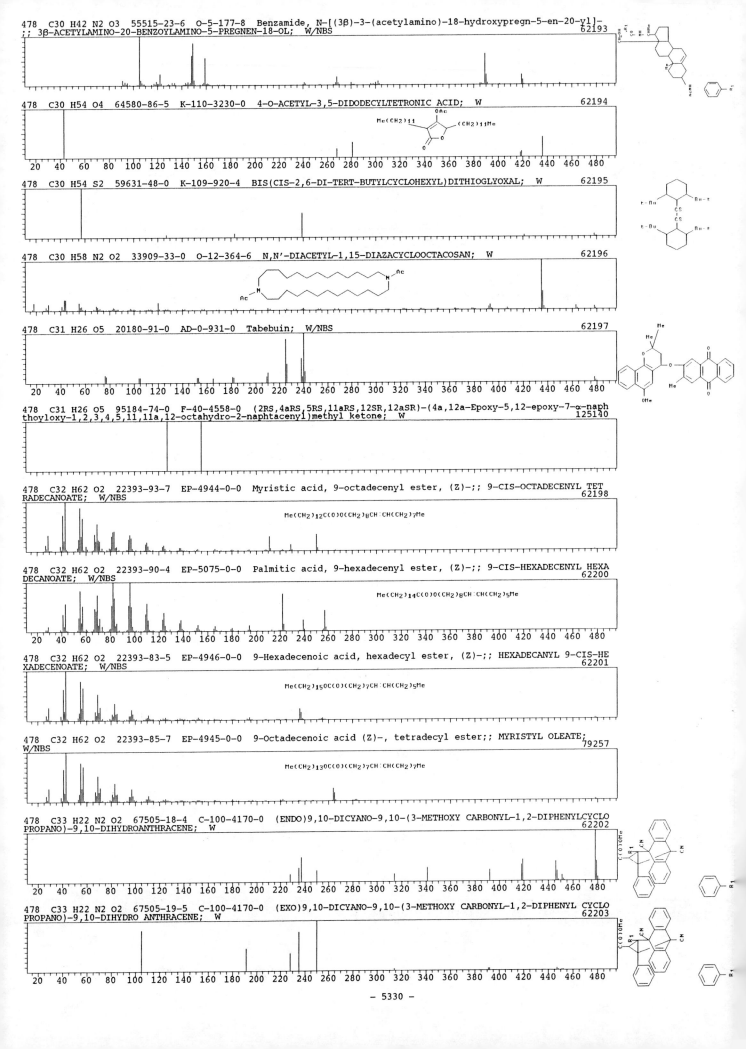

478　C30 H42 N2 O3　55515-23-6　O-5-177-8　Benzamide, N-[(3β)-3-(acetylamino)-18-hydroxypregn-5-en-20-yl]-
;;　3β-ACETYLAMINO-20-BENZOYLAMINO-5-PREGNEN-18-OL;　W/NBS
62193

478　C30 H54 O4　64580-86-5　K-110-3230-0　4-O-ACETYL-3,5-DIDODECYLTETRONIC ACID;　W
62194

478　C30 H54 S2　59631-48-0　K-109-920-4　BIS(CIS-2,6-DI-TERT-BUTYLCYCLOHEXYL)DITHIOGLYOXAL;　W
62195

478　C30 H58 N2 O2　33909-33-0　O-12-364-6　N,N'-DIACETYL-1,15-DIAZACYCLOOCTACOSAN;　W
62196

478　C31 H26 O5　20180-91-0　AD-0-931-0　Tabebuin;　W/NBS
62197

478　C31 H26 O5　95184-74-0　F-40-4558-0　(2RS,4aRS,5RS,11aRS,12SR,12aSR)-(4a,12a-Epoxy-5,12-epoxy-7-α-naph
thoyloxy-1,2,3,4,5,11,11a,12-octahydro-2-naphtacenyl)methyl ketone;　W
125140

478　C32 H62 O2　22393-93-7　EP-4944-0-0　Myristic acid, 9-octadecenyl ester, (Z)-;;　9-CIS-OCTADECENYL TET
RADECANOATE;　W/NBS
62198

478　C32 H62 O2　22393-90-4　EP-5075-0-0　Palmitic acid, 9-hexadecenyl ester, (Z)-;;　9-CIS-HEXADECENYL HEXA
DECANOATE;　W/NBS
62200

478　C32 H62 O2　22393-83-5　EP-4946-0-0　9-Hexadecenoic acid, hexadecyl ester, (Z)-;;　HEXADECANYL 9-CIS-HE
XADECENOATE;　W/NBS
62201

478　C32 H62 O2　22393-85-7　EP-4945-0-0　9-Octadecenoic acid (Z)-, tetradecyl ester;;　MYRISTYL OLEATE;
W/NBS
79257

478　C33 H22 N2 O2　67505-18-4　C-100-4170-0　(ENDO)9,10-DICYANO-9,10-(3-METHOXY CARBONYL-1,2-DIPHENYLCYCLO
PROPANO)-9,10-DIHYDROANTHRACENE;　W
62202

478　C33 H22 N2 O2　67505-19-5　C-100-4170-0　(EXO)9,10-DICYANO-9,10-(3-METHOXY CARBONYL-1,2-DIPHENYL CYCLO
PROPANO)-9,10-DIHYDRO ANTHRACENE;　W
62203

478   C33 H26 Fe   HE-1982-0-0   IRON, (TETRAPHENYLCYCLOBUTADIEN)-CYCLOPENTADIEN;   W   62204

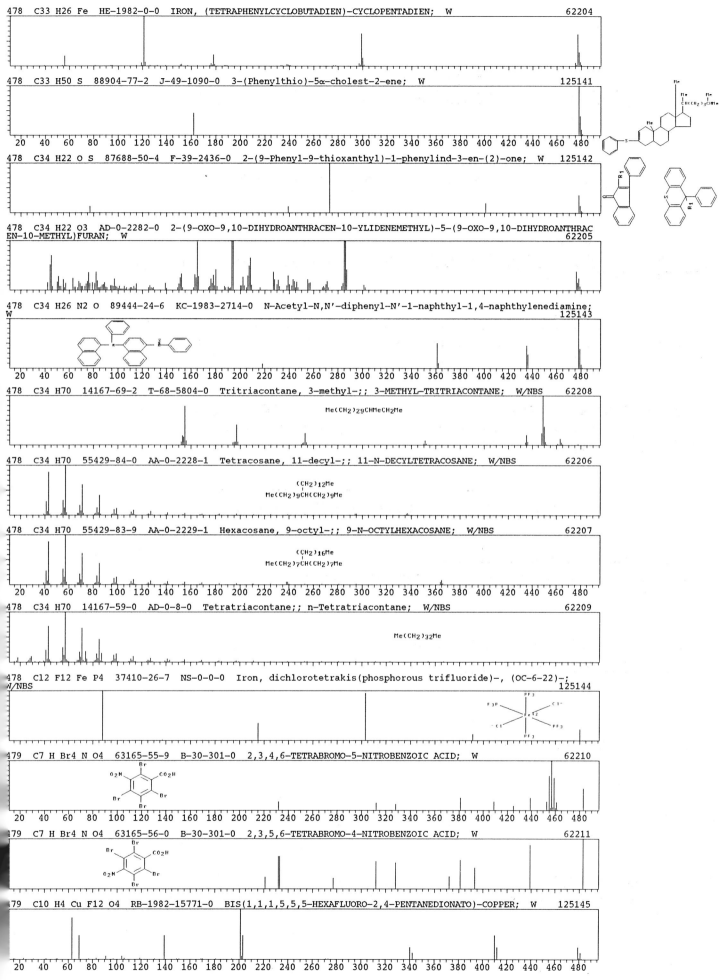

478   C33 H50 S   88904-77-2   J-49-1090-0   3-(Phenylthio)-5α-cholest-2-ene;   W   125141

478   C34 H22 O S   87688-50-4   F-39-2436-0   2-(9-Phenyl-9-thioxanthyl)-1-phenylind-3-en-(2)-one;   W   125142

478   C34 H22 O3   AD-0-2282-0   2-(9-OXO-9,10-DIHYDROANTHRACEN-10-YLIDENEMETHYL)-5-(9-OXO-9,10-DIHYDROANTHRAC
EN-10-METHYL)FURAN;   W   62205

478   C34 H26 N2 O   89444-24-6   KC-1983-2714-0   N-Acetyl-N,N'-diphenyl-N'-1-naphthyl-1,4-naphthylenediamine;
W   125143

478   C34 H70   14167-69-2   T-68-5804-0   Tritriacontane, 3-methyl-;; 3-METHYL-TRITRIACONTANE;   W/NBS   62208

478   C34 H70   55429-84-0   AA-0-2228-1   Tetracosane, 11-decyl-;; 11-N-DECYLTETRACOSANE;   W/NBS   62206

478   C34 H70   55429-83-9   AA-0-2229-1   Hexacosane, 9-octyl-;; 9-N-OCTYLHEXACOSANE;   W/NBS   62207

478   C34 H70   14167-59-0   AD-0-8-0   Tetratriacontane;; n-Tetratriacontane;   W/NBS   62209

478   Cl2 F12 Fe P4   37410-26-7   NS-0-0-0   Iron, dichlorotetrakis(phosphorous trifluoride)-, (OC-6-22)-;
W/NBS   125144

479   C7 H Br4 N O4   63165-55-9   B-30-301-0   2,3,4,6-TETRABROMO-5-NITROBENZOIC ACID;   W   62210

479   C7 H Br4 N O4   63165-56-0   B-30-301-0   2,3,5,6-TETRABROMO-4-NITROBENZOIC ACID;   W   62211

479   C10 H4 Cu F12 O4   RB-1982-15771-0   BIS(1,1,1,5,5,5-HEXAFLUORO-2,4-PENTANEDIONATO)-COPPER;   W   125145

479   C14 H11 F10 N O4 S   87613-20-5   J-49-4782-0   Diethyl 3,5-Bis(pentafluoroethyl)-2H-1,4-thiazine-2,6-dicarboxylate;   W
125146

479   C16 H13 F12 N O2   80360-42-5   J-47-1024-0   1,4-Dihydro-8-[1-hydroxy-1-(trifluoromethyl)-2,2,2-trifluoroethyl]-6-methyl-2,2-dimethyl-4,4-bis(trifluoromethyl)-2H-[3,1]benzoxazine;   W
125147

479   C17 H15 Br2 N5 O2   IC-1215-0-0   N-(2-Cyanoethyl)-4-(2,6-dibromo-4-nitrophenylazo)-N-ethylaniline;   W
125148

479   C18 H29 N3 O10 S   21026-92-6   T-68-4091-0   Glutamine, N2-carboxy-N-[1-[(carboxymethyl)carbamoyl]-2-mercaptoethyl]-, N2-ethyl dimethyl ester, ethyl carbonate (ester), L-;; ETHOXYCARBONYL-γ-GLU-OME-5-ETHOXYCARBONYL-CYS-GLY-OME;   W/NBS
62212

479   C19 H45 N O5 Si4   HE-1982-0-0   INOSOSE, 2-DESOXY-, O-METHYLOXIM, TETRAKIS-O-(TRIMETHYLSILYL)-;   W
62213

479   C21 H22 Br N O7   68244-13-3   Y-19-1322-0   1-(3',4'-Dimethoxy-6'-bromobenzyl)-1,6,7-trimethoxyisoquinoline-3,4-dione;   W
125149

479   C21 H45 D8 N3 O3 Si3   61043-80-9   KO-2-329-1   O-TRIMETHYLSILYLATED THR-SER-ALA-D8;   W
62214

479   C22 H41 N5 O3 Si2   51549-32-7   O-15-209-1   3',5'-BIS-O-TERT-BUTYLDIMETHYLSILYL-2-DEOXYADENOSINE;   W
62215

479   C24 H21 N3 O8   37922-00-2   NS-0-0-0   6H-Pyridazino[1,6-a]quinoxaline-2,3,4-tricarboxylic acid, 4a,5-dihydro-4a-hydroxy-5-oxo-6-(phenylmethyl)-, trimethyl ester;   W/NBS
125150

479   C24 H37 N3 O7   65564-03-6   KO-4-142-13   DIHYDRO-AM-TOXIN I MEOH;   W
62216

479   C25 H37 N O8   63533-75-5   J-44-4574-1   (3α,5β,7α,12α)-7,12-Bis(acetyloxy)-3-(nitrooxy)pregnan-20-one;   W
125151

479   C25 H37 N O8   63533-73-3   J-44-4574-1   (3α,5β,7α,12α)-3,7-Bis(acetyloxy)-12-(nitrooxy)pregnan-20-one;   W
125152

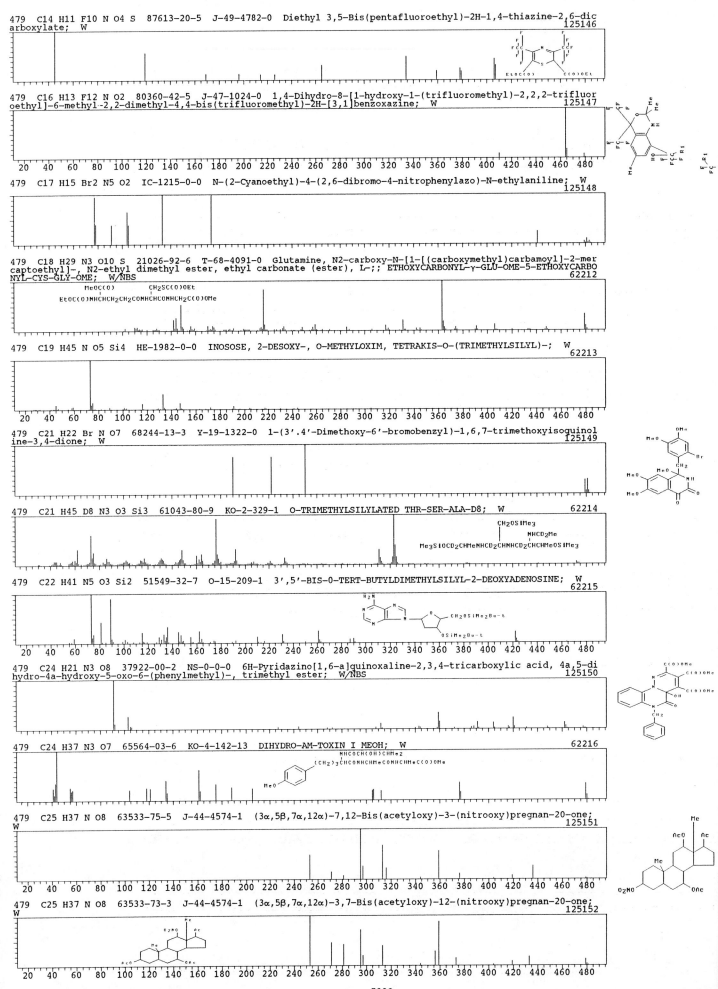

479 C26 H28 Cl N3 P2 42998-71-0 NS-0-0-0 Phosphinimidic amide, N'-(aminochlorodiphenylphosphoranyl)-N-ethyl-P,P-diphenyl-; W/NBS
125153

479 C26 H29 N3 O6 73442-93-0 K-113-785-0 Ethylester of 1-Benzyloxycarbonyl-2-isopropyl-β-(4-nitrobenzyliden)-3-imidazolin-4-propionic acid; W
125154

479 C26 H41 N O7 OM-1981-317-0 PGE2 ME ESTER MOX DIACETATE ISOMER 1; W
62217

479 C26 H41 N O7 OM-1981-318-0 PGE2 ME ESTER MOX DIACETATE ISOMER 2; W
62218

479 C26 H41 N O7 CD-279-0-0 14-Acetylneoline; W
125155

479 C26 H45 N3 O3 S 72636-91-0 H-62-1941-0 13-(8'-Amino-4'-tosyl-4'-azaoctyl)-12-dodecanlactam; W
125156

479 C26 H49 N O3 Si2 32221-29-7 EP-8223-0-0 5α-Androstan-17-one, 3α,11β-bis(trimethylsiloxy)-,O-methyloxime;; 5α-ANDROSTANE-3α,11β-DIOL-17-ONE MO TMS; W/NBS
62219

479 C26 H49 N O3 Si2 32206-64-7 SW-0-30-0 5β-Androstan-17-one, 3α,11β-bis(trimethylsiloxy)-, O-methyloxime;; 5β-ANDROSTANE-3α,11β-DIOL-17-ONE MO TMS; W
79259

479 C26 H49 N O3 Si2 39780-65-9 EP-8471-0-0 Androstan-17-one, 3,11-bis[(trimethylsilyl)oxy]-, O-methyloxime, (3β,5α,11b)-;; 5α-ANDROSTANE-3β,11β-DIOL-17-ONE MO TMS; W/NBS
62221

479 C26 H49 N O3 Si2 SH-1981-0-0 BISTRIMETHYLSILYL 3β,16α-DIHYDROXY-5α-ANDROSTANE-17-ONE METHOXIME; W
62222

479 C26 H49 N O3 Si2 SW-0-175-0 BISTRIMETHYLSILYL 3α,16α-DIHYDROXY-5α-ANDROSTANE-17-ONE METHOXIME; W
62223

479 C26 H49 N O3 Si2 SW-0-172-0 BISTRIMETHYLSILYL 3α,16α-DIHYDROXY-5β-ANDROSTANE-17-ONE METHOXIME; W
62224

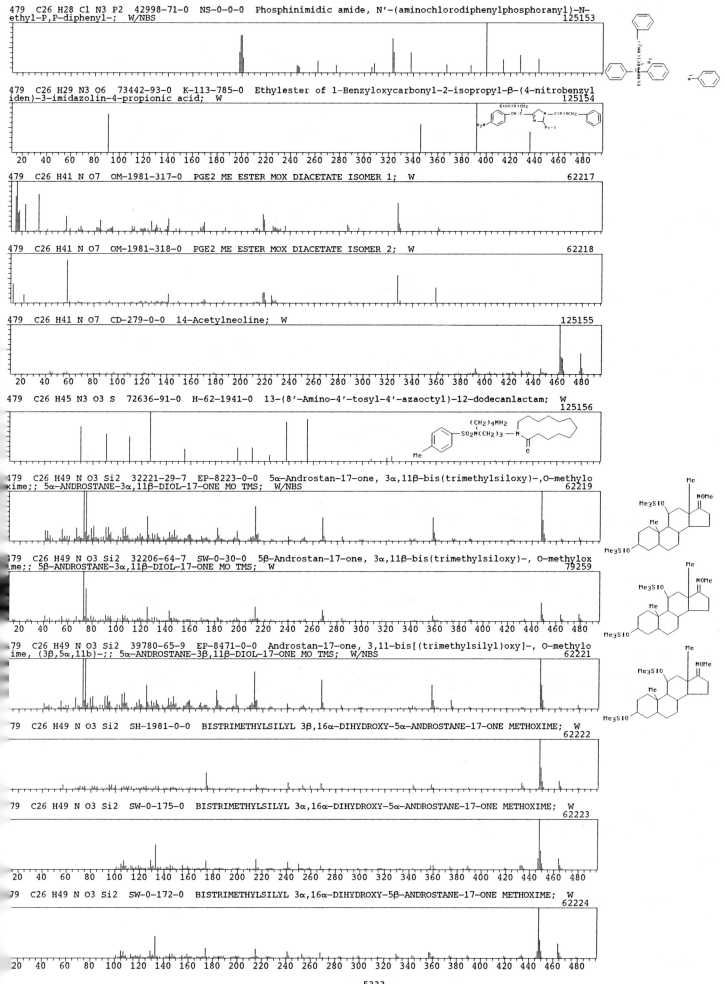

479 C26 H49 N O3 Si2 SH-1981-0-0 3α,18-DIHYDROXY-5α-ANDROSTAN-17-ONE MO TMS; W   62225

479 C26 H49 N O3 Si2 SH-1981-0-0 3β,15α-DIHYDROXY-5α-ANDROSTAN-17-ONE MO TMS; W   62226

20 40 60 80 100 120 140 160 180 200 220 240 260 280 300 320 340 360 380 400 420 440 460 480

479 C26 H49 N O3 Si2 SH-1981-0-0 16β-HYDROXYANDROSTENE MO TMS; W   62227

479 C26 H49 N O3 Si2 SH-1981-0-0 16α-HYDROXYANDROSTENE MO TMS OR 3α,16-DIHYDROXY-5α-ANDROSTAN-17-ONE MO
TMS; W   62228

479 C26 H49 N O3 Si2 SH-1981-0-0 3β,17β-DIHYDROXY-5α-ANDROSTAN-16-ONE MO TMS; W   62229

20 40 60 80 100 120 140 160 180 200 220 240 260 280 300 320 340 360 380 400 420 440 460 480

479 C26 H49 N O3 Si2 SH-1981-0-0 Methyloxime, trimethylsilyl- 16α-Hydroxyandrosterone or Methyloxime, tr
imethylsilyl- 3α,16-Dihydroxy-5α-androst-17-one; W   125157

479 C26 H49 N O3 Si2 SH-1981-0-0 Monomethyloxime,bis(trimethylsilyl)-16α-Hydroxy-5α-dihydrotestosterone;
W   125158

479 C27 H29 N O7 56335-87-6 SD-1981-0-0 2H,8H-Benzo[1,2-b:5,4-b']dipyran-10-propanol, 5-methoxy-2,2,8,
8-tetramethyl-, 4-nitrobenzoate;; W/NBS   62230

20 40 60 80 100 120 140 160 180 200 220 240 260 280 300 320 340 360 380 400 420 440 460 480

479 C27 H29 N O7 86467-72-3 H-66-793-0 3,4,6-tri-O-benzyl-1-deoxy-1-nitro-D-manno-pyranose; W 125159

479 C27 H33 N3 O5 12626-17-4 NS--29223-0-0 5H,14H-Pyrrolo[1'',2'':4',5']pyrazino[1',2':1,6]pyrido[3,4-
b]indole-5,14-dione, 1,2,3,5a,6,11,12,14a-octahydro-5a,6-dihydroxy-9-methoxy-11-(3-methyl-2-butenyl)-12-(2-
methyl-1-propenyl)-, [5aR-(5aα,6α,12β,14aα)]-;; 5H,14H-Pyrrolo[1'',2'':4',5']pyrazino[1',2':1,6]pyrido[3,4-
b]indole-5,14-dione, 1,2,3,5a,6,11,12,14a-octahydro-5a,6-dihydroxy-9-methoxy-11-(3-me; NBS   62231

479 C27 H33 N3 O5 40451-43-2 MY-1981-0-0 LANOSULIN; W   62232

20 40 60 80 100 120 140 160 180 200 220 240 260 280 300 320 340 360 380 400 420 440 460 480

479 C27 H45 N O6 BS-5-81-0 Methyl glycocholate; W   125160

479 C27 H45 N O6 BS-5-82-0 Methyl glycoallocholate; W   125161

20 40 60 80 100 120 140 160 180 200 220 240 260 280 300 320 340 360 380 400 420 440 460 480

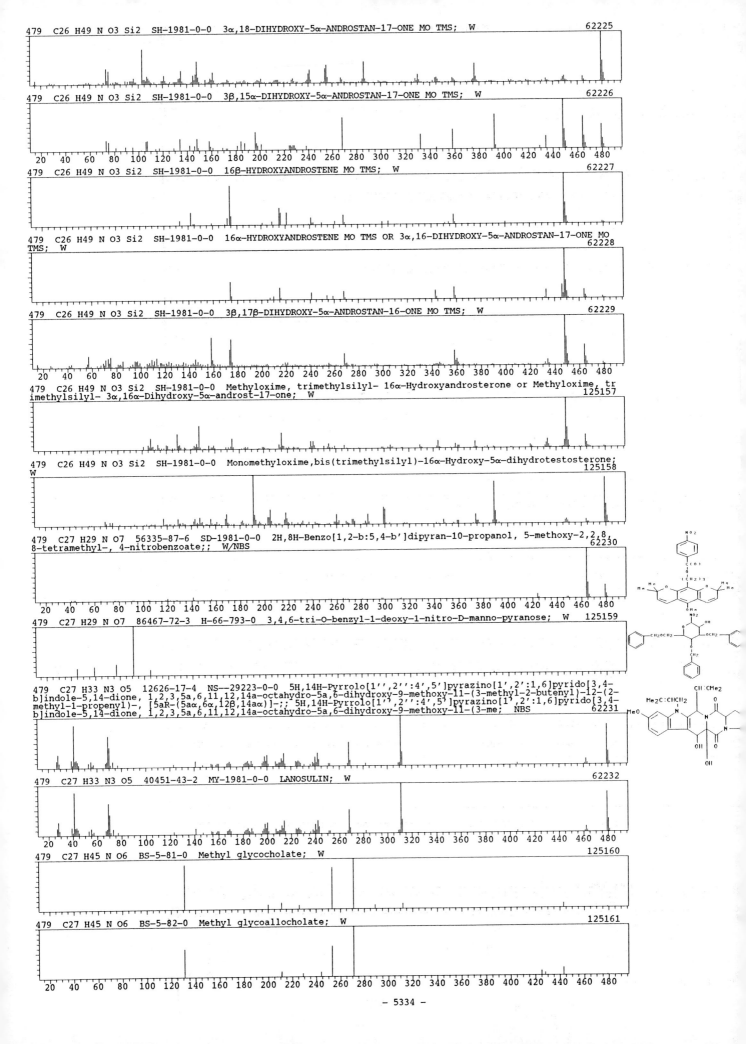

479 C27 H53 N3 O2 Si 89578-79-0 C-106-3244-0 Lactam derivative of [2α,6β(S*)]-N-[4-[(methylsulfonyl)oxy]butyl]-1-[3[[(4-methylphenyl)sulfonyl]aminopropyl]-6-[2-[[(1,1-dimethylethyl)dimethylsilyl]oxy]heptyl]-1,2,3,6-tetrahydropyridine acetamide; W
125162

479 C28 H49 N O5 20238-89-5 T-68-5444-0 OCTAHYDROPIERICIDIN β ACETATE; W/NBS
62233

20 40 60 80 100 120 140 160 180 200 220 240 260 280 300 320 340 360 380 400 420 440 460 480

479 C29 H21 N O4 S 61164-99-6 K-112-274-4 DIMETHYL ESTER OF 3,4,7-TRIPHENYL-2,1-BENZISOTHIAZOL-5,6-DICARBOXYLIC ACID; W
62234

479 C29 H37 N O5 14930-96-2 H-53-699-3 PHOMINE; W/NBS
62235

479 C29 H41 N O3 Si 57325-70-9 EP-8313-0-0 Androst-5-ene-11,17-dione, 3-[(trimethylsilyl)oxy]-, 17-[O-(phenylmethyl)oxime], (3β)-;; 5-ANDROSTEN-3β-OL-11,17-DIONE BO TMS; W/NBS
62236

20 40 60 80 100 120 140 160 180 200 220 240 260 280 300 320 340 360 380 400 420 440 460 480

479 C31 H33 N3 O2 72553-51-6 K-114-2152-0 6'',7''-Dimethyl-2'''-phenyltrispiro[biscyclohexan-1,1'':1',2''(2''H)-[1H]cyclopenta[b]quinoxalin-2'',4'''(5''H)-oxazol]-5'''-one; W
125163

79 C33 H53 N O 62115-00-8 C-99-185-0 3-(CYANO-BUTOXYMETHYLENE)-4-CHOLESTENE; W
62237

79 C34 H29 N Si 56805-08-4 EP-7410-0-0 7-Silabicyclo[2.2.1]hept-5-ene-2-carbonitrile, 7-ethenyl-7-methl-1,4,5,6-tetraphenyl-;; 3-CYANO-1,4,5,6-TETRAPHENYL-7-METHYL-7-VINYL-7-SILABICYCLO(2.2.1) HEPT-5-ENE; W/NBS
62238

20 40 60 80 100 120 140 160 180 200 220 240 260 280 300 320 340 360 380 400 420 440 460 480

30 C8 H7 Br2 O2 Re 55839-76-4 AG-94-432-8 (η5-Methylcyclopentadienyl)(dicarbonyl)rheniumdibromide; W
125164

30 C8 H14 Br2 Pd2 OS-5-136-0 Bis-.pi.-methallyl-palladium bromide; W
125165

0 C8 H14 Br2 Pd2 OS-5-137-0 Bis-.pi.-crotyl-palladium bromide; W
125166

20 40 60 80 100 120 140 160 180 200 220 240 260 280 300 320 340 360 380 400 420 440 460 480

0 C10 H5 Cl2 F11 N2 O3 73997-89-4 KO-6-386-18 2,2-BIS(CHLORODIFLUOROMETHYL)-4-HEPTAFLUOROPROPYLCARBONAMINOMETHYL-1,3-OXAZOLIDINONE; W
62239

0 40 60 80 100 120 140 160 180 200 220 240 260 280 300 320 340 360 380 400 420 440 460 480

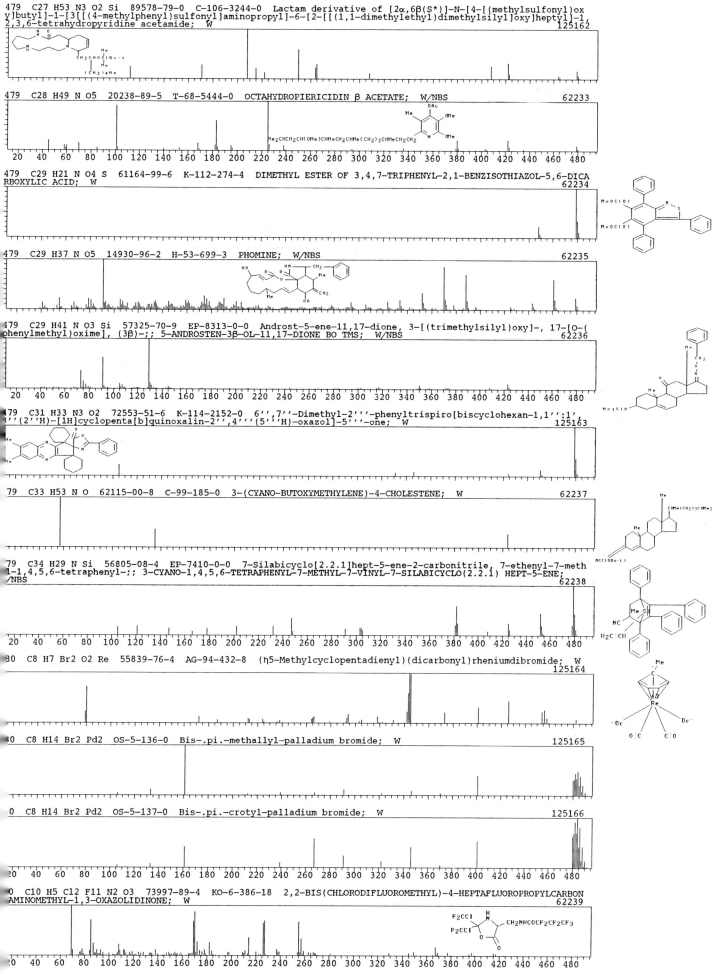

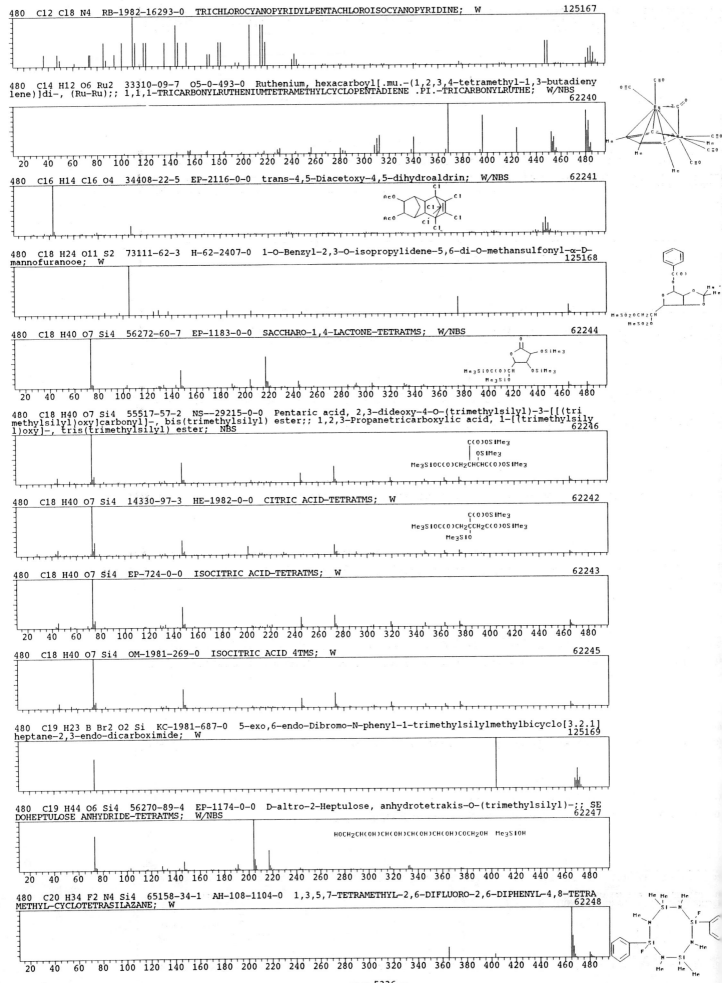

480　C12 Cl8 N4　RB-1982-16293-0　TRICHLOROCYANOPYRIDYLPENTACHLOROISOCYANOPYRIDINE;　W　　　125167

480　C14 H12 O6 Ru2　33310-09-7　O5-0-493-0　Ruthenium, hexacarboyl[.mu.-(1,2,3,4-tetramethyl-1,3-butadieny lene)]di-, (Ru-Ru);; 1,1,1-TRICARBONYLRUTHENIUMTETRAMETHYLCYCLOPENTADIENE .PI.-TRICARBONYLRUTHE;　W/NBS
62240

480　C16 H14 Cl6 O4　34408-22-5　EP-2116-0-0　trans-4,5-Diacetoxy-4,5-dihydroaldrin;　W/NBS　　62241

480　C18 H24 O11 S2　73111-62-3　H-62-2407-0　1-O-Benzyl-2,3-O-isopropylidene-5,6-di-O-methansulfonyl-α-D-mannofuranooe;　W
125168

480　C18 H40 O7 Si4　56272-60-7　EP-1183-0-0　SACCHARO-1,4-LACTONE-TETRATMS;　W/NBS　　62244

480　C18 H40 O7 Si4　55517-57-2　NS--29215-0-0　Pentaric acid, 2,3-dideoxy-4-O-(trimethylsilyl)-3-[[(tri methylsilyl)oxy]carbonyl]-, bis(trimethylsilyl) ester;; 1,2,3-Propanetricarboxylic acid, 1-[[trimethylsily l)oxy]-, tris(trimethylsilyl) ester;　NBS
62246

480　C18 H40 O7 Si4　14330-97-3　HE-1982-0-0　CITRIC ACID-TETRATMS;　W　　62242

480　C18 H40 O7 Si4　EP-724-0-0　ISOCITRIC ACID-TETRATMS;　W　　62243

480　C18 H40 O7 Si4　OM-1981-269-0　ISOCITRIC ACID 4TMS;　W　　62245

480　C19 H23 B Br2 O2 Si　KC-1981-687-0　5-exo,6-endo-Dibromo-N-phenyl-1-trimethylsilylmethylbicyclo[3.2.1] heptane-2,3-endo-dicarboximide;　W
125169

480　C19 H44 O6 Si4　56270-89-4　EP-1174-0-0　D-altro-2-Heptulose, anhydrotetrakis-O-(trimethylsilyl)-;; SE DOHEPTULOSE ANHYDRIDE-TETRATMS;　W/NBS
62247

480　C20 H34 F2 N4 Si4　65158-34-1　AH-108-1104-0　1,3,5,7-TETRAMETHYL-2,6-DIFLUORO-2,6-DIPHENYL-4,8-TETRA METHYL-CYCLOTETRASILAZANE;　W
62248

480 C21 H13 F9 O Si 58102-00-4 H-59-1236-18 TRIS(M-TRIFLUOROMETHYLPHENYL)SILANOL; W   62249

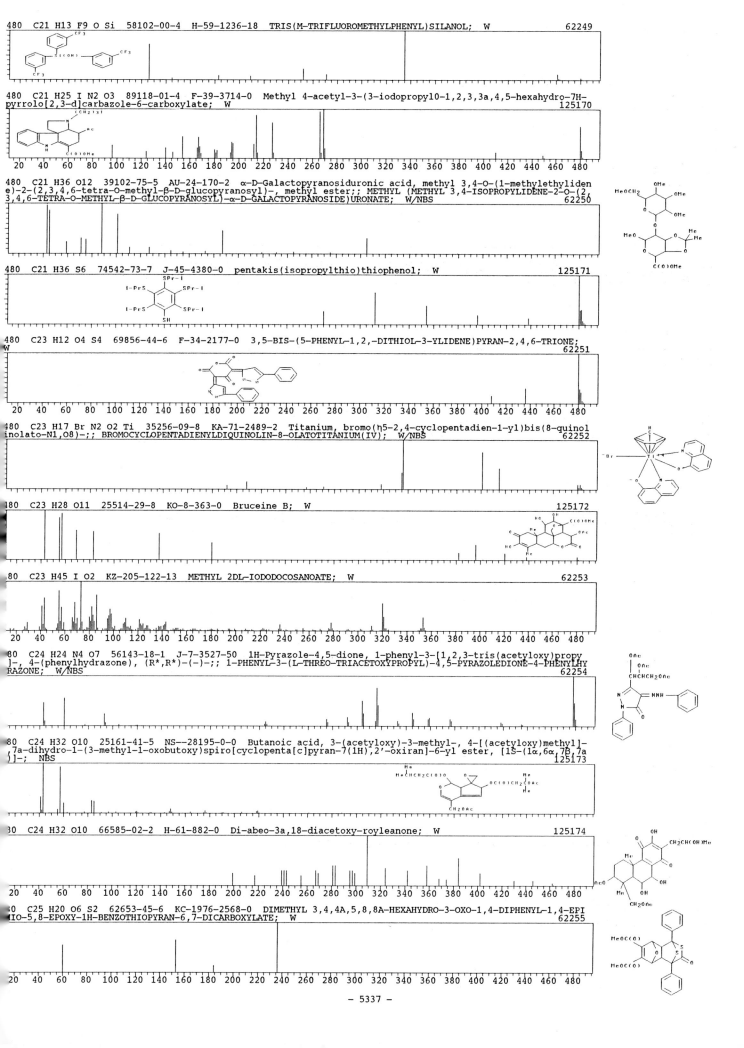

480 C21 H25 I N2 O3 89118-01-4 F-39-3714-0 Methyl 4-acetyl-3-(3-iodopropyl0-1,2,3,3a,4,5-hexahydro-7H-
pyrrolo[2,3-d]carbazole-6-carboxylate; W   125170

480 C21 H36 O12 39102-75-5 AU-24-170-2 α-D-Galactopyranosiduronic acid, methyl 3,4-O-(1-methylethyliden
e)-2-(2,3,4,6-tetra-O-methyl-β-D-glucopyranosyl)-, methyl ester;; METHYL (METHYL 3,4-ISOPROPYLIDENE-2-O-(2,
3,4,6-TETRA-O-METHYL-β-D-GLUCOPYRANOSYL)-α-D-GALACTOPYRANOSIDE)URONATE; W/NBS   62250

480 C21 H36 S6 74542-73-7 J-45-4380-0 pentakis(isopropylthio)thiophenol; W   125171

480 C23 H12 O4 S4 69856-44-6 F-34-2177-0 3,5-BIS-(5-PHENYL-1,2,-DITHIOL-3-YLIDENE)PYRAN-2,4,6-TRIONE;
W   62251

480 C23 H17 Br N2 O2 Ti 35256-09-8 KA-71-2489-2 Titanium, bromo(η5-2,4-cyclopentadien-1-yl)bis(8-quinol
inolato-N1,O8)-;; BROMOCYCLOPENTADIENYLDIQUINOLIN-8-OLATOTITANIUM(IV); W/NBS   62252

180 C23 H28 O11 25514-29-8 KO-8-363-0 Bruceine B; W   125172

180 C23 H45 I O2 KZ-205-122-13 METHYL 2DL-IODODOCOSANOATE; W   62253

80 C24 H24 N4 O7 56143-18-1 J-7-3527-50 1H-Pyrazole-4,5-dione, 1-phenyl-3-[1,2,3-tris(acetyloxy)propy
]-, 4-(phenylhydrazone), (R*,R*)-(-)-;; 1-PHENYL-3-(L-THRÉO-TRIACETOXYPROPYL)-4,5-PYRAZOLEDIONE-4-PHENYLHY
RAZONE; W/NBS   62254

80 C24 H32 O10 25161-41-5 NS--28195-0-0 Butanoic acid, 3-(acetyloxy)-3-methyl-, 4-[(acetyloxy)methyl]-
7a-dihydro-1-(3-methyl-1-oxobutoxy)spiro[cyclopenta[c]pyran-7(1H),2'-oxiran]-6-yl ester, [1S-(1α,6α,7β,7a
)]-; NBS   125173

30 C24 H32 O10 66585-02-2 H-61-882-0 Di-abeo-3a,18-diacetoxy-royleanone; W   125174

0 C25 H20 O6 S2 62653-45-6 KC-1976-2568-0 DIMETHYL 3,4,4A,5,8,8A-HEXAHYDRO-3-OXO-1,4-DIPHENYL-1,4-EPI
IO-5,8-EPOXY-1H-BENZOTHIOPYRAN-6,7-DICARBOXYLATE; W   62255

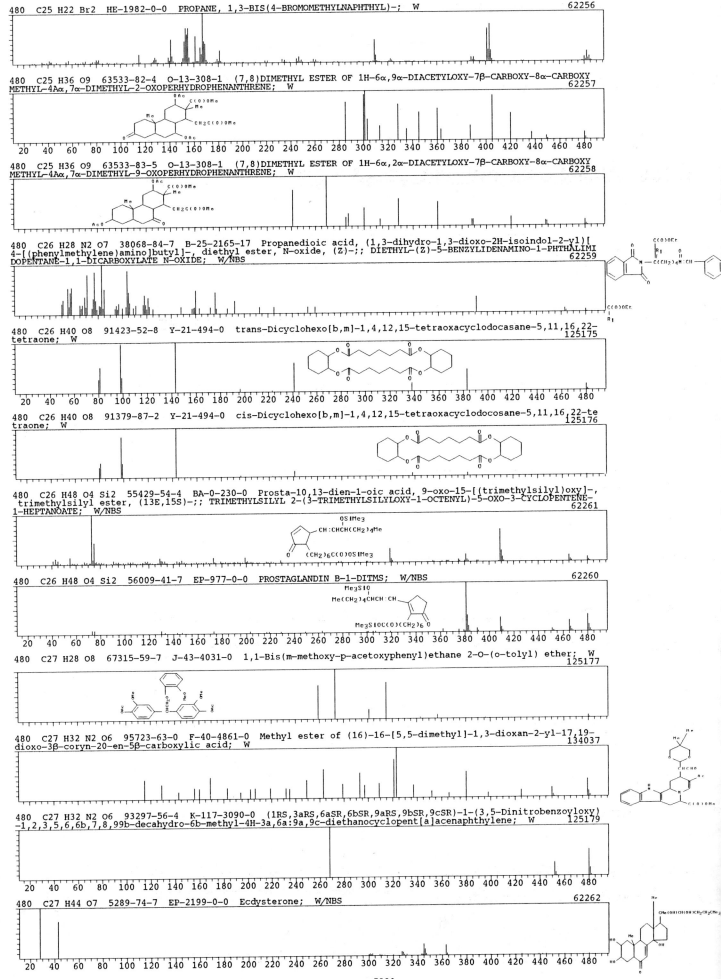

480 C25 H22 Br2 HE-1982-0-0 PROPANE, 1,3-BIS(4-BROMOMETHYLNAPHTHYL)-; W 62256

480 C25 H36 O9 63533-82-4 O-13-308-1 (7,8)DIMETHYL ESTER OF 1H-6α,9α-DIACETYLOXY-7β-CARBOXY-8α-CARBOXY METHYL-4Aα,7α-DIMETHYL-2-OXOPERHYDROPHENANTHRENE; W 62257

480 C25 H36 O9 63533-83-5 O-13-308-1 (7,8)DIMETHYL ESTER OF 1H-6α,2α-DIACETYLOXY-7β-CARBOXY-8α-CARBOXY METHYL-4Aα,7α-DIMETHYL-9-OXOPERHYDROPHENANTHRENE; W 62258

480 C26 H28 N2 O7 38068-84-7 B-25-2165-17 Propanedioic acid, (1,3-dihydro-1,3-dioxo-2H-isoindol-2-yl)[4-[(phenylmethylene)amino]butyl]-, diethyl ester, N-oxide, (Z)-;; DIETHYL-(Z)-5-BENZYLIDENAMINO-1-PHTHALIMI DOPENTANE-1,1-DICARBOXYLATE N-OXIDE; W/NBS 62259

480 C26 H40 O8 91423-52-8 Y-21-494-0 trans-Dicyclohexo[b,m]-1,4,12,15-tetraoxacyclodocasane-5,11,16,22- tetraone; W 125175

480 C26 H40 O8 91379-87-2 Y-21-494-0 cis-Dicyclohexo[b,m]-1,4,12,15-tetraoxacyclodocosane-5,11,16,22-te traone; W 125176

480 C26 H48 O4 Si2 55429-54-4 BA-0-230-0 Prosta-10,13-dien-1-oic acid, 9-oxo-15-[(trimethylsilyl)oxy]-, trimethylsilyl ester, (13E,15S)-;; TRIMETHYLSILYL 2-(3-TRIMETHYLSILYLOXY-1-OCTENYL)-5-OXO-3-CYCLOPENTENE- 1-HEPTANOATE; W/NBS 62261

480 C26 H48 O4 Si2 56009-41-7 EP-977-0-0 PROSTAGLANDIN B-1-DITMS; W/NBS 62260

480 C27 H28 O8 67315-59-7 J-43-4031-0 1,1-Bis(m-methoxy-p-acetoxyphenyl)ethane 2-O-(o-tolyl) ether; W 125177

480 C27 H32 N2 O6 95723-63-0 F-40-4861-0 Methyl ester of (16)-16-[5,5-dimethyl]-1,3-dioxan-2-yl-17,19- dioxo-3β-coryn-20-en-5β-carboxylic acid; W 134037

480 C27 H32 N2 O6 93297-56-4 K-117-3090-0 (1RS,3aRS,6aSR,6bSR,9aRS,9bSR,9cSR)-1-(3,5-Dinitrobenzoyloxy) -1,2,3,5,6,6b,7,8,99b-decahydro-6b-methyl-4H-3a,6a:9a,9c-diethanocyclopent[a]acenaphthylene; W 125179

480 C27 H44 O7 5289-74-7 EP-2199-0-0 Ecdysterone; W/NBS 62262

480   C27 H44 O7   52717-49-4   NS-0-0-0   Cholest-7-en-6-one, 2,3,14,22,25,26-hexahydroxy-, (2β,3β,5β,22R)-;
W/NBS
125180

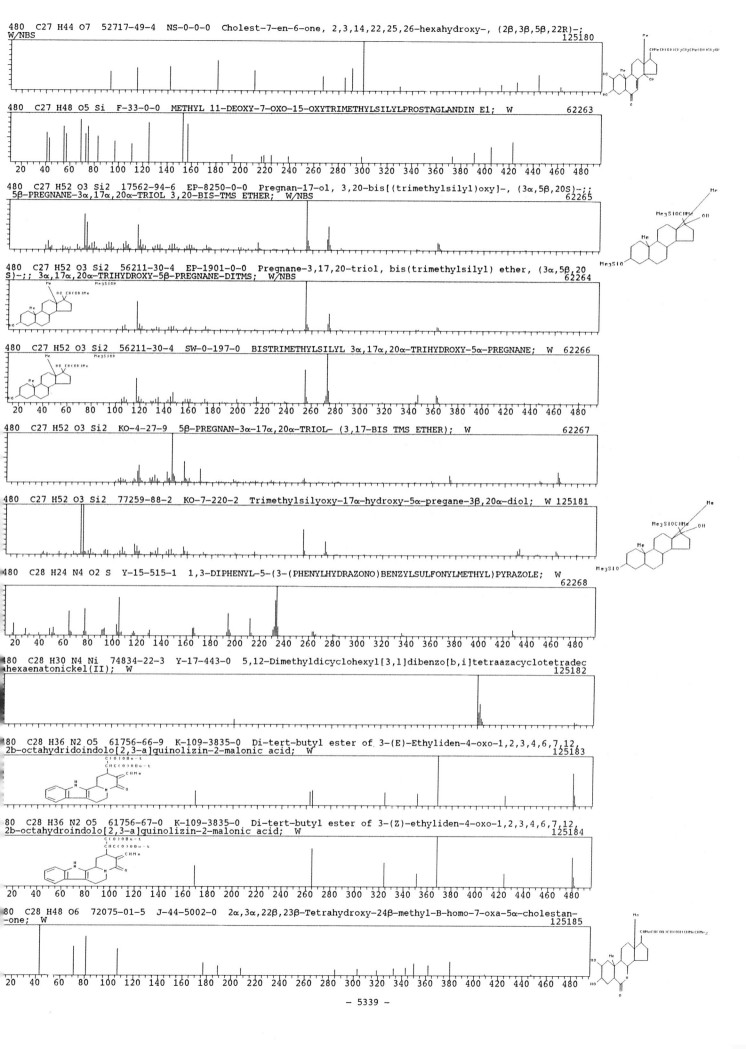

480   C27 H48 O5 Si   F-33-0-0   METHYL 11-DEOXY-7-OXO-15-OXYTRIMETHYLSILYLPROSTAGLANDIN E1;   W   62263

480   C27 H52 O3 Si2   17562-94-6   EP-8250-0-0   Pregnan-17-ol, 3,20-bis[(trimethylsilyl)oxy]-, (3α,5β,20S)-;;
5β-PREGNANE-3α,17α,20α-TRIOL 3,20-BIS-TMS ETHER;   W/NBS   62265

480   C27 H52 O3 Si2   56211-30-4   EP-1901-0-0   Pregnane-3,17,20-triol, bis(trimethylsilyl) ether, (3α,5β,20
S)-;; 3α,17α,20α-TRIHYDROXY-5β-PREGNANE-DITMS;   W/NBS   62264

480   C27 H52 O3 Si2   56211-30-4   SW-0-197-0   BISTRIMETHYLSILYL 3α,17α,20α-TRIHYDROXY-5α-PREGNANE;   W   62266

480   C27 H52 O3 Si2   KO-4-27-9   5β-PREGNAN-3α-17α,20α-TRIOL- (3,17-BIS TMS ETHER);   W   62267

480   C27 H52 O3 Si2   77259-88-2   KO-7-220-2   Trimethylsilyloxy-17α-hydroxy-5α-pregane-3β,20α-diol;   W 125181

480   C28 H24 N4 O2 S   Y-15-515-1   1,3-DIPHENYL-5-(3-(PHENYLHYDRAZONO)BENZYLSULFONYLMETHYL)PYRAZOLE;   W   62268

480   C28 H30 N4 Ni   74834-22-3   Y-17-443-0   5,12-Dimethyldicyclohexyl[3,1]dibenzo[b,i]tetraazacyclotetradec
ahexaenatonickel(II);   W   125182

480   C28 H36 N2 O5   61756-66-9   K-109-3835-0   Di-tert-butyl ester of 3-(E)-Ethyliden-4-oxo-1,2,3,4,6,7,12,
2b-octahydridoindolo[2,3-a]quinolizin-2-malonic acid;   W   125183

480   C28 H36 N2 O5   61756-67-0   K-109-3835-0   Di-tert-butyl ester of 3-(Z)-ethyliden-4-oxo-1,2,3,4,6,7,12,
2b-octahydroindolo[2,3-a]quinolizin-2-malonic acid;   W   125184

480   C28 H48 O6   72075-01-5   J-44-5002-0   2α,3α,22β,23β-Tetrahydroxy-24β-methyl-B-homo-7-oxa-5α-cholestan-
-one;   W   125185

480   C29 H27 F3 O3   77507-36-9   J-46-3245-0   1-Phenyl-2,2-dimethylprop-1-yl α-[1-(9-anthryl)-2,2,2-trifluor
oethoxy]acetate;   W
125186

480   C30 H28 N2 O4   65591-24-4   K-111-238-0   ETHYL N,N'-(4,4'-BIPHENYLYLEX)DI(ANTHRANILCARBOXYLIC ACID EST
ER);   W
62271

480   C30 H28 N2 O4   86168-91-4   B-36-766-0   (6aβ,7β,15β)-(+,-)-6,6,9,15,18-Pentamethyl-6,6a,7,9,15,18-hexa
hydro-7,15-methano-8H,17H-quino[3'',4'':5',6']pyrano[3',4':5,6]oxocino[3,2-c]quinoline-8,17-dione;   W
125188

480   C30 H36 N6   89703-50-4   K-117-546-0   (c-4a,c-8a)-4a,5,8,8a-Tetrahydro-r-5,c-8-ethanophthalazine;   W
125189

480   C30 H40 O5   73246-54-5   KC-1979-3130-0   21-HYDROXYPRISTIMERIN;   W
62272

480   C30 H40 O5   71638-22-7   K-116-770-0   1,2-Ethandiyl-1-(4-benzoylbenzoato)-2-tetradecanoate;   W   125190

480   C30 H56 O4   56256-46-3   NS-0-0-0   1,3-Dioxolane, 4-[[(2-methoxy-4-tricosynyl)oxy]methyl]-2,2-dimethyl-
;   W/NBS
125191

480   C30 H57 O2 P   HE-1986-1753-0   DIMENTHOXY-MENTHYL-PHOSPHINE;   W
125192

480   C31 H44 O4   86547-43-5   H-66-767-0   3β-Benzyloxy-12β-(3,4,5,6-tetrahydro-2H-pyran-2-yl)oxy-9β,13β-etha
no-9β-podocarpan-15-one;   W
125193

480   C31 H60 O3   52262-75-6   G-39-469-3   14,16-Hentriacontanedione, 25-hydroxy-;;   25-Hydroxy-14,16-hentriac
ontanedione;   W/NBS
62273

480   C32 H32 O4   68463-60-5   K-111-3145-0   12-(DEHYDROCACALOHASTIN-14-YL)CACALOHASTIN;   W
62274

480   C32 H48 O3   75808-33-2   B-33-2084-0   (23S)-16,23-Epoxy-14(15 to 30)abeo-20.xi.-dammara-13,15(30)16-tri
en-3β-yl acetate;   W
125194

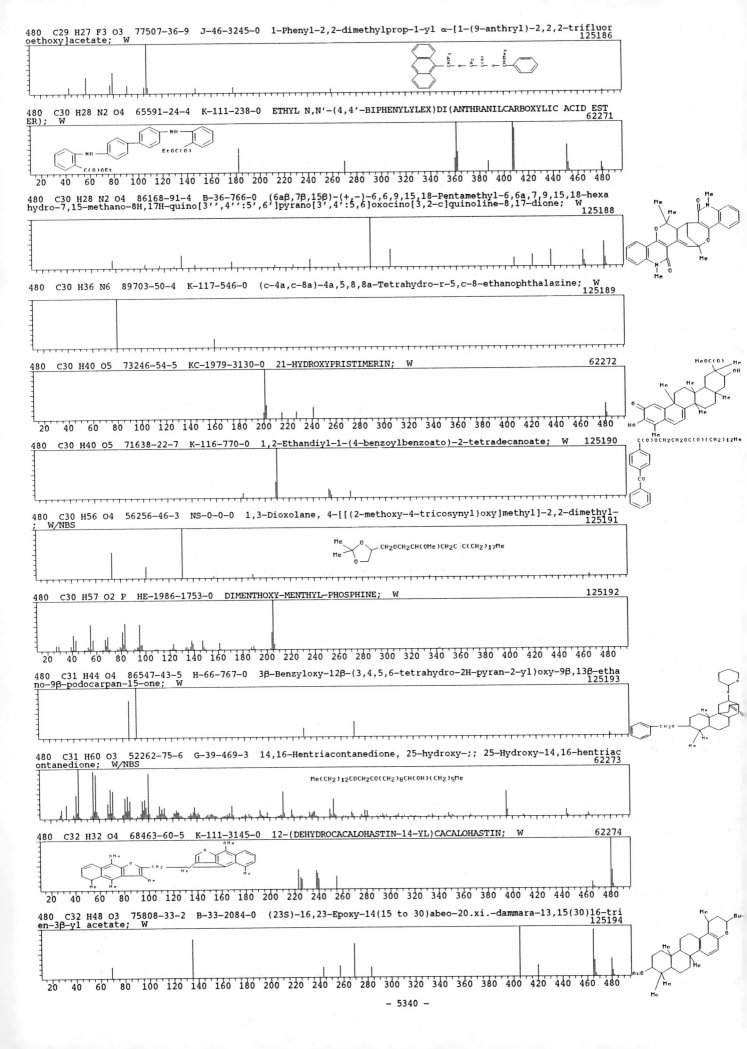

480 C32 H48 O3 80239-46-9 H-64-1880-0 25-(Tetrahydro-2H-pyran-2-yl)oxy-1,4,6-cholestatrien-3-one; W
125195

480 C32 H48 O3 KC-1981-2726-0 3,3-Ethylenedioxylup-20(29)-en-30-al; W
125196

480 C32 H64 O2 540-10-3 EP-5065-0-0 Hexadecanoic acid, hexadecyl ester;; HEXADECANYL HEXADECANOATE;
W/NBS
62275

Me(CH2)15OC(O)(CH2)14Me

480 C32 H62 D2 O2 34689-05-9 EP-5171-0-0 2,2-DIDEUTERIO-TETRADECANYL OCTADECANOATE; W
62277

Me(CH2)11CD2CH2OC(O)(CH2)16Me

480 C32 H64 O2 3234-81-9 EP-5178-0-0 Tetradecanoic acid, octadecyl ester;; Stearyl myristate; W/NBS
62278

Me(CH2)17OC(O)(CH2)12Me

480 C32 H62 D2 O2 34689-01-5 EP-5174-0-0 OCTADECANYL 2,2-DIDEUTERIO-TETRADECANOATE; W
62279

Me(CH2)17OC(O)CD2(CH2)11Me

480 C32 H64 O2 17661-50-6 EP-5179-0-0 Octadecanoic acid, tetradecyl ester;; Myristyl stearate; W 79271

Me(CH2)13OC(O)(CH2)16Me

480 C33 H25 Co 1278-02-0 HE-1982-0-0 Cobalt, [1,1',1'',1'''-(η4-1,3-cyclobutadiene-1,2,3,4-tetrayl)tet
rakis[benzene]](η5-2,4-cyclopentadien-1-yl)-;; COBALT, CYCLOPENTADIENYL-TETRAPHENYLCYCLOBUTADIENE; W 62280

480 C34 H24 O3 55429-86-2 AD-0-1501-0 [1,1'-Biphenyl]-2-carboxaldehyde, 4',4'''-(3,4-furandiyldi-2,1-et
henediyl)bis-;; 3,4-BIS(2-(2'-FORMYL-2-BIPHENYL)-VINYL)-FURAN (IMPURITY ?); W/NBS
62281

480 C34 H25 O P 68457-27-2 K-115-589-0 Phenyl[2-(trans-2-phenylethenyl)phenyl][2-(2-phenylethinyl)phen
yl]phosphineoxide; W
125197

480 C34 H25 O P 68457-29-4 K-115-589-0 1,3-Diphenyl-6'-(phenylethenyliden)spiro[1-benzophosphorin-2(
1H),1'-cyclohexa-2',4'-dien]-1-oxide; W
125198

480 C35 H60 AC-1977-169-0 29-n-Pentyl-(17αH,21βH)-hop-22(29)-ene; W
125199

480 C35 H60 AC-1977-169-0 29-Isopentyl-(17αH,21βH)-hop-22(29)-ene; W
125200

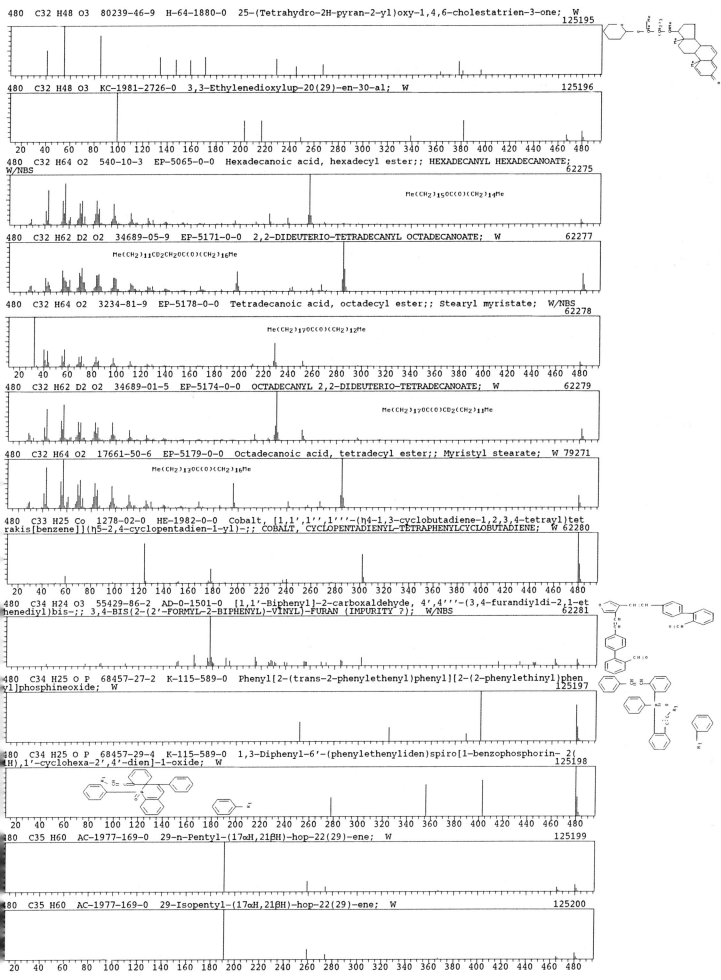

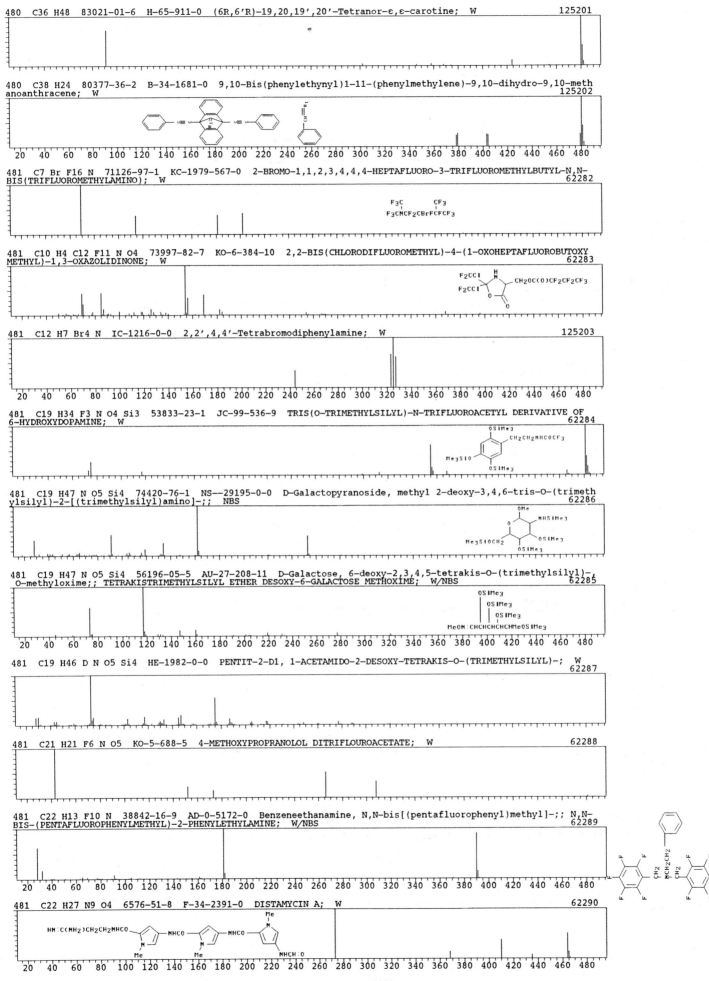

480   C36 H48   83021-01-6   H-65-911-0   (6R,6'R)-19,20,19',20'-Tetranor-ε,ε-carotine;   W   125201

480   C38 H24   80377-36-2   B-34-1681-0   9,10-Bis(phenylethynyl)1-11-(phenylmethylene)-9,10-dihydro-9,10-meth
anoanthracene;   W   125202

481   C7 Br F16 N   71126-97-1   KC-1979-567-0   2-BROMO-1,1,2,3,4,4,4-HEPTAFLUORO-3-TRIFLUOROMETHYLBUTYL-N,N-
BIS(TRIFLUOROMETHYLAMINO);   W   62282

F3C   CF3
F3CNCF2CBrFCFCF3

481   C10 H4 Cl2 F11 N O4   73997-82-7   KO-6-384-10   2,2-BIS(CHLORODIFLUOROMETHYL)-4-(1-OXOHEPTAFLUOROBUTOXY
METHYL)-1,3-OXAZOLIDINONE;   W   62283

481   C12 H7 Br4 N   IC-1216-0-0   2,2',4,4'-Tetrabromodiphenylamine;   W   125203

481   C19 H34 F3 N O4 Si3   53833-23-1   JC-99-536-9   TRIS(O-TRIMETHYLSILYL)-N-TRIFLUOROACETYL DERIVATIVE OF
6-HYDROXYDOPAMINE;   W   62284

481   C19 H47 N O5 Si4   74420-76-1   NS--29195-0-0   D-Galactopyranoside, methyl 2-deoxy-3,4,6-tris-O-(trimeth
ylsilyl)-2-[(trimethylsilyl)amino]-;;   NBS   62286

481   C19 H47 N O5 Si4   56196-05-5   AU-27-208-11   D-Galactose, 6-deoxy-2,3,4,5-tetrakis-O-(trimethylsilyl)-,
O-methyloxime;; TETRAKISTRIMETHYLSILYL ETHER DESOXY-6-GALACTOSE METHOXIME;   W/NBS   62285

481   C19 H46 D N O5 Si4   HE-1982-0-0   PENTIT-2-D1, 1-ACETAMIDO-2-DESOXY-TETRAKIS-O-(TRIMETHYLSILYL)-;   W
          62287

481   C21 H21 F6 N O5   KO-5-688-5   4-METHOXYPROPRANOLOL DITRIFLOUROACETATE;   W   62288

481   C22 H13 F10 N   38842-16-9   AD-0-5172-0   Benzeneethanamine, N,N-bis[(pentafluorophenyl)methyl]-;;   N,N-
BIS-(PENTAFLUOROPHENYLMETHYL)-2-PHENYLETHYLAMINE;   W/NBS   62289

481   C22 H27 N9 O4   6576-51-8   F-34-2391-0   DISTAMYCIN A;   W   62290

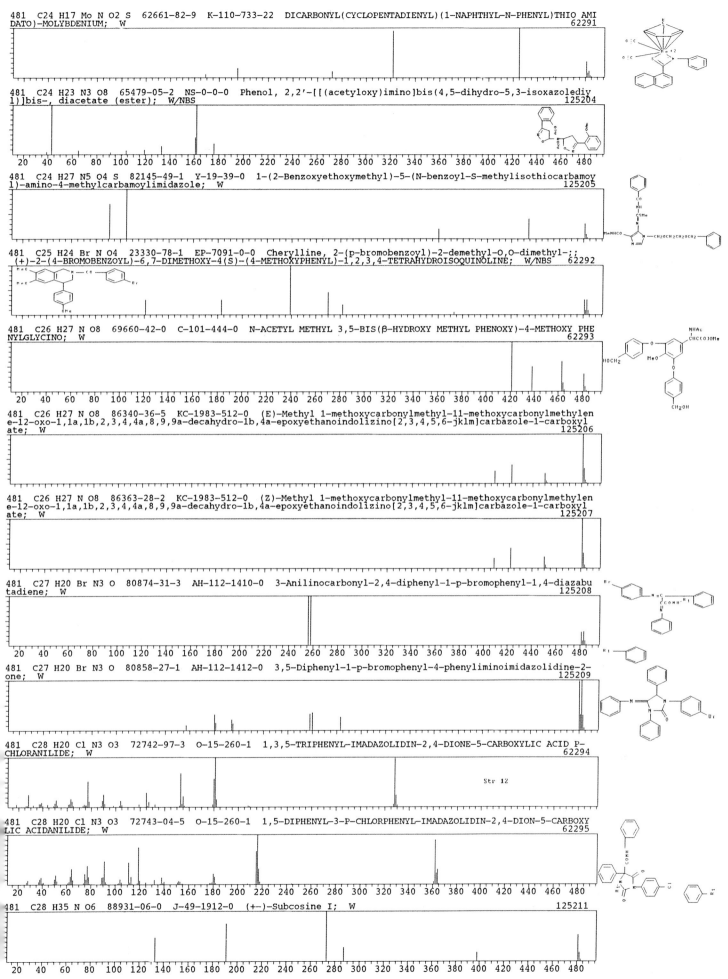

481   C24 H17 Mo N O2 S   62661-82-9   K-110-733-22   DICARBONYL(CYCLOPENTADIENYL)(1-NAPHTHYL-N-PHENYL)THIO AMI
DATO)-MOLYBDENIUM;   W
62291

481   C24 H23 N3 O8   65479-05-2   NS-0-0-0   Phenol, 2,2'-[[(acetyloxy)imino]bis(4,5-dihydro-5,3-isoxazolediy
1)]bis-, diacetate (ester);   W/NBS
125204

20 40 60 80 100 120 140 160 180 200 220 240 260 280 300 320 340 360 380 400 420 440 460 480

481   C24 H27 N5 O4 S   82145-49-1   Y-19-39-0   1-(2-Benzoxyethoxymethyl)-5-(N-benzoyl-S-methylisothiocarbamoy
1)-amino-4-methylcarbamoylimidazole;   W
125205

481   C25 H24 Br N O4   23330-78-1   EP-7091-0-0   Cherylline, 2-(p-bromobenzoyl)-2-demethyl-O,O-dimethyl-;;
(+)-2-(4-BROMOBENZOYL)-6,7-DIMETHOXY-4(S)-(4-METHOXYPHENYL)-1,2,3,4-TETRAHYDROISOQUINOLINE;   W/NBS   62292

481   C26 H27 N O8   69660-42-0   C-101-444-0   N-ACETYL METHYL 3,5-BIS(β-HYDROXY METHYL PHENOXY)-4-METHOXY PHE
NYLGLYCINO;   W
62293

20 40 60 80 100 120 140 160 180 200 220 240 260 280 300 320 340 360 380 400 420 440 460 480

481   C26 H27 N O8   86340-36-5   KC-1983-512-0   (E)-Methyl 1-methoxycarbonylmethyl-11-methoxycarbonylmethylen
e-12-oxo-1,1a,1b,2,3,4,4a,8,9,9a-decahydro-1b,4a-epoxyethanoindolizino[2,3,4,5,6-jklm]carbazole-1-carboxyl
ate;   W
125206

481   C26 H27 N O8   86363-28-2   KC-1983-512-0   (Z)-Methyl 1-methoxycarbonylmethyl-11-methoxycarbonylmethylen
e-12-oxo-1,1a,1b,2,3,4,4a,8,9,9a-decahydro-1b,4a-epoxyethanoindolizino[2,3,4,5,6-jklm]carbazole-1-carboxyl
ate;   W
125207

481   C27 H20 Br N3 O   80874-31-3   AH-112-1410-0   3-Anilinocarbonyl-2,4-diphenyl-1-p-bromophenyl-1,4-diazabu
tadiene;   W
125208

20 40 60 80 100 120 140 160 180 200 220 240 260 280 300 320 340 360 380 400 420 440 460 480

481   C27 H20 Br N3 O   80858-27-1   AH-112-1412-0   3,5-Diphenyl-1-p-bromophenyl-4-phenyliminoimidazolidine-2-
one;   W
125209

481   C28 H20 Cl N3 O3   72742-97-3   O-15-260-1   1,3,5-TRIPHENYL-IMADAZOLIDIN-2,4-DIONE-5-CARBOXYLIC ACID P-
CHLORANILIDE;   W
62294

Str 12

481   C28 H20 Cl N3 O3   72743-04-5   O-15-260-1   1,5-DIPHENYL-3-P-CHLORPHENYL-IMADAZOLIDIN-2,4-DION-5-CARBOXY
LIC ACIDANILIDE;   W
62295

20 40 60 80 100 120 140 160 180 200 220 240 260 280 300 320 340 360 380 400 420 440 460 480

481   C28 H35 N O6   88931-06-0   J-49-1912-0   (+-)-Subcosine I;   W
125211

20 40 60 80 100 120 140 160 180 200 220 240 260 280 300 320 340 360 380 400 420 440 460 480

481   C28 H35 N O6   89771-50-6   J-49-1912-0   (+-)-Episubcosine II;   W   125212

481   C28 H51 N O5   69121-68-2   NS--29192-0-0   Tetradecanamide, N-[3-[4-(acetyloxy)-3-methyl-7-oxabicyclo[4.
1.0]hept-1-yl]propyl]-7-methoxy-N-methyl-;;   NBS   62296

481   C29 H23 N O6   75233-35-1   J-45-4901-0   3-(Methoxalyl)-3-(carbomethoxy)-2-p-tolyl-5,6-diphenyl-4(3h)-py
ridone;   W   125213

481   C29 H27 N3 O4   71522-28-6   I-57-1702-0   N-(Z-(1'-BENZYL-3'-(2''-CARBOXYINDOL-7''-YL)OXINDOL-3'-YL)ETH
YL)-N-METHYLACETAMIDE;   W   62297

481   C29 H39 N O5   IC-1216-0-0   1-Benzyl-3a,5-dihydroxy-7-methyl-4[4-methyl-7-(5-oxo-tetrahydrofuran-2-yl)]
-6-methylene-3-oxo-3a,4,5,6,7,7a-hexahydroisoindoline;   W   125214

481   C29 H39 N O5   IC-1216-0-0   1-Benzyl-8,17-dihydroxy12,19-dimethyl-3,5-dioxo-8-methylene-4-oxa-14-ene-pe
rhydrocyclotetradeca[2,3-d]isoindole;   W   125215

481   C29 H43 N O3 Si   57305-11-0   EP-8225-0-0   Androstane-11,17-dione, 3-[(trimethylsilyl)oxy]-, 17-[O-(phe
nylmethyl)oxime], (3α,5α)-;;   5α-ANDROSTAN-3α-OL-11,17-DIONE BO TMS;   W/NBS   62298

481   C30 H31 N O3 Si   79139-18-7   K-114-1023-0   3-(Diethylamino)-2-[3-(triphenylsiloxy)-3-butenoyl]-2-cyclo
buten-1-one;   W   125216

481   C31 H47 N O3   HE-1986-1225-0   METHYL ESTER OF 1,7,8-TRI-TERT-BUTYL-3-NEOPENTYL-5-PHENYL-2-OXA-4-AZABIC
YCLO[4.2.0]OCTA-3,7-DIEN-6-CARBOXYLIC ACID;   W   125217

481   C32 H35 N O3   79632-06-7   J-46-5388-0   2,3β-Dibenzyl-5α,6-dimethyl-8α-[(benzyloxy)methyl]-9-β-hydroxy-
2,3,4β,5β,8β,9β-hexahydro-1H-isoindole-1-one;   W   125218

482   C7 H3 Br5   87-83-2   B-30-302-0   Benzene, pentabromomethyl-;;   Flammex 5BT;   W   62299

482   C7 H3 Br5   IC-1217-0-0   Pentabromomethyl-benzene;   W   125219

- 5344 -

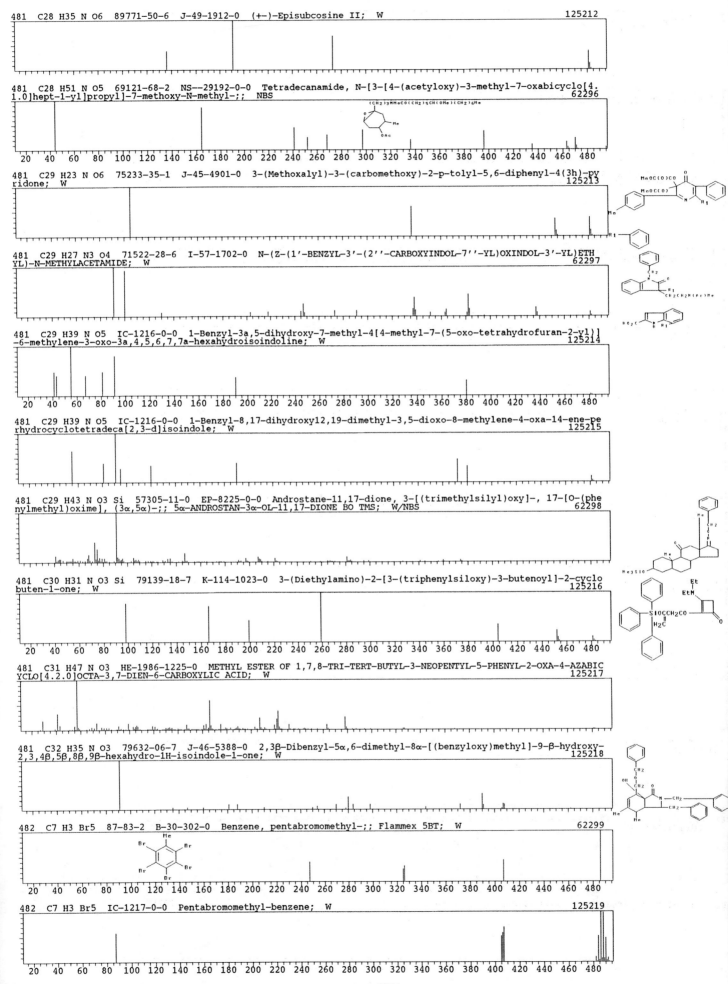

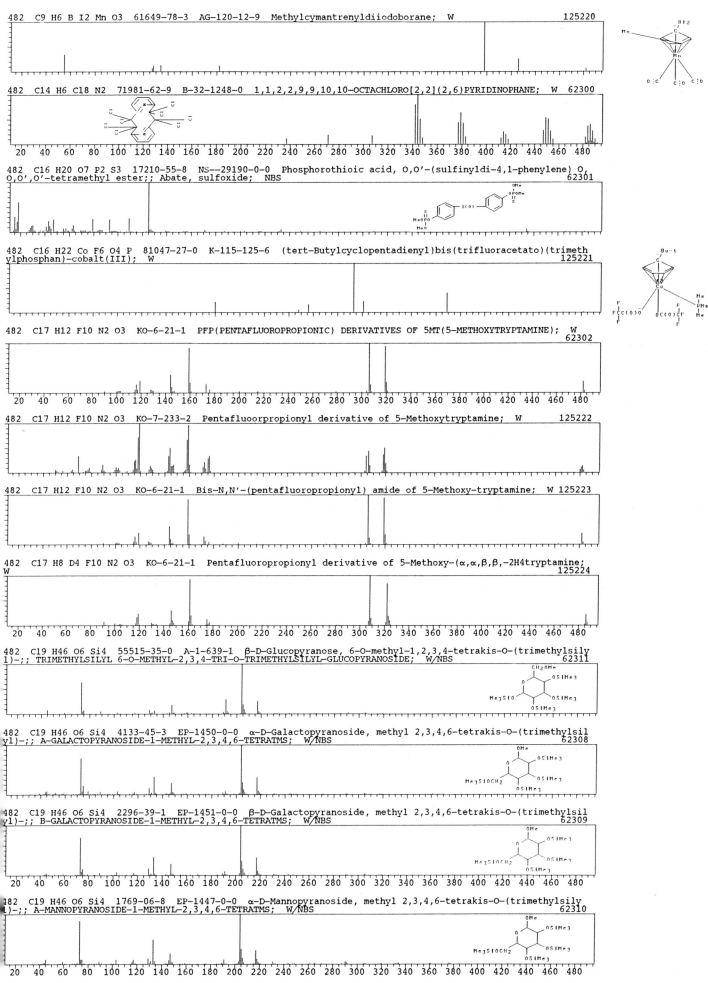

482   C9 H6 B I2 Mn O3   61649-78-3   AG-120-12-9   Methylcymantrenyldiiodoborane;   W   125220

482   C14 H6 Cl8 N2   71981-62-9   B-32-1248-0   1,1,2,2,9,9,10,10-OCTACHLORO[2,2](2,6)PYRIDINOPHANE;   W   62300

482   C16 H20 O7 P2 S3   17210-55-8   NS--29190-0-0   Phosphorothioic acid, O,O'-(sulfinyldi-4,1-phenylene) O,
O,O',O'-tetramethyl ester;; Abate, sulfoxide;   NBS   62301

482   C16 H22 Co F6 O4 P   81047-27-0   K-115-125-6   (tert-Butylcyclopentadienyl)bis(trifluoracetato)(trimeth
ylphosphan)-cobalt(III);   W   125221

482   C17 H12 F10 N2 O3   KO-6-21-1   PFP(PENTAFLUOROPROPIONIC) DERIVATIVES OF 5MT(5-METHOXYTRYPTAMINE);   W
                                                                                                            62302

482   C17 H12 F10 N2 O3   KO-7-233-2   Pentafluoorpropionyl derivative of 5-Methoxytryptamine;   W   125222

482   C17 H12 F10 N2 O3   KO-6-21-1   Bis-N,N'-(pentafluoropropionyl) amide of 5-Methoxy-tryptamine;   W 125223

482   C17 H8 D4 F10 N2 O3   KO-6-21-1   Pentafluoropropionyl derivative of 5-Methoxy-(α,α,β,β,-2H4tryptamine;
W   125224

482   C19 H46 O6 Si4   55515-35-0   A-1-639-1   β-D-Glucopyranose, 6-O-methyl-1,2,3,4-tetrakis-O-(trimethylsily
l)-;; TRIMETHYLSILYL 6-O-METHYL-2,3,4-TRI-O-TRIMETHYLSILYL-GLUCOPYRANOSIDE;   W/NBS   62311

482   C19 H46 O6 Si4   4133-45-3   EP-1450-0-0   α-D-Galactopyranoside, methyl 2,3,4,6-tetrakis-O-(trimethylsil
yl)-;; A-GALACTOPYRANOSIDE-1-METHYL-2,3,4,6-TETRATMS;   W/NBS   62308

482   C19 H46 O6 Si4   2296-39-1   EP-1451-0-0   β-D-Galactopyranoside, methyl 2,3,4,6-tetrakis-O-(trimethylsil
yl)-;; B-GALACTOPYRANOSIDE-1-METHYL-2,3,4,6-TETRATMS;   W/NBS   62309

482   C19 H46 O6 Si4   1769-06-8   EP-1447-0-0   α-D-Mannopyranoside, methyl 2,3,4,6-tetrakis-O-(trimethylsily
l)-;; A-MANNOPYRANOSIDE-1-METHYL-2,3,4,6-TETRATMS;   W/NBS   62310

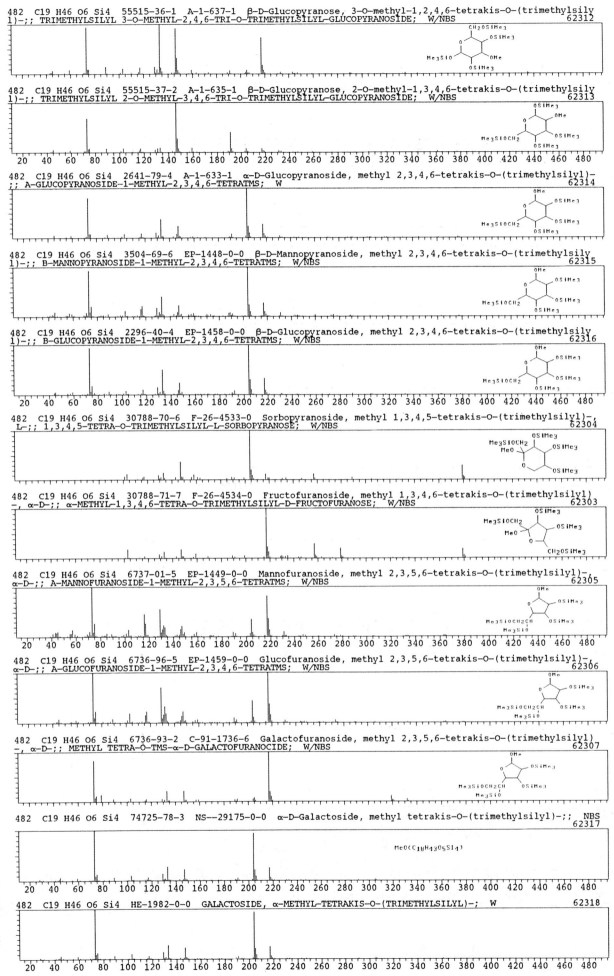

482 C19 H46 O6 Si4 55515-36-1 A-1-637-1 β-D-Glucopyranose, 3-O-methyl-1,2,4,6-tetrakis-O-(trimethylsily l)-;; TRIMETHYLSILYL 3-O-METHYL-2,4,6-TRI-O-TRIMETHYLSILYL-GLUCOPYRANOSIDE; W/NBS
62312

482 C19 H46 O6 Si4 55515-37-2 A-1-635-1 β-D-Glucopyranose, 2-O-methyl-1,3,4,6-tetrakis-O-(trimethylsily l)-;; TRIMETHYLSILYL 2-O-METHYL-3,4,6-TRI-O-TRIMETHYLSILYL-GLUCOPYRANOSIDE; W/NBS
62313

482 C19 H46 O6 Si4 2641-79-4 A-1-633-1 α-D-Glucopyranoside, methyl 2,3,4,6-tetrakis-O-(trimethylsilyl)- ;; A-GLUCOPYRANOSIDE-1-METHYL-2,3,4,6-TETRATMS; W
62314

482 C19 H46 O6 Si4 3504-69-6 EP-1448-0-0 β-D-Mannopyranoside, methyl 2,3,4,6-tetrakis-O-(trimethylsily l)-;; B-MANNOPYRANOSIDE-1-METHYL-2,3,4,6-TETRATMS; W/NBS
62315

482 C19 H46 O6 Si4 2296-40-4 EP-1458-0-0 β-D-Glucopyranoside, methyl 2,3,4,6-tetrakis-O-(trimethylsily l)-;; B-GLUCOPYRANOSIDE-1-METHYL-2,3,4,6-TETRATMS; W/NBS
62316

482 C19 H46 O6 Si4 30788-70-6 F-26-4533-0 Sorbopyranoside, methyl 1,3,4,5-tetrakis-O-(trimethylsilyl)-, L-;; 1,3,4,5-TETRA-O-TRIMETHYLSILYL-L-SORBOPYRANOSE; W/NBS
62304

482 C19 H46 O6 Si4 30788-71-7 F-26-4534-0 Fructofuranoside, methyl 1,3,4,6-tetrakis-O-(trimethylsilyl) -, α-D-;; α-METHYL-1,3,4,6-TETRA-O-TRIMETHYLSILYL-D-FRUCTOFURANOSE; W/NBS
62303

482 C19 H46 O6 Si4 6737-01-5 EP-1449-0-0 Mannofuranoside, methyl 2,3,5,6-tetrakis-O-(trimethylsilyl)- α-D-;; A-MANNOFURANOSIDE-1-METHYL-2,3,5,6-TETRATMS; W/NBS
62305

482 C19 H46 O6 Si4 6736-96-5 EP-1459-0-0 Glucofuranoside, methyl 2,3,5,6-tetrakis-O-(trimethylsilyl)- α-D-;; A-GLUCOFURANOSIDE-1-METHYL-2,3,4,6-TETRATMS; W/NBS
62306

482 C19 H46 O6 Si4 6736-93-2 C-91-1736-6 Galactofuranoside, methyl 2,3,5,6-tetrakis-O-(trimethylsilyl) -, α-D-;; METHYL TETRA-O-TMS-α-D-GALACTOFURANOCIDE; W/NBS
62307

482 C19 H46 O6 Si4 74725-78-3 NS--29175-0-0 α-D-Galactoside, methyl tetrakis-O-(trimethylsilyl)-;; NBS
62317

482 C19 H46 O6 Si4 HE-1982-0-0 GALACTOSIDE, α-METHYL-TETRAKIS-O-(TRIMETHYLSILYL)-; W
62318

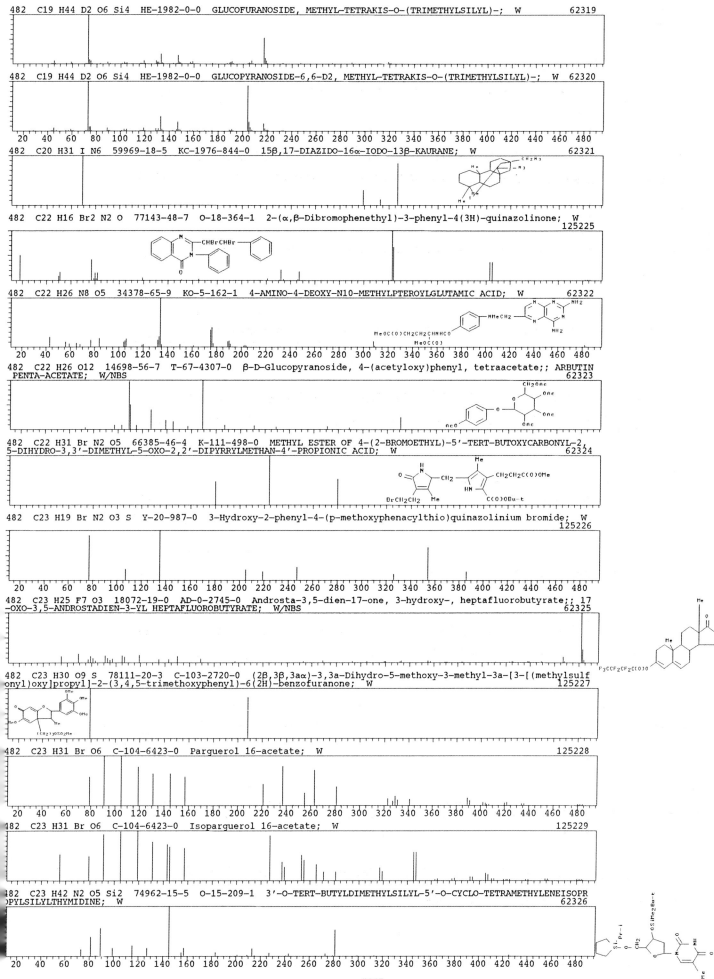

482 C19 H44 D2 O6 Si4 HE-1982-0-0 GLUCOFURANOSIDE, METHYL-TETRAKIS-O-(TRIMETHYLSILYL)-; W   62319

482 C19 H44 D2 O6 Si4 HE-1982-0-0 GLUCOPYRANOSIDE-6,6-D2, METHYL-TETRAKIS-O-(TRIMETHYLSILYL)-; W 62320

482 C20 H31 I N6 59969-18-5 KC-1976-844-0 15β,17-DIAZIDO-16α-IODO-13β-KAURANE; W   62321

482 C22 H16 Br2 N2 O 77143-48-7 O-18-364-1 2-(α,β-Dibromophenethyl)-3-phenyl-4(3H)-quinazolinone; W
                                                                                      125225

482 C22 H26 N8 O5 34378-65-9 KO-5-162-1 4-AMINO-4-DEOXY-N10-METHYLPTEROYLGLUTAMIC ACID; W   62322

482 C22 H26 O12 14698-56-7 T-67-4307-0 β-D-Glucopyranoside, 4-(acetyloxy)phenyl, tetraacetate;; ARBUTIN
   PENTA-ACETATE; W/NBS   62323

482 C22 H31 Br N2 O5 66385-46-4 K-111-498-0 METHYL ESTER OF 4-(2-BROMOETHYL)-5'-TERT-BUTOXYCARBONYL-2,
5-DIHYDRO-3,3'-DIMETHYL-5-OXO-2,2'-DIPYRRYLMETHAN-4'-PROPIONIC ACID; W   62324

482 C23 H19 Br N2 O3 S Y-20-987-0 3-Hydroxy-2-phenyl-4-(p-methoxyphenacylthio)quinazolinium bromide; W
                                                                                      125226

482 C23 H25 F7 O3 18072-19-0 AD-0-2745-0 Androsta-3,5-dien-17-one, 3-hydroxy-, heptafluorobutyrate;; 17
-OXO-3,5-ANDROSTADIEN-3-YL HEPTAFLUOROBUTYRATE; W/NBS   62325

482 C23 H30 O9 S 78111-20-3 C-103-2720-0 (2β,3β,3aα)-3,3a-Dihydro-5-methoxy-3-methyl-3a-[3-[(methylsulf
onyl)oxy]propyl]-2-(3,4,5-trimethoxyphenyl)-6(2H)-benzofuranone; W   125227

482 C23 H31 Br O6 C-104-6423-0 Parguerol 16-acetate; W   125228

482 C23 H31 Br O6 C-104-6423-0 Isoparguerol 16-acetate; W   125229

482 C23 H42 N2 O5 Si2 74962-15-5 O-15-209-1 3'-O-TERT-BUTYLDIMETHYLSILYL-5'-O-CYCLO-TETRAMETHYLENEISOPR
OPYLSILYLTHYMIDINE; W   62326

482   C23 H42 N2 O5 Si2   O-15-209-1   5'-O-TERT-BUTYLDIMETHYLSILYL-3'-O-CYCLO-TETRAMETHYLENEISOPROPYLSILYLTHY
MIDINE;   W   62327

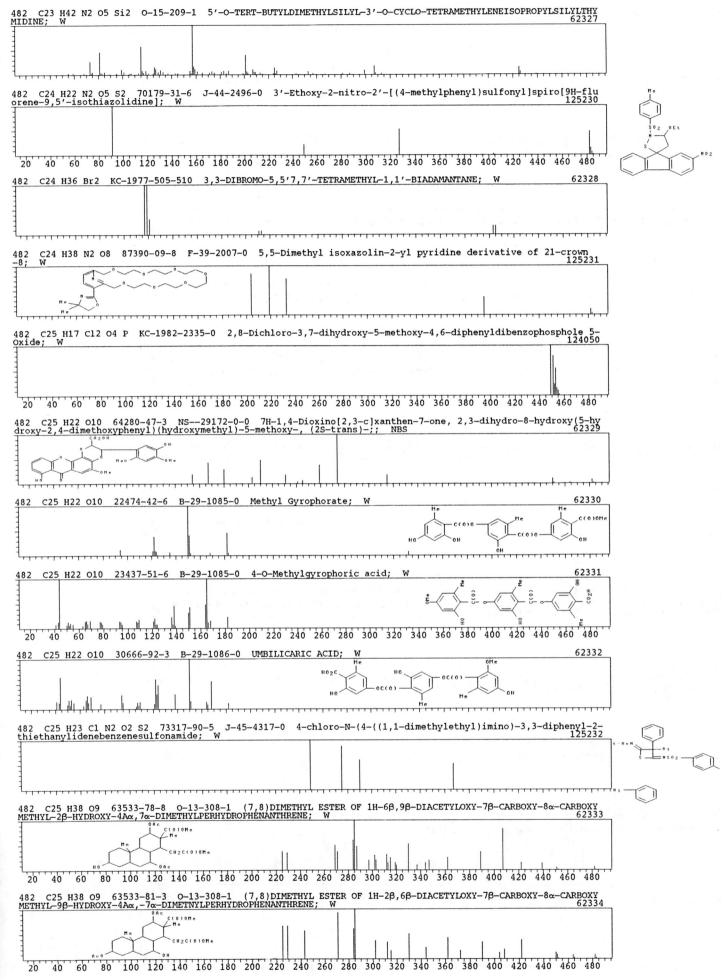

482   C24 H22 N2 O5 S2   70179-31-6   J-44-2496-0   3'-Ethoxy-2-nitro-2'-[(4-methylphenyl)sulfonyl]spiro[9H-flu
orene-9,5'-isothiazolidine];   W   125230

482   C24 H36 Br2   KC-1977-505-510   3,3-DIBROMO-5,5'7,7'-TETRAMETHYL-1,1'-BIADAMANTANE;   W   62328

482   C24 H38 N2 O8   87390-09-8   F-39-2007-0   5,5-Dimethyl isoxazolin-2-yl pyridine derivative of 21-crown
-8;   W   125231

482   C25 H17 Cl2 O4 P   KC-1982-2335-0   2,8-Dichloro-3,7-dihydroxy-5-methoxy-4,6-diphenyldibenzophosphole 5-
Oxide;   W   124050

482   C25 H22 O10   64280-47-3   NS--29172-0-0   7H-1,4-Dioxino[2,3-c]xanthen-7-one, 2,3-dihydro-8-hydroxy(5-hy
droxy-2,4-dimethoxyphenyl)(hydroxymethyl)-5-methoxy-, (2S-trans)-;;   NBS   62329

482   C25 H22 O10   22474-42-6   B-29-1085-0   Methyl Gyrophorate;   W   62330

482   C25 H22 O10   23437-51-6   B-29-1085-0   4-O-Methylgyrophoric acid;   W   62331

482   C25 H22 O10   30666-92-3   B-29-1086-0   UMBILICARIC ACID;   W   62332

482   C25 H23 Cl N2 O2 S2   73317-90-5   J-45-4317-0   4-chloro-N-(4-((1,1-dimethylethyl)imino)-3,3-diphenyl-2-
thiethanylidenebenzenesulfonamide;   W   125232

482   C25 H38 O9   63533-78-8   O-13-308-1   (7,8)DIMETHYL ESTER OF 1H-6β,9β-DIACETYLOXY-7β-CARBOXY-8α-CARBOXY
METHYL-2β-HYDROXY-4Aα,7α-DIMETHYLPERHYDROPHENANTHRENE;   W   62333

482   C25 H38 O9   63533-81-3   O-13-308-1   (7,8)DIMETHYL ESTER OF 1H-2β,6β-DIACETYLOXY-7β-CARBOXY-8α-CARBOXY
METHYL-9β-HYDROXY-4Aα,-7α-DIMETNYLPERHYDROPHENANTHRENE;   W   62334

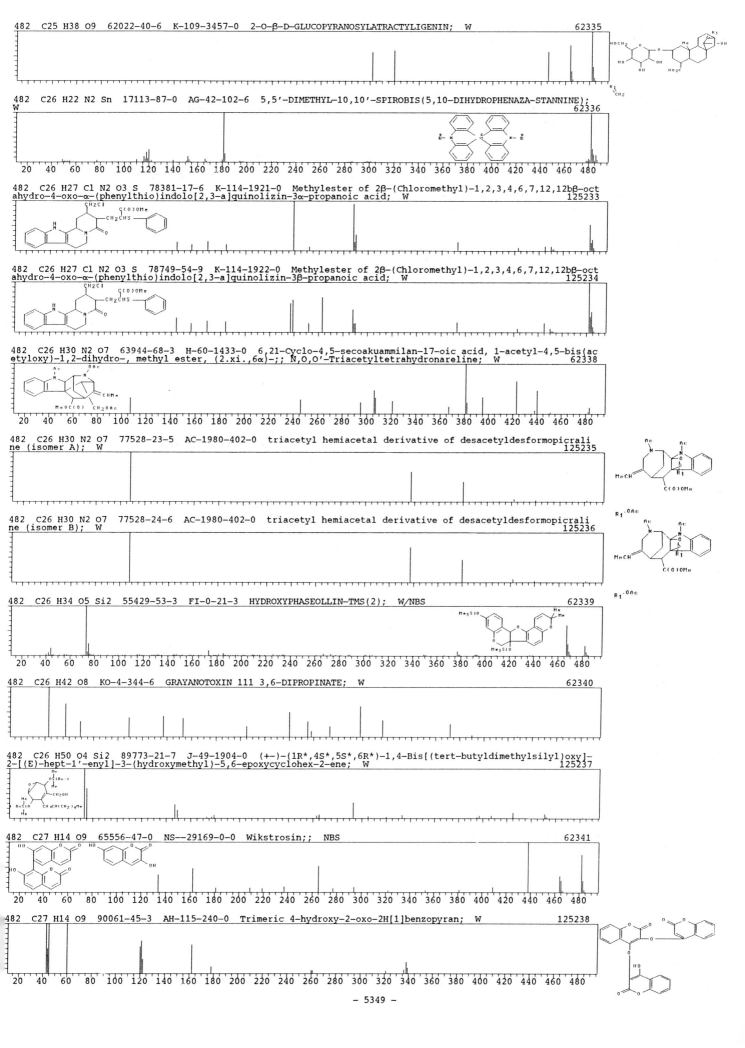

482 C25 H38 O9 62022-40-6 K-109-3457-0 2-O-β-D-GLUCOPYRANOSYLATRACTYLIGENIN; W   62335

482 C26 H22 N2 Sn 17113-87-0 AG-42-102-6 5,5'-DIMETHYL-10,10'-SPIROBIS(5,10-DIHYDROPHENAZA-STANNINE);
W   62336

482 C26 H27 Cl N2 O3 S 78381-17-6 K-114-1921-0 Methylester of 2β-(Chloromethyl)-1,2,3,4,6,7,12,12bβ-oct
ahydro-4-oxo-α-(phenylthio)indolo[2,3-a]quinolizin-3α-propanoic acid; W   125233

482 C26 H27 Cl N2 O3 S 78749-54-9 K-114-1922-0 Methylester of 2β-(Chloromethyl)-1,2,3,4,6,7,12,12bβ-oct
ahydro-4-oxo-α-(phenylthio)indolo[2,3-a]quinolizin-3β-propanoic acid; W   125234

482 C26 H30 N2 O7 63944-68-3 H-60-1433-0 6,21-Cyclo-4,5-secoakuammilan-17-oic acid, 1-acetyl-4,5-bis(ac
etyloxy)-1,2-dihydro-, methyl ester, (2.xi.,6α)-;; N,O,O'-Triacetyltetrahydronareline; W   62338

482 C26 H30 N2 O7 77528-23-5 AC-1980-402-0 triacetyl hemiacetal derivative of desacetyldesformopicrali
ne (isomer A); W   125235

482 C26 H30 N2 O7 77528-24-6 AC-1980-402-0 triacetyl hemiacetal derivative of desacetyldesformopicrali
ne (isomer B); W   125236

482 C26 H34 O5 Si2 55429-53-3 FI-0-21-3 HYDROXYPHASEOLLIN-TMS(2); W/NBS   62339

482 C26 H42 O8 KO-4-344-6 GRAYANOTOXIN 111 3,6-DIPROPINATE; W   62340

482 C26 H50 O4 Si2 89773-21-7 J-49-1904-0 (+-)-(1R*,4S*,5S*,6R*)-1,4-Bis[(tert-butyldimethylsilyl)oxy]-
2-[(E)-hept-1'-enyl]-3-(hydroxymethyl)-5,6-epoxycyclohex-2-ene; W   125237

482 C27 H14 O9 65556-47-0 NS--29169-0-0 Wikstrosin;; NBS   62341

482 C27 H14 O9 90061-45-3 AH-115-240-0 Trimeric 4-hydroxy-2-oxo-2H[1]benzopyran; W   125238

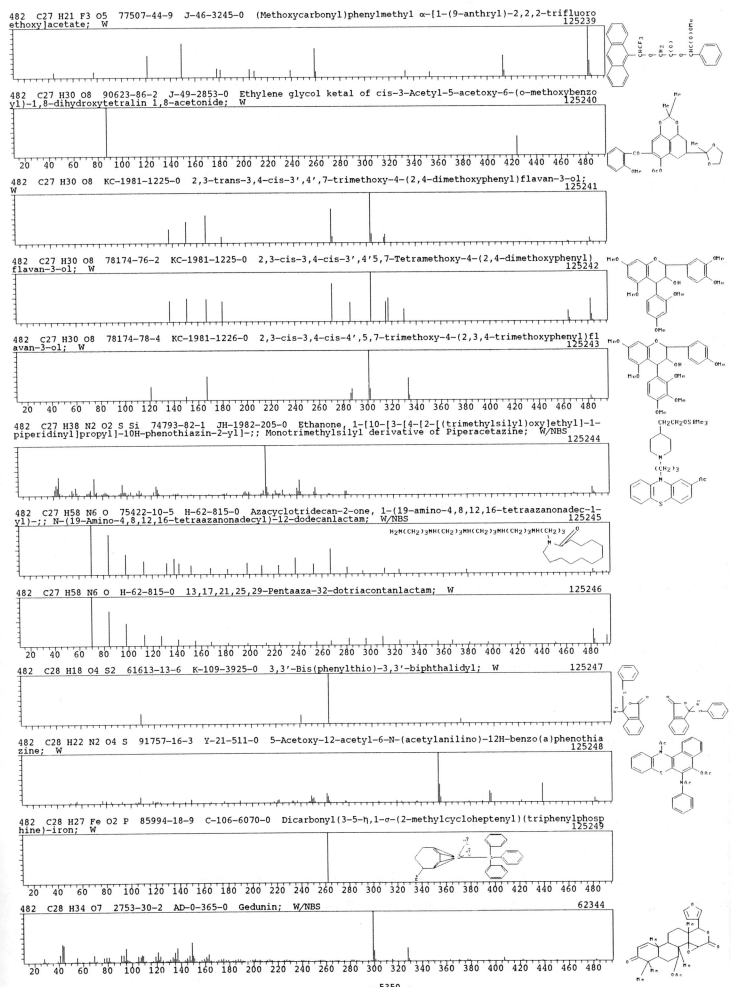

482   C27 H21 F3 O5   77507-44-9   J-46-3245-0   (Methoxycarbonyl)phenylmethyl α-[1-(9-anthryl)-2,2,2-trifluoro
ethoxy]acetate;   W
125239

482   C27 H30 O8   90623-86-2   J-49-2853-0   Ethylene glycol ketal of cis-3-Acetyl-5-acetoxy-6-(o-methoxybenzo
yl)-1,8-dihydroxytetralin 1,8-acetonide;   W
125240

482   C27 H30 O8   KC-1981-1225-0   2,3-trans-3,4-cis-3',4',7-trimethoxy-4-(2,4-dimethoxyphenyl)flavan-3-ol;
W
125241

482   C27 H30 O8   78174-76-2   KC-1981-1225-0   2,3-cis-3,4-cis-3',4'5,7-Tetramethoxy-4-(2,4-dimethoxyphenyl)
flavan-3-ol;   W
125242

482   C27 H30 O8   78174-78-4   KC-1981-1226-0   2,3-cis-3,4-cis-4',5,7-trimethoxy-4-(2,3,4-trimethoxyphenyl)fl
avan-3-ol;   W
125243

482   C27 H38 N2 O2 S Si   74793-82-1   JH-1982-205-0   Ethanone, 1-[10-[3-[4-[2-[(trimethylsilyl)oxy]ethyl]-1-
piperidinyl]propyl]-10H-phenothiazin-2-yl]-;;   Monotrimethylsilyl derivative of Piperacetazine;   W/NBS
125244

482   C27 H58 N6 O   75422-10-5   H-62-815-0   Azacyclotridecan-2-one, 1-(19-amino-4,8,12,16-tetraazanonadec-1-
yl)-;;   N-(19-Amino-4,8,12,16-tetraazanonadecyl)-12-dodecanlactam;   W/NBS
125245

482   C27 H58 N6 O   H-62-815-0   13,17,21,25,29-Pentaaza-32-dotriacontanlactam;   W
125246

482   C28 H18 O4 S2   61613-13-6   K-109-3925-0   3,3'-Bis(phenylthio)-3,3'-biphthalidyl;   W
125247

482   C28 H22 N2 O4 S   91757-16-3   Y-21-511-0   5-Acetoxy-12-acetyl-6-N-(acetylanilino)-12H-benzo(a)phenothia
zine;   W
125248

482   C28 H27 Fe O2 P   85994-18-9   C-106-6070-0   Dicarbonyl(3-5-η,1-σ-(2-methylcycloheptenyl)(triphenylphosp
hine)-iron;   W
125249

482   C28 H34 O7   2753-30-2   AD-0-365-0   Gedunin;   W/NBS
62344

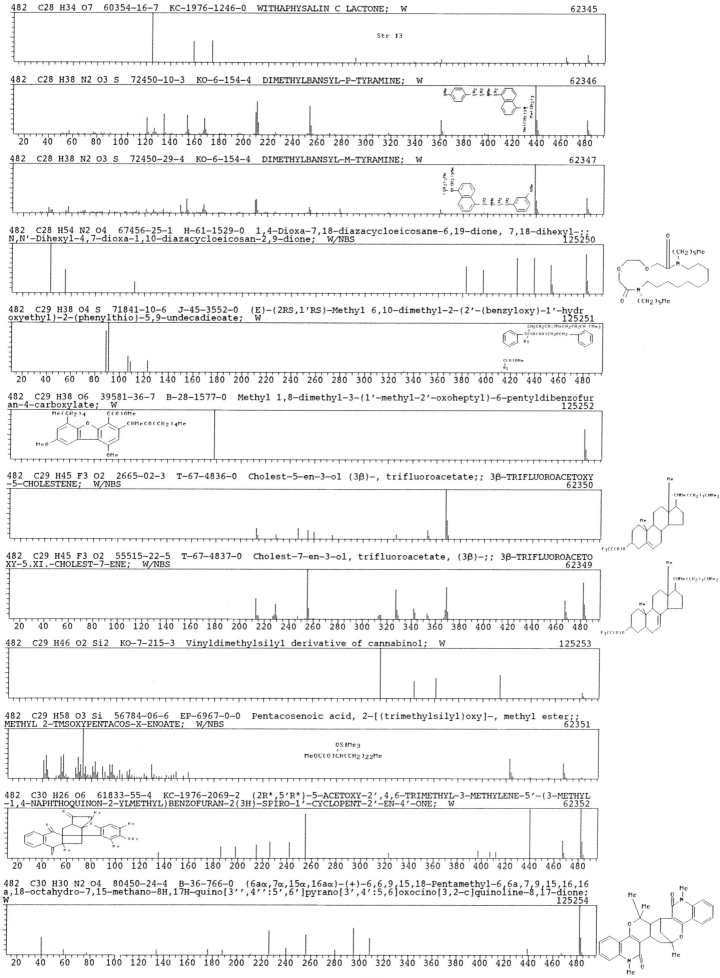

482 C28 H34 O7 60354-16-7 KC-1976-1246-0 WITHAPHYSALIN C LACTONE; W   62345

Str 13

482 C28 H38 N2 O3 S 72450-10-3 KO-6-154-4 DIMETHYLBANSYL-P-TYRAMINE; W   62346

20 40 60 80 100 120 140 160 180 200 220 240 260 280 300 320 340 360 380 400 420 440 460 480

482 C28 H38 N2 O3 S 72450-29-4 KO-6-154-4 DIMETHYLBANSYL-M-TYRAMINE; W   62347

482 C28 H54 N2 O4 67456-25-1 H-61-1529-0 1,4-Dioxa-7,18-diazacycloeicosane-6,19-dione, 7,18-dihexyl-;;
N,N'-Dihexyl-4,7-dioxa-1,10-diazacycloeicosan-2,9-dione; W/NBS   125250

482 C29 H38 O4 S 71841-10-6 J-45-3552-0 (E)-(2RS,1'RS)-Methyl 6,10-dimethyl-2-(2'-(benzyloxy)-1'-hydr
oxyethyl)-2-(phenylthio)-5,9-undecadieoate; W   125251

20 40 60 80 100 120 140 160 180 200 220 240 260 280 300 320 340 360 380 400 420 440 460 480

482 C29 H38 O6 39581-36-7 B-28-1577-0 Methyl 1,8-dimethyl-3-(1'-methyl-2'-oxoheptyl)-6-pentyldibenzofur
an-4-carboxylate; W   125252

482 C29 H45 F3 O2 2665-02-3 T-67-4836-0 Cholest-5-en-3-ol (3β)-, trifluoroacetate;; 3β-TRIFLUOROACETOXY
-5-CHOLESTENE; W/NBS   62350

482 C29 H45 F3 O2 55515-22-5 T-67-4837-0 Cholest-7-en-3-ol, trifluoroacetate, (3β)-;; 3β-TRIFLUOROACETO
XY-5.XI.-CHOLEST-7-ENE; W/NBS   62349

20 40 60 80 100 120 140 160 180 200 220 240 260 280 300 320 340 360 380 400 420 440 460 480

482 C29 H46 O2 Si2 KO-7-215-3 Vinyldimethylsilyl derivative of cannabinol; W   125253

482 C29 H58 O3 Si 56784-06-6 EP-6967-0-0 Pentacosenoic acid, 2-[(trimethylsilyl)oxy]-, methyl ester;;
METHYL 2-TMSOXYPENTACOS-X-ENOATE; W/NBS   62351

482 C30 H26 O6 61833-55-4 KC-1976-2069-2 (2R*,5'R*)-5-ACETOXY-2',4,6-TRIMETHYL-3-METHYLENE-5'-(3-METHYL
-1,4-NAPHTHOQUINON-2-YLMETHYL)BENZOFURAN-2(3H)-SPIRO-1'-CYCLOPENT-2'-EN-4'-ONE; W   62352

20 40 60 80 100 120 140 160 180 200 220 240 260 280 300 320 340 360 380 400 420 440 460 480

482 C30 H30 N2 O4 80450-24-4 B-36-766-0 (6aα,7α,15α,16aα)-(+)-6,6,9,15,18-Pentamethyl-6,6a,7,9,15,16,16
a,18-octahydro-7,15-methano-8H,17H-quino[3'',4''':5',6']pyrano[3',4':5,6]oxocino[3,2-c]quinoline-8,17-dione;
W   125254

20 40 60 80 100 120 140 160 180 200 220 240 260 280 300 320 340 360 380 400 420 440 460 480

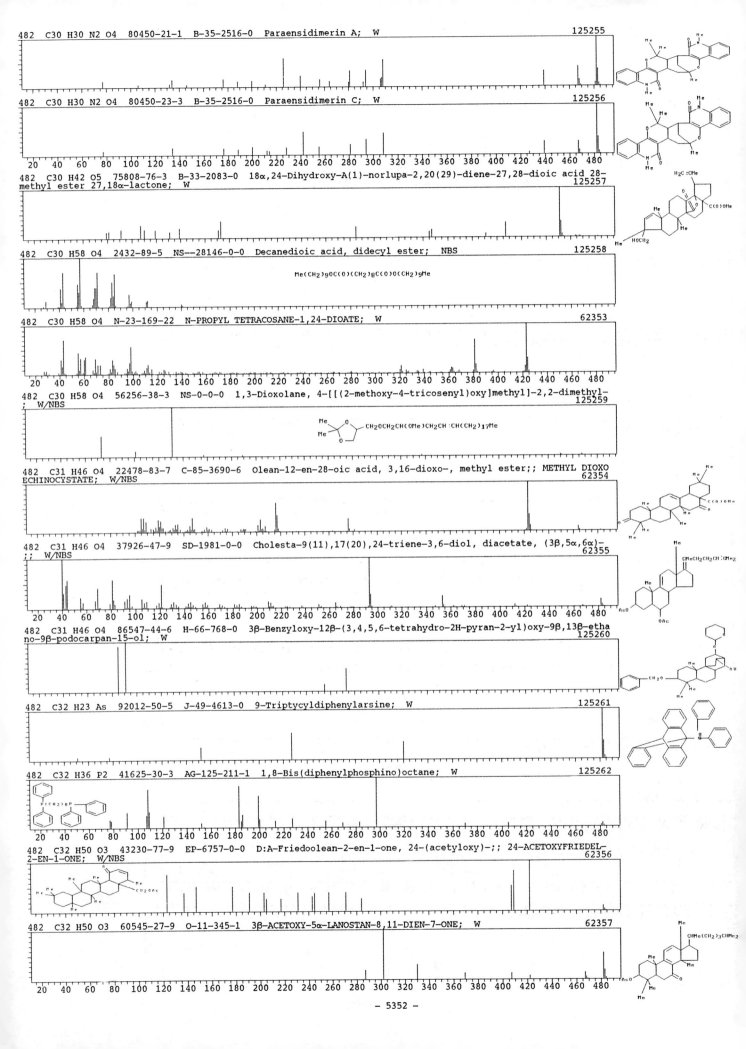

482 C30 H30 N2 O4 80450-21-1 B-35-2516-0 Paraensidimerin A; W   125255

482 C30 H30 N2 O4 80450-23-3 B-35-2516-0 Paraensidimerin C; W   125256

482 C30 H42 O5 75808-76-3 B-33-2083-0 18α,24-Dihydroxy-A(1)-norlupa-2,20(29)-diene-27,28-dioic acid 28-
methyl ester 27,18α-lactone; W   125257

482 C30 H58 O4 2432-89-5 NS--28146-0-0 Decanedioic acid, didecyl ester; NBS   125258

Me(CH2)9OC(O)(CH2)8C(O)O(CH2)9Me

482 C30 H58 O4 N-23-169-22 N-PROPYL TETRACOSANE-1,24-DIOATE; W   62353

482 C30 H58 O4 56256-38-3 NS-0-0-0 1,3-Dioxolane, 4-[[(2-methoxy-4-tricosenyl)oxy]methyl]-2,2-dimethyl-
; W/NBS   125259

Me O CH2OCH2CH(OMe)CH2CH:CH(CH2)17Me
Me O

482 C31 H46 O4 22478-83-7 C-85-3690-6 Olean-12-en-28-oic acid, 3,16-dioxo-, methyl ester;; METHYL DIOXO
ECHINOCYSTATE; W/NBS   62354

482 C31 H46 O4 37926-47-9 SD-1981-0-0 Cholesta-9(11),17(20),24-triene-3,6-diol, diacetate, (3β,5α,6α)-
;; W/NBS   62355

482 C31 H46 O4 86547-44-6 H-66-768-0 3β-Benzyloxy-12β-(3,4,5,6-tetrahydro-2H-pyran-2-yl)oxy-9β,13β-etha
no-9β-podocarpan-15-ol; W   125260

482 C32 H23 As 92012-50-5 J-49-4613-0 9-Triptycyldiphenylarsine; W   125261

482 C32 H36 P2 41625-30-3 AG-125-211-1 1,8-Bis(diphenylphosphino)octane; W   125262

482 C32 H50 O3 43230-77-9 EP-6757-0-0 D:A-Friedoolean-2-en-1-one, 24-(acetyloxy)-;; 24-ACETOXYFRIEDEL-
2-EN-1-ONE; W/NBS   62356

482 C32 H50 O3 60545-27-9 O-11-345-1 3β-ACETOXY-5α-LANOSTAN-8,11-DIEN-7-ONE; W   62357

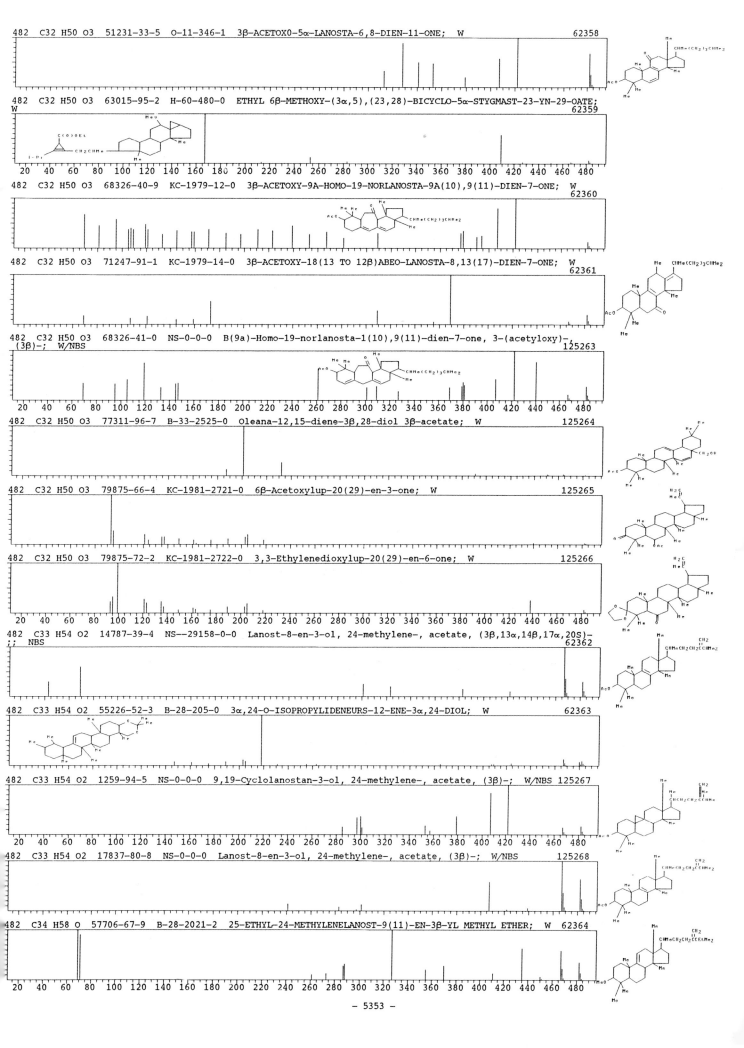

482 C32 H50 O3 51231-33-5 O-11-346-1 3β-ACETOXO-5α-LANOSTA-6,8-DIEN-11-ONE; W   62358

482 C32 H50 O3 63015-95-2 H-60-480-0 ETHYL 6β-METHOXY-(3α,5),(23,28)-BICYCLO-5α-STYGMAST-23-YN-29-OATE;
W   62359

20 40 60 80 100 120 140 160 180 200 220 240 260 280 300 320 340 360 380 400 420 440 460 480

482 C32 H50 O3 68326-40-9 KC-1979-12-0 3β-ACETOXY-9A-HOMO-19-NORLANOSTA-9A(10),9(11)-DIEN-7-ONE; W   62360

482 C32 H50 O3 71247-91-1 KC-1979-14-0 3β-ACETOXY-18(13 TO 12β)ABEO-LANOSTA-8,13(17)-DIEN-7-ONE; W   62361

482 C32 H50 O3 68326-41-0 NS-0-0-0 B(9a)-Homo-19-norlanosta-1(10),9(11)-dien-7-one, 3-(acetyloxy)-
(3β)-; W/NBS   125263

20 40 60 80 100 120 140 160 180 200 220 240 260 280 300 320 340 360 380 400 420 440 460 480

482 C32 H50 O3 77311-96-7 B-33-2525-0 Oleana-12,15-diene-3β,28-diol 3β-acetate; W   125264

482 C32 H50 O3 79875-66-4 KC-1981-2721-0 6β-Acetoxylup-20(29)-en-3-one; W   125265

482 C32 H50 O3 79875-72-2 KC-1981-2722-0 3,3-Ethylenedioxylup-20(29)-en-6-one; W   125266

20 40 60 80 100 120 140 160 180 200 220 240 260 280 300 320 340 360 380 400 420 440 460 480

482 C33 H54 O2 14787-39-4 NS--29158-0-0 Lanost-8-en-3-ol, 24-methylene-, acetate, (3β,13α,14β,17α,20S)-
;; NBS   62362

482 C33 H54 O2 55226-52-3 B-28-205-0 3α,24-O-ISOPROPYLIDENEURS-12-ENE-3α,24-DIOL; W   62363

482 C33 H54 O2 1259-94-5 NS-0-0-0 9,19-Cyclolanostan-3-ol, 24-methylene-, acetate, (3β)-; W/NBS 125267

20 40 60 80 100 120 140 160 180 200 220 240 260 280 300 320 340 360 380 400 420 440 460 480

482 C33 H54 O2 17837-80-8 NS-0-0-0 Lanost-8-en-3-ol, 24-methylene-, acetate, (3β)-; W/NBS   125268

482 C34 H58 O 57706-67-9 B-28-2021-2 25-ETHYL-24-METHYLENELANOST-9(11)-EN-3β-YL METHYL ETHER; W 62364

20 40 60 80 100 120 140 160 180 200 220 240 260 280 300 320 340 360 380 400 420 440 460 480

- 5353 -

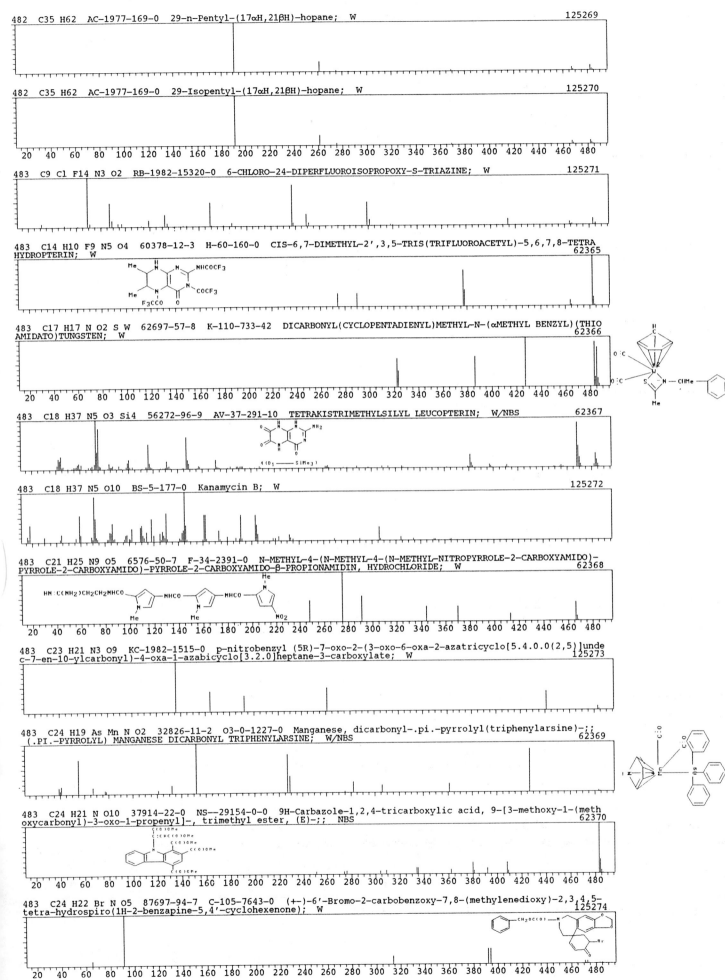

482   C35 H62   AC-1977-169-0   29-n-Pentyl-(17αH,21βH)-hopane;   W   125269

482   C35 H62   AC-1977-169-0   29-Isopentyl-(17αH,21βH)-hopane;   W   125270

483   C9 Cl F14 N3 O2   RB-1982-15320-0   6-CHLORO-24-DIPERFLUOROISOPROPOXY-S-TRIAZINE;   W   125271

483   C14 H10 F9 N5 O4   60378-12-3   H-60-160-0   CIS-6,7-DIMETHYL-2',3,5-TRIS(TRIFLUOROACETYL)-5,6,7,8-TETRA
HYDROPTERIN;   W   62365

483   C17 H17 N O2 S W   62697-57-8   K-110-733-42   DICARBONYL(CYCLOPENTADIENYL)METHYL-N-(αMETHYL BENZYL)(THIO
AMIDATO)TUNGSTEN;   W   62366

483   C18 H37 N5 O3 Si4   56272-96-9   AV-37-291-10   TETRAKISTRIMETHYLSILYL LEUCOPTERIN;   W/NBS   62367

483   C18 H37 N5 O10   BS-5-177-0   Kanamycin B;   W   125272

483   C21 H25 N9 O5   6576-50-7   F-34-2391-0   N-METHYL-4-(N-METHYL-4-(N-METHYL-NITROPYRROLE-2-CARBOXYAMIDO)-
PYRROLE-2-CARBOXYAMIDO)-PYRROLE-2-CARBOXYAMIDO-β-PROPIONAMIDIN, HYDROCHLORIDE;   W   62368

483   C23 H21 N3 O9   KC-1982-1515-0   p-nitrobenzyl (5R)-7-oxo-2-(3-oxo-6-oxa-2-azatricyclo[5.4.0.0(2,5)]unde
c-7-en-10-ylcarbonyl)-4-oxa-1-azabicyclo[3.2.0]heptane-3-carboxylate;   W   125273

483   C24 H19 As Mn N O2   32826-11-2   O3-0-1227-0   Manganese, dicarbonyl-.pi.-pyrrolyl(triphenylarsine)-;;
(.PI.-PYRROLYL) MANGANESE DICARBONYL TRIPHENYLARSINE;   W/NBS   62369

483   C24 H21 N O10   37914-22-0   NS--29154-0-0   9H-Carbazole-1,2,4-tricarboxylic acid, 9-[3-methoxy-1-(meth
oxycarbonyl)-3-oxo-1-propenyl]-, trimethyl ester, (E)-;;   NBS   62370

483   C24 H22 Br N O5   87697-94-7   C-105-7643-0   (+-)-6'-Bromo-2-carbobenzoxy-7,8-(methylenedioxy)-2,3,4,5-
tetra-hydrospiro(1H-2-benzapine-5,4'-cyclohexenone);   W   125274

483   C25 H29 N3 O7   KC-1982-1170-0   L-α-Hydroxyisovaleryl-L-phenylanyl-L-proline p-nitrophenyl ester; W
125275

483   C25 H41 N O8   CD-283-0-0   Pseudaconine; W
125276

483   C26 H29 N O8   86340-44-5   KC-1983-522-0   (Z,Z)-Dimethyl 6,7-bis(methoxycarbonyl)-5-methyl-5,9,10,11,12,12a-hexahydrocyclonona[b]indol-12a-ylmaleate; W
125277

483   C26 H29 N O8   86340-43-4   KC-1983-522-0   Tetramethyl 5-methyl-5,7,8,9,10,10a-hexahydrocyclohepta[b]indol-6,10a-diyldimaleate; W
125278

483   C26 H45 N O5 S   BS-5-95-0   Taurolithocholic acid; W
125279

483   C26 H45 N O5 S   BS-5-96-0   Tauroallolithocholic acid; W
125280

483   C26 H53 N O3 Si2   55429-51-1   AD-0-5898-0   Acetamide, N-[2-[(trimethylsilyl)oxy]-1-[[(trimethylsilyl)oxy]methyl]-3,7-heptadecadienyl]-;; N-ACETYL-DI-O-TMS-SPHINGA-4,8-DIENINE; W/NBS
62371

483   C26 H53 N O3 Si2   SW-0-226-0   BISTRIMETHYLSILYL N-ACETYL SPHINGA-4,14-DIENINE; W
62372

483   C27 H37 N3 O5   37580-37-3   O7-0-817-0   L-Alanine, N-[N-[N-[(2-hydroxy-1-naphthalenyl)methylene]-L-valyl]-L-isoleucyl]-, ethyl ester;; 2-HYDROXY-1-NAPHTHYLIDENE VALYLISOLEUCYLALANINE ETHYL ESTER; W/NBS   62373

483   C27 H41 N O3 Si2   74299-04-0   NS--29151-0-0   19-Norpregna-1,3,5,7,9-pentaen-21-al, 3,17-bis[(trimethylsilyl)oxy]-, O-methyloxime, (17α)-;; NBS
62374

483   C28 H30 Fe N O P   59568-04-6   K-109-1059-0   CARBONYL(CYCLOPENTADIENYL)METHYL(((S)-N-METHYL-1-PHENYLETHYLAMINO)DIPHENYLPHOSPHIN-P)IRON; W
62375

483   C28 H41 N O4 Si   69588-03-0   E-51-270-1   3-TRIMETHYLSILYLOXY-ETORPHINE-TMS; W
62376

483   C29 H25 N O6   22628-23-5   N-30-369-2   RHIZOCARPIC ACID METHYL ETHER; W/NBS
62377

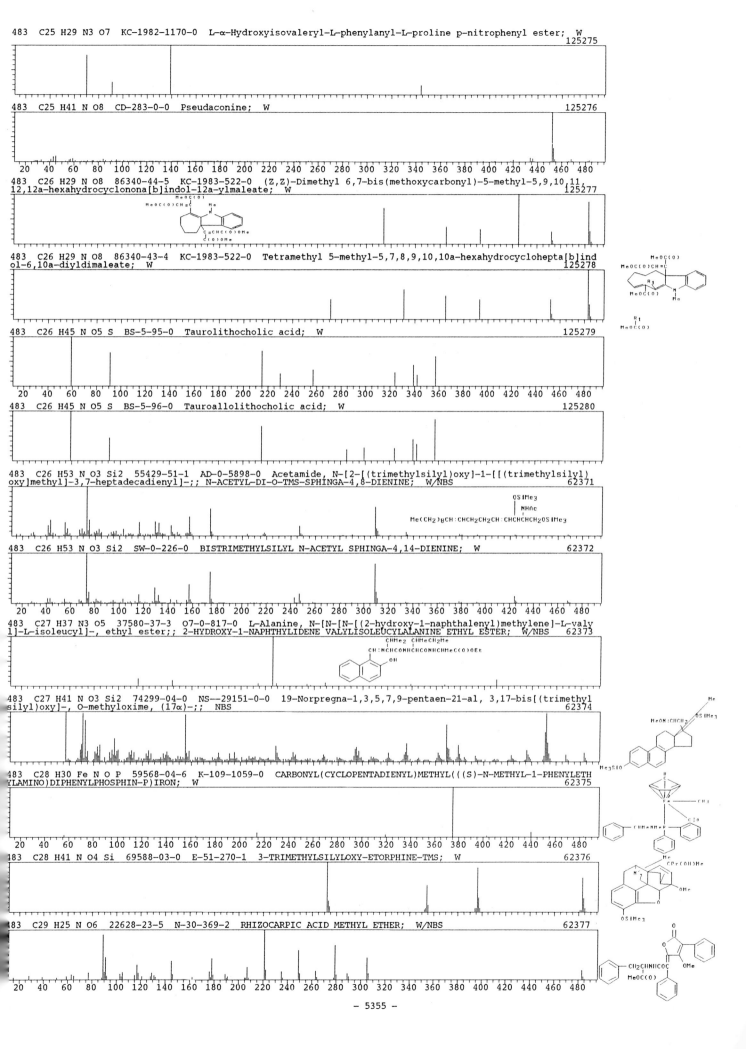

483 C29 H25 N O6 76447-16-0 J-46-1370-0 Benzyl 3,5-bis(benzyloxy)-4-methyl-2-nitrobenzoate; W 125282

483 C30 H45 N O4 58545-19-0 B-28-2666-0 Methyl 3-nitrilo-4-hydroxy-11-oxo-23-nor-3,4-seco-18β-olean-12-
en-30-oate; W 62378

483 C30 H45 N O4 HE-1986-1224-0 TERT-BUTYL ESTER OF 1,5,6-TRI-TERT-BUTYL-2-METHOXYCARBONYL-3-PHENYL-2-AZ
ABICYCLO[2.2.0]HEX-5-EN-4-CARBOXYLIC ACID; W 125283

483 C30 H49 N3 O2 89647-79-0 I-62-300-0 Spirolucidine; W 125284

483 C31 H33 N O4 56847-11-1 EP-5754-0-0 Benzenemethanol, α-[[[[4-methoxy-3-(phenylmethoxy)phenyl]meth
yl]methylamino]methyl]-4-(phenylmethoxy)-;; P-BENZOXY-α-(((3-BENZYLOXY-4-METHOXYBENZYL)METHYLAMINO)-METHYL)
BENZYLALCOHOL; W/NBS 62379

Str 14

483 C31 H49 N O3 F-32-1291-0 3β-ACETOXY-CHOLEST-6-ENO-(7,6-D)2'-METHOXYOXAZOL; W 62380

483 C31 H49 N O3 72204-13-8 B-32-813-0 (22S,25R)-22,26-ACETYLEPIMINOCHOLEST-5-EN-3β-YL ACETATE; W 62381

483 C31 H49 N O3 80925-45-7 J-47-1723-0 (25R,S)-3β-Acetoxy-N,N-dimethylcholesta-5,22-dien-26-amide; W 125285

483 C32 H21 N O2 S 87612-95-1 Y-20-972-0 1,3,4,7-Tetraphenylthieno[3,4-c]pyridine-6-carboxylic acid; W 125286

483 C35 H21 N3 85731-55-1 Y-19-1488-0 8,10,11-Triphenylacenaphtho[1,2-b]pyrido[4,3-e]pyridine; W 125287

484 C2 H4 F10 Hg S2 85362-82-9 K-116-655-0 Bis(pentafluorosulfanylmethyl)mercury; W 125288

484 C6 H Br5 O 608-71-9 RB-1982-14919-0 Phenol, pentabromo-;; Flammex 5BP; W 125289

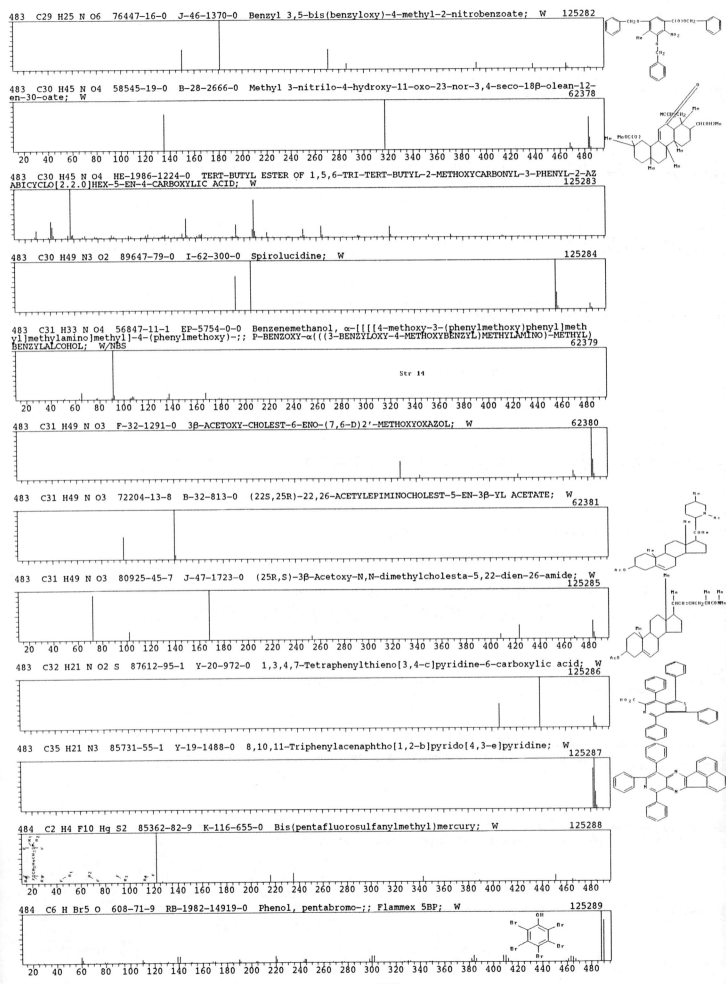

484   C8 H5 Cl Hg 96Mo O3   AD-0-2870-0   .PI.-CYCLOPENTADIENYL-TRICARBONYL-MOLYBDENUM MERCURY-MONOCHLORIDE;
W   62383

484   C8 H5 Cl Hg Mo O3   12079-83-3   NS--29109-0-0   Molybdenum, tricarbonyl(chloromercury)(η5-2,4-cyclopenta
dien-1-yl)-, (Hg-Mo);;   NBS   62384

20   40   60   80   100   120   140   160   180   200   220   240   260   280   300   320   340   360   380   400   420   440   460   480

484   C9 Fe3 O9 S2   TJ-29-1533-0   TRIS-IRON-(NONA-CARBONYL)DI-SULPHUR MONOISOTOPIC;   W   62385

484   C9 Fe3 O9 S2   22309-04-2   AG-124-39-1   (Nonacarbonyl)iron sulfide;   W   125290

                                                                            Str 15

484   C9 H22 N4 O5 P Re   K-113-646-0   Tricarbonyl(1,2-ethanediamine-N,N')1,2-ethanediamine-N-rhenium(I)-di
methylphosphinate;   W   125291

20   40   60   80   100   120   140   160   180   200   220   240   260   280   300   320   340   360   380   400   420   440   460   480

484   C11 H27 Fe O11 P3   HE-1982-0-0   IRON, DICARBONYL-TRIS(TRIMETHYLPHOSPHITE)-;   W   62386

484   C12 H13 Cl2 F2 N2 O8 P S   RB-1982-15531-0   O,O-DIETHYL O-(2,6-DINITRO-4-(2,2-DICHLORO-1,1-DIFLUOROETHO
XY)PHENYL)PHOSPHOR-;   W   125292

484   C12 H41 F N6 Si7   64639-38-9   AG-135-171-0   Fluorosilyl-bis(hexamethylcyclotrisilazane);   W   125293

20   40   60   80   100   120   140   160   180   200   220   240   260   280   300   320   340   360   380   400   420   440   460   480

484   C14 H6 F10 Sn   801-79-6   NS--29147-0-0   Stannane, dimethylbis(pentafluorophenyl)-;;   NBS   62387

484   C14 H15 Br3 O4   23103-18-6   EP-6055-0-0   Altropyranoside, methyl 2,3,6-tribromo-2,3,6-trideoxy-, benzo
ate, α-D-;; METHYL-4-O-BENZOYL-2,3,6-TRIBROMO-2-3-6-TRIDEOXY α D ALTROPYRANDOXID;   W/NBS   62388

484   C15 F16   25078-75-5   O7-0-11-0   Benzene, 1,1'-(1,1,2,2,3,3-hexafluoro-1,3-propanediyl)bis[2,3,4,5,6-pe
ntafluoro-;; PERFLUORO-1,3-DIPHENYLPROPANE;   W/NBS   62389

20   40   60   80   100   120   140   160   180   200   220   240   260   280   300   320   340   360   380   400   420   440   460   480

484   C18 H22 Co2 O8   83479-82-7   F-38-1507-0   Hexacarbonyl-.mu.-(η-dineopentyloxyethyne)-dicobalt;   W
                                                                                                          125294

484   C18 H36 N4 O11   BS-5-176-0   Kanamycin A;   W   125295

20   40   60   80   100   120   140   160   180   200   220   240   260   280   300   320   340   360   380   400   420   440   460   480

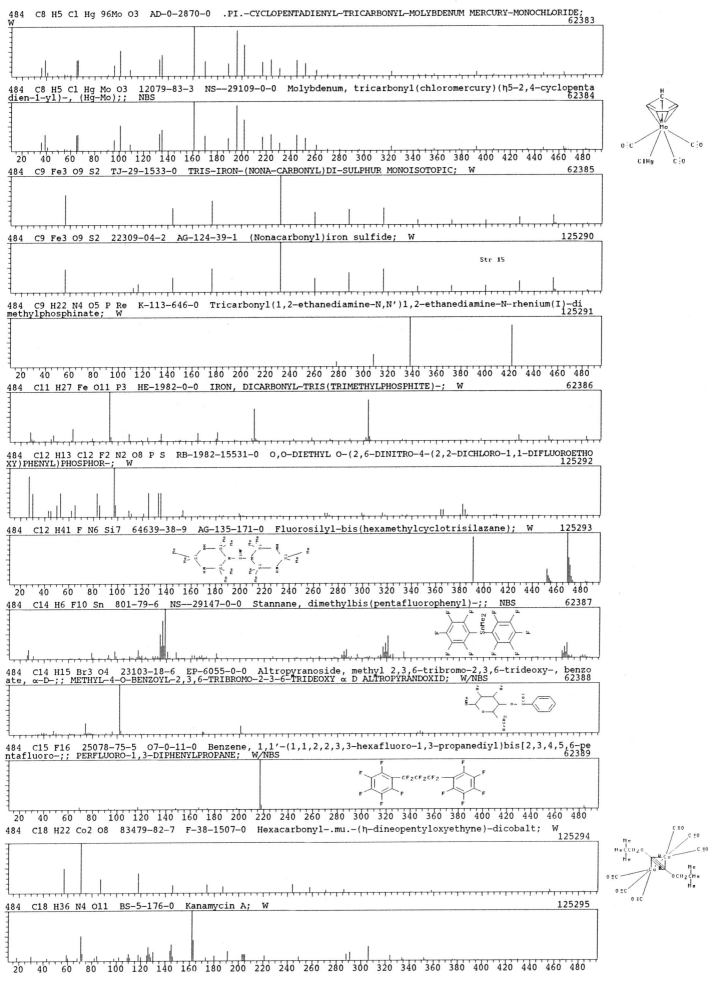

484   C19 H16 O2 Pb   RB-1982-15772-0   LEAD(FORMYLOXY)TRIPHENYL;   W   125296

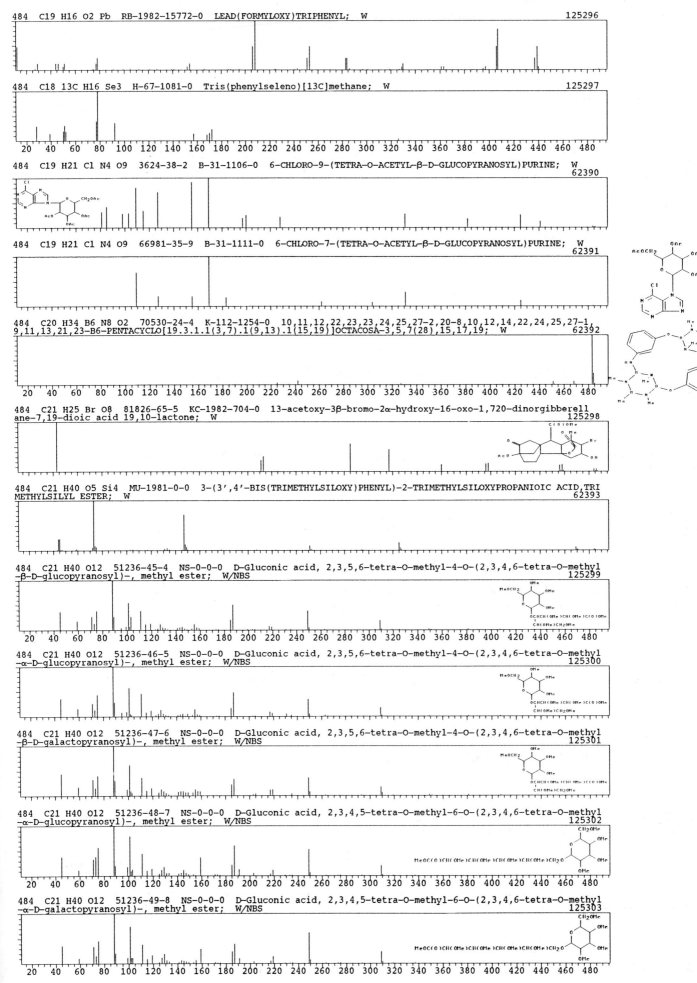

484   C18 13C H16 Se3   H-67-1081-0   Tris(phenylseleno)[13C]methane;   W   125297

484   C19 H21 Cl N4 O9   3624-38-2   B-31-1106-0   6-CHLORO-9-(TETRA-O-ACETYL-β-D-GLUCOPYRANOSYL)PURINE;   W
                                                                                            62390

484   C19 H21 Cl N4 O9   66981-35-9   B-31-1111-0   6-CHLORO-7-(TETRA-O-ACETYL-β-D-GLUCOPYRANOSYL)PURINE;   W
                                                                                            62391

484   C20 H34 B6 N8 O2   70530-24-4   K-112-1254-0   10,11,12,22,23,23,24,25,27-2,20-8,10,12,14,22,24,25,27-1,
9,11,13,21,23-B6-PENTACYCLO[19.3.1.1(3,7).1(9,13).1(15,19)]OCTACOSA-3,5,7(28),15,17,19;   W   62392

484   C21 H25 Br O8   81826-65-5   KC-1982-704-0   13-acetoxy-3β-bromo-2α-hydroxy-16-oxo-1,720-dinorgibberell
ane-7,19-dioic acid 19,10-lactone;   W   125298

484   C21 H40 O5 Si4   MU-1981-0-0   3-(3',4'-BIS(TRIMETHYLSILOXY)PHENYL)-2-TRIMETHYLSILOXYPROPANIOIC ACID,TRI
METHYLSILYL ESTER;   W   62393

484   C21 H40 O12   51236-45-4   NS-0-0-0   D-Gluconic acid, 2,3,5,6-tetra-O-methyl-4-O-(2,3,4,6-tetra-O-methyl
-β-D-glucopyranosyl)-, methyl ester;   W/NBS   125299

484   C21 H40 O12   51236-46-5   NS-0-0-0   D-Gluconic acid, 2,3,5,6-tetra-O-methyl-4-O-(2,3,4,6-tetra-O-methyl
-α-D-glucopyranosyl)-, methyl ester;   W/NBS   125300

484   C21 H40 O12   51236-47-6   NS-0-0-0   D-Gluconic acid, 2,3,5,6-tetra-O-methyl-4-O-(2,3,4,6-tetra-O-methyl
-β-D-galactopyranosyl)-, methyl ester;   W/NBS   125301

484   C21 H40 O12   51236-48-7   NS-0-0-0   D-Gluconic acid, 2,3,4,5-tetra-O-methyl-6-O-(2,3,4,6-tetra-O-methyl
-α-D-glucopyranosyl)-, methyl ester;   W/NBS   125302

484   C21 H40 O12   51236-49-8   NS-0-0-0   D-Gluconic acid, 2,3,4,5-tetra-O-methyl-6-O-(2,3,4,6-tetra-O-methyl
-α-D-galactopyranosyl)-, methyl ester;   W/NBS   125303

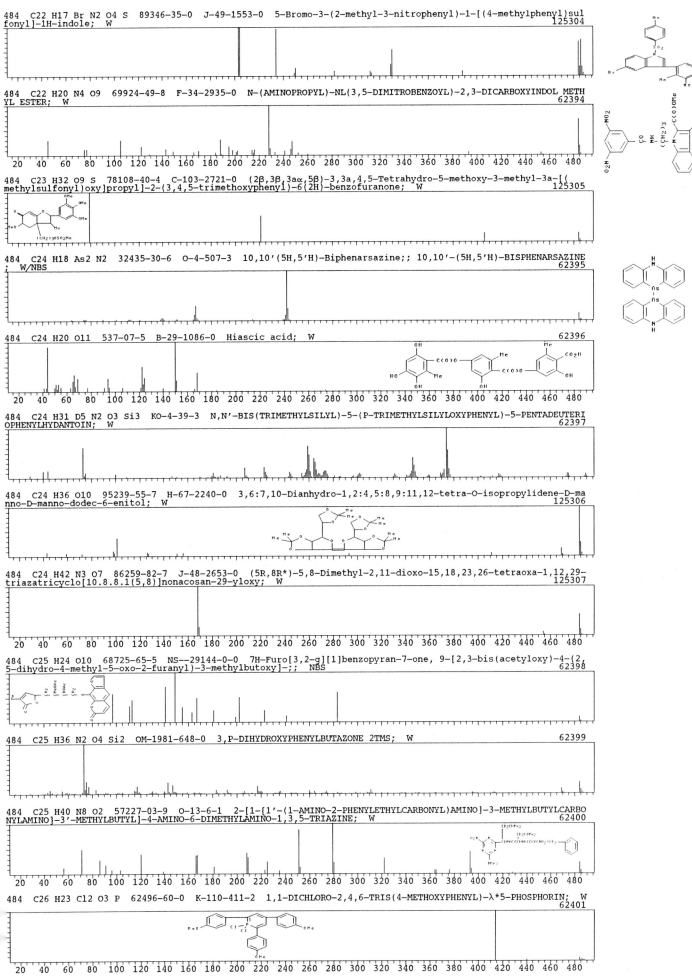

484 C22 H17 Br N2 O4 S 89346-35-0 J-49-1553-0 5-Bromo-3-(2-methyl-3-nitrophenyl)-1-[(4-methylphenyl)sul
fonyl]-1H-indole; W
125304

484 C22 H20 N4 O9 69924-49-8 F-34-2935-0 N-(AMINOPROPYL)-NL(3,5-DIMITROBENZOYL)-2,3-DICARBOXYINDOL METH
YL ESTER; W
62394

20 40 60 80 100 120 140 160 180 200 220 240 260 280 300 320 340 360 380 400 420 440 460 480

484 C23 H32 O9 S 78108-40-4 C-103-2721-0 (2β,3β,3aα,5β)-3,3a,4,5-Tetrahydro-5-methoxy-3-methyl-3a-[(
methylsulfonyl)oxy]propyl]-2-(3,4,5-trimethoxyphenyl)-6(2H)-benzofuranone; W
125305

484 C24 H18 As2 N2 32435-30-6 O-4-507-3 10,10'(5H,5'H)-Biphenarsazine;; 10,10'-(5H,5'H)-BISPHENARSAZINE
; W/NBS
62395

484 C24 H20 O11 537-07-5 B-29-1086-0 Hiascic acid; W
62396

20 40 60 80 100 120 140 160 180 200 220 240 260 280 300 320 340 360 380 400 420 440 460 480

484 C24 H31 D5 N2 O3 Si3 KO-4-39-3 N,N'-BIS(TRIMETHYLSILYL)-5-(P-TRIMETHYLSILYLOXYPHENYL)-5-PENTADEUTERI
OPHENYLHYDANTOIN; W
62397

484 C24 H36 O10 95239-55-7 H-67-2240-0 3,6:7,10-Dianhydro-1,2:4,5:8,9:11,12-tetra-O-isopropylidene-D-ma
nno-D-manno-dodec-6-enitol; W
125306

484 C24 H42 N3 O7 86259-82-7 J-48-2653-0 (5R,8R*)-5,8-Dimethyl-2,11-dioxo-15,18,23,26-tetraoxa-1,12,29-
triazatricyclo[10.8.8.1(5,8)]nonacosan-29-yloxy; W
125307

20 40 60 80 100 120 140 160 180 200 220 240 260 280 300 320 340 360 380 400 420 440 460 480

484 C25 H24 O10 68725-65-5 NS--29144-0-0 7H-Furo[3,2-g][1]benzopyran-7-one, 9-[2,3-bis(acetyloxy)-4-(2,
5-dihydro-4-methyl-5-oxo-2-furanyl)-3-methylbutoxy]-;; NBS
62398

484 C25 H36 N2 O4 Si2 OM-1981-648-0 3,P-DIHYDROXYPHENYLBUTAZONE 2TMS; W
62399

484 C25 H40 N8 O2 57227-03-9 O-13-6-1 2-[1-{1'-(1-AMINO-2-PHENYLETHYLCARBONYL)AMINO]-3-METHYLBUTYLCARBO
NYLAMINO]-3'-METHYLBUTYL]-4-AMINO-6-DIMETHYLAMINO-1,3,5-TRIAZINE; W
62400

20 40 60 80 100 120 140 160 180 200 220 240 260 280 300 320 340 360 380 400 420 440 460 480

484 C26 H23 Cl2 O3 P 62496-60-0 K-110-411-2 1,1-DICHLORO-2,4,6-TRIS(4-METHOXYPHENYL)-λ*5-PHOSPHORIN; W
62401

20 40 60 80 100 120 140 160 180 200 220 240 260 280 300 320 340 360 380 400 420 440 460 480

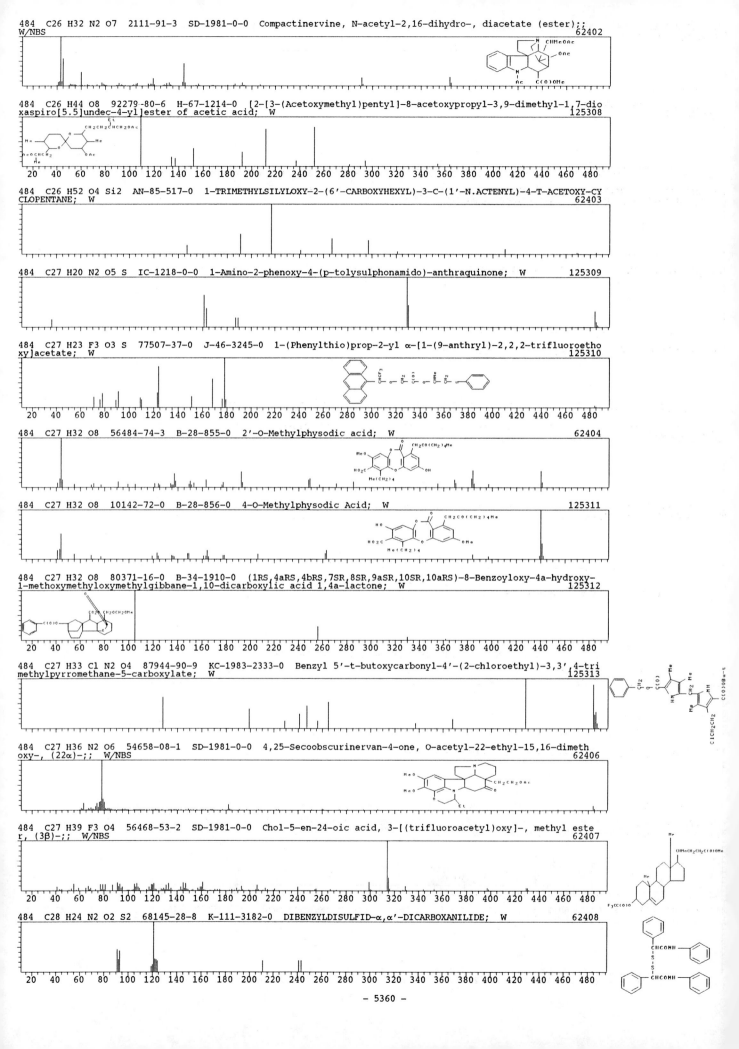

484 C26 H32 N2 O7   2111-91-3   SD-1981-0-0   Compactinervine, N-acetyl-2,16-dihydro-, diacetate (ester);;
W/NBS   62402

484 C26 H44 O8   92279-80-6   H-67-1214-0   [2-[3-(Acetoxymethyl)pentyl]-8-acetoxypropyl-3,9-dimethyl-1,7-dio
xaspiro[5.5]undec-4-yl]ester of acetic acid; W   125308

484 C26 H52 O4 Si2   AN-85-517-0   1-TRIMETHYLSILYLOXY-2-(6'-CARBOXYHEXYL)-3-C-(1'-N.ACTENYL)-4-T-ACETOXY-CY
CLOPENTANE; W   62403

484 C27 H20 N2 O5 S   IC-1218-0-0   1-Amino-2-phenoxy-4-(p-tolysulphonamido)-anthraquinone; W   125309

484 C27 H23 F3 O3 S   77507-37-0   J-46-3245-0   1-(Phenylthio)prop-2-yl α-[1-(9-anthryl)-2,2,2-trifluoroetho
xy]acetate; W   125310

484 C27 H32 O8   56484-74-3   B-28-855-0   2'-O-Methylphysodic acid; W   62404

484 C27 H32 O8   10142-72-0   B-28-856-0   4-O-Methylphysodic Acid; W   125311

484 C27 H32 O8   80371-16-0   B-34-1910-0   (1RS,4aRS,4bRS,7SR,8SR,9aSR,10SR,10aRS)-8-Benzoyloxy-4a-hydroxy-
1-methoxymethyloxymethylgibbane-1,10-dicarboxylic acid 1,4a-lactone; W   125312

484 C27 H33 Cl N2 O4   87944-90-9   KC-1983-2333-0   Benzyl 5'-t-butoxycarbonyl-4'-(2-chloroethyl)-3,3',4-tri
methylpyrromethane-5-carboxylate; W   125313

484 C27 H36 N2 O6   54658-08-1   SD-1981-0-0   4,25-Secoobscurinervan-4-one, O-acetyl-22-ethyl-15,16-dimeth
oxy-, (22α)-;; W/NBS   62406

484 C27 H39 F3 O4   56468-53-2   SD-1981-0-0   Chol-5-en-24-oic acid, 3-[(trifluoroacetyl)oxy]-, methyl este
r, (3β)-;; W/NBS   62407

484 C28 H24 N2 O2 S2   68145-28-8   K-111-3182-0   DIBENZYLDISULFID-α,α'-DICARBOXANILIDE; W   62408

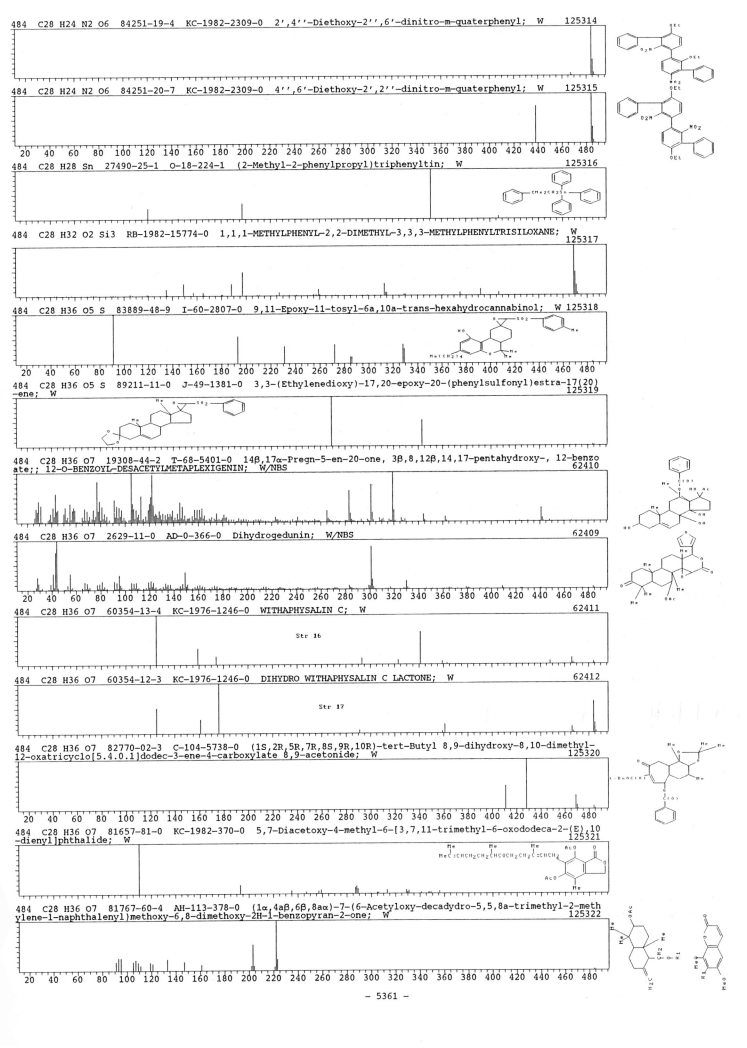

484   C28 H24 N2 O6   84251-19-4   KC-1982-2309-0   2',4''-Diethoxy-2'',6'-dinitro-m-quaterphenyl;   W   125314

484   C28 H24 N2 O6   84251-20-7   KC-1982-2309-0   4'',6'-Diethoxy-2',2''-dinitro-m-quaterphenyl;   W   125315

484   C28 H28 Sn   27490-25-1   O-18-224-1   (2-Methyl-2-phenylpropyl)triphenyltin;   W   125316

484   C28 H32 O2 Si3   RB-1982-15774-0   1,1,1-METHYLPHENYL-2,2-DIMETHYL-3,3,3-METHYLPHENYLTRISILOXANE;   W   125317

484   C28 H36 O5 S   83889-48-9   I-60-2807-0   9,11-Epoxy-11-tosyl-6a,10a-trans-hexahydrocannabinol;   W   125318

484   C28 H36 O5 S   89211-11-0   J-49-1381-0   3,3-(Ethylenedioxy)-17,20-epoxy-20-(phenylsulfonyl)estra-17(20)
-ene;   W   125319

484   C28 H36 O7   19308-44-2   T-68-5401-0   14β,17α-Pregn-5-en-20-one, 3β,8,12β,14,17-pentahydroxy-, 12-benzo
ate;;   12-O-BENZOYL-DESACETYLMETAPLEXIGENIN;   W/NBS   62410

484   C28 H36 O7   2629-11-0   AD-0-366-0   Dihydrogedunin;   W/NBS   62409

484   C28 H36 O7   60354-13-4   KC-1976-1246-0   WITHAPHYSALIN C;   W   62411

Str 16

484   C28 H36 O7   60354-12-3   KC-1976-1246-0   DIHYDRO WITHAPHYSALIN C LACTONE;   W   62412

Str 17

484   C28 H36 O7   82770-02-3   C-104-5738-0   (1S,2R,5R,7R,8S,9R,10R)-tert-Butyl 8,9-dihydroxy-8,10-dimethyl-
12-oxatricyclo[5.4.0.1]dodec-3-ene-4-carboxylate 8,9-acetonide;   W   125320

484   C28 H36 O7   81657-81-0   KC-1982-370-0   5,7-Diacetoxy-4-methyl-6-[3,7,11-trimethyl-6-oxododeca-2-(E),10
-dienyl]phthalide;   W   125321

484   C28 H36 O7   81767-60-4   AH-113-378-0   (1α,4aβ,6β,8aα)-7-(6-Acetyloxy-decadydro-5,5,8a-trimethyl-2-meth
ylene-1-naphthalenyl)methoxy-6,8-dimethoxy-2H-1-benzopyran-2-one;   W   125322

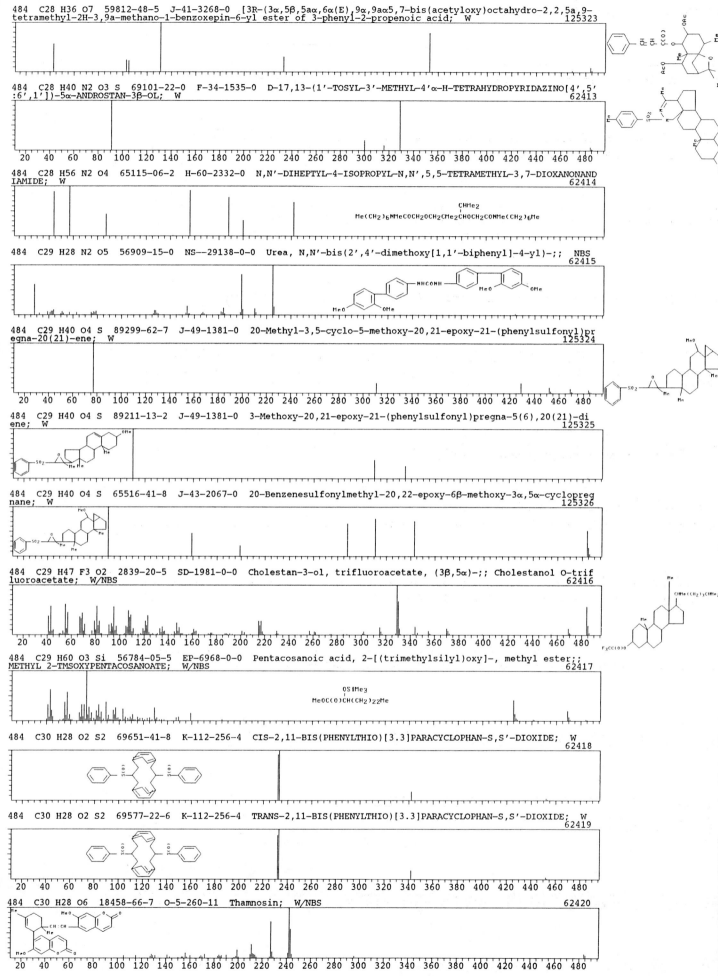

484   C28 H36 O7   59812-48-5   J-41-3268-0   [3R-(3α,5β,5aα,6α(E),9α,9aα5,7-bis(acetyloxy)octahydro-2,2,5a,9-tetramethyl-2H-3,9a-methano-1-benzoxepin-6-yl ester of 3-phenyl-2-propenoic acid;   W   125323

484   C28 H40 N2 O3 S   69101-22-0   F-34-1535-0   D-17,13-(1'-TOSYL-3'-METHYL-4'α-H-TETRAHYDROPYRIDAZINO[4',5':6',1'])-5α-ANDROSTAN-3β-OL;   W   62413

484   C28 H56 N2 O4   65115-06-2   H-60-2332-0   N,N'-DIHEPTYL-4-ISOPROPYL-N,N',5,5-TETRAMETHYL-3,7-DIOXANONANDIAMIDE;   W   62414

Me(CH₂)₆NMeCOCH₂OCH₂CMe₂CHOCH₂CONMe(CH₂)₆Me
CHMe₂

484   C29 H28 N2 O5   56909-15-0   NS--29138-0-0   Urea, N,N'-bis(2',4'-dimethoxy[1,1'-biphenyl]-4-yl)-;;   NBS   62415

484   C29 H40 O4 S   89299-62-7   J-49-1381-0   20-Methyl-3,5-cyclo-5-methoxy-20,21-epoxy-21-(phenylsulfonyl)pregna-20(21)-ene;   W   125324

484   C29 H40 O4 S   89211-13-2   J-49-1381-0   3-Methoxy-20,21-epoxy-21-(phenylsulfonyl)pregna-5(6),20(21)-diene;   W   125325

484   C29 H40 O4 S   65516-41-8   J-43-2067-0   20-Benzenesulfonylmethyl-20,22-epoxy-6β-methoxy-3α,5α-cyclopregnane;   W   125326

484   C29 H47 F3 O2   2839-20-5   SD-1981-0-0   Cholestan-3-ol, trifluoroacetate, (3β,5α)-;; Cholestanol O-trifluoroacetate;   W/NBS   62416

484   C29 H60 O3 Si   56784-05-5   EP-6968-0-0   Pentacosanoic acid, 2-[(trimethylsilyl)oxy]-, methyl ester;;
METHYL 2-TMSOXYPENTACOSANOATE;   W/NBS   62417

OSiMe₃
MeOC(O)CH(CH₂)₂₂Me

484   C30 H28 O2 S2   69651-41-8   K-112-256-4   CIS-2,11-BIS(PHENYLTHIO)[3.3]PARACYCLOPHAN-S,S'-DIOXIDE;   W   62418

484   C30 H28 O2 S2   69577-22-6   K-112-256-4   TRANS-2,11-BIS(PHENYLTHIO)[3.3]PARACYCLOPHAN-S,S'-DIOXIDE;   W   62419

484   C30 H28 O6   18458-66-7   O-5-260-11   Thamnosin;   W/NBS   62420

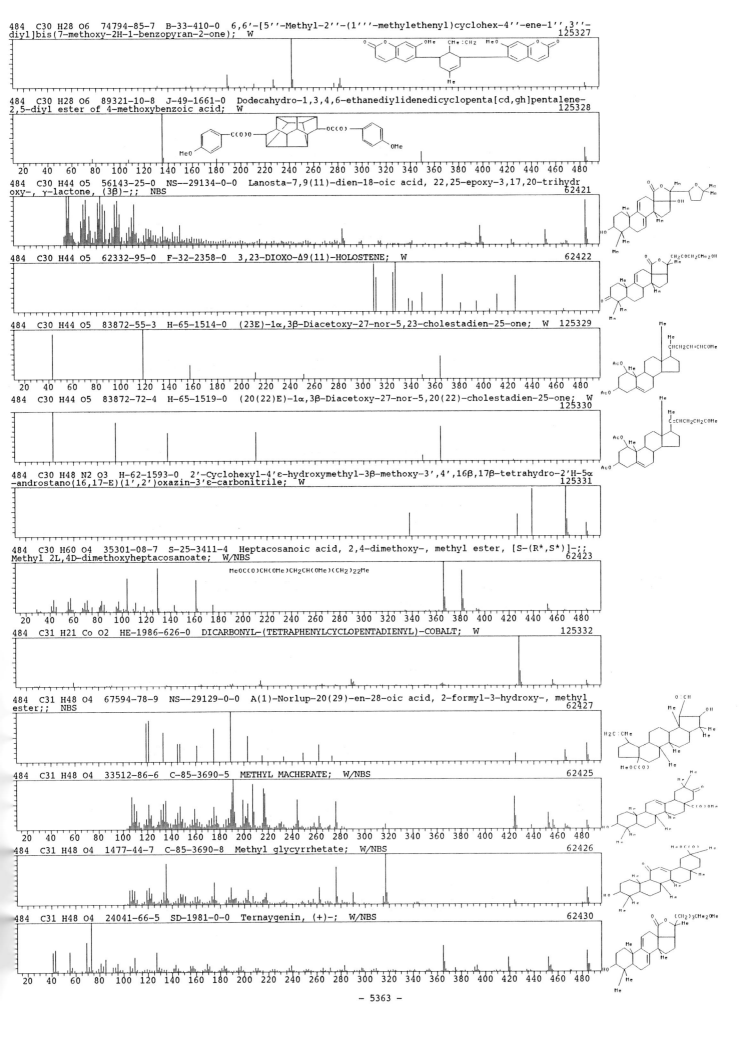

484 C30 H28 O6 74794-85-7 B-33-410-0 6,6'-[5''-Methyl-2''-(1'''-methylethenyl)cyclohex-4''-ene-1'',3''-diyl]bis(7-methoxy-2H-1-benzopyran-2-one); W 125327

484 C30 H28 O6 89321-10-8 J-49-1661-0 Dodecahydro-1,3,4,6-ethanediylidenedicyclopenta[cd,gh]pentalene-2,5-diyl ester of 4-methoxybenzoic acid; W 125328

484 C30 H44 O5 56143-25-0 NS--29134-0-0 Lanosta-7,9(11)-dien-18-oic acid, 22,25-epoxy-3,17,20-trihydroxy-, γ-lactone, (3β)-;; NBS 62421

484 C30 H44 O5 62332-95-0 F-32-2358-0 3,23-DIOXO-Δ9(11)-HOLOSTENE; W 62422

484 C30 H44 O5 83872-55-3 H-65-1514-0 (23E)-1α,3β-Diacetoxy-27-nor-5,23-cholestadien-25-one; W 125329

484 C30 H44 O5 83872-72-4 H-65-1519-0 (20(22)E)-1α,3β-Diacetoxy-27-nor-5,20(22)-cholestadien-25-one; W 125330

484 C30 H48 N2 O3 H-62-1593-0 2'-Cyclohexyl-4'ε-hydroxymethyl-3β-methoxy-3',4',16β,17β-tetrahydro-2'H-5α-androstano(16,17-E)(1',2')oxazin-3'ε-carbonitrile; W 125331

484 C30 H60 O4 35301-08-7 S-25-3411-4 Heptacosanoic acid, 2,4-dimethoxy-, methyl ester, [S-(R*,S*)]-;; Methyl 2L,4D-dimethoxyheptacosanoate; W/NBS 62423

MeOC(O)CH(OMe)CH2CH(OMe)(CH2)22Me

484 C31 H21 Co O2 HE-1986-626-0 DICARBONYL-(TETRAPHENYLCYCLOPENTADIENYL)-COBALT; W 125332

484 C31 H48 O4 67594-78-9 NS--29129-0-0 A(1)-Norlup-20(29)-en-28-oic acid, 2-formyl-3-hydroxy-, methyl ester;; NBS 62427

484 C31 H48 O4 33512-86-6 C-85-3690-5 METHYL MACHERATE; W/NBS 62425

484 C31 H48 O4 1477-44-7 C-85-3690-8 Methyl glycyrrhetate; W/NBS 62426

484 C31 H48 O4 24041-66-5 SD-1981-0-0 Ternaygenin, (+)-; W/NBS 62430

- 5363 -

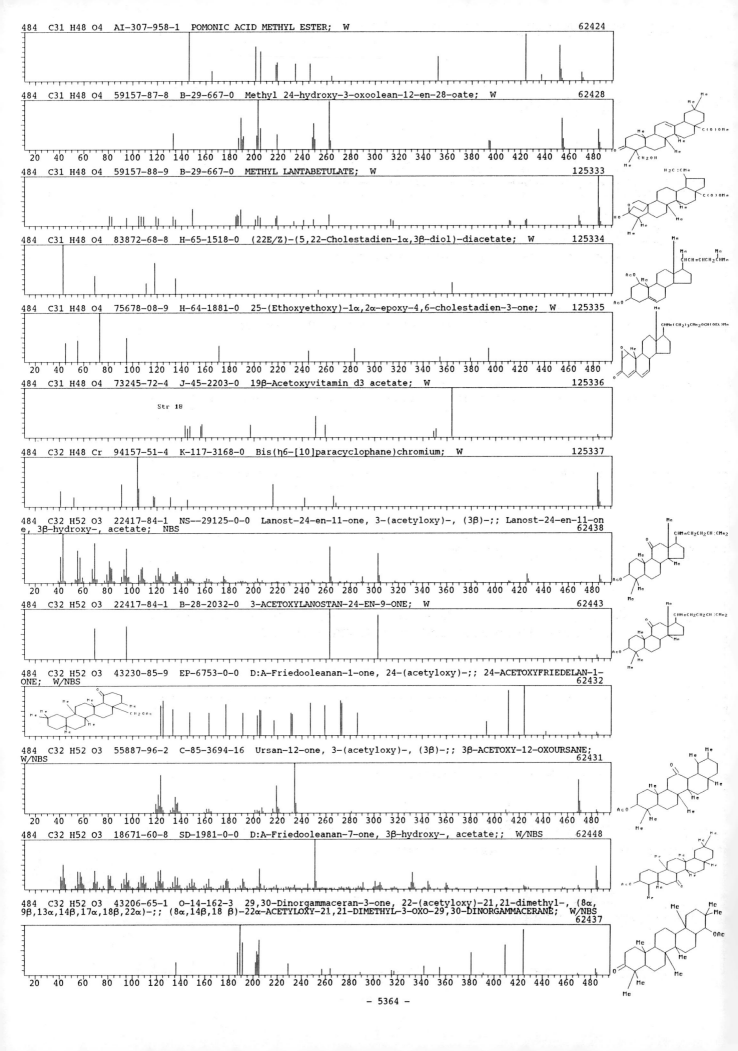

484 C31 H48 O4 AI-307-958-1 POMONIC ACID METHYL ESTER; W   62424

484 C31 H48 O4 59157-87-8 B-29-667-0 Methyl 24-hydroxy-3-oxoolean-12-en-28-oate; W   62428

484 C31 H48 O4 59157-88-9 B-29-667-0 METHYL LANTABETULATE; W   125333

484 C31 H48 O4 83872-68-8 H-65-1518-0 (22E/Z)-(5,22-Cholestadien-1α,3β-diol)-diacetate; W   125334

484 C31 H48 O4 75678-08-9 H-64-1881-0 25-(Ethoxyethoxy)-1α,2α-epoxy-4,6-cholestadien-3-one; W 125335

484 C31 H48 O4 73245-72-4 J-45-2203-0 19β-Acetoxyvitamin d3 acetate; W   125336

Str 1B

484 C32 H48 Cr 94157-51-4 K-117-3168-0 Bis(η6-[10]paracyclophane)chromium; W   125337

484 C32 H52 O3 22417-84-1 NS--29125-0-0 Lanost-24-en-11-one, 3-(acetyloxy)-, (3β)-;; Lanost-24-en-11-on
e, 3β-hydroxy-, acetate; NBS   62438

484 C32 H52 O3 22417-84-1 B-28-2032-0 3-ACETOXYLANOSTAN-24-EN-9-ONE; W   62443

484 C32 H52 O3 43230-85-9 EP-6753-0-0 D:A-Friedooleanan-1-one, 24-(acetyloxy)-;; 24-ACETOXYFRIEDELAN-1-
ONE; W/NBS   62432

484 C32 H52 O3 55887-96-2 C-85-3694-16 Ursan-12-one, 3-(acetyloxy)-, (3β)-;; 3β-ACETOXY-12-OXOURSANE;
W/NBS   62431

484 C32 H52 O3 18671-60-8 SD-1981-0-0 D:A-Friedooleanan-7-one, 3β-hydroxy-, acetate;; W/NBS   62448

484 C32 H52 O3 43206-65-1 O-14-162-3 29,30-Dinorgammaceran-3-one, 22-(acetyloxy)-21,21-dimethyl-, (8α,
9β,13α,14β,17α,18β,22α)-;; (8α,14β,18 β)-22α-ACETYLOXY-21,21-DIMETHYL-3-OXO-29,30-DINORGAMMACERANE; W/NBS
62437

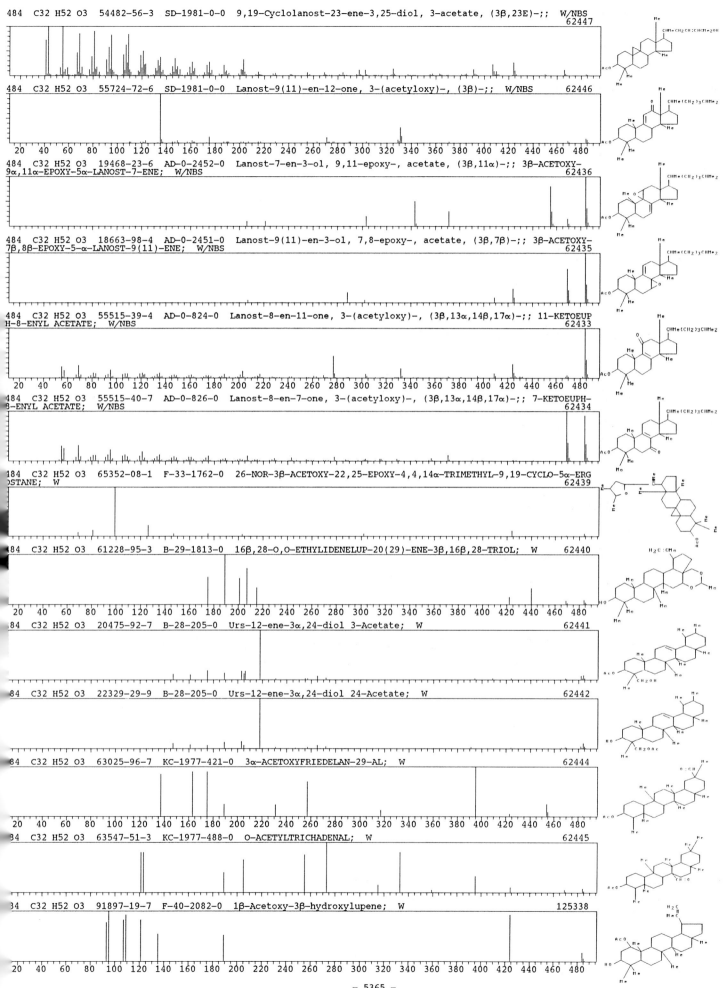

484   C32 H52 O3   54482-56-3   SD-1981-0-0   9,19-Cyclolanost-23-ene-3,25-diol, 3-acetate, (3β,23E)-;;   W/NBS
62447

484   C32 H52 O3   55724-72-6   SD-1981-0-0   Lanost-9(11)-en-12-one, 3-(acetyloxy)-, (3β)-;;   W/NBS   62446

484   C32 H52 O3   19468-23-6   AD-0-2452-0   Lanost-7-en-3-ol, 9,11-epoxy-, acetate, (3β,11α)-;; 3β-ACETOXY-
9α,11α-EPOXY-5α-LANOST-7-ENE;   W/NBS   62436

484   C32 H52 O3   18663-98-4   AD-0-2451-0   Lanost-9(11)-en-3-ol, 7,8-epoxy-, acetate, (3β,7β)-;; 3β-ACETOXY-
7β,8β-EPOXY-5-α-LANOST-9(11)-ENE;   W/NBS   62435

484   C32 H52 O3   55515-39-4   AD-0-824-0   Lanost-8-en-11-one, 3-(acetyloxy)-, (3β,13α,14β,17α)-;; 11-KETOEUP
H-8-ENYL ACETATE;   W/NBS   62433

484   C32 H52 O3   55515-40-7   AD-0-826-0   Lanost-8-en-7-one, 3-(acetyloxy)-, (3β,13α,14β,17α)-;; 7-KETOEUPH-
8-ENYL ACETATE;   W/NBS   62434

484   C32 H52 O3   65352-08-1   F-33-1762-0   26-NOR-3β-ACETOXY-22,25-EPOXY-4,4,14α-TRIMETHYL-9,19-CYCLO-5α-ERG
OSTANE;   W   62439

484   C32 H52 O3   61228-95-3   B-29-1813-0   16β,28-O,O-ETHYLIDENELUP-20(29)-ENE-3β,16β,28-TRIOL;   W   62440

484   C32 H52 O3   20475-92-7   B-28-205-0   Urs-12-ene-3α,24-diol 3-Acetate;   W   62441

484   C32 H52 O3   22329-29-9   B-28-205-0   Urs-12-ene-3α,24-diol 24-Acetate;   W   62442

484   C32 H52 O3   63025-96-7   KC-1977-421-0   3α-ACETOXYFRIEDELAN-29-AL;   W   62444

484   C32 H52 O3   63547-51-3   KC-1977-488-0   O-ACETYLTRICHADENAL;   W   62445

484   C32 H52 O3   91897-19-7   F-40-2082-0   1β-Acetoxy-3β-hydroxylupene;   W   125338

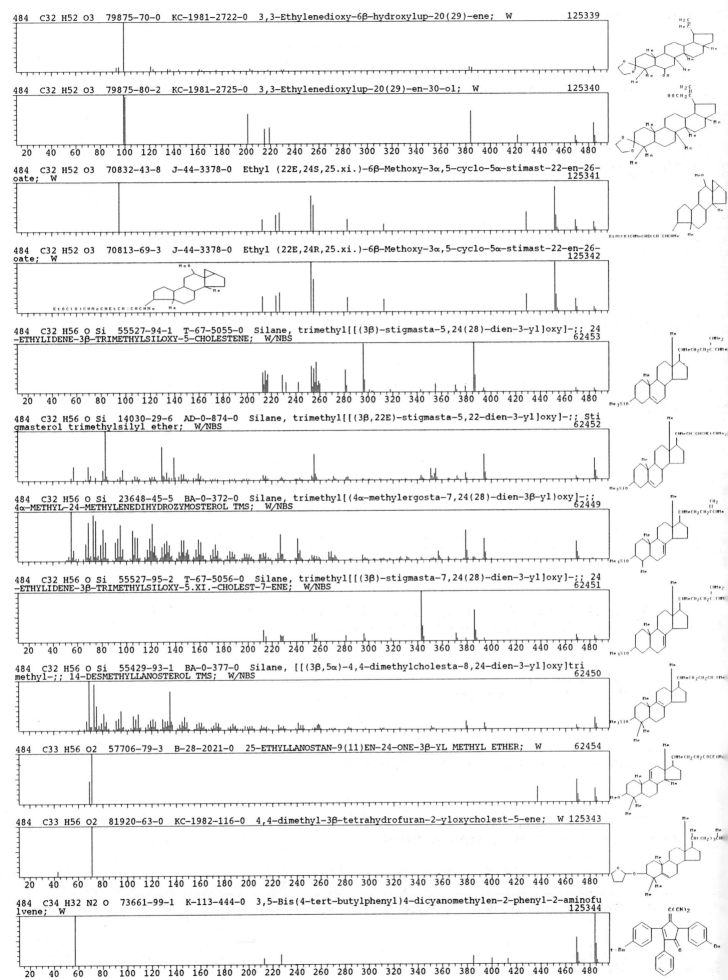

484 C32 H52 O3 79875-70-0 KC-1981-2722-0 3,3-Ethylenedioxy-6β-hydroxylup-20(29)-ene; W 125339

484 C32 H52 O3 79875-80-2 KC-1981-2725-0 3,3-Ethylenedioxylup-20(29)-en-30-ol; W 125340

484 C32 H52 O3 70832-43-8 J-44-3378-0 Ethyl (22E,24S,25.xi.)-6β-Methoxy-3α,5-cyclo-5α-stimast-22-en-26-oate; W 125341

484 C32 H52 O3 70813-69-3 J-44-3378-0 Ethyl (22E,24R,25.xi.)-6β-Methoxy-3α,5-cyclo-5α-stimast-22-en-26-oate; W 125342

484 C32 H56 O Si 55527-94-1 T-67-5055-0 Silane, trimethyl[[(3β)-stigmasta-5,24(28)-dien-3-yl]oxy]-;; 24-ETHYLIDENE-3β-TRIMETHYLSILOXY-5-CHOLESTENE; W/NBS 62453

484 C32 H56 O Si 14030-29-6 AD-0-874-0 Silane, trimethyl[[(3β,22E)-stigmasta-5,22-dien-3-yl]oxy]-;; Stigmasterol trimethylsilyl ether; W/NBS 62452

484 C32 H56 O Si 23648-45-5 BA-0-372-0 Silane, trimethyl[(4α-methylergosta-7,24(28)-dien-3β-yl)oxy]-;; 4α-METHYL-24-METHYLENEDIHYDROZYMOSTEROL TMS; W/NBS 62449

484 C32 H56 O Si 55527-95-2 T-67-5056-0 Silane, trimethyl[[(3β)-stigmasta-7,24(28)-dien-3-yl]oxy]-;; 24-ETHYLIDENE-3β-TRIMETHYLSILOXY-5.XI.-CHOLEST-7-ENE; W/NBS 62451

484 C32 H56 O Si 55429-93-1 BA-0-377-0 Silane, [[(3β,5α)-4,4-dimethylcholesta-8,24-dien-3-yl]oxy]trimethyl-;; 14-DESMETHYLLANOSTEROL TMS; W/NBS 62450

484 C33 H56 O2 57706-79-3 B-28-2021-0 25-ETHYLLANOSTAN-9(11)EN-24-ONE-3β-YL METHYL ETHER; W 62454

484 C33 H56 O2 81920-63-0 KC-1982-116-0 4,4-dimethyl-3β-tetrahydrofuran-2-yloxycholest-5-ene; W 125343

484 C34 H32 N2 O 73661-99-1 K-113-444-0 3,5-Bis(4-tert-butylphenyl)4-dicyanomethylen-2-phenyl-2-aminofulvene; W 125344

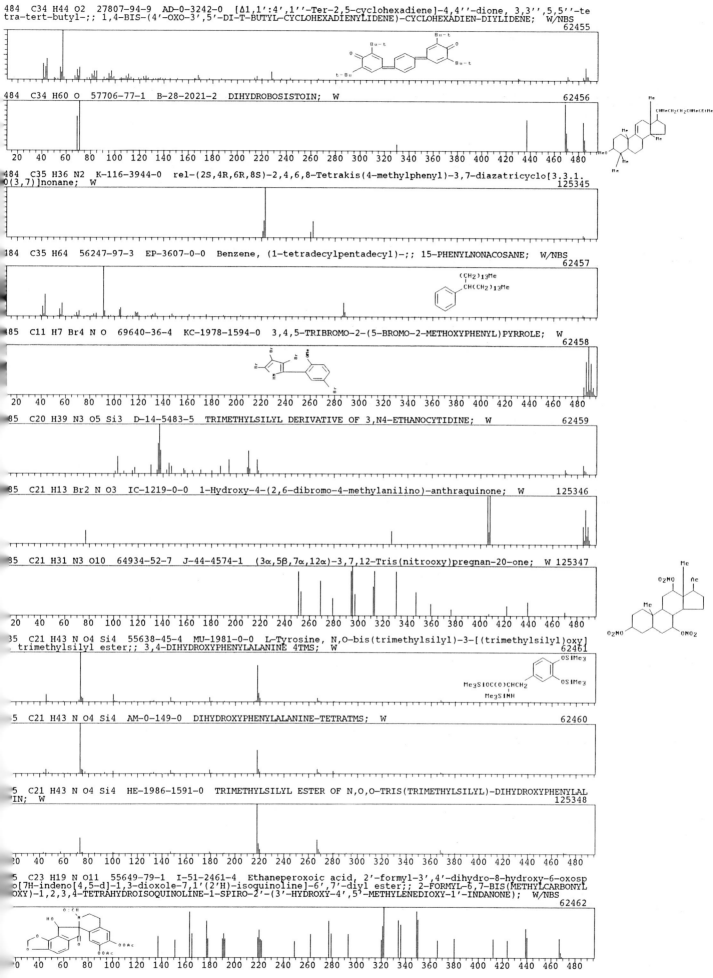

484 C34 H44 O2 27807-94-9 AD-0-3242-0 [Δ1,1':4',1''-Ter-2,5-cyclohexadiene]-4,4''-dione, 3,3'',5,5''-tetra-tert-butyl-;; 1,4-BIS-(4'-OXO-3',5'-DI-T-BUTYL-CYCLOHEXADIENYLIDENE)-CYCLOHEXADIEN-DIYLIDENE; W/NBS
62455

484 C34 H60 O 57706-77-1 B-28-2021-2 DIHYDROBOSISTOIN; W
62456

484 C35 H36 N2 K-116-3944-0 rel-(2S,4R,6R,8S)-2,4,6,8-Tetrakis(4-methylphenyl)-3,7-diazatricyclo[3.3.1.0(3,7)]nonane; W
125345

484 C35 H64 56247-97-3 EP-3607-0-0 Benzene, (1-tetradecylpentadecyl)-;; 15-PHENYLNONACOSANE; W/NBS
62457

485 C11 H7 Br4 N O 69640-36-4 KC-1978-1594-0 3,4,5-TRIBROMO-2-(5-BROMO-2-METHOXYPHENYL)PYRROLE; W
62458

485 C20 H39 N3 O5 Si3 D-14-5483-5 TRIMETHYLSILYL DERIVATIVE OF 3,N4-ETHANOCYTIDINE; W
62459

485 C21 H13 Br2 N O3 IC-1219-0-0 1-Hydroxy-4-(2,6-dibromo-4-methylanilino)-anthraquinone; W
125346

485 C21 H31 N3 O10 64934-52-7 J-44-4574-1 (3α,5β,7α,12α)-3,7,12-Tris(nitrooxy)pregnan-20-one; W 125347

485 C21 H43 N O4 Si4 55638-45-4 MU-1981-0-0 L-Tyrosine, N,O-bis(trimethylsilyl)-3-[(trimethylsilyl)oxy]-, trimethylsilyl ester;; 3,4-DIHYDROXYPHENYLALANINE 4TMS; W
62461

485 C21 H43 N O4 Si4 AM-0-149-0 DIHYDROXYPHENYLALANINE-TETRATMS; W
62460

485 C21 H43 N O4 Si4 HE-1986-1591-0 TRIMETHYLSILYL ESTER OF N,O,O-TRIS(TRIMETHYLSILYL)-DIHYDROXYPHENYLALANIN; W
125348

485 C23 H19 N O11 55649-79-1 I-51-2461-4 Ethaneperoxoic acid, 2'-formyl-3',4'-dihydro-8-hydroxy-6-oxospiro[7H-indeno[4,5-d]-1,3-dioxole-7,1'(2'H)-isoquinoline]-6',7'-diyl ester;; 2-FORMYL-6,7-BIS(METHYLCARBONYLOXY)-1,2,3,4-TETRAHYDROISOQUINOLINE-1-SPIRO-2'-(3'-HYDROXY-4',5'-METHYLENEDIOXY-1'-INDANONE); W/NBS
62462

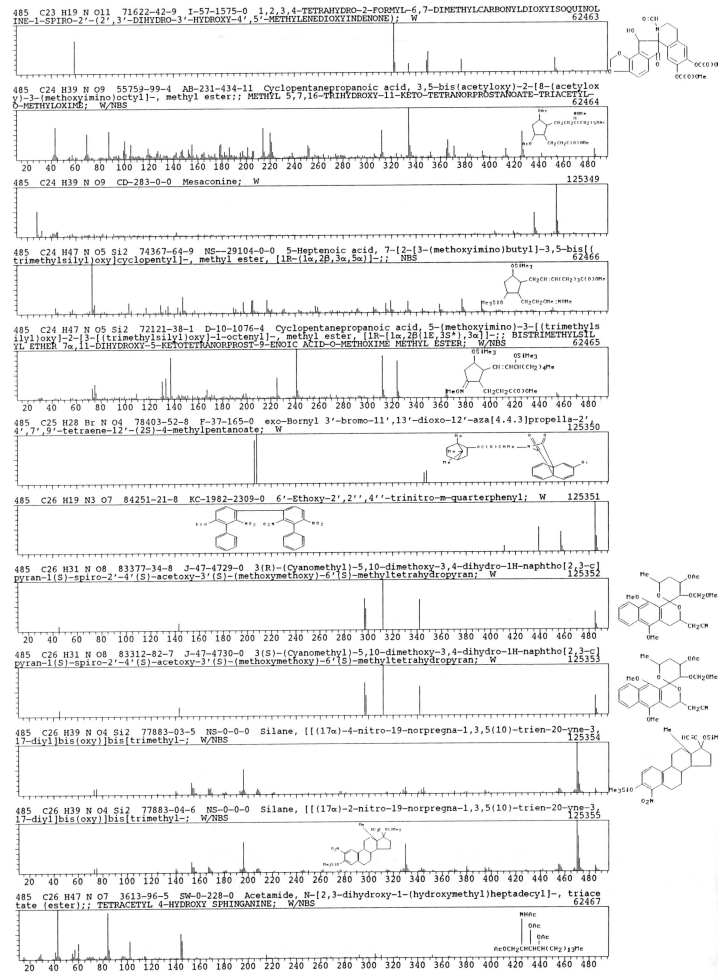

485 C23 H19 N O11 71622-42-9 I-57-1575-0 1,2,3,4-TETRAHYDRO-2-FORMYL-6,7-DIMETHYLCARBONYLDIOXYISOQUINOL
INE-1-SPIRO-2'-(2',3'-DIHYDRO-3'-HYDROXY-4',5'-METHYLENEDIOXYINDENONE); W
62463

485 C24 H39 N O9 55759-99-4 AB-231-434-11 Cyclopentanepropanoic acid, 3,5-bis(acetyloxy)-2-[8-(acetylox
y)-3-(methoxyimino)octyl]-, methyl ester;; METHYL 5,7,16-TRIHYDROXY-11-KETO-TETRANORPROSTANOATE-TRIACETYL-
O-METHYLOXIME; W/NBS
62464

485 C24 H39 N O9 CD-283-0-0 Mesaconine; W
125349

485 C24 H47 N O5 Si2 74367-64-9 NS--29104-0-0 5-Heptenoic acid, 7-[2-[3-(methoxyimino)butyl]-3,5-bis[(
trimethylsilyl)oxy]cyclopentyl]-, methyl ester, [1R-(1α,2β,3α,5α)]-;; NBS
62466

485 C24 H47 N O5 Si2 72121-38-1 D-10-1076-4 Cyclopentanepropanoic acid, 5-(methoxyimino)-3-[(trimethyls
ilyl)oxy]-2-[3-[(trimethylsilyl)oxy]-1-octenyl]-, methyl ester, [1R-[1α,2β(1E,3S*),3α]-;; BISTRIMETHYLSIL
YL ETHER 7α,11-DIHYDROXY-5-KETOTETRANORPROST-9-ENOIC ACID-O-METHOXIME METHYL ESTER; W/NBS
62465

485 C25 H28 Br N O4 78403-52-8 F-37-165-0 exo-Bornyl 3'-bromo-11',13'-dioxo-12'-aza[4.4.3]propella-2',
4',7',9'-tetraene-12'-(2S)-4-methylpentanoate; W
125350

485 C26 H19 N3 O7 84251-21-8 KC-1982-2309-0 6'-Ethoxy-2',2'',4''-trinitro-m-quarterphenyl; W 125351

485 C26 H31 N O8 83377-34-8 J-47-4729-0 3(R)-(Cyanomethyl)-5,10-dimethoxy-3,4-dihydro-1H-naphtho[2,3-c]
pyran-1(S)-spiro-2'-4'(S)-acetoxy-3'(S)-(methoxymethoxy)-6'(S)-methyltetrahydropyran; W 125352

485 C26 H31 N O8 83312-82-7 J-47-4730-0 3(S)-(Cyanomethyl)-5,10-dimethoxy-3,4-dihydro-1H-naphtho[2,3-c]
pyran-1(S)-spiro-2'-4'(S)-acetoxy-3'(S)-(methoxymethoxy)-6'(S)-methyltetrahydropyran; W 125353

485 C26 H39 N O4 Si2 77883-03-5 NS-0-0-0 Silane, [[(17α)-4-nitro-19-norpregna-1,3,5(10)-trien-20-yne-3,
17-diyl]bis(oxy)]bis[trimethyl-; W/NBS
125354

485 C26 H39 N O4 Si2 77883-04-6 NS-0-0-0 Silane, [[(17α)-2-nitro-19-norpregna-1,3,5(10)-trien-20-yne-3,
17-diyl]bis(oxy)]bis[trimethyl-; W/NBS
125355

485 C26 H47 N O7 3613-96-5 SW-0-228-0 Acetamide, N-[2,3-dihydroxy-1-(hydroxymethyl)heptadecyl]-, triace
tate (ester);; TETRACETYL 4-HYDROXY SPHINGANINE; W/NBS
62467

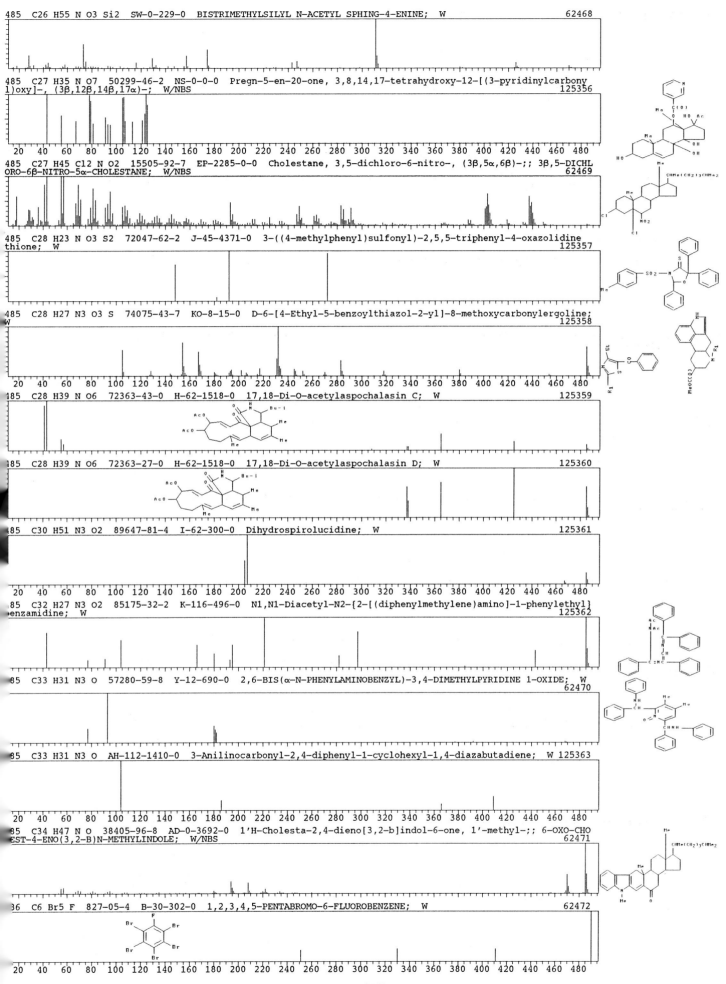

485   C26 H55 N O3 Si2   SW-0-229-0   BISTRIMETHYLSILYL N-ACETYL SPHING-4-ENINE;   W   62468

485   C27 H35 N O7   50299-46-2   NS-0-0-0   Pregn-5-en-20-one, 3,8,14,17-tetrahydroxy-12-[(3-pyridinylcarbonyl)oxy]-, (3β,12β,14β,17α)-;   W/NBS   125356

485   C27 H45 Cl2 N O2   15505-92-7   EP-2285-0-0   Cholestane, 3,5-dichloro-6-nitro-, (3β,5α,6β)-;; 3β,5-DICHLORO-6β-NITRO-5α-CHOLESTANE;   W/NBS   62469

485   C28 H23 N O3 S2   72047-62-2   J-45-4371-0   3-((4-methylphenyl)sulfonyl)-2,5,5-triphenyl-4-oxazolidinethione;   W   125357

485   C28 H27 N3 O3 S   74075-43-7   KO-8-15-0   D-6-[4-Ethyl-5-benzoylthiazol-2-yl]-8-methoxycarbonylergoline;   W   125358

485   C28 H39 N O6   72363-43-0   H-62-1518-0   17,18-Di-O-acetylaspochalasin C;   W   125359

485   C28 H39 N O6   72363-27-0   H-62-1518-0   17,18-Di-O-acetylaspochalasin D;   W   125360

485   C30 H51 N3 O2   89647-81-4   I-62-300-0   Dihydrospirolucidine;   W   125361

485   C32 H27 N3 O2   85175-32-2   K-116-496-0   N1,N1-Diacetyl-N2-{2-[(diphenylmethylene)amino]-1-phenylethyl}benzamidine;   W   125362

485   C33 H31 N3 O   57280-59-8   Y-12-690-0   2,6-BIS(α-N-PHENYLAMINOBENZYL)-3,4-DIMETHYLPYRIDINE 1-OXIDE;   W   62470

485   C33 H31 N3 O   AH-112-1410-0   3-Anilinocarbonyl-2,4-diphenyl-1-cyclohexyl-1,4-diazabutadiene;   W 125363

485   C34 H47 N O   38405-96-8   AD-0-3692-0   1'H-Cholesta-2,4-dieno[3,2-b]indol-6-one, 1'-methyl-;; 6-OXO-CHOLEST-4-ENO(3,2-B)N-METHYLINDOLE;   W/NBS   62471

486   C6 Br5 F   827-05-4   B-30-302-0   1,2,3,4,5-PENTABROMO-6-FLUOROBENZENE;   W   62472

- 5369 -

486   C10 Cl10 O   143-50-0   PG-1982-1026-0   Kepone;   W   62473

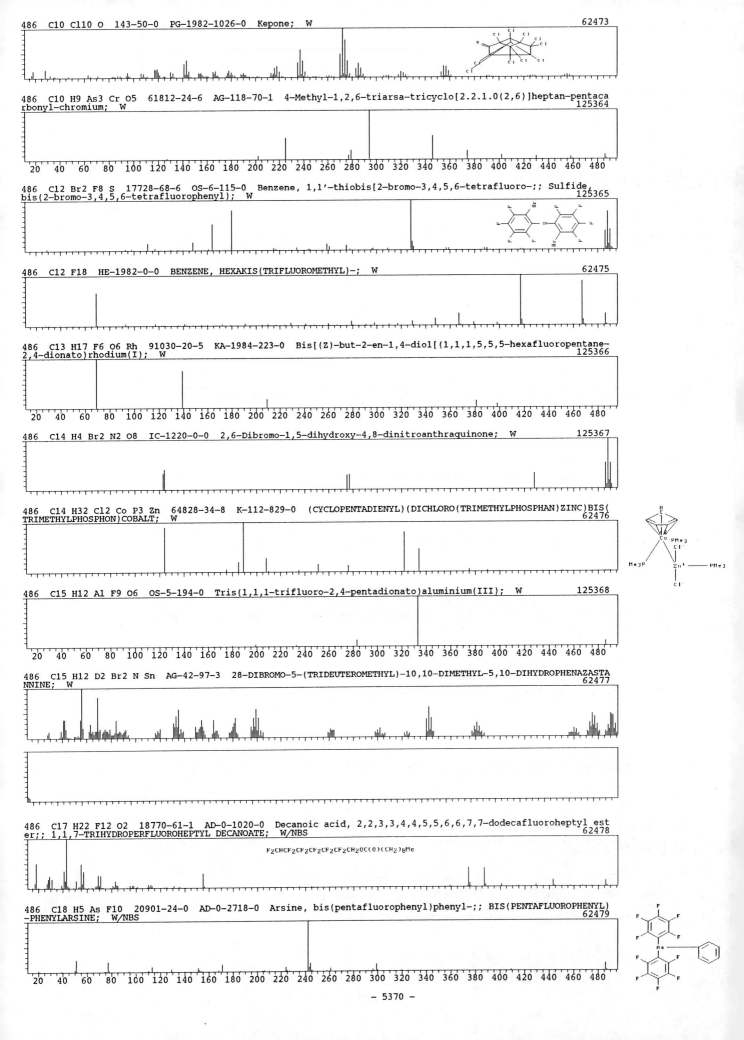

486   C10 H9 As3 Cr O5   61812-24-6   AG-118-70-1   4-Methyl-1,2,6-triarsa-tricyclo[2.2.1.0(2,6)]heptan-pentaca
rbonyl-chromium;   W   125364

486   C12 Br2 F8 S   17728-68-6   OS-6-115-0   Benzene, 1,1'-thiobis[2-bromo-3,4,5,6-tetrafluoro-;; Sulfide,
bis(2-bromo-3,4,5,6-tetrafluorophenyl);   W   125365

486   C12 F18   HE-1982-0-0   BENZENE, HEXAKIS(TRIFLUOROMETHYL)-;   W   62475

486   C13 H17 F6 O6 Rh   91030-20-5   KA-1984-223-0   Bis[(Z)-but-2-en-1,4-diol[(1,1,1,5,5,5-hexafluoropentane-
2,4-dionato)rhodium(I);   W   125366

486   C14 H4 Br2 N2 O8   IC-1220-0-0   2,6-Dibromo-1,5-dihydroxy-4,8-dinitroanthraquinone;   W   125367

486   C14 H32 Cl2 Co P3 Zn   64828-34-8   K-112-829-0   (CYCLOPENTADIENYL)(DICHLORO(TRIMETHYLPHOSPHAN)ZINC)BIS(
TRIMETHYLPHOSPHON)COBALT;   W   62476

486   C15 H12 Al F9 O6   OS-5-194-0   Tris(1,1,1-trifluoro-2,4-pentadionato)aluminium(III);   W   125368

486   C15 H12 D2 Br2 N Sn   AG-42-97-3   28-DIBROMO-5-(TRIDEUTEROMETHYL)-10,10-DIMETHYL-5,10-DIHYDROPHENAZASTA
NNINE;   W   62477

486   C17 H22 F12 O2   18770-61-1   AD-0-1020-0   Decanoic acid, 2,2,3,3,4,4,5,5,6,6,7,7-dodecafluoroheptyl est
er;; 1,1,7-TRIHYDROPERFLUOROHEPTYL DECANOATE;   W/NBS   62478

F2CHCF2CF2CF2CF2CF2CH2OC(O)(CH2)8Me

486   C18 H5 As F10   20901-24-0   AD-0-2718-0   Arsine, bis(pentafluorophenyl)phenyl-;; BIS(PENTAFLUOROPHENYL)
-PHENYLARSINE;   W/NBS   62479

486 C19 H8 F6 N4 O S2 57459-30-0 J-41-624-0 Bis[3-(3-Trifluoromethylphenyl-1,2,4-thiadiazol-5-yl] keto
ne; W   134042

486 C19 H17 I N4 Ni B-29-2278-0 16,17-Dihydro-5H-dibenzo[f,m][1,4,8,12]tetraazacyclopentadecinatonickel(
II) Iodide; W   125369

486 C20 H13 Cl3 O8 69987-57-1 F-34-2500-1 2,4,5-TRICHLORO-1,3,6-TRI-O-ACETYLNORLICHEXANTHON; W   62480

486 C20 H22 O14 91758-61-1 I-62-1190-0 Hexamethyl cis-3,7-dioxocyclo[3.3.0]octane-1,2,4,5,6,8-hexacarbo
xylate; W   125370

486 C20 H28 O8 Zr 17501-44-9 T-66-1827-0 Zirconium, tetrakis(2,4-pentanedionato-O,O')-;; Zirconium acet
ylacetonate; W/NBS   62481

486 C20 H32 B6 N6 O4 70530-21-1 K-112-1254-0 10,11,12,22,23,24,25,27-ME8-2,8,14,20-O4-10,12,22,24,25,27
-N6-1,9,11,13,21,23-B6-PENTACYCLO[19.3.1.1(3,7).1(9,13).1(15,19)]OCTACOSA-3,5,7(28),15,; W   62482

486 C21 H42 O5 Si4 MU-1981-0-0 1-TRIMETHYLSILYLOXYCARBONYL-1-TRIMETHYLSILYLOXY-2-(3',4'-BIS(TRIMETHYLSIL
YLOXY)PHENYL)-ETHANE; W   62483

486 C22 H13 F6 O4 P 81044-40-8 C-104-2500-0 3,4:8,9-Dibenzo-5,7-bis(trifluoromethyl)-1-phenyl-2,6,10,11
-tetraoxa-1-phospha(V)tricyclo[5.3.1.0]undecane; W   125371

486 C22 H16 Cl2 N4 O S2 23651-57-2 Y-14-877-1 N-(4-P-CHLOROPHENYL-2-THIAZOLYL)-β-(4'-CHLORO-PHENYL-2'-
THIAZOLYLAMINO)-CROTONAMIDE; W   62484

486 C22 H23 Cl N6 O5 IC-1221-0-0 N-Ethyl-N-(2-(2,5-dioxopyrrolidin-1-ylethyl))-3-acetamido-4-(2-chloro-
-nitrophenylazo)aniline; W   125372

486 C23 H22 N2 O8 S 35837-30-0 AU-21-399-8 Uridine, 2',3'-O-(phenylmethylene)-, 5'-(4-methylbenzenesulf
onate);; 2',3'-O-BENZYLIDENE-5'-O-P-TOLYLSULFONYLURIDINE; W/NBS   62485

486 C23 H35 I O3 38037-12-6 F-34-1533-0 3β-ACETOXY-18-IODO-5α,17β-PREGNAN-20-ONE; W   62486

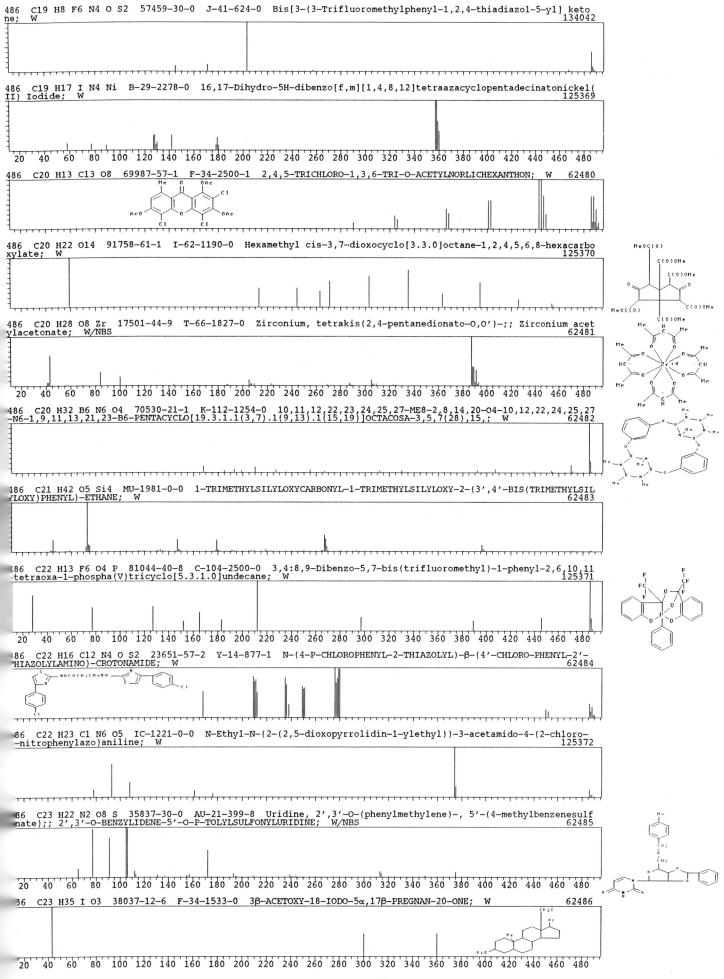

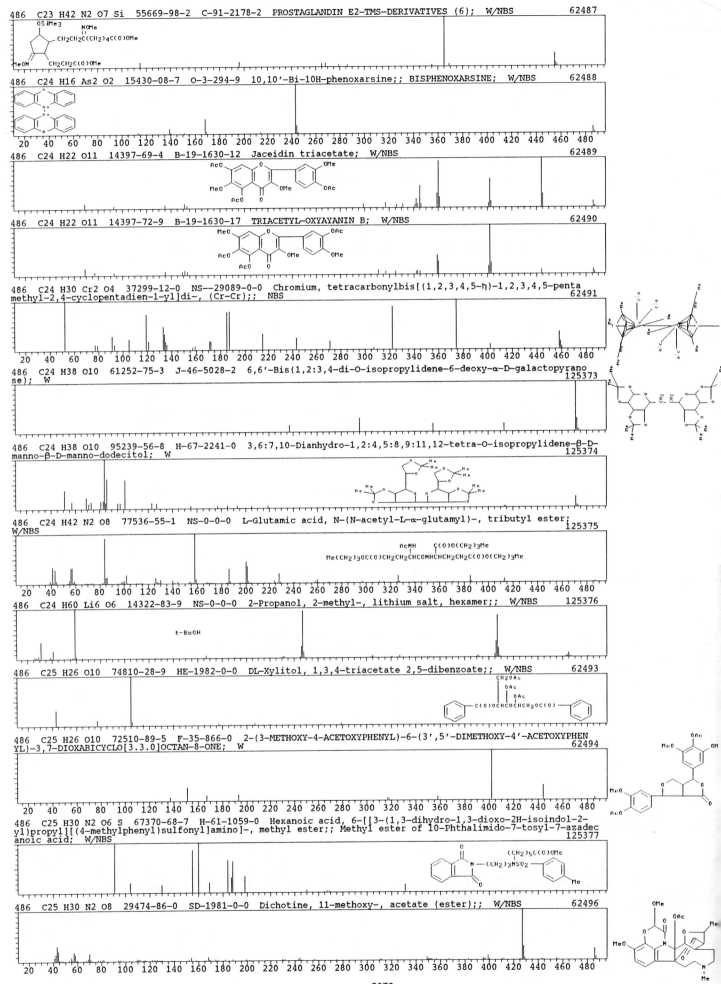

486   C23 H42 N2 O7 Si   55669-98-2   C-91-2178-2   PROSTAGLANDIN E2-TMS-DERIVATIVES (6);   W/NBS   62487

486   C24 H16 As2 O2   15430-08-7   O-3-294-9   10,10'-Bi-10H-phenoxarsine;;   BISPHENOXARSINE;   W/NBS   62488

486   C24 H22 O11   14397-69-4   B-19-1630-12   Jaceidin triacetate;   W/NBS   62489

486   C24 H22 O11   14397-72-9   B-19-1630-17   TRIACETYL-OXYAYANIN B;   W/NBS   62490

486   C24 H30 Cr2 O4   37299-12-0   NS--29089-0-0   Chromium, tetracarbonylbis[(1,2,3,4,5-η)-1,2,3,4,5-penta
methyl-2,4-cyclopentadien-1-yl]di-, (Cr-Cr);;   NBS   62491

486   C24 H38 O10   61252-75-3   J-46-5028-2   6,6'-Bis(1,2:3,4-di-O-isopropylidene-6-deoxy-α-D-galactopyrano
se);   W   125373

486   C24 H38 O10   95239-56-8   H-67-2241-0   3,6:7,10-Dianhydro-1,2:4,5:8,9:11,12-tetra-O-isopropylidene-β-D-
manno-β-D-manno-dodecitol;   W   125374

486   C24 H42 N2 O8   77536-55-1   NS-0-0-0   L-Glutamic acid, N-(N-acetyl-L-α-glutamyl)-, tributyl ester;
W/NBS   125375

486   C24 H60 Li6 O6   14322-83-9   NS-0-0-0   2-Propanol, 2-methyl-, lithium salt, hexamer;;   W/NBS   125376

486   C25 H26 O10   74810-28-9   HE-1982-0-0   DL-Xylitol, 1,3,4-triacetate 2,5-dibenzoate;;   W/NBS   62493

486   C25 H26 O10   72510-89-5   F-35-866-0   2-(3-METHOXY-4-ACETOXYPHENYL)-6-(3',5'-DIMETHOXY-4'-ACETOXYPHEN
YL)-3,7-DIOXABICYCLO[3.3.0]OCTAN-8-ONE;   W   62494

486   C25 H30 N2 O6 S   67370-68-7   H-61-1059-0   Hexanoic acid, 6-[[3-(1,3-dihydro-1,3-dioxo-2H-isoindol-2-
yl)propyl][(4-methylphenyl)sulfonyl]amino]-, methyl ester;;   Methyl ester of 10-Phthalimido-7-tosyl-7-azadec
anoic acid;   W/NBS   125377

486   C25 H30 N2 O8   29474-86-0   SD-1981-0-0   Dichotine, 11-methoxy-, acetate (ester);;   W/NBS   62496

486 C25 H50 O5 Si2 55760-04-8 AB-280-468-2 Cyclopentanepentanoic acid, 2-(3-oxooctyl)-3,5-bis[(trimeth ylsilyl)oxy]-, methyl ester, [1R-(1α,2β,3α,5α)]-;; METHYL 7α,9α-DIHYDROXY-13-KETODINORPROSTANOATE BISTRI METHYLSILYL ETHER; W/NBS
62497

486 C26 H30 N6 O2 Si 68950-24-3 J-44-1321-0 9-[5-O-(tert-Butyldiphenylsilyl)-2,3-epimino-2,3-dideoxy-β-D-lyxofuranosyl]adenine; W
125378

486 C26 H30 O9 751-49-5 OS-6-109-0 Limonoic acid, 19-deoxy-6,19-epoxy-, 16,17-lactone;; Limonilic acid; W
125379

Str 19

486 C26 H30 O9 53899-46-0 B-28-857-0 3-Hydroxyphysodic Acid; W
62499

486 C26 H30 O9 79742-36-2 B-34-1132-0 (aR,5R,6S,7R)-5,10-Diacetoxy-1,9,11-trimethoxy-6,7-dimethyl-2,3-methylenedioxy-5,6,7,8-tetrahydrodibenzo[a,c]cyclooctene; W
125380

486 C26 H38 N4 O5 U-74-1920-21 N-ACETYLABYSSENIN-C; W
62500

486 C26 H35 D3 N4 O5 U-74-1920-22 N-TRIDEUTERIOACETYLABYSSENIN-C; W
62501

486 C27 H28 B F5 N2 72886-63-6 K-112-3826-0 1,3-DI-TERT-BUTYL-2-(PENTAFLUOROPHENYL)-4,4-DIPHENYL-1,3,2-DIAZABORETIDIN; W
62502

486 C27 H34 O8 19314-71-7 SD-1981-0-0 β-Resorcylic acid, 6-(2-oxoheptyl)-, 4-ester with methyl 6-pentyl-β-resorcylate;; W/NBS
62504

486 C27 H34 O8 B-31-1051-0 4-O-METHYLOLIVETORIC ACID; W
62503

486 C27 H34 O8 80141-16-8 B-34-816-0 (aR,5S,6S,7R)-5-Acetoxy-12-acetyl-1,2,3,9,11-pentamethoxy-6,7-dimethyl-5,6,7,8-tetrahydrodibenzo[a,c]cyclooctene; W
125381

486 C27 H38 N2 O6 72101-38-3 NS--29082-0-0 4,25-Secoobscurinervan-4-ol, 25-ethyl-15,16-dimethoxy-, 25-acetate, (4β,22α)-;; NBS
62506

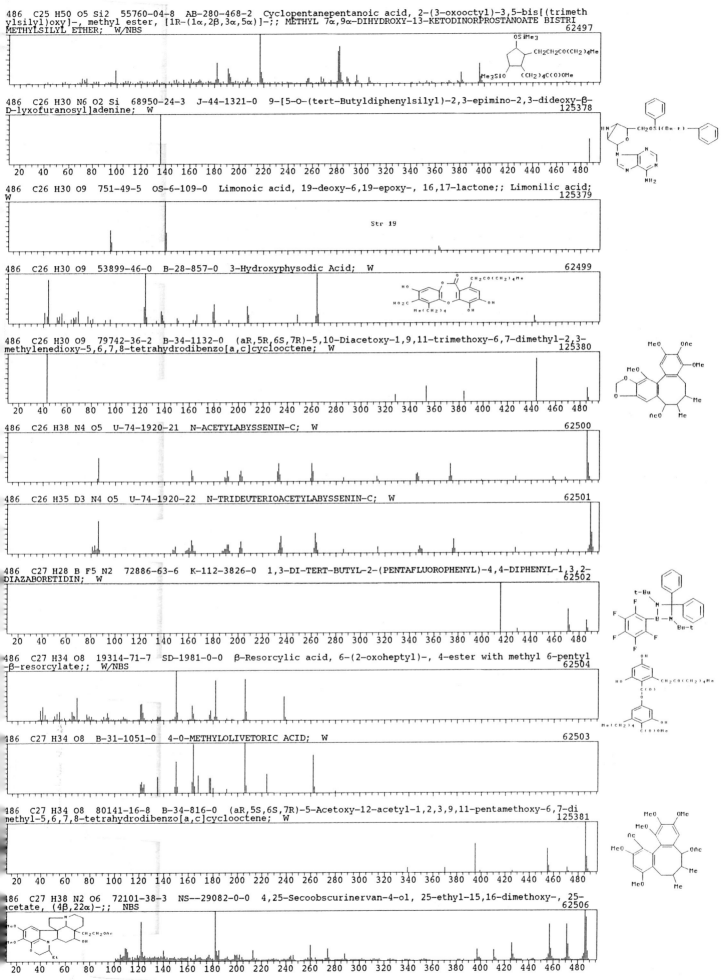

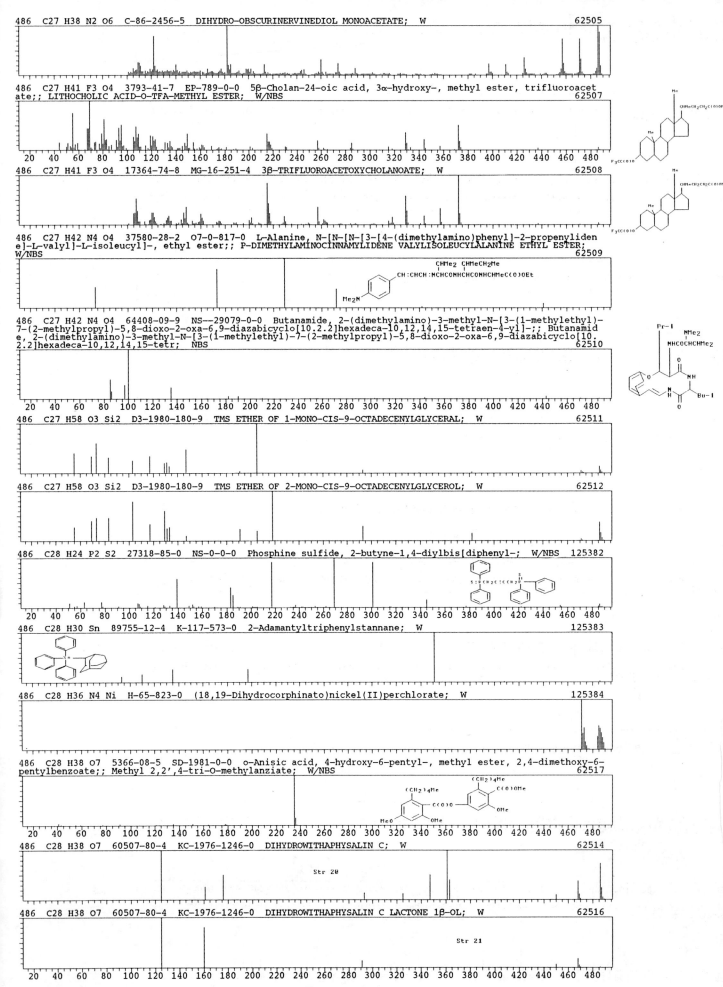

486   C27 H38 N2 O6   C-86-2456-5   DIHYDRO-OBSCURINERVINEDIOL MONOACETATE;   W
62505

486   C27 H41 F3 O4   3793-41-7   EP-789-0-0   5β-Cholan-24-oic acid, 3α-hydroxy-, methyl ester, trifluoroacet
ate;;   LITHOCHOLIC ACID-O-TFA-METHYL ESTER;   W/NBS
62507

486   C27 H41 F3 O4   17364-74-8   MG-16-251-4   3β-TRIFLUOROACETOXYCHOLANOATE;   W
62508

486   C27 H42 N4 O4   37580-28-2   O7-0-817-0   L-Alanine, N-[N-[N-[3-[4-(dimethylamino)phenyl]-2-propyliden
e]-L-valyl]-L-isoleucyl]-, ethyl ester;;   P-DIMETHYLAMINOCINNAMYLIDENE VALYLISOLEUCYLALANINE ETHYL ESTER;
W/NBS
62509

486   C27 H42 N4 O4   64408-09-9   NS--29079-0-0   Butanamide, 2-(dimethylamino)-3-methyl-N-[3-(1-methylethyl)-
7-(2-methylpropyl)-5,8-dioxo-2-oxa-6,9-diazabicyclo[10.2.2]hexadeca-10,12,14,15-tetraen-4-yl]-;;   Butanamid
e, 2-(dimethylamino)-3-methyl-N-[3-(1-methylethyl)-7-(2-methylpropyl)-5,8-dioxo-2-oxa-6,9-diazabicyclo[10.
2.2]hexadeca-10,12,14,15-tetr;   NBS
62510

486   C27 H58 O3 Si2   D3-1980-180-9   TMS ETHER OF 1-MONO-CIS-9-OCTADECENYLGLYCERAL;   W
62511

486   C27 H58 O3 Si2   D3-1980-180-9   TMS ETHER OF 2-MONO-CIS-9-OCTADECENYLGLYCEROL;   W
62512

486   C28 H24 P2 S2   27318-85-0   NS-0-0-0   Phosphine sulfide, 2-butyne-1,4-diylbis[diphenyl-;   W/NBS   125382

486   C28 H30 Sn   89755-12-4   K-117-573-0   2-Adamantyltriphenylstannane;   W
125383

486   C28 H36 N4 Ni   H-65-823-0   (18,19-Dihydrocorphinato)nickel(II)perchlorate;   W
125384

486   C28 H38 O7   5366-08-5   SD-1981-0-0   o-Anisic acid, 4-hydroxy-6-pentyl-, methyl ester, 2,4-dimethoxy-6-
pentylbenzoate;;   Methyl 2,2',4-tri-O-methylanziate;   W/NBS
62517

486   C28 H38 O7   60507-80-4   KC-1976-1246-0   DIHYDROWITHAPHYSALIN C;   W
62514

Str 20

486   C28 H38 O7   60507-80-4   KC-1976-1246-0   DIHYDROWITHAPHYSALIN C LACTONE 1β-OL;   W
62516

Str 21

486   C28 H38 O7   KC-1982-2838-0   (22R)-6β,14α,17β,20α(F)-tetrahydroxy-1-oxowitha-2,4,24-trienolide;   W
125385

486   C28 H54 O4 S   IC-1221-0-0   Dihendecyl thiodipropionate;   W
125386

486   C28 H58 O4 Si   87515-19-3   F-39-2214-0   (R,S)-Methyl-3-(tert-butyldimethylsilyloxy)-2-(1-octadecycloxy)propionate;   W
125387

MeOC(O) Me
Me(CH2)17OCHCH2OSiBu-t
Me

486   C28 H62 O2 Si2   22396-30-1   BA-0-360-0   3,26-Dioxa-2,27-disilaoctacosane, 2,2,27,27-tetramethyl-;; 1,22-BIS(TRIMETHYLSILYLOXY)DOCOSANE;   W/NBS
62518

Me3SiO(CH2)22OSiMe3

486   C29 H26 Fe2   77793-00-1   AH-112-102-0   1,3-Diferrocenyl-3-phenyl-propene;   W
125388

486   C29 H26 O5 S   52516-85-5   C-97-833-0   MIXTURE OF CIS AND TRANS-1,2-DI-(P-METHOXYPHENYL)-2-PHENYLVINYL P-TOLUENESULFONATE;   W
62519

486   C29 H28 O3 P2   87953-00-2   K-116-3149-0   [1,3-Bis(diphenylphosphoryl)-1-butenyl]-methyl-ether;   W
125389

486   C29 H28 O3 P2   88017-77-0   K-116-3160-0   (1-Diphenylphosphoryl-3-methyl-2-butenyl)-diphenylphosphinate;   W
125390

486   C29 H30 N2 O5   95689-19-3   F-40-4859-0   Methyl ester of (17R,20E)-4-Benzyl-17,21-epoxy-17-hydroxy-16-methylene-19-oxo-3β-4,21-secocorn-20-en-5β-carboxylic acid;   W
125391

486   C29 H30 N2 O5   90140-54-8   K-117-1483-0   3-Benzoyl-1-(3,4-dimethoxyphenyl)-5-ethyl-7,8-dimethoxy-4-methyl-3H-2,3-benzodiazepine;   W
125392

486   C29 H42 O4 S   83872-45-1   H-65-1510-0   (20S)S-(20-Methyl-5-pregnen-3β,21-diol)-21-p-toluolsulfonate;   W
125393

486   C29 H42 O6   21902-97-6   J-4-3863-2   27-DEOXY-2,3-DIHYDRO-3-METHOXYWITHAFERIN;   W
62520

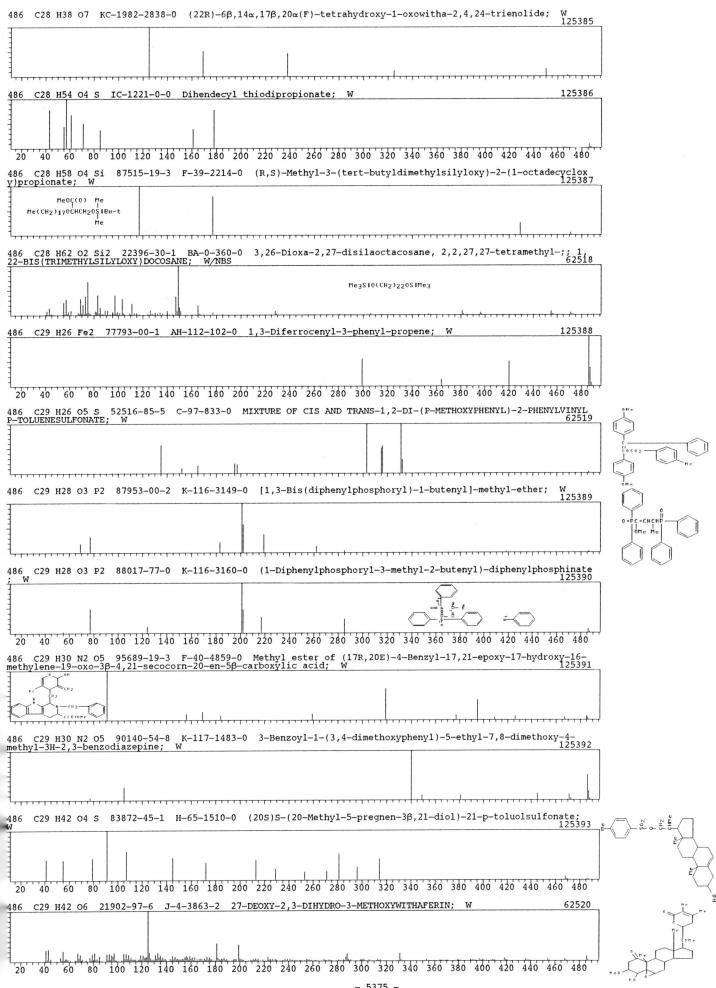

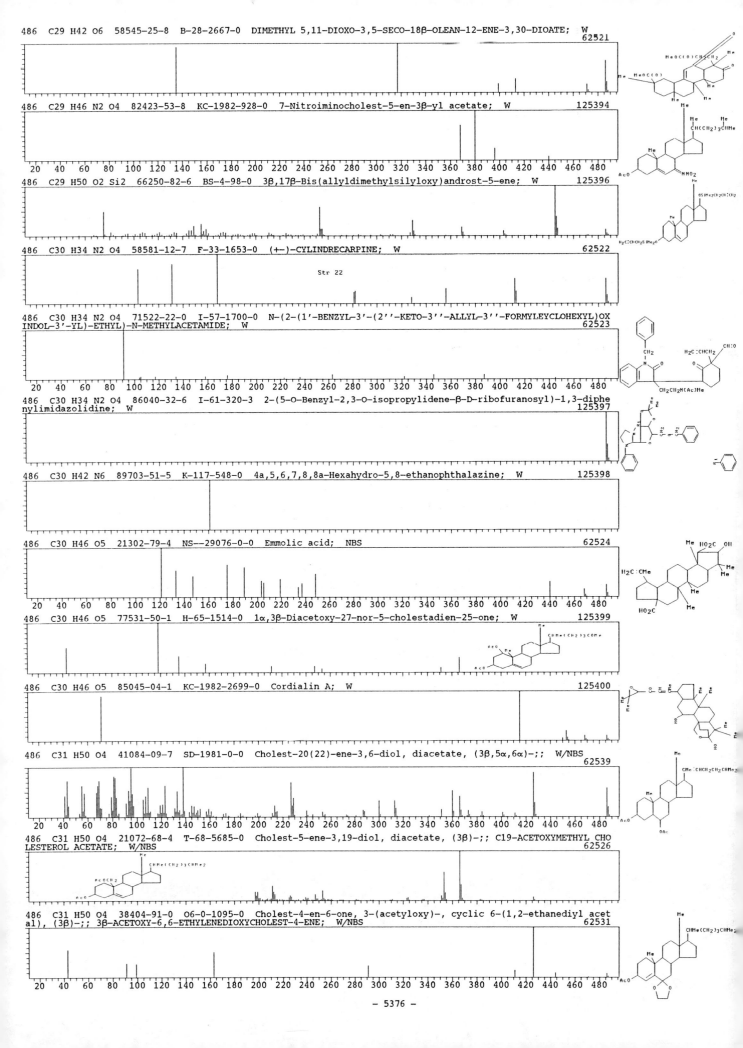

486   C29 H42 O6   58545-25-8   B-28-2667-0   DIMETHYL 5,11-DIOXO-3,5-SECO-18β-OLEAN-12-ENE-3,30-DIOATE;   W
62521

486   C29 H46 N2 O4   82423-53-8   KC-1982-928-0   7-Nitroiminocholest-5-en-3β-yl acetate;   W   125394

486   C29 H50 O2 Si2   66250-82-6   BS-4-98-0   3β,17β-Bis(allyldimethylsilyloxy)androst-5-ene;   W   125396

486   C30 H34 N2 O4   58581-12-7   F-33-1653-0   (+-)-CYLINDRECARPINE;   W   62522

Str 22

486   C30 H34 N2 O4   71522-22-0   I-57-1700-0   N-(2-(1'-BENZYL-3'-(2''-KETO-3''-ALLYL-3''-FORMYLEYCLOHEXYL)OX INDOL-3'-YL)-ETHYL)-N-METHYLACETAMIDE;   W   62523

486   C30 H34 N2 O4   86040-32-6   I-61-320-3   2-(5-O-Benzyl-2,3-O-isopropylidene-β-D-ribofuranosyl)-1,3-diphe nylimidazolidine;   W   125397

486   C30 H42 N6   89703-51-5   K-117-548-0   4a,5,6,7,8,8a-Hexahydro-5,8-ethanophthalazine;   W   125398

486   C30 H46 O5   21302-79-4   NS--29076-0-0   Emmolic acid;   NBS   62524

486   C30 H46 O5   77531-50-1   H-65-1514-0   1α,3β-Diacetoxy-27-nor-5-cholestadien-25-one;   W   125399

486   C30 H46 O5   85045-04-1   KC-1982-2699-0   Cordialin A;   W   125400

486   C31 H50 O4   41084-09-7   SD-1981-0-0   Cholest-20(22)-ene-3,6-diol, diacetate, (3β,5α,6α)-;;   W/NBS   62539

486   C31 H50 O4   21072-68-4   T-68-5685-0   Cholest-5-ene-3,19-diol, diacetate, (3β)-;; C19-ACETOXYMETHYL CHO LESTEROL ACETATE;   W/NBS   62526

486   C31 H50 O4   38404-91-0   O6-0-1095-0   Cholest-4-en-6-one, 3-(acetyloxy)-, cyclic 6-(1,2-ethanediyl acet al), (3β)-;; 3β-ACETOXY-6,6-ETHYLENEDIOXYCHOLEST-4-ENE;   W/NBS   62531

486 C31 H50 O4 6831-09-0 NS--29074-0-0 Lup-20(29)-en-28-oic acid, 2,3-dihydroxy-, methyl ester, (2α, 3β)-;; NBS   62532

486 C31 H50 O4 26563-64-4 EP-7510-0-0 Olean-12-en-28-oic acid, 2,3-dihydroxy-, methyl ester, (2α,3α)-;; Methyl 3-epimaslinate; W/NBS   62527

486 C31 H50 O4 14511-72-9 T-67-4992-0 Urs-12-en-28-oic acid, 3,19-dihydroxy-, methyl ester, (3β)-;; METHYL POMOLATE; W/NBS   62530

486 C31 H50 O4 4891-77-4 AD-0-1246-0 Olean-12-en-28-oic acid, 3,29-dihydroxy-, methyl ester, (3β,20α)- ;; METHYL-MESEMBRYANTHEMOIDIGENATE; W/NBS   62529

486 C31 H50 O4 74420-84-1 NS--29066-0-0 Cholest-8(14)-ene-3,7-diol, diacetate, (3β,5α,7β)-;; NBS 62535

486 C31 H50 O4 69140-09-6 NS--29067-0-0 Cholest-8(14)-ene-3,15-diol, diacetate, (3β,5α,15β)-;; 5α-Cholest-8(14)-ene-3β,15β-diacetate; NBS   62534

486 C31 H50 O4 AT-64-3182-4 METHYL COMMATE D; W   62525

486 C31 H50 O4 AT-64-3182-3 METHYL COMMATE C; W   62528

486 C31 H50 O4 59157-93-6 B-29-670-0 Methyl 22α,25-dihydroxyolean-12-en-28-oate; W   62536

Str 23

486 C31 H50 O4 59157-93-6 B-29-670-0 METHYL 22β,25-DIHYDROXYOLEAN-12-EN-28-OATE; W   62537

Str 24

486 C31 H50 O4 B-31-2743-0 METHYL ECHINOCYSTATE; W   62538

486 C31 H50 O4 38242-20-5 NS-0-0-0 D:C-Friedo-B':A'-neogammacer-9(11)-en-23-oic acid, 2,3-dihydroxy-, methyl ester, (2α,3β,4α)-; W/NBS   125402

486 C31 H50 O4 58028-17-4 KC-1981-249-0 2-Acetyl-4-oxa-5-acetoxymethylcholest-2-ene; W   125403

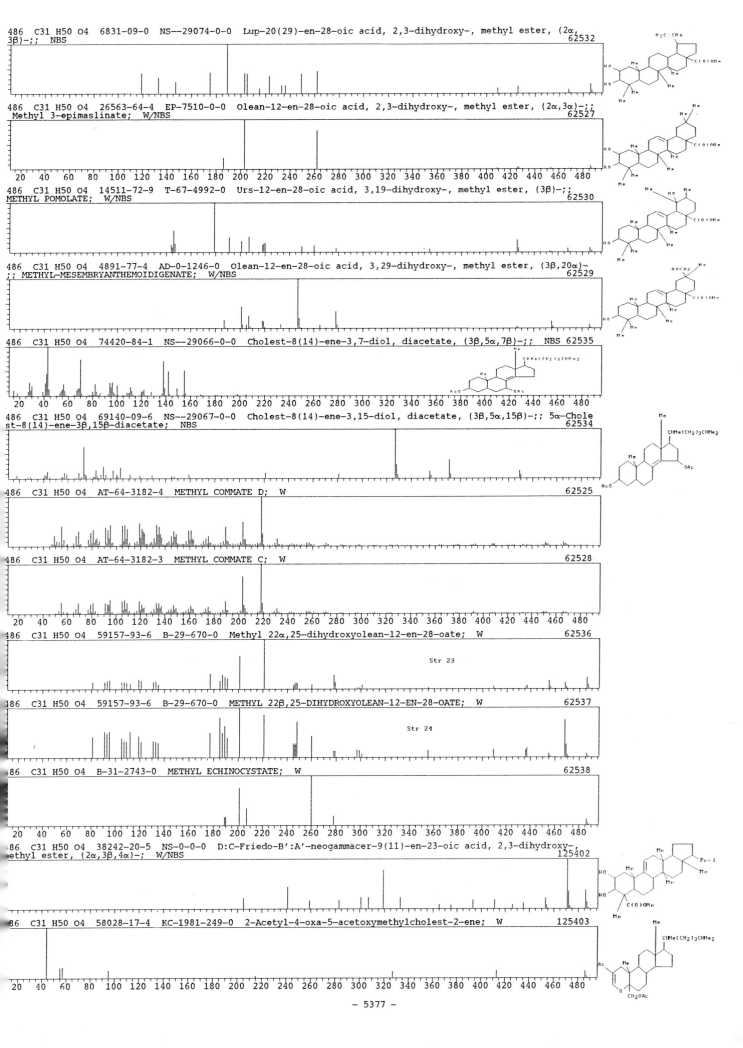

- 5377 -

486   C31 H50 O4   69140-08-5   NS--29068-0-0   Cholest-8(14)-ene-3,15-diol, diacetate, (3β,5α,15α)-;; 5α-Chole
st-8(14)-ene-3β,15α-diacetate;   NBS
62533

486   C32 H18 N6   63948-24-3   K-115-456-0   Bis(benzimidazo)[1',2':1,8:2'',1'':4,5][1,5]diazocino[3,2-b:7,6-
b']diquinolinehydrate;   W
125404

486   C32 H26 N2 O3   92670-79-6   J-49-4654-0   Methyl ester of (Z)-α-(2-Phenoxy-2-phenylethenyl)-3,5-diphenyl
-1H-pyrazole-1-acetic acid;   W
125405

486   C32 H26 N2 O3   92670-75-2   J-49-4654-0   (Z)-4-Methyl-α-(2-phenoxy-2-phenylethenyl)-3,5-diphenyl-1H-pyr
azole-1-acetic acid;   W
125406

486   C32 H38 O4   26537-41-7   O-2-971-8   DIPHENYLTETRAHYDROERIOSTANOL;   W/NBS
62540

486   C32 H38 O4   56335-88-7   SD-1981-0-0   2H,8H-Benzo[1,2-b:3,4-b']dipyran-6-propanol, 3,4,9,10-tetrahydro-
5-methoxy-2,2,8,8-tetramethyl-α,α-diphenyl-;;   W/NBS
62541

486   C32 H54 O3   54411-53-9   SD-1981-0-0   Lanostan-11-one, 18-(acetyloxy)-;;   W/NBS
62550

486   C32 H54 O3   25116-68-1   SD-1981-0-0   Lanostan-3β-ol, 11β,18-epoxy-, acetate;;   W/NBS
62552

486   C32 H54 O3   55700-22-6   AD-0-821-0   Lanostan-11-one, 3-(acetyloxy)-, (13α,14β,17α)-;;   11-KETOEUPHANYL
ACETATE;   W/NBS
62544

486   C32 H54 O3   10049-93-1   SD-1981-0-0   Lanostan-11-one, 3-(acetyloxy)-, (3β)-;; Lanostan-11-one, 3β-hydr
oxy-, acetate;   W/NBS
62551

486   C32 H54 O3   55515-38-3   T-68-5736-0   D:A-Friedooleanane-3,7-diol, 3-acetate;;   3-ACETOXY-7-HYDROXYFRIED
ELANE;   W/NBS
62542

486   C32 H54 O3   43206-38-8   O-14-161-2   29,30-Dinorgammacerane-3,22-diol, 21,21-dimethyl-, 3-acetate, (3β,
8α,9β,13α,14α,17α,18β,22α)-;;   (8α,14β,18β)-3β-ACETYLOXY-22α-HYDROXY-21,21-DIMETHYL-29,30-DINORGAMMACERANE;
W/NBS
62546

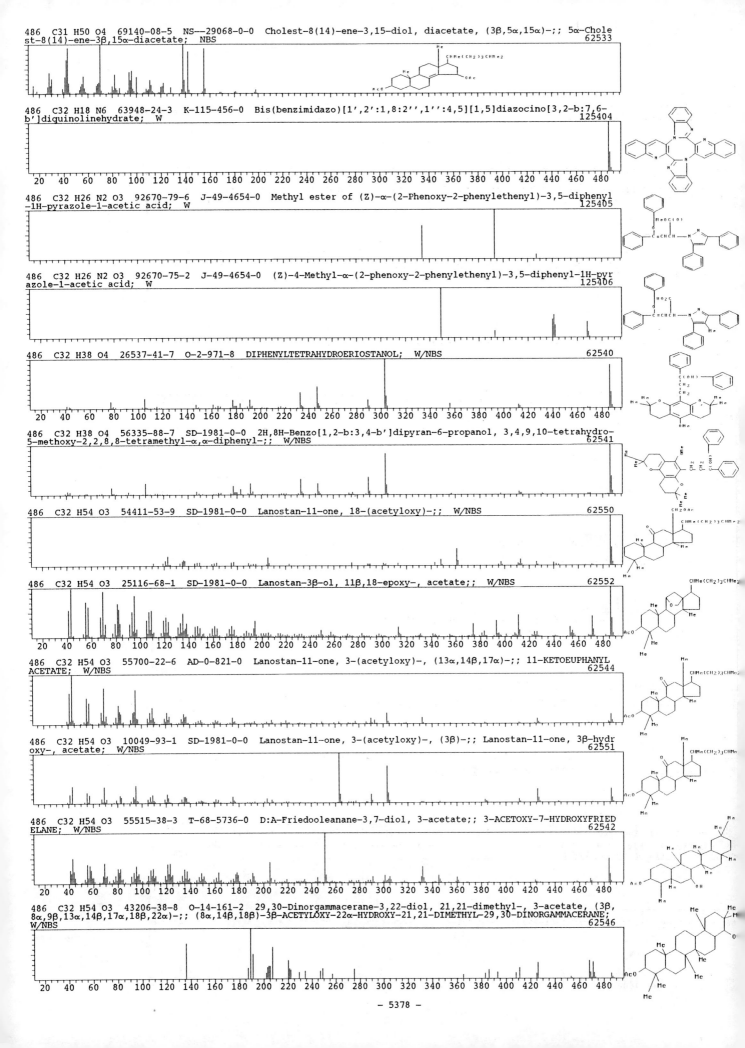

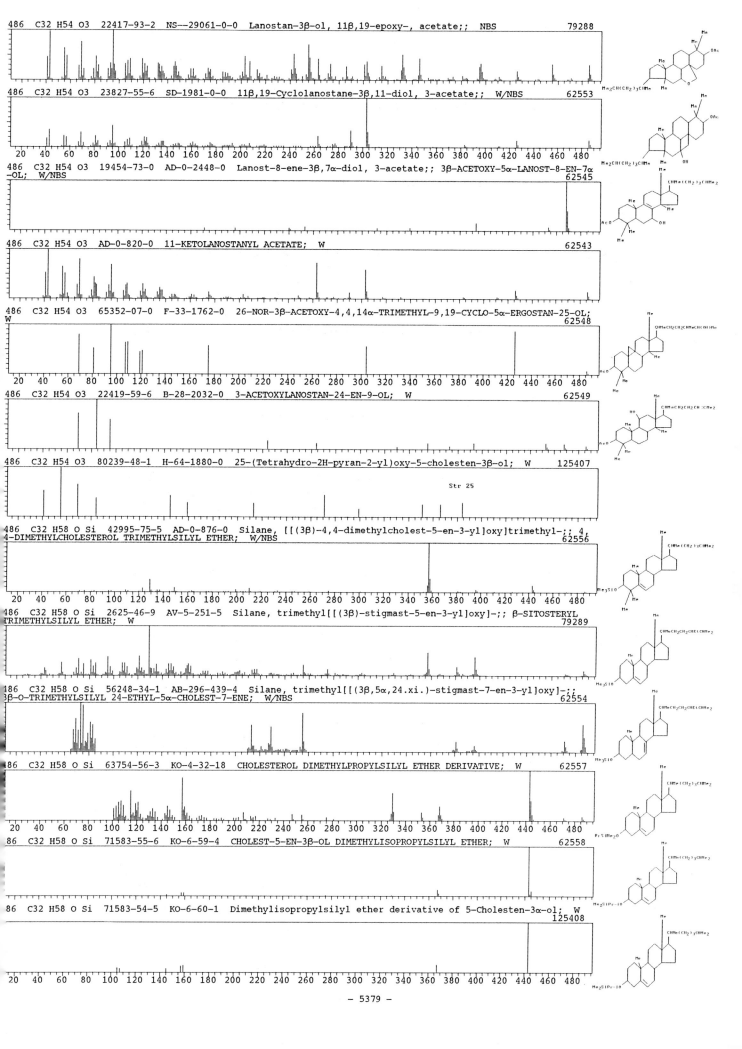

486   C32 H54 O3   22417-93-2   NS--29061-0-0   Lanostan-3β-ol, 11β,19-epoxy-, acetate;;   NBS   79288

486   C32 H54 O3   23827-55-6   SD-1981-0-0   11β,19-Cyclolanostane-3β,11-diol, 3-acetate;;   W/NBS   62553

486   C32 H54 O3   19454-73-0   AD-0-2448-0   Lanost-8-ene-3β,7α-diol, 3-acetate;;   3β-ACETOXY-5α-LANOST-8-EN-7α
-OL;   W/NBS   62545

486   C32 H54 O3   AD-0-820-0   11-KETOLANOSTANYL ACETATE;   W   62543

486   C32 H54 O3   65352-07-0   F-33-1762-0   26-NOR-3β-ACETOXY-4,4,14α-TRIMETHYL-9,19-CYCLO-5α-ERGOSTAN-25-OL;
W   62548

486   C32 H54 O3   22419-59-6   B-28-2032-0   3-ACETOXYLANOSTAN-24-EN-9-OL;   W   62549

486   C32 H54 O3   80239-48-1   H-64-1880-0   25-(Tetrahydro-2H-pyran-2-yl)oxy-5-cholesten-3β-ol;   W   125407

486   C32 H58 O Si   42995-75-5   AD-0-876-0   Silane, [[(3β)-4,4-dimethylcholest-5-en-3-yl]oxy]trimethyl-;;   4,
4-DIMETHYLCHOLESTEROL TRIMETHYLSILYL ETHER;   W/NBS   62556

486   C32 H58 O Si   2625-46-9   AV-5-251-5   Silane, trimethyl[[(3β)-stigmast-5-en-3-yl]oxy]-;;   β-SITOSTERYL
TRIMETHYLSILYL ETHER;   W   79289

486   C32 H58 O Si   56248-34-1   AB-296-439-4   Silane, trimethyl[[(3β,5α,24.xi.)-stigmast-7-en-3-yl]oxy]-;;
3β-O-TRIMETHYLSILYL 24-ETHYL-5α-CHOLEST-7-ENE;   W/NBS   62554

486   C32 H58 O Si   63754-56-3   KO-4-32-18   CHOLESTEROL DIMETHYLPROPYLSILYL ETHER DERIVATIVE;   W   62557

486   C32 H58 O Si   71583-55-6   KO-6-59-4   CHOLEST-5-EN-3β-OL DIMETHYLISOPROPYLSILYL ETHER;   W   62558

486   C32 H58 O Si   71583-54-5   KO-6-60-1   Dimethylisopropylsilyl ether derivative of 5-Cholesten-3α-ol;   W   125408

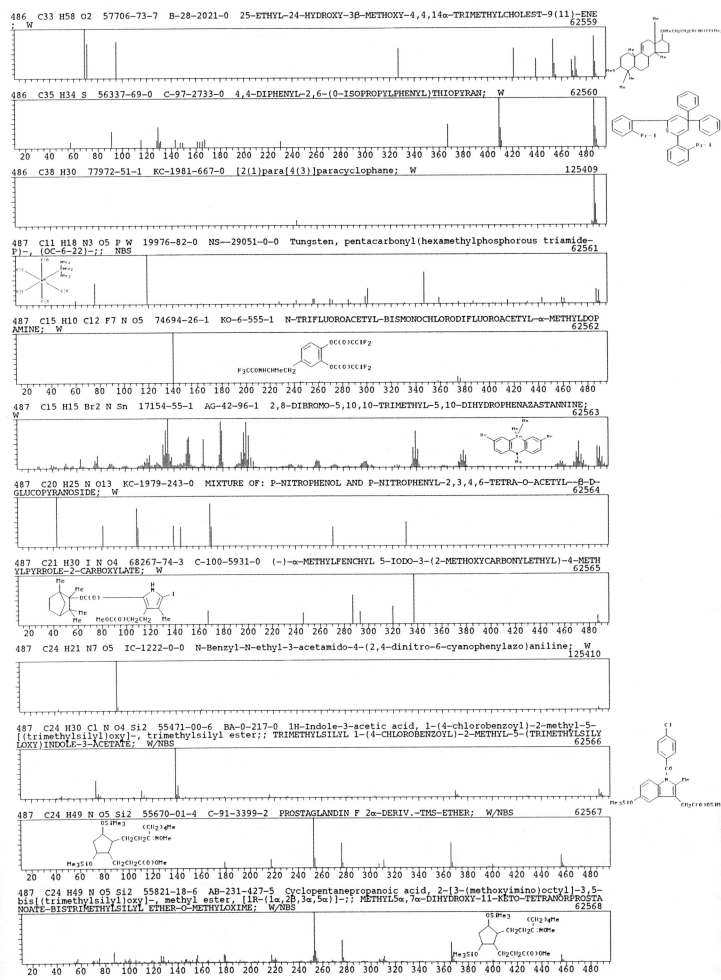

486   C33 H58 O2   57706-73-7   B-28-2021-0   25-ETHYL-24-HYDROXY-3β-METHOXY-4,4,14α-TRIMETHYLCHOLEST-9(11)-ENE
; W   62559

486   C35 H34 S   56337-69-0   C-97-2733-0   4,4-DIPHENYL-2,6-(O-ISOPROPYLPHENYL)THIOPYRAN; W   62560

486   C38 H30   77972-51-1   KC-1981-667-0   [2(1)para[4(3)]paracyclophane; W   125409

487   C11 H18 N3 O5 P W   19976-82-0   NS--29051-0-0   Tungsten, pentacarbonyl(hexamethylphosphorous triamide-
P)-, (OC-6-22)-;;   NBS   62561

487   C15 H10 Cl2 F7 N O5   74694-26-1   KO-6-555-1   N-TRIFLUOROACETYL-BISMONOCHLORODIFLUOROACETYL-α-METHYLDOP
AMINE; W   62562

487   C15 H15 Br2 N Sn   17154-55-1   AG-42-96-1   2,8-DIBROMO-5,10,10-TRIMETHYL-5,10-DIHYDROPHENAZASTANNINE;
W   62563

487   C20 H25 N O13   KC-1979-243-0   MIXTURE OF: P-NITROPHENOL AND P-NITROPHENYL-2,3,4,6-TETRA-O-ACETYL--β-D-
GLUCOPYRANOSIDE; W   62564

487   C21 H30 I N O4   68267-74-3   C-100-5931-0   (-)-α-METHYLFENCHYL 5-IODO-3-(2-METHOXYCARBONYLETHYL)-4-METH
YLPYRROLE-2-CARBOXYLATE; W   62565

487   C24 H21 N7 O5   IC-1222-0-0   N-Benzyl-N-ethyl-3-acetamido-4-(2,4-dinitro-6-cyanophenylazo)aniline; W
                                                                                                            125410

487   C24 H30 Cl N O4 Si2   55471-00-6   BA-0-217-0   1H-Indole-3-acetic acid, 1-(4-chlorobenzoyl)-2-methyl-5-
[(trimethylsilyl)oxy]-, trimethylsilyl ester;; TRIMETHYLSILYL 1-(4-CHLOROBENZOYL)-2-METHYL-5-(TRIMETHYLSILY
LOXY)INDOLE-3-ACETATE; W/NBS   62566

487   C24 H49 N O5 Si2   55670-01-4   C-91-3399-2   PROSTAGLANDIN F 2α-DERIV.-TMS-ETHER; W/NBS   62567

487   C24 H49 N O5 Si2   55821-18-6   AB-231-427-5   Cyclopentanepropanoic acid, 2-[3-(methoxyimino)octyl]-3,5-
bis[(trimethylsilyl)oxy]-, methyl ester, [1R-(1α,2β,3α,5α)]-;; METHYL5α,7α-DIHYDROXY-11-KETO-TETRANORPROSTA
NOATE-BISTRIMETHYLSILYL ETHER-O-METHYLOXIME; W/NBS   62568

487 C25 H18 Br N3 O3 Y-20-381-0 3-(p-Nitro-phenyl)-5,7-diphenyl-2H-pyrido[2,1-f]-1,3,4-oxadiazinium brom
ides; W   125411

487 C26 H21 N3 O5 S 86317-96-6 K-116-1829-0 N-[2-(4-Methylphenylsulfonyloxy)phenyl]-N'-(4-nitrophenyl)
benzamidine; W   125412

20 40 60 80 100 120 140 160 180 200 220 240 260 280 300 320 340 360 380 400 420 440 460 480

487 C26 H21 N3 O5 S 86317-97-7 K-116-1829-0 N-[4-(4-Methylphenylsulfonyloxy)phenyl]-N'-(4-nitrophenyl)
benzamidine; W   125413

487 C26 H33 N O6 S 82769-86-6 C-104-5736-0 (1'S,1S,2S,3R,4R,6S)-2,6-Dimethyl-3-9(((1'-phenylethyl)amin
o)-carbonyl)oxy)-4-(phenylsulfonyl)cyclohexane-1,2-diol 1,2-acetonide; W   125414

487 C26 H37 N3 O6 77536-56-2 NS-0-0-0 L-Tryptophan, N-(N-acetyl-L-α-glutamyl)-, dibutyl ester; W/NBS
                                                                                               125415

20 40 60 80 100 120 140 160 180 200 220 240 260 280 300 320 340 360 380 400 420 440 460 480

487 C26 H57 N O3 Si2 SW-0-230-0 BISTRIMETHYLSILYL N-ACETYL SPHINGANINE; W   62569

487 C26 H57 N O3 Si2 D3-1980-190-9 BIS-O-TRIMETHYLSILYL-N-ACETYLSPHINGANINE; W   62570

487 C27 H34 Co N5 T-68-5595-0 DICYANIDE OF COBALT COMPLEX WITH CORRIN-TYPE LIGAND; W   62571

20 40 60 80 100 120 140 160 180 200 220 240 260 280 300 320 340 360 380 400 420 440 460 480

487 C28 H33 N O3 Si2 73593-57-4 K-113-320-0 4-Oxo-2,3,4-triphenyl-2,3-bis(trimethylsiloxy)butanenitrile
; W   125416

487 C29 H29 N O6 83953-26-8 KC-1982-2146-0 N-Acetoxy-2,3-O-isopropylidene-5-O-trityl-D-ribonimido-1,4-
lactone; W   125417

487 C30 H24 F3 N O2 77507-23-4 J-46-3244-0 N-[1-(1-Naphthyl)ethyl]-α-[1-(9-anthryl)-2,2,2-trifluoroetho
xy]acetamide; W   125418

20 40 60 80 100 120 140 160 180 200 220 240 260 280 300 320 340 360 380 400 420 440 460 480

487 C30 H24 F3 N O2 77507-24-5 J-46-3244-0 N-[1-(2-Naphthyl)ethyl]-α-[1-(9-anthryl)-2,2,2-trifluoroetho
xy]acetamide; W   125419

20 40 60 80 100 120 140 160 180 200 220 240 260 280 300 320 340 360 380 400 420 440 460 480

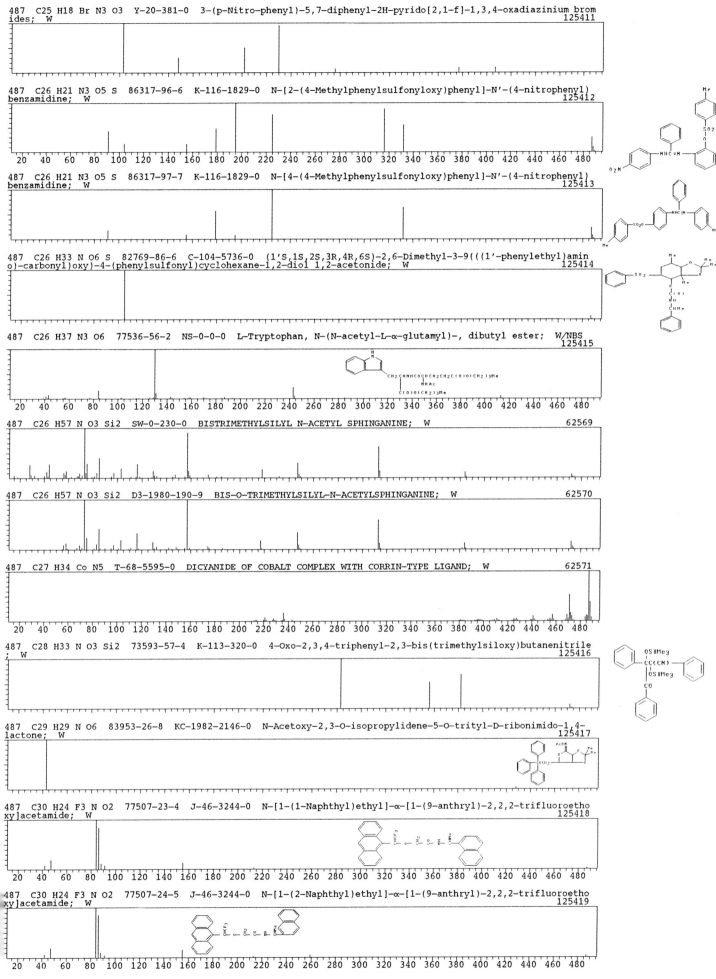

487 C30 H29 N3 Si2 77123-68-3 J-46-2285-0 10-[(Trimethylsilyl)ethynyl]-6-[3-[(trimethylsilyl)ethynyl]ph
enyl]benzimidazoquinazoline; W   125420

487 C32 H29 N3 O2 73661-65-1 K-113-436-0 6,6-Dicyano-2-diethylamino-1,4-bis(4-methoxyphenyl)-3-phenyl-
2-aminofulvene; W   125421

487 C33 H29 N O3 73476-16-1 K-114-955-0 10-[4-(4-Cyanobenzyloxy)pentyloxy]-10-phenyl-9(10H)-anthraceno
ne; W   125422

487 C34 H49 N O 34535-60-9 AD-0-3746-0 1'H-Cholest-3-eno[3,4-b]indol-6-one, 1'-methyl-, (5β)-;; 6-OXO-
5β-CHOLESTANO(3,4-B)N-METHYLINDOLE; W/NBS   62573

487 C34 H49 N O 38389-03-6 AD-0-3731-0 1'H-Cholest-2-eno[3,2-b]indol-6-one, 1'-methyl-, (5α)-;; 6-OXO-
5α-CHOLESTANO(3,2-B)N-METHYLINDOLE; W/NBS   62574

487 C34 H49 N O 34535-61-0 O-5-1192-2 1'H-Cholest-2-eno[3,2-b]indol-6-one, 1'-methyl-, (5β)-;; 6-OXO-5β
-CHOLESTANO-(3,2-B)-N-METHYLINDOLE; W/NBS   62575

487 C34 H49 N O 38389-06-9 AD-0-3734-0 1'H-Cholest-2-eno[3,2-b]indol-6-one, 5'-methyl-, (5α)-;; 6-OXO-
5α-CHOLESTANO(3,2-B)5'-METHYLINDOLE; W/NBS   62572

488 C3 H5 F9 Fe I P3 42911-96-6 NS-0-0-0 Iron, iodotris(phosphorous trifluoride)(η3-2-propenyl)-;
W/NBS   125423

488 C6 H6 C19 O4 P RB-1982-15532-0 TRIS-(2,2,2-TRICHLOROETHYL)PHOSPHATE; W   125424

488 C10 H2 Cl10 O 1034-41-9 NS-0-0-0 KEPOL; W/NBS   125425

488 C10 H2 Cl10 O RB-1982-14401-0 DECACHLOROPENTACYCLODECAN-10-OL; W   125426

488 C10 H19 As3 N2 S3 51678-03-6 TX-0-253-0 Arsinothious acid, dimethyl-, 2,4,6-pyrimidinetriyl ester;;
2,4,6-TRIS(DIMETHYLARSINO)THIO)PYRIMIDINE; W/NBS   62577

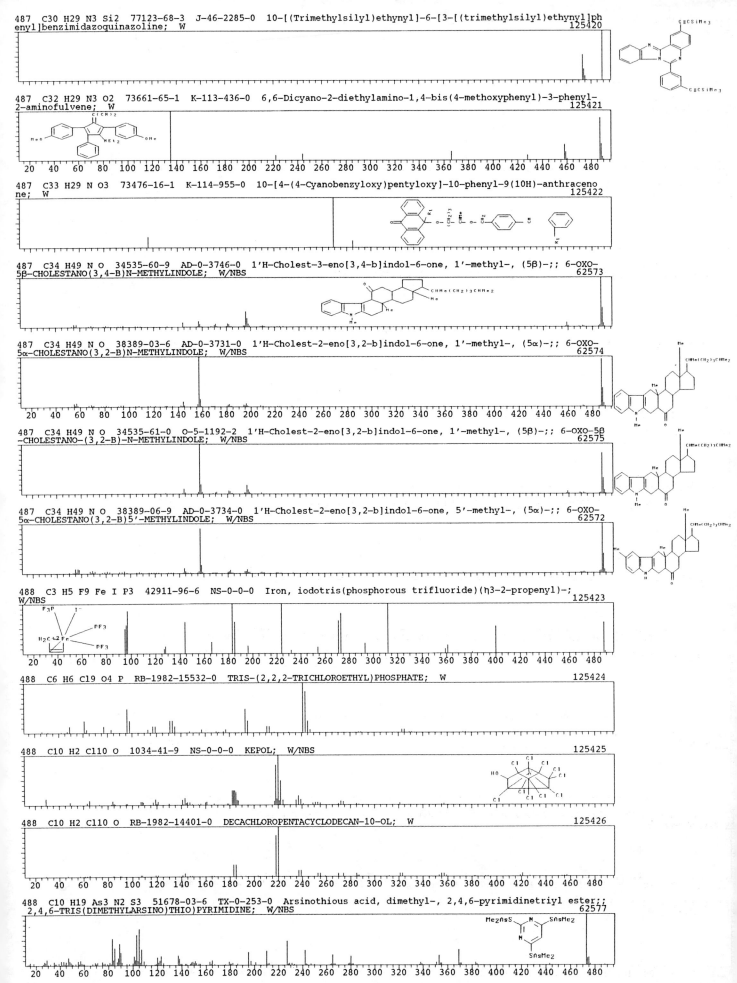

- 5382 -

488   C12 H4 Cl7 O2 P S   RB-1982-15533-0   O,O-BIS-(2,4,5-TRICHLOROPHENYL)ESTER OF PHOSPHOROCHLORIDOTHIOIC ACID;   W   125427

488   C13 H Cl9 O   PG-1982-255-0   NONACHLORO-BENZOPHENONE;   W   62578

488   C16 H10 F10 O6   55683-25-5   NS-0-0-0   Benzenepropanoic acid, α,4-bis(2,2,3,3,3-pentafluoro-1-oxopropoxy)-, methyl ester;; PARA-HYDROXYPHENYLLACTIC ACID METHYL ESTER PENTAFLUOROPROPIONATE;   W/NBS   125428

488   C16 H41 O7 P Si4   55723-94-9   BA-0-308-0   Phosphoric acid, 4-oxo-2,3-bis[(trimethylsilyl)oxy]butyl bis(trimethylsilyl)ester, [R-(R*,R*)]-;; O-TETRAKIS(TRIMETHYLSILYL)-D-ERYTHROSE-4-PHOSPHATE (ISOMER B);   W   62581

488   C16 H41 O7 P Si4   BA-0-307-0   O-TETRAKIS(TRIMETHYLSILYL)-D-ERYTHROSE-4-PHOSPHATE (ISOMER A);   W   62580

488   C20 H14 Br2 N2 O S   80746-27-6   KC-1981-2950-0   N'-benzoyl-N'-(2,4-dibromophenyl)benzothiohydrazide;   W   125429

488   C20 H40 O6 Si4   OM-1981-731-0   FUMARYLACETOACETIC ACID 4TMS;   W   62582

488   C21 H20 N4 O10   F-41-1759-0   4-(3,4,5-Trimethoxybenzyl)pyridine;   W   125430

488   C22 H32 O12   51885-42-8   AU-32-13-8   α-D-Galactopyranose, 6-O-(2,4-di-O-acetyl-3,6-anhydro-β-D-galactopyranosyl)-1,2:3,4-bis-O-(1-methylethylidene)-;; 6-O-(2,4-DI-O-ACETYL-3,6-ANHYDRO-β-D-GALACTOPYRANOSYL)-1,2:3,4-DI-O-ISOPROPYLIDENE-α-D-GALACTOPYRANOSE;   W/NBS   62583

488   C22 H38 Ni S4   33915-69-4   B-29-261-2   BIS(2,2,6,6-TETRAMETHYLHEPTANE-3,5-DITHONATO)NICKEL(II);   W   125431

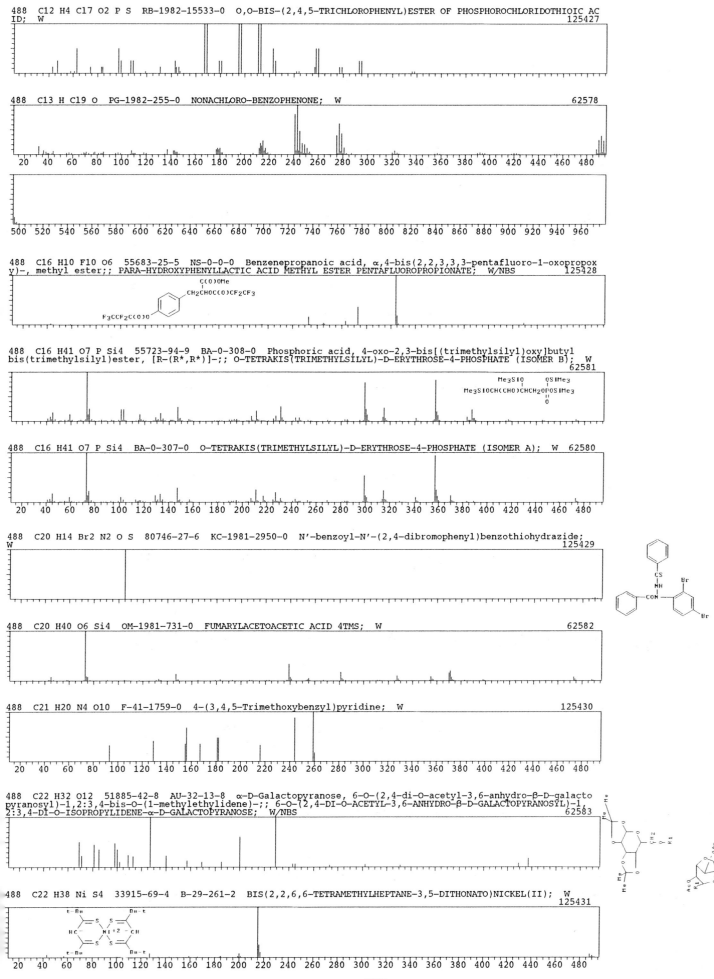

488 C23 H23 D5 N4 O8 55521-21-6 T-68-5001-0 D5-BENZOYL-GLU-OME-N-ETHOXYCARBONYL-HIS; W   62586

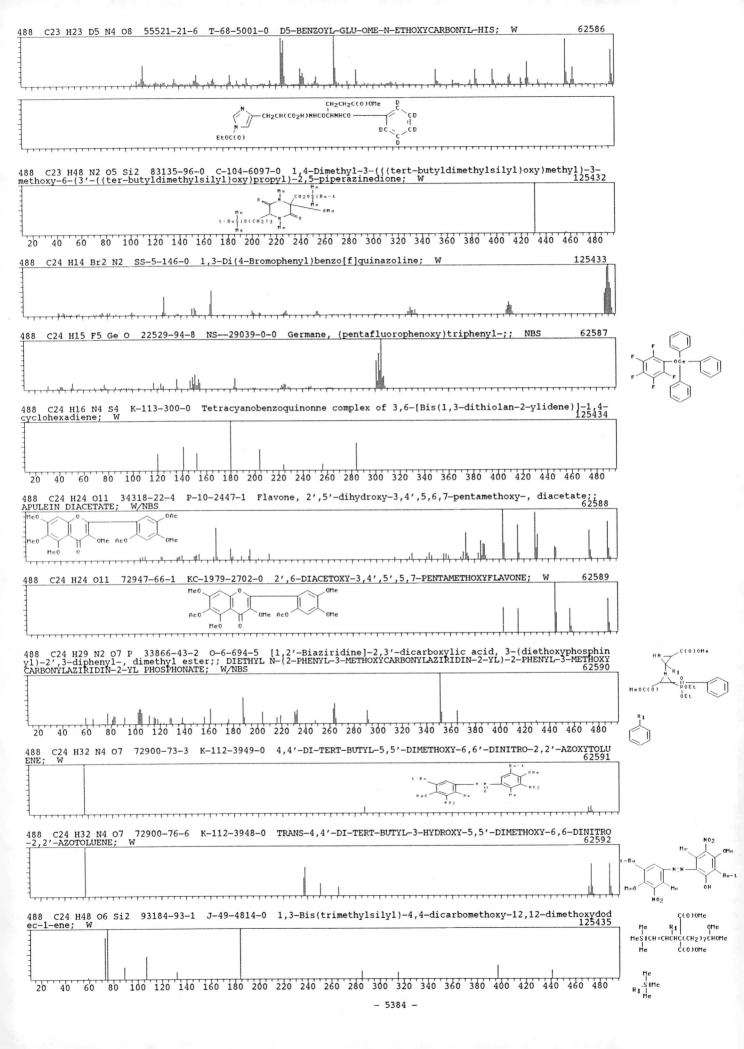

488 C23 H48 N2 O5 Si2 83135-96-0 C-104-6097-0 1,4-Dimethyl-3-(((tert-butyldimethylsilyl)oxy)methyl)-3-methoxy-6-(3'-((ter-butyldimethylsilyl)oxy)propyl)-2,5-piperazinedione; W   125432

488 C24 H14 Br2 N2 SS-5-146-0 1,3-Di(4-Bromophenyl)benzo[f]quinazoline; W   125433

488 C24 H15 F5 Ge O 22529-94-8 NS--29039-0-0 Germane, (pentafluorophenoxy)triphenyl-;; NBS   62587

488 C24 H16 N4 S4 K-113-300-0 Tetracyanobenzoquinonne complex of 3,6-[Bis(1,3-dithiolan-2-ylidene)]-1,4-cyclohexadiene; W   125434

488 C24 H24 O11 34318-22-4 P-10-2447-1 Flavone, 2',5'-dihydroxy-3,4',5,6,7-pentamethoxy-, diacetate;; APULEIN DIACETATE; W/NBS   62588

488 C24 H24 O11 72947-66-1 KC-1979-2702-0 2',6-DIACETOXY-3,4',5',5,7-PENTAMETHOXYFLAVONE; W   62589

488 C24 H29 N2 O7 P 33866-43-2 O-6-694-5 [1,2'-Biaziridine]-2,3'-dicarboxylic acid, 3-(diethoxyphosphinyl)-2',3-diphenyl-, dimethyl ester;; DIETHYL N-(2-PHENYL-3-METHOXYCARBONYLAZIRIDIN-2-YL)-2-PHENYL-3-METHOXYCARBONYLAZIRIDIN-2-YL PHOSPHONATE; W/NBS   62590

488 C24 H32 N4 O7 72900-73-3 K-112-3949-0 4,4'-DI-TERT-BUTYL-5,5'-DIMETHOXY-6,6'-DINITRO-2,2'-AZOXYTOLUENE; W   62591

488 C24 H32 N4 O7 72900-76-6 K-112-3948-0 TRANS-4,4'-DI-TERT-BUTYL-3-HYDROXY-5,5'-DIMETHOXY-6,6-DINITRO-2,2'-AZOTOLUENE; W   62592

488 C24 H48 O6 Si2 93184-93-1 J-49-4814-0 1,3-Bis(trimethylsilyl)-4,4-dicarbomethoxy-12,12-dimethoxydodec-1-ene; W   125435

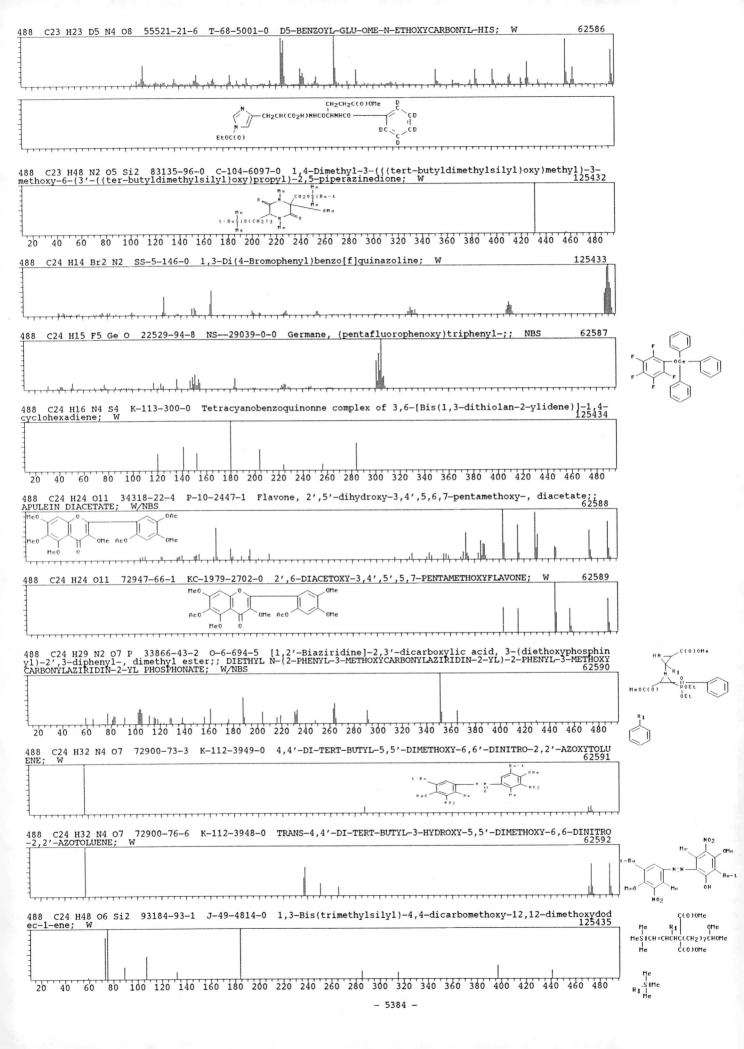

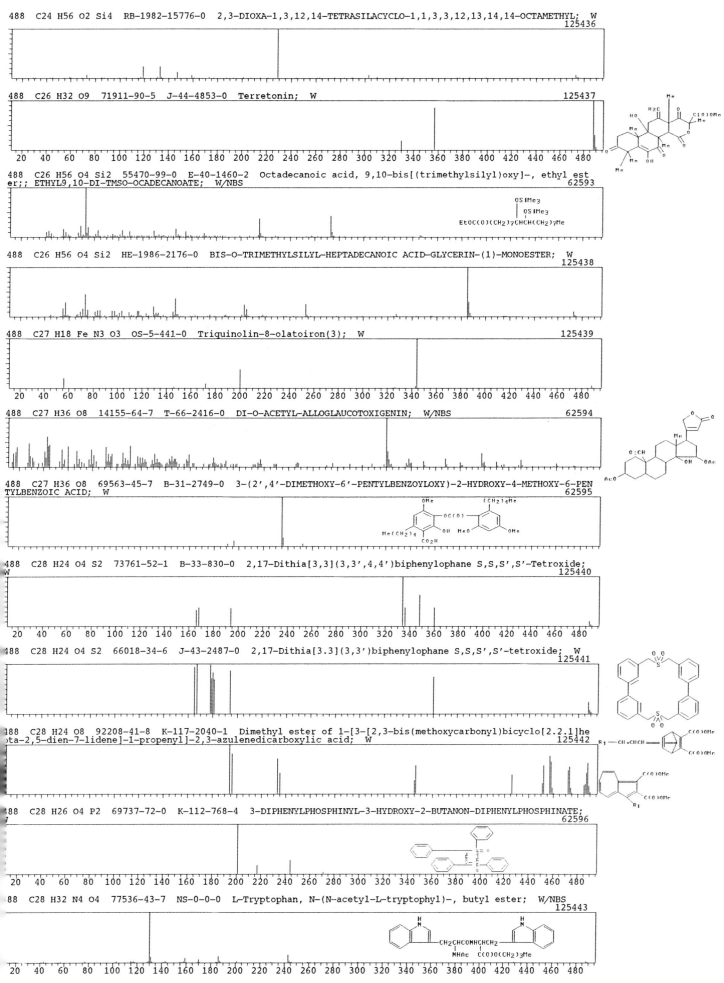

488 C24 H56 O2 Si4 RB-1982-15776-0 2,3-DIOXA-1,3,12,14-TETRASILACYCLO-1,1,3,3,12,13,14,14-OCTAMETHYL; W
125436

488 C26 H32 O9 71911-90-5 J-44-4853-0 Terretonin; W
125437

20 40 60 80 100 120 140 160 180 200 220 240 260 280 300 320 340 360 380 400 420 440 460 480

488 C26 H56 O4 Si2 55470-99-0 E-40-1460-2 Octadecanoic acid, 9,10-bis[(trimethylsilyl)oxy]-, ethyl est
er;; ETHYL9,10-DI-TMSO-OCADECANOATE; W/NBS
62593

488 C26 H56 O4 Si2 HE-1986-2176-0 BIS-O-TRIMETHYLSILYL-HEPTADECANOIC ACID-GLYCERIN-(1)-MONOESTER; W
125438

488 C27 H18 Fe N3 O3 OS-5-441-0 Triquinolin-8-olatoiron(3); W
125439

20 40 60 80 100 120 140 160 180 200 220 240 260 280 300 320 340 360 380 400 420 440 460 480

488 C27 H36 O8 14155-64-7 T-66-2416-0 DI-O-ACETYL-ALLOGLAUCOTOXIGENIN; W/NBS
62594

488 C27 H36 O8 69563-45-7 B-31-2749-0 3-(2',4'-DIMETHOXY-6'-PENTYLBENZOYLOXY)-2-HYDROXY-4-METHOXY-6-PEN
TYLBENZOIC ACID; W
62595

488 C28 H24 O4 S2 73761-52-1 B-33-830-0 2,17-Dithia[3,3](3,3',4,4')biphenylophane S,S,S',S'-Tetroxide;
W
125440

20 40 60 80 100 120 140 160 180 200 220 240 260 280 300 320 340 360 380 400 420 440 460 480

488 C28 H24 O4 S2 66018-34-6 J-43-2487-0 2,17-Dithia[3.3](3,3')biphenylophane S,S,S',S'-tetroxide; W
125441

488 C28 H24 O8 92208-41-8 K-117-2040-1 Dimethyl ester of 1-[3-[2,3-bis(methoxycarbonyl)bicyclo[2.2.1]he
pta-2,5-dien-7-lidene]-1-propenyl]-2,3-azulenedicarboxylic acid; W
125442

488 C28 H26 O4 P2 69737-72-0 K-112-768-4 3-DIPHENYLPHOSPHINYL-3-HYDROXY-2-BUTANON-DIPHENYLPHOSPHINATE;
62596

20 40 60 80 100 120 140 160 180 200 220 240 260 280 300 320 340 360 380 400 420 440 460 480

488 C28 H32 N4 O4 77536-43-7 NS-0-0-0 L-Tryptophan, N-(N-acetyl-L-tryptophyl)-, butyl ester; W/NBS
125443

20 40 60 80 100 120 140 160 180 200 220 240 260 280 300 320 340 360 380 400 420 440 460 480

488   C28 H36 N6 O2   75812-74-7   J-46-1086-0   1,4-Bis[N-[[(dimethylamido)amino]cyclopropyl]phenylamino]-2-butyne;   W
125444

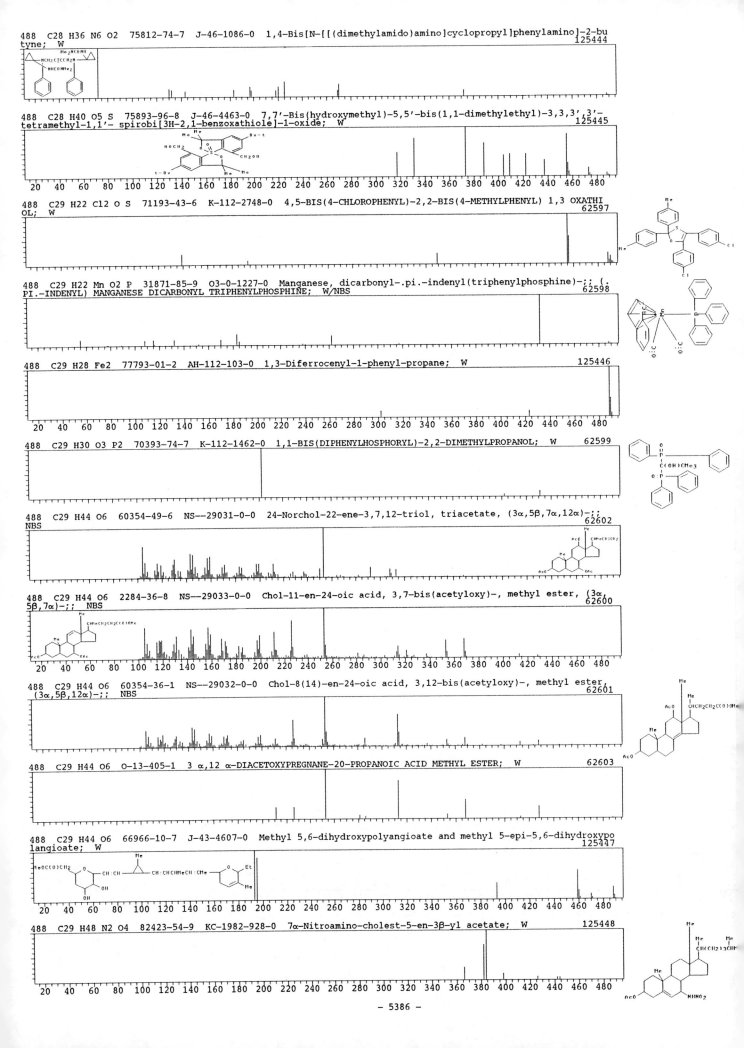

488   C28 H40 O5 S   75893-96-8   J-46-4463-0   7,7'-Bis(hydroxymethyl)-5,5'-bis(1,1-dimethylethyl)-3,3,3',3'-tetramethyl-1,1'- spirobi[3H-2,1-benzoxathiole]-1-oxide;   W
125445

488   C29 H22 Cl2 O S   71193-43-6   K-112-2748-0   4,5-BIS(4-CHLOROPHENYL)-2,2-BIS(4-METHYLPHENYL) 1,3 OXATHIOL;   W
62597

488   C29 H22 Mn O2 P   31871-85-9   O3-0-1227-0   Manganese, dicarbonyl-.pi.-indenyl(triphenylphosphine)-;; (.PI.-INDENYL) MANGANESE DICARBONYL TRIPHENYLPHOSPHINE;   W/NBS
62598

488   C29 H28 Fe2   77793-01-2   AH-112-103-0   1,3-Diferrocenyl-1-phenyl-propane;   W
125446

488   C29 H30 O3 P2   70393-74-7   K-112-1462-0   1,1-BIS(DIPHENYLHOSPHORYL)-2,2-DIMETHYLPROPANOL;   W
62599

488   C29 H44 O6   60354-49-6   NS--29031-0-0   24-Norchol-22-ene-3,7,12-triol, triacetate, (3α,5β,7α,12α)-;; NBS
62602

488   C29 H44 O6   2284-36-8   NS--29033-0-0   Chol-11-en-24-oic acid, 3,7-bis(acetyloxy)-, methyl ester, (3α,5β,7α)-;;   NBS
62600

488   C29 H44 O6   60354-36-1   NS--29032-0-0   Chol-8(14)-en-24-oic acid, 3,12-bis(acetyloxy)-, methyl ester, (3α,5β,12α)-;;   NBS
62601

488   C29 H44 O6   O-13-405-1   3 α,12 α-DIACETOXYPREGNANE-20-PROPANOIC ACID METHYL ESTER;   W
62603

488   C29 H44 O6   66966-10-7   J-43-4607-0   Methyl 5,6-dihydroxypolyangioate and methyl 5-epi-5,6-dihydroxypolangioate;   W
125447

488   C29 H48 N2 O4   82423-54-9   KC-1982-928-0   7α-Nitroamino-cholest-5-en-3β-yl acetate;   W
125448

488 C29 H48 N2 O4 82423-55-0 KC-1982-928-0 7β-Nitroamino-cholest-5-en-3βyl acetate; W   125449

488 C29 H52 O2 Si2 66250-80-4 BS-4-95-0 Silane, [[(3α,5α,17β)-androstane-3,17-diyl]bis(oxy)]bis[dimeth
yl-2-propenyl-; W   125450

488 C29 H52 O2 Si2 66250-81-5 BS-4-96-0 Silane, [[(3β,5α,17β)-androstane-3,17-diyl]bis(oxy)]bis[dimeth
yl-2-propenyl-; W   125451

488 C29 H52 O2 Si2 66293-36-5 BS-4-97-0 Silane, [[(3α,5β,17β)-androstane-3,17-diyl]bis(oxy)]bis[dimeth
yl-2-propenyl-; W   125452

488 C29 H52 O2 Si2 77259-74-6 KO-7-213-1 Vinyldimetlsilyl derivative of 5α-Pregane-3β,20α-diol; W
                                                                                        125453

488 C29 H52 O2 Si2 77259-75-7 KO-7-213-1 Vinyldimetlsilyl derivative of 5α-Pregane-3β,20β-diol; W
                                                                                        125454

488 C30 H48 O5 56143-30-7 SD-1981-0-0 Ergostane-3,6-dione, 25-(acetyloxy)-5-hydroxy-, (5α)-;; W/NBS
                                                                                        62608

488 C30 H48 O5 60877-02-3 F-32-1079-0 3β,11α,15α-TRIHYDROXYCYCLOART-24-EN-26-OIC ACID; W   62604

188 C30 H48 O5 F-32-2357-0 Δ9(11)-HOLOSTEBE-3β,23ε,25-TRIOL; W   62605

188 C30 H48 O5 64340-23-4 B-30-1399-0 15α,16α,22α,28-TETRAHYDROXYALEAN-12-EN-3-ONE; W   62606

Str 26

188 C30 H48 O5 64285-74-1 B-30-1318-0 (16α,21β,22α,28-TETRAHYDROXYOLEAN-12-EN-3-ONE; W   62607

188 C30 H48 O5 86632-20-4 J-48-3761-0 Orthosphenic acid; W   125455

88 C30 H52 O3 Si 55759-94-9 D-10-2802-6 9,10-Secocholesta-5,7,10(19)-triene-1,3-diol, 25-[(trimethylsi
yl)oxy]-, (3β,5Z,7E)-;; 1,25-DIHYDROXYCHOLECALCIFEROL-25-TRIMETHYLSILYL ETHER; W/NBS   62609

- 5387 -

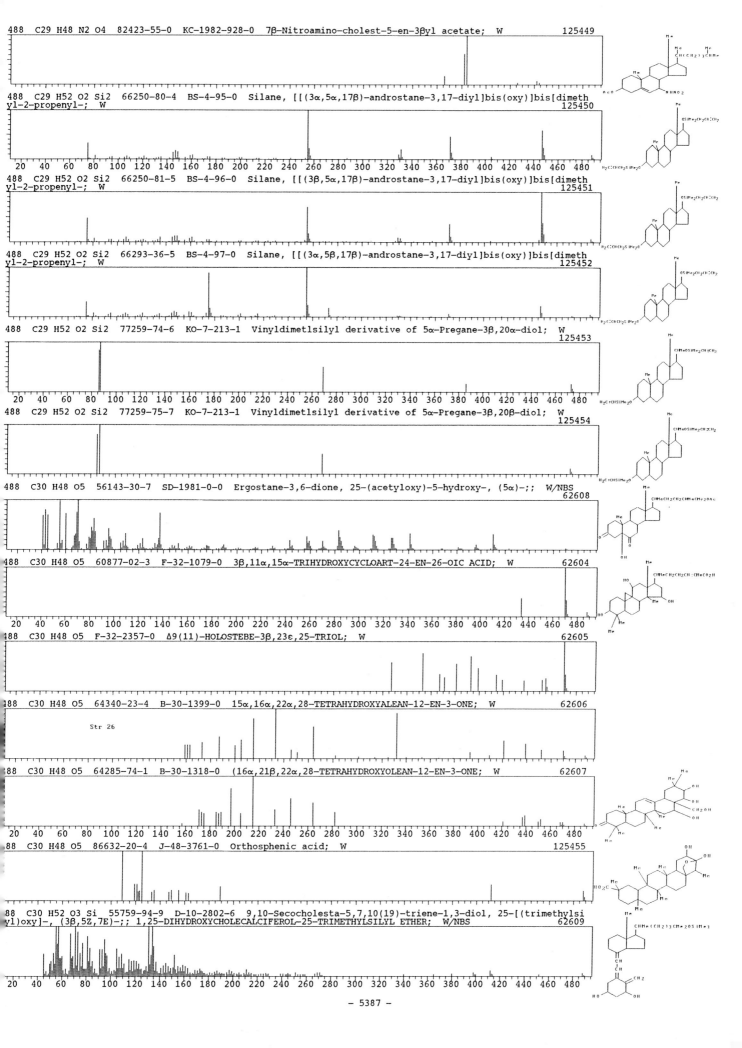

488 C31 H24 N2 O4 B-36-1137-0 N,N'-Bis(2'-Benzoylphenyl)cyclopropane-1,1-dicarboxamide; W   125475

488 C31 H52 O4 54498-63-4 NS--29023-0-0 Cholestane-2-carboxylic acid, 3,3-[1,2-ethanediylbis(oxy)]-,
methyl ester, (2β,5α)-;; NBS   62614

488 C31 H52 O4 3514-28-1 SD-1981-0-0 Cholestane-3,6-diol, diacetate, (3β,5α,6α)-;; 5α-Cholestane-3β,6α-
diol, diacetate; W/NBS   62617

488 C31 H52 O4 28809-56-5 O-4-377-1 Cholestane-2,19-diol, diacetate, (2α,5α)-;; 2α-ACETOXY-19-ACETOXY
METHYL-5α-CHOLESTANE; W/NBS   62610

488 C31 H52 O4 28809-57-6 O-4-377-2 Cholestane-1,19-diol, diacetate, (1α,5α)-;; 1α-ACETOXY-19-ACETOXY
METHYL-5α-CHOLESTANE; W/NBS   62611

488 C31 H52 O4 19518-70-8 O6-0-1095-0 Cholestan-6-one, 3-(acetyloxy)-, cyclic 1,2-ethanediyl acetal,
(3β,5α)-;; 3β-ACETOXY-6,6-ETHYLENEDIOXY-5α-CHOLESTANE; W/NBS   62612

488 C31 H52 O4 21671-01-2 O-10-982-1 3,4-Secodammar-4(28)-en-3-oic acid, 20,24-epoxy-25-hydroxy-, meth
yl ester, (24S)-;; Methyl shorate; W/NBS   62613

488 C31 H52 O4 63727-16-2 C-99-4483-0 3β,14α-DIHYDROXY-5α-STIGMAST-15-ONE 3-ACETATE; W   62615

488 C31 H52 O4 57030-54-3 AH-106-690-9 THE DIACETATE DERIVATIVE OF CHOLESTANE-2α-3α-DIOL; W   62616

488 C31 H56 O2 Si 65599-00-0 NS-0-0-0 Androstan-17-one, 3-[(tributylsilyl)oxy]-, (3β,5α)-; W/NBS   125456

488 C31 H56 O2 Si 65599-01-1 NS-0-0-0 Androstan-17-one, 3-[(tributylsilyl)oxy]-, (3α,5β)-; W/NBS   125457

488 C31 H56 O2 Si 65599-02-2 NS-0-0-0 Androstan-17-one, 3-[(tributylsilyl)oxy]-, (3β,5β)-; W/NBS   125458

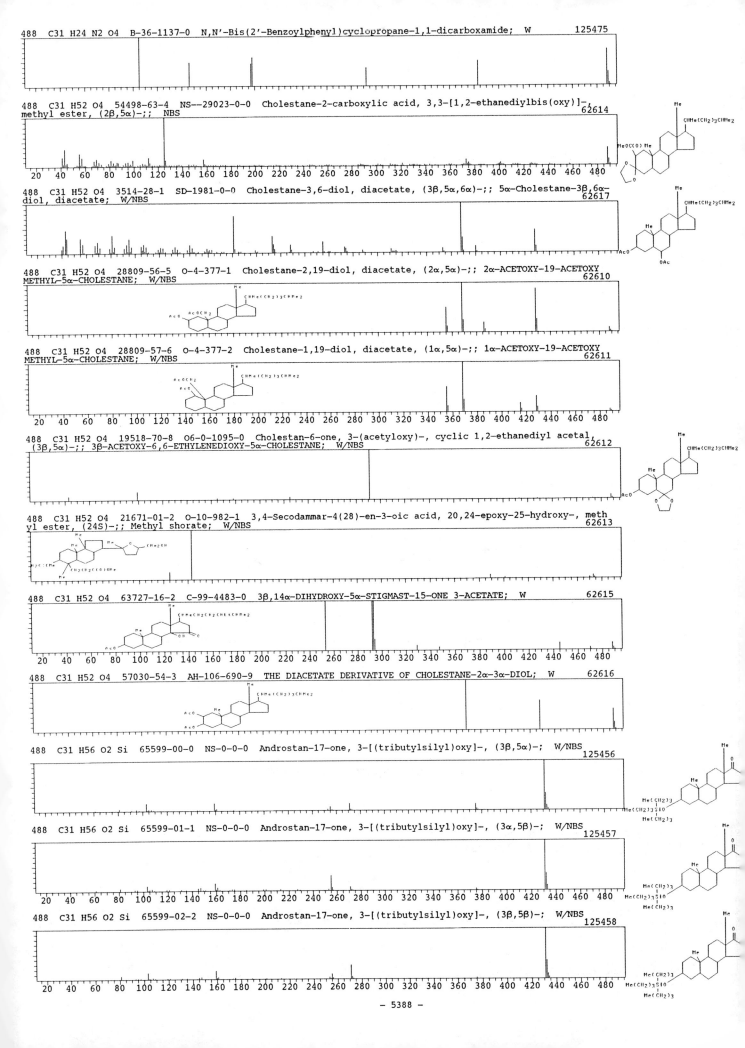

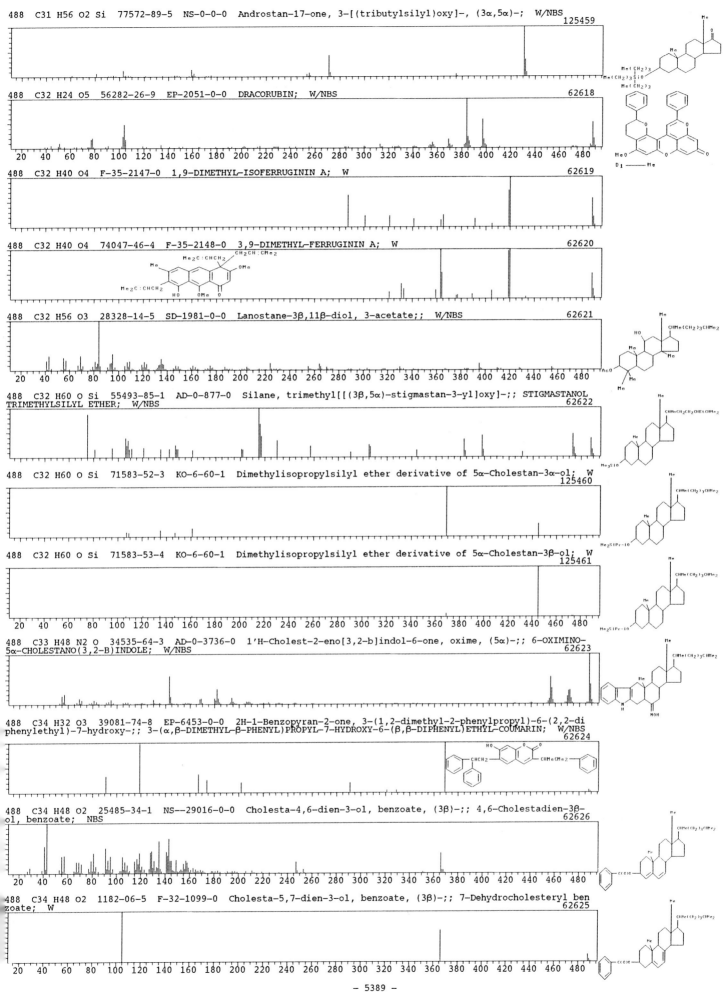

488 C31 H56 O2 Si 77572-89-5 NS-0-0-0 Androstan-17-one, 3-[(tributylsilyl)oxy]-, (3α,5α)-; W/NBS
125459

488 C32 H24 O5 56282-26-9 EP-2051-0-0 DRACORUBIN; W/NBS
62618

488 C32 H40 O4 F-35-2147-0 1,9-DIMETHYL-ISOFERRUGININ A; W
62619

488 C32 H40 O4 74047-46-4 F-35-2148-0 3,9-DIMETHYL-FERRUGININ A; W
62620

488 C32 H56 O3 28328-14-5 SD-1981-0-0 Lanostane-3β,11β-diol, 3-acetate;; W/NBS
62621

488 C32 H60 O Si 55493-85-1 AD-0-877-0 Silane, trimethyl[[(3β,5α)-stigmastan-3-yl]oxy]-;; STIGMASTANOL
TRIMETHYLSILYL ETHER; W/NBS
62622

488 C32 H60 O Si 71583-52-3 KO-6-60-1 Dimethylisopropylsilyl ether derivative of 5α-Cholestan-3α-ol; W
125460

488 C32 H60 O Si 71583-53-4 KO-6-60-1 Dimethylisopropylsilyl ether derivative of 5α-Cholestan-3β-ol; W
125461

488 C33 H48 N2 O 34535-64-3 AD-0-3736-0 1'H-Cholest-2-eno[3,2-b]indol-6-one, oxime, (5α)-;; 6-OXIMINO-
5α-CHOLESTANO(3,2-B)INDOLE; W/NBS
62623

488 C34 H32 O3 39081-74-8 EP-6453-0-0 2H-1-Benzopyran-2-one, 3-(1,2-dimethyl-2-phenylpropyl)-6-(2,2-di
phenylethyl)-7-hydroxy-;; 3-(α,β-DIMETHYL-β-PHENYL)PROPYL-7-HYDROXY-6-(β,β-DIPHENYL)ETHYL-COUMARIN; W/NBS
62624

488 C34 H48 O2 25485-34-1 NS--29016-0-0 Cholesta-4,6-dien-3-ol, benzoate, (3β)-;; 4,6-Cholestadien-3β-
ol, benzoate; NBS
62626

488 C34 H48 O2 1182-06-5 F-32-1099-0 Cholesta-5,7-dien-3-ol, benzoate, (3β)-;; 7-Dehydrocholesteryl ben
zoate; W
62625

488   C37 H28 O   HE-1986-1092-0   1,5-BIS(1-CHRYSYL)-3-PENTANOL;   W   125462

488   C37 H28 O   70456-58-5   J-44-2486-0   5-Acetyl-1,2,3,4,5-pentaphenyl-1,3-cyclopentadiene;   W   125463

489   C11 H3 Cl8 N3 S   24478-10-2   AD-0-3134-0   1,3,5-Triazine, 2-[(3,4-dichlorophenyl)thio]-4,6-bis(trichloromethyl)-;;   2-(3',4'-DICHLOROPHENYLTHIO)-4,6-BIS(TRICHLOROMETHYL)-S-TRIAZINE;   W/NBS   62627

489   C15 H21 O6 Os   15635-86-6   T-66-1402-0   Osmium, tris(2,4-pentanedionato-O,O')-, (OC-6-11)-;;   TRIS(ACETYLACETONATO)OSMIUM(III);   W/NBS   62628

489   C18 H39 N3 O6 S3   56221-19-3   O-11-1222-1   1,3,5,-TRIS(N-PENTYLSULPHONYL) HEXAHYDRO-1,3,5,-TRIAZINE;   W   62629

489   C19 H29 N3 O8 P2   58276-64-5   O-12-121-1   PHTHALIMIDOMETHYL(ETHOXY)PHOSPHINYLAMINOMETHYL(ETHOXY)PHOSPHINYLALANINEETHYL ESTER;   W   62630

489   C21 H33 Br2 N O2   40572-24-5   EP-6532-0-0   Benzene, 1,3-dibromo-2,4,6-tris(2,2-dimethylpropyl)-5-nitro-;;   1,3-DIBROMO-2,4,6-TRINEOPENTYL-5-NITROBENZENE;   W/NBS   62631

489   C21 H39 N5 O8   55649-82-6   D-13-600-1   D-Streptamine, 4-O-[3-(acetylamino)-6-(aminomethyl)-3,4-dihydro-2H-pyran-2-yl]-2-deoxy-6-O-[3-deoxy-4-C-methyl-3-(methylamino)-β-L-arabinopyranosyl]-, (2S-cis)-;;   2'-N-ACETYLSISOMICIN;   W/NBS   62632

489   C22 H35 N O11   84036-66-8   H-65-1139-0   (t-Butyl) ester of (3S,3aR,5R,6S,6aR)- and (3S,3aR,5S,6S,6aR)-5,6-Dihydroxy-2-(2,3:5,6-di-O-isopropylidene-α-D-mannofuranosyl)-perhydrofuro[2,3-d]isoxazol-3-carboxylic acid;   W   125464

489   C24 H31 N3 O8   70582-71-7   SB-33-132-0   [1-(Ethoxycarbonyl)methyl-4-formylpyrrol-2-ylmethyl][1-(ethoxycarbonyl)methyl-5-formylpyrrol-2-ylmethyl]amine;   W   125465

489   C24 H31 N3 O8   70582-72-8   SB-33-131-0   Di-[1-(ethoxycarbonyl)methyl-5-formylpyrrol-2-ylmethyl]ethoxycarbonylmethylamine;   W   125466

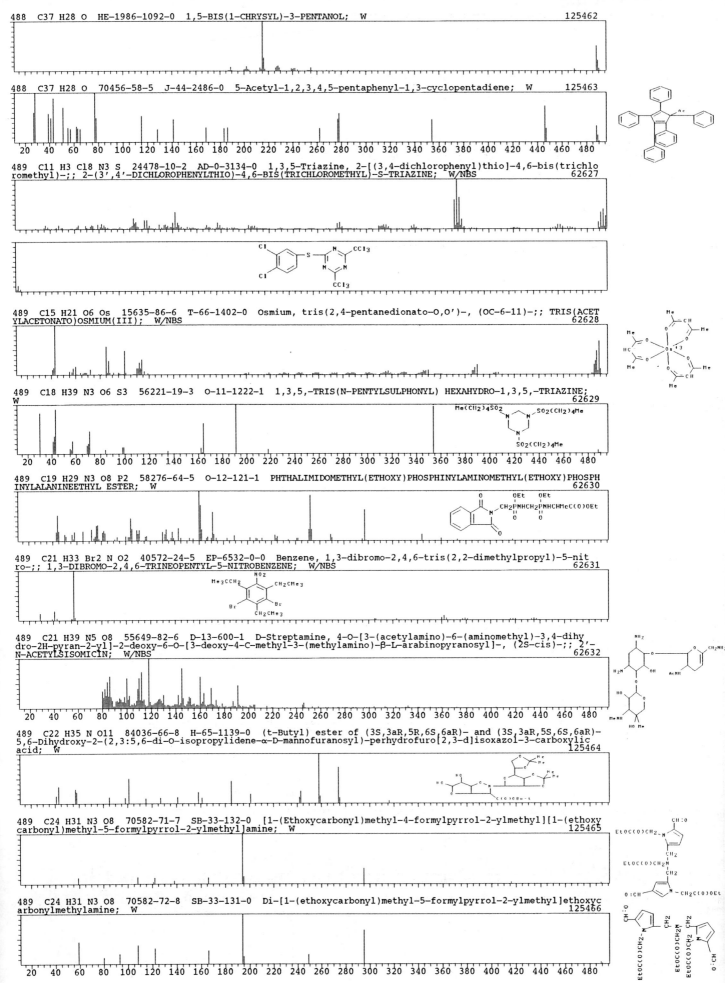

489   C25 H35 N O7 Si   77883-01-3   NS-0-0-0   Estra-1,3,5(10)-triene-16,17-diol, 4-nitro-3-[(trimethylsilyl)
oxy]-, diacetate (ester), (16α,17β)-;   W/NBS   125467

489   C25 H35 N O7 Si   77883-02-4   NS-0-0-0   Estra-1,3,5(10)-triene-16,17-diol, 2-nitro-3-[(trimethylsilyl)
oxy]-, diacetate (ester), (16α,17β)-;   W/NBS   125468

489   C26 H48 Cl N O5   RB-1982-10192-0   NONYLPHENOXYTETRAETHOXYPROPYLAMINE HYDROCHLORIDE;   W   125469

489   C28 H31 N O3 S Si   89817-09-4   Y-20-1463-0   trans-8,9,10,11-tetrahydro-8-(4-methyl)benzenesulfonamido-
9-[(trimethylsilyl)oxy]benz[a]anthracene;   W   125470

489   C28 H39 N4 Ni   H-65-823-0   (cis-4,18,19,20-Tetrahydrocorphinato)nickel(II)perchlorate;   W   125471

489   C28 H39 N4 Ni   H-65-823-0   (1,18,19,20-Tetrahydrocorphinato)nickel(II)tetraphenyl borate;   W   125472

489   C28 H43 N O6   72363-29-2   H-62-1518-0   17,18-Di-O-acetyl-6,7,19,20-tetrahydroaspochalasin D;   W 125473

489   C28 H43 N O6   72401-78-6   H-62-1519-0   17,18-Di-O-acetyl-6,7,19,20-tetrahydro-6-epi-aspochalasin D;   W
                                                                                  125474

489   C30 H39 N3 O3   5263-31-0   SD-1981-0-0   Tubulosan, 8',10,11-trimethoxy-;; Tubulosine, O-methyl-;   W/NBS
                                                                                  62633

489   C31 H52 Cl N O   38759-54-5   EP-6903-0-0   Butanamide, 4-chloro-N-[(3β)-cholest-5-en-3-yl]-;; 3β-(4'CHLO
RO)BUTYRAMIDOCHOLEST-5-ENE;   W/NBS   62634

489   C32 H28 N O2 P   87886-04-2   Y-20-901-0   2-Pivaloyl-3-triphenyl phosphazene inden-1-one;   W   125476

489   C32 H29 N P2   42998-69-6   NS-0-0-0   Phosphinous amide, N-(diphenylphosphino)-P,P-diphenyl-N-(1-phenyl
ethyl)-;   W/NBS   125477

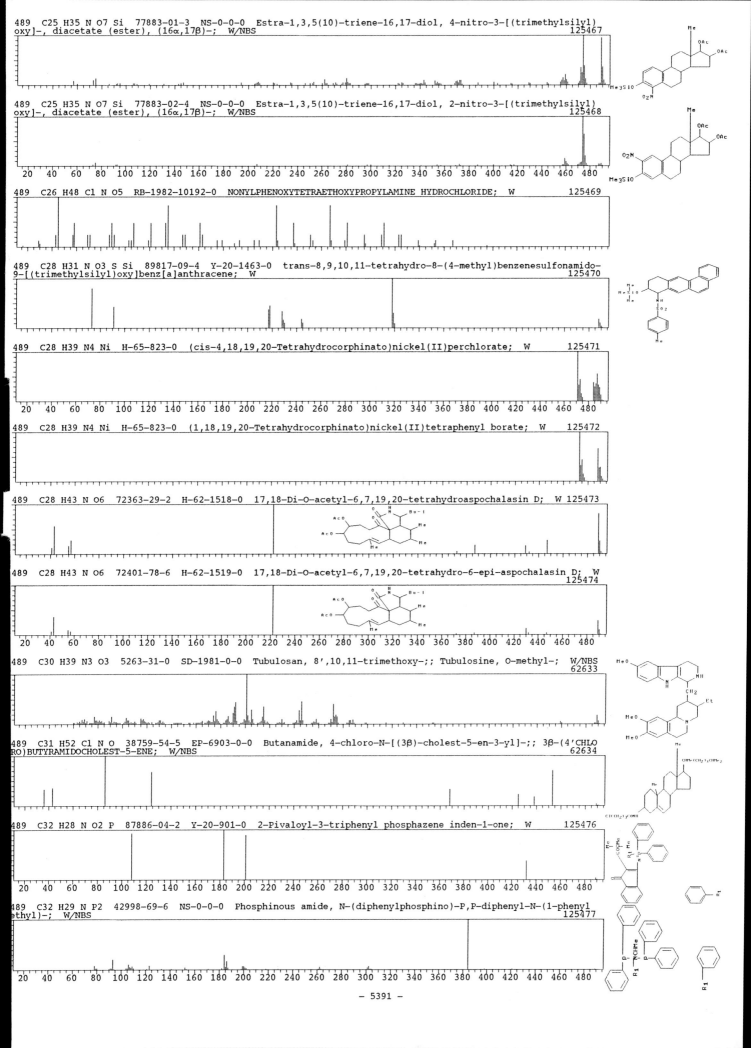

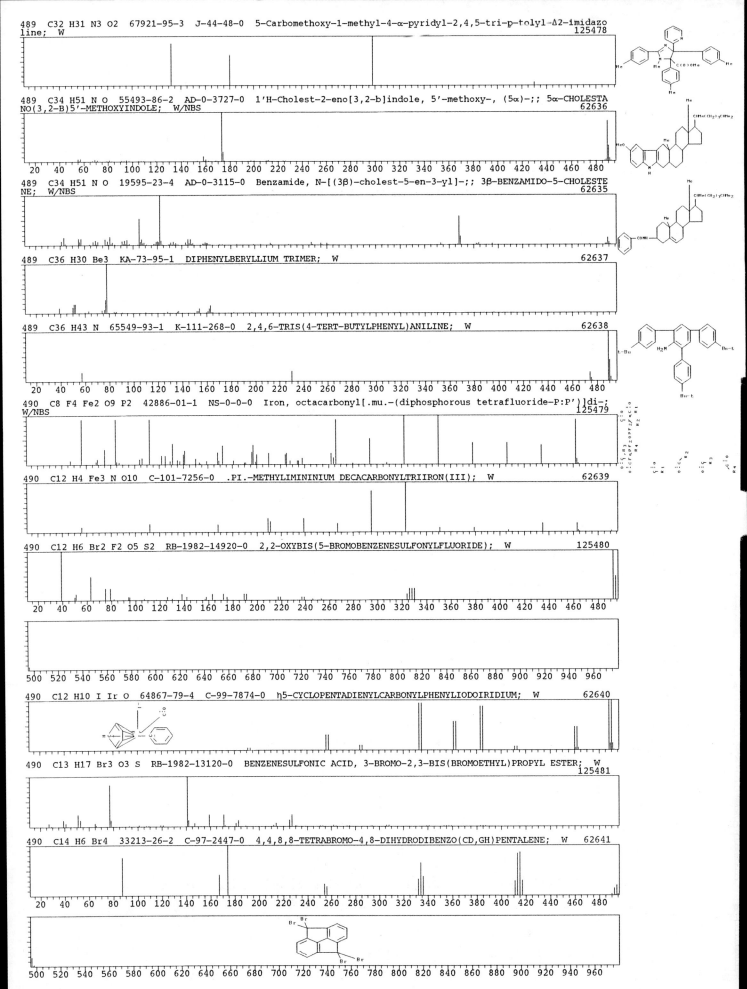

489   C32 H31 N3 O2   67921-95-3   J-44-48-0   5-Carbomethoxy-1-methyl-4-α-pyridyl-2,4,5-tri-p-tolyl-Δ2-imidazo
line;   W   125478

489   C34 H51 N O   55493-86-2   AD-0-3727-0   1'H-Cholest-2-eno[3,2-b]indole, 5'-methoxy-, (5α)-;; 5α-CHOLESTA
NO(3,2-B)5'-METHOXYINDOLE;   W/NBS   62636

489   C34 H51 N O   19595-23-4   AD-0-3115-0   Benzamide, N-[(3β)-cholest-5-en-3-yl]-;; 3β-BENZAMIDO-5-CHOLESTE
NE;   W/NBS   62635

489   C36 H30 Be3   KA-73-95-1   DIPHENYLBERYLLIUM TRIMER;   W   62637

489   C36 H43 N   65549-93-1   K-111-268-0   2,4,6-TRIS(4-TERT-BUTYLPHENYL)ANILINE;   W   62638

490   C8 F4 Fe2 O9 P2   42886-01-1   NS-0-0-0   Iron, octacarbonyl[.mu.-(diphosphorous tetrafluoride-P:P')]di-;
W/NBS   125479

490   C12 H4 Fe3 N O10   C-101-7256-0   .PI.-METHYLIMININIUM DECACARBONYLTRIIRON(III);   W   62639

490   C12 H6 Br2 F2 O5 S2   RB-1982-14920-0   2,2-OXYBIS(5-BROMOBENZENESULFONYLFLUORIDE);   W   125480

490   C12 H10 I Ir O   64867-79-4   C-99-7874-0   η5-CYCLOPENTADIENYLCARBONYLPHENYLIODOIRIDIUM;   W 62640

490   C13 H17 Br3 O3 S   RB-1982-13120-0   BENZENESULFONIC ACID, 3-BROMO-2,3-BIS(BROMOETHYL)PROPYL ESTER;   W
                                                                                                    125481

490   C14 H6 Br4   33213-26-2   C-97-2447-0   4,4,8,8-TETRABROMO-4,8-DIHYDRODIBENZO(CD,GH)PENTALENE;   W   62641

490 C30 H51 Cl O Si 69688-10-4 NS--28984-0-0 Silane, (chloromethyl)dimethyl[[(3β,6E)-9,10-secocholesta-
5(10),6,8(14)-trien-3-yl]oxy]-;; NBS
62681

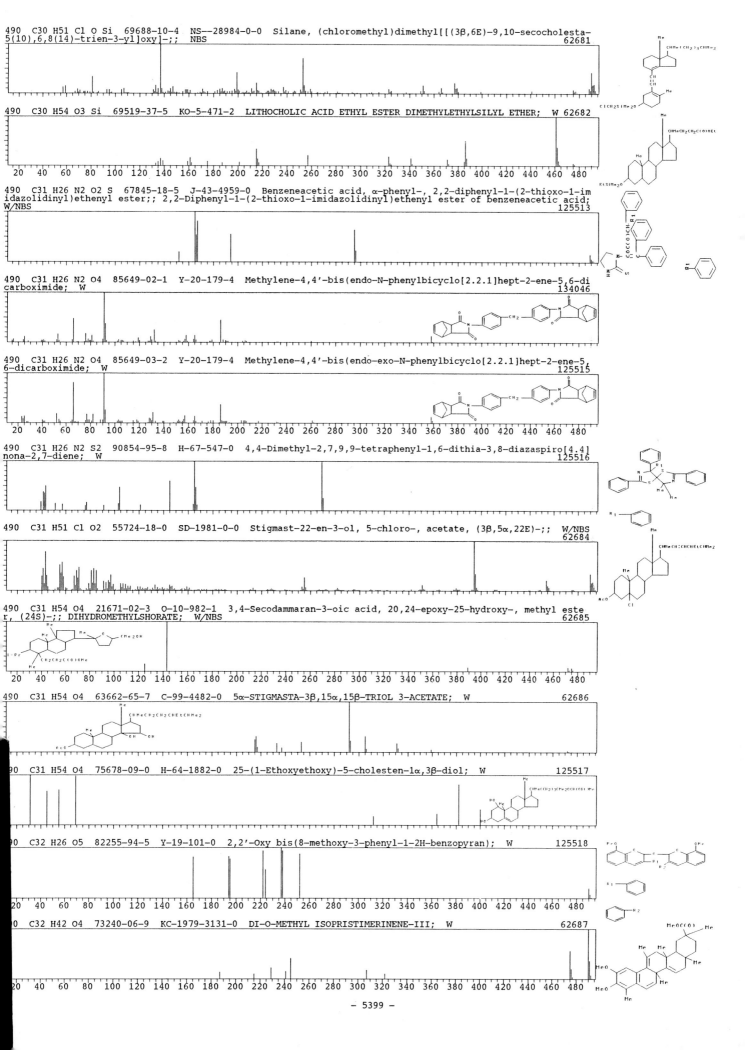

490 C30 H54 O3 Si 69519-37-5 KO-5-471-2 LITHOCHOLIC ACID ETHYL ESTER DIMETHYLETHYLSILYL ETHER; W 62682

490 C31 H26 N2 O2 S 67845-18-5 J-43-4959-0 Benzeneacetic acid, α-phenyl-, 2,2-diphenyl-1-(2-thioxo-1-im
idazolidinyl)ethenyl ester;; 2,2-Diphenyl-1-(2-thioxo-1-imidazolidinyl)ethenyl ester of benzeneacetic acid;
W/NBS
125513

490 C31 H26 N2 O4 85649-02-1 Y-20-179-4 Methylene-4,4'-bis(endo-N-phenylbicyclo[2.2.1]hept-2-ene-5,6-di
carboximide; W
134046

490 C31 H26 N2 O4 85649-03-2 Y-20-179-4 Methylene-4,4'-bis(endo-exo-N-phenylbicyclo[2.2.1]hept-2-ene-5,
6-dicarboximide; W
125515

490 C31 H26 N2 S2 90854-95-8 H-67-547-0 4,4-Dimethyl-2,7,9,9-tetraphenyl-1,6-dithia-3,8-diazaspiro[4.4]
nona-2,7-diene; W
125516

490 C31 H51 Cl O2 55724-18-0 SD-1981-0-0 Stigmast-22-en-3-ol, 5-chloro-, acetate, (3β,5α,22E)-;; W/NBS
62684

490 C31 H54 O4 21671-02-3 O-10-982-1 3,4-Secodammaran-3-oic acid, 20,24-epoxy-25-hydroxy-, methyl este
r, (24S)-;; DIHYDROMETHYLSHORATE; W/NBS
62685

490 C31 H54 O4 63662-65-7 C-99-4482-0 5α-STIGMASTA-3β,15α,15β-TRIOL 3-ACETATE; W
62686

90 C31 H54 O4 75678-09-0 H-64-1882-0 25-(1-Ethoxyethoxy)-5-cholesten-1α,3β-diol; W
125517

90 C32 H26 O5 82255-94-5 Y-19-101-0 2,2'-Oxy bis(8-methoxy-3-phenyl-1-2H-benzopyran); W
125518

0 C32 H42 O4 73240-06-9 KC-1979-3131-0 DI-O-METHYL ISOPRISTIMERINENE-III; W
62687

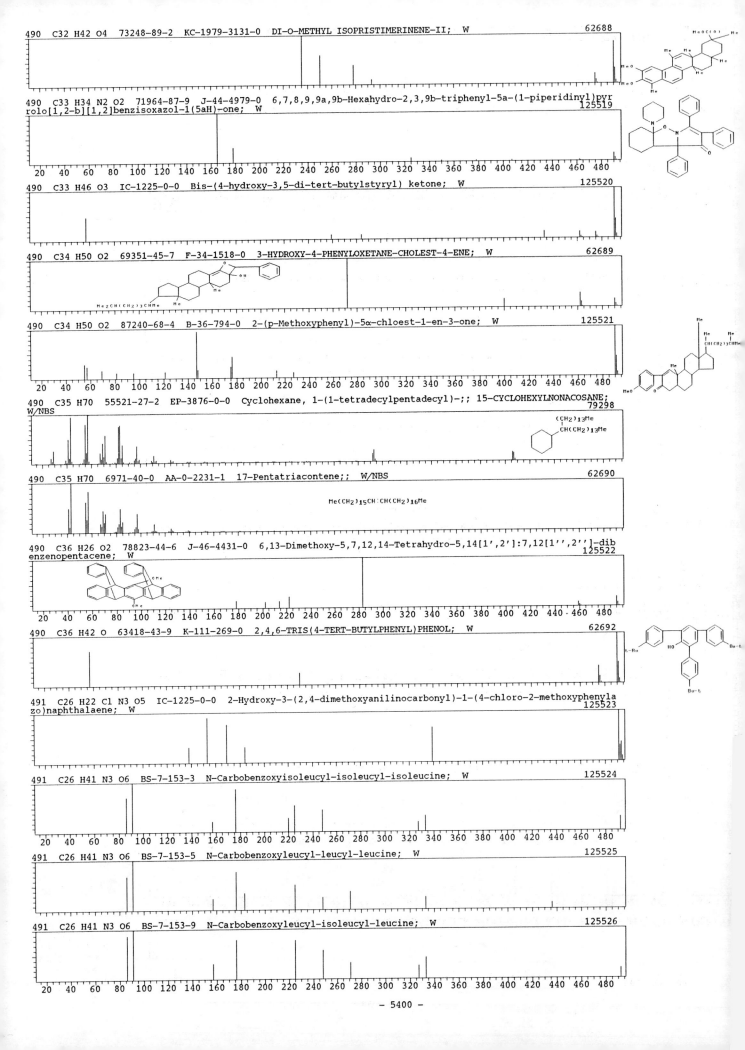

490　C32 H42 O4　73248-89-2　KC-1979-3131-0　DI-O-METHYL ISOPRISTIMERINENE-II;　W　　　62688

490　C33 H34 N2 O2　71964-87-9　J-44-4979-0　6,7,8,9,9a,9b-Hexahydro-2,3,9b-triphenyl-5a-(1-piperidinyl)pyr
rolo[1,2-b][1,2]benzisoxazol-1(5aH)-one;　W　　　125519

490　C33 H46 O3　IC-1225-0-0　Bis-(4-hydroxy-3,5-di-tert-butylstyryl) ketone;　W　　　125520

490　C34 H50 O2　69351-45-7　F-34-1518-0　3-HYDROXY-4-PHENYLOXETANE-CHOLEST-4-ENE;　W　　　62689

490　C34 H50 O2　87240-68-4　B-36-794-0　2-(p-Methoxyphenyl)-5α-chloest-1-en-3-one;　W　　　125521

490　C35 H70　55521-27-2　EP-3876-0-0　Cyclohexane, 1-(1-tetradecylpentadecyl)-;; 15-CYCLOHEXYLNONACOSANE;
W/NBS　　　79298

490　C35 H70　6971-40-0　AA-0-2231-1　17-Pentatriacontene;;　W/NBS　　　62690

490　C36 H26 O2　78823-44-6　J-46-4431-0　6,13-Dimethoxy-5,7,12,14-Tetrahydro-5,14[1',2']:7,12[1'',2'']-dib
enzenopentacene;　W　　　125522

490　C36 H42 O　63418-43-9　K-111-269-0　2,4,6-TRIS(4-TERT-BUTYLPHENYL)PHENOL;　W　　　62692

491　C26 H22 Cl N3 O5　IC-1225-0-0　2-Hydroxy-3-(2,4-dimethoxyanilinocarbonyl)-1-(4-chloro-2-methoxyphenyla
zo)naphthalaene;　W　　　125523

491　C26 H41 N3 O6　BS-7-153-3　N-Carbobenzoxyisoleucyl-isoleucyl-isoleucine;　W　　　125524

491　C26 H41 N3 O6　BS-7-153-5　N-Carbobenzoxyleucyl-leucyl-leucine;　W　　　125525

491　C26 H41 N3 O6　BS-7-153-9　N-Carbobenzoxyleucyl-isoleucyl-leucine;　W　　　125526

491 C27 H29 N3 O4 S 92098-02-7 C-107-3260-0 1,4-Dibenzyl-3-(2'-thiopyridyl)-6-(1''-(hydroxymethyl)-3''-(hydroxypropyl))-2,5-piperazinedione; W   125527

491 C27 H29 N3 O4 S 95782-30-2 C-107-3260-0 1,4-Dibenzyl-3-(2'-thiopyridyl)-6-(1''-(hydroxymethyl)-3''-(hydroxypropyl)-2,5-piperazinedione; W   125528

20 40 60 80 100 120 140 160 180 200 220 240 260 280 300 320 340 360 380 400 420 440 460 480 500

491 C27 H41 N O7 B-28-655-0 N-ACETYL 7β-ACETOXYNORERYTHROSTACHALDINE; W   125529

491 C27 H41 N O7 CD-279-0-0 Monoacetylcondelphine; W   125530

491 C28 H18 Br N3 O IC-1225-0-0 7-Oxo-2-phenyl-4-bromo-6-(4-toluidino)-7H-benzo[e]pyrimidine; W   125531

20 40 60 80 100 120 140 160 180 200 220 240 260 280 300 320 340 360 380 400 420 440 460 480 500

491 C28 H29 N O7 86340-59-2 KC-1983-525-0 Trimethyl 9-oxo-5,6,7a,8,9,10-hexahydro-7a,10-etheno-4H-pyrido[3,2,1-ij]carbazole-8-spirocyclohexane-11,12,13-tricarboxylate; W   125532

491 C29 H49 N O5 IC-1225-0-0 1-Cyclohexylmethyl-3a,5-dihydroxy-6,7-dimethyl-4-(4-methyl-7-(5-oxotetrahydrofuran-2-yl)heptyl)-3-oxo-3a,4-5,6,7,7a-hexahydro-isoindoline; W   125533

491 C31 H45 N O2 Si 57325-71-0 EP-8314-0-0 Pregna-5,16-dien-20-one, 3-[(trimethylsilyl)oxy]-, O-(phenylmethyl)oxime, (3β)-;; 5,16-PREGNADIEN-3β-OL-20-ONE BO TMS; W/NBS   62694

20 40 60 80 100 120 140 160 180 200 220 240 260 280 300 320 340 360 380 400 420 440 460 480 500

491 C33 H49 N O2 17373-01-2 T-68-5761-0 6-Azacholest-4-en-7-one, 6-benzyl-3α-hydroxy-;; 3α-HYDROXY-N-BENZYL-6-AZA-4-CHOLESTEN-7-ONE; W/NBS   62695

491 C34 H53 N O 38253-24-6 KC-1983-578-0 2-(N-Methylanilino)-5α-cholestan-3-one; W   125535

Str 27

91 C34 H53 N O 38257-51-1 KC-1983-579-0 Cholestano-phenylazetidinol; W   125536

20 40 60 80 100 120 140 160 180 200 220 240 260 280 300 320 340 360 380 400 420 440 460 480 500

91 C35 H25 N O2 55975-69-4 J-40-2877-0 4a,5,10,10a-Tetrahydro-2-methyl-1,3-diphenyl-5,10[1',2']benzeno-H-naphth[2,3-f]isoindol-4,11-dione; W   125537

20 40 60 80 100 120 140 160 180 200 220 240 260 280 300 320 340 360 380 400 420 440 460 480 500

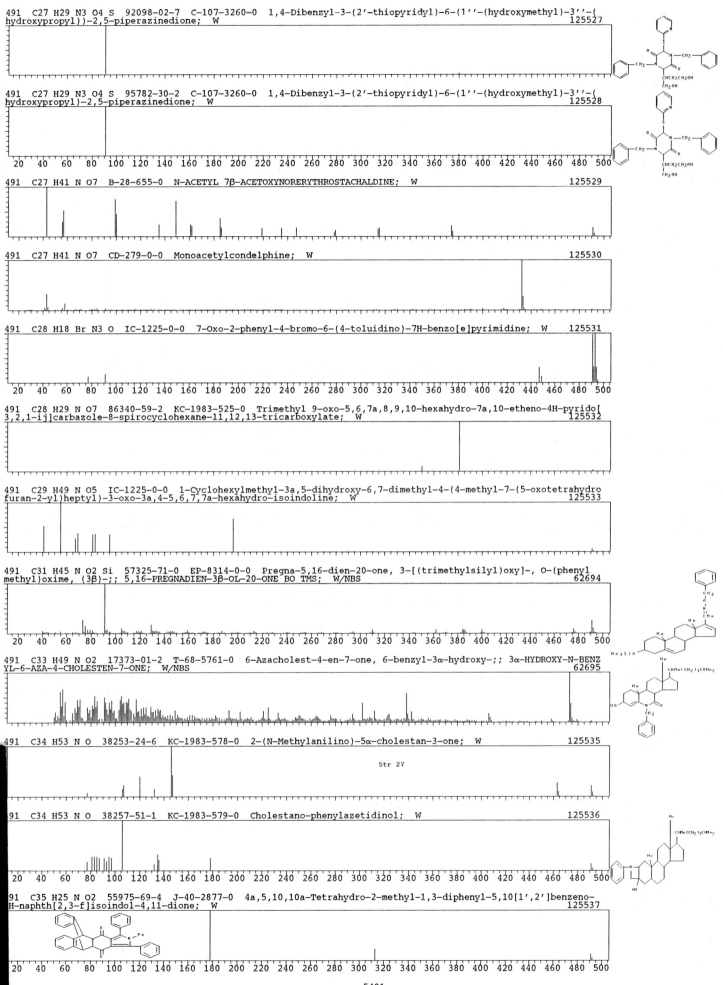

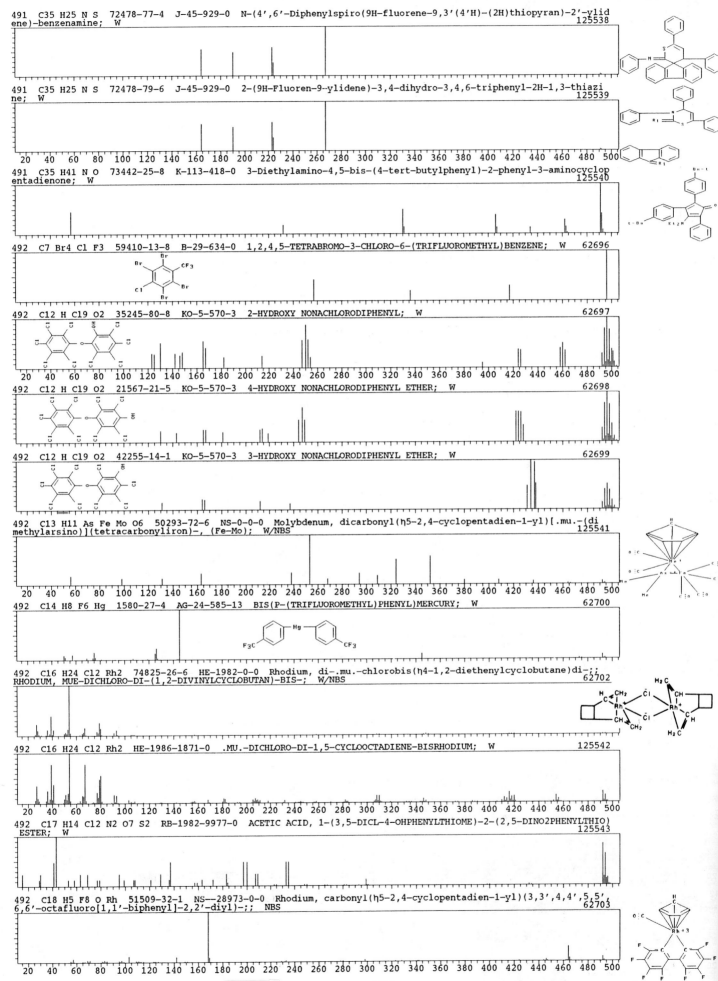

491   C35 H25 N S   72478-77-4   J-45-929-0   N-(4',6'-Diphenylspiro(9H-fluorene-9,3'(4'H)-(2H)thiopyran)-2'-ylid
ene)-benzenamine;   W   125538

491   C35 H25 N S   72478-79-6   J-45-929-0   2-(9H-Fluoren-9-ylidene)-3,4-dihydro-3,4,6-triphenyl-2H-1,3-thiazi
ne;   W   125539

491   C35 H41 N O   73442-25-8   K-113-418-0   3-Diethylamino-4,5-bis-(4-tert-butylphenyl)-2-phenyl-3-aminocyclop
entadienone;   W   125540

492   C7 Br4 Cl F3   59410-13-8   B-29-634-0   1,2,4,5-TETRABROMO-3-CHLORO-6-(TRIFLUOROMETHYL)BENZENE;   W   62696

492   C12 H Cl9 O2   35245-80-8   KO-5-570-3   2-HYDROXY NONACHLORODIPHENYL;   W   62697

492   C12 H Cl9 O2   21567-21-5   KO-5-570-3   4-HYDROXY NONACHLORODIPHENYL ETHER;   W   62698

492   C12 H Cl9 O2   42255-14-1   KO-5-570-3   3-HYDROXY NONACHLORODIPHENYL ETHER;   W   62699

492   C13 H11 As Fe Mo O6   50293-72-6   NS-0-0-0   Molybdenum, dicarbonyl(η5-2,4-cyclopentadien-1-yl)[.mu.-(di
methylarsino)](tetracarbonyliron)-, (Fe-Mo);   W/NBS   125541

492   C14 H8 F6 Hg   1580-27-4   AG-24-585-13   BIS(P-(TRIFLUOROMETHYL)PHENYL)MERCURY;   W   62700

492   C16 H24 Cl2 Rh2   74825-26-6   HE-1982-0-0   Rhodium, di-.mu.-chlorobis(η4-1,2-diethenylcyclobutane)di-;;
RHODIUM, MUE-DICHLORO-DI-(1,2-DIVINYLCYCLOBUTAN)-BIS-;   W/NBS   62702

492   C16 H24 Cl2 Rh2   HE-1986-1871-0   .MU.-DICHLORO-DI-1,5-CYCLOOCTADIENE-BISRHODIUM;   W   125542

492   C17 H14 Cl2 N2 O7 S2   RB-1982-9977-0   ACETIC ACID, 1-(3,5-DICL-4-OHPHENYLTHIOME)-2-(2,5-DINO2PHENYLTHIO)
ESTER;   W   125543

492   C18 H5 F8 O Rh   51509-32-1   NS--28973-0-0   Rhodium, carbonyl(η5-2,4-cyclopentadien-1-yl)(3,3',4,4',5,5'
6,6'-octafluoro[1,1'-biphenyl]-2,2'-diyl)-;;   NBS   62703

- 5402 -

492   C18 H20 O16   AN-87-84-0   6H,8H-Benzopyrano(3,4-b)benzopyrano(4,3-e)pyran-6,7,8-trione;   W   125544

492   C18 H42 Ga2 O6   KA-70-2347-1   GALLIUM ISOPROPOXIDE;   W   62704

20   40   60   80   100   120   140   160   180   200   220   240   260   280   300   320   340   360   380   400   420   440   460   480   500

492   C20 H28 O14   55493-88-4   AD-0-3207-0   Propanoic acid, 3-(acetyloxy)-2-[(2,3,4,6-tetra-O-acetyl-α-D-manno
pyranosyl)oxy]-, methyl ester, (R)-;; METHYL 3-O-ACETYL-2-O-(2',3'',4',6'-TETRA-O-ACETYL-α-D-MANNOPYRANOSYL)-
D-GLYCERATE;   W/NBS   62705

492   C21 H34 Br I   25347-09-5   EP-6539-0-0   Benzene, 2-bromo-1,3,5-tris(2,2-dimethylpropyl)-4-iodo-;   2-BROMO
-4-IODO-1,3,5-TRINEOPENTYLBENZENE;   W/NBS   62706

492   C22 H24 N2 O11   29742-55-0   F-26-4024-1   1H-Indole, 5-nitro-1-(2,3,4,6-tetra-O-acetyl-β-D-glucopyranosy
l)-;; 1-(2,3,4,6-TETRA-O-ACETYL-β-D-GLUCOPYRANOSYL)-5-NITROINDOLE;   W/NBS   62707

20   40   60   80   100   120   140   160   180   200   220   240   260   280   300   320   340   360   380   400   420   440   460   480   500

492   C22 H45 Cl O4 Si3   60164-90-1   C-98-3643-0   TERT-BUTYLDIMETHYLSILYL 2,4-BIS(TERT-BUTYLDIMETHYLSILOXY)-2-
CHLOROBUT-3-YNOATE;   W   62708

492   C24 H12 S6   56598-45-9   K-111-1334-0   HEXATHIENYLENE;   W   62709

492   C24 H28 O7 S2   81825-48-1   H-65-184-0   5exo,6exo-Epoxybicyclo[2.2.2]octane-2endo,3endo-dimethyl bis(p-to
luenesulfonate);   W   125545

20   40   60   80   100   120   140   160   180   200   220   240   260   280   300   320   340   360   380   400   420   440   460   480   500

492   C24 H32 N2 O5 S2   71190-62-0   K-112-2251-0   METHYL ESTER OF 5'-T-BUTOXYCARBONYL-1,5-DIHYDRO-3,3'-DIMETH
YL-4-(2-METHYL-4-(2-METHYL-1,3-DITHIOLAN-2-YL)-5-OXO-2,2'-PYRROMETHEN-4'-PROPIONIC;   W   62710

492   C25 H32 O10   71155-66-3   J-44-3403-0   Repandin A;   W   125546

92   C25 H32 O10   71138-44-8   J-44-3404-0   Repandin C;   W   125547

20   40   60   80   100   120   140   160   180   200   220   240   260   280   300   320   340   360   380   400   420   440   460   480   500

92   C25 H33 Br O5   90164-48-0   AH-114-1262-0   Methyl ester, cyclic 13,14-anhydride derivative of [4α,8α,12α]
16-(1-bromo-1-methylethyl)-17,19-dinoratis-15-ene-4,13,14-tricarboxylic acid;   W   125548

20   40   60   80   100   120   140   160   180   200   220   240   260   280   300   320   340   360   380   400   420   440   460   480   500

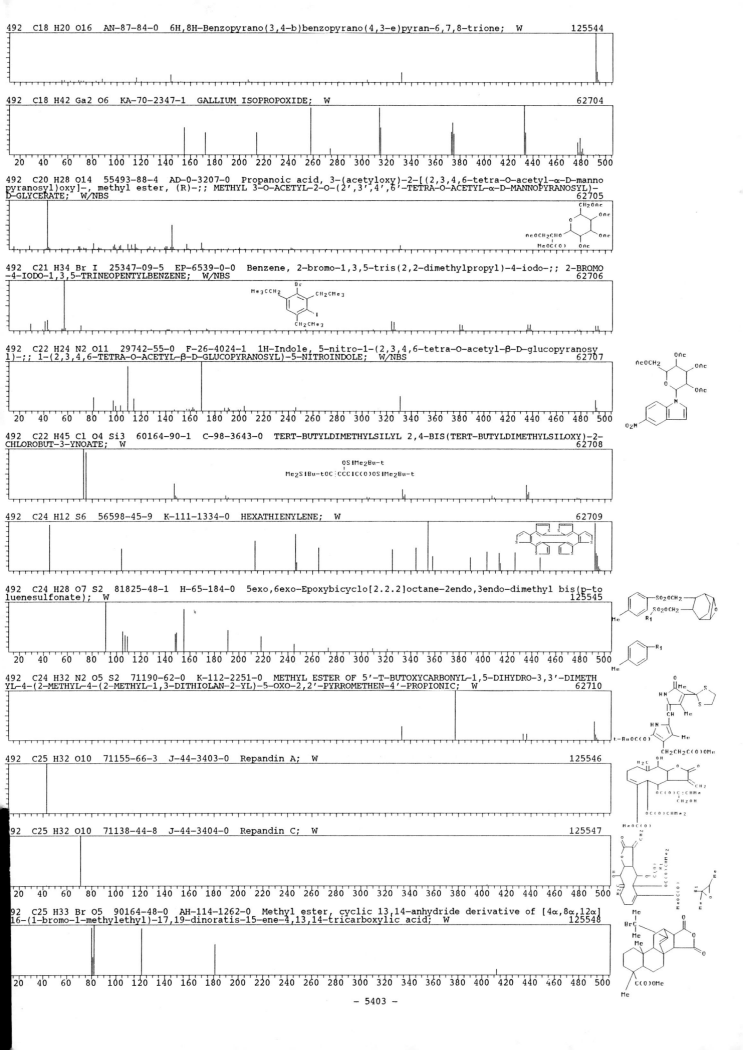

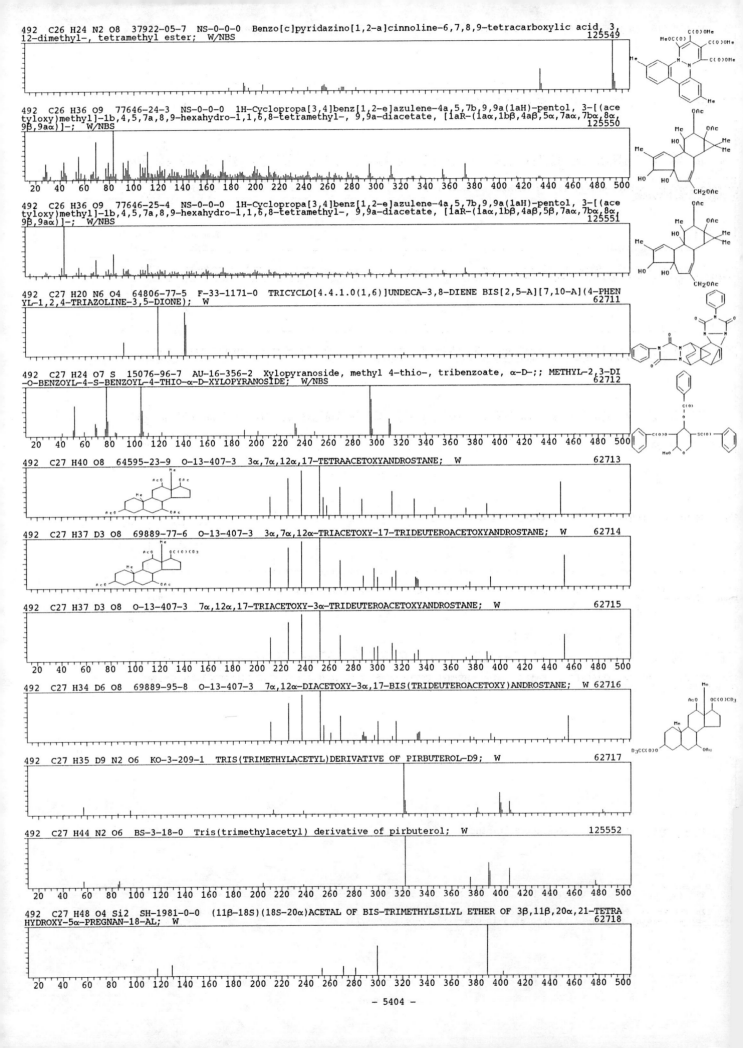

492 C26 H24 N2 O8 37922-05-7 NS-0-0-0 Benzo[c]pyridazino[1,2-a]cinnoline-6,7,8,9-tetracarboxylic acid, 3,
12-dimethyl-, tetramethyl ester; W/NBS   125549

492 C26 H36 O9 77646-24-3 NS-0-0-0 1H-Cyclopropa[3,4]benz[1,2-e]azulene-4a,5,7b,9,9a(1aH)-pentol, 3-[(ace
tyloxy)methyl]-1b,4,5,7a,8,9-hexahydro-1,1,6,8-tetramethyl-, 9,9a-diacetate, [1aR-(1aα,1bβ,4aβ,5α,7aα,7bα,8α,
9β,9aα)]-; W/NBS   125550

492 C26 H36 O9 77646-25-4 NS-0-0-0 1H-Cyclopropa[3,4]benz[1,2-e]azulene-4a,5,7b,9,9a(1aH)-pentol, 3-[(ace
tyloxy)methyl]-1b,4,5,7a,8,9-hexahydro-1,1,6,8-tetramethyl-, 9,9a-diacetate, [1aR-(1aα,1bβ,4aβ,5β,7aα,7bα,8α,
9β,9aα)]-; W/NBS   125551

492 C27 H20 N6 O4 64806-77-5 F-33-1171-0 TRICYCLO[4.4.1.0(1,6)]UNDECA-3,8-DIENE BIS[2,5-A][7,10-A](4-PHEN
YL-1,2,4-TRIAZOLINE-3,5-DIONE); W   62711

492 C27 H24 O7 S 15076-96-7 AU-16-356-2 Xylopyranoside, methyl 4-thio-, tribenzoate, α-D-;; METHYL-2,3-DI
-O-BENZOYL-4-S-BENZOYL-4-THIO-α-D-XYLOPYRANOSIDE; W/NBS   62712

492 C27 H40 O8 64595-23-9 O-13-407-3 3α,7α,12α,17-TETRAACETOXYANDROSTANE; W   62713

492 C27 H37 D3 O8 69889-77-6 O-13-407-3 3α,7α,12α-TRIACETOXY-17-TRIDEUTEROACETOXYANDROSTANE; W   62714

492 C27 H37 D3 O8 O-13-407-3 7α,12α,17-TRIACETOXY-3α-TRIDEUTEROACETOXYANDROSTANE; W   62715

492 C27 H34 D6 O8 69889-95-8 O-13-407-3 7α,12α-DIACETOXY-3α,17-BIS(TRIDEUTEROACETOXY)ANDROSTANE; W 62716

492 C27 H35 D9 N2 O6 KO-3-209-1 TRIS(TRIMETHYLACETYL)DERIVATIVE OF PIRBUTEROL-D9; W   62717

492 C27 H44 N2 O6 BS-3-18-0 Tris(trimethylacetyl) derivative of pirbuterol; W   125552

492 C27 H48 O4 Si2 SH-1981-0-0 (11β-18S)(18S-20α)ACETAL OF BIS-TRIMETHYLSILYL ETHER OF 3β,11β,20α,21-TETRA
HYDROXY-5α-PREGNAN-18-AL; W   62718

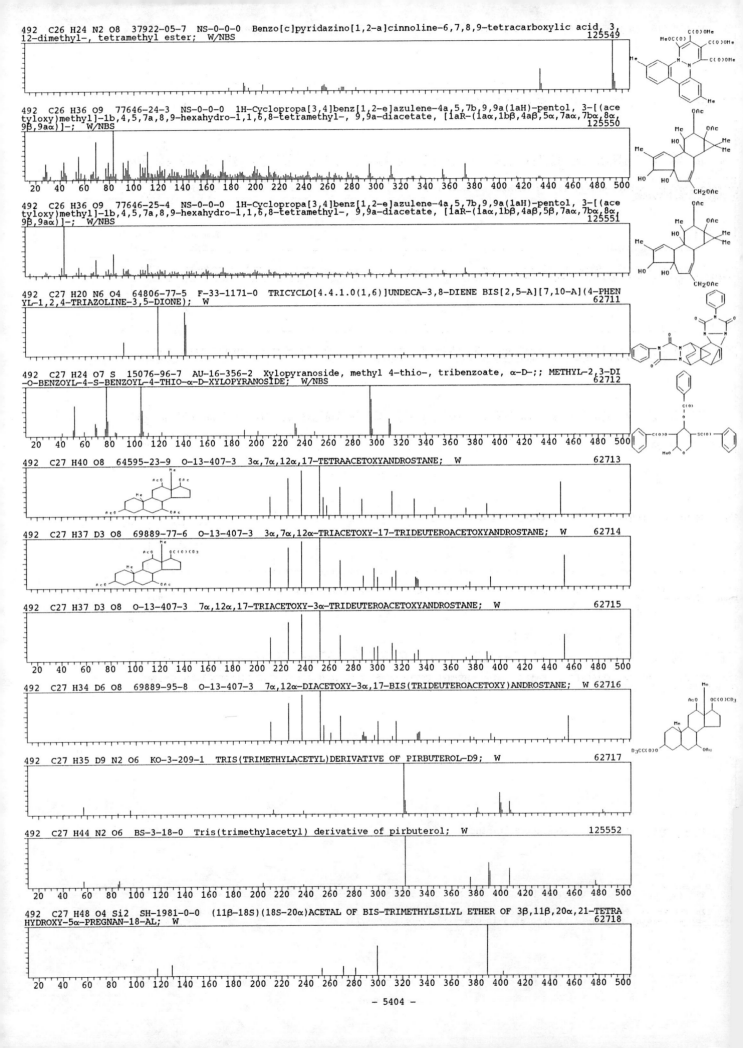

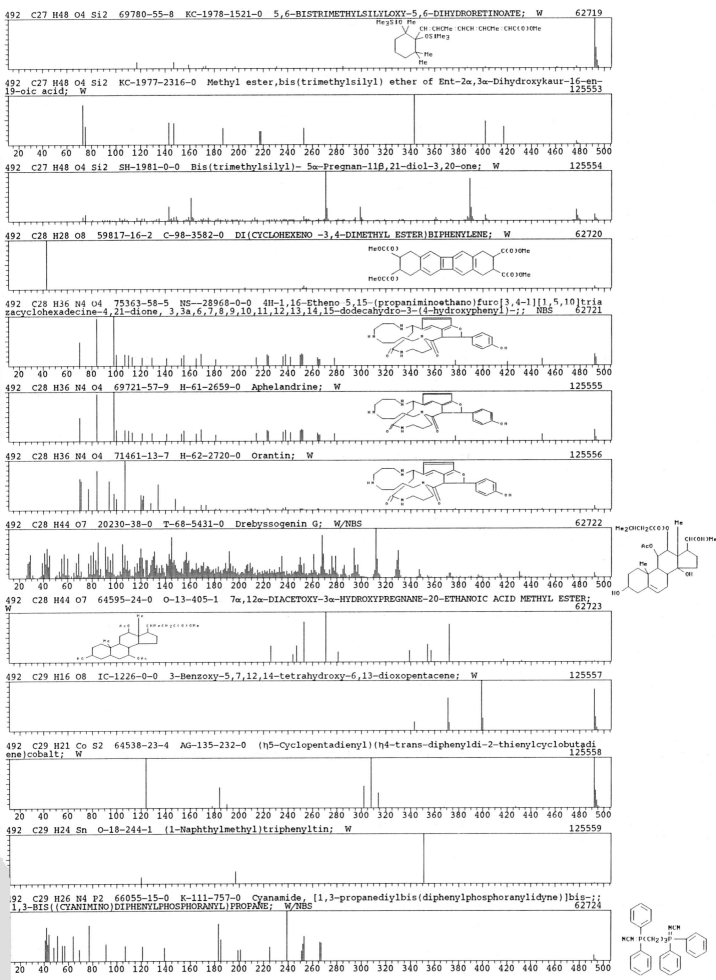

492   C27 H48 O4 Si2   69780-55-8   KC-1978-1521-0   5,6-BISTRIMETHYLSILYLOXY-5,6-DIHYDRORETINOATE;   W   62719

492   C27 H48 O4 Si2   KC-1977-2316-0   Methyl ester,bis(trimethylsilyl) ether of Ent-2α,3α-Dihydroxykaur-16-en-19-oic acid;   W   125553

492   C27 H48 O4 Si2   SH-1981-0-0   Bis(trimethylsilyl)- 5α-Pregnan-11β,21-diol-3,20-one;   W   125554

492   C28 H28 O8   59817-16-2   C-98-3582-0   DI(CYCLOHEXENO -3,4-DIMETHYL ESTER)BIPHENYLENE;   W   62720

492   C28 H36 N4 O4   75363-58-5   NS--28968-0-0   4H-1,16-Etheno 5,15-(propaniminoethano)furo[3,4-1][1,5,10]triazacyclohexadecine-4,21-dione, 3,3a,6,7,8,9,10,11,12,13,14,15-dodecahydro-3-(4-hydroxyphenyl)-;;   NBS   62721

492   C28 H36 N4 O4   69721-57-9   H-61-2659-0   Aphelandrine;   W   125555

492   C28 H36 N4 O4   71461-13-7   H-62-2720-0   Orantin;   W   125556

492   C28 H44 O7   20230-38-0   T-68-5431-0   Drebyssogenin G;   W/NBS   62722

492   C28 H44 O7   64595-24-0   O-13-405-1   7α,12α-DIACETOXY-3α-HYDROXYPREGNANE-20-ETHANOIC ACID METHYL ESTER;   W   62723

492   C29 H16 O8   IC-1226-0-0   3-Benzoxy-5,7,12,14-tetrahydroxy-6,13-dioxopentacene;   W   125557

492   C29 H21 Co S2   64538-23-4   AG-135-232-0   (η5-Cyclopentadienyl)(η4-trans-diphenyldi-2-thienylcyclobutadiene)cobalt;   W   125558

492   C29 H24 Sn   O-18-244-1   (1-Naphthylmethyl)triphenyltin;   W   125559

492   C29 H26 N4 P2   66055-15-0   K-111-757-0   Cyanamide, [1,3-propanediylbis(diphenylphosphoranylidyne)]bis-;;
1,3-BIS((CYANIMINO)DIPHENYLPHOSPHORANYL)PROPANE;   W/NBS   62724

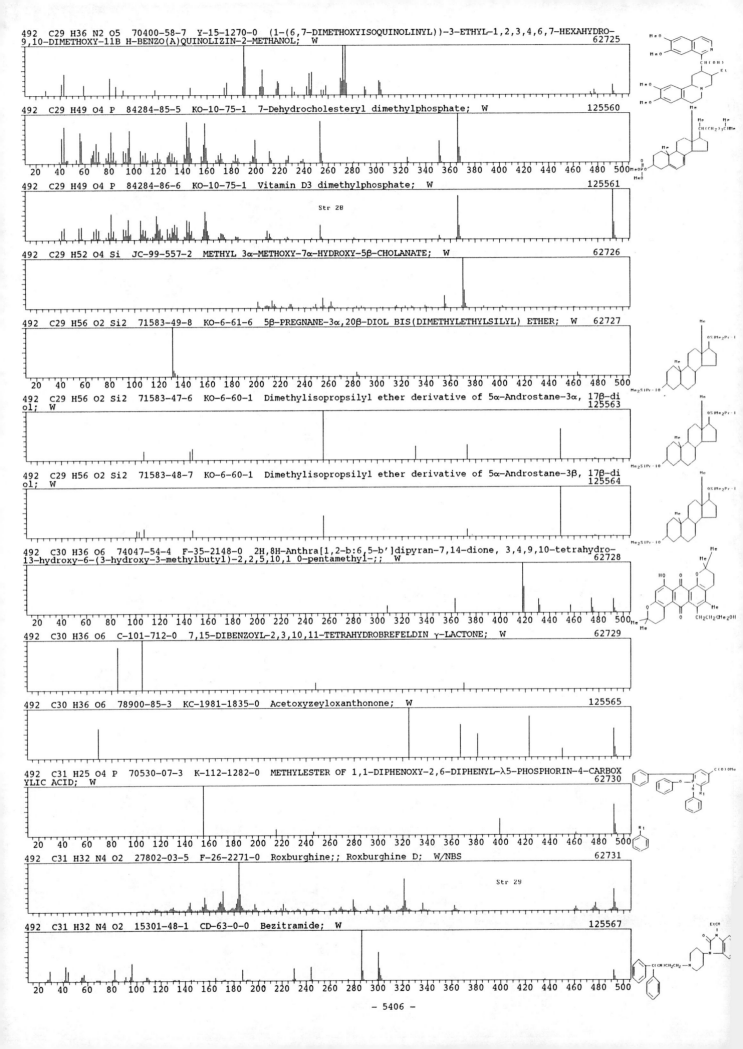

492 C29 H36 N2 O5 70400-58-7 Y-15-1270-0 (1-(6,7-DIMETHOXYISOQUINOLINYL))-3-ETHYL-1,2,3,4,6,7-HEXAHYDRO-
9,10-DIMETHOXY-11B H-BENZO(A)QUINOLIZIN-2-METHANOL; W
62725

492 C29 H49 O4 P 84284-85-5 KO-10-75-1 7-Dehydrocholesteryl dimethylphosphate; W
125560

492 C29 H49 O4 P 84284-86-6 KO-10-75-1 Vitamin D3 dimethylphosphate; W
125561

Str 28

492 C29 H52 O4 Si JC-99-557-2 METHYL 3α-METHOXY-7α-HYDROXY-5β-CHOLANATE; W
62726

492 C29 H56 O2 Si2 71583-49-8 KO-6-61-6 5β-PREGNANE-3α,20β-DIOL BIS(DIMETHYLETHYLSILYL) ETHER; W 62727

492 C29 H56 O2 Si2 71583-47-6 KO-6-60-1 Dimethylisopropsilyl ether derivative of 5α-Androstane-3α, 17β-di
ol; W
125563

492 C29 H56 O2 Si2 71583-48-7 KO-6-60-1 Dimethylisopropsilyl ether derivative of 5α-Androstane-3β, 17β-di
ol; W
125564

492 C30 H36 O6 74047-54-4 F-35-2148-0 2H,8H-Anthra[1,2-b:6,5-b']dipyran-7,14-dione, 3,4,9,10-tetrahydro-
13-hydroxy-6-(3-hydroxy-3-methylbutyl)-2,2,5,10,1 0-pentamethyl-;; W
62728

492 C30 H36 O6 C-101-712-0 7,15-DIBENZOYL-2,3,10,11-TETRAHYDROBREFELDIN γ-LACTONE; W
62729

492 C30 H36 O6 78900-85-3 KC-1981-1835-0 Acetoxyzeyloxanthonone; W
125565

492 C31 H25 O4 P 70530-07-3 K-112-1282-0 METHYLESTER OF 1,1-DIPHENOXY-2,6-DIPHENYL-λ5-PHOSPHORIN-4-CARBOX
YLIC ACID; W
62730

492 C31 H32 N4 O2 27802-03-5 F-26-2271-0 Roxburghine;; Roxburghine D; W/NBS
62731

Str 29

492 C31 H32 N4 O2 15301-48-1 CD-63-0-0 Bezitramide; W
125567

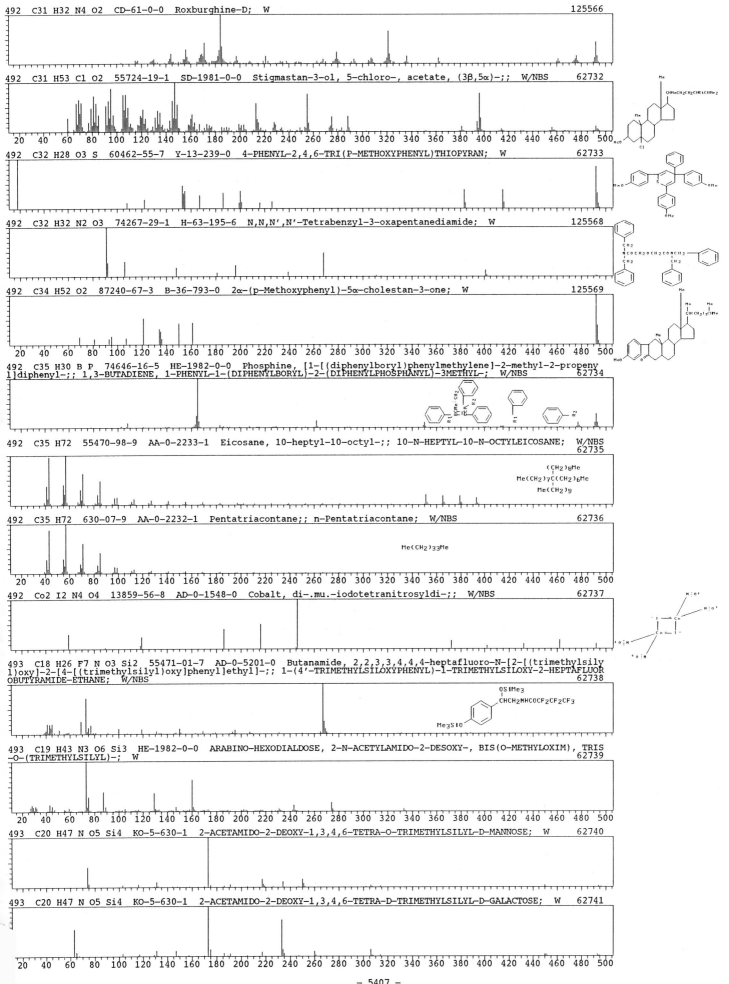

492   C31 H32 N4 O2   CD-61-0-0   Roxburghine-D; W   125566

492   C31 H53 Cl O2   55724-19-1   SD-1981-0-0   Stigmastan-3-ol, 5-chloro-, acetate, (3β,5α)-;; W/NBS   62732

492   C32 H28 O3 S   60462-55-7   Y-13-239-0   4-PHENYL-2,4,6-TRI(P-METHOXYPHENYL)THIOPYRAN; W   62733

492   C32 H32 N2 O3   74267-29-1   H-63-195-6   N,N,N',N'-Tetrabenzyl-3-oxapentanediamide; W   125568

492   C34 H52 O2   87240-67-3   B-36-793-0   2α-(p-Methoxyphenyl)-5α-cholestan-3-one; W   125569

492   C35 H30 B P   74646-16-5   HE-1982-0-0   Phosphine, [1-[(diphenylboryl)phenylmethylene]-2-methyl-2-propeny l]diphenyl-;; 1,3-BUTADIENE, 1-PHENYL-1-(DIPHENYLBORYL)-2-(DIPHENYLPHOSPHANYL)-3METHYL-; W/NBS   62734

492   C35 H72   55470-98-9   AA-0-2233-1   Eicosane, 10-heptyl-10-octyl-;; 10-N-HEPTYL-10-N-OCTYLEICOSANE; W/NBS   62735

492   C35 H72   630-07-9   AA-0-2232-1   Pentatriacontane;; n-Pentatriacontane; W/NBS   62736

492   Co2 I2 N4 O4   13859-56-8   AD-0-1548-0   Cobalt, di-.mu.-iodotetranitrosyldi-;; W/NBS   62737

493   C18 H26 F7 N O3 Si2   55471-01-7   AD-0-5201-0   Butanamide, 2,2,3,3,4,4,4-heptafluoro-N-[2-[(trimethylsily l)oxy]-2-[4-[(trimethylsilyl)oxy]phenyl]ethyl]-;; 1-(4'-TRIMETHYLSILOXYPHENYL)-1-TRIMETHYLSILOXY-2-HEPTAFLUOR OBUTYRAMIDE-ETHANE; W/NBS   62738

493   C19 H43 N3 O6 Si3   HE-1982-0-0   ARABINO-HEXODIALDOSE, 2-N-ACETYLAMIDO-2-DESOXY-, BIS(O-METHYLOXIM), TRIS -O-(TRIMETHYLSILYL)-; W   62739

493   C20 H47 N O5 Si4   KO-5-630-1   2-ACETAMIDO-2-DEOXY-1,3,4,6-TETRA-O-TRIMETHYLSILYL-D-MANNOSE; W   62740

493   C20 H47 N O5 Si4   KO-5-630-1   2-ACETAMIDO-2-DEOXY-1,3,4,6-TETRA-D-TRIMETHYLSILYL-D-GALACTOSE; W   62741

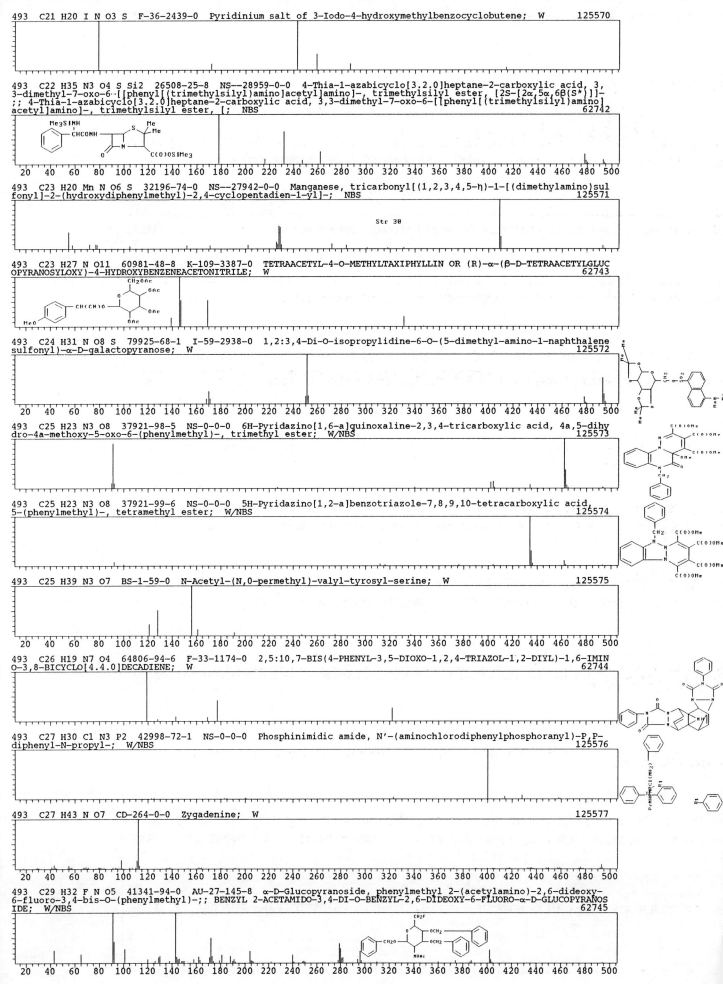

493   C21 H20 I N O3 S   F-36-2439-0   Pyridinium salt of 3-Iodo-4-hydroxymethylbenzocyclobutene;   W   125570

493   C22 H35 N3 O4 S Si2   26508-25-8   NS--28959-0-0   4-Thia-1-azabicyclo[3.2.0]heptane-2-carboxylic acid, 3,
3-dimethyl-7-oxo-6-[[phenyl[(trimethylsilyl)amino]acetyl]amino]-, trimethylsilyl ester, [2S-[2α,5α,6β(S*)]]-
;; 4-Thia-1-azabicyclo[3.2.0]heptane-2-carboxylic acid, 3,3-dimethyl-7-oxo-6-[[phenyl[(trimethylsilyl)amino]
acetyl]amino]-, trimethylsilyl ester, [;   NBS   62742

493   C23 H20 Mn N O6 S   32196-74-0   NS--27942-0-0   Manganese, tricarbonyl[(1,2,3,4,5-η)-1-[(dimethylamino)sul
fonyl]-2-(hydroxydiphenylmethyl)-2,4-cyclopentadien-1-yl]-;   NBS   125571

Str 30

493   C23 H27 N O11   60981-48-8   K-109-3387-0   TETRAACETYL-4-O-METHYLTAXIPHYLLIN OR (R)-α-(β-D-TETRAACETYLGLUC
OPYRANOSYLOXY)-4-HYDROXYBENZENEACETONITRILE;   W   62743

493   C24 H31 N O8 S   79925-68-1   I-59-2938-0   1,2:3,4-Di-O-isopropylidine-6-O-(5-dimethyl-amino-1-naphthalene
sulfonyl)-α-D-galactopyranose;   W   125572

493   C25 H23 N3 O8   37921-98-5   NS-0-0-0   6H-Pyridazino[1,6-a]quinoxaline-2,3,4-tricarboxylic acid, 4a,5-dihy
dro-4a-methoxy-5-oxo-6-(phenylmethyl)-, trimethyl ester;   W/NBS   125573

493   C25 H23 N3 O8   37921-99-6   NS-0-0-0   5H-Pyridazino[1,2-a]benzotriazole-7,8,9,10-tetracarboxylic acid,
5-(phenylmethyl)-, tetramethyl ester;   W/NBS   125574

493   C25 H39 N3 O7   BS-1-59-0   N-Acetyl-(N,O-permethyl)-valyl-tyrosyl-serine;   W   125575

493   C26 H19 N7 O4   64806-94-6   F-33-1174-0   2,5:10,7-BIS(4-PHENYL-3,5-DIOXO-1,2,4-TRIAZOL-1,2-DIYL)-1,6-IMIN
O-3,8-BICYCLO[4.4.0]DECADIENE;   W   62744

493   C27 H30 Cl N3 P2   42998-72-1   NS-0-0-0   Phosphinimidic amide, N'-(aminochlorodiphenylphosphoranyl)-P,P-
diphenyl-N-propyl-;   W/NBS   125576

493   C27 H43 N O7   CD-264-0-0   Zygadenine;   W   125577

493   C29 H32 F N O5   41341-94-0   AU-27-145-8   α-D-Glucopyranoside, phenylmethyl 2-(acetylamino)-2,6-dideoxy-
6-fluoro-3,4-bis-O-(phenylmethyl)-;; BENZYL 2-ACETAMIDO-3,4-DI-O-BENZYL-2,6-DIDEOXY-6-FLUORO-α-D-GLUCOPYRANOS
IDE;   W/NBS   62745

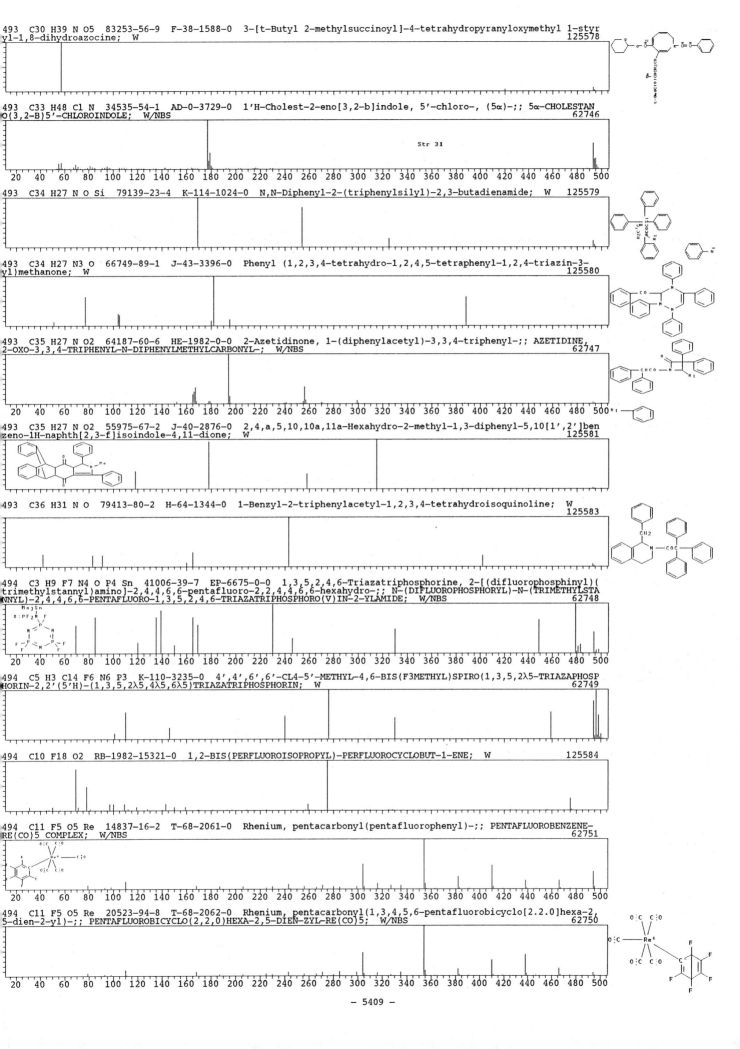

493 C30 H39 N O5 83253-56-9 F-38-1588-0 3-[t-Butyl 2-methylsuccinoyl]-4-tetrahydropyranyloxymethyl 1-styryl-1,8-dihydroazocine; W
125578

493 C33 H48 Cl N 34535-54-1 AD-0-3729-0 1'H-Cholest-2-eno[3,2-b]indole, 5'-chloro-, (5α)-;; 5α-CHOLESTANO(3,2-B)5'-CHLOROINDOLE; W/NBS
62746

Str 31

493 C34 H27 N O Si 79139-23-4 K-114-1024-0 N,N-Diphenyl-2-(triphenylsilyl)-2,3-butadienamide; W 125579

493 C34 H27 N3 O 66749-89-1 J-43-3396-0 Phenyl (1,2,3,4-tetrahydro-1,2,4,5-tetraphenyl-1,2,4-triazin-3-yl)methanone; W
125580

493 C35 H27 N O2 64187-60-6 HE-1982-0-0 2-Azetidinone, 1-(diphenylacetyl)-3,3,4-triphenyl-;; AZETIDINE, 2-OXO-3,3,4-TRIPHENYL-N-DIPHENYLMETHYLCARBONYL-; W/NBS
62747

493 C35 H27 N O2 55975-67-2 J-40-2876-0 2,4,a,5,10,10a,11a-Hexahydro-2-methyl-1,3-diphenyl-5,10[1',2']benzeno-1H-naphth[2,3-f]isoindole-4,11-dione; W
125581

493 C36 H31 N O 79413-80-2 H-64-1344-0 1-Benzyl-2-triphenylacetyl-1,2,3,4-tetrahydroisoquinoline; W
125583

494 C3 H9 F7 N4 O P4 Sn 41006-39-7 EP-6675-0-0 1,3,5,2,4,6-Triazatriphosphorine, 2-[(difluorophosphinyl)(trimethylstannyl)amino]-2,4,4,6,6-pentafluoro-2,2,4,4,6,6-hexahydro-;; N-(DIFLUOROPHOSPHORYL)-N-(TRIMETHYLSTANNYL)-2,4,4,6,6-PENTAFLUORO-1,3,5,2,4,6-TRIAZATRIPHOSPHORO(V)IN-2-YLAMIDE; W/NBS
62748

494 C5 H3 Cl4 F6 N6 P3 K-110-3235-0 4',4',6',6'-CL4-5'-METHYL-4,6-BIS(F3METHYL)SPIRO(1,3,5,2λ5-TRIAZAPHOSPHORIN-2,2'(5'H)-(1,3,5,2λ5,4λ5,6λ5)TRIAZATRIPHOSPHORIN; W
62749

494 C10 F18 O2 RB-1982-15321-0 1,2-BIS(PERFLUOROISOPROPYL)-PERFLUOROCYCLOBUT-1-ENE; W 125584

494 C11 F5 O5 Re 14837-16-2 T-68-2061-0 Rhenium, pentacarbonyl(pentafluorophenyl)-;; PENTAFLUOROBENZENE-RE(CO)5 COMPLEX; W/NBS
62751

494 C11 F5 O5 Re 20523-94-8 T-68-2062-0 Rhenium, pentacarbonyl(1,3,4,5,6-pentafluorobicyclo[2.2.0]hexa-2,5-dien-2-yl)-;; PENTAFLUOROBICYCLO(2,2,0)HEXA-2,5-DIEN-ZYL-RE(CO)5; W/NBS
62750

494   C11 H7 Cl2 F11 N2 O3   73997-90-7   KO-6-386-19   2,2-BIS(CHLORODIFLUOROMETHYL)-4-(2-(N-HEPTAFLUOROPROPYLCA
RBONYLAMINO)ETHYL)-1,3-OXAZOLIDINONE;   W
62752

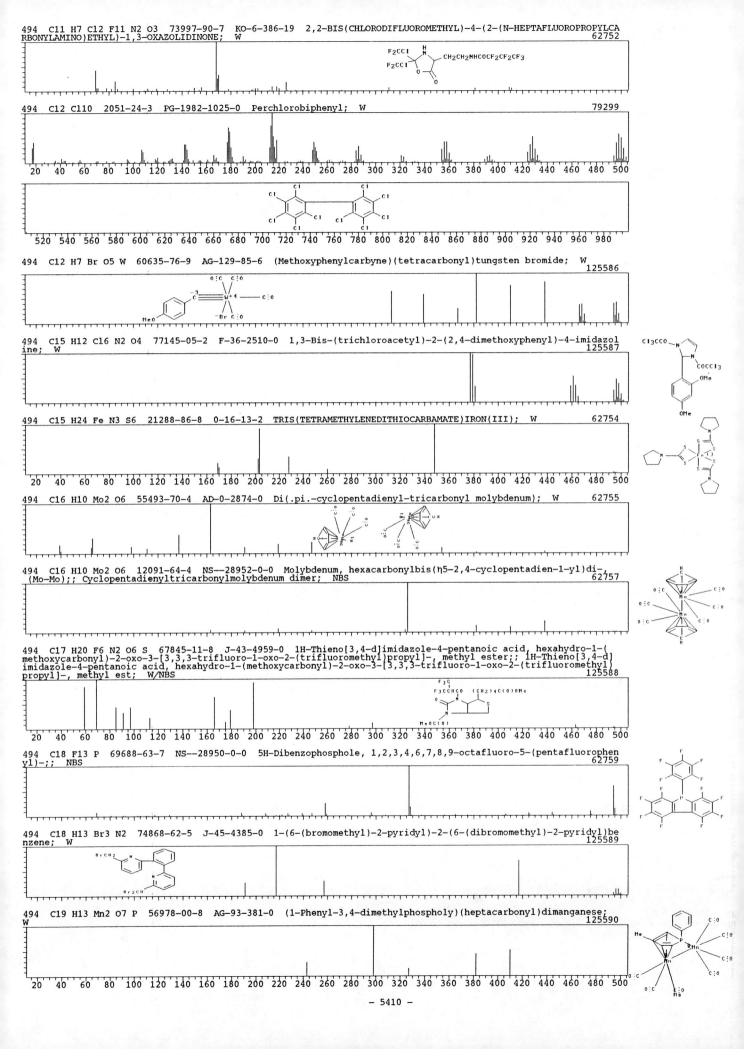

494   C12 Cl10   2051-24-3   PG-1982-1025-0   Perchlorobiphenyl;   W
79299

494   C12 H7 Br O5 W   60635-76-9   AG-129-85-6   (Methoxyphenylcarbyne)(tetracarbonyl)tungsten bromide;   W
125586

494   C15 H12 Cl6 N2 O4   77145-05-2   F-36-2510-0   1,3-Bis-(trichloroacetyl)-2-(2,4-dimethoxyphenyl)-4-imidazol
ine;   W
125587

494   C15 H24 Fe N3 S6   21288-86-8   0-16-13-2   TRIS(TETRAMETHYLENEDITHIOCARBAMATE)IRON(III);   W
62754

494   C16 H10 Mo2 O6   55493-70-4   AD-0-2874-0   Di(.pi.-cyclopentadienyl-tricarbonyl molybdenum);   W
62755

494   C16 H10 Mo2 O6   12091-64-4   NS--28952-0-0   Molybdenum, hexacarbonylbis(η5-2,4-cyclopentadien-1-yl)di-
(Mo-Mo);;   Cyclopentadienyltricarbonylmolybdenum dimer;   NBS
62757

494   C17 H20 F6 N2 O6 S   67845-11-8   J-43-4959-0   1H-Thieno[3,4-d]imidazole-4-pentanoic acid, hexahydro-1-(
methoxycarbonyl)-2-oxo-3-[3,3,3-trifluoro-1-oxo-2-(trifluoromethyl)propyl]-,   methyl ester;;   1H-Thieno[3,4-d]
imidazole-4-pentanoic acid, hexahydro-1-(methoxycarbonyl)-2-oxo-3-[3,3,3-trifluoro-1-oxo-2-(trifluoromethyl)
propyl]-,   methyl est;   W/NBS
125588

494   C18 F13 P   69688-63-7   NS--28950-0-0   5H-Dibenzophosphole, 1,2,3,4,6,7,8,9-octafluoro-5-(pentafluorophen
yl)-;;   NBS
62759

494   C18 H13 Br3 N2   74868-62-5   J-45-4385-0   1-(6-(bromomethyl)-2-pyridyl)-2-(6-(dibromomethyl)-2-pyridyl)be
nzene;   W
125589

494   C19 H13 Mn2 O7 P   56978-00-8   AG-93-381-0   (1-Phenyl-3,4-dimethylphospholy)(heptacarbonyl)dimanganese;
W
125590

494 C19 H16 Nb2 O4 77110-77-1 K-117-91-0 Carbonylbis(η5-cyclopentadienyl)-.mu.-hydrido-[tricarbonyl(η5-cyclopentadienyl)niobium]niobium; W
125591

494 C19 H42 O7 Si4 OM-1981-651-0 METHYLCITRIC ACID 4TMS; W
62760

494 C20 H22 Cl2 O2 Te 78733-66-1 C-103-5199-0 Bis[2-(5-methyl-2,3-dihydrobenzofuranyl)methyl]tellurium dichloride; W
125592

494 C20 H34 B4 O11 HE-1982-0-0 O-(2',3',:5',6'-DI-O-ETHYLBORANDIYL-D-MANNOFURANOSYL)-(1-1)-2,3:5,6-DI-O-ETHYLBORANDIYL-D-MANNOFURANOSIDE; W
62761

494 C20 H39 B5 O10 74792-96-4 HE-1982-0-0 DL-Xylitol, cyclic 1,4:2,3-bis(ethylboronate) 5,5'-(ethylboronate);; 1,1'-O-ETHYLBORANDIYL-DI-(2,5:3,4-DI-OETHYLBORANDIYL-D,L-XYLITOL); W/NBS
62762

494 C20 H46 O6 Si4 50459-27-3 C-95-7828-8 Cyclohexanecarboxylic acid, 1,3,4,5-tetrakis[(trimethylsilyl)oxy]-, methyl ester, [1S-(1α,3α,4β,5β)]-;; METHYL EPI-QUINATE TRIMETHYLSILYL ETHER; W/NBS
62763

494 C22 H26 N2 O11 26386-09-4 F-26-4023-3 1H-Indole, 2,3-dihydro-5-nitro-1-(2,3,4,6-tetra-O-acetyl-β-D-glucopyranosyl)-;; 2,3-DIHYDRO-5-NITROINDOLYL N-PERACETYL-β-D-GLUCOPYRANOSIDE; W/NBS
62764

494 C23 H24 F6 O5 34210-15-6 SD-1981-0-0 Estra-1,3,5(10)-triene-16α,17β-diol, 3-methoxy-, bis(trifluoroacetate);; 3-Methoxyestriol bis(trifluoroacetate); W/NBS
62766

494 C23 H24 F6 O5 56588-10-4 SD-1981-0-0 Estra-1,3,5(10)-triene-11,17-diol, 3-methoxy-, bis(trifluoroacetate), (11α,17β)-;; W/NBS
62765

494 C23 H26 O6 S3 SB-32-586-0 1-(4-ACETOXY-3-METHOXYPHENYL)-ETHYL-TRANS-3-(4-ACETOXY-3-METHOXYPHENYL)-2-PROPENYL OLIGOSULFIDE (NO. OF S ATOMS UNKNOWN); W
62767

494 C24 H25 F7 O3 FI-0-46-6 ETHISTERONE-HFB; W
62768

494 C24 H30 O7 S2 81825-47-0 H-65-184-0 5exo-Hydroxybicyclo[2.2.2]octane-2endo,3endo-dimethyl bis(p-toluenesulfonate); W
125593

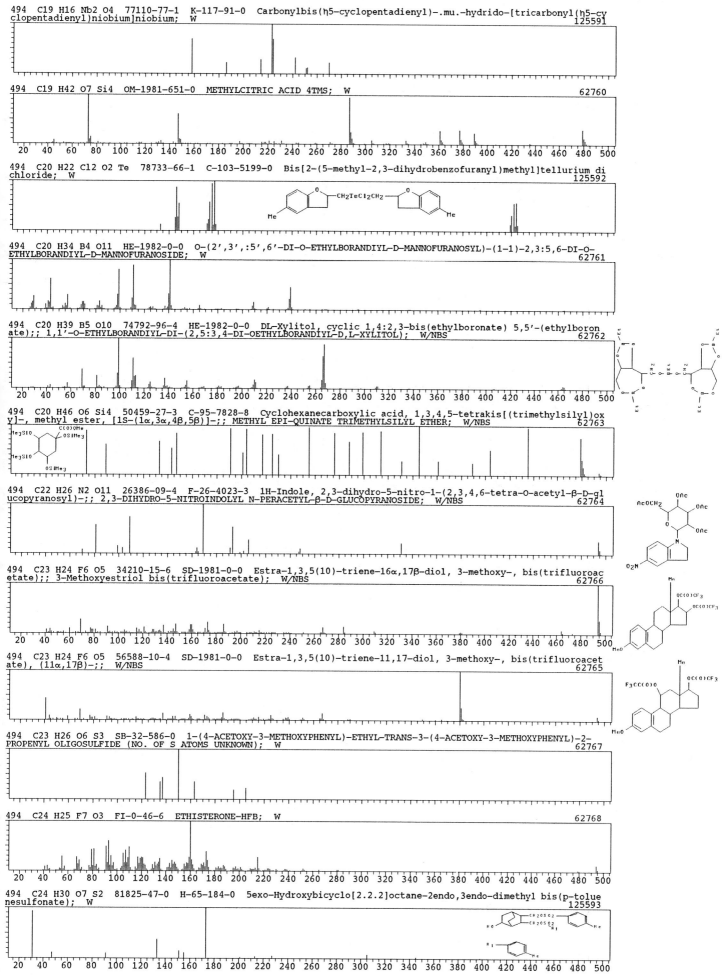

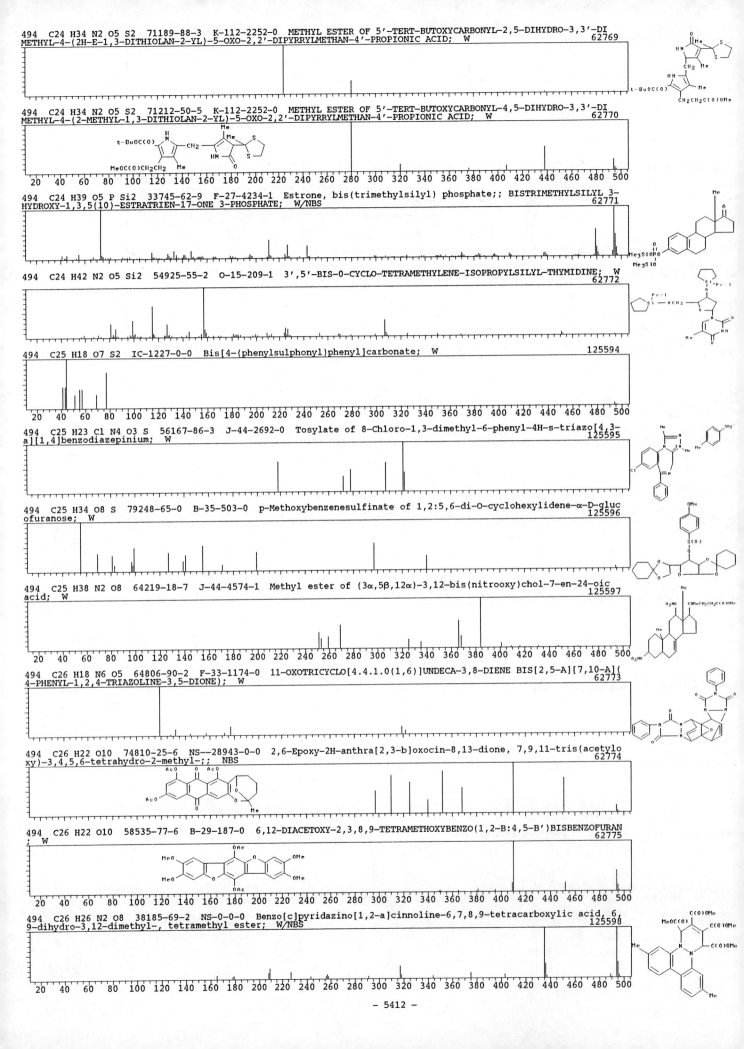

494   C24 H34 N2 O5 S2   71189-88-3   K-112-2252-0   METHYL ESTER OF 5'-TERT-BUTOXYCARBONYL-2,5-DIHYDRO-3,3'-DI
METHYL-4-(2H-E-1,3-DITHIOLAN-2-YL)-5-OXO-2,2'-DIPYRRYLMETHAN-4'-PROPIONIC ACID;   W   62769

494   C24 H34 N2 O5 S2   71212-50-5   K-112-2252-0   METHYL ESTER OF 5'-TERT-BUTOXYCARBONYL-4,5-DIHYDRO-3,3'-DI
METHYL-4-(2-METHYL-1,3-DITHIOLAN-2-YL)-5-OXO-2,2'-DIPYRRYLMETHAN-4'-PROPIONIC ACID;   W   62770

494   C24 H39 O5 P Si2   33745-62-9   F-27-4234-1   Estrone, bis(trimethylsilyl) phosphate;;   BISTRIMETHYLSILYL 3-
HYDROXY-1,3,5(10)-ESTRATRIEN-17-ONE 3-PHOSPHATE;   W/NBS   62771

494   C24 H42 N2 O5 Si2   54925-55-2   O-15-209-1   3',5'-BIS-O-CYCLO-TETRAMETHYLENE-ISOPROPYLSILYL-THYMIDINE;   W
                                                                                                                      62772

494   C25 H18 O7 S2   IC-1227-0-0   Bis[4-(phenylsulphonyl)phenyl]carbonate;   W   125594

494   C25 H23 Cl N4 O3 S   56167-86-3   J-44-2692-0   Tosylate of 8-Chloro-1,3-dimethyl-6-phenyl-4H-s-triazo[4,3-
a][1,4]benzodiazepinium;   W   125595

494   C25 H34 O8 S   79248-65-0   B-35-503-0   p-Methoxybenzenesulfinate of 1,2:5,6-di-O-cyclohexylidene-α-D-gluc
ofuranose;   W   125596

494   C25 H38 N2 O8   64219-18-7   J-44-4574-1   Methyl ester of (3α,5β,12α)-3,12-bis(nitrooxy)chol-7-en-24-oic
acid;   W   125597

494   C26 H18 N6 O5   64806-90-2   F-33-1174-0   11-OXOTRICYCLO[4.4.1.0(1,6)]UNDECA-3,8-DIENE BIS[2,5-A][7,10-A](
4-PHENYL-1,2,4-TRIAZOLINE-3,5-DIONE);   W   62773

494   C26 H22 O10   74810-25-6   NS--28943-0-0   2,6-Epoxy-2H-anthra[2,3-b]oxocin-8,13-dione, 7,9,11-tris(acetylo
xy)-3,4,5,6-tetrahydro-2-methyl-;;   NBS   62774

494   C26 H22 O10   58535-77-6   B-29-187-0   6,12-DIACETOXY-2,3,8,9-TETRAMETHOXYBENZO(1,2-B:4,5-B')BISBENZOFURAN
;   W   62775

494   C26 H26 N2 O8   38185-69-2   NS-0-0-0   Benzo[c]pyridazino[1,2-a]cinnoline-6,7,8,9-tetracarboxylic acid, 6,
9-dihydro-3,12-dimethyl-, tetramethyl ester;   W/NBS   125598

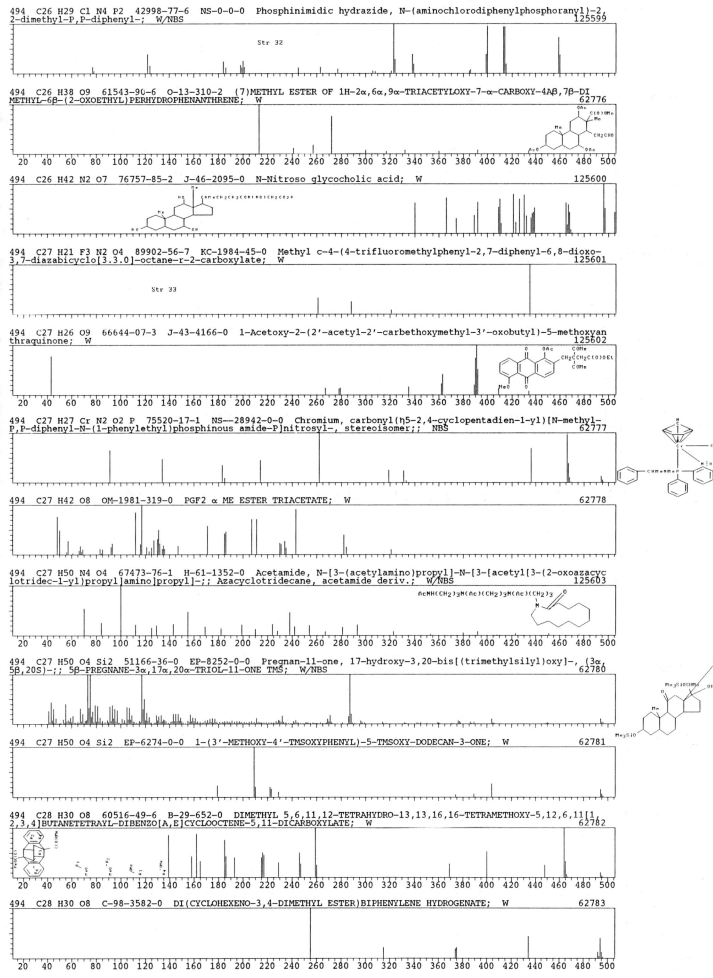

494 C26 H29 Cl N4 P2 42998-77-6 NS-0-0-0 Phosphinimidic hydrazide, N-(aminochlorodiphenylphosphoranyl)-2,2-dimethyl-P,P-diphenyl-; W/NBS
125599

Str 32

494 C26 H38 O9 61543-90-6 O-13-310-2 (7)METHYL ESTER OF 1H-2α,6α,9α-TRIACETYLOXY-7-α-CARBOXY-4Aβ,7β-DIMETHYL-6β-(2-OXOETHYL)PERHYDROPHENANTHRENE; W
62776

494 C26 H42 N2 O7 76757-85-2 J-46-2095-0 N-Nitroso glycocholic acid; W
125600

494 C27 H21 F3 N2 O4 89902-56-7 KC-1984-45-0 Methyl c-4-(4-trifluoromethylphenyl-2,7-diphenyl-6,8-dioxo-3,7-diazabicyclo[3.3.0]-octane-r-2-carboxylate; W
125601

Str 33

494 C27 H26 O9 66644-07-3 J-43-4166-0 1-Acetoxy-2-(2'-acetyl-2'-carbethoxymethyl-3'-oxobutyl)-5-methoxyanthraquinone; W
125602

494 C27 H27 Cr N2 O2 P 75520-17-1 NS--28942-0-0 Chromium, carbonyl(η5-2,4-cyclopentadien-1-yl)[N-methyl-P,P-diphenyl-N-(1-phenylethyl)phosphinous amide-P]nitrosyl-, stereoisomer;; NBS
62777

494 C27 H42 O8 OM-1981-319-0 PGF2 α ME ESTER TRIACETATE; W
62778

494 C27 H50 N4 O4 67473-76-1 H-61-1352-0 Acetamide, N-[3-(acetylamino)propyl]-N-[3-[acetyl[3-(2-oxoazacyclotridec-1-yl)propyl]amino]propyl]-;; Azacyclotridecane, acetamide deriv.; W/NBS
125603

494 C27 H50 O4 Si2 51166-36-0 EP-8252-0-0 Pregnan-11-one, 17-hydroxy-3,20-bis[(trimethylsilyl)oxy]-, (3α,5β,20S)-;; 5β-PREGNANE-3α,17α,20α-TRIOL-11-ONE TMS; W/NBS
62780

494 C27 H50 O4 Si2 EP-6274-0-0 1-(3'-METHOXY-4'-TMSOXYPHENYL)-5-TMSOXY-DODECAN-3-ONE; W
62781

494 C28 H30 O8 60516-49-6 B-29-652-0 DIMETHYL 5,6,11,12-TETRAHYDRO-13,13,16,16-TETRAMETHOXY-5,12,6,11[1,2,3,4]BUTANETETRAYL-DIBENZO[A,E]CYCLOOCTENE-5,11-DICARBOXYLATE; W
62782

494 C28 H30 O8 C-98-3582-0 DI(CYCLOHEXENO-3,4-DIMETHYL ESTER)BIPHENYLENE HYDROGENATE; W
62783

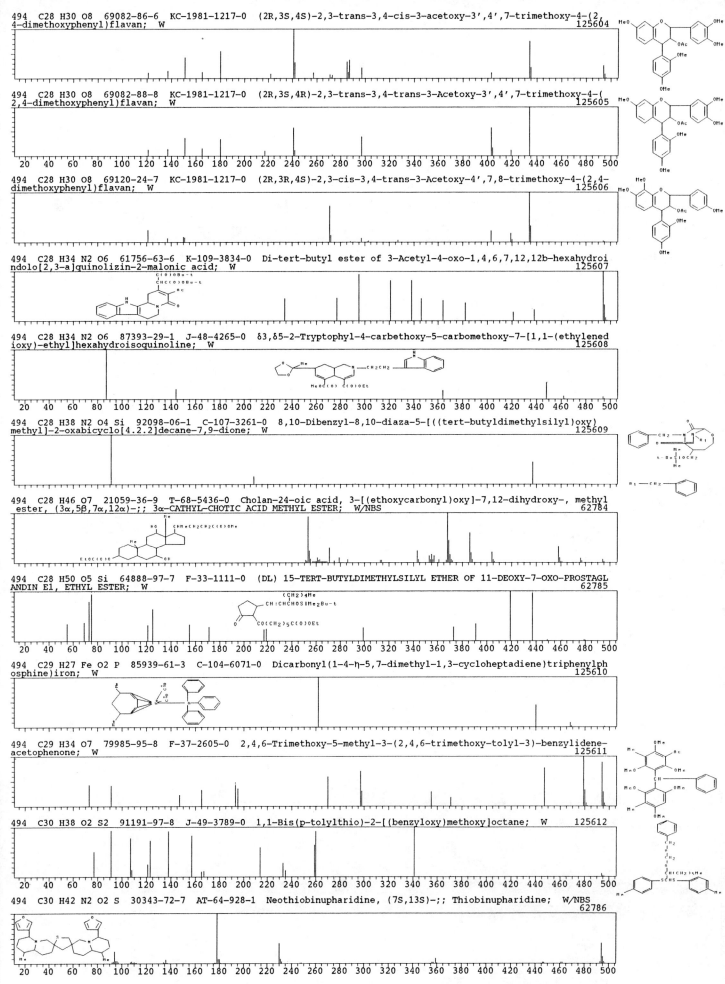

494 C28 H30 O8 69082-86-6 KC-1981-1217-0 (2R,3S,4S)-2,3-trans-3,4-cis-3-acetoxy-3',4',7-trimethoxy-4-(2,4-dimethoxyphenyl)flavan; W
125604

494 C28 H30 O8 69082-88-8 KC-1981-1217-0 (2R,3S,4R)-2,3-trans-3,4-trans-3-Acetoxy-3',4',7-trimethoxy-4-(2,4-dimethoxyphenyl)flavan; W
125605

494 C28 H30 O8 69120-24-7 KC-1981-1217-0 (2R,3R,4S)-2,3-cis-3,4-trans-3-Acetoxy-4',7,8-trimethoxy-4-(2,4-dimethoxyphenyl)flavan; W
125606

494 C28 H34 N2 O6 61756-63-6 K-109-3834-0 Di-tert-butyl ester of 3-Acetyl-4-oxo-1,4,6,7,12,12b-hexahydroindolo[2,3-a]quinolizin-2-malonic acid; W
125607

494 C28 H34 N2 O6 87393-29-1 J-48-4265-0 δ3,δ5-2-Tryptophyl-4-carbethoxy-5-carbomethoxy-7-[1,1-(ethylenedioxy)-ethyl]hexahydroisoquinoline; W
125608

494 C28 H38 N2 O4 Si 92098-06-1 C-107-3261-0 8,10-Dibenzyl-8,10-diaza-5-[((tert-butyldimethylsilyl)oxy)methyl]-2-oxabicyclo[4.2.2]decane-7,9-dione; W
125609

494 C28 H46 O7 21059-36-9 T-68-5436-0 Cholan-24-oic acid, 3-[(ethoxycarbonyl)oxy]-7,12-dihydroxy-, methyl ester, (3α,5β,7α,12α)-;; 3α-CATHYL-CHOTIC ACID METHYL ESTER; W/NBS
62784

494 C28 H50 O5 Si 64888-97-7 F-33-1111-0 (DL) 15-TERT-BUTYLDIMETHYLSILYL ETHER OF 11-DEOXY-7-OXO-PROSTAGLANDIN E1, ETHYL ESTER; W
62785

494 C29 H27 Fe O2 P 85939-61-3 C-104-6071-0 Dicarbonyl(1-4-η-5,7-dimethyl-1,3-cycloheptadiene)triphenylphosphine)iron; W
125610

494 C29 H34 O7 79985-95-8 F-37-2605-0 2,4,6-Trimethoxy-5-methyl-3-(2,4,6-trimethoxy-tolyl-3)-benzylidene-acetophenone; W
125611

494 C30 H38 O2 S2 91191-97-8 J-49-3789-0 1,1-Bis(p-tolylthio)-2-[(benzyloxy)methoxy]octane; W 125612

494 C30 H42 N2 O2 S 30343-72-7 AT-64-928-1 Neothiobinupharidine, (7S,13S)-;; Thiobinupharidine; W/NBS
62786

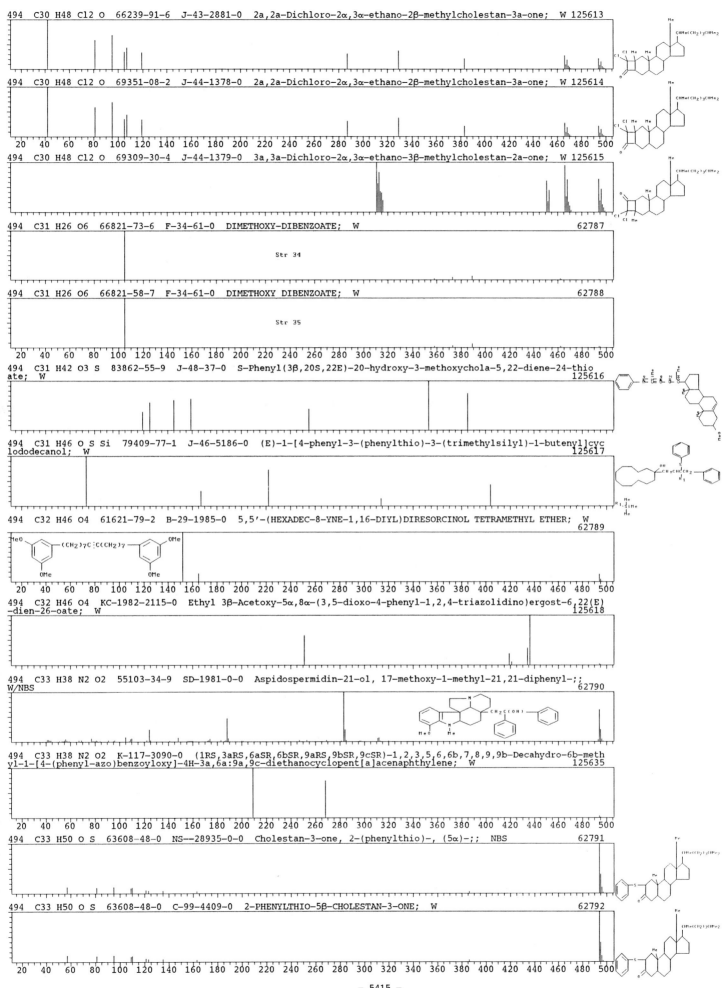

494   C30 H48 Cl2 O   66239-91-6   J-43-2881-0   2a,2a-Dichloro-2α,3α-ethano-2β-methylcholestan-3a-one;   W 125613

494   C30 H48 Cl2 O   69351-08-2   J-44-1378-0   2a,2a-Dichloro-2α,3α-ethano-2β-methylcholestan-3a-one;   W 125614

494   C30 H48 Cl2 O   69309-30-4   J-44-1379-0   3a,3a-Dichloro-2α,3α-ethano-3β-methylcholestan-2a-one;   W 125615

494   C31 H26 O6   66821-73-6   F-34-61-0   DIMETHOXY-DIBENZOATE;   W   62787

Str 34

494   C31 H26 O6   66821-58-7   F-34-61-0   DIMETHOXY DIBENZOATE;   W   62788

Str 35

494   C31 H42 O3 S   83862-55-9   J-48-37-0   S-Phenyl(3β,20S,22E)-20-hydroxy-3-methoxychola-5,22-diene-24-thioate;   W   125616

494   C31 H46 O S Si   79409-77-1   J-46-5186-0   (E)-1-[4-phenyl-3-(phenylthio)-3-(trimethylsilyl)-1-butenyl]cyclododecanol;   W   125617

494   C32 H46 O4   61621-79-2   B-29-1985-0   5,5'-(HEXADEC-8-YNE-1,16-DIYL)DIRESORCINOL TETRAMETHYL ETHER;   W   62789

494   C32 H46 O4   KC-1982-2115-0   Ethyl 3β-Acetoxy-5α,8α-(3,5-dioxo-4-phenyl-1,2,4-triazolidino)ergost-6,22(E)-dien-26-oate;   W   125618

494   C33 H38 N2 O2   55103-34-9   SD-1981-0-0   Aspidospermidin-21-ol, 17-methoxy-1-methyl-21,21-diphenyl-;;   W/NBS   62790

494   C33 H38 N2 O2   K-117-3090-0   (1RS,3aRS,6aSR,6bSR,9aRS,9bSR,9cSR)-1,2,3,5,6,6b,7,8,9,9b-Decahydro-6b-methyl-1-[4-(phenyl-azo)benzoyloxy]-4H-3a,6a:9a,9c-diethanocyclopent[a]acenaphthylene;   W   125635

494   C33 H50 O S   63608-48-0   NS--28935-0-0   Cholestan-3-one, 2-(phenylthio)-, (5α)-;;   NBS   62791

494   C33 H50 O S   63608-48-0   C-99-4409-0   2-PHENYLTHIO-5β-CHOLESTAN-3-ONE;   W   62792

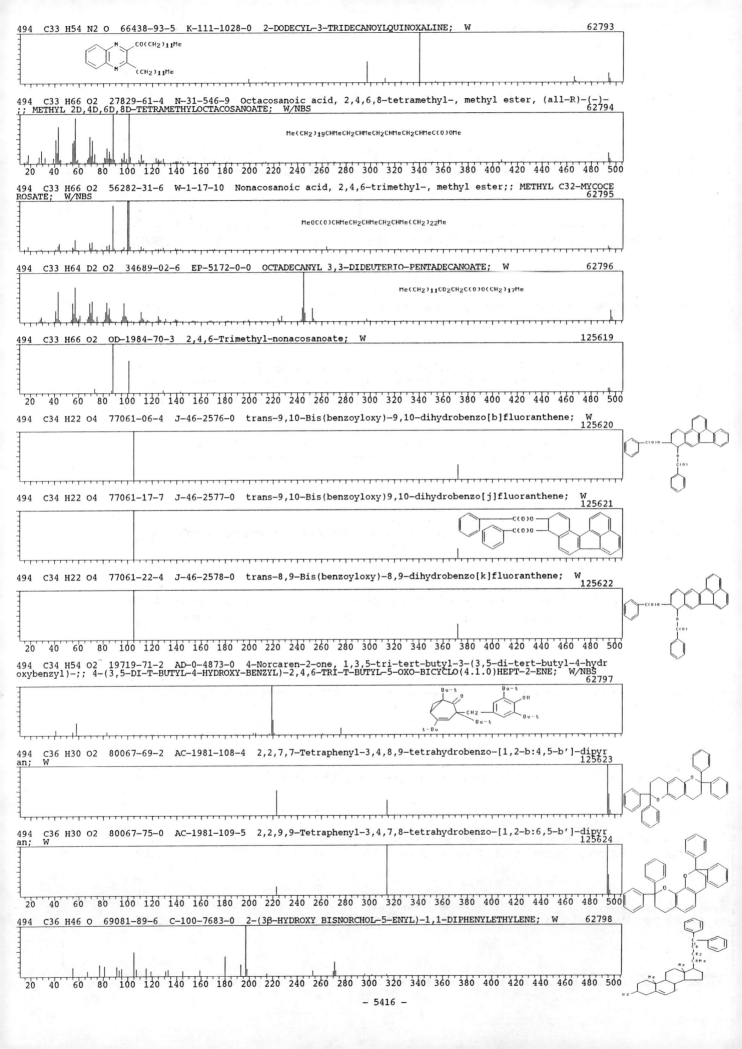

494   C33 H54 N2 O   66438-93-5   K-111-1028-0   2-DODECYL-3-TRIDECANOYLQUINOXALINE;   W   62793

494   C33 H66 O2   27829-61-4   N-31-546-9   Octacosanoic acid, 2,4,6,8-tetramethyl-, methyl ester, (all-R)-(-)-
;; METHYL 2D,4D,6D,8D-TETRAMETHYLOCTACOSANOATE;   W/NBS   62794

Me(CH2)19CHMeCH2CHMeCH2CHMeCH2CHMeC(O)OMe

494   C33 H66 O2   56282-31-6   W-1-17-10   Nonacosanoic acid, 2,4,6-trimethyl-, methyl ester;; METHYL C32-MYCOCE
ROSATE;   W/NBS   62795

MeOC(O)CHMeCH2CHMeCH2CHMe(CH2)22Me

494   C33 H64 D2 O2   34689-02-6   EP-5172-0-0   OCTADECANYL 3,3-DIDEUTERIO-PENTADECANOATE;   W   62796

Me(CH2)11CD2CH2C(O)O(CH2)17Me

494   C33 H66 O2   OD-1984-70-3   2,4,6-Trimethyl-nonacosanoate;   W   125619

494   C34 H22 O4   77061-06-4   J-46-2576-0   trans-9,10-Bis(benzoyloxy)-9,10-dihydrobenzo[b]fluoranthene;   W
                                                                                                                   125620

494   C34 H22 O4   77061-17-7   J-46-2577-0   trans-9,10-Bis(benzoyloxy)9,10-dihydrobenzo[j]fluoranthene;   W
                                                                                                                   125621

494   C34 H22 O4   77061-22-4   J-46-2578-0   trans-8,9-Bis(benzoyloxy)-8,9-dihydrobenzo[k]fluoranthene;   W
                                                                                                                   125622

494   C34 H54 O2   19719-71-2   AD-0-4873-0   4-Norcaren-2-one, 1,3,5-tri-tert-butyl-3-(3,5-di-tert-butyl-4-hydr
oxybenzyl)-;; 4-(3,5-DI-T-BUTYL-4-HYDROXY-BENZYL)-2,4,6-TRI-T-BUTYL-5-OXO-BICYCLO(4.1.0)HEPT-2-ENE;   W/NBS
                                                                                                                    62797

494   C36 H30 O2   80067-69-2   AC-1981-108-4   2,2,7,7-Tetraphenyl-3,4,8,9-tetrahydrobenzo-[1,2-b:4,5-b']-dipyr
an;   W   125623

494   C36 H30 O2   80067-75-0   AC-1981-109-5   2,2,9,9-Tetraphenyl-3,4,7,8-tetrahydrobenzo-[1,2-b:6,5-b']-dipyr
an;   W   125624

494   C36 H46 O   69081-89-6   C-100-7683-0   2-(3β-HYDROXY BISNORCHOL-5-ENYL)-1,1-DIPHENYLETHYLENE;   W   62798

495   C11 H6 Cl2 F11 N O4   73997-83-8   KO-6-384-11   2,2-BIS(CHLORODIFLUOROMETHYL)-4-(1-(1-OXOHEPTAFLUOROBUTOX
Y)ETHYL)-1,3-OXAZOLIDINONE;   W
62799

495   C19 H37 N5 O3 Si4   55649-41-7   AV-37-291-11   6-Pteridinecarboxylic acid, 2-[bis(trimethylsilyl)amino]-1,
4-dihydro-4-oxo-1-(trimethylsilyl)-, trimethylsilyl ester;; TETRAKISTRIMETHYLSILYL PTERIN-6-CARBOXYLIC ACID;
W/NBS
62800

495   C20 H35 F2 N3 Si5   62978-26-1   K-110-1282-0   1,3-BIS(FLUOROMETHYLPHENYLSILYL)-2,2,4,4,6,6-HEXAMETHYLCYCL
OTRISILAZANE;   W
62801

495   C20 H47 D2 N O5 Si4   HE-1982-0-0   HEXIT-1,4-D2, 2-ACETAMIDO-2,3-DIDESOXY-, O-METHYLOXIM, TRIS-O-(TRIMETH
YLSILYL)-;   W
62802

495   C22 H41 N5 O4 Si2   51549-35-0   O-15-209-1   3',5'-BIS-O-TERT-BUTYLDIMETHYLSILYL-2'-DEOXYGUANOSINE;   W
62803

495   C24 H25 F3 N O5 P   65138-47-8   K-113-64-0   Dimethylester of 5,5,7-Trimethyl2,2-diphenyl-7-(trifluormeth
yl)-6-oxa-1-aza-5λ(5)-phosphabicyclo(3.2.0)hept-3-ene-3,4-dicarboxylic acid;   W
125625

495   C25 H19 Mo N O2 S   62865-39-8   K-110-733-23   DICARBONYL(CYCLOPENTADIENYL)(1-NAPHTHYL-N-BENZYL-THIOAMIDAT
O)-MOLYBDENUM;   W
62804

495   C25 H25 N3 O4 S2   72955-30-7   KC-1979-2391-0   1,2,4,5-TETRAHYDRO-5-METHYL-5-P-TOLYLSULPHONYLAMINO-4-P-TO
LYSULPHONYLIMINOPYTTOLO(3,2,1-HI)INDOLE;   W
62805

495   C26 H25 N O9   83640-65-7   B-35-1222-0   Tetramethyl 1-(4-methoxyphenyl)-6-phenyl-1,6-dihydropyridine-2,3,
4,5-tetracarboxylate;   W
125626

495   C26 H25 N O9   83640-71-5   B-35-1224-0   Tetramethyl ester of 6-(4-methoxyphenyl)-1-phenyl-1,6-dihydropyri
dine-2,3,4,5-tetracarboxylic acid;   W
125627

495   C26 H41 N O8   50676-21-6   NS--28930-0-0   Aconitane-1,7,8,14-tetrol, 20-ethyl-6,16-dimethoxy-4-(methoxy
methyl)-, 14-acetate, (1α,6β,14α,16β)-;;   NBS
62806

495   C27 H23 Cl2 N S2   75931-73-6   O-15-25-1   N,N-BIS(4'-P-CHLOROPHENYLTHIO-2'-BUTYNYL)-4-METHYLANILINE;   W
62807

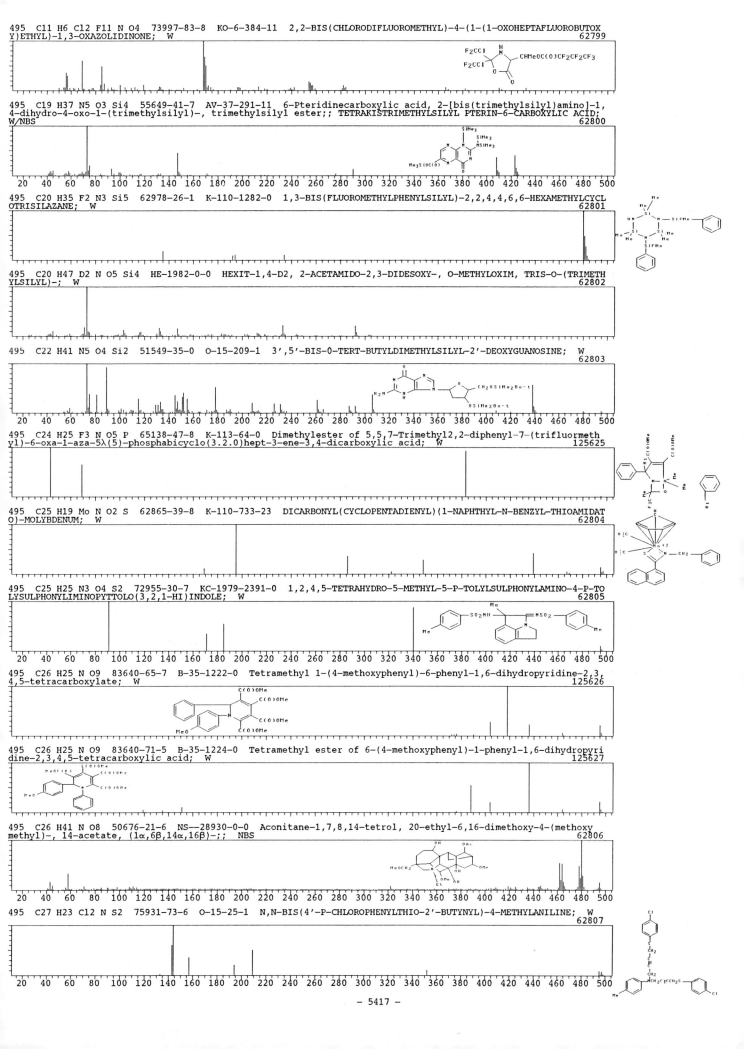

495 C27 H26 Cl N O6 93295-63-7 H-67-1570-0 2,3-O-Isopropylidene-1-nitro-5-O-trityl-β-D-ribofuranosylchlor
ide; W   125628

495 C27 H29 N O8 86363-29-3 KC-1983-523-0 Dimethyl 3-methoxycarbonylmethyl-2-oxo-2,4,5,6,7,7a,12,13-octa
hydro-11H-furo[2'',3'':2',3']cyclohepta[1',2':4,5]pyrrolo[3,2,1-ij]quinolin-7a-ylmaleate; W   125629

495 C27 H29 N O8 86340-46-7 KC-1983-523-0 Dimethyl 3-methoxycarbonylmethyl-2-oxo-2,4,5,6,7,7a,12,13-octa
hydro-11H-furo[2'',3'':2',3',]cyclohepta[1',2':4,5]pyrrolo[3,2,1-ij]quinolin-7a-ylfumarate; W 125630

495 C28 H45 N7 O RB-1982-9481-0 N,N'-BIS(4-DIETHYLAMINO-2-ETHOXYBENZYLIDENEAMINO)GUANIDINE; W 125631

495 C28 H53 N3 O4 53602-40-7 H-57-426-7 N(6),N(4')-DIACETYL-NEOONCINOTINIC ACID METHYL ESTER; W/NBS
                                                                                             62808

495 C28 H53 N3 O4 D3-1980-80-24 N,N'-DIACETYL-ONCINOTINE; W   62809

495 C29 H25 N3 O5 639-48-5 NS--28928-0-0 Nicomorphine; NBS   62810

495 C29 H53 N O5 IC-1228-0-0 1-Cyclohexylmethyl-3a,5-dihydroxy-4-(8,11-dihydroxy-4-methyl-undecyl)-6,7-di
methyl-3-oxo-3a,4,5,6,7,7a-hexahydroisoindoline; W   125633

495 C30 H29 N3 O4 71522-29-7 I-57-1701-0 N-(2-(1'-BENZYL-3'-(2''-METHOXYCARBONYLINDOL-7''-YL)OXINDOL-3'-
YL)ETHYL)-N-METHYLACETAMIDE; W   62811

495 C32 H27 Cl2 N 88365-75-7 K-116-3944-0 rel-(1R,4S,5S,6S,8R)-2,4-Bis(4-chlorophenyl)-6,8-diphenyl-3-aza
bicyclo[3.3.1]non-2-ene; W   125634

495 C32 H49 N O3 26216-84-2 AD-0-3500-0 2,4-Cyclohexadien-1-one, 4,6-di-tert-butyl-2-(3,5-di-tert-butyl-
2-hydroxyphenyl)-6-morpholino-;; 2-(3,5-DI-T-BUTYL-2-HYDROXYPHENYL)-4,6-DI-T-BUTYL-6-MORPHOLINOCYCLOHEXADIENO
NE; W/NBS   62812

495 C34 H41 N O2 55493-87-3 AD-0-3720-0 1'H-Androst-16-eno[17,16-b]indol-3-ol, 1'-(phenylmethyl)-, acet
ate (ester), (3β,5α)-;; 3β-ACETOXY-5α-ANDROSTANO(17,16-B)N-BENZYLINDOLE; W/NBS   62813

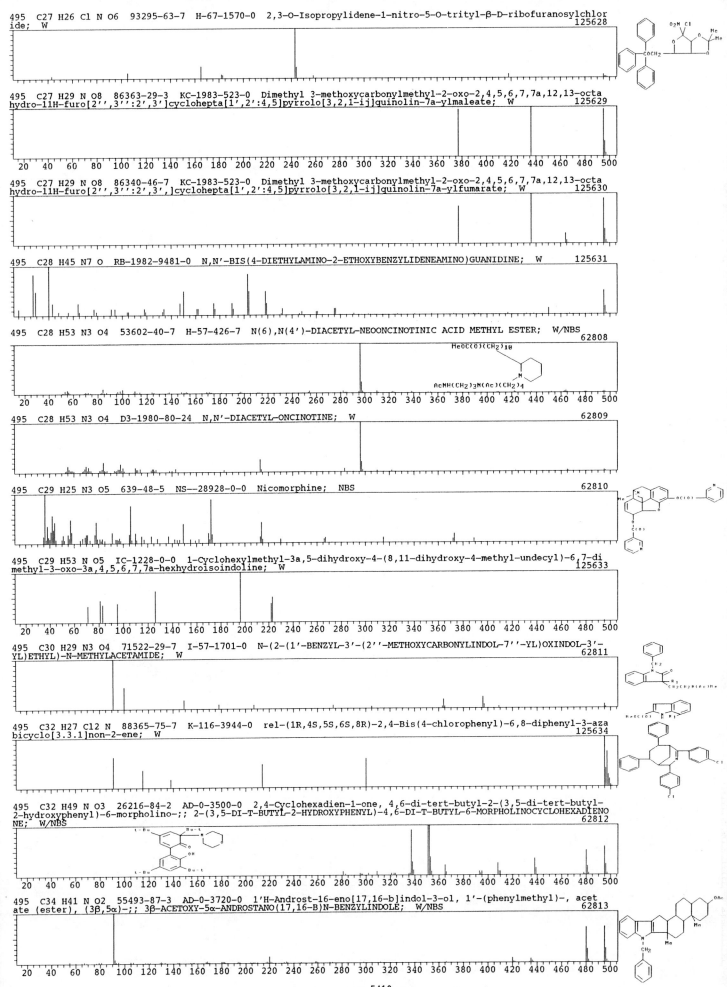

495 C35 H45 N O KC-1984-273-0 4-Tetradecyloxy-N-(9,10-dihydro-2-phenanthrylmethylene)aniline; W   125636

495 F9 O Re Sb KA-1981-1213-0 Tetrafluoro oxorhenium pentafluoro antimony adduct; W   125637

496 C8 H5 Br5 85-22-3 NS--28904-0-0 Benzene, pentabromoethyl-;; Pentabromoethylbenzene; NBS   62814

496 C8 H5 Br5 RB-1982-14912-0 α,α,α',2,5-PENTABROMO-P-XYLENE; W   125638

496 C14 H12 Br4 68185-77-3 J-44-8-0 2,3,6,7-Tetrabromo-1,4,5,8-tetramethylnapthalene; W   125639

496 C16 H22 Cl2 Pd2 B-33-2267-0 Di-.mu.-chloro-bis[α-2-η-(2-methylenebicyclo[2.2.1]heptane)palladium((I)];
W   125640

496 C16 H28 Cl2 Rh2 12308-60-0 NS--28925-0-0 Rhodium, tetrakis[(1,2,3-η)-2-butenyl]di-.mu.-chlorodi-;;
NBS   62815

496 C16 H28 Cl2 Rh2 HE-1982-0-0 RHODIUM, MUE-DICHLORO-TETRACROTYL-BIS-; W   62816

496 C17 H14 F6 N6 O5 69145-31-9 KO-6-3-17 N-PROPYL-2,2'-AZOXYBIS-(α,α,α-TRIFLUORO-6-NITRO-P-TOLUIDINE);
W   62817

496 C17 H14 F6 N6 O5 69145-30-8 KO-6-3-18 N'-PROPYL-2,2'-AZOXYBIS-(α,α,α-TRIFLUORO-6-NITRO-P-TOLUIDINE);
W   62818

496 C18 H44 N2 O4 S Si4 AM-0-155-0 LANTHIONINE-TETRATMS; W   62819

496 C19 H28 O3 Os HE-1986-794-0 TRICARBONYL-BIS(CYCLOOCTENE)OSMIUM; W   125641

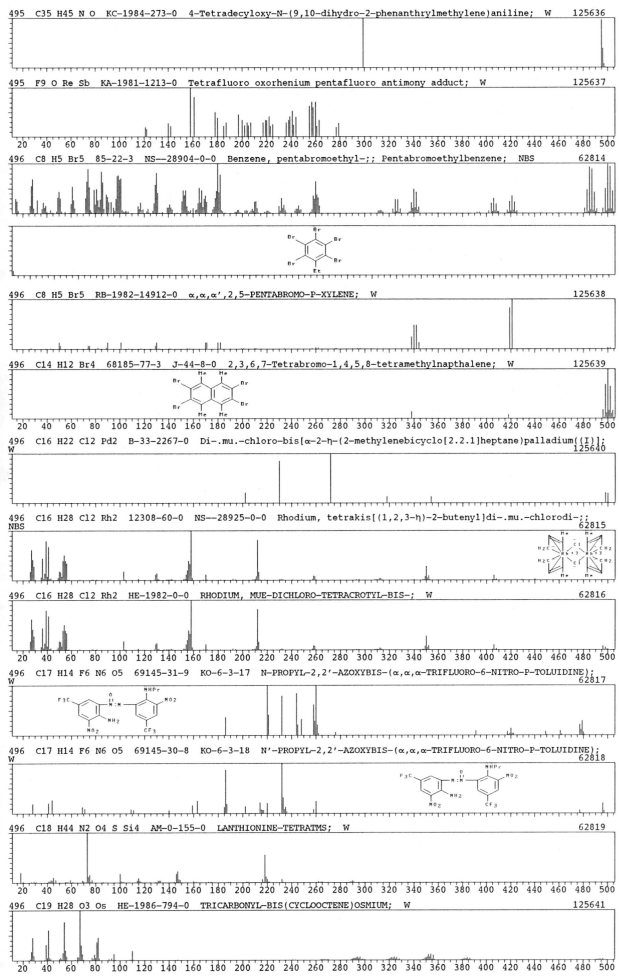

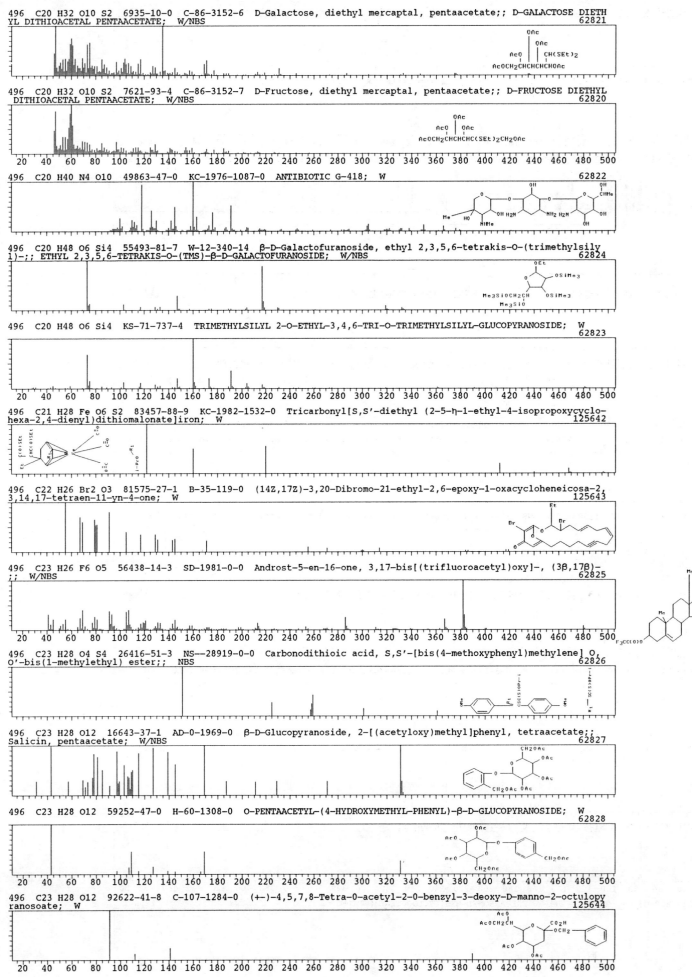

496   C20 H32 O10 S2   6935-10-0   C-86-3152-6   D-Galactose, diethyl mercaptal, pentaacetate;; D-GALACTOSE DIETH
YL DITHIOACETAL PENTAACETATE;   W/NBS   62821

496   C20 H32 O10 S2   7621-93-4   C-86-3152-7   D-Fructose, diethyl mercaptal, pentaacetate;; D-FRUCTOSE DIETHYL
DITHIOACETAL PENTAACETATE;   W/NBS   62820

496   C20 H40 N4 O10   49863-47-0   KC-1976-1087-0   ANTIBIOTIC G-418;   W   62822

496   C20 H48 O6 Si4   55493-81-7   W-12-340-14   β-D-Galactofuranoside, ethyl 2,3,5,6-tetrakis-O-(trimethylsily
l)-;; ETHYL 2,3,5,6-TETRAKIS-O-(TMS)-β-D-GALACTOFURANOSIDE;   W/NBS   62824

496   C20 H48 O6 Si4   KS-71-737-4   TRIMETHYLSILYL 2-O-ETHYL-3,4,6-TRI-O-TRIMETHYLSILYL-GLUCOPYRANOSIDE;   W
                                                                                                          62823

496   C21 H28 Fe O6 S2   83457-88-9   KC-1982-1532-0   Tricarbonyl[S,S'-diethyl (2-5-η-1-ethyl-4-isopropoxycyclo-
hexa-2,4-dienyl)dithiomalonate]iron;   W   125642

496   C22 H26 Br2 O3   81575-27-1   B-35-119-0   (14Z,17Z)-3,20-Dibromo-21-ethyl-2,6-epoxy-1-oxacycloheneicosa-2,
3,14,17-tetraen-11-yn-4-one;   W   125643

496   C23 H26 F6 O5   56438-14-3   SD-1981-0-0   Androst-5-en-16-one, 3,17-bis[(trifluoroacetyl)oxy]-, (3β,17β)-
;;   W/NBS   62825

496   C23 H28 O4 S4   26416-51-3   NS--28919-0-0   Carbonodithioic acid, S,S'-[bis(4-methoxyphenyl)methylene] O,
O'-bis(1-methylethyl) ester;;   NBS   62826

496   C23 H28 O12   16643-37-1   AD-0-1969-0   β-D-Glucopyranoside, 2-[(acetyloxy)methyl]phenyl, tetraacetate;;
Salicin, pentaacetate;   W/NBS   62827

496   C23 H28 O12   59252-47-0   H-60-1308-0   O-PENTAACETYL-(4-HYDROXYMETHYL-PHENYL)-β-D-GLUCOPYRANOSIDE;   W
                                                                                                          62828

496   C23 H28 O12   92622-41-8   C-107-1284-0   (+-)-4,5,7,8-Tetra-O-acetyl-2-O-benzyl-3-deoxy-D-manno-2-octulopy
ranosoate;   W   125644

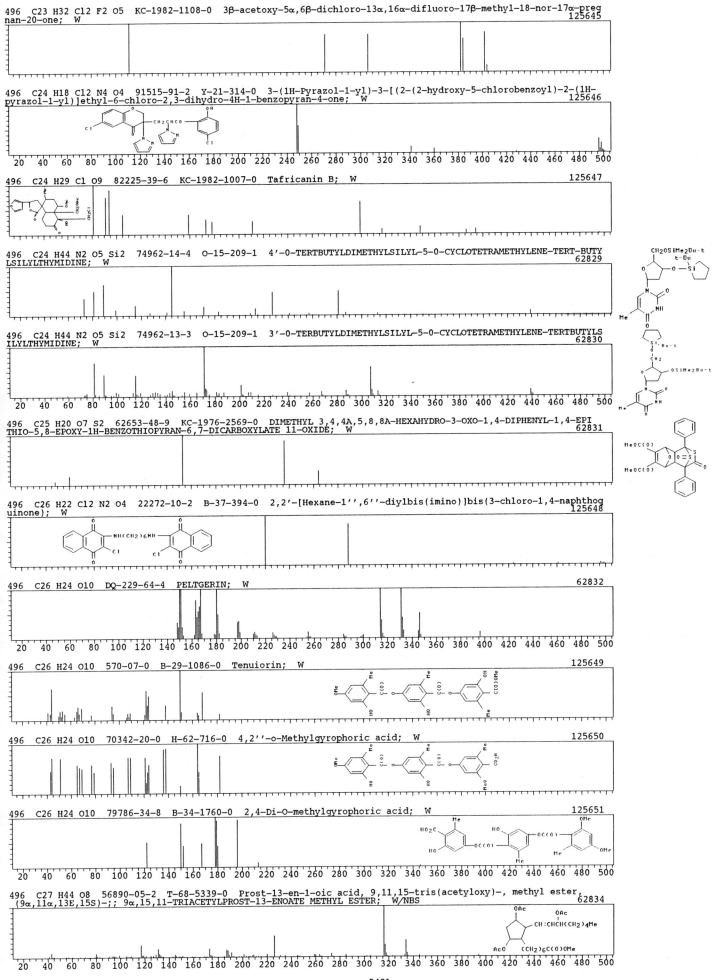

496 C23 H32 Cl2 F2 O5 KC-1982-1108-0 3β-acetoxy-5α,6β-dichloro-13α,16α-difluoro-17β-methyl-18-nor-17α-preg
nan-20-one; W   125645

496 C24 H18 Cl2 N4 O4 91515-91-2 Y-21-314-0 3-(1H-Pyrazol-1-yl)-3-[(2-(2-hydroxy-5-chlorobenzoyl)-2-(1H-
pyrazol-1-yl)]ethyl-6-chloro-2,3-dihydro-4H-1-benzopyran-4-one; W   125646

496 C24 H29 Cl O9 82225-39-6 KC-1982-1007-0 Tafricanin B; W   125647

496 C24 H44 N2 O5 Si2 74962-14-4 O-15-209-1 4'-O-TERTBUTYLDIMETHYLSILYL-5-O-CYCLOTETRAMETHYLENE-TERT-BUTY
LSILYLTHYMIDINE; W   62829

496 C24 H44 N2 O5 Si2 74962-13-3 O-15-209-1 3'-O-TERBUTYLDIMETHYLSILYL-5-O-CYCLOTETRAMETHYLENE-TERTBUTYLS
ILYLTHYMIDINE; W   62830

496 C25 H20 O7 S2 62653-48-9 KC-1976-2569-0 DIMETHYL 3,4,4A,5,8,8A-HEXAHYDRO-3-OXO-1,4-DIPHENYL-1,4-EPI
THIO-5,8-EPOXY-1H-BENZOTHIOPYRAN-6,7-DICARBOXYLATE 11-OXIDE; W   62831

496 C26 H22 Cl2 N2 O4 22272-10-2 B-37-394-0 2,2'-[Hexane-1'',6''-diylbis(imino)]bis(3-chloro-1,4-naphthog
uinone); W   125648

496 C26 H24 O10 DQ-229-64-4 PELTGERIN; W   62832

496 C26 H24 O10 570-07-0 B-29-1086-0 Tenuiorin; W   125649

496 C26 H24 O10 70342-20-0 H-62-716-0 4,2''-o-Methylgyrophoric acid; W   125650

496 C26 H24 O10 79786-34-8 B-34-1760-0 2,4-Di-O-methylgyrophoric acid; W   125651

496 C27 H44 O8 56890-05-2 T-68-5339-0 Prost-13-en-1-oic acid, 9,11,15-tris(acetyloxy)-, methyl ester,
(9α,11α,13E,15S)-;; 9α,15,11-TRIACETYLPROST-13-ENOATE METHYL ESTER; W/NBS   62834

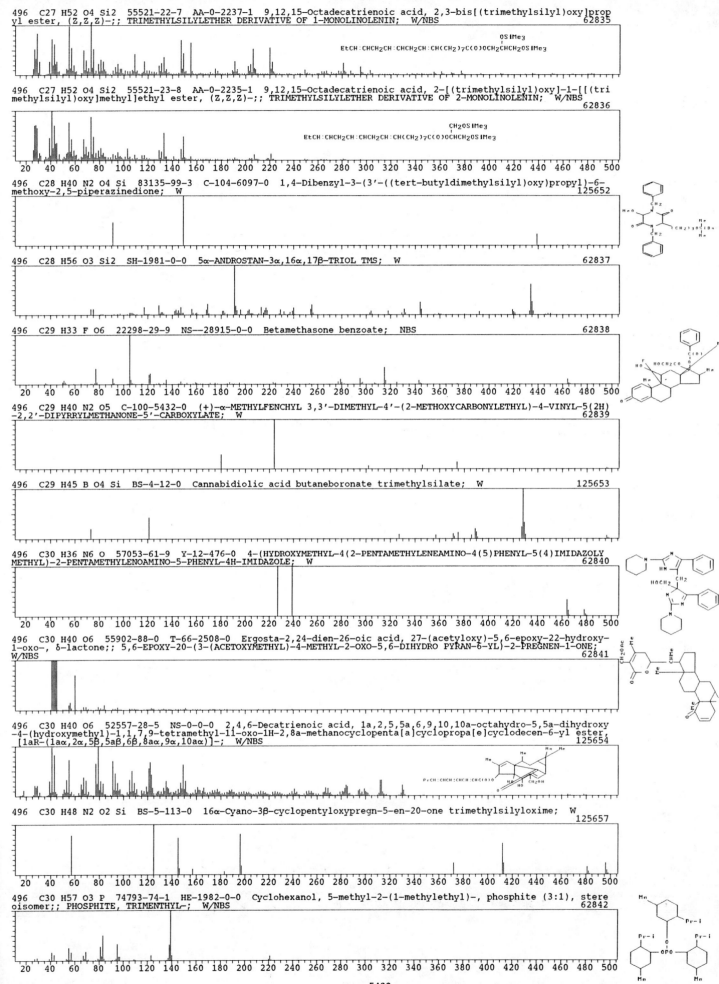

496　C27 H52 O4 Si2　55521-22-7　AA-0-2237-1　9,12,15-Octadecatrienoic acid, 2,3-bis[(trimethylsilyl)oxy]propyl ester, (Z,Z,Z)-;; TRIMETHYLSILYLETHER DERIVATIVE OF 1-MONOLINOLENIN;　W/NBS　62835

OSIMe₃
EtCH:CHCH₂CH:CHCH₂CH:CH(CH₂)₇C(O)OCH₂CHCH₂OSIMe₃

496　C27 H52 O4 Si2　55521-23-8　AA-0-2235-1　9,12,15-Octadecatrienoic acid, 2-[(trimethylsilyl)oxy]-1-[[(trimethylsilyl)oxy]methyl]ethyl ester, (Z,Z,Z)-;; TRIMETHYLSILYLETHER DERIVATIVE OF 2-MONOLINOLENIN;　W/NBS　62836

CH₂OSIMe₃
EtCH:CHCH₂CH:CHCH₂CH:CH(CH₂)₇C(O)OCH₂OSIMe₃

496　C28 H40 N2 O4 Si　83135-99-3　C-104-6097-0　1,4-Dibenzyl-3-(3'-((tert-butyldimethylsilyl)oxy)propyl)-6-methoxy-2,5-piperazinedione;　W　125652

496　C28 H56 O3 Si2　SH-1981-0-0　5α-ANDROSTAN-3α,16α,17β-TRIOL TMS;　W　62837

496　C29 H33 F O6　22298-29-9　NS--28915-0-0　Betamethasone benzoate;　NBS　62838

496　C29 H40 N2 O5　C-100-5432-0　(+)-α-METHYLFENCHYL 3,3'-DIMETHYL-4'-(2-METHOXYCARBONYLETHYL)-4-VINYL-5(2H)-2,2'-DIPYRRYLMETHANONE-5'-CARBOXYLATE;　W　62839

496　C29 H45 B O4 Si　BS-4-12-0　Cannabidiolic acid butaneboronate trimethylsilate;　W　125653

496　C30 H36 N6 O　57053-61-9　Y-12-476-0　4-(HYDROXYMETHYL-4(2-PENTAMETHYLENEAMINO-4(5)PHENYL-5(4)IMIDAZOLYMETHYL)-2-PENTAMETHYLENOAMINO-5-PHENYL-4H-IMIDAZOLE;　W　62840

496　C30 H40 O6　55902-88-0　T-66-2508-0　Ergosta-2,24-dien-26-oic acid, 27-(acetyloxy)-5,6-epoxy-22-hydroxy-1-oxo-, δ-lactone;; 5,6-EPOXY-20-(3-(ACETOXYMETHYL)-4-METHYL-2-OXO-5,6-DIHYDRO PYRAN-6-YL)-2-PREGNEN-1-ONE;　W/NBS　62841

496　C30 H40 O6　52557-28-5　NS-0-0-0　2,4,6-Decatrienoic acid, 1a,2,5,5a,6,9,10,10a-octahydro-5,5a-dihydroxy-4-(hydroxymethyl)-1,1,7,9-tetramethyl-11-oxo-1H-2,8a-methanocyclopenta[a]cyclopropa[e]cyclodecen-6-yl ester, [1aR-(1aα,2α,5β,5aβ,6β,8aα,9α,10aα)]-;　W/NBS　125654

496　C30 H48 N2 O2 Si　BS-5-113-0　16α-Cyano-3β-cyclopentyloxypregn-5-en-20-one trimethylsilyloxime;　W　125657

496　C30 H57 O3 P　74793-74-1　HE-1982-0-0　Cyclohexanol, 5-methyl-2-(1-methylethyl)-, phosphite (3:1), stereoisomer;; PHOSPHITE, TRIMENTHYL-;　W/NBS　62842

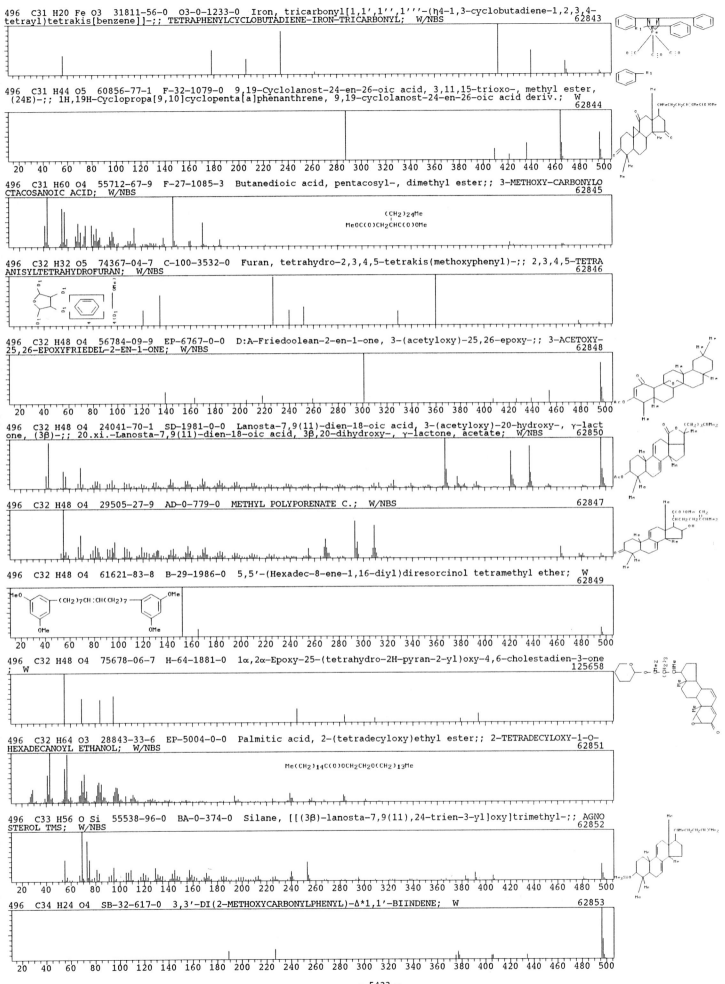

496   C31 H20 Fe O3   31811-56-0   O3-0-1233-0   Iron, tricarbonyl[1,1',1'',1'''-(η4-1,3-cyclobutadiene-1,2,3,4-tetrayl)tetrakis[benzene]]-;;   TETRAPHENYLCYCLOBUTADIENE-IRON-TRICARBONYL;   W/NBS
62843

496   C31 H44 O5   60856-77-1   F-32-1079-0   9,19-Cyclolanost-24-en-26-oic acid, 3,11,15-trioxo-, methyl ester, (24E)-;;   1H,19H-Cyclopropa[9,10]cyclopenta[a]phenanthrene, 9,19-cyclolanost-24-en-26-oic acid deriv.;   W
62844

496   C31 H60 O4   55712-67-9   F-27-1085-3   Butanedioic acid, pentacosyl-, dimethyl ester;;   3-METHOXY-CARBONYLOCTACOSANOIC ACID;   W/NBS
62845

496   C32 H32 O5   74367-04-7   C-100-3532-0   Furan, tetrahydro-2,3,4,5-tetrakis(methoxyphenyl)-;;   2,3,4,5-TETRAANISYLTETRAHYDROFURAN;   W/NBS
62846

496   C32 H48 O4   56784-09-9   EP-6767-0-0   D:A-Friedoolean-2-en-1-one, 3-(acetyloxy)-25,26-epoxy-;;   3-ACETOXY-25,26-EPOXYFRIEDEL-2-EN-1-ONE;   W/NBS
62848

496   C32 H48 O4   24041-70-1   SD-1981-0-0   Lanosta-7,9(11)-dien-18-oic acid, 3-(acetyloxy)-20-hydroxy-, γ-lactone, (3β)-;;   20.xi.-Lanosta-7,9(11)-dien-18-oic acid, 3β,20-dihydroxy-, γ-lactone, acetate;   W/NBS
62850

496   C32 H48 O4   29505-27-9   AD-0-779-0   METHYL POLYPORENATE C.;   W/NBS
62847

496   C32 H48 O4   61621-83-8   B-29-1986-0   5,5'-(Hexadec-8-ene-1,16-diyl)diresorcinol tetramethyl ether;   W
62849

496   C32 H48 O4   75678-06-7   H-64-1881-0   1α,2α-Epoxy-25-(tetrahydro-2H-pyran-2-yl)oxy-4,6-cholestadien-3-one;   W
125658

496   C32 H64 O3   28843-33-6   EP-5004-0-0   Palmitic acid, 2-(tetradecyloxy)ethyl ester;;   2-TETRADECYLOXY-1-O-HEXADECANOYL ETHANOL;   W/NBS
62851

496   C33 H56 O Si   55538-96-0   BA-0-374-0   Silane, [[(3β)-lanosta-7,9(11),24-trien-3-yl]oxy]trimethyl-;;   AGNOSTEROL TMS;   W/NBS
62852

496   C34 H24 O4   SB-32-617-0   3,3'-DI(2-METHOXYCARBONYLPHENYL)-Δ*1,1'-BIINDENE;   W
62853

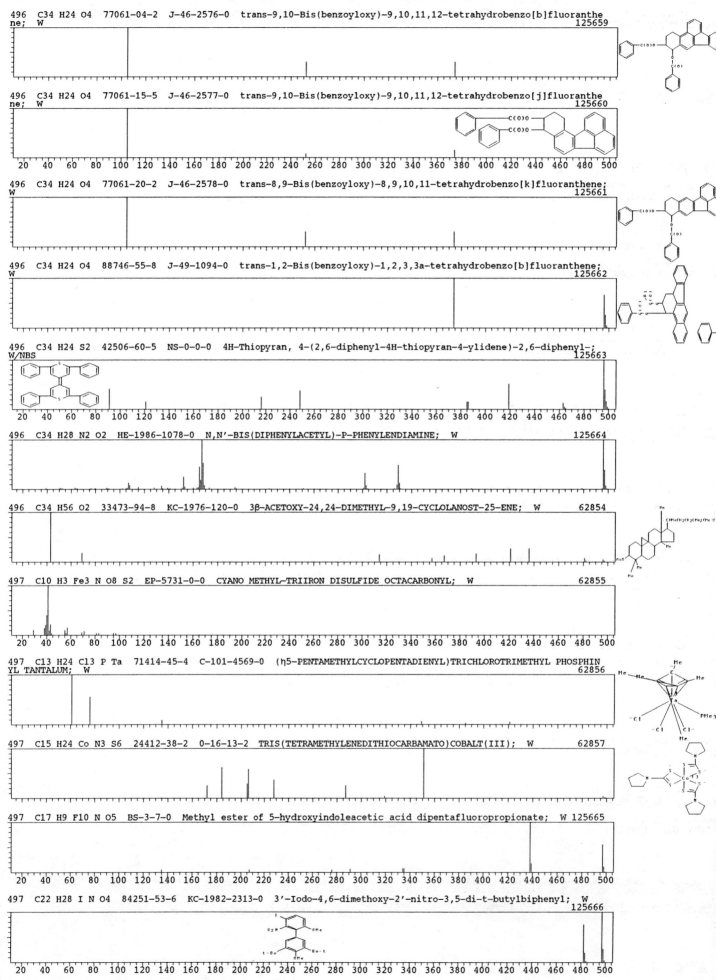

496 C34 H24 O4 77061-04-2 J-46-2576-0 trans-9,10-Bis(benzoyloxy)-9,10,11,12-tetrahydrobenzo[b]fluoranthene; W
125659

496 C34 H24 O4 77061-15-5 J-46-2577-0 trans-9,10-Bis(benzoyloxy)-9,10,11,12-tetrahydrobenzo[j]fluoranthene; W
125660

496 C34 H24 O4 77061-20-2 J-46-2578-0 trans-8,9-Bis(benzoyloxy)-8,9,10,11-tetrahydrobenzo[k]fluoranthene; W
125661

496 C34 H24 O4 88746-55-8 J-49-1094-0 trans-1,2-Bis(benzoyloxy)-1,2,3,3a-tetrahydrobenzo[b]fluoranthene; W
125662

496 C34 H24 S2 42506-60-5 NS-0-0-0 4H-Thiopyran, 4-(2,6-diphenyl-4H-thiopyran-4-ylidene)-2,6-diphenyl-; W/NBS
125663

496 C34 H28 N2 O2 HE-1986-1078-0 N,N'-BIS(DIPHENYLACETYL)-P-PHENYLENDIAMINE; W
125664

496 C34 H56 O2 33473-94-8 KC-1976-120-0 3β-ACETOXY-24,24-DIMETHYL-9,19-CYCLOLANOST-25-ENE; W
62854

497 C10 H3 Fe3 N O8 S2 EP-5731-0-0 CYANO METHYL-TRIIRON DISULFIDE OCTACARBONYL; W
62855

497 C13 H24 Cl3 P Ta 71414-45-4 C-101-4569-0 (η5-PENTAMETHYLCYCLOPENTADIENYL)TRICHLOROTRIMETHYL PHOSPHINYL TANTALUM; W
62856

497 C15 H24 Co N3 S6 24412-38-2 0-16-13-2 TRIS(TETRAMETHYLENEDITHIOCARBAMATO)COBALT(III); W
62857

497 C17 H9 F10 N O5 BS-3-7-0 Methyl ester of 5-hydroxyindoleacetic acid dipentafluoropropionate; W 125665

497 C22 H28 I N O4 84251-53-6 KC-1982-2313-0 3'-Iodo-4,6-dimethoxy-2'-nitro-3,5-di-t-butylbiphenyl; W
125666

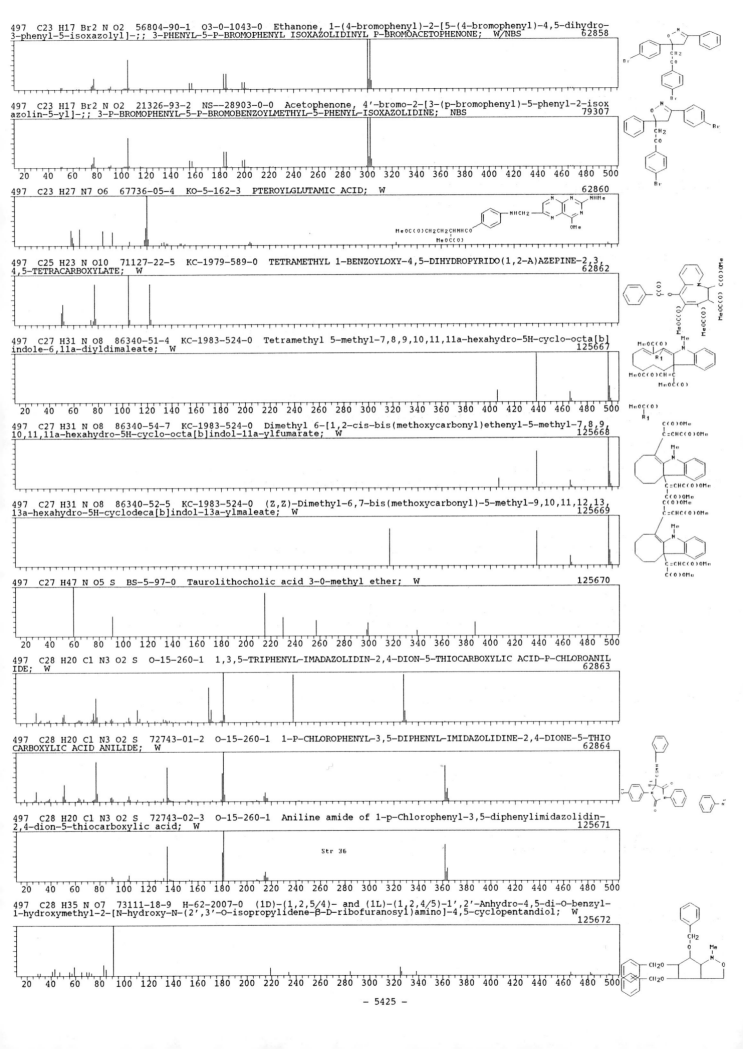

497 C23 H17 Br2 N O2 56804-90-1 O3-0-1043-0 Ethanone, 1-(4-bromophenyl)-2-[5-(4-bromophenyl)-4,5-dihydro-3-phenyl-5-isoxazolyl]-;; 3-PHENYL-5-P-BROMOPHENYL ISOXAZOLIDINYL P-BROMOACETOPHENONE; W/NBS   62858

497 C23 H17 Br2 N O2 21326-93-2 NS--28903-0-0 Acetophenone, 4'-bromo-2-[3-(p-bromophenyl)-5-phenyl-2-isoxazolin-5-yl]-;; 3-P-BROMOPHENYL-5-P-BROMOBENZOYLMETHYL-5-PHENYL-ISOXAZOLIDINE; NBS   79307

497 C23 H27 N7 O6 67736-05-4 KO-5-162-3 PTEROYLGLUTAMIC ACID; W   62860

497 C25 H23 N O10 71127-22-5 KC-1979-589-0 TETRAMETHYL 1-BENZOYLOXY-4,5-DIHYDROPYRIDO(1,2-A)AZEPINE-2,3,4,5-TETRACARBOXYLATE; W   62862

497 C27 H31 N O8 86340-51-4 KC-1983-524-0 Tetramethyl 5-methyl-7,8,9,10,11,11a-hexahydro-5H-cyclo-octa[b]indole-6,11a-diyldimaleate; W   125667

497 C27 H31 N O8 86340-54-7 KC-1983-524-0 Dimethyl 6-[1,2-cis-bis(methoxycarbonyl)ethenyl-5-methyl-7,8,9,10,11,11a-hexahydro-5H-cyclo-octa[b]indol-11a-ylfumarate; W   125668

497 C27 H31 N O8 86340-52-5 KC-1983-524-0 (Z,Z)-Dimethyl-6,7-bis(methoxycarbonyl)-5-methyl-9,10,11,12,13,13a-hexahydro-5H-cyclodeca[b]indol-13a-ylmaleate; W   125669

497 C27 H47 N O5 S BS-5-97-0 Taurolithocholic acid 3-O-methyl ether; W   125670

497 C28 H20 Cl N3 O2 S O-15-260-1 1,3,5-TRIPHENYL-IMADAZOLIDIN-2,4-DION-5-THIOCARBOXYLIC ACID-P-CHLOROANILIDE; W   62863

497 C28 H20 Cl N3 O2 S 72743-01-2 O-15-260-1 1-P-CHLOROPHENYL-3,5-DIPHENYL-IMIDAZOLIDINE-2,4-DIONE-5-THIOCARBOXYLIC ACID ANILIDE; W   62864

497 C28 H20 Cl N3 O2 S 72743-02-3 O-15-260-1 Aniline amide of 1-p-Chlorophenyl-3,5-diphenylimidazolidin-2,4-dion-5-thiocarboxylic acid; W   125671

Str 36

497 C28 H35 N O7 73111-18-9 H-62-2007-0 (1D)-(1,2,5/4)- and (1L)-(1,2,4/5)-1',2'-Anhydro-4,5-di-O-benzyl-1-hydroxymethyl-2-[N-hydroxy-N-(2',3'-O-isopropylidene-β-D-ribofuranosyl)amino]-4,5-cyclopentandiol; W   125672

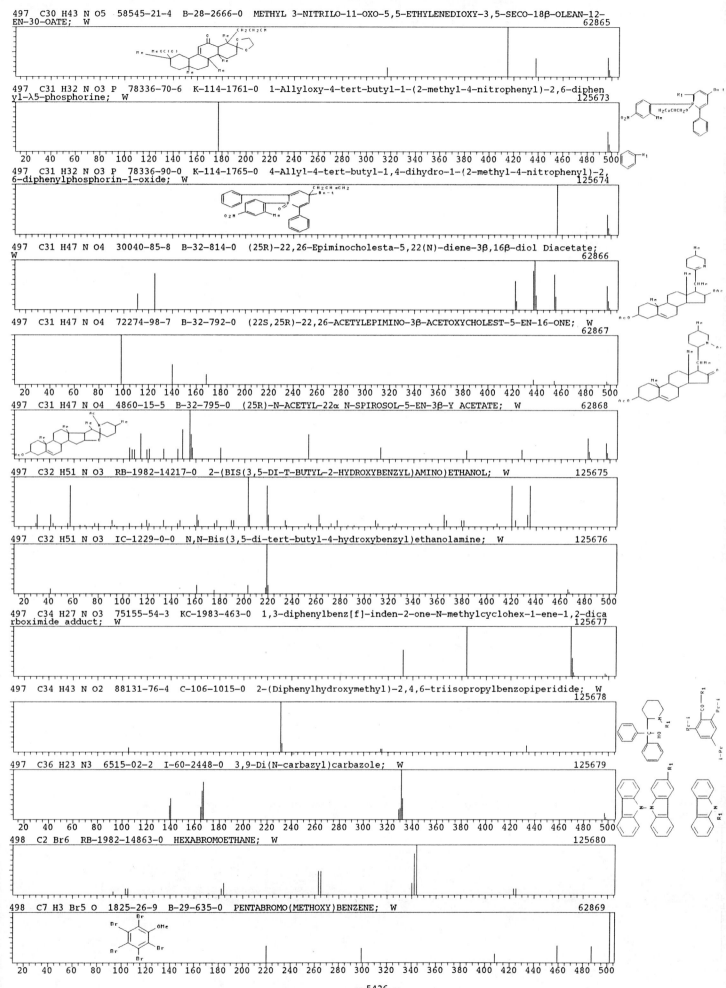

497 C30 H43 N O5 58545-21-4 B-28-2666-0 METHYL 3-NITRILO-11-OXO-5,5-ETHYLENEDIOXY-3,5-SECO-18β-OLEAN-12-EN-30-OATE; W
62865

497 C31 H32 N O3 P 78336-70-6 K-114-1761-0 1-Allyloxy-4-tert-butyl-1-(2-methyl-4-nitrophenyl)-2,6-diphenyl-λ5-phosphorine; W
125673

497 C31 H32 N O3 P 78336-90-0 K-114-1765-0 4-Allyl-4-tert-butyl-1,4-dihydro-1-(2-methyl-4-nitrophenyl)-2,6-diphenylphosphorin-1-oxide; W
125674

497 C31 H47 N O4 30040-85-8 B-32-814-0 (25R)-22,26-Epiminocholesta-5,22(N)-diene-3β,16β-diol Diacetate; W
62866

497 C31 H47 N O4 72274-98-7 B-32-792-0 (22S,25R)-22,26-ACETYLEPIMINO-3β-ACETOXYCHOLEST-5-EN-16-ONE; W
62867

497 C31 H47 N O4 4860-15-5 B-32-795-0 (25R)-N-ACETYL-22α N-SPIROSOL-5-EN-3β-Y ACETATE; W
62868

497 C32 H51 N O3 RB-1982-14217-0 2-(BIS(3,5-DI-T-BUTYL-2-HYDROXYBENZYL)AMINO)ETHANOL; W
125675

497 C32 H51 N O3 IC-1229-0-0 N,N-Bis(3,5-di-tert-butyl-4-hydroxybenzyl)ethanolamine; W
125676

497 C34 H27 N O3 75155-54-3 KC-1983-463-0 1,3-diphenylbenz[f]inden-2-one-N-methylcyclohex-1-ene-1,2-dicarboximide adduct; W
125677

497 C34 H43 N O2 88131-76-4 C-106-1015-0 2-(Diphenylhydroxymethyl)-2,4,6-triisopropylbenzopiperidide; W
125678

497 C36 H23 N3 6515-02-2 I-60-2448-0 3,9-Di(N-carbazyl)carbazole; W
125679

498 C2 Br6 RB-1982-14863-0 HEXABROMOETHANE; W
125680

498 C7 H3 Br5 O 1825-26-9 B-29-635-0 PENTABROMO(METHOXY)BENZENE; W
62869

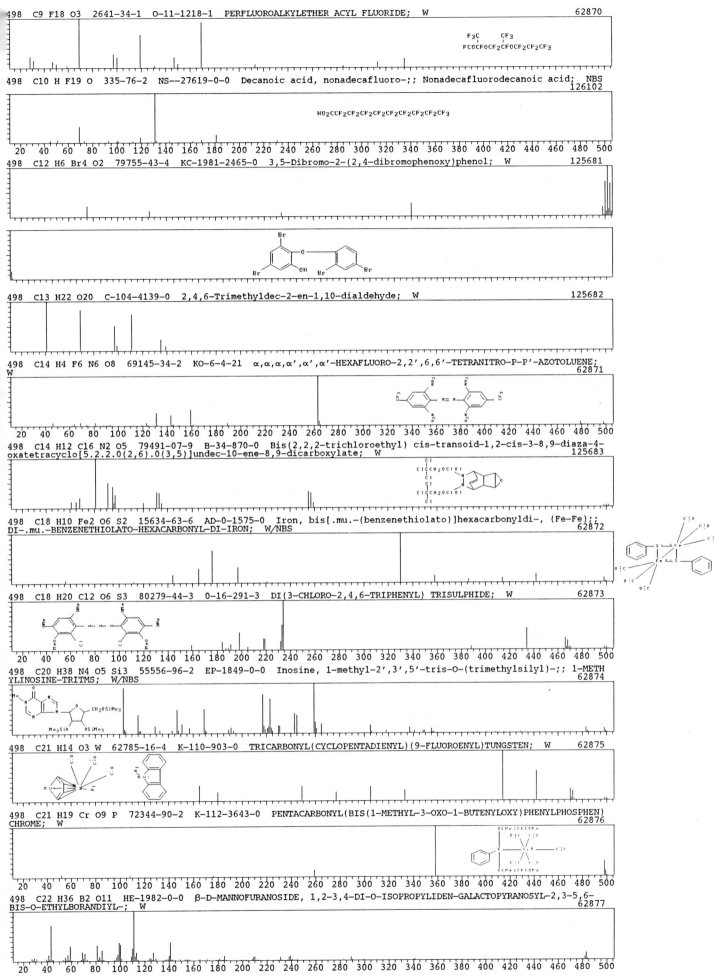

498   C9 F18 O3   2641-34-1   O-11-1218-1   PERFLUOROALKYLETHER ACYL FLUORIDE;   W   62870

498   C10 H F19 O   335-76-2   NS--27619-0-0   Decanoic acid, nonadecafluoro-;; Nonadecafluorodecanoic acid;   NBS
126102

498   C12 H6 Br4 O2   79755-43-4   KC-1981-2465-0   3,5-Dibromo-2-(2,4-dibromophenoxy)phenol;   W   125681

498   C13 H22 O20   C-104-4139-0   2,4,6-Trimethyldec-2-en-1,10-dialdehyde;   W   125682

498   C14 H4 F6 N6 O8   69145-34-2   KO-6-4-21   α,α,α,α',α',α'-HEXAFLUORO-2,2',6,6'-TETRANITRO-P-P'-AZOTOLUENE;
W   62871

498   C14 H12 Cl6 N2 O5   79491-07-9   B-34-870-0   Bis(2,2,2-trichloroethyl) cis-transoid-1,2-cis-3-8,9-diaza-4-
oxatetracyclo[5.2.2.0(2,6).0(3,5)]undec-10-ene-8,9-dicarboxylate;   W   125683

498   C18 H10 Fe2 O6 S2   15634-63-6   AD-0-1575-0   Iron, bis[.mu.-(benzenethiolato)]hexacarbonyldi-, (Fe-Fe);;
DI-.mu.-BENZENETHIOLATO-HEXACARBONYL-DI-IRON;   W/NBS   62872

498   C18 H20 Cl2 O6 S3   80279-44-3   O-16-291-3   DI(3-CHLORO-2,4,6-TRIPHENYL) TRISULPHIDE;   W   62873

498   C20 H38 N4 O5 Si3   55556-96-2   EP-1849-0-0   Inosine, 1-methyl-2',3',5'-tris-O-(trimethylsilyl)-;; 1-METH
YLINOSINE-TRITMS;   W/NBS   62874

498   C21 H14 O3 W   62785-16-4   K-110-903-0   TRICARBONYL(CYCLOPENTADIENYL)(9-FLUOROENYL)TUNGSTEN;   W   62875

498   C21 H19 Cr O9 P   72344-90-2   K-112-3643-0   PENTACARBONYL(BIS(1-METHYL-3-OXO-1-BUTENYLOXY)PHENYLPHOSPHEN)
CHROME;   W   62876

498   C22 H36 B2 O11   HE-1982-0-0   β-D-MANNOFURANOSIDE, 1,2-3,4-DI-O-ISOPROPYLIDEN-GALACTOPYRANOSYL-2,3-5,6-
BIS-O-ETHYLBORANDIYL-;   W   62877

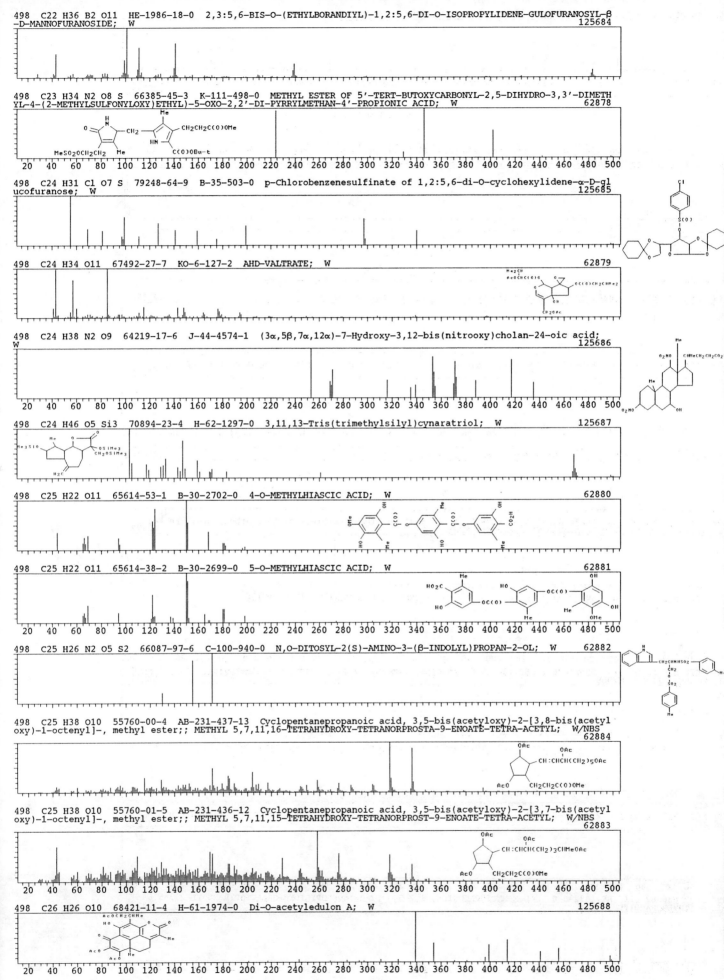

498   C22 H36 B2 O11   HE-1986-18-0   2,3:5,6-BIS-O-(ETHYLBORANDIYL)-1,2:5,6-DI-O-ISOPROPYLIDENE-GULOFURANOSYL-β
-D-MANNOFURANOSIDE;   W   125684

498   C23 H34 N2 O8 S   66385-45-3   K-111-498-0   METHYL ESTER OF 5'-TERT-BUTOXYCARBONYL-2,5-DIHYDRO-3,3'-DIMETH
YL-4-(2-METHYLSULFONYLOXY)ETHYL)-5-OXO-2,2'-DI-PYRRYLMETHAN-4'-PROPIONIC ACID;   W   62878

498   C24 H31 Cl O7 S   79248-64-9   B-35-503-0   p-Chlorobenzenesulfinate of 1,2:5,6-di-O-cyclohexylidene-α-D-gl
ucofuranose;   W   125685

498   C24 H34 O11   67492-27-7   KO-6-127-2   AHD-VALTRATE;   W   62879

498   C24 H38 N2 O9   64219-17-6   J-44-4574-1   (3α,5β,7α,12α)-7-Hydroxy-3,12-bis(nitrooxy)cholan-24-oic acid;
W   125686

498   C24 H46 O5 Si3   70894-23-4   H-62-1297-0   3,11,13-Tris(trimethylsilyl)cynaratriol;   W   125687

498   C25 H22 O11   65614-53-1   B-30-2702-0   4-O-METHYLHIASCIC ACID;   W   62880

498   C25 H22 O11   65614-38-2   B-30-2699-0   5-O-METHYLHIASCIC ACID;   W   62881

498   C25 H26 N2 O5 S2   66087-97-6   C-100-940-0   N,O-DITOSYL-2(S)-AMINO-3-(β-INDOLYL)PROPAN-2-OL;   W   62882

498   C25 H38 O10   55760-00-4   AB-231-437-13   Cyclopentanepropanoic acid, 3,5-bis(acetyloxy)-2-[3,8-bis(acetyl
oxy)-1-octenyl]-, methyl ester;; METHYL 5,7,11,16-TETRAHYDROXY-TETRANORPROSTA-9-ENOATE-TETRA-ACETYL;   W/NBS
62884

498   C25 H38 O10   55760-01-5   AB-231-436-12   Cyclopentanepropanoic acid, 3,5-bis(acetyloxy)-2-[3,7-bis(acetyl
oxy)-1-octenyl]-, methyl ester;; METHYL 5,7,11,15-TETRAHYDROXY-TETRANORPROST-9-ENOATE-TETRA-ACETYL;   W/NBS
62883

498   C26 H26 O10   68421-11-4   H-61-1974-0   Di-O-acetyledulon A;   W   125688

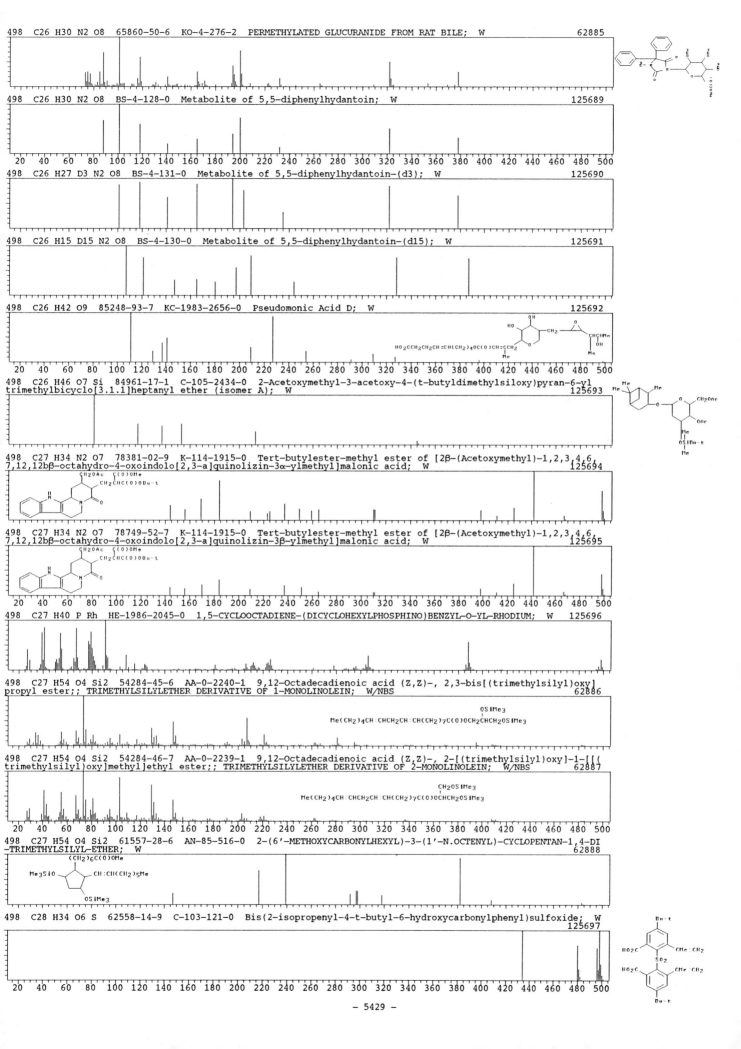

498   C26 H30 N2 O8   65860-50-6   KO-4-276-2   PERMETHYLATED GLUCURANIDE FROM RAT BILE;   W   62885

498   C26 H30 N2 O8   BS-4-128-0   Metabolite of 5,5-diphenylhydantoin;   W   125689

498   C26 H27 D3 N2 O8   BS-4-131-0   Metabolite of 5,5-diphenylhydantoin-(d3);   W   125690

498   C26 H15 D15 N2 O8   BS-4-130-0   Metabolite of 5,5-diphenylhydantoin-(d15);   W   125691

498   C26 H42 O9   85248-93-7   KC-1983-2656-0   Pseudomonic Acid D;   W   125692

498   C26 H46 O7 Si   84961-17-1   C-105-2434-0   2-Acetoxymethyl-3-acetoxy-4-(t-butyldimethylsiloxy)pyran-6-yl
trimethylbicyclo[3.1.1]heptanyl ether (isomer A);   W   125693

498   C27 H34 N2 O7   78381-02-9   K-114-1915-0   Tert-butylester-methyl ester of [2β-(Acetoxymethyl)-1,2,3,4,6,
7,12,12bβ-octahydro-4-oxoindolo[2,3-a]quinolizin-3α-ylmethyl]malonic acid;   W   125694

498   C27 H34 N2 O7   78749-52-7   K-114-1915-0   Tert-butylester-methyl ester of [2β-(Acetoxymethyl)-1,2,3,4,6,
7,12,12bβ-octahydro-4-oxoindolo[2,3-a]quinolizin-3β-ylmethyl]malonic acid;   W   125695

498   C27 H40 P Rh   HE-1986-2045-0   1,5-CYCLOOCTADIENE-(DICYCLOHEXYLPHOSPHINO)BENZYL-O-YL-RHODIUM;   W   125696

498   C27 H54 O4 Si2   54284-45-6   AA-0-2240-1   9,12-Octadecadienoic acid (Z,Z)-, 2,3-bis[(trimethylsilyl)oxy]
propyl ester;;   TRIMETHYLSILYLETHER DERIVATIVE OF 1-MONOLINOLEIN;   W/NBS   62886

498   C27 H54 O4 Si2   54284-46-7   AA-0-2239-1   9,12-Octadecadienoic acid (Z,Z)-, 2-[(trimethylsilyl)oxy]-1-[[(
trimethylsilyl)oxy]methyl]ethyl ester;;   TRIMETHYLSILYLETHER DERIVATIVE OF 2-MONOLINOLEIN;   W/NBS   62887

498   C27 H54 O4 Si2   61557-28-6   AN-85-516-0   2-(6'-METHOXYCARBONYLHEXYL)-3-(1'-N.OCTENYL)-CYCLOPENTAN-1,4-DI
-TRIMETHYLSILYL-ETHER;   W   62888

498   C28 H34 O6 S   62558-14-9   C-103-121-0   Bis(2-isopropenyl-4-t-butyl-6-hydroxycarbonylphenyl)sulfoxide;   W
                                                                                                                 125697

- 5429 -

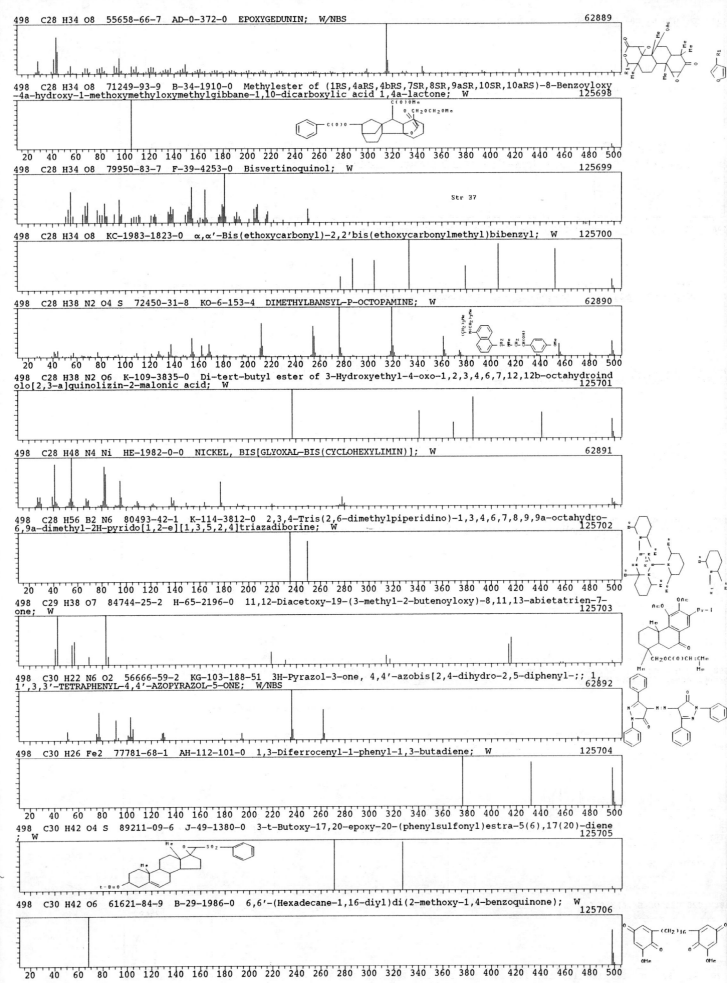

498 C28 H34 O8 55658-66-7 AD-0-372-0 EPOXYGEDUNIN; W/NBS   62889

498 C28 H34 O8 71249-93-9 B-34-1910-0 Methylester of (1RS,4aRS,4bRS,7SR,8SR,9aSR,10SR,10aRS)-8-Benzoyloxy
-4a-hydroxy-1-methoxymethyloxymethylgibbane-1,10-dicarboxylic acid 1,4a-lactone; W   125698

498 C28 H34 O8 79950-83-7 F-39-4253-0 Bisvertinoquinol; W   125699

Str 37

498 C28 H34 O8 KC-1983-1823-0 α,α'-Bis(ethoxycarbonyl)-2,2'bis(ethoxycarbonylmethyl)bibenzyl; W   125700

498 C28 H38 N2 O4 S 72450-31-8 KO-6-153-4 DIMETHYLBANSYL-P-OCTOPAMINE; W   62890

498 C28 H38 N2 O6 K-109-3835-0 Di-tert-butyl ester of 3-Hydroxyethyl-4-oxo-1,2,3,4,6,7,12,12b-octahydroind
olo[2,3-a]quinolizin-2-malonic acid; W   125701

498 C28 H48 N4 Ni HE-1982-0-0 NICKEL, BIS[GLYOXAL-BIS(CYCLOHEXYLIMIN)]; W   62891

498 C28 H56 B2 N6 80493-42-1 K-114-3812-0 2,3,4-Tris(2,6-dimethylpiperidino)-1,3,4,6,7,8,9,9a-octahydro-
6,9a-dimethyl-2H-pyrido[1,2-e][1,3,5,2,4]triazadiborine; W   125702

498 C29 H38 O7 84744-25-2 H-65-2196-0 11,12-Diacetoxy-19-(3-methyl-2-butenoyloxy)-8,11,13-abietatrien-7-
one; W   125703

498 C30 H22 N6 O2 56666-59-2 KG-103-188-51 3H-Pyrazol-3-one, 4,4'-azobis[2,4-dihydro-2,5-diphenyl-;; 1,
1',3,3'-TETRAPHENYL-4,4'-AZOPYRAZOL-5-ONE; W/NBS   62892

498 C30 H26 Fe2 77781-68-1 AH-112-101-0 1,3-Diferrocenyl-1-phenyl-1,3-butadiene; W   125704

498 C30 H42 O4 S 89211-09-6 J-49-1380-0 3-t-Butoxy-17,20-epoxy-20-(phenylsulfonyl)estra-5(6),17(20)-diene
; W   125705

498 C30 H42 O6 61621-84-9 B-29-1986-0 6,6'-(Hexadecane-1,16-diyl)di(2-methoxy-1,4-benzoquinone); W   125706

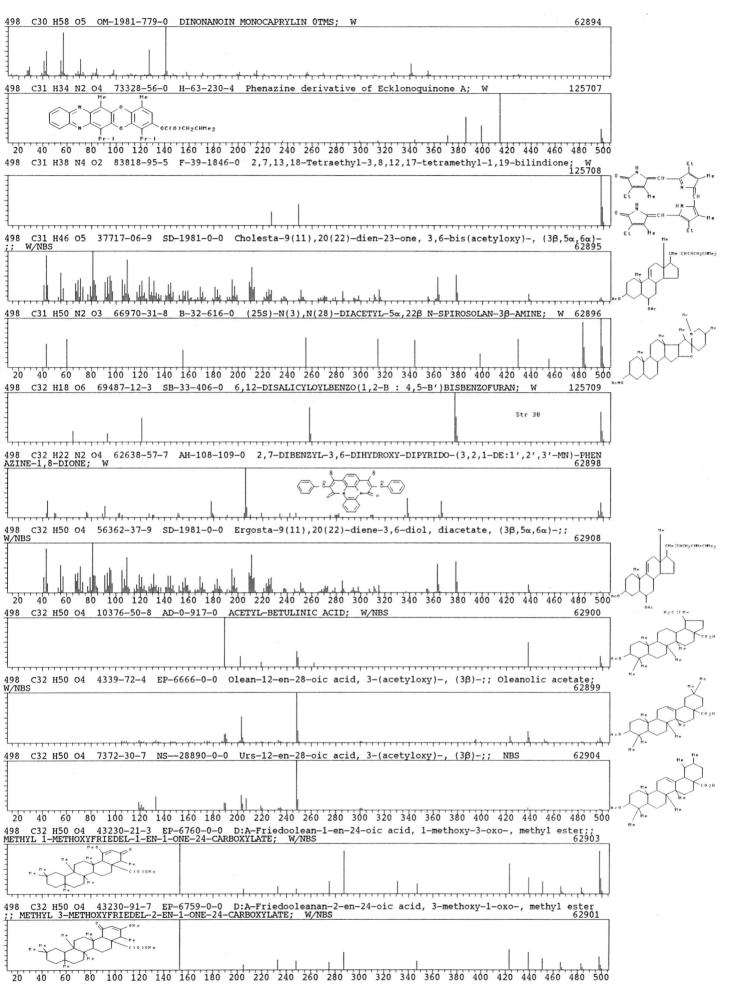

498   C30 H58 O5   OM-1981-779-0   DINONANOIN MONOCAPRYLIN OTMS;   W   62894

498   C31 H34 N2 O4   73328-56-0   H-63-230-4   Phenazine derivative of Ecklonoquinone A;   W   125707

498   C31 H38 N4 O2   83818-95-5   F-39-1846-0   2,7,13,18-Tetraethyl-3,8,12,17-tetramethyl-1,19-bilindione;   W
                                                                                                         125708

498   C31 H46 O5   37717-06-9   SD-1981-0-0   Cholesta-9(11),20(22)-dien-23-one, 3,6-bis(acetyloxy)-, (3β,5α,6α)-
;;   W/NBS   62895

498   C31 H50 N2 O3   66970-31-8   B-32-616-0   (25S)-N(3),N(28)-DIACETYL-5α,22β N-SPIROSOLAN-3β-AMINE;   W   62896

498   C32 H18 O6   69487-12-3   SB-33-406-0   6,12-DISALICYLOYLBENZO(1,2-B : 4,5-B')BISBENZOFURAN;   W   125709
                                                                                      Str 38

498   C32 H22 N2 O4   62638-57-7   AH-108-109-0   2,7-DIBENZYL-3,6-DIHYDROXY-DIPYRIDO-(3,2,1-DE:1',2',3'-MN)-PHEN
AZINE-1,8-DIONE;   W   62898

498   C32 H50 O4   56362-37-9   SD-1981-0-0   Ergosta-9(11),20(22)-diene-3,6-diol, diacetate, (3β,5α,6α)-;;
W/NBS   62908

498   C32 H50 O4   10376-50-8   AD-0-917-0   ACETYL-BETULINIC ACID;   W/NBS   62900

498   C32 H50 O4   4339-72-4   EP-6666-0-0   Olean-12-en-28-oic acid, 3-(acetyloxy)-, (3β)-;;   Oleanolic acetate;
W/NBS   62899

498   C32 H50 O4   7372-30-7   NS--28890-0-0   Urs-12-en-28-oic acid, 3-(acetyloxy)-, (3β)-;;   NBS   62904

498   C32 H50 O4   43230-21-3   EP-6760-0-0   D:A-Friedoolean-1-en-24-oic acid, 1-methoxy-3-oxo-, methyl ester;;
METHYL 1-METHOXYFRIEDEL-1-EN-1-ONE-24-CARBOXYLATE;   W/NBS   62903

498   C32 H50 O4   43230-91-7   EP-6759-0-0   D:A-Friedooleanan-2-en-24-oic acid, 3-methoxy-1-oxo-, methyl ester
;;   METHYL 3-METHOXYFRIEDEL-2-EN-1-ONE-24-CARBOXYLATE;   W/NBS   62901

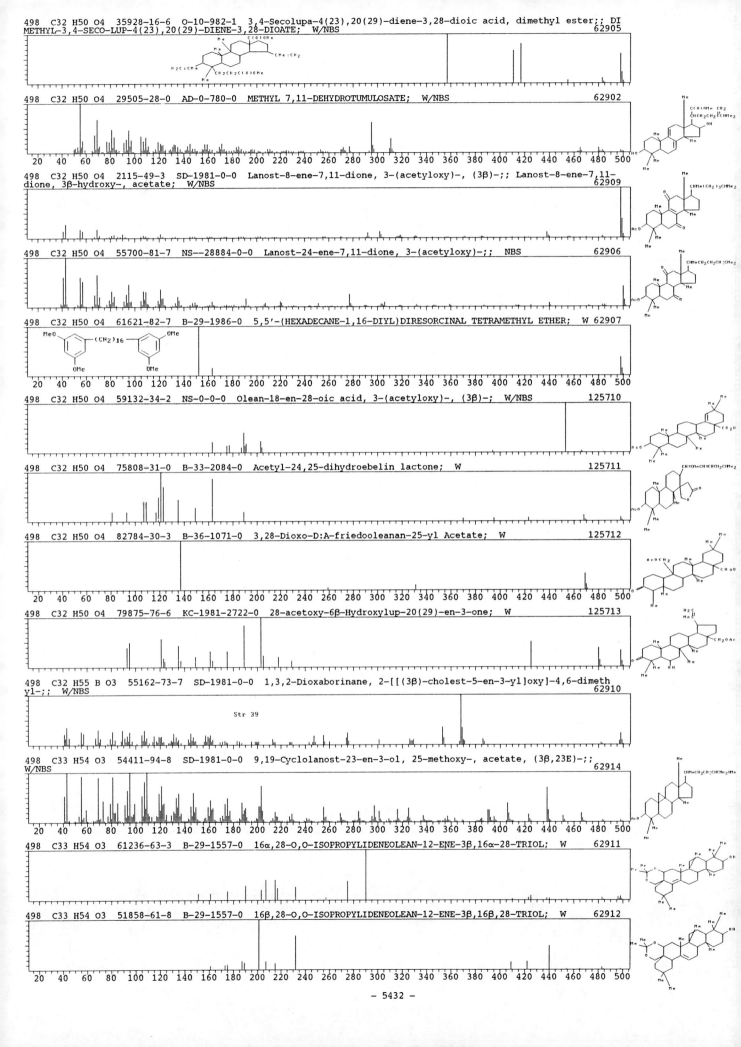

498 C32 H50 O4 35928-16-6 O-10-982-1 3,4-Secolupa-4(23),20(29)-diene-3,28-dioic acid, dimethyl ester;; DI
METHYL-3,4-SECO-LUP-4(23),20(29)-DIENE-3,28-DIOATE; W/NBS   62905

498 C32 H50 O4 29505-28-0 AD-0-780-0 METHYL 7,11-DEHYDROTUMULOSATE; W/NBS   62902

498 C32 H50 O4 2115-49-3 SD-1981-0-0 Lanost-8-ene-7,11-dione, 3-(acetyloxy)-, (3β)-;; Lanost-8-ene-7,11-
dione, 3β-hydroxy-, acetate; W/NBS   62909

498 C32 H50 O4 55700-81-7 NS--28884-0-0 Lanost-24-ene-7,11-dione, 3-(acetyloxy)-;; NBS   62906

498 C32 H50 O4 61621-82-7 B-29-1986-0 5,5'-(HEXADECANE-1,16-DIYL)DIRESORCINAL TETRAMETHYL ETHER; W 62907

498 C32 H50 O4 59132-34-2 NS-0-0-0 Olean-18-en-28-oic acid, 3-(acetyloxy)-, (3β)-; W/NBS   125710

498 C32 H50 O4 75808-31-0 B-33-2084-0 Acetyl-24,25-dihydroebelin lactone; W   125711

498 C32 H50 O4 82784-30-3 B-36-1071-0 3,28-Dioxo-D:A-friedooleanan-25-yl Acetate; W   125712

498 C32 H50 O4 79875-76-6 KC-1981-2722-0 28-acetoxy-6β-Hydroxylup-20(29)-en-3-one; W   125713

498 C32 H55 B O3 55162-73-7 SD-1981-0-0 1,3,2-Dioxaborinane, 2-[[(3β)-cholest-5-en-3-yl]oxy]-4,6-dimeth
yl-;; W/NBS   62910

Str 39

498 C33 H54 O3 54411-94-8 SD-1981-0-0 9,19-Cyclolanost-23-en-3-ol, 25-methoxy-, acetate, (3β,23E)-;;
W/NBS   62914

498 C33 H54 O3 61236-63-3 B-29-1557-0 16α,28-O,O-ISOPROPYLIDENEOLEAN-12-ENE-3β,16α-28-TRIOL; W   62911

498 C33 H54 O3 51858-61-8 B-29-1557-0 16β,28-O,O-ISOPROPYLIDENEOLEAN-12-ENE-3β,16β,28-TRIOL; W   62912

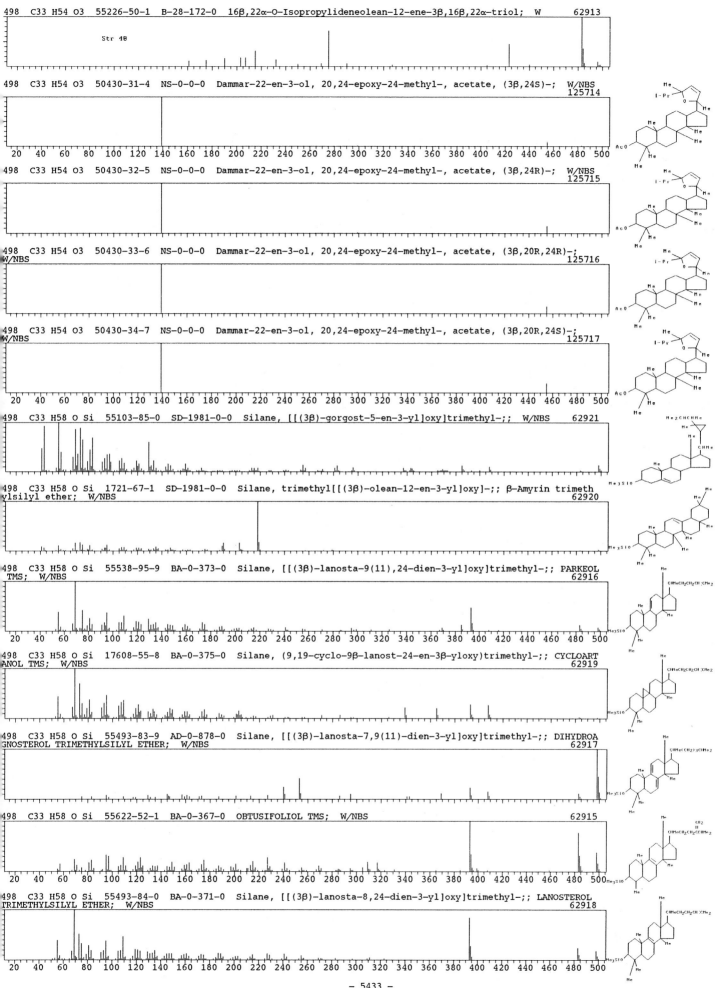

498   C33 H54 O3   55226-50-1   B-28-172-0   16β,22α-O-Isopropylideneolean-12-ene-3β,16β,22α-triol;   W   62913

Str 40

498   C33 H54 O3   50430-31-4   NS-0-0-0   Dammar-22-en-3-ol, 20,24-epoxy-24-methyl-, acetate, (3β,24S)-;   W/NBS
                                                                                                                            125714

498   C33 H54 O3   50430-32-5   NS-0-0-0   Dammar-22-en-3-ol, 20,24-epoxy-24-methyl-, acetate, (3β,24R)-;   W/NBS
                                                                                                                            125715

498   C33 H54 O3   50430-33-6   NS-0-0-0   Dammar-22-en-3-ol, 20,24-epoxy-24-methyl-, acetate, (3β,20R,24R)-;
W/NBS   125716

498   C33 H54 O3   50430-34-7   NS-0-0-0   Dammar-22-en-3-ol, 20,24-epoxy-24-methyl-, acetate, (3β,20R,24S)-;
W/NBS   125717

498   C33 H58 O Si   55103-85-0   SD-1981-0-0   Silane, [[(3β)-gorgost-5-en-3-yl]oxy]trimethyl-;;   W/NBS   62921

498   C33 H58 O Si   1721-67-1   SD-1981-0-0   Silane, trimethyl[[(3β)-olean-12-en-3-yl]oxy]-;;   β-Amyrin trimeth
ylsilyl ether;   W/NBS   62920

498   C33 H58 O Si   55538-95-9   BA-0-373-0   Silane, [[(3β)-lanosta-9(11),24-dien-3-yl]oxy]trimethyl-;;   PARKEOL
TMS;   W/NBS   62916

498   C33 H58 O Si   17608-55-8   BA-0-375-0   Silane, (9,19-cyclo-9β-lanost-24-en-3β-yloxy)trimethyl-;;   CYCLOART
ANOL TMS;   W/NBS   62919

498   C33 H58 O Si   55493-83-9   AD-0-878-0   Silane, [[(3β)-lanosta-7,9(11)-dien-3-yl]oxy]trimethyl-;;   DIHYDROA
GNOSTEROL TRIMETHYLSILYL ETHER;   W/NBS   62917

498   C33 H58 O Si   55622-52-1   BA-0-367-0   OBTUSIFOLIOL TMS;   W/NBS   62915

498   C33 H58 O Si   55493-84-0   BA-0-371-0   Silane, [[(3β)-lanosta-8,24-dien-3-yl]oxy]trimethyl-;;   LANOSTEROL
TRIMETHYLSILYL ETHER;   W/NBS   62918

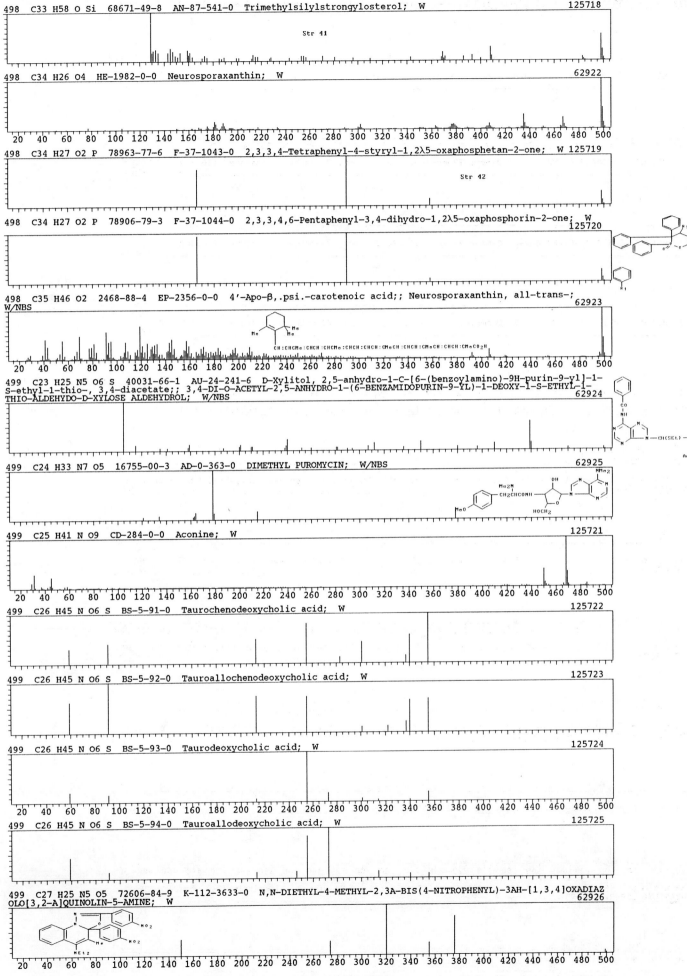

498   C33 H58 O Si   68671-49-8   AN-87-541-0   Trimethylsilylstrongylosterol;   W   125718

Str 41

498   C34 H26 O4   HE-1982-0-0   Neurosporaxanthin;   W   62922

498   C34 H27 O2 P   78963-77-6   F-37-1043-0   2,3,3,4-Tetraphenyl-4-styryl-1,2λ5-oxaphosphetan-2-one;   W 125719

Str 42

498   C34 H27 O2 P   78906-79-3   F-37-1044-0   2,3,3,4,6-Pentaphenyl-3,4-dihydro-1,2λ5-oxaphosphorin-2-one;   W
125720

498   C35 H46 O2   2468-88-4   EP-2356-0-0   4'-Apo-β,.psi.-carotenoic acid;; Neurosporaxanthin, all-trans-;
W/NBS   62923

499   C23 H25 N5 O6 S   40031-66-1   AU-24-241-6   D-Xylitol, 2,5-anhydro-1-C-[6-(benzoylamino)-9H-purin-9-yl]-1-
S-ethyl-1-thio-, 3,4-diacetate;; 3,4-DI-O-ACETYL-2,5-ANHYDRO-1-(6-BENZAMIDOPURIN-9-YL)-1-DEOXY-1-S-ETHYL-1-
THIO-ALDEHYDO-D-XYLOSE ALDEHYDROL;   W/NBS   62924

499   C24 H33 N7 O5   16755-00-3   AD-0-363-0   DIMETHYL PUROMYCIN;   W/NBS   62925

499   C25 H41 N O9   CD-284-0-0   Aconine;   W   125721

499   C26 H45 N O6 S   BS-5-91-0   Taurochenodeoxycholic acid;   W   125722

499   C26 H45 N O6 S   BS-5-92-0   Tauroallochenodeoxycholic acid;   W   125723

499   C26 H45 N O6 S   BS-5-93-0   Taurodeoxycholic acid;   W   125724

499   C26 H45 N O6 S   BS-5-94-0   Tauroallodeoxycholic acid;   W   125725

499   C27 H25 N5 O5   72606-84-9   K-112-3633-0   N,N-DIETHYL-4-METHYL-2,3A-BIS(4-NITROPHENYL)-3AH-[1,3,4]OXADIAZ
OLO[3,2-A]QUINOLIN-5-AMINE;   W   62926

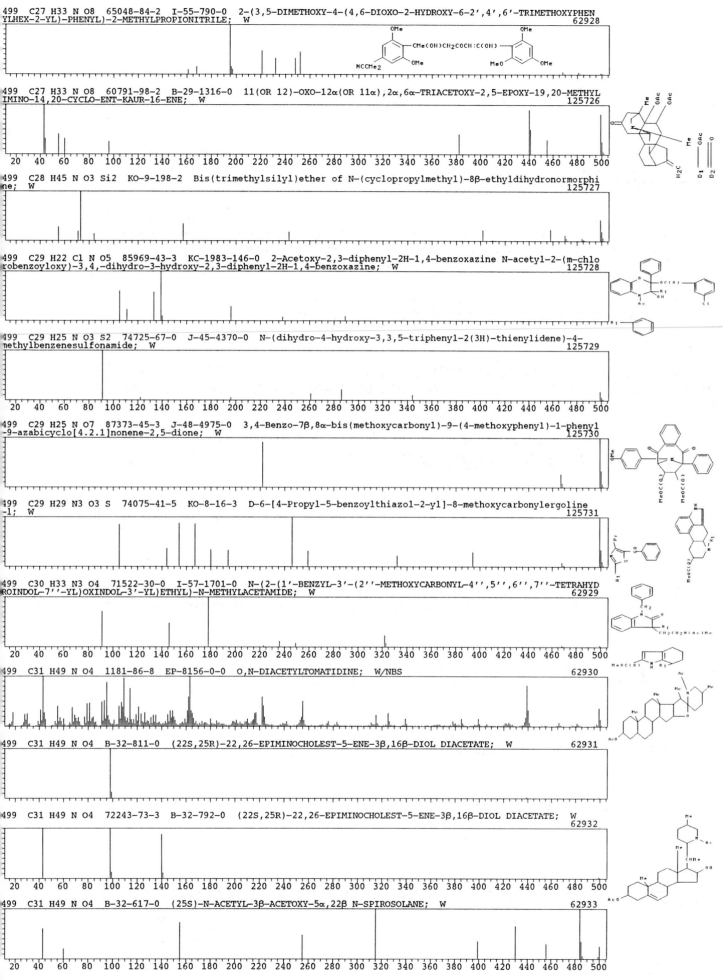

499 C27 H33 N O8 65048-84-2 I-55-790-0 2-(3,5-DIMETHOXY-4-(4,6-DIOXO-2-HYDROXY-6-2',4',6'-TRIMETHOXYPHEN
YLHEX-2-YL)-PHENYL)-2-METHYLPROPIONITRILE; W
62928

499 C27 H33 N O8 60791-98-2 B-29-1316-0 11(OR 12)-OXO-12α(OR 11α),2α,6α-TRIACETOXY-2,5-EPOXY-19,20-METHYL
IMINO-14,20-CYCLO-ENT-KAUR-16-ENE; W
125726

499 C28 H45 N O3 Si2 KO-9-198-2 Bis(trimethylsilyl)ether of N-(cyclopropylmethyl)-8β-ethyldihydronormorphi
ne; W
125727

499 C29 H22 Cl N O5 85969-43-3 KC-1983-146-0 2-Acetoxy-2,3-diphenyl-2H-1,4-benzoxazine N-acetyl-2-(m-chlo
robenzoyloxy)-3,4,-dihydro-3-hydroxy-2,3-diphenyl-2H-1,4-benzoxazine; W
125728

499 C29 H25 N O3 S2 74725-67-0 J-45-4370-0 N-(dihydro-4-hydroxy-3,3,5-triphenyl-2(3H)-thienylidene)-4-
methylbenzenesulfonamide; W
125729

499 C29 H25 N O7 87373-45-3 J-48-4975-0 3,4-Benzo-7β,8α-bis(methoxycarbonyl)-9-(4-methoxyphenyl)-1-phenyl
-9-azabicyclo[4.2.1]nonene-2,5-dione; W
125730

499 C29 H29 N3 O3 S 74075-41-5 KO-8-16-3 D-6-[4-Propyl-5-benzoylthiazol-2-yl]-8-methoxycarbonylergoline
-1; W
125731

499 C30 H33 N3 O4 71522-30-0 I-57-1701-0 N-(2-(1'-BENZYL-3'-(2''-METHOXYCARBONYL-4'',5'',6'',7''-TETRAHYD
ROINDOL-7''-YL)OXINDOL-3'-YL)ETHYL)-N-METHYLACETAMIDE; W
62929

499 C31 H49 N O4 1181-86-8 EP-8156-0-0 O,N-DIACETYLTOMATIDINE; W/NBS
62930

499 C31 H49 N O4 B-32-811-0 (22S,25R)-22,26-EPIMINOCHOLEST-5-ENE-3β,16β-DIOL DIACETATE; W
62931

499 C31 H49 N O4 72243-73-3 B-32-792-0 (22S,25R)-22,26-EPIMINOCHOLEST-5-ENE-3β,16β-DIOL DIACETATE; W
62932

499 C31 H49 N O4 B-32-617-0 (25S)-N-ACETYL-3β-ACETOXY-5α,22β N-SPIROSOLANE; W
62933

- 5435 -

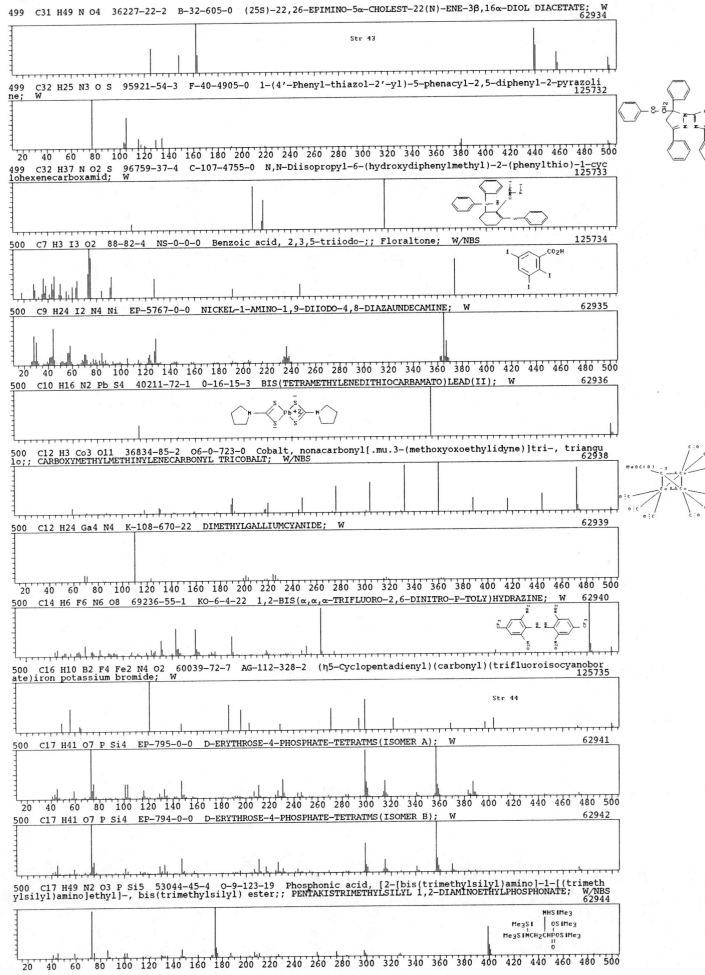

499   C31 H49 N O4   36227-22-2   B-32-605-0   (25S)-22,26-EPIMINO-5α-CHOLEST-22(N)-ENE-3β,16α-DIOL DIACETATE;   W
62934

Str 43

499   C32 H25 N3 O S   95921-54-3   F-40-4905-0   1-(4'-Phenyl-thiazol-2'-yl)-5-phenacyl-2,5-diphenyl-2-pyrazoline;   W
125732

499   C32 H37 N O2 S   96759-37-4   C-107-4755-0   N,N-Diisopropyl-6-(hydroxydiphenylmethyl)-2-(phenylthio)-1-cyclohexenecarboxamid;   W
125733

500   C7 H3 I3 O2   88-82-4   NS-0-0-0   Benzoic acid, 2,3,5-triiodo-;; Floraltone;   W/NBS
125734

500   C9 H24 I2 N4 Ni   EP-5767-0-0   NICKEL-1-AMINO-1,9-DIIODO-4,8-DIAZAUNDECAMINE;   W
62935

500   C10 H16 N2 Pb S4   40211-72-1   0-16-15-3   BIS(TETRAMETHYLENEDITHIOCARBAMATO)LEAD(II);   W
62936

500   C12 H3 Co3 O11   36834-85-2   O6-0-723-0   Cobalt, nonacarbonyl[.mu.3-(methoxyoxoethylidyne)]tri-, triangulo;;   CARBOXYMETHYLMETHINYLENECARBONYL TRICOBALT;   W/NBS
62938

500   C12 H24 Ga4 N4   K-108-670-22   DIMETHYLGALLIUMCYANIDE;   W
62939

500   C14 H6 F6 N6 O8   69236-55-1   KO-6-4-22   1,2-BIS(α,α,α-TRIFLUORO-2,6-DINITRO-P-TOLY)HYDRAZINE;   W
62940

500   C16 H10 B2 F4 Fe2 N4 O2   60039-72-7   AG-112-328-2   (η5-Cyclopentadienyl)(carbonyl)(trifluoroisocyanoborate)iron potassium bromide;   W
125735

Str 44

500   C17 H41 O7 P Si4   EP-795-0-0   D-ERYTHROSE-4-PHOSPHATE-TETRATMS(ISOMER A);   W
62941

500   C17 H41 O7 P Si4   EP-794-0-0   D-ERYTHROSE-4-PHOSPHATE-TETRATMS(ISOMER B);   W
62942

500   C17 H49 N2 O3 P Si5   53044-45-4   O-9-123-19   Phosphonic acid, [2-[bis(trimethylsilyl)amino]-1-[(trimethylsilyl)amino]ethyl]-, bis(trimethylsilyl) ester;; PENTAKISTRIMETHYLSILYL 1,2-DIAMINOETHYLPHOSPHONATE;   W/NBS
62944

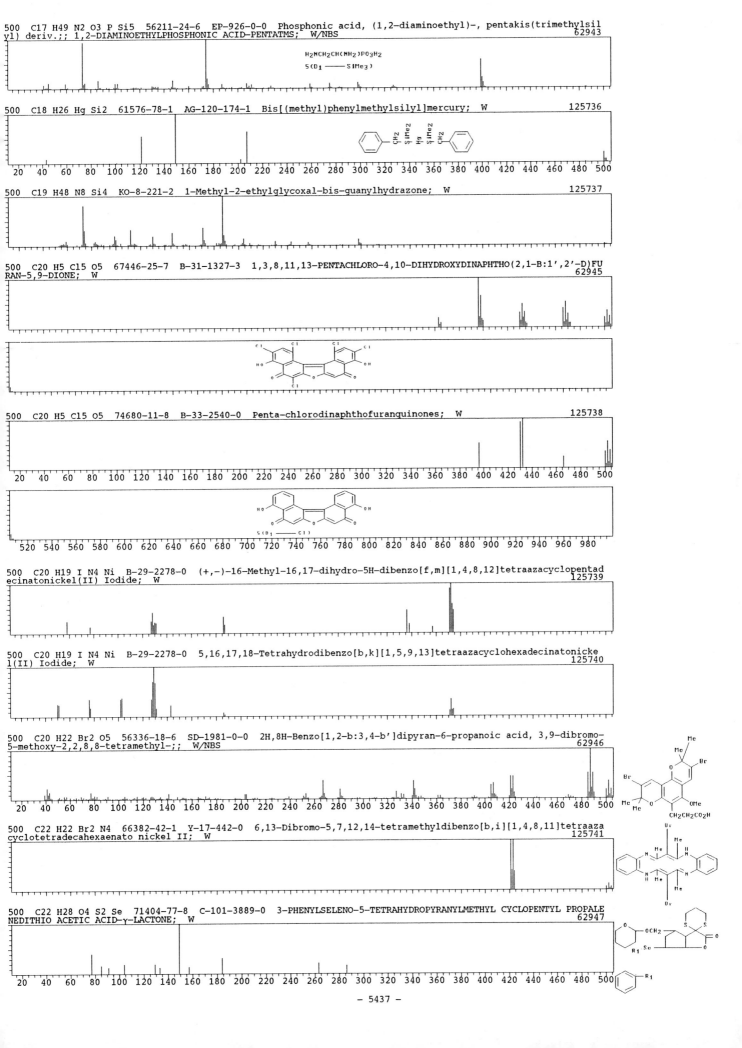

500   C17 H49 N2 O3 P Si5   56211-24-6   EP-926-0-0   Phosphonic acid, (1,2-diaminoethyl)-, pentakis(trimethylsil yl) deriv.;; 1,2-DIAMINOETHYLPHOSPHONIC ACID-PENTATMS;   W/NBS
62943

H2NCH2CH(NH2)PO3H2
5(D1 ——— SiMe3)

500   C18 H26 Hg Si2   61576-78-1   AG-120-174-1   Bis[(methyl)phenylmethylsilyl]mercury;   W
125736

500   C19 H48 N8 Si4   KO-8-221-2   1-Methyl-2-ethylglyoxal-bis-guanylhydrazone;   W
125737

500   C20 H5 Cl5 O5   67446-25-7   B-31-1327-3   1,3,8,11,13-PENTACHLORO-4,10-DIHYDROXYDINAPHTHO(2,1-B:1',2'-D)FU RAN-5,9-DIONE;   W
62945

500   C20 H5 Cl5 O5   74680-11-8   B-33-2540-0   Penta-chlorodinaphthofuranquinones;   W
125738

5(D1 ——— Cl1)

500   C20 H19 I N4 Ni   B-29-2278-0   (+,-)-16-Methyl-16,17-dihydro-5H-dibenzo[f,m][1,4,8,12]tetraazacyclopentad ecinatonickel(II) Iodide;   W
125739

500   C20 H19 I N4 Ni   B-29-2278-0   5,16,17,18-Tetrahydrodibenzo[b,k][1,5,9,13]tetraazacyclohexadecinatonicke l(II) Iodide;   W
125740

500   C20 H22 Br2 O5   56336-18-6   SD-1981-0-0   2H,8H-Benzo[1,2-b:3,4-b']dipyran-6-propanoic acid, 3,9-dibromo- 5-methoxy-2,2,8,8-tetramethyl-;;   W/NBS
62946

500   C22 H22 Br2 N4   66382-42-1   Y-17-442-0   6,13-Dibromo-5,7,12,14-tetramethyldibenzo[b,i][1,4,8,11]tetraaza cyclotetradecahexaenato nickel II;   W
125741

500   C22 H28 O4 S2 Se   71404-77-8   C-101-3889-0   3-PHENYLSELENO-5-TETRAHYDROPYRANYLMETHYL CYCLOPENTYL PROPALE NEDITHIO ACETIC ACID-γ-LACTONE;   W
62947

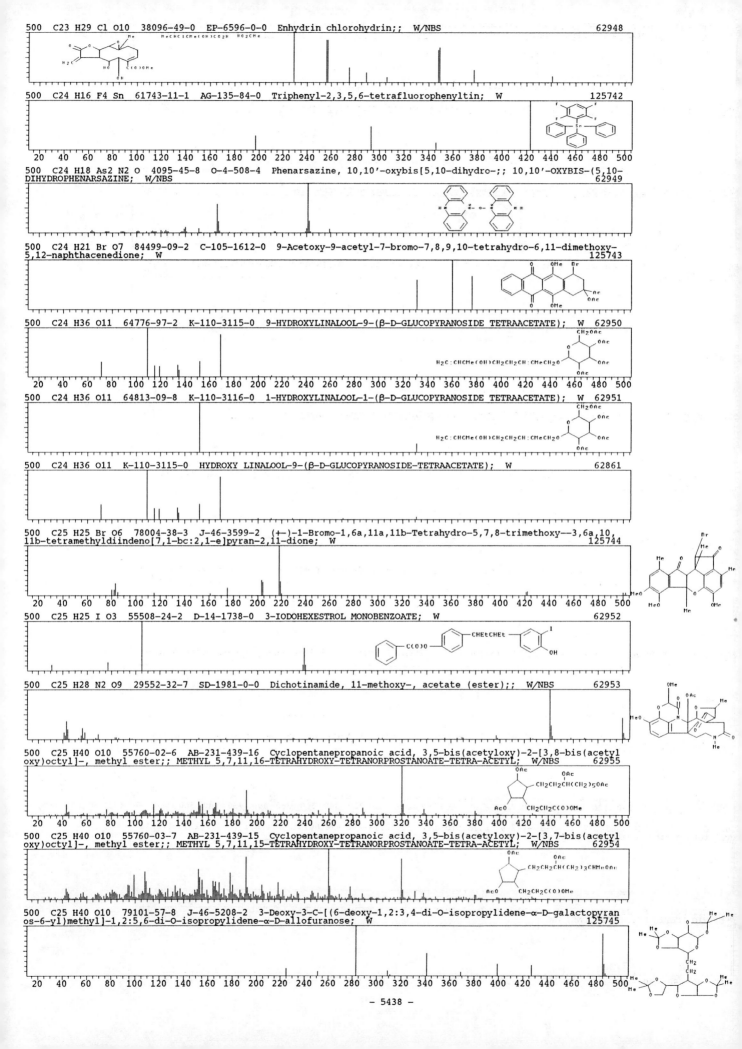

500   C23 H29 Cl O10   38096-49-0   EP-6596-0-0   Enhydrin chlorohydrin;;   W/NBS   62948

500   C24 H16 F4 Sn   61743-11-1   AG-135-84-0   Triphenyl-2,3,5,6-tetrafluorophenyltin;   W   125742

20   40   60   80   100 120 140 160 180 200 220 240 260 280 300 320 340 360 380 400 420 440 460 480 500

500   C24 H18 As2 N2 O   4095-45-8   O-4-508-4   Phenarsazine, 10,10'-oxybis[5,10-dihydro-;;   10,10'-OXYBIS-(5,10-
DIHYDROPHENARSAZINE;   W/NBS   62949

500   C24 H21 Br O7   84499-09-2   C-105-1612-0   9-Acetoxy-9-acetyl-7-bromo-7,8,9,10-tetrahydro-6,11-dimethoxy-
5,12-naphthacenedione;   W   125743

500   C24 H36 O11   64776-97-2   K-110-3115-0   9-HYDROXYLINALOOL-9-(β-D-GLUCOPYRANOSIDE TETRAACETATE);   W   62950

20   40   60   80   100 120 140 160 180 200 220 240 260 280 300 320 340 360 380 400 420 440 460 480 500

500   C24 H36 O11   64813-09-8   K-110-3116-0   1-HYDROXYLINALOOL-1-(β-D-GLUCOPYRANOSIDE TETRAACETATE);   W   62951

500   C24 H36 O11   K-110-3115-0   HYDROXY LINALOOL-9-(β-D-GLUCOPYRANOSIDE-TETRAACETATE);   W   62861

500   C25 H25 Br O6   78004-38-3   J-46-3599-2   (+-)-1-Bromo-1,6a,11a,11b-Tetrahydro-5,7,8-trimethoxy--3,6a,10,
11b-tetramethyldiindeno[7,1-bc:2,1-e]pyran-2,11-dione;   W   125744

20   40   60   80   100 120 140 160 180 200 220 240 260 280 300 320 340 360 380 400 420 440 460 480 500

500   C25 H25 I O3   55508-24-2   D-14-1738-0   3-IODOHEXESTROL MONOBENZOATE;   W   62952

500   C25 H28 N2 O9   29552-32-7   SD-1981-0-0   Dichotinamide, 11-methoxy-, acetate (ester);;   W/NBS   62953

500   C25 H40 O10   55760-02-6   AB-231-439-16   Cyclopentanepropanoic acid, 3,5-bis(acetyloxy)-2-[3,8-bis(acetyl
oxy)octyl]-, methyl ester;;   METHYL 5,7,11,16-TETRAHYDROXY-TETRANORPROSTANOATE-TETRA-ACETYL;   W/NBS   62955

20   40   60   80   100 120 140 160 180 200 220 240 260 280 300 320 340 360 380 400 420 440 460 480 500

500   C25 H40 O10   55760-03-7   AB-231-439-15   Cyclopentanepropanoic acid, 3,5-bis(acetyloxy)-2-[3,7-bis(acetyl
oxy)octyl]-, methyl ester;;   METHYL 5,7,11,15-TETRAHYDROXY-TETRANORPROSTANOATE-TETRA-ACETYL;   W/NBS   62954

500   C25 H40 O10   79101-57-8   J-46-5208-2   3-Deoxy-3-C-[(6-deoxy-1,2:3,4-di-O-isopropylidene-α-D-galactopyran
os-6-yl)methyl]-1,2:5,6-di-O-isopropylidene-α-D-allofuranose;   W   125745

20   40   60   80   100 120 140 160 180 200 220 240 260 280 300 320 340 360 380 400 420 440 460 480 500

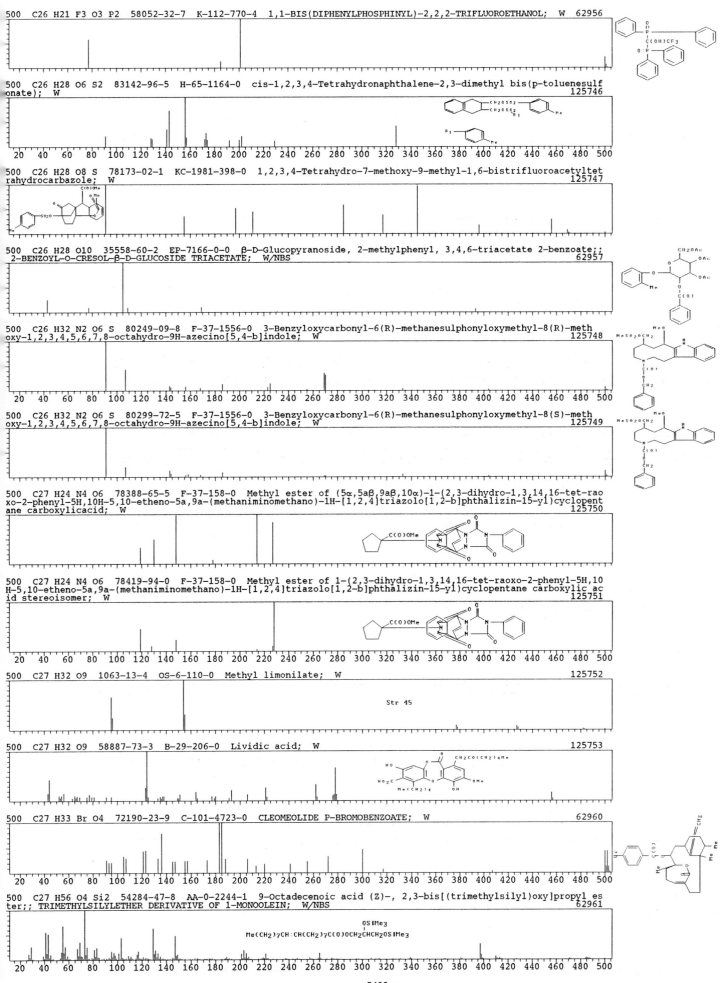

500 C26 H21 F3 O3 P2 58052-32-7 K-112-770-4 1,1-BIS(DIPHENYLPHOSPHINYL)-2,2,2-TRIFLUOROETHANOL; W 62956

500 C26 H28 O6 S2 83142-96-5 H-65-1164-0 cis-1,2,3,4-Tetrahydronaphthalene-2,3-dimethyl bis(p-toluenesulf
onate); W   125746

500 C26 H28 O8 S 78173-02-1 KC-1981-398-0 1,2,3,4-Tetrahydro-7-methoxy-9-methyl-1,6-bistrifluoroacetyltet
rahydrocarbazole; W   125747

500 C26 H28 O10 35558-60-2 EP-7166-0-0 β-D-Glucopyranoside, 2-methylphenyl, 3,4,6-triacetate 2-benzoate;;
2-BENZOYL-O-CRESOL-β-D-GLUCOSIDE TRIACETATE; W/NBS   62957

500 C26 H32 N2 O6 S 80249-09-8 F-37-1556-0 3-Benzyloxycarbonyl-6(R)-methanesulphonyloxymethyl-8(R)-meth
oxy-1,2,3,4,5,6,7,8-octahydro-9H-azecino[5,4-b]indole; W   125748

500 C26 H32 N2 O6 S 80299-72-5 F-37-1556-0 3-Benzyloxycarbonyl-6(R)-methanesulphonyloxymethyl-8(S)-meth
oxy-1,2,3,4,5,6,7,8-octahydro-9H-azecino[5,4-b]indole; W   125749

500 C27 H24 N4 O6 78388-65-5 F-37-158-0 Methyl ester of (5α,5aβ,9aβ,10α)-1-(2,3-dihydro-1,3,14,16-tet-rao
xo-2-phenyl-5H,10H-5,10-etheno-5a,9a-(methaniminomethano)-1H-[1,2,4]triazolo[1,2-b]phthalizin-15-yl)cyclopent
ane carboxylicacid; W   125750

500 C27 H24 N4 O6 78419-94-0 F-37-158-0 Methyl ester of 1-(2,3-dihydro-1,3,14,16-tet-raoxo-2-phenyl-5H,10
H-5,10-etheno-5a,9a-(methaniminomethano)-1H-[1,2,4]triazolo[1,2-b]phthalizin-15-yl)cyclopentane carboxylic ac
id stereoisomer; W   125751

500 C27 H32 O9 1063-13-4 OS-6-110-0 Methyl limonilate; W   125752

Str 45

500 C27 H32 O9 58887-73-3 B-29-206-0 Lividic acid; W   125753

500 C27 H33 Br O4 72190-23-9 C-101-4723-0 CLEOMEOLIDE P-BROMOBENZOATE; W   62960

500 C27 H56 O4 Si2 54284-47-8 AA-0-2244-1 9-Octadecenoic acid (Z)-, 2,3-bis[(trimethylsilyl)oxy]propyl es
ter;; TRIMETHYLSILYLETHER DERIVATIVE OF 1-MONOOLEIN; W/NBS   62961

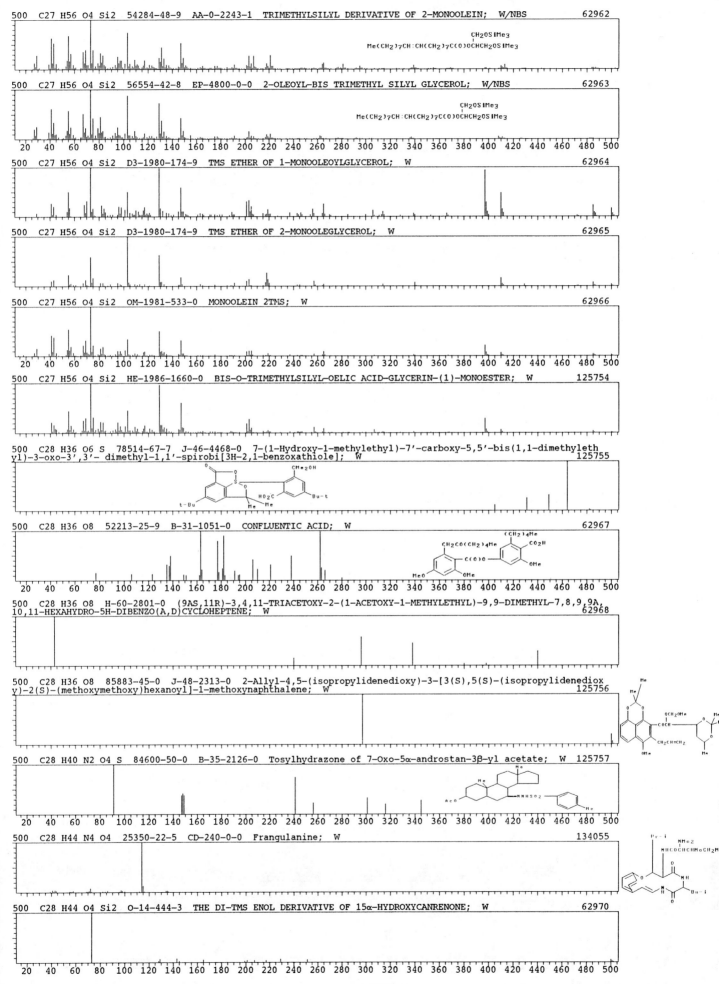

500   C27 H56 O4 Si2   54284-48-9   AA-0-2243-1   TRIMETHYLSILYL DERIVATIVE OF 2-MONOOLEIN;   W/NBS   62962

CH₂OSiMe₃
Me(CH₂)₇CH:CH(CH₂)₇C(O)OCHCH₂OSiMe₃

500   C27 H56 O4 Si2   56554-42-8   EP-4800-0-0   2-OLEOYL-BIS TRIMETHYL SILYL GLYCEROL;   W/NBS   62963

CH₂OSiMe₃
Me(CH₂)₇CH:CH(CH₂)₇C(O)OCHCH₂OSiMe₃

500   C27 H56 O4 Si2   D3-1980-174-9   TMS ETHER OF 1-MONOOLEOYLGLYCEROL;   W   62964

500   C27 H56 O4 Si2   D3-1980-174-9   TMS ETHER OF 2-MONOOLEGLYCEROL;   W   62965

500   C27 H56 O4 Si2   OM-1981-533-0   MONOOLEIN 2TMS;   W   62966

500   C27 H56 O4 Si2   HE-1986-1660-0   BIS-O-TRIMETHYLSILYL-OELIC ACID-GLYCERIN-(1)-MONOESTER;   W   125754

500   C28 H36 O6 S   78514-67-7   J-46-4468-0   7-(1-Hydroxy-1-methylethyl)-7'-carboxy-5,5'-bis(1,1-dimethylethyl)-3-oxo-3',3'- dimethyl-1,1'-spirobi[3H-2,1-benzoxathiole];   W   125755

500   C28 H36 O8   52213-25-9   B-31-1051-0   CONFLUENTIC ACID;   W   62967

500   C28 H36 O8   H-60-2801-0   (9AS,11R)-3,4,11-TRIACETOXY-2-(1-ACETOXY-1-METHYLETHYL)-9,9-DIMETHYL-7,8,9,9A,10,11-HEXAHYDRO-5H-DIBENZO(A,D)CYCLOHEPTENE;   W   62968

500   C28 H36 O8   85883-45-0   J-48-2313-0   2-Allyl-4,5-(isopropylidenedioxy)-3-[3(S),5(S)-(isopropylidenedioxy)-2(S)-(methoxymethoxy)hexanoyl]-1-methoxynaphthalene;   W   125756

500   C28 H40 N2 O4 S   84600-50-0   B-35-2126-0   Tosylhydrazone of 7-Oxo-5α-androstan-3β-yl acetate;   W   125757

500   C28 H44 N4 O4   25350-22-5   CD-240-0-0   Frangulanine;   W   134055

500   C28 H44 O4 Si2   O-14-444-3   THE DI-TMS ENOL DERIVATIVE OF 15α-HYDROXYCANRENONE;   W   62970

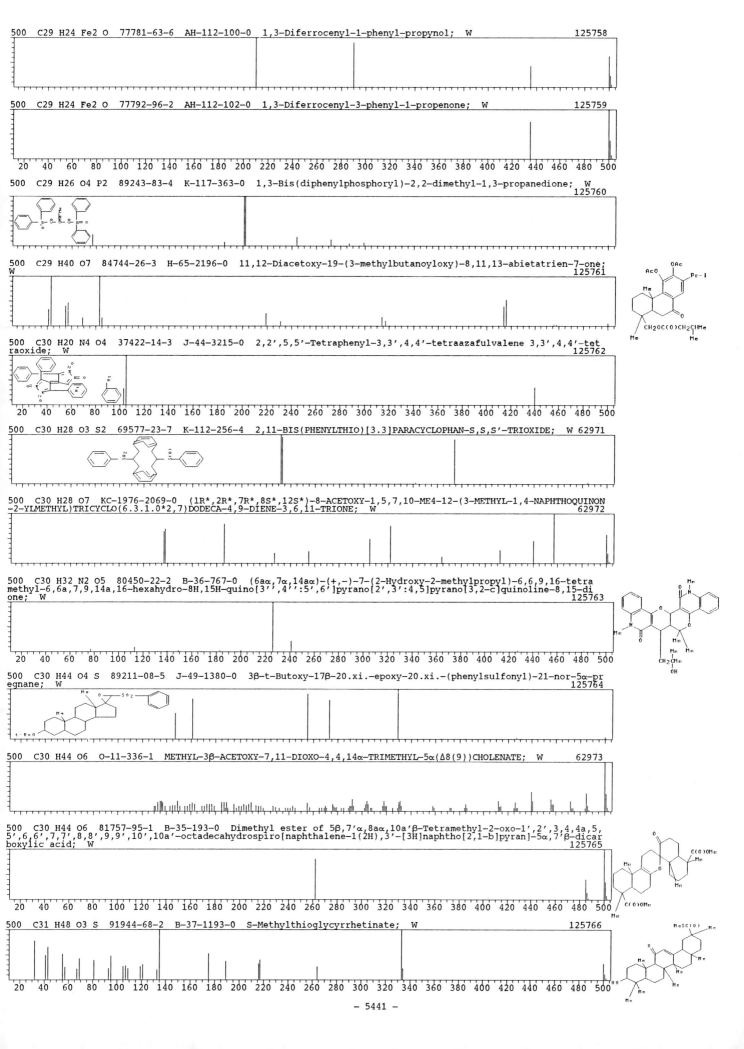

500 C29 H24 Fe2 O 77781-63-6 AH-112-100-0 1,3-Diferrocenyl-1-phenyl-propynol; W 125758

500 C29 H24 Fe2 O 77792-96-2 AH-112-102-0 1,3-Diferrocenyl-3-phenyl-1-propenone; W 125759

500 C29 H26 O4 P2 89243-83-4 K-117-363-0 1,3-Bis(diphenylphosphoryl)-2,2-dimethyl-1,3-propanedione; W
125760

500 C29 H40 O7 84744-26-3 H-65-2196-0 11,12-Diacetoxy-19-(3-methylbutanoyloxy)-8,11,13-abietatrien-7-one;
W
125761

500 C30 H20 N4 O4 37422-14-3 J-44-3215-0 2,2',5,5'-Tetraphenyl-3,3',4,4'-tetraazafulvalene 3,3',4,4'-tet
raoxide; W
125762

500 C30 H28 O3 S2 69577-23-7 K-112-256-4 2,11-BIS(PHENYLTHIO)[3.3]PARACYCLOPHAN-S,S,S'-TRIOXIDE; W 62971

500 C30 H28 O7 KC-1976-2069-0 (1R*,2R*,7R*,8S*,12S*)-8-ACETOXY-1,5,7,10-ME4-12-(3-METHYL-1,4-NAPHTHOQUINON
-2-YLMETHYL)TRICYCLO(6.3.1.0*2,7)DODECA-4,9-DIENE-3,6,11-TRIONE; W
62972

500 C30 H32 N2 O5 80450-22-2 B-36-767-0 (6aα,7α,14aα)-(+,-)-7-(2-Hydroxy-2-methylpropyl)-6,6,9,16-tetra
methyl-6,6a,7,9,14a,16-hexahydro-8H,15H-quino[3'',4'':5',6']pyrano[2',3':4,5]pyrano[3,2-c]quinoline-8,15-di
one; W
125763

500 C30 H44 O4 S 89211-08-5 J-49-1380-0 3β-t-Butoxy-17β-20.xi.-epoxy-20.xi.-(phenylsulfonyl)-21-nor-5α-pr
egnane; W
125764

500 C30 H44 O6 O-11-336-1 METHYL-3β-ACETOXY-7,11-DIOXO-4,4,14α-TRIMETHYL-5α(Δ8(9))CHOLENATE; W 62973

500 C30 H44 O6 81757-95-1 B-35-193-0 Dimethyl ester of 5β,7'α,8aα,10a'β-Tetramethyl-2-oxo-1',2',3,4,4a,5,
5',6,6',7,7',8,8',9,9',10',10a'-octadecahydrospiro[naphthalene-1(2H),3'-[3H]naphtho[2,1-b]pyran]-5α,7'β-dicar
boxylic acid; W
125765

500 C31 H48 O3 S 91944-68-2 B-37-1193-0 S-Methylthioglycyrrhetinate; W 125766

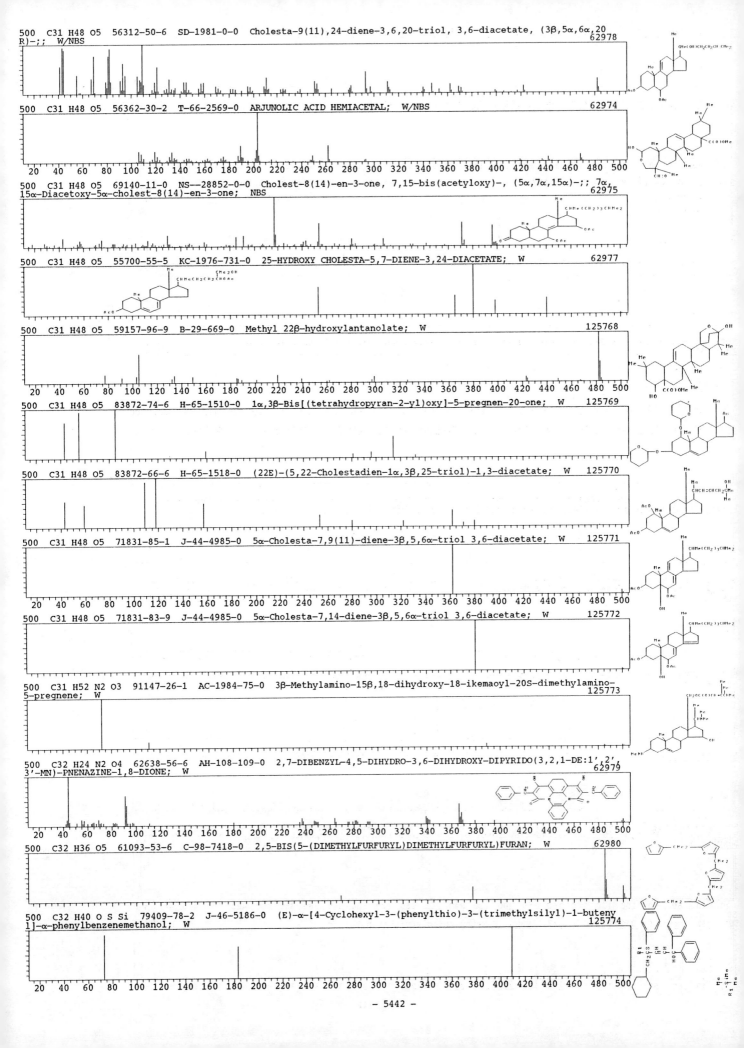

500 C31 H48 O5 56312-50-6 SD-1981-0-0 Cholesta-9(11),24-diene-3,6,20-triol, 3,6-diacetate, (3β,5α,6α,20 R)-;; W/NBS 62978

500 C31 H48 O5 56362-30-2 T-66-2569-0 ARJUNOLIC ACID HEMIACETAL; W/NBS 62974

500 C31 H48 O5 69140-11-0 NS--28852-0-0 Cholest-8(14)-en-3-one, 7,15-bis(acetyloxy)-, (5α,7α,15α)-;; 7α, 15α-Diacetoxy-5α-cholest-8(14)-en-3-one; NBS 62975

500 C31 H48 O5 55700-55-5 KC-1976-731-0 25-HYDROXY CHOLESTA-5,7-DIENE-3,24-DIACETATE; W 62977

500 C31 H48 O5 59157-96-9 B-29-669-0 Methyl 22β-hydroxylantanolate; W 125768

500 C31 H48 O5 83872-74-6 H-65-1510-0 1α,3β-Bis[(tetrahydropyran-2-yl)oxy]-5-pregnen-20-one; W 125769

500 C31 H48 O5 83872-66-6 H-65-1518-0 (22E)-(5,22-Cholestadien-1α,3β,25-triol)-1,3-diacetate; W 125770

500 C31 H48 O5 71831-85-1 J-44-4985-0 5α-Cholesta-7,9(11)-diene-3β,5,6α-triol 3,6-diacetate; W 125771

500 C31 H48 O5 71831-83-9 J-44-4985-0 5α-Cholesta-7,14-diene-3β,5,6α-triol 3,6-diacetate; W 125772

500 C31 H52 N2 O3 91147-26-1 AC-1984-75-0 3β-Methylamino-15β,18-dihydroxy-18-ikemaoyl-20S-dimethylamino-5-pregnene; W 125773

500 C32 H24 N2 O4 62638-56-6 AH-108-109-0 2,7-DIBENZYL-4,5-DIHYDRO-3,6-DIHYDROXY-DIPYRIDO(3,2,1-DE:1',2',3'-MN)-PNENAZINE-1,8-DIONE; W 62979

500 C32 H36 O5 61093-53-6 C-98-7418-0 2,5-BIS(5-(DIMETHYLFURFURYL)DIMETHYLFURFURYL)FURAN; W 62980

500 C32 H40 O S Si 79409-78-2 J-46-5186-0 (E)-α-[4-Cyclohexyl-3-(phenylthio)-3-(trimethylsilyl)-1-butenyl]-α-phenylbenzenemethanol; W 125774

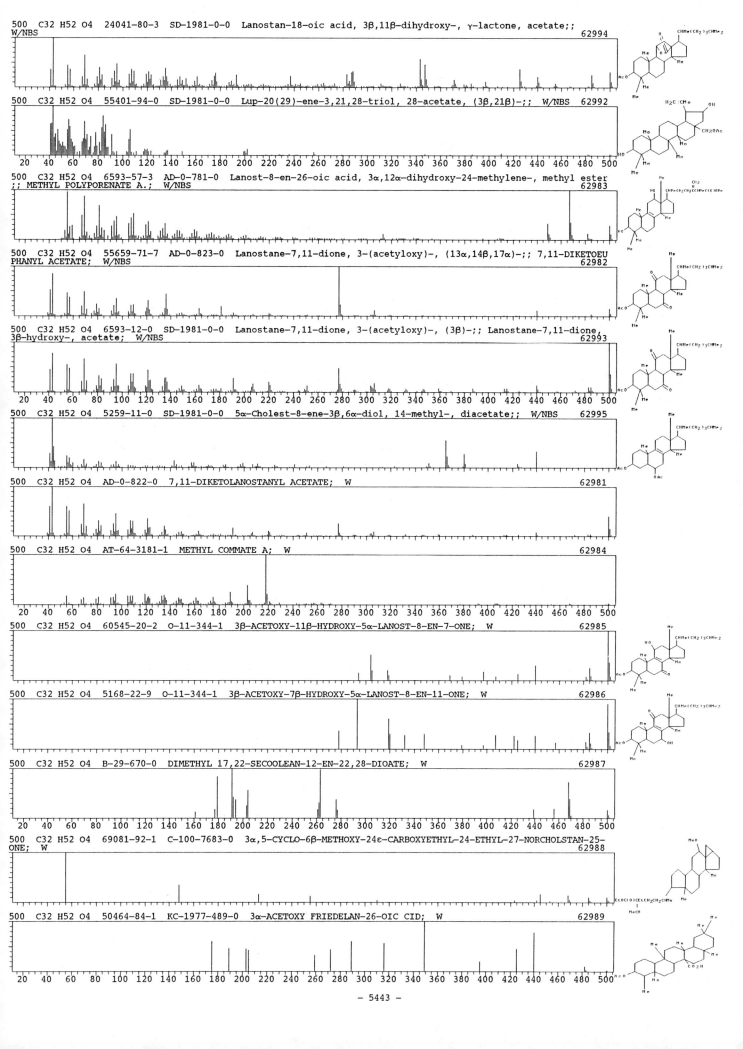

500 C32 H52 O4 24041-80-3 SD-1981-0-0 Lanostan-18-oic acid, 3β,11β-dihydroxy-, γ-lactone, acetate;;
W/NBS   62994

500 C32 H52 O4 55401-94-0 SD-1981-0-0 Lup-20(29)-ene-3,21,28-triol, 28-acetate, (3β,21β)-;; W/NBS 62992

500 C32 H52 O4 6593-57-3 AD-0-781-0 Lanost-8-en-26-oic acid, 3α,12α-dihydroxy-24-methylene-, methyl ester
;; METHYL POLYPORENATE A.; W/NBS   62983

500 C32 H52 O4 55659-71-7 AD-0-823-0 Lanostane-7,11-dione, 3-(acetyloxy)-, (13α,14β,17α)-;; 7,11-DIKETOEU
PHANYL ACETATE; W/NBS   62982

500 C32 H52 O4 6593-12-0 SD-1981-0-0 Lanostane-7,11-dione, 3-(acetyloxy)-, (3β)-;; Lanostane-7,11-dione,
3β-hydroxy-, acetate; W/NBS   62993

500 C32 H52 O4 5259-11-0 SD-1981-0-0 5α-Cholest-8-ene-3β,6α-diol, 14-methyl-, diacetate;; W/NBS 62995

500 C32 H52 O4 AD-0-822-0 7,11-DIKETOLANOSTANYL ACETATE; W   62981

500 C32 H52 O4 AT-64-3181-1 METHYL COMMATE A; W   62984

500 C32 H52 O4 60545-20-2 O-11-344-1 3β-ACETOXY-11β-HYDROXY-5α-LANOST-8-EN-7-ONE; W   62985

500 C32 H52 O4 5168-22-9 O-11-344-1 3β-ACETOXY-7β-HYDROXY-5α-LANOST-8-EN-11-ONE; W   62986

500 C32 H52 O4 B-29-670-0 DIMETHYL 17,22-SECOOLEAN-12-EN-22,28-DIOATE; W   62987

500 C32 H52 O4 69081-92-1 C-100-7683-0 3α,5-CYCLO-6β-METHOXY-24ε-CARBOXYETHYL-24-ETHYL-27-NORCHOLSTAN-25-
ONE; W   62988

500 C32 H52 O4 50464-84-1 KC-1977-489-0 3α-ACETOXY FRIEDELAN-26-OIC CID; W   62989

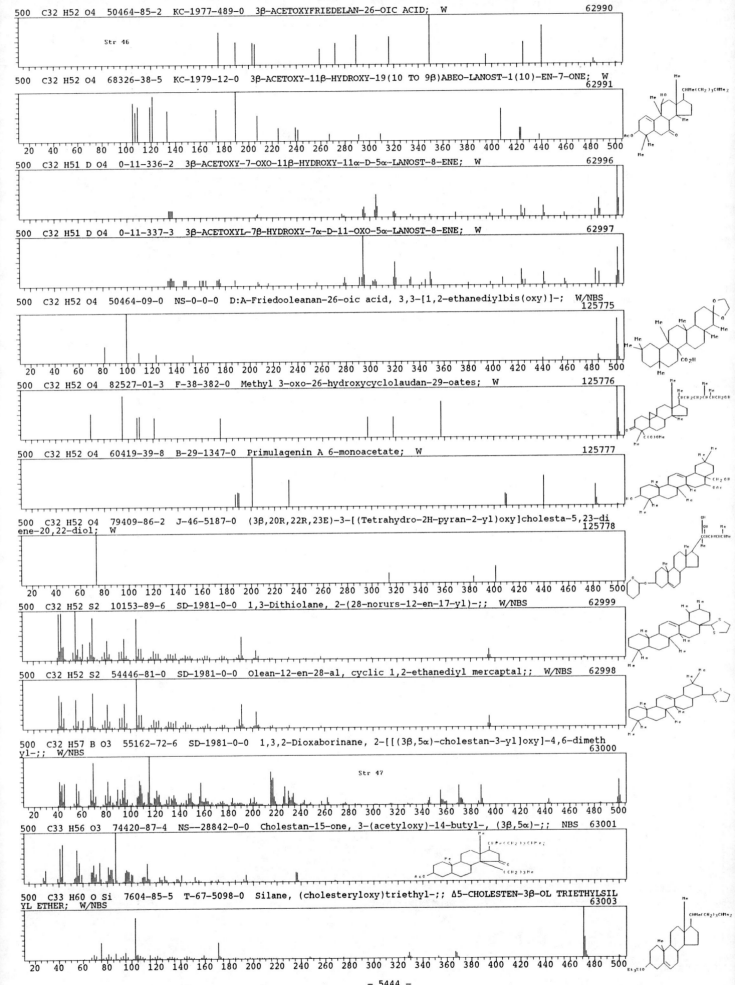

500   C32 H52 O4   50464-85-2   KC-1977-489-0   3β-ACETOXYFRIEDELAN-26-OIC ACID;   W   62990

Str 46

500   C32 H52 O4   68326-38-5   KC-1979-12-0   3β-ACETOXY-11β-HYDROXY-19(10 TO 9β)ABEO-LANOST-1(10)-EN-7-ONE;   W   62991

20 40 60 80 100 120 140 160 180 200 220 240 260 280 300 320 340 360 380 400 420 440 460 480 500

500   C32 H51 D O4   0-11-336-2   3β-ACETOXY-7-OXO-11β-HYDROXY-11α-D-5α-LANOST-8-ENE;   W   62996

500   C32 H51 D O4   0-11-337-3   3β-ACETOXYL-7β-HYDROXY-7α-D-11-OXO-5α-LANOST-8-ENE;   W   62997

500   C32 H52 O4   50464-09-0   NS-0-0-0   D:A-Friedooleanan-26-oic acid, 3,3-[1,2-ethanediylbis(oxy)]-;   W/NBS
125775

20 40 60 80 100 120 140 160 180 200 220 240 260 280 300 320 340 360 380 400 420 440 460 480 500

500   C32 H52 O4   82527-01-3   F-38-382-0   Methyl 3-oxo-26-hydroxycyclolaudan-29-oates;   W   125776

500   C32 H52 O4   60419-39-8   B-29-1347-0   Primulagenin A 6-monoacetate;   W   125777

500   C32 H52 O4   79409-86-2   J-46-5187-0   (3β,20R,22R,23E)-3-[(Tetrahydro-2H-pyran-2-yl)oxy]cholesta-5,23-di
ene-20,22-diol;   W   125778

20 40 60 80 100 120 140 160 180 200 220 240 260 280 300 320 340 360 380 400 420 440 460 480 500

500   C32 H52 S2   10153-89-6   SD-1981-0-0   1,3-Dithiolane, 2-(28-norurs-12-en-17-yl)-;;   W/NBS   62999

500   C32 H52 S2   54446-81-0   SD-1981-0-0   Olean-12-en-28-al, cyclic 1,2-ethanediyl mercaptal;;   W/NBS   62998

500   C32 H57 B O3   55162-72-6   SD-1981-0-0   1,3,2-Dioxaborinane, 2-[[(3β,5α)-cholestan-3-yl]oxy]-4,6-dimeth
yl-;;   W/NBS   63000

Str 47

20 40 60 80 100 120 140 160 180 200 220 240 260 280 300 320 340 360 380 400 420 440 460 480 500

500   C33 H56 O3   74420-87-4   NS--28842-0-0   Cholestan-15-one, 3-(acetyloxy)-14-butyl-, (3β,5α)-;;   NBS   63001

500   C33 H60 O Si   7604-85-5   T-67-5098-0   Silane, (cholesteryloxy)triethyl-;;   Δ5-CHOLESTEN-3β-OL TRIETHYLSIL
YL ETHER;   W/NBS   63003

20 40 60 80 100 120 140 160 180 200 220 240 260 280 300 320 340 360 380 400 420 440 460 480 500

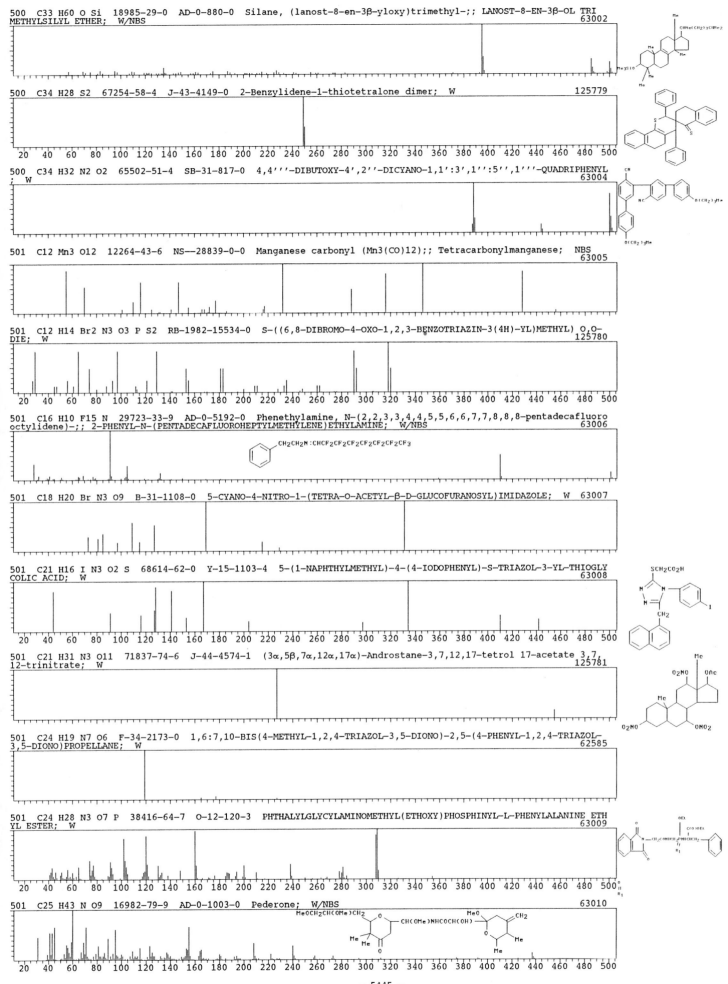

500   C33 H60 O Si   18985-29-0   AD-0-880-0   Silane, (lanost-8-en-3β-yloxy)trimethyl-;; LANOST-8-EN-3β-OL TRI
METHYLSILYL ETHER;   W/NBS   63002

500   C34 H28 S2   67254-58-4   J-43-4149-0   2-Benzylidene-1-thiotetralone dimer;   W   125779

500   C34 H32 N2 O2   65502-51-4   SB-31-817-0   4,4'''-DIBUTOXY-4',2''-DICYANO-1,1':3',1'':5'',1'''-QUADRIPHENYL
;   W   63004

501   C12 Mn3 O12   12264-43-6   NS--28839-0-0   Manganese carbonyl (Mn3(CO)12);; Tetracarbonylmanganese;   NBS
                                                                                                   63005

501   C12 H14 Br2 N3 O3 P S2   RB-1982-15534-0   S-((6,8-DIBROMO-4-OXO-1,2,3-BENZOTRIAZIN-3(4H)-YL)METHYL) O,O-
DIE;   W   125780

501   C16 H10 F15 N   29723-33-9   AD-0-5192-0   Phenethylamine, N-(2,2,3,3,4,4,5,5,6,6,7,7,8,8,8-pentadecafluoro
octylidene)-;; 2-PHENYL-N-(PENTADECAFLUOROHEPTYLMETHYLENE)ETHYLAMINE;   W/NBS   63006

501   C18 H20 Br N3 O9   B-31-1108-0   5-CYANO-4-NITRO-1-(TETRA-O-ACETYL-β-D-GLUCOFURANOSYL)IMIDAZOLE;   W   63007

501   C21 H16 I N3 O2 S   68614-62-0   Y-15-1103-4   5-(1-NAPHTHYLMETHYL)-4-(4-IODOPHENYL)-S-TRIAZOL-3-YL-THIOGLY
COLIC ACID;   W   63008

501   C21 H31 N3 O11   71837-74-6   J-44-4574-1   (3α,5β,7α,12α,17α)-Androstane-3,7,12,17-tetrol 17-acetate 3,7,
12-trinitrate;   W   125781

501   C24 H19 N7 O6   F-34-2173-0   1,6:7,10-BIS(4-METHYL-1,2,4-TRIAZOL-3,5-DIONO)-2,5-(4-PHENYL-1,2,4-TRIAZOL-
3,5-DIONO)PROPELLANE;   W   62585

501   C24 H28 N3 O7 P   38416-64-7   O-12-120-3   PHTHALYLGLYCYLAMINOMETHYL(ETHOXY)PHOSPHINYL-L-PHENYLALANINE ETH
YL ESTER;   W   63009

501   C25 H43 N O9   16982-79-9   AD-0-1003-0   Pederone;   W/NBS   63010

501 C27 H18 Ga N3 O3 OS-5-435-0 Triquinolin-8-olatogallium(3); W   125782

501 C27 H35 N O8 60783-09-7 B-29-1314-0 2,20-EPOXY-19,20-METHYLIMINO-14,20-CYCLO-ENT-KAUR-16-ENE-2α,11,6α
-TRIACETOXY-5-OL; W   63011

20 40 60 80 100 120 140 160 180 200 220 240 260 280 300 320 340 360 380 400 420 440 460 480 500

501 C27 H35 N O8 CD-171-0-0 O-Demethyldeoxyharringtonine; W   125783

501 C27 H43 N O4 Si2 JH-1982-190-0 Bis(trimethylsilyl) derivative of Nalbuphine; W   125784

501 C28 H24 N O2 P S2 KC-1976-1407-0 1-ETHYLTHIO-2-(NITROPHENYL)-2-(TRIPHENYLPHOSPHONIO)ETHENETHIOLATE; W
                                                                                            63013

501 C31 H23 N3 O4 IC-1232-0-0 2-(2,3-Diphenylpropyl)-1,3,5,10-tetrahydro-4,11-diamino-1,3,5,10-tetroxo-2H-
naphtha[2,3-f]isoindole; W   125785

501 C31 H51 N O4 71473-02-4 K-112-2689-0 (25R)-26-AMINO-5-CHOLESTEN-3β,16β-DIOL-DIACETATE; W   63014

501 C31 H51 N O4 58028-19-6 KC-1981-249-0 3-Deutero-2-acetyl-4-oxa-5-acetoxymethylcholest-2-ene; W
                                                                                            125786

20 40 60 80 100 120 140 160 180 200 220 240 260 280 300 320 340 360 380 400 420 440 460 480 500

501 C32 H23 N O5 59631-67-3 K-109-1112-0 2-BENZOYLOXY-1-(N-BENZOYL-N-(BENZOYLOXY)AMINOMETHYL)NAPHTHALENE;
W   63015

501 C33 H27 N O4 39656-25-2 KC-1977-16-1854 DIMETHYL2',5'-DIHYDRO-1'-METHYL-5',5'-DIPHENYLFLUORENE-9-SPIR
O-2'-PYRROLE-3',4'-DICARBOXYLATE; W   63016

502 C6 Br5 Cl 13075-05-3 B-30-302-0 1,2,3,4,5-PENTABROMO-6-CHLOROBENZENE; W   63017

20 40 60 80 100 120 140 160 180 200 220 240 260 280 300 320 340 360 380 400 420 440 460 480 500

520 540 560 580 600 620 640 660 680 700 720 740 760 780 800 820 840 860 880 900 920 940 960 980

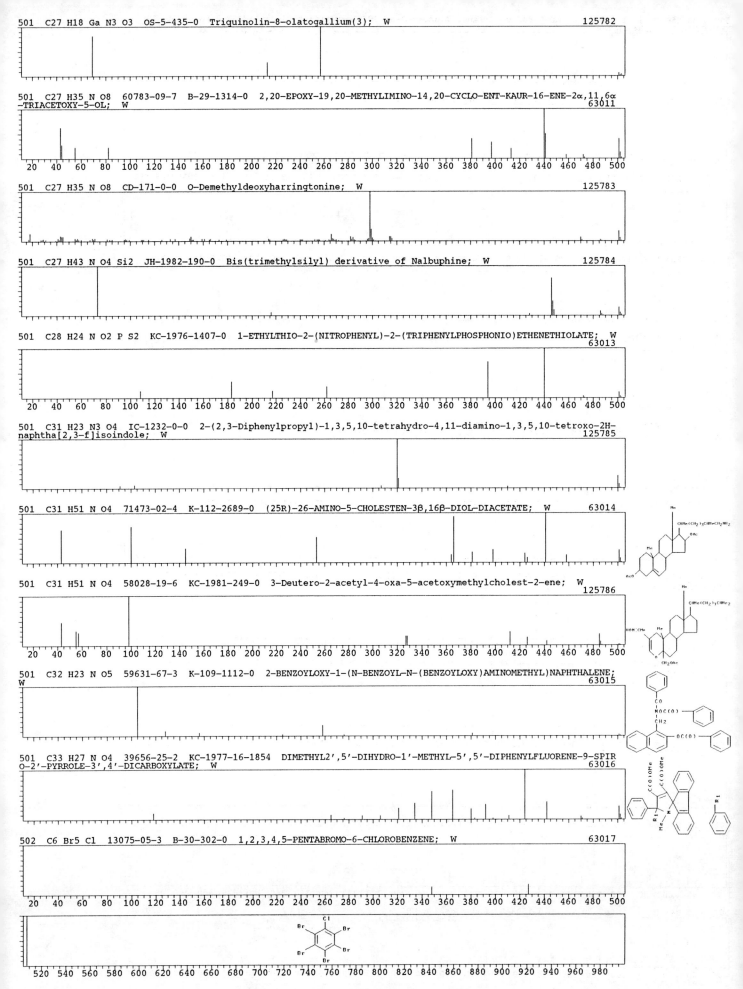

502   C8 H Br4 F3 O2   63165-57-1   B-30-302-0   2,3,5,6-TETRABROMO-4-TRIFLUOROMETHYLBENZOIC ACID;   W   63018

502   C8 H Br4 F3 O2   63165-58-2   B-30-302-0   2,3,4,6-TETRABROMO-5-TRIFLUOROMETHYLBENZOIC ACID;   W   63019

502   C10 H2 Br4 S2   88687-03-0   J-49-1029-0   1,3,6,8-Tetrabromobenzo[1,2-c:3,4-c]dithiophene;   W   125787

502   C12 H40 F2 N6 Si7   62978-31-8   K-110-1283-0   DIFLUOROBIS(2,2,4,4,6,6-HEXAMETHYLCYCLOTRISILAZAN-1YL)SIL
ANE;   W   63020

502   C13 H5 Co F14 O   38531-08-7   NS-0-0-0   Cobalt, carbonyl(η5-2,4-cyclopentadien-1-yl)(heptafluoropropyl)[
2,3,3,3-tetrafluoro-1-(trifluoromethyl)-1-propenyl]-;   W/NBS   125788

502   C15 H30 I P2 Rh   86225-07-2   K-116-1461-0   Iodo(methyl)(pentamethylcyclopentadienyl)(tetramethyldiphosph
an)rhodium(III);   W   125789

502   C15 H27 D3 I P2 Rh   86225-08-3   K-116-1461-0   Iodo(methyl)(pentamethyl cyclopentadienyl)(tetramethyldiph
osphan)rhodium;   W   125790

502   C17 H25 Br2 Cl O5   72926-50-2   J-45-1527-0   (2α(2S*,4R*,5R*),3α,5β)-2-(3-(Acetyloxy)-5-bromotetrahydro-
2,6,6-trimethyl-2H-pyran-2-yl)-4-bromo-5-chloro-2-hydroxy-5-methyl-cyclohexanone;   W   125791

502   C17 H43 O7 P Si4   69688-43-3   BA-0-310-0   D-erythro-Pentofuranose, 2-deoxy-1,3-bis-O-(trimethylsilyl)-,
bis(trimethylsilyl) phosphate;;   O-TETRAKIS(TRIMETHYLSILYL)-2-DEOXY-D-RIBOFURANOSE-5-PHOSPHATE;   W/NBS   63021

502   C18 H12 Br2 Fe N2 O2   OS-5-437-0   Bromodiquinolin-8-ato iron(3);   W   125792

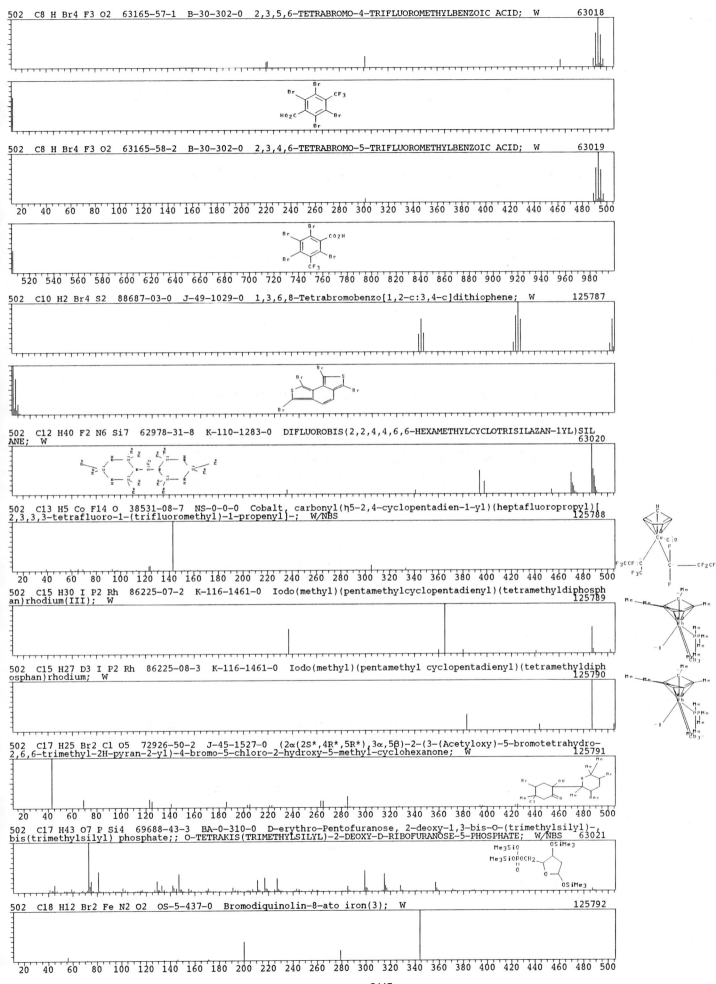

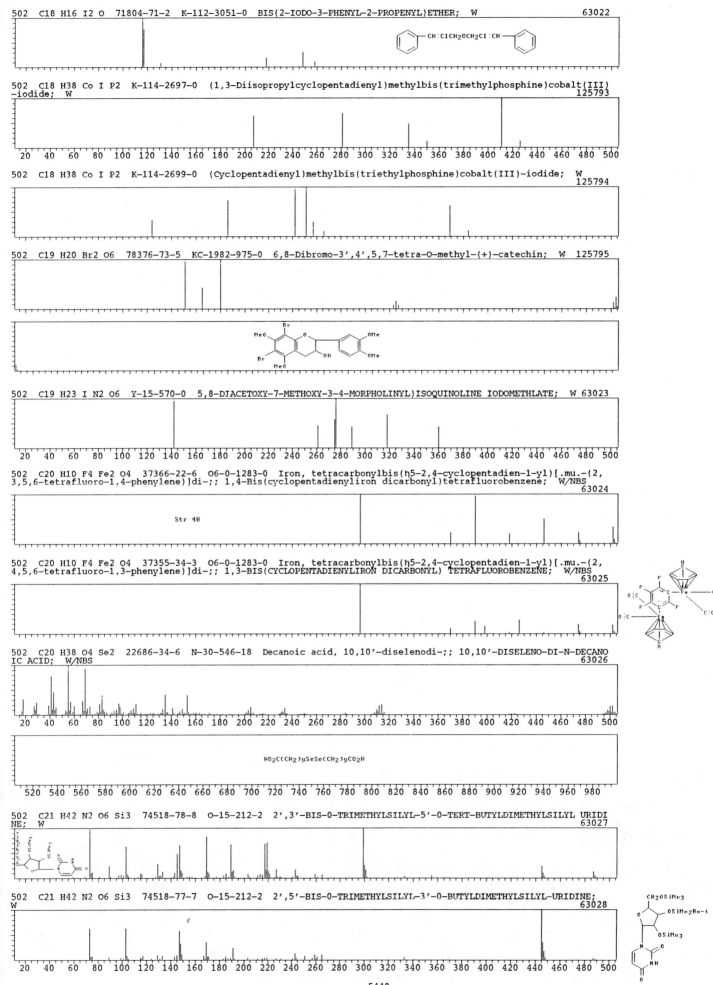

502   C18 H16 I2 O   71804-71-2   K-112-3051-0   BIS(2-IODO-3-PHENYL-2-PROPENYL)ETHER;   W   63022

502   C18 H38 Co I P2   K-114-2697-0   (1,3-Diisopropylcyclopentadienyl)methylbis(trimethylphosphine)cobalt(III)
-iodide;   W   125793

502   C18 H38 Co I P2   K-114-2699-0   (Cyclopentadienyl)methylbis(triethylphosphine)cobalt(III)-iodide;   W
125794

502   C19 H20 Br2 O6   78376-73-5   KC-1982-975-0   6,8-Dibromo-3',4',5,7-tetra-O-methyl-(+)-catechin;   W   125795

502   C19 H23 I N2 O6   Y-15-570-0   5,8-DIACETOXY-7-METHOXY-3-4-MORPHOLINYL)ISOQUINOLINE IODOMETHLATE;   W 63023

502   C20 H10 F4 Fe2 O4   37366-22-6   O6-0-1283-0   Iron, tetracarbonylbis(η5-2,4-cyclopentadien-1-yl)[.mu.-(2,
3,5,6-tetrafluoro-1,4-phenylene)]di-;; 1,4-Bis(cyclopentadienyliron dicarbonyl)tetrafluorobenzene;   W/NBS
63024

Str 48

502   C20 H10 F4 Fe2 O4   37355-34-3   O6-0-1283-0   Iron, tetracarbonylbis(η5-2,4-cyclopentadien-1-yl)[.mu.-(2,
4,5,6-tetrafluoro-1,3-phenylene)]di-;; 1,3-BIS(CYCLOPENTADIENYLIRON DICARBONYL) TETRAFLUOROBENZENE;   W/NBS
63025

502   C20 H38 O4 Se2   22686-34-6   N-30-546-18   Decanoic acid, 10,10'-diselenodi-;; 10,10'-DISELENO-DI-N-DECANO
IC ACID;   W/NBS   63026

HO2C((CH2)9SeSe(CH2)9CO2H

502   C21 H42 N2 O6 Si3   74518-78-8   O-15-212-2   2',3'-BIS-O-TRIMETHYLSILYL-5'-O-TERT-BUTYLDIMETHYLSILYL URIDI
NE;   W   63027

502   C21 H42 N2 O6 Si3   74518-77-7   O-15-212-2   2',5'-BIS-O-TRIMETHYLSILYL-3'-O-BUTYLDIMETHYLSILYL-URIDINE;
W   63028

- 5448 -

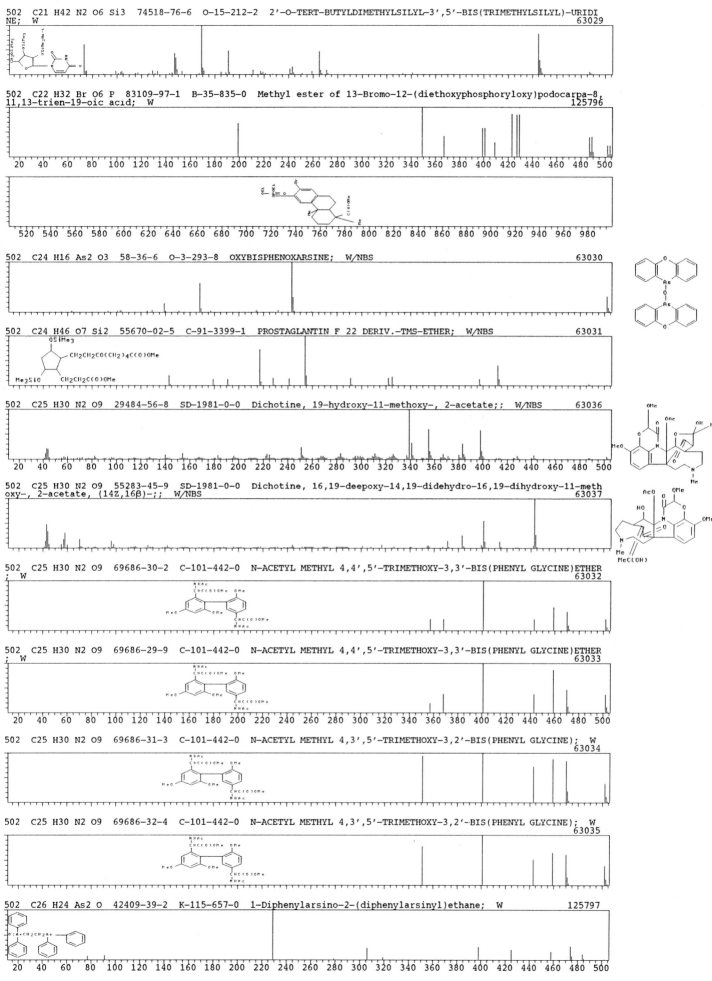

502 C21 H42 N2 O6 Si3 74518-76-6 O-15-212-2 2'-O-TERT-BUTYLDIMETHYLSILYL-3',5'-BIS(TRIMETHYLSILYL)-URIDINE; W
63029

502 C22 H32 Br O6 P 83109-97-1 B-35-835-0 Methyl ester of 13-Bromo-12-(diethoxyphosphoryloxy)podocarpa-8,11,13-trien-19-oic acid; W
125796

502 C24 H16 As2 O3 58-36-6 O-3-293-8 OXYBISPHENOXARSINE; W/NBS
63030

502 C24 H46 O7 Si2 55670-02-5 C-91-3399-1 PROSTAGLANTIN F 22 DERIV.-TMS-ETHER; W/NBS
63031

502 C25 H30 N2 O9 29484-56-8 SD-1981-0-0 Dichotine, 19-hydroxy-11-methoxy-, 2-acetate;; W/NBS
63036

502 C25 H30 N2 O9 55283-45-9 SD-1981-0-0 Dichotine, 16,19-deepoxy-14,19-didehydro-16,19-dihydroxy-11-methoxy-, 2-acetate, (14Z,16β)-;; W/NBS
63037

502 C25 H30 N2 O9 69686-30-2 C-101-442-0 N-ACETYL METHYL 4,4',5'-TRIMETHOXY-3,3'-BIS(PHENYL GLYCINE)ETHER; W
63032

502 C25 H30 N2 O9 69686-29-9 C-101-442-0 N-ACETYL METHYL 4,4',5'-TRIMETHOXY-3,3'-BIS(PHENYL GLYCINE)ETHER; W
63033

502 C25 H30 N2 O9 69686-31-3 C-101-442-0 N-ACETYL METHYL 4,3',5'-TRIMETHOXY-3,2'-BIS(PHENYL GLYCINE); W
63034

502 C25 H30 N2 O9 69686-32-4 C-101-442-0 N-ACETYL METHYL 4,3',5'-TRIMETHOXY-3,2'-BIS(PHENYL GLYCINE); W
63035

502 C26 H24 As2 O 42409-39-2 K-115-657-0 1-Diphenylarsino-2-(diphenylarsinyl)ethane; W
125797

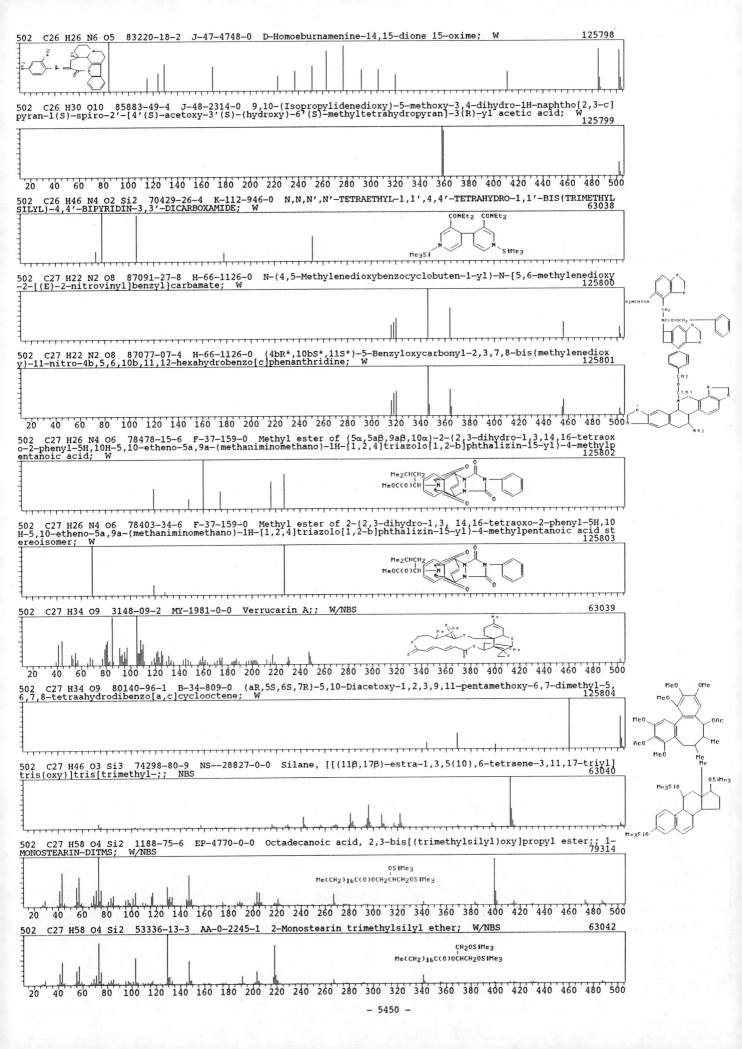

502 C26 H26 N6 O5 83220-18-2 J-47-4748-0 D-Homoeburnamenine-14,15-dione 15-oxime; W   125798

502 C26 H30 O10 85883-49-4 J-48-2314-0 9,10-(Isopropylidenedioxy)-5-methoxy-3,4-dihydro-1H-naphtho[2,3-c]
pyran-1(S)-spiro-2'-[4'(S)-acetoxy-3'(S)-(hydroxy)-6'(S)-methyltetrahydropyran]-3(R)-yl acetic acid; W   125799

502 C26 H46 N4 O2 Si2 70429-26-4 K-112-946-0 N,N,N',N'-TETRAETHYL-1,1',4,4'-TETRAHYDRO-1,1'-BIS(TRIMETHYL
SILYL)-4,4'-BIPYRIDIN-3,3'-DICARBOXAMIDE; W   63038

502 C27 H22 N2 O8 87091-27-8 H-66-1126-0 N-(4,5-Methylenedioxybenzocyclobuten-1-yl)-N-{5,6-methylenedioxy
-2-[(E)-2-nitrovinyl]benzyl}carbamate; W   125800

502 C27 H22 N2 O8 87077-07-4 H-66-1126-0 (4bR*,10bS*,11S*)-5-Benzyloxycarbonyl-2,3,7,8-bis(methylenediox
y)-11-nitro-4b,5,6,10b,11,12-hexahydrobenzo[c]phenanthridine; W   125801

502 C27 H26 N4 O6 78478-15-6 F-37-159-0 Methyl ester of (5α,5aβ,9aβ,10α)-2-(2,3-dihydro-1,3,14,16-tetraox
o-2-phenyl-5H,10H-5,10-etheno-5a,9a-(methaniminomethano)-1H-[1,2,4]triazolo[1,2-b]phthalizin-15-yl)-4-methylp
entanoic acid; W   125802

502 C27 H26 N4 O6 78403-34-6 F-37-159-0 Methyl ester of 2-(2,3-dihydro-1,3,14,16-tetraoxo-2-phenyl-5H,10
H-5,10-etheno-5a,9a-(methaniminomethano)-1H-[1,2,4]triazolo[1,2-b]phthalizin-15-yl)-4-methylpentanoic acid st
ereoisomer; W   125803

502 C27 H34 O9 3148-09-2 MY-1981-0-0 Verrucarin A;; W/NBS   63039

502 C27 H34 O9 80140-96-1 B-34-809-0 (aR,5S,6S,7R)-5,10-Diacetoxy-1,2,3,9,11-pentamethoxy-6,7-dimethyl-5,
6,7,8-tetrahydrodibenzo[a,c]cyclooctene; W   125804

502 C27 H46 O3 Si3 74298-80-9 NS--28827-0-0 Silane, [[(11β,17β)-estra-1,3,5(10),6-tetraene-3,11,17-triyl]
tris(oxy)]tris[trimethyl-;; NBS   63040

502 C27 H58 O4 Si2 1188-75-6 EP-4770-0-0 Octadecanoic acid, 2,3-bis[(trimethylsilyl)oxy]propyl ester;; 1-
MONOSTEARIN-DITMS; W/NBS   79314

502 C27 H58 O4 Si2 53336-13-3 AA-0-2245-1 2-Monostearin trimethylsilyl ether; W/NBS   63042

- 5450 -

502   C27 H58 O4 Si2   HE-1986-2173-0   BIS-O-TRIMETHYLSILYL-STEARINIC ACID-GLYCERIN-(1)-MONOESTER;   W   125805

502   C27 H58 O4 Si2   HE-1986-2364-0   BIS-TRIMETHYLSILYL-STEARINIC ACID-GLYCERIN-(2)-MONOESTER;   W   125806

502   C27 H58 O4 Si2   IC-1232-0-0   O,O''-Trimethylsilyl-O'-stearoylglycerol;   W   125807

502   C28 H24 O5 P2   RB-1982-15535-0   10,10'-OXYBIS-(2,8-DIMETHYL-10,10'-DIOXIDE)PHENOXAPHOSPHINE;   W   125808

502   C28 H25 P3 Ti   37299-20-0   EP-2423-0-0   Titanium, bis(η5-2,4-cyclopentadien-1-yl)[1,2,3-triphenyltriphos
phinato(2-)-P1,P3]-;; BIS(.PI.-CYCLOPENTADIENYL)(1,2,3-TRIPHENYLTRIPHOSPHANATO-P1,P3)-TITANIUM;   W/NBS   63043

Str 49

502   C28 H27 N2 O5 P   72502-65-9   K-112-3476-0   DIMETHYLESTER OF 8,8-DIMETHYL-3,5,5-TRIPHENYL-5H-[1,2λ*5]AZAP
HOSPHOL-6,7-DICARBOXYLIC ACID;   W   63044

502   C28 H38 O6 S   62558-13-8   C-103-120-0   Bis[2-(1-hydroxy-1-methylethyl)-4-(1,1-dimethylethyl)-6-carboxyph
enyl]sulfide;   W   125809

502   C28 H38 O8   61621-88-3   B-29-1987-0   2,2'-(Tetradecane-1,14-diyl)di(3-hydroxy-6-methoxy-1,4-benzoquino
ne);   W   63045

502   C28 H46 O4 Si2   O-11-745-3   Δ9-TETRAHYDROCANNABIVARINIC ACID (OR Δ9-TETRAHYDROCANNABINOLACID B), BIS(TRI
METHYLSILYL) ETHER;   W   63046

502   C28 H46 O4 Si2   O-11-742-1   Δ9-TETRAHYDROCANNABINOL ACID, BIS(TRIMETHYLSILYL) ETHER;   W   63047

502   C28 H46 O4 Si2   70284-90-1   KO-5-527-2   3β,15α-DIHYDROXY-3-DEOXO-CANRENONE BIS-TRIMETHYLSILYLOXY ETHER;
W   63048

502   C28 H28 D18 O4 Si2   O-11-743-2   D9-TRIMETHYLSILYL DERIVATIVE OF Δ9-TETRAHYDROCANNABINOL ACID A;   W 63050

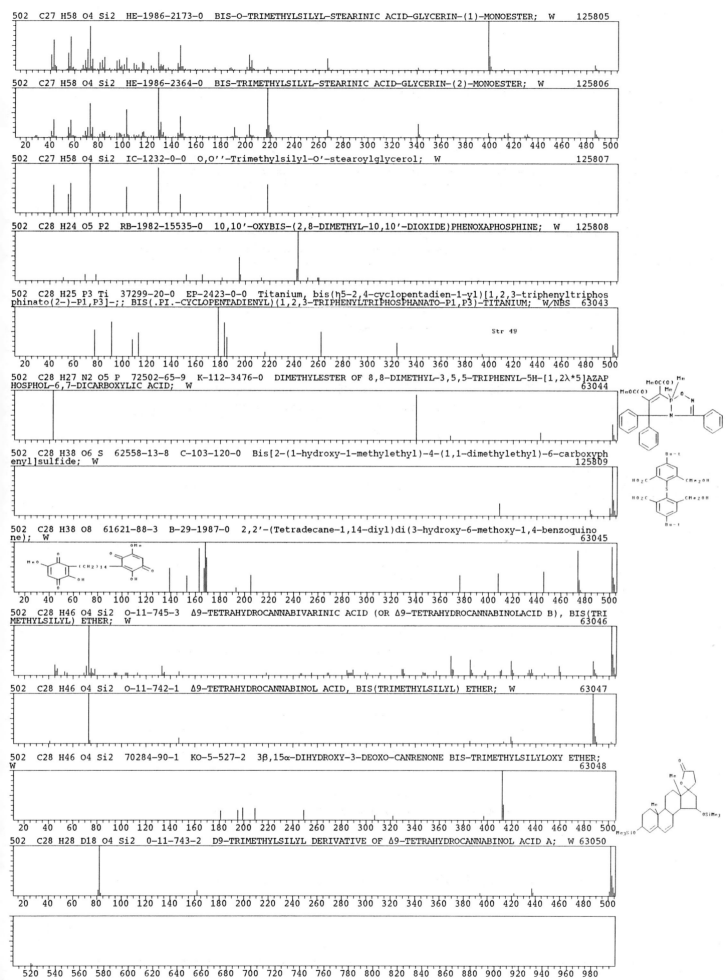

502　C28 H46 O4 Si2　BS-4-6-0　δ(1)-Tetrahydrocannabinolic acid trimethylsilate;　W　　125810

502　C28 H46 O4 Si2　86668-44-2　J-48-3261-0　1-Hydroxy-10(tert-butyldimethylsiloxy)-8,9-(di-tert-butylsilyle
nedioxy)-1(R*),4,4a(S*),9-(R*),9a(S*),10(S*)-hexahydroanthracene;　W　　125811

502　C28 H58 O5 Si　84961-22-8　C-105-2434-0　2-Methyl-4,5-dimethoxy-3-(t-butyldimethylsiloxy)pyran-6-yl tet
radecyl ether (isomer A);　W　　125812

502　C28 H58 O5 Si　84961-21-7　C-105-2434-0　2-Methyl-4,5-dimethoxy-3-(t-butyldimethylsiloxy)pyran-6-yl tet
radecyl ether (isomer B);　W　　125813

502　C29 H26 Fe2 O　77781-65-8　AH-112-100-0　1,3-Diferrocenyl-1-phenyl-propenol;　W　　125814

502　C29 H26 O8　53831-02-0　K-107-3324-49　6,8-DIPHENYLTRICYCLO[4.3.0.0(2,5)]NONA-3,8-DIEN-1,2,3,9-TETRA(
METHYL CARBOXYLATE);　W　　63051

502　C29 H26 O8　53831-04-2　K-107-3324-52　1,8-DIPHENYLPENTACYCLO[4.3.0.0(2,5).0(3,8).0(4,7)]NONAN-2,3,4,7-
TETRA(METHYLCARBOXYLATE);　W　　63052

502　C29 H42 O5 S　80164-09-6　H-65-1509-0　(20S)-(20-Methyl-5-pregnen-1α,3β,21-triol)-21-p-toluolsulfonate;
W　　125815

502　C29 H42 O7　69889-90-3　O-13-406-2　7α,12α-DIACETOXY-3-OXOPREGN-4-EN-20-PROPANOIC ACID METHYL ESTER;　W
63053

502　C29 H42 O7　64219-22-3　J-44-4574-0　Methyl ester of (3α,5β,12α)-3,12-bis(acetyloxy)-6-oxochol-7-en-24-
oic acid;　W　　125816

502　C29 H50 O3 Si2　KO-7-213-1　Bis(vinyldimethylsilyl) derivative of δ(5)-pregane-3β,17α,20α-triol;　W
125817

502　C30 H14 O8　74798-77-9　Y-17-229-0　1,4-Bis[3-(2,5-dioxo-2H,5H-pyrano[3,2-c][1]benzopyranyl)]benzene;　W
125818

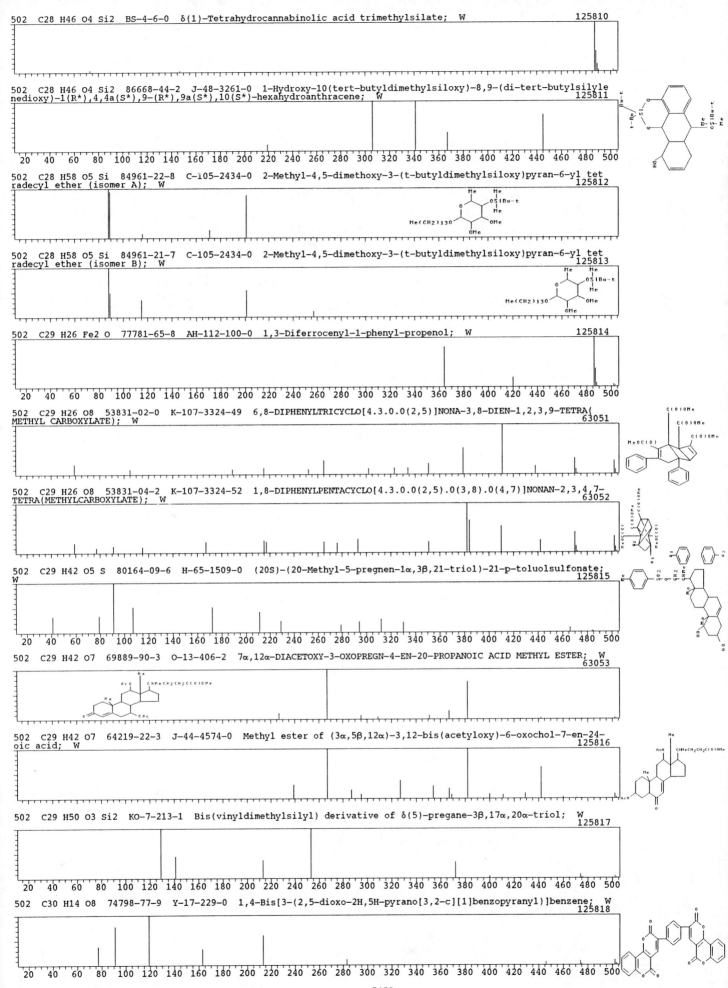

502 C30 H22 N4 O4 3474-99-5 O-10-608-1 [4,4'-Bipyrazolidine]-3,3',5,5'-tetrone, 1,1',2,2'-tetraphenyl-;;
1,1',2,2'-TETRAPHENYL-(4,4'-BIPYRAZOLIDINE)-3,3',5,5'-TETRAONE; W/NBS
63054

502 C30 H30 O7 28753-30-2 O-5-263-13 Dihydrothamnosinoxide; W/NBS
63055

502 C30 H34 N2 O5 56143-43-2 SD-1981-0-0 Aspidospermidin-21-oic acid, 20-hydroxy-17-methoxy-1-(1-oxo-3-ph
enyl-2-propenyl)-, methyl ester, [1(E)]-;; W/NBS
63056

502 C30 H46 O6 60545-21-3 O-11-345-1 METHYL-3β-ACETOXY-7-OXO-11β-HYDROXY-4,4,14-α-TRIMETHYL-5α-(Δ8(9))CHO
LENATE; W
63057

502 C30 H46 O6 70000-05-4 KC-1978-1469-0 1,2-BIS-(5Aβ,9Bβ-DIMETHYL-6-OXO-9AαH)-PERHYDRONAPTHO(2,1-B)-FURA
N-3A-YLOXY)ETHANE; W
63058

502 C30 H45 D O6 0-11-337-4 METHYL 3β-ACETOXY-7β-HYDROXY-7α-D-11-OXO-4,4,14α-TRIMETHYL-5α-Δ9,9-CHOLENATE;
W
63059

502 C30 H46 O6 89766-93-8 KC-1983-2762-0 (20S)-3β-Acetoxy-16-methoxy-16α,30-epoxy-24,25,26,27-tetranor-5α
-dammarane-23,20-carbolactone; W
125819

502 C30 H46 O6 CD-463-0-0 Medicagenic acid; W
125820

502 C31 H50 O5 56052-67-6 SD-1981-0-0 Cholestan-24-one, 3,12-bis(acetyloxy)-, (3α,5β,12α)-;; W/NBS 63069

502 C31 H50 O5 56312-51-7 SD-1981-0-0 Cholest-9(11)-ene-3,6,20-triol, 3,6-diacetate, (3β,5α,6α,20R)-;;
W/NBS
63067

502 C31 H50 O5 55401-91-7 SD-1981-0-0 30-Norlupan-28-oic acid, 3-hydroxy-21-methoxy-20-oxo-, methyl este
r, (3β)-;; W/NBS
63068

502 C31 H50 O5 13850-15-2 T-66-2578-0 METHYL TORMENTATE; W/NBS
63061

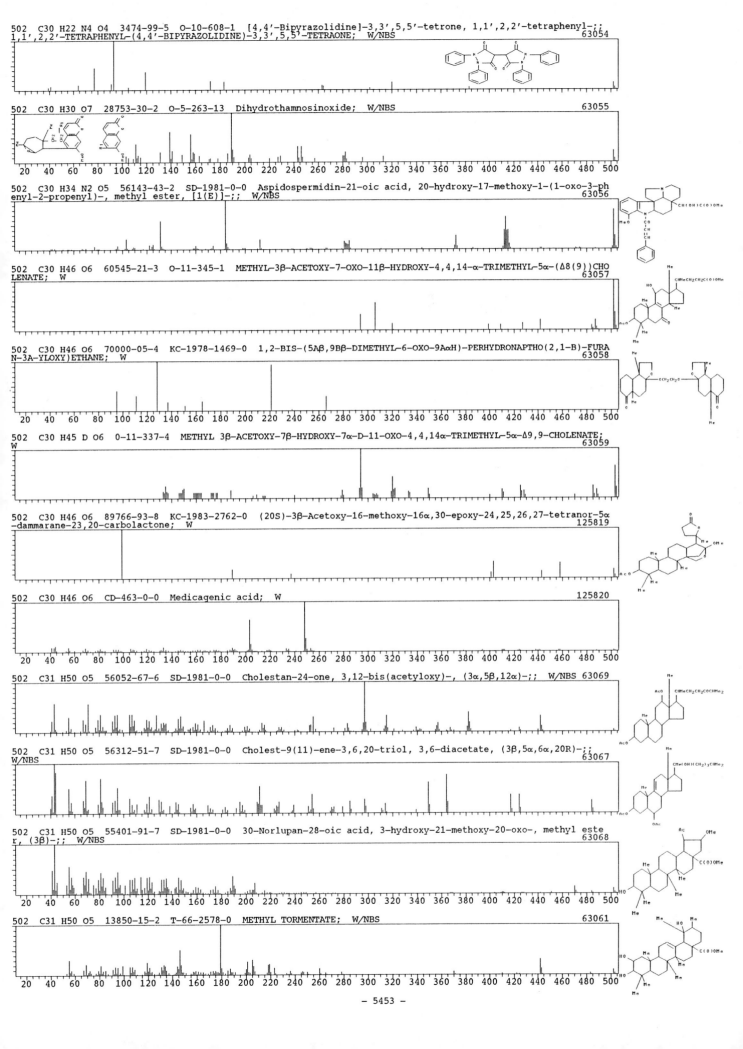

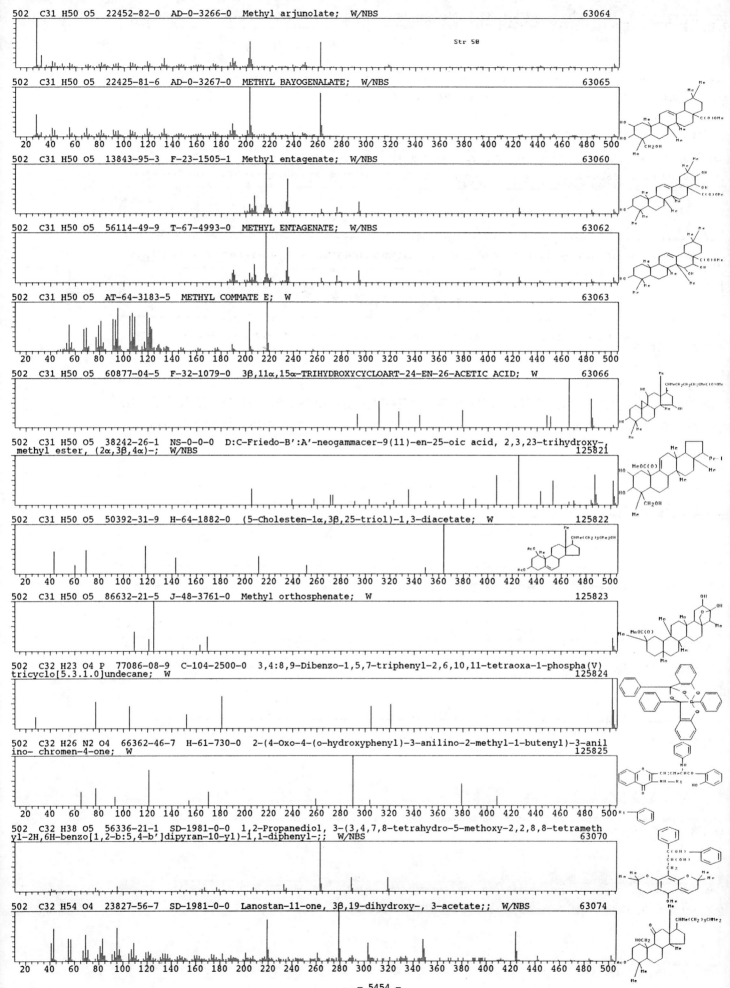

502 C31 H50 O5 22452-82-0 AD-0-3266-0 Methyl arjunolate; W/NBS   63064

Str 50

502 C31 H50 O5 22425-81-6 AD-0-3267-0 METHYL BAYOGENALATE; W/NBS   63065

502 C31 H50 O5 13843-95-3 F-23-1505-1 Methyl entagenate; W/NBS   63060

502 C31 H50 O5 56114-49-9 T-67-4993-0 METHYL ENTAGENATE; W/NBS   63062

502 C31 H50 O5 AT-64-3183-5 METHYL COMMATE E; W   63063

502 C31 H50 O5 60877-04-5 F-32-1079-0 3β,11α,15α-TRIHYDROXYCYCLOART-24-EN-26-ACETIC ACID; W   63066

502 C31 H50 O5 38242-26-1 NS-0-0-0 D:C-Friedo-B':A'-neogammacer-9(11)-en-25-oic acid, 2,3,23-trihydroxy-
     methyl ester, (2α,3β,4α)-; W/NBS   125821

502 C31 H50 O5 50392-31-9 H-64-1882-0 (5-Cholesten-1α,3β,25-triol)-1,3-diacetate; W   125822

502 C31 H50 O5 86632-21-5 J-48-3761-0 Methyl orthosphenate; W   125823

502 C32 H23 O4 P 77086-08-9 C-104-2500-0 3,4:8,9-Dibenzo-1,5,7-triphenyl-2,6,10,11-tetraoxa-1-phospha(V)
     tricyclo[5.3.1.0]undecane; W   125824

502 C32 H26 N2 O4 66362-46-7 H-61-730-0 2-(4-Oxo-4-(o-hydroxyphenyl)-3-anilino-2-methyl-1-butenyl)-3-anil
     ino- chromen-4-one; W   125825

502 C32 H38 O5 56336-21-1 SD-1981-0-0 1,2-Propanediol, 3-(3,4,7,8-tetrahydro-5-methoxy-2,2,8,8-tetrameth
     yl-2H,6H-benzo[1,2-b:5,4-b']dipyran-10-yl)-1,1-diphenyl-;; W/NBS   63070

502 C32 H54 O4 23827-56-7 SD-1981-0-0 Lanostan-11-one, 3β,19-dihydroxy-, 3-acetate;; W/NBS   63074

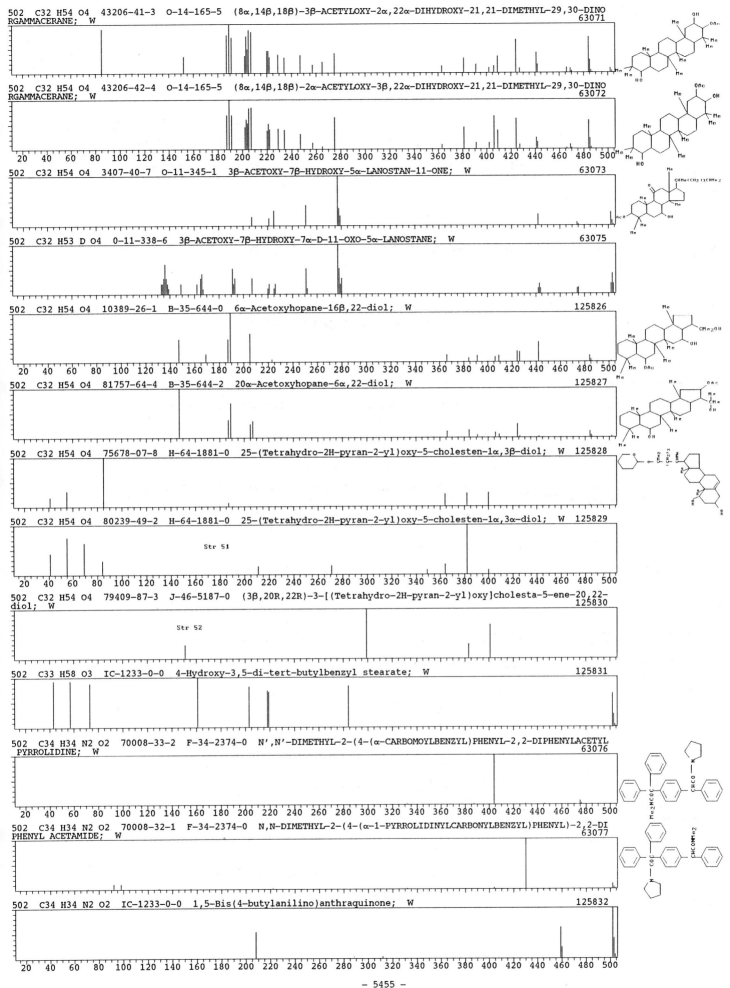

502   C32 H54 O4   43206-41-3   O-14-165-5   (8α,14β,18β)-3β-ACETYLOXY-2α,22α-DIHYDROXY-21,21-DIMETHYL-29,30-DINO
RGAMMACERANE;   W   63071

502   C32 H54 O4   43206-42-4   O-14-165-5   (8α,14β,18β)-2α-ACETYLOXY-3β,22α-DIHYDROXY-21,21-DIMETHYL-29,30-DINO
RGAMMACERANE;   W   63072

502   C32 H54 O4   3407-40-7   O-11-345-1   3β-ACETOXY-7β-HYDROXY-5α-LANOSTAN-11-ONE;   W   63073

502   C32 H53 D O4   0-11-338-6   3β-ACETOXY-7β-HYDROXY-7α-D-11-OXO-5α-LANOSTANE;   W   63075

502   C32 H54 O4   10389-26-1   B-35-644-0   6α-Acetoxyhopane-16β,22-diol;   W   125826

502   C32 H54 O4   81757-64-4   B-35-644-2   20α-Acetoxyhopane-6α,22-diol;   W   125827

502   C32 H54 O4   75678-07-8   H-64-1881-0   25-(Tetrahydro-2H-pyran-2-yl)oxy-5-cholesten-1α,3β-diol;   W   125828

502   C32 H54 O4   80239-49-2   H-64-1881-0   25-(Tetrahydro-2H-pyran-2-yl)oxy-5-cholesten-1α,3α-diol;   W   125829

Str 51

502   C32 H54 O4   79409-87-3   J-46-5187-0   (3β,20R,22R)-3-[(Tetrahydro-2H-pyran-2-yl)oxy]cholesta-5-ene-20,22-
diol;   W   125830

Str 52

502   C33 H58 O3   IC-1233-0-0   4-Hydroxy-3,5-di-tert-butylbenzyl stearate;   W   125831

502   C34 H34 N2 O2   70008-33-2   F-34-2374-0   N',N'-DIMETHYL-2-(4-(α-CARBOMOYLBENZYL)PHENYL-2,2-DIPHENYLACETYL
PYRROLIDINE;   W   63076

502   C34 H34 N2 O2   70008-32-1   F-34-2374-0   N,N-DIMETHYL-2-(4-(α-1-PYRROLIDINYLCARBONYLBENZYL)PHENYL)-2,2-DI
PHENYL ACETAMIDE;   W   63077

502   C34 H34 N2 O2   IC-1233-0-0   1,5-Bis(4-butylanilino)anthraquinone;   W   125832

502 C34 H38 N4 76915-42-9 J-46-2192-0 6,7-dipropyl-1,3,5,8-tetramethyl-2,4-divinylporphyrin; W 125833

502 C36 H38 O2 88811-90-9 K-117-340-0 2,4-Dibenzyloxy-5,7-di-tert-butylphenanthrene; W 125834

502 C36 H42 N2 HE-1986-1028-0 6,13-BIS(2,5-DIMETHYLPHENYL)-5,6,7,12,13,14-HEXAHYDRO-1,4,8,11-TETRAMETHYL-
DIBENZO[c,h]DIAZECINE; W 125835

503 C15 H8 F15 N O 77970-67-3 O-15-614-2 N-BENZYL-N-PENTADECAFLUORO-OCTANAMIDE; W 63078

503 C15 H10 Cl3 F6 N O5 74694-27-2 KO-6-555-1 TRISMONOCHLORODIFLUOROACETYL-α-METHYLDOPAMINE; W 63079

503 C18 H29 F4 N3 Si5 K-110-1243-0 1,3-BIS(DIFLUOROPHENYLSILYL)-2,2,4,4,6,6-HEXAMETHYLCYCLOTRISILAZANE; W
63080

503 C21 H30 Br N O3 Se 71041-97-9 J-44-3466-0 (2α,4aβ,6α,8aα)-(+/-)-6-Bromooctahydro-2,5,5,8a-tetramethyl
-2-[-[(2-nitrophenyl)seleno]ethyl]-2H-1-benzopyran; W 125836

503 C21 H30 Br N O3 Se 71075-19-9 J-44-3466-0 (2α,4aα,6β,8aβ)-(+/-)-6-Bromooctahydro-2,5,5,8a-tetramethyl
-2-[-[(2-nitrophenyl)seleno]ethyl]-2H-1-benzopyran; W 125837

503 C22 H18 Br N O8 49616-73-1 NS--28810-0-0 2,4-Cyclopentadiene-1,2,3,4-tetracarboxylic acid, 5-(4-bromo
-2-quinolinyl)-, tetramethyl ester;; NBS 63081

503 C26 H49 N O8 79647-12-4 I-59-1731-0 2R,3R-(+)-N-Tetradecyl-2-carboxamido-3-carboxyl-1,4,7,10,13-penta
oxacyclopentadecane; W 125838

503 C27 H25 N3 O7 68195-91-5 F-34-1160-0 5,5-BIS ETHOXY CARBONYL -1,2α-DIPHENYL-3-P-NITRO-PHENYL-4-IMIDAZ
OLIDONE; W 63082

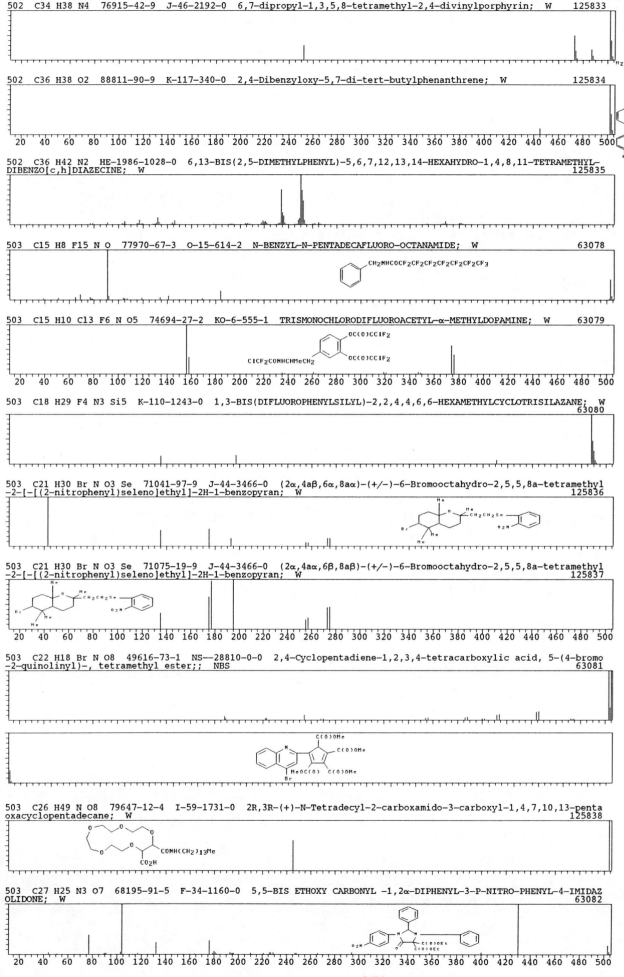

503 C27 H37 N O8 60783-14-4 B-29-1315-0 6β,11α,12α-TRIACETOXY-19,20-METHYLIMINO-14,20-CYCLO-ENT-KAUR-16-
ENE-2α,5β-DIOL; W
63083

503 C27 H37 N O8 60812-05-7 B-29-1316-0 2α,11α,12α-TRIACETOXY-19,20-METHYLIMINO-14,20-CYCLO-ENT-KAUR-16-
ENE-5β,6α-DIOL; W
125839

503 C27 H45 N3 O4 Si SH-1981-0-0 Tris(methyloxime),monotrimethylsilyl-21-Hydroxy-4-pregnene-3,19,20-trione
; W
125840

503 C27 H49 N O2 Si3 77883-14-8 NS-0-0-0 Silanamine, N-[(17β)-3,17-bis[(trimethylsilyl)oxy]estra-1,3,5(
10)-trien-2-yl]-1,1,1-trimethyl-; W/NBS
125841

503 C27 H49 N O2 Si3 77883-25-1 NS-0-0-0 Silanamine, N-[(17β)-3,17-bis[(trimethylsilyl)oxy]estra-1,3,5(
10)-trien-4-yl]-1,1,1-trimethyl-; W/NBS
125842

503 C28 H45 N O5 Si 74312-90-6 NS--28808-0-0 Pregn-4-ene-3,20-dione, 17-(acetyloxy)-6-methyl-21-[(trimeth
ylsilyl)oxy]-, 3-(O-methyloxime), (6α)-;; NBS
63085

503 C28 H45 N O5 Si 69833-70-1 NS--28809-0-0 Pregn-4-ene-3,20-dione, 17-(acetyloxy)-6-methyl-6-[(trimeth
ylsilyl)oxy]-, 3-(O-methyloxime), (6β)-;; NBS
63086

503 C29 H45 N O6 79156-49-3 B-34-620-0 (20S,22R,25R)-26-Acetylamino-2β,3β,14α-trihydroxy-5β-furost-7-en-
6-one; W
125843

503 C30 H33 N O6 CD-274-0-0 Episcopalidine; W
125844

503 C30 H37 N O4 Si 88761-51-7 J-49-4797-0 2',3'-Dihydro-6'-[[(tert-butyldiphenylsilyl)oxy]methyl]5'-oxo-
7'-propylspiro[1,3-dioxolane-2,1'(5'H)-indolizine]; W
125845

503 C30 H37 N O4 Si 88761-50-6 J-49-4797-0 2',3'-Dihydro-7'-[[(tert-butyldiphenylsilyl)oxy]methyl]5'-oxo-
6'-propylspiro[1,3-dioxolane-2,1'(5'H)-indolizine]; W
125846

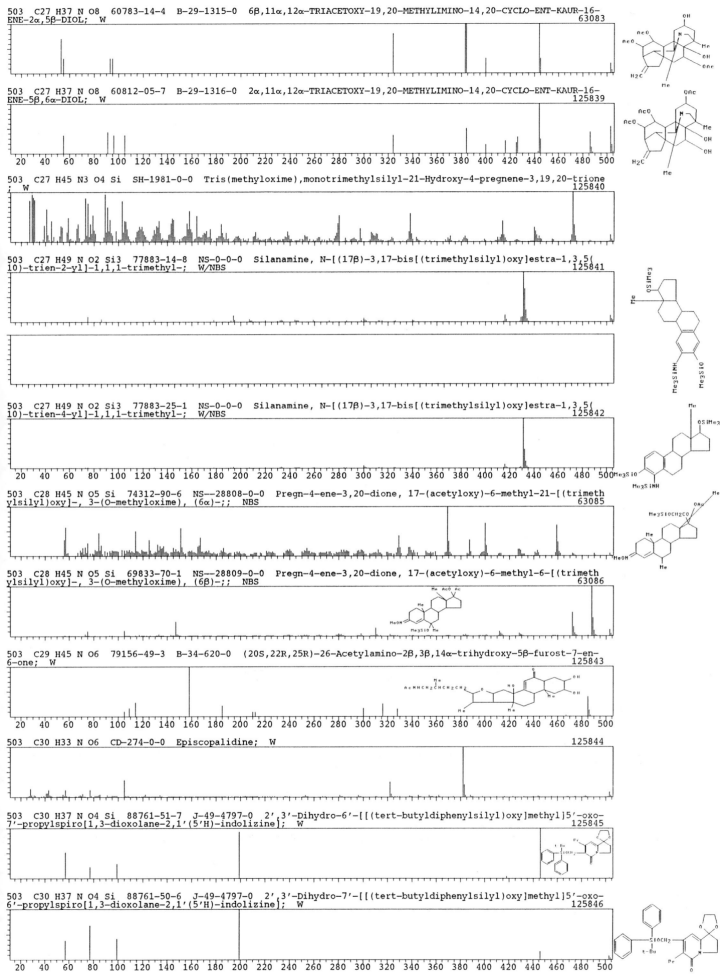

- 5457 -

503 C31 H37 N O5 73111-38-3 H-62-2013-0 (1L)-(1,2,4/3,5)-1-Acetoxymethyl-3,4,5-tri-O-benzyl-2-dimethylamino-3,4,5-cyclopentantriol; W
125847

Str 53

503 C32 H29 N3 O3 65179-71-7 K-110-3865-0 1,2,3-TRIBENZOYL-1-BUTYL-3-PHENYLGUANIDINE; W
63087

503 C33 H29 N O4 65273-11-2 KC-1977-16-1854 DIMETHYL 1'-DIPHENYLMETHYLFLUORENE-9-SPIRO-2'-PYRROLIDINE-3'4'-DICARBOXYLATE; W
63088

504 C9 H21 I O P2 Pd S2 72969-26-7 K-113-272-0 Acetyl(trimethylphosphine)[(trimethylphosphonio)dithioformiato-S,S']palladiumiodide; W
125848

504 C10 H2 Cl10 O2 4715-22-4 KO-5-236-0 Kepone hydrate; W
63089

504 C10 H2 F12 Ni O2 S2 31541-97-6 EP-7206-0-0 Nickel, bis(1,1,1,5,5,5-hexafluoro-4-thioxo-2-pentanonato-O,S)-;; BIS-(HEXAFLUORO-MONOTHIO-ACETYLACETONE)-NICKEL (II); W/NBS
63090

504 C10 H15 F6 Ir P2 34822-30-5 EP-6844-0-0 Iridium, [(1,2,3,4,5-η)-1,2,3,4,5-pentamethyl-2,4-cyclopentadien-1-yl]bis(phosphorous trifluoride)-;; PENTAMETHYLCYCLOPENTADIENYLIRIDIUM-BIS-TRIFLUOROPHOSPHINE; W/NBS
63091

504 C12 H3 Mn3 O12 18444-56-9 AD-0-1542-0 Manganese, dodecacarbonyltrihydrotri-, triangulo;; HYDRIDOMANGANESE TETRACARBONYL TRIMER; W/NBS
63092

504 C14 H44 O6 Si7 HE-1986-2387-0 1,1,3,3,5,5,7,7,9,9,11,11,13,13-TETRADECAMETHYL-HEPTASILOXANE; W 125849

504 C15 H17 F9 N4 O5 JD-8-184-3 (TFA)3-ANG ISOPROPYL ESTER; W
63093

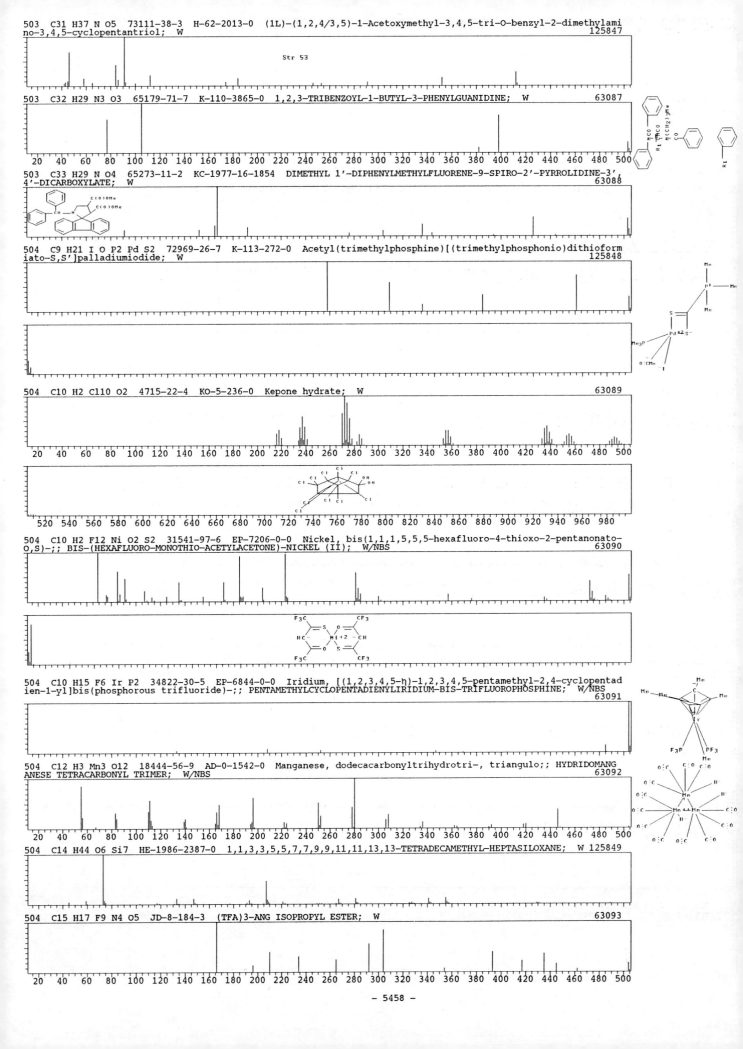

504   C15 H23 O6 Tl   54156-78-4   C-104-3985-0   2-((Diacetato)thallio)-3-acetoxycyclononene;   W   125850

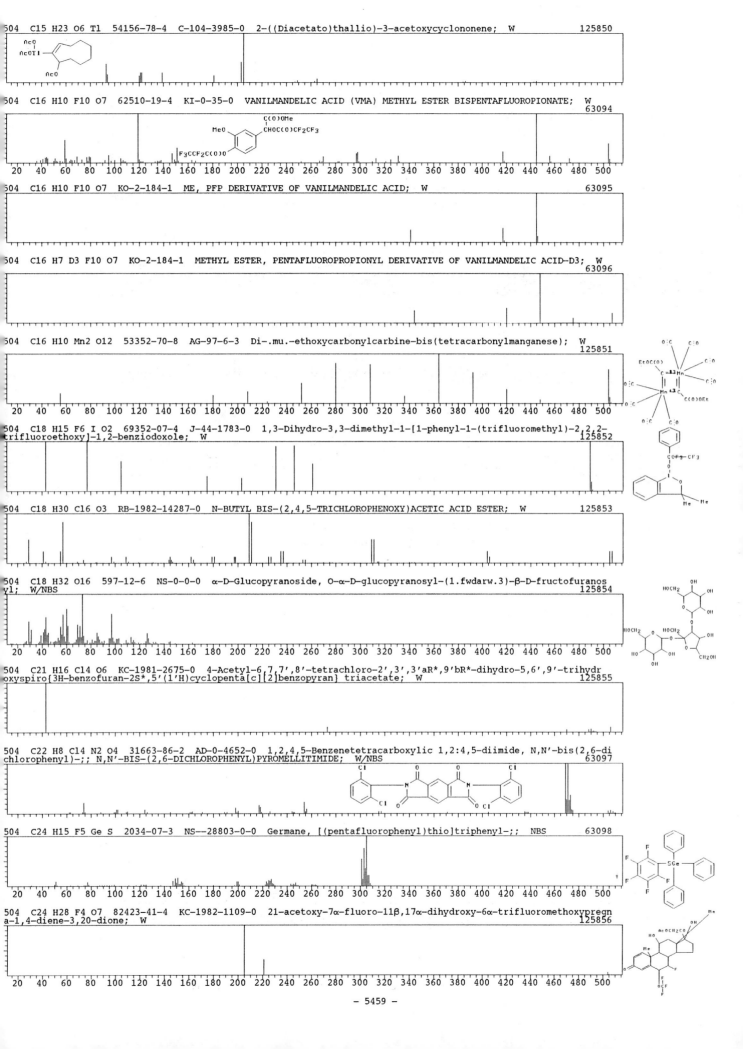

504   C16 H10 F10 O7   62510-19-4   KI-0-35-0   VANILMANDELIC ACID (VMA) METHYL ESTER BISPENTAFLUOROPIONATE;   W
63094

504   C16 H10 F10 O7   KO-2-184-1   ME, PFP DERIVATIVE OF VANILMANDELIC ACID;   W   63095

504   C16 H7 D3 F10 O7   KO-2-184-1   METHYL ESTER, PENTAFLUOROPROPIONYL DERIVATIVE OF VANILMANDELIC ACID-D3;   W
63096

504   C16 H10 Mn2 O12   53352-70-8   AG-97-6-3   Di-.mu.-ethoxycarbonylcarbine-bis(tetracarbonylmanganese);   W
125851

504   C18 H15 F6 I O2   69352-07-4   J-44-1783-0   1,3-Dihydro-3,3-dimethyl-1-[1-phenyl-1-(trifluoromethyl)-2,2,2-trifluoroethoxy]-1,2-benziodoxole;   W
125852

504   C18 H30 Cl6 O3   RB-1982-14287-0   N-BUTYL BIS-(2,4,5-TRICHLOROPHENOXY)ACETIC ACID ESTER;   W   125853

504   C18 H32 O16   597-12-6   NS-0-0-0   α-D-Glucopyranoside, O-α-D-glucopyranosyl-(1.fwdarw.3)-β-D-fructofuranosyl;   W/NBS
125854

504   C21 H16 Cl4 O6   KC-1981-2675-0   4-Acetyl-6,7,7',8'-tetrachloro-2',3',3'aR*,9'bR*-dihydro-5,6',9'-trihydroxyspiro[3H-benzofuran-2S*,5'(1'H)cyclopenta[c][2]benzopyran} triacetate;   W   125855

504   C22 H8 Cl4 N2 O4   31663-86-2   AD-0-4652-0   1,2,4,5-Benzenetetracarboxylic 1,2:4,5-diimide, N,N'-bis(2,6-dichlorophenyl)-;; N,N'-BIS-(2,6-DICHLOROPHENYL)PYROMELLITIMIDE;   W/NBS   63097

504   C24 H15 F5 Ge S   2034-07-3   NS--28803-0-0   Germane, [(pentafluorophenyl)thio]triphenyl-;;   NBS   63098

504   C24 H28 F4 O7   82423-41-4   KC-1982-1109-0   21-acetoxy-7α-fluoro-11β,17α-dihydroxy-6α-trifluoromethoxypregna-1,4-diene-3,20-dione;   W   125856

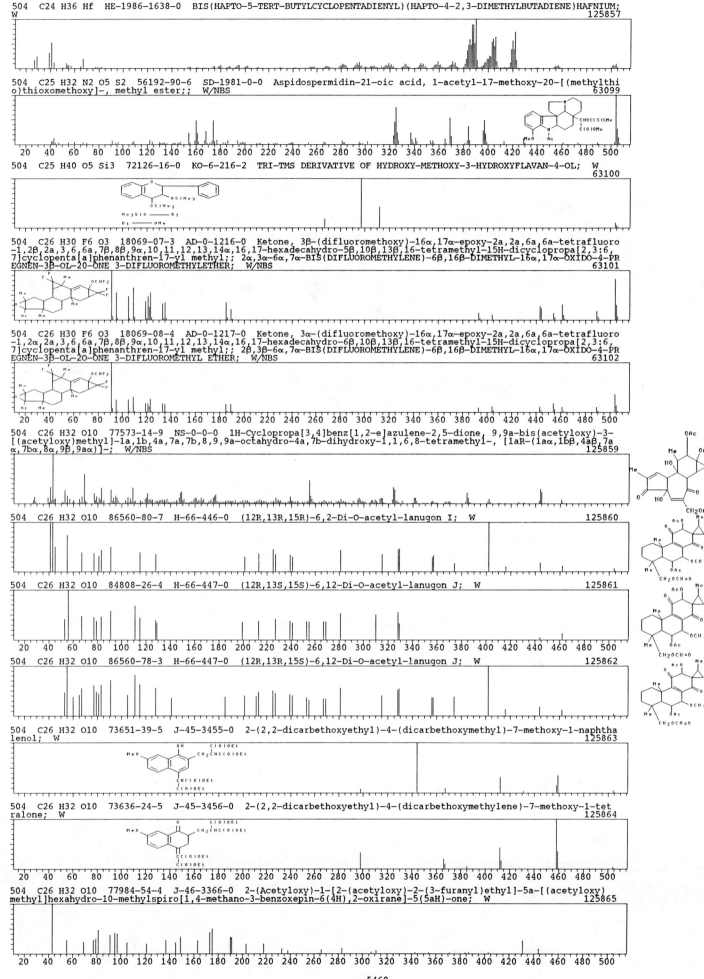

504   C24 H36 Hf   HE-1986-1638-0   BIS(HAPTO-5-TERT-BUTYLCYCLOPENTADIENYL)(HAPTO-4-2,3-DIMETHYLBUTADIENE)HAFNIUM;
W   125857

504   C25 H32 N2 O5 S2   56192-90-6   SD-1981-0-0   Aspidospermidin-21-oic acid, 1-acetyl-17-methoxy-20-[(methylthi
o)thioxomethoxy]-, methyl ester;;   W/NBS   63099

504   C25 H40 O5 Si3   72126-16-0   KO-6-216-2   TRI-TMS DERIVATIVE OF HYDROXY-METHOXY-3-HYDROXYFLAVAN-4-OL;   W
                                                                                             63100

504   C26 H30 F6 O3   18069-07-3   AD-0-1216-0   Ketone, 3β-(difluoromethoxy)-16α,17α-epoxy-2a,2a,6a,6a-tetrafluoro
-1,2β,2a,3,6,6a,7β,8β,9α,10,11,12,13,14α,16,17-hexadecahydro-5β,10β,13β,16-tetramethyl-15H-dicyclopropa[2,3:6,
7]cyclopenta[a]phenanthren-17-yl methyl;;   2α,3α-6α,7α-BIS(DIFLUOROMETHYLENE)-6β,16β-DIMETHYL-16α,17α-OXIDO-4-PR
EGNEN-3β-OL-20-ONE 3-DIFLUOROMETHYLETHER;   W/NBS   63101

504   C26 H30 F6 O3   18069-08-4   AD-0-1217-0   Ketone, 3α-(difluoromethoxy)-16α,17α-epoxy-2a,2a,6a,6a-tetrafluoro
-1,2α,2a,3,6,6a,7β,8β,9α,10,11,12,13,14α,16,17-hexadecahydro-6β,10β,13β,16-tetramethyl-15H-dicyclopropa[2,3:6,
7]cyclopenta[a]phenanthren-17-yl methyl;;   2β,3β-6α,7α-BIS(DIFLUOROMETHYLENE)-6β,16β-DIMETHYL-16α,17α-OXIDO-4-PR
EGNEN-3β-OL-20-ONE 3-DIFLUOROMETHYL ETHER;   W/NBS   63102

504   C26 H32 O10   77573-14-9   NS-0-0-0   1H-Cyclopropa[3,4]benz[1,2-e]azulene-2,5-dione, 9,9a-bis(acetyloxy)-3-
[(acetyloxy)methyl]-1a,1b,4a,7a,7b,8,9,9a-octahydro-4a,7b-dihydroxy-1,1,6,8-tetramethyl-, [1aR-(1aα,1bβ,4aβ,7a
α,7bα,8α,9β,9aα)]-;   W/NBS   125859

504   C26 H32 O10   86560-80-7   H-66-446-0   (12R,13R,15R)-6,2-Di-O-acetyl-lanugon I;   W   125860

504   C26 H32 O10   84808-26-4   H-66-447-0   (12R,13S,15S)-6,12-Di-O-acetyl-lanugon J;   W   125861

504   C26 H32 O10   86560-78-3   H-66-447-0   (12R,13R,15S)-6,12-Di-O-acetyl-lanugon J;   W   125862

504   C26 H32 O10   73651-39-5   J-45-3455-0   2-(2,2-dicarbethoxyethyl)-4-(dicarbethoxymethyl)-7-methoxy-1-naphtha
lenol;   W   125863

504   C26 H32 O10   73636-24-5   J-45-3456-0   2-(2,2-dicarbethoxyethyl)-4-(dicarbethoxymethylene)-7-methoxy-1-tet
ralone;   W   125864

504   C26 H32 O10   77984-54-4   J-46-3366-0   2-(Acetyloxy)-1-[2-(acetyloxy)-2-(3-furanyl)ethyl]-5a-[(acetyloxy)
methyl]hexahydro-10-methylspiro[1,4-methano-3-benzoxepin-6(4H),2-oxirane]-5(5aH)-one;   W   125865

- 5460 -

504   C26 H32 O10   85883-48-3   J-48-2313-0   9,10-(Isopropylidenedioxy)-5-methoxy-3,4-dihydro-1H-naphtho[2,3-c]py ran-1(S)-spiro-2'-[4'(S)-hydroxy-3'(S)-(methoxymethoxy)-6'(S)-methyltetrahydropyran]-3(R)-yl acetic acid; W
125866

504   C27 H44 O5 Si2   69833-66-5   JD-7-413-3   ALDOSTERONE DI-TMS; W/NBS
63103

504   C27 H48 N6 O3   67969-51-1   NS--28798-0-0   Triimidazo[1,5-a:1',5'-c:1'',5''-e][1,3,5]triazine-1,5,9(2H,6H, 10H)-trione, 2,6,10-tris(1,1-dimethylethyl)hexahydro-3,3,7,7,11,11-hexamethyl-;; NBS
63104

504   C27 H48 O3 Si3   33287-47-7   EP-8484-0-0   Silane, (estra-1,3,5(10)-trien-3,6α,17β-triyltrioxy)tris[trimeth yl-;; 1,3,5(10)-ESTRATRIENE-3,6α,17β-TRIOL TMS; W/NBS
63107

504   C27 H48 O3 Si3   69688-19-3   NS--28790-0-0   Silane, [[(6β,17β)-estra-1,3,5(10)-triene-3,6,17-triyl]tris(ox y)]tris[trimethyl-;; NBS
63112

504   C27 H48 O3 Si3   69688-37-5   NS--28787-0-0   Silane, [[(7α,17β)-estra-1,3,5(10)-triene-3,7,17-triyl]tris(ox y)]tris[trimethyl-;; NBS
63115

504   C27 H48 O3 Si3   57305-21-2   EP-8231-0-0   Silane, [[(16β,17β)-estra-1,3,5(10)-triene-3,16,17-triyl]tris(ox y)]tris[trimethyl-;; 1,3,5(10)-ESTRATRIENE-3,16β,17β-TRIOL TMS; W/NBS
63106

504   C27 H48 O3 Si3   57305-23-4   EP-8233-0-0   Silane, [[(16α,17α)-estra-1,3,5(10)-triene-3,16,17-triyl]tris(ox y)]tris[trimethyl-;; 1,3,5(10)-ESTRATRIENE-3,16α,17α-TRIOL TMS; W/NBS
63109

504   C27 H48 O3 Si3   57305-22-3   EP-8232-0-0   Silane, [[(16β,17α)-estra-1,3,5(10)-triene-3,16,17-triyl]tris(ox y)]tris[trimethyl-;; 3,16β,17α-Trihydroxy-1,3,5(10)-estratriene 3,16,17-tms; W/NBS
63110

504   C27 H48 O3 Si3   69833-61-0   NS--28786-0-0   Silane, [[(15α,17β)-estra-1,3,5(10)-triene-3,15,17-triyl]tris(oxy)]tris[trimethyl-;; NBS
63116

504   C27 H48 O3 Si3   69833-74-5   NS--28785-0-0   Silane, [[(15β,17β)-estra-1,3,5(10)-triene-3,15,17-triyl]tris(oxy)]tris[trimethyl-;; NBS
63117

504   C27 H48 O3 Si3   18888-17-0   SH-1981-0-0   Silane, [[(16α,17β)-estra-1,3,5(10)-triene-3,16,17-triyl]tris(ox y)]tris[trimethyl-;; ESTRIOL TRIMETHYLSILYL ETHER; W
125867

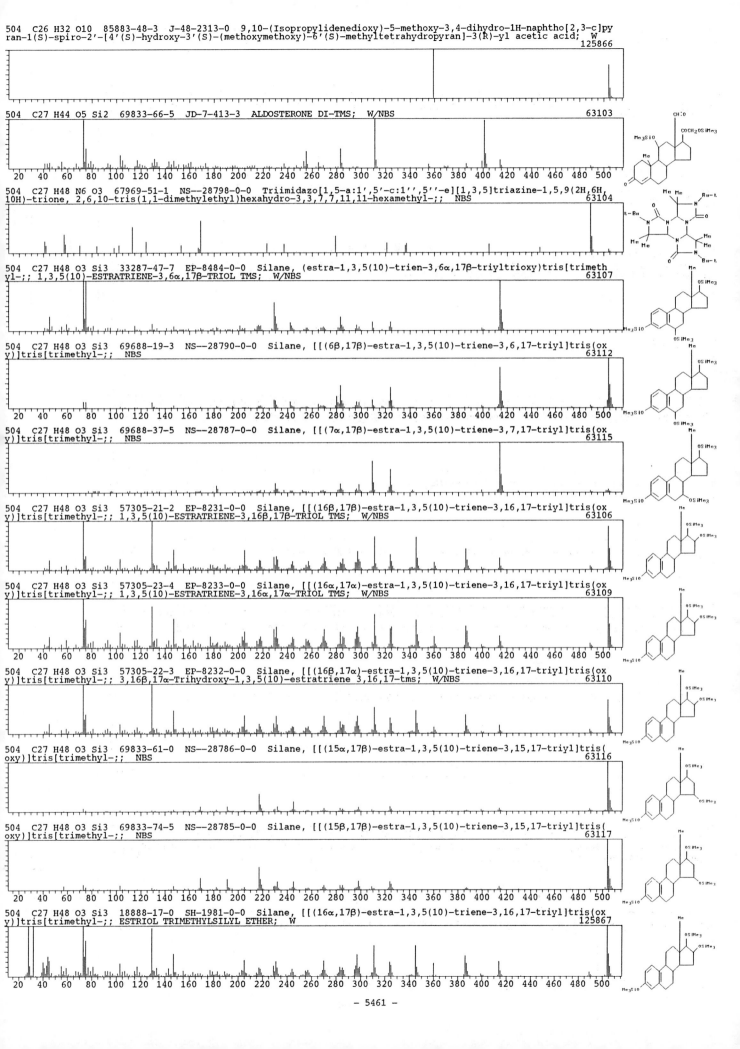

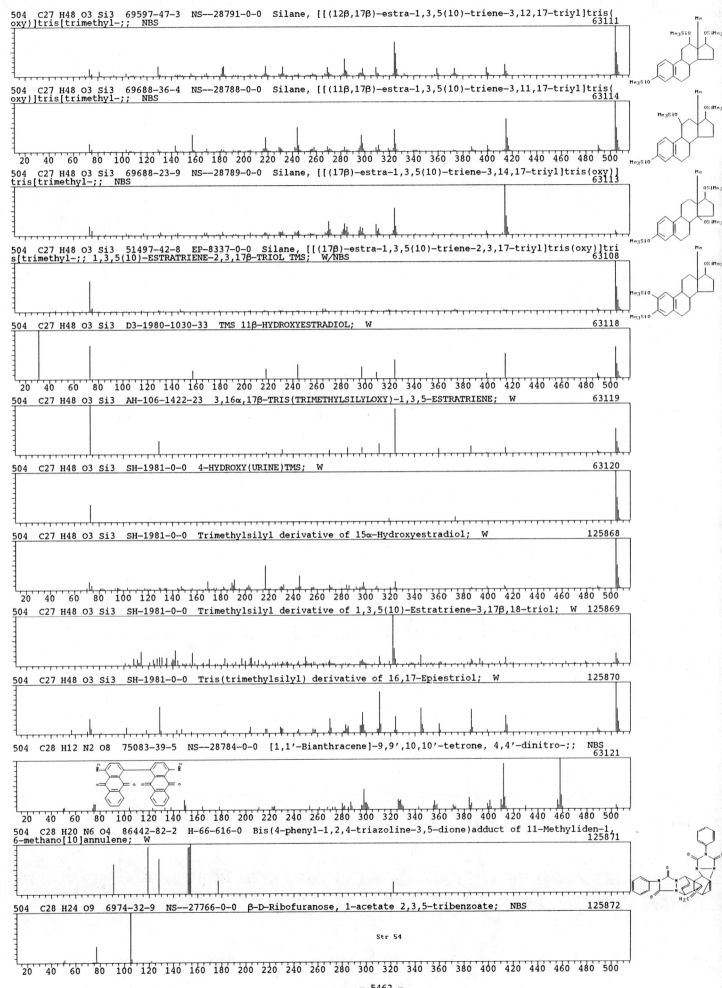

504   C27 H48 O3 Si3   69597-47-3   NS--28791-0-0   Silane, [[(12β,17β)-estra-1,3,5(10)-triene-3,12,17-triyl]tris(oxy)]tris[trimethyl-;;   NBS
63111

504   C27 H48 O3 Si3   69688-36-4   NS--28788-0-0   Silane, [[(11β,17β)-estra-1,3,5(10)-triene-3,11,17-triyl]tris(oxy)]tris[trimethyl-;;   NBS
63114

504   C27 H48 O3 Si3   69688-23-9   NS--28789-0-0   Silane, [[(17β)-estra-1,3,5(10)-triene-3,14,17-triyl]tris(oxy)]tris[trimethyl-;;   NBS
63113

504   C27 H48 O3 Si3   51497-42-8   EP-8337-0-0   Silane, [[(17β)-estra-1,3,5(10)-triene-2,3,17-triyl]tris(oxy)]tris[trimethyl-;; 1,3,5(10)-ESTRATRIENE-2,3,17β-TRIOL TMS;   W/NBS
63108

504   C27 H48 O3 Si3   D3-1980-1030-33   TMS 11β-HYDROXYESTRADIOL;   W
63118

504   C27 H48 O3 Si3   AH-106-1422-23   3,16α,17β-TRIS(TRIMETHYLSILYLOXY)-1,3,5-ESTRATRIENE;   W
63119

504   C27 H48 O3 Si3   SH-1981-0-0   4-HYDROXY(URINE)TMS;   W
63120

504   C27 H48 O3 Si3   SH-1981-0-0   Trimethylsilyl derivative of 15α-Hydroxyestradiol;   W
125868

504   C27 H48 O3 Si3   SH-1981-0-0   Trimethylsilyl derivative of 1,3,5(10)-Estratriene-3,17β,18-triol;   W   125869

504   C27 H48 O3 Si3   SH-1981-0-0   Tris(trimethylsilyl) derivative of 16,17-Epiestriol;   W
125870

504   C28 H12 N2 O8   75083-39-5   NS--28784-0-0   [1,1'-Bianthracene]-9,9',10,10'-tetrone, 4,4'-dinitro-;;   NBS
63121

504   C28 H20 N6 O4   86442-82-2   H-66-616-0   Bis(4-phenyl-1,2,4-triazoline-3,5-dione)adduct of 11-Methyliden-1,6-methano[10]annulene;   W
125871

504   C28 H24 O9   6974-32-9   NS--27766-0-0   β-D-Ribofuranose, 1-acetate 2,3,5-tribenzoate;   NBS
125872

Str 54

504 C28 H24 O9 62350-97-4 K-110-1056-0 FASCICULIN-A-TETRAMETHYLETHER; W   63122

504 C28 H24 O9 74695-00-4 B-33-2539-0 Dimethyl 4,5,9,10-tetramethoxydinaphtho[2,1-b:1',2'-d]furan-2,12-dica
rboxylate; W   125873

504 C28 H40 O4 S2 K-116-3845-0 9,12,22,25-Tetramethoxy-2,19-dithia[7.7]paracyclophane; W   125874

504 C28 H40 O8 KC-1977-214-6 12,14,27,29-DIBENZO-1,4,7,10,16,19,22,25-OCTAOXATRIACONTA-12,27-DIENE; W 63123

504 C28 H40 O8 KC-1977-214-6 12,13,26,27-DIBENZO-1,4,7,10,15,18,21,24-OCTAOXAOCTACOSA-12,26-DIENE; W 63124

504 C28 H40 O8 78708-78-8 AN-90-495-0 2β,3α,9.xi.,13.xi.-Tetraacetoxy-1(15),8(19)-trinervitadiene; W
                                                                                                   125875

504 C28 H40 O8 85573-21-3 B-36-129-0 3-[3',3',6',6'-Di(ethylenedioxy)-3a'-methylperhydroinden-1'-yl]-1,1,5,
5-di(ethylenedioxy)-7a-methyl-2,4,5,6,7,7a-hexahydro-1H-indene; W   125876

504 C29 H28 O6 S 70145-43-6 J-44-2580-0 4'-Phenylthio-rot-2'-enonic acid; W   125877

504 C29 H29 Cl N2 O4 62786-73-6 KC-1976-2503-0 DIBENZYL 3-(2-CHLOROETHYL)-3',4-DIMETHYLPYRROMETHANE-5,5'-DI
CARBOXYLATE; W   63125

504 C29 H29 O6 P 84251-41-2 KC-1982-2311-0 Methyl Bis-(4',6-dimethoxybiphenyl-2-yl)phosphinate; W 125878

Str 55

504 C29 H36 N4 O4 52617-26-2 CD-252-0-0 Amphibine F; W   125879

504 C29 H44 O7 4947-65-3 NS--28782-0-0 Cholan-24-oic acid, 7,12-bis(acetyloxy)-3-oxo-, methyl ester, (5β,
7α,12α)-;; 5β-Cholan-24-oic acid, 7α,12α-dihydroxy-3-oxo-, methyl ester, diacetate; NBS   63128

504 C29 H44 O7 21066-20-6 NS--28781-0-0 Cholan-24-oic acid, 3,12-bis(acetyloxy)-7-oxo-, methyl ester, (3α,
5β,12α)-;; NBS   63129

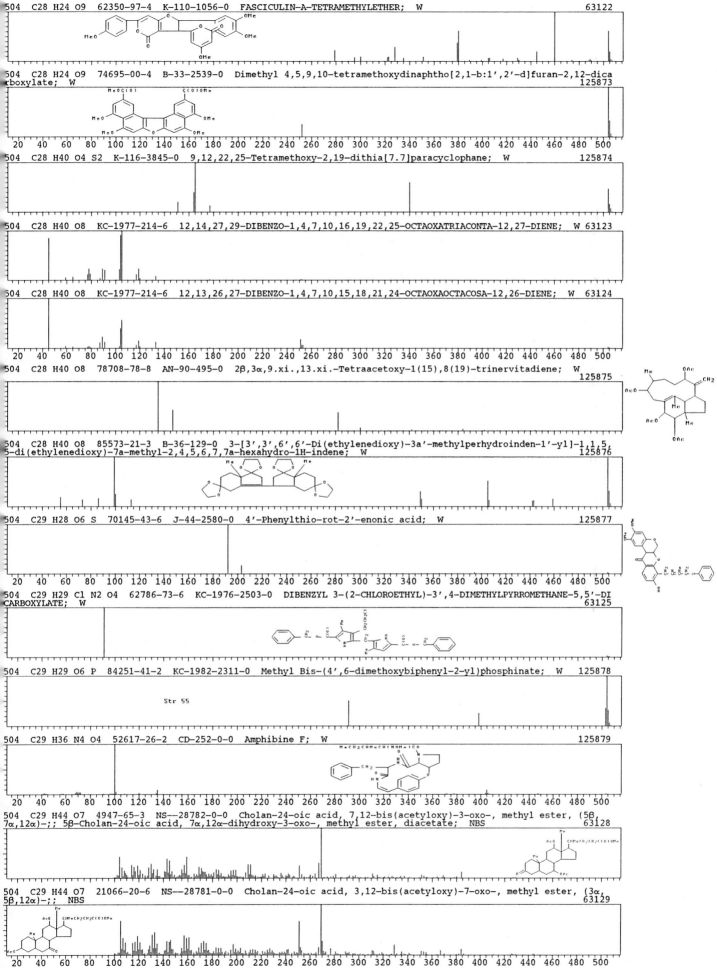

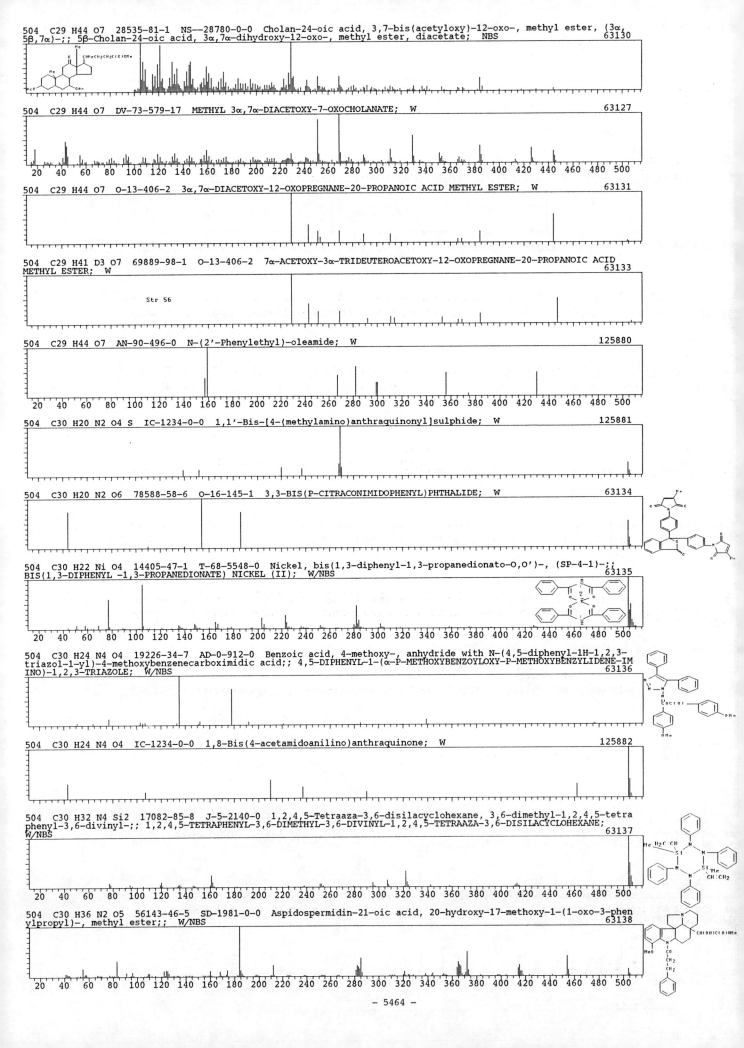

504   C29 H44 O7   28535-81-1   NS--28780-0-0   Cholan-24-oic acid, 3,7-bis(acetyloxy)-12-oxo-, methyl ester, (3α, 5β,7α)-;; 5β-Cholan-24-oic acid, 3α,7α-dihydroxy-12-oxo-, methyl ester, diacetate; NBS   63130

504   C29 H44 O7   DV-73-579-17   METHYL 3α,7α-DIACETOXY-7-OXOCHOLANATE;   W   63127

504   C29 H44 O7   O-13-406-2   3α,7α-DIACETOXY-12-OXOPREGNANE-20-PROPANOIC ACID METHYL ESTER;   W   63131

504   C29 H41 D3 O7   69889-98-1   O-13-406-2   7α-ACETOXY-3α-TRIDEUTEROACETOXY-12-OXOPREGNANE-20-PROPANOIC ACID
METHYL ESTER;   W   63133

Str 56

504   C29 H44 O7   AN-90-496-0   N-(2'-Phenylethyl)-oleamide;   W   125880

504   C30 H20 N2 O4 S   IC-1234-0-0   1,1'-Bis-[4-(methylamino)anthraquinonyl]sulphide;   W   125881

504   C30 H20 N2 O6   78588-58-6   O-16-145-1   3,3-BIS(P-CITRACONIMIDOPHENYL)PHTHALIDE;   W   63134

504   C30 H22 Ni O4   14405-47-1   T-68-5548-0   Nickel, bis(1,3-diphenyl-1,3-propanedionato-O,O')-, (SP-4-1)-;;
BIS(1,3-DIPHENYL -1,3-PROPANEDIONATE) NICKEL (II);   W/NBS   63135

504   C30 H24 N4 O4   19226-34-7   AD-0-912-0   Benzoic acid, 4-methoxy-, anhydride with N-(4,5-diphenyl-1H-1,2,3-
triazol-1-yl)-4-methoxybenzenecarboximidic acid;;   4,5-DIPHENYL-1-(α-P-METHOXYBENZOYLOXY-P-METHOXYBENZYLIDENE-IM
INO)-1,2,3-TRIAZOLE;   W/NBS   63136

504   C30 H24 N4 O4   IC-1234-0-0   1,8-Bis(4-acetamidoanilino)anthraquinone;   W   125882

504   C30 H32 N4 Si2   17082-85-8   J-5-2140-0   1,2,4,5-Tetraaza-3,6-disilacyclohexane, 3,6-dimethyl-1,2,4,5-tetra
phenyl-3,6-divinyl-;;   1,2,4,5-TETRAPHENYL-3,6-DIMETHYL-3,6-DIVINYL-1,2,4,5-TETRAAZA-3,6-DISILACYCLOHEXANE;
W/NBS   63137

504   C30 H36 N2 O5   56143-46-5   SD-1981-0-0   Aspidospermidin-21-oic acid, 20-hydroxy-17-methoxy-1-(1-oxo-3-phen
ylpropyl)-, methyl ester;;   W/NBS   63138

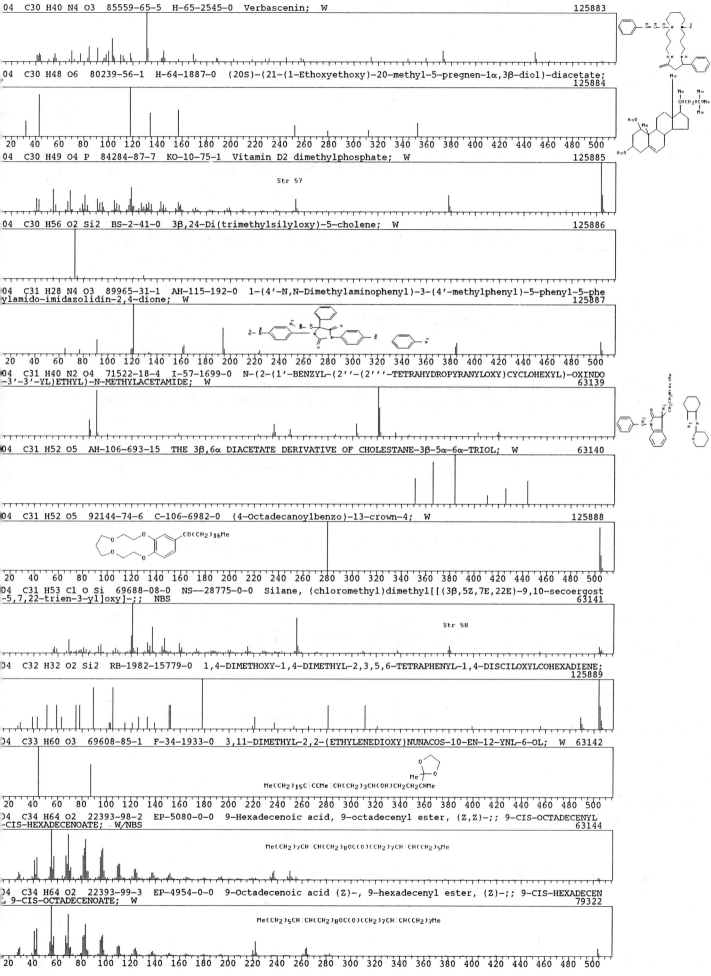

04 C30 H40 N4 O3 85559-65-5 H-65-2545-0 Verbascenin; W   125883

04 C30 H48 O6 80239-56-1 H-64-1887-0 (20S)-(21-(1-Ethoxyethoxy)-20-methyl-5-pregnen-1α,3β-diol)-diacetate;
                                                                                     125884

04 C30 H49 O4 P 84284-87-7 KO-10-75-1 Vitamin D2 dimethylphosphate; W   125885

Str 57

04 C30 H56 O2 Si2 BS-2-41-0 3β,24-Di(trimethylsilyloxy)-5-cholene; W   125886

04 C31 H28 N4 O3 89965-31-1 AH-115-192-0 1-(4'-N,N-Dimethylaminophenyl)-3-(4'-methylphenyl)-5-phenyl-5-phe
nylamido-imidazolidin-2,4-dione; W   125887

04 C31 H40 N2 O4 71522-18-4 I-57-1699-0 N-(2-(1'-BENZYL-(2''-(2'''-TETRAHYDROPYRANYLOXY)CYCLOHEXYL)-OXINDO
-3'-3'-YL)ETHYL)-N-METHYLACETAMIDE; W   63139

04 C31 H52 O5 AH-106-693-15 THE 3β,6α DIACETATE DERIVATIVE OF CHOLESTANE-3β-5α-6α-TRIOL; W   63140

04 C31 H52 O5 92144-74-6 C-106-6982-0 (4-Octadecanoylbenzo)-13-crown-4; W   125888

04 C31 H53 Cl O Si 69688-08-0 NS--28775-0-0 Silane, (chloromethyl)dimethyl[[(3β,5Z,7E,22E)-9,10-secoergost
-5,7,22-trien-3-yl]oxy]-;; NBS   63141

Str 58

04 C32 H32 O2 Si2 RB-1982-15779-0 1,4-DIMETHOXY-1,4-DIMETHYL-2,3,5,6-TETRAPHENYL-1,4-DISCILOXYLCOHEXADIENE;
                                                                                     125889

04 C33 H60 O3 69608-85-1 F-34-1933-0 3,11-DIMETHYL-2,2-(ETHYLENEDIOXY)NUNACOS-10-EN-12-YNL-6-OL; W 63142

04 C34 H64 O2 22393-98-2 EP-5080-0-0 9-Hexadecenoic acid, 9-octadecenyl ester, (Z,Z)-;; 9-CIS-OCTADECENYL
-CIS-HEXADECENOATE; W/NBS   63144

04 C34 H64 O2 22393-99-3 EP-4954-0-0 9-Octadecenoic acid (Z)-, 9-hexadecenyl ester, (Z)-;; 9-CIS-HEXADECEN
 9-CIS-OCTADECENOATE; W   79322

504   C36 H24 O3   92012-80-1   J-49-4171-0   (4α,9α,9aα)-9,9a-Dihydro-3,4,9,9a-tetraphenyl-4,9-epoxynaphtho[2,3-c]furan-1(4H)-one;   W   125890

504   C36 H24 O3   92012-83-4   J-49-4171-0   [3-(Benxoyloxy)-1,4-diphenyl-2-naphthalenyl]phenylmethanone;   W   125891

504   C37 H44 O   63418-46-2   K-111-269-0   METHYL-(2,4,6-TRIS(4-TERT-BUTYLPHENYL)PHENYL)-ETHER;   W   63145

505   C18 H16 Br2 Co N4   65230-35-5   B-30-2507-0   2,11-DIBROMO-5,6,7,8,15,16-HEXAHYDRODIBENZO[3,M][1,4,8,11]TETRAAZACYCLOTETRADECINATOCOBALT (II);   W   63146

505   C19 H44 N3 O P S Si4   63698-94-2   K-110-2381-0   (N-(PHENYLTHIOMETHYL)TRIMETHYLSILYLAMINO)(TRIMETHYLSILYLAMINO)(TRIMETHYLSILYLAMINO)(TRIMETHYLSILOXY)PHOSPHORANE;   W   63147

505   C20 H40 B N O9 Si2   EP-1567-0-0   B-N-ACETYLNEURAMINIC ME ESTER-2-ME-8,9-ME-BORONATE-3,7-DITMS;   W   63148

505   C20 H40 B N O9 Si2   EP-1568-0-0   B-N-ACETYLNEURAMINIC ME ESTER-2-ME-7,9-ME-BORONATE-3,8-DITMS;   W   63149

505   C20 H44 N O4 P Si4   53044-43-2   O-9-120-16   Phosphonic acid, [1-[(trimethylsilyl)amino]-2-[4-[(trimethylsilyl)oxy]phenyl]ethyl]-, bis(trimethylsilyl) ester;; 1-AMINO-2-(4-HYDROXYPHENYL)ETHYLPHOSPHONIC ACID-TETRATMS;   W   79323

505   C21 H31 N O13   56323-64-9   AU-27-6-1   D-glycero-D-galacto-2-Nonulopyranosidonic acid, methyl 5-(acetylamino)-3,5-dideoxy-, methyl ester, 4,7,8,9-tetraacetate;; PERACETYLATED METHYL ESTER METHYL GLYCOSIDE OF N-ACETYLNEURAMINIC ACID;   W/NBS   63151

505   C21 H25 D6 N O13   56323-63-8   AU-27-6-2   PERACETYLATED TRIDEUTERIOMETHYL ESTER TRIDEUTERIOMETHYL GLYCOSIDE OF N-ACETYLNEURAMINIC ACID;   W   63152

505   C22 H28 F5 N O3 Si2   55517-87-8   AD-0-5179-0   Benzeneethanamine, 3-methoxy-N-[(pentafluorophenyl)methylene]-β,4-bis[(trimethylsilyl)oxy]-;; 2-(3'-METHOXY-4'-TRIMETHYLSILOXYPHENYL)-2-TRIMETHYLSILOXY-N-PENTAFLUOROPHENYLMETHYLENE-ETHYLAMINE;   W/NBS   63153

505   C24 H30 Cl N O2 Pd   78067-80-8   B-33-2763-0   Chloro[α-4-6-η-(3,17-dioxoandrostenyl)](pyridine)palladium(II);   W   125892

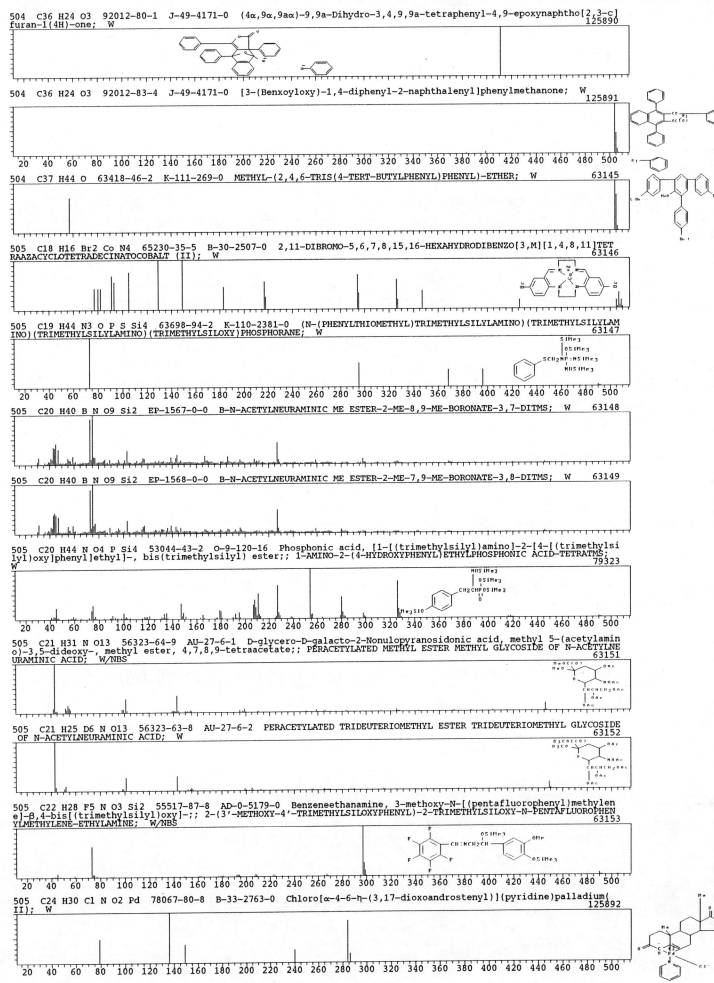

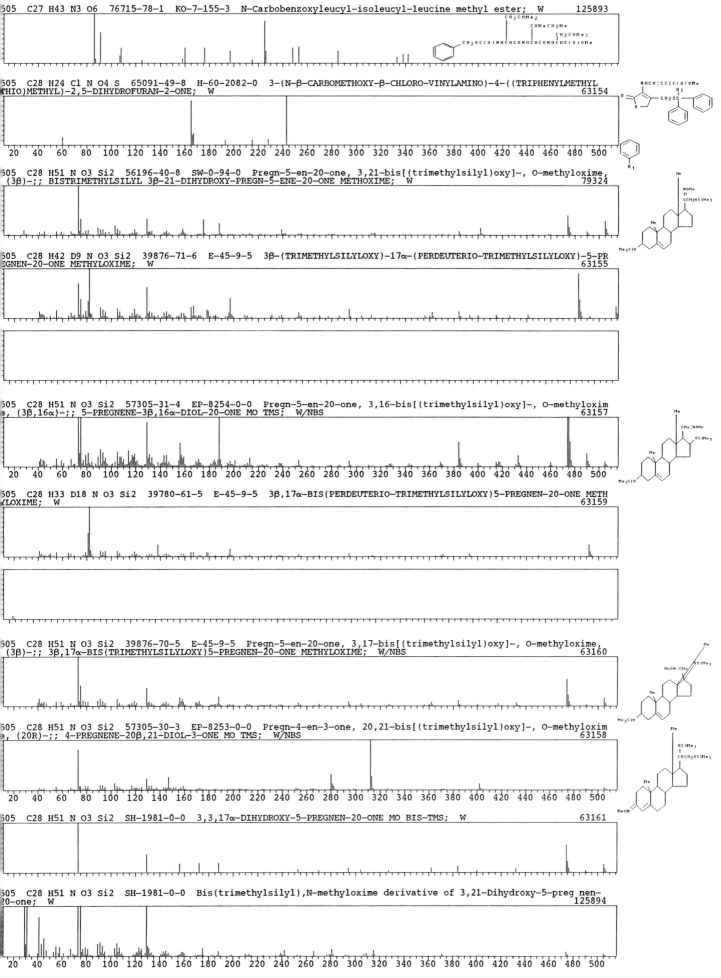

505   C27 H43 N3 O6   76715-78-1   KO-7-155-3   N-Carbobenzoxyleucyl-isoleucyl-leucine methyl ester;   W   125893

505   C28 H24 Cl N O4 S   65091-49-8   H-60-2082-0   3-(N-β-CARBOMETHOXY-β-CHLORO-VINYLAMINO)-4-((TRIPHENYLMETHYL THIO)METHYL)-2,5-DIHYDROFURAN-2-ONE;   W   63154

505   C28 H51 N O3 Si2   56196-40-8   SW-0-94-0   Pregn-5-en-20-one, 3,21-bis[(trimethylsilyl)oxy]-, O-methyloxime, (3β)-;;   BISTRIMETHYLSILYL 3β-21-DIHYDROXY-PREGN-5-ENE-20-ONE METHOXIME;   W   79324

505   C28 H42 D9 N O3 Si2   39876-71-6   E-45-9-5   3β-(TRIMETHYLSILYLOXY)-17α-(PERDEUTERIO-TRIMETHYLSILYLOXY)-5-PREGNEN-20-ONE METHYLOXIME;   W   63155

505   C28 H51 N O3 Si2   57305-31-4   EP-8254-0-0   Pregn-5-en-20-one, 3,16-bis[(trimethylsilyl)oxy]-, O-methyloxime, (3β,16α)-;;   5-PREGNENE-3β,16α-DIOL-20-ONE MO TMS;   W/NBS   63157

505   C28 H33 D18 N O3 Si2   39780-61-5   E-45-9-5   3β,17α-BIS(PERDEUTERIO-TRIMETHYLSILYLOXY)5-PREGNEN-20-ONE METHYLOXIME;   W   63159

505   C28 H51 N O3 Si2   39876-70-5   E-45-9-5   Pregn-5-en-20-one, 3,17-bis[(trimethylsilyl)oxy]-, O-methyloxime, (3β)-;;   3β,17α-BIS(TRIMETHYLSILYLOXY)5-PREGNEN-20-ONE METHYLOXIME;   W/NBS   63160

505   C28 H51 N O3 Si2   57305-30-3   EP-8253-0-0   Pregn-4-en-3-one, 20,21-bis[(trimethylsilyl)oxy]-, O-methyloxime, (20R)-;;   4-PREGNENE-20β,21-DIOL-3-ONE MO TMS;   W/NBS   63158

505   C28 H51 N O3 Si2   SH-1981-0-0   3,3,17α-DIHYDROXY-5-PREGNEN-20-ONE MO BIS-TMS;   W   63161

505   C28 H51 N O3 Si2   SH-1981-0-0   Bis(trimethylsilyl),N-methyloxime derivative of 3,21-Dihydroxy-5-pregnen-20-one;   W   125894

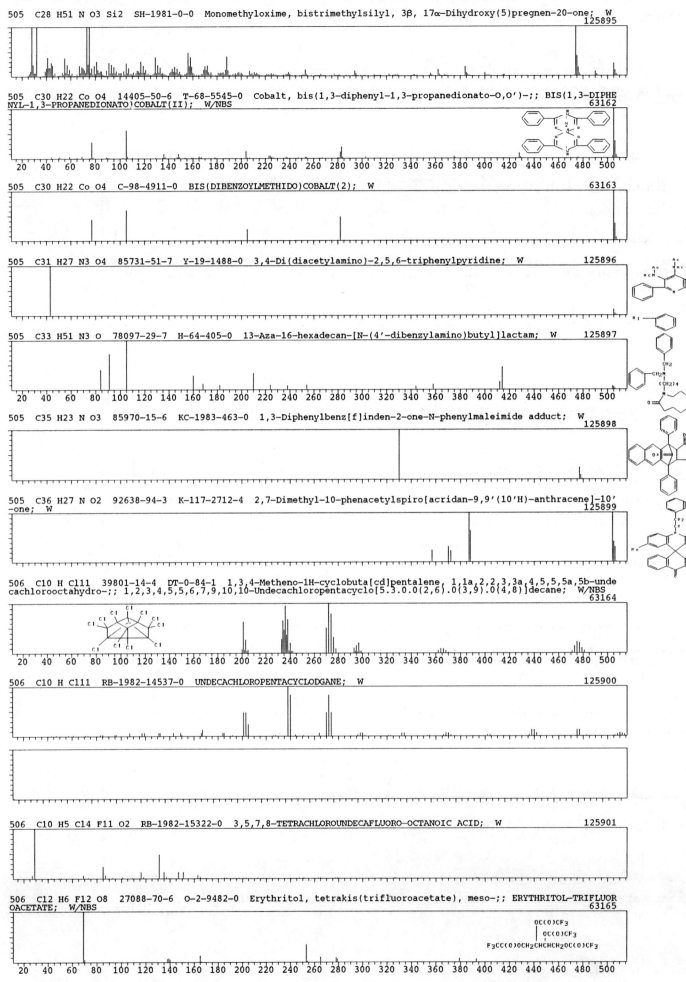

505 C28 H51 N O3 Si2 SH-1981-0-0 Monomethyloxime, bistrimethylsilyl, 3β, 17α-Dihydroxy(5)pregnen-20-one; W
125895

505 C30 H22 Co O4 14405-50-6 T-68-5545-0 Cobalt, bis(1,3-diphenyl-1,3-propanedionato-O,O')-;; BIS(1,3-DIPHE
NYL-1,3-PROPANEDIONATO)COBALT(II); W/NBS
63162

505 C30 H22 Co O4 C-98-4911-0 BIS(DIBENZOYLMETHIDO)COBALT(2); W
63163

505 C31 H27 N3 O4 85731-51-7 Y-19-1488-0 3,4-Di(diacetylamino)-2,5,6-triphenylpyridine; W
125896

505 C33 H51 N3 O 78097-29-7 H-64-405-0 13-Aza-16-hexadecan-[N-(4'-dibenzylamino)butyl]lactam; W 125897

505 C35 H23 N O3 85970-15-6 KC-1983-463-0 1,3-Diphenylbenz[f]inden-2-one-N-phenylmaleimide adduct; W
125898

505 C36 H27 N O2 92638-94-3 K-117-2712-4 2,7-Dimethyl-10-phenacetylspiro[acridan-9,9'(10'H)-anthracene]-10'
-one; W
125899

506 C10 H Cl11 39801-14-4 DT-0-84-1 1,3,4-Metheno-1H-cyclobuta[cd]pentalene, 1,1a,2,2,3,3a,4,5,5,5a,5b-unde
cachlorooctahydro-;; 1,2,3,4,5,5,6,7,9,10,10-Undecachloropentacyclo[5.3.0.0(2,6).0(3,9).0(4,8)]decane; W/NBS
63164

506 C10 H Cl11 RB-1982-14537-0 UNDECACHLOROPENTACYCLODGANE; W
125900

506 C10 H5 Cl4 F11 O2 RB-1982-15322-0 3,5,7,8-TETRACHLOROUNDECAFLUORO-OCTANOIC ACID; W
125901

506 C12 H6 F12 O8 27088-70-6 O-2-9482-0 Erythritol, tetrakis(trifluoroacetate), meso-;; ERYTHRITOL-TRIFLUOR
OACETATE; W/NBS
63165

506 C13 H3 C19 O2 50544-04-2 KO-5-568-2 2-METHOXY NONACHLORODIPHENYL ETHER; W   63166

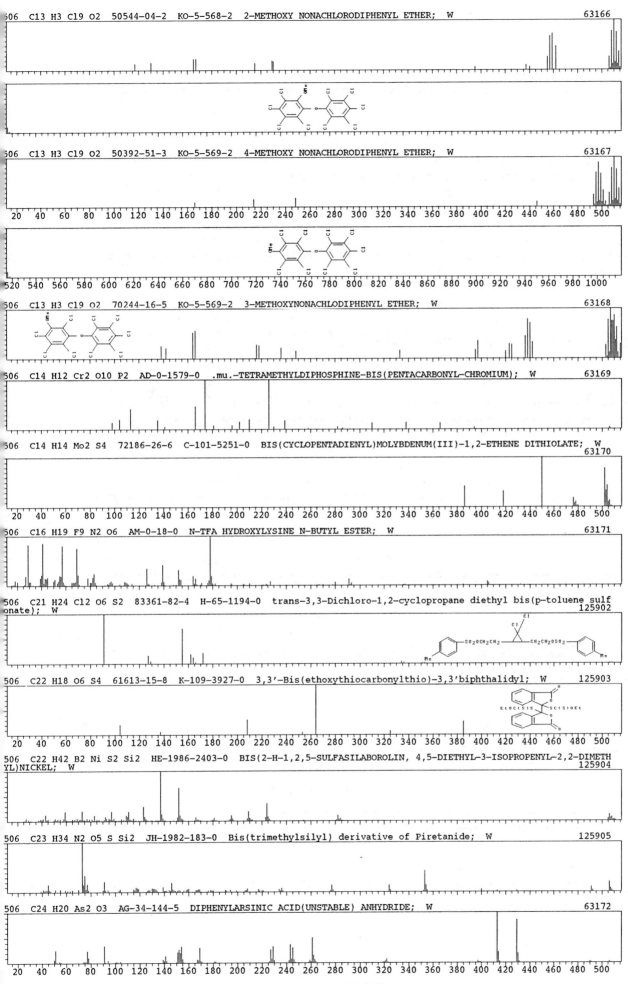

506 C13 H3 C19 O2 50392-51-3 KO-5-569-2 4-METHOXY NONACHLORODIPHENYL ETHER; W   63167

506 C13 H3 C19 O2 70244-16-5 KO-5-569-2 3-METHOXYNONACHLODIPHENYL ETHER; W   63168

506 C14 H12 Cr2 O10 P2 AD-0-1579-0 .mu.-TETRAMETHYLDIPHOSPHINE-BIS(PENTACARBONYL-CHROMIUM); W   63169

506 C14 H14 Mo2 S4 72186-26-6 C-101-5251-0 BIS(CYCLOPENTADIENYL)MOLYBDENUM(III)-1,2-ETHENE DITHIOLATE; W   63170

506 C16 H19 F9 N2 O6 AM-0-18-0 N-TFA HYDROXYLYSINE N-BUTYL ESTER; W   63171

506 C21 H24 Cl2 O6 S2 83361-82-4 H-65-1194-0 trans-3,3-Dichloro-1,2-cyclopropane diethyl bis(p-toluene sulfonate); W   125902

506 C22 H18 O6 S4 61613-15-8 K-109-3927-0 3,3'-Bis(ethoxythiocarbonylthio)-3,3'biphthalidyl; W   125903

506 C22 H42 B2 Ni S2 Si2 HE-1986-2403-0 BIS(2-H-1,2,5-SULFASILABOROLIN, 4,5-DIETHYL-3-ISOPROPENYL-2,2-DIMETHYL)NICKEL; W   125904

506 C23 H34 N2 O5 S Si2 JH-1982-183-0 Bis(trimethylsilyl) derivative of Piretanide; W   125905

506 C24 H20 As2 O3 AG-34-144-5 DIPHENYLARSINIC ACID(UNSTABLE) ANHYDRIDE; W   63172

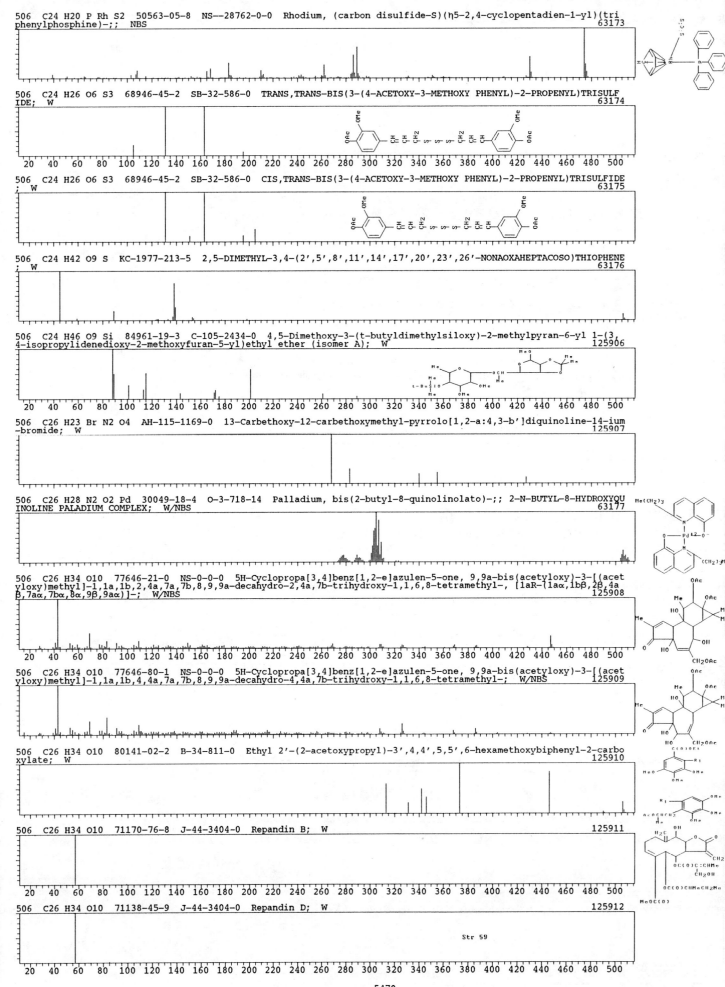

506 C24 H20 P Rh S2 50563-05-8 NS--28762-0-0 Rhodium, (carbon disulfide-S)(η5-2,4-cyclopentadien-1-yl)(triphenylphosphine)-;; NBS
63173

506 C24 H26 O6 S3 68946-45-2 SB-32-586-0 TRANS,TRANS-BIS(3-(4-ACETOXY-3-METHOXY PHENYL)-2-PROPENYL)TRISULFIDE; W
63174

506 C24 H26 O6 S3 68946-45-2 SB-32-586-0 CIS,TRANS-BIS(3-(4-ACETOXY-3-METHOXY PHENYL)-2-PROPENYL)TRISULFIDE; W
63175

506 C24 H42 O9 S KC-1977-213-5 2,5-DIMETHYL-3,4-(2',5',8',11',14',17',20',23',26'-NONAOXAHEPTACOSO)THIOPHENE; W
63176

506 C24 H46 O9 Si 84961-19-3 C-105-2434-0 4,5-Dimethoxy-3-(t-butyldimethylsiloxy)-2-methylpyran-6-yl 1-(3,4-isopropylidenedioxy-2-methoxyfuran-5-yl)ethyl ether (isomer A); W
125906

506 C26 H23 Br N2 O4 AH-115-1169-0 13-Carbethoxy-12-carbethoxymethyl-pyrrolo[1,2-a:4,3-b']diquinoline-14-ium-bromide; W
125907

506 C26 H28 N2 O2 Pd 30049-18-4 O-3-718-14 Palladium, bis(2-butyl-8-quinolinolato)-;; 2-N-BUTYL-8-HYDROXYQUINOLINE PALADIUM COMPLEX; W/NBS
63177

506 C26 H34 O10 77646-21-0 NS-0-0-0 5H-Cyclopropa[3,4]benz[1,2-e]azulen-5-one, 9,9a-bis(acetyloxy)-3-[(acetyloxy)methyl]-1,1a,1b,2,4a,7a,7b,8,9,9a-decahydro-2,4a,7b-trihydroxy-1,1,6,8-tetramethyl-, [1aR-(1aα,1bβ,2β,4aβ,7aα,7bα,8α,9β,9aα)]-; W/NBS
125908

506 C26 H34 O10 77646-80-1 NS-0-0-0 5H-Cyclopropa[3,4]benz[1,2-e]azulen-5-one, 9,9a-bis(acetyloxy)-3-[(acetyloxy)methyl]-1,1a,1b,4,4a,7a,7b,8,9,9a-decahydro-4,4a,7b-trihydroxy-1,1,6,8-tetramethyl-; W/NBS
125909

506 C26 H34 O10 80141-02-2 B-34-811-0 Ethyl 2'-(2-acetoxypropyl)-3',4,4',5,5',6-hexamethoxybiphenyl-2-carboxylate; W
125910

506 C26 H34 O10 71170-76-8 J-44-3404-0 Repandin B; W
125911

506 C26 H34 O10 71138-45-9 J-44-3404-0 Repandin D; W
125912

Str 59

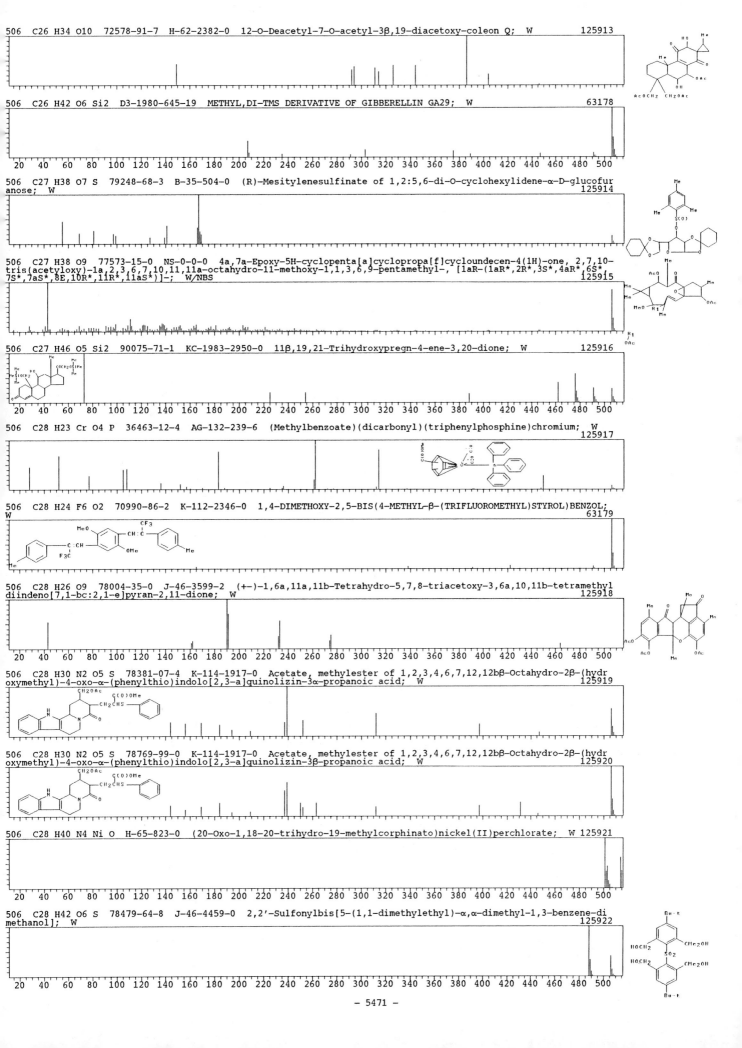

506   C26 H34 O10   72578-91-7   H-62-2382-0   12-O-Deacetyl-7-O-acetyl-3β,19-diacetoxy-coleon Q;   W   125913

506   C26 H42 O6 Si2   D3-1980-645-19   METHYL,DI-TMS DERIVATIVE OF GIBBERELLIN GA29;   W   63178

506   C27 H38 O7 S   79248-68-3   B-35-504-0   (R)-Mesitylenesulfinate of 1,2:5,6-di-O-cyclohexylidene-α-D-glucofuranose;   W   125914

506   C27 H38 O9   77573-15-0   NS-0-0-0   4a,7a-Epoxy-5H-cyclopenta[a]cyclopropa[f]cycloundecen-4(1H)-one, 2,7,10-tris(acetyloxy)-1a,2,3,6,7,10,11,11a-octahydro-11-methoxy-1,1,3,6,9-pentamethyl-, [1aR-(1aR*,2R*,3S*,4aR*,6S*,7S*,7aS*,8E,10R*,11R*,11aS*)]-;   W/NBS   125915

506   C27 H46 O5 Si2   90075-71-1   KC-1983-2950-0   11β,19,21-Trihydroxypregn-4-ene-3,20-dione;   W   125916

506   C28 H23 Cr O4 P   36463-12-4   AG-132-239-6   (Methylbenzoate)(dicarbonyl)(triphenylphosphine)chromium;   W   125917

506   C28 H24 F6 O2   70990-86-2   K-112-2346-0   1,4-DIMETHOXY-2,5-BIS(4-METHYL-β-(TRIFLUOROMETHYL)STYROL)BENZOL;   W   63179

506   C28 H26 O9   78004-35-0   J-46-3599-2   (+-)-1,6a,11a,11b-Tetrahydro-5,7,8-triacetoxy-3,6a,10,11b-tetramethyl diindeno[7,1-bc:2,1-e]pyran-2,11-dione;   W   125918

506   C28 H30 N2 O5 S   78381-07-4   K-114-1917-0   Acetate, methylester of 1,2,3,4,6,7,12,12bβ-Octahydro-2β-(hydroxymethyl)-4-oxo-α-(phenylthio)indolo[2,3-a]quinolizin-3α-propanoic acid;   W   125919

506   C28 H30 N2 O5 S   78769-99-0   K-114-1917-0   Acetate, methylester of 1,2,3,4,6,7,12,12bβ-Octahydro-2β-(hydroxymethyl)-4-oxo-α-(phenylthio)indolo[2,3-a]quinolizin-3β-propanoic acid;   W   125920

506   C28 H40 N4 Ni O   H-65-823-0   (20-Oxo-1,18-20-trihydro-19-methylcorphinato)nickel(II)perchlorate;   W 125921

506   C28 H42 O6 S   78479-64-8   J-46-4459-0   2,2'-Sulfonylbis[5-(1,1-dimethylethyl)-α,α-dimethyl-1,3-benzene-di methanol];   W   125922

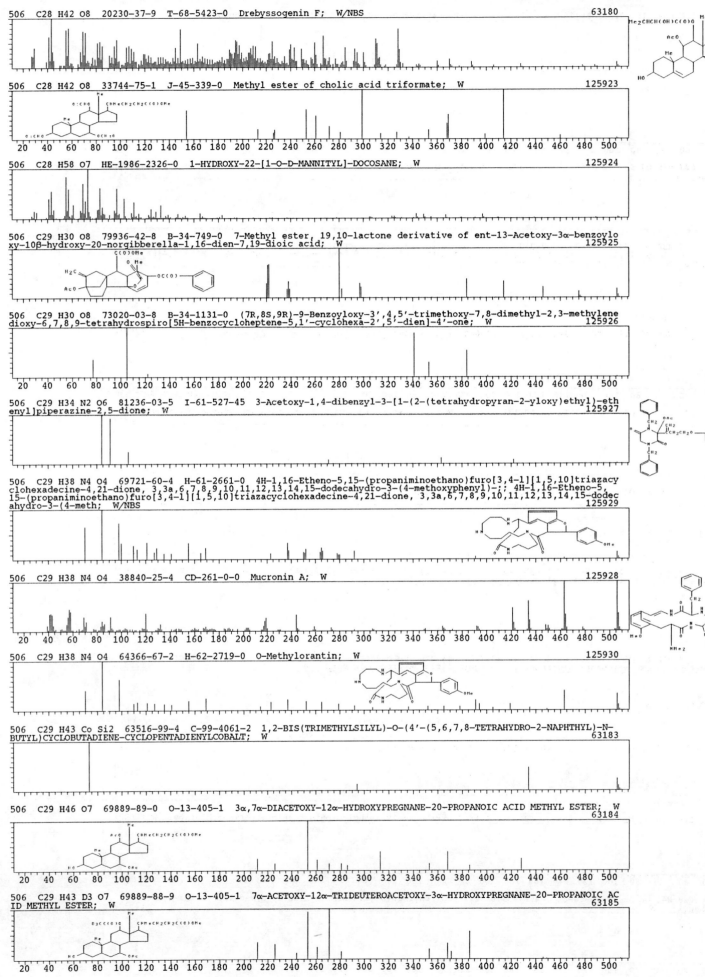

506   C28 H42 O8   20230-37-9   T-68-5423-0   Drebyssogenin F;   W/NBS   63180

506   C28 H42 O8   33744-75-1   J-45-339-0   Methyl ester of cholic acid triformate;   W   125923

506   C28 H58 O7   HE-1986-2326-0   1-HYDROXY-22-[1-O-D-MANNITYL]-DOCOSANE;   W   125924

506   C29 H30 O8   79936-42-8   B-34-749-0   7-Methyl ester, 19,10-lactone derivative of ent-13-Acetoxy-3α-benzoylo
xy-10β-hydroxy-20-norgibberella-1,16-dien-7,19-dioic acid;   W   125925

506   C29 H30 O8   73020-03-8   B-34-1131-0   (7R,8S,9R)-9-Benzoyloxy-3',4,5'-trimethoxy-7,8-dimethyl-2,3-methylene
dioxy-6,7,8,9-tetrahydrospiro[5H-benzocycloheptene-5,1'-cyclohexa-2',5'-dien]-4'-one;   W   125926

506   C29 H34 N2 O6   81236-03-5   I-61-527-45   3-Acetoxy-1,4-dibenzyl-3-[1-(2-(tetrahydropyran-2-yloxy)ethyl)-eth
enyl]piperazine-2,5-dione;   W   125927

506   C29 H38 N4 O4   69721-60-4   H-61-2661-0   4H-1,16-Etheno-5,15-(propaniminoethano)furo[3,4-1][1,5,10]triazacy
clohexadecine-4,21-dione, 3,3a,6,7,8,9,10,11,12,13,14,15-dodecahydro-3-(4-methoxyphenyl)-;; 4H-1,16-Etheno-5,
15-(propaniminoethano)furo[3,4-1][1,5,10]triazacyclohexadecine-4,21-dione, 3,3a,6,7,8,9,10,11,12,13,14,15-dodec
ahydro-3-(4-meth;   W/NBS   125929

506   C29 H38 N4 O4   38840-25-4   CD-261-0-0   Mucronin A;   W   125928

506   C29 H38 N4 O4   64366-67-2   H-62-2719-0   O-Methylorantin;   W   125930

506   C29 H43 Co Si2   63516-99-4   C-99-4061-2   1,2-BIS(TRIMETHYLSILYL)-O-(4'-(5,6,7,8-TETRAHYDRO-2-NAPHTHYL)-N-
BUTYL)CYCLOBUTADIENE-CYCLOPENTADIENYLCOBALT;   W   63183

506   C29 H46 O7   69889-89-0   O-13-405-1   3α,7α-DIACETOXY-12α-HYDROXYPREGNANE-20-PROPANOIC ACID METHYL ESTER;   W
                                                                                                                             63184

506   C29 H43 D3 O7   69889-88-9   O-13-405-1   7α-ACETOXY-12α-TRIDEUTEROACETOXY-3α-HYDROXYPREGNANE-20-PROPANOIC AC
ID METHYL ESTER;   W   63185

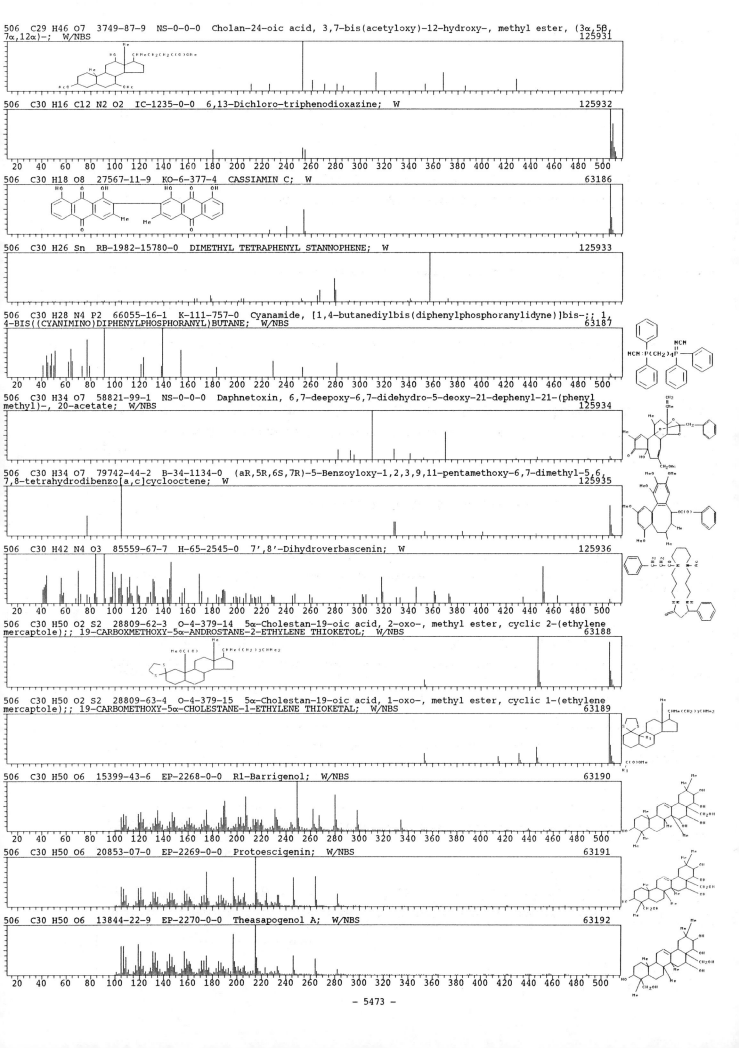

506 C29 H46 O7 3749-87-9 NS-0-0-0 Cholan-24-oic acid, 3,7-bis(acetyloxy)-12-hydroxy-, methyl ester, (3α,5β, 7α,12α)-; W/NBS   125931

506 C30 H16 Cl2 N2 O2 IC-1235-0-0 6,13-Dichloro-triphenodioxazine; W   125932

506 C30 H18 O8 27567-11-9 KO-6-377-4 CASSIAMIN C; W   63186

506 C30 H26 Sn RB-1982-15780-0 DIMETHYL TETRAPHENYL STANNOPHENE; W   125933

506 C30 H28 N4 P2 66055-16-1 K-111-757-0 Cyanamide, [1,4-butanediylbis(diphenylphosphoranylidyne)]bis-;; 1, 4-BIS((CYANIMINO)DIPHENYLPHOSPHORANYL)BUTANE; W/NBS   63187

506 C30 H34 O7 58821-99-1 NS-0-0-0 Daphnetoxin, 6,7-deepoxy-6,7-didehydro-5-deoxy-21-dephenyl-21-(phenyl methyl)-, 20-acetate; W/NBS   125934

506 C30 H34 O7 79742-44-2 B-34-1134-0 (aR,5R,6S,7R)-5-Benzoyloxy-1,2,3,9,11-pentamethoxy-6,7-dimethyl-5,6, 7,8-tetrahydrodibenzo[a,c]cyclooctene; W   125935

506 C30 H42 N4 O3 85559-67-7 H-65-2545-0 7',8'-Dihydroverbascenin; W   125936

506 C30 H50 O2 S2 28809-62-3 O-4-379-14 5α-Cholestan-19-oic acid, 2-oxo-, methyl ester, cyclic 2-(ethylene mercaptole);; 19-CARBOXMETHOXY-5α-ANDROSTANE-2-ETHYLENE THIOKETOL; W/NBS   63188

506 C30 H50 O2 S2 28809-63-4 O-4-379-15 5α-Cholestan-19-oic acid, 1-oxo-, methyl ester, cyclic 1-(ethylene mercaptole);; 19-CARBOMETHOXY-5α-CHOLESTANE-1-ETHYLENE THIOKETAL; W/NBS   63189

506 C30 H50 O6 15399-43-6 EP-2268-0-0 R1-Barrigenol; W/NBS   63190

506 C30 H50 O6 20853-07-0 EP-2269-0-0 Protoescigenin; W/NBS   63191

506 C30 H50 O6 13844-22-9 EP-2270-0-0 Theasapogenol A; W/NBS   63192

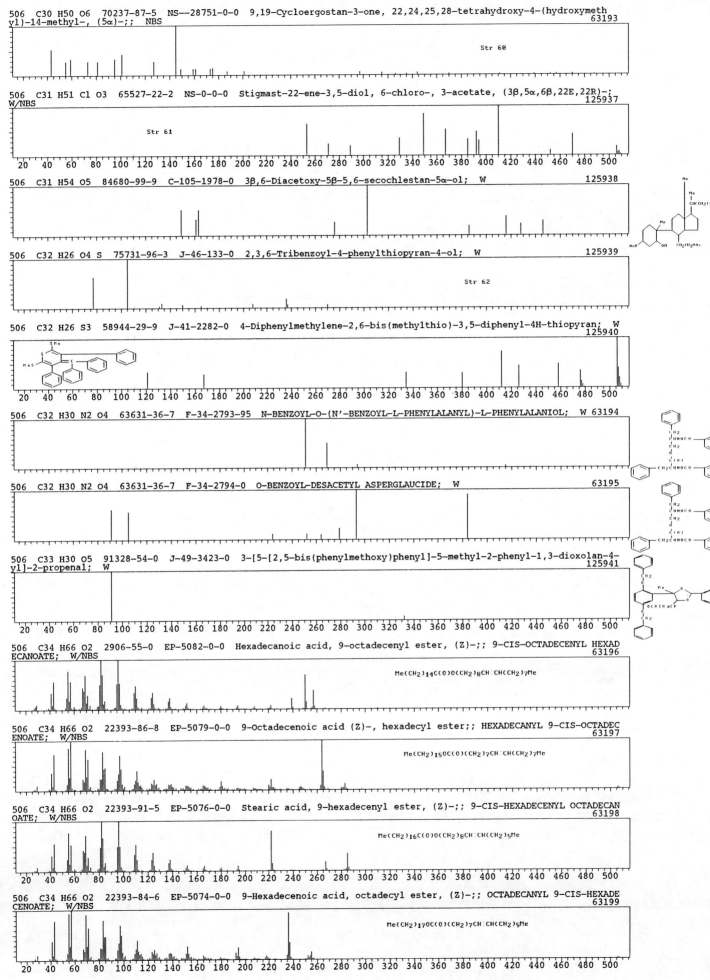

506   C30 H50 O6   70237-87-5   NS--28751-0-0   9,19-Cycloergostan-3-one, 22,24,25,28-tetrahydroxy-4-(hydroxymethyl)-14-methyl-, (5α)-;;   NBS
63193

Str 60

506   C31 H51 Cl O3   65527-22-2   NS-0-0-0   Stigmast-22-ene-3,5-diol, 6-chloro-, 3-acetate, (3β,5α,6β,22E,22R)-; W/NBS
125937

Str 61

506   C31 H54 O5   84680-99-9   C-105-1978-0   3β,6-Diacetoxy-5β-5,6-secochlestan-5α-ol; W
125938

506   C32 H26 O4 S   75731-96-3   J-46-133-0   2,3,6-Tribenzoyl-4-phenylthiopyran-4-ol; W
125939

Str 62

506   C32 H26 S3   58944-29-9   J-41-2282-0   4-Diphenylmethylene-2,6-bis(methylthio)-3,5-diphenyl-4H-thiopyran; W
125940

506   C32 H30 N2 O4   63631-36-7   F-34-2793-95   N-BENZOYL-O-(N'-BENZOYL-L-PHENYLALANYL)-L-PHENYLALANIOL;   W 63194

506   C32 H30 N2 O4   63631-36-7   F-34-2794-0   O-BENZOYL-DESACETYL ASPERGLAUCIDE;   W
63195

506   C33 H30 O5   91328-54-0   J-49-3423-0   3-[5-[2,5-bis(phenylmethoxy)phenyl]-5-methyl-2-phenyl-1,3-dioxolan-4-yl]-2-propenal;   W
125941

506   C34 H66 O2   2906-55-0   EP-5082-0-0   Hexadecanoic acid, 9-octadecenyl ester, (Z)-;; 9-CIS-OCTADECENYL HEXADECANOATE;   W/NBS
63196

Me(CH2)14C(O)O(CH2)8CH:CH(CH2)7Me

506   C34 H66 O2   22393-86-8   EP-5079-0-0   9-Octadecenoic acid (Z)-, hexadecyl ester;; HEXADECANYL 9-CIS-OCTADECENOATE;   W/NBS
63197

Me(CH2)15OC(O)(CH2)7CH:CH(CH2)7Me

506   C34 H66 O2   22393-91-5   EP-5076-0-0   Stearic acid, 9-hexadecenyl ester, (Z)-;; 9-CIS-HEXADECENYL OCTADECANOATE;   W/NBS
63198

Me(CH2)16C(O)O(CH2)8CH:CH(CH2)5Me

506   C34 H66 O2   22393-84-6   EP-5074-0-0   9-Hexadecenoic acid, octadecyl ester, (Z)-;; OCTADECANYL 9-CIS-HEXADECENOATE;   W/NBS
63199

Me(CH2)17OC(O)(CH2)7CH:CH(CH2)5Me

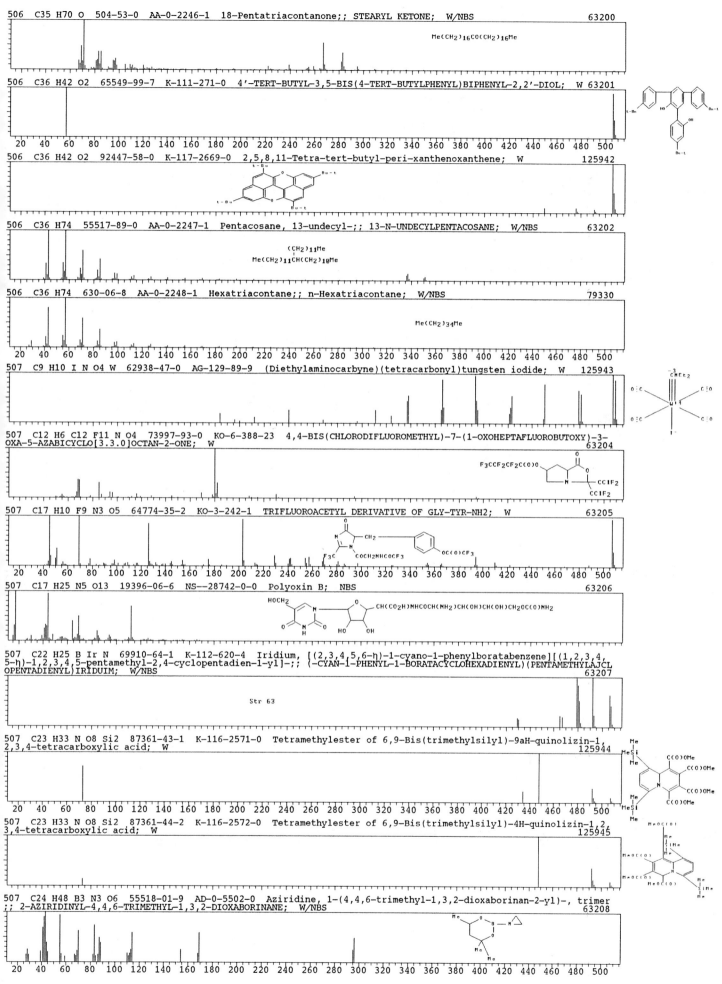

506 C35 H70 O 504-53-0 AA-0-2246-1 18-Pentatriacontanone;; STEARYL KETONE; W/NBS   63200

Me(CH2)16CO(CH2)16Me

506 C36 H42 O2 65549-99-7 K-111-271-0 4'-TERT-BUTYL-3,5-BIS(4-TERT-BUTYLPHENYL)BIPHENYL-2,2'-DIOL; W 63201

506 C36 H42 O2 92447-58-0 K-117-2669-0 2,5,8,11-Tetra-tert-butyl-peri-xanthenoxanthene; W   125942

506 C36 H74 55517-89-0 AA-0-2247-1 Pentacosane, 13-undecyl-;; 13-N-UNDECYLPENTACOSANE; W/NBS   63202

(CH2)11Me
Me(CH2)11CH(CH2)10Me

506 C36 H74 630-06-8 AA-0-2248-1 Hexatriacontane;; n-Hexatriacontane; W/NBS   79330

Me(CH2)34Me

507 C9 H10 I N O4 W 62938-47-0 AG-129-89-9 (Diethylaminocarbyne)(tetracarbonyl)tungsten iodide; W 125943

507 C12 H6 Cl2 F11 N O4 73997-93-0 KO-6-388-23 4,4-BIS(CHLORODIFLUOROMETHYL)-7-(1-OXOHEPTAFLUOROBUTOXY)-3-OXA-5-AZABICYCLO[3.3.0]OCTAN-2-ONE; W 63204

F3CCF2CF2C(O)O

507 C17 H10 F9 N3 O5 64774-35-2 KO-3-242-1 TRIFLUOROACETYL DERIVATIVE OF GLY-TYR-NH2; W 63205

507 C17 H25 N5 O13 19396-06-6 NS--28742-0-0 Polyoxin B; NBS 63206

507 C22 H25 B Ir N 69910-64-1 K-112-620-4 Iridium, [(2,3,4,5,6-η)-1-cyano-1-phenylboratabenzene][(1,2,3,4,5-η)-1,2,3,4,5-pentamethyl-2,4-cyclopentadien-1-yl]-;; (-CYAN-1-PHENYL-1-BORATACYCLOHEXADIENYL)(PENTAMETHYLAJCLOPENTADIENYL)IRIDUIM; W/NBS 63207

Str 63

507 C23 H33 N O8 Si2 87361-43-1 K-116-2571-0 Tetramethylester of 6,9-Bis(trimethylsilyl)-9aH-quinolizin-1,2,3,4-tetracarboxylic acid; W 125944

507 C23 H33 N O8 Si2 87361-44-2 K-116-2572-0 Tetramethylester of 6,9-Bis(trimethylsilyl)-4H-quinolizin-1,2,3,4-tetracarboxylic acid; W 125945

507 C24 H48 B3 N3 O6 55518-01-9 AD-0-5502-0 Aziridine, 1-(4,4,6-trimethyl-1,3,2-dioxaborinan-2-yl)-, trimer;; 2-AZIRIDINYL-4,4,6-TRIMETHYL-1,3,2-DIOXABORINANE; W/NBS 63208

507 C25 H25 Fe N O7 72602-54-1 K-112-3585-0 DI-TERT-BUTYLESTER OF (AZEPINO(2,1,7-CD)INDOLIZEN-4,5-DICARBOXY
LIC ACID)TRICARBONYLIROX; W
63209

507 C27 H49 N O4 Si2 56009-42-8 EP-981-0-0 PROSTAGLANDIN B-2-METHOXIME-DITMS; W/NBS
63210

507 C28 H32 Cl N3 P2 42998-73-2 NS-0-0-0 Phosphinimidic amide, N'-(aminochlorodiphenylphosphoranyl)-N-butyl
-P,P-diphenyl-; W/NBS
125946

507 C28 H53 N O3 Si2 57305-33-6 EP-8259-0-0 Pregnan-20-one, 3,21-bis[(trimethylsilyl)oxy]-, O-methyloxime
(3α,5α)-;; 5α-PREGNANE-3α,21-DIOL-20-ONE MO TMS; W/NBS
63211

507 C28 H53 N O3 Si2 56196-41-9 EP-8258-0-0 Pregnan-20-one, 3,21-bis[(trimethylsilyl)oxy]-, O-methyloxime
(3α,5β)-;; BISTRIMETHYLSILYL 3α,21-DIHYDROXY-5β-PREGNANE-20-ONE METHOXIME; W/NBS
63212

507 C28 H53 N O3 Si2 33287-44-4 EP-8256-0-0 5β-Pregnan-20-one, 3α,6α-bis(trimethylsiloxy)-, O-methyl oxime
;; 5β-PREGNANE-3α,6α-DIOL-20-ONE MO TMS; W/NBS
63216

507 C28 H53 N O3 Si2 39780-68-2 EP-8260-0-0 Pregnan-20-one, 3,17-bis[(trimethylsilyl)oxy]-, O-methyloxime
(3β,5α)-;; 5α-PREGNANE-3β,17α-DIOL-20-ONE MO TMS; W/NBS
63214

507 C28 H35 D18 N O3 Si2 57397-23-6 EP-8447-0-0 5α-PREGNANE-3α,17α-DIOL-20-ONE MO TMS D9; W
63215

507 C28 H53 N O3 Si2 33287-42-2 EP-8291-0-0 5α-Pregnan-20-one, 3α,17-bis(trimethylsiloxy)-, O-methyl oxime
;; 5α-PREGNANE-3α,17α-DIOL-20-ONE MO TMS; W/NBS
63219

507 C28 H53 N O3 Si2 57305-32-5 EP-8257-0-0 Pregnan-20-one, 3,17-bis[(trimethylsilyl)oxy]-, O-methyloxime
(3α,5β)-;; 5β-PREGNANE-3α,17α-DIOL-20-ONE MO TMS; W/NBS
63220

507 C28 H53 N O3 Si2 SW-0-169-0 BISTRIMETHYLSILYL 3β,21-DIHYDROXY-5α-PREGNANE-20-ONE METHOXIME; W
63213

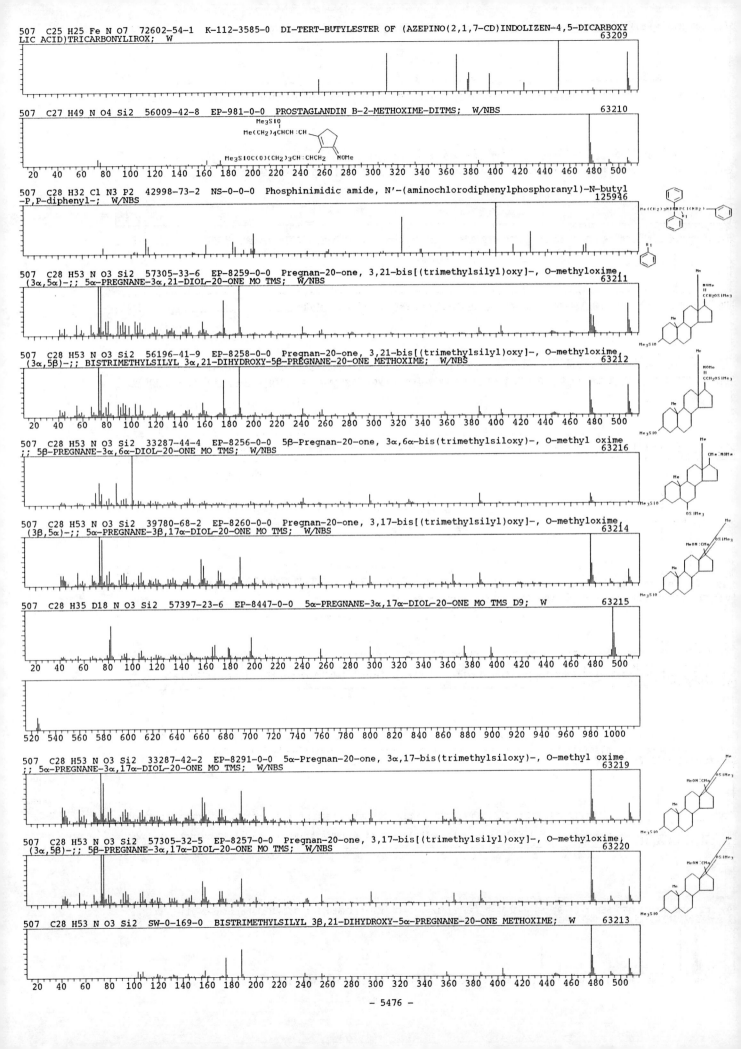

- 5476 -

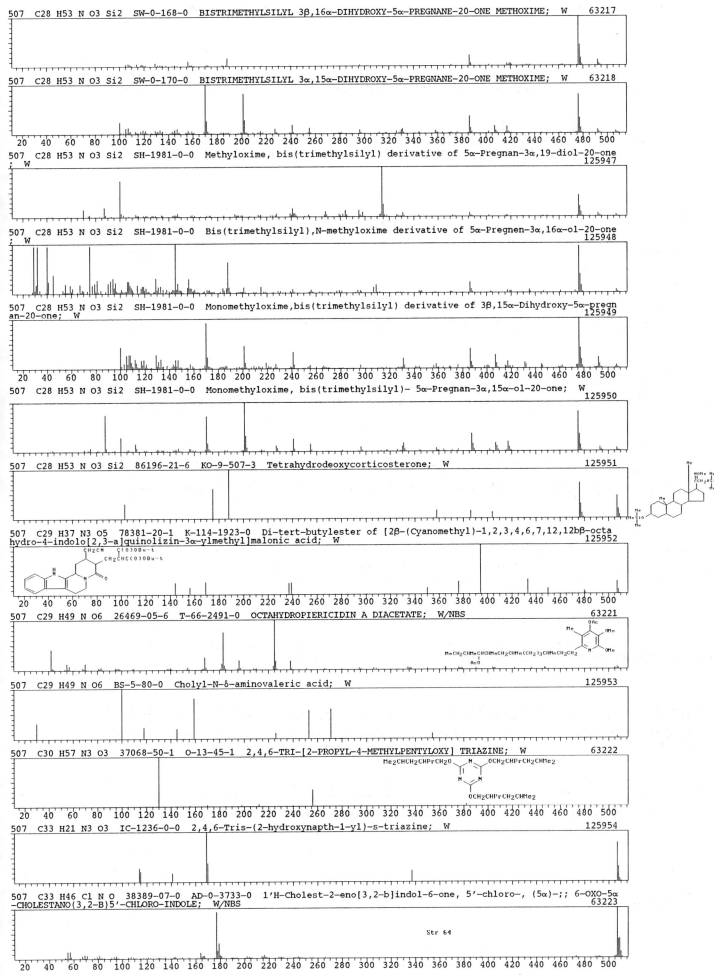

507 C28 H53 N O3 Si2 SW-0-168-0 BISTRIMETHYLSILYL 3β,16α-DIHYDROXY-5α-PREGNANE-20-ONE METHOXIME; W 63217

507 C28 H53 N O3 Si2 SW-0-170-0 BISTRIMETHYLSILYL 3α,15α-DIHYDROXY-5α-PREGNANE-20-ONE METHOXIME; W 63218

507 C28 H53 N O3 Si2 SH-1981-0-0 Methyloxime, bis(trimethylsilyl) derivative of 5α-Pregnan-3α,19-diol-20-one ; W 125947

507 C28 H53 N O3 Si2 SH-1981-0-0 Bis(trimethylsilyl),N-methyloxime derivative of 5α-Pregnen-3α,16α-ol-20-one ; W 125948

507 C28 H53 N O3 Si2 SH-1981-0-0 Monomethyloxime,bis(trimethylsilyl) derivative of 3β,15α-Dihydroxy-5α-pregnan-20-one; W 125949

507 C28 H53 N O3 Si2 SH-1981-0-0 Monomethyloxime, bis(trimethylsilyl)- 5α-Pregnan-3α,15α-ol-20-one; W 125950

507 C28 H53 N O3 Si2 86196-21-6 KO-9-507-3 Tetrahydrodeoxycorticosterone; W 125951

507 C29 H37 N3 O5 78381-20-1 K-114-1923-0 Di-tert-butylester of [2β-(Cyanomethyl)-1,2,3,4,6,7,12,12bβ-octahydro-4-indolo[2,3-a]quinolizin-3α-ylmethyl]malonic acid; W 125952

507 C29 H49 N O6 26469-05-6 T-66-2491-0 OCTAHYDROPIERICIDIN A DIACETATE; W/NBS 63221

507 C29 H49 N O6 BS-5-80-0 Cholyl-N-δ-aminovaleric acid; W 125953

507 C30 H57 N3 O3 37068-50-1 O-13-45-1 2,4,6-TRI-[2-PROPYL-4-METHYLPENTYLOXY] TRIAZINE; W 63222

507 C33 H21 N3 O3 IC-1236-0-0 2,4,6-Tris-(2-hydroxynapth-1-yl)-s-triazine; W 125954

507 C33 H46 Cl N O 38389-07-0 AD-0-3733-0 1'H-Cholest-2-eno[3,2-b]indol-6-one, 5'-chloro-, (5α)-;; 6-OXO-5α-CHOLESTANO(3,2-B)5'-CHLORO-INDOLE; W/NBS 63223

Str 64

507   C34 H25 N3 O2   85731-56-2   Y-19-1488-0   Ethyl 2,3,5,8-Tetraphenylpyrido[3,4-b]pyridine-7-carboxylate;   W
125955

508   C6 H6 Br5 Cl   87-84-3   RB-1982-11374-0   PENTABROMOCHLOROCYCLOHEXANE;   W
125956

20 40 60 80 100 120 140 160 180 200 220 240 260 280 300 320 340 360 380 400 420 440 460 480 500

520 540 560 580 600 620 640 660 680 700 720 740 760 780 800 820 840 860 880 900 920 940 960 980 1000

508   C7 H3 F6 Mn N2 O8 S3   82389-60-4   K-115-1466-0   [Bis(trifluoromethylsulfonyl)sulfurdiimido]tetracarbonyl
methylmanganese;   W
125957

508   C9 F7 O5 Re   38317-69-0   NS-0-0-0   Rhenium, pentacarbonyl[2,3,3-tetrafluoro-1-(trifluoromethyl)-1-propeny
l]-, (OC-6-21)-;   W/NBS
125958

508   C10 H28 Sn3   56177-42-5   O-10-19-2   Stannane, methylidynetris[trimethyl-;;   TRIS(TRIMETHYLSTANNYL)METHANE;
W/NBS
63224

508   C12 H9 Cl2 F11 N2 O3   73997-91-8   KO-6-386-20   2,2-BIS(CHLORODIFLUOROMETHYL)-4-(3-(N-HEPTAFLUOROPROPYLCARB
ONYL)AMINOPROPYL)-1,3-OXAZOLIDINONE;   W
63225

508   C12 H16 Br4 O2   61605-37-6   H-59-1324-20   4,4,8,8-TETRABROMO-1,3,5,7-TETRAMETHYL-ANTI-TRICYCLO[5.1.0.0(3,
5)]OCTANE-2,6-TRANS-DIOL;   W
63226

508   C12 H27 Cr O12 P3   62413-59-6   HE-1982-0-0   Chromium, tricarbonyltris(trimethyl phosphite-P)-;;   CHROMIUM,
TRICARBONYL-TRIS(TRIMETHYLPHOSPHIT);   W/NBS
63227

20 40 60 80 100 120 140 160 180 200 220 240 260 280 300 320 340 360 380 400 420 440 460 480 500

508   C16 H6 F10 Sn   1247-12-7   NS--28729-0-0   Stannane, bis(pentafluorophenyl)divinyl-;;   Tin, bis(pentafluoroph
enyl)divinyl-;   NBS
63228

508   C19 H21 Br N6 O6   IC-1237-0-0   Aniline-5-acetamide, N,N-diethyl-2-methoxy-4-(2,4-dinitro-6-bromophenylaz
o)-;   W
125959

508   C21 H25 Cl N6 O7   IC-1237-0-0   N-(1,3-Dimethyl-3-hydroxybutyl)-2-methoxy-4-(2,4-dinitro-6-chlorophenylazo)
-aniline-5-acetamide;   W
125960

20 40 60 80 100 120 140 160 180 200 220 240 260 280 300 320 340 360 380 400 420 440 460 480 500

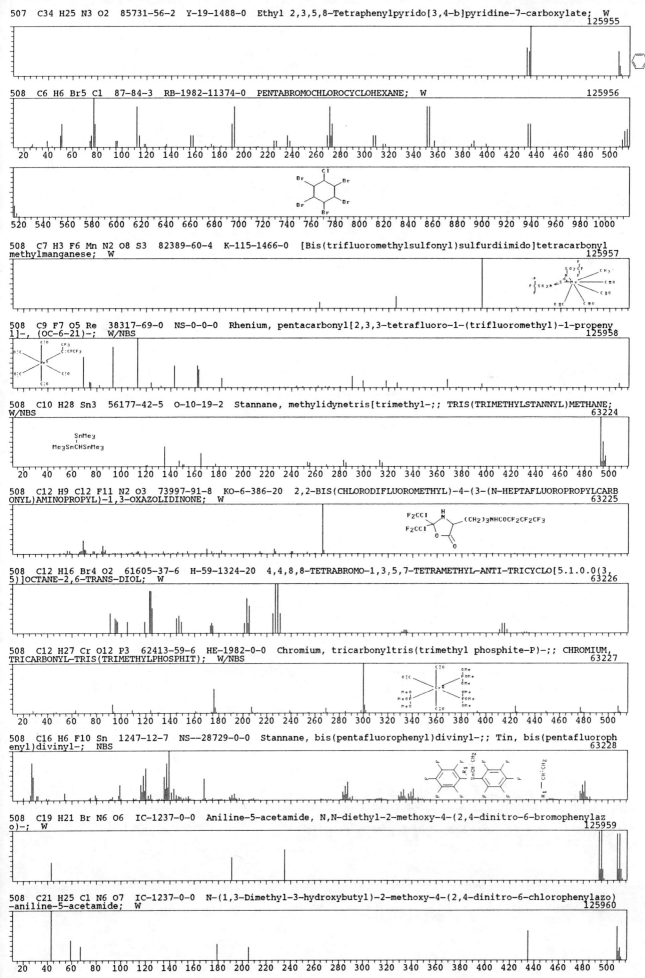

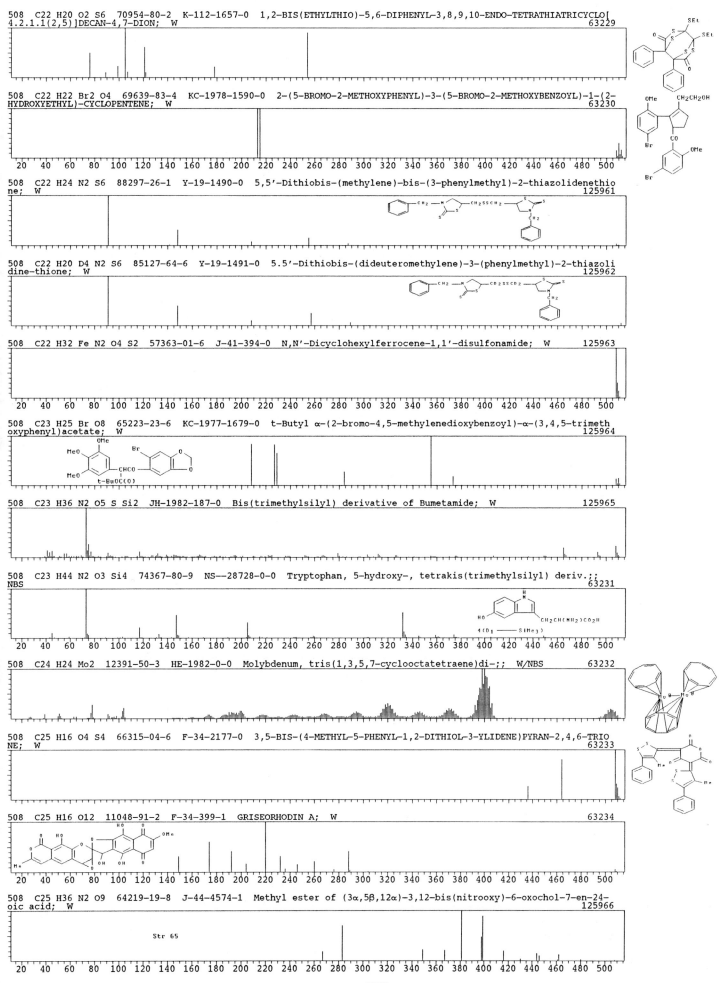

508 C22 H20 O2 S6 70954-80-2 K-112-1657-0 1,2-BIS(ETHYLTHIO)-5,6-DIPHENYL-3,8,9,10-ENDO-TETRATHIATRICYCLO[
4.2.1.1(2,5)]DECAN-4,7-DION; W   63229

508 C22 H22 Br2 O4 69639-83-4 KC-1978-1590-0 2-(5-BROMO-2-METHOXYPHENYL)-3-(5-BROMO-2-METHOXYBENZOYL)-1-(2-
HYDROXYETHYL)-CYCLOPENTENE; W   63230

20 40 60 80 100 120 140 160 180 200 220 240 260 280 300 320 340 360 380 400 420 440 460 480 500

508 C22 H24 N2 S6 88297-26-1 Y-19-1490-0 5,5'-Dithiobis-(methylene)-bis-(3-phenylmethyl)-2-thiazolidenethio
ne; W   125961

508 C22 H20 D4 N2 S6 85127-64-6 Y-19-1491-0 5.5'-Dithiobis-(dideuteromethylene)-3-(phenylmethyl)-2-thiazoli
dine-thione; W   125962

508 C22 H32 Fe N2 O4 S2 57363-01-6 J-41-394-0 N,N'-Dicyclohexylferrocene-1,1'-disulfonamide; W   125963

20 40 60 80 100 120 140 160 180 200 220 240 260 280 300 320 340 360 380 400 420 440 460 480 500

508 C23 H25 Br O8 65223-23-6 KC-1977-1679-0 t-Butyl α-(2-bromo-4,5-methylenedioxybenzoyl)-α-(3,4,5-trimeth
oxyphenyl)acetate; W   125964

508 C23 H36 N2 O5 S Si2 JH-1982-187-0 Bis(trimethylsilyl) derivative of Bumetamide; W   125965

508 C23 H44 N2 O3 Si4 74367-80-9 NS--28728-0-0 Tryptophan, 5-hydroxy-, tetrakis(trimethylsilyl) deriv.;;
NBS   63231

20 40 60 80 100 120 140 160 180 200 220 240 260 280 300 320 340 360 380 400 420 440 460 480 500

508 C24 H24 Mo2 12391-50-3 HE-1982-0-0 Molybdenum, tris(1,3,5,7-cyclooctatetraene)di-;; W/NBS   63232

508 C25 H16 O4 S4 66315-04-6 F-34-2177-0 3,5-BIS-(4-METHYL-5-PHENYL-1,2-DITHIOL-3-YLIDENE)PYRAN-2,4,6-TRIO
NE; W   63233

508 C25 H16 O12 11048-91-2 F-34-399-1 GRISEORHODIN A; W   63234

20 40 60 80 100 120 140 160 180 200 220 240 260 280 300 320 340 360 380 400 420 440 460 480 500

508 C25 H36 N2 O9 64219-19-8 J-44-4574-1 Methyl ester of (3α,5β,12α)-3,12-bis(nitrooxy)-6-oxochol-7-en-24-
oic acid; W   125966

Str 65

20 40 60 80 100 120 140 160 180 200 220 240 260 280 300 320 340 360 380 400 420 440 460 480 500

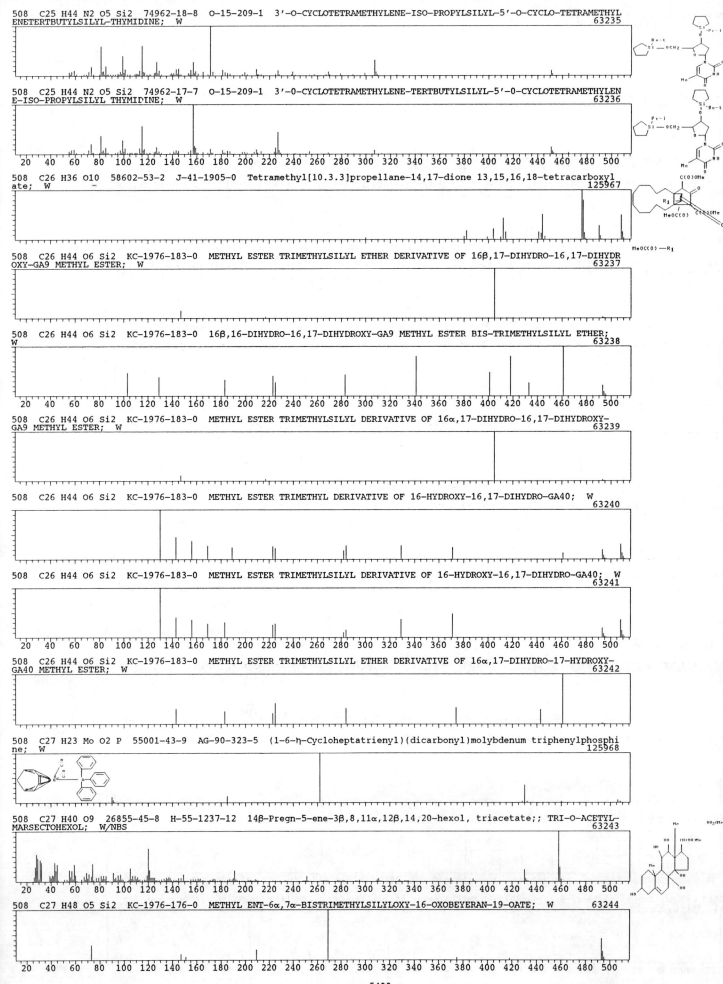

508   C25 H44 N2 O5 Si2   74962-18-8   O-15-209-1   3'-O-CYCLOTETRAMETHYLENE-ISO-PROPYLSILYL-5'-O-CYCLO-TETRAMETHYL
ENETERTBUTYLSILYL-THYMIDINE;   W   63235

508   C25 H44 N2 O5 Si2   74962-17-7   O-15-209-1   3'-O-CYCLOTETRAMETHYLENE-TERTBUTYLSILYL-5'-O-CYCLOTETRAMETHYLEN
E-ISO-PROPYLSILYL THYMIDINE;   W   63236

508   C26 H36 O10   58602-53-2   J-41-1905-0   Tetramethyl[10.3.3]propellane-14,17-dione 13,15,16,18-tetracarboxyl
ate;   W   125967

508   C26 H44 O6 Si2   KC-1976-183-0   METHYL ESTER TRIMETHYLSILYL ETHER DERIVATIVE OF 16β,17-DIHYDRO-16,17-DIHYDR
OXY-GA9 METHYL ESTER;   W   63237

508   C26 H44 O6 Si2   KC-1976-183-0   16β,16-DIHYDRO-16,17-DIHYDROXY-GA9 METHYL ESTER BIS-TRIMETHYLSILYL ETHER;
W   63238

508   C26 H44 O6 Si2   KC-1976-183-0   METHYL ESTER TRIMETHYLSILYL DERIVATIVE OF 16α,17-DIHYDRO-16,17-DIHYDROXY-
GA9 METHYL ESTER;   W   63239

508   C26 H44 O6 Si2   KC-1976-183-0   METHYL ESTER TRIMETHYL DERIVATIVE OF 16-HYDROXY-16,17-DIHYDRO-GA40;   W
                                                                                                                 63240

508   C26 H44 O6 Si2   KC-1976-183-0   METHYL ESTER TRIMETHYLSILYL DERIVATIVE OF 16-HYDROXY-16,17-DIHYDRO-GA40;   W
                                                                                                                 63241

508   C26 H44 O6 Si2   KC-1976-183-0   METHYL ESTER TRIMETHYLSILYL ETHER DERIVATIVE OF 16α,17-DIHYDRO-17-HYDROXY-
GA40 METHYL ESTER;   W   63242

508   C27 H23 Mo O2 P   55001-43-9   AG-90-323-5   (1-6-η-Cycloheptatrienyl)(dicarbonyl)molybdenum triphenylphosphi
ne;   W   125968

508   C27 H40 O9   26855-45-8   H-55-1237-12   14β-Pregn-5-ene-3β,8,11α,12β,14,20-hexol, triacetate;; TRI-O-ACETYL-
MARSECTOHEXOL;   W/NBS   63243

508   C27 H48 O5 Si2   KC-1976-176-0   METHYL ENT-6α,7α-BISTRIMETHYLSILYLOXY-16-OXOBEYERAN-19-OATE;   W   63244

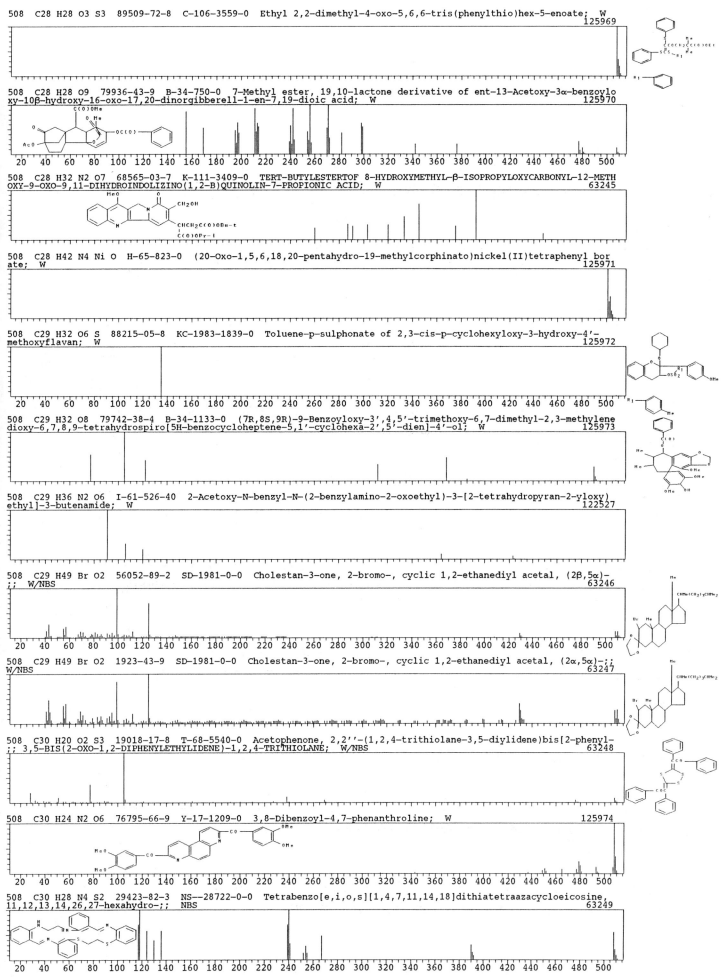

508   C28 H28 O3 S3   89509-72-8   C-106-3559-0   Ethyl 2,2-dimethyl-4-oxo-5,6,6-tris(phenylthio)hex-5-enoate;   W
125969

508   C28 H28 O9   79936-43-9   B-34-750-0   7-Methyl ester, 19,10-lactone derivative of ent-13-Acetoxy-3α-benzoylo
xy-10β-hydroxy-16-oxo-17,20-dinorgibberell-1-en-7,19-dioic acid;   W
125970

508   C28 H32 N2 O7   68565-03-7   K-111-3409-0   TERT-BUTYLESTERTOF 8-HYDROXYMETHYL-β-ISOPROPYLOXYCARBONYL-12-METH
OXY-9-OXO-9,11-DIHYDROINDOLIZINO(1,2-B)QUINOLIN-7-PROPIONIC ACID;   W
63245

508   C28 H42 N4 Ni O   H-65-823-0   (20-Oxo-1,5,6,18,20-pentahydro-19-methylcorphinato)nickel(II)tetraphenyl bor
ate;   W
125971

508   C29 H32 O6 S   88215-05-8   KC-1983-1839-0   Toluene-p-sulphonate of 2,3-cis-p-cyclohexyloxy-3-hydroxy-4'-
methoxyflavan;   W
125972

508   C29 H32 O8   79742-38-4   B-34-1133-0   (7R,8S,9R)-9-Benzoyloxy-3',4,5'-trimethoxy-6,7-dimethyl-2,3-methylene
dioxy-6,7,8,9-tetrahydrospiro[5H-benzocycloheptene-5,1'-cyclohexa-2',5'-dien]-4'-ol;   W
125973

508   C29 H36 N2 O6   I-61-526-40   2-Acetoxy-N-benzyl-N-(2-benzylamino-2-oxoethyl)-3-[2-tetrahydropyran-2-yloxy)
ethyl]-3-butenamide;   W
122527

508   C29 H49 Br O2   56052-89-2   SD-1981-0-0   Cholestan-3-one, 2-bromo-, cyclic 1,2-ethanediyl acetal, (2β,5α)-
;;   W/NBS
63246

508   C29 H49 Br O2   1923-43-9   SD-1981-0-0   Cholestan-3-one, 2-bromo-, cyclic 1,2-ethanediyl acetal, (2α,5α)-;;
W/NBS
63247

508   C30 H20 O2 S3   19018-17-8   T-68-5540-0   Acetophenone, 2,2''-(1,2,4-trithiolane-3,5-diylidene)bis[2-phenyl-
;;   3,5-BIS(2-OXO-1,2-DIPHENYLETHYLIDENE)-1,2,4-TRITHIOLANE;   W/NBS
63248

508   C30 H24 N2 O6   76795-66-9   Y-17-1209-0   3,8-Dibenzoyl-4,7-phenanthroline;   W
125974

508   C30 H28 N4 S2   29423-82-3   NS--28722-0-0   Tetrabenzo[e,i,o,s][1,4,7,11,14,18]dithiatetraazacycloeicosine,
11,12,13,14,26,27-hexahydro-;;   NBS
63249

508 C30 H29 Fe O2 P 91550-61-7 C-106-6070-0 Dicarbonyl(3-5-η,1-σ-(2-(3-propenyl)cycloheptenyl))(triphenylphosphine)iron; W
125975

Str 66

508 C30 H52 O6 56053-00-0 SD-1981-0-0 Ergostane-3,5,6,12,25-pentol, 25-acetate, (3β,5α,6β,12β)-;; W/NBS
63250

Str 67

508 C31 H26 O3 P2 22400-41-5 O-10-242-1 Phosphonic acid, [(triphenylphosphoranylidene)methyl]-, diphenyl es
ter;; DIPHENYL-TRIPHENYL PHOSPHORANYLIDENE METHYL- PHOSPHONATE; W/NBS
63251

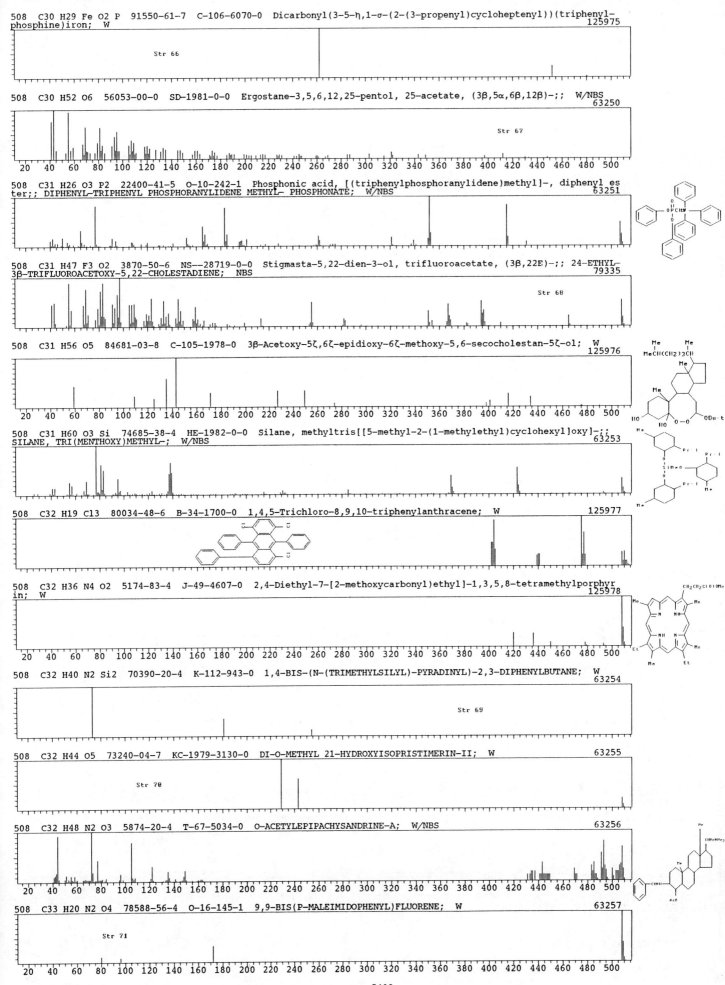

508 C31 H47 F3 O2 3870-50-6 NS--28719-0-0 Stigmasta-5,22-dien-3-ol, trifluoroacetate, (3β,22E)-;; 24-ETHYL-
3β-TRIFLUOROACETOXY-5,22-CHOLESTADIENE; NBS
79335

Str 68

508 C31 H56 O5 84681-03-8 C-105-1978-0 3β-Acetoxy-5ζ,6ζ-epidioxy-6ζ-methoxy-5,6-secocholestan-5ζ-ol; W
125976

508 C31 H60 O3 Si 74685-38-4 HE-1982-0-0 Silane, methyltris[[5-methyl-2-(1-methylethyl)cyclohexyl]oxy]-;;
SILANE, TRI(MENTHOXY)METHYL-; W/NBS
63253

508 C32 H19 Cl3 80034-48-6 B-34-1700-0 1,4,5-Trichloro-8,9,10-triphenylanthracene; W
125977

508 C32 H36 N4 O2 5174-83-4 J-49-4607-0 2,4-Diethyl-7-[2-methoxycarbonyl)ethyl]-1,3,5,8-tetramethylporphyr
in; W
125978

508 C32 H40 N2 Si2 70390-20-4 K-112-943-0 1,4-BIS-(N-(TRIMETHYLSILYL)-PYRADINYL)-2,3-DIPHENYLBUTANE; W
63254

Str 69

508 C32 H44 O5 73240-04-7 KC-1979-3130-0 DI-O-METHYL 21-HYDROXYISOPRISTIMERIN-II; W
63255

Str 70

508 C32 H48 N2 O3 5874-20-4 T-67-5034-0 O-ACETYLEPIPACHYSANDRINE-A; W/NBS
63256

508 C33 H20 N2 O4 78588-56-4 O-16-145-1 9,9-BIS(P-MALEIMIDOPHENYL)FLUORENE; W
63257

Str 71

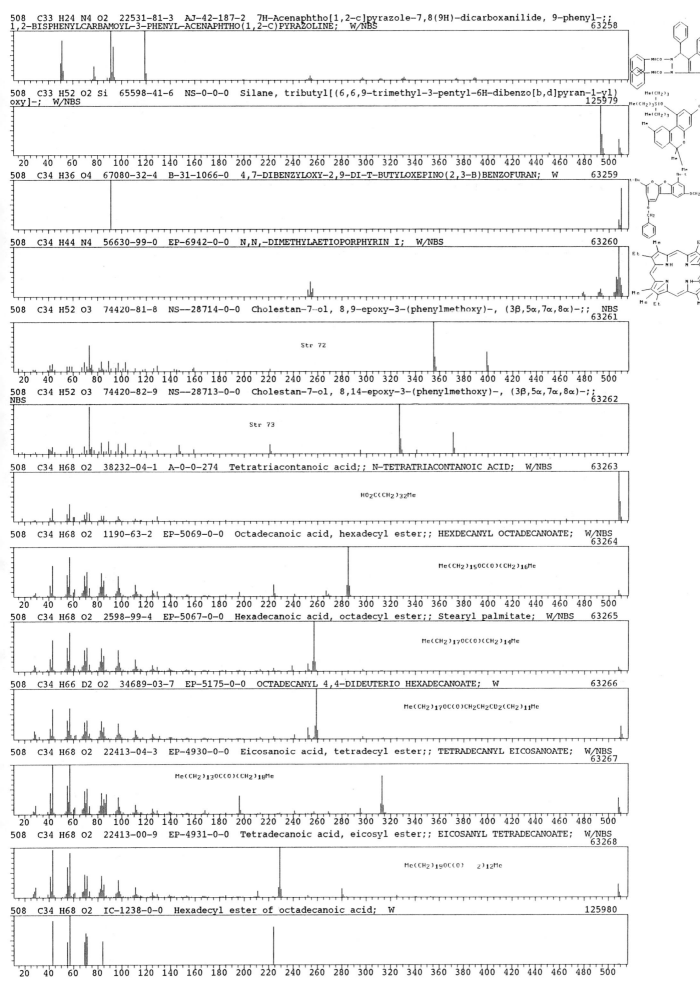

508   C33 H24 N4 O2   22531-81-3   AJ-42-187-2   7H-Acenaphtho[1,2-c]pyrazole-7,8(9H)-dicarboxanilide, 9-phenyl-;;
1,2-BISPHENYLCARBAMOYL-3-PHENYL-ACENAPHTHO(1,2-C)PYRAZOLINE;   W/NBS   63258

508   C33 H52 O2 Si   65598-41-6   NS-0-0-0   Silane, tributyl[(6,6,9-trimethyl-3-pentyl-6H-dibenzo[b,d]pyran-1-yl)
oxy]-;   W/NBS   125979

508   C34 H36 O4   67080-32-4   B-31-1066-0   4,7-DIBENZYLOXY-2,9-DI-T-BUTYLOXEPINO(2,3-B)BENZOFURAN;   W   63259

508   C34 H44 N4   56630-99-0   EP-6942-0-0   N,N,-DIMETHYLAETIOPORPHYRIN I;   W/NBS   63260

508   C34 H52 O3   74420-81-8   NS--28714-0-0   Cholestan-7-ol, 8,9-epoxy-3-(phenylmethoxy)-, (3β,5α,7α,8α)-;;   NBS
63261

Str 72

508   C34 H52 O3   74420-82-9   NS--28713-0-0   Cholestan-7-ol, 8,14-epoxy-3-(phenylmethoxy)-, (3β,5α,7α,8α)-;;
NBS   63262

Str 73

508   C34 H68 O2   38232-04-1   A-0-0-274   Tetratriacontanoic acid;;   N-TETRATRIACONTANOIC ACID;   W/NBS   63263

HO2C(CH2)32Me

508   C34 H68 O2   1190-63-2   EP-5069-0-0   Octadecanoic acid, hexadecyl ester;;   HEXDECANYL OCTADECANOATE;   W/NBS
63264

Me(CH2)15OC(O)(CH2)16Me

508   C34 H68 O2   2598-99-4   EP-5067-0-0   Hexadecanoic acid, octadecyl ester;;   Stearyl palmitate;   W/NBS   63265

Me(CH2)17OC(O)(CH2)14Me

508   C34 H66 D2 O2   34689-03-7   EP-5175-0-0   OCTADECANYL 4,4-DIDEUTERIO HEXADECANOATE;   W   63266

Me(CH2)17OC(O)CH2CH2CD2(CH2)11Me

508   C34 H68 O2   22413-04-3   EP-4930-0-0   Eicosanoic acid, tetradecyl ester;;   TETRADECANYL EICOSANOATE;   W/NBS
63267

Me(CH2)13OC(O)(CH2)18Me

508   C34 H68 O2   22413-00-9   EP-4931-0-0   Tetradecanoic acid, eicosyl ester;;   EICOSANYL TETRADECANOATE;   W/NBS
63268

Me(CH2)19OC(O)   2)12Me

508   C34 H68 O2   IC-1238-0-0   Hexadecyl ester of octadecanoic acid;   W   125980

508   C35 H28 N2 O2   88313-00-2   F-39-2083-0   6-Methyl-3-oxa-3,4-dihydro-2,3-diphenyl-4-(4'-benzamido-1-tol-3-yl)quinoline;   W

125981

508   C35 H29 Co   55518-03-1   O-3-27-3   Cobalt, [1,1',1'',1'''-[(1,2,3,4-η)-1,3-cyclohexadiene-1,2,3,4-tetrayl]tetrakis[benzene]](η5-2,4-cyclopentadien-1-yl)-;; .PI.-CYCLOPENTADIENYL-1,2,3,4-TETRAPHENYLCYCLOHEXA-1,3-DIENE COBALT;   W/NBS

63269

508   C35 H29 Co   HE-1986-837-0   CYCLOPENTADIENYL-1,2,3,4-TETRAPHENYLCYCLOHEXADIENECOBALT;   W

125982

508   C35 H72 O   55517-90-3   AA-0-2249-2   1-Pentatriacontanol;; PENTATRIACONTANOL-1;   W/NBS

63270

Me(CH₂)₃₄OH

508   C36 H28 Ti   39475-28-0   AG-55-311-1   CYCLOOCTATETRAENE(TETRAPHENYLCYCLOBUTADIENE)TITANIUM;   W

63271

508   C36 H44 O2   92447-59-1   K-117-2671-0   2,5,8,11-Tetra-tert-butyldibenzo[a,kl]xanthen-1-ol;   W

125983

508   C37 H48 O   69081-90-9   C-100-7683-0   2-(3α,5-CYCLO-6β-METHOXYBISNORCHOLYL)-1,1-DIPHENYLETHYLENE;   W   63272

Str 74

508   C37 H48 O   CD-500-0-0   β-Apo-2'-carotenol;   W

125984

508   C40 H28   72862-25-0   SB-33-446-0   [2](1,4)NAPHTHALENO[2]PARACYCLO[2](1,4)NAPHTHALENO[2]PARACYCLOPHANETETRAENE;   W

63273

508   C40 H28   72862-26-1   SB-33-447-0   [2](1,5)NAPHTHALENO(2)PARACYCLO[2](1,5)NAPHTHOLENO[2]PARACYCLOPHANETETRAENE;   W

63274

508   C40 H28   72862-23-8   SB-33-447-0   [2](1,5)Naphthaleno[2]paracyclo[2](1,5)-naphthaleno[2]paracyclophanetetraene;   W

125985

509   C10 H3 F12 N O S4   90594-23-3   K-117-1898-0   1-Acetyl-2,3,4,5-tetrakis(trifluoromethylthio)pyrrole;   W

125986

509   C12 H6 F15 N3 O2   82644-64-2   AC-1982-94-0   1,5-Dimethyl-4-nitro-3-perfluoroheptyl-pyrazole;   W

125987

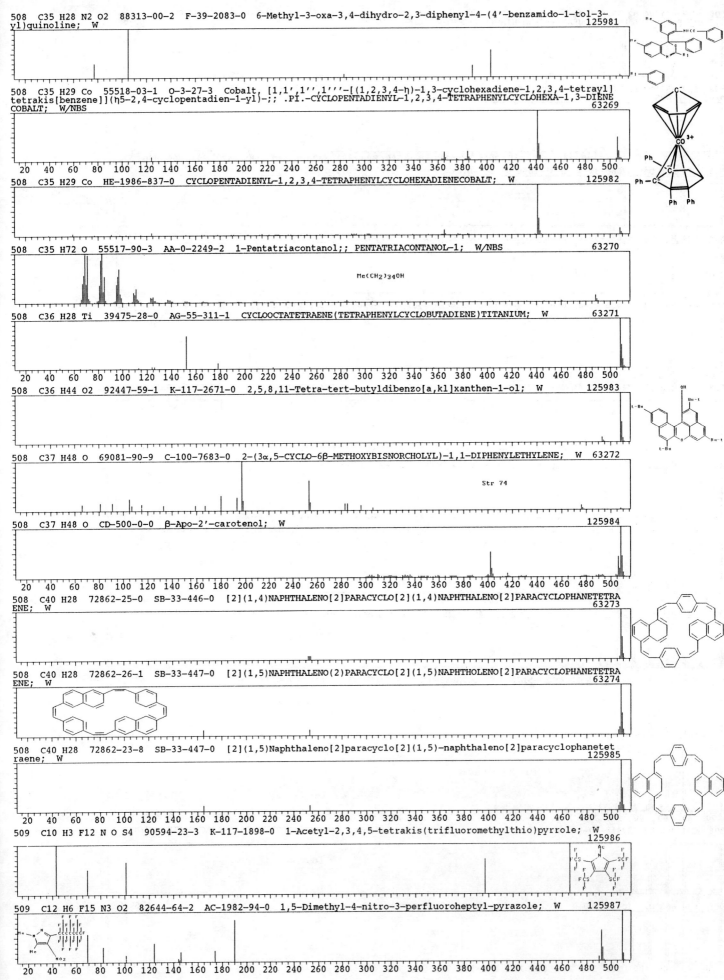

509   C13 F17 N O   RB-1982-15323-0   PERFLUORO-4-(1-BUTOXYETHYL)BENZONITRILE;   W   125988

509   C16 H19 Cl4 N O9   18422-35-0   AD-0-1002-0   β-D-Glucopyranose, 2-deoxy-2-[(tetrachloroethylidene)amino]-,
1,3,4,6-tetraacetate;;   1,3,4,6-TETRA-O-ACETYL-2-DEOXY-2-TETRACHLOROETHYLIDENEAMINO-β-D-GLUCOPYRANOSE;   W/NBS
                                                                                                                        63275

509   C18 H16 Br2 Cu N4   65230-36-6   B-30-2508-0   2,11-DIBROMO-5,6,7,8,15,16-HEXAHYDRODIBENZO[E,M][1,4,8,11]TET
RAAZACYCLOTETRADECINATOCOPPER (II);   W   63276

509   C20 H47 N O6 Si4   55721-25-0   AU-10-117-10   D-Galactose, 2-(acetylamino)-2-deoxy-3,4,5,6-tetrakis-O-(tri
methylsilyl)-;;   2-ACETAMIDO-2-DEOXY-TETRA-O-(TRIMETHYLSILYL)-D-GALACTOSE;   W/NBS   63282

509   C20 H47 N O6 Si4   55529-74-3   C-91-1738-7   D-Glucose, 2-(acetylamino)-2-deoxy-3,4,5,6-tetrakis-O-(trimeth
ylsilyl)-;;   2-ACETAMIDO-2-DEOXY-D-GLUCOSE-TMS-ETHER;   W/NBS   63283

509   C20 H47 N O6 Si4   55517-99-2   AM-0-72-0   Galactose, 2-(acetylamino)-2-deoxy-3,4,5,6-tetrakis-O-(trimethyls
ilyl)-;;   N-ACETYLGALACTOSAMINE-TETRATMS;   W/NBS   63285

509   C20 H47 N O6 Si4   53110-70-6   EP-1453-0-0   α-D-Mannopyranose, 2-(acetylamino)-2-deoxy-1,3,4,6-tetrakis-O-(
trimethylsilyl)-;;   2-ACETAMIDO-A-MANNOPYRANOSE-1,3,4,6-TETRATMS;   W/NBS   63277

509   C20 H47 N O6 Si4   53110-71-7   EP-1454-0-0   β-D-Mannopyranose, 2-(acetylamino)-2-deoxy-1,3,4,6-tetrakis-O-(
trimethylsilyl)-;;   2-ACETAMIDO-B-MANNOPYRANOSE-1,3,4,6-TETRATMS;   W/NBS   63278

509   C20 H47 N O6 Si4   53110-73-9   EP-1456-0-0   β-D-Galactopyranose, 2-(acetylamino)-2-deoxy-1,3,4,6-tetrakis-
O-(trimethylsilyl)-;;   2-ACETAMIDO-B-GALACTOPYRANOSE-1,3,4,6-TETRATMS;   W/NBS   63279

509   C20 H47 N O6 Si4   53110-72-8   EP-1455-0-0   α-D-Galactopyranose, 2-(acetylamino)-2-deoxy-1,3,4,6-tetrakis-
O-(trimethylsilyl)-;;   2-ACETAMIDO-A-GALACTOPYRANOSE-1,3,4,6-TETRATMS;   W/NBS   63280

509   C20 H47 N O6 Si4   31980-72-0   AM-0-71-0   Glucopyranose, 2-acetamido-2-deoxy-1,3,4,6-tetrakis-O-(trimethyls
ilyl)-, D-;;   N-ACETYLGLUCOSAMINE-TETRATMS;   W/NBS   63281

509   C20 H47 N O6 Si4   53110-68-2   NS--28703-0-0   α-D-Glucopyranose, 2-(acetylamino)-2-deoxy-1,3,4,6-tetrakis-
O-(trimethylsilyl)-;;   NBS   63286

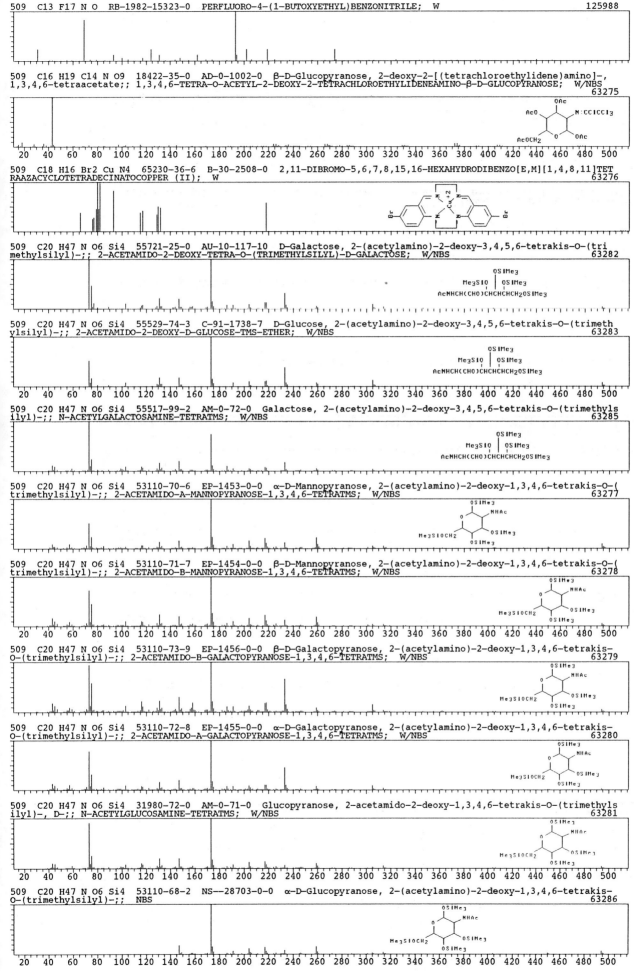

- 5485 -

509 C20 H47 N O6 Si4 53110-69-3 NS--28702-0-0 β-D-Glucopyranose, 2-(acetylamino)-2-deoxy-1,3,4,6-tetrakis-O-(trimethylsilyl)-;; NBS
63287

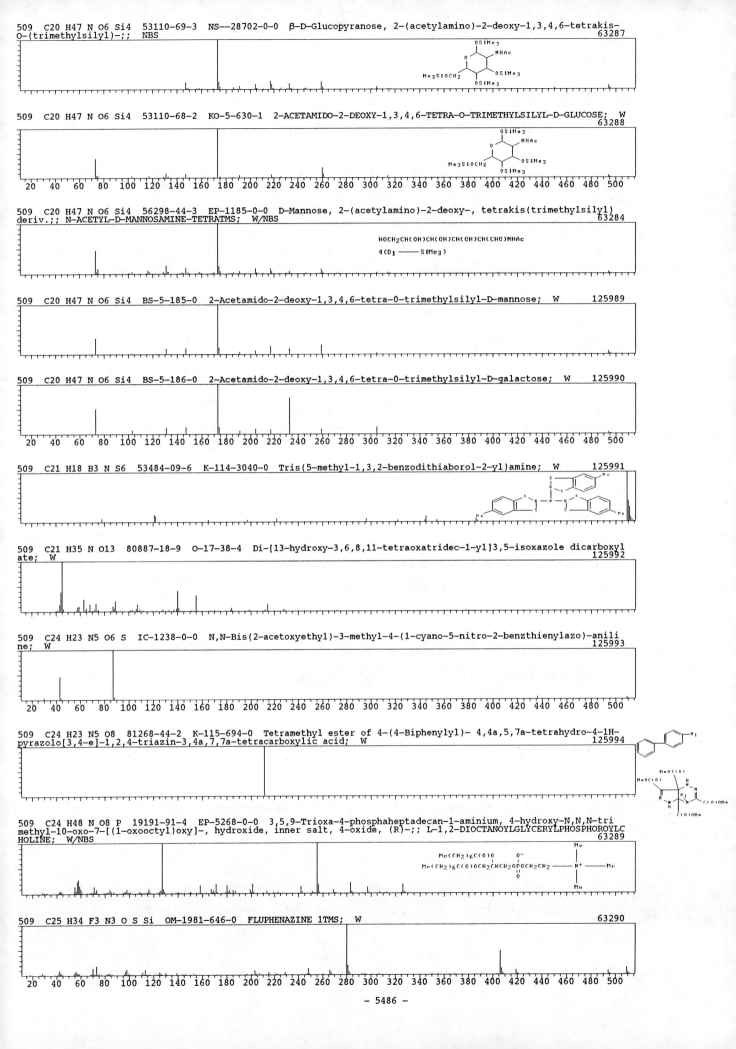

509 C20 H47 N O6 Si4 53110-68-2 KO-5-630-1 2-ACETAMIDO-2-DEOXY-1,3,4,6-TETRA-O-TRIMETHYLSILYL-D-GLUCOSE; W
63288

509 C20 H47 N O6 Si4 56298-44-3 EP-1185-0-0 D-Mannose, 2-(acetylamino)-2-deoxy-, tetrakis(trimethylsilyl) deriv.;; N-ACETYL-D-MANNOSAMINE-TETRATMS; W/NBS
63284

509 C20 H47 N O6 Si4 BS-5-185-0 2-Acetamido-2-deoxy-1,3,4,6-tetra-0-trimethylsilyl-D-mannose; W   125989

509 C20 H47 N O6 Si4 BS-5-186-0 2-Acetamido-2-deoxy-1,3,4,6-tetra-0-trimethylsilyl-D-galactose; W   125990

509 C21 H18 B3 N S6 53484-09-6 K-114-3040-0 Tris(5-methyl-1,3,2-benzodithiaborol-2-yl)amine; W   125991

509 C21 H35 N O13 80887-18-9 O-17-38-4 Di-[13-hydroxy-3,6,8,11-tetraoxatridec-1-yl]3,5-isoxazole dicarboxylate; W   125992

509 C24 H23 N5 O6 S IC-1238-0-0 N,N-Bis(2-acetoxyethyl)-3-methyl-4-(1-cyano-5-nitro-2-benzthienylazo)-aniline; W   125993

509 C24 H23 N5 O8 81268-44-2 K-115-694-0 Tetramethyl ester of 4-(4-Biphenylyl)- 4,4a,5,7a-tetrahydro-4-1H-pyrazolo[3,4-e]-1,2,4-triazin-3,4a,7,7a-tetracarboxylic acid; W   125994

509 C24 H48 N O8 P 19191-91-4 EP-5268-0-0 3,5,9-Trioxa-4-phosphaheptadecan-1-aminium, 4-hydroxy-N,N,N-trimethyl-10-oxo-7-[(1-oxooctyl)oxy]-, hydroxide, inner salt, 4-oxide, (R)-;; L-1,2-DIOCTANOYLGLYCERYLPHOSPHOROYLCHOLINE; W/NBS
63289

509 C25 H34 F3 N3 O S Si OM-1981-646-0 FLUPHENAZINE 1TMS; W   63290

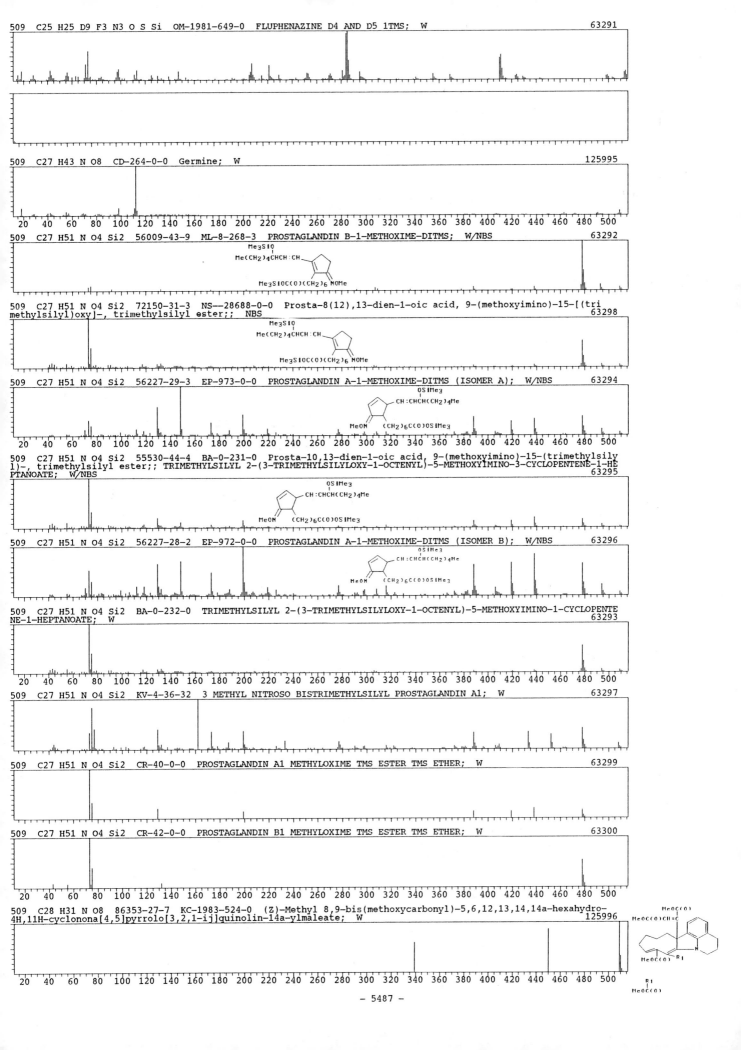

509   C25 H25 D9 F3 N3 O S Si   OM-1981-649-0   FLUPHENAZINE D4 AND D5 1TMS;   W   63291

509   C27 H43 N O8   CD-264-0-0   Germine;   W   125995

509   C27 H51 N O4 Si2   56009-43-9   ML-8-268-3   PROSTAGLANDIN B-1-METHOXIME-DITMS;   W/NBS   63292

509   C27 H51 N O4 Si2   72150-31-3   NS--28688-0-0   Prosta-8(12),13-dien-1-oic acid, 9-(methoxyimino)-15-[(tri
methylsilyl)oxy]-, trimethylsilyl ester;;   NBS   63298

509   C27 H51 N O4 Si2   56227-29-3   EP-973-0-0   PROSTAGLANDIN A-1-METHOXIME-DITMS (ISOMER A);   W/NBS   63294

509   C27 H51 N O4 Si2   55530-44-4   BA-0-231-0   Prosta-10,13-dien-1-oic acid, 9-(methoxyimino)-15-(trimethylsily
l)-, trimethylsilyl ester;; TRIMETHYLSILYL 2-(3-TRIMETHYLSILYLOXY-1-OCTENYL)-5-METHOXYIMINO-3-CYCLOPENTENE-1-HE
PTANOATE;   W/NBS   63295

509   C27 H51 N O4 Si2   56227-28-2   EP-972-0-0   PROSTAGLANDIN A-1-METHOXIME-DITMS (ISOMER B);   W/NBS   63296

509   C27 H51 N O4 Si2   BA-0-232-0   TRIMETHYLSILYL 2-(3-TRIMETHYLSILYLOXY-1-OCTENYL)-5-METHOXYIMINO-1-CYCLOPENTE
NE-1-HEPTANOATE;   W   63293

509   C27 H51 N O4 Si2   KV-4-36-32   3 METHYL NITROSO BISTRIMETHYLSILYL PROSTAGLANDIN A1;   W   63297

509   C27 H51 N O4 Si2   CR-40-0-0   PROSTAGLANDIN A1 METHYLOXIME TMS ESTER TMS ETHER;   W   63299

509   C27 H51 N O4 Si2   CR-42-0-0   PROSTAGLANDIN B1 METHYLOXIME TMS ESTER TMS ETHER;   W   63300

509   C28 H31 N O8   86353-27-7   KC-1983-524-0   (Z)-Methyl 8,9-bis(methoxycarbonyl)-5,6,12,13,14,14a-hexahydro-
4H,11H-cyclonona[4,5]pyrrolo[3,2,1-ij]quinolin-14a-ylmaleate;   W   125996

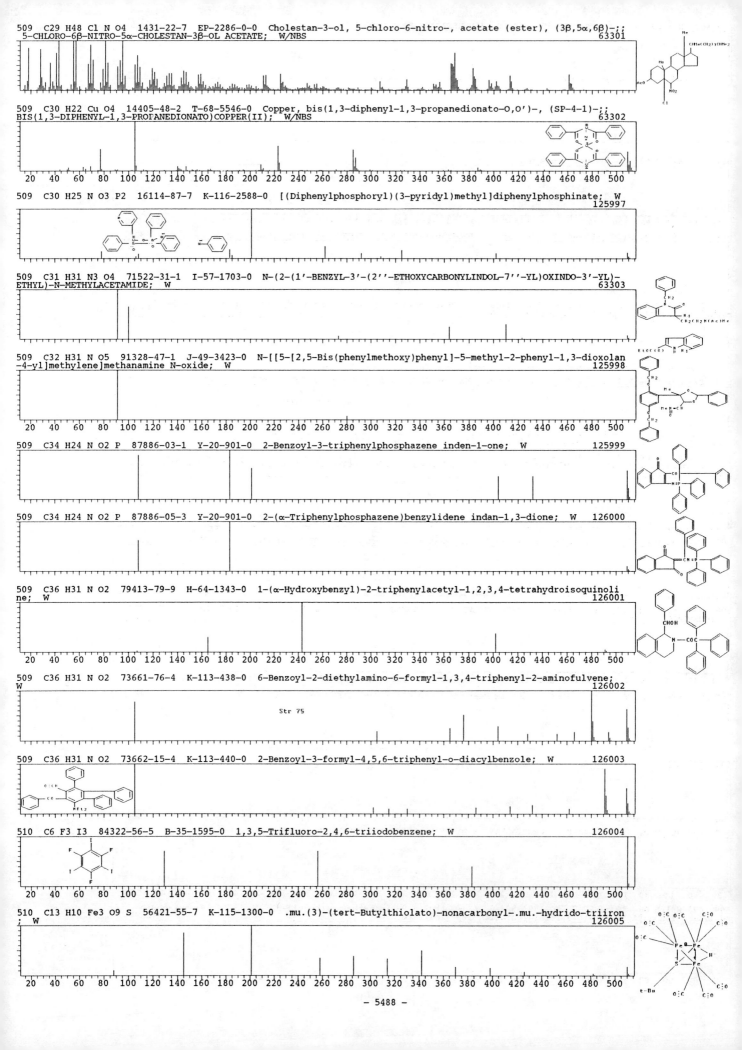

509 C29 H48 Cl N O4 1431-22-7 EP-2286-0-0 Cholestan-3-ol, 5-chloro-6-nitro-, acetate (ester), (3β,5α,6β)-;;
5-CHLORO-6β-NITRO-5α-CHOLESTAN-3β-OL ACETATE; W/NBS   63301

509 C30 H22 Cu O4 14405-48-2 T-68-5546-0 Copper, bis(1,3-diphenyl-1,3-propanedionato-O,O')-, (SP-4-1)-;;
BIS(1,3-DIPHENYL-1,3-PROPANEDIONATO)COPPER(II); W/NBS   63302

509 C30 H25 N O3 P2 16114-87-7 K-116-2588-0 [(Diphenylphosphoryl)(3-pyridyl)methyl]diphenylphosphinate; W
125997

509 C31 H31 N3 O4 71522-31-1 I-57-1703-0 N-(2-(1'-BENZYL-3'-(2''-ETHOXYCARBONYLINDOL-7''-YL)OXINDO-3'-YL)-
ETHYL)-N-METHYLACETAMIDE; W   63303

509 C32 H31 N O5 91328-47-1 J-49-3423-0 N-[[5-[2,5-Bis(phenylmethoxy)phenyl]-5-methyl-2-phenyl-1,3-dioxolan
-4-yl]methylene]methanamine N-oxide; W   125998

509 C34 H24 N O2 P 87886-03-1 Y-20-901-0 2-Benzoyl-3-triphenylphosphazene inden-1-one; W   125999

509 C34 H24 N O2 P 87886-05-3 Y-20-901-0 2-(α-Triphenylphosphazene)benzylidene indan-1,3-dione; W   126000

509 C36 H31 N O2 79413-79-9 H-64-1343-0 1-(α-Hydroxybenzyl)-2-triphenylacetyl-1,2,3,4-tetrahydroisoquinoli
ne; W   126001

509 C36 H31 N O2 73661-76-4 K-113-438-0 6-Benzoyl-2-diethylamino-6-formyl-1,3,4-triphenyl-2-aminofulvene;
W   126002

Str 75

509 C36 H31 N O2 73662-15-4 K-113-440-0 2-Benzoyl-3-formyl-4,5,6-triphenyl-o-diacylbenzole; W   126003

510 C6 F3 I3 84322-56-5 B-35-1595-0 1,3,5-Trifluoro-2,4,6-triiodobenzene; W   126004

510 C13 H10 Fe3 O9 S 56421-55-7 K-115-1300-0 .mu.(3)-(tert-Butylthiolato)-nonacarbonyl-.mu.-hydrido-triiron
; W   126005

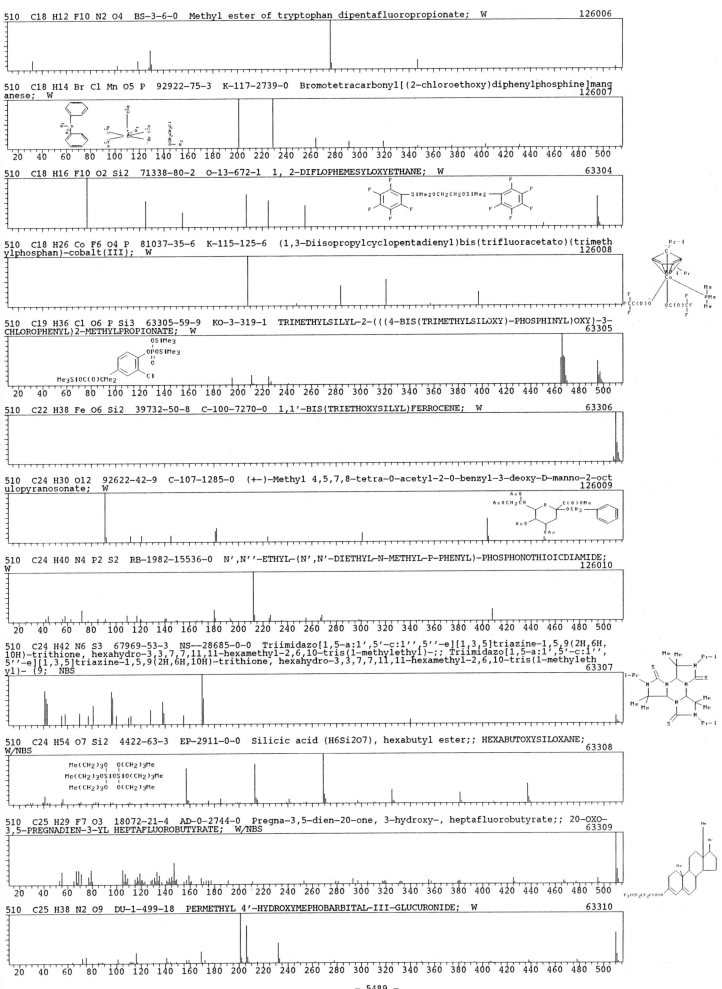

510   C18 H12 F10 N2 O4   BS-3-6-0   Methyl ester of tryptophan dipentafluoropropionate;   W   126006

510   C18 H14 Br Cl Mn O5 P   92922-75-3   K-117-2739-0   Bromotetracarbonyl[(2-chloroethoxy)diphenylphosphine]manganese;   W   126007

510   C18 H16 F10 O2 Si2   71338-80-2   O-13-672-1   1,2-DIFLOPHEMESYLOXYETHANE;   W   63304

510   C18 H26 Co F6 O4 P   81037-35-6   K-115-125-6   (1,3-Diisopropylcyclopentadienyl)bis(trifluoracetato)(trimethylphosphan)-cobalt(III);   W   126008

510   C19 H36 Cl O6 P Si3   63305-59-9   KO-3-319-1   TRIMETHYLSILYL-2-(((4-BIS(TRIMETHYLSILOXY)-PHOSPHINYL)OXY)-3-CHLOROPHENYL)2-METHYLPROPIONATE;   W   63305

510   C22 H38 Fe O6 Si2   39732-50-8   C-100-7270-0   1,1'-BIS(TRIETHOXYSILYL)FERROCENE;   W   63306

510   C24 H30 O12   92622-42-9   C-107-1285-0   (+-)-Methyl 4,5,7,8-tetra-0-acetyl-2-0-benzyl-3-deoxy-D-manno-2-octulopyranosonate;   W   126009

510   C24 H40 N4 P2 S2   RB-1982-15536-0   N',N''-ETHYL-(N',N'-DIETHYL-N-METHYL-P-PHENYL)-PHOSPHONOTHIOICDIAMIDE;   W   126010

510   C24 H42 N6 S3   67969-53-3   NS--28685-0-0   Triimidazo[1,5-a:1',5'-c:1'',5''-e][1,3,5]triazine-1,5,9(2H,6H,10H)-trithione, hexahydro-3,3,7,7,11,11-hexamethyl-2,6,10-tris(1-methylethyl)-;;   Triimidazo[1,5-a:1',5'-c:1'',5''-e][1,3,5]triazine-1,5,9(2H,6H,10H)-trithione, hexahydro-3,3,7,7,11,11-hexamethyl-2,6,10-tris(1-methylethyl)- (9;   NBS   63307

510   C24 H54 O7 Si2   4422-63-3   EP-2911-0-0   Silicic acid (H6Si2O7), hexabutyl ester;;   HEXABUTOXYSILOXANE;   W/NBS   63308

510   C25 H29 F7 O3   18072-21-4   AD-0-2744-0   Pregna-3,5-dien-20-one, 3-hydroxy-, heptafluorobutyrate;;   20-OXO-3,5-PREGNADIEN-3-YL HEPTAFLUOROBUTYRATE;   W/NBS   63309

510   C25 H38 N2 O9   DU-1-499-18   PERMETHYL 4'-HYDROXYMEPHOBARBITAL-III-GLUCURONIDE;   W   63310

510   C26 H22 O11   C-105-1612-0   Triacetate derivative of 4-demethoxydaunomycinone (isomer B);   W   126011

510   C26 H38 O10   61543-94-0   O-13-308-1   (8)METHYL ESTER OF 1H-2α,6α,9α-TRIACETYLOXY-7α-CARBOXY-8β-CARBOXYMETH
YL-4Aβ,7β-DIMETHYLPERHYDROPHENANTHRENE;   W   63311

510   C26 H42 N2 O8   86534-19-2   F-40-2969-0   N,N'-Dimethyl diaza-crown ether;   W   126012

510   C27 H26 O10   80144-98-5   B-34-2009-0   2''-O-Methyltenuiorin;   W   126013

510   C27 H26 O10   80144-97-4   B-34-2009-0   3-Hydroxy-4-methoxycarbonyl-5-methylphenyl 4-(2-hydroxy-4-methoxy-6-
methylbenzoyloxy)-2-methoxy-6-methylbenzoate;   W   126014

510   C27 H26 O10   80145-00-2   B-34-2011-0   3-Hydroxy-4-methoxycarbonyl-5-methylphenyl 2-hydroxy-4-(2,4-dimeth
oxy-6-methylbenzoyloxy)-6-methylbenzoate;   W   126015

510   C27 H27 Br O5   93286-38-5   H-67-1339-0   2,3,4-Tri-O-benzyl-6-bromo-6-deoxy-D-mannono-1,5-lactone;   W   126016

510   C27 H27 Cl N2 O2 S2   72047-60-0   J-45-4371-0   5-((4-chlorophenyl)sulfonyl)-2-(diphenylmethylene)-N,N-dieth
yl-2,5-dihydro-5-methyl-4-thiazolamine;   W   126017

510   C27 H50 O5 Si2   56248-51-2   D-10-3660-3   Prosta-7,13-dien-1-oic acid, 6,9-epoxy-11,15-bis[(trimethylsilyl)
oxy]-, methyl ester, (13E,15S)-;; BISTRIMETHYLSILYL ETHER 6(9)-OXY-11,15-DIHYDROXYPROSTA-7,13-DIENOIC ACID METH
YLESTER;   W/NBS   63312

510   C27 H50 O5 Si2   74985-64-1   NS--28681-0-0   Prosta-7,13-dien-1-oic acid, 6,9-epoxy-11,15-bis[(trimethylsily
l)oxy]-, methyl ester;;   NBS   63313

510   C27 H50 O5 Si2   C-100-769-0   PROSTACYCLIN METHYL ESTER BIS(TRIMETHYLSILY)ETHER;   W   63314

510   C27 H50 O5 Si2   OM-1981-549-0   6(9)-OXY-PGF-1A ME 2TMS;   W   63315

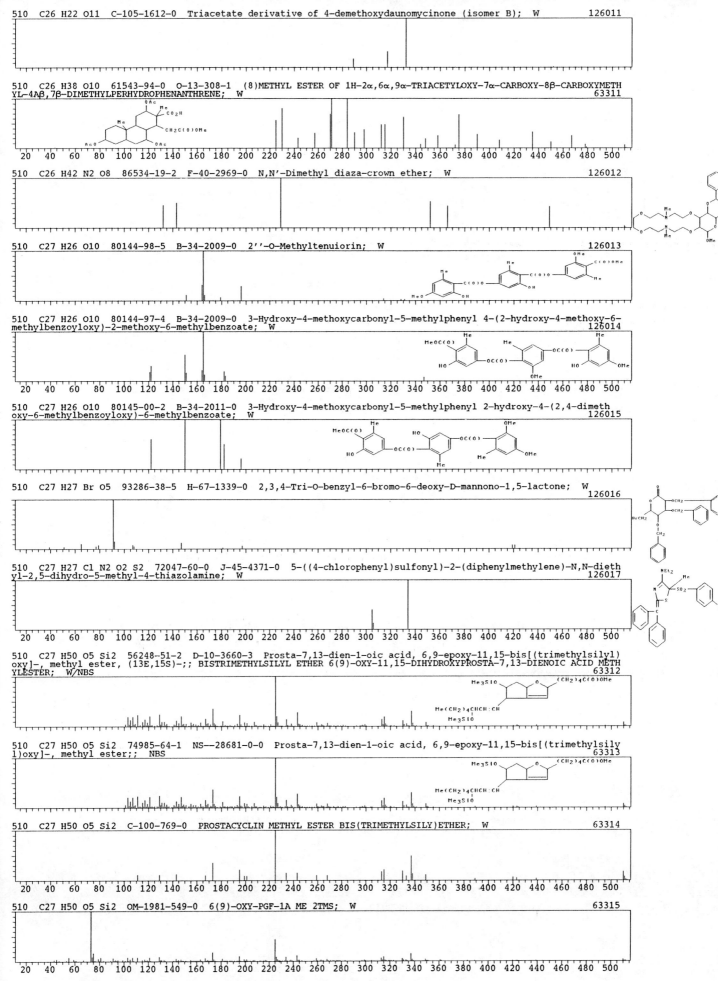

510 C28 H38 N4 O5 L2-1975-73-4 O,N-PERMETHYLATED PHE-GLY-PHE-GLY; W   63316

510 C28 H46 O8 F-21-1807-7 DESACYL-KONDURENGOGENIN-A-MONOSIDS; W   63317

510 C28 H34 O8 30576-19-3 O-3-330-6 3-Cyclohexene-1-acetic acid, 4-[1α-(3-furyl-3,7,8,8a-tetrahydro-8aα-
methyl-3-oxo-1H-2-benzopyran-5-yl)methyl]-3-hydroxy-2α,6,6-trimethyl-5-oxo-, methyl ester, acetate;; C.O.D. ACE
TATE; W/NBS   63318

510 C29 H42 N4 O4 57237-94-2 O7-0-817-0 4-DIMETHYLAMINO-1-NAPHTHYLIDENE VALYLISOLEUCYLALANINE ETHYL ESTER;
W/NBS   63319

510 C30 H22 O4 S2 61613-14-7 K-109-3925-0 3,3'-Bis(p-tolylthio)-3,3'-biphthalidyl; W   126018

510 C30 H22 O4 Zn 21333-45-9 T-68-5549-0 Zinc, bis(1,3-diphenyl-1,3-propanedionato-O,O')-, (T-4)-;; BIS(1,
3-DIPHENYL-1,3-PROPANEDIONATO)ZINC(II); W/NBS   63320

510 C30 H26 N2 O2 S2 74725-64-7 J-45-4370-0 1-methyl-1H-indol-3-yl ester of N-((4-methylphenyl)sulfonyl)-α-
phenyl-benzeneethanimidothioic acid; W   126019

510 C30 H26 N2 O6 40732-55-6 O7-0-317-0 3'-O-ACETYL-5'-O-TRITYL-2,2'-ANHYDROURIDINE; W/NBS   63321

510 C30 H30 N4 O4 76013-29-1 F-36-1837-0 12,18-Dimethyl-13,17-di(2'-methoxycarbonylethyl)porphin; W 126020

510 C30 H30 N4 O4 76003-91-3 F-36-1837-0 12,17-Dimethyl-13,18-di(2'-methoxycarbonylethyl)porphin; W 126021

510 C30 H30 N4 O4 76003-90-2 F-36-1837-0 12,18-Di(2'-methoxycarbonylethyl)-13,17-dimethylporphin; W 126022

510 C30 H38 O7 78012-25-6 KC-1983-2847-0 Zeylasterone; W   126023

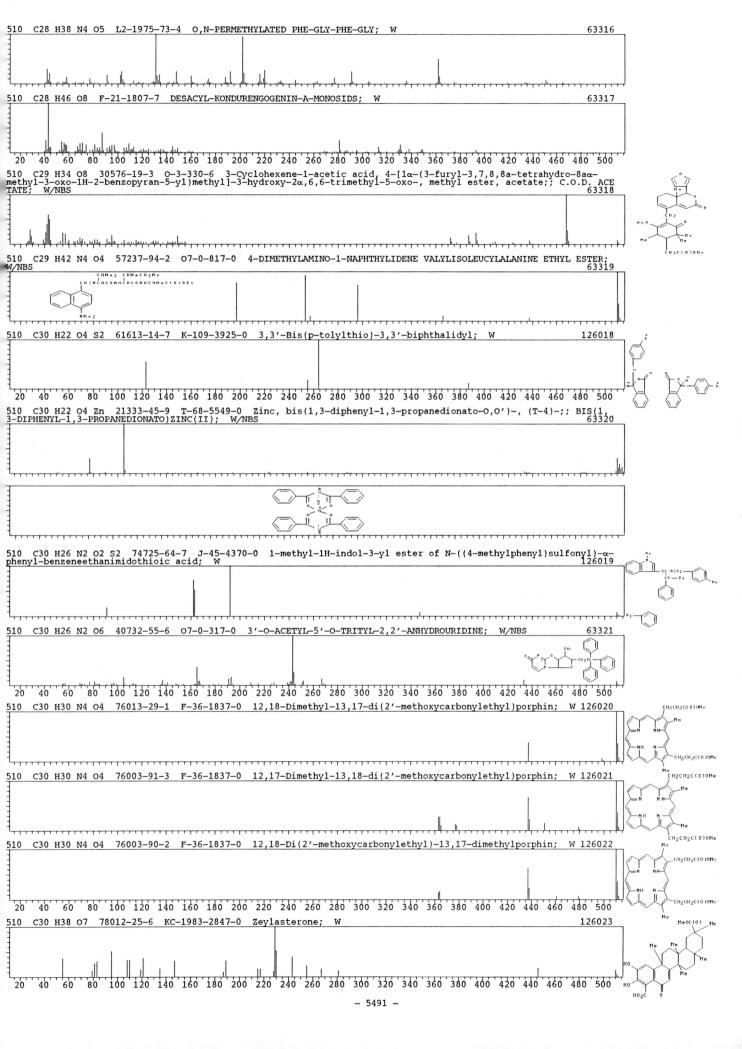

- 5491 -

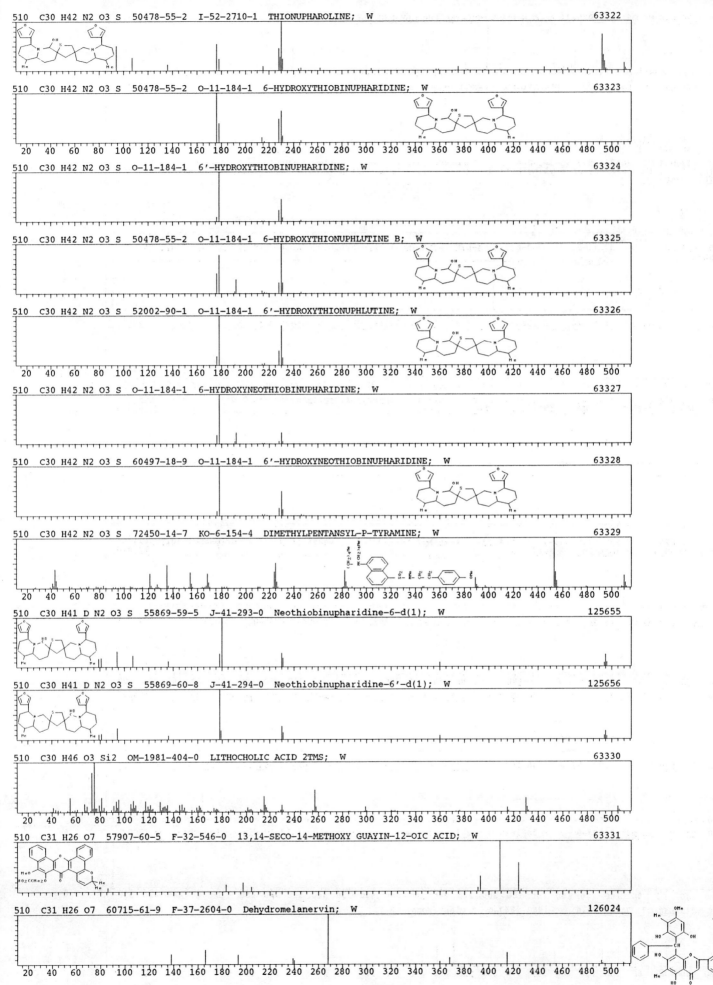

510   C30 H42 N2 O3 S   50478-55-2   I-52-2710-1   THIONUPHAROLINE;   W   63322

510   C30 H42 N2 O3 S   50478-55-2   O-11-184-1   6-HYDROXYTHIOBINUPHARIDINE;   W   63323

510   C30 H42 N2 O3 S   O-11-184-1   6'-HYDROXYTHIOBINUPHARIDINE;   W   63324

510   C30 H42 N2 O3 S   50478-55-2   O-11-184-1   6-HYDROXYTHIONUPHLUTINE B;   W   63325

510   C30 H42 N2 O3 S   52002-90-1   O-11-184-1   6'-HYDROXYTHIONUPHLUTINE;   W   63326

510   C30 H42 N2 O3 S   O-11-184-1   6-HYDROXYNEOTHIOBINUPHARIDINE;   W   63327

510   C30 H42 N2 O3 S   60497-18-9   O-11-184-1   6'-HYDROXYNEOTHIOBINUPHARIDINE;   W   63328

510   C30 H42 N2 O3 S   72450-14-7   KO-6-154-4   DIMETHYLPENTANSYL-P-TYRAMINE;   W   63329

510   C30 H41 D N2 O3 S   55869-59-5   J-41-293-0   Neothiobinupharidine-6-d(1);   W   125655

510   C30 H41 D N2 O3 S   55869-60-8   J-41-294-0   Neothiobinupharidine-6'-d(1);   W   125656

510   C30 H46 O3 Si2   OM-1981-404-0   LITHOCHOLIC ACID 2TMS;   W   63330

510   C31 H26 O7   57907-60-5   F-32-546-0   13,14-SECO-14-METHOXY GUAYIN-12-OIC ACID;   W   63331

510   C31 H26 O7   60715-61-9   F-37-2604-0   Dehydromelanervin;   W   126024

510   C31 H49 F3 O2   55517-92-5   T-67-4984-0   Stigmast-7-en-3-ol, trifluoroacetate, (3β,24.xi.)-;; 24-ETHYL-3β-
TRIFLUOROACETOXY-5.XI.-CHOLEST-7-ENE;   W/NBS   63332

Str 76

510   C32 H31 O2 P S   64894-34-4   KC-1977-1462-0   2-DIPHENYLPHOSPHINOYL-1-PHENYL-2-(5-PHENYLTHIOCYCLO-PENT-1-ENY
L)PROPAN-1-OL;   W   63333

Str 77

20   40   60   80   100   120   140   160   180   200   220   240   260   280   300   320   340   360   380   400   420   440   460   480   500

510   C32 H38 N4 O2   78668-16-3   F-37-765-0   (Z,Z,Z)-1,19-Dioxo-3,8,12,17-tetraethyl-2,7,13,18-tetramethyl-21,24
-methylen-1,19,22,24-tetrahydro-21H-biline;   W   126025

510   C32 H38 N4 O2   78668-17-4   F-37-765-0   (Z,Z,Z)-1,19-Dioxo-3,8,12,17-tetraethyl-2,7,13,18-tetramethyl-21,22
-methylen-1,19,22,24-tetrahydro-21H-biline;   W   126026

510   C32 H38 N4 O2   78668-18-5   F-37-765-0   (Z,Z,Z)-1,19-Dioxo-2,3,12,17-tetraethyl-2,7,13,18-tetramethyl-22,23
-methylen-1,19,22,23-tetrahydro-21H-biline;   W   126027

20   40   60   80   100   120   140   160   180   200   220   240   260   280   300   320   340   360   380   400   420   440   460   480   500

510   C32 H46 O5   56052-65-4   SD-1981-0-0   Lanosta-9(11),25-dien-18-oic acid, 3-(acetyloxy)-20-hydroxy-16-oxo-
γ-lactone, (3β)-;;   W/NBS   63334

510   C33 H50 O4   39701-82-1   AQ-6-585-23   Oleana-12,15-dien-28-oic acid, 3-(acetyloxy)-, methyl ester, (3β)-;;
METHYL 15-DEHYDRO-OLEANOLATE-3-ACETATE;   W/NBS   63335

510   C33 H66 O3   18951-36-5   T-67-5099-0   METHYL CORYNOMYLATE;   W/NBS   63336

Me(CH2)14CH(OH)CHC(O)OMe
(CH2)13Me

20   40   60   80   100   120   140   160   180   200   220   240   260   280   300   320   340   360   380   400   420   440   460   480   500

510   C33 H66 O3   17369-87-8   EP-4829-0-0   Octadecanoic acid, 3-hydroxy-2-tetradecyl-, methyl ester;; METHYL-2-
TETRADECANYL-3-HYDROXY OCTADECANOATE (α);   W/NBS   63337

Me(CH2)14CH(OH)CHC(O)OMe
(CH2)13Me

510   C33 H66 O3   28808-36-8   T-67-5100-0   PHTHIODIOLONE A;   W/NBS   63338

Me(CH2)28CH(OH)CH2CH(OH)(CH2)4CHMeCOEt

510   C34 H40 P2   27721-03-5   AG-125-210-1   1,10-Bis(diphenylphosphino)decane;   W   126028

P(CH2)10P

510   C34 H70 O2   17367-09-8   EP-5201-0-0   Hexadecane, 1,1'-[1,2-ethanediylbis(oxy)]bis-;; 1,2-DIHEXADECYLOXY
ETHANE;   W/NBS   63339

Me(CH2)15OCH2CH2O(CH2)15Me

20   40   60   80   100   120   140   160   180   200   220   240   260   280   300   320   340   360   380   400   420   440   460   480   500

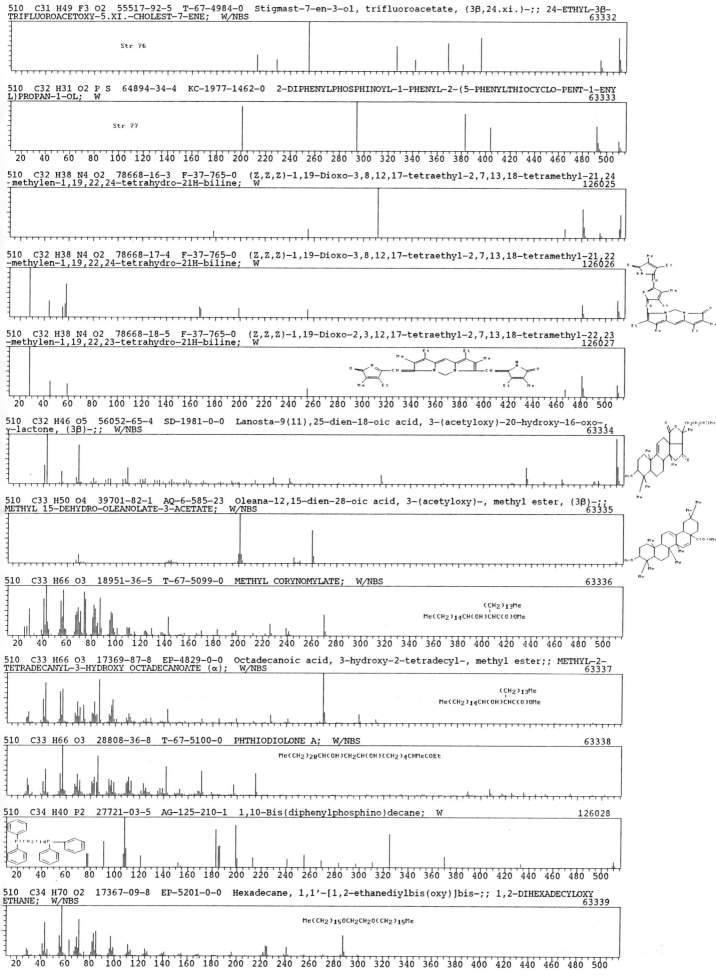

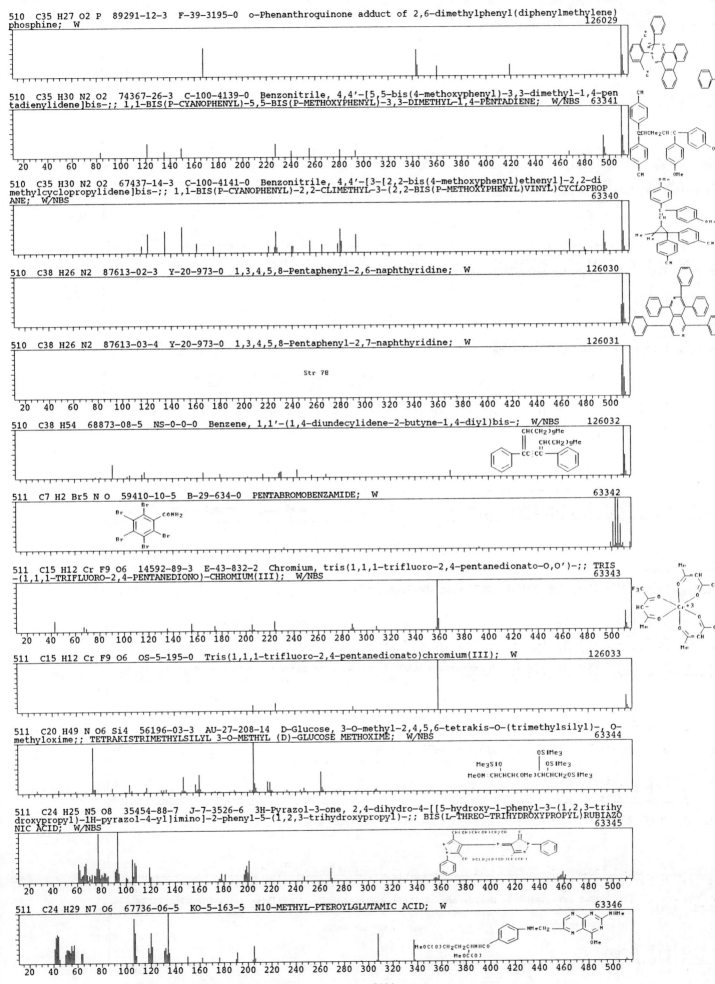

510 C35 H27 O2 P 89291-12-3 F-39-3195-0 o-Phenanthroquinone adduct of 2,6-dimethylphenyl(diphenylmethylene)phosphine; W
126029

510 C35 H30 N2 O2 74367-26-3 C-100-4139-0 Benzonitrile, 4,4'-[5,5-bis(4-methoxyphenyl)-3,3-dimethyl-1,4-pentadienylidene]bis-;; 1,1-BIS(P-CYANOPHENYL)-5,5-BIS(P-METHOXYPHENYL)-3,3-DIMETHYL-1,4-PENTADIENE; W/NBS 63341

510 C35 H30 N2 O2 67437-14-3 C-100-4141-0 Benzonitrile, 4,4'-[3-[2,2-bis(4-methoxyphenyl)ethenyl]-2,2-dimethylcyclopropylidene]bis-;; 1,1-BIS(P-CYANOPHENYL)-2,2-DIMETHYL-3-(2,2-BIS(P-METHOXYPHENYL)VINYL)CYCLOPROPANE; W/NBS
63340

510 C38 H26 N2 87613-02-3 Y-20-973-0 1,3,4,5,8-Pentaphenyl-2,6-naphthyridine; W 126030

510 C38 H26 N2 87613-03-4 Y-20-973-0 1,3,4,5,8-Pentaphenyl-2,7-naphthyridine; W 126031

Str 78

510 C38 H54 68873-08-5 NS-0-0-0 Benzene, 1,1'-(1,4-diundecylidene-2-butyne-1,4-diyl)bis-; W/NBS 126032

511 C7 H2 Br5 N O 59410-10-5 B-29-634-0 PENTABROMOBENZAMIDE; W 63342

511 C15 H12 Cr F9 O6 14592-89-3 E-43-832-2 Chromium, tris(1,1,1-trifluoro-2,4-pentanedionato-O,O')-;; TRIS-(1,1,1-TRIFLUORO-2,4-PENTANEDIONO)-CHROMIUM(III); W/NBS
63343

511 C15 H12 Cr F9 O6 OS-5-195-0 Tris(1,1,1-trifluoro-2,4-pentanedionato)chromium(III); W 126033

511 C20 H49 N O6 Si4 56196-03-3 AU-27-208-14 D-Glucose, 3-O-methyl-2,4,5,6-tetrakis-O-(trimethylsilyl)-, O-methyloxime;; TETRAKISTRIMETHYLSILYL 3-O-METHYL (D)-GLUCOSE METHOXIME; W/NBS 63344

511 C24 H25 N5 O8 35454-88-7 J-7-3526-6 3H-Pyrazol-3-one, 2,4-dihydro-4-[[5-hydroxy-1-phenyl-3-(1,2,3-trihydroxypropyl)-1H-pyrazol-4-yl]imino]-2-phenyl-5-(1,2,3-trihydroxypropyl)-;; BIS(L-THREO-TRIHYDROXYPROPYL)RUBIAZONIC ACID; W/NBS
63345

511 C24 H29 N7 O6 67736-06-5 KO-5-163-5 N10-METHYL-PTEROYLGLUTAMIC ACID; W 63346

511   C27 H33 N3 O7 12771-72-1 NS-0-0-0   VERRUCULOGEN (IMPURE);   W/NBS   126034

511   C28 H24 Co N2 S2 53575-86-3 B-29-1420-0   Bis[2-{(p-tolylimino)methyl}benzenethiolato-S,N]cobalt(II); W
                                                                                              126035

20   40   60   80   100 120 140 160 180 200 220 240 260 280 300 320 340 360 380 400 420 440 460 480 500

511   C28 H37 N O6 Si C-106-4185-0   Benzocyclobutenedione;   W   126036

511   C28 H37 N O6 Si 90696-89-2 C-106-4185-0   2,8-[(2',8'-Dioxa-7-oxo-8-methyl)nonamethylene]-3-[(2-t-butyldi
methylsiloxy)-3-cyano]propylnaphthoquinone;   W   126037

511   C28 H37 N3 O6 37059-21-5 J-49-2668-0   Triethyl-4,4',4''-Methylidynetris(3,5-dimethyl-1H-pyrrole-2-carbox
ylate;   W   126038

20   40   60   80   100 120 140 160 180 200 220 240 260 280 300 320 340 360 380 400 420 440 460 480 500

511   C28 H57 N O3 Si2 SW-0-231-0   BISTRIMETHYLSILYL N-ACETYL EICOSASPHINGA-4,11-DIENINE;   W   63347

511   C29 H28 F3 N O4 K-107-2210-21   1-(2-(α-CARBOXYBENZYL)-CIS-1,2-DIPHENYLCYCLOPROPYL)PYRROLIDINIUM-TFA; W
                                                                                               63348

511   C30 H29 N3 O5 18732-52-0 AD-0-1328-0   Propiophenone, 3,3''-(hydroxyimino)bis[2'-hydroxy-3-phenyl-, dioxi
me;; N,N-BIS-(1-PHENYL-3-HYDROXYIMINO-3-O-HYDROXYPHENYL-PROPYL)HYDROXYL-AMINE;   W/NBS   63349

20   40   60   80   100 120 140 160 180 200 220 240 260 280 300 320 340 360 380 400 420 440 460 480 500

511   C31 H33 N3 O4 IC-1240-0-0   3-(4-Ethoxyanilinocarbonyl)-1-(4-hexyloxyphenylazo)-2-naphthalenol; W 126039

511   C31 H49 N3 O3 89647-80-3 I-62-300-0   N-Formylspirolucidine;   W   126040

511   C37 H25 N3 85731-54-0 Y-19-1488-0   2,3,5,7,8-Pentaphenylpyrido[3,4-b]pyridine;   W   126042

Str 79

20   40   60   80   100 120 140 160 180 200 220 240 260 280 300 320 340 360 380 400 420 440 460 480 500

512   C5 Br5 F3 RB-1982-15324-0   PENTABROMOTRIFLUOROCYCLOPENTENE;   W   126043

20   40   60   80   100 120 140 160 180 200 220 240 260 280 300 320 340 360 380 400 420 440 460 480 500

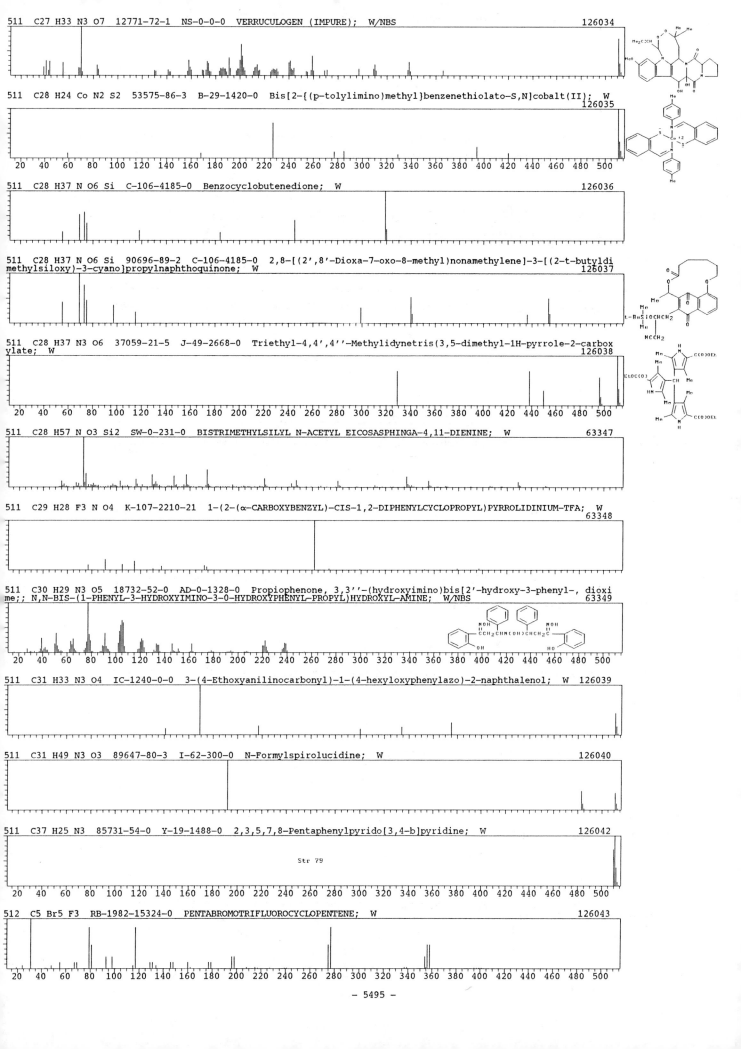

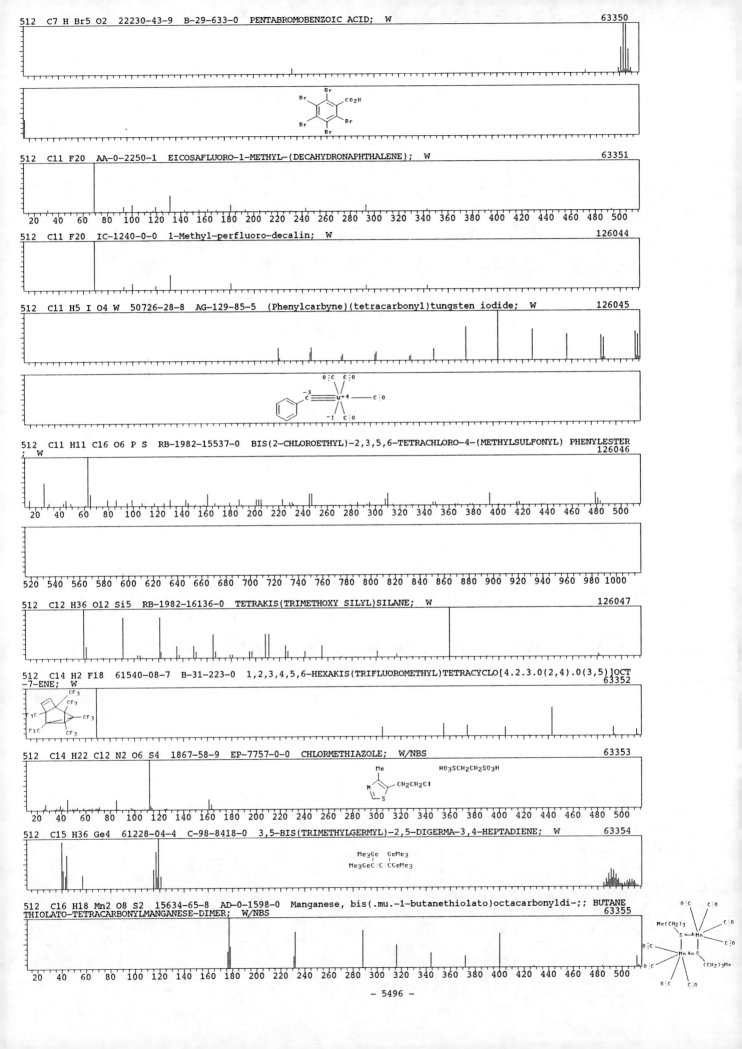

512 C7 H Br5 O2 22230-43-9 B-29-633-0 PENTABROMOBENZOIC ACID; W   63350

512 C11 F20 AA-0-2250-1 EICOSAFLUORO-1-METHYL-(DECAHYDRONAPHTHALENE); W   63351

512 C11 F20 IC-1240-0-0 1-Methyl-perfluoro-decalin; W   126044

512 C11 H5 I O4 W 50726-28-8 AG-129-85-5 (Phenylcarbyne)(tetracarbonyl)tungsten iodide; W   126045

512 C11 H11 Cl6 O6 P S RB-1982-15537-0 BIS(2-CHLOROETHYL)-2,3,5,6-TETRACHLORO-4-(METHYLSULFONYL) PHENYLESTER
; W   126046

512 C12 H36 O12 Si5 RB-1982-16136-0 TETRAKIS(TRIMETHOXY SILYL)SILANE; W   126047

512 C14 H2 F18 61540-08-7 B-31-223-0 1,2,3,4,5,6-HEXAKIS(TRIFLUOROMETHYL)TETRACYCLO[4.2.3.0(2,4).0(3,5)]OCT
-7-ENE; W   63352

512 C14 H22 Cl2 N2 O6 S4 1867-58-9 EP-7757-0-0 CHLORMETHIAZOLE; W/NBS   63353

512 C15 H36 Ge4 61228-04-4 C-98-8418-0 3,5-BIS(TRIMETHYLGERMYL)-2,5-DIGERMA-3,4-HEPTADIENE; W   63354

512 C16 H18 Mn2 O8 S2 15634-65-8 AD-0-1598-0 Manganese, bis(.mu.-1-butanethiolato)octacarbonyldi-;; BUTANE
THIOLATO-TETRACARBONYLMANGANESE-DIMER; W/NBS   63355

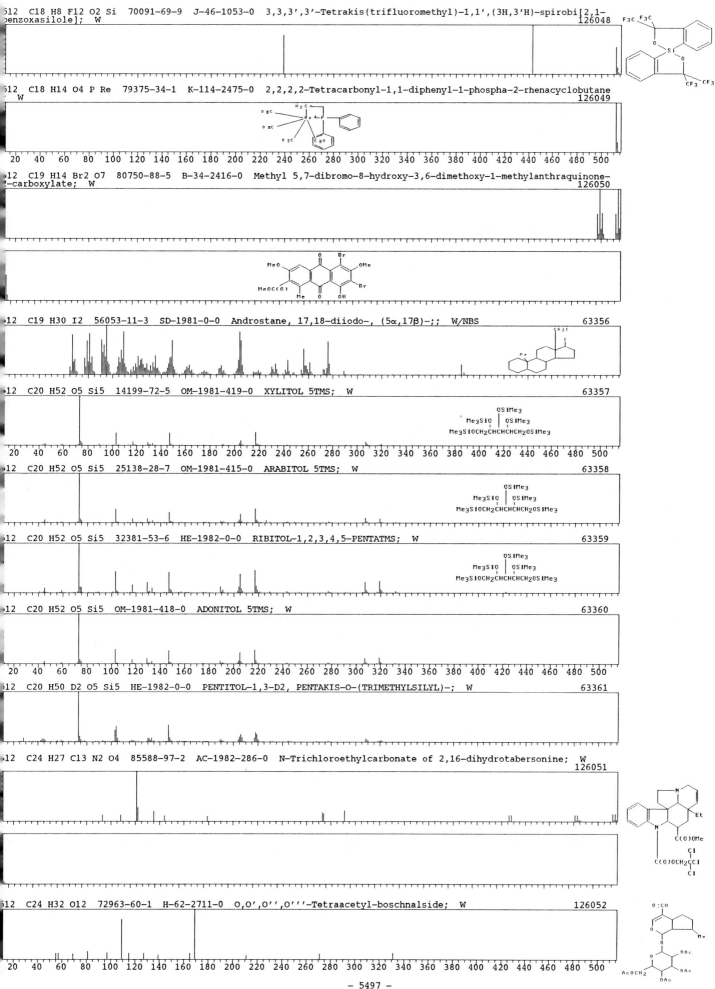

512 C18 H8 F12 O2 Si 70091-69-9 J-46-1053-0 3,3,3',3'-Tetrakis(trifluoromethyl)-1,1',(3H,3'H)-spirobi[2,1-benzoxasilole]; W
126048

512 C18 H14 O4 P Re 79375-34-1 K-114-2475-0 2,2,2,2-Tetracarbonyl-1,1-diphenyl-1-phospha-2-rhenacyclobutane; W
126049

512 C19 H14 Br2 O7 80750-88-5 B-34-2416-0 Methyl 5,7-dibromo-8-hydroxy-3,6-dimethoxy-1-methylanthraquinone-2-carboxylate; W
126050

512 C19 H30 I2 56053-11-3 SD-1981-0-0 Androstane, 17,18-diiodo-, (5α,17β)-;; W/NBS
63356

512 C20 H52 O5 Si5 14199-72-5 OM-1981-419-0 XYLITOL 5TMS; W
63357

512 C20 H52 O5 Si5 25138-28-7 OM-1981-415-0 ARABITOL 5TMS; W
63358

512 C20 H52 O5 Si5 32381-53-6 HE-1982-0-0 RIBITOL-1,2,3,4,5-PENTATMS; W
63359

512 C20 H52 O5 Si5 OM-1981-418-0 ADONITOL 5TMS; W
63360

512 C20 H50 D2 O5 Si5 HE-1982-0-0 PENTITOL-1,3-D2, PENTAKIS-O-(TRIMETHYLSILYL)-; W
63361

512 C24 H27 Cl3 N2 O4 85588-97-2 AC-1982-286-0 N-Trichloroethylcarbonate of 2,16-dihydrotabersonine; W
126051

512 C24 H32 O12 72963-60-1 H-62-2711-0 O,O',O'',O'''-Tetraacetyl-boschnalside; W
126052

512 C24 H20 D12 O12 H-62-2711-0 O,O',O'',O'''-Tetra(acetyl-D3)-boschnalside; W   126053

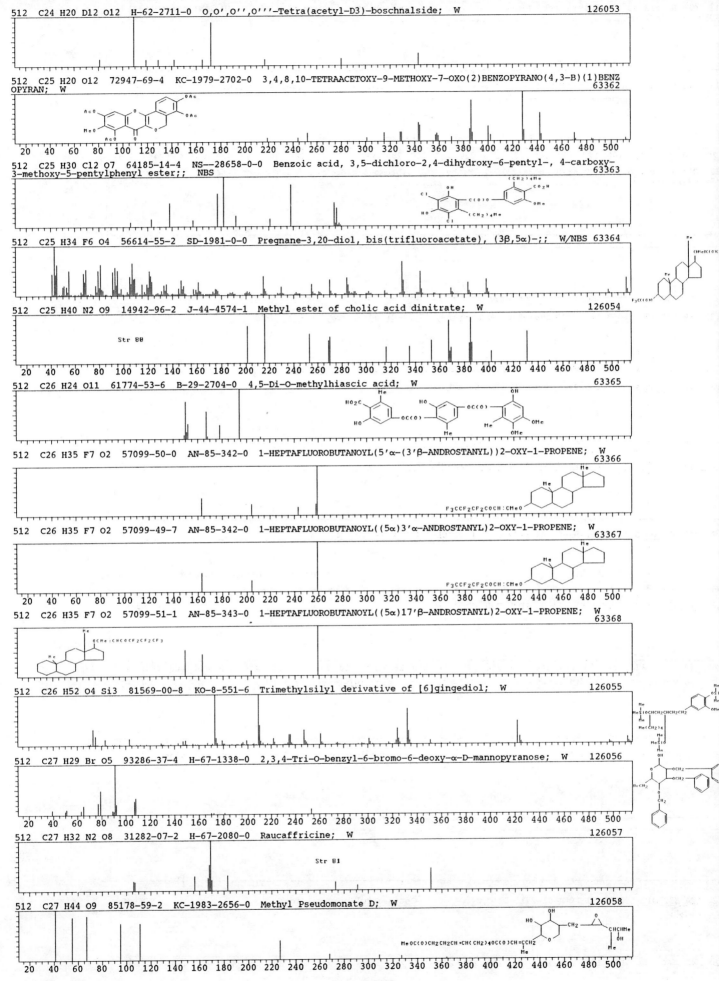

512 C25 H20 O12 72947-69-4 KC-1979-2702-0 3,4,8,10-TETRAACETOXY-9-METHOXY-7-OXO(2)BENZOPYRANO(4,3-B)(1)BENZ
OPYRAN; W   63362

512 C25 H30 Cl2 O7 64185-14-4 NS--28658-0-0 Benzoic acid, 3,5-dichloro-2,4-dihydroxy-6-pentyl-, 4-carboxy-
3-methoxy-5-pentylphenyl ester;; NBS   63363

512 C25 H34 F6 O4 56614-55-2 SD-1981-0-0 Pregnane-3,20-diol, bis(trifluoroacetate), (3β,5α)-;; W/NBS 63364

512 C25 H40 N2 O9 14942-96-2 J-44-4574-1 Methyl ester of cholic acid dinitrate; W   126054

Str 80

512 C26 H24 O11 61774-53-6 B-29-2704-0 4,5-Di-O-methylhiascic acid; W   63365

512 C26 H35 F7 O2 57099-50-0 AN-85-342-0 1-HEPTAFLUOROBUTANOYL(5'α-(3'β-ANDROSTANYL))2-OXY-1-PROPENE; W
                                                                                                       63366

512 C26 H35 F7 O2 57099-49-7 AN-85-342-0 1-HEPTAFLUOROBUTANOYL((5α)3'α-ANDROSTANYL)2-OXY-1-PROPENE; W
                                                                                                       63367

512 C26 H35 F7 O2 57099-51-1 AN-85-343-0 1-HEPTAFLUOROBUTANOYL((5α)17'β-ANDROSTANYL)2-OXY-1-PROPENE; W
                                                                                                       63368

512 C26 H52 O4 Si3 81569-00-8 KO-8-551-6 Trimethylsilyl derivative of [6]gingediol; W   126055

512 C27 H29 Br O5 93286-37-4 H-67-1338-0 2,3,4-Tri-O-benzyl-6-bromo-6-deoxy-α-D-mannopyranose; W 126056

512 C27 H32 N2 O8 31282-07-2 H-67-2080-0 Raucaffricine; W   126057

Str 81

512 C27 H44 O9 85178-59-2 KC-1983-2656-0 Methyl Pseudomonate D; W   126058

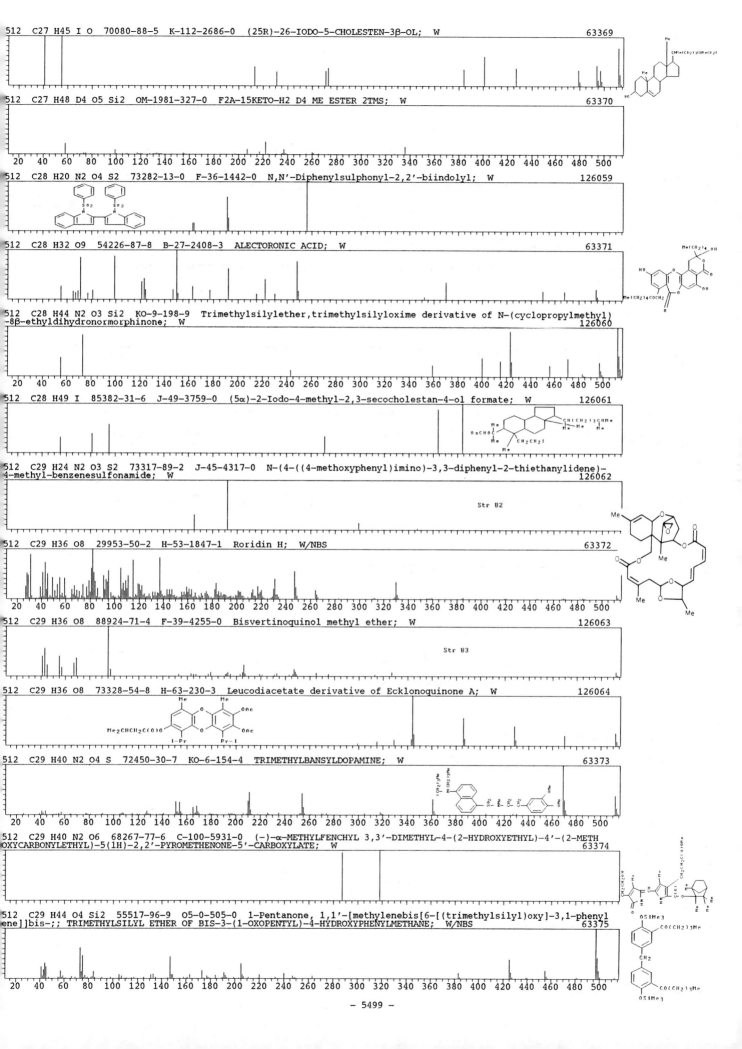

512　C27 H45 I O　70080-88-5　K-112-2686-0　(25R)-26-IODO-5-CHOLESTEN-3β-OL;　W　63369

512　C27 H48 D4 O5 Si2　OM-1981-327-0　F2A-15KETO-H2 D4 ME ESTER 2TMS;　W　63370

512　C28 H20 N2 O4 S2　73282-13-0　F-36-1442-0　N,N'-Diphenylsulphonyl-2,2'-biindolyl;　W　126059

512　C28 H32 O9　54226-87-8　B-27-2408-3　ALECTORONIC ACID;　W　63371

512　C28 H44 N2 O3 Si2　KO-9-198-9　Trimethylsilylether,trimethylsilyloxime derivative of N-(cyclopropylmethyl)-8β-ethyldihydronormorphinone;　W　126060

512　C28 H49 I　85382-31-6　J-49-3759-0　(5α)-2-Iodo-4-methyl-2,3-secocholestan-4-ol formate;　W　126061

512　C29 H24 N2 O3 S2　73317-89-2　J-45-4317-0　N-(4-((4-methoxyphenyl)imino)-3,3-diphenyl-2-thiethanylidene)-4-methyl-benzenesulfonamide;　W　126062

Str 82

512　C29 H36 O8　29953-50-2　H-53-1847-1　Roridin H;　W/NBS　63372

512　C29 H36 O8　88924-71-4　F-39-4255-0　Bisvertinoquinol methyl ether;　W　126063

Str 83

512　C29 H36 O8　73328-54-8　H-63-230-3　Leucodiacetate derivative of Ecklonoquinone A;　W　126064

512　C29 H40 N2 O4 S　72450-30-7　KO-6-154-4　TRIMETHYLBANSYLDOPAMINE;　W　63373

512　C29 H40 N2 O6　68267-77-6　C-100-5931-0　(-)-α-METHYLFENCHYL 3,3'-DIMETHYL-4-(2-HYDROXYETHYL)-4'-(2-METHOXYCARBONYLETHYL)-5(1H)-2,2'-PYROMETHENONE-5'-CARBOXYLATE;　W　63374

512　C29 H44 O4 Si2　55517-96-9　O5-0-505-0　1-Pentanone, 1,1'-[methylenebis[6-[(trimethylsilyl)oxy]-3,1-phenylene]]bis-;; TRIMETHYLSILYL ETHER OF BIS-3-(1-OXOPENTYL)-4-HYDROXYPHENYLMETHANE;　W/NBS　63375

512 C29 H26 D18 O4 Si2 O-5-516-10 BIS-4-HYDROXY-3-(1-OXOPENTYL) PHENYLMETHANE DI-TMS D9-ETHER; W   63376

512 C30 H32 N4 O4 27934-21-0 O-3-344-3 5,10,15,20(22H,24H)-Porphinetetrone, 2,7,12-triethyl-3,8,13,18-tetra
methyl-;; 2,4,8-TRIETHYL-1,3,5,7-TETRAMETHYL-α,β,γ,Δ-TETRAOXOPORPHIN; W/NBS   63377

512 C30 H33 Fe O2 P C-106-6070-0 Dicarbonyl(5-(but-2-enyl)-1,3-cycloheptadiene)(triphenylphosphine)iron; W
                                                                                                   126065

512 C30 H36 N2 O2 Si2 70390-36-2 K-112-946-0 3,3'-(1,2-DIPHENYL-1,2-BIS(TRIMETHYLSILOXY)-1,2-ETHANDIYL)BIS(
PYRIDINE); W   63378

512 C30 H38 N4 Ni 74834-13-2 Y-17-441-0 5,7,12,14-Tetra-n-propyldibenzo[b,l][1,4,8,11]tetraazacyclotetradec
ahexaenatonickel(II); W   126066

512 C30 H40 O Ti2 HE-1986-1018-0 .MU.-OXO-BIS(CYCLOPENTADIENYL-PENTAMETHYLCYCLOPENTADIENYL)BIS-TITANIUM; W
                                                                                                   126067

512 C30 H40 O7 21902-99-8 J-4-3863-3 Withacnistin; W/NBS   63379

Str 84

512 C30 H40 O7 KC-1982-2839-0 (22R)-4β-Acetoxy-5β,6β-epoxy-27-hydroxy-1-oxowitha-2,24-dienolide; W   126068

512 C30 H56 O6 56554-53-1 EP-5056-0-0 Hexanoic acid, 3,5,5-trimethyl-, 1,2,3-propanetriyl ester;; TRI (3,
5,5 TRIMETHYLHEXANOYL) GLYCEROL; W/NBS   63380

OC(O)CH2CHMeCH2CMe3
Me3CCH2CHMeCH2C(O)OCH2CHCH2OC(O)CH2CHMeCH2CMe3

512 C30 H56 O6 OM-1981-777-0 TRINONANOIN OTMS; W   63381

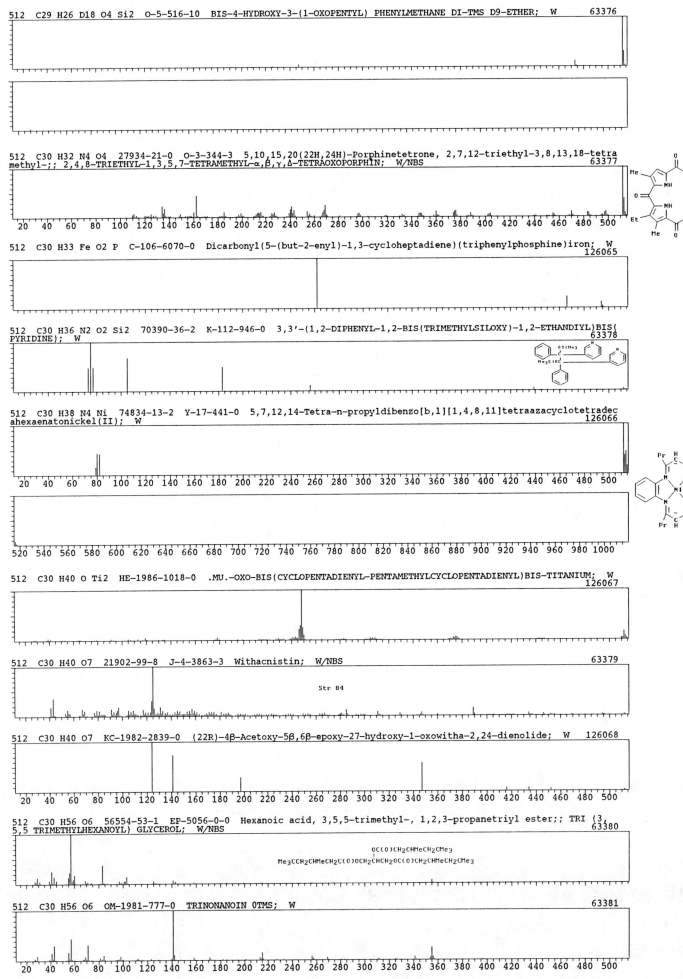

512 C30 H60 N2 O4 67456-22-8 H-61-1529-0 Acetamide, 2,2'-[1,2-ethanediylbis(oxy)]bis[N,N-dihexyl-;; N,N,N',
N'-Tetrahexyl-3,6-dioxaoctandiamide; W/NBS   126069

Me(CH2)5   (CH2)5Me
Me(CH2)5NCOCH2OCH2CH2OCH2CON(CH2)5Me

512 C31 H28 N8 85738-50-7 KC-1983-15-0 5,7-Dimethyl-3-phenyl-2-[3-(5,7-dimethyl-3-phenylpyrazolo[1,5-a]pyri
midin-2-ylamino)prop-2-enylideneamino]pyrazolo[1,5-a]pyrimidine; W   126070

512 C31 H28 O7 82571-09-3 F-38-136-0 (2,4-Dihydroxy-6-methoxy)-tolyl-3-(5,7-dihydroxy-6-methyl)-flavanonyl-
8-phenylmethane; W   126071

512 C31 H28 O7 60715-58-4 F-37-2603-0 2,4-Dihydroxy-6-methoxy-tolyl-3-(5,7-dihydroxy-6-methyl)-flavanonyl-
8-phenylmethane; W   126072

512 C31 H48 N2 O4 H-62-1594-0 3β-Actoxy-4'ε-hydroxymethyl-2'-cyclohexyl-3',4',16β,17β-tetrahydro-2'H-5α-andr
ostano(16,17-E)(1',2')oxazin-3'ε-carbonitrile; W   126073

512 C31 H48 O4 Si 56051-70-8 T-66-2568-0 AZAFRIN METHYLESTER TRIMETHYLSILYL ETHER; W/NBS   63383

Me3SiO Me
CH:CHCMe:CHCH:CHCMe:CHCH:CHCH:CMeCH:CHC(O)OMe
OH
Me
Me

512 C31 H48 O4 Si H-65-362-0 Methyl ester of (+)-(5S,6S)-6-Hydroxy-5-trimethylsilyloxy-5,6-dihydro-10'-apo-β
-carotin-10'-oic acid; W   126464

512 C31 H60 O5 HE-1982-0-0 GLYCERINE-1,3-DIMYRISTATE; W   63384

512 C32 H32 O6 68313-25-7 B-31-1746-0 3,3,6,6-TETRA(4-METHOXYPHENYL)-1,2-DIOXANE; W   63385

512 C32 H32 O6 IC-1240-0-0 Diamyl binaphthalene dioxide quinone; W   126074

512 C32 H33 Br O 65549-94-2 K-111-269-0 4-(4-BROMOPHENYL)-2,6-BIS(4-TERT-BUTYLPHENYL)PHENOL; W   63386

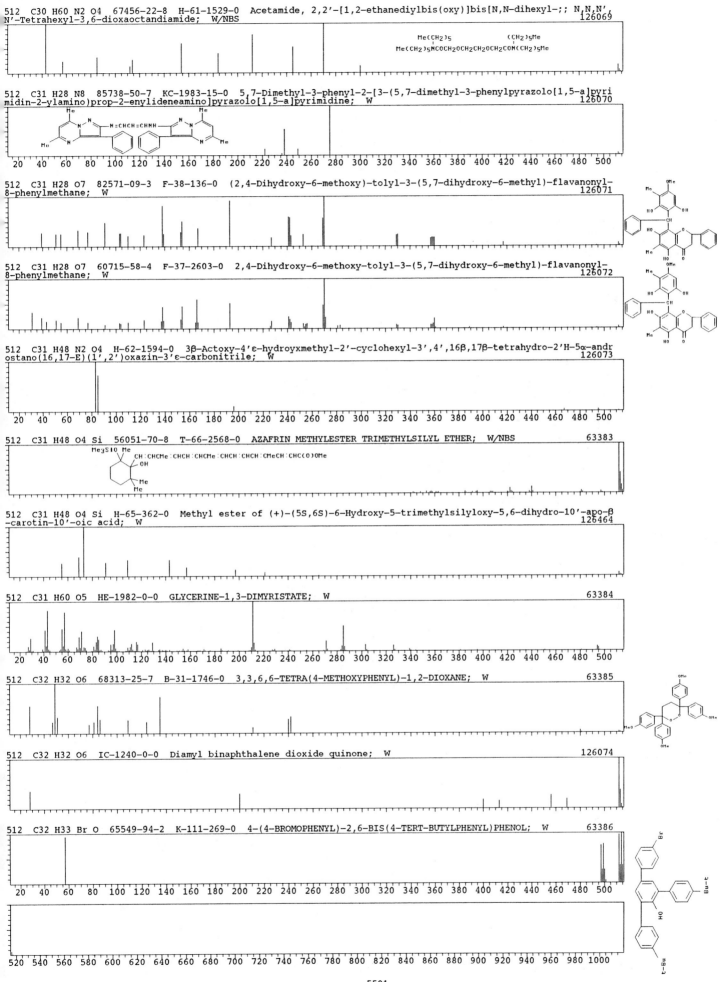

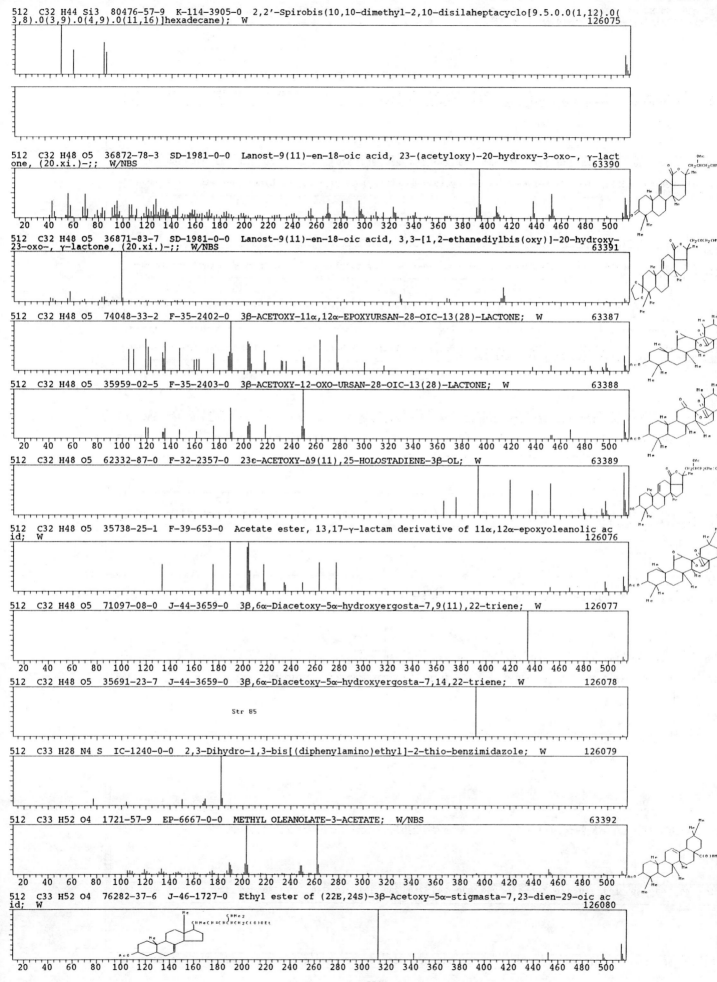

512 C32 H44 Si3 80476-57-9 K-114-3905-0 2,2'-Spirobis(10,10-dimethyl-2,10-disilaheptacyclo[9.5.0.0(1,12).0(3,8).0(3,9).0(4,9).0(11,16)]hexadecane); W
126075

512 C32 H48 O5 36872-78-3 SD-1981-0-0 Lanost-9(11)-en-18-oic acid, 23-(acetyloxy)-20-hydroxy-3-oxo-, γ-lactone, (20.xi.)-;; W/NBS
63390

20 40 60 80 100 120 140 160 180 200 220 240 260 280 300 320 340 360 380 400 420 440 460 480 500

512 C32 H48 O5 36871-83-7 SD-1981-0-0 Lanost-9(11)-en-18-oic acid, 3,3-[1,2-ethanediylbis(oxy)]-20-hydroxy-23-oxo-, γ-lactone, (20.xi.)-;; W/NBS
63391

512 C32 H48 O5 74048-33-2 F-35-2402-0 3β-ACETOXY-11α,12α-EPOXYURSAN-28-OIC-13(28)-LACTONE; W
63387

512 C32 H48 O5 35959-02-5 F-35-2403-0 3β-ACETOXY-12-OXO-URSAN-28-OIC-13(28)-LACTONE; W
63388

20 40 60 80 100 120 140 160 180 200 220 240 260 280 300 320 340 360 380 400 420 440 460 480 500

512 C32 H48 O5 62332-87-0 F-32-2357-0 23ε-ACETOXY-Δ9(11),25-HOLOSTADIENE-3β-OL; W
63389

512 C32 H48 O5 35738-25-1 F-39-653-0 Acetate ester, 13,17-γ-lactam derivative of 11α,12α-epoxyoleanolic acid; W
126076

512 C32 H48 O5 71097-08-0 J-44-3659-0 3β,6α-Diacetoxy-5α-hydroxyergosta-7,9(11),22-triene; W
126077

20 40 60 80 100 120 140 160 180 200 220 240 260 280 300 320 340 360 380 400 420 440 460 480 500

512 C32 H48 O5 35691-23-7 J-44-3659-0 3β,6α-Diacetoxy-5α-hydroxyergosta-7,14,22-triene; W
126078

Str 85

512 C33 H28 N4 S IC-1240-0-0 2,3-Dihydro-1,3-bis[(diphenylamino)ethyl]-2-thio-benzimidazole; W
126079

512 C33 H52 O4 1721-57-9 EP-6667-0-0 METHYL OLEANOLATE-3-ACETATE; W/NBS
63392

20 40 60 80 100 120 140 160 180 200 220 240 260 280 300 320 340 360 380 400 420 440 460 480 500

512 C33 H52 O4 76282-37-6 J-46-1727-0 Ethyl ester of (22E,24S)-3β-Acetoxy-5α-stigmasta-7,23-dien-29-oic acid; W
126080

20 40 60 80 100 120 140 160 180 200 220 240 260 280 300 320 340 360 380 400 420 440 460 480 500

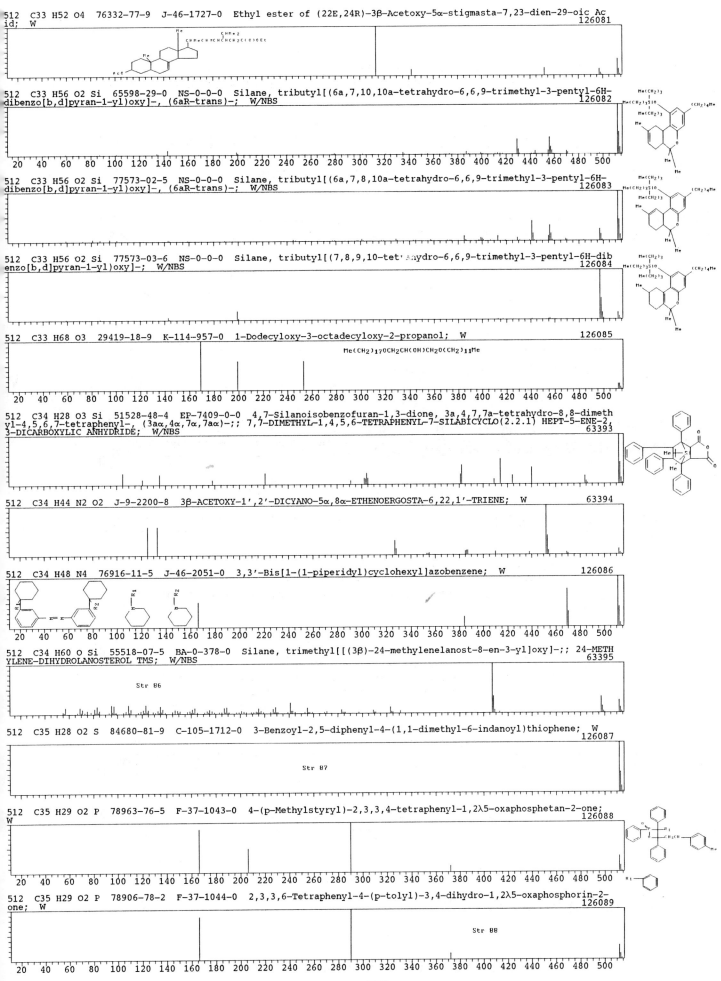

512 C33 H52 O4 76332-77-9 J-46-1727-0 Ethyl ester of (22E,24R)-3β-Acetoxy-5α-stigmasta-7,23-dien-29-oic Ac id; W
126081

512 C33 H56 O2 Si 65598-29-0 NS-0-0-0 Silane, tributyl[(6a,7,10,10a-tetrahydro-6,6,9-trimethyl-3-pentyl-6H-dibenzo[b,d]pyran-1-yl)oxy]-, (6aR-trans)-; W/NBS
126082

512 C33 H56 O2 Si 77573-02-5 NS-0-0-0 Silane, tributyl[(6a,7,8,10a-tetrahydro-6,6,9-trimethyl-3-pentyl-6H-dibenzo[b,d]pyran-1-yl)oxy]-, (6aR-trans)-; W/NBS
126083

512 C33 H56 O2 Si 77573-03-6 NS-0-0-0 Silane, tributyl[(7,8,9,10-tetrahydro-6,6,9-trimethyl-3-pentyl-6H-dibenzo[b,d]pyran-1-yl)oxy]-; W/NBS
126084

512 C33 H68 O3 29419-18-9 K-114-957-0 1-Dodecyloxy-3-octadecyloxy-2-propanol; W
126085

Me(CH₂)₁₇OCH₂CH(OH)CH₂O(CH₂)₁₁Me

512 C34 H28 O3 Si 51528-48-4 EP-7409-0-0 4,7-Silanoisobenzofuran-1,3-dione, 3a,4,7,7a-tetrahydro-8,8-dimethyl-4,5,6,7-tetraphenyl-, (3aα,4α,7α,7aα)-;; 7,7-DIMETHYL-1,4,5,6-TETRAPHENYL-7-SILABICYCLO(2.2.1) HEPT-5-ENE-2,3-DICARBOXYLIC ANHYDRIDE; W/NBS
63393

512 C34 H44 N2 O2 J-9-2200-8 3β-ACETOXY-1',2'-DICYANO-5α,8α-ETHENOERGOSTA-6,22,1'-TRIENE; W
63394

512 C34 H48 N4 76916-11-5 J-46-2051-0 3,3'-Bis[1-(1-piperidyl)cyclohexyl]azobenzene; W
126086

512 C34 H60 O Si 55518-07-5 BA-0-378-0 Silane, trimethyl[[(3β)-24-methylenelanost-8-en-3-yl]oxy]-;; 24-METHYLENE-DIHYDROLANOSTEROL TMS; W/NBS
63395

Str 86

512 C35 H28 O2 S 84680-81-9 C-105-1712-0 3-Benzoyl-2,5-diphenyl-4-(1,1-dimethyl-6-indanoyl)thiophene; W
126087

Str 87

512 C35 H29 O2 P 78963-76-5 F-37-1043-0 4-(p-Methylstyryl)-2,3,3,4-tetraphenyl-1,2λ5-oxaphosphetan-2-one; W
126088

512 C35 H29 O2 P 78906-78-2 F-37-1044-0 2,3,3,6-Tetraphenyl-4-(p-tolyl)-3,4-dihydro-1,2λ5-oxaphosphorin-2-one; W
126089

Str 88

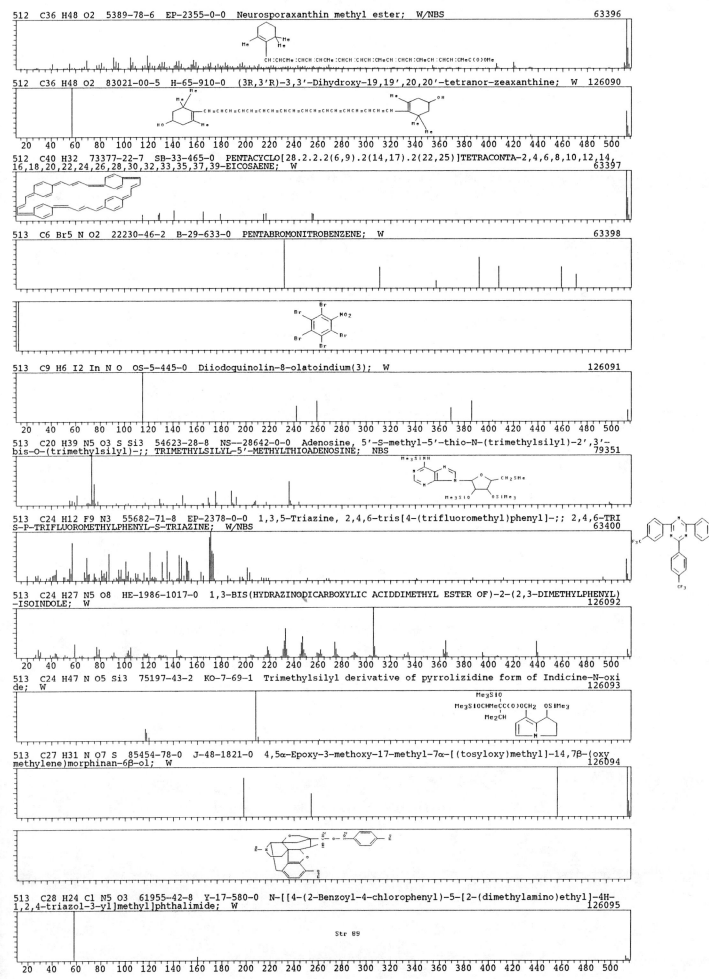

512   C36 H48 O2   5389-78-6   EP-2355-0-0   Neurosporaxanthin methyl ester;   W/NBS   63396

CH:CHCMe:CHCH:CHCMe:CHCH:CHCH:CMeCH:CHCH:CHeCH:CHCH:CMeC(O)OMe

512   C36 H48 O2   83021-00-5   H-65-910-0   (3R,3'R)-3,3'-Dihydroxy-19,19',20,20'-tetranor-zeaxanthine;   W   126090

512   C40 H32   73377-22-7   SB-33-465-0   PENTACYCLO[28.2.2.2(6,9).2(14,17).2(22,25)]TETRACONTA-2,4,6,8,10,12,14,
16,18,20,22,24,26,28,30,32,33,35,37,39-EICOSAENE;   W   63397

513   C6 Br5 N O2   22230-46-2   B-29-633-0   PENTABROMONITROBENZENE;   W   63398

513   C9 H6 I2 In N O   OS-5-445-0   Diiodoquinolin-8-olatoindium(3);   W   126091

513   C20 H39 N5 O3 S Si3   54623-28-8   NS--28642-0-0   Adenosine, 5'-S-methyl-5'-thio-N-(trimethylsilyl)-2',3'-
bis-O-(trimethylsilyl)-;;   TRIMETHYLSILYL-5'-METHYLTHIOADENOSINE;   NBS   79351

513   C24 H12 F9 N3   55682-71-8   EP-2378-0-0   1,3,5-Triazine, 2,4,6-tris[4-(trifluoromethyl)phenyl]-;;   2,4,6-TRI
S-P-TRIFLUOROMETHYLPHENYL-S-TRIAZINE;   W/NBS   63400

513   C24 H27 N5 O8   HE-1986-1017-0   1,3-BIS(HYDRAZINODICARBOXYLIC ACIDDIMETHYL ESTER OF)-2-(2,3-DIMETHYLPHENYL)
-ISOINDOLE;   W   126092

513   C24 H47 N O5 Si3   75197-43-2   KO-7-69-1   Trimethylsilyl derivative of pyrrolizidine form of Indicine-N-oxi
de;   W   126093

513   C27 H31 N O7 S   85454-78-0   J-48-1821-0   4,5α-Epoxy-3-methoxy-17-methyl-7α-[(tosyloxy)methyl]-14,7β-(oxy
methylene)morphinan-6β-ol;   W   126094

513   C28 H24 Cl N5 O3   61955-42-8   Y-17-580-0   N-[[4-(2-Benzoyl-4-chlorophenyl)-5-[2-(dimethylamino)ethyl]-4H-
1,2,4-triazol-3-yl]methyl]phthalimide;   W   126095

Str 89

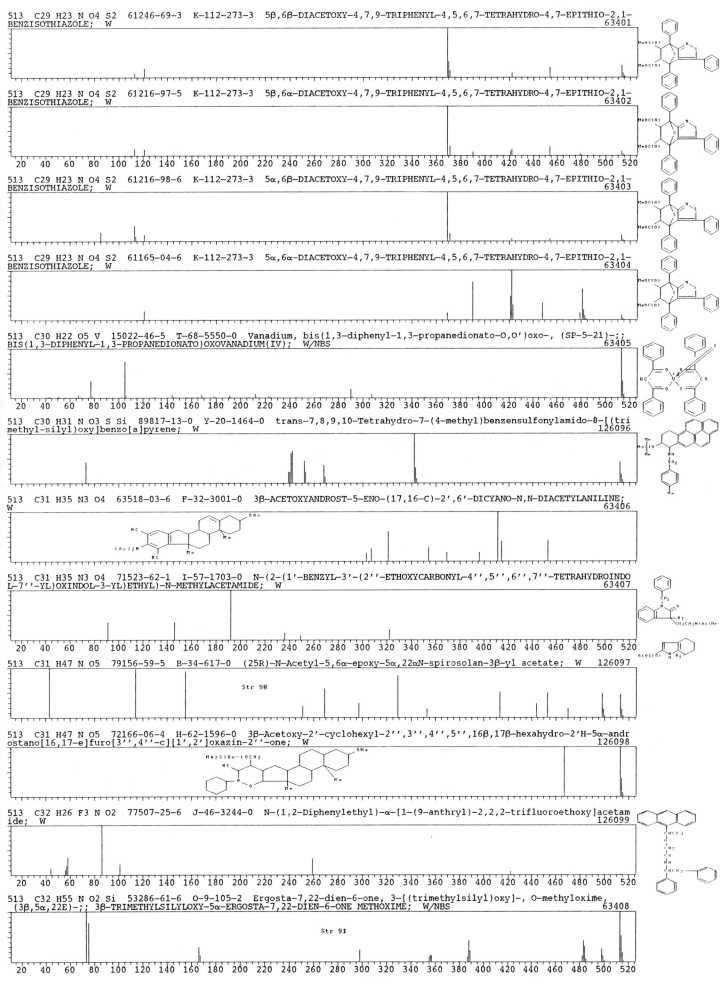

513 C29 H23 N O4 S2 61246-69-3 K-112-273-3 5β,6β-DIACETOXY-4,7,9-TRIPHENYL-4,5,6,7-TETRAHYDRO-4,7-EPITHIO-2,1-
BENZISOTHIAZOLE; W   63401

513 C29 H23 N O4 S2 61216-97-5 K-112-273-3 5β,6α-DIACETOXY-4,7,9-TRIPHENYL-4,5,6,7-TETRAHYDRO-4,7-EPITHIO-2,1-
BENZISOTHIAZOLE; W   63402

513 C29 H23 N O4 S2 61216-98-6 K-112-273-3 5α,6β-DIACETOXY-4,7,9-TRIPHENYL-4,5,6,7-TETRAHYDRO-4,7-EPITHIO-2,1-
BENZISOTHIAZOLE; W   63403

513 C29 H23 N O4 S2 61165-04-6 K-112-273-3 5α,6α-DIACETOXY-4,7,9-TRIPHENYL-4,5,6,7-TETRAHYDRO-4,7-EPITHIO-2,1-
BENZISOTHIAZOLE; W   63404

513 C30 H22 O5 V 15022-46-5 T-68-5550-0 Vanadium, bis(1,3-diphenyl-1,3-propanedionato-O,O')oxo-, (SP-5-21)-;;
BIS(1,3-DIPHENYL-1,3-PROPANEDIONATO)OXOVANADIUM(IV); W/NBS   63405

513 C30 H31 N O3 S Si 89817-13-0 Y-20-1464-0 trans-7,8,9,10-Tetrahydro-7-(4-methyl)benzensulfonylamido-8-[(tri
methyl-silyl)oxy]benzo[a]pyrene; W   126096

513 C31 H35 N3 O4 63518-03-6 F-32-3001-0 3β-ACETOXYANDROST-5-ENO-(17,16-C)-2',6'-DICYANO-N,N-DIACETYLANILINE;
W   63406

513 C31 H35 N3 O4 71523-62-1 I-57-1703-0 N-(2-(1'-BENZYL-3'-(2''-ETHOXYCARBONYL-4'',5'',6'',7''-TETRAHYDROINDO
L-7''-YL)OXINDOL-3-YL)ETHYL)-N-METHYLACETAMIDE; W   63407

513 C31 H47 N O5 79156-59-5 B-34-617-0 (25R)-N-Acetyl-5,6α-epoxy-5α,22αN-spirosolan-3β-yl acetate; W 126097

Str 90

513 C31 H47 N O5 72166-06-4 H-62-1596-0 3β-Acetoxy-2'-cyclohexyl-2'',3'',4'',5'',16β,17β-hexahydro-2'H-5α-andr
ostano[16,17-e]furo[3'',4''-c][1',2']oxazin-2''-one; W   126098

513 C32 H26 F3 N O2 77507-25-6 J-46-3244-0 N-(1,2-Diphenylethyl)-α-[1-(9-anthryl)-2,2,2-trifluoroethoxy]acetam
ide; W   126099

513 C32 H55 N O2 Si 53286-61-6 O-9-105-2 Ergosta-7,22-dien-6-one, 3-[(trimethylsilyl)oxy]-, O-methyloxime,
(3β,5α,22E)-;; 3β-TRIMETHYLSILYLOXY-5α-ERGOSTA-7,22-DIEN-6-ONE METHOXIME; W/NBS   63408

Str 91

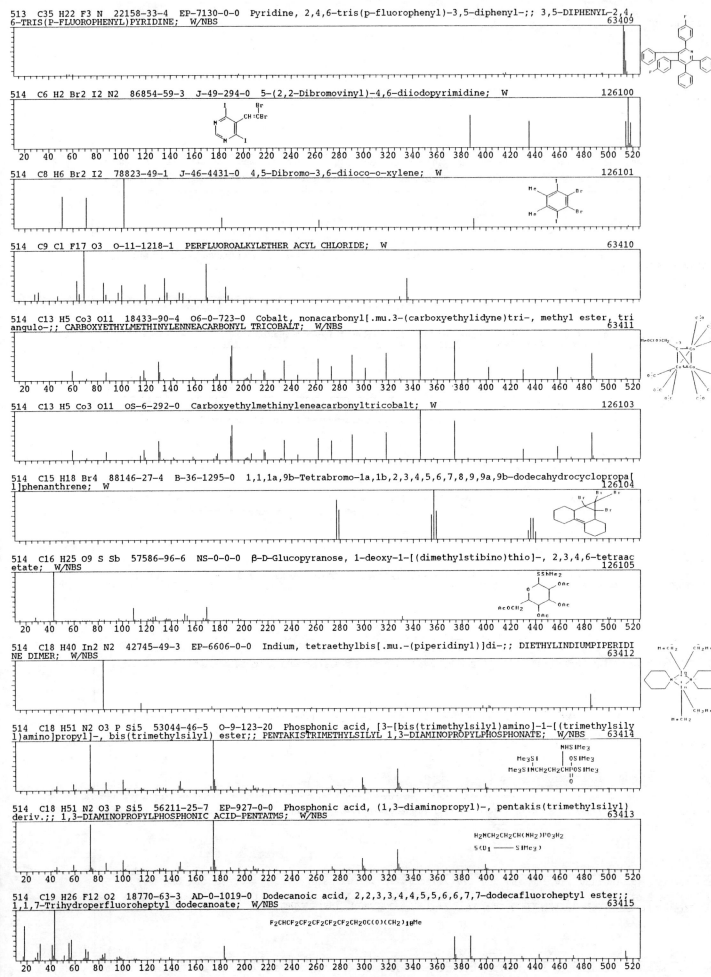

513 C35 H22 F3 N 22158-33-4 EP-7130-0-0 Pyridine, 2,4,6-tris(p-fluorophenyl)-3,5-diphenyl-;; 3,5-DIPHENYL-2,4,6-TRIS(P-FLUOROPHENYL)PYRIDINE; W/NBS
63409

514 C6 H2 Br2 I2 N2 86854-59-3 J-49-294-0 5-(2,2-Dibromovinyl)-4,6-diiodopyrimidine; W
126100

514 C8 H6 Br2 I2 78823-49-1 J-46-4431-0 4,5-Dibromo-3,6-diioco-o-xylene; W
126101

514 C9 Cl F17 O3 O-11-1218-1 PERFLUOROALKYLETHER ACYL CHLORIDE; W
63410

514 C13 H5 Co3 O11 18433-90-4 O6-0-723-0 Cobalt, nonacarbonyl[.mu.3-(carboxyethylidyne)tri-, methyl ester, triangulo-;; CARBOXYETHYLMETHINYLENNEACARBONYL TRICOBALT; W/NBS
63411

514 C13 H5 Co3 O11 OS-6-292-0 Carboxyethylmethinyleneacarbonyltricobalt; W
126103

514 C15 H18 Br4 88146-27-4 B-36-1295-0 1,1,1a,9b-Tetrabromo-1a,1b,2,3,4,5,6,7,8,9,9a,9b-dodecahydrocyclopropa[1]phenanthrene; W
126104

514 C16 H25 O9 S Sb 57586-96-6 NS-0-0-0 β-D-Glucopyranose, 1-deoxy-1-[(dimethylstibino)thio]-, 2,3,4,6-tetraacetate; W/NBS
126105

514 C18 H40 In2 N2 42745-49-3 EP-6606-0-0 Indium, tetraethylbis[.mu.-(piperidinyl)]di-;; DIETHYLINDIUMPIPERIDINE DIMER; W/NBS
63412

514 C18 H51 N2 O3 P Si5 53044-46-5 O-9-123-20 Phosphonic acid, [3-[bis(trimethylsilyl)amino]-1-[(trimethylsilyl)amino]propyl]-, bis(trimethylsilyl) ester;; PENTAKISTRIMETHYLSILYL 1,3-DIAMINOPROPYLPHOSPHONATE; W/NBS
63414

514 C18 H51 N2 O3 P Si5 56211-25-7 EP-927-0-0 Phosphonic acid, (1,3-diaminopropyl)-, pentakis(trimethylsilyl) deriv.;; 1,3-DIAMINOPROPYLPHOSPHONIC ACID-PENTATMS; W/NBS
63413

514 C19 H26 F12 O2 18770-63-3 AD-0-1019-0 Dodecanoic acid, 2,2,3,3,4,4,5,5,6,6,7,7-dodecafluoroheptyl ester;; 1,1,7-Trihydroperfluoroheptyl dodecanoate; W/NBS
63415

514   C20 H14 Te2   32294-58-9   AC-71-926-3   Ditelluride, di-1-naphthalenyl;; DI-(α,α'-NAPHTHYL)DITELLURITE;   W/NBS
63416

514   C20 H14 Te2   1666-12-2   AC-71-926-4   Ditelluride, di-2-naphthalenyl;; DI-(β,β'-NAPHTHYL)DITELLURITE;   W/NBS
63417

514   C20 H20 Cl2 N4 O8   IC-1241-0-0   N,N-Bis(2-carbomethoxyethoxyethoxy)-4-(2,6-dichloro-4-nitrophenylazo)-aniline
; W
126106

514   C21 H20 F6 O6 S   92817-06-6   J-49-4904-0   16β-(Trifluoroacetoxy)-3-[[(trifluoromethyl)sulonyl]oxy]estra-1,3,5(10)-trien-17-one;   W
126107
Str 92

514   C21 H24 Br2 O5   56336-17-5   SD-1981-0-0   2H,8H-Benzo[1,2-b:5,4-b']dipyran-10-propanoic acid, 3,7-dibromo-5-methoxy-2,2,8,8-tetramethyl-, methyl ester;;   W/NBS
63418

514   C22 H36 B6 N6 O4   70530-22-2   K-112-1254-0   5,10,11,12,17,22,23,24,25,27-ME10-2,8,14,20-O4-10,12,22,24,25,27-N6-1,9,11,13,21,23-B6-PENTACYCLO[19.3.1.1(3,7).1(9,13).1(15,19)]OCTACOSA-3,5,7(28),15,17,19(26)HEXAENE;   W   63419

514   C24 H18 O13   56691-89-5   B-28-1797-0   Triacetylsalazinic acid;   W
126108

514   C25 H30 N4 O8   85738-53-0   KC-1983-15-0   Ethyl 2-(2,2-Bisethoxycarbonyl-1-ethoxyethylamino)-4,7-dihydro-7-oxo-3-phenylpyrazolo[1,5-a]pyrimidine-6-carboxylate;   W
126109

514   C26 H20 Cl2 O7   73322-27-7   K-113-481-0   2,3,4-Tri-O-benzoyl-2-C-chloro-β-D-xylopyranosyl chloride;   W 126110

514   C26 H28 Cd N2 O2   15685-80-0   O-3-712-6   Cadmium, bis(2-butyl-8-quinolinolato-N1,O8)-, (T-4)-;; 2-N-BUTYL-8-HYDROXYQUINOLINE CADNIUM COMPLEX;   W/NBS
63420

514   C26 H30 N2 O9   78381-10-9   K-114-1918-0   Dimethylester of [1,2,3,4,6,7,12,12bβ-Octahydro-2β-[(methoxycarbonyl)acetoxymethyl]indolo[2,3-a]quinolizin-3α-ylmethyl]malonic acid;   W
126111

514   C26 H42 O10   79068-98-7   J-46-5208-2   1,2-Bis(3-deoxy-1,2:5,6-di-O-isopropylidene-α-D-allofuranos-3-yl)ethane
; W
126112

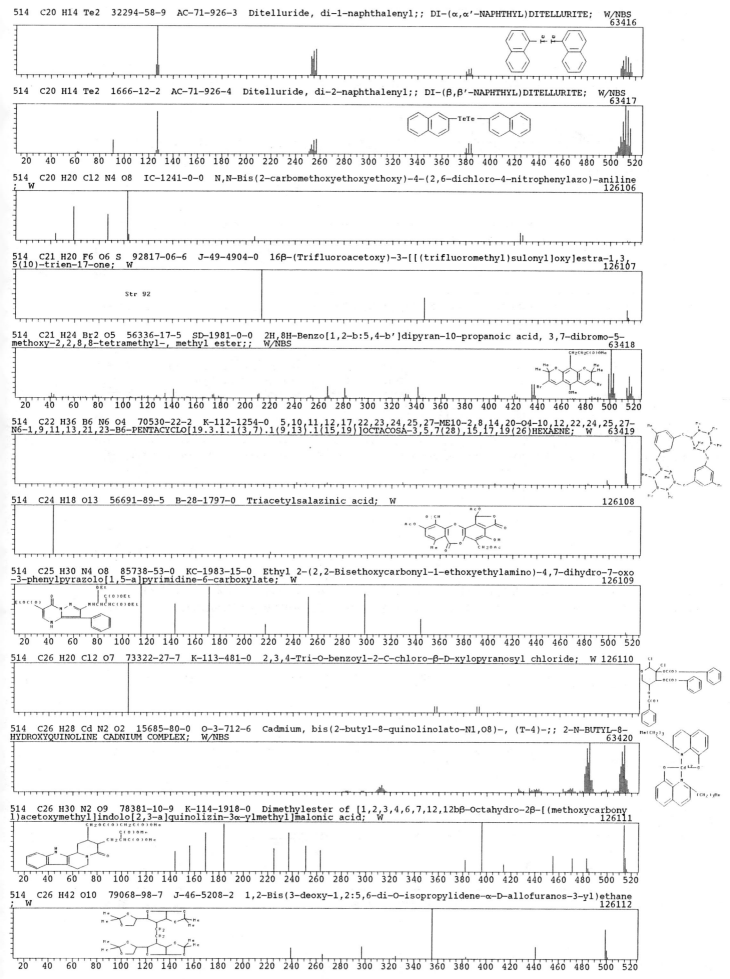

514   C27 H46 O9   40980-52-7   KC-1978-564-0   METHYL ISOPSEUDOMONATE A;   W   63421

514   C27 H54 O5 Si2   56248-52-3   D-10-3661-5   Prostan-1-oic acid, 6,9-epoxy-11,15-bis[(trimethylsilyl)oxy]-, meth
yl ester, (15S)-;; BISTRIMETHYLSILYL ETHER 6(9)-OXY-11,15-DIHYDROXYPROSTANOIC ACID METHYL ESTER;   W/NBS   63422

514   C27 H54 O5 Si2   74842-21-0   NS--28627-0-0   Prostan-1-oic acid, 6,9-epoxy-11,15-bis[(trimethylsilyl)oxy]-,
methyl ester, (9α)-;;   NBS   63423

514   C28 H18 O10   3692-07-7   MY-1981-0-0   USTILAGINOIDIN A;   W   63424

514   C28 H20 Br2   HE-1982-0-0   CYCLOBUTENE, 3,4-DIBROMO-1,2,3,4-TETRAPHENYL-;   W   63425

514   C28 H34 O7 S   79248-66-1   B-35-503-0   1-Naphthalenesulfinate of 1,2:5,6-di-O-cyclohexylidene-α-D-glucofurano
se;   W   126113

514   C28 H34 O7 S   79248-67-2   B-35-503-0   2-Napthalenesulfinate of 1,2:5,6-di-O-cyclohexylidene-α-D-glucofuranose
;   W   126114

514   C28 H34 O9   CD-542-0-0   Gomisin B;   W   126115

514   C28 H46 O3 Si3   80249-63-4   K-114-3371-0   2,2,4,6-Tetra-tert-butyl-4,6-diphenylcyclotrisiloxane;   W   126116

514   C29 H38 O4 S2   80115-34-0   H-64-1403-0   3β,19-Epoxy-17-(3'-furyl)-3α-methoxy-4,4-dimethyl-16,16-trimethylendi
thio-16,17-seco-13α-5,15-androstadien-17-one;   W   126117

514   C29 H38 O8   19314-73-9   SD-1981-0-0   o-Anisic acid, 4-hydroxy-6-pentyl-, methyl ester, ester with 2-hydroxy-
6-(2-oxoheptyl)-p-anisic acid;;   W/NBS   63429

514   C29 H38 O8   21596-60-1   T-68-5478-0   DEMETHYLARDISIAQUINONE B;   W/NBS   63426

514   C29 H38 O8   67093-46-3   B-31-1051-0   2-0-METHYLCONFLUENTIC ACID;   W   63427

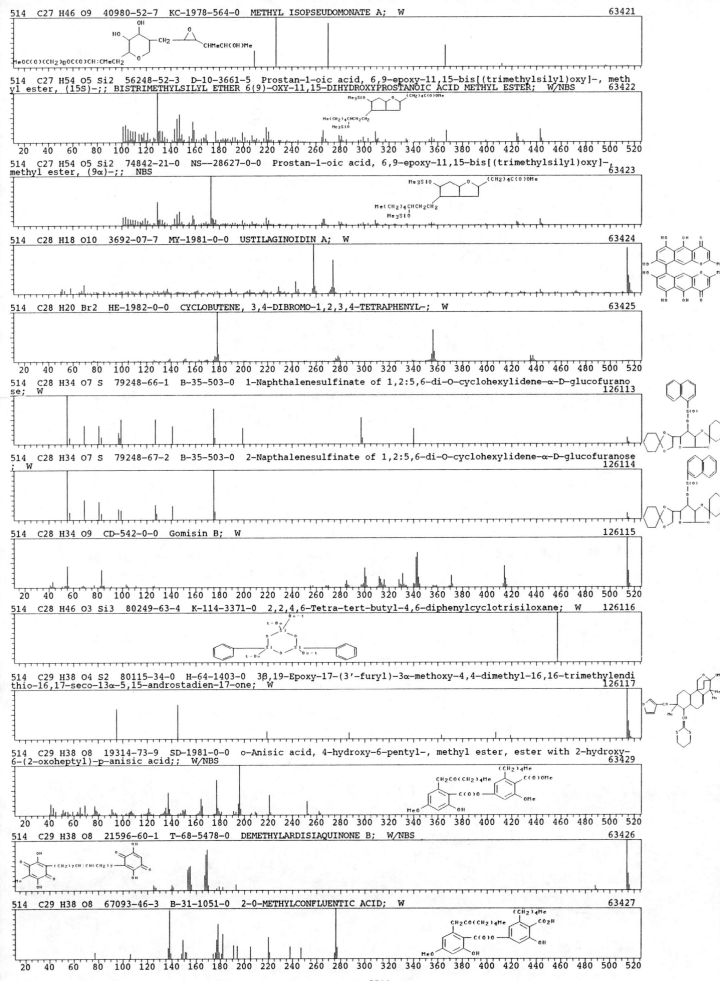

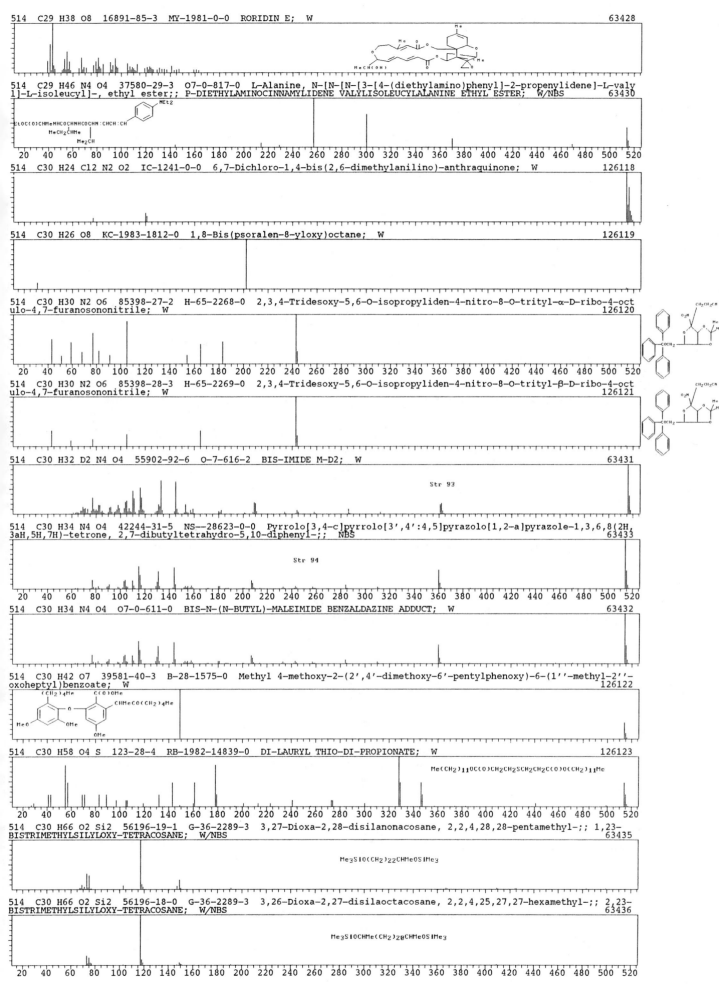

514   C29 H38 O8   16891-85-3   MY-1981-0-0   RORIDIN E;   W   63428

514   C29 H46 N4 O4   37580-29-3   O7-0-817-0   L-Alanine, N-[N-[N-[3-[4-(diethylamino)phenyl]-2-propenylidene]-L-valy
l]-L-isoleucyl]-, ethyl ester;;   P-DIETHYLAMINOCINNAMYLIDENE VALYLISOLEUCYLALANINE ETHYL ESTER;   W/NBS   63430

514   C30 H24 Cl2 N2 O2   IC-1241-0-0   6,7-Dichloro-1,4-bis(2,6-dimethylanilino)-anthraquinone;   W   126118

514   C30 H26 O8   KC-1983-1812-0   1,8-Bis(psoralen-8-yloxy)octane;   W   126119

514   C30 H30 N2 O6   85398-27-2   H-65-2268-0   2,3,4-Tridesoxy-5,6-O-isopropyliden-4-nitro-8-O-trityl-α-D-ribo-4-oct
ulo-4,7-furanosononitrile;   W   126120

514   C30 H30 N2 O6   85398-28-3   H-65-2269-0   2,3,4-Tridesoxy-5,6-O-isopropyliden-4-nitro-8-O-trityl-β-D-ribo-4-oct
ulo-4,7-furanosononitrile;   W   126121

514   C30 H32 D2 N4 O4   55902-92-6   O-7-616-2   BIS-IMIDE M-D2;   W   63431

Str 93

514   C30 H34 N4 O4   42244-31-5   NS--28623-0-0   Pyrrolo[3,4-c]pyrrolo[3',4':4,5]pyrazolo[1,2-a]pyrazole-1,3,6,8(2H,
3aH,5H,7H)-tetrone, 2,7-dibutyltetrahydro-5,10-diphenyl-;;   NBS   63433

Str 94

514   C30 H34 N4 O4   O7-0-611-0   BIS-N-(N-BUTYL)-MALEIMIDE BENZALDAZINE ADDUCT;   W   63432

514   C30 H42 O7   39581-40-3   B-28-1575-0   Methyl 4-methoxy-2-(2',4'-dimethoxy-6'-pentylphenoxy)-6-(1''-methyl-2''-
oxoheptyl)benzoate;   W   126122

514   C30 H58 O4 S   123-28-4   RB-1982-14839-0   DI-LAURYL THIO-DI-PROPIONATE;   W   126123

Me(CH2)11OC(O)CH2CH2SCH2CH2C(O)O(CH2)11Me

514   C30 H66 O2 Si2   56196-19-1   G-36-2289-3   3,27-Dioxa-2,28-disilanonacosane, 2,2,4,28,28-pentamethyl-;;   1,23-
BISTRIMETHYLSILYLOXY-TETRACOSANE;   W/NBS   63435

Me3SiO(CH2)22CHMeOSiMe3

514   C30 H66 O2 Si2   56196-18-0   G-36-2289-3   3,26-Dioxa-2,27-disilaoctacosane, 2,2,4,25,27,27-hexamethyl-;;   2,23-
BISTRIMETHYLSILYLOXY-TETRACOSANE;   W/NBS   63436

Me3SiOCHMe(CH2)20CHMeOSiMe3

514   C31 H46 O4 S   69081-91-0   C-100-7683-0   3α,5-CYCLO-6β-METHOXY-23-HYDROXYNORCHOLANYL TOSYLATE;   W   63437

Str 95

514   C31 H46 O6   56196-24-8   K-102-2075-2   Furost-5-ene-3,26-diol, 22,25-epoxy-, diacetate, (3β,22α,25S)-;;   NUATIG
ENIN-DIOCETATE;   W/NBS   63439

514   C31 H46 O6   7554-95-2   AD-0-3948-0   Kryptogenin-3,26-diacetate;   W   63438

Str 96

514   C31 H46 O6   F-37-602-0   1α,3β-Diacetoxyspirost-5-ene;   W   126124

514   C31 H46 O6   88972-28-5   KC-1983-2824-0   (22S,24S)-16,24:22,25-Dianhydro-16,22,22,24,25-pentahydroxy-2β-meth
oxy-3,11-dioxocucurbit-5-ene;   W   126125

514   C31 H54 O2 Si2   65598-45-0   NS-0-0-0   Silane, [[2-[3-methyl-6-(1-methylethenyl)-2-cyclohexen-1-yl]-5-pentyl-
1,3-phenylene]bis(oxy)]bis[dimethylpropyl-, (1R-trans)-;   W/NBS   126126

514   C32 H18 N8   IC-1242-0-0   Phthalocyanine;   W   126127

514   C32 H26 N4 O3   60026-92-8   F-32-581-0   2,4,10,11-TETRAZA-3,5,12-TRIOXO-2,4,11-TRIPHENYLTHRICYCLO[4.4.2.0(2,
6)]DODECANE;   W   63440

Str 97

514   C32 H47 Cl O3   66556-33-0   K-111-1162-0   3β-ACETOXYOLEANA-9(11),12-DIEN-28-ACID CHLOIDE;   W   63441

Str 98

514   C32 H50 O5   36872-76-1   SD-1981-0-0   Lanost-9(11)-en-18-oic acid, 23-(acetyloxy)-3,20-dihydroxy-, γ-lactone,
(3β,20.xi.)-;;   W/NBS   63447

Str 99

514   C32 H50 O5   36871-82-6   SD-1981-0-0   Lanost-9(11)-en-18-oic acid, 3,3-[1,2-ethanediylbis(oxy)]-20,23-dihydr
oxy-, γ-lactone, (20.xi.)-;;   W/NBS   63448

Str 100

514   C32 H50 O5   26339-85-5   T-66-2597-0   PRIVEROGENIN A-16-ACETATE;   W/NBS   63442

514   C32 H50 O5   74048-38-7   F-35-2402-0   3β-ACETOXY-12α-HYDROXY-URSAN-28-OIC-13(28)-LACTONE;   W   63443

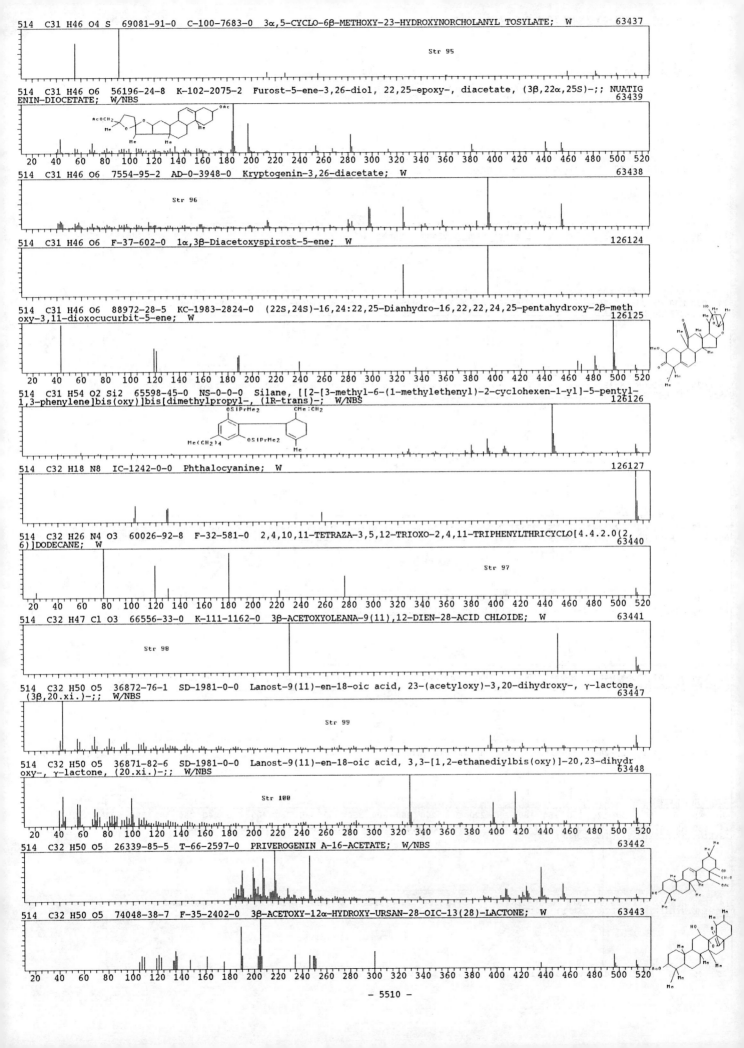

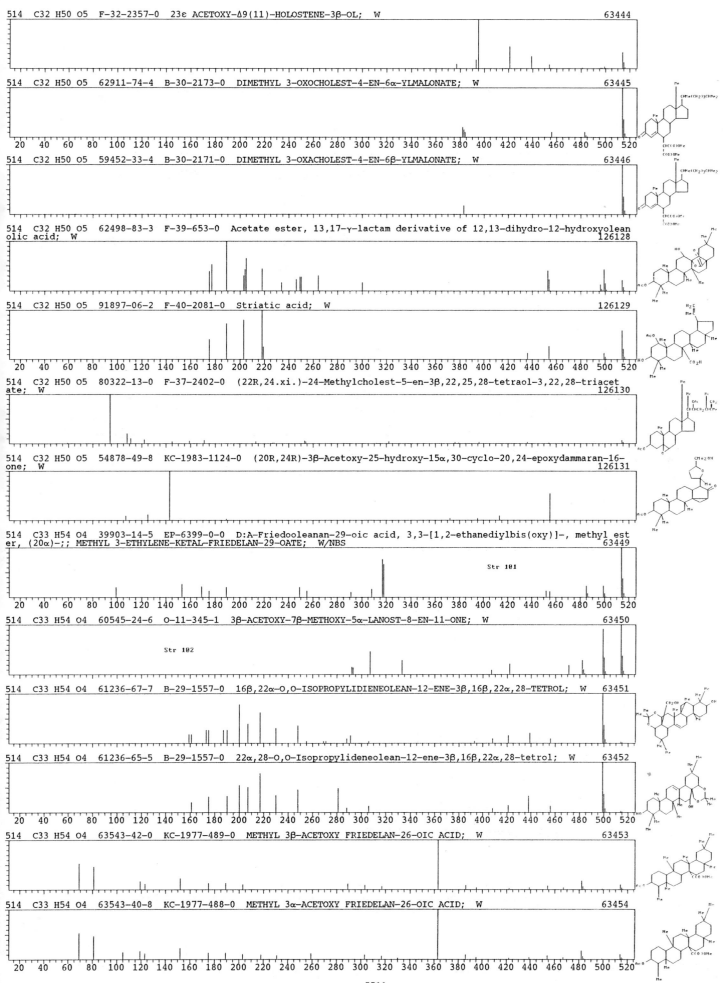

514　C32 H50 O5　F-32-2357-0　23ε ACETOXY-Δ9(11)-HOLOSTENE-3β-OL;　W　　　　63444

514　C32 H50 O5　62911-74-4　B-30-2173-0　DIMETHYL 3-OXOCHOLEST-4-EN-6α-YLMALONATE;　W　　　63445

514　C32 H50 O5　59452-33-4　B-30-2171-0　DIMETHYL 3-OXACHOLEST-4-EN-6β-YLMALONATE;　W　　　63446

514　C32 H50 O5　62498-83-3　F-39-653-0　Acetate ester, 13,17-γ-lactam derivative of 12,13-dihydro-12-hydroxyolean
olic acid;　W　　　　126128

514　C32 H50 O5　91897-06-2　F-40-2081-0　Striatic acid;　W　　　　126129

514　C32 H50 O5　80322-13-0　F-37-2402-0　(22R,24.xi.)-24-Methylcholest-5-en-3β,22,25,28-tetraol-3,22,28-triacet
ate;　W　　　　126130

514　C32 H50 O5　54878-49-8　KC-1983-1124-0　(20R,24R)-3β-Acetoxy-25-hydroxy-15α,30-cyclo-20,24-epoxydammaran-16-
one;　W　　　　126131

514　C33 H54 O4　39903-14-5　EP-6399-0-0　D:A-Friedooleanan-29-oic acid, 3,3-[1,2-ethanediylbis(oxy)]-, methyl est
er, (20α)-;; METHYL 3-ETHYLENE-KETAL-FRIEDELAN-29-OATE;　W/NBS　　　　63449

Str 101

514　C33 H54 O4　60545-24-6　O-11-345-1　3β-ACETOXY-7β-METHOXY-5α-LANOST-8-EN-11-ONE;　W　　　63450

Str 102

514　C33 H54 O4　61236-67-7　B-29-1557-0　16β,22α-O,O-ISOPROPYLIDENEOLEAN-12-ENE-3β,16β,22α,28-TETROL;　W　63451

514　C33 H54 O4　61236-65-5　B-29-1557-0　22α,28-O,O-Isopropylideneolean-12-ene-3β,16β,22α,28-tetrol;　W　63452

514　C33 H54 O4　63543-42-0　KC-1977-489-0　METHYL 3β-ACETOXY FRIEDELAN-26-OIC ACID;　W　　　63453

514　C33 H54 O4　63543-40-8　KC-1977-488-0　METHYL 3α-ACETOXY FRIEDELAN-26-OIC ACID;　W　　　63454

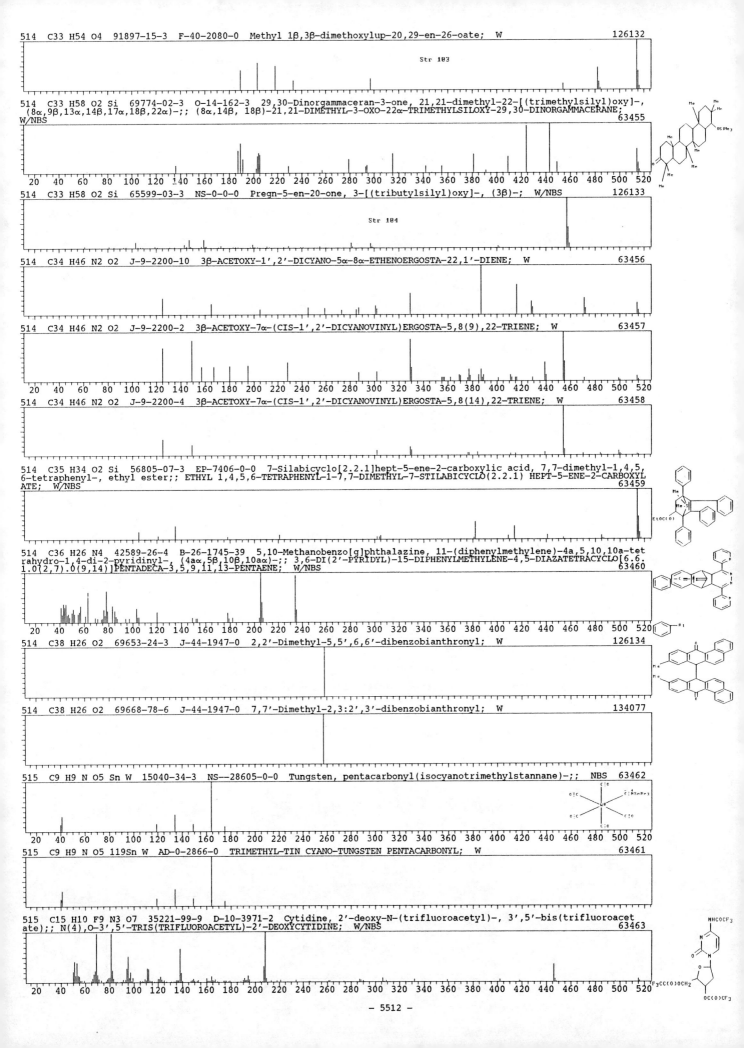

514 C33 H54 O4 91897-15-3 F-40-2080-0 Methyl 1β,3β-dimethoxylup-20,29-en-26-oate; W   126132

Str 103

514 C33 H58 O2 Si 69774-02-3 O-14-162-3 29,30-Dinorgammaceran-3-one, 21,21-dimethyl-22-[(trimethylsilyl)oxy]-, (8α,9β,13α,14β,17α,18β,22α)-;; (8α,14β, 18β)-21,21-DIMETHYL-3-OXO-22α-TRIMETHYLSILOXY-29,30-DINORGAMMACERANE; W/NBS   63455

514 C33 H58 O2 Si 65599-03-3 NS-0-0-0 Pregn-5-en-20-one, 3-[(tributylsilyl)oxy]-, (3β)-; W/NBS   126133

Str 104

514 C34 H46 N2 O2 J-9-2200-10 3β-ACETOXY-1',2'-DICYANO-5α-8α-ETHENOERGOSTA-22,1'-DIENE; W   63456

514 C34 H46 N2 O2 J-9-2200-2 3β-ACETOXY-7α-(CIS-1',2'-DICYANOVINYL)ERGOSTA-5,8(9),22-TRIENE; W   63457

514 C34 H46 N2 O2 J-9-2200-4 3β-ACETOXY-7α-(CIS-1',2'-DICYANOVINYL)ERGOSTA-5,8(14),22-TRIENE; W   63458

514 C35 H34 O2 Si 56805-07-3 EP-7406-0-0 7-Silabicyclo[2.2.1]hept-5-ene-2-carboxylic acid, 7,7-dimethyl-1,4,5,6-tetraphenyl-, ethyl ester;; ETHYL 1,4,5,6-TETRAPHENYL-1-7,7-DIMETHYL-7-STILABICYCLO(2.2.1) HEPT-5-ENE-2-CARBOXYLATE; W/NBS   63459

514 C36 H26 N4 42589-26-4 B-26-1745-39 5,10-Methanobenzo[g]phthalazine, 11-(diphenylmethylene)-4a,5,10,10a-tetrahydro-1,4-di-2-pyridinyl-, (4aα,5β,10β,10aα)-;; 3,6-DI(2'-PYRIDYL)-15-DIPHENYLMETHYLENE-4,5-DIAZATETRACYCLO[6.6.1.0(2,7).0(9,14)]PENTADECA-3,5,9,11,13-PENTAENE; W/NBS   63460

514 C38 H26 O2 69653-24-3 J-44-1947-0 2,2'-Dimethyl-5,5',6,6'-dibenzobianthronyl; W   126134

514 C38 H26 O2 69668-78-6 J-44-1947-0 7,7'-Dimethyl-2,3:2',3'-dibenzobianthronyl; W   134077

515 C9 H9 N O5 Sn W 15040-34-3 NS--28605-0-0 Tungsten, pentacarbonyl(isocyanotrimethylstannane)-;; NBS 63462

515 C9 H9 N O5 119Sn W AD-0-2866-0 TRIMETHYL-TIN CYANO-TUNGSTEN PENTACARBONYL; W   63461

515 C15 H10 F9 N3 O7 35221-99-9 D-10-3971-2 Cytidine, 2'-deoxy-N-(trifluoroacetyl)-, 3',5'-bis(trifluoroacetate);; N(4),O-3',5'-TRIS(TRIFLUOROACETYL)-2'-DEOXYCYTIDINE; W/NBS   63463

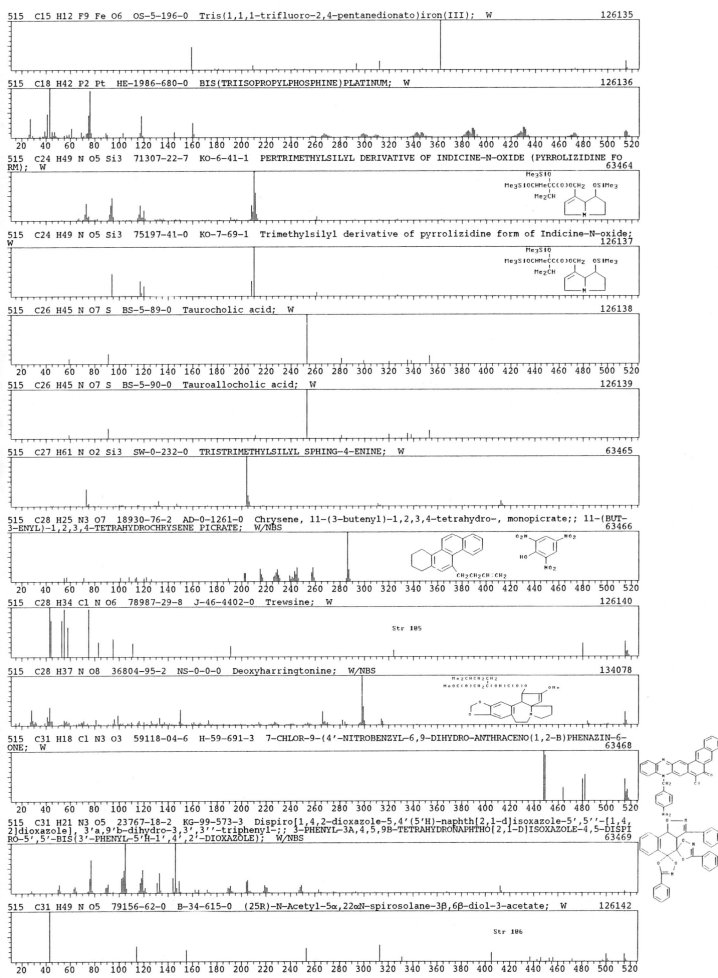

515   C15 H12 F9 Fe O6   OS-5-196-0   Tris(1,1,1-trifluoro-2,4-pentanedionato)iron(III);   W   126135

515   C18 H42 P2 Pt   HE-1986-680-0   BIS(TRIISOPROPYLPHOSPHINE)PLATINUM;   W   126136

20   40   60   80   100   120   140   160   180   200   220   240   260   280   300   320   340   360   380   400   420   440   460   480   500   520

515   C24 H49 N O5 Si3   71307-22-7   KO-6-41-1   PERTRIMETHYLSILYL DERIVATIVE OF INDICINE-N-OXIDE (PYRROLIZIDINE FORM);   W   63464

Me3SiO
Me3SiOCHMeCC(O)OCH2 OSiMe3
Me2CH

515   C24 H49 N O5 Si3   75197-41-0   KO-7-69-1   Trimethylsilyl derivative of pyrrolizidine form of Indicine-N-oxide;   W   126137

Me3SiO
Me3SiOCHMeCC(O)OCH2 OSiMe3
Me2CH

515   C26 H45 N O7 S   BS-5-89-0   Taurocholic acid;   W   126138

20   40   60   80   100   120   140   160   180   200   220   240   260   280   300   320   340   360   380   400   420   440   460   480   500   520

515   C26 H45 N O7 S   BS-5-90-0   Tauroallocholic acid;   W   126139

515   C27 H61 N O2 Si3   SW-0-232-0   TRISTRIMETHYLSILYL SPHING-4-ENINE;   W   63465

515   C28 H25 N3 O7   18930-76-2   AD-0-1261-0   Chrysene, 11-(3-butenyl)-1,2,3,4-tetrahydro-, monopicrate;; 11-(BUT-3-ENYL)-1,2,3,4-TETRAHYDROCHRYSENE PICRATE;   W/NBS   63466

O2N   NO2
HO
NO2
CH2CH2CH:CH2

20   40   60   80   100   120   140   160   180   200   220   240   260   280   300   320   340   360   380   400   420   440   460   480   500   520

515   C28 H34 Cl N O6   78987-29-8   J-46-4402-0   Trewsine;   W   126140

Str 105

515   C28 H37 N O8   36804-95-2   NS-0-0-0   Deoxyharringtonine;   W/NBS   134078

Me2CHCH2CH2
MeOC(O)CH2C(OH)C(O)O   OMe
O
O

515   C31 H18 Cl N3 O3   59118-04-6   H-59-691-3   7-CHLOR-9-(4'-NITROBENZYL-6,9-DIHYDRO-ANTHRACENO(1,2-B)PHENAZIN-6-ONE;   W   63468

20   40   60   80   100   120   140   160   180   200   220   240   260   280   300   320   340   360   380   400   420   440   460   480   500   520

515   C31 H21 N3 O5   23767-18-2   KG-99-573-3   Dispiro[1,4,2-dioxazole-5,4'(5'H)-naphth[2,1-d]isoxazole-5',5''-[1,4,2]dioxazole], 3'a,9'b-dihydro-3,3',3''-triphenyl-;; 3-PHENYL-3A,4,5,9B-TETRAHYDRONAPHTHO[2,1-D]ISOXAZOLE-4,5-DISPIRO-5',5'-BIS(3'-PHENYL-5'H-1',4',2'-DIOXAZOLE);   W/NBS   63469

515   C31 H49 N O5   79156-62-0   B-34-615-0   (25R)-N-Acetyl-5α,22αN-spirosolane-3β,6β-diol-3-acetate;   W   126142

Str 106

20   40   60   80   100   120   140   160   180   200   220   240   260   280   300   320   340   360   380   400   420   440   460   480   500   520

515　C31 H49 N O5　79156-61-9　B-34-616-0　(25R)-N-Acetyl-5β,22αN-spirosolane-3β,5β-diol-3-acetate;　W　126143

Str 107

515　C31 H49 N O5　79156-60-8　B-34-616-0　(25R)-N-Acetyl-22αN-spirosolane-3β,5α-diol-3-acetate;　W　126144

Str 108

515　C31 H49 N O5　79156-57-3　B-34-617-0　(20S,22R,25R)-26-Acetylamino-6-oxo-5α-furostan-3β-yl acetate;　W 126145

515　C32 H53 N O4　19454-74-1　AD-0-2447-0　Lanost-8-ene-3β,7α-diol, 3-acetate nitrite;; 3β-ACETOXY-5α-LANOST-8-EN
-7α-YL NITRITE ESTER;　W/NBS　63470

Str 109

515　C33 H29 N O3 Si　79139-21-2　K-114-1024-0　3-(N-Methylanilino)-2-[3-(triphenylsiloxy)-3-butenoyl]-2-cyclobute
n-1-one;　W　126146

515　C35 H18 Cl N3　IC-1242-0-0　6-Chloro-2,4-bis(pyren-1-yl)-s-triazine;　W　126147

515　C35 H65 N O　54725-10-9　SD-1981-0-0　Cholestan-3-ol, 2-(dibutylamino)-, (2β,3α,5α)-;;　W/NBS　63471

515　C37 H25 N O2　87612-97-3　Y-20-973-0　3,4-Dibenzoyl-2,5,6-triphenylpyridine;　W　126148

Str 110

515　C37 H25 N S　87612-94-0　Y-20-972-0　1,3,4,6,7-Pentaphenylthieno[3,4-c]pyridine;　W　126149

Str 111

516　C8 H20 F10 N2 Ni P4　53079-16-6　NS-0-0-0　Nickel, bis(diethylphosphoramidous difluoride-P)bis(phosphorous tr
ifluoride)-, (T-4)-;　W/NBS　126150

516　C16 H10 Cd Cr2 O6　55518-02-0　AD-0-5768-0　Chromium, (cadmium)hexacarbonylbis(η5-2,4-cyclopentadien-1-yl)di
-, (2Cd-Cr);; BIS-(CYCLOPENTADIENYLTRICARBONYLCHROMIUM)CADMIUM;　W/NBS　63472

516　C17 H11 Br3 O4　66178-14-1　B-31-302-0　BROMOBIS(4'-BROMOBENZOYL)METHYL ACETATE;　W　63473

516　C18 H8 F12 O2 S　70091-68-8　J-46-1052-0　3,3,3',3'-Tetrakis(trifluoromethyl)-1,1'-spiro[3H-2,1-benzoxathio
le];　W　126152

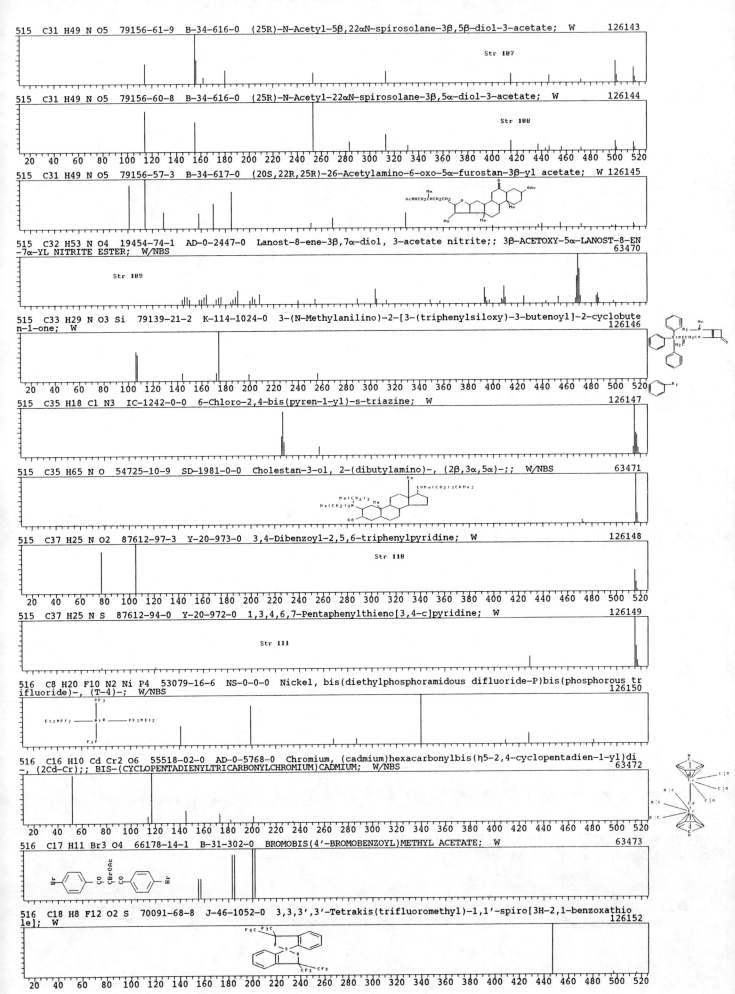

516 C18 H9 F12 O2 P 77121-88-1 C-103-1235-0 Bis(cyclic) derivative of bis(phenyl-bis(trifluoromethyl)methoxy)phosphine; W
126153

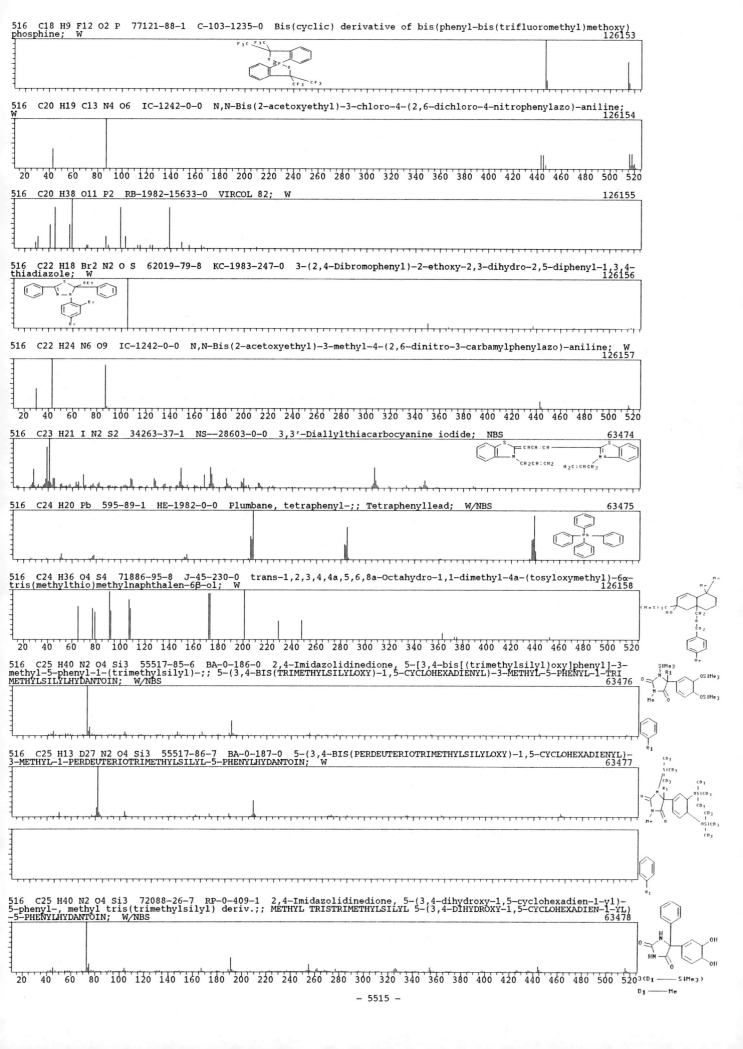

516 C20 H19 Cl3 N4 O6 IC-1242-0-0 N,N-Bis(2-acetoxyethyl)-3-chloro-4-(2,6-dichloro-4-nitrophenylazo)-aniline; W
126154

516 C20 H38 O11 P2 RB-1982-15633-0 VIRCOL 82; W
126155

516 C22 H18 Br2 N2 O S 62019-79-8 KC-1983-247-0 3-(2,4-Dibromophenyl)-2-ethoxy-2,3-dihydro-2,5-diphenyl-1,3,4-thiadiazole; W
126156

516 C22 H24 N6 O9 IC-1242-0-0 N,N-Bis(2-acetoxyethyl)-3-methyl-4-(2,6-dinitro-3-carbamylphenylazo)-aniline; W
126157

516 C23 H21 I N2 S2 34263-37-1 NS--28603-0-0 3,3'-Diallylthiacarbocyanine iodide; NBS
63474

516 C24 H20 Pb 595-89-1 HE-1982-0-0 Plumbane, tetraphenyl-;; Tetraphenyllead; W/NBS
63475

516 C24 H36 O4 S4 71886-95-8 J-45-230-0 trans-1,2,3,4,4a,5,6,8a-Octahydro-1,1-dimethyl-4a-(tosyloxymethyl)-6α-tris(methylthio)methylnaphthalen-6β-ol; W
126158

516 C25 H40 N2 O4 Si3 55517-85-6 BA-0-186-0 2,4-Imidazolidinedione, 5-[3,4-bis[(trimethylsilyl)oxy]phenyl]-3-methyl-5-phenyl-1-(trimethylsilyl)-;; 5-(3,4-BIS(TRIMETHYLSILYLOXY)-1,5-CYCLOHEXADIENYL)-3-METHYL-5-PHENYL-1-TRIMETHYLSILYLHYDANTOIN; W/NBS
63476

516 C25 H13 D27 N2 O4 Si3 55517-86-7 BA-0-187-0 5-(3,4-BIS(PERDEUTERIOTRIMETHYLSILYLOXY)-1,5-CYCLOHEXADIENYL)-3-METHYL-1-PERDEUTERIOTRIMETHYLSILYL-5-PHENYLHYDANTOIN; W
63477

516 C25 H40 N2 O4 Si3 72088-26-7 RP-0-409-1 2,4-Imidazolidinedione, 5-(3,4-dihydroxy-1,5-cyclohexadien-1-yl)-5-phenyl-, methyl tris(trimethylsilyl) deriv.;; METHYL TRISTRIMETHYLSILYL 5-(3,4-DIHYDROXY-1,5-CYCLOHEXADIEN-1-YL)-5-PHENYLHYDANTOIN; W/NBS
63478

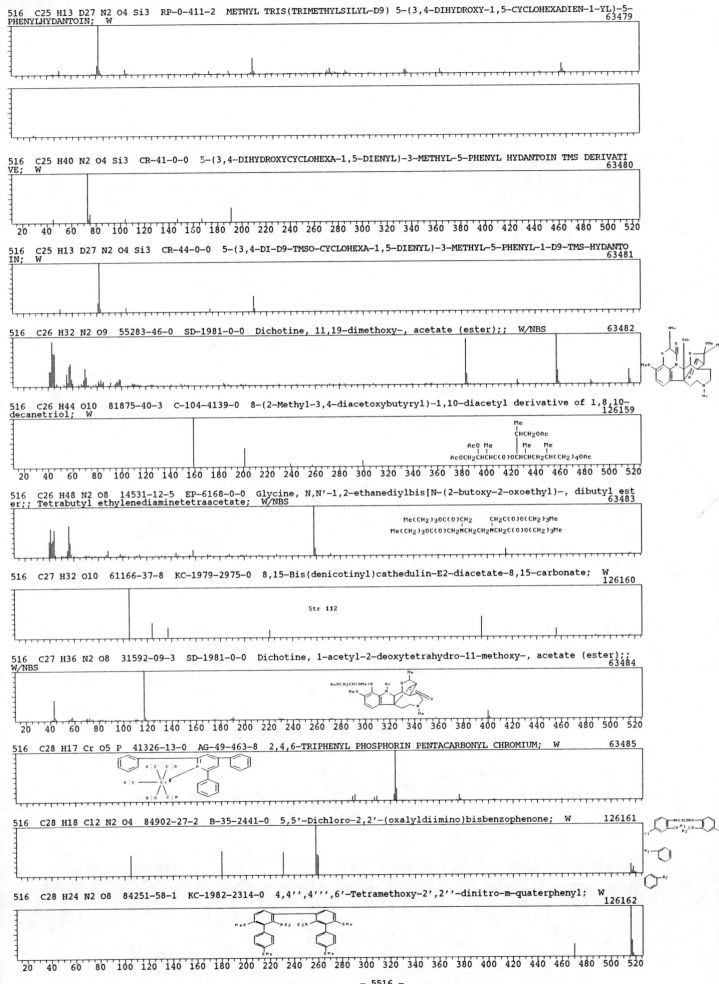

516   C25 H13 D27 N2 O4 Si3   RP-0-411-2   METHYL TRIS(TRIMETHYLSILYL-D9) 5-(3,4-DIHYDROXY-1,5-CYCLOHEXADIEN-1-YL)-5-
PHENYLHYDANTOIN;   W
63479

516   C25 H40 N2 O4 Si3   CR-41-0-0   5-(3,4-DIHYDROXYCYCLOHEXA-1,5-DIENYL)-3-METHYL-5-PHENYL HYDANTOIN TMS DERIVATI
VE;   W
63480

516   C25 H13 D27 N2 O4 Si3   CR-44-0-0   5-(3,4-DI-D9-TMSO-CYCLOHEXA-1,5-DIENYL)-3-METHYL-5-PHENYL-1-D9-TMS-HYDANTO
IN;   W
63481

516   C26 H32 N2 O9   55283-46-0   SD-1981-0-0   Dichotine, 11,19-dimethoxy-, acetate (ester);;   W/NBS
63482

516   C26 H44 O10   81875-40-3   C-104-4139-0   8-(2-Methyl-3,4-diacetoxybutyryl)-1,10-diacetyl derivative of 1,8,10-
decanetriol;   W
126159

516   C26 H48 N2 O8   14531-12-5   EP-6168-0-0   Glycine, N,N'-1,2-ethanediylbis[N-(2-butoxy-2-oxoethyl)-, dibutyl est
er;;   Tetrabutyl ethylenediaminetetraacetate;   W/NBS
63483

516   C27 H32 O10   61166-37-8   KC-1979-2975-0   8,15-Bis(denicotinyl)cathedulin-E2-diacetate-8,15-carbonate;   W
126160

Str 112

516   C27 H36 N2 O8   31592-09-3   SD-1981-0-0   Dichotine, 1-acetyl-2-deoxytetrahydro-11-methoxy-, acetate (ester);;
W/NBS
63484

516   C28 H17 Cr O5 P   41326-13-0   AG-49-463-8   2,4,6-TRIPHENYL PHOSPHORIN PENTACARBONYL CHROMIUM;   W
63485

516   C28 H18 Cl2 N2 O4   84902-27-2   B-35-2441-0   5,5'-Dichloro-2,2'-(oxalyldiimino)bisbenzophenone;   W
126161

516   C28 H24 N2 O8   84251-58-1   KC-1982-2314-0   4,4'',4''',6'-Tetramethoxy-2',2''-dinitro-m-quaterphenyl;   W
126162

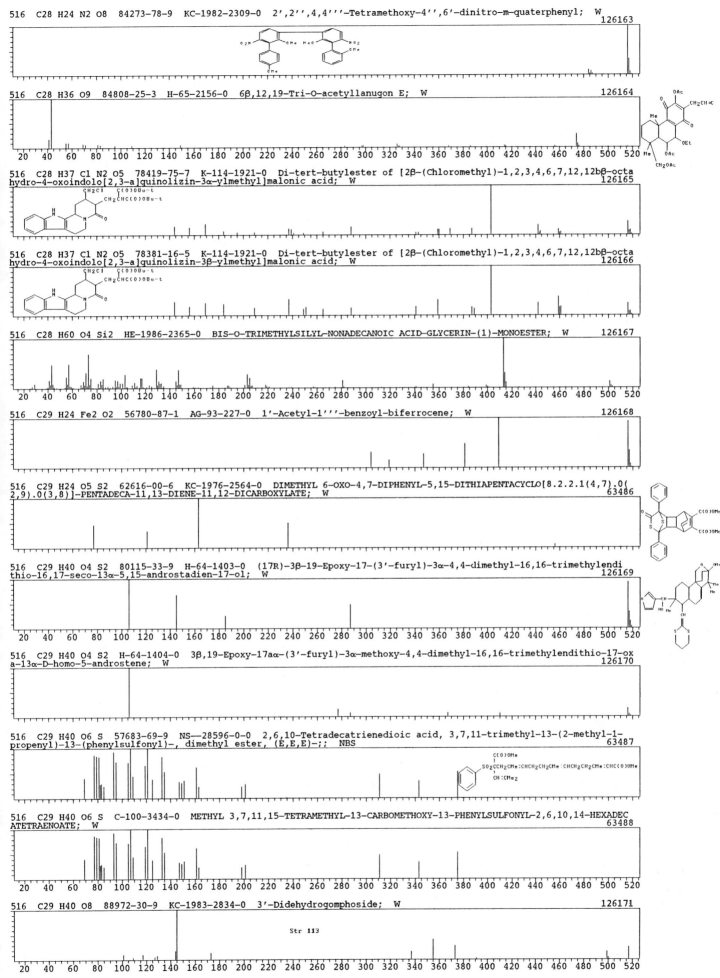

516   C28 H24 N2 O8   84273-78-9   KC-1982-2309-0   2',2'',4,4'''-Tetramethoxy-4'',6'-dinitro-m-quaterphenyl;   W
126163

516   C28 H36 O9   84808-25-3   H-65-2156-0   6β,12,19-Tri-O-acetyllanugon E;   W
126164

516   C28 H37 Cl N2 O5   78419-75-7   K-114-1921-0   Di-tert-butylester of [2β-(Chloromethyl)-1,2,3,4,6,7,12,12bβ-octa
hydro-4-oxoindolo[2,3-a]quinolizin-3α-ylmethyl]malonic acid;   W
126165

516   C28 H37 Cl N2 O5   78381-16-5   K-114-1921-0   Di-tert-butylester of [2β-(Chloromethyl)-1,2,3,4,6,7,12,12bβ-octa
hydro-4-oxoindolo[2,3-a]quinolizin-3β-ylmethyl]malonic acid;   W
126166

516   C28 H60 O4 Si2   HE-1986-2365-0   BIS-O-TRIMETHYLSILYL-NONADECANOIC ACID-GLYCERIN-(1)-MONOESTER;   W   126167

516   C29 H24 Fe2 O2   56780-87-1   AG-93-227-0   1'-Acetyl-1'''-benzoyl-biferrocene;   W
126168

516   C29 H24 O5 S2   62616-00-6   KC-1976-2564-0   DIMETHYL 6-OXO-4,7-DIPHENYL-5,15-DITHIAPENTACYCLO[8.2.1(4,7).0(
2,9).0(3,8)]-PENTADECA-11,13-DIENE-11,12-DICARBOXYLATE;   W
63486

516   C29 H40 O4 S2   80115-33-9   H-64-1403-0   (17R)-3β-19-Epoxy-17-(3'-furyl)-3α-4,4-dimethyl-16,16-trimethylendi
thio-16,17-seco-13α-5,15-androstadien-17-ol;   W
126169

516   C29 H40 O4 S2   H-64-1404-0   3β,19-Epoxy-17aα-(3'-furyl)-3α-methoxy-4,4-dimethyl-16,16-trimethylendithio-17-ox
a-13α-D-homo-5-androstene;   W
126170

516   C29 H40 O6 S   57683-69-9   NS--28596-0-0   2,6,10-Tetradecatrienedioic acid, 3,7,11-trimethyl-13-(2-methyl-1-
propenyl)-13-(phenylsulfonyl)-, dimethyl ester, (E,E,E)-;;   NBS
63487

516   C29 H40 O6 S   C-100-3434-0   METHYL 3,7,11,15-TETRAMETHYL-13-CARBOMETHOXY-13-PHENYLSULFONYL-2,6,10,14-HEXADEC
ATETRAENOATE;   W
63488

516   C29 H40 O8   88972-30-9   KC-1983-2834-0   3'-Didehydrogomphoside;   W
126171

Str 113

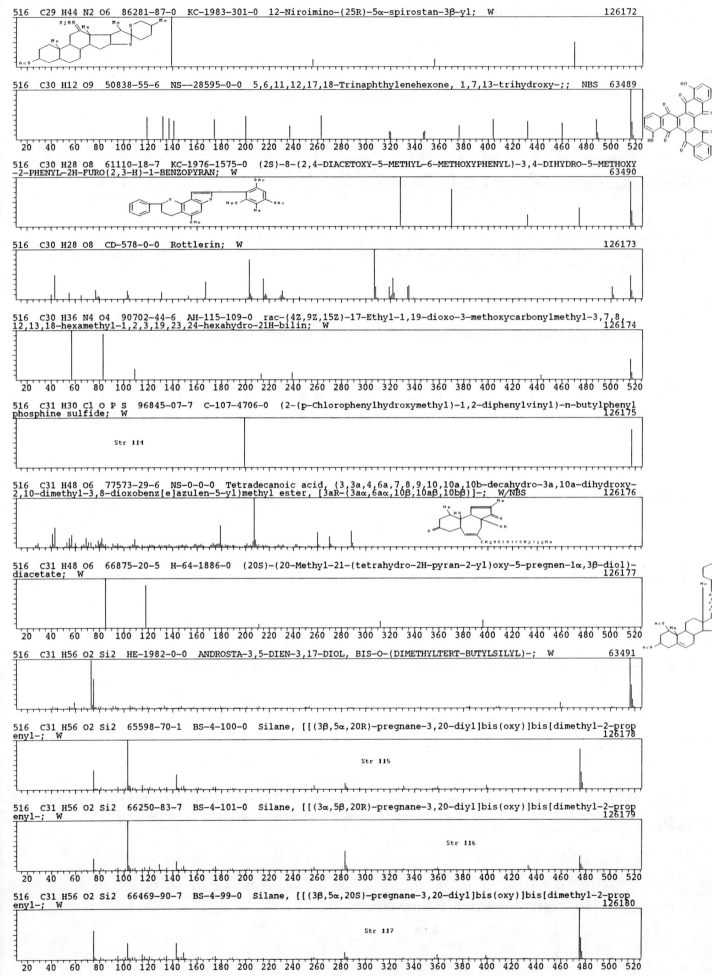

516 C29 H44 N2 O6 86281-87-0 KC-1983-301-0 12-Niroimino-(25R)-5α-spirostan-3β-yl; W 126172

516 C30 H12 O9 50838-55-6 NS--28595-0-0 5,6,11,12,17,18-Trinaphthylenehexone, 1,7,13-trihydroxy-;; NBS 63489

516 C30 H28 O8 61110-18-7 KC-1976-1575-0 (2S)-8-(2,4-DIACETOXY-5-METHYL-6-METHOXYPHENYL)-3,4-DIHYDRO-5-METHOXY
-2-PHENYL-2H-FURO(2,3-H)-1-BENZOPYRAN; W 63490

516 C30 H28 O8 CD-578-0-0 Rottlerin; W 126173

516 C30 H36 N4 O4 90702-44-6 AH-115-109-0 rac-(4Z,9Z,15Z)-17-Ethyl-1,19-dioxo-3-methoxycarbonylmethyl-3,7,8,
12,13,18-hexamethyl-1,2,3,19,23,24-hexahydro-21H-bilin; W 126174

516 C31 H30 Cl O P S 96845-07-7 C-107-4706-0 (2-(p-Chlorophenylhydroxymethyl)-1,2-diphenylvinyl)-n-butylphenyl
phosphine sulfide; W 126175

Str 114

516 C31 H48 O6 77573-29-6 NS-0-0-0 Tetradecanoic acid, (3,3a,4,6a,7,8,9,10,10a,10b-decahydro-3a,10a-dihydroxy-
2,10-dimethyl-3,8-dioxobenz[e]azulen-5-yl)methyl ester, [3aR-(3aα,6aα,10β,10aβ,10bβ)]-; W/NBS 126176

516 C31 H48 O6 66875-20-5 H-64-1886-0 (20S)-(20-Methyl-21-(tetrahydro-2H-pyran-2-yl)oxy-5-pregnen-1α,3β-diol)-
diacetate; W 126177

516 C31 H56 O2 Si2 HE-1982-0-0 ANDROSTA-3,5-DIEN-3,17-DIOL, BIS-O-(DIMETHYLTERT-BUTYLSILYL)-; W 63491

516 C31 H56 O2 Si2 65598-70-1 BS-4-100-0 Silane, [[(3β,5α,20R)-pregnane-3,20-diyl]bis(oxy)]bis[dimethyl-2-prop
enyl-; W 126178

Str 115

516 C31 H56 O2 Si2 66250-83-7 BS-4-101-0 Silane, [[(3α,5β,20R)-pregnane-3,20-diyl]bis(oxy)]bis[dimethyl-2-prop
enyl-; W 126179

Str 116

516 C31 H56 O2 Si2 66469-90-7 BS-4-99-0 Silane, [[(3β,5α,20S)-pregnane-3,20-diyl]bis(oxy)]bis[dimethyl-2-prop
enyl-; W 126180

Str 117

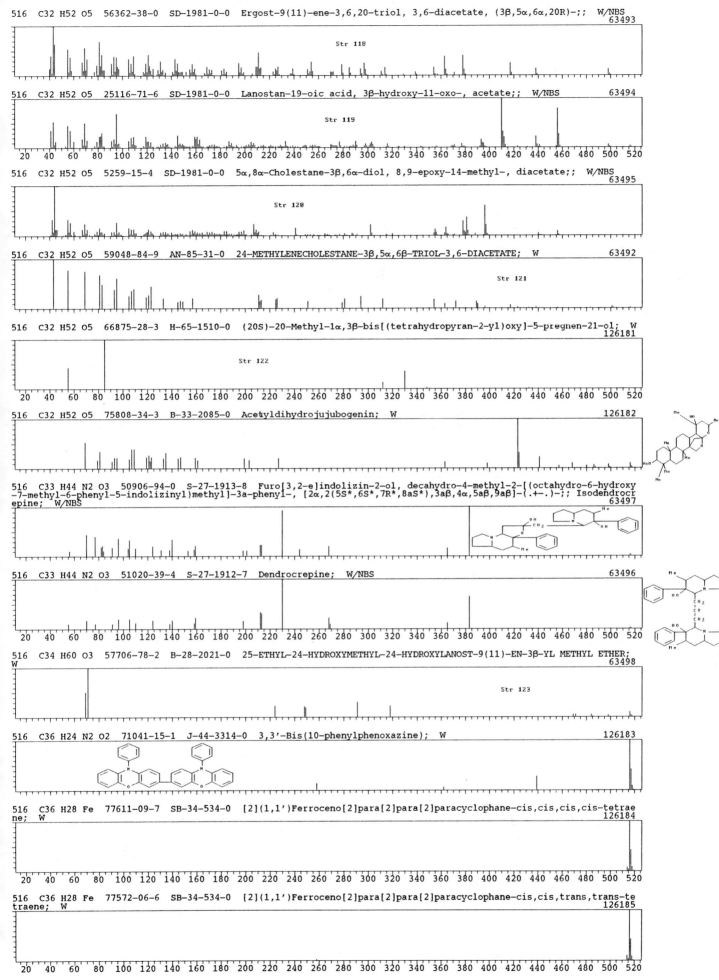

516   C32 H52 O5   56362-38-0   SD-1981-0-0   Ergost-9(11)-ene-3,6,20-triol, 3,6-diacetate, (3β,5α,6α,20R)-;;   W/NBS
63493

Str 118

516   C32 H52 O5   25116-71-6   SD-1981-0-0   Lanostan-19-oic acid, 3β-hydroxy-11-oxo-, acetate;;   W/NBS   63494

Str 119

516   C32 H52 O5   5259-15-4   SD-1981-0-0   5α,8α-Cholestane-3β,6α-diol, 8,9-epoxy-14-methyl-, diacetate;;   W/NBS
63495

Str 120

516   C32 H52 O5   59048-84-9   AN-85-31-0   24-METHYLENECHOLESTANE-3β,5α,6β-TRIOL-3,6-DIACETATE;   W   63492

Str 121

516   C32 H52 O5   66875-28-3   H-65-1510-0   (20S)-20-Methyl-1α,3β-bis[(tetrahydropyran-2-yl)oxy]-5-pregnen-21-ol;   W
126181

Str 122

516   C32 H52 O5   75808-34-3   B-33-2085-0   Acetyldihydrojujubogenin;   W   126182

516   C33 H44 N2 O3   50906-94-0   S-27-1913-8   Furo[3,2-e]indolizin-2-ol, decahydro-4-methyl-2-[(octahydro-6-hydroxy
-7-methyl-6-phenyl-5-indolizinyl)methyl]-3a-phenyl-, [2α,2(5S*,6S*,7R*,8aS*),3aβ,4α,5aβ,9aβ]-(.+-.)-;; Isodendrocr
epine;   W/NBS
63497

516   C33 H44 N2 O3   51020-39-4   S-27-1912-7   Dendrocrepine;   W/NBS   63496

516   C34 H60 O3   57706-78-2   B-28-2021-0   25-ETHYL-24-HYDROXYMETHYL-24-HYDROXYLANOST-9(11)-EN-3β-YL METHYL ETHER;
W
63498

Str 123

516   C36 H24 N2 O2   71041-15-1   J-44-3314-0   3,3'-Bis(10-phenylphenoxazine);   W   126183

516   C36 H28 Fe   77611-09-7   SB-34-534-0   [2](1,1')Ferroceno[2]para[2]para[2]paracyclophane-cis,cis,cis,cis-tetrae
ne;   W
126184

516   C36 H28 Fe   77572-06-6   SB-34-534-0   [2](1,1')Ferroceno[2]para[2]para[2]paracyclophane-cis,cis,trans,trans-te
traene;   W
126185

516 C36 H28 Fe 77611-10-0 SB-34-534-0 [2](1,1')Ferroceno[2]para[2]para[2]paracyclophane-cis,cis,cis,trans-tet
raene; W   126186

516 C36 H52 O2 1990-11-0 NS-0-0-0 Cholest-5-en-3-ol (3β)-, 3-phenyl-2-propenoate; W/NBS   126187

Str 124

516 C36 H52 O2 6020-45-7 EP-2201-0-0 Stigmasta-5,7-dien-3-ol, benzoate, (3β)-;; 7-DEHYDROSITOSTERYL BENZOATE;
W/NBS   63499

Str 125

516 C37 H28 N2 O 2873-76-9 B-36-413-0 4,4'-Bis(diphenylamino)benzophenone; W   126188

516 C40 H36 72862-24-9 SB-33-447-0 (2)(1,4)NAPHTHALENO(2)PARACYCLO(2)(1,4)NAPHTHOLENO(2)PARACYCLOPHANE; W
63500

516 C40 H36 72862-27-2 SB-33-447-0 [2](2,6)Naphthaleno[2]paracyclo[2](2,6)naphthaleno[2]paracyclophane; W
63501

517 C16 H10 F15 N O 55521-09-0 AD-0-5170-0 Octanamide, 2,2,3,3,4,4,5,5,6,6,7,7,8,8,8-pentadecafluoro-N-(2-phen
ylethyl)-;; N-PENTADECAFLUOROOCTANOYL-2-PHENYLETHYLAMINE; W/NBS   63502

CH2CH2NHCOCF2CF2CF2CF2CF2CF2CF3

517 C16 H10 F15 N O O-15-614-2 N(PHENYL-2-ETHYL)N-PENTADECAFLUOROOCTANAMIDE; W   63503

517 C17 H44 N O7 P Si4 55517-61-8 BA-0-309-0 Phosphorimidic acid, methoxy-, 4-oxo-2,3-bis[(trimethylsilyl)oxy]
butyl bis(trimethylsilyl) ester, [R-(R*,R*)]-;; O-TETRAKIS(TRIMETHYLSILYL)-D-ERYTHROSE-4-PHOSPHATE METHYLOXIME;
W/NBS   63504

Me3SiO   OSiMe3
Me3SiOCH(CHO)CHCH2OP:NOMe
OSiMe3

517 C21 H28 Br N O7 S 81925-36-2 F-37-3980-0 N-Tosyl benzylamine acetal; W   126189

OMe
CH2CHOMe
MeO   CH2NSO2   Me
MeO   Br
OMe

517 C21 H45 F2 N3 Si5 64639-37-8 AG-135-171-0 Bis(fluorodimethylsilyl)-hexamethylcyclotrisilazane; W 126190

Me   H   Me
Me Si   Si Me
SiFMe N   N SiF(Bu-1)2
Si
Me Me

517 C23 H17 F6 N O6 GC-13-104-5 Trifluoroacetyl haloxone; W   126191

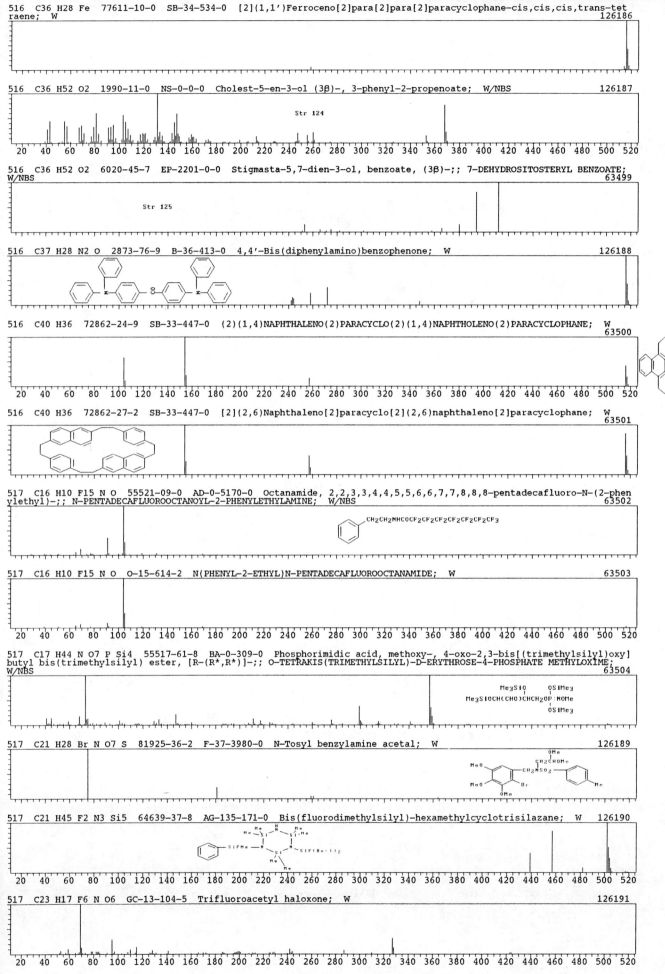

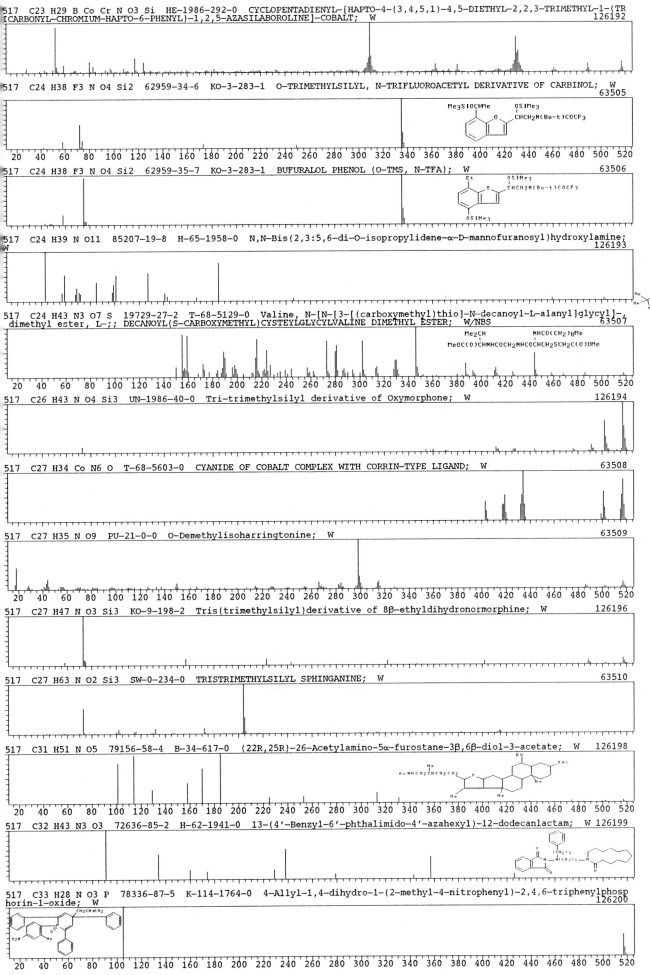

517   C23 H29 B Co Cr N O3 Si   HE-1986-292-0   CYCLOPENTADIENYL-[HAPTO-4-(3,4,5,1)-4,5-DIETHYL-2,2,3-TRIMETHYL-1-(TR
ICARBONYL-CHROMIUM-HAPTO-6-PHENYL)-1,2,5-AZASILABOROLINE]-COBALT;   W   126192

517   C24 H38 F3 N O4 Si2   62959-34-6   KO-3-283-1   O-TRIMETHYLSILYL, N-TRIFLUOROACETYL DERIVATIVE OF CARBINOL;   W
                                                                                                              63505

517   C24 H38 F3 N O4 Si2   62959-35-7   KO-3-283-1   BUFURALOL PHENOL (O-TMS, N-TFA);   W
                                                                                                              63506

517   C24 H39 N O11   85207-19-8   H-65-1958-0   N,N-Bis(2,3:5,6-di-O-isopropylidene-α-D-mannofuranosyl)hydroxylamine;
W   126193

517   C24 H43 N3 O7 S   19729-27-2   T-68-5129-0   Valine, N-[N-[3-[(carboxymethyl)thio]-N-decanoyl-L-alanyl]glycyl]-,
dimethyl ester, L-;;   DECANOYL(S-CARBOXYMETHYL)CYSTEYLGLYCYLVALINE DIMETHYL ESTER;   W/NBS   63507

517   C26 H43 N O4 Si3   UN-1986-40-0   Tri-trimethylsilyl derivative of Oxymorphone;   W   126194

517   C27 H34 Co N6 O   T-68-5603-0   CYANIDE OF COBALT COMPLEX WITH CORRIN-TYPE LIGAND;   W   63508

517   C27 H35 N O9   PU-21-0-0   O-Demethylisoharringtonine;   W   63509

517   C27 H47 N O3 Si3   KO-9-198-2   Tris(trimethylsilyl)derivative of 8β-ethyldihydronormorphine;   W   126196

517   C27 H63 N O2 Si3   SW-0-234-0   TRISTRIMETHYLSILYL SPHINGANINE;   W   63510

517   C31 H51 N O5   79156-58-4   B-34-617-0   (22R,25R)-26-Acetylamino-5α-furostane-3β,6β-diol-3-acetate;   W   126198

517   C32 H43 N3 O3   72636-85-2   H-62-1941-0   13-(4'-Benzyl-6'-phthalimido-4'-azahexyl)-12-dodecanlactam;   W 126199

517   C33 H28 N O3 P   78336-87-5   K-114-1764-0   4-Allyl-1,4-dihydro-1-(2-methyl-4-nitrophenyl)-2,4,6-triphenylphosp
horin-1-oxide;   W   126200

517 C33 H28 N O3 P 78336-88-6 K-114-1764-0 2-Allyl-1,2-dihydro-1-(2-methyl-4-nitrophenyl)-2,4,6-triphenylphosphorin-1-oxide; W
126201

517 C33 H28 N O3 P 78336-89-7 K-114-1764-0 6-(2-Methyl-4-nitrophenyl)-3,5,7-triphenyl-6-phosphatricyclo[3.3.1.0(2,7)]non-3-en-6-oxide; W
126202

517 C33 H59 N O3 81920-84-5 KC-1982-115-0 3,3-Bisisopropoxy-3,4-secocholest-5-en-4-one-oxime; W
126203

517 C37 H27 N O2 73078-84-9 J-45-1671-0 1'-(3-Tolyl)dispiro(anthrone-10,2'-pyrrolidine-3',10''-anthrone); W
126204

517 C37 H27 N S 72478-81-0 J-45-929-0 N-(5',6'-Dihydro-4'-phenylspiro(9H-fluorene-9,3'(4'H)-(2H)naphtho-(1,2-b)thiopyran)-2'-ylidene-benzenenamine; W
126205

517 C37 H27 N S 72478-84-3 J-45-929-0 2-(9H-fluoren-9-ylidene)-3,4,5,6-tetrahydro-3,4-diphenyl-2H-naphtho(2,1-e)-1,3-thiazine; W
126206

518 C11 H4 Cl10 O2 67000-51-5 KO-5-236-0 KEPONE HEMIKETAL; W
63511

518 C14 H12 Cl4 F4 N2 O2 P2 94445-98-4 AH-115-892-0 2,2,4,4-Tetrafluoro-1,3-dimethyl-2,4-bis(2,4-dichlorophenoxy)-1,3,2λ5,4λ5-diazadiphosphetidine; W
126207

518 C14 H42 O7 Si7 107-50-6 HE-1986-2361-0 TETRADECAMETHYLCYCLOHEPTASILOXANE; W
126208

518 C15 H12 Co F9 O6 OS-5-197-0 Tris(1,1,1-trifluoro-2,4-pentanedionato)cobalt(III); W
126209

518 C16 H19 F9 N4 O5 AM-0-8-0 N-TFA ARGININE N-BU ESTER; W
63513

518 C17 H31 Cl5 N2 O3 S RB-1982-13997-0 CARBANILIC ACID, 3,5-DICHLORO-, 2,3,4-TRICHLORO-6-(DIETHYLSULFAMOYL)PHENYL ESTER; W
126211

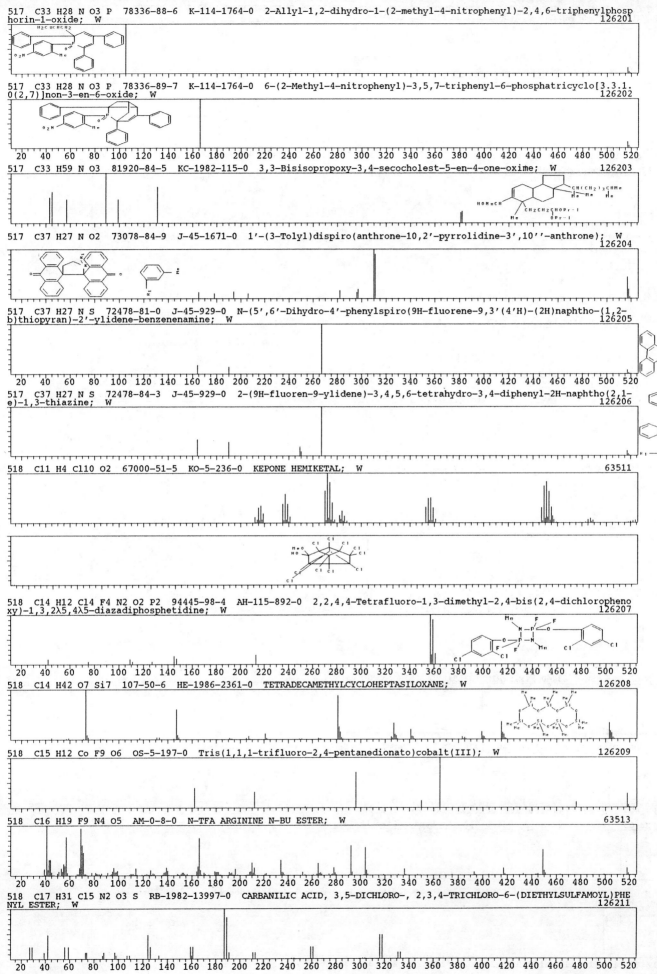

518 C19 H18 Br2 N4 Ni 65230-30-0 B-30-2508-0 2,11-DIBROMO-6,7,8,15,16,17-HEXAHYDRO-5H-DIBENZO[E,M][1,4,8,12]TETRAAZACYCLOPENTADECINATONICKEL (II); W
63514

518 C19 H18 Br2 N4 Ni 59448-31-6 B-30-2508-0 2,12-DIBROMO-6,7,8,15,16,17-HEXAHYDRO-5H-DIBENZO[F,M][1,4,8,12]TETRAAZACYCLOHEXADECINATONICKEL(II); W
63515

518 C19 H18 Br2 N4 Ni B-30-2508-0 2,12-DIBROMO-6,7,8,9,16,17-HEXAHYDRO-5H-DIBENZO[F,M][1,4,8,12]TETRAAZACYCLOPENTADECINATONICKEL (II); W
63516

518 C19 H21 Cl Hg O2 36794-29-3 O7-0-1019-0 Mercury, chloro(3,17-dioxoandrosta-1,4,6-trien-2-yl)-;; 2-CHLOROMERCURI-1,4,6-ANDROSTATRIENE-3,17-DIONE; W/NBS
63517

518 C22 H8 F10 Fe 55517-82-3 O-2-9982-0 Ferrocene, 1,1'-bis(pentafluorophenyl)-;; DI-PENTAFLUOROPHENYLFERROCENE; W/NBS
63518

518 C23 H34 O13 HE-1986-526-0 6-O-(2,3,5-TRI-O-ACETYL-β-D-LYXOFURANOSYL)-1,2:3,4-DI-O-ISOPROPYLIDENE-α-D-GALACTOPYRANOSE; W
126212

518 C23 H34 O13 HE-1986-1757-0 3-O-(2,3,4-TRI-O-ACETYL-α-D-RIBOPYRANOSYL)-1,2-5,6-DI-O-ISOPROPYLIDENE-α-D-GLUCOFURANOSE; W
126213

518 C24 H15 F5 Sn 1058-08-8 NS--28583-0-0 Stannane, (pentafluorophenyl)triphenyl-;; NBS
63519

518 C24 H16 As2 O2 S RB-1982-15781-0 10,10'-THIOBISPHENOXYARSINE; W
126214

518 C24 H16 Cl2 O5 S2 IC-1244-0-0 Bis[(4-chlorophenylsulphonyl)phenyl] ether; W
126215

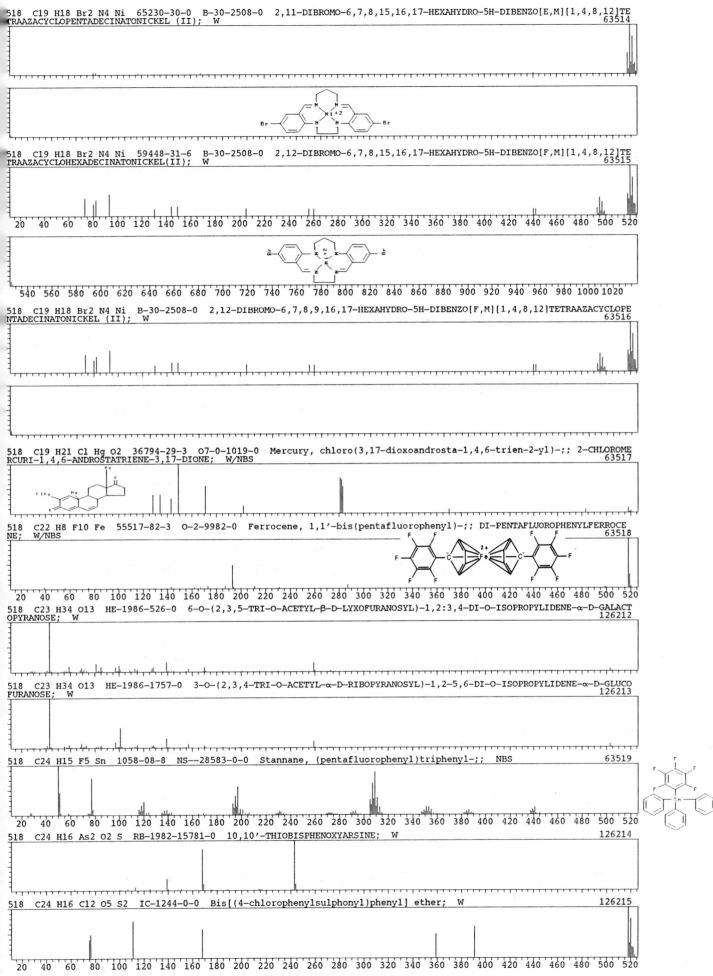

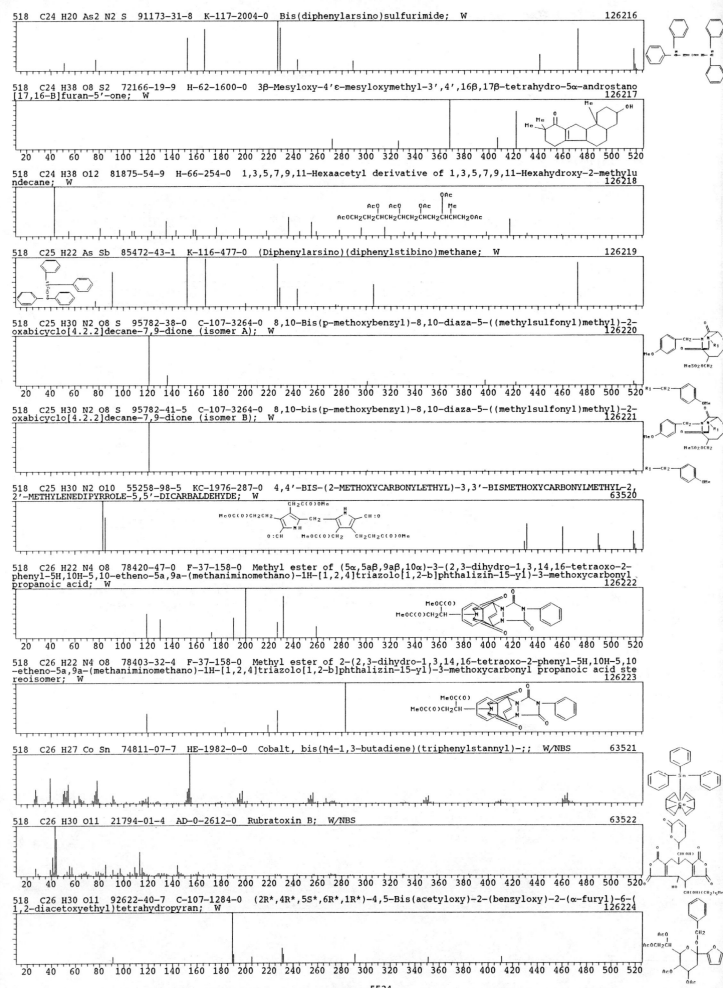

518   C24 H20 As2 N2 S   91173-31-8   K-117-2004-0   Bis(diphenylarsino)sulfurimide;   W
126216

518   C24 H38 O8 S2   72166-19-9   H-62-1600-0   3β-Mesyloxy-4'ε-mesyloxymethyl-3',4',16β,17β-tetrahydro-5α-androstano
[17,16-B]furan-5'-one;   W
126217

518   C24 H38 O12   81875-54-9   H-66-254-0   1,3,5,7,9,11-Hexaacetyl derivative of 1,3,5,7,9,11-Hexahydroxy-2-methylu
ndecane;   W
126218

518   C25 H22 As Sb   85472-43-1   K-116-477-0   (Diphenylarsino)(diphenylstibino)methane;   W
126219

518   C25 H30 N2 O8 S   95782-38-0   C-107-3264-0   8,10-Bis(p-methoxybenzyl)-8,10-diaza-5-((methylsulfonyl)methyl)-2-
oxabicyclo[4.2.2]decane-7,9-dione (isomer A);   W
126220

518   C25 H30 N2 O8 S   95782-41-5   C-107-3264-0   8,10-bis(p-methoxybenzyl)-8,10-diaza-5-((methylsulfonyl)methyl)-2-
oxabicyclo[4.2.2]decane-7,9-dione (isomer B);   W
126221

518   C25 H30 N2 O10   55258-98-5   KC-1976-287-0   4,4'-BIS-(2-METHOXYCARBONYLETHYL)-3,3'-BISMETHOXYCARBONYLMETHYL-2,
2'-METHYLENEDIPYRROLE-5,5'-DICARBALDEHYDE;   W
63520

518   C26 H22 N4 O8   78420-47-0   F-37-158-0   Methyl ester of (5α,5aβ,9aβ,10α)-3-(2,3-dihydro-1,3,14,16-tetraoxo-2-
phenyl-5H,10H-5,10-etheno-5a,9a-(methaniminomethano)-1H-[1,2,4]triazolo[1,2-b]phthalizin-15-yl)-3-methoxycarbonyl
propanoic acid;   W
126222

518   C26 H22 N4 O8   78403-32-4   F-37-158-0   Methyl ester of 2-(2,3-dihydro-1,3,14,16-tetraoxo-2-phenyl-5H,10H-5,10
-etheno-5a,9a-(methaniminomethano)-1H-[1,2,4]triazolo[1,2-b]phthalizin-15-yl)-3-methoxycarbonyl propanoic acid ste
reoisomer;   W
126223

518   C26 H27 Co Sn   74811-07-7   HE-1982-0-0   Cobalt, bis(η4-1,3-butadiene)(triphenylstannyl)-;;   W/NBS
63521

518   C26 H30 O11   21794-01-4   AD-0-2612-0   Rubratoxin B;   W/NBS
63522

518   C26 H30 O11   92622-40-7   C-107-1284-0   (2R*,4R*,5S*,6R*,1R*)-4,5-Bis(acetyloxy)-2-(benzyloxy)-2-(α-furyl)-6-(
1,2-diacetoxyethyl)tetrahydropyran;   W
126224

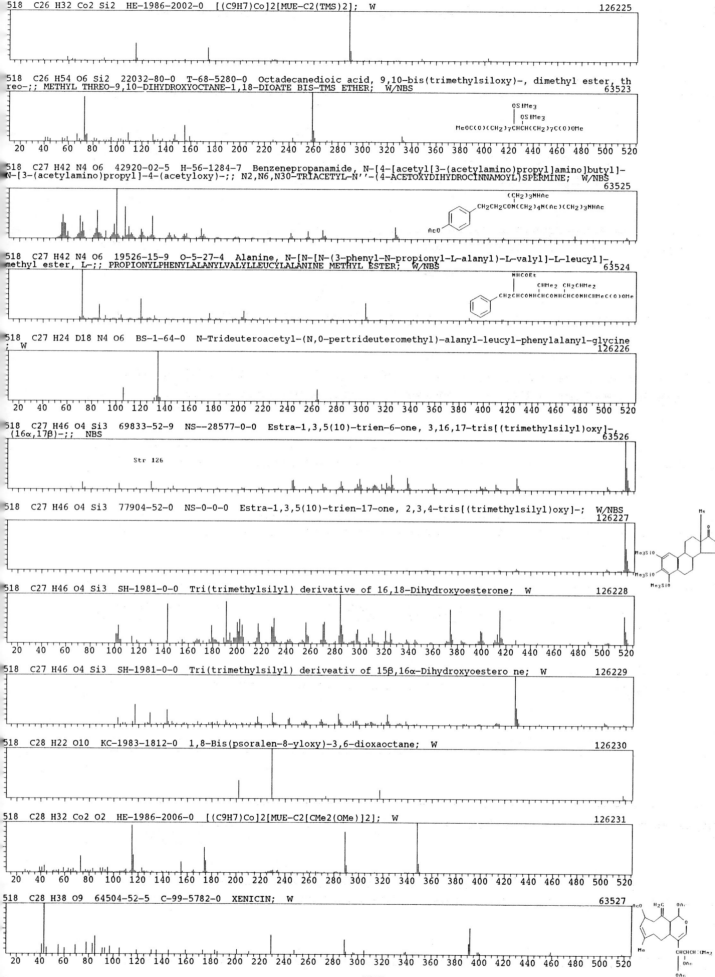

518   C26 H32 Co2 Si2   HE-1986-2002-0   [(C9H7)Co]2[MUE-C2(TMS)2];   W   126225

518   C26 H54 O6 Si2   22032-80-0   T-68-5280-0   Octadecanedioic acid, 9,10-bis(trimethylsiloxy)-, dimethyl ester, th
reo-;; METHYL THREO-9,10-DIHYDROXYOCTANE-1,18-DIOATE BIS-TMS ETHER;   W/NBS   63523

OSIMe3
OSIMe3
MeOC(O)(CH2)7CHCH(CH2)7C(O)OMe

518   C27 H42 N4 O6   42920-02-5   H-56-1284-7   Benzenepropanamide, N-[4-[acetyl[3-(acetylamino)propyl]amino]butyl]-
N-[3-(acetylamino)propyl]-4-(acetyloxy)-;; N2,N6,N30-TRIACETYL-N''-(4-ACETOXYDIHYDROCINNAMOYL)SPERMINE;   W/NBS
                                                                                                    63525

(CH2)3NHAc
CH2CH2CON(CH2)4N(Ac)(CH2)3NHAc
AcO

518   C27 H42 N4 O6   19526-15-9   O-5-27-4   Alanine, N-[N-[N-(3-phenyl-N-propionyl-L-alanyl)-L-valyl]-L-leucyl]-,
methyl ester, L-;; PROPIONYLPHENYLALANYLVALYLLEUCYLALANINE METHYL ESTER;   W/NBS   63524

NHCOEt
CHMe2 CH2CHMe2
CH2CHCONHCHCONHCHCONHCHMeC(O)OMe

518   C27 H24 D18 N4 O6   BS-1-64-0   N-Trideuteroacetyl-(N,O-pertrideuteromethyl)-alanyl-leucyl-phenylalanyl-glycine
;   W   126226

518   C27 H46 O4 Si3   69833-52-9   NS--28577-0-0   Estra-1,3,5(10)-trien-6-one, 3,16,17-tris[(trimethylsilyl)oxy]-,
(16α,17β)-;;   NBS   63526

Str 126

518   C27 H46 O4 Si3   77904-52-0   NS-0-0-0   Estra-1,3,5(10)-trien-17-one, 2,3,4-tris[(trimethylsilyl)oxy]-;   W/NBS
                                                                                                    126227

Me
Me3SiO
Me3SiO
Me3SiO

518   C27 H46 O4 Si3   SH-1981-0-0   Tri(trimethylsilyl) derivative of 16,18-Dihydroxyoesterone;   W   126228

518   C27 H46 O4 Si3   SH-1981-0-0   Tri(trimethylsilyl) deriveativ of 15β,16α-Dihydroxyoestero ne;   W   126229

518   C28 H22 O10   KC-1983-1812-0   1,8-Bis(psoralen-8-yloxy)-3,6-dioxaoctane;   W   126230

518   C28 H32 Co2 O2   HE-1986-2006-0   [(C9H7)Co]2[MUE-C2[CMe2(OMe)]2];   W   126231

518   C28 H38 O9   64504-52-5   C-99-5782-0   XENICIN;   W   63527

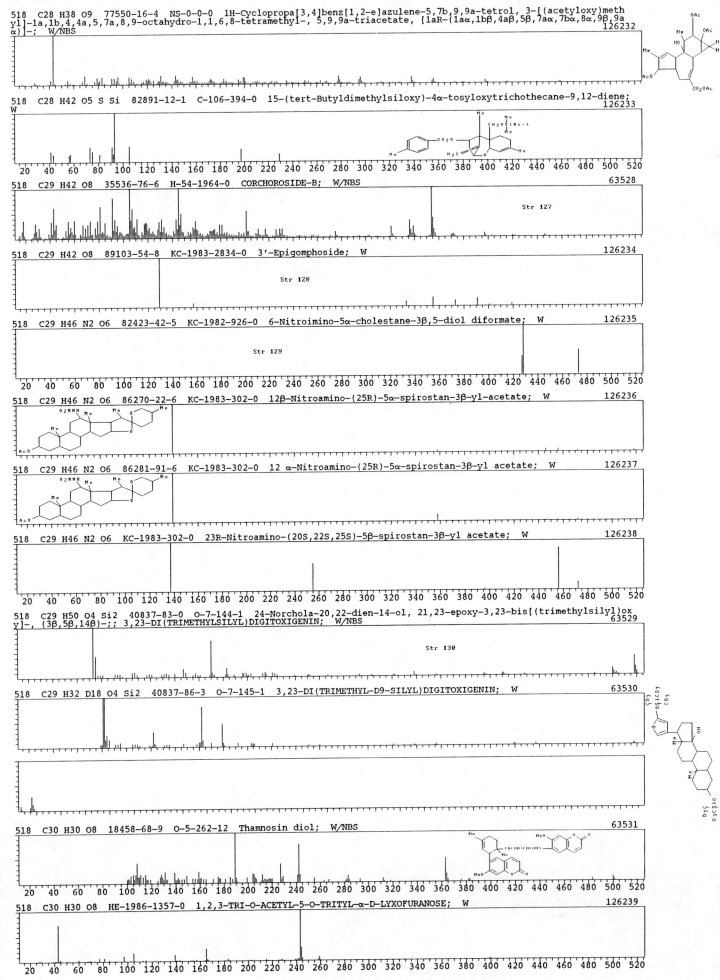

518　C28 H38 O9　77550-16-4　NS-0-0-0　1H-Cyclopropa[3,4]benz[1,2-e]azulene-5,7b,9,9a-tetrol, 3-[(acetyloxy)meth
yl]-1a,1b,4,4a,5,7a,8,9-octahydro-1,1,6,8-tetramethyl-, 5,9,9a-triacetate, [1aR-(1aα,1bβ,4aβ,5β,7aα,7bα,8α,9β,9a
α)]-;　W/NBS　　　　　　　　　　　　　　　　　　　　　　　　　　　　　　　　　　　　126232

518　C28 H42 O5 S Si　82891-12-1　C-106-394-0　15-(tert-Butyldimethylsiloxy)-4α-tosyloxytrichothecane-9,12-diene;
W　　126233

20　40　60　80　100　120　140　160　180　200　220　240　260　280　300　320　340　360　380　400　420　440　460　480　500　520

518　C29 H42 O8　35536-76-6　H-54-1964-0　CORCHOROSIDE-B;　W/NBS　　　　　　　　　63528

Str 127

518　C29 H42 O8　89103-54-8　KC-1983-2834-0　3'-Epigomphoside;　W　　　　　　　　126234

Str 128

518　C29 H46 N2 O6　82423-42-5　KC-1982-926-0　6-Nitroimino-5α-cholestane-3β,5-diol diformate;　W　　126235

Str 129

20　40　60　80　100　120　140　160　180　200　220　240　260　280　300　320　340　360　380　400　420　440　460　480　500　520

518　C29 H46 N2 O6　86270-22-6　KC-1983-302-0　12β-Nitroamino-(25R)-5α-spirostan-3β-yl-acetate;　W　126236

518　C29 H46 N2 O6　86281-91-6　KC-1983-302-0　12 α-Nitroamino-(25R)-5α-spirostan-3β-yl acetate;　W　126237

518　C29 H46 N2 O6　KC-1983-302-0　23R-Nitroamino-(20S,22S,25S)-5β-spirostan-3β-yl acetate;　W　126238

20　40　60　80　100　120　140　160　180　200　220　240　260　280　300　320　340　360　380　400　420　440　460　480　500　520

518　C29 H50 O4 Si2　40837-83-0　O-7-144-1　24-Norchola-20,22-dien-14-ol, 21,23-epoxy-3,23-bis[(trimethylsilyl)ox
y]-, (3β,5β,14β)-;;　3,23-DI(TRIMETHYLSILYL)DIGITOXIGENIN;　W/NBS　　　　　　　　63529

Str 130

518　C29 H32 D18 O4 Si2　40837-86-3　O-7-145-1　3,23-DI(TRIMETHYL-D9-SILYL)DIGITOXIGENIN;　W　63530

518　C30 H30 O8　18458-68-9　O-5-262-12　Thamnosin diol;　W/NBS　　　　　　　　　63531

20　40　60　80　100　120　140　160　180　200　220　240　260　280　300　320　340　360　380　400　420　440　460　480　500　520

518　C30 H30 O8　HE-1986-1357-0　1,2,3-TRI-O-ACETYL-5-O-TRITYL-α-D-LYXOFURANOSE;　W　126239

20　40　60　80　100　120　140　160　180　200　220　240　260　280　300　320　340　360　380　400　420　440　460　480　500　520

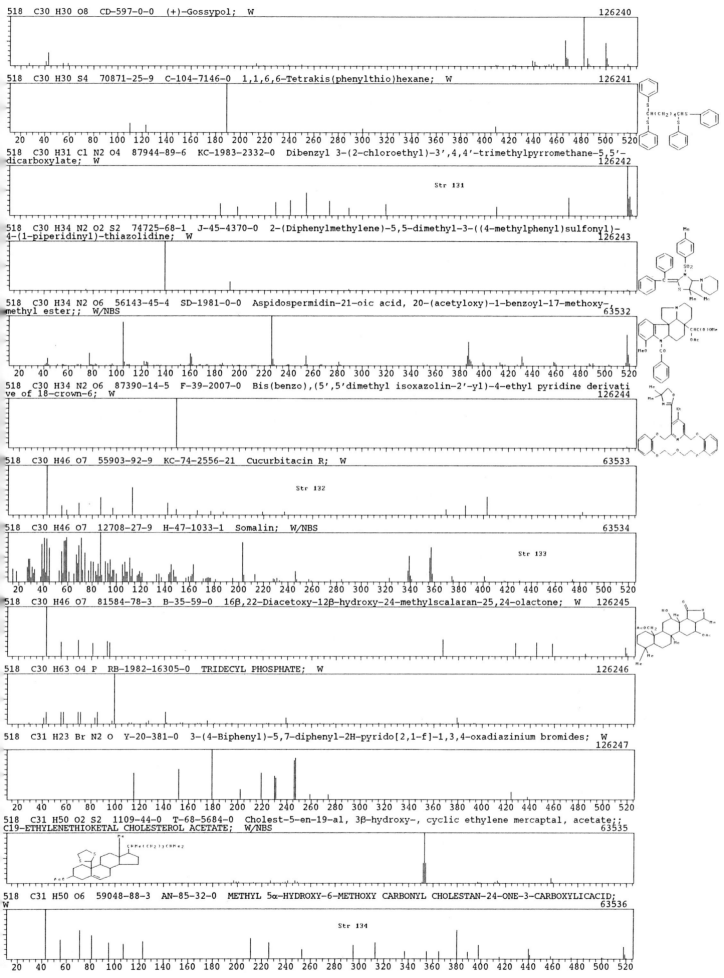

518　C30 H30 O8　CD-597-0-0　(+)-Gossypol;　W　126240

518　C30 H30 S4　70871-25-9　C-104-7146-0　1,1,6,6-Tetrakis(phenylthio)hexane;　W　126241

20　40　60　80　100　120　140　160　180　200　220　240　260　280　300　320　340　360　380　400　420　440　460　480　500　520

518　C30 H31 Cl N2 O4　87944-89-6　KC-1983-2332-0　Dibenzyl 3-(2-chloroethyl)-3',4,4'-trimethylpyrromethane-5,5'-dicarboxylate;　W　126242

Str 131

518　C30 H34 N2 O2 S2　74725-68-1　J-45-4370-0　2-(Diphenylmethylene)-5,5-dimethyl-3-((4-methylphenyl)sulfonyl)-4-(1-piperidinyl)-thiazolidine;　W　126243

518　C30 H34 N2 O6　56143-45-4　SD-1981-0-0　Aspidospermidin-21-oic acid, 20-(acetyloxy)-1-benzoyl-17-methoxy-,methyl ester;;　W/NBS　63532

20　40　60　80　100　120　140　160　180　200　220　240　260　280　300　320　340　360　380　400　420　440　460　480　500　520

518　C30 H34 N2 O6　87390-14-5　F-39-2007-0　Bis(benzo),(5',5'dimethyl isoxazolin-2'-yl)-4-ethyl pyridine derivative of 18-crown-6;　W　126244

518　C30 H46 O7　55903-92-9　KC-74-2556-21　Cucurbitacin R;　W　63533

Str 132

518　C30 H46 O7　12708-27-9　H-47-1033-1　Somalin;　W/NBS　63534

Str 133

20　40　60　80　100　120　140　160　180　200　220　240　260　280　300　320　340　360　380　400　420　440　460　480　500　520

518　C30 H46 O7　81584-78-3　B-35-59-0　16β,22-Diacetoxy-12β-hydroxy-24-methylscalaran-25,24-olactone;　W　126245

518　C30 H63 O4 P　RB-1982-16305-0　TRIDECYL PHOSPHATE;　W　126246

518　C31 H23 Br N2 O　Y-20-381-0　3-(4-Biphenyl)-5,7-diphenyl-2H-pyrido[2,1-f]-1,3,4-oxadiazinium bromides;　W　126247

20　40　60　80　100　120　140　160　180　200　220　240　260　280　300　320　340　360　380　400　420　440　460　480　500　520

518　C31 H50 O2 S2　1109-44-0　T-68-5684-0　Cholest-5-en-19-al, 3β-hydroxy-, cyclic ethylene mercaptal, acetate;;
C19-ETHYLENETHIOKETAL CHOLESTEROL ACETATE;　W/NBS　63535

518　C31 H50 O6　59048-88-3　AN-85-32-0　METHYL 5α-HYDROXY-6-METHOXY CARBONYL CHOLESTAN-24-ONE-3-CARBOXYLICACID;
W　63536

Str 134

20　40　60　80　100　120　140　160　180　200　220　240　260　280　300　320　340　360　380　400　420　440　460　480　500　520

518   C31 H50 O6   77058-95-8   NS-0-0-0   Tetradecanoic acid, 3,3a,4,6a,7,8,9,10,10a,10b-decahydro-3a,10a-dihydroxy-
5-(hydroxymethyl)-2,10-dimethyl-3-oxobenz[e]azulen-8-yl ester, [3aR-(3aα,6aα,8α,10β,10aβ,10bβ)]-; W/NBS   126248

518   C31 H50 O6   77573-28-5   NS-0-0-0   Tetradecanoic acid, (3,3a,4,6a,7,8,9,10,10a,10b-decahydro-3a,8,10a-trihydr
oxy-2,10-dimethyl-3-oxobenz[e]azulen-5-yl)methyl ester, [3aR-(3aα,6aα,8β,10β,10aβ,10bβ)]-; W/NBS   126249

518   C31 H50 O6   77646-22-1   NS-0-0-0   Tetradecanoic acid, (3,3a,4,6a,7,8,9,10,10a,10b-decahydro-3a,8,10a-trihydr
oxy-2,10-dimethyl-3-oxobenz[e]azulen-5-yl)methyl ester, [3aR-(3aα,6aα,8α,10β,10aβ,10bβ)]-; W/NBS   126250

518   C31 H50 O6   84681-07-2   C-105-1978-0   6,6-Diacetoxy-10-hydroxy-5,6:5,10-disecocholest-3-en-5-oic; W   126251

518   C31 H58 O2 Si2   HE-1982-0-0   TESTOSTERONE, BIS-O-(DIMETHYLTERT-BUTYLSILYL)-; W   63537

518   C31 H58 O2 Si2   42151-22-4   NS-0-0-0   Silane, [[(3β,17β)-androst-5-ene-3,17-diyl]bis(oxy)]bis[(1,1-dimethyl
ethyl)dimethyl-; W/NBS   126252
Str 135

518   C31 H58 O2 Si2   65598-85-8   NS-0-0-0   Silane, [[(3β,17β)-androst-5-ene-3,17-diyl]bis(oxy)]bis[triethyl-;
W/NBS   126253

518   C31 H58 O2 Si2   BS-4-110-0   3-α,17-β-Bis(tert-butyldimethylsilyloxy)androst-5-ene; W   126254

518   C32 H54 O5   56259-25-7   SD-1981-0-0   Lanost-9(11)-ene-3,18,20,23-tetrol, 23-acetate, (3β,20.xi.)-;; W/NBS   63539
Str 136

518   C32 H54 O5   59048-83-8   AN-85-31-0   24.XI.-METHYLCHOLESTANE-3β,5α,6β-TRIOL-3,6-DIACETATE; W   63538
Str 137

518   C32 H54 O5   92279-81-7   H-67-1215-0   12-Ethyl-14-[4-hydroxy-8-(2-hydroxypropyl)-3,9-dimethyl-1,7-dioxaspiro[
5.5]undec-2-yl]-4,6-dimethyl-3-tetradecanone; W   126255

518   C33 H38 N6   18210-71-4   NS--28569-0-0   Hodgkinsine;; 3a,3'a(1H,1'H):7',3''a(1''H)-Terpyrrolo[2,3-b]indole, 2,
2',2'',3,3',3'',8,8',8'',8a,8'a,8''a-dodecahydro-1,1',1''-trimethyl-; NBS   63540

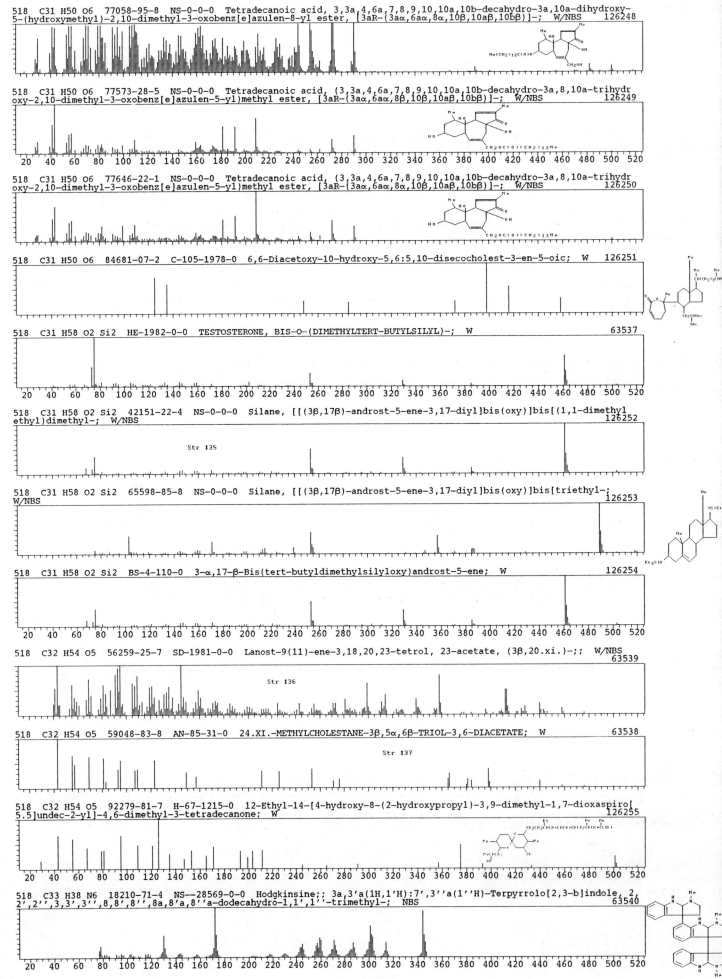

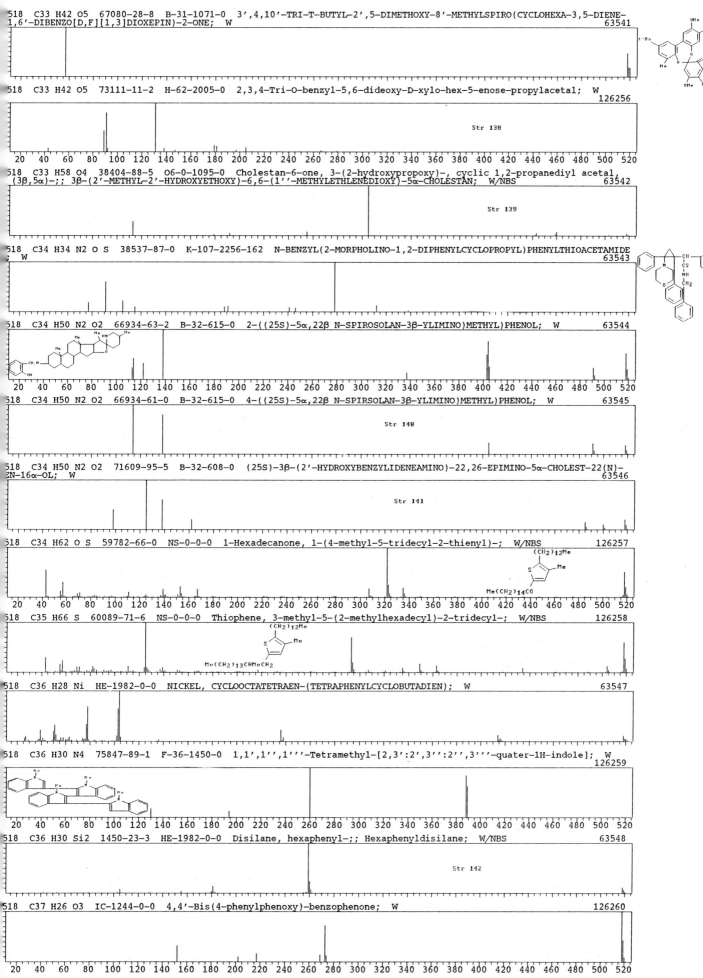

518 C33 H42 O5 67080-28-8 B-31-1071-0 3',4,10'-TRI-T-BUTYL-2',5-DIMETHOXY-8'-METHYLSPIRO(CYCLOHEXA-3,5-DIENE-1,6'-DIBENZO[D,F][1,3]DIOXEPIN)-2-ONE; W 63541

518 C33 H42 O5 73111-11-2 H-62-2005-0 2,3,4-Tri-O-benzyl-5,6-dideoxy-D-xylo-hex-5-enose-propylacetal; W 126256

Str 138

518 C33 H58 O4 38404-88-5 O6-0-1095-0 Cholestan-6-one, 3-(2-hydroxypropoxy)-, cyclic 1,2-propanediyl acetal, (3β,5α)-;; 3β-(2'-METHYL-2'-HYDROXYETHOXY)-6,6-(1''-METHYLETHLENEDIOXY)-5α-CHOLESTAN; W/NBS 63542

Str 139

518 C34 H34 N2 O S 38537-87-0 K-107-2256-162 N-BENZYL(2-MORPHOLINO-1,2-DIPHENYLCYCLOPROPYL)PHENYLTHIOACETAMIDE ; W 63543

518 C34 H50 N2 O2 66934-63-2 B-32-615-0 2-((25S)-5α,22β N-SPIROSOLAN-3β-YLIMINO)METHYL)PHENOL; W 63544

518 C34 H50 N2 O2 66934-61-0 B-32-615-0 4-((25S)-5α,22β N-SPIRSOLAN-3β-YLIMINO)METHYL)PHENOL; W 63545

Str 140

518 C34 H50 N2 O2 71609-95-5 B-32-608-0 (25S)-3β-(2'-HYDROXYBENZYLIDENEAMINO)-22,26-EPIMINO-5α-CHOLEST-22(N)-EN-16α-OL; W 63546

Str 141

518 C34 H62 O S 59782-66-0 NS-0-0-0 1-Hexadecanone, 1-(4-methyl-5-tridecyl-2-thienyl)-; W/NBS 126257

518 C35 H66 S 60089-71-6 NS-0-0-0 Thiophene, 3-methyl-5-(2-methylhexadecyl)-2-tridecyl-; W/NBS 126258

518 C36 H28 Ni HE-1982-0-0 NICKEL, CYCLOOCTATETRAEN-(TETRAPHENYLCYCLOBUTADIEN); W 63547

518 C36 H30 N4 75847-89-1 F-36-1450-0 1,1',1'',1'''-Tetramethyl-[2,3':2',3'':2'',3'''-quater-1H-indole]; W 126259

518 C36 H30 Si2 1450-23-3 HE-1982-0-0 Disilane, hexaphenyl-;; Hexaphenyldisilane; W/NBS 63548

Str 142

518 C37 H26 O3 IC-1244-0-0 4,4'-Bis(4-phenylphenoxy)-benzophenone; W 126260

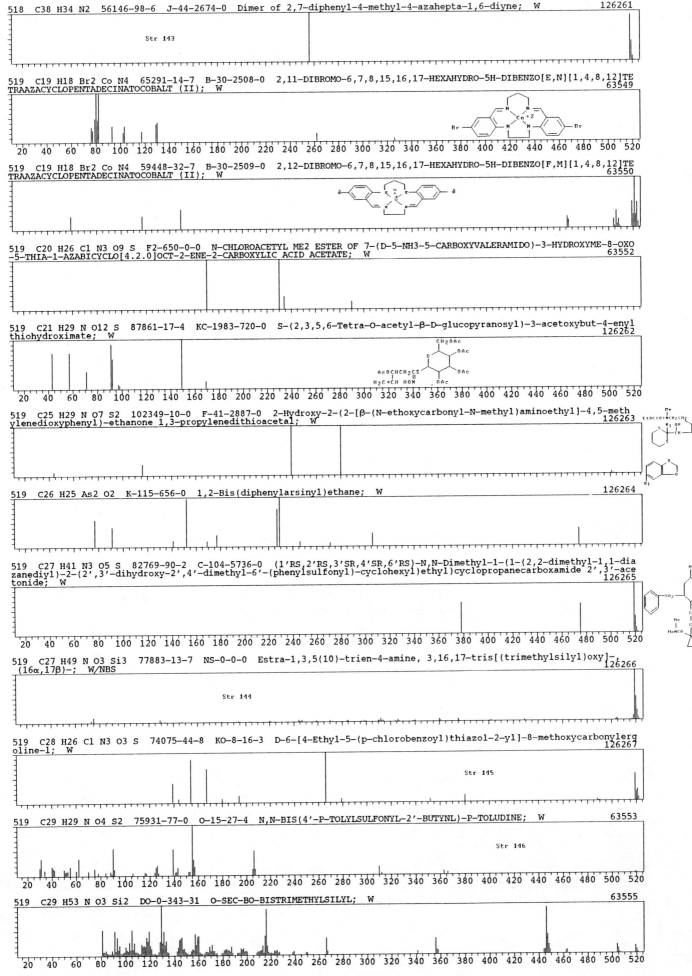

518   C38 H34 N2   56146-98-6   J-44-2674-0   Dimer of 2,7-diphenyl-4-methyl-4-azahepta-1,6-diyne;   W   126261

Str 143

519   C19 H18 Br2 Co N4   65291-14-7   B-30-2508-0   2,11-DIBROMO-6,7,8,15,16,17-HEXAHYDRO-5H-DIBENZO[E,N][1,4,8,12]TE
TRAAZACYCLOPENTADECINATOCOBALT (II);   W   63549

519   C19 H18 Br2 Co N4   59448-32-7   B-30-2509-0   2,12-DIBROMO-6,7,8,15,16,17-HEXAHYDRO-5H-DIBENZO[F,M][1,4,8,12]TE
TRAAZACYCLOPENTADECINATOCOBALT (II);   W   63550

519   C20 H26 Cl N3 O9 S   F2-650-0-0   N-CHLOROACETYL ME2 ESTER OF 7-(D-5-NH3-5-CARBOXYVALERAMIDO)-3-HYDROXYME-8-OXO
-5-THIA-1-AZABICYCLO[4.2.0]OCT-2-ENE-2-CARBOXYLIC ACID ACETATE;   W   63552

519   C21 H29 N O12 S   87861-17-4   KC-1983-720-0   S-(2,3,5,6-Tetra-O-acetyl-β-D-glucopyranosyl)-3-acetoxybut-4-enyl
thiohydroximate;   W   126262

519   C25 H29 N O7 S2   102349-10-0   F-41-2887-0   2-Hydroxy-2-(2-[β-(N-ethoxycarbonyl-N-methyl)aminoethyl]-4,5-meth
ylenedioxyphenyl)-ethanone 1,3-propylenedithioacetal;   W   126263

519   C26 H25 As2 O2   K-115-656-0   1,2-Bis(diphenylarsinyl)ethane;   W   126264

519   C27 H41 N3 O5 S   82769-90-2   C-104-5736-0   (1'RS,2'RS,3'SR,4'SR,6'RS)-N,N-Dimethyl-1-(1-(2,2-dimethyl-1,1-dia
zanediyl)-2-(2',3'-dihydroxy-2',4'-dimethyl-6'-(phenylsulfonyl)-cyclohexyl)ethyl)cyclopropanecarboxamide 2',3'-ace
tonide;   W   126265

519   C27 H49 N O3 Si3   77883-13-7   NS-0-0-0   Estra-1,3,5(10)-trien-4-amine, 3,16,17-tris[(trimethylsilyl)oxy]-,
(16α,17β)-;   W/NBS   126266

Str 144

519   C28 H26 Cl N3 O3 S   74075-44-8   KO-8-16-3   D-6-[4-Ethyl-5-(p-chlorobenzoyl)thiazol-2-yl]-8-methoxycarbonylerg
oline-1;   W   126267

Str 145

519   C29 H29 N O4 S2   75931-77-0   O-15-27-4   N,N-BIS(4'-P-TOLYLSULFONYL-2'-BUTYNL)-P-TOLUDINE;   W   63553

Str 146

519   C29 H53 N O3 Si2   DO-0-343-31   O-SEC-BO-BISTRIMETHYLSILYL;   W   63555

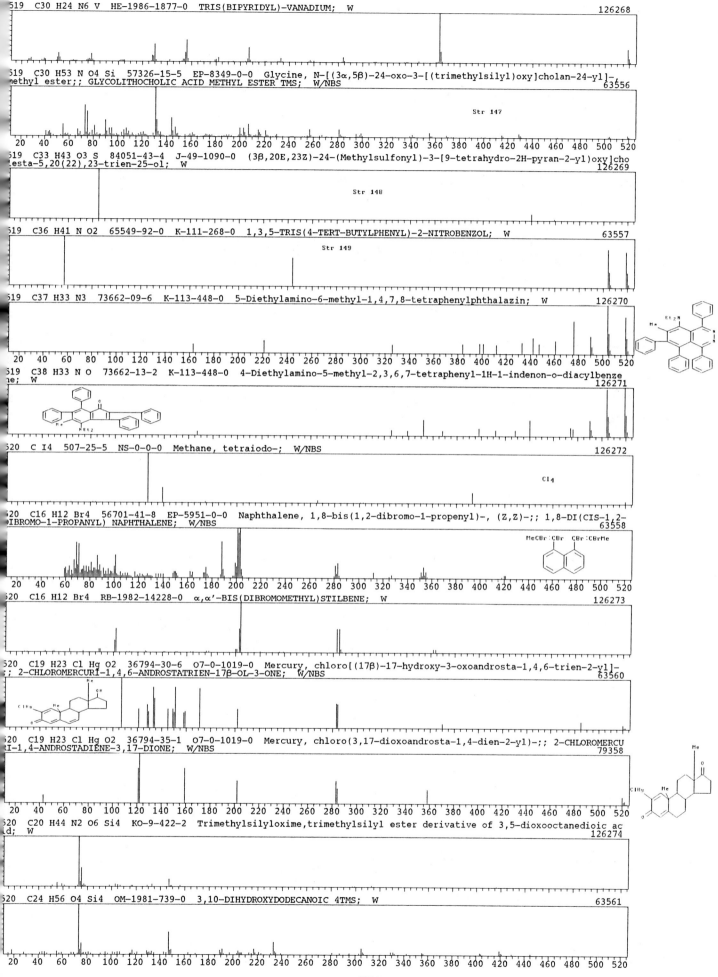

519　C30 H24 N6 V　HE-1986-1877-0　TRIS(BIPYRIDYL)-VANADIUM;　W
126268

519　C30 H53 N O4 Si　57326-15-5　EP-8349-0-0　Glycine, N-[(3α,5β)-24-oxo-3-[(trimethylsilyl)oxy]cholan-24-yl]-, methyl ester;; GLYCOLITHOCHOLIC ACID METHYL ESTER TMS;　W/NBS
63556

Str 147

519　C33 H43 O3 S　84051-43-4　J-49-1090-0　(3β,20E,23Z)-24-(Methylsulfonyl)-3-[9-tetrahydro-2H-pyran-2-yl)oxy]cholesta-5,20(22),23-trien-25-ol;　W
126269

Str 148

519　C36 H41 N O2　65549-92-0　K-111-268-0　1,3,5-TRIS(4-TERT-BUTYLPHENYL)-2-NITROBENZOL;　W
63557

Str 149

519　C37 H33 N3　73662-09-6　K-113-448-0　5-Diethylamino-6-methyl-1,4,7,8-tetraphenylphthalazin;　W
126270

519　C38 H33 N O　73662-13-2　K-113-448-0　4-Diethylamino-5-methyl-2,3,6,7-tetraphenyl-1H-1-indenon-o-diacylbenzene;　W
126271

520　C I4　507-25-5　NS-0-0-0　Methane, tetraiodo-;　W/NBS
126272

CI4

520　C16 H12 Br4　56701-41-8　EP-5951-0-0　Naphthalene, 1,8-bis(1,2-dibromo-1-propenyl)-, (Z,Z)-;; 1,8-DI(CIS-1,2-DIBROMO-1-PROPANYL) NAPHTHALENE;　W/NBS
63558

MeCBr:CBr　CBr:CBrMe

520　C16 H12 Br4　RB-1982-14228-0　α,α'-BIS(DIBROMOMETHYL)STILBENE;　W
126273

520　C19 H23 Cl Hg O2　36794-30-6　O7-0-1019-0　Mercury, chloro[(17β)-17-hydroxy-3-oxoandrosta-1,4,6-trien-2-yl]-;; 2-CHLOROMERCURI-1,4,6-ANDROSTATRIEN-17β-OL-3-ONE;　W/NBS
63560

520　C19 H23 Cl Hg O2　36794-35-1　O7-0-1019-0　Mercury, chloro(3,17-dioxoandrosta-1,4-dien-2-yl)-;; 2-CHLOROMERCURI-1,4-ANDROSTADIENE-3,17-DIONE;　W/NBS
79358

520　C20 H44 N2 O6 Si4　KO-9-422-2　Trimethylsilyloxime,trimethylsilyl ester derivative of 3,5-dioxooctanedioic acid;　W
126274

520　C24 H56 O4 Si4　OM-1981-739-0　3,10-DIHYDROXYDODECANOIC 4TMS;　W
63561

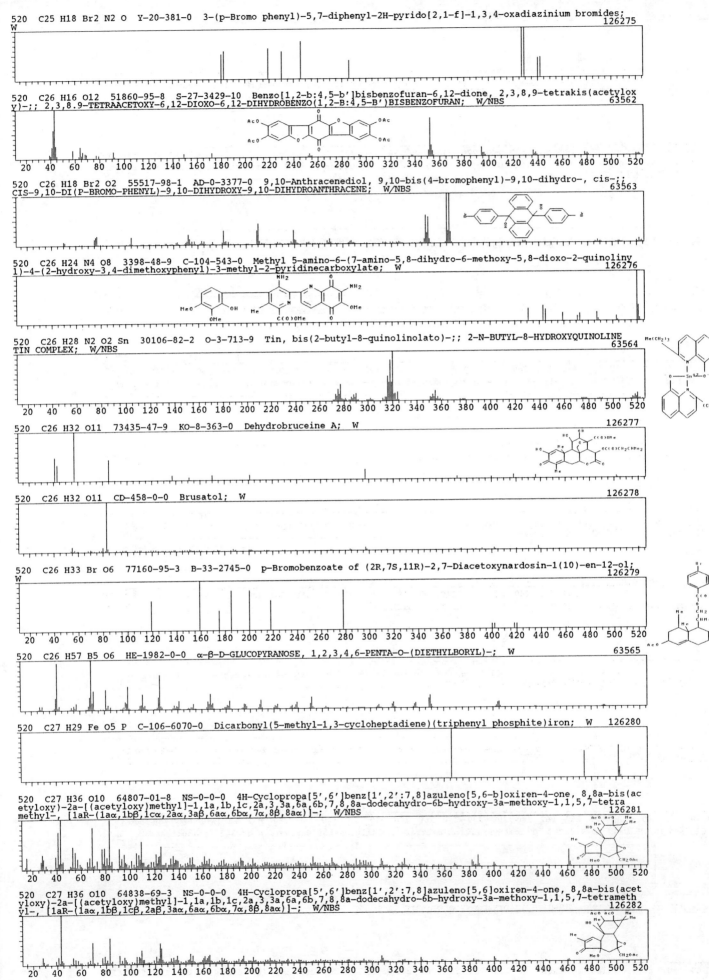

520   C25 H18 Br2 N2 O   Y-20-381-0   3-(p-Bromo phenyl)-5,7-diphenyl-2H-pyrido[2,1-f]-1,3,4-oxadiazinium bromides;
W   126275

520   C26 H16 O12   51860-95-8   S-27-3429-10   Benzo[1,2-b:4,5-b']bisbenzofuran-6,12-dione, 2,3,8,9-tetrakis(acetylox
y)-;; 2,3,8,9-TETRAACETOXY-6,12-DIOXO-6,12-DIHYDROBENZO(1,2-B:4,5-B')BISBENZOFURAN;   W/NBS   63562

520   C26 H18 Br2 O2   55517-98-1   AD-0-3377-0   9,10-Anthracenediol, 9,10-bis(4-bromophenyl)-9,10-dihydro-, cis-;;
CIS-9,10-DI(P-BROMO-PHENYL)-9,10-DIHYDROXY-9,10-DIHYDROANTHRACENE;   W/NBS   63563

520   C26 H24 N4 O8   3398-48-9   C-104-543-0   Methyl 5-amino-6-(7-amino-5,8-dihydro-6-methoxy-5,8-dioxo-2-quinoliny
l)-4-(2-hydroxy-3,4-dimethoxyphenyl)-3-methyl-2-pyridinecarboxylate;   W   126276

520   C26 H28 N2 O2 Sn   30106-82-2   O-3-713-9   Tin, bis(2-butyl-8-quinolinolato)-;;   2-N-BUTYL-8-HYDROXYQUINOLINE
TIN COMPLEX;   W/NBS   63564

520   C26 H32 O11   73435-47-9   KO-8-363-0   Dehydrobruceine A;   W   126277

520   C26 H32 O11   CD-458-0-0   Brusatol;   W   126278

520   C26 H33 Br O6   77160-95-3   B-33-2745-0   p-Bromobenzoate of (2R,7S,11R)-2,7-Diacetoxynardosin-1(10)-en-12-ol;
W   126279

520   C26 H57 B5 O6   HE-1982-0-0   α-β-D-GLUCOPYRANOSE, 1,2,3,4,6-PENTA-O-(DIETHYLBORYL)-;   W   63565

520   C27 H29 Fe O5 P   C-106-6070-0   Dicarbonyl(5-methyl-1,3-cycloheptadiene)(triphenyl phosphite)iron;   W   126280

520   C27 H36 O10   64807-01-8   NS-0-0-0   4H-Cyclopropa[5',6']benz[1',2':7,8]azuleno[5,6-b]oxiren-4-one, 8,8a-bis(ac
etyloxy)-2a-[(acetyloxy)methyl]-1,1a,1b,1c,2a,3,3a,6a,6b,7,8,8a-dodecahydro-6b-hydroxy-3a-methoxy-1,1,5,7-tetra
methyl-, [1aR-(1aα,1bβ,1cα,2aα,3aβ,6aα,6bα,7α,8β,8aα)]-;   W/NBS   126281

520   C27 H36 O10   64838-69-3   NS-0-0-0   4H-Cyclopropa[5',6']benz[1',2':7,8]azuleno[5,6]oxiren-4-one, 8,8a-bis(acet
yloxy)-2a-[(acetyloxy)methyl]-1,1a,1b,1c,2a,3,3a,6a,6b,7,8,8a-dodecahydro-6b-hydroxy-3a-methoxy-1,1,5,7-tetrameth
yl-, [1aR-(1aα,1bβ,1cβ,2aβ,3aα,6aα,6bα,7α,8β,8aα)]-;   W/NBS   126282

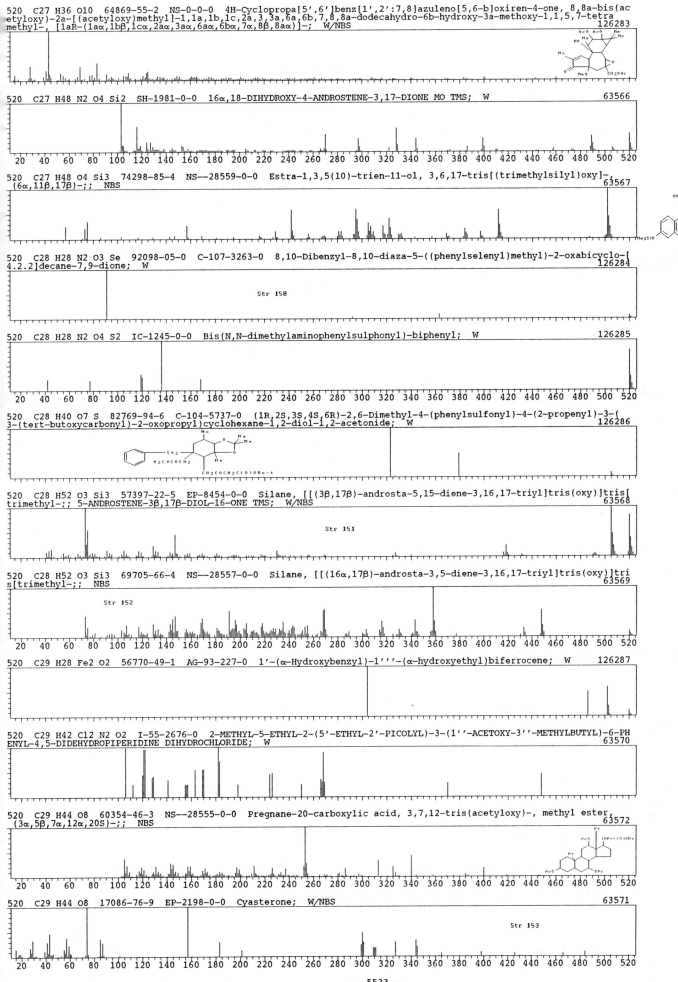

520 C27 H36 O10 64869-55-2 NS-0-0-0 4H-Cyclopropa[5',6']benz[1',2':7,8]azuleno[5,6-b]oxiren-4-one, 8,8a-bis(acetyloxy)-2a-[(acetyloxy)methyl]-1,1a,1b,1c,2a,3,3a,6a,6b,7,8,8a-dodecahydro-6b-hydroxy-3a-methoxy-1,1,5,7-tetramethyl-, [1aR-(1aα,1bβ,1cα,2aα,3aα,6aα,6bα,7α,8β,8aα)]-; W/NBS 126283

520 C27 H48 N2 O4 Si2 SH-1981-0-0 16α,18-DIHYDROXY-4-ANDROSTENE-3,17-DIONE MO TMS; W 63566

520 C27 H48 O4 Si3 74298-85-4 NS--28559-0-0 Estra-1,3,5(10)-trien-11-ol, 3,6,17-tris[(trimethylsilyl)oxy]-(6α,11β,17β)-;; NBS 63567

520 C28 H28 N2 O3 Se 92098-05-0 C-107-3263-0 8,10-Dibenzyl-8,10-diaza-5-((phenylselenyl)methyl)-2-oxabicyclo-[4.2.2]decane-7,9-dione; W 126284
Str 158

520 C28 H28 N2 O4 S2 IC-1245-0-0 Bis(N,N-dimethylaminophenylsulphonyl)-biphenyl; W 126285

520 C28 H40 O7 S 82769-94-6 C-104-5737-0 (1R,2S,3S,4S,6R)-2,6-Dimethyl-4-(phenylsulfonyl)-4-(2-propenyl)-3-(3-(tert-butoxycarbonyl)-2-oxopropyl)cyclohexane-1,2-diol-1,2-acetonide; W 126286

520 C28 H52 O3 Si3 57397-22-5 EP-8454-0-0 Silane, [[(3β,17β)-androsta-5,15-diene-3,16,17-triyl]tris(oxy)]tris[trimethyl-;; 5-ANDROSTENE-3β,17β-DIOL-16-ONE TMS; W/NBS 63568
Str 151

520 C28 H52 O3 Si3 69705-66-4 NS--28557-0-0 Silane, [[(16α,17β)-androsta-3,5-diene-3,16,17-triyl]tris(oxy)]tris[trimethyl-;; NBS 63569
Str 152

520 C29 H28 Fe2 O2 56770-49-1 AG-93-227-0 1'-(α-Hydroxybenzyl)-1'''-(α-hydroxyethyl)biferrocene; W 126287

520 C29 H42 Cl2 N2 O2 I-55-2676-0 2-METHYL-5-ETHYL-2-(5'-ETHYL-2'-PICOLYL)-3-(1''-ACETOXY-3''-METHYLBUTYL)-6-PHENYL-4,5-DIDEHYDROPIPERIDINE DIHYDROCHLORIDE; W 63570

520 C29 H44 O8 60354-46-3 NS--28555-0-0 Pregnane-20-carboxylic acid, 3,7,12-tris(acetyloxy)-, methyl ester, (3α,5β,7α,12α,20S)-;; NBS 63572

520 C29 H44 O8 17086-76-9 EP-2198-0-0 Cyasterone; W/NBS 63571
Str 153

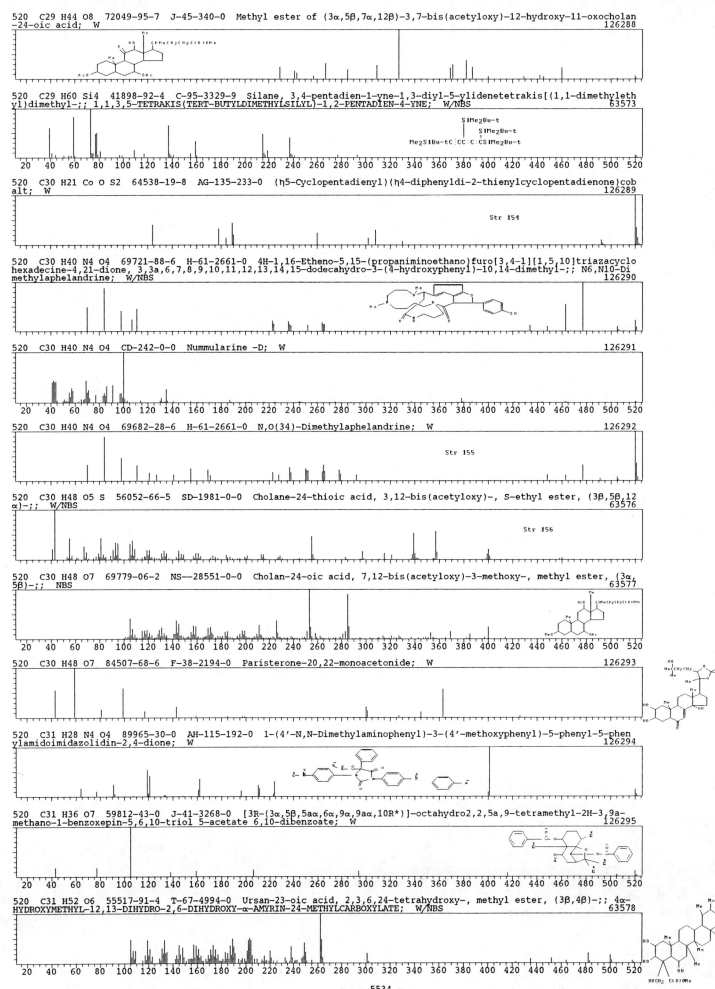

520   C29 H44 O8   72049-95-7   J-45-340-0   Methyl ester of (3α,5β,7α,12β)-3,7-bis(acetyloxy)-12-hydroxy-11-oxocholan-24-oic acid;   W
126288

520   C29 H60 Si4   41898-92-4   C-95-3329-9   Silane, 3,4-pentadien-1-yne-1,3-diyl-5-ylidenetetrakis[(1,1-dimethylethyl)dimethyl-;;   1,1,3,5-TETRAKIS(TERT-BUTYLDIMETHYLSILYL)-1,2-PENTADIEN-4-YNE;   W/NBS
63573

520   C30 H21 Co O S2   64538-19-8   AG-135-233-0   (η5-Cyclopentadienyl)(η4-diphenyldi-2-thienylcyclopentadienone)cobalt;   W
126289

Str 154

520   C30 H40 N4 O4   69721-88-6   H-61-2661-0   4H-1,16-Etheno-5,15-(propaniminoethano)furo[3,4-1][1,5,10]triazacyclohexadecine-4,21-dione, 3,3a,6,7,8,9,10,11,12,13,14,15-dodecahydro-3-(4-hydroxyphenyl)-10,14-dimethyl-;;   N6,N10-Dimethylaphelandrine;   W/NBS
126290

520   C30 H40 N4 O4   CD-242-0-0   Nummularine -D;   W
126291

520   C30 H40 N4 O4   69682-28-6   H-61-2661-0   N,O(34)-Dimethylaphelandrine;   W
126292

Str 155

520   C30 H48 O5 S   56052-66-5   SD-1981-0-0   Cholane-24-thioic acid, 3,12-bis(acetyloxy)-, S-ethyl ester, (3β,5β,12α)-;;   W/NBS
63576

Str 156

520   C30 H48 O7   69779-06-2   NS--28551-0-0   Cholan-24-oic acid, 7,12-bis(acetyloxy)-3-methoxy-, methyl ester, (3α,5β)-;;   NBS
63577

520   C30 H48 O7   84507-68-6   F-38-2194-0   Paristerone-20,22-monoacetonide;   W
126293

520   C31 H28 N4 O4   89965-30-0   AH-115-192-0   1-(4'-N,N-Dimethylaminophenyl)-3-(4'-methoxyphenyl)-5-phenyl-5-phenylamidoimidazolidin-2,4-dione;   W
126294

520   C31 H36 O7   59812-43-0   J-41-3268-0   [3R-(3α,5β,5aα,6α,9α,9aα,10R*)]-octahydro2,2,5a,9-tetramethyl-2H-3,9a-methano-1-benzoxepin-5,6,10-triol 5-acetate 6,10-dibenzoate;   W
126295

520   C31 H52 O6   55517-91-4   T-67-4994-0   Ursan-23-oic acid, 2,3,6,24-tetrahydroxy-, methyl ester, (3β,4β)-;;   4α-HYDROXYMETHYL-12,13-DIHYDRO-2,6-DIHYDROXY-α-AMYRIN-24-METHYLCARBOXYLATE;   W/NBS
63578

520   C31 H60 O2 Si2   71583-60-3   KO-6-61-6   5β-PREGNANE-3α,20β-DIOL BIS(DIMETHYL-N-PROPYLSILYL) ETHER;   W   63579

Str 157

520   C31 H60 O2 Si2   71611-63-7   KO-6-61-6   5β-PREGNANE-3α,20β-DIOL BIS(DIMETHYLISOPROPYLSILYL) ETHER;   W   63580

Str 158

20   40   60   80   100 120 140 160 180 200 220 240 260 280 300 320 340 360 380 400 420 440 460 480 500 520

520   C31 H60 O2 Si2   71611-64-8   KO-6-79-1   5α-ANDROSTANE-3α,17α-DIOL BIS-T-BUTYLDIMETHYLSILYL ETHER;   W   63581

Str 159

520   C31 H60 O2 Si2   57711-47-4   BS-4-108-0   Silane, [[(3β,5α,17β)-androstane-3,17-diyl]bis(oxy)]bis[(1,1-dimethyl ethyl)dimethyl-;   W   126296

Str 160

520   C31 H60 O2 Si2   57711-48-5   BS-4-107-0   Silane, [[(3α,5α,17β)-androstane-3,17-diyl]bis(oxy)]bis[(1,1-dimethyl ethyl)dimethyl-;   W   126297

Str 161

20   40   60   80   100 120 140 160 180 200 220 240 260 280 300 320 340 360 380 400 420 440 460 480 500 520

520   C31 H60 O2 Si2   63754-63-2   NS-0-0-0   Silane, [[(3β,5α,20R)-pregnane-3,20-diyl]bis(oxy)]bis[dimethylpropyl-;   W/NBS   126298

Str 162

520   C31 H60 O2 Si2   65598-75-6   BS-4-109-0   Silane, [[(3α,5β,17β)-androstane-3,17-diyl]bis(oxy)]bis[(1,1-dimethyl ethyl)dimethyl-;   W   126299

Str 163

520   C31 H60 O2 Si2   65598-82-5   NS-0-0-0   Silane, [[(3α,5α,17β)-androstane-3,17-diyl]bis(oxy)]bis[triethyl-;   W/NBS   126300

20   40   60   80   100 120 140 160 180 200 220 240 260 280 300 320 340 360 380 400 420 440 460 480 500 520

520   C31 H60 O2 Si2   65598-83-6   NS-0-0-0   Silane, [[(3β,5α,17β)-androstane-3,17-diyl]bis(oxy)]bis[triethyl-;   W/NBS   126301

520   C31 H60 O2 Si2   65598-84-7   NS-0-0-0   Silane, [[(3α,5β,17β)-androstane-3,17-diyl]bis(oxy)]bis[triethyl-;   W/NBS   126302

520   C31 H60 O2 Si2   71583-58-9   KO-6-60-1   Dimethylisopropylsilyl ether derivative of 5α-Pregnane-3α,20α-diol;   W   126303

Str 164

20   40   60   80   100 120 140 160 180 200 220 240 260 280 300 320 340 360 380 400 420 440 460 480 500 520

520   C31 H60 O2 Si2   71583-57-8   KO-6-60-1   Dimethylisopropylsilyl ether derivative of 5α-Pregnane-3α, 20β-diol;   W   126304

Str 165

20   40   60   80   100 120 140 160 180 200 220 240 260 280 300 320 340 360 380 400 420 440 460 480 500 520

- 5535 -

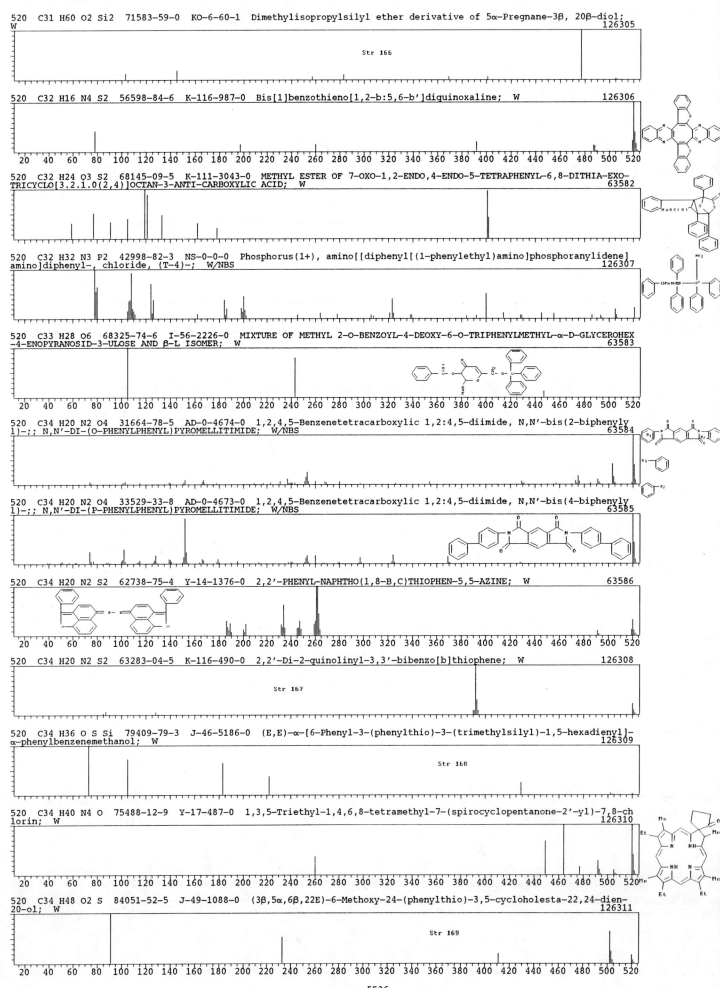

520 C31 H60 O2 Si2 71583-59-0 KO-6-60-1 Dimethylisopropylsilyl ether derivative of 5α-Pregnane-3β, 20β-diol;
W   126305

Str 166

520 C32 H16 N4 S2 56598-84-6 K-116-987-0 Bis[1]benzothieno[1,2-b:5,6-b']diquinoxaline; W   126306

520 C32 H24 O3 S2 68145-09-5 K-111-3043-0 METHYL ESTER OF 7-OXO-1,2-ENDO,4-ENDO-5-TETRAPHENYL-6,8-DITHIA-EXO-
TRICYCLO[3.2.1.0(2,4)]OCTAN-3-ANTI-CARBOXYLIC ACID; W   63582

520 C32 H32 N3 P2 42998-82-3 NS-0-0-0 Phosphorus(1+), amino[[diphenyl[(1-phenylethyl)amino]phosphoranylidene]
amino]diphenyl-, chloride, (T-4)-; W/NBS   126307

520 C33 H28 O6 68325-74-6 I-56-2226-0 MIXTURE OF METHYL 2-O-BENZOYL-4-DEOXY-6-O-TRIPHENYLMETHYL-α-D-GLYCEROHEX
-4-ENOPYRANOSID-3-ULOSE AND β-L ISOMER; W   63583

520 C34 H20 N2 O4 31664-78-5 AD-0-4674-0 1,2,4,5-Benzenetetracarboxylic 1,2:4,5-diimide, N,N'-bis(2-biphenyl
l)-;; N,N'-DI-(O-PHENYLPHENYL)PYROMELLITIMIDE; W/NBS   63584

520 C34 H20 N2 O4 33529-33-8 AD-0-4673-0 1,2,4,5-Benzenetetracarboxylic 1,2:4,5-diimide, N,N'-bis(4-biphenyl
l)-;; N,N'-DI-(P-PHENYLPHENYL)PYROMELLITIMIDE; W/NBS   63585

520 C34 H20 N2 S2 62738-75-4 Y-14-1376-0 2,2'-PHENYL-NAPHTHO(1,8-B,C)THIOPHEN-5,5-AZINE; W   63586

520 C34 H20 N2 S2 63283-04-5 K-116-490-0 2,2'-Di-2-quinolinyl-3,3'-bibenzo[b]thiophene; W   126308

Str 167

520 C34 H36 O S Si 79409-79-3 J-46-5186-0 (E,E)-α-[6-Phenyl-3-(phenylthio)-3-(trimethylsilyl)-1,5-hexadienyl]-
α-phenylbenzenemethanol; W   126309

Str 168

520 C34 H40 N4 O 75488-12-9 Y-17-487-0 1,3,5-Triethyl-1,4,6,8-tetramethyl-7-(spirocyclopentanone-2'-yl)-7,8-ch
lorin; W   126310

520 C34 H48 O2 S 84051-52-5 J-49-1088-0 (3β,5α,6β,22E)-6-Methoxy-24-(phenylthio)-3,5-cycloholesta-22,24-dien-
20-ol; W   126311

Str 169

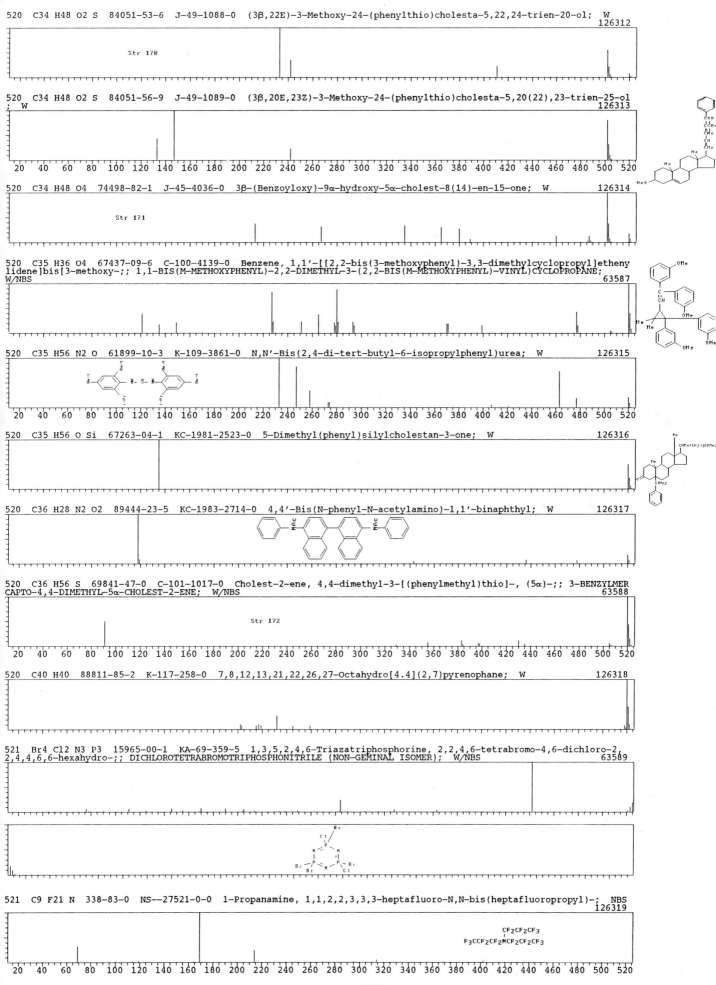

520   C34 H48 O2 S   84051-53-6   J-49-1088-0   (3β,22E)-3-Methoxy-24-(phenylthio)cholesta-5,22,24-trien-20-ol;   W
126312

Str 170

520   C34 H48 O2 S   84051-56-9   J-49-1089-0   (3β,20E,23Z)-3-Methoxy-24-(phenylthio)cholesta-5,20(22),23-trien-25-ol
;   W
126313

520   C34 H48 O4   74498-82-1   J-45-4036-0   3β-(Benzoyloxy)-9α-hydroxy-5α-cholest-8(14)-en-15-one;   W
126314

Str 171

520   C35 H36 O4   67437-09-6   C-100-4139-0   Benzene, 1,1'-[[2,2-bis(3-methoxyphenyl)-3,3-dimethylcyclopropyl]etheny
lidene]bis[3-methoxy-;;   1,1-BIS(M-METHOXYPHENYL)-2,2-DIMETHYL-3-(2,2-BIS(M-METHOXYPHENYL)-VINYL)CYCLOPROPANE;
W/NBS
63587

520   C35 H56 N2 O   61899-10-3   K-109-3861-0   N,N'-Bis(2,4-di-tert-butyl-6-isopropylphenyl)urea;   W
126315

520   C35 H56 O Si   67263-04-1   KC-1981-2523-0   5-Dimethyl(phenyl)silylcholestan-3-one;   W
126316

520   C36 H28 N2 O2   89444-23-5   KC-1983-2714-0   4,4'-Bis(N-phenyl-N-acetylamino)-1,1'-binaphthyl;   W
126317

520   C36 H56 S   69841-47-0   C-101-1017-0   Cholest-2-ene, 4,4-dimethyl-3-[(phenylmethyl)thio]-, (5α)-;;   3-BENZYLMER
CAPTO-4,4-DIMETHYL-5α-CHOLEST-2-ENE;   W/NBS
63588

Str 172

520   C40 H40   88811-85-2   K-117-258-0   7,8,12,13,21,22,26,27-Octahydro[4.4](2,7)pyrenophane;   W
126318

521   Br4 Cl2 N3 P3   15965-00-1   KA-69-359-5   1,3,5,2,4,6-Triazatriphosphorine, 2,2,4,6-tetrabromo-4,6-dichloro-2,
2,4,4,6,6-hexahydro-;;   DICHLOROTETRABROMOTRIPHOSPHONITRILE (NON-GEMINAL ISOMER);   W/NBS
63589

521   C9 F21 N   338-83-0   NS--27521-0-0   1-Propanamine, 1,1,2,2,3,3,3-heptafluoro-N,N-bis(heptafluoropropyl)-;   NBS
126319

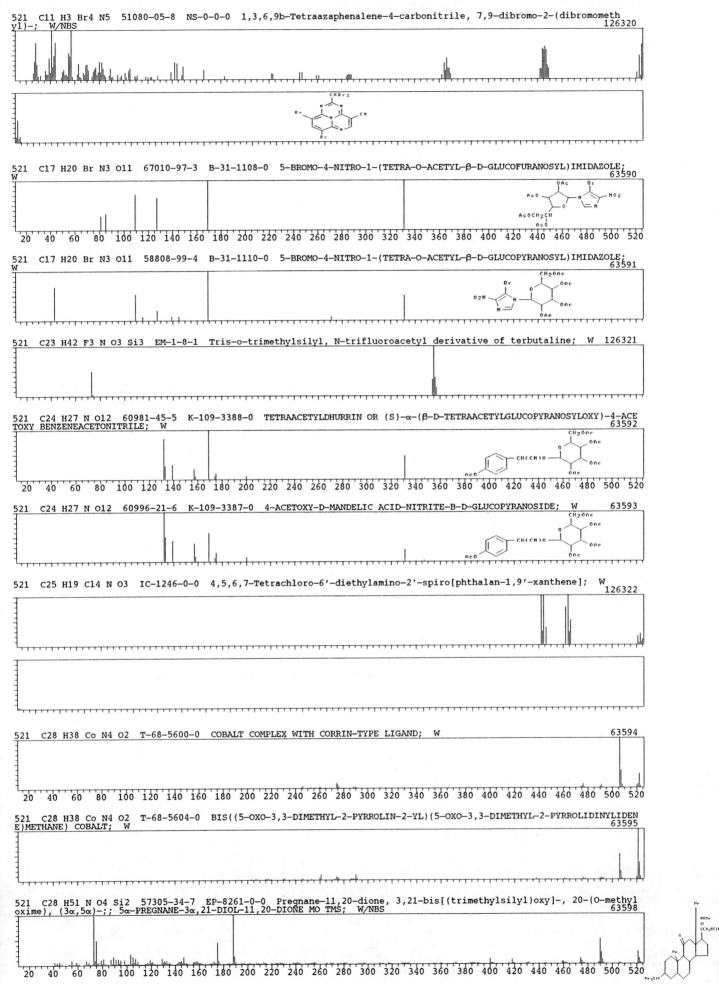

521 C11 H3 Br4 N5 51080-05-8 NS-0-0-0 1,3,6,9b-Tetraazaphenalene-4-carbonitrile, 7,9-dibromo-2-(dibromometh
yl)-; W/NBS   126320

521 C17 H20 Br N3 O11 67010-97-3 B-31-1108-0 5-BROMO-4-NITRO-1-(TETRA-O-ACETYL-β-D-GLUCOFURANOSYL)IMIDAZOLE;
W   63590

521 C17 H20 Br N3 O11 58808-99-4 B-31-1110-0 5-BROMO-4-NITRO-1-(TETRA-O-ACETYL-β-D-GLUCOPYRANOSYL)IMIDAZOLE;
W   63591

521 C23 H42 F3 N O3 Si3 EM-1-8-1 Tris-o-trimethylsilyl, N-trifluoroacetyl derivative of terbutaline; W 126321

521 C24 H27 N O12 60981-45-5 K-109-3388-0 TETRAACETYLDHURRIN OR (S)-α-(β-D-TETRAACETYLGLUCOPYRANOSYLOXY)-4-ACE
TOXY BENZENEACETONITRILE; W   63592

521 C24 H27 N O12 60996-21-6 K-109-3387-0 4-ACETOXY-D-MANDELIC ACID-NITRITE-B-D-GLUCOPYRANOSIDE; W   63593

521 C25 H19 Cl4 N O3 IC-1246-0-0 4,5,6,7-Tetrachloro-6'-diethylamino-2'-spiro[phthalan-1,9'-xanthene]; W
                                                                                                                126322

521 C28 H38 Co N4 O2 T-68-5600-0 COBALT COMPLEX WITH CORRIN-TYPE LIGAND; W   63594

521 C28 H38 Co N4 O2 T-68-5604-0 BIS((5-OXO-3,3-DIMETHYL-2-PYRROLIN-2-YL)(5-OXO-3,3-DIMETHYL-2-PYRROLIDINYLIDEN
E)METHANE) COBALT; W   63595

521 C28 H51 N O4 Si2 57305-34-7 EP-8261-0-0 Pregnane-11,20-dione, 3,21-bis[(trimethylsilyl)oxy]-, 20-(O-methyl
oxime), (3α,5α)-;; 5α-PREGNANE-3α,21-DIOL-11,20-DIONE MO TMS; W/NBS   63598

521 C28 H51 N O4 Si2 57305-35-8 EP-8262-0-0 Pregnane-11,20-dione, 3,21-bis[(trimethylsilyl)oxy]-, 20-(O-methyl oxime), (3α,5β)-;; 5β-PREGNANE-3α,21-DIOL-11,20-DIONE MO TMS; W/NBS
63599

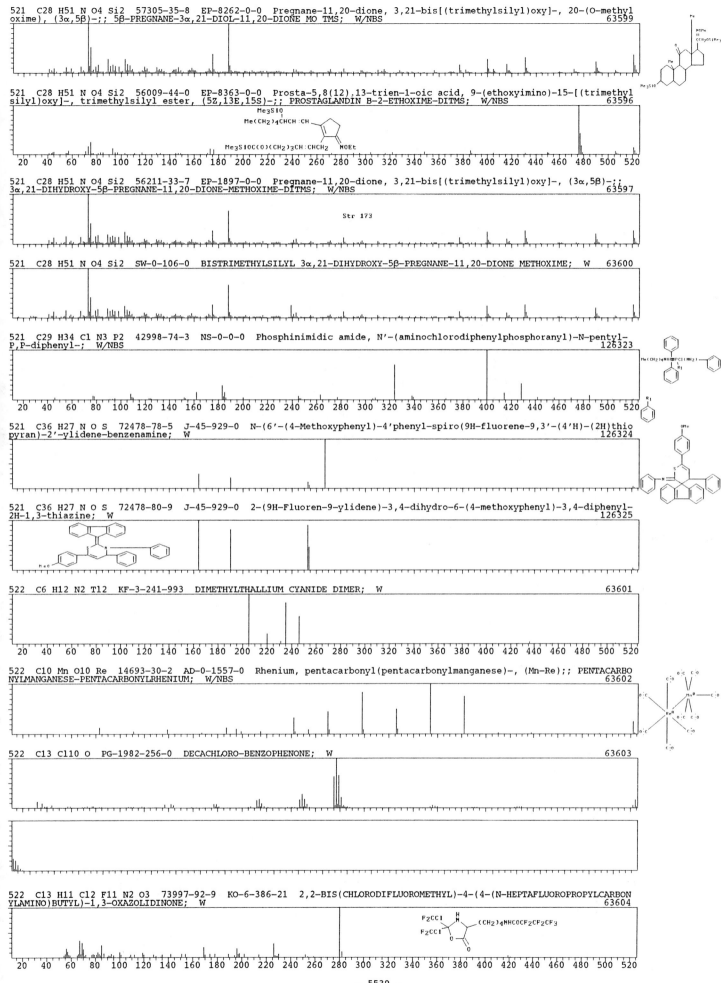

521 C28 H51 N O4 Si2 56009-44-0 EP-8363-0-0 Prosta-5,8(12),13-trien-1-oic acid, 9-(ethoxyimino)-15-[(trimethyl silyl)oxy]-, trimethylsilyl ester, (5Z,13E,15S)-;; PROSTAGLANDIN B-2-ETHOXIME-DITMS; W/NBS
63596

Me₃SiO
Me(CH₂)₄CHCH:CH
Me₃SiOC(O)(CH₂)₃CH:CHCH₂   NOEt

| 20 | 40 | 60 | 80 | 100 | 120 | 140 | 160 | 180 | 200 | 220 | 240 | 260 | 280 | 300 | 320 | 340 | 360 | 380 | 400 | 420 | 440 | 460 | 480 | 500 | 520 |

521 C28 H51 N O4 Si2 56211-33-7 EP-1897-0-0 Pregnane-11,20-dione, 3,21-bis[(trimethylsilyl)oxy]-, (3α,5β)-;; 3α,21-DIHYDROXY-5β-PREGNANE-11,20-DIONE-METHOXIME-DITMS; W/NBS
63597

Str 173

521 C28 H51 N O4 Si2 SW-0-106-0 BISTRIMETHYLSILYL 3α,21-DIHYDROXY-5β-PREGNANE-11,20-DIONE METHOXIME; W   63600

521 C29 H34 Cl N3 P2 42998-74-3 NS-0-0-0 Phosphinimidic amide, N'-(aminochlorodiphenylphosphoranyl)-N-pentyl-P,P-diphenyl-; W/NBS
126323

Me(CH₂)₄NHP PCl(NH₂)
R₁
R₁

521 C36 H27 N O S 72478-78-5 J-45-929-0 N-(6'-(4-Methoxyphenyl)-4'phenyl-spiro(9H-fluorene-9,3'-(4'H)-(2H)thio pyran)-2'-ylidene-benzenamine; W
126324

OMe

521 C36 H27 N O S 72478-80-9 J-45-929-0 2-(9H-Fluoren-9-ylidene)-3,4-dihydro-6-(4-methoxyphenyl)-3,4-diphenyl-2H-1,3-thiazine; W
126325

MeO

522 C6 H12 N2 Tl2 KF-3-241-993 DIMETHYLTHALLIUM CYANIDE DIMER; W   63601

| 20 | 40 | 60 | 80 | 100 | 120 | 140 | 160 | 180 | 200 | 220 | 240 | 260 | 280 | 300 | 320 | 340 | 360 | 380 | 400 | 420 | 440 | 460 | 480 | 500 | 520 |

522 C10 Mn O10 Re 14693-30-2 AD-0-1557-0 Rhenium, pentacarbonyl(pentacarbonylmanganese)-, (Mn-Re);; PENTACARBO NYLMANGANESE-PENTACARBONYLRHENIUM; W/NBS
63602

522 C13 Cl10 O PG-1982-256-0 DECACHLORO-BENZOPHENONE; W   63603

522 C13 H11 Cl2 F11 N2 O3 73997-92-9 KO-6-386-21 2,2-BIS(CHLORODIFLUOROMETHYL)-4-(4-(N-HEPTAFLUOROPROPYLCARBON YLAMINO)BUTYL)-1,3-OXAZOLIDINONE; W
63604

F₂CCl   H
F₂CCl   (CH₂)₄NHCOCF₂CF₂CF₃
O

| 20 | 40 | 60 | 80 | 100 | 120 | 140 | 160 | 180 | 200 | 220 | 240 | 260 | 280 | 300 | 320 | 340 | 360 | 380 | 400 | 420 | 440 | 460 | 480 | 500 | 520 |

522 C13 H12 B F4 O5 Re K-117-3172-0 Pentacarbonyl(1,2,3-trimethyl-4-methylene-1-cyclobutene)rhenium-tetrafluoro borate; W
126326

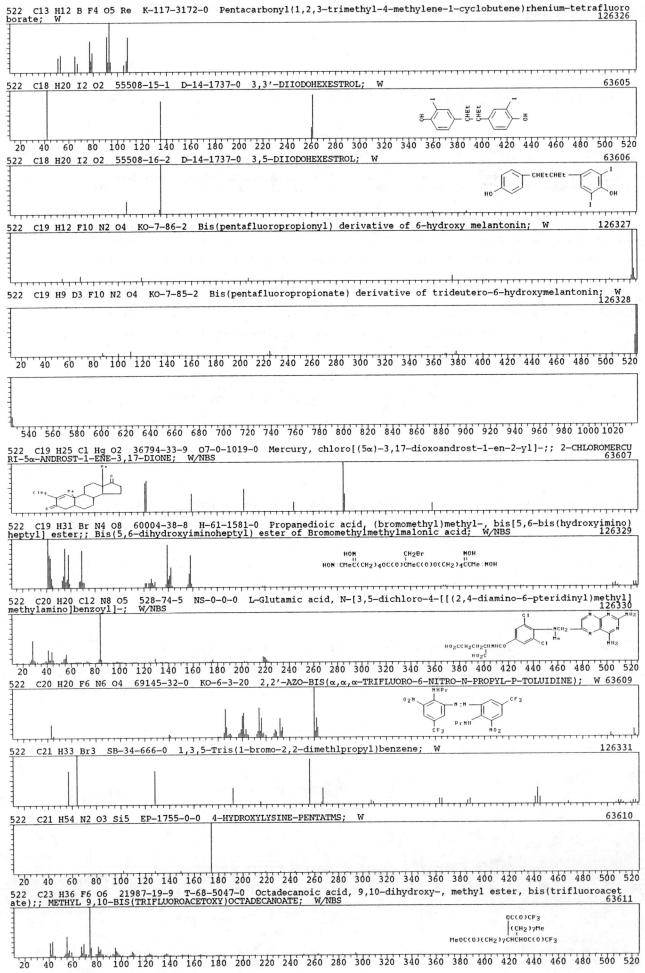

522 C18 H20 I2 O2 55508-15-1 D-14-1737-0 3,3'-DIIODOHEXESTROL; W
63605

522 C18 H20 I2 O2 55508-16-2 D-14-1737-0 3,5-DIIODOHEXESTROL; W
63606

522 C19 H12 F10 N2 O4 KO-7-86-2 Bis(pentafluoropropionyl) derivative of 6-hydroxy melantonin; W
126327

522 C19 H9 D3 F10 N2 O4 KO-7-85-2 Bis(pentafluoropropionate) derivative of trideutero-6-hydroxymelantonin; W
126328

522 C19 H25 Cl Hg O2 36794-33-9 O7-0-1019-0 Mercury, chloro[(5α)-3,17-dioxoandrost-1-en-2-yl]-;; 2-CHLOROMERCURI-5α-ANDROST-1-ENE-3,17-DIONE; W/NBS
63607

522 C19 H31 Br N4 O8 60004-38-8 H-61-1581-0 Propanedioic acid, (bromomethyl)methyl-, bis[5,6-bis(hydroxyimino) heptyl] ester;; Bis(5,6-dihydroxyiminoheptyl) ester of Bromomethylmethylmalonic acid; W/NBS
126329

522 C20 H20 Cl2 N8 O5 528-74-5 NS-0-0-0 L-Glutamic acid, N-[3,5-dichloro-4-[[(2,4-diamino-6-pteridinyl)methyl]methylamino]benzoyl]-; W/NBS
126330

522 C20 H20 F6 N6 O4 69145-32-0 KO-6-3-20 2,2'-AZO-BIS(α,α,α-TRIFLUORO-6-NITRO-N-PROPYL-P-TOLUIDINE); W 63609

522 C21 H33 Br3 SB-34-666-0 1,3,5-Tris(1-bromo-2,2-dimethlpropyl)benzene; W
126331

522 C21 H54 N2 O3 Si5 EP-1755-0-0 4-HYDROXYLYSINE-PENTATMS; W
63610

522 C23 H36 F6 O6 21987-19-9 T-68-5047-0 Octadecanoic acid, 9,10-dihydroxy-, methyl ester, bis(trifluoroacetate);; METHYL 9,10-BIS(TRIFLUOROACETOXY)OCTADECANOATE; W/NBS
63611

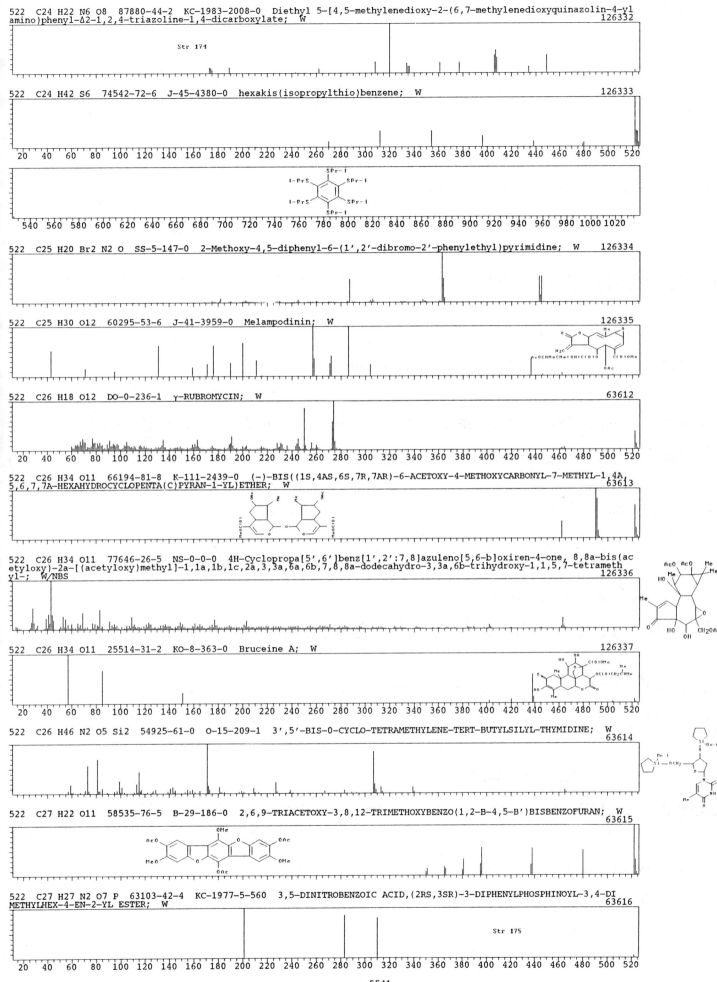

522 C24 H22 N6 O8 87880-44-2 KC-1983-2008-0 Diethyl 5-[4,5-methylenedioxy-2-(6,7-methylenedioxyquinazolin-4-yl
amino)phenyl-Δ2-1,2,4-triazoline-1,4-dicarboxylate; W   126332

Str 174

522 C24 H42 S6 74542-72-6 J-45-4380-0 hexakis(isopropylthio)benzene; W   126333

522 C25 H20 Br2 N2 O SS-5-147-0 2-Methoxy-4,5-diphenyl-6-(1',2'-dibromo-2'-phenylethyl)pyrimidine; W 126334

522 C25 H30 O12 60295-53-6 J-41-3959-0 Melampodinin; W   126335

522 C26 H18 O12 DO-0-236-1 γ-RUBROMYCIN; W   63612

522 C26 H34 O11 66194-81-8 K-111-2439-0 (-)-BIS((1S,4AS,6S,7R,7AR)-6-ACETOXY-4-METHOXYCARBONYL-7-METHYL-1,4A,
5,6,7,7A-HEXAHYDROCYCLOPENTA(C)PYRAN-1-YL)ETHER; W   63613

522 C26 H34 O11 77646-26-5 NS-0-0-0 4H-Cyclopropa[5',6']benz[1',2':7,8]azuleno[5,6-b]oxiren-4-one, 8,8a-bis(ac
etyloxy)-2a-[(acetyloxy)methyl]-1,1a,1b,1c,2a,3,3a,6a,6b,7,8,8a-dodecahydro-3,3a,6b-trihydroxy-1,1,5,7-tetrameth
yl-; W/NBS   126336

522 C26 H34 O11 25514-31-2 KO-8-363-0 Bruceine A; W   126337

522 C26 H46 N2 O5 Si2 54925-61-0 O-15-209-1 3',5'-BIS-O-CYCLO-TETRAMETHYLENE-TERT-BUTYLSILYL-THYMIDINE; W
                                                                                                      63614

522 C27 H22 O11 58535-76-5 B-29-186-0 2,6,9-TRIACETOXY-3,8,12-TRIMETHOXYBENZO(1,2-B-4,5-B')BISBENZOFURAN; W
                                                                                                      63615

522 C27 H27 N2 O7 P 63103-42-4 KC-1977-5-560 3,5-DINITROBENZOIC ACID,(2RS,3SR)-3-DIPHENYLPHOSPHINOYL-3,4-DI
METHYLHEX-4-EN-2-YL ESTER; W   63616

Str 175

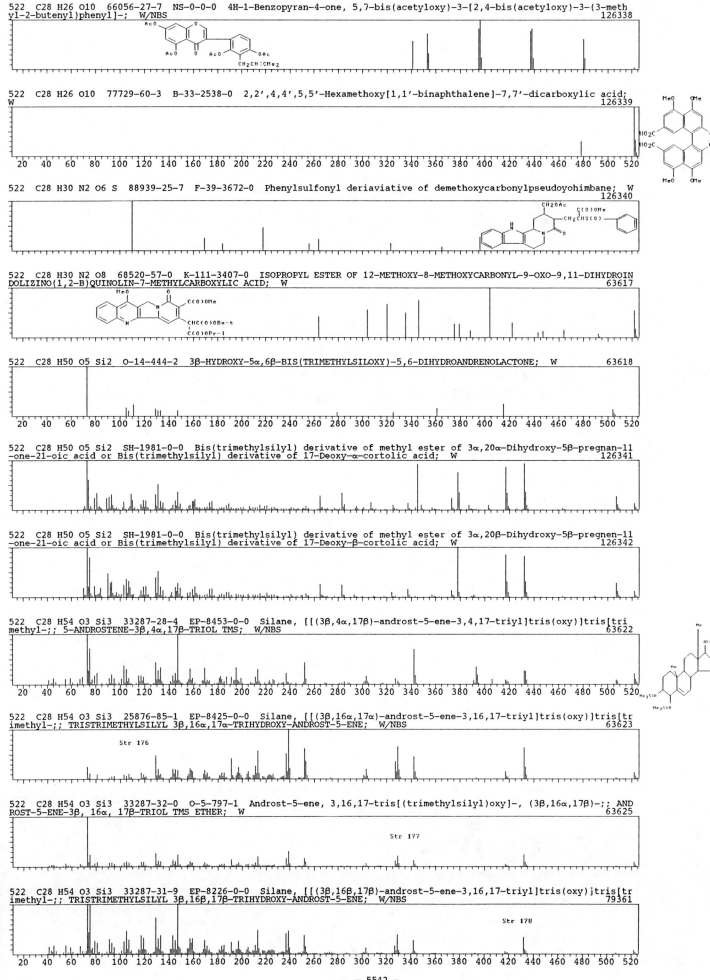

522 C28 H26 O10 66056-27-7 NS-0-0-0 4H-1-Benzopyran-4-one, 5,7-bis(acetyloxy)-3-[2,4-bis(acetyloxy)-3-(3-meth yl-2-butenyl)phenyl]-; W/NBS
126338

522 C28 H26 O10 77729-60-3 B-33-2538-0 2,2',4,4',5,5'-Hexamethoxy[1,1'-binaphthalene]-7,7'-dicarboxylic acid; W
126339

522 C28 H30 N2 O6 S 88939-25-7 F-39-3672-0 Phenylsulfonyl deriaviative of demethoxycarbonylpseudoyohimbane; W
126340

522 C28 H30 N2 O8 68520-57-0 K-111-3407-0 ISOPROPYL ESTER OF 12-METHOXY-8-METHOXYCARBONYL-9-OXO-9,11-DIHYDROIN DOLIZINO(1,2-B)QUINOLIN-7-METHYLCARBOXYLIC ACID; W
63617

522 C28 H50 O5 Si2 O-14-444-2 3β-HYDROXY-5α,6β-BIS(TRIMETHYLSILOXY)-5,6-DIHYDROANDRENOLACTONE; W
63618

522 C28 H50 O5 Si2 SH-1981-0-0 Bis(trimethylsilyl) derivative of methyl ester of 3α,20α-Dihydroxy-5β-pregnan-11 -one-21-oic acid or Bis(trimethylsilyl) derivative of 17-Deoxy-α-cortolic acid; W
126341

522 C28 H50 O5 Si2 SH-1981-0-0 Bis(trimethylsilyl) derivative of methyl ester of 3α,20β-Dihydroxy-5β-pregnen-11 -one-21-oic acid or Bis(trimethylsilyl) derivative of 17-Deoxy-β-cortolic acid; W
126342

522 C28 H54 O3 Si3 33287-28-4 EP-8453-0-0 Silane, [[(3β,4α,17β)-androst-5-ene-3,4,17-triyl]tris(oxy)]tris[tri methyl-;; 5-ANDROSTENE-3β,4α,17β-TRIOL TMS; W/NBS
63622

522 C28 H54 O3 Si3 25876-85-1 EP-8425-0-0 Silane, [[(3β,16α,17α)-androst-5-ene-3,16,17-triyl]tris(oxy)]tris[tr imethyl-;; TRISTRIMETHYLSILYL 3β,16α,17α-TRIHYDROXY-ANDROST-5-ENE; W/NBS
63623

Str 176

522 C28 H54 O3 Si3 33287-32-0 O-5-797-1 Androst-5-ene, 3,16,17-tris[(trimethylsilyl)oxy]-, (3β,16α,17β)-;; AND ROST-5-ENE-3β, 16α, 17β-TRIOL TMS ETHER; W
63625

Str 177

522 C28 H54 O3 Si3 33287-31-9 EP-8226-0-0 Silane, [[(3β,16β,17β)-androst-5-ene-3,16,17-triyl]tris(oxy)]tris[tr imethyl-;; TRISTRIMETHYLSILYL 3β,16β,17β-TRIHYDROXY-ANDROST-5-ENE; W/NBS
79361

Str 178

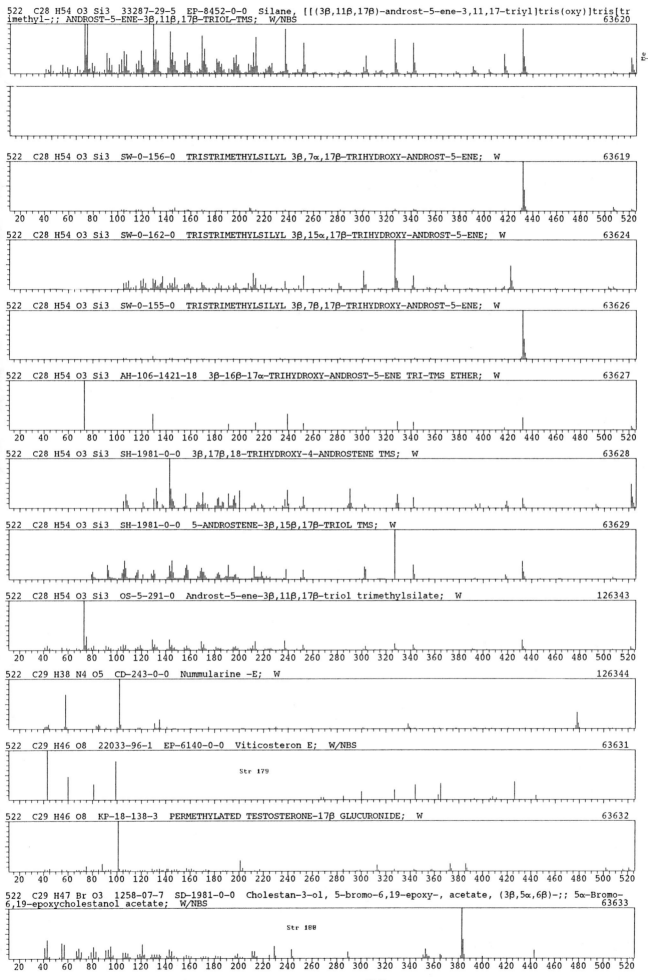

522 C28 H54 O3 Si3 33287-29-5 EP-8452-0-0 Silane, [[(3β,11β,17β)-androst-5-ene-3,11,17-triyl]tris(oxy)]tris[tr imethyl-;; ANDROST-5-ENE-3β,11β,17β-TRIOL-TMS; W/NBS   63620

522 C28 H54 O3 Si3 SW-0-156-0 TRISTRIMETHYLSILYL 3β,7α,17β-TRIHYDROXY-ANDROST-5-ENE; W   63619

522 C28 H54 O3 Si3 SW-0-162-0 TRISTRIMETHYLSILYL 3β,15α,17β-TRIHYDROXY-ANDROST-5-ENE; W   63624

522 C28 H54 O3 Si3 SW-0-155-0 TRISTRIMETHYLSILYL 3β,7β,17β-TRIHYDROXY-ANDROST-5-ENE; W   63626

522 C28 H54 O3 Si3 AH-106-1421-18 3β-16β-17α-TRIHYDROXY-ANDROST-5-ENE TRI-TMS ETHER; W   63627

522 C28 H54 O3 Si3 SH-1981-0-0 3β,17β,18-TRIHYDROXY-4-ANDROSTENE TMS; W   63628

522 C28 H54 O3 Si3 SH-1981-0-0 5-ANDROSTENE-3β,15β,17β-TRIOL TMS; W   63629

522 C28 H54 O3 Si3 OS-5-291-0 Androst-5-ene-3β,11β,17β-triol trimethylsilate; W   126343

522 C29 H38 N4 O5 CD-243-0-0 Nummularine -E; W   126344

522 C29 H46 O8 22033-96-1 EP-6140-0-0 Viticosteron E; W/NBS   63631

Str 179

522 C29 H46 O8 KP-18-138-3 PERMETHYLATED TESTOSTERONE-17β GLUCURONIDE; W   63632

522 C29 H47 Br O3 1258-07-7 SD-1981-0-0 Cholestan-3-ol, 5-bromo-6,19-epoxy-, acetate, (3β,5α,6β)-;; 5α-Bromo-6,19-epoxycholestanol acetate; W/NBS   63633

Str 188

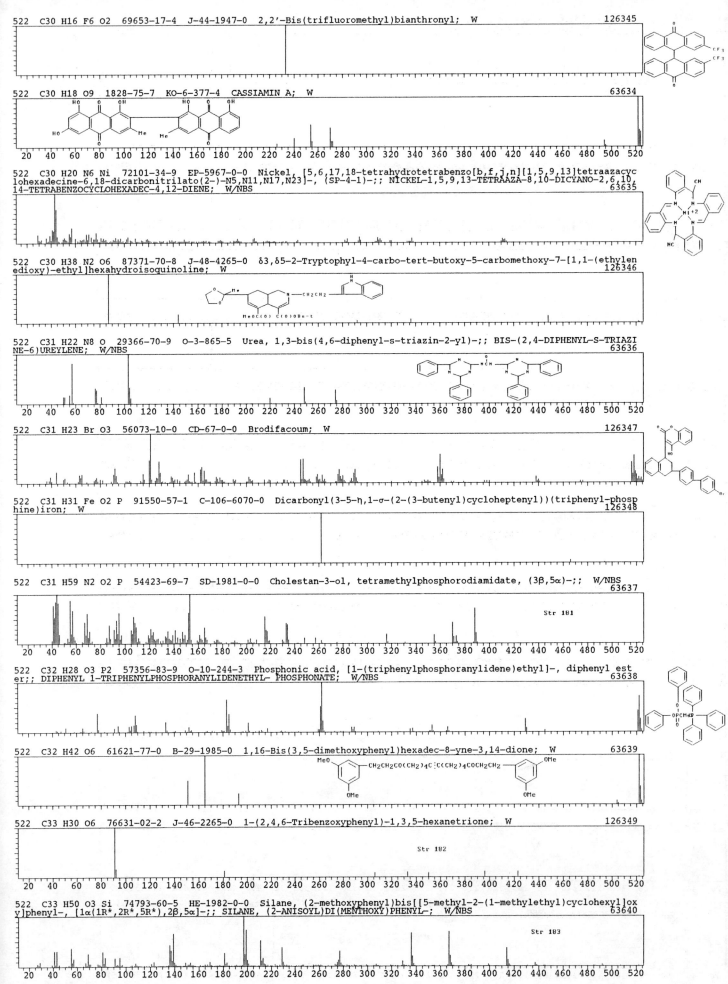

522 C30 H16 F6 O2 69653-17-4 J-44-1947-0 2,2'-Bis(trifluoromethyl)bianthronyl; W 126345

522 C30 H18 O9 1828-75-7 KO-6-377-4 CASSIAMIN A; W 63634

522 C30 H20 N6 Ni 72101-34-9 EP-5967-0-0 Nickel, [5,6,17,18-tetrahydrotetrabenzo[b,f,j,n][1,5,9,13]tetraazacyc
lohexadecine-6,18-dicarbonitrilato(2-)-N5,N11,N17,N23]-, (SP-4-1)-;; NICKEL-1,5,9,13-TETRAAZA-8,10-DICYANO-2,6,10,
14-TETRABENZOCYCLOHEXADEC-4,12-DIENE; W/NBS 63635

522 C30 H38 N2 O6 87371-70-8 J-48-4265-0 δ3,δ5-2-Tryptophyl-4-carbo-tert-butoxy-5-carbomethoxy-7-[1,1-(ethylen
edioxy)-ethyl]hexahydroisoquinoline; W 126346

522 C31 H22 N8 O 29366-70-9 O-3-865-5 Urea, 1,3-bis(4,6-diphenyl-s-triazin-2-yl)-;; BIS-(2,4-DIPHENYL-S-TRIAZI
NE-6)UREYLENE; W/NBS 63636

522 C31 H23 Br O3 56073-10-0 CD-67-0-0 Brodifacoum; W 126347

522 C31 H31 Fe O2 P 91550-57-1 C-106-6070-0 Dicarbonyl(3-5-η,1-σ-(2-(3-butenyl)cycloheptenyl))(triphenyl-phosp
hine)iron; W 126348

522 C31 H59 N2 O2 P 54423-69-7 SD-1981-0-0 Cholestan-3-ol, tetramethylphosphorodiamidate, (3β,5α)-;; W/NBS
63637
Str 181

522 C32 H28 O3 P2 57356-83-9 O-10-244-3 Phosphonic acid, [1-(triphenylphosphoranylidene)ethyl]-, diphenyl est
er;; DIPHENYL 1-TRIPHENYLPHOSPHORANYLIDENETHYL- PHOSPHONATE; W/NBS 63638

522 C32 H42 O6 61621-77-0 B-29-1985-0 1,16-Bis(3,5-dimethoxyphenyl)hexadec-8-yne-3,14-dione; W 63639

522 C33 H30 O6 76631-02-2 J-46-2265-0 1-(2,4,6-Tribenzoxyphenyl)-1,3,5-hexanetrione; W 126349
Str 182

522 C33 H50 O3 Si 74793-60-5 HE-1982-0-0 Silane, (2-methoxyphenyl)bis[[5-methyl-2-(1-methylethyl)cyclohexyl]ox
y]phenyl-, [1α(1R*,2R*,5R*),2β,5α]-;; SILANE, (2-ANISOYL)DI(MENTHOXY)PHENYL-; W/NBS 63640
Str 183

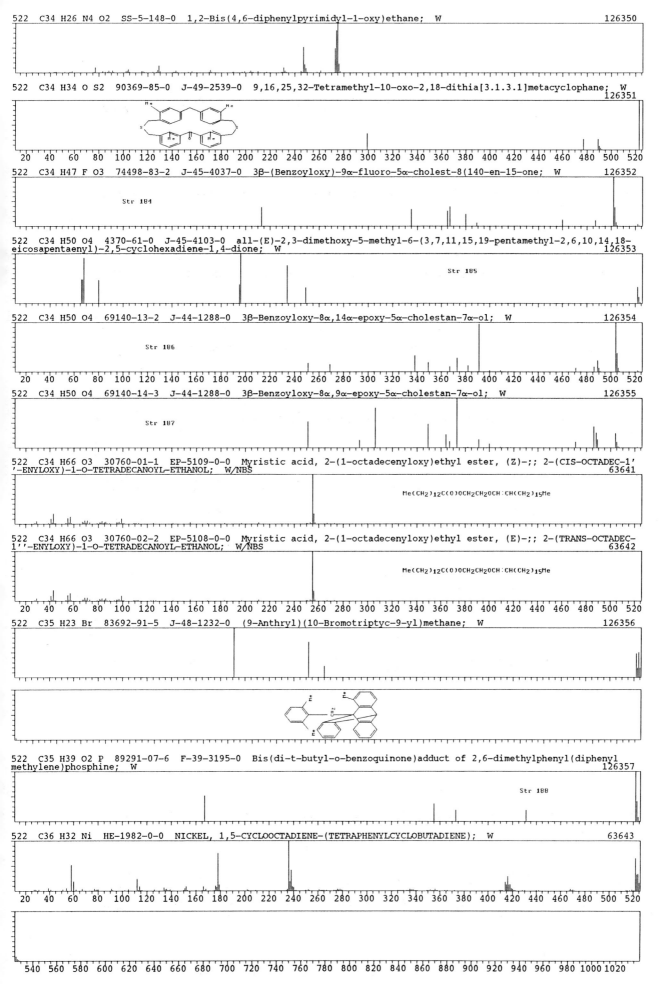

522　C34 H26 N4 O2　SS-5-148-0　1,2-Bis(4,6-diphenylpyrimidyl-1-oxy)ethane;　W　126350

522　C34 H34 O S2　90369-85-0　J-49-2539-0　9,16,25,32-Tetramethyl-10-oxo-2,18-dithia[3.1.3.1]metacyclophane;　W
126351

522　C34 H47 F O3　74498-83-2　J-45-4037-0　3β-(Benzoyloxy)-9α-fluoro-5α-cholest-8(140-en-15-one;　W　126352

Str 184

522　C34 H50 O4　4370-61-0　J-45-4103-0　all-(E)-2,3-dimethoxy-5-methyl-6-(3,7,11,15,19-pentamethyl-2,6,10,14,18-eicosapentaenyl)-2,5-cyclohexadiene-1,4-dione;　W
126353

Str 185

522　C34 H50 O4　69140-13-2　J-44-1288-0　3β-Benzoyloxy-8α,14α-epoxy-5α-cholestan-7α-ol;　W　126354

Str 186

522　C34 H50 O4　69140-14-3　J-44-1288-0　3β-Benzoyloxy-8α,9α-epoxy-5α-cholestan-7α-ol;　W　126355

Str 187

522　C34 H66 O3　30760-01-1　EP-5109-0-0　Myristic acid, 2-(1-octadecenyloxy)ethyl ester, (Z)-;; 2-(CIS-OCTADEC-1'-ENYLOXY)-1-O-TETRADECANOYL-ETHANOL;　W/NBS　63641

Me(CH2)12C(O)OCH2CH2OCH:CH(CH2)15Me

522　C34 H66 O3　30760-02-2　EP-5108-0-0　Myristic acid, 2-(1-octadecenyloxy)ethyl ester, (E)-;; 2-(TRANS-OCTADEC-1''-ENYLOXY)-1-O-TETRADECANOYL-ETHANOL;　W/NBS　63642

Me(CH2)12C(O)OCH2CH2OCH:CH(CH2)15Me

522　C35 H23 Br　83692-91-5　J-48-1232-0　(9-Anthryl)(10-Bromotriptyc-9-yl)methane;　W　126356

522　C35 H39 O2 P　89291-07-6　F-39-3195-0　Bis(di-t-butyl-o-benzoquinone)adduct of 2,6-dimethylphenyl(diphenylmethylene)phosphine;　W
126357

Str 188

522　C36 H32 Ni　HE-1982-0-0　NICKEL, 1,5-CYCLOOCTADIENE-(TETRAPHENYLCYCLOBUTADIENE);　W　63643

522 C36 H58 O2 73599-16-3 K-113-388-0 1-tert-Butoxy-4,4-bis(4,8-dimethyl-3,7-nonadienyl)-1,2,46,7,7a-hexahy
dro-7a-methyl-5H-inden-5-one; W   126358

Str 189

522 C40 H26 O 74065-76-2 K-113-1439-0 13,18-Dihydro-13,18-diphenyl-13,18-epoxybenzo[b]tetraphenylene; W
                                                                                                      126359

522 C41 H30 20168-15-4 K-114-1672-0 4-(2,3,4,5-Tetraphenyl-2,4-cyclopentadien-1-yliden)bicyclo[5.4.1]dodeca-2,
5,7,9,11-pentaene; W   126360

523 C18 H6 F12 N3 P 70758-18-8 I-57-1016-0 TRIS(4-AMINO-2,3,5,6-TETRAFLUOROPHENYL)PHOSPHINE; W   63644

523 C19 H18 Br2 Cu N4 59448-33-8 B-30-2509-0 2,12-DIBROMO-6,7,8,9,16,17-HEXAHYDRO-5H-DIBENZO[F,M][1,4,8,12]TET
RAAZACYCLOPETADECINATOCOPPER (II); W   63645

523 C19 H20 Cl3 N3 O6 S 68908-00-9 KC-1978-821-0 (3R,8R,8AR)-2,2,2-TRICHLOROETHYL 2,2-DIMETHYL-5,7-DIOXO-8-PHE
NOXYACETAMIDOPERHYDROTHIAZOLO(3,2-C)PYRIMIDINE-3-CARBOXYLATE; W   63646

523 C19 H20 Cl3 N3 O6 S 68897-23-4 KC-1978-821-0 (3R,8S,8AR)-2,2,2-TRICHLOROETHYL 2,2-DIMETHYL-5,7-DIOXO-8-PHE
NOXYACETAMIDO-PERHYDRO THIAZOLE-(3,2-C)PYRIMIDINE-3-CARBOXYLATE; W   63647

523 C20 H21 N5 O8 S2 42362-45-8 J-8-4341-5 2-Propanamine, N-[2,2-bis[(2,4-dinitrophenyl)thio]ethenyl]-N-(1-
methylethyl)-;; 1,1-BIS(2,4-DINITROPHENYLTHIO)-2-DI-ISOPROPYLAMINOETHENE; W/NBS   63648

523 C21 H23 Br2 N3 O S H-62-169-0 5,5-DIMETHYL-2-(4'-BROMOPHENACYLTHIO)-3-PHENYL-Δ'-IMIDAZOLIN-4-DIMETHYLIMINI
UM BROMIDE; W   63649

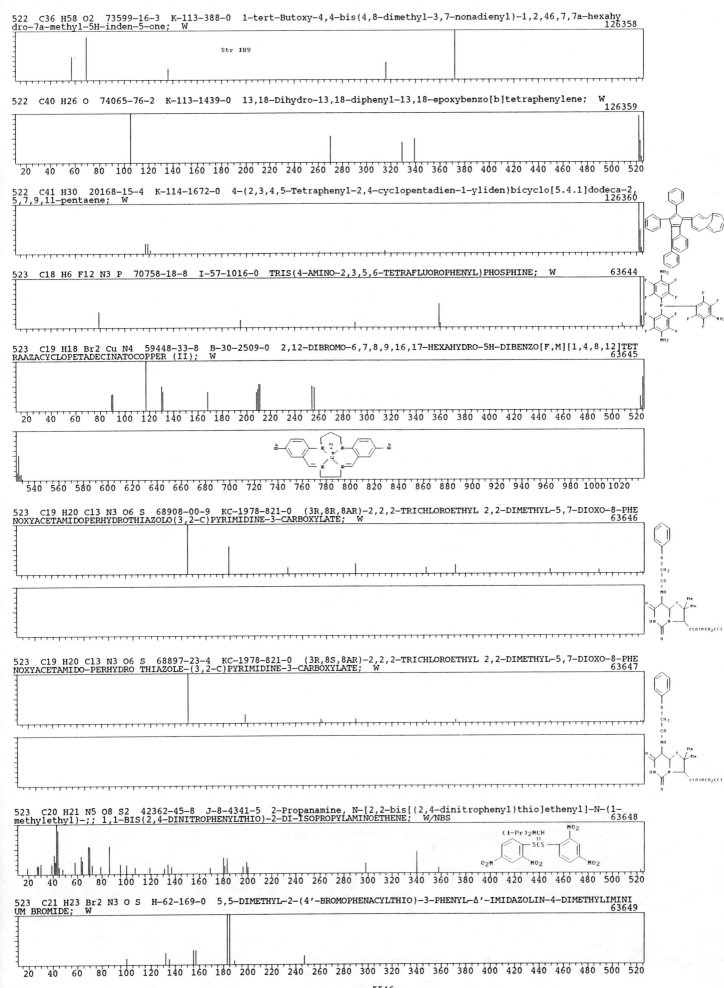

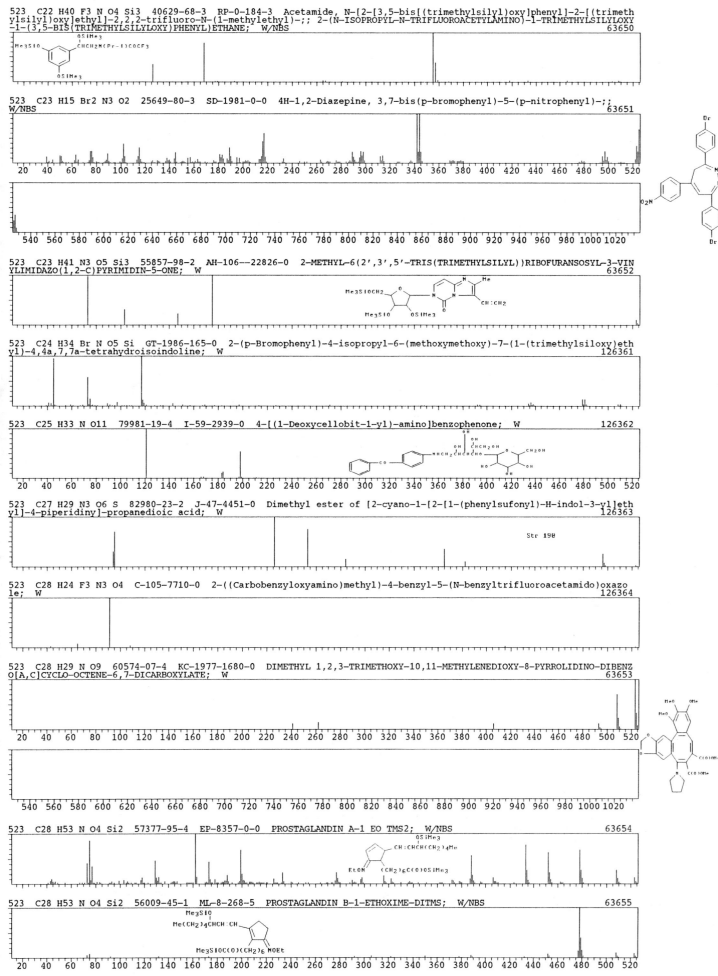

523　C22 H40 F3 N O4 Si3　40629-68-3　RP-0-184-3　Acetamide, N-[2-[3,5-bis[(trimethylsilyl)oxy]phenyl]-2-[(trimethylsilyl)oxy]ethyl]-2,2,2-trifluoro-N-(1-methylethyl)-;; 2-(N-ISOPROPYL-N-TRIFLUOROACETYLAMINO)-1-TRIMETHYLSILYLOXY-1-(3,5-BIS(TRIMETHYLSILYLOXY)PHENYL)ETHANE;　W/NBS
63650

523　C23 H15 Br2 N3 O2　25649-80-3　SD-1981-0-0　4H-1,2-Diazepine, 3,7-bis(p-bromophenyl)-5-(p-nitrophenyl)-;;
W/NBS
63651

523　C23 H41 N3 O5 Si3　55857-98-2　AH-106--22826-0　2-METHYL-6(2',3',5'-TRIS(TRIMETHYLSILYL))RIBOFURANSOSYL-3-VINYLIMIDAZO(1,2-C)PYRIMIDIN-5-ONE;　W
63652

523　C24 H34 Br N O5 Si　GT-1986-165-0　2-(p-Bromophenyl)-4-isopropyl-6-(methoxymethoxy)-7-(1-(trimethylsiloxy)ethyl)-4,4a,7,7a-tetrahydroisoindoline;　W
126361

523　C25 H33 N O11　79981-19-4　I-59-2939-0　4-[(1-Deoxycellobit-1-yl)-amino]benzophenone;　W
126362

523　C27 H29 N3 O6 S　82980-23-2　J-47-4451-0　Dimethyl ester of [2-cyano-1-[2-[1-(phenylsufonyl)-H-indol-3-yl]ethyl]-4-piperidinyl]-propanedioic acid;　W
126363
Str 198

523　C28 H24 F3 N3 O4　C-105-7710-0　2-((Carbobenzyloxyamino)methyl)-4-benzyl-5-(N-benzyltrifluoroacetamido)oxazole;　W
126364

523　C28 H29 N O9　60574-07-4　KC-1977-1680-0　DIMETHYL 1,2,3-TRIMETHOXY-10,11-METHYLENEDIOXY-8-PYRROLIDINO-DIBENZO[A,C]CYCLO-OCTENE-6,7-DICARBOXYLATE;　W
63653

523　C28 H53 N O4 Si2　57377-95-4　EP-8357-0-0　PROSTAGLANDIN A-1 EO TMS2;　W/NBS
63654

523　C28 H53 N O4 Si2　56009-45-1　ML-8-268-5　PROSTAGLANDIN B-1-ETHOXIME-DITMS;　W/NBS
63655

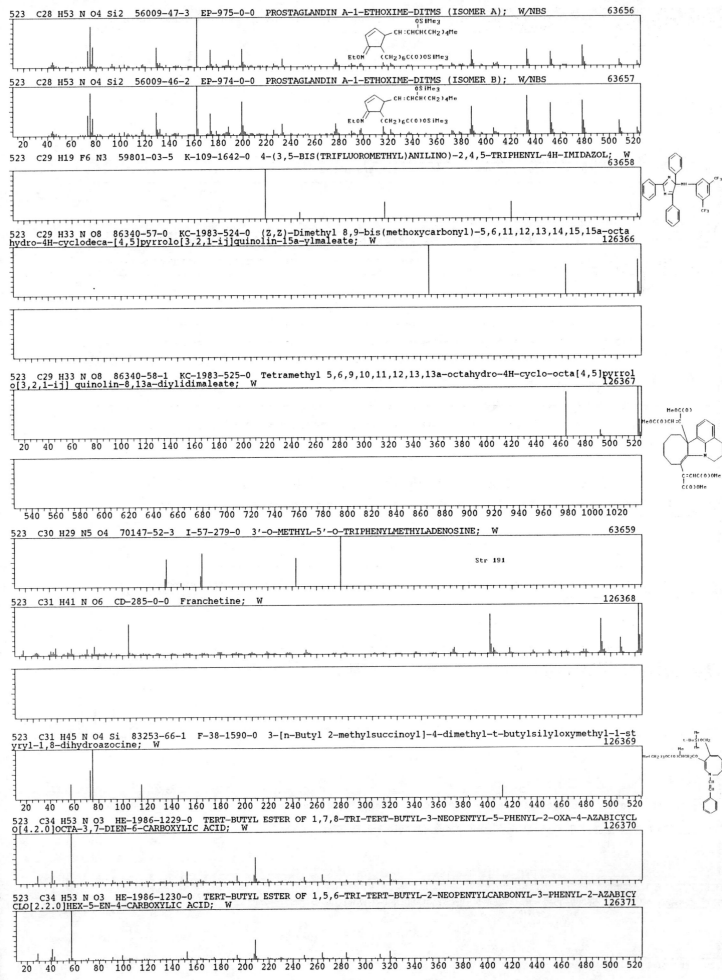

523   C28 H53 N O4 Si2   56009-47-3   EP-975-0-0   PROSTAGLANDIN A-1-ETHOXIME-DITMS (ISOMER A);   W/NBS   63656

523   C28 H53 N O4 Si2   56009-46-2   EP-974-0-0   PROSTAGLANDIN A-1-ETHOXIME-DITMS (ISOMER B);   W/NBS   63657

523   C29 H19 F6 N3   59801-03-5   K-109-1642-0   4-(3,5-BIS(TRIFLUOROMETHYL)ANILINO)-2,4,5-TRIPHENYL-4H-IMIDAZOL;   W   63658

523   C29 H33 N O8   86340-57-0   KC-1983-524-0   (Z,Z)-Dimethyl 8,9-bis(methoxycarbonyl)-5,6,11,12,13,14,15,15a-octa hydro-4H-cyclodeca-[4,5]pyrrolo[3,2,1-ij]quinolin-15a-ylmaleate;   W   126366

523   C29 H33 N O8   86340-58-1   KC-1983-525-0   Tetramethyl 5,6,9,10,11,12,13,13a-octahydro-4H-cyclo-octa[4,5]pyrrol o[3,2,1-ij] quinolin-8,13a-diylidimaleate;   W   126367

523   C30 H29 N5 O4   70147-52-3   I-57-279-0   3'-O-METHYL-5'-O-TRIPHENYLMETHYLADENOSINE;   W   63659

Str 191

523   C31 H41 N O6   CD-285-0-0   Franchetine;   W   126368

523   C31 H45 N O4 Si   83253-66-1   F-38-1590-0   3-[n-Butyl 2-methylsuccinoyl]-4-dimethyl-t-butylsilyloxymethyl-1-st yryl-1,8-dihydroazocine;   W   126369

523   C34 H53 N O3   HE-1986-1229-0   TERT-BUTYL ESTER OF 1,7,8-TRI-TERT-BUTYL-3-NEOPENTYL-5-PHENYL-2-OXA-4-AZABICYCL O[4.2.0]OCTA-3,7-DIEN-6-CARBOXYLIC ACID;   W   126370

523   C34 H53 N O3   HE-1986-1230-0   TERT-BUTYL ESTER OF 1,5,6-TRI-TERT-BUTYL-2-NEOPENTYLCARBONYL-3-PHENYL-2-AZABICY CLO[2.2.0]HEX-5-EN-4-CARBOXYLIC ACID;   W   126371

524   C8 H24 F5 N4 O3 P3 Si4   61565-37-5   K-109-3962-0   2,2,4,4,6,6,8,8-Octamethyl-7-(pentafluoro-1,3,5,2λ5,4λ5,6λ5-triazatriphosphorin-2-yl)-1,3,5-trioxa-5-aza-2,4,6,8-tetrasilacyclooctane;   W   126372

Str 192

524   C9 H5 Br5 O   3555-11-1   NS-0-0-0   Benzene, pentabromo(2-propenyloxy)-;   W/NBS   126373

524   C9 H7 Br2 O4 Re   57376-74-6   AG-94-433-9   (η5-Acetylcyclopentadienyl)(dicarbonyl)rhenium dibromide;   W   126374

524   C12 F20   36481-20-6   J-43-4982-0   Tricyclo[3.3.1.1(3,7)]decane, 1,2,2,3,4,4,6,6,8,8,9,9,10,10-tetradecafluoro-5,7-bis(trifluoromethyl)-;; Perfluoro-1,3-dimethyladamantane;   W/NBS   126375

524   C12 H2 Co3 F3 O9   15663-91-9   T-68-2338-0   Cobalt, nonacarbonyl[.mu.3-(3,3,3-trifluoropropylidyne)]tri-, triangulo-;; .MU.3-(3,3,3-TRIFLUOROPROPYL)-TRIS(TRICARBONYLCOBALT);   W/NBS   63661

524   C18 H48 Ge2 Si4   AG-116-206-0   1,1,3,3-Tetramethyl-2,2,4,4-tetrakis(trimethylsilyl)-1,3-digermacyclobutane;   W   126376

524   C20 H48 N2 O6 Si4   62108-13-8   NS--28513-0-0   xylo-Hexos-5-ulose, 2,3,4,6-tetrakis-O-(trimethylsilyl)-, bis(O-methyloxime);;   NBS   63662

524   C20 H48 N2 O6 Si4   62108-14-9   NS--28512-0-0   ribo-Hexos-3-ulose, 2,4,5,6-tetrakis-O-(trimethylsilyl)-, bis(O-methyloxime);;   NBS   63663

524   C20 H48 N2 O6 Si4   62108-39-8   NS--28511-0-0   D-gluco-Hexodialdose, 2,3,4,5-tetrakis-O-(trimethylsilyl)-, bis(O-methyloxime);;   NBS   63664

524   C20 H48 N2 O6 Si4   62181-81-1   HE-1982-0-0   galacto-Hexodialdose, 2,3,4,5-tetrakis-O-(trimethylsilyl)-, bis(O-methyloxime);; GALACTO-HEXADIALDOSE, BIS(O-METHYLOXIM), TETRAKIS-O-(TRIMETHYLSILYL)-;   W/NBS   63667

524   C20 H48 N2 O6 Si4   HE-1982-0-0   5-KETO-GLUCOSE, BIS(O-METHYLOXIME), TETRAKIS-O-(TRIMETHYLSILYL)-;   W   63665

524   C20 H48 N2 O6 Si4   HE-1982-0-0   3-KETO-GLUCOSE, BIS(O-METHYLOXIME), TETRAKIS-O-(TRIMETHYLSILYL)-;   W   63666

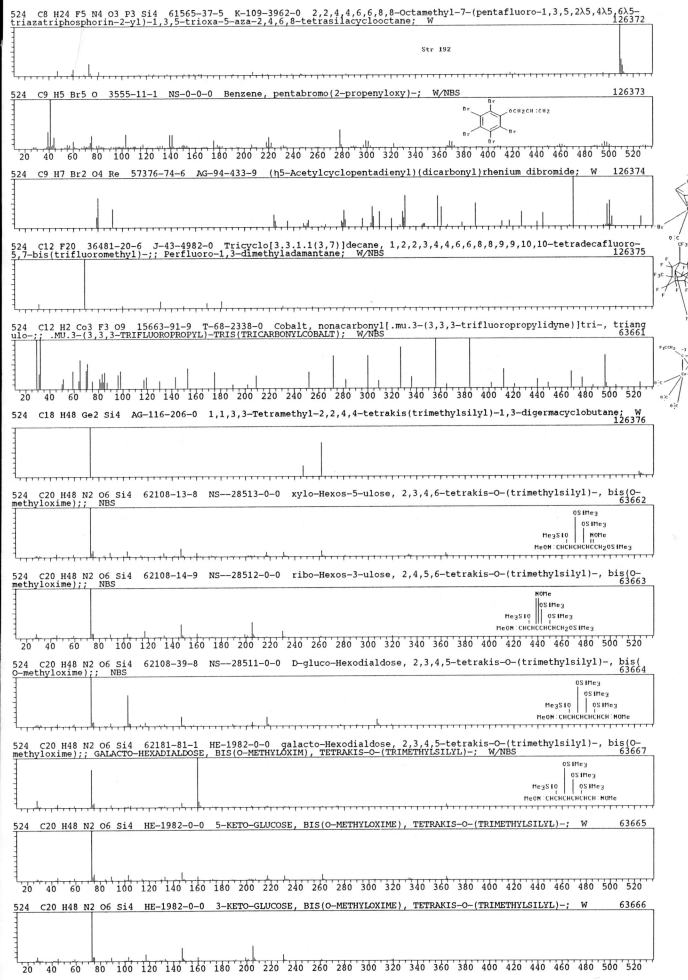

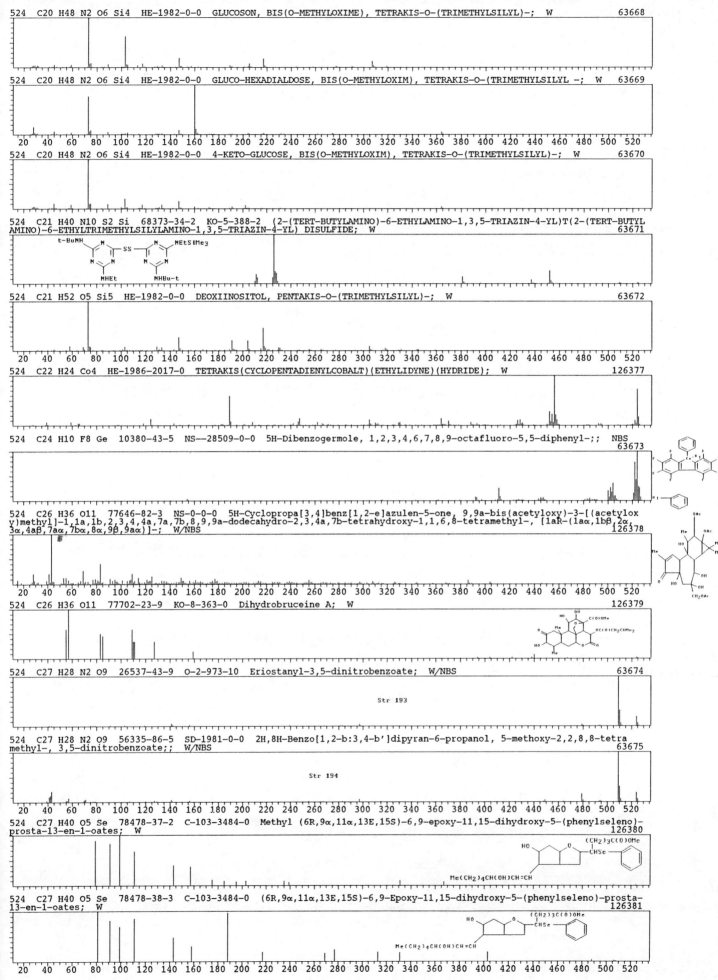

524 C20 H48 N2 O6 Si4 HE-1982-0-0 GLUCOSON, BIS(O-METHYLOXIME), TETRAKIS-O-(TRIMETHYLSILYL)-; W 63668

524 C20 H48 N2 O6 Si4 HE-1982-0-0 GLUCO-HEXADIALDOSE, BIS(O-METHYLOXIM), TETRAKIS-O-(TRIMETHYLSILYL -; W 63669

524 C20 H48 N2 O6 Si4 HE-1982-0-0 4-KETO-GLUCOSE, BIS(O-METHYLOXIM), TETRAKIS-O-(TRIMETHYLSILYL)-; W 63670

524 C21 H40 N10 S2 Si 68373-34-2 KO-5-388-2 (2-(TERT-BUTYLAMINO)-6-ETHYLAMINO-1,3,5-TRIAZIN-4-YL)T(2-(TERT-BUTYL AMINO)-6-ETHYLTRIMETHYLSILYLAMINO-1,3,5-TRIAZIN-4-YL) DISULFIDE; W 63671

524 C21 H52 O5 Si5 HE-1982-0-0 DEOXIINOSITOL, PENTAKIS-O-(TRIMETHYLSILYL)-; W 63672

524 C22 H24 Co4 HE-1986-2017-0 TETRAKIS(CYCLOPENTADIENYLCOBALT)(ETHYLIDYNE)(HYDRIDE); W 126377

524 C24 H10 F8 Ge 10380-43-5 NS--28509-0-0 5H-Dibenzogermole, 1,2,3,4,6,7,8,9-octafluoro-5,5-diphenyl-;; NBS 63673

524 C26 H36 O11 77646-82-3 NS-0-0-0 5H-Cyclopropa[3,4]benz[1,2-e]azulen-5-one, 9,9a-bis(acetyloxy)-3-[(acetylox y)methyl]-1,1a,1b,2,3,4,4a,7a,7b,8,9,9a-dodecahydro-2,3,4a,7b-tetrahydroxy-1,1,6,8-tetramethyl-, [1aR-(1aα,1bβ,2α, 3α,4aβ,7aα,7bα,8α,9β,9aα)]-; W/NBS 126378

524 C26 H36 O11 77702-23-9 KO-8-363-0 Dihydrobruceine A; W 126379

524 C27 H28 N2 O9 26537-43-9 O-2-973-10 Eriostanyl-3,5-dinitrobenzoate; W/NBS 63674

Str 193

524 C27 H28 N2 O9 56335-86-5 SD-1981-0-0 2H,8H-Benzo[1,2-b:3,4-b']dipyran-6-propanol, 5-methoxy-2,2,8,8-tetra methyl-, 3,5-dinitrobenzoate;; W/NBS 63675

Str 194

524 C27 H40 O5 Se 78478-37-2 C-103-3484-0 Methyl (6R,9α,11α,13E,15S)-6,9-epoxy-11,15-dihydroxy-5-(phenylseleno)- prosta-13-en-1-oates; W 126380

524 C27 H40 O5 Se 78478-38-3 C-103-3484-0 (6R,9α,11α,13E,15S)-6,9-Epoxy-11,15-dihydroxy-5-(phenylseleno)-prosta- 13-en-1-oates; W 126381

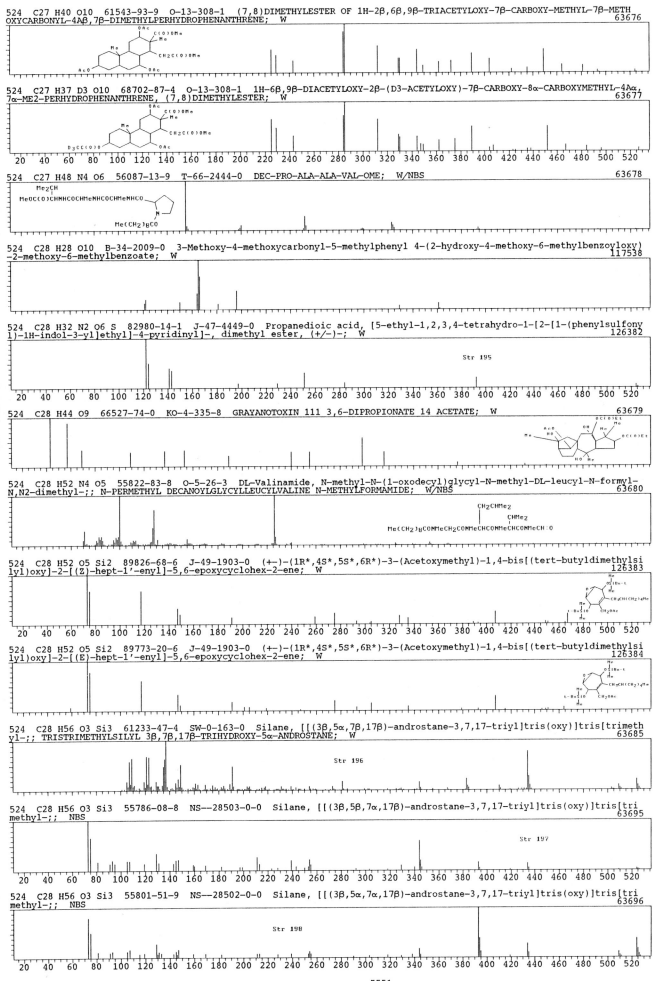

524   C27 H40 O10   61543-93-9   O-13-308-1   (7,8)DIMETHYLESTER OF 1H-2β,6β,9β-TRIACETYLOXY-7β-CARBOXY-METHYL-7β-METH
OXYCARBONYL-4Aβ,7β-DIMETHYLPERHYDROPHENANTHRENE;   W
63676

524   C27 H37 D3 O10   68702-87-4   O-13-308-1   1H-6β,9β-DIACETYLOXY-2β-(D3-ACETYLOXY)-7β-CARBOXY-8α-CARBOXYMETHYL-4Aα,
7α-ME2-PERHYDROPHENANTHRENE, (7,8)DIMETHYLESTER;   W
63677

524   C27 H48 N4 O6   56087-13-9   T-66-2444-0   DEC-PRO-ALA-ALA-VAL-OME;   W/NBS
63678

524   C28 H28 O10   B-34-2009-0   3-Methoxy-4-methoxycarbonyl-5-methylphenyl 4-(2-hydroxy-4-methoxy-6-methylbenzoyloxy)
-2-methoxy-6-methylbenzoate;   W
117538

524   C28 H32 N2 O6 S   82980-14-1   J-47-4449-0   Propanedioic acid, [5-ethyl-1,2,3,4-tetrahydro-1-[2-[1-(phenylsulfony
l)-1H-indol-3-yl]ethyl]-4-pyridinyl]-, dimethyl ester, (+/-)-;   W
126382
Str 195

524   C28 H44 O9   66527-74-0   KO-4-335-8   GRAYANOTOXIN III 3,6-DIPROPIONATE 14 ACETATE;   W
63679

524   C28 H52 N4 O5   55822-83-8   O-5-26-3   DL-Valinamide, N-methyl-N-(1-oxodecyl)glycyl-N-methyl-DL-leucyl-N-formyl-
N,N2-dimethyl-;;   N-PERMETHYL DECANOYLGLYCYLLEUCYLVALINE N-METHYLFORMAMIDE;   W/NBS
63680

524   C28 H52 O5 Si2   89826-68-6   J-49-1903-0   (+-)-(1R*,4S*,5S*,6R*)-3-(Acetoxymethyl)-1,4-bis[(tert-butyldimethylsi
lyl)oxy]-2-[(Z)-hept-1'-enyl]-5,6-epoxycyclohex-2-ene;   W
126383

524   C28 H52 O5 Si2   89773-20-6   J-49-1903-0   (+-)-(1R*,4S*,5S*,6R*)-3-(Acetoxymethyl)-1,4-bis[(tert-butyldimethylsi
lyl)oxy]-2-[(E)-hept-1'-enyl]-5,6-epoxycyclohex-2-ene;   W
126384

524   C28 H56 O3 Si3   61233-47-4   SW-0-163-0   Silane, [[(3β,5α,7β,17β)-androstane-3,7,17-triyl]tris(oxy)]tris[trimeth
yl-;;   TRISTRIMETHYLSILYL 3β,7β,17β-TRIHYDROXY-5α-ANDROSTANE;   W
63685
Str 196

524   C28 H56 O3 Si3   55786-08-8   NS--28503-0-0   Silane, [[(3β,5β,7α,17β)-androstane-3,7,17-triyl]tris(oxy)]tris[tri
methyl-;;   NBS
63695
Str 197

524   C28 H56 O3 Si3   55801-51-9   NS--28502-0-0   Silane, [[(3β,5α,7α,17β)-androstane-3,7,17-triyl]tris(oxy)]tris[tri
methyl-;;   NBS
63696
Str 198

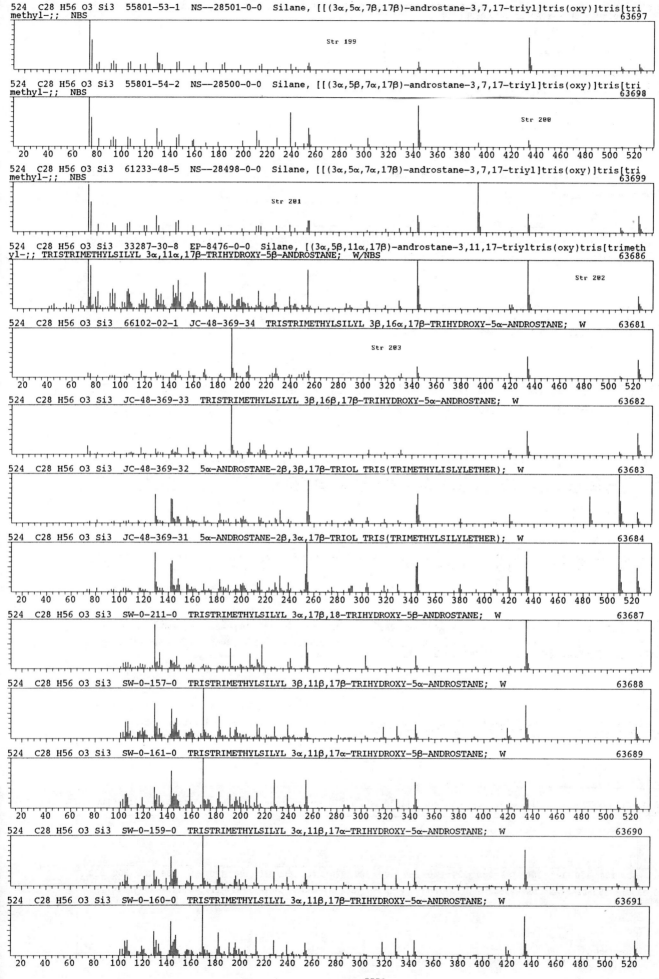

524 C28 H56 O3 Si3 55801-53-1 NS--28501-0-0 Silane, [[(3α,5α,7β,17β)-androstane-3,7,17-triyl]tris(oxy)]tris[tri
methyl-;; NBS   63697
Str 199

524 C28 H56 O3 Si3 55801-54-2 NS--28500-0-0 Silane, [[(3α,5β,7α,17β)-androstane-3,7,17-triyl]tris(oxy)]tris[tri
methyl-;; NBS   63698
Str 200

524 C28 H56 O3 Si3 61233-48-5 NS--28498-0-0 Silane, [[(3α,5α,7α,17β)-androstane-3,7,17-triyl]tris(oxy)]tris[tri
methyl-;; NBS   63699
Str 201

524 C28 H56 O3 Si3 33287-30-8 EP-8476-0-0 Silane, [(3α,5β,11α,17β)-androstane-3,11,17-triyltris(oxy)tris[trimeth
yl-;; TRISTRIMETHYLSILYL 3α,11α,17β-TRIHYDROXY-5β-ANDROSTANE; W/NBS   63686
Str 202

524 C28 H56 O3 Si3 66102-02-1 JC-48-369-34 TRISTRIMETHYLSILYL 3β,16α,17β-TRIHYDROXY-5α-ANDROSTANE; W   63681
Str 203

524 C28 H56 O3 Si3 JC-48-369-33 TRISTRIMETHYLSILYL 3β,16β,17β-TRIHYDROXY-5α-ANDROSTANE; W   63682

524 C28 H56 O3 Si3 JC-48-369-32 5α-ANDROSTANE-2β,3β,17β-TRIOL TRIS(TRIMETHYLISLYLETHER); W   63683

524 C28 H56 O3 Si3 JC-48-369-31 5α-ANDROSTANE-2β,3α,17β-TRIOL TRIS(TRIMETHYLSILYLETHER); W   63684

524 C28 H56 O3 Si3 SW-0-211-0 TRISTRIMETHYLSILYL 3α,17β,18-TRIHYDROXY-5β-ANDROSTANE; W   63687

524 C28 H56 O3 Si3 SW-0-157-0 TRISTRIMETHYLSILYL 3β,11β,17β-TRIHYDROXY-5α-ANDROSTANE; W   63688

524 C28 H56 O3 Si3 SW-0-161-0 TRISTRIMETHYLSILYL 3α,11β,17α-TRIHYDROXY-5β-ANDROSTANE; W   63689

524 C28 H56 O3 Si3 SW-0-159-0 TRISTRIMETHYLSILYL 3α,11β,17α-TRIHYDROXY-5α-ANDROSTANE; W   63690

524 C28 H56 O3 Si3 SW-0-160-0 TRISTRIMETHYLSILYL 3α,11β,17β-TRIHYDROXY-5α-ANDROSTANE; W   63691

- 5552 -

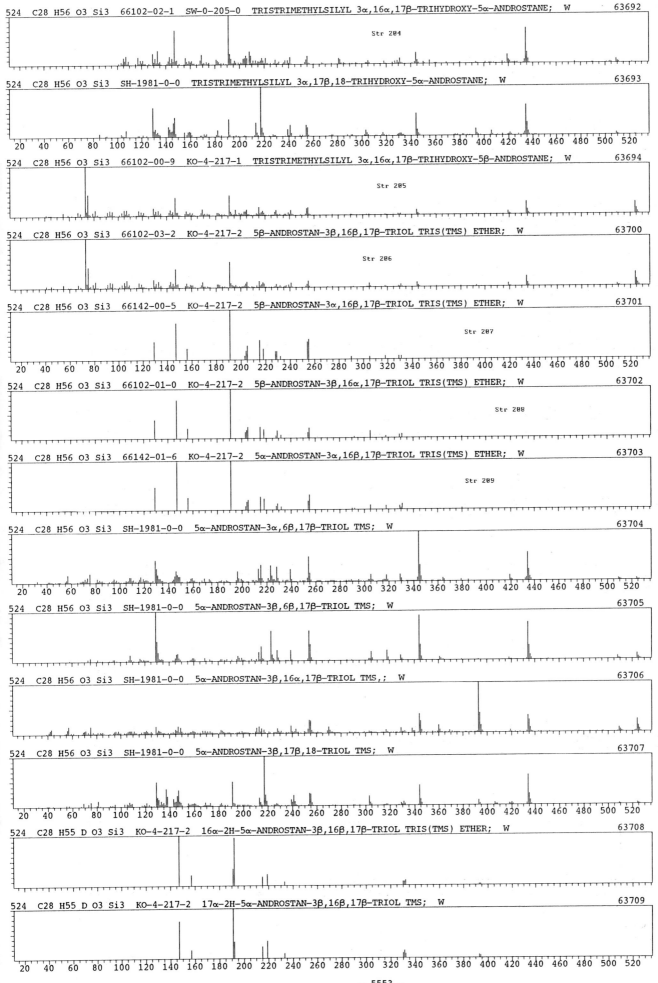

524 C28 H56 O3 Si3 66102-02-1 SW-0-205-0 TRISTRIMETHYLSILYL 3α,16α,17β-TRIHYDROXY-5α-ANDROSTANE; W 63692

Str 204

524 C28 H56 O3 Si3 SH-1981-0-0 TRISTRIMETHYLSILYL 3α,17β,18-TRIHYDROXY-5α-ANDROSTANE; W 63693

524 C28 H56 O3 Si3 66102-00-9 KO-4-217-1 TRISTRIMETHYLSILYL 3α,16α,17β-TRIHYDROXY-5β-ANDROSTANE; W 63694

Str 205

524 C28 H56 O3 Si3 66102-03-2 KO-4-217-2 5β-ANDROSTAN-3β,16β,17β-TRIOL TRIS(TMS) ETHER; W 63700

Str 206

524 C28 H56 O3 Si3 66142-00-5 KO-4-217-2 5β-ANDROSTAN-3α,16β,17β-TRIOL TRIS(TMS) ETHER; W 63701

Str 207

524 C28 H56 O3 Si3 66102-01-0 KO-4-217-2 5β-ANDROSTAN-3β,16α,17β-TRIOL TRIS(TMS) ETHER; W 63702

Str 208

524 C28 H56 O3 Si3 66142-01-6 KO-4-217-2 5α-ANDROSTAN-3α,16β,17β-TRIOL TRIS(TMS) ETHER; W 63703

Str 209

524 C28 H56 O3 Si3 SH-1981-0-0 5α-ANDROSTAN-3α,6β,17β-TRIOL TMS; W 63704

524 C28 H56 O3 Si3 SH-1981-0-0 5α-ANDROSTAN-3β,6β,17β-TRIOL TMS; W 63705

524 C28 H56 O3 Si3 SH-1981-0-0 5α-ANDROSTAN-3β,16α,17β-TRIOL TMS,; W 63706

524 C28 H56 O3 Si3 SH-1981-0-0 5α-ANDROSTAN-3β,17β,18-TRIOL TMS; W 63707

524 C28 H55 D O3 Si3 KO-4-217-2 16α-2H-5α-ANDROSTAN-3β,16β,17β-TRIOL TRIS(TMS) ETHER; W 63708

524 C28 H55 D O3 Si3 KO-4-217-2 17α-2H-5α-ANDROSTAN-3β,16β,17β-TRIOL TMS; W 63709

- 5553 -

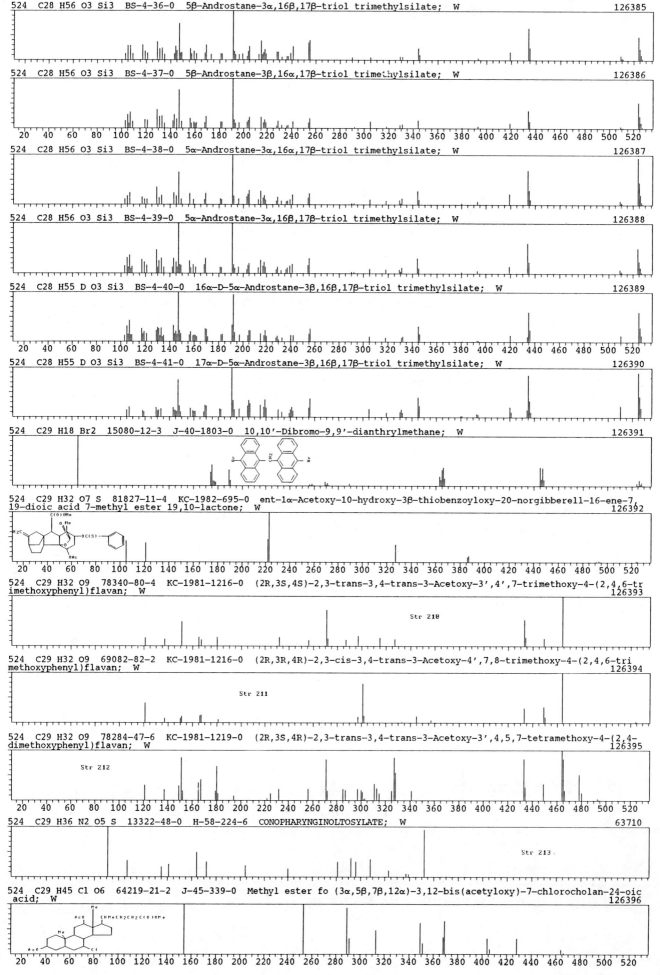

524   C28 H56 O3 Si3   BS-4-36-0   5β-Androstane-3α,16β,17β-triol trimethylsilate;   W   126385

524   C28 H56 O3 Si3   BS-4-37-0   5β-Androstane-3β,16α,17β-triol trimethylsilate;   W   126386

524   C28 H56 O3 Si3   BS-4-38-0   5α-Androstane-3α,16α,17β-triol trimethylsilate;   W   126387

524   C28 H56 O3 Si3   BS-4-39-0   5α-Androstane-3α,16β,17β-triol trimethylsilate;   W   126388

524   C28 H55 D O3 Si3   BS-4-40-0   16α-D-5α-Androstane-3β,16β,17β-triol trimethylsilate;   W   126389

524   C28 H55 D O3 Si3   BS-4-41-0   17α-D-5α-Androstane-3β,16β,17β-triol trimethylsilate;   W   126390

524   C29 H18 Br2   15080-12-3   J-40-1803-0   10,10'-Dibromo-9,9'-dianthrylmethane;   W   126391

524   C29 H32 O7 S   81827-11-4   KC-1982-695-0   ent-1α-Acetoxy-10-hydroxy-3β-thiobenzoyloxy-20-norgibberell-16-ene-7,
19-dioic acid 7-methyl ester 19,10-lactone;   W   126392

524   C29 H32 O9   78340-80-4   KC-1981-1216-0   (2R,3S,4S)-2,3-trans-3,4-trans-3-Acetoxy-3',4',7-trimethoxy-4-(2,4,6-tr
imethoxyphenyl)flavan;   W   126393

Str 210

524   C29 H32 O9   69082-82-2   KC-1981-1216-0   (2R,3R,4R)-2,3-cis-3,4-trans-3-Acetoxy-4',7,8-trimethoxy-4-(2,4,6-tri
methoxyphenyl)flavan;   W   126394

Str 211

524   C29 H32 O9   78284-47-6   KC-1981-1219-0   (2R,3S,4R)-2,3-trans-3,4-trans-3-Acetoxy-3',4,5,7-tetramethoxy-4-(2,4-
dimethoxyphenyl)flavan;   W   126395

Str 212

524   C29 H36 N2 O5 S   13322-48-0   H-58-224-6   CONOPHARYNGINOLTOSYLATE;   W   63710

Str 213

524   C29 H45 Cl O6   64219-21-2   J-45-339-0   Methyl ester fo (3α,5β,7β,12α)-3,12-bis(acetyloxy)-7-chlorocholan-24-oic
acid;   W   126396

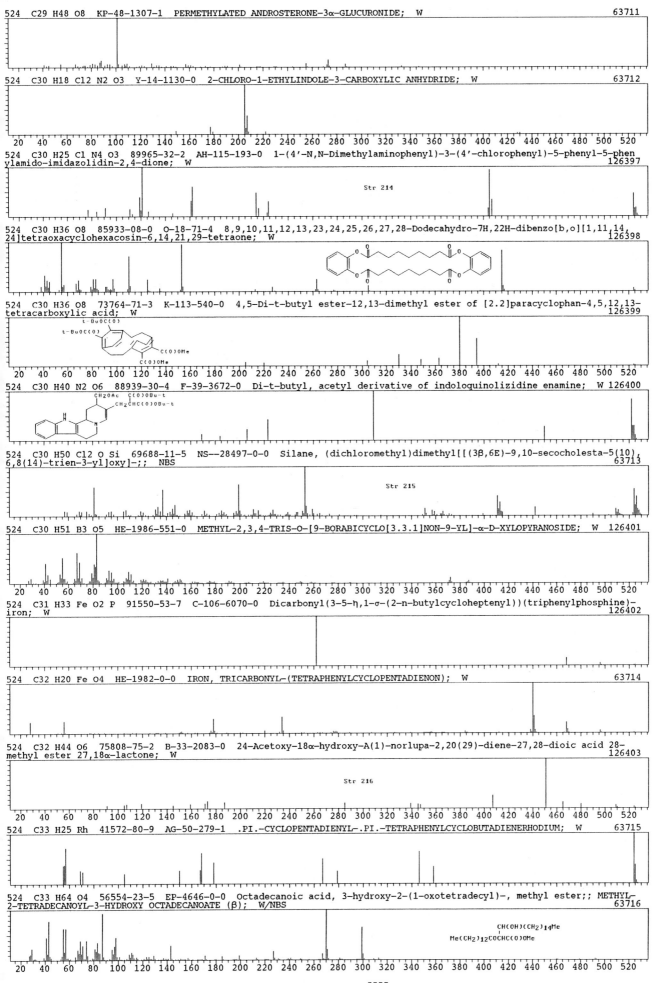

524   C29 H48 O8   KP-48-1307-1   PERMETHYLATED ANDROSTERONE-3α-GLUCURONIDE;   W   63711

524   C30 H18 Cl2 N2 O3   Y-14-1130-0   2-CHLORO-1-ETHYLINDOLE-3-CARBOXYLIC ANHYDRIDE;   W   63712

524   C30 H25 Cl N4 O3   89965-32-2   AH-115-193-0   1-(4'-N,N-Dimethylaminophenyl)-3-(4'-chlorophenyl)-5-phenyl-5-phen
ylamido-imidazolidin-2,4-dione;   W   126397

Str 214

524   C30 H36 O8   85933-08-0   O-18-71-4   8,9,10,11,12,13,23,24,25,26,27,28-Dodecahydro-7H,22H-dibenzo[b,o][1,11,14,
24]tetraoxacyclohexacosin-6,14,21,29-tetraone;   W   126398

524   C30 H36 O8   73764-71-3   K-113-540-0   4,5-Di-t-butyl ester-12,13-dimethyl ester of [2.2]paracyclophan-4,5,12,13-
tetracarboxylic acid;   W   126399

524   C30 H40 N2 O6   88939-30-4   F-39-3672-0   Di-t-butyl, acetyl derivative of indoloquinolizidine enamine;   W 126400

524   C30 H50 Cl2 O Si   69688-11-5   NS--28497-0-0   Silane, (dichloromethyl)dimethyl[[(3β,6E)-9,10-secocholesta-5(10),
6,8(14)-trien-3-yl]oxy]-;;   NBS   63713

Str 215

524   C30 H51 B3 O5   HE-1986-551-0   METHYL-2,3,4-TRIS-O-[9-BORABICYCLO[3.3.1]NON-9-YL]-α-D-XYLOPYRANOSIDE;   W 126401

524   C31 H33 Fe O2 P   91550-53-7   C-106-6070-0   Dicarbonyl(3-5-η,1-σ-(2-n-butylcycloheptenyl))(triphenylphosphine)-
iron;   W   126402

524   C32 H20 Fe O4   HE-1982-0-0   IRON, TRICARBONYL-(TETRAPHENYLCYCLOPENTADIENON);   W   63714

524   C32 H44 O6   75808-75-2   B-33-2083-0   24-Acetoxy-18α-hydroxy-A(1)-norlupa-2,20(29)-diene-27,28-dioic acid 28-
methyl ester 27,18α-lactone;   W   126403

Str 216

524   C33 H25 Rh   41572-80-9   AG-50-279-1   .PI.-CYCLOPENTADIENYL-.PI.-TETRAPHENYLCYCLOBUTADIENERHODIUM;   W   63715

524   C33 H64 O4   56554-23-5   EP-4646-0-0   Octadecanoic acid, 3-hydroxy-2-(1-oxotetradecyl)-, methyl ester;; METHYL-
2-TETRADECANOYL-3-HYDROXY OCTADECANOATE (β);   W/NBS   63716

- 5555 -

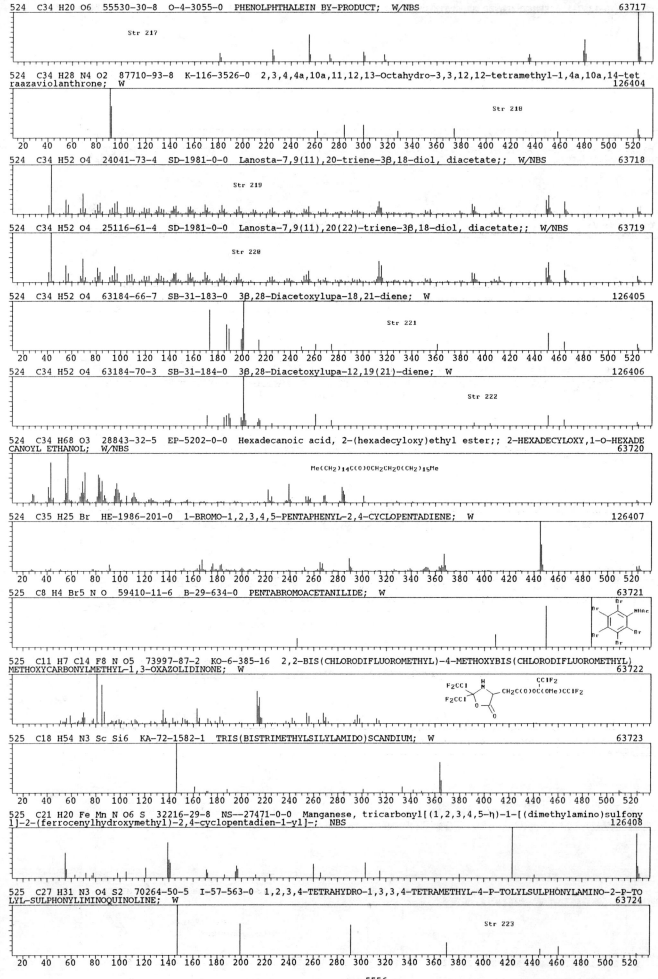

524　C34 H20 O6　55530-30-8　O-4-3055-0　PHENOLPHTHALEIN BY-PRODUCT;　W/NBS　63717

Str 217

524　C34 H28 N4 O2　87710-93-8　K-116-3526-0　2,3,4,4a,10a,11,12,13-Octahydro-3,3,12,12-tetramethyl-1,4a,10a,14-tet raazaviolanthrone;　W　126404

Str 218

20　40　60　80　100　120　140　160　180　200　220　240　260　280　300　320　340　360　380　400　420　440　460　480　500　520

524　C34 H52 O4　24041-73-4　SD-1981-0-0　Lanosta-7,9(11),20-triene-3β,18-diol, diacetate;;　W/NBS　63718

Str 219

524　C34 H52 O4　25116-61-4　SD-1981-0-0　Lanosta-7,9(11),20(22)-triene-3β,18-diol, diacetate;;　W/NBS　63719

Str 220

524　C34 H52 O4　63184-66-7　SB-31-183-0　3β,28-Diacetoxylupa-18,21-diene;　W　126405

Str 221

524　C34 H52 O4　63184-70-3　SB-31-184-0　3β,28-Diacetoxylupa-12,19(21)-diene;　W　126406

Str 222

524　C34 H68 O3　28843-32-5　EP-5202-0-0　Hexadecanoic acid, 2-(hexadecyloxy)ethyl ester;;　2-HEXADECYLOXY,1-O-HEXADE CANOYL ETHANOL;　W/NBS　63720

Me(CH2)14C(O)OCH2CH2O(CH2)15Me

524　C35 H25 Br　HE-1986-201-0　1-BROMO-1,2,3,4,5-PENTAPHENYL-2,4-CYCLOPENTADIENE;　W　126407

20　40　60　80　100　120　140　160　180　200　220　240　260　280　300　320　340　360　380　400　420　440　460　480　500　520

525　C8 H4 Br5 N O　59410-11-6　B-29-634-0　PENTABROMOACETANILIDE;　W　63721

525　C11 H7 Cl4 F8 N O5　73997-87-2　KO-6-385-16　2,2-BIS(CHLORODIFLUOROMETHYL)-4-METHOXYBIS(CHLORODIFLUOROMETHYL) METHOXYCARBONYLMETHYL-1,3-OXAZOLIDINONE;　W　63722

525　C18 H54 N3 Sc Si6　KA-72-1582-1　TRIS(BISTRIMETHYLSILYLAMIDO)SCANDIUM;　W　63723

20　40　60　80　100　120　140　160　180　200　220　240　260　280　300　320　340　360　380　400　420　440　460　480　500　520

525　C21 H20 Fe Mn N O6 S　32216-29-8　NS--27471-0-0　Manganese, tricarbonyl[(1,2,3,4,5-η)-1-[(dimethylamino)sulfony l]-2-(ferrocenylhydroxymethyl)-2,4-cyclopentadien-1-yl]-;　NBS　126408

525　C27 H31 N3 O4 S2　70264-50-5　I-57-563-0　1,2,3,4-TETRAHYDRO-1,3,3,4-TETRAMETHYL-4-P-TOLYLSULPHONYLAMINO-2-P-TO LYL-SULPHONYLIMINOQUINOLINE;　W　63724

Str 223

20　40　60　80　100　120　140　160　180　200　220　240　260　280　300　320　340　360　380　400　420　440　460　480　500　520

525　C27 H35 N5 S Zn　T-68-5526-0　ZINC COMPLEX WITH CORRIN-TYPE LIGAND;　W　63725

525　C28 H20 Br N3 O3　72743-00-1　O-15-260-1　1-P-BROMOPHENYL-3,5-DIPHENYL-IMIDAZOLIDIN-2,4-DIONE-5-CARBOXYLICACID ANILIDE;　W　63726

Str 224

525　C28 H44 Cl N O6　69121-74-0　NS--28491-0-0　4-Tetradecenamide, N-[3-[4-(acetyloxy)-3-methyl-2-oxo-7-oxabicyclo[4.1.0]hept-1-yl]-2-chloro-2-propenyl]-7-methoxy-N-methyl-;;　NBS　63727

525　C29 H51 N O7　O-17-36-3　Di-[13-hydroxy-3,6,8,11-tetraoxatridec-1-yl]3,5-isoxazole dicarboxylate;　W　126409

525　C30 H14 Co N2 S2　B-29-1420-5　2,2'-[1,2-Ethanediylbis(nitrilomethylidyne)]bis(benzenethio)cobalt(II);　W　126410

525　C30 H47 N O3 Si2　87202-22-0　F-39-910-0　5,6,13,13a-Tetrahydro-2,3-dimethoxy-10-neopentoxy-11,12-bis(trimethylsilyl)-8H-dibenzo(a,g)quinolizine;　W　126411

Str 225

526　C8 H7 Br5 N2　RB-1982-14915-0　N-(PENTABROMOPHENYL)ETHYLENEDIAMINE;　W　126412

526　C10 Cl10 Fe　11121-63-4　O-4-392-12　Ferrocene, decachloro-;;　1,1',2,2',3,3',4,4',5,5'-DECACHLOROFERROCENE;　W/NBS　63728

526　C14 H22 O6 Si2 W　73382-50-0　KA-1980-94-5　Pentacarbonyl[2,2-bis(trimethylsilyl)-1-methoxyethylidene]tungsten;　W　126413

526　C17 H41 Cl Co Cu P4　64828-35-9　K-112-830-0　(CHLOROBIS(TRIMETHYLPHOSPHAN)COPPER)(CYCLOPENTADIENYL)BIS(TRIMETHYLPHOSPHAN)COBALT;　W　63729

526　C18 H41 F3 N2 O Si6　80249-61-2　K-114-3370-0　1-[Bis(trimethylsilyl)amino]-1-[(difluorophenylsilyl)(trimethylsilyl)amino]-1-fluoro-3,3,3-trimethyldisiloxane;　W　126414

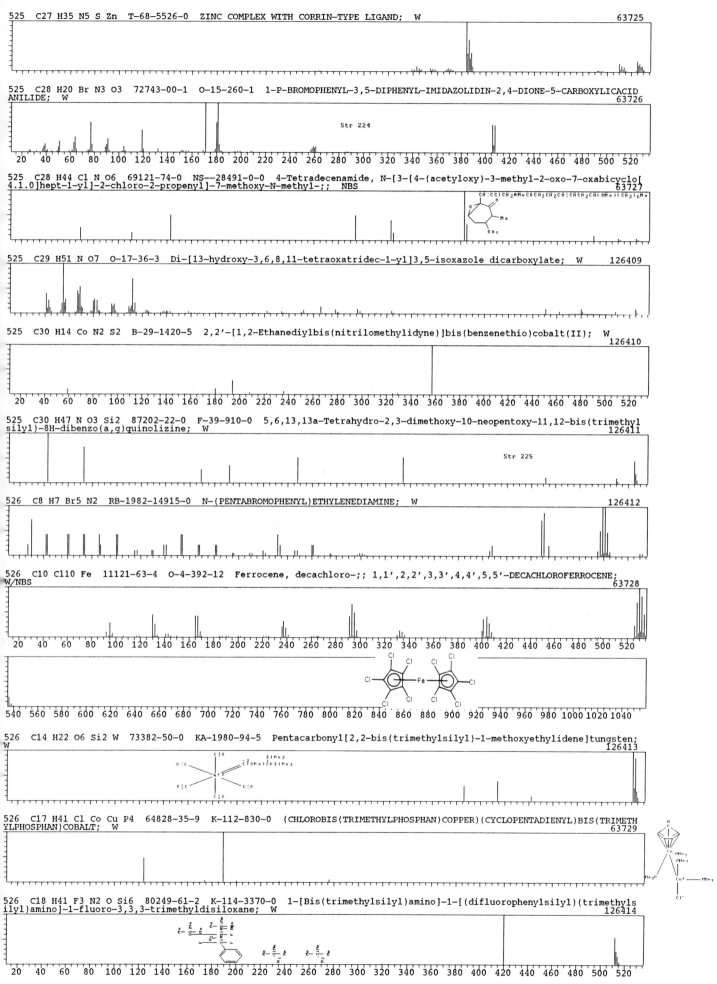

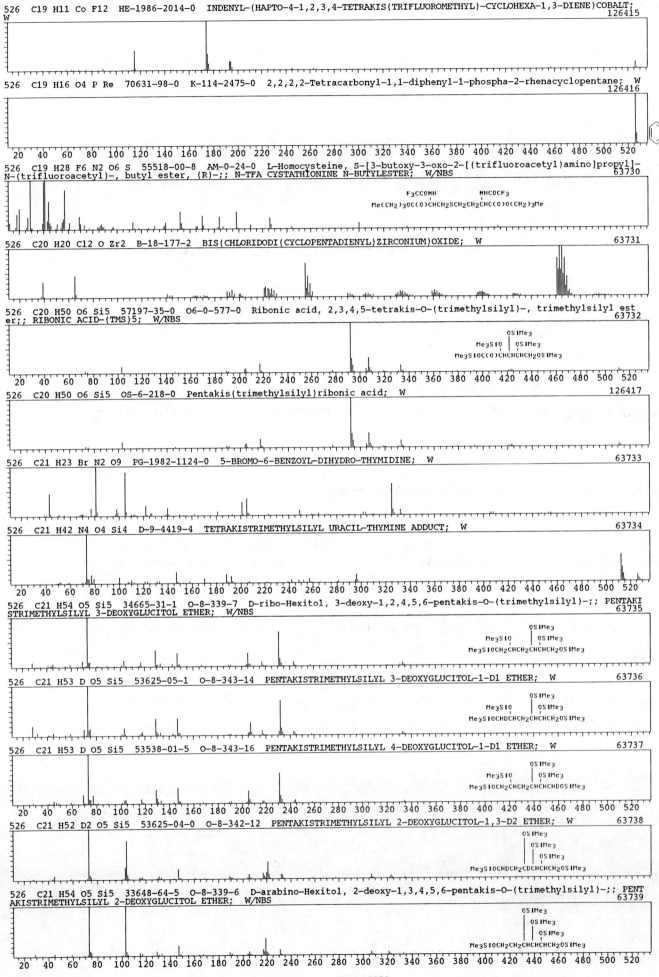

526   C19 H11 Co F12   HE-1986-2014-0   INDENYL-(HAPTO-4-1,2,3,4-TETRAKIS(TRIFLUOROMETHYL)-CYCLOHEXA-1,3-DIENE)COBALT;
W   126415

526   C19 H16 O4 P Re   70631-98-0   K-114-2475-0   2,2,2,2-Tetracarbonyl-1,1-diphenyl-1-phospha-2-rhenacyclopentane;   W
                                                                                                                    126416

526   C19 H28 F6 N2 O6 S   55518-00-8   AM-0-24-0   L-Homocysteine, S-[3-butoxy-3-oxo-2-[(trifluoroacetyl)amino]propyl]-
N-(trifluoroacetyl)-, butyl ester, (R)-;; N-TFA CYSTATHIONINE N-BUTYLESTER;   W/NBS   63730

F3CCONH   NHCOCF3
Me(CH2)3OC(O)CHCH2SCH2CH2CHC(O)O(CH2)3Me

526   C20 H20 Cl2 O Zr2   B-18-177-2   BIS(CHLORIDODI(CYCLOPENTADIENYL)ZIRCONIUM)OXIDE;   W   63731

526   C20 H50 O6 Si5   57197-35-0   O6-0-577-0   Ribonic acid, 2,3,4,5-tetrakis-O-(trimethylsilyl)-, trimethylsilyl est
er;; RIBONIC ACID-(TMS)5;   W/NBS   63732

OSiMe3
Me3SiO   OSiMe3
Me3SiOC(O)CHCHCHCH2OSiMe3

526   C20 H50 O6 Si5   OS-6-218-0   Pentakis(trimethylsilyl)ribonic acid;   W   126417

526   C21 H23 Br N2 O9   PG-1982-1124-0   5-BROMO-6-BENZOYL-DIHYDRO-THYMIDINE;   W   63733

526   C21 H42 N4 O4 Si4   D-9-4419-4   TETRAKISTRIMETHYLSILYL URACIL-THYMINE ADDUCT;   W   63734

526   C21 H54 O5 Si5   34665-31-1   O-8-339-7   D-ribo-Hexitol, 3-deoxy-1,2,4,5,6-pentakis-O-(trimethylsilyl)-;; PENTAKI
STRIMETHYLSILYL 3-DEOXYGLUCITOL ETHER;   W/NBS   63735

OSiMe3
Me3SiO   OSiMe3
Me3SiOCH2CHCH2CHCHCH2OSiMe3

526   C21 H53 D O5 Si5   53625-05-1   O-8-343-14   PENTAKISTRIMETHYLSILYL 3-DEOXYGLUCITOL-1-D1 ETHER;   W   63736

OSiMe3
Me3SiO   OSiMe3
Me3SiOCHDCHCH2CHCHCH2OSiMe3

526   C21 H53 D O5 Si5   53538-01-5   O-8-343-16   PENTAKISTRIMETHYLSILYL 4-DEOXYGLUCITOL-1-D1 ETHER;   W   63737

OSiMe3
Me3SiO   OSiMe3
Me3SiOCH2CHCH2CHCHCHDOSiMe3

526   C21 H52 D2 O5 Si5   53625-04-0   O-8-342-12   PENTAKISTRIMETHYLSILYL 2-DEOXYGLUCITOL-1,3-D2 ETHER;   W   63738

OSiMe3
OSiMe3
OSiMe3
Me3SiOCHDCH2CDCHCHCH2OSiMe3

526   C21 H54 O5 Si5   33648-64-5   O-8-339-6   D-arabino-Hexitol, 2-deoxy-1,3,4,5,6-pentakis-O-(trimethylsilyl)-;; PENT
AKISTRIMETHYLSILYL 2-DEOXYGLUCITOL ETHER;   W/NBS   63739

OSiMe3
OSiMe3
OSiMe3
Me3SiOCH2CH2CHCHCHCH2OSiMe3

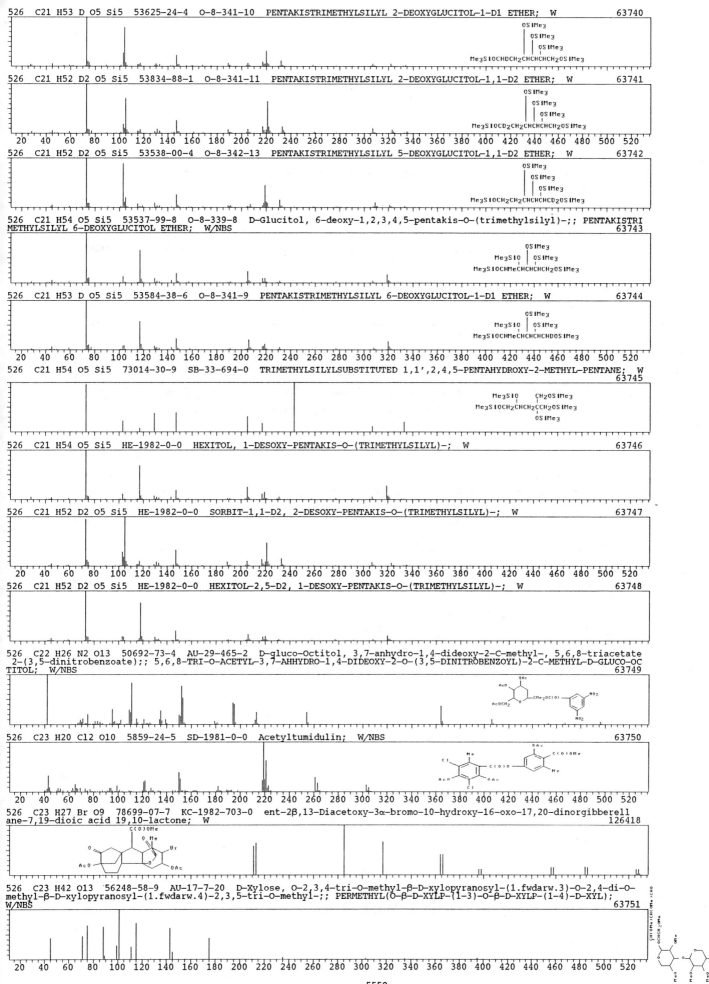

526 C21 H53 D O5 Si5   53625-24-4   O-8-341-10   PENTAKISTRIMETHYLSILYL 2-DEOXYGLUCITOL-1-D1 ETHER;   W   63740

OS IMe₃
OS IMe₃
OS IMe₃
Me₃SIOCHDCH₂CHCHCHCH₂OS IMe₃

526 C21 H52 D2 O5 Si5   53834-88-1   O-8-341-11   PENTAKISTRIMETHYLSILYL 2-DEOXYGLUCITOL-1,1-D2 ETHER;   W   63741

OS IMe₃
OS IMe₃
OS IMe₃
Me₃SIOCD₂CH₂CHCHCHCH₂OS IMe₃

526 C21 H52 D2 O5 Si5   53538-00-4   O-8-342-13   PENTAKISTRIMETHYLSILYL 5-DEOXYGLUCITOL-1,1-D2 ETHER;   W   63742

OS IMe₃
OS IMe₃
OS IMe₃
Me₃SIOCH₂CH₂CHCHCHCD₂OS IMe₃

526 C21 H54 O5 Si5   53537-99-8   O-8-339-8   D-Glucitol, 6-deoxy-1,2,3,4,5-pentakis-O-(trimethylsilyl)-;; PENTAKISTRI
METHYLSILYL 6-DEOXYGLUCITOL ETHER;   W/NBS   63743

OS IMe₃
Me₃SIO   OS IMe₃
Me₃SIOCHMeCHCHCHCH₂OS IMe₃

526 C21 H53 D O5 Si5   53584-38-6   O-8-341-9   PENTAKISTRIMETHYLSILYL 6-DEOXYGLUCITOL-1-D1 ETHER;   W   63744

OS IMe₃
Me₃SIO   OS IMe₃
Me₃SIOCHMeCHCHCHCHDOS IMe₃

526 C21 H54 O5 Si5   73014-30-9   SB-33-694-0   TRIMETHYLSILYLSUBSTITUTED 1,1',2,4,5-PENTAHYDROXY-2-METHYL-PENTANE;   W
63745

Me₃SIO   CH₂OS IMe₃
Me₃SIOCH₂CHCH₂CCH₂OS IMe₃
OS IMe₃

526 C21 H54 O5 Si5   HE-1982-0-0   HEXITOL, 1-DESOXY-PENTAKIS-O-(TRIMETHYLSILYL)-;   W   63746

526 C21 H52 D2 O5 Si5   HE-1982-0-0   SORBIT-1,1-D2, 2-DESOXY-PENTAKIS-O-(TRIMETHYLSILYL)-;   W   63747

526 C21 H52 D2 O5 Si5   HE-1982-0-0   HEXITOL-2,5-D2, 1-DESOXY-PENTAKIS-O-(TRIMETHYLSILYL)-;   W   63748

526 C22 H26 N2 O13   50692-73-4   AU-29-465-2   D-gluco-Octitol, 3,7-anhydro-1,4-dideoxy-2-C-methyl-, 5,6,8-triacetate
2-(3,5-dinitrobenzoate);; 5,6,8-TRI-O-ACETYL-3,7-AHHYDRO-1,4-DIDEOXY-2-O-(3,5-DINITROBENZOYL)-2-C-METHYL-D-GLUCO-OC
TITOL;   W/NBS   63749

526 C23 H20 Cl2 O10   5859-24-5   SD-1981-0-0   Acetyltumidulin;   W/NBS   63750

526 C23 H27 Br O9   78699-07-7   KC-1982-703-0   ent-2β,13-Diacetoxy-3α-bromo-10-hydroxy-16-oxo-17,20-dinorgibberell
ane-7,19-dioic acid 19,10-lactone;   W   126418

526 C23 H42 O13   56248-58-9   AU-17-7-20   D-Xylose, O-2,3,4-tri-O-methyl-β-D-xylopyranosyl-(1.fwdarw.3)-O-2,4-di-O-
methyl-β-D-xylopyranosyl-(1.fwdarw.4)-2,3,5-tri-O-methyl-;; PERMETHYL(O-β-D-XYLP-(1-3)-O-β-D-XYLP-(1-4)-D-XYL);
W/NBS   63751

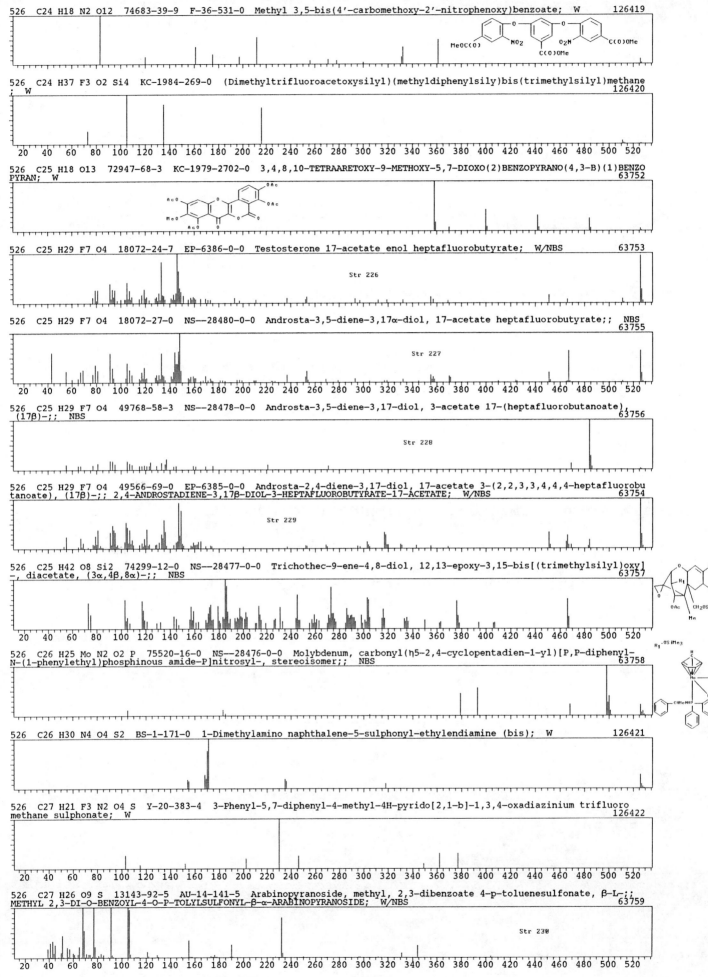

526   C24 H18 N2 O12   74683-39-9   F-36-531-0   Methyl 3,5-bis(4'-carbomethoxy-2'-nitrophenoxy)benzoate;   W   126419

526   C24 H37 F3 O2 Si4   KC-1984-269-0   (Dimethyltrifluoroacetoxysilyl)(methyldiphenylsily)bis(trimethylsilyl)methane
; W   126420

526   C25 H18 O13   72947-68-3   KC-1979-2702-0   3,4,8,10-TETRAARETOXY-9-METHOXY-5,7-DIOXO(2)BENZOPYRANO(4,3-B)(1)BENZO
PYRAN; W   63752

526   C25 H29 F7 O4   18072-24-7   EP-6386-0-0   Testosterone 17-acetate enol heptafluorobutyrate;   W/NBS   63753

Str 226

526   C25 H29 F7 O4   18072-27-0   NS--28480-0-0   Androsta-3,5-diene-3,17α-diol, 17-acetate heptafluorobutyrate;;   NBS   63755

Str 227

526   C25 H29 F7 O4   49768-58-3   NS--28478-0-0   Androsta-3,5-diene-3,17-diol, 3-acetate 17-(heptafluorobutanoate),
(17β)-;;   NBS   63756

Str 228

526   C25 H29 F7 O4   49566-69-0   EP-6385-0-0   Androsta-2,4-diene-3,17-diol, 17-acetate 3-(2,2,3,3,4,4,4-heptafluorobu
tanoate), (17β)-;; 2,4-ANDROSTADIENE-3,17β-DIOL-3-HEPTAFLUOROBUTYRATE-17-ACETATE;   W/NBS   63754

Str 229

526   C25 H42 O8 Si2   74299-12-0   NS--28477-0-0   Trichothec-9-ene-4,8-diol, 12,13-epoxy-3,15-bis[(trimethylsilyl)oxy]
-, diacetate, (3α,4β,8α)-;;   NBS   63757

526   C26 H25 Mo N2 O2 P   75520-16-0   NS--28476-0-0   Molybdenum, carbonyl(η5-2,4-cyclopentadien-1-yl)[P,P-diphenyl-
N-(1-phenylethyl)phosphinous amide-P]nitrosyl-, stereoisomer;;   NBS   63758

526   C26 H30 N4 O4 S2   BS-1-171-0   1-Dimethylamino naphthalene-5-sulphonyl-ethylendiamine (bis);   W   126421

526   C27 H21 F3 N2 O4 S   Y-20-383-4   3-Phenyl-5,7-diphenyl-4-methyl-4H-pyrido[2,1-b]-1,3,4-oxadiazinium trifluoro
methane sulphonate;   W   126422

526   C27 H26 O9 S   13143-92-5   AU-14-141-5   Arabinopyranoside, methyl, 2,3-dibenzoate 4-p-toluenesulfonate, β-L-;;
METHYL 2,3-DI-O-BENZOYL-4-O-P-TOLYLSULFONYL-β-α-ARABINOPYRANOSIDE;   W/NBS   63759

Str 230

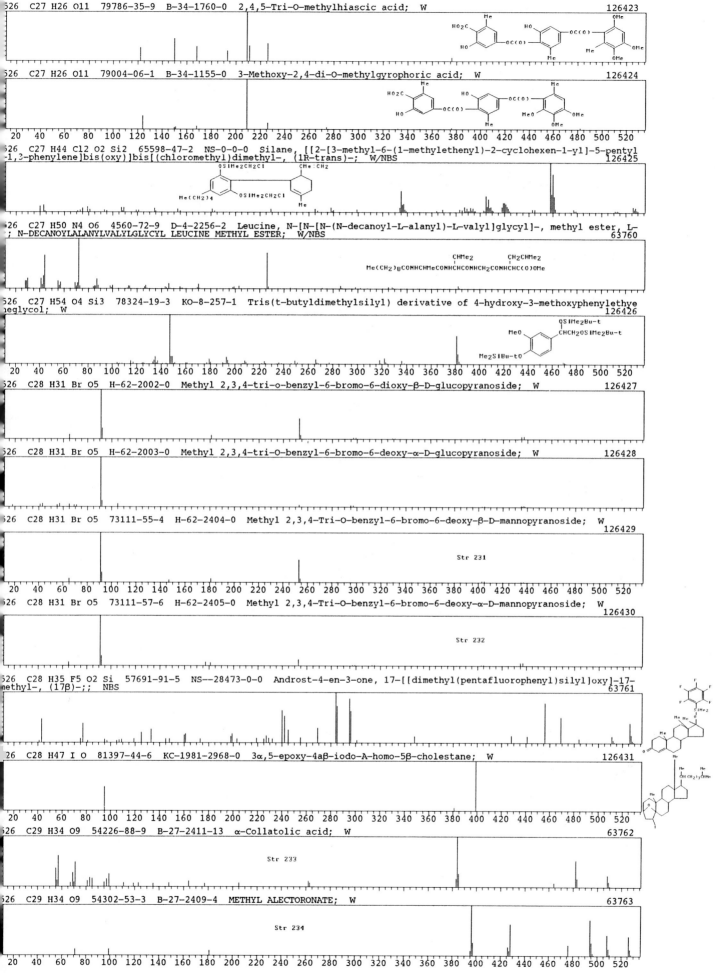

526   C27 H26 O11   79786-35-9   B-34-1760-0   2,4,5-Tri-O-methylhiascic acid;   W   126423

526   C27 H26 O11   79004-06-1   B-34-1155-0   3-Methoxy-2,4-di-O-methylgyrophoric acid;   W   126424

526   C27 H44 Cl2 O2 Si2   65598-47-2   NS-0-0-0   Silane, [[2-[3-methyl-6-(1-methylethenyl)-2-cyclohexen-1-yl]-5-pentyl
-1,3-phenylene]bis(oxy)]bis[(chloromethyl)dimethyl-, (1R-trans)-;   W/NBS   126425

526   C27 H50 N4 O6   4560-72-9   D-4-2256-2   Leucine, N-[N-[N-(N-decanoyl-L-alanyl)-L-valyl]glycyl]-, methyl ester, L-
; N-DECANOYLALANYLVALYLGLYCYL LEUCINE METHYL ESTER;   W/NBS   63760

526   C27 H54 O4 Si3   78324-19-3   KO-8-257-1   Tris(t-butyldimethylsilyl) derivative of 4-hydroxy-3-methoxyphenylethye
neglycol;   W   126426

526   C28 H31 Br O5   H-62-2002-0   Methyl 2,3,4-tri-o-benzyl-6-bromo-6-dioxy-β-D-glucopyranoside;   W   126427

526   C28 H31 Br O5   H-62-2003-0   Methyl 2,3,4-tri-O-benzyl-6-bromo-6-deoxy-α-D-glucopyranoside;   W   126428

526   C28 H31 Br O5   73111-55-4   H-62-2404-0   Methyl 2,3,4-Tri-O-benzyl-6-bromo-6-deoxy-β-D-mannopyranoside;   W   126429

Str 231

526   C28 H31 Br O5   73111-57-6   H-62-2405-0   Methyl 2,3,4-Tri-O-benzyl-6-bromo-6-deoxy-α-D-mannopyranoside;   W   126430

Str 232

526   C28 H35 F5 O2 Si   57691-91-5   NS--28473-0-0   Androst-4-en-3-one, 17-[[dimethyl(pentafluorophenyl)silyl]oxy]-17-
methyl-, (17β)-;;   NBS   63761

526   C28 H47 I O   81397-44-6   KC-1981-2968-0   3α,5-epoxy-4aβ-iodo-A-homo-5β-cholestane;   W   126431

526   C29 H34 O9   54226-88-9   B-27-2411-13   α-Collatolic acid;   W   63762

Str 233

526   C29 H34 O9   54302-53-3   B-27-2409-4   METHYL ALECTORONATE;   W   63763

Str 234

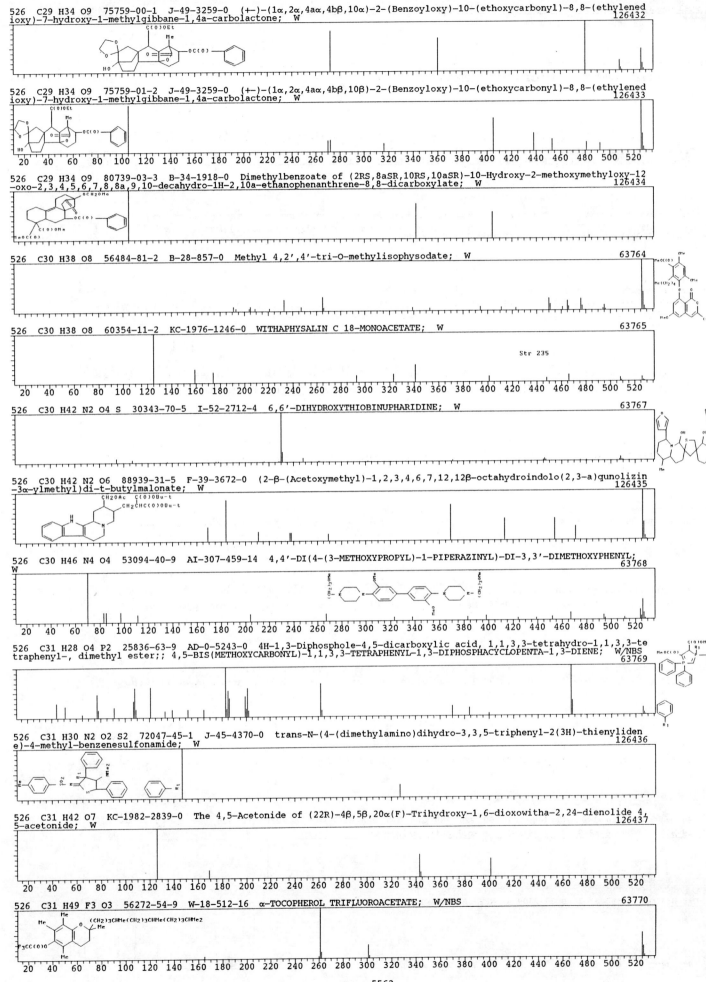

526 C29 H34 O9 75759-00-1 J-49-3259-0 (+-)-(1α,2α,4aα,4bβ,10α)-2-(Benzoyloxy)-10-(ethoxycarbonyl)-8,8-(ethylenedioxy)-7-hydroxy-1-methylgibbane-1,4a-carbolactone; W
126432

526 C29 H34 O9 75759-01-2 J-49-3259-0 (+-)-(1α,2α,4aα,4bβ,10β)-2-(Benzoyloxy)-10-(ethoxycarbonyl)-8,8-(ethylenedioxy)-7-hydroxy-1-methylgibbane-1,4a-carbolactone; W
126433

526 C29 H34 O9 80739-03-3 B-34-1918-0 Dimethylbenzoate of (2RS,8aSR,10RS,10aSR)-10-Hydroxy-2-methoxymethyloxy-12-oxo-2,3,4,5,6,7,8,8a,9,10-decahydro-1H-2,10a-ethanophenanthrene-8,8-dicarboxylate; W
126434

526 C30 H38 O8 56484-81-2 B-28-857-0 Methyl 4,2',4'-tri-O-methylisophysodate; W
63764

526 C30 H38 O8 60354-11-2 KC-1976-1246-0 WITHAPHYSALIN C 18-MONOACETATE; W
63765

Str 235

526 C30 H42 N2 O4 S 30343-70-5 I-52-2712-4 6,6'-DIHYDROXYTHIOBINUPHARIDINE; W
63767

526 C30 H42 N2 O6 88939-31-5 F-39-3672-0 (2-β-(Acetoxymethyl)-1,2,3,4,6,7,12,12β-octahydroindolo(2,3-a)qunolizin-3α-ylmethyl)di-t-butylmalonate; W
126435

526 C30 H46 N4 O4 53094-40-9 AI-307-459-14 4,4'-DI(4-(3-METHOXYPROPYL)-1-PIPERAZINYL)-DI-3,3'-DIMETHOXYPHENYL; W
63768

526 C31 H28 O4 P2 25836-63-9 AD-0-5243-0 4H-1,3-Diphosphole-4,5-dicarboxylic acid, 1,1,3,3-tetrahydro-1,1,3,3-tetraphenyl-, dimethyl ester;; 4,5-BIS(METHOXYCARBONYL)-1,1,3,3-TETRAPHENYL-1,3-DIPHOSPHACYCLOPENTA-1,3-DIENE; W/NBS
63769

526 C31 H30 N2 O2 S2 72047-45-1 J-45-4370-0 trans-N-(4-(dimethylamino)dihydro-3,3,5-triphenyl-2(3H)-thenylidene)-4-methyl-benzenesulfonamide; W
126436

526 C31 H42 O7 KC-1982-2839-0 The 4,5-Acetonide of (22R)-4β,5β,20α(F)-Trihydroxy-1,6-dioxowitha-2,24-dienolide 4,5-acetonide; W
126437

526 C31 H49 F3 O3 56272-54-9 W-18-512-16 α-TOCOPHEROL TRIFLUOROACETATE; W/NBS
63770

526 C31 H50 O3 Si2 65598-56-3 BS-4-121-0 Silane, [[9-[[(dimethyl-2-propenylsilyl)oxy]methyl]-6a,7,10,10a-tetrahy
dro-6,6-dimethyl-3-pentyl-6H-dibenzo[b,d]pyran-1-yl]oxy]dimethyl-2-propenyl-, (6aR-trans)-; W   126438

Str 236

526 C31 H50 O3 Si2 66250-97-3 BS-4-122-0 Silane, [[9-[[(dimethyl-2-propenylsilyl)oxy]methyl]-6a,7,8,10a-tetrahy
dro-6,6-dimethyl-3-pentyl-6H-dibenzo[b,d]pyran-1-yl]oxy]dimethyl-2-propenyl-, (6aR-trans)-; W   126439

Str 237

20 40 60 80 100 120 140 160 180 200 220 240 260 280 300 320 340 360 380 400 420 440 460 480 500 520

526 C31 H50 O3 Si2 66251-00-1 BS-4-120-0 Silane, [(6a,7,8,10a-tetrahydro-6,6,9-trimethyl-3-pentyl-6H-dibenzo[b,
d]pyran-1,8-diyl)bis(oxy)]bis[dimethyl-2-propenyl-, [6aR-(6aα,8β,10aβ)]-; W   126440

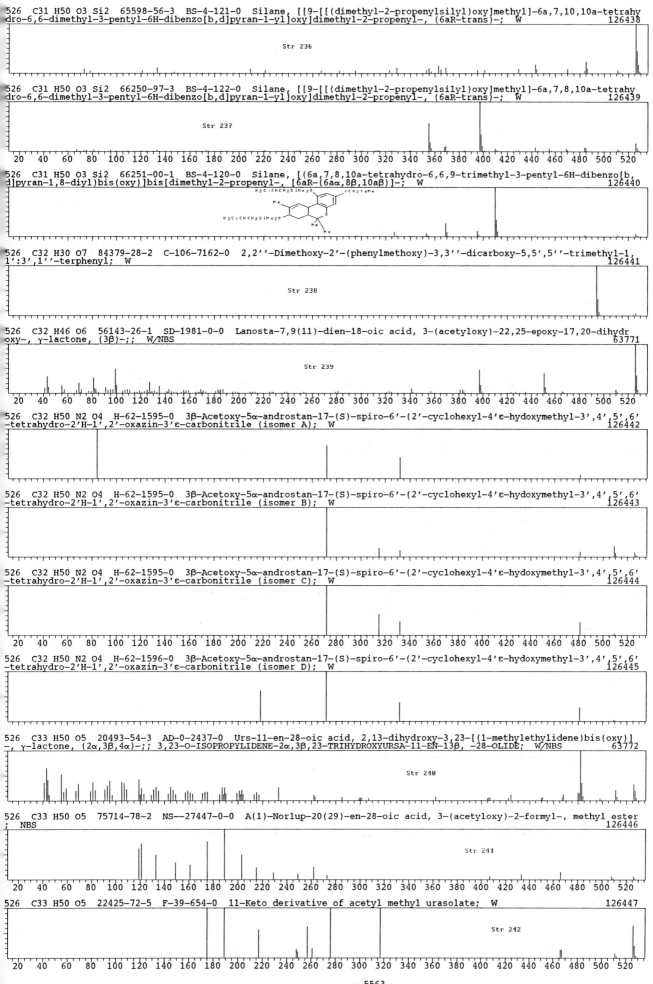

526 C32 H30 O7 84379-28-2 C-106-7162-0 2,2''-Dimethoxy-2'-(phenylmethoxy)-3,3''-dicarboxy-5,5',5''-trimethyl-1,
1':3',1''-terphenyl; W   126441

Str 238

526 C32 H46 O6 56143-26-1 SD-1981-0-0 Lanosta-7,9(11)-dien-18-oic acid, 3-(acetyloxy)-22,25-epoxy-17,20-dihydr
oxy-, γ-lactone, (3β)-;; W/NBS   63771

Str 239

20 40 60 80 100 120 140 160 180 200 220 240 260 280 300 320 340 360 380 400 420 440 460 480 500 520

526 C32 H50 N2 O4 H-62-1595-0 3β-Acetoxy-5α-androstan-17-(S)-spiro-6'-(2'-cyclohexyl-4'ε-hydoxymethyl-3',4',5',6'
-tetrahydro-2'H-1',2'-oxazin-3'ε-carbonitrile (isomer A); W   126442

526 C32 H50 N2 O4 H-62-1595-0 3β-Acetoxy-5α-androstan-17-(S)-spiro-6'-(2'-cyclohexyl-4'ε-hydoxymethyl-3',4',5',6'
-tetrahydro-2'H-1',2'-oxazin-3'ε-carbonitrile (isomer B); W   126443

526 C32 H50 N2 O4 H-62-1595-0 3β-Acetoxy-5α-androstan-17-(S)-spiro-6'-(2'-cyclohexyl-4'ε-hydoxymethyl-3',4',5',6'
-tetrahydro-2'H-1',2'-oxazin-3'ε-carbonitrile (isomer C); W   126444

20 40 60 80 100 120 140 160 180 200 220 240 260 280 300 320 340 360 380 400 420 440 460 480 500 520

526 C32 H50 N2 O4 H-62-1596-0 3β-Acetoxy-5α-androstan-17-(S)-spiro-6'-(2'-cyclohexyl-4'ε-hydoxymethyl-3',4',5',6'
-tetrahydro-2'H-1',2'-oxazin-3'ε-carbonitrile (isomer D); W   126445

526 C33 H50 O5 20493-54-3 AD-0-2437-0 Urs-11-en-28-oic acid, 2,13-dihydroxy-3,23-[(1-methylethylidene)bis(oxy)]
-, γ-lactone, (2α,3β,4α)-;; 3,23-O-ISOPROPYLIDENE-2α,3β,23-TRIHYDROXYURSA-11-EN-13β, -28-OLIDE; W/NBS   63772

Str 240

526 C33 H50 O5 75714-78-2 NS--27447-0-0 A(1)-Norlup-20(29)-en-28-oic acid, 3-(acetyloxy)-2-formyl-, methyl ester
; NBS   126446

Str 241

20 40 60 80 100 120 140 160 180 200 220 240 260 280 300 320 340 360 380 400 420 440 460 480 500 520

526 C33 H50 O5 22425-72-5 F-39-654-0 11-Keto derivative of acetyl methyl urasolate; W   126447

Str 242

20 40 60 80 100 120 140 160 180 200 220 240 260 280 300 320 340 360 380 400 420 440 460 480 500 520

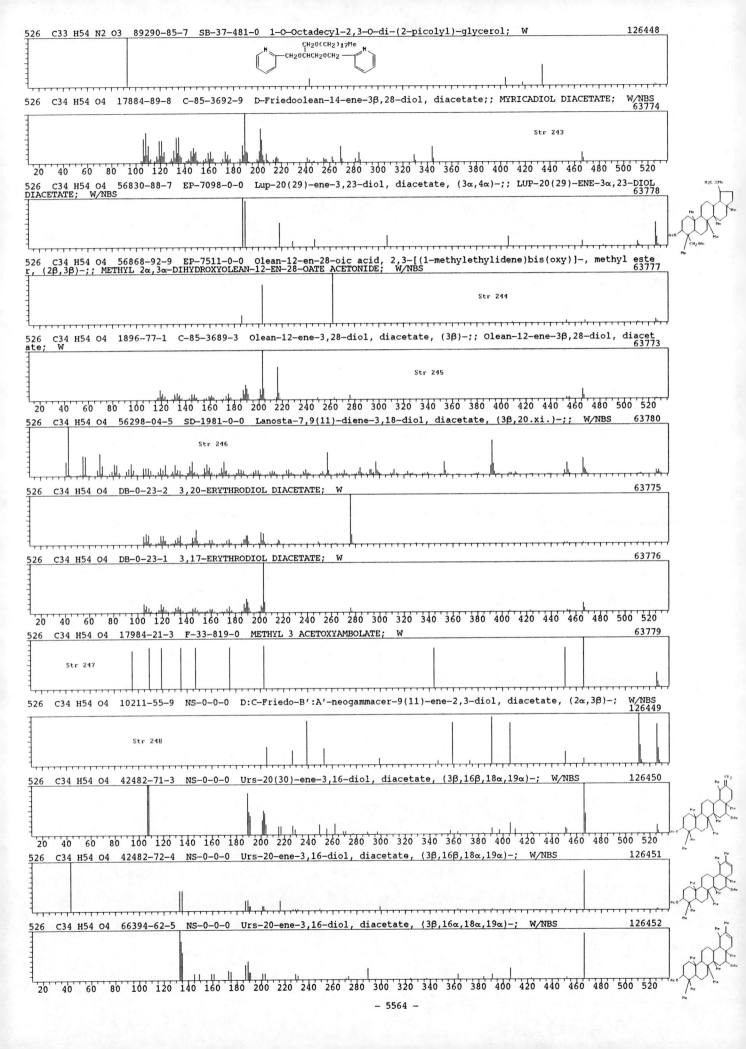

526　C33 H54 N2 O3　89290-85-7　SB-37-481-0　1-O-Octadecyl-2,3-O-di-(2-picolyl)-glycerol;　W　　　　126448

526　C34 H54 O4　17884-89-8　C-85-3692-9　D-Friedoolean-14-ene-3β,28-diol, diacetate;;　MYRICADIOL DIACETATE;　W/NBS
63774

Str 243

526　C34 H54 O4　56830-88-7　EP-7098-0-0　Lup-20(29)-ene-3,23-diol, diacetate, (3α,4α)-;;　LUP-20(29)-ENE-3α,23-DIOL
DIACETATE;　W/NBS　　　　63778

526　C34 H54 O4　56868-92-9　EP-7511-0-0　Olean-12-en-28-oic acid, 2,3-[(1-methylethylidene)bis(oxy)]-, methyl este
r, (2β,3β)-;;　METHYL 2α,3α-DIHYDROXYOLEAN-12-EN-28-OATE ACETONIDE;　W/NBS　　　　63777

Str 244

526　C34 H54 O4　1896-77-1　C-85-3689-3　Olean-12-ene-3,28-diol, diacetate, (3β)-;;　Olean-12-ene-3β,28-diol, diacet
ate;　W　　　　63773

Str 245

526　C34 H54 O4　56298-04-5　SD-1981-0-0　Lanosta-7,9(11)-diene-3,18-diol, diacetate, (3β,20.xi.)-;;　W/NBS　　　63780

Str 246

526　C34 H54 O4　DB-0-23-2　3,20-ERYTHRODIOL DIACETATE;　W　　　　63775

526　C34 H54 O4　DB-0-23-1　3,17-ERYTHRODIOL DIACETATE;　W　　　　63776

526　C34 H54 O4　17984-21-3　F-33-819-0　METHYL 3 ACETOXYAMBOLATE;　W　　　　63779

Str 247

526　C34 H54 O4　10211-55-9　NS-0-0-0　D:C-Friedo-B':A'-neogammacer-9(11)-ene-2,3-diol, diacetate, (2α,3β)-;　W/NBS
126449

Str 248

526　C34 H54 O4　42482-71-3　NS-0-0-0　Urs-20(30)-ene-3,16-diol, diacetate, (3β,16β,18α,19α)-;　W/NBS　　　126450

526　C34 H54 O4　42482-72-4　NS-0-0-0　Urs-20-ene-3,16-diol, diacetate, (3β,16β,18α,19α)-;　W/NBS　　　126451

526　C34 H54 O4　66394-62-5　NS-0-0-0　Urs-20-ene-3,16-diol, diacetate, (3β,16α,18α,19α)-;　W/NBS　　　126452

- 5564 -

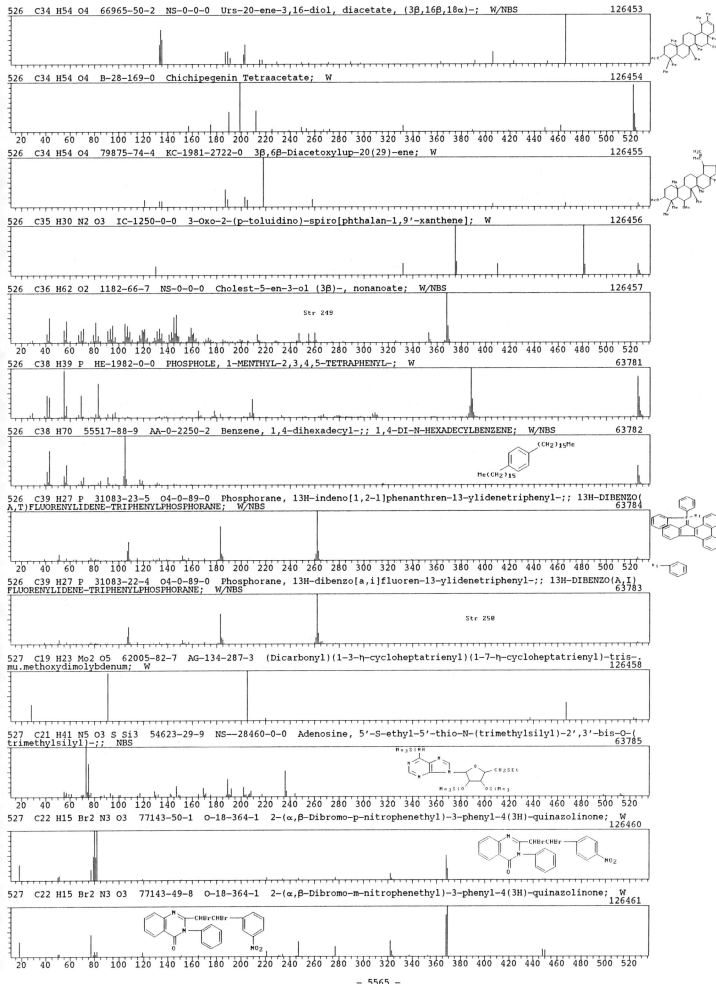

526   C34 H54 O4   66965-50-2   NS-0-0-0   Urs-20-ene-3,16-diol, diacetate, (3β,16β,18α)-;   W/NBS   126453

526   C34 H54 O4   B-28-169-0   Chichipegenin Tetraacetate;   W   126454

526   C34 H54 O4   79875-74-4   KC-1981-2722-0   3β,6β-Diacetoxylup-20(29)-ene;   W   126455

526   C35 H30 N2 O3   IC-1250-0-0   3-Oxo-2-(p-toluidino)-spiro[phthalan-1,9'-xanthene];   W   126456

526   C36 H62 O2   1182-66-7   NS-0-0-0   Cholest-5-en-3-ol (3β)-, nonanoate;   W/NBS   126457

Str 249

526   C38 H39 P   HE-1982-0-0   PHOSPHOLE, 1-MENTHYL-2,3,4,5-TETRAPHENYL-;   W   63781

526   C38 H70   55517-88-9   AA-0-2250-2   Benzene, 1,4-dihexadecyl-;;   1,4-DI-N-HEXADECYLBENZENE;   W/NBS   63782

526   C39 H27 P   31083-23-5   O4-0-89-0   Phosphorane, 13H-indeno[1,2-l]phenanthren-13-ylidenetriphenyl-;;   13H-DIBENZO(
A,T)FLUORENYLIDENE-TRIPHENYLPHOSPHORANE;   W/NBS   63784

526   C39 H27 P   31083-22-4   O4-0-89-0   Phosphorane, 13H-dibenzo[a,i]fluoren-13-ylidenetriphenyl-;;   13H-DIBENZO(A,I)
FLUORENYLIDENE-TRIPHENYLPHOSPHORANE;   W/NBS   63783

Str 250

527   C19 H23 Mo2 O5   62005-82-7   AG-134-287-3   (Dicarbonyl)(1-3-η-cycloheptatrienyl)(1-7-η-cycloheptatrienyl)-tris-.
mu.methoxydimolybdenum;   W   126458

527   C21 H41 N5 O3 S Si3   54623-29-9   NS--28460-0-0   Adenosine, 5'-S-ethyl-5'-thio-N-(trimethylsilyl)-2',3'-bis-O-(
trimethylsilyl)-;;   NBS   63785

527   C22 H15 Br2 N3 O3   77143-50-1   O-18-364-1   2-(α,β-Dibromo-p-nitrophenethyl)-3-phenyl-4(3H)-quinazolinone;   W   126460

527   C22 H15 Br2 N3 O3   77143-49-8   O-18-364-1   2-(α,β-Dibromo-m-nitrophenethyl)-3-phenyl-4(3H)-quinazolinone;   W   126461

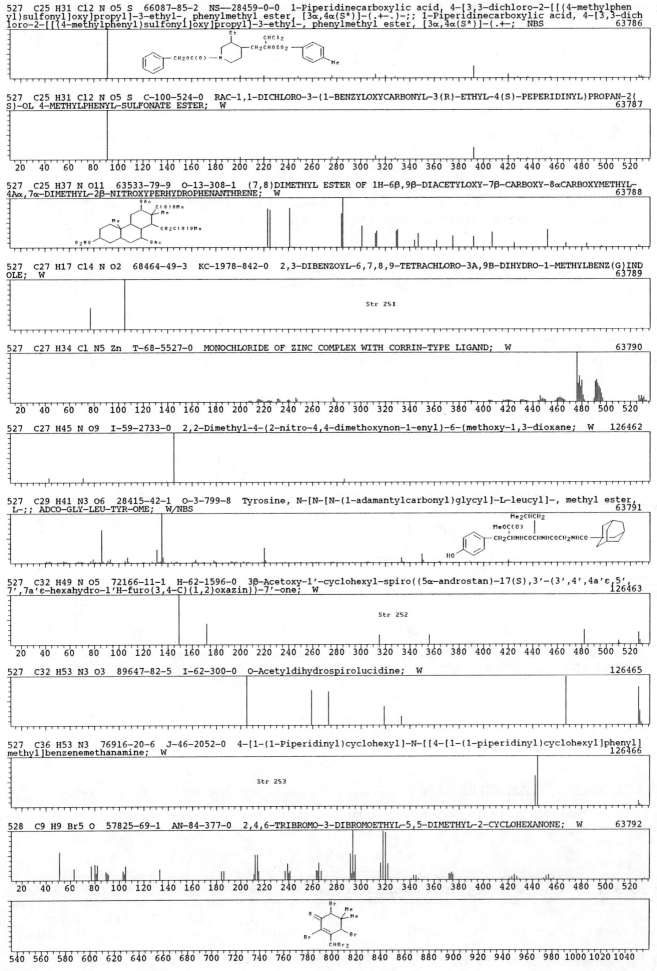

527 C25 H31 Cl2 N O5 S 66087-85-2 NS--28459-0-0 1-Piperidinecarboxylic acid, 4-[3,3-dichloro-2-[[(4-methylphenyl)sulfonyl]oxy]propyl]-3-ethyl-, phenylmethyl ester, [3α,4α(S*)]-(.+-.)-;; 1-Piperidinecarboxylic acid, 4-[3,3-dichloro-2-[[(4-methylphenyl)sulfonyl]oxy]propyl]-3-ethyl-, phenylmethyl ester, [3α,4α(S*)]-(.+-; NBS   63786

527 C25 H31 Cl2 N O5 S C-100-524-0 RAC-1,1-DICHLORO-3-(1-BENZYLOXYCARBONYL-3(R)-ETHYL-4(S)-PEPERIDINYL)PROPAN-2(S)-OL 4-METHYLPHENYL-SULFONATE ESTER; W   63787

527 C25 H37 N O11 63533-79-9 O-13-308-1 (7,8)DIMETHYL ESTER OF 1H-6β,9β-DIACETYLOXY-7β-CARBOXY-8αCARBOXYMETHYL-4Aα,7α-DIMETHYL-2β-NITROXYPERHYDROPHENANTHRENE; W   63788

527 C27 H17 Cl4 N O2 68464-49-3 KC-1978-842-0 2,3-DIBENZOYL-6,7,8,9-TETRACHLORO-3A,9B-DIHYDRO-1-METHYLBENZ(G)INDOLE; W   63789

Str 251

527 C27 H34 Cl N5 Zn T-68-5527-0 MONOCHLORIDE OF ZINC COMPLEX WITH CORRIN-TYPE LIGAND; W   63790

527 C27 H45 N O9 I-59-2733-0 2,2-Dimethyl-4-(2-nitro-4,4-dimethoxynon-1-enyl)-6-(methoxy-1,3-dioxane; W   126462

527 C29 H41 N3 O6 28415-42-1 O-3-799-8 Tyrosine, N-[N-[N-(1-adamantylcarbonyl)glycyl]-L-leucyl]-, methyl ester, L-;; ADCO-GLY-LEU-TYR-OME; W/NBS   63791

527 C32 H49 N O5 72166-11-1 H-62-1596-0 3β-Acetoxy-1'-cyclohexyl-spiro((5α-androstan)-17(S),3'-(3',4',4a'ε,5',7',7a'ε-hexahydro-1'H-furo(3,4-C)(1,2)oxazin))-7'-one; W   126463

Str 252

527 C32 H53 N3 O3 89647-82-5 I-62-300-0 O-Acetyldihydrospirolucidine; W   126465

527 C36 H53 N3 76916-20-6 J-46-2052-0 4-[1-(1-Piperidinyl)cyclohexyl]-N-[[4-[1-(1-piperidinyl)cyclohexyl]phenyl]methyl]benzenemethanamine; W   126466

Str 253

528 C9 H9 Br5 O 57825-69-1 AN-84-377-0 2,4,6-TRIBROMO-3-DIBROMOETHYL-5,5-DIMETHYL-2-CYCLOHEXANONE; W   63792

528 C9 H9 Br5 O 57825-68-0 AN-84-377-0 2,6,6-TRIBROMO-3-DIBROMOMETHYL-5,5-DIMETHYL-2-CYCLOHEXANONE; W   63793

528 C9 H9 Br5 O 57825-67-9 AN-84-376-0 2,4,6,6-TETRABROMO-3-BROMOETHYL-5,5-DIMETHYL-2-CYCLOHEXANONE; W   63794

528 C12 H20 Br2 Rh2 12307-73-2 HE-1982-0-0 Rhodium, di-.mu.-bromotetrakis(η3-2-propenyl)di-;; Tetraallyldirhodi
um dibromide; W/NBS   63795

528 C18 H44 N2 O4 S2 Si4 69688-44-4 AM-0-144-0 CYSTINE-TETRATMS; W/NBS   63796

528 C18 H54 N3 Si6 Ti KA-72-1582-2 TRIS(BISTRIMETHYLSILYLAMIDO)TITANIUM; W   63797

528 C18 H54 N4 P2 Si5 67277-83-2 K-111-1820-0 1,3,2,4-Diazadiphosphetidine, 4-[bis(trimethylsilyl)amino]-2,2,4,
4-tetrahydro-2,2,2-trimethyl-1,3-bis(trimethylsilyl)-4-[(trimethylsilyl)imino]-;; 4-(BIS(TRIMETHYLSILYL)AMINO)-2,2,
2-TRIMETHYL-1,3-BIS(TRIMETHYLSILYL)-4-(TRIMETHYLSILYLIMINO)-1,3,2λ*5,4λ*5-DIAZOPHOSPHETIDINE; W/NBS   63798

528 C18 H57 N3 Sc Si6 37512-28-0 NS--28440-0-0 Silanamine, 1,1,1-trimethyl-N-(trimethylsilyl)-, scandium(3+) sa
lt;; NBS   63799

528 C19 H53 N2 O3 P Si5 53044-47-6 O-9-123-21 Phosphonic acid, [4-[bis(trimethylsilyl)amino]-1-[(trimethylsilyl)
amino]butyl]-, bis(trimethylsilyl) ester;; PENTAKISTRIMETHYLSILYL 1,4-DIAMINOBUTYLPHOSPHONATE; W/NBS   63801

528 C19 H53 N2 O3 P Si5 56211-26-8 EP-928-0-0 Phosphonic acid, (1,4-diaminobutyl)-, pentakis(trimethylsilyl) der
iv.;; 1,4-DIAMINOBUTYLPHOSPHONIC ACID-PENTATMS; W/NBS   63800

528 C22 H14 Br2 N2 S2 KC-1981-2954-0 3-(2,4-Dibromophenyl)-5-phenyl-2-thiobenzoylmethylene-2H-1,3,4-thiadiazolene
; W   126459

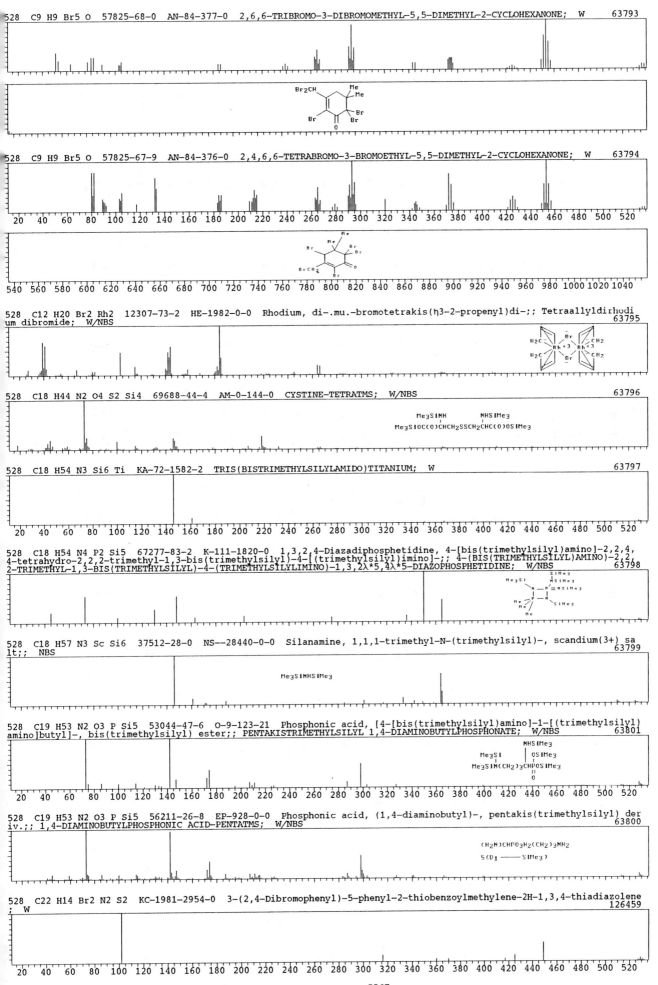

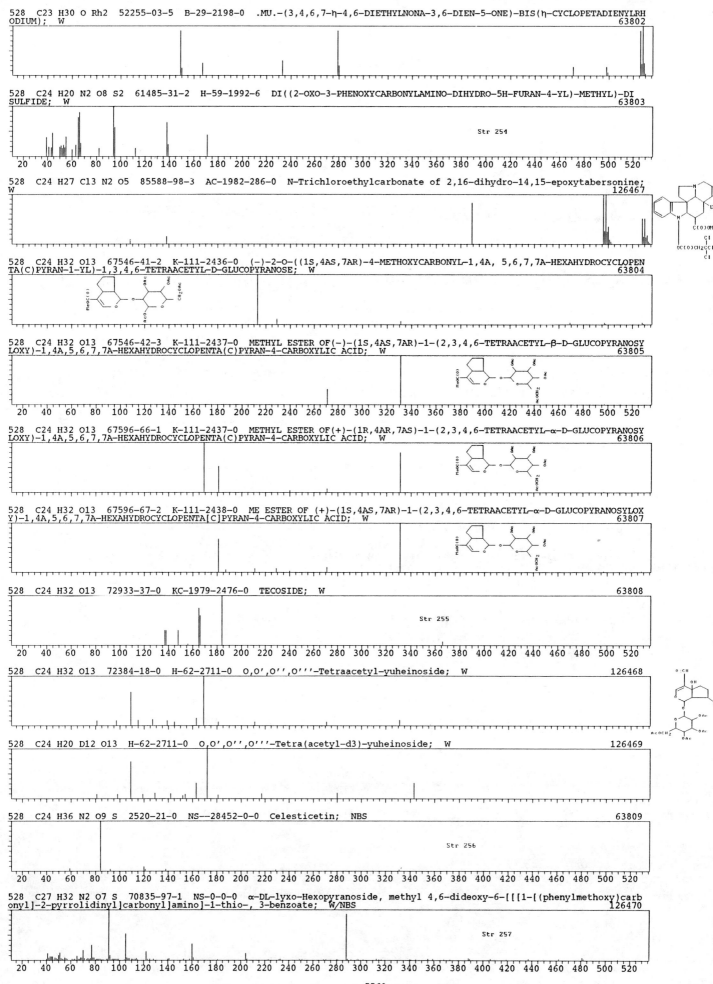

528   C23 H30 O Rh2   52255-03-5   B-29-2198-0   .MU.-(3,4,6,7-η-4,6-DIETHYLNONA-3,6-DIEN-5-ONE)-BIS(η-CYCLOPETADIENYLRH
ODIUM);   W
63802

528   C24 H20 N2 O8 S2   61485-31-2   H-59-1992-6   DI((2-OXO-3-PHENOXYCARBONYLAMINO-DIHYDRO-5H-FURAN-4-YL)-METHYL)-DI
SULFIDE;   W
63803

Str 254

528   C24 H27 Cl3 N2 O5   85588-98-3   AC-1982-286-0   N-Trichloroethylcarbonate of 2,16-dihydro-14,15-epoxytabersonine;
W
126467

528   C24 H32 O13   67546-41-2   K-111-2436-0   (-)-2-O-((1S,4AS,7AR)-4-METHOXYCARBONYL-1,4A, 5,6,7,7A-HEXAHYDROCYCLOPEN
TA(C)PYRAN-1-YL)-1,3,4,6-TETRAACETYL-D-GLUCOPYRANOSE;   W
63804

528   C24 H32 O13   67546-42-3   K-111-2437-0   METHYL ESTER OF(-)-(1S,4AS,7AR)-1-(2,3,4,6-TETRAACETYL-β-D-GLUCOPYRANOSY
LOXY)-1,4A,5,6,7,7A-HEXAHYDROCYCLOPENTA(C)PYRAN-4-CARBOXYLIC ACID;   W
63805

528   C24 H32 O13   67596-66-1   K-111-2437-0   METHYL ESTER OF(+)-(1R,4AR,7AS)-1-(2,3,4,6-TETRAACETYL-α-D-GLUCOPYRANOSY
LOXY)-1,4A,5,6,7,7A-HEXAHYDROCYCLOPENTA(C)PYRAN-4-CARBOXYLIC ACID;   W
63806

528   C24 H32 O13   67596-67-2   K-111-2438-0   ME ESTER OF (+)-(1S,4AS,7AR)-1-(2,3,4,6-TETRAACETYL-α-D-GLUCOPYRANOSYLOX
Y)-1,4A,5,6,7,7A-HEXAHYDROCYCLOPENTA[C]PYRAN-4-CARBOXYLIC ACID;   W
63807

528   C24 H32 O13   72933-37-0   KC-1979-2476-0   TECOSIDE;   W
63808

Str 255

528   C24 H32 O13   72384-18-0   H-62-2711-0   O,O',O'',O'''-Tetraacetyl-yuheinoside;   W
126468

528   C24 H20 D12 O13   H-62-2711-0   O,O',O'',O'''-Tetra(acetyl-d3)-yuheinoside;   W
126469

528   C24 H36 N2 O9 S   2520-21-0   NS--28452-0-0   Celesticetin;   NBS
63809

Str 256

528   C27 H32 N2 O7 S   70835-97-1   NS-0-0-0   α-DL-lyxo-Hexopyranoside, methyl 4,6-dideoxy-6-[[[1-[(phenylmethoxy)carb
onyl]-2-pyrrolidinyl]carbonyl]amino]-1-thio-, 3-benzoate;   W/NBS
126470

Str 257

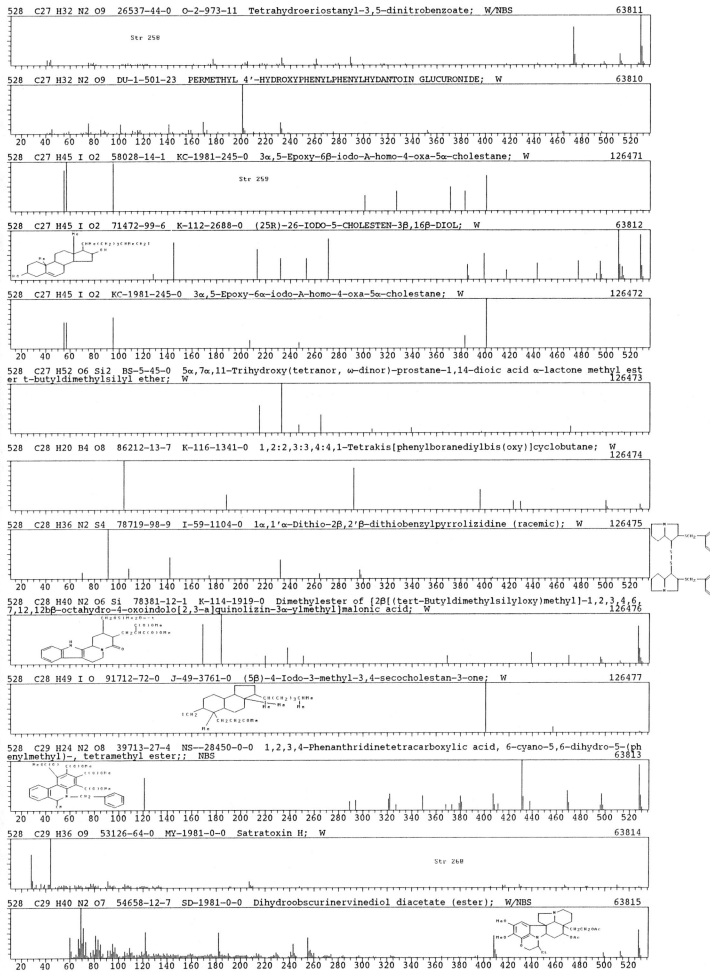

528 C27 H32 N2 O9 26537-44-0 O-2-973-11 Tetrahydroeriostanyl-3,5-dinitrobenzoate; W/NBS   63811

Str 258

528 C27 H32 N2 O9 DU-1-501-23 PERMETHYL 4'-HYDROXYPHENYLPHENYLHYDANTOIN GLUCURONIDE; W   63810

528 C27 H45 I O2 58028-14-1 KC-1981-245-0 3α,5-Epoxy-6β-iodo-A-homo-4-oxa-5α-cholestane; W   126471

Str 259

528 C27 H45 I O2 71472-99-6 K-112-2688-0 (25R)-26-IODO-5-CHOLESTEN-3β,16β-DIOL; W   63812

528 C27 H45 I O2 KC-1981-245-0 3α,5-Epoxy-6α-iodo-A-homo-4-oxa-5α-cholestane; W   126472

528 C27 H52 O6 Si2 BS-5-45-0 5α,7α,11-Trihydroxy(tetranor, ω-dinor)-prostane-1,14-dioic acid α-lactone methyl ester t-butyldimethylsilyl ether; W   126473

528 C28 H20 B4 O8 86212-13-7 K-116-1341-0 1,2:2,3:3,4:4,1-Tetrakis[phenylboranediylbis(oxy)]cyclobutane; W   126474

528 C28 H36 N2 S4 78719-98-9 I-59-1104-0 1α,1'α-Dithio-2β,2'β-dithiobenzylpyrrolizidine (racemic); W   126475

528 C28 H40 N2 O6 Si 78381-12-1 K-114-1919-0 Dimethylester of {2β[(tert-Butyldimethylsilyloxy)methyl]-1,2,3,4,6,7,12,12bβ-octahydro-4-oxoindolo[2,3-a]quinolizin-3α-ylmethyl}malonic acid; W   126476

528 C28 H49 I O 91712-72-0 J-49-3761-0 (5β)-4-Iodo-3-methyl-3,4-secocholestan-3-one; W   126477

528 C29 H24 N2 O8 39713-27-4 NS--28450-0-0 1,2,3,4-Phenanthridinetetracarboxylic acid, 6-cyano-5,6-dihydro-5-(phenylmethyl)-, tetramethyl ester;; NBS   63813

528 C29 H36 O9 53126-64-0 MY-1981-0-0 Satratoxin H; W   63814

Str 260

528 C29 H40 N2 O7 54658-12-7 SD-1981-0-0 Dihydroobscurinervinediol diacetate (ester); W/NBS   63815

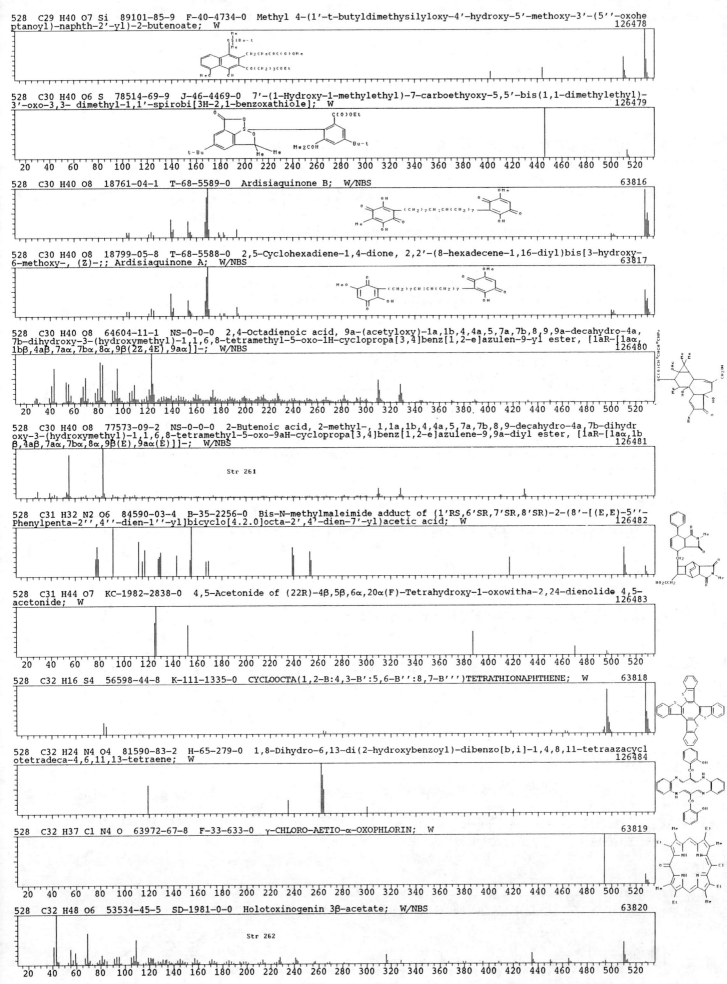

528 C29 H40 O7 Si 89101-85-9 F-40-4734-0 Methyl 4-(1'-t-butyldimethysilyloxy-4'-hydroxy-5'-methoxy-3'-(5''-oxohe ptanoyl)-naphth-2'-yl)-2-butenoate; W
126478

528 C30 H40 O6 S 78514-69-9 J-46-4469-0 7'-(1-Hydroxy-1-methylethyl)-7-carboethyoxy-5,5'-bis(1,1-dimethylethyl)- 3'-oxo-3,3- dimethyl-1,1'-spirobi[3H-2,1-benzoxathiole]; W
126479

528 C30 H40 O8 18761-04-1 T-68-5589-0 Ardisiaquinone B; W/NBS
63816

528 C30 H40 O8 18799-05-8 T-68-5588-0 2,5-Cyclohexadiene-1,4-dione, 2,2'-(8-hexadecene-1,16-diyl)bis[3-hydroxy- 6-methoxy-, (Z)-;; Ardisiaquinone A; W/NBS
63817

528 C30 H40 O8 64604-11-1 NS-0-0-0 2,4-Octadienoic acid, 9a-(acetyloxy)-1a,1b,4,4a,5,7a,7b,8,9,9a-decahydro-4a, 7b-dihydroxy-3-(hydroxymethyl)-1,1,6,8-tetramethyl-5-oxo-1H-cyclopropa[3,4]benz[1,2-e]azulen-9-yl ester, [1aR-[1aα, 1bβ,4aβ,7aα,7bα,8α,9β(2Z,4E),9aα]]-; W/NBS
126480

528 C30 H40 O8 77573-09-2 NS-0-0-0 2-Butenoic acid, 2-methyl-, 1,1a,1b,4,4a,5,7a,7b,8,9-decahydro-4a,7b-dihydr oxy-3-(hydroxymethyl)-1,1,6,8-tetramethyl-5-oxo-9aH-cyclopropa[3,4]benz[1,2-e]azulene-9,9a-diyl ester, [1aR-[1aα,1b β,4aβ,7aα,7bα,8α,9β(E),9aα(E)]]-; W/NBS
126481
Str 261

528 C31 H32 N2 O6 84590-03-4 B-35-2256-0 Bis-N-methylmaleimide adduct of (1'RS,6'SR,7'SR,8'SR)-2-(8'-[(E,E)-5''- Phenylpenta-2'',4''-dien-1''-yl]bicyclo[4.2.0]octa-2',4'-dien-7'-yl)acetic acid; W
126482

528 C31 H44 O7 KC-1982-2838-0 4,5-Acetonide of (22R)-4β,5β,6α,20α(F)-Tetrahydroxy-1-oxowitha-2,24-dienolide 4,5- acetonide; W
126483

528 C32 H16 S4 56598-44-8 K-111-1335-0 CYCLOOCTA(1,2-B:4,3-B':5,6-B'':8,7-B''')TETRATHIONAPHTHENE; W
63818

528 C32 H24 N4 O4 81590-83-2 H-65-279-0 1,8-Dihydro-6,13-di(2-hydroxybenzoyl)-dibenzo[b,i]-1,4,8,11-tetraazacycl otetradeca-4,6,11,13-tetraene; W
126484

528 C32 H37 Cl N4 O 63972-67-8 F-33-633-0 γ-CHLORO-AETIO-α-OXOPHLORIN; W
63819

528 C32 H48 O6 53534-45-5 SD-1981-0-0 Holotoxinogenin 3β-acetate; W/NBS
63820
Str 262

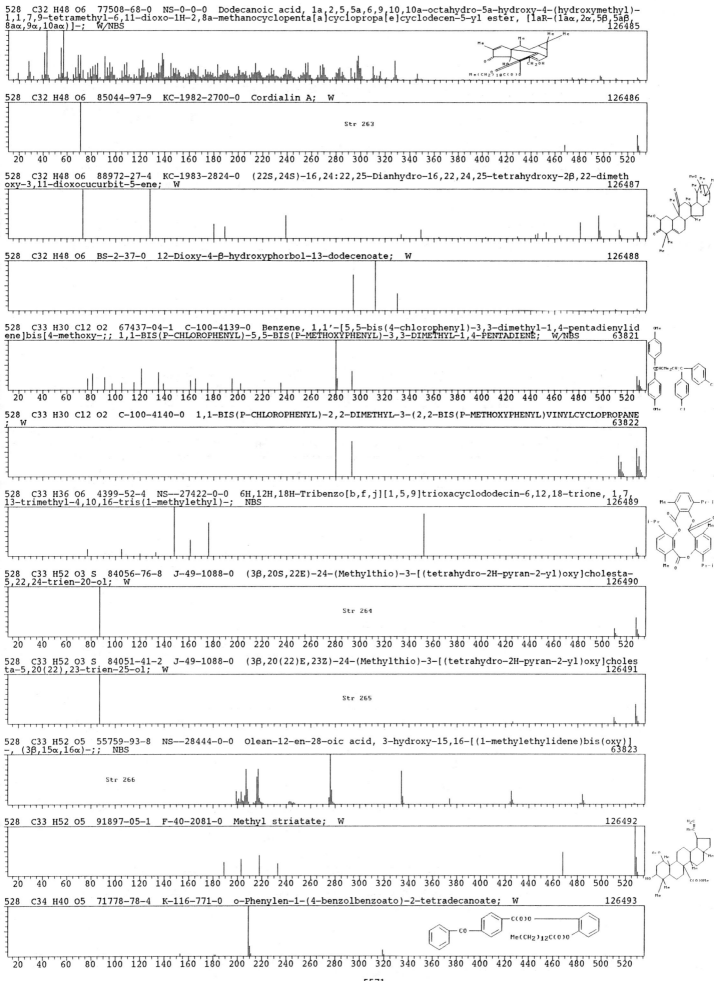

528 C32 H48 O6 77508-68-0 NS-0-0-0 Dodecanoic acid, 1a,2,5,5a,6,9,10,10a-octahydro-5a-hydroxy-4-(hydroxymethyl)-1,1,7,9-tetramethyl-6,11-dioxo-1H-2,8a-methanocyclopenta[a]cyclopropa[e]cyclodecen-5-yl ester, [1aR-(1aα,2α,5β,5aβ,8aα,9α,10aα)]-; W/NBS
126485

528 C32 H48 O6 85044-97-9 KC-1982-2700-0 Cordialin A; W
126486
Str 263

528 C32 H48 O6 88972-27-4 KC-1983-2824-0 (22S,24S)-16,24:22,25-Dianhydro-16,22,24,25-tetrahydroxy-2β,22-dimethoxy-3,11-dioxocucurbit-5-ene; W
126487

528 C32 H48 O6 BS-2-37-0 12-Dioxy-4-β-hydroxyphorbol-13-dodecenoate; W
126488

528 C33 H30 Cl2 O2 67437-04-1 C-100-4139-0 Benzene, 1,1'-[5,5-bis(4-chlorophenyl)-3,3-dimethyl-1,4-pentadienylidene]bis[4-methoxy-;; 1,1-BIS(P-CHLOROPHENYL)-5,5-BIS(P-METHOXYPHENYL)-3,3-DIMETHYL-1,4-PENTADIENE; W/NBS
63821

528 C33 H30 Cl2 O2 C-100-4140-0 1,1-BIS(P-CHLOROPHENYL)-2,2-DIMETHYL-3-(2,2-BIS(P-METHOXYPHENYL)VINYLCYCLOPROPANE ; W
63822

528 C33 H36 O6 4399-52-4 NS--27422-0-0 6H,12H,18H-Tribenzo[b,f,j][1,5,9]trioxacyclododecin-6,12,18-trione, 1,7,13-trimethyl-4,10,16-tris(1-methylethyl)-; NBS
126489

528 C33 H52 O3 S 84056-76-8 J-49-1088-0 (3β,20S,22E)-24-(Methylthio)-3-[(tetrahydro-2H-pyran-2-yl)oxy]cholesta-5,22,24-trien-20-ol; W
126490
Str 264

528 C33 H52 O3 S 84051-41-2 J-49-1088-0 (3β,20(22)E,23Z)-24-(Methylthio)-3-[(tetrahydro-2H-pyran-2-yl)oxy]cholesta-5,20(22),23-trien-25-ol; W
126491
Str 265

528 C33 H52 O5 55759-93-8 NS--28444-0-0 Olean-12-en-28-oic acid, 3-hydroxy-15,16-[(1-methylethylidene)bis(oxy)]-, (3β,15α,16α)-;; NBS
63823
Str 266

528 C33 H52 O5 91897-05-1 F-40-2081-0 Methyl striatate; W
126492

528 C34 H40 O5 71778-78-4 K-116-771-0 o-Phenylen-1-(4-benzolbenzoato)-2-tetradecanoate; W
126493

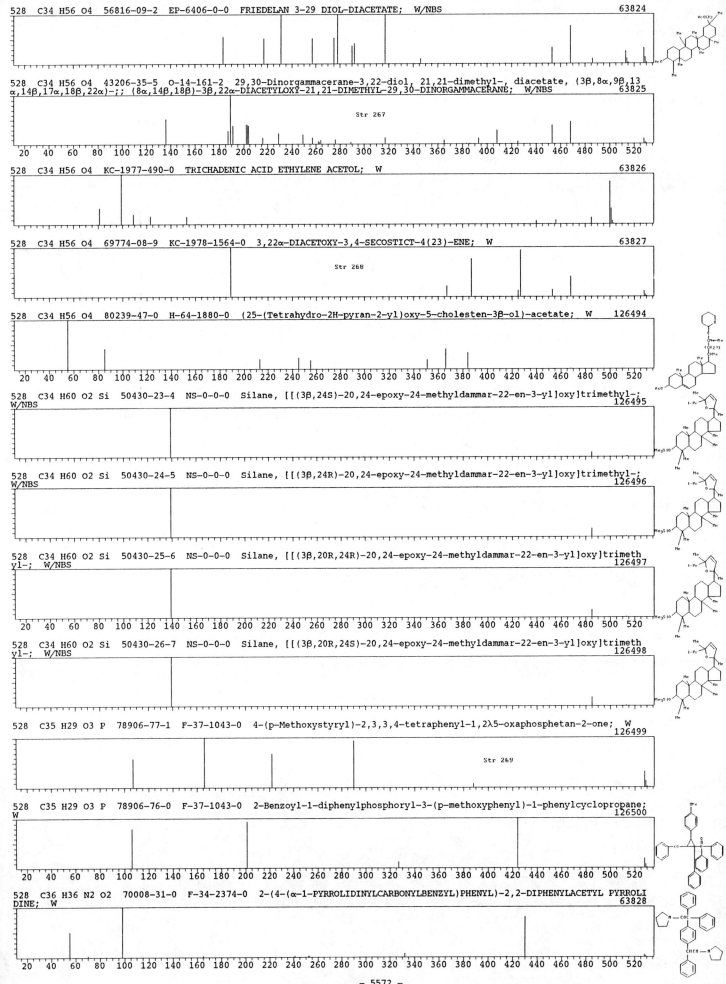

528　C34 H56 O4　56816-09-2　EP-6406-0-0　FRIEDELAN 3-29 DIOL-DIACETATE;　W/NBS　63824

528　C34 H56 O4　43206-35-5　O-14-161-2　29,30-Dinorgammacerane-3,22-diol, 21,21-dimethyl-, diacetate, (3β,8α,9β,13α,14β,17α,18β,22α)-;; (8α,14β,18β)-3β,22α-DIACETYLOXY-21,21-DIMETHYL-29,30-DINORGAMMACERANE;　W/NBS　63825

Str 267

528　C34 H56 O4　KC-1977-490-0　TRICHADENIC ACID ETHYLENE ACETOL;　W　63826

528　C34 H56 O4　69774-08-9　KC-1978-1564-0　3,22α-DIACETOXY-3,4-SECOSTICT-4(23)-ENE;　W　63827

Str 268

528　C34 H56 O4　80239-47-0　H-64-1880-0　(25-(Tetrahydro-2H-pyran-2-yl)oxy-5-cholesten-3β-ol)-acetate;　W　126494

528　C34 H60 O2 Si　50430-23-4　NS-0-0-0　Silane, [[(3β,24S)-20,24-epoxy-24-methyldammar-22-en-3-yl]oxy]trimethyl-;　W/NBS　126495

528　C34 H60 O2 Si　50430-24-5　NS-0-0-0　Silane, [[(3β,24R)-20,24-epoxy-24-methyldammar-22-en-3-yl]oxy]trimethyl-;　W/NBS　126496

528　C34 H60 O2 Si　50430-25-6　NS-0-0-0　Silane, [[(3β,20R,24R)-20,24-epoxy-24-methyldammar-22-en-3-yl]oxy]trimethyl-;　W/NBS　126497

528　C34 H60 O2 Si　50430-26-7　NS-0-0-0　Silane, [[(3β,20R,24S)-20,24-epoxy-24-methyldammar-22-en-3-yl]oxy]trimethyl-;　W/NBS　126498

528　C35 H29 O3 P　78906-77-1　F-37-1043-0　4-(p-Methoxystyryl)-2,3,3,4-tetraphenyl-1,2λ5-oxaphosphetan-2-one;　W　126499

Str 269

528　C35 H29 O3 P　78906-76-0　F-37-1043-0　2-Benzoyl-1-diphenylphosphoryl-3-(p-methoxyphenyl)-1-phenylcyclopropane;　W　126500

528　C36 H36 N2 O2　70008-31-0　F-34-2374-0　2-(4-(α-1-PYRROLIDINYLCARBONYLBENZYL)PHENYL)-2,2-DIPHENYLACETYL PYRROLIDINE;　W　63828

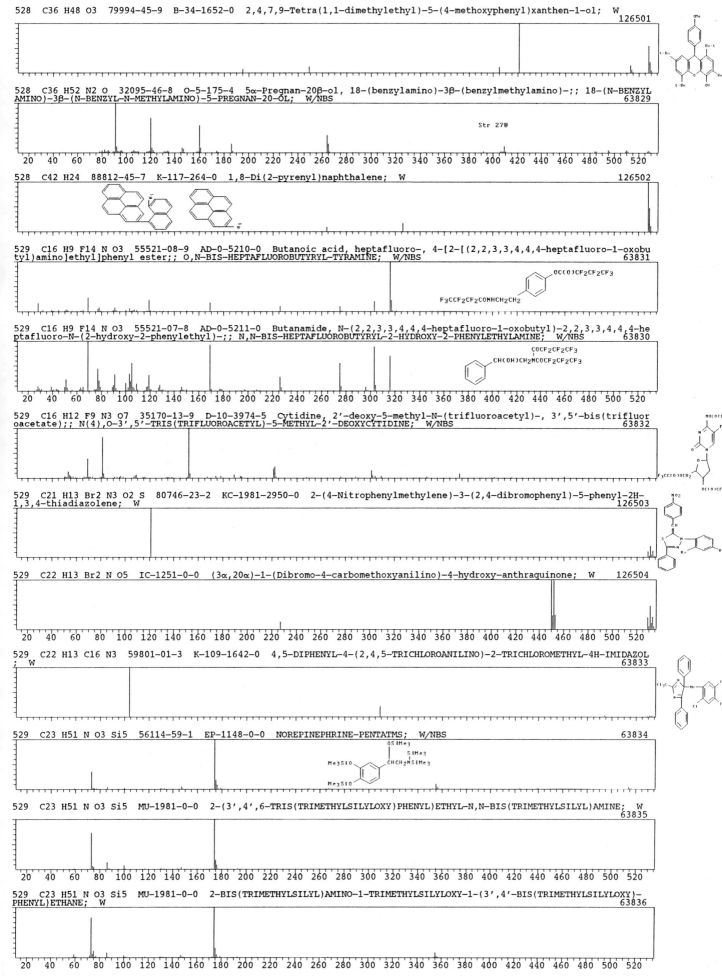

528 C36 H48 O3 79994-45-9 B-34-1652-0 2,4,7,9-Tetra(1,1-dimethylethyl)-5-(4-methoxyphenyl)xanthen-1-ol; W
126501

528 C36 H52 N2 O 32095-46-8 O-5-175-4 5α-Pregnan-20β-ol, 18-(benzylamino)-3β-(benzylmethylamino)-;; 18-(N-BENZYL
AMINO)-3β-(N-BENZYL-N-METHYLAMINO)-5-PREGNAN-20-OL; W/NBS
63829

Str 270

528 C42 H24 88812-45-7 K-117-264-0 1,8-Di(2-pyrenyl)naphthalene; W
126502

529 C16 H9 F14 N O3 55521-08-9 AD-0-5210-0 Butanoic acid, heptafluoro-, 4-[2-[(2,2,3,3,4,4,4-heptafluoro-1-oxobu
tyl)amino]ethyl]phenyl ester;; O,N-BIS-HEPTAFLUOROBUTYRYL-TYRAMINE; W/NBS
63831

529 C16 H9 F14 N O3 55521-07-8 AD-0-5211-0 Butanamide, N-(2,2,3,3,4,4,4-heptafluoro-1-oxobutyl)-2,2,3,3,4,4,4-he
ptafluoro-N-(2-hydroxy-2-phenylethyl)-;; N,N-BIS-HEPTAFLUOROBUTYRYL-2-HYDROXY-2-PHENYLETHYLAMINE; W/NBS
63830

529 C16 H12 F9 N3 O7 35170-13-9 D-10-3974-5 Cytidine, 2'-deoxy-5-methyl-N-(trifluoroacetyl)-, 3',5'-bis(trifluor
oacetate);; N(4),O-3',5'-TRIS(TRIFLUOROACETYL)-5-METHYL-2'-DEOXYCYTIDINE; W/NBS
63832

529 C21 H13 Br2 N3 O2 S 80746-23-2 KC-1981-2950-0 2-(4-Nitrophenylmethylene)-3-(2,4-dibromophenyl)-5-phenyl-2H-
1,3,4-thiadiazolene; W
126503

529 C22 H13 Br2 N O5 IC-1251-0-0 (3α,20α)-1-(Dibromo-4-carbomethoxyanilino)-4-hydroxy-anthraquinone; W 126504

529 C22 H13 Cl6 N3 59801-01-3 K-109-1642-0 4,5-DIPHENYL-4-(2,4,5-TRICHLOROANILINO)-2-TRICHLOROMETHYL-4H-IMIDAZOL
; W
63833

529 C23 H51 N O3 Si5 56114-59-1 EP-1148-0-0 NOREPINEPHRINE-PENTATMS; W/NBS
63834

529 C23 H51 N O3 Si5 MU-1981-0-0 2-(3',4',6-TRIS(TRIMETHYLSILYLOXY)PHENYL)ETHYL-N,N-BIS(TRIMETHYLSILYL)AMINE; W
63835

529 C23 H51 N O3 Si5 MU-1981-0-0 2-BIS(TRIMETHYLSILYL)AMINO-1-TRIMETHYLSILYLOXY-1-(3',4'-BIS(TRIMETHYLSILYLOXY)-
PHENYL)ETHANE; W
63836

529 C24 H19 Mn N O2 Sb 32826-12-3 O3-0-1227-0 Manganese, dicarbonyl-.pi.-pyrrolyl(triphenylstibine)-;; (.PI.-PYR
ROLYL) MANGANESE DICARBONYL TRIPHENYLSTIBINE; W/NBS   63837

529 C25 H39 N O11 84036-68-0 H-65-1140-0 (t-Butyl) ester of (3S,3aR,4aR,7aR,7bS)-2-(2,3:5,6-Di-O-isopropylidene-
α-D-mannofuranosyl)-6,6-dimethyl-perhydro[1,3]dioxolo[4',5':4,5]furo[2,3-d]isoxazol-3-carboxylic acid; W 126505

529 C27 H35 N3 O8 55517-95-8 O-3-73-3 L-Tyrosine, N-[N-[N-[(phenylmethoxy)carbonyl]-L-seryl]-L-leucyl]-, methyl
ester;; BENZYLOXYCARBONYL-SE-LEU-TYR-OME; W/NBS   63838

529 C28 H23 N3 O8 73510-64-2 K-113-621-0 Tetramethyl ester of 4-(2-Cyano-1-methylethenyl)-7b,11a-dihydrobenzo[a]
pyrrolo[1',2':3,4]pyrimido[6,1,2cd]pyrrolizin-8,9,10,11-tetracarboxylic acid; W   126506

529 C29 H39 N O8 39007-91-5 B-27-921-1 Rostratine; W   63839
Str 271

529 C29 H56 F N3 O4 55823-08-0 H-57-433-33 Piperidinium, 1-[4-[acetyl[3-(acetylamino)propyl]amino]butyl]-2-(11-
methoxy-11-oxoundecyl)-1-methyl-, fluoride, (R)-;; N(6),N(4')-DIACETYL-NEOONCINOTINIC ACID-METHYL ESTER-N(1)-METHYL
FLUORIDE; W/NBS   63840

529 C31 H47 N O6 79156-66-4 B-34-614-0 (25R)-N-Acetyl-6-oxo-5α,22αN-spirosolane-3β,5-diol-3-acetate; W 126507
Str 272

529 C31 H47 N O6 IC-1251-0-0 1-Benzyl-3a5-dihydroxy-4-(8,11-dihydroxy-9-ethoxy-4-methyl-undec-1-enyl)-7-methyl-6-
methylene-3-oxo3a,4,5,6,7,7a-hexahydro-isoindoline; W   126508

529 C32 H35 N O6 63535-56-8 H-60-441-0 3ε-HYDROXY-3ε-METHYL-1-(2',3'-0-ISOPROPYLIDEN-5'-0-TRITYL-B-D-RIBOFURANOS
YL)-2-PYRROLIDINONE; W   63841
Str 273

529 C36 H35 N O3 72511-95-6 B-32-1325-0 N,N-DIBENZYL(2-(3',5'-DIBENZYLOXYPHENYL)-2-HYDROXYETHYL)AMINE; W 63842
Str 274

529 C36 H35 N O3 64090-42-2 H-60-1469-0 CIS-1,2-DIBENZYL-N-((2-HYDROXY-2,2-DIPHENYL)-ETHYL)-CYCLOHEXANE-1,2-DICA
RBOXIMIDE; W   63843
Str 275

530 C9 H10 B2 Fe I2 O3 S 63338-96-5 AG-131-10-5 3,4-Diethyl-2,5-diiodo-1,2,5-thiadiborolene-tricarbonyliron; W
126509

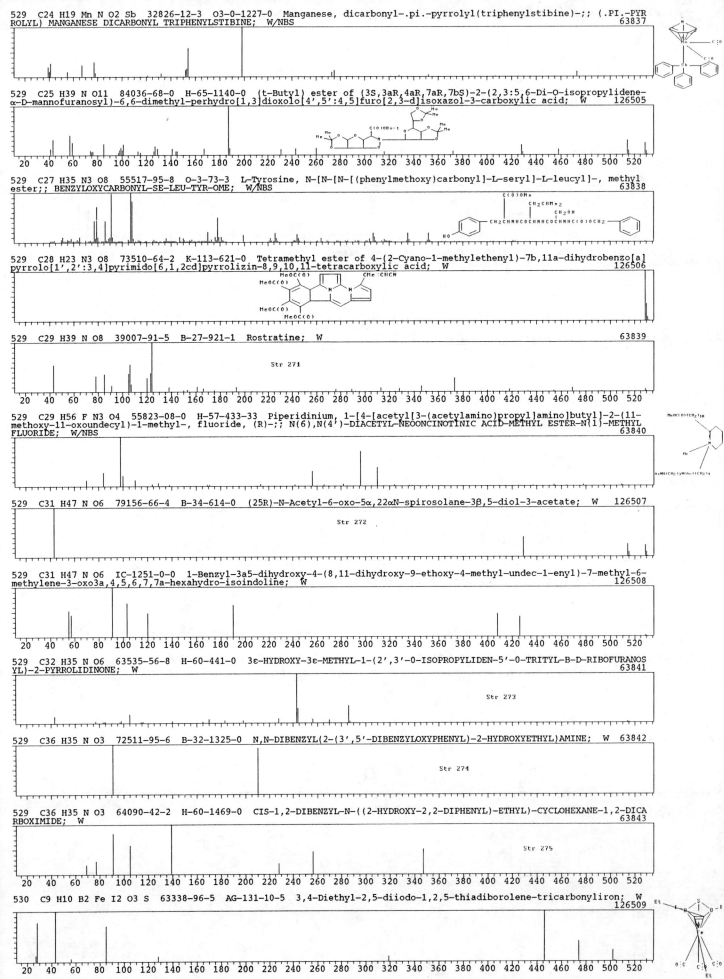

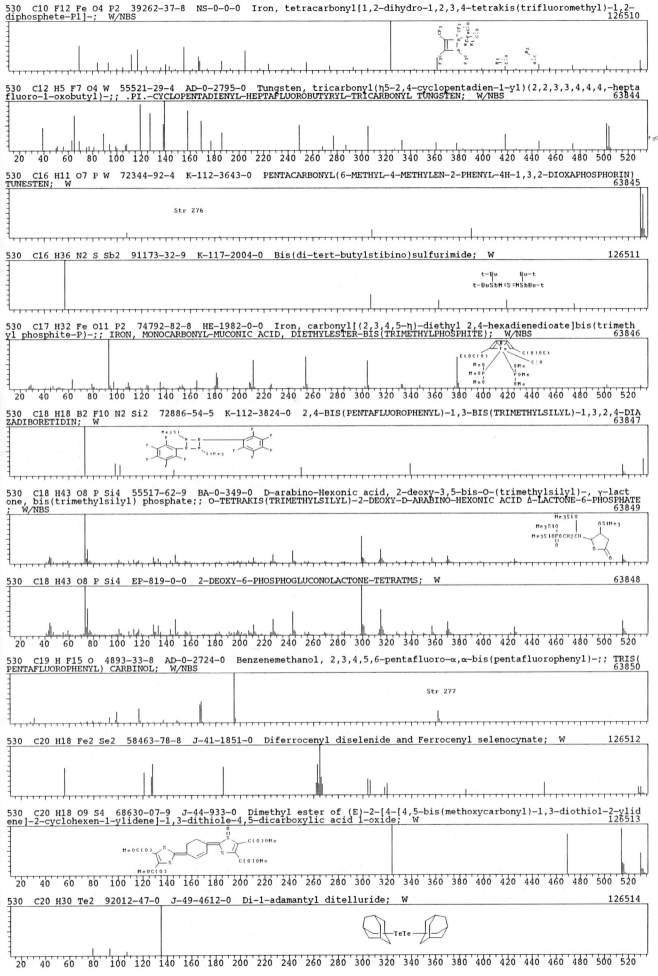

530 C10 F12 Fe O4 P2 39262-37-8 NS-0-0-0 Iron, tetracarbonyl[1,2-dihydro-1,2,3,4-tetrakis(trifluoromethyl)-1,2-diphosphete-P1]-; W/NBS
126510

530 C12 H5 F7 O4 W 55521-29-4 AD-0-2795-0 Tungsten, tricarbonyl(η5-2,4-cyclopentadien-1-yl)(2,2,3,3,4,4,4,-heptafluoro-1-oxobutyl)-;; .PI.-CYCLOPENTADIENYL-HEPTAFLUOROBUTYRYL-TRICARBONYL TUNGSTEN; W/NBS
63844

20 40 60 80 100 120 140 160 180 200 220 240 260 280 300 320 340 360 380 400 420 440 460 480 500 520

530 C16 H11 O7 P W 72344-92-4 K-112-3643-0 PENTACARBONYL(6-METHYL-4-METHYLEN-2-PHENYL-4H-1,3,2-DIOXAPHOSPHORIN)TUNESTEN; W
63845

Str 276

530 C16 H36 N2 S Sb2 91173-32-9 K-117-2004-0 Bis(di-tert-butylstibino)sulfurimide; W
126511

530 C17 H32 Fe O11 P2 74792-82-8 HE-1982-0-0 Iron, carbonyl[(2,3,4,5-η)-diethyl 2,4-hexadienedioate]bis(trimethyl phosphite-P)-;; IRON, MONOCARBONYL-MUCONIC ACID, DIETHYLESTER-BIS(TRIMETHYLPHOSPHITE); W/NBS
63846

530 C18 H18 B2 F10 N2 Si2 72886-54-5 K-112-3824-0 2,4-BIS(PENTAFLUOROPHENYL)-1,3-BIS(TRIMETHYLSILYL)-1,3,2,4-DIAZADIBORETIDIN; W
63847

530 C18 H43 O8 P Si4 55517-62-9 BA-0-349-0 D-arabino-Hexonic acid, 2-deoxy-3,5-bis-O-(trimethylsilyl)-, γ-lactone, bis(trimethylsilyl) phosphate;; O-TETRAKIS(TRIMETHYLSILYL)-2-DEOXY-D-ARABINO-HEXONIC ACID Δ-LACTONE-6-PHOSPHATE; W/NBS
63849

530 C18 H43 O8 P Si4 EP-819-0-0 2-DEOXY-6-PHOSPHOGLUCONOLACTONE-TETRATMS; W
63848

20 40 60 80 100 120 140 160 180 200 220 240 260 280 300 320 340 360 380 400 420 440 460 480 500 520

530 C19 H F15 O 4893-33-8 AD-0-2724-0 Benzenemethanol, 2,3,4,5,6-pentafluoro-α,α-bis(pentafluorophenyl)-;; TRIS(PENTAFLUOROPHENYL) CARBINOL; W/NBS
63850

Str 277

530 C20 H18 Fe2 Se2 58463-78-8 J-41-1851-0 Diferrocenyl diselenide and Ferrocenyl selenocynate; W
126512

530 C20 H18 O9 S4 68630-07-9 J-44-933-0 Dimethyl ester of (E)-2-[4-[4,5-bis(methoxycarbonyl)-1,3-diothiol-2-ylidene]-2-cyclohexen-1-ylidene]-1,3-dithiole-4,5-dicarboxylic acid 1-oxide; W
126513

20 40 60 80 100 120 140 160 180 200 220 240 260 280 300 320 340 360 380 400 420 440 460 480 500 520

530 C20 H30 Te2 92012-47-0 J-49-4612-0 Di-1-adamantyl ditelluride; W
126514

20 40 60 80 100 120 140 160 180 200 220 240 260 280 300 320 340 360 380 400 420 440 460 480 500 520

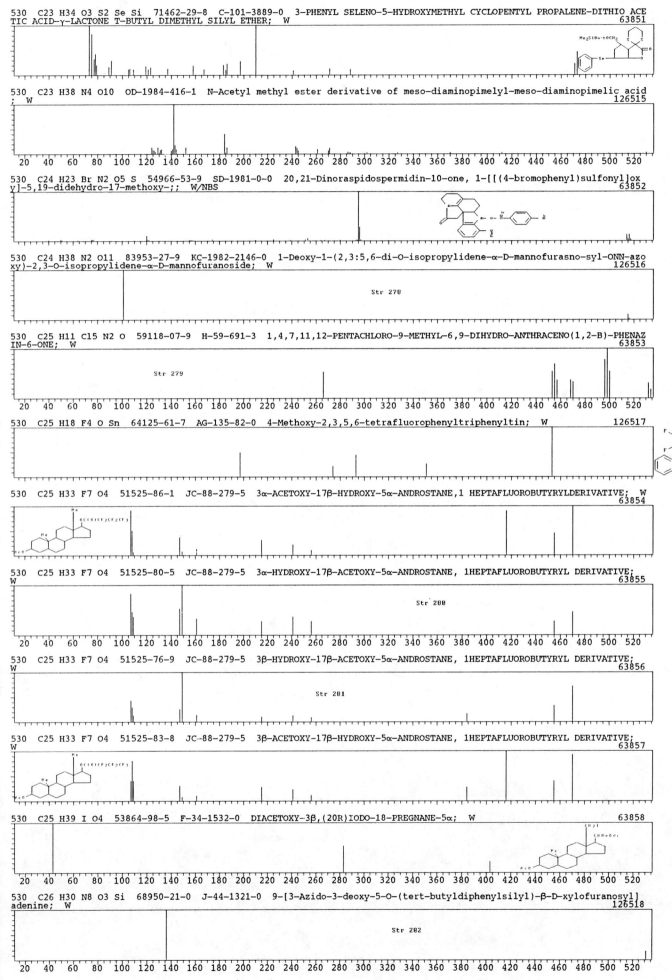

530 C23 H34 O3 S2 Se Si   71462-29-8   C-101-3889-0   3-PHENYL SELENO-5-HYDROXYMETHYL CYCLOPENTYL PROPALENE-DITHIO ACE
TIC ACID-γ-LACTONE T-BUTYL DIMETHYL SILYL ETHER;   W
63851

530 C23 H38 N4 O10   OD-1984-416-1   N-Acetyl methyl ester derivative of meso-diaminopimelyl-meso-diaminopimelic acid
; W
126515

530 C24 H23 Br N2 O5 S   54966-53-9   SD-1981-0-0   20,21-Dinoraspidospermidin-10-one, 1-[[(4-bromophenyl)sulfonyl]ox
y]-5,19-didehydro-17-methoxy-;; W/NBS
63852

530 C24 H38 N2 O11   83953-27-9   KC-1982-2146-0   1-Deoxy-1-(2,3:5,6-di-O-isopropylidene-α-D-mannofurasno-syl-ONN-azo
xy)-2,3-O-isopropylidene-α-D-mannofuranoside; W
126516

Str 278

530 C25 H11 Cl5 N2 O   59118-07-9   H-59-691-3   1,4,7,11,12-PENTACHLORO-9-METHYL-6,9-DIHYDRO-ANTHRACENO(1,2-B)-PHENAZ
IN-6-ONE; W
63853

Str 279

530 C25 H18 F4 O Sn   64125-61-7   AG-135-82-0   4-Methoxy-2,3,5,6-tetrafluorophenyltriphenyltin; W
126517

530 C25 H33 F7 O4   51525-86-1   JC-88-279-5   3α-ACETOXY-17β-HYDROXY-5α-ANDROSTANE,1 HEPTAFLUOROBUTYRYLDERIVATIVE; W
63854

530 C25 H33 F7 O4   51525-80-5   JC-88-279-5   3α-HYDROXY-17β-ACETOXY-5α-ANDROSTANE, 1HEPTAFLUOROBUTYRYL DERIVATIVE;
W
63855

Str 280

530 C25 H33 F7 O4   51525-76-9   JC-88-279-5   3β-HYDROXY-17β-ACETOXY-5α-ANDROSTANE, 1HEPTAFLUOROBUTYRYL DERIVATIVE;
W
63856

Str 281

530 C25 H33 F7 O4   51525-83-8   JC-88-279-5   3β-ACETOXY-17β-HYDROXY-5α-ANDROSTANE, 1HEPTAFLUOROBUTYRYL DERIVATIVE;
W
63857

530 C25 H39 I O4   53864-98-5   F-34-1532-0   DIACETOXY-3β,(20R)IODO-18-PREGNANE-5α; W
63858

530 C26 H30 N8 O3 Si   68950-21-0   J-44-1321-0   9-[3-Azido-3-deoxy-5-O-(tert-butyldiphenylsilyl)-β-D-xylofuranosyl]
adenine; W
126518

Str 282

530 C26 H34 N4 O4 Zn 69782-63-4 T-68-5605-0 Zinc, bis[[5,5'-methylenebis[3,4-dihydro-4,4-dimethyl-2H-pyrrol-2-on
ato]](1-)-N1,N1']-, (T-4)-;; BIS((5-OXO-3,3-DIMETHYL-2-PYRROLIN-2-YL)(5-OXO-3,3-DIMETHYL-2-PYRROLIDINYLIDENE)METH
ANE) ZINC(II); W/NBS   63859

530 C26 H42 O11 85888-08-0 H-66-256-0 3,7-Anhydro-3,7-dihydroxy-5,9,13-triacetyl-4,8,12-trimethylpentadecandi-
oic acid-dimethylester; W   126519

Str 283

20 40 60 80 100 120 140 160 180 200 220 240 260 280 300 320 340 360 380 400 420 440 460 480 500 520

530 C26 H42 O11 81875-51-6 C-104-4140-0 Peracetyl derivative of 2-(2,6,8-trihydroxy-2-nonyl)-1-oxa-3,5-dimethyl-
4,6-dihydroxycyclohexane; W   126520

530 C26 H54 O5 Si3 55759-98-3 AB-231-430-8 Cyclopentanepropanoic acid, 3,5-bis[(trimethylsilyl)oxy]-2-[3-[(tri
methylsilyl)oxy]-1-octenyl]-, methyl ester, [1R-[1α,2β(1E,3S*),3α,5α]]-;; METHYL 5α,7α,11-TRIHYDROXY-TETRANORPROST-
9-ENOATE-TRISTRIMETHYLSILYL ETHER; W/NBS   63860

530 C27 H46 N2 O3 Si3 KO-9-198-2 Trimethylsilylether,trimethylsilylamino,trimethylsilyloxime derivative of 8β-eth
yldihydronormorphinone; W   126521

20 40 60 80 100 120 140 160 180 200 220 240 260 280 300 320 340 360 380 400 420 440 460 480 500 520

530 C27 H47 I O2 91712-67-3 J-49-3760-0 (5β,10.xi.)-2-Iodo-19-nor-1,2-secocholestan-10-ol formate; W 126522

530 C28 H22 N2 O5 S2 72047-63-3 J-45-4371-0 3-((4-methylphenyl)sulfonyl)-2-(3-nitrophenyl)-5,5-diphenyl-4-oxazol
idinethione; W   126523

Str 284

530 C28 H26 N4 O7 92220-51-4 J-49-4343-0 2,6-Diphenyl-9β-(2,3,5-tri-O-acetyl-D-ribofuranosyl)purine; W 126524

Str 285

20 40 60 80 100 120 140 160 180 200 220 240 260 280 300 320 340 360 380 400 420 440 460 480 500 520

530 C28 H34 O10 75379-00-9 NS--28425-0-0 9(1H)-Phenanthrenone, 5,6,8-tris(acetyloxy)-7-[2-(acetyloxy)-1-methyl
ethyl]-2,3,4,4a-tetrahydro-10-hydroxy-1,1,4a-trimethyl-, [R-(R*,R*)]-;; NBS   63861

530 C28 H34 O10 66584-88-1 H-61-882-0 11,12,14,16-Tetra-o-acetyl-coleon C; W   126525

530 C29 H38 O9 69993-66-4 K-112-434-0 8β-(ANGELOYLOXY)PROUSTIANOL-ANGELICATE; W   63862

20 40 60 80 100 120 140 160 180 200 220 240 260 280 300 320 340 360 380 400 420 440 460 480 500 520

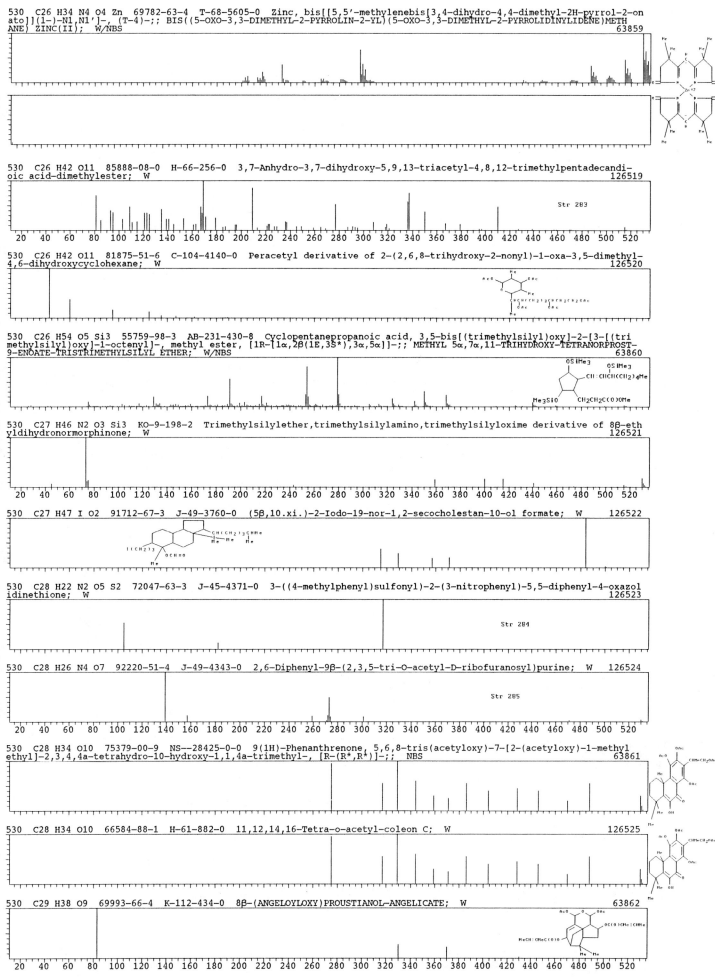

530 C29 H62 O4 Si2 55517-94-7 AA-0-2252-1 Eicosanoic acid, 2,3-bis[(trimethylsilyl)oxy]propyl ester;; TRIMETHYLS
ILYLETHER DERIVATIVE OF 1-MONOARACHIDIN; W/NBS   63863

OSiMe3
Me(CH2)18C(O)OCH2CHCH2OSiMe3

530 C29 H62 O4 Si2 55517-93-6 AA-0-2251-1 Eicosanoic acid, 2-[(trimethylsilyl)oxy]-1-[[(trimethylsilyl)oxy]meth
yl]ethyl ester;; TRIMETHYLSILYLETHER DERIVATIVE OF 2-MONOARACHIDIN; W/NBS   63864

CH2OSiMe3
Me(CH2)18C(O)OCHCH2OSiMe3

20   40   60   80   100 120 140 160 180 200 220 240 260 280 300 320 340 360 380 400 420 440 460 480 500 520

530 C29 H62 O4 Si2 HE-1986-2366-0 BIS-O-TRIMETHYLSILYL-EICOSANOIC ACID-GLYCERIN-(1)-MONOESTER; W   126526

530 C30 H18 N4 O6 71041-18-4 J-44-3315-0 3,3'-Dinitro-10-phenyl-7,10'-biphenoxazine; W   126527

Str 286

530 C30 H26 O9 62350-96-3 K-110-1055-0 HYPHOLOMIN-A-TETRAMETHYLETHER; W   63865

530 C30 H30 N2 O5 S 23145-64-4 O-2-339-7 p-Benzotoluidide, 2-(N-veratryl-p-toluenesulfonamido)-;; 2-(N-TOSYL-N-(
3,4-DIMETHOXYBENZYL))ANTHRANILIC ACID-P-TOLVIDIDE; W/NBS   63866

530 C30 H34 N2 O3 S Si 73454-06-5 J-45-2630-0 1,4-Dimethyl-3-(((tert-butyldiphenylsilyl)oxy)-methylene)-6-mercap
to-6-benzyl-2,5-piperazinedione; W   126528

530 C30 H42 O8 21551-67-7 B-29-1986-0 Dihydroardisiaquinone A; W   63867

20   40   60   80   100 120 140 160 180 200 220 240 260 280 300 320 340 360 380 400 420 440 460 480 500 520

530 C30 H54 N2 O2 Si2 SH-1981-0-0 STANOZOLOL TMS; W   63868

530 C31 H38 N4 O4 90702-40-2 AH-115-110-0 rac-(4Z,10Z,15Z)- and rac-(4E,10Z,15Z)-17-Ethyl-1,19-dioxo-3-methoxyca
rbonylmethyl-3,7,8,12,13,18,21-heptamethyl-1,2,3,19,22,24-hexahydro-21H-bilin; W   126529

Str 287

530 C31 H54 O3 Si2 77590-72-8 NS-0-0-0 Silane, [[9-[[(dimethylpropylsilyl)oxy]methyl]-6a,7,8,10a-tetrahydro-6,6-
dimethyl-3-pentyl-6H-dibenzo[b,d]pyran-1-yl]oxy]dimethylpropyl-, (6aR-trans)-; W/NBS   126530

20   40   60   80   100 120 140 160 180 200 220 240 260 280 300 320 340 360 380 400 420 440 460 480 500 520

530 C32 H32 N4 Ni HE-1986-132-0 BIS[GLYOXAL-BIS(P-TOLYLIMINO)]NICKEL; W   126531

20   40   60   80   100 120 140 160 180 200 220 240 260 280 300 320 340 360 380 400 420 440 460 480 500 520

540 560 580 600 620 640 660 680 700 720 740 760 780 800 820 840 860 880 900 920 940 960 980 1000 1020 1040

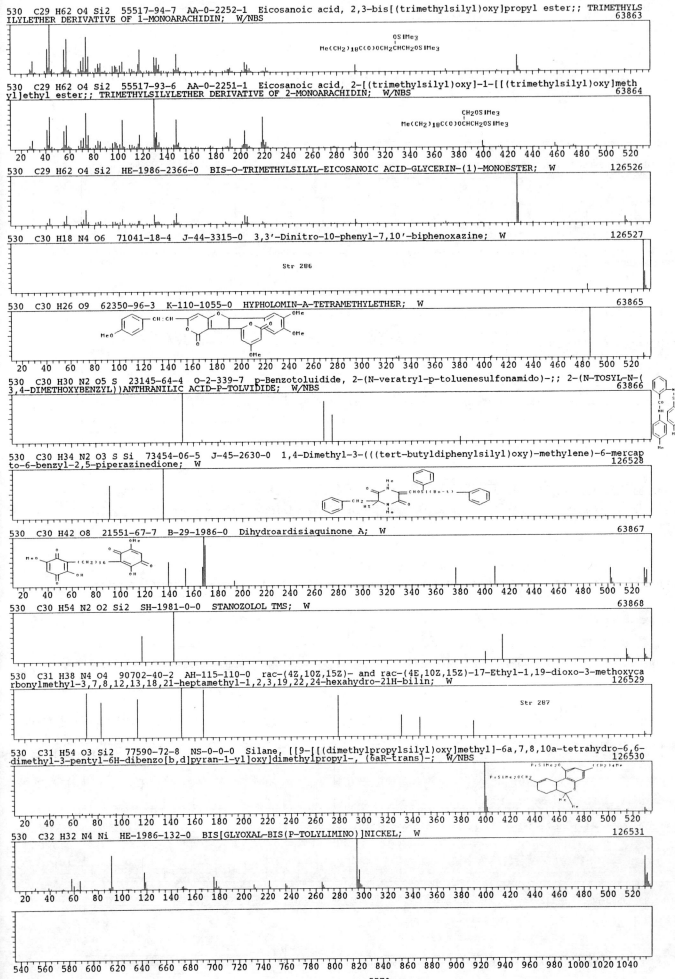

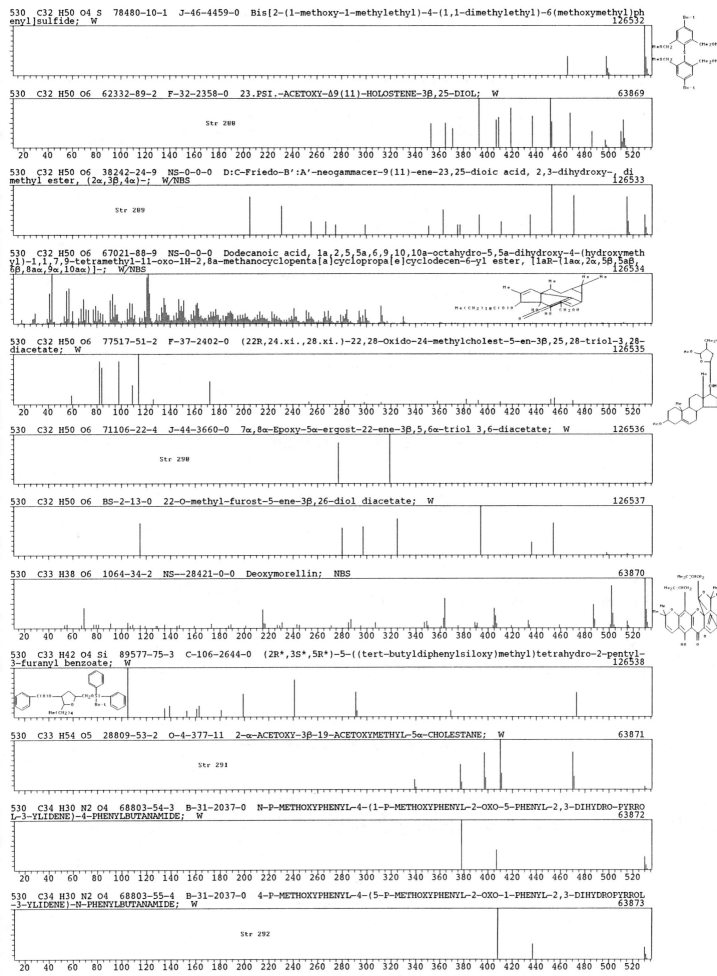

530　C32 H50 O4 S　78480-10-1　J-46-4459-0　Bis[2-(1-methoxy-1-methylethyl)-4-(1,1-dimethylethyl)-6(methoxymethyl)phenyl]sulfide;　W
126532

530　C32 H50 O6　62332-89-2　F-32-2358-0　23.PSI.-ACETOXY-Δ9(11)-HOLOSTENE-3β,25-DIOL;　W
63869

Str 288

530　C32 H50 O6　38242-24-9　NS-0-0-0　D:C-Friedo-B':A'-neogammacer-9(11)-ene-23,25-dioic acid, 2,3-dihydroxy-, dimethyl ester, (2α,3β,4α)-;　W/NBS
126533

Str 289

530　C32 H50 O6　67021-88-9　NS-0-0-0　Dodecanoic acid, 1a,2,5,5a,6,9,10,10a-octahydro-5,5a-dihydroxy-4-(hydroxymethyl)-1,1,7,9-tetramethyl-11-oxo-1H-2,8a-methanocyclopenta[a]cyclopropa[e]cyclodecen-6-yl ester, [1aR-(1aα,2α,5β,5aβ,6β,8aα,9α,10aα)]-;　W/NBS
126534

530　C32 H50 O6　77517-51-2　F-37-2402-0　(22R,24.xi.,28.xi.)-22,28-Oxido-24-methylcholest-5-en-3β,25,28-triol-3,28-diacetate;　W
126535

530　C32 H50 O6　71106-22-4　J-44-3660-0　7α,8α-Epoxy-5α-ergost-22-ene-3β,5,6α-triol 3,6-diacetate;　W
126536

Str 290

530　C32 H50 O6　BS-2-13-0　22-O-methyl-furost-5-ene-3β,26-diol diacetate;　W
126537

530　C33 H38 O6　1064-34-2　NS--28421-0-0　Deoxymorellin;　NBS
63870

530　C33 H42 O4 Si　89577-75-3　C-106-2644-0　(2R*,3S*,5R*)-5-((tert-butyldiphenylsiloxy)methyl)tetrahydro-2-pentyl-3-furanyl benzoate;　W
126538

530　C33 H54 O5　28809-53-2　O-4-377-11　2-α-ACETOXY-3β-19-ACETOXYMETHYL-5α-CHOLESTANE;　W
63871

Str 291

530　C34 H30 N2 O4　68803-54-3　B-31-2037-0　N-P-METHOXYPHENYL-4-(1-P-METHOXYPHENYL-2-OXO-5-PHENYL-2,3-DIHYDRO-PYRROL-3-YLIDENE)-4-PHENYLBUTANAMIDE;　W
63872

530　C34 H30 N2 O4　68803-55-4　B-31-2037-0　4-P-METHOXYPHENYL-4-(5-P-METHOXYPHENYL-2-OXO-1-PHENYL-2,3-DIHYDROPYRROL-3-YLIDENE)-N-PHENYLBUTANAMIDE;　W
63873

Str 292

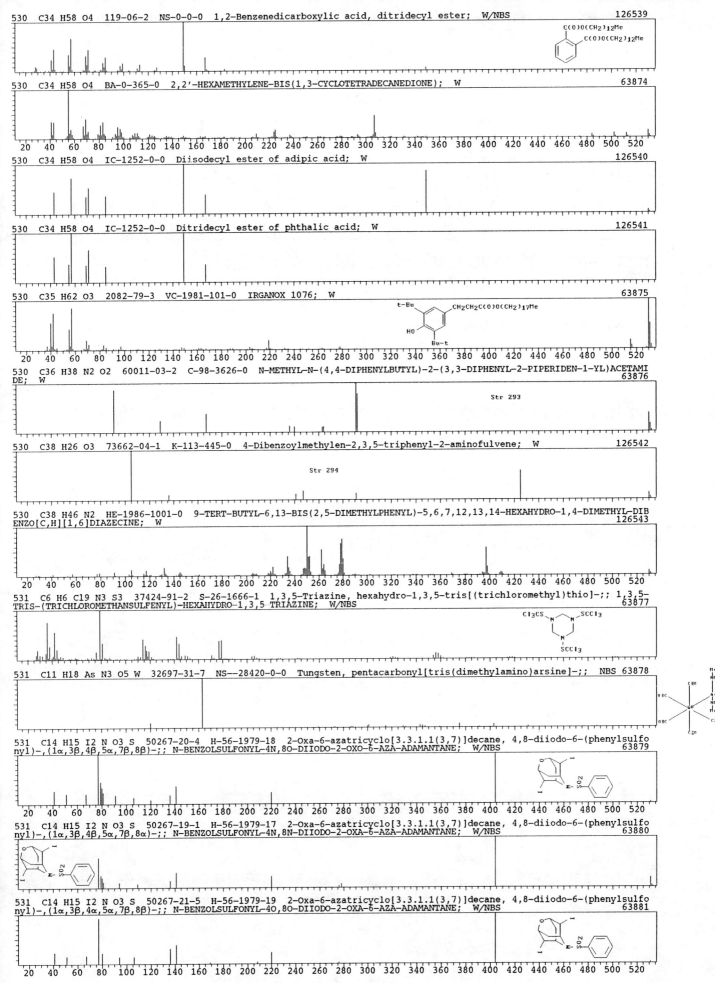

530 C34 H58 O4 119-06-2 NS-0-0-0 1,2-Benzenedicarboxylic acid, ditridecyl ester; W/NBS   126539

530 C34 H58 O4 BA-0-365-0 2,2'-HEXAMETHYLENE-BIS(1,3-CYCLOTETRADECANEDIONE); W   63874

530 C34 H58 O4 IC-1252-0-0 Diisodecyl ester of adipic acid; W   126540

530 C34 H58 O4 IC-1252-0-0 Ditridecyl ester of phthalic acid; W   126541

530 C35 H62 O3 2082-79-3 VC-1981-101-0 IRGANOX 1076; W   63875

530 C36 H38 N2 O2 60011-03-2 C-98-3626-0 N-METHYL-N-(4,4-DIPHENYLBUTYL)-2-(3,3-DIPHENYL-2-PIPERIDEN-1-YL)ACETAMIDE; W   63876

Str 293

530 C38 H26 O3 73662-04-1 K-113-445-0 4-Dibenzoylmethylen-2,3,5-triphenyl-2-aminofulvene; W   126542

Str 294

530 C38 H46 N2 HE-1986-1001-0 9-TERT-BUTYL-6,13-BIS(2,5-DIMETHYLPHENYL)-5,6,7,12,13,14-HEXAHYDRO-1,4-DIMETHYL-DIBENZO[C,H][1,6]DIAZECINE; W   126543

531 C6 H6 Cl9 N3 S3 37424-91-2 S-26-1666-1 1,3,5-Triazine, hexahydro-1,3,5-tris[(trichloromethyl)thio]-;; 1,3,5-TRIS-(TRICHLOROMETHANSULFENYL)-HEXAHYDRO-1,3,5 TRIAZINE; W/NBS   63877

531 C11 H18 As N3 O5 W 32697-31-7 NS--28420-0-0 Tungsten, pentacarbonyl[tris(dimethylamino)arsine]-;; NBS 63878

531 C14 H15 I2 N O3 S 50267-20-4 H-56-1979-18 2-Oxa-6-azatricyclo[3.3.1.1(3,7)]decane, 4,8-diiodo-6-(phenylsulfonyl)-,(1α,3β,4β,5α,7β,8β)-;; N-BENZOLSULFONYL-4N,8O-DIIODO-2-OXO-6-AZA-ADAMANTANE; W/NBS   63879

531 C14 H15 I2 N O3 S 50267-19-1 H-56-1979-17 2-Oxa-6-azatricyclo[3.3.1.1(3,7)]decane, 4,8-diiodo-6-(phenylsulfonyl)-,(1α,3β,4β,5α,7β,8α)-;; N-BENZOLSULFONYL-4N,8N-DIIODO-2-OXA-6-AZA-ADAMANTANE; W/NBS   63880

531 C14 H15 I2 N O3 S 50267-21-5 H-56-1979-19 2-Oxa-6-azatricyclo[3.3.1.1(3,7)]decane, 4,8-diiodo-6-(phenylsulfonyl)-,(1α,3β,4α,5α,7β,8β)-;; N-BENZOLSULFONYL-4O,8O-DIIODO-2-OXA-6-AZA-ADAMANTANE; W/NBS   63881

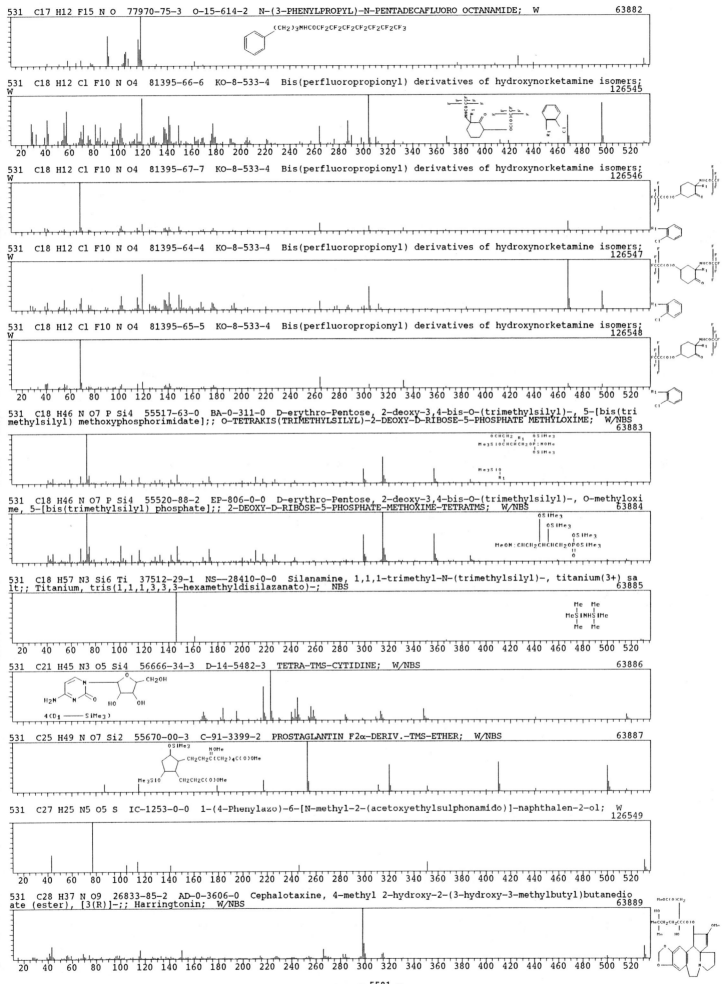

531   C17 H12 F15 N O   77970-75-3   O-15-614-2   N-(3-PHENYLPROPYL)-N-PENTADECAFLUORO OCTANAMIDE;   W   63882

(CH2)3NHCOCF2CF2CF2CF2CF2CF2CF3

531   C18 H12 Cl F10 N O4   81395-66-6   KO-8-533-4   Bis(perfluoropropionyl) derivatives of hydroxynorketamine isomers;
W   126545

531   C18 H12 Cl F10 N O4   81395-67-7   KO-8-533-4   Bis(perfluoropropionyl) derivatives of hydroxynorketamine isomers;
W   126546

531   C18 H12 Cl F10 N O4   81395-64-4   KO-8-533-4   Bis(perfluoropropionyl) derivatives of hydroxynorketamine isomers;
W   126547

531   C18 H12 Cl F10 N O4   81395-65-5   KO-8-533-4   Bis(perfluoropropionyl) derivatives of hydroxynorketamine isomers;
W   126548

531   C18 H46 N O7 P Si4   55517-63-0   BA-0-311-0   D-erythro-Pentose, 2-deoxy-3,4-bis-O-(trimethylsilyl)-, 5-[bis(tri
methylsilyl) methoxyphosphorimidate];;   O-TETRAKIS(TRIMETHYLSILYL)-2-DEOXY-D-RIBOSE-5-PHOSPHATE METHYLOXIME;   W/NBS
                                                                                                                 63883

531   C18 H46 N O7 P Si4   55520-88-2   EP-806-0-0   D-erythro-Pentose, 2-deoxy-3,4-bis-O-(trimethylsilyl)-, O-methyloxi
me, 5-[bis(trimethylsilyl) phosphate];;   2-DEOXY-D-RIBOSE-5-PHOSPHATE-METHOXIME-TETRATMS;   W/NBS   63884

531   C18 H57 N3 Si6 Ti   37512-29-1   NS--28410-0-0   Silanamine, 1,1,1-trimethyl-N-(trimethylsilyl)-, titanium(3+) sa
lt;;   Titanium, tris(1,1,1,3,3,3-hexamethyldisilazanato)-;   NBS   63885

531   C21 H45 N3 O5 Si4   56666-34-3   D-14-5482-3   TETRA-TMS-CYTIDINE;   W/NBS   63886

531   C25 H49 N O7 Si2   55670-00-3   C-91-3399-2   PROSTAGLANTIN F2α-DERIV.-TMS-ETHER;   W/NBS   63887

531   C27 H25 N5 O5 S   IC-1253-0-0   1-(4-Phenylazo)-6-[N-methyl-2-(acetoxyethylsulphonamido)]-naphthalen-2-ol;   W
                                                                                                                 126549

531   C28 H37 N O9   26833-85-2   AD-0-3606-0   Cephalotaxine, 4-methyl 2-hydroxy-2-(3-hydroxy-3-methylbutyl)butanedio
ate (ester), [3(R)]-;;   Harringtonin;   W/NBS   63889

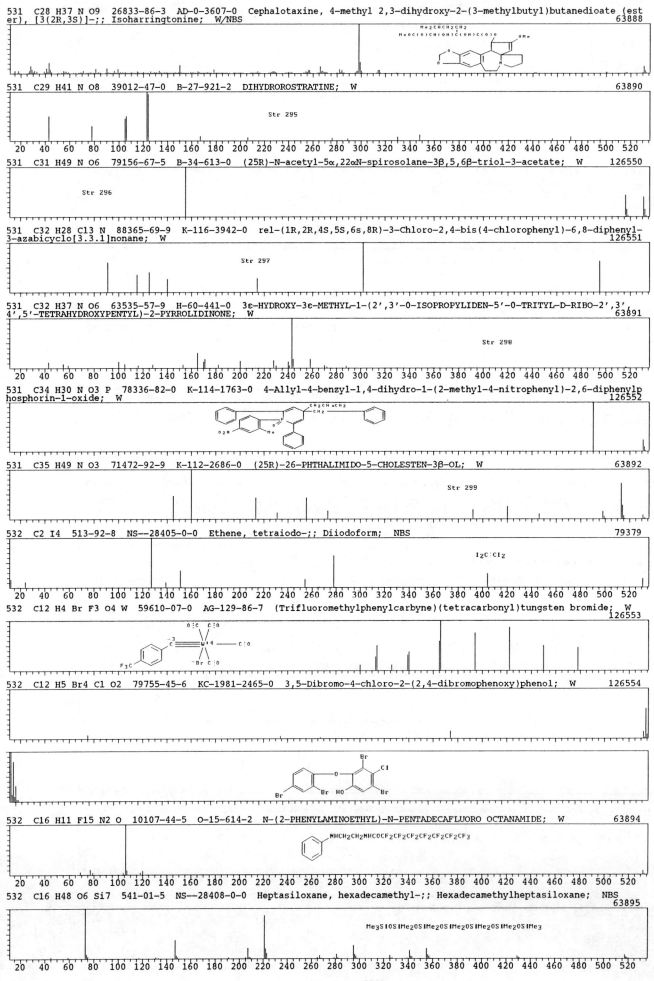

531 C28 H37 N O9 26833-86-3 AD-0-3607-0 Cephalotaxine, 4-methyl 2,3-dihydroxy-2-(3-methylbutyl)butanedioate (ester), [3(2R,3S)]-;; Isoharringtonine; W/NBS   63888

531 C29 H41 N O8 39012-47-0 B-27-921-2 DIHYDROROSTRATINE; W   63890

Str 295

531 C31 H49 N O6 79156-67-5 B-34-613-0 (25R)-N-acetyl-5α,22αN-spirosolane-3β,5,6β-triol-3-acetate; W   126550

Str 296

531 C32 H28 Cl3 N 88365-69-9 K-116-3942-0 rel-(1R,2R,4S,5S,6s,8R)-3-Chloro-2,4-bis(4-chlorophenyl)-6,8-diphenyl-3-azabicyclo[3.3.1]nonane; W   126551

Str 297

531 C32 H37 N O6 63535-57-9 H-60-441-0 3ε-HYDROXY-3ε-METHYL-1-(2',3'-0-ISOPROPYLIDEN-5'-0-TRITYL-D-RIBO-2',3',4',5'-TETRAHYDROXYPENTYL)-2-PYRROLIDINONE; W   63891

Str 298

531 C34 H30 N O3 P 78336-82-0 K-114-1763-0 4-Allyl-4-benzyl-1,4-dihydro-1-(2-methyl-4-nitrophenyl)-2,6-diphenylphosphorin-1-oxide; W   126552

531 C35 H49 N O3 71472-92-9 K-112-2686-0 (25R)-26-PHTHALIMIDO-5-CHOLESTEN-3β-OL; W   63892

Str 299

532 C2 I4 513-92-8 NS--28405-0-0 Ethene, tetraiodo-;; Diiodoform; NBS   79379

I₂C:CI₂

532 C12 H4 Br F3 O4 W 59610-07-0 AG-129-86-7 (Trifluoromethylphenylcarbyne)(tetracarbonyl)tungsten bromide; W   126553

532 C12 H5 Br4 Cl O2 79755-45-6 KC-1981-2465-0 3,5-Dibromo-4-chloro-2-(2,4-dibromophenoxy)phenol; W   126554

532 C16 H11 F15 N2 O 10107-44-5 O-15-614-2 N-(2-PHENYLAMINOETHYL)-N-PENTADECAFLUORO OCTANAMIDE; W   63894

532 C16 H48 O6 Si7 541-01-5 NS--28408-0-0 Heptasiloxane, hexadecamethyl-;; Hexadecamethylheptasiloxane; NBS   63895

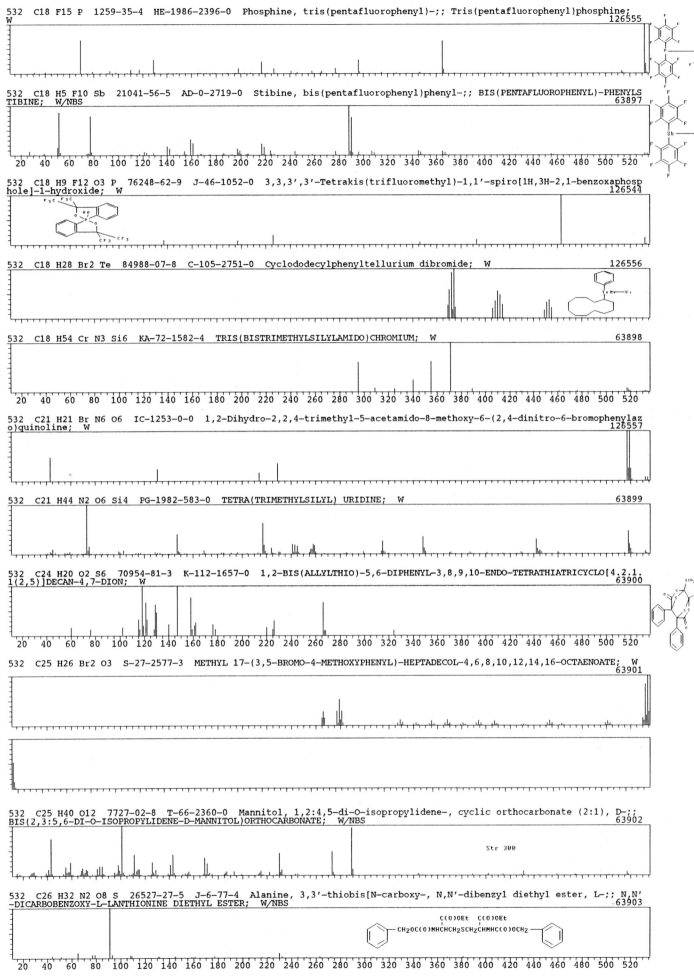

532   C18 F15 P   1259-35-4   HE-1986-2396-0   Phosphine, tris(pentafluorophenyl)-;; Tris(pentafluorophenyl)phosphine;
W
126555

532   C18 H5 F10 Sb   21041-56-5   AD-0-2719-0   Stibine, bis(pentafluorophenyl)phenyl-;; BIS(PENTAFLUOROPHENYL)-PHENYLS
TIBINE;   W/NBS
63897

532   C18 H9 F12 O3 P   76248-62-9   J-46-1052-0   3,3,3',3'-Tetrakis(trifluoromethyl)-1,1'-spiro[1H,3H-2,1-benzoxaphosp
hole]-1-hydroxide;   W
126544

532   C18 H28 Br2 Te   84988-07-8   C-105-2751-0   Cyclododecylphenyltellurium dibromide;   W
126556

532   C18 H54 Cr N3 Si6   KA-72-1582-4   TRIS(BISTRIMETHYLSILYLAMIDO)CHROMIUM;   W
63898

532   C21 H21 Br N6 O6   IC-1253-0-0   1,2-Dihydro-2,2,4-trimethyl-5-acetamido-8-methoxy-6-(2,4-dinitro-6-bromophenylaz
o)quinoline;   W
126557

532   C21 H44 N2 O6 Si4   PG-1982-583-0   TETRA(TRIMETHYLSILYL) URIDINE;   W
63899

532   C24 H20 O2 S6   70954-81-3   K-112-1657-0   1,2-BIS(ALLYLTHIO)-5,6-DIPHENYL-3,8,9,10-ENDO-TETRATHIATRICYCLO[4.2.1.
1(2,5)]DECAN-4,7-DION;   W
63900

532   C25 H26 Br2 O3   S-27-2577-3   METHYL 17-(3,5-BROMO-4-METHOXYPHENYL)-HEPTADECOL-4,6,8,10,12,14,16-OCTAENOATE;   W
63901

532   C25 H40 O12   7727-02-8   T-66-2360-0   Mannitol, 1,2:4,5-di-O-isopropylidene-, cyclic orthocarbonate (2:1), D-;;
BIS(2,3:5,6-DI-O-ISOPROPYLIDENE-D-MANNITOL)ORTHOCARBONATE;   W/NBS
63902

Str 300

532   C26 H32 N2 O8 S   26527-27-5   J-6-77-4   Alanine, 3,3'-thiobis[N-carboxy-, N,N'-dibenzyl diethyl ester, L-;; N,N'
-DICARBOBENZOXY-L-LANTHIONINE DIETHYL ESTER;   W/NBS
63903

532   C27 H29 Cl O7 S   78259-44-6   KC-1981-676-0   3α-Chloro-13-toluene-p-sulphonate;   W   126558

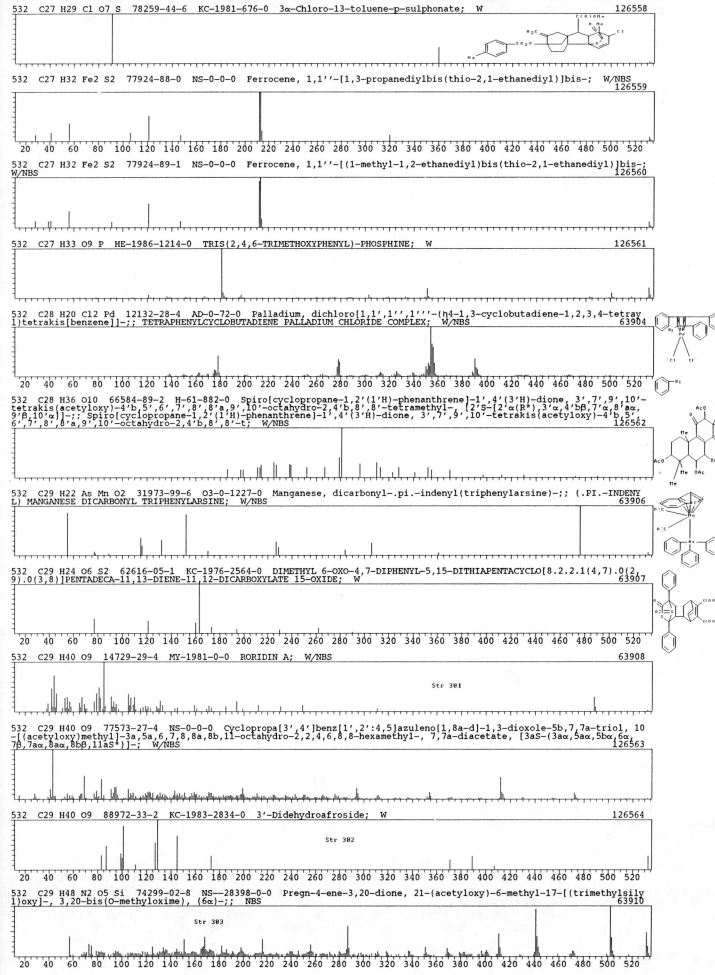

532   C27 H32 Fe2 S2   77924-88-0   NS-0-0-0   Ferrocene, 1,1''-[1,3-propanediylbis(thio-2,1-ethanediyl)]bis-;   W/NBS
                                                                                                                      126559

532   C27 H32 Fe2 S2   77924-89-1   NS-0-0-0   Ferrocene, 1,1''-[(1-methyl-1,2-ethanediyl)bis(thio-2,1-ethanediyl)]bis-;
W/NBS   126560

532   C27 H33 O9 P   HE-1986-1214-0   TRIS(2,4,6-TRIMETHOXYPHENYL)-PHOSPHINE;   W   126561

532   C28 H20 Cl2 Pd   12132-28-4   AD-0-72-0   Palladium, dichloro[1,1',1'',1'''-(η4-1,3-cyclobutadiene-1,2,3,4-tetray
l)tetrakis[benzene]]-;; TETRAPHENYLCYCLOBUTADIENE PALLADIUM CHLORIDE COMPLEX;   W/NBS   63904

532   C28 H36 O10   66584-89-2   H-61-882-0   Spiro[cyclopropane-1,2'(1'H)-phenanthrene]-1',4'(3'H)-dione, 3',7',9',10'-
tetrakis(acetyloxy)-4'b,5',6',7',8',8'a,9',10'-octahydro-2,4'b,8',8'-tetramethyl-, [2'S-[2'α(R*),3'α,4'bβ,7'α,8'aα,
9'β,10'α]]-;; Spiro[cyclopropane-1,2'(1'H)-phenanthrene]-1',4'(3'H)-dione, 3',7',9',10'-tetrakis(acetyloxy)-4'b,5',
6',7',8',8'a,9',10'-octahydro-2,4'b,8',8'-t;   W/NBS   126562

532   C29 H22 As Mn O2   31973-99-6   O3-0-1227-0   Manganese, dicarbonyl-.pi.-indenyl(triphenylarsine)-;; (.PI.-INDENY
L) MANGANESE DICARBONYL TRIPHENYLARSINE;   W/NBS   63906

532   C29 H24 O6 S2   62616-05-1   KC-1976-2564-0   DIMETHYL 6-OXO-4,7-DIPHENYL-5,15-DITHIAPENTACYCLO[8.2.2.1(4,7).0(2,
9).0(3,8)]PENTADECA-11,13-DIENE-11,12-DICARBOXYLATE 15-OXIDE;   W   63907

532   C29 H40 O9   14729-29-4   MY-1981-0-0   RORIDIN A;   W/NBS   63908

Str 301

532   C29 H40 O9   77573-27-4   NS-0-0-0   Cyclopropa[3',4']benz[1',2':4,5]azuleno[1,8a-d]-1,3-dioxole-5b,7,7a-triol, 10
-[(acetyloxy)methyl]-3a,5a,6,7,8,8a,8b,11-octahydro-2,2,4,6,8,8-hexamethyl-, 7,7a-diacetate, [3aS-(3aα,5aα,5bα,6α,
7β,7aα,8aα,8bβ,11aS*)]-;   W/NBS   126563

532   C29 H40 O9   88972-33-2   KC-1983-2834-0   3'-Didehydroafroside;   W   126564

Str 302

532   C29 H48 N2 O5 Si   74299-02-8   NS--28398-0-0   Pregn-4-ene-3,20-dione, 21-(acetyloxy)-6-methyl-17-[(trimethylsily
l)oxy]-, 3,20-bis(O-methyloxime), (6α)-;;   NBS   63910

Str 303

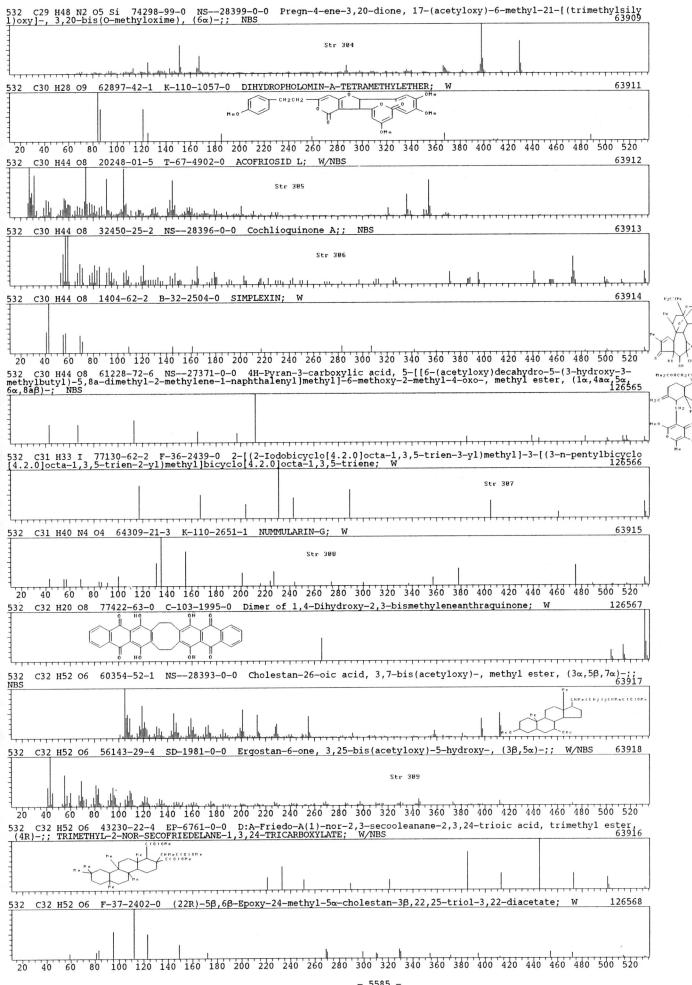

532   C29 H48 N2 O5 Si   74298-99-0   NS--28399-0-0   Pregn-4-ene-3,20-dione, 17-(acetyloxy)-6-methyl-21-[(trimethylsilyl)oxy]-, 3,20-bis(O-methyloxime), (6α)-;;   NBS
63909
Str 304

532   C30 H28 O9   62897-42-1   K-110-1057-0   DIHYDROPHOLOMIN-A-TETRAMETHYLETHER;   W
63911

532   C30 H44 O8   20248-01-5   T-67-4902-0   ACOFRIOSID L;   W/NBS
63912
Str 305

532   C30 H44 O8   32450-25-2   NS--28396-0-0   Cochlioquinone A;;   NBS
63913
Str 306

532   C30 H44 O8   1404-62-2   B-32-2504-0   SIMPLEXIN;   W
63914

532   C30 H44 O8   61228-72-6   NS--27371-0-0   4H-Pyran-3-carboxylic acid, 5-[[6-(acetyloxy)decahydro-5-(3-hydroxy-3-methylbutyl)-5,8a-dimethyl-2-methylene-1-naphthalenyl]methyl]-6-methoxy-2-methyl-4-oxo-, methyl ester, (1α,4aα,5α,6α,8aβ)-;   NBS
126565

532   C31 H33 I   77130-62-2   F-36-2439-0   2-[(2-Iodobicyclo[4.2.0]octa-1,3,5-trien-3-yl)methyl]-3-[(3-n-pentylbicyclo[4.2.0]octa-1,3,5-trien-2-yl)methyl]bicyclo[4.2.0]octa-1,3,5-triene;   W
126566
Str 307

532   C31 H40 N4 O4   64309-21-3   K-110-2651-1   NUMMULARIN-G;   W
63915
Str 308

532   C32 H20 O8   77422-63-0   C-103-1995-0   Dimer of 1,4-Dihydroxy-2,3-bismethyleneanthraquinone;   W
126567

532   C32 H52 O6   60354-52-1   NS--28393-0-0   Cholestan-26-oic acid, 3,7-bis(acetyloxy)-, methyl ester, (3α,5β,7α)-;;   NBS
63917

532   C32 H52 O6   56143-29-4   SD-1981-0-0   Ergostan-6-one, 3,25-bis(acetyloxy)-5-hydroxy-, (3β,5α)-;;   W/NBS
63918
Str 309

532   C32 H52 O6   43230-22-4   EP-6761-0-0   D:A-Friedo-A(1)-nor-2,3-secooleanane-2,3,24-trioic acid, trimethyl ester, (4R)-;;   TRIMETHYL-2-NOR-SECOFRIEDELANE-1,3,24-TRICARBOXYLATE;   W/NBS
63916

532   C32 H52 O6   F-37-2402-0   (22R)-5β,6β-Epoxy-24-methyl-5α-cholestan-3β,22,25-triol-3,22-diacetate;   W
126568

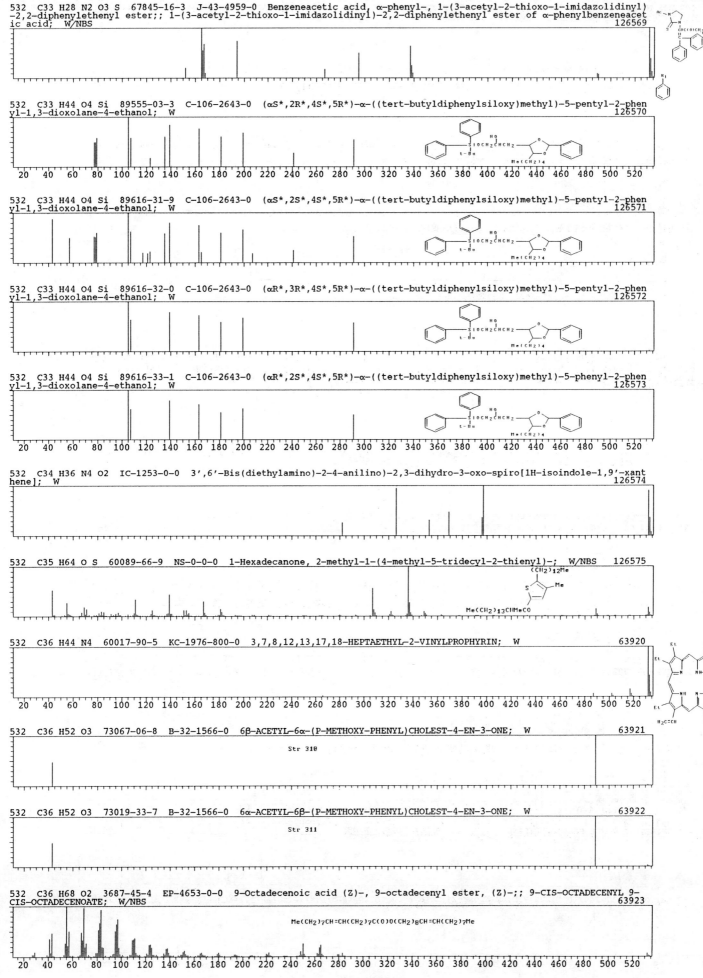

532 C33 H28 N2 O3 S 67845-16-3 J-43-4959-0 Benzeneacetic acid, α-phenyl-, 1-(3-acetyl-2-thioxo-1-imidazolidinyl)
-2,2-diphenylethenyl ester;; 1-(3-acetyl-2-thioxo-1-imidazolidinyl)-2,2-diphenylethenyl ester of α-phenylbenzeneacet
ic acid; W/NBS   126569

532 C33 H44 O4 Si 89555-03-3 C-106-2643-0 (αS*,2R*,4S*,5R*)-α-((tert-butyldiphenylsiloxy)methyl)-5-pentyl-2-phen
yl-1,3-dioxolane-4-ethanol; W   126570

532 C33 H44 O4 Si 89616-31-9 C-106-2643-0 (αS*,2S*,4S*,5R*)-α-((tert-butyldiphenylsiloxy)methyl)-5-pentyl-2-phen
yl-1,3-dioxolane-4-ethanol; W   126571

532 C33 H44 O4 Si 89616-32-0 C-106-2643-0 (αR*,3R*,4S*,5R*)-α-((tert-butyldiphenylsiloxy)methyl)-5-pentyl-2-phen
yl-1,3-dioxolane-4-ethanol; W   126572

532 C33 H44 O4 Si 89616-33-1 C-106-2643-0 (αR*,2S*,4S*,5R*)-α-((tert-butyldiphenylsiloxy)methyl)-5-phenyl-2-phen
yl-1,3-dioxolane-4-ethanol; W   126573

532 C34 H36 N4 O2 IC-1253-0-0 3',6'-Bis(diethylamino)-2-4-anilino)-2,3-dihydro-3-oxo-spiro[1H-isoindole-1,9'-xant
hene]; W   126574

532 C35 H64 O S 60089-66-9 NS-0-0-0 1-Hexadecanone, 2-methyl-1-(4-methyl-5-tridecyl-2-thienyl)-; W/NBS 126575

532 C36 H44 N4 60017-90-5 KC-1976-800-0 3,7,8,12,13,17,18-HEPTAETHYL-2-VINYLPROPHYRIN; W   63920

532 C36 H52 O3 73067-06-8 B-32-1566-0 6β-ACETYL-6α-(P-METHOXY-PHENYL)CHOLEST-4-EN-3-ONE; W   63921

Str 310

532 C36 H52 O3 73019-33-7 B-32-1566-0 6α-ACETYL-6β-(P-METHOXY-PHENYL)CHOLEST-4-EN-3-ONE; W   63922

Str 311

532 C36 H68 O2 3687-45-4 EP-4653-0-0 9-Octadecenoic acid (Z)-, 9-octadecenyl ester, (Z)-;; 9-CIS-OCTADECENYL 9-
CIS-OCTADECENOATE; W/NBS   63923

Me(CH2)7CH=CH(CH2)7C(O)O(CH2)8CH=CH(CH2)7Me

532   C36 H68 S   59782-73-9   NS-0-0-0   Thiophene, 3-methyl-5-octadecyl-2-tridecyl-;   W/NBS   126576

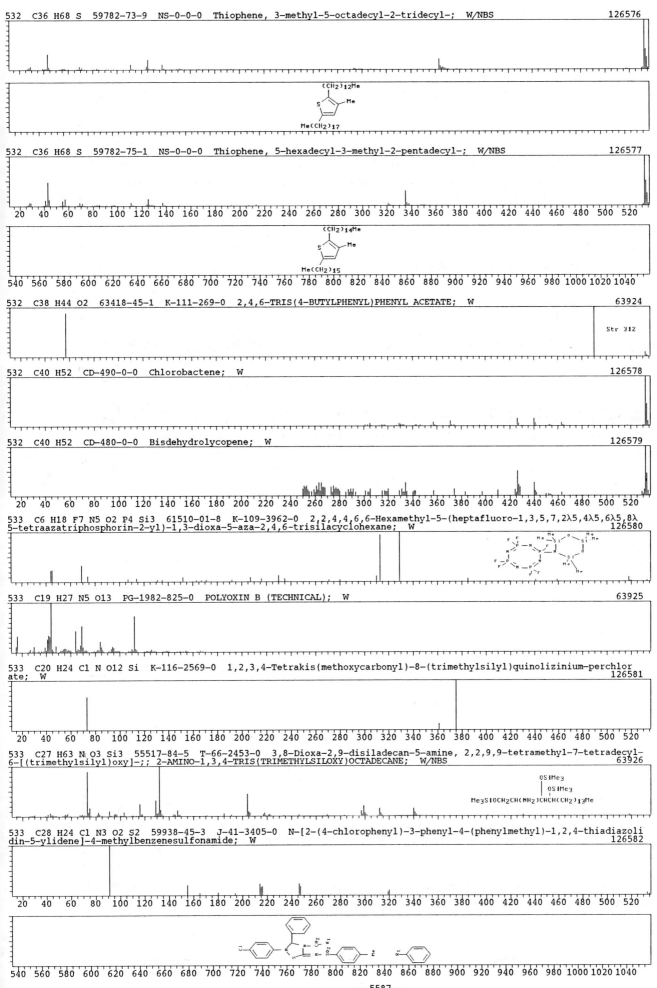

532   C36 H68 S   59782-75-1   NS-0-0-0   Thiophene, 5-hexadecyl-3-methyl-2-pentadecyl-;   W/NBS   126577

532   C38 H44 O2   63418-45-1   K-111-269-0   2,4,6-TRIS(4-BUTYLPHENYL)PHENYL ACETATE;   W   63924

Str 312

532   C40 H52   CD-490-0-0   Chlorobactene;   W   126578

532   C40 H52   CD-480-0-0   Bisdehydrolycopene;   W   126579

533   C6 H18 F7 N5 O2 P4 Si3   61510-01-8   K-109-3962-0   2,2,4,4,6,6-Hexamethyl-5-(heptafluoro-1,3,5,7,2λ5,4λ5,6λ5,8λ
5-tetraazatriphosphorin-2-yl)-1,3-dioxa-5-aza-2,4,6-trisilacyclohexane;   W   126580

533   C19 H27 N5 O13   PG-1982-825-0   POLYOXIN B (TECHNICAL);   W   63925

533   C20 H24 Cl N O12 Si   K-116-2569-0   1,2,3,4-Tetrakis(methoxycarbonyl)-8-(trimethylsilyl)quinolizinium-perchlor
ate;   W   126581

533   C27 H63 N O3 Si3   55517-84-5   T-66-2453-0   3,8-Dioxa-2,9-disiladecan-5-amine, 2,2,9,9-tetramethyl-7-tetradecyl-
6-[(trimethylsilyl)oxy]-;; 2-AMINO-1,3,4-TRIS(TRIMETHYLSILOXY)OCTADECANE;   W/NBS   63926

533   C28 H24 Cl N3 O2 S2   59938-45-3   J-41-3405-0   N-[2-(4-chlorophenyl)-3-phenyl-4-(phenylmethyl)-1,2,4-thiadiazoli
din-5-ylidene]-4-methylbenzenesulfonamide;   W   126582

533 C29 H27 N O9 F-41-2888-0 2-Benzoyloxy-2-(2-[β-(N-ethoxycarbonyl-N-methyl)aminoethyl]-4,5-methylenedioxphen
yl)-1-(3,4-methylenedioxyphenyl)-ethanone; W
126583

533 C30 H31 N O8 35398-13-6 H-65-2264-0 1-O-Acetyl-2-desoxy-3,4-O-isopropyliden-2-nitro-6-O-trityl-β-D-psicofura
nose; W
126584

533 C30 H31 N O8 81601-80-1 KC-1982-524-0 (+-)-5,6,7,8-Tetrahydro-1,2,3-trimethoxy-10,11-methylene-dioxy-8-oxodi
benzo[a,c]; W
126585

Str 313

533 C30 H55 N O3 Si2 DO-0-343-32 O-IPO-BISTRIMETHYLSILYL; W
63927

533 C31 H35 N O7 64018-66-2 H-60-1293-0 β-Alanine, N-hydroxy-N-[2,3-O-(1-methylethylidene)-5-O-(triphenylmethyl)
-β-D-ribofuranosyl]-, methyl ester;; N-HYDROXY-N-(2,3-O-ISOPROPYLIDEN-5-O-TRITYL-β-D-RIBOFURANOSYL)-3-AMINOPROPIONIC
ACID METHYL ESTER; W/NBS
63928

Str 314

533 C32 H27 N3 O3 S 74075-40-4 KO-8-16-3 D-6-[4-Phenyl-5-benzoylthiazol-2-yl]-8-methoxycarbonylergoline-1; W
126586

533 C33 H47 N O3 Si 91384-97-3 J-49-3843-0 17β-Hydroxy-3α-phthalimido-4-androstene tert-butyldimethylsilyl ether
; W
126587

Str 315

533 C33 H47 N O3 Si 91384-98-4 J-49-3843-0 17β-Hydroxy-3β-phthalimido-4-androstene tert-butyldimethylsilylether;
W
126588

Str 316

533 C34 H47 N O4 BS-5-169-0 O-(p-Nitrobenzoyl)-vitamin D[3]; W
126589

533 C34 H47 N O4 BS-5-170-0 O-(p-Nitrobenzoyl)-epi-vitamin D[3]; W
126590

533 C34 H47 N O4 BS-5-171-0 O-(p-Nitrobenzoyl)-trans-vitamin D[3]; W
126591

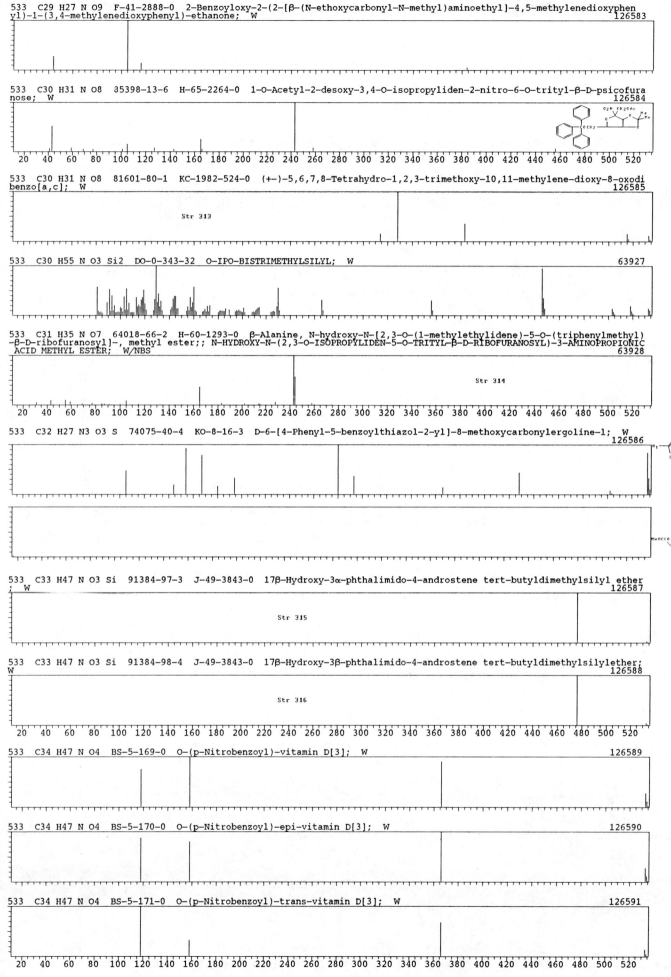

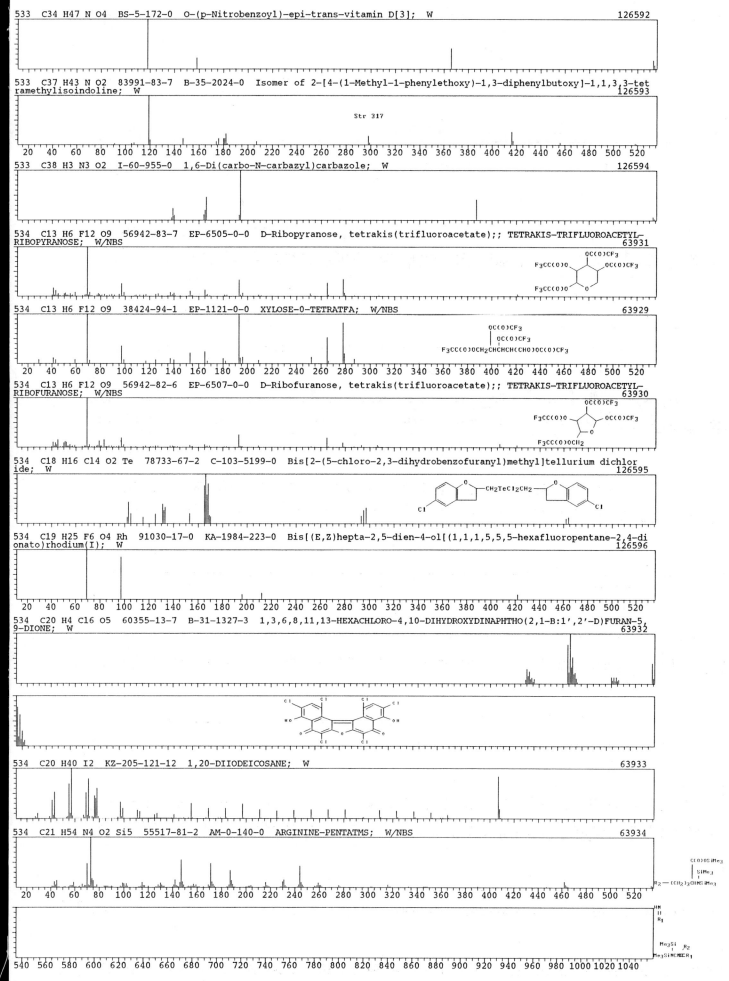

533   C34 H47 N O4   BS-5-172-0   O-(p-Nitrobenzoyl)-epi-trans-vitamin D[3];   W   126592

533   C37 H43 N O2   83991-83-7   B-35-2024-0   Isomer of 2-[4-(1-Methyl-1-phenylethoxy)-1,3-diphenylbutoxy]-1,1,3,3-tet
ramethylisoindoline;   W   126593

Str 317

533   C38 H3 N3 O2   I-60-955-0   1,6-Di(carbo-N-carbazyl)carbazole;   W   126594

534   C13 H6 F12 O9   56942-83-7   EP-6505-0-0   D-Ribopyranose, tetrakis(trifluoroacetate);;   TETRAKIS-TRIFLUOROACETYL-
RIBOPYRANOSE;   W/NBS   63931

534   C13 H6 F12 O9   38424-94-1   EP-1121-0-0   XYLOSE-0-TETRATFA;   W/NBS   63929

534   C13 H6 F12 O9   56942-82-6   EP-6507-0-0   D-Ribofuranose, tetrakis(trifluoroacetate);;   TETRAKIS-TRIFLUOROACETYL-
RIBOFURANOSE;   W/NBS   63930

534   C18 H16 Cl4 O2 Te   78733-67-2   C-103-5199-0   Bis[2-(5-chloro-2,3-dihydrobenzofuranyl)methyl]tellurium dichlor
ide;   W   126595

534   C19 H25 F6 O4 Rh   91030-17-0   KA-1984-223-0   Bis[(E,Z)hepta-2,5-dien-4-ol[(1,1,1,5,5,5-hexafluoropentane-2,4-di
onato)rhodium(I)];   W   126596

534   C20 H4 Cl6 O5   60355-13-7   B-31-1327-3   1,3,6,8,11,13-HEXACHLORO-4,10-DIHYDROXYDINAPHTHO(2,1-B:1',2'-D)FURAN-5,
9-DIONE;   W   63932

534   C20 H40 I2   KZ-205-121-12   1,20-DIIODEICOSANE;   W   63933

534   C21 H54 N4 O2 Si5   55517-81-2   AM-0-140-0   ARGININE-PENTATMS;   W/NBS   63934

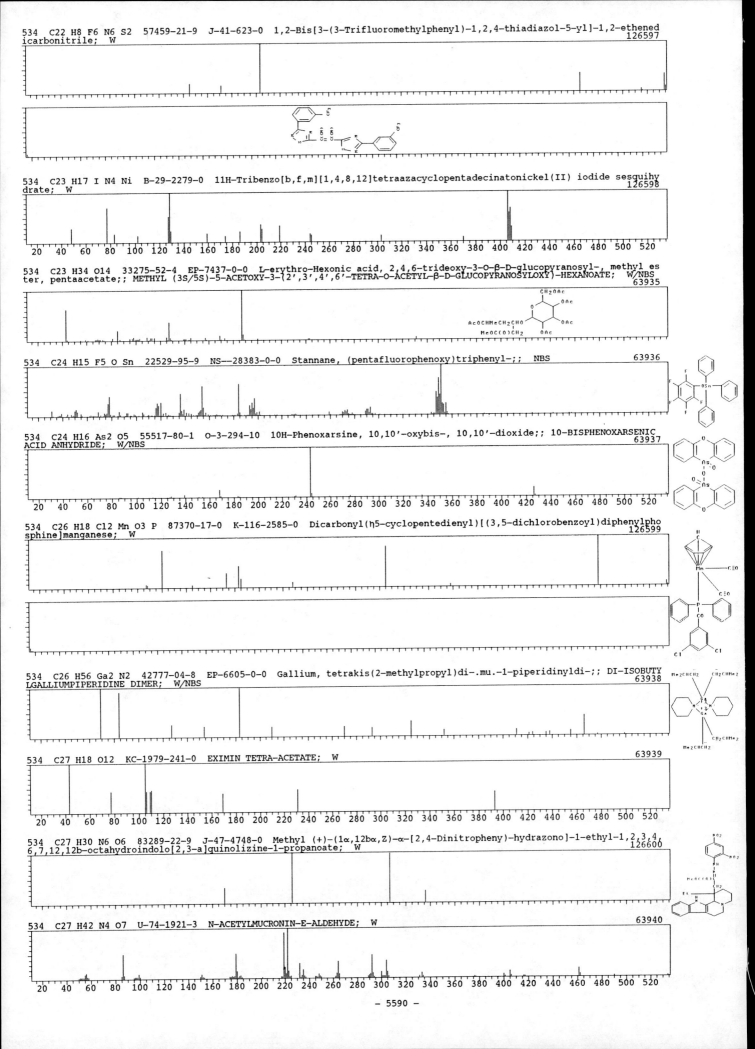

534 C28 H38 O10 51906-02-6 NS-0-0-0 4a,7a-Epoxy-5H-cyclopenta[a]cyclopropa[f]cycloundecen-4(1H)-one, 2,7,10,11-tetrakis(acetyloxy)-1a,2,3,6,7,10,11,11a-octahydro-1,1,3,6,9-pentamethyl-, [1aR-(1aR*,2R*,3S*,4aR*,6S*,7S*,7aS*,8E,10R*,11R*,11aS*)]-; W/NBS
126601

Str 318

534 C28 H38 O10 77698-37-4 NS-0-0-0 1H-Cyclopropa[3,4]benz[1,2-e]azulene-4a,5,7b,9,9a(1aH)-pentol, 3-[(acetyloxy)methyl]-1b,4,5,7a,8,9-hexahydro-1,1,6,8-tetramethyl-, 5,9,9a-triacetate, [1aR-(1aα,1bβ,4aβ,5β,7aα,7bα,8α,9β,9aα)]-; W/NBS
126602

Str 319

534 C28 H38 O10 85170-94-1 H-65-2177-0 (2'S)-6β,19-Dihydroxy-12-O-acetyl-7α-methoxy-13-(2'methoxypropyl)-13-desisopropylroyleanone; W
126603

Str 320

534 C28 H50 O4 Si3 51497-48-4 NS--28380-0-0 Silane, [[(16α,17β)-2-methoxyestra-1,3,5(10)-triene-3,16,17-triyl]tris(oxy)]tris[trimethyl-;; NBS
63941

Str 321

534 C28 H50 O4 Si3 77882-88-3 NS-0-0-0 Silane, [[(17β)-4-methoxyestra-1,3,5(10)-triene-2,3,17-triyl]tris(oxy)]tris[trimethyl-; W/NBS
126604

Str 322

534 C28 H50 O4 Si3 77882-89-4 NS-0-0-0 Silane, [[(17β)-3-methoxyestra-1,3,5(10)-triene-2,4-17-triyl]tris(oxy)]tris[trimethyl-; W/NBS
126605

Str 323

534 C28 H50 O4 Si3 77882-92-9 NS-0-0-0 Silane, [[(17β)-2-methoxyestra-1,3,5(10)-triene-3,4,17-triyl]tris(oxy)]tris[trimethyl-; W/NBS
126606

Str 324

534 C28 H50 O4 Si3 77883-30-8 NS-0-0-0 Silane, [[(6α,11β,17β)-3-methoxyestra-1,3,5(10)-triene-6,11,17-triyl]tris(oxy)]tris[trimethyl-; W/NBS
126607

Str 325

534 C29 H26 O10 35082-49-6 NS--28379-0-0 Perylo[1,12-def]-1,3-dioxepin-5,11-dione, 6,12-dihydroxy-8,9-bis(2-hydroxypropyl)-7,10-dimethoxy-, stereoisomer;; NBS
63942

Str 326

534 C29 H26 O10 62350-99-6 K-110-1056-0 FASCICULIN-B-PENTAMETHYLETHER; W
63943

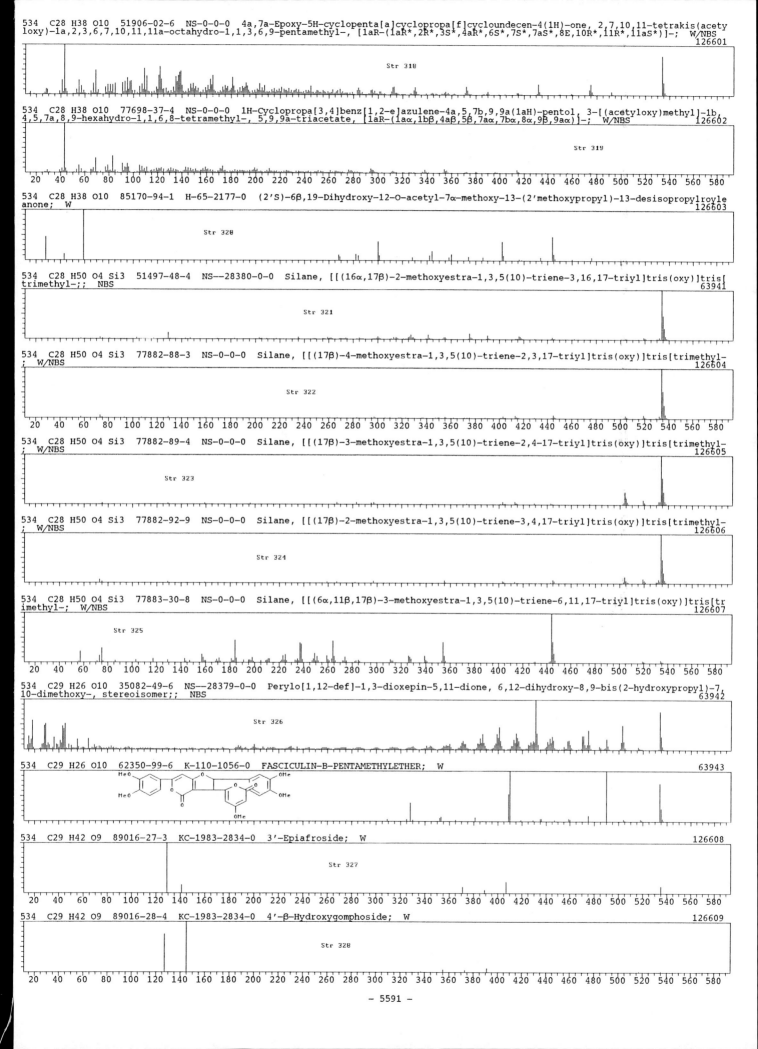

534 C29 H42 O9 89016-27-3 KC-1983-2834-0 3'-Epiafroside; W
126608

Str 327

534 C29 H42 O9 89016-28-4 KC-1983-2834-0 4'-β-Hydroxygomphoside; W
126609

Str 328

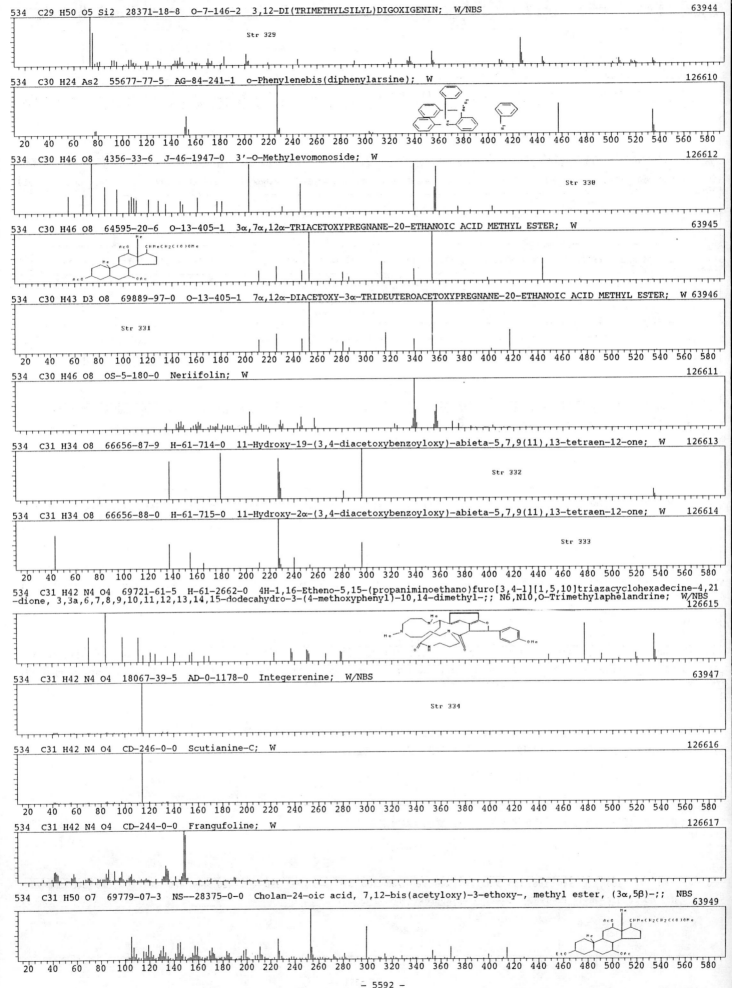

534 C29 H50 O5 Si2 28371-18-8 O-7-146-2 3,12-DI(TRIMETHYLSILYL)DIGOXIGENIN; W/NBS   63944

Str 329

534 C30 H24 As2 55677-77-5 AG-84-241-1 o-Phenylenebis(diphenylarsine); W   126610

534 C30 H46 O8 4356-33-6 J-46-1947-0 3'-O-Methylevomonoside; W   126612

Str 330

534 C30 H46 O8 64595-20-6 O-13-405-1 3α,7α,12α-TRIACETOXYPREGNANE-20-ETHANOIC ACID METHYL ESTER; W   63945

534 C30 H43 D3 O8 69889-97-0 O-13-405-1 7α,12α-DIACETOXY-3α-TRIDEUTEROACETOXYPREGNANE-20-ETHANOIC ACID METHYL ESTER; W 63946

Str 331

534 C30 H46 O8 OS-5-180-0 Neriifolin; W   126611

534 C31 H34 O8 66656-87-9 H-61-714-0 11-Hydroxy-19-(3,4-diacetoxybenzoyloxy)-abieta-5,7,9(11),13-tetraen-12-one; W   126613

Str 332

534 C31 H34 O8 66656-88-0 H-61-715-0 11-Hydroxy-2α-(3,4-diacetoxybenzoyloxy)-abieta-5,7,9(11),13-tetraen-12-one; W   126614

Str 333

534 C31 H42 N4 O4 69721-61-5 H-61-2662-0 4H-1,16-Etheno-5,15-(propaniminoethano)furo[3,4-1][1,5,10]triazacyclohexadecine-4,21
-dione, 3,3a,6,7,8,9,10,11,12,13,14,15-dodecahydro-3-(4-methoxyphenyl)-10,14-dimethyl-;; N6,N10,O-Trimethylaphelandrine; W/NBS
                                                                                                       126615

534 C31 H42 N4 O4 18067-39-5 AD-0-1178-0 Integerrenine; W/NBS   63947

Str 334

534 C31 H42 N4 O4 CD-246-0-0 Scutianine-C; W   126616

534 C31 H42 N4 O4 CD-244-0-0 Frangufoline; W   126617

534 C31 H50 O7 69779-07-3 NS--28375-0-0 Cholan-24-oic acid, 7,12-bis(acetyloxy)-3-ethoxy-, methyl ester, (3α,5β)-;; NBS
                                                                                                        63949

534   C31 H50 O7   55106-11-1   J-45-339-0   Methyl ester of (3α,5β,7α,12α)-12-hydroxy-3,7-bis(1-oxopropoxy)cholan-24-oic acid;   W
126618

534   C32 H23 I   89265-22-5   SB-37-593-0   Iodo[2(4)]Paracyclphanetetraene;   W
126619

Str 335

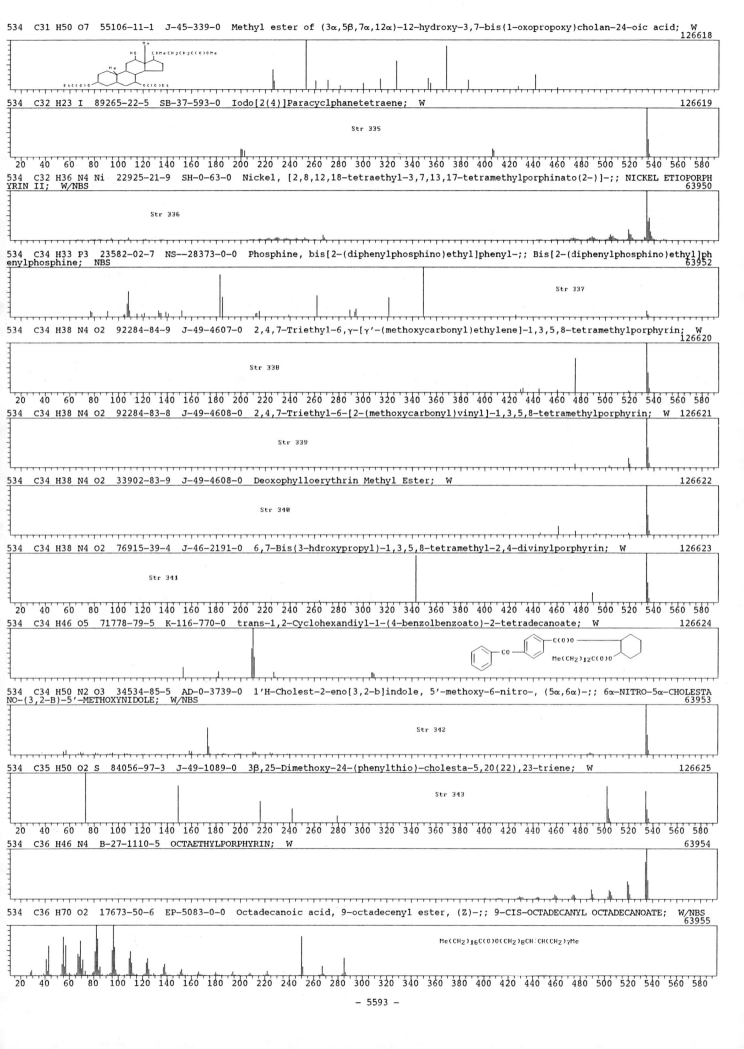

534   C32 H36 N4 Ni   22925-21-9   SH-0-63-0   Nickel, [2,8,12,18-tetraethyl-3,7,13,17-tetramethylporphinato(2-)]-;; NICKEL ETIOPORPH
YRIN II;   W/NBS
63950

Str 336

534   C34 H33 P3   23582-02-7   NS--28373-0-0   Phosphine, bis[2-(diphenylphosphino)ethyl]phenyl-;; Bis[2-(diphenylphosphino)ethyl]ph
enylphosphine;   NBS
63952

Str 337

534   C34 H38 N4 O2   92284-84-9   J-49-4607-0   2,4,7-Triethyl-6,γ-[γ'-(methoxycarbonyl)ethylene]-1,3,5,8-tetramethylporphyrin;   W
126620

Str 338

534   C34 H38 N4 O2   92284-83-8   J-49-4608-0   2,4,7-Triethyl-6-[2-(methoxycarbonyl)vinyl]-1,3,5,8-tetramethylporphyrin;   W   126621

Str 339

534   C34 H38 N4 O2   33902-83-9   J-49-4608-0   Deoxophylloerythrin Methyl Ester;   W
126622

Str 340

534   C34 H38 N4 O2   76915-39-4   J-46-2191-0   6,7-Bis(3-hdroxypropyl)-1,3,5,8-tetramethyl-2,4-divinylporphyrin;   W
126623

Str 341

534   C34 H46 O5   71778-79-5   K-116-770-0   trans-1,2-Cyclohexandiyl-1-(4-benzolbenzoato)-2-tetradecanoate;   W
126624

534   C34 H50 N2 O3   34534-85-5   AD-0-3739-0   1'H-Cholest-2-eno[3,2-b]indole, 5'-methoxy-6-nitro-, (5α,6α)-;; 6α-NITRO-5α-CHOLESTA
NO-(3,2-B)-5'-METHOXYNIDOLE;   W/NBS
63953

Str 342

534   C35 H50 O2 S   84056-97-3   J-49-1089-0   3β,25-Dimethoxy-24-(phenylthio)-cholesta-5,20(22),23-triene;   W
126625

Str 343

534   C36 H46 N4   B-27-1110-5   OCTAETHYLPORPHYRIN;   W
63954

534   C36 H70 O2   17673-50-6   EP-5083-0-0   Octadecanoic acid, 9-octadecenyl ester, (Z)-;; 9-CIS-OCTADECANYL OCTADECANOATE;   W/NBS
63955

Me(CH2)16C(O)O(CH2)8CH:CH(CH2)7Me

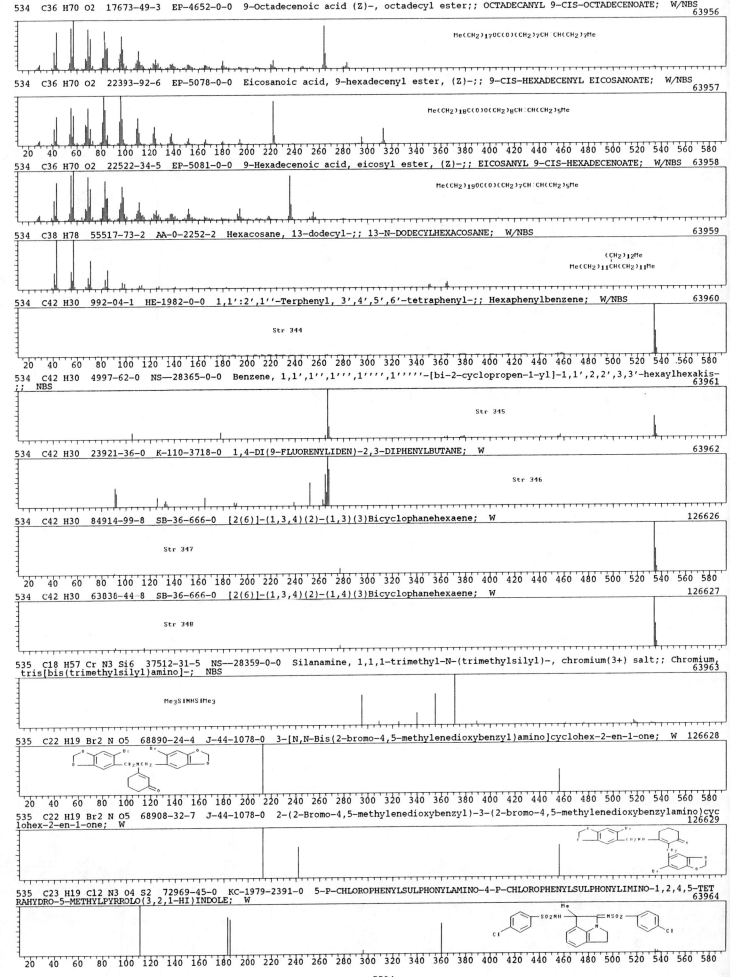

534　C36 H70 O2　17673-49-3　EP-4652-0-0　9-Octadecenoic acid (Z)-, octadecyl ester;; OCTADECANYL 9-CIS-OCTADECENOATE;　W/NBS
63956

Me(CH2)17OC(O)(CH2)7CH:CH(CH2)7Me

534　C36 H70 O2　22393-92-6　EP-5078-0-0　Eicosanoic acid, 9-hexadecenyl ester, (Z)-;; 9-CIS-HEXADECENYL EICOSANOATE;　W/NBS
63957

Me(CH2)18C(O)O(CH2)8CH:CH(CH2)5Me

534　C36 H70 O2　22522-34-5　EP-5081-0-0　9-Hexadecenoic acid, eicosyl ester, (Z)-;; EICOSANYL 9-CIS-HEXADECENOATE;　W/NBS　63958

Me(CH2)19OC(O)(CH2)7CH:CH(CH2)5Me

534　C38 H78　55517-73-2　AA-0-2252-2　Hexacosane, 13-dodecyl-;; 13-N-DODECYLHEXACOSANE;　W/NBS
63959

(CH2)12Me
Me(CH2)11CH(CH2)11Me

534　C42 H30　992-04-1　HE-1982-0-0　1,1':2',1''-Terphenyl, 3',4',5',6'-tetraphenyl-;; Hexaphenylbenzene;　W/NBS
63960

Str 344

534　C42 H30　4997-62-0　NS--28365-0-0　Benzene, 1,1',1'',1''',1'''',1'''''-[bi-2-cyclopropen-1-yl]-1,1',2,2',3,3'-hexaylhexakis-
;;　NBS
63961

Str 345

534　C42 H30　23921-36-0　K-110-3718-0　1,4-DI(9-FLUORENYLIDEN)-2,3-DIPHENYLBUTANE;　W
63962

Str 346

534　C42 H30　84914-99-8　SB-36-666-0　[2(6)]-(1,3,4)(2)-(1,3)(3)Bicyclophanehexaene;　W
126626

Str 347

534　C42 H30　63838-44-8　SB-36-666-0　[2(6)]-(1,3,4)(2)-(1,4)(3)Bicyclophanehexaene;　W
126627

Str 348

535　C18 H57 Cr N3 Si6　37512-31-5　NS--28359-0-0　Silanamine, 1,1,1-trimethyl-N-(trimethylsilyl)-, chromium(3+) salt;; Chromium,
tris[bis(trimethylsilyl)amino]-;　NBS
63963

Me3SiNHSiMe3

535　C22 H19 Br2 N O5　68890-24-4　J-44-1078-0　3-[N,N-Bis(2-bromo-4,5-methylenedioxybenzyl)amino]cyclohex-2-en-1-one;　W　126628

535　C22 H19 Br2 N O5　68908-32-7　J-44-1078-0　2-(2-Bromo-4,5-methylenedioxybenzyl)-3-(2-bromo-4,5-methylenedioxybenzylamino)cyc
lohex-2-en-1-one;　W
126629

535　C23 H19 Cl2 N3 O4 S2　72969-45-0　KC-1979-2391-0　5-P-CHLOROPHENYLSULPHONYLAMINO-4-P-CHLOROPHENYLSULPHONYLIMINO-1,2,4,5-TET
RAHYDRO-5-METHYLPYRROLO(3,2,1-HI)INDOLE;　W
63964

535 C23 H53 N O5 Si4 72021-24-0 HE-1982-0-0 Piperidine, 1-[2,3,4,6-tetrakis-O-(trimethylsilyl)-β-D-glucopyranosyl]-;; N-(β-D-GLUCOPYRANOSYL)-PIPERIDIN, TETRAKIS-O-(TRIMETHYLSILYL)-; W/NBS
63965

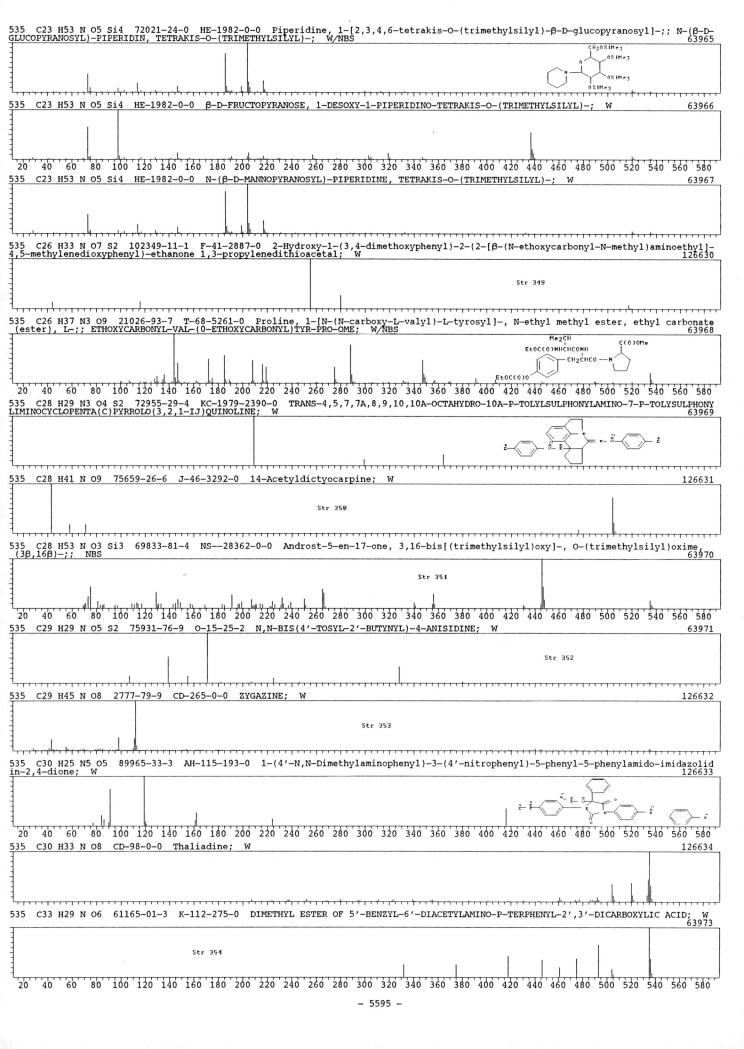

535 C23 H53 N O5 Si4 HE-1982-0-0 β-D-FRUCTOPYRANOSE, 1-DESOXY-1-PIPERIDINO-TETRAKIS-O-(TRIMETHYLSILYL)-; W
63966

535 C23 H53 N O5 Si4 HE-1982-0-0 N-(β-D-MANNOPYRANOSYL)-PIPERIDINE, TETRAKIS-O-(TRIMETHYLSILYL)-; W
63967

535 C26 H33 N O7 S2 102349-11-1 F-41-2887-0 2-Hydroxy-1-(3,4-dimethoxyphenyl)-2-(2-[β-(N-ethoxycarbonyl-N-methyl)aminoethyl]-4,5-methylenedioxyphenyl)-ethanone 1,3-propylenedithioacetal; W
126630
Str 349

535 C26 H37 N3 O9 21026-93-7 T-68-5261-0 Proline, 1-[N-(N-carboxy-L-valyl)-L-tyrosyl]-, N-ethyl methyl ester, ethyl carbonate (ester), L-;; ETHOXYCARBONYL-VAL-(0-ETHOXYCARBONYL)TYR-PRO-OME; W/NBS
63968

535 C28 H29 N3 O4 S2 72955-29-4 KC-1979-2390-0 TRANS-4,5,7,7A,8,9,10,10A-OCTAHYDRO-10A-P-TOLYLSULPHONYLAMINO-7-P-TOLYSULPHONYLIMINOCYCLOPENTA(C)PYRROLO(3,2,1-IJ)QUINOLINE; W
63969

535 C28 H41 N O9 75659-26-6 J-46-3292-0 14-Acetyldictyocarpine; W
126631
Str 350

535 C28 H53 N O3 Si3 69833-81-4 NS--28362-0-0 Androst-5-en-17-one, 3,16-bis[(trimethylsilyl)oxy]-, O-(trimethylsilyl)oxime, (3β,16β)-;; NBS
63970
Str 351

535 C29 H29 N O5 S2 75931-76-9 O-15-25-2 N,N-BIS(4'-TOSYL-2'-BUTYNYL)-4-ANISIDINE; W
63971
Str 352

535 C29 H45 N O8 2777-79-9 CD-265-0-0 ZYGAZINE; W
126632
Str 353

535 C30 H25 N5 O5 89965-33-3 AH-115-193-0 1-(4'-N,N-Dimethylaminophenyl)-3-(4'-nitrophenyl)-5-phenyl-5-phenylamido-imidazolidin-2,4-dione; W
126633

535 C30 H33 N O8 CD-98-0-0 Thaliadine; W
126634

535 C33 H29 N O6 61165-01-3 K-112-275-0 DIMETHYL ESTER OF 5'-BENZYL-6'-DIACETYLAMINO-P-TERPHENYL-2',3'-DICARBOXYLIC ACID; W
63973
Str 354

535   C37 H29 N O3   74764-45-7   HE-1982-0-0   2-Azetidinone, 1-(1,3-dioxo-2,2-diphenylbutyl)-3,3,4-triphenyl-;; AZETIDIN, 2-OXO-3,
3,4-TRIPHENYL-N-ACETYL(DIPHENYLMETHYL)CARBONYL-;   W/NBS   63974

Str 355

536   C7 Br5 F3   5360-57-6   B-30-302-0   1,2,3,4,5-PENTABROMO-6-(TRIFLUOROMETHYL)BENZENE;   W   63975

536   C10 H10 F8 I2   74779-66-1   HE-1982-0-0   1-Decene, 7,7,8,8,9,9,10,10-octafluoro-5,10-diiodo-;; 9-DECENE, 1,1,2,2,3,3,4,4-OCTA
FLUORO-1,6-DIIODO-;   W/NBS   63976

H2C:CHCH2CH2CHICH2CF2CF2CF2CF2I

536   C15 H12 Fe3 O9 S   81877-18-1   K-115-1300-0   Nonacarbonyl-.mu.(3)-(cyclohexylthiolato)-.mu.-hydrido-triiron;   W   126635

Str 356

536   C15 H14 F10 N4 O2 S2   90598-06-4   K-117-1720-0   1,1'-(Methylenedi-4,1-phenylene)bis[3-(pentafluorosulfanyl)urea];   W   126636

Str 357

536   C16 H36 Mo N6 O4 P2   27342-90-1   NS--28357-0-0   Molybdenum, tetracarbonylbis(hexamethylphosphorous triamide-P)-;; Molybdenu
m, tetracarbonylbis(tris(dimethylamine)phosphine)-;   NBS   63976

536   C16 H36 96Mo N6 O4 P2   AD-0-2755-0   BIS(TRIS(DIMETHYLAMINO)PHOSPHINE)-TETRACARBONYL MOLYBDENUM;   W   63977

536   C17 H36 Ge4   41898-96-8   C-95-3329-12   Germane, 1,3-pentadiyn-1-yl-5-ylidynetetrakis[trimethyl-;; 1,5,5,5-TETRAKIS(TRIMETHYL
GERMYL)-1,3-PENTADIYNE;   W/NBS   63979

536   C17 H36 Ge4   41898-95-7   C-95-3329-11   Germane, 3,4-pentadien-1-yne-1,3-diyl 5-ylidenetetrakis[trimethyl-;; 1,1,3,5-TETRAKI
S(TRIMETHYLGERMYL)-1,2-PENTADIEN-4-YNE;   W/NBS   63980

536   C18 H54 Fe N3 Si6   KA-72-1582-5   TRIS(BISTRIMETHYLSILYLAMIDO)IRON;   W   63981

536   C19 H10 F3 Mo O7 P   78803-79-9   K-114-2282-0   Pentacarbonyl[diphenyl(trifluoracetoxy)phosphine]molybdenum;   W   126637

Str 358

536   C20 H16 O6 W   TR-67-1636-0   trans-Tetracarbonylbis[(1R,2R,3S,4S)-2,3-η-(5,6-dimethylidene-7-oxabicyclo[2.2.1]hept-2-ene)]tun
gsten;   W   126638

536   C24 H39 F3 N4 O6   C-99-8477-8   N-TRIFLUOROACETYL PROLYLVALYL-α-AMINOISOBUTYL-α-AMINOISOBUTYRIC ACID N-BUTYL ESTER;   W   63982

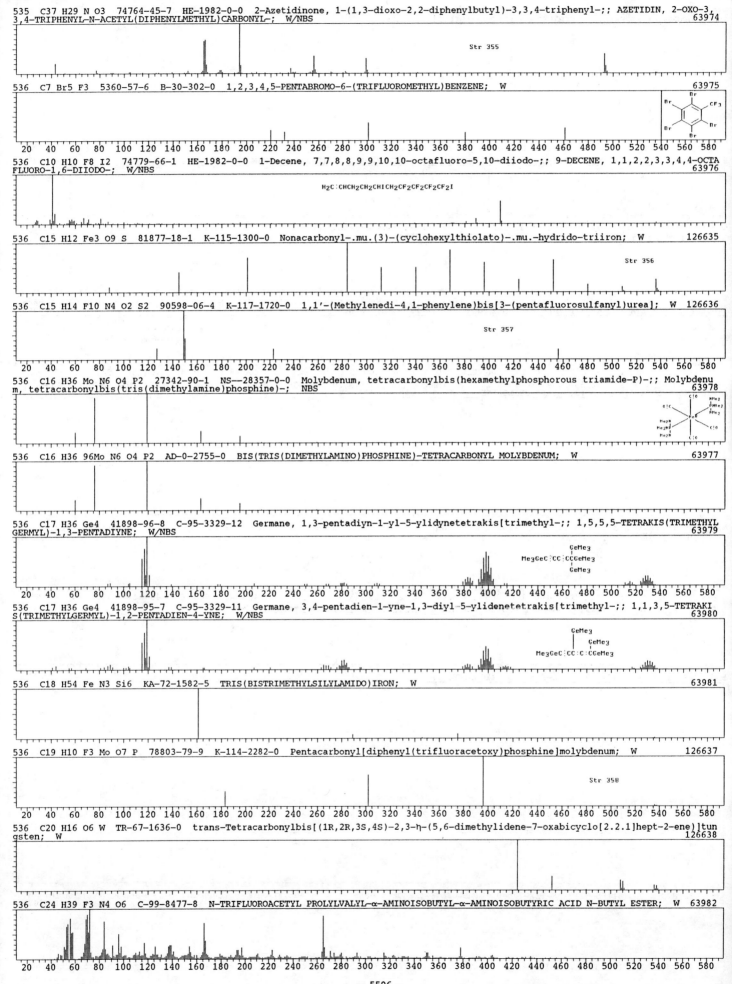

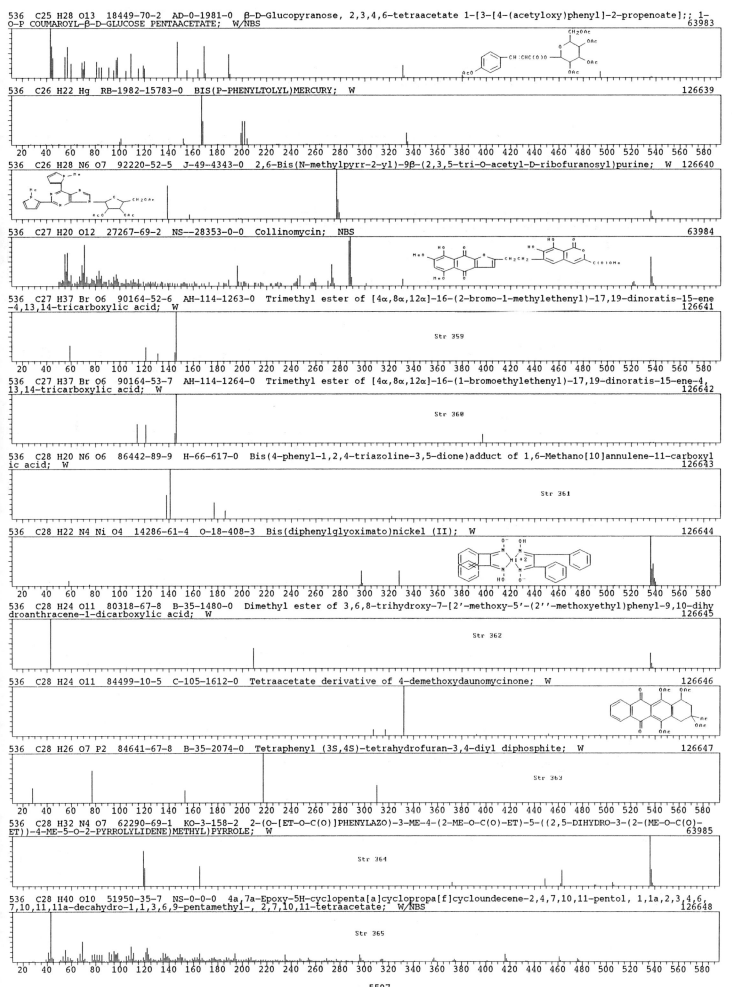

536   C25 H28 O13   18449-70-2   AD-0-1981-0   β-D-Glucopyranose, 2,3,4,6-tetraacetate 1-[3-[4-(acetyloxy)phenyl]-2-propenoate];; 1-
O-P COUMAROYL-β-D-GLUCOSE PENTAACETATE;   W/NBS
63983

536   C26 H22 Hg   RB-1982-15783-0   BIS(P-PHENYLTOLYL)MERCURY;   W
126639

20   40   60   80   100   120   140   160   180   200   220   240   260   280   300   320   340   360   380   400   420   440   460   480   500   520   540   560   580

536   C26 H28 N6 O7   92220-52-5   J-49-4343-0   2,6-Bis(N-methylpyrr-2-yl)-9β-(2,3,5-tri-O-acetyl-D-ribofuranosyl)purine;   W   126640

536   C27 H20 O12   27267-69-2   NS--28353-0-0   Collinomycin;   NBS
63984

536   C27 H37 Br O6   90164-52-6   AH-114-1263-0   Trimethyl ester of [4α,8α,12α]-16-(2-bromo-1-methylethenyl)-17,19-dinoratis-15-ene
-4,13,14-tricarboxylic acid;   W
126641

Str 359

20   40   60   80   100   120   140   160   180   200   220   240   260   280   300   320   340   360   380   400   420   440   460   480   500   520   540   560   580

536   C27 H37 Br O6   90164-53-7   AH-114-1264-0   Trimethyl ester of [4α,8α,12α]-16-(1-bromoethylethenyl)-17,19-dinoratis-15-ene-4,
13,14-tricarboxylic acid;   W
126642

Str 360

536   C28 H20 N6 O6   86442-89-9   H-66-617-0   Bis(4-phenyl-1,2,4-triazoline-3,5-dione)adduct of 1,6-Methano[10]annulene-11-carboxyl
ic acid;   W
126643

Str 361

536   C28 H22 N4 Ni O4   14286-61-4   O-18-408-3   Bis(diphenylglyoximato)nickel (II);   W
126644

536   C28 H24 O11   80318-67-8   B-35-1480-0   Dimethyl ester of 3,6,8-trihydroxy-7-[2'-methoxy-5'-(2''-methoxyethyl)phenyl-9,10-dihy
droanthracene-1-dicarboxylic acid;   W
126645

Str 362

536   C28 H24 O11   84499-10-5   C-105-1612-0   Tetraacetate derivative of 4-demethoxydaunomycinone;   W
126646

536   C28 H26 O7 P2   84641-67-8   B-35-2074-0   Tetraphenyl (3S,4S)-tetrahydrofuran-3,4-diyl diphosphite;   W
126647

Str 363

536   C28 H32 N4 O7   62290-69-1   KO-3-158-2   2-(O-[ET-O-C(O)]PHENYLAZO)-3-ME-4-(2-ME-O-C(O)-ET)-5-((2,5-DIHYDRO-3-(2-(ME-O-C(O)-
ET))-4-ME-5-O-2-PYRROLYLIDENE)METHYL)PYRROLE;   W
63985

Str 364

536   C28 H40 O10   51950-35-7   NS-0-0-0   4a,7a-Epoxy-5H-cyclopenta[a]cyclopropa[f]cycloundecene-2,4,7,10,11-pentol, 1,1a,2,3,4,6,
7,10,11,11a-decahydro-1,1,3,6,9-pentamethyl-, 2,7,10,11-tetraacetate;   W/NBS
126648

Str 365

20   40   60   80   100   120   140   160   180   200   220   240   260   280   300   320   340   360   380   400   420   440   460   480   500   520   540   560   580

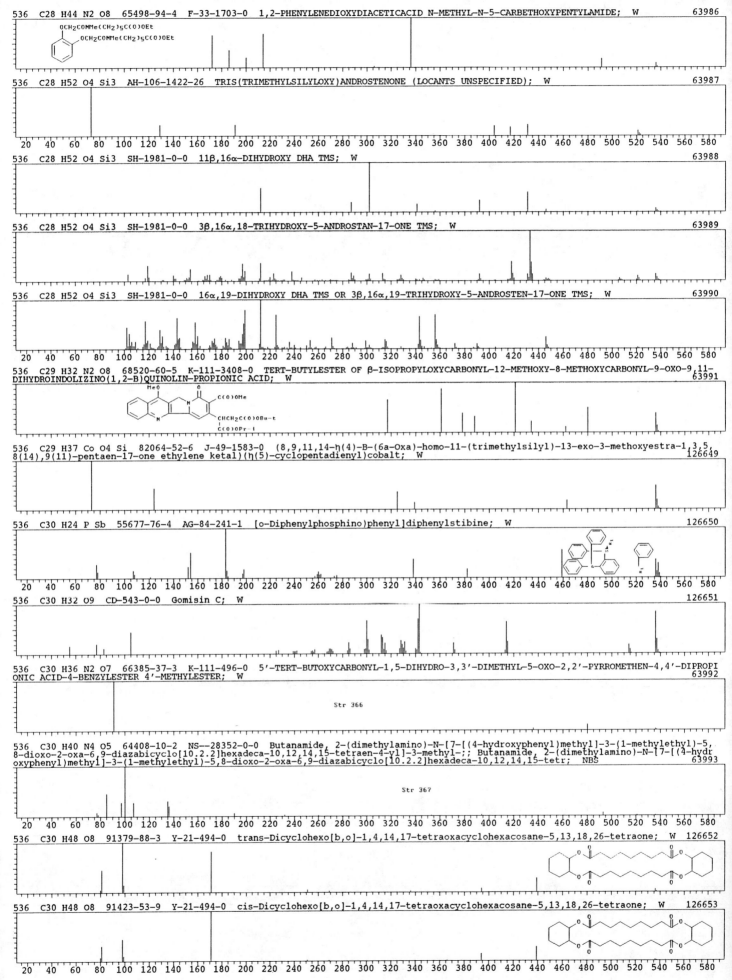

536   C28 H44 N2 O8   65498-94-4   F-33-1703-0   1,2-PHENYLENEDIOXYDIACETICACID N-METHYL-N-5-CARBETHOXYPENTYLAMIDE;   W   63986

536   C28 H52 O4 Si3   AH-106-1422-26   TRIS(TRIMETHYLSILYLOXY)ANDROSTENONE (LOCANTS UNSPECIFIED);   W   63987

536   C28 H52 O4 Si3   SH-1981-0-0   11β,16α-DIHYDROXY DHA TMS;   W   63988

536   C28 H52 O4 Si3   SH-1981-0-0   3β,16α,18-TRIHYDROXY-5-ANDROSTAN-17-ONE TMS;   W   63989

536   C28 H52 O4 Si3   SH-1981-0-0   16α,19-DIHYDROXY DHA TMS OR 3β,16α,19-TRIHYDROXY-5-ANDROSTEN-17-ONE TMS;   W   63990

536   C29 H32 N2 O8   68520-60-5   K-111-3408-0   TERT-BUTYLESTER OF β-ISOPROPYLOXYCARBONYL-12-METHOXY-8-METHOXYCARBONYL-9-OXO-9,11-DIHYDROINDOLIZINO(1,2-B)QUINOLIN-PROPIONIC ACID;   W   63991

536   C29 H37 Co O4 Si   82064-52-6   J-49-1583-0   (8,9,11,14-η(4)-B-(6a-Oxa)-homo-11-(trimethylsilyl)-13-exo-3-methoxyestra-1,3,5,8(14),9(11)-pentaen-17-one ethylene ketal)(η(5)-cyclopentadienyl)cobalt;   W   126649

536   C30 H24 P Sb   55677-76-4   AG-84-241-1   [o-Diphenylphosphino)phenyl]diphenylstibine;   W   126650

536   C30 H32 O9   CD-543-0-0   Gomisin C;   W   126651

536   C30 H36 N2 O7   66385-37-3   K-111-496-0   5'-TERT-BUTOXYCARBONYL-1,5-DIHYDRO-3,3'-DIMETHYL-5-OXO-2,2'-PYRROMETHEN-4,4'-DIPROPIONIC ACID-4-BENZYLESTER 4'-METHYLESTER;   W   63992

Str 366

536   C30 H40 N4 O5   64408-10-2   NS--28352-0-0   Butanamide, 2-(dimethylamino)-N-[7-[(4-hydroxyphenyl)methyl]-3-(1-methylethyl)-5,8-dioxo-2-oxa-6,9-diazabicyclo[10.2.2]hexadeca-10,12,14,15-tetraen-4-yl]-3-methyl-;; Butanamide, 2-(dimethylamino)-N-[7-[(4-hydroxyphenyl)methyl]-3-(1-methylethyl)-5,8-dioxo-2-oxa-6,9-diazabicyclo[10.2.2]hexadeca-10,12,14,15-tetr;   NBS   63993

Str 367

536   C30 H48 O8   91379-88-3   Y-21-494-0   trans-Dicyclohexo[b,o]-1,4,14,17-tetraoxacyclohexacosane-5,13,18,26-tetraone;   W   126652

536   C30 H48 O8   91423-53-9   Y-21-494-0   cis-Dicyclohexo[b,o]-1,4,14,17-tetraoxacyclohexacosane-5,13,18,26-tetraone;   W   126653

536   C31 H40 N2 O6   75266-56-7   J-45-4816-0   Benzyl (Z,E)-4-methyl-3-(4'-β-(((n'-((tert-butyloxy)-carbonyl)-1-prolyl)amino)ethyl)
phenoxy)-2-pentenoate;   W   126654

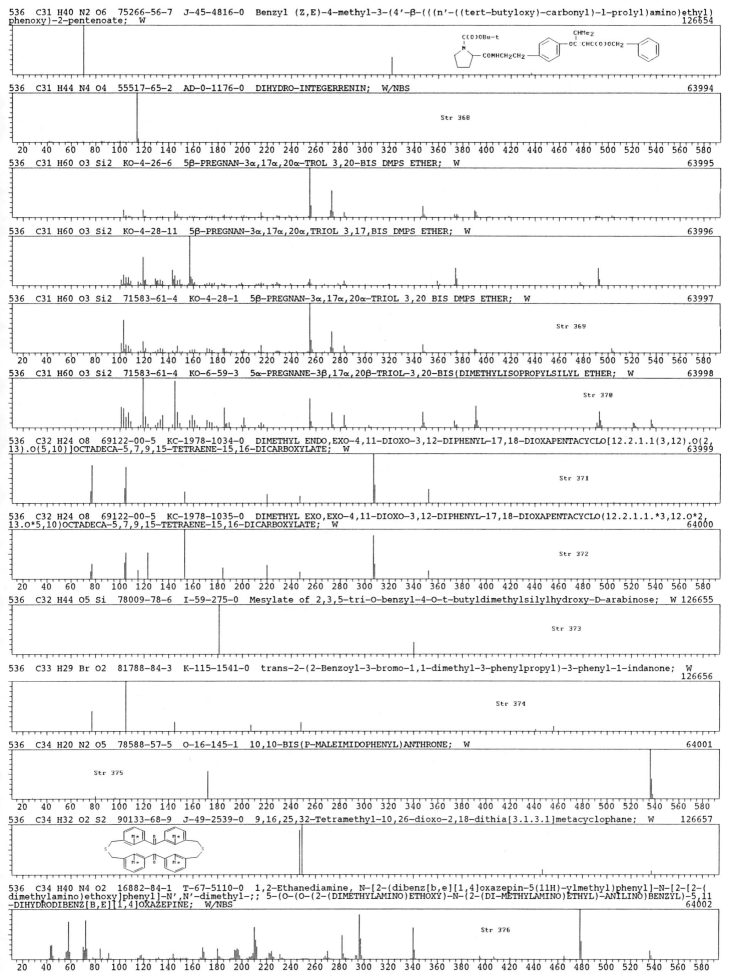

536   C31 H44 N4 O4   55517-65-2   AD-0-1176-0   DIHYDRO-INTEGERRENIN;   W/NBS   63994

Str 368

536   C31 H60 O3 Si2   KO-4-26-6   5β-PREGNAN-3α,17α,20α-TROL 3,20-BIS DMPS ETHER;   W   63995

536   C31 H60 O3 Si2   KO-4-28-11   5β-PREGNAN-3α,17α,20α,TRIOL 3,17,BIS DMPS ETHER;   W   63996

536   C31 H60 O3 Si2   71583-61-4   KO-4-28-1   5β-PREGNAN-3α,17α,20α-TRIOL 3,20 BIS DMPS ETHER;   W   63997

Str 369

536   C31 H60 O3 Si2   71583-61-4   KO-6-59-3   5α-PREGNANE-3β,17α,20β-TRIOL-3,20-BIS(DIMETHYLISOPROPYLSILYL ETHER;   W   63998

Str 370

536   C32 H24 O8   69122-00-5   KC-1978-1034-0   DIMETHYL ENDO,EXO-4,11-DIOXO-3,12-DIPHENYL-17,18-DIOXAPENTACYCLO[12.2.1.1(3,12).O(2,
13).O(5,10)]OCTADECA-5,7,9,15-TETRAENE-15,16-DICARBOXYLATE;   W   63999

Str 371

536   C32 H24 O8   69122-00-5   KC-1978-1035-0   DIMETHYL EXO,EXO-4,11-DIOXO-3,12-DIPHENYL-17,18-DIOXAPENTACYCLO(12.2.1.1.*3,12.0*2,
13.O*5,10)OCTADECA-5,7,9,15-TETRAENE-15,16-DICARBOXYLATE;   W   64000

Str 372

536   C32 H44 O5 Si   78009-78-6   I-59-275-0   Mesylate of 2,3,5-tri-O-benzyl-4-O-t-butyldimethylsilylhydroxy-D-arabinose;   W 126655

Str 373

536   C33 H29 Br O2   81788-84-3   K-115-1541-0   trans-2-(2-Benzoyl-3-bromo-1,1-dimethyl-3-phenylpropyl)-3-phenyl-1-indanone;   W
                                                                                                                                         126656

Str 374

536   C34 H20 N2 O5   78588-57-5   O-16-145-1   10,10-BIS(P-MALEIMIDOPHENYL)ANTHRONE;   W   64001

Str 375

536   C34 H32 O2 S2   90133-68-9   J-49-2539-0   9,16,25,32-Tetramethyl-10,26-dioxo-2,18-dithia[3.1.3.1]metacyclophane;   W   126657

536   C34 H40 N4 O2   16882-84-1   T-67-5110-0   1,2-Ethanediamine, N-[2-(dibenz[b,e][1,4]oxazepin-5(11H)-ylmethyl)phenyl]-N-[2-[2-(
dimethylamino)ethoxy]phenyl]-N',N-dimethyl-;; 5-(O-(O-(2-(DIMETHYLAMINO)ETHOXY)-N-(2-(DI-METHYLAMINO)ETHYL)-ANILINO)BENZYL)-5,11
-DIHYDRODIBENZ[B,E][1,4]OXAZEPINE;   W/NBS   64002

Str 376

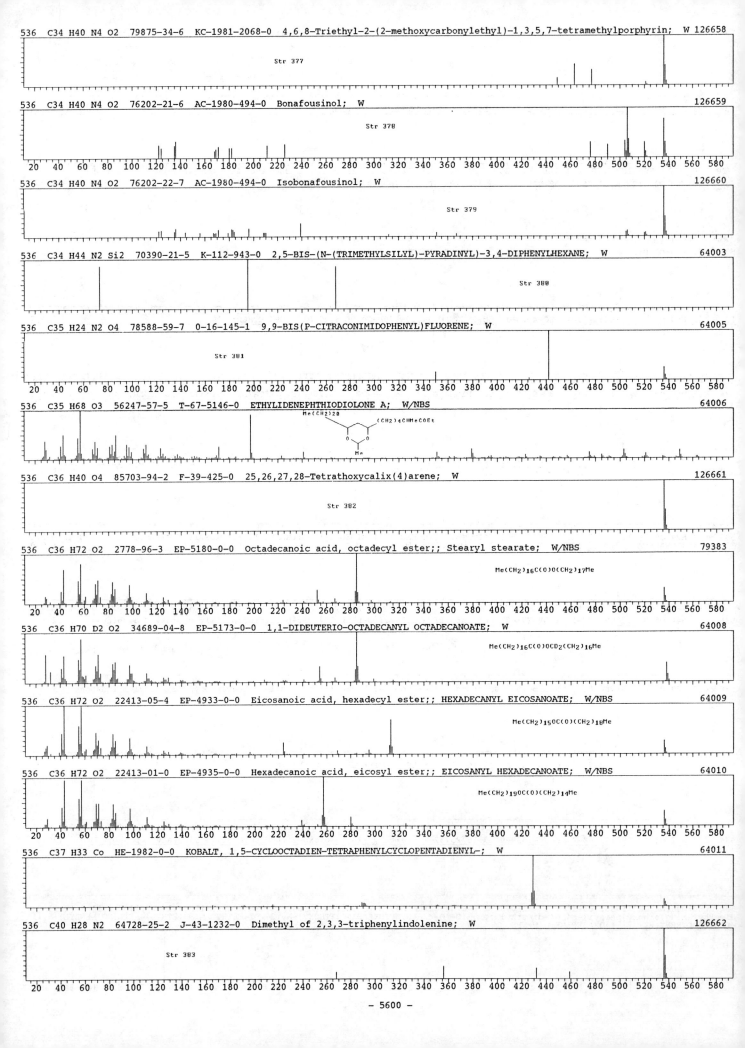

536   C34 H40 N4 O2   79875-34-6   KC-1981-2068-0   4,6,8-Triethyl-2-(2-methoxycarbonylethyl)-1,3,5,7-tetramethylporphyrin;   W 126658

Str 377

536   C34 H40 N4 O2   76202-21-6   AC-1980-494-0   Bonafousinol;   W   126659

Str 378

536   C34 H40 N4 O2   76202-22-7   AC-1980-494-0   Isobonafousinol;   W   126660

Str 379

536   C34 H44 N2 Si2   70390-21-5   K-112-943-0   2,5-BIS-(N-(TRIMETHYLSILYL)-PYRADINYL)-3,4-DIPHENYLHEXANE;   W   64003

Str 380

536   C35 H24 N2 O4   78588-59-7   0-16-145-1   9,9-BIS(P-CITRACONIMIDOPHENYL)FLUORENE;   W   64005

Str 381

536   C35 H68 O3   56247-57-5   T-67-5146-0   ETHYLIDENEPHTHIODIOLONE A;   W/NBS   64006

Me(CH2)20   (CH2)4CHMeCOEt

Me

536   C36 H40 O4   85703-94-2   F-39-425-0   25,26,27,28-Tetrathoxycalix(4)arene;   W   126661

Str 382

536   C36 H72 O2   2778-96-3   EP-5180-0-0   Octadecanoic acid, octadecyl ester;; Stearyl stearate;   W/NBS   79383

Me(CH2)16C(O)O(CH2)17Me

536   C36 H70 D2 O2   34689-04-8   EP-5173-0-0   1,1-DIDEUTERIO-OCTADECANYL OCTADECANOATE;   W   64008

Me(CH2)16C(O)OCD2(CH2)16Me

536   C36 H72 O2   22413-05-4   EP-4933-0-0   Eicosanoic acid, hexadecyl ester;; HEXADECANYL EICOSANOATE;   W/NBS   64009

Me(CH2)15OC(O)(CH2)18Me

536   C36 H72 O2   22413-01-0   EP-4935-0-0   Hexadecanoic acid, eicosyl ester;; EICOSANYL HEXADECANOATE;   W/NBS   64010

Me(CH2)19OC(O)(CH2)14Me

536   C37 H33 Co   HE-1982-0-0   KOBALT, 1,5-CYCLOOCTADIEN-TETRAPHENYLCYCLOPENTADIENYL-;   W   64011

536   C40 H28 N2   64728-25-2   J-43-1232-0   Dimethyl of 2,3,3-triphenylindolenine;   W   126662

Str 383

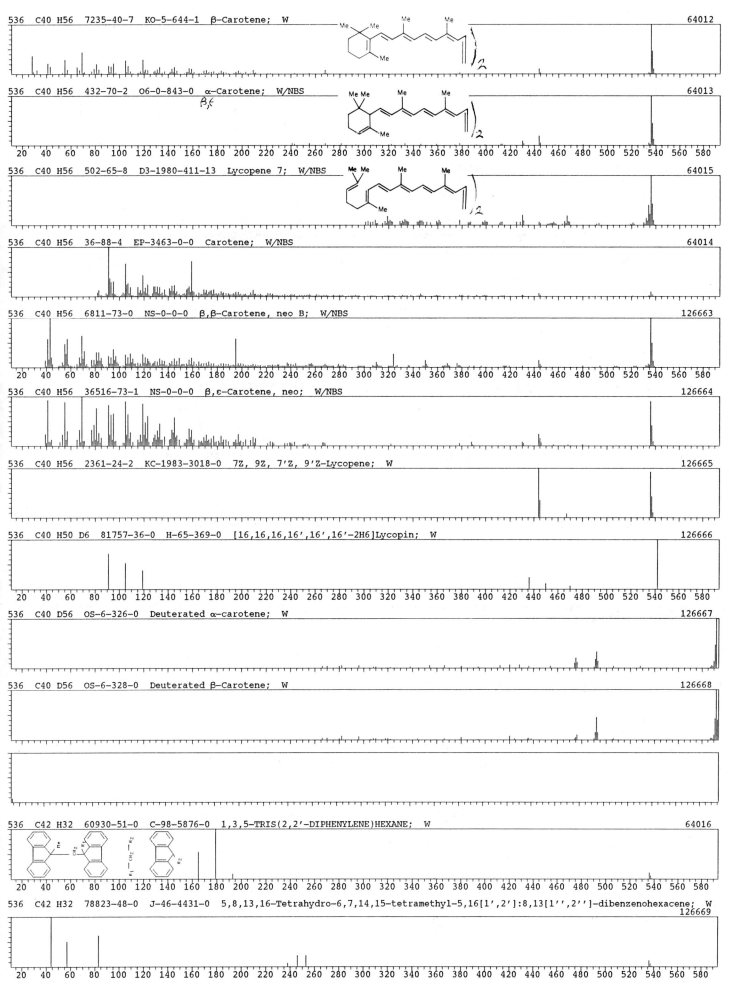

536　C40 H56　7235-40-7　KO-5-644-1　β-Carotene;　W　　　　　64012

536　C40 H56　432-70-2　O6-0-843-0　α-Carotene;　W/NBS　　　　64013
β,ε

536　C40 H56　502-65-8　D3-1980-411-13　Lycopene 7;　W/NBS　　64015

536　C40 H56　36-88-4　EP-3463-0-0　Carotene;　W/NBS　　　　　64014

536　C40 H56　6811-73-0　NS-0-0-0　β,β-Carotene, neo B;　W/NBS　126663

536　C40 H56　36516-73-1　NS-0-0-0　β,ε-Carotene, neo;　W/NBS　126664

536　C40 H56　2361-24-2　KC-1983-3018-0　7Z, 9Z, 7'Z, 9'Z-Lycopene;　W　126665

536　C40 H50 D6　81757-36-0　H-65-369-0　[16,16,16,16',16',16'-2H6]Lycopin;　W　126666

536　C40 D56　OS-6-326-0　Deuterated α-carotene;　W　126667

536　C40 D56　OS-6-328-0　Deuterated β-Carotene;　W　126668

536　C42 H32　60930-51-0　C-98-5876-0　1,3,5-TRIS(2,2'-DIPHENYLENE)HEXANE;　W　64016

536　C42 H32　78823-48-0　J-46-4431-0　5,8,13,16-Tetrahydro-6,7,14,15-tetramethyl-5,16[1',2']:8,13[1'',2'']-dibenzenohexacene;　W
126669

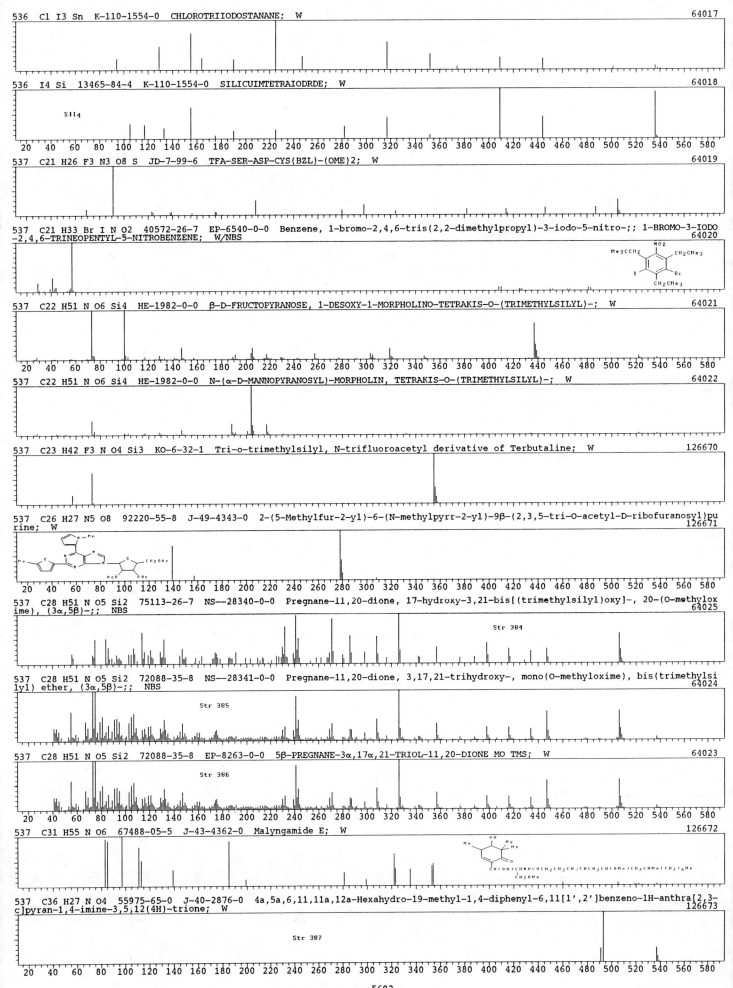

536   Cl I3 Sn   K-110-1554-0   CHLOROTRIIODOSTANANE;   W   64017

536   I4 Si   13465-84-4   K-110-1554-0   SILICUIMTETRAIODRDE;   W   64018

SiI₄

537   C21 H26 F3 N3 O8 S   JD-7-99-6   TFA-SER-ASP-CYS(BZL)-(OME)2;   W   64019

537   C21 H33 Br I N O2   40572-26-7   EP-6540-0-0   Benzene, 1-bromo-2,4,6-tris(2,2-dimethylpropyl)-3-iodo-5-nitro-;; 1-BROMO-3-IODO
-2,4,6-TRINEOPENTYL-5-NITROBENZENE;   W/NBS   64020

537   C22 H51 N O6 Si4   HE-1982-0-0   β-D-FRUCTOPYRANOSE, 1-DESOXY-1-MORPHOLINO-TETRAKIS-O-(TRIMETHYLSILYL)-;   W   64021

537   C22 H51 N O6 Si4   HE-1982-0-0   N-(α-D-MANNOPYRANOSYL)-MORPHOLIN, TETRAKIS-O-(TRIMETHYLSILYL)-;   W   64022

537   C23 H42 F3 N O4 Si3   KO-6-32-1   Tri-o-trimethylsilyl, N-trifluoroacetyl derivative of Terbutaline;   W   126670

537   C26 H27 N5 O8   92220-55-8   J-49-4343-0   2-(5-Methylfur-2-yl)-6-(N-methylpyrr-2-yl)-9β-(2,3,5-tri-O-acetyl-D-ribofuranosyl)pu
rine;   W   126671

537   C28 H51 N O5 Si2   75113-26-7   NS--28340-0-0   Pregnane-11,20-dione, 17-hydroxy-3,21-bis[(trimethylsilyl)oxy]-, 20-(O-methylox
ime), (3α,5β)-;;   NBS   64025

Str 384

537   C28 H51 N O5 Si2   72088-35-8   NS--28341-0-0   Pregnane-11,20-dione, 3,17,21-trihydroxy-, mono(O-methyloxime), bis(trimethylsi
lyl) ether, (3α,5β)-;;   NBS   64024

Str 385

537   C28 H51 N O5 Si2   72088-35-8   EP-8263-0-0   5β-PREGNANE-3α,17α,21-TRIOL-11,20-DIONE MO TMS;   W   64023

Str 386

537   C31 H55 N O6   67488-05-5   J-43-4362-0   Malyngamide E;   W   126672

537   C36 H27 N O4   55975-65-0   J-40-2876-0   4a,5a,6,11,11a,12a-Hexahydro-19-methyl-1,4-diphenyl-6,11[1',2']benzeno-1H-anthra[2,3-
c]pyran-1,4-imine-3,5,12(4H)-trione;   W   126673

Str 387

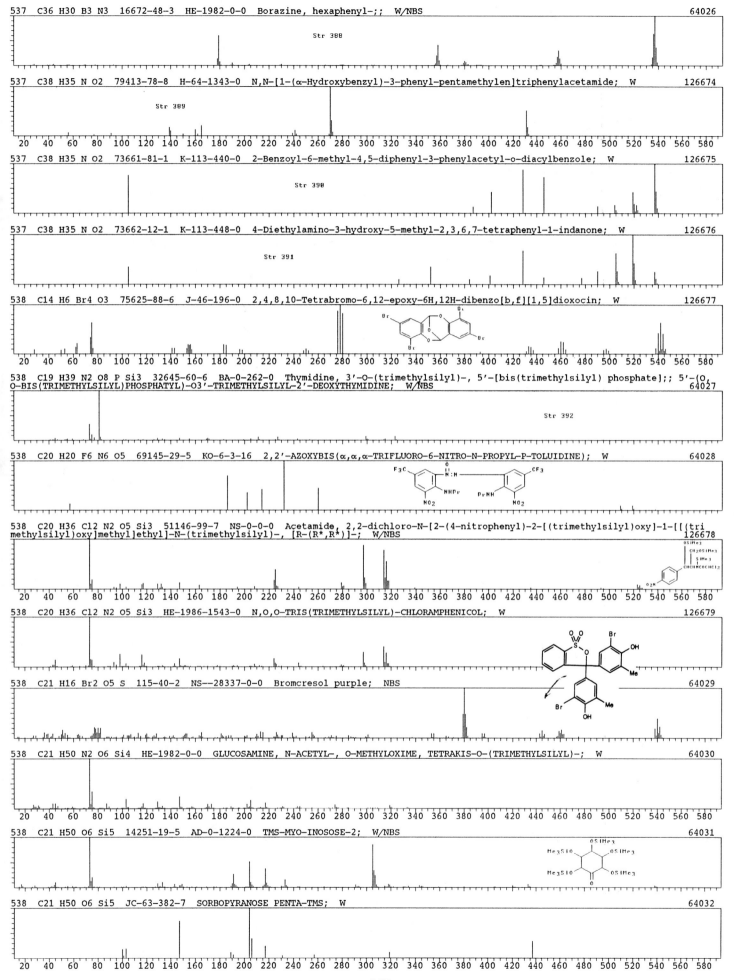

537　C36 H30 B3 N3　16672-48-3　HE-1982-0-0　Borazine, hexaphenyl-;;　W/NBS　64026

Str 388

537　C38 H35 N O2　79413-78-8　H-64-1343-0　N,N-[1-(α-Hydroxybenzyl)-3-phenyl-pentamethylen]triphenylacetamide;　W　126674

Str 389

537　C38 H35 N O2　73661-81-1　K-113-440-0　2-Benzoyl-6-methyl-4,5-diphenyl-3-phenylacetyl-o-diacylbenzole;　W　126675

Str 390

537　C38 H35 N O2　73662-12-1　K-113-448-0　4-Diethylamino-3-hydroxy-5-methyl-2,3,6,7-tetraphenyl-1-indanone;　W　126676

Str 391

538　C14 H6 Br4 O3　75625-88-6　J-46-196-0　2,4,8,10-Tetrabromo-6,12-epoxy-6H,12H-dibenzo[b,f][1,5]dioxocin;　W　126677

538　C19 H39 N2 O8 P Si3　32645-60-6　BA-0-262-0　Thymidine, 3'-O-(trimethylsilyl)-, 5'-[bis(trimethylsilyl) phosphate];; 5'-(O,O-BIS(TRIMETHYLSILYL)PHOSPHATYL)-O3'-TRIMETHYLSILYL-2'-DEOXYTHYMIDINE;　W/NBS　64027

Str 392

538　C20 H20 F6 N6 O5　69145-29-5　KO-6-3-16　2,2'-AZOXYBIS(α,α,α-TRIFLUORO-6-NITRO-N-PROPYL-P-TOLUIDINE);　W　64028

538　C20 H36 Cl2 N2 O5 Si3　51146-99-7　NS-0-0-0　Acetamide, 2,2-dichloro-N-[2-(4-nitrophenyl)-2-[(trimethylsilyl)oxy]-1-[[(trimethylsilyl)oxy]methyl]ethyl]-N-(trimethylsilyl)-, [R-(R*,R*)]-;　W/NBS　126678

538　C20 H36 Cl2 N2 O5 Si3　HE-1986-1543-0　N,O,O-TRIS(TRIMETHYLSILYL)-CHLORAMPHENICOL;　W　126679

538　C21 H16 Br2 O5 S　115-40-2　NS--28337-0-0　Bromcresol purple;　NBS　64029

538　C21 H50 N2 O6 Si4　HE-1982-0-0　GLUCOSAMINE, N-ACETYL-, O-METHYLOXIME, TETRAKIS-O-(TRIMETHYLSILYL)-;　W　64030

538　C21 H50 O6 Si5　14251-19-5　AD-0-1224-0　TMS-MYO-INOSOSE-2;　W/NBS　64031

538　C21 H50 O6 Si5　JC-63-382-7　SORBOPYRANOSE PENTA-TMS;　W　64032

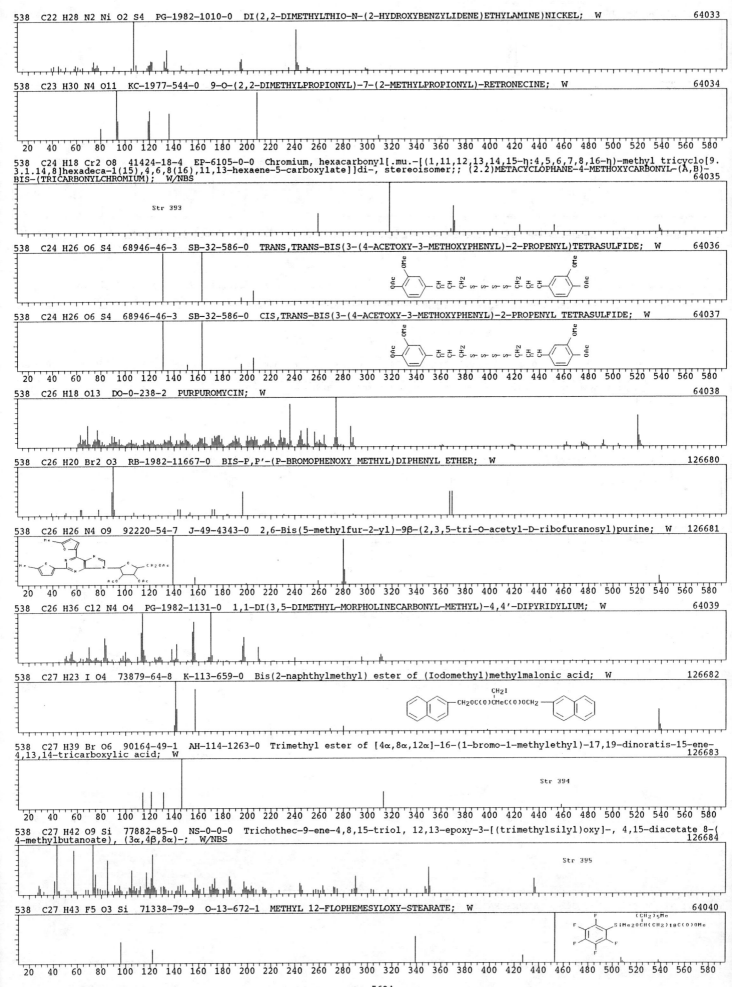

538 C22 H28 N2 Ni O2 S4   PG-1982-1010-0   DI(2,2-DIMETHYLTHIO-N-(2-HYDROXYBENZYLIDENE)ETHYLAMINE)NICKEL;   W   64033

538 C23 H30 N4 O11   KC-1977-544-0   9-O-(2,2-DIMETHYLPROPIONYL)-7-(2-METHYLPROPIONYL)-RETRONECINE;   W   64034

538 C24 H18 Cr2 O8   41424-18-4   EP-6105-0-0   Chromium, hexacarbonyl[.mu.-[(1,11,12,13,14,15-η:4,5,6,7,8,16-η)-methyl tricyclo[9.3.1.14,8]hexadeca-1(15),4,6,8(16),11,13-hexaene-5-carboxylate]]di-, stereoisomer;; (2.2)METACYCLOPHANE-4-METHOXYCARBONYL-(A,B)-BIS-(TRICARBONYLCHROMIUM);   W/NBS   64035

Str 393

538 C24 H26 O6 S4   68946-46-3   SB-32-586-0   TRANS,TRANS-BIS(3-(4-ACETOXY-3-METHOXYPHENYL)-2-PROPENYL)TETRASULFIDE;   W   64036

538 C24 H26 O6 S4   68946-46-3   SB-32-586-0   CIS,TRANS-BIS(3-(4-ACETOXY-3-METHOXYPHENYL)-2-PROPENYL TETRASULFIDE;   W   64037

538 C26 H18 O13   DO-0-238-2   PURPUROMYCIN;   W   64038

538 C26 H20 Br2 O3   RB-1982-11667-0   BIS-P,P'-(P-BROMOPHENOXY METHYL)DIPHENYL ETHER;   W   126680

538 C26 H26 N4 O9   92220-54-7   J-49-4343-0   2,6-Bis(5-methylfur-2-yl)-9β-(2,3,5-tri-O-acetyl-D-ribofuranosyl)purine;   W   126681

538 C26 H36 Cl2 N4 O4   PG-1982-1131-0   1,1-DI(3,5-DIMETHYL-MORPHOLINECARBONYL-METHYL)-4,4'-DIPYRIDYLIUM;   W   64039

538 C27 H23 I O4   73879-64-8   K-113-659-0   Bis(2-naphthylmethyl) ester of (Iodomethyl)methylmalonic acid;   W   126682

538 C27 H39 Br O6   90164-49-1   AH-114-1263-0   Trimethyl ester of [4α,8α,12α]-16-(1-bromo-1-methylethyl)-17,19-dinoratis-15-ene-4,13,14-tricarboxylic acid;   W   126683

Str 394

538 C27 H42 O9 Si   77882-85-0   NS-0-0-0   Trichothec-9-ene-4,8,15-triol, 12,13-epoxy-3-[(trimethylsilyl)oxy]-, 4,15-diacetate 8-(4-methylbutanoate), (3α,4β,8α)-;   W/NBS   126684

Str 395

538 C27 H43 F5 O3 Si   71338-79-9   O-13-672-1   METHYL 12-FLOPHEMESYLOXY-STEARATE;   W   64040

538　C28 H26 O11　61035-69-6　B-29-1150-0　2-O-ACETYLTENUIORIN;　W
64041

538　C28 H42 O10　61543-96-2　O-13-308-1　(8)ETHYL ESTER OF 1H-2β,6β,9β-TRIACETYLOXY-8α-CARBOXYMETHYL-7β-METHOXYCARBONYL-4Aβ,7β-
DIMETHYLPERHYDRO-;　W
64042

20　40　60　80　100　120　140　160　180　200　220　240　260　280　300　320　340　360　380　400　420　440　460　480　500　520　540　560　580

538　C28 H50 O6 Si2　D3-1980-645-19　DIMETHYL,DI-TRIMETHYLSILYL DERIVATIVE OF GIBBERELLIN CA42;　W
64043

538　C28 H54 O4 Si3　SH-1981-0-0　3α,16α,17β-TRIHYDROXY-5β-ANDROSTAN-11-ONE TMS;　W
64044

538　C28 H54 O4 Si3　SH-1981-0-0　3α,16β,17β-TRIHYDROXY-5β-ANDROSTAN-11-ONE TMS;　W
64045

20　40　60　80　100　120　140　160　180　200　220　240　260　280　300　320　340　360　380　400　420　440　460　480　500　520　540　560　580

538　C28 H54 O4 Si3　KC-1977-2323-0　Tris(trimethylsilyl) ether of Ent-2β,3α,19-trihydroxy-17-norkauran-16-one;　W
126685

538　C28 H54 O4 Si3　KC-1977-2323-0　Tris(trimethylsilyl) ether of Ent-2α,3β,19-trihydroxy-17-norkauran-16-one;　W
126686

538　C28 H45 D13 N4 O2 Si2　61075-60-3　KO-2-333-2　O-TRIMETHYSILYLATED(D13)-PHE-GLU-ARG;　W
64046

20　40　60　80　100　120　140　160　180　200　220　240　260　280　300　320　340　360　380　400　420　440　460　480　500　520　540　560　580

538　C29 H30 O10　19314-74-0　SD-1981-0-0　4,2-Cresotic acid, 6-methoxy-, bimol. ester, methyl ester, 4,6-dimethoxy-o-toluate;;
W/NBS
64047

538　C29 H34 N2 O8　87377-74-0　F-39-1854-0　Benzyl 3-(methoxycarbonylmethyl)-4'-(ethoxycarbonylmethyl-3',4-dimethyl-5'-ethoxycar
bonylpyrromethane-5-carboxylate;　W
126687

Str 396

538　C29 H38 N4 O6　74974-43-9　KC-1982-313-0　(S)-trans-5-Amino-6-(4-hydroxyphenyl)hex-3-enoylglycyl-(L)-phenylalanyl-(L)-leuci
ne;　W
126688

20　40　60　80　100　120　140　160　180　200　220　240　260　280　300　320　340　360　380　400　420　440　460　480　500　520　540　560　580

538　C30 H18 O10　568-42-3　MY-1981-0-0　Iridoskyrin;　W/NBS
64049

Str 397

538　C30 H18 O10　602-06-2　MY-1981-0-0　Skyrin;　W/NBS
64050

Str 398

20　40　60　80　100　120　140　160　180　200　220　240　260　280　300　320　340　360　380　400　420　440　460　480　500　520　540　560　580

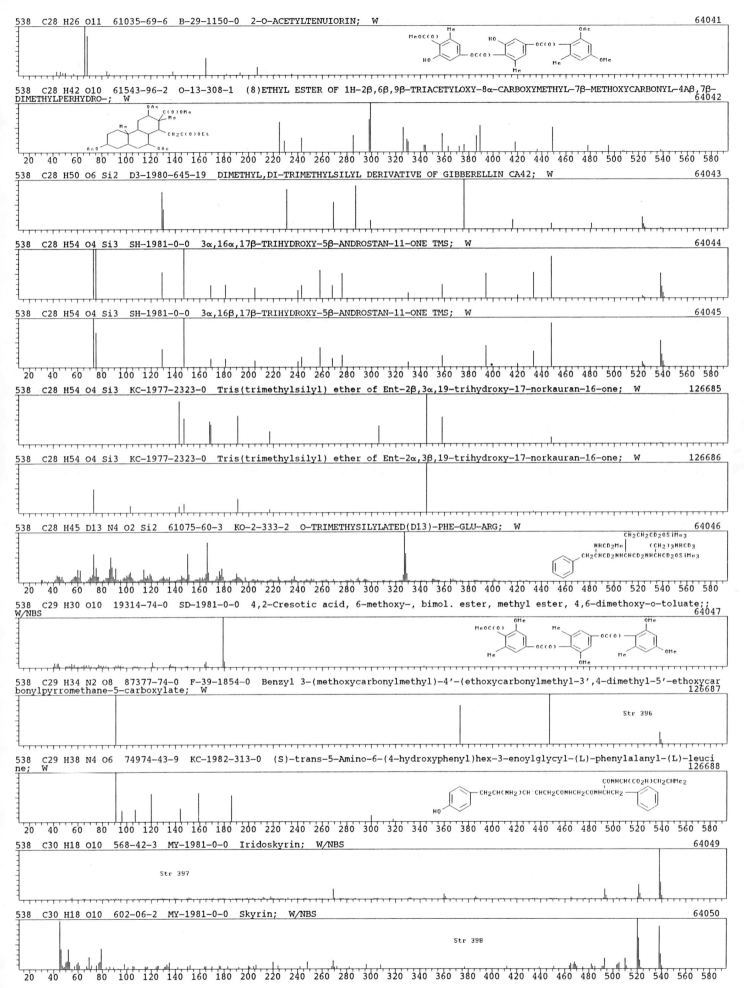

538　C30 H18 O10　27090-20-6　F-26-1413-0　3,3-BIAPIGENINYL;　W/NBS
64048

Str 399

538　C30 H34 O9　91127-24-1　J-49-3257-0　(+-)-(4bβ,10α)-Benzyl 8,8-(Ethylenedioxy)-2-methoxy-1-(methoxycarbonyl)-7-[(methoxymethyl)oxy]gibba-1,3,4a(10a)-triene-10-carboxylate;　W
126689

Str 400

20　40　60　80　100　120　140　160　180　200　220　240　260　280　300　320　340　360　380　400　420　440　460　480　500　520　540　560　580

538　C30 H38 N2 Si4　21116-67-6　NS--28330-0-0　Cyclodisilazane, 2,2,4,4-tetramethyl-1,3-bis(methyldiphenylsilyl)-;;　NBS
64051

Str 401

538　C31 H52 Cl2 O Si　69688-09-1　NS--28329-0-0　Silane, (dichloromethyl)dimethyl[[(3β,5Z,7E,22E)-9,10-secoergosta-5,7,22-trien-3-yl]oxy]-;;　NBS
64052

Str 402

538　C32 H26 O8　71641-34-4　KO-6-118-1　MESO-ERYTHRITOL-PERBENZOATE;　W
64053

Str 403

20　40　60　80　100　120　140　160　180　200　220　240　260　280　300　320　340　360　380　400　420　440　460　480　500　520　540　560　580

538　C32 H26 O8　71669-15-3　KO-6-118-1　DL-THREITOL-PERBENZOATE;　W
64054

Str 404

538　C32 H28 O4 P2　25836-64-0　AD-0-5242-0　1,4-Diphosphorin-2,3-dicarboxylic acid, 1,1,4,4-tetrahydro-1,1,4,4-tetraphenyl-, dimethyl ester;; 2,3-BIS(METHOXYCARBONYL)-1,1,4,4-TETRAPHENYL-1,4-DIPHOSPHABENZENE;　W/NBS
64055

Str 405

538　C32 H28 O4 P2　16115-02-9　K-116-2588-0　[α-(Diphenylphosphoryl)-4-methoxybenzyl]diphenylphosphinate;　W
126690

Str 406

20　40　60　80　100　120　140　160　180　200　220　240　260　280　300　320　340　360　380　400　420　440　460　480　500　520　540　560　580

538　C32 H38 N6 O2　SS-5-149-0　2,2'-Di(4-hexyloxylphenyl)-5,5'-azo-pyrimidine;　W
126691

538　C32 H42 O7　55529-71-0　AD-0-229-0　Pregnan-20-one, 12-(acetyloxy)-3,14-dihydroxy-11-[(1-oxo-3-phenyl-2-propenyl)oxy]-, (3β,11α,12β,14β)-;; 3,14-DIHYDROXY-12-ACETOXY-11-CINNAMOYLOXYPREGNAN-20-ONE;　W/NBS
64056

Str 407

538　C32 H42 O7　F-21-1805-6　KONDURANGOGENINS A;　W
64057

538　C32 H42 O7　58262-71-8　O-11-474-1　LEUKOTRIDENTOQUINONTRIACETATE;　W
64058

Str 408

538　C32 H42 O7　27825-38-3　NS-0-0-0　2-Oxatricyclo[20.2.2.1(3,7)]heptacosa-3,5,7(27),22,24,25-hexaene-5,24,25-triol, triacetate;;　W/NBS
126692

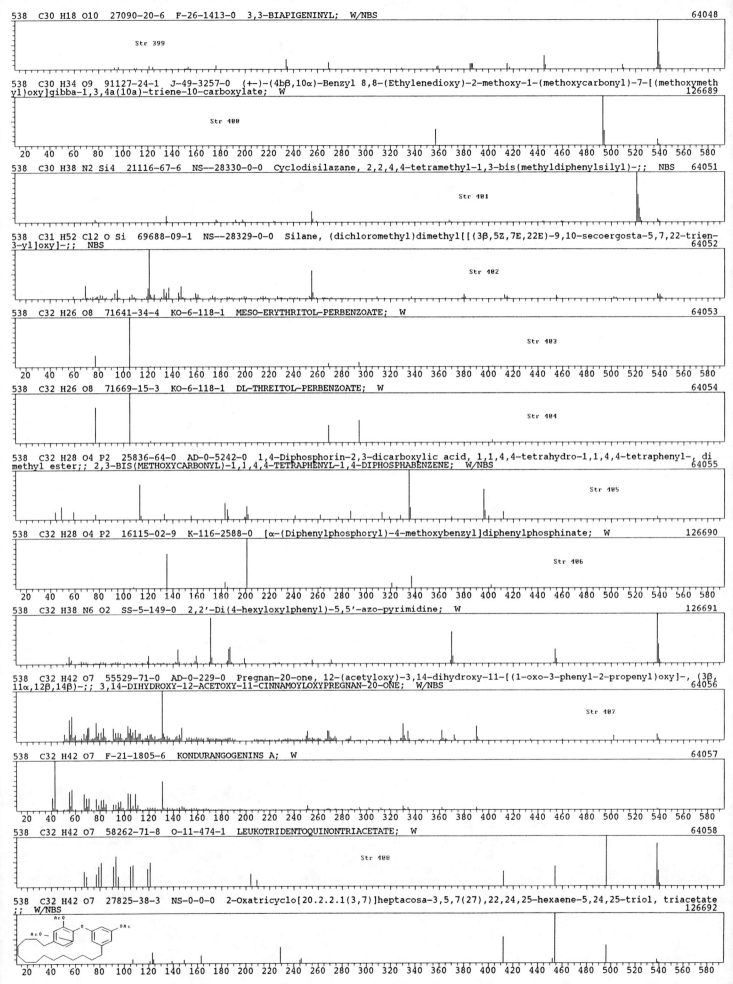

20　40　60　80　100　120　140　160　180　200　220　240　260　280　300　320　340　360　380　400　420　440　460　480　500　520　540　560　580

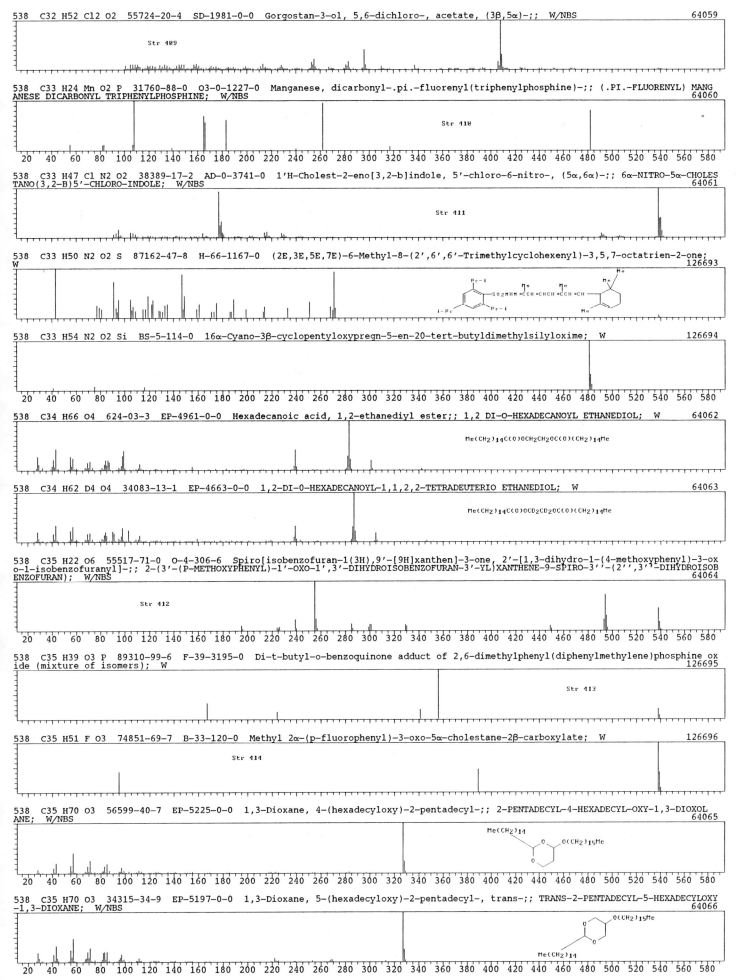

538   C32 H52 Cl2 O2 55724-20-4 SD-1981-0-0 Gorgostan-3-ol, 5,6-dichloro-, acetate, (3β,5α)-;; W/NBS   64059

Str 409

538   C33 H24 Mn O2 P 31760-88-0 O3-0-1227-0 Manganese, dicarbonyl-.pi.-fluorenyl(triphenylphosphine)-;; (.PI.-FLUORENYL) MANGANESE DICARBONYL TRIPHENYLPHOSPHINE; W/NBS   64060

Str 410

538   C33 H47 Cl N2 O2 38389-17-2 AD-0-3741-0 1'H-Cholest-2-eno[3,2-b]indole, 5'-chloro-6-nitro-, (5α,6α)-;; 6α-NITRO-5α-CHOLESTANO(3,2-B)5'-CHLORO-INDOLE; W/NBS   64061

Str 411

538   C33 H50 N2 O2 S 87162-47-8 H-66-1167-0 (2E,3E,5E,7E)-6-Methyl-8-(2',6',6'-Trimethylcyclohexenyl)-3,5,7-octatrien-2-one; W   126693

538   C33 H54 N2 O2 Si BS-5-114-0 16α-Cyano-3β-cyclopentyloxypregn-5-en-20-tert-butyldimethylsilyloxime; W   126694

538   C34 H66 O4 624-03-3 EP-4961-0-0 Hexadecanoic acid, 1,2-ethanediyl ester;; 1,2 DI-O-HEXADECANOYL ETHANEDIOL; W   64062

$Me(CH_2)_{14}C(O)OCH_2CH_2OC(O)(CH_2)_{14}Me$

538   C34 H62 D4 O4 34083-13-1 EP-4663-0-0 1,2-DI-0-HEXADECANOYL-1,1,2,2-TETRADEUTERIO ETHANEDIOL; W   64063

$Me(CH_2)_{14}C(O)OCD_2CD_2OC(O)(CH_2)_{14}Me$

538   C35 H22 O6 55517-71-0 O-4-306-6 Spiro[isobenzofuran-1(3H),9'-[9H]xanthen]-3-one, 2'-[1,3-dihydro-1-(4-methoxyphenyl)-3-oxo-1-isobenzofuranyl]-;; 2-(3'-(P-METHOXYPHENYL)-1'-OXO-1',3'-DIHYDROISOBENZOFURAN-3'-YL)XANTHENE-9-SPIRO-3''-(2'',3''-DIHYDROISOBENZOFURAN); W/NBS   64064

Str 412

538   C35 H39 O3 P 89310-99-6 F-39-3195-0 Di-t-butyl-o-benzoquinone adduct of 2,6-dimethylphenyl(diphenylmethylene)phosphine oxide (mixture of isomers); W   126695

Str 413

538   C35 H51 F O3 74851-69-7 B-33-120-0 Methyl 2α-(p-fluorophenyl)-3-oxo-5α-cholestane-2β-carboxylate; W   126696

Str 414

538   C35 H70 O3 56599-40-7 EP-5225-0-0 1,3-Dioxane, 4-(hexadecyloxy)-2-pentadecyl-;; 2-PENTADECYL-4-HEXADECYL-OXY-1,3-DIOXOLANE; W/NBS   64065

538   C35 H70 O3 34315-34-9 EP-5197-0-0 1,3-Dioxane, 5-(hexadecyloxy)-2-pentadecyl-, trans-;; TRANS-2-PENTADECYL-5-HEXADECYLOXY-1,3-DIOXANE; W/NBS   64066

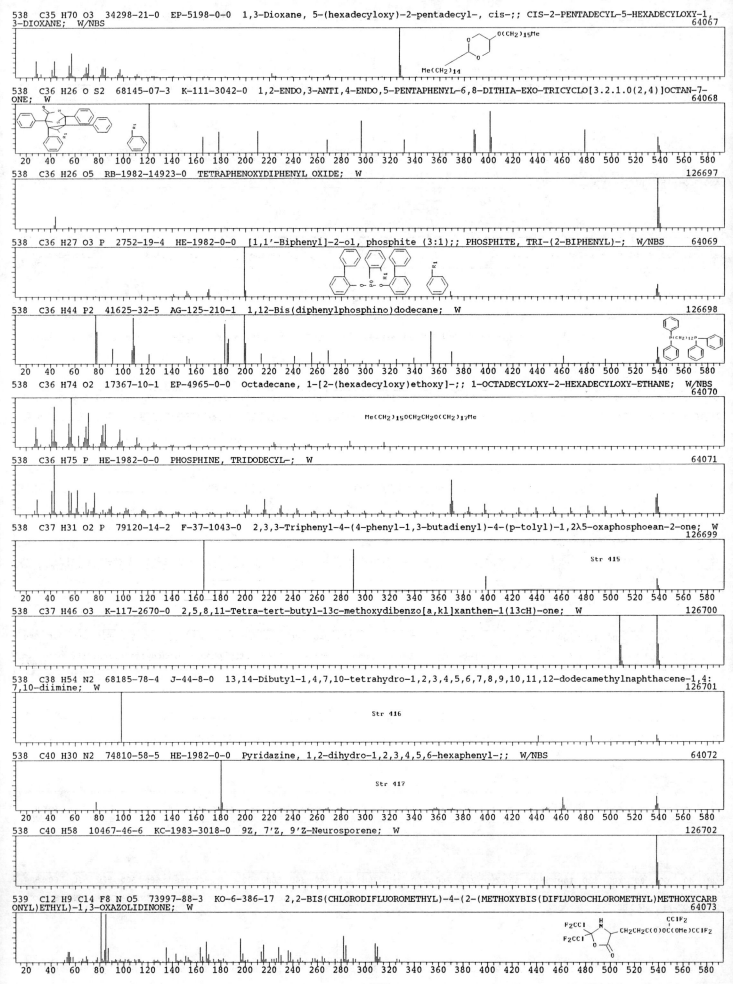

538 C35 H70 O3 34298-21-0 EP-5198-0-0 1,3-Dioxane, 5-(hexadecyloxy)-2-pentadecyl-, cis-;; CIS-2-PENTADECYL-5-HEXADECYLOXY-1,3-DIOXANE; W/NBS
64067

538 C36 H26 O S2 68145-07-3 K-111-3042-0 1,2-ENDO,3-ANTI,4-ENDO,5-PENTAPHENYL-6,8-DITHIA-EXO-TRICYCLO[3.2.1.0(2,4)]OCTAN-7-ONE; W
64068

538 C36 H26 O5 RB-1982-14923-0 TETRAPHENOXYDIPHENYL OXIDE; W
126697

538 C36 H27 O3 P 2752-19-4 HE-1982-0-0 [1,1'-Biphenyl]-2-ol, phosphite (3:1);; PHOSPHITE, TRI-(2-BIPHENYL)-; W/NBS
64069

538 C36 H44 P2 41625-32-5 AG-125-210-1 1,12-Bis(diphenylphosphino)dodecane; W
126698

538 C36 H74 O2 17367-10-1 EP-4965-0-0 Octadecane, 1-[2-(hexadecyloxy)ethoxy]-;; 1-OCTADECYLOXY-2-HEXADECYLOXY-ETHANE; W/NBS
64070

538 C36 H75 P HE-1982-0-0 PHOSPHINE, TRIDODECYL-; W
64071

538 C37 H31 O2 P 79120-14-2 F-37-1043-0 2,3,3-Triphenyl-4-(4-phenyl-1,3-butadienyl)-4-(p-tolyl)-1,2λ5-oxaphosphoean-2-one; W
126699

538 C37 H46 O3 K-117-2670-0 2,5,8,11-Tetra-tert-butyl-13c-methoxydibenzo[a,kl]xanthen-1(13cH)-one; W
126700

538 C38 H54 N2 68185-78-4 J-44-8-0 13,14-Dibutyl-1,4,7,10-tetrahydro-1,2,3,4,5,6,7,8,9,10,11,12-dodecamethylnaphthacene-1,4:7,10-diimine; W
126701

538 C40 H30 N2 74810-58-5 HE-1982-0-0 Pyridazine, 1,2-dihydro-1,2,3,4,5,6-hexaphenyl-;; W/NBS
64072

538 C40 H58 10467-46-6 KC-1983-3018-0 9Z, 7'Z, 9'Z-Neurosporene; W
126702

539 C12 H9 Cl4 F8 N O5 73997-88-3 KO-6-386-17 2,2-BIS(CHLORODIFLUOROMETHYL)-4-(2-(METHOXYBIS(DIFLUOROCHLOROMETHYL)METHOXYCARBONYL)ETHYL)-1,3-OXAZOLIDINONE; W
64073

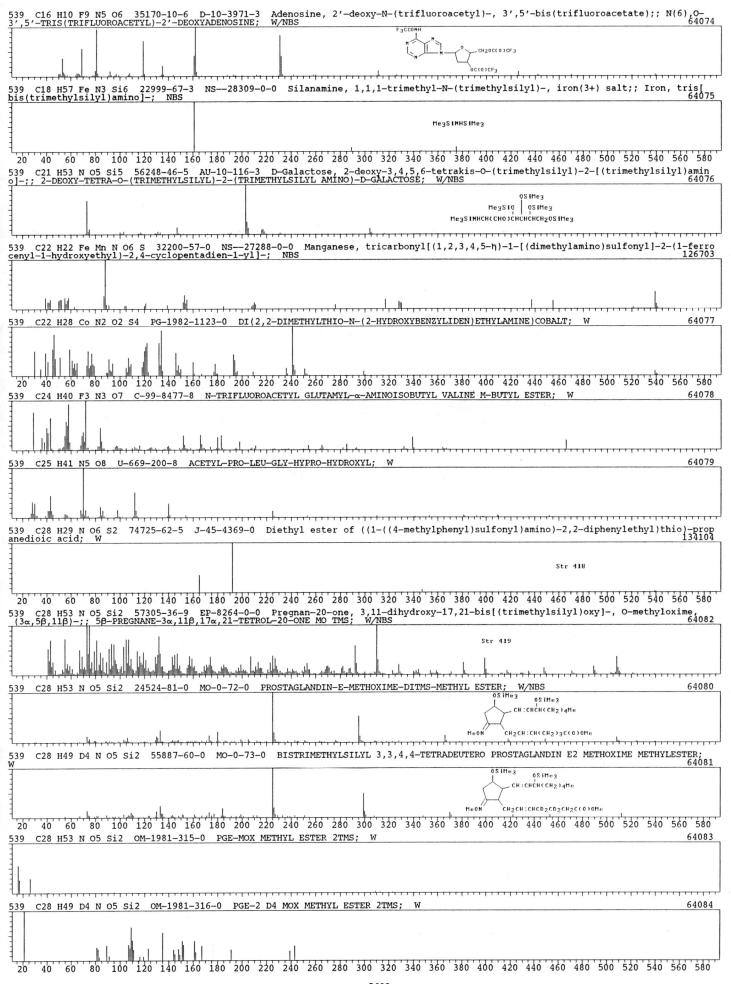

539 C16 H10 F9 N5 O6 35170-10-6 D-10-3971-3 Adenosine, 2'-deoxy-N-(trifluoroacetyl)-, 3',5'-bis(trifluoroacetate);; N(6),O-
3',5'-TRIS(TRIFLUOROACETYL)-2'-DEOXYADENOSINE; W/NBS
64074

539 C18 H57 Fe N3 Si6 22999-67-3 NS--28309-0-0 Silanamine, 1,1,1-trimethyl-N-(trimethylsilyl)-, iron(3+) salt;; Iron, tris[
bis(trimethylsilyl)amino]-; NBS
64075

Me3SiNHSiMe3

539 C21 H53 N O5 Si5 56248-46-5 AU-10-116-3 D-Galactose, 2-deoxy-3,4,5,6-tetrakis-O-(trimethylsilyl)-2-[(trimethylsilyl)amin
o]-;; 2-DEOXY-TETRA-O-(TRIMETHYLSILYL)-2-(TRIMETHYLSILYL AMINO)-D-GALACTOSE; W/NBS
64076

539 C22 H22 Fe Mn N O6 S 32200-57-0 NS--27288-0-0 Manganese, tricarbonyl[(1,2,3,4,5-η)-1-[(dimethylamino)sulfonyl]-2-(1-ferro
cenyl-1-hydroxyethyl)-2,4-cyclopentadien-1-yl]-; NBS
126703

539 C22 H28 Co N2 O2 S4 PG-1982-1123-0 DI(2,2-DIMETHYLTHIO-N-(2-HYDROXYBENZYLIDEN)ETHYLAMINE)COBALT; W
64077

539 C24 H40 F3 N3 O7 C-99-8477-8 N-TRIFLUOROACETYL GLUTAMYL-α-AMINOISOBUTYL VALINE M-BUTYL ESTER; W
64078

539 C25 H41 N5 O8 U-669-200-8 ACETYL-PRO-LEU-GLY-HYPRO-HYDROXYL; W
64079

539 C28 H29 N O6 S2 74725-62-5 J-45-4369-0 Diethyl ester of ((1-((4-methylphenyl)sulfonyl)amino)-2,2-diphenylethyl)thio)-prop
anedioic acid; W
134104

Str 418

539 C28 H53 N O5 Si2 57305-36-9 EP-8264-0-0 Pregnan-20-one, 3,11-dihydroxy-17,21-bis[(trimethylsilyl)oxy]-, O-methyloxime,
(3α,5β,11β)-;; 5β-PREGNANE-3α,11β,17α,21-TETROL-20-ONE MO TMS; W/NBS
64082

Str 419

539 C28 H53 N O5 Si2 24524-81-0 MO-0-72-0 PROSTAGLANDIN-E-METHOXIME-DITMS-METHYL ESTER; W/NBS
64080

539 C28 H49 D4 N O5 Si2 55887-60-0 MO-0-73-0 BISTRIMETHYLSILYL 3,3,4,4-TETRADEUTERO PROSTAGLANDIN E2 METHOXIME METHYLESTER;
W
64081

539 C28 H53 N O5 Si2 OM-1981-315-0 PGE-MOX METHYL ESTER 2TMS; W
64083

539 C28 H49 D4 N O5 Si2 OM-1981-316-0 PGE-2 D4 MOX METHYL ESTER 2TMS; W
64084

539   C30 H27 Fe O6   14323-17-2   T-66-2504-0   Iron, tris(1-phenyl-1,3-butanedionato-O,O')-;; TRIS(BENZOYLACETONATO)IRON(III);
W/NBS   64085

Str 420

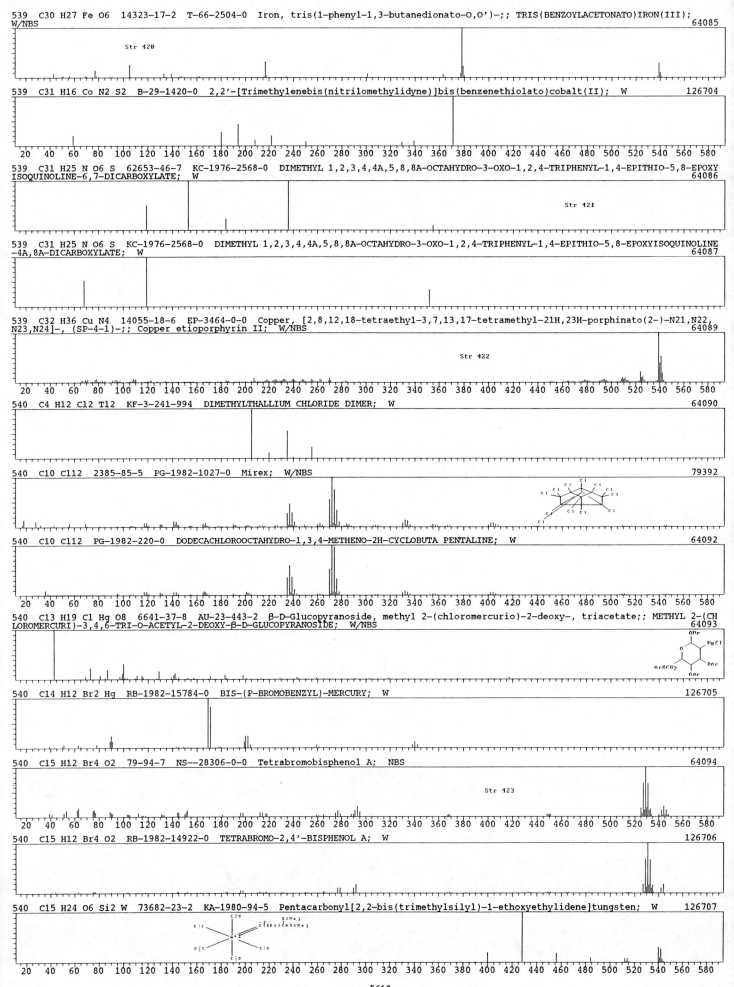

539   C31 H16 Co N2 S2   B-29-1420-0   2,2'-[Trimethylenebis(nitrilomethylidyne)]bis(benzenethiolato)cobalt(II);   W   126704

539   C31 H25 N O6 S   62653-46-7   KC-1976-2568-0   DIMETHYL 1,2,3,4,4A,5,8,8A-OCTAHYDRO-3-OXO-1,2,4-TRIPHENYL-1,4-EPITHIO-5,8-EPOXY
ISOQUINOLINE-6,7-DICARBOXYLATE;   W   64086

Str 421

539   C31 H25 N O6 S   KC-1976-2568-0   DIMETHYL 1,2,3,4,4A,5,8,8A-OCTAHYDRO-3-OXO-1,2,4-TRIPHENYL-1,4-EPITHIO-5,8-EPOXYISOQUINOLINE
-4A,8A-DICARBOXYLATE;   W   64087

539   C32 H36 Cu N4   14055-18-6   EP-3464-0-0   Copper, [2,8,12,18-tetraethyl-3,7,13,17-tetramethyl-21H,23H-porphinato(2-)-N21,N22,
N23,N24]-, (SP-4-1)-;; Copper etioporphyrin II;   W/NBS   64089

Str 422

540   C4 H12 Cl2 Tl2   KF-3-241-994   DIMETHYLTHALLIUM CHLORIDE DIMER;   W   64090

540   C10 Cl12   2385-85-5   PG-1982-1027-0   Mirex;   W/NBS   79392

540   C10 Cl12   PG-1982-220-0   DODECACHLOROOCTAHYDRO-1,3,4-METHENO-2H-CYCLOBUTA PENTALINE;   W   64092

540   C13 H19 Cl Hg O8   6641-37-8   AU-23-443-2   β-D-Glucopyranoside, methyl 2-(chloromercurio)-2-deoxy-, triacetate;; METHYL 2-(CH
LOROMERCURI)-3,4,6-TRI-O-ACETYL-2-DEOXY-β-D-GLUCOPYRANOSIDE;   W/NBS   64093

540   C14 H12 Br2 Hg   RB-1982-15784-0   BIS-(P-BROMOBENZYL)-MERCURY;   W   126705

540   C15 H12 Br4 O2   79-94-7   NS--28306-0-0   Tetrabromobisphenol A;   NBS   64094

Str 423

540   C15 H12 Br4 O2   RB-1982-14922-0   TETRABROMO-2,4'-BISPHENOL A;   W   126706

540   C15 H24 O6 Si2 W   73682-23-2   KA-1980-94-5   Pentacarbonyl[2,2-bis(trimethylsilyl)-1-ethoxyethylidene]tungsten;   W   126707

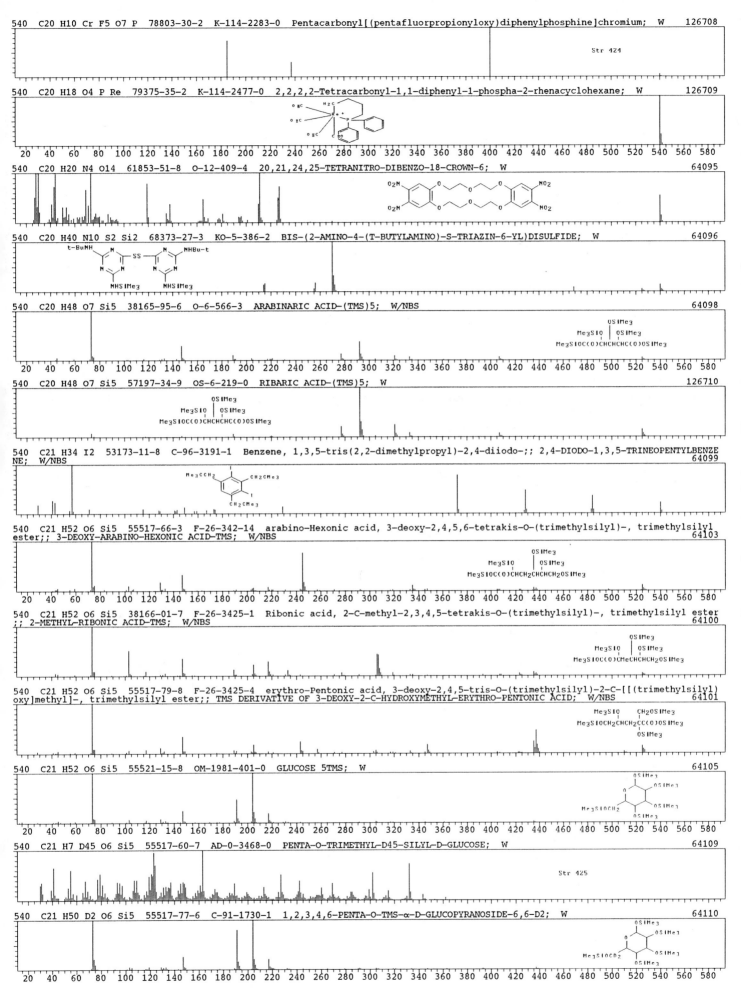

540   C20 H10 Cr F5 O7 P   78803-30-2   K-114-2283-0   Pentacarbonyl[(pentafluorpropionyloxy)diphenylphosphine]chromium;   W   126708

Str 424

540   C20 H18 O4 P Re   79375-35-2   K-114-2477-0   2,2,2,2-Tetracarbonyl-1,1-diphenyl-1-phospha-2-rhenacyclohexane;   W   126709

540   C20 H20 N4 O14   61853-51-8   O-12-409-4   20,21,24,25-TETRANITRO-DIBENZO-18-CROWN-6;   W   64095

540   C20 H40 N10 S2 Si2   68373-27-3   KO-5-386-2   BIS-(2-AMINO-4-(T-BUTYLAMINO)-S-TRIAZIN-6-YL)DISULFIDE;   W   64096

540   C20 H48 O7 Si5   38165-95-6   O-6-566-3   ARABINARIC ACID-(TMS)5;   W/NBS   64098

540   C20 H48 O7 Si5   57197-34-9   OS-6-219-0   RIBARIC ACID-(TMS)5;   W   126710

540   C21 H34 I2   53173-11-8   C-96-3191-1   Benzene, 1,3,5-tris(2,2-dimethylpropyl)-2,4-diiodo-;; 2,4-DIODO-1,3,5-TRINEOPENTYLBENZENE;   W/NBS   64099

540   C21 H52 O6 Si5   55517-66-3   F-26-342-14   arabino-Hexonic acid, 3-deoxy-2,4,5,6-tetrakis-O-(trimethylsilyl)-, trimethylsilyl ester;; 3-DEOXY-ARABINO-HEXONIC ACID-TMS;   W/NBS   64103

540   C21 H52 O6 Si5   38166-01-7   F-26-3425-1   Ribonic acid, 2-C-methyl-2,3,4,5-tetrakis-O-(trimethylsilyl)-, trimethylsilyl ester;; 2-METHYL-RIBONIC ACID-TMS;   W/NBS   64100

540   C21 H52 O6 Si5   55517-79-8   F-26-3425-4   erythro-Pentonic acid, 3-deoxy-2,4,5-tris-O-(trimethylsilyl)-2-C-[[(trimethylsilyl)oxy]methyl]-, trimethylsilyl ester;; TMS DERIVATIVE OF 3-DEOXY-2-C-HYDROXYMETHYL-ERYTHRO-PENTONIC ACID;   W/NBS   64101

540   C21 H52 O6 Si5   55521-15-8   OM-1981-401-0   GLUCOSE 5TMS;   W   64105

540   C21 H7 D45 O6 Si5   55517-60-7   AD-0-3468-0   PENTA-O-TRIMETHYL-D45-SILYL-D-GLUCOSE;   W   64109

Str 425

540   C21 H50 D2 O6 Si5   55517-77-6   C-91-1730-1   1,2,3,4,6-PENTA-O-TMS-α-D-GLUCOPYRANOSIDE-6,6-D2;   W   64110

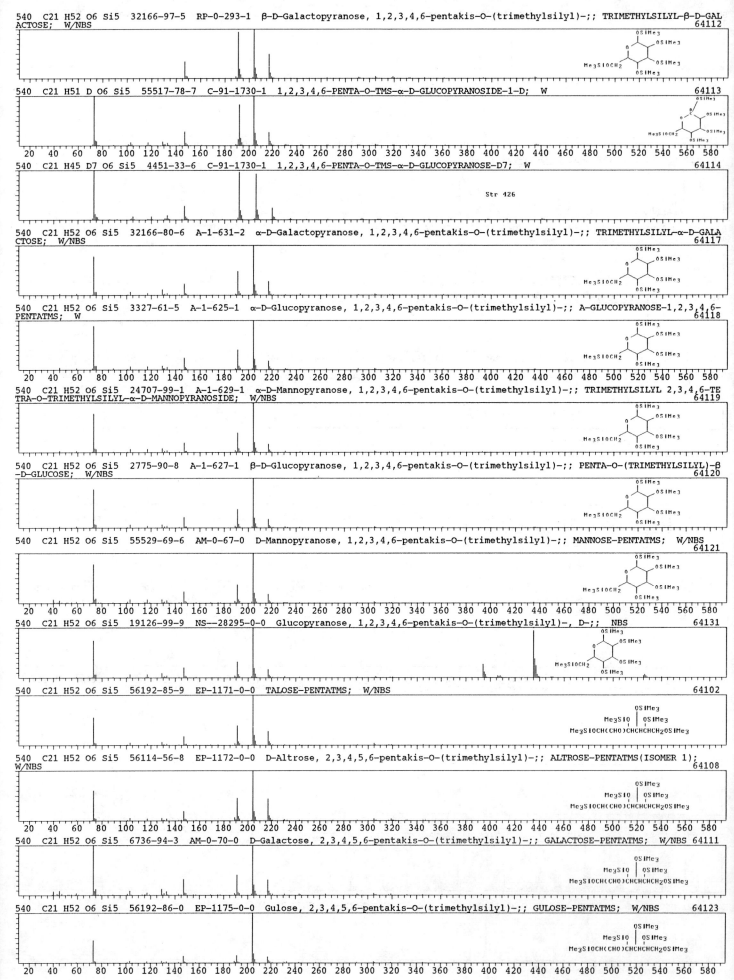

540　C21 H52 O6 Si5　32166-97-5　RP-0-293-1　β-D-Galactopyranose, 1,2,3,4,6-pentakis-O-(trimethylsilyl)-;; TRIMETHYLSILYL-β-D-GALACTOSE; W/NBS
64112

540　C21 H51 D O6 Si5　55517-78-7　C-91-1730-1　1,2,3,4,6-PENTA-O-TMS-α-D-GLUCOPYRANOSIDE-1-D; W
64113

540　C21 H45 D7 O6 Si5　4451-33-6　C-91-1730-1　1,2,3,4,6-PENTA-O-TMS-α-D-GLUCOPYRANOSE-D7; W
64114

Str 426

540　C21 H52 O6 Si5　32166-80-6　A-1-631-2　α-D-Galactopyranose, 1,2,3,4,6-pentakis-O-(trimethylsilyl)-;; TRIMETHYLSILYL-α-D-GALACTOSE; W/NBS
64117

540　C21 H52 O6 Si5　3327-61-5　A-1-625-1　α-D-Glucopyranose, 1,2,3,4,6-pentakis-O-(trimethylsilyl)-;; A-GLUCOPYRANOSE-1,2,3,4,6-PENTATMS; W
64118

540　C21 H52 O6 Si5　24707-99-1　A-1-629-1　α-D-Mannopyranose, 1,2,3,4,6-pentakis-O-(trimethylsilyl)-;; TRIMETHYLSILYL 2,3,4,6-TETRA-O-TRIMETHYLSILYL-α-D-MANNOPYRANOSIDE; W/NBS
64119

540　C21 H52 O6 Si5　2775-90-8　A-1-627-1　β-D-Glucopyranose, 1,2,3,4,6-pentakis-O-(trimethylsilyl)-;; PENTA-O-(TRIMETHYLSILYL)-β-D-GLUCOSE; W/NBS
64120

540　C21 H52 O6 Si5　55529-69-6　AM-0-67-0　D-Mannopyranose, 1,2,3,4,6-pentakis-O-(trimethylsilyl)-;; MANNOSE-PENTATMS; W/NBS
64121

540　C21 H52 O6 Si5　19126-99-9　NS--28295-0-0　Glucopyranose, 1,2,3,4,6-pentakis-O-(trimethylsilyl)-, D-;; NBS
64131

540　C21 H52 O6 Si5　56192-85-9　EP-1171-0-0　TALOSE-PENTATMS; W/NBS
64102

540　C21 H52 O6 Si5　56114-56-8　EP-1172-0-0　D-Altrose, 2,3,4,5,6-pentakis-O-(trimethylsilyl)-;; ALTROSE-PENTATMS(ISOMER 1); W/NBS
64108

540　C21 H52 O6 Si5　6736-94-3　AM-0-70-0　D-Galactose, 2,3,4,5,6-pentakis-O-(trimethylsilyl)-;; GALACTOSE-PENTATMS; W/NBS 64111

540　C21 H52 O6 Si5　56192-86-0　EP-1175-0-0　Gulose, 2,3,4,5,6-pentakis-O-(trimethylsilyl)-;; GULOSE-PENTATMS; W/NBS
64123

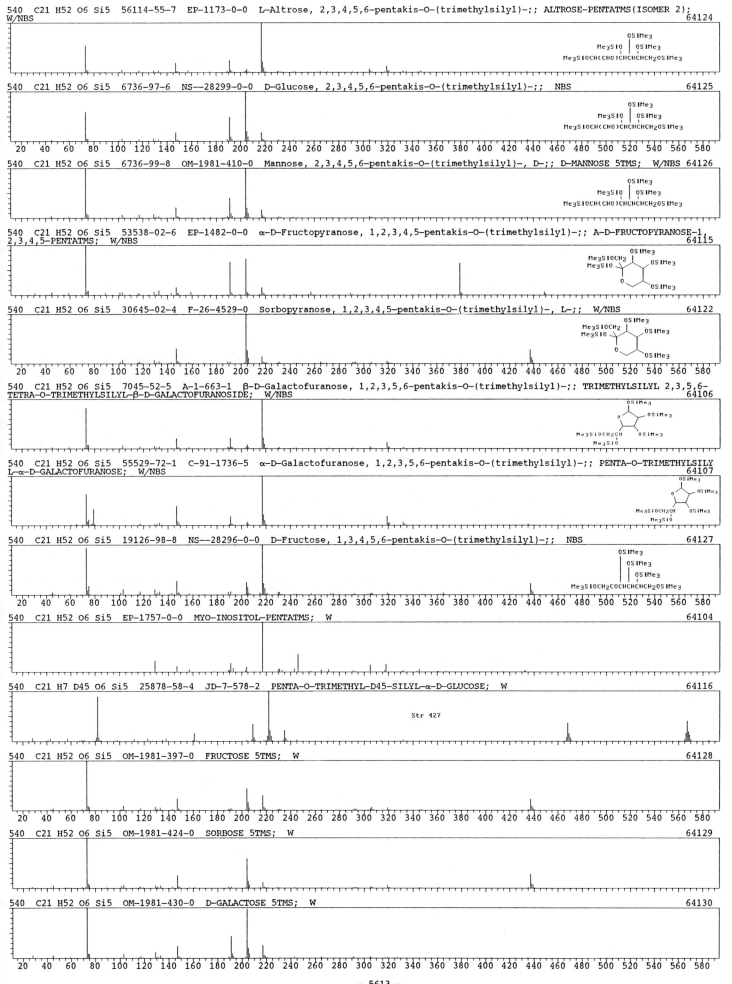

540　C21 H52 O6 Si5　56114-55-7　EP-1173-0-0　L-Altrose, 2,3,4,5,6-pentakis-O-(trimethylsilyl)-;; ALTROSE-PENTATMS(ISOMER 2);
W/NBS　　64124

OSiMe3
Me3SiO │ OSiMe3
Me3SiOCH(CHO)CHCHCHCH2OSiMe3

540　C21 H52 O6 Si5　6736-97-6　NS--28299-0-0　D-Glucose, 2,3,4,5,6-pentakis-O-(trimethylsilyl)-;;　NBS　　64125

OSiMe3
Me3SiO │ OSiMe3
Me3SiOCH(CHO)CHCHCHCH2OSiMe3

540　C21 H52 O6 Si5　6736-99-8　OM-1981-410-0　Mannose, 2,3,4,5,6-pentakis-O-(trimethylsilyl)-, D-;; D-MANNOSE 5TMS;　W/NBS　64126

OSiMe3
Me3SiO │ OSiMe3
Me3SiOCH(CHO)CHCHCHCH2OSiMe3

540　C21 H52 O6 Si5　53538-02-6　EP-1482-0-0　α-D-Fructopyranose, 1,2,3,4,5-pentakis-O-(trimethylsilyl)-;; A-D-FRUCTOPYRANOSE-1
2,3,4,5-PENTATMS;　W/NBS　　64115

540　C21 H52 O6 Si5　30645-02-4　F-26-4529-0　Sorbopyranose, 1,2,3,4,5-pentakis-O-(trimethylsilyl)-, L-;;　W/NBS　　64122

540　C21 H52 O6 Si5　7045-52-5　A-1-663-1　β-D-Galactofuranose, 1,2,3,5,6-pentakis-O-(trimethylsilyl)-;; TRIMETHYLSILYL 2,3,5,6-
TETRA-O-TRIMETHYLSILYL-β-D-GALACTOFURANOSIDE;　W/NBS　　64106

540　C21 H52 O6 Si5　55529-72-1　C-91-1736-5　α-D-Galactofuranose, 1,2,3,5,6-pentakis-O-(trimethylsilyl)-;; PENTA-O-TRIMETHYLSILY
L-α-D-GALACTOFURANOSE;　W/NBS　　64107

540　C21 H52 O6 Si5　19126-98-8　NS--28296-0-0　D-Fructose, 1,3,4,5,6-pentakis-O-(trimethylsilyl)-;;　NBS　　64127

OSiMe3
│
OSiMe3
│
OSiMe3
│
Me3SiOCH2COCHCHCHCH2OSiMe3

540　C21 H52 O6 Si5　EP-1757-0-0　MYO-INOSITOL-PENTATMS;　W　　64104

540　C21 H7 D45 O6 Si5　25878-58-4　JD-7-578-2　PENTA-O-TRIMETHYL-D45-SILYL-α-D-GLUCOSE;　W　　64116

Str 427

540　C21 H52 O6 Si5　OM-1981-397-0　FRUCTOSE 5TMS;　W　　64128

540　C21 H52 O6 Si5　OM-1981-424-0　SORBOSE 5TMS;　W　　64129

540　C21 H52 O6 Si5　OM-1981-430-0　D-GALACTOSE 5TMS;　W　　64130

- 5613 -

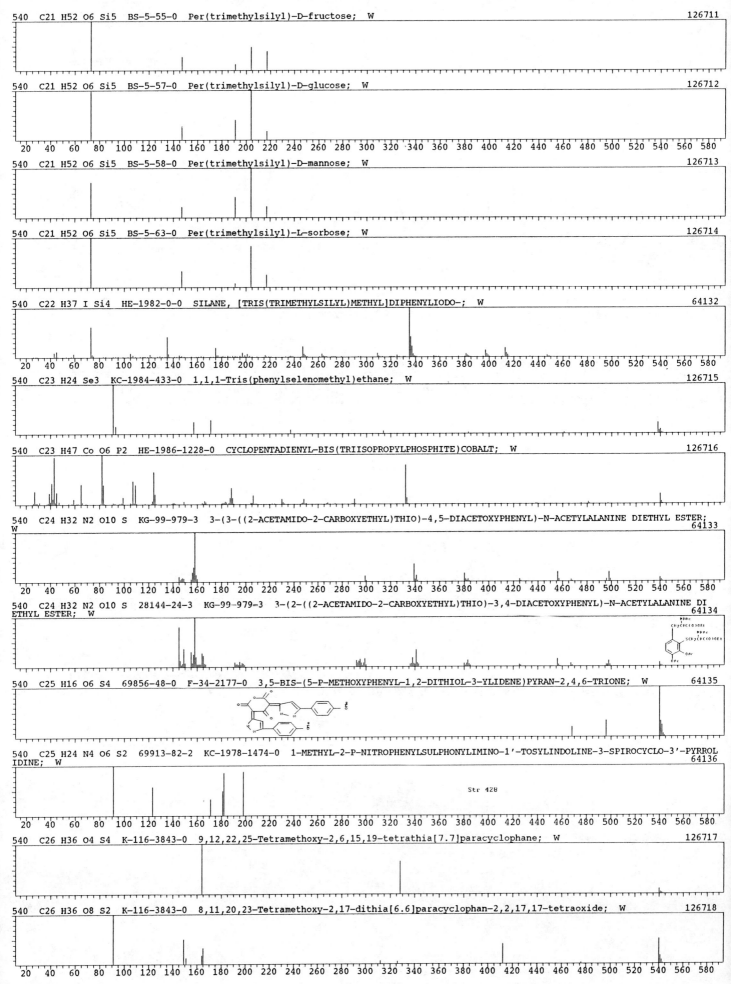

540   C21 H52 O6 Si5   BS-5-55-0   Per(trimethylsilyl)-D-fructose;   W   126711

540   C21 H52 O6 Si5   BS-5-57-0   Per(trimethylsilyl)-D-glucose;   W   126712

540   C21 H52 O6 Si5   BS-5-58-0   Per(trimethylsilyl)-D-mannose;   W   126713

540   C21 H52 O6 Si5   BS-5-63-0   Per(trimethylsilyl)-L-sorbose;   W   126714

540   C22 H37 I Si4   HE-1982-0-0   SILANE, [TRIS(TRIMETHYLSILYL)METHYL]DIPHENYLIODO-;   W   64132

540   C23 H24 Se3   KC-1984-433-0   1,1,1-Tris(phenylselenomethyl)ethane;   W   126715

540   C23 H47 Co O6 P2   HE-1986-1228-0   CYCLOPENTADIENYL-BIS(TRIISOPROPYLPHOSPHITE)COBALT;   W   126716

540   C24 H32 N2 O10 S   KG-99-979-3   3-(3-((2-ACETAMIDO-2-CARBOXYETHYL)THIO)-4,5-DIACETOXYPHENYL)-N-ACETYLALANINE DIETHYL ESTER;
W   64133

540   C24 H32 N2 O10 S   28144-24-3   KG-99-979-3   3-(2-((2-ACETAMIDO-2-CARBOXYETHYL)THIO)-3,4-DIACETOXYPHENYL)-N-ACETYLALANINE DI
ETHYL ESTER;   W   64134

540   C25 H16 O6 S4   69856-48-0   F-34-2177-0   3,5-BIS-(5-P-METHOXYPHENYL-1,2-DITHIOL-3-YLIDENE)PYRAN-2,4,6-TRIONE;   W   64135

540   C25 H24 N4 O6 S2   69913-82-2   KC-1978-1474-0   1-METHYL-2-P-NITROPHENYLSULPHONYLIMINO-1'-TOSYLINDOLINE-3-SPIROCYCLO-3'-PYRROL
IDINE;   W   64136

Str 428

540   C26 H36 O4 S4   K-116-3843-0   9,12,22,25-Tetramethoxy-2,6,15,19-tetrathia[7.7]paracyclophane;   W   126717

540   C26 H36 O8 S2   K-116-3843-0   8,11,20,23-Tetramethoxy-2,17-dithia[6.6]paracyclophan-2,2,17,17-tetraoxide;   W   126718

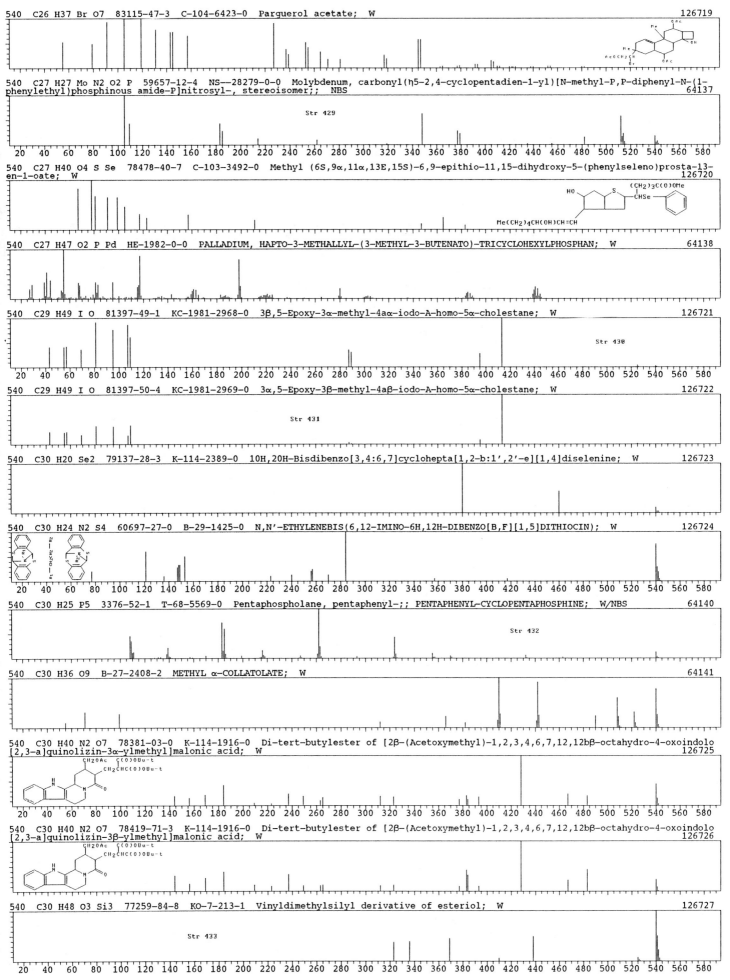

540　C26 H37 Br O7　83115-47-3　C-104-6423-0　Parguerol acetate;　W　126719

540　C27 H27 Mo N2 O2 P　59657-12-4　NS--28279-0-0　Molybdenum, carbonyl(η5-2,4-cyclopentadien-1-yl)[N-methyl-P,P-diphenyl-N-(1-phenylethyl)phosphinous amide-P]nitrosyl-, stereoisomer;;　NBS　64137

Str 429

540　C27 H40 O4 S Se　78478-40-7　C-103-3492-0　Methyl (6S,9α,11α,13E,15S)-6,9-epithio-11,15-dihydroxy-5-(phenylseleno)prosta-13-en-1-oate;　W　126720

540　C27 H47 O2 P Pd　HE-1982-0-0　PALLADIUM, HAPTO-3-METHALLYL-(3-METHYL-3-BUTENATO)-TRICYCLOHEXYLPHOSPHAN;　W　64138

540　C29 H49 I O　81397-49-1　KC-1981-2968-0　3β,5-Epoxy-3α-methyl-4aα-iodo-A-homo-5α-cholestane;　W　126721

Str 430

540　C29 H49 I O　81397-50-4　KC-1981-2969-0　3α,5-Epoxy-3β-methyl-4aβ-iodo-A-homo-5α-cholestane;　W　126722

Str 431

540　C30 H20 Se2　79137-28-3　K-114-2389-0　10H,20H-Bisdibenzo[3,4:6,7]cyclohepta[1,2-b:1',2'-e][1,4]diselenine;　W　126723

540　C30 H24 N2 S4　60697-27-0　B-29-1425-0　N,N'-ETHYLENEBIS(6,12-IMINO-6H,12H-DIBENZO[B,F][1,5]DITHIOCIN);　W　126724

540　C30 H25 P5　3376-52-1　T-68-5569-0　Pentaphospholane, pentaphenyl-;;　PENTAPHENYL-CYCLOPENTAPHOSPHINE;　W/NBS　64140

Str 432

540　C30 H36 O9　B-27-2408-2　METHYL α-COLLATOLATE;　W　64141

540　C30 H40 N2 O7　78381-03-0　K-114-1916-0　Di-tert-butylester of [2β-(Acetoxymethyl)-1,2,3,4,6,7,12,12bβ-octahydro-4-oxoindolo[2,3-a]quinolizin-3α-ylmethyl]malonic acid;　W　126725

540　C30 H40 N2 O7　78419-71-3　K-114-1916-0　Di-tert-butylester of [2β-(Acetoxymethyl)-1,2,3,4,6,7,12,12bβ-octahydro-4-oxoindolo[2,3-a]quinolizin-3β-ylmethyl]malonic acid;　W　126726

540　C30 H48 O3 Si3　77259-84-8　KO-7-213-1　Vinyldimethylsilyl derivative of esteriol;　W　126727

Str 433

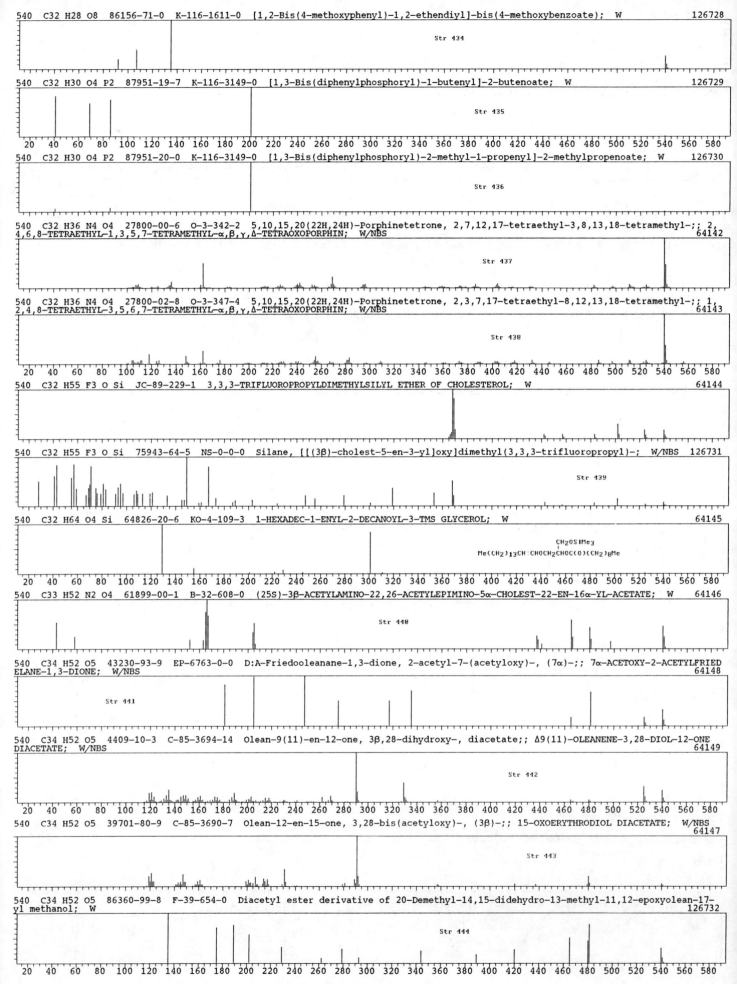

540　C32 H28 O8　86156-71-0　K-116-1611-0　[1,2-Bis(4-methoxyphenyl)-1,2-ethendiyl]-bis(4-methoxybenzoate);　W　　126728

Str 434

540　C32 H30 O4 P2　87951-19-7　K-116-3149-0　[1,3-Bis(diphenylphosphoryl)-1-butenyl]-2-butenoate;　W　　126729

Str 435

540　C32 H30 O4 P2　87951-20-0　K-116-3149-0　[1,3-Bis(diphenylphosphoryl)-2-methyl-1-propenyl]-2-methylpropenoate;　W　　126730

Str 436

540　C32 H36 N4 O4　27800-00-6　O-3-342-2　5,10,15,20(22H,24H)-Porphinetetrone, 2,7,12,17-tetraethyl-3,8,13,18-tetramethyl-;; 2,4,6,8-TETRAETHYL-1,3,5,7-TETRAMETHYL-α,β,γ,Δ-TETRAOXOPORPHIN;　W/NBS　　64142

Str 437

540　C32 H36 N4 O4　27800-02-8　O-3-347-4　5,10,15,20(22H,24H)-Porphinetetrone, 2,3,7,17-tetraethyl-8,12,13,18-tetramethyl-;; 1,2,4,8-TETRAETHYL-3,5,6,7-TETRAMETHYL-α,β,γ,Δ-TETRAOXOPORPHIN;　W/NBS　　64143

Str 438

540　C32 H55 F3 O Si　JC-89-229-1　3,3,3-TRIFLUOROPROPYLDIMETHYLSILYL ETHER OF CHOLESTEROL;　W　　64144

540　C32 H55 F3 O Si　75943-64-5　NS-0-0-0　Silane, [[(3β)-cholest-5-en-3-yl]oxy]dimethyl(3,3,3-trifluoropropyl)-;　W/NBS　126731

Str 439

540　C32 H64 O4 Si　64826-20-6　KO-4-109-3　1-HEXADEC-1-ENYL-2-DECANOYL-3-TMS GLYCEROL;　W　　64145

CH2OSiMe3
|
Me(CH2)13CH:CHOCH2CHOC(O)(CH2)8Me

540　C33 H52 N2 O4　61899-00-1　B-32-608-0　(25S)-3β-ACETYLAMINO-22,26-ACETYLEPIMINO-5α-CHOLEST-22-EN-16α-YL-ACETATE;　W　　64146

Str 440

540　C34 H52 O5　43230-93-9　EP-6763-0-0　D:A-Friedooleanane-1,3-dione, 2-acetyl-7-(acetyloxy)-, (7α)-;; 7α-ACETOXY-2-ACETYLFRIEDELANE-1,3-DIONE;　W/NBS　　64148

Str 441

540　C34 H52 O5　4409-10-3　C-85-3694-14　Olean-9(11)-en-12-one, 3β,28-dihydroxy-, diacetate;; Δ9(11)-OLEANENE-3,28-DIOL-12-ONE DIACETATE;　W/NBS　　64149

Str 442

540　C34 H52 O5　39701-80-9　C-85-3690-7　Olean-12-en-15-one, 3,28-bis(acetyloxy)-, (3β)-;; 15-OXOERYTHRODIOL DIACETATE;　W/NBS　　64147

Str 443

540　C34 H52 O5　86360-99-8　F-39-654-0　Diacetyl ester derivative of 20-Demethyl-14,15-didehydro-13-methyl-11,12-epoxyolean-17-yl methanol;　W　　126732

Str 444

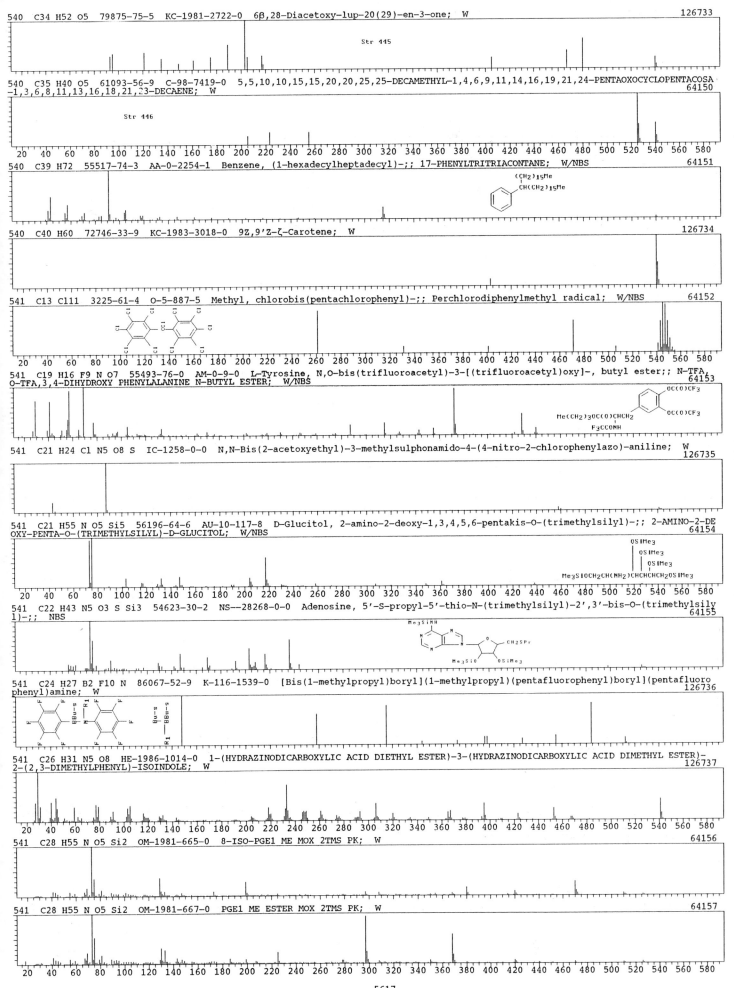

540　C34 H52 O5　79875-75-5　KC-1981-2722-0　6β,28-Diacetoxy-lup-20(29)-en-3-one;　W　　　　126733

Str 445

540　C35 H40 O5　61093-56-9　C-98-7419-0　5,5,10,10,15,15,20,20,25,25-DECAMETHYL-1,4,6,9,11,14,16,19,21,24-PENTAOXOCYCLOPENTACOSA
-1,3,6,8,11,13,16,18,21,23-DECAENE;　W　　　　64150

Str 446

540　C39 H72　55517-74-3　AA-0-2254-1　Benzene, (1-hexadecylheptadecyl)-;; 17-PHENYLTRITRIACONTANE;　W/NBS　　　64151

540　C40 H60　72746-33-9　KC-1983-3018-0　9Z,9'Z-ζ-Carotene;　W　　　　126734

541　C13 Cl11　3225-61-4　O-5-887-5　Methyl, chlorobis(pentachlorophenyl)-;; Perchlorodiphenylmethyl radical;　W/NBS　　　64152

541　C19 H16 F9 N O7　55493-76-0　AM-0-9-0　L-Tyrosine, N,O-bis(trifluoroacetyl)-3-[(trifluoroacetyl)oxy]-, butyl ester;; N-TFA,
O-TFA,3,4-DIHYDROXY PHENYLALANINE N-BUTYL ESTER;　W/NBS　　　64153

541　C21 H24 Cl N5 O8 S　IC-1258-0-0　N,N-Bis(2-acetoxyethyl)-3-methylsulphonamido-4-(4-nitro-2-chlorophenylazo)-aniline;　W
126735

541　C21 H55 N O5 Si5　56196-64-6　AU-10-117-8　D-Glucitol, 2-amino-2-deoxy-1,3,4,5,6-pentakis-O-(trimethylsilyl)-;; 2-AMINO-2-DE
OXY-PENTA-O-(TRIMETHYLSILYL)-D-GLUCITOL;　W/NBS　　　64154

541　C22 H43 N5 O3 S Si3　54623-30-2　NS--28268-0-0　Adenosine, 5'-S-propyl-5'-thio-N-(trimethylsilyl)-2',3'-bis-O-(trimethylsily
l)-;;　NBS　　　64155

541　C24 H27 B2 F10 N　86067-52-9　K-116-1539-0　[Bis(1-methylpropyl)boryl](1-methylpropyl)(pentafluorophenyl)boryl](pentafluoro
phenyl)amine;　W　　　126736

541　C26 H31 N5 O8　HE-1986-1014-0　1-(HYDRAZINODICARBOXYLIC ACID DIETHYL ESTER)-3-(HYDRAZINODICARBOXYLIC ACID DIMETHYL ESTER)-
2-(2,3-DIMETHYLPHENYL)-ISOINDOLE;　W　　　126737

541　C28 H55 N O5 Si2　OM-1981-665-0　8-ISO-PGE1 ME MOX 2TMS PK;　W　　　64156

541　C28 H55 N O5 Si2　OM-1981-667-0　PGE1 ME ESTER MOX 2TMS PK;　W　　　64157

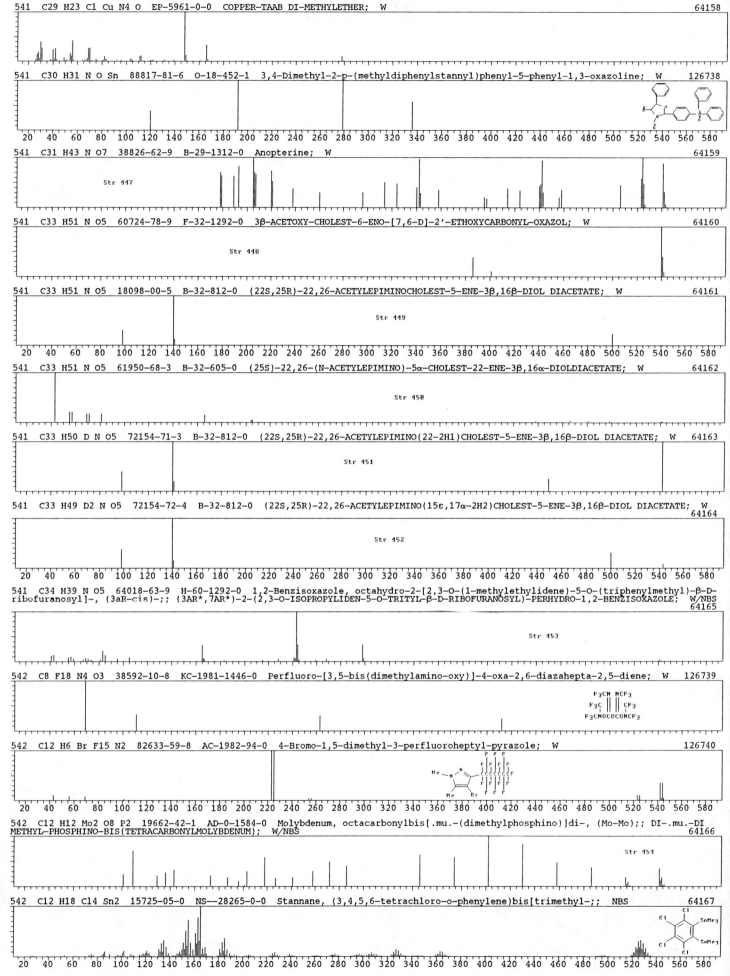

541　C29 H23 Cl Cu N4 O　EP-5961-0-0　COPPER-TAAB DI-METHYLETHER;　W　64158

541　C30 H31 N O Sn　88817-81-6　O-18-452-1　3,4-Dimethyl-2-p-(methyldiphenylstannyl)phenyl-5-phenyl-1,3-oxazoline;　W　126738

541　C31 H43 N O7　38826-62-9　B-29-1312-0　Anopterine;　W　64159
Str 447

541　C33 H51 N O5　60724-78-9　F-32-1292-0　3β-ACETOXY-CHOLEST-6-ENO-[7,6-D]-2'-ETHOXYCARBONYL-OXAZOL;　W　64160
Str 448

541　C33 H51 N O5　18098-00-5　B-32-812-0　(22S,25R)-22,26-ACETYLEPIMINOCHOLEST-5-ENE-3β,16β-DIOL DIACETATE;　W　64161
Str 449

541　C33 H51 N O5　61950-68-3　B-32-605-0　(25S)-22,26-(N-ACETYLEPIMINO)-5α-CHOLEST-22-ENE-3β,16α-DIOLDIACETATE;　W　64162
Str 450

541　C33 H50 D N O5　72154-71-3　B-32-812-0　(22S,25R)-22,26-ACETYLEPIMINO(22-2H1)CHOLEST-5-ENE-3β,16β-DIOL DIACETATE;　W　64163
Str 451

541　C33 H49 D2 N O5　72154-72-4　B-32-812-0　(22S,25R)-22,26-ACETYLEPIMINO(15ε,17α-2H2)CHOLEST-5-ENE-3β,16β-DIOL DIACETATE;　W　64164
Str 452

541　C34 H39 N O5　64018-63-9　H-60-1292-0　1,2-Benzisoxazole, octahydro-2-[2,3-O-(1-methylethylidene)-5-O-(triphenylmethyl)-β-D-ribofuranosyl]-, (3aR-cis)-;; (3AR*,7AR*)-2-(2,3-O-ISOPROPYLIDEN-5-O-TRITYL-β-D-RIBOFURANOSYL)-PERHYDRO-1,2-BENZISOXAZOLE;　W/NBS　64165
Str 453

542　C8 F18 N4 O3　38592-10-8　KC-1981-1446-0　Perfluoro-[3,5-bis(dimethylamino-oxy)]-4-oxa-2,6-diazahepta-2,5-diene;　W　126739

542　C12 H6 Br F15 N2　82633-59-8　AC-1982-94-0　4-Bromo-1,5-dimethyl-3-perfluoroheptyl-pyrazole;　W　126740

542　C12 H12 Mo2 O8 P2　19662-42-1　AD-0-1584-0　Molybdenum, octacarbonylbis[.mu.-(dimethylphosphino)]di-, (Mo-Mo);; DI-.mu.-DI METHYL-PHOSPHINO-BIS(TETRACARBONYLMOLYBDENUM);　W/NBS　64166
Str 454

542　C12 H18 Cl4 Sn2　15725-05-0　NS--28265-0-0　Stannane, (3,4,5,6-tetrachloro-o-phenylene)bis[trimethyl-;;　NBS　64167

542 C12 H18 Cl4 Sn2 18689-05-9 NS--28264-0-0 Stannane, (2,3,5,6-tetrachloro-p-phenylene)bis[trimethyl-;; NBS   64168

542 C16 H14 Pb S4 68409-47-2 J-44-571-0 Lead dithiophenylacetate; W   126741

542 C17 H11 Br2 Cu N3 O4 55494-17-2 AD-0-3218-0 Copper, bis(4-bromo-3,5-cyclohexadiene-1,2-dione i-oximato-N2,O1)(pyridine)-
;; BIS(4-BROMO-1-QUINONE-2-OXIMATO)COPPER(II)-PYRIDINE 1:1 ADDUCT; W/NBS   64169

Str 455

542 C20 H4 F14 O2 38795-55-0 OS-6-338-0 Benzene, 1,2,4,5-tetrafluoro-3,6-bis[(pentafluorophenoxy)methyl]-;; Bis(pentafluoroph
enoxy)-2,3,5,6-tetrafluoroxylylene; W   126742

542 C22 H30 Co I P2 K-110-3491-0 (CYCLOPENTADIENYL)BIS(DIMETHYLPHENYLPHOSPHIN)METHYLCOBALT(III)-IODIDE; W   64171

542 C23 H33 Co3 Si2 HE-1986-2033-0 TRIS(CYCLOPENTADIENYLCOBALT).MU.-(TRIMETHYLSILYLMETHYLIDYNE); W   126743

542 C24 H14 F8 Fe O2 55494-18-3 O-2-9988-0 Ferrocene, 1,1'-bis(2,3,4,5-tetrafluoro-6-methoxyphenyl)-;; DI-O-METHOXYTETRAFLUOR
OFERROCENE; W/NBS   64172

542 C24 H22 Cl4 N2 O2 S KC-1977-706-710 3,5-BIS-(3,5-DICHLOR-2,4,6-TRIMETHYLPHENYL)-3A,4A,7A,7B-TETRAHYDROTHIENO(2,3-D:5,4-D')
DI-ISOXAZOLE; W   64173

542 C24 H22 Cl4 N2 O2 S KC-1977-706-710 3,7-BIS-(3,5-DICHLORO-2,4,6-TRIMETHYLPHENYL)-3A,4A,7A,7BTETRAHYDROTHIENO(2,3-D:4,5-DI-
ISOXAZOLE; W   64174

542 C24 H44 As2 N2 S 91173-30-7 K-117-2004-0 Bis(dicyclohexylarsino)sulfurimide; W   126744

542 C24 H46 O13 32581-27-4 AU-17-15-20 Xylitol, O-2,3,4-tri-O-methyl-β-D-xylopyranosyl-(1.fwdarw.3)-O-2,4-di-O-methyl-β-D-xyl
opyranosyl-(1.fwdarw.3)-1,2,4,5-tetra-O-methyl-, L-;; PERMETHYL(O-β-D-XYLP-(1-3)-O-β-D-XYLP-(1-4)-D-XYL)ALDITOL; W/NBS   64175

Str 456

542 C26 H35 Cl O10 77573-33-2 NS-0-0-0 5H-Cyclopropa[3,4]benz[1,2-e]azulen-5-one, 9,9a-bis(acetyloxy)-3-[(acetyloxy)methyl]-
2-chloro-1,1a,1b,2,3,4,4a,7a,7b,8,9,9a-dodecahydro-3,4a,7b-trihydroxy-1,1,6,8-tetramethyl-, [1aR-(1aα,1bβ,2α,3α,4aβ,7aα,7bα,8α,
9β,9aα)]-; W/NBS   126745

Str 457

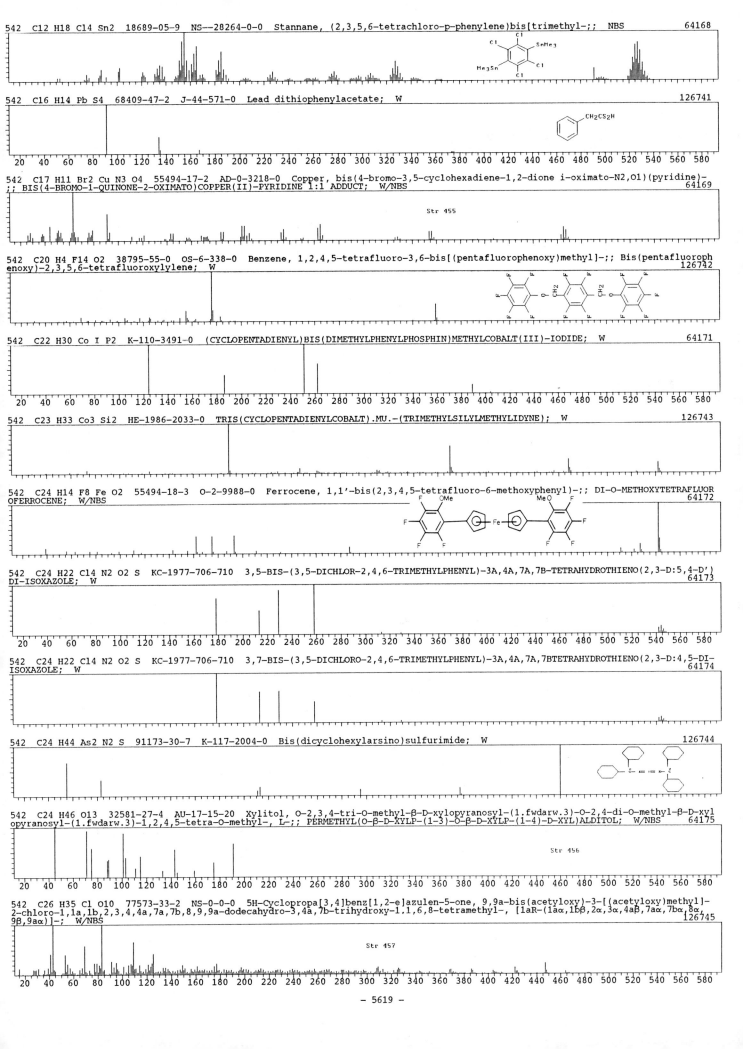

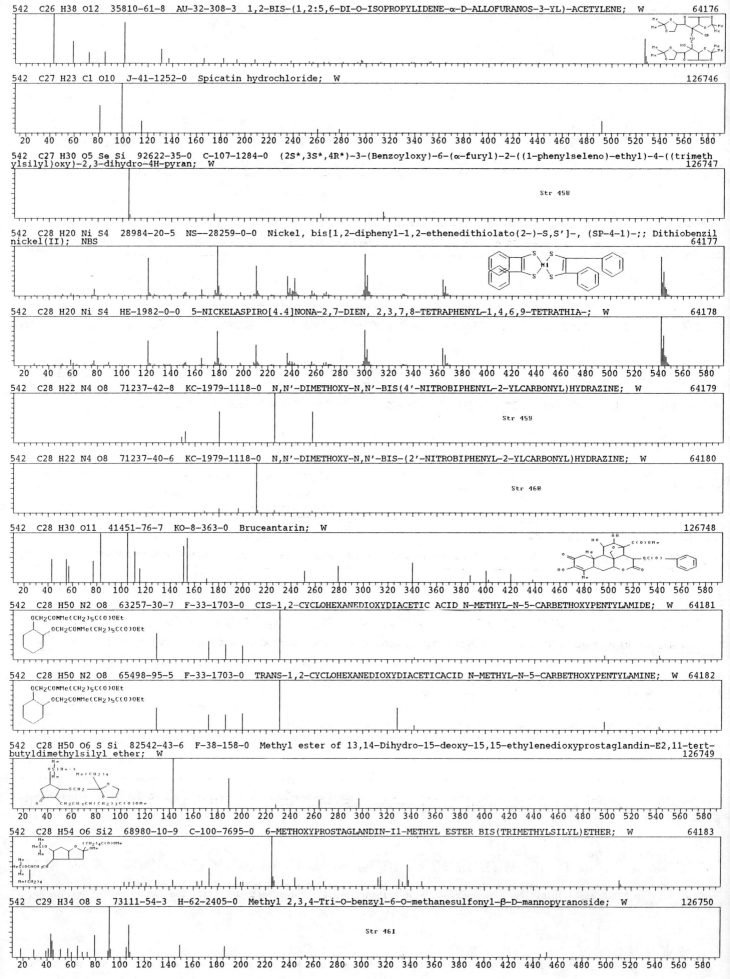

542   C26 H38 O12   35810-61-8   AU-32-308-3   1,2-BIS-(1,2:5,6-DI-O-ISOPROPYLIDENE-α-D-ALLOFURANOS-3-YL)-ACETYLENE;   W   64176

542   C27 H23 Cl O10   J-41-1252-0   Spicatin hydrochloride;   W   126746

542   C27 H30 O5 Se Si   92622-35-0   C-107-1284-0   (2S*,3S*,4R*)-3-(Benzoyloxy)-6-(α-furyl)-2-((1-phenylseleno)-ethyl)-4-((trimethylsilyl)oxy)-2,3-dihydro-4H-pyran;   W   126747

Str 458

542   C28 H20 Ni S4   28984-20-5   NS--28259-0-0   Nickel, bis[1,2-diphenyl-1,2-ethenedithiolato(2-)-S,S']-, (SP-4-1)-;; Dithiobenzil nickel(II);   NBS   64177

542   C28 H20 Ni S4   HE-1982-0-0   5-NICKELASPIRO[4.4]NONA-2,7-DIEN, 2,3,7,8-TETRAPHENYL-1,4,6,9-TETRATHIA-;   W   64178

542   C28 H22 N4 O8   71237-42-8   KC-1979-1118-0   N,N'-DIMETHOXY-N,N'-BIS(4'-NITROBIPHENYL-2-YLCARBONYL)HYDRAZINE;   W   64179

Str 459

542   C28 H22 N4 O8   71237-40-6   KC-1979-1118-0   N,N'-DIMETHOXY-N,N'-BIS-(2'-NITROBIPHENYL-2-YLCARBONYL)HYDRAZINE;   W   64180

Str 460

542   C28 H30 O11   41451-76-7   KO-8-363-0   Bruceantarin;   W   126748

542   C28 H50 N2 O8   63257-30-7   F-33-1703-0   CIS-1,2-CYCLOHEXANEDIOXYDIACETIC ACID N-METHYL-N-5-CARBETHOXYPENTYLAMIDE;   W   64181

OCH2CONMe(CH2)5C(O)OEt
OCH2CONMe(CH2)5C(O)OEt

542   C28 H50 N2 O8   65498-95-5   F-33-1703-0   TRANS-1,2-CYCLOHEXANEDIOXYDIACETICACID N-METHYL-N-5-CARBETHOXYPENTYLAMINE;   W   64182

OCH2CONMe(CH2)5C(O)OEt
OCH2CONMe(CH2)5C(O)OEt

542   C28 H50 O6 S Si   82542-43-6   F-38-158-0   Methyl ester of 13,14-Dihydro-15-deoxy-15,15-ethylenedioxyprostaglandin-E2,11-tert-butyldimethylsilyl ether;   W   126749

542   C28 H54 O6 Si2   68980-10-9   C-100-7695-0   6-METHOXYPROSTAGLANDIN-I1-METHYL ESTER BIS(TRIMETHYLSILYL)ETHER;   W   64183

542   C29 H34 O8 S   73111-54-3   H-62-2405-0   Methyl 2,3,4-Tri-O-benzyl-6-O-methanesulfonyl-β-D-mannopyranoside;   W   126750

Str 461

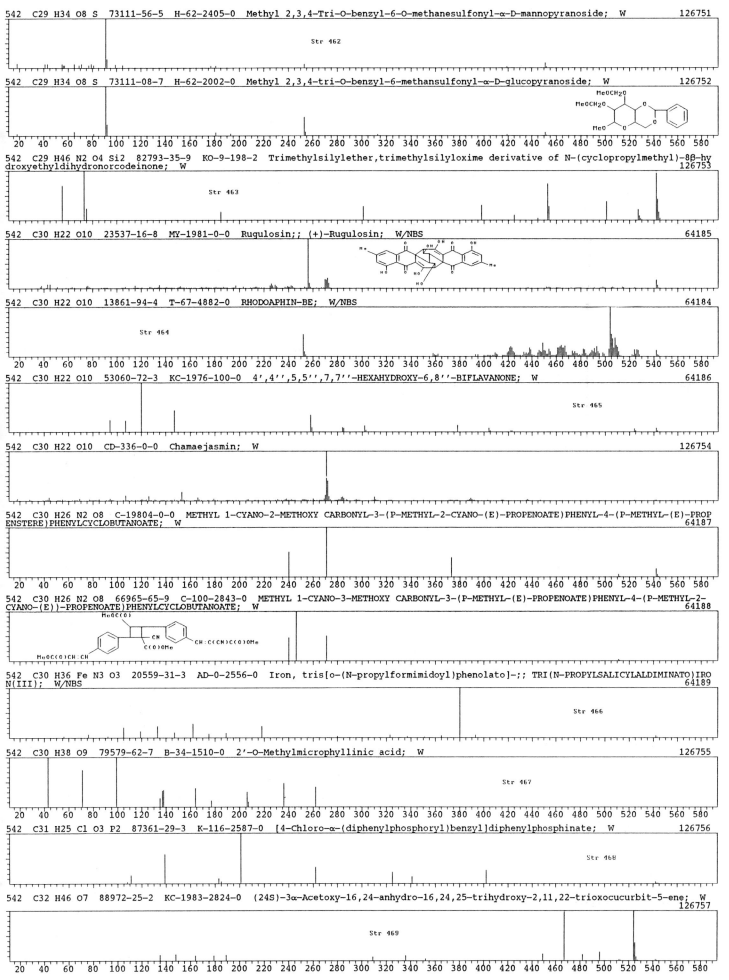

542 C29 H34 O8 S 73111-56-5 H-62-2405-0 Methyl 2,3,4-Tri-O-benzyl-6-O-methanesulfonyl-α-D-mannopyranoside; W   126751

Str 462

542 C29 H34 O8 S 73111-08-7 H-62-2002-0 Methyl 2,3,4-tri-O-benzyl-6-methansulfonyl-α-D-glucopyranoside; W   126752

542 C29 H46 N2 O4 Si2 82793-35-9 KO-9-198-2 Trimethylsilylether,trimethylsilyloxime derivative of N-(cyclopropylmethyl)-8β-hydroxyethyldihydronorcodeinone; W   126753

Str 463

542 C30 H22 O10 23537-16-8 MY-1981-0-0 Rugulosin;; (+)-Rugulosin; W/NBS   64185

542 C30 H22 O10 13861-94-4 T-67-4882-0 RHODOAPHIN-BE; W/NBS   64184

Str 464

542 C30 H22 O10 53060-72-3 KC-1976-100-0 4',4'',5,5'',7,7''-HEXAHYDROXY-6,8''-BIFLAVANONE; W   64186

Str 465

542 C30 H22 O10 CD-336-0-0 Chamaejasmin; W   126754

542 C30 H26 N2 O8 C-19804-0-0 METHYL 1-CYANO-2-METHOXY CARBONYL-3-(P-METHYL-2-CYANO-(E)-PROPENOATE)PHENYL-4-(P-METHYL-(E)-PROPENSTERE)PHENYLCYCLOBUTANOATE; W   64187

542 C30 H26 N2 O8 66965-65-9 C-100-2843-0 METHYL 1-CYANO-3-METHOXY CARBONYL-3-(P-METHYL-(E)-PROPENOATE)PHENYL-4-(P-METHYL-2-CYANO-(E))-PROPENOATE)PHENYLCYCLOBUTANOATE; W   64188

542 C30 H36 Fe N3 O3 20559-31-3 AD-0-2556-0 Iron, tris[o-(N-propylformimidoyl)phenolato]-;; TRI(N-PROPYLSALICYLALDIMINATO)IRON(III); W/NBS   64189

Str 466

542 C30 H38 O9 79579-62-7 B-34-1510-0 2'-O-Methylmicrophyllinic acid; W   126755

Str 467

542 C31 H25 Cl O3 P2 87361-29-3 K-116-2587-0 [4-Chloro-α-(diphenylphosphoryl)benzyl]diphenylphosphinate; W   126756

Str 468

542 C32 H46 O7 88972-25-2 KC-1983-2824-0 (24S)-3α-Acetoxy-16,24-anhydro-16,24,25-trihydroxy-2,11,22-trioxocucurbit-5-ene; W   126757

Str 469

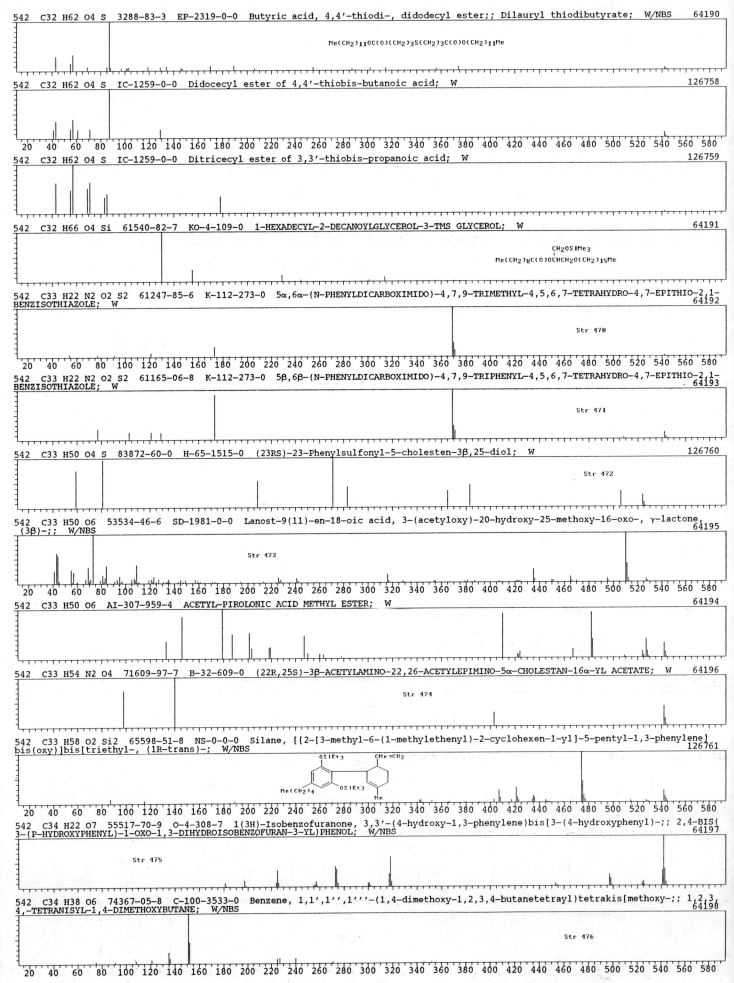

542 C32 H62 O4 S 3288-83-3 EP-2319-0-0 Butyric acid, 4,4'-thiodi-, didodecyl ester;; Dilauryl thiodibutyrate; W/NBS 64190

Me(CH2)11OC(O)(CH2)3S(CH2)3C(O)O(CH2)11Me

542 C32 H62 O4 S IC-1259-0-0 Didocecyl ester of 4,4'-thiobis-butanoic acid; W 126758

542 C32 H62 O4 S IC-1259-0-0 Ditricecyl ester of 3,3'-thiobis-propanoic acid; W 126759

542 C32 H66 O4 Si 61540-82-7 KO-4-109-0 1-HEXADECYL-2-DECANOYLGLYCEROL-3-TMS GLYCEROL; W 64191

CH2OSiMe3
|
Me(CH2)8C(O)OCHCH2O(CH2)15Me

542 C33 H22 N2 O2 S2 61247-85-6 K-112-273-0 5α,6α-(N-PHENYLDICARBOXIMIDO)-4,7,9-TRIMETHYL-4,5,6,7-TETRAHYDRO-4,7-EPITHIO-2,1-BENZISOTHIAZOLE; W 64192

Str 470

542 C33 H22 N2 O2 S2 61165-06-8 K-112-273-0 5β,6β-(N-PHENYLDICARBOXIMIDO)-4,7,9-TRIPHENYL-4,5,6,7-TETRAHYDRO-4,7-EPITHIO-2,1-BENZISOTHIAZOLE; W 64193

Str 471

542 C33 H50 O4 S 83872-60-0 H-65-1515-0 (23RS)-23-Phenylsulfonyl-5-cholesten-3β,25-diol; W 126760

Str 472

542 C33 H50 O6 53534-46-6 SD-1981-0-0 Lanost-9(11)-en-18-oic acid, 3-(acetyloxy)-20-hydroxy-25-methoxy-16-oxo-, γ-lactone, (3β)-;; W/NBS 64195

Str 473

542 C33 H50 O6 AI-307-959-4 ACETYL-PIROLONIC ACID METHYL ESTER; W 64194

542 C33 H54 N2 O4 71609-97-7 B-32-609-0 (22R,25S)-3β-ACETYLAMINO-22,26-ACETYLEPIMINO-5α-CHOLESTAN-16α-YL ACETATE; W 64196

Str 474

542 C33 H58 O2 Si2 65598-51-8 NS-0-0-0 Silane, [[2-[3-methyl-6-(1-methylethenyl)-2-cyclohexen-1-yl]-5-pentyl-1,3-phenylene]bis(oxy)]bis[triethyl-, (1R-trans)-; W/NBS 126761

542 C34 H22 O7 55517-70-9 O-4-308-7 1(3H)-Isobenzofuranone, 3,3'-(4-hydroxy-1,3-phenylene)bis[3-(4-hydroxyphenyl)-;; 2,4-BIS(3-(P-HYDROXYPHENYL)-1-OXO-1,3-DIHYDROISOBENZOFURAN-3-YL)PHENOL; W/NBS 64197

Str 475

542 C34 H38 O6 74367-05-8 C-100-3533-0 Benzene, 1,1',1'',1'''-(1,4-dimethoxy-1,2,3,4-butanetetrayl)tetrakis[methoxy-;; 1,2,3,4,-TETRANISYL-1,4-DIMETHOXYBUTANE; W/NBS 64198

Str 476

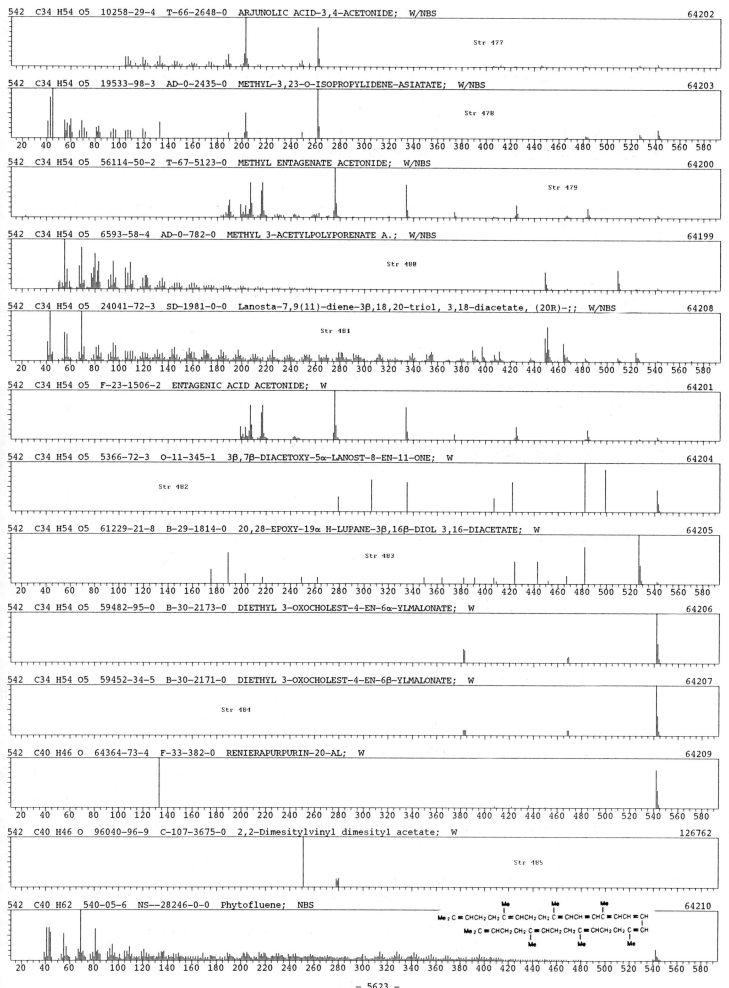

542 C34 H54 O5 10258-29-4 T-66-2648-0 ARJUNOLIC ACID-3,4-ACETONIDE; W/NBS   64202

Str 477

542 C34 H54 O5 19533-98-3 AD-0-2435-0 METHYL-3,23-O-ISOPROPYLIDENE-ASIATATE; W/NBS   64203

Str 478

542 C34 H54 O5 56114-50-2 T-67-5123-0 METHYL ENTAGENATE ACETONIDE; W/NBS   64200

Str 479

542 C34 H54 O5 6593-58-4 AD-0-782-0 METHYL 3-ACETYLPOLYPORENATE A.; W/NBS   64199

Str 480

542 C34 H54 O5 24041-72-3 SD-1981-0-0 Lanosta-7,9(11)-diene-3β,18,20-triol, 3,18-diacetate, (20R)-;; W/NBS   64208

Str 481

542 C34 H54 O5 F-23-1506-2 ENTAGENIC ACID ACETONIDE; W   64201

542 C34 H54 O5 5366-72-3 O-11-345-1 3β,7β-DIACETOXY-5α-LANOST-8-EN-11-ONE; W   64204

Str 482

542 C34 H54 O5 61229-21-8 B-29-1814-0 20,28-EPOXY-19α H-LUPANE-3β,16β-DIOL 3,16-DIACETATE; W   64205

Str 483

542 C34 H54 O5 59482-95-0 B-30-2173-0 DIETHYL 3-OXOCHOLEST-4-EN-6α-YLMALONATE; W   64206

542 C34 H54 O5 59452-34-5 B-30-2171-0 DIETHYL 3-OXOCHOLEST-4-EN-6β-YLMALONATE; W   64207

Str 484

542 C40 H46 O 64364-73-4 F-33-382-0 RENIERAPURPURIN-20-AL; W   64209

542 C40 H46 O 96040-96-9 C-107-3675-0 2,2-Dimesitylvinyl dimesityl acetate; W   126762

Str 485

542 C40 H62 540-05-6 NS--28246-0-0 Phytofluene; NBS   64210

- 5623 -

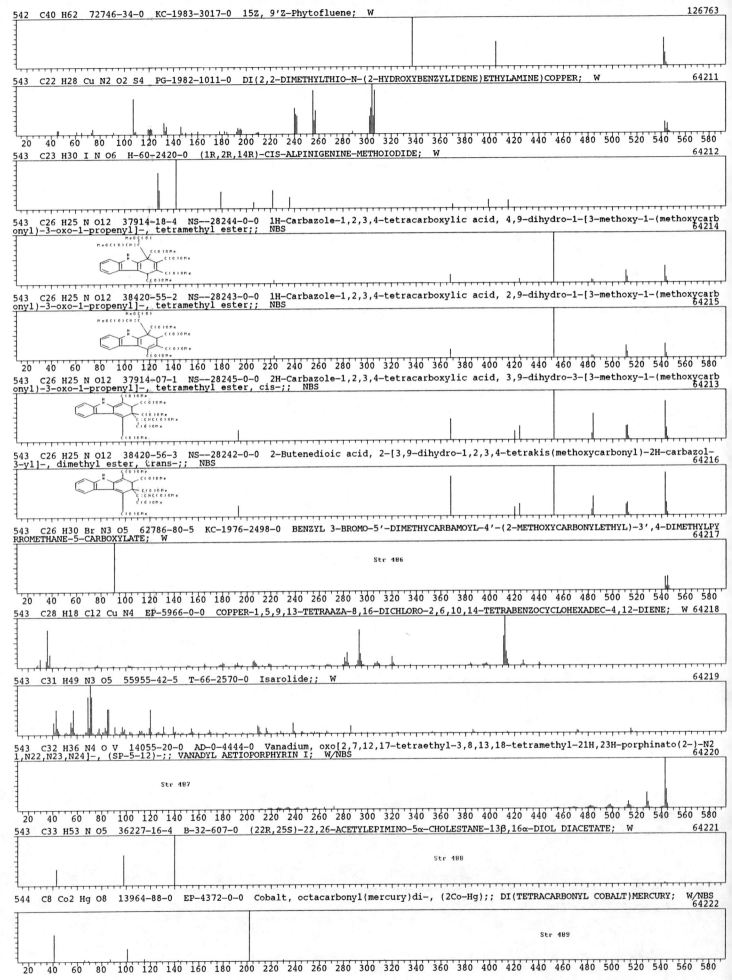

542   C40 H62   72746-34-0   KC-1983-3017-0   15Z, 9'Z-Phytofluene;   W   126763

543   C22 H28 Cu N2 O2 S4   PG-1982-1011-0   DI(2,2-DIMETHYLTHIO-N-(2-HYDROXYBENZYLIDENE)ETHYLAMINE)COPPER;   W   64211

543   C23 H30 I N O6   H-60-2420-0   (1R,2R,14R)-CIS-ALPINIGENINE-METHOIODIDE;   W   64212

543   C26 H25 N O12   37914-18-4   NS--28244-0-0   1H-Carbazole-1,2,3,4-tetracarboxylic acid, 4,9-dihydro-1-[3-methoxy-1-(methoxycarb
onyl)-3-oxo-1-propenyl]-, tetramethyl ester;;   NBS   64214

543   C26 H25 N O12   38420-55-2   NS--28243-0-0   1H-Carbazole-1,2,3,4-tetracarboxylic acid, 2,9-dihydro-1-[3-methoxy-1-(methoxycarb
onyl)-3-oxo-1-propenyl]-, tetramethyl ester;;   NBS   64215

543   C26 H25 N O12   37914-07-1   NS--28245-0-0   2H-Carbazole-1,2,3,4-tetracarboxylic acid, 3,9-dihydro-3-[3-methoxy-1-(methoxycarb
onyl)-3-oxo-1-propenyl]-, tetramethyl ester, cis-;;   NBS   64213

543   C26 H25 N O12   38420-56-3   NS--28242-0-0   2-Butenedioic acid, 2-[3,9-dihydro-1,2,3,4-tetrakis(methoxycarbonyl)-2H-carbazol-
3-yl]-, dimethyl ester, trans-;;   NBS   64216

543   C26 H30 Br N3 O5   62786-80-5   KC-1976-2498-0   BENZYL 3-BROMO-5'-DIMETHYCARBAMOYL-4'-(2-METHOXYCARBONYLETHYL)-3',4-DIMETHYLPY
RROMETHANE-5-CARBOXYLATE;   W   64217

Str 486

543   C28 H18 Cl2 Cu N4   EP-5966-0-0   COPPER-1,5,9,13-TETRAAZA-8,16-DICHLORO-2,6,10,14-TETRABENZOCYCLOHEXADEC-4,12-DIENE;   W 64218

543   C31 H49 N3 O5   55955-42-5   T-66-2570-0   Isarolide;;   W   64219

543   C32 H36 N4 O V   14055-20-0   AD-0-4444-0   Vanadium, oxo[2,7,12,17-tetraethyl-3,8,13,18-tetramethyl-21H,23H-porphinato(2-)-N2
1,N22,N23,N24]-, (SP-5-12)-;; VANADYL AETIOPORPHYRIN I;   W/NBS   64220

Str 487

543   C33 H53 N O5   36227-16-4   B-32-607-0   (22R,25S)-22,26-ACETYLEPIMINO-5α-CHOLESTANE-13β,16α-DIOL DIACETATE;   W   64221

Str 488

544   C8 Co2 Hg O8   13964-88-0   EP-4372-0-0   Cobalt, octacarbonyl(mercury)di-, (2Co-Hg);;   DI(TETRACARBONYL COBALT)MERCURY;   W/NBS   64222

Str 489

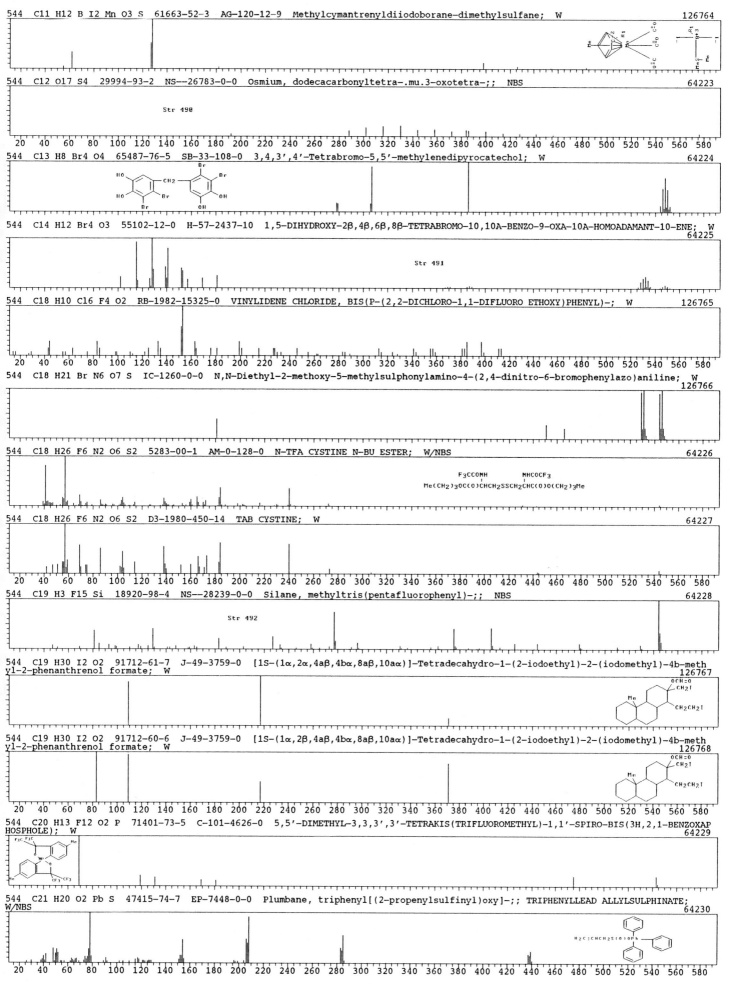

544   C11 H12 B I2 Mn O3 S   61663-52-3   AG-120-12-9   Methylcymantrenyldiiodoborane-dimethylsulfane;   W
126764

544   C12 O17 S4   29994-93-2   NS--26783-0-0   Osmium, dodecacarbonyltetra-.mu.3-oxotetra-;;   NBS
64223

Str 490

20 40 60 80 100 120 140 160 180 200 220 240 260 280 300 320 340 360 380 400 420 440 460 480 500 520 540 560 580

544   C13 H8 Br4 O4   65487-76-5   SB-33-108-0   3,4,3',4'-Tetrabromo-5,5'-methylenedipyrocatechol;   W
64224

544   C14 H12 Br4 O3   55102-12-0   H-57-2437-10   1,5-DIHYDROXY-2β,4β,6β,8β-TETRABROMO-10,10A-BENZO-9-OXA-10A-HOMOADAMANT-10-ENE;   W
64225

Str 491

544   C18 H10 Cl6 F4 O2   RB-1982-15325-0   VINYLIDENE CHLORIDE, BIS(P-(2,2-DICHLORO-1,1-DIFLUORO ETHOXY)PHENYL)-;   W
126765

544   C18 H21 Br N6 O7 S   IC-1260-0-0   N,N-Diethyl-2-methoxy-5-methylsulphonylamino-4-(2,4-dinitro-6-bromophenylazo)aniline;   W
126766

544   C18 H26 F6 N2 O6 S2   5283-00-1   AM-0-128-0   N-TFA CYSTINE N-BU ESTER;   W/NBS
64226

F3CCONH   NHCOCF3
Me(CH2)3OC(O)CHCH2SSCH2CHC(O)O(CH2)3Me

544   C18 H26 F6 N2 O6 S2   D3-1980-450-14   TAB CYSTINE;   W
64227

20 40 60 80 100 120 140 160 180 200 220 240 260 280 300 320 340 360 380 400 420 440 460 480 500 520 540 560 580

544   C19 H3 F15 Si   18920-98-4   NS--28239-0-0   Silane, methyltris(pentafluorophenyl)-;;   NBS
64228

Str 492

544   C19 H30 I2 O2   91712-61-7   J-49-3759-0   [1S-(1α,2α,4aβ,4bα,8aβ,10aα)]-Tetradecahydro-1-(2-iodoethyl)-2-(iodomethyl)-4b-meth
yl-2-phenanthrenol formate;   W
126767

544   C19 H30 I2 O2   91712-60-6   J-49-3759-0   [1S-(1α,2β,4aβ,4bα,8aβ,10aα)]-Tetradecahydro-1-(2-iodoethyl)-2-(iodomethyl)-4b-meth
yl-2-phenanthrenol formate;   W
126768

20 40 60 80 100 120 140 160 180 200 220 240 260 280 300 320 340 360 380 400 420 440 460 480 500 520 540 560 580

544   C20 H13 F12 O2 P   71401-73-5   C-101-4626-0   5,5'-DIMETHYL-3,3,3',3'-TETRAKIS(TRIFLUOROMETHYL)-1,1'-SPIRO-BIS(3H,2,1-BENZOXAP
HOSPHOLE);   W
64229

544   C21 H20 O2 Pb S   47415-74-7   EP-7448-0-0   Plumbane, triphenyl[(2-propenylsulfinyl)oxy]-;;   TRIPHENYLLEAD ALLYLSULPHINATE;
W/NBS
64230

20 40 60 80 100 120 140 160 180 200 220 240 260 280 300 320 340 360 380 400 420 440 460 480 500 520 540 560 580

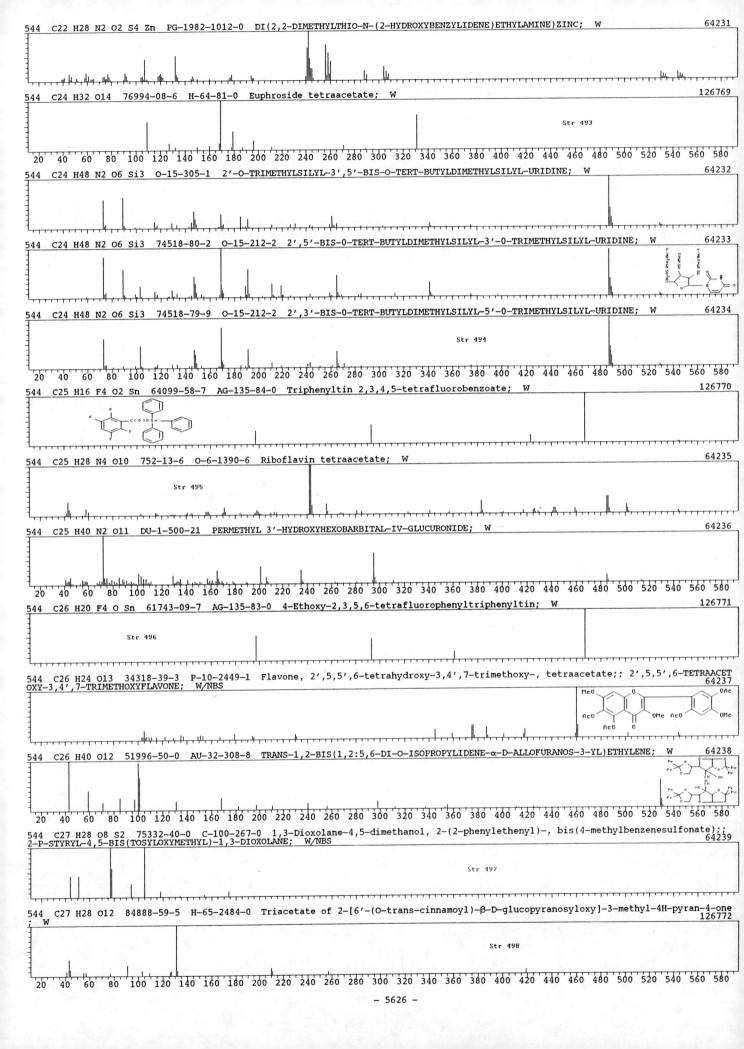

544 C22 H28 N2 O2 S4 Zn PG-1982-1012-0 DI(2,2-DIMETHYLTHIO-N-(2-HYDROXYBENZYLIDENE)ETHYLAMINE)ZINC; W   64231

544 C24 H32 O14 76994-08-6 H-64-81-0 Euphroside tetraacetate; W   126769

Str 493

544 C24 H48 N2 O6 Si3 O-15-305-1 2'-O-TRIMETHYLSILYL-3',5'-BIS-O-TERT-BUTYLDIMETHYLSILYL-URIDINE; W   64232

544 C24 H48 N2 O6 Si3 74518-80-2 O-15-212-2 2',5'-BIS-O-TERT-BUTYLDIMETHYLSILYL-3'-O-TRIMETHYLSILYL-URIDINE; W   64233

544 C24 H48 N2 O6 Si3 74518-79-9 O-15-212-2 2',3'-BIS-O-TERT-BUTYLDIMETHYLSILYL-5'-O-TRIMETHYLSILYL-URIDINE; W   64234

Str 494

544 C25 H16 F4 O2 Sn 64099-58-7 AG-135-84-0 Triphenyltin 2,3,4,5-tetrafluorobenzoate; W   126770

544 C25 H28 N4 O10 752-13-6 O-6-1390-6 Riboflavin tetraacetate; W   64235

Str 495

544 C25 H40 N2 O11 DU-1-500-21 PERMETHYL 3'-HYDROXYHEXOBARBITAL-IV-GLUCURONIDE; W   64236

544 C26 H20 F4 O Sn 61743-09-7 AG-135-83-0 4-Ethoxy-2,3,5,6-tetrafluorophenyltriphenyltin; W   126771

Str 496

544 C26 H24 O13 34318-39-3 P-10-2449-1 Flavone, 2',5,5',6-tetrahydroxy-3,4',7-trimethoxy-, tetraacetate;; 2',5,5',6-TETRAACET OXY-3,4',7-TRIMETHOXYFLAVONE; W/NBS   64237

544 C26 H40 O12 51996-50-0 AU-32-308-8 TRANS-1,2-BIS(1,2:5,6-DI-O-ISOPROPYLIDENE-α-D-ALLOFURANOS-3-YL)ETHYLENE; W   64238

544 C27 H28 O8 S2 75332-40-0 C-100-267-0 1,3-Dioxolane-4,5-dimethanol, 2-(2-phenylethenyl)-, bis(4-methylbenzenesulfonate);; 2-P-STYRYL-4,5-BIS(TOSYLOXYMETHYL)-1,3-DIOXOLANE; W/NBS   64239

Str 497

544 C27 H28 O12 84888-59-5 H-65-2484-0 Triacetate of 2-[6'-(O-trans-cinnamoyl)-β-D-glucopyranosyloxy]-3-methyl-4H-pyran-4-one ; W   126772

Str 498

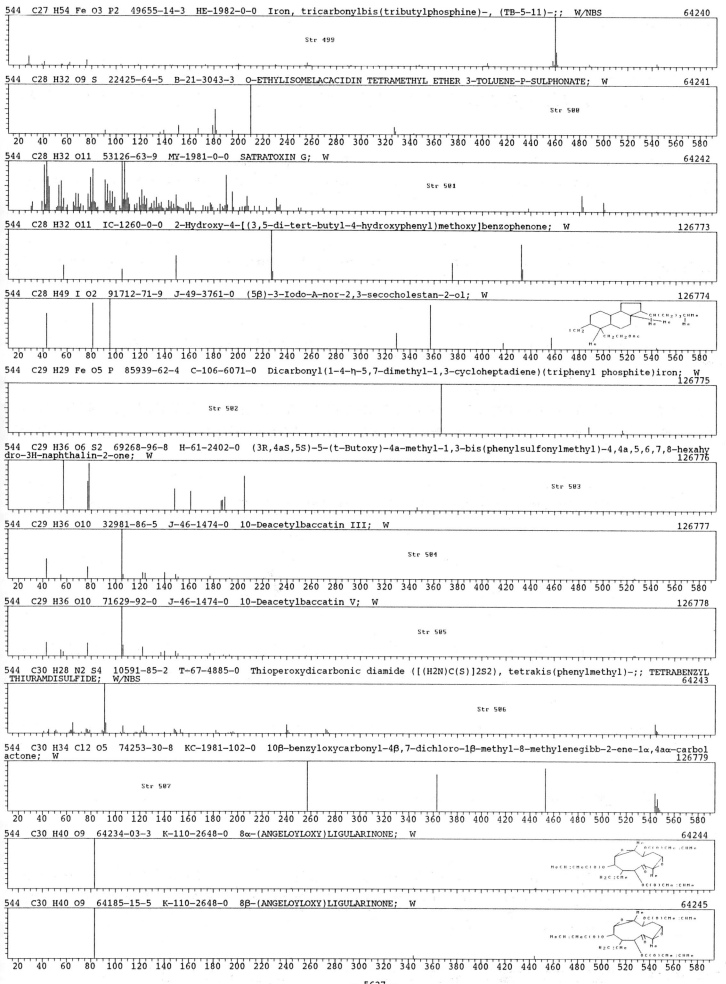

544 C27 H54 Fe O3 P2 49655-14-3 HE-1982-0-0 Iron, tricarbonylbis(tributylphosphine)-, (TB-5-11)-;; W/NBS   64240

Str 499

544 C28 H32 O9 S 22425-64-5 B-21-3043-3 O-ETHYLISOMELACACIDIN TETRAMETHYL ETHER 3-TOLUENE-P-SULPHONATE; W   64241

Str 500

20 40 60 80 100 120 140 160 180 200 220 240 260 280 300 320 340 360 380 400 420 440 460 480 500 520 540 560 580

544 C28 H32 O11 53126-63-9 MY-1981-0-0 SATRATOXIN G; W   64242

Str 501

544 C28 H32 O11 IC-1260-0-0 2-Hydroxy-4-[(3,5-di-tert-butyl-4-hydroxyphenyl)methoxy]benzophenone; W   126773

544 C28 H49 I O2 91712-71-9 J-49-3761-0 (5β)-3-Iodo-A-nor-2,3-secocholestan-2-ol; W   126774

544 C29 H29 Fe O5 P 85939-62-4 C-106-6071-0 Dicarbonyl(1-4-η-5,7-dimethyl-1,3-cycloheptadiene)(triphenyl phosphite)iron; W
126775

Str 502

544 C29 H36 O6 S2 69268-96-8 H-61-2402-0 (3R,4aS,5S)-5-(t-Butoxy)-4a-methyl-1,3-bis(phenylsulfonylmethyl)-4,4a,5,6,7,8-hexahy
dro-3H-naphthalin-2-one; W   126776

Str 503

544 C29 H36 O10 32981-86-5 J-46-1474-0 10-Deacetylbaccatin III; W   126777

Str 504

20 40 60 80 100 120 140 160 180 200 220 240 260 280 300 320 340 360 380 400 420 440 460 480 500 520 540 560 580

544 C29 H36 O10 71629-92-0 J-46-1474-0 10-Deacetylbaccatin V; W   126778

Str 505

544 C30 H28 N2 S4 10591-85-2 T-67-4885-0 Thioperoxydicarbonic diamide ([(H2N)C(S)]2S2), tetrakis(phenylmethyl)-;; TETRABENZYL
THIURAMDISULFIDE; W/NBS   64243

Str 506

544 C30 H34 Cl2 O5 74253-30-8 KC-1981-102-0 10β-benzyloxycarbonyl-4β,7-dichloro-1β-methyl-8-methylenegibb-2-ene-1α,4aα-carbol
actone; W   126779

Str 507

20 40 60 80 100 120 140 160 180 200 220 240 260 280 300 320 340 360 380 400 420 440 460 480 500 520 540 560 580

544 C30 H40 O9 64234-03-3 K-110-2648-0 8α-(ANGELOYLOXY)LIGULARINONE; W   64244

544 C30 H40 O9 64185-15-5 K-110-2648-0 8β-(ANGELOYLOXY)LIGULARINONE; W   64245

20 40 60 80 100 120 140 160 180 200 220 240 260 280 300 320 340 360 380 400 420 440 460 480 500 520 540 560 580

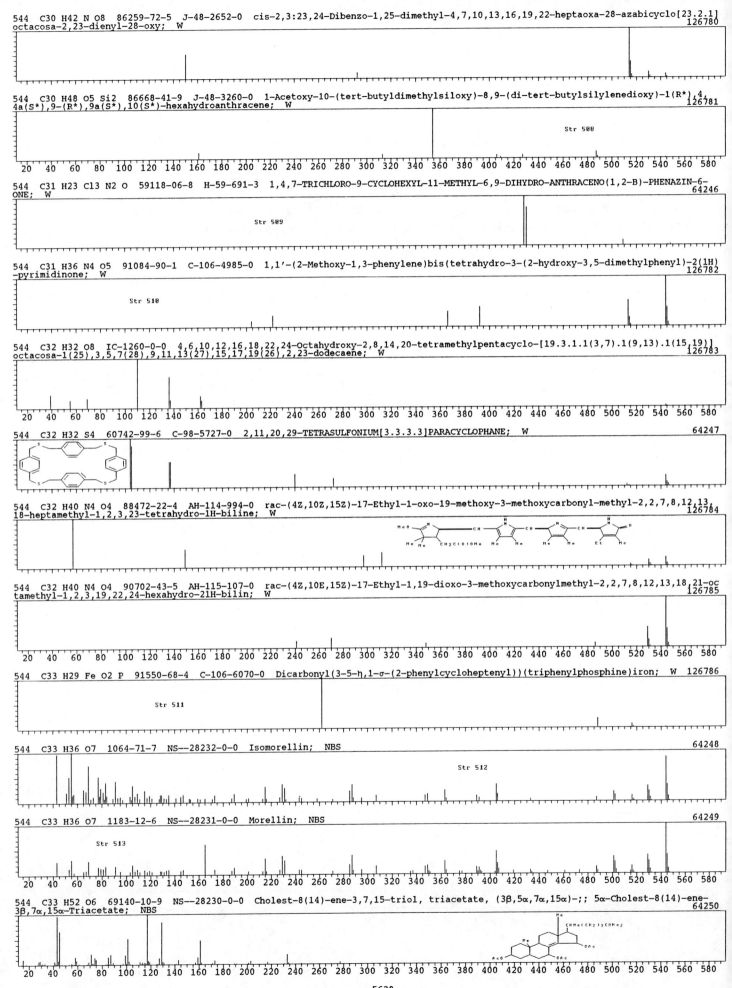

544   C30 H42 N O8   86259-72-5   J-48-2652-0   cis-2,3:23,24-Dibenzo-1,25-dimethyl-4,7,10,13,16,19,22-heptaoxa-28-azabicyclo[23.2.1]octacosa-2,23-dienyl-28-oxy;   W
126780

544   C30 H48 O5 Si2   86668-41-9   J-48-3260-0   1-Acetoxy-10-(tert-butyldimethylsiloxy)-8,9-(di-tert-butylsilylenedioxy)-1(R*),4,4a(S*),9-(R*),9a(S*),10(S*)-hexahydroanthracene;   W
126781

Str 508

544   C31 H23 Cl3 N2 O   59118-06-8   H-59-691-3   1,4,7-TRICHLORO-9-CYCLOHEXYL-11-METHYL-6,9-DIHYDRO-ANTHRACENO(1,2-B)-PHENAZIN-6-ONE;   W
64246

Str 509

544   C31 H36 N4 O5   91084-90-1   C-106-4985-0   1,1'-(2-Methoxy-1,3-phenylene)bis(tetrahydro-3-(2-hydroxy-3,5-dimethylphenyl)-2(1H)-pyrimidinone;   W
126782

Str 510

544   C32 H32 O8   IC-1260-0-0   4,6,10,12,16,18,22,24-Octahydroxy-2,8,14,20-tetramethylpentacyclo-[19.3.1.1(3,7).1(9,13).1(15,19)]octacosa-1(25),3,5,7(28),9,11,13(27),15,17,19(26),2,23-dodecaene;   W
126783

544   C32 H32 S4   60742-99-6   C-98-5727-0   2,11,20,29-TETRASULFONIUM[3.3.3.3]PARACYCLOPHANE;   W
64247

544   C32 H40 N4 O4   88472-22-4   AH-114-994-0   rac-(4Z,10Z,15Z)-17-Ethyl-1-oxo-19-methoxy-3-methoxycarbonyl-methyl-2,2,7,8,12,13,18-heptamethyl-1,2,3,23-tetrahydro-1H-biline;   W
126784

544   C32 H40 N4 O4   90702-43-5   AH-115-107-0   rac-(4Z,10E,15Z)-17-Ethyl-1,19-dioxo-3-methoxycarbonylmethyl-2,2,7,8,12,13,18,21-octamethyl-1,2,3,19,22,24-hexahydro-21H-bilin;   W
126785

544   C33 H29 Fe O2 P   91550-68-4   C-106-6070-0   Dicarbonyl(3-5-η,1-σ-(2-phenylcycloheptenyl))(triphenylphosphine)iron;   W   126786

Str 511

544   C33 H36 O7   1064-71-7   NS--28232-0-0   Isomorellin;   NBS
64248

Str 512

544   C33 H36 O7   1183-12-6   NS--28231-0-0   Morellin;   NBS
64249

Str 513

544   C33 H52 O6   69140-10-9   NS--28230-0-0   Cholest-8(14)-ene-3,7,15-triol, triacetate, (3β,5α,7α,15α)-;; 5α-Cholest-8(14)-ene-3β,7α,15α-Triacetate;   NBS
64250

544   C33 H52 O6   66875-27-2   H-65-1510-0   Methyl ester of (20S)-1α,3β-Bis[(tetrahydropyran-2-yl)oxy]-5-pregnen-20-carboxylic acid
; W
126788

Str 514

544   C33 H52 O6   79037-05-1   F-37-603-0   1β,3β,16β-Triacetoxycholest-5-ene; W
126789

Str 515

20 40 60 80 100 120 140 160 180 200 220 240 260 280 300 320 340 360 380 400 420 440 460 480 500 520 540 560 580

544   C33 H52 O6   39783-14-7   H-64-1883-0   (5-Cholesten-1α,3β,25-triol)-triacetate; W
126790

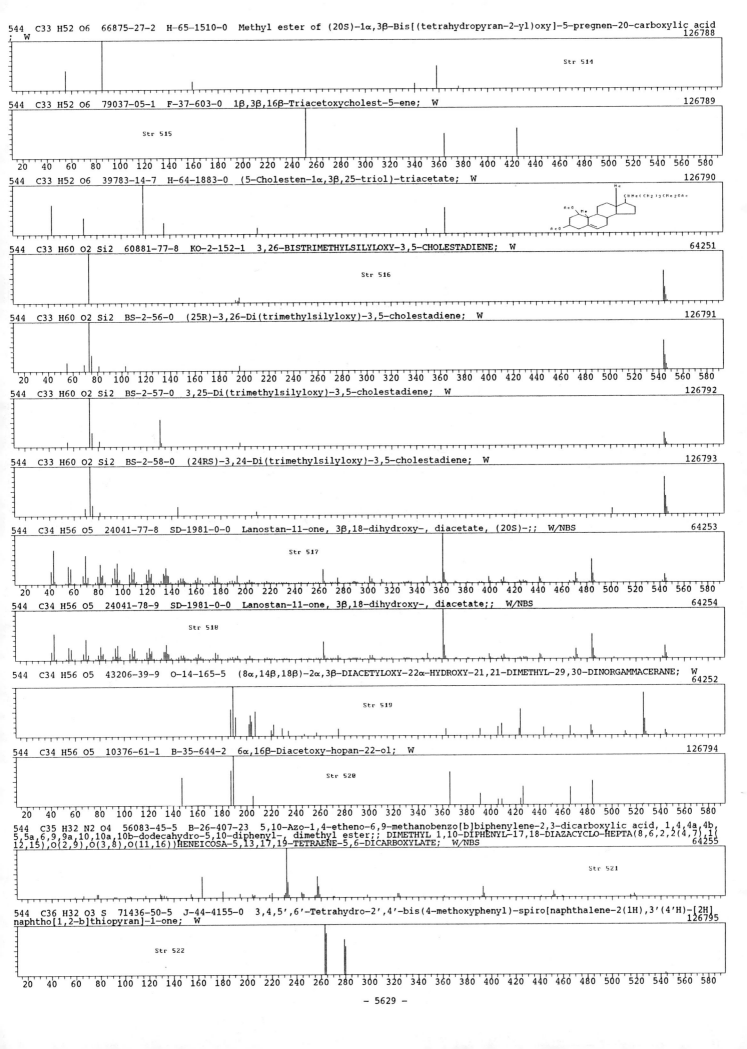

544   C33 H60 O2 Si2   60881-77-8   KO-2-152-1   3,26-BISTRIMETHYLSILYLOXY-3,5-CHOLESTADIENE; W
64251

Str 516

544   C33 H60 O2 Si2   BS-2-56-0   (25R)-3,26-Di(trimethylsilyloxy)-3,5-cholestadiene; W
126791

544   C33 H60 O2 Si2   BS-2-57-0   3,25-Di(trimethylsilyloxy)-3,5-cholestadiene; W
126792

20 40 60 80 100 120 140 160 180 200 220 240 260 280 300 320 340 360 380 400 420 440 460 480 500 520 540 560 580

544   C33 H60 O2 Si2   BS-2-58-0   (24RS)-3,24-Di(trimethylsilyloxy)-3,5-cholestadiene; W
126793

544   C34 H56 O5   24041-77-8   SD-1981-0-0   Lanostan-11-one, 3β,18-dihydroxy-, diacetate, (20S)-;; W/NBS
64253

Str 517

544   C34 H56 O5   24041-78-9   SD-1981-0-0   Lanostan-11-one, 3β,18-dihydroxy-, diacetate;; W/NBS
64254

Str 518

544   C34 H56 O5   43206-39-9   O-14-165-5   (8α,14β,18β)-2α,3β-DIACETYLOXY-22α-HYDROXY-21,21-DIMETHYL-29,30-DINORGAMMACERANE; W
64252

Str 519

544   C34 H56 O5   10376-61-1   B-35-644-2   6α,16β-Diacetoxy-hopan-22-ol; W
126794

Str 520

20 40 60 80 100 120 140 160 180 200 220 240 260 280 300 320 340 360 380 400 420 440 460 480 500 520 540 560 580

544   C35 H32 N2 O4   56083-45-5   B-26-407-23   5,10-Azo-1,4-etheno-6,9-methanobenzo[b]biphenylene-2,3-dicarboxylic acid, 1,4,4a,4b,
5,5a,6,9,9a,10,10a,10b-dodecahydro-5,10-diphenyl- dimethyl ester;; DIMETHYL 1,10-DIPHENYL-17,18-DIAZACYCLO-HEPTA(8,6,2,2(4,7),1(
12,15),O(2,9),O(3,8),O(11,16))HENEICOSA-5,13,17,19-TETRAENE-5,6-DICARBOXYLATE; W/NBS
64255

Str 521

544   C36 H32 O3 S   71436-50-5   J-44-4155-0   3,4,5',6'-Tetrahydro-2',4'-bis(4-methoxyphenyl)-spiro[naphthalene-2(1H),3'(4'H)-[2H]
naphtho[1,2-b]thiopyran]-1-one; W
126795

Str 522

20 40 60 80 100 120 140 160 180 200 220 240 260 280 300 320 340 360 380 400 420 440 460 480 500 520 540 560 580

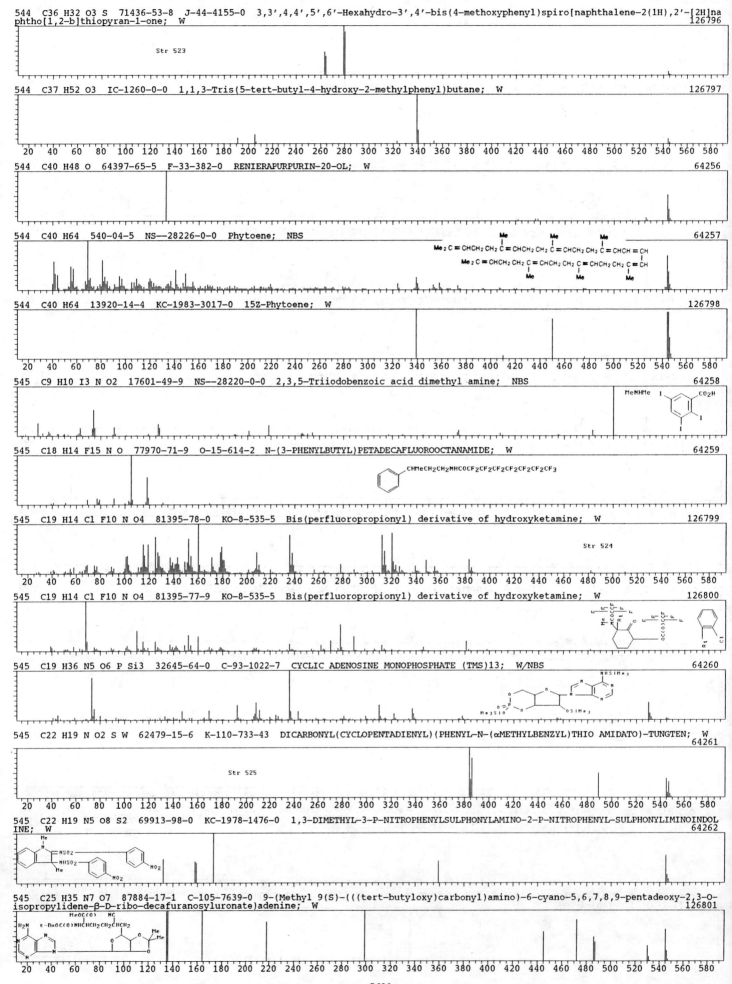

544   C36 H32 O3 S   71436-53-8   J-44-4155-0   3,3',4,4',5',6'-Hexahydro-3',4'-bis(4-methoxyphenyl)spiro[naphthalene-2(1H),2'-[2H]naphtho[1,2-b]thiopyran-1-one;   W   126796

Str 523

544   C37 H52 O3   IC-1260-0-0   1,1,3-Tris(5-tert-butyl-4-hydroxy-2-methylphenyl)butane;   W   126797

544   C40 H48 O   64397-65-5   F-33-382-0   RENIERAPURPURIN-20-OL;   W   64256

544   C40 H64   540-04-5   NS--28226-0-0   Phytoene;   NBS   64257

544   C40 H64   13920-14-4   KC-1983-3017-0   15Z-Phytoene;   W   126798

545   C9 H10 I3 N O2   17601-49-9   NS--28220-0-0   2,3,5-Triiodobenzoic acid dimethyl amine;   NBS   64258

545   C18 H14 F15 N O   77970-71-9   O-15-614-2   N-(3-PHENYLBUTYL)PETADECAFLUOROOCTANAMIDE;   W   64259

545   C19 H14 Cl F10 N O4   81395-78-0   KO-8-535-5   Bis(perfluoropropionyl) derivative of hydroxyketamine;   W   126799

Str 524

545   C19 H14 Cl F10 N O4   81395-77-9   KO-8-535-5   Bis(perfluoropropionyl) derivative of hydroxyketamine;   W   126800

545   C19 H36 N5 O6 P Si3   32645-64-0   C-93-1022-7   CYCLIC ADENOSINE MONOPHOSPHATE (TMS)13;   W/NBS   64260

545   C22 H19 N O2 S W   62479-15-6   K-110-733-43   DICARBONYL(CYCLOPENTADIENYL)(PHENYL-N-(αMETHYLBENZYL)THIO AMIDATO)-TUNGTEN;   W   64261

Str 525

545   C22 H19 N5 O8 S2   69913-98-0   KC-1978-1476-0   1,3-DIMETHYL-3-P-NITROPHENYLSULPHONYLAMINO-2-P-NITROPHENYL-SULPHONYLIMINOINDOLINE;   W   64262

545   C25 H35 N7 O7   87884-17-1   C-105-7639-0   9-(Methyl 9(S)-(((tert-butyloxy)carbonyl)amino)-6-cyano-5,6,7,8,9-pentadeoxy-2,3-O-isopropylidene-β-D-ribo-decafuranosyluronate)adenine;   W   126801

545 C28 H20 Cl2 Cu N4 69815-52-7 NS--28223-0-0 Copper, [6,18-dichloro-5,6,17,18-tetrahydrotetrabenzo[b,f,j,n][1,5,9,13]tetraa
zacyclohexadecinato(2-)-N5,N11,N17,N23]-, (SP-4-1)-;; NBS   64263

Str 526

545 C29 H39 N O9 26833-87-4 AD-0-3608-0 Homoharringtonine; W/NBS   64264

Str 527

545 C29 H39 N O9 38826-64-1 B-29-1315-0 2α,6α,11α,12α-TETRAACETOXY-19,20-METHYLIMINO-14,20-CYCLO-ENT-KAUR-16-ENE-5β-OL; W
126802

Str 528

545 C31 H35 N3 O4 S CD-51-0-0 Benethamine; W   126803

545 C32 H35 N O7 64018-64-0 H-60-1293-0 5-Isoxazolidinecarboxylic acid, 2-[2,3-O-(1-methylethylidene)-5-O-(triphenylmethyl)-β
-D-ribofuranosyl]-, methyl ester, (S)-;; (5S)-2-(2,3-O-ISOPROPYLIDEN-5-O-TRITYL-β-D-RIBOFURANOSYL)-ISOXAZOLIDIN-5-CARBOXYLIC ACID
METHYL ESTER; W/NBS   64266

Str 529

545 C35 H51 N3 O2 58581-86-5 J-41-2101-0 [1R-(1α(R*),3aβ,4(8S*,10R*),7aα]]-10-[[1-(1,5-dimethylhexyl)octahydro-7a-methyl-4H-
inden-4-ylidene]methyl]-2,3,5,6,7,8,9,10-octahydro-8-hydroxy-2-phenyl-1H-[1,2,4]triazolo[1,2-b]phthalazin-1-one; W   126804

Str 530

545 C36 H51 N O3 60724-76-7 F-32-1291-0 3β-ACETOXY-CHOLEST-6-ENO-[7,6-D]-2'-PHENYLOXAZOL; W   64267

546 C6 Br6 87-82-1 NS--28203-0-0 Benzene, hexabromo-;; Hexabromobenzene; NBS   64268

546 C6 H18 N4 O2 S4 Sn2 70411-78-8 K-112-1378-0 2,8-BIS(TRIMETHYLSTANNYL)-1,3,5λ4,7,2,4,6,8-TETRATHIATETRAZOCIN-1,1-DIOXIDE;
W   64269

546 C6 H18 S3 Sn3 16892-64-1 NS--27176-0-0 1,3,5-Trithia-2,4,6-tristannacyclohexane, 2,2,4,4,6,6-hexamethyl-;; Hexamethylcycl
otristannathiane; NBS   126806

546 C8 H2 F12 N2 O2 S5 90614-18-9 K-117-1899-0 2,3,4,5-Tetrakis(trifluoromethylthio)pyrrole; W   126807

Str 531

546 C16 H15 F6 Ir O2 68586-44-7 B-31-1951-0 1-3-η-(CARBONYL(1,2-BIS(TRIFLUOROMETHYL)-3-OXO-PROP-1-ENE-1,3-DIYL))-(η-PENTAMETH
YLCYCLOPENTADIENYL)IRIDIUM; W   64270

Str 532

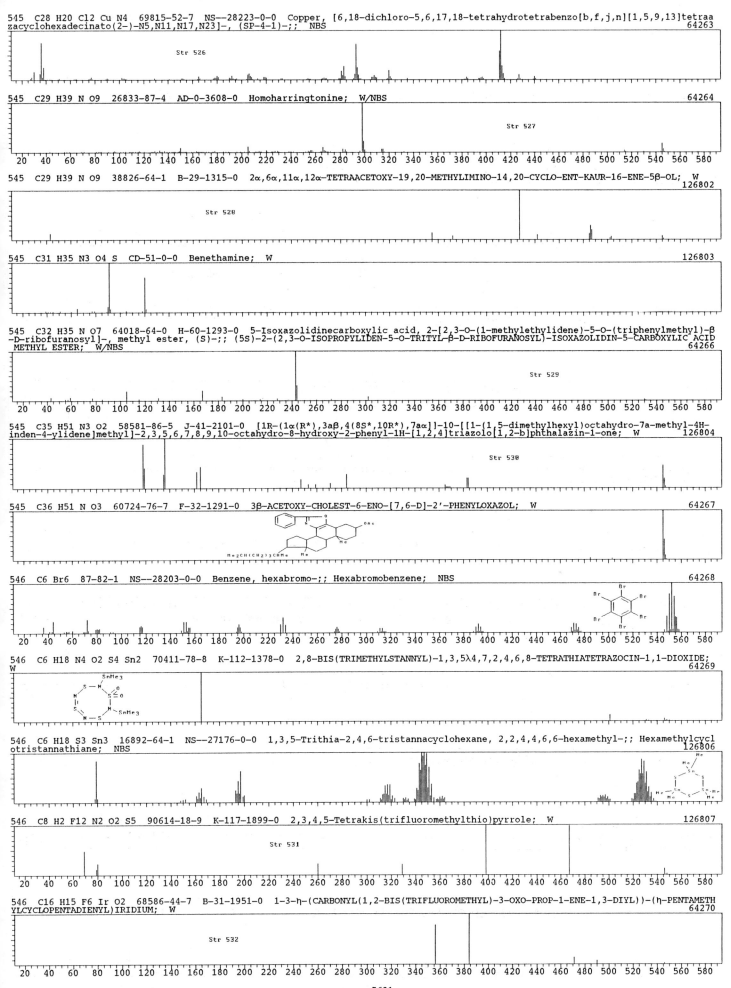

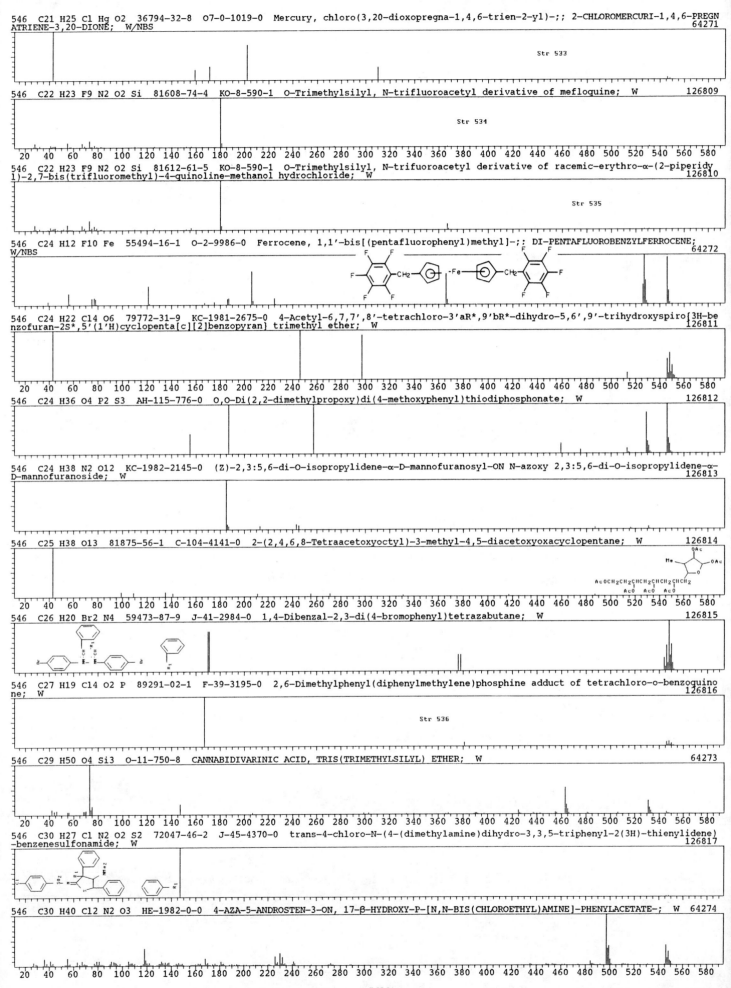

546   C21 H25 Cl Hg O2   36794-32-8   O7-0-1019-0   Mercury, chloro(3,20-dioxopregna-1,4,6-trien-2-yl)-;;   2-CHLOROMERCURI-1,4,6-PREGN
ATRIENE-3,20-DIONE;   W/NBS   64271

Str 533

546   C22 H23 F9 N2 O2 Si   81608-74-4   KO-8-590-1   O-Trimethylsilyl, N-trifluoroacetyl derivative of mefloquine;   W   126809

Str 534

20 40 60 80 100 120 140 160 180 200 220 240 260 280 300 320 340 360 380 400 420 440 460 480 500 520 540 560 580

546   C22 H23 F9 N2 O2 Si   81612-61-5   KO-8-590-1   O-Trimethylsilyl, N-trifuoroacetyl derivative of racemic-erythro-α-(2-piperidy
1)-2,7-bis(trifluoromethyl)-4-quinoline-methanol hydrochloride;   W   126810

Str 535

546   C24 H12 F10 Fe   55494-16-1   O-2-9986-0   Ferrocene, 1,1'-bis[(pentafluorophenyl)methyl]-;;   DI-PENTAFLUOROBENZYLFERROCENE;
W/NBS   64272

546   C24 H22 Cl4 O6   79772-31-9   KC-1981-2675-0   4-Acetyl-6,7,7',8'-tetrachloro-3'aR*,9'bR*-dihydro-5,6',9'-trihydroxyspiro[3H-be
nzofuran-2S*,5'(1'H)cyclopenta[c][2]benzopyran] trimethyl ether;   W   126811

546   C24 H36 O4 P2 S3   AH-115-776-0   O,O-Di(2,2-dimethylpropoxy)di(4-methoxyphenyl)thiodiphosphonate;   W   126812

546   C24 H38 N2 O12   KC-1982-2145-0   (Z)-2,3:5,6-di-O-isopropylidene-α-D-mannofuranosyl-ON N-azoxy 2,3:5,6-di-O-isopropylidene-α-
D-mannofuranoside;   W   126813

546   C25 H38 O13   81875-56-1   C-104-4141-0   2-(2,4,6,8-Tetraacetoxyoctyl)-3-methyl-4,5-diacetoxyoxacyclopentane;   W   126814

546   C26 H20 Br2 N4   59473-87-9   J-41-2984-0   1,4-Dibenzal-2,3-di(4-bromophenyl)tetrazabutane;   W   126815

546   C27 H19 Cl4 O2 P   89291-02-1   F-39-3195-0   2,6-Dimethylphenyl(diphenylmethylene)phosphine adduct of tetrachloro-o-benzoquino
ne;   W   126816

Str 536

546   C29 H50 O4 Si3   O-11-750-8   CANNABIDIVARINIC ACID, TRIS(TRIMETHYLSILYL) ETHER;   W   64273

20 40 60 80 100 120 140 160 180 200 220 240 260 280 300 320 340 360 380 400 420 440 460 480 500 520 540 560 580

546   C30 H27 Cl N2 O2 S2   72047-46-2   J-45-4370-0   trans-4-chloro-N-(4-(dimethylamine)dihydro-3,3,5-triphenyl-2(3H)-thienylidene)
-benzenesulfonamide;   W   126817

546   C30 H40 Cl2 N2 O3   HE-1982-0-0   4-AZA-5-ANDROSTEN-3-ON, 17-β-HYDROXY-P-[N,N-BIS(CHLOROETHYL)AMINE]-PHENYLACETATE-;   W   64274

20 40 60 80 100 120 140 160 180 200 220 240 260 280 300 320 340 360 380 400 420 440 460 480 500 520 540 560 580

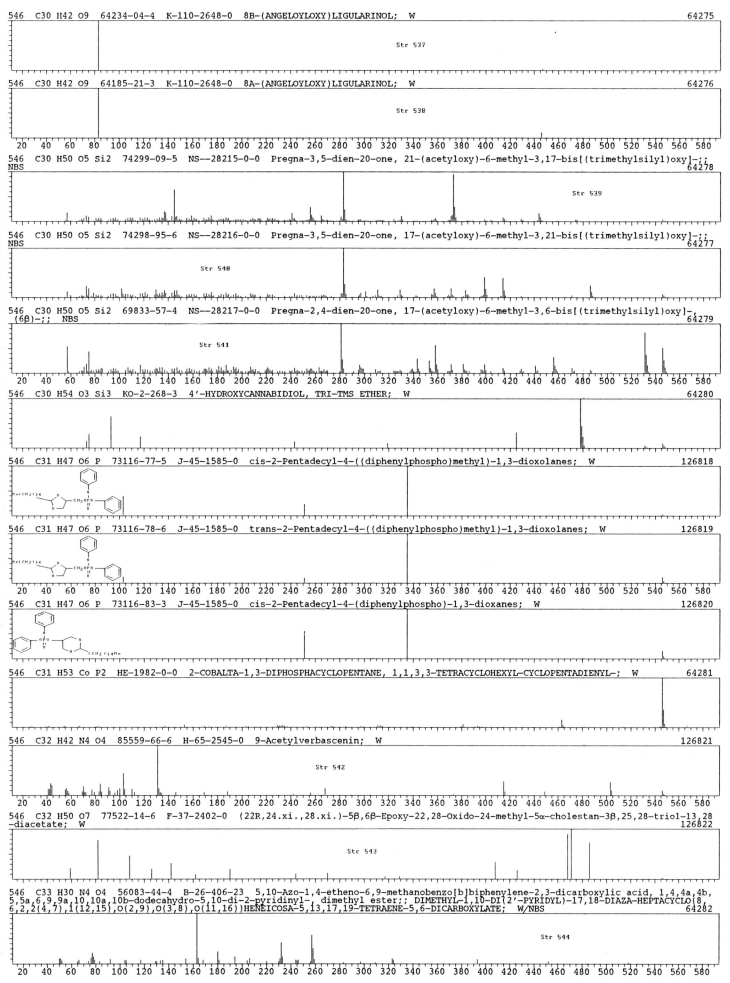

546   C30 H42 O9   64234-04-4   K-110-2648-0   8B-(ANGELOYLOXY)LIGULARINOL;   W   64275

Str 537

546   C30 H42 O9   64185-21-3   K-110-2648-0   8A-(ANGELOYLOXY)LIGULARINOL;   W   64276

Str 538

20   40   60   80   100 120 140 160 180 200 220 240 260 280 300 320 340 360 380 400 420 440 460 480 500 520 540 560 580

546   C30 H50 O5 Si2   74299-09-5   NS--28215-0-0   Pregna-3,5-dien-20-one, 21-(acetyloxy)-6-methyl-3,17-bis[(trimethylsilyl)oxy]-;;
NBS   64278

Str 539

546   C30 H50 O5 Si2   74298-95-6   NS--28216-0-0   Pregna-3,5-dien-20-one, 17-(acetyloxy)-6-methyl-3,21-bis[(trimethylsilyl)oxy]-;;
NBS   64277

Str 540

546   C30 H50 O5 Si2   69833-57-4   NS--28217-0-0   Pregna-2,4-dien-20-one, 17-(acetyloxy)-6-methyl-3,6-bis[(trimethylsilyl)oxy]-,
(6β)-;;   NBS   64279

Str 541

546   C30 H54 O3 Si3   KO-2-268-3   4'-HYDROXYCANNABIDIOL, TRI-TMS ETHER;   W   64280

546   C31 H47 O6 P   73116-77-5   J-45-1585-0   cis-2-Pentadecyl-4-((diphenylphospho)methyl)-1,3-dioxolanes;   W   126818

546   C31 H47 O6 P   73116-78-6   J-45-1585-0   trans-2-Pentadecyl-4-((diphenylphospho)methyl)-1,3-dioxolanes;   W   126819

20   40   60   80   100 120 140 160 180 200 220 240 260 280 300 320 340 360 380 400 420 440 460 480 500 520 540 560 580

546   C31 H47 O6 P   73116-83-3   J-45-1585-0   cis-2-Pentadecyl-4-(diphenylphospho)-1,3-dioxanes;   W   126820

546   C31 H53 Co P2   HE-1982-0-0   2-COBALTA-1,3-DIPHOSPHACYCLOPENTANE, 1,1,3,3-TETRACYCLOHEXYL-CYCLOPENTADIENYL-;   W   64281

546   C32 H42 N4 O4   85559-66-6   H-65-2545-0   9-Acetylverbascenin;   W   126821

Str 542

20   40   60   80   100 120 140 160 180 200 220 240 260 280 300 320 340 360 380 400 420 440 460 480 500 520 540 560 580

546   C32 H50 O7   77522-14-6   F-37-2402-0   (22R,24.xi.,28.xi.)-5β,6β-Epoxy-22,28-Oxido-24-methyl-5α-cholestan-3β,25,28-triol-13,28
-diacetate;   W   126822

Str 543

546   C33 H30 N4 O4   56083-44-4   B-26-406-23   5,10-Azo-1,4-etheno-6,9-methanobenzo[b]biphenylene-2,3-dicarboxylic acid, 1,4,4a,4b,
5,5a,6,9,9a,10,10a,10b-dodecahydro-5,10-di-2-pyridinyl-, dimethyl ester;; DIMETHYL-1,10-DI(2'-PYRIDYL)-17,18-DIAZA-HEPTACYCLO(8,
6,2,2(4,7),1(12,15),O(2,9),O(3,8),O(11,16))HENEICOSA-5,13,17,19-TETRAENE-5,6-DICARBOXYLATE;   W/NBS   64282

Str 544

20   40   60   80   100 120 140 160 180 200 220 240 260 280 300 320 340 360 380 400 420 440 460 480 500 520 540 560 580

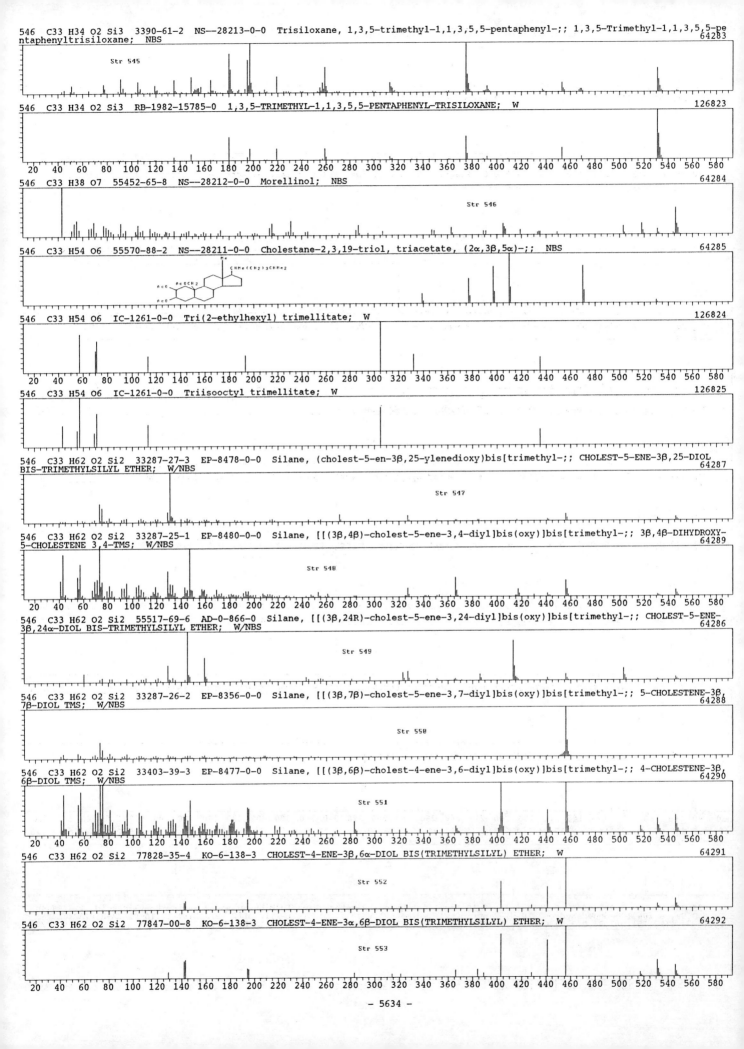

546 C33 H34 O2 Si3 3390-61-2 NS--28213-0-0 Trisiloxane, 1,3,5-trimethyl-1,1,3,5,5-pentaphenyl-;; 1,3,5-Trimethyl-1,1,3,5,5-pentaphenyltrisiloxane; NBS
64283

Str 545

546 C33 H34 O2 Si3 RB-1982-15785-0 1,3,5-TRIMETHYL-1,1,3,5,5-PENTAPHENYL-TRISILOXANE; W
126823

546 C33 H38 O7 55452-65-8 NS--28212-0-0 Morellinol; NBS
64284

Str 546

546 C33 H54 O6 55570-88-2 NS--28211-0-0 Cholestane-2,3,19-triol, triacetate, (2α,3β,5α)-;; NBS
64285

546 C33 H54 O6 IC-1261-0-0 Tri(2-ethylhexyl) trimellitate; W
126824

546 C33 H54 O6 IC-1261-0-0 Triisooctyl trimellitate; W
126825

546 C33 H62 O2 Si2 33287-27-3 EP-8478-0-0 Silane, (cholest-5-en-3β,25-ylenedioxy)bis[trimethyl-;; CHOLEST-5-ENE-3β,25-DIOL BIS-TRIMETHYLSILYL ETHER; W/NBS
64287

Str 547

546 C33 H62 O2 Si2 33287-25-1 EP-8480-0-0 Silane, [[(3β,4β)-cholest-5-ene-3,4-diyl]bis(oxy)]bis[trimethyl-;; 3β,4β-DIHYDROXY-5-CHOLESTENE 3,4-TMS; W/NBS
64289

Str 548

546 C33 H62 O2 Si2 55517-69-6 AD-0-866-0 Silane, [[(3β,24R)-cholest-5-ene-3,24-diyl]bis(oxy)]bis[trimethyl-;; CHOLEST-5-ENE-3β,24α-DIOL BIS-TRIMETHYLSILYL ETHER; W/NBS
64286

Str 549

546 C33 H62 O2 Si2 33287-26-2 EP-8356-0-0 Silane, [[(3β,7β)-cholest-5-ene-3,7-diyl]bis(oxy)]bis[trimethyl-;; 5-CHOLESTENE-3β,7β-DIOL TMS; W/NBS
64288

Str 550

546 C33 H62 O2 Si2 33403-39-3 EP-8477-0-0 Silane, [[(3β,6β)-cholest-4-ene-3,6-diyl]bis(oxy)]bis[trimethyl-;; 4-CHOLESTENE-3β,6β-DIOL TMS; W/NBS
64290

Str 551

546 C33 H62 O2 Si2 77828-35-4 KO-6-138-3 CHOLEST-4-ENE-3β,6α-DIOL BIS(TRIMETHYLSILYL) ETHER; W
64291

Str 552

546 C33 H62 O2 Si2 77847-00-8 KO-6-138-3 CHOLEST-4-ENE-3α,6β-DIOL BIS(TRIMETHYLSILYL) ETHER; W
64292

Str 553

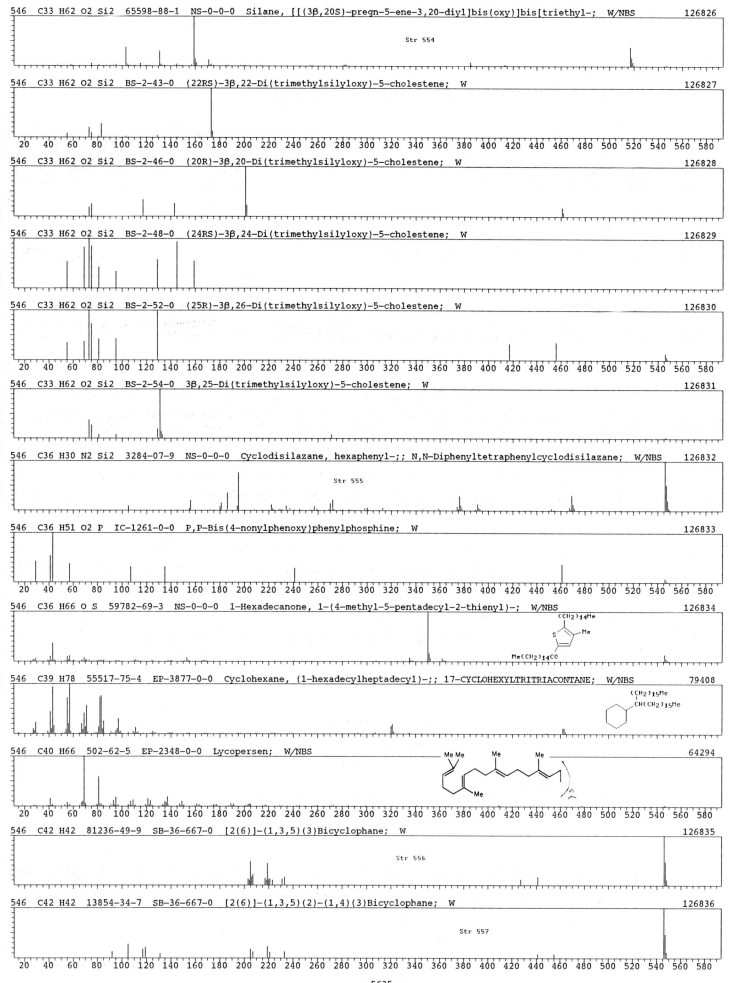

546   C33 H62 O2 Si2   65598-88-1   NS-0-0-0   Silane, [[(3β,20S)-pregn-5-ene-3,20-diyl]bis(oxy)]bis[triethyl-;   W/NBS   126826

Str 554

546   C33 H62 O2 Si2   BS-2-43-0   (22RS)-3β,22-Di(trimethylsilyloxy)-5-cholestene;   W   126827

546   C33 H62 O2 Si2   BS-2-46-0   (20R)-3β,20-Di(trimethylsilyloxy)-5-cholestene;   W   126828

546   C33 H62 O2 Si2   BS-2-48-0   (24RS)-3β,24-Di(trimethylsilyloxy)-5-cholestene;   W   126829

546   C33 H62 O2 Si2   BS-2-52-0   (25R)-3β,26-Di(trimethylsilyloxy)-5-cholestene;   W   126830

546   C33 H62 O2 Si2   BS-2-54-0   3β,25-Di(trimethylsilyloxy)-5-cholestene;   W   126831

546   C36 H30 N2 Si2   3284-07-9   NS-0-0-0   Cyclodisilazane, hexaphenyl-;;   N,N-Diphenyltetraphenylcyclodisilazane;   W/NBS   126832

Str 555

546   C36 H51 O2 P   IC-1261-0-0   P,P-Bis(4-nonylphenoxy)phenylphosphine;   W   126833

546   C36 H66 O S   59782-69-3   NS-0-0-0   1-Hexadecanone, 1-(4-methyl-5-pentadecyl-2-thienyl)-;   W/NBS   126834

546   C39 H78   55517-75-4   EP-3877-0-0   Cyclohexane, (1-hexadecylheptadecyl)-;;   17-CYCLOHEXYLTRITRIACONTANE;   W/NBS   79408

546   C40 H66   502-62-5   EP-2348-0-0   Lycopersen;   W/NBS   64294

546   C42 H42   81236-49-9   SB-36-667-0   [2(6)]-(1,3,5)(3)Bicyclophane;   W   126835

Str 556

546   C42 H42   13854-34-7   SB-36-667-0   [2(6)]-(1,3,5)(2)-(1,4)(3)Bicyclophane;   W   126836

Str 557

547   C16 H11 F14 N O4   67412-36-6   KO-5-178-2   RETRONECINE DIHEPTAFLUOROBUTYRATE;   W   64295

547   C24 H31 Cl2 N O5 S2   66087-87-4   C-100-584-0   RAC-1,1-DICHLORO-3-(3(R)-ETHYL-1((4-METHYLPHENYL)SULFONYLOXY)-4(S)-PIPERIDINY
L)PROPAN-2(S)-OL 4-METHYLPHENYL-SULFONATE ESTER;   W   64296

547   C24 H37 N O13   85398-19-2   H-65-2266-0   7-Desoxy-1,2:3,4:8,9:11,12-tetra-O-isopropyliden-7-nitro-β-D-manno-D-glycero-α-D-gal
acto-dodeco-1,5-pyranos-7-ulo-7,10-furanose;   W   126837

Str 558

547   C27 H18 In N3 O3   OS-5-432-0   Triquinolin-8-olatoindium(III);   W   126838

547   C27 H25 N5 O8   75314-21-5   H-62-693-0   Benzamide, N-[6-[2-O-acetyl-5-O-benzoyl-3-C-(hydroxymethyl)-β-D-xylofuranosyl]-9H-pur
in-9-yl]-;; O-ACETYL-2-O-BENZOYL-5'-C-HYDROXYMETHYL-3'-B-D-XYLOFURANOSYL-9-BENZAMIDO-6-PURINE;   W/NBS   64297

Str 559

547   C28 H49 N O4 Si3   69855-54-5   NS--28201-0-0   Estra-1,3,5(10)-trien-6-one, 3,16,17-tris[(trimethylsilyl)oxy]-, O-methyloxime,
(16α,17β)-;;   NBS   64298

Str 560

547   C28 H49 N O4 Si3   SH-1981-0-0   Methyloxime, trimethylsilyl-15β,16α-Dihydroxyesterone;   W   126839

547   C28 H53 N O9   76871-54-0   I-59-1731-0   2R,3R-(+)-N-Tetradecyl-2-carboxamido-3-carboxyl-1,4,7,10,13,16-hexaoxacyclooctadecane
;   W   126840

547   C30 H33 N3 O5 S   83933-12-4   KC-1982-2020-0   5-Benzyl 3-(6,7,8,14-Tetrahydro-6,14-endo-ethenothebain 7α-ylcarbonyl)thiocarba
zate;   W   126841

547   C32 H41 N3 O5   77248-17-0   J-46-3425-0   3β-(Methoxymethoxy)-5α,8α-(4-phenyl-1,2-urazolo)bis-norchola-6-ene-22-al;   W   126844

Str 561

547   C35 H49 N O4   71473-00-2   K-112-2688-0   (25R)-26-PHTHALIMIDO-5-CHOLESTEN-3β,16β-DIOL;   W   64299

Str 562

547   C38 H29 N O S   72478-82-1   J-45-929-0   N-(5',6'-Dihydro-4'-(4-methoxyphenyl)spiro(9H-fluorene-9,3'(4'H)-(2H)-naphtho(1,2-b)
thiopyran)-2'-ylidene-benzenamine;   W   126845

Str 563

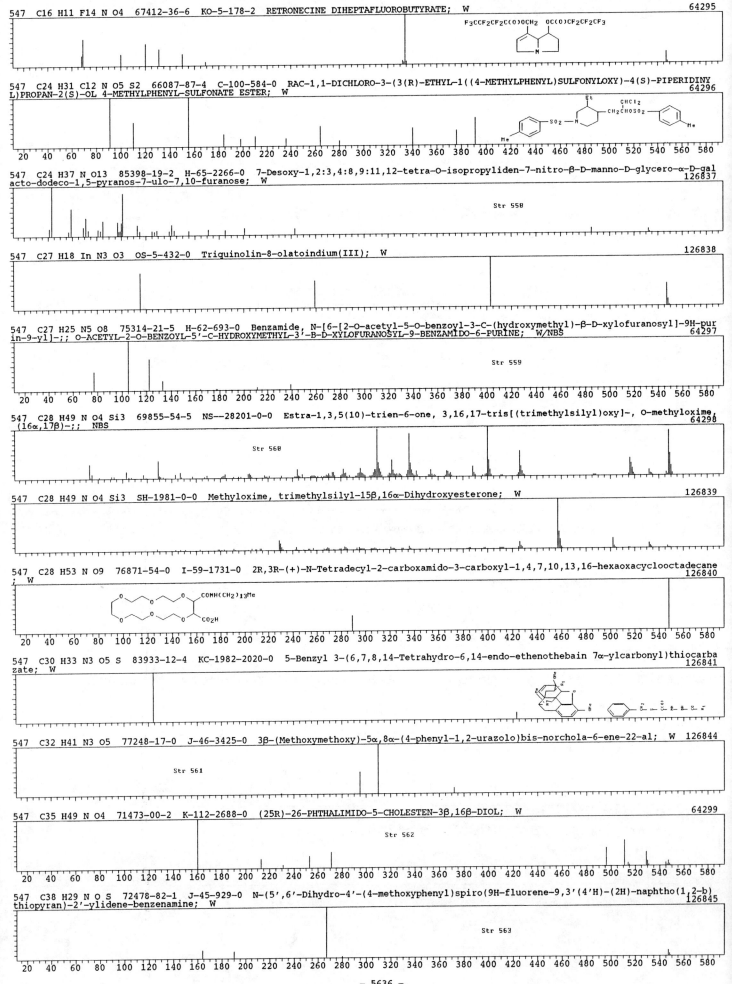

547 C38 H29 N O S 72478-85-4 J-45-929-0 2-(9H-Fluorena-ylidene)-3,4,5,6-tetrahydro-4-(4-methoxyphenyl)-3-phenyl-2H-naphtho(2,1-e)-1,3-thiazine; W
126846

Str 564

547 C40 H53 N 55517-76-5 AD-0-3723-0 1'H-Cholesta-2,5-dieno[3,2-b]indole, 1'-(phenylmethyl)-;; CHOLEST-5-ENO(3,2-B)N-BENZYLINDOLE; W/NBS
64300

Str 565

548 C13 H5 F12 I O2 80360-45-8 J-47-1025-0 10-Methyl-3,3,7,7-tetrakis(trifluoromethyl)-4,6-benzo-1-ioda-2,8-dioxabicyclo[3.3.1]octane; W
126847

548 C14 H8 F12 O9 49561-09-3 EP-6506-0-0 α-L-Mannopyranose, 6-deoxy-, tetrakis(trifluoroacetate);; TETRAKIS-TRIFLUOROACETYL-RHAMNOPYRANOSE; W/NBS
64301

548 C18 F15 O P 2729-11-5 AD-0-2727-0 Phosphine oxide, tris(pentafluorophenyl)-;; TRIS(PENTAFLUOROPHENYL) PHOSPHINE OXIDE; W/NBS
64302

Str 566

548 C18 H12 Se4 67497-85-2 O-13-125-4 DIPHENYLTETRASELENAFULVALENE; W
64303

548 C18 H19 Br3 N2 O3 HE-1982-0-0 5H-INDENO[1,2-B]-1,2,3,4-TETRAHYDROPYRAZIN-5-ON, 9-ETHOXYCARBONYL-N,N'-DIETHYL-6,7,8-TRIBROMO-; W
64304

548 C19 H23 N2 O4 Tl 58410-75-6 B-29-54-0 (4,4'-DIETHOXYCARBONYL-3,3',5,5'-TETRAMETHYLDIPYRROMETHENATO)THALLIUM; W
64305

548 C20 H32 Cl2 Rh2 HE-1982-0-0 RHODIUM, MUE-DICHLORO-DI-(CIS-1,2-DIVINYLCYCLOHEXAN)-BIS-; W
64308

548 C20 H32 Cl2 Rh2 HE-1982-0-0 RHODIUM, MUE-DICHLORO-DI-(TRANS-1,2-DIVINYLCYCLOHEXAN)-BIS-; W
64309

548 C20 H32 Cl2 Rh2 74811-01-1 NS--28197-0-0 Rhodium, di-.mu.-chlorobis(η4-1,2-diethenylcyclohexane)di-, stereoisomer;; NBS
64306

548 C20 H32 Cl2 Rh2 74842-27-6 NS--28196-0-0 Rhodium, bis[1,2-bis(η2-ethenyl)cyclohexane]di-.mu.-chlorodi-, stereoisomer;; NBS
64307

Str 567

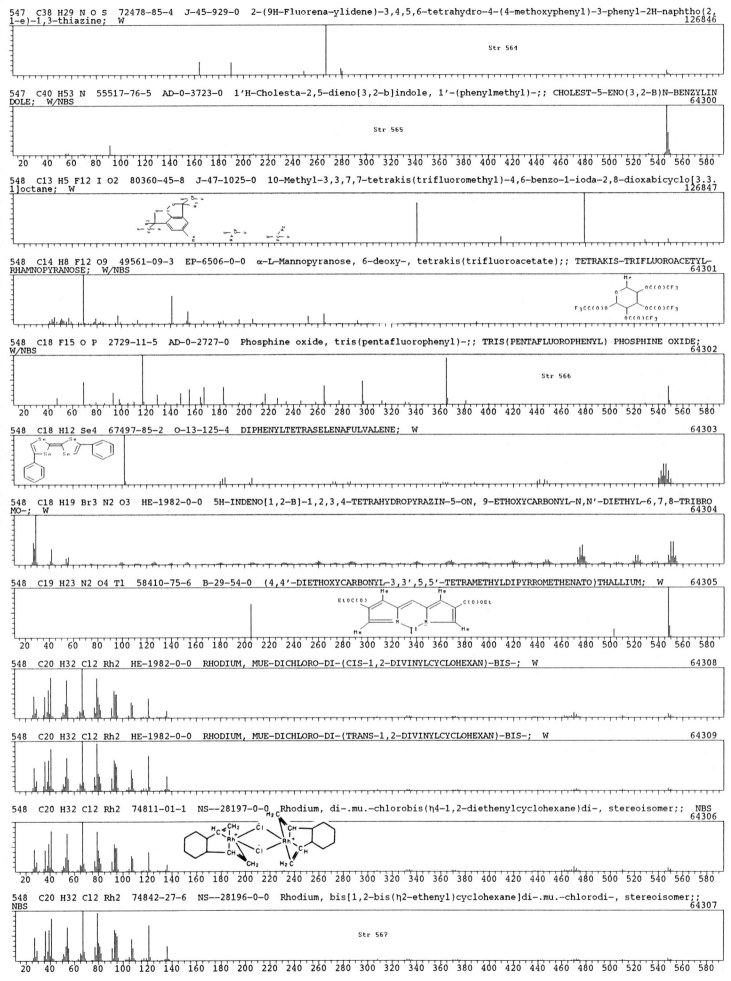

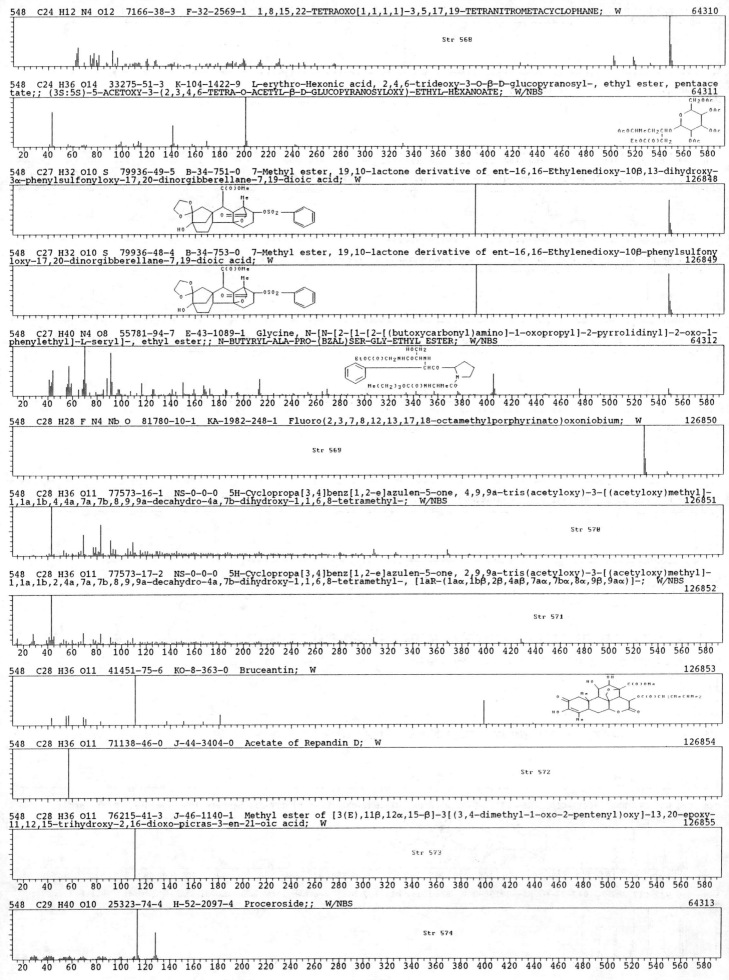

548 C24 H12 N4 O12 7166-38-3 F-32-2569-1 1,8,15,22-TETRAOXO[1,1,1,1]-3,5,17,19-TETRANITROMETACYCLOPHANE; W   64310

Str 568

548 C24 H36 O14 33275-51-3 K-104-1422-9 L-erythro-Hexonic acid, 2,4,6-trideoxy-3-O-β-D-glucopyranosyl-, ethyl ester, pentaace
tate;; (3S:5S)-5-ACETOXY-3-(2,3,4,6-TETRA-O-ACETYL-β-D-GLUCOPYRANOSYLOXY)-ETHYL-HEXANOATE; W/NBS   64311

548 C27 H32 O10 S 79936-49-5 B-34-751-0 7-Methyl ester, 19,10-lactone derivative of ent-16,16-Ethylenedioxy-10β,13-dihydroxy-
3α-phenylsulfonyloxy-17,20-dinorgibberellane-7,19-dioic acid; W   126848

548 C27 H32 O10 S 79936-48-4 B-34-753-0 7-Methyl ester, 19,10-lactone derivative of ent-16,16-Ethylenedioxy-10β-phenylsulfony
loxy-17,20-dinorgibberellane-7,19-dioic acid; W   126849

548 C27 H40 N4 O8 55781-94-7 E-43-1089-1 Glycine, N-[N-[2-[1-[2-[(butoxycarbonyl)amino]-1-oxopropyl]-2-pyrrolidinyl]-2-oxo-1-
phenylethyl]-L-seryl]-, ethyl ester;; N-BUTYRYL-ALA-PRO-(BZAL)SER-GLY-ETHYL ESTER; W/NBS   64312

548 C28 H28 F N4 Nb O 81780-10-1 KA-1982-248-1 Fluoro(2,3,7,8,12,13,17,18-octamethylporphyrinato)oxoniobium; W   126850

Str 569

548 C28 H36 O11 77573-16-1 NS-0-0-0 5H-Cyclopropa[3,4]benz[1,2-e]azulen-5-one, 4,9,9a-tris(acetyloxy)-3-[(acetyloxy)methyl]-
1,1a,1b,4,4a,7a,7b,8,9,9a-decahydro-4a,7b-dihydroxy-1,1,6,8-tetramethyl-; W/NBS   126851

Str 570

548 C28 H36 O11 77573-17-2 NS-0-0-0 5H-Cyclopropa[3,4]benz[1,2-e]azulen-5-one, 2,9,9a-tris(acetyloxy)-3-[(acetyloxy)methyl]-
1,1a,1b,2,4a,7a,7b,8,9,9a-decahydro-4a,7b-dihydroxy-1,1,6,8-tetramethyl-, [1aR-(1aα,1bβ,2β,4aβ,7aα,7bα,8α,9β,9aα)]-; W/NBS   126852

Str 571

548 C28 H36 O11 41451-75-6 KO-8-363-0 Bruceantin; W   126853

548 C28 H36 O11 71138-46-0 J-44-3404-0 Acetate of Repandin D; W   126854

Str 572

548 C28 H36 O11 76215-41-3 J-46-1140-1 Methyl ester of [3(E),11β,12α,15-β]-3[(3,4-dimethyl-1-oxo-2-pentenyl)oxy]-13,20-epoxy-
11,12,15-trihydroxy-2,16-dioxo-picras-3-en-21-oic acid; W   126855

Str 573

548 C29 H40 O10 25323-74-4 H-52-2097-4 Proceroside;; W/NBS   64313

Str 574

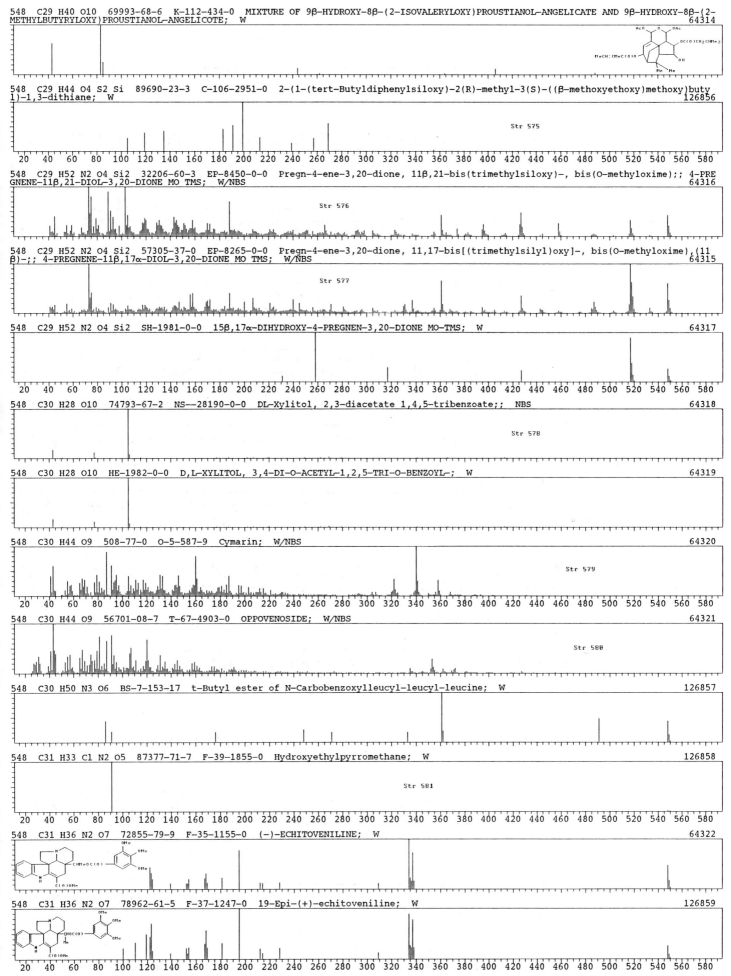

548 C29 H40 O10 69993-68-6 K-112-434-0 MIXTURE OF 9β-HYDROXY-8β-(2-ISOVALERYLOXY)PROUSTIANOL-ANGELICATE AND 9β-HYDROXY-8β-(2-METHYLBUTYRYLOXY)PROUSTIANOL-ANGELICOTE; W
64314

548 C29 H44 O4 S2 Si 89690-23-3 C-106-2951-0 2-(1-(tert-Butyldiphenylsiloxy)-2(R)-methyl-3(S)-((β-methoxyethoxy)methoxy)butyl)-1,3-dithiane; W
126856

Str 575

548 C29 H52 N2 O4 Si2 32206-60-3 EP-8450-0-0 Pregn-4-ene-3,20-dione, 11β,21-bis(trimethylsiloxy)-, bis(O-methyloxime);; 4-PREGNENE-11β,21-DIOL-3,20-DIONE MO TMS; W/NBS
64316

Str 576

548 C29 H52 N2 O4 Si2 57305-37-0 EP-8265-0-0 Pregn-4-ene-3,20-dione, 11,17-bis[(trimethylsilyl)oxy]-, bis(O-methyloxime),(11β)-;; 4-PREGNENE-11β,17α-DIOL-3,20-DIONE MO TMS; W/NBS
64315

Str 577

548 C29 H52 N2 O4 Si2 SH-1981-0-0 15β,17α-DIHYDROXY-4-PREGNEN-3,20-DIONE MO-TMS; W
64317

548 C30 H28 O10 74793-67-2 NS--28190-0-0 DL-Xylitol, 2,3-diacetate 1,4,5-tribenzoate;; NBS
64318

Str 578

548 C30 H28 O10 HE-1982-0-0 D,L-XYLITOL, 3,4-DI-O-ACETYL-1,2,5-TRI-O-BENZOYL-; W
64319

548 C30 H44 O9 508-77-0 O-5-587-9 Cymarin; W/NBS
64320

Str 579

548 C30 H44 O9 56701-08-7 T-67-4903-0 OPPOVENOSIDE; W/NBS
64321

Str 580

548 C30 H50 N3 O6 BS-7-153-17 t-Butyl ester of N-Carbobenzoxylleucyl-leucyl-leucine; W
126857

548 C31 H33 Cl N2 O5 87377-71-7 F-39-1855-0 Hydroxyethylpyrromethane; W
126858

Str 581

548 C31 H36 N2 O7 72855-79-9 F-35-1155-0 (-)-ECHITOVENILINE; W
64322

548 C31 H36 N2 O7 78962-61-5 F-37-1247-0 19-Epi-(+)-echitoveniline; W
126859

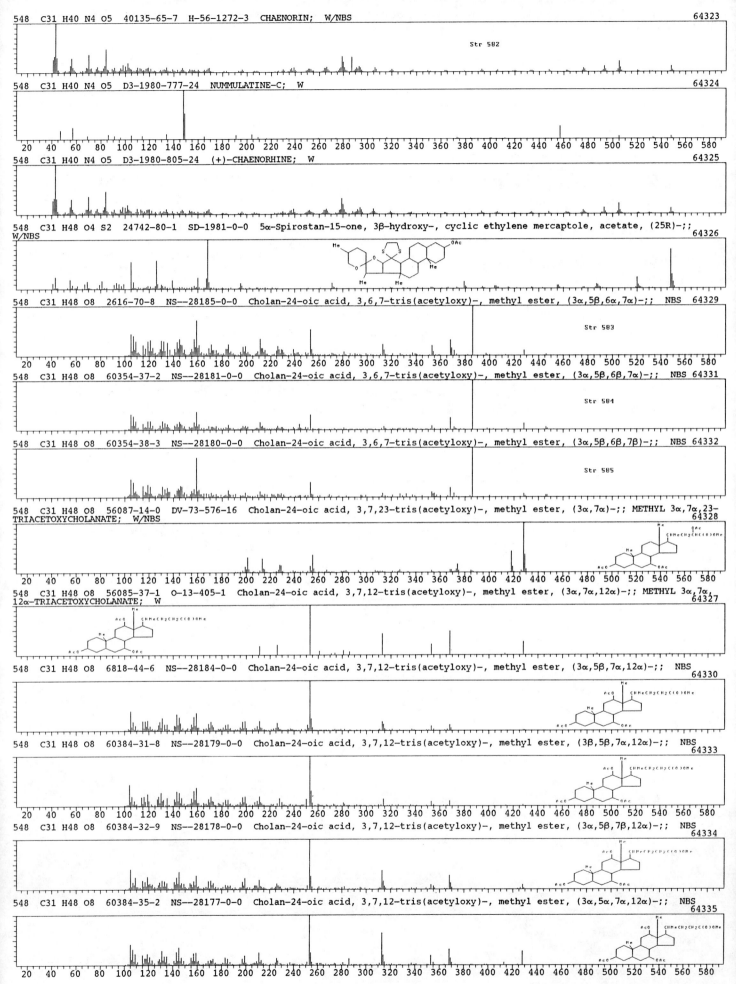

548 C31 H40 N4 O5 40135-65-7 H-56-1272-3 CHAENORIN; W/NBS   64323

Str 582

548 C31 H40 N4 O5 D3-1980-777-24 NUMMULATINE-C; W   64324

548 C31 H40 N4 O5 D3-1980-805-24 (+)-CHAENORHINE; W   64325

548 C31 H48 O4 S2 24742-80-1 SD-1981-0-0 5α-Spirostan-15-one, 3β-hydroxy-, cyclic ethylene mercaptole, acetate, (25R)-;;
W/NBS   64326

548 C31 H48 O8 2616-70-8 NS--28185-0-0 Cholan-24-oic acid, 3,6,7-tris(acetyloxy)-, methyl ester, (3α,5β,6α,7α)-;; NBS 64329

Str 583

548 C31 H48 O8 60354-37-2 NS--28181-0-0 Cholan-24-oic acid, 3,6,7-tris(acetyloxy)-, methyl ester, (3α,5β,6β,7α)-;; NBS 64331

Str 584

548 C31 H48 O8 60354-38-3 NS--28180-0-0 Cholan-24-oic acid, 3,6,7-tris(acetyloxy)-, methyl ester, (3α,5β,6β,7β)-;; NBS 64332

Str 585

548 C31 H48 O8 56087-14-0 DV-73-576-16 Cholan-24-oic acid, 3,7,23-tris(acetyloxy)-, methyl ester, (3α,7α)-;; METHYL 3α,7α,23-
TRIACETOXYCHOLANATE; W/NBS   64328

548 C31 H48 O8 56085-37-1 O-13-405-1 Cholan-24-oic acid, 3,7,12-tris(acetyloxy)-, methyl ester, (3α,7α,12α)-;; METHYL 3α,7α,
12α-TRIACETOXYCHOLANATE; W   64327

548 C31 H48 O8 6818-44-6 NS--28184-0-0 Cholan-24-oic acid, 3,7,12-tris(acetyloxy)-, methyl ester, (3α,5β,7α,12α)-;; NBS
64330

548 C31 H48 O8 60384-31-8 NS--28179-0-0 Cholan-24-oic acid, 3,7,12-tris(acetyloxy)-, methyl ester, (3β,5β,7α,12α)-;; NBS
64333

548 C31 H48 O8 60384-32-9 NS--28178-0-0 Cholan-24-oic acid, 3,7,12-tris(acetyloxy)-, methyl ester, (3α,5β,7β,12α)-;; NBS
64334

548 C31 H48 O8 60384-35-2 NS--28177-0-0 Cholan-24-oic acid, 3,7,12-tris(acetyloxy)-, methyl ester, (3α,5α,7α,12α)-;; NBS
64335

548   C31 H45 D3 O8   69889-83-4   O-13-405-1   3α,7α-DIACETOXY-12α-TRIDEUTEROACETOXYPREGNANE-20-PROPANOIC ACID METHYL ESTER;   W
64336

548   C31 H42 D6 O8   70049-15-9   O-13-405-1   7α-ACETOXY-3α,12α-BIS(TRIDEUTEROACETOXY)PREGNANE-20-PROPANOIC ACID METHYL ESTER;   W
64337

Str 586

548   C31 H42 D6 O8   69889-84-5   O-13-404-1   3α-ACETOXY-7α,12α-BIS(TRIDEUTEROACETOXY)PREGNANE-20-PROPANOIC ACID METHYL ESTER;   W
64338

Str 587

548   C31 H56 O4 Si2   91860-82-1   C-107-1263-0   (2S*,1'S*,6R*)-2-((1'-Benzyloxy)propyl)-4-((triethylsilyl)oxy)-6-(4'-((ter-butyldi
methylsilyl)oxy)butyl)-2,3-dihydro-6H-pyran;   W
126860

548   C32 H44 N4 O4   69721-90-0   H-61-2667-0   4-(3'-Buten-1'-yl)-3,9-dioxo-11-[N-methyl-N-(3''-dimethylaminopropyl)amino]-15-hydr
oxy-2-[(p-methoxyphenyl)methylene]-4,8-diazabicyclo[10.3.1]hexadeca-12,14,1(16)-triene;   W
126861

Str 588

548   C32 H44 N4 O4   85559-68-8   H-65-2546-0   9-Acetyl-7',8'-dihydroverbascenin;   W
126862

Str 589

548   C32 H32 D12 N4 O4   H-61-2667-0   4-(3'-Buten-1'-yl)-3,9-dioxo-11-[N-methyl-d3-N-(3''-di(methyl-d3)aminopropyl)amino]-15-hydr
oxy-2-[(p-(methoxy-d3)phenyl)methylene]-4,8-diazabicyclo[10.3.1]hexadeca-12,14,1(16)-triene;   W
126863

548   C33 H64 O2 Si2   33283-15-7   EP-8482-0-0   Silane, [[(3β,5α)-cholestane-3,5-diyl]bis(oxy)]bis[trimethyl-;;   CHOLESTANE-3β,5α-DI
OL TMS;   W/NBS
64340

Str 590

548   C33 H64 O2 Si2   33403-38-2   EP-8481-0-0   Silane, (5β-cholestan-3β,6β-ylenedioxy)bis[trimethyl-;;   5β-CHOLESTANE-3β,6β-DIOL
TMS;   W/NBS
64339

Str 591

548   C33 H64 O2 Si2   77828-32-1   KO-6-138-3   5α-CHOLESTANE-3β,6α-DIOL TMS;   W
64341

Str 592

548   C33 H64 O2 Si2   40272-80-8   KO-6-138-3   5α-CHOLESTANE-3β,6β-DIOL BIS(TRIMETHLYSILYL)ETHER;   W
64342

Str 593

548   C33 H64 O2 Si2   77828-33-2   KO-6-138-3   5α-CHOESTANE-3α,6β-DIOL BIS(TRIMETHLYSILYL) ETHER;   W
64343

Str 594

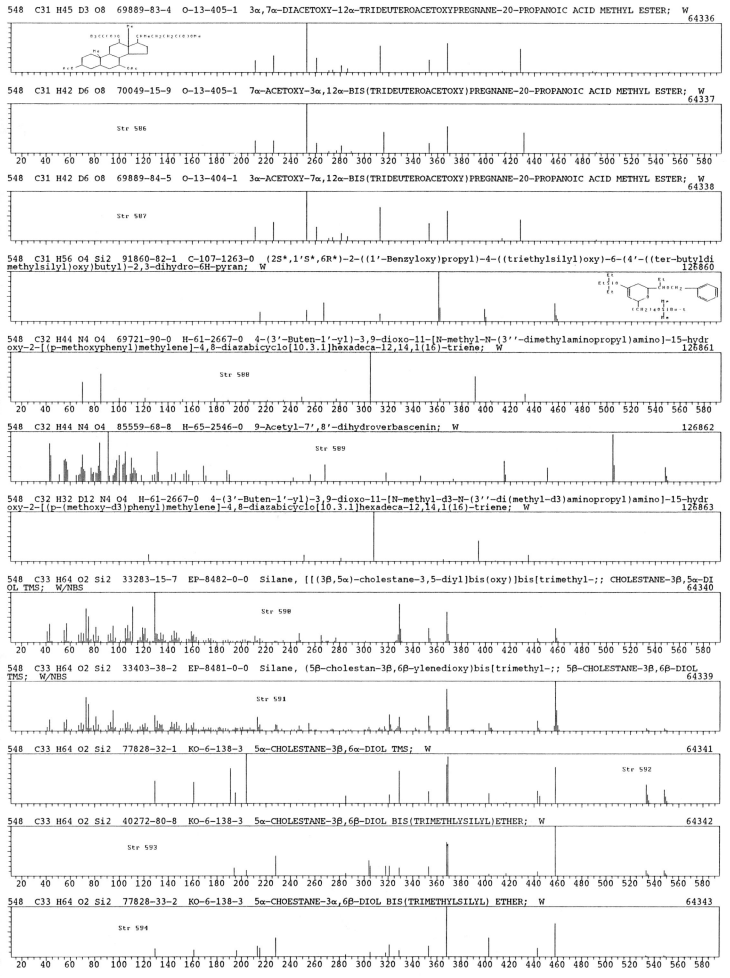

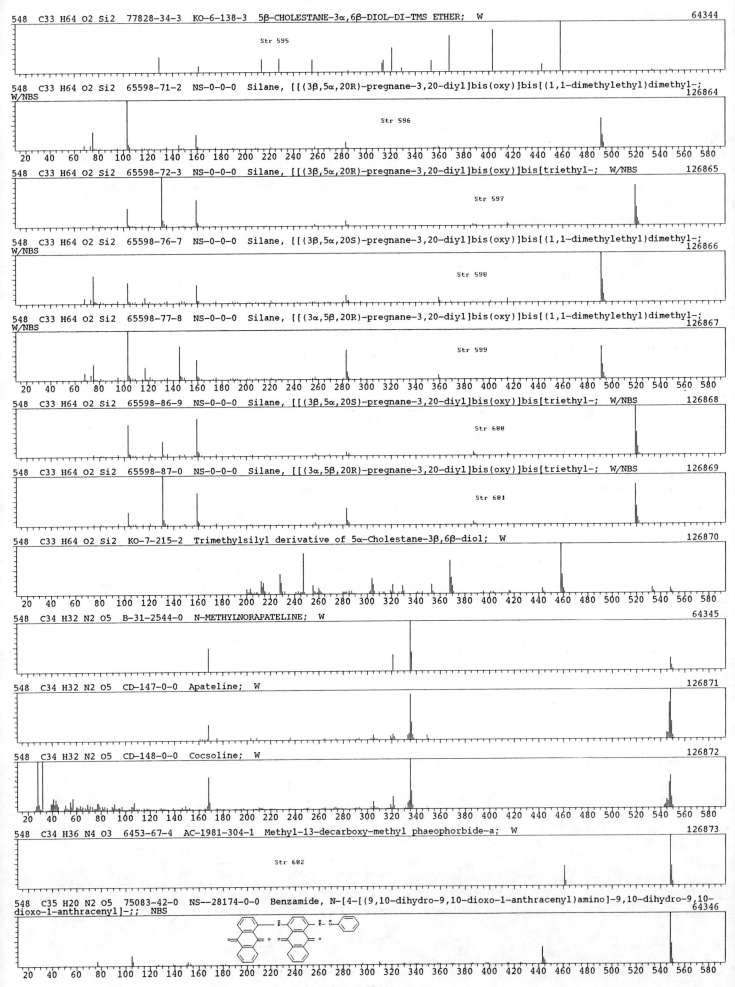

548 C33 H64 O2 Si2 77828-34-3 KO-6-138-3 5β-CHOLESTANE-3α,6β-DIOL-DI-TMS ETHER; W   64344

Str 595

548 C33 H64 O2 Si2 65598-71-2 NS-0-0-0 Silane, [[(3β,5α,20R)-pregnane-3,20-diyl]bis(oxy)]bis[(1,1-dimethylethyl)dimethyl-; W/NBS   126864

Str 596

548 C33 H64 O2 Si2 65598-72-3 NS-0-0-0 Silane, [[(3β,5α,20R)-pregnane-3,20-diyl]bis(oxy)]bis[triethyl-; W/NBS   126865

Str 597

548 C33 H64 O2 Si2 65598-76-7 NS-0-0-0 Silane, [[(3β,5α,20S)-pregnane-3,20-diyl]bis(oxy)]bis[(1,1-dimethylethyl)dimethyl-; W/NBS   126866

Str 598

548 C33 H64 O2 Si2 65598-77-8 NS-0-0-0 Silane, [[(3α,5β,20R)-pregnane-3,20-diyl]bis(oxy)]bis[(1,1-dimethylethyl)dimethyl-; W/NBS   126867

Str 599

548 C33 H64 O2 Si2 65598-86-9 NS-0-0-0 Silane, [[(3β,5α,20S)-pregnane-3,20-diyl]bis(oxy)]bis[triethyl-; W/NBS   126868

Str 600

548 C33 H64 O2 Si2 65598-87-0 NS-0-0-0 Silane, [[(3α,5β,20R)-pregnane-3,20-diyl]bis(oxy)]bis[triethyl-; W/NBS   126869

Str 601

548 C33 H64 O2 Si2 KO-7-215-2 Trimethylsilyl derivative of 5α-Cholestane-3β,6β-diol; W   126870

548 C34 H32 N2 O5 B-31-2544-0 N-METHYLNORAPATELINE; W   64345

548 C34 H32 N2 O5 CD-147-0-0 Apateline; W   126871

548 C34 H32 N2 O5 CD-148-0-0 Cocsoline; W   126872

548 C34 H36 N4 O3 6453-67-4 AC-1981-304-1 Methyl-13-decarboxy-methyl phaeophorbide-a; W   126873

Str 602

548 C35 H20 N2 O5 75083-42-0 NS--28174-0-0 Benzamide, N-[4-[(9,10-dihydro-9,10-dioxo-1-anthracenyl)amino]-9,10-dihydro-9,10-dioxo-1-anthracenyl]-;; NBS   64346

548   C35 H48 O5   89766-99-4   KC-1983-2763-0   Spirolactone,phenylketone,dihydro derivative of (20R)-3β-acetoxy-16-methoxy-16α,30-epoxy-24,25,26,27-tetranor-5α-dammarane-23,20-carbolactone (isomer A);   W
126874

Str 603

548   C35 H48 O5   89766-98-3   KC-1983-2763-0   Spirolactone,phenylketone,dihydro derivative of (20R)-3β-acetoxy-16-methoxy-16α,30-epoxy-24,25,26,27-tetranor-5α-dammarane-23,20-carbolactone (isomer B);   W
126875

Str 604

548   C36 H24 N2 O4   31663-80-6   AD-0-4659-0   1,2,4,5-Benzenetetracarboxylic 1,2:4,5-diimide, N,N'-bis(diphenylmethyl)-;; N,N'-BIS(DIPHENYLMETHYL)PYROMELLITIMIDE;   W/NBS
64347

548   C36 H52 O2 S   84051-57-0   J-49-1089-0   (3β,20E,23Z)-25-Ethoxy-3-methoxy-24-(phenylthio)cholesta-5,20(22),23-triene;   W
126876

Str 605

548   C37 H72 O2   55517-72-1   T-67-5188-0   Cyclopropanepentadecanoic acid, 2-octadecyl-, methyl ester;; METHYL 16,17-METHYLENEPENTATRIACONTANOATE;   W/NBS
64348

Me(CH₂)₁₇ (CH₂)₁₄C(O)OMe

548   C38 H36 Si2   80476-55-7   K-114-3904-0   2,2,10,10-Tetraphenyl-2,10-disilaheptacyclo[9.5.0.0(1,12).0(3,8).0(3,9).0(4,9).0(11,16)]hexadecane;   W
126877

548   C40 H52 O   H-60-2786-0   (3R)-β,.PSI.-CAROTEN-3-OL;   W
64349

549   C21 H28 Cl N3 O10 S   F2-650-0-0   N-CHLOROACETYL ME2 ESTER OF 7-(D-5-NH2-5-CARBOXYVALERAMIDO)-3-HYDROXYME-7-MEO-8-OXO-5-THIA-1-AZABICYCLO[4.2.0]OCT-2-ENE-2-CARBOXYLIC ACID ACETAT;   W
64350

549   C22 H23 N5 O8 S2   42362-44-7   J-8-4341-4   Cyclohexanamine, N-[2,2-bis[(2,4-dinitrophenyl)thio]ethenyl]-N-ethyl-;; 1,1-BIS(2,4-DINITROPHENYLTHIO)-2-N-ETHYLCYCLOHEXYLAMINOETHENE;   W/NBS
64351

Str 606

549   C24 H21 Cl2 N3 O4 S2   72987-66-7   KC-1979-2391-0   4-P-CHLOROPHENYLSULPHONYLIMINO-5-(N-METHYL-P-CHLOROPHENYLSULPHONYLAMINO)-1,2,4,5-TETRAHYDRO-5-METHYLPYRROLO(3,2,1-HI)INDOLE;   W
64352

549   C24 H27 N3 O6 S3   52082-68-5   O-11-1222-1   1,3,5,-TRIS(O-METHYLPHENYLSULPHONYL) HEXAHYDRO-1,3,5,-TRIAZINE;   W
64353

549   C24 H27 N3 O6 S3   23865-44-3   O-11-1222-1   1,3,5,-TRIS(P-METHYLPHENYLSULPHONYL) HEXAHYDRO-1,3,5,-TRIAZINE;   W
64354

Str 607

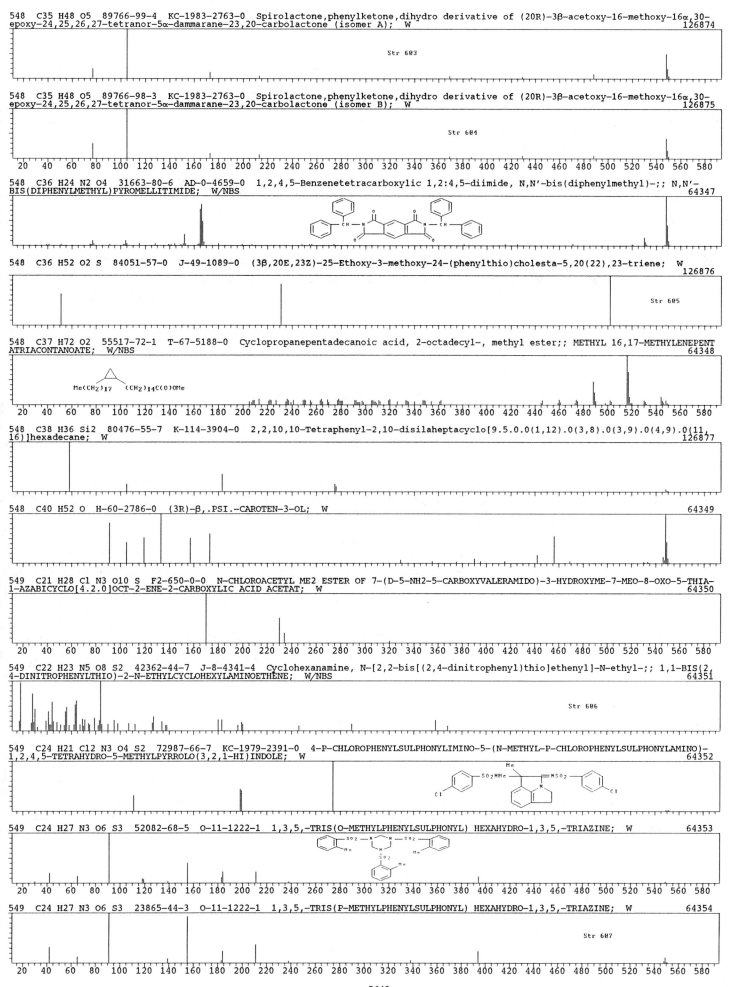

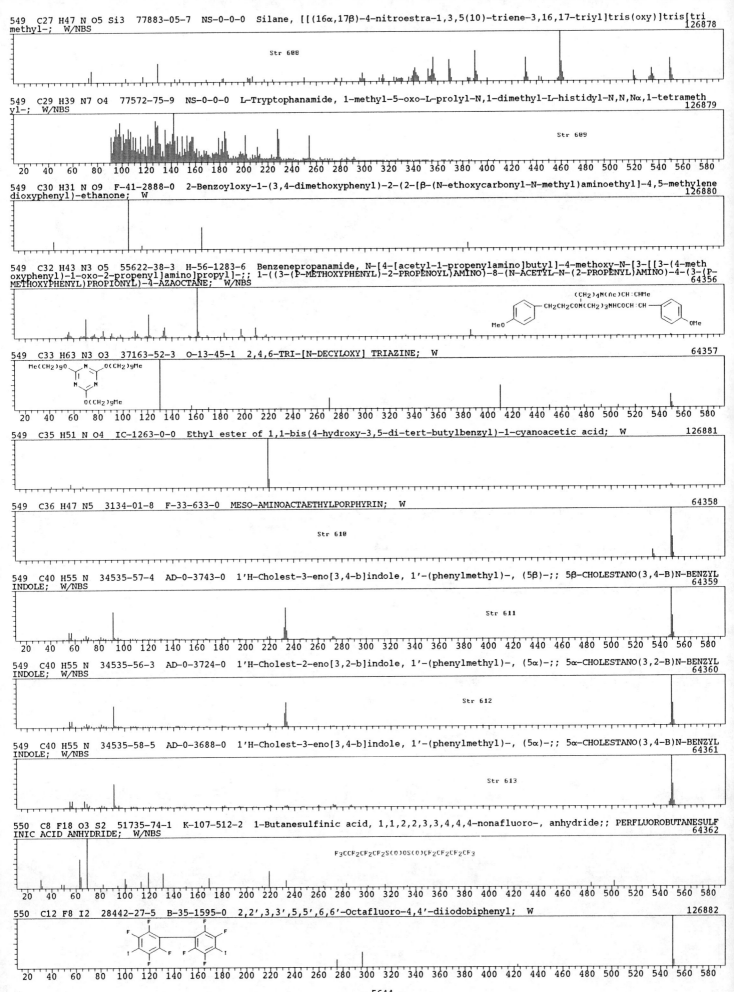

549   C27 H47 N O5 Si3   77883-05-7   NS-0-0-0   Silane, [[(16α,17β)-4-nitroestra-1,3,5(10)-triene-3,16,17-triyl]tris(oxy)]tris[trimethyl-;   W/NBS   126878

Str 608

549   C29 H39 N7 O4   77572-75-9   NS-0-0-0   L-Tryptophanamide, 1-methyl-5-oxo-L-prolyl-N,1-dimethyl-L-histidyl-N,N,Nα,1-tetramethyl-;   W/NBS   126879

Str 609

549   C30 H31 N O9   F-41-2888-0   2-Benzoyloxy-1-(3,4-dimethoxyphenyl)-2-(2-[β-(N-ethoxycarbonyl-N-methyl)aminoethyl]-4,5-methylenedioxyphenyl)-ethanone;   W   126880

549   C32 H43 N3 O5   55622-38-3   H-56-1283-6   Benzenepropanamide, N-[4-[acetyl-1-propenylamino]butyl]-4-methoxy-N-[3-[[3-(4-methoxyphenyl)-1-oxo-2-propenyl]amino]propyl]-;;   1-((3-(P-METHOXYPHENYL)-2-PROPENOYL)AMINO)-8-(N-ACETYL-N-(2-PROPENYL)AMINO)-4-(3-(P-METHOXYPHENYL)PROPIONYL)-4-AZAOCTANE;   W/NBS   64356

549   C33 H63 N3 O3   37163-52-3   O-13-45-1   2,4,6-TRI-[N-DECYLOXY] TRIAZINE;   W   64357

549   C35 H51 N O4   IC-1263-0-0   Ethyl ester of 1,1-bis(4-hydroxy-3,5-di-tert-butylbenzyl)-1-cyanoacetic acid;   W   126881

549   C36 H47 N5   3134-01-8   F-33-633-0   MESO-AMINOACTAETHYLPORPHYRIN;   W   64358

Str 610

549   C40 H55 N   34535-57-4   AD-0-3743-0   1'H-Cholest-3-eno[3,4-b]indole, 1'-(phenylmethyl)-, (5β)-;;   5β-CHOLESTANO(3,4-B)N-BENZYLINDOLE;   W/NBS   64359

Str 611

549   C40 H55 N   34535-56-3   AD-0-3724-0   1'H-Cholest-2-eno[3,2-b]indole, 1'-(phenylmethyl)-, (5α)-;;   5α-CHOLESTANO(3,2-B)N-BENZYLINDOLE;   W/NBS   64360

Str 612

549   C40 H55 N   34535-58-5   AD-0-3688-0   1'H-Cholest-3-eno[3,4-b]indole, 1'-(phenylmethyl)-, (5α)-;;   5α-CHOLESTANO(3,4-B)N-BENZYLINDOLE;   W/NBS   64361

Str 613

550   C8 F18 O3 S2   51735-74-1   K-107-512-2   1-Butanesulfinic acid, 1,1,2,2,3,3,4,4,4-nonafluoro-, anhydride;;   PERFLUOROBUTANESULFINIC ACID ANHYDRIDE;   W/NBS   64362

F3CCF2CF2CF2S(O)OS(O)CF2CF2CF2CF3

550   C12 F8 I2   28442-27-5   B-35-1595-0   2,2',3,3',5,5',6,6'-Octafluoro-4,4'-diiodobiphenyl;   W   126882

- 5644 -

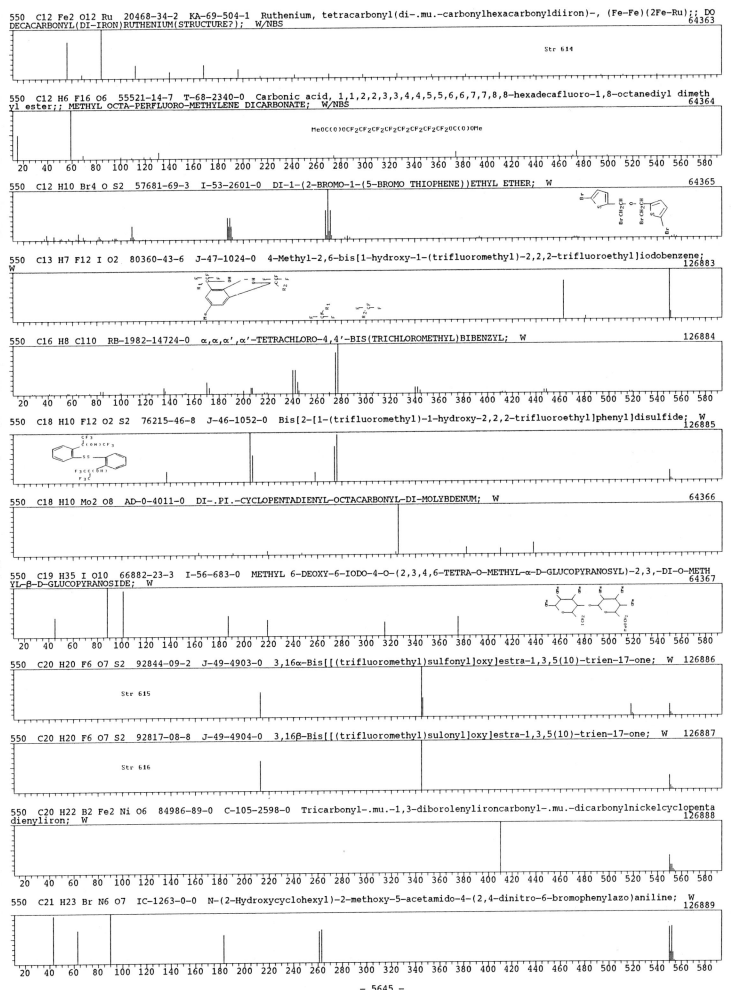

550 C12 Fe2 O12 Ru 20468-34-2 KA-69-504-1 Ruthenium, tetracarbonyl(di-.mu.-carbonylhexacarbonyldiiron)-, (Fe-Fe)(2Fe-Ru);; DO
DECACARBONYL(DI-IRON)RUTHENIUM(STRUCTURE?); W/NBS
64363

Str 614

550 C12 H6 F16 O6 55521-14-7 T-68-2340-0 Carbonic acid, 1,1,2,2,3,3,4,4,5,5,6,6,7,7,8,8-hexadecafluoro-1,8-octanediyl dimeth
yl ester;; METHYL OCTA-PERFLUORO-METHYLENE DICARBONATE; W/NBS
64364

MeOC(O)OCF2CF2CF2CF2CF2CF2CF2CF2OC(O)OMe

550 C12 H10 Br4 O S2 57681-69-3 I-53-2601-0 DI-1-(2-BROMO-1-(5-BROMO THIOPHENE))ETHYL ETHER; W
64365

550 C13 H7 F12 I O2 80360-43-6 J-47-1024-0 4-Methyl-2,6-bis[1-hydroxy-1-(trifluoromethyl)-2,2,2-trifluoroethyl]iodobenzene;
W
126883

550 C16 H8 Cl10 RB-1982-14724-0 α,α,α',α'-TETRACHLORO-4,4'-BIS(TRICHLOROMETHYL)BIBENZYL; W
126884

550 C18 H10 F12 O2 S2 76215-46-8 J-46-1052-0 Bis[2-[1-(trifluoromethyl)-1-hydroxy-2,2,2-trifluoroethyl]phenyl]disulfide; W
126885

550 C18 H10 Mo2 O8 AD-0-4011-0 DI-.PI.-CYCLOPENTADIENYL-OCTACARBONYL-DI-MOLYBDENUM; W
64366

550 C19 H35 I O10 66882-23-3 I-56-683-0 METHYL 6-DEOXY-6-IODO-4-O-(2,3,4,6-TETRA-O-METHYL-α-D-GLUCOPYRANOSYL)-2,3,-DI-O-METH
YL-β-D-GLUCOPYRANOSIDE; W
64367

550 C20 H20 F6 O7 S2 92844-09-2 J-49-4903-0 3,16α-Bis[[(trifluoromethyl)sulfonyl]oxy]estra-1,3,5(10)-trien-17-one; W 126886

Str 615

550 C20 H20 F6 O7 S2 92817-08-8 J-49-4904-0 3,16β-Bis[[(trifluoromethyl)sulonyl]oxy]estra-1,3,5(10)-trien-17-one; W 126887

Str 616

550 C20 H22 B2 Fe2 Ni O6 84986-89-0 C-105-2598-0 Tricarbonyl-.mu.-1,3-diborolenyliironcarbonyl-.mu.-dicarbonylnickelcyclopenta
dienyliron; W
126888

550 C21 H23 Br N6 O7 IC-1263-0-0 N-(2-Hydroxycyclohexyl)-2-methoxy-5-acetamido-4-(2,4-dinitro-6-bromophenylazo)aniline; W
126889

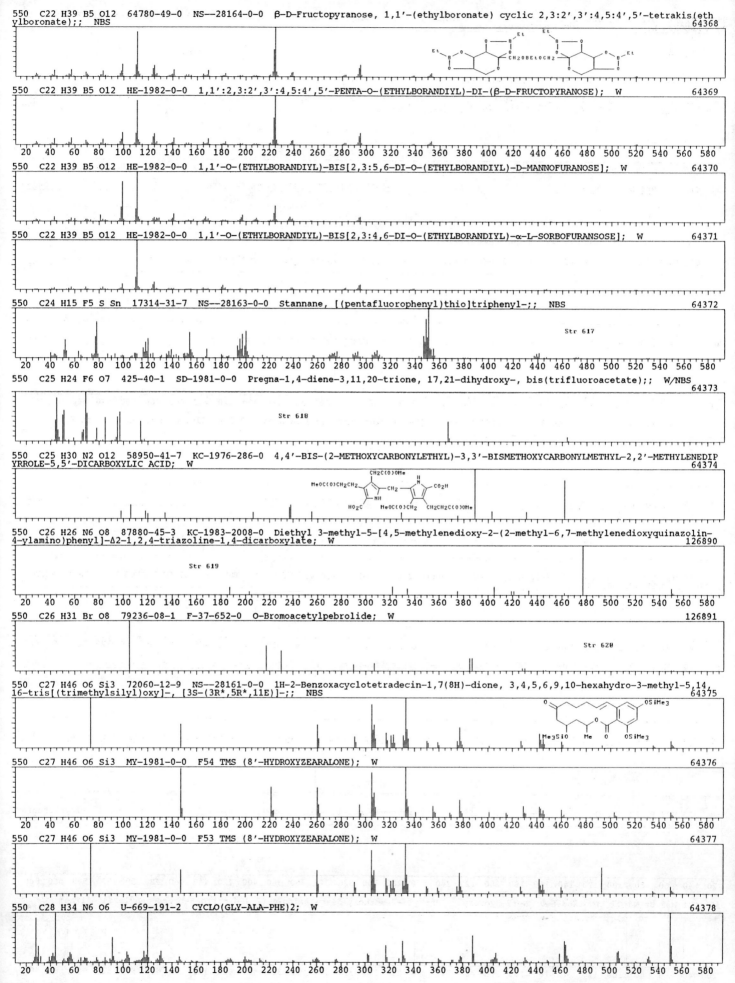

550 C22 H39 B5 O12   64780-49-0   NS--28164-0-0   β-D-Fructopyranose, 1,1'-(ethylboronate) cyclic 2,3:2',3':4,5:4',5'-tetrakis(ethylboronate);;   NBS
64368

550 C22 H39 B5 O12   HE-1982-0-0   1,1':2,3:2',3':4,5:4',5'-PENTA-O-(ETHYLBORANDIYL)-DI-(β-D-FRUCTOPYRANOSE);   W
64369

550 C22 H39 B5 O12   HE-1982-0-0   1,1'-O-(ETHYLBORANDIYL)-BIS[2,3:5,6-DI-O-(ETHYLBORANDIYL)-D-MANNOFURANOSE];   W
64370

550 C22 H39 B5 O12   HE-1982-0-0   1,1'-O-(ETHYLBORANDIYL)-BIS[2,3:4,6-DI-O-(ETHYLBORANDIYL)-α-L-SORBOFURANSOSE];   W
64371

550 C24 H15 F5 S Sn   17314-31-7   NS--28163-0-0   Stannane, [(pentafluorophenyl)thio]triphenyl-;;   NBS
64372
Str 617

550 C25 H24 F6 O7   425-40-1   SD-1981-0-0   Pregna-1,4-diene-3,11,20-trione, 17,21-dihydroxy-, bis(trifluoroacetate);;   W/NBS
64373
Str 618

550 C25 H30 N2 O12   58950-41-7   KC-1976-286-0   4,4'-BIS-(2-METHOXYCARBONYLETHYL)-3,3'-BISMETHOXYCARBONYLMETHYL-2,2'-METHYLENEDIPYRROLE-5,5'-DICARBOXYLIC ACID;   W
64374

550 C26 H26 N6 O8   87880-45-3   KC-1983-2008-0   Diethyl 3-methyl-5-[4,5-methylenedioxy-2-(2-methyl-6,7-methylenedioxyquinazolin-4-ylamino)phenyl]-Δ2-1,2,4-triazoline-1,4-dicarboxylate;   W
126890
Str 619

550 C26 H31 Br O8   79236-08-1   F-37-652-0   O-Bromoacetylpebrolide;   W
126891
Str 620

550 C27 H46 O6 Si3   72060-12-9   NS--28161-0-0   1H-2-Benzoxacyclotetradecin-1,7(8H)-dione, 3,4,5,6,9,10-hexahydro-3-methyl-5,14,16-tris[(trimethylsilyl)oxy]-, [3S-(3R*,5R*,11E)]-;;   NBS
64375

550 C27 H46 O6 Si3   MY-1981-0-0   F54 TMS (8'-HYDROXYZEARALONE);   W
64376

550 C27 H46 O6 Si3   MY-1981-0-0   F53 TMS (8'-HYDROXYZEARALONE);   W
64377

550 C28 H34 N6 O6   U-669-191-2   CYCLO(GLY-ALA-PHE)2;   W
64378

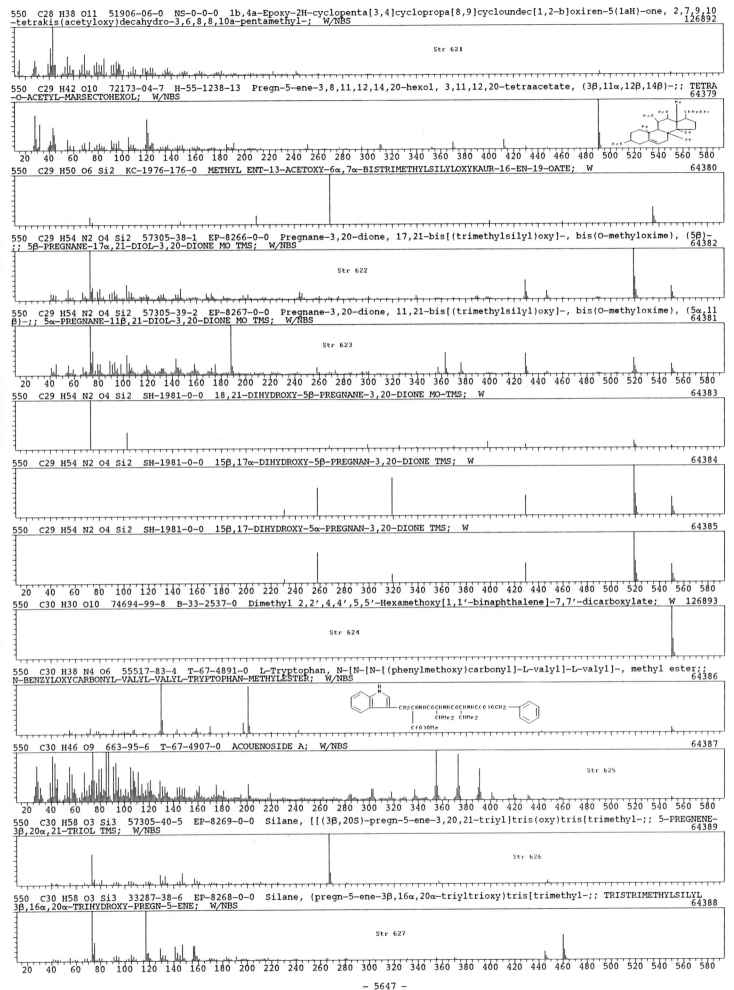

550 C28 H38 O11 51906-06-0 NS-0-0-0 1b,4a-Epoxy-2H-cyclopenta[3,4]cyclopropa[8,9]cycloundec[1,2-b]oxiren-5(1aH)-one, 2,7,9,10
-tetrakis(acetyloxy)decahydro-3,6,8,8,10a-pentamethyl-; W/NBS   126892

Str 621

550 C29 H42 O10 72173-04-7 H-55-1238-13 Pregn-5-ene-3,8,11,12,14,20-hexol, 3,11,12,20-tetraacetate, (3β,11α,12β,14β)-;; TETRA
-O-ACETYL-MARSECTOHEXOL; W/NBS   64379

550 C29 H50 O6 Si2 KC-1976-176-0 METHYL ENT-13-ACETOXY-6α,7α-BISTRIMETHYLSILYLOXYKAUR-16-EN-19-OATE; W   64380

550 C29 H54 N2 O4 Si2 57305-38-1 EP-8266-0-0 Pregnane-3,20-dione, 17,21-bis[(trimethylsilyl)oxy]-, bis(O-methyloxime), (5β)-
;; 5β-PREGNANE-17α,21-DIOL-3,20-DIONE MO TMS; W/NBS   64382

Str 622

550 C29 H54 N2 O4 Si2 57305-39-2 EP-8267-0-0 Pregnane-3,20-dione, 11,21-bis[(trimethylsilyl)oxy]-, bis(O-methyloxime), (5α,11
β)-;; 5α-PREGNANE-11β,21-DIOL-3,20-DIONE MO TMS; W/NBS   64381

Str 623

550 C29 H54 N2 O4 Si2 SH-1981-0-0 18,21-DIHYDROXY-5β-PREGNANE-3,20-DIONE MO-TMS; W   64383

550 C29 H54 N2 O4 Si2 SH-1981-0-0 15β,17α-DIHYDROXY-5β-PREGNAN-3,20-DIONE TMS; W   64384

550 C29 H54 N2 O4 Si2 SH-1981-0-0 15β,17-DIHYDROXY-5α-PREGNAN-3,20-DIONE TMS; W   64385

550 C30 H30 O10 74694-99-8 B-33-2537-0 Dimethyl 2,2',4,4',5,5'-Hexamethoxy[1,1'-binaphthalene]-7,7'-dicarboxylate; W 126893

Str 624

550 C30 H38 N4 O6 55517-83-4 T-67-4891-0 L-Tryptophan, N-[N-[N-[(phenylmethoxy)carbonyl]-L-valyl]-L-valyl]-, methyl ester;;
N-BENZYLOXYCARBONYL-VALYL-VALYL-TRYPTOPHAN-METHYLESTER; W/NBS   64386

550 C30 H46 O9 663-95-6 T-67-4907-0 ACOUENOSIDE A; W/NBS   64387

Str 625

550 C30 H58 O3 Si3 57305-40-5 EP-8269-0-0 Silane, [[(3β,20S)-pregn-5-ene-3,20,21-triyl]tris(oxy)tris[trimethyl-;; 5-PREGNENE-
3β,20α,21-TRIOL TMS; W/NBS   64389

Str 626

550 C30 H58 O3 Si3 33287-38-6 EP-8268-0-0 Silane, (pregn-5-ene-3β,16α,20α-triyltrioxy)tris[trimethyl-;; TRISTRIMETHYLSILYL
3β,16α,20α-TRIHYDROXY-PREGN-5-ENE; W/NBS   64388

Str 627

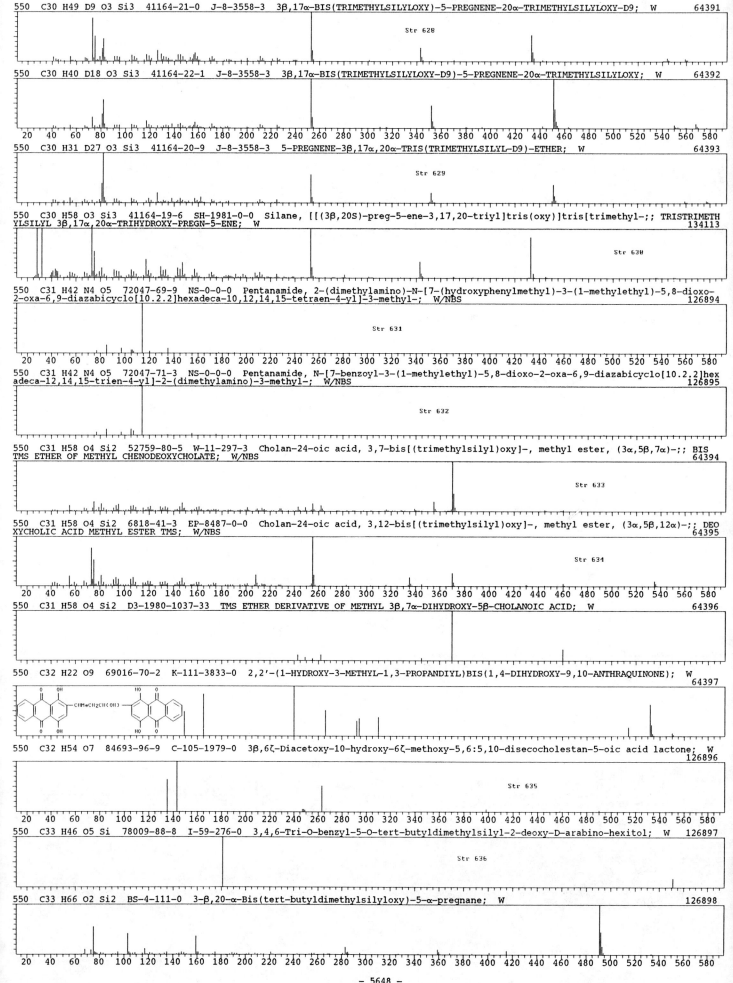

550 C30 H49 D9 O3 Si3 41164-21-0 J-8-3558-3 3β,17α-BIS(TRIMETHYLSILYLOXY)-5-PREGNENE-20α-TRIMETHYLSILYLOXY-D9; W 64391

Str 628

550 C30 H40 D18 O3 Si3 41164-22-1 J-8-3558-3 3β,17α-BIS(TRIMETHYLSILYLOXY-D9)-5-PREGNENE-20α-TRIMETHYLSILYLOXY; W 64392

550 C30 H31 D27 O3 Si3 41164-20-9 J-8-3558-3 5-PREGNENE-3β,17α,20α-TRIS(TRIMETHYLSILYL-D9)-ETHER; W 64393

Str 629

550 C30 H58 O3 Si3 41164-19-6 SH-1981-0-0 Silane, [[(3β,20S)-preg-5-ene-3,17,20-triyl]tris(oxy)]tris[trimethyl-;; TRISTRIMETH
YLSILYL 3β,17α,20α-TRIHYDROXY-PREGN-5-ENE; W 134113

Str 630

550 C31 H42 N4 O5 72047-69-9 NS-0-0-0 Pentanamide, 2-(dimethylamino)-N-[7-(hydroxyphenylmethyl)-3-(1-methylethyl)-5,8-dioxo-
2-oxa-6,9-diazabicyclo[10.2.2]hexadeca-10,12,14,15-tetraen-4-yl]-3-methyl-; W/NBS 126894

Str 631

550 C31 H42 N4 O5 72047-71-3 NS-0-0-0 Pentanamide, N-[7-benzoyl-3-(1-methylethyl)-5,8-dioxo-2-oxa-6,9-diazabicyclo[10.2.2]hex
adeca-12,14,15-trien-4-yl]-2-(dimethylamino)-3-methyl-; W/NBS 126895

Str 632

550 C31 H58 O4 Si2 52759-80-5 W-11-297-3 Cholan-24-oic acid, 3,7-bis[(trimethylsilyl)oxy]-, methyl ester, (3α,5β,7α)-;; BIS
TMS ETHER OF METHYL CHENODEOXYCHOLATE; W/NBS 64394

Str 633

550 C31 H58 O4 Si2 6818-41-3 EP-8487-0-0 Cholan-24-oic acid, 3,12-bis[(trimethylsilyl)oxy]-, methyl ester, (3α,5β,12α)-;; DEO
XYCHOLIC ACID METHYL ESTER TMS; W/NBS 64395

Str 634

550 C31 H58 O4 Si2 D3-1980-1037-33 TMS ETHER DERIVATIVE OF METHYL 3β,7α-DIHYDROXY-5β-CHOLANOIC ACID; W 64396

550 C32 H22 O9 69016-70-2 K-111-3833-0 2,2'-(1-HYDROXY-3-METHYL-1,3-PROPANDIYL)BIS(1,4-DIHYDROXY-9,10-ANTHRAQUINONE); W 64397

550 C32 H54 O7 84693-96-9 C-105-1979-0 3β,6ζ-Diacetoxy-10-hydroxy-6ζ-methoxy-5,6:5,10-disecocholestan-5-oic acid lactone; W 126896

Str 635

550 C33 H46 O5 Si 78009-88-8 I-59-276-0 3,4,6-Tri-O-benzyl-5-O-tert-butyldimethylsilyl-2-deoxy-D-arabino-hexitol; W 126897

Str 636

550 C33 H66 O2 Si2 BS-4-111-0 3-β,20-α-Bis(tert-butyldimethylsilyloxy)-5-α-pregnane; W 126898

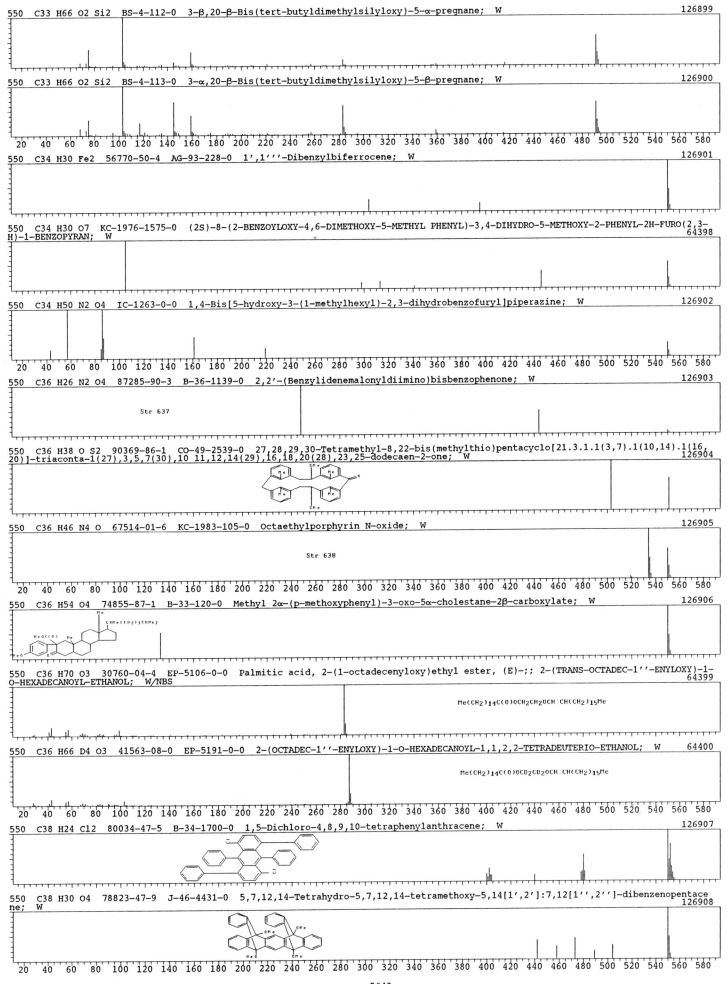

550   C33 H66 O2 Si2   BS-4-112-0   3-β,20-β-Bis(tert-butyldimethylsilyloxy)-5-α-pregnane;   W   126899

550   C33 H66 O2 Si2   BS-4-113-0   3-α,20-β-Bis(tert-butyldimethylsilyloxy)-5-β-pregnane;   W   126900

550   C34 H30 Fe2   56770-50-4   AG-93-228-0   1',1'''-Dibenzylbiferrocene;   W   126901

550   C34 H30 O7   KC-1976-1575-0   (2S)-8-(2-BENZOYLOXY-4,6-DIMETHOXY-5-METHYL PHENYL)-3,4-DIHYDRO-5-METHOXY-2-PHENYL-2H-FURO(2,3-H)-1-BENZOPYRAN;   W   64398

550   C34 H50 N2 O4   IC-1263-0-0   1,4-Bis[5-hydroxy-3-(1-methylhexyl)-2,3-dihydrobenzofuryl]piperazine;   W   126902

550   C36 H26 N2 O4   87285-90-3   B-36-1139-0   2,2'-(Benzylidenemalonyldiimino)bisbenzophenone;   W   126903
Str 637

550   C36 H38 O S2   90369-86-1   CO-49-2539-0   27,28,29,30-Tetramethyl-8,22-bis(methylthio)pentacyclo[21.3.1.1(3,7).1(10,14).1(16,20)]-triaconta-1(27),3,5,7(30),10 11,12,14(29),16,18,20(28),23,25-dodecaen-2-one;   W   126904

550   C36 H46 N4 O   67514-01-6   KC-1983-105-0   Octaethylporphyrin N-oxide;   W   126905
Str 638

550   C36 H54 O4   74855-87-1   B-33-120-0   Methyl 2α-(p-methoxyphenyl)-3-oxo-5α-cholestane-2β-carboxylate;   W   126906

550   C36 H70 O3   30760-04-4   EP-5106-0-0   Palmitic acid, 2-(1-octadecenyloxy)ethyl ester, (E)-;; 2-(TRANS-OCTADEC-1''-ENYLOXY)-1-O-HEXADECANOYL-ETHANOL;   W/NBS   64399
Me(CH2)14C(O)OCH2CH2OCH:CH(CH2)15Me

550   C36 H66 D4 O3   41563-08-0   EP-5191-0-0   2-(OCTADEC-1''-ENYLOXY)-1-O-HEXADECANOYL-1,1,2,2-TETRADEUTERIO-ETHANOL;   W   64400
Me(CH2)14C(O)OCD2CD2OCH:CH(CH2)15Me

550   C38 H24 Cl2   80034-47-5   B-34-1700-0   1,5-Dichloro-4,8,9,10-tetraphenylanthracene;   W   126907

550   C38 H30 O4   78823-47-9   J-46-4431-0   5,7,12,14-Tetrahydro-5,7,12,14-tetramethoxy-5,14[1',2']:7,12[1'',2'']-dibenzenopentacene;   W   126908

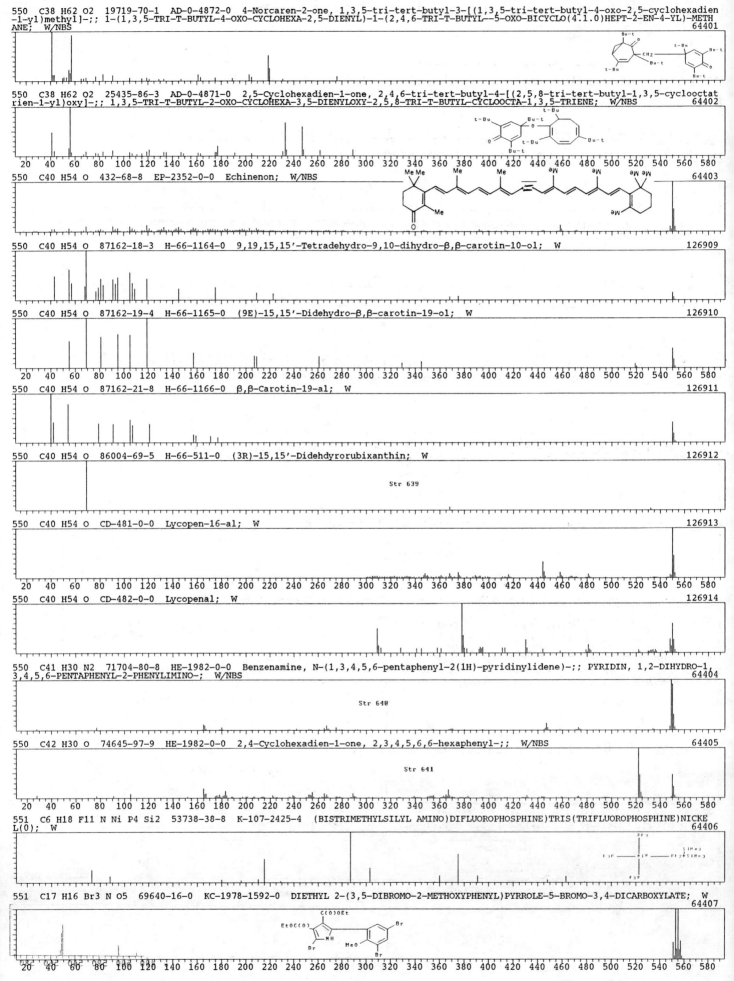

550　C38 H62 O2　19719-70-1　AD-0-4872-0　4-Norcaren-2-one, 1,3,5-tri-tert-butyl-3-[(1,3,5-tri-tert-butyl-4-oxo-2,5-cyclohexadien-1-yl)methyl]-;; 1-(1,3,5-TRI-T-BUTYL-4-OXO-CYCLOHEXA-2,5-DIENYL)-1-(2,4,6-TRI-T-BUTYL--5-OXO-BICYCLO(4.1.0)HEPT-2-EN-4-YL)-METHANE; W/NBS　64401

550　C38 H62 O2　25435-86-3　AD-0-4871-0　2,5-Cyclohexadien-1-one, 2,4,6-tri-tert-butyl-4-[(2,5,8-tri-tert-butyl-1,3,5-cyclooctatrien-1-yl)oxy]-;; 1,3,5-TRI-T-BUTYL-2-OXO-CYCLOHEXA-3,5-DIENYLOXY-2,5,8-TRI-T-BUTYL-CYCLOOCTA-1,3,5-TRIENE; W/NBS　64402

550　C40 H54 O　432-68-8　EP-2352-0-0　Echinenon; W/NBS　64403

550　C40 H54 O　87162-18-3　H-66-1164-0　9,19,15,15'-Tetradehydro-9,10-dihydro-β,β-carotin-10-ol; W　126909

550　C40 H54 O　87162-19-4　H-66-1165-0　(9E)-15,15'-Didehydro-β,β-carotin-19-ol; W　126910

550　C40 H54 O　87162-21-8　H-66-1166-0　β,β-Carotin-19-al; W　126911

550　C40 H54 O　86004-69-5　H-66-511-0　(3R)-15,15'-Didehdyrorubixanthin; W　126912

Str 639

550　C40 H54 O　CD-481-0-0　Lycopen-16-al; W　126913

550　C40 H54 O　CD-482-0-0　Lycopenal; W　126914

550　C41 H30 N2　71704-80-8　HE-1982-0-0　Benzenamine, N-(1,3,4,5,6-pentaphenyl-2(1H)-pyridinylidene)-;; PYRIDIN, 1,2-DIHYDRO-1,3,4,5,6-PENTAPHENYL-2-PHENYLIMINO-; W/NBS　64404

Str 640

550　C42 H30 O　74645-97-9　HE-1982-0-0　2,4-Cyclohexadien-1-one, 2,3,4,5,6,6-hexaphenyl-;; W/NBS　64405

Str 641

551　C6 H18 F11 N Ni P4 Si2　53738-38-8　K-107-2425-4　(BISTRIMETHYLSILYL AMINO)DIFLUOROPHOSPHINE)TRIS(TRIFLUOROPHOSPHINE)NICKEL(0); W　64406

551　C17 H16 Br3 N O5　69640-16-0　KC-1978-1592-0　DIETHYL 2-(3,5-DIBROMO-2-METHOXYPHENYL)PYRROLE-5-BROMO-3,4-DICARBOXYLATE; W　64407

- 5650 -

551 C24 H45 N5 O4 Si3 32352-58-2 AV-40-427-3 Adenosine, N-(3-methyl-3-butenyl)-2',3',5'-tris-O-(trimethylsilyl)-;; TRISTRI
METHYLSILYL N6-(Δ3-ISOPENTENYL)ADENOSINE; W/NBS
64408

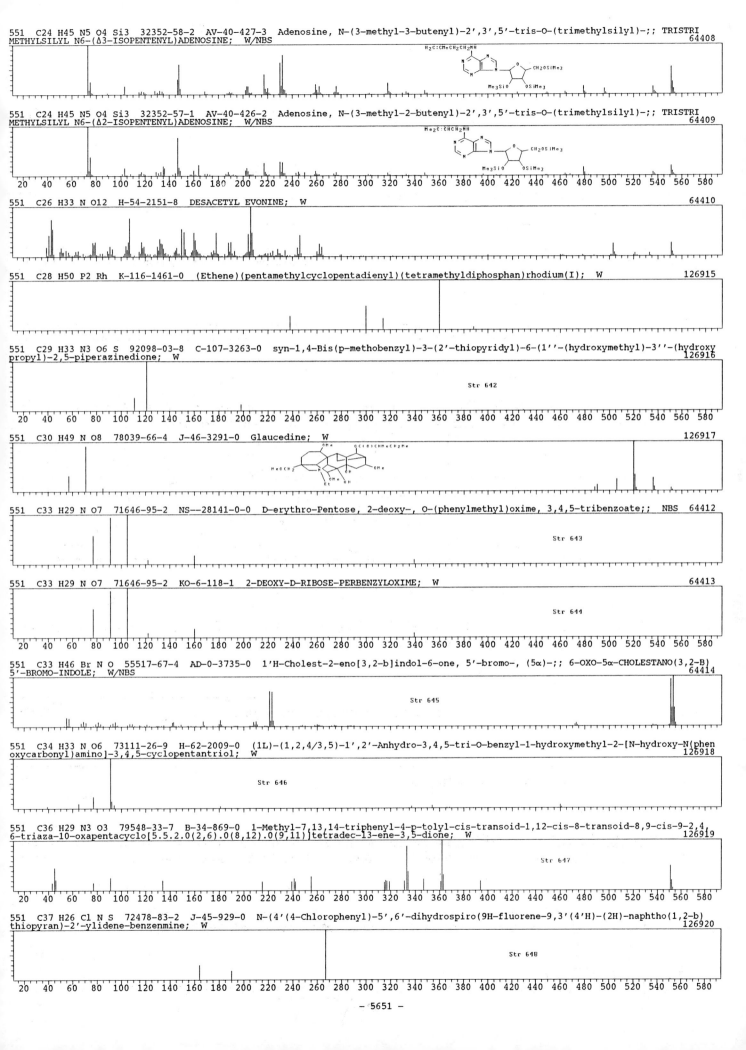

551 C24 H45 N5 O4 Si3 32352-57-1 AV-40-426-2 Adenosine, N-(3-methyl-2-butenyl)-2',3',5'-tris-O-(trimethylsilyl)-;; TRISTRI
METHYLSILYL N6-(Δ2-ISOPENTENYL)ADENOSINE; W/NBS
64409

551 C26 H33 N O12 H-54-2151-8 DESACETYL EVONINE; W
64410

551 C28 H50 P2 Rh K-116-1461-0 (Ethene)(pentamethylcyclopentadienyl)(tetramethyldiphosphan)rhodium(I); W
126915

551 C29 H33 N3 O6 S 92098-03-8 C-107-3263-0 syn-1,4-Bis(p-methobenzyl)-3-(2'-thiopyridyl)-6-(1''-(hydroxymethyl)-3''-(hydroxy
propyl)-2,5-piperazinedione; W
126916
Str 642

551 C30 H49 N O8 78039-66-4 J-46-3291-0 Glaucedine; W
126917

551 C33 H29 N O7 71646-95-2 NS--28141-0-0 D-erythro-Pentose, 2-deoxy-, O-(phenylmethyl)oxime, 3,4,5-tribenzoate;; NBS 64412
Str 643

551 C33 H29 N O7 71646-95-2 KO-6-118-1 2-DEOXY-D-RIBOSE-PERBENZYLOXIME; W
64413
Str 644

551 C33 H46 Br N O 55517-67-4 AD-0-3735-0 1'H-Cholest-2-eno[3,2-b]indol-6-one, 5'-bromo-, (5α)-;; 6-OXO-5α-CHOLESTANO(3,2-B)
5'-BROMO-INDOLE; W/NBS
64414
Str 645

551 C34 H33 N O6 73111-26-9 H-62-2009-0 (1L)-(1,2,4/3,5)-1',2'-Anhydro-3,4,5-tri-O-benzyl-1-hydroxymethyl-2-[N-hydroxy-N(phen
oxycarbonyl)amino]-3,4,5-cyclopentantriol; W
126918
Str 646

551 C36 H29 N3 O3 79548-33-7 B-34-869-0 1-Methyl-7,13,14-triphenyl-4-p-tolyl-cis-transoid-1,12-cis-8-transoid-8,9-cis-9-2,4,
6-triaza-10-oxapentacyclo[5.5.2.0(2,6).0(8,12).0(9,11)]tetradec-13-ene-3,5-dione; W
126919
Str 647

551 C37 H26 Cl N S 72478-83-2 J-45-929-0 N-(4'(4-Chlorophenyl)-5',6'-dihydrospiro(9H-fluorene-9,3'(4'H)-(2H)-naphtho(1,2-b)
thiopyran)-2'-ylidene-benzenmine; W
126920
Str 648

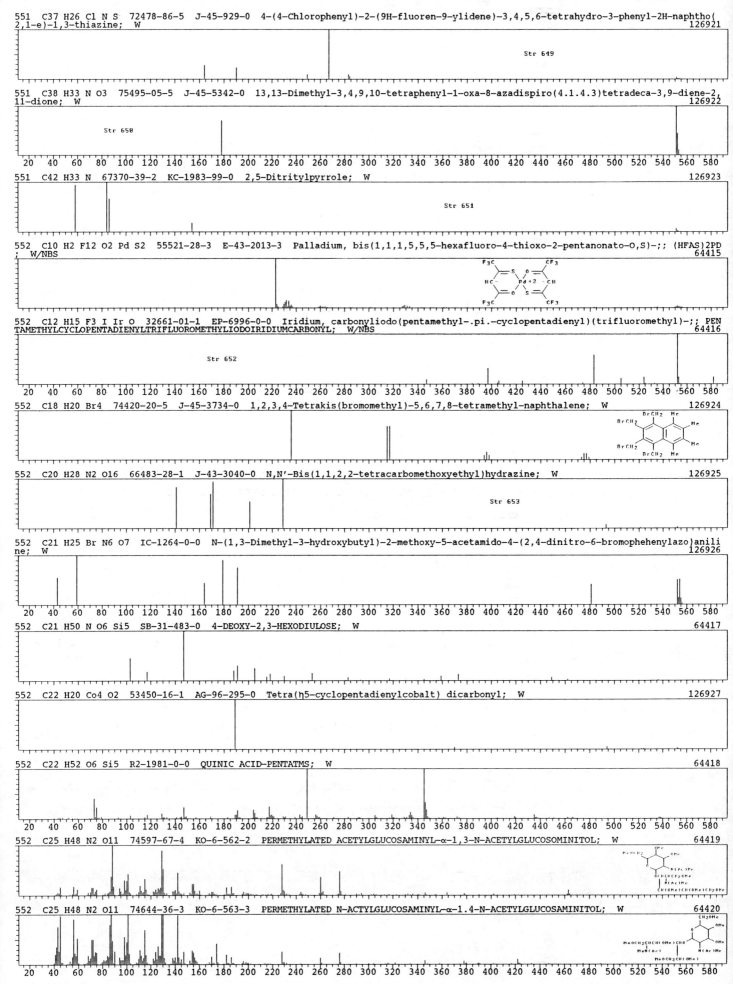

551 C37 H26 Cl N S 72478-86-5 J-45-929-0 4-(4-Chlorophenyl)-2-(9H-fluoren-9-ylidene)-3,4,5,6-tetrahydro-3-phenyl-2H-naphtho(2,1-e)-1,3-thiazine; W 126921

Str 649

551 C38 H33 N O3 75495-05-5 J-45-5342-0 13,13-Dimethyl-3,4,9,10-tetraphenyl-1-oxa-8-azadispiro(4.1.4.3)tetradeca-3,9-diene-2,11-dione; W 126922

Str 650

551 C42 H33 N 67370-39-2 KC-1983-99-0 2,5-Ditritylpyrrole; W 126923

Str 651

552 C10 H2 F12 O2 Pd S2 55521-28-3 E-43-2013-3 Palladium, bis(1,1,1,5,5,5-hexafluoro-4-thioxo-2-pentanonato-O,S)-;; (HFAS)2PD; W/NBS 64415

552 C12 H15 F3 I Ir O 32661-01-1 EP-6996-0-0 Iridium, carbonyliodo(pentamethyl-.pi.-cyclopentadienyl)(trifluoromethyl)-;; PENTAMETHYLCYCLOPENTADIENYLTRIFLUOROMETHYLIODOIRIDIUMCARBONYL; W/NBS 64416

Str 652

552 C18 H20 Br4 74420-20-5 J-45-3734-0 1,2,3,4-Tetrakis(bromomethyl)-5,6,7,8-tetramethyl-naphthalene; W 126924

552 C20 H28 N2 O16 66483-28-1 J-43-3040-0 N,N'-Bis(1,1,2,2-tetracarbomethoxyethyl)hydrazine; W 126925

Str 653

552 C21 H25 Br N6 O7 IC-1264-0-0 N-(1,3-Dimethyl-3-hydroxybutyl)-2-methoxy-5-acetamido-4-(2,4-dinitro-6-bromophehenylazo)aniline; W 126926

552 C21 H50 N O6 Si5 SB-31-483-0 4-DEOXY-2,3-HEXODIULOSE; W 64417

552 C22 H20 Co4 O2 53450-16-1 AG-96-295-0 Tetra(η5-cyclopentadienylcobalt) dicarbonyl; W 126927

552 C22 H52 O6 Si5 R2-1981-0-0 QUINIC ACID-PENTATMS; W 64418

552 C25 H48 N2 O11 74597-67-4 KO-6-562-2 PERMETHYLATED ACETYLGLUCOSAMINYL-α-1,3-N-ACETYLGLUCOSOMINITOL; W 64419

552 C25 H48 N2 O11 74644-36-3 KO-6-563-3 PERMETHYLATED N-ACTYLGLUCOSAMINYL-α-1.4-N-ACETYLGLUCOSAMINITOL; W 64420

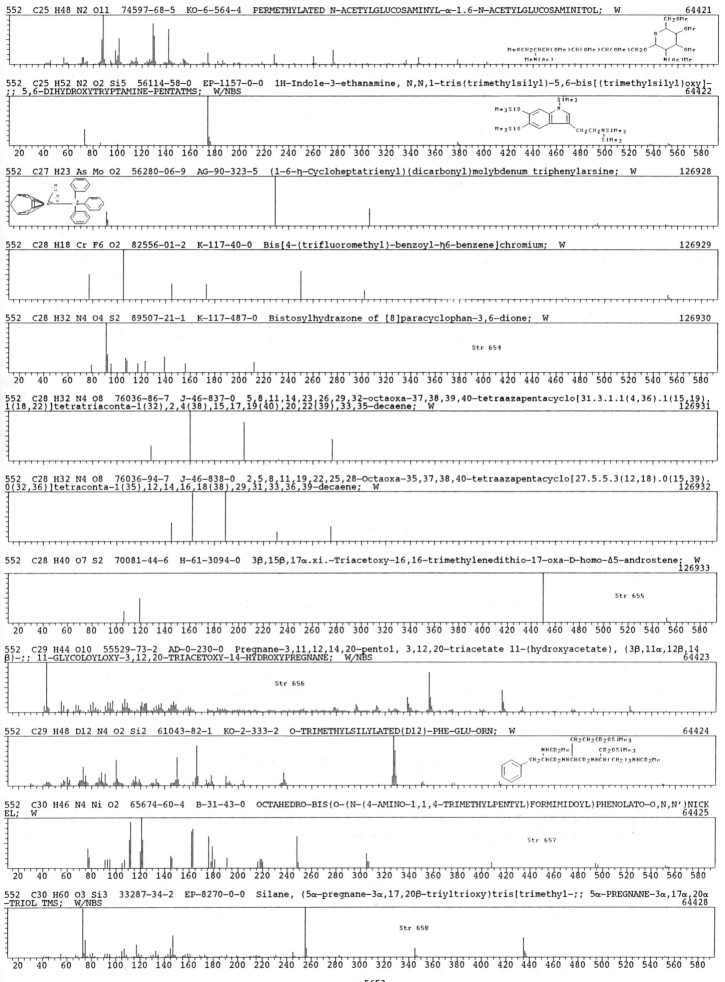

552   C25 H48 N2 O11   74597-68-5   KO-6-564-4   PERMETHYLATED N-ACETYLGLUCOSAMINYL-α-1.6-N-ACETYLGLUCOSAMINITOL; W   64421

552   C25 H52 N2 O2 Si5   56114-58-0   EP-1157-0-0   1H-Indole-3-ethanamine, N,N,1-tris(trimethylsilyl)-5,6-bis[(trimethylsilyl)oxy]-
-;; 5,6-DIHYDROXYTRYPTAMINE-PENTATMS; W/NBS   64422

552   C27 H23 As Mo O2   56280-06-9   AG-90-323-5   (1-6-η-Cycloheptatrienyl)(dicarbonyl)molybdenum triphenylarsine; W   126928

552   C28 H18 Cr F6 O2   82556-01-2   K-117-40-0   Bis[4-(trifluoromethyl)-benzoyl-η6-benzene]chromium; W   126929

552   C28 H32 N4 O4 S2   89507-21-1   K-117-487-0   Bistosylhydrazone of [8]paracyclophan-3,6-dione; W   126930
Str 654

552   C28 H32 N4 O8   76036-86-7   J-46-837-0   5,8,11,14,23,26,29,32-octaoxa-37,38,39,40-tetraazapentacyclo[31.3.1.1(4,36).1(15,19).
1(18,22)]tetratriaconta-1(32),2,4(38),15,17,19(40),20,22(39),33,35-decaene; W   126931

552   C28 H32 N4 O8   76036-94-7   J-46-838-0   2,5,8,11,19,22,25,28-Octaoxa-35,37,38,40-tetraazapentacyclo[27.5.5.3(12,18).0(15,39).
0(32,36)]tetraconta-1(35),12,14,16,18(38),29,31,33,36,39-decaene; W   126932

552   C28 H40 O7 S2   70081-44-6   H-61-3094-0   3β,15β,17α.xi.-Triacetoxy-16,16-trimethylenedithio-17-oxa-D-homo-Δ5-androstene; W   126933
Str 655

552   C29 H44 O10   55529-73-2   AD-0-230-0   Pregnane-3,11,12,14,20-pentol, 3,12,20-triacetate 11-(hydroxyacetate), (3β,11α,12β,14
β)-;; 11-GLYCOLOYLOXY-3,12,20-TRIACETOXY-14-HYDROXYPREGNANE; W/NBS   64423
Str 656

552   C29 H48 D12 N4 O2 Si2   61043-82-1   KO-2-333-2   O-TRIMETHYLSILYLATED(D12)-PHE-GLU-ORN; W   64424

552   C30 H46 N4 Ni O2   65674-60-4   B-31-43-0   OCTAHEDRO-BIS(O-(N-(4-AMINO-1,1,4-TRIMETHYLPENTYL)FORMIMIDOYL)PHENOLATO-O,N,N')NICK
EL; W   64425
Str 657

552   C30 H60 O3 Si3   33287-34-2   EP-8270-0-0   Silane, (5α-pregnane-3α,17,20β-triyltrioxy)tris[trimethyl-;; 5α-PREGNANE-3α,17α,20α
-TRIOL TMS; W/NBS   64428
Str 658

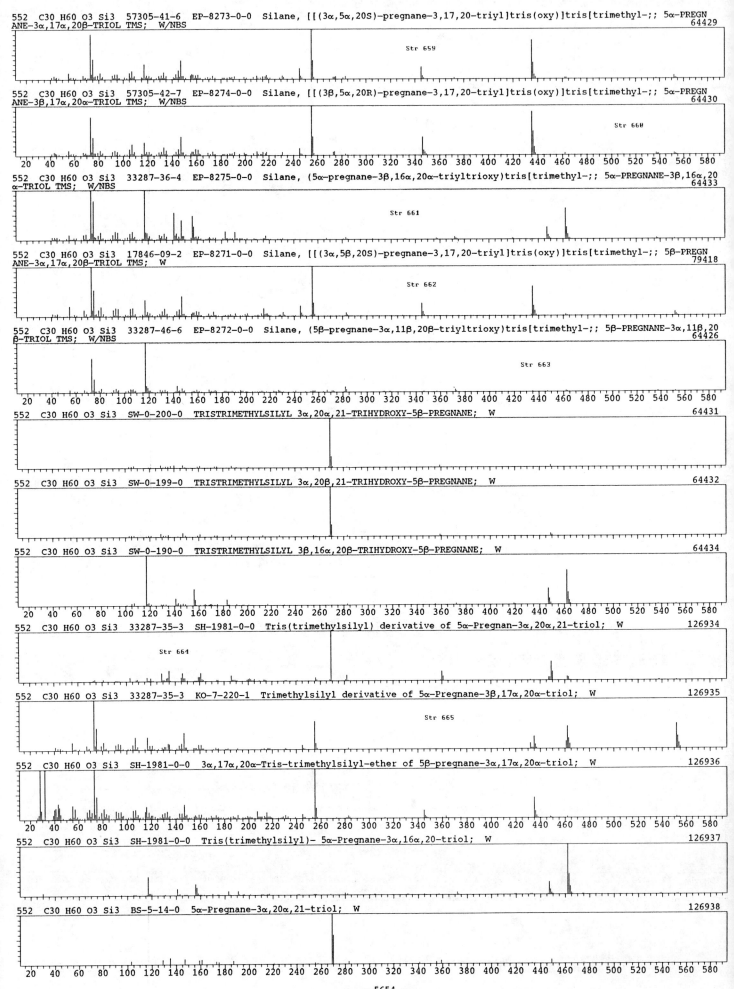

552 C30 H60 O3 Si3 57305-41-6 EP-8273-0-0 Silane, [[(3α,5α,20S)-pregnane-3,17,20-triyl]tris(oxy)]tris[trimethyl-;; 5α-PREGN
ANE-3α,17α,20β-TRIOL TMS; W/NBS   64429

Str 659

552 C30 H60 O3 Si3 57305-42-7 EP-8274-0-0 Silane, [[(3β,5α,20R)-pregnane-3,17,20-triyl]tris(oxy)]tris[trimethyl-;; 5α-PREGN
ANE-3β,17α,20α-TRIOL TMS; W/NBS   64430

Str 660

552 C30 H60 O3 Si3 33287-36-4 EP-8275-0-0 Silane, (5α-pregnane-3β,16α,20-triyltrioxy)tris[trimethyl-;; 5α-PREGNANE-3β,16α,20
α-TRIOL TMS; W/NBS   64433

Str 661

552 C30 H60 O3 Si3 17846-09-2 EP-8271-0-0 Silane, [[(3α,5β,20S)-pregnane-3,17,20-triyl]tris(oxy)]tris[trimethyl-;; 5β-PREGN
ANE-3α,17α,20β-TRIOL TMS; W   79418

Str 662

552 C30 H60 O3 Si3 33287-46-6 EP-8272-0-0 Silane, (5β-pregnane-3α,11β,20β-triyltrioxy)tris[trimethyl-;; 5β-PREGNANE-3α,11β,20
β-TRIOL TMS; W/NBS   64426

Str 663

552 C30 H60 O3 Si3 SW-0-200-0 TRISTRIMETHYLSILYL 3α,20α,21-TRIHYDROXY-5β-PREGNANE; W   64431

552 C30 H60 O3 Si3 SW-0-199-0 TRISTRIMETHYLSILYL 3α,20β,21-TRIHYDROXY-5β-PREGNANE; W   64432

552 C30 H60 O3 Si3 SW-0-190-0 TRISTRIMETHYLSILYL 3β,16α,20β-TRIHYDROXY-5β-PREGNANE; W   64434

552 C30 H60 O3 Si3 33287-35-3 SH-1981-0-0 Tris(trimethylsilyl) derivative of 5α-Pregnan-3α,20α,21-triol; W   126934

Str 664

552 C30 H60 O3 Si3 33287-35-3 KO-7-220-1 Trimethylsilyl derivative of 5α-Pregnane-3β,17α,20α-triol; W   126935

Str 665

552 C30 H60 O3 Si3 SH-1981-0-0 3α,17α,20α-Tris-trimethylsilyl-ether of 5β-pregnane-3α,17α,20α-triol; W   126936

552 C30 H60 O3 Si3 SH-1981-0-0 Tris(trimethylsilyl)- 5α-Pregnane-3α,16α,20-triol; W   126937

552 C30 H60 O3 Si3 BS-5-14-0 5α-Pregnane-3α,20α,21-triol; W   126938

552   C30 H60 O3 Si3   BS-5-15-0   5α-Pregnane-3α,20β,21-triol;   W   126939

552   C31 H36 N8 O2   57227-01-7   O-13-6-1   2-AMINO-4-DIMETHYLAMINO-6-{1-[1-(1-AMINO-2-PHENYLETHYLCARBONYLAMINO)-2-PHENYLETHYLCARBO
NYLAMINO]-2-PHENYLETHYL}-1,3,5-TRIAZINE;   W
                                                                        Str 666   64435

552   C31 H36 O9   77508-84-0   NS-0-0-0   Benzenepropanol, 4-(acetyloxy)-3-methoxy-β-(2-methoxyphenoxy)-γ-(2-methoxy-4-propylphenox
y)-, acetate; W/NBS
                                                                        Str 667   126940

552   C31 H44 N4 O5   72047-70-2   NS-0-0-0   Pentanamide, 2-(dimethylamino)-N-[7-(hydroxyphenylmethyl)-3-(1-methylethyl)-5,8-dioxo-
2-oxa-6,9-diazabicyclo[10.2.2]hexadeca-12,14,15-trien-4-yl]-3-methyl-; W/NBS
                                                                        Str 668   126941

552   C32 H40 O8   85933-09-1   O-18-71-5   8,9,10,11,12,13,14,24,25,26,27,28,29,30-Tetradecahydro-7H,23H-dibenzo[b,p][1,12,15,26]tet
raoxycyclooctacosin-6,15,22,31-tetraone; W   126942

552   C33 H44 O7   55452-66-9   NS--28129-0-0   1,5-Methano-1H,3H,8H-furo[3,4-g]pyrano[3,2-b]xanthene-1-butanal, 3a,4,5,6,6a,7,9,10-
octahydro-8-hydroxy-α,3,3,11,11-pentamethyl-13-(3-methylbutyl)-7,15-dioxo-;; 1,5-Methano-1H,3H,8H-furo[3,4-g]pyrano[3,2-b]xanthen
e-1-butanal, 3a,4,5,6,6a,7,9,10-octahydro-8-hydroxy-α,3,3,11,11-pentamethyl-13-(3-meth;   NBS   64436

552   C34 H25 O Rh   31851-08-8   AD-0-5748-0   Rhodium, (η5-2,4-cyclopentadien-1-yl)[(2,3,4,5-η)-2,3,4,5-tetraphenyl-2,4-cyclopentad
ien-1-one]-;; TETRAPHENYLCYCLOPENTADIENONE-.PI.-CYCLOPENTADIENYLRHODIUM; W/NBS
                                                                        Str 669   64437

552   C34 H30 N4 Ni   74834-14-3   Y-17-442-0   5,12-Dibenzyl-7,14-dimethyldibenzo[b,i][1,4,8,11]tetradecahexaenatonickel(II); W
                                                                        Str 670   126943

552   C34 H32 N8   18711-04-1   EP-7023-0-0   Pyrimido[5,4-d]pyrimidine, 2,4,6,8-tetra-o-toluidino-;; 2,4,6,8-TETRA(O-METHYLANILINO)
PYRIMIDO(5,4-D)PYRIMIDINE; W/NBS
                   Str 671   64439

552   C34 H32 N8   18711-05-2   EP-7022-0-0   Pyrimido[5,4-d]pyrimidine, 2,4,6,8-tetra-m-toluidino-;; 2,4,6,8-TETRA(M-METHYLANILINO)
PYRIMIDO(5,4-D)PYRIMIDINE; W/NBS
                                                                        Str 672   64438

552   C34 H52 O2 S Si   88930-43-2   J-49-1087-0   (6β,20S,23Z0-20-Hydroxy-6-methoxy-24-(phenylthio)-24-(trimethylsilyl)-3α,5α-cycloc
hol-23-ene; W
                                                                        Str 673   126944

552   C35 H52 O5   467-81-2   B-31-1321-0   22β-((Z)-2-METHYLBUT-2-ENOYLOXY)-3-OXOOLEANO-13β,28-LACTONE; W   64440

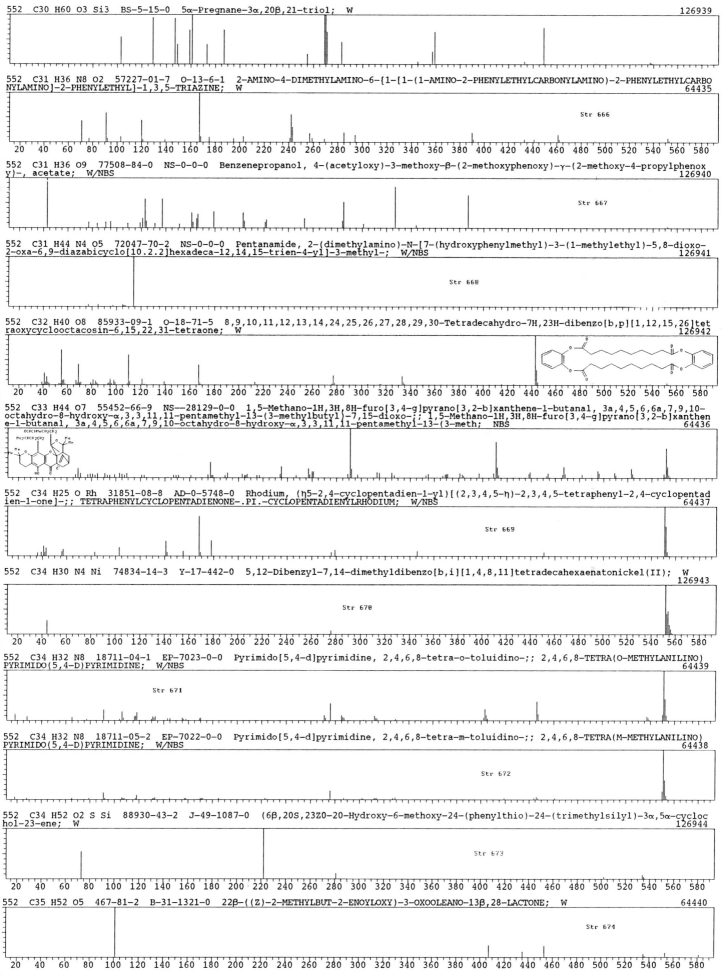

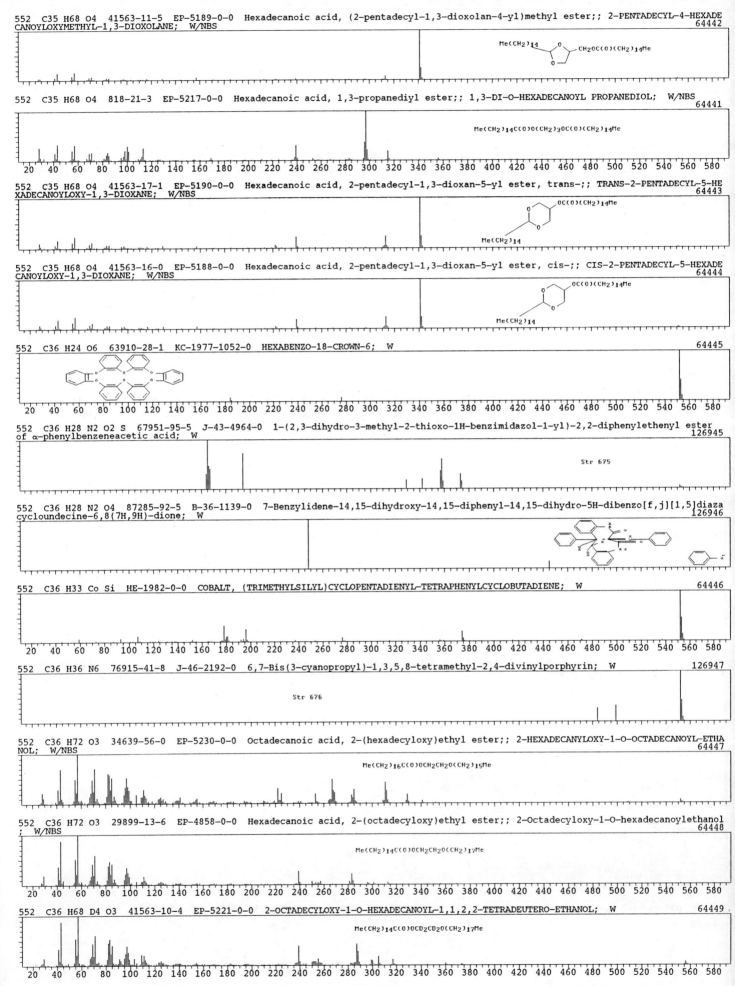

552 C35 H68 O4 41563-11-5 EP-5189-0-0 Hexadecanoic acid, (2-pentadecyl-1,3-dioxolan-4-yl)methyl ester;; 2-PENTADECYL-4-HEXADE
CANOYLOXYMETHYL-1,3-DIOXOLANE; W/NBS
64442

Me(CH2)14 O CH2OC(O)(CH2)14Me

552 C35 H68 O4 818-21-3 EP-5217-0-0 Hexadecanoic acid, 1,3-propanediyl ester;; 1,3-DI-O-HEXADECANOYL PROPANEDIOL; W/NBS
64441

Me(CH2)14C(O)O(CH2)3OC(O)(CH2)14Me

552 C35 H68 O4 41563-17-1 EP-5190-0-0 Hexadecanoic acid, 2-pentadecyl-1,3-dioxan-5-yl ester, trans-;; TRANS-2-PENTADECYL-5-HE
XADECANOYLOXY-1,3-DIOXANE; W/NBS
64443

OC(O)(CH2)14Me

Me(CH2)14

552 C35 H68 O4 41563-16-0 EP-5188-0-0 Hexadecanoic acid, 2-pentadecyl-1,3-dioxan-5-yl ester, cis-;; CIS-2-PENTADECYL-5-HEXADE
CANOYLOXY-1,3-DIOXANE; W/NBS
64444

OC(O)(CH2)14Me

Me(CH2)14

552 C36 H24 O6 63910-28-1 KC-1977-1052-0 HEXABENZO-18-CROWN-6; W
64445

552 C36 H28 N2 O2 S 67951-95-5 J-43-4964-0 1-(2,3-dihydro-3-methyl-2-thioxo-1H-benzimidazol-1-yl)-2,2-diphenylethenyl ester
of α-phenylbenzeneacetic acid; W
126945

Str 675

552 C36 H28 N2 O4 87285-92-5 B-36-1139-0 7-Benzylidene-14,15-dihydroxy-14,15-diphenyl-14,15-dihydro-5H-dibenzo[f,j][1,5]diaza
cycloundecine-6,8(7H,9H)-dione; W
126946

552 C36 H33 Co Si HE-1982-0-0 COBALT, (TRIMETHYLSILYL)CYCLOPENTADIENYL-TETRAPHENYLCYCLOBUTADIENE; W
64446

552 C36 H36 N6 76915-41-8 J-46-2192-0 6,7-Bis(3-cyanopropyl)-1,3,5,8-tetramethyl-2,4-divinylporphyrin; W
126947

Str 676

552 C36 H72 O3 34639-56-0 EP-5230-0-0 Octadecanoic acid, 2-(hexadecyloxy)ethyl ester;; 2-HEXADECANYLOXY-1-O-OCTADECANOYL-ETHA
NOL; W/NBS
64447

Me(CH2)16C(O)OCH2CH2O(CH2)15Me

552 C36 H72 O3 29899-13-6 EP-4858-0-0 Hexadecanoic acid, 2-(octadecyloxy)ethyl ester;; 2-Octadecyloxy-1-O-hexadecanoylethanol
; W/NBS
64448

Me(CH2)14C(O)OCH2CH2O(CH2)17Me

552 C36 H68 D4 O3 41563-10-4 EP-5221-0-0 2-OCTADECYLOXY-1-O-HEXADECANOYL-1,1,2,2-TETRADEUTERO-ETHANOL; W
64449

Me(CH2)14C(O)OCD2CD2O(CH2)17Me

552   C40 H40 O2   83095-79-8   J-47-4372-0   8,9,20,21-Tetrahydro-6,6,11,11,18,18,23,23-octamethyl-5,24:12,17-diepoxy-5,24:12,17-die
thenodibenzo[a,k]cycloeicosene-7,10,19,22(6H,11H,18H,23H)-tetrone;   W   126948

552   C40 H56 O   19891-74-8   EP-2343-0-0   Lycoxanthin;   W/NBS   64451

552   C40 H56 O   1923-89-3   NS--28119-0-0   β,β-Carotene, 5,6-epoxy-5,6-dihydro-;;   NBS   64452

552   C40 H56 O   S-27-2501-9   ALEURIAXANTHIN;   W   64450

552   C40 H56 O   87162-12-7   H-66-1165-0   β,β-Carotin-19-ol;   W   126949

552   C40 H56 O   73553-27-2   H-62-2538-0   (5S,6R)-5,6-Epoxy-5,6-dihydro-β,.psi.-carotin;   W   126950

552   C40 H56 O   73574-04-6   H-62-2538-0   (5S,6R,6'R)-5,6-Epoxy-5,6-dihydro-β,ε-carotin;   W   126951

552   C40 H56 O   CD-490-0-0   Rubixanthin;   W   126952

552   C40 H56 O   66609-71-0   H-61-831-0   5,6-Epoxy-5,6-dihydro-β,β-carotin;   W   126953

552   C44 H24   61670-61-9   SB-30-689-0   PROPELLICENE;   W   64453

553   C10 H8 Br5 N O   90490-46-3   B-36-2104-0   4-Bromo-N-bromoacetyl-N-methyl-2-tribromomethylaniline;   W   126954

553   C16 H7 F12 N O7   53833-22-0   JC-99-534-5   6-HYDROXYDOPAMINE-TRIFLUOROACETYL;   W   64455

553   C17 H12 F9 N5 O6   35170-12-8   D-10-3974-6   Adenosine, 2'-deoxy-N-methyl-N-(trifluoroacetyl)-, 3',5'-bis(trifluoroacetate);;
N(6),O-3',5'-TRIS(TRIFLUOROACETYL)-N(6)-METHYL-2'-DEOXYADENOSINE;   W/NBS   64457

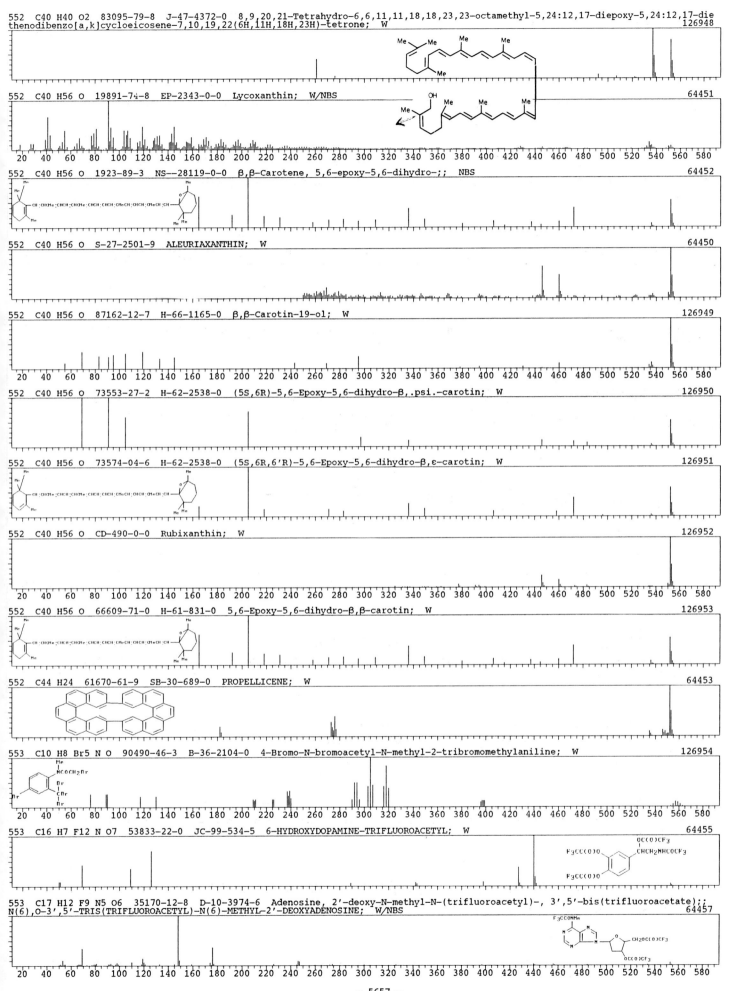

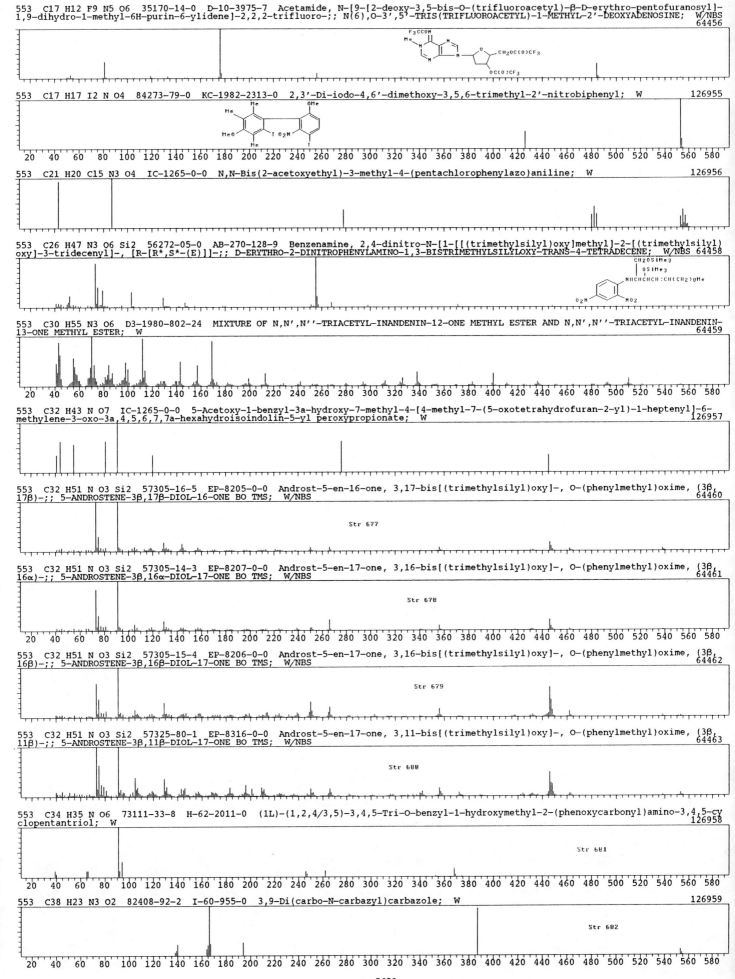

553 C17 H12 F9 N5 O6 35170-14-0 D-10-3975-7 Acetamide, N-[9-[2-deoxy-3,5-bis-O-(trifluoroacetyl)-β-D-erythro-pentofuranosyl]-1,9-dihydro-1-methyl-6H-purin-6-ylidene]-2,2,2-trifluoro-;; N(6),O-3',5'-TRIS(TRIFLUOROACETYL)-1-METHYL-2'-DEOXYADENOSINE; W/NBS
64456

553 C17 H17 I2 N O4 84273-79-0 KC-1982-2313-0 2,3'-Di-iodo-4,6'-dimethoxy-3,5,6-trimethyl-2'-nitrobiphenyl; W
126955

553 C21 H20 Cl5 N3 O4 IC-1265-0-0 N,N-Bis(2-acetoxyethyl)-3-methyl-4-(pentachlorophenylazo)aniline; W
126956

553 C26 H47 N3 O6 Si2 56272-05-0 AB-270-128-9 Benzenamine, 2,4-dinitro-N-[1-[[(trimethylsilyl)oxy]methyl]-2-[(trimethylsilyl)oxy]-3-tridecenyl]-, [R-[R*,S*-(E)]]-;; D-ERYTHRO-2-DINITROPHENYLAMINO-1,3-BISTRIMETHYLSILYLOXY-TRANS-4-TETRADECENE; W/NBS 64458

553 C30 H55 N3 O6 D3-1980-802-24 MIXTURE OF N,N',N''-TRIACETYL-INANDENIN-12-ONE METHYL ESTER AND N,N',N''-TRIACETYL-INANDENIN-13-ONE METHYL ESTER; W
64459

553 C32 H43 N O7 IC-1265-0-0 5-Acetoxy-1-benzyl-3a-hydroxy-7-methyl-4-[4-methyl-7-(5-oxotetrahydrofuran-2-yl)-1-heptenyl]-6-methylene-3-oxo-3a,4,5,6,7,7a-hexahydroisoindolin-5-yl peroxypropionate; W
126957

553 C32 H51 N O3 Si2 57305-16-5 EP-8205-0-0 Androst-5-en-16-one, 3,17-bis[(trimethylsilyl)oxy]-, O-(phenylmethyl)oxime, (3β,17β)-;; 5-ANDROSTENE-3β,17β-DIOL-16-ONE BO TMS; W/NBS
64460

Str 677

553 C32 H51 N O3 Si2 57305-14-3 EP-8207-0-0 Androst-5-en-17-one, 3,16-bis[(trimethylsilyl)oxy]-, O-(phenylmethyl)oxime, (3β,16α)-;; 5-ANDROSTENE-3β,16α-DIOL-17-ONE BO TMS; W/NBS
64461

Str 678

553 C32 H51 N O3 Si2 57305-15-4 EP-8206-0-0 Androst-5-en-17-one, 3,16-bis[(trimethylsilyl)oxy]-, O-(phenylmethyl)oxime, (3β,16β)-;; 5-ANDROSTENE-3β,16β-DIOL-17-ONE BO TMS; W/NBS
64462

Str 679

553 C32 H51 N O3 Si2 57325-80-1 EP-8316-0-0 Androst-5-en-17-one, 3,11-bis[(trimethylsilyl)oxy]-, O-(phenylmethyl)oxime, (3β,11β)-;; 5-ANDROSTENE-3β,11β-DIOL-17-ONE BO TMS; W/NBS
64463

Str 680

553 C34 H35 N O6 73111-33-8 H-62-2011-0 (1L)-(1,2,4/3,5)-3,4,5-Tri-O-benzyl-1-hydroxymethyl-2-(phenoxycarbonyl)amino-3,4,5-cyclopentantriol; W
126958

Str 681

553 C38 H23 N3 O2 82408-92-2 I-60-955-0 3,9-Di(carbo-N-carbazyl)carbazole; W
126959

Str 682

553   C38 H51 N O2   88131-77-5   C-106-1015-0   2-(Diphenylhydroxymethyl)-4-tert-bityl-2,4,6-triisopropylbenzopiperidide;   W   126960

Str 683

554   C12 H27 Mo O12 P3   22261-14-9   HE-1982-0-0   Molybdenum, tricarbonyltris(trimethyl phosphite-P)-;; MOLYBDENUM, TRICARBONYL-TR
IS(TRIMETHYLPHOSPHIT);   W/NBS   64464

554   C21 H50 O7 Si5   52842-25-8   DP-7-48-1   α-D-Glucopyranuronic acid, 1,2,3,4-tetrakis-O-(trimethylsilyl)-, trimethylsilyl ester
;; TRIMETHYLSILYL (TRIMETHYLSILYL 2,3,4-TRI-O-TRIMETHYLSILYL-α-D-GLUCOPYRANOSIDE)URONATE;   W   64470

554   C21 H50 O7 Si5   52842-24-7   DP-7-49-2   β-D-Glucopyranuronic acid, 1,2,3,4-tetrakis-O-(trimethylsilyl)-, trimethylsilyl ester
;; PER-O-TRIMETHYLSILYL-β-D-GLUCURONIC ACID;   W   64471

554   C21 H50 O7 Si5   55530-80-8   EP-768-0-0   D-Glucuronic acid, 2,3,4,5-tetrakis-O-(trimethylsilyl)-, trimethylsilyl ester;; GLUC
URONIC ACID-PENTATMS;   W/NBS   64469

554   C21 H50 O7 Si5   56192-87-1   EP-1177-0-0   Galacturonic acid, 2,3,4,5-tetrakis-O-(trimethylsilyl)-, trimethylsilyl ester;; GAL
ACTURONIC ACID-PENTATMS;   W/NBS   64473

554   C21 H50 O7 Si5   52783-57-0   DP-7-50-3   D-Glucofuranuronic acid, 1,2,3,5-tetrakis-O-(trimethylsilyl)-, trimethylsilyl ester;;
TRIMETHYLSILYL (TRIMETHYLSILYL 2,3-TRI-O-TRIMETHYLSILYL-D-GLUCOFURANOSIDE)URONATE;   W   64468

554   C21 H50 O7 Si5   38166-08-4   O6-0-577-0   ribo-5-Hexulosonic acid, 2,3,4,6-tetrakis-O-(trimethylsilyl)-, trimethylsilyl ester
;; RIBO-5-HEXULOSONIC ACID-(TMS)5;   W/NBS   64467

554   C21 H50 O7 Si5   38165-99-0   O-6-573-9   arabino-Hexaric acid, 3-deoxy-2,4,5-tris-O-(trimethylsilyl)-, bis(trimethylsilyl) est
er;; 3-DEOXY-ARABINO-HEXARIC ACID-(TMS)5;   W/NBS   64465

554   C21 H50 O7 Si5   38165-98-9   O-6-573-7   arabino-Hexaric acid, 2-deoxy-3,4,5-tris-O-(trimethylsilyl)-, bis(trimethylsilyl) est
er;; 2-DEOXY-ARABINO-HEXARIC ACID-(TMS)5;   W/NBS   64466

554   C21 H50 O7 Si5   JC-63-381-6   2-KETO-L-GULONIC ACID (β-PYRANOSE);   W   64472

554   C21 H50 O7 Si5   JC-63-380-5   2-KETO-L-GULONIC ACID (α-PYRANOSE);   W   64474

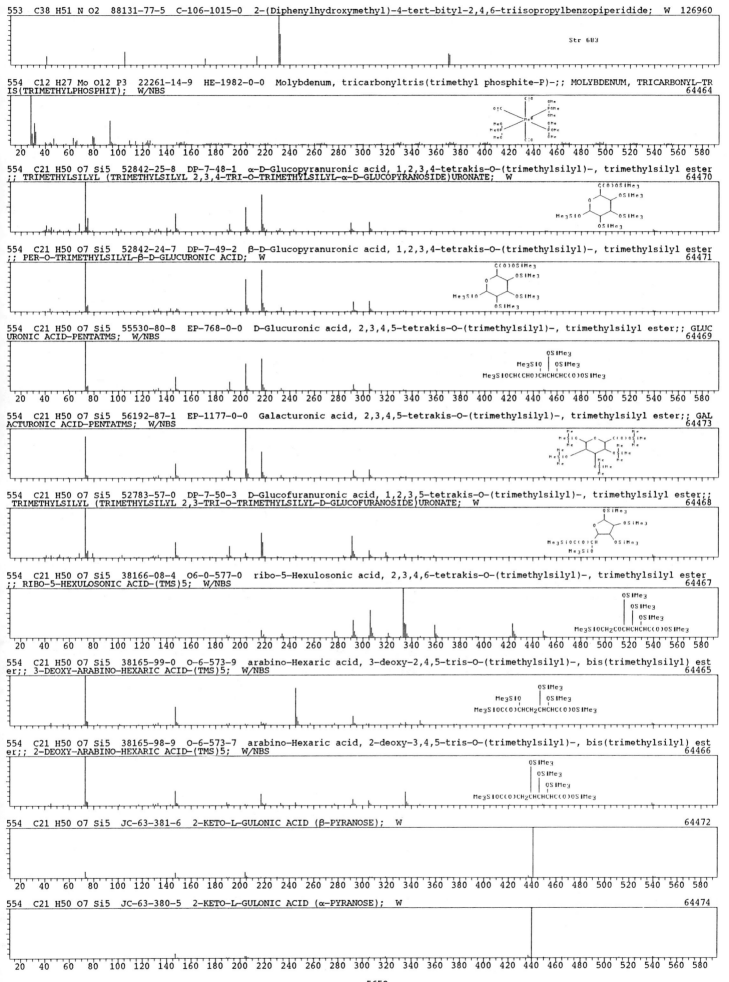

- 5659 -

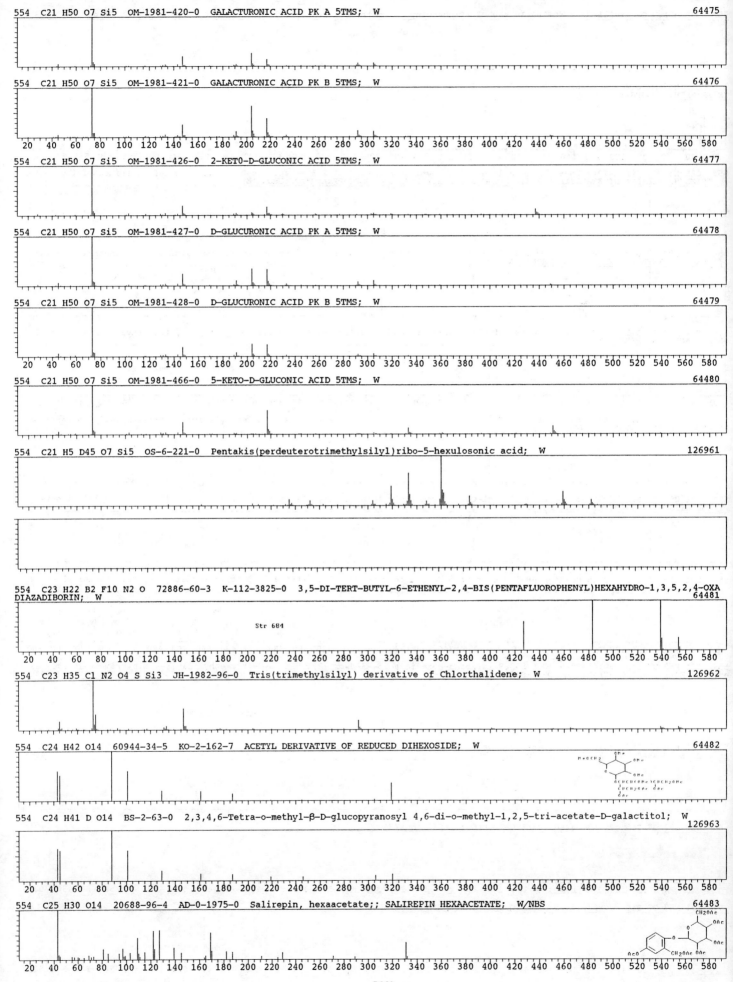

554   C21 H50 O7 Si5   OM-1981-420-0   GALACTURONIC ACID PK A 5TMS;   W   64475

554   C21 H50 O7 Si5   OM-1981-421-0   GALACTURONIC ACID PK B 5TMS;   W   64476

554   C21 H50 O7 Si5   OM-1981-426-0   2-KETO-D-GLUCONIC ACID 5TMS;   W   64477

554   C21 H50 O7 Si5   OM-1981-427-0   D-GLUCURONIC ACID PK A 5TMS;   W   64478

554   C21 H50 O7 Si5   OM-1981-428-0   D-GLUCURONIC ACID PK B 5TMS;   W   64479

554   C21 H50 O7 Si5   OM-1981-466-0   5-KETO-D-GLUCONIC ACID 5TMS;   W   64480

554   C21 H5 D45 O7 Si5   OS-6-221-0   Pentakis(perdeuterotrimethylsilyl)ribo-5-hexulosonic acid;   W   126961

554   C23 H22 B2 F10 N2 O   72886-60-3   K-112-3825-0   3,5-DI-TERT-BUTYL-6-ETHENYL-2,4-BIS(PENTAFLUOROPHENYL)HEXAHYDRO-1,3,5,2,4-OXA
DIAZADIBORIN;   W   64481

Str 684

554   C23 H35 Cl N2 O4 S Si3   JH-1982-96-0   Tris(trimethylsilyl) derivative of Chlorthalidene;   W   126962

554   C24 H42 O14   60944-34-5   KO-2-162-7   ACETYL DERIVATIVE OF REDUCED DIHEXOSIDE;   W   64482

554   C24 H41 D O14   BS-2-63-0   2,3,4,6-Tetra-o-methyl-β-D-glucopyranosyl 4,6-di-o-methyl-1,2,5-tri-acetate-D-galactitol;   W
126963

554   C25 H30 O14   20688-96-4   AD-0-1975-0   Salirepin, hexaacetate;; SALIREPIN HEXAACETATE;   W/NBS   64483

554   C26 H20 Al2 Cl4 N2   31390-30-4   KA-70-2636-1   Aluminum, tetrachlorobis[.mu.-(1,1-diphenylmethyleniminato)]di-;; DIPHENYLMETH
YLENEAMINOALUMINIUM DICHLORIDE DIMER;   W/NBS   64484

Str 685

554   C26 H38 N2 O11   DU-1-499-19   PERMETHYL 4-HYDROXYMEPHOBARBITAL-IV-GLUCURONIDE;   W   64485

554   C27 H42 N2 O10   14942-97-3   J-44-4574-1   Methyl ester of cholic acid 7-acetate 3,12-dinitrate;   W   126964

Str 686

554   C28 H23 Cr O7 P   63928-77-8   AG-132-239-6   (Methylbenzoate)(dicarbonyl)(triphenoxyphosphine)chromium;   W   126965

Str 687

554   C28 H26 O4 S4   IC-1266-0-0   Bis[4-(2-phenylsulphonylethyl)phenyl] disulphide;   W   126966

554   C28 H29 Mo N2 O2 P   75657-54-4   NS--28099-0-0   Molybdenum, carbonyl(η5-2,4-cyclopentadien-1-yl)[N-ethyl-P,P-diphenyl-N-(1-ph
enylethyl)phosphinous amide-P]nitrosyl-, stereoisomer;;   NBS   64486

Str 688

554   C28 H34 N4 O4 S2   13285-10-4   EP-790-0-0   PUTRESCINE-DIDANSYL;   W/NBS   64487

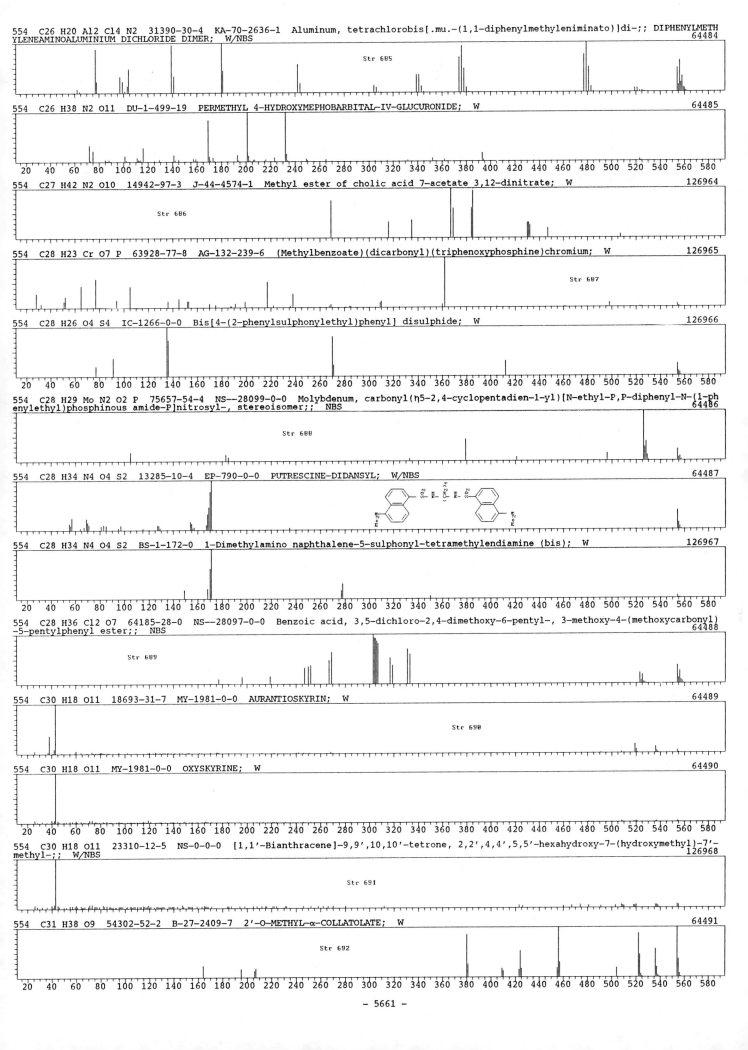

554   C28 H34 N4 O4 S2   BS-1-172-0   1-Dimethylamino naphthalene-5-sulphonyl-tetramethylendiamine (bis);   W   126967

554   C28 H36 Cl2 O7   64185-28-0   NS--28097-0-0   Benzoic acid, 3,5-dichloro-2,4-dimethoxy-6-pentyl-, 3-methoxy-4-(methoxycarbonyl)
-5-pentylphenyl ester;;   NBS   64488

Str 689

554   C30 H18 O11   18693-31-7   MY-1981-0-0   AURANTIOSKYRIN;   W   64489

Str 690

554   C30 H18 O11   MY-1981-0-0   OXYSKYRINE;   W   64490

554   C30 H18 O11   23310-12-5   NS-0-0-0   [1,1'-Bianthracene]-9,9',10,10'-tetrone, 2,2',4,4',5,5'-hexahydroxy-7-(hydroxymethyl)-7'-
methyl-;;   W/NBS   126968

Str 691

554   C31 H38 O9   54302-52-2   B-27-2409-7   2'-O-METHYL-α-COLLATOLATE;   W   64491

Str 692

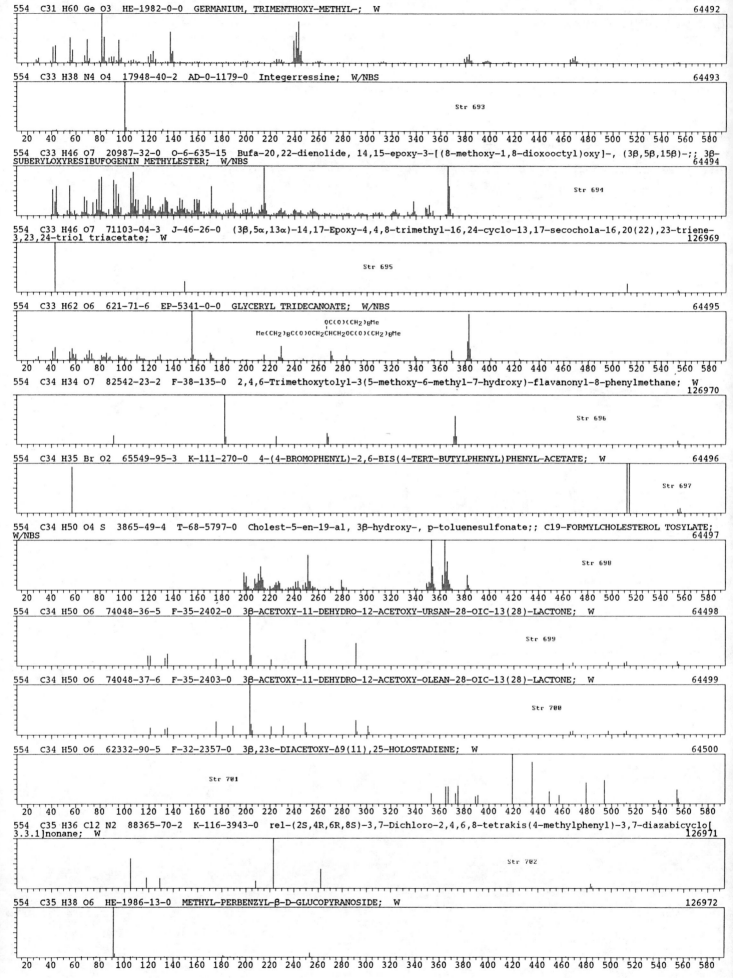

554   C31 H60 Ge O3   HE-1982-0-0   GERMANIUM, TRIMENTHOXY-METHYL-;   W   64492

554   C33 H38 N4 O4   17948-40-2   AD-0-1179-0   Integerressine;   W/NBS   64493

Str 693

554   C33 H46 O7   20987-32-0   O-6-635-15   Bufa-20,22-dienolide, 14,15-epoxy-3-[(8-methoxy-1,8-dioxooctyl)oxy]-, (3β,5β,15β)-;; 3β-
SUBERYLOXYRESIBUFOGENIN METHYLESTER;   W/NBS   64494

Str 694

554   C33 H46 O7   71103-04-3   J-46-26-0   (3β,5α,13α)-14,17-Epoxy-4,4,8-trimethyl-16,24-cyclo-13,17-secochola-16,20(22),23-triene-
3,23,24-triol triacetate;   W   126969

Str 695

554   C33 H62 O6   621-71-6   EP-5341-0-0   GLYCERYL TRIDECANOATE;   W/NBS   64495

OC(O)(CH2)8Me
Me(CH2)8C(O)OCH2CHCH2OC(O)(CH2)8Me

554   C34 H34 O7   82542-23-2   F-38-135-0   2,4,6-Trimethoxytolyl-3(5-methoxy-6-methyl-7-hydroxy)-flavanonyl-8-phenylmethane;   W
126970

Str 696

554   C34 H35 Br O2   65549-95-3   K-111-270-0   4-(4-BROMOPHENYL)-2,6-BIS(4-TERT-BUTYLPHENYL)PHENYL-ACETATE;   W   64496

Str 697

554   C34 H50 O4 S   3865-49-4   T-68-5797-0   Cholest-5-en-19-al, 3β-hydroxy-, p-toluenesulfonate;; C19-FORMYLCHOLESTEROL TOSYLATE;
W/NBS   64497

Str 698

554   C34 H50 O6   74048-36-5   F-35-2402-0   3β-ACETOXY-11-DEHYDRO-12-ACETOXY-URSAN-28-OIC-13(28)-LACTONE;   W   64498

Str 699

554   C34 H50 O6   74048-37-6   F-35-2403-0   3β-ACETOXY-11-DEHYDRO-12-ACETOXY-OLEAN-28-OIC-13(28)-LACTONE;   W   64499

Str 700

554   C34 H50 O6   62332-90-5   F-32-2357-0   3β,23ε-DIACETOXY-Δ9(11),25-HOLOSTADIENE;   W   64500

Str 701

554   C35 H36 Cl2 N2   88365-70-2   K-116-3943-0   rel-(2S,4R,6R,8S)-3,7-Dichloro-2,4,6,8-tetrakis(4-methylphenyl)-3,7-diazabicyclo[
3.3.1]nonane;   W   126971

Str 702

554   C35 H38 O6   HE-1986-13-0   METHYL-PERBENZYL-β-D-GLUCOPYRANOSIDE;   W   126972

554   C35 H46 N4 O2   35050-46-5   KC-1981-324-0   2,3,7,8,12,13,17,18-Octaethyl-21H,24H-bilin-1,19-dione;   W   126973

Str 703

554   C35 H54 O5   61236-70-2   B-29-1558-0   22α,28-O,O-ISOPROPYLIDENE-16-OXOOLEAN-12-ENE-3β,22α,28-TRIOL, 3-ACETATE;   W   64501

Str 704

20 40 60 80 100 120 140 160 180 200 220 240 260 280 300 320 340 360 380 400 420 440 460 480 500 520 540 560 580

554   C36 H26 O6   72193-15-8   SB-33-406-0   6,12-BIS(2-HYDROXY-5-METHYLBENZOYL)-2,8-DIMETHYLBENZO(1,2 : 4,5-B')BISBENZOFURAN;   W   126974

Str 705

554   C36 H27 O4 P   HE-1986-863-0   TRIS(2-BIPHENYL)ESTER OF PHOSPHORIC ACID;   W   126975

554   C36 H30 N2 O4   90552-94-6   J-49-2826-0   (3aR*,3bS*,8aS*)-6,8a-Diethyl-3a,8a-dihydro-2,3a,3b,5-tetraphenyl-3H-dipyrrol[1,2-b: 3',2'-d]isoxazole-3,4(3bH)-dione-1-oxide;   W   126976

Str 706

20 40 60 80 100 120 140 160 180 200 220 240 260 280 300 320 340 360 380 400 420 440 460 480 500 520 540 560 580

554   C36 H36 N2 Ni   HE-1982-0-0   NICKEL, GLYOXAL-BIS(ISOPROPYLIMIN)-TETRAPHENYLCYCLOBUTADIEN;   W   64503

554   C36 H44 B2 N4   86067-33-6   K-116-1536-0   1-[Bis(2,4,6-timethylphenyl)boryl]-4,5-bis(2,4,6-trimethylphenyl)-2-tetrazaborol; W   126977

Str 707

554   C38 H34 O4   55508-26-4   D-14-1738-0   3-PHENYLHEXESTROL DIBENZOATE;   W   64504

20 40 60 80 100 120 140 160 180 200 220 240 260 280 300 320 340 360 380 400 420 440 460 480 500 520 540 560 580

554   C40 H58 O   105-92-0   EP-2344-0-0   Rhodopin;   W/NBS   64505

555   C16 H10 F9 N5 O7   35170-11-7   D-10-3971-4   Guanosine, 2'-deoxy-N-(trifluoroacetyl)-, 3',5'-bis(trifluoroacetate);;   N(2),O-3',5'-TRIS(TRIFLUOROACETYL)-2'-DEOXYGUANOSINE;   W/NBS   64506

555   C24 H33 N3 O12   39945-18-1   NS-0-0-0   2,4,6-TRIS(DIETHOXYCARBONYLMETHYLENE)HEXAHYDRO-S-TRIAZINE;   W/NBS   126978

20 40 60 80 100 120 140 160 180 200 220 240 260 280 300 320 340 360 380 400 420 440 460 480 500 520 540 560 580

555   C26 H33 N7 O7   57237-91-9   O7-0-817-0   Glycine, N-[N-[N-[N-[N-[N-[[N-(dimethylamino)-1-naphthalenyl]methylene]glycyl]glycyl]glycyl]glycyl]glycyl]-, methyl ester;;   4-DIMETHYLAMINONAPHTHYLIDENE GLYCYLGLYCYLGLYCYLGLYCYL-GLYCYLGLYCINE METH;   W/NBS   64508

555   C31 H25 N O7 S   62615-98-9   KC-1976-2569-0   DIMETHYL 1,2,3,4,4A,5,8,8A-OCTAHYDRO-3-OXO-1,2,4-TRIPHENYL-1,4-EPITHIO-5,8-EPOXY ISOQUINOLINE-6,7-DICARBOXYLATE 11-OXIDE;   W   64509

Str 708

20 40 60 80 100 120 140 160 180 200 220 240 260 280 300 320 340 360 380 400 420 440 460 480 500 520 540 560 580

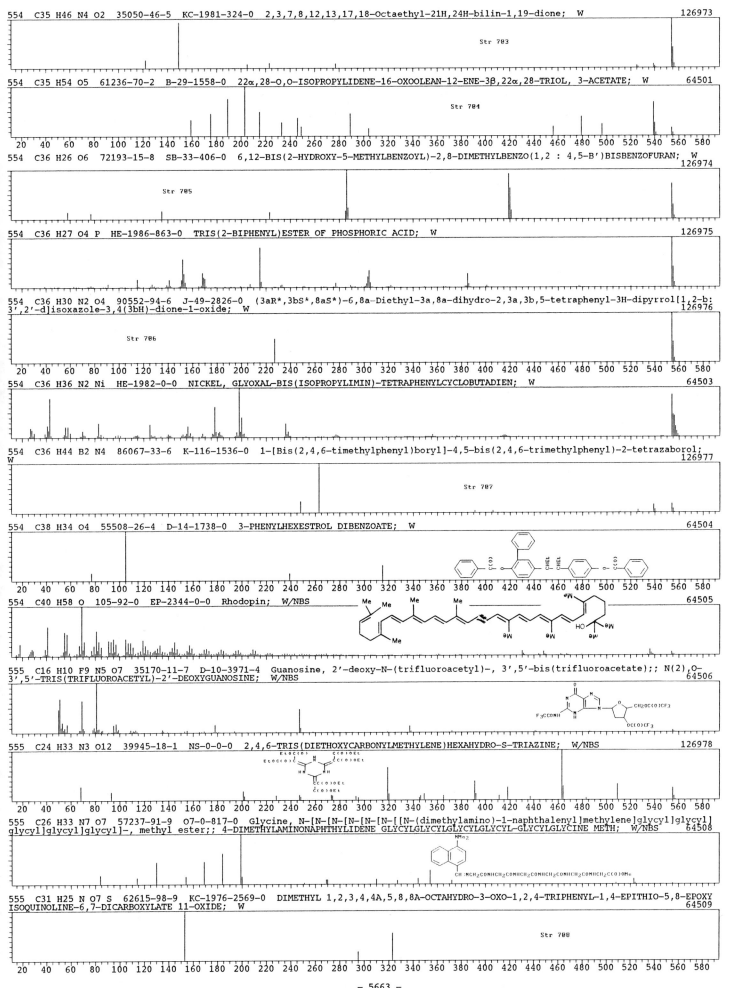

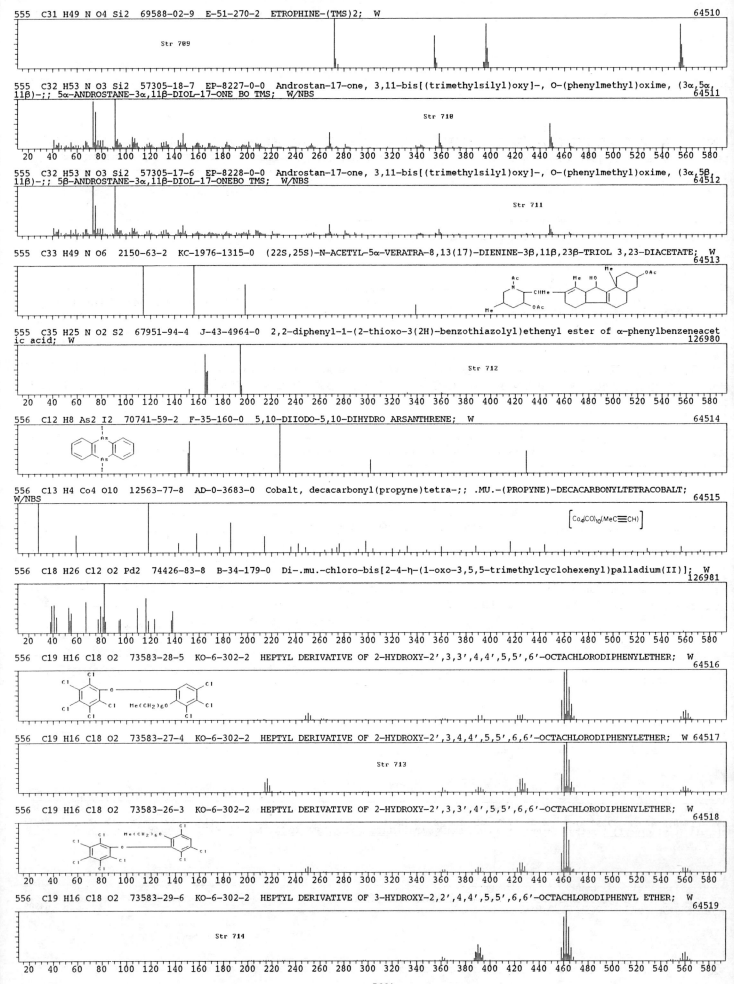

555   C31 H49 N O4 Si2   69588-02-9   E-51-270-2   ETROPHINE-(TMS)2;   W
64510

Str 709

555   C32 H53 N O3 Si2   57305-18-7   EP-8227-0-0   Androstan-17-one, 3,11-bis[(trimethylsilyl)oxy]-, O-(phenylmethyl)oxime, (3α,5α,11β)-;; 5α-ANDROSTANE-3α,11β-DIOL-17-ONE BO TMS;   W/NBS
64511

Str 710

555   C32 H53 N O3 Si2   57305-17-6   EP-8228-0-0   Androstan-17-one, 3,11-bis[(trimethylsilyl)oxy]-, O-(phenylmethyl)oxime, (3α,5β,11β)-;; 5β-ANDROSTANE-3α,11β-DIOL-17-ONEBO TMS;   W/NBS
64512

Str 711

555   C33 H49 N O6   2150-63-2   KC-1976-1315-0   (22S,25S)-N-ACETYL-5α-VERATRA-8,13(17)-DIENINE-3β,11β,23β-TRIOL 3,23-DIACETATE;   W
64513

555   C35 H25 N O2 S2   67951-94-4   J-43-4964-0   2,2-diphenyl-1-(2-thioxo-3(2H)-benzothiazolyl)ethenyl ester of α-phenylbenzeneacetic acid;   W
126980

Str 712

556   C12 H8 As2 I2   70741-59-2   F-35-160-0   5,10-DIIODO-5,10-DIHYDRO ARSANTHRENE;   W
64514

556   C13 H4 Co4 O10   12563-77-8   AD-0-3683-0   Cobalt, decacarbonyl(propyne)tetra-;; .MU.-(PROPYNE)-DECACARBONYLTETRACOBALT;   W/NBS
64515

556   C18 H26 Cl2 O2 Pd2   74426-83-8   B-34-179-0   Di-.mu.-chloro-bis[2-4-η-(1-oxo-3,5,5-trimethylcyclohexenyl)palladium(II)];   W
126981

556   C19 H16 Cl8 O2   73583-28-5   KO-6-302-2   HEPTYL DERIVATIVE OF 2-HYDROXY-2',3,3',4,4',5,5',6'-OCTACHLORODIPHENYLETHER;   W
64516

556   C19 H16 Cl8 O2   73583-27-4   KO-6-302-2   HEPTYL DERIVATIVE OF 2-HYDROXY-2',3,4,4',5,5',6,6'-OCTACHLORODIPHENYLETHER;   W 64517

Str 713

556   C19 H16 Cl8 O2   73583-26-3   KO-6-302-2   HEPTYL DERIVATIVE OF 2-HYDROXY-2',3,3',4',5,5',6,6'-OCTACHLORODIPHENYLETHER;   W
64518

556   C19 H16 Cl8 O2   73583-29-6   KO-6-302-2   HEPTYL DERIVATIVE OF 3-HYDROXY-2,2',4,4',5,5',6,6'-OCTACHLORODIPHENYL ETHER;   W
64519

Str 714

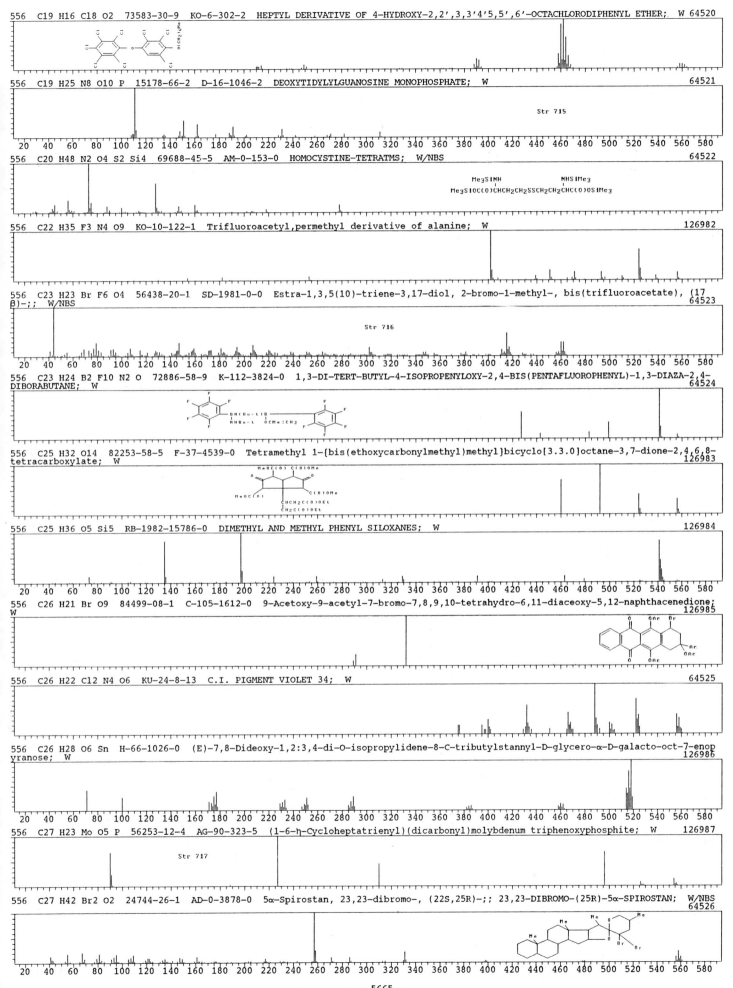

556   C19 H16 C18 O2   73583-30-9   KO-6-302-2   HEPTYL DERIVATIVE OF 4-HYDROXY-2,2',3,3'4'5,5',6'-OCTACHLORODIPHENYL ETHER;   W 64520

556   C19 H25 N8 O10 P   15178-66-2   D-16-1046-2   DEOXYTIDYLYLGUANOSINE MONOPHOSPHATE;   W   64521

Str 715

556   C20 H48 N2 O4 S2 Si4   69688-45-5   AM-0-153-0   HOMOCYSTINE-TETRATMS;   W/NBS   64522

Me3SiNH   NHSiMe3
| |
Me3SiOC(O)CHCH2CH2SSCH2CH2CHC(O)OSiMe3

556   C22 H35 F3 N4 O9   KO-10-122-1   Trifluoroacetyl,permethyl derivative of alanine;   W   126982

556   C23 H23 Br F6 O4   56438-20-1   SD-1981-0-0   Estra-1,3,5(10)-triene-3,17-diol, 2-bromo-1-methyl-, bis(trifluoroacetate), (17β)-;;   W/NBS   64523

Str 716

556   C23 H24 B2 F10 N2 O   72886-58-9   K-112-3824-0   1,3-DI-TERT-BUTYL-4-ISOPROPENYLOXY-2,4-BIS(PENTAFLUOROPHENYL)-1,3-DIAZA-2,4-DIBORABUTANE;   W   64524

556   C25 H32 O14   82253-58-5   F-37-4539-0   Tetramethyl 1-{bis(ethoxycarbonylmethyl)methyl}bicyclo[3.3.0]octane-3,7-dione-2,4,6,8-tetracarboxylate;   W   126983

556   C25 H36 O5 Si5   RB-1982-15786-0   DIMETHYL AND METHYL PHENYL SILOXANES;   W   126984

556   C26 H21 Br O9   84499-08-1   C-105-1612-0   9-Acetoxy-9-acetyl-7-bromo-7,8,9,10-tetrahydro-6,11-diaceoxy-5,12-naphthacenedione;   W   126985

556   C26 H22 Cl2 N4 O6   KU-24-8-13   C.I. PIGMENT VIOLET 34;   W   64525

556   C26 H28 O6 Sn   H-66-1026-0   (E)-7,8-Dideoxy-1,2:3,4-di-O-isopropylidene-8-C-tributylstannyl-D-glycero-α-D-galacto-oct-7-enopyranose;   W   126986

556   C27 H23 Mo O5 P   56253-12-4   AG-90-323-5   (1-6-η-Cycloheptatrienyl)(dicarbonyl)molybdenum triphenoxyphosphite;   W   126987

Str 717

556   C27 H42 Br2 O2   24744-26-1   AD-0-3878-0   5α-Spirostan, 23,23-dibromo-, (22S,25R)-;; 23,23-DIBROMO-(25R)-5α-SPIROSTAN;   W/NBS   64526

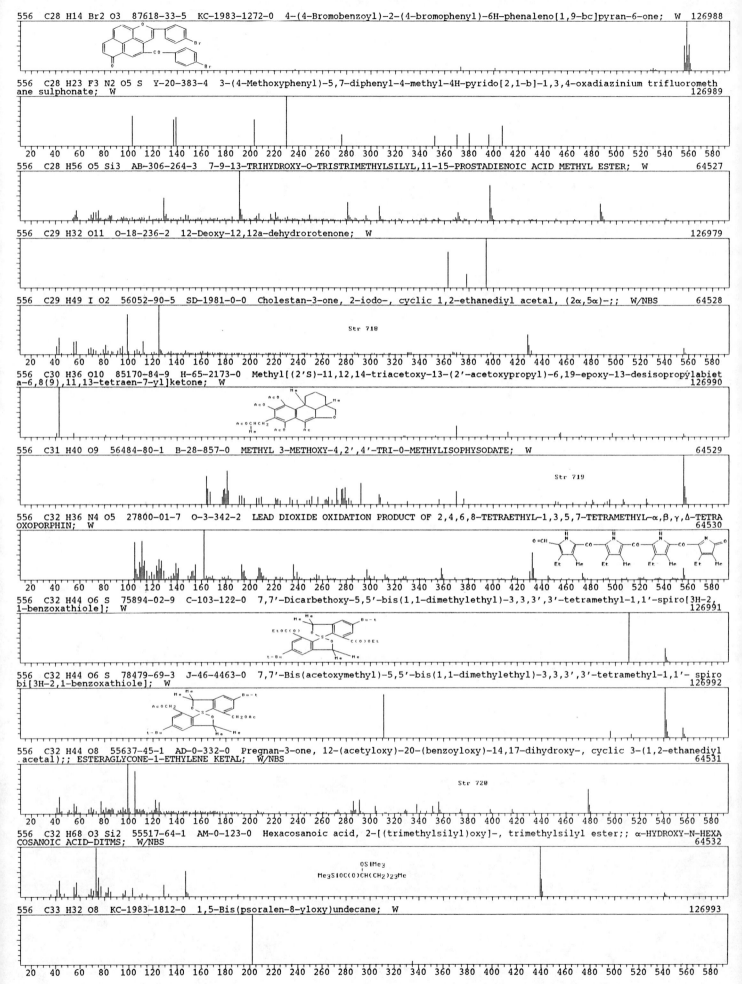

556   C28 H14 Br2 O3   87618-33-5   KC-1983-1272-0   4-(4-Bromobenzoyl)-2-(4-bromophenyl)-6H-phenaleno[1,9-bc]pyran-6-one;   W   126988

556   C28 H23 F3 N2 O5 S   Y-20-383-4   3-(4-Methoxyphenyl)-5,7-diphenyl-4-methyl-4H-pyrido[2,1-b]-1,3,4-oxadiazinium trifluoromethane sulphonate;   W   126989

556   C28 H56 O5 Si3   AB-306-264-3   7-9-13-TRIHYDROXY-O-TRISTRIMETHYLSILYL,11-15-PROSTADIENOIC ACID METHYL ESTER;   W   64527

556   C29 H32 O11   O-18-236-2   12-Deoxy-12,12a-dehydrorotenone;   W   126979

556   C29 H49 I O2   56052-90-5   SD-1981-0-0   Cholestan-3-one, 2-iodo-, cyclic 1,2-ethanediyl acetal, (2α,5α)-;;   W/NBS   64528

Str 718

556   C30 H36 O10   85170-84-9   H-65-2173-0   Methyl[(2'S)-11,12,14-triacetoxy-13-(2'-acetoxypropyl)-6,19-epoxy-13-desisopropylabieta-6,8(9),11,13-tetraen-7-yl]ketone;   W   126990

556   C31 H40 O9   56484-80-1   B-28-857-0   METHYL 3-METHOXY-4,2',4'-TRI-O-METHYLISOPHYSODATE;   W   64529

Str 719

556   C32 H36 N4 O5   27800-01-7   O-3-342-2   LEAD DIOXIDE OXIDATION PRODUCT OF 2,4,6,8-TETRAETHYL-1,3,5,7-TETRAMETHYL-α,β,γ,Δ-TETRAOXOPORPHIN;   W   64530

556   C32 H44 O6 S   75894-02-9   C-103-122-0   7,7'-Dicarbethoxy-5,5'-bis(1,1-dimethylethyl)-3,3,3',3'-tetramethyl-1,1'-spiro[3H-2,1-benzoxathiole];   W   126991

556   C32 H44 O6 S   78479-69-3   J-46-4463-0   7,7'-Bis(acetoxymethyl)-5,5'-bis(1,1-dimethylethyl)-3,3,3',3'-tetramethyl-1,1'- spirobi[3H-2,1-benzoxathiole];   W   126992

556   C32 H44 O8   55637-45-1   AD-0-332-0   Pregnan-3-one, 12-(acetyloxy)-20-(benzoyloxy)-14,17-dihydroxy-, cyclic 3-(1,2-ethanediyl acetal);;   ESTERAGLYCONE-1-ETHYLENE KETAL;   W/NBS   64531

Str 720

556   C32 H68 O3 Si2   55517-64-1   AM-0-123-0   Hexacosanoic acid, 2-[(trimethylsilyl)oxy]-, trimethylsilyl ester;;   α-HYDROXY-N-HEXACOSANOIC ACID-DITMS;   W/NBS   64532

556   C33 H32 O8   KC-1983-1812-0   1,5-Bis(psoralen-8-yloxy)undecane;   W   126993

556   C33 H40 N4 O4   55517-68-5   AD-0-1177-0   DIHYDRO-INTEGERRESSIN;   W/NBS   64533

Str 721

556   C33 H40 N4 O4   66314-76-9   F-33-2963-0   16-ACETYL-1,15,19-BILINTRIONE;   W   64534

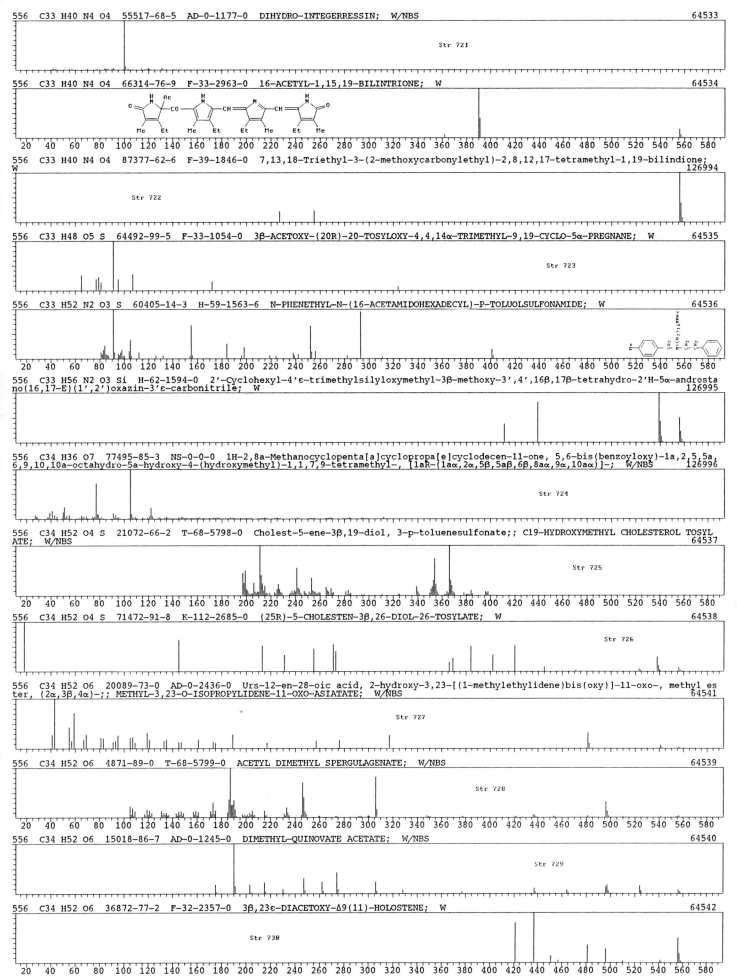

20 40 60 80 100 120 140 160 180 200 220 240 260 280 300 320 340 360 380 400 420 440 460 480 500 520 540 560 580

556   C33 H40 N4 O4   87377-62-6   F-39-1846-0   7,13,18-Triethyl-3-(2-methoxycarbonylethyl)-2,8,12,17-tetramethyl-1,19-bilindione;
W   126994

Str 722

556   C33 H48 O5 S   64492-99-5   F-33-1054-0   3β-ACETOXY-(20R)-20-TOSYLOXY-4,4,14α-TRIMETHYL-9,19-CYCLO-5α-PREGNANE;   W   64535

Str 723

556   C33 H52 N2 O3 S   60405-14-3   H-59-1563-6   N-PHENETHYL-N-(16-ACETAMIDOHEXADECYL)-P-TOLUOLSULFONAMIDE;   W   64536

20 40 60 80 100 120 140 160 180 200 220 240 260 280 300 320 340 360 380 400 420 440 460 480 500 520 540 560 580

556   C33 H56 N2 O3 Si   H-62-1594-0   2'-Cyclohexyl-4'ε-trimethylsilyloxymethyl-3β-methoxy-3',4',16β,17β-tetrahydro-2'H-5α-androsta
no(16,17-E)(1',2')oxazin-3'ε-carbonitrile;   W   126995

556   C34 H36 O7   77495-85-3   NS-0-0-0   1H-2,8a-Methanocyclopenta[a]cyclopropa[e]cyclodecen-11-one, 5,6-bis(benzoyloxy)-1a,2,5,5a,
6,9,10,10a-octahydro-5a-hydroxy-4-(hydroxymethyl)-1,1,7,9-tetramethyl-, [1aR-(1aα,2α,5β,5aβ,6β,8aα,9α,10aα)]-;   W/NBS   126996

Str 724

556   C34 H52 O4 S   21072-66-2   T-68-5798-0   Cholest-5-ene-3β,19-diol, 3-p-toluenesulfonate;;   C19-HYDROXYMETHYL CHOLESTEROL TOSYL
ATE;   W/NBS   64537

Str 725

556   C34 H52 O4 S   71472-91-8   K-112-2685-0   (25R)-5-CHOLESTEN-3β,26-DIOL-26-TOSYLATE;   W   64538

Str 726

556   C34 H52 O6   20089-73-0   AD-0-2436-0   Urs-12-en-28-oic acid, 2-hydroxy-3,23-[(1-methylethylidene)bis(oxy)]-11-oxo-, methyl es
ter, (2α,3β,4α)-;;   METHYL-3,23-O-ISOPROPYLIDENE-11-OXO-ASIATATE;   W/NBS   64541

Str 727

556   C34 H52 O6   4871-89-0   T-68-5799-0   ACETYL DIMETHYL SPERGULAGENATE;   W/NBS   64539

Str 728

556   C34 H52 O6   15018-86-7   AD-0-1245-0   DIMETHYL-QUINOVATE ACETATE;   W/NBS   64540

Str 729

556   C34 H52 O6   36872-77-2   F-32-2357-0   3β,23ε-DIACETOXY-Δ9(11)-HOLOSTENE;   W   64542

Str 730

20 40 60 80 100 120 140 160 180 200 220 240 260 280 300 320 340 360 380 400 420 440 460 480 500 520 540 560 580

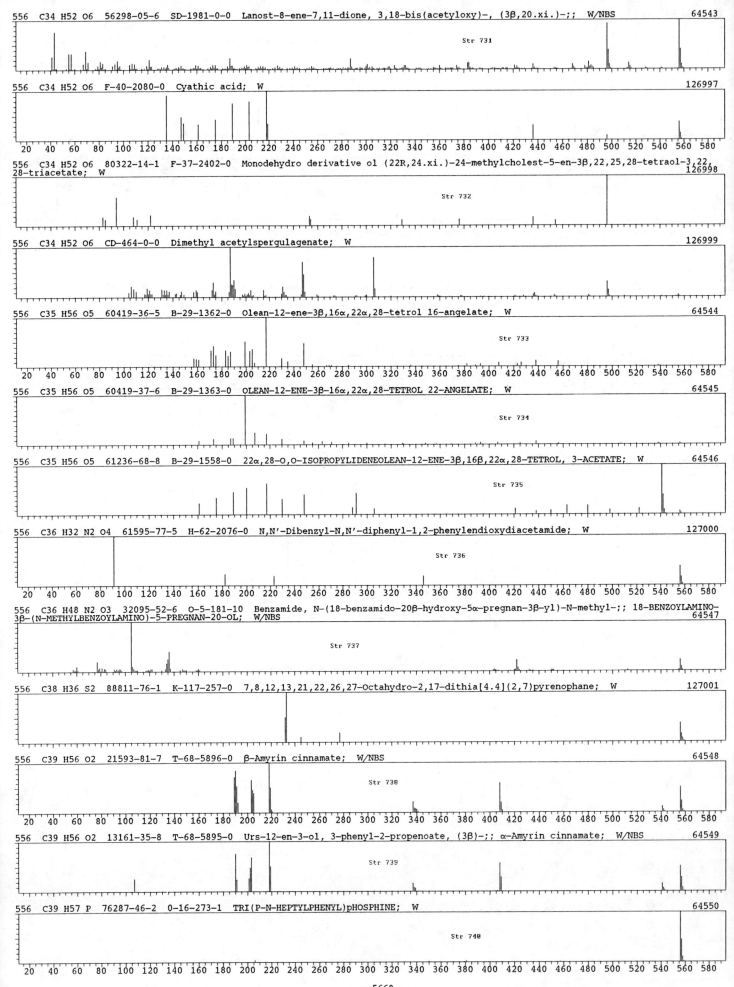

556 C34 H52 O6 56298-05-6 SD-1981-0-0 Lanost-8-ene-7,11-dione, 3,18-bis(acetyloxy)-, (3β,20.xi.)-;; W/NBS   64543

Str 731

556 C34 H52 O6 F-40-2080-0 Cyathic acid; W   126997

556 C34 H52 O6 80322-14-1 F-37-2402-0 Monodehydro derivative ol (22R,24.xi.)-24-methylcholest-5-en-3β,22,25,28-tetraol-3,22,
28-triacetate; W   126998

Str 732

556 C34 H52 O6 CD-464-0-0 Dimethyl acetylspergulagenate; W   126999

556 C35 H56 O5 60419-36-5 B-29-1362-0 Olean-12-ene-3β,16α,22α,28-tetrol 16-angelate; W   64544

Str 733

556 C35 H56 O5 60419-37-6 B-29-1363-0 OLEAN-12-ENE-3β-16α,22α,28-TETROL 22-ANGELATE; W   64545

Str 734

556 C35 H56 O5 61236-68-8 B-29-1558-0 22α,28-O,O-ISOPROPYLIDENEOLEAN-12-ENE-3β,16β,22α,28-TETROL, 3-ACETATE; W   64546

Str 735

556 C36 H32 N2 O4 61595-77-5 H-62-2076-0 N,N'-Dibenzyl-N,N'-diphenyl-1,2-phenylendioxydiacetamide; W   127000

Str 736

556 C36 H48 N2 O3 32095-52-6 O-5-181-10 Benzamide, N-(18-benzamido-20β-hydroxy-5α-pregnan-3β-yl)-N-methyl-;; 18-BENZOYLAMINO-
3β-(N-METHYLBENZOYLAMINO)-5-PREGNAN-20-OL; W/NBS   64547

Str 737

556 C38 H36 S2 88811-76-1 K-117-257-0 7,8,12,13,21,22,26,27-Octahydro-2,17-dithia[4.4](2,7)pyrenophane; W   127001

556 C39 H56 O2 21593-81-7 T-68-5896-0 β-Amyrin cinnamate; W/NBS   64548

Str 738

556 C39 H56 O2 13161-35-8 T-68-5895-0 Urs-12-en-3-ol, 3-phenyl-2-propenoate, (3β)-;; α-Amyrin cinnamate; W/NBS   64549

Str 739

556 C39 H57 P 76287-46-2 0-16-273-1 TRI(P-N-HEPTYLPHENYL)pHOSPHINE; W   64550

Str 740

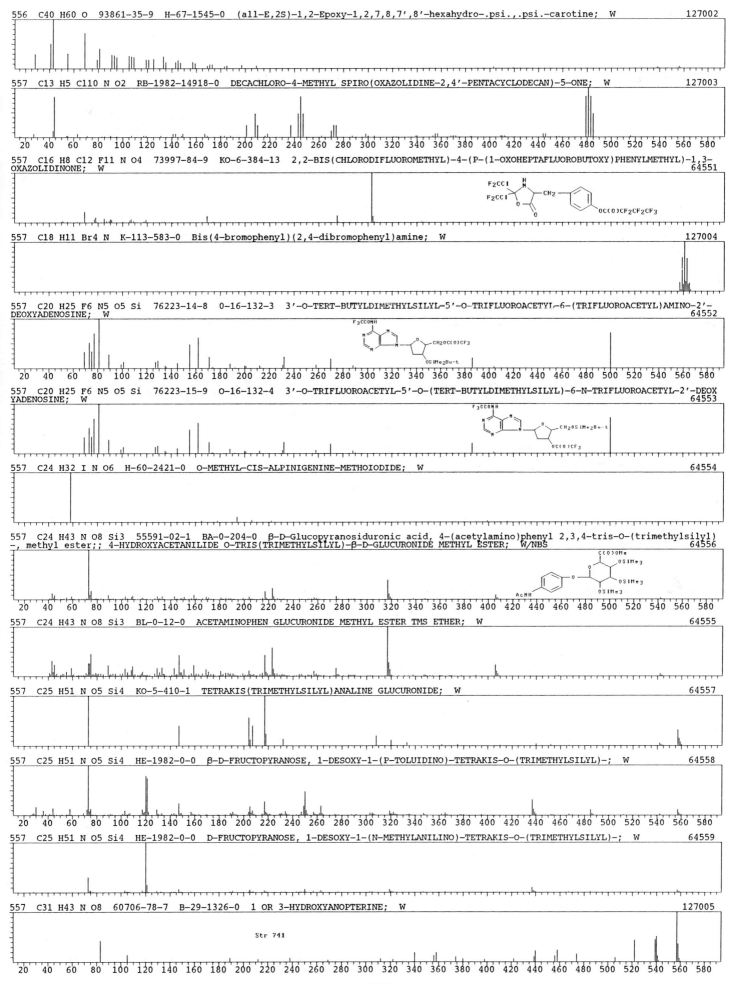

556   C40 H60 O   93861-35-9   H-67-1545-0   (all-E,2S)-1,2-Epoxy-1,2,7,8,7',8'-hexahydro-.psi.,.psi.-carotine;   W   127002

557   C13 H5 Cl10 N O2   RB-1982-14918-0   DECACHLORO-4-METHYL SPIRO(OXAZOLIDINE-2,4'-PENTACYCLODECAN)-5-ONE;   W   127003

557   C16 H8 Cl2 F11 N O4   73997-84-9   KO-6-384-13   2,2-BIS(CHLORODIFLUOROMETHYL)-4-(P-(1-OXOHEPTAFLUOROBUTOXY)PHENYLMETHYL)-1,3-OXAZOLIDINONE;   W   64551

557   C18 H11 Br4 N   K-113-583-0   Bis(4-bromophenyl)(2,4-dibromophenyl)amine;   W   127004

557   C20 H25 F6 N5 O5 Si   76223-14-8   O-16-132-3   3'-O-TERT-BUTYLDIMETHYLSILYL-5'-O-TRIFLUOROACETYL-6-(TRIFLUOROACETYL)AMINO-2'-DEOXYADENOSINE;   W   64552

557   C20 H25 F6 N5 O5 Si   76223-15-9   O-16-132-4   3'-O-TRIFLUOROACETYL-5'-O-(TERT-BUTYLDIMETHYLSILYL)-6-N-TRIFLUOROACETYL-2'-DEOXYADENOSINE;   W   64553

557   C24 H32 I N O6   H-60-2421-0   O-METHYL-CIS-ALPINIGENINE-METHOIODIDE;   W   64554

557   C24 H43 N O8 Si3   55591-02-1   BA-0-204-0   β-D-Glucopyranosiduronic acid, 4-(acetylamino)phenyl 2,3,4-tris-O-(trimethylsilyl)-, methyl ester;;   4-HYDROXYACETANILIDE O-TRIS(TRIMETHYLSILYL)-β-D-GLUCURONIDE METHYL ESTER;   W/NBS   64556

557   C24 H43 N O8 Si3   BL-0-12-0   ACETAMINOPHEN GLUCURONIDE METHYL ESTER TMS ETHER;   W   64555

557   C25 H51 N O5 Si4   KO-5-410-1   TETRAKIS(TRIMETHYLSILYL)ANALINE GLUCURONIDE;   W   64557

557   C25 H51 N O5 Si4   HE-1982-0-0   β-D-FRUCTOPYRANOSE, 1-DESOXY-1-(P-TOLUIDINO)-TETRAKIS-O-(TRIMETHYLSILYL)-;   W   64558

557   C25 H51 N O5 Si4   HE-1982-0-0   D-FRUCTOPYRANOSE, 1-DESOXY-1-(N-METHYLANILINO)-TETRAKIS-O-(TRIMETHYLSILYL)-;   W   64559

557   C31 H43 N O8   60706-78-7   B-29-1326-0   1 OR 3-HYDROXYANOPTERINE;   W   127005

Str 741

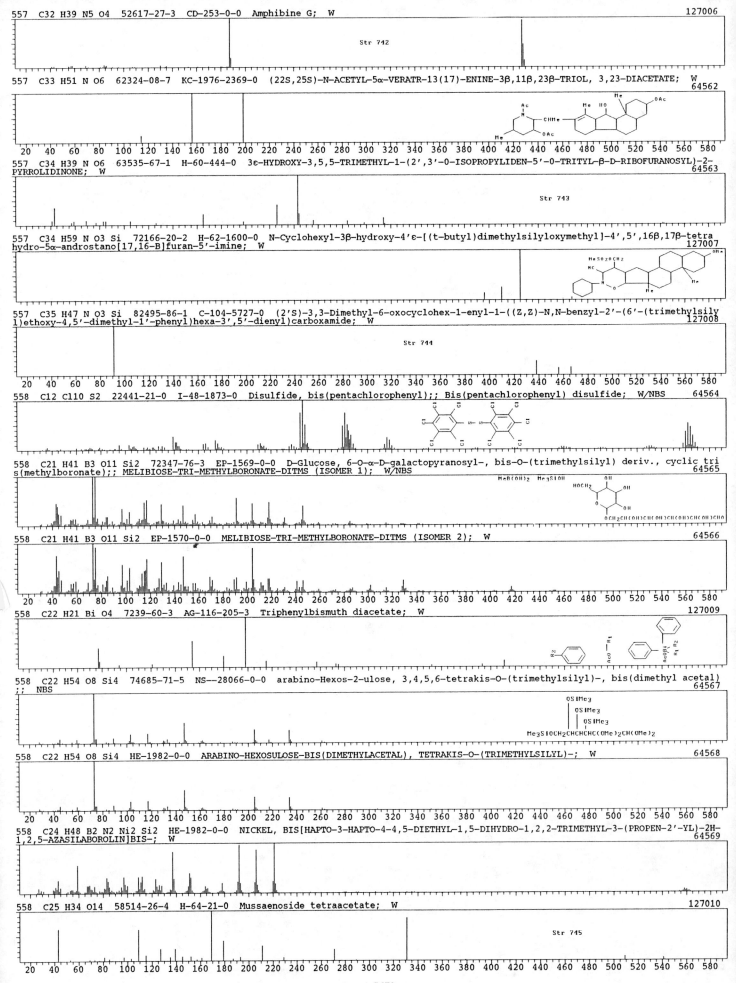

557　C32 H39 N5 O4　52617-27-3　CD-253-0-0　Amphibine G;　W　127006

Str 742

557　C33 H51 N O6　62324-08-7　KC-1976-2369-0　(22S,25S)-N-ACETYL-5α-VERATR-13(17)-ENINE-3β,11β,23β-TRIOL, 3,23-DIACETATE;　W
64562

557　C34 H39 N O6　63535-67-1　H-60-444-0　3ε-HYDROXY-3,5,5-TRIMETHYL-1-(2',3'-0-ISOPROPYLIDEN-5'-0-TRITYL-β-D-RIBOFURANOSYL)-2-PYRROLIDINONE;　W
64563

Str 743

557　C34 H59 N O3 Si　72166-20-2　H-62-1600-0　N-Cyclohexyl-3β-hydroxy-4'ε-[(t-butyl)dimethylsilyloxymethyl]-4',5',16β,17β-tetra
hydro-5α-androstano[17,16-B]furan-5'-imine;　W
127007

557　C35 H47 N O3 Si　82495-86-1　C-104-5727-0　(2'S)-3,3-Dimethyl-6-oxocyclohex-1-enyl-1-((Z,Z)-N,N-benzyl-2'-(6'-(trimethylsily
l)ethoxy-4,5'-dimethyl-1'-phenyl)hexa-3',5'-dienyl)carboxamide;　W
127008

Str 744

558　C12 Cl10 S2　22441-21-0　I-48-1873-0　Disulfide, bis(pentachlorophenyl);; Bis(pentachlorophenyl) disulfide;　W/NBS
64564

558　C21 H41 B3 O11 Si2　72347-76-3　EP-1569-0-0　D-Glucose, 6-O-α-D-galactopyranosyl-, bis-O-(trimethylsilyl) deriv., cyclic tri
s(methylboronate);; MELIBIOSE-TRI-METHYLBORONATE-DITMS (ISOMER 1);　W/NBS
64565

558　C21 H41 B3 O11 Si2　EP-1570-0-0　MELIBIOSE-TRI-METHYLBORONATE-DITMS (ISOMER 2);　W
64566

558　C22 H21 Bi O4　7239-60-3　AG-116-205-3　Triphenylbismuth diacetate;　W
127009

558　C22 H54 O8 Si4　74685-71-5　NS--28066-0-0　arabino-Hexos-2-ulose, 3,4,5,6-tetrakis-O-(trimethylsilyl)-, bis(dimethyl acetal)
;;　NBS
64567

558　C22 H54 O8 Si4　HE-1982-0-0　ARABINO-HEXOSULOSE-BIS(DIMETHYLACETAL), TETRAKIS-O-(TRIMETHYLSILYL)-;　W
64568

558　C24 H48 B2 N2 Ni2 Si2　HE-1982-0-0　NICKEL, BIS[HAPTO-3-HAPTO-4-4,5-DIETHYL-1,5-DIHYDRO-1,2,2-TRIMETHYL-3-(PROPEN-2'-YL)-2H-
1,2,5-AZASILABOROLIN]BIS-;　W
64569

558　C25 H34 O14　58514-26-4　H-64-21-0　Mussaenoside tetraacetate;　W
127010

Str 745

558 C26 H26 N2 O12 88295-13-0 KC-1983-1617-0 Hemiacetal dimer of 2-benzoylamino-2-deoxy-2-hydroxymethyl-D-mannono-1,4-lactone
; W
127011

Str 746

558 C26 H30 N4 O10 21066-33-1 O-6-1393-7 3-Methylriboflavin tetraacetate; W/NBS
64570

Str 747

20 40 60 80 100 120 140 160 180 200 220 240 260 280 300 320 340 360 380 400 420 440 460 480 500 520 540 560 580

558 C27 H31 N2 O9 P 79953-21-2 C-104-542-0 Methyl 5-amino-4-(3,4-dimethoxy-2-(phenylmethoxy)phenyl)-6-((dimethoxyphosphinyl)
acetyl)-3-methyl-2-pyridinecarboxylate; W
127012

Str 748

558 C28 H30 O7 Se 92622-37-2 C-107-1284-0 (2R*,4R*,5S*,6S*)-4,5-Bis(acetyloxy)-2-(benzyloxy)-2-(α-furyl)-6-((pheneleno)ethyl)
tetrahydropyran; W
127013

Str 749

558 C28 H30 O12 17019-76-0 AD-0-1971-0 Populin, tetraacetate;; POPULIN TETRAACETATE; W/NBS
64572

Str 750

20 40 60 80 100 120 140 160 180 200 220 240 260 280 300 320 340 360 380 400 420 440 460 480 500 520 540 560 580

558 C28 H30 O12 18449-66-6 AD-0-1972-0 Tremuloidin, tetraacetate;; TREMULOIDIN TETRAACETATE; W/NBS
64571

558 C28 H30 O12 74741-51-8 HE-1982-0-0 Hexitol, 1,3,4,6-tetraacetate 2,5-dibenzoate;; HEXAN, 2,5-DIBENZOYLOXY-1,3,4,6-TETRAAC
ETOXY-; W/NBS
64573

558 C28 H30 O12 F-33-142-0 2α-(3,4β-METHYLENEDIOXYPHENYL)-3β-ACETOXYMETHYL-4-(α-ACETOXY-3,4α-METHYLENEDIOXYBENZYL)-4-ACETOXYTE
TRAHYDROFURAN; W
64574

20 40 60 80 100 120 140 160 180 200 220 240 260 280 300 320 340 360 380 400 420 440 460 480 500 520 540 560 580

558 C28 H54 O7 Si2 BS-5-44-0 5α,7α-Dihydroxy-11-oxo(tetranor, ω-dinor)-prostane-1,14-dioic acid methyl ester t-butyldimethylsi
lyl ether; W
127014

558 C28 H58 O5 Si3 55759-96-1 AB-231-428-6 Cyclopentanepentanoic acid, 3,5-bis[(trimethylsilyl)oxy]-2-[3-[(trimethylsilyl)ox
y]-1-octenyl]-, methyl ester;; METHYL 7α,9α,13-TRIHYDROXY-DINORPROST-11-ENOATE-TRISTRIMETHYLSILYL ETHER; W/NBS
64575

558 C29 H34 O9 S 78173-17-8 KC-1981-398-0 Methyl ent-2β-acetoxy-10β-hydroxy-7-methoxycarbonyl-13-toluene-p-sulphonyloxy-20-no
rgibberell-16-en-19-oate 19,10-lactone; W
127015

558 C29 H51 I O2 91712-64-0 J-49-3760-0 (5β)-4-Iodo-2-methyl-3,4-secocholestan-2-ol formate; W
127016

20 40 60 80 100 120 140 160 180 200 220 240 260 280 300 320 340 360 380 400 420 440 460 480 500 520 540 560 580

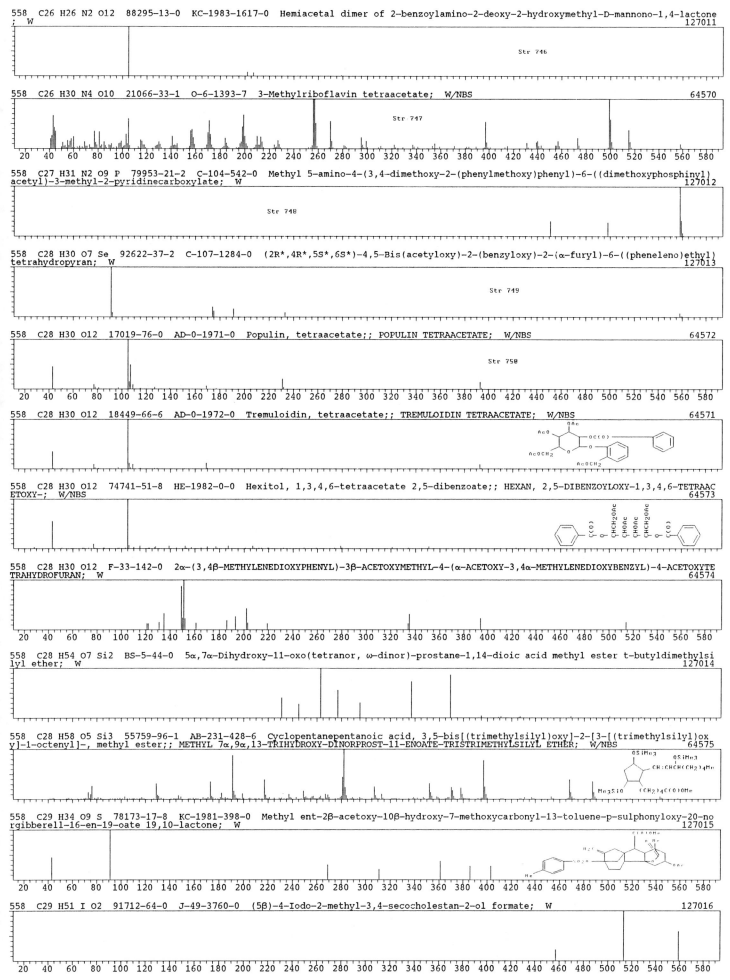

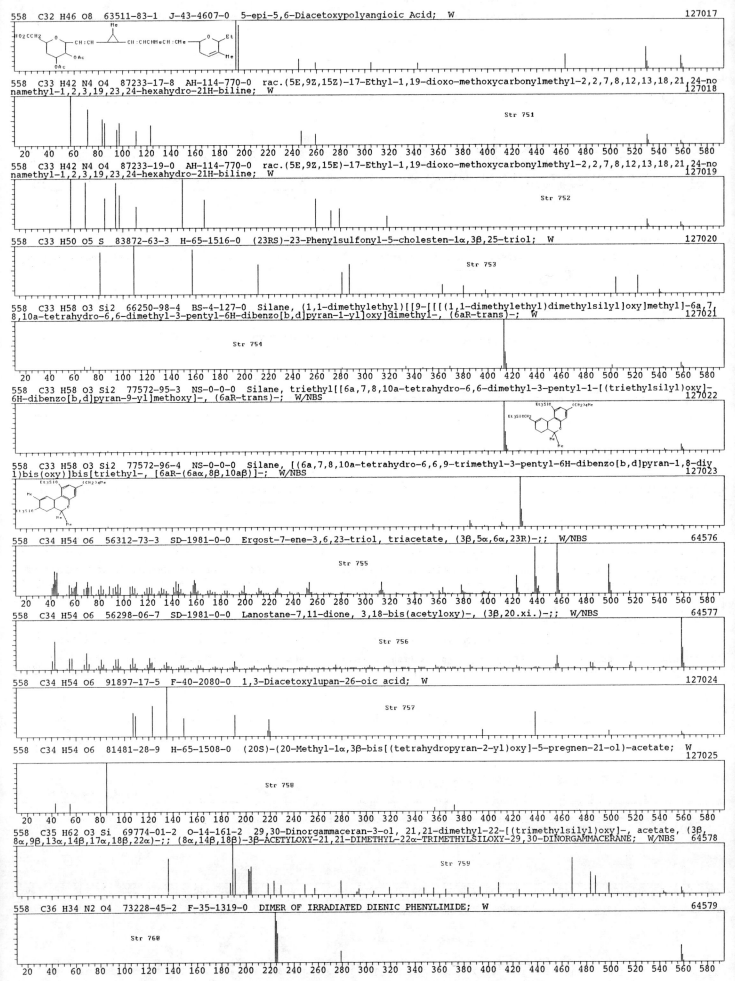

558 C32 H46 O8   63511-83-1 J-43-4607-0   5-epi-5,6-Diacetoxypolyangioic Acid; W   127017

558 C33 H42 N4 O4   87233-17-8 AH-114-770-0   rac.(5E,9Z,15Z)-17-Ethyl-1,19-dioxo-methoxycarbonylmethyl-2,2,7,8,12,13,18,21,24-no
namethyl-1,2,3,19,23,24-hexahydro-21H-biline;   W   127018
Str 751

558 C33 H42 N4 O4   87233-19-0 AH-114-770-0   rac.(5E,9Z,15E)-17-Ethyl-1,19-dioxo-methoxycarbonylmethyl-2,2,7,8,12,13,18,21,24-no
namethyl-1,2,3,19,23,24-hexahydro-21H-biline;   W   127019
Str 752

558 C33 H50 O5 S   83872-63-3 H-65-1516-0   (23RS)-23-Phenylsulfonyl-5-cholesten-1α,3β,25-triol; W   127020
Str 753

558 C33 H58 O3 Si2   66250-98-4 BS-4-127-0   Silane, (1,1-dimethylethyl)[[9-[[[(1,1-dimethylethyl)dimethylsilyl]oxy]methyl]-6a,7,
8,10a-tetrahydro-6,6-dimethyl-3-pentyl-6H-dibenzo[b,d]pyran-1-yl]oxy]dimethyl-, (6aR-trans)-; W   127021
Str 754

558 C33 H58 O3 Si2   77572-95-3 NS-0-0-0   Silane, triethyl[[6a,7,8,10a-tetrahydro-6,6-dimethyl-3-pentyl-1-[(triethylsilyl)oxy]-
6H-dibenzo[b,d]pyran-9-yl]methoxy]-, (6aR-trans)-;   W/NBS   127022

558 C33 H58 O3 Si2   77572-96-4 NS-0-0-0   Silane, [(6a,7,8,10a-tetrahydro-6,6,9-trimethyl-3-pentyl-6H-dibenzo[b,d]pyran-1,8-diy
l)bis(oxy)]bis[triethyl-, [6aR-(6aα,8β,10aβ)]-;   W/NBS   127023

558 C34 H54 O6   56312-73-3 SD-1981-0-0   Ergost-7-ene-3,6,23-triol, triacetate, (3β,5α,6α,23R)-;;   W/NBS   64576
Str 755

558 C34 H54 O6   56298-06-7 SD-1981-0-0   Lanostane-7,11-dione, 3,18-bis(acetyloxy)-, (3β,20.xi.)-;;   W/NBS   64577
Str 756

558 C34 H54 O6   91897-17-5 F-40-2080-0   1,3-Diacetoxylupan-26-oic acid; W   127024
Str 757

558 C34 H54 O6   81481-28-9 H-65-1508-0   (20S)-(20-Methyl-1α,3β-bis[(tetrahydropyran-2-yl)oxy]-5-pregnen-21-ol)-acetate; W   127025
Str 758

558 C35 H62 O3 Si   69774-01-2 O-14-161-2   29,30-Dinorgammaceran-3-ol, 21,21-dimethyl-22-[(trimethylsilyl)oxy]-, acetate, (3β,
8α,9β,13α,14β,17α,18β,22α)-;; (8α,14β,18β)-3β-ACETYLOXY-21,21-DIMETHYL-22α-TRIMETHYLSILOXY-29,30-DINORGAMMACERANE; W/NBS   64578
Str 759

558 C36 H34 N2 O4   73228-45-2 F-35-1319-0   DIMER OF IRRADIATED DIENIC PHENYLIMIDE; W   64579
Str 760

- 5672 -

558　C39 H31 Co　HE-1986-2019-0　INDENYL-(HAPTO-4-1,2,3,4-TETRAPHENYLCYCLOHEXA-1,3-DIENE)COBALT;　W　　　127026

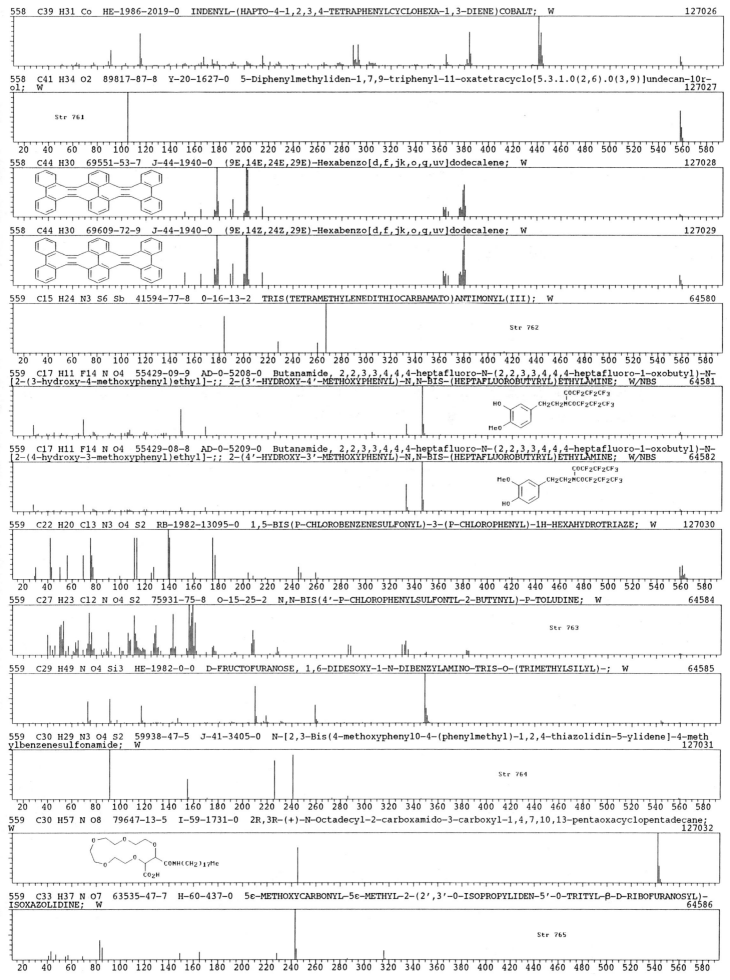

558　C41 H34 O2　89817-87-8　Y-20-1627-0　5-Diphenylmethyliden-1,7,9-triphenyl-11-oxatetracyclo[5.3.1.0(2,6).0(3,9)]undecan-10r-ol;　W　　　127027

Str 761

558　C44 H30　69551-53-7　J-44-1940-0　(9E,14E,24E,29E)-Hexabenzo[d,f,jk,o,q,uv]dodecalene;　W　　　127028

558　C44 H30　69609-72-9　J-44-1940-0　(9E,14Z,24Z,29E)-Hexabenzo[d,f,jk,o,q,uv]dodecalene;　W　　　127029

559　C15 H24 N3 S6 Sb　41594-77-8　0-16-13-2　TRIS(TETRAMETHYLENEDITHIOCARBAMATO)ANTIMONYL(III);　W　　　64580

Str 762

559　C17 H11 F14 N O4　55429-09-9　AD-0-5208-0　Butanamide, 2,2,3,3,4,4,4-heptafluoro-N-(2,2,3,3,4,4,4-heptafluoro-1-oxobutyl)-N-[2-(3-hydroxy-4-methoxyphenyl)ethyl]-;; 2-(3'-HYDROXY-4'-METHOXYPHENYL)-N,N-BIS-(HEPTAFLUOROBUTYRYL)ETHYLAMINE;　W/NBS　　　64581

559　C17 H11 F14 N O4　55429-08-8　AD-0-5209-0　Butanamide, 2,2,3,3,4,4,4-heptafluoro-N-(2,2,3,3,4,4,4-heptafluoro-1-oxobutyl)-N-[2-(4-hydroxy-3-methoxyphenyl)ethyl]-;; 2-(4'-HYDROXY-3'-METHOXYPHENYL)-N,N-BIS-(HEPTAFLUOROBUTYRYL)ETHYLAMINE;　W/NBS　　　64582

559　C22 H20 Cl3 N3 O4 S2　RB-1982-13095-0　1,5-BIS(P-CHLOROBENZENESULFONYL)-3-(P-CHLOROPHENYL)-1H-HEXAHYDROTRIAZE;　W　　　127030

559　C27 H23 Cl2 N O4 S2　75931-75-8　O-15-25-2　N,N-BIS(4'-P-CHLOROPHENYLSULFONTL-2-BUTYNYL)-P-TOLUDINE;　W　　　64584

Str 763

559　C29 H49 N O4 Si3　HE-1982-0-0　D-FRUCTOFURANOSE, 1,6-DIDESOXY-1-N-DIBENZYLAMINO-TRIS-O-(TRIMETHYLSILYL)-;　W　　　64585

559　C30 H29 N3 O4 S2　59938-47-5　J-41-3405-0　N-[2,3-Bis(4-methoxyphenyl0-4-(phenylmethyl)-1,2,4-thiazolidin-5-ylidene]-4-methylbenzenesulfonamide;　W　　　127031

Str 764

559　C30 H57 N O8　79647-13-5　I-59-1731-0　2R,3R-(+)-N-Octadecyl-2-carboxamido-3-carboxyl-1,4,7,10,13-pentaoxacyclopentadecane;　W　　　127032

559　C33 H37 N O7　63535-47-7　H-60-437-0　5ε-METHOXYCARBONYL-5ε-METHYL-2-(2',3'-0-ISOPROPYLIDEN-5'-0-TRITYL-β-D-RIBOFURANOSYL)-ISOXAZOLIDINE;　W　　　64586

Str 765

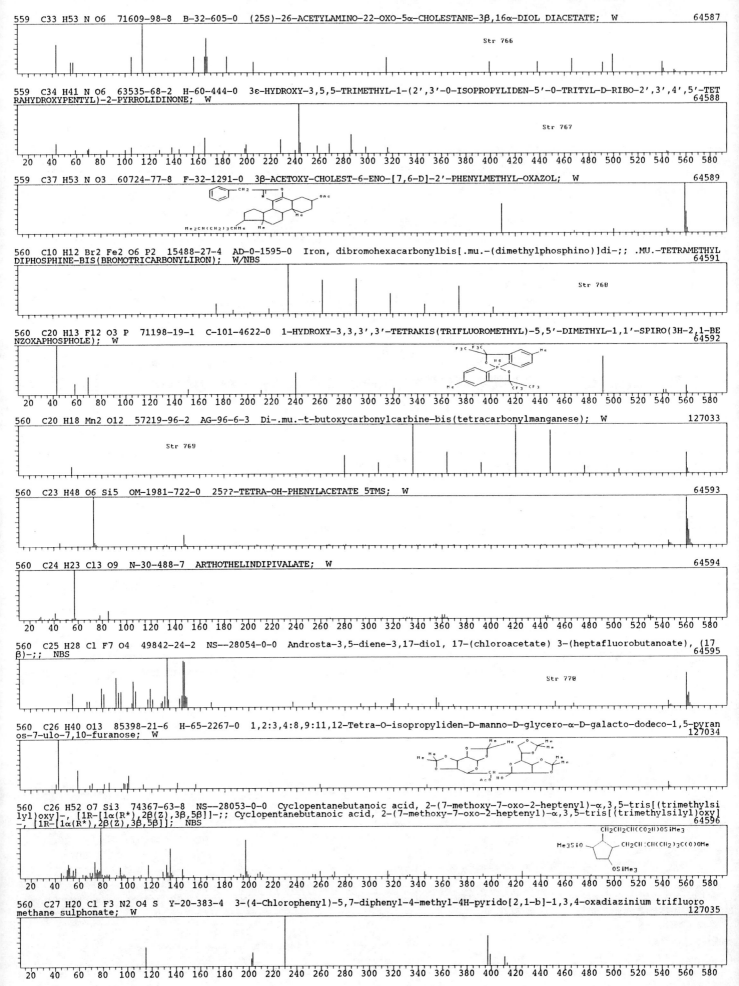

559   C33 H53 N O6   71609-98-8   B-32-605-0   (25S)-26-ACETYLAMINO-22-OXO-5α-CHOLESTANE-3β,16α-DIOL DIACETATE;   W
64587

Str 766

559   C34 H41 N O6   63535-68-2   H-60-444-0   3ε-HYDROXY-3,5,5-TRIMETHYL-1-(2',3'-O-ISOPROPYLIDEN-5'-O-TRITYL-D-RIBO-2',3',4',5'-TET
RAHYDROXYPENTYL)-2-PYRROLIDINONE;   W
64588

Str 767

559   C37 H53 N O3   60724-77-8   F-32-1291-0   3β-ACETOXY-CHOLEST-6-ENO-[7,6-D]-2'-PHENYLMETHYL-OXAZOL;   W
64589

560   C10 H12 Br2 Fe2 O6 P2   15488-27-4   AD-0-1595-0   Iron, dibromohexacarbonylbis[.mu.-(dimethylphosphino)]di-;;   .MU.-TETRAMETHYL
DIPHOSPHINE-BIS(BROMOTRICARBONYLIRON);   W/NBS
64591

Str 768

560   C20 H13 F12 O3 P   71198-19-1   C-101-4622-0   1-HYDROXY-3,3,3',3'-TETRAKIS(TRIFLUOROMETHYL)-5,5'-DIMETHYL-1,1'-SPIRO(3H-2,1-BE
NZOXAPHOSPHOLE);   W
64592

560   C20 H18 Mn2 O12   57219-96-2   AG-96-6-3   Di-.mu.-t-butoxycarbonylcarbine-bis(tetracarbonylmanganese);   W
127033

Str 769

560   C23 H48 O6 Si5   OM-1981-722-0   25??-TETRA-OH-PHENYLACETATE 5TMS;   W
64593

560   C24 H23 Cl3 O9   N-30-488-7   ARTHOTHELINDIPIVALATE;   W
64594

560   C25 H28 Cl F7 O4   49842-24-2   NS--28054-0-0   Androsta-3,5-diene-3,17-diol, 17-(chloroacetate) 3-(heptafluorobutanoate), (17
β)-;;   NBS
64595

Str 770

560   C26 H40 O13   85398-21-6   H-65-2267-0   1,2:3,4:8,9:11,12-Tetra-O-isopropyliden-D-manno-D-glycero-α-D-galacto-dodeco-1,5-pyran
os-7-ulo-7,10-furanose;   W
127034

560   C26 H52 O7 Si3   74367-63-8   NS--28053-0-0   Cyclopentanebutanoic acid, 2-(7-methoxy-7-oxo-2-heptenyl)-α,3,5-tris[(trimethylsi
lyl)oxy]-, [1R-[1α(R*),2β(Z),3β,5β]]-;; Cyclopentanebutanoic acid, 2-(7-methoxy-7-oxo-2-heptenyl)-α,3,5-tris[(trimethylsilyl)oxy]
-, [1R-[1α(R*),2β(Z),3β,5β]];   NBS
64596

560   C27 H20 Cl F3 N2 O4 S   Y-20-383-4   3-(4-Chlorophenyl)-5,7-diphenyl-4-methyl-4H-pyrido[2,1-b]-1,3,4-oxadiazinium trifluoro
methane sulphonate;   W
127035

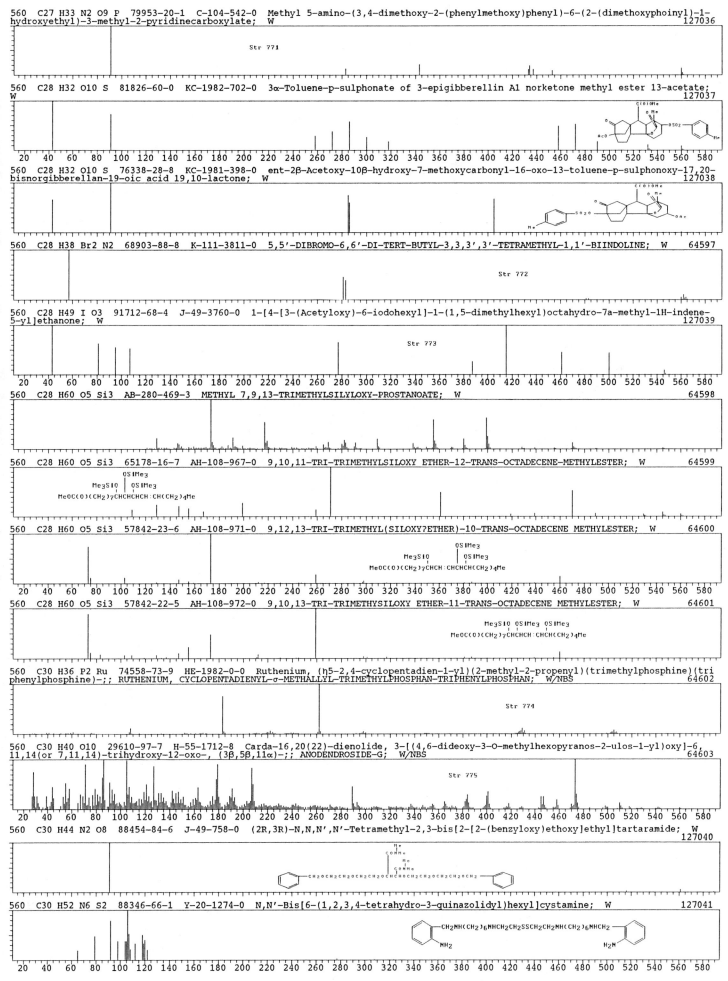

560   C27 H33 N2 O9 P   79953-20-1   C-104-542-0   Methyl 5-amino-(3,4-dimethoxy-2-(phenylmethoxy)phenyl)-6-(2-(dimethoxyphoinyl)-1-hydroxyethyl)-3-methyl-2-pyridinecarboxylate;   W
127036

Str 771

560   C28 H32 O10 S   81826-60-0   KC-1982-702-0   3α-Toluene-p-sulphonate of 3-epigibberellin A1 norketone methyl ester 13-acetate;   W
127037

560   C28 H32 O10 S   76338-28-8   KC-1981-398-0   ent-2β-Acetoxy-10β-hydroxy-7-methoxycarbonyl-16-oxo-13-toluene-p-sulphonoxy-17,20-bisnorgibberellan-19-oic acid 19,10-lactone;   W
127038

560   C28 H38 Br2 N2   68903-88-8   K-111-3811-0   5,5'-DIBROMO-6,6'-DI-TERT-BUTYL-3,3,3',3'-TETRAMETHYL-1,1'-BIINDOLINE;   W
64597

Str 772

560   C28 H49 I O3   91712-68-4   J-49-3760-0   1-[4-[3-(Acetyloxy)-6-iodohexyl]-1-(1,5-dimethylhexyl)octahydro-7a-methyl-1H-indene-5-yl]ethanone;   W
127039

Str 773

560   C28 H60 O5 Si3   AB-280-469-3   METHYL 7,9,13-TRIMETHYLSILYLOXY-PROSTANOATE;   W
64598

560   C28 H60 O5 Si3   65178-16-7   AH-108-967-0   9,10,11-TRI-TRIMETHYLSILOXY ETHER-12-TRANS-OCTADECENE-METHYLESTER;   W
64599

560   C28 H60 O5 Si3   57842-23-6   AH-108-971-0   9,12,13-TRI-TRIMETHYL(SILOXY?ETHER)-10-TRANS-OCTADECENE METHYLESTER;   W
64600

560   C28 H60 O5 Si3   57842-22-5   AH-108-972-0   9,10,13-TRI-TRIMETHYSILOXY ETHER-11-TRANS-OCTADECENE METHYLESTER;   W
64601

560   C30 H36 P2 Ru   74558-73-9   HE-1982-0-0   Ruthenium, (η5-2,4-cyclopentadien-1-yl)(2-methyl-2-propenyl)(trimethylphosphine)(triphenylphosphine)-;;   RUTHENIUM, CYCLOPENTADIENYL-σ-METHALLYL-TRIMETHYLPHOSPHAN-TRIPHENYLPHOSPHAN;   W/NBS
64602

Str 774

560   C30 H40 O10   29610-97-7   H-55-1712-8   Carda-16,20(22)-dienolide, 3-[(4,6-dideoxy-3-O-methylhexopyranos-2-ulos-1-yl)oxy]-6,11,14(or 7,11,14)-trihydroxy-12-oxo-, (3β,5β,11α)-;;   ANODENDROSIDE-G;   W/NBS
64603

Str 775

560   C30 H44 N2 O8   88454-84-6   J-49-758-0   (2R,3R)-N,N,N',N'-Tetramethyl-2,3-bis[2-[2-(benzyloxy)ethoxy]ethyl]tartaramide;   W
127040

560   C30 H52 N6 S2   88346-66-1   Y-20-1274-0   N,N'-Bis[6-(1,2,3,4-tetrahydro-3-quinazolidyl)hexyl]cystamine;   W
127041

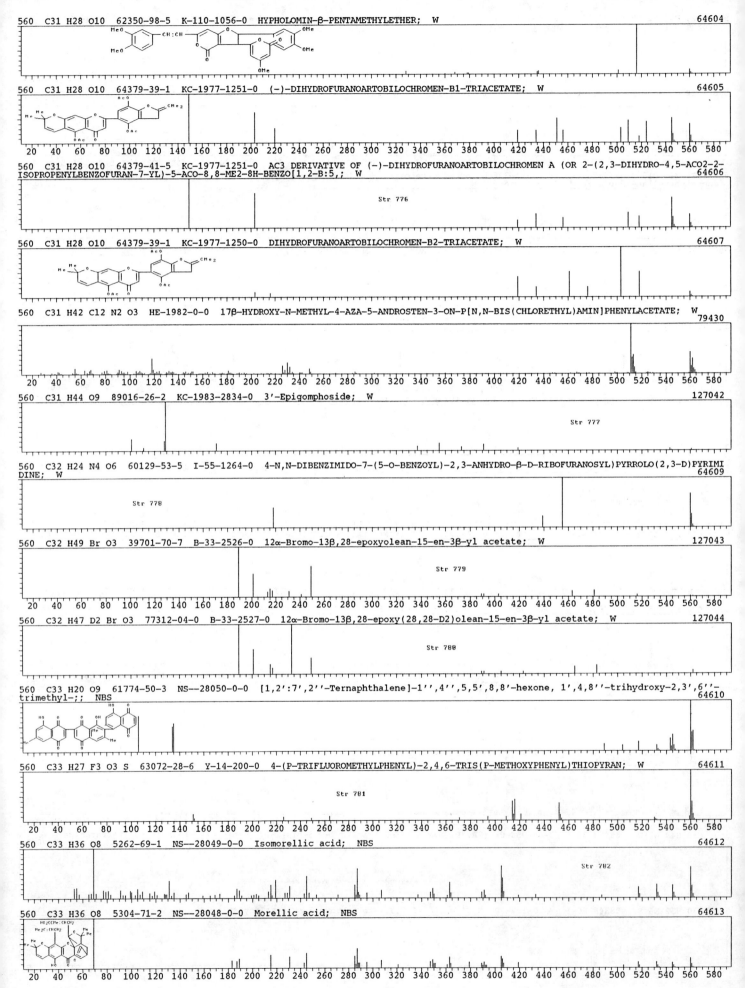

560   C31 H28 O10   62350-98-5   K-110-1056-0   HYPHOLOMIN-β-PENTAMETHYLETHER;   W   64604

560   C31 H28 O10   64379-39-1   KC-1977-1251-0   (-)-DIHYDROFURANOARTOBILOCHROMEN-B1-TRIACETATE;   W   64605

560   C31 H28 O10   64379-41-5   KC-1977-1251-0   AC3 DERIVATIVE OF (-)-DIHYDROFURANOARTOBILOCHROMEN A (OR 2-(2,3-DIHYDRO-4,5-ACO2-2-
ISOPROPENYLBENZOFURAN-7-YL)-5-ACO-8,8-ME2-8H-BENZO[1,2-B:5,;   W   64606

Str 776

560   C31 H28 O10   64379-39-1   KC-1977-1250-0   DIHYDROFURANOARTOBILOCHROMEN-B2-TRIACETATE;   W   64607

560   C31 H42 Cl2 N2 O3   HE-1982-0-0   17β-HYDROXY-N-METHYL-4-AZA-5-ANDROSTEN-3-ON-P[N,N-BIS(CHLORETHYL)AMIN]PHENYLACETATE;   W   79430

560   C31 H44 O9   89016-26-2   KC-1983-2834-0   3'-Epigomphoside;   W   127042

Str 777

560   C32 H24 N4 O6   60129-53-5   I-55-1264-0   4-N,N-DIBENZIMIDO-7-(5-O-BENZOYL)-2,3-ANHYDRO-β-D-RIBOFURANOSYL)PYRROLO(2,3-D)PYRIMI
DINE;   W   64609

Str 778

560   C32 H49 Br O3   39701-70-7   B-33-2526-0   12α-Bromo-13β,28-epoxyolean-15-en-3β-yl acetate;   W   127043

Str 779

560   C32 H47 D2 Br O3   77312-04-0   B-33-2527-0   12α-Bromo-13β,28-epoxy(28,28-D2)olean-15-en-3β-yl acetate;   W   127044

Str 780

560   C33 H20 O9   61774-50-3   NS--28050-0-0   [1,2':7,2''-Ternaphthalene]-1'',4'',5,5',8,8'-hexone, 1',4,8''-trihydroxy-2,3',6''-
trimethyl-;;   NBS   64610

560   C33 H27 F3 O3 S   63072-28-6   Y-14-200-0   4-(P-TRIFLUOROMETHYLPHENYL)-2,4,6-TRIS(P-METHOXYPHENYL)THIOPYRAN;   W   64611

Str 781

560   C33 H36 O8   5262-69-1   NS--28049-0-0   Isomorellic acid;   NBS   64612

Str 782

560   C33 H36 O8   5304-71-2   NS--28048-0-0   Morellic acid;   NBS   64613

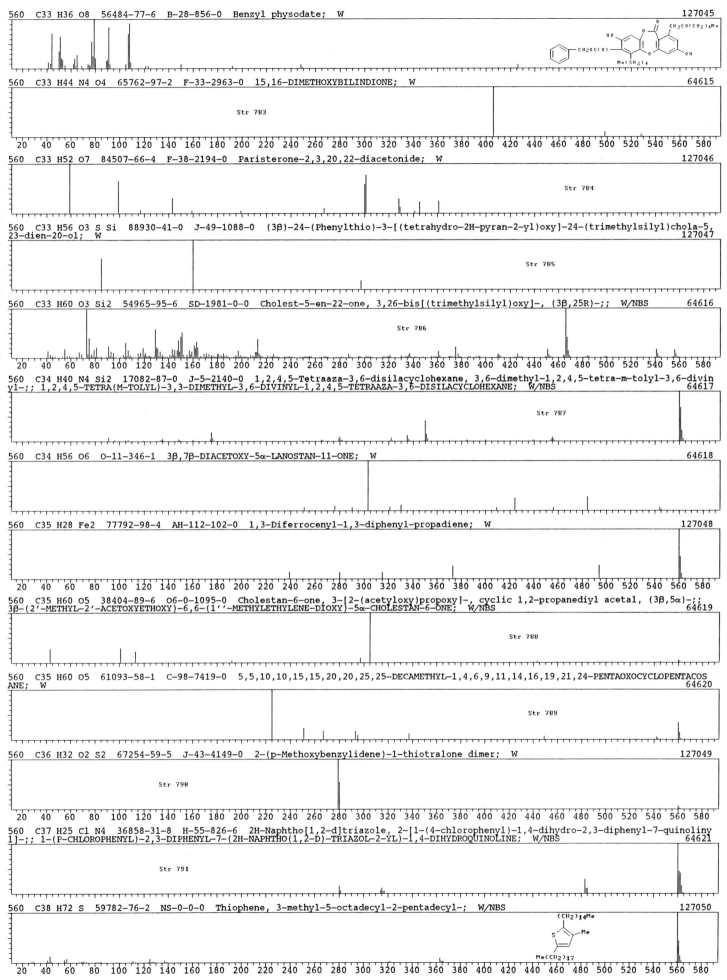

560   C33 H36 O8   56484-77-6   B-28-856-0   Benzyl physodate;   W   127045

560   C33 H44 N4 O4   65762-97-2   F-33-2963-0   15,16-DIMETHOXYBILINDIONE;   W   64615

Str 783

560   C33 H52 O7   84507-66-4   F-38-2194-0   Paristerone-2,3,20,22-diacetonide;   W   127046

Str 784

560   C33 H56 O3 S Si   88930-41-0   J-49-1088-0   (3β)-24-(Phenylthio)-3-[(tetrahydro-2H-pyran-2-yl)oxy]-24-(trimethylsilyl)chola-5,
23-dien-20-ol;   W   127047

Str 785

560   C33 H60 O3 Si2   54965-95-6   SD-1981-0-0   Cholest-5-en-22-one, 3,26-bis[(trimethylsilyl)oxy]-, (3β,25R)-;;   W/NBS   64616

Str 786

560   C34 H40 N4 Si2   17082-87-0   J-5-2140-0   1,2,4,5-Tetraaza-3,6-disilacyclohexane, 3,6-dimethyl-1,2,4,5-tetra-m-tolyl-3,6-divin
yl-;;   1,2,4,5-TETRA(M-TOLYL)-3,3-DIMETHYL-3,6-DIVINYL-1,2,4,5-TETRAAZA-3,6-DISILACYCLOHEXANE;   W/NBS   64617

Str 787

560   C34 H56 O6   O-11-346-1   3β,7β-DIACETOXY-5α-LANOSTAN-11-ONE;   W   64618

560   C35 H28 Fe2   77792-98-4   AH-112-102-0   1,3-Diferrocenyl-1,3-diphenyl-propadiene;   W   127048

560   C35 H60 O5   38404-89-6   O6-0-1095-0   Cholestan-6-one, 3-[2-(acetyloxy)propoxy]-, cyclic 1,2-propanediyl acetal, (3β,5α)-;;
3β-(2'-METHYL-2'-ACETOXYETHOXY)-6,6-(1''-METHYLETHYLENE-DIOXY)-5α-CHOLESTAN-6-ONE;   W/NBS   64619

Str 788

560   C35 H60 O5   61093-58-1   C-98-7419-0   5,5,10,10,15,15,20,20,25,25-DECAMETHYL-1,4,6,9,11,14,16,19,21,24-PENTAOXOCYCLOPENTACOS
ANE;   W   64620

Str 789

560   C36 H32 O2 S2   67254-59-5   J-43-4149-0   2-(p-Methoxybenzylidene)-1-thiotralone dimer;   W   127049

Str 790

560   C37 H25 Cl N4   36858-31-8   H-55-826-6   2H-Naphtho[1,2-d]triazole, 2-[1-(4-chlorophenyl)-1,4-dihydro-2,3-diphenyl-7-quinoliny
l]-;;   1-(P-CHLOROPHENYL)-2,3-DIPHENYL-7-(2H-NAPHTHO(1,2-D)-TRIAZOL-2-YL)-1,4-DIHYDROQUINOLINE;   W/NBS   64621

Str 791

560   C38 H72 S   59782-76-2   NS-0-0-0   Thiophene, 3-methyl-5-octadecyl-2-pentadecyl-;   W/NBS   127050

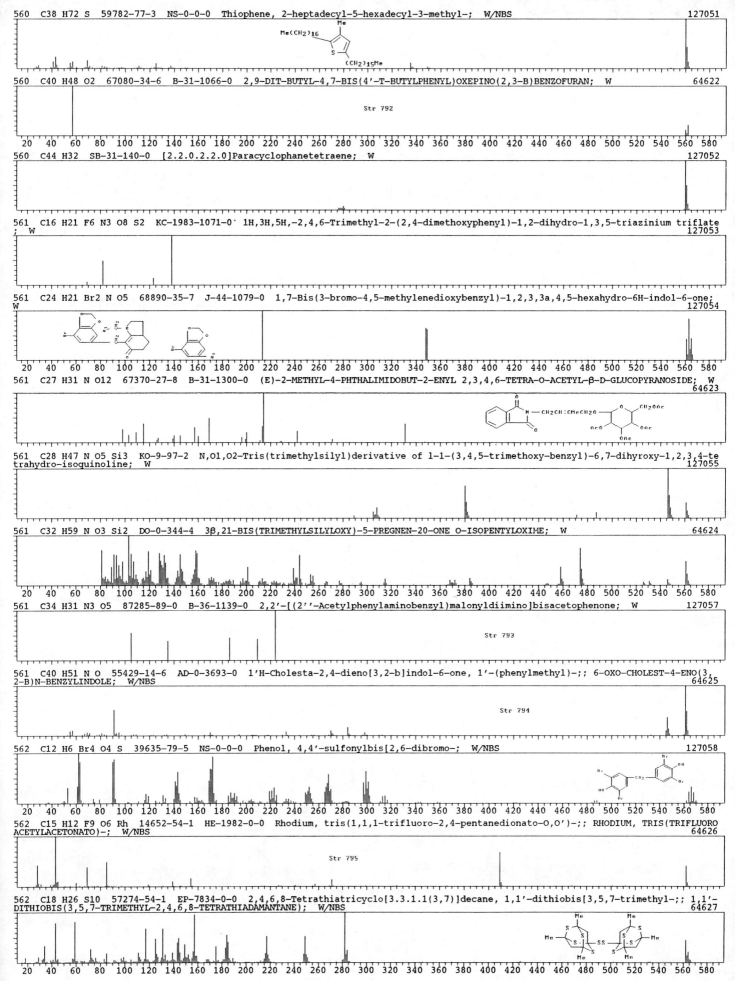

560   C38 H72 S   59782-77-3   NS-0-0-0   Thiophene, 2-heptadecyl-5-hexadecyl-3-methyl-; W/NBS   127051

560   C40 H48 O2   67080-34-6   B-31-1066-0   2,9-DIT-BUTYL-4,7-BIS(4'-T-BUTYLPHENYL)OXEPINO(2,3-B)BENZOFURAN; W   64622

Str 792

560   C44 H32   SB-31-140-0   [2.2.0.2.2.0]Paracyclophanetetraene; W   127052

561   C16 H21 F6 N3 O8 S2   KC-1983-1071-0·   1H,3H,5H,-2,4,6-Trimethyl-2-(2,4-dimethoxyphenyl)-1,2-dihydro-1,3,5-triazinium triflate
; W   127053

561   C24 H21 Br2 N O5   68890-35-7   J-44-1079-0   1,7-Bis(3-bromo-4,5-methylenedioxybenzyl)-1,2,3,3a,4,5-hexahydro-6H-indol-6-one;
W   127054

561   C27 H31 N O12   67370-27-8   B-31-1300-0   (E)-2-METHYL-4-PHTHALIMIDOBUT-2-ENYL 2,3,4,6-TETRA-O-ACETYL-β-D-GLUCOPYRANOSIDE; W
       64623

561   C28 H47 N O5 Si3   KO-9-97-2   N,O1,O2-Tris(trimethylsilyl)derivative of 1-1-(3,4,5-trimethoxy-benzyl)-6,7-dihyroxy-1,2,3,4-te
trahydro-isoquinoline; W   127055

561   C32 H59 N O3 Si2   DO-0-344-4   3β,21-BIS(TRIMETHYLSILYLOXY)-5-PREGNEN-20-ONE O-ISOPENTYLOXIME; W   64624

561   C34 H31 N3 O5   87285-89-0   B-36-1139-0   2,2'-[(2''-Acetylphenylaminobenzyl)malonyldiimino]bisacetophenone; W   127057

Str 793

561   C40 H51 N O   55429-14-6   AD-0-3693-0   1'H-Cholesta-2,4-dieno[3,2-b]indol-6-one, 1'-(phenylmethyl)-;; 6-OXO-CHOLEST-4-ENO(3,
2-B)N-BENZYLINDOLE; W/NBS   64625

Str 794

562   C12 H6 Br4 O4 S   39635-79-5   NS-0-0-0   Phenol, 4,4'-sulfonylbis[2,6-dibromo-; W/NBS   127058

562   C15 H12 F9 O6 Rh   14652-54-1   HE-1982-0-0   Rhodium, tris(1,1,1-trifluoro-2,4-pentanedionato-O,O')-;; RHODIUM, TRIS(TRIFLUORO
ACETYLACETONATO)-; W/NBS   64626

Str 795

562   C18 H26 S10   57274-54-1   EP-7834-0-0   2,4,6,8-Tetrathiatricyclo[3.3.1.1(3,7)]decane, 1,1'-dithiobis[3,5,7-trimethyl-;; 1,1'-
DITHIOBIS(3,5,7-TRIMETHYL-2,4,6,8-TETRATHIADAMANTANE); W/NBS   64627

562 C22 H20 Co2 Se2 HE-1986-2039-0 DICYCLOPENTADIENYL-2,4-DIPHENYL-1,3-DICOBALTA-2,4-DITHIACYCLOBUTANE; W   127059

562 C23 H26 N6 O11 IC-1269-0-0 N-(2-Methoxyethoxycarbonylethyl)-3-acetylamino-6-methoxy-4-(3,5-dinitro-2-methoxycarbonylphenyl
azo)aniline; W   127061

20 40 60 80 100 120 140 160 180 200 220 240 260 280 300 320 340 360 380 400 420 440 460 480 500 520 540 560 580

562 C23 H34 N2 O14 78300-72-8 F-37-109-0 Methyl-O-(2,3,4-tri-O-acetyl-β-D-galactopyranosiduronamide)-(1 to 4)-2-acetamido-3-
O-acetyl-2,6-didesoxy-β-D-glucopyranoside; W   127062

Str 796

562 C24 H30 Be4 O13 56377-88-9 EP-4132-0-0 Beryllium, hexakis[.mu.-(3-butenoato-O:O')]-.mu.4-oxotetra-;; OXOHEXAKIS(VINYLACET
ATO)TETRABERYLLIUM(II); W/NBS   64628

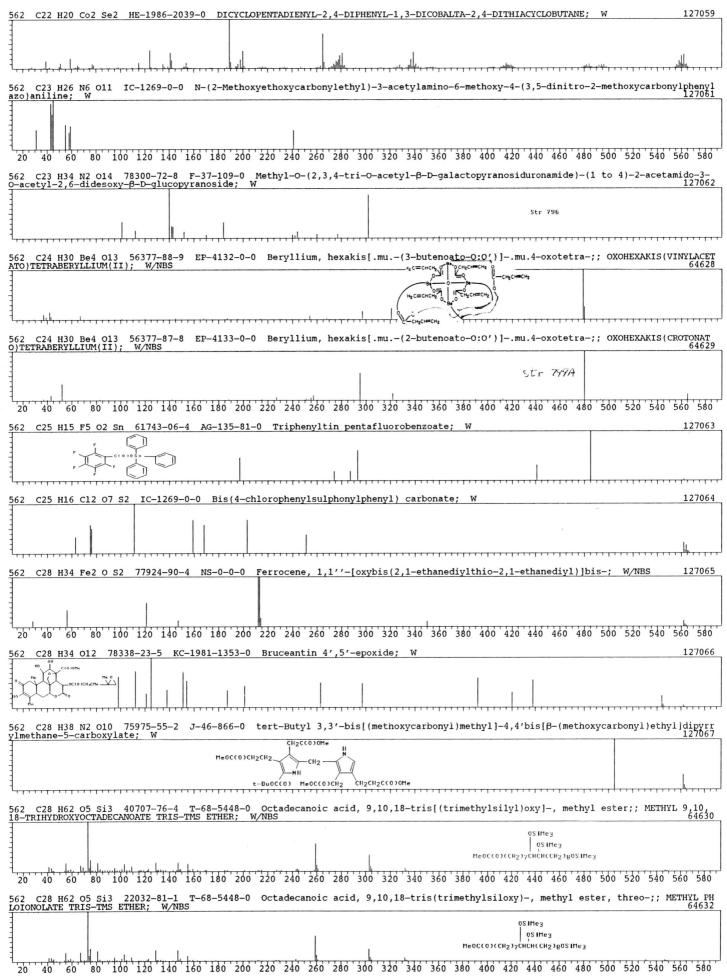

562 C24 H30 Be4 O13 56377-87-8 EP-4133-0-0 Beryllium, hexakis[.mu.-(2-butenoato-O:O')]-.mu.4-oxotetra-;; OXOHEXAKIS(CROTONAT
O)TETRABERYLLIUM(II); W/NBS   64629

Str 777A

20 40 60 80 100 120 140 160 180 200 220 240 260 280 300 320 340 360 380 400 420 440 460 480 500 520 540 560 580

562 C25 H15 F5 O2 Sn 61743-06-4 AG-135-81-0 Triphenyltin pentafluorobenzoate; W   127063

562 C25 H16 Cl2 O7 S2 IC-1269-0-0 Bis(4-chlorophenylsulphonylphenyl) carbonate; W   127064

562 C28 H34 Fe2 O S2 77924-90-4 NS-0-0-0 Ferrocene, 1,1''-[oxybis(2,1-ethanediylthio-2,1-ethanediyl)]bis-; W/NBS 127065

20 40 60 80 100 120 140 160 180 200 220 240 260 280 300 320 340 360 380 400 420 440 460 480 500 520 540 560 580

562 C28 H34 O12 78338-23-5 KC-1981-1353-0 Bruceantin 4',5'-epoxide; W   127066

562 C28 H38 N2 O10 75975-55-2 J-46-866-0 tert-Butyl 3,3'-bis[(methoxycarbonyl)methyl]-4,4'bis[β-(methoxycarbonyl)ethyl]dipyrr
ylmethane-5-carboxylate; W   127067

562 C28 H62 O5 Si3 40707-76-4 T-68-5448-0 Octadecanoic acid, 9,10,18-tris[(trimethylsilyl)oxy]-, methyl ester;; METHYL 9,10,
18-TRIHYDROXYOCTADECANOATE TRIS-TMS ETHER; W/NBS   64630

20 40 60 80 100 120 140 160 180 200 220 240 260 280 300 320 340 360 380 400 420 440 460 480 500 520 540 560 580

562 C28 H62 O5 Si3 22032-81-1 T-68-5448-0 Octadecanoic acid, 9,10,18-tris(trimethylsiloxy)-, methyl ester, threo-;; METHYL PH
LOIONOLATE TRIS-TMS ETHER; W/NBS   64632

20 40 60 80 100 120 140 160 180 200 220 240 260 280 300 320 340 360 380 400 420 440 460 480 500 520 540 560 580

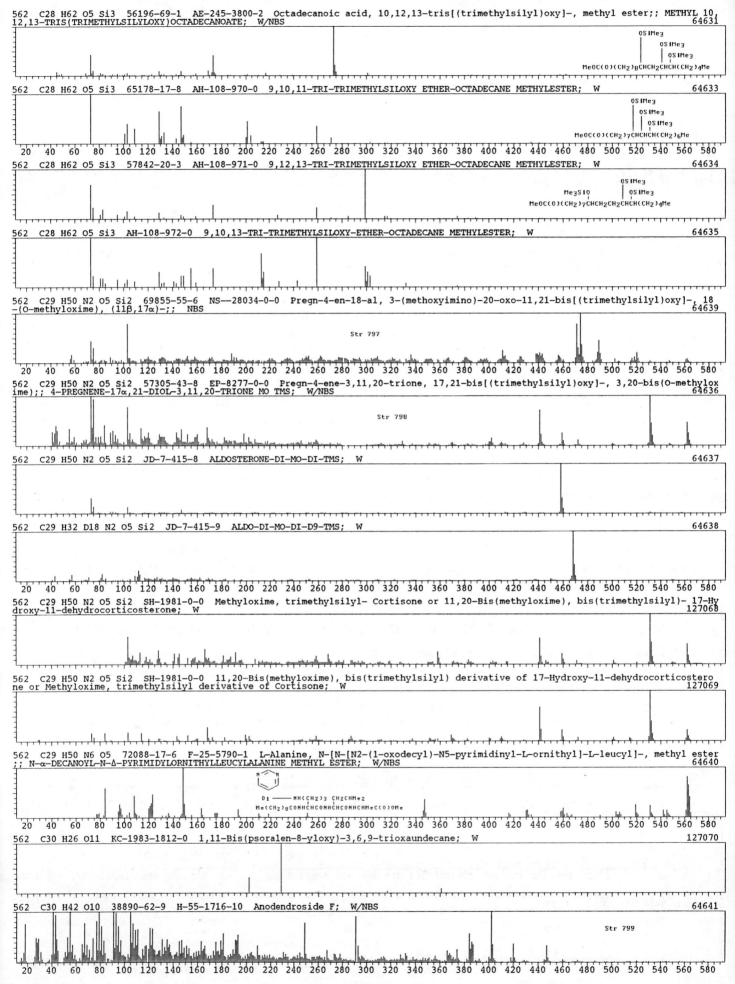

562 C28 H62 O5 Si3 56196-69-1 AE-245-3800-2 Octadecanoic acid, 10,12,13-tris[(trimethylsilyl)oxy]-, methyl ester;; METHYL 10,
12,13-TRIS(TRIMETHYLSILYLOXY)OCTADECANOATE; W/NBS   64631

562 C28 H62 O5 Si3 65178-17-8 AH-108-970-0 9,10,11-TRI-TRIMETHYLSILOXY ETHER-OCTADECANE METHYLESTER; W   64633

562 C28 H62 O5 Si3 57842-20-3 AH-108-971-0 9,12,13-TRI-TRIMETHYLSILOXY ETHER-OCTADECANE METHYLESTER; W   64634

562 C28 H62 O5 Si3 AH-108-972-0 9,10,13-TRI-TRIMETHYLSILOXY-ETHER-OCTADECANE METHYLESTER; W   64635

562 C29 H50 N2 O5 Si2 69855-55-6 NS--28034-0-0 Pregn-4-en-18-al, 3-(methoxyimino)-20-oxo-11,21-bis[(trimethylsilyl)oxy]-, 18
-(O-methyloxime), (11β,17α)-;; NBS   64639

Str 797

562 C29 H50 N2 O5 Si2 57305-43-8 EP-8277-0-0 Pregn-4-ene-3,11,20-trione, 17,21-bis[(trimethylsilyl)oxy]-, 3,20-bis(O-methylox
ime);; 4-PREGNENE-17α,21-DIOL-3,11,20-TRIONE MO TMS; W/NBS   64636

Str 798

562 C29 H50 N2 O5 Si2 JD-7-415-8 ALDOSTERONE-DI-MO-DI-TMS; W   64637

562 C29 H32 D18 N2 O5 Si2 JD-7-415-9 ALDO-DI-MO-DI-D9-TMS; W   64638

562 C29 H50 N2 O5 Si2 SH-1981-0-0 Methyloxime, trimethylsilyl- Cortisone or 11,20-Bis(methyloxime), bis(trimethylsilyl)- 17-Hy
droxy-11-dehydrocorticosterone; W   127068

562 C29 H50 N2 O5 Si2 SH-1981-0-0 11,20-Bis(methyloxime), bis(trimethylsilyl) derivative of 17-Hydroxy-11-dehydrocorticostero
ne or Methyloxime, trimethylsilyl derivative of Cortisone; W   127069

562 C29 H50 N6 O5 72088-17-6 F-25-5790-1 L-Alanine, N-[N-[N2-(1-oxodecyl)-N5-pyrimidinyl-L-ornithyl]-L-leucyl]-, methyl ester
;; N-α-DECANOYL-N-Δ-PYRIMIDYLORNITHYLLEUCYLALANINE METHYL ESTER; W/NBS   64640

562 C30 H26 O11 KC-1983-1812-0 1,11-Bis(psoralen-8-yloxy)-3,6,9-trioxaundecane; W   127070

562 C30 H42 O10 38890-62-9 H-55-1716-10 Anodendroside F; W/NBS   64641

Str 799

562 C30 H50 O6 Si2 69833-78-9 NS--28031-0-0 Pregn-4-ene-3,20-dione, 17-(acetyloxy)-6-methyl-6,21-bis[(trimethylsilyl)oxy]-, (6β)-;; NBS
64642

Str 800

562 C30 H54 N2 O4 Si2 74299-03-9 NS--28030-0-0 Pregn-4-ene-3,20-dione, 6-methyl-17,21-bis[(trimethylsilyl)oxy]-, bis(O-methyl oxime), (6α)-;; NBS
64643

Str 801

562 C30 H54 O4 Si3 B-31-1805-0 3',11'-BIS(TRIMETHYLSILYLOXY)-1-0-TRIMETHYLSILYL-Δ9-6A,10A-TRANS-TETRAHYDROCANNABINOL; W 64644

562 C31 H30 O10 64379-42-6 KC-1977-1251-0 CHROMANOARTOBILOCHROMENE B TRIACETATE; W
64645

562 C31 H30 O10 64379-56-2 KC-1977-1250-0 ACETYLATION CHROMANOARTOBILOCHROMENE A; W
64646

Str 802

562 C31 H34 N2 O8 55870-82-1 F-30-2764-1 Aspidospermidine-3-carboxylic acid, 2,3-didehydro-16-methoxy-20-[[(7-methoxy-1,3-benzodioxol-5-yl)carbonyl]oxy]-, methyl ester, (5α,12β,19α,20R)-;; Echitoserpine; W
64647

562 C31 H46 O9 69889-93-6 O-13-406-2 3α,7α,12β-TRIACETOXY-11-OXOPREGNANE-20-PROPANOIC ACID METHYL ESTER; W
64648

562 C31 H46 O9 69889-92-5 O-13-406-2 3α,7α,11α-TRIACETOXY-12-OXOPREGNANE-20-PROPANOIC ACID METHYL ESTER; W
64649

562 C31 H43 D3 O9 70341-61-6 O-13-406-2 3α,7α-DIACETOXY-12β-TRIDEUTEROACETOXY-11-OXOPREGNANE-20-PROPANOIC ACID METHYL ESTER; W
64650

562 C31 H50 N2 O5 S H-62-1597-1 2'-Cyclohexyl-4'ε-mesyloxymethyl-3β-methoxy-3',4',16β,17β-tetrahydro-2'H-5α-androstano[16,17-e][1',2']oxazin-3'ε-carbonitrile; W
127071

562 C31 H50 N2 O5 S H-62-1597-0 2'-Cyclohexyl-4'ε-mesyloxymethyl-3β-methoxy-3',4',16β,17β-tetrahydro-2'H-5α-androstano[17,16-e][1',2']oxazin-3'ε-carbonitrile; W
127072

562 C32 H42 N4 O5 69744-53-2 H-61-2661-0 N,O(34)-Dimethyl-N'-acetylaphelandrine; W
127073

Str 803

562 C32 H50 O6 S 78480-12-3 J-46-4459-0 1,1'-Sulfonylbis[4-(1,1-dimethylethyl)-2-(methoxymethyl)-6-(1-methoxy-1-methylethyl)benzene]; W
127074

Str 804

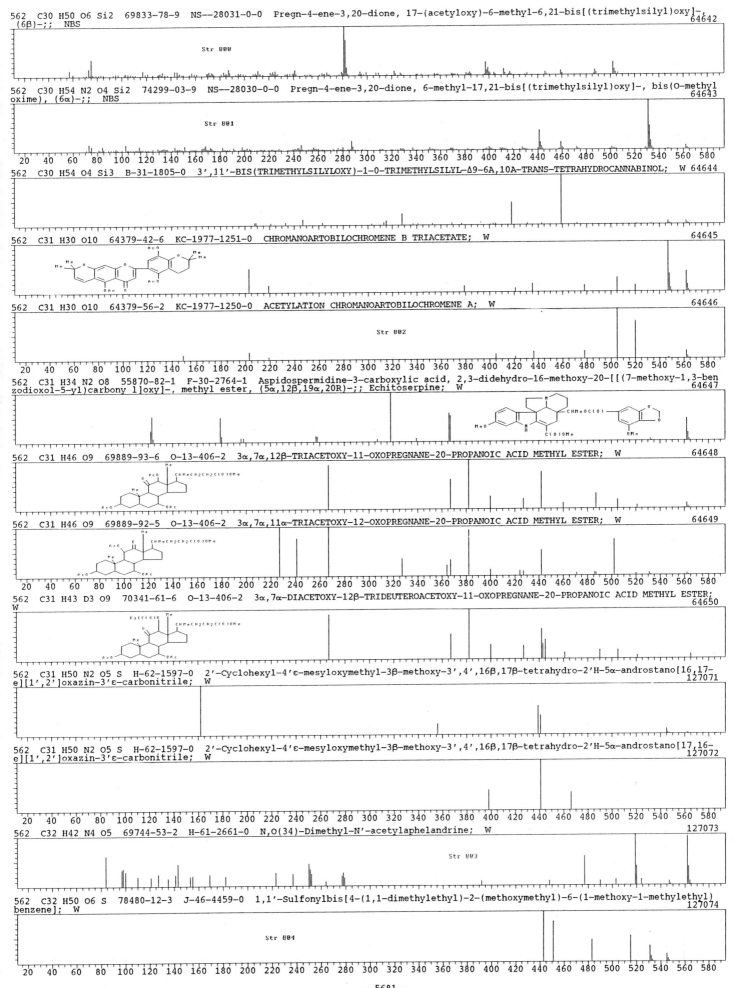

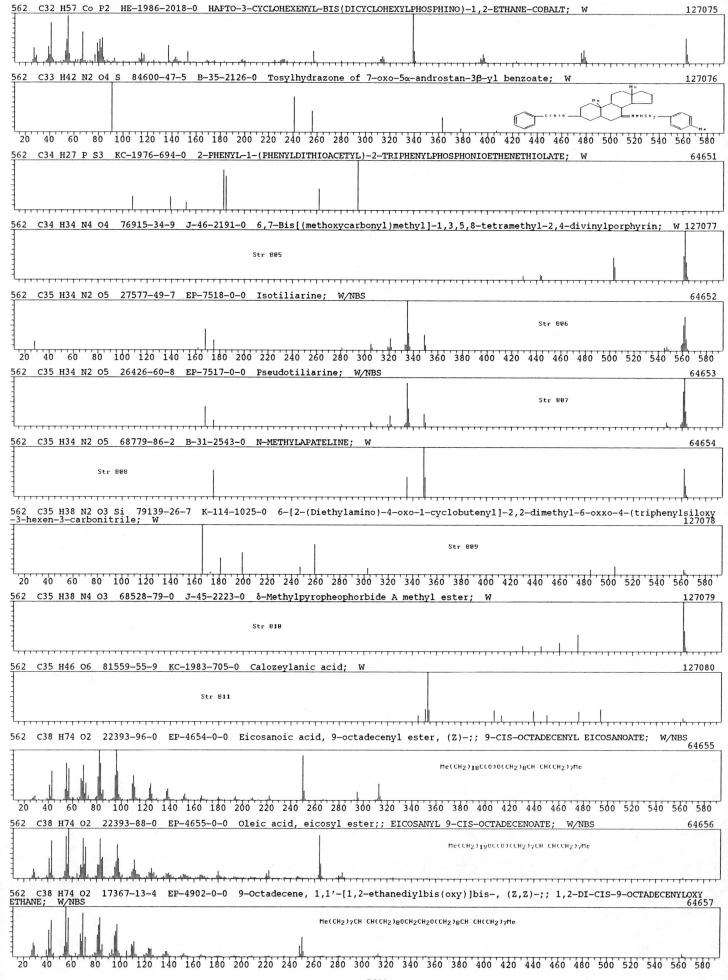

562　C32 H57 Co P2　HE-1986-2018-0　HAPTO-3-CYCLOHEXENYL-BIS(DICYCLOHEXYLPHOSPHINO)-1,2-ETHANE-COBALT;　W　127075

562　C33 H42 N2 O4 S　84600-47-5　B-35-2126-0　Tosylhydrazone of 7-oxo-5α-androstan-3β-yl benzoate;　W　127076

562　C34 H27 P S3　KC-1976-694-0　2-PHENYL-1-(PHENYLDITHIOACETYL)-2-TRIPHENYLPHOSPHONIOETHENETHIOLATE;　W　64651

562　C34 H34 N4 O4　76915-34-9　J-46-2191-0　6,7-Bis[(methoxycarbonyl)methyl]-1,3,5,8-tetramethyl-2,4-divinylporphyrin;　W 127077

Str 805

562　C35 H34 N2 O5　27577-49-7　EP-7518-0-0　Isotiliarine;　W/NBS　64652

Str 806

562　C35 H34 N2 O5　26426-60-8　EP-7517-0-0　Pseudotiliarine;　W/NBS　64653

Str 807

562　C35 H34 N2 O5　68779-86-2　B-31-2543-0　N-METHYLAPATELINE;　W　64654

Str 808

562　C35 H38 N2 O3 Si　79139-26-7　K-114-1025-0　6-[2-(Diethylamino)-4-oxo-1-cyclobutenyl]-2,2-dimethyl-6-oxxo-4-(triphenylsiloxy)-3-hexen-3-carbonitrile;　W　127078

Str 809

562　C35 H38 N4 O3　68528-79-0　J-45-2223-0　δ-Methylpyropheophorbide A methyl ester;　W　127079

Str 810

562　C35 H46 O6　81559-55-9　KC-1983-705-0　Calozeylanic acid;　W　127080

Str 811

562　C38 H74 O2　22393-96-0　EP-4654-0-0　Eicosanoic acid, 9-octadecenyl ester, (Z)-;; 9-CIS-OCTADECENYL EICOSANOATE;　W/NBS　64655

Me(CH2)18C(O)O(CH2)8CH:CH(CH2)7Me

562　C38 H74 O2　22393-88-0　EP-4655-0-0　Oleic acid, eicosyl ester;; EICOSANYL 9-CIS-OCTADECENOATE;　W/NBS　64656

Me(CH2)19O(CO)(CH2)7CH:CH(CH2)7Me

562　C38 H74 O2　17367-13-4　EP-4902-0-0　9-Octadecene, 1,1'-[1,2-ethanediylbis(oxy)]bis-, (Z,Z)-;; 1,2-DI-CIS-9-OCTADECENYLOXY
ETHANE;　W/NBS　64657

Me(CH2)7CH:CH(CH2)8OCH2CH2O(CH2)8CH:CH(CH2)7Me

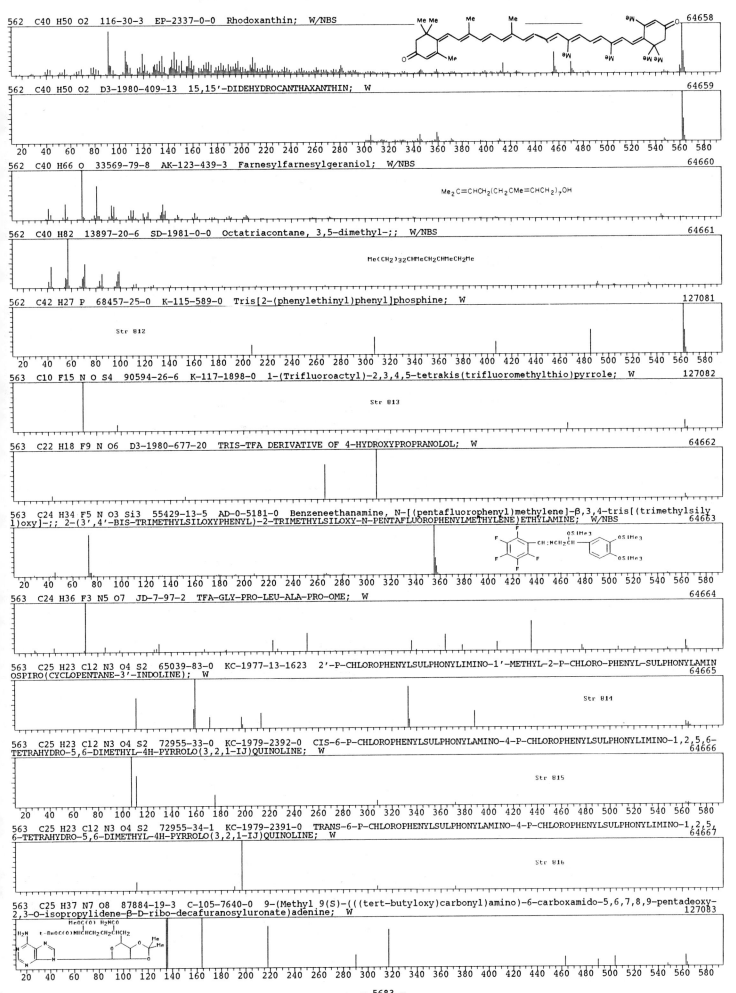

562   C40 H50 O2   116-30-3   EP-2337-0-0   Rhodoxanthin;   W/NBS   64658

562   C40 H50 O2   D3-1980-409-13   15,15'-DIDEHYDROCANTHAXANTHIN;   W   64659

562   C40 H66 O   33569-79-8   AK-123-439-3   Farnesylfarnesylgeraniol;   W/NBS   64660

Me2C=CHCH2(CH2CMe=CHCH2)7OH

562   C40 H82   13897-20-6   SD-1981-0-0   Octatriacontane, 3,5-dimethyl-;;   W/NBS   64661

Me(CH2)32CHMeCH2CHMeCH2Me

562   C42 H27 P   68457-25-0   K-115-589-0   Tris[2-(phenylethinyl)phenyl]phosphine;   W   127081

Str 812

563   C10 F15 N O S4   90594-26-6   K-117-1898-0   1-(Trifluoroactyl)-2,3,4,5-tetrakis(trifluoromethylthio)pyrrole;   W   127082

Str 813

563   C22 H18 F9 N O6   D3-1980-677-20   TRIS-TFA DERIVATIVE OF 4-HYDROXYPROPRANOLOL;   W   64662

563   C24 H34 F5 N O3 Si3   55429-13-5   AD-0-5181-0   Benzeneethanamine, N-[(pentafluorophenyl)methylene]-β,3,4-tris[(trimethylsilyl)oxy]-;;   2-(3',4'-BIS-TRIMETHYLSILOXYPHENYL)-2-TRIMETHYLSILOXY-N-PENTAFLUOROPHENYLMETHYLENE)ETHYLAMINE;   W/NBS   64663

563   C24 H36 F3 N5 O7   JD-7-97-2   TFA-GLY-PRO-LEU-ALA-PRO-OME;   W   64664

563   C25 H23 Cl2 N3 O4 S2   65039-83-0   KC-1977-13-1623   2'-P-CHLOROPHENYLSULPHONYLIMINO-1'-METHYL-2-P-CHLORO-PHENYL-SULPHONYLAMINOSPIRO(CYCLOPENTANE-3'-INDOLINE);   W   64665

Str 814

563   C25 H23 Cl2 N3 O4 S2   72955-33-0   KC-1979-2392-0   CIS-6-P-CHLOROPHENYLSULPHONYLAMINO-4-P-CHLOROPHENYLSULPHONYLIMINO-1,2,5,6-TETRAHYDRO-5,6-DIMETHYL-4H-PYRROLO(3,2,1-IJ)QUINOLINE;   W   64666

Str 815

563   C25 H23 Cl2 N3 O4 S2   72955-34-1   KC-1979-2391-0   TRANS-6-P-CHLOROPHENYLSULPHONYLAMINO-4-P-CHLOROPHENYLSULPHONYLIMINO-1,2,5,6-TETRAHYDRO-5,6-DIMETHYL-4H-PYRROLO(3,2,1-IJ)QUINOLINE;   W   64667

Str 816

563   C25 H37 N7 O8   87884-19-3   C-105-7640-0   9-(Methyl 9(S)-(((tert-butyloxy)carbonyl)amino)-6-carboxamido-5,6,7,8,9-pentadeoxy-2,3-O-isopropylidene-β-D-ribo-decafuranosyluronate)adenine;   W   127083

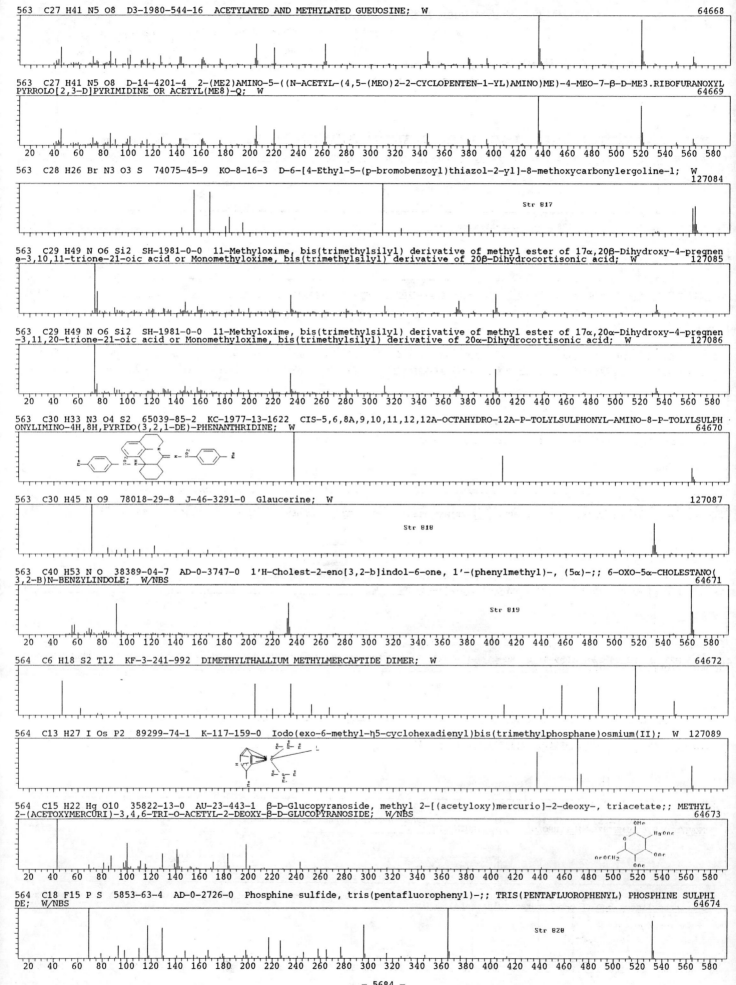

563 C27 H41 N5 O8 D3-1980-544-16 ACETYLATED AND METHYLATED GUEUOSINE; W 64668

563 C27 H41 N5 O8 D-14-4201-4 2-(ME2)AMINO-5-((N-ACETYL-(4,5-(MEO)2-2-CYCLOPENTEN-1-YL)AMINO)ME)-4-MEO-7-β-D-ME3.RIBOFURANOXYL PYRROLO[2,3-D]PYRIMIDINE OR ACETYL(ME8)-Q; W 64669

563 C28 H26 Br N3 O3 S 74075-45-9 KO-8-16-3 D-6-[4-Ethyl-5-(p-bromobenzoyl)thiazol-2-yl]-8-methoxycarbonylergoline-1; W 127084

Str 817

563 C29 H49 N O6 Si2 SH-1981-0-0 11-Methyloxime, bis(trimethylsilyl) derivative of methyl ester of 17α,20β-Dihydroxy-4-pregnene-3,10,11-trione-21-oic acid or Monomethyloxime, bis(trimethylsilyl) derivative of 20β-Dihydrocortisonic acid; W 127085

563 C29 H49 N O6 Si2 SH-1981-0-0 11-Methyloxime, bis(trimethylsilyl) derivative of methyl ester of 17α,20α-Dihydroxy-4-pregnen-3,11,20-trione-21-oic acid or Monomethyloxime, bis(trimethylsilyl) derivative of 20α-Dihydrocortisonic acid; W 127086

563 C30 H33 N3 O4 S2 65039-85-2 KC-1977-13-1622 CIS-5,6,8A,9,10,11,12,12A-OCTAHYDRO-12A-P-TOLYLSULPHONYL-AMINO-8-P-TOLYLSULPHONYLIMINO-4H,8H,PYRIDO(3,2,1-DE)-PHENANTHRIDINE; W 64670

563 C30 H45 N O9 78018-29-8 J-46-3291-0 Glaucerine; W 127087

Str 818

563 C40 H53 N O 38389-04-7 AD-0-3747-0 1'H-Cholest-2-eno[3,2-b]indol-6-one, 1'-(phenylmethyl)-, (5α)-;; 6-OXO-5α-CHOLESTANO(3,2-B)N-BENZYLINDOLE; W/NBS 64671

Str 819

564 C6 H18 S2 T12 KF-3-241-992 DIMETHYLTHALLIUM METHYLMERCAPTIDE DIMER; W 64672

564 C13 H27 I Os P2 89299-74-1 K-117-159-0 Iodo(exo-6-methyl-η5-cyclohexadienyl)bis(trimethylphosphane)osmium(II); W 127089

564 C15 H22 Hg O10 35822-13-0 AU-23-443-1 β-D-Glucopyranoside, methyl 2-[(acetyloxy)mercurio]-2-deoxy-, triacetate;; METHYL 2-(ACETOXYMERCURI)-3,4,6-TRI-O-ACETYL-2-DEOXY-β-D-GLUCOPYRANOSIDE; W/NBS 64673

564 C18 F15 P S 5853-63-4 AD-0-2726-0 Phosphine sulfide, tris(pentafluorophenyl)-;; TRIS(PENTAFLUOROPHENYL) PHOSPHINE SULPHIDE; W/NBS 64674

Str 820

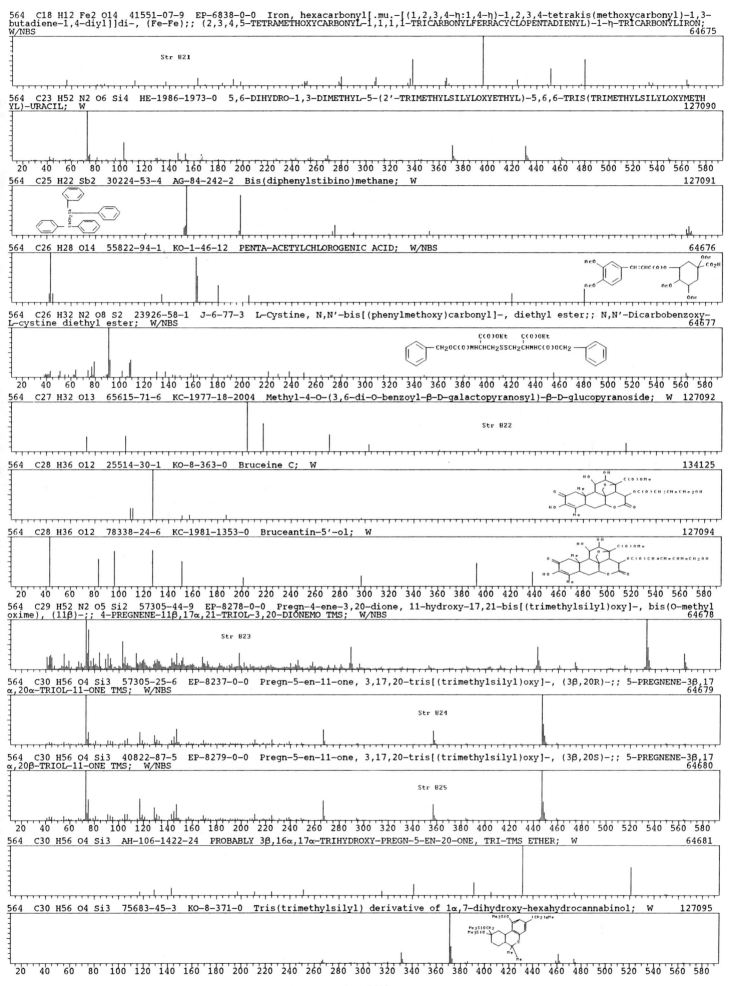

564   C18 H12 Fe2 O14   41551-07-9   EP-6838-0-0   Iron, hexacarbonyl[.mu.-[(1,2,3,4-η:1,4-η)-1,2,3,4-tetrakis(methoxycarbonyl)-1,3-butadiene-1,4-diyl]]di-, (Fe-Fe);; (2,3,4,5-TETRAMETHOXYCARBONYL-1,1,1,1-TRICARBONYLFERRACYCLOPENTADIENYL)-1-η-TRICARBONYLIRON; W/NBS
64675

Str 821

564   C23 H52 N2 O6 Si4   HE-1986-1973-0   5,6-DIHYDRO-1,3-DIMETHYL-5-(2'-TRIMETHYLSILYLOXYETHYL)-5,6,6-TRIS(TRIMETHYLSILYLOXYMETHYL)-URACIL; W
127090

564   C25 H22 Sb2   30224-53-4   AG-84-242-2   Bis(diphenylstibino)methane; W
127091

564   C26 H28 O14   55822-94-1   KO-1-46-12   PENTA-ACETYLCHLOROGENIC ACID; W/NBS
64676

564   C26 H32 N2 O8 S2   23926-58-1   J-6-77-3   L-Cystine, N,N'-bis[(phenylmethoxy)carbonyl]-, diethyl ester;; N,N'-Dicarbobenzoxy-L-cystine diethyl ester; W/NBS
64677

564   C27 H32 O13   65615-71-6   KC-1977-18-2004   Methyl-4-O-(3,6-di-O-benzoyl-β-D-galactopyranosyl)-β-D-glucopyranoside; W   127092

Str 822

564   C28 H36 O12   25514-30-1   KO-8-363-0   Bruceine C; W
134125

564   C28 H36 O12   78338-24-6   KC-1981-1353-0   Bruceantin-5'-ol; W
127094

564   C29 H52 N2 O5 Si2   57305-44-9   EP-8278-0-0   Pregn-4-ene-3,20-dione, 11-hydroxy-17,21-bis[(trimethylsilyl)oxy]-, bis(O-methyl oxime), (11β)-;; 4-PREGNENE-11β,17α,21-TRIOL-3,20-DIONEMO TMS; W/NBS
64678

Str 823

564   C30 H56 O4 Si3   57305-25-6   EP-8237-0-0   Pregn-5-en-11-one, 3,17,20-tris[(trimethylsilyl)oxy]-, (3β,20R)-;; 5-PREGNENE-3β,17α,20α-TRIOL-11-ONE TMS; W/NBS
64679

Str 824

564   C30 H56 O4 Si3   40822-87-5   EP-8279-0-0   Pregn-5-en-11-one, 3,17,20-tris[(trimethylsilyl)oxy]-, (3β,20S)-;; 5-PREGNENE-3β,17α,20β-TRIOL-11-ONE TMS; W/NBS
64680

Str 825

564   C30 H56 O4 Si3   AH-106-1422-24   PROBABLY 3β,16α,17α-TRIHYDROXY-PREGN-5-EN-20-ONE, TRI-TMS ETHER; W
64681

564   C30 H56 O4 Si3   75683-45-3   KO-8-371-0   Tris(trimethylsilyl) derivative of 1α,7-dihydroxy-hexahydrocannabinol; W   127095

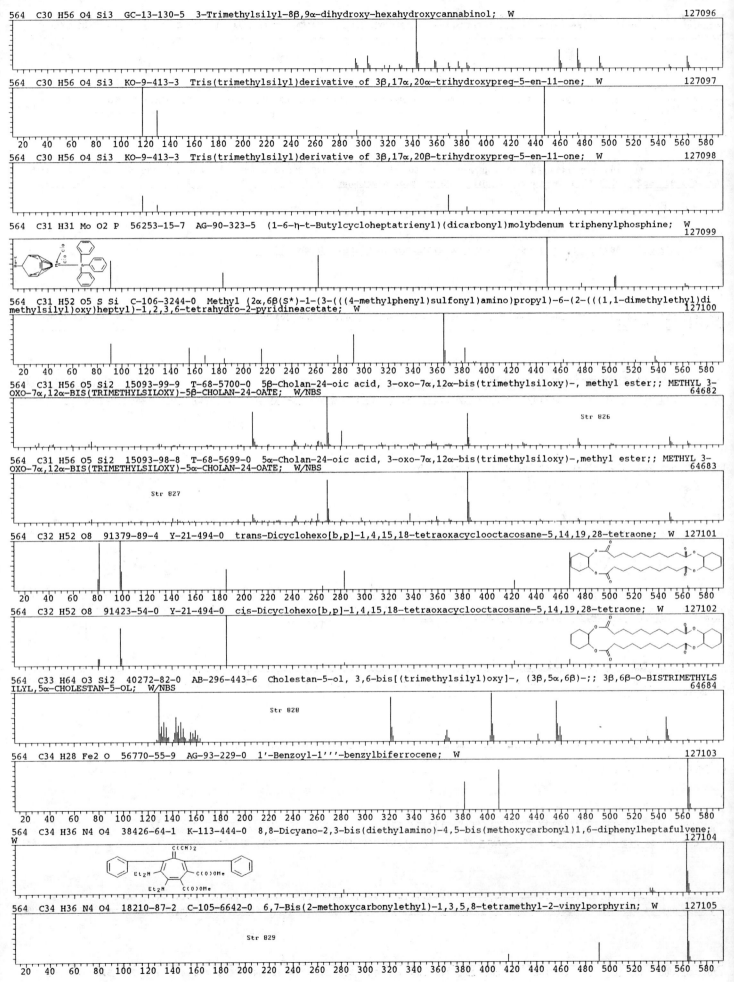

564   C30 H56 O4 Si3   GC-13-130-5   3-Trimethylsilyl-8β,9α-dihydroxy-hexahydroxycannabinol; W   127096

564   C30 H56 O4 Si3   KO-9-413-3   Tris(trimethylsilyl)derivative of 3β,17α,20α-trihydroxypreg-5-en-11-one; W   127097

564   C30 H56 O4 Si3   KO-9-413-3   Tris(trimethylsilyl)derivative of 3β,17α,20β-trihydroxypreg-5-en-11-one; W   127098

564   C31 H31 Mo O2 P   56253-15-7   AG-90-323-5   (1-6-η-t-Butylcycloheptatrienyl)(dicarbonyl)molybdenum triphenylphosphine; W
127099

564   C31 H52 O5 S Si   C-106-3244-0   Methyl (2α,6β(S*)-1-(3-(((4-methylphenyl)sulfonyl)amino)propyl)-6-(2-(((1,1-dimethylethyl)di
methylsilyl)oxy)heptyl)-1,2,3,6-tetrahydro-2-pyridineacetate; W   127100

564   C31 H56 O5 Si2   15093-99-9   T-68-5700-0   5β-Cholan-24-oic acid, 3-oxo-7α,12α-bis(trimethylsiloxy)-, methyl ester;; METHYL 3-
OXO-7α,12α-BIS(TRIMETHYLSILOXY)-5β-CHOLAN-24-OATE;   W/NBS   64682
Str 826

564   C31 H56 O5 Si2   15093-98-8   T-68-5699-0   5α-Cholan-24-oic acid, 3-oxo-7α,12α-bis(trimethylsiloxy)-,methyl ester;; METHYL 3-
OXO-7α,12α-BIS(TRIMETHYLSILOXY)-5α-CHOLAN-24-OATE;   W/NBS   64683
Str 827

564   C32 H52 O8   91379-89-4   Y-21-494-0   trans-Dicyclohexo[b,p]-1,4,15,18-tetraoxacyclooctacosane-5,14,19,28-tetraone; W   127101

564   C32 H52 O8   91423-54-0   Y-21-494-0   cis-Dicyclohexo[b,p]-1,4,15,18-tetraoxacyclooctacosane-5,14,19,28-tetraone; W   127102

564   C33 H64 O3 Si2   40272-82-0   AB-296-443-6   Cholestan-5-ol, 3,6-bis[(trimethylsilyl)oxy]-, (3β,5α,6β)-;; 3β,6β-O-BISTRIMETHYLS
ILYL,5α-CHOLESTAN-5-OL;   W/NBS   64684
Str 828

564   C34 H28 Fe2 O   56770-55-9   AG-93-229-0   1'-Benzoyl-1'''-benzylbiferrocene; W   127103

564   C34 H36 N4 O4   38426-64-1   K-113-444-0   8,8-Dicyano-2,3-bis(diethylamino)-4,5-bis(methoxycarbonyl)1,6-diphenylheptafulvene;
W   127104

564   C34 H36 N4 O4   18210-87-2   C-105-6642-0   6,7-Bis(2-methoxycarbonylethyl)-1,3,5,8-tetramethyl-2-vinylporphyrin; W   127105
Str 829

564   C34 H36 N4 O4   17467-73-1   C-105-6643-0   6,7-Bis(2-methoxycarbonylethyl)-1,3,5,8-tetramethyl-4-vinylporphyrin;   W   127106

Str 830

564   C35 H32 Fe2   77792-99-5   AH-112-102-0   1,3-Diferrocenyl-1,3-diphenyl-propane;   W   127107

564   C35 H32 O7   72327-93-6   J-46-2266-0   1-(2,4,6-Tribenzoxyphenyl)-1,3,5,7-octanetetraone;   W   127108

Str 831

564   C35 H40 N4 O3   68464-68-6   C-107-4952-0   Methyl mesobacteriopheophorbide C(Et,Me);   W   127109

Str 832

564   C35 H40 N4 O3   60820-65-7   AC-1980-490-1   Bonafousine;   W   127110

Str 833

564   C35 H40 N4 O3   76202-19-2   AC-1980-494-3   Isobonafousine;   W   127111

Str 834

564   C36 H24 N2 O5   78588-60-0   O-16-145-1   10,10-BIS(P-CITRACONIMIDOPHENYL)ANTHRONE;   W   64685

Str 835

564   C36 H36 O6   90340-80-0   B-37-575-0   Diethyl 2,2'-[t-2,t-4-di(p-methoxyphenyl)cyclobutane-t-1,c-3-diyl]dibenzoate;   W   127112

Str 836

564   C38 H44 O4   22453-06-1   S-23-3297-3   Violerythrin;   W/NBS   64686

564   C38 H52 N4   56630-82-1   EP-6943-0-0   N,N,-DIMETHYLOCTAETHYLPORPHYRIN;   W/NBS   64687

Str 837

564   C38 H76 O2   22432-79-7   EP-5070-0-0   Eicosanoic acid, octadecyl ester;;   OCTADECANYL EICOSANOATE;   W/NBS   64688

Me(CH2)17OC(O)(CH2)18Me

564   C38 H76 O2   22413-02-1   EP-4650-0-0   Octadecanoic acid, eicosyl ester;;   EICOSANYL OCTADECANOATE;   W/NBS   64689

Me(CH2)16C(O)O(CH2)19Me

564   C38 H76 O2   56599-41-8   EP-5233-0-0   9-Octadecene, 1-[2-(octadecyloxy)ethoxy]-;;   2-OCTADECYLOXY-1-CIS-9-OCTADECENYLOXY ETHANE;   W/NBS   64690

Me(CH2)17OCH2CH2O(CH2)8CH:CH(CH2)7Me

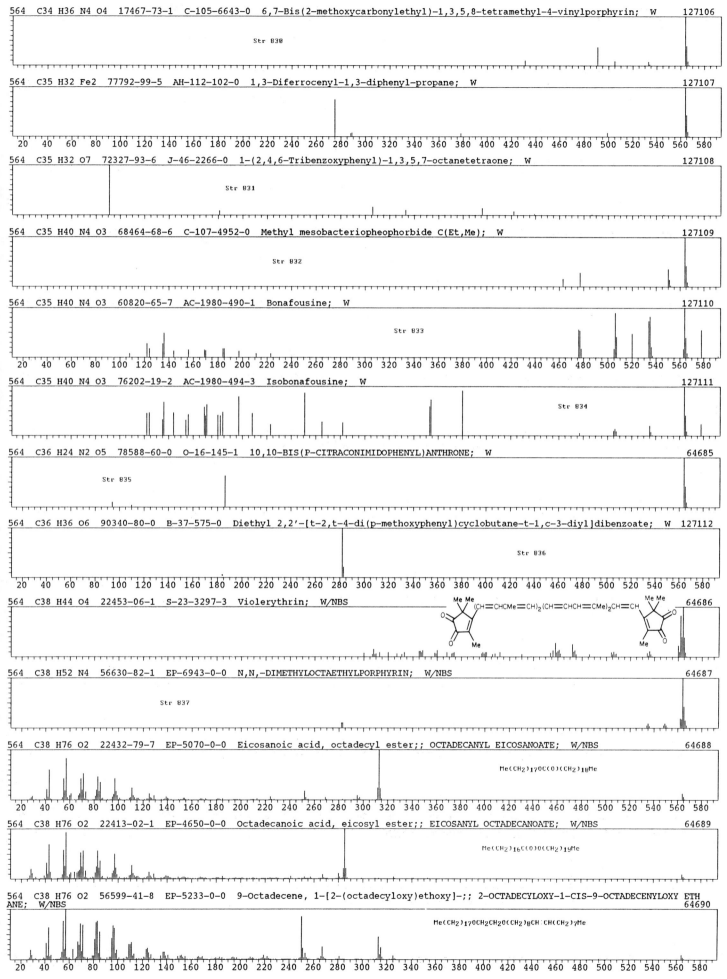

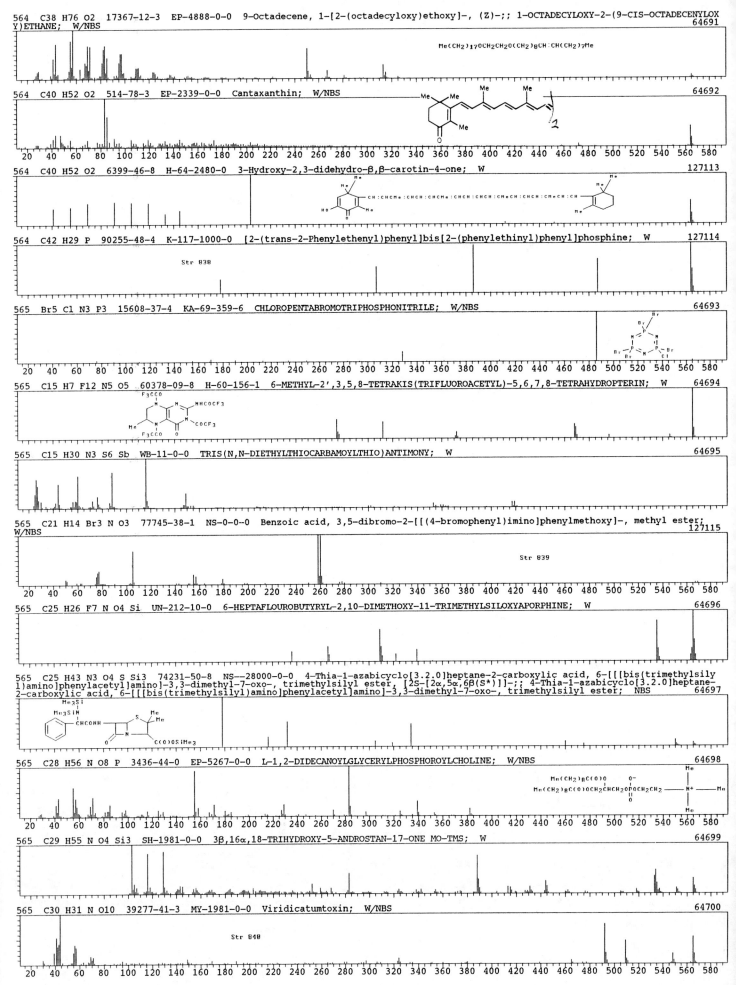

564   C38 H76 O2   17367-12-3   EP-4888-0-0   9-Octadecene, 1-[2-(octadecyloxy)ethoxy]-, (Z)-;; 1-OCTADECYLOXY-2-(9-CIS-OCTADECENYLOXY)ETHANE; W/NBS
64691

Me(CH2)17OCH2CH2O(CH2)8CH:CH(CH2)7Me

564   C40 H52 O2   514-78-3   EP-2339-0-0   Cantaxanthin; W/NBS
64692

564   C40 H52 O2   6399-46-8   H-64-2480-0   3-Hydroxy-2,3-didehydro-β,β-carotin-4-one; W
127113

564   C42 H29 P   90255-48-4   K-117-1000-0   [2-(trans-2-Phenylethenyl)phenyl]bis[2-(phenylethinyl)phenyl]phosphine; W
127114

Str 838

565   Br5 Cl N3 P3   15608-37-4   KA-69-359-6   CHLOROPENTABROMOTRIPHOSPHONITRILE; W/NBS
64693

565   C15 H7 F12 N5 O5   60378-09-8   H-60-156-1   6-METHYL-2',3,5,8-TETRAKIS(TRIFLUOROACETYL)-5,6,7,8-TETRAHYDROPTERIN; W
64694

565   C15 H30 N3 S6 Sb   WB-11-0-0   TRIS(N,N-DIETHYLTHIOCARBAMOYLTHIO)ANTIMONY; W
64695

565   C21 H14 Br3 N O3   77745-38-1   NS-0-0-0   Benzoic acid, 3,5-dibromo-2-[[(4-bromophenyl)imino]phenylmethoxy]-, methyl ester; W/NBS
127115

Str 839

565   C25 H26 F7 N O4 Si   UN-212-10-0   6-HEPTAFLOUROBUTYRYL-2,10-DIMETHOXY-11-TRIMETHYLSILOXYAPORPHINE; W
64696

565   C25 H43 N3 O4 S Si3   74231-50-8   NS--28000-0-0   4-Thia-1-azabicyclo[3.2.0]heptane-2-carboxylic acid, 6-[[[bis(trimethylsilyl)amino]phenylacetyl]amino]-3,3-dimethyl-7-oxo-, trimethylsilyl ester, [2S-[2α,5α,6β(S*)]]-;; 4-Thia-1-azabicyclo[3.2.0]heptane-2-carboxylic acid, 6-[[[bis(trimethylsilyl)amino]phenylacetyl]amino]-3,3-dimethyl-7-oxo-, trimethylsilyl ester; NBS
64697

565   C28 H56 N O8 P   3436-44-0   EP-5267-0-0   L-1,2-DIDECANOYLGLYCERYLPHOSPHOROYLCHOLINE; W/NBS
64698

565   C29 H55 N O4 Si3   SH-1981-0-0   3β,16α,18-TRIHYDROXY-5-ANDROSTAN-17-ONE MO-TMS; W
64699

565   C30 H31 N O10   39277-41-3   MY-1981-0-0   Viridicatumtoxin; W/NBS
64700

Str 840

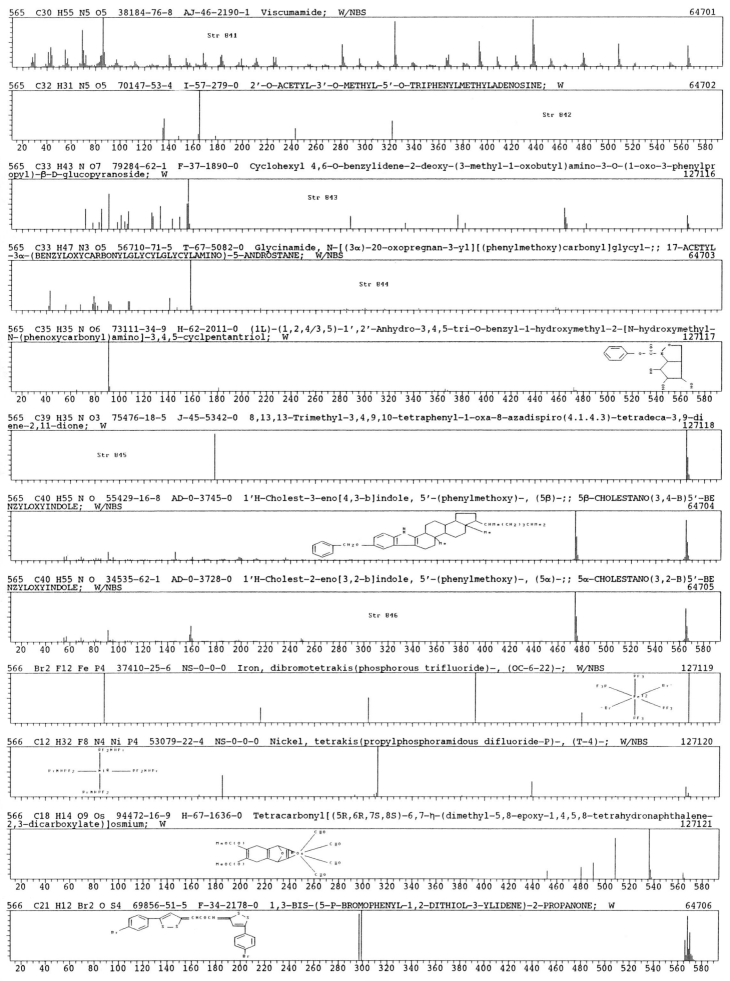

565   C30 H55 N5 O5   38184-76-8   AJ-46-2190-1   Viscumamide;   W/NBS   64701

Str 841

565   C32 H31 N5 O5   70147-53-4   I-57-279-0   2'-O-ACETYL-3'-O-METHYL-5'-O-TRIPHENYLMETHYLADENOSINE;   W   64702

Str 842

565   C33 H43 N O7   79284-62-1   F-37-1890-0   Cyclohexyl 4,6-O-benzylidene-2-deoxy-(3-methyl-1-oxobutyl)amino-3-O-(1-oxo-3-phenylpropyl)-β-D-glucopyranoside;   W   127116

Str 843

565   C33 H47 N3 O5   56710-71-5   T-67-5082-0   Glycinamide, N-[(3α)-20-oxopregnan-3-yl][(phenylmethoxy)carbonyl]glycyl-;;   17-ACETYL-3α-(BENZYLOXYCARBONYLGLYCYLGLYCYLAMINO)-5-ANDROSTANE;   W/NBS   64703

Str 844

565   C35 H35 N O6   73111-34-9   H-62-2011-0   (1L)-(1,2,4/3,5)-1',2'-Anhydro-3,4,5-tri-O-benzyl-1-hydroxymethyl-2-[N-hydroxymethyl-N-(phenoxycarbonyl)amino]-3,4,5-cyclpentantriol;   W   127117

565   C39 H35 N O3   75476-18-5   J-45-5342-0   8,13,13-Trimethyl-3,4,9,10-tetraphenyl-1-oxa-8-azadispiro(4.1.4.3)-tetradeca-3,9-diene-2,11-dione;   W   127118

Str 845

565   C40 H55 N O   55429-16-8   AD-0-3745-0   1'H-Cholest-3-eno[4,3-b]indole, 5'-(phenylmethoxy)-, (5β)-;;   5β-CHOLESTANO(3,4-B)5'-BENZYLOXYINDOLE;   W/NBS   64704

565   C40 H55 N O   34535-62-1   AD-0-3728-0   1'H-Cholest-2-eno[3,2-b]indole, 5'-(phenylmethoxy)-, (5α)-;;   5α-CHOLESTANO(3,2-B)5'-BENZYLOXYINDOLE;   W/NBS   64705

Str 846

566   Br2 F12 Fe P4   37410-25-6   NS-0-0-0   Iron, dibromotetrakis(phosphorous trifluoride)-, (OC-6-22)-;   W/NBS   127119

566   C12 H32 F8 N4 Ni P4   53079-22-4   NS-0-0-0   Nickel, tetrakis(propylphosphoramidous difluoride-P)-, (T-4)-;   W/NBS   127120

566   C18 H14 O9 Os   94472-16-9   H-67-1636-0   Tetracarbonyl[(5R,6R,7S,8S)-6,7-η-(dimethyl-5,8-epoxy-1,4,5,8-tetrahydronaphthalene-2,3-dicarboxylate)]osmium;   W   127121

566   C21 H12 Br2 O S4   69856-51-5   F-34-2178-0   1,3-BIS-(5-P-BROMOPHENYL-1,2-DITHIOL-3-YLIDENE)-2-PROPANONE;   W   64706

566 C22 H54 N2 O5 Si5 56247-64-4 AB-286-109-1 Hexanedioic acid, 5-[bis(trimethylsilyl)amino]-2-[[bis(trimethylsilyl)amino]
methyl]-2-[(trimethylsilyl)oxy]-;; PENTAKISTRIMETHYLSILYL TABTOXININE; W/NBS   64707

Me3Si   OSiMe3   SiMe3
Me3SiNCH(CO2H)CH2CH2C(CO2H)CH2NSiMe3

566 C24 H20 O Sb2 7065-22-7 AG-87-173-2 Bis(diphenylstibine)oxide; W   127122

566 C26 H30 O8 S3 SB-32-586-0 3-(1-ACETOXY-1-(4-ACETOXY-3-METHOXY PHENYL))-PROPYL-TRANS-3-(4-ACETOXY-3-METHOXYPHENYL)-2-PROPEN
YLTRISULFIDE; W   64708

566 C29 H26 O12 74978-22-6 NS--27992-0-0 5,12-Naphthacenedione, 6,8,11-tris(acetyloxy)-8-[(acetyloxy)acetyl]-7,8,9,10-tetrahy
dro-1-methoxy-, (R)-;; NBS   64709

566 C29 H53 P2 Rh HE-1986-2042-0 HAPTO-3-ALLYL-BIS(DICYCLOHEXYLPHOSPHINO)-1,2-ETHANE-RHODIUM; W   127123

566 C30 H30 O11 61035-70-9 B-29-1150-0 Methyl 2-O-acetyl-2',2''-di-O-methylgyrophorate; W   127124

566 C30 H34 N2 O9 77495-90-0 NS-0-0-0 Benzo[c]pyridazino[1,2-a]cinnoline-6,7,8,9-tetracarboxylic acid, 6(or 7)-ethoxy-6,7-di
hydro-, tetraethyl ester; W/NBS   127125

Str 847

566 C30 H58 O4 Si3 57397-20-3 EP-8441-0-0 Pregnan-20-one, 3,17,21-tris[(trimethylsilyl)oxy]-, (3α,5β)-;; 5β-PREGNANE-3α,17α,
21-TRIOL-20-ONE TMS; W/NBS   64713

Str 848

566 C30 H58 O4 Si3 56248-33-0 E-45-1095-6 Pregn-20-en-11-ol, 3,20,21-tris[(trimethylsilyl)oxy]-, (3α,5β,11β)-;; 3α,20,21-TRIS
TRIMETHYLSILYLYLOXY-11β-HYDROXY-5β-PREGN-20-ENE; W/NBS   64715

Str 849

566 C30 H58 O4 Si3 56196-42-0 SW-0-198-0 Pregnan-20-one, 3,11,21-tris[(trimethylsilyl)oxy]-, (3α,5β,11β)-;; 3α,11β,21-TRIHYDR
OXY-5β-PREGNANE-20-ONE-TRITMS; W/NBS   64714

Str 850

566 C30 H58 O4 Si3 57397-18-9 NS--27987-0-0 Pregnan-20-one, 3,11,21-tris[(trimethylsilyl)oxy]-, (3α,5α,11β)-;; NBS   64716

Str 851

566 C30 H58 O4 Si3 57377-54-5 EP-8280-0-0 Pregnan-11-one, 3,20,21-tris[(trimethylsilyl)oxy]-, (3α,5β,20S)-;; 5β-PREGNANE-3α,
20α,21-TRIOL-11-ONE TMS; W/NBS   64712

Str 852

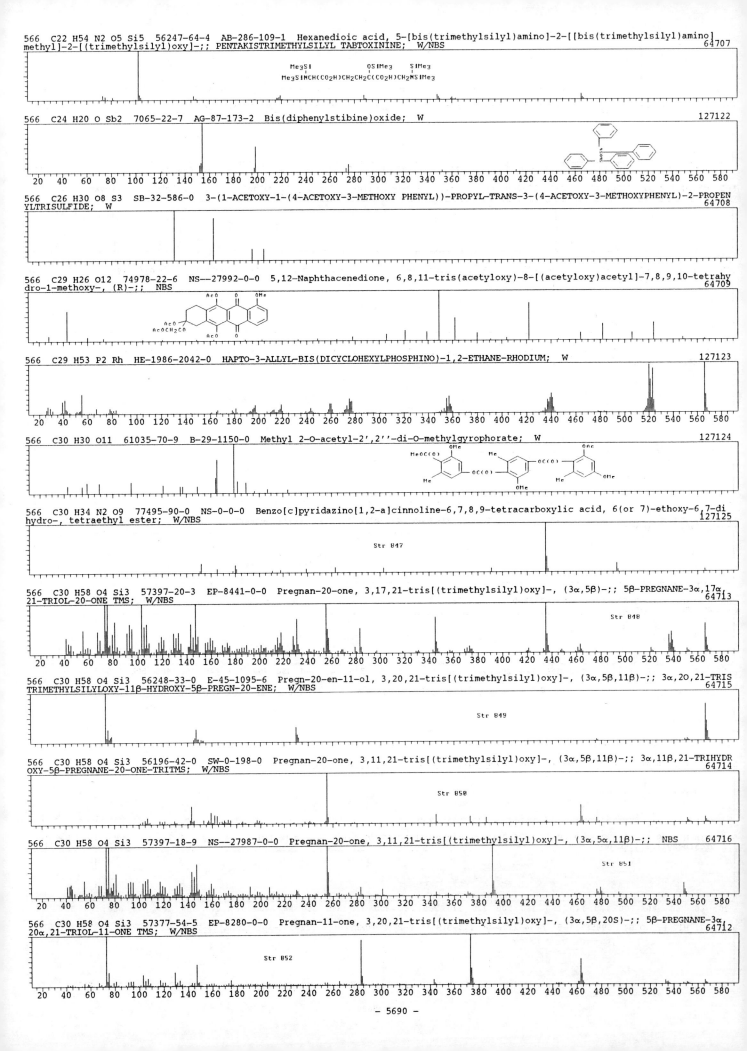

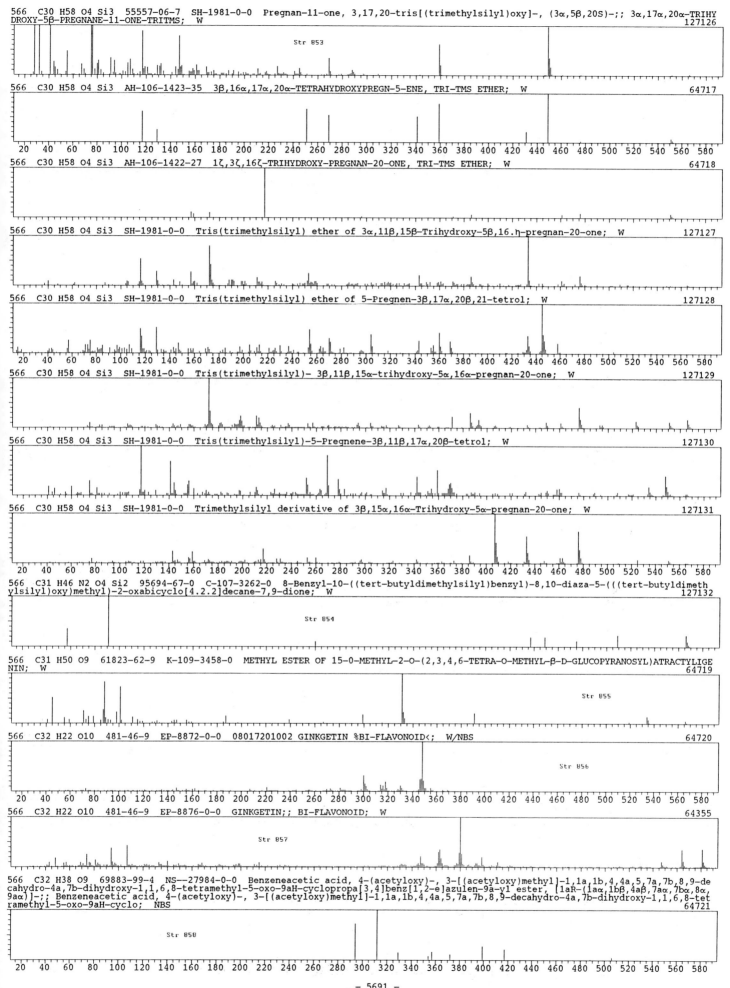

566 C30 H58 O4 Si3 55557-06-7 SH-1981-0-0 Pregnan-11-one, 3,17,20-tris[(trimethylsilyl)oxy]-, (3α,5β,20S)-;; 3α,17α,20α-TRIHYDROXY-5β-PREGNANE-11-ONE-TRITMS; W   127126

Str 853

566 C30 H58 O4 Si3 AH-106-1423-35 3β,16α,17α,20α-TETRAHYDROXYPREGN-5-ENE, TRI-TMS ETHER; W   64717

20 40 60 80 100 120 140 160 180 200 220 240 260 280 300 320 340 360 380 400 420 440 460 480 500 520 540 560 580

566 C30 H58 O4 Si3 AH-106-1422-27 1ζ,3ζ,16ζ-TRIHYDROXY-PREGNAN-20-ONE, TRI-TMS ETHER; W   64718

566 C30 H58 O4 Si3 SH-1981-0-0 Tris(trimethylsilyl) ether of 3α,11β,15β-Trihydroxy-5β,16.η-pregnan-20-one; W   127127

566 C30 H58 O4 Si3 SH-1981-0-0 Tris(trimethylsilyl) ether of 5-Pregnen-3β,17α,20β,21-tetrol; W   127128

20 40 60 80 100 120 140 160 180 200 220 240 260 280 300 320 340 360 380 400 420 440 460 480 500 520 540 560 580

566 C30 H58 O4 Si3 SH-1981-0-0 Tris(trimethylsilyl)- 3β,11β,15α-trihydroxy-5α,16α-pregnan-20-one; W   127129

566 C30 H58 O4 Si3 SH-1981-0-0 Tris(trimethylsilyl)-5-Pregnene-3β,11β,17α,20β-tetrol; W   127130

566 C30 H58 O4 Si3 SH-1981-0-0 Trimethylsilyl derivative of 3β,15α,16α-Trihydroxy-5α-pregnan-20-one; W   127131

20 40 60 80 100 120 140 160 180 200 220 240 260 280 300 320 340 360 380 400 420 440 460 480 500 520 540 560 580

566 C31 H46 N2 O4 Si2 95694-67-0 C-107-3262-0 8-Benzyl-10-((tert-butyldimethylsilyl)benzyl)-8,10-diaza-5-(((tert-butyldimethylsilyl)oxy)methyl)-2-oxabicyclo[4.2.2]decane-7,9-dione; W   127132

Str 854

566 C31 H50 O9 61823-62-9 K-109-3458-0 METHYL ESTER OF 15-O-METHYL-2-O-(2,3,4,6-TETRA-O-METHYL-β-D-GLUCOPYRANOSYL)ATRACTYLIGENIN; W   64719

Str 855

566 C32 H22 O10 481-46-9 EP-8872-0-0 08017201002 GINKGETIN %BI-FLAVONOID<; W/NBS   64720

Str 856

566 C32 H22 O10 481-46-9 EP-8876-0-0 GINKGETIN;; BI-FLAVONOID; W   64355

Str 857

566 C32 H38 O9 69883-99-4 NS--27984-0-0 Benzeneacetic acid, 4-(acetyloxy)-, 3-[(acetyloxy)methyl]-1,1a,1b,4,4a,5,7a,7b,8,9-decahydro-4a,7b-dihydroxy-1,1,6,8-tetramethyl-5-oxo-9aH-cyclopropa[3,4]benz[1,2-e]azulen-9a-yl ester, [1aR-(1aα,1bβ,4aβ,7aα,7bα,8α,9aα)]-;; Benzeneacetic acid, 4-(acetyloxy)-, 3-[(acetyloxy)methyl]-1,1a,1b,4,4a,5,7a,7b,8,9-decahydro-4a,7b-dihydroxy-1,1,6,8-tetramethyl-5-oxo-9aH-cyclo; NBS   64721

Str 858

20 40 60 80 100 120 140 160 180 200 220 240 260 280 300 320 340 360 380 400 420 440 460 480 500 520 540 560 580

566　C33 H34 N4 O5　15295-20-2　KC-1983-443-0　1:1 Mixture of 2- and 4-formyl-6,7-bis-(2-methoxycarbonylethyl)-1,3,5,8-tetrameth
ylporphyrin;　W

Str 859

127133

566　C33 H66 O3 Si2　EM-1-54-1　3α,20α-Bis-tert-butyldimethylsilyl-ether of 5β-pregnane-3α,17α,20α-triol;　W

127134

566　C34 H28 Cl2 N2 O2　IC-1271-0-0　1,4-Bis(5,6,7,8-tetrahydronaphth-2-ylamino)-6,7-dichloroanthraquinone;　W

127135

566　C34 H30 Fe2 O　56770-51-5　AG-93-228-0　1'-Benzyl-1'''-(α-hydroxybenzyl)biferrocene;　W

127136

566　C34 H30 O8　61375-95-9　B-29-1862-0　5,6-Di-O-benzoyl-1,3(R):2,4(S)-di-O-benzylidene-D-glucitol;　W

Str 860

127137

566　C34 H38 N4 O4　59954-18-6　J-45-2223-0　2-(1-hydroxyethyl)-2-devinylpyropheophorbide A methyl ester;　W

Str 861

127138

566　C34 H38 N4 O4　61665-26-7　J-45-2223-0　Methyl ester of (3S-(3α,4β,9(R*)))-14-ethyl-9-(1-hydroxyethyl)-4,8,13,18-teramethyl-
20-oxo-3-phorbinepropanoic acid;　W

Str 862

127139

566　C34 H38 N4 O4　66230-00-0　J-45-2223-0　2-(2-Hydroxyethyl)-2-devinylpyropheophorbide A methyl ester;　W

Str 863

127140

566　C34 H46 O7　78798-25-1　KC-1983-1530-0　32-O-Methylkijanolide;　W

Str 864

127141

566　C35 H50 O6　81559-56-0　KC-1983-705-0　Tetrahydrocalozeylanic Acid;　W

Str 865

127142

566　C36 H50 N6　KC-1978-1680-0　3-(3'-(3''-DIMETHYLAMINOETHYLINDOLIN-7''-YL)-3'-DIMETHYLAMINOETHYLINDOLIN-7'-YL)-3-DIMETHYLAMINO
ETHYLINDOLINE;　W

64723

566　C36 H70 O4　55429-46-4　BA-0-363-0　Tetratriacontanedioic acid, dimethyl ester;; DIMETHYL TETRATRIACONTANEDIOATE;　W/NBS

64726

MeOC(O)(CH2)32C(O)OMe

566　C36 H70 O4　26158-81-6　EP-4950-0-0　Octadecanoic acid, 2-[(1-oxohexadecyl)oxy]ethyl ester;; 1-O-HEXADECANOYL-2-O-OCTADECANO
YL ETHANEDIOL;　W

79436

Me(CH2)14C(O)OCH2CH2OC(O)(CH2)16Me

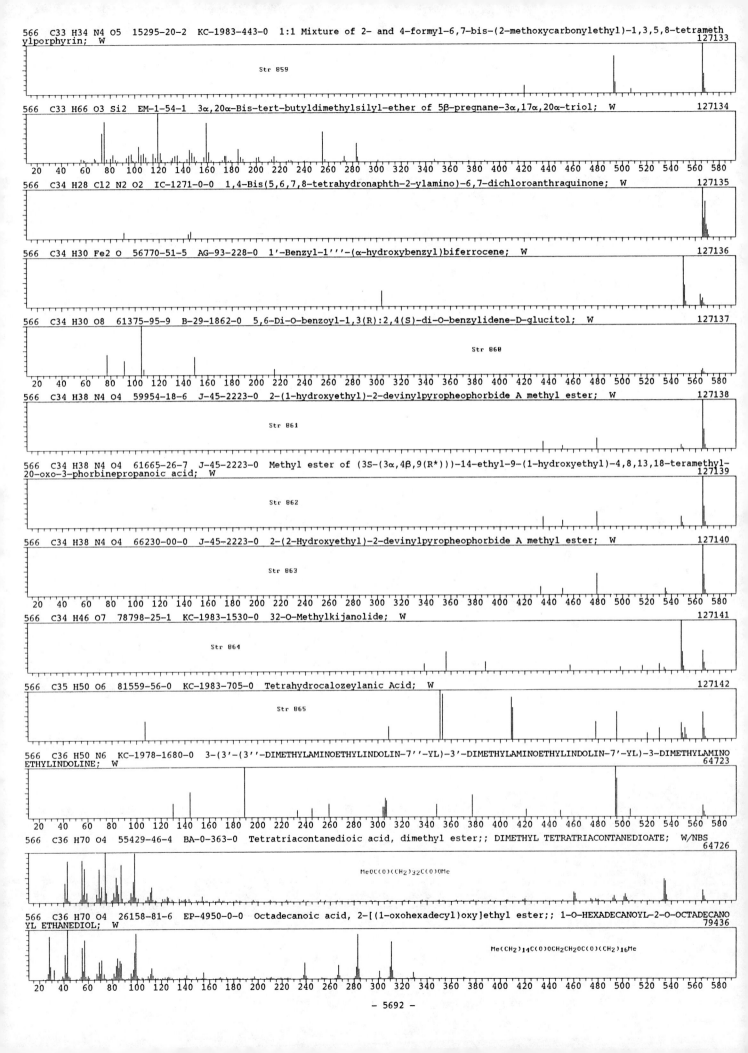

- 5692 -

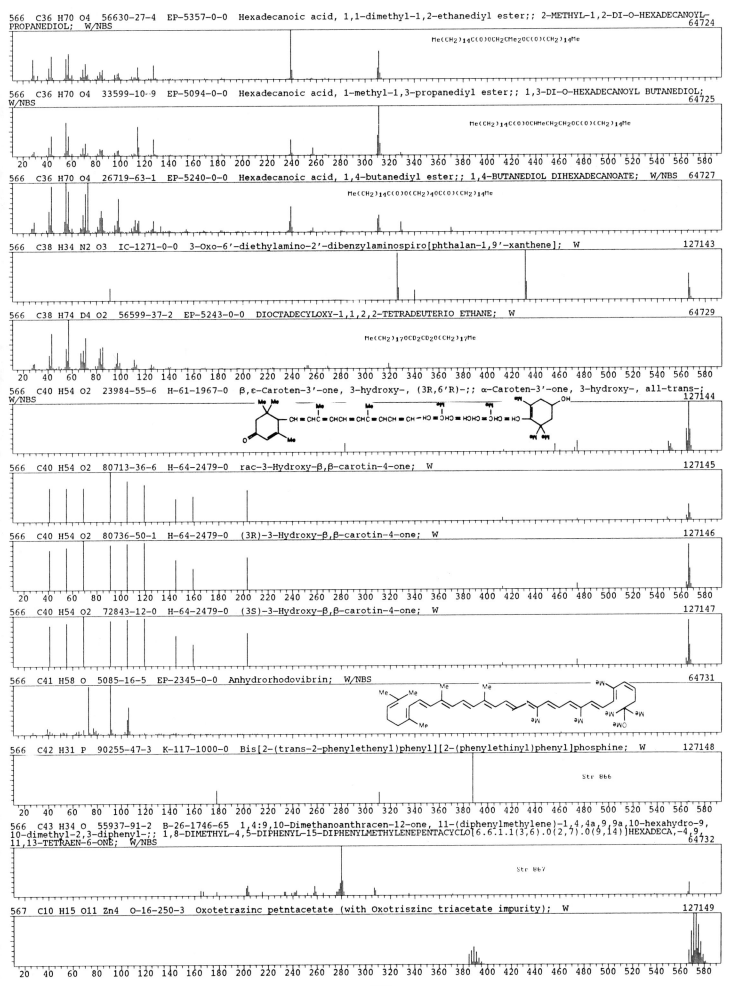

566 C36 H70 O4 56630-27-4 EP-5357-0-0 Hexadecanoic acid, 1,1-dimethyl-1,2-ethanediyl ester;; 2-METHYL-1,2-DI-O-HEXADECANOYL-PROPANEDIOL; W/NBS 64724

Me(CH2)14C(O)OCH2CMe2OC(O)(CH2)14Me

566 C36 H70 O4 33599-10-9 EP-5094-0-0 Hexadecanoic acid, 1-methyl-1,3-propanediyl ester;; 1,3-DI-O-HEXADECANOYL BUTANEDIOL; W/NBS 64725

Me(CH2)14C(O)OCHMeCH2CH2OC(O)(CH2)14Me

566 C36 H70 O4 26719-63-1 EP-5240-0-0 Hexadecanoic acid, 1,4-butanediyl ester;; 1,4-BUTANEDIOL DIHEXADECANOATE; W/NBS 64727

Me(CH2)14C(O)O(CH2)4OC(O)(CH2)14Me

566 C38 H34 N2 O3 IC-1271-0-0 3-Oxo-6'-diethylamino-2'-dibenzylaminospiro[phthalan-1,9'-xanthene]; W 127143

566 C38 H74 D4 O2 56599-37-2 EP-5243-0-0 DIOCTADECYLOXY-1,1,2,2-TETRADEUTERIO ETHANE; W 64729

Me(CH2)17OCD2CD2O(CH2)17Me

566 C40 H54 O2 23984-55-6 H-61-1967-0 β,ε-Caroten-3'-one, 3-hydroxy-, (3R,6'R)-;; α-Caroten-3'-one, 3-hydroxy-, all-trans-; W/NBS 127144

566 C40 H54 O2 80713-36-6 H-64-2479-0 rac-3-Hydroxy-β,β-carotin-4-one; W 127145

566 C40 H54 O2 80736-50-1 H-64-2479-0 (3R)-3-Hydroxy-β,β-carotin-4-one; W 127146

566 C40 H54 O2 72843-12-0 H-64-2479-0 (3S)-3-Hydroxy-β,β-carotin-4-one; W 127147

566 C41 H58 O 5085-16-5 EP-2345-0-0 Anhydrorhodovibrin; W/NBS 64731

566 C42 H31 P 90255-47-3 K-117-1000-0 Bis[2-(trans-2-phenylethenyl)phenyl][2-(phenylethinyl)phenyl]phosphine; W 127148

Str 866

566 C43 H34 O 55937-91-2 B-26-1746-65 1,4:9,10-Dimethanoanthracen-12-one, 11-(diphenylmethylene)-1,4,4a,9,9a,10-hexahydro-9,10-dimethyl-2,3-diphenyl-;; 1,8-DIMETHYL-4,5-DIPHENYL-15-DIPHENYLMETHYLENEPENTACYCLO[6.6.1.1(3,6).0(2,7).0(9,14)]HEXADECA,-4,9,11,13-TETRAEN-6-ONE; W/NBS 64732

Str 867

567 C10 H15 O11 Zn4 O-16-250-3 Oxotetrazinc petntacetate (with Oxotriszinc triacetate impurity); W 127149

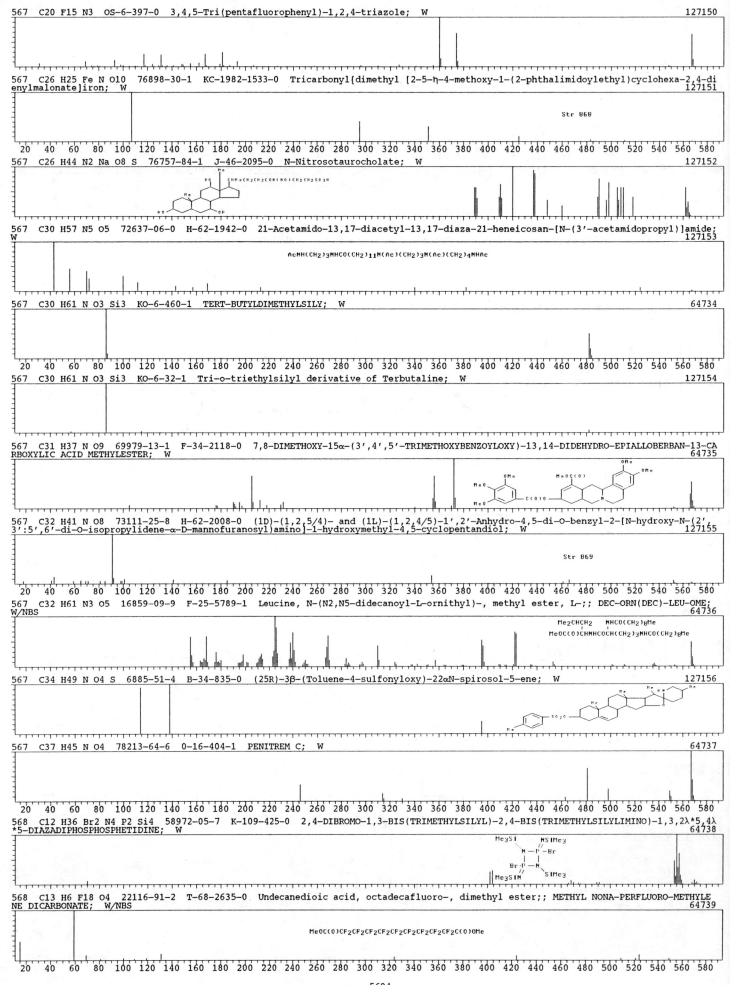

567   C20 F15 N3 OS-6-397-0   3,4,5-Tri(pentafluorophenyl)-1,2,4-triazole;   W   127150

567   C26 H25 Fe N O10   76898-30-1   KC-1982-1533-0   Tricarbonyl{dimethyl [2-5-η-4-methoxy-1-(2-phthalimidoylethyl)cyclohexa-2,4-di
enylmalonate]iron;   W   127151

Str 868

567   C26 H44 N2 Na O8 S   76757-84-1   J-46-2095-0   N-Nitrosotaurocholate;   W   127152

567   C30 H57 N5 O5   72637-06-0   H-62-1942-0   21-Acetamido-13,17-diacetyl-13,17-diaza-21-heneicosan-[N-(3'-acetamidopropyl)]amide;
W   127153

AcNH(CH2)3NHCO(CH2)11N(Ac)(CH2)3N(Ac)(CH2)4NHAc

567   C30 H61 N O3 Si3   KO-6-460-1   TERT-BUTYLDIMETHYLSILY;   W   64734

567   C30 H61 N O3 Si3   KO-6-32-1   Tri-o-triethylsilyl derivative of Terbutaline;   W   127154

567   C31 H37 N O9   69979-13-1   F-34-2118-0   7,8-DIMETHOXY-15α-(3',4',5'-TRIMETHOXYBENZOYLOXY)-13,14-DIDEHYDRO-EPIALLOBERBAN-13-CA
RBOXYLIC ACID METHYLESTER;   W   64735

567   C32 H41 N O8   73111-25-8   H-62-2008-0   (1D)-(1,2,5/4)- and (1L)-(1,2,4/5)-1',2'-Anhydro-4,5-di-O-benzyl-2-[N-hydroxy-N-(2'
3':5',6'-di-O-isopropylidene-α-D-mannofuranosyl)amino]-1-hydroxymethyl-4,5-cyclopentandiol;   W   127155

Str 869

567   C32 H61 N3 O5   16859-09-9   F-25-5789-1   Leucine, N-(N2,N5-didecanoyl-L-ornithyl)-, methyl ester, L-;; DEC-ORN(DEC)-LEU-OME;
W/NBS   64736

567   C34 H49 N O4 S   6885-51-4   B-34-835-0   (25R)-3β-(Toluene-4-sulfonyloxy)-22αN-spirosol-5-ene;   W   127156

567   C37 H45 N O4   78213-64-6   O-16-404-1   PENITREM C;   W   64737

568   C12 H36 Br2 N4 P2 Si4   58972-05-7   K-109-425-0   2,4-DIBROMO-1,3-BIS(TRIMETHYLSILYL)-2,4-BIS(TRIMETHYLSILYLIMINO)-1,3,2λ*5,4λ
*5-DIAZADIPHOSPHETIDINE;   W   64738

568   C13 H6 F18 O4   22116-91-2   T-68-2635-0   Undecanedioic acid, octadecafluoro-, dimethyl ester;; METHYL NONA-PERFLUORO-METHYLE
NE DICARBONATE;   W/NBS   64739

MeOC(O)CF2CF2CF2CF2CF2CF2CF2CF2C(O)OMe

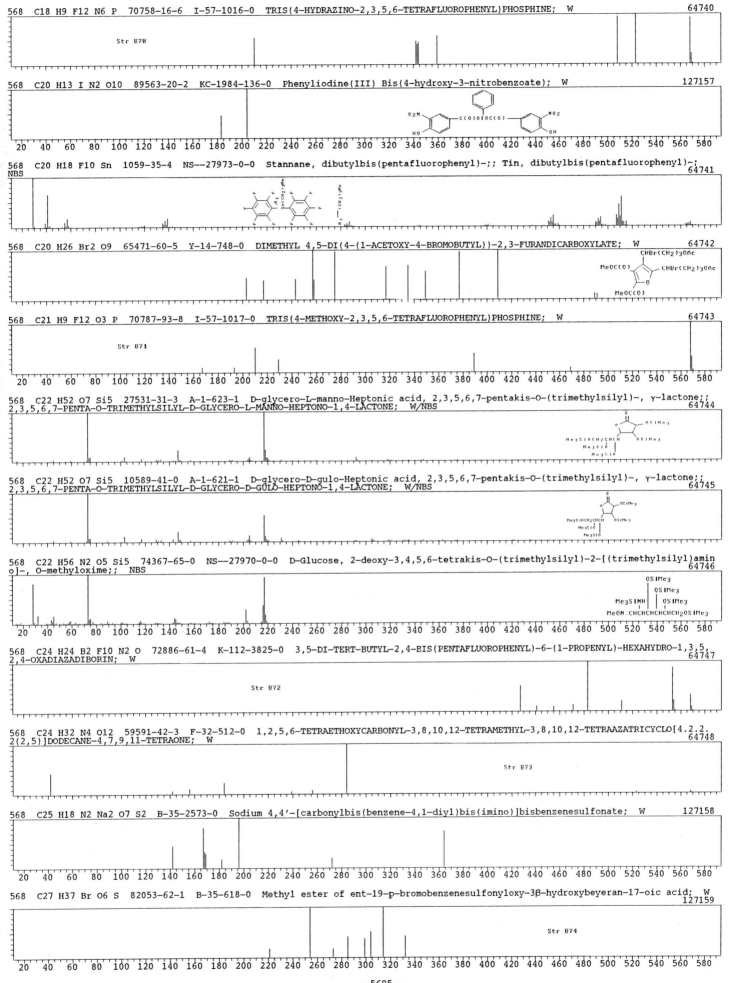

568 C18 H9 F12 N6 P 70758-16-6 I-57-1016-0 TRIS(4-HYDRAZINO-2,3,5,6-TETRAFLUOROPHENYL)PHOSPHINE; W 64740

Str 870

568 C20 H13 I N2 O10 89563-20-2 KC-1984-136-0 Phenyliodine(III) Bis(4-hydroxy-3-nitrobenzoate); W 127157

568 C20 H18 F10 Sn 1059-35-4 NS--27973-0-0 Stannane, dibutylbis(pentafluorophenyl)-;; Tin, dibutylbis(pentafluorophenyl)-; 64741
NBS

568 C20 H26 Br2 O9 65471-60-5 Y-14-748-0 DIMETHYL 4,5-DI(4-(1-ACETOXY-4-BROMOBUTYL))-2,3-FURANDICARBOXYLATE; W 64742

568 C21 H9 F12 O3 P 70787-93-8 I-57-1017-0 TRIS(4-METHOXY-2,3,5,6-TETRAFLUOROPHENYL)PHOSPHINE; W 64743

Str 871

568 C22 H52 O7 Si5 27531-31-3 A-1-623-1 D-glycero-L-manno-Heptonic acid, 2,3,5,6,7-pentakis-O-(trimethylsilyl)-, γ-lactone;;
2,3,5,6,7-PENTA-O-TRIMETHYLSILYL-D-GLYCERO-L-MANNO-HEPTONO-1,4-LACTONE; W/NBS 64744

568 C22 H52 O7 Si5 10589-41-0 A-1-621-1 D-glycero-D-gulo-Heptonic acid, 2,3,5,6,7-pentakis-O-(trimethylsilyl)-, γ-lactone;;
2,3,5,6,7-PENTA-O-TRIMETHYLSILYL-D-GLYCERO-D-GULO-HEPTONO-1,4-LACTONE; W/NBS 64745

568 C22 H56 N2 O5 Si5 74367-65-0 NS--27970-0-0 D-Glucose, 2-deoxy-3,4,5,6-tetrakis-O-(trimethylsilyl)-2-[(trimethylsilyl)amin
o]-, O-methyloxime;; NBS 64746

568 C24 H24 B2 F10 N2 O 72886-61-4 K-112-3825-0 3,5-DI-TERT-BUTYL-2,4-BIS(PENTAFLUOROPHENYL)-6-(1-PROPENYL)-HEXAHYDRO-1,3,5,
2,4-OXADIAZADIBORIN; W 64747

Str 872

568 C24 H32 N4 O12 59591-42-3 F-32-512-0 1,2,5,6-TETRAETHOXYCARBONYL-3,8,10,12-TETRAMETHYL-3,8,10,12-TETRAAZATRICYCLO[4.2.2.
2(2,5)]DODECANE-4,7,9,11-TETRAONE; W 64748

Str 873

568 C25 H18 N2 Na2 O7 S2 B-35-2573-0 Sodium 4,4'-[carbonylbis(benzene-4,1-diyl)bis(imino)]bisbenzenesulfonate; W 127158

568 C27 H37 Br O6 S 82053-62-1 B-35-618-0 Methyl ester of ent-19-p-bromobenzenesulfonyloxy-3β-hydroxybeyeran-17-oic acid; W 127159

Str 874

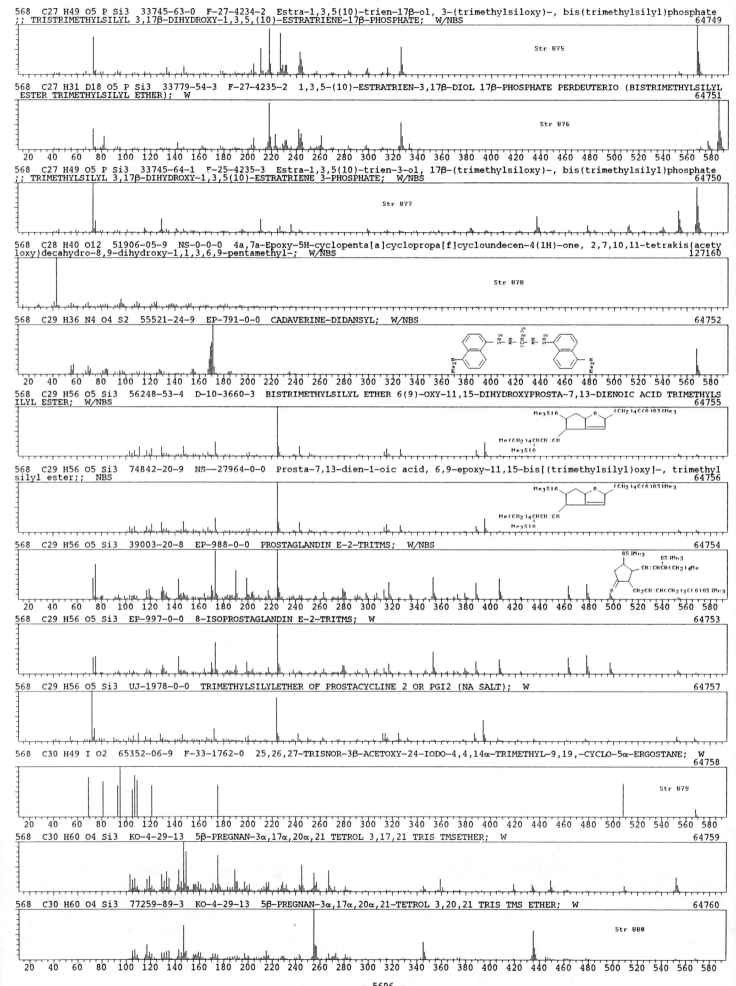

568   C27 H49 O5 P Si3   33745-63-0   F-27-4234-2   Estra-1,3,5(10)-trien-17β-ol, 3-(trimethylsiloxy)-, bis(trimethylsilyl)phosphate
;; TRISTRIMETHYLSILYL 3,17β-DIHYDROXY-1,3,5,(10)-ESTRATRIENE-17β-PHOSPHATE; W/NBS   64749

Str 875

568   C27 H31 D18 O5 P Si3   33779-54-3   F-27-4235-2   1,3,5-(10)-ESTRATRIEN-3,17β-DIOL 17β-PHOSPHATE PERDEUTERIO (BISTRIMETHYLSILYL
ESTER TRIMETHYLSILYL ETHER); W   64751

Str 876

568   C27 H49 O5 P Si3   33745-64-1   F-25-4235-3   Estra-1,3,5(10)-trien-3-ol, 17β-(trimethylsiloxy)-, bis(trimethylsilyl)phosphate
;; TRIMETHYLSILYL 3,17β-DIHYDROXY-1,3,5,(10)-ESTRATRIENE 3-PHOSPHATE; W/NBS   64750

Str 877

568   C28 H40 O12   51906-05-9   NS-0-0-0   4a,7a-Epoxy-5H-cyclopenta[a]cyclopropa[f]cycloundecen-4(1H)-one, 2,7,10,11-tetrakis(acety
loxy)decahydro-8,9-dihydroxy-1,1,3,6,9-pentamethyl-; W/NBS   127160

Str 878

568   C29 H36 N4 O4 S2   55521-24-9   EP-791-0-0   CADAVERINE-DIDANSYL; W/NBS   64752

568   C29 H56 O5 Si3   56248-53-4   D-10-3660-3   BISTRIMETHYLSILYL ETHER 6(9)-OXY-11,15-DIHYDROXYPROSTA-7,13-DIENOIC ACID TRIMETHYLS
ILYL ESTER; W/NBS   64755

568   C29 H56 O5 Si3   74842-20-9   NS--27964-0-0   Prosta-7,13-dien-1-oic acid, 6,9-epoxy-11,15-bis[(trimethylsilyl)oxy]-, trimethyl
silyl ester;; NBS   64756

568   C29 H56 O5 Si3   39003-20-8   EP-988-0-0   PROSTAGLANDIN E-2-TRITMS; W/NBS   64754

568   C29 H56 O5 Si3   EP-997-0-0   8-ISOPROSTAGLANDIN E-2-TRITMS; W   64753

568   C29 H56 O5 Si3   UJ-1978-0-0   TRIMETHYLSILYLETHER OF PROSTACYCLINE 2 OR PGI2 (NA SALT); W   64757

568   C30 H49 I O2   65352-06-9   F-33-1762-0   25,26,27-TRISNOR-3β-ACETOXY-24-IODO-4,4,14α-TRIMETHYL-9,19,-CYCLO-5α-ERGOSTANE; W
                                                                                                                         64758

Str 879

568   C30 H60 O4 Si3   KO-4-29-13   5β-PREGNAN-3α,17α,20α,21 TETROL 3,17,21 TRIS TMSETHER; W   64759

568   C30 H60 O4 Si3   77259-89-3   KO-4-29-13   5β-PREGNAN-3α,17α,20α,21-TETROL 3,20,21 TRIS TMS ETHER; W   64760

Str 880

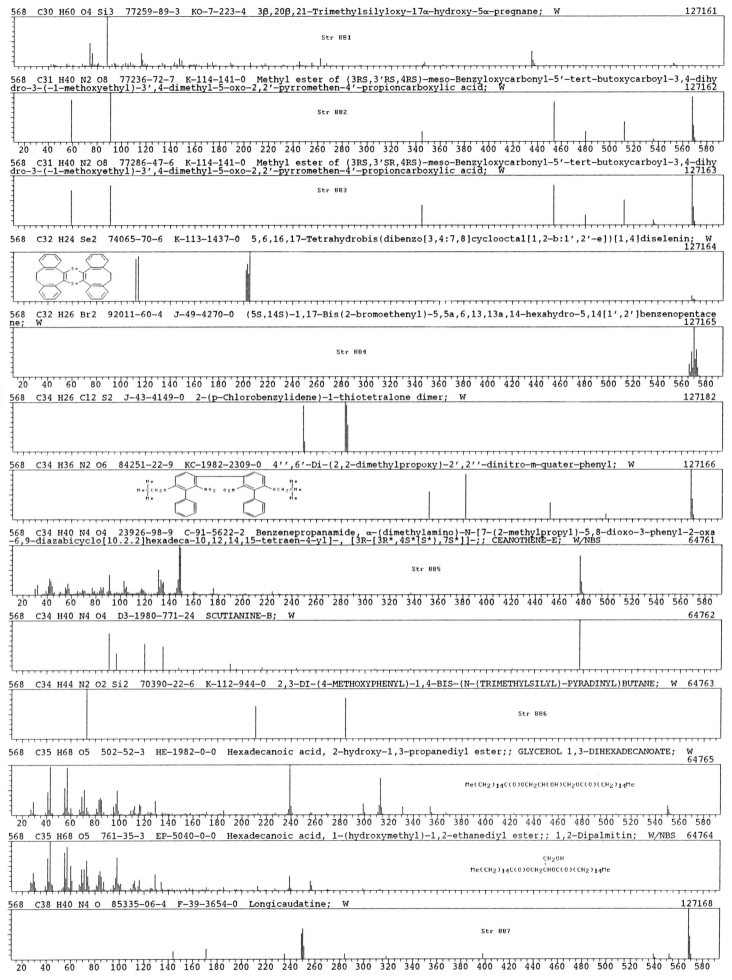

568   C30 H60 O4 Si3   77259-89-3   KO-7-223-4   3β,20β,21-Trimethylsilyloxy-17α-hydroxy-5α-pregnane;   W   127161

Str 881

568   C31 H40 N2 O8   77236-72-7   K-114-141-0   Methyl ester of (3RS,3'RS,4RS)-meso-Benzyloxycarbonyl-5'-tert-butoxycarboyl-3,4-dihy
dro-3-(-1-methoxyethyl)-3',4-dimethyl-5-oxo-2,2'-pyrromethen-4'-propioncarboxylic acid;   W   127162

Str 882

568   C31 H40 N2 O8   77286-47-6   K-114-141-0   Methyl ester of (3RS,3'SR,4RS)-meso-Benzyloxycarbonyl-5'-tert-butoxycarboyl-3,4-dihy
dro-3-(-1-methoxyethyl)-3',4-dimethyl-5-oxo-2,2'-pyrromethen-4'-propioncarboxylic acid;   W   127163

Str 883

568   C32 H24 Se2   74065-70-6   K-113-1437-0   5,6,16,17-Tetrahydrobis(dibenzo[3,4:7,8]cyclooctal[1,2-b:1',2'-e])[1,4]diselenin;   W
127164

568   C32 H26 Br2   92011-60-4   J-49-4270-0   (5S,14S)-1,17-Bis(2-bromoethenyl)-5,5a,6,13,13a,14-hexahydro-5,14[1',2']benzopentace
ne;   W   127165

Str 884

568   C34 H26 Cl2 S2   J-43-4149-0   2-(p-Chlorobenzylidene)-1-thiotetralone dimer;   W   127182

568   C34 H36 N2 O6   84251-22-9   KC-1982-2309-0   4'',6'-Di-(2,2-dimethylpropoxy)-2',2''-dinitro-m-quater-phenyl;   W   127166

568   C34 H40 N4 O4   23926-98-9   C-91-5622-2   Benzenepropanamide, α-(dimethylamino)-N-[7-(2-methylpropyl)-5,8-dioxo-3-phenyl-2-oxa
-6,9-diazabicyclo[10.2.2]hexadeca-10,12,14,15-tetraen-4-yl]-, [3R-[3R*,4S*(S*),7S*]]-;;   CEANOTHINE-E;   W/NBS   64761

Str 885

568   C34 H40 N4 O4   D3-1980-771-24   SCUTIANINE-B;   W   64762

568   C34 H44 N2 O2 Si2   70390-22-6   K-112-944-0   2,3-DI-(4-METHOXYPHENYL)-1,4-BIS-(N-(TRIMETHYLSILYL)-PYRADINYL)BUTANE;   W   64763

Str 886

568   C35 H68 O5   502-52-3   HE-1982-0-0   Hexadecanoic acid, 2-hydroxy-1,3-propanediyl ester;;   GLYCEROL 1,3-DIHEXADECANOATE;   W
64765

Me(CH2)14C(O)OCH2CH(OH)CH2OC(O)(CH2)14Me

568   C35 H68 O5   761-35-3   EP-5040-0-0   Hexadecanoic acid, 1-(hydroxymethyl)-1,2-ethanediyl ester;;   1,2-Dipalmitin;   W/NBS   64764

CH2OH
Me(CH2)14C(O)OCH2CHOC(O)(CH2)14Me

568   C38 H40 N4 O   85335-06-4   F-39-3654-0   Longicaudatine;   W   127168

Str 887

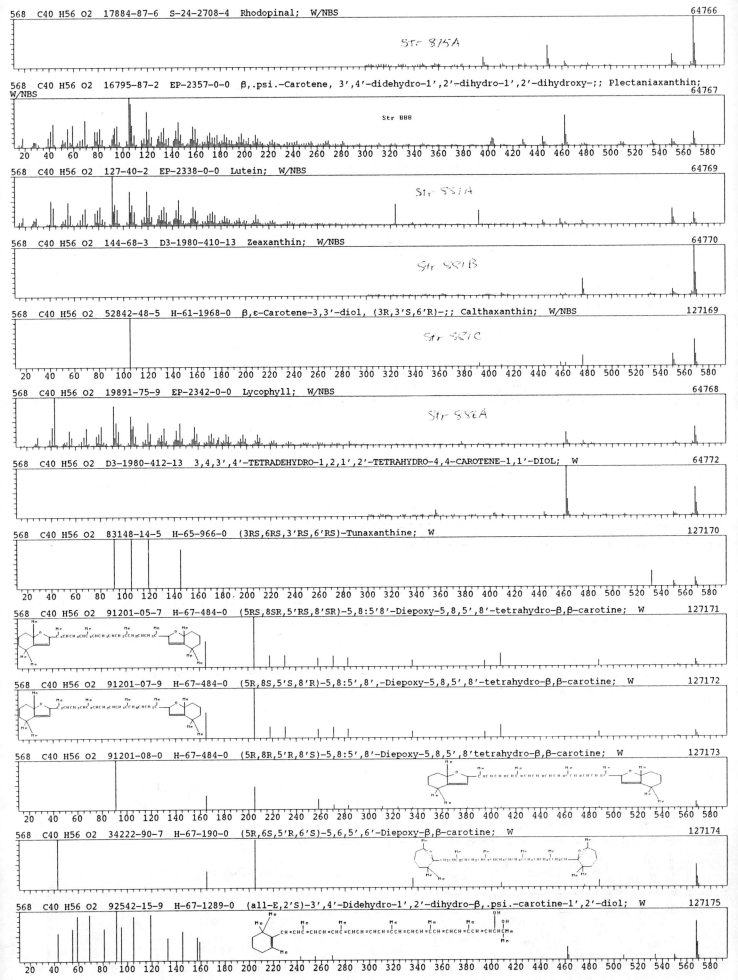

568   C40 H56 O2   17884-87-6   S-24-2708-4   Rhodopinal;   W/NBS   64766

Str 875A

568   C40 H56 O2   16795-87-2   EP-2357-0-0   β,.psi.-Carotene, 3',4'-didehydro-1',2'-dihydro-1',2'-dihydroxy-;; Plectaniaxanthin;
W/NBS   64767

Str 888

20   40   60   80   100 120 140 160 180 200 220 240 260 280 300 320 340 360 380 400 420 440 460 480 500 520 540 560 580

568   C40 H56 O2   127-40-2   EP-2338-0-0   Lutein;   W/NBS   64769

Str 881A

568   C40 H56 O2   144-68-3   D3-1980-410-13   Zeaxanthin;   W/NBS   64770

Str 881B

568   C40 H56 O2   52842-48-5   H-61-1968-0   β,ε-Carotene-3,3'-diol, (3R,3'S,6'R)-;; Calthaxanthin;   W/NBS   127169

Str 881C

20   40   60   80   100 120 140 160 180 200 220 240 260 280 300 320 340 360 380 400 420 440 460 480 500 520 540 560 580

568   C40 H56 O2   19891-75-9   EP-2342-0-0   Lycophyll;   W/NBS   64768

Str 882A

568   C40 H56 O2   D3-1980-412-13   3,4,3',4'-TETRADEHYDRO-1,2,1',2'-TETRAHYDRO-4,4-CAROTENE-1,1'-DIOL;   W   64772

568   C40 H56 O2   83148-14-5   H-65-966-0   (3RS,6RS,3'RS,6'RS)-Tunaxanthine;   W   127170

20   40   60   80   100 120 140 160 180 200 220 240 260 280 300 320 340 360 380 400 420 440 460 480 500 520 540 560 580

568   C40 H56 O2   91201-05-7   H-67-484-0   (5RS,8SR,5'RS,8'SR)-5,8:5'8'-Diepoxy-5,8,5',8'-tetrahydro-β,β-carotine;   W   127171

568   C40 H56 O2   91201-07-9   H-67-484-0   (5R,8S,5'S,8'R)-5,8:5',8',-Diepoxy-5,8,5',8'-tetrahydro-β,β-carotine;   W   127172

568   C40 H56 O2   91201-08-0   H-67-484-0   (5R,8R,5'R,8'S)-5,8:5',8'-Diepoxy-5,8,5',8'tetrahydro-β,β-carotine;   W   127173

20   40   60   80   100 120 140 160 180 200 220 240 260 280 300 320 340 360 380 400 420 440 460 480 500 520 540 560 580

568   C40 H56 O2   34222-90-7   H-67-190-0   (5R,6S,5'R,6'S)-5,6,5',6'-Diepoxy-β,β-carotine;   W   127174

568   C40 H56 O2   92542-15-9   H-67-1289-0   (all-E,2'S)-3',4'-Didehydro-1',2'-dihydro-β,.psi.-carotine-1',2'-diol;   W   127175

20   40   60   80   100 120 140 160 180 200 220 240 260 280 300 320 340 360 380 400 420 440 460 480 500 520 540 560 580

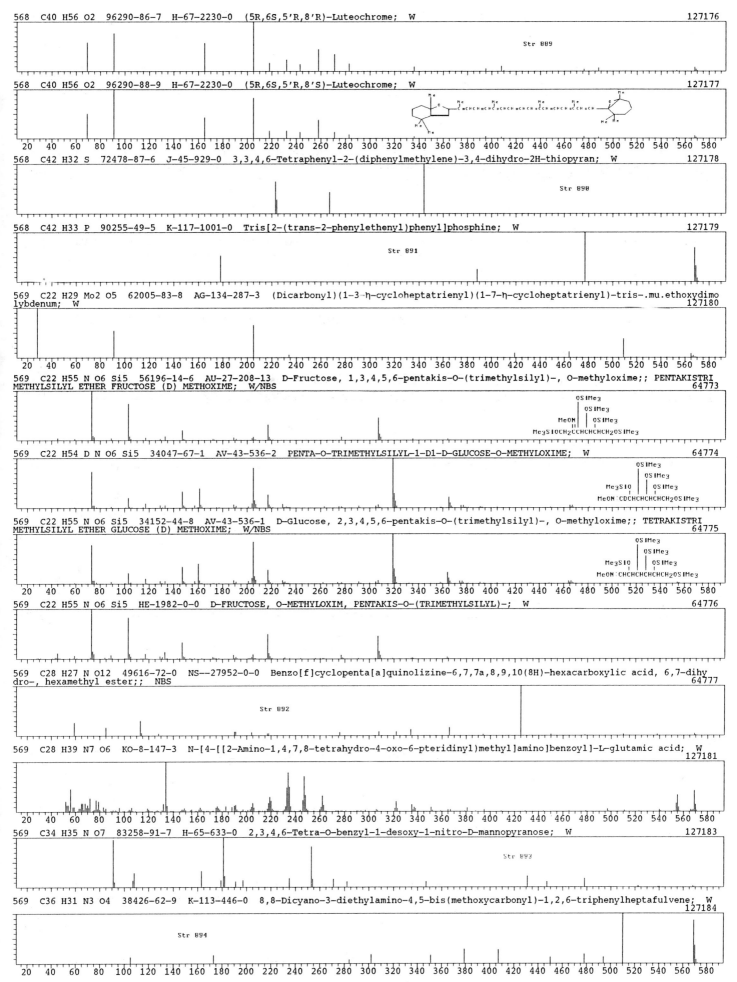

568　C40 H56 O2　96290-86-7　H-67-2230-0　(5R,6S,5'R,8'R)-Luteochrome;　W　　　　127176

Str 889

568　C40 H56 O2　96290-88-9　H-67-2230-0　(5R,6S,5'R,8'S)-Luteochrome;　W　　　　127177

568　C42 H32 S　72478-87-6　J-45-929-0　3,3,4,6-Tetraphenyl-2-(diphenylmethylene)-3,4-dihydro-2H-thiopyran;　W　　　127178

Str 890

568　C42 H33 P　90255-49-5　K-117-1001-0　Tris[2-(trans-2-phenylethenyl)phenyl]phosphine;　W　　　127179

Str 891

569　C22 H29 Mo2 O5　62005-83-8　AG-134-287-3　(Dicarbonyl)(1-3-η-cycloheptatrienyl)(1-7-η-cycloheptatrienyl)-tris-.mu.ethoxydimo lybdenum;　W　　　127180

569　C22 H55 N O6 Si5　56196-14-6　AU-27-208-13　D-Fructose, 1,3,4,5,6-pentakis-O-(trimethylsilyl)-, O-methyloxime;;　PENTAKISTRI METHYLSILYL ETHER FRUCTOSE (D) METHOXIME;　W/NBS　　　64773

569　C22 H54 D N O6 Si5　34047-67-1　AV-43-536-2　PENTA-O-TRIMETHYLSILYL-1-D1-D-GLUCOSE-O-METHYLOXIME;　W　　　64774

569　C22 H55 N O6 Si5　34152-44-8　AV-43-536-1　D-Glucose, 2,3,4,5,6-pentakis-O-(trimethylsilyl)-, O-methyloxime;;　TETRAKISTRI METHYLSILYL ETHER GLUCOSE (D) METHOXIME;　W/NBS　　　64775

569　C22 H55 N O6 Si5　HE-1982-0-0　D-FRUCTOSE, O-METHYLOXIM, PENTAKIS-O-(TRIMETHYLSILYL)-;　W　　　64776

569　C28 H27 N O12　49616-72-0　NS--27952-0-0　Benzo[f]cyclopenta[a]quinolizine-6,7,7a,8,9,10(8H)-hexacarboxylic acid, 6,7-dihy dro-, hexamethyl ester;;　NBS　　　64777

Str 892

569　C28 H39 N7 O6　KO-8-147-3　N-[4-[[2-Amino-1,4,7,8-tetrahydro-4-oxo-6-pteridinyl)methyl]amino]benzoyl]-L-glutamic acid;　W　　　127181

569　C34 H35 N O7　83258-91-7　H-65-633-0　2,3,4,6-Tetra-O-benzyl-1-desoxy-1-nitro-D-mannopyranose;　W　　　127183

Str 893

569　C36 H31 N3 O4　38426-62-9　K-113-446-0　8,8-Dicyano-3-diethylamino-4,5-bis(methoxycarbonyl)-1,2,6-triphenylheptafulvene;　W　　　127184

Str 894

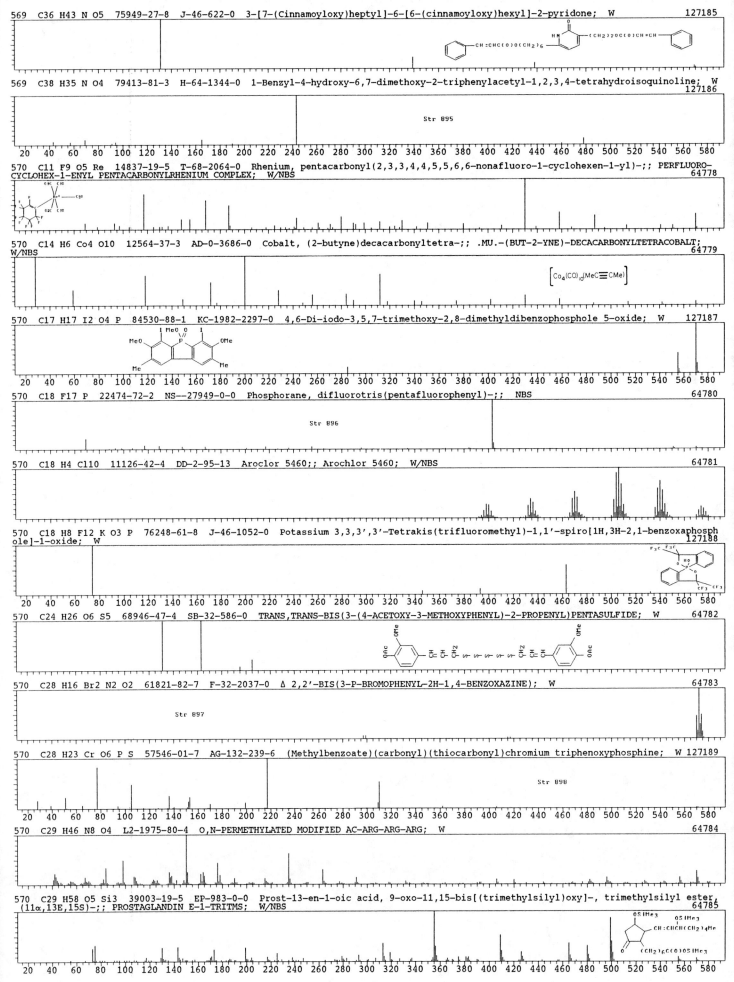

569　C36 H43 N O5　75949-27-8　J-46-622-0　3-[7-(Cinnamoyloxy)heptyl]-6-[6-(cinnamoyloxy)hexyl]-2-pyridone;　W
127185

569　C38 H35 N O4　79413-81-3　H-64-1344-0　1-Benzyl-4-hydroxy-6,7-dimethoxy-2-triphenylacetyl-1,2,3,4-tetrahydroisoquinoline;　W
127186

Str 895

20　40　60　80　100　120　140　160　180　200　220　240　260　280　300　320　340　360　380　400　420　440　460　480　500　520　540　560　580

570　C11 F9 O5 Re　14837-19-5　T-68-2064-0　Rhenium, pentacarbonyl(2,3,3,4,4,5,5,6,6-nonafluoro-1-cyclohexen-1-yl)-;; PERFLUORO-
CYCLOHEX-1-ENYL PENTACARBONYLRHENIUM COMPLEX;　W/NBS
64778

570　C14 H6 Co4 O10　12564-37-3　AD-0-3686-0　Cobalt, (2-butyne)decacarbonyltetra-;; .MU.-(BUT-2-YNE)-DECACARBONYLTETRACOBALT;
W/NBS
64779

570　C17 H17 I2 O4 P　84530-88-1　KC-1982-2297-0　4,6-Di-iodo-3,5,7-trimethoxy-2,8-dimethyldibenzophosphole 5-oxide;　W　127187

20　40　60　80　100　120　140　160　180　200　220　240　260　280　300　320　340　360　380　400　420　440　460　480　500　520　540　560　580

570　C18 F17 P　22474-72-2　NS--27949-0-0　Phosphorane, difluorotris(pentafluorophenyl)-;;　NBS
64780

Str 896

570　C18 H4 Cl10　11126-42-4　DD-2-95-13　Aroclor 5460;; Arochlor 5460;　W/NBS
64781

570　C18 H8 F12 K O3 P　76248-61-8　J-46-1052-0　Potassium 3,3,3',3'-Tetrakis(trifluoromethyl)-1,1'-spiro[1H,3H-2,1-benzoxaphosph
ole]-1-oxide;　W
127188

20　40　60　80　100　120　140　160　180　200　220　240　260　280　300　320　340　360　380　400　420　440　460　480　500　520　540　560　580

570　C24 H26 O6 S5　68946-47-4　SB-32-586-0　TRANS,TRANS-BIS(3-(4-ACETOXY-3-METHOXYPHENYL)-2-PROPENYL)PENTASULFIDE;　W
64782

570　C28 H16 Br2 N2 O2　61821-82-7　F-32-2037-0　Δ 2,2'-BIS(3-P-BROMOPHENYL-2H-1,4-BENZOXAZINE);　W
64783

Str 897

570　C28 H23 Cr O6 P S　57546-01-7　AG-132-239-6　(Methylbenzoate)(carbonyl)(thiocarbonyl)chromium triphenoxyphosphine;　W 127189

Str 898

570　C29 H46 N8 O4　L2-1975-80-4　O,N-PERMETHYLATED MODIFIED AC-ARG-ARG-ARG;　W
64784

570　C29 H58 O5 Si3　39003-19-5　EP-983-0-0　Prost-13-en-1-oic acid, 9-oxo-11,15-bis[(trimethylsilyl)oxy]-, trimethylsilyl ester,
(11α,13E,15S)-;; PROSTAGLANDIN E-1-TRITMS;　W/NBS
64785

20　40　60　80　100　120　140　160　180　200　220　240　260　280　300　320　340　360　380　400　420　440　460　480　500　520　540　560　580

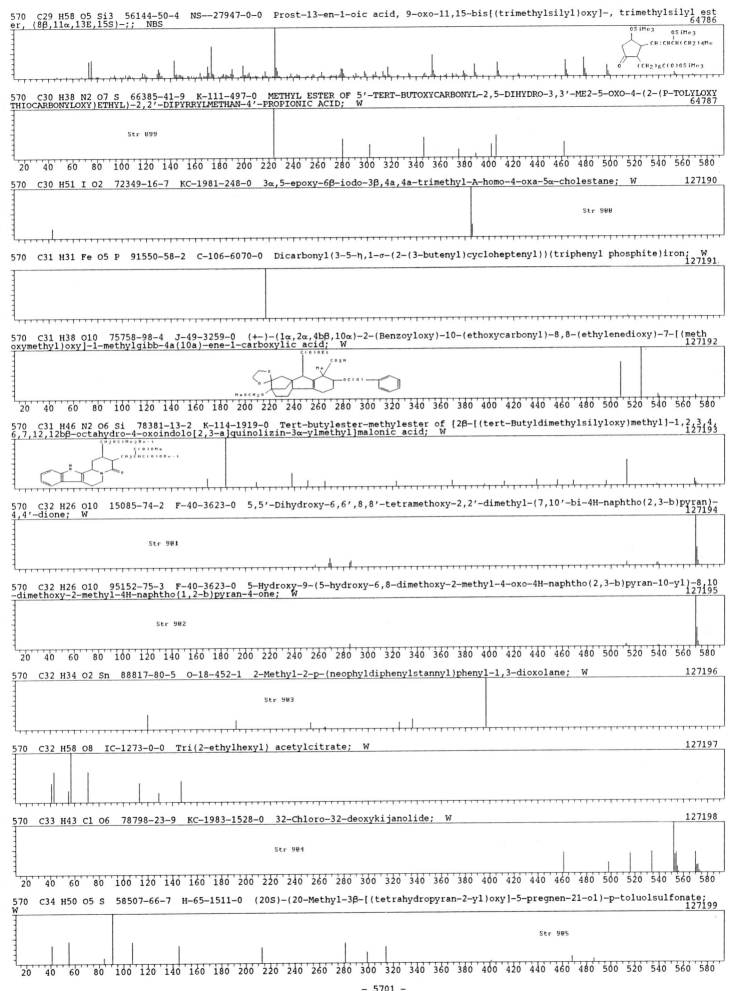

570 C29 H58 O5 Si3 56144-50-4 NS--27947-0-0 Prost-13-en-1-oic acid, 9-oxo-11,15-bis[(trimethylsilyl)oxy]-, trimethylsilyl ester, (8β,11α,13E,15S)-;; NBS
64786

570 C30 H38 N2 O7 S 66385-41-9 K-111-497-0 METHYL ESTER OF 5'-TERT-BUTOXYCARBONYL-2,5-DIHYDRO-3,3'-ME2-5-OXO-4-(2-(P-TOLYLOXY THIOCARBONYLOXY)ETHYL)-2,2'-DIPYRRYLMETHAN-4'-PROPIONIC ACID; W
64787

Str 899

570 C30 H51 I O2 72349-16-7 KC-1981-248-0 3α,5-epoxy-6β-iodo-3β,4a,4a-trimethyl-A-homo-4-oxa-5α-cholestane; W
127190

Str 900

570 C31 H31 Fe O5 P 91550-58-2 C-106-6070-0 Dicarbonyl(3-5-η,1-σ-(2-(3-butenyl)cycloheptenyl))(triphenyl phosphite)iron; W
127191

570 C31 H38 O10 75758-98-4 J-49-3259-0 (+-)-(1α,2α,4bβ,10α)-2-(Benzoyloxy)-10-(ethoxycarbonyl)-8,8-(ethylenedioxy)-7-[(methoxymethyl)oxy]-1-methylgibb-4a(10a)-ene-1-carboxylic acid; W
127192

570 C31 H46 N2 O6 Si 78381-13-2 K-114-1919-0 Tert-butylester-methylester of {2β-[(tert-Butyldimethylsilyloxy)methyl]-1,2,3,4,6,7,12,12bβ-octahydro-4-oxoindolo[2,3-a]quinolizin-3α-ylmethyl}malonic acid; W
127193

570 C32 H26 O10 15085-74-2 F-40-3623-0 5,5'-Dihydroxy-6,6',8,8'-tetramethoxy-2,2'-dimethyl-(7,10'-bi-4H-naphtho(2,3-b)pyran)-4,4'-dione; W
127194

Str 901

570 C32 H26 O10 95152-75-3 F-40-3623-0 5-Hydroxy-9-(5-hydroxy-6,8-dimethoxy-2-methyl-4-oxo-4H-naphtho(2,3-b)pyran-10-yl)-8,10-dimethoxy-2-methyl-4H-naphtho(1,2-b)pyran-4-one; W
127195

Str 902

570 C32 H34 O2 Sn 88817-80-5 O-18-452-1 2-Methyl-2-p-(neophyldiphenylstannyl)phenyl-1,3-dioxolane; W
127196

Str 903

570 C32 H58 O8 IC-1273-0-0 Tri(2-ethylhexyl) acetylcitrate; W
127197

570 C33 H43 Cl O6 78798-23-9 KC-1983-1528-0 32-Chloro-32-deoxykijanolide; W
127198

Str 904

570 C34 H50 O5 S 58507-66-7 H-65-1511-0 (20S)-(20-Methyl-3β-[(tetrahydropyran-2-yl)oxy]-5-pregnen-21-ol)-p-toluolsulfonate; W
127199

Str 905

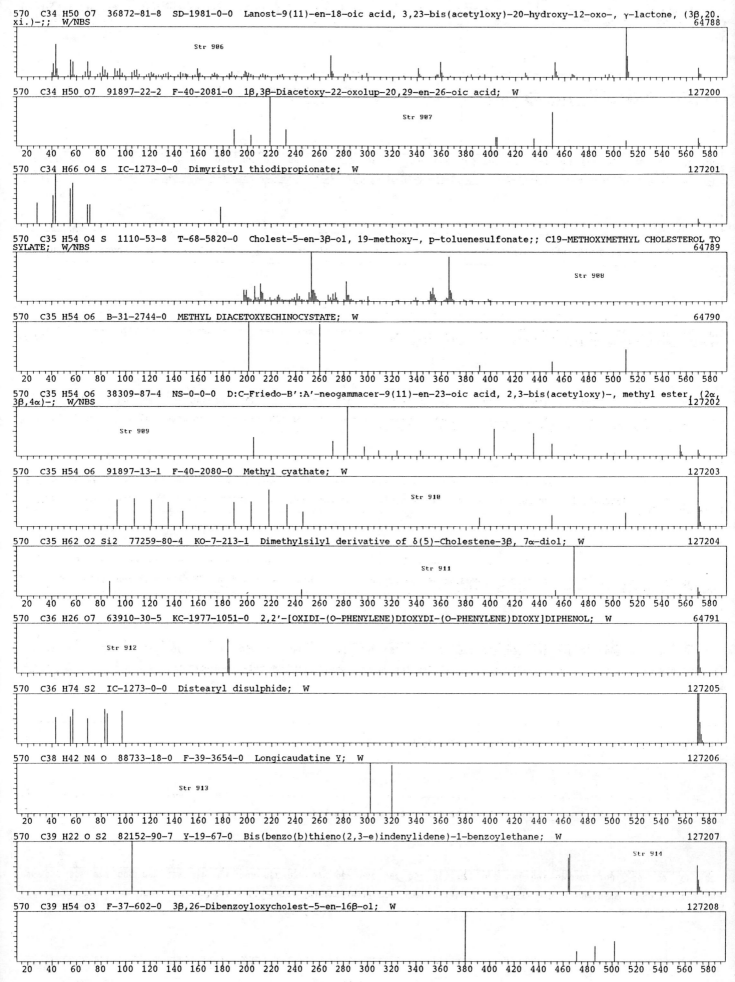

570   C34 H50 O7   36872-81-8   SD-1981-0-0   Lanost-9(11)-en-18-oic acid, 3,23-bis(acetyloxy)-20-hydroxy-12-oxo-, γ-lactone, (3β,20. xi.)-;; W/NBS
64788

Str 906

570   C34 H50 O7   91897-22-2   F-40-2081-0   1β,3β-Diacetoxy-22-oxolup-20,29-en-26-oic acid; W
127200

Str 907

570   C34 H66 O4 S   IC-1273-0-0   Dimyristyl thiodipropionate; W
127201

570   C35 H54 O4 S   1110-53-8   T-68-5820-0   Cholest-5-en-3β-ol, 19-methoxy-, p-toluenesulfonate;; C19-METHOXYMETHYL CHOLESTEROL TOSYLATE; W/NBS
64789

Str 908

570   C35 H54 O6   B-31-2744-0   METHYL DIACETOXYECHINOCYSTATE; W
64790

570   C35 H54 O6   38309-87-4   NS-0-0-0   D:C-Friedo-B':A'-neogammacer-9(11)-en-23-oic acid, 2,3-bis(acetyloxy)-, methyl ester, (2α,3β,4α)-; W/NBS
127202

Str 909

570   C35 H54 O6   91897-13-1   F-40-2080-0   Methyl cyathate; W
127203

Str 910

570   C35 H62 O2 Si2   77259-80-4   KO-7-213-1   Dimethylsilyl derivative of δ(5)-Cholestene-3β, 7α-diol; W
127204

Str 911

570   C36 H26 O7   63910-30-5   KC-1977-1051-0   2,2'-[OXIDI-(O-PHENYLENE)DIOXYDI-(O-PHENYLENE)DIOXY]DIPHENOL; W
64791

Str 912

570   C36 H74 S2   IC-1273-0-0   Distearyl disulphide; W
127205

570   C38 H42 N4 O   88733-18-0   F-39-3654-0   Longicaudatine Y; W
127206

Str 913

570   C39 H22 O S2   82152-90-7   Y-19-67-0   Bis(benzo(b)thieno(2,3-e)indenylidene)-1-benzoylethane; W
127207

Str 914

570   C39 H54 O3   F-37-602-0   3β,26-Dibenzoyloxycholest-5-en-16β-ol; W
127208

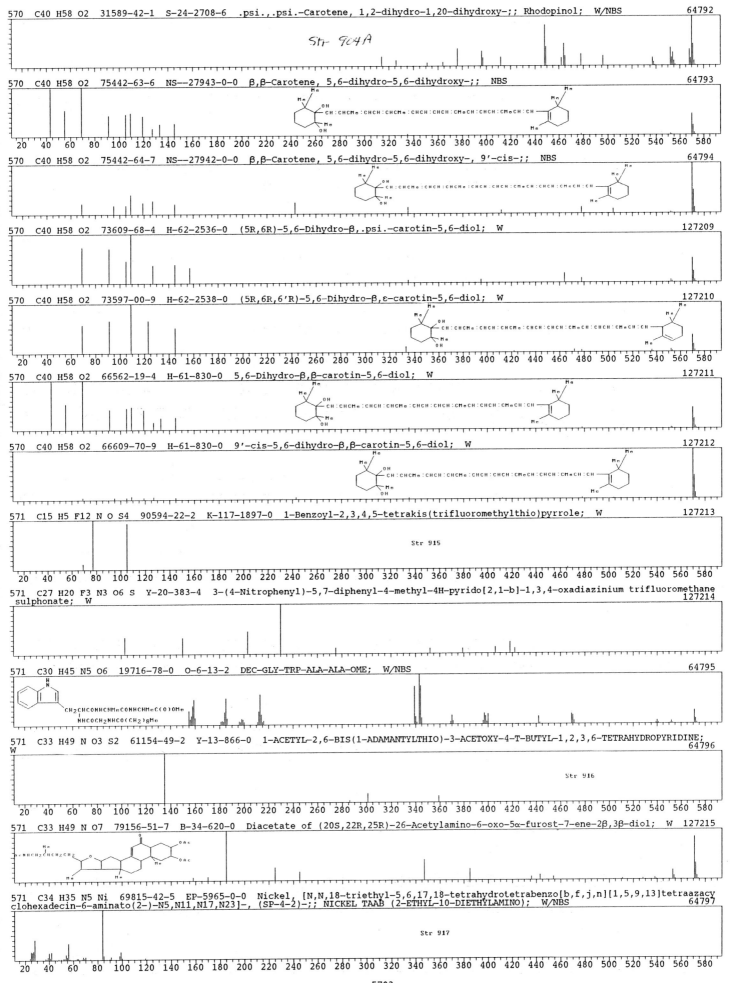

570　C40 H58 O2　31589-42-1　S-24-2708-6　.psi.,.psi.-Carotene, 1,2-dihydro-1,20-dihydroxy-;; Rhodopinol;　W/NBS　64792

Str 904A

570　C40 H58 O2　75442-63-6　NS--27943-0-0　β,β-Carotene, 5,6-dihydro-5,6-dihydroxy-;;　NBS　64793

570　C40 H58 O2　75442-64-7　NS--27942-0-0　β,β-Carotene, 5,6-dihydro-5,6-dihydroxy-, 9'-cis-;;　NBS　64794

570　C40 H58 O2　73609-68-4　H-62-2536-0　(5R,6R)-5,6-Dihydro-β,.psi.-carotin-5,6-diol;　W　127209

570　C40 H58 O2　73597-00-9　H-62-2538-0　(5R,6R,6'R)-5,6-Dihydro-β,ε-carotin-5,6-diol;　W　127210

570　C40 H58 O2　66562-19-4　H-61-830-0　5,6-Dihydro-β,β-carotin-5,6-diol;　W　127211

570　C40 H58 O2　66609-70-9　H-61-830-0　9'-cis-5,6-dihydro-β,β-carotin-5,6-diol;　W　127212

571　C15 H5 F12 N O S4　90594-22-2　K-117-1897-0　1-Benzoyl-2,3,4,5-tetrakis(trifluoromethylthio)pyrrole;　W　127213

Str 915

571　C27 H20 F3 N3 O6 S　Y-20-383-4　3-(4-Nitrophenyl)-5,7-diphenyl-4-methyl-4H-pyrido[2,1-b]-1,3,4-oxadiazinium trifluoromethane sulphonate;　W　127214

571　C30 H45 N5 O6　19716-78-0　O-6-13-2　DEC-GLY-TRP-ALA-ALA-OME;　W/NBS　64795

571　C33 H49 N O3 S2　61154-49-2　Y-13-866-0　1-ACETYL-2,6-BIS(1-ADAMANTYLTHIO)-3-ACETOXY-4-T-BUTYL-1,2,3,6-TETRAHYDROPYRIDINE;　W　64796

Str 916

571　C33 H49 N O7　79156-51-7　B-34-620-0　Diacetate of (20S,22R,25R)-26-Acetylamino-6-oxo-5α-furost-7-ene-2β,3β-diol;　W　127215

571　C34 H35 N5 Ni　69815-42-5　EP-5965-0-0　Nickel, [N,N,18-triethyl-5,6,17,18-tetrahydrotetrabenzo[b,f,j,n][1,5,9,13]tetraazacyclohexadecin-6-aminato(2-)-N5,N11,N17,N23]-, (SP-4-2)-;; NICKEL TAAB (2-ETHYL-10-DIETHYLAMINO);　W/NBS　64797

Str 917

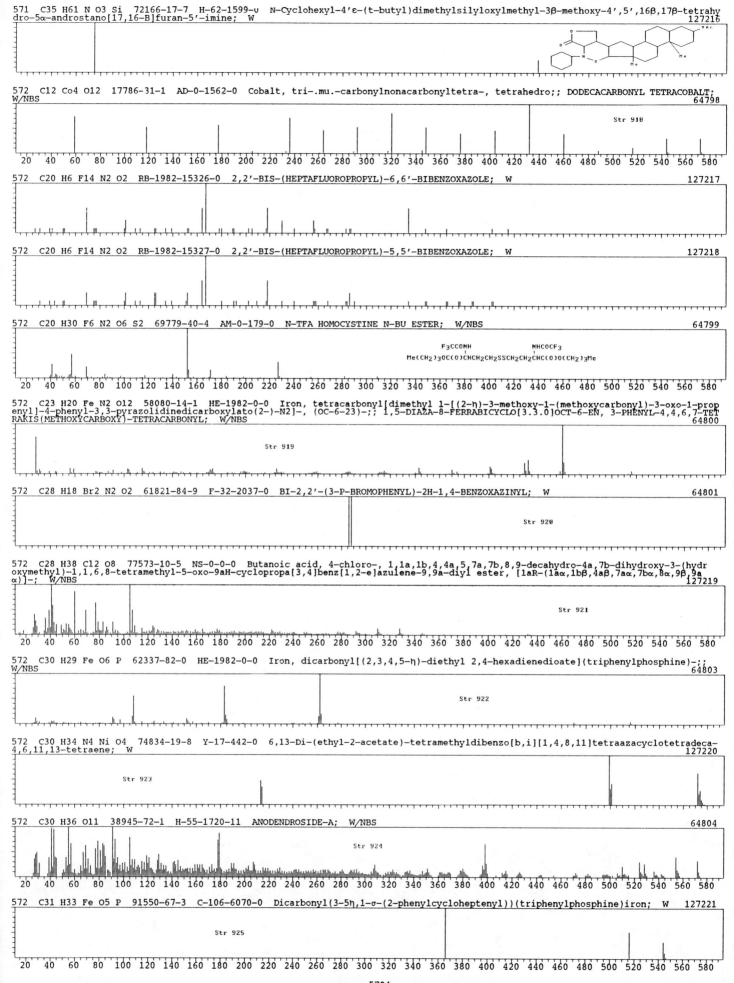

571 C35 H61 N O3 Si 72166-17-7 H-62-1599-υ N-Cyclohexyl-4'ε-(t-butyl)dimethylsilyloxylmethyl-3β-methoxy-4',5',16β,17β-tetrahy
dro-5α-androstano[17,16-B]furan-5'-imine; W
127216

572 C12 Co4 O12 17786-31-1 AD-0-1562-0 Cobalt, tri-.mu.-carbonylnonacarbonyltetra-, tetrahedro;; DODECACARBONYL TETRACOBALT;
W/NBS
64798

Str 918

20 40 60 80 100 120 140 160 180 200 220 240 260 280 300 320 340 360 380 400 420 440 460 480 500 520 540 560 580

572 C20 H6 F14 N2 O2 RB-1982-15326-0 2,2'-BIS-(HEPTAFLUOROPROPYL)-6,6'-BIBENZOXAZOLE; W
127217

572 C20 H6 F14 N2 O2 RB-1982-15327-0 2,2'-BIS-(HEPTAFLUOROPROPYL)-5,5'-BIBENZOXAZOLE; W
127218

572 C20 H30 F6 N2 O6 S2 69779-40-4 AM-0-179-0 N-TFA HOMOCYSTINE N-BU ESTER; W/NBS
64799

F₃CCONH   NHCOCF₃
Me(CH₂)₃OC(O)CHCH₂CH₂SSCH₂CH₂CHC(O)O(CH₂)₃Me

20 40 60 80 100 120 140 160 180 200 220 240 260 280 300 320 340 360 380 400 420 440 460 480 500 520 540 560 580

572 C23 H20 Fe N2 O12 58080-14-1 HE-1982-0-0 Iron, tetracarbonyl[dimethyl 1-[(2-η)-3-methoxy-1-(methoxycarbonyl)-3-oxo-1-prop
enyl]-4-phenyl-3,3-pyrazolidinedicarboxylato(2-)-N2]-, (OC-6-23)-;; 1,5-DIAZA-8-FERRABICYCLO[3.3.0]OCT-6-EN, 3-PHENYL-4,4,6,7-TET
RAKIS(METHOXYCARBOXY)-TETRACARBONYL; W/NBS
64800

Str 919

572 C28 H18 Br2 N2 O2 61821-84-9 F-32-2037-0 BI-2,2'-(3-P-BROMOPHENYL)-2H-1,4-BENZOXAZINYL; W
64801

Str 920

572 C28 H38 Cl2 O8 77573-10-5 NS-0-0-0 Butanoic acid, 4-chloro-, 1,1a,1b,4,4a,5,7a,7b,8,9-decahydro-4a,7b-dihydroxy-3-(hydr
oxymethyl)-1,1,6,8-tetramethyl-5-oxo-9aH-cyclopropa[3,4]benz[1,2-e]azulene-9,9a-diyl ester, [1aR-(1aα,1bβ,4aβ,7aα,7bα,8α,9β,9a
α)]-; W/NBS
127219

Str 921

20 40 60 80 100 120 140 160 180 200 220 240 260 280 300 320 340 360 380 400 420 440 460 480 500 520 540 560 580

572 C30 H29 Fe O6 P 62337-82-0 HE-1982-0-0 Iron, dicarbonyl[(2,3,4,5-η)-diethyl 2,4-hexadienedioate](triphenylphosphine)-;;
W/NBS
64803

Str 922

572 C30 H34 N4 Ni O4 74834-19-8 Y-17-442-0 6,13-Di-(ethyl-2-acetate)-tetramethyldibenzo[b,i][1,4,8,11]tetraazacyclotetradeca-
4,6,11,13-tetraene; W
127220

Str 923

572 C30 H36 O11 38945-72-1 H-55-1720-11 ANODENDROSIDE-A; W/NBS
64804

Str 924

20 40 60 80 100 120 140 160 180 200 220 240 260 280 300 320 340 360 380 400 420 440 460 480 500 520 540 560 580

572 C31 H33 Fe O5 P 91550-67-3 C-106-6070-0 Dicarbonyl(3-5η,1-σ-(2-phenylcycloheptenyl))(triphenylphosphine)iron; W 127221

Str 925

20 40 60 80 100 120 140 160 180 200 220 240 260 280 300 320 340 360 380 400 420 440 460 480 500 520 540 560 580

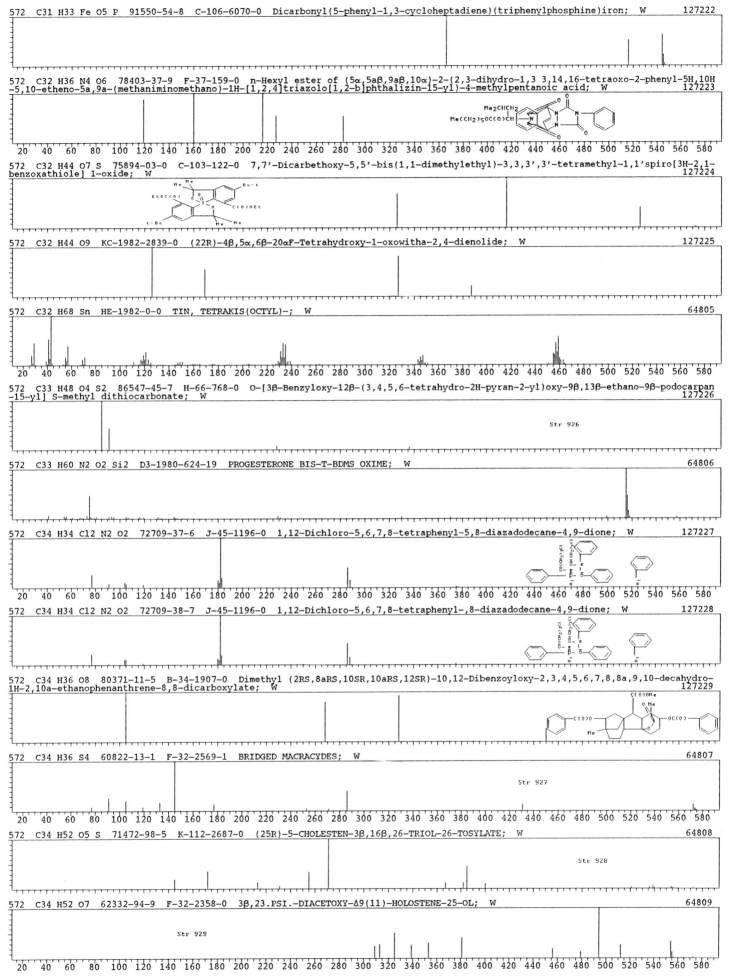

572　C31 H33 Fe O5 P　91550-54-8　C-106-6070-0　Dicarbonyl(5-phenyl-1,3-cycloheptadiene)(triphenylphosphine)iron;　W　127222

572　C32 H36 N4 O6　78403-37-9　F-37-159-0　n-Hexyl ester of (5α,5aβ,9aβ,10α)-2-(2,3-dihydro-1,3 3,14,16-tetraoxo-2-phenyl-5H,10H -5,10-etheno-5a,9a-(methaniminomethano)-1H-[1,2,4]triazolo[1,2-b]phthalizin-15-yl)-4-methylpentanoic acid;　W　127223

572　C32 H44 O7 S　75894-03-0　C-103-122-0　7,7'-Dicarbethoxy-5,5'-bis(1,1-dimethylethyl)-3,3,3',3'-tetramethyl-1,1'spiro[3H-2,1-benzoxathiole] 1-oxide;　W　127224

572　C32 H44 O9　KC-1982-2839-0　(22R)-4β,5α,6β-20αF-Tetrahydroxy-1-oxowitha-2,4-dienolide;　W　127225

572　C32 H68 Sn　HE-1982-0-0　TIN, TETRAKIS(OCTYL)-;　W　64805

572　C33 H48 O4 S2　86547-45-7　H-66-768-0　O-[3β-Benzyloxy-12β-(3,4,5,6-tetrahydro-2H-pyran-2-yl)oxy-9β,13β-ethano-9β-podocarpan -15-yl] S-methyl dithiocarbonate;　W　127226

Str 926

572　C33 H60 N2 O2 Si2　D3-1980-624-19　PROGESTERONE BIS-T-BDMS OXIME;　W　64806

572　C34 H34 Cl2 N2 O2　72709-37-6　J-45-1196-0　1,12-Dichloro-5,6,7,8-tetraphenyl-5,8-diazadodecane-4,9-dione;　W　127227

572　C34 H34 Cl2 N2 O2　72709-38-7　J-45-1196-0　1,12-Dichloro-5,6,7,8-tetraphenyl-,8-diazadodecane-4,9-dione;　W　127228

572　C34 H36 O8　80371-11-5　B-34-1907-0　Dimethyl (2RS,8aRS,10SR,10aRS,12SR)-10,12-Dibenzoyloxy-2,3,4,5,6,7,8,8a,9,10-decahydro-1H-2,10a-ethanophenanthrene-8,8-dicarboxylate;　W　127229

572　C34 H36 S4　60822-13-1　F-32-2569-1　BRIDGED MACRACYDES;　W　64807

Str 927

572　C34 H52 O5 S　71472-98-5　K-112-2687-0　(25R)-5-CHOLESTEN-3β,16β,26-TRIOL-26-TOSYLATE;　W　64808

Str 928

572　C34 H52 O7　62332-94-9　F-32-2358-0　3β,23.PSI.-DIACETOXY-Δ9(11)-HOLOSTENE-25-OL;　W　64809

Str 929

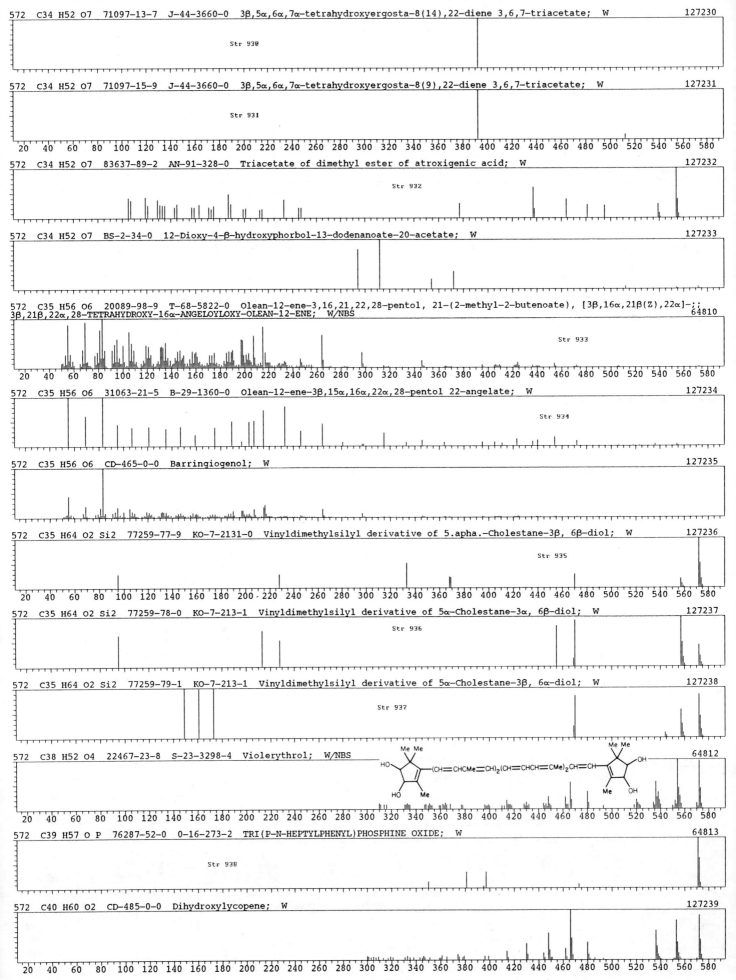

572 C34 H52 O7 71097-13-7 J-44-3660-0 3β,5α,6α,7α-tetrahydroxyergosta-8(14),22-diene 3,6,7-triacetate; W   127230

Str 930

572 C34 H52 O7 71097-15-9 J-44-3660-0 3β,5α,6α,7α-tetrahydroxyergosta-8(9),22-diene 3,6,7-triacetate; W   127231

Str 931

572 C34 H52 O7 83637-89-2 AN-91-328-0 Triacetate of dimethyl ester of atroxigenic acid; W   127232

Str 932

572 C34 H52 O7 BS-2-34-0 12-Dioxy-4-β-hydroxyphorbol-13-dodenanoate-20-acetate; W   127233

572 C35 H56 O6 20089-98-9 T-68-5822-0 Olean-12-ene-3,16,21,22,28-pentol, 21-(2-methyl-2-butenoate), [3β,16α,21β(Z),22α]-;;
3β,21β,22α,28-TETRAHYDROXY-16α-ANGELOYLOXY-OLEAN-12-ENE; W/NBS   64810

Str 933

572 C35 H56 O6 31063-21-5 B-29-1360-0 Olean-12-ene-3β,15α,16α,22α,28-pentol 22-angelate; W   127234

Str 934

572 C35 H56 O6 CD-465-0-0 Barringiogenol; W   127235

572 C35 H64 O2 Si2 77259-77-9 KO-7-2131-0 Vinyldimethylsilyl derivative of 5.apha.-Cholestane-3β, 6β-diol; W   127236

Str 935

572 C35 H64 O2 Si2 77259-78-0 KO-7-213-1 Vinyldimethylsilyl derivative of 5α-Cholestane-3α, 6β-diol; W   127237

Str 936

572 C35 H64 O2 Si2 77259-79-1 KO-7-213-1 Vinyldimethylsilyl derivative of 5α-Cholestane-3β, 6α-diol; W   127238

Str 937

572 C38 H52 O4 22467-23-8 S-23-3298-4 Violerythrol; W/NBS   64812

572 C39 H57 O P 76287-52-0 0-16-273-2 TRI(P-N-HEPTYLPHENYL)PHOSPHINE OXIDE; W   64813

Str 938

572 C40 H60 O2 CD-485-0-0 Dihydroxylycopene; W   127239

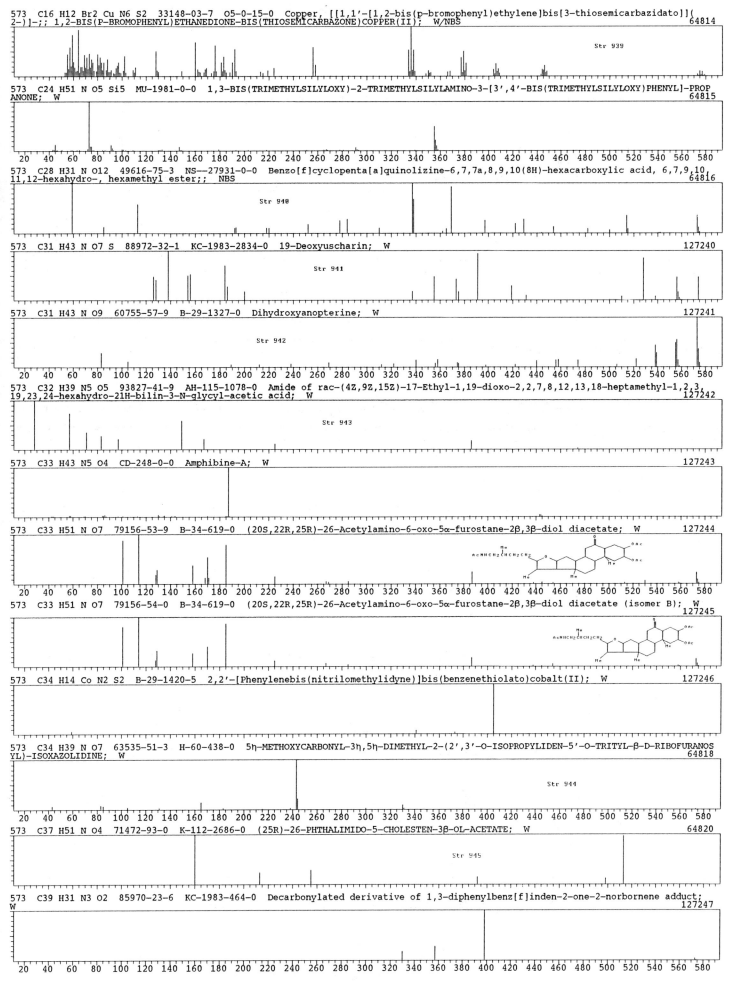

573　C16 H12 Br2 Cu N6 S2　33148-03-7　O5-0-15-0　Copper, [[1,1'-[1,2-bis(p-bromophenyl)ethylene]bis[3-thiosemicarbazidato]](2-)]-;; 1,2-BIS(P-BROMOPHENYL)ETHANEDIONE-BIS(THIOSEMICARBAZONE)COPPER(II);　W/NBS
64814
Str 939

573　C24 H51 N O5 Si5　MU-1981-0-0　1,3-BIS(TRIMETHYLSILYLOXY)-2-TRIMETHYLSILYLAMINO-3-[3',4'-BIS(TRIMETHYLSILYLOXY)PHENYL]-PROP ANONE;　W
64815

573　C28 H31 N O12　49616-75-3　NS--27931-0-0　Benzo[f]cyclopenta[a]quinolizine-6,7,7a,8,9,10(8H)-hexacarboxylic acid, 6,7,9,10, 11,12-hexahydro-, hexamethyl ester;;　NBS
64816
Str 940

573　C31 H43 N O7 S　88972-32-1　KC-1983-2834-0　19-Deoxyuscharin;　W
127240
Str 941

573　C31 H43 N O9　60755-57-9　B-29-1327-0　Dihydroxyanopterine;　W
127241
Str 942

573　C32 H39 N5 O5　93827-41-9　AH-115-1078-0　Amide of rac-(4Z,9Z,15Z)-17-Ethyl-1,19-dioxo-2,2,7,8,12,13,18-heptamethyl-1,2,3, 19,23,24-hexahydro-21H-bilin-3-N-glycyl-acetic acid;　W
127242
Str 943

573　C33 H43 N5 O4　CD-248-0-0　Amphibine-A;　W
127243

573　C33 H51 N O7　79156-53-9　B-34-619-0　(20S,22R,25R)-26-Acetylamino-6-oxo-5α-furostane-2β,3β-diol diacetate;　W
127244

573　C33 H51 N O7　79156-54-0　B-34-619-0　(20S,22R,25R)-26-Acetylamino-6-oxo-5α-furostane-2β,3β-diol diacetate (isomer B);　W
127245

573　C34 H14 Co N2 S2　B-29-1420-5　2,2'-[Phenylenebis(nitrilomethylidyne)]bis(benzenethiolato)cobalt(II);　W
127246

573　C34 H39 N O7　63535-51-3　H-60-438-0　5η-METHOXYCARBONYL-3η,5η-DIMETHYL-2-(2',3'-O-ISOPROPYLIDEN-5'-O-TRITYL-β-D-RIBOFURANOS YL)-ISOXAZOLIDINE;　W
64818
Str 944

573　C37 H51 N O4　71472-93-0　K-112-2686-0　(25R)-26-PHTHALIMIDO-5-CHOLESTEN-3β-OL-ACETATE;　W
64820
Str 945

573　C39 H31 N3 O2　85970-23-6　KC-1983-464-0　Decarbonylated derivative of 1,3-diphenylbenz[f]inden-2-one-2-norbornene adduct; W
127247

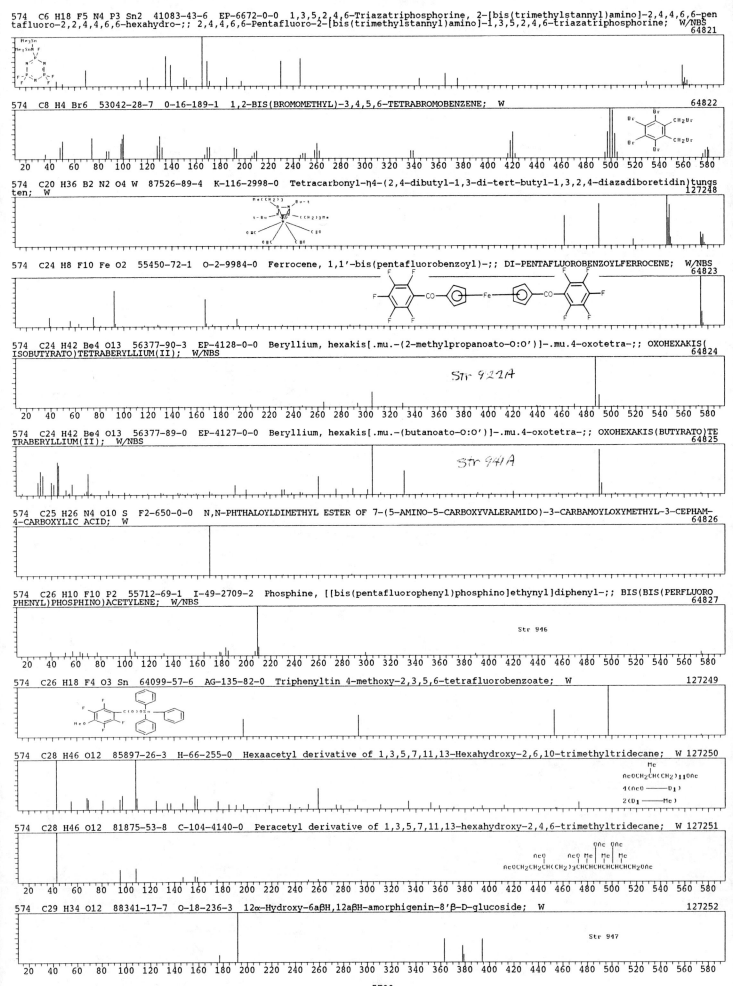

574   C6 H18 F5 N4 P3 Sn2   41083-43-6   EP-6672-0-0   1,3,5,2,4,6-Triazatriphosphorine, 2-[bis(trimethylstannyl)amino]-2,4,4,6,6-pentafluoro-2,2,4,4,6,6-hexahydro-;; 2,4,4,6,6-Pentafluoro-2-[bis(trimethylstannyl)amino]-1,3,5,2,4,6-triazatriphosphorine;   W/NBS
64821

574   C8 H4 Br6   53042-28-7   0-16-189-1   1,2-BIS(BROMOMETHYL)-3,4,5,6-TETRABROMOBENZENE;   W
64822

574   C20 H36 B2 N2 O4 W   87526-89-4   K-116-2998-0   Tetracarbonyl-η4-(2,4-dibutyl-1,3-di-tert-butyl-1,3,2,4-diazadiboretidin)tungsten;   W
127248

574   C24 H8 F10 Fe O2   55450-72-1   O-2-9984-0   Ferrocene, 1,1'-bis(pentafluorobenzoyl)-;; DI-PENTAFLUOROBENZOYLFERROCENE;   W/NBS
64823

574   C24 H42 Be4 O13   56377-90-3   EP-4128-0-0   Beryllium, hexakis[.mu.-(2-methylpropanoato-O:O')]-.mu.4-oxotetra-;; OXOHEXAKIS(ISOBUTYRATO)TETRABERYLLIUM(II);   W/NBS
64824

574   C24 H42 Be4 O13   56377-89-0   EP-4127-0-0   Beryllium, hexakis[.mu.-(butanoato-O:O')]-.mu.4-oxotetra-;; OXOHEXAKIS(BUTYRATO)TETRABERYLLIUM(II);   W/NBS
64825

574   C25 H26 N4 O10 S   F2-650-0-0   N,N-PHTHALOYLDIMETHYL ESTER OF 7-(5-AMINO-5-CARBOXYVALERAMIDO)-3-CARBAMOYLOXYMETHYL-3-CEPHAM-4-CARBOXYLIC ACID;   W
64826

574   C26 H10 F10 P2   55712-69-1   I-49-2709-2   Phosphine, [[bis(pentafluorophenyl)phosphino]ethynyl]diphenyl-;; BIS(BIS(PERFLUOROPHENYL)PHOSPHINO)ACETYLENE;   W/NBS
64827

574   C26 H18 F4 O3 Sn   64099-57-6   AG-135-82-0   Triphenyltin 4-methoxy-2,3,5,6-tetrafluorobenzoate;   W
127249

574   C28 H46 O12   85897-26-3   H-66-255-0   Hexaacetyl derivative of 1,3,5,7,11,13-Hexahydroxy-2,6,10-trimethyltridecane;   W 127250

574   C28 H46 O12   81875-53-8   C-104-4140-0   Peracetyl derivative of 1,3,5,7,11,13-hexahydroxy-2,4,6-trimethyltridecane;   W 127251

574   C29 H34 O12   88341-17-7   O-18-236-3   12α-Hydroxy-6aβH,12aβH-amorphigenin-8'β-D-glucoside;   W
127252

574 C30 H22 O12 18913-18-3 T-67-4884-0 [3,8'-Bi-4H-1-benzopyran]-4,4'-dione, 2'-(3,4-dihydroxyphenyl)-2,2',3,3'-tetrahydro-3',5',7,7'-pentahydroxy-2-(4-hydroxyphenyl)-;; 3-(3,3',4',5,7-PENTAHYDROXYFLAVANON-8-YL)-4',5,7-TRIHYDROXYFLAVANONE; W/NBS
64828

Str 948

574 C30 H22 O12 21884-44-6 MY-1981-0-0 Rugulosin, 8,8'-dihydroxy-, (1S,1'S,2R,2'R,3S,3'S,9aR,9'aR)-;; Luteoskyrin; W
64830

574 C30 H22 O12 1685-91-2 MY-1981-0-0 [8,8'-Bi-1H-naphtho[2,3-c]pyran]-1,1',6,6',9,9'-hexone, 3,3',4,4'-tetrahydro-10,10'-dihydroxy-7,7'-dimethoxy-3,3'-dimethyl-, [R-(R*,R*)]-;; Xanthomegnin; W/NBS
64829

574 C30 H22 O12 MY-1981-0-0 RUBROSKYRINE; W
64831

574 C30 H22 O12 21884-47-9 NS-0-0-0 9,17-Methanonaphtho[2'3':5,6]cyclohept[1,2-d]anthracene-5,11,14,16,18(17H)-pentone, 7,8,8a,9-tetrahydro-1,4,6,8,10,15,19-heptahydroxy-2,12-dimethyl-, [8S-(8α,8aα,9α,17α,17aS*,19S*)]-; W/NBS
127253

Str 949

574 C30 H24 O8 P2 IC-1274-0-0 1,3-Bis(diphenylphosphono)benzene; W
127254

574 C30 H38 O11 29428-87-3 H-55-1707-6 ANODENDROSIDE-E2; W/NBS
64832

Str 950

574 C30 H38 O11 CD-479-0-0 Toosendanin; W
127255

574 C31 H54 O4 Si3 O-11-748-6 CANNABIDIOLIC ACID, TRIS(TRIMETHYLSILYL) ETHER; W
64833

574 C31 H54 O4 Si3 BS-4-7-0 Cannabidiolic acid trimethylsilate; W
127256

574 C32 H30 O10 62682-07-9 K-110-1062-0 3,14'-BIHISPIDINYL-HEXAMETHYLETHER; W
64834

574 C32 H46 O7 S 75894-01-8 C-103-122-0 Bis[2-(1-hydroxy-1-methyethyl)-4-(1,1-dimethyethyl-6-(carbethoxy)phenyl] sulfide; W
127257

Str 951

574 C33 H50 O8 55429-17-9 AD-0-3133-0 Spirostan-2,3,27-triol, triacetate, (2α,25s)-3β,5α,25s)-;; 5α,20βH,22α O,25 O-SPIROSTAN-2α,3β,27-TRIACETATE; W/NBS
64835

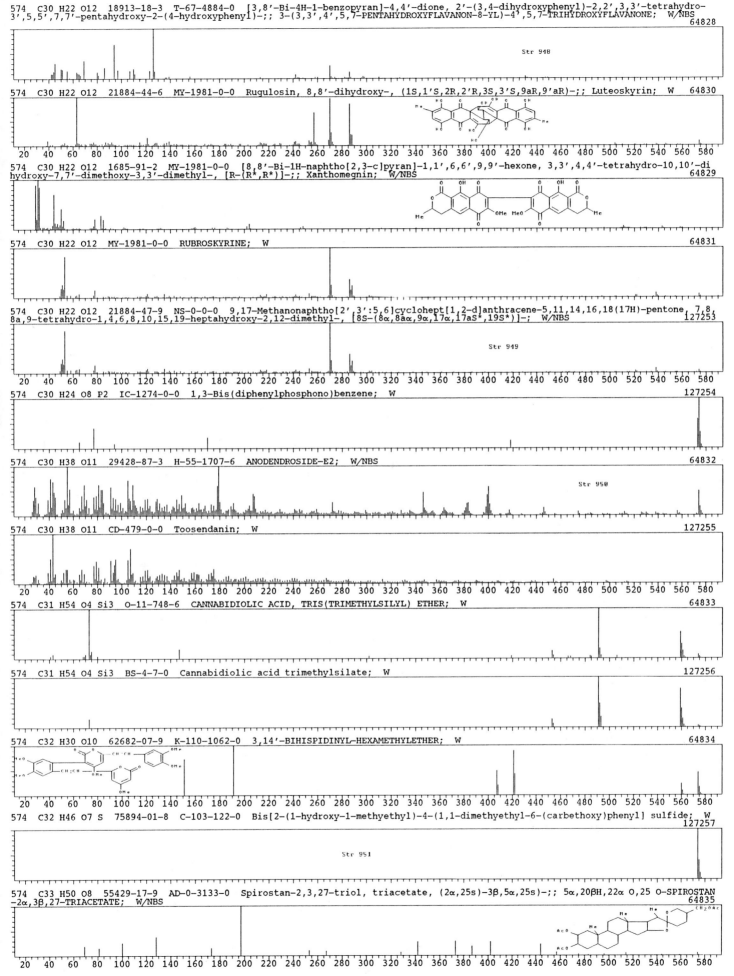

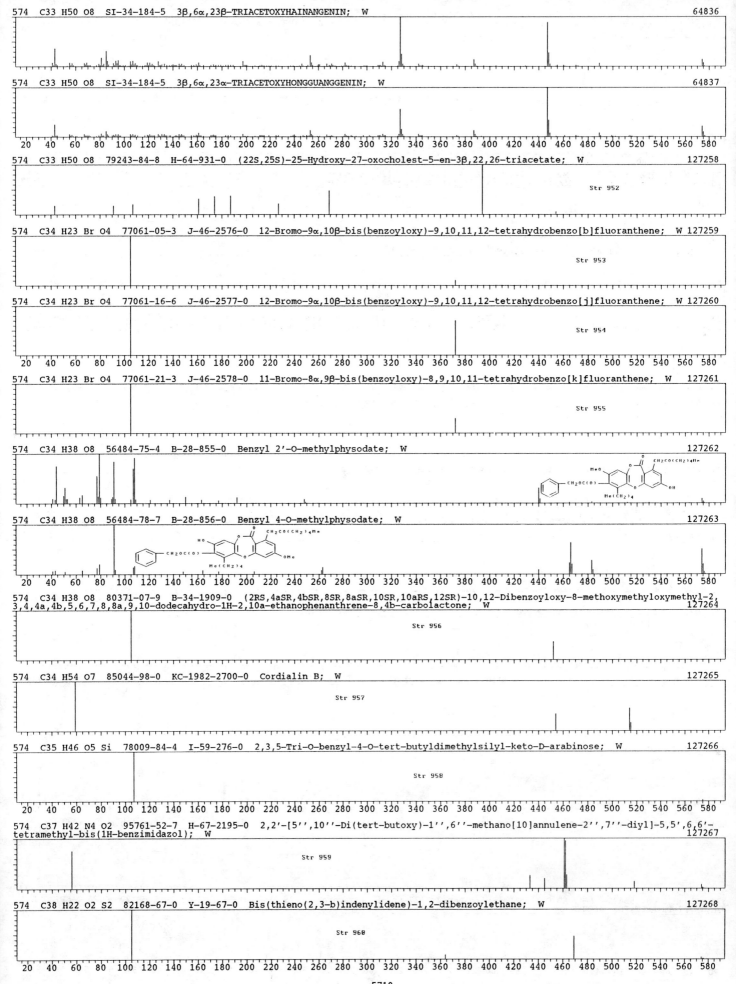

574　C33 H50 O8　SI-34-184-5　3β,6α,23β-TRIACETOXYHAINANGENIN;　W　64836

574　C33 H50 O8　SI-34-184-5　3β,6α,23α-TRIACETOXYHONGGUANGGENIN;　W　64837

574　C33 H50 O8　79243-84-8　H-64-931-0　(22S,25S)-25-Hydroxy-27-oxocholest-5-en-3β,22,26-triacetate;　W　127258
Str 952

574　C34 H23 Br O4　77061-05-3　J-46-2576-0　12-Bromo-9α,10β-bis(benzoyloxy)-9,10,11,12-tetrahydrobenzo[b]fluoranthene;　W 127259
Str 953

574　C34 H23 Br O4　77061-16-6　J-46-2577-0　12-Bromo-9α,10β-bis(benzoyloxy)-9,10,11,12-tetrahydrobenzo[j]fluoranthene;　W 127260
Str 954

574　C34 H23 Br O4　77061-21-3　J-46-2578-0　11-Bromo-8α,9β-bis(benzoyloxy)-8,9,10,11-tetrahydrobenzo[k]fluoranthene;　W　127261
Str 955

574　C34 H38 O8　56484-75-4　B-28-855-0　Benzyl 2'-O-methylphysodate;　W　127262

574　C34 H38 O8　56484-78-7　B-28-856-0　Benzyl 4-O-methylphysodate;　W　127263

574　C34 H38 O8　80371-07-9　B-34-1909-0　(2RS,4aSR,4bSR,8SR,8aSR,10SR,10aRS,12SR)-10,12-Dibenzoyloxy-8-methoxymethyloxymethyl-2,3,4,4a,4b,5,6,7,8,8a,9,10-dodecahydro-1H-2,10a-ethanophenanthrene-8,4b-carbolactone;　W　127264
Str 956

574　C34 H54 O7　85044-98-0　KC-1982-2700-0　Cordialin B;　W　127265
Str 957

574　C35 H46 O5 Si　78009-84-4　I-59-276-0　2,3,5-Tri-O-benzyl-4-O-tert-butyldimethylsilyl-keto-D-arabinose;　W　127266
Str 958

574　C37 H42 N4 O2　95761-52-7　H-67-2195-0　2,2'-[5'',10''-Di(tert-butoxy)-1'',6''-methano[10]annulene-2'',7''-diyl]-5,5',6,6'-tetramethyl-bis(1H-benzimidazol);　W　127267
Str 959

574　C38 H22 O2 S2　82168-67-0　Y-19-67-0　Bis(thieno(2,3-b)indenylidene)-1,2-dibenzoylethane;　W　127268
Str 960

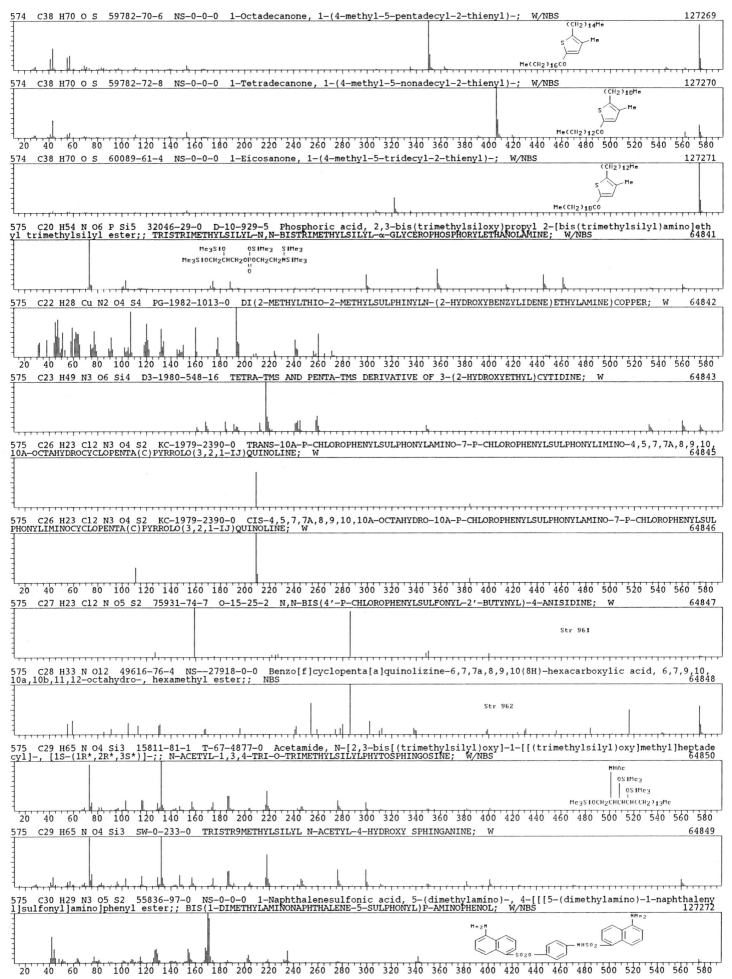

574   C38 H70 O S   59782-70-6   NS-0-0-0   1-Octadecanone, 1-(4-methyl-5-pentadecyl-2-thienyl)-;   W/NBS   127269

574   C38 H70 O S   59782-72-8   NS-0-0-0   1-Tetradecanone, 1-(4-methyl-5-nonadecyl-2-thienyl)-;   W/NBS   127270

574   C38 H70 O S   60089-61-4   NS-0-0-0   1-Eicosanone, 1-(4-methyl-5-tridecyl-2-thienyl)-;   W/NBS   127271

575   C20 H54 N O6 P Si5   32046-29-0   D-10-929-5   Phosphoric acid, 2,3-bis(trimethylsiloxy)propyl 2-[bis(trimethylsilyl)amino]eth yl trimethylsilyl ester;; TRISTRIMETHYLSILYL-N,N-BISTRIMETHYLSILYL-α-GLYCEROPHOSPHORYLETHANOLAMINE; W/NBS   64841

575   C22 H28 Cu N2 O4 S4   PG-1982-1013-0   DI(2-METHYLTHIO-2-METHYLSULPHINYLN-(2-HYDROXYBENZYLIDENE)ETHYLAMINE)COPPER;   W   64842

575   C23 H49 N3 O6 Si4   D3-1980-548-16   TETRA-TMS AND PENTA-TMS DERIVATIVE OF 3-(2-HYDROXYETHYL)CYTIDINE;   W   64843

575   C26 H23 Cl2 N3 O4 S2   KC-1979-2390-0   TRANS-10A-P-CHLOROPHENYLSULPHONYLAMINO-7-P-CHLOROPHENYLSULPHONYLIMINO-4,5,7,7A,8,9,10, 10A-OCTAHYDROCYCLOPENTA(C)PYRROLO(3,2,1-IJ)QUINOLINE;   W   64845

575   C26 H23 Cl2 N3 O4 S2   KC-1979-2390-0   CIS-4,5,7,7A,8,9,10,10A-OCTAHYDRO-10A-P-CHLOROPHENYLSULPHONYLAMINO-7-P-CHLOROPHENYLSUL PHONYLIMINOCYCLOPENTA(C)PYRROLO(3,2,1-IJ)QUINOLINE;   W   64846

575   C27 H23 Cl2 N O5 S2   75931-74-7   O-15-25-2   N,N-BIS(4'-P-CHLOROPHENYLSULFONYL-2'-BUTYNYL)-4-ANISIDINE;   W   64847

Str 961

575   C28 H33 N O12   49616-76-4   NS--27918-0-0   Benzo[f]cyclopenta[a]quinolizine-6,7,7a,8,9,10(8H)-hexacarboxylic acid, 6,7,9,10, 10a,10b,11,12-octahydro-, hexamethyl ester;;   NBS   64848

Str 962

575   C29 H65 N O4 Si3   15811-81-1   T-67-4877-0   Acetamide, N-[2,3-bis[(trimethylsilyl)oxy]-1-[[(trimethylsilyl)oxy]methyl]heptade cyl]-, [1S-(1R*,2R*,3S*)]-;; N-ACETYL-1,3,4-TRI-O-TRIMETHYLSILYLPHYTOSPHINGOSINE; W/NBS   64850

575   C29 H65 N O4 Si3   SW-0-233-0   TRISTR9METHYLSILYL N-ACETYL-4-HYDROXY SPHINGANINE;   W   64849

575   C30 H29 N3 O5 S2   55836-97-0   NS-0-0-0   1-Naphthalenesulfonic acid, 5-(dimethylamino)-, 4-[[[5-(dimethylamino)-1-naphthaleny l]sulfonyl]amino]phenyl ester;; BIS(1-DIMETHYLAMINONAPHTHALENE-5-SULPHONYL)P-AMINOPHENOL; W/NBS   127272

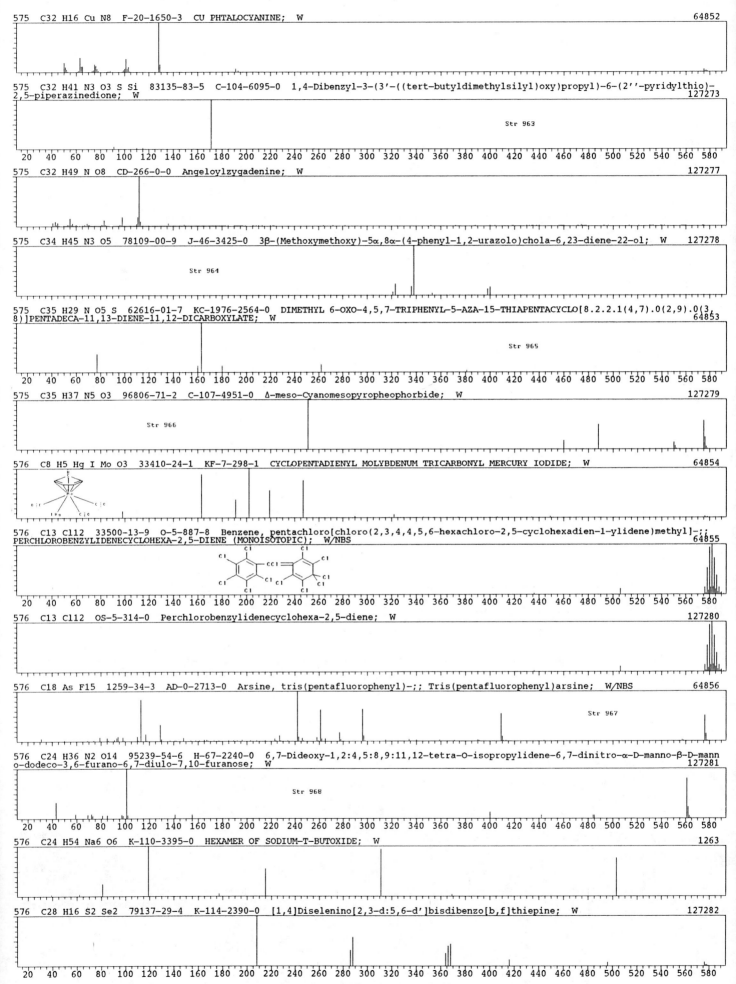

575 C32 H16 Cu N8 F-20-1650-3 CU PHTALOCYANINE; W   64852

575 C32 H41 N3 O3 S Si 83135-83-5 C-104-6095-0 1,4-Dibenzyl-3-(3'-((tert-butyldimethylsilyl)oxy)propyl)-6-(2''-pyridylthio)-
2,5-piperazinedione; W   127273

Str 963

575 C32 H49 N O8 CD-266-0-0 Angeloylzygadenine; W   127277

575 C34 H45 N3 O5 78109-00-9 J-46-3425-0 3β-(Methoxymethoxy)-5α,8α-(4-phenyl-1,2-urazolo)chola-6,23-diene-22-ol; W   127278

Str 964

575 C35 H29 N O5 S 62616-01-7 KC-1976-2564-0 DIMETHYL 6-OXO-4,5,7-TRIPHENYL-5-AZA-15-THIAPENTACYCLO[8.2.2.1(4,7).0(2,9).0(3,
8)]PENTADECA-11,13-DIENE-11,12-DICARBOXYLATE; W   64853

Str 965

575 C35 H37 N5 O3 96806-71-2 C-107-4951-0 Δ-meso-Cyanomesopyropheophorbide; W   127279

Str 966

576 C8 H5 Hg I Mo O3 33410-24-1 KF-7-298-1 CYCLOPENTADIENYL MOLYBDENUM TRICARBONYL MERCURY IODIDE; W   64854

576 C13 Cl12 33500-13-9 O-5-887-8 Benzene, pentachloro[chloro(2,3,4,4,5,6-hexachloro-2,5-cyclohexadien-1-ylidene)methyl]-;;
PERCHLOROBENZYLIDENECYCLOHEXA-2,5-DIENE (MONOISOTOPIC); W/NBS   64855

576 C13 Cl12 OS-5-314-0 Perchlorobenzylidenecyclohexa-2,5-diene; W   127280

576 C18 As F15 1259-34-3 AD-0-2713-0 Arsine, tris(pentafluorophenyl)-;; Tris(pentafluorophenyl)arsine; W/NBS   64856

Str 967

576 C24 H36 N2 O14 95239-54-6 H-67-2240-0 6,7-Dideoxy-1,2:4,5:8,9:11,12-tetra-O-isopropylidene-6,7-dinitro-α-D-manno-β-D-mann
o-dodeco-3,6-furano-6,7-diulo-7,10-furanose; W   127281

Str 968

576 C24 H54 Na6 O6 K-110-3395-0 HEXAMER OF SODIUM-T-BUTOXIDE; W   1263

576 C28 H16 S2 Se2 79137-29-4 K-114-2390-0 [1,4]Diselenino[2,3-d:5,6-d']bisdibenzo[b,f]thiepine; W   127282

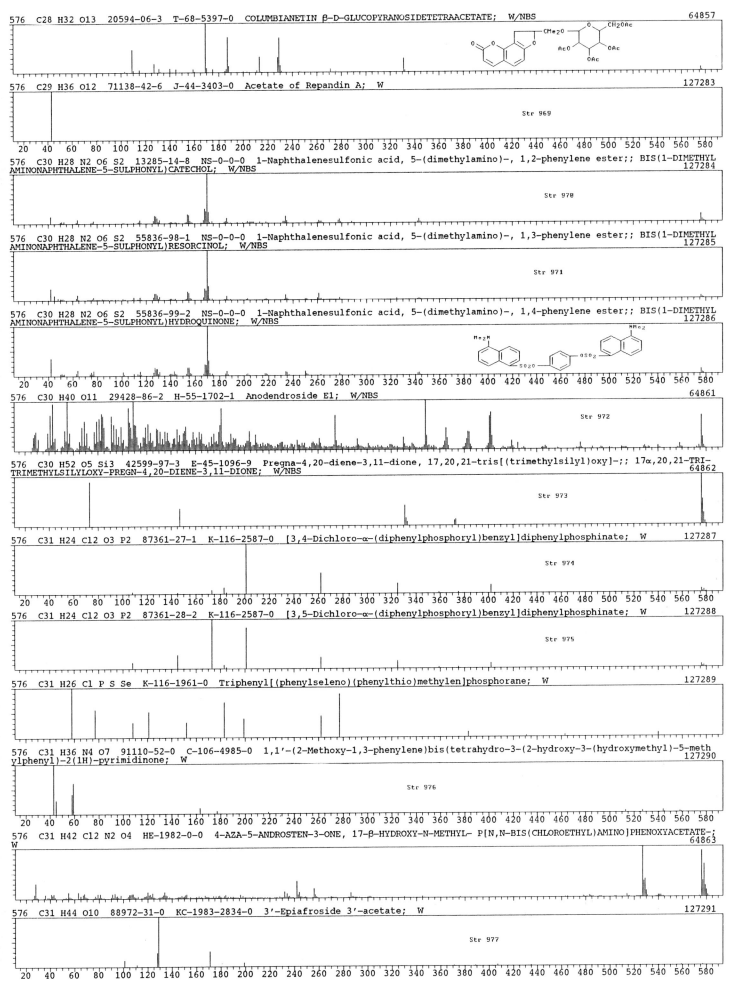

576   C28 H32 O13   20594-06-3   T-68-5397-0   COLUMBIANETIN β-D-GLUCOPYRANOSIDETETRAACETATE;   W/NBS   64857

576   C29 H36 O12   71138-42-6   J-44-3403-0   Acetate of Repandin A;   W   127283

Str 969

576   C30 H28 N2 O6 S2   13285-14-8   NS-0-0-0   1-Naphthalenesulfonic acid, 5-(dimethylamino)-, 1,2-phenylene ester;; BIS(1-DIMETHYL
AMINONAPHTHALENE-5-SULPHONYL)CATECHOL;   W/NBS   127284

Str 970

576   C30 H28 N2 O6 S2   55836-98-1   NS-0-0-0   1-Naphthalenesulfonic acid, 5-(dimethylamino)-, 1,3-phenylene ester;; BIS(1-DIMETHYL
AMINONAPHTHALENE-5-SULPHONYL)RESORCINOL;   W/NBS   127285

Str 971

576   C30 H28 N2 O6 S2   55836-99-2   NS-0-0-0   1-Naphthalenesulfonic acid, 5-(dimethylamino)-, 1,4-phenylene ester;; BIS(1-DIMETHYL
AMINONAPHTHALENE-5-SULPHONYL)HYDROQUINONE;   W/NBS   127286

576   C30 H40 O11   29428-86-2   H-55-1702-1   Anodendroside E1;   W/NBS   64861

Str 972

576   C30 H52 O5 Si3   42599-97-3   E-45-1096-9   Pregna-4,20-diene-3,11-dione, 17,20,21-tris[(trimethylsilyl)oxy]-;; 17α,20,21-TRI-
TRIMETHYLSILYLOXY-PREGN-4,20-DIENE-3,11-DIONE;   W/NBS   64862

Str 973

576   C31 H24 Cl2 O3 P2   87361-27-1   K-116-2587-0   [3,4-Dichloro-α-(diphenylphosphoryl)benzyl]diphenylphosphinate;   W   127287

Str 974

576   C31 H24 Cl2 O3 P2   87361-28-2   K-116-2587-0   [3,5-Dichloro-α-(diphenylphosphoryl)benzyl]diphenylphosphinate;   W   127288

Str 975

576   C31 H26 Cl P S Se   K-116-1961-0   Triphenyl[(phenylseleno)(phenylthio)methylen]phosphorane;   W   127289

576   C31 H36 N4 O7   91110-52-0   C-106-4985-0   1,1'-(2-Methoxy-1,3-phenylene)bis(tetrahydro-3-(2-hydroxy-3-(hydroxymethyl)-5-meth
ylphenyl)-2(1H)-pyrimidinone;   W   127290

Str 976

576   C31 H42 Cl2 N2 O4   HE-1982-0-0   4-AZA-5-ANDROSTEN-3-ONE, 17-β-HYDROXY-N-METHYL- P[N,N-BIS(CHLOROETHYL)AMINO]PHENOXYACETATE-;
W   64863

576   C31 H44 O10   88972-31-0   KC-1983-2834-0   3'-Epiafroside 3'-acetate;   W   127291

Str 977

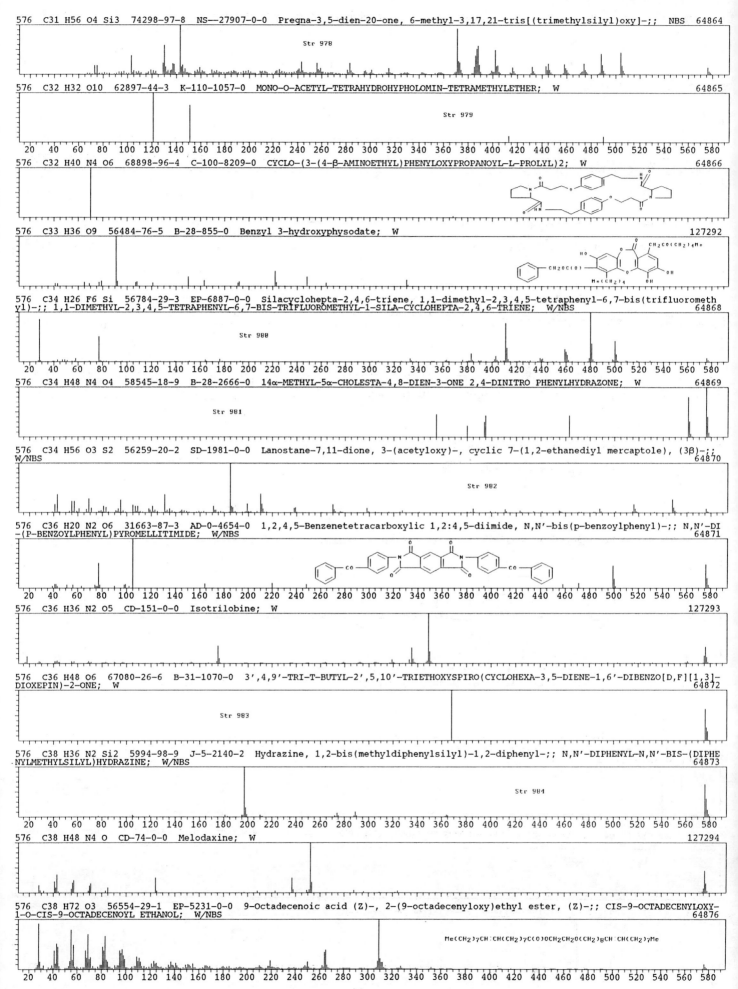

576   C31 H56 O4 Si3   74298-97-8   NS--27907-0-0   Pregna-3,5-dien-20-one, 6-methyl-3,17,21-tris[(trimethylsilyl)oxy]-;;   NBS   64864

Str 978

576   C32 H32 O10   62897-44-3   K-110-1057-0   MONO-O-ACETYL-TETRAHYDROHYPHOLOMIN-TETRAMETHYLETHER;   W   64865

Str 979

576   C32 H40 N4 O6   68898-96-4   C-100-8209-0   CYCLO-(3-(4-β-AMINOETHYL)PHENYLOXYPROPANOYL-L-PROLYL)2;   W   64866

576   C33 H36 O9   56484-76-5   B-28-855-0   Benzyl 3-hydroxyphysodate;   W   127292

576   C34 H26 F6 Si   56784-29-3   EP-6887-0-0   Silacyclohepta-2,4,6-triene, 1,1-dimethyl-2,3,4,5-tetraphenyl-6,7-bis(trifluoromethyl)-;; 1,1-DIMETHYL-2,3,4,5-TETRAPHENYL-6,7-BIS-TRIFLUOROMETHYL-1-SILA-CYCLOHEPTA-2,4,6-TRIENE;   W/NBS   64868

Str 980

576   C34 H48 N4 O4   58545-18-9   B-28-2666-0   14α-METHYL-5α-CHOLESTA-4,8-DIEN-3-ONE 2,4-DINITRO PHENYLHYDRAZONE;   W   64869

Str 981

576   C34 H56 O3 S2   56259-20-2   SD-1981-0-0   Lanostane-7,11-dione, 3-(acetyloxy)-, cyclic 7-(1,2-ethanediyl mercaptole), (3β)-;; W/NBS   64870

Str 982

576   C36 H20 N2 O6   31663-87-3   AD-0-4654-0   1,2,4,5-Benzenetetracarboxylic 1,2:4,5-diimide, N,N'-bis(p-benzoylphenyl)-;; N,N'-DI-(P-BENZOYLPHENYL)PYROMELLITIMIDE;   W/NBS   64871

576   C36 H36 N2 O5   CD-151-0-0   Isotrilobine;   W   127293

576   C36 H48 O6   67080-26-6   B-31-1070-0   3',4,9'-TRI-T-BUTYL-2',5,10'-TRIETHOXYSPIRO(CYCLOHEXA-3,5-DIENE-1,6'-DIBENZO[D,F][1,3]-DIOXEPIN)-2-ONE;   W   64872

Str 983

576   C38 H36 N2 Si2   5994-98-9   J-5-2140-2   Hydrazine, 1,2-bis(methyldiphenylsilyl)-1,2-diphenyl-;; N,N'-DIPHENYL-N,N'-BIS-(DIPHENYLMETHYLSILYL)HYDRAZINE;   W/NBS   64873

Str 984

576   C38 H48 N4 O   CD-74-0-0   Melodaxine;   W   127294

576   C38 H72 O3   56554-29-1   EP-5231-0-0   9-Octadecenoic acid (Z)-, 2-(9-octadecenyloxy)ethyl ester, (Z)-;; CIS-9-OCTADECENYLOXY-1-O-CIS-9-OCTADECENOYL ETHANOL;   W/NBS   64876

Me(CH₂)₇CH:CH(CH₂)₇C(O)OCH₂CH₂O(CH₂)₈CH:CH(CH₂)₇Me

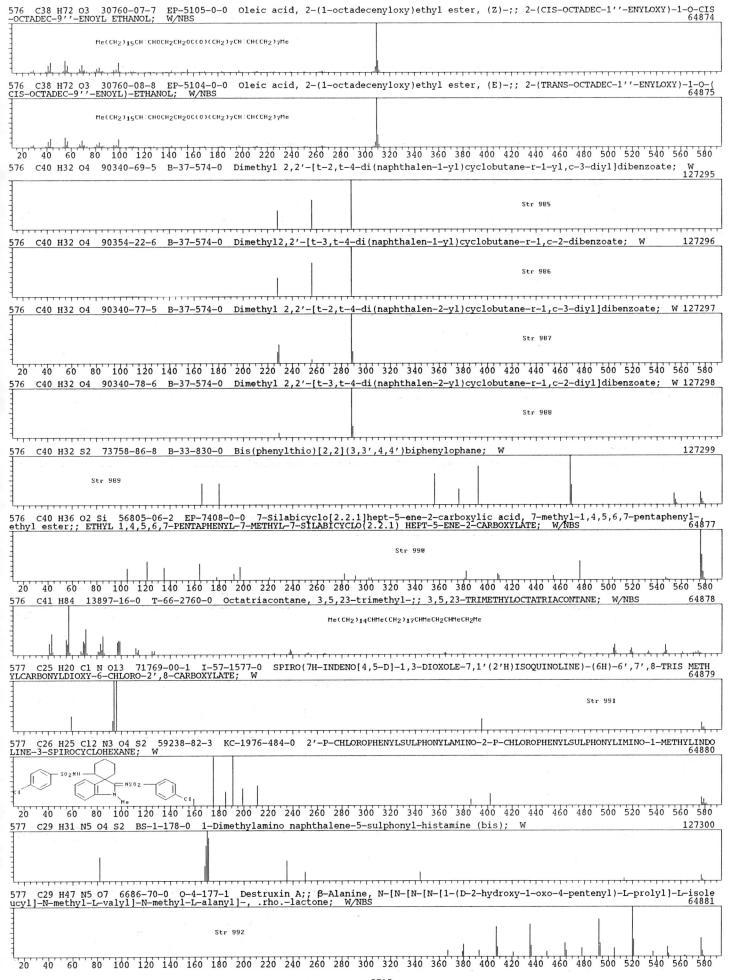

576 C38 H72 O3 30760-07-7 EP-5105-0-0 Oleic acid, 2-(1-octadecenyloxy)ethyl ester, (Z)-;; 2-(CIS-OCTADEC-1''-ENYLOXY)-1-O-CIS-OCTADEC-9''-ENOYL ETHANOL; W/NBS
64874

Me(CH₂)₁₅CH:CHOCH₂CH₂OC(O)(CH₂)₇CH:CH(CH₂)₇Me

576 C38 H72 O3 30760-08-8 EP-5104-0-0 Oleic acid, 2-(1-octadecenyloxy)ethyl ester, (E)-;; 2-(TRANS-OCTADEC-1''-ENYLOXY)-1-O-(CIS-OCTADEC-9''-ENOYL)-ETHANOL; W/NBS
64875

Me(CH₂)₁₅CH:CHOCH₂CH₂OC(O)(CH₂)₇CH:CH(CH₂)₇Me

576 C40 H32 O4 90340-69-5 B-37-574-0 Dimethyl 2,2'-[t-2,t-4-di(naphthalen-1-yl)cyclobutane-r-1-yl,c-3-diyl]dibenzoate; W
127295

Str 985

576 C40 H32 O4 90354-22-6 B-37-574-0 Dimethyl2,2'-[t-3,t-4-di(naphthalen-1-yl)cyclobutane-r-1,c-2-dibenzoate; W 127296

Str 986

576 C40 H32 O4 90340-77-5 B-37-574-0 Dimethyl 2,2'-[t-2,t-4-di(naphthalen-2-yl)cyclobutane-r-1,c-3-diyl]dibenzoate; W 127297

Str 987

576 C40 H32 O4 90340-78-6 B-37-574-0 Dimethyl 2,2'-[t-3,t-4-di(naphthalen-2-yl)cyclobutane-r-1,c-2-diyl]dibenzoate; W 127298

Str 988

576 C40 H32 S2 73758-86-8 B-33-830-0 Bis(phenylthio)[2,2](3,3',4,4')biphenylophane; W
127299

Str 989

576 C40 H36 O2 Si 56805-06-2 EP-7408-0-0 7-Silabicyclo[2.2.1]hept-5-ene-2-carboxylic acid, 7-methyl-1,4,5,6,7-pentaphenyl-ethyl ester;; ETHYL 1,4,5,6,7-PENTAPHENYL-7-METHYL-7-SILABICYCLO(2.2.1) HEPT-5-ENE-2-CARBOXYLATE; W/NBS
64877

Str 990

576 C41 H84 13897-16-0 T-66-2760-0 Octatriacontane, 3,5,23-trimethyl-;; 3,5,23-TRIMETHYLOCTATRIACONTANE; W/NBS
64878

Me(CH₂)₁₄CHMe(CH₂)₁₇CHMeCH₂CHMeCH₂Me

577 C25 H20 Cl N O13 71769-00-1 I-57-1577-0 SPIRO(7H-INDENO[4,5-D]-1,3-DIOXOLE-7,1'(2'H)ISOQUINOLINE)-(6H)-6',7',8-TRIS METHYLCARBONYLDIOXY-6-CHLORO-2',8-CARBOXYLATE; W
64879

Str 991

577 C26 H25 Cl2 N3 O4 S2 59238-82-3 KC-1976-484-0 2'-P-CHLOROPHENYLSULPHONYLAMINO-2-P-CHLOROPHENYLSULPHONYLIMINO-1-METHYLINDOLINE-3-SPIROCYCLOHEXANE; W
64880

577 C29 H31 N5 O4 S2 BS-1-178-0 1-Dimethylamino naphthalene-5-sulphonyl-histamine (bis); W
127300

577 C29 H47 N5 O7 6686-70-0 O-4-177-1 Destruxin A;; β-Alanine, N-[N-[N-[N-[1-(D-2-hydroxy-1-oxo-4-pentenyl)-L-prolyl]-L-isoleucyl]-N-methyl-L-valyl]-N-methyl-L-alanyl]-, .rho.-lactone; W/NBS
64881

Str 992

577　C31 H47 N O9　78018-27-6　J-46-3290-0　Glaucenine;　W　　　　　　　　　　　　127301

Str 993

577　C32 H39 N3 O7　60283-69-4　KC-1976-730-0　3β-ACETOXY-3',5'-DIOXO-4'-PHENYL-5α,8α-(1',2')-1',2',4'-TRIAAZOLIDINO-6,7-SECO-23,24-DINOR-CHLOLANE-6,7,22-TRIAL;　W　　　　　　　　　　64882

Str 994

20　40　60　80　100　120　140　160　180　200　220　240　260　280　300　320　340　360　380　400　420　440　460　480　500　520　540　560　580

578　C6 H5 I3 Sn　14532-07-1　K-110-1554-0　IODOPHENYLSTANNANEN;　W　　　　　　　　　　64883

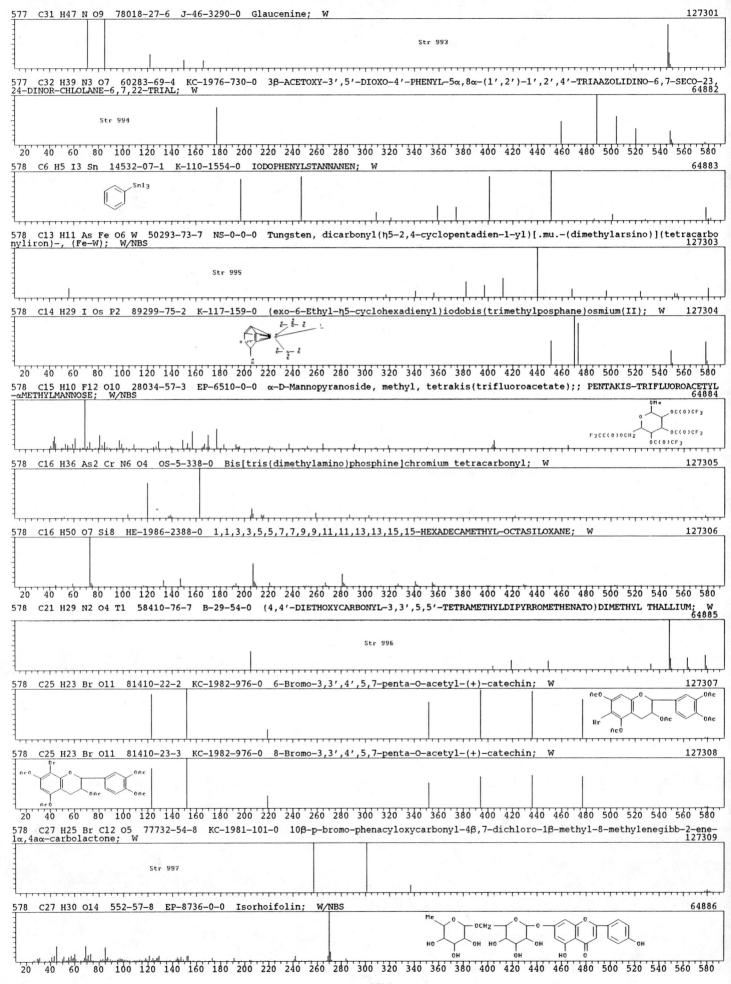

578　C13 H11 As Fe O6 W　50293-73-7　NS-0-0-0　Tungsten, dicarbonyl(η5-2,4-cyclopentadien-1-yl)[.mu.-(dimethylarsino)](tetracarbonyliron)-, (Fe-W);　W/NBS　　　　　　　　127303

Str 995

578　C14 H29 I Os P2　89299-75-2　K-117-159-0　(exo-6-Ethyl-η5-cyclohexadienyl)iodobis(trimethylposphane)osmium(II);　W　　127304

20　40　60　80　100　120　140　160　180　200　220　240　260　280　300　320　340　360　380　400　420　440　460　480　500　520　540　560　580

578　C15 H10 F12 O10　28034-57-3　EP-6510-0-0　α-D-Mannopyranoside, methyl, tetrakis(trifluoroacetate);; PENTAKIS-TRIFLUOROACETYL-αMETHYLMANNOSE;　W/NBS　　　　　　　　　64884

578　C16 H36 As2 Cr N6 O4　OS-5-338-0　Bis[tris(dimethylamino)phosphine]chromium tetracarbonyl;　W　　　127305

578　C16 H50 O7 Si8　HE-1986-2388-0　1,1,3,3,5,5,7,7,9,9,11,11,13,13,15,15-HEXADECAMETHYL-OCTASILOXANE;　W　　127306

20　40　60　80　100　120　140　160　180　200　220　240　260　280　300　320　340　360　380　400　420　440　460　480　500　520　540　560　580

578　C21 H29 N2 O4 Tl　58410-76-7　B-29-54-0　(4,4'-DIETHOXYCARBONYL-3,3',5,5'-TETRAMETHYLDIPYRROMETHENATO)DIMETHYL THALLIUM;　W　　64885

Str 996

578　C25 H23 Br O11　81410-22-2　KC-1982-976-0　6-Bromo-3,3',4',5,7-penta-O-acetyl-(+)-catechin;　W　　127307

578　C25 H23 Br O11　81410-23-3　KC-1982-976-0　8-Bromo-3,3',4',5,7-penta-O-acetyl-(+)-catechin;　W　　127308

20　40　60　80　100　120　140　160　180　200　220　240　260　280　300　320　340　360　380　400　420　440　460　480　500　520　540　560　580

578　C27 H25 Br Cl2 O5　77732-54-8　KC-1981-101-0　10β-p-bromo-phenacyloxycarbonyl-4β,7-dichloro-1β-methyl-8-methylenegibb-2-ene-1α,4aα-carbolactone;　W　　127309

Str 997

578　C27 H30 O14　552-57-8　EP-8736-0-0　Isorhoifolin;　W/NBS　　　　　64886

20　40　60　80　100　120　140　160　180　200　220　240　260　280　300　320　340　360　380　400　420　440　460　480　500　520　540　560　580

578 C27 H30 O14 EP-8738-0-0 5167202003 FLAVONE 5,7 OH, 6C GLYCOSID; W   64887

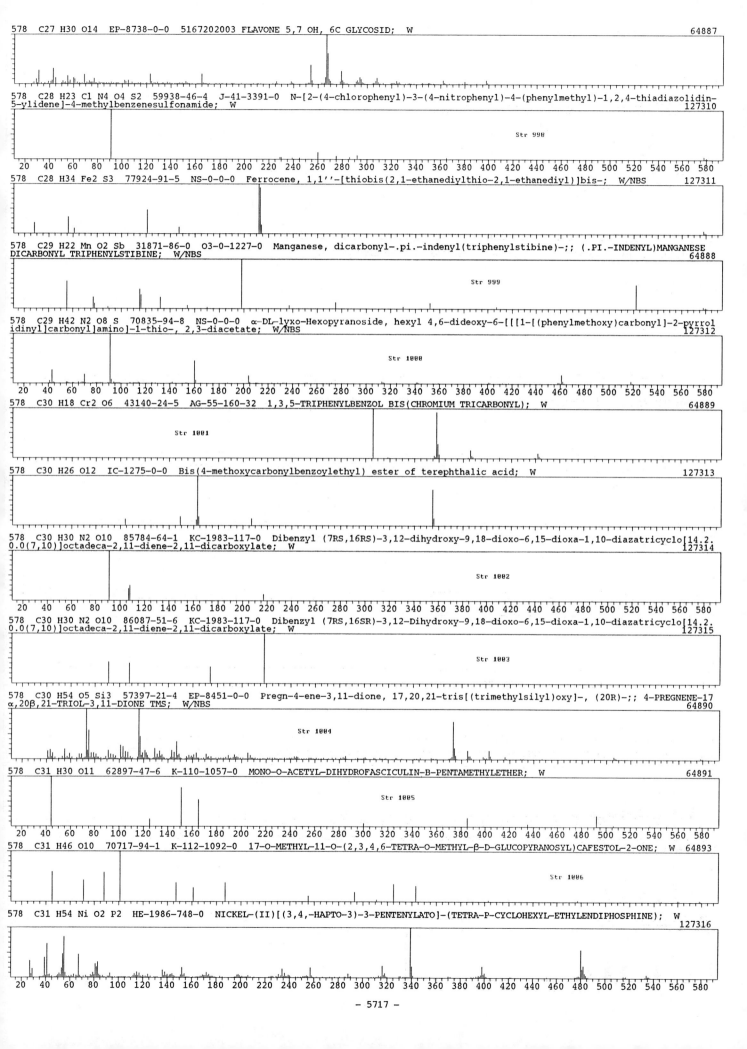

578 C28 H23 Cl N4 O4 S2 59938-46-4 J-41-3391-0 N-[2-(4-chlorophenyl)-3-(4-nitrophenyl)-4-(phenylmethyl)-1,2,4-thiadiazolidin-
5-ylidene]-4-methylbenzenesulfonamide; W   127310

Str 998

578 C28 H34 Fe2 S3 77924-91-5 NS-0-0-0 Ferrocene, 1,1''-[thiobis(2,1-ethanediylthio-2,1-ethanediyl)]bis-; W/NBS   127311

578 C29 H22 Mn O2 Sb 31871-86-0 O3-0-1227-0 Manganese, dicarbonyl-.pi.-indenyl(triphenylstibine)-;; (.PI.-INDENYL)MANGANESE
DICARBONYL TRIPHENYLSTIBINE; W/NBS   64888

Str 999

578 C29 H42 N2 O8 S 70835-94-8 NS-0-0-0 α-DL-lyxo-Hexopyranoside, hexyl 4,6-dideoxy-6-[[[1-[(phenylmethoxy)carbonyl]-2-pyrrol
idinyl]carbonyl]amino]-1-thio-, 2,3-diacetate; W/NBS   127312

Str 1000

578 C30 H18 Cr2 O6 43140-24-5 AG-55-160-32 1,3,5-TRIPHENYLBENZOL BIS(CHROMIUM TRICARBONYL); W   64889

Str 1001

578 C30 H26 O12 IC-1275-0-0 Bis(4-methoxycarbonylbenzoylethyl) ester of terephthalic acid; W   127313

578 C30 H30 N2 O10 85784-64-1 KC-1983-117-0 Dibenzyl (7RS,16RS)-3,12-dihydroxy-9,18-dioxo-6,15-dioxa-1,10-diazatricyclo[14.2.
0.0(7,10)]octadeca-2,11-diene-2,11-dicarboxylate; W   127314

Str 1002

578 C30 H30 N2 O10 86087-51-6 KC-1983-117-0 Dibenzyl (7RS,16SR)-3,12-Dihydroxy-9,18-dioxo-6,15-dioxa-1,10-diazatricyclo[14.2.
0.0(7,10)]octadeca-2,11-diene-2,11-dicarboxylate; W   127315

Str 1003

578 C30 H54 O5 Si3 57397-21-4 EP-8451-0-0 Pregn-4-ene-3,11-dione, 17,20,21-tris[(trimethylsilyl)oxy]-, (20R)-;; 4-PREGNENE-17
α,20β,21-TRIOL-3,11-DIONE TMS; W/NBS   64890

Str 1004

578 C31 H30 O11 62897-47-6 K-110-1057-0 MONO-O-ACETYL-DIHYDROFASCICULIN-B-PENTAMETHYLETHER; W   64891

Str 1005

578 C31 H46 O10 70717-94-1 K-112-1092-0 17-O-METHYL-11-O-(2,3,4,6-TETRA-O-METHYL-β-D-GLUCOPYRANOSYL)CAFESTOL-2-ONE; W 64893

Str 1006

578 C31 H54 Ni O2 P2 HE-1986-748-0 NICKEL-(II)[(3,4,-HAPTO-3)-3-PENTENYLATO]-(TETRA-P-CYCLOHEXYL-ETHYLENDIPHOSPHINE); W   127316

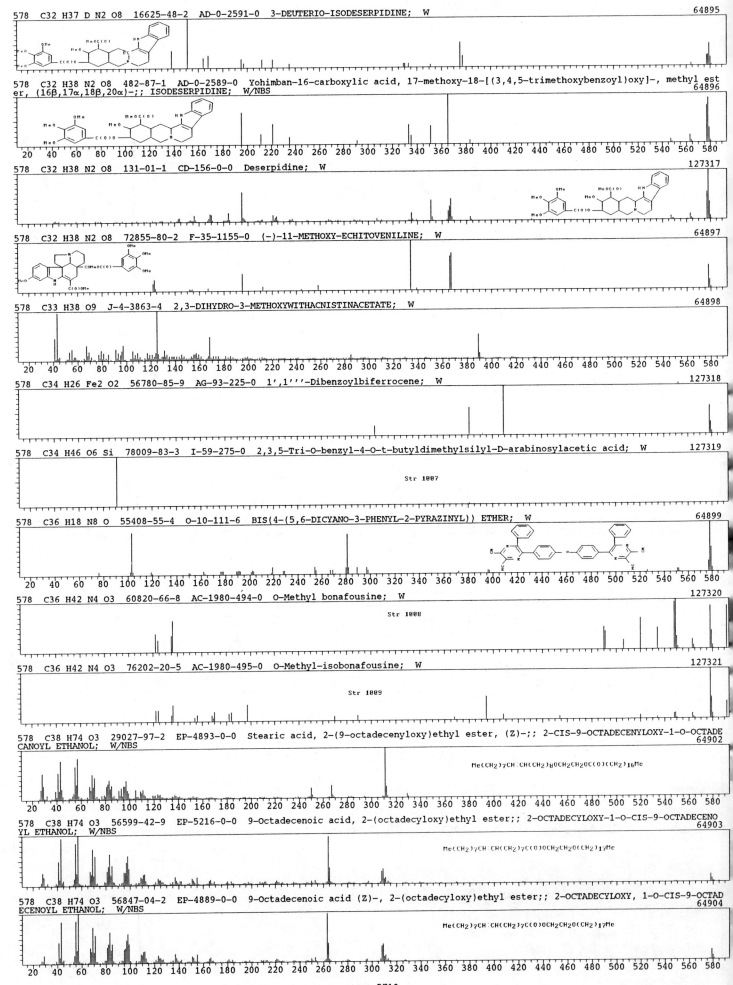

578   C32 H37 D N2 O8   16625-48-2   AD-0-2591-0   3-DEUTERIO-ISODESERPIDINE;   W   64895

578   C32 H38 N2 O8   482-87-1   AD-0-2589-0   Yohimban-16-carboxylic acid, 17-methoxy-18-[(3,4,5-trimethoxybenzoyl)oxy]-, methyl ester, (16β,17α,18β,20α)-;; ISODESERPIDINE;   W/NBS   64896

578   C32 H38 N2 O8   131-01-1   CD-156-0-0   Deserpidine;   W   127317

578   C32 H38 N2 O8   72855-80-2   F-35-1155-0   (-)-11-METHOXY-ECHITOVENILINE;   W   64897

578   C33 H38 O9   J-4-3863-4   2,3-DIHYDRO-3-METHOXYWITHACNISTINACETATE;   W   64898

578   C34 H26 Fe2 O2   56780-85-9   AG-93-225-0   1',1'''-Dibenzoylbiferrocene;   W   127318

578   C34 H46 O6 Si   78009-83-3   I-59-275-0   2,3,5-Tri-O-benzyl-4-O-t-butyldimethylsilyl-D-arabinosylacetic acid;   W   127319
Str 1007

578   C36 H18 N8 O   55408-55-4   O-10-111-6   BIS(4-(5,6-DICYANO-3-PHENYL-2-PYRAZINYL)) ETHER;   W   64899

578   C36 H42 N4 O3   60820-66-8   AC-1980-494-0   O-Methyl bonafousine;   W   127320
Str 1008

578   C36 H42 N4 O3   76202-20-5   AC-1980-495-0   O-Methyl-isobonafousine;   W   127321
Str 1009

578   C38 H74 O3   29027-97-2   EP-4893-0-0   Stearic acid, 2-(9-octadecenyloxy)ethyl ester, (Z)-;; 2-CIS-9-OCTADECENYLOXY-1-O-OCTADECANOYL ETHANOL;   W/NBS   64902
Me(CH2)7CH:CH(CH2)8OCH2CH2OC(O)(CH2)16Me

578   C38 H74 O3   56599-42-9   EP-5216-0-0   9-Octadecenoic acid, 2-(octadecyloxy)ethyl ester;; 2-OCTADECYLOXY-1-O-CIS-9-OCTADECENOYL ETHANOL;   W/NBS   64903
Me(CH2)7CH:CH(CH2)7C(O)OCH2CH2O(CH2)17Me

578   C38 H74 O3   56847-04-2   EP-4889-0-0   9-Octadecenoic acid (Z)-, 2-(octadecyloxy)ethyl ester;; 2-OCTADECYLOXY, 1-O-CIS-9-OCTADECENOYL ETHANOL;   W/NBS   64904
Me(CH2)7CH:CH(CH2)7C(O)OCH2CH2O(CH2)17Me

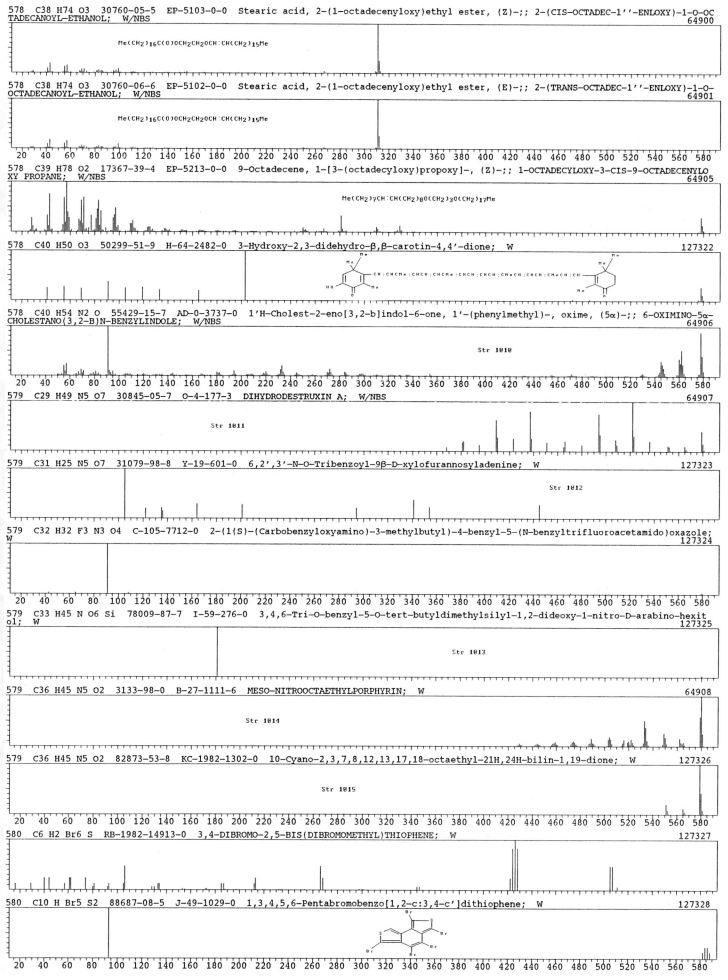

578 C38 H74 O3 30760-05-5 EP-5103-0-0 Stearic acid, 2-(1-octadecenyloxy)ethyl ester, (Z)-;; 2-(CIS-OCTADEC-1''-ENLOXY)-1-O-OCTADECANOYL-ETHANOL; W/NBS
64900

Me(CH2)16C(O)OCH2CH2OCH:CH(CH2)15Me

578 C38 H74 O3 30760-06-6 EP-5102-0-0 Stearic acid, 2-(1-octadecenyloxy)ethyl ester, (E)-;; 2-(TRANS-OCTADEC-1''-ENLOXY)-1-O-OCTADECANOYL-ETHANOL; W/NBS
64901

Me(CH2)16C(O)OCH2CH2OCH:CH(CH2)15Me

578 C39 H78 O2 17367-39-4 EP-5213-0-0 9-Octadecene, 1-[3-(octadecyloxy)propoxy]-, (Z)-;; 1-OCTADECYLOXY-3-CIS-9-OCTADECENYLOXY PROPANE; W/NBS
64905

Me(CH2)7CH:CH(CH2)8O(CH2)3O(CH2)17Me

578 C40 H50 O3 50299-51-9 H-64-2482-0 3-Hydroxy-2,3-didehydro-β,β-carotin-4,4'-dione; W
127322

578 C40 H54 N2 O 55429-15-7 AD-0-3737-0 1'H-Cholest-2-eno[3,2-b]indol-6-one, 1'-(phenylmethyl)-, oxime, (5α)-;; 6-OXIMINO-5α-CHOLESTANO(3,2-B)N-BENZYLINDOLE; W/NBS
64906

Str 1010

579 C29 H49 N5 O7 30845-05-7 O-4-177-3 DIHYDRODESTRUXIN A; W/NBS
64907

Str 1011

579 C31 H25 N5 O7 31079-98-8 Y-19-601-0 6,2',3'-N-O-Tribenzoyl-9β-D-xylofurannosyladenine; W
127323

Str 1012

579 C32 H32 F3 N3 O4 C-105-7712-0 2-(1(S)-(Carbobenzyloxyamino)-3-methylbutyl)-4-benzyl-5-(N-benzyltrifluoroacetamido)oxazole; W
127324

579 C33 H45 N O6 Si 78009-87-7 I-59-276-0 3,4,6-Tri-O-benzyl-5-O-tert-butyldimethylsilyl-1,2-dideoxy-1-nitro-D-arabino-hexitol; W
127325

Str 1013

579 C36 H45 N5 O2 3133-98-0 B-27-1111-6 MESO-NITROOCTAETHYLPORPHYRIN; W
64908

Str 1014

579 C36 H45 N5 O2 82873-53-8 KC-1982-1302-0 10-Cyano-2,3,7,8,12,13,17,18-octaethyl-21H,24H-bilin-1,19-dione; W
127326

Str 1015

580 C6 H2 Br6 S RB-1982-14913-0 3,4-DIBROMO-2,5-BIS(DIBROMOMETHYL)THIOPHENE; W
127327

580 C10 H Br5 S2 88687-08-5 J-49-1029-0 1,3,4,5,6-Pentabromobenzo[1,2-c:3,4-c']dithiophene; W
127328

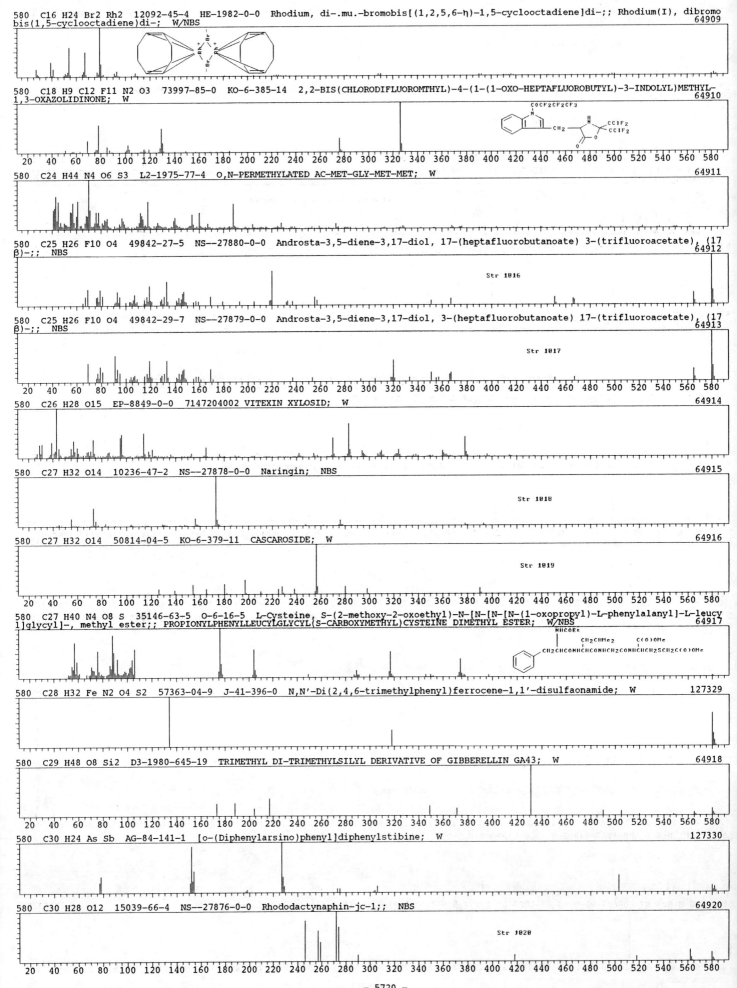

580 C16 H24 Br2 Rh2 12092-45-4 HE-1982-0-0 Rhodium, di-.mu.-bromobis[(1,2,5,6-η)-1,5-cyclooctadiene]di-;; Rhodium(I), dibromo bis(1,5-cyclooctadiene)di-; W/NBS
64909

580 C18 H9 Cl2 F11 N2 O3 73997-85-0 KO-6-385-14 2,2-BIS(CHLORODIFLUOROMTHYL)-4-(1-(1-OXO-HEPTAFLUOROBUTYL)-3-INDOLYL)METHYL-1,3-OXAZOLIDINONE; W
64910

580 C24 H44 N4 O6 S3 L2-1975-77-4 O,N-PERMETHYLATED AC-MET-GLY-MET-MET; W
64911

580 C25 H26 F10 O4 49842-27-5 NS--27880-0-0 Androsta-3,5-diene-3,17-diol, 17-(heptafluorobutanoate) 3-(trifluoroacetate), (17β)-;; NBS
64912

Str 1016

580 C25 H26 F10 O4 49842-29-7 NS--27879-0-0 Androsta-3,5-diene-3,17-diol, 3-(heptafluorobutanoate) 17-(trifluoroacetate), (17β)-;; NBS
64913

Str 1017

580 C26 H28 O15 EP-8849-0-0 7147204002 VITEXIN XYLOSID; W
64914

580 C27 H32 O14 10236-47-2 NS--27878-0-0 Naringin; NBS
64915

Str 1018

580 C27 H32 O14 50814-04-5 KO-6-379-11 CASCAROSIDE; W
64916

Str 1019

580 C27 H40 N4 O8 S 35146-63-5 O-6-16-5 L-Cysteine, S-(2-methoxy-2-oxoethyl)-N-[N-[N-[N-(1-oxopropyl)-L-phenylalanyl]-L-leucyl]glycyl]-, methyl ester;; PROPIONYLPHENYLLEUCYLGLYCYL(S-CARBOXYMETHYL)CYSTEINE DIMETHYL ESTER; W/NBS
64917

580 C28 H32 Fe N2 O4 S2 57363-04-9 J-41-396-0 N,N'-Di(2,4,6-trimethylphenyl)ferrocene-1,1'-disulfaonamide; W
127329

580 C29 H48 O8 Si2 D3-1980-645-19 TRIMETHYL DI-TRIMETHYLSILYL DERIVATIVE OF GIBBERELLIN GA43; W
64918

580 C30 H24 As Sb AG-84-141-1 [o-(Diphenylarsino)phenyl]diphenylstibine; W
127330

580 C30 H28 O12 15039-66-4 NS--27876-0-0 Rhododactynaphin-jc-1;; NBS
64920

Str 1020

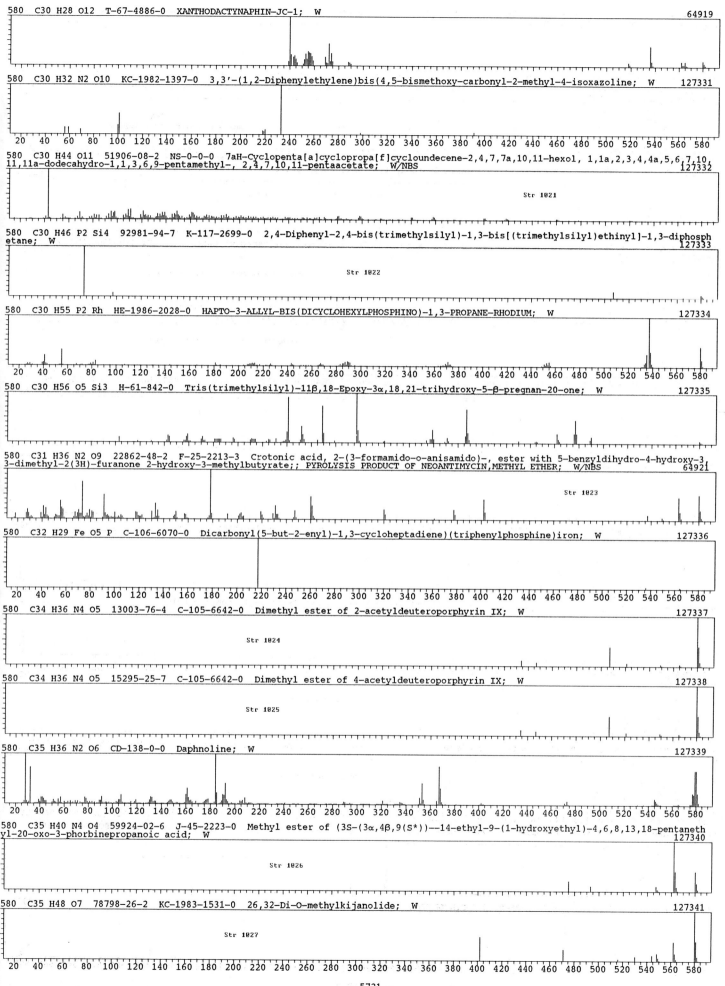

580   C30 H28 O12   T-67-4886-0   XANTHODACTYNAPHIN-JC-1;   W   64919

580   C30 H32 N2 O10   KC-1982-1397-0   3,3'-(1,2-Diphenylethylene)bis(4,5-bismethoxy-carbonyl-2-methyl-4-isoxazoline;   W   127331

580   C30 H44 O11   51906-08-2   NS-0-0-0   7aH-Cyclopenta[a]cyclopropa[f]cycloundecene-2,4,7,7a,10,11-hexol, 1,1a,2,3,4,4a,5,6,7,10,
11,11a-dodecahydro-1,1,3,6,9-pentamethyl-, 2,4,7,10,11-pentaacetate;   W/NBS   127332
                                                        Str 1021

580   C30 H46 P2 Si4   92981-94-7   K-117-2699-0   2,4-Diphenyl-2,4-bis(trimethylsilyl)-1,3-bis[(trimethylsilyl)ethinyl]-1,3-diphosph
etane;   W   127333
                                           Str 1022

580   C30 H55 P2 Rh   HE-1986-2028-0   HAPTO-3-ALLYL-BIS(DICYCLOHEXYLPHOSPHINO)-1,3-PROPANE-RHODIUM;   W   127334

580   C30 H56 O5 Si3   H-61-842-0   Tris(trimethylsilyl)-11β,18-Epoxy-3α,18,21-trihydroxy-5-β-pregnan-20-one;   W   127335

580   C31 H36 N2 O9   22862-48-2   F-25-2213-3   Crotonic acid, 2-(3-formamido-o-anisamido)-, ester with 5-benzyldihydro-4-hydroxy-3,
3-dimethyl-2(3H)-furanone 2-hydroxy-3-methylbutyrate;;   PYROLYSIS PRODUCT OF NEOANTIMYCIN,METHYL ETHER;   W/NBS   64921
                                                        Str 1023

580   C32 H29 Fe O5 P   C-106-6070-0   Dicarbonyl(5-but-2-enyl)-1,3-cycloheptadiene)(triphenylphosphine)iron;   W   127336

580   C34 H36 N4 O5   13003-76-4   C-105-6642-0   Dimethyl ester of 2-acetyldeuteroporphyrin IX;   W   127337
                                           Str 1024

580   C34 H36 N4 O5   15295-25-7   C-105-6642-0   Dimethyl ester of 4-acetyldeuteroporphyrin IX;   W   127338
                                           Str 1025

580   C35 H36 N2 O6   CD-138-0-0   Daphnoline;   W   127339

580   C35 H40 N4 O4   59924-02-6   J-45-2223-0   Methyl ester of (3S-(3α,4β,9(S*))--14-ethyl-9-(1-hydroxyethyl)-4,6,8,13,18-pentaneth
yl-20-oxo-3-phorbinepropanoic acid;   W   127340
                                           Str 1026

580   C35 H48 O7   78798-26-2   KC-1983-1531-0   26,32-Di-O-methylkijanolide;   W   127341
                                           Str 1027

580 C36 H52 O6 89766-96-1 KC-1983-2762-0 Acetyl derivative of phenylketo-(20S)-3β-Acetoxy-16-methoxy-16α,30-epoxy-24,25,26,27
-tetranor-5α-dammarane-23,20-carbolactone; W
127342

Str 1828

580 C37 H72 O4 56599-93-0 EP-5352-0-0 Hexadecanoic acid, 1-(1-methylethyl)-1,2-ethanediyl ester;; 3-METHYL-1,2-O-DIHEXADECANO
YL-1,2-BUTANEDIOL; W/NBS
64927

CHMe₂
Me(CH₂)₁₄C(O)OCH₂CHOC(O)(CH₂)₁₄Me

20 40 60 80 100 120 140 160 180 200 220 240 260 280 300 320 340 360 380 400 420 440 460 480 500 520 540 560 580

580 C37 H72 O4 26933-79-9 EP-5241-0-0 Hexadecanoic acid, 1,5-pentanediyl ester;; 1,5-PENTANEDIOL DIHEXADECANOATE; W/NBS
64926

Me(CH₂)₁₄C(O)O(CH₂)₅OC(O)(CH₂)₁₄Me

580 C37 H72 O4 56599-99-6 EP-5365-0-0 Hexadecanoic acid, 1,1,2-trimethyl-1,2-ethanediyl ester;; 2,3-DI-O-HEXADECANOYL-2-METH
YL-BUTANEDIOL; W/NBS
64922

Me(CH₂)₁₄C(O)OCHMeCMe₂OC(O)(CH₂)₁₄Me

580 C37 H71 D O4 56599-91-8 EP-5349-0-0 2,3-DI-O-HEXADECANOYL-2-METHYL-3-DEUTERO-BUTANEDIOL; W
64923

Me(CH₂)₁₄C(O)OCDMeCMe₂OC(O)(CH₂)₁₄Me

20 40 60 80 100 120 140 160 180 200 220 240 260 280 300 320 340 360 380 400 420 440 460 480 500 520 540 560 580

580 C37 H72 O4 EP-5351-0-0 3-METHYL-1,3-O-DIHEXADECANOYL-BUTANEDIOL; W
64924

580 C37 H72 O4 EP-5313-0-0 ISOPENTANEDIOL DIHEXADECANOATE; W
64925

580 C38 H19 Cl3 80034-30-6 B-34-1715-0 1,2,3-Trichloro-4,9,10-tris(phenylethynyl)anthracene; W
127343

Str 1829

20 40 60 80 100 120 140 160 180 200 220 240 260 280 300 320 340 360 380 400 420 440 460 480 500 520 540 560 580

580 C38 H36 N4 O2 IC-1276-0-0 1,2-Bis[4-[2-(2-tert-butylphenyl)-1,3,4-oxadiazol-5-yl]phenyl]ethylene; W
127344

580 C38 H76 O3 28843-25-6 EP-4857-0-0 Octadecanoic acid, 2-(octadecyloxy)ethyl ester;; 2-OCTADECYLOXY,1-O-OCTADECANOYL ETHA
NOL; W/NBS
64928

Me(CH₂)₁₆C(O)OCH₂CH₂O(CH₂)₁₇Me

580 C39 H80 O2 35545-51-8 EP-5005-0-0 Octadecane, 1,1'-[(1-methyl-1,2-ethanediyl)bis(oxy)]bis-;; 1,2-DIOCTADECYLOXY PROPANE;
W/NBS
64929

Me(CH₂)₁₇OCHMeCH₂O(CH₂)₁₇Me

20 40 60 80 100 120 140 160 180 200 220 240 260 280 300 320 340 360 380 400 420 440 460 480 500 520 540 560 580

580 C39 H80 O2 17367-38-3 EP-5237-0-0 Octadecane, 1,1'-[1,3-propanediylbis(oxy)]bis-;; 1,3-DIOCTADECYLOXY-PROPANE; W/NBS
64930

Me(CH₂)₁₇O(CH₂)₃O(CH₂)₁₇Me

580 C40 H52 O3 55906-76-8 H-60-2786-0 3,8-DIHYDROXY-K,.CHI.-CAROTEN-6-ONE; W
64931

20 40 60 80 100 120 140 160 180 200 220 240 260 280 300 320 340 360 380 400 420 440 460 480 500 520 540 560 580

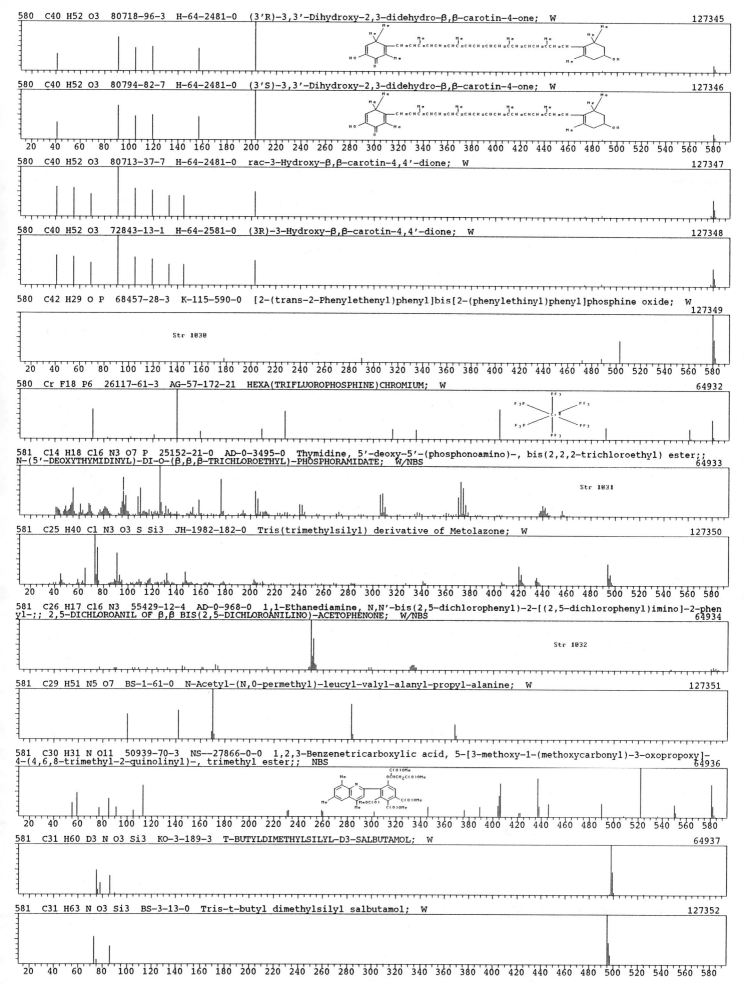

580   C40 H52 O3   80718-96-3   H-64-2481-0   (3'R)-3,3'-Dihydroxy-2,3-didehydro-β,β-carotin-4-one;   W

127345

580   C40 H52 O3   80794-82-7   H-64-2481-0   (3'S)-3,3'-Dihydroxy-2,3-didehydro-β,β-carotin-4-one;   W

127346

580   C40 H52 O3   80713-37-7   H-64-2481-0   rac-3-Hydroxy-β,β-carotin-4,4'-dione;   W

127347

580   C40 H52 O3   72843-13-1   H-64-2581-0   (3R)-3-Hydroxy-β,β-carotin-4,4'-dione;   W

127348

580   C42 H29 O P   68457-28-3   K-115-590-0   [2-(trans-2-Phenylethenyl)phenyl]bis[2-(phenylethinyl)phenyl]phosphine oxide;   W

127349

Str 1030

580   Cr F18 P6   26117-61-3   AG-57-172-21   HEXA(TRIFLUOROPHOSPHINE)CHROMIUM;   W

64932

581   C14 H18 Cl6 N3 O7 P   25152-21-0   AD-0-3495-0   Thymidine, 5'-deoxy-5'-(phosphonoamino)-, bis(2,2,2-trichloroethyl) ester;;
N-(5'-DEOXYTHYMIDINYL)-DI-O-(β,β,β-TRICHLOROETHYL)-PHOSPHORAMIDATE;   W/NBS

64933

Str 1031

581   C25 H40 Cl N3 O3 S Si3   JH-1982-182-0   Tris(trimethylsilyl) derivative of Metolazone;   W

127350

581   C26 H17 Cl6 N3   55429-12-4   AD-0-968-0   1,1-Ethanediamine, N,N'-bis(2,5-dichlorophenyl)-2-[(2,5-dichlorophenyl)imino]-2-phen
yl-;; 2,5-DICHLOROANIL OF β,β BIS(2,5-DICHLOROANILINO)-ACETOPHENONE;   W/NBS

64934

Str 1032

581   C29 H51 N5 O7   BS-1-61-0   N-Acetyl-(N,0-permethyl)-leucyl-valyl-alanyl-propyl-alanine;   W

127351

581   C30 H31 N O11   50939-70-3   NS--27866-0-0   1,2,3-Benzenetricarboxylic acid, 5-[3-methoxy-1-(methoxycarbonyl)-3-oxopropoxy]-
4-(4,6,8-trimethyl-2-quinolinyl)-, trimethyl ester;;   NBS

64936

581   C31 H60 D3 N O3 Si3   KO-3-189-3   T-BUTYLDIMETHYLSILYL-D3-SALBUTAMOL;   W

64937

581   C31 H63 N O3 Si3   BS-3-13-0   Tris-t-butyl dimethylsilyl salbutamol;   W

127352

581　C31 H60 D3 N O3 Si3　BS-3-14-0　Tris-t-butyl-d3-dimethylsilyl salbutamol;　W　　　127353

581　C33 H43 N O8　CD-277-0-0　Diacetylepiscopalisine;　W　　　127355

581　C34 H55 N O3 Si2　57325-86-7　EP-8318-0-0　Pregn-5-en-20-one, 3,21-bis[(trimethylsilyl)oxy]-, O-(phenylmethyl)oxime, (3β)-;;
5-PREGNENE-3β,21-DIOL-20-ONE BO TMS;　W/NBS　　64940

Str 1033

581　C34 H55 N O3 Si2　57325-87-8　EP-8317-0-0　Pregn-5-en-20-one, 3,16-bis[(trimethylsilyl)oxy]-, O-(phenylmethyl)oxime, (3β,16α)-;;　5-PREGNENE-3β,16α-DIOL-20-ONE BO TMS;　W/NBS　　64938

Str 1034

581　C34 H55 N O3 Si2　57326-12-2　EP-8345-0-0　Pregn-5-en-20-one, 3,17-bis[(trimethylsilyl)oxy]-, O-(phenylmethyl)oxime, (3β)-;;
5-PREGNENE-3β,17α-DIOL-20-ONE BO TMS;　W/NBS　　64939

Str 1035

582　C11 F22 O2　RB-1982-15328-0　2-PERFLUOROOCTOXY-PERFLUOROPROPIONYL FLUORIDE;　W　　　127356

582　C12 F8 I2 S　19638-37-0　NS--27862-0-0　Benzene, 1,1'-thiobis[2,3,4,5-tetrafluoro-6-iodo-;; Bis(2-iodotetrafluorophenyl) sulphide;　NBS　　79444

582　C18 B3 F15 O3　HE-1982-0-0　BOROXINE, TRI-(PERFLUOROPHENYL)-;　W　　　64942

582　C18 H5 F8 Ir O　51509-33-2　NS--27861-0-0　Iridium, carbonyl(η5-2,4-cyclopentaldien-1-yl)(3,3',4,4',5,5',6,6'-octafluoro[1,1'-biphenyl]-2,2'-diyl)-;;　NBS　　64943

Str 1036

582　C26 H34 N2 O13　25644-83-1　AU-20-280-3　L-Serine, N-[(phenylmethoxy)carbonyl]-O-[3,4,6-tri-O-acetyl-2-(acetylamino)-2-deoxy-β-D-glucopyranosyl]-, methyl ester;; O-(2-ACETAMIDO-3,4,6-TRI-O-ACETYL-2-DEOXY-β-D-GLUCOPYRANOSYL)-N-BENZYLOXYCARBONYL-L-SERINE METHYLESTER;　W/NBS　　64944

582　C27 H34 O14　CD-591-0-0　Syringin peracetate;　W　　　127358

582　C28 H38 N8 O6　KO-8-146-1　Permethylated methotrexate-related component B (human urine);　W　　　127359

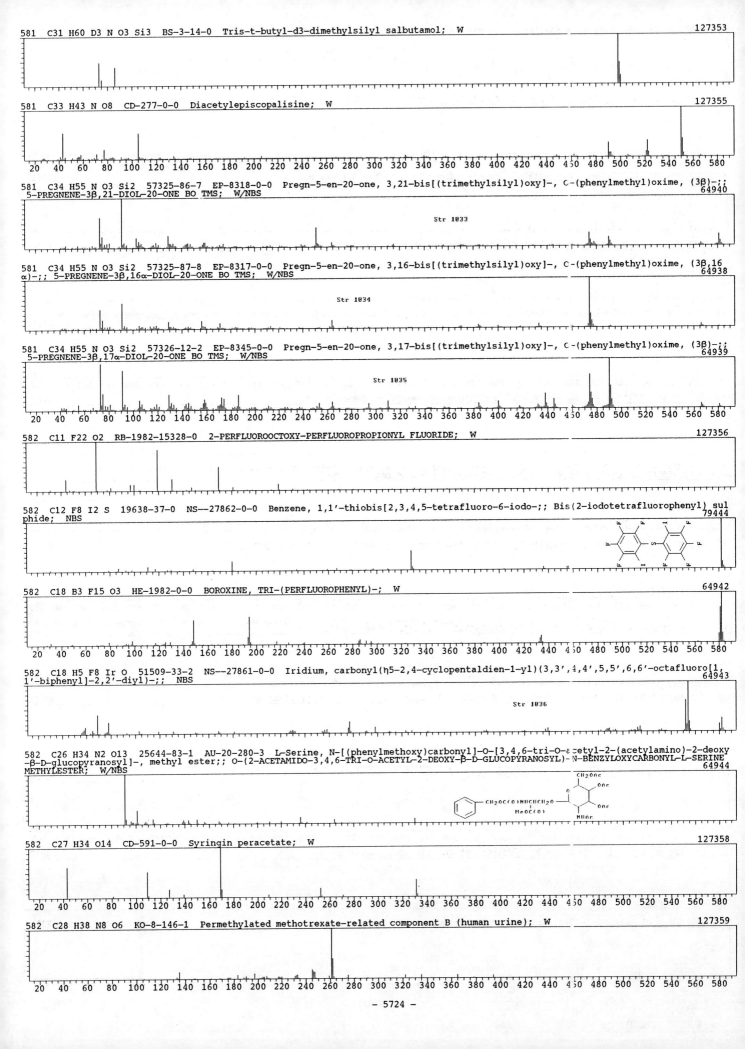

582   C28 H46 O9 Si2   KC-1977-2323-0   Bis(trimethylsilyl) ether of Methyl ent-2α,3β-dihydroxy-16-oxo-17-norgibberellane-7,19,20-tr
ioate;   W   127360

582   C29 H50 O8 Si2   D3-1980-645-19   TRIMETHYL, DI-TMS DERIVATIVE OF GIBBERELLIN GA41;   W   64945

20   40   60   80   100   120   140   160   180   200   220   240   260   280   300   320   340   360   380   400   420   440   460   480   500   520   540   560   580

582   C30 H38 N4 O4 S2   BS-1-177-0   1-Dimethylamino naphthalene-5-sulphonyl-hexamethylendiamine (bis);   W   127361

582   C30 H58 O5 Si3   17563-00-7   EP-8444-0-0   5β-Pregnan-20-one, 17-hydroxy-3α,11β,21-tris(trimethylsiloxy)-;;   5β-PREGNANE-3α,11
β,17α,21-TETROL-20-ONE TMS;   W/NBS   64946

Str 1037

582   C31 H34 O9 S   81827-02-3   KC-1982-693-0   ent-1α,13-Diacetoxy-10-hydroxy-3β-thiobenzoyloxy-20-norgibberell-16-ene-7,19-dioic
acid 7-methyl ester 19,10-lactone;   W   127362

582   C31 H38 N2 O9   92098-17-4   C-107-6265-0   N,N'-Bis(p-methoxybenzyl)-2',3'-O-isoprolidenebicyclomycin;   W   127363

Str 1038

582   C31 H45 F7 O2   18003-78-6   JC-90-380-1   CHOLESTEROL HEPTAFLUOROBUTYRATE;   W   64947

Str 1039

582   C32 H58 O7 Si   68216-40-0   C-100-6215-0   METHYL(+)-8-EPI-9-OXO-11α-TETRAHYDROPYRANYLOXY-15β-HYDROXY-19α/β-TERT-BUTYLDIMETHYL
SILYOXYPROST-13-CIS-ENOATE;   W   64948

Str 1040

582   C32 H58 O7 Si   68216-30-8   C-100-6215-0   METHYL(+,-)-9-OXO-11α-TETRAHYDROPYRANYLOXY-15β-HYDROXY-19α/β-TERT-BUTYLDIMETHYLSILY
OXYPROST-13-CIS-ENOATE;   W   64949

Str 1041

582   C32 H58 O7 Si   68216-29-5   C-100-6215-0   METHYL(+,-)-9-OXO-11-TETRAHYDDROPYRANYLOXY-15α-HYDROXY-19α/β-TERT-BUTYLDIMETHYL-SIL
YOXYPROST-13-CIS-ENOATE;   W   64950

Str 1042

582   C32 H58 O7 Si   C-100-6216-0   METHYL(+,-)-9-OXO-11α-TETRAHYDROPYRANYLOXY-15α-HYDROXY-19α/β-BUTYLDIMETHYLSILYOXYPROST-13-TRANS
-ENOATE;   W   64951

20   40   60   80   100   120   140   160   180   200   220   240   260   280   300   320   340   360   380   400   420   440   460   480   500   520   540   560   580

582   C33 H54 N6 O3   67969-52-2   NS--27858-0-0   Triimidazo[1,5-a:1',5'-c:1'',5''-e][1,3,5]triazine-1,5,9(2H,6H,10H)-trione, 2,6,10
-tricyclohexylhexahydro-3,3,7,7,11,11-hexamethyl-;;   NBS   64952

Str 1043

20   40   60   80   100   120   140   160   180   200   220   240   260   280   300   320   340   360   380   400   420   440   460   480   500   520   540   560   580

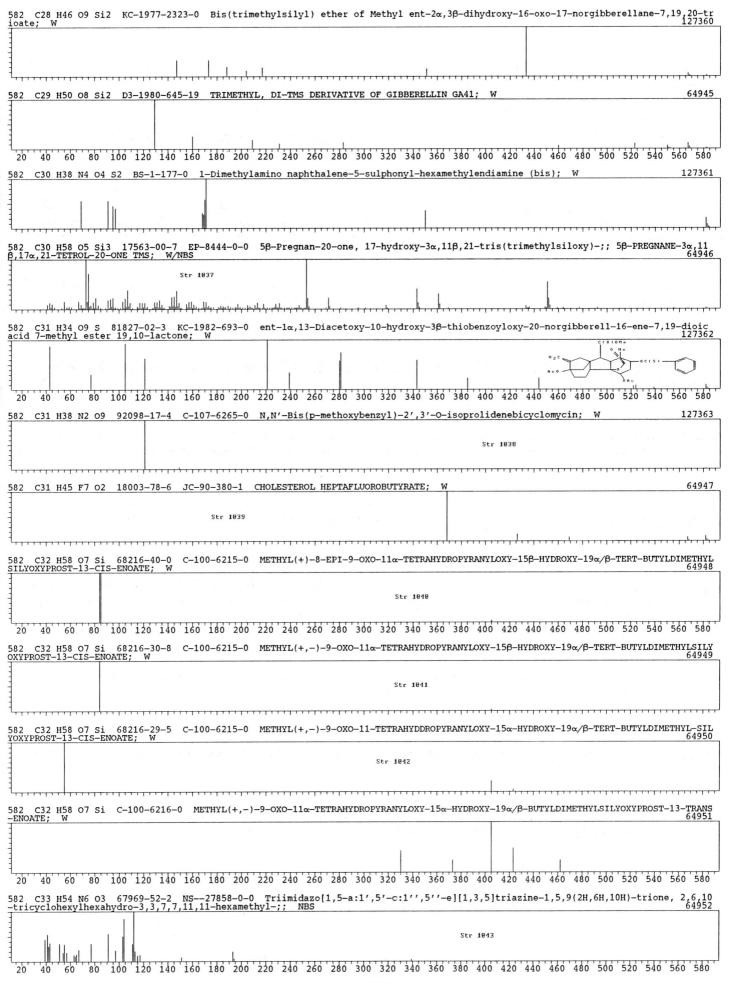

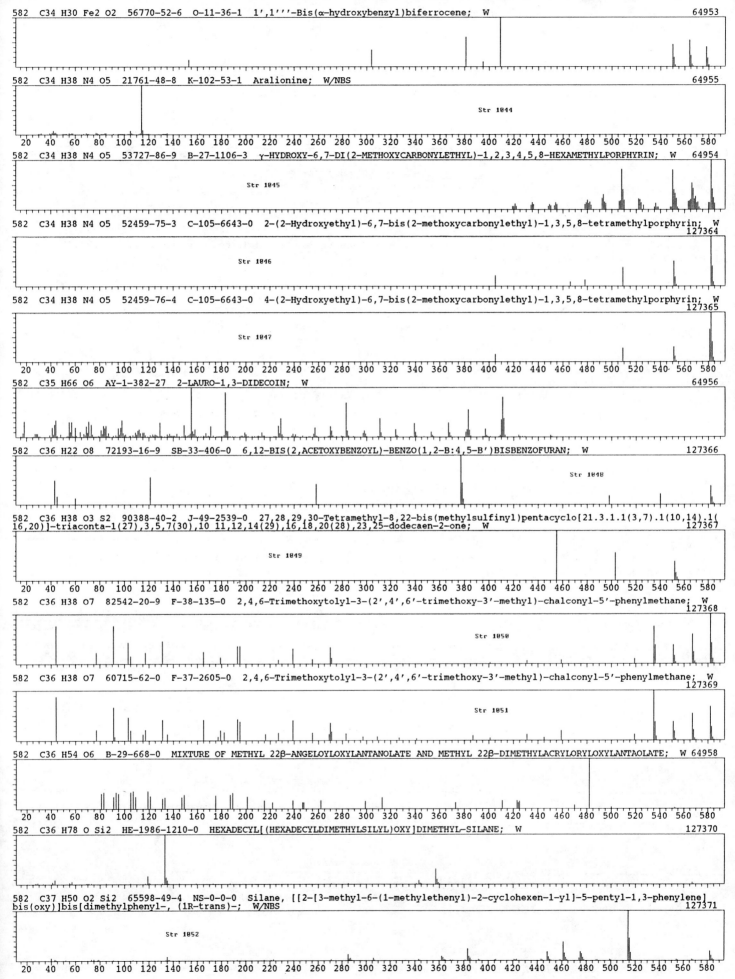

582　C34 H30 Fe2 O2　56770-52-6　O-11-36-1　1',1'''-Bis(α-hydroxybenzyl)biferrocene;　W　　64953

582　C34 H38 N4 O5　21761-48-8　K-102-53-1　Aralionine;　W/NBS　　64955

Str 1044

582　C34 H38 N4 O5　53727-86-9　B-27-1106-3　γ-HYDROXY-6,7-DI(2-METHOXYCARBONYLETHYL)-1,2,3,4,5,8-HEXAMETHYLPORPHYRIN;　W　　64954

Str 1045

582　C34 H38 N4 O5　52459-75-3　C-105-6643-0　2-(2-Hydroxyethyl)-6,7-bis(2-methoxycarbonylethyl)-1,3,5,8-tetramethylporphyrin;　W　　127364

Str 1046

582　C34 H38 N4 O5　52459-76-4　C-105-6643-0　4-(2-Hydroxyethyl)-6,7-bis(2-methoxycarbonylethyl)-1,3,5,8-tetramethylporphyrin;　W　　127365

Str 1047

582　C35 H66 O6　AY-1-382-27　2-LAURO-1,3-DIDECOIN;　W　　64956

582　C36 H22 O8　72193-16-9　SB-33-406-0　6,12-BIS(2,ACETOXYBENZOYL)-BENZO(1,2-B:4,5-B')BISBENZOFURAN;　W　　127366

Str 1048

582　C36 H38 O3 S2　90388-40-2　J-49-2539-0　27,28,29,30-Tetramethyl-8,22-bis(methylsulfinyl)pentacyclo[21.3.1.1(3,7).1(10,14).1(16,20)]-triaconta-1(27),3,5,7(30),10　11,12,14(29),16,18,20(28),23,25-dodecaen-2-one;　W　　127367

Str 1049

582　C36 H38 O7　82542-20-9　F-38-135-0　2,4,6-Trimethoxytolyl-3-(2',4',6'-trimethoxy-3'-methyl)-chalconyl-5'-phenylmethane;　W　　127368

Str 1050

582　C36 H38 O7　60715-62-0　F-37-2605-0　2,4,6-Trimethoxytolyl-3-(2',4',6'-trimethoxy-3'-methyl)-chalconyl-5'-phenylmethane;　W　　127369

Str 1051

582　C36 H54 O6　B-29-668-0　MIXTURE OF METHYL 22β-ANGELOYLOXYLANTANOLATE AND METHYL 22β-DIMETHYLACRYLORYLOXYLANTAOLATE;　W 64958

582　C36 H78 O Si2　HE-1986-1210-0　HEXADECYL[(HEXADECYLDIMETHYLSILYL)OXY]DIMETHYL-SILANE;　W　　127370

582　C37 H50 O2 Si2　65598-49-4　NS-0-0-0　Silane, [[2-[3-methyl-6-(1-methylethenyl)-2-cyclohexen-1-yl]-5-pentyl-1,3-phenylene]bis(oxy)]bis[dimethylphenyl-, (1R-trans)-;　W/NBS　　127371

Str 1052

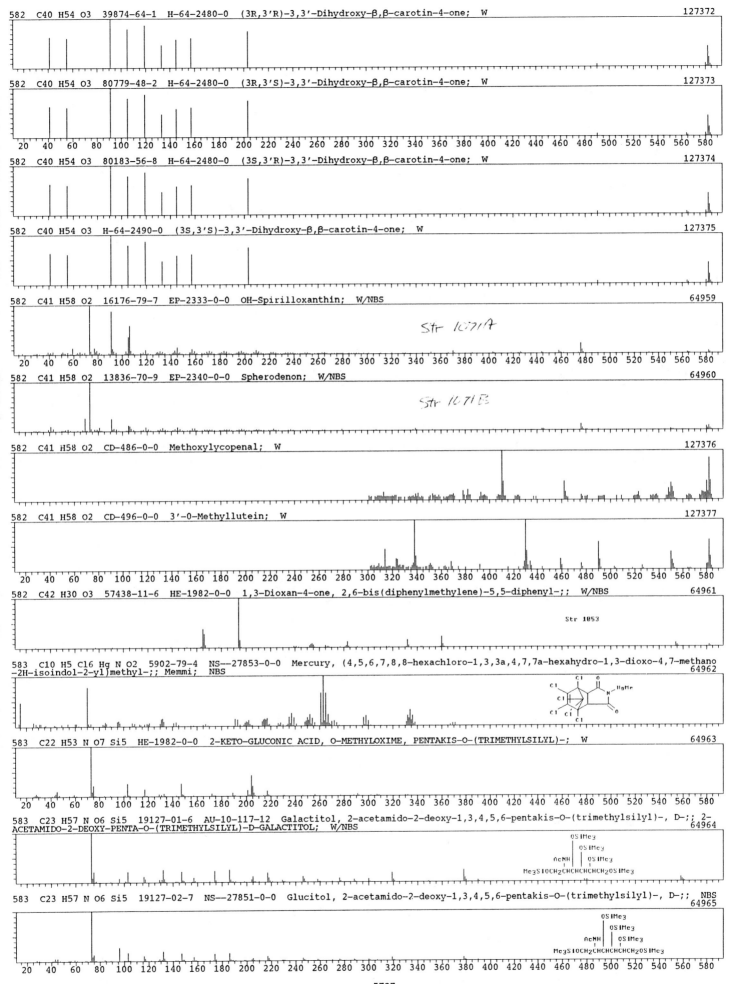

582 C40 H54 O3 39874-64-1 H-64-2480-0 (3R,3'R)-3,3'-Dihydroxy-β,β-carotin-4-one; W 127372

582 C40 H54 O3 80779-48-2 H-64-2480-0 (3R,3'S)-3,3'-Dihydroxy-β,β-carotin-4-one; W 127373

582 C40 H54 O3 80183-56-8 H-64-2480-0 (3S,3'R)-3,3'-Dihydroxy-β,β-carotin-4-one; W 127374

582 C40 H54 O3 H-64-2490-0 (3S,3'S)-3,3'-Dihydroxy-β,β-carotin-4-one; W 127375

582 C41 H58 O2 16176-79-7 EP-2333-0-0 OH-Spirilloxanthin; W/NBS 64959
Str 1071A

582 C41 H58 O2 13836-70-9 EP-2340-0-0 Spherodenon; W/NBS 64960
Str 1071B

582 C41 H58 O2 CD-486-0-0 Methoxylycopenal; W 127376

582 C41 H58 O2 CD-496-0-0 3'-O-Methyllutein; W 127377

582 C42 H30 O3 57438-11-6 HE-1982-0-0 1,3-Dioxan-4-one, 2,6-bis(diphenylmethylene)-5,5-diphenyl-;; W/NBS 64961
Str 1053

583 C10 H5 Cl6 Hg N O2 5902-79-4 NS--27853-0-0 Mercury, (4,5,6,7,8,8-hexachloro-1,3,3a,4,7,7a-hexahydro-1,3-dioxo-4,7-methano-2H-isoindol-2-yl)methyl-;; Memmi; NBS 64962

583 C22 H53 N O7 Si5 HE-1982-0-0 2-KETO-GLUCONIC ACID, O-METHYLOXIME, PENTAKIS-O-(TRIMETHYLSILYL)-; W 64963

583 C23 H57 N O6 Si5 19127-01-6 AU-10-117-12 Galactitol, 2-acetamido-2-deoxy-1,3,4,5,6-pentakis-O-(trimethylsilyl)-, D-;; 2-ACETAMIDO-2-DEOXY-PENTA-O-(TRIMETHYLSILYL)-D-GALACTITOL; W/NBS 64964

583 C23 H57 N O6 Si5 19127-02-7 NS--27851-0-0 Glucitol, 2-acetamido-2-deoxy-1,3,4,5,6-pentakis-O-(trimethylsilyl)-, D-;; NBS 64965

583   C23 H57 N O6 Si5   74464-41-8   NS--27850-0-0   D-Mannitol, 2-(acetylamino)-2-deoxy-1,3,4,5,6-pentakis-O-(trimethylsilyl)-;;
NBS   64966

583   C23 H57 N O6 Si5   HE-1982-0-0   SORBITOL, 2-ACETAMIDO-2-DESOXY-PENTAKIS-O-(TRIMETHYLSILYL)-; W   64967

583   C23 H56 D N O6 Si5   HE-1982-0-0   SORBITOL-1-D1, 2-ACETAMIDO-2-DESOXY-PENTAKIS-O-(TRIMETHYLSILYL9-; W   64968

583   C23 H55 D2 N O6 Si5   HE-1982-0-0   ARABINO-HEXITOL-1,6-D2, 2-ACETAMIDO-2-DESOXY-PENTAKIS-O-(TRIMETHYLSILYL)-; W   64969

583   C23 H55 D2 N O6 Si5   HE-1982-0-0   ARABINO-HEXITOL-1,1-D2, 2-ACETAMIDO-2-DESOXY-PENTAKIS-O-(TRIMETHYLSILYL)-; W   64970

583   C26 H33 N O14   14364-68-2   AU-20-279-1   L-Alanine, N-[(phenylmethoxy)carbonyl]-3-[(2,3,4,6-tetra-O-acetyl-β-D-galactopyranos yl)oxy]-, methyl ester;; N-BENZYLOXYCARBONYL-O-(TETRA-O-ACETYL-β-D-GALACTOPYRANOSYL)-L-SERINE METHYL ESTER; W/NBS   64971

583   C27 H23 Br2 N S2   75931-71-4   O-15-25-1   N,N-BIS(4'-P-BROMOPHENYLTHIO-2'-BUTYNYL)-4-METHYLANILINE; W   64972

Str 1054

583   C28 H37 N7 O7   55515-17-8   AD-0-364-0   DIMETHYL PUROMYCIN DIACETATE; W/NBS   64973

583   C29 H29 N O12   49616-80-0   NS--27847-0-0   Cyclobut[4,5]azepino[1,2-a]quinoline-7,7a,8,9,9a,10-hexacarboxylic acid, 10,11-di hydro-1-methyl-, hexamethyl ester;; NBS   64974

Str 1055

583   C29 H29 N O12   49679-09-6   NS--27846-0-0   Benzo[f]cyclopenta[a]quinolizine-6,7,7a,8,9,10(8H)-hexacarboxylic acid, 6,7-dihy dro-4-methyl-, hexamethyl ester;; NBS   64975

Str 1056

583   C29 H33 N3 O8 S   92216-29-0   C-107-3263-0   anti-1,4-Bis(p-methoxybenzyl)-3-(2'-thiopyridyl)-6-(1''-(hydroxymethyl)-3''-(hydr oxypropyl))-2,5-piperazinedione; W   127378

Str 1057

583   C29 H33 N3 O8 S   95782-44-8   C-107-3263-0   syn-1,4-Bis(p-methoxybenzyl)-3-(2'-thiopyridyl)-6-(1''-(hydroxymethyl)-3''-(ydrox ypropyl))-2,5-piperazinedione; W   127379

Str 1058

583   C31 H37 N O10   69979-02-8   F-34-2118-0   7,8-DIMETHOXY-14-OXO-15-(3',4',5')-TRIMETHOXYBENZOYLOXY EPIALLOBERBAN-13-CARBOXYLIC ACID METHYLESTER; W   64976

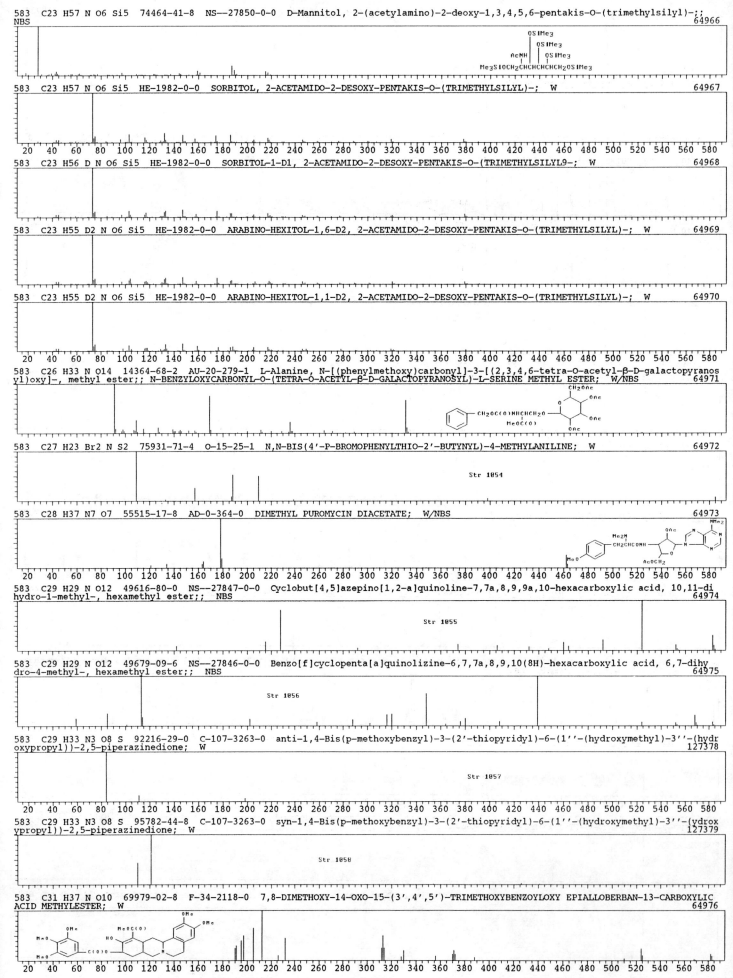

- 5728 -

583   C31 H37 N O10   69915-48-6   F-34-2118-0   7,8-DIMETHOXY-15-OXO-14β-(3',4',5'-TRIMETHOXYBENZOYLOXY)-EPIALLOBERBAN-13α-CARBOXYL
IC ACID-METHYLESTER;   W
                                                                                                                    64977

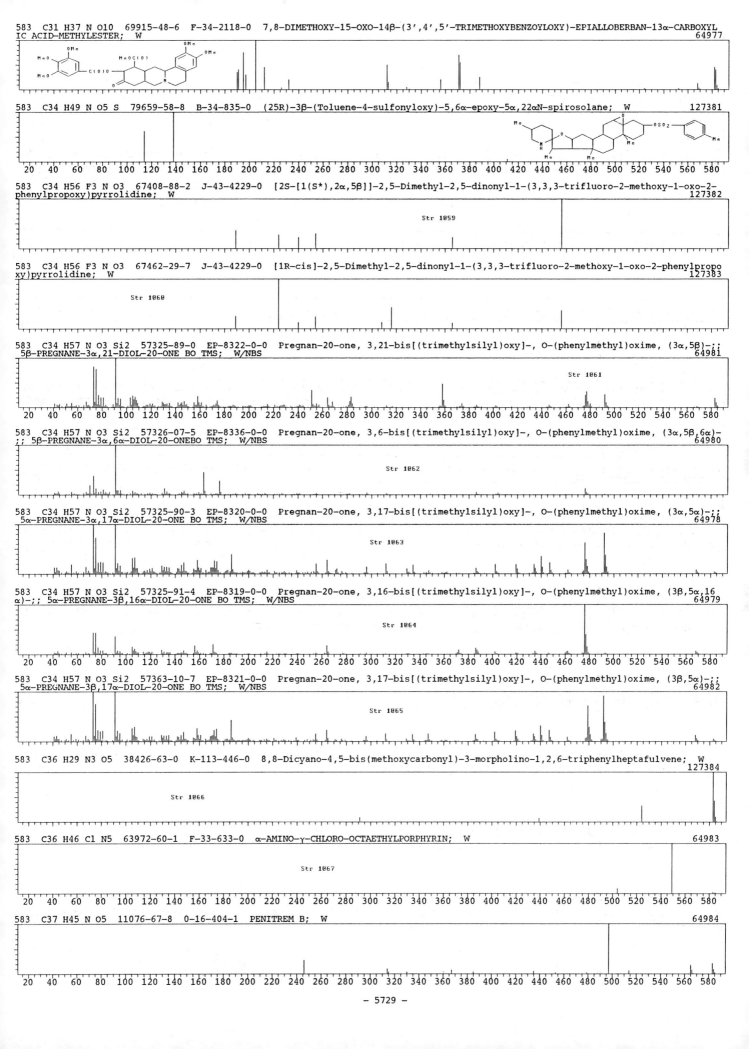

583   C34 H49 N O5 S   79659-58-8   B-34-835-0   (25R)-3β-(Toluene-4-sulfonyloxy)-5,6α-epoxy-5α,22αN-spirosolane;   W   127381

583   C34 H56 F3 N O3   67408-88-2   J-43-4229-0   [2S-[1(S*),2α,5β]]-2,5-Dimethyl-2,5-dinonyl-1-(3,3,3-trifluoro-2-methoxy-1-oxo-2-
phenylpropoxy)pyrrolidine;   W
                                                                                                                    127382
Str 1059

583   C34 H56 F3 N O3   67462-29-7   J-43-4229-0   [1R-cis]-2,5-Dimethyl-2,5-dinonyl-1-(3,3,3-trifluoro-2-methoxy-1-oxo-2-phenylpropo
xy)pyrrolidine;   W
                                                                                                                    127383
Str 1060

583   C34 H57 N O3 Si2   57325-89-0   EP-8322-0-0   Pregnan-20-one, 3,21-bis[(trimethylsilyl)oxy]-, O-(phenylmethyl)oxime, (3α,5β)-;;
5β-PREGNANE-3α,21-DIOL-20-ONE BO TMS;   W/NBS   64981
Str 1061

583   C34 H57 N O3 Si2   57326-07-5   EP-8336-0-0   Pregnan-20-one, 3,6-bis[(trimethylsilyl)oxy]-, O-(phenylmethyl)oxime, (3α,5β,6α)-
;; 5β-PREGNANE-3α,6α-DIOL-20-ONEBO TMS;   W/NBS   64980
Str 1062

583   C34 H57 N O3 Si2   57325-90-3   EP-8320-0-0   Pregnan-20-one, 3,17-bis[(trimethylsilyl)oxy]-, O-(phenylmethyl)oxime, (3α,5α)-;;
5α-PREGNANE-3α,17α-DIOL-20-ONE BO TMS;   W/NBS   64978
Str 1063

583   C34 H57 N O3 Si2   57325-91-4   EP-8319-0-0   Pregnan-20-one, 3,16-bis[(trimethylsilyl)oxy]-, O-(phenylmethyl)oxime, (3β,5α,16
α)-;; 5α-PREGNANE-3β,16α-DIOL-20-ONE BO TMS;   W/NBS   64979
Str 1064

583   C34 H57 N O3 Si2   57363-10-7   EP-8321-0-0   Pregnan-20-one, 3,17-bis[(trimethylsilyl)oxy]-, O-(phenylmethyl)oxime, (3β,5α)-;;
5α-PREGNANE-3β,17α-DIOL-20-ONE BO TMS;   W/NBS   64982
Str 1065

583   C36 H29 N3 O5   38426-63-0   K-113-446-0   8,8-Dicyano-4,5-bis(methoxycarbonyl)-3-morpholino-1,2,6-triphenylheptafulvene;   W
                                                                                                                    127384
Str 1066

583   C36 H46 Cl N5   63972-60-1   F-33-633-0   α-AMINO-γ-CHLORO-OCTAETHYLPORPHYRIN;   W   64983
Str 1067

583   C37 H45 N O5   11076-67-8   O-16-404-1   PENITREM B;   W   64984

583　C37 H45 N O5　75949-33-6　J-46-622-0　2-Methoxy-3-[7-(Cinnamoyloxy)heptyl]-6-[6-(cinnamoyloxy)hexyl]pyridine;　W　127385

584　C15 H12 Br F15 N2　82633-57-6　AC-1982-94-0　4-Bromo-1-methyl-3-perfluoroheptyl-5-(t-butyl)-pyrazole;　W　127386

584　C16 H28 O6 Si3 W　73382-52-2　KA-1980-94-5　Pentacarbonyl[1-(trimethylsilyloxy)-2,2-bis(trimethylsilyl)ethylidene]tungsten;　W　127387

Str 1068

584　C24 H39 F3 N4 O9　KO-10-122-1　Trifluoroacetyl, permethyl derivative of valine;　W　127388

584　C28 H37 Cl O11　77573-36-5　NS-0-0-0　5H-Cyclopropa[3,4]benz[1,2-e]azulen-5-one, 3,9,9a-tris(acetyloxy)-3-[(acetyloxy)methyl]-2-chloro-1,1a,1b,2,3,4,4a,7a,7b,8,9,9a-dodecahydro-4a,7b-dihydroxy-1,1,6,8-tetramethyl-, [1aR-(1aα,1bβ,2α,3β,4aβ,7aα,7bα,8α,9β,9aα)]-;　W/NBS　127389

Str 1069

584　C28 H38 F6 O6　56438-17-6　SD-1981-0-0　Cholan-24-oic acid, 3,12-bis[(trifluoroacetyl)oxy]-, (3α,5β,12α)-;;　W/NBS　64985

Str 1070

584　C29 H37 F5 N2 O3 Si　57691-90-4　NS--27839-0-0　Androst-4-ene-3,17-dione, 11-[[dimethyl(pentafluorophenyl)silyl]oxy]-, bis(O-methyloxime), (11β)-;;　NBS　64986

Str 1071

584　C30 H60 O5 Si3　56247-88-2　NS--27837-0-0　Prosta-11,15-dien-1-oic acid, 7,9,13-tris[(trimethylsilyl)oxy], methyl ester;;　NBS　64989

584　C30 H60 O5 Si3　52058-16-9　OM-1981-407-0　PGF2α METHYL ESTER TRIMETHYLSILYL ETHER;　W　64987

584　C30 H60 O5 Si3　57325-69-6　EP-8282-0-0　Pregnane-3,11,17,20,21-pentol, tris(trimethylsilyl) ether, (3α,5β,11β,20R)-;;　5β-PREGNANE-3α,11β,17α,20β,21-PENTOL TMS;　W/NBS　64988

Str 1072

584　C30 H60 O5 Si3　OM-1981-495-0　F2A 1415-1418 ME 3TMS POWELL;　W　64990

584　C30 H60 O5 Si3　BM-1978-69-4　Tris(trimethylsilyl),methylester derivative of prostaglandin F2α;　W　127391

584　C30 H56 D4 O5 Si3　BM-1978-69-4　Tris(trimethylsilyl),methylester derivative of 3,3,4,4-[2H4]-prostaglandin F2α (with H/D exchange products);　W　127392

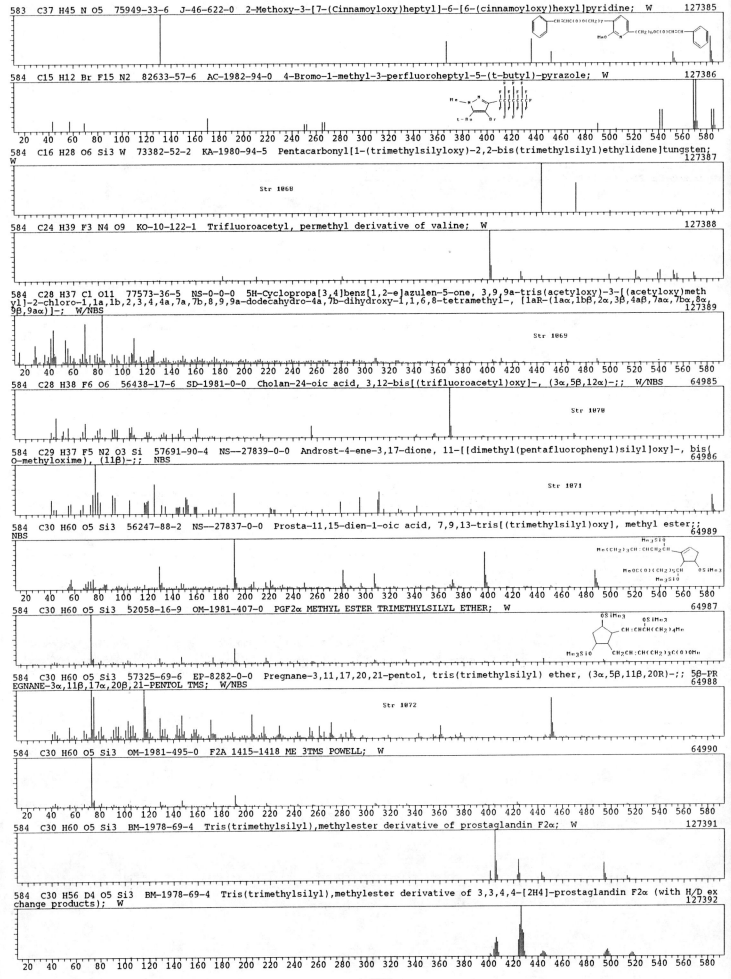

584   C30 H56 D4 O5 Si3   BM-1978-69-4   Tris(trimethylsilyl),methylester derivative of 3,3,4,4-[2H4]-prostaglandin F2α;   W   127393

584   C31 H36 O11   32981-89-8   J-46-1474-0   5β,20-Epoxy-1,2α,4,7β,10β-pentahydroxy-tax-11-ene-9,13-dione-4,10-diacetate-2-benzoate
; W   127394

Str 1073

20   40   60   80   100 120 140 160 180 200 220 240 260 280 300 320 340 360 380 400 420 440 460 480 500 520 540 560 580

584   C31 H36 O11   76497-09-1   J-46-1474-0   [2aR-(2aα,4α,4aβ,6β,11β,12α,12aα,12b-α)]-6,12b-bis(acetyloxy)-12-(benzoyloxy)-2a,3,4,
4a,11,12,12a,12b-octahydro-4,11-dihydroxy-4a,8,13,13-tetramethyl-7,11-methan-1H-cyclodeca[3,4]benz[1,2-b]oxete-5,9(6H,10H)-dione;
W   127395

Str 1074

584   C31 H40 N2 O7 S   79228-23-2   KC-1981-2911-0   3-(2,2-Diethoxyethyl-N-p-tolylsulphonyl)-5,6,7-trimethoxy-1,4-dimethylcarbazole
; W   127396

584   C32 H40 O10   75758-97-3   J-49-3258-0   (+-)-(1α,2α,4bβ,10α)-Ethyl-2-(benzoyloxy)-8,8-(ethylenedioxy)-1-(methoxycarbonyl)-7-[(
methoxymethyl)oxy]-1-methyl-2-oxogibb-4a(10a)-ene-10-carboxylate;   W   127397

20   40   60   80   100 120 140 160 180 200 220 240 260 280 300 320 340 360 380 400 420 440 460 480 500 520 540 560 580

584   C33 H36 N4 O6   635-65-4   NS--27835-0-0   Bilirubin;   NBS   64991

584   C33 H41 B F4 N2 O2   K-116-1995-0   4-(2,2-Dimethoxyethyl)-1,7-bis(1,3,3-trimethyl-2-indolinyl)heptamethinium-tetrafluorobor
ate;   W   127398

584   C34 H40 N4 O5   53766-27-1   NS-0-0-0   Benzenepropanamide, α-(dimethylamino)-N-[7-(hydroxyphenylmethyl)-3-(1-methylethyl)-5,8-
dioxo-2-oxa-6,9-diazabicyclo[10.2.2]hexadeca-10,12,14,15-tetraen-4-yl]-, [3S-[3R*,4R*(R*),7R*(S*)]]-;   W/NBS   127399

Str 1075

20   40   60   80   100 120 140 160 180 200 220 240 260 280 300 320 340 360 380 400 420 440 460 480 500 520 540 560 580

584   C34 H40 N4 O5   53797-27-6   NS-0-0-0   Benzenepropanamide, α-(dimethylamino)-N-[7-(hydroxyphenylmethyl)-3-(1-methylethyl)-5,8-
dioxo-2-oxa-6,9-diazabicyclo[10.2.2]hexadeca-10,12,14,15-tetraen-4-yl]-, [3R-[3R*,4R*(S*),7R*(S*)]]-;   W/NBS   127400

Str 1076

584   C34 H48 O8   71725-97-8   B-32-782-0   PONASTERONE C 22α4-BENZYLIDENE ACETAL;   W   64992

Str 1077

584   C34 H48 O8   71725-98-9   B-32-782-0   PONASTERONE C 20,22-BENZYLIDENE ACETAL;   W   64993

Str 1078

20   40   60   80   100 120 140 160 180 200 220 240 260 280 300 320 340 360 380 400 420 440 460 480 500 520 540 560 580

584   C34 H48 O8   33465-16-6   B-32-2505-0   12-DEOXY-6α,7α-EPOXY-5β-HYDROXYPHORBOL 13-TETRADECA-2,4-DIENOATE;   W   64994

Str 1079

20   40   60   80   100 120 140 160 180 200 220 240 260 280 300 320 340 360 380 400 420 440 460 480 500 520 540 560 580

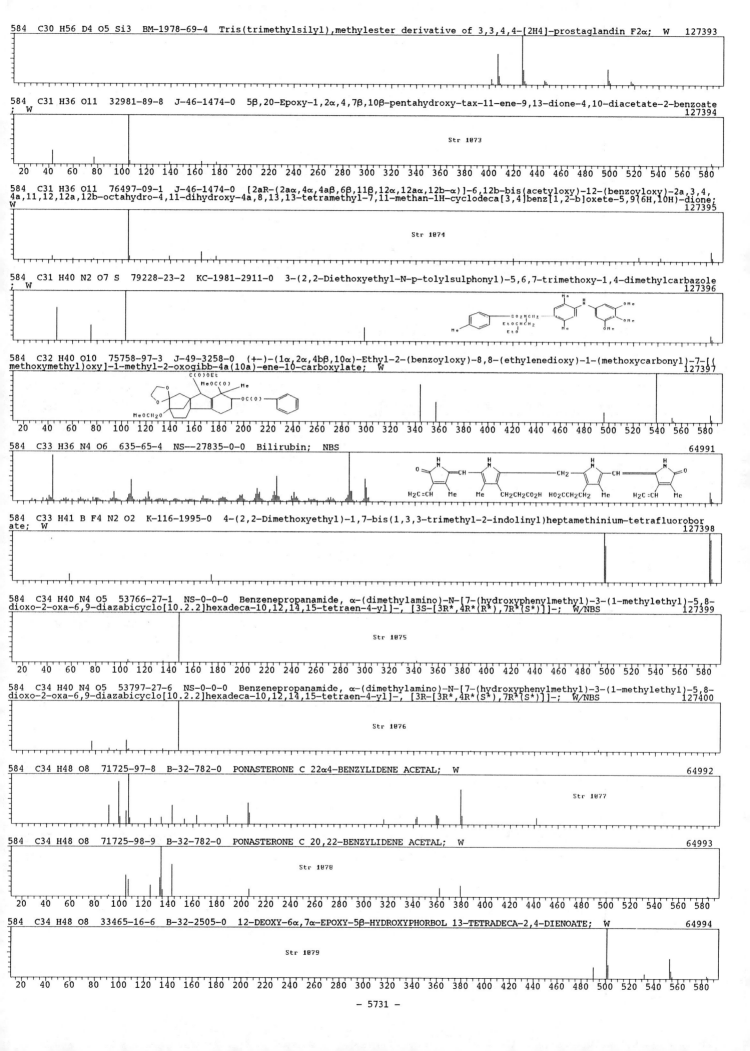

584 C34 H48 O8 F-39-637-0 (2E,6E,14E)-1-(1'-Hydroxy-4'-methoxy-6'-methyl-phenyl)-5,13-dihydroxy-12-one-3,7,11,15-tetramethylhe
xadeca-2,6,14-triene (isomer B); W
127401

584 C35 H52 O5 S 83872-46-2 H-65-1511-0 (23RS)-25,25-Ethylendioxy-6β-methoxy-23-phenylsulfonyl-27-nor-3β,5-cyclo-5α-cholest
ane; W
127402

Str 1000

584 C36 H28 P4 38234-91-2 EP-6193-0-0 Dibenzo[c,g][1,2,5,6]tetraphosphocin, 5,6,11,12-tetrahydro-5,6,11,12-tetraphenyl-;; 5,
6,11,12-TETRAHYDRO-5,6,11,12-TETRAPHENYLDIBENZOTETRAPHOSPHOCIN; W/NBS
64995

Str 1001

584 C36 H45 Cl N4 O 63972-65-6 F-33-633-0 γ-CHLORO-OCTAETHYL-α-OXOPHLORIN; W
64996

Str 1002

584 C36 H56 O6 55226-49-8 B-28-171-0 Olean-12-ene-3β,16β,22α-triol triacetate; W
64997

Str 1003

584 C36 H56 O6 77508-69-1 NS-0-0-0 Hexadecanoic acid, 1a,2,5,5a,6,9,10,10a-octahydro-5a-hydroxy-4-(hydroxymethyl)-1,1,7,9-tet
ramethyl-6,11-dioxo-1H-2,8a-methanocyclopenta[a]cyclopropa[e]cyclodecen-5-yl ester, [1aR-(1aα,2α,5β,5aβ,8aα,9α,10aα)]-; W/NBS
127403

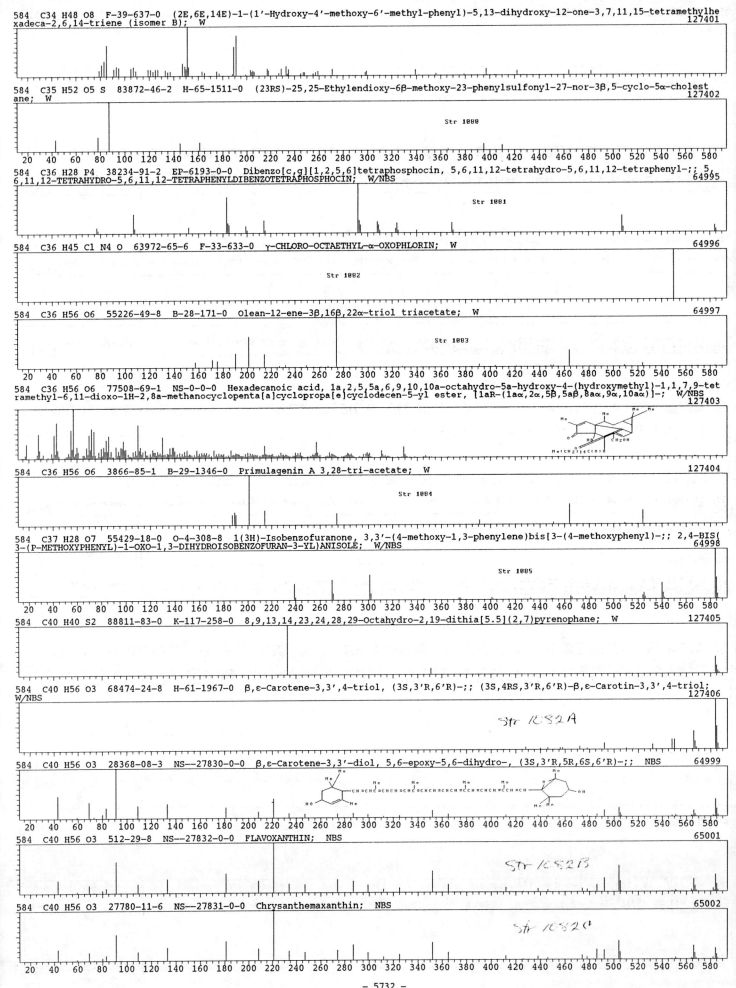

584 C36 H56 O6 3866-85-1 B-29-1346-0 Primulagenin A 3,28-tri-acetate; W
127404

Str 1004

584 C37 H28 O7 55429-18-0 O-4-308-8 1(3H)-Isobenzofuranone, 3,3'-(4-methoxy-1,3-phenylene)bis[3-(4-methoxyphenyl)-;; 2,4-BIS(
3-(P-METHOXYPHENYL)-1-OXO-1,3-DIHYDROISOBENZOFURAN-3-YL)ANISOLE; W/NBS
64998

Str 1005

584 C40 H40 S2 88811-83-0 K-117-258-0 8,9,13,14,23,24,28,29-Octahydro-2,19-dithia[5.5](2,7)pyrenophane; W
127405

584 C40 H56 O3 68474-24-8 H-61-1967-0 β,ε-Carotene-3,3',4-triol, (3S,3'R,6'R)-;; (3S,4RS,3'R,6'R)-β,ε-Carotin-3,3',4-triol;
W/NBS
127406

Str 1082A

584 C40 H56 O3 28368-08-3 NS--27830-0-0 β,ε-Carotene-3,3'-diol, 5,6-epoxy-5,6-dihydro-, (3S,3'R,5R,6S,6'R)-;; NBS
64999

584 C40 H56 O3 512-29-8 NS--27832-0-0 FLAVOXANTHIN; NBS
65001

Str 1082B

584 C40 H56 O3 27780-11-6 NS--27831-0-0 Chrysanthemaxanthin; NBS
65002

Str 1082C

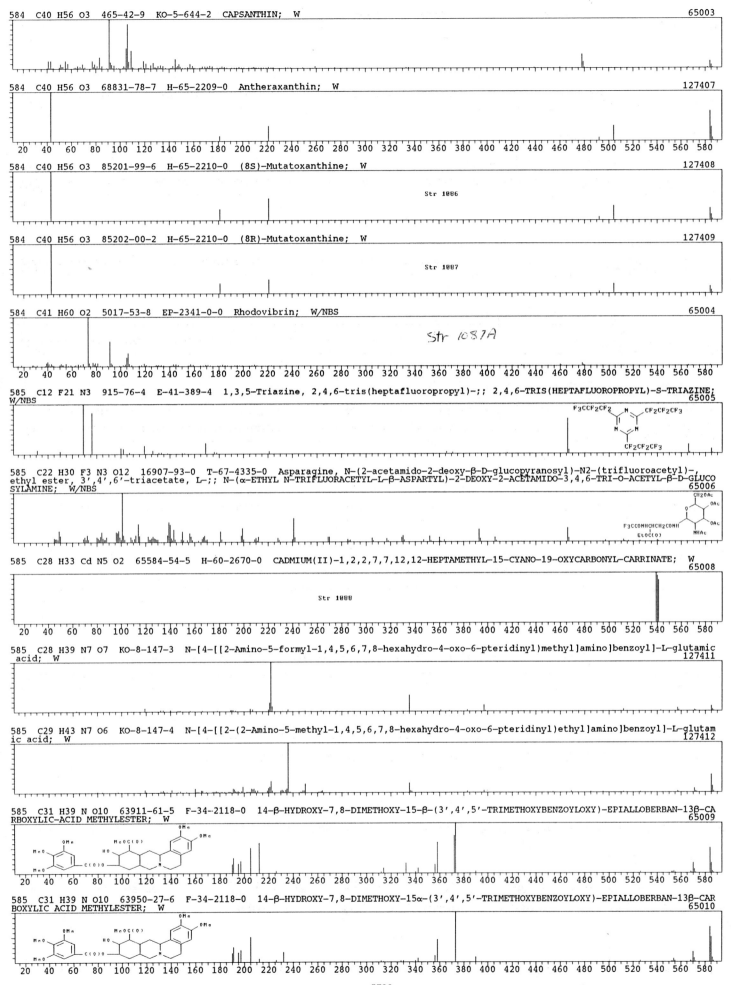

584   C40 H56 O3   465-42-9   KO-5-644-2   CAPSANTHIN;   W   65003

584   C40 H56 O3   68831-78-7   H-65-2209-0   Antheraxanthin;   W   127407

584   C40 H56 O3   85201-99-6   H-65-2210-0   (8S)-Mutatoxanthine;   W   127408

Str 1886

584   C40 H56 O3   85202-00-2   H-65-2210-0   (8R)-Mutatoxanthine;   W   127409

Str 1887

584   C41 H60 O2   5017-53-8   EP-2341-0-0   Rhodovibrin;   W/NBS   65004

Str 1087A

585   C12 F21 N3   915-76-4   E-41-389-4   1,3,5-Triazine, 2,4,6-tris(heptafluoropropyl)-;; 2,4,6-TRIS(HEPTAFLUOROPROPYL)-S-TRIAZINE;
W/NBS   65005

585   C22 H30 F3 N3 O12   16907-93-0   T-67-4335-0   Asparagine, N-(2-acetamido-2-deoxy-β-D-glucopyranosyl)-N2-(trifluoroacetyl)-,
ethyl ester, 3',4',6'-triacetate, L-;; N-(α-ETHYL N-TRIFLUORACETYL-L-β-ASPARTYL)-2-DEOXY-2-ACETAMIDO-3,4,6-TRI-O-ACETYL-β-D-GLUCO
SYLAMINE;   W/NBS   65006

585   C28 H33 Cd N5 O2   65584-54-5   H-60-2670-0   CADMIUM(II)-1,2,2,7,7,12,12-HEPTAMETHYL-15-CYANO-19-OXYCARBONYL-CARRINATE;   W
65008

Str 1088

585   C28 H39 N7 O7   KO-8-147-3   N-[4-[[2-Amino-5-formyl-1,4,5,6,7,8-hexahydro-4-oxo-6-pteridinyl)methyl]amino]benzoyl]-L-glutamic
acid;   W   127411

585   C29 H43 N7 O6   KO-8-147-4   N-[4-[[2-(2-Amino-5-methyl-1,4,5,6,7,8-hexahydro-4-oxo-6-pteridinyl)ethyl]amino]benzoyl]-L-glutam
ic acid;   W   127412

585   C31 H39 N O10   63911-61-5   F-34-2118-0   14-β-HYDROXY-7,8-DIMETHOXY-15-β-(3',4',5'-TRIMETHOXYBENZOYLOXY)-EPIALLOBERBAN-13β-CA
RBOXYLIC-ACID METHYLESTER;   W   65009

585   C31 H39 N O10   63950-27-6   F-34-2118-0   14-β-HYDROXY-7,8-DIMETHOXY-15α-(3',4',5'-TRIMETHOXYBENZOYLOXY)-EPIALLOBERBAN-13β-CAR
BOXYLIC ACID METHYLESTER;   W   65010

585　C31 H39 N O10　69979-24-4　F-34-2121-0　14-β-HYDROXY-7,8-DIMETHOXY-15α-(3',4',5'-TRIMETHOXYBENZOYLOXY)-ALLOBERBAN-13α-CARBOXYLICACID METHYLESTER;　W
65011

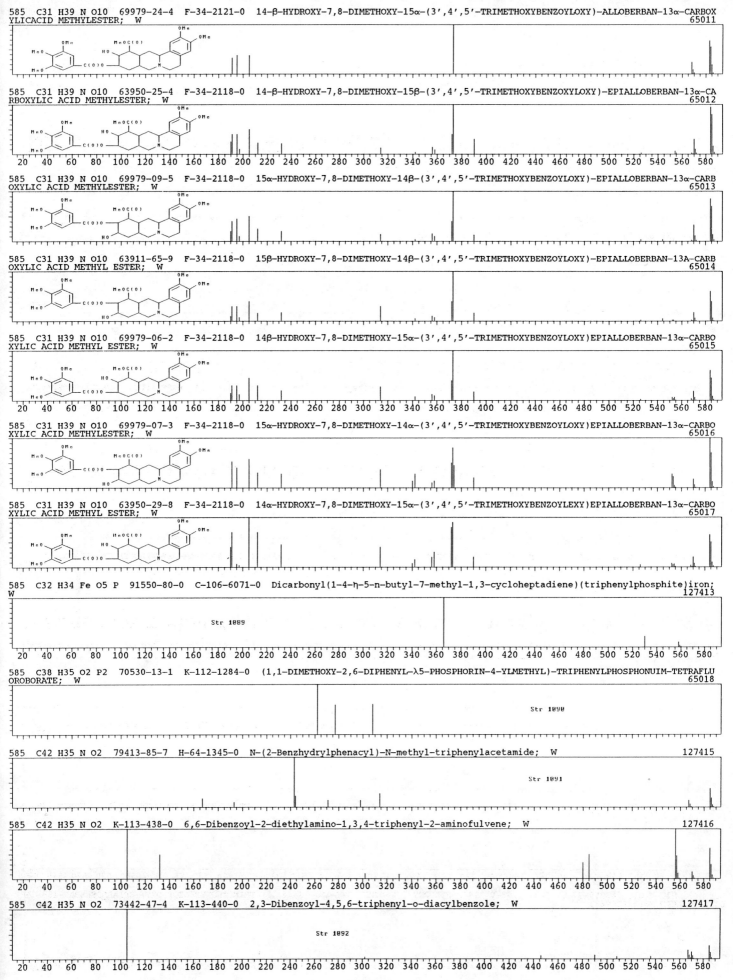

585　C31 H39 N O10　63950-25-4　F-34-2118-0　14-β-HYDROXY-7,8-DIMETHOXY-15β-(3',4',5'-TRIMETHOXYBENZOXYLOXY)-EPIALLOBERBAN-13α-CARBOXYLIC ACID METHYLESTER;　W
65012

585　C31 H39 N O10　69979-09-5　F-34-2118-0　15α-HYDROXY-7,8-DIMETHOXY-14β-(3',4',5'-TRIMETHOXYBENZOYLOXY)-EPIALLOBERBAN-13α-CARBOXYLIC ACID METHYLESTER;　W
65013

585　C31 H39 N O10　63911-65-9　F-34-2118-0　15β-HYDROXY-7,8-DIMETHOXY-14β-(3',4',5'-TRIMETHOXYBENZOYLOXY)-EPIALLOBERBAN-13A-CARBOXYLIC ACID METHYL ESTER;　W
65014

585　C31 H39 N O10　69979-06-2　F-34-2118-0　14β-HYDROXY-7,8-DIMETHOXY-15α-(3',4',5'-TRIMETHOXYBENZOYLOXY)EPIALLOBERBAN-13α-CARBOXYLIC ACID METHYL ESTER;　W
65015

585　C31 H39 N O10　69979-07-3　F-34-2118-0　15α-HYDROXY-7,8-DIMETHOXY-14α-(3',4',5'-TRIMETHOXYBENZOYLOXY)EPIALLOBERBAN-13α-CARBOXYLIC ACID METHYLESTER;　W
65016

585　C31 H39 N O10　63950-29-8　F-34-2118-0　14α-HYDROXY-7,8-DIMETHOXY-15α-(3',4',5'-TRIMETHOXYBENZOYLEXY)EPIALLOBERBAN-13α-CARBOXYLIC ACID METHYL ESTER;　W
65017

585　C32 H34 Fe O5 P　91550-80-0　C-106-6071-0　Dicarbonyl(1-4-η-5-n-butyl-7-methyl-1,3-cycloheptadiene)(triphenylphosphite)iron;　W
127413

Str 1089

585　C38 H35 O2 P2　70530-13-1　K-112-1284-0　(1,1-DIMETHOXY-2,6-DIPHENYL-λ5-PHOSPHORIN-4-YLMETHYL)-TRIPHENYLPHOSPHONUIM-TETRAFLUOROBORATE;　W
65018

Str 1090

585　C42 H35 N O2　79413-85-7　H-64-1345-0　N-(2-Benzhydrylphenacyl)-N-methyl-triphenylacetamide;　W
127415

Str 1091

585　C42 H35 N O2　K-113-438-0　6,6-Dibenzoyl-2-diethylamino-1,3,4-triphenyl-2-aminofulvene;　W
127416

585　C42 H35 N O2　73442-47-4　K-113-440-0　2,3-Dibenzoyl-4,5,6-triphenyl-o-diacylbenzole;　W
127417

Str 1092

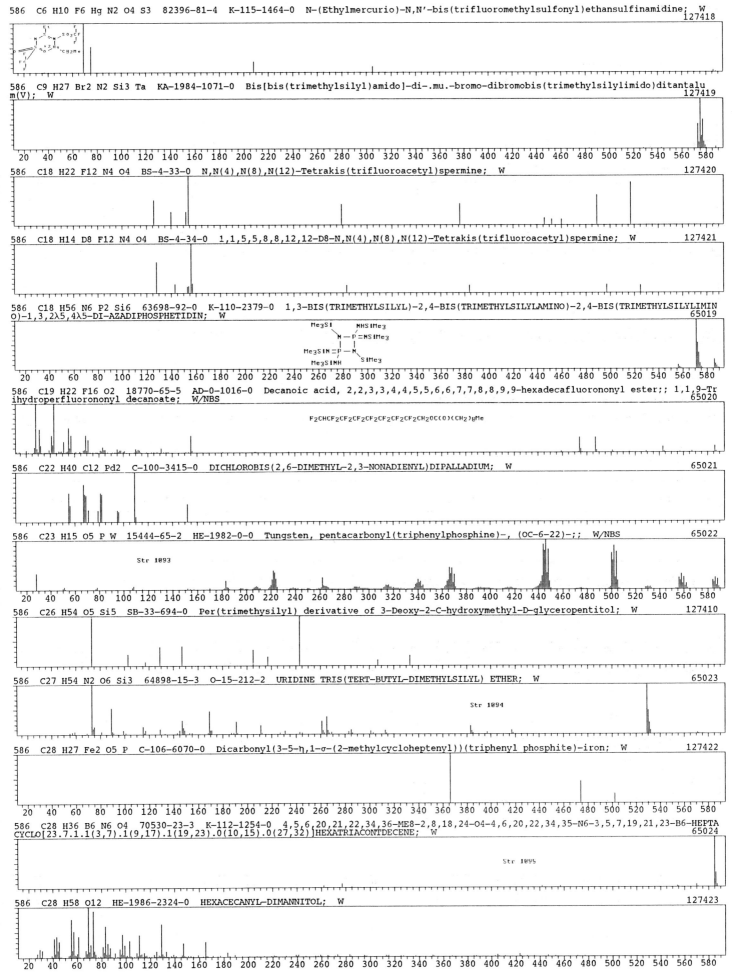

586   C6 H10 F6 Hg N2 O4 S3   82396-81-4   K-115-1464-0   N-(Ethylmercurio)-N,N'-bis(trifluoromethylsulfonyl)ethansulfinamidine;   W
127418

586   C9 H27 Br2 N2 Si3 Ta   KA-1984-1071-0   Bis[bis(trimethylsilyl)amido]-di-.mu.-bromo-dibromobis(trimethylsilylimido)ditantalu
m(V);   W
127419

586   C18 H22 F12 N4 O4   BS-4-33-0   N,N(4),N(8),N(12)-Tetrakis(trifluoroacetyl)spermine;   W
127420

586   C18 H14 D8 F12 N4 O4   BS-4-34-0   1,1,5,5,8,8,12,12-D8-N,N(4),N(8),N(12)-Tetrakis(trifluoroacetyl)spermine;   W
127421

586   C18 H56 N6 P2 Si6   63698-92-0   K-110-2379-0   1,3-BIS(TRIMETHYLSILYL)-2,4-BIS(TRIMETHYLSILYLAMINO)-2,4-BIS(TRIMETHYLSILYLIMIN
O)-1,3,2λ5,4λ5-DI-AZADIPHOSPHETIDIN;   W
65019

586   C19 H22 F16 O2   18770-65-5   AD-0-1016-0   Decanoic acid, 2,2,3,3,4,4,5,5,6,6,7,7,8,8,9,9-hexadecafluorononyl ester;; 1,1,9-Tr
ihydroperfluorononyl decanoate;   W/NBS
65020

586   C22 H40 Cl2 Pd2   C-100-3415-0   DICHLOROBIS(2,6-DIMETHYL-2,3-NONADIENYL)DIPALLADIUM;   W
65021

586   C23 H15 O5 P W   15444-65-2   HE-1982-0-0   Tungsten, pentacarbonyl(triphenylphosphine)-, (OC-6-22)-;;   W/NBS
65022

586   C26 H54 O5 Si5   SB-33-694-0   Per(trimethysilyl) derivative of 3-Deoxy-2-C-hydroxymethyl-D-glyceropentitol;   W
127410

586   C27 H54 N2 O6 Si3   64898-15-3   O-15-212-2   URIDINE TRIS(TERT-BUTYL-DIMETHYLSILYL) ETHER;   W
65023

586   C28 H27 Fe2 O5 P   C-106-6070-0   Dicarbonyl(3-5-η,1-σ-(2-methylcycloheptenyl))(triphenyl phosphite)-iron;   W
127422

586   C28 H36 B6 N6 O4   70530-23-3   K-112-1254-0   4,5,6,20,21,22,34,36-ME8-2,8,18,24-O4-4,6,20,22,34,35-N6-3,5,7,19,21,23-B6-HEPTA
CYCLO[23.7.1.1(3,7).1(9,17).1(19,23).0(10,15).0(27,32)]HEXATRIACONTDECENE;   W
65024

586   C28 H58 O12   HE-1986-2324-0   HEXACECANYL-DIMANNITOL;   W
127423

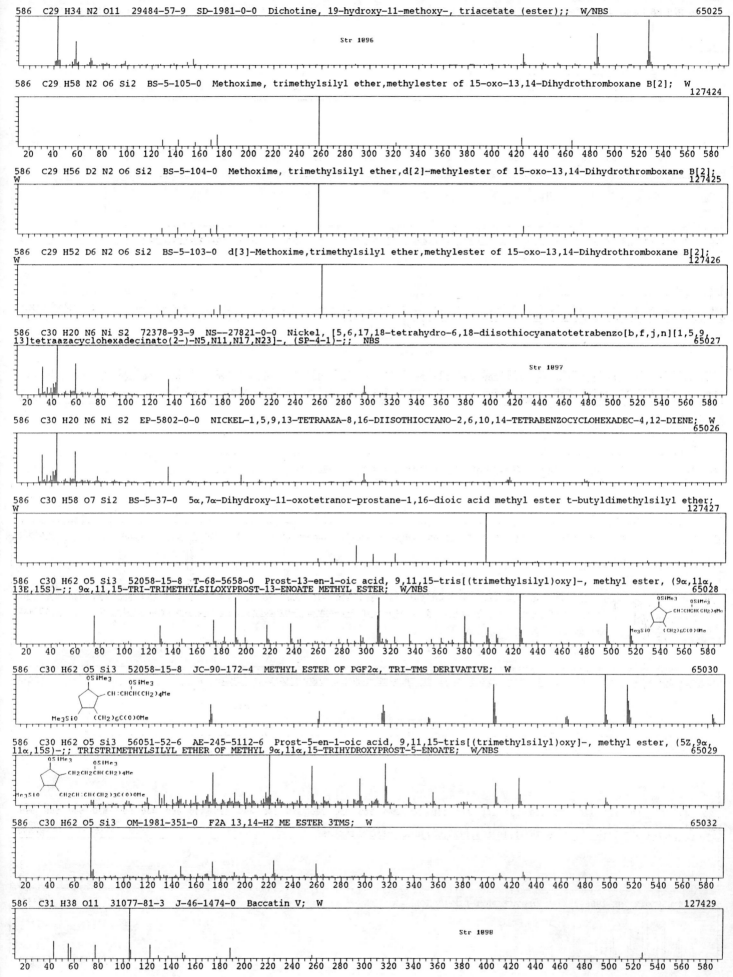

586　C29 H34 N2 O11　29484-57-9　SD-1981-0-0　Dichotine, 19-hydroxy-11-methoxy-, triacetate (ester);;　W/NBS　65025

Str 1096

586　C29 H58 N2 O6 Si2　BS-5-105-0　Methoxime, trimethylsilyl ether,methylester of 15-oxo-13,14-Dihydrothromboxane B[2];　W　127424

586　C29 H56 D2 N2 O6 Si2　BS-5-104-0　Methoxime, trimethylsilyl ether,d[2]-methylester of 15-oxo-13,14-Dihydrothromboxane B[2];　W　127425

586　C29 H52 D6 N2 O6 Si2　BS-5-103-0　d[3]-Methoxime,trimethylsilyl ether,methylester of 15-oxo-13,14-Dihydrothromboxane B[2];　W　127426

586　C30 H20 N6 Ni S2　72378-93-9　NS--27821-0-0　Nickel, [5,6,17,18-tetrahydro-6,18-diisothiocyanatotetrabenzo[b,f,j,n][1,5,9,13]tetraazacyclohexadecinato(2-)-N5,N11,N17,N23]-, (SP-4-1)-;;　NBS　65027

Str 1097

586　C30 H20 N6 Ni S2　EP-5802-0-0　NICKEL-1,5,9,13-TETRAAZA-8,16-DIISOTHIOCYANO-2,6,10,14-TETRABENZOCYCLOHEXADEC-4,12-DIENE;　W　65026

586　C30 H58 O7 Si2　BS-5-37-0　5α,7α-Dihydroxy-11-oxotetranor-prostane-1,16-dioic acid methyl ester t-butyldimethylsilyl ether;　W　127427

586　C30 H62 O5 Si3　52058-15-8　T-68-5658-0　Prost-13-en-1-oic acid, 9,11,15-tris[(trimethylsilyl)oxy]-, methyl ester, (9α,11α,13E,15S)-;;　9α,11,15-TRI-TRIMETHYLSILOXYPROST-13-ENOATE METHYL ESTER;　W/NBS　65028

586　C30 H62 O5 Si3　52058-15-8　JC-90-172-4　METHYL ESTER OF PGF2α, TRI-TMS DERIVATIVE;　W　65030

586　C30 H62 O5 Si3　56051-52-6　AE-245-5112-6　Prost-5-en-1-oic acid, 9,11,15-tris[(trimethylsilyl)oxy]-, methyl ester, (5Z,9α,11α,15S)-;;　TRISTRIMETHYLSILYL ETHER OF METHYL 9α,11α,15-TRIHYDROXYPROST-5-ENOATE;　W/NBS　65029

586　C30 H62 O5 Si3　OM-1981-351-0　F2A 13,14-H2 ME ESTER 3TMS;　W　65032

586　C31 H38 O11　31077-81-3　J-46-1474-0　Baccatin V;　W　127429

Str 1098

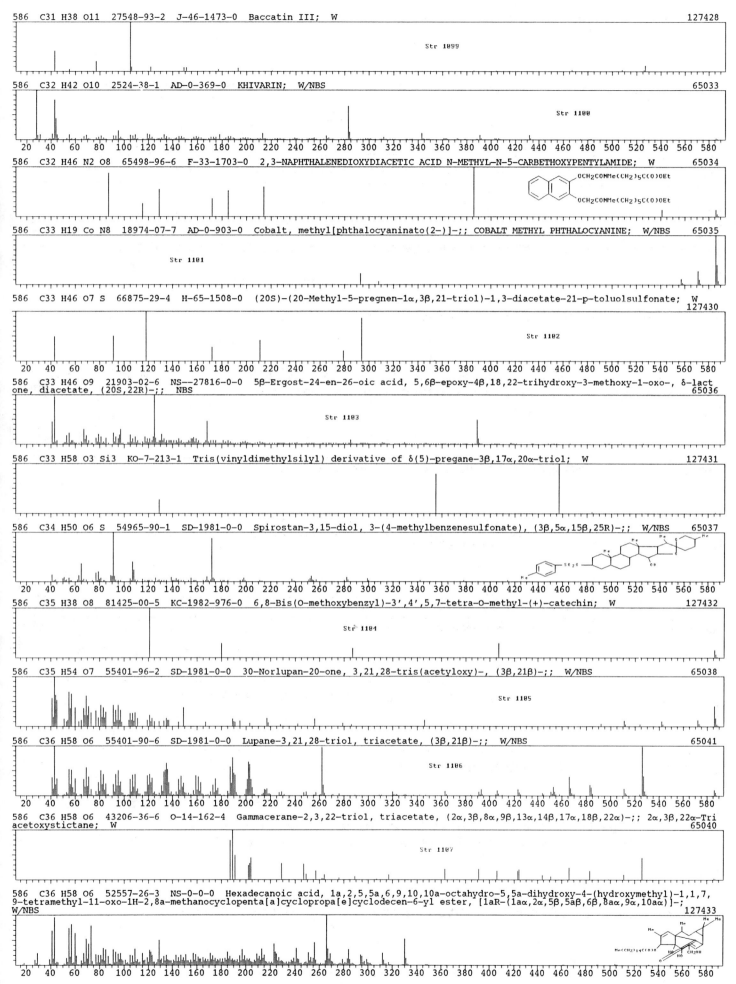

586   C31 H38 O11   27548-93-2   J-46-1473-0   Baccatin III;   W   127428

Str 1099

586   C32 H42 O10   2524-38-1   AD-0-369-0   KHIVARIN;   W/NBS   65033

Str 1100

586   C32 H46 N2 O8   65498-96-6   F-33-1703-0   2,3-NAPHTHALENEDIOXYDIACETIC ACID N-METHYL-N-5-CARBETHOXYPENTYLAMIDE;   W   65034

586   C33 H19 Co N8   18974-07-7   AD-0-903-0   Cobalt, methyl[phthalocyaninato(2-)]-;; COBALT METHYL PHTHALOCYANINE;   W/NBS   65035

Str 1101

586   C33 H46 O7 S   66875-29-4   H-65-1508-0   (20S)-(20-Methyl-5-pregnen-1α,3β,21-triol)-1,3-diacetate-21-p-toluolsulfonate;   W   127430

Str 1102

586   C33 H46 O9   21903-02-6   NS--27816-0-0   5β-Ergost-24-en-26-oic acid, 5,6β-epoxy-4β,18,22-trihydroxy-3-methoxy-1-oxo-, δ-lact one, diacetate, (20S,22R)-;;   NBS   65036

Str 1103

586   C33 H58 O3 Si3   KO-7-213-1   Tris(vinyldimethylsilyl) derivative of δ(5)-pregane-3β,17α,20α-triol;   W   127431

586   C34 H50 O6 S   54965-90-1   SD-1981-0-0   Spirostan-3,15-diol, 3-(4-methylbenzenesulfonate), (3β,5α,15β,25R)-;;   W/NBS   65037

586   C35 H38 O8   81425-00-5   KC-1982-976-0   6,8-Bis(O-methoxybenzyl)-3',4',5,7-tetra-O-methyl-(+)-catechin;   W   127432

Str 1104

586   C35 H54 O7   55401-96-2   SD-1981-0-0   30-Norlupan-20-one, 3,21,28-tris(acetyloxy)-, (3β,21β)-;;   W/NBS   65038

Str 1105

586   C36 H58 O6   55401-90-6   SD-1981-0-0   Lupane-3,21,28-triol, triacetate, (3β,21β)-;;   W/NBS   65041

Str 1106

586   C36 H58 O6   43206-36-6   O-14-162-4   Gammacerane-2,3,22-triol, triacetate, (2α,3β,8α,9β,13α,14β,17α,18β,22α)-;; 2α,3β,22α-Tri acetoxystictane;   W   65040

Str 1107

586   C36 H58 O6   52557-26-3   NS-0-0-0   Hexadecanoic acid, 1a,2,5,5a,6,9,10,10a-octahydro-5,5a-dihydroxy-4-(hydroxymethyl)-1,1,7, 9-tetramethyl-11-oxo-1H-2,8a-methanocyclopenta[a]cyclopropa[e]cyclodecen-6-yl ester, [1aR-(1aα,2α,5β,5aβ,6β,8aα,9α,10aα)]-; W/NBS   127433

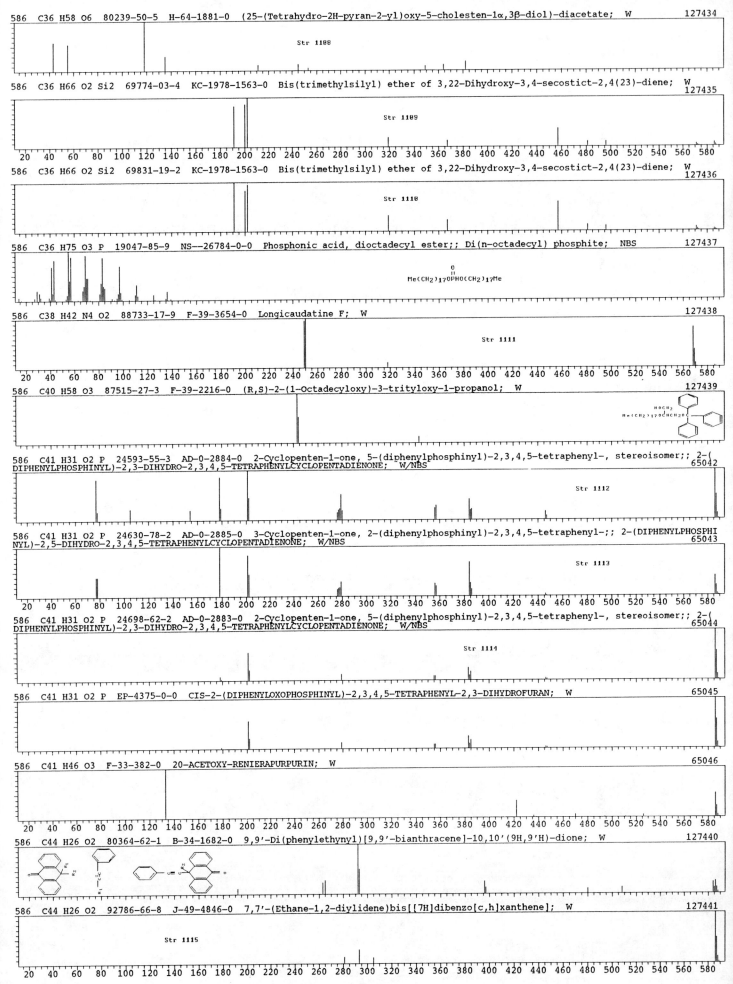

586   C36 H58 O6   80239-50-5   H-64-1881-0   (25-(Tetrahydro-2H-pyran-2-yl)oxy-5-cholesten-1α,3β-diol)-diacetate;   W   127434

Str 1108

586   C36 H66 O2 Si2   69774-03-4   KC-1978-1563-0   Bis(trimethylsilyl) ether of 3,22-Dihydroxy-3,4-secostict-2,4(23)-diene;   W   127435

Str 1109

586   C36 H66 O2 Si2   69831-19-2   KC-1978-1563-0   Bis(trimethylsilyl) ether of 3,22-Dihydroxy-3,4-secostict-2,4(23)-diene;   W   127436

Str 1110

586   C36 H75 O3 P   19047-85-9   NS--26784-0-0   Phosphonic acid, dioctadecyl ester;; Di(n-octadecyl) phosphite;   NBS   127437

Me(CH2)17OPHO(CH2)17Me

586   C38 H42 N4 O2   88733-17-9   F-39-3654-0   Longicaudatine F;   W   127438

Str 1111

586   C40 H58 O3   87515-27-3   F-39-2216-0   (R,S)-2-(1-Octadecyloxy)-3-trityloxy-1-propanol;   W   127439

586   C41 H31 O2 P   24593-55-3   AD-0-2884-0   2-Cyclopenten-1-one, 5-(diphenylphosphinyl)-2,3,4,5-tetraphenyl-, stereoisomer;; 2-(DIPHENYLPHOSPHINYL)-2,3-DIHYDRO-2,3,4,5-TETRAPHENYLCYCLOPENTADIENONE;   W/NBS   65042

Str 1112

586   C41 H31 O2 P   24630-78-2   AD-0-2885-0   3-Cyclopenten-1-one, 2-(diphenylphosphinyl)-2,3,4,5-tetraphenyl-;; 2-(DIPHENYLPHOSPHINYL)-2,5-DIHYDRO-2,3,4,5-TETRAPHENYLCYCLOPENTADIENONE;   W/NBS   65043

Str 1113

586   C41 H31 O2 P   24698-62-2   AD-0-2883-0   2-Cyclopenten-1-one, 5-(diphenylphosphinyl)-2,3,4,5-tetraphenyl-, stereoisomer;; 2-(DIPHENYLPHOSPHINYL)-2,3-DIHYDRO-2,3,4,5-TETRAPHENYLCYCLOPENTADIENONE;   W/NBS   65044

Str 1114

586   C41 H31 O2 P   EP-4375-0-0   CIS-2-(DIPHENYLOXOPHOSPHINYL)-2,3,4,5-TETRAPHENYL-2,3-DIHYDROFURAN;   W   65045

586   C41 H46 O3   F-33-382-0   20-ACETOXY-RENIERAPURPURIN;   W   65046

586   C44 H26 O2   80364-62-1   B-34-1682-0   9,9'-Di(phenylethynyl)[9,9'-bianthracene]-10,10'(9H,9'H)-dione;   W   127440

586   C44 H26 O2   92786-66-8   J-49-4846-0   7,7'-(Ethane-1,2-diylidene)bis[[7H]dibenzo[c,h]xanthene];   W   127441

Str 1115

586 C44 H30 N2 87613-00-1 Y-20-973-0 1,3,4,5,7,8-Hexaphenyl-2,6-naphthyridine; W 127442

Str 1116

586 C44 H30 N2 87613-01-2 Y-20-973-0 1,3,4,5,6,8-Hexaphenyl-2,7-naphthyridine; W 127443

Str 1117

587 C16 H16 Br3 N O8 66653-29-0 J-43-3729-0 Tetramethyl 1,6,7-Tribromo-3a,6,7,7a-tetrahydroindole-2,3,3a,4-tetracarboxylate;
W 127444

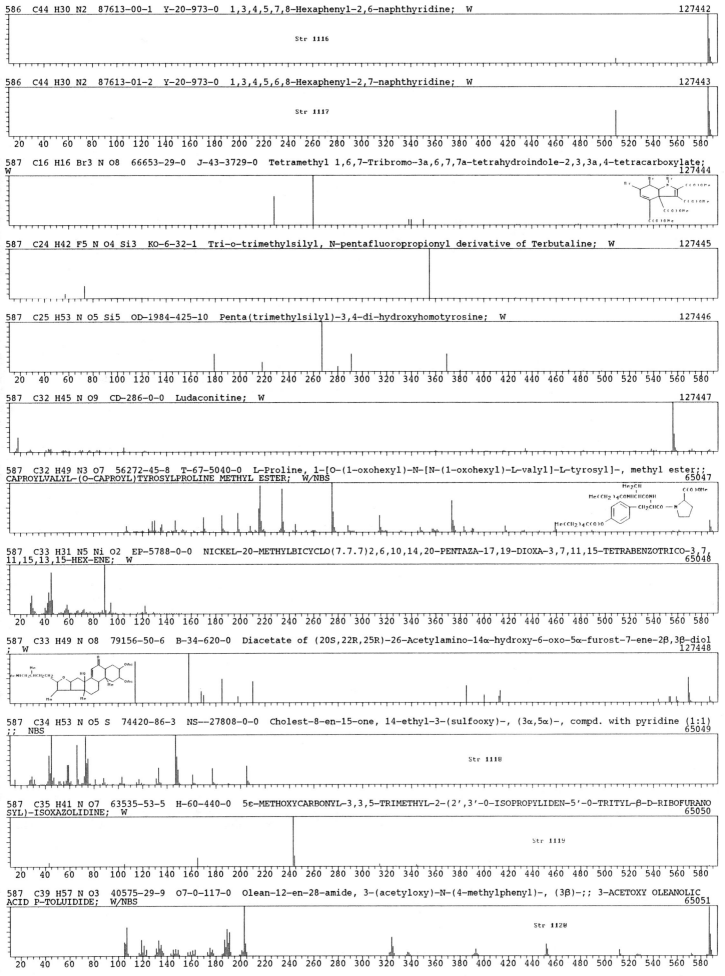

587 C24 H42 F5 N O4 Si3 KO-6-32-1 Tri-o-trimethylsilyl, N-pentafluoropropionyl derivative of Terbutaline; W 127445

587 C25 H53 N O5 Si5 OD-1984-425-10 Penta(trimethylsilyl)-3,4-di-hydroxyhomotyrosine; W 127446

587 C32 H45 N O9 CD-286-0-0 Ludaconitine; W 127447

587 C32 H49 N3 O7 56272-45-8 T-67-5040-0 L-Proline, 1-[O-(1-oxohexyl)-N-[N-(1-oxohexyl)-L-valyl]-L-tyrosyl]-, methyl ester;;
CAPROYLVALYL-(O-CAPROYL)TYROSYLPROLINE METHYL ESTER; W/NBS 65047

587 C33 H31 N5 Ni O2 EP-5788-0-0 NICKEL-20-METHYLBICYCLO(7.7.7)2,6,10,14,20-PENTAZA-17,19-DIOXA-3,7,11,15-TETRABENZOTRICO-3,7,
11,15,13,15-HEX-ENE; W 65048

587 C33 H49 N O8 79156-50-6 B-34-620-0 Diacetate of (20S,22R,25R)-26-Acetylamino-14α-hydroxy-6-oxo-5α-furost-7-ene-2β,3β-diol
; W 127448

587 C34 H53 N O5 S 74420-86-3 NS--27808-0-0 Cholest-8-en-15-one, 14-ethyl-3-(sulfooxy)-, (3α,5α)-, compd. with pyridine (1:1)
;; NBS 65049

Str 1118

587 C35 H41 N O7 63535-53-5 H-60-440-0 5ε-METHOXYCARBONYL-3,3,5-TRIMETHYL-2-(2',3'-0-ISOPROPYLIDEN-5'-0-TRITYL-β-D-RIBOFURANO
SYL)-ISOXAZOLIDINE; W 65050

Str 1119

587 C39 H57 N O3 40575-29-9 O7-0-117-0 Olean-12-en-28-amide, 3-(acetyloxy)-N-(4-methylphenyl)-, (3β)-;; 3-ACETOXY OLEANOLIC
ACID P-TOLUIDIDE; W/NBS 65051

Str 1120

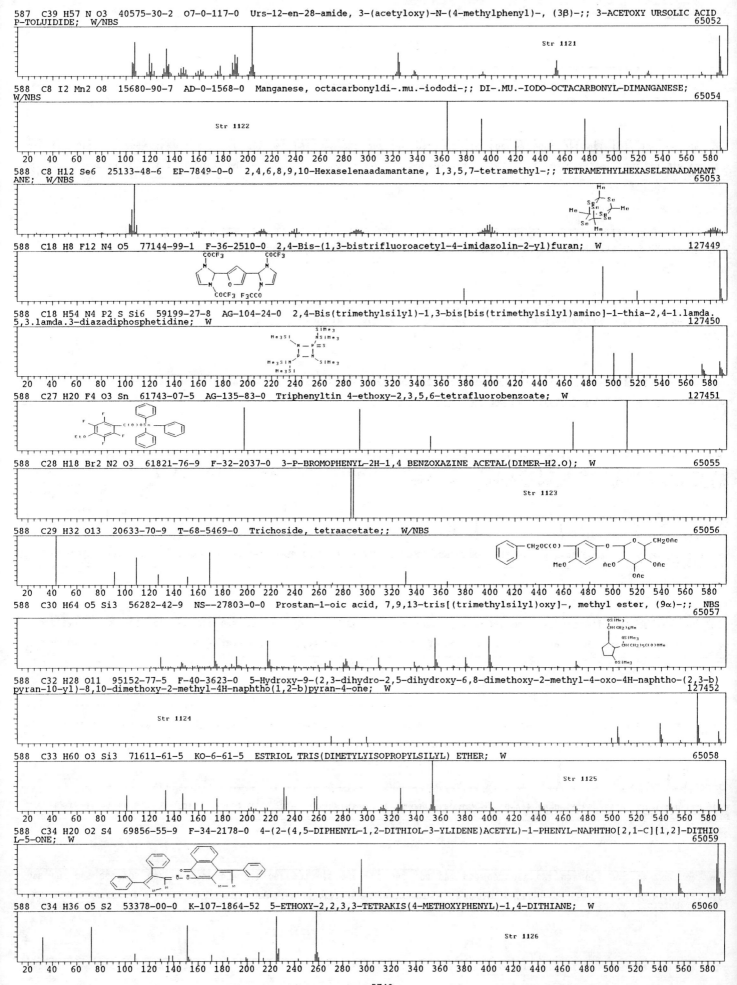

587   C39 H57 N O3   40575-30-2   O7-0-117-0   Urs-12-en-28-amide, 3-(acetyloxy)-N-(4-methylphenyl)-, (3β)-;; 3-ACETOXY URSOLIC ACID P-TOLUIDIDE;   W/NBS
Str 1121
65052

588   C8 I2 Mn2 O8   15680-90-7   AD-0-1568-0   Manganese, octacarbonyldi-.mu.-iododi-;; DI-.MU.-IODO-OCTACARBONYL-DIMANGANESE; W/NBS
Str 1122
65054

588   C8 H12 Se6   25133-48-6   EP-7849-0-0   2,4,6,8,9,10-Hexaselenaadamantane, 1,3,5,7-tetramethyl-;; TETRAMETHYLHEXASELENAADAMANTANE;   W/NBS
65053

588   C18 H8 F12 N4 O5   77144-99-1   F-36-2510-0   2,4-Bis-(1,3-bistrifluoroacetyl-4-imidazolin-2-yl)furan;   W
127449

588   C18 H54 N4 P2 S Si6   59199-27-8   AG-104-24-0   2,4-Bis(trimethylsilyl)-1,3-bis[bis(trimethylsilyl)amino]-1-thia-2,4-1.lamda.5,3.lamda.3-diazadiphosphetidine;   W
127450

588   C27 H20 F4 O3 Sn   61743-07-5   AG-135-83-0   Triphenyltin 4-ethoxy-2,3,5,6-tetrafluorobenzoate;   W
127451

588   C28 H18 Br2 N2 O3   61821-76-9   F-32-2037-0   3-P-BROMOPHENYL-2H-1,4 BENZOXAZINE ACETAL(DIMER-H2.O);   W
Str 1123
65055

588   C29 H32 O13   20633-70-9   T-68-5469-0   Trichoside, tetraacetate;;   W/NBS
65056

588   C30 H64 O5 Si3   56282-42-9   NS--27803-0-0   Prostan-1-oic acid, 7,9,13-tris[(trimethylsilyl)oxy]-, methyl ester, (9α)-;;   NBS
65057

588   C32 H28 O11   95152-77-5   F-40-3623-0   5-Hydroxy-9-(2,3-dihydro-2,5-dihydroxy-6,8-dimethoxy-2-methyl-4-oxo-4H-naphtho-(2,3-b)pyran-10-yl)-8,10-dimethoxy-2-methyl-4H-naphtho(1,2-b)pyran-4-one;   W
Str 1124
127452

588   C33 H60 O3 Si3   71611-61-5   KO-6-61-5   ESTRIOL TRIS(DIMETYLYISOPROPYLSILYL) ETHER;   W
Str 1125
65058

588   C34 H20 O2 S4   69856-55-9   F-34-2178-0   4-(2-(4,5-DIPHENYL-1,2-DITHIOL-3-YLIDENE)ACETYL)-1-PHENYL-NAPHTHO[2,1-C][1,2]-DITHIOL-5-ONE;   W
65059

588   C34 H36 O5 S2   53378-00-0   K-107-1864-52   5-ETHOXY-2,2,3,3-TETRAKIS(4-METHOXYPHENYL)-1,4-DITHIANE;   W
Str 1126
65060

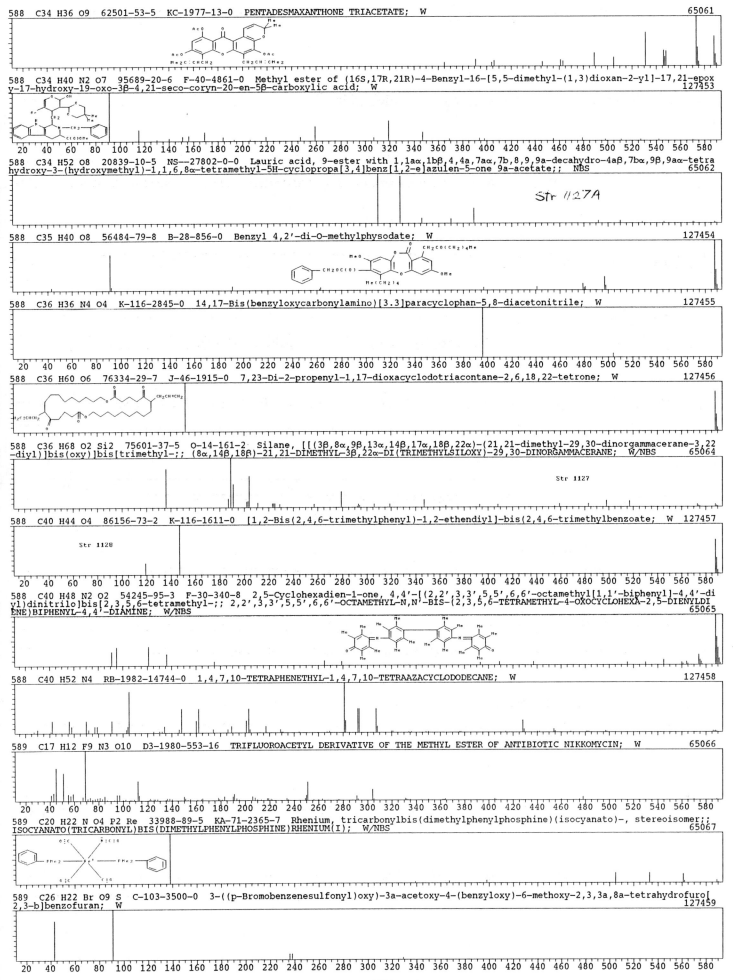

588　C34 H36 O9　62501-53-5　KC-1977-13-0　PENTADESMAXANTHONE TRIACETATE;　W　　　　　　　　　65061

588　C34 H40 N2 O7　95689-20-6　F-40-4861-0　Methyl ester of (16S,17R,21R)-4-Benzyl-16-[5,5-dimethyl-(1,3)dioxan-2-yl]-17,21-epoxy-17-hydroxy-19-oxo-3β-4,21-seco-coryn-20-en-5β-carboxylic acid;　W　　　　　127453

588　C34 H52 O8　20839-10-5　NS--27802-0-0　Lauric acid, 9-ester with 1,1aα,1bβ,4,4a,7aα,7b,8,9,9a-decahydro-4aβ,7bα,9β,9aα-tetrahydroxy-3-(hydroxymethyl)-1,1,6,8α-tetramethyl-5H-cyclopropa[3,4]benz[1,2-e]azulen-5-one 9a-acetate;;　NBS　　65062

Str 1127A

588　C35 H40 O8　56484-79-8　B-28-856-0　Benzyl 4,2'-di-O-methylphysodate;　W　　　　　　　127454

588　C36 H36 N4 O4　K-116-2845-0　14,17-Bis(benzyloxycarbonylamino)[3.3]paracyclophan-5,8-diacetonitrile;　W　　　127455

588　C36 H60 O6　76334-29-7　J-46-1915-0　7,23-Di-2-propenyl-1,17-dioxacyclodotriacontane-2,6,18,22-tetrone;　W　　　127456

588　C36 H68 O2 Si2　75601-37-5　O-14-161-2　Silane, [[(3β,8α,9β,13α,14β,17α,18β,22α)-(21,21-dimethyl-29,30-dinorgammacerane-3,22-diyl)]bis(oxy)]bis[trimethyl-;;　(8α,14β,18β)-21,21-DIMETHYL-3β,22α-DI(TRIMETHYLSILOXY)-29,30-DINORGAMMACERANE;　W/NBS　　65064

Str 1127

588　C40 H44 O4　86156-73-2　K-116-1611-0　[1,2-Bis(2,4,6-trimethylphenyl)-1,2-ethendiyl]-bis(2,4,6-trimethylbenzoate;　W　127457

Str 1128

588　C40 H48 N2 O2　54245-95-3　F-30-340-8　2,5-Cyclohexadien-1-one, 4,4'-[(2,2',3,3',5,5',6,6'-octamethyl[1,1'-biphenyl]-4,4'-diyl)dinitrilo]bis[2,3,5,6-tetramethyl-;;　2,2',3,3',5,5',6,6'-OCTAMETHYL-N,N'-BIS-(2,3,5,6-TETRAMETHYL-4-OXOCYCLOHEXA-2,5-DIENYLIDENE)BIPHENYL-4,4'-DIAMINE;　W/NBS　　65065

588　C40 H52 N4　RB-1982-14744-0　1,4,7,10-TETRAPHENETHYL-1,4,7,10-TETRAAZACYCLODODECANE;　W　　127458

589　C17 H12 F9 N3 O10　D3-1980-553-16　TRIFLUOROACETYL DERIVATIVE OF THE METHYL ESTER OF ANTIBIOTIC NIKKOMYCIN;　W　　65066

589　C20 H22 N O4 P2 Re　33988-89-5　KA-71-2365-7　Rhenium, tricarbonylbis(dimethylphenylphosphine)(isocyanato)-, stereoisomer;;　ISOCYANATO(TRICARBONYL)BIS(DIMETHYLPHENYLPHOSPHINE)RHENIUM(I);　W/NBS　　65067

589　C26 H22 Br O9 S　C-103-3500-0　3-((p-Bromobenzenesulfonyl)oxy)-3a-acetoxy-4-(benzyloxy)-6-methoxy-2,3,3a,8a-tetrahydrofuro[2,3-b]benzofuran;　W　　127459

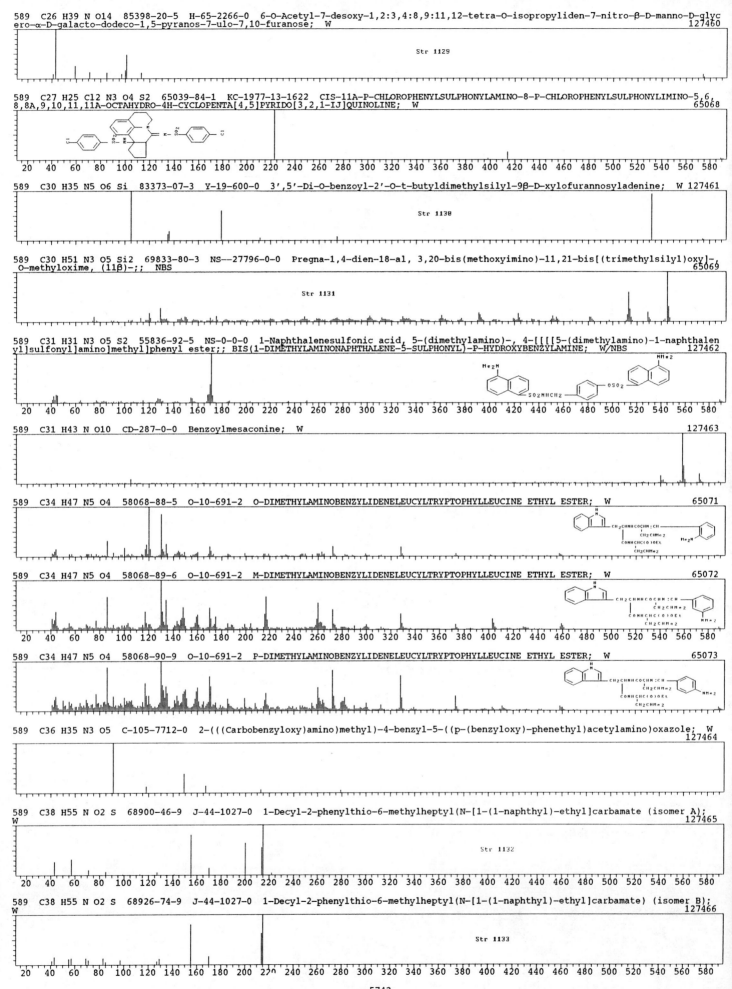

589 C26 H39 N O14   85398-20-5   H-65-2266-0   6-O-Acetyl-7-desoxy-1,2:3,4:8,9:11,12-tetra-O-isopropyliden-7-nitro-β-D-manno-D-glyc ero-α-D-galacto-dodeco-1,5-pyranos-7-ulo-7,10-furanose;   W   127460

Str 1129

589 C27 H25 Cl2 N3 O4 S2   65039-84-1   KC-1977-13-1622   CIS-11A-P-CHLOROPHENYLSULPHONYLAMINO-8-P-CHLOROPHENYLSULPHONYLIMINO-5,6, 8,8A,9,10,11,11A-OCTAHYDRO-4H-CYCLOPENTA[4,5]PYRIDO[3,2,1-IJ]QUINOLINE;   W   65068

589 C30 H35 N5 O6 Si   83373-07-3   Y-19-600-0   3',5'-Di-O-benzoyl-2'-O-t-butyldimethylsilyl-9β-D-xylofurannosyladenine;   W 127461

Str 1130

589 C30 H51 N3 O5 Si2   69833-80-3   NS--27796-0-0   Pregna-1,4-dien-18-al, 3,20-bis(methoxyimino)-11,21-bis[(trimethylsilyl)oxy]-, O-methyloxime, (11β)-;;   NBS   65069

Str 1131

589 C31 H31 N3 O5 S2   55836-92-5   NS-0-0-0   1-Naphthalenesulfonic acid, 5-(dimethylamino)-, 4-[[[[5-(dimethylamino)-1-naphthalen yl]sulfonyl]amino]methyl]phenyl ester;; BIS(1-DIMETHYLAMINONAPHTHALENE-5-SULPHONYL)-P-HYDROXYBENZYLAMINE;   W/NBS   127462

589 C31 H43 N O10   CD-287-0-0   Benzoylmesaconine;   W   127463

589 C34 H47 N5 O4   58068-88-5   O-10-691-2   O-DIMETHYLAMINOBENZYLIDENELEUCYLTRYPTOPHYLLEUCINE ETHYL ESTER;   W   65071

589 C34 H47 N5 O4   58068-89-6   O-10-691-2   M-DIMETHYLAMINOBENZYLIDENELEUCYLTRYPTOPHYLLEUCINE ETHYL ESTER;   W   65072

589 C34 H47 N5 O4   58068-90-9   O-10-691-2   P-DIMETHYLAMINOBENZYLIDENELEUCYLTRYPTOPHYLLEUCINE ETHYL ESTER;   W   65073

589 C36 H35 N3 O5   C-105-7712-0   2-(((Carbobenzyloxy)amino)methyl)-4-benzyl-5-((p-(benzyloxy)-phenethyl)acetylamino)oxazole;   W   127464

589 C38 H55 N O2 S   68900-46-9   J-44-1027-0   1-Decyl-2-phenylthio-6-methylheptyl(N-[1-(1-naphthyl)-ethyl]carbamate (isomer A); W   127465

Str 1132

589 C38 H55 N O2 S   68926-74-9   J-44-1027-0   1-Decyl-2-phenylthio-6-methylheptyl(N-[1-(1-naphthyl)-ethyl]carbamate) (isomer B); W   127466

Str 1133

590 C20 H23 F9 N2 O8 56196-63-5 AK-135-660-2 5-Undecenedioic acid, 2,10-bis[(trifluoroacetyl)amino]-5-[[(trifluoroacetyl)oxy]methyl]-, dimethyl ester;; 2,10-BIS(N-TRIFLUOROACETYLAMINO)-5-TRIFLUOROACETOXYMETHYL-5-UNDECENEDIOIC ACID DIMETHYL ESTER; W/NBS
65074

590 C20 H51 O8 P Si5 69688-47-7 BA-0-317-0 D-Xylopyranose, 2,3,4-tris-O-(trimethylsilyl)-, bis(trimethylsilyl) phosphate;; O-PENTAKIS(TRIMETHYLSILYL)-D-XYLOPYRANOSE-1-PHOSPHATE; W/NBS
65078

590 C20 H51 O8 P Si5 55520-86-0 EP-797-0-0 D-Ribose, 2,3,4-tris-O-(trimethylsilyl)-, 5-[bis(trimethylsilyl) phosphate];; D-RIBOSE-5-PHOSPHATE-PENTATMS; W/NBS
65080

590 C20 H51 O8 P Si5 69688-46-6 BA-0-312-0 D-Ribofuranose, 2,3,5-tris-O-(trimethylsilyl)-, bis(trimethylsilyl) phosphate;; O-PENTAKIS(TRIMETHYLSILYL)-D-RIBOFURANOSE-1-PHOSPHATE; W/NBS
65076

590 C20 H51 O8 P Si5 69744-64-5 BA-0-313-0 D-Ribofuranose, 1,2,3-tris-O-(trimethylsilyl)-, bis(trimethylsilyl) phosphate;; O-PENTAKIS(TRIMETHYLSILYL)-D-RIBOFURANOSE-5-PHOSPHATE; W/NBS
65077

590 C20 H51 O8 P Si5 55520-85-9 BA-0-318-0 D-threo-2-Pentulose, 1,3,4-tris-O-(trimethylsilyl)-, 5-[bis(trimethylsilyl) phosphate];; O-PENTAKIS(TRIMETHYLSILYL)-D-XYLULOSE-5-PHOSPHATE; W/NBS
65075

590 C20 H51 O8 P Si5 55520-87-1 BA-0-315-0 D-erythro-2-Pentulose, 1,3,4-tris-O-(trimethylsilyl)-, 5-[bis(trimethylsilyl) phosphate];; O-PENTAKIS(TRIMETHYLSILYL)-D-RIBULOSE-5-PHOSPHATE; W/NBS
65079

590 C20 H51 O8 P Si5 56192-89-3 EP-1140-0-0 erythro-2-Pentulose, 1,3,4-tris-O-(trimethylsilyl)-, 5-[bis(trimethylsilyl) phosphate];; RIBULOSE-5-PHOSPHATE-PENTATMS; W/NBS
65081

590 C20 H51 O8 P Si5 EP-796-0-0 D-RIBOSE-1-PHOSPHATE-PENTATMS; W
65082

590 C20 H51 O8 P Si5 EP-798-0-0 D-XYLOSE-1-PHOSPHATE-PENTATMS; W
65083

590 C24 H19 F9 O7 56438-13-2 SD-1981-0-0 Estra-1,3,5(10)-trien-17-one, 3,6,7-tris[(trifluoroacetyl)oxy]-, (6α,7α)-;; W/NBS
65084

590 C24 H34 N2 O15 78300-73-9 F-37-110-0 O-(2,3,4-Tri-O-acetyl-β-D-galactopyranosiduronamide)-(1 to 4)-2-acetamido-1,3-di-O-acetyl-2,6-didesoxy-α-D-glucopyranoside; W
127467

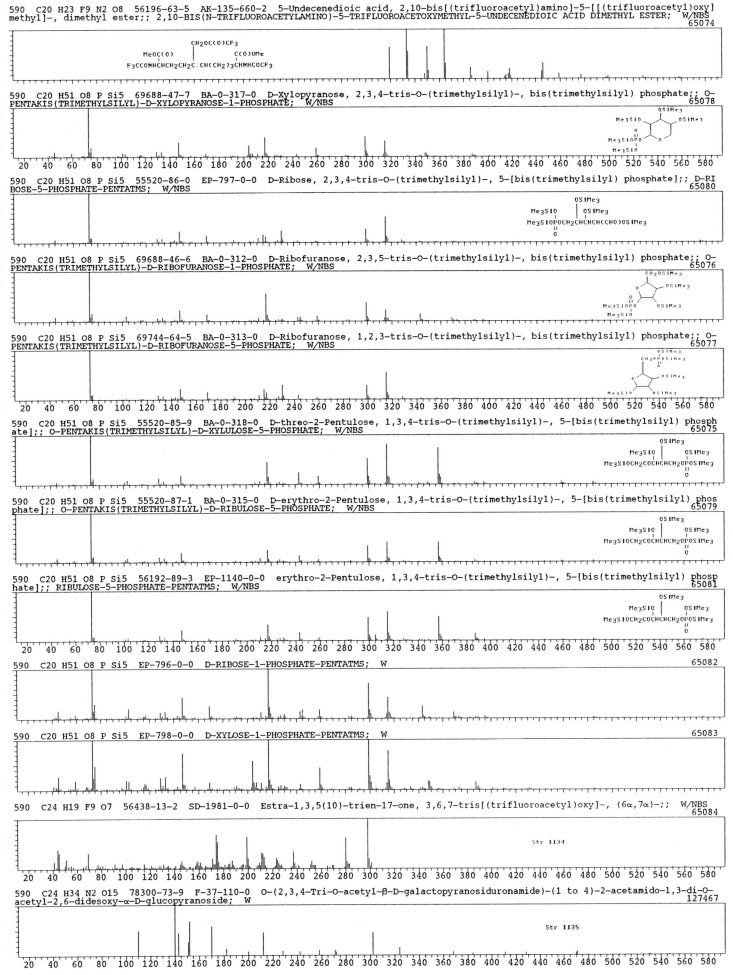

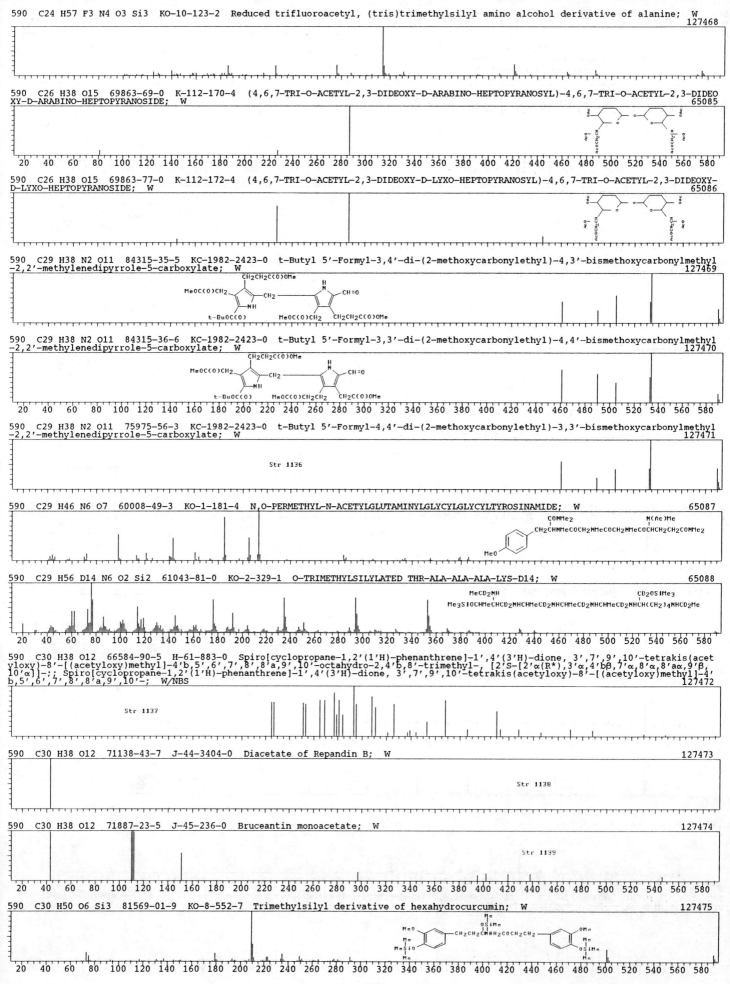

590 C24 H57 F3 N4 O3 Si3 KO-10-123-2 Reduced trifluoroacetyl, (tris)trimethylsilyl amino alcohol derivative of alanine; W
127468

590 C26 H38 O15 69863-69-0 K-112-170-4 (4,6,7-TRI-O-ACETYL-2,3-DIDEOXY-D-ARABINO-HEPTOPYRANOSYL)-4,6,7-TRI-O-ACETYL-2,3-DIDEOXY-D-ARABINO-HEPTOPYRANOSIDE; W
65085

590 C26 H38 O15 69863-77-0 K-112-172-4 (4,6,7-TRI-O-ACETYL-2,3-DIDEOXY-D-LYXO-HEPTOPYRANOSYL)-4,6,7-TRI-O-ACETYL-2,3-DIDEOXY-D-LYXO-HEPTOPYRANOSIDE; W
65086

590 C29 H38 N2 O11 84315-35-5 KC-1982-2423-0 t-Butyl 5'-Formyl-3,4'-di-(2-methoxycarbonylethyl)-4,3'-bismethoxycarbonylmethyl-2,2'-methylenedipyrrole-5-carboxylate; W
127469

590 C29 H38 N2 O11 84315-36-6 KC-1982-2423-0 t-Butyl 5'-Formyl-3,3'-di-(2-methoxycarbonylethyl)-4,4'-bismethoxycarbonylmethyl-2,2'-methylenedipyrrole-5-carboxylate; W
127470

590 C29 H38 N2 O11 75975-56-3 KC-1982-2423-0 t-Butyl 5'-Formyl-4,4'-di-(2-methoxycarbonylethyl)-3,3'-bismethoxycarbonylmethyl-2,2'-methylenedipyrrole-5-carboxylate; W
127471

Str 1136

590 C29 H46 N6 O7 60008-49-3 KO-1-181-4 N,O-PERMETHYL-N-ACETYLGLUTAMINYLGLYCYLGLYCYLTYROSINAMIDE; W
65087

590 C29 H56 D14 N6 O2 Si2 61043-81-0 KO-2-329-1 O-TRIMETHYLSILYLATED THR-ALA-ALA-ALA-LYS-D14; W
65088

590 C30 H38 O12 66584-90-5 H-61-883-0 Spiro[cyclopropane-1,2'(1'H)-phenanthrene]-1',4'(3'H)-dione, 3',7',9',10'-tetrakis(acetyloxy)-8'-[(acetyloxy)methyl]-4'b,5',6',7',8',8'a,9',10'-octahydro-2,4'b,8'-trimethyl-, [2'S-[2'α(R*),3'α,4'bβ,7'α,8'α,8'aα,9'β,10'α]]-; Spiro[cyclopropane-1,2'(1'H)-phenanthrene]-1',4'(3'H)-dione, 3',7',9',10'-tetrakis(acetyloxy)-8'-[(acetyloxy)methyl]-4'b,5',6',7',8',8'a,9',10'-; W/NBS
127472

Str 1137

590 C30 H38 O12 71138-43-7 J-44-3404-0 Diacetate of Repandin B; W
127473

Str 1138

590 C30 H38 O12 71887-23-5 J-45-236-0 Bruceantin monoacetate; W
127474

Str 1139

590 C30 H50 O6 Si3 81569-01-9 KO-8-552-7 Trimethylsilyl derivative of hexahydrocurcumin; W
127475

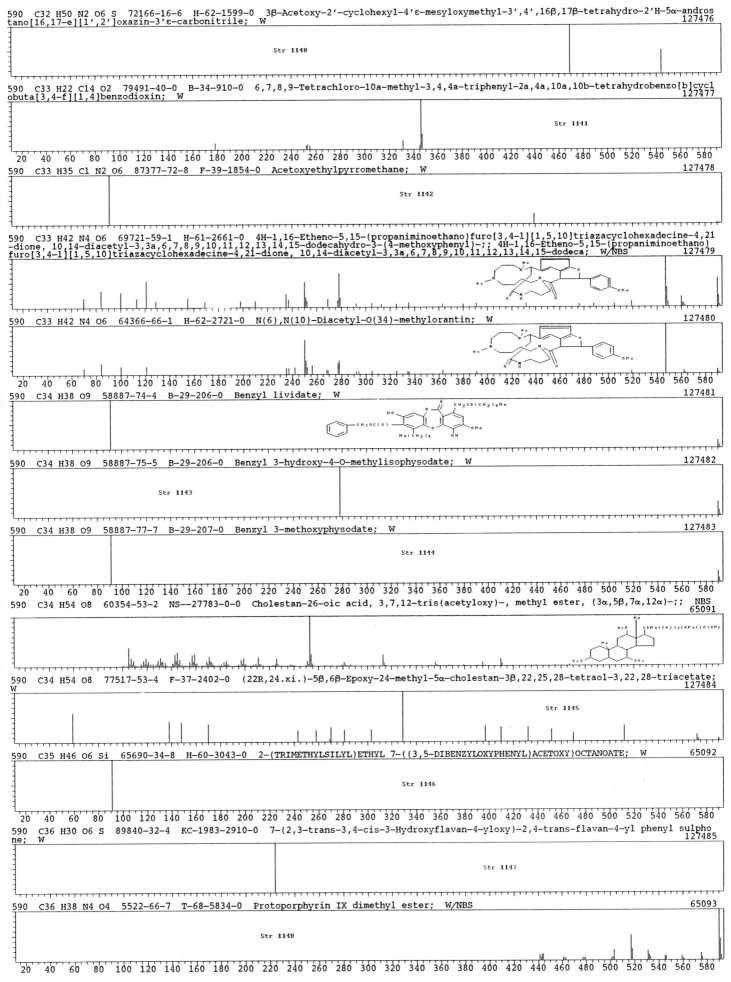

590   C32 H50 N2 O6 S   72166-16-6   H-62-1599-0   3β-Acetoxy-2'-cyclohexyl-4'ε-mesyloxymethyl-3',4',16β,17β-tetrahydro-2'H-5α-andros
tano[16,17-e][1',2']oxazin-3'ε-carbonitrile;   W   127476

Str 1140

590   C33 H22 Cl4 O2   79491-40-0   B-34-910-0   6,7,8,9-Tetrachloro-10a-methyl-3,4,4a-triphenyl-2a,4a,10a,10b-tetrahydrobenzo[b]cycl
obuta[3,4-f][1,4]benzodioxin;   W   127477

Str 1141

590   C33 H35 Cl N2 O6   87377-72-8   F-39-1854-0   Acetoxyethylpyrromethane;   W   127478

Str 1142

590   C33 H42 N4 O6   69721-59-1   H-61-2661-0   4H-1,16-Etheno-5,15-(propaniminoethano)furo[3,4-1][1,5,10]triazacyclohexadecine-4,21
-dione, 10,14-diacetyl-3,3a,6,7,8,9,10,11,12,13,14,15-dodecahydro-3-(4-methoxyphenyl)-;;   4H-1,16-Etheno-5,15-(propaniminoethano)
furo[3,4-1][1,5,10]triazacyclohexadecine-4,21-dione, 10,14-diacetyl-3,3a,6,7,8,9,10,11,12,13,14,15-dodeca;   W/NBS   127479

590   C33 H42 N4 O6   64366-66-1   H-62-2721-0   N(6),N(10)-Diacetyl-O(34)-methylorantin;   W   127480

590   C34 H38 O9   58887-74-4   B-29-206-0   Benzyl lividate;   W   127481

590   C34 H38 O9   58887-75-5   B-29-206-0   Benzyl 3-hydroxy-4-O-methylisophysodate;   W   127482

Str 1143

590   C34 H38 O9   58887-77-7   B-29-207-0   Benzyl 3-methoxyphysodate;   W   127483

Str 1144

590   C34 H54 O8   60354-53-2   NS--27783-0-0   Cholestan-26-oic acid, 3,7,12-tris(acetyloxy)-, methyl ester, (3α,5β,7α,12α)-;;   NBS
65091

590   C34 H54 O8   77517-53-4   F-37-2402-0   (22R,24.xi.)-5β,6β-Epoxy-24-methyl-5α-cholestan-3β,22,25,28-tetraol-3,22,28-triacetate;
W   127484

Str 1145

590   C35 H46 O6 Si   65690-34-8   H-60-3043-0   2-(TRIMETHYLSILYL)ETHYL 7-((3,5-DIBENZYLOXYPHENYL)ACETOXY)OCTANOATE;   W   65092

Str 1146

590   C36 H30 O6 S   89840-32-4   KC-1983-2910-0   7-(2,3-trans-3,4-cis-3-Hydroxyflavan-4-yloxy)-2,4-trans-flavan-4-yl phenyl sulpho
ne;   W   127485

Str 1147

590   C36 H38 N4 O4   5522-66-7   T-68-5834-0   Protoporphyrin IX dimethyl ester;   W/NBS   65093

Str 1148

590   C38 H46 N4 O2   76643-56-6   KC-1982-585-0   α,β-Diformyloctaethylporphyrin;   W   127486

590   C38 H54 O3 S   84051-51-4   J-49-1088-0   (3β,22E)-24-(Phenylthio)-3-[(tetrahydro-2H-pyran-2-yl)oxy]chole
sta-5,22,24-trien-20-ol;   W   127487

590   C38 H54 O3 S   84051-55-8   J-49-1089-0   (3β,20E,23Z)-24-(Phenylthio)-3-[(tetrahydro-2H-pyran-2-yl)oxy]
cholesta-5,20(22),23-trien-25-ol;   W   127488

590   C38 H70 O4   928-24-5   EP-4870-0-0   9-Octadecenoic acid (Z)-, 1,2-ethanediyl ester;; 1,2-DI-O-CIS-9-OCT
ADECENOYL ETHANEDIOL;   W/NBS   65094

Me(CH2)7CH:CH(CH2)7C(O)OCH2CH2OC(O)(CH2)7CH:CH(CH2)7Me

590   C38 H74 O2 Si   OM-1981-177-0   PALMITELAIDIC ACID 1TMS;   W   65095

590   C39 H38 N2 Si2   15951-44-7   J-5-2140-2   Hydrazobenzene, 4-methyl-N,N'-bis(methyldiphenylsilyl)-;; N-PH
ENYL-N'-(P-TOLYL)-N,N'-BIS-(DIPHENYLMETHYLSILYL)HYDRAZINE;   W/NBS   65096

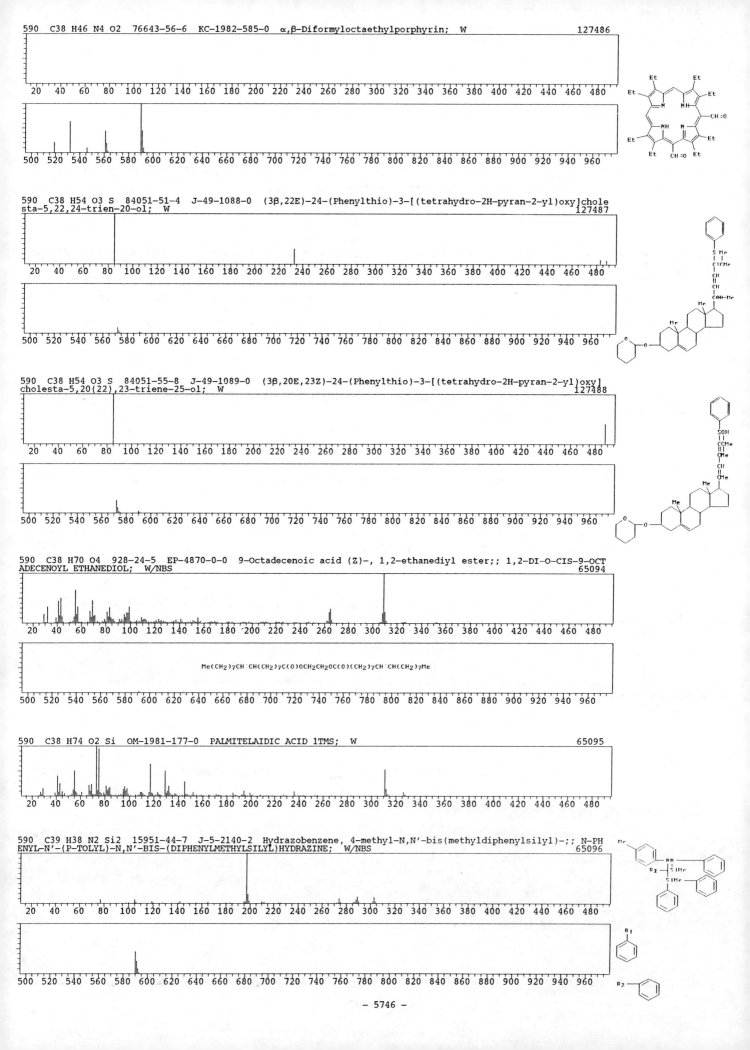

590 C39 H58 O4 1065-31-2 J-45-4103-0 all-(E)-2-(3,7,11,15,19,23-hexamethyl-2,6,10,14,18,22-tetracosahex
aenyl)-5,6-dimethoxy-3-methyl-2,5-cyclohexadiene-1,4-dione; W   127489

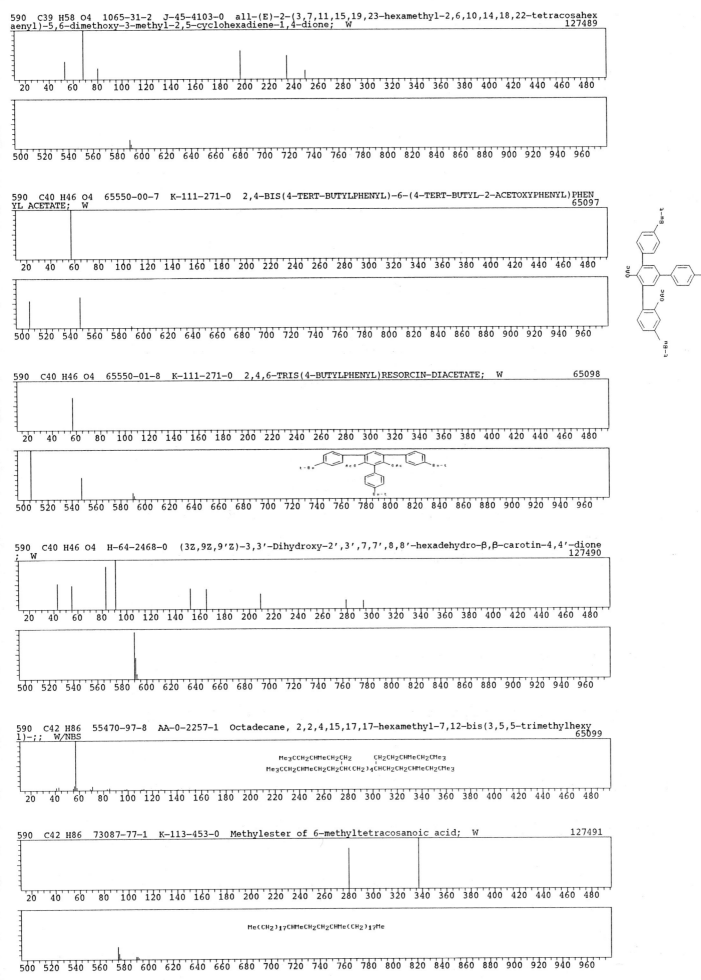

590 C40 H46 O4 65550-00-7 K-111-271-0 2,4-BIS(4-TERT-BUTYLPHENYL)-6-(4-TERT-BUTYL-2-ACETOXYPHENYL)PHEN
YL ACETATE; W   65097

590 C40 H46 O4 65550-01-8 K-111-271-0 2,4,6-TRIS(4-BUTYLPHENYL)RESORCIN-DIACETATE; W   65098

590 C40 H46 O4 H-64-2468-0 (3Z,9Z,9'Z)-3,3'-Dihydroxy-2',3',7,7',8,8'-hexadehydro-β,β-carotin-4,4'-dione
; W   127490

590 C42 H86 55470-97-8 AA-0-2257-1 Octadecane, 2,2,4,15,17,17-hexamethyl-7,12-bis(3,5,5-trimethylhexy
l)-;; W/NBS   65099

Me₃CCH₂CHMeCH₂CH₂   CH₂CH₂CHMeCH₂CMe₃
Me₃CCH₂CHMeCH₂CH₂CH(CH₂)₄CHCH₂CH₂CHMeCH₂CMe₃

590 C42 H86 73087-77-1 K-113-453-0 Methylester of 6-methyltetracosanoic acid; W   127491

Me(CH₂)₁₇CHMeCH₂CH₂CHMe(CH₂)₁₇Me

591   C17 H8 F15 N O5   59785-38-5   NS-0-0-0   Propanoic acid, pentafluoro-, 4-[2-[(2,2,3,3,3-pentafluoro-1-ox
opropyl)amino]ethyl]-1,2-phenylene ester;;   W/NBS   127492

591   C17 H8 F15 N O5   52118-49-7   FI-0-32-6   DOPAMINE-PFP;   W   65100

591   C17 H8 F15 N O5   KO-9-303-1   Tris(pentafluoropropionyl) derivative of dopamine;   W   127493

591   C24 H53 N3 O6 Si4   78146-55-1   F-37-195-0   Tetra-(trimethylsilyl)-nicotianamine;   W   127494

591   C27 H27 Cl2 N3 O4 S2   62256-21-7   KC-1976-2258-0   7A,11A-BIS-(P-CHLOROPHENYLSULPHONYLAMINO)-4,5,7A,8,
9,10,11,11A-OCTAHYDRO-7H-PYRROLO(3,2,1-DE)PHENANTHRIDINE;   W   65102

591   C30 H49 N5 O7   55466-29-0   NS--27776-0-0   Roseotoxin B;;   NBS   65103

591   C30 H53 N3 O5 Si2   69833-79-0   NS--27775-0-0   Pregn-4-en-18-al, 3,20-bis(methoxyimino)-11,21-bis[(tri
methylsilyl)oxy]-, O-methyloxime, (11β)-;;   NBS   65104

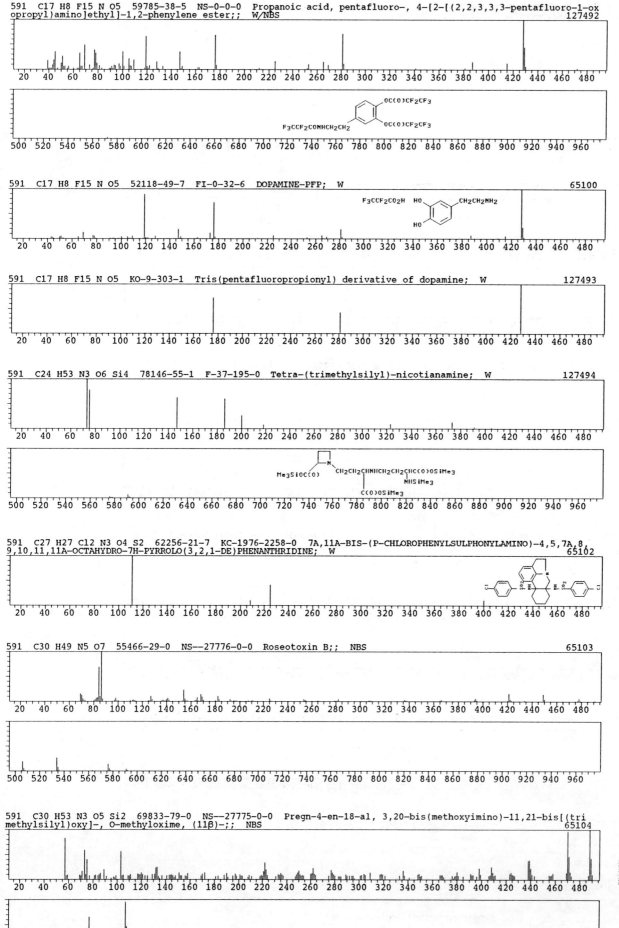

591 C30 H53 N3 O5 Si2 69854-81-5 NS--27774-0-0 Pregn-4-en-18-al, 3,20-bis(methoxyimino)-11,21-bis[(tri
methylsilyl)oxy]-, O-methyloxime, (11β,17α)-;; NBS
65105

591 C30 H57 N O3 Si4 77883-12-6 NS-0-0-0 Estra-1,3,5(10)-trien-4-amine, N-(trimethylsilyl)-3,16,17-tris
[(trimethylsilyl)oxy]-, (16α,17β)-; W/NBS
127495

591 C30 H57 N O3 Si4 77883-15-9 NS-0-0-0 Silanamine, 1,1,1-trimethyl-N-[(16α,17β)-3,16,17-tris[(trimeth
ylsilyl)oxy]estra-1,3,5(10)-trien-2-yl]-; W/NBS
127496

591 C35 H29 N O6 S 62616-06-2 KC-1976-2564-0 DIMETHYL 6-OXO-4,5,7-TRIPHENYL-5-AZA-15-THIAPENTACYCLO[8.
2.2.1(4,7).0(2,9).0(3,8)]PENTADECA-11,13-DIENE-11,12-DICARBOXYLATE 15-OXIDE; W
65106

591 C35 H49 N3 O5 60283-68-3 KC-1976-731-0 3',5'-DIOXO-4'-PHENYL-5α,8α-(1',2')-1',2',4'-TRIAZOLIDINOCHO
LEST-6-ENE-3β,24ε,25-TRIOL; W
65107

591 C35 H49 N3 O5 60283-62-7 KC-1976-735-0 3',5'-DIOXO-4'-PHENYL-5α,8α-(1',2')-1',2',4'-TRIAZOLIDINOCHO
LEST-6-ENE-3β,25ε,26-TRIOL; W
65108

591 C35 H49 N3 O5 AH-113-444-0 (5,10)-(7,8)-Dioxirane derivative of α-4-phenyl-1,2,4-triazolin-3,5-dione
adduct of vitamin D3 (isomer A); W
127497

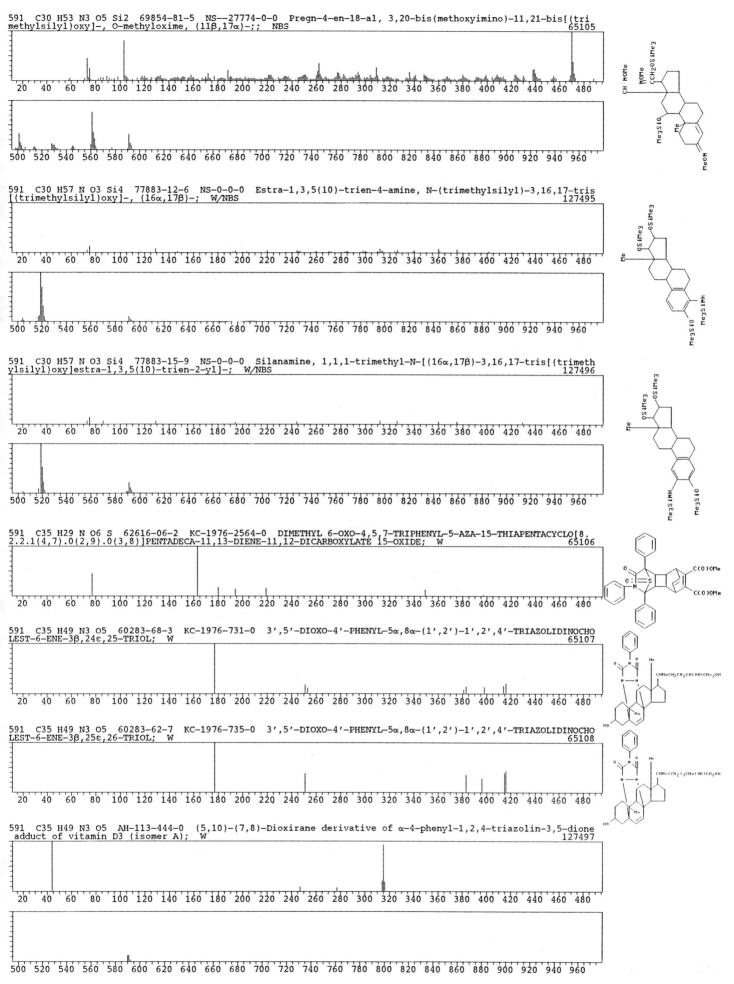

591   C35 H49 N3 O5   83148-22-5   AH-113-444-0   (5,10)-(7,8)-Dioxirane derivative of α-4-phenyl-1,2,4-triazol
in-3,5-dione adduct of vitamin D3 (isomer B);   W   127498

591   C35 H49 N3 O5   82858-70-6   AH-113-444-0   (5,10)-(7,8)-Dioxirane derivative of α-4-phenyl-1,2,4-triazol
in-3,5-dione adduct of vitamin D3 (isomer C);   W   127499

591   C35 H49 N3 O5   AH-113-445-0   (5,10)-(7,8)-Dioxirane derivative of β-4-phenyl-1,2,4-triazolin-3,5-dione
adduct of vitamin D3 (isomer A);   W   127500

591   C35 H49 N3 O5   AH-113-446-0   (5,10)-(7,8)-Dioxirane derivative of β-4-phenyl-1,2,4-triazolin-3,5-dione
adduct of vitamin D3 (isomer B);   W   127501

591   C36 H69 N3 O3   37163-53-4   O-13-45-1   2,4,6-TRI-[N-UNDECYCLOXY] TRIAZINE;   W   65109

592   C7 H4 Br2 I O2 Re   55839-77-5   AG-94-435-10   (η5-Iodocyclopentadienyl)(dicarbonyl)rhenium dibromide;
W   127502

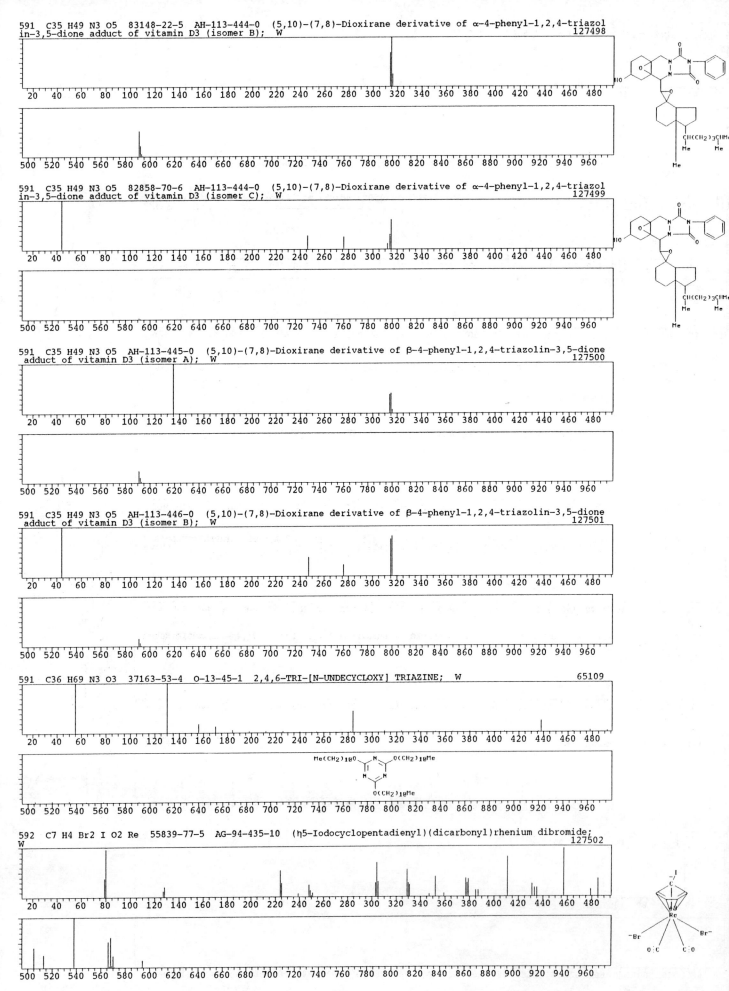

592 C10 Hg Mn2 O10 15525-07-2 AD-0-2867-0 Manganese, decacarbonyl(mercury)di-, (2Hg-Mn);; DI(PENTACARBO
NYL MANGANESE)MERCURY; W/NBS   65110

592 C15 H31 I Os P2 89299-76-3 K-117-159-0 Iodo(exo-6-propyl-η5-cyclohexadienyl)bis(trimethylphosphane)
osmium(II); W   127503

592 C16 H48 O8 Si8 556-68-3 HE-1986-2362-0 HEXADECAMETHYLCYCLOOCTASILOXANE; W   127504

592 C17 H7 F15 O6 82807-46-3 KO-9-149-1 Tri(pentafluoropropionyl)derivative of O-hydroxyphenylglycol(De
rivative of 1-(2-Hydroxyphenyl)ethane-1,2-diol); W   127505

592 C23 H25 Br N6 O8 IC-1280-0-0 N-(2-Acetoxycyclohexyl)-2-methoxy-5-acetamido-4-(2,4-dinitro-]D6-bromo
phenylazo)aniline; W   127506

592 C27 H26 Sb2 38611-86-8 AG-84-242-2 1,3-Bis(diphenylstibino)propane; W   127507

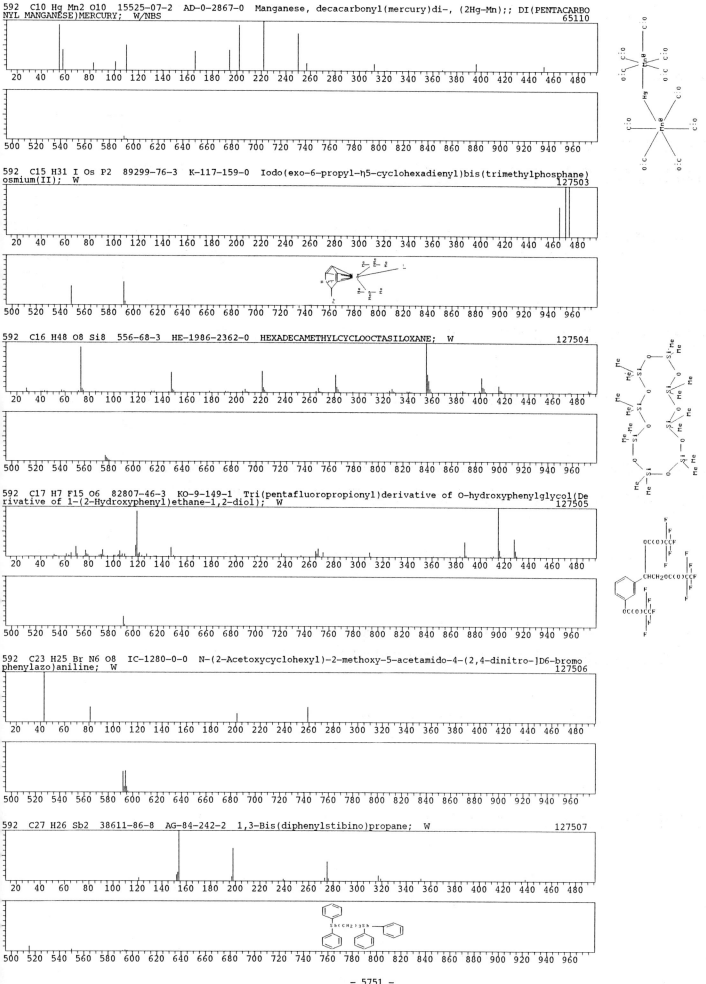

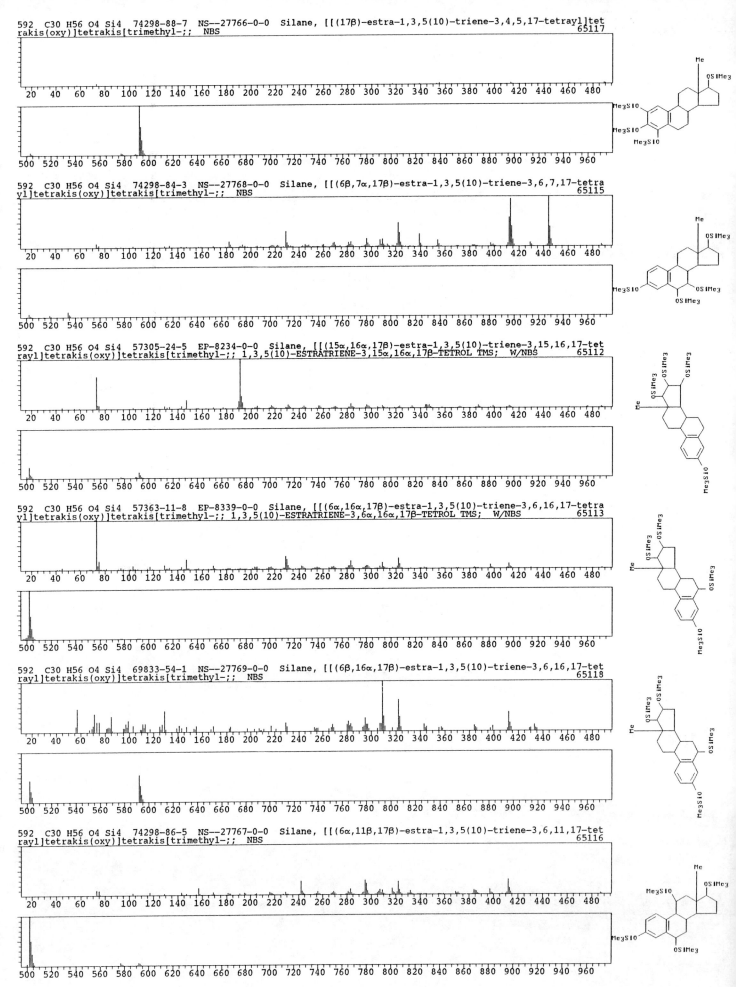

592 C30 H56 O4 Si4 74298-88-7 NS--27766-0-0 Silane, [[(17β)-estra-1,3,5(10)-triene-3,4,5,17-tetrayl]tetrakis(oxy)]tetrakis[trimethyl-;; NBS
65117

592 C30 H56 O4 Si4 74298-84-3 NS--27768-0-0 Silane, [[(6β,7α,17β)-estra-1,3,5(10)-triene-3,6,7,17-tetrayl]tetrakis(oxy)]tetrakis[trimethyl-;; NBS
65115

592 C30 H56 O4 Si4 57305-24-5 EP-8234-0-0 Silane, [[(15α,16α,17β)-estra-1,3,5(10)-triene-3,15,16,17-tetrayl]tetrakis(oxy)]tetrakis[trimethyl-;; 1,3,5(10)-ESTRATRIENE-3,15α,16α,17β-TETROL TMS; W/NBS
65112

592 C30 H56 O4 Si4 57363-11-8 EP-8339-0-0 Silane, [[(6α,16α,17β)-estra-1,3,5(10)-triene-3,6,16,17-tetrayl]tetrakis(oxy)]tetrakis[trimethyl-;; 1,3,5(10)-ESTRATRIENE-3,6α,16α,17β-TETROL TMS; W/NBS
65113

592 C30 H56 O4 Si4 69833-54-1 NS--27769-0-0 Silane, [[(6β,16α,17β)-estra-1,3,5(10)-triene-3,6,16,17-tetrayl]tetrakis(oxy)]tetrakis[trimethyl-;; NBS
65118

592 C30 H56 O4 Si4 74298-86-5 NS--27767-0-0 Silane, [[(6α,11β,17β)-estra-1,3,5(10)-triene-3,6,11,17-tetrayl]tetrakis(oxy)]tetrakis[trimethyl-;; NBS
65116

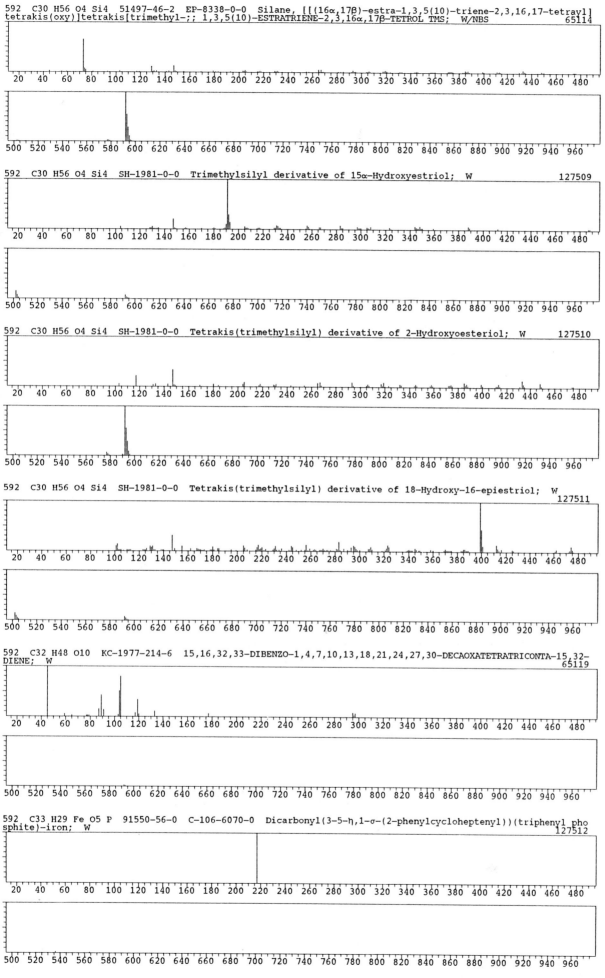

592 C30 H56 O4 Si4 51497-46-2 EP-8338-0-0 Silane, [[(16α,17β)-estra-1,3,5(10)-triene-2,3,16,17-tetrayl]
tetrakis(oxy)]tetrakis[trimethyl-;; 1,3,5(10)-ESTRATRIENE-2,3,16α,17β-TETROL TMS; W/NBS   65114

592 C30 H56 O4 Si4 SH-1981-0-0 Trimethylsilyl derivative of 15α-Hydroxyestriol; W   127509

592 C30 H56 O4 Si4 SH-1981-0-0 Tetrakis(trimethylsilyl) derivative of 2-Hydroxyoesteriol; W   127510

592 C30 H56 O4 Si4 SH-1981-0-0 Tetrakis(trimethylsilyl) derivative of 18-Hydroxy-16-epiestriol; W
                                                                                                          127511

592 C32 H48 O10 KC-1977-214-6 15,16,32,33-DIBENZO-1,4,7,10,13,18,21,24,27,30-DECAOXATETRATRICONTA-15,32-
DIENE; W   65119

592 C33 H29 Fe O5 P 91550-56-0 C-106-6070-0 Dicarbonyl(3-5-η,1-σ-(2-phenylcycloheptenyl))(triphenyl pho
sphite)-iron; W   127512

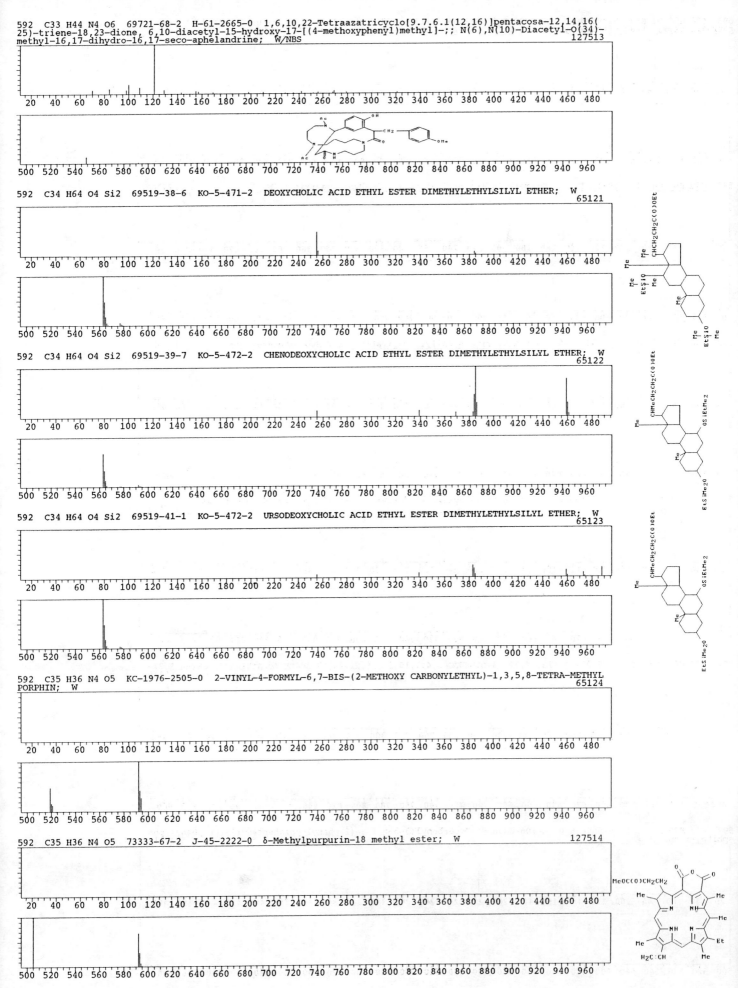

592   C33 H44 N4 O6   69721-68-2   H-61-2665-0   1,6,10,22-Tetraazatricyclo[9.7.6.1(12,16)]pentacosa-12,14,16(25)-triene-18,23-dione, 6,10-diacetyl-15-hydroxy-17-[(4-methoxyphenyl)methyl]-;; N(6),N(10)-Diacetyl-O(34)-methyl-16,17-dihydro-16,17-seco-aphelandrine;   W/NBS   127513

592   C34 H64 O4 Si2   69519-38-6   KO-5-471-2   DEOXYCHOLIC ACID ETHYL ESTER DIMETHYLETHYLSILYL ETHER;   W   65121

592   C34 H64 O4 Si2   69519-39-7   KO-5-472-2   CHENODEOXYCHOLIC ACID ETHYL ESTER DIMETHYLETHYLSILYL ETHER;   W   65122

592   C34 H64 O4 Si2   69519-41-1   KO-5-472-2   URSODEOXYCHOLIC ACID ETHYL ESTER DIMETHYLETHYLSILYL ETHER;   W   65123

592   C35 H36 N4 O5   KC-1976-2505-0   2-VINYL-4-FORMYL-6,7-BIS-(2-METHOXY CARBONYLETHYL)-1,3,5,8-TETRA-METHYL PORPHIN;   W   65124

592   C35 H36 N4 O5   73333-67-2   J-45-2222-0   δ-Methylpurpurin-18 methyl ester;   W   127514

592 C35 H48 O6 Si CO-59-275-0 Methyl ester of 2,3,5-tri-O-benzyl-4-O-t-butyldimethylsilyl-D-arabinosylac
etic acid; W
127515

592 C36 H40 N4 O4 92284-86-1 J-49-4608-0 2,4-Diethyl-7-[2-(methoxycarbonyl)ethyl]-6,γ-[γ'-(methoxycarbo
nyl)ethylene]-1,3,5,8-tetramethylporphyrin; W
127516

592 C36 H40 N4 O4 92284-87-2 J-49-4608-0 2,4-Diethyl-6-[2-(methoxycarbonyl)vinyl]-1,3,5,8-tetramethylpo
rphyrin; W
127517

592 C37 H28 N4 O4 2652-72-4 O-10-609-1 3,5-Pyrazolidinedione, 4,4'-(phenylmethylene)bis[1,2-diphenyl-;;
1,1',2,2',TETRAPHENYL-(4,4'-BIPYRAZOLIDINE-3,3',5,5'-TETRAONYL-1-PHENYLMETHANE; W/NBS
65125

592 C38 H6 D6 N2 O6 57288-11-6 J-41-318-0 O,O-Bis(trideuteriomethyl)krukovine; W
127518

592 C38 H48 N4 O2 14287-38-8 KC-1983-106-0 5-Acetoxyoctaethylporphyrin; W
127519

592 C38 H56 O3 S 79409-83-9 J-46-5187-0 (3β,20S,22E,24S)-24-(Phenylthio)-3-[(tetrahydro-2H-pyran-2-yl)
oxy]cholesta-5,22-dien-20-ol; W
127520

592 C38 H72 O4 26291-64-5 EP-4863-0-0 Oleic acid, 2-hydroxyethyl ester stearate;; 1-O-OCTADECANOYL-2-O-
CIS-9-OCTADECENOYL ETHANEDIOL; W/NBS
65126

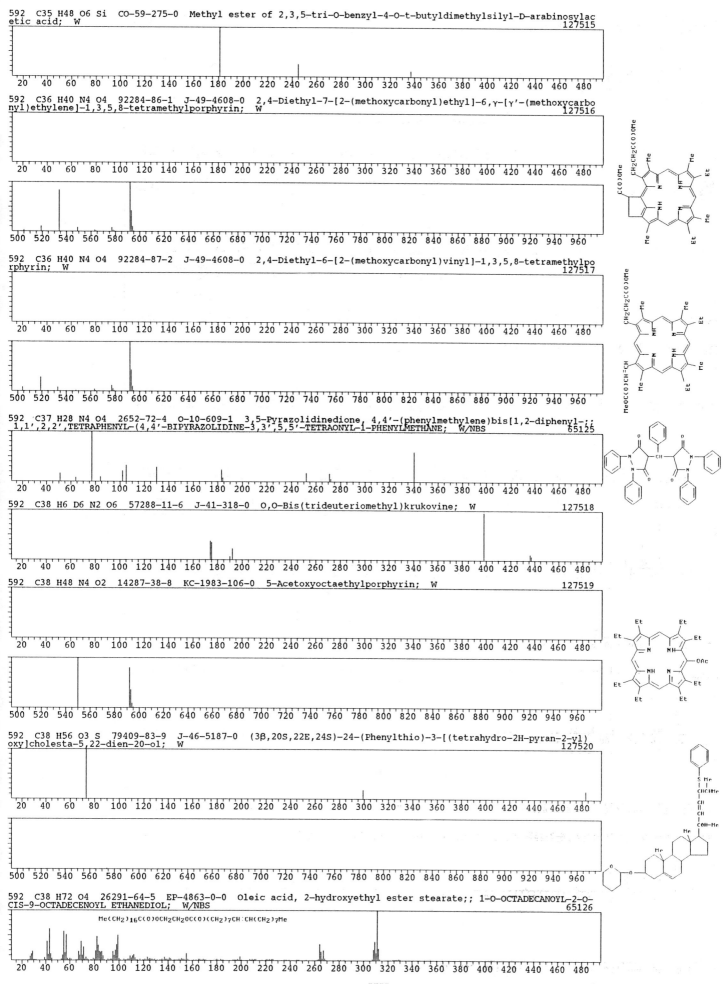

- 5755 -

592 C39 H76 O3 17367-41-8 EP-4967-0-0 Oleic acid, 3-(octadecyloxy)propyl ester;; 3-OCTADECYLOXY-1-O-CIS
-OCTADEC-9-EN-OYL-PROPANOL; W/NBS   65127

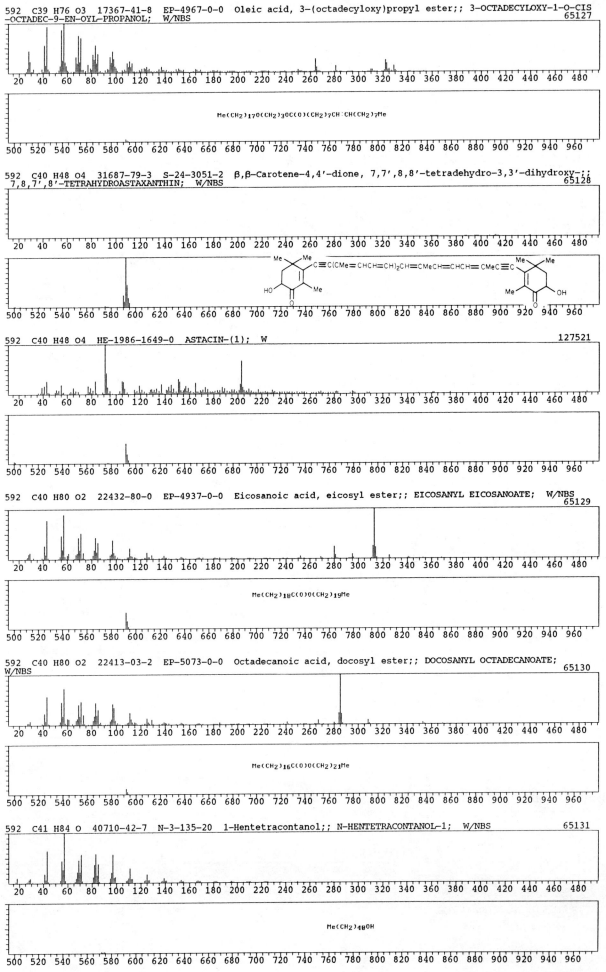

Me(CH₂)₁₇O(CH₂)₃OC(O)(CH₂)₇CH:CH(CH₂)₇Me

592 C40 H48 O4 31687-79-3 S-24-3051-2 β,β-Carotene-4,4'-dione, 7,7',8,8'-tetradehydro-3,3'-dihydroxy-;;
7,8,7',8'-TETRAHYDROASTAXANTHIN; W/NBS   65128

592 C40 H48 O4 HE-1986-1649-0 ASTACIN-(1); W   127521

592 C40 H80 O2 22432-80-0 EP-4937-0-0 Eicosanoic acid, eicosyl ester;; EICOSANYL EICOSANOATE; W/NBS
                                                                                                     65129

Me(CH₂)₁₈C(O)O(CH₂)₁₉Me

592 C40 H80 O2 22413-03-2 EP-5073-0-0 Octadecanoic acid, docosyl ester;; DOCOSANYL OCTADECANOATE;
W/NBS   65130

Me(CH₂)₁₆C(O)O(CH₂)₂₁Me

592 C41 H84 O 40710-42-7 N-3-135-20 1-Hentetracontanol;; N-HENTETRACONTANOL-1; W/NBS   65131

Me(CH₂)₄₀OH

593   C28 H29 Cl2 N9 O2   IC-1281-0-0   N-[4-(2-Chloroanilino)-triazin-2-ylaminobutyl]-N-ethyl-3-methyl-4-(4-nitrophenylazo)aniline;   W   127522

593   C30 H51 N5 O7   2503-26-6   O-4-177-2   Destruxin B;; β-Alanine, N-[N-[N-[N-[1-(D-2-hydroxy-4-methyl-1-oxopentyl)-L-prolyl]-L-isoleucyl]-N-methyl-L-valyl]-N-methyl-L-alanyl]-, .rho.-lactone;   W/NBS   65132

593   C31 H48 Br N O5   79156-63-1   B-34-615-0   (25R)-N-Acetyl-5α-bromo-5α,22αN-spirosolane-3β,6β-diol-3-acetate;   W   127523

593   C31 H59 N O4 Si3   57325-74-3   EP-8283-0-0   Pregn-4-en-3-one, 17,20,21-tris[(trimethylsilyl)oxy]-, O-methyloxime, (20S)-;;   4-PREGNENE-17α,20α,21-TRIOL-3-ONE MO TMS;   W/NBS   65133

593   C31 H59 N O4 Si3   35554-02-0   EP-8284-0-0   Pregn-4-en-3-one, 17,20,21-tris[(trimethylsilyl)oxy]-, O-methyloxime, (20R)-;;   4-PREGNENE-17α,20β,21-TRIOL-3-ONE MO TMS;   W/NBS   65134

593   C31 H59 N O4 Si3   SH-1981-0-0   3β,15β,17-TRIHYDROXY-5-PREGNEN-20-ONE MO-TMS;   W   65135

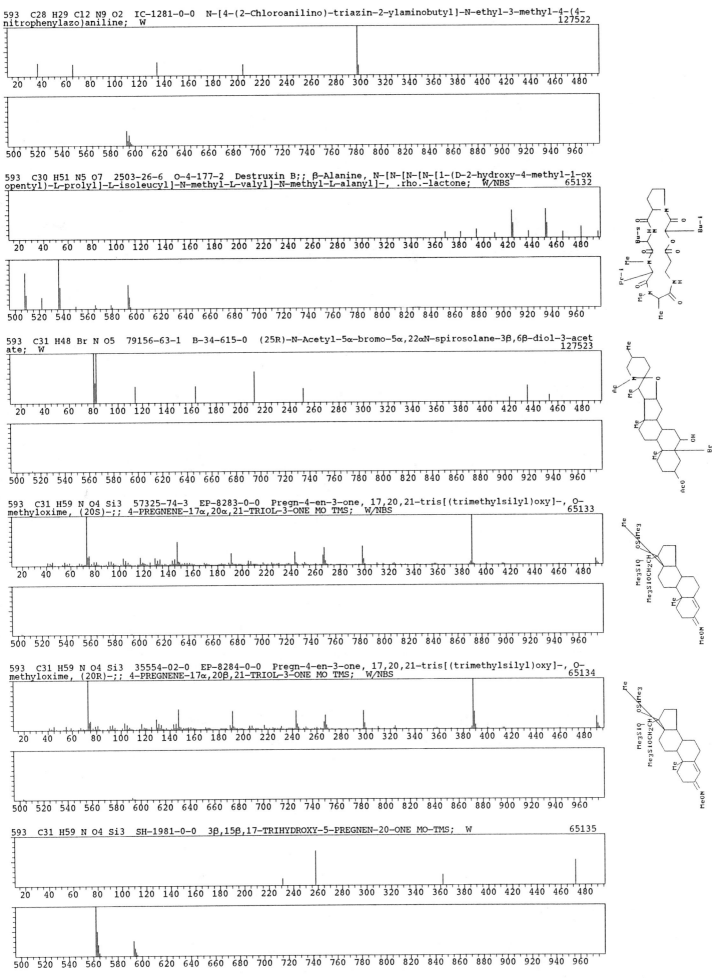

593 C35 H39 N5 O4 18397-13-2 AD-0-1174-0 Integerrine; W/NBS   65136

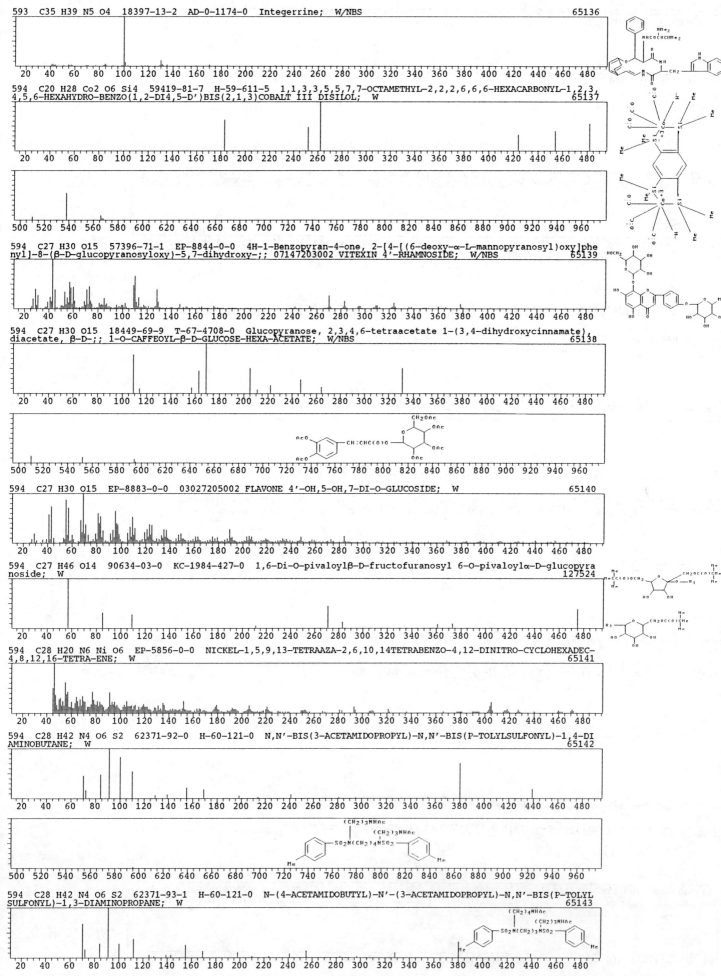

594 C20 H28 Co2 O6 Si4 59419-81-7 H-59-611-5 1,1,3,3,5,5,7,7-OCTAMETHYL-2,2,2,6,6,-HEXACARBONYL-1,2,3,
4,5,6-HEXAHYDRO-BENZO(1,2-DI4,5-D')BIS(2,1,3)COBALT III DISILOL; W   65137

594 C27 H30 O15 57396-71-1 EP-8844-0-0 4H-1-Benzopyran-4-one, 2-[4-[(6-deoxy-α-L-mannopyranosyl)oxy]phe
nyl]-8-(β-D-glucopyranosyloxy)-5,7-dihydroxy-;; 07147203002 VITEXIN 4'-RHAMNOSIDE; W/NBS   65139

594 C27 H30 O15 18449-69-9 T-67-4708-0 Glucopyranose, 2,3,4,6-tetraacetate 1-(3,4-dihydroxycinnamate)
diacetate, β-D-;; 1-O-CAFFEOYL-β-D-GLUCOSE-HEXA-ACETATE; W/NBS   65138

594 C27 H30 O15 EP-8883-0-0 03027205002 FLAVONE 4'-OH,5-OH,7-DI-O-GLUCOSIDE; W   65140

594 C27 H46 O14 90634-03-0 KC-1984-427-0 1,6-Di-O-pivaloylβ-D-fructofuranosyl 6-O-pivaloylα-D-glucopyra
noside; W   127524

594 C28 H20 N6 Ni O6 EP-5856-0-0 NICKEL-1,5,9,13-TETRAAZA-2,6,10,14TETRABENZO-4,12-DINITRO-CYCLOHEXADEC-
4,8,12,16-TETRA-ENE; W   65141

594 C28 H42 N4 O6 S2 62371-92-0 H-60-121-0 N,N'-BIS(3-ACETAMIDOPROPYL)-N,N'-BIS(P-TOLYLSULFONYL)-1,4-DI
AMINOBUTANE; W   65142

594 C28 H42 N4 O6 S2 62371-93-1 H-60-121-0 N-(4-ACETAMIDOBUTYL)-N'-(3-ACETAMIDOPROPYL)-N,N'-BIS(P-TOLYL
SULFONYL)-1,3-DIAMINOPROPANE; W   65143

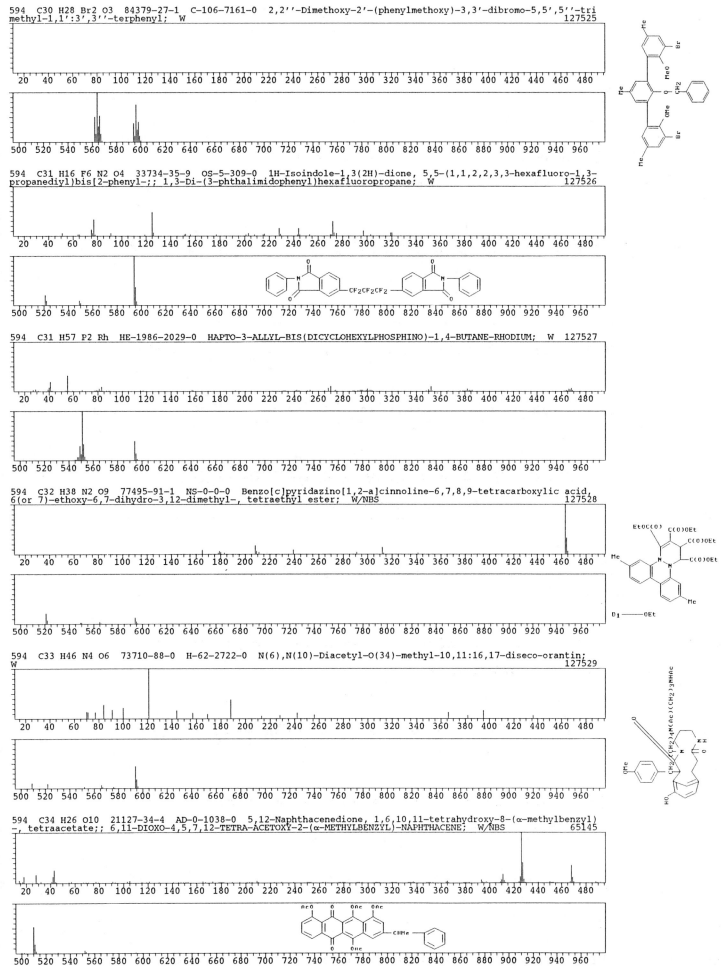

594   C30 H28 Br2 O3   84379-27-1   C-106-7161-0   2,2''-Dimethoxy-2'-(phenylmethoxy)-3,3'-dibromo-5,5',5''-tri
methyl-1,1':3',3''-terphenyl;   W   127525

594   C31 H16 F6 N2 O4   33734-35-9   OS-5-309-0   1H-Isoindole-1,3(2H)-dione, 5,5-(1,1,2,2,3,3-hexafluoro-1,3-
propanediyl)bis[2-phenyl-;; 1,3-Di-(3-phthalimidophenyl)hexafluoropropane;   W   127526

594   C31 H57 P2 Rh   HE-1986-2029-0   HAPTO-3-ALLYL-BIS(DICYCLOHEXYLPHOSPHINO)-1,4-BUTANE-RHODIUM;   W   127527

594   C32 H38 N2 O9   77495-91-1   NS-0-0-0   Benzo[c]pyridazino[1,2-a]cinnoline-6,7,8,9-tetracarboxylic acid,
6(or 7)-ethoxy-6,7-dihydro-3,12-dimethyl-, tetraethyl ester;   W/NBS   127528

594   C33 H46 N4 O6   73710-88-0   H-62-2722-0   N(6),N(10)-Diacetyl-O(34)-methyl-10,11:16,17-diseco-orantin;
W   127529

594   C34 H26 O10   21127-34-4   AD-0-1038-0   5,12-Naphthacenedione, 1,6,10,11-tetrahydroxy-8-(α-methylbenzyl)
-, tetraacetate;; 6,11-DIOXO-4,5,7,12-TETRA-ACETOXY-2-(α-METHYLBENZYL)-NAPHTHACENE;   W/NBS   65145

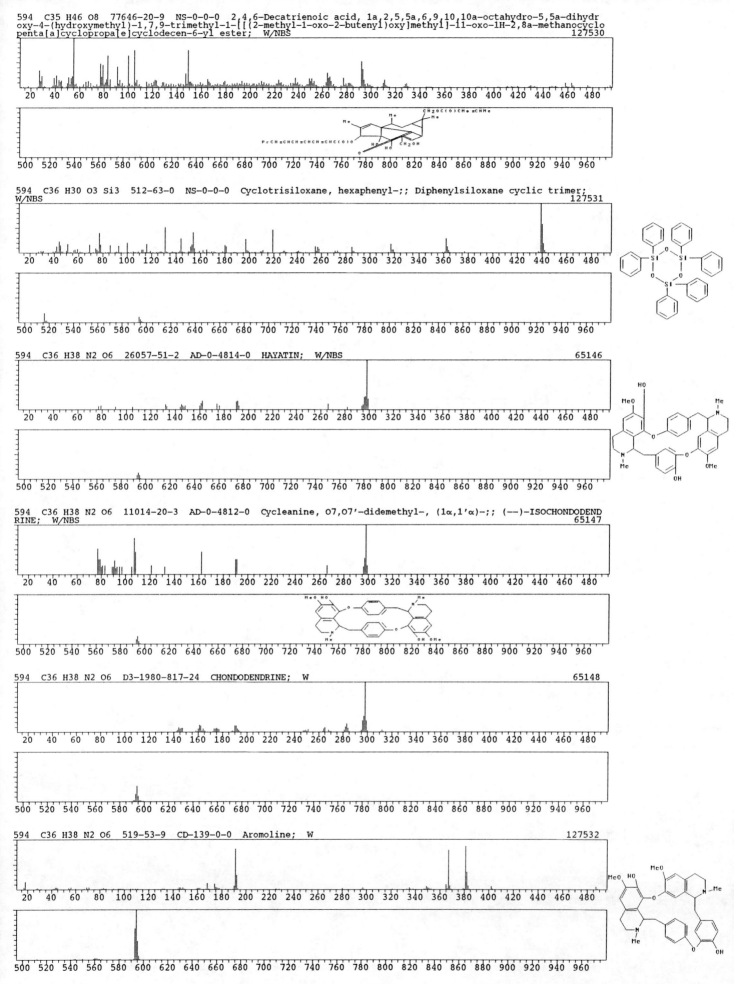

594   C35 H46 O8   77646-20-9   NS-0-0-0   2,4,6-Decatrienoic acid, 1a,2,5,5a,6,9,10,10a-octahydro-5,5a-dihydr
oxy-4-(hydroxymethyl)-1,7,9-trimethyl-1-[[[(2-methyl-1-oxo-2-butenyl)oxy]methyl]-11-oxo-1H-2,8a-methanocyclo
penta[a]cyclopropa[e]cyclodecen-6-yl ester;   W/NBS   127530

594   C36 H30 O3 Si3   512-63-0   NS-0-0-0   Cyclotrisiloxane, hexaphenyl-;; Diphenylsiloxane cyclic trimer;
W/NBS   127531

594   C36 H38 N2 O6   26057-51-2   AD-0-4814-0   HAYATIN; W/NBS   65146

594   C36 H38 N2 O6   11014-20-3   AD-0-4812-0   Cycleanine, O7,O7'-didemethyl-, (1α,1'α)-;; (--)-ISOCHONDODEND
RINE; W/NBS   65147

594   C36 H38 N2 O6   D3-1980-817-24   CHONDODENDRINE; W   65148

594   C36 H38 N2 O6   519-53-9   CD-139-0-0   Aromoline; W   127532

594   C36 H38 N2 O6   55702-00-6   J-40-2648-0   Norpanursine;   W   127533

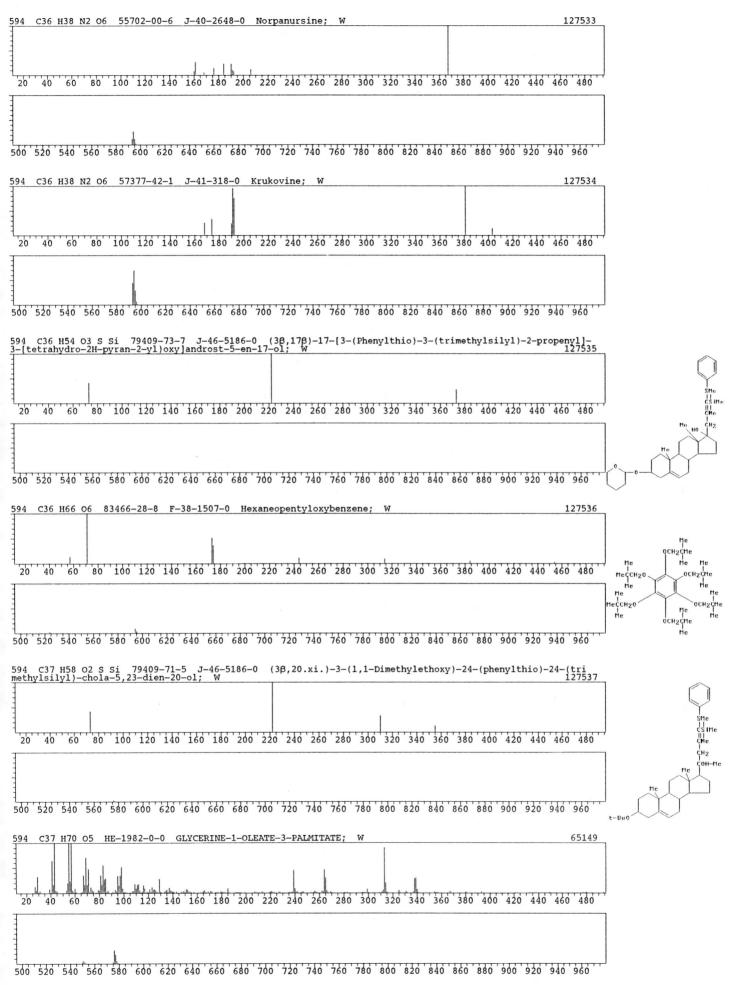

594   C36 H38 N2 O6   57377-42-1   J-41-318-0   Krukovine;   W   127534

594   C36 H54 O3 S Si   79409-73-7   J-46-5186-0   (3β,17β)-17-[3-(Phenylthio)-3-(trimethylsilyl)-2-propenyl]-
3-[tetrahydro-2H-pyran-2-yl)oxy]androst-5-en-17-ol;   W   127535

594   C36 H66 O6   83466-28-8   F-38-1507-0   Hexaneopentyloxybenzene;   W   127536

594   C37 H58 O2 S Si   79409-71-5   J-46-5186-0   (3β,20.xi.)-3-(1,1-Dimethylethoxy)-24-(phenylthio)-24-(tri
methylsilyl)-chola-5,23-dien-20-ol;   W   127537

594   C37 H70 O5   HE-1982-0-0   GLYCERINE-1-OLEATE-3-PALMITATE;   W   65149

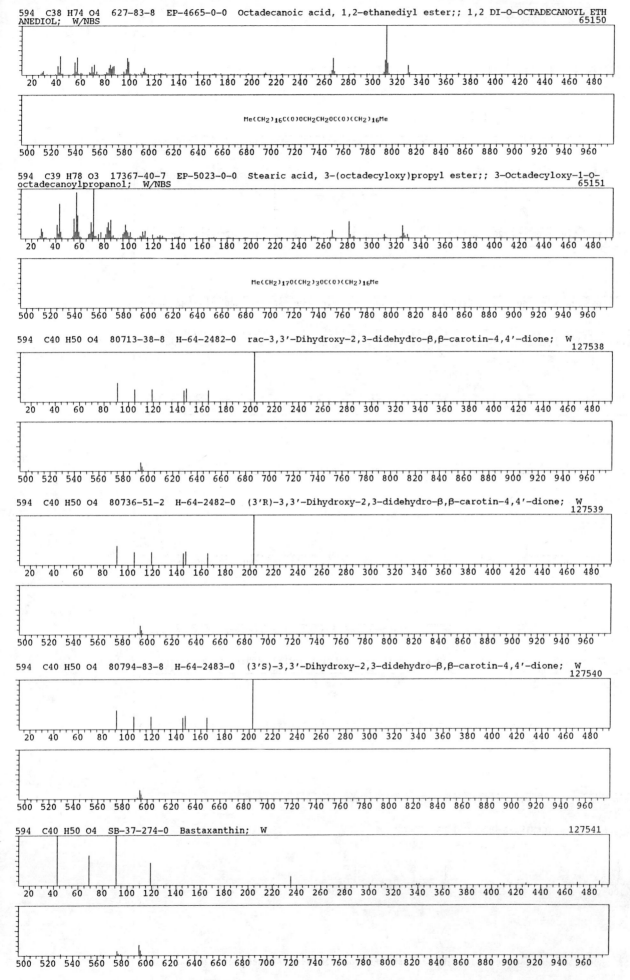

594 C38 H74 O4 627-83-8 EP-4665-0-0 Octadecanoic acid, 1,2-ethanediyl ester;; 1,2 DI-O-OCTADECANOYL ETH
ANEDIOL; W/NBS
65150

Me(CH2)16C(O)OCH2CH2OC(O)(CH2)16Me

594 C39 H78 O3 17367-40-7 EP-5023-0-0 Stearic acid, 3-(octadecyloxy)propyl ester;; 3-Octadecyloxy-1-O-
octadecanoylpropanol; W/NBS
65151

Me(CH2)17O(CH2)3OC(O)(CH2)16Me

594 C40 H50 O4 80713-38-8 H-64-2482-0 rac-3,3'-Dihydroxy-2,3-didehydro-β,β-carotin-4,4'-dione; W
127538

594 C40 H50 O4 80736-51-2 H-64-2482-0 (3'R)-3,3'-Dihydroxy-2,3-didehydro-β,β-carotin-4,4'-dione; W
127539

594 C40 H50 O4 80794-83-8 H-64-2483-0 (3'S)-3,3'-Dihydroxy-2,3-didehydro-β,β-carotin-4,4'-dione; W
127540

594 C40 H50 O4 SB-37-274-0 Bastaxanthin; W
127541

594   C40 H54 N2 O2   38389-14-9   AD-0-3738-0   1'H-Cholest-2-eno[3,2-b]indole, 6-nitro-1'-(phenylmethyl)-
(5α,6α)-;;   6α-NITRO-5α-CHOLESTANO(3,2-B)N-BENZYLINDOLE;   W/NBS
65152

594   C40 H82 O2   56554-64-4   EP-4923-0-0   Hexadecane, 1,1-bis(dodecyloxy)-;;   DIDODECYLOXY HEXADECANE;
W/NBS
65153

(CH2)14Me
|
Me(CH2)11OCHO(CH2)11Me

594   C42 H58 O2   S-27-2501-8   ALEURIAXANTHIN ACETATE;   W
65154

594   C42 H58 O2   EP-2346-0-0   RUBIXANTHIN ACETATE;   W
65155

594   C42 H58 O2   73841-13-1   SB-33-740-0   5,6-DI-T-BUTYL-2,3'-DI-(4-T-BUTYLBENZOYL-1,4,6,1',4',5',6'-OCTAHY
DROPHENYL;   W
65156

594   C44 H34 S   72478-89-8   J-45-929-0   3,3,4-Triphenyl-2-(diphenylmethylene)-5,6-dihydrobenzo(H)thiochrom
an;   W
127542

595   C19 H13 F12 N O7   40629-67-2   RP-0-182-22   Acetic acid, trifluoro-, 4-[2-[(1-methylethyl)(trifluoroace
tyl)amino]-1-[(trifluoroacetyl)oxy]ethyl]-1,2-phenylene ester;;   2-(N-ISOPROPYL-N-TRIFLUOROACETYLAMINO)-1-TR
IFLUOROACETOXY-1-(3,4-TRIFLUOROACETOXYPHENYL)ETHANE;   W/NBS
65157

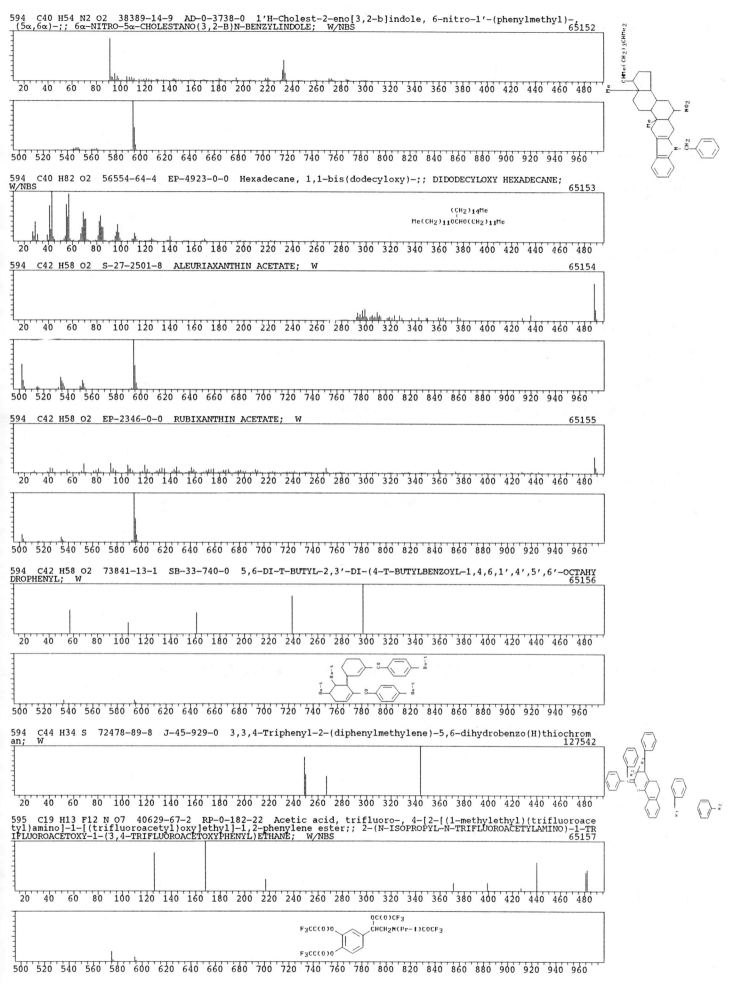

595   C20 H12 Br3 N3 O2 S   57139-16-9   O-12-628-1   2,5-BIS(4-BROMPHENYL)-H4-BROMPHENYLIMINO)-1λ*4,2,5-THIADI
AZOLIDINE-3,4-DIONE;   W   65158

595   C21 H46 N3 O7 P Si4   32645-72-0   BA-0-261-0   Cytidine, 2'-deoxy-N-(trimethylsilyl)-3'-O-(trimethylsily
l)-, 5'-[bis(trimethylsilyl) phosphate];; 5'-(O,O-BIS(TRIMETHYLSILYL)PHOSPHATYL)-N4,O3'-BIS(TRIMETHYLSILYL)
-2'-DEOXYCYTIDINE;   W/NBS   65159

595   C28 H37 N O13   73257-63-3   KC-1979-2981-0   6-Acetylevonyminyl ester of Evonic acid;   W   127543

595   C31 H23 Br2 N3   91545-76-5   AH-115-322-0   2,8-Bis(4-bromophenyl)-4,6-diphenyl-1,6-dihydro-4H-pyrimido[
1,2-a]pyrimidine;   W   127544

595   C31 H61 N O4 Si3   56196-43-1   EP-8287-0-0   Pregnan-20-one, 3,17,21-tris[(trimethylsilyl)oxy]-, O-meth
yloxime, (3α,5β)-;; TRISTRIMETHYLSILYL 3α,17α,21-TRIHYDROXY-5β-PREGNANE-20-ONE METHOXIME;   W/NBS   65163

595   C31 H61 N O4 Si3   32206-69-2   EP-8290-0-0   Pregnan-20-one, 3,11,21-tris[(trimethylsilyl)oxy]-, O-meth
yloxime, (3α,5α,11β)-;; 5α-PREGNANE-3α,11β,21-TRIOL-20-ONEMO TMS;   W/NBS   65161

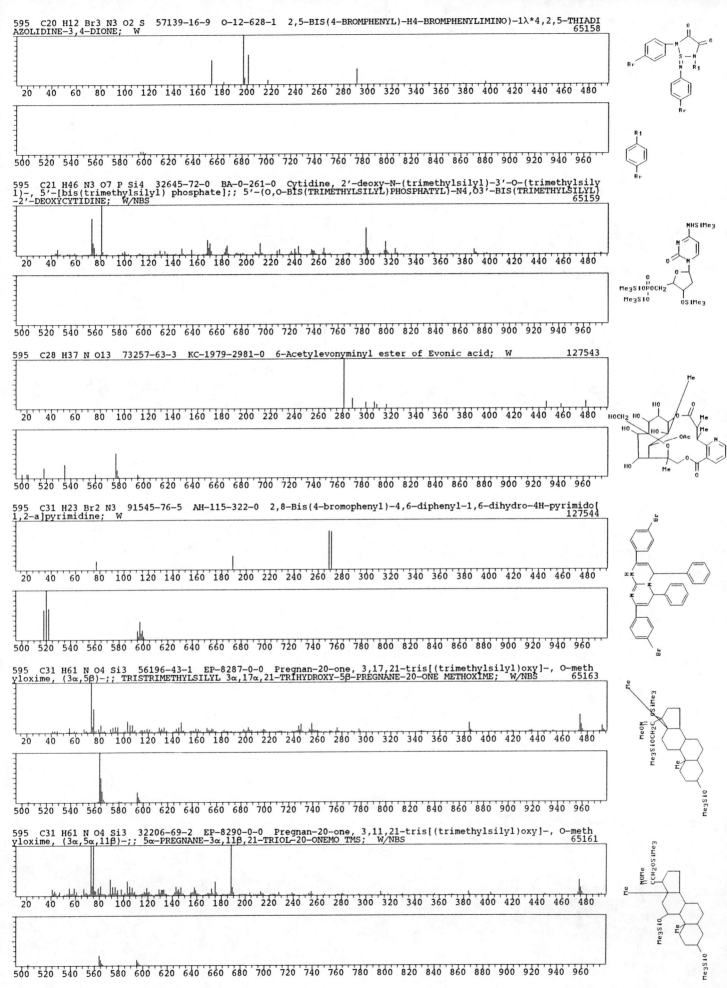

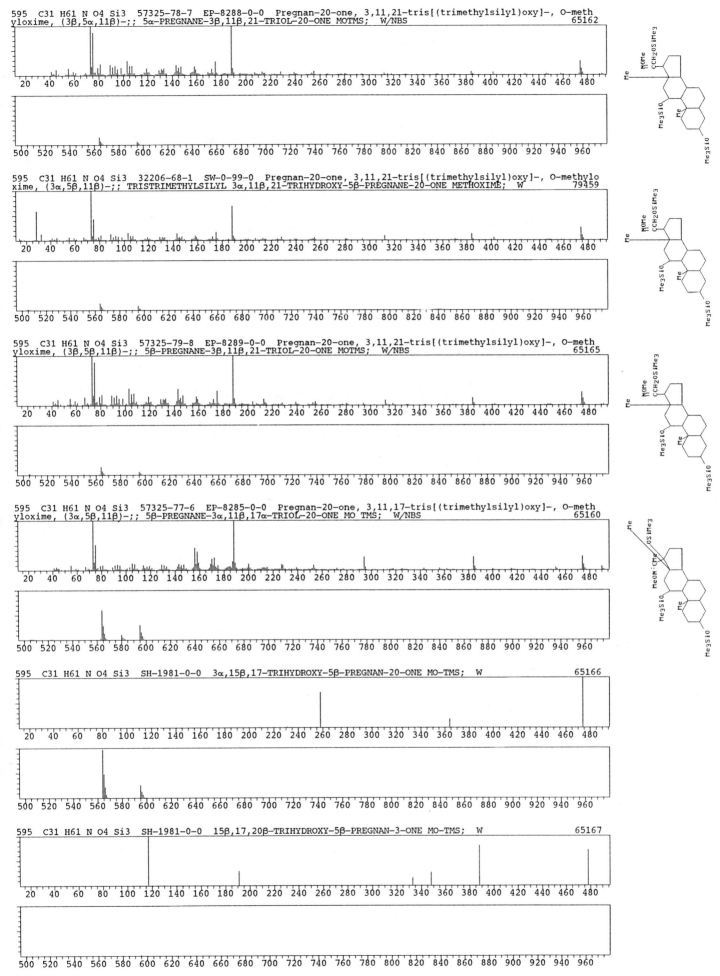

595 C31 H61 N O4 Si3 57325-78-7 EP-8288-0-0 Pregnan-20-one, 3,11,21-tris[(trimethylsilyl)oxy]-, O-meth
yloxime, (3β,5α,11β)-;; 5α-PREGNANE-3β,11β,21-TRIOL-20-ONE MOTMS; W/NBS   65162

595 C31 H61 N O4 Si3 32206-68-1 SW-0-99-0 Pregnan-20-one, 3,11,21-tris[(trimethylsilyl)oxy]-, O-methylo
xime, (3α,5β,11β)-;; TRISTRIMETHYLSILYL 3α,11β,21-TRIHYDROXY-5β-PREGNANE-20-ONE METHOXIME; W   79459

595 C31 H61 N O4 Si3 57325-79-8 EP-8289-0-0 Pregnan-20-one, 3,11,21-tris[(trimethylsilyl)oxy]-, O-meth
yloxime, (3β,5β,11β)-;; 5β-PREGNANE-3β,11β,21-TRIOL-20-ONE MOTMS; W/NBS   65165

595 C31 H61 N O4 Si3 57325-77-6 EP-8285-0-0 Pregnan-20-one, 3,11,17-tris[(trimethylsilyl)oxy]-, O-meth
yloxime, (3α,5β,11β)-;; 5β-PREGNANE-3α,11β,17α-TRIOL-20-ONE MO TMS; W/NBS   65160

595 C31 H61 N O4 Si3 SH-1981-0-0 3α,15β,17-TRIHYDROXY-5β-PREGNAN-20-ONE MO-TMS; W   65166

595 C31 H61 N O4 Si3 SH-1981-0-0 15β,17,20β-TRIHYDROXY-5β-PREGNAN-3-ONE MO-TMS; W   65167

595   C31 H61 N O4 Si3   SH-1981-0-0   15β,17α,20β-TRIHYDROXY-5α-PREGNAN-3-ONE MO-TMS;   W   65168

595   C31 H61 N O4 Si3   SH-1981-0-0   3β,15β,17α-TRIHYDROXY-5α-PREGNAN-20-ONE MO TMS;   W   65169

595   C32 H37 N O10   61166-34-5   KC-1979-2974-0   8-DINICOTINYLCATHEDULIN;   W   65170

595   C33 H29 N3 O8   74615-12-6   C-104-543-0   Methyl 5-amino-6-(5,8-dihydro-6-methoxy-5,8-dioxo-2-quinolinyl
-4-(3,4-dimethoxy-2-(phenylmethoxy)phenyl)-3-methyl-2-pyridinecarboxylate;   W   127545

595   C34 H36 Cu N4 O2   76703-06-5   KC-1982-585-0   α,β-diformyletio-porphyrin-I;   W   127546

595   C34 H36 Cu N4 O2   76703-05-4   KC-1982-585-0   α,γ-diformyletioporphyrin-I;   W   127547

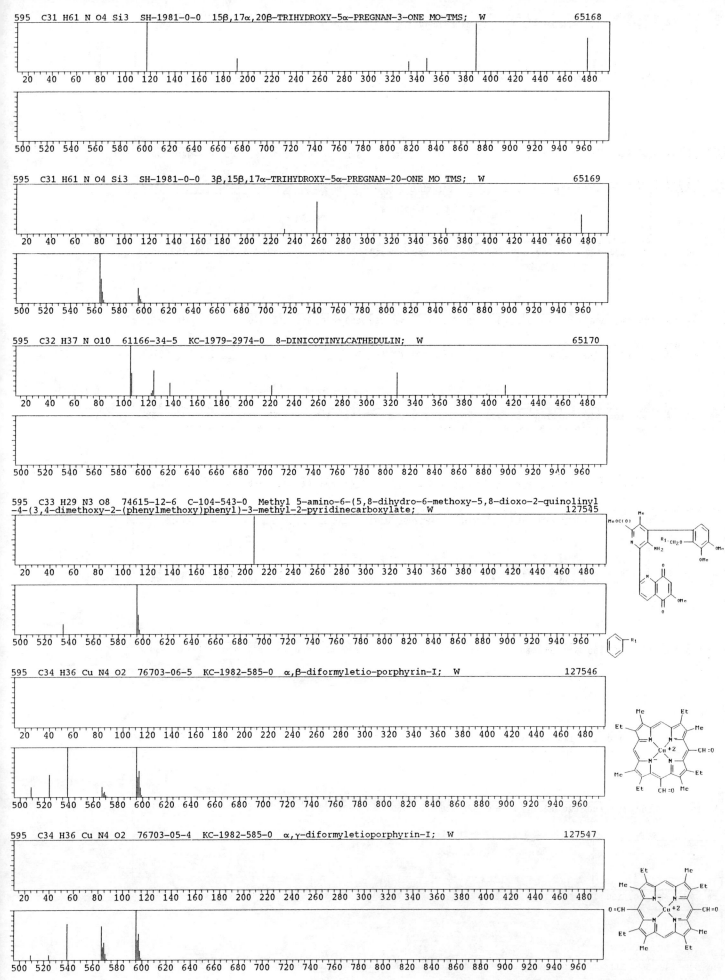

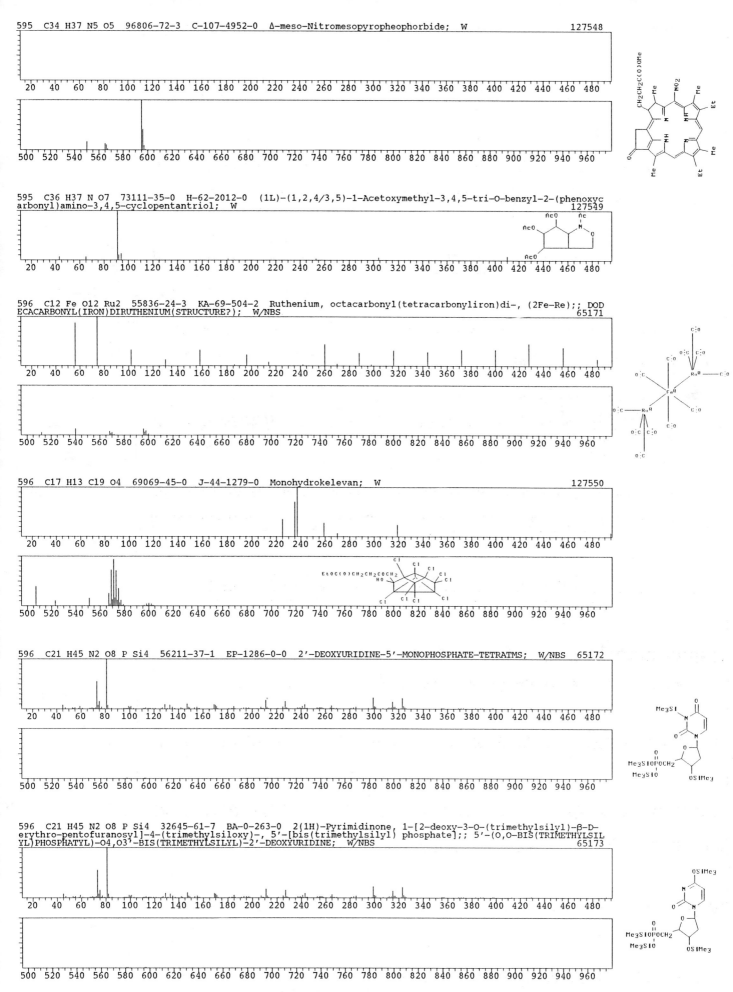

595 C34 H37 N5 O5 96806-72-3 C-107-4952-0 Δ-meso-Nitromesopyropheophorbide; W   127548

595 C36 H37 N O7 73111-35-0 H-62-2012-0 (1L)-(1,2,4/3,5)-1-Acetoxymethyl-3,4,5-tri-O-benzyl-2-(phenoxyc
arbonyl)amino-3,4,5-cyclopentantriol; W   127549

596 C12 Fe O12 Ru2 55836-24-3 KA-69-504-2 Ruthenium, octacarbonyl(tetracarbonyliron)di-, (2Fe-Re);; DOD
ECACARBONYL(IRON)DIRUTHENIUM(STRUCTURE?); W/NBS   65171

596 C17 H13 Cl9 O4 69069-45-0 J-44-1279-0 Monohydrokelevan; W   127550

596 C21 H45 N2 O8 P Si4 56211-37-1 EP-1286-0-0 2'-DEOXYURIDINE-5'-MONOPHOSPHATE-TETRATMS; W/NBS 65172

596 C21 H45 N2 O8 P Si4 32645-61-7 BA-0-263-0 2(1H)-Pyrimidinone, 1-[2-deoxy-3-O-(trimethylsilyl)-β-D-
erythro-pentofuranosyl]-4-(trimethylsiloxy)-, 5'-[bis(trimethylsilyl) phosphate];; 5'-(O,O-BIS(TRIMETHYLSIL
YL)PHOSPHATYL)-O4,O3'-BIS(TRIMETHYLSILYL)-2'-DEOXYURIDINE; W/NBS   65173

596 C22 H19 Br3 N2 O3 HE-1982-0-0 5H-INDENO[1,2-B]-1,2,3,4-TETRAHYDROPYRAZIN-5-ON, 7,8,9-TRIBROMO-N,N'-DIETHYL-6-PHENOXYCARBONYL-; W   65174

596 C24 H20 Fe4 O4 12203-87-1 AD-0-2771-0 Iron, tetra-.mu.3-carbonyltetrakis(η5-2,4-cyclopentadien-1-yl)tetra-, tetrahedro;; CYCLOPENTADIENYLCARBONYLIRON TETRAMER; W/NBS   65175

596 C25 H26 Cl F9 O4 49842-33-3 NS--27731-0-0 Androsta-3,5-diene-3,17-diol, 3-(chlorodifluoroacetate) 17-(heptafluorobutanoate), (17β)-;; NBS   65176

596 C25 H26 Cl F9 O4 74825-21-1 NS--27730-0-0 Androsta-3,5-diene-3,17-diol, 17-(chlorodifluoroacetate) 3-(heptafluorobutanoate), (17β)-;; NBS   65177

596 C26 H26 Hg Si2 59466-93-2 AG-120-175-1 Bis(methyldiphenylsilyl)mercury; W   127551

596 C26 H28 B2 F10 N2 O 72886-59-0 K-112-3825-0 1,3-DI-TERT-BUTYL-4-(1-CYCLOHEXENYLOXY)-2,4-BIS(PENTAFLUOROPHENYL)-1,3-DIAZA-2,4-DIBORABUTANE; W   65178

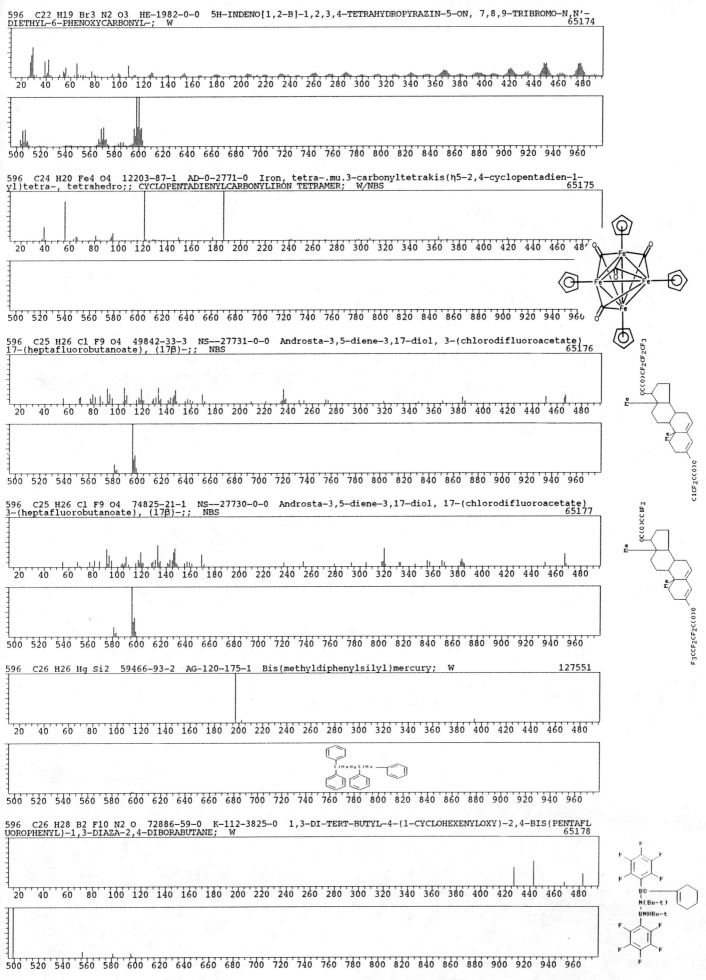

596 C27 H36 N2 O13 23141-51-7 AU-20-280-4 L-Threonine, N-[(phenylmethoxy)carbonyl]-O-[3,4,6-tri-O-acety
l-2-(acetylamino)-2-deoxy-β-D-glucopyranosyl]-, methyl ester;; O-(2-ACETAMIDO-3,4,6-TRI-O-ACETYL-2-DEOXY-β-
D-GLUCOPYRANOSYL)-N-BENZYLOXYCARBONYL-L-THREONINE METHYL ESTER; W/NBS
65179

596 C28 H20 F6 O4 P2 69737-31-1 K-112-769-4 3-DIPHENYLPHOSPHINYL-1,1,1,4,4,4-HEXAFLUORO-3-HYDROXY-2-BUT
ANON-DIPHENYLPHOSPHINATE; W
65180

596 C28 H20 F6 O4 P2 69737-29-7 K-112-774-0 1,1-BIS(DIPHENYLPHOSPHINYL)-2,2,2-TRIFLUOROETHYL-TRIFLUOROA
CETATE; W
65181

596 C30 H28 O13 54725-40-5 SD-1981-0-0 1-Naphthacenecarboxylic acid, 5,7,10,12-tetrakis(acetyloxy)-2-
ethyl-1,2,3,4,6,11-hexahydro-2-hydroxy-6,11-dioxo-, methyl ester;; W/NBS
65182

596 C30 H48 N2 O10 Y-15-1270-0 α-(1-(6,7-DIMETHOXYISOQUINOLINYL))-3-ETHYL-1,2,3,4,6,7-HEXAHYDRO-9,10-DI
METHOXY-11B H-BENZO(A)QUINOLIZIN-2-METHANOL BENZOATE WITH 4 H2O; W
65183

596 C31 H36 N2 O10 58950-48-4 KC-1976-289-0 BENZYL 3,4'-BIS-(2-METHOXYCARBONYL)-3',4-BISMETHOXYCARBONYL
METHYL-2,2'-METHYLENEDIPYRROLE-5-CARBOXYLATE; W
65184

596 C31 H60 O5 Si3 SH-1981-0-0 Tris(trimethylsilyl) derivative of 3α,17α,20α-Trihydroxy-5-β-pregnen-21-
oic acid or Trimethylsilyl derivative of 11-Deoxy-α-cortolic acid; W
127552

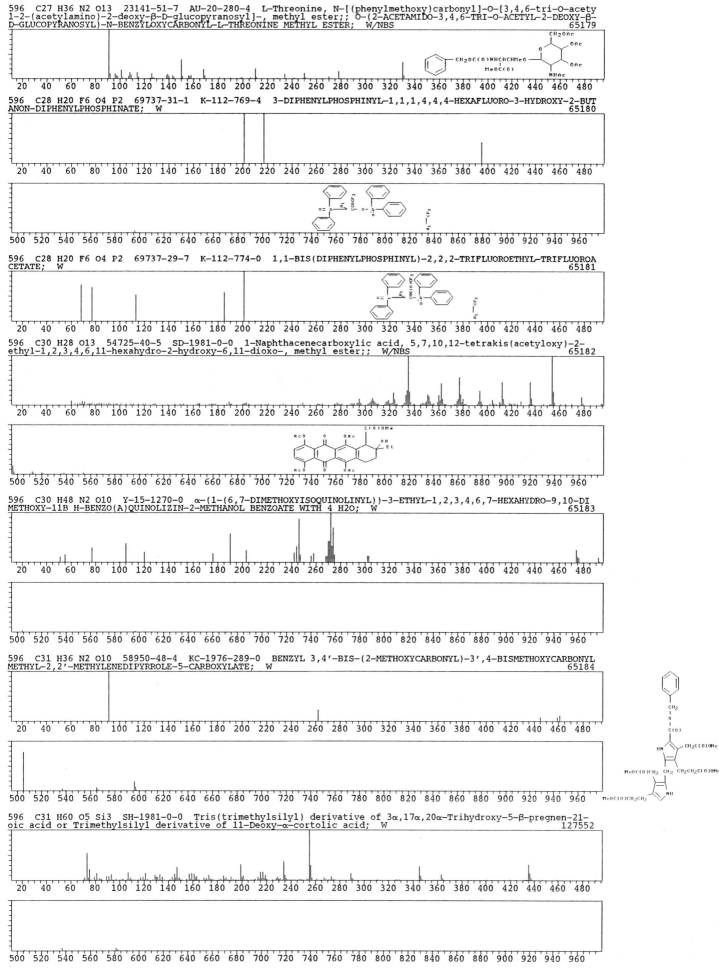

596   C32 H36 O11   85950-08-9   KC-1983-236-0   Tetra-acetate derivative of 1,1'-(1,3,7,9-tetrahydroxy-2,6-di
methyldibenzofuran-4,8-diyl)-3,3'-dimethyldibutan-1-one; W
127553

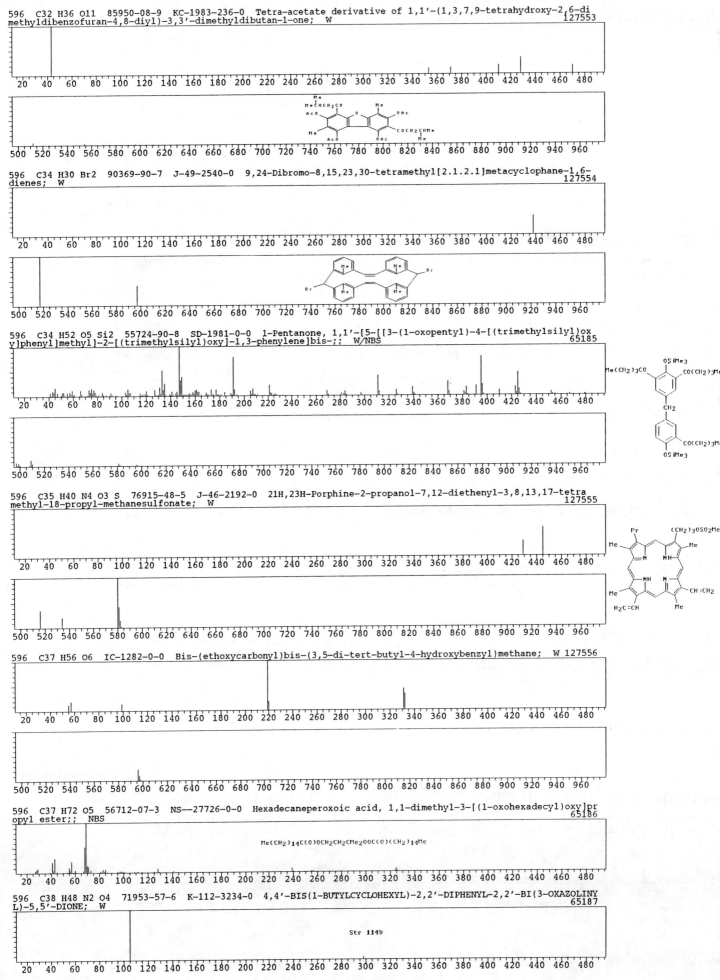

596   C34 H30 Br2   90369-90-7   J-49-2540-0   9,24-Dibromo-8,15,23,30-tetramethyl[2.1.2.1]metacyclophane-1,6-
dienes; W
127554

596   C34 H52 O5 Si2   55724-90-8   SD-1981-0-0   1-Pentanone, 1,1'-[5-[[3-(1-oxopentyl)-4-[(trimethylsilyl)ox
y]phenyl]methyl]-2-[(trimethylsilyl)oxy]-1,3-phenylene]bis-;; W/NBS
65185

596   C35 H40 N4 O3 S   76915-48-5   J-46-2192-0   21H,23H-Porphine-2-propanol-7,12-diethenyl-3,8,13,17-tetra
methyl-18-propyl-methanesulfonate; W
127555

596   C37 H56 O6   IC-1282-0-0   Bis-(ethoxycarbonyl)bis-(3,5-di-tert-butyl-4-hydroxybenzyl)methane; W 127556

596   C37 H72 O5   56712-07-3   NS--27726-0-0   Hexadecaneperoxoic acid, 1,1-dimethyl-3-[(1-oxohexadecyl)oxy]pr
opyl ester;; NBS
65186

596   C38 H48 N2 O4   71953-57-6   K-112-3234-0   4,4'-BIS(1-BUTYLCYCLOHEXYL)-2,2'-DIPHENYL-2,2'-BI(3-OXAZOLINY
L)-5,5'-DIONE; W
65187

Str 1149

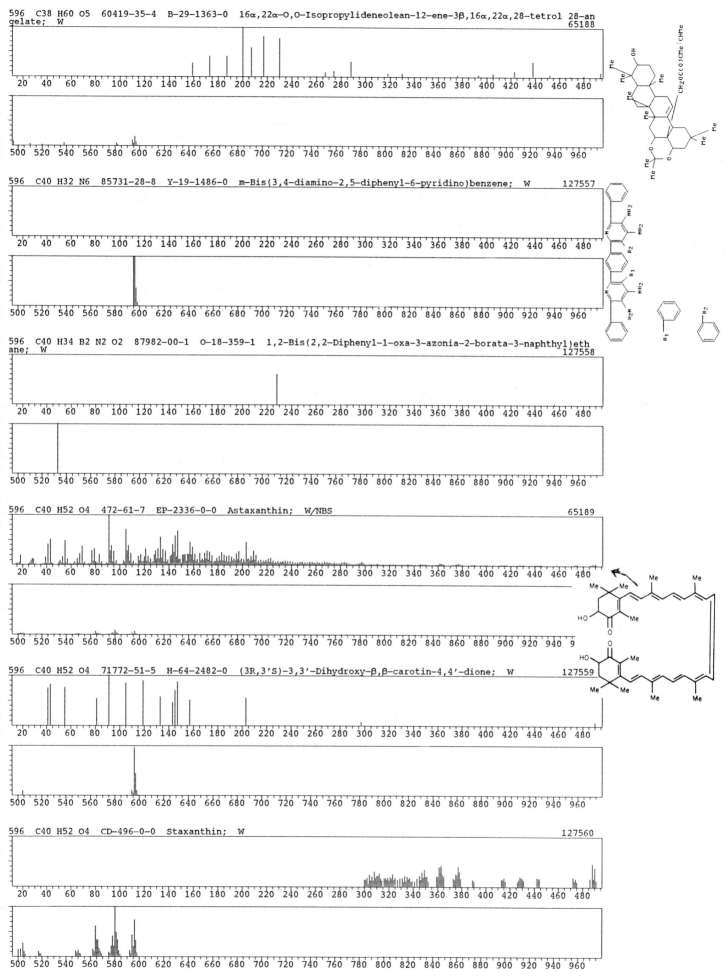

596   C38 H60 O5   60419-35-4   B-29-1363-0   16α,22α-O,O-Isopropylideneolean-12-ene-3β,16α,22α,28-tetrol 28-an
gelate;   W
65188

596   C40 H32 N6   85731-28-8   Y-19-1486-0   m-Bis(3,4-diamino-2,5-diphenyl-6-pyridino)benzene;   W   127557

596   C40 H34 B2 N2 O2   87982-00-1   O-18-359-1   1,2-Bis(2,2-Diphenyl-1-oxa-3-azonia-2-borata-3-naphthyl)eth
ane;   W
127558

596   C40 H52 O4   472-61-7   EP-2336-0-0   Astaxanthin;   W/NBS
65189

596   C40 H52 O4   71772-51-5   H-64-2482-0   (3R,3'S)-3,3'-Dihydroxy-β,β-carotin-4,4'-dione;   W   127559

596   C40 H52 O4   CD-496-0-0   Staxanthin;   W
127560

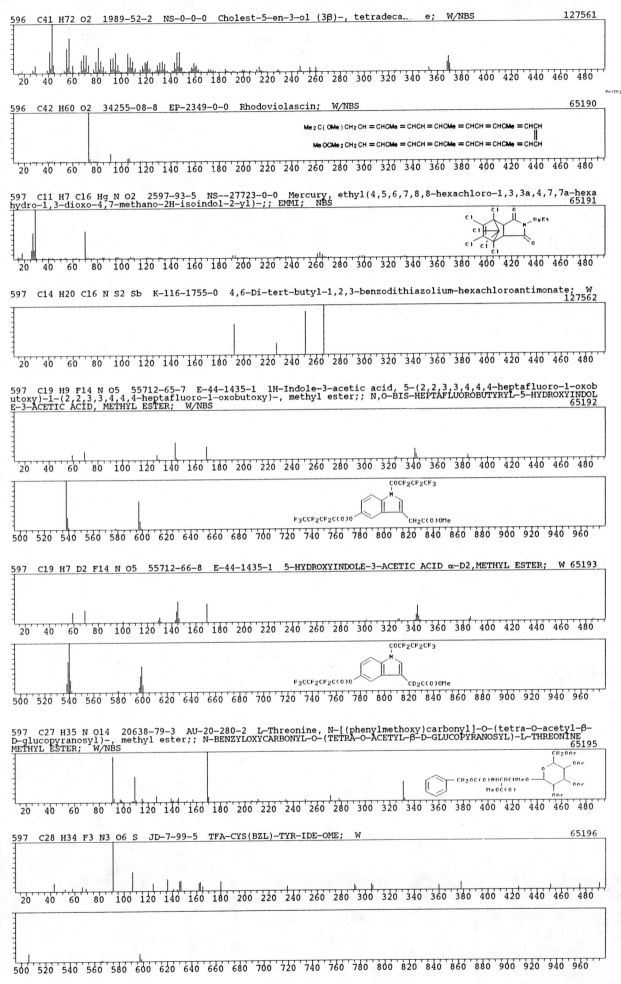

596 C41 H72 O2 1989-52-2 NS-0-0-0 Cholest-5-en-3-ol (3β)-, tetradeca... e; W/NBS   127561

596 C42 H60 O2 34255-08-8 EP-2349-0-0 Rhodoviolascin; W/NBS   65190

Me2C(OMe)CH2CH=CHCMe=CHCH=CHCMe=CHCH=CHCMe=CHCH

MeOCMe2CH2CH=CHCMe=CHCH=CHCMe=CHCH=CHCMe=CHCH

597 C11 H7 Cl6 Hg N O2 2597-93-5 NS--27723-0-0 Mercury, ethyl(4,5,6,7,8,8-hexachloro-1,3,3a,4,7,7a-hexa
hydro-1,3-dioxo-4,7-methano-2H-isoindol-2-yl)-;; EMMI; NBS   65191

597 C14 H20 Cl6 N S2 Sb K-116-1755-0 4,6-Di-tert-butyl-1,2,3-benzodithiazolium-hexachloroantimonate; W
127562

597 C19 H9 F14 N O5 55712-65-7 E-44-1435-1 1H-Indole-3-acetic acid, 5-(2,2,3,3,4,4,4-heptafluoro-1-oxob
utoxy)-1-(2,2,3,3,4,4,4-heptafluoro-1-oxobutoxy)-, methyl ester;; N,O-BIS-HEPTAFLUOROBUTYRYL-5-HYDROXYINDOL
E-3-ACETIC ACID, METHYL ESTER; W/NBS   65192

597 C19 H7 D2 F14 N O5 55712-66-8 E-44-1435-1 5-HYDROXYINDOLE-3-ACETIC ACID α-D2,METHYL ESTER; W 65193

597 C27 H35 N O14 20638-79-3 AU-20-280-2 L-Threonine, N-[(phenylmethoxy)carbonyl]-O-(tetra-O-acetyl-β-
D-glucopyranosyl)-, methyl ester;; N-BENZYLOXYCARBONYL-O-(TETRA-O-ACETYL-β-D-GLUCOPYRANOSYL)-L-THREONINE
METHYL ESTER; W/NBS   65195

597 C28 H34 F3 N3 O6 S JD-7-99-5 TFA-CYS(BZL)-TYR-IDE-OME; W   65196

597   C30 H31 N O12   49616-84-4   NS--27720-0-0   Cyclobut[4,5]azepino[1,2-a]quinoline-7,7a,8,9,9a,10-hexacarb
oxylic acid, 10,11-dihydro-1,3-dimethyl-, hexamethyl ester;;   NBS   65197

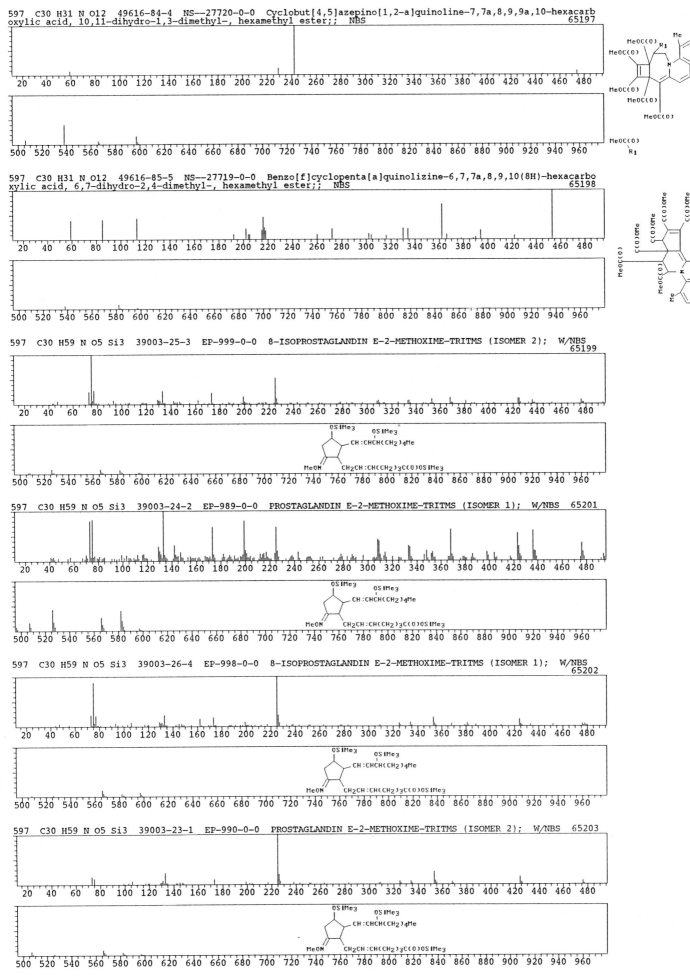

597   C30 H31 N O12   49616-85-5   NS--27719-0-0   Benzo[f]cyclopenta[a]quinolizine-6,7,7a,8,9,10(8H)-hexacarbo
xylic acid, 6,7-dihydro-2,4-dimethyl-, hexamethyl ester;;   NBS   65198

597   C30 H59 N O5 Si3   39003-25-3   EP-999-0-0   8-ISOPROSTAGLANDIN E-2-METHOXIME-TRITMS (ISOMER 2);   W/NBS
                                                                                                            65199

597   C30 H59 N O5 Si3   39003-24-2   EP-989-0-0   PROSTAGLANDIN E-2-METHOXIME-TRITMS (ISOMER 1);   W/NBS   65201

597   C30 H59 N O5 Si3   39003-26-4   EP-998-0-0   8-ISOPROSTAGLANDIN E-2-METHOXIME-TRITMS (ISOMER 1);   W/NBS
                                                                                                            65202

597   C30 H59 N O5 Si3   39003-23-1   EP-990-0-0   PROSTAGLANDIN E-2-METHOXIME-TRITMS (ISOMER 2);   W/NBS   65203

597   C30 H59 N O5 Si3   72377-56-1   NS--27714-0-0   Prosta-5,13-dien-1-oic acid, 9-(methoxyimino)-11,15-bis[(trimethylsilyl)oxy]-, trimethylsilyl ester, (8.xi.,12.xi.)-;;   NBS
65204

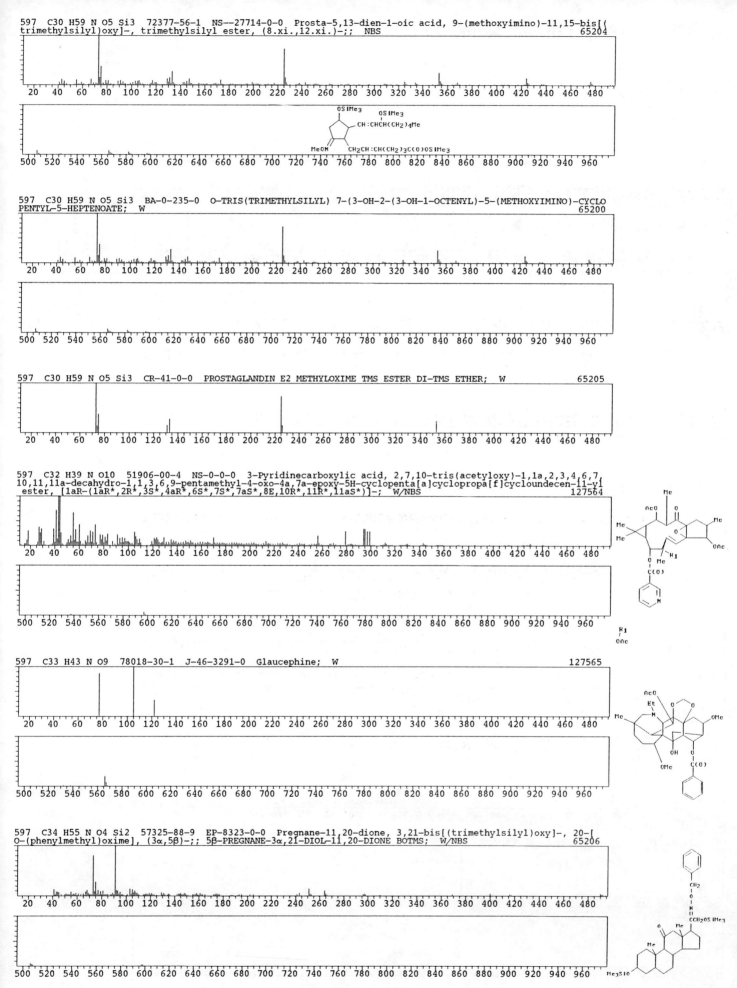

597   C30 H59 N O5 Si3   BA-0-235-0   O-TRIS(TRIMETHYLSILYL) 7-(3-OH-2-(3-OH-1-OCTENYL)-5-(METHOXYIMINO)-CYCLOPENTYL-5-HEPTENOATE; W
65200

597   C30 H59 N O5 Si3   CR-41-0-0   PROSTAGLANDIN E2 METHYLOXIME TMS ESTER DI-TMS ETHER; W
65205

597   C32 H39 N O10   51906-00-4   NS-0-0-0   3-Pyridinecarboxylic acid, 2,7,10-tris(acetyloxy)-1,1a,2,3,4,6,7,10,11,11a-decahydro-1,1,3,6,9-pentamethyl-4-oxo-4a,7a-epoxy-5H-cyclopenta[a]cyclopropa[f]cycloundecen-11-yl ester, [1aR-(1aR*,2R*,3S*,4aR*,6S*,7S*,7aS*,8E,10R*,11R*,11aS*)]-; W/NBS
127564

597   C33 H43 N O9   78018-30-1   J-46-3291-0   Glaucephine; W
127565

597   C34 H55 N O4 Si2   57325-88-9   EP-8323-0-0   Pregnane-11,20-dione, 3,21-bis[(trimethylsilyl)oxy]-, 20-[O-(phenylmethyl)oxime], (3α,5β)-;; 5β-PREGNANE-3α,21-DIOL-11,20-DIONE BOTMS; W/NBS
65206

597 C35 H55 N3 O5 28417-15-4 O-3-798-7 Leucine, N-[N2,N6-bis(1-adamantylcarbonyl)-L-lysyl]-, methyl est
er, L-;; DI-ADCO-LYS-LEU-OME; W/NBS   65207

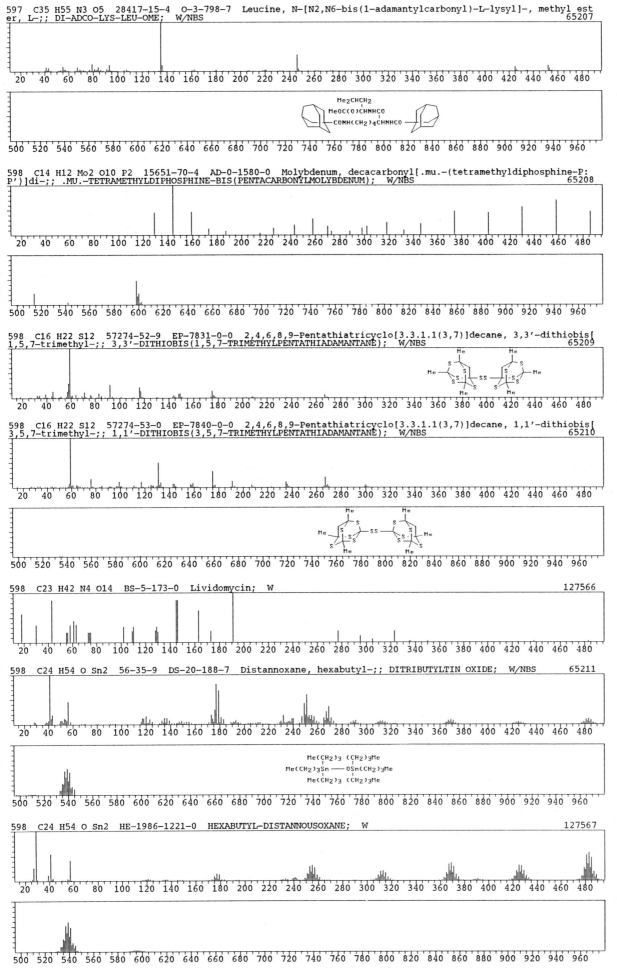

598 C14 H12 Mo2 O10 P2 15651-70-4 AD-0-1580-0 Molybdenum, decacarbonyl[.mu.-(tetramethyldiphosphine-P:
P')]di-;; .MU.-TETRAMETHYLDIPHOSPHINE-BIS(PENTACARBONYLMOLYBDENUM); W/NBS   65208

598 C16 H22 S12 57274-52-9 EP-7831-0-0 2,4,6,8,9-Pentathiatricyclo[3.3.1.1(3,7)]decane, 3,3'-dithiobis[
1,5,7-trimethyl-;; 3,3'-DITHIOBIS(1,5,7-TRIMETHYLPENTATHIADAMANTANE); W/NBS   65209

598 C16 H22 S12 57274-53-0 EP-7840-0-0 2,4,6,8,9-Pentathiatricyclo[3.3.1.1(3,7)]decane, 1,1'-dithiobis[
3,5,7-trimethyl-;; 1,1'-DITHIOBIS(3,5,7-TRIMETHYLPENTATHIADAMANTANE); W/NBS   65210

598 C23 H42 N4 O14 BS-5-173-0 Lividomycin; W   127566

598 C24 H54 O Sn2 56-35-9 DS-20-188-7 Distannoxane, hexabutyl-;; DITRIBUTYLTIN OXIDE; W/NBS   65211

598 C24 H54 O Sn2 HE-1986-1221-0 HEXABUTYL-DISTANNOUSOXANE; W   127567

598   C26 H30 O16   EP-8857-0-0   08217205002   FLAVONOL 3',4',5,7-OH,3-O-ARAGLUCOSIDE;   W   65194

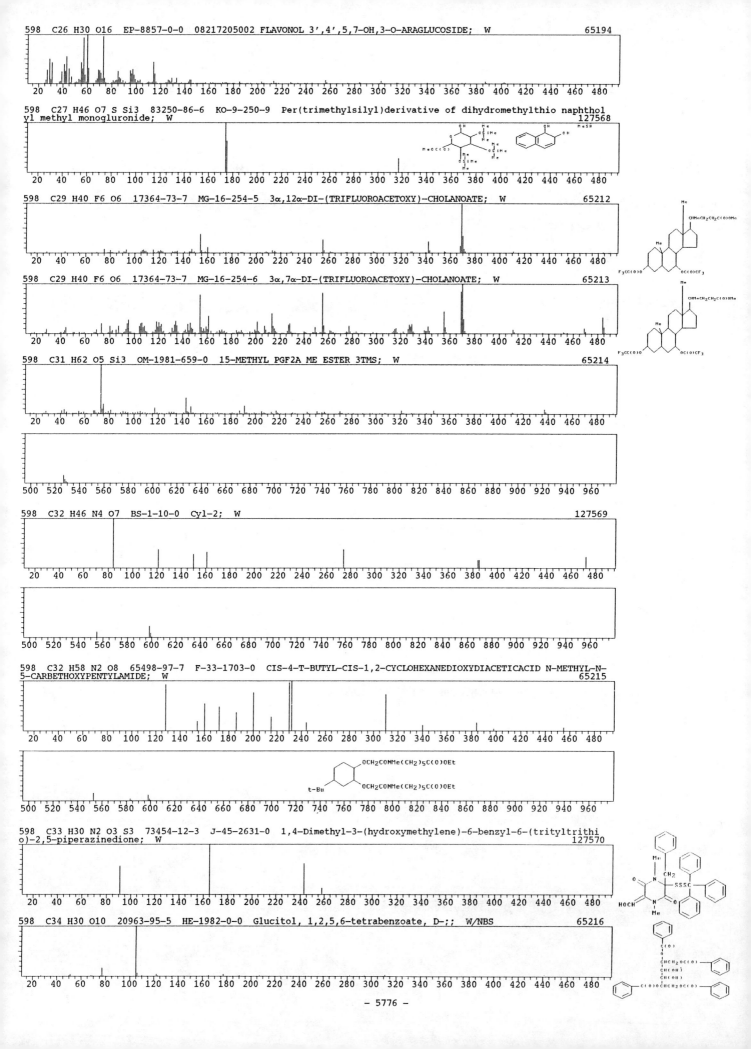

598   C27 H46 O7 S Si3   83250-86-6   KO-9-250-9   Per(trimethylsilyl)derivative of dihydromethylthio naphthol
yl methyl monogluronide;   W   127568

598   C29 H40 F6 O6   17364-73-7   MG-16-254-5   3α,12α-DI-(TRIFLUOROACETOXY)-CHOLANOATE;   W   65212

598   C29 H40 F6 O6   17364-73-7   MG-16-254-6   3α,7α-DI-(TRIFLUOROACETOXY)-CHOLANOATE;   W   65213

598   C31 H62 O5 Si3   OM-1981-659-0   15-METHYL PGF2A ME ESTER 3TMS;   W   65214

598   C32 H46 N4 O7   BS-1-10-0   Cyl-2;   W   127569

598   C32 H58 N2 O8   65498-97-7   F-33-1703-0   CIS-4-T-BUTYL-CIS-1,2-CYCLOHEXANEDIOXYDIACETICACID N-METHYL-N-
5-CARBETHOXYPENTYLAMIDE;   W   65215

598   C33 H30 N2 O3 S3   73454-12-3   J-45-2631-0   1,4-Dimethyl-3-(hydroxymethylene)-6-benzyl-6-(trityltrithi
o)-2,5-piperazinedione;   W   127570

598   C34 H30 O10   20963-95-5   HE-1982-0-0   Glucitol, 1,2,5,6-tetrabenzoate, D-;;   W/NBS   65216

- 5776 -

598　C34 H30 O10　58965-07-4　KC-1976-100-0　4',4'',7,7''-TETRAMETHOXY-5,5''-DIHYDROXY-6,8''-BILFLAVANONE;
W
65217

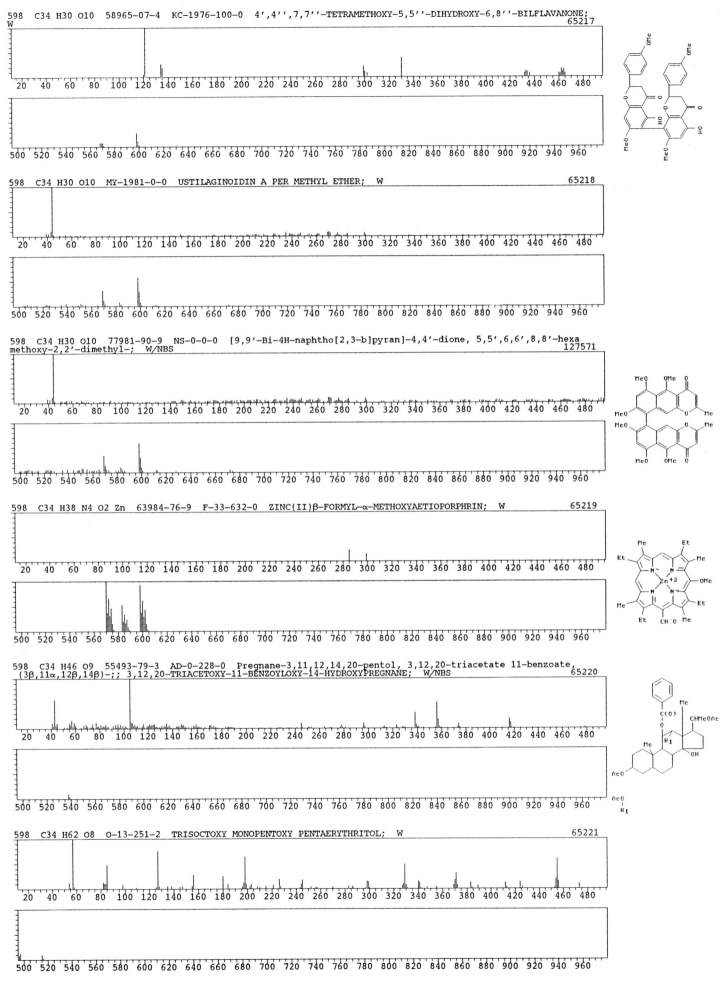

598　C34 H30 O10　MY-1981-0-0　USTILAGINOIDIN A PER METHYL ETHER;　W
65218

598　C34 H30 O10　77981-90-9　NS-0-0-0　[9,9'-Bi-4H-naphtho[2,3-b]pyran]-4,4'-dione, 5,5',6,6',8,8'-hexa
methoxy-2,2'-dimethyl-;　W/NBS
127571

598　C34 H38 N4 O2 Zn　63984-76-9　F-33-632-0　ZINC(II)β-FORMYL-α-METHOXYAETIOPORPHRIN;　W
65219

598　C34 H46 O9　55493-79-3　AD-0-228-0　Pregnane-3,11,12,14,20-pentol, 3,12,20-triacetate 11-benzoate,
(3β,11α,12β,14β)-;;　3,12,20-TRIACETOXY-11-BENZOYLOXY-14-HYDROXYPREGNANE;　W/NBS
65220

598　C34 H62 O8　O-13-251-2　TRISOCTOXY MONOPENTOXY PENTAERYTHRITOL;　W
65221

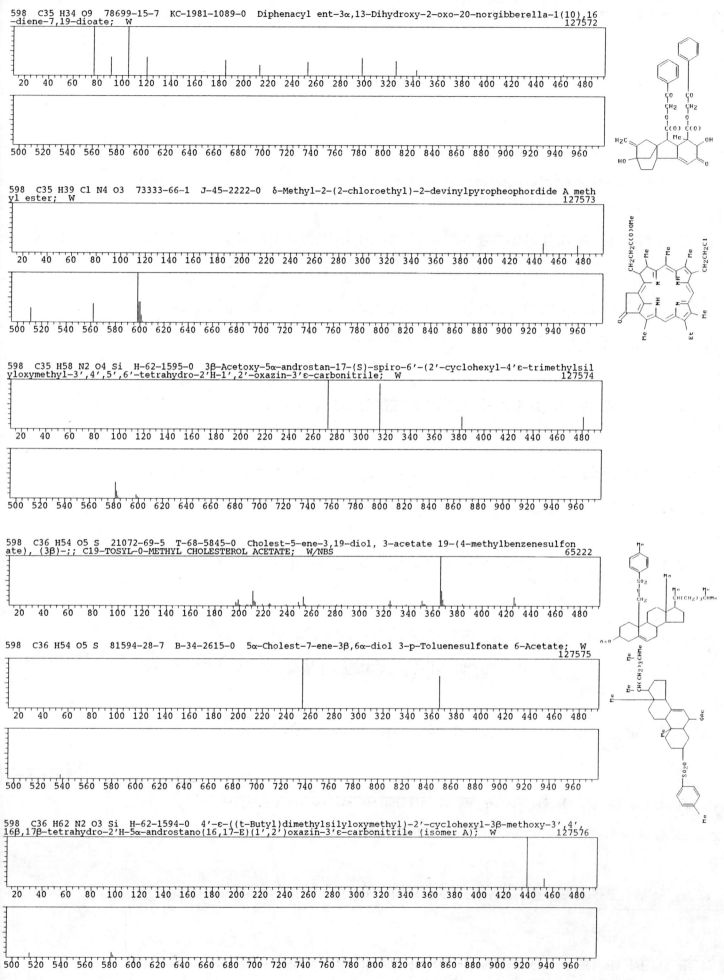

598 C35 H34 O9 78699-15-7 KC-1981-1089-0 Diphenacyl ent-3α,13-Dihydroxy-2-oxo-20-norgibberella-1(10),16
-diene-7,19-dioate; W
127572

598 C35 H39 Cl N4 O3 73333-66-1 J-45-2222-0 δ-Methyl-2-(2-chloroethyl)-2-devinylpyropheophordide A meth
yl ester; W
127573

598 C35 H58 N2 O4 Si H-62-1595-0 3β-Acetoxy-5α-androstan-17-(S)-spiro-6'-(2'-cyclohexyl-4'ε-trimethylsil
yloxymethyl-3',4',5',6'-tetrahydro-2'H-1',2'-oxazin-3'ε-carbonitrile; W
127574

598 C36 H54 O5 S 21072-69-5 T-68-5845-0 Cholest-5-ene-3,19-diol, 3-acetate 19-(4-methylbenzenesulfon
ate), (3β)-;; C19-TOSYL-0-METHYL CHOLESTEROL ACETATE; W/NBS
65222

598 C36 H54 O5 S 81594-28-7 B-34-2615-0 5α-Cholest-7-ene-3β,6α-diol 3-p-Toluenesulfonate 6-Acetate; W
127575

598 C36 H62 N2 O3 Si H-62-1594-0 4'-ε-((t-Butyl)dimethylsilyloxymethyl)-2'-cyclohexyl-3β-methoxy-3',4',
16β,17β-tetrahydro-2'H-5α-androstano(16,17-E)(1',2')oxazin-3'ε-carbonitrile (isomer A); W
127576

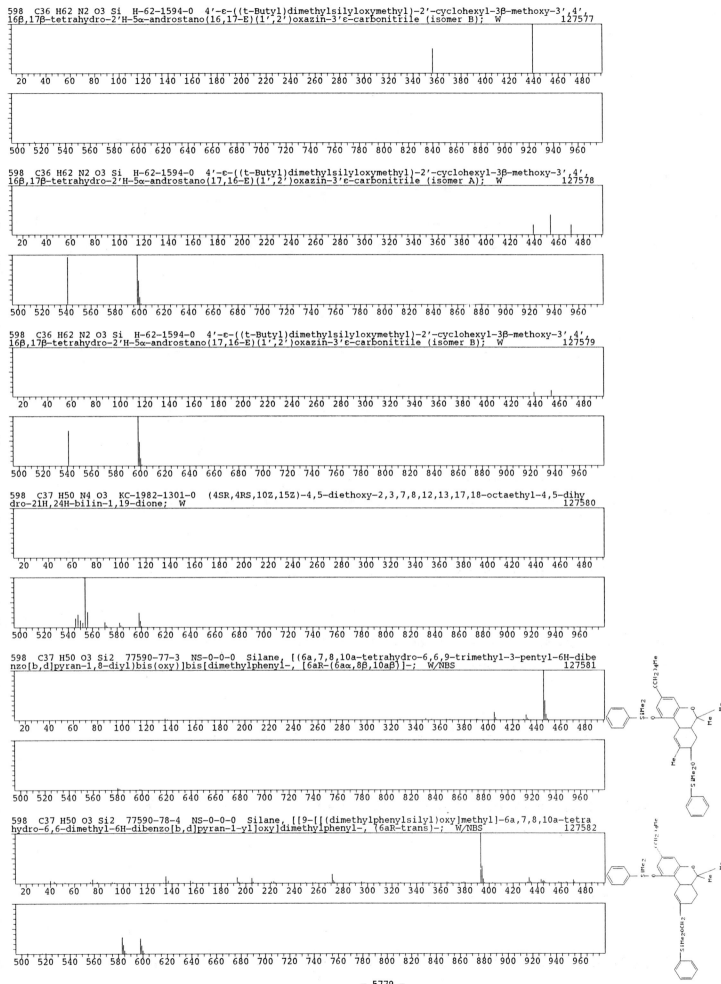

598 C36 H62 N2 O3 Si H-62-1594-0   4'-ε-((t-Butyl)dimethylsilyloxymethyl)-2'-cyclohexyl-3β-methoxy-3',4',
16β,17β-tetrahydro-2'H-5α-androstano(16,17-E)(1',2')oxazin-3'ε-carbonitrile (isomer B); W   127577

598 C36 H62 N2 O3 Si H-62-1594-0   4'-ε-((t-Butyl)dimethylsilyloxymethyl)-2'-cyclohexyl-3β-methoxy-3',4',
16β,17β-tetrahydro-2'H-5α-androstano(17,16-E)(1',2')oxazin-3'ε-carbonitrile (isomer A); W   127578

598 C36 H62 N2 O3 Si H-62-1594-0   4'-ε-((t-Butyl)dimethylsilyloxymethyl)-2'-cyclohexyl-3β-methoxy-3',4',
16β,17β-tetrahydro-2'H-5α-androstano(17,16-E)(1',2')oxazin-3'ε-carbonitrile (isomer B); W   127579

598 C37 H50 N4 O3 KC-1982-1301-0   (4SR,4RS,10Z,15Z)-4,5-diethoxy-2,3,7,8,12,13,17,18-octaethyl-4,5-dihy
dro-21H,24H-bilin-1,19-dione; W   127580

598 C37 H50 O3 Si2 77590-77-3 NS-0-0-0 Silane, [(6a,7,8,10a-tetrahydro-6,6,9-trimethyl-3-pentyl-6H-dibe
nzo[b,d]pyran-1,8-diyl)bis(oxy)]bis[dimethylphenyl-, [6aR-(6aα,8β,10aβ)]-; W/NBS   127581

598 C37 H50 O3 Si2 77590-78-4 NS-0-0-0 Silane, [[9-[[(dimethylphenylsilyl)oxy]methyl]-6a,7,8,10a-tetra
hydro-6,6-dimethyl-6H-dibenzo[b,d]pyran-1-yl]oxy]dimethylphenyl-, (6aR-trans)-; W/NBS   127582

- 5779 -

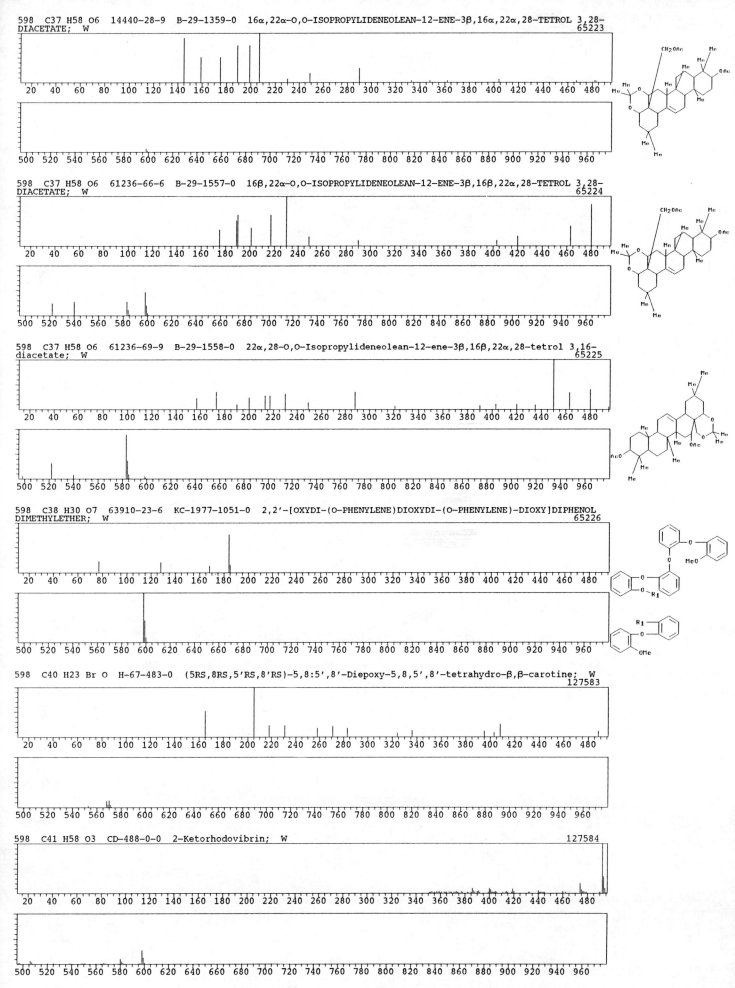

598   C37 H58 O6   14440-28-9   B-29-1359-0   16α,22α-O,O-ISOPROPYLIDENEOLEAN-12-ENE-3β,16α,22α,28-TETROL 3,28-DIACETATE;   W
65223

598   C37 H58 O6   61236-66-6   B-29-1557-0   16β,22α-O,O-ISOPROPYLIDENEOLEAN-12-ENE-3β,16β,22α,28-TETROL 3,28-DIACETATE;   W
65224

598   C37 H58 O6   61236-69-9   B-29-1558-0   22α,28-O,O-Isopropylideneolean-12-ene-3β,16β,22α,28-tetrol 3,16-diacetate;   W
65225

598   C38 H30 O7   63910-23-6   KC-1977-1051-0   2,2'-[OXYDI-(O-PHENYLENE)DIOXYDI-(O-PHENYLENE)-DIOXY]DIPHENOL DIMETHYLETHER;   W
65226

598   C40 H23 Br O   H-67-483-0   (5RS,8RS,5'RS,8'RS)-5,8:5',8'-Diepoxy-5,8,5',8'-tetrahydro-β,β-carotine;   W
127583

598   C41 H58 O3   CD-488-0-0   2-Ketorhodovibrin;   W
127584

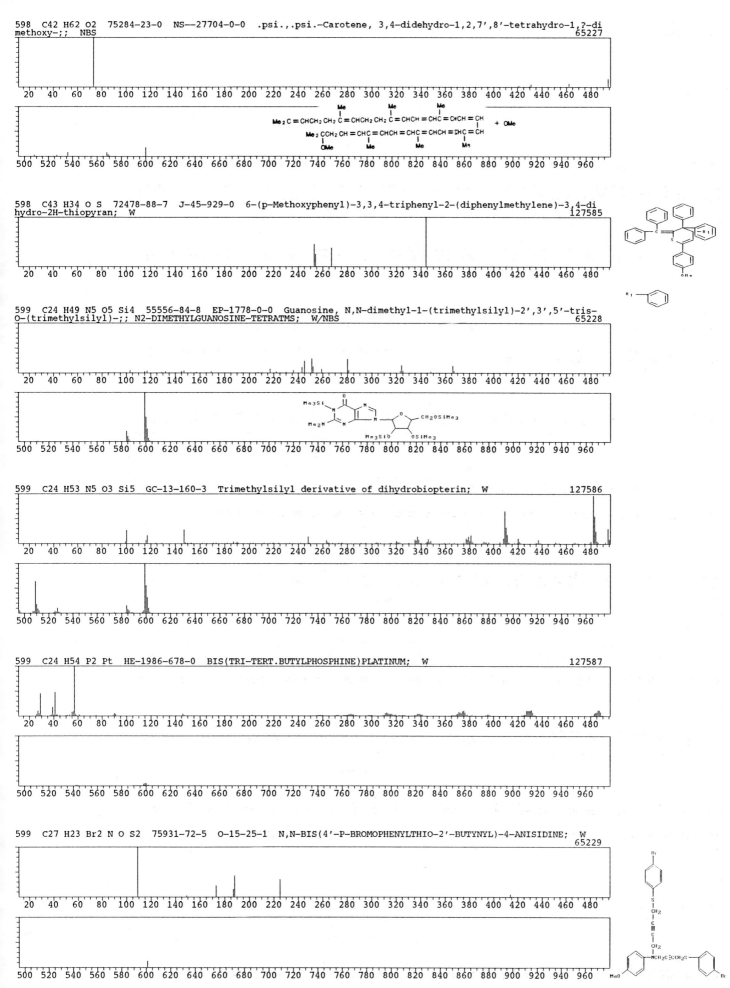

598 C42 H62 O2 75284-23-0 NS--27704-0-0 .psi.,.psi.-Carotene, 3,4-didehydro-1,2,7',8'-tetrahydro-1,2-dimethoxy-;; NBS 65227

598 C43 H34 O S 72478-88-7 J-45-929-0 6-(p-Methoxyphenyl)-3,3,4-triphenyl-2-(diphenylmethylene)-3,4-dihydro-2H-thiopyran; W 127585

599 C24 H49 N5 O5 Si4 55556-84-8 EP-1778-0-0 Guanosine, N,N-dimethyl-1-(trimethylsilyl)-2',3',5'-tris-O-(trimethylsilyl)-;; N2-DIMETHYLGUANOSINE-TETRATMS; W/NBS 65228

599 C24 H53 N5 O3 Si5 GC-13-160-3 Trimethylsilyl derivative of dihydrobiopterin; W 127586

599 C24 H54 P2 Pt HE-1986-678-0 BIS(TRI-TERT.BUTYLPHOSPHINE)PLATINUM; W 127587

599 C27 H23 Br2 N O S2 75931-72-5 O-15-25-1 N,N-BIS(4'-P-BROMOPHENYLTHIO-2'-BUTYNYL)-4-ANISIDINE; W 65229

599   C27 H26 Br N3 O8   I-57-2327-0   (2S)-N-((1E)-5,6,7-TRIACETOXYSTYR-1-YL)-2-ACETYLAMINO-3-(6-BROMOINDOL-
3YL)PROPIONAMIDE;   W   65230

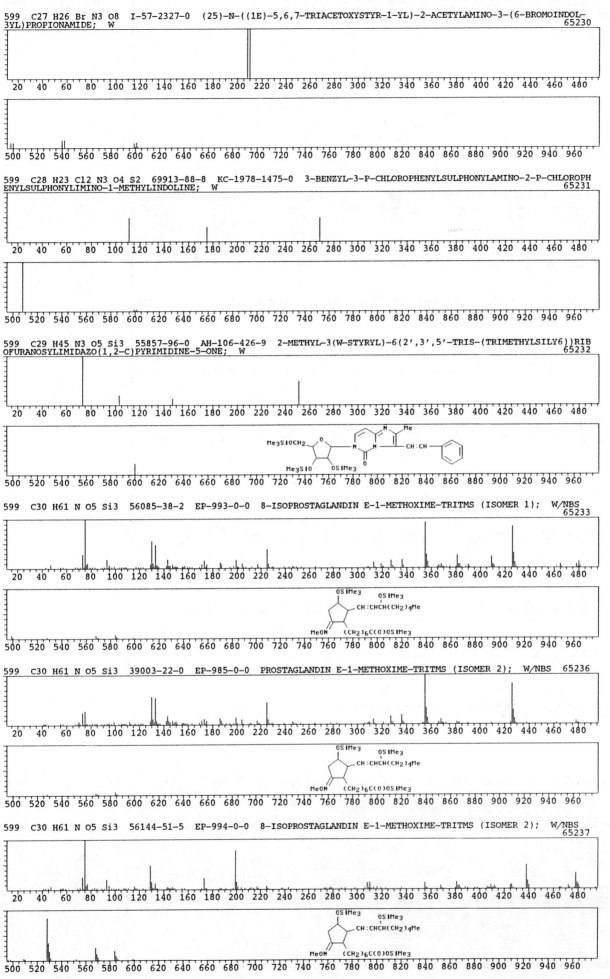

599   C28 H23 Cl2 N3 O4 S2   69913-88-8   KC-1978-1475-0   3-BENZYL-3-P-CHLOROPHENYLSULPHONYLAMINO-2-P-CHLOROPH
ENYLSULPHONYLIMINO-1-METHYLINDOLINE;   W   65231

599   C29 H45 N3 O5 Si3   55857-96-0   AH-106-426-9   2-METHYL-3(W-STYRYL)-6(2',3',5'-TRIS-(TRIMETHYLSILY6))RIB
OFURANOSYLIMIDAZO(1,2-C)PYRIMIDINE-5-ONE;   W   65232

599   C30 H61 N O5 Si3   56085-38-2   EP-993-0-0   8-ISOPROSTAGLANDIN E-1-METHOXIME-TRITMS (ISOMER 1);   W/NBS
                                                                                                              65233

599   C30 H61 N O5 Si3   39003-22-0   EP-985-0-0   PROSTAGLANDIN E-1-METHOXIME-TRITMS (ISOMER 2);   W/NBS   65236

599   C30 H61 N O5 Si3   56144-51-5   EP-994-0-0   8-ISOPROSTAGLANDIN E-1-METHOXIME-TRITMS (ISOMER 2);   W/NBS
                                                                                                              65237

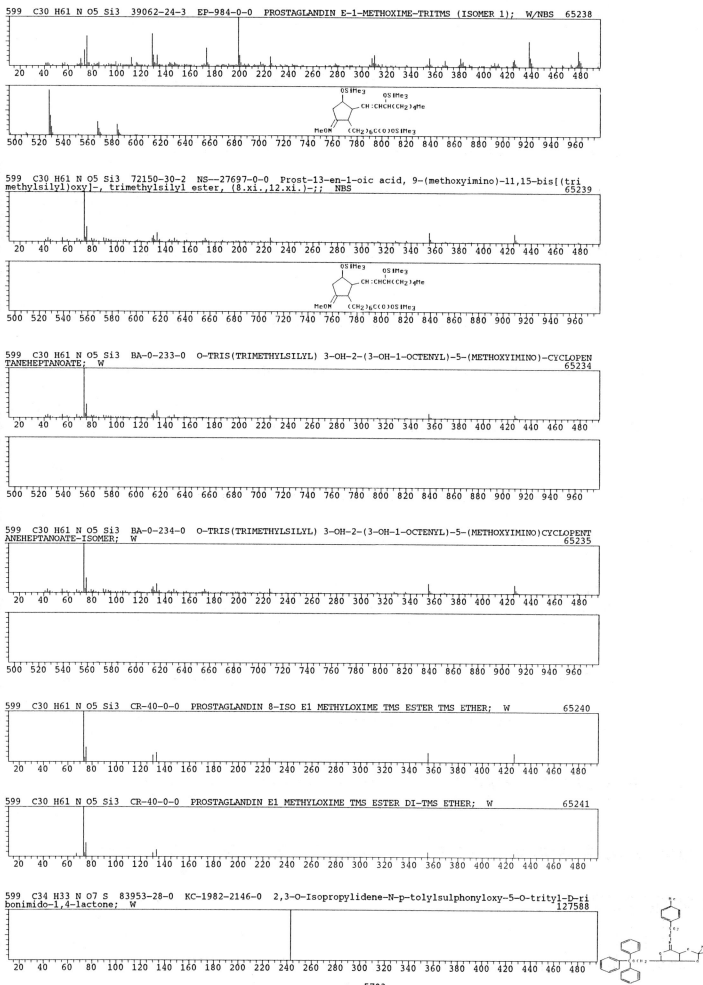

599 C30 H61 N O5 Si3 39062-24-3 EP-984-0-0 PROSTAGLANDIN E-1-METHOXIME-TRITMS (ISOMER 1); W/NBS 65238

599 C30 H61 N O5 Si3 72150-30-2 NS--27697-0-0 Prost-13-en-1-oic acid, 9-(methoxyimino)-11,15-bis[(tri methylsilyl)oxy]-, trimethylsilyl ester, (8.xi.,12.xi.)-;; NBS 65239

599 C30 H61 N O5 Si3 BA-0-233-0 O-TRIS(TRIMETHYLSILYL) 3-OH-2-(3-OH-1-OCTENYL)-5-(METHOXYIMINO)-CYCLOPEN TANEHEPTANOATE; W 65234

599 C30 H61 N O5 Si3 BA-0-234-0 O-TRIS(TRIMETHYLSILYL) 3-OH-2-(3-OH-1-OCTENYL)-5-(METHOXYIMINO)CYCLOPENT ANEHEPTANOATE-ISOMER; W 65235

599 C30 H61 N O5 Si3 CR-40-0-0 PROSTAGLANDIN 8-ISO E1 METHYLOXIME TMS ESTER TMS ETHER; W 65240

599 C30 H61 N O5 Si3 CR-40-0-0 PROSTAGLANDIN E1 METHYLOXIME TMS ESTER DI-TMS ETHER; W 65241

599 C34 H33 N O7 S 83953-28-0 KC-1982-2146-0 2,3-O-Isopropylidene-N-p-tolylsulphonyloxy-5-O-trityl-D-ri bonimido-1,4-lactone; W 127588

- 5783 -

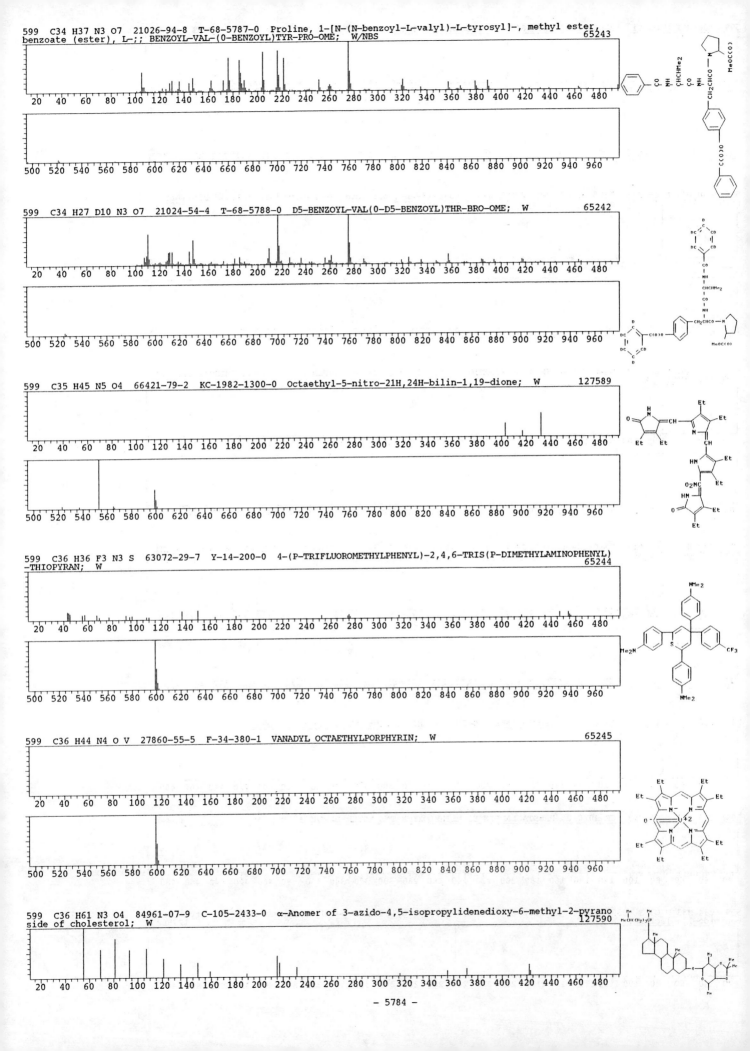

599   C34 H37 N3 O7   21026-94-8   T-68-5787-0   Proline, 1-[N-(N-benzoyl-L-valyl)-L-tyrosyl]-, methyl ester, benzoate (ester), L-;; BENZOYL-VAL-(0-BENZOYL)TYR-PRO-OME; W/NBS   65243

599   C34 H27 D10 N3 O7   21024-54-4   T-68-5788-0   D5-BENZOYL-VAL(0-D5-BENZOYL)THR-BRO-OME; W   65242

599   C35 H45 N5 O4   66421-79-2   KC-1982-1300-0   Octaethyl-5-nitro-21H,24H-bilin-1,19-dione; W   127589

599   C36 H36 F3 N3 S   63072-29-7   Y-14-200-0   4-(P-TRIFLUOROMETHYLPHENYL)-2,4,6-TRIS(P-DIMETHYLAMINOPHENYL)-THIOPYRAN; W   65244

599   C36 H44 N4 O V   27860-55-5   F-34-380-1   VANADYL OCTAETHYLPORPHYRIN; W   65245

599   C36 H61 N3 O4   84961-07-9   C-105-2433-0   α-Anomer of 3-azido-4,5-isopropylidenedioxy-6-methyl-2-pyrano side of cholesterol; W   127590

599   C37 H45 N O6   78213-66-8   0-16-404-1   PENITREM E;   W   65246

599   C37 H49 N3 O4   60283-61-6   KC-1976-734-0   3',5'-DIOXO-4'-PHENYL-5α,8α-(1',2')-1',2',4'-TRIAZOLIDINOCHO
LESTA-6,25-DIEN-3β-YL ACETATE;   W   65247

599   C19 N7 P6   33992-37-9   EP-6338-0-0   3aH,6aH,9aH-1,3,4,6,7,9,9b-Heptaaza-2,3a,5,6a,8,9a-hexaphosphaphen
alene, 2,2,3a,5,5,6a,8,8,9a-nonachloro-2,2,5,5,8,8-hexahydro-;; Nitrilohexaphosphonitrilic chloride;   W/NBS
                                                                                          65248

600   C12 F10 Hg S2   37612-94-5   O-8-149-2   Mercury, bis[(pentafluorophenyl)thio]-;; BIS(PENTAFLUOROPHENYL
THIO)MERCURY;   W/NBS   65249

600   C12 H8 Cl4 F8 N2 O4 S2   73997-96-3   KO-6-388-26   BIS(2,2-BIS(CHLORODIFLUOROMETHYL)-5-OXO-1,3-OXAZOLIDI
N-4-YLMETHYL)-DISULFIDE;   W   65250

600   C17 H16 Br4 O4   65487-78-7   SB-33-108-0   3,3',4,4'-TETRAMETHOXY-5,5',6,6'-TETRABROMODIPHENYLMETHANE;
W   65251

600   C24 H24 O6 S6   72036-86-3   KC-1979-1713-1   4,6,12,14,20,22-HEXAMETHOXY-1,2,9,10,17,18-HEXATHIA[2.2.2]
METACYCLOPHANE;   W   65252

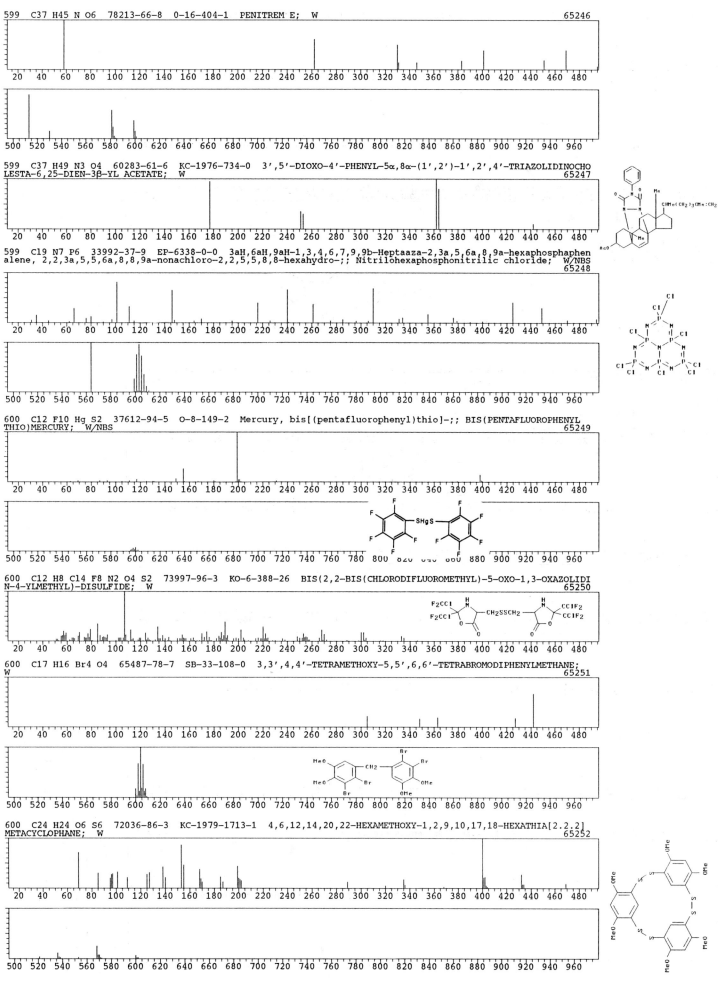

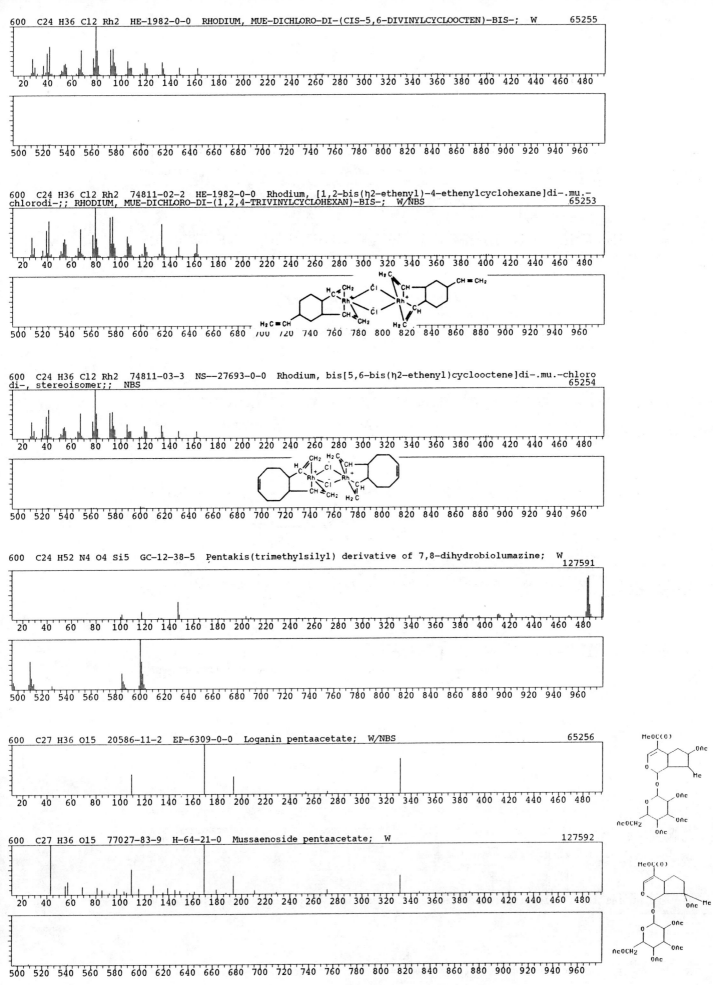

600 C24 H36 Cl2 Rh2 HE-1982-0-0 RHODIUM, MUE-DICHLORO-DI-(CIS-5,6-DIVINYLCYCLOOCTEN)-BIS-; W 65255

600 C24 H36 Cl2 Rh2 74811-02-2 HE-1982-0-0 Rhodium, [1,2-bis(η2-ethenyl)-4-ethenylcyclohexane]di-.mu.-chlorodi-;; RHODIUM, MUE-DICHLORO-DI-(1,2,4-TRIVINYLCYCLOHEXAN)-BIS-; W/NBS 65253

600 C24 H36 Cl2 Rh2 74811-03-3 NS--27693-0-0 Rhodium, bis[5,6-bis(η2-ethenyl)cyclooctene]di-.mu.-chloro di-, stereoisomer;; NBS 65254

600 C24 H52 N4 O4 Si5 GC-12-38-5 Pentakis(trimethylsilyl) derivative of 7,8-dihydrobiolumazine; W 127591

600 C27 H36 O15 20586-11-2 EP-6309-0-0 Loganin pentaacetate; W/NBS 65256

600 C27 H36 O15 77027-83-9 H-64-21-0 Mussaenoside pentaacetate; W 127592

- 5786 -

600 C28 H44 B4 Co2 Ni 84959-87-5 C-105-2598-0 Bis[cyclopentadienyl-.mu.-1,3-diborolenylcobalt]nickel;
W
127593

600 C28 H49 Cl O6 Si3 KO-7-229-3 Trimethylsilyl derivative of methyl ester of tetranorcloprostenol with
Trimethylethylsilyl derivative of δ-lactone of tetranorcloprosternol; W
127594

600 C28 H56 N2 O6 Si3 64911-27-9 O-15-212-2 2',3',5'-TRIS-O-TERT-BUTYLDIMETHYLSILYL-5-METHYL URIDINE;
W
65257

600 C29 H36 N4 O10 7652-80-4 O-6-1396-8 Riboflavin, 2',3',4',5'-tetrapropanoate;; Riboflavin tetrapropi
onate; W/NBS
65258

600 C30 H36 N2 O11 28215-67-0 I-48-1852-2 Hexaacetyladrepine; W/NBS
65259

600 C30 H60 O6 Si3 35275-54-8 D-10-3666-1 5-Octenoic acid, 8-[tetrahydro-4-[(trimethylsilyl)oxy]-5-[3-
[(trimethylsilyl)oxy]-1-octenyl]-2-furanyl]-8-[(trimethylsilyl)oxy]-, methyl ester;; TRISTRIMETHYLSILYL ETH
ER 9(12)-OXY-8,11,15-TRIHYDROXYEICOSA-5,13-DIENOIC ACID METHYL ESTER; W/NBS
65260

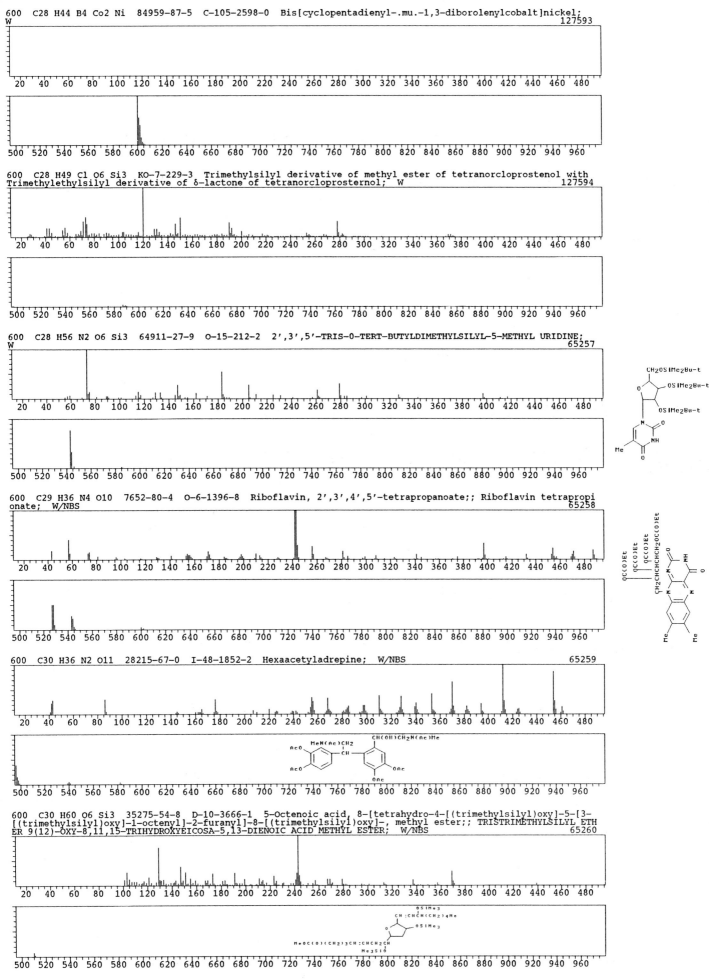

600   C30 H60 O6 Si3   OM-1981-478-0   THROMBOXANE B2 ME ESTER 3TMS;   W   65261

600   C30 H60 O6 Si3   OM-1981-551-0   6-OXO-PGF-1A ME 3TMS;   W   65262

600   C30 H52 D8 O6 Si3   OM-1981-431-0   D8-THROMBOXANE ME ESTER 3TMS;   W   65263

600   C32 H40 N4 O4 Si2   17082-89-2   J-5-2140-0   1,2,4,5-Tetraaza-3,6-disilacyclohexane, 1,2,4,5-tetrakis(m-
methoxyphenyl)-3,3,6,6-tetramethyl-;; 1,2,4,5-TETRA(M-ANISYL)-3,3,6,6-TETRAMETHYL-1,2,4,5-TETRAAZA-3,6-DISI
LACYCLOHEXANE;   W/NBS   65264

600   C34 H36 N2 O8   87377-77-3   F-39-1855-0   Dibenzyl 3-ethoxycarbonylmethyl-4'-methyoxycarbonylmethyl-4,3'
-dimethylpyrromethane-5,5'-dicarboxylate;   W   127596

600   C34 H37 Cl N4 O4   52459-78-6   KC-1976-2504-0   2-(2-CHLOROETHYL)-6,7-BIS-(2-METHOXY CARBONYLETHYL)-1,3,
5,8-TETRAMETHYL-PORPHIN;   W   65265

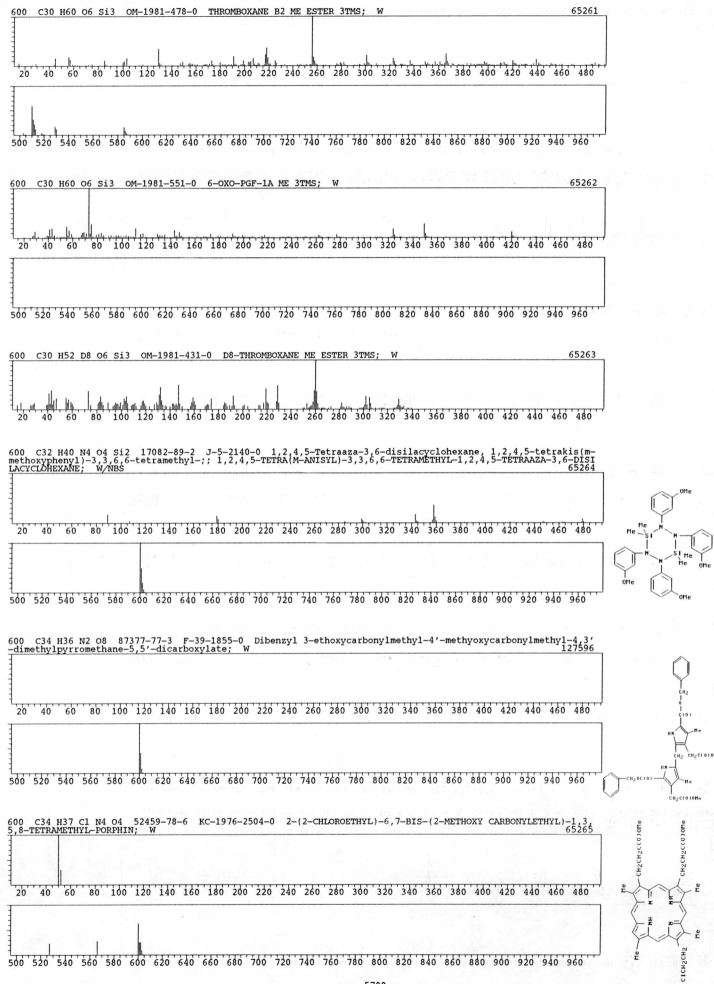

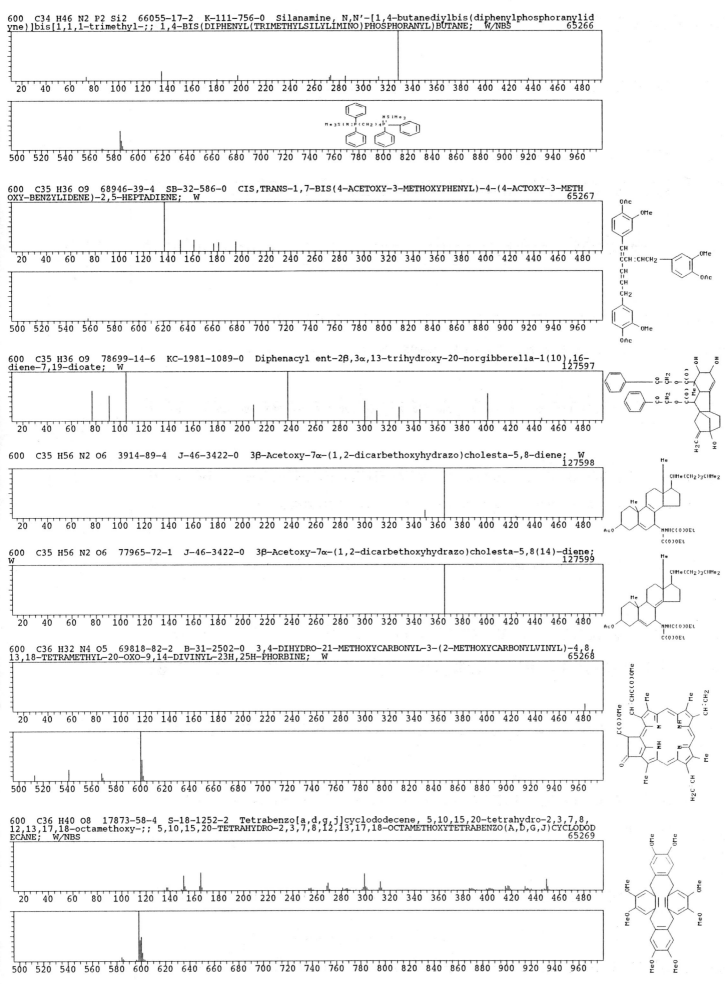

600 C34 H46 N2 P2 Si2 66055-17-2 K-111-756-0 Silanamine, N,N'-[1,4-butanediylbis(diphenylphosphoranylid yne)]bis[1,1,1-trimethyl-;; 1,4-BIS(DIPHENYL(TRIMETHYLSILYLIMINO)PHOSPHORANYL)BUTANE; W/NBS   65266

600 C35 H36 O9 68946-39-4 SB-32-586-0 CIS,TRANS-1,7-BIS(4-ACETOXY-3-METHOXYPHENYL)-4-(4-ACTOXY-3-METH OXY-BENZYLIDENE)-2,5-HEPTADIENE; W   65267

600 C35 H36 O9 78699-14-6 KC-1981-1089-0 Diphenacyl ent-2β,3α,13-trihydroxy-20-norgibberella-1(10),16-diene-7,19-dioate; W   127597

600 C35 H56 N2 O6 3914-89-4 J-46-3422-0 3β-Acetoxy-7α-(1,2-dicarbethoxyhydrazo)cholesta-5,8-diene; W   127598

600 C35 H56 N2 O6 77965-72-1 J-46-3422-0 3β-Acetoxy-7α-(1,2-dicarbethoxyhydrazo)cholesta-5,8(14)-diene; W   127599

600 C36 H32 N4 O5 69818-82-2 B-31-2502-0 3,4-DIHYDRO-21-METHOXYCARBONYL-3-(2-METHOXYCARBONYLVINYL)-4,8, 13,18-TETRAMETHYL-20-OXO-9,14-DIVINYL-23H,25H-PHORBINE; W   65268

600 C36 H40 O8 17873-58-4 S-18-1252-2 Tetrabenzo[a,d,g,j]cyclododecene, 5,10,15,20-tetrahydro-2,3,7,8, 12,13,17,18-octamethoxy-;; 5,10,15,20-TETRAHYDRO-2,3,7,8,12,13,17,18-OCTAMETHOXYTETRABENZO(A,D,G,J)CYCLODOD ECANE; W/NBS   65269

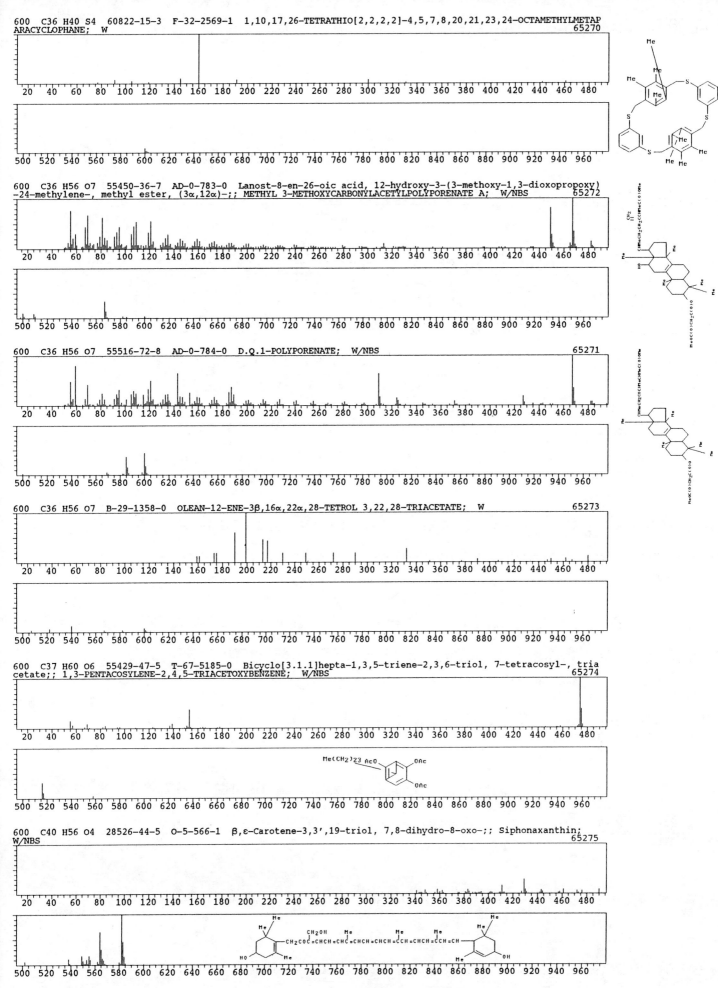

600 C36 H40 S4 60822-15-3 F-32-2569-1 1,10,17,26-TETRATHIO[2,2,2,2]-4,5,7,8,20,21,23,24-OCTAMETHYLMETAP
ARACYCLOPHANE; W   65270

600 C36 H56 O7 55450-36-7 AD-0-783-0 Lanost-8-en-26-oic acid, 12-hydroxy-3-(3-methoxy-1,3-dioxopropoxy)
-24-methylene-, methyl ester, (3α,12α)-;; METHYL 3-METHOXYCARBONYLACETYLPOLYPORENATE A; W/NBS   65272

600 C36 H56 O7 55516-72-8 AD-0-784-0 D.Q.1-POLYPORENATE; W/NBS   65271

600 C36 H56 O7 B-29-1358-0 OLEAN-12-ENE-3β,16α,22α,28-TETROL 3,22,28-TRIACETATE; W   65273

600 C37 H60 O6 55429-47-5 T-67-5185-0 Bicyclo[3.1.1]hepta-1,3,5-triene-2,3,6-triol, 7-tetracosyl-, tria
cetate;; 1,3-PENTACOSYLENE-2,4,5-TRIACETOXYBENZENE; W/NBS   65274

600 C40 H56 O4 28526-44-5 O-5-566-1 β,ε-Carotene-3,3′,19-triol, 7,8-dihydro-8-oxo-;; Siphonaxanthin;
W/NBS   65275

- 5790 -

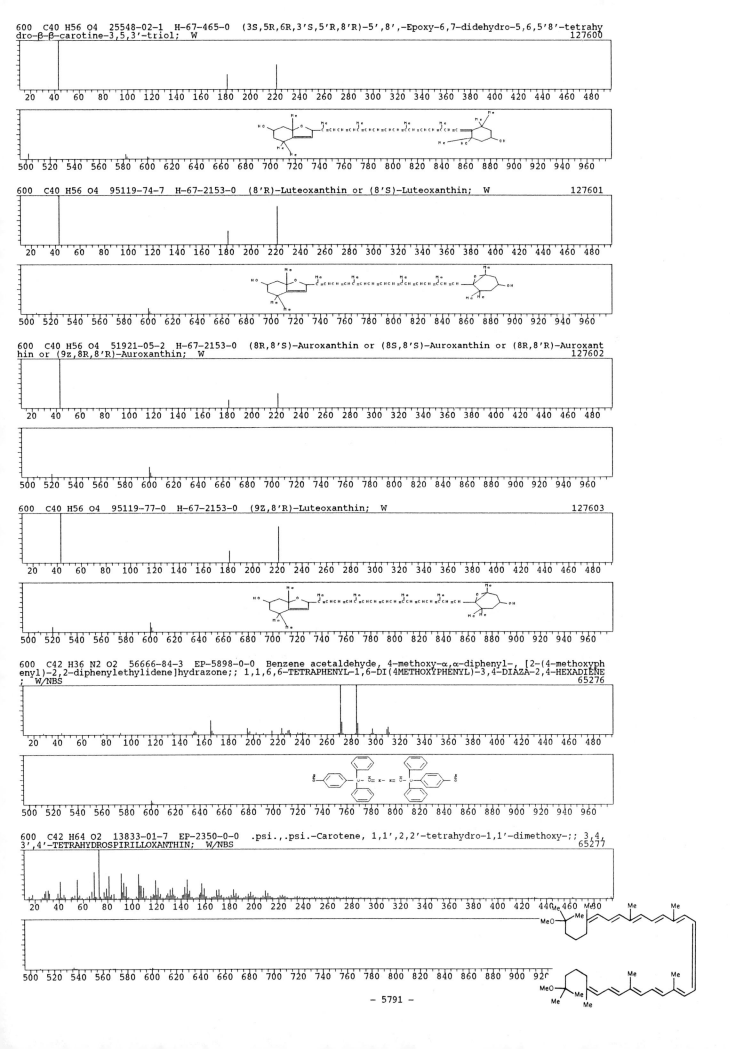

600   C40 H56 O4   25548-02-1   H-67-465-0   (3S,5R,6R,3'S,5'R,8'R)-5',8',-Epoxy-6,7-didehydro-5,6,5'8'-tetrahydro-β-β-carotine-3,5,3'-triol;   W   127600

600   C40 H56 O4   95119-74-7   H-67-2153-0   (8'R)-Luteoxanthin or (8'S)-Luteoxanthin;   W   127601

600   C40 H56 O4   51921-05-2   H-67-2153-0   (8R,8'S)-Auroxanthin or (8S,8'S)-Auroxanthin or (8R,8'R)-Auroxanthin or (9z,8R,8'R)-Auroxanthin;   W   127602

600   C40 H56 O4   95119-77-0   H-67-2153-0   (9Z,8'R)-Luteoxanthin;   W   127603

600   C42 H36 N2 O2   56666-84-3   EP-5898-0-0   Benzene acetaldehyde, 4-methoxy-α,α-diphenyl-, [2-(4-methoxyphenyl)-2,2-diphenylethylidene]hydrazone;; 1,1,6,6-TETRAPHENYL-1,6-DI(4METHOXYPHENYL)-3,4-DIAZA-2,4-HEXADIENE ;   W/NBS   65276

600   C42 H64 O2   13833-01-7   EP-2350-0-0   .psi.,.psi.-Carotene, 1,1',2,2'-tetrahydro-1,1'-dimethoxy-;; 3,4, 3',4'-TETRAHYDROSPIRILLOXANTHIN;   W/NBS   65277

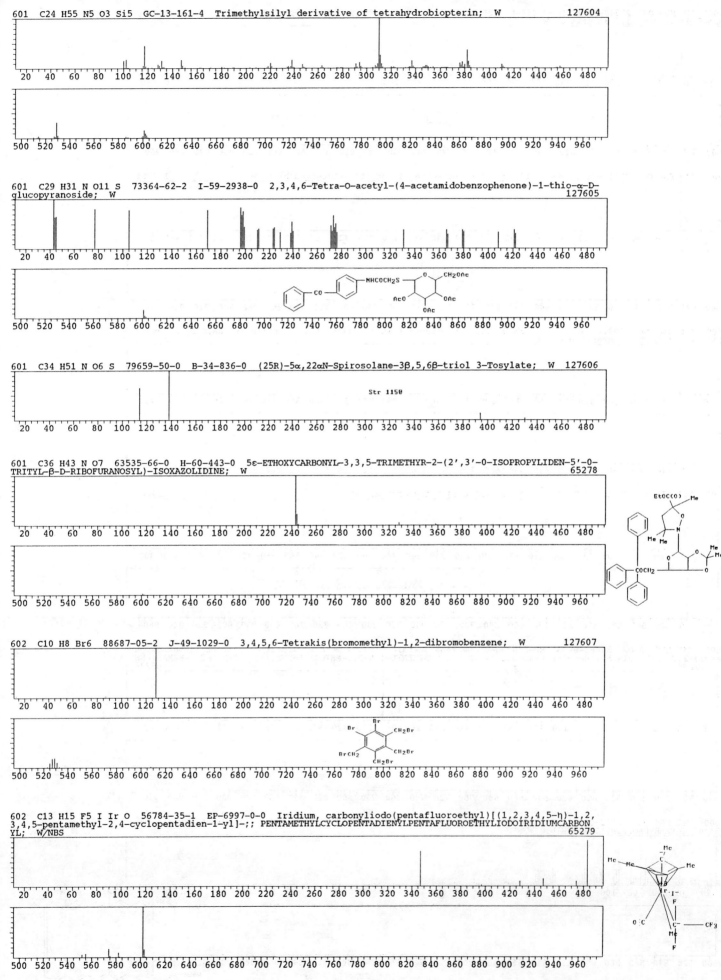

601   C24 H55 N5 O3 Si5   GC-13-161-4   Trimethylsilyl derivative of tetrahydrobiopterin;   W   127604

601   C29 H31 N O11 S   73364-62-2   I-59-2938-0   2,3,4,6-Tetra-O-acetyl-(4-acetamidobenzophenone)-1-thio-α-D-glucopyranoside;   W   127605

601   C34 H51 N O6 S   79659-50-0   B-34-836-0   (25R)-5α,22αN-Spirosolane-3β,5,6β-triol 3-Tosylate;   W   127606

Str 1150

601   C36 H43 N O7   63535-66-0   H-60-443-0   5ε-ETHOXYCARBONYL-3,3,5-TRIMETHYR-2-(2',3'-O-ISOPROPYLIDEN-5'-O-TRITYL-β-D-RIBOFURANOSYL)-ISOXAZOLIDINE;   W   65278

602   C10 H8 Br6   88687-05-2   J-49-1029-0   3,4,5,6-Tetrakis(bromomethyl)-1,2-dibromobenzene;   W   127607

602   C13 H15 F5 I Ir O   56784-35-1   EP-6997-0-0   Iridium, carbonyliodo(pentafluoroethyl)[(1,2,3,4,5-η)-1,2,3,4,5-pentamethyl-2,4-cyclopentadien-1-yl]-;;   PENTAMETHYLCYCLOPENTADIENYLPENTAFLUOROETHYLIODOIRIDIUMCARBONYL;   W/NBS   65279

602　C16 H5 F15 N4 O4　82633-73-6　AC-1982-94-0　5-nitrophenyl-4-nitro-3-perfluoroheptyl-pyrazole;　W
127608

602　C16 H20 Hg N2 O4 S3　82396-78-9　K-115-1465-0　N-(Methylmercurio)-N,N'-bis(4-methylphenylsulfonyl)meth
anesulfinamidine;　W
127609

602　C18 Cl2 F15 P　5864-22-2　AD-0-2728-0　Phosphorane, dichlorotris(pentafluorophenyl)-;; TRIS(PENTAFLUOR
OPHENYL) DICHLOROPHOSPHORANE;　W/NBS
65280

602　C24 H14 As2 Cl2 O5　55429-19-1　O-3-297-11　10H-Phenoxarsine, 10,10'-oxybis[2-chloro-, 10,10'-dioxide
;; 2-CHLORO-10-PHENOXARSENIC ACID ANHYDRIDE;　W/NBS
65281

602　C30 H62 O6 Si3　35275-56-0　D-10-3667-3　2-Furanoctanoic acid, tetrahydro-η,4-bis[(trimethylsilyl)oxy]
-5-[3-[(trimethylsilyl)oxy]-1-octenyl]-, methyl ester;; TRISTRIMETHYLSILYL ETHER 9(12)-OXY-8,11,15-TRIHYDR
OXYEICOSA-13-ENOIC ACID METHYL ESTER;　W/NBS
65282

602　C31 H38 O12　H-65-2177-0　6β,7ε,11,12,14-Pentaacetoxy-13-allyl-19-formyloxy-13-desi-isopropylbieta-8(
9),11,13-triene;　W
127563

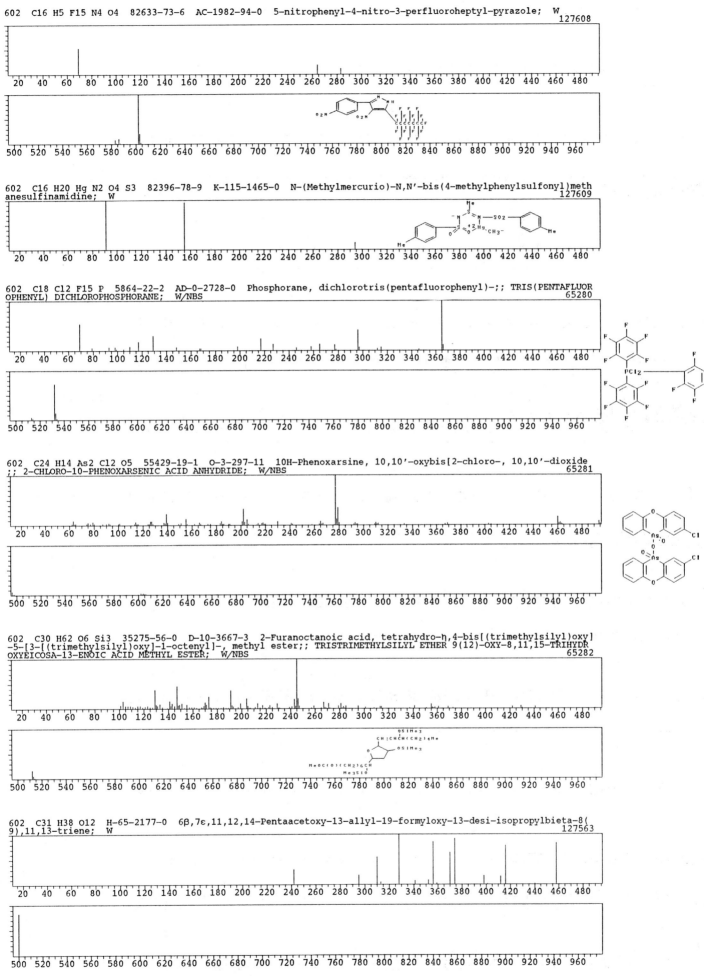

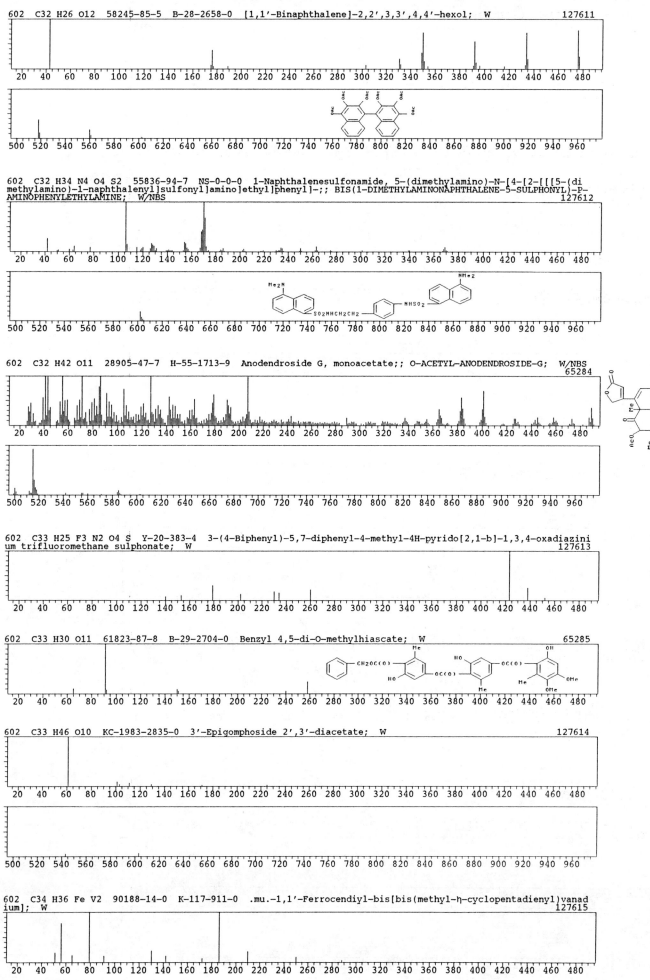

602   C32 H26 O12   58245-85-5   B-28-2658-0   [1,1'-Binaphthalene]-2,2',3,3',4,4'-hexol;   W   127611

602   C32 H34 N4 O4 S2   55836-94-7   NS-0-0-0   1-Naphthalenesulfonamide, 5-(dimethylamino)-N-[4-[2-[[[5-(di
methylamino)-1-naphthalenyl]sulfonyl]amino]ethyl]phenyl]-;; BIS(1-DIMETHYLAMINONAPHTHALENE-5-SULPHONYL)-P-
AMINOPHENYLETHYLAMINE;   W/NBS   127612

602   C32 H42 O11   28905-47-7   H-55-1713-9   Anodendroside G, monoacetate;; O-ACETYL-ANODENDROSIDE-G;   W/NBS
                                                                                                                                 65284

602   C33 H25 F3 N2 O4 S   Y-20-383-4   3-(4-Biphenyl)-5,7-diphenyl-4-methyl-4H-pyrido[2,1-b]-1,3,4-oxadiazini
um trifluoromethane sulphonate;   W   127613

602   C33 H30 O11   61823-87-8   B-29-2704-0   Benzyl 4,5-di-O-methylhiascate;   W   65285

602   C33 H46 O10   KC-1983-2835-0   3'-Epigomphoside 2',3'-diacetate;   W   127614

602   C34 H36 Fe V2   90188-14-0   K-117-911-0   .mu.-1,1'-Ferrocendiyl-bis[bis(methyl-η-cyclopentadienyl)vanad
ium];   W   127615

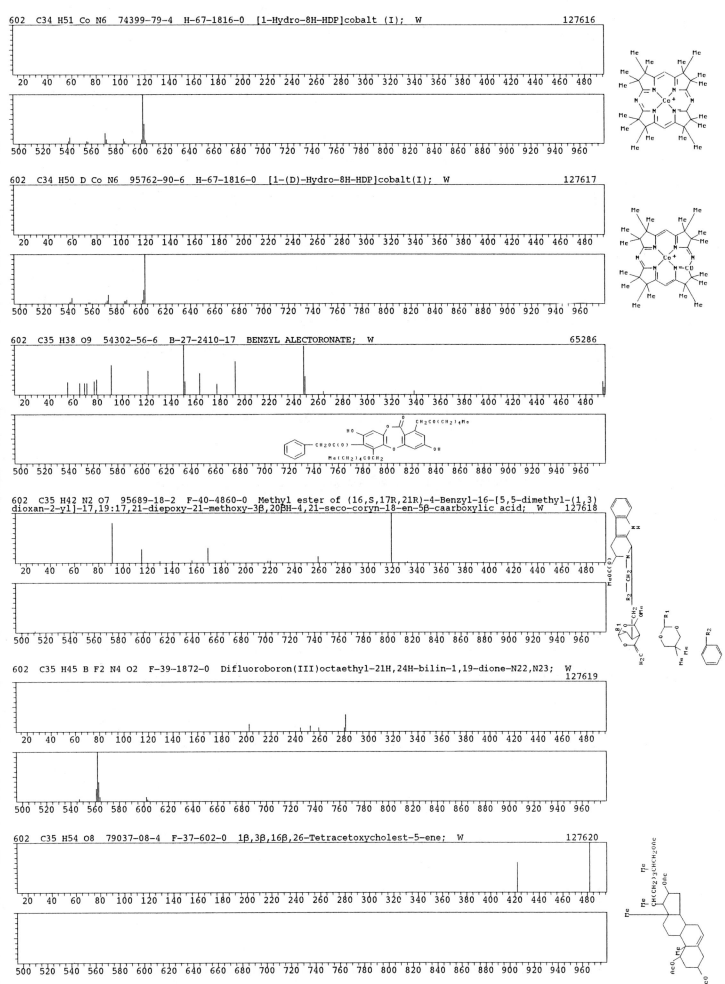

602   C34 H51 Co N6   74399-79-4   H-67-1816-0   [1-Hydro-8H-HDP]cobalt (I);   W   127616

602   C34 H50 D Co N6   95762-90-6   H-67-1816-0   [1-(D)-Hydro-8H-HDP]cobalt(I);   W   127617

602   C35 H38 O9   54302-56-6   B-27-2410-17   BENZYL ALECTORONATE;   W   65286

602   C35 H42 N2 O7   95689-18-2   F-40-4860-0   Methyl ester of (16,S,17R,21R)-4-Benzyl-16-[5,5-dimethyl-(1,3)
dioxan-2-yl]-17,19:17,21-diepoxy-21-methoxy-3β,20βH-4,21-seco-coryn-18-en-5β-caarboxylic acid;   W   127618

602   C35 H45 B F2 N4 O2   F-39-1872-0   Difluoroboron(III)octaethyl-21H,24H-bilin-1,19-dione-N22,N23;   W
                                                                                                    127619

602   C35 H54 O8   79037-08-4   F-37-602-0   1β,3β,16β,26-Tetracetoxycholest-5-ene;   W   127620

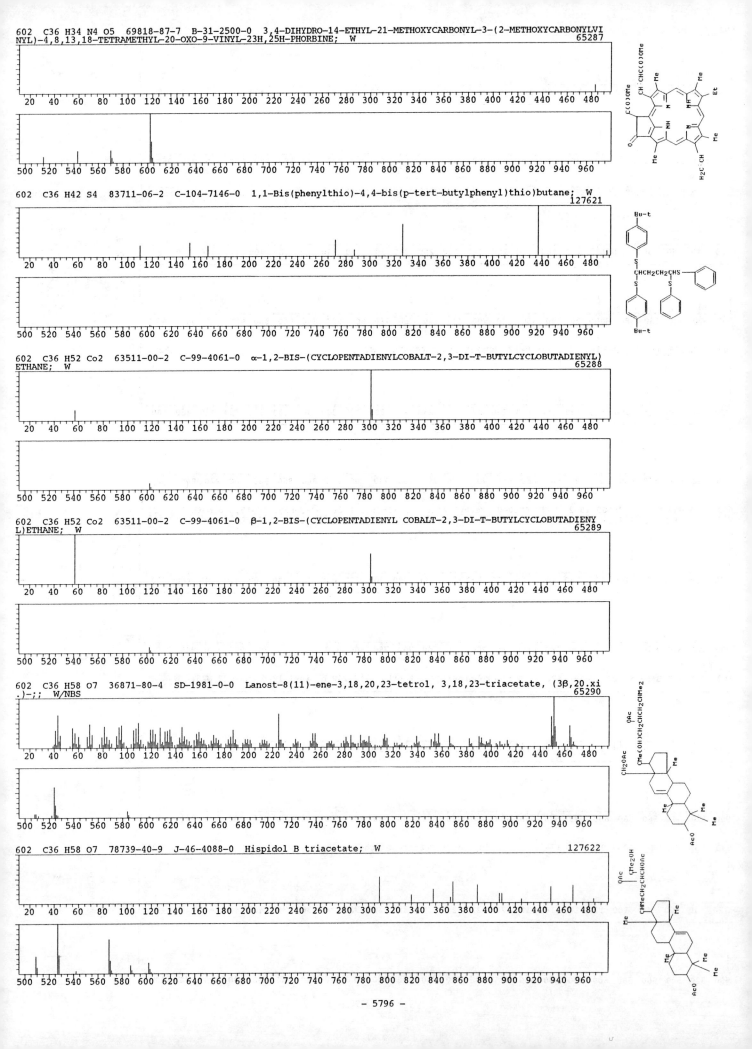

602   C36 H34 N4 O5   69818-87-7   B-31-2500-0   3,4-DIHYDRO-14-ETHYL-21-METHOXYCARBONYL-3-(2-METHOXYCARBONYLVI
NYL)-4,8,13,18-TETRAMETHYL-20-OXO-9-VINYL-23H,25H-PHORBINE;   W
65287

602   C36 H42 S4   83711-06-2   C-104-7146-0   1,1-Bis(phenylthio)-4,4-bis(p-tert-butylphenyl)thio)butane;   W
127621

602   C36 H52 Co2   63511-00-2   C-99-4061-0   α-1,2-BIS-(CYCLOPENTADIENYLCOBALT-2,3-DI-T-BUTYLCYCLOBUTADIENYL)
ETHANE;   W
65288

602   C36 H52 Co2   63511-00-2   C-99-4061-0   β-1,2-BIS-(CYCLOPENTADIENYL COBALT-2,3-DI-T-BUTYLCYCLOBUTADIENY
L)ETHANE;   W
65289

602   C36 H58 O7   36871-80-4   SD-1981-0-0   Lanost-8(11)-ene-3,18,20,23-tetrol, 3,18,23-triacetate, (3β,20.xi
.)-;;   W/NBS
65290

602   C36 H58 O7   78739-40-9   J-46-4088-0   Hispidol B triacetate;   W
127622

602　C37 H70 O2 Si2　65598-95-0　NS-0-0-0　Silane, [[(3β,17β)-androst-5-ene-3,17-diyl]bis(oxy)]bis[tripropyl-;　W/NBS　127623

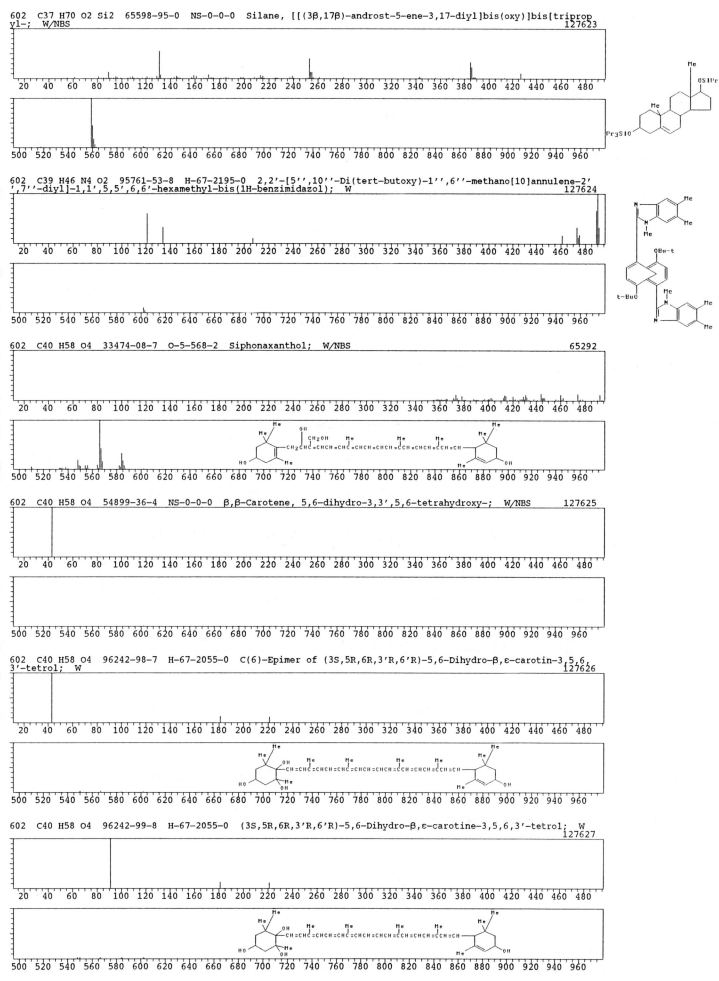

602　C39 H46 N4 O2　95761-53-8　H-67-2195-0　2,2'-[5'',10''-Di(tert-butoxy)-1'',6''-methano[10]annulene-2',7''-diyl]-1,1',5,5',6,6'-hexamethyl-bis(1H-benzimidazol);　W　127624

602　C40 H58 O4　33474-08-7　O-5-568-2　Siphonaxanthol;　W/NBS　65292

602　C40 H58 O4　54899-36-4　NS-0-0-0　β,β-Carotene, 5,6-dihydro-3,3',5,6-tetrahydroxy-;　W/NBS　127625

602　C40 H58 O4　96242-98-7　H-67-2055-0　C(6)-Epimer of (3S,5R,6R,3'R,6'R)-5,6-Dihydro-β,ε-carotin-3,5,6,3'-tetrol;　W　127626

602　C40 H58 O4　96242-99-8　H-67-2055-0　(3S,5R,6R,3'R,6'R)-5,6-Dihydro-β,ε-carotine-3,5,6,3'-tetrol;　W　127627

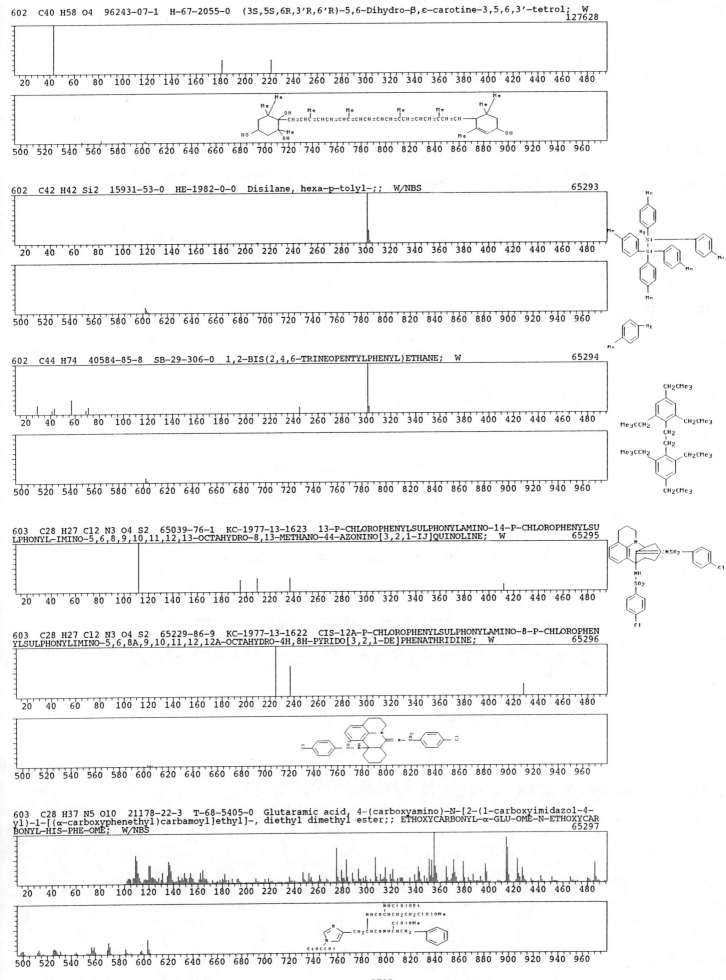

602 C40 H58 O4 96243-07-1 H-67-2055-0 (3S,5S,6R,3'R,6'R)-5,6-Dihydro-β,ε-carotine-3,5,6,3'-tetrol; W
127628

602 C42 H42 Si2 15931-53-0 HE-1982-0-0 Disilane, hexa-p-tolyl-;; W/NBS
65293

602 C44 H74 40584-85-8 SB-29-306-0 1,2-BIS(2,4,6-TRINEOPENTYLPHENYL)ETHANE; W
65294

603 C28 H27 Cl2 N3 O4 S2 65039-76-1 KC-1977-13-1623 13-P-CHLOROPHENYLSULPHONYLAMINO-14-P-CHLOROPHENYLSU
LPHONYL-IMINO-5,6,8,9,10,11,12,13-OCTAHYDRO-8,13-METHANO-44-AZONINO[3,2,1-IJ]QUINOLINE; W
65295

603 C28 H27 Cl2 N3 O4 S2 65229-86-9 KC-1977-13-1622 CIS-12A-P-CHLOROPHENYLSULPHONYLAMINO-8-P-CHLOROPHEN
YLSULPHONYLIMINO-5,6,8A,9,10,11,12,12A-OCTAHYDRO-4H,8H-PYRIDO[3,2,1-DE]PHENATHRIDINE; W
65296

603 C28 H37 N5 O10 21178-22-3 T-68-5405-0 Glutaramic acid, 4-(carboxyamino)-N-[2-(1-carboxyimidazol-4-
yl)-1-[(α-carboxyphenethyl)carbamoyl]ethyl]-, diethyl dimethyl ester;; ETHOXYCARBONYL-α-GLU-OME-N-ETHOXYCAR
BONYL-HIS-PHE-OME; W/NBS
65297

603 C31 H38 Cl N O9   69915-49-7   F-34-2118-0   14β-CHLORO-7,8-DIMETHOXY-15α-(3',4',5'-TRIMETHOXYBENZOYLOXY)
-EPIALLOBERBAN-13α-CARBOXYLIC ACID METHYLESTER; W   65298

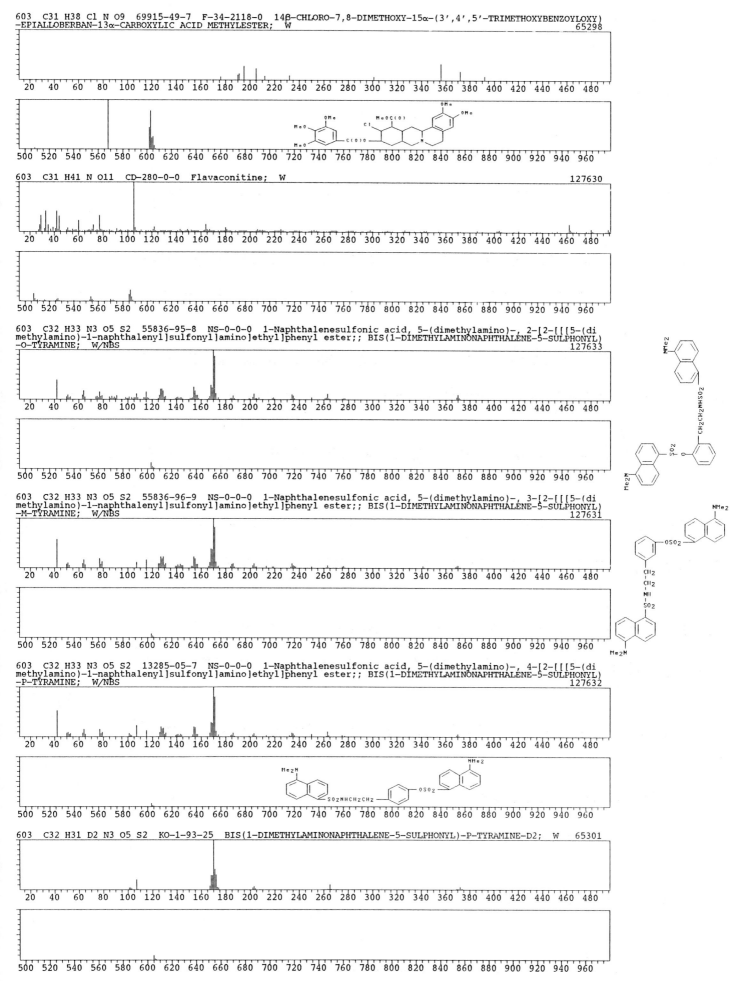

603   C31 H41 N O11   CD-280-0-0   Flavaconitine; W   127630

603   C32 H33 N3 O5 S2   55836-95-8   NS-0-0-0   1-Naphthalenesulfonic acid, 5-(dimethylamino)-, 2-[2-[[[5-(di
methylamino)-1-naphthalenyl]sulfonyl]amino]ethyl]phenyl ester;; BIS(1-DIMETHYLAMINONAPHTHALENE-5-SULPHONYL)
-O-TYRAMINE; W/NBS   127633

603   C32 H33 N3 O5 S2   55836-96-9   NS-0-0-0   1-Naphthalenesulfonic acid, 5-(dimethylamino)-, 3-[2-[[[5-(di
methylamino)-1-naphthalenyl]sulfonyl]amino]ethyl]phenyl ester;; BIS(1-DIMETHYLAMINONAPHTHALENE-5-SULPHONYL)
-M-TYRAMINE; W/NBS   127631

603   C32 H33 N3 O5 S2   13285-05-7   NS-0-0-0   1-Naphthalenesulfonic acid, 5-(dimethylamino)-, 4-[2-[[[5-(di
methylamino)-1-naphthalenyl]sulfonyl]amino]ethyl]phenyl ester;; BIS(1-DIMETHYLAMINONAPHTHALENE-5-SULPHONYL)
-P-TYRAMINE; W/NBS   127632

603   C32 H31 D2 N3 O5 S2   KO-1-93-25   BIS(1-DIMETHYLAMINONAPHTHALENE-5-SULPHONYL)-P-TYRAMINE-D2; W   65301

603 C32 H61 N O9 76871-55-1 I-59-1731-0 2R,3R-(+)-N-Octadecyl-2-carboxamido-3-carboxyl-1,4,7,10,13,16-hexaoxacyclooctadecane; W   127634

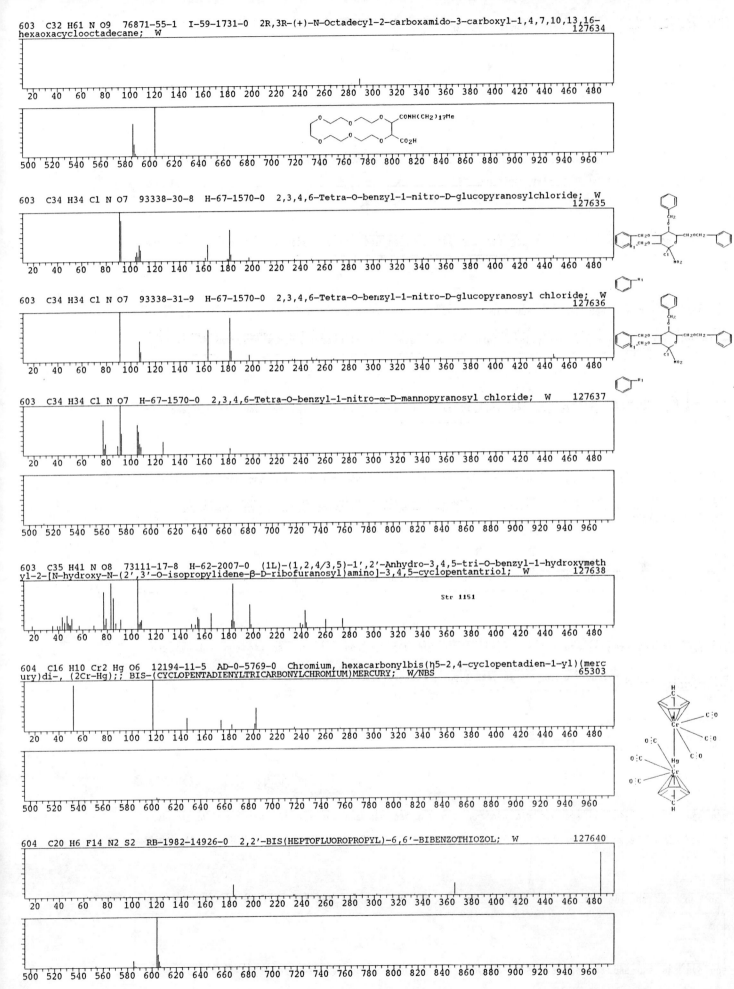

603 C34 H34 Cl N O7 93338-30-8 H-67-1570-0 2,3,4,6-Tetra-O-benzyl-1-nitro-D-glucopyranosylchloride; W   127635

603 C34 H34 Cl N O7 93338-31-9 H-67-1570-0 2,3,4,6-Tetra-O-benzyl-1-nitro-D-glucopyranosyl chloride; W   127636

603 C34 H34 Cl N O7 H-67-1570-0 2,3,4,6-Tetra-O-benzyl-1-nitro-α-D-mannopyranosyl chloride; W 127637

603 C35 H41 N O8 73111-17-8 H-62-2007-0 (1L)-(1,2,4/3,5)-1',2'-Anhydro-3,4,5-tri-O-benzyl-1-hydroxymethyl-2-[N-hydroxy-N-(2',3'-O-isopropylidene-β-D-ribofuranosyl)amino]-3,4,5-cyclopentantriol; W   127638

Str 1151

604 C16 H10 Cr2 Hg O6 12194-11-5 AD-0-5769-0 Chromium, hexacarbonylbis(η5-2,4-cyclopentadien-1-yl)(mercury)di-, (2Cr-Hg);; BIS-(CYCLOPENTADIENYLTRICARBONYLCHROMIUM)MERCURY; W/NBS   65303

604 C20 H6 F14 N2 S2 RB-1982-14926-0 2,2'-BIS(HEPTOFLUOROPROPYL)-6,6'-BIBENZOTHIOZOL; W 127640

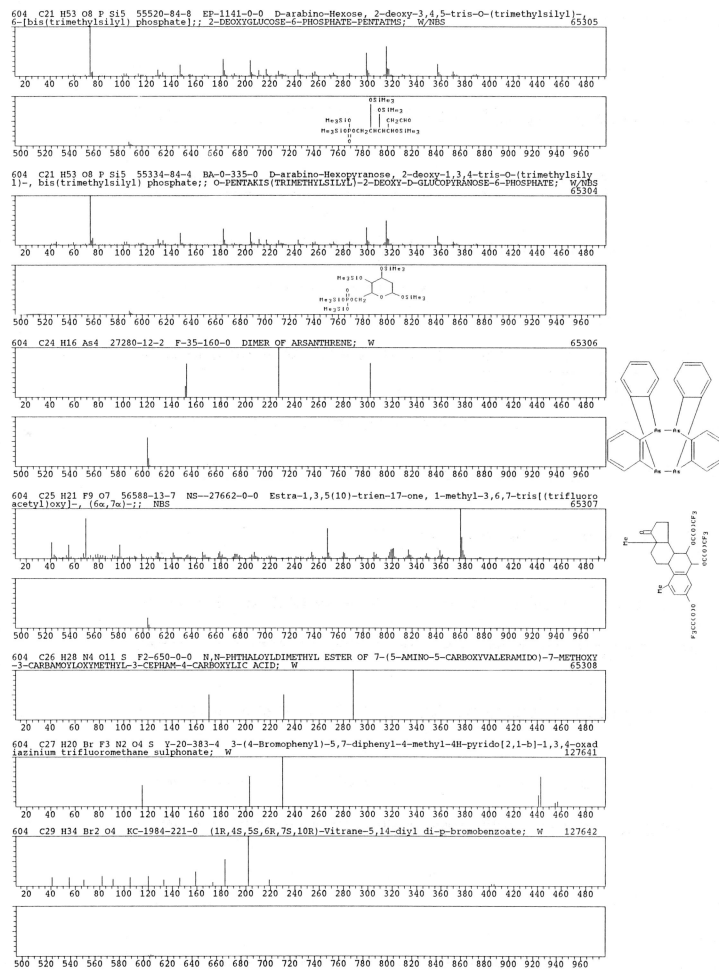

604   C21 H53 O8 P Si5   55520-84-8   EP-1141-0-0   D-arabino-Hexose, 2-deoxy-3,4,5-tris-O-(trimethylsilyl)-,
6-[bis(trimethylsilyl) phosphate];; 2-DEOXYGLUCOSE-6-PHOSPHATE-PENTATMS; W/NBS
65305

604   C21 H53 O8 P Si5   55334-84-4   BA-0-335-0   D-arabino-Hexopyranose, 2-deoxy-1,3,4-tris-O-(trimethylsilyl)-, bis(trimethylsilyl) phosphate;; O-PENTAKIS(TRIMETHYLSILYL)-2-DEOXY-D-GLUCOPYRANOSE-6-PHOSPHATE; W/NBS
65304

604   C24 H16 As4   27280-12-2   F-35-160-0   DIMER OF ARSANTHRENE; W
65306

604   C25 H21 F9 O7   56588-13-7   NS--27662-0-0   Estra-1,3,5(10)-trien-17-one, 1-methyl-3,6,7-tris[(trifluoro
acetyl)oxy]-, (6α,7α)-;; NBS
65307

604   C26 H28 N4 O11 S   F2-650-0-0   N,N-PHTHALOYLDIMETHYL ESTER OF 7-(5-AMINO-5-CARBOXYVALERAMIDO)-7-METHOXY
-3-CARBAMOYLOXYMETHYL-3-CEPHAM-4-CARBOXYLIC ACID; W
65308

604   C27 H20 Br F3 N2 O4 S   Y-20-383-4   3-(4-Bromophenyl)-5,7-diphenyl-4-methyl-4H-pyrido[2,1-b]-1,3,4-oxad
iazinium trifluoromethane sulphonate; W
127641

604   C29 H34 Br2 O4   KC-1984-221-0   (1R,4S,5S,6R,7S,10R)-Vitrane-5,14-diyl di-p-bromobenzoate; W   127642

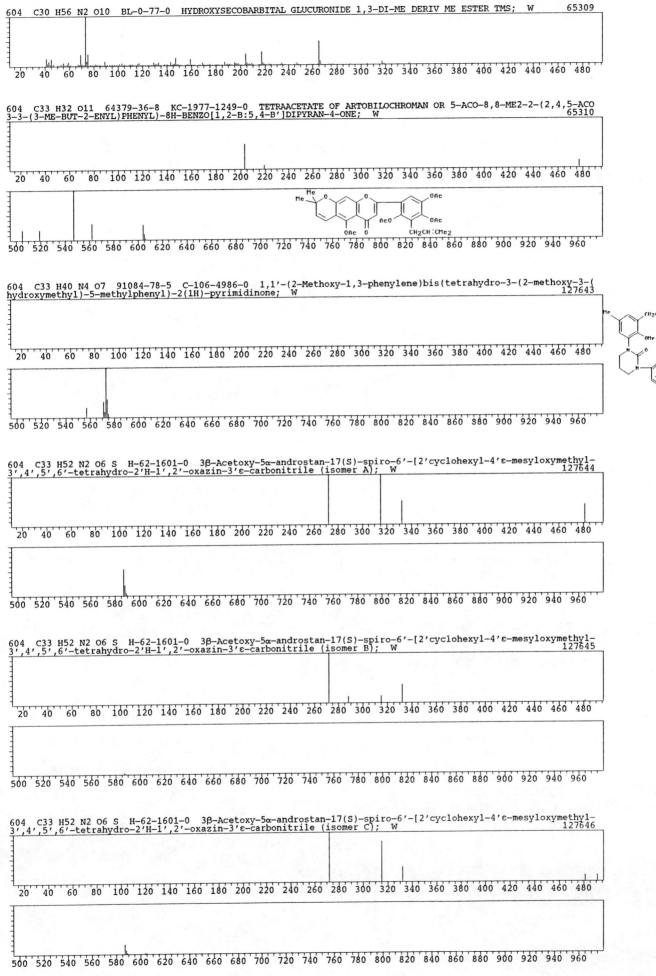

604   C30 H56 N2 O10   BL-0-77-0   HYDROXYSECOBARBITAL GLUCURONIDE 1,3-DI-ME DERIV ME ESTER TMS;   W   65309

604   C33 H32 O11   64379-36-8   KC-1977-1249-0   TETRAACETATE OF ARTOBILOCHROMAN OR 5-ACO-8,8-ME2-2-(2,4,5-ACO
3-3-(3-ME-BUT-2-ENYL)PHENYL)-8H-BENZO[1,2-B:5,4-B']DIPYRAN-4-ONE;   W   65310

604   C33 H40 N4 O7   91084-78-5   C-106-4986-0   1,1'-(2-Methoxy-1,3-phenylene)bis(tetrahydro-3-(2-methoxy-3-(
hydroxymethyl)-5-methylphenyl)-2(1H)-pyrimidinone;   W   127643

604   C33 H52 N2 O6 S   H-62-1601-0   3β-Acetoxy-5α-androstan-17(S)-spiro-6'-[2'cyclohexyl-4'ε-mesyloxymethyl-
3',4',5',6'-tetrahydro-2'H-1',2'-oxazin-3'ε-carbonitrile (isomer A);   W   127644

604   C33 H52 N2 O6 S   H-62-1601-0   3β-Acetoxy-5α-androstan-17(S)-spiro-6'-[2'cyclohexyl-4'ε-mesyloxymethyl-
3',4',5',6'-tetrahydro-2'H-1',2'-oxazin-3'ε-carbonitrile (isomer B);   W   127645

604   C33 H52 N2 O6 S   H-62-1601-0   3β-Acetoxy-5α-androstan-17(S)-spiro-6'-[2'cyclohexyl-4'ε-mesyloxymethyl-
3',4',5',6'-tetrahydro-2'H-1',2'-oxazin-3'ε-carbonitrile (isomer C);   W   127646

604   C33 H52 N2 O6 S   H-62-1601-0   3β-Acetoxy-5α-androstan-17(S)-spiro-6'-[2'cyclohexyl-4'ε-mesyloxymethyl-3',4',5',6'-tetrahydro-2'H-1',2'-oxazin-3'ε-carbonitrile (isomer D);   W   127647

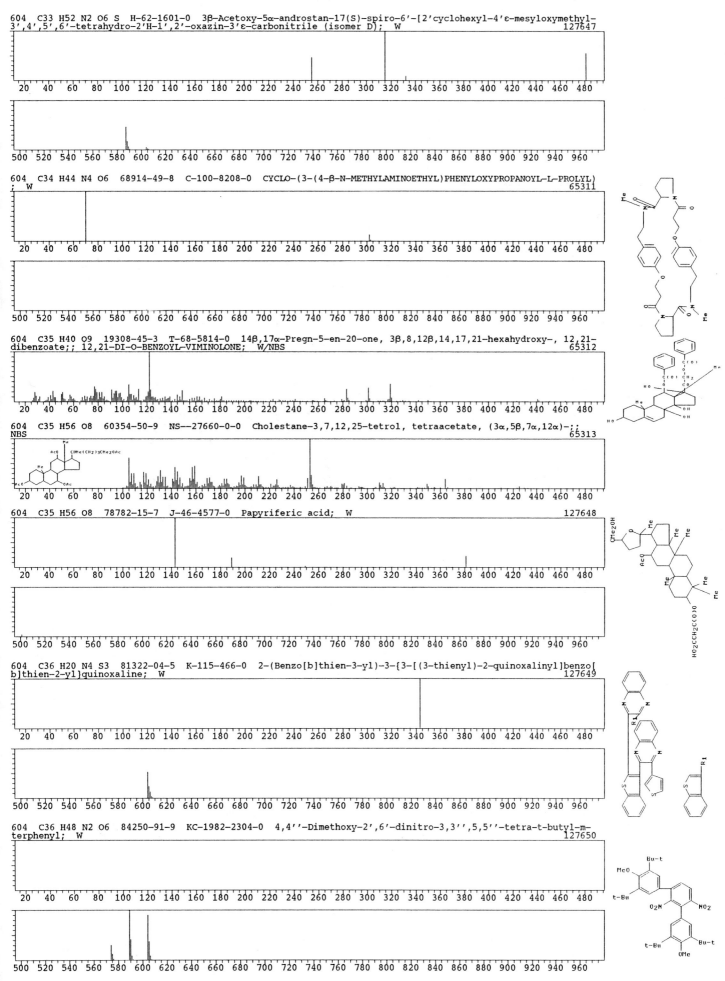

604   C34 H44 N4 O6   68914-49-8   C-100-8208-0   CYCLO-(3-(4-β-N-METHYLAMINOETHYL)PHENYLOXYPROPANOYL-L-PROLYL); W   65311

604   C35 H40 O9   19308-45-3   T-68-5814-0   14β,17α-Pregn-5-en-20-one, 3β,8,12β,14,17,21-hexahydroxy-, 12,21-dibenzoate;;   12,21-DI-O-BENZOYL-VIMINOLONE;   W/NBS   65312

604   C35 H56 O8   60354-50-9   NS--27660-0-0   Cholestane-3,7,12,25-tetrol, tetraacetate, (3α,5β,7α,12α)-;; NBS   65313

604   C35 H56 O8   78782-15-7   J-46-4577-0   Papyriferic acid;   W   127648

604   C36 H20 N4 S3   81322-04-5   K-115-466-0   2-(Benzo[b]thien-3-yl)-3-{3-[(3-thienyl)-2-quinoxalinyl]benzo[b]thien-2-yl}quinoxaline;   W   127649

604   C36 H48 N2 O6   84250-91-9   KC-1982-2304-0   4,4''-Dimethoxy-2',6'-dinitro-3,3'',5,5''-tetra-t-butyl-m-terphenyl;   W   127650

- 5803 -

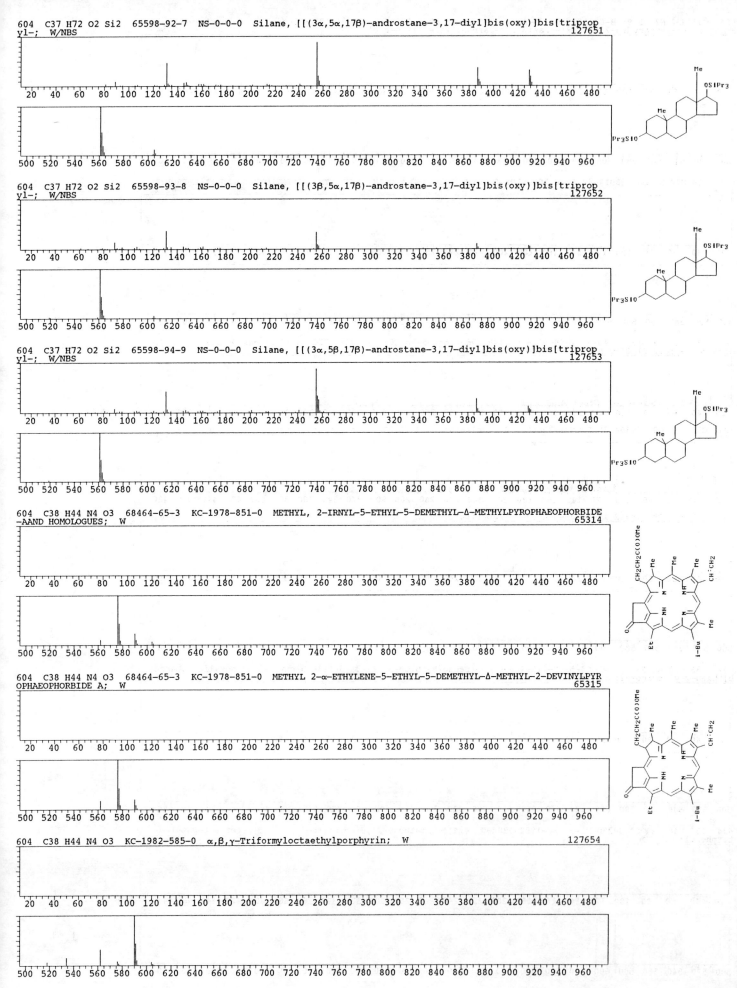

604   C37 H72 O2 Si2   65598-92-7   NS-0-0-0   Silane, [[(3α,5α,17β)-androstane-3,17-diyl]bis(oxy)]bis[tripropyl-;   W/NBS   127651

604   C37 H72 O2 Si2   65598-93-8   NS-0-0-0   Silane, [[(3β,5α,17β)-androstane-3,17-diyl]bis(oxy)]bis[tripropyl-;   W/NBS   127652

604   C37 H72 O2 Si2   65598-94-9   NS-0-0-0   Silane, [[(3α,5β,17β)-androstane-3,17-diyl]bis(oxy)]bis[tripropyl-;   W/NBS   127653

604   C38 H44 N4 O3   68464-65-3   KC-1978-851-0   METHYL, 2-IRNYL-5-ETHYL-5-DEMETHYL-Δ-METHYLPYROPHAEOPHORBIDE
-AAND HOMOLOGUES;   W   65314

604   C38 H44 N4 O3   68464-65-3   KC-1978-851-0   METHYL 2-α-ETHYLENE-5-ETHYL-5-DEMETHYL-Δ-METHYL-2-DEVINYLPYR
OPHAEOPHORBIDE A;   W   65315

604   C38 H44 N4 O3   KC-1982-585-0   α,β,γ-Triformyloctaethylporphyrin;   W   127654

604 C40 H32 N2 O4 IC-1285-0-0 [(6-Anilino-2,7-dimethylpropyl)phenoxy]-2,3-dihydro-7H-dibenz(f,ij)isoquin
oline; W   127655

604 C41 H32 O5 82542-14-1 F-38-135-0 6,8-Bis-diphenylmethyl-naringenin; W   127656

604 C41 H56 N4 80056-64-0 KC-1981-2629-0 meso-n-Pentyloctaethylporphyrin; W   127657

604 C42 H36 O4 90340-65-1 B-37-574-0 Diethyl 2,2'-[t-2,t-4-di(naphthalen-1-yl)-cyclobutane-r-1,c-3-dyl]
dibenzoate; W   127658

604 C42 H36 O4 90340-66-2 B-37-574-0 Diethyl 2,2'-[t-3,t-4-di(naphthalen-1-yl)cyclobutane-r-1,c-2-diyl]
dibenzoate; W   127659

604 C42 H36 O4 90340-72-0 B-37-574-0 Diethyl 2,2'-[t-2,t-4,di(naphthalen-2-yl)cyclobutane-r-1,c-3diyl]
benzoate; W   127660

604 C42 H36 O4 90340-73-1 B-37-574-0 Diethyl 2,2'[t-3,t-4-di(naphthalen-2-yl)cyclobutane-r-1,c-2-diyl]
dibenzoate; W   127661

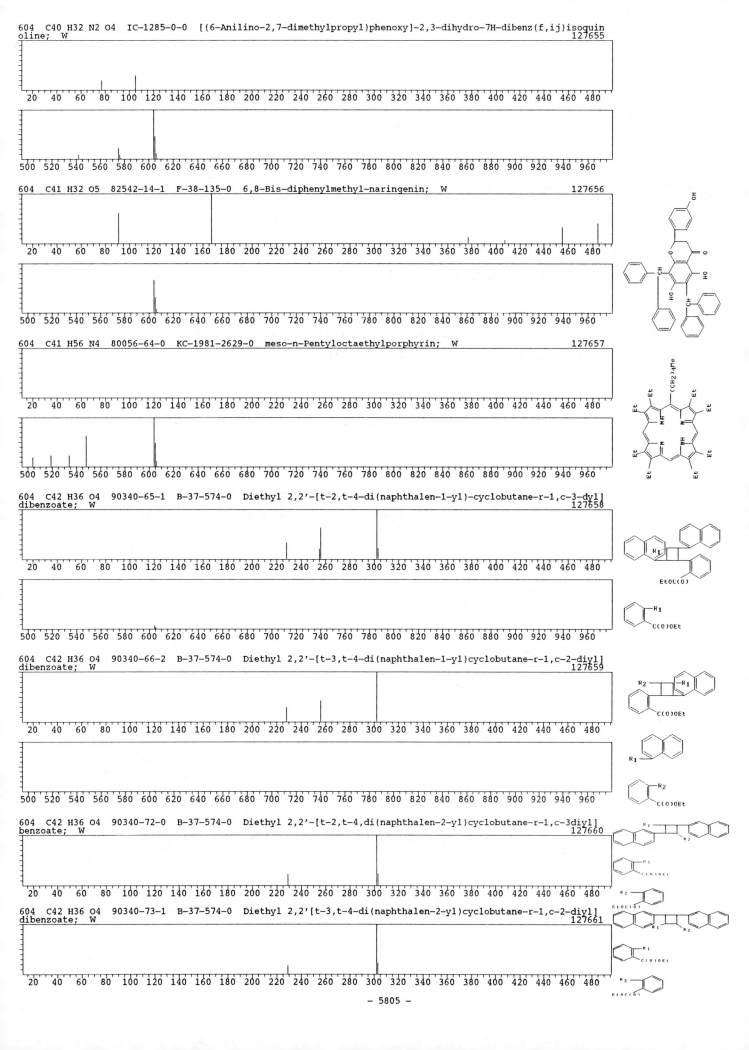

604   C43 H88   55162-61-3   SD-1981-0-0   Tetracontane, 3,5,24-trimethyl-;;   W/NBS   65318

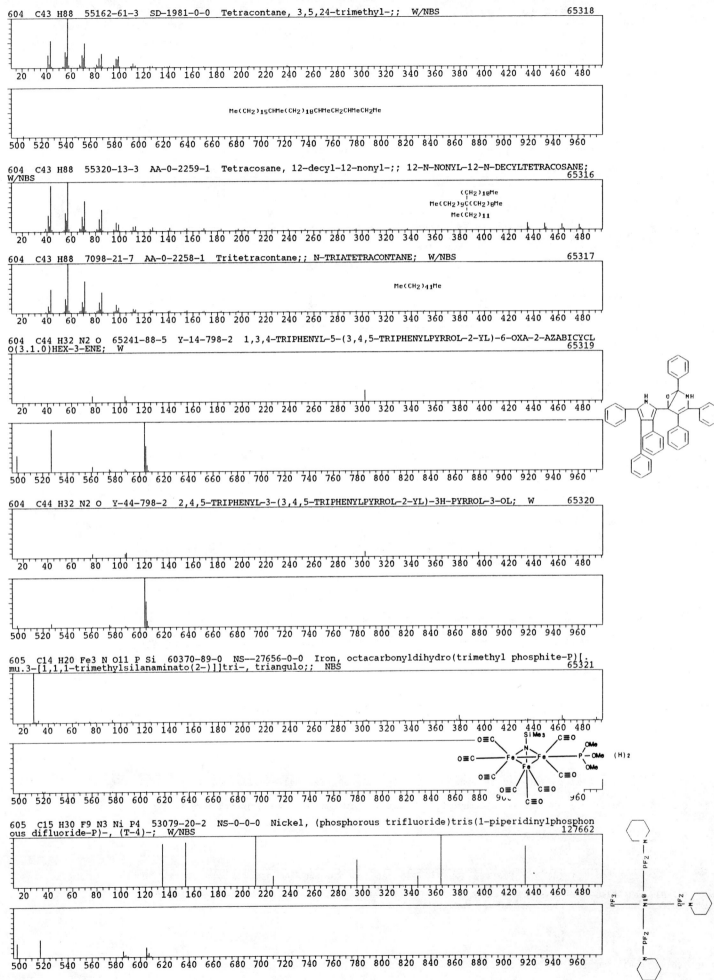

Me(CH2)15CHMe(CH2)18CHMeCH2CHMeCH2Me

604   C43 H88   55320-13-3   AA-0-2259-1   Tetracosane, 12-decyl-12-nonyl-;;   12-N-NONYL-12-N-DECYLTETRACOSANE;
W/NBS   65316

                                                                        (CH2)18Me
                                                                Me(CH2)9C(CH2)8Me
                                                                    Me(CH2)11

604   C43 H88   7098-21-7   AA-0-2258-1   Tritetracontane;;   N-TRIATETRACONTANE;   W/NBS   65317

                                                        Me(CH2)41Me

604   C44 H32 N2 O   65241-88-5   Y-14-798-2   1,3,4-TRIPHENYL-5-(3,4,5-TRIPHENYLPYRROL-2-YL)-6-OXA-2-AZABICYCL
O(3.1.0)HEX-3-ENE;   W   65319

604   C44 H32 N2 O   Y-44-798-2   2,4,5-TRIPHENYL-3-(3,4,5-TRIPHENYLPYRROL-2-YL)-3H-PYRROL-3-OL;   W   65320

605   C14 H20 Fe3 N O11 P Si   60370-89-0   NS--27656-0-0   Iron, octacarbonyldihydro(trimethyl phosphite-P)[.
mu.3-[1,1,1-trimethylsilanaminato(2-)]]tri-, triangulo;;   NBS   65321

605   C15 H30 F9 N3 Ni P4   53079-20-2   NS-0-0-0   Nickel, (phosphorous trifluoride)tris(1-piperidinylphosphon
ous difluoride-P)-, (T-4)-;   W/NBS   127662

605　C18 H10 F15 N O5　74231-47-3　NS--27654-0-0　Propanoic acid, pentafluoro-, 4-[2-[(2,2,3,3,3-pentafluor
o-1-oxopropyl)amino]propyl]-1,2-phenylene ester;;　NBS　65323

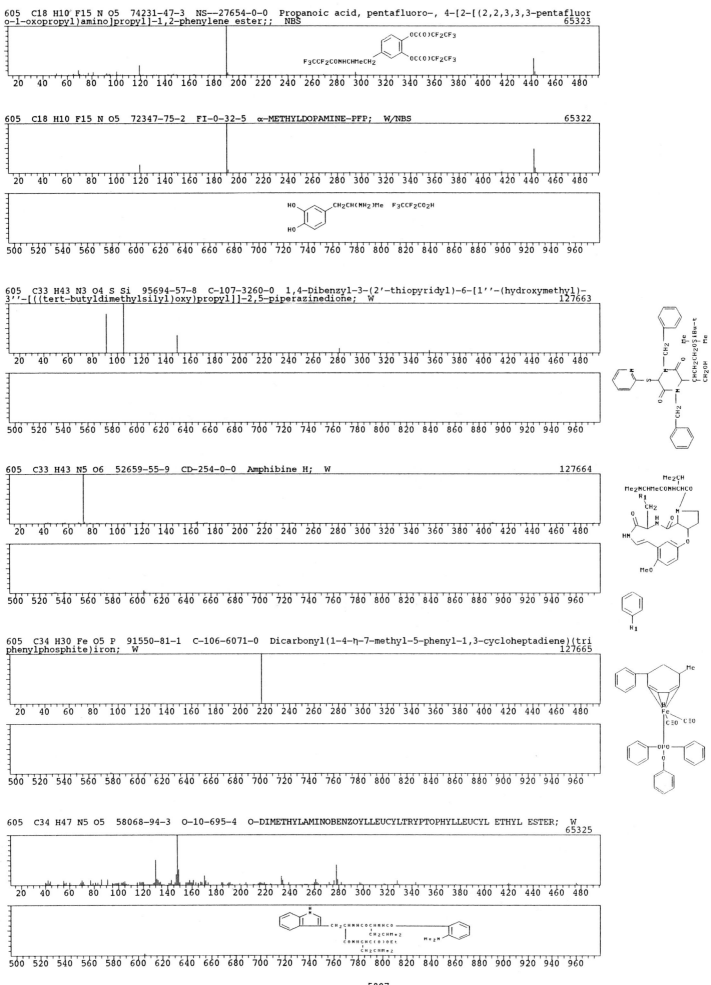

OC(O)CF₂CF₃

F₃CCF₂CONHCHMeCH₂ ... OC(O)CF₂CF₃

20　40　60　80　100　120　140　160　180　200　220　240　260　280　300　320　340　360　380　400　420　440　460　480

605　C18 H10 F15 N O5　72347-75-2　FI-0-32-5　α-METHYLDOPAMINE-PFP;　W/NBS　65322

20　40　60　80　100　120　140　160　180　200　220　240　260　280　300　320　340　360　380　400　420　440　460　480

HO ... CH₂CH(NH₂)Me　F₃CCF₂CO₂H
HO

500　520　540　560　580　600　620　640　660　680　700　720　740　760　780　800　820　840　860　880　900　920　940　960

605　C33 H43 N3 O4 S Si　95694-57-8　C-107-3260-0　1,4-Dibenzyl-3-(2'-thiopyridyl)-6-[1''-(hydroxymethyl)-
3''-[(((tert-butyldimethylsilyl)oxy)propyl]]-2,5-piperazinedione;　W　127663

20　40　60　80　100　120　140　160　180　200　220　240　260　280　300　320　340　360　380　400　420　440　460　480

500　520　540　560　580　600　620　640　660　680　700　720　740　760　780　800　820　840　860　880　900　920　940　960

605　C33 H43 N5 O6　52659-55-9　CD-254-0-0　Amphibine H;　W　127664

20　40　60　80　100　120　140　160　180　200　220　240　260　280　300　320　340　360　380　400　420　440　460　480

500　520　540　560　580　600　620　640　660　680　700　720　740　760　780　800　820　840　860　880　900　920　940　960

605　C34 H30 Fe O5 P　91550-81-1　C-106-6071-0　Dicarbonyl(1-4-η-7-methyl-5-phenyl-1,3-cycloheptadiene)(tri
phenylphosphite)iron;　W　127665

20　40　60　80　100　120　140　160　180　200　220　240　260　280　300　320　340　360　380　400　420　440　460　480

500　520　540　560　580　600　620　640　660　680　700　720　740　760　780　800　820　840　860　880　900　920　940　960

605　C34 H47 N5 O5　58068-94-3　O-10-695-4　O-DIMETHYLAMINOBENZOYLLEUCYLTRYPTOPHYLLEUCYL ETHYL ESTER;　W　65325

20　40　60　80　100　120　140　160　180　200　220　240　260　280　300　320　340　360　380　400　420　440　460　480

500　520　540　560　580　600　620　640　660　680　700　720　740　760　780　800　820　840　860　880　900　920　940　960

- 5807 -

605   C34 H47 N5 O5   58068-95-4   O-10-695-4   M-DIMETHYLAMINOBENZOYLLEUCYLTRYPTOPHYLLEUCYL ETHYL ESTER;   W
                                                                                                    65326

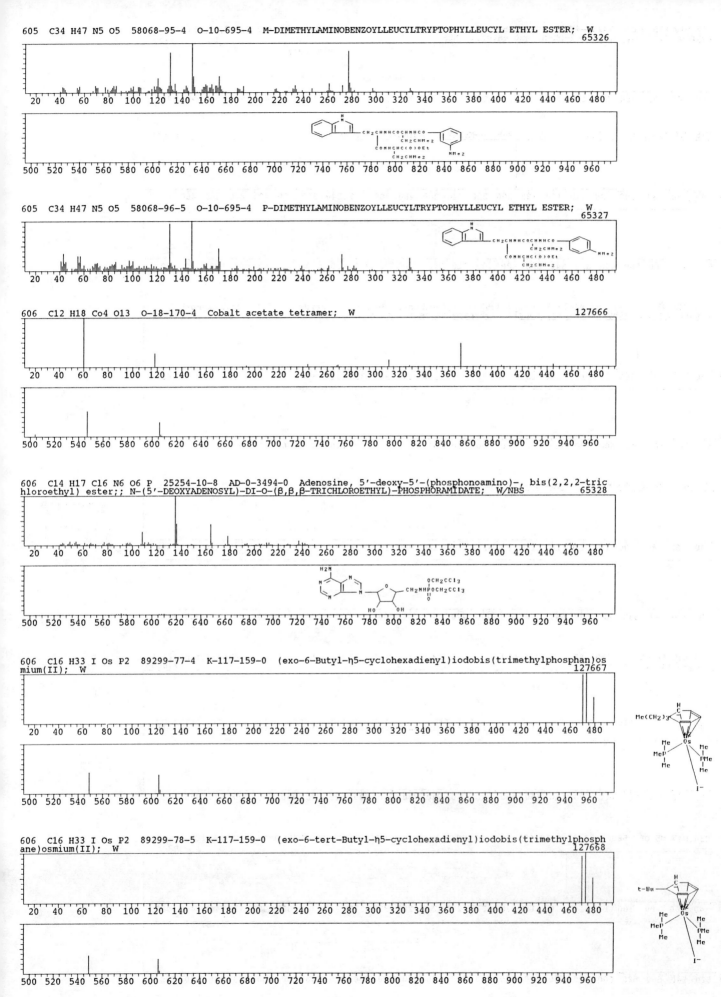

605   C34 H47 N5 O5   58068-96-5   O-10-695-4   P-DIMETHYLAMINOBENZOYLLEUCYLTRYPTOPHYLLEUCYL ETHYL ESTER;   W
                                                                                                    65327

606   C12 H18 Co4 O13   O-18-170-4   Cobalt acetate tetramer;   W   127666

606   C14 H17 Cl6 N6 O6 P   25254-10-8   AD-0-3494-0   Adenosine, 5'-deoxy-5'-(phosphonoamino)-, bis(2,2,2-tric
hloroethyl) ester;; N-(5'-DEOXYADENOSYL)-DI-O-(β,β,β-TRICHLOROETHYL)-PHOSPHORAMIDATE;   W/NBS   65328

606   C16 H33 I Os P2   89299-77-4   K-117-159-0   (exo-6-Butyl-η5-cyclohexadienyl)iodobis(trimethylphosphan)os
mium(II);   W   127667

606   C16 H33 I Os P2   89299-78-5   K-117-159-0   (exo-6-tert-Butyl-η5-cyclohexadienyl)iodobis(trimethylphosph
ane)osmium(II);   W   127668

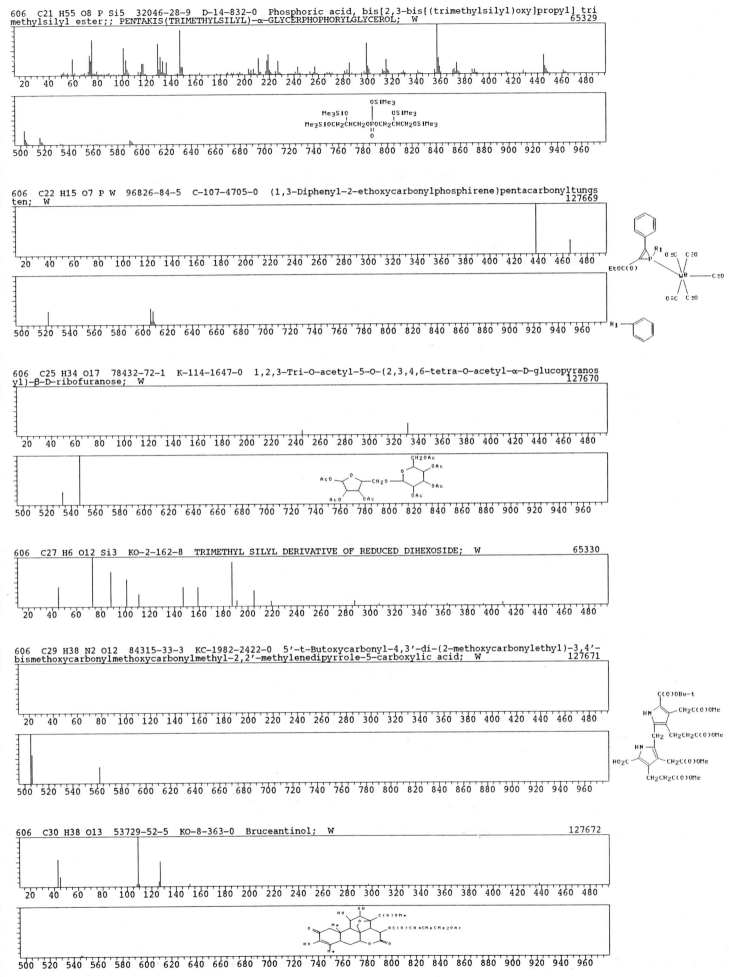

606   C21 H55 O8 P Si5   32046-28-9   D-14-832-0   Phosphoric acid, bis[2,3-bis[(trimethylsilyl)oxy]propyl] tri
methylsilyl ester;; PENTAKIS(TRIMETHYLSILYL)-α-GLYCERPHOPHORYLGLYCEROL;   W   65329

606   C22 H15 O7 P W   96826-84-5   C-107-4705-0   (1,3-Diphenyl-2-ethoxycarbonylphosphirene)pentacarbonyltungs
ten;   W   127669

606   C25 H34 O17   78432-72-1   K-114-1647-0   1,2,3-Tri-O-acetyl-5-O-(2,3,4,6-tetra-O-acetyl-α-D-glucopyranos
yl)-β-D-ribofuranose;   W   127670

606   C27 H6 O12 Si3   KO-2-162-8   TRIMETHYL SILYL DERIVATIVE OF REDUCED DIHEXOSIDE;   W   65330

606   C29 H38 N2 O12   84315-33-3   KC-1982-2422-0   5'-t-Butoxycarbonyl-4,3'-di-(2-methoxycarbonylethyl)-3,4'-
bismethoxycarbonylmethoxycarbonylmethyl-2,2'-methylenedipyrrole-5-carboxylic acid;   W   127671

606   C30 H38 O13   53729-52-5   KO-8-363-0   Bruceantinol;   W   127672

606 C31 H31 Br N2 O6 62786-93-0 KC-1976-2498-0 DIBENZYL 3-BROMO-4'-(2-METHOXYCARBONYLETHYL)-3',4-DIMETH
YL-PYRROMETHANE-5,5'-DICARBOXYLATE; W   65331

606 C32 H34 N2 O10 85784-65-2 KC-1983-117-0 Dibenzyl (7RS,16Rs)-3,12-Dimethoxy-9,18-dioxo-6,15-dioxa-1,
10-diazatricyclo[14.2.0.0(7,10)]octadeca-2,11-diene-2,11-dicarboxylate; W   127673

606 C32 H34 N2 O10 85848-93-7 KC-1983-117-0 Dibenzyl (7RS,16SR)-3,12-Dimethoxy-9,18-dioxo-6,15-dioxa-1,
10-diazatricyclo[14.2.0.0(7,10)]octadeca-2,11-diene-2,11-dicarboxylate; W   127674

606 C32 H58 O5 Si3 40837-85-2 O-7-146-2 24-Norchola-20,22-dien-14-ol, 21,23-epoxy-3,12,23-tris[(trimeth
ylsilyl)oxy]-, (3β,5β,12β,14β)-;; 3,12,23-TRI(TRIMETHYLSILYL)DIGOXIGENIN; W/NBS   65332

606 C33 H22 Cl4 O3 79491-39-7 B-34-910-0 5,6,7,8-Tetrachloro-9a-methyl-2,3,3a-triphenyl-1a,1b,3a,9a,9b,
9c-hexahydrobenzo[b]oxireno[2',3':1,2]cyclobutan[3,4-f]benzodioxin; W   127675

606 C33 H50 O10 84507-62-0 F-38-2193-0 Paristerone-2,3,22-triacetate; W   127676

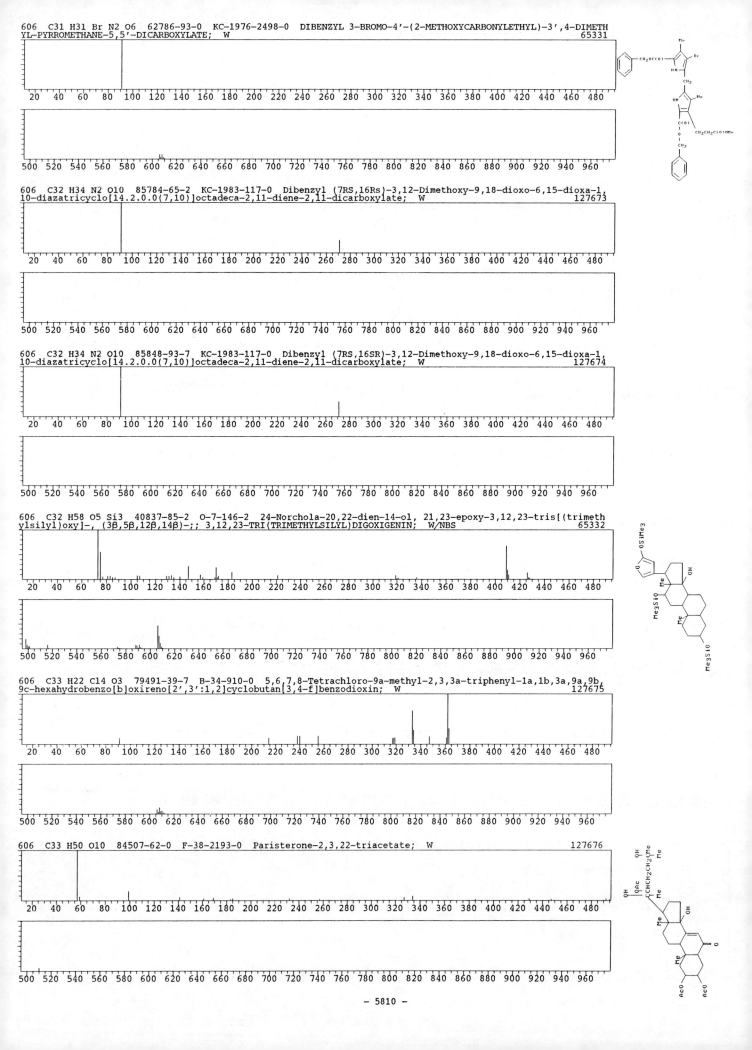

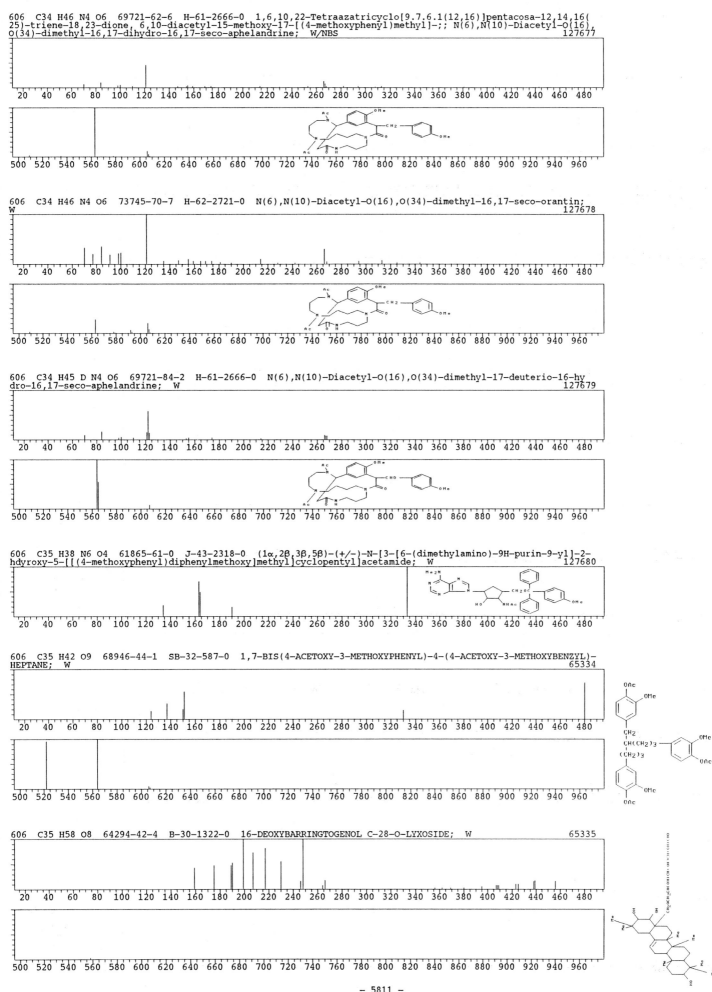

606   C34 H46 N4 O6   69721-62-6   H-61-2666-0   1,6,10,22-Tetraazatricyclo[9.7.6.1(12,16)]pentacosa-12,14,16(25)-triene-18,23-dione, 6,10-diacetyl-15-methoxy-17-[(4-methoxyphenyl)methyl]-;; N(6),N(10)-Diacetyl-O(16),O(34)-dimethyl-16,17-dihydro-16,17-seco-aphelandrine; W/NBS
127677

606   C34 H46 N4 O6   73745-70-7   H-62-2721-0   N(6),N(10)-Diacetyl-O(16),O(34)-dimethyl-16,17-seco-orantin; W
127678

606   C34 H45 D N4 O6   69721-84-2   H-61-2666-0   N(6),N(10)-Diacetyl-O(16),O(34)-dimethyl-17-deuterio-16-hydro-16,17-seco-aphelandrine; W
127679

606   C35 H38 N6 O4   61865-61-0   J-43-2318-0   (1α,2β,3β,5β)-(+/-)-N-[3-[6-(dimethylamino)-9H-purin-9-yl]-2-hdyroxy-5-[[(4-methoxyphenyl)diphenylmethoxy]methyl]cyclopentyl]acetamide; W
127680

606   C35 H42 O9   68946-44-1   SB-32-587-0   1,7-BIS(4-ACETOXY-3-METHOXYPHENYL)-4-(4-ACETOXY-3-METHOXYBENZYL)-HEPTANE; W
65334

606   C35 H58 O8   64294-42-4   B-30-1322-0   16-DEOXYBARRINGTOGENOL C-28-O-LYXOSIDE; W
65335

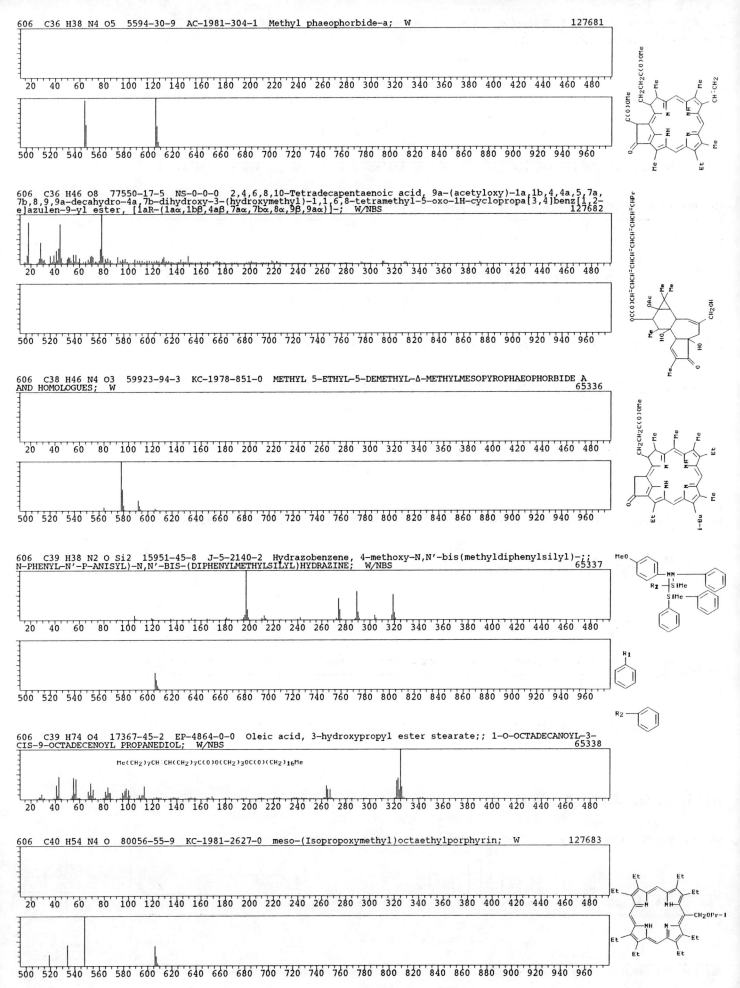

606   C36 H38 N4 O5   5594-30-9   AC-1981-304-1   Methyl phaeophorbide-a;   W   127681

606   C36 H46 O8   77550-17-5   NS-0-0-0   2,4,6,8,10-Tetradecapentaenoic acid, 9a-(acetyloxy)-1a,1b,4,4a,5,7a,
7b,8,9,9a-decahydro-4a,7b-dihydroxy-3-(hydroxymethyl)-1,1,6,8-tetramethyl-5-oxo-1H-cyclopropa[3,4]benz[1,2-
e]azulen-9-yl ester, [1aR-(1aα,1bβ,4aβ,7aα,7bα,8α,9β,9aα)]-;   W/NBS   127682

606   C38 H46 N4 O3   59923-94-3   KC-1978-851-0   METHYL 5-ETHYL-5-DEMETHYL-Δ-METHYLMESOPYROPHAEOPHORBIDE A
AND HOMOLOGUES;   W   65336

606   C39 H38 N2 O Si2   15951-45-8   J-5-2140-2   Hydrazobenzene, 4-methoxy-N,N'-bis(methyldiphenylsilyl)-;;
N-PHENYL-N'-P-ANISYL)-N,N'-BIS-(DIPHENYLMETHYLSILYL)HYDRAZINE;   W/NBS   65337

606   C39 H74 O4   17367-45-2   EP-4864-0-0   Oleic acid, 3-hydroxypropyl ester stearate;; 1-O-OCTADECANOYL-3-
CIS-9-OCTADECENOYL PROPANEDIOL;   W/NBS   65338

606   C40 H54 N4 O   80056-55-9   KC-1981-2627-0   meso-(Isopropoxymethyl)octaethylporphyrin;   W   127683

606   C41 H82 O2   40710-29-0   N-3-128-5   Hentetracontanoic acid;; N-HENTETRACONTANOIC ACID;   W/NBS   65339

HO2C(CH2)39Me

606   C41 H82 O2   18082-12-7   SD-1981-0-0   Octatriacontanoic acid, 34,36-dimethyl-, methyl ester;;   W/NBS
65340

MeOC(O)(CH2)32CHMeCH2CHMeCH2Me

606   C48 H30   HE-1982-0-0   BENZENE, 1,2,3-TRIPHENYL-3,4,6-TRI-(PHENYLETHYNYL)-;   W   65341

606   C48 H30   84907-58-4   SB-36-659-0   [2(6)]Paracyclophanetrienetriyne;   W   127684

606   C48 H30   25911-58-4   J-46-4432-0   5,6,11,12,17,18-Hexahydro-5,18[1',2']6,11[1'',2'']:12,17[1''',2''']-
tribenzenotrinaphthylene;   W   127685

607   C24 H18 F12 N3 P   70758-14-4   I-57-1015-0   TRIS(4-(N,N-DIMETHYLAMINO)-2,3,5,6-TETRAFLUOROPHENYL)PHOSPH
INE;   W   65342

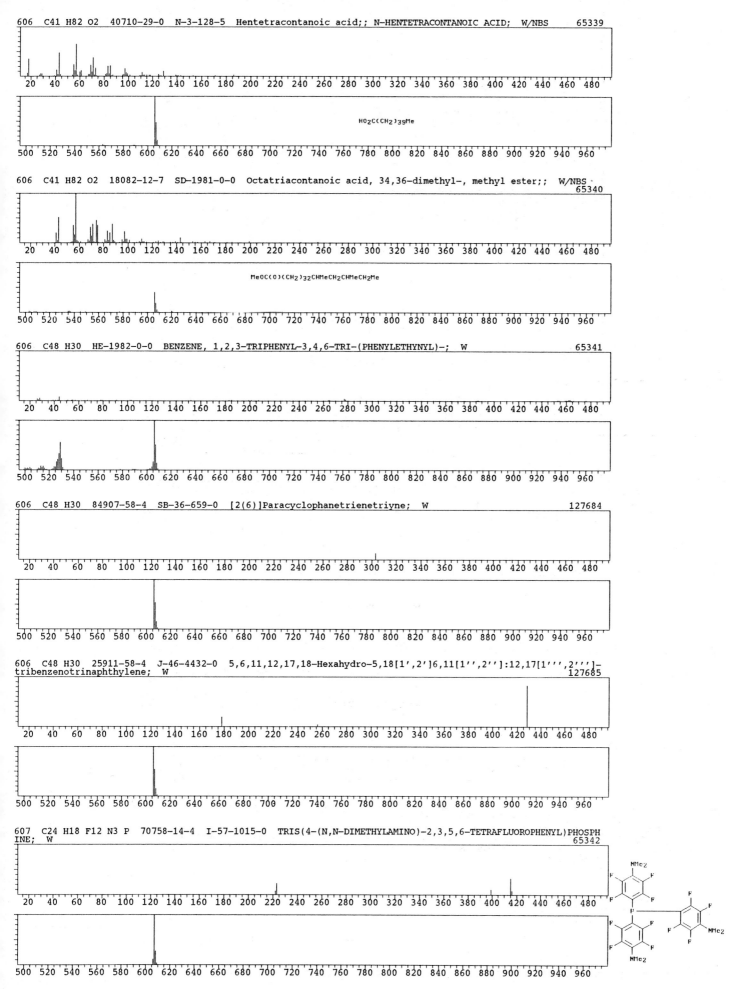

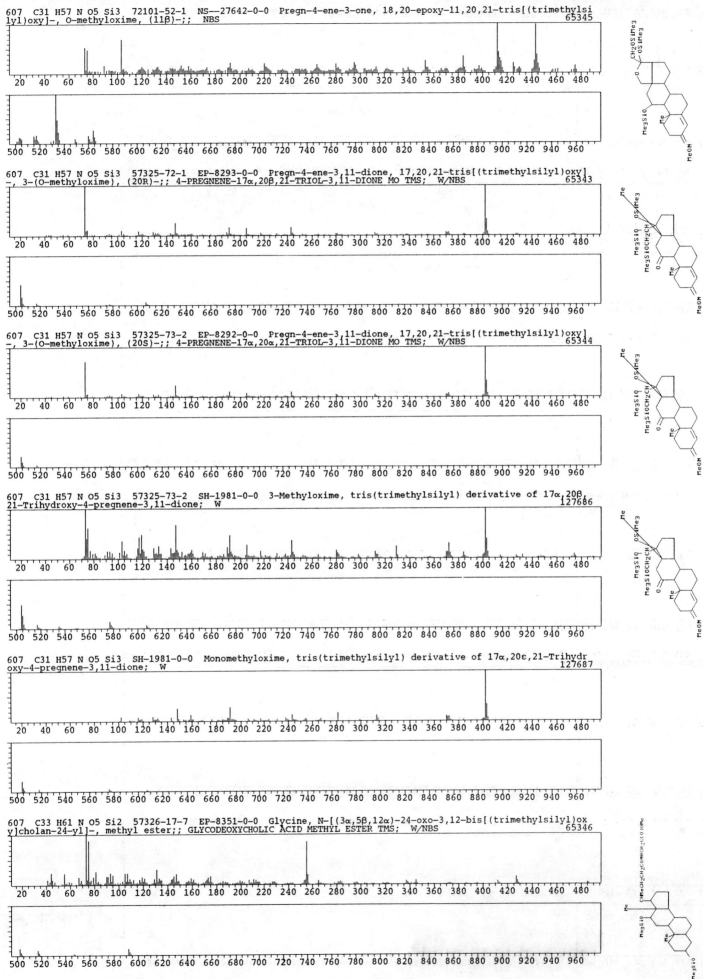

607 C31 H57 N O5 Si3 72101-52-1 NS--27642-0-0 Pregn-4-ene-3-one, 18,20-epoxy-11,20,21-tris[(trimethylsi
lyl)oxy]-, O-methyloxime, (11β)-;; NBS   65345

607 C31 H57 N O5 Si3 57325-72-1 EP-8293-0-0 Pregn-4-ene-3,11-dione, 17,20,21-tris[(trimethylsilyl)oxy]
-, 3-(O-methyloxime), (20R)-;; 4-PREGNENE-17α,20β,21-TRIOL-3,11-DIONE MO TMS; W/NBS   65343

607 C31 H57 N O5 Si3 57325-73-2 EP-8292-0-0 Pregn-4-ene-3,11-dione, 17,20,21-tris[(trimethylsilyl)oxy]
-, 3-(O-methyloxime), (20S)-;; 4-PREGNENE-17α,20α,21-TRIOL-3,11-DIONE MO TMS; W/NBS   65344

607 C31 H57 N O5 Si3 57325-73-2 SH-1981-0-0 3-Methyloxime, tris(trimethylsilyl) derivative of 17α,20β,
21-Trihydroxy-4-pregnene-3,11-dione; W   127686

607 C31 H57 N O5 Si3 SH-1981-0-0 Monomethyloxime, tris(trimethylsilyl) derivative of 17α,20ε,21-Trihydr
oxy-4-pregnene-3,11-dione; W   127687

607 C33 H61 N O5 Si2 57326-17-7 EP-8351-0-0 Glycine, N-[(3α,5β,12α)-24-oxo-3,12-bis[(trimethylsilyl)ox
y]cholan-24-yl]-, methyl ester;; GLYCODEOXYCHOLIC ACID METHYL ESTER TMS; W/NBS   65346

607 C35 H49 N3 O6 78109-10-1 J-46-3427-0 3β-Hydroxy-5α,8α-(4-phenyl-1,2-urazolo)cholesta-6-en-25-ol-26,
23-lactone; W
127688

607 C35 H49 N3 O6 78183-84-3 J-46-3427-0 3β-Hydroxy-5α,8α-(4-phenyl-1,2-urazolo)cholesta-6-en-25-ol-26,
23-lactone; W
127689

607 C35 H49 N3 O6 78183-85-4 J-46-3427-0 3β-Hydroxy-5α,8α-(4-phenyl-1,2-urazolo)cholesta-6-en-25-ol-26,
23-lactone; W
127690

607 C35 H49 N3 O6 78183-86-5 J-46-3427-0 3β-Hydroxy-5α,8α-(4-phenyl-1,2-urazolo)cholesta-6-en-25-ol-26,
23-lactone; W
127691

607 C37 H45 N5 O3 79900-13-3 KC-1981-2068-0 [2-(2-Carbonylethyl)-4,6,8-triethyl-1,3,5,7-tetramethyl-por
phyrinyl]-glycine ethyl ester; W
127692

607 C39 H77 N O3 82280-27-1 H-65-60-0 N,N-Dioctadecyl carbamoyl acetic acid; W
127693

607 C40 H57 N5 50530-25-1 K-110-655-0 3-CYCLOHEXYL-4-CYCLOHEXYLIMINO-1-(N,N′-DICYCLOHEXYLAMIDINO)-2,
5-D IPHENYLIMIDAZOLIDINE; W
65347

608 C19 H26 N4 O5 P Re K-113-646-0 Tricarbonyl(1,2-ethanediamine-N,N′)1,2-ethanediamine-N-rhenium(I)-di
phenylphosphinate; W
127694

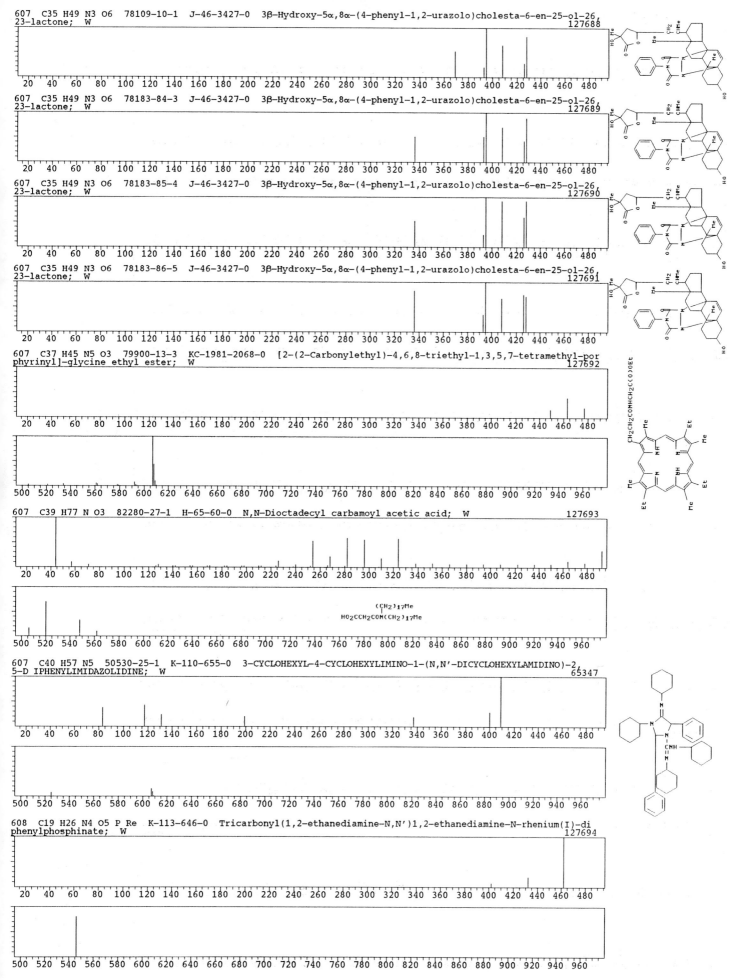

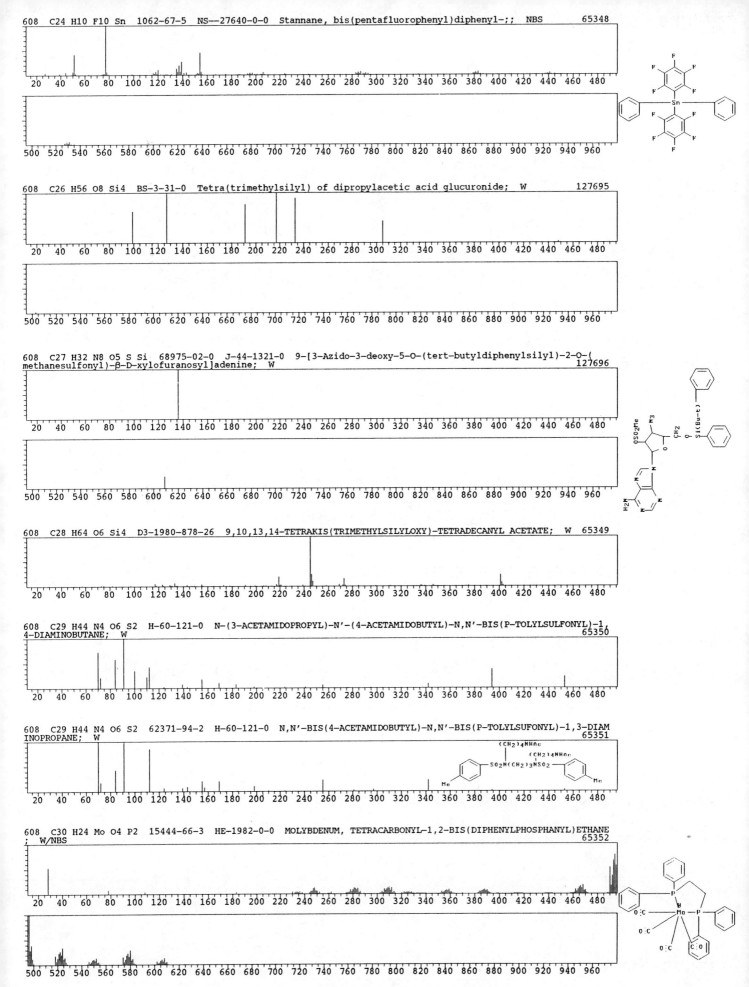

608 C24 H10 F10 Sn 1062-67-5 NS--27640-0-0 Stannane, bis(pentafluorophenyl)diphenyl-;; NBS 65348

608 C26 H56 O8 Si4 BS-3-31-0 Tetra(trimethylsilyl) of dipropylacetic acid glucuronide; W 127695

608 C27 H32 N8 O5 S Si 68975-02-0 J-44-1321-0 9-[3-Azido-3-deoxy-5-O-(tert-butyldiphenylsilyl)-2-O-(methanesulfonyl)-β-D-xylofuranosyl]adenine; W 127696

608 C28 H64 O6 Si4 D3-1980-878-26 9,10,13,14-TETRAKIS(TRIMETHYLSILYLOXY)-TETRADECANYL ACETATE; W 65349

608 C29 H44 N4 O6 S2 H-60-121-0 N-(3-ACETAMIDOPROPYL)-N'-(4-ACETAMIDOBUTYL)-N,N'-BIS(P-TOLYLSULFONYL)-1,4-DIAMINOBUTANE; W 65350

608 C29 H44 N4 O6 S2 62371-94-2 H-60-121-0 N,N'-BIS(4-ACETAMIDOBUTYL)-N,N'-BIS(P-TOLYLSUFONYL)-1,3-DIAMINOPROPANE; W 65351

608 C30 H24 Mo O4 P2 15444-66-3 HE-1982-0-0 MOLYBDENUM, TETRACARBONYL-1,2-BIS(DIPHENYLPHOSPHANYL)ETHANE; W/NBS 65352

608   C30 H28 N2 S4 Zn   14726-36-4   NS--27638-0-0   Zinc, bis[bis(phenylmethyl)carbamodithioato-S,S']-, (T-4)-;; Zinc dibenzyldithiocarbonate;   NBS   65353

608   C30 H40 O13   77573-25-2   NS-0-0-0   5H-Cyclopropa[3,4]benz[1,2-e]azulen-5-one, 2,4a,9,9a-tetrakis(acetyloxy)-3,[(acetyloxy)methyl]-1,1a,1b,2,3,4,4a,7b,8,9,9a-dodecahydro-2,7b-dihydroxy-1,1,6,8-tetramethyl-[1aR-(1aα,1bβ,2α,3β,4aβ,7aα,7bα,8α,9β,9aα)]-;   W/NBS   127697

608   C32 H32 O12   86381-53-5   H-66-32-0   Hexamethyl pentacyclo[6.6.6.0.0.0]icosa-2(7),4,9(14),11,15(20)-17-hexaene-4,5,11,12,17,18-hexacarboxylate;   W   127698

608   C32 H36 N2 O8 S   29484-58-0   SD-1981-0-0   Dichotine, 19-(benzylthio)-11-methoxy-, acetate (ester);;
W/NBS   65354

608   C33 H39 D N2 O9   16625-47-1   AD-0-2590-0   3-DEUTERIO-ISORESERPINE;   W   65355

608   C33 H40 N2 O9   50-55-5   CD-62-0-0   Methyl reserpate 3,4,5-trimethoxybenzoic acid ester;   W   134143

608   C33 H40 N2 O9   482-85-9   AD-0-2585-0   Isoreserpine;   W/NBS   65357

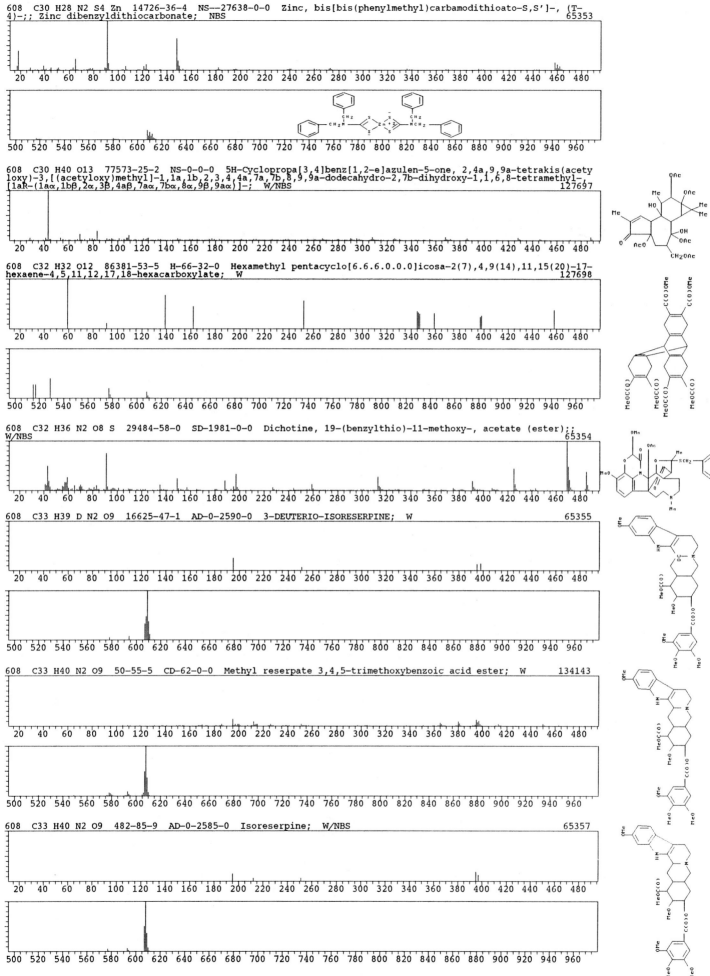

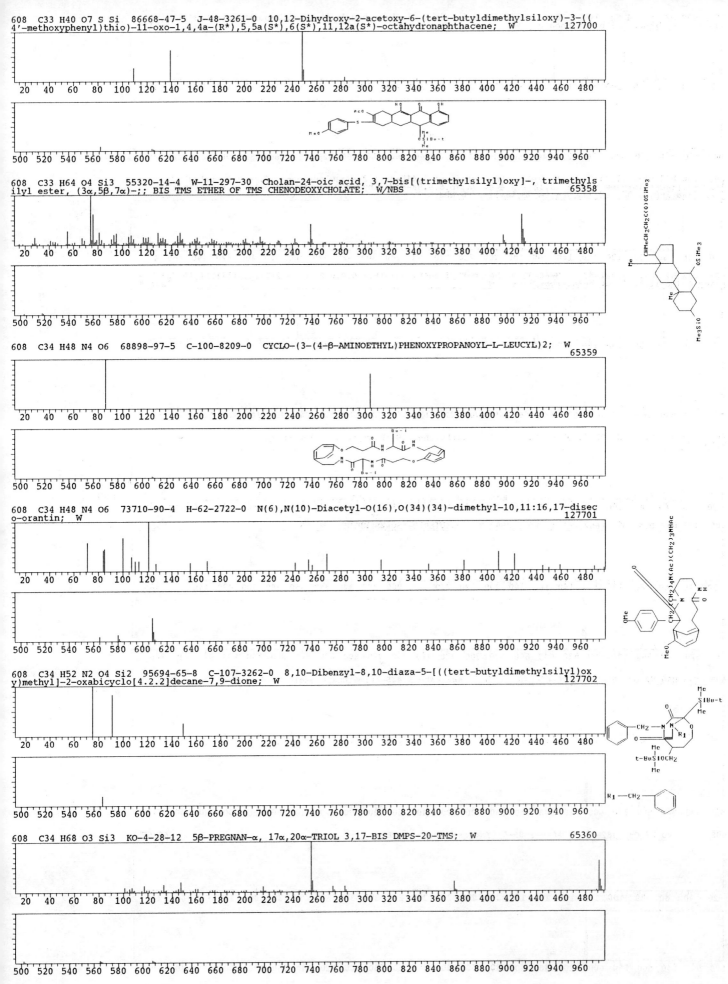

608 C33 H40 O7 S Si 86668-47-5 J-48-3261-0 10,12-Dihydroxy-2-acetoxy-6-(tert-butyldimethylsiloxy)-3-((4'-methoxyphenyl)thio)-11-oxo-1,4,4a-(R*),5,5a(S*),6(S*),11,12a(S*)-octahydronaphthacene; W 127700

608 C33 H64 O4 Si3 55320-14-4 W-11-297-30 Cholan-24-oic acid, 3,7-bis[(trimethylsilyl)oxy]-, trimethylsilyl ester, (3α,5β,7α)-;; BIS TMS ETHER OF TMS CHENODEOXYCHOLATE; W/NBS 65358

608 C34 H48 N4 O6 68898-97-5 C-100-8209-0 CYCLO-(3-(4-β-AMINOETHYL)PHENOXYPROPANOYL-L-LEUCYL)2; W 65359

608 C34 H48 N4 O6 73710-90-4 H-62-2722-0 N(6),N(10)-Diacetyl-O(16),O(34)(34)-dimethyl-10,11:16,17-diseco-orantin; W 127701

608 C34 H52 N2 O4 Si2 95694-65-8 C-107-3262-0 8,10-Dibenzyl-8,10-diaza-5-[((tert-butyldimethylsilyl)oxy)methyl]-2-oxabicyclo[4.2.2]decane-7,9-dione; W 127702

608 C34 H68 O3 Si3 KO-4-28-12 5β-PREGNAN-α, 17α,20α-TRIOL 3,17-BIS DMPS-20-TMS; W 65360

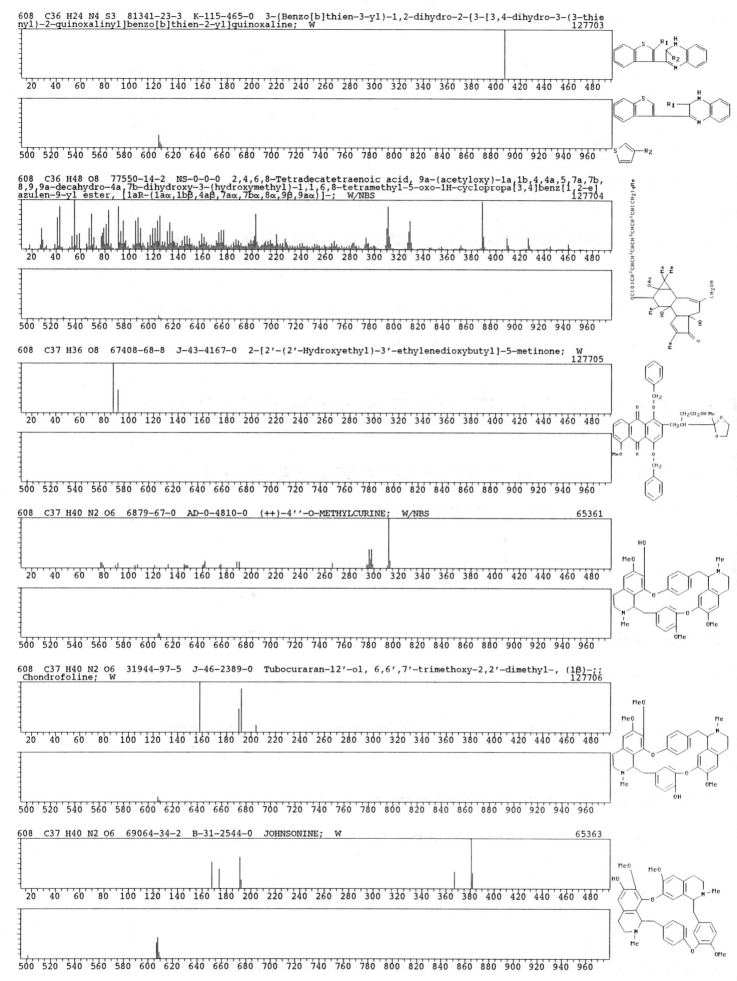

608   C36 H24 N4 S3   81341-23-3   K-115-465-0   3-(Benzo[b]thien-3-yl)-1,2-dihydro-2-{3-[3,4-dihydro-3-(3-thie
nyl)-2-quinoxalinyl]benzo[b]thien-2-yl}quinoxaline;   W
127703

608   C36 H48 O8   77550-14-2   NS-0-0-0   2,4,6,8-Tetradecatetraenoic acid, 9a-(acetyloxy)-1a,1b,4,4a,5,7a,7b,
8,9,9a-decahydro-4a,7b-dihydroxy-3-(hydroxymethyl)-1,1,6,8-tetramethyl-5-oxo-1H-cyclopropa[3,4]benz[1,2-e]
azulen-9-yl ester, [1aR-(1aα,1bβ,4aβ,7aα,7bα,8α,9β,9aα)]-;   W/NBS
127704

608   C37 H36 O8   67408-68-8   J-43-4167-0   2-[2'-(2'-Hydroxyethyl)-3'-ethylenedioxybutyl]-5-metinone;   W
127705

608   C37 H40 N2 O6   6879-67-0   AD-0-4810-0   (++)-4''-O-METHYLCURINE;   W/NBS
65361

608   C37 H40 N2 O6   31944-97-5   J-46-2389-0   Tubocuraran-12'-ol, 6,6',7'-trimethoxy-2,2'-dimethyl-, (1β)-;;
Chondrofoline;   W
127706

608   C37 H40 N2 O6   69064-34-2   B-31-2544-0   JOHNSONINE;   W
65363

608   C37 H40 N2 O6   D3-1980-815-24   Oxyacanthine;   W   65364

608   C37 H40 N2 O6   CD-140-0-0   Berbamine;   W   127707

608   C37 H40 N2 O6   J-47-900-0   Trihydro derivative of colorflammine;   W   127708

608   C37 H40 N2 O6   J-47-900-0   Trihydro derivative of berbacolorflammine;   W   127709

608   C37 H40 N2 O6   55701-99-0   J-40-2648-0   Panurensine;   W   127710

608   C37 H40 N2 O6   55702-02-8   J-40-2469-0   O-Methylnorpanurensine;   W   127711

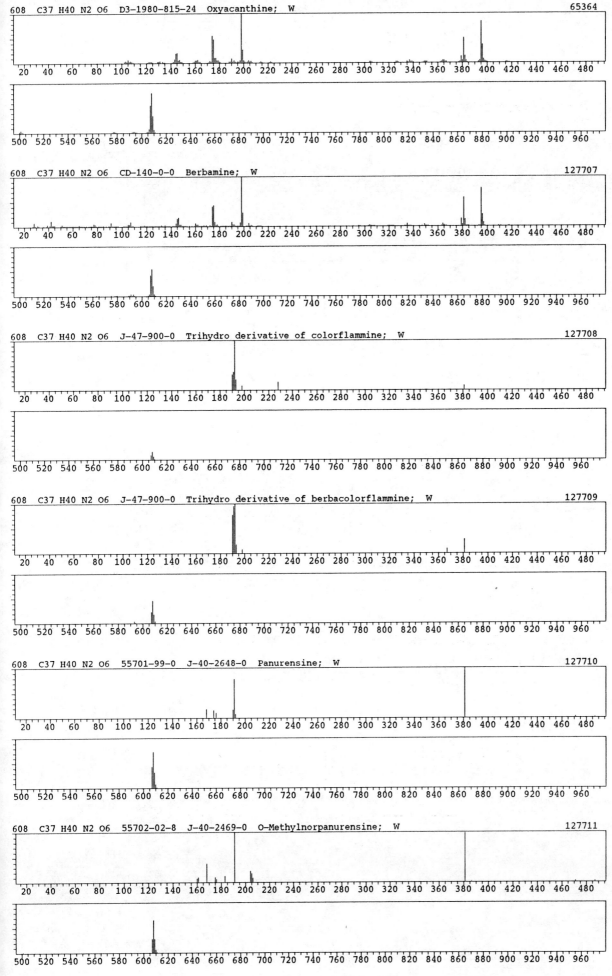

608   C37 H40 N2 O6   17132-74-0   J-41-318-0   Homoaromoline;   W   127712

608   C37 H37 D3 N2 O6   J-47-900-0   Trideuterio derivative of colorflammine;   W   127713

608   C38 H56 O4 S   79409-84-0   J-46-5187-0   (3β,20S,22E,24S)-24-(Phenylsulfinyl)-3-[(tetrahydro-2H-pyran-2-yl)oxy]cholesta-5,22-dien-20-ol;   W   127714

608   C38 H56 O4 S   79409-85-1   J-46-5187-0   (3β,23E)-24-(Phenylsulfinyl)-3-[(tetrahydro-2H-pyran-2-yl)oxy]cholesta-5,23-dien-20-ol;   W   127715

608   C39 H44 O6   TC-7-4-3   D.G.E.B.A.;   W   65365

608   C39 H44 O6   61093-54-7   C-98-7418-0   2,2-1SOPROPYLIDENEBIS((5-(DIMETHYLFURFURYL)-DIMETHYLFURFURYL)FURAN);   W   65366

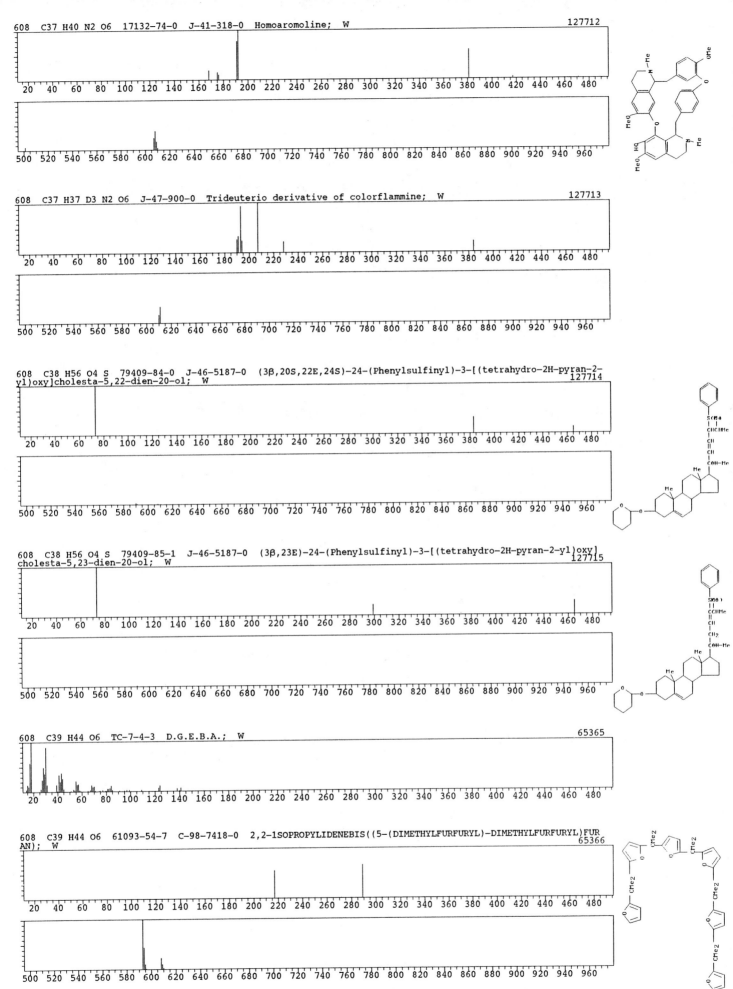

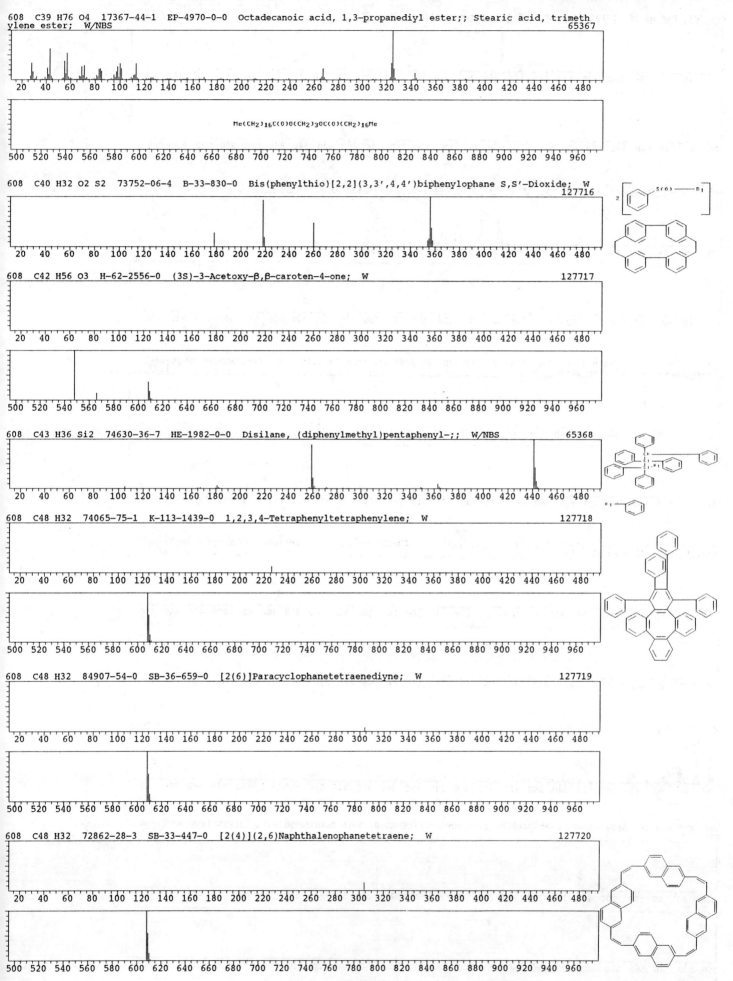

608   C39 H76 O4   17367-44-1   EP-4970-0-0   Octadecanoic acid, 1,3-propanediyl ester;; Stearic acid, trimeth
ylene ester;   W/NBS   65367

Me(CH2)16C(O)O(CH2)3OC(O)(CH2)16Me

608   C40 H32 O2 S2   73752-06-4   B-33-830-0   Bis(phenylthio)[2,2](3,3',4,4')biphenylophane S,S'-Dioxide;   W
                                                                                                        127716

608   C42 H56 O3   H-62-2556-0   (3S)-3-Acetoxy-β,β-caroten-4-one;   W   127717

608   C43 H36 Si2   74630-36-7   HE-1982-0-0   Disilane, (diphenylmethyl)pentaphenyl-;;   W/NBS   65368

608   C48 H32   74065-75-1   K-113-1439-0   1,2,3,4-Tetraphenyltetraphenylene;   W   127718

608   C48 H32   84907-54-0   SB-36-659-0   [2(6)]Paracyclophanetetraenediyne;   W   127719

608   C48 H32   72862-28-3   SB-33-447-0   [2(4)](2,6)Naphthalenophanetetraene;   W   127720

609 Br6 N3 P3 13701-85-4 AD-0-1012-0 1,3,5,2,4,6-Triazatriphosphorine, 2,2,4,4,6,6-hexabromo-2,2,4,4,6,6-hexahydro-;; HEXABROMOTRIPHOSPHONITRILE; W   65369

609 C5 H5 Co Ge I3 N O 52124-71-7 NS-0-0-0 Cobalt, (η5-2,4-cyclopentadien-1-yl)nitrosyl(triiodogermyl)-; W/NBS   127721

609 C20 H15 F12 N O7 KO-6-32-1 Tetra-trifluoroacetyl derivative of Terbutaline; W   127722

609 C21 H18 Cl3 N3 O6 S3 52082-69-6 O-11-1222-1 1,3,5,-TRIS(P-CHLOROPHENYLSULPHONYL) HEXAHYDRO-1,3,5,-TRIAZINE; W   65370

609 C28 H55 N5 O4 Si3 64911-28-0 O-15-212-2 2',3',5'-TRIS-O--TERT-BUTYLDIMETHYLSILYL-ADENOSINE; W   65371

609 C31 H59 N O5 Si3 57325-75-4 EP-8295-0-0 Pregnane-11,20-dione, 3,17,21-tris[(trimethylsilyl)oxy]-20-(O-methyloxime), (3α,5α)-;; 5α-PREGNANE-3α,17α,21-TRIOL-11,20-DIONE MO TMS; W/NBS   65373

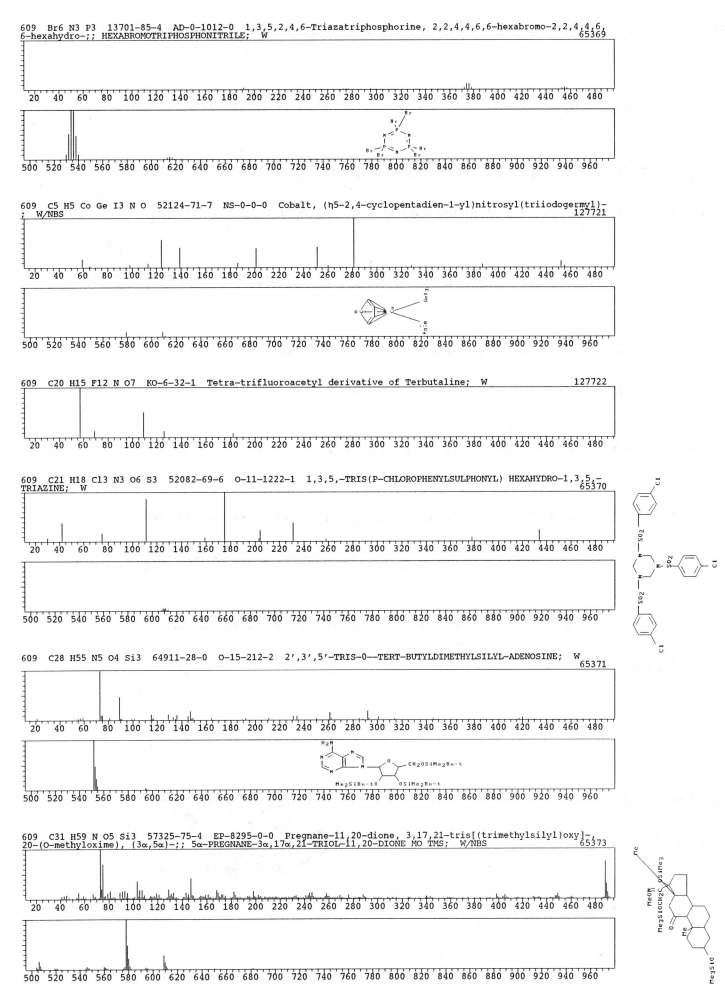

- 5823 -

609   C31 H59 N O5 Si3   57325-76-5   SW-0-12-0   Pregnane-11,20-dione, 3,17,21-tris[(trimethylsilyl)oxy]-, 20
-(O-methyloxime), (3α,5β)-;; TRISTRIMETHYLSILYL 3α,17α,21-TRIHYDROXY-5β-PREGNANE-11,20-DIONE METHOXIME; W
                                                                                                      79467

609   C31 H59 N O5 Si3   86196-22-7   KO-9-508-5   Tris(trimethylsilyl) ether of tetrahydrocortisone methyloxi
me; W   127723

609   C32 H15 Cl Cu N8   F-20-1360-4   COPPER MONOCHLOROPHTHALOCYANINE; W   65374

609   C34 H47 N3 O5 S   72636-90-9   H-62-1941-0   13-(8'-Phthalimido-4'-tosyl-4'-azaoctyl)-12-dodecanlactam;
W   127724

609   C37 H43 N O5 Si   91328-51-7   J-49-3423-0   3-[5-[2,5-Bis(phenylmethoxy)phenyl]-5-methyl-2-phenyl-1,3-di
oxolan-4-yl]-2-methyl-5-(trimethylsilyl)isoxazolidine; W   127726

610   C12 H12 Be4 Cl6 O13   12560-98-4   EP-4136-0-0   Beryllium, hexakis(chloroacetato)oxotetra-;; OXOHEXAKIS(
CHLOROACETATO)TETRABERYLLIUM(II); W/NBS   65375

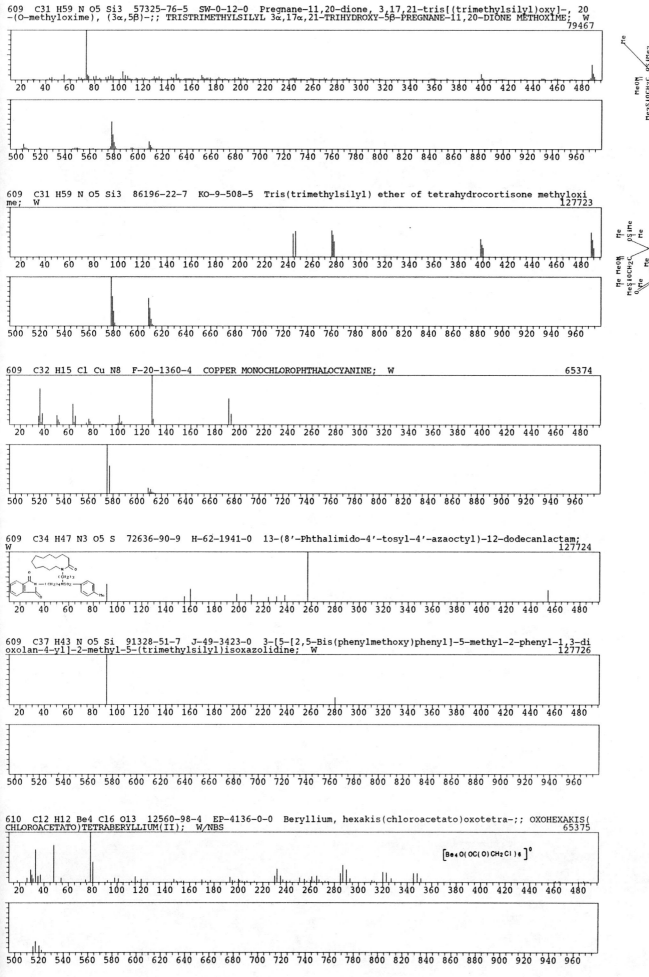

610   C14 H5 F12 Ir O   73207-43-9   B-32-2157-0   1-CARBONYL-N-CYCLOPENTADIENYL-2,3,4,5-TETRA(TRIFLOUROMETHYL)
IRIDIOCYCLOBUTADIENE;   W   65376

610   C15 H27 F12 N2 O2 P Si3   68222-37-7   K-111-3110-0   2-(BIS(TRIMETHYLSILYL)AMINO)-4,4,5,5-TETRAKIS(TRIFL
UOROMETHYL)-2-(TRIMETHYLSILYLIMINO)-1,3,2λ5-DIOXAPHOSPHOLANE;   W   65377

610   C17 H6 F16 O6   KO-9-303-1   Bis(pentafluoropropionyl),hexafluoro-2-propylester, derivative of dihydroxy
phenylacetic acid;   W   127727

610   C27 H30 O16   29428-58-8   NS--27625-0-0   Lucenin 2;   NBS   65378

610   C27 H46 N2 O8 Si3   KO-4-157-7   4-HYDROXYANTIPYRINEGLUCURONIC METHYLESTER (TMS DERIVATIVE);   W   65379

610   C27 H46 N2 O8 Si3   KO-4-158-8   3-HYDROXYMETHYLANTIPYRINEGLUCURONIC METHYLESTER (TMS DERIVATIVE);   W   65380

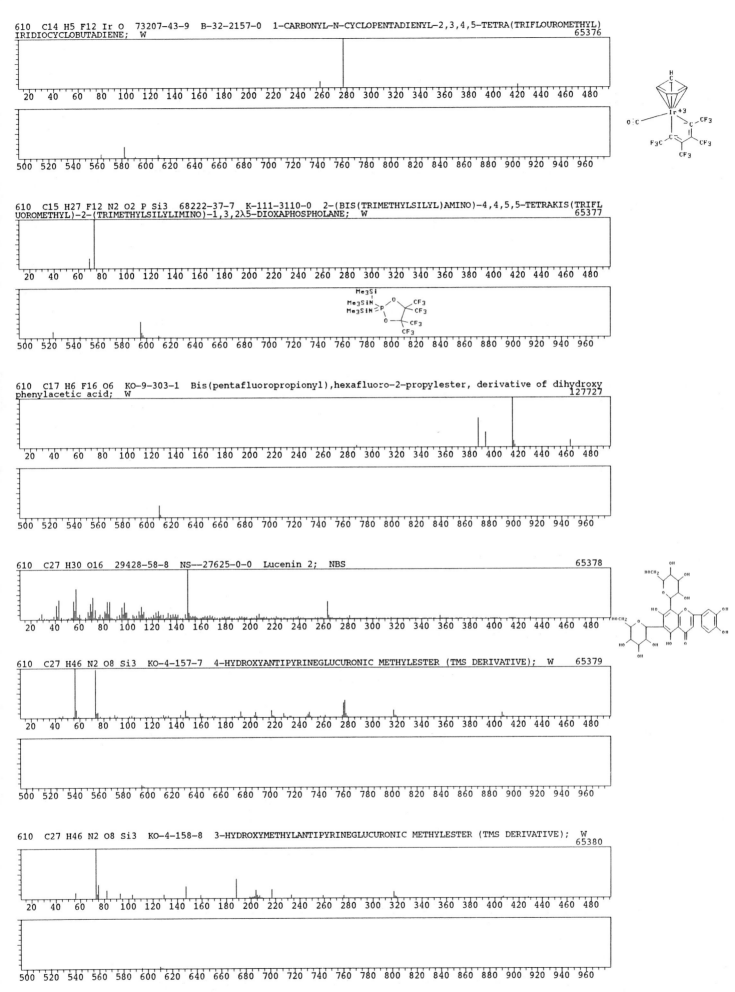

610   C29 H38 O14   78416-55-4   K-114-1645-0   Benzyl-2,3-O-isopropyliden-5-O-(2,3,4,6-tetra-O-acetyl-β-D-gluc
opyranosyl)-β-D-ribofuranoside;   W   127728

| 20 | 40 | 60 | 80 | 100 | 120 | 140 | 160 | 180 | 200 | 220 | 240 | 260 | 280 | 300 | 320 | 340 | 360 | 380 | 400 | 420 | 440 | 460 | 480 |

| 500 | 520 | 540 | 560 | 580 | 600 | 620 | 640 | 660 | 680 | 700 | 720 | 740 | 760 | 780 | 800 | 820 | 840 | 860 | 880 | 900 | 920 | 940 | 960 |

610   C30 H37 F3 N2 O8   51067-80-2   I-58-524-0   [2S,3S,4R,5R,6(S*),7(R*)]-N-[7-[5-[7-(1,2-Dihydro-4-methoxy-
2-oxo-3-pyridinyl)-6-methyl-7-oxo-1(E),3(E),5(E)-hepta-trienyl]tetrahydro-3,4-dihydroxyfuran-2-yl]-6-meth
oxy-5,7-dimethyl-2(E),4(E)-heptadienyl]trifluoroacetamide;   W   127729

| 20 | 40 | 60 | 80 | 100 | 120 | 140 | 160 | 180 | 200 | 220 | 240 | 260 | 280 | 300 | 320 | 340 | 360 | 380 | 400 | 420 | 440 | 460 | 480 |

610   C31 H58 N2 O2 Sn   77256-69-0   J-46-2372-0   1-(Tributylstannyl)ethyl ester of 2,6-bis(dimethylamino)-3
5-bis(1-methylethyl)benzoic acid;   W   127730

| 20 | 40 | 60 | 80 | 100 | 120 | 140 | 160 | 180 | 200 | 220 | 240 | 260 | 280 | 300 | 320 | 340 | 360 | 380 | 400 | 420 | 440 | 460 | 480 |

| 500 | 520 | 540 | 560 | 580 | 600 | 620 | 640 | 660 | 680 | 700 | 720 | 740 | 760 | 780 | 800 | 820 | 840 | 860 | 880 | 900 | 920 | 940 | 960 |

610   C31 H58 O6 Si3   SH-1981-0-0   Tris(trimethylsilyl), methyl ester derivative of 3α,17α,20β-Trihydroxy-11
-oxo-5β-pregnen-21-oic acid;   W   127731

| 20 | 40 | 60 | 80 | 100 | 120 | 140 | 160 | 180 | 200 | 220 | 240 | 260 | 280 | 300 | 320 | 340 | 360 | 380 | 400 | 420 | 440 | 460 | 480 |

| 500 | 520 | 540 | 560 | 580 | 600 | 620 | 640 | 660 | 680 | 700 | 720 | 740 | 760 | 780 | 800 | 820 | 840 | 860 | 880 | 900 | 920 | 940 | 960 |

610   C31 H58 O6 Si3   SH-1981-0-0   Trimethylsilyl derivative of methyl ester of 3α,17α,20α,-Trihydroxy-11-ox
o-5β-pregnan-21-oic acid or Trimethylsilyl derivative of α-Cortolonic acid;   W   127732

| 20 | 40 | 60 | 80 | 100 | 120 | 140 | 160 | 180 | 200 | 220 | 240 | 260 | 280 | 300 | 320 | 340 | 360 | 380 | 400 | 420 | 440 | 460 | 480 |

| 500 | 520 | 540 | 560 | 580 | 600 | 620 | 640 | 660 | 680 | 700 | 720 | 740 | 760 | 780 | 800 | 820 | 840 | 860 | 880 | 900 | 920 | 940 | 960 |

610   C31 H62 O4 Si4   AH-106-1422-28   TETRA(TRIMETHYLSILOXY)ANDROSTENE (LOCANTS UNSPECIFIED);   W   65381

| 20 | 40 | 60 | 80 | 100 | 120 | 140 | 160 | 180 | 200 | 220 | 240 | 260 | 280 | 300 | 320 | 340 | 360 | 380 | 400 | 420 | 440 | 460 | 480 |

| 500 | 520 | 540 | 560 | 580 | 600 | 620 | 640 | 660 | 680 | 700 | 720 | 740 | 760 | 780 | 800 | 820 | 840 | 860 | 880 | 900 | 920 | 940 | 960 |

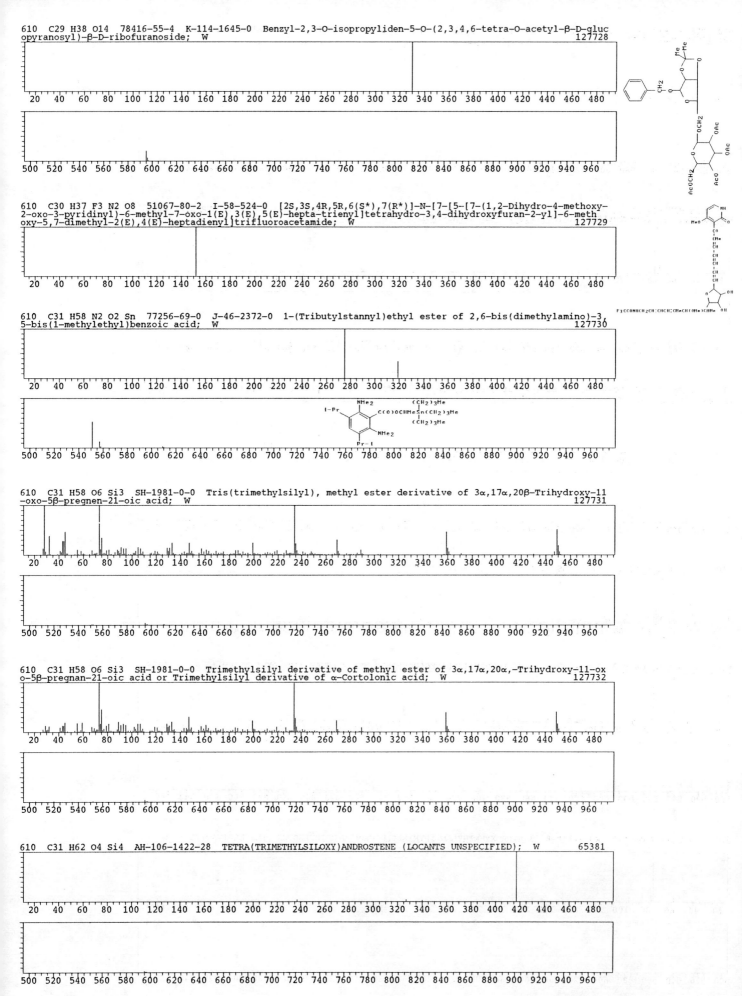

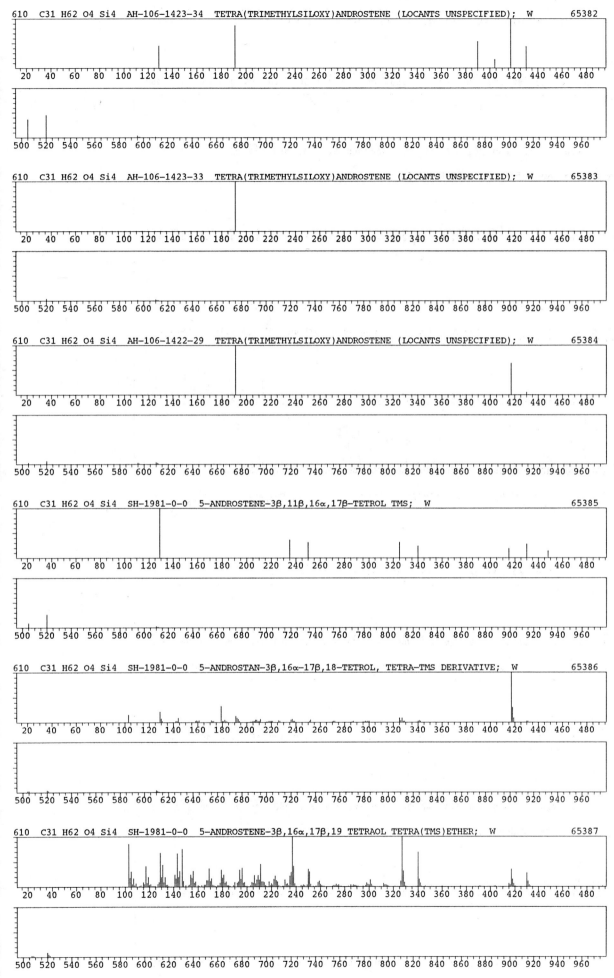

610   C31 H62 O4 Si4   AH-106-1423-34   TETRA(TRIMETHYLSILOXY)ANDROSTENE (LOCANTS UNSPECIFIED);   W   65382

610   C31 H62 O4 Si4   AH-106-1423-33   TETRA(TRIMETHYLSILOXY)ANDROSTENE (LOCANTS UNSPECIFIED);   W   65383

610   C31 H62 O4 Si4   AH-106-1422-29   TETRA(TRIMETHYLSILOXY)ANDROSTENE (LOCANTS UNSPECIFIED);   W   65384

610   C31 H62 O4 Si4   SH-1981-0-0   5-ANDROSTENE-3β,11β,16α,17β-TETROL TMS;   W   65385

610   C31 H62 O4 Si4   SH-1981-0-0   5-ANDROSTAN-3β,16α-17β,18-TETROL, TETRA-TMS DERIVATIVE;   W   65386

610   C31 H62 O4 Si4   SH-1981-0-0   5-ANDROSTENE-3β,16α,17β,19 TETRAOL TETRA(TMS)ETHER;   W   65387

610 C33 H30 N4 O8 74615-13-7 C-104-543-0 Methyl 5-amino-6-(7-amino-5,8-dihydro-6-methoxy-5,8-dioxo-2-qu
inolinyl)-4-(3,4-dimethoxy-2-(phenylmethoxy)phenyl-3-methyl-2-pyridinecarboxylate; W   127733

610 C33 H42 N2 O9 16625-52-8 AD-0-2592-0 DIHYDRORESERPINE; W/NBS   65388

610 C33 H42 N2 O9 83705-24-2 J-48-47-0 3-Ethoxy-3,4-seco-4-(isopropoxycarbonyl)-1,3,5,6,14,21-hexahy
dro-17,18-bis(isopropoxycarbonyl)benz[g]indolo[2,3-a]quinolizine; W   127734

610 C33 H42 O7 S Si 81418-10-2 J-48-3261-0 10,11-Dihydroxy-2-acetoxy-6-(tert-butyldimethylsiloxy)-3-((
4'-methoxyphenyl)thio)-12-oxo-1,4,4a-(R*),5,5a(S*),6(S*),11(R*),11a(R*),12,12a(S*)-decahydronaphthacene; W
                                                                                                     127735

610 C33 H62 O6 Si2 64888-96-6 F-33-1110-0 7-OXOPROSTAGLANDINE 1 METHYL ESTER 15-TERT-BUTYLDIMETHYLSILYL
ETHER; W   65389

610 C33 H66 O4 Si3 KO-4-26-7 5β-PREGNAN-3α,17α,70α,21-TETROL 3,20,21-TRIS DMES ETHER; W   65390

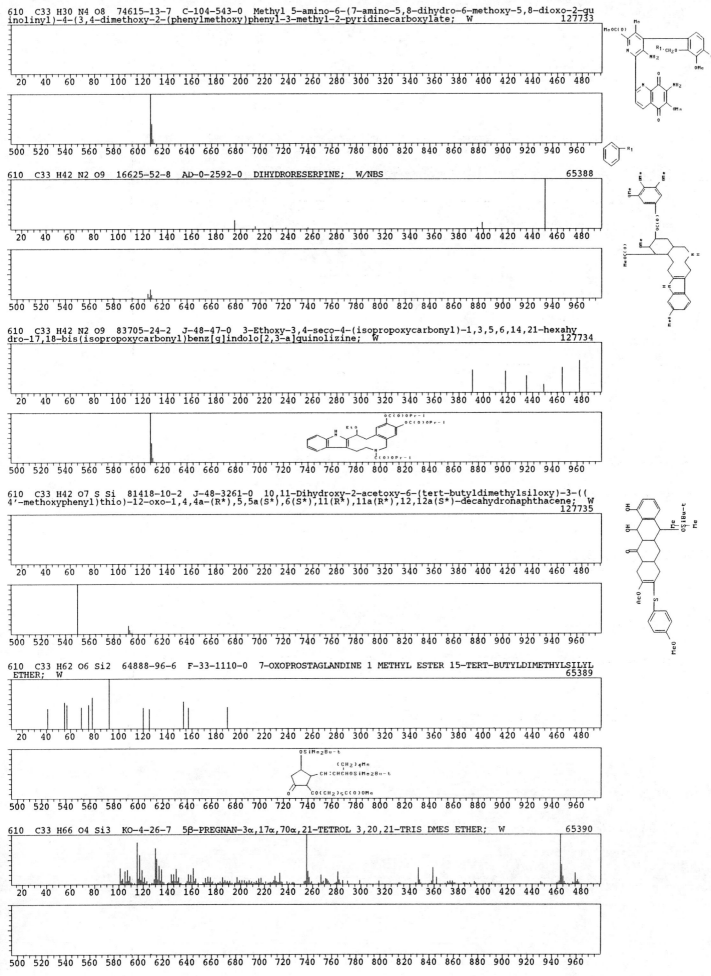

- 5828 -

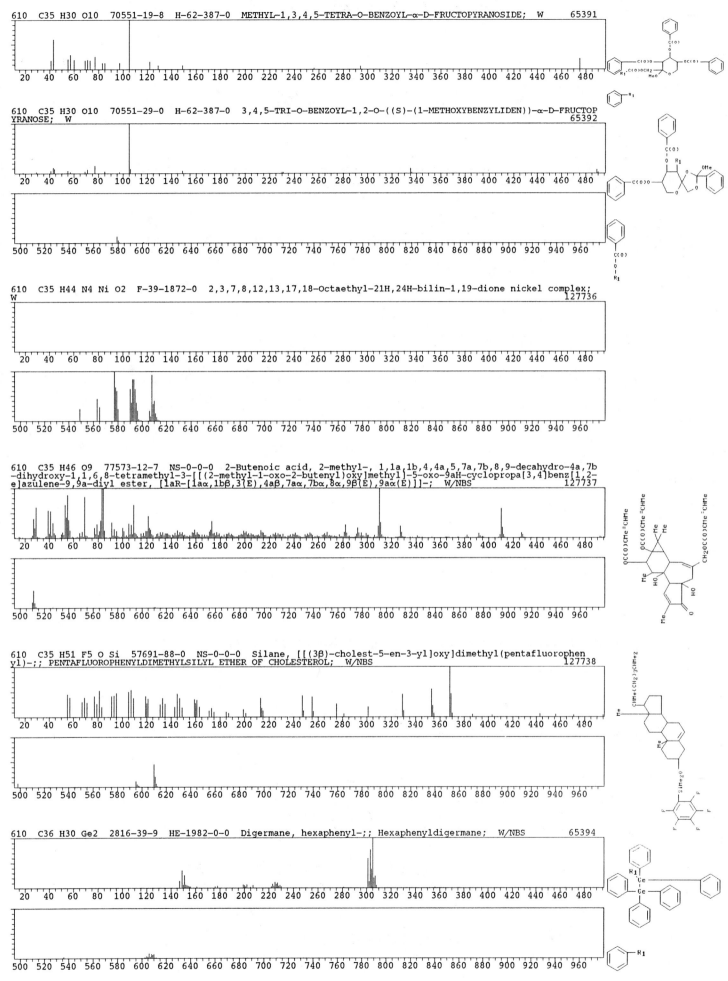

610 C35 H30 O10   70551-19-8   H-62-387-0   METHYL-1,3,4,5-TETRA-O-BENZOYL-α-D-FRUCTOPYRANOSIDE; W   65391

610 C35 H30 O10   70551-29-0   H-62-387-0   3,4,5-TRI-O-BENZOYL-1,2-O-((S)-(1-METHOXYBENZYLIDEN))-α-D-FRUCTOP
YRANOSE; W   65392

610 C35 H44 N4 Ni O2   F-39-1872-0   2,3,7,8,12,13,17,18-Octaethyl-21H,24H-bilin-1,19-dione nickel complex;
W   127736

610 C35 H46 O9   77573-12-7   NS-0-0-0   2-Butenoic acid, 2-methyl-, 1,1a,1b,4,4a,5,7a,7b,8,9-decahydro-4a,7b
-dihydroxy-1,1,6,8-tetramethyl-3-[[(2-methyl-1-oxo-2-butenyl)oxy]methyl]-5-oxo-9aH-cyclopropa[3,4]benz[1,2-
e]azulene-9,9a-diyl ester, [1aR-[1aα,1bβ,3(E),4aβ,7aα,7bα,8α,9β(E),9aα(E)]]-; W/NBS   127737

610 C35 H51 F5 O Si   57691-88-0   NS-0-0-0   Silane, [[(3β)-cholest-5-en-3-yl]oxy]dimethyl(pentafluorophen
yl)-;; PENTAFLUOROPHENYLDIMETHYLSILYL ETHER OF CHOLESTEROL; W/NBS   127738

610 C36 H30 Ge2   2816-39-9   HE-1982-0-0   Digermane, hexaphenyl-;; Hexaphenyldigermane; W/NBS   65394

610   C36 H34 Fe2 O2   56770-53-7   AG-93-229-0   1',1'''-Bis(α-methoxybenzyl)biferrocene;   W   127739

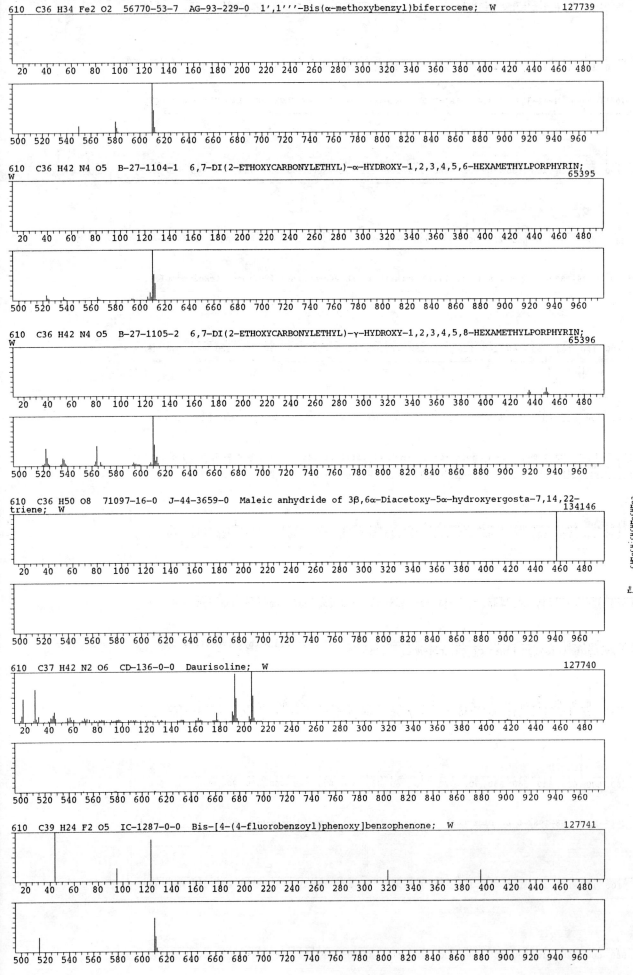

610   C36 H42 N4 O5   B-27-1104-1   6,7-DI(2-ETHOXYCARBONYLETHYL)-α-HYDROXY-1,2,3,4,5,6-HEXAMETHYLPORPHYRIN;
W   65395

610   C36 H42 N4 O5   B-27-1105-2   6,7-DI(2-ETHOXYCARBONYLETHYL)-γ-HYDROXY-1,2,3,4,5,8-HEXAMETHYLPORPHYRIN;
W   65396

610   C36 H50 O8   71097-16-0   J-44-3659-0   Maleic anhydride of 3β,6α-Diacetoxy-5α-hydroxyergosta-7,14,22-
triene;   W   134146

610   C37 H42 N2 O6   CD-136-0-0   Daurisoline;   W   127740

610   C39 H24 F2 O5   IC-1287-0-0   Bis-[4-(4-fluorobenzoyl)phenoxy]benzophenone;   W   127741

610 C39 H54 O2 Si2 65598-50-7 NS-0-0-0 Silane, [[2-[3-methyl-6-(1-methylethenyl)-2-cyclohexen-1-yl]-5-pentyl-1,3-phenylene]bis(oxy)]bis[dimethyl(phenylmethyl)-, (1R-trans)-; W/NBS 127742

610 C40 H54 N2 O3 38389-13-8 AD-0-3740-0 1'H-Cholest-2-eno[3,2-b]indole, 6-nitro-, 5'-(phenylmethoxy)-;; 6α-NITRO-5α-CHOLESTANO(3,2-B)5'-BENZYLOXYINDOLE; W/NBS 65398

610 C41 H36 B2 N2 O2 87982-05-6 O-18-359-1 1,2-Bis(2,2-Diphenyl-1-oxa-3-azonia-2-borata-3-naphthyl)propane; W 127743

610 C41 H36 B2 N2 O2 87982-01-2 O-18-359-1 1,3-Bis(2,2-Diphenyl-1-oxa-3-azonia-2-borata-3-naphthyl)propane; W 127744

610 C41 H54 O4 F-37-603-0 1β,3β-Dibenzoyloxycholest-5-ene-16β-ol; W 127745

610 C42 H37 B Ni HE-1986-67-0 1,5-CYCLOOCTADIENE-PENTAPHENYLBOROLNICKEL; W 127746

610   C42 H58 O3   92466-44-9   H-67-1289-0   (all-E,2'S)-2'-Acetoxy-3',4'-didehydro-1',2'-dihydro-β,.psi.-caro
tin-1'-ol;   W
127747

610   C43 H78 O   22032-48-0   T-68-5947-0   Cholest-5-ene, 3-(hexadecyloxy)-, (3β)-;;   CHOLESTERYL HEXADECYL ET
HER;   W/NBS
65399

610   C48 H34   16716-11-3   EP-3474-0-0   1,1':3',1'':3''',1''':3'''',1'''':3''''',1''''':3'''''',1'''''':3''''''
1'''''''-Octiphenyl;;   m-Octiphenyl;   W/NBS
65400

611   C10 Cl3 F12 N O S4   90594-24-4   K-117-1898-0   1-(Trichloroacetyl)-2,3,4,5-tetrakis(trifluoromethylthi
o)pyrrole;   W
127748

611   C19 H27 N5 O14 P2   2147-10-6   D-16-1047-4   5'-O-PHOSPHORYLTHYMIDYLYL-(3'-5')-2'-DEOXYCYTIDINE;   W
65401

611   C24 H23 Br2 N O8   74965-16-5   J-45-4583-0   Tetramethyl 2,3-dibromo-1-(2,6-dimethylphenyl)-indoline-4,
5,6,7-tetracarboxylate;   W
127749

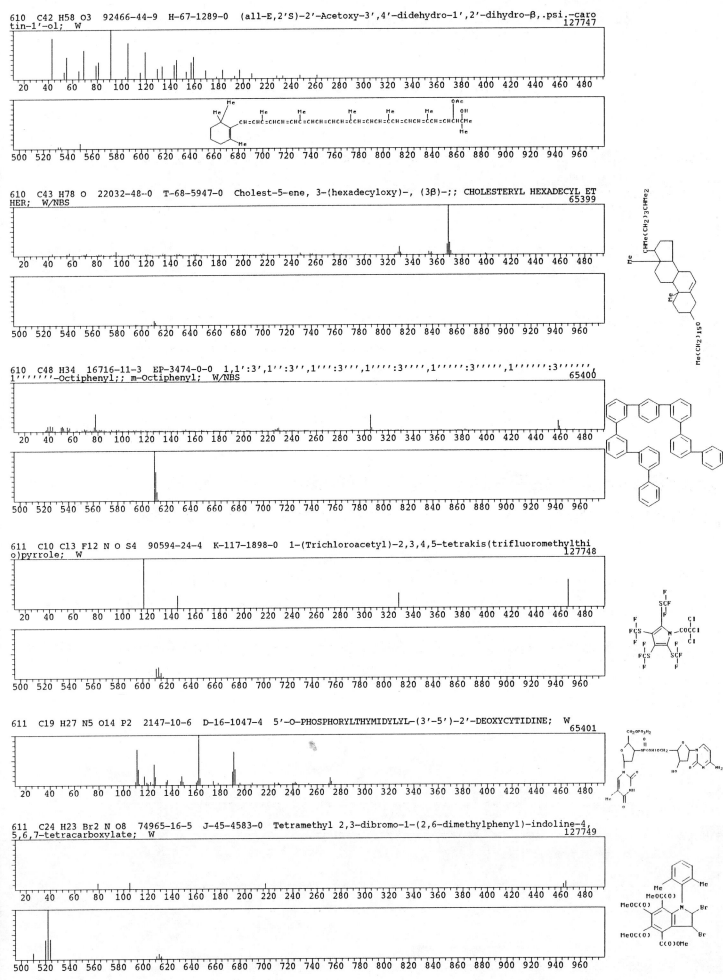

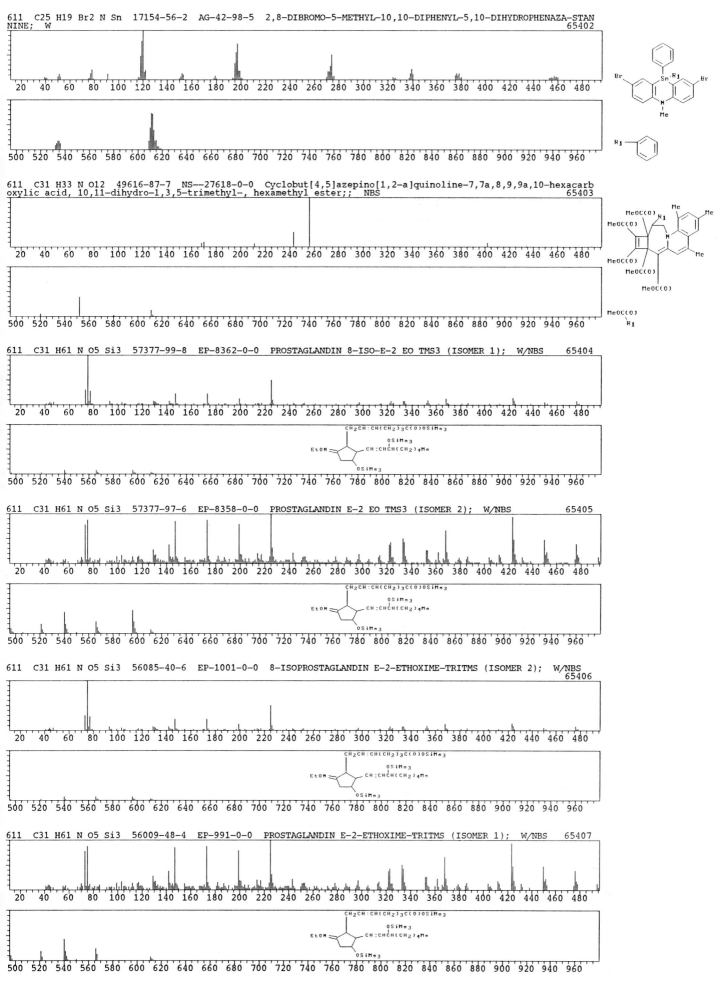

611　C25 H19 Br2 N Sn　17154-56-2　AG-42-98-5　2,8-DIBROMO-5-METHYL-10,10-DIPHENYL-5,10-DIHYDROPHENAZA-STAN
NINE;　W
65402

611　C31 H33 N O12　49616-87-7　NS--27618-0-0　Cyclobut[4,5]azepino[1,2-a]quinoline-7,7a,8,9,9a,10-hexacarb
oxylic acid, 10,11-dihydro-1,3,5-trimethyl-, hexamethyl ester;;　NBS
65403

611　C31 H61 N O5 Si3　57377-99-8　EP-8362-0-0　PROSTAGLANDIN 8-ISO-E-2 EO TMS3 (ISOMER 1);　W/NBS　65404

611　C31 H61 N O5 Si3　57377-97-6　EP-8358-0-0　PROSTAGLANDIN E-2 EO TMS3 (ISOMER 2);　W/NBS　65405

611　C31 H61 N O5 Si3　56085-40-6　EP-1001-0-0　8-ISOPROSTAGLANDIN E-2-ETHOXIME-TRITMS (ISOMER 2);　W/NBS
65406

611　C31 H61 N O5 Si3　56009-48-4　EP-991-0-0　PROSTAGLANDIN E-2-ETHOXIME-TRITMS (ISOMER 1);　W/NBS　65407

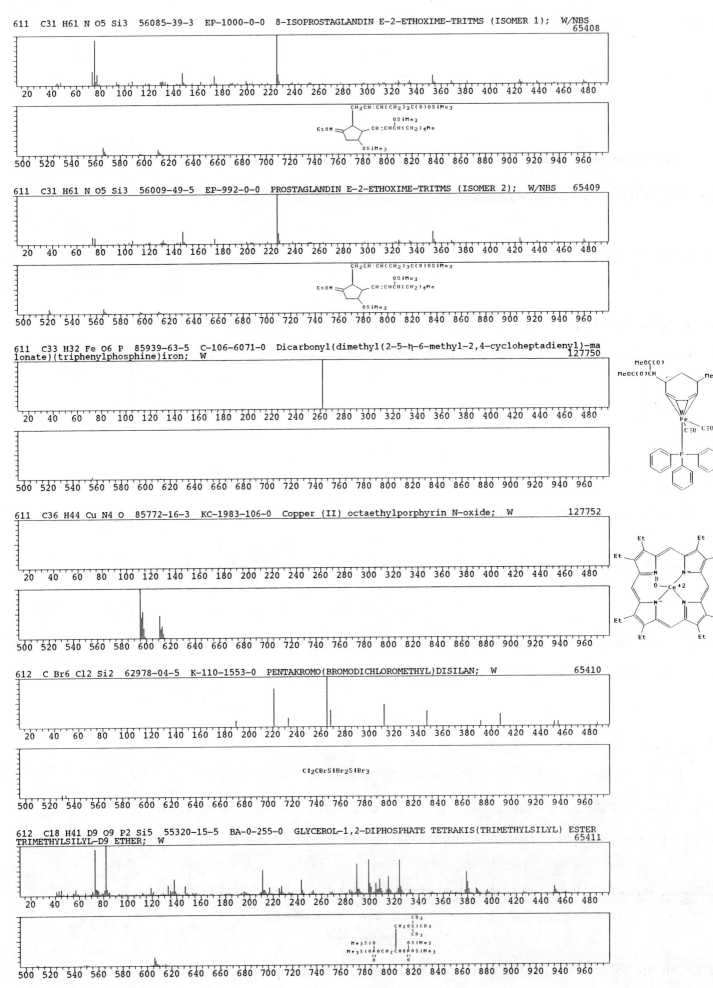

611 C31 H61 N O5 Si3   56085-39-3   EP-1000-0-0   8-ISOPROSTAGLANDIN E-2-ETHOXIME-TRITMS (ISOMER 1);   W/NBS
65408

611 C31 H61 N O5 Si3   56009-49-5   EP-992-0-0   PROSTAGLANDIN E-2-ETHOXIME-TRITMS (ISOMER 2);   W/NBS   65409

611 C33 H32 Fe O6 P   85939-63-5   C-106-6071-0   Dicarbonyl(dimethyl(2-5-η-6-methyl-2,4-cycloheptadienyl)-ma
lonate)(triphenylphosphine)iron;   W
127750

611 C36 H44 Cu N4 O   85772-16-3   KC-1983-106-0   Copper (II) octaethylporphyrin N-oxide;   W   127752

612 C Br6 Cl2 Si2   62978-04-5   K-110-1553-0   PENTAKROMO(BROMODICHLOROMETHYL)DISILAN;   W   65410

612 C18 H41 D9 O9 P2 Si5   55320-15-5   BA-0-255-0   GLYCEROL-1,2-DIPHOSPHATE TETRAKIS(TRIMETHYLSILYL) ESTER
TRIMETHYLSILYL-D9 ETHER;   W
65411

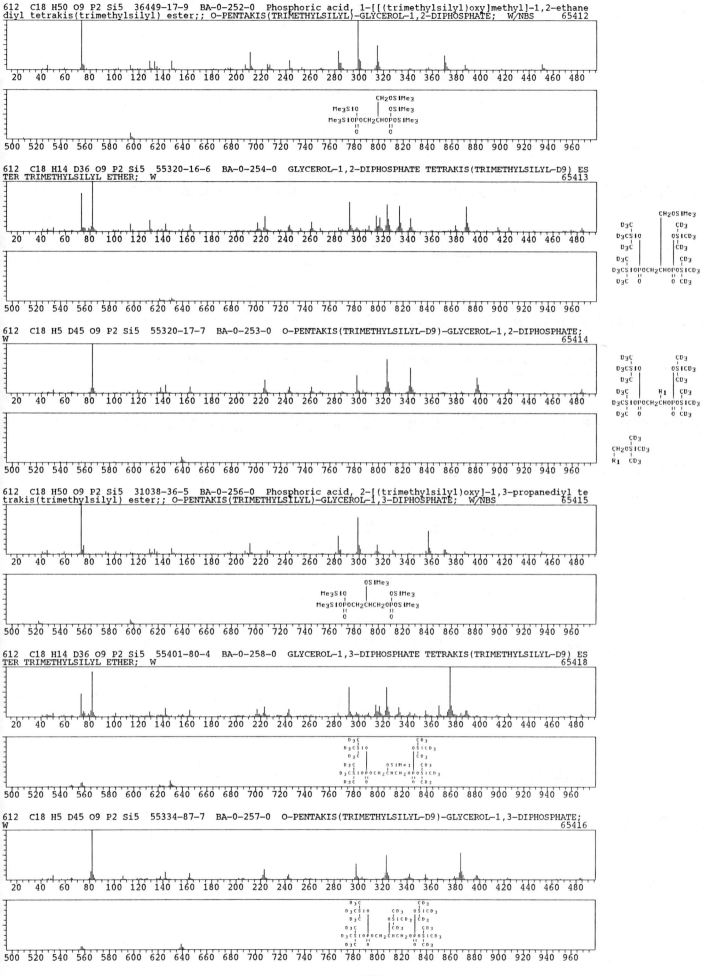

612 C18 H50 O9 P2 Si5 36449-17-9 BA-0-252-0 Phosphoric acid, 1-[[(trimethylsilyl)oxy]methyl]-1,2-ethane diyl tetrakis(trimethylsilyl) ester;; O-PENTAKIS(TRIMETHYLSILYL)-GLYCEROL-1,2-DIPHOSPHATE; W/NBS   65412

612 C18 H14 D36 O9 P2 Si5 55320-16-6 BA-0-254-0 GLYCEROL-1,2-DIPHOSPHATE TETRAKIS(TRIMETHYLSILYL-D9) ESTER TRIMETHYLSILYL ETHER; W   65413

612 C18 H5 D45 O9 P2 Si5 55320-17-7 BA-0-253-0 O-PENTAKIS(TRIMETHYLSILYL-D9)-GLYCEROL-1,2-DIPHOSPHATE; W   65414

612 C18 H50 O9 P2 Si5 31038-36-5 BA-0-256-0 Phosphoric acid, 2-[(trimethylsilyl)oxy]-1,3-propanediyl tetrakis(trimethylsilyl) ester;; O-PENTAKIS(TRIMETHYLSILYL)-GLYCEROL-1,3-DIPHOSPHATE; W/NBS   65415

612 C18 H14 D36 O9 P2 Si5 55401-80-4 BA-0-258-0 GLYCEROL-1,3-DIPHOSPHATE TETRAKIS(TRIMETHYLSILYL-D9) ESTER TRIMETHYLSILYL ETHER; W   65418

612 C18 H5 D45 O9 P2 Si5 55334-87-7 BA-0-257-0 O-PENTAKIS(TRIMETHYLSILYL-D9)-GLYCEROL-1,3-DIPHOSPHATE; W   65416

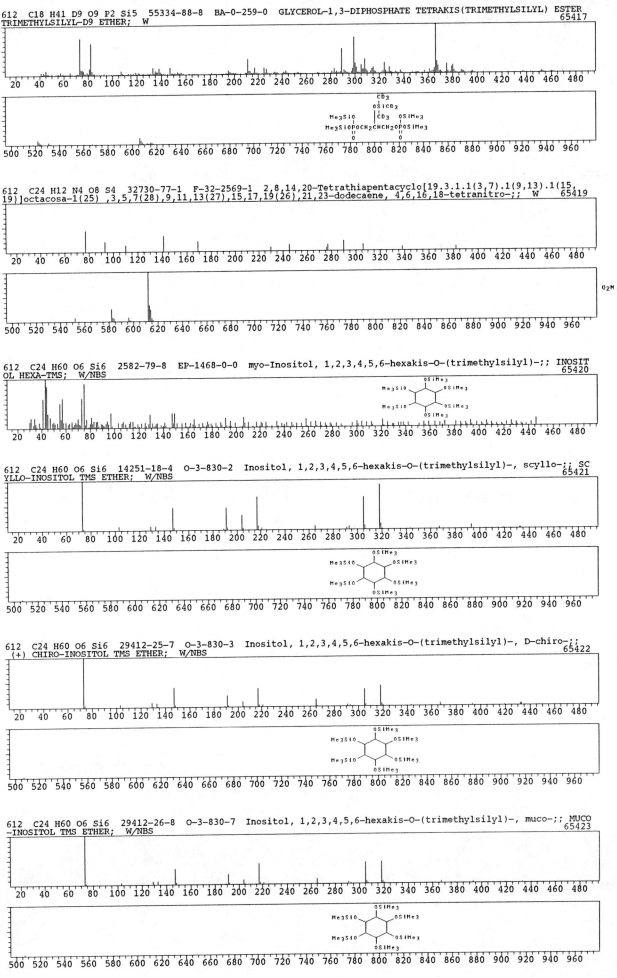

612 C18 H41 D9 O9 P2 Si5 55334-88-8 BA-0-259-0 GLYCEROL-1,3-DIPHOSPHATE TETRAKIS(TRIMETHYLSILYL) ESTER
TRIMETHYLSILYL-D9 ETHER; W   65417

612 C24 H12 N4 O8 S4 32730-77-1 F-32-2569-1 2,8,14,20-Tetrathiapentacyclo[19.3.1.1(3,7).1(9,13).1(15,
19)]octacosa-1(25),3,5,7(28),9,11,13(27),15,17,19(26),21,23-dodecaene, 4,6,16,18-tetranitro-;; W   65419

612 C24 H60 O6 Si6 2582-79-8 EP-1468-0-0 myo-Inositol, 1,2,3,4,5,6-hexakis-O-(trimethylsilyl)-;; INOSIT
OL HEXA-TMS; W/NBS   65420

612 C24 H60 O6 Si6 14251-18-4 O-3-830-2 Inositol, 1,2,3,4,5,6-hexakis-O-(trimethylsilyl)-, scyllo-;; SC
YLLO-INOSITOL TMS ETHER; W/NBS   65421

612 C24 H60 O6 Si6 29412-25-7 O-3-830-3 Inositol, 1,2,3,4,5,6-hexakis-O-(trimethylsilyl)-, D-chiro-;;
(+) CHIRO-INOSITOL TMS ETHER; W/NBS   65422

612 C24 H60 O6 Si6 29412-26-8 O-3-830-7 Inositol, 1,2,3,4,5,6-hexakis-O-(trimethylsilyl)-, muco-;; MUCO
-INOSITOL TMS ETHER; W/NBS   65423

C24 H60 O6 Si6 29267-01-4 O-3-830-5 Inositol, 1,2,3,4,5,6-hexakis-O-(trimethylsilyl)-, epi-;; EPI-
INOSITOL TMS ETHER; W/NBS   65424

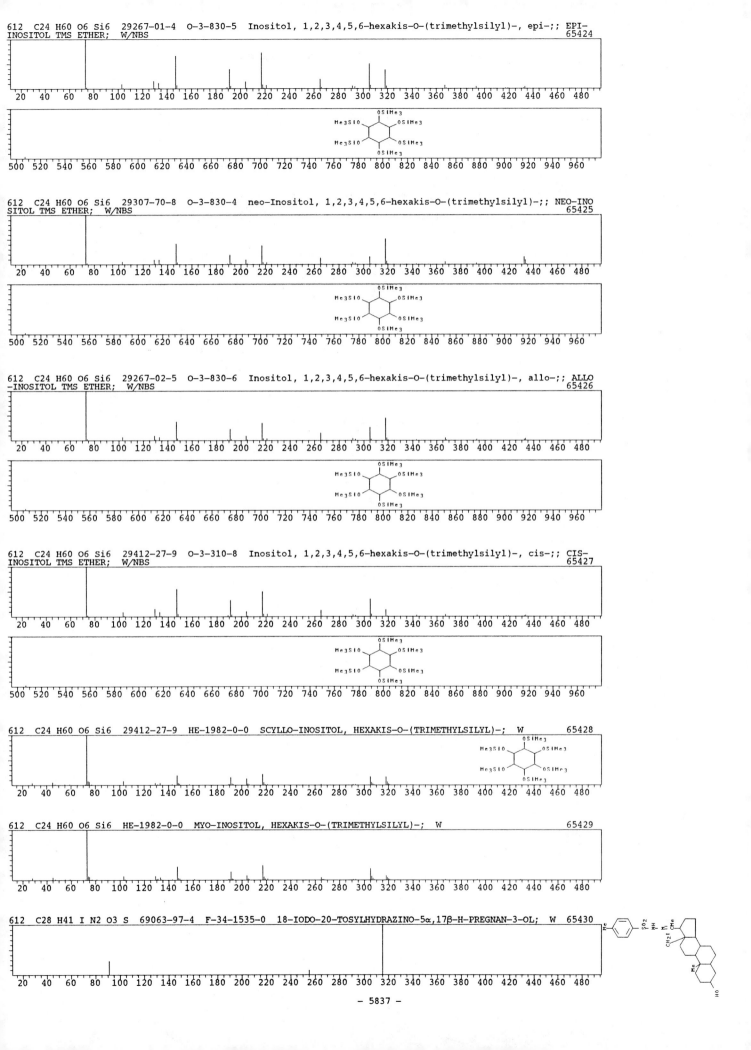

612 C24 H60 O6 Si6 29307-70-8 O-3-830-4 neo-Inositol, 1,2,3,4,5,6-hexakis-O-(trimethylsilyl)-;; NEO-INO
SITOL TMS ETHER; W/NBS   65425

612 C24 H60 O6 Si6 29267-02-5 O-3-830-6 Inositol, 1,2,3,4,5,6-hexakis-O-(trimethylsilyl)-, allo-;; ALLO
-INOSITOL TMS ETHER; W/NBS   65426

612 C24 H60 O6 Si6 29412-27-9 O-3-310-8 Inositol, 1,2,3,4,5,6-hexakis-O-(trimethylsilyl)-, cis-;; CIS-
INOSITOL TMS ETHER; W/NBS   65427

612 C24 H60 O6 Si6 29412-27-9 HE-1982-0-0 SCYLLO-INOSITOL, HEXAKIS-O-(TRIMETHYLSILYL)-; W   65428

612 C24 H60 O6 Si6 HE-1982-0-0 MYO-INOSITOL, HEXAKIS-O-(TRIMETHYLSILYL)-; W   65429

612 C28 H41 I N2 O3 S 69063-97-4 F-34-1535-0 18-IODO-20-TOSYLHYDRAZINO-5α,17β-H-PREGNAN-3-OL; W 65430

612   C29 H42 O Rh2   68913-96-2   B-31-1949-0   BIS(η-PENTAMETHYLCYCLOPENTADIENYL)(1-5-η-(1,2,4,5-TETRAMETHYL-
3-OXO-PENTA-1,4-DIENE-1,5-DIYL))DIRHODIUM;   W
65431

612   C31 H41 F5 N2 O3 Si   57691-93-7   NS--27601-0-0   Pregn-4-ene-3,20-dione, 17-[[dimethyl(pentafluorophen
yl)silyl]oxy]-, bis(O-methyloxime);;   NBS
65432

612   C32 H53 I O3   28328-13-4   SD-1981-0-0   Lanostan-3β-ol, 11β,18-epoxy-19-iodo-, acetate;;   W/NBS   65433

612   C32 H64 O5 Si3   OM-1981-658-0   16,16-DIME F2A ME ESTER 3TMS;   W
65434

612   C33 H40 O11   60538-74-1   MY-1981-0-0   SATRATOXIN H DIACETATE;   W
65435

612   C33 H45 Br O4 Si   78009-90-2   I-59-276-0   3,4,6-Tri-O-benzyl-1-bromo-5-O-tert-butyldimethylsilyl-1,2-
dideoxy-D-arabino-hexitol;   W
127753

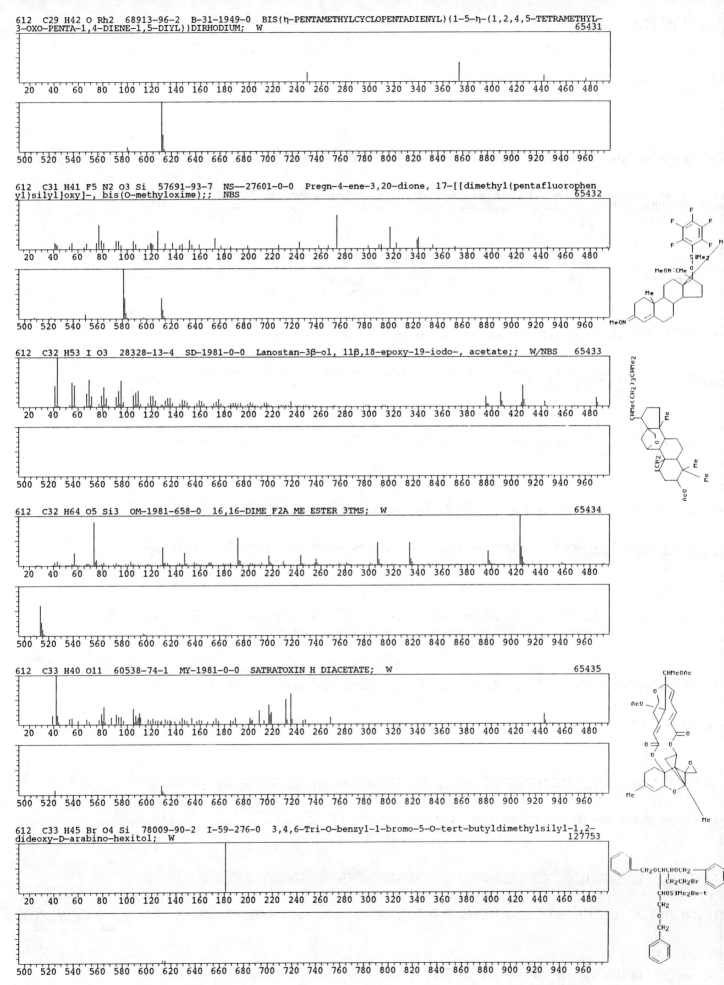

612   C34 H52 N2 O6 Si   78381-14-3   K-114-1920-0   Di-tert-butylester of {2β-[(tert-Butyldimethylsilyloxy)
methyl]-1,2,3,4,6,7,12,12bβ-octahydro-4-oxoindolo[2,3-a]quinolizin-3α-ylmethyl}malonic acid; W   127754

612   C35 H32 O10   58965-08-5   KC-1976-100-0   4',4'',5'',7,7''-PENTAMETHOXY-5-HYDROXY-6,8''-BILFLAVANONE; W
                                                                                                      65436

612   C35 H40 N4 O6   73948-32-0   K-113-1609-0   Methyl ester of 3(E)-ethylidene-1,2,3,19,21,24-hexahydro-2,7,
13,17-tetramethyl-1,19-dioxo-18-vinyl-22H-bilin-8,12-dipropionic acid; W   127755

612   C35 H53 F5 O Si   57691-87-9   NS-0-0-0   Silane, [[(3β,5α)-cholestan-3-yl]oxy]dimethyl(pentafluorophen
yl)-;; W/NBS   127756

612   C36 H56 N2 O4 S   IC-1288-0-0   N,N'-Dilauryl-4,4'-diaminodiphenylsulphone; W   127757

612   C36 H56 N2 O6   79307-94-1   KC-1981-2126-0   7α-[N,N'-bis(ethoxycarbonyl)hydrazino]ergosta-5,8(9),22-tri
en-3β-ylacetate; W   127758

612   C36 H56 N2 O6   79307-95-2   KC-1981-2126-0   7α[NN'-bis(ethoxycarbonyl)hydrazino]-ergosta-5,8(14),22-tri
en-3β-yl acetate; W   127759

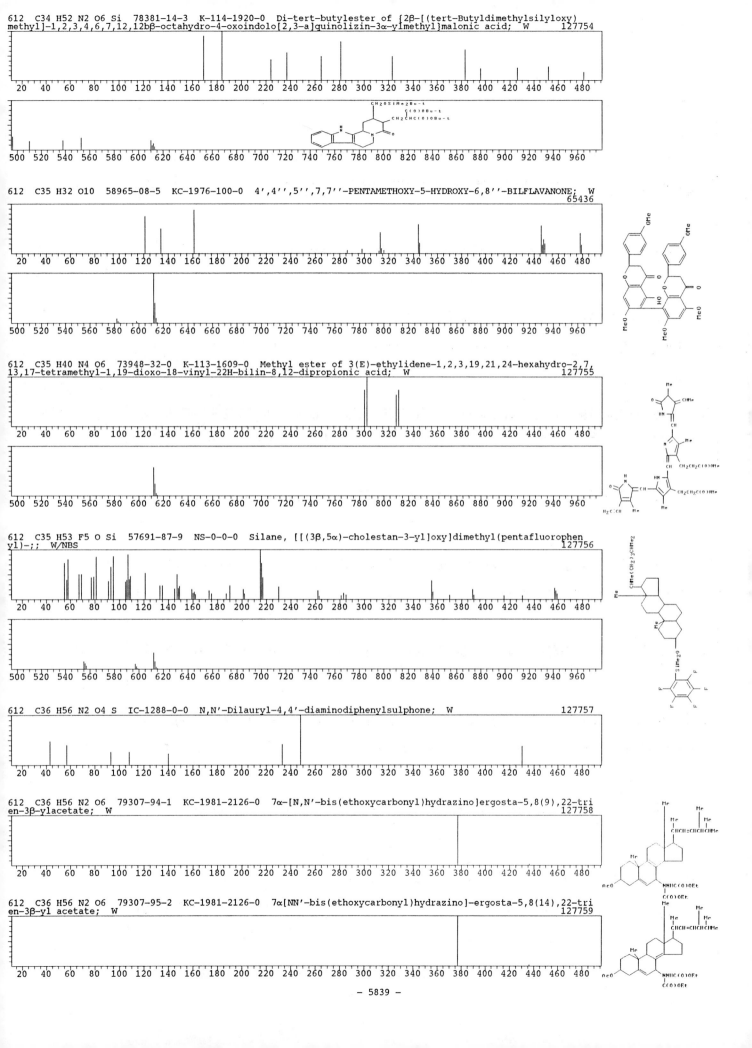

612 C38 H60 O6 B-29-1358-0 15α,16α:22α,28-DI-O,O-ISOPROPYLIDENEOLEAN-12-ENE-3β,15α,22α,28-PENTOL-3-ACET
ATE; W
65438

612 C38 H60 O6 60419-31-0 B-29-1361-0 15α,16α-O,O-ISOPROPYLIDENEOLEAN-12-ENE-3β,15α,16α,22α,28-PENTOL
22-ANGELATE; W
65439

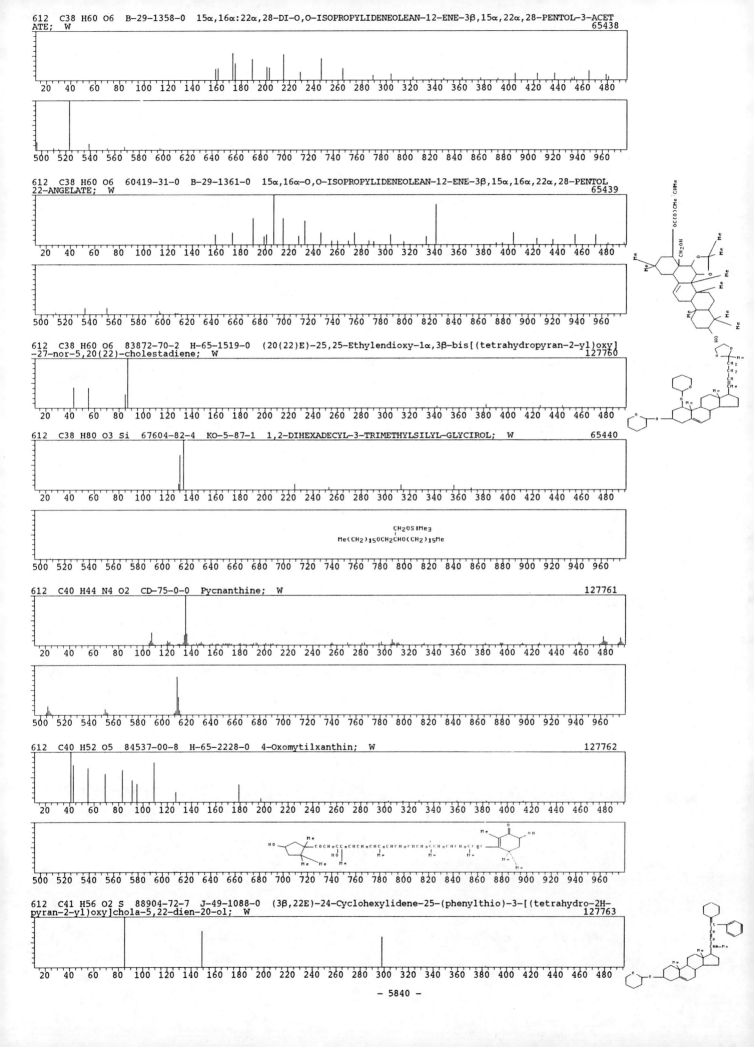

612 C38 H60 O6 83872-70-2 H-65-1519-0 (20(22)E)-25,25-Ethylendioxy-1α,3β-bis[(tetrahydropyran-2-yl)oxy]
-27-nor-5,20(22)-cholestadiene; W
127760

612 C38 H80 O3 Si 67604-82-4 KO-5-87-1 1,2-DIHEXADECYL-3-TRIMETHYLSILYL-GLYCIROL; W
65440

CH2OSiMe3
|
Me(CH2)15OCH2CHO(CH2)15Me

612 C40 H44 N4 O2 CD-75-0-0 Pycnanthine; W
127761

612 C40 H52 O5 84537-00-8 H-65-2228-0 4-Oxomytilxanthin; W
127762

612 C41 H56 O2 S 88904-72-7 J-49-1088-0 (3β,22E)-24-Cyclohexylidene-25-(phenylthio)-3-[(tetrahydro-2H-
pyran-2-yl)oxy]chola-5,22-dien-20-ol; W
127763

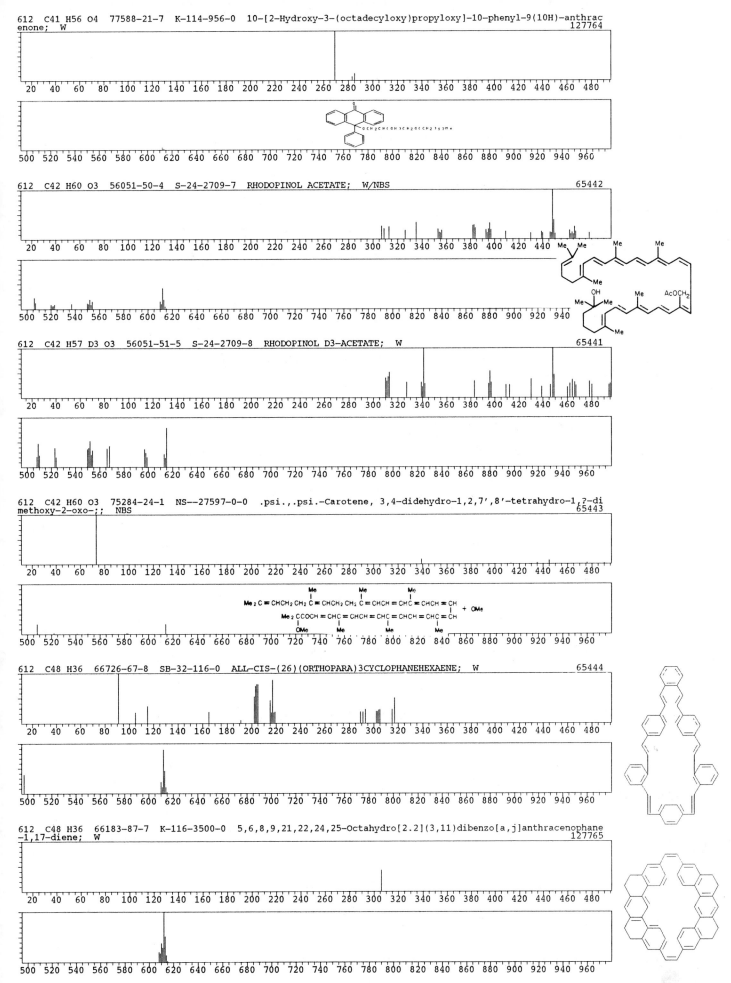

612   C41 H56 O4   77588-21-7   K-114-956-0   10-[2-Hydroxy-3-(octadecyloxy)propyloxy]-10-phenyl-9(10H)-anthracenone;   W
127764

612   C42 H60 O3   56051-50-4   S-24-2709-7   RHODOPINOL ACETATE;   W/NBS
65442

612   C42 H57 D3 O3   56051-51-5   S-24-2709-8   RHODOPINOL D3-ACETATE;   W
65441

612   C42 H60 O3   75284-24-1   NS--27597-0-0   .psi.,.psi.-Carotene, 3,4-didehydro-1,2,7',8'-tetrahydro-1,?-dimethoxy-2-oxo-;;   NBS
65443

612   C48 H36   66726-67-8   SB-32-116-0   ALL-CIS-(26)(ORTHOPARA)3CYCLOPHANEHEXAENE;   W
65444

612   C48 H36   66183-87-7   K-116-3500-0   5,6,8,9,21,22,24,25-Octahydro[2.2](3,11)dibenzo[a,j]anthracenophane-1,17-diene;   W
127765

- 5841 -

612   C48 H36   84907-52-8   SB-36-658-0   cis,cis,trans,cis,cis,trans-[2(6)]-Paracyclophanehexaene;   W   127766

613   C24 H51 N5 O4 Si5   GC-12-37-2   Pentakis(trimethylsilyl) derivative of d-erythro neopterin;   W   127767

613   C24 H63 N O5 Si6   56282-41-8   AU-10-117-9   D-Glucitol, 2-deoxy-1,3,4,5,6-pentakis-O-(trimethylsilyl)-2-[(trimethylsilyl)amino]-;; 2-DEOXY-PENTA-O-(TRIMETHYLSILYL)-2-(TRIMETHYLSILYL-AMINO)-D-GLUCITOL;   W/NBS
65445

613   C24 H63 N O5 Si6   74420-75-0   NS--27595-0-0   D-Mannitol, 2-deoxy-1,3,4,5,6-pentakis-O-(trimethylsilyl)-2-[(trimethylsilyl)amino]-;;   NBS
65446

613   C28 H34 Cl N5 Ni O5   H-60-2670-0   NICKEL(II)-1,2,2,7,7,12,12-HEPTAMETHYL-15-CYANO-19-FORMYL-CARRINATE-PERCHLOROATE;   W
65447

613   C31 H47 N7 O6   KO-8-147-3   N-[4-[[2-Amino-1,4,5,6,7,8-hexahydro-4-oxo-6-pteridinyl)methyl]amino]benzoyl]-L-glutamic acid;   W
127768

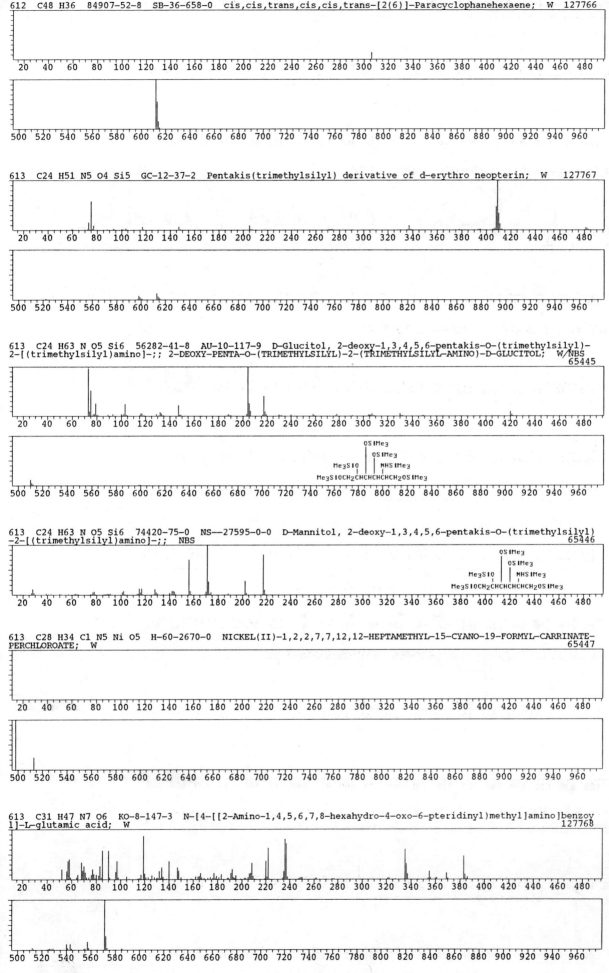

613 C31 H63 N O5 Si3 57377-98-7 EP-8360-0-0 PROSTAGLANDIN 8-ISO-E-1 EO TMS3 (ISOMER 2); W/NBS   65448

613 C31 H63 N O5 Si3 56085-41-7 EP-995-0-0 8-ISOPROSTAGLANDIN E-1-ETHOXIME-TRITMS (ISOMER 1); W/NBS
65449

613 C31 H63 N O5 Si3 56144-49-1 EP-986-0-0 PROSTAGLANDIN E-1-ETHOXIME-TRITMS (ISOMER 1); W/NBS   65450

613 C31 H63 N O5 Si3 56085-42-8 EP-996-0-0 8-ISOPROSTAGLANDIN E-1-ETHOXIME-TRITMS (ISOMER 2); W/NBS
65451

613 C31 H63 N O5 Si3 56009-50-8 EP-987-0-0 PROSTAGLANDIN E-1-ETHOXIME-TRITMS (ISOMER 2); W/NBS   65452

613 C34 H47 N O9 CD-288-0-0 Chasmaconitine; W   127769

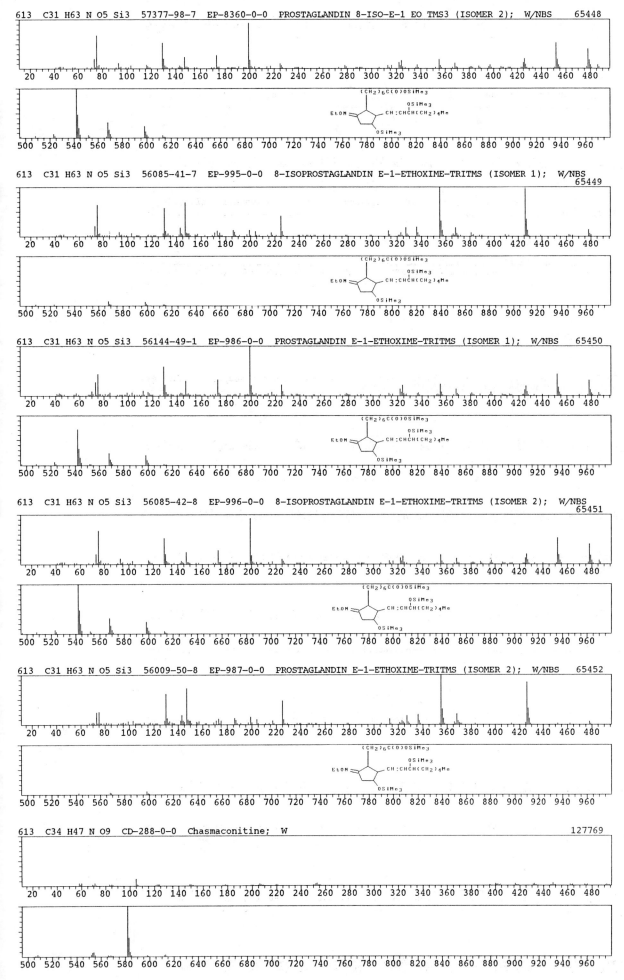

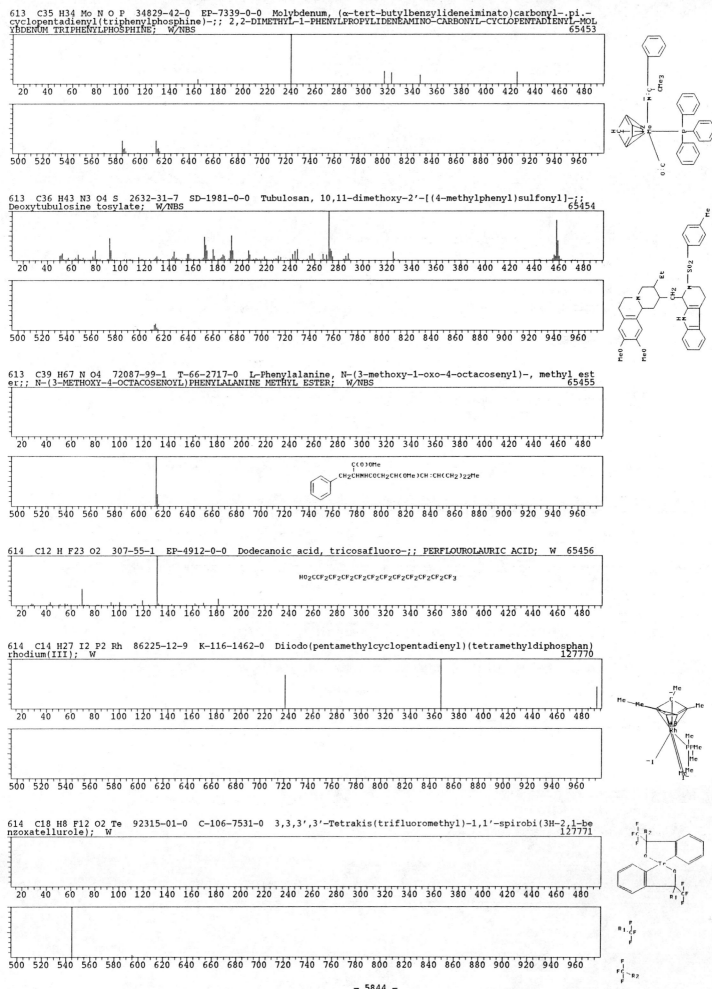

613   C35 H34 Mo N O P   34829-42-0   EP-7339-0-0   Molybdenum, (α-tert-butylbenzylideneiminato)carbonyl-.pi.-
cyclopentadienyl(triphenylphosphine)-;; 2,2-DIMETHYL-1-PHENYLPROPYLIDENEAMINO-CARBONYL-CYCLOPENTADIENYL-MOL
YBDENUM TRIPHENYLPHOSPHINE; W/NBS   65453

613   C36 H43 N3 O4 S   2632-31-7   SD-1981-0-0   Tubulosan, 10,11-dimethoxy-2'-[(4-methylphenyl)sulfonyl]-;;
Deoxytubulosine tosylate; W/NBS   65454

613   C39 H67 N O4   72087-99-1   T-66-2717-0   L-Phenylalanine, N-(3-methoxy-1-oxo-4-octacosenyl)-, methyl est
er;; N-(3-METHOXY-4-OCTACOSENOYL)PHENYLALANINE METHYL ESTER; W/NBS   65455

C(O)OMe
|
CH2CHNHCOCH2CH(OMe)CH:CH(CH2)22Me

614   C12 H F23 O2   307-55-1   EP-4912-0-0   Dodecanoic acid, tricosafluoro-;; PERFLOUROLAURIC ACID; W   65456

HO2CCF2CF2CF2CF2CF2CF2CF2CF2CF2CF2CF3

614   C14 H27 I2 P2 Rh   86225-12-9   K-116-1462-0   Diiodo(pentamethylcyclopentadienyl)(tetramethyldiphosphan)
rhodium(III); W   127770

614   C18 H8 F12 O2 Te   92315-01-0   C-106-7531-0   3,3,3',3'-Tetrakis(trifluoromethyl)-1,1'-spirobi(3H-2,1-be
nzoxatellurole); W   127771

614   C19 H9 F15 N2 O4   BS-3-8-0   5-hydroxytryptamine tripentafluoropropionate;   W   127772

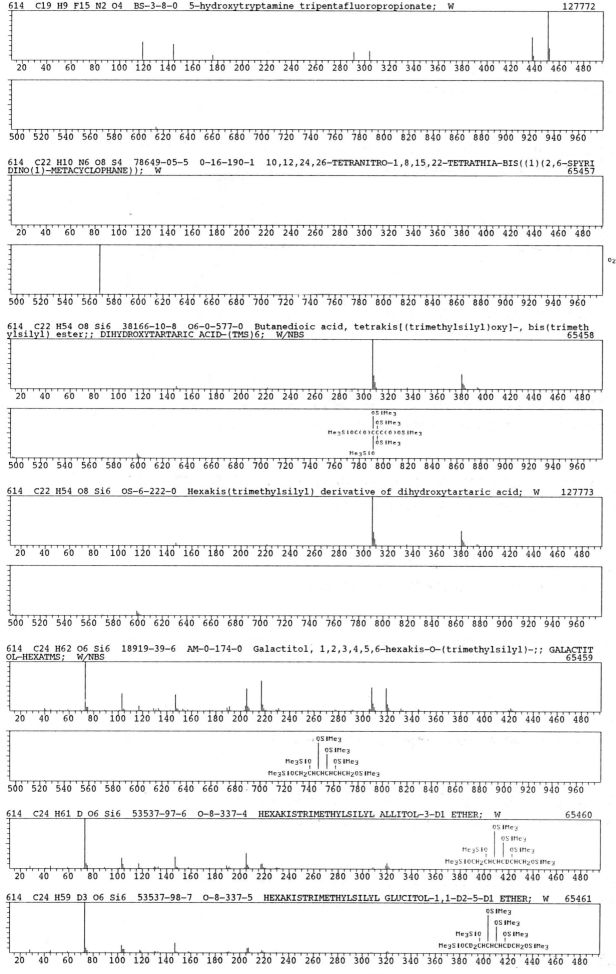

614   C22 H10 N6 O8 S4   78649-05-5   0-16-190-1   10,12,24,26-TETRANITRO-1,8,15,22-TETRATHIA-BIS((1)(2,6-SPYRI
DINO(1)-METACYCLOPHANE));   W   65457

614   C22 H54 O8 Si6   38166-10-8   06-0-577-0   Butanedioic acid, tetrakis[(trimethylsilyl)oxy]-, bis(trimeth
ylsilyl) ester;;   DIHYDROXYTARTARIC ACID-(TMS)6;   W/NBS   65458

OSiMe3
|
OSiMe3
||
Me3SIOC(O)CCC(O)OSiMe3
||
OSiMe3
|
Me3SiO

614   C22 H54 O8 Si6   OS-6-222-0   Hexakis(trimethylsilyl) derivative of dihydroxytartaric acid;   W   127773

614   C24 H62 O6 Si6   18919-39-6   AM-0-174-0   Galactitol, 1,2,3,4,5,6-hexakis-O-(trimethylsilyl)-;;   GALACTIT
OL-HEXATMS;   W/NBS   65459

OSiMe3
|
OSiMe3
|
Me3SIO   OSiMe3
| |
Me3SIOCH2CHCHCHCHCH2OSiMe3

614   C24 H61 D O6 Si6   53537-97-6   0-8-337-4   HEXAKISTRIMETHYLSILYL ALLITOL-3-D1 ETHER;   W   65460

OSiMe3
|
OSiMe3
|
Me3SIO   OSiMe3
| |
Me3SIOCH2CHCHCDCHCH2OSiMe3

614   C24 H59 D3 O6 Si6   53537-98-7   0-8-337-5   HEXAKISTRIMETHYLSILYL GLUCITOL-1,1-D2-5-D1 ETHER;   W   65461

OSiMe3
|
OSiMe3
|
Me3SIO   OSiMe3
| |
Me3SIOCD2CHCHCDCHCH2OSiMe3

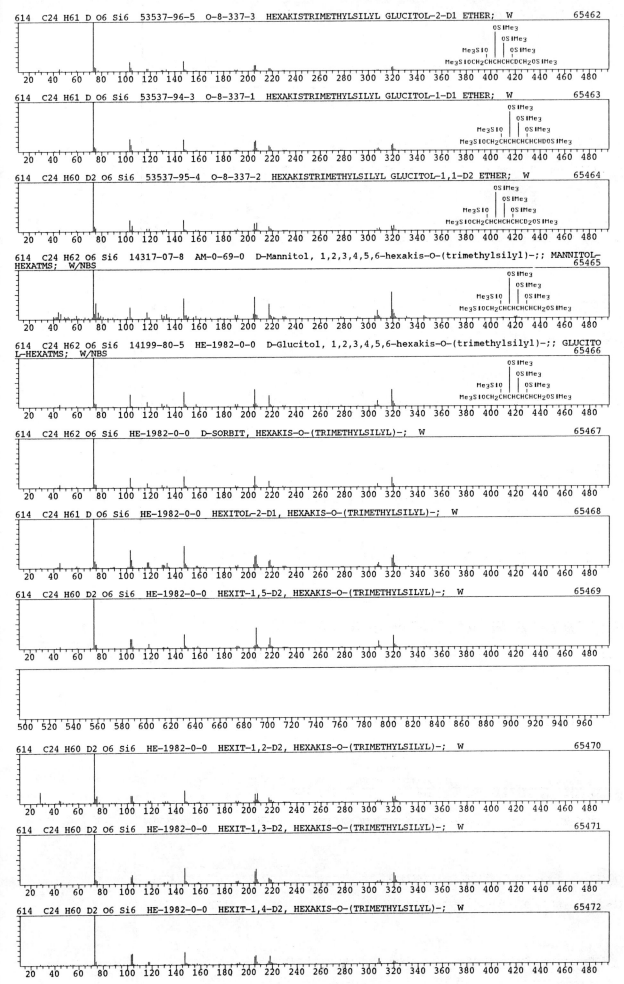

614　C24 H61 D O6 Si6　53537-96-5　O-8-337-3　HEXAKISTRIMETHYLSILYL GLUCITOL-2-D1 ETHER;　W　　　65462

OS Me₃
OS Me₃
Me₃S O　　OS Me₃
Me₃S OCH₂CHCHCHCDCH₂OS Me₃

20 40 60 80 100 120 140 160 180 200 220 240 260 280 300 320 340 360 380 400 420 440 460 480

614　C24 H61 D O6 Si6　53537-94-3　O-8-337-1　HEXAKISTRIMETHYLSILYL GLUCITOL-1-D1 ETHER;　W　　　65463

OS Me₃
OS Me₃
Me₃S O　　OS Me₃
Me₃S OCH₂CHCHCHCHCHDOS Me₃

20 40 60 80 100 120 140 160 180 200 220 240 260 280 300 320 340 360 380 400 420 440 460 480

614　C24 H60 D2 O6 Si6　53537-95-4　O-8-337-2　HEXAKISTRIMETHYLSILYL GLUCITOL-1,1-D2 ETHER;　W　　　65464

OS Me₃
OS Me₃
Me₃S O　　OS Me₃
Me₃S OCH₂CHCHCHCHCD₂OS Me₃

20 40 60 80 100 120 140 160 180 200 220 240 260 280 300 320 340 360 380 400 420 440 460 480

614　C24 H62 O6 Si6　14317-07-8　AM-0-69-0　D-Mannitol, 1,2,3,4,5,6-hexakis-O-(trimethylsilyl)-;; MANNITOL-
HEXATMS;　W/NBS　　　65465

OS Me₃
OS Me₃
Me₃S O　　OS Me₃
Me₃S OCH₂CHCHCHCHCH₂OS Me₃

20 40 60 80 100 120 140 160 180 200 220 240 260 280 300 320 340 360 380 400 420 440 460 480

614　C24 H62 O6 Si6　14199-80-5　HE-1982-0-0　D-Glucitol, 1,2,3,4,5,6-hexakis-O-(trimethylsilyl)-;; GLUCITO
L-HEXATMS;　W/NBS　　　65466

OS Me₃
OS Me₃
Me₃S O　　OS Me₃
Me₃S OCH₂CHCHCHCHCH₂OS Me₃

20 40 60 80 100 120 140 160 180 200 220 240 260 280 300 320 340 360 380 400 420 440 460 480

614　C24 H62 O6 Si6　HE-1982-0-0　D-SORBIT, HEXAKIS-O-(TRIMETHYLSILYL)-;　W　　　65467

20 40 60 80 100 120 140 160 180 200 220 240 260 280 300 320 340 360 380 400 420 440 460 480

614　C24 H61 D O6 Si6　HE-1982-0-0　HEXITOL-2-D1, HEXAKIS-O-(TRIMETHYLSILYL)-;　W　　　65468

20 40 60 80 100 120 140 160 180 200 220 240 260 280 300 320 340 360 380 400 420 440 460 480

614　C24 H60 D2 O6 Si6　HE-1982-0-0　HEXIT-1,5-D2, HEXAKIS-O-(TRIMETHYLSILYL)-;　W　　　65469

20 40 60 80 100 120 140 160 180 200 220 240 260 280 300 320 340 360 380 400 420 440 460 480

500 520 540 560 580 600 620 640 660 680 700 720 740 760 780 800 820 840 860 880 900 920 940 960

614　C24 H60 D2 O6 Si6　HE-1982-0-0　HEXIT-1,2-D2, HEXAKIS-O-(TRIMETHYLSILYL)-;　W　　　65470

20 40 60 80 100 120 140 160 180 200 220 240 260 280 300 320 340 360 380 400 420 440 460 480

614　C24 H60 D2 O6 Si6　HE-1982-0-0　HEXIT-1,3-D2, HEXAKIS-O-(TRIMETHYLSILYL)-;　W　　　65471

20 40 60 80 100 120 140 160 180 200 220 240 260 280 300 320 340 360 380 400 420 440 460 480

614　C24 H60 D2 O6 Si6　HE-1982-0-0　HEXIT-1,4-D2, HEXAKIS-O-(TRIMETHYLSILYL)-;　W　　　65472

20 40 60 80 100 120 140 160 180 200 220 240 260 280 300 320 340 360 380 400 420 440 460 480

614 C24 H60 D2 O6 Si6 HE-1982-0-0 MANNITOL-1,1-D2, HEXAKIS-O-(TRIMETHYLSILYL)-; W   65598

614 C27 H50 O15 56248-59-0 AU-17-7-19 L-Arabinose, O-2,3,4,6-tetra-O-methyl-β-D-galactopyranosyl-(1.fwd
arw.6)-O-2,3,4-tri-O-methyl-β-D-galactopyranosyl-(1.fwdarw.3)-2,4,5-tri-O-methyl-;; PERMETHYL(O-β-D-GALP-(
1-6)-O-β-D-GALP-(1-3)-L-ARA); W/NBS   65473

614 C27 H50 O15 60944-32-3 KO-2-158-2 PERMETHYLATED TRISACCHARIDE; W   65474

614 C28 H24 Br2 O2 S2 71095-54-0 KC-1979-837-0 1,6-BIS(2-BROMOPHENYL)-1,6-DIPHENYL-3,4-DITHIAHEXANE-1,
6-DIOL; W   65475

614 C29 H40 F6 O7 72049-91-3 J-45-339-0 Methyl ester of (3α,5β,7α,12α)-3-hydroxy-7,12-bis[(trifluoroace
tyl)oxy]cholan-24-oic acid; W   127774

614 C31 H48 Cr O5 P2 56557-14-3 AG-94-59-0 (Dicyclohexylphosphine)(pentacyclocarbonyl)chromium; W
                                                                                      127775

614 C34 H27 Cl O9 70551-23-4 H-62-386-0 1,3,4,5-TETRA-O-BENZOYL-β-D-FRUCTOPYRANOSYL CHLORIDE; W 65476

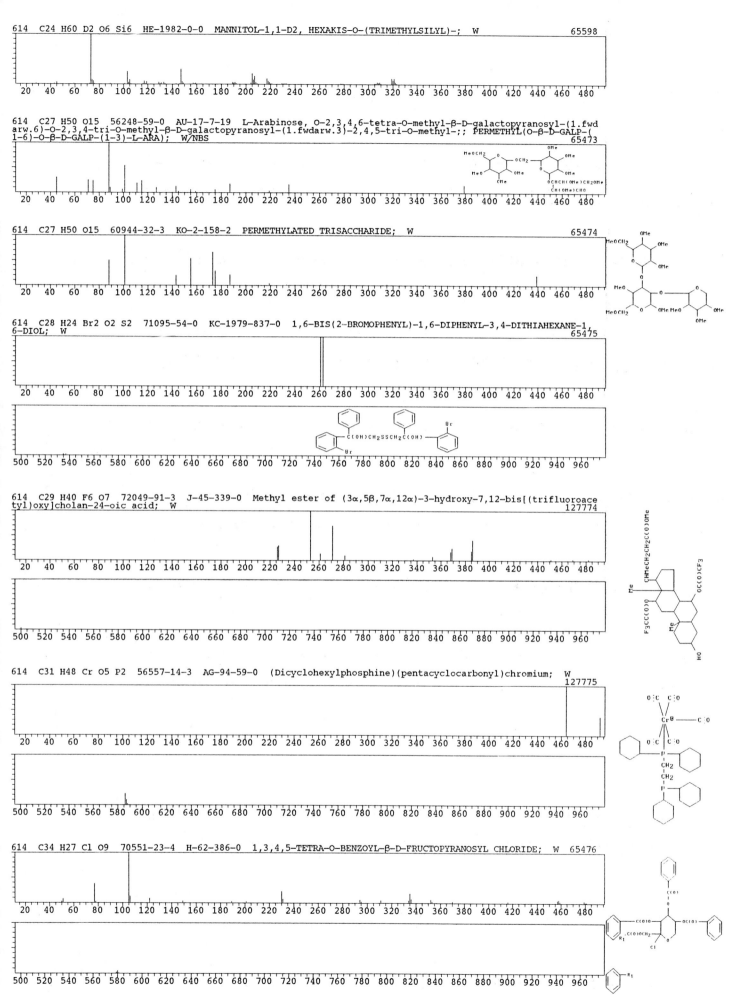

614   C34 H30 O11   49586-44-9   EP-6800-0-0   Candicanin acetate;   W/NBS   65477

614   C35 H39 Cl N4 O4   Y-17-485-0   2-(2-Chloroethyl)-4-ethyl-6-methoxycarbonyl-7-(3-Methoxycarbonylpropyl)-
1,3,5,8-tetramethylporphyrin;   W   127776

614   C35 H39 Cl N4 O4   87944-95-4   KC-1983-2334-0   2-(2-Chloroethyl)-6,7-bis(2-methoxycarbonylethyl)-1,3,4,
5,8-pentamethylporphyrin;   W   127777

614   C35 H42 N4 O6   71189-93-0   K-112-2254-0   (+)-(2R,16R)-PHYCOERY THROBILIN DIMETHYL ESTER;   W   65478

614   C35 H42 N4 O6   68330-93-8   C-100-5932-0   (+)-(25,16R)-PHYCOERYTHROBILIN DIMETHYL ESTER;   W   65479

614   C35 H42 N4 O6   21059-15-4   NS-0-0-0   21H-Biline-8,12-dipropanoic acid, 3,18-diethyl-1,19,22,24-tetrahy
dro-2,7,13,17-tetramethyl-1,19-dioxo-, dimethyl ester;   W/NBS   127778

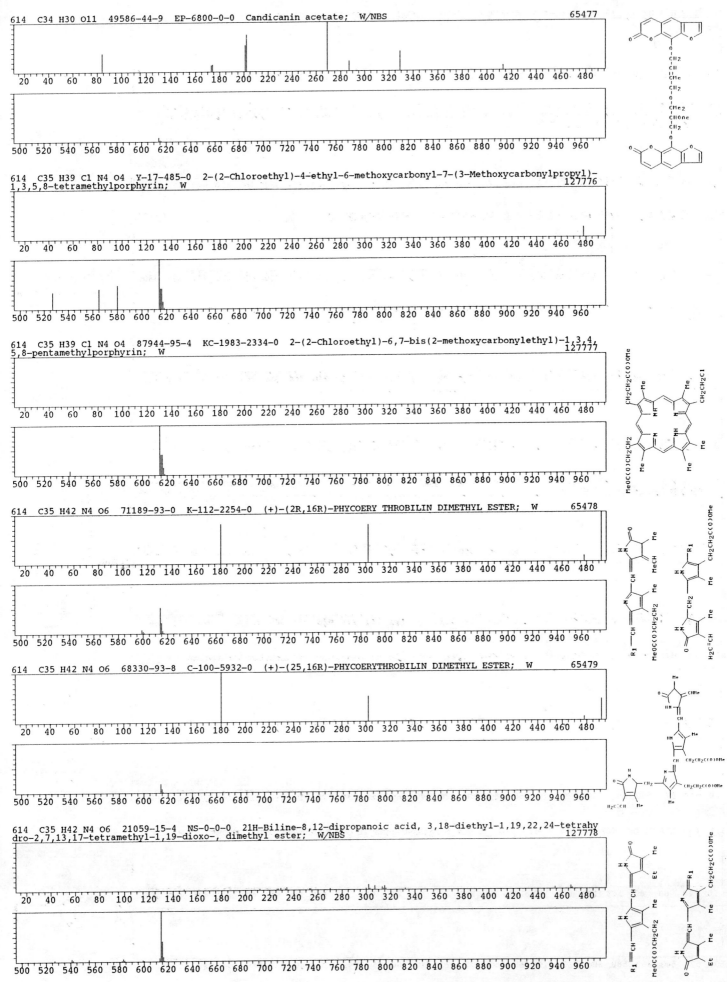

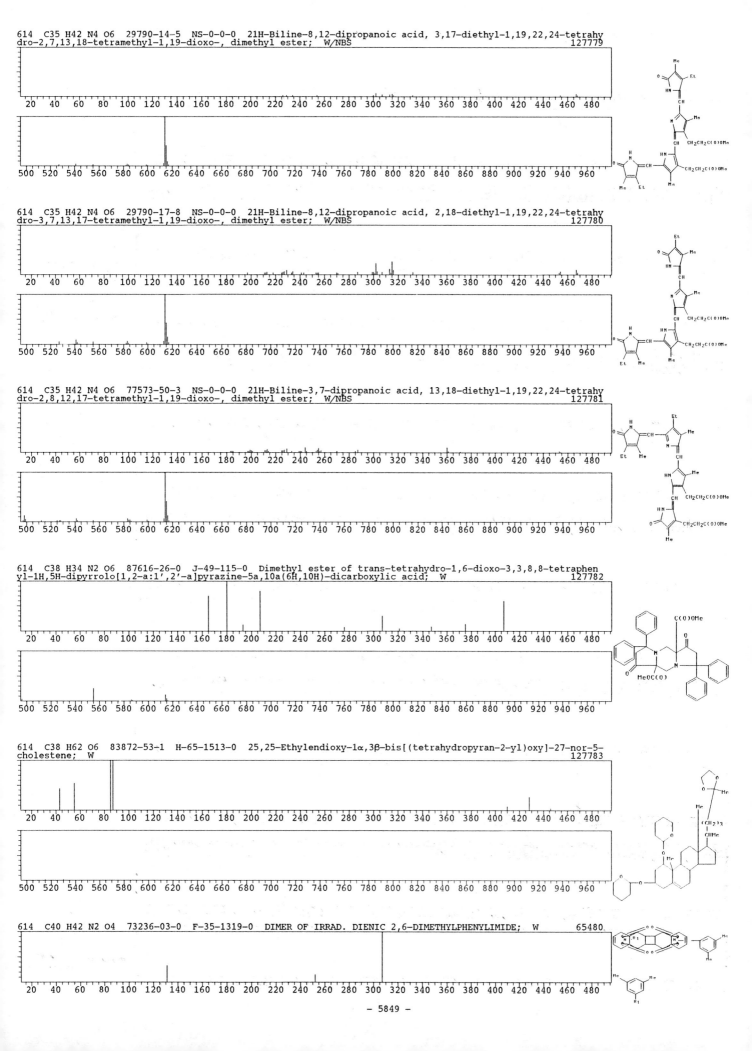

614 C35 H42 N4 O6 29790-14-5 NS-0-0-0 21H-Biline-8,12-dipropanoic acid, 3,17-diethyl-1,19,22,24-tetrahy
dro-2,7,13,18-tetramethyl-1,19-dioxo-, dimethyl ester; W/NBS   127779

614 C35 H42 N4 O6 29790-17-8 NS-0-0-0 21H-Biline-8,12-dipropanoic acid, 2,18-diethyl-1,19,22,24-tetrahy
dro-3,7,13,17-tetramethyl-1,19-dioxo-, dimethyl ester; W/NBS   127780

614 C35 H42 N4 O6 77573-50-3 NS-0-0-0 21H-Biline-3,7-dipropanoic acid, 13,18-diethyl-1,19,22,24-tetrahy
dro-2,8,12,17-tetramethyl-1,19-dioxo-, dimethyl ester; W/NBS   127781

614 C38 H34 N2 O6 87616-26-0 J-49-115-0 Dimethyl ester of trans-tetrahydro-1,6-dioxo-3,3,8,8-tetraphen
yl-1H,5H-dipyrrolo[1,2-a:1',2'-a]pyrazine-5a,10a(6H,10H)-dicarboxylic acid; W   127782

614 C38 H62 O6 83872-53-1 H-65-1513-0 25,25-Ethylendioxy-1α,3β-bis[(tetrahydropyran-2-yl)oxy]-27-nor-5-
cholestene; W   127783

614 C40 H42 N2 O4 73236-03-0 F-35-1319-0 DIMER OF IRRAD. DIENIC 2,6-DIMETHYLPHENYLIMIDE; W   65480

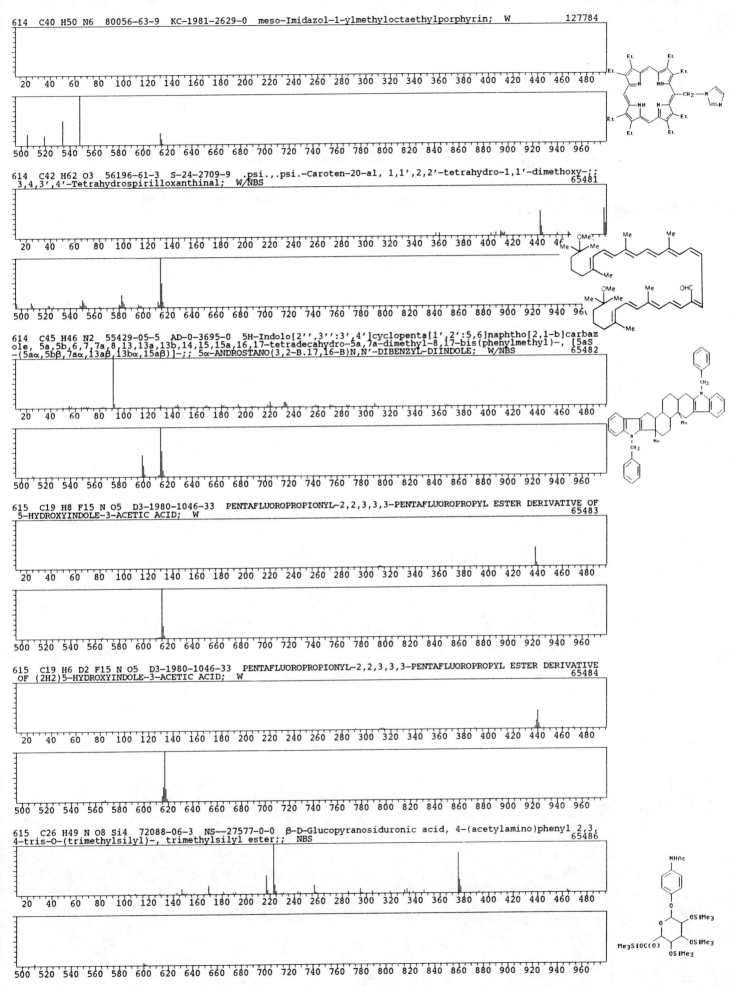

614   C40 H50 N6   80056-63-9   KC-1981-2629-0   meso-Imidazol-1-ylmethyloctaethylporphyrin;   W   127784

614   C42 H62 O3   56196-61-3   S-24-2709-9   .psi.,.psi.-Caroten-20-al, 1,1',2,2'-tetrahydro-1,1'-dimethoxy-;;
3,4,3',4'-Tetrahydrospirilloxanthinal;   W/NBS   65481

614   C45 H46 N2   55429-05-5   AD-0-3695-0   5H-Indolo[2'',3'':3',4']cyclopenta[1',2':5,6]naphtho[2,1-b]carbaz
ole, 5a,5b,6,7,7a,8,13,13a,13b,14,15,15a,16,17-tetradecahydro-5a,7a-dimethyl-8,17-bis(phenylmethyl)-, [5aS
-(5aα,5bβ,7aα,13aβ,13bα,15aβ)]-;; 5α-ANDROSTANO(3,2-B.17,16-B)N,N'-DIBENZYL-DIINDOLE;   W/NBS   65482

615   C19 H8 F15 N O5   D3-1980-1046-33   PENTAFLUOROPROPIONYL-2,2,3,3,3-PENTAFLUOROPROPYL ESTER DERIVATIVE OF
5-HYDROXYINDOLE-3-ACETIC ACID;   W   65483

615   C19 H6 D2 F15 N O5   D3-1980-1046-33   PENTAFLUOROPROPIONYL-2,2,3,3,3-PENTAFLUOROPROPYL ESTER DERIVATIVE
OF (2H2)5-HYDROXYINDOLE-3-ACETIC ACID;   W   65484

615   C26 H49 N O8 Si4   72088-06-3   NS--27577-0-0   β-D-Glucopyranosiduronic acid, 4-(acetylamino)phenyl 2,3,
4-tris-O-(trimethylsilyl)-, trimethylsilyl ester;;   NBS   65486

615   C26 H49 N O8 Si4   EP-1750-0-0   4-GLUCURONOSIDO-N-ACETYLANILINE-TETRATMS;   W   65485

615   C28 H53 N3 O8 Si2   27089-69-6   AB-270-129-10   Tridecylamine, N-(2,4-dinitrophenyl)-2-methoxy-1-(meth
oxymethyl)-3,4-bis(trimethylsiloxy)-, erythro-;; D-ERYTHRO-2-DINITROPHENYLAMINO-1,3-DIMETHOXY-4,5-BISTRI
METHYLSILYLOXYTETRADECANE;   W/NBS   65487

615   C33 H33 N3 O9   79087-35-7   C-103-5135-0   1,3,5-Tris(N-methyl-N-(2,3-dihydroxybenzoyl)aminomethyl)benze
ne;   W   127785

615   C33 H45 N O10   CD-289-0-0   Hypaconitine;   W   127786

615   C34 H57 N O5 Si2   OM-1981-323-0   PGE2 ME ESTER BENZYL OXIME 2TMS;   W   65488

615   C34 H57 N O5 Si2   OM-1981-328-0   PGD-2 BZ OXIME METHYL ESTER 2TMS;   W   65489

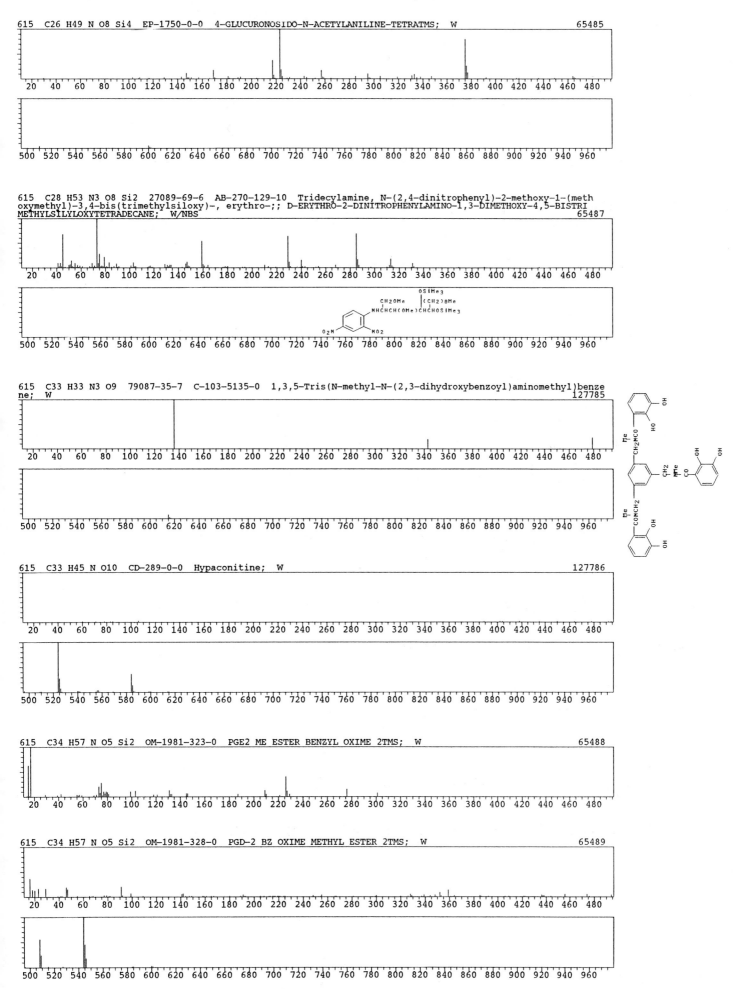

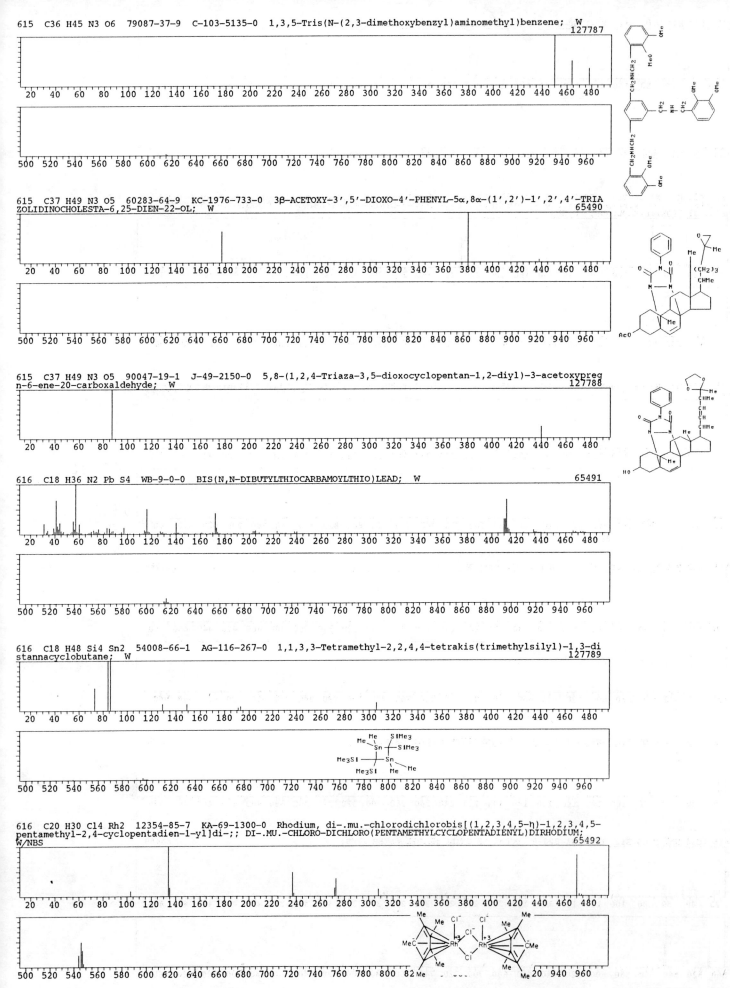

615　C36 H45 N3 O6　79087-37-9　C-103-5135-0　1,3,5-Tris(N-(2,3-dimethoxybenzyl)aminomethyl)benzene;　W
127787

615　C37 H49 N3 O5　60283-64-9　KC-1976-733-0　3β-ACETOXY-3',5'-DIOXO-4'-PHENYL-5α,8α-(1',2')-1',2',4'-TRIA
ZOLIDINOCHOLESTA-6,25-DIEN-22-OL;　W
65490

615　C37 H49 N3 O5　90047-19-1　J-49-2150-0　5,8-(1,2,4-Triaza-3,5-dioxocyclopentan-1,2-diyl)-3-acetoxypreg
n-6-ene-20-carboxaldehyde;　W
127788

616　C18 H36 N2 Pb S4　WB-9-0-0　BIS(N,N-DIBUTYLTHIOCARBAMOYLTHIO)LEAD;　W
65491

616　C18 H48 Si4 Sn2　54008-66-1　AG-116-267-0　1,1,3,3-Tetramethyl-2,2,4,4-tetrakis(trimethylsilyl)-1,3-di
stannacyclobutane;　W
127789

616　C20 H30 Cl4 Rh2　12354-85-7　KA-69-1300-0　Rhodium, di-.mu.-chlorodichlorobis[(1,2,3,4,5-η)-1,2,3,4,5-
pentamethyl-2,4-cyclopentadien-1-yl]di-;; DI-.MU.-CHLORO-DICHLORO(PENTAMETHYLCYCLOPENTADIENYL)DIRHODIUM;
W/NBS
65492

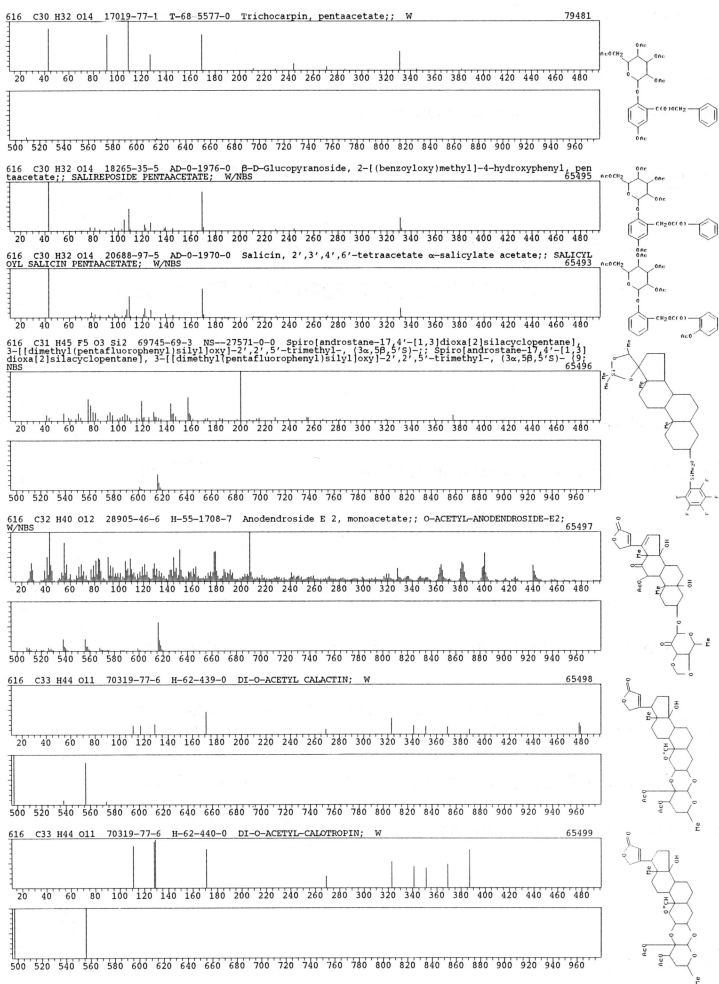

616   C30 H32 O14   17019-77-1   T-68-5577-0   Trichocarpin, pentaacetate;;   W   79481

616   C30 H32 O14   18265-35-5   AD-0-1976-0   β-D-Glucopyranoside, 2-[(benzoyloxy)methyl]-4-hydroxyphenyl, pen
taacetate;;   SALIREPOSIDE PENTAACETATE;   W/NBS   65495

616   C30 H32 O14   20688-97-5   AD-0-1970-0   Salicin, 2',3',4',6'-tetraacetate α-salicylate acetate;;   SALICYL
OYL SALICIN PENTAACETATE;   W/NBS   65493

616   C31 H45 F5 O3 Si2   69745-69-3   NS--27571-0-0   Spiro[androstane-17,4'-[1,3]dioxa[2]silacyclopentane],
3-[[dimethyl(pentafluorophenyl)silyl]oxy]-2',2',5'-trimethyl-, (3α,5β,5'S)-;; Spiro[androstane-17,4'-[1,3]
dioxa[2]silacyclopentane], 3-[[dimethyl(pentafluorophenyl)silyl]oxy]-2',2',5'-trimethyl-, (3α,5β,5'S)- (9;
NBS   65496

616   C32 H40 O12   28905-46-6   H-55-1708-7   Anodendroside E 2, monoacetate;;   O-ACETYL-ANODENDROSIDE-E2;
W/NBS   65497

616   C33 H44 O11   70319-77-6   H-62-439-0   DI-O-ACETYL CALACTIN;   W   65498

616   C33 H44 O11   70319-77-6   H-62-440-0   DI-O-ACETYL-CALOTROPIN;   W   65499

616   C34 H32 O11   79004-03-8   B-34-1155-0   Benzyl 3-methoxy-2,4-di-O-methylgyrophorate;   W   127790

616   C34 H48 O10   73025-04-4   B-32-2504-0   SIMPLEXIN DIACETATE;   W   65500

616   C35 H44 N4 O6   66385-51-1   K-111-500-0   DIMETHYL ESTER OF (4R,16R)-3-ETHYL-2,7,13,17-TETRAMETHYL-1,19-DIOXO-18-VINYL-1,4,5,15,16,19,21,24-OCTAHYDRO-22H-BILIN-8,12-DIPROPIONIC ACID;   W   65501

616   C35 H44 N4 O6   66385-52-2   K-111-501-0   DIMETHYL ESTER OF (14R,16S)-3-ETHYL-2,7,13,17-TETRAMETHYL-1,19-DIOXO-18-VINYL-1,4,5,15,16,19,21,24-OCTAHYDRO-22H-BILIN-8,12-DIPROPIONIC ACID;   W   65502

616   C35 H44 N4 O6   29789-71-7   NS-0-0-0   21H-Biline-8,12-dipropanoic acid, 3,17-diethyl-1,4,5,19,23,24-hexahydro-2,7,13,18-tetramethyl-1,19-dioxo-, dimethyl ester, (4R)-;   W/NBS   127791

616   C35 H44 N4 O6   29789-73-9   NS-0-0-0   21H-Biline-8,12-dipropanoic acid, 3,18-diethyl-1,4,5,19,23,24-hexahydro-2,7,13,17-tetramethyl-1,19-dioxo-, dimethyl ester, (4R)-;   W/NBS   127792

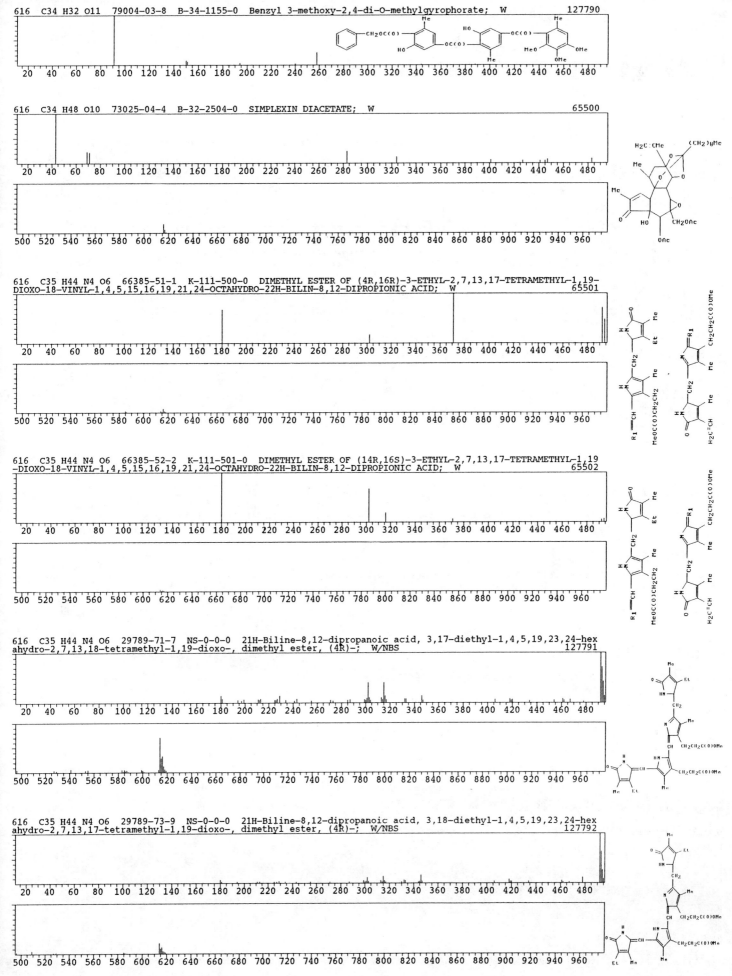

- 5854 -

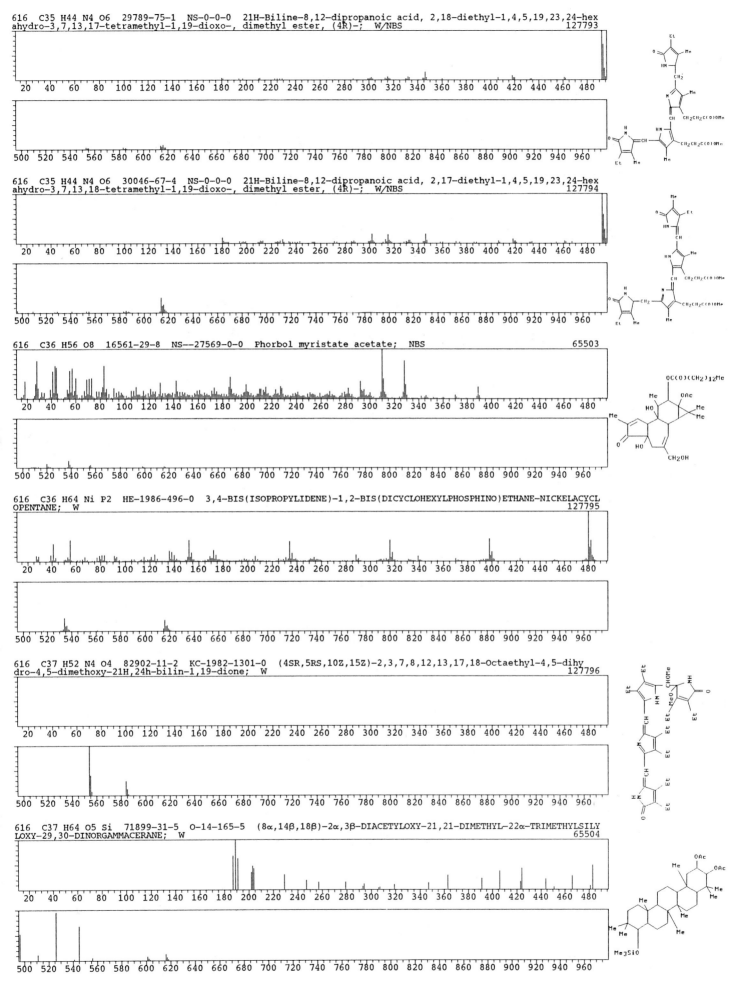

616   C35 H44 N4 O6   29789-75-1   NS-0-0-0   21H-Biline-8,12-dipropanoic acid, 2,18-diethyl-1,4,5,19,23,24-hex
ahydro-3,7,13,17-tetramethyl-1,19-dioxo-, dimethyl ester, (4R)-;   W/NBS   127793

616   C35 H44 N4 O6   30046-67-4   NS-0-0-0   21H-Biline-8,12-dipropanoic acid, 2,17-diethyl-1,4,5,19,23,24-hex
ahydro-3,7,13,18-tetramethyl-1,19-dioxo-, dimethyl ester, (4R)-;   W/NBS   127794

616   C36 H56 O8   16561-29-8   NS--27569-0-0   Phorbol myristate acetate;   NBS   65503

616   C36 H64 Ni P2   HE-1986-496-0   3,4-BIS(ISOPROPYLIDENE)-1,2-BIS(DICYCLOHEXYLPHOSPHINO)ETHANE-NICKELACYCL
OPENTANE;   W   127795

616   C37 H52 N4 O4   82902-11-2   KC-1982-1301-0   (4SR,5RS,10Z,15Z)-2,3,7,8,12,13,17,18-Octaethyl-4,5-dihy
dro-4,5-dimethoxy-21H,24h-bilin-1,19-dione;   W   127796

616   C37 H64 O5 Si   71899-31-5   O-14-165-5   (8α,14β,18β)-2α,3β-DIACETYLOXY-21,21-DIMETHYL-22α-TRIMETHYLSILY
LOXY-29,30-DINORGAMMACERANE;   W   65504

616 C40 H48 N4 O2 CD-76-0-0 Kopsoffine; W   127797

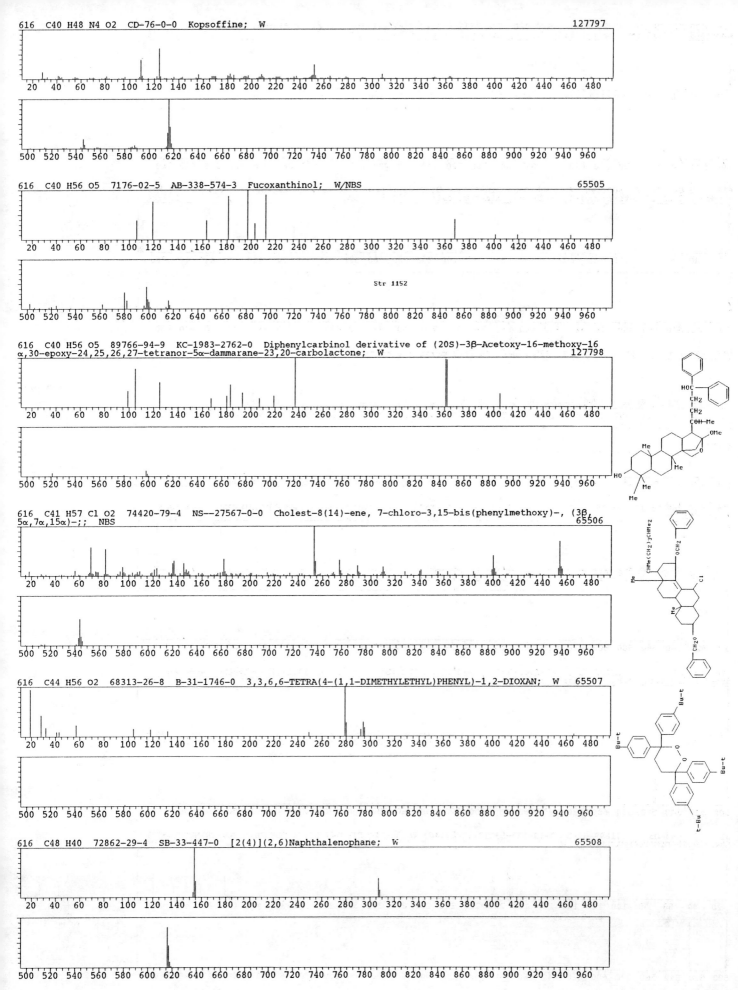

616 C40 H56 O5 7176-02-5 AB-338-574-3 Fucoxanthinol; W/NBS   65505

Str 1152

616 C40 H56 O5 89766-94-9 KC-1983-2762-0 Diphenylcarbinol derivative of (20S)-3β-Acetoxy-16-methoxy-16
α,30-epoxy-24,25,26,27-tetranor-5α-dammarane-23,20-carbolactone; W   127798

616 C41 H57 Cl O2 74420-79-4 NS--27567-0-0 Cholest-8(14)-ene, 7-chloro-3,15-bis(phenylmethoxy)-, (3β,
5α,7α,15α)-;; NBS   65506

616 C44 H56 O2 68313-26-8 B-31-1746-0 3,3,6,6-TETRA(4-(1,1-DIMETHYLETHYL)PHENYL)-1,2-DIOXAN; W   65507

616 C48 H40 72862-29-4 SB-33-447-0 [2(4)](2,6)Naphthalenophane; W   65508

616 C48 H40 19576-05-7 K-116-3499-0 5,6,8,9,21,22,24,25-Octahydro[2.2](3,11)dibenzo[a,j]anthracenophane
; W   127799

617 C19 H10 F15 N O5 KI-0-31-0 TRI-PENTAFLUOROPROPIONATE; W   65509

617 C19 H10 F15 N O5 KI-0-39-0 TRI-PENTAFLUOROPROPIONATE; W   65510

617 C33 H35 N3 O5 S2 55836-93-6 KO-1-93-26 1-Naphthalenesulfonic acid, 5-(dimethylamino)-, 4-[2-[[[5-(
dimethylamino)-1-naphthalenyl]sulfonyl]amino]propyl]phenyl ester, (.+-.)-;; BIS(1-DIMETHYLAMINONAPHTHALENE-
5-SULPHONYL)-P-HYDROXYAMPHETAMINE; W/NBS   65511

617 C33 H35 N3 O5 S2 36600-88-1 NS-0-0-0 1-Naphthalenesulfonic acid, 5-(dimethylamino)-, 4-[2-[[[5-(di
methylamino)-1-naphthalenyl]sulfonyl]amino]propyl]phenyl ester; W/NBS   127800

617 C37 H44 Cl N O5 0-16-404-1 PENITREM D; W   65512

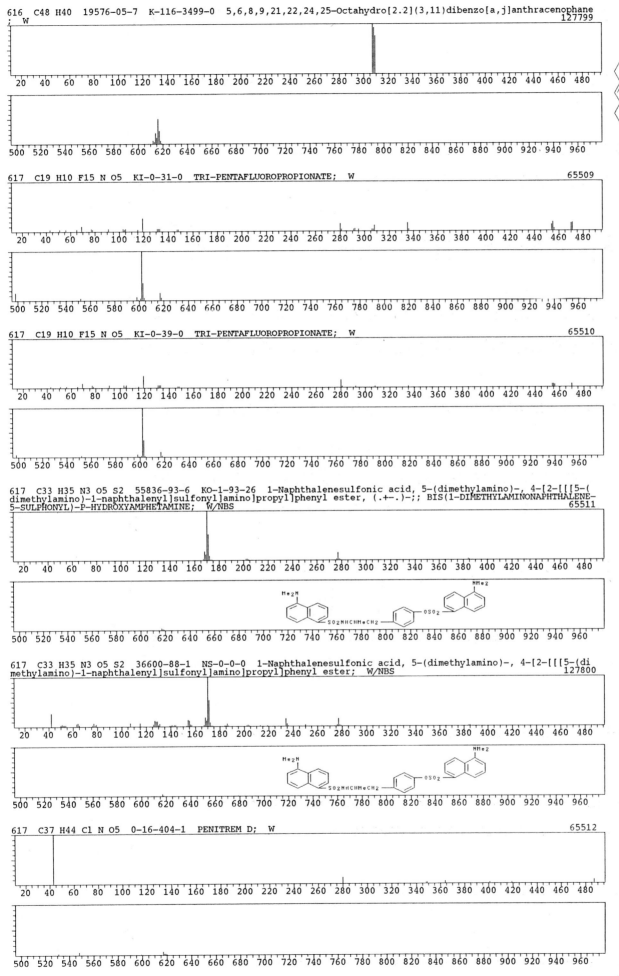

618   C16 H50 B F3 N6 Si8   K-114-2291-0   Fluorobis[3-fluorodimethylsilyl-2,2,4,4,6,6-hexamethyl-1,3,5-triaza
-2,4,6-trisilacyclohexyl]borane;   W   127801

618   C18 H6 Co4 O10   12568-53-5   AD-0-3682-0   Cobalt, decacarbonyl(ethynylbenzene)tetra-;; .MU.-(PHENYLACET
YLENE)-DECACARBONYLTETRACOBALT;   W/NBS   65513

[Co₄(CO)₁₀(PhC≡CH)]

618   C22 H34 S10   17525-43-8   EP-7814-0-0   2,4,6,8-Tetrathiatricyclo[3.3.1.1(3,7)]decane, 1,1'-dithiobis[3,
5,7,10,10-pentamethyl-;; 1,1'-DITHIOBIS(3,5,7,10,10-PENTAMETHYL-2,4,6,8-TETRATHIADAMANTANE);   W/NBS   65514

618   C26 H20 I2 O2   84389-87-7   KC-1982-2320-0   2',2''-Di-iodo-4'',6'-dimethoxy-m-quaterphenyl;   W   127802

618   C26 H61 F3 N4 O3 Si3   KO-10-123-2   Reduced trifluoroacetyl, (tris)trimethylsilyl amino alcohol derivat
ive of valine;   W   127803

618   C30 H62 O7 Si3   OM-1981-660-0   17-PHE-18,19,20-TRINOR F2A ME 3T 618;   W   65515

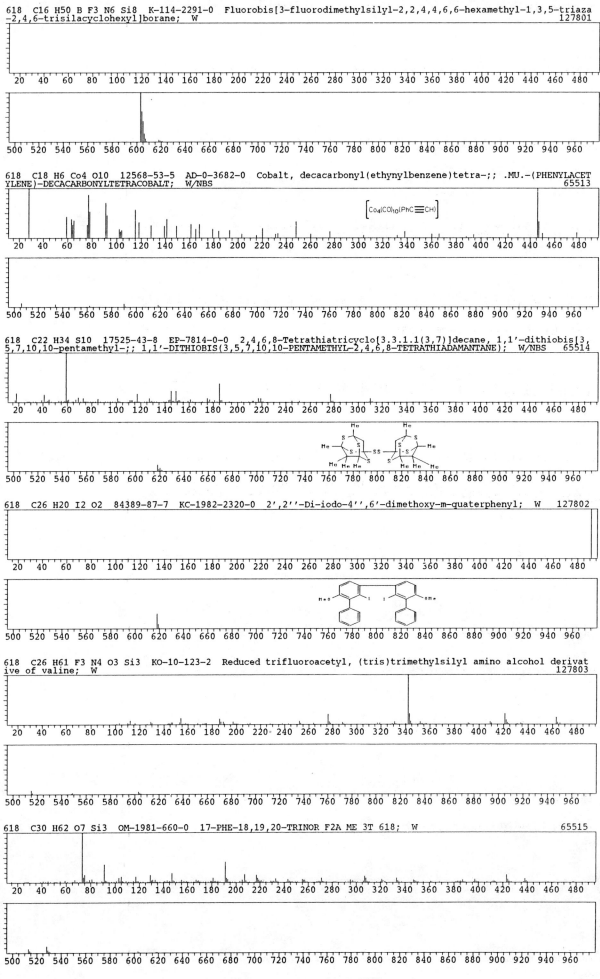

618   C32 H16 Ge N8 S   K-116-1864-0   (Phthalocyaninato)-.mu.-thio-germanium; W   127804

618   C32 H42 O12 56051-65-1 H-55-1705-5 Card-20(22)-enolide, 11-(acetyloxy)-3-[(6-deoxy-3,4-O-methyleneh
exopyranos-2-ulos-1-yl)oxy]-5,14-dihydroxy-12-oxo-, (3β,5α,11α)-;; O-ACETYL-ANODENDROSIDE-E1; W/NBS 65516

618   C33 H46 O11 29685-54-9 KC-1976-1360-0   17-OXOANDROST-5-EN-3β-YL TETRA-O-ACETYL-β-D-GLUCOPYRANOSIDE;
W   65517

618   C34 H42 N4 O7 69755-30-2 H-61-2660-0 4H-1,16-Etheno-5,15-(propaniminoethano)furo[3,4-1][1,5,10]tria
zacyclohexadecine-4,21-dione, 10,14-diacetyl-3-[4-(acetyloxy)phenyl]-3,3a,6,7,8,9,10,11,12,13,14,15-dodeca
hydro-;; 4H-1,16-Etheno-5,15-(propaniminoethano)furo[3,4-1][1,5,10]triazacyclohexadecine-4,21-dione, 10,14-
diacetyl-3-[4-(acetyloxy)phenyl]-3,3a,6,7,8,9,; W/NBS   127805

618   C34 H33 D9 N4 O7 69721-82-0 H-61-2660-0 N(6),N(10),O(34)-Tris(trideuterioacetyl)aphelandrine; W
                                                                                                  127807

618   C34 H42 N4 O7 71657-56-2 H-62-2720-0 N(6),N(10),O(34)-Triacetylorantin; W   127806

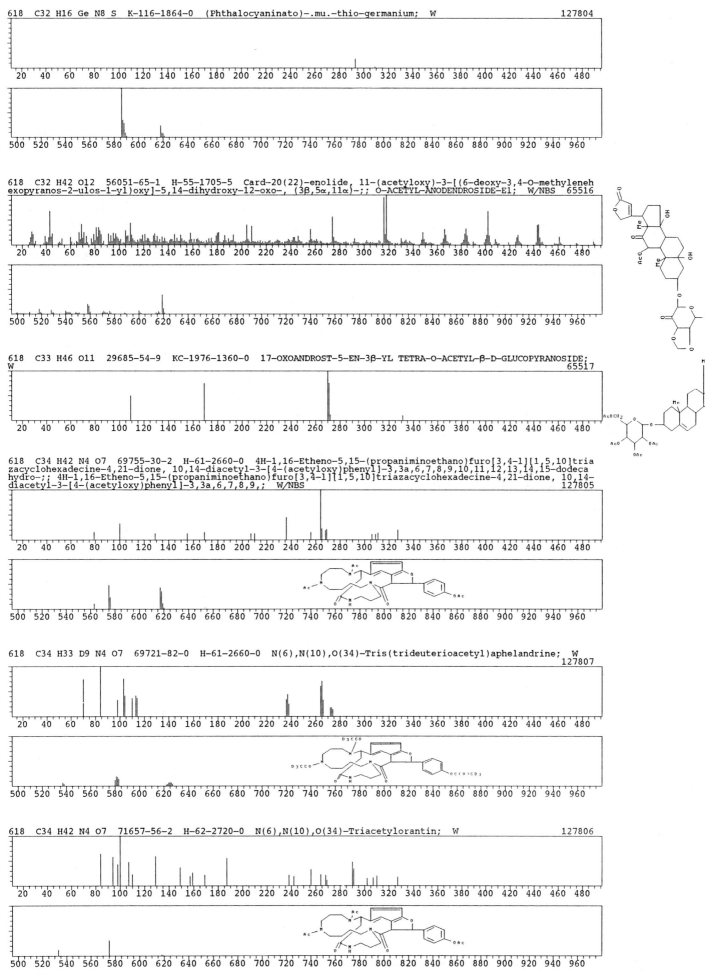

- 5859 -

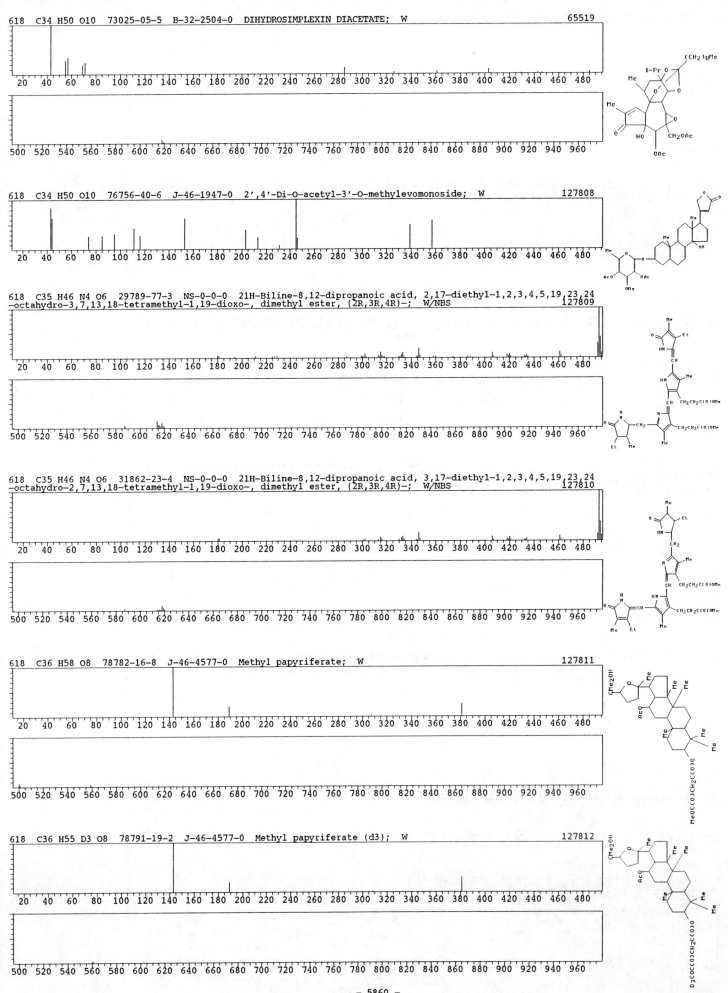

618   C34 H50 O10   73025-05-5   B-32-2504-0   DIHYDROSIMPLEXIN DIACETATE;   W   65519

618   C34 H50 O10   76756-40-6   J-46-1947-0   2',4'-Di-O-acetyl-3'-O-methylevomonoside;   W   127808

618   C35 H46 N4 O6   29789-77-3   NS-0-0-0   21H-Biline-8,12-dipropanoic acid, 2,17-diethyl-1,2,3,4,5,19,23,24
-octahydro-3,7,13,18-tetramethyl-1,19-dioxo-, dimethyl ester, (2R,3R,4R)-;   W/NBS   127809

618   C35 H46 N4 O6   31862-23-4   NS-0-0-0   21H-Biline-8,12-dipropanoic acid, 3,17-diethyl-1,2,3,4,5,19,23,24
-octahydro-2,7,13,18-tetramethyl-1,19-dioxo-, dimethyl ester, (2R,3R,4R)-;   W/NBS   127810

618   C36 H58 O8   78782-16-8   J-46-4577-0   Methyl papyriferate;   W   127811

618   C36 H55 D3 O8   78791-19-2   J-46-4577-0   Methyl papyriferate (d3);   W   127812

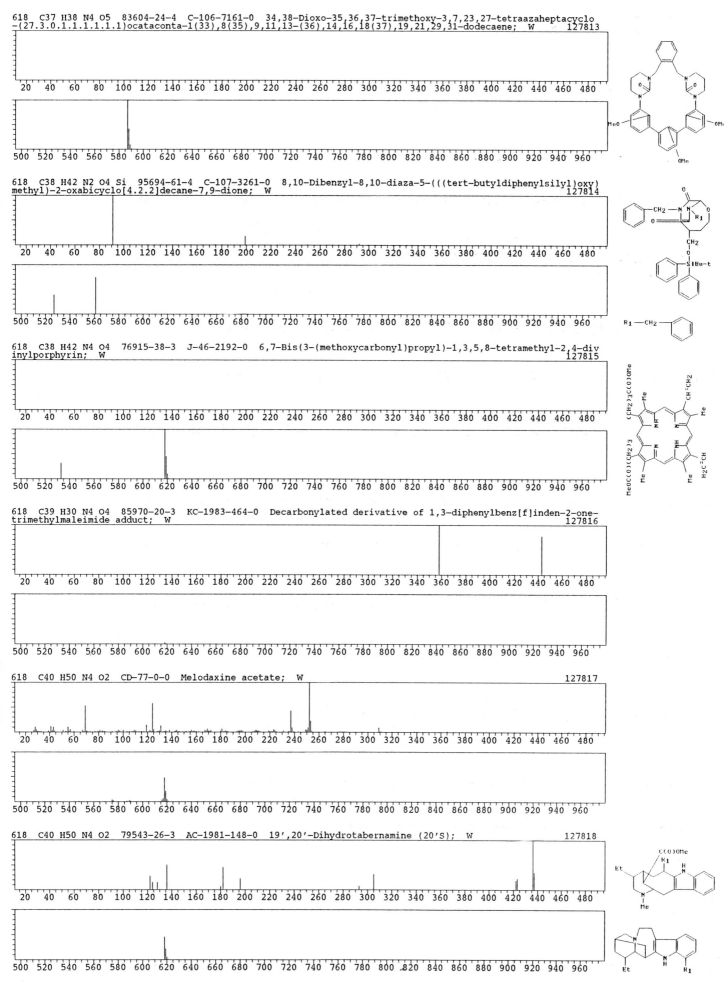

618 C37 H38 N4 O5 83604-24-4 C-106-7161-0 34,38-Dioxo-35,36,37-trimethoxy-3,7,23,27-tetraazaheptacyclo
-(27.3.0.1.1.1.1.1.1)ocataconta-1(33),8(35),9,11,13-(36),14,16,18(37),19,21,29,31-dodecaene; W   127813

618 C38 H42 N2 O4 Si 95694-61-4 C-107-3261-0 8,10-Dibenzyl-8,10-diaza-5-(((tert-butyldiphenylsilyl)oxy)
methyl)-2-oxabicyclo[4.2.2]decane-7,9-dione; W   127814

618 C38 H42 N4 O4 76915-38-3 J-46-2192-0 6,7-Bis(3-(methoxycarbonyl)propyl)-1,3,5,8-tetramethyl2,4-div
inylporphyrin; W   127815

618 C39 H30 N4 O4 85970-20-3 KC-1983-464-0 Decarbonylated derivative of 1,3-diphenylbenz[f]inden-2-one-
trimethylmaleimide adduct; W   127816

618 C40 H50 N4 O2 CD-77-0-0 Melodaxine acetate; W   127817

618 C40 H50 N4 O2 79543-26-3 AC-1981-148-0 19',20'-Dihydrotabernamine (20'S); W   127818

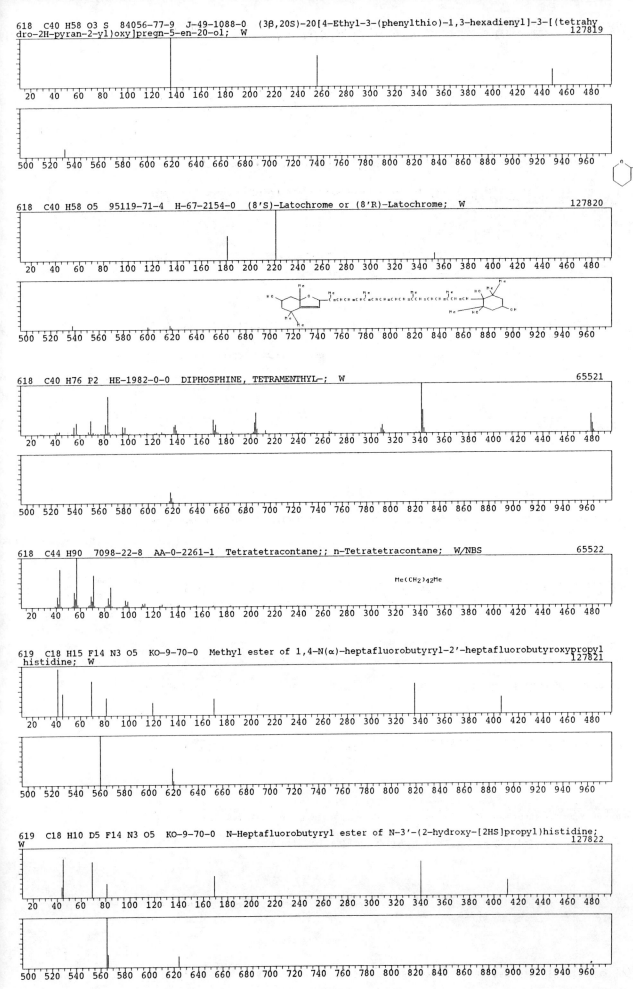

618   C40 H58 O3 S   84056-77-9   J-49-1088-0   (3β,20S)-20[4-Ethyl-3-(phenylthio)-1,3-hexadienyl]-3-[(tetrahy
dro-2H-pyran-2-yl)oxy]pregn-5-en-20-ol;   W   127819

618   C40 H58 O5   95119-71-4   H-67-2154-0   (8'S)-Latochrome or (8'R)-Latochrome;   W   127820

618   C40 H76 P2   HE-1982-0-0   DIPHOSPHINE, TETRAMENTHYL-;   W   65521

618   C44 H90   7098-22-8   AA-0-2261-1   Tetratetracontane;; n-Tetratetracontane;   W/NBS   65522

Me(CH2)42Me

619   C18 H15 F14 N3 O5   KO-9-70-0   Methyl ester of 1,4-N(α)-heptafluorobutyryl-2'-heptafluorobutyroxypropyl
histidine;   W   127821

619   C18 H10 D5 F14 N3 O5   KO-9-70-0   N-Heptafluorobutyryl ester of N-3'-(2-hydroxy-[2HS]propyl)histidine;
W   127822

619 C21 H54 N O8 P Si5 32204-75-4 D-10-929-3 Serine, N-(trimethylsilyl)-, L-, trimethylsilyl ester, 2,
3-bis(trimethylsiloxy)propyl trimethylsilyl phosphate (ester);; TETRAKISTRIMETHYLSILYL-N-TRIMETHYLSILYL-α-
GLYCEROPHOSPHORYLSERINE; W/NBS   65523

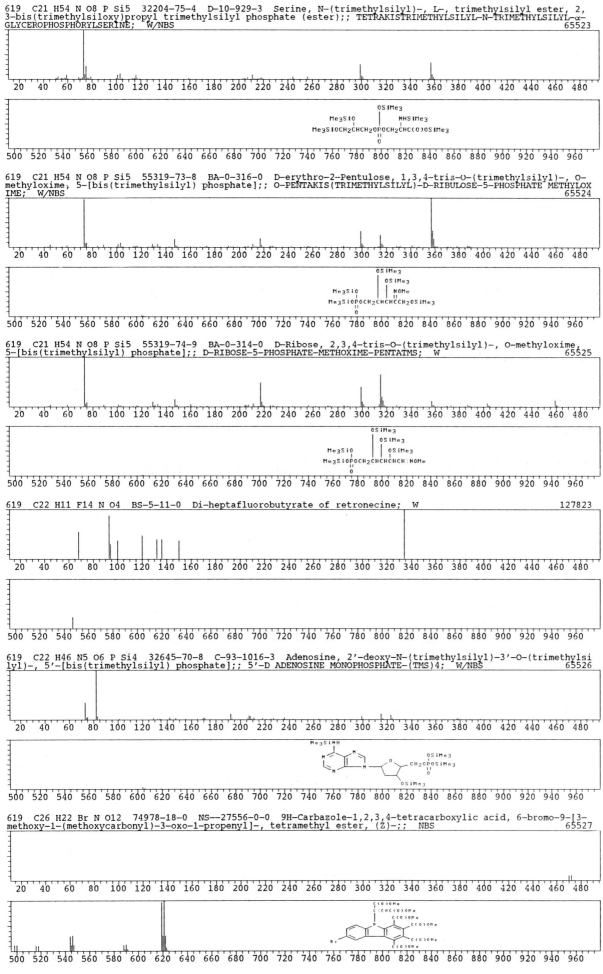

619 C21 H54 N O8 P Si5 55319-73-8 BA-0-316-0 D-erythro-2-Pentulose, 1,3,4-tris-O-(trimethylsilyl)-, O-
methyloxime, 5-[bis(trimethylsilyl) phosphate];; O-PENTAKIS(TRIMETHYLSILYL)-D-RIBULOSE-5-PHOSPHATE METHYLOX
IME; W/NBS   65524

619 C21 H54 N O8 P Si5 55319-74-9 BA-0-314-0 D-Ribose, 2,3,4-tris-O-(trimethylsilyl)-, O-methyloxime,
5-[bis(trimethylsilyl) phosphate];; D-RIBOSE-5-PHOSPHATE-METHOXIME-PENTATMS; W   65525

619 C22 H11 F14 N O4 BS-5-11-0 Di-heptafluorobutyrate of retronecine; W   127823

619 C22 H46 N5 O6 P Si4 32645-70-8 C-93-1016-3 Adenosine, 2'-deoxy-N-(trimethylsilyl)-3'-O-(trimethylsi
lyl)-, 5'-[bis(trimethylsilyl) phosphate];; 5'-D ADENOSINE MONOPHOSPHATE-(TMS)4; W/NBS   65526

619 C26 H22 Br N O12 74978-18-0 NS--27556-0-0 9H-Carbazole-1,2,3,4-tetracarboxylic acid, 6-bromo-9-[3-
methoxy-1-(methoxycarbonyl)-3-oxo-1-propenyl]-, tetramethyl ester, (Z)-;; NBS   65527

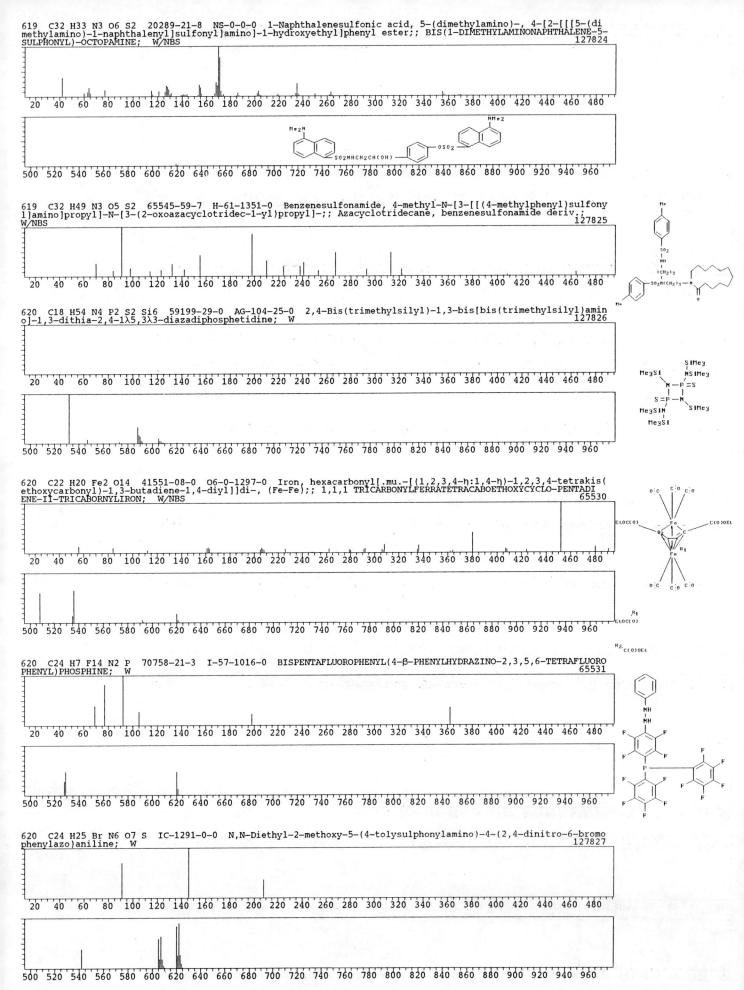

619 C32 H33 N3 O6 S2 20289-21-8 NS-0-0-0 1-Naphthalenesulfonic acid, 5-(dimethylamino)-, 4-[2-[[[5-(di methylamino)-1-naphthalenyl]sulfonyl]amino]-1-hydroxyethyl]phenyl ester;; BIS(1-DIMETHYLAMINONAPHTHALENE-5-SULPHONYL)-OCTOPAMINE; W/NBS   127824

619 C32 H49 N3 O5 S2 65545-59-7 H-61-1351-0 Benzenesulfonamide, 4-methyl-N-[3-[[(4-methylphenyl)sulfony l]amino]propyl]-N-[3-(2-oxoazacyclotridec-1-yl)propyl]-;; Azacyclotridecane, benzenesulfonamide deriv.; W/NBS   127825

620 C18 H54 N4 P2 S2 Si6 59199-29-0 AG-104-25-0 2,4-Bis(trimethylsilyl)-1,3-bis[bis(trimethylsilyl)amin o]-1,3-dithia-2,4-1λ5,3λ3-diazadiphosphetidine; W   127826

620 C22 H20 Fe2 O14 41551-08-0 O6-0-1297-0 Iron, hexacarbonyl[.mu.-[(1,2,3,4-η:1,4-η)-1,2,3,4-tetrakis(ethoxycarbonyl)-1,3-butadiene-1,4-diyl]]di-, (Fe-Fe);; 1,1,1 TRICARBONYLFERRATETRACABOETHOXYCYCLO-PENTADI ENE-II-TRICABORNYLIRON; W/NBS   65530

620 C24 H7 F14 N2 P 70758-21-3 I-57-1016-0 BISPENTAFLUOROPHENYL(4-β-PHENYLHYDRAZINO-2,3,5,6-TETRAFLUORO PHENYL)PHOSPHINE; W   65531

620 C24 H25 Br N6 O7 S IC-1291-0-0 N,N-Diethyl-2-methoxy-5-(4-tolysulphonylamino)-4-(2,4-dinitro-6-bromo phenylazo)aniline; W   127827

620 C26 H45 Cl O9 Si3 55373-82-5 BA-0-208-0 D-Glucuronic acid, 2,3,4-tris-O-(trimethylsilyl)-, methyl ester, 2-(4-chlorophenoxy)-2-methylpropanoate;; 2-(4-CHLOROPHENOXY)-2-METHYLPROPIONIC ESTER OF METHYL O2, O3,O4-TRIS(TRIMETHYLSILYL)GLUCURONATE; W/NBS   65532

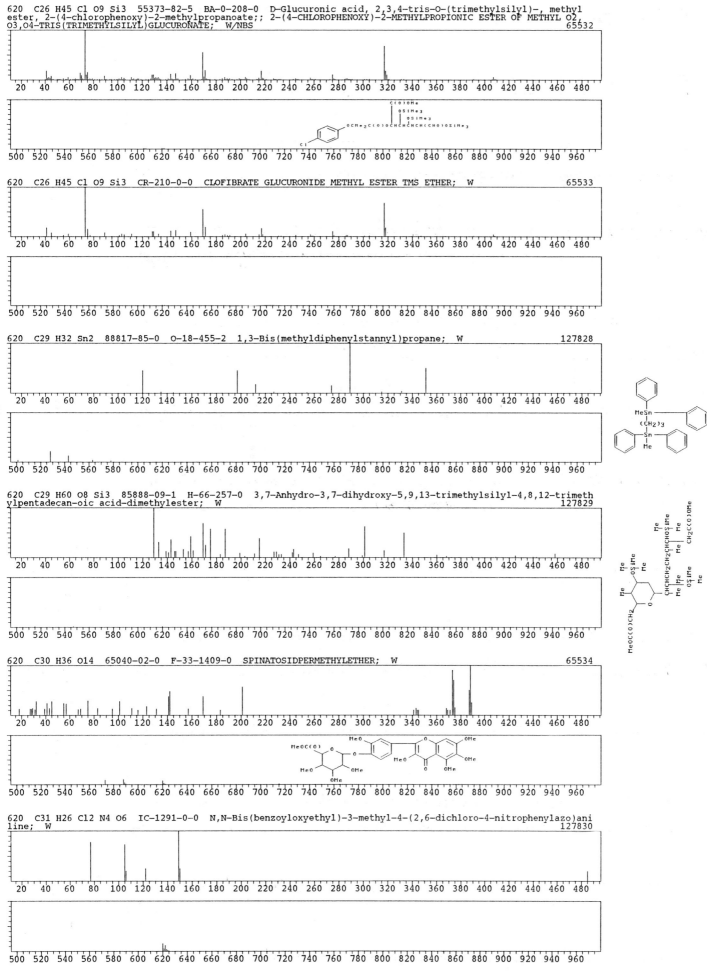

620 C26 H45 Cl O9 Si3 CR-210-0-0 CLOFIBRATE GLUCURONIDE METHYL ESTER TMS ETHER; W   65533

620 C29 H32 Sn2 88817-85-0 O-18-455-2 1,3-Bis(methyldiphenylstannyl)propane; W   127828

620 C29 H60 O8 Si3 85888-09-1 H-66-257-0 3,7-Anhydro-3,7-dihydroxy-5,9,13-trimethylsilyl-4,8,12-trimethylpentadecan-oic acid-dimethylester; W   127829

620 C30 H36 O14 65040-02-0 F-33-1409-0 SPINATOSIDPERMETHYLETHER; W   65534

620 C31 H26 Cl2 N4 O6 IC-1291-0-0 N,N-Bis(benzoyloxyethyl)-3-methyl-4-(2,6-dichloro-4-nitrophenylazo)aniline; W   127830

620   C31 H40 O11 S 91127-32-1 J-49-3259-0 (+/-)-(1α,2α,4bβ,10α)-8,8-(ethylenedioxy)-2-hydroxy-1-(methoxy carbonyl)-7-[(methoxymethyl)oxy]-1-methylgibb-4a(10a)-ene-10-carboxylic acid 2α-benzenesulfonate; W 127831

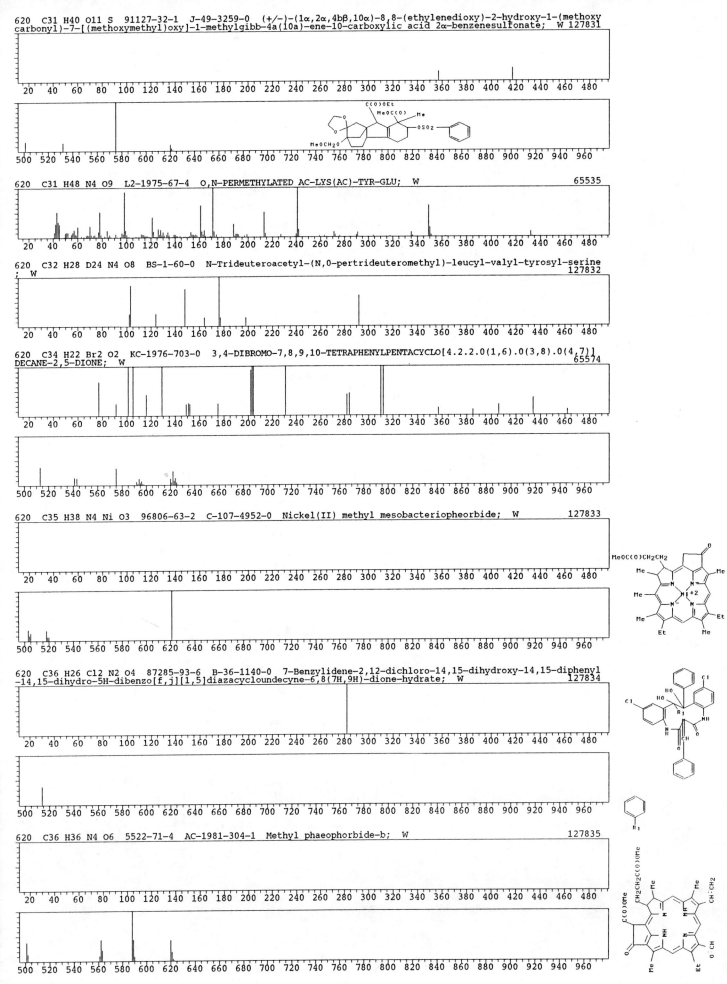

620   C31 H48 N4 O9 L2-1975-67-4 O,N-PERMETHYLATED AC-LYS(AC)-TYR-GLU; W   65535

620   C32 H28 D24 N4 O8 BS-1-60-0 N-Trideuteroacetyl-(N,O-pertrideuteromethyl)-leucyl-valyl-tyrosyl-serine ; W   127832

620   C34 H22 Br2 O2 KC-1976-703-0 3,4-DIBROMO-7,8,9,10-TETRAPHENYLPENTACYCLO[4.2.2.0(1,6).0(3,8).0(4,7)] DECANE-2,5-DIONE; W   65574

620   C35 H38 N4 Ni O3 96806-63-2 C-107-4952-0 Nickel(II) methyl mesobacteriopheorbide; W   127833

620   C36 H26 Cl2 N2 O4 87285-93-6 B-36-1140-0 7-Benzylidene-2,12-dichloro-14,15-dihydroxy-14,15-diphenyl -14,15-dihydro-5H-dibenzo[f,j][1,5]diazacycloundecyne-6,8(7H,9H)-dione-hydrate; W   127834

620   C36 H36 N4 O6 5522-71-4 AC-1981-304-1 Methyl phaeophorbide-b; W   127835

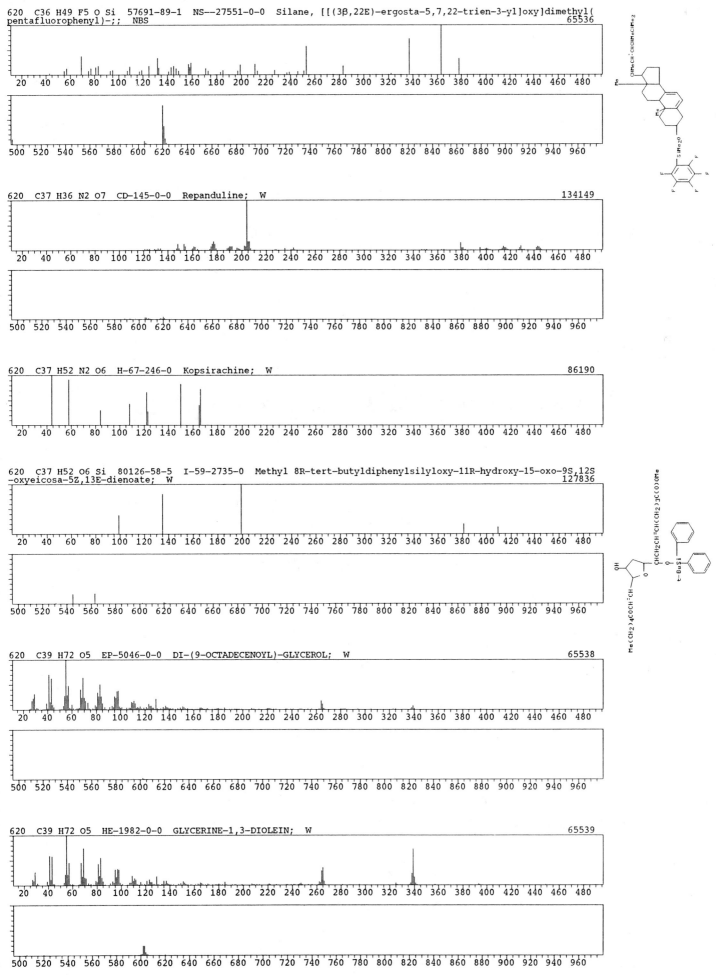

620   C36 H49 F5 O Si   57691-89-1   NS--27551-0-0   Silane, [[(3β,22E)-ergosta-5,7,22-trien-3-yl]oxy]dimethyl(
pentafluorophenyl)-;;   NBS   65536

620   C37 H36 N2 O7   CD-145-0-0   Repanduline;   W   134149

620   C37 H52 N2 O6   H-67-246-0   Kopsirachine;   W   86190

620   C37 H52 O6 Si   80126-58-5   I-59-2735-0   Methyl 8R-tert-butyldiphenylsilyloxy-11R-hydroxy-15-oxo-9S,12S
-oxyeicosa-5Z,13E-dienoate;   W   127836

620   C39 H72 O5   EP-5046-0-0   DI-(9-OCTADECENOYL)-GLYCEROL;   W   65538

620   C39 H72 O5   HE-1982-0-0   GLYCERINE-1,3-DIOLEIN;   W   65539

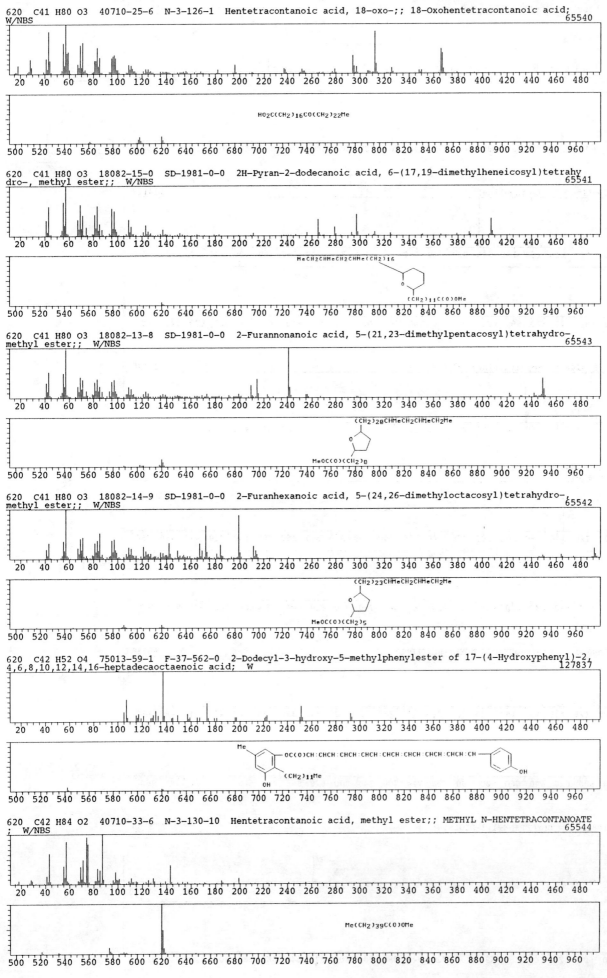

620 C41 H80 O3 40710-25-6 N-3-126-1 Hentetracontanoic acid, 18-oxo-;; 18-Oxohentetracontanoic acid;
W/NBS
65540

HO2C(CH2)16CO(CH2)22Me

620 C41 H80 O3 18082-15-0 SD-1981-0-0 2H-Pyran-2-dodecanoic acid, 6-(17,19-dimethylheneicosyl)tetrahy
dro-, methyl ester;; W/NBS
65541

MeCH2CHMeCH2CHMe(CH2)16

(CH2)11C(O)OMe

620 C41 H80 O3 18082-13-8 SD-1981-0-0 2-Furannonanoic acid, 5-(21,23-dimethylpentacosyl)tetrahydro-
methyl ester;; W/NBS
65543

(CH2)28CHMeCH2CHMeCH2Me

MeOC(O)(CH2)8

620 C41 H80 O3 18082-14-9 SD-1981-0-0 2-Furanhexanoic acid, 5-(24,26-dimethyloctacosyl)tetrahydro-
methyl ester;; W/NBS
65542

(CH2)23CHMeCH2CHMeCH2Me

MeOC(O)(CH2)5

620 C42 H52 O4 75013-59-1 F-37-562-0 2-Dodecyl-3-hydroxy-5-methylphenylester of 17-(4-Hydroxyphenyl)-2,
4,6,8,10,12,14,16-heptadecaoctaenoic acid; W
127837

Me   OC(O)CH:CHCH:CHCH:CHCH:CHCH:CHCH:CHCH:CHCH:CH   OH
(CH2)11Me
OH

620 C42 H84 O2 40710-33-6 N-3-130-10 Hentetracontanoic acid, methyl ester;; METHYL N-HENTETRACONTANOATE
; W/NBS
65544

Me(CH2)39C(O)OMe

620   C44 H28 O4   J-47-144-0   trans-1,2-Bis(1-naphthyl)-1,2-ethenediol di-1-naphthoate;   W   127838

620   C44 H28 O4   J-47-144-0   trans-1,2-Bis(2-naphthyl)-1,2-ethenediol di-2-naphthoate;   W   127839

620   C44 H28 O4   80364-58-5   B-34-1681-0   10-Hydroxy-10-[10'-hydroxy-9',10'-bis(phenylethynyl)-9',10'-dihyd
roanthracen-9'-yloxy]anthrone and 10-[9',10'-Dihydroxy-10'-(phenylethynyl)-9',10'-dihydroanthracen-9'-ylox
y]-10-(phenylethynyl)anthrone;   W   127840

621   C18 H10 F15 N O6   66582-82-9   NS--27545-0-0   Propanoic acid, pentafluoro-, 1-[3-methoxy-4-(2,2,3,3,3-
pentafluoro-1-oxopropoxy)phenyl]-2-[(2,2,3,3,3-pentafluoro-1-oxopropyl)amino]ethyl ester;;   NBS   65545

621   C18 H7 D3 F15 N O6   BS-5-212-0   Pentafluoropropionyl-α-d[2]-β-d[1]-normetanephrine;   W   127841

621   C30 H47 N5 O9   27545-13-7   O-3-74-4   Glycine, N-[N-[N-[N-(N-carboxy-L-threonyl)-L-alanyl]-L-leucyl]-L-
leucyl]-, N-benzyl methyl ester;;   BENZYLOXYCARBONYL-THR-ALA-LEU-LEU-GLY-OME;   W/NBS   65546

621   C30 H71 N O4 Si4   72088-27-8   NS--27543-0-0   1,3,4,5-Octadecanetetrol, 2-amino-, tetrakis(trimethylsil
yl) deriv.;;   NBS   65547

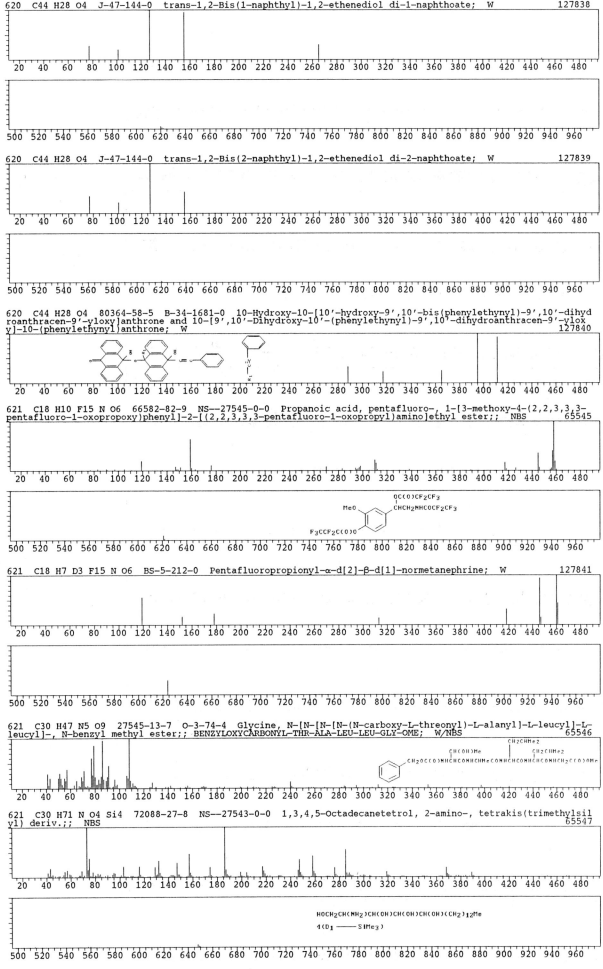

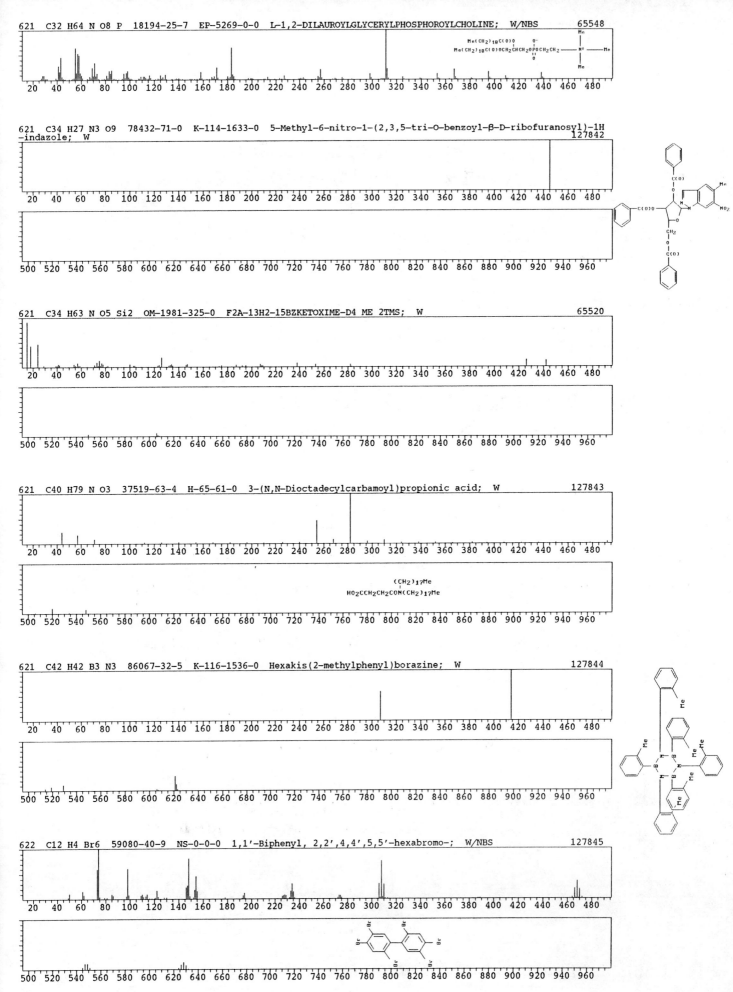

621 C32 H64 N O8 P 18194-25-7 EP-5269-0-0 L-1,2-DILAUROYLGLYCERYLPHOSPHOROYLCHOLINE; W/NBS 65548

621 C34 H27 N3 O9 78432-71-0 K-114-1633-0 5-Methyl-6-nitro-1-(2,3,5-tri-O-benzoyl-β-D-ribofuranosyl)-1H
-indazole; W 127842

621 C34 H63 N O5 Si2 OM-1981-325-0 F2A-13H2-15BZKETOXIME-D4 ME 2TMS; W 65520

621 C40 H79 N O3 37519-63-4 H-65-61-0 3-(N,N-Dioctadecylcarbamoyl)propionic acid; W 127843

621 C42 H42 B3 N3 86067-32-5 K-116-1536-0 Hexakis(2-methylphenyl)borazine; W 127844

622 C12 H4 Br6 59080-40-9 NS-0-0-0 1,1'-Biphenyl, 2,2',4,4',5,5'-hexabromo-; W/NBS 127845

622   C12 H4 Br6   59261-08-4   NS-0-0-0   2,2',4,4',6,6'-Hexabromobiphenyl;   W/NBS   127846

622   C12 H4 Br6   60044-26-0   NS-0-0-0   3,3',4,4',5,5'-Hexabromobiphenyl;   W/NBS   127847

622   C14 H42 N2 O2 Si4 Sn2   61509-99-7   K-109-3961-0   2,2,4,4,6,6,8,8-Octamethyl-3,7-bis(trimethylstannyl)-
1,5-dioxa-3,7-diaza-2,4,6,8-tetrasilacyclooctane;   W   127848

622   C16 H36 N6 O4 P2 W   19976-86-4   AD-0-2756-0   Tungsten, tetracarbonylbis(hexamethylphosphorous triamide
-P)-;;   BIS(TRIS(DIMETHYLAMINO)PHOSPHINE)-TETRACARBONYL TUNGSTEN;   W/NBS   65549

622   C18 F15 Sb   3910-39-2   AD-0-2714-0   Stibine, tris(pentafluorophenyl)-;;   Tris(pentafluorophenyl)stibine
;   W/NBS   65550

622   C18 H9 F15 O7   56728-12-2   NS-0-0-0   Propanoic acid, pentafluoro-, 1-[3-methoxy-4-(2,2,3,3,3-pentafluo
ro-1-oxopropoxy)phenyl]-1,2-ethanediyl ester;;   3-METHOXY-4-HYDROXYPHENYLGLYCOL TRIS(PENTAFLUOROPROPIONYL)
ETHER;   W/NBS   127849

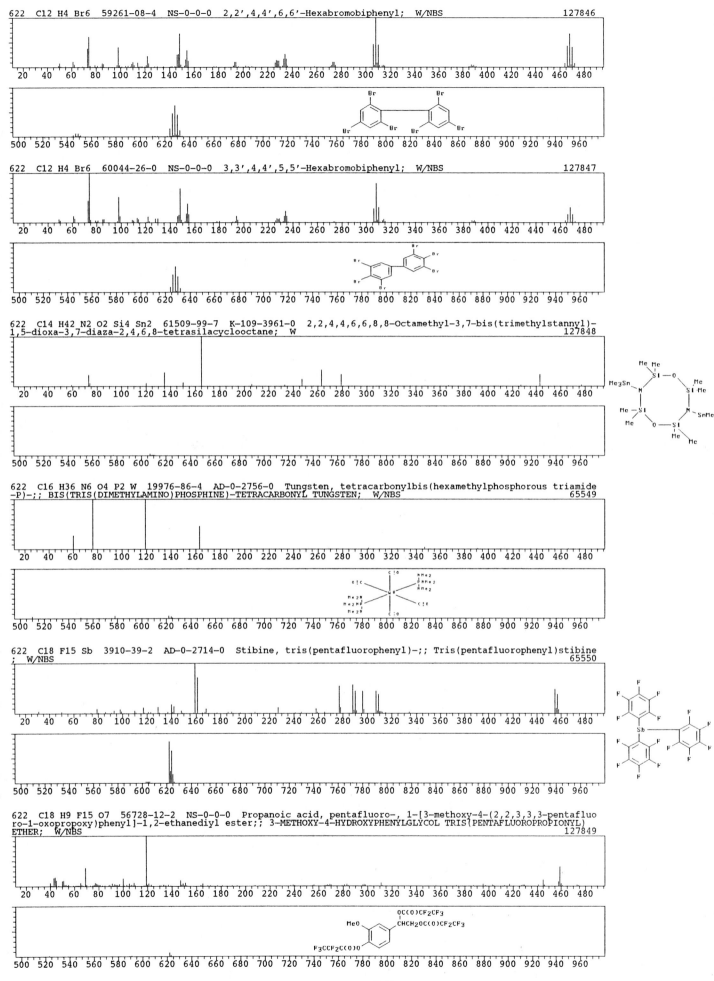

622   C18 H6 D3 F15 O7   D3-1980-1046-33   PENTAFLUOROPROPIONYL DERIVATIVE OF 3-TRIDEUTEROMETHOXY-4-HYDROXY-PH
ENYLETHYLENE GLYCOL;   W
65552

622   C18 H9 F15 O7   KO-9-149-1   Tri-(pentafluoropropionyl)derivative of 3-methoxy-4-hydroxyphenyl-glycol;
W
127850

622   C23 H15 F9 O10   54638-67-4   JC-101-208-2   METHYL-TRIFLUOROACETYL DERIVATIVE OF 1-NAPHTHOL GLUCURONIDE;
W
65553

622   C27 H28 Br2 O5 S   76-59-5   NS--27538-0-0   Bromthymol Blue;   NBS
65554

622   C30 H54 N2 O6 Si3   64933-63-7   O-15-212-2   2',3',5'-TRIS-O-CYCLO-TETRAMETHYLENE-ISO-PROPYLSILYL-URIDI
NE;   W
65555

622   C35 H58 O9   64285-76-3   B-30-1320-0   BARRINGTOZENOL C-28-O-LYXOSIDE;   W
65556

Str 1153

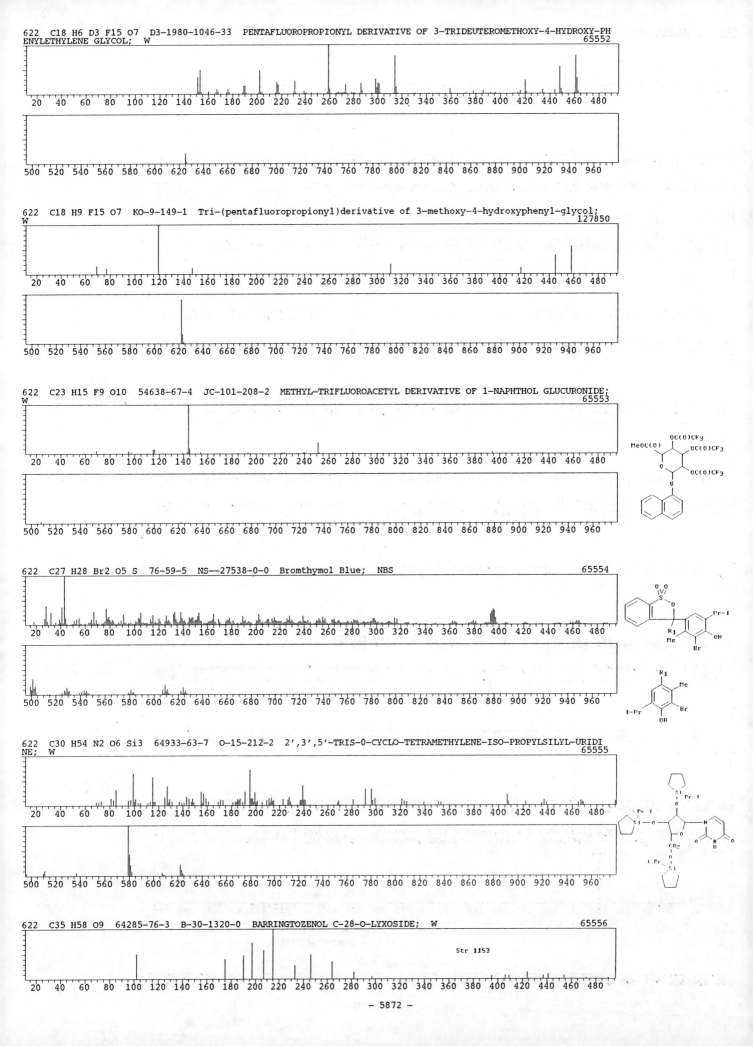

622　C36 H30 O10　27090-23-9　KC-1981-557-0　Hexa-O-methyltaiwaniaflavone;　W　127851

622　C36 H38 N4 O6　53358-14-8　AC-1981-304-1　Methyl-13-hydroxy-phaeophorbide-a;　W　127852

622　C37 H34 O9　67408-67-7　J-43-4167-0　2-(2'-Carboxymethyl-3'-ethylenedioxybutyl)-5-methoxy-1,4-dibenzyl oxyanthraquinone;　W　127853

622　C37 H38 N2 O7　CD-146-0-0　Repandulinol;　W　127854

622　C38 H42 N2 O6　2233-44-5　AD-0-4811-0　(--)-CURINE DIMETHYL ETHER;　W/NBS　65558

622　C38 H42 N2 O6　KQ-8-619-3　OXYACANTHIN METHYLETHER;　W　65557

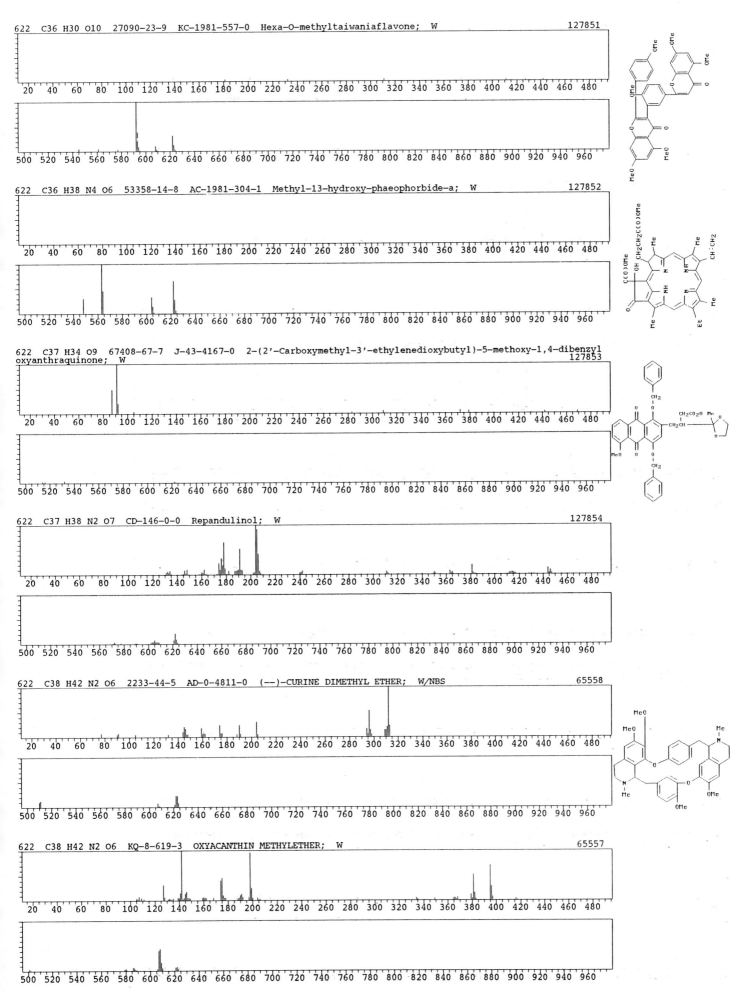

622　C38 H42 N2 O6　CD-142-0-0　Isotetrandrine;　W　　　　　　　　　127855

622　C38 H42 N2 O6　CD-143-0-0　Thalmidine;　W　　　　　　　　　127856

622　C38 H42 N2 O6　55702-01-7　J-40-2468-0　O-Methylpanurensine;　W　　　127857

622　C38 H42 N2 O6　1263-79-2　J-41-318-0　O,O-Dimethylkrukovine;　W　　127858

622　C38 H46 N4 O4　59923-98-7　KC-1978-850-0　METHYL 5-ETHYL-2-(1-HYDROXY ETHYL)-5-DEMETHYL-Δ-METHYL-2-DEV
INYLPYRO PHAEOPHORBIDE A AND HOMOLOGUES;　W　　　　　　　　　　　65559

622　C38 H58 O3 S Si　79409-72-6　J-46-5186-0　(3β,23Z)-3-[(Tetrahydro-2H-pyran-2-yl)oxy]-24-(phenylthio)-
24-(trimethylsilyl)-chola-5,23-dien-20-ol;　W　　　　　　　　　127859

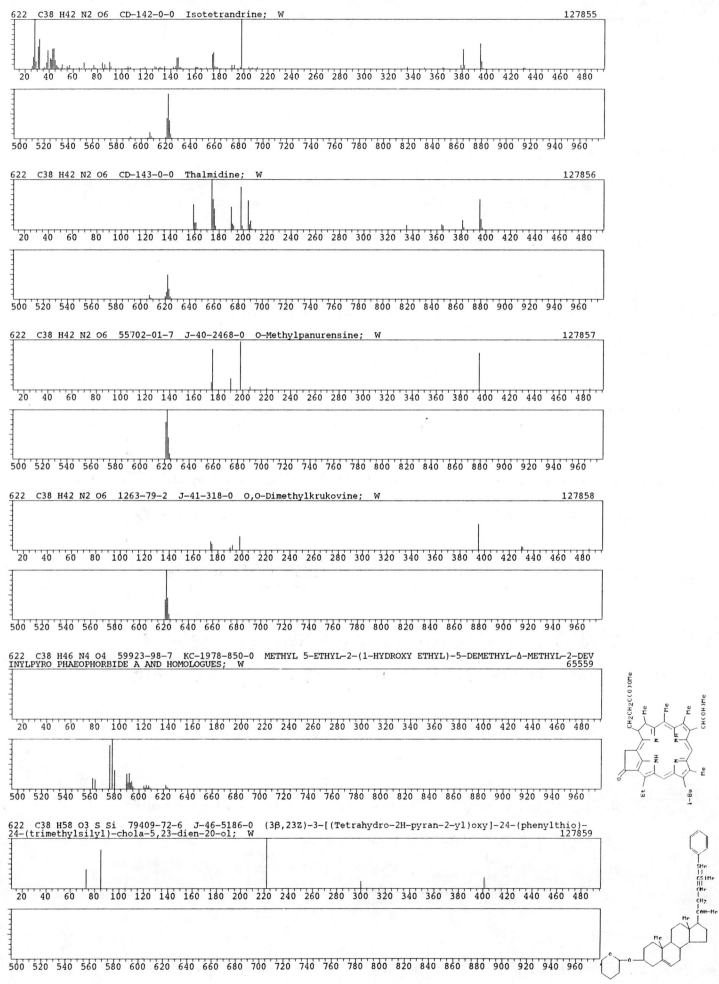

622 C39 H74 O5 HE-1982-0-0 GLYCERIN-1-OLEAT-3-STEARAT; W   65560

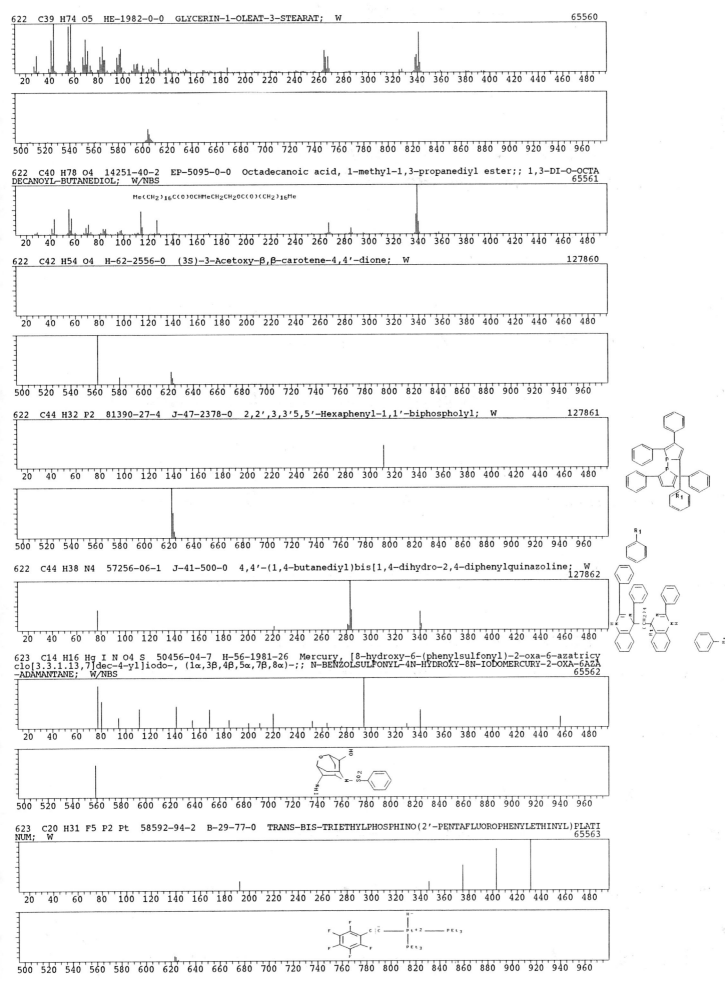

622 C40 H78 O4 14251-40-2 EP-5095-0-0 Octadecanoic acid, 1-methyl-1,3-propanediyl ester;; 1,3-DI-O-OCTA
DECANOYL-BUTANEDIOL; W/NBS   65561

Me(CH₂)₁₆C(O)OCHMeCH₂CH₂OC(O)(CH₂)₁₆Me

622 C42 H54 O4 H-62-2556-0 (3S)-3-Acetoxy-β,β-carotene-4,4'-dione; W   127860

622 C44 H32 P2 81390-27-4 J-47-2378-0 2,2',3,3'5,5'-Hexaphenyl-1,1'-biphospholyl; W   127861

622 C44 H38 N4 57256-06-1 J-41-500-0 4,4'-(1,4-butanediyl)bis[1,4-dihydro-2,4-diphenylquinazoline; W
                                                                                                 127862

623 C14 H16 Hg I N O4 S 50456-04-7 H-56-1981-26 Mercury, [8-hydroxy-6-(phenylsulfonyl)-2-oxa-6-azatricy
clo[3.3.1.13,7]dec-4-yl]iodo-, (1α,3β,4β,5α,7β,8α)-;; N-BENZOLSULFONYL-4N-HYDROXY-8N-IODOMERCURY-2-OXA-6AZA
-ADAMANTANE; W/NBS   65562

623 C20 H31 F5 P2 Pt 58592-94-2 B-29-77-0 TRANS-BIS-TRIETHYLPHOSPHINO(2'-PENTAFLUOROPHENYLETHINYL)PLATI
NUM; W   65563

- 5875 -

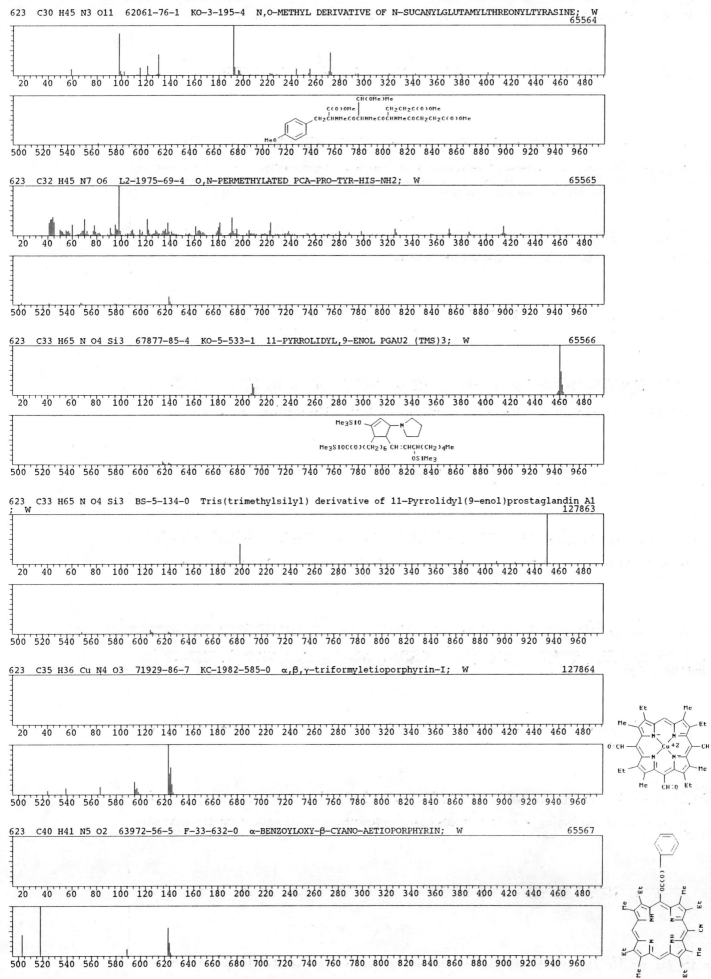

623   C30 H45 N3 O11   62061-76-1   KO-3-195-4   N,O-METHYL DERIVATIVE OF N-SUCANYLGLUTAMYLTHREONYLTYRASINE;   W
65564

623   C32 H45 N7 O6   L2-1975-69-4   O,N-PERMETHYLATED PCA-PRO-TYR-HIS-NH2;   W
65565

623   C33 H65 N O4 Si3   67877-85-4   KO-5-533-1   11-PYRROLIDYL,9-ENOL PGAU2 (TMS)3;   W
65566

623   C33 H65 N O4 Si3   BS-5-134-0   Tris(trimethylsilyl) derivative of 11-Pyrrolidyl(9-enol)prostaglandin A1
; W
127863

623   C35 H36 Cu N4 O3   71929-86-7   KC-1982-585-0   α,β,γ-triformyletioporphyrin-I;   W
127864

623   C40 H41 N5 O2   63972-56-5   F-33-632-0   α-BENZOYLOXY-β-CYANO-AETIOPORPHYRIN;   W
65567

- 5876 -

624   C12 H16 F12 Fe2 P4   HE-1982-0-0   IRON, BIS(CYCLOHEXEN)-TETRAKIS(TRIFLUORPHOSPHAN)-BIS-;   W   65568

624   C12 H20 I2 Rh2   12307-77-6   HE-1982-0-0   Rhodium, di-.mu.-iodotetrakis(η3-2-propenyl)di-;; RHODIUM, DI
-MUE-IODO-TETRA-ALLYL-BIS-;   W/NBS   65569

624   C14 F24   SJ-1986-4-0   Perfluorophenanthrene;   W   127865

624   C14 F24   67700-17-8   J-43-4982-0   Perfluorotetramethyladamantane;   W   127866

624   C22 H18 F14 N2 O3   JC-90-193-1   N,N-Bis(heptafluorobutyryl)-2,6-pipecoloxylidide;   W   127867

624   C30 H24 Bi P   62336-23-6   AG-122-354-1   (o-Diphenylphosphinophenyl)diphenylbismuthine;   W   127868

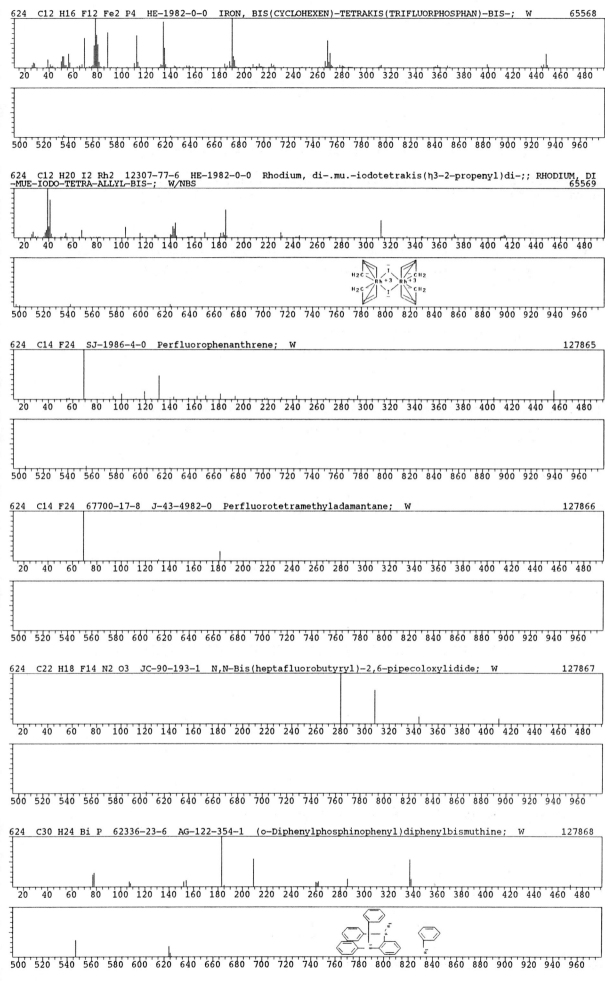

624   C31 H28 O14   24385-09-9   NS--27533-0-0   5,12-Naphthacenedione, 6,8,10,11-tetrakis(acetyloxy)-8-[(acety
loxy)acetyl]-7,8,9,10-tetrahydro-1-methoxy-, (8S-cis)-;;   NBS
65570

624   C31 H39 F3 N2 O8   51067-79-9   I-58-524-0   [2S,3S,4R,6(S*),7(R*)]-N-{7-[5-(7-(1,2-Dihydro-4-methoxy-1-
methyl-2-oxo-3-pyridinyl)-6-methyl-7-oxo-1(E),3(E),5(E)-heptatrienyl]tetrahydro-3,4-dihydroxyfuran-2-yl]-6-
methoxy-5,7-dimethyl-2(E),4(E)-heptadienyl]trifluoroacetamide;   W
127869

624   C32 H36 N2 O11   58950-47-3   KC-1976-289-0   BENZYL 3,4'-BIS-(2-METHOXYCARBONYLETHYL)-3',4-BISMETHOXYCAR
BONYLMETHYL-5'-FORMYL-2,2'-METHYLENEDIPYRROLE-5-CARBOXYLATE;   W
65571

624   C32 H34 D2 N2 O11   KC-1976-289-0   (13C2)BENZYL 3,4'-BIS-(2-METHOXYCARBONYLETHYL)-3',4-BISMETHOXYCARBON
YLMETHYL-5'-FORMYL-2,2'-METHYLENEDIPYRROLE-5-CARBOXYLATE;   W
65572

624   C32 H56 N4 O8   AT-64-2760-1   SPORIDESMOLIDE III;   W
65573

624   C34 H26 Br2 O2   59614-63-0   KC-1976-703-0   3,4-DIBROMO-7,8,9,10-TETRAPHENYLPENTACYCLO[4.2.2.0(1,6).0(
3,8).0(4,7)]DECAN-2,5-DIOL;   W
65575

624   C34 H64 N4 O6   33281-02-6   O6-0-1-0   2,B-DIDECANOYL-LYS-GLY-VAL-OME;   W/NBS
65576

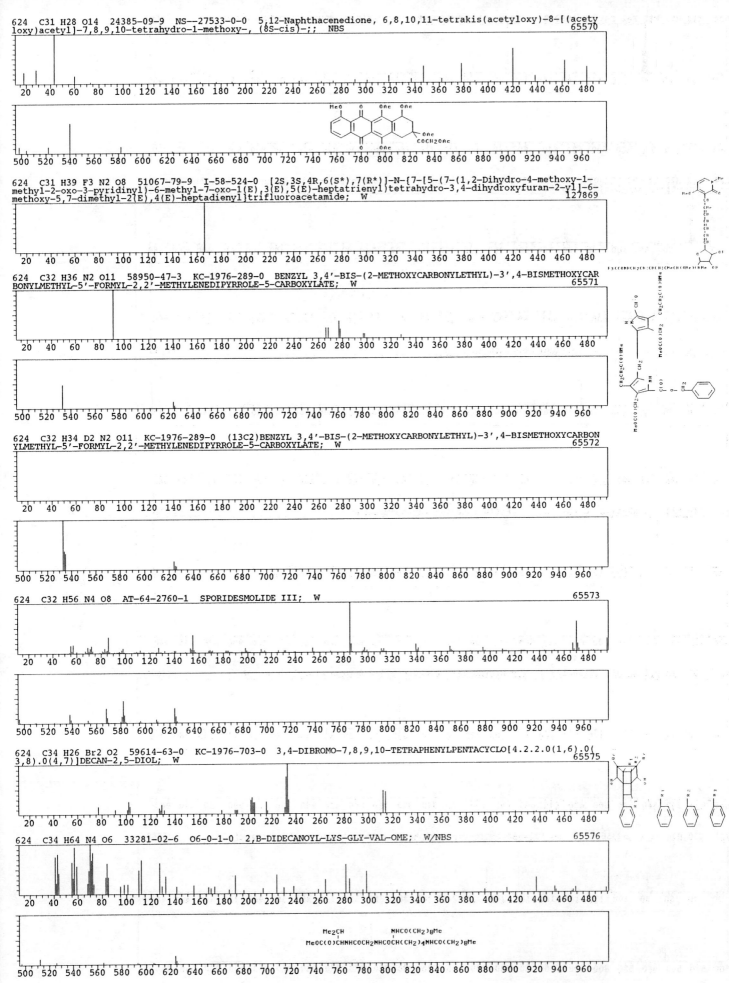

– 5878 –

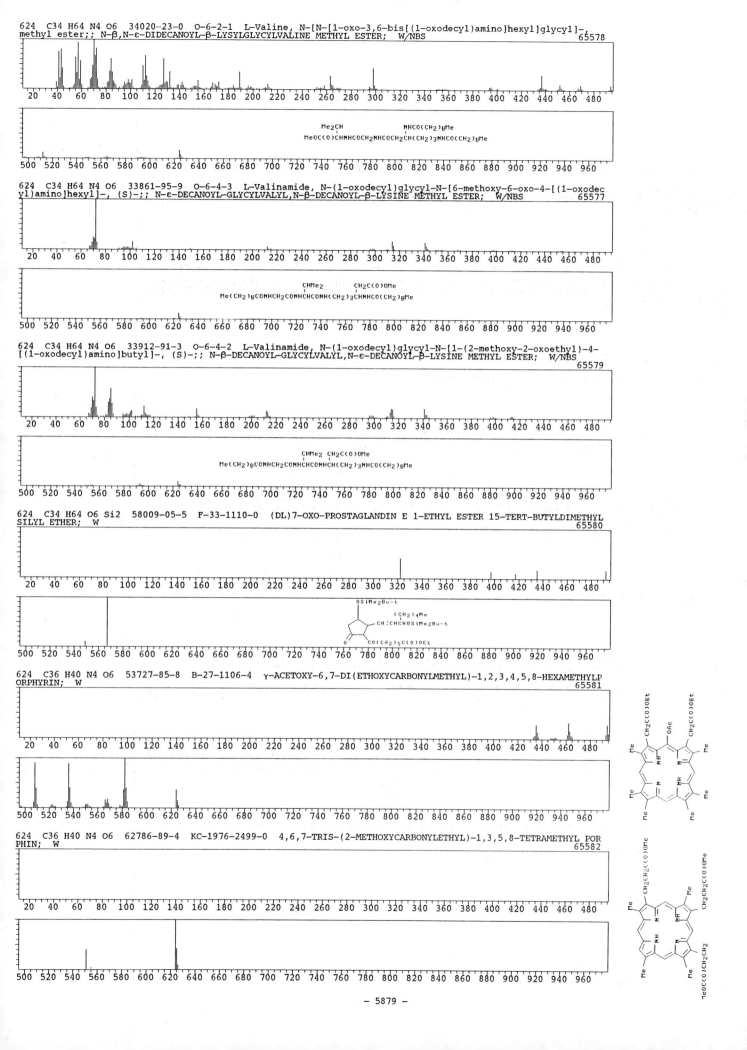

624　C34 H64 N4 O6　34020-23-0　O-6-2-1　L-Valine, N-[N-[1-oxo-3,6-bis[(1-oxodecyl)amino]hexyl]glycyl]-,
methyl ester;; N-β,N-ε-DIDECANOYL-β-LYSYLGLYCYLVALINE METHYL ESTER;　W/NBS　　65578

Me₂CH　　　　　NHCO(CH₂)₈Me
|　　　　　　　|
MeOC(O)CHNHCOCH₂NHCOCH₂CH(CH₂)₃NHCO(CH₂)₈Me

624　C34 H64 N4 O6　33861-95-9　O-6-4-3　L-Valinamide, N-(1-oxodecyl)glycyl-N-[6-methoxy-6-oxo-4-[(1-oxodec
yl)amino]hexyl]-, (S)-;; N-ε-DECANOYL-GLYCYLVALYL,N-β-DECANOYL-β-LYSINE METHYL ESTER;　W/NBS　　65577

CHMe₂　　　CH₂C(O)OMe
|　　　　　　|
Me(CH₂)₈CONHCH₂CONHCHCONH(CH₂)₃CHNHCO(CH₂)₈Me

624　C34 H64 N4 O6　33912-91-3　O-6-4-2　L-Valinamide, N-(1-oxodecyl)glycyl-N-[1-(2-methoxy-2-oxoethyl)-4-
[(1-oxodecyl)amino]butyl]-, (S)-;; N-β-DECANOYL-GLYCYLVALYL,N-ε-DECANOYL-β-LYSINE METHYL ESTER;　W/NBS　　65579

CHMe₂　CH₂C(O)OMe
|　　　|
Me(CH₂)₈CONHCH₂CONHCHCONHCH(CH₂)₃NHCO(CH₂)₈Me

624　C34 H64 O6 Si2　58009-05-5　F-33-1110-0　(DL)7-OXO-PROSTAGLANDIN E 1-ETHYL ESTER 15-TERT-BUTYLDIMETHYL
SILYL ETHER;　W　　65580

OSiMe₂Bu-t
|
(CH₂)₄Me
CH:CHCHOSiMe₂Bu-t
CO(CH₂)₅C(O)OEt
O=

624　C36 H40 N4 O6　53727-85-8　B-27-1106-4　γ-ACETOXY-6,7-DI(ETHOXYCARBONYLMETHYL)-1,2,3,4,5,8-HEXAMETHYLP
ORPHYRIN;　W　　65581

624　C36 H40 N4 O6　62786-89-4　KC-1976-2499-0　4,6,7-TRIS-(2-METHOXYCARBONYLETHYL)-1,3,5,8-TETRAMETHYL POR
PHIN;　W　　65582

624 C36 H40 N4 O6 62786-94-1 KC-1976-2500-0 2,6,7-TRIS-(2-METHOXYCARBONYLETHYL)-1,3,5,8-TETRAMETHYLPORP
HIN; W   65583

20 40 60 80 100 120 140 160 180 200 220 240 260 280 300 320 340 360 380 400 420 440 460 480

500 520 540 560 580 600 620 640 660 680 700 720 740 760 780 800 820 840 860 880 900 920 940 960

624 C36 H40 N4 O6 78403-41-5 F-37-160-0 exobornyl ester of (5α,5aβ,9aβ,10α)-2-(2,3-dihydro-1,3,14,16-te
traoxo-2-phenyl-5H,10H-5,10-etheno-5a,9a-(methaniminomethano)-1H-[1,2,4]triazolo[1,2-b]phthalizin-15-yl)-4-
methylpentanoic acid; W   127870

20 40 60 80 100 120 140 160 180 200 220 240 260 280 300 320 340 360 380 400 420 440 460 480

624 C36 H40 N4 O6 78420-52-7 F-37-160-0 exoBornyl ester of 2-(2,3-dihydro-1,3, 14,16-tetraoxo-2-phenyl-
5H,10H-5,10-etheno-5a,9a-(methaniminomethano)-1H-[1,2,4]triazolo[1,2-b]phthalizin-15-yl)-4-methylpentanoic
acid stereoisomer; W   127871

20 40 60 80 100 120 140 160 180 200 220 240 260 280 300 320 340 360 380 400 420 440 460 480

624 C37 H40 N2 O7 64252-82-0 J-43-583-0 Thaligosidine; W   127872

20 40 60 80 100 120 140 160 180 200 220 240 260 280 300 320 340 360 380 400 420 440 460 480

500 520 540 560 580 600 620 640 660 680 700 720 740 760 780 800 820 840 860 880 900 920 940 960

624 C37 H52 O8 26122-89-4 S-23-2608-3 METHYL 1-HEXOSYL-1,2-DIHYDRO-3,4-DIDEHYDRO-APO-8'-LYCOPENOATE; W
                                                                                           65584

20 40 60 80 100 120 140 160 180 200 220 240 260 280 300 320 340 360 380 400 420 440 460 480

500 520 540 560 580 600 620 640 660 680 700 720 740 760 780 800 820 840 860 880 900 920 940 960

624 C38 H44 N2 O2 S2 79250-29-6 F-37-2186-0 N,N'-Di-t-butyl-2,2,2',2'-tetraphenyl-3,3'-dithiodipropiona
mide; W   127873

20 40 60 80 100 120 140 160 180 200 220 240 260 280 300 320 340 360 380 400 420 440 460 480

500 520 540 560 580 600 620 640 660 680 700 720 740 760 780 800 820 840 860 880 900 920 940 960

624 C38 H44 N2 O6 CD-137-0-0 Dauricine; W   127874

20 40 60 80 100 120 140 160 180 200 220 240 260 280 300 320 340 360 380 400 420 440 460 480

500 520 540 560 580 600 620 640 660 680 700 720 740 760 780 800 820 840 860 880 900 920 940 960

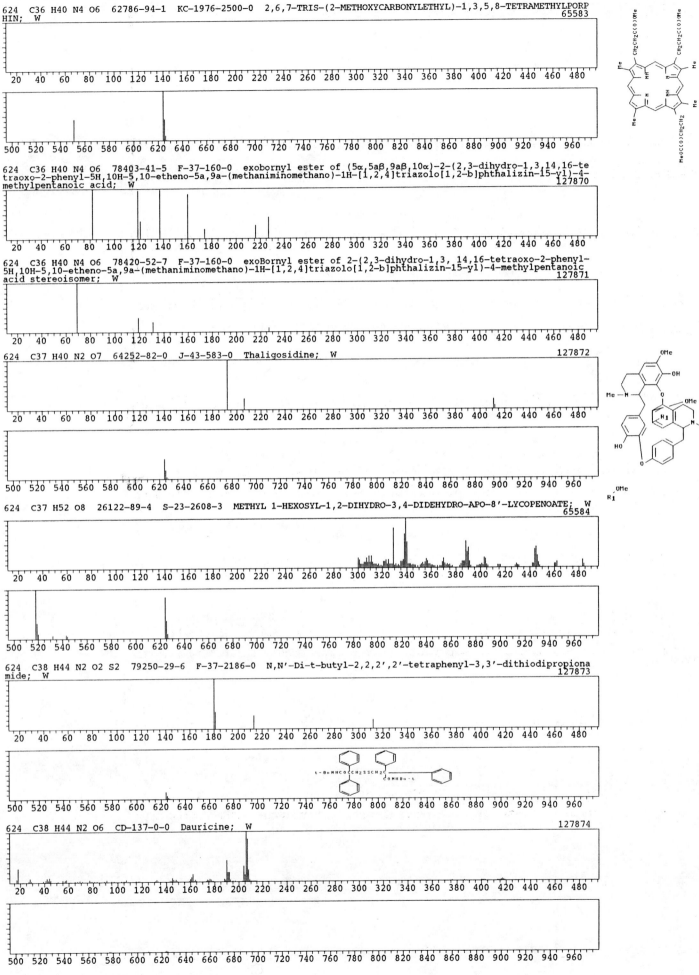

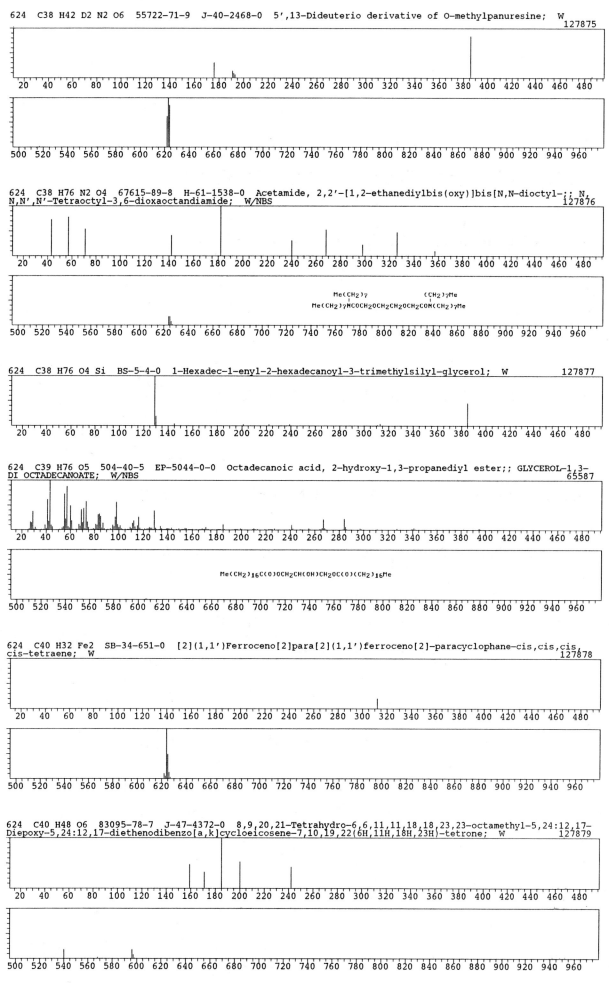

624 C38 H42 D2 N2 O6 55722-71-9 J-40-2468-0 5',13-Dideuterio derivative of O-methylpanuresine; W
127875

624 C38 H76 N2 O4 67615-89-8 H-61-1538-0 Acetamide, 2,2'-[1,2-ethanediylbis(oxy)]bis[N,N-dioctyl-;; N, N,N',N'-Tetraoctyl-3,6-dioxaoctandiamide; W/NBS
127876

Me(CH₂)₇   (CH₂)₇Me
                                          | |
                             Me(CH₂)₇NCOCH₂OCH₂CH₂OCH₂CON(CH₂)₇Me

624 C38 H76 O4 Si BS-5-4-0 1-Hexadec-1-enyl-2-hexadecanoyl-3-trimethylsilyl-glycerol; W
127877

624 C39 H76 O5 504-40-5 EP-5044-0-0 Octadecanoic acid, 2-hydroxy-1,3-propanediyl ester;; GLYCEROL-1,3-DI OCTADECANOATE; W/NBS
65587

Me(CH₂)₁₆C(O)OCH₂CH(OH)CH₂OC(O)(CH₂)₁₆Me

624 C40 H32 Fe2 SB-34-651-0 [2](1,1')Ferroceno[2]para[2](1,1')ferroceno[2]-paracyclophane-cis,cis,cis, cis-tetraene; W
127878

624 C40 H48 O6 83095-78-7 J-47-4372-0 8,9,20,21-Tetrahydro-6,6,11,11,18,18,23,23-octamethyl-5,24:12,17-Diepoxy-5,24:12,17-diethenodibenzo[a,k]cycloeicosene-7,10,19,22(6H,11H,18H,23H)-tetrone; W
127879

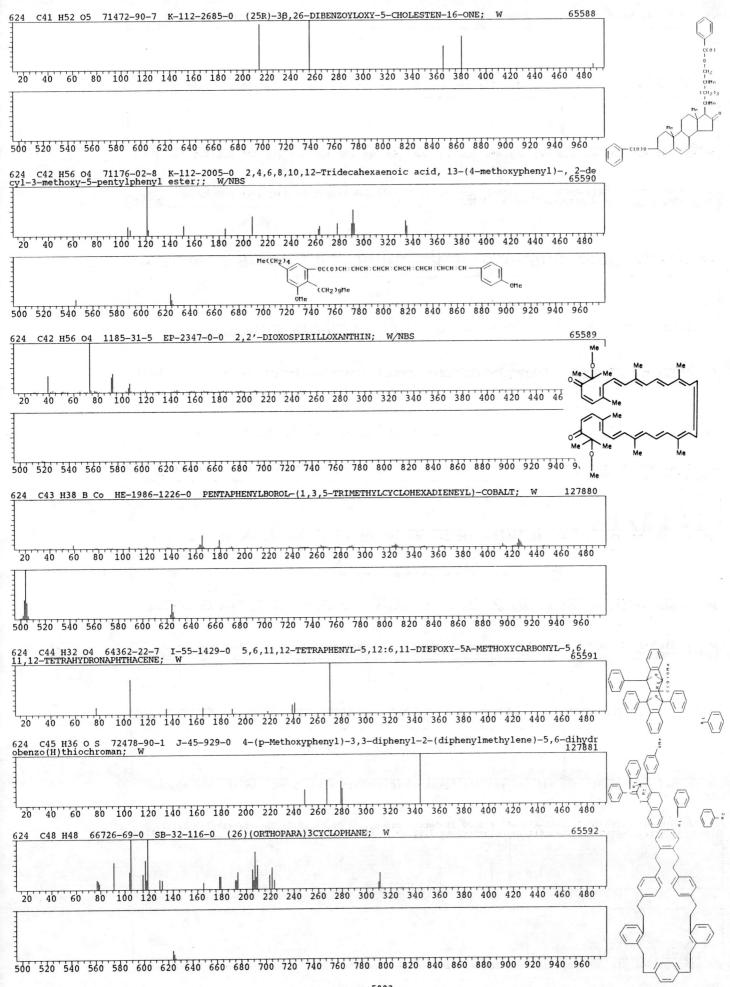

624   C41 H52 O5   71472-90-7   K-112-2685-0   (25R)-3β,26-DIBENZOYLOXY-5-CHOLESTEN-16-ONE;   W   65588

624   C42 H56 O4   71176-02-8   K-112-2005-0   2,4,6,8,10,12-Tridecahexaenoic acid, 13-(4-methoxyphenyl)-, 2-de
cyl-3-methoxy-5-pentylphenyl ester;;   W/NBS   65590

624   C42 H56 O4   1185-31-5   EP-2347-0-0   2,2'-DIOXOSPIRILLOXANTHIN;   W/NBS   65589

624   C43 H38 B Co   HE-1986-1226-0   PENTAPHENYLBOROL-(1,3,5-TRIMETHYLCYCLOHEXADIENEYL)-COBALT;   W   127880

624   C44 H32 O4   64362-22-7   I-55-1429-0   5,6,11,12-TETRAPHENYL-5,12:6,11-DIEPOXY-5A-METHOXYCARBONYL-5,6,
11,12-TETRAHYDRONAPHTHACENE;   W   65591

624   C45 H36 O S   72478-90-1   J-45-929-0   4-(p-Methoxyphenyl)-3,3-diphenyl-2-(diphenylmethylene)-5,6-dihydr
obenzo(H)thiochroman;   W   127881

624   C48 H48   66726-69-0   SB-32-116-0   (26)(ORTHOPARA)3CYCLOPHANE;   W   65592

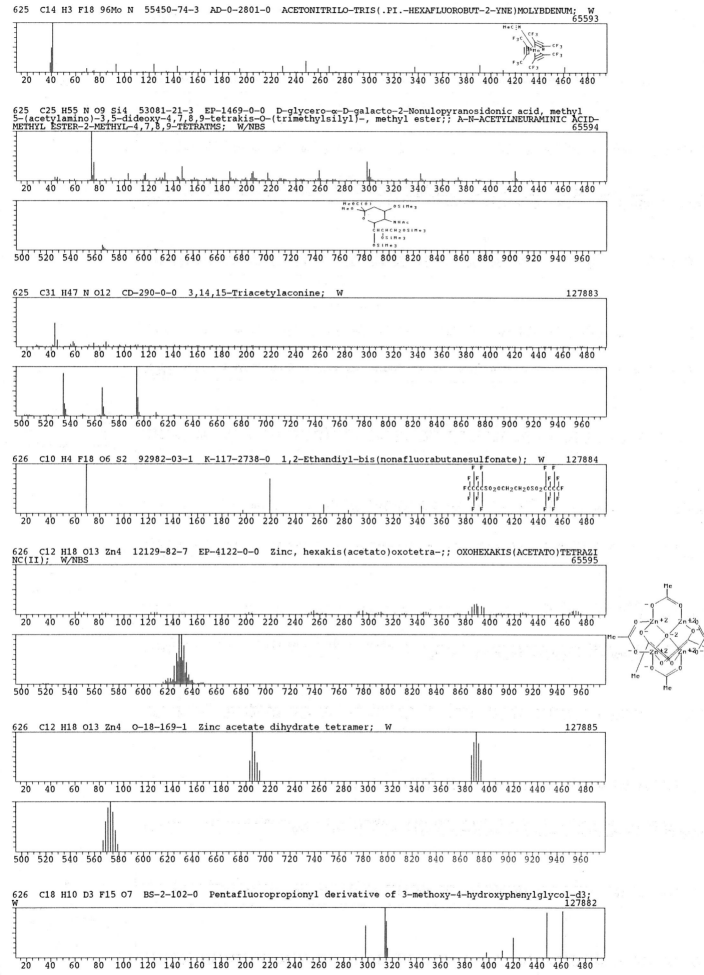

625   C14 H3 F18 96Mo N   55450-74-3   AD-0-2801-0   ACETONITRILO-TRIS(.PI.-HEXAFLUOROBUT-2-YNE)MOLYBDENUM;   W
65593

625   C25 H55 N O9 Si4   53081-21-3   EP-1469-0-0   D-glycero-α-D-galacto-2-Nonulopyranosidonic acid, methyl
5-(acetylamino)-3,5-dideoxy-4,7,8,9-tetrakis-O-(trimethylsilyl)-, methyl ester;; A-N-ACETYLNEURAMINIC ACID-
METHYL ESTER-2-METHYL-4,7,8,9-TETRATMS;   W/NBS
65594

625   C31 H47 N O12   CD-290-0-0   3,14,15-Triacetylaconine;   W
127883

626   C10 H4 F18 O6 S2   92982-03-1   K-117-2738-0   1,2-Ethandiyl-bis(nonafluorabutanesulfonate);   W   127884

626   C12 H18 O13 Zn4   12129-82-7   EP-4122-0-0   Zinc, hexakis(acetato)oxotetra-;; OXOHEXAKIS(ACETATO)TETRAZI
NC(II);   W/NBS
65595

626   C12 H18 O13 Zn4   O-18-169-1   Zinc acetate dihydrate tetramer;   W
127885

626   C18 H10 D3 F15 O7   BS-2-102-0   Pentafluoropropionyl derivative of 3-methoxy-4-hydroxyphenylglycol-d3;
W
127882

- 5883 -

626   C18 H29 I Os P2   89215-80-5   K-117-159-0   Iodo(exo-6-phenyl-η5-cyclohexadienyl)bis(trimethylphosphane)
osmium(II);   W   127886

626   C18 H48 O10 P2 Si5   55334-89-9   BA-0-300-0   3,5-Dioxa-4-phospha-2-silaoctan-8-oic acid, 7-[[bis[(tri
methylsilyl)oxy]phosphinyl]oxy]-2,2-dimethyl-4-[(trimethylsilyl)oxy]-, trimethylsilyl ester, 4-oxide, (R)-
;; PENTAKIS(TRIMETHYLSILYL) D-GLYCERATE-2,3-DIPHOSPHATE;   W/NBS   65596

626   C18 H3 D45 O10 P2 Si5   55334-90-2   BA-0-301-0   PENTAKIS(TRIMETHYLSILYL-D9) D-GLYCERATE-2,3-DIPHOSPHATE
; W   65597

626   C26 H36 Br2 N4 Zn   91278-36-3   H-67-932-0   Dibromobis[2-dimethylamino-3,3-dimethyl-4-phenyl-1-azetine-
N(1)]-zinc;   W   127887

626   C27 H26 N6 O12   56051-46-8   T-66-2413-0   L-Valine, N-[N,O-bis(2,4-dinitrophenyl)-L-tyrosyl]-, methyl
ester;; O,N-DIDINITROPHENYLTYROSYLVALINE METHYL;   W/NBS   65599

626   C27 H27 N2 O2 P W   75520-18-2   NS--27520-0-0   Tungsten, carbonyl(η5-2,4-cyclopentadien-1-yl)[N-methyl-
P,P-diphenyl-N-(1-phenylethyl)phosphinous amide-P]nitrosyl-, stereoisomer;;   NBS   65600

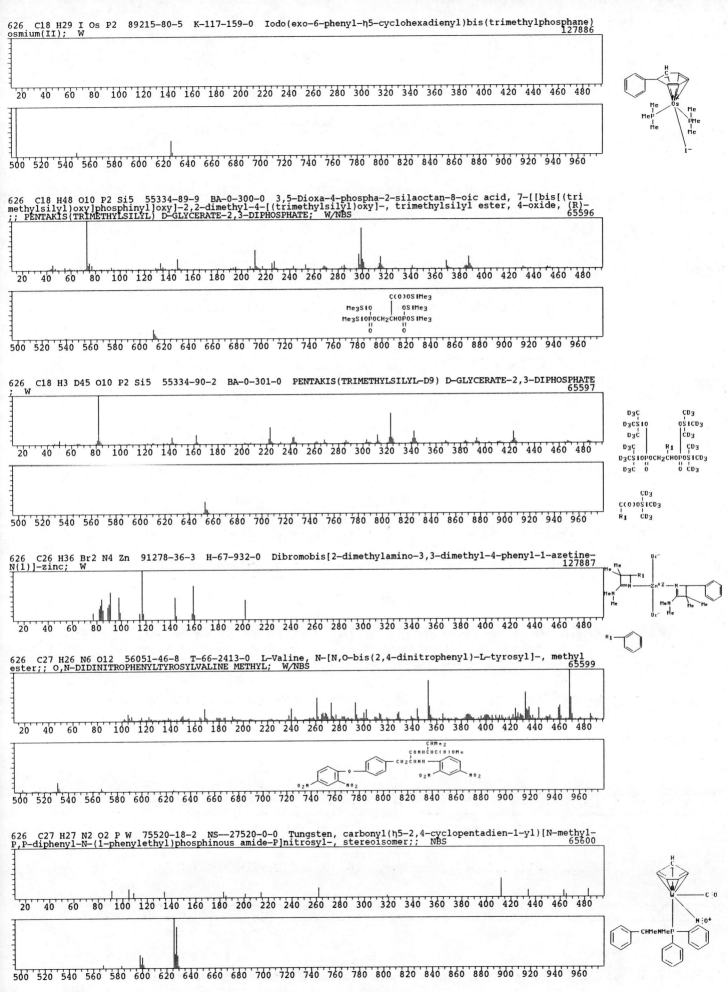

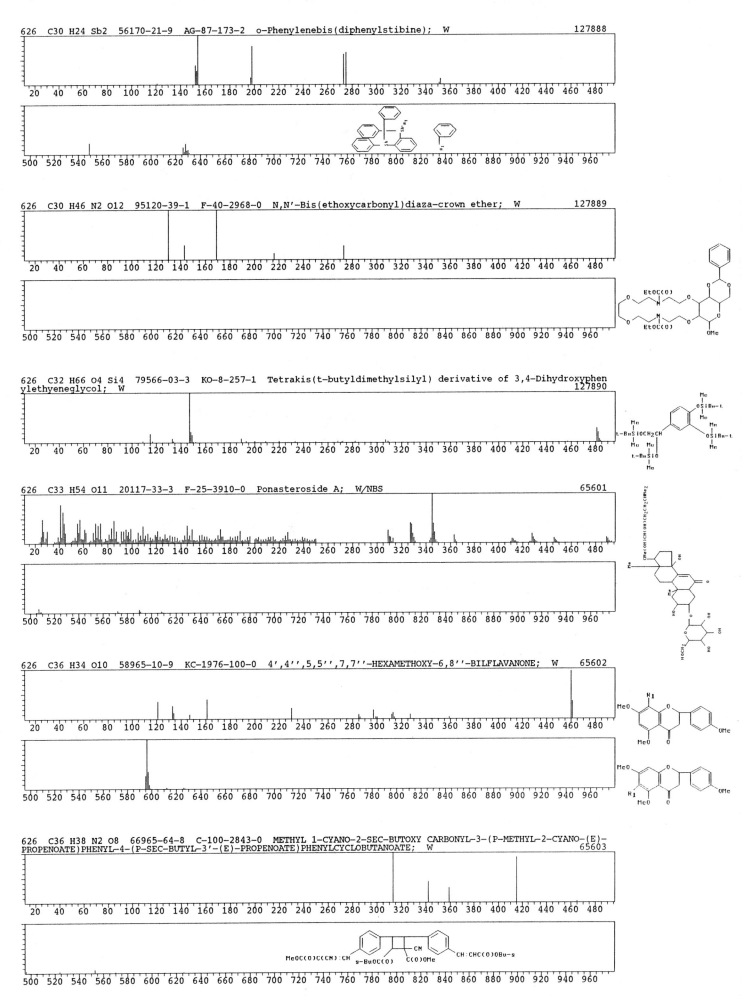

626   C30 H24 Sb2   56170-21-9   AG-87-173-2   o-Phenylenebis(diphenylstibine);   W   127888

626   C30 H46 N2 O12   95120-39-1   F-40-2968-0   N,N'-Bis(ethoxycarbonyl)diaza-crown ether;   W   127889

626   C32 H66 O4 Si4   79566-03-3   KO-8-257-1   Tetrakis(t-butyldimethylsilyl) derivative of 3,4-Dihydroxyphen
ylethyeneglycol;   W   127890

626   C33 H54 O11   20117-33-3   F-25-3910-0   Ponasteroside A;   W/NBS   65601

626   C36 H34 O10   58965-10-9   KC-1976-100-0   4',4'',5,5'',7,7''-HEXAMETHOXY-6,8''-BILFLAVANONE;   W   65602

626   C36 H38 N2 O8   66965-64-8   C-100-2843-0   METHYL 1-CYANO-2-SEC-BUTOXY CARBONYL-3-(P-METHYL-2-CYANO-(E)-
PROPENOATE)PHENYL-4-(P-SEC-BUTYL-3'-(E)-PROPENOATE)PHENYLCYCLOBUTANOATE;   W   65603

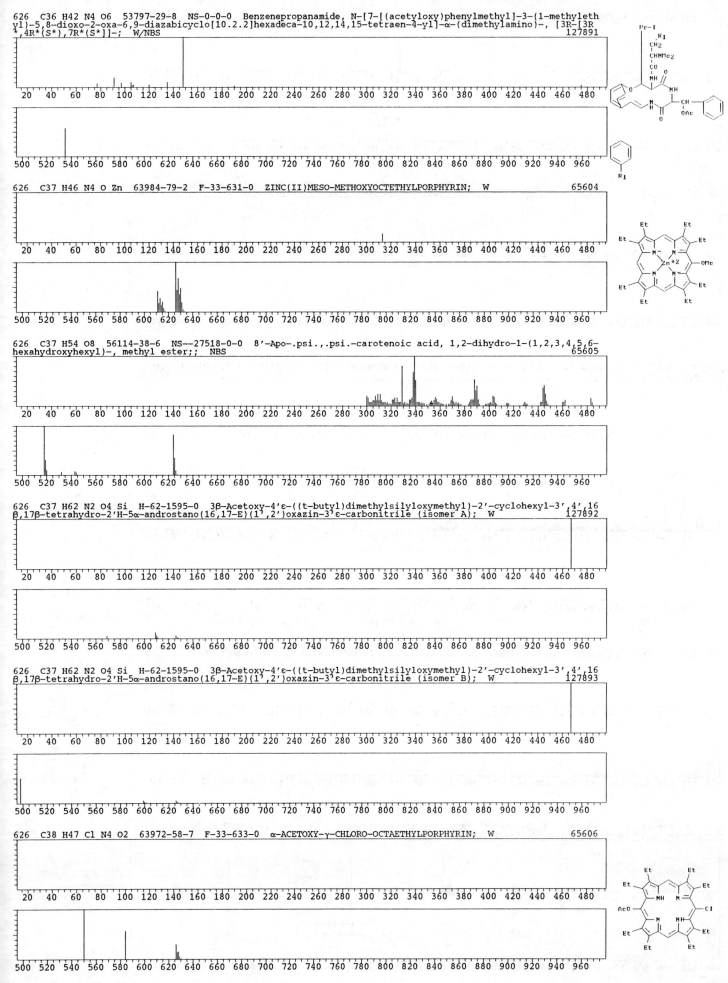

626 C36 H42 N4 O6 53797-29-8 NS-0-0-0 Benzenepropanamide, N-[7-[(acetyloxy)phenylmethyl]-3-(1-methyleth yl)-5,8-dioxo-2-oxa-6,9-diazabicyclo[10.2.2]hexadeca-10,12,14,15-tetraen-4-yl]-α-(dimethylamino)-, [3R-[3R *,4R*(S*),7R*(S*)]]-; W/NBS   127891

626 C37 H46 N4 O Zn 63984-79-2 F-33-631-0 ZINC(II)MESO-METHOXYOCTETHYLPORPHYRIN; W   65604

626 C37 H54 O8 56114-38-6 NS--27518-0-0 8'-Apo-.psi.,.psi.-carotenoic acid, 1,2-dihydro-1-(1,2,3,4,5,6-hexahydroxyhexyl)-, methyl ester;; NBS   65605

626 C37 H62 N2 O4 Si H-62-1595-0 3β-Acetoxy-4'ε-((t-butyl)dimethylsilyloxymethyl)-2'-cyclohexyl-3',4',16 β,17β-tetrahydro-2'H-5α-androstano(16,17-E)(1',2')oxazin-3'ε-carbonitrile (isomer A); W   127892

626 C37 H62 N2 O4 Si H-62-1595-0 3β-Acetoxy-4'ε-((t-butyl)dimethylsilyloxymethyl)-2'-cyclohexyl-3',4',16 β,17β-tetrahydro-2'H-5α-androstano(16,17-E)(1',2')oxazin-3'ε-carbonitrile (isomer B); W   127893

626 C38 H47 Cl N4 O2 63972-58-7 F-33-633-0 α-ACETOXY-γ-CHLORO-OCTAETHYLPORPHYRIN; W   65606

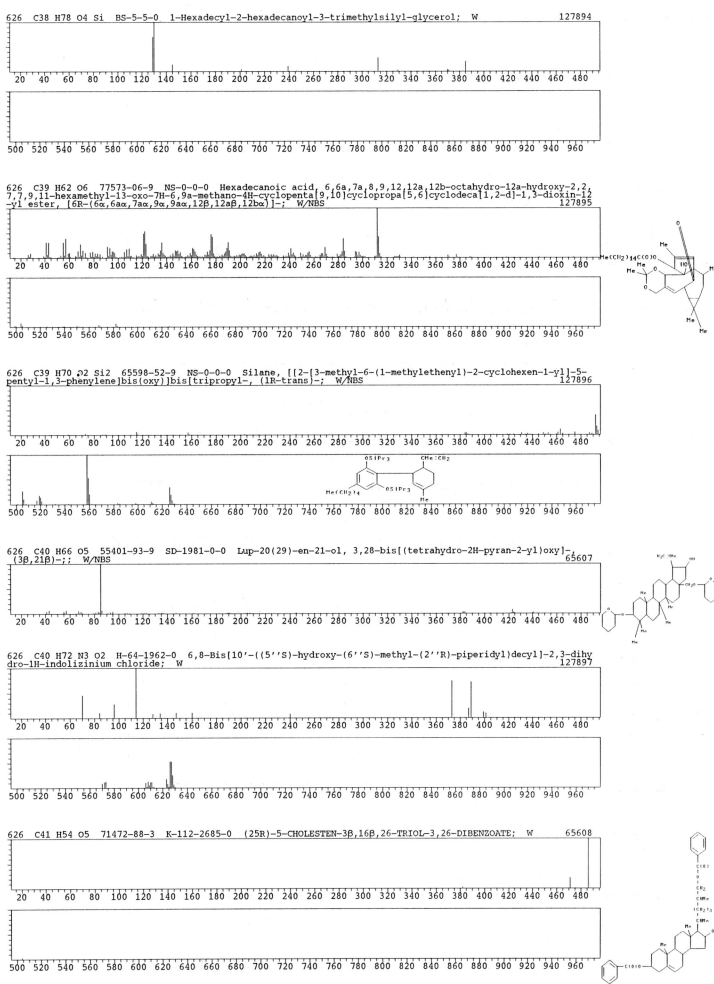

626   C38 H78 O4 Si   BS-5-5-0   1-Hexadecyl-2-hexadecanoyl-3-trimethylsilyl-glycerol;   W   127894

626   C39 H62 O6   77573-06-9   NS-0-0-0   Hexadecanoic acid, 6,6a,7a,8,9,12,12a,12b-octahydro-12a-hydroxy-2,2,
7,7,9,11-hexamethyl-13-oxo-7H-6,9a-methano-4H-cyclopenta[9,10]cyclopropa[5,6]cyclodeca[1,2-d]-1,3-dioxin-12
-yl ester, [6R-(6α,6aα,7aα,9α,9aα,12β,12aβ,12bα)]-;   W/NBS   127895

626   C39 H70 O2 Si2   65598-52-9   NS-0-0-0   Silane, [[2-[3-methyl-6-(1-methylethenyl)-2-cyclohexen-1-yl]-5-
pentyl-1,3-phenylene]bis(oxy)]bis[tripropyl-, (1R-trans)-;   W/NBS   127896

626   C40 H66 O5   55401-93-9   SD-1981-0-0   Lup-20(29)-en-21-ol, 3,28-bis[(tetrahydro-2H-pyran-2-yl)oxy]-,
(3β,21β)-;;   W/NBS   65607

626   C40 H72 N3 O2   H-64-1962-0   6,8-Bis[10'-((5''S)-hydroxy-(6''S)-methyl-(2''R)-piperidyl)decyl]-2,3-dihy
dro-1H-indolizinium chloride;   W   127897

626   C41 H54 O5   71472-88-3   K-112-2685-0   (25R)-5-CHOLESTEN-3β,16β,26-TRIOL-3,26-DIBENZOATE;   W   65608

626 F18 Mo P6 15339-46-5 AG-57-172-22 HEXA(TRIFLUOROPHOSPHINE)MOLYBDENUM; W   65609

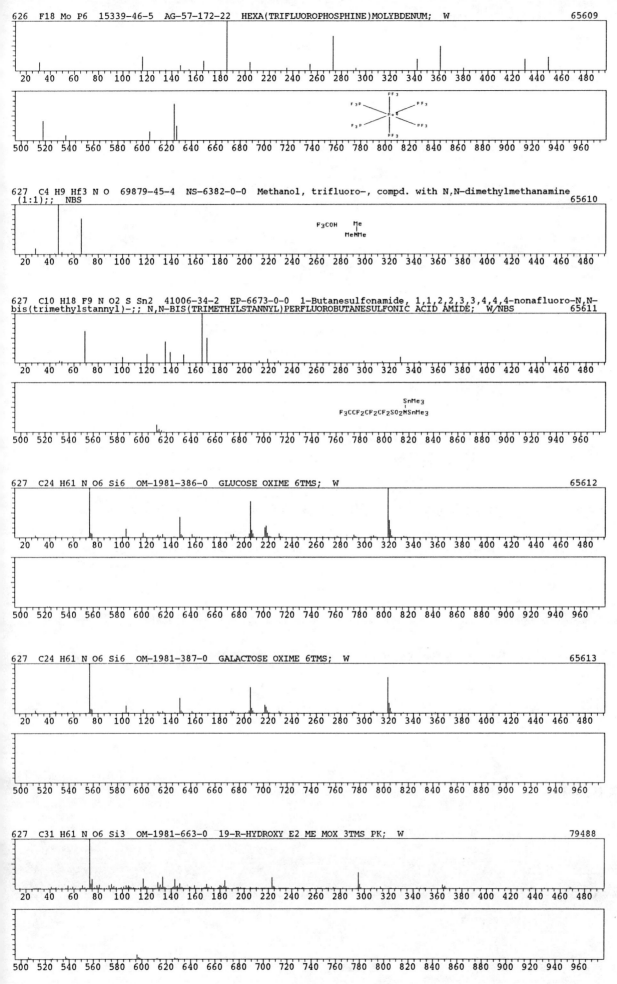

627 C4 H9 Hf3 N O 69879-45-4 NS-6382-0-0 Methanol, trifluoro-, compd. with N,N-dimethylmethanamine
  (1:1);; NBS   65610

627 C10 H18 F9 N O2 S Sn2 41006-34-2 EP-6673-0-0 1-Butanesulfonamide, 1,1,2,2,3,3,4,4,4-nonafluoro-N,N-
bis(trimethylstannyl)-;; N,N-BIS(TRIMETHYLSTANNYL)PERFLUOROBUTANESULFONIC ACID AMIDE; W/NBS   65611

627 C24 H61 N O6 Si6 OM-1981-386-0 GLUCOSE OXIME 6TMS; W   65612

627 C24 H61 N O6 Si6 OM-1981-387-0 GALACTOSE OXIME 6TMS; W   65613

627 C31 H61 N O6 Si3 OM-1981-663-0 19-R-HYDROXY E2 ME MOX 3TMS PK; W   79488

627   C41 H73 N O3   77588-24-0   K-114-957-0   2-(4-Cyanobenzyloxy)-1-dodecyloxy-3-(octadecyloxy)propane;   W
127898

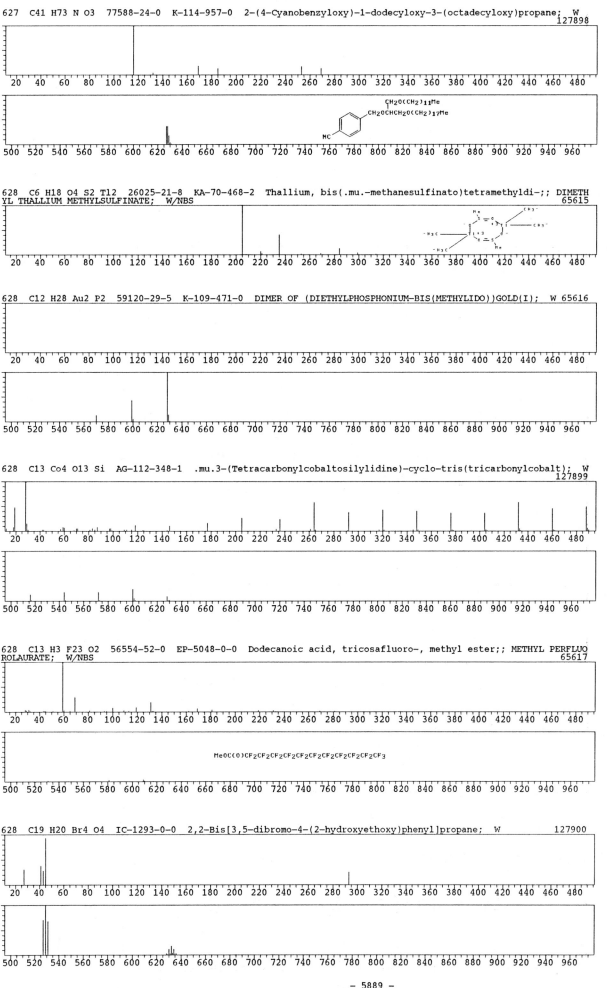

628   C6 H18 O4 S2 Tl2   26025-21-8   KA-70-468-2   Thallium, bis(.mu.-methanesulfinato)tetramethyldi-;; DIMETH
YL THALLIUM METHYLSULFINATE;   W/NBS
65615

628   C12 H28 Au2 P2   59120-29-5   K-109-471-0   DIMER OF (DIETHYLPHOSPHONIUM-BIS(METHYLIDO))GOLD(I);   W 65616

628   C13 Co4 O13 Si   AG-112-348-1   .mu.3-(Tetracarbonylcobaltosilylidine)-cyclo-tris(tricarbonylcobalt);   W
127899

628   C13 H3 F23 O2   56554-52-0   EP-5048-0-0   Dodecanoic acid, tricosafluoro-, methyl ester;; METHYL PERFLUO
ROLAURATE;   W/NBS
65617

628   C19 H20 Br4 O4   IC-1293-0-0   2,2-Bis[3,5-dibromo-4-(2-hydroxyethoxy)phenyl]propane;   W   127900

628 C20 H28 O8 Th 17499-48-8 T-66-1826-0 Thorium, tetrakis(2,4-pentanedionato-O,O')-,;; Tetrakis(acetyl acetonato)thorium(IV); W/NBS   65618

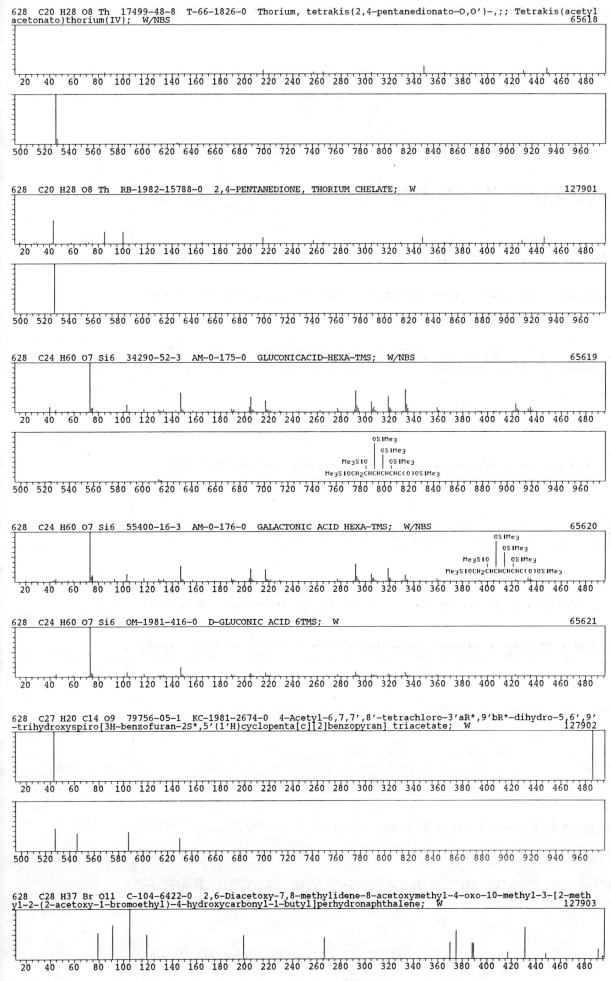

628 C20 H28 O8 Th RB-1982-15788-0 2,4-PENTANEDIONE, THORIUM CHELATE; W   127901

628 C24 H60 O7 Si6 34290-52-3 AM-0-175-0 GLUCONICACID-HEXA-TMS; W/NBS   65619

OS iMe3
OS iMe3
Me3S iO   OS iMe3
Me3S iOCH2CHCHCHCHCC(O)OS iMe3

628 C24 H60 O7 Si6 55400-16-3 AM-0-176-0 GALACTONIC ACID HEXA-TMS; W/NBS   65620

OS iMe3
OS iMe3
Me3S iO   OS iMe3
Me3S iOCH2CHCHCHCHCC(O)OS iMe3

628 C24 H60 O7 Si6 OM-1981-416-0 D-GLUCONIC ACID 6TMS; W   65621

628 C27 H20 Cl4 O9 79756-05-1 KC-1981-2674-0 4-Acetyl-6,7,7',8'-tetrachloro-3'aR*,9'bR*-dihydro-5,6',9' -trihydroxyspiro{3H-benzofuran-2S*,5'(1'H)cyclopenta[c][2]benzopyran} triacetate; W   127902

628 C28 H37 Br O11 C-104-6422-0 2,6-Diacetoxy-7,8-methylidene-8-acetoxymethyl-4-oxo-10-methyl-3-[2-meth yl-2-(2-acetoxy-1-bromoethyl)-4-hydroxycarbonyl-1-butyl]perhydronaphthalene; W   127903

628　C31 H32 O14　82891-64-3　J-47-4310-0　7β-(Tetraacetoglucosyl)-6,11-dihydroxyzantho[2,3-g]tetralin;　W

127904

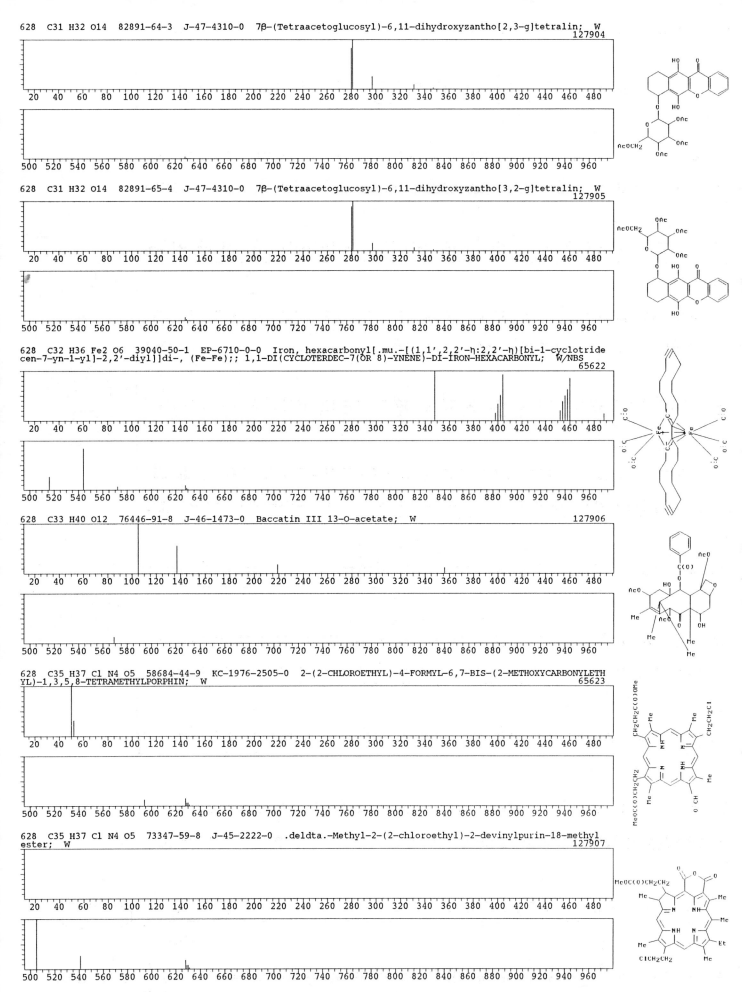

628　C31 H32 O14　82891-65-4　J-47-4310-0　7β-(Tetraacetoglucosyl)-6,11-dihydroxyzantho[3,2-g]tetralin;　W

127905

628　C32 H36 Fe2 O6　39040-50-1　EP-6710-0-0　Iron, hexacarbonyl[.mu.-[(1,1',2,2'-η:2,2'-η)[bi-1-cyclotride
cen-7-yn-1-yl]-2,2'-diyl]]di-, (Fe-Fe);;　1,1-DI(CYCLOTERDEC-7(OR 8)-YNENE)-DI-IRON-HEXACARBONYL;　W/NBS

65622

628　C33 H40 O12　76446-91-8　J-46-1473-0　Baccatin III 13-O-acetate;　W

127906

628　C35 H37 Cl N4 O5　58684-44-9　KC-1976-2505-0　2-(2-CHLOROETHYL)-4-FORMYL-6,7-BIS-(2-METHOXYCARBONYLETH
YL)-1,3,5,8-TETRAMETHYLPORPHIN;　W

65623

628　C35 H37 Cl N4 O5　73347-59-8　J-45-2222-0　.deldta.-Methyl-2-(2-chloroethyl)-2-devinylpurin-18-methyl
ester;　W

127907

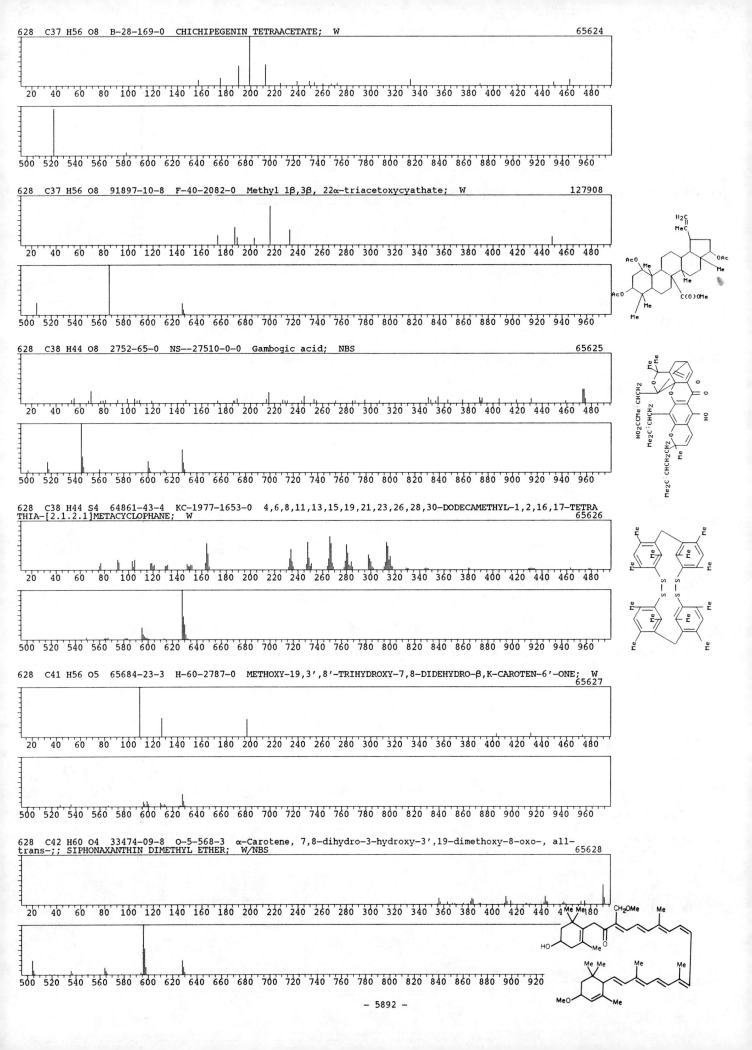

628   C37 H56 O8   B-28-169-0   CHICHIPEGENIN TETRAACETATE;   W   65624

628   C37 H56 O8   91897-10-8   F-40-2082-0   Methyl 1β,3β, 22α-triacetoxycyathate;   W   127908

628   C38 H44 O8   2752-65-0   NS--27510-0-0   Gambogic acid;   NBS   65625

628   C38 H44 S4   64861-43-4   KC-1977-1653-0   4,6,8,11,13,15,19,21,23,26,28,30-DODECAMETHYL-1,2,16,17-TETRA
THIA-[2.1.2.1]METACYCLOPHANE;   W   65626

628   C41 H56 O5   65684-23-3   H-60-2787-0   METHOXY-19,3',8'-TRIHYDROXY-7,8-DIDEHYDRO-β,κ-CAROTEN-6'-ONE;   W
65627

628   C42 H60 O4   33474-09-8   O-5-568-3   α-Carotene, 7,8-dihydro-3-hydroxy-3',19-dimethoxy-8-oxo-, all-
trans-;; SIPHONAXANTHIN DIMETHYL ETHER;   W/NBS   65628

628   C44 H33 Cl S   72478-91-2   J-45-929-0   4-(p-Chlorophenyl)-3,3-diphenyl-2-(diphenylmethylene)-5,6-dihydr
obenzo(H)thiochroman;   W   127909

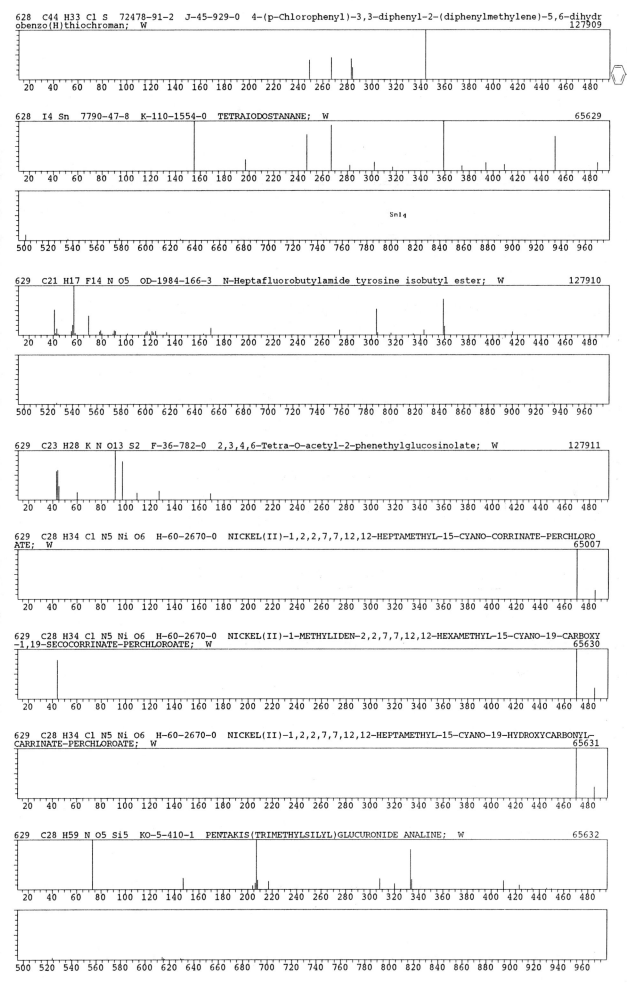

628   I4 Sn   7790-47-8   K-110-1554-0   TETRAIODOSTANANE;   W   65629

SnI4

629   C21 H17 F14 N O5   OD-1984-166-3   N-Heptafluorobutylamide tyrosine isobutyl ester;   W   127910

629   C23 H28 K N O13 S2   F-36-782-0   2,3,4,6-Tetra-O-acetyl-2-phenethylglucosinolate;   W   127911

629   C28 H34 Cl N5 Ni O6   H-60-2670-0   NICKEL(II)-1,2,2,7,7,12,12-HEPTAMETHYL-15-CYANO-CORRINATE-PERCHLORO
ATE;   W   65007

629   C28 H34 Cl N5 Ni O6   H-60-2670-0   NICKEL(II)-1-METHYLIDEN-2,2,7,7,12,12-HEXAMETHYL-15-CYANO-19-CARBOXY
-1,19-SECOCORRINATE-PERCHLOROATE;   W   65630

629   C28 H34 Cl N5 Ni O6   H-60-2670-0   NICKEL(II)-1,2,2,7,7,12,12-HEPTAMETHYL-15-CYANO-19-HYDROXYCARBONYL-
CARRINATE-PERCHLOROATE;   W   65631

629   C28 H59 N O5 Si5   KO-5-410-1   PENTAKIS(TRIMETHYLSILYL)GLUCURONIDE ANALINE;   W   65632

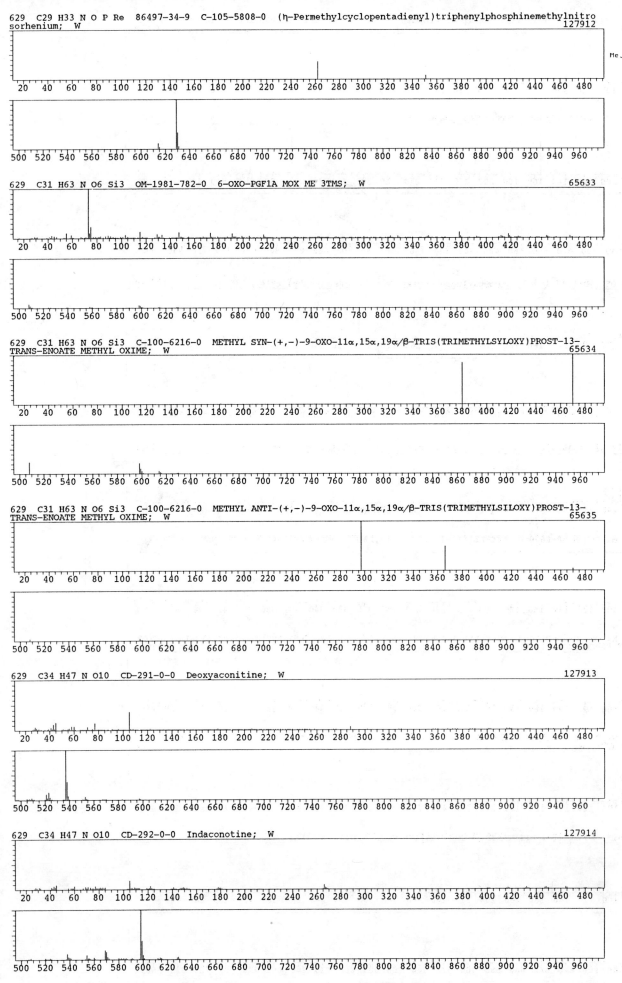

629   C29 H33 N O P Re   86497-34-9   C-105-5808-0   (η-Permethylcyclopentadienyl)triphenylphosphinemethylnitro
sorhenium; W   127912

629   C31 H63 N O6 Si3   OM-1981-782-0   6-OXO-PGF1A MOX ME 3TMS; W   65633

629   C31 H63 N O6 Si3   C-100-6216-0   METHYL SYN-(+,-)-9-OXO-11α,15α,19α/β-TRIS(TRIMETHYLSYLOXY)PROST-13-
TRANS-ENOATE METHYL OXIME; W   65634

629   C31 H63 N O6 Si3   C-100-6216-0   METHYL ANTI-(+,-)-9-OXO-11α,15α,19α/β-TRIS(TRIMETHYLSILOXY)PROST-13-
TRANS-ENOATE METHYL OXIME; W   65635

629   C34 H47 N O10   CD-291-0-0   Deoxyaconitine; W   127913

629   C34 H47 N O10   CD-292-0-0   Indaconotine; W   127914

629   C36 H47 N5 O5   93827-40-8   AH-115-1078-0   Amide of rac-(4Z,9Z,15Z)-17-Ethyl-1,19-dioxo-2,2,7,8,12,13,
18-heptamethyl-1,2,3,19,23,24-hexahydro-21H-bilin-3-N-tert-butoxycarbonylmethylacetic acid;   W   127915

629   C37 H47 N3 O6   60283-67-2   KC-1976-730-0   3β-ACETOXY-25-HYDROXY-4'-PHENYL-5α,8α-(1',2')-1',2',4'-TRIAZ
OLIDINOCHOLESTA-6,22-DIENE-3',5',24-TRIONE;   W   65636

629   C37 H47 N3 O6   78109-11-2   J-46-3428-0   (23R)-3β-(methoxymethoxy)-5α,8α-(4-phenyl-1,2-urazolo)cholesta
-6,25-diene-26,23-lactone;   W   127916

629   C37 H47 N3 O6   78183-87-6   J-46-3428-0   (23S)-3β-(methoxymethoxy)-5α,8α-(4-phenyl-1,2-urazolo)cholesta
-6,25-diene-26,23-lactone;   W   127917

630   C17 H12 Cl10 O4   4234-79-1   NS--26479-0-0   1,3,4-Metheno-1H-cyclobuta[cd]pentalene-2-pentanoic acid,
1,1a,3,3a,4,5,5,5a,5b,6-decachlorooctahydro-2-hydroxy-γ-oxo-, ethyl ester;   NBS   127918

630   C20 H10 Br4 O4   76-62-0   NS-0-0-0   1(3H)-Isobenzofuranone, 3,3-bis(3,5-dibromo-4-hydroxyphenyl)-;
W/NBS   127919

630   C21 H19 O9 P W   72344-91-3   K-112-3643-0   PENTACARBONYL(BIS(1-METHYL-3-OXO-1-BUTENYLOXY)PHENYLPHOSPHA
N)TUNQSTEN;   W   65637

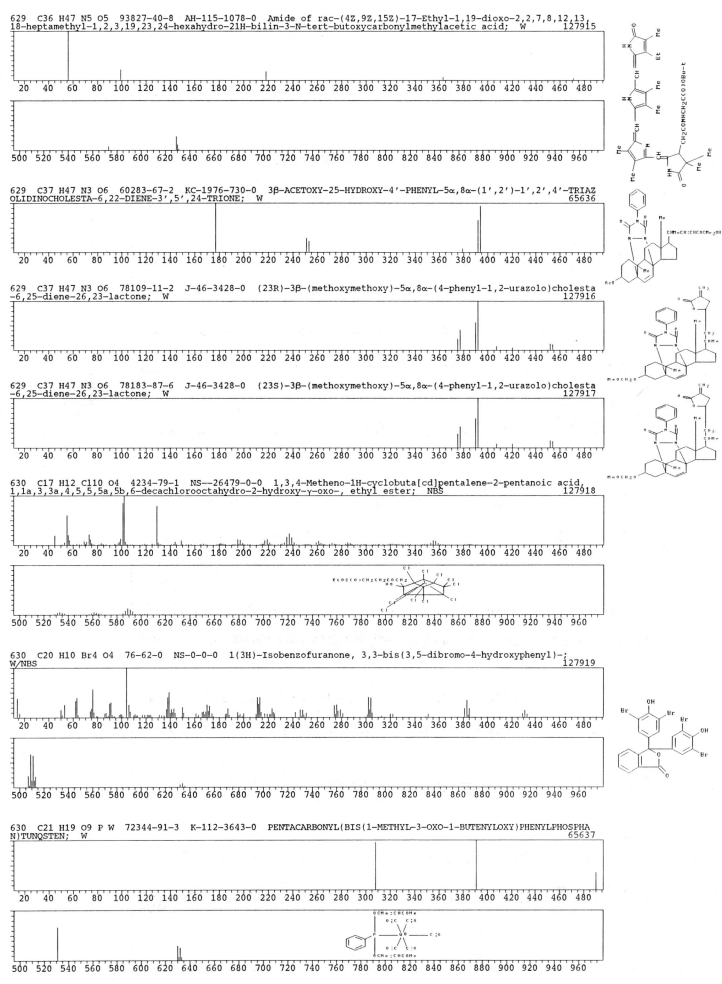

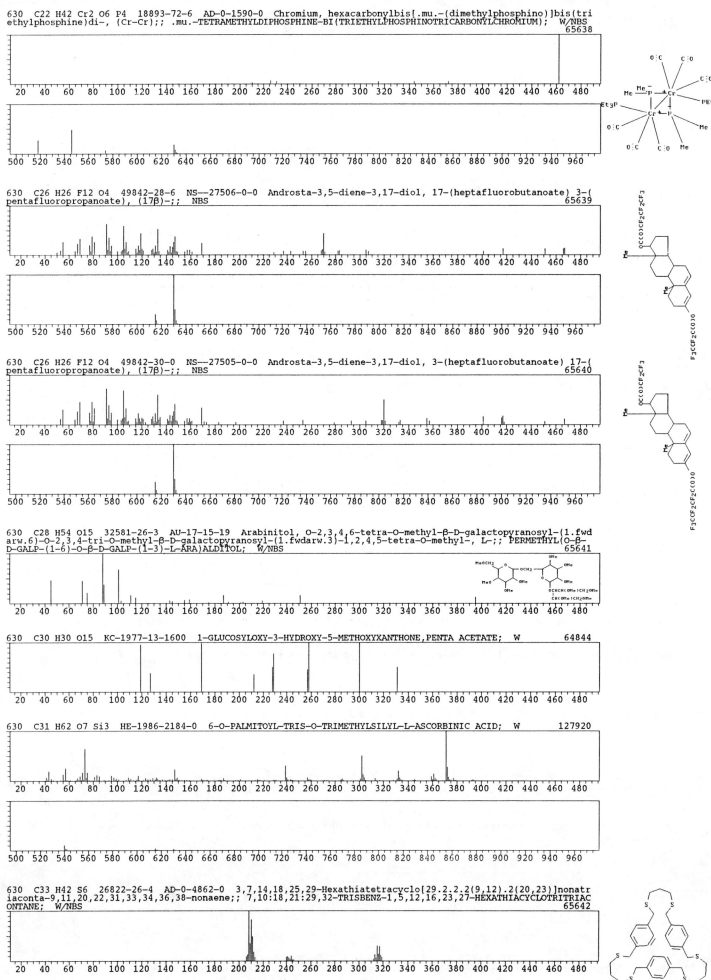

630   C22 H42 Cr2 O6 P4   18893-72-6   AD-0-1590-0   Chromium, hexacarbonylbis[.mu.-(dimethylphosphino)]bis(tri
ethylphosphine)di-, (Cr-Cr);; .mu.-TETRAMETHYLDIPHOSPHINE-BI(TRIETHYLPHOSPHINOTRICARBONYLCHROMIUM);   W/NBS
65638

630   C26 H26 F12 O4   49842-28-6   NS--27506-0-0   Androsta-3,5-diene-3,17-diol, 17-(heptafluorobutanoate) 3-(
pentafluoropropanoate), (17β)-;;   NBS
65639

630   C26 H26 F12 O4   49842-30-0   NS--27505-0-0   Androsta-3,5-diene-3,17-diol, 3-(heptafluorobutanoate) 17-(
pentafluoropropanoate), (17β)-;;   NBS
65640

630   C28 H54 O15   32581-26-3   AU-17-15-19   Arabinitol, O-2,3,4,6-tetra-O-methyl-β-D-galactopyranosyl-(1.fwd
arw.6)-O-2,3,4-tri-O-methyl-β-D-galactopyranosyl-(1.fwdarw.3)-1,2,4,5-tetra-O-methyl-, L-;;   PERMETHYL(O-β-
D-GALP-(1-6)-O-β-D-GALP-(1-3)-L-ARA)ALDITOL;   W/NBS
65641

630   C30 H30 O15   KC-1977-13-1600   1-GLUCOSYLOXY-3-HYDROXY-5-METHOXYXANTHONE,PENTA ACETATE;   W
64844

630   C31 H62 O7 Si3   HE-1986-2184-0   6-O-PALMITOYL-TRIS-O-TRIMETHYLSILYL-L-ASCORBIC ACID;   W
127920

630   C33 H42 S6   26822-26-4   AD-0-4862-0   3,7,14,18,25,29-Hexathiatetracyclo[29.2.2.2(9,12).2(20,23)]nonatr
iaconta-9,11,20,22,31,33,34,36,38-nonaene;; 7,10:18,21:29,32-TRISBENZ-1,5,12,16,23,27-HEXATHIACYCLOTRITRIAC
ONTANE;   W/NBS
65642

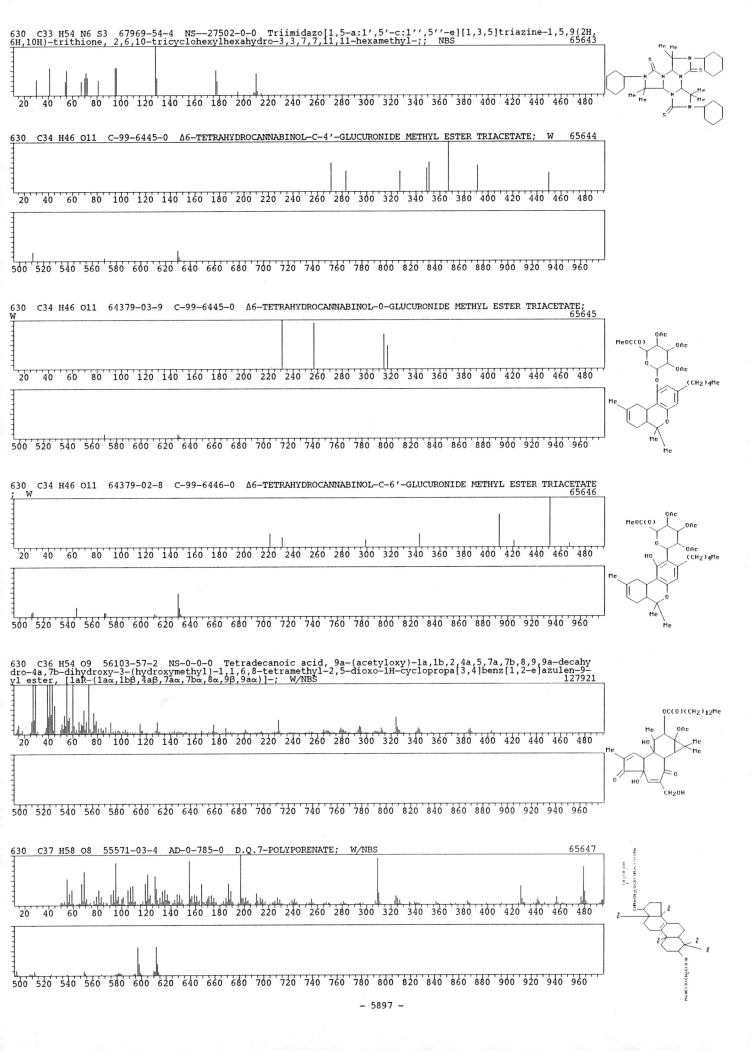

630   C33 H54 N6 S3   67969-54-4   NS--27502-0-0   Triimidazo[1,5-a:1',5'-c:1'',5''-e][1,3,5]triazine-1,5,9(2H, 6H,10H)-trithione, 2,6,10-tricyclohexylhexahydro-3,3,7,7,11,11-hexamethyl-;;   NBS   65643

630   C34 H46 O11   C-99-6445-0   Δ6-TETRAHYDROCANNABINOL-C-4'-GLUCURONIDE METHYL ESTER TRIACETATE;   W   65644

630   C34 H46 O11   64379-03-9   C-99-6445-0   Δ6-TETRAHYDROCANNABINOL-0-GLUCURONIDE METHYL ESTER TRIACETATE;
W   65645

630   C34 H46 O11   64379-02-8   C-99-6446-0   Δ6-TETRAHYDROCANNABINOL-C-6'-GLUCURONIDE METHYL ESTER TRIACETATE
; W   65646

630   C36 H54 O9   56103-57-2   NS-0-0-0   Tetradecanoic acid, 9a-(acetyloxy)-1a,1b,2,4a,5,7a,7b,8,9,9a-decahy
dro-4a,7b-dihydroxy-3-(hydroxymethyl)-1,1,6,8-tetramethyl-2,5-dioxo-1H-cyclopropa[3,4]benz[1,2-e]azulen-9-
yl ester, [1aR-(1aα,1bβ,4aβ,7aα,7bα,8α,9β,9aα)]-;   W/NBS   127921

630   C37 H58 O8   55571-03-4   AD-0-785-0   D.Q.7-POLYPORENATE;   W/NBS   65647

- 5897 -

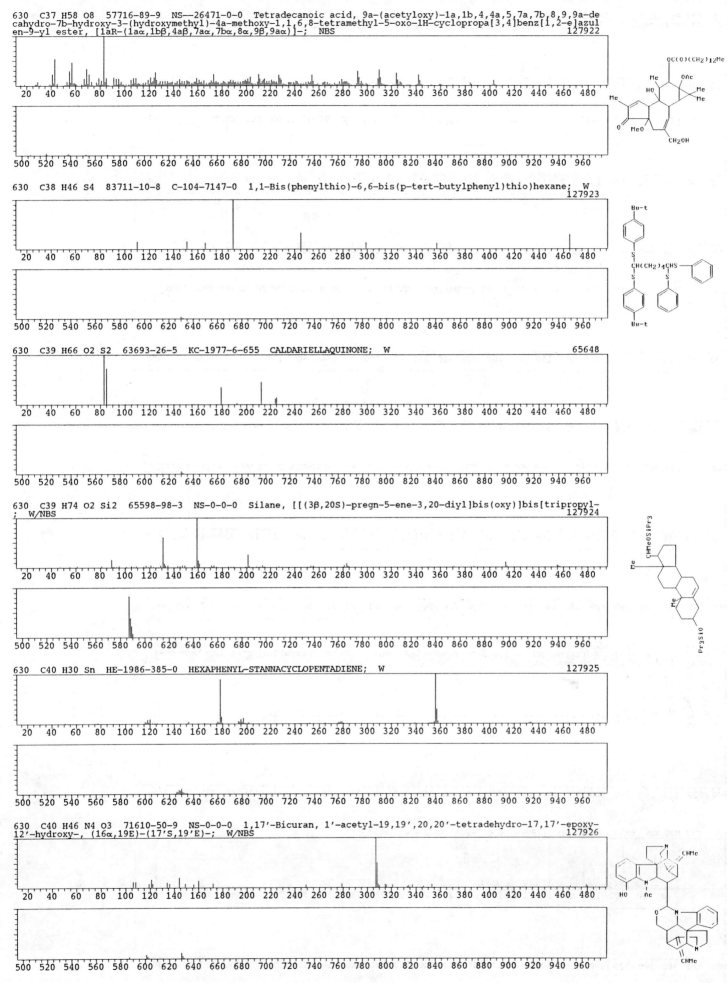

630   C37 H58 O8   57716-89-9   NS--26471-0-0   Tetradecanoic acid, 9a-(acetyloxy)-1a,1b,4,4a,5,7a,7b,8,9,9a-de
cahydro-7b-hydroxy-3-(hydroxymethyl)-4a-methoxy-1,1,6,8-tetramethyl-5-oxo-1H-cyclopropa[3,4]benz[1,2-e]azul
en-9-yl ester, [1aR-(1aα,1bβ,4aβ,7aα,7bα,8α,9β,9aα)]-;   NBS   127922

630   C38 H46 S4   83711-10-8   C-104-7147-0   1,1-Bis(phenylthio)-6,6-bis(p-tert-butylphenyl)thio)hexane;   W
                                                                                                127923

630   C39 H66 O2 S2   63693-26-5   KC-1977-6-655   CALDARIELLAQUINONE;   W   65648

630   C39 H74 O2 Si2   65598-98-3   NS-0-0-0   Silane, [[(3β,20S)-pregn-5-ene-3,20-diyl]bis(oxy)]bis[tripropyl-
;   W/NBS   127924

630   C40 H30 Sn   HE-1986-385-0   HEXAPHENYL-STANNACYCLOPENTADIENE;   W   127925

630   C40 H46 N4 O3   71610-50-9   NS-0-0-0   1,17'-Bicuran, 1'-acetyl-19,19',20,20'-tetradehydro-17,17'-epoxy-
12'-hydroxy-, (16α,19E)-(17'S,19'E)-;   W/NBS   127926

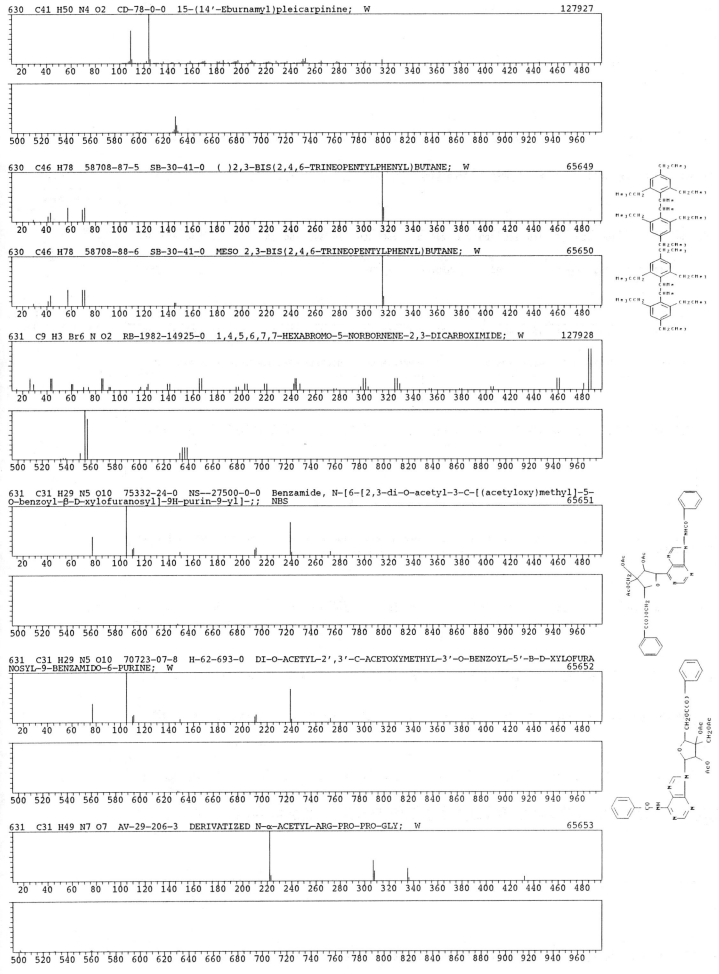

630   C41 H50 N4 O2   CD-78-0-0   15-(14'-Eburnamyl)pleicarpinine;   W   127927

630   C46 H78   58708-87-5   SB-30-41-0   ( )2,3-BIS(2,4,6-TRINEOPENTYLPHENYL)BUTANE;   W   65649

630   C46 H78   58708-88-6   SB-30-41-0   MESO 2,3-BIS(2,4,6-TRINEOPENTYLPHENYL)BUTANE;   W   65650

631   C9 H3 Br6 N O2   RB-1982-14925-0   1,4,5,6,7,7-HEXABROMO-5-NORBORNENE-2,3-DICARBOXIMIDE;   W   127928

631   C31 H29 N5 O10   75332-24-0   NS--27500-0-0   Benzamide, N-[6-[2,3-di-O-acetyl-3-C-[(acetyloxy)methyl]-5-
O-benzoyl-β-D-xylofuranosyl]-9H-purin-9-yl]-;;   NBS   65651

631   C31 H29 N5 O10   70723-07-8   H-62-693-0   DI-O-ACETYL-2',3'-C-ACETOXYMETHYL-3'-O-BENZOYL-5'-B-D-XYLOFURA
NOSYL-9-BENZAMIDO-6-PURINE;   W   65652

631   C31 H49 N7 O7   AV-29-206-3   DERIVATIZED N-α-ACETYL-ARG-PRO-PRO-GLY;   W   65653

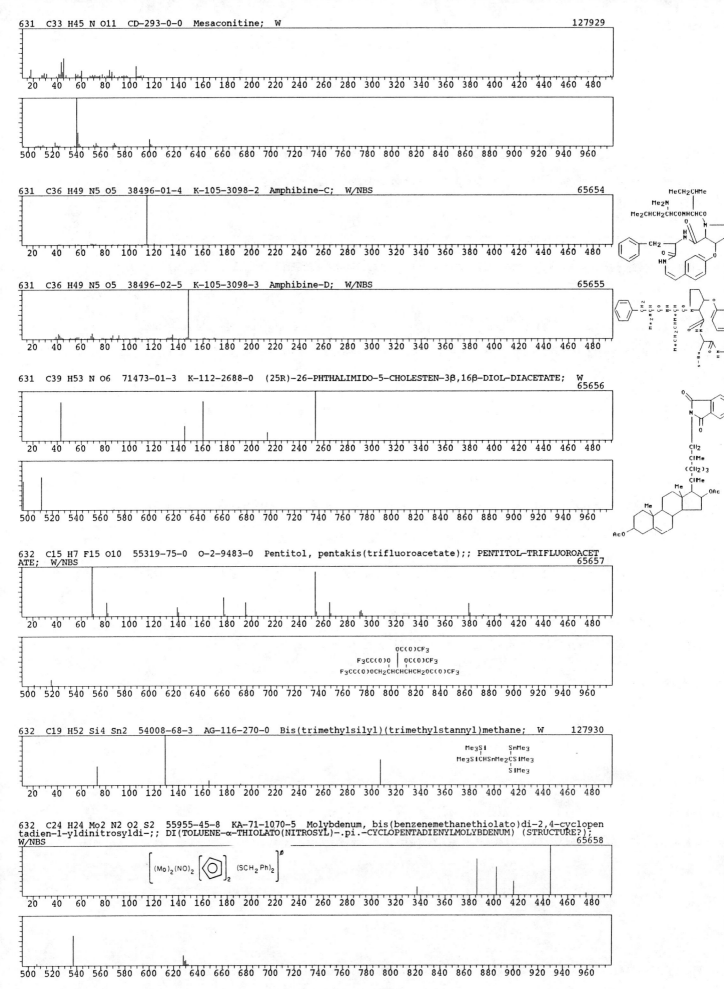

631   C33 H45 N O11   CD-293-0-0   Mesaconitine;   W   127929

631   C36 H49 N5 O5   38496-01-4   K-105-3098-2   Amphibine-C;   W/NBS   65654

631   C36 H49 N5 O5   38496-02-5   K-105-3098-3   Amphibine-D;   W/NBS   65655

631   C39 H53 N O6   71473-01-3   K-112-2688-0   (25R)-26-PHTHALIMIDO-5-CHOLESTEN-3β,16β-DIOL-DIACETATE;   W
                                                                                        65656

632   C15 H7 F15 O10   55319-75-0   O-2-9483-0   Pentitol, pentakis(trifluoroacetate);;   PENTITOL-TRIFLUOROACET
ATE;   W/NBS   65657

632   C19 H52 Si4 Sn2   54008-68-3   AG-116-270-0   Bis(trimethylsilyl)(trimethylstannyl)methane;   W   127930

632   C24 H24 Mo2 N2 O2 S2   55955-45-8   KA-71-1070-5   Molybdenum, bis(benzenemethanethiolato)di-2,4-cyclopen
tadien-1-yldinitrosyldi-;;   DI(TOLUENE-α-THIOLATO(NITROSYL)-.pi.-CYCLOPENTADIENYLMOLYBDENUM) (STRUCTURE?);;
W/NBS   65658

632   C24 H34 B4 Fe2 Ni O6   84959-90-0   C-105-2598-0   Bis[tricarbonyl-.mu.-(1,3-diborolenyl)iron]nickel;   W
127931

632   C26 H37 B O17   HE-1982-0-0   SUCROSE, 1',2,3,3',4',6'-HEXA-O-ACETYL-4,6-O-BORANDIYL-;   W
65659

632   C30 H20 Cl4 N8   18710-95-7   EP-7021-0-0   Pyrimido[5,4-d]pyrimidine, 2,4,6,8-tetrakis(m-chloroanilino)-
;; 2,4,6,8-TETRA(M-CHLOROANILINO)PYRIMIDO(5,4-D)PYRIMIDINE;   W/NBS
65660

632   C30 H32 O11 S2   73111-61-2   H-62-2407-0   1-O-Benzyl-2,3-O-isopropylidene-5,6-di-O-p-toluolsulfonyl-α-
D-mannofuranose;   W
127932

632   C30 H48 Mo O4 P2   56557-44-9   AG-94-60-0   cis-(Dicyclohexylphosphine)(tetracarbonyl)molybdenum;   W
127933

632   C31 H28 Cl4 N2 O4   8027-00-7   NS--27496-0-0   Dilan;   NBS
65661

632   C34 H36 N2 O8 S   70835-98-2   NS-0-0-0   α-DL-lyxo-Hexopyranoside, methyl 4,6-dideoxy-6-[[[1-[(phenyl
methoxy)carbonyl]-2-pyrrolidinyl]carbonyl]amino]-1-thio-, 2,3-dibenzoate;   W/NBS
127934

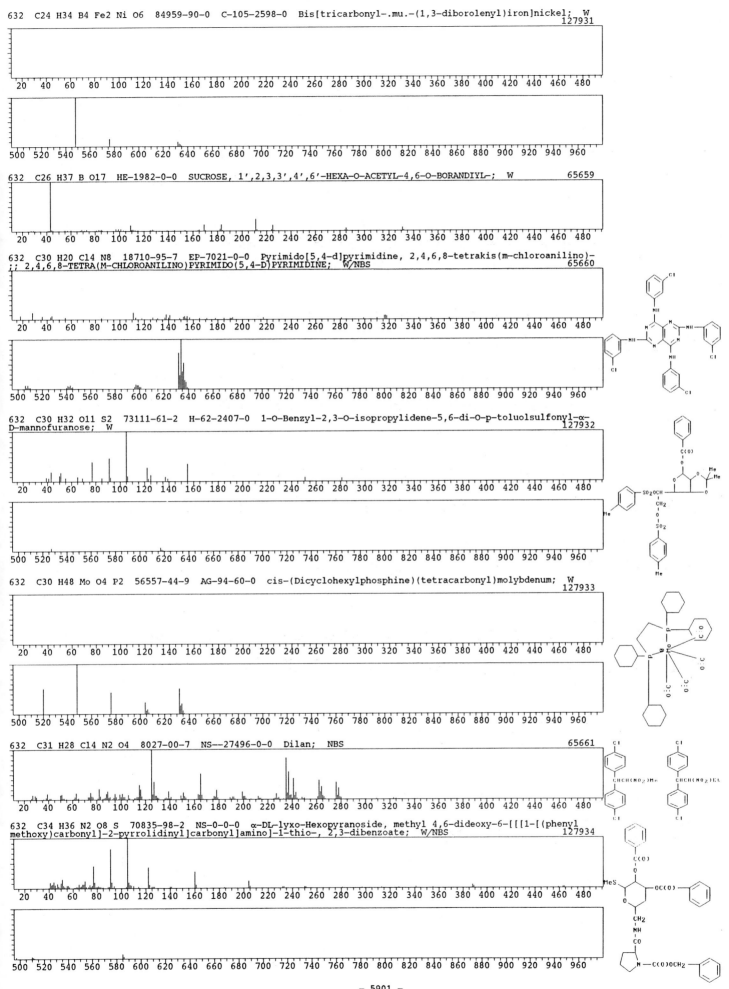

632   C34 H64 O10   88454-79-9   J-49-759-0   (2R,3R)-11,12-Didecyl-1,4,7,10,13,16-hexaoxacyclooctadecane-2,3-
dicarboxylic acid; W   127935

632   C35 H20 O4 S4   69856-47-9   F-34-2177-0   3,5-BIS-(4,5-DIPHENYL-1,2-DITHIOL-3-YLIDENE)PYRAN-2,4,6-TRIONE
; W   65662

632   C35 H44 N4 O7   91084-88-7   C-106-4985-0   1,1'-(2-Methoxy-1,3-phenylene)bis(tetrahydro-3-(2-methoxymeth
oxy)-3,5-dimethylphenyl)-2(1H)-pyrimidinone; W   127936

632   C36 H68 O3 Si3   56009-12-2   D-9-2921-8   Silane, [[(3β,5Z,7E)-9,10-secocholesta-5,7,10(19)-triene-3,21,
25-triyl]tris(oxy)]tris[trimethyl-;; 21,25-DIHYDROXYCHOLECALCIFEROL TRISTRIMETHYLSILYL ETHER; W/NBS   65664

632   C36 H68 O3 Si3   55759-95-0   D-10-2803-8   Silane, [[(3β,5Z,7E)-9,10-secocholesta-5,7,10(19)-triene-1,3,
25-triyl]tris(oxy)tris[trimethyl-;; 1,25-DIHYDROXYCHOLECALCIFEROL-TRISTRIMETHYLSILYL ETHER; W/NBS   65665

632   C36 H68 O3 Si3   56009-11-1   D-9-4779-2   Silane, [[(3β,5Z,7E)-9,10-secocholesta-5,7,10(19)-triene-3,25,
26-triyl]tris(oxy)]tris[trimethyl-;; 25,26-DIHYDROXYCHOLECALCIFEROL TRISTRIMETHYLSILYL ETHER; W/NBS   65663

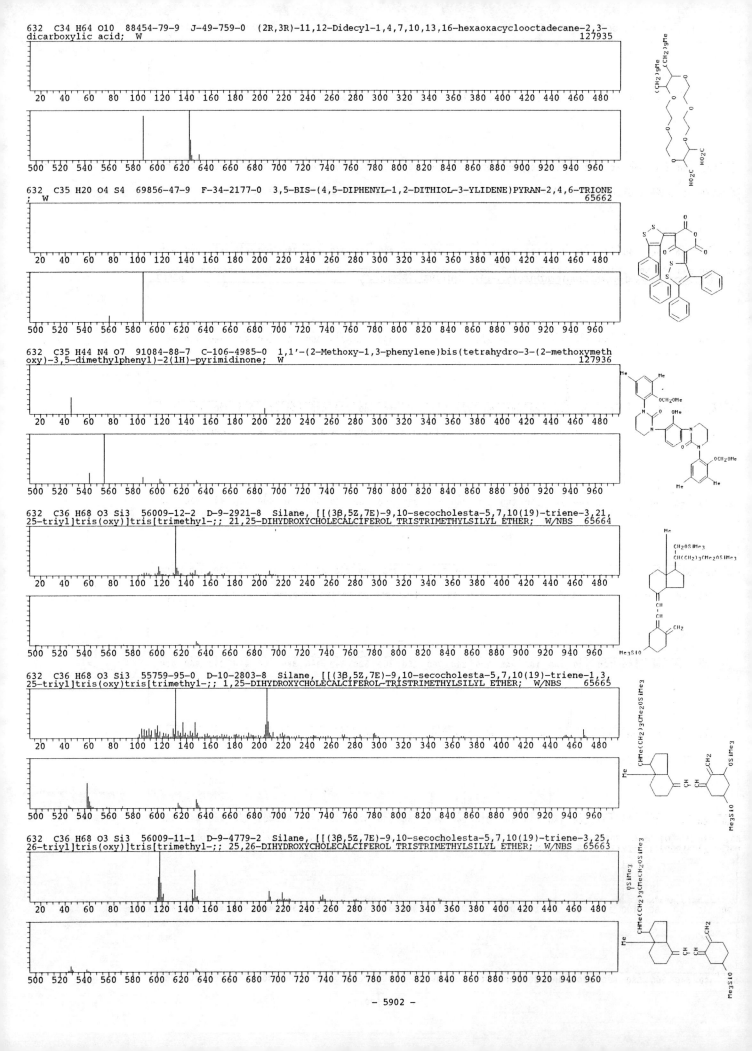

632   C36 H68 O3 Si3   KC-1976-734-0   3β,25ε,26-TRISTRIMETHYLSILYL-CHOLESTA-5,7-DIENE;   W   65666

632   C38 H36 N2 O7   50816-65-4   NS-0-0-0   Oxyacanthan-6,12'-diol, 6',7-epoxy-2,2'-dimethyl-, diacetate (ester), (1'α)-;   W/NBS   127937

632   C39 H41 Cl N4 O2   63972-59-8   F-33-633-0   α-BENZOYLOXY-γ-CHLORO-AETIOPHORPHYRIN-I;   W   65667

632   C39 H76 O2 Si2   65598-96-1   NS-0-0-0   Silane, [[(3β,5α,20S)-pregnane-3,20-diyl]bis(oxy)]bis[tripropyl-;   W/NBS   127938

632   C39 H76 O2 Si2   65598-97-2   NS-0-0-0   Silane, [[(3α,5β,20R)-pregnane-3,20-diyl]bis(oxy)]bis[tripropyl-;   W/NBS   127939

632   C39 H76 O2 Si2   77572-88-4   NS-0-0-0   Silane, [[(3β,5α,20R)-pregnane-3,20-diyl]bis(oxy)]bis[tripropyl-;   W/NBS   127940

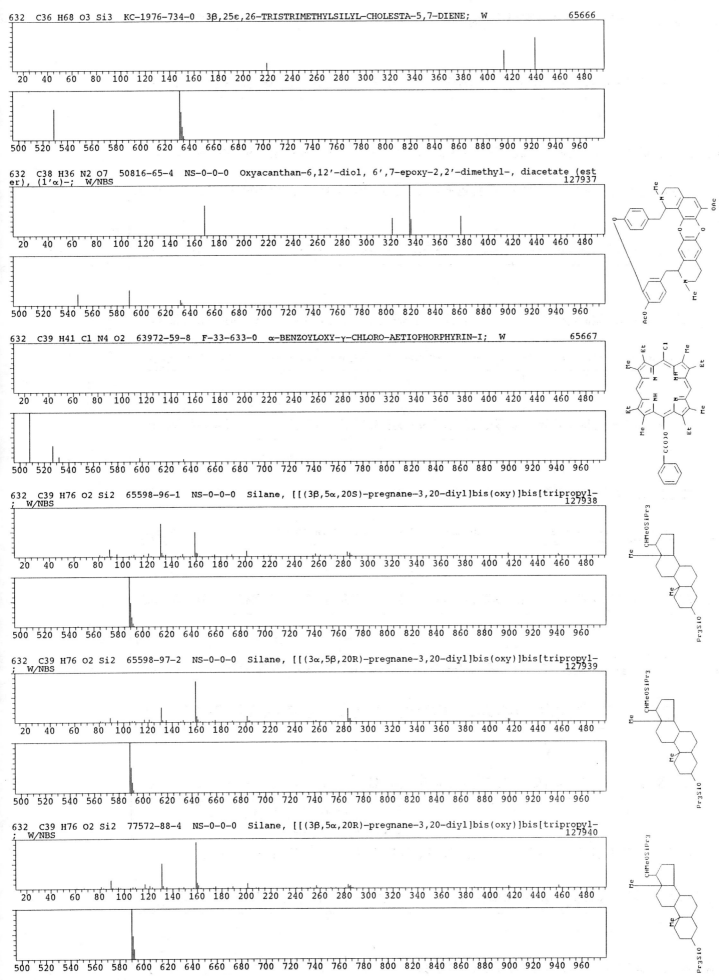

632   C40 H56 O6   75363-15-4   NS--27490-0-0   Urs-12-en-28-oic acid, 2-hydroxy-3-[[3-(4-hydroxyphenyl)-1-oxo-2-propenyl]oxy]-, methyl ester, (2α,3β)-;;   NBS
65668

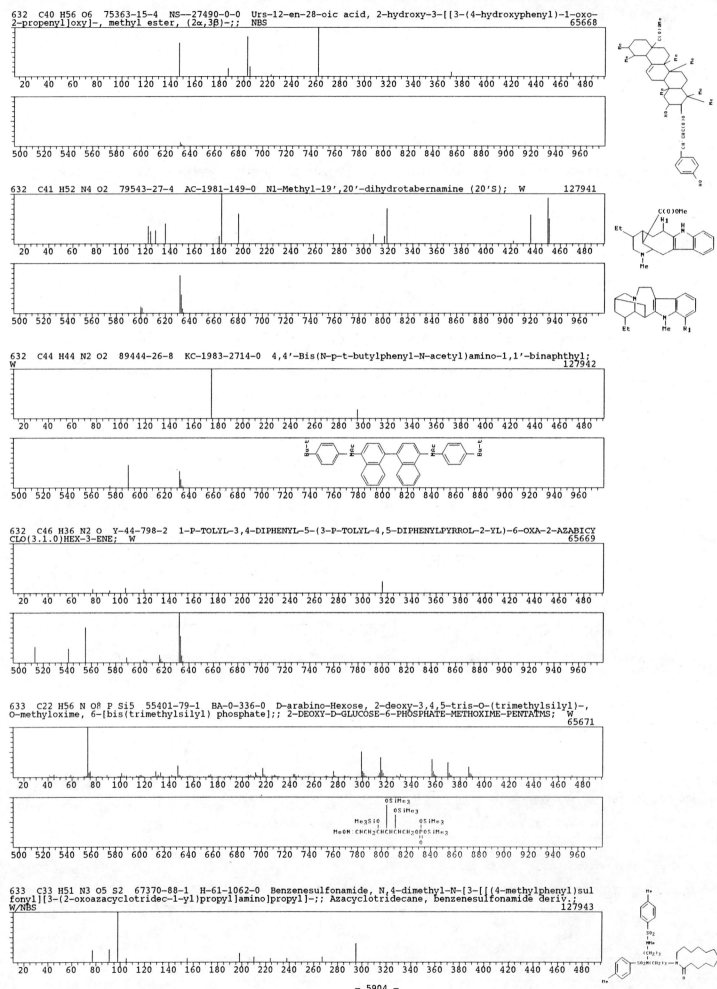

632   C41 H52 N4 O2   79543-27-4   AC-1981-149-0   N1-Methyl-19',20'-dihydrotabernamine (20'S);   W   127941

632   C44 H44 N2 O2   89444-26-8   KC-1983-2714-0   4,4'-Bis(N-p-t-butylphenyl-N-acetyl)amino-1,1'-binaphthyl;
W   127942

632   C46 H36 N2 O   Y-44-798-2   1-P-TOLYL-3,4-DIPHENYL-5-(3-P-TOLYL-4,5-DIPHENYLPYRROL-2-YL)-6-OXA-2-AZABICYCLO(3.1.0)HEX-3-ENE;   W   65669

633   C22 H56 N O8 P Si5   55401-79-1   BA-0-336-0   D-arabino-Hexose, 2-deoxy-3,4,5-tris-O-(trimethylsilyl)-, O-methyloxime, 6-[bis(trimethylsilyl) phosphate];; 2-DEOXY-D-GLUCOSE-6-PHOSPHATE-METHOXIME-PENTATMS;   W
65671

633   C33 H51 N3 O5 S2   67370-88-1   H-61-1062-0   Benzenesulfonamide, N,4-dimethyl-N-[3-[[(4-methylphenyl)sulfonyl][3-(2-oxoazacyclotridec-1-yl)propyl]amino]propyl]-;; Azacyclotridecane, benzenesulfonamide deriv.;
W/NBS   127943

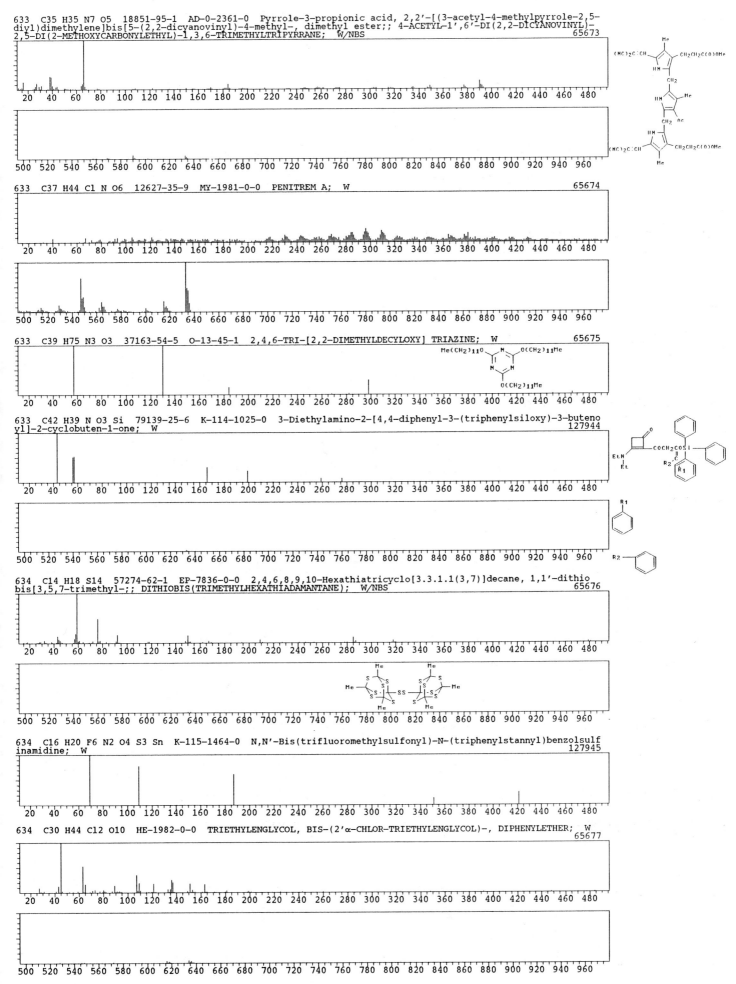

633 C35 H35 N7 O5 18851-95-1 AD-0-2361-0 Pyrrole-3-propionic acid, 2,2'-[(3-acetyl-4-methylpyrrole-2,5-diyl)dimethylene]bis[5-(2,2-dicyanovinyl)-4-methyl-, dimethyl ester;; 4-ACETYL-1',6'-DI(2,2-DICYANOVINYL)-2,5-DI(2-METHOXYCARBONYLETHYL)-1,3,6-TRIMETHYLTRIPYRRANE; W/NBS 65673

633 C37 H44 Cl N O6 12627-35-9 MY-1981-0-0 PENITREM A; W 65674

633 C39 H75 N3 O3 37163-54-5 O-13-45-1 2,4,6-TRI-[2,2-DIMETHYLDECYLOXY] TRIAZINE; W 65675

633 C42 H39 N O3 Si 79139-25-6 K-114-1025-0 3-Diethylamino-2-[4,4-diphenyl-3-(triphenylsiloxy)-3-butenoyl]-2-cyclobuten-1-one; W 127944

634 C14 H18 S14 57274-62-1 EP-7836-0-0 2,4,6,8,9,10-Hexathiatricyclo[3.3.1.1(3,7)]decane, 1,1'-dithiobis[3,5,7-trimethyl-;; DITHIOBIS(TRIMETHYLHEXATHIADAMANTANE); W/NBS 65676

634 C16 H20 F6 N2 O4 S3 Sn K-115-1464-0 N,N'-Bis(trifluoromethylsulfonyl)-N-(triphenylstannyl)benzolsulfinamidine; W 127945

634 C30 H44 Cl2 O10 HE-1982-0-0 TRIETHYLENGLYCOL, BIS-(2'α-CHLOR-TRIETHYLENGLYCOL)-, DIPHENYLETHER; W 65677

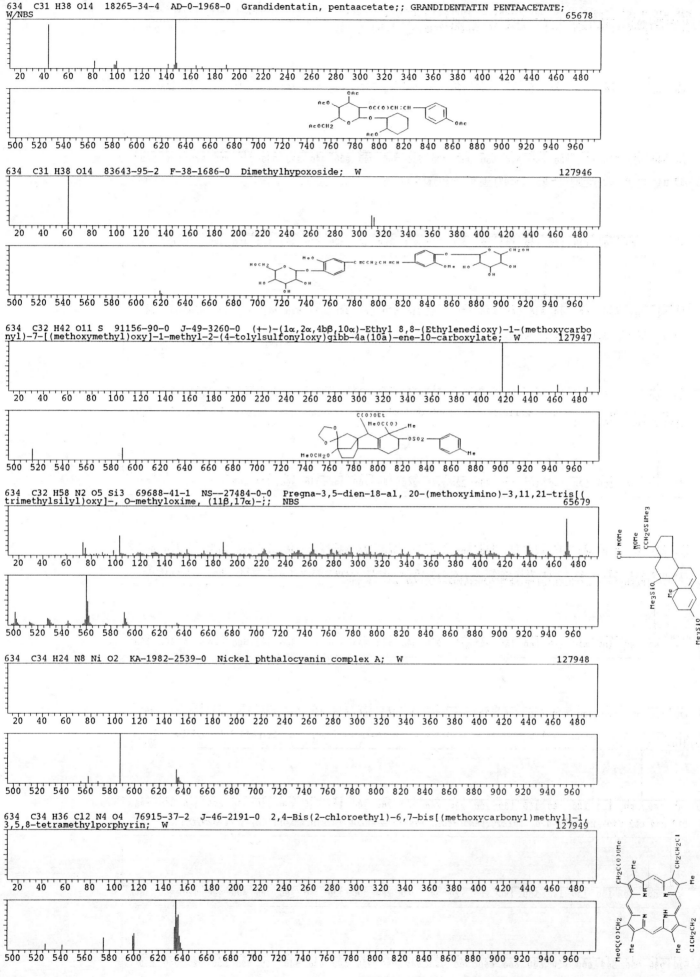

634   C31 H38 O14   18265-34-4   AD-0-1968-0   Grandidentatin, pentaacetate;; GRANDIDENTATIN PENTAACETATE;
W/NBS
65678

634   C31 H38 O14   83643-95-2   F-38-1686-0   Dimethylhypoxoside; W
127946

634   C32 H42 O11 S   91156-90-0   J-49-3260-0   (+-)-(1α,2α,4bβ,10α)-Ethyl 8,8-(Ethylenedioxy)-1-(methoxycarbo
nyl)-7-[(methoxymethyl)oxy]-1-methyl-2-(4-tolylsulfonyloxy)gibb-4a(10a)-ene-10-carboxylate; W   127947

634   C32 H58 N2 O5 Si3   69688-41-1   NS--27484-0-0   Pregna-3,5-dien-18-al, 20-(methoxyimino)-3,11,21-tris[(
trimethylsilyl)oxy]-, O-methyloxime, (11β,17α)-;;   NBS
65679

634   C34 H24 N8 Ni O2   KA-1982-2539-0   Nickel phthalocyanin complex A; W
127948

634   C34 H36 Cl2 N4 O4   76915-37-2   J-46-2191-0   2,4-Bis(2-chloroethyl)-6,7-bis[(methoxycarbonyl)methyl]-1,
3,5,8-tetramethylporphyrin; W
127949

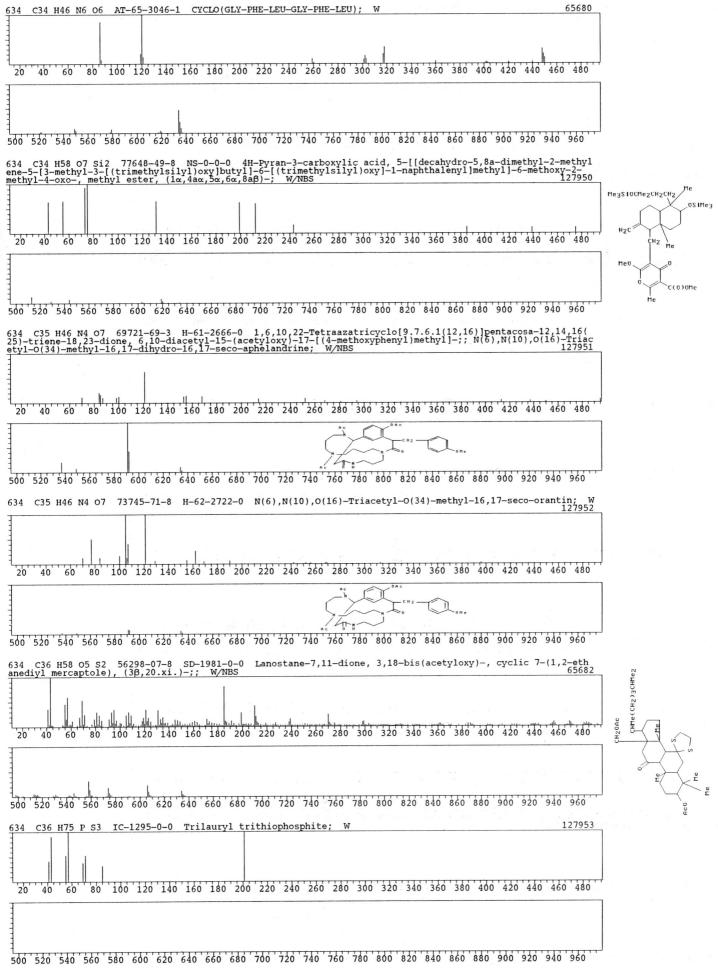

634   C34 H46 N6 O6   AT-65-3046-1   CYCLO(GLY-PHE-LEU-GLY-PHE-LEU);   W   65680

634   C34 H58 O7 Si2   77648-49-8   NS-0-0-0   4H-Pyran-3-carboxylic acid, 5-[[decahydro-5,8a-dimethyl-2-methyl
ene-5-[3-methyl-3-[(trimethylsilyl)oxy]butyl]-6-[(trimethylsilyl)oxy]-1-naphthalenyl]methyl]-6-methoxy-2-
methyl-4-oxo-, methyl ester, (1α,4aα,5α,6α,8aβ)-;   W/NBS   127950

634   C35 H46 N4 O7   69721-69-3   H-61-2666-0   1,6,10,22-Tetraazatricyclo[9.7.6.1(12,16)]pentacosa-12,14,16(
25)-triene-18,23-dione, 6,10-diacetyl-15-(acetyloxy)-17-[(4-methoxyphenyl)methyl]-;; N(6),N(10),O(16)-Triac
etyl-O(34)-methyl-16,17-dihydro-16,17-seco-aphelandrine;   W/NBS   127951

634   C35 H46 N4 O7   73745-71-8   H-62-2722-0   N(6),N(10),O(16)-Triacetyl-O(34)-methyl-16,17-seco-orantin;   W
                                                                                                127952

634   C36 H58 O5 S2   56298-07-8   SD-1981-0-0   Lanostane-7,11-dione, 3,18-bis(acetyloxy)-, cyclic 7-(1,2-eth
anediyl mercaptole), (3β,20.xi.)-;;   W/NBS   65682

634   C36 H75 P S3   IC-1295-0-0   Trilauryl trithiophosphite;   W   127953

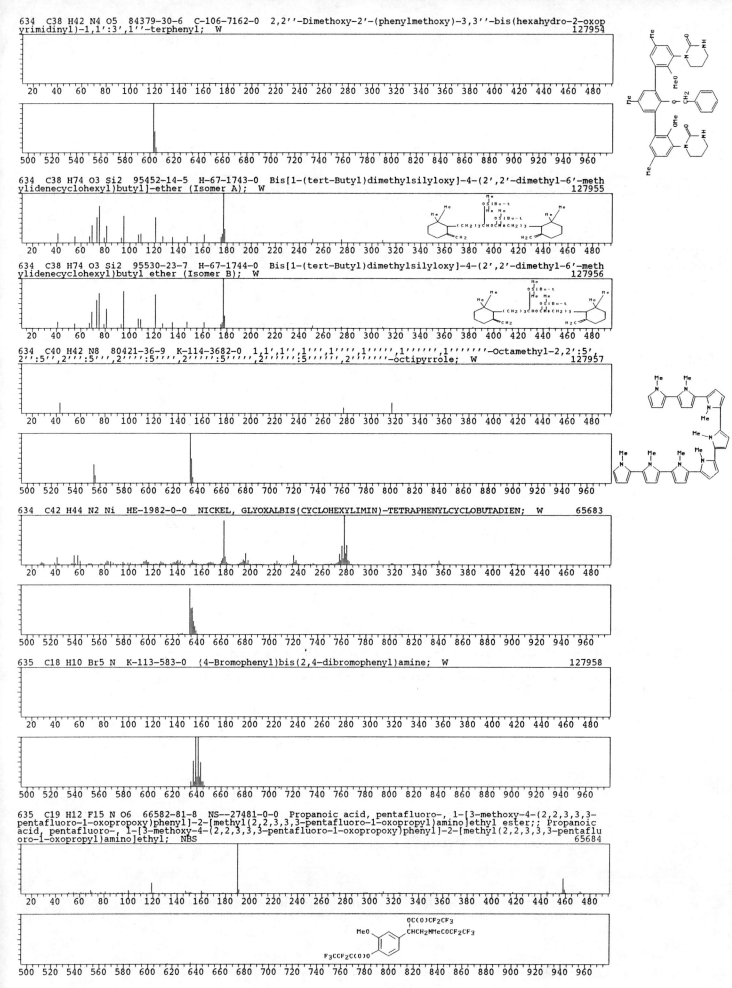

634　C38 H42 N4 O5　84379-30-6　C-106-7162-0　2,2''-Dimethoxy-2'-(phenylmethoxy)-3,3''-bis(hexahydro-2-oxop
yrimidinyl)-1,1':3',1''-terphenyl;　W　　　　127954

634　C38 H74 O3 Si2　95452-14-5　H-67-1743-0　Bis[1-(tert-Butyl)dimethylsilyloxy]-4-(2',2'-dimethyl-6'-meth
ylidenecyclohexyl)butyl]-ether (Isomer A);　W　　　　127955

634　C38 H74 O3 Si2　95530-23-7　H-67-1744-0　Bis[1-(tert-Butyl)dimethylsilyloxy]-4-(2',2'-dimethyl-6'-meth
ylidenecyclohexyl)butyl ether (Isomer B);　W　　　　127956

634　C40 H42 N8　80421-36-9　K-114-3682-0　1,1',1'',1''',1'''',1''''',1'''''',1'''''''-Octamethyl-2,2':5'
2'':5'',2''':5''',2'''':5'''',2''''':5''''',2'''''':5'''''',2'''''''-octipyrrole;　W　　　　127957

634　C42 H44 N2 Ni　HE-1982-0-0　NICKEL, GLYOXALBIS(CYCLOHEXYLIMIN)-TETRAPHENYLCYCLOBUTADIEN;　W　　65683

635　C18 H10 Br5 N　K-113-583-0　(4-Bromophenyl)bis(2,4-dibromophenyl)amine;　W　　　　127958

635　C19 H12 F15 N O6　66582-81-8　NS--27481-0-0　Propanoic acid, pentafluoro-, 1-[3-methoxy-4-(2,2,3,3,3-
pentafluoro-1-oxopropoxy)phenyl]-2-[methyl(2,2,3,3,3-pentafluoro-1-oxopropyl)amino]ethyl ester;; Propanoic
acid, pentafluoro-, 1-[3-methoxy-4-(2,2,3,3,3-pentafluoro-1-oxopropoxy)phenyl]-2-[methyl(2,2,3,3,3-pentaflu
oro-1-oxopropyl)amino]ethyl;　NBS　　　　65684

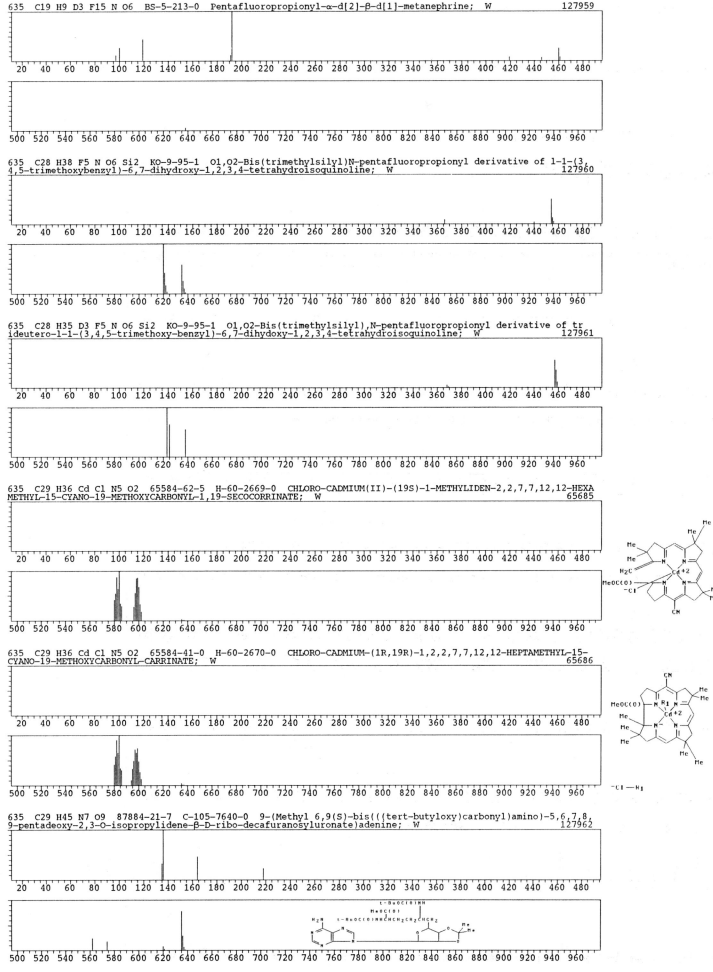

635 C19 H9 D3 F15 N O6 BS-5-213-0 Pentafluoropropionyl-α-d[2]-β-d[1]-metanephrine; W 127959

635 C28 H38 F5 N O6 Si2 KO-9-95-1 O1,O2-Bis(trimethylsilyl)N-pentafluoropropionyl derivative of 1-1-(3,4,5-trimethoxybenzyl)-6,7-dihydroxy-1,2,3,4-tetrahydroisoquinoline; W 127960

635 C28 H35 D3 F5 N O6 Si2 KO-9-95-1 O1,O2-Bis(trimethylsilyl),N-pentafluoropropionyl derivative of trideutero-1-1-(3,4,5-trimethoxy-benzyl)-6,7-dihydoxy-1,2,3,4-tetrahydroisoquinoline; W 127961

635 C29 H36 Cd Cl N5 O2 65584-62-5 H-60-2669-0 CHLORO-CADMIUM(II)-(19S)-1-METHYLIDEN-2,2,7,7,12,12-HEXAMETHYL-15-CYANO-19-METHOXYCARBONYL-1,19-SECOCORRINATE; W 65685

635 C29 H36 Cd Cl N5 O2 65584-41-0 H-60-2670-0 CHLORO-CADMIUM-(1R,19R)-1,2,2,7,7,12,12-HEPTAMETHYL-15-CYANO-19-METHOXYCARBONYL-CARRINATE; W 65686

635 C29 H45 N7 O9 87884-21-7 C-105-7640-0 9-(Methyl 6,9(S)-bis(((tert-butyloxy)carbonyl)amino)-5,6,7,8,9-pentadeoxy-2,3-O-isopropylidene-β-D-ribo-decafuranosyluronate)adenine; W 127962

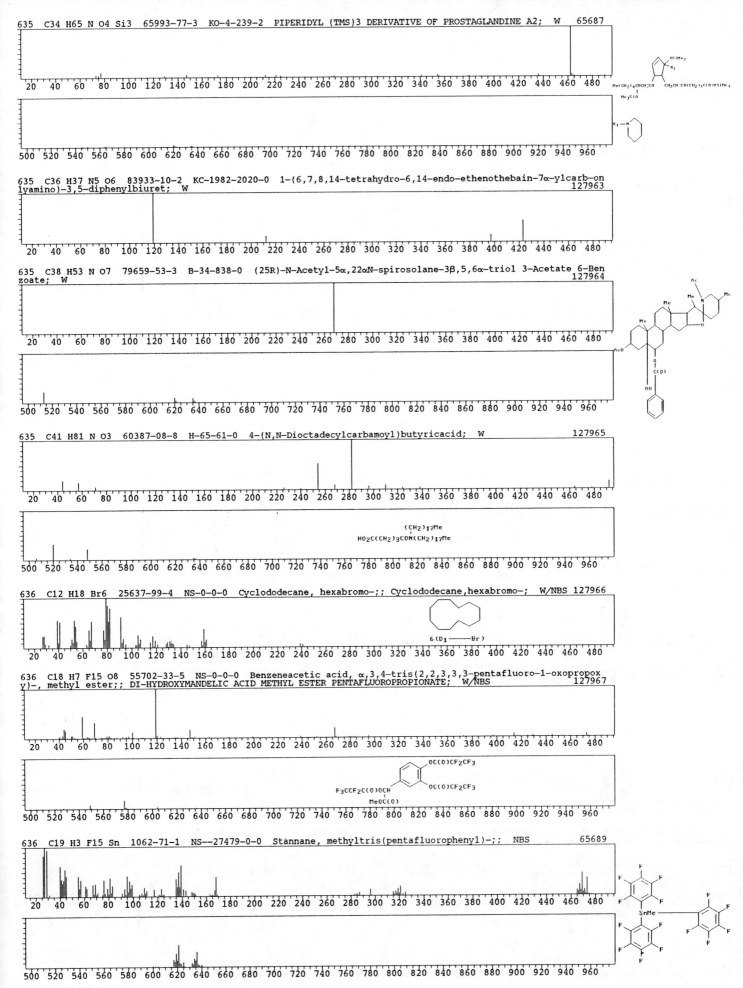

635 C34 H65 N O4 Si3 65993-77-3 KO-4-239-2 PIPERIDYL (TMS)3 DERIVATIVE OF PROSTAGLANDINE A2; W   65687

635 C36 H37 N5 O6 83933-10-2 KC-1982-2020-0 1-(6,7,8,14-tetrahydro-6,14-endo-ethenothebain-7α-ylcarb-on lyamino)-3,5-diphenylbiuret; W   127963

635 C38 H53 N O7 79659-53-3 B-34-838-0 (25R)-N-Acetyl-5α,22αN-spirosolane-3β,5,6α-triol 3-Acetate 6-Benzoate; W   127964

635 C41 H81 N O3 60387-08-8 H-65-61-0 4-(N,N-Dioctadecylcarbamoyl)butyricacid; W   127965

636 C12 H18 Br6 25637-99-4 NS-0-0-0 Cyclododecane, hexabromo-;; Cyclododecane,hexabromo-; W/NBS 127966

636 C18 H7 F15 O8 55702-33-5 NS-0-0-0 Benzeneacetic acid, α,3,4-tris(2,2,3,3,3-pentafluoro-1-oxopropox y)-, methyl ester;; DI-HYDROXYMANDELIC ACID METHYL ESTER PENTAFLUOROPROPIONATE; W/NBS 127967

636 C19 H3 F15 Sn 1062-71-1 NS--27479-0-0 Stannane, methyltris(pentafluorophenyl)-;; NBS   65689

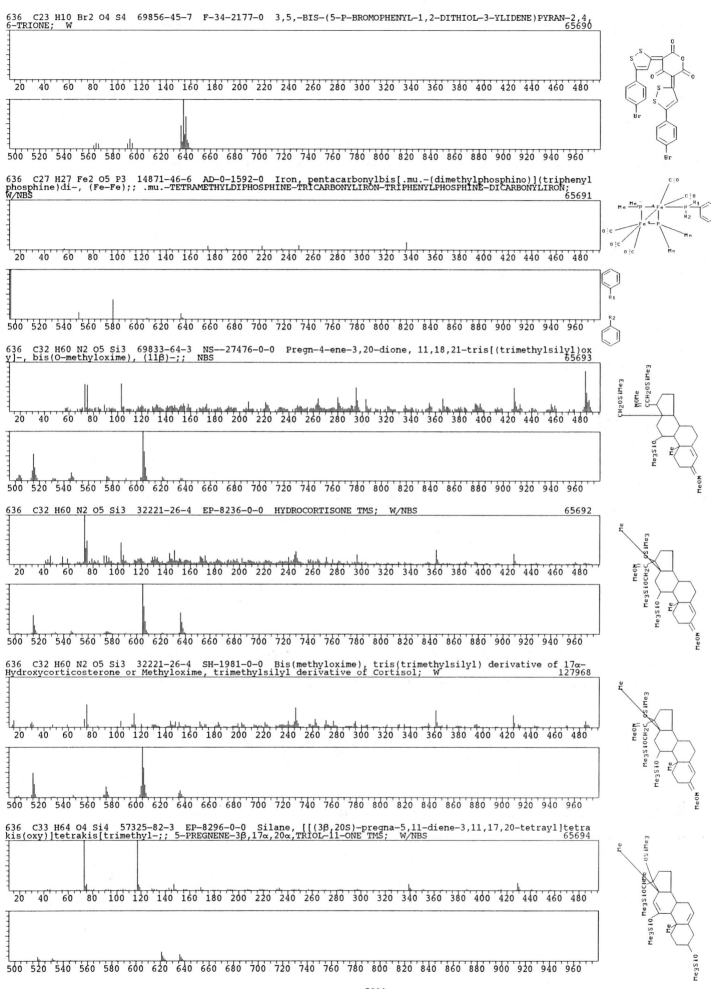

636   C23 H10 Br2 O4 S4   69856-45-7   F-34-2177-0   3,5,-BIS-(5-P-BROMOPHENYL-1,2-DITHIOL-3-YLIDENE)PYRAN-2,4,
6-TRIONE;   W   65690

636   C27 H27 Fe2 O5 P3   14871-46-6   AD-0-1592-0   Iron, pentacarbonylbis[.mu.-(dimethylphosphino)](triphenyl
phosphine)di-, (Fe-Fe);;   .mu.-TETRAMETHYLDIPHOSPHINE-TRICARBONYLIRON-TRIPHENYLPHOSPHINE-DICARBONYLIRON;
W/NBS   65691

636   C32 H60 N2 O5 Si3   69833-64-3   NS--27476-0-0   Pregn-4-ene-3,20-dione, 11,18,21-tris[(trimethylsilyl)ox
y]-, bis(O-methyloxime), (11β)-;;   NBS   65693

636   C32 H60 N2 O5 Si3   32221-26-4   EP-8236-0-0   HYDROCORTISONE TMS;   W/NBS   65692

636   C32 H60 N2 O5 Si3   32221-26-4   SH-1981-0-0   Bis(methyloxime), tris(trimethylsilyl) derivative of 17α-
Hydroxycorticosterone or Methyloxime, trimethylsilyl derivative of Cortisol;   W   127968

636   C33 H64 O4 Si4   57325-82-3   EP-8296-0-0   Silane, [[(3β,20S)-pregna-5,11-diene-3,11,17,20-tetrayl]tetra
kis(oxy)]tetrakis[trimethyl-;;   5-PREGNENE-3β,17α,20α,TRIOL-11-ONE TMS;   W/NBS   65694

636   C34 H20 Fe2 O6   33310-05-3   O5-0-493-0   Iron, hexacarbonyl[.mu.-[(1,2,3,4-η:1,4-η)-1,2,3,4-tetraphenyl
-1,3-butadiene-1,4-diyl]]di-, (Fe-Fe);; 1,1,1 TRICARBONYLFERRATETRAPHENYLCYCLOPENTADIENE-.PI.-TRICARBONYLIR
ON; W/NBS   65695

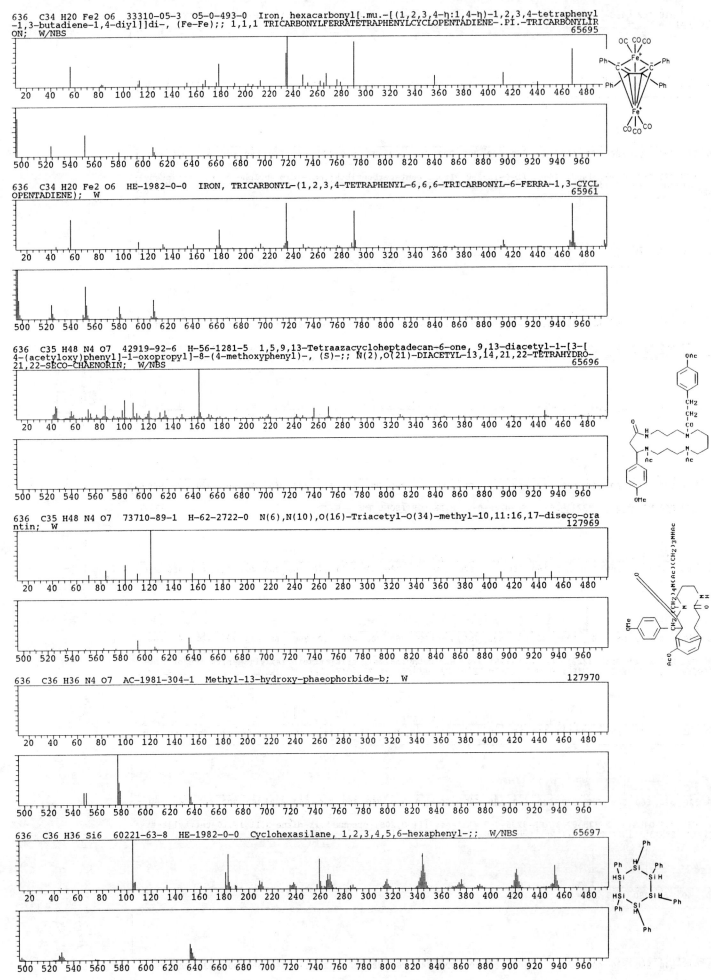

636   C34 H20 Fe2 O6   HE-1982-0-0   IRON, TRICARBONYL-(1,2,3,4-TETRAPHENYL-6,6,6-TRICARBONYL-6-FERRA-1,3-CYCL
OPENTADIENE); W   65961

636   C35 H48 N4 O7   42919-92-6   H-56-1281-5   1,5,9,13-Tetraazacycloheptadecan-6-one, 9,13-diacetyl-1-[3-[
4-(acetyloxy)phenyl]-1-oxopropyl]-8-(4-methoxyphenyl)-, (S)-;; N(2),O(21)-DIACETYL-13,14,21,22-TETRAHYDRO-
21,22-SECO-CHAENORIN; W/NBS   65696

636   C35 H48 N4 O7   73710-89-1   H-62-2722-0   N(6),N(10),O(16)-Triacetyl-O(34)-methyl-10,11:16,17-diseco-ora
ntin; W   127969

636   C36 H36 N4 O7   AC-1981-304-1   Methyl-13-hydroxy-phaeophorbide-b; W   127970

636   C36 H36 Si6   60221-63-8   HE-1982-0-0   Cyclohexasilane, 1,2,3,4,5,6-hexaphenyl-;; W/NBS   65697

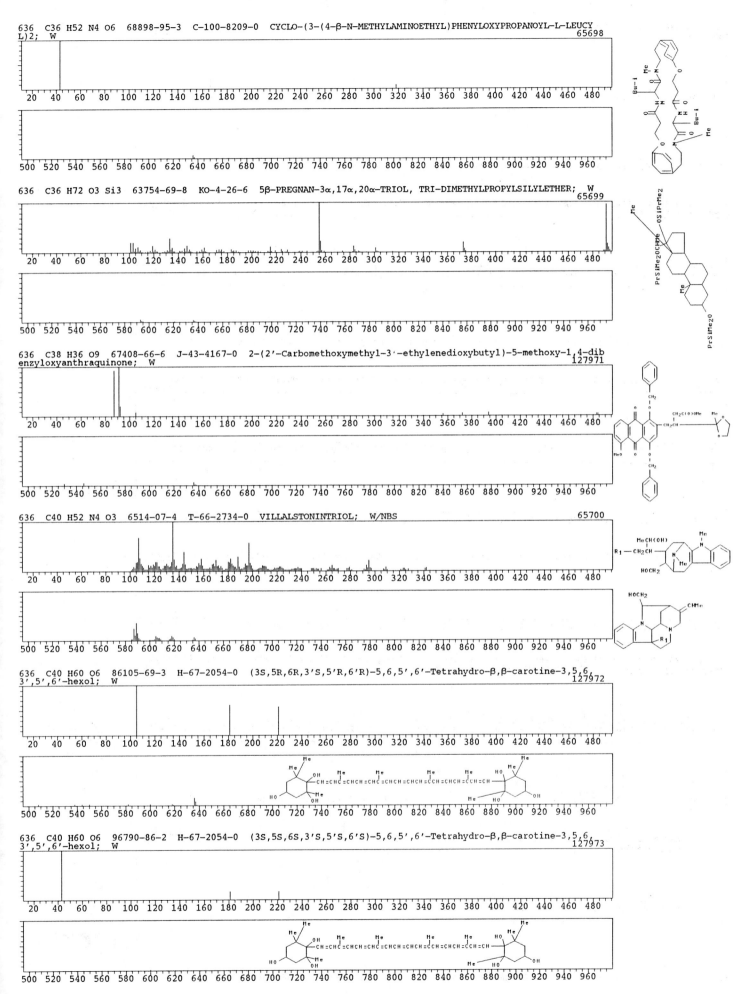

636   C36 H52 N4 O6   68898-95-3   C-100-8209-0   CYCLO-(3-(4-β-N-METHYLAMINOETHYL)PHENYLOXYPROPANOYL-L-LEUCY
L)2; W   65698

636   C36 H72 O3 Si3   63754-69-8   KO-4-26-6   5β-PREGNAN-3α,17α,20α-TRIOL, TRI-DIMETHYLPROPYLSILYLETHER; W
                                                                                           65699

636   C38 H36 O9   67408-66-6   J-43-4167-0   2-(2'-Carbomethoxymethyl-3'-ethylenedioxybutyl)-5-methoxy-1,4-dib
enzyloxyanthraquinone; W   127971

636   C40 H52 N4 O3   6514-07-4   T-66-2734-0   VILLALSTONINTRIOL; W/NBS   65700

636   C40 H60 O6   86105-69-3   H-67-2054-0   (3S,5R,6R,3'S,5'R,6'R)-5,6,5',6'-Tetrahydro-β,β-carotine-3,5,6,
3',5',6'-hexol; W   127972

636   C40 H60 O6   96790-86-2   H-67-2054-0   (3S,5S,6S,3'S,5'S,6'S)-5,6,5',6'-Tetrahydro-β,β-carotine-3,5,6,
3',5',6'-hexol; W   127973

636   C42 H38 O2 P2   62556-15-4   KC-1976-2561-0   1,4-BIS-(DIPHENYLPHOSPHINOYL)-2,3-DIPHENYL-4-METHYLCYCLOPEN
TANE; W   65701

636   C42 H68 O4   71142-40-0   NS--27470-0-0   Benzenetridecanoic acid, 4-methoxy-, 2-decyl-3-methoxy-5-pentyl
phenyl ester;;   NBS   65702

636   C42 H68 O4   K-112-2338-0   2-DECYL-3-METHOXY-5-PENTYL PHENYLESTER OF 13-(4-METHYOXYPHENYL)TRIDECANE CAR
BOXYLIC ACID; W   65703

637   C25 H42 F7 N O4 Si3   KO-6-32-1   Tri-o-trimethylsilyl, N-heptafluorobutyryl derivative of Terbutaline;
W   127974

637   C33 H59 N5 O7   10409-85-5   T-66-2627-0   L-Valine, N-[N-[N-[N-(3-hydroxy-1-oxododecyl)glycyl]-L-valy
l]-D-leucyl]-L-alanyl]-, .rho.-lactone, (R)-;; Isariin; W/NBS   65704

637   C34 H43 N3 O9   79072-70-1   J-46-5236-0   N(1),N(10)-Diisopropyl-N(1),N(5),N(10)-tris(2,3-dihydroxybenzo
yl)-1,5,10-triazadecane; W   127975

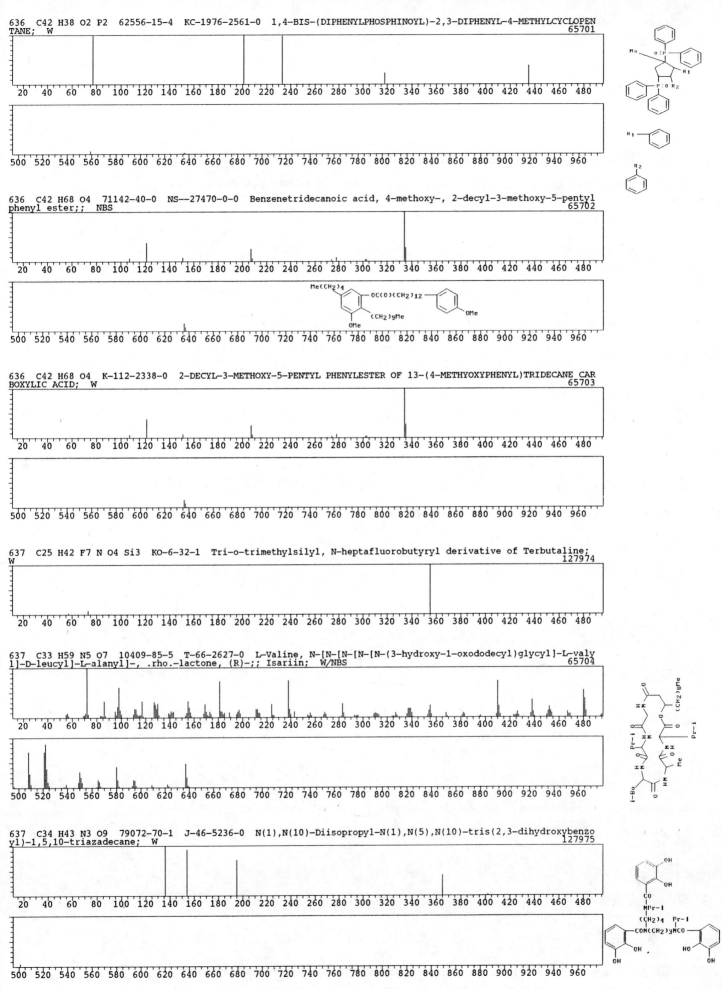

637   C34 H67 N O4 Si3   67204-54-0   KO-5-533-1   11-PIPERIDYL,9-ENOL PGA1 (TMS)3;   W   65705

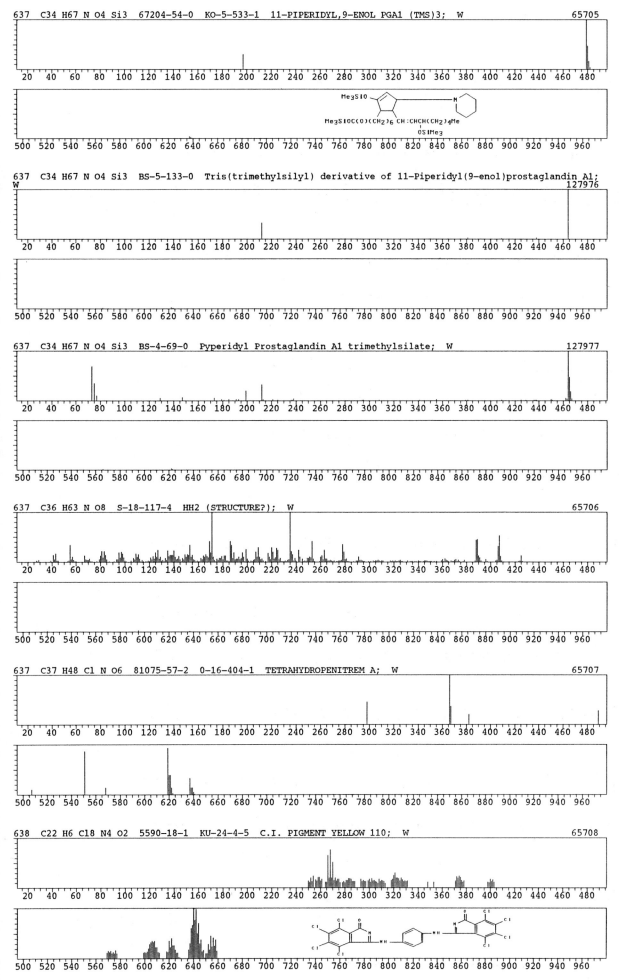

637   C34 H67 N O4 Si3   BS-5-133-0   Tris(trimethylsilyl) derivative of 11-Piperidyl(9-enol)prostaglandin A1;
W   127976

637   C34 H67 N O4 Si3   BS-4-69-0   Pyperidyl Prostaglandin A1 trimethylsilate;   W   127977

637   C36 H63 N O8   S-18-117-4   HH2 (STRUCTURE?);   W   65706

637   C37 H48 Cl N O6   81075-57-2   0-16-404-1   TETRAHYDROPENITREM A;   W   65707

638   C22 H6 Cl8 N4 O2   5590-18-1   KU-24-4-5   C.I. PIGMENT YELLOW 110;   W   65708

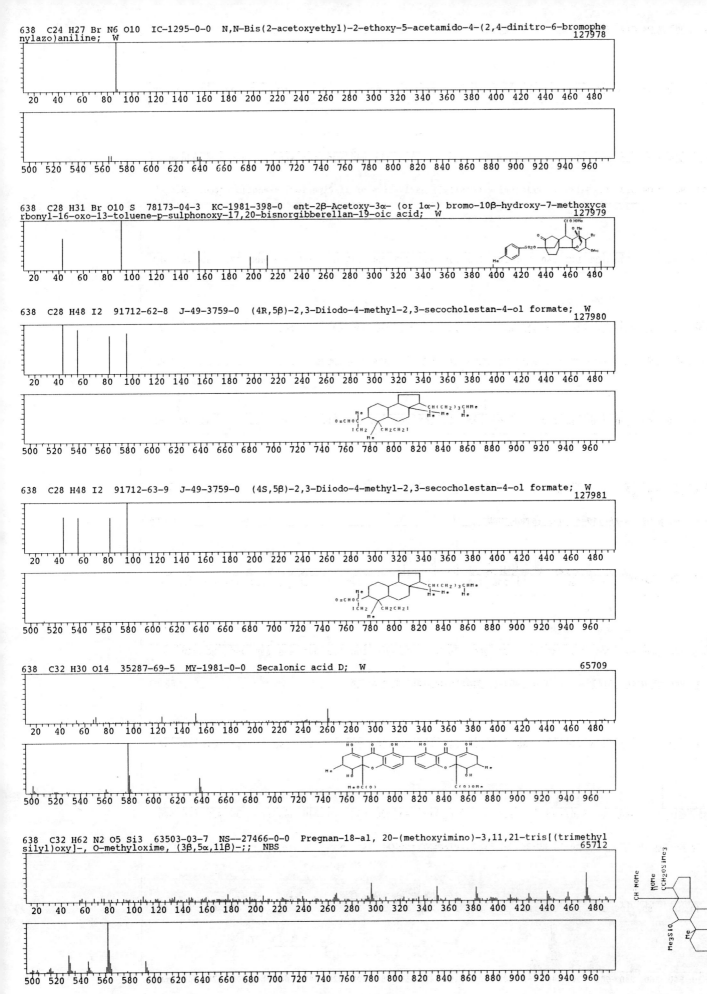

638   C24 H27 Br N6 O10   IC-1295-0-0   N,N-Bis(2-acetoxyethyl)-2-ethoxy-5-acetamido-4-(2,4-dinitro-6-bromophe
nylazo)aniline;   W
127978

638   C28 H31 Br O10 S   78173-04-3   KC-1981-398-0   ent-2β-Acetoxy-3α- (or 1α-) bromo-10β-hydroxy-7-methoxyca
rbonyl-16-oxo-13-toluene-p-sulphonoxy-17,20-bisnorgibberellan-19-oic acid;   W
127979

638   C28 H48 I2   91712-62-8   J-49-3759-0   (4R,5β)-2,3-Diiodo-4-methyl-2,3-secocholestan-4-ol formate;   W
127980

638   C28 H48 I2   91712-63-9   J-49-3759-0   (4S,5β)-2,3-Diiodo-4-methyl-2,3-secocholestan-4-ol formate;   W
127981

638   C32 H30 O14   35287-69-5   MY-1981-0-0   Secalonic acid D;   W
65709

638   C32 H62 N2 O5 Si3   63503-03-7   NS--27466-0-0   Pregnan-18-al, 20-(methoxyimino)-3,11,21-tris[(trimethyl
silyl)oxy]-, O-methyloxime, (3β,5α,11β)-;;   NBS
65712

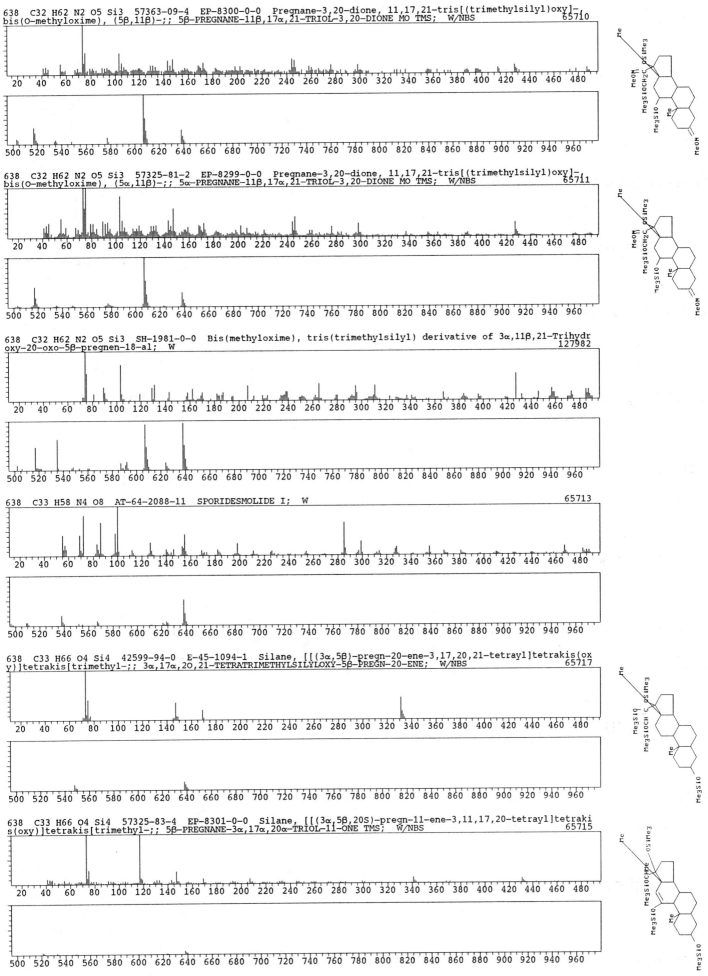

638   C32 H62 N2 O5 Si3   57363-09-4   EP-8300-0-0   Pregnane-3,20-dione, 11,17,21-tris[(trimethylsilyl)oxy]-
bis(O-methyloxime), (5β,11β)-;; 5β-PREGNANE-11β,17α,21-TRIOL-3,20-DIONE MO TMS;   W/NBS   65710

638   C32 H62 N2 O5 Si3   57325-81-2   EP-8299-0-0   Pregnane-3,20-dione, 11,17,21-tris[(trimethylsilyl)oxy]-
bis(O-methyloxime), (5α,11β)-;; 5α-PREGNANE-11β,17α,21-TRIOL-3,20-DIONE MO TMS;   W/NBS   65711

638   C32 H62 N2 O5 Si3   SH-1981-0-0   Bis(methyloxime), tris(trimethylsilyl) derivative of 3α,11β,21-Trihydr
oxy-20-oxo-5β-pregnen-18-al; W   127982

638   C33 H58 N4 O8   AT-64-2088-11   SPORIDESMOLIDE I; W   65713

638   C33 H66 O4 Si4   42599-94-0   E-45-1094-1   Silane, [[(3α,5β)-pregn-20-ene-3,17,20,21-tetrayl]tetrakis(ox
y)]tetrakis[trimethyl-;; 3α,17α,20,21-TETRATRIMETHYLSILYLOXY-5β-PREGN-20-ENE;   W/NBS   65717

638   C33 H66 O4 Si4   57325-83-4   EP-8301-0-0   Silane, [[(3α,5β,20S)-pregn-11-ene-3,11,17,20-tetrayl]tetraki
s(oxy)]tetrakis[trimethyl-;; 5β-PREGNANE-3α,17α,20α-TRIOL-11-ONE TMS;   W/NBS   65715

- 5917 -

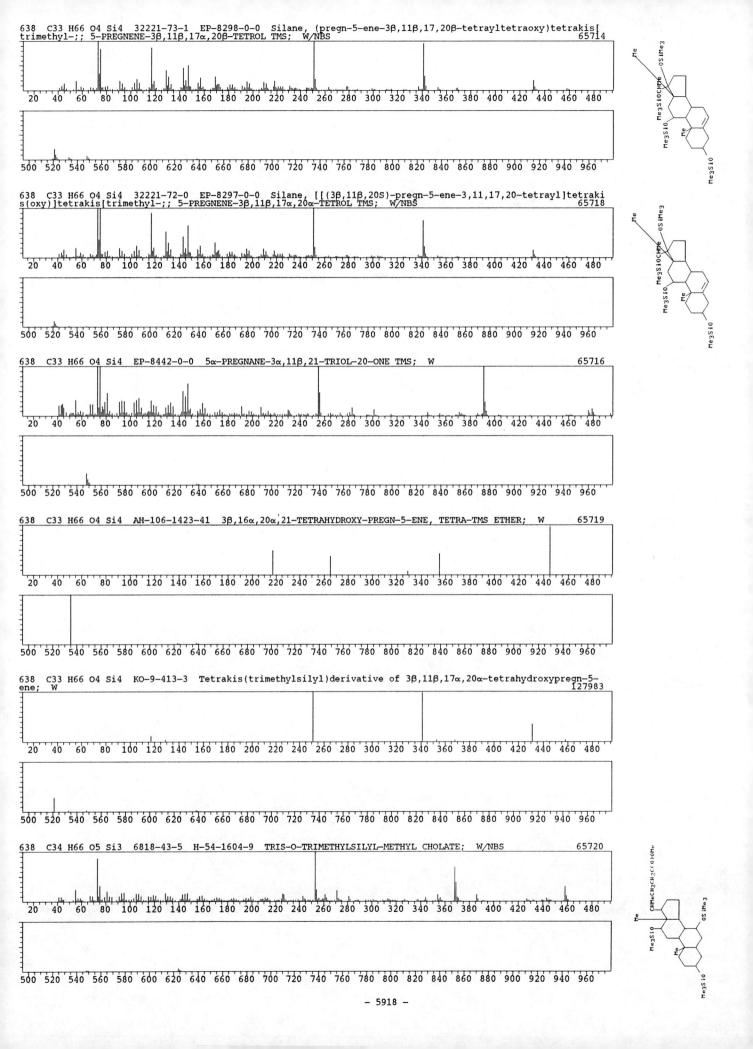

638 C33 H66 O4 Si4 32221-73-1 EP-8298-0-0 Silane, (pregn-5-ene-3β,11β,17,20β-tetrayltetraoxy)tetrakis[
trimethyl-;; 5-PREGNENE-3β,11β,17α,20β-TETROL TMS; W/NBS   65714

638 C33 H66 O4 Si4 32221-72-0 EP-8297-0-0 Silane, [[(3β,11β,20S)-pregn-5-ene-3,11,17,20-tetrayl]tetraki
s(oxy)]tetrakis[trimethyl-;; 5-PREGNENE-3β,11β,17α,20α-TETROL TMS; W/NBS   65718

638 C33 H66 O4 Si4 EP-8442-0-0 5α-PREGNANE-3α,11β,21-TRIOL-20-ONE TMS; W   65716

638 C33 H66 O4 Si4 AH-106-1423-41 3β,16α,20α,21-TETRAHYDROXY-PREGN-5-ENE, TETRA-TMS ETHER; W   65719

638 C33 H66 O4 Si4 KO-9-413-3 Tetrakis(trimethylsilyl)derivative of 3β,11β,17α,20α-tetrahydroxypregn-5-
ene; W   127983

638 C34 H66 O5 Si3 6818-43-5 H-54-1604-9 TRIS-O-TRIMETHYLSILYL-METHYL CHOLATE; W/NBS   65720

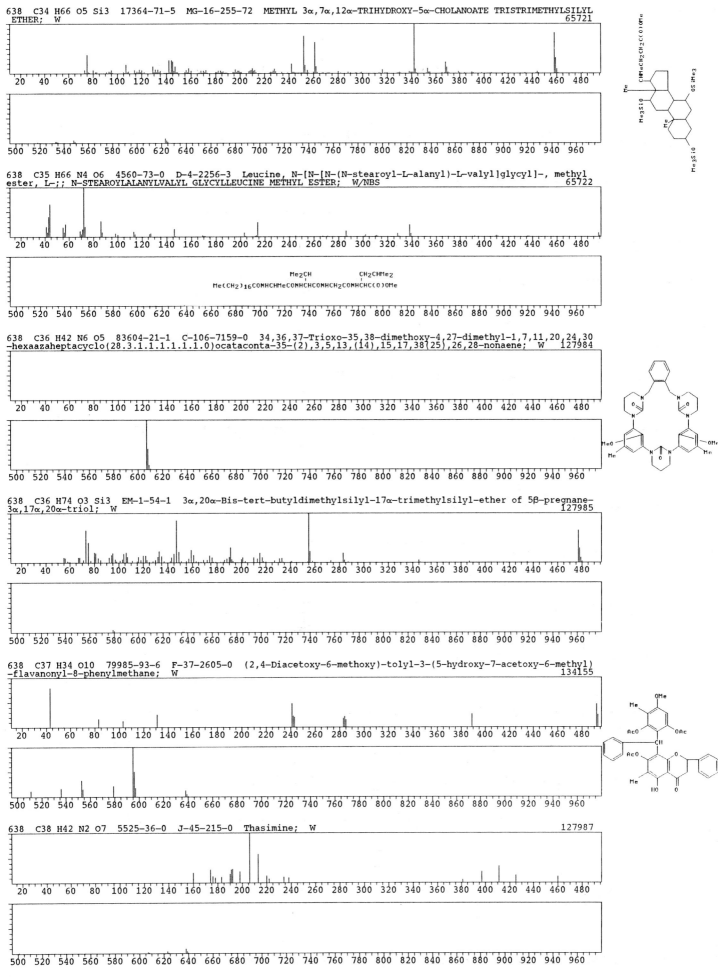

638   C34 H66 O5 Si3   17364-71-5   MG-16-255-72   METHYL 3α,7α,12α-TRIHYDROXY-5α-CHOLANOATE TRISTRIMETHYLSILYL
ETHER; W   65721

638   C35 H66 N4 O6   4560-73-0   D-4-2256-3   Leucine, N-[N-[N-(N-stearoyl-L-alanyl)-L-valyl]glycyl]-, methyl
ester, L-;; N-STEAROYLALANYLVALYL GLYCYLLEUCINE METHYL ESTER; W/NBS   65722

638   C36 H42 N6 O5   83604-21-1   C-106-7159-0   34,36,37-Trioxo-35,38-dimethoxy-4,27-dimethyl-1,7,11,20,24,30
-hexaazaheptacyclo(28.3.1.1.1.1.1.1.0)ocataconta-35-(2),3,5,13,(14),15,17,38(25),26,28-nonaene; W   127984

638   C36 H74 O3 Si3   EM-1-54-1   3α,20α-Bis-tert-butyldimethylsilyl-17α-trimethylsilyl-ether of 5β-pregnane-
3α,17α,20α-triol; W   127985

638   C37 H34 O10   79985-93-6   F-37-2605-0   (2,4-Diacetoxy-6-methoxy)-tolyl-3-(5-hydroxy-7-acetoxy-6-methyl)
-flavanonyl-8-phenylmethane; W   134155

638   C38 H42 N2 O7   5525-36-0   J-45-215-0   Thasimine; W   127987

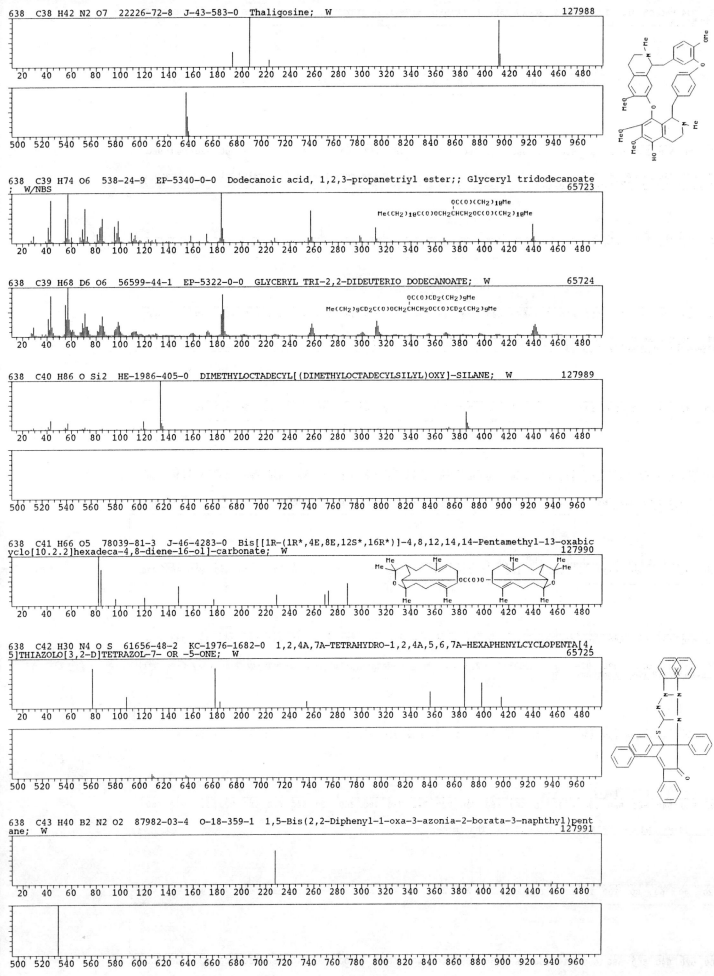

638   C38 H42 N2 O7   22226-72-8   J-43-583-0   Thaligosine;   W   127988

638   C39 H74 O6   538-24-9   EP-5340-0-0   Dodecanoic acid, 1,2,3-propanetriyl ester;; Glyceryl tridodecanoate
;   W/NBS   65723

OC(O)(CH2)10Me
Me(CH2)10C(O)OCH2CHCH2OC(O)(CH2)10Me

638   C39 H68 D6 O6   56599-44-1   EP-5322-0-0   GLYCERYL TRI-2,2-DIDEUTERIO DODECANOATE;   W   65724

OC(O)CD2(CH2)9Me
Me(CH2)9CD2C(O)OCH2CHCH2OC(O)CD2(CH2)9Me

638   C40 H86 O Si2   HE-1986-405-0   DIMETHYLOCTADECYL[(DIMETHYLOCTADECYLSILYL)OXY]-SILANE;   W   127989

638   C41 H66 O5   78039-81-3   J-46-4283-0   Bis[[1R-(1R*,4E,8E,12S*,16R*)]-4,8,12,14,14-Pentamethyl-13-oxabic
yclo[10.2.2]hexadeca-4,8-diene-16-ol]-carbonate;   W   127990

638   C42 H30 N4 O S   61656-48-2   KC-1976-1682-0   1,2,4A,7A-TETRAHYDRO-1,2,4A,5,6,7A-HEXAPHENYLCYCLOPENTA[4,
5]THIAZOLO[3,2-D]TETRAZOL-7- OR -5-ONE;   W   65725

638   C43 H40 B2 N2 O2   87982-03-4   O-18-359-1   1,5-Bis(2,2-Diphenyl-1-oxa-3-azonia-2-borata-3-naphthyl)pent
ane;   W   127991

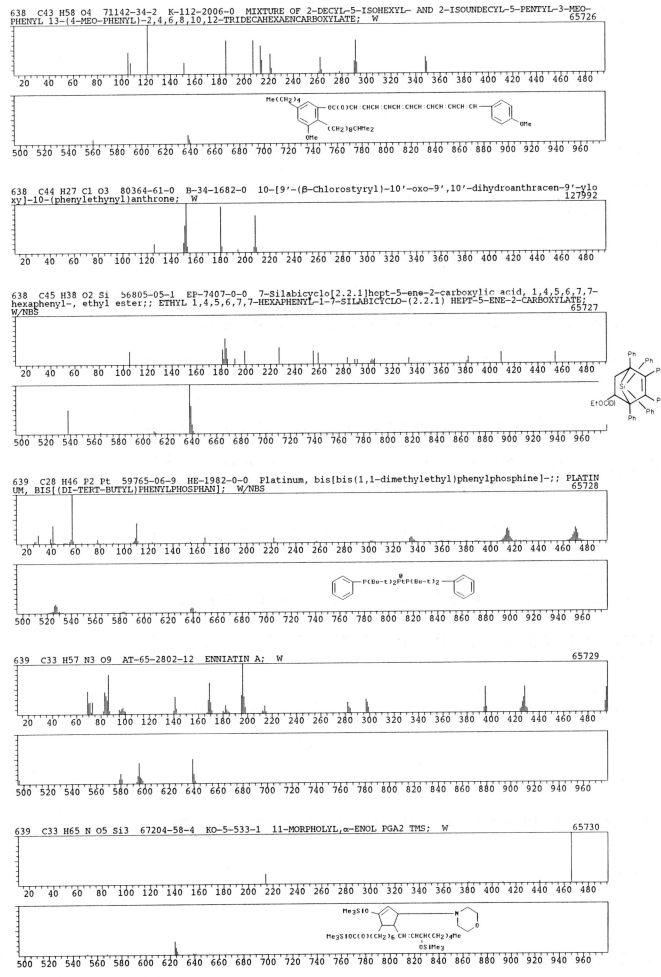

638 C43 H58 O4 71142-34-2 K-112-2006-0 MIXTURE OF 2-DECYL-5-ISOHEXYL- AND 2-ISOUNDECYL-5-PENTYL-3-MEO-
PHENYL 13-(4-MEO-PHENYL)-2,4,6,8,10,12-TRIDECAHEXAENCARBOXYLATE; W   65726

638 C44 H27 Cl O3 80364-61-0 B-34-1682-0 10-[9'-(β-Chlorostyryl)-10'-oxo-9',10'-dihydroanthracen-9'-ylo
xy]-10-(phenylethynyl)anthrone; W   127992

638 C45 H38 O2 Si 56805-05-1 EP-7407-0-0 7-Silabicyclo[2.2.1]hept-5-ene-2-carboxylic acid, 1,4,5,6,7,7-
hexaphenyl-, ethyl ester;; ETHYL 1,4,5,6,7,7-HEXAPHENYL-1-7-SILABICYCLO-(2.2.1) HEPT-5-ENE-2-CARBOXYLATE;
W/NBS   65727

639 C28 H46 P2 Pt 59765-06-9 HE-1982-0-0 Platinum, bis[bis(1,1-dimethylethyl)phenylphosphine]-;; PLATIN
UM, BIS[(DI-TERT-BUTYL)PHENYLPHOSPHAN]; W/NBS   65728

639 C33 H57 N3 O9 AT-65-2802-12 ENNIATIN A; W   65729

639 C33 H65 N O5 Si3 67204-58-4 KO-5-533-1 11-MORPHOLYL,α-ENOL PGA2 TMS; W   65730

639   C35 H53 N5 O6   55822-84-9   T-67-5141-0   DL-Leucine, N-[N-[1-[N-(1-oxodecyl)glycyl]-DL-prolyl]-DL-tryptophyl]-, methyl ester;; DECANOYLGLYCYLPROLYLTRYPTOPHYLLEUCINE METHYL ESTER;   W/NBS   65731

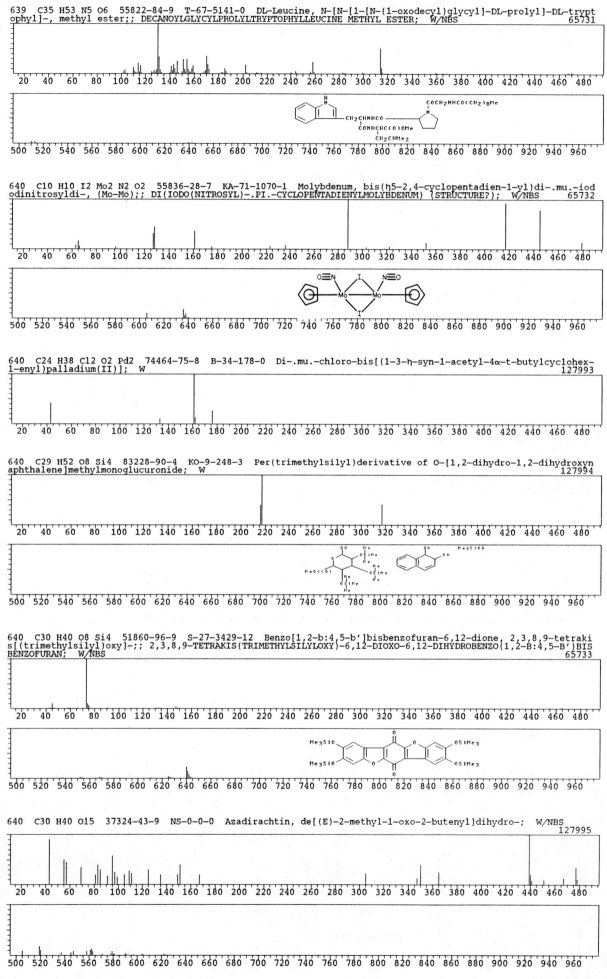

640   C10 H10 I2 Mo2 N2 O2   55836-28-7   KA-71-1070-1   Molybdenum, bis(η5-2,4-cyclopentadien-1-yl)di-.mu.-iododinitrosyldi-, (Mo-Mo);; DI(IODO(NITROSYL)-.PI.-CYCLOPENTADIENYLMOLYBDENUM) (STRUCTURE?);   W/NBS   65732

640   C24 H38 Cl2 O2 Pd2   74464-75-8   B-34-178-0   Di-.mu.-chloro-bis[(1-3-η-syn-1-acetyl-4α-t-butylcyclohex-1-enyl)palladium(II)];   W   127993

640   C29 H52 O8 Si4   83228-90-4   KO-9-248-3   Per(trimethylsilyl)derivative of O-[1,2-dihydro-1,2-dihydroxynaphthalene]methylmonoglucuronide;   W   127994

640   C30 H40 O8 Si4   51860-96-9   S-27-3429-12   Benzo[1,2-b:4,5-b']bisbenzofuran-6,12-dione, 2,3,8,9-tetrakis[(trimethylsilyl)oxy]-;; 2,3,8,9-TETRAKIS(TRIMETHYLSILYLOXY)-6,12-DIOXO-6,12-DIHYDROBENZO(1,2-B:4,5-B')BISBENZOFURAN;   W/NBS   65733

640   C30 H40 O15   37324-43-9   NS-0-0-0   Azadirachtin, de[(E)-2-methyl-1-oxo-2-butenyl]dihydro-;   W/NBS   127995

- 5922 -

640 C32 H40 N4 O10 76036-85-6 J-46-837-0 5,8,11,14,17,26,29,32,35,38-Decaoxa-43,44,45,46-tetraazapentac
yclo[37.3.1.1(4,42).1(18,22).1(21,25)]hexatetraconta-1(43),2,4(44),18,20,22(46),23,25(45),39,41-decaene; W
127996

640 C32 H40 N4 O10 76036-93-6 J-46-838-0 2,5,8,11,14,22,25,28,31,34-Decaoxa-41,43,44,46-tetraazapentacy
clo[33.5.3.3(15,21).0(18,45).0(38,42)]hexatetraconta-1(41),15,17,19,21(44),35,37,39,42,45-decaene; W
127997

640 C32 H64 O5 Si4 BA-0-236-0 O-TETRAKIS(TRIMETHYLSILYL) 3,5-DIHYDROXY-2-(3-HYDROXY-1-OCTENYL)-CYCLOPENT
ANEHEPTANOATE; W
65734

640 C32 H64 O5 Si4 65963-04-4 KO-4-238-1 PROSTAGLANDIN E2 9-ENOL TETRAKIS(TMS) DERIVATIVE; W   65735

640 C33 H60 N4 O8 56272-49-2 T-66-2628-0 L-Leucine, N-[N-[N-[N-(1-oxodecyl)-L-α-aspartyl]-L-valyl]-L-al
anyl]-, 1-butyl 4-methyl ester;; DECANOYL(O-METHYL)ASPARTYLVALYLALANYLLEUCINE BUTYL ESTER; W/NBS   65736

640 C33 H68 O4 Si4 57325-84-5 EP-8303-0-0 Silane, [[(3β,5α,20R)-pregnane-3,17,20,21-tetrayl]tetrakis(ox
y)]tetrakis[trimethyl-;; 5α-PREGNANE-3β,17α,20β,21-TETROL TMS; W/NBS   65739

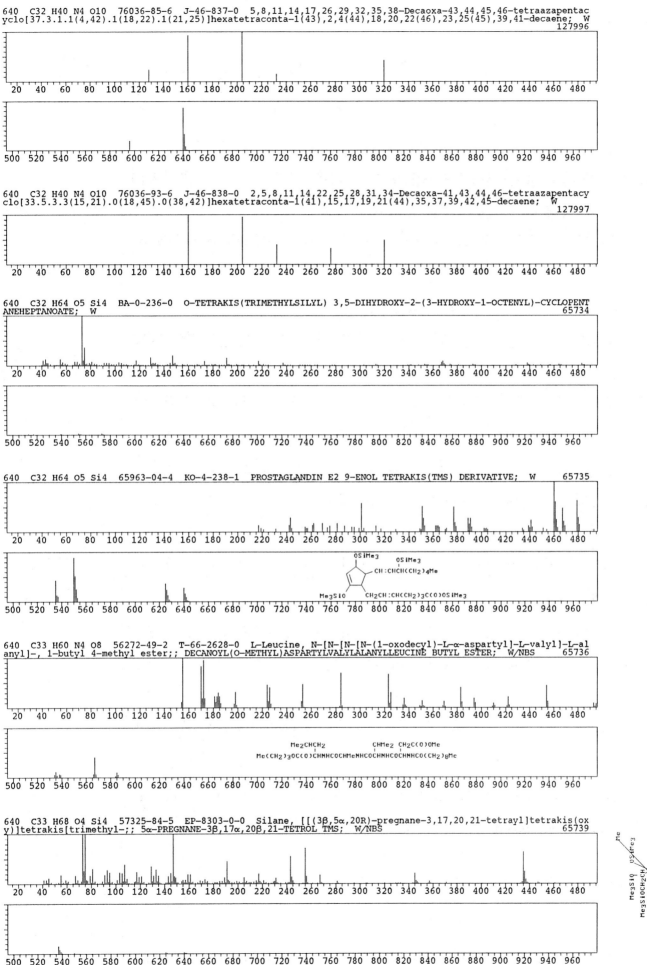

- 5923 -

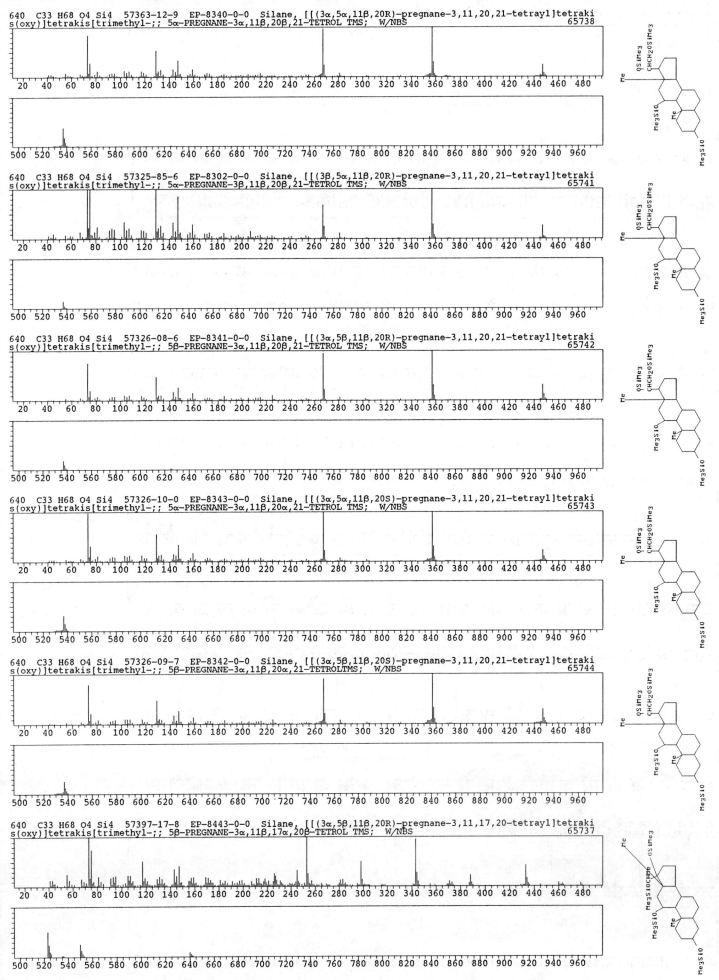

640   C33 H68 O4 Si4   57363-12-9   EP-8340-0-0   Silane, [[(3α,5α,11β,20R)-pregnane-3,11,20,21-tetrayl]tetraki
s(oxy)]tetrakis[trimethyl-;;   5α-PREGNANE-3α,11β,20β,21-TETROL TMS;   W/NBS   65738

640   C33 H68 O4 Si4   57325-85-6   EP-8302-0-0   Silane, [[(3β,5α,11β,20R)-pregnane-3,11,20,21-tetrayl]tetraki
s(oxy)]tetrakis[trimethyl-;;   5α-PREGNANE-3β,11β,20β,21-TETROL TMS;   W/NBS   65741

640   C33 H68 O4 Si4   57326-08-6   EP-8341-0-0   Silane, [[(3α,5β,11β,20R)-pregnane-3,11,20,21-tetrayl]tetraki
s(oxy)]tetrakis[trimethyl-;;   5β-PREGNANE-3α,11β,20β,21-TETROL TMS;   W/NBS   65742

640   C33 H68 O4 Si4   57326-10-0   EP-8343-0-0   Silane, [[(3α,5α,11β,20S)-pregnane-3,11,20,21-tetrayl]tetraki
s(oxy)]tetrakis[trimethyl-;;   5α-PREGNANE-3α,11β,20α,21-TETROL TMS;   W/NBS   65743

640   C33 H68 O4 Si4   57326-09-7   EP-8342-0-0   Silane, [[(3α,5β,11β,20S)-pregnane-3,11,20,21-tetrayl]tetraki
s(oxy)]tetrakis[trimethyl-;;   5β-PREGNANE-3α,11β,20α,21-TETROLTMS;   W/NBS   65744

640   C33 H68 O4 Si4   57397-17-8   EP-8443-0-0   Silane, [[(3α,5β,11β,20R)-pregnane-3,11,17,20-tetrayl]tetraki
s(oxy)]tetrakis[trimethyl-;;   5β-PREGNANE-3α,11β,17α,20β-TETROL TMS;   W/NBS   65737

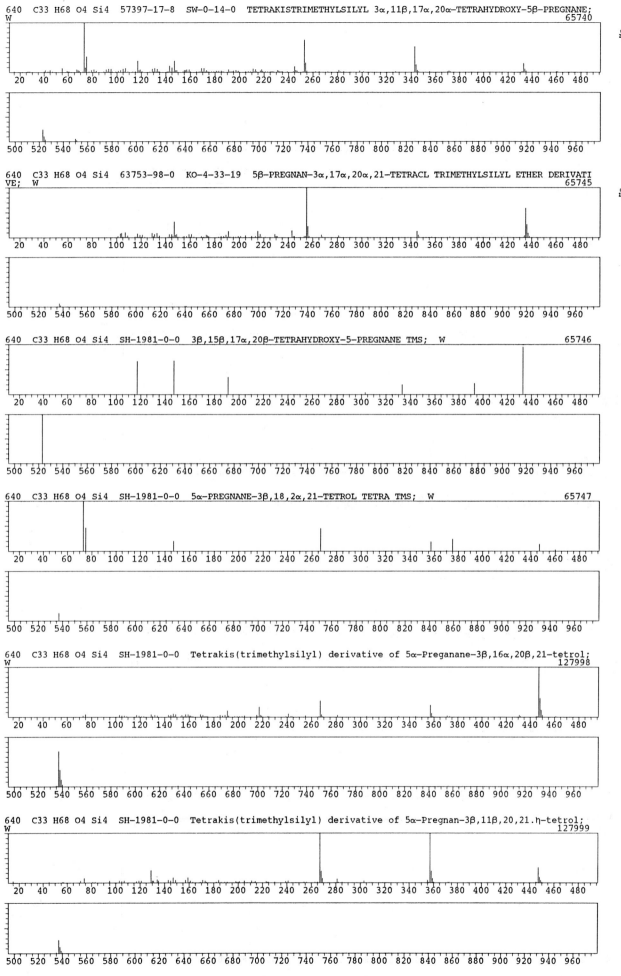

640   C33 H68 O4 Si4   57397-17-8   SW-0-14-0   TETRAKISTRIMETHYLSILYL 3α,11β,17α,20α-TETRAHYDROXY-5β-PREGNANE;
W   65740

640   C33 H68 O4 Si4   63753-98-0   KO-4-33-19   5β-PREGNAN-3α,17α,20α,21-TETRACL TRIMETHYLSILYL ETHER DERIVATI
VE; W   65745

640   C33 H68 O4 Si4   SH-1981-0-0   3β,15β,17α,20β-TETRAHYDROXY-5-PREGNANE TMS;   W   65746

640   C33 H68 O4 Si4   SH-1981-0-0   5α-PREGNANE-3β,18,2α,21-TETROL TETRA TMS;   W   65747

640   C33 H68 O4 Si4   SH-1981-0-0   Tetrakis(trimethylsilyl) derivative of 5α-Preganane-3β,16α,20β,21-tetrol;
W   127998

640   C33 H68 O4 Si4   SH-1981-0-0   Tetrakis(trimethylsilyl) derivative of 5α-Pregnan-3β,11β,20,21.η-tetrol;
W   127999

640   C33 H68 O4 Si4   SH-1981-0-0   Tetrakis(trimethylsilyl) ether of 5β-Pjregnen-3α,17α,20α,21-tetrol;   W
128000

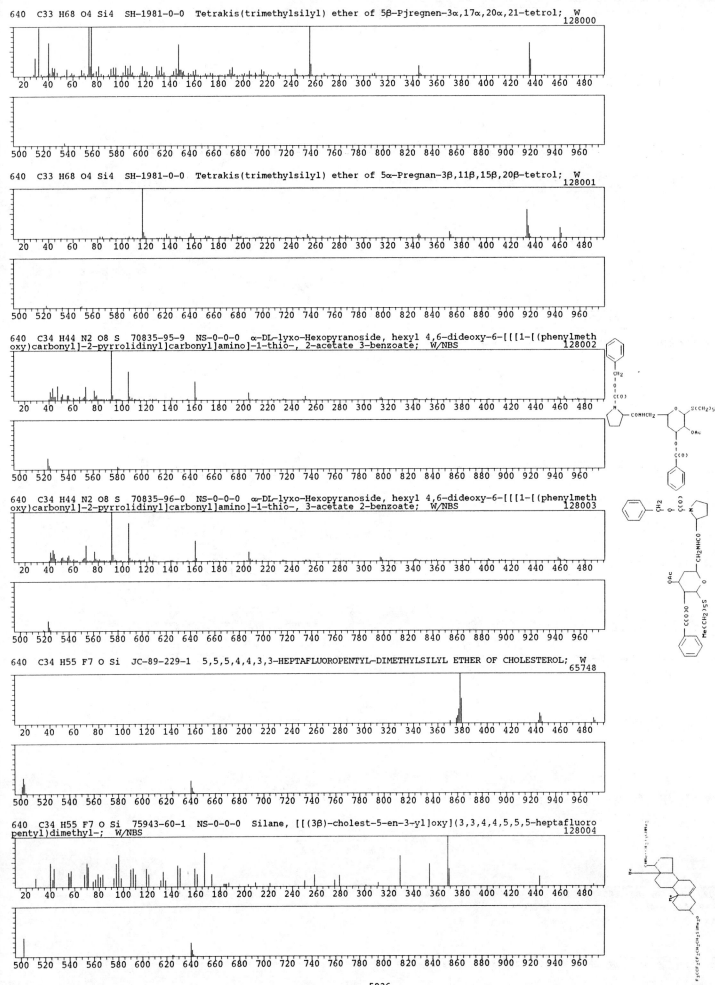

640   C33 H68 O4 Si4   SH-1981-0-0   Tetrakis(trimethylsilyl) ether of 5α-Pregnan-3β,11β,15β,20β-tetrol;   W
128001

640   C34 H44 N2 O8 S   70835-95-9   NS-0-0-0   α-DL-lyxo-Hexopyranoside, hexyl 4,6-dideoxy-6-[[[1-[(phenylmeth
oxy)carbonyl]-2-pyrrolidinyl]carbonyl]amino]-1-thio-, 2-acetate 3-benzoate;   W/NBS   128002

640   C34 H44 N2 O8 S   70835-96-0   NS-0-0-0   α-DL-lyxo-Hexopyranoside, hexyl 4,6-dideoxy-6-[[[1-[(phenylmeth
oxy)carbonyl]-2-pyrrolidinyl]carbonyl]amino]-1-thio-, 3-acetate 2-benzoate;   W/NBS   128003

640   C34 H55 F7 O Si   JC-89-229-1   5,5,5,4,4,3,3-HEPTAFLUOROPENTYL-DIMETHYLSILYL ETHER OF CHOLESTEROL;   W
65748

640   C34 H55 F7 O Si   75943-60-1   NS-0-0-0   Silane, [[(3β)-cholest-5-en-3-yl]oxy](3,3,4,4,5,5,5-heptafluoro
pentyl)dimethyl-;   W/NBS   128004

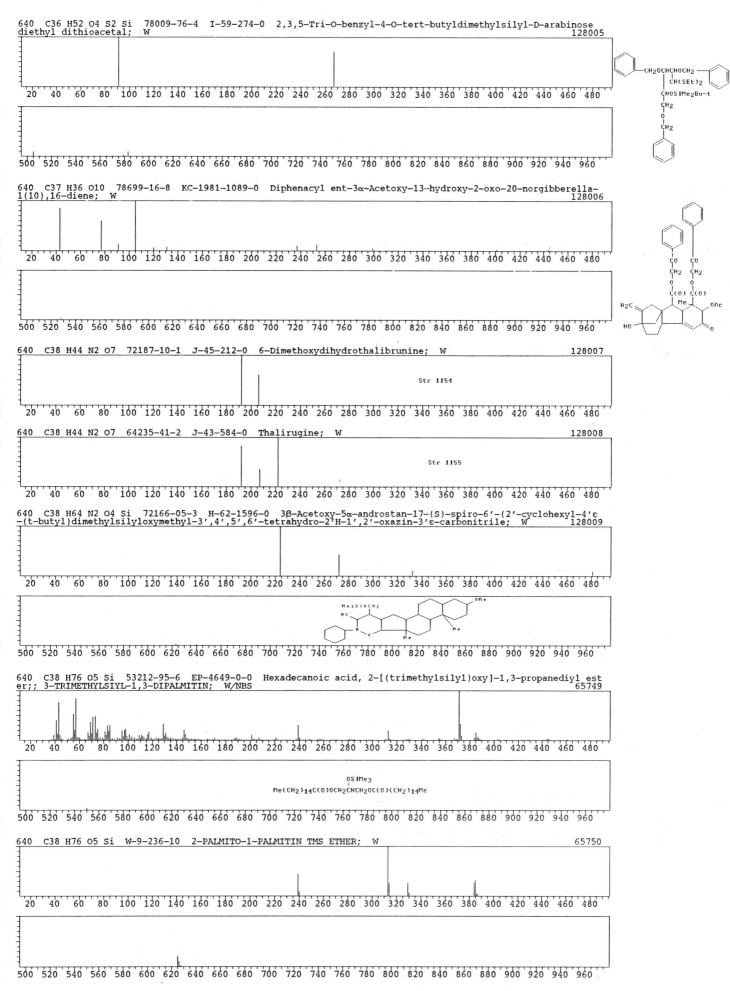

640 C36 H52 O4 S2 Si 78009-76-4 I-59-274-0 2,3,5-Tri-O-benzyl-4-O-tert-butyldimethylsilyl-D-arabinose
diethyl dithioacetal; W   128005

640 C37 H36 O10 78699-16-8 KC-1981-1089-0 Diphenacyl ent-3α-Acetoxy-13-hydroxy-2-oxo-20-norgibberella-
1(10),16-diene; W   128006

640 C38 H44 N2 O7 72187-10-1 J-45-212-0 6-Dimethoxydihydrothalibrunine; W   128007

Str 1154

640 C38 H44 N2 O7 64235-41-2 J-43-584-0 Thalirugine; W   128008

Str 1155

640 C38 H64 N2 O4 Si 72166-05-3 H-62-1596-0 3β-Acetoxy-5α-androstan-17-(S)-spiro-6'-(2'-cyclohexyl-4'ε
-(t-butyl)dimethylsilyloxymethyl-3',4',5',6'-tetrahydro-2'H-1',2'-oxazin-3'ε-carbonitrile; W 128009

640 C38 H76 O5 Si 53212-95-6 EP-4649-0-0 Hexadecanoic acid, 2-[(trimethylsilyl)oxy]-1,3-propanediyl est
er;; 3-TRIMETHYLSIYL-1,3-DIPALMITIN; W/NBS   65749

640 C38 H76 O5 Si W-9-236-10 2-PALMITO-1-PALMITIN TMS ETHER; W   65750

- 5927 -

640   C38 H76 O5 Si   W-9-236-10   3-PALMITO-1-PALMITIN TMS ETHER;   W   65751

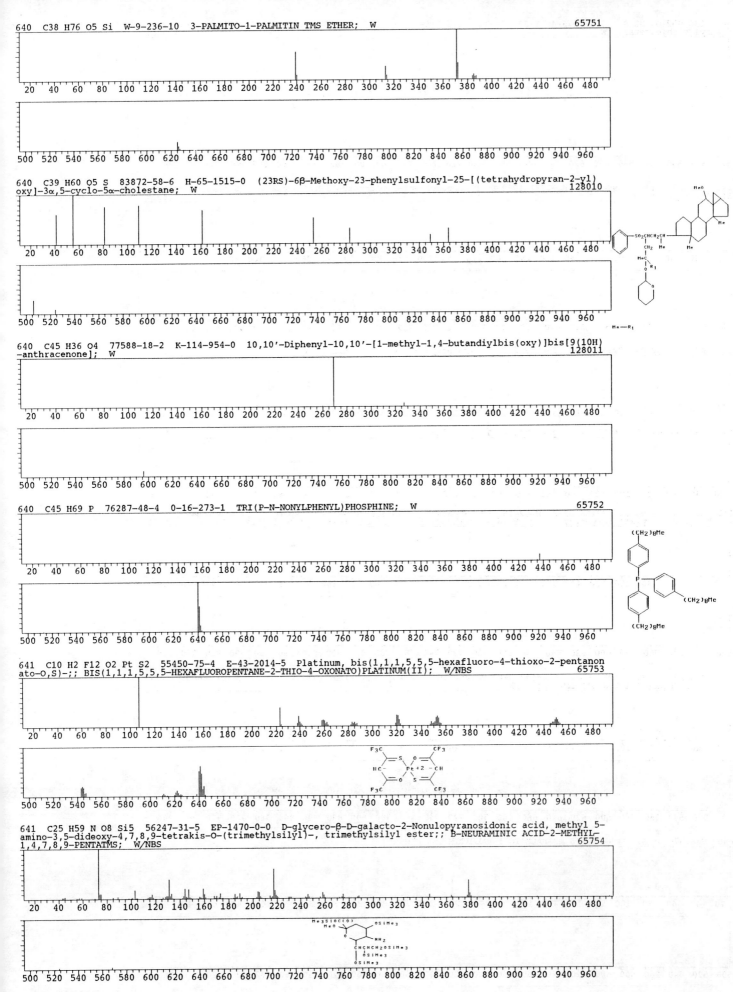

640   C39 H60 O5 S   83872-58-6   H-65-1515-0   (23RS)-6β-Methoxy-23-phenylsulfonyl-25-[(tetrahydropyran-2-yl)
oxy]-3α,5-cyclo-5α-cholestane;   W   128010

640   C45 H36 O4   77588-18-2   K-114-954-0   10,10'-Diphenyl-10,10'-[1-methyl-1,4-butandiylbis(oxy)]bis[9(10H)
-anthracenone];   W   128011

640   C45 H69 P   76287-48-4   0-16-273-1   TRI(P-N-NONYLPHENYL)PHOSPHINE;   W   65752

641   C10 H2 F12 O2 Pt S2   55450-75-4   E-43-2014-5   Platinum, bis(1,1,1,5,5,5-hexafluoro-4-thioxo-2-pentanon
ato-O,S)-;;   BIS(1,1,1,5,5,5-HEXAFLUOROPENTANE-2-THIO-4-OXONATO)PLATINUM(II);   W/NBS   65753

641   C25 H59 N O8 Si5   56247-31-5   EP-1470-0-0   D-glycero-β-D-galacto-2-Nonulopyranosidonic acid, methyl 5-
amino-3,5-dideoxy-4,7,8,9-tetrakis-O-(trimethylsilyl)-, trimethylsilyl ester;;   B-NEURAMINIC ACID-2-METHYL-
1,4,7,8,9-PENTATMS;   W/NBS   65754

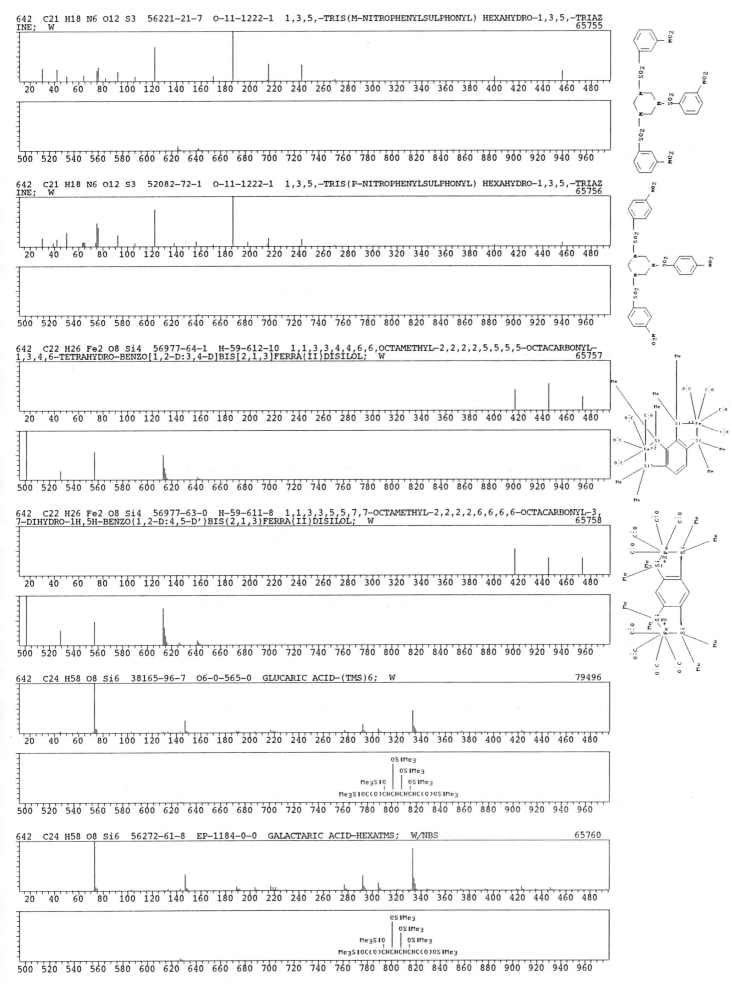

642 C21 H18 N6 O12 S3   56221-21-7   O-11-1222-1   1,3,5,-TRIS(M-NITROPHENYLSULPHONYL) HEXAHYDRO-1,3,5,-TRIAZINE; W   65755

642 C21 H18 N6 O12 S3   52082-72-1   O-11-1222-1   1,3,5,-TRIS(P-NITROPHENYLSULPHONYL) HEXAHYDRO-1,3,5,-TRIAZINE; W   65756

642 C22 H26 Fe2 O8 Si4 56977-64-1 H-59-612-10 1,1,3,3,4,4,6,6,OCTAMETHYL-2,2,2,2,5,5,5,5-OCTACARBONYL-1,3,4,6-TETRAHYDRO-BENZO[1,2-D:3,4-D]BIS[2,1,3]FERRA(II)DISILOL; W   65757

642 C22 H26 Fe2 O8 Si4 56977-63-0 H-59-611-8 1,1,3,3,5,5,7,7-OCTAMETHYL-2,2,2,2,6,6,6,6-OCTACARBONYL-3,7-DIHYDRO-1H,5H-BENZO(1,2-D:4,5-D')BIS(2,1,3)FERRA(II)DISILOL; W   65758

642 C24 H58 O8 Si6   38165-96-7   O6-0-565-0   GLUCARIC ACID-(TMS)6; W   79496

642 C24 H58 O8 Si6   56272-61-8   EP-1184-0-0   GALACTARIC ACID-HEXATMS;   W/NBS   65760

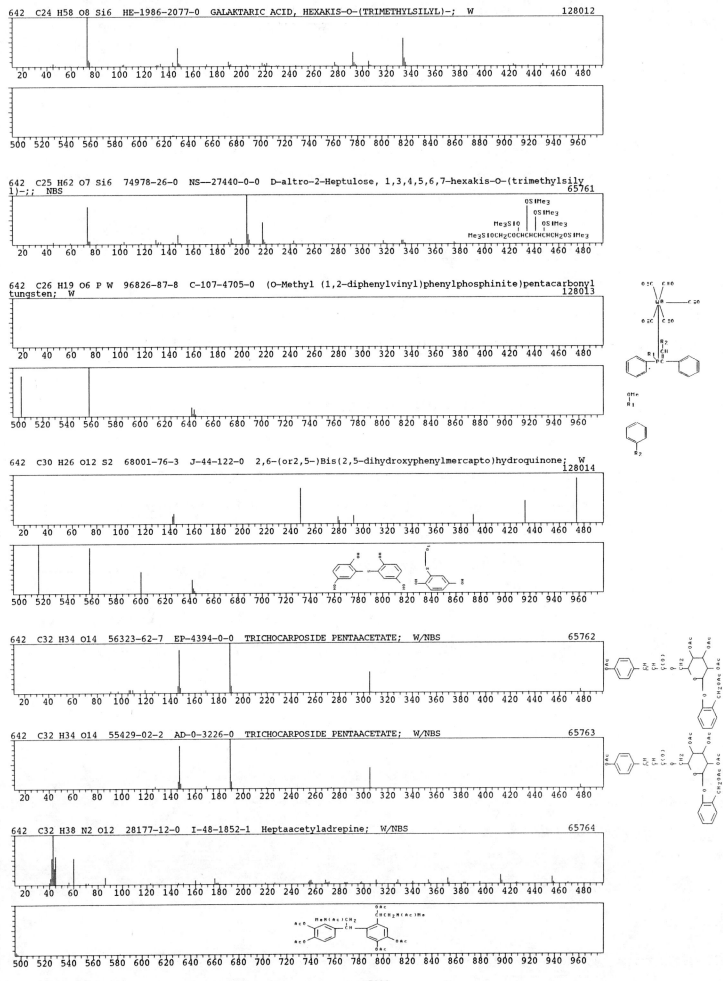

642   C24 H58 O8 Si6   HE-1986-2077-0   GALAKTARIC ACID, HEXAKIS-O-(TRIMETHYLSILYL)-;   W   128012

642   C25 H62 O7 Si6   74978-26-0   NS--27440-0-0   D-altro-2-Heptulose, 1,3,4,5,6,7-hexakis-O-(trimethylsily
l)-;;   NBS   65761

OSiMe3
OSiMe3
Me3SiO   OSiMe3
Me3SiOCH2COCHCHCHCHCH2OSiMe3

642   C26 H19 O6 P W   96826-87-8   C-107-4705-0   (O-Methyl (1,2-diphenylvinyl)phenylphosphinite)pentacarbonyl
tungsten;   W   128013

642   C30 H26 O12 S2   68001-76-3   J-44-122-0   2,6-(or2,5-)Bis(2,5-dihydroxyphenylmercapto)hydroquinone;   W   128014

642   C32 H34 O14   56323-62-7   EP-4394-0-0   TRICHOCARPOSIDE PENTAACETATE;   W/NBS   65762

642   C32 H34 O14   55429-02-2   AD-0-3226-0   TRICHOCARPOSIDE PENTAACETATE;   W/NBS   65763

642   C32 H38 N2 O12   28177-12-0   I-48-1852-1   Heptaacetyladrepine;   W/NBS   65764

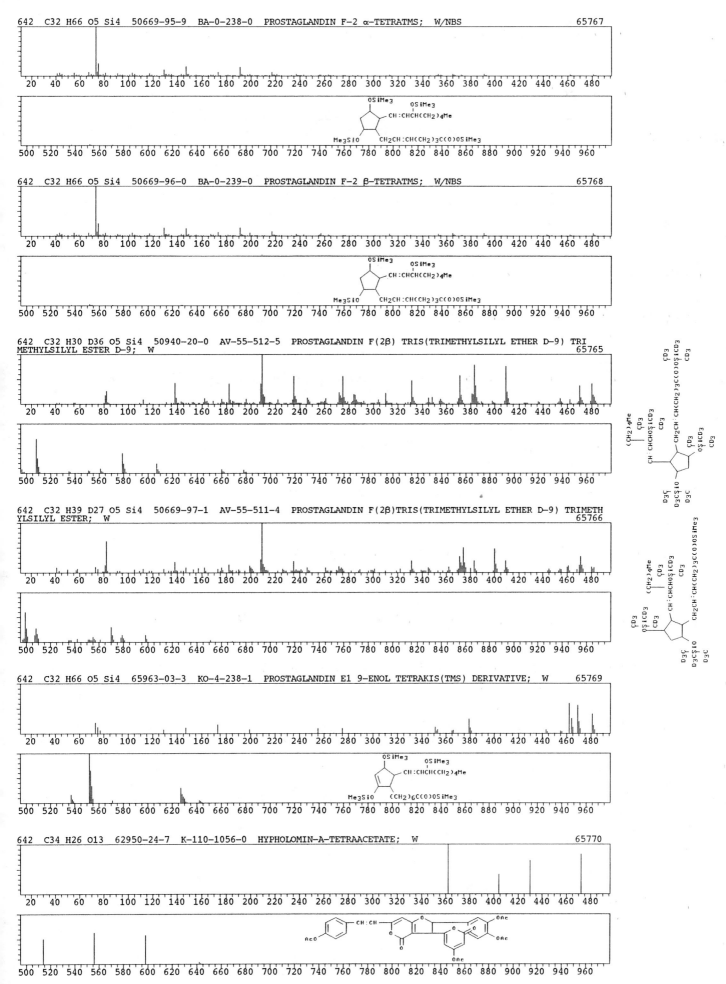

642   C32 H66 O5 Si4   50669-95-9   BA-0-238-0   PROSTAGLANDIN F-2 α-TETRATMS; W/NBS   65767

642   C32 H66 O5 Si4   50669-96-0   BA-0-239-0   PROSTAGLANDIN F-2 β-TETRATMS; W/NBS   65768

642   C32 H30 D36 O5 Si4   50940-20-0   AV-55-512-5   PROSTAGLANDIN F(2β) TRIS(TRIMETHYLSILYL ETHER D-9) TRI
METHYLSILYL ESTER D-9; W   65765

642   C32 H39 D27 O5 Si4   50669-97-1   AV-55-511-4   PROSTAGLANDIN F(2β)TRIS(TRIMETHYLSILYL ETHER D-9) TRIMETH
YLSILYL ESTER; W   65766

642   C32 H66 O5 Si4   65963-03-3   KO-4-238-1   PROSTAGLANDIN E1 9-ENOL TETRAKIS(TMS) DERIVATIVE; W   65769

642   C34 H26 O13   62950-24-7   K-110-1056-0   HYPHOLOMIN-A-TETRAACETATE; W   65770

642   C34 H38 N6 O5 S   65942-44-1   J-43-2318-0   (1α,2β,3β,5β)-(+/-)-N-[3-[6-(dimethylamino)-9H-purin-9-yl]-
2-hdyroxy-5-[[(4-methoxyphenyl)diphenylmethoxy]methyl]cyclopentyl]methanesulfonamide;   W   128015

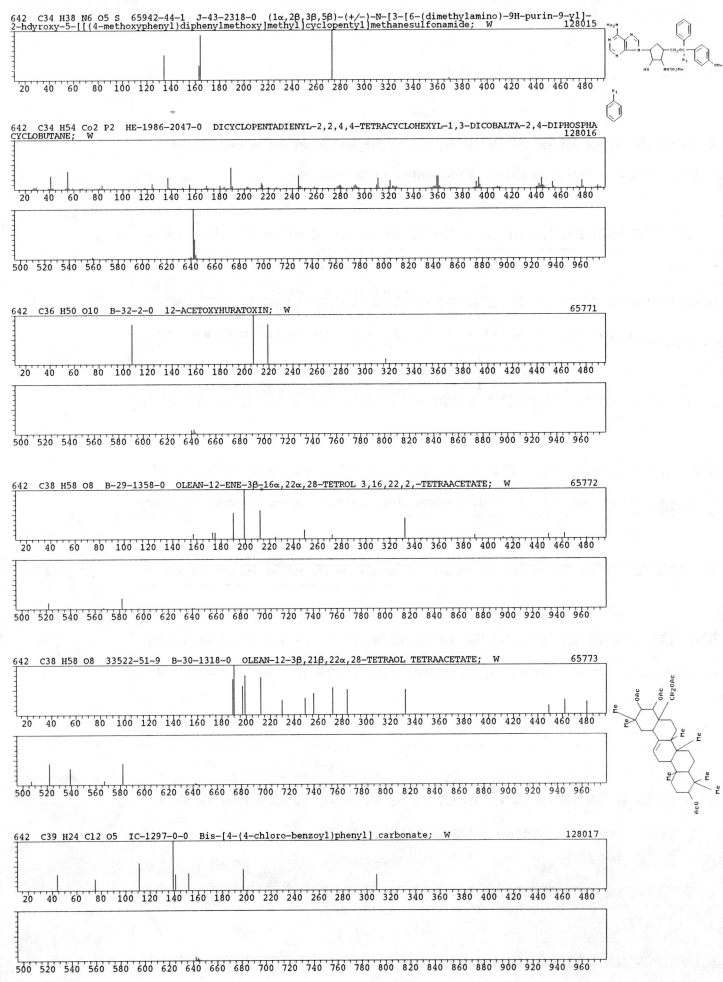

642   C34 H54 Co2 P2   HE-1986-2047-0   DICYCLOPENTADIENYL-2,2,4,4-TETRACYCLOHEXYL-1,3-DICOBALTA-2,4-DIPHOSPHA
CYCLOBUTANE;   W   128016

642   C36 H50 O10   B-32-2-0   12-ACETOXYHURATOXIN;   W   65771

642   C38 H58 O8   B-29-1358-0   OLEAN-12-ENE-3β-16α,22α,28-TETROL 3,16,22,2,-TETRAACETATE;   W   65772

642   C38 H58 O8   33522-51-9   B-30-1318-0   OLEAN-12-3β,21β,22α,28-TETRAOL TETRAACETATE;   W   65773

642   C39 H24 Cl2 O5   IC-1297-0-0   Bis-[4-(4-chloro-benzoyl)phenyl] carbonate;   W   128017

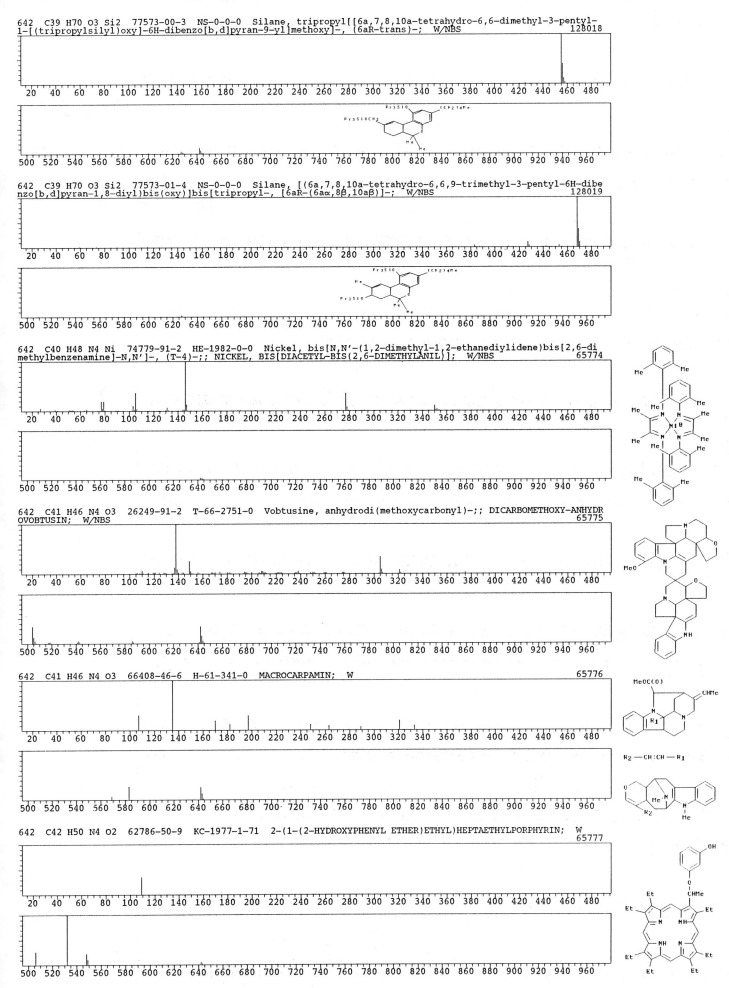

642 C39 H70 O3 Si2 77573-00-3 NS-0-0-0 Silane, tripropyl[[6a,7,8,10a-tetrahydro-6,6-dimethyl-3-pentyl-1-[(tripropylsilyl)oxy]-6H-dibenzo[b,d]pyran-9-yl]methoxy]-, (6aR-trans)-; W/NBS   128018

642 C39 H70 O3 Si2 77573-01-4 NS-0-0-0 Silane, [(6a,7,8,10a-tetrahydro-6,6,9-trimethyl-3-pentyl-6H-dibenzo[b,d]pyran-1,8-diyl)bis(oxy)]bis[tripropyl-, [6aR-(6aα,8β,10aβ)]-; W/NBS   128019

642 C40 H48 N4 Ni 74779-91-2 HE-1982-0-0 Nickel, bis[N,N'-(1,2-dimethyl-1,2-ethanediylidene)bis[2,6-dimethylbenzenamine]-N,N']-, (T-4)-;; NICKEL, BIS[DIACETYL-BIS(2,6-DIMETHYLANIL)]; W/NBS   65774

642 C41 H46 N4 O3 26249-91-2 T-66-2751-0 Vobtusine, anhydrodi(methoxycarbonyl)-;; DICARBOMETHOXY-ANHYDROVOBTUSIN; W/NBS   65775

642 C41 H46 N4 O3 66408-46-6 H-61-341-0 MACROCARPAMIN; W   65776

642 C42 H50 N4 O2 62786-50-9 KC-1977-1-71 2-(1-(2-HYDROXYPHENYL ETHER)ETHYL)HEPTAETHYLPORPHYRIN; W   65777

642   C42 H50 N4 O2   62786-51-0   KC-1977-1-71   2-(1-(2,4-DIHYDROXYPHENYL)ETHYL)HEPTAETHYLPORPHYRIN;   W 65778

642   C42 H18 D32 N4 O2   KC-1977-1-71   3-(1-RESOCINYLETHYL)DEUTEROPORPHYRIN;   W   65779

642   C42 H18 D32 N4 O2   KC-1977-1-71   8-(1-RESOCINYLETHYL)DEUTEROPORPHYRIN;   W   65780

642   C44 H30 N6   IC-1297-0-0   4,4'-Bis-(2,4-diphenyltriazine-6-yl)-stilbene;   W   128020

642   C44 H66 O3   IC-1297-0-0   2,6-Di-tert-butyl-4,4-bis-(3,5-di-tert-butyl-4-hydroxy-benzyl)-cyclohexa-2,5-
dienone;   W   128021

643   C12 F23 N O3   51445-03-5   O-11-1218-1   PERFLUOROALKYLETHER NITRILE;   W   65781

643   C25 H53 N5 O5 Si5   53294-38-5   C-90-4183-1   GUANOSINE-PENTATMS;   W/NBS   65782

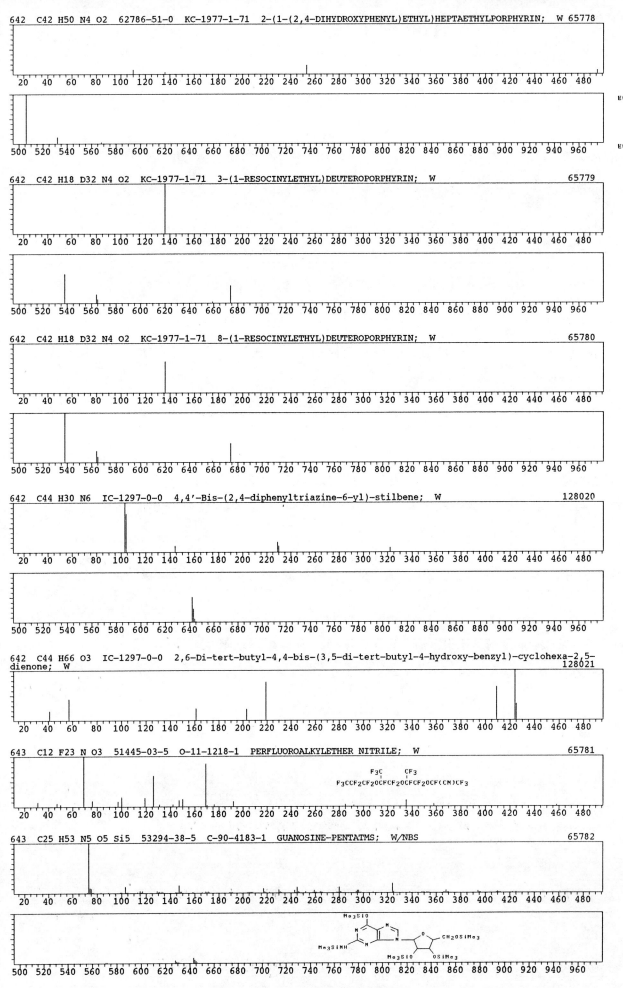

643   C34 H53 N5 O7   BS-1-63-0   N-Acetyl-(N,O-permethyl)-prolyl-leucyl-phenylalanyl-valyl-glycine;   W 128022

643   C35 H49 N O10   CD-294-0-0   Crassicauline A;   W   128023

643   C39 H59 Cl2 N O2   3546-10-9   NS-0-0-0   Cholest-5-en-3-ol (3β)-, 4-[bis(2-chloroethyl)amino]benzeneacet
ate;   W/NBS   128024

Str 1156

644   C25 H18 N2 O5 P Re   73132-90-8   K-113-647-0   2,2'-Bipyridine(tricarbonyl)diphenylphosphinato-O-rheniu
m(I);   W   128025

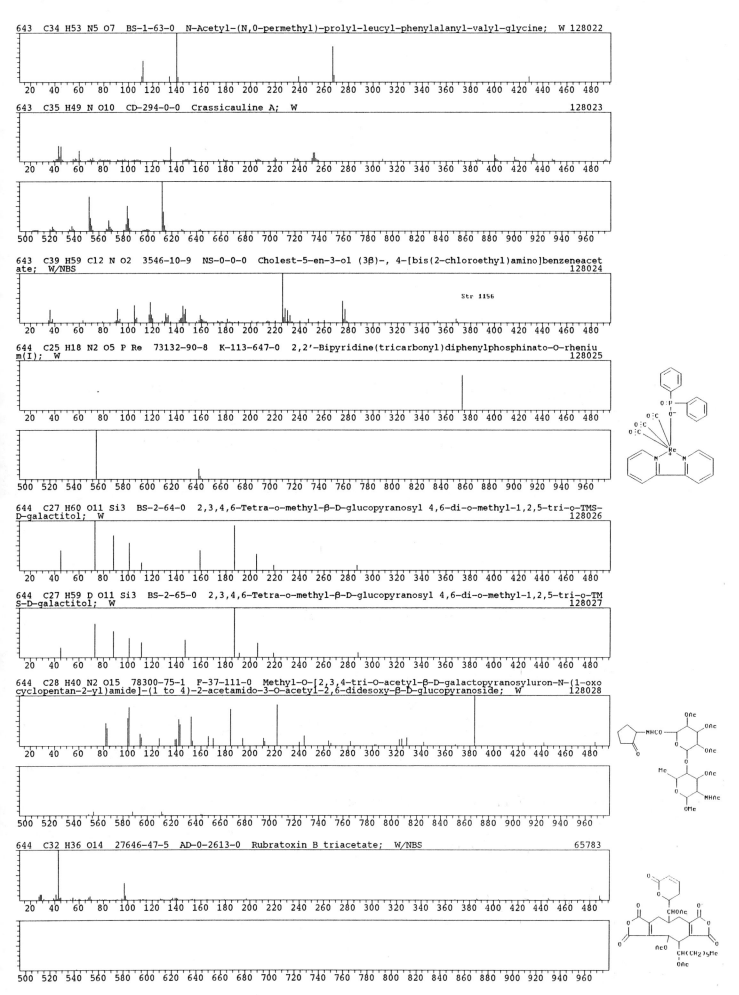

644   C27 H60 O11 Si3   BS-2-64-0   2,3,4,6-Tetra-o-methyl-β-D-glucopyranosyl 4,6-di-o-methyl-1,2,5-tri-o-TMS-
D-galactitol;   W   128026

644   C27 H59 D O11 Si3   BS-2-65-0   2,3,4,6-Tetra-o-methyl-β-D-glucopyranosyl 4,6-di-o-methyl-1,2,5-tri-o-TM
S-D-galactitol;   W   128027

644   C28 H40 N2 O15   78300-75-1   F-37-111-0   Methyl-O-[2,3,4-tri-O-acetyl-β-D-galactopyranosyluron-N-(1-oxo
cyclopentan-2-yl)amide]-(1 to 4)-2-acetamido-3-O-acetyl-2,6-didesoxy-β-D-glucopyranoside;   W   128028

644   C32 H36 O14   27646-47-5   AD-0-2613-0   Rubratoxin B triacetate;   W/NBS   65783

644   C32 H68 O5 Si4   50669-94-8   BA-0-237-0   PROSTAGLANDIN F-1 α-TETRATMS;   W/NBS   65784

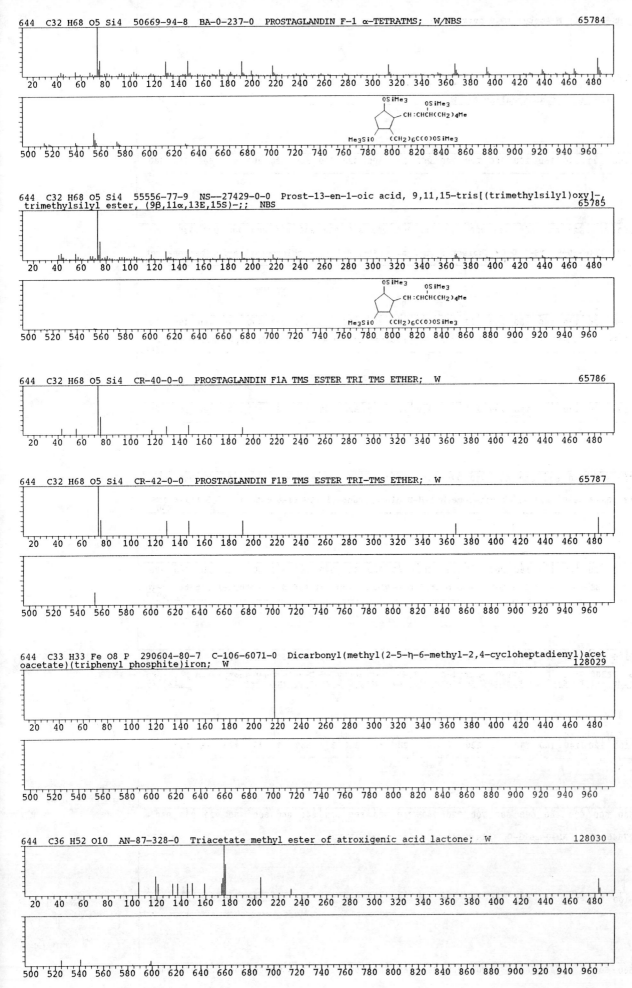

644   C32 H68 O5 Si4   55556-77-9   NS--27429-0-0   Prost-13-en-1-oic acid, 9,11,15-tris[(trimethylsilyl)oxy]-,
   trimethylsilyl ester, (9β,11α,13E,15S)-;;   NBS   65785

644   C32 H68 O5 Si4   CR-40-0-0   PROSTAGLANDIN F1A TMS ESTER TRI TMS ETHER;   W   65786

644   C32 H68 O5 Si4   CR-42-0-0   PROSTAGLANDIN F1B TMS ESTER TRI-TMS ETHER;   W   65787

644   C33 H33 Fe O8 P   290604-80-7   C-106-6071-0   Dicarbonyl(methyl(2-5-η-6-methyl-2,4-cycloheptadienyl)acet
oacetate)(triphenyl phosphite)iron;   W   128029

644   C36 H52 O10   AN-87-328-0   Triacetate methyl ester of atroxigenic acid lactone;   W   128030

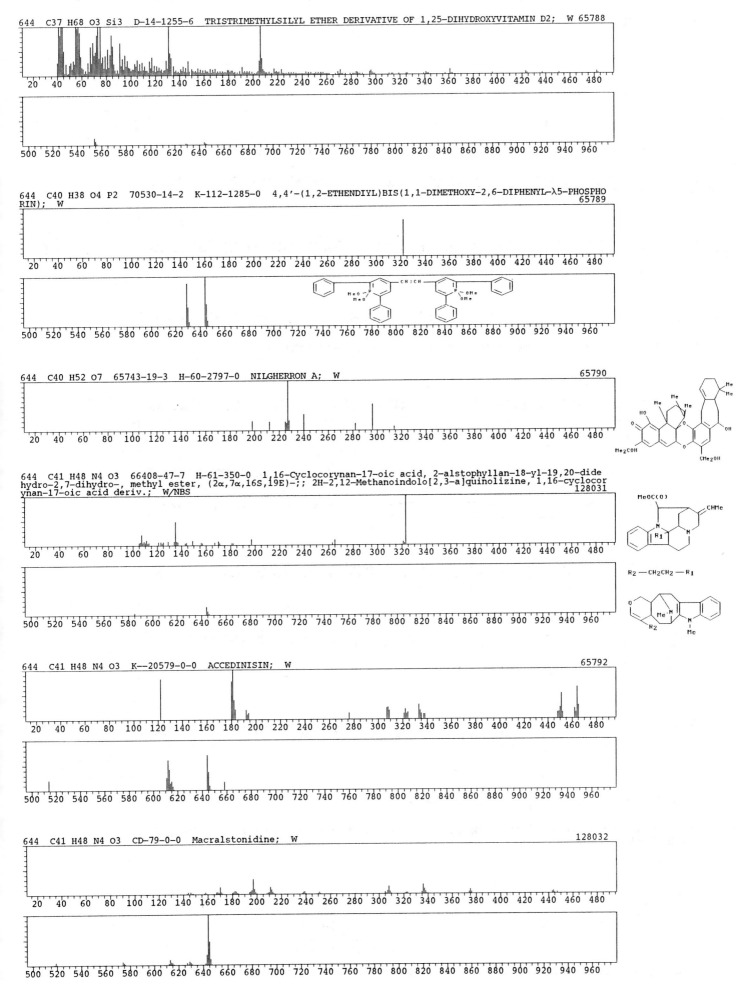

644   C37 H68 O3 Si3   D-14-1255-6   TRISTRIMETHYLSILYL ETHER DERIVATIVE OF 1,25-DIHYDROXYVITAMIN D2;   W 65788

644   C40 H38 O4 P2   70530-14-2   K-112-1285-0   4,4'-(1,2-ETHENDIYL)BIS(1,1-DIMETHOXY-2,6-DIPHENYL-λ5-PHOSPHO
RIN);   W
65789

644   C40 H52 O7   65743-19-3   H-60-2797-0   NILGHERRON A;   W
65790

644   C41 H48 N4 O3   66408-47-7   H-61-350-0   1,16-Cyclocorynan-17-oic acid, 2-alstophyllan-18-yl-19,20-dide
hydro-2,7-dihydro-, methyl ester, (2α,7α,16S,19E)-;; 2H-2,12-Methanoindolo[2,3-a]quinolizine, 1,16-cyclocor
ynan-17-oic acid deriv.;   W/NBS
128031

644   C41 H48 N4 O3   K--20579-0-0   ACCEDINISIN;   W
65792

644   C41 H48 N4 O3   CD-79-0-0   Macralstonidine;   W
128032

644   C42 H28 O7   63910-29-2   KC-1977-1052-0   HEPTABENZO-21-CROWN-7;   W   65793

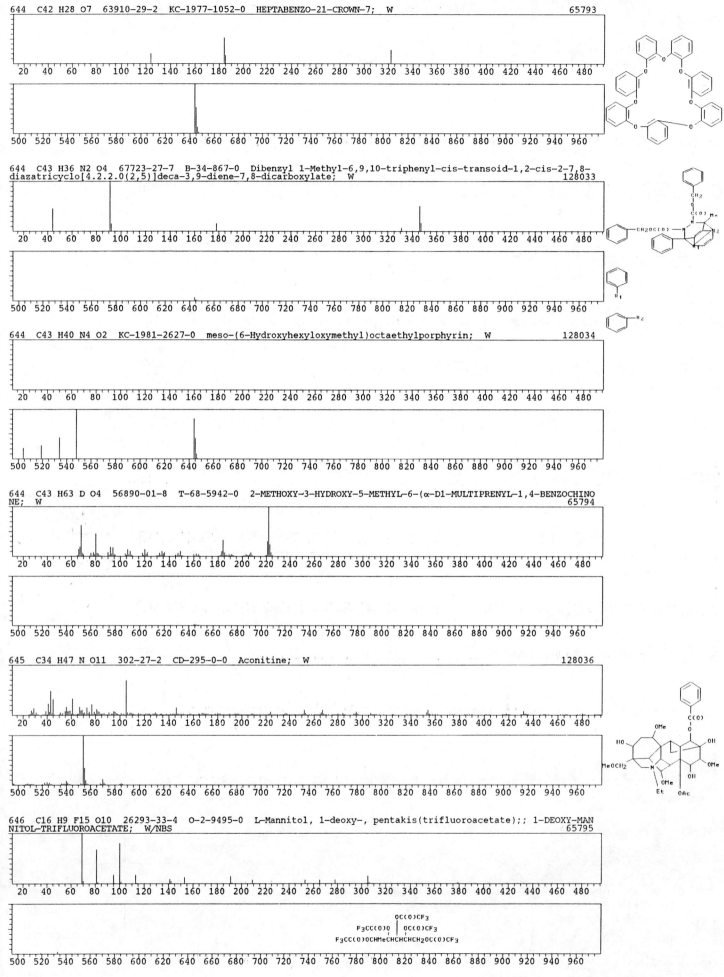

644   C43 H36 N2 O4   67723-27-7   B-34-867-0   Dibenzyl 1-Methyl-6,9,10-triphenyl-cis-transoid-1,2-cis-2-7,8-
diazatricyclo[4.2.2.0(2,5)]deca-3,9-diene-7,8-dicarboxylate;   W   128033

644   C43 H40 N4 O2   KC-1981-2627-0   meso-(6-Hydroxyhexyloxymethyl)octaethylporphyrin;   W   128034

644   C43 H63 D O4   56890-01-8   T-68-5942-0   2-METHOXY-3-HYDROXY-5-METHYL-6-(α-D1-MULTIPRENYL-1,4-BENZOCHINO
NE;   W   65794

645   C34 H47 N O11   302-27-2   CD-295-0-0   Aconitine;   W   128036

646   C16 H9 F15 O10   26293-33-4   O-2-9495-0   L-Mannitol, 1-deoxy-, pentakis(trifluoroacetate);; 1-DEOXY-MAN
NITOL-TRIFLUOROACETATE;   W/NBS   65795

646 C16 H9 F15 O10 26388-40-9 O-2-9506-0 D-arabino-Hexitol, 2-deoxy-, pentakis(trifluoroacetate);; 2-DE
OXY-D-ARAKINO-HEXITOL-TRIFLUOROACETATE; W/NBS
65796

646 C20 H14 Br F15 N2 82633-64-5 AC-1982-94-0 4-Bromo-1-phenyl-3-perfluoroheptyl-5-(t-butyl)-pyrazole;
W
128037

646 C21 H6 F12 O10 62751-02-4 K-110-1068-0 LEUKOHYMENOQUINONE-TETRAKIS(TRIFLUOROACETATE); W 65797

646 C31 H66 O6 Si4 KO-5-442-1 TETRAKIS(TRIMETHYLSILYL)ETHER OF 10,11,12,13-TETRAHYDROXY-TRANS-10-TRANS-
12-OCTADECADIENOIC ACID METHYL ESTER; W
65798

646 C31 H66 O6 Si4 KO-5-442-1 TETRAKIS(TRIMETHYLSILYL)ETHER OF 9,10,11,12-TETRAHYDROXY-TRANS-9-TRANS-11-
OCTADECADIENOIC ACID METHYL ESTER; W
65799

646 C32 H31 Fe O9 P 85939-52-2 C-106-6071-0 dicarbonyl(dimethyl(2-5-η-cyclohepta-2,4-dienyl)malonate)(
malonate)(triphenylphosphite)iron; W
128038

646 C34 H30 O13 31449-13-5 B-29-1086-0 Aphthosin; W 65800

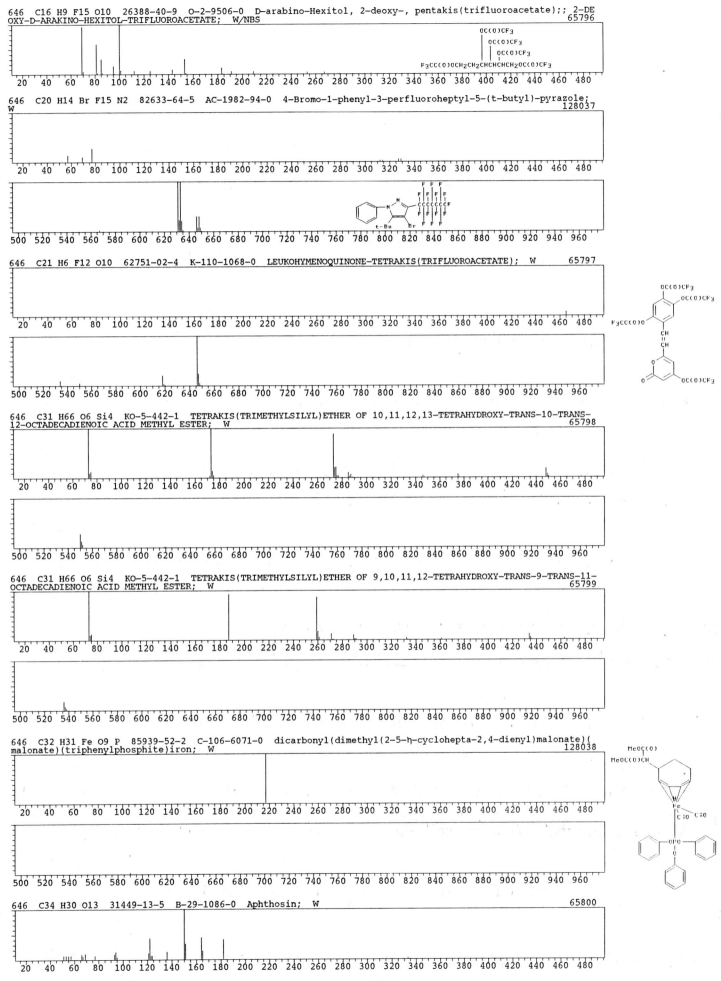

646 C34 H50 N2 O10 88454-99-3 J-49-759-0 (2R,3R,11R,12R)-N,N,N',N'-Tetramethyl-11,12-bis[(benzyloxy)
methyl)]-1,4,7,10,13,16-hexaoxacyclooctadecane-2,3-dicarboxamide; W   128039

646 C36 H46 N4 O7 77286-48-7 K-114-142-0 Methyl ester of (2RS,3RS,3'RS)-18-Ethyl-3-(1-methoxyethyl)-1,
2,3,19,22,24-hexahydro-2,7,13,17-tetramethyl-1,19-dioxo-21H-bilin-8,12-dipropionic acid; W   128040

646 C36 H46 N4 O7 77236-74-9 K-114-143-0 Methyl ester of (2RS,3RS,3'SR)-18-Ethyl-3-(1-methoxyethyl)-1,
2,3,19,22,24-hexahydro-2,7,13,17-tetramethyl-1,19-dioxo-21H-bilin-8,12-dipropionic acid; W   128041

646 C38 H70 O4 Si2 71899-30-4 O-14-165-5 (8α,14β,18β)-3β-ACETYLOXY-21,21-DIMETHYL-2α,22α-DI(TRIMETHYLSI
LYLOXY)-29,30-DINORGAMMACERANE; W   65801

646 C38 H70 O4 Si2 71899-29-1 O-14-165-5 (8 α,14 β,18 β)-2 α-ACETYLOXY-21,21-DIMETHYL-3 β,22 α-DI(TRI
METHYLSILYLOXY)-29,30-DINORGAMMACERANE; W   65802

646 C41 H50 N4 O3 66408-48-8 H-61-350-0 1,16-Cyclocorynan-17-oic acid, 19,20-didehydro-2-(20,21-dihydro
alstophyllan-18-yl)-2,7-dihydro-, methyl ester, (2α,7α,16S,19E)-;; 1,16-Cyclocorynan-17-oic acid, 19,20-did
ehydro-2-(20,21-dihydroalstophyllan-18-yl)-2,7-dihydro-, methyl ester, (2α,7α,16S,19E)- (9CI; W/NBS 128042

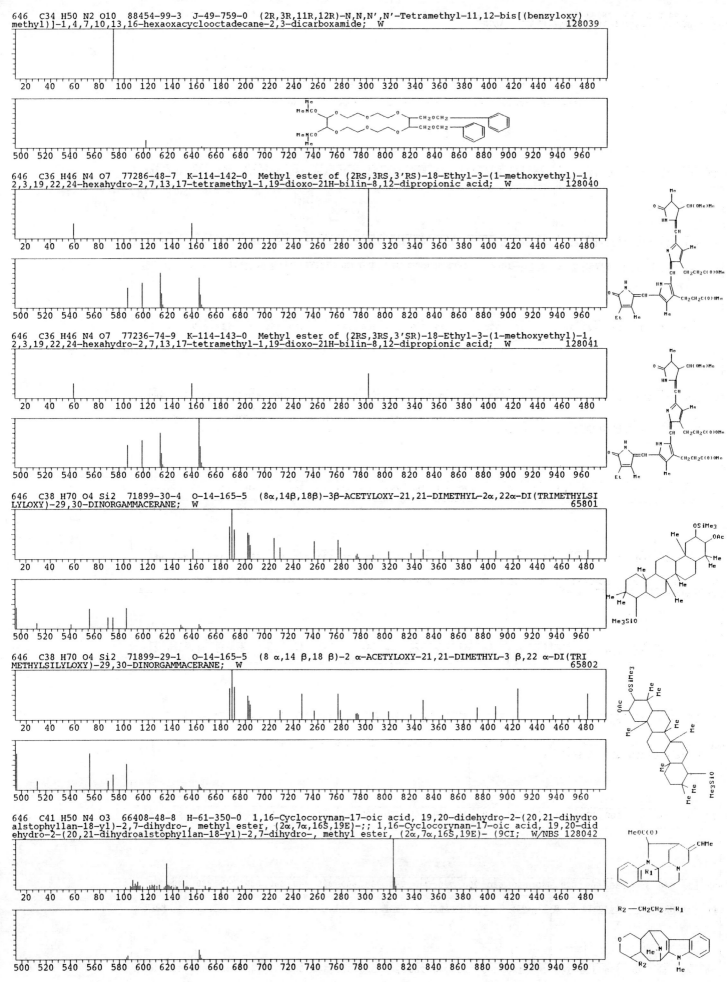

646 C44 H62 N4   69912-06-7   J-44-2079-0   1,3,5,7-Tetramethyl-2,4,6,8-tetra(n-pentyl)porphyrin;   W   128043

20 40 60 80 100 120 140 160 180 200 220 240 260 280 300 320 340 360 380 400 420 440 460 480

500 520 540 560 580 600 620 640 660 680 700 720 740 760 780 800 820 840 860 880 900 920 940 960

646 C46 H34 N2 O2   65241-99-8   Y-14-801-0   N-(4-BENZOYL-8-METHYL-2,3-DIPHENYL-9B-P-TOLYL-9BH-BENZ(G)INDOLE
-5-YL)BENZAMIDE;   W   65804

20 40 60 80 100 120 140 160 180 200 220 240 260 280 300 320 340 360 380 400 420 440 460 480

500 520 540 560 580 600 620 640 660 680 700 720 740 760 780 800 820 840 860 880 900 920 940 960

647 C15 H24 Bi N3 S6   40211-73-2   O-16-13-2   TRIS(TETRAMETHYLENEDITHIOCARBAMATO)BIMUTH(III);   W   65805

20 40 60 80 100 120 140 160 180 200 220 240 260 280 300 320 340 360 380 400 420 440 460 480

500 520 540 560 580 600 620 640 660 680 700 720 740 760 780 800 820 840 860 880 900 920 940 960

647 C26 H57 N3 O6 Si5   D-14-5482-4   A MIXTURE OF THE TETRA- AND PENTA-TRIMETHYLSILYL DERIVATIVES OF 3-β-HY
DROXYETHYLCYTIDINE;   W   65806

20 40 60 80 100 120 140 160 180 200 220 240 260 280 300 320 340 360 380 400 420 440 460 480

500 520 540 560 580 600 620 640 660 680 700 720 740 760 780 800 820 840 860 880 900 920 940 960

647 C27 H23 Br2 N O4 S2   75931-79-2   O-15-25-2   N,N-BIS(4'-P-BROMOPHENYLSULFONYL-2'-BUTYNYL)-4-METHYLANILI
NE;   W   65807

20 40 60 80 100 120 140 160 180 200 220 240 260 280 300 320 340 360 380 400 420 440 460 480

500 520 540 560 580 600 620 640 660 680 700 720 740 760 780 800 820 840 860 880 900 920 940 960

647 C32 H37 N7 O8   87884-15-9   C-105-7639-0   9-(Methyl 9(S)-(((tert-butyloxy)carbonyl)amino)-6-cyano-5,6,
7,8,9-pentadeoxy-2,3-O-isopropylidene-β-D-ribo-deca-6-enofuranosyluronate)-adenine;   W   128044

20 40 60 80 100 120 140 160 180 200 220 240 260 280 300 320 340 360 380 400 420 440 460 480

500 520 540 560 580 600 620 640 660 680 700 720 740 760 780 800 820 840 860 880 900 920 940 960

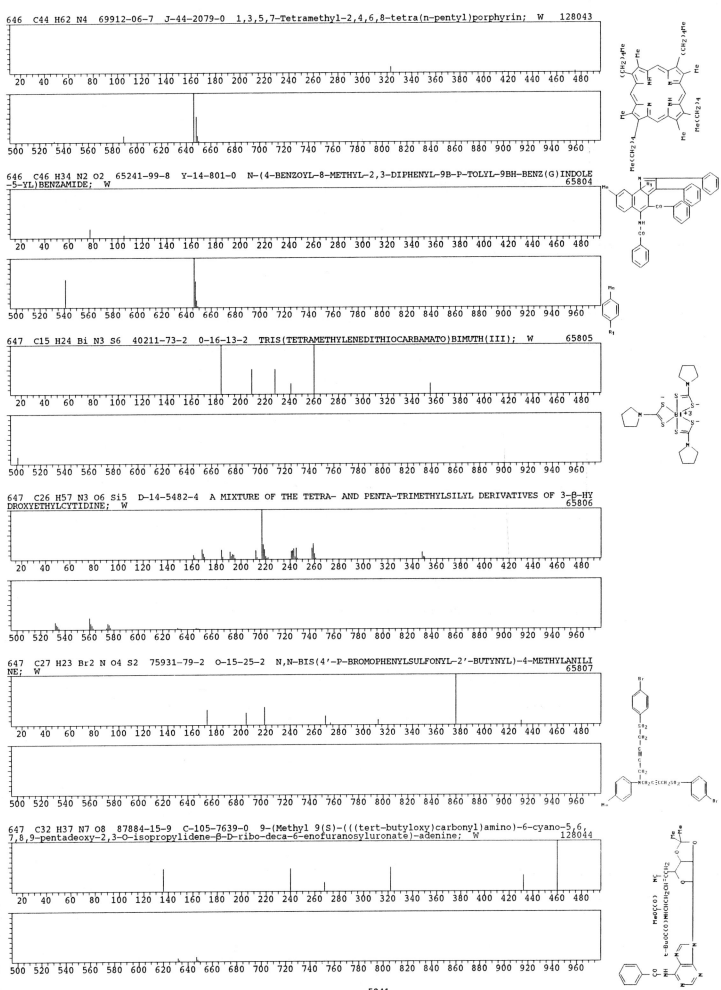

647 C32 H57 N O5 Si4 HE-1982-0-0 β-D-FRUCTOPYRANOSE, 1-N-DIBENZYLAMINO-1-DESOXY-TETRAKIS-O-(TRIMETHYLSIL
YL)-; W   65808

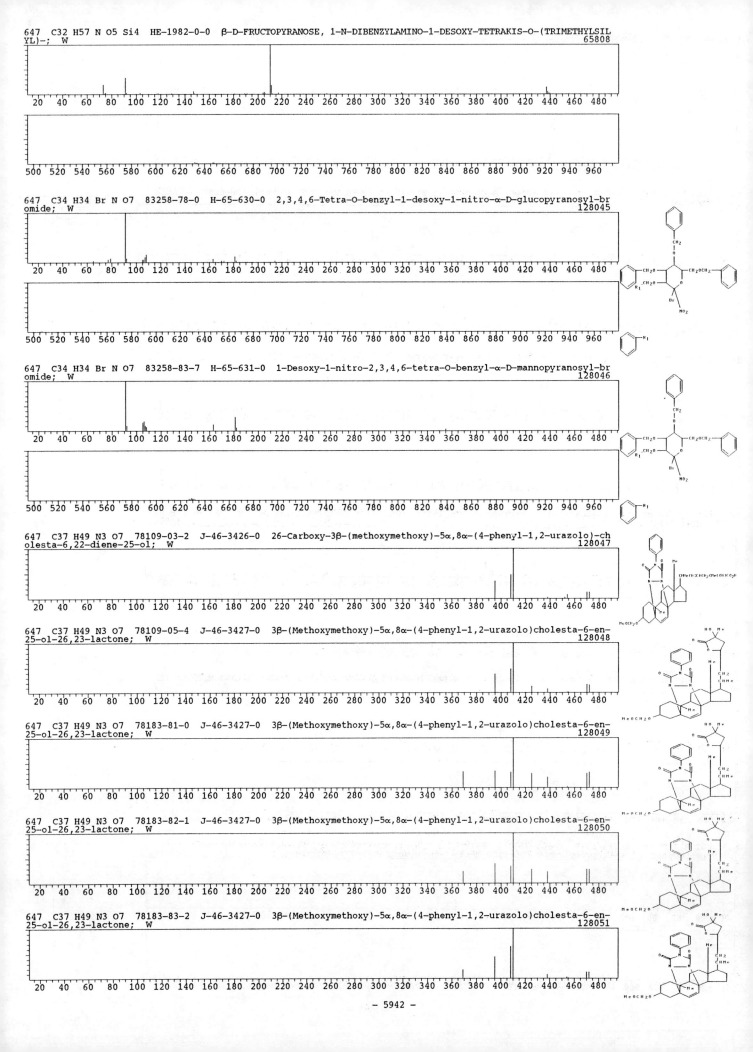

647 C34 H34 Br N O7 83258-78-0 H-65-630-0 2,3,4,6-Tetra-O-benzyl-1-desoxy-1-nitro-α-D-glucopyranosyl-br
omide; W   128045

647 C34 H34 Br N O7 83258-83-7 H-65-631-0 1-Desoxy-1-nitro-2,3,4,6-tetra-O-benzyl-α-D-mannopyranosyl-br
omide; W   128046

647 C37 H49 N3 O7 78109-03-2 J-46-3426-0 26-Carboxy-3β-(methoxymethoxy)-5α,8α-(4-phenyl-1,2-urazolo)-ch
olesta-6,22-diene-25-ol; W   128047

647 C37 H49 N3 O7 78109-05-4 J-46-3427-0 3β-(Methoxymethoxy)-5α,8α-(4-phenyl-1,2-urazolo)cholesta-6-en-
25-ol-26,23-lactone; W   128048

647 C37 H49 N3 O7 78183-81-0 J-46-3427-0 3β-(Methoxymethoxy)-5α,8α-(4-phenyl-1,2-urazolo)cholesta-6-en-
25-ol-26,23-lactone; W   128049

647 C37 H49 N3 O7 78183-82-1 J-46-3427-0 3β-(Methoxymethoxy)-5α,8α-(4-phenyl-1,2-urazolo)cholesta-6-en-
25-ol-26,23-lactone; W   128050

647 C37 H49 N3 O7 78183-83-2 J-46-3427-0 3β-(Methoxymethoxy)-5α,8α-(4-phenyl-1,2-urazolo)cholesta-6-en-
25-ol-26,23-lactone; W   128051

- 5942 -

647 C37 H49 N3 O7   78109-07-6   J-46-3427-0   3β-(Methoxymethoxy)-5α,8α-(4-phenyl-1,2-urazolo)cholesta-6-en-
25-ol-26,22-lactone;   W   128052

647 C38 H41 N5 O5   91084-75-2   C-106-4985-0   7,8,17,18-Tetrahydro-35-methoxy-1,3,21,23-tetramethyl-16H,31H
-5,9,15,19-dimethano-10,14-metheno-26,30-nitrilo-6H,25H-dibenzo(b,s)(1,21,4,8,14,18)dioxatetraazacyclooctac
osine;   W   128053

648 As F15 O3 Se3   63044-19-9   K-110-1474-0   ARSEN-TRIS(PENTAFLUOROORTHOSELENATE);   W   65809

648 C15 H3 Al F18 O6   OS-5-202-0   Tris-(1,1,1,5,5,5,-hexafluoro-2,4-pentanedionato aluminium(III);   W 128054

648 C18 H14 Ir2 O2   64867-80-7   C-99-7874-0   BIS(η5-CYCLOPENTADIENYL)BIS(CARBONYL)-.MU.-(O-PHENYLENE)-DIIR
IDIUM(R-R);   W   65810

648 C18 H19 I3 O2   55508-17-3   D-14-1737-0   3,3',5-TRIIODOHEXESTROL;   W   65811

648 C23 H57 O9 P Si5   55334-86-6   BA-0-392-0   L-chiro-Inositol, 1,2,3,5,6-pentakis-O-(trimethylsilyl)-, di
methyl phosphate;; PENTAKIS(TRIMETHYLSILYL)-(-)-CHIRO-INOSITOL-3-PHOSPHATE DIMETHYL ESTER;   W/NBS   65812

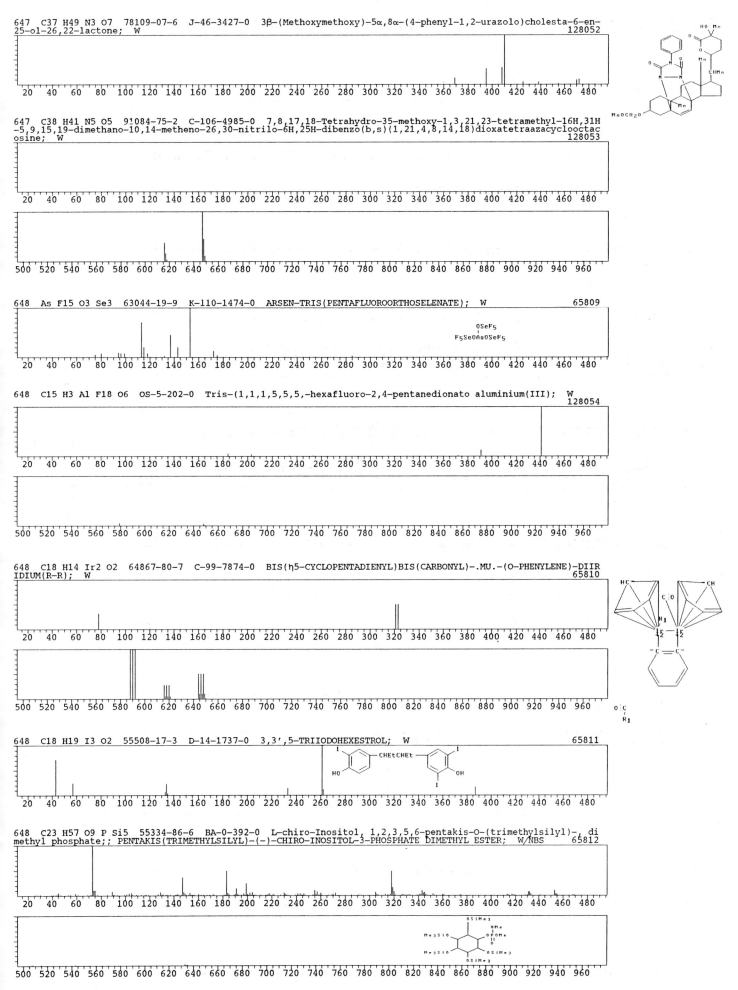

- 5943 -

648 C23 H57 O9 P Si5 55569-48-7 BA-0-388-0 myo-Inositol, 1,3,4,5,6-pentakis-O-(trimethylsilyl)-, dimeth
yl phosphate;; PENTAKIS(TRIMETHYLSILYL)MYO-INOSITOL-2-PHOSPHATE DIMETHYL ESTER; W/NBS 65813

648 C24 H60 Al4 O12 KA-70-2347-3 ALUMINUM ETHOXIDE; W 65814

648 C27 H52 N2 O10 Si3 55556-81-5 NS--27421-0-0 α-D-Glucopyranosiduronic acid, 3-(5-ethylhexahydro-2,4,
6-trioxo-5-pyrimidinyl)-1,1-dimethylpropyl 2,3,4-tris-O-(trimethylsilyl)-, methyl ester;; α-D-Glucopyranosi
duronic acid, 3-(5-ethylhexahydro-2,4,6-trioxo-5-pyrimidinyl)-1,1-dimethylpropyl 2,3,4-tris-O-(trimethylsil
yl)-, methyl e; NBS 65815

648 C27 H52 N2 O10 Si3 55556-80-4 NS--27422-0-0 β-D-Glucopyranosiduronic acid, 3-(5-ethylhexahydro-2,4,
6-trioxo-5-pyrimidinyl)-1,1-dimethylpropyl 2,3,4-tris-O-(trimethylsilyl)-, methyl ester;; β-D-Glucopyranosi
duronic acid, 3-(5-ethylhexahydro-2,4,6-trioxo-5-pyrimidinyl)-1,1-dimethylpropyl 2,3,4-tris-O-(trimethylsil
yl)-, methyl es; NBS 65816

648 C31 H37 Br O10 77882-84-9 MY-1981-0-0 P-BROMOBENZOYL-T2; W 65817

648 C31 H37 Br O10 75758-99-5 J-49-3259-0 (+-)-(1α,2α,4aα,4bβ,10α)-2-(Benzoyloxy)-10a-bromo-10-(ethoxyc
arbonyl)-8,8-(ethylenedioxy)-7-[(methoxymethyl)oxy]-1-methylgibbane-1,4a-carbolactone; W 128055

648 C32 H48 N4 O6 S2 67370-93-8 H-61-1059-0 7,18-Ditosyl-12-oxo-7,11,18-triaza-21-heneicosanlactam; W 128056

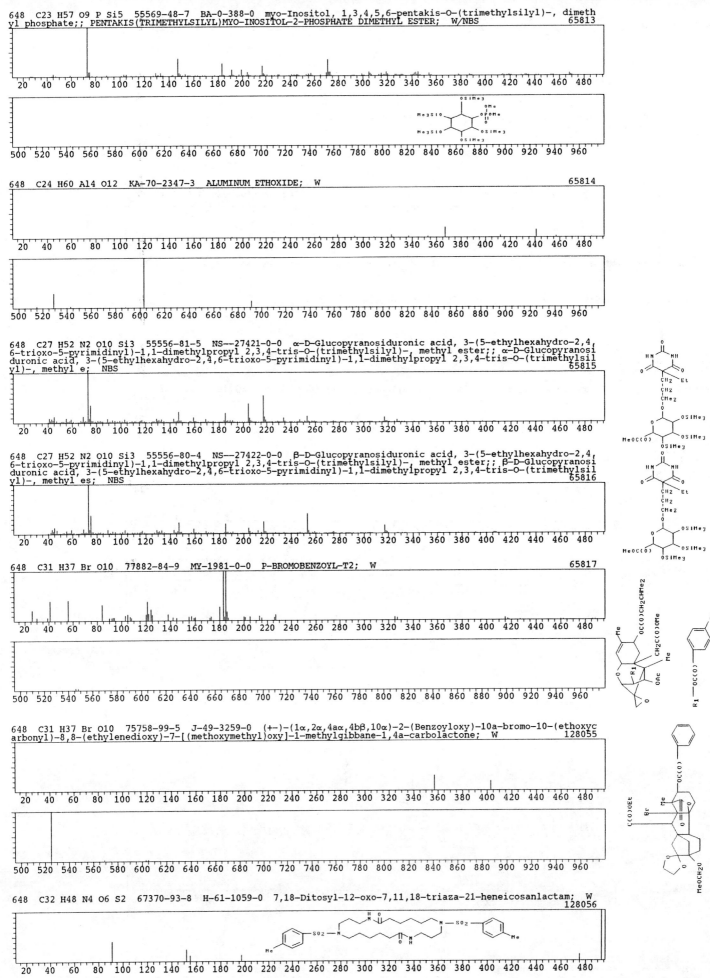

648 C32 H68 O7 Si3 67689-99-0 KO-5-147-1 METHYL-2,3,4-TRIS-O-TRIMETHYLSILYL-6-O-PALMITOYL-α-D-GLUPYRANO
SIDE; W   65818

648 C32 H68 O7 Si3 67721-70-4 KO-5-148-2 METHYL-2,3,6-TRIS-O-TMS-4-O-PALMITOYL-α-D-GLUCOPYRANOSIDE; W
                                                                                                            65819

648 C32 H68 O7 Si3 67690-00-0 KO-5-148-3 METHYL-2,4,6-TRIS-TMS-3-O-PALMITOYL-α-D-GLUCOPYRAMOSIDE; W
                                                                                                            65820

648 C32 H68 O7 Si3 67690-01-1 KO-5-148-4 METHYL-3,4,6-TRIS-O-TMS-2-O-PALMITOYL-α-D-GLUCOPYRANOSIDE; W
                                                                                                            65821

648 C34 H32 O13 81424-99-9 KC-1982-976-0 6-(p-Acetoxybenzyl)-3,3',4',5,7-penta-O-acetyl-(+)-catechin;
W   128057

648 C34 H32 O13 81410-27-7 KC-1982-977-0 6-(o-Acetoxybenzyl)-3,3',4',5,7-penta-O-acetyl-(+)-catechin;
W   128058

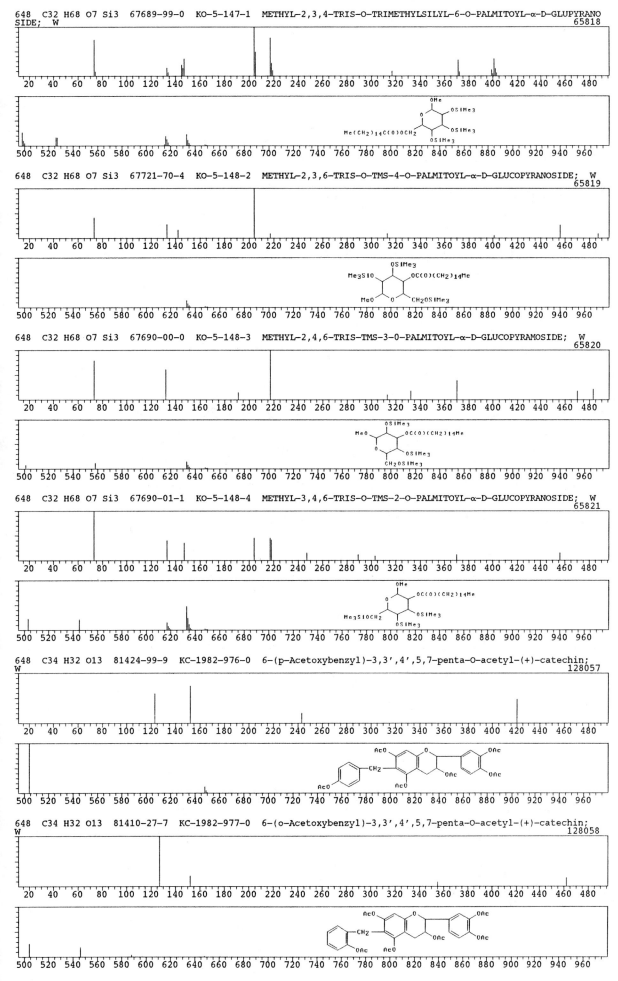

648   C34 H32 O13   81410-29-9   KC-1982-977-0   8-(o-Acetoxybenzyl)-3,3',4',5,7-penta-O-acetyl-(+)-catechin;
W   128059

648   C34 H56 O8 Si2   64623-57-0   SB-31-503-0   2-(1'-(2-Methoxy-4-methylphenoxy)-2'-(t-butyldimethylsilylox
y)ethyl)-5-(2-methoxy-4-methylphenoxy)-6-(t-butyldimethylsilyloxy)-tetrahydro-1,3-dioxin;   W   128060

648   C35 H52 O11   84507-61-9   F-38-2193-0   Paristerone-2,3,22,25-tetraacetate;   W   128061

648   C36 H28 P4 S2   EP-6195-0-0   TETRAHYDROTETRAPHENYLDIBENZO-1,2,5,6-TETRAPHOSPHOCIN DISULPHIDE;   W   65822

648   C36 D30 P6   HE-1982-0-0   HEXAPHOSPHACYCLOHEXAN, HEXAKIS(D5-PHENYL)-;   W   65823

648   C36 H40 O11   52748-45-5   H-56-2140-2   Pentherin II;;   PENTHERIN-II;   W/NBS   65824

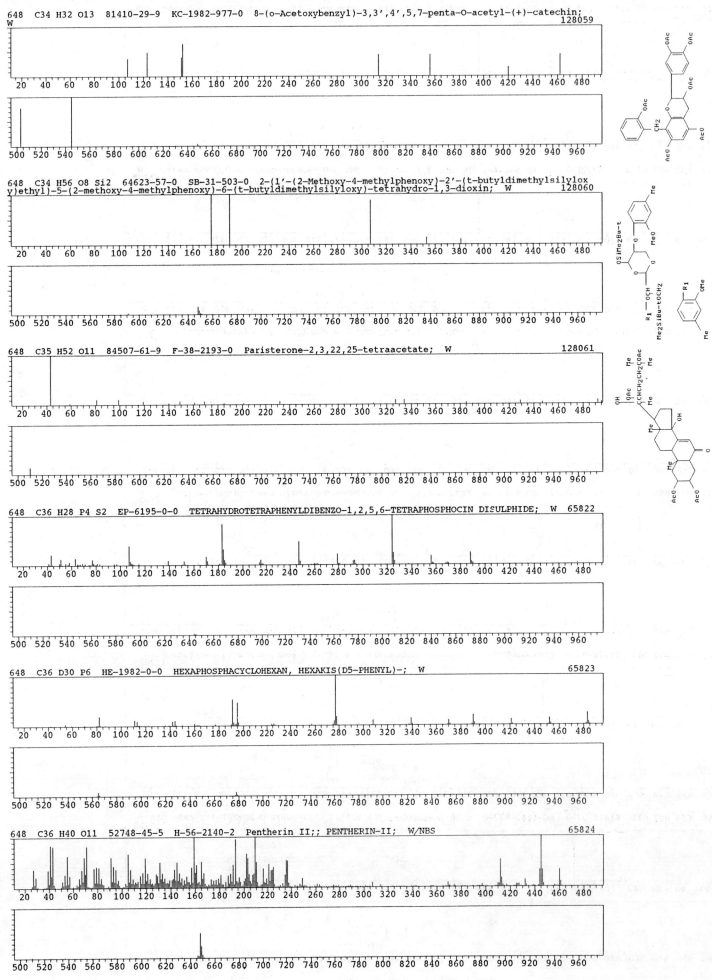

648 C36 H56 O10 60354-51-0 NS--27419-0-0 Cholestan-26-oic acid, 3,7,12,24-tetrakis(acetyloxy)-, methyl ester, (3α,5β,7α,12α)-;; NBS 65825

648 C39 H56 N2 O6 89759-09-1 H-67-246-0 O,O'-Dimethylkopsirachin; W 128062

648 C40 H40 O8 78076-85-4 K-114-769-0 7,7',8,8'-Tetrahydro-2,2,2',2',6,6,6',6',10,10'-decamethylcyclobuta[1,2-c:4,3-c']-cis-7,8,7'.8'-trans-7,7',8,8'-syn-bis[2H,6H,12H-benzo[1,2-b:3,4-b':5,6-b'']tripyran]-12,12'-dione; W 128063

648 C40 H41 I 77130-63-3 F-36-2439-0 2-[[2-[(2-Iodobicyclo[4.2.0]octa-1,3,5-trien-3-yl)methyl]bicyclo[4.2.0]octa-1,3,5-trien-2-yl]methyl]-3-[(3-pentylbicyclo[4.2.0]octa-1,3,5-trien-2-yl)methyl]bi- cyclo[4.2.0]octa-1,3,5-triene; W 128064

648 C41 H52 N4 O3 61542-15-2 K-109-3536-0 TETRAHYDRO-ACCEDINISIN; W 65826

648 C41 H52 N4 O3 56691-97-5 B-28-1820-0 16-Demethoxycarbonyldihydrovoacamine; W 128065

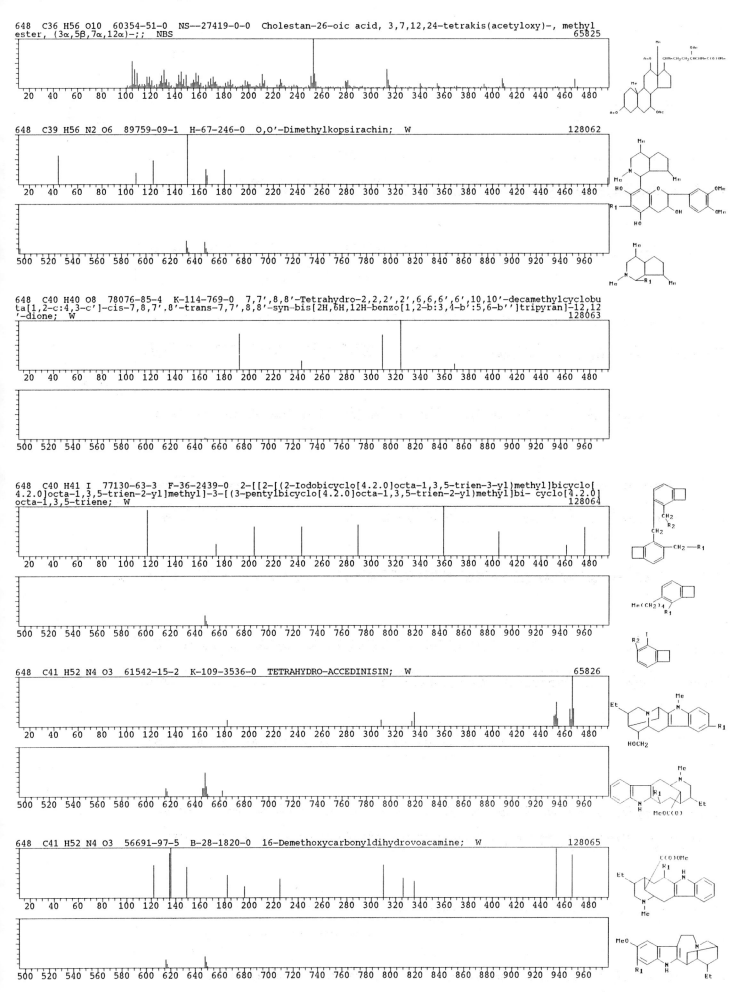

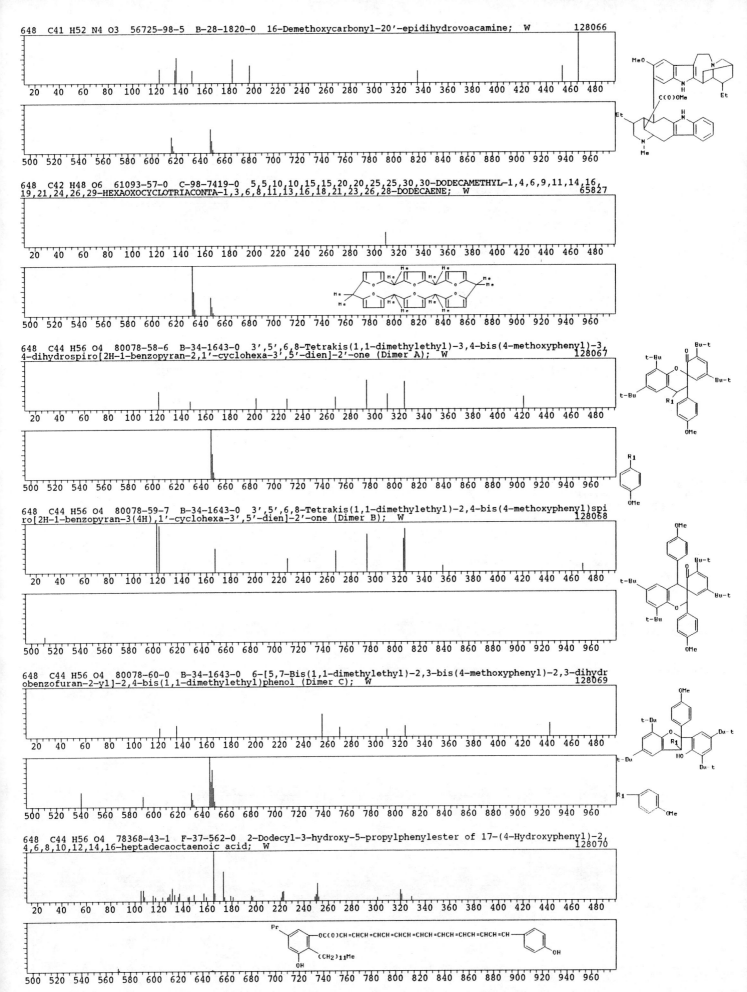

648   C41 H52 N4 O3   56725-98-5   B-28-1820-0   16-Demethoxycarbonyl-20'-epidihydrovoacamine;   W   128066

648   C42 H48 O6   61093-57-0   C-98-7419-0   5,5,10,10,15,15,20,20,25,25,30,30-DODECAMETHYL-1,4,6,9,11,14,16,19,21,24,26,29-HEXAOXOCYCLOTRIACONTA-1,3,6,8,11,13,16,18,21,23,26,28-DODECAENE;   W   65827

648   C44 H56 O4   80078-58-6   B-34-1643-0   3',5',6,8-Tetrakis(1,1-dimethylethyl)-3,4-bis(4-methoxyphenyl)-3,4-dihydrospiro[2H-1-benzopyran-2,1'-cyclohexa-3',5'-dien]-2'-one (Dimer A);   W   128067

648   C44 H56 O4   80078-59-7   B-34-1643-0   3',5',6,8-Tetrakis(1,1-dimethylethyl)-2,4-bis(4-methoxyphenyl)spiro[2H-1-benzopyran-3(4H),1'-cyclohexa-3',5'-dien]-2'-one (Dimer B);   W   128068

648   C44 H56 O4   80078-60-0   B-34-1643-0   6-[5,7-Bis(1,1-dimethylethyl)-2,3-bis(4-methoxyphenyl)-2,3-dihydrobenzofuran-2-yl]-2,4-bis(1,1-dimethylethyl)phenol (Dimer C);   W   128069

648   C44 H56 O4   78368-43-1   F-37-562-0   2-Dodecyl-3-hydroxy-5-propylphenylester of 17-(4-Hydroxyphenyl)-2,4,6,8,10,12,14,16-heptadecaoctaenoic acid;   W   128070

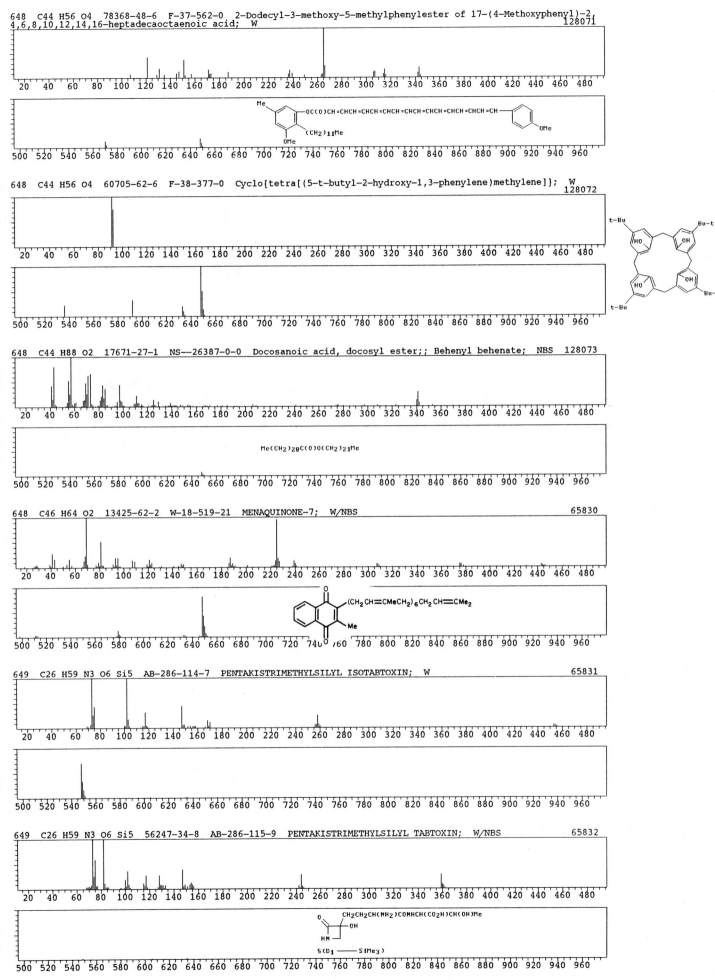

648   C44 H56 O4   78368-48-6   F-37-562-0   2-Dodecyl-3-methoxy-5-methylphenylester of 17-(4-Methoxyphenyl)-2, 4,6,8,10,12,14,16-heptadecaoctaenoic acid;   W   128071

Me
〔OC(O)CH=CHCH=CH=CHCH=CH=CHCH=CH=CHCH=CHCH=CH〕⟨4-C6H4-OMe⟩
(CH2)11Me
OMe

648   C44 H56 O4   60705-62-6   F-38-377-0   Cyclo{tetra[(5-t-butyl-2-hydroxy-1,3-phenylene)methylene]};   W
                                                                                                128072

648   C44 H88 O2   17671-27-1   NS--26387-0-0   Docosanoic acid, docosyl ester;;   Behenyl behenate;   NBS   128073

Me(CH2)28C(O)O(CH2)21Me

648   C46 H64 O2   13425-62-2   W-18-519-21   MENAQUINONE-7;   W/NBS   65830

(CH2CH=CMeCH2)6CH2CH=CMe2
Me
O

649   C26 H59 N3 O6 Si5   AB-286-114-7   PENTAKISTRIMETHYLSILYL ISOTABTOXIN;   W   65831

649   C26 H59 N3 O6 Si5   56247-34-8   AB-286-115-9   PENTAKISTRIMETHYLSILYL TABTOXIN;   W/NBS   65832

O   CH2CH2CH(NH2)CONHCH(CO2H)CH(OH)Me
                 OH
HN
5(D1——SIMe3)

649   C33 H35 N3 O7 S2   36600-91-6   NS-0-0-0   1-Naphthalenesulfonic acid, 5-(dimethylamino)-, 4-[2-[[[5-(di methylamino)-1-naphthalenyl]sulfonyl]amino]-1-hydroxyethyl]-2-methoxyphenyl ester;; BIS(1-DIMETHYLAMINONAPH THALENE-5-SULPHONYL)NORMETANEPHRINE;   W/NBS
128074

649   C33 H48 Br N O7   B-34-619-0   (20S,22R,25R)-26-Acetylamino-7α-bromo-6-oxo-5α-furostane-2β,3β-diol-diace tate;   W
128078

650   C16 H8 O9 Ru3   94472-19-2   H-67-1644-0   Octacarbonyl-.mu.-[1,3-η:4,5,C-η-(5,6-dimethylidene-1,3-cycloh exadiene)]-.mu.2-oxo-triangulo-triruthenium;   W
128075

650   C27 H38 O18   55887-84-8   AU-13-334-2   α-D-Glucopyranose, 3-O-methyl-, 1,2,4-triacetate, anhydride with α-D-glucopyranose 1,2,4,6-tetraacetate;; 3'-O-METHYL-DI-α-GLUCOSE-3,6'-ANHYDRID-HEPTA-ACETATE;   W/NBS
65836

650   C30 H20 Fe2 O6 P2   15024-71-2   AD-0-1589-0   Iron, hexacarbonylbis[.mu.-(diphenylphosphino)]di-;; DI-. MU.-DIPHENYL-PHOSPHINO-BIS(TRICARBONYLIRON);   W/NBS
65837

650   C31 H70 O6 Si4   35437-08-2   NS--27408-0-0   Octadecanoic acid, 14,15,17,18-tetrakis[(trimethylsilyl)ox y]-, methyl ester;;   NBS
65844

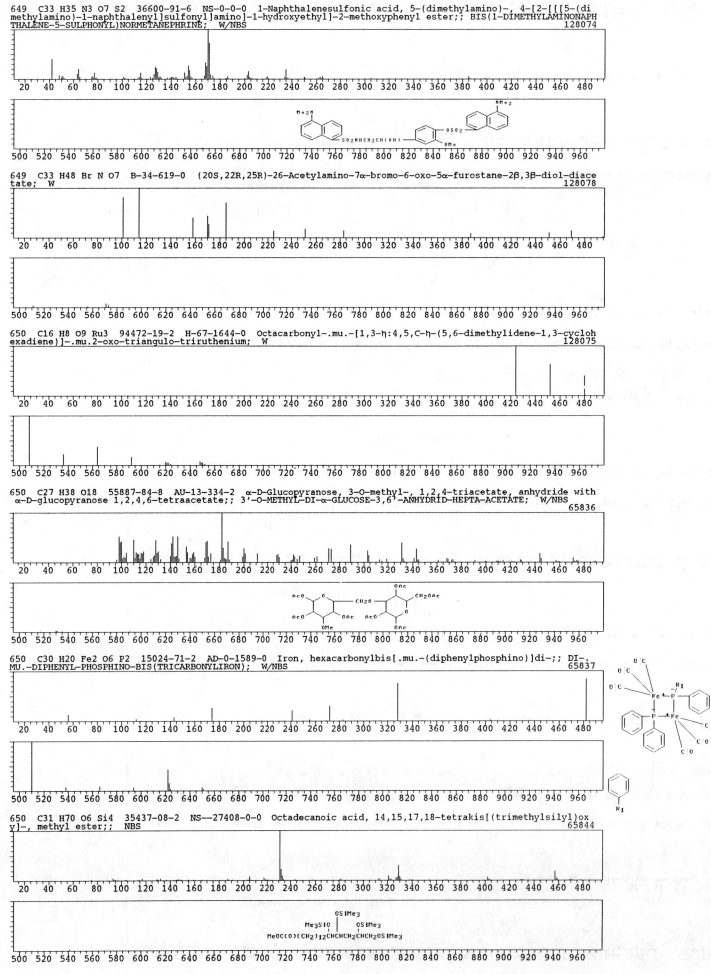

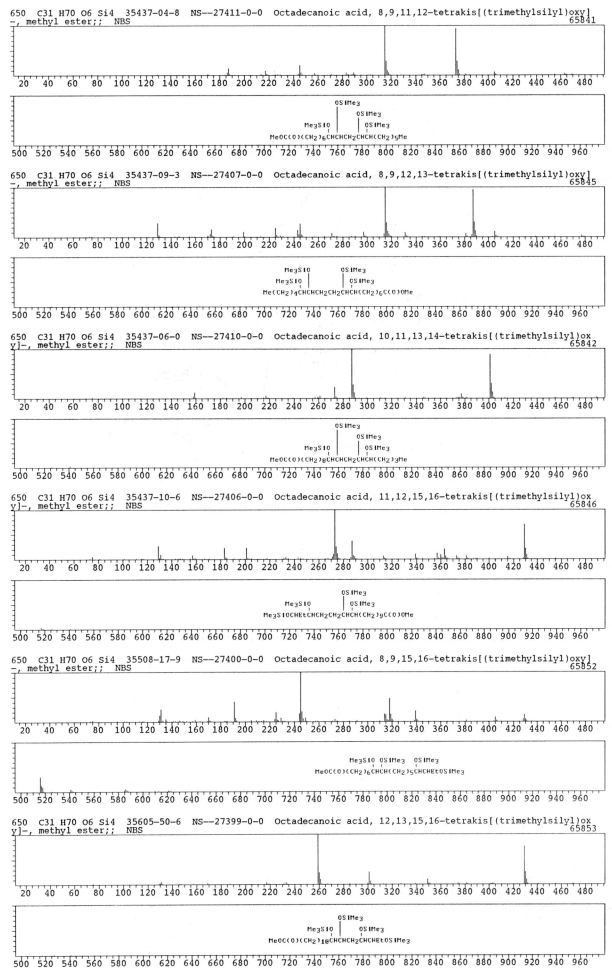

650 C31 H70 O6 Si4 35437-04-8 NS--27411-0-0 Octadecanoic acid, 8,9,11,12-tetrakis[(trimethylsilyl)oxy]
-, methyl ester;; NBS   65841

OSiMe3
Me3SiO | OSiMe3
MeOC(O)(CH2)6CHCHCH2CHCH(CH2)5Me

650 C31 H70 O6 Si4 35437-09-3 NS--27407-0-0 Octadecanoic acid, 8,9,12,13-tetrakis[(trimethylsilyl)oxy]
-, methyl ester;; NBS   65845

Me3SiO   OSiMe3
Me3SiO | OSiMe3
Me(CH2)4CHCHCH2CH2CHCH(CH2)6C(O)OMe

650 C31 H70 O6 Si4 35437-06-0 NS--27410-0-0 Octadecanoic acid, 10,11,13,14-tetrakis[(trimethylsilyl)ox
y]-, methyl ester;; NBS   65842

OSiMe3
OSiMe3
Me3SiO | OSiMe3
MeOC(O)(CH2)8CHCHCH2CHCH(CH2)3Me

650 C31 H70 O6 Si4 35437-10-6 NS--27406-0-0 Octadecanoic acid, 11,12,15,16-tetrakis[(trimethylsilyl)ox
y]-, methyl ester;; NBS   65846

OSiMe3
Me3SiO | OSiMe3
Me3SiOCHEtCHCH2CH2CHCH(CH2)9C(O)OMe

650 C31 H70 O6 Si4 35508-17-9 NS--27400-0-0 Octadecanoic acid, 8,9,15,16-tetrakis[(trimethylsilyl)oxy]
-, methyl ester;; NBS   65852

Me3SiO OSiMe3 OSiMe3
MeOC(O)(CH2)6CHCH(CH2)5CHCHEtOSiMe3

650 C31 H70 O6 Si4 35605-50-6 NS--27399-0-0 Octadecanoic acid, 12,13,15,16-tetrakis[(trimethylsilyl)ox
y]-, methyl ester;; NBS   65853

OSiMe3
Me3SiO | OSiMe3
MeOC(O)(CH2)10CHCHCH2CHCHEtOSiMe3

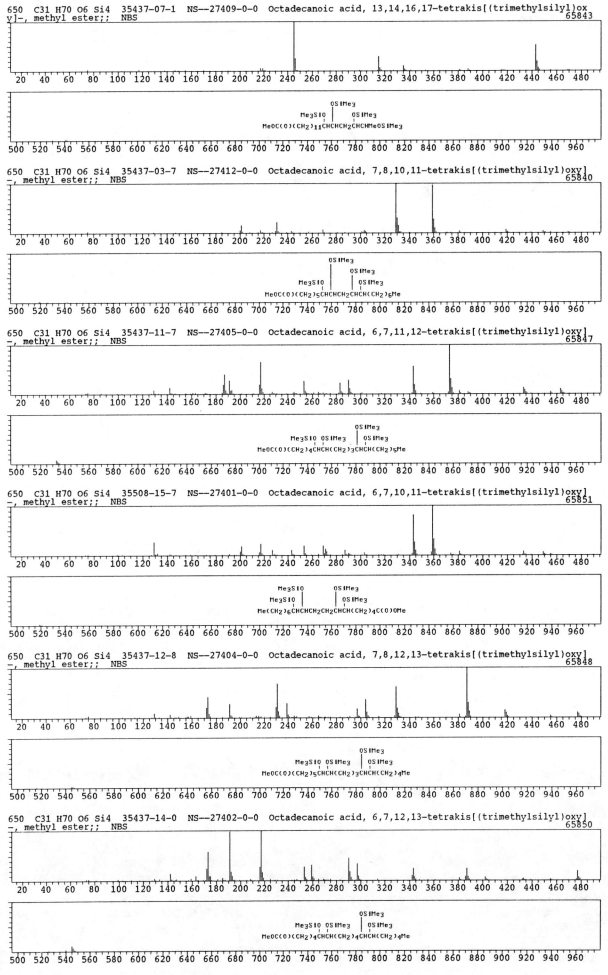

650 C31 H70 O6 Si4 35437-07-1 NS--27409-0-0 Octadecanoic acid, 13,14,16,17-tetrakis[(trimethylsilyl)oxy]-, methyl ester;; NBS   65843

OSiMe3
Me3SiO | OSiMe3
MeOC(O)(CH2)11CHCHCH2CHCHMeOSiMe3

650 C31 H70 O6 Si4 35437-03-7 NS--27412-0-0 Octadecanoic acid, 7,8,10,11-tetrakis[(trimethylsilyl)oxy]-, methyl ester;; NBS   65840

OSiMe3
Me3SiO | OSiMe3
MeOC(O)(CH2)5CHCHCH2CHCH(CH2)6Me

650 C31 H70 O6 Si4 35437-11-7 NS--27405-0-0 Octadecanoic acid, 6,7,11,12-tetrakis[(trimethylsilyl)oxy]-, methyl ester;; NBS   65847

OSiMe3
Me3SiO OSiMe3 | OSiMe3
MeOC(O)(CH2)4CHCH(CH2)3CHCH(CH2)5Me

650 C31 H70 O6 Si4 35508-15-7 NS--27401-0-0 Octadecanoic acid, 6,7,10,11-tetrakis[(trimethylsilyl)oxy]-, methyl ester;; NBS   65851

Me3SiO   OSiMe3
Me3SiO | OSiMe3
Me(CH2)6CHCHCH2CH2CHCH(CH2)4C(O)OMe

650 C31 H70 O6 Si4 35437-12-8 NS--27404-0-0 Octadecanoic acid, 7,8,12,13-tetrakis[(trimethylsilyl)oxy]-, methyl ester;; NBS   65848

OSiMe3
Me3SiO OSiMe3 | OSiMe3
MeOC(O)(CH2)5CHCH(CH2)3CHCH(CH2)4Me

650 C31 H70 O6 Si4 35437-14-0 NS--27402-0-0 Octadecanoic acid, 6,7,12,13-tetrakis[(trimethylsilyl)oxy]-, methyl ester;; NBS   65850

OSiMe3
Me3SiO OSiMe3 | OSiMe3
MeOC(O)(CH2)4CHCH(CH2)4CHCH(CH2)4Me

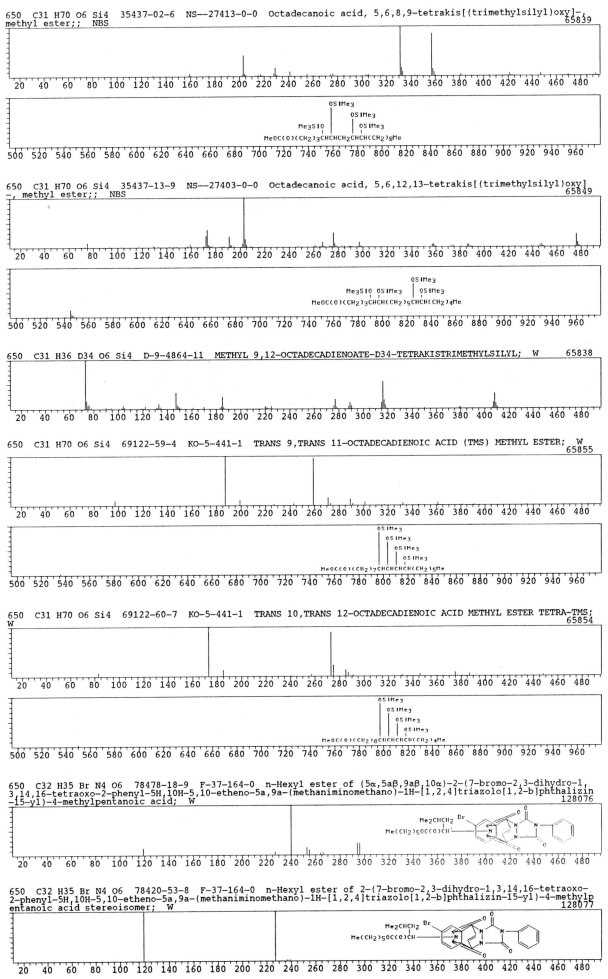

650 C31 H70 O6 Si4   35437-02-6   NS--27413-0-0   Octadecanoic acid, 5,6,8,9-tetrakis[(trimethylsilyl)oxy]-
methyl ester;; NBS   65839

OSIMe3
                                OSIMe3
                Me3SIO   OSIMe3
                    | |
        MeOC(O)(CH2)3CHCHCH2CHCH(CH2)8Me

650 C31 H70 O6 Si4   35437-13-9   NS--27403-0-0   Octadecanoic acid, 5,6,12,13-tetrakis[(trimethylsilyl)oxy]
-, methyl ester;; NBS   65849

                                        OSIMe3
                Me3SIO OSIMe3   OSIMe3
                    | | |
        MeOC(O)(CH2)3CHCH(CH2)5CHCH(CH2)4Me

650 C31 H36 D34 O6 Si4   D-9-4864-11   METHYL 9,12-OCTADECADIENOATE-D34-TETRAKISTRIMETHYLSILYL; W   65838

650 C31 H70 O6 Si4   69122-59-4   KO-5-441-1   TRANS 9,TRANS 11-OCTADECADIENOIC ACID (TMS) METHYL ESTER; W
                                                                                                 65855

                                OSIMe3
                                OSIMe3
                                 OSIMe3
                                  OSIMe3
                                   |
            MeOC(O)(CH2)7CHCHCHCH(CH2)5Me

650 C31 H70 O6 Si4   69122-60-7   KO-5-441-1   TRANS 10,TRANS 12-OCTADECADIENOIC ACID METHYL ESTER TETRA-TMS;
W   65854

                                OSIMe3
                                OSIMe3
                                 OSIMe3
                                  OSIMe3
                                   |
            MeOC(O)(CH2)0CHCHCHCH(CH2)4Me

650 C32 H35 Br N4 O6   78478-18-9   F-37-164-0   n-Hexyl ester of (5α,5aβ,9aβ,10α)-2-(7-bromo-2,3-dihydro-1,
3,14,16-tetraoxo-2-phenyl-5H,10H-5,10-etheno-5a,9a-(methaniminomethano)-1H-[1,2,4]triazolo[1,2-b]phthalizin
-15-yl)-4-methylpentanoic acid; W   128076

650 C32 H35 Br N4 O6   78420-53-8   F-37-164-0   n-Hexyl ester of 2-(7-bromo-2,3-dihydro-1,3,14,16-tetraoxo-
2-phenyl-5H,10H-5,10-etheno-5a,9a-(methaniminomethano)-1H-[1,2,4]triazolo[1,2-b]phthalizin-15-yl)-4-methylp
entanoic acid stereoisomer; W   128077

650   C32 H35 Br N4 O6   F-35-1326-0   (SYN)-12-AZA-3-BROMO-7,10-(2,5-DIOXO-1-PHENYL-1,3,4-TRIAZOLIDIN-3,4-DIYL)-12-(4-METHYL-1-N-HEXYLOXEYCARBONYLBUTYL)(4.4.3)PROPELLA2,4,8-TRIENE;   W   65833

650   C32 H35 Br N4 O6   F-35-1326-0   (ANTI)-12-AZA-3-BROMO-7,10-(2,5-DIOXO-1-PHENYL-1,3,4-TRIAZOLIDIN-3,4-DIYL)-12-(4-METHYL-1-N-HEXYLOXYCARBONYLBUTYL)[4.9.3]PROPELLA-2,4,8-TRIENE;   W   65834

650   C32 H58 N2 O6 Si3   57326-06-4   EP-8335-0-0   Pregn-4-ene-3,11,20-trione, 6,17,21-tris[(trimethylsilyl)oxy]-, 3,20-bis(O-methyloxime), (6β)-;;   4-PREGNENE-6β,17α,21-TRIOL-3,11,20-TRIONEMO TMS;   W/NBS   65858

650   C33 H46 O13   72101-43-0   NS--27397-0-0   β-D-Glucopyranoside, [5-[2-(2,5-dihydro-2-oxo-3-furanyl)ethyl]decahydro-1,4a-dimethyl-6-oxo-1-naphthalenyl]methyl, tetraacetate, [1R-(1α,4aα,5α,8aβ)]-;;   β-D-Glucopyranoside, [5-[2-(2,5-dihydro-2-oxo-3-furanyl)ethyl]decahydro-1,4a-dimethyl-6-oxo-1-naphthalenyl]methyl, tetraacetate, [1R-(1.alp;   NBS   65860

650   C33 H46 O13   34918-06-4   F-27-5086-6   NEOANDROGRAPHOLIDE NORKETONE ACETATE;   W   65859

650   C35 H22 Fe2 O6   74421-53-7   NS--27396-0-0   Iron, hexacarbonyl[.mu.-[(1,2,3,4-η:1,5-η)-1,2,3,4-tetraphenyl-1,3-pentadiene-1,5-diyl]]di-, (Fe-Fe);;   NBS   65861

650   C37 H62 O9   BS-2-21-0   Tokoronin permethylate;   W   128079

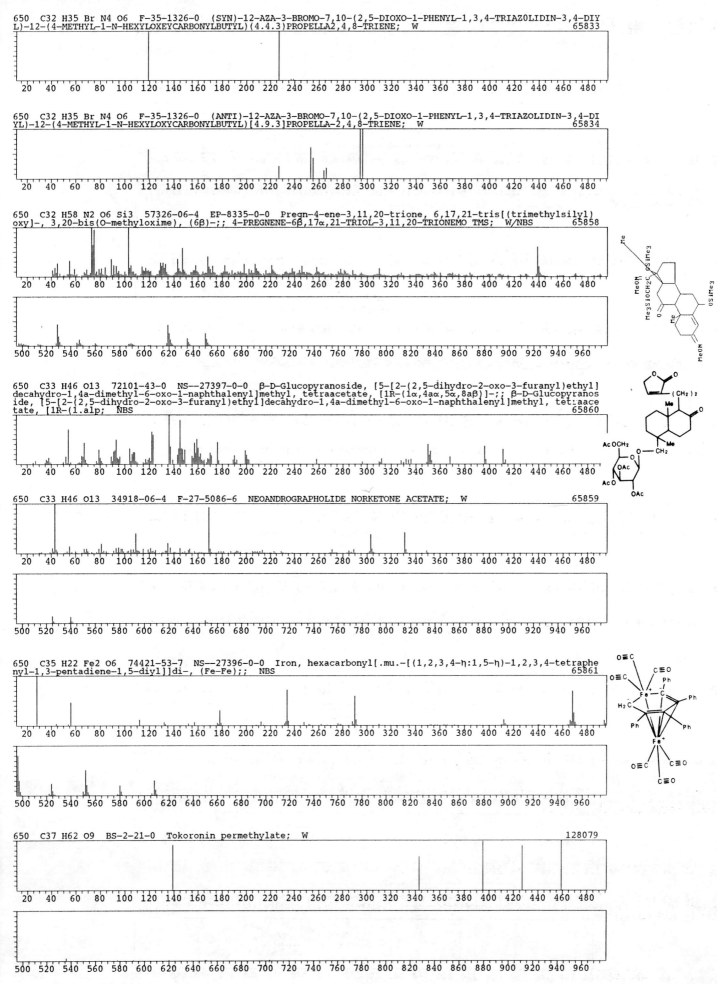

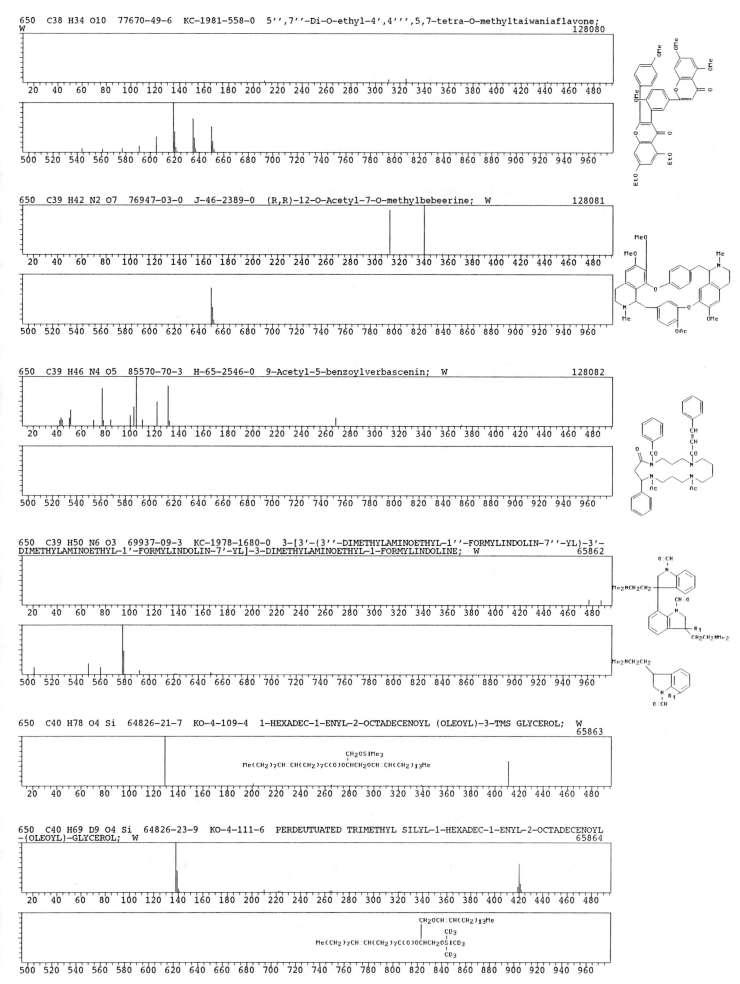

650   C38 H34 O10   77670-49-6   KC-1981-558-0   5'',7''-Di-O-ethyl-4',4''',5,7-tetra-O-methyltaiwaniaflavone;
W   128080

650   C39 H42 N2 O7   76947-03-0   J-46-2389-0   (R,R)-12-O-Acetyl-7-O-methylbebeerine;   W
                                                                                           128081

650   C39 H46 N4 O5   85570-70-3   H-65-2546-0   9-Acetyl-5-benzoylverbascenin;   W
                                                                                           128082

650   C39 H50 N6 O3   69937-09-3   KC-1978-1680-0   3-[3'-(3''-DIMETHYLAMINOETHYL-1''-FORMYLINDOLIN-7''-YL)-3'-
DIMETHYLAMINOETHYL-1'-FORMYLINDOLIN-7'-YL]-3-DIMETHYLAMINOETHYL-1-FORMYLINDOLINE;   W   65862

650   C40 H78 O4 Si   64826-21-7   KO-4-109-4   1-HEXADEC-1-ENYL-2-OCTADECENOYL (OLEOYL)-3-TMS GLYCEROL;   W
                                                                                           65863

650   C40 H69 D9 O4 Si   64826-23-9   KO-4-111-6   PERDEUTUATED TRIMETHYL SILYL-1-HEXADEC-1-ENYL-2-OCTADECENOYL
-(OLEOYL)-GLYCEROL;   W   65864

650   C42 H22 N2 O6   IC-1299-0-0   1,1',5,1''-Trianthrimide;   W   128083

650   C43 H38 O6   C-98-5384-0   2,4,4',6-TETRAKIS(BENZYLOXY)-2'-METHYL-6'-METHYL-BENZOPHENONE;   W   65865

650   C43 H70 O4   K-112-2008-0   MIX OF 2-DECYL-3-MEO-5-(4-ME PENTYL)PHESTER OF AND 3-MEO-2-(9-ME DECYL)-5-PE
NTYL PHESTER OF 13-(4-MEOPH)TRIDECANCARBOXYLIC ACID;   W   65866

650   C44 H36 N2 Ni   HE-1982-0-0   NICKEL, (TETRAPHENYLCYCLOBUTADIEN)[DIACETYLBIS(ANIL)];   W   65867

650   C44 H58 O4   71246-59-8   K-112-2006-0   2,4,6,8,10,12,14-Pentadecaheptaenoic acid, 15-(4-methoxyphenyl)
-, 2-decyl-3-methoxy-5-pentylphenyl ester;; 2-DECYL-3-METHOXY-5-PENTYLPHENYLESTER OF 15-(4-METHOXYPHENYL)-
2,4,6,8,10,12,14-PENTADECAHEPTAENCARBOXYLIC ACID;   W/NBS   65868

650   C45 H78 O2   303-43-5   NS--26362-0-0   Cholest-5-en-3-ol (3β)-, 9-octadecenoate, (Z)-;   NBS   128084

Str 1157

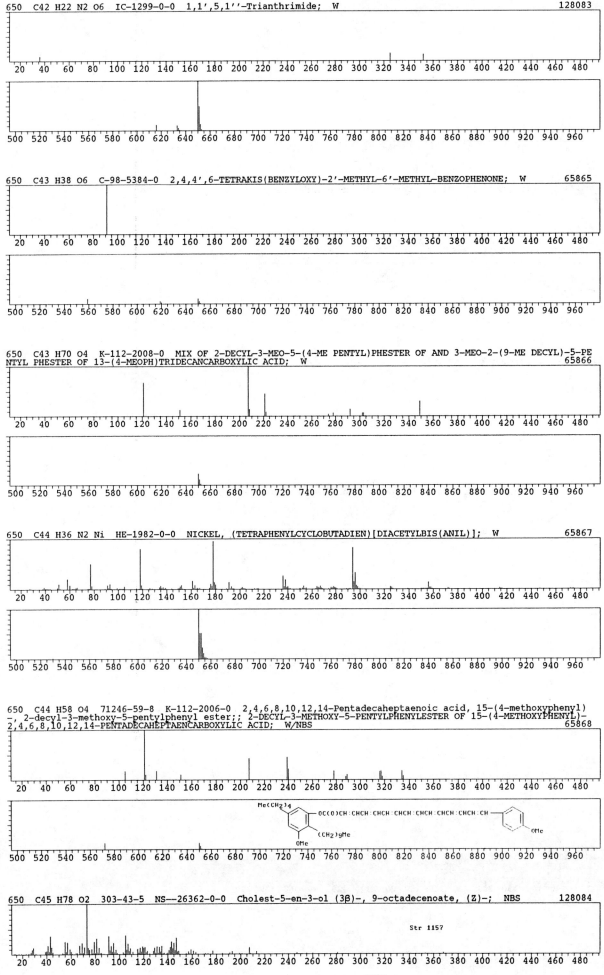

651 C22 H9 Br4 N O3 IC-1299-0-0 Tetrabromo-2-(3-hydroxy-1,2-dihydro-quinol-2-ylidene)-2,3-dihydro-1H-ben
z[f]indene; W   128085

651 C29 H37 N3 O14 83239-13-8 KO-9-258-2 (S)-N,N'-Bis-[2-acetoxy-1-(acetoxymethyl)ethyl]-5-[(2-acetoxy-
1-amino]-1,3-benzenedicarboxyamide; W   128086

651 C33 H49 N O12 CD-284-0-0 Tetraacetylpseudaconine; W   128087

651 C34 H61 N3 O5 S Si 89578-76-7 C-106-3244-0 (2α,6β(S*))-N-(4-Hydroxybutyl)-1-(3-((4-methylphenyl)sul
fonyl)-amino)propyl)-6-(2-((1,1-dimethylethyl)dimethylsilyl)heptyl)-1,2,3,6-tetrahydro-2-pyridineacetamide;
W   128088

651 C35 H65 N5 O6 55556-78-0 F-25-5786-0 Glycine, N-[N5-[imino[(1-oxodecyl)amino]methyl]-N2-(1-oxodecy
l)-N-(1-oxohexyl)-L-ornithyl]-, methyl ester;; N(α)-HEXANOYL-N,N(ε)BISDECANOYLARGINYLGLYCINE METHYL ESTER;
W/NBS   65869

651 C35 H69 N O4 Si3 67877-87-6 KO-5-533-1 11-HEXAMETHYLENIMINAL,9-ENOL PGA1 (TMS)3; W   65870

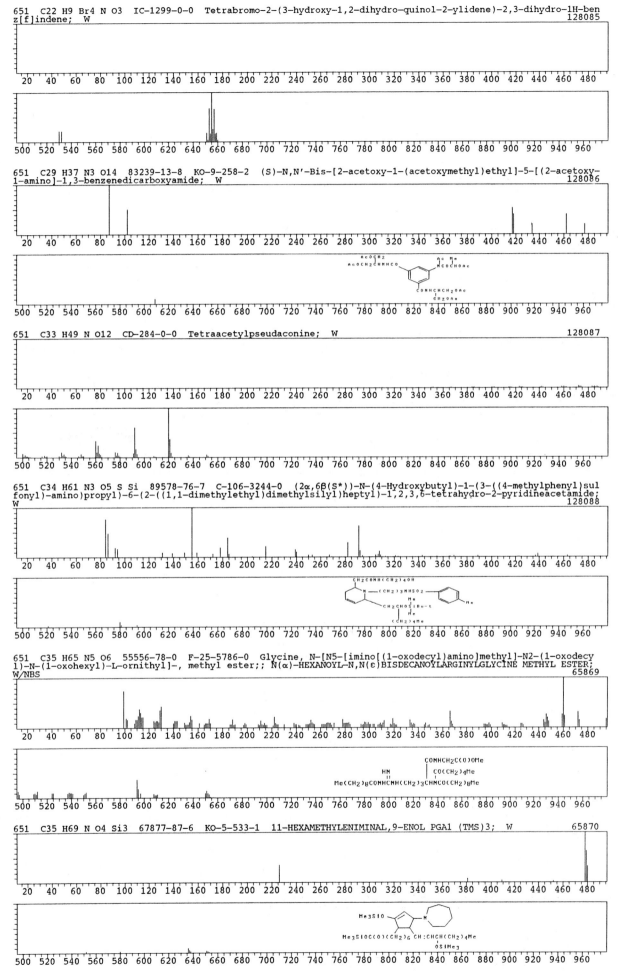

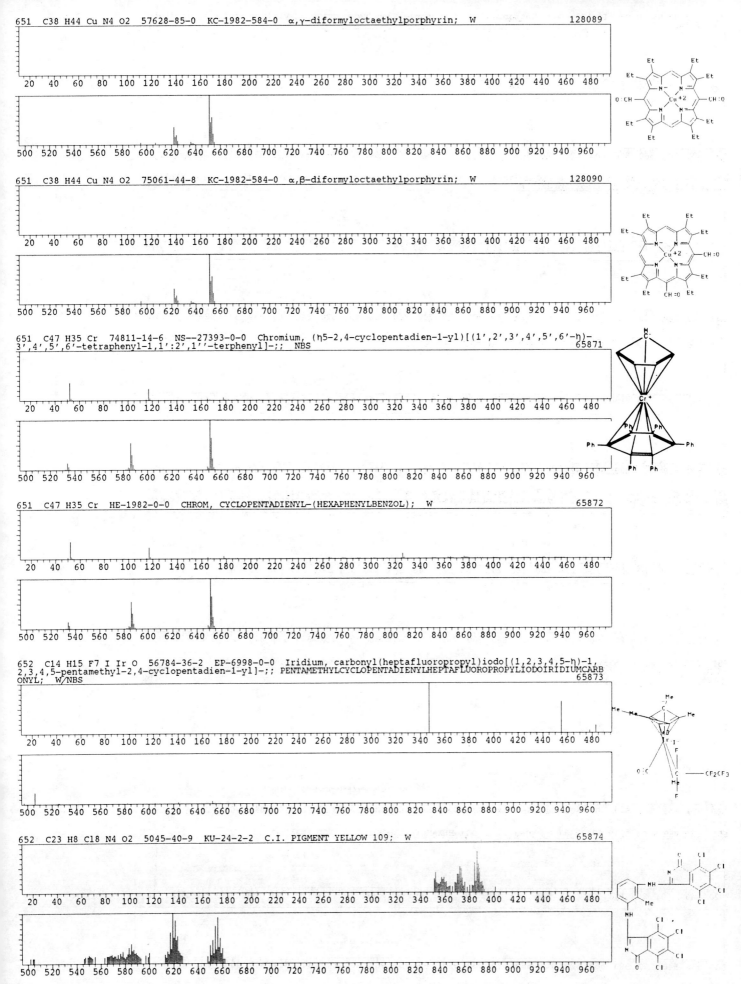

651 C38 H44 Cu N4 O2 57628-85-0 KC-1982-584-0 α,γ-diformyloctaethylporphyrin; W   128089

651 C38 H44 Cu N4 O2 75061-44-8 KC-1982-584-0 α,β-diformyloctaethylporphyrin; W   128090

651 C47 H35 Cr 74811-14-6 NS--27393-0-0 Chromium, (η5-2,4-cyclopentadien-1-yl)[(1',2',3',4',5',6'-η)-
3',4',5',6'-tetraphenyl-1,1':2',1''-terphenyl]-;; NBS   65871

651 C47 H35 Cr HE-1982-0-0 CHROM, CYCLOPENTADIENYL-(HEXAPHENYLBENZOL); W   65872

652 C14 H15 F7 I Ir O 56784-36-2 EP-6998-0-0 Iridium, carbonyl(heptafluoropropyl)iodo[(1,2,3,4,5-η)-1,
2,3,4,5-pentamethyl-2,4-cyclopentadien-1-yl]-;; PENTAMETHYLCYCLOPENTADIENYLHEPTAFLUOROPROPYLIODOIRIDIUMCARB
ONYL; W/NBS   65873

652 C23 H8 Cl8 N4 O2 5045-40-9 KU-24-2-2 C.I. PIGMENT YELLOW 109; W   65874

652   C24 H21 F12 N6 P   70758-17-7   I-57-1016-0   TRIS(4-N,N-DIMETHYLHYDRAZINO-2,3,5,6-TETRAFLUOROPHENYL)PHOS
PHINE; W   65875

652   C27 H32 N4 O15   59957-65-2   K-114-1619-0   1,5-Dihydro-1,5-bis(2,3,5-tri-O-acetyl-β-D-ribofuranosyl)-4H
-pyrazolo[3,4-d]pyrimidin-4-one; W   128091

652   C28 H20 Cl4 N2 O8   84251-56-9   KC-1982-2314-0   3,3''',5,5'''-Tetrachloro-4,4'',4''',6'-tetramethoxy-
2',2''-dinitro-m-quaterphenyl; W   128092

652   C30 H37 Br O11   C-104-6422-0   2,6-Diacetoxy-7,8-methylidene-8-acetoxymethyl-4-oxo-10-methyl-3-(2-meth
yl-2-(2-acetoxy-1-bromoethyl)-4,4-dimethoxy-1-butyl)perhydronaphthalene; W   128093

652   C33 H64 O5 Si4   AH-106-1423-37   3α,17α,21-TRIHYDROXY-5β-PREGNAN-11,12-DIONE, TETRA-TMS (NOTIDENTIFIED)
; W   65876

652   C34 H60 N4 O8   AT-64-2088-12   SPORIDESMOLIDE II; W   65877

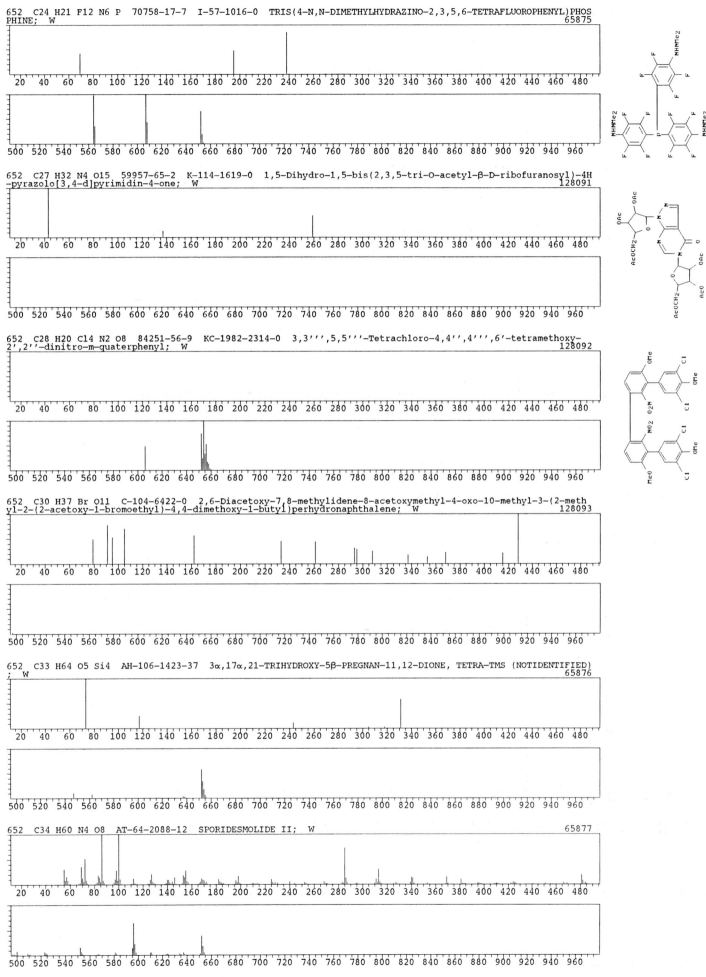

652   C34 H60 N4 O8   AT-65-2802-21   SPORIDESMOLIDE ANALOG; W   65878

652   C36 H72 O4 Si3   KO-4-26-8   5β-PREGNAN-3α,17α,20α,21-TETROL 3,20,21 TRIS DIMETHYLPROPYLSILYL ETHER; W
                                                                                                              65879

652   C36 H72 O4 Si3   KO-4-29-14   5β-PREGNAN-3α,17α,20α,21-TETROL-3,17,21-TRIS(DIMETHYL-N-PROPYLSILYL)ETHER;
W   65880

652   C37 H40 N4 O7   62786-90-7   KC-1976-2500-0   2-FORMYL-4,6,7-TRIS-(2-METHOXYCARBONYLETHYL)-1,3,5,8-TETRA
METHYLPORPHIN; W   65881

652   C37 H40 N4 O7   62786-91-8   KC-1976-2500-0   4-FORMYL-2,6,7-TRIS-(2-METHOXYCARBONYLETHYL)-1,3,5,8-TETRA
METHYLPORPHIN; W   65882

652   C38 H24 Cl4 O2   79491-41-1   B-34-910-0   6,7,8,9-Tetrachloro-3,4,4a,10a-tetraphenyl-2a,4a,10a,10b-tetra
hydrobenzo[b]cyclobuta[3,4-f][1,4]benzodioxin; W   128094

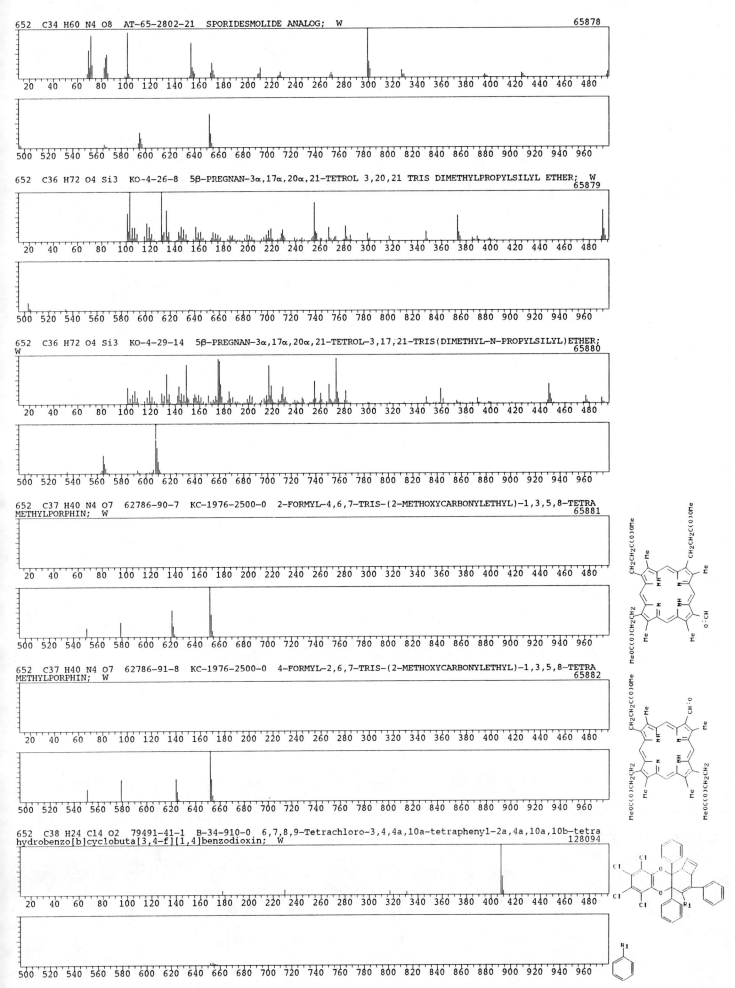

- 5960 -

652   C40 H80 O4 Si 61540-87-2 D3-1980-182-9 TMS ETHER OF 1-HEXADECYL-2-OLEOYLGLYCEROL; W   65883

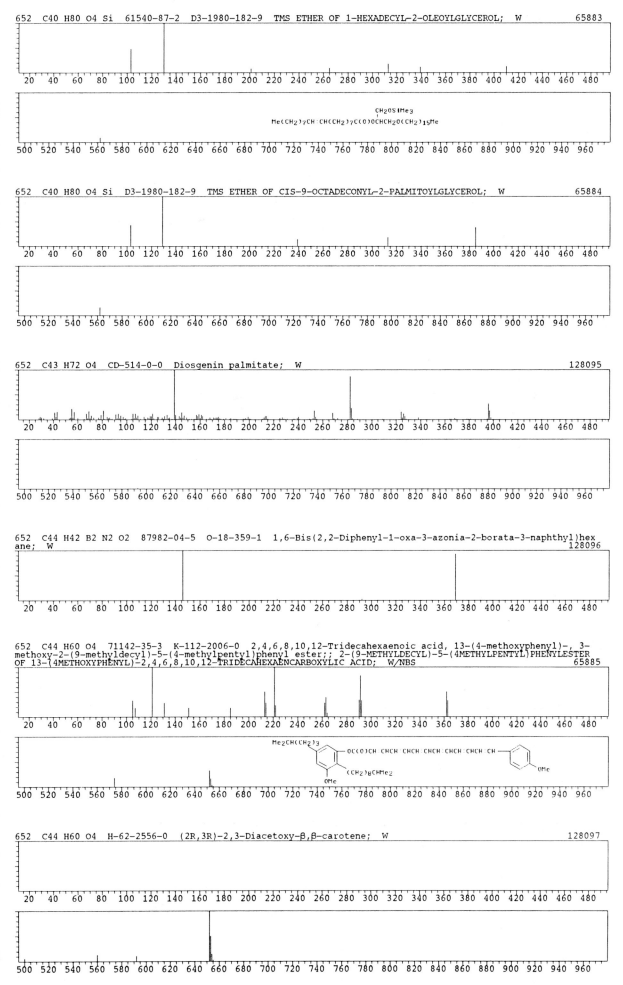

CH2OSiMe3
|
Me(CH2)7CH:CH(CH2)7C(O)OCHCH2O(CH2)15Me

652   C40 H80 O4 Si D3-1980-182-9 TMS ETHER OF CIS-9-OCTADECONYL-2-PALMITOYLGLYCEROL; W   65884

652   C43 H72 O4 CD-514-0-0 Diosgenin palmitate; W   128095

652   C44 H42 B2 N2 O2 87982-04-5 O-18-359-1 1,6-Bis(2,2-Diphenyl-1-oxa-3-azonia-2-borata-3-naphthyl)hex
ane; W   128096

652   C44 H60 O4 71142-35-3 K-112-2006-0 2,4,6,8,10,12-Tridecahexaenoic acid, 13-(4-methoxyphenyl)-, 3-
methoxy-2-(9-methyldecyl)-5-(4-methylpentyl)phenyl ester;; 2-(9-METHYLDECYL)-5-(4METHYLPENTYL)PHENYLESTER
OF 13-(4METHOXYPHENYL)-2,4,6,8,10,12-TRIDECAHEXAENCARBOXYLIC ACID; W/NBS   65885

Me2CH(CH2)3
OC(O)CH:CHCH:CHCH:CHCH:CHCH:CHCH:CH——OMe
(CH2)8CHMe2
OMe

652   C44 H60 O4 H-62-2556-0 (2R,3R)-2,3-Diacetoxy-β,β-carotene; W   128097

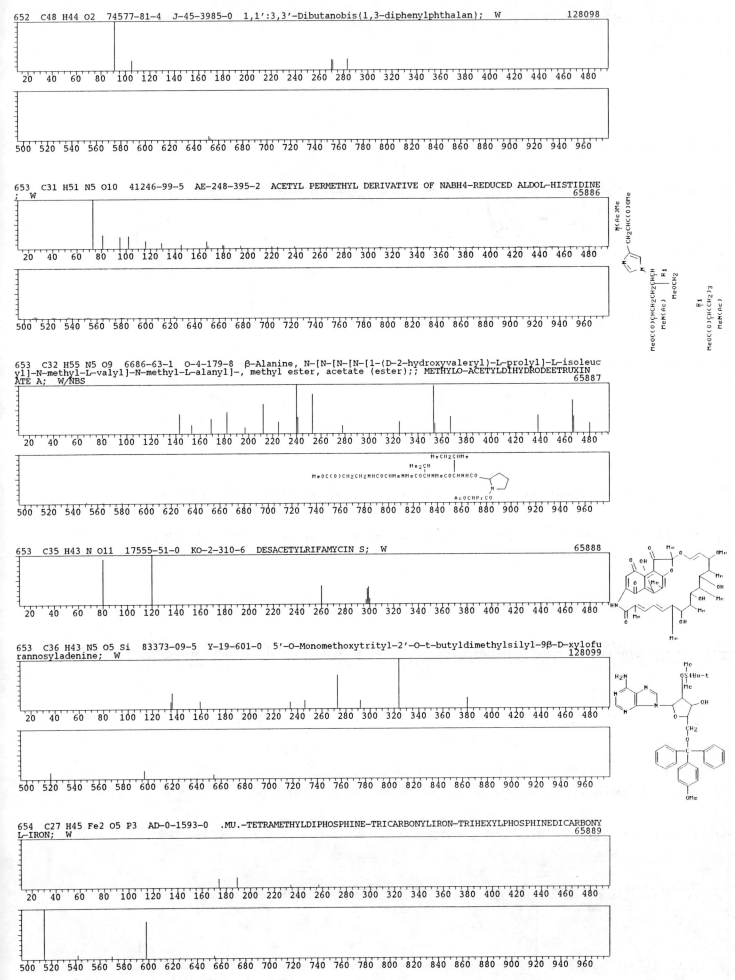

652   C48 H44 O2   74577-81-4   J-45-3985-0   1,1':3,3'-Dibutanobis(1,3-diphenylphthalan);   W   128098

653   C31 H51 N5 O10   41246-99-5   AE-248-395-2   ACETYL PERMETHYL DERIVATIVE OF NABH4-REDUCED ALDOL-HISTIDINE
;   W   65886

653   C32 H55 N5 O9   6686-63-1   O-4-179-8   β-Alanine, N-[N-[N-[N-[1-(D-2-hydroxyvaleryl)-L-prolyl]-L-isoleuc
yl]-N-methyl-L-valyl]-N-methyl-L-alanyl]-, methyl ester, acetate (ester);; METHYLO-ACETYLDIHYDRODEETRUXIN
ATE A;   W/NBS   65887

653   C35 H43 N O11   17555-51-0   KO-2-310-6   DESACETYLRIFAMYCIN S;   W   65888

653   C36 H43 N5 O5 Si   83373-09-5   Y-19-601-0   5'-O-Monomethoxytrityl-2'-O-t-butyldimethylsilyl-9β-D-xylofu
rannosyladenine;   W   128099

654   C27 H45 Fe2 O5 P3   AD-0-1593-0   .MU.-TETRAMETHYLDIPHOSPHINE-TRICARBONYLIRON-TRIHEXYLPHOSPHINEDICARBONY
L-IRON;   W   65889

654 C28 H50 P2 Rh2 86244-24-8 K-116-1461-0 .mu.-(Tetramethyldiphosphan)-bis[(ethene)(pentamethylcyclope
ntadienyl)rhodium(I)]; W
128100

654 C29 H54 N2 O7 Si4 77590-68-2 NS-0-0-0 D-Glucose, 3,4,5,6-tetrakis-O-(trimethylsilyl)-, O-methyloxim
e, 2-(1H-indole-3-acetate); W/NBS
128101

654 C29 H54 N2 O7 Si4 77590-69-3 NS-0-0-0 D-Glucose, 2,3,5,6-tetrakis-O-(trimethylsilyl)-, O-methyloxim
e, 4-(1H-indole-3-acetate); W/NBS
128102

654 C29 H54 N2 O7 Si4 77590-70-6 NS-0-0-0 D-Glucose, 2,3,4,5-tetrakis-O-(trimethylsilyl)-, O-methyloxim
e, 6-(1H-indole-3-acetate); W/NBS
128103

654 C30 H38 O16 78420-93-6 K-114-1645-0 Benzyl-2,3-di-O-acetyl-5-O-(2,3,4,6-tetra-O-acetyl-β-D-glucopyr
anosyl)-β-D-ribofuranoside; W
128104

654 C32 H26 N6 O6 S2 IC-1299-0-0 4,4'-Bis-(4-phenyl-2H-1,2,3-triazol-2-yl)-2-methylsulpho-stilbene; W
128105

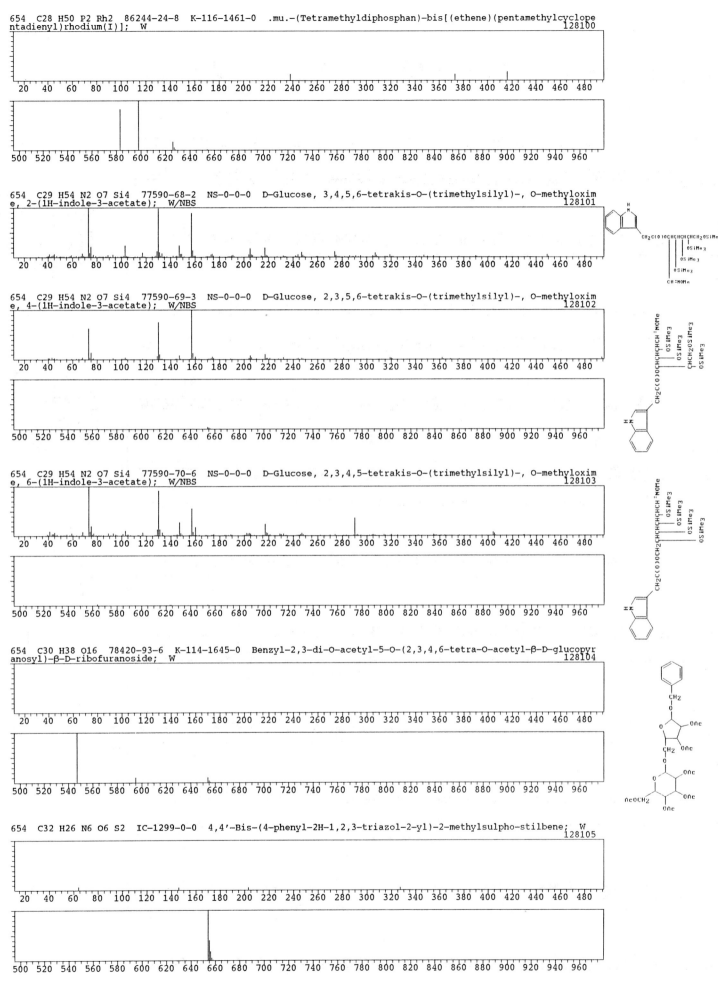

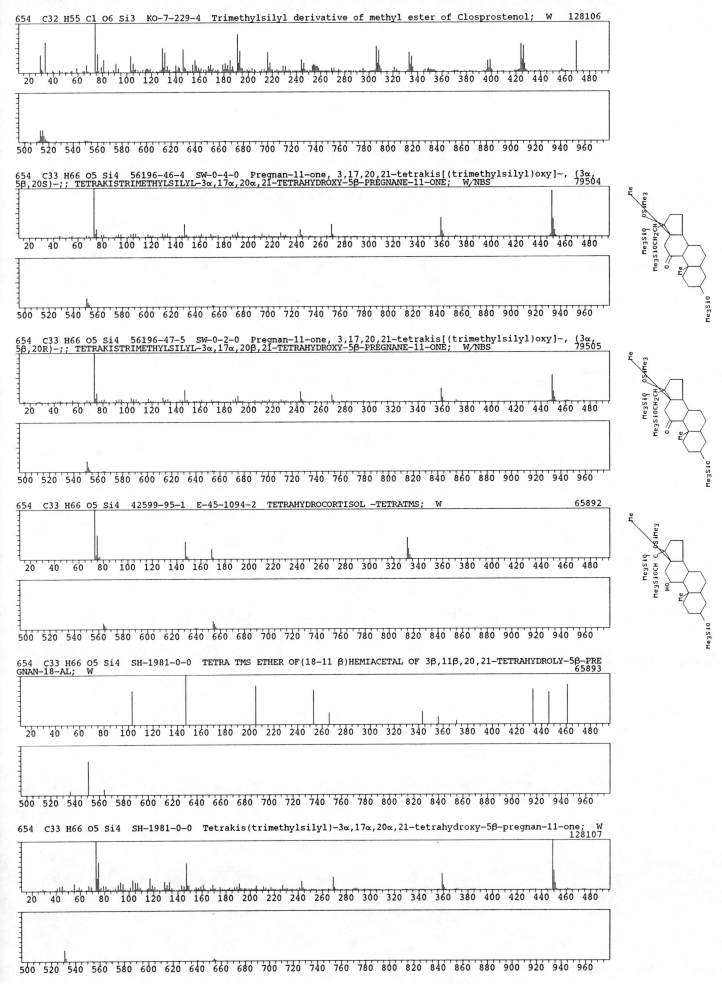

654   C32 H55 Cl O6 Si3   KO-7-229-4   Trimethylsilyl derivative of methyl ester of Closprostenol;   W   128106

654   C33 H66 O5 Si4   56196-46-4   SW-0-4-0   Pregnan-11-one, 3,17,20,21-tetrakis[(trimethylsilyl)oxy]-, (3α, 5β,20S)-;; TETRAKISTRIMETHYLSILYL-3α,17α,20α,21-TETRAHYDROXY-5β-PREGNANE-11-ONE;   W/NBS   79504

654   C33 H66 O5 Si4   56196-47-5   SW-0-2-0   Pregnan-11-one, 3,17,20,21-tetrakis[(trimethylsilyl)oxy]-, (3α, 5β,20R)-;; TETRAKISTRIMETHYLSILYL-3α,17α,20β,21-TETRAHYDROXY-5β-PREGNANE-11-ONE;   W/NBS   79505

654   C33 H66 O5 Si4   42599-95-1   E-45-1094-2   TETRAHYDROCORTISOL -TETRATMS;   W   65892

654   C33 H66 O5 Si4   SH-1981-0-0   TETRA TMS ETHER OF(18-11 β)HEMIACETAL OF 3β,11β,20,21-TETRAHYDROLY-5β-PREGNAN-18-AL;   W   65893

654   C33 H66 O5 Si4   SH-1981-0-0   Tetrakis(trimethylsilyl)-3α,17α,20α,21-tetrahydroxy-5β-pregnan-11-one;   W   128107

654   C35 H42 O12   CD-578-0-0   Trisflavaspidic acid; W   128108

654   C35 H42 O12   CD-579-0-0   Trisdesaspidin; W   128109

654   C37 H42 N4 O7   78437-53-3   I-59-783-0   2-Formyl-4-(2,2-dimethoxyethyl)deuteroporphyrin dimethyl ester;
W   128110

654   C37 H42 N4 O7   78422-42-1   I-59-783-0   2-(2,2-Dimethoxyethyl)-4-formyldeuteroporphyrin dimethyl ester;
W   128111

654   C37 H58 O6 Si2   86668-50-0   J-48-3262-0   6-(tert-Butyldimethylsiloxy)-10,11-(di-tert-butylsilylenediox
y)-12,12-(2',2'-dimethylpropane-1',3'-diyldioxy)-2-oxo-1,2,4a(R*),5,5a(S*),6(S*),11(R*),11a-(R*),12,12a(S*)
-decahydronaphthacene; W   128112

654   C38 H42 N2 O8   66965-62-6   C-100-2843-0   ETHYL 1-CYANO-3-SEC-BUTYOXY CARBONYL -3-(P-SEC-BUTYL-3'-(E)-
PROPENOATE)PHENYL-4-(P-ETHYL-2-CYANO-(E)-PROPENOATE)PHENYLCYCLOBUTANOATE; W   65894

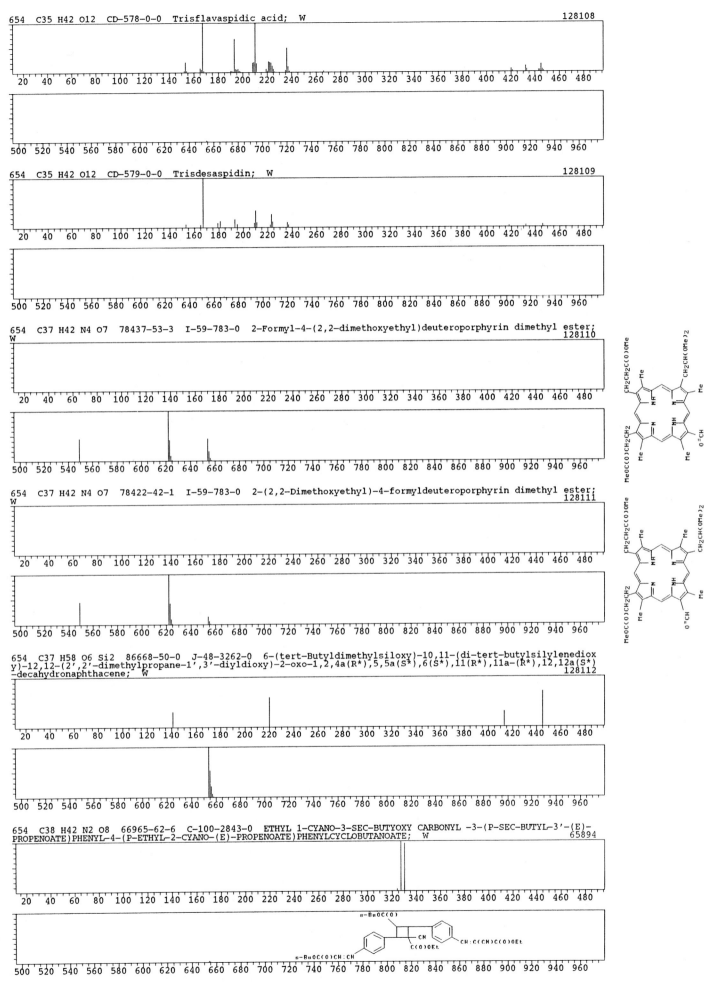

- 5965 -

654   C38 H42 N2 O8   59553-88-7   J-45-215-0   N'-Northalibrunine;   W   128113

654   C38 H42 N2 O8   72205-62-0   J-45-215-0   Epinorthalibrunine;   W   128114

654   C38 H45 Cl N5 O3   Y-17-485-0   2-(2-Chloroethyl)-4-ethyl-6-methoxycarbonyl-7-(3-pyrrolidinylcarbonylpro
pyl)-1,3,5,8-tetramethylporphyrin;   W   128115

654   C38 H46 N4 O6   KC-1978-851-0   1,3,8-TRIMETHYL-4,5-DIETHYL-2-(1-HYDROXYETHYL)-6-C-(1-OXO ETHYLENE)-7,8-
DIHYDRO-7-(2-METHOXYCARBONYLETHYL-8'-ACETYLBIBLITRIENE AND HOMOLOGUES;   W   65895

654   C39 H46 N2 O7   64215-93-6   J-43-585-0   Thaliruginine;   W   128116

654   C39 H58 O6 S   83872-47-3   H-65-1511-0   (23RS)-25,25-Ethylendioxy-23-phenylsulfonyl-3β-[(tetrahydropyra
n-2-yl)oxy]-27-nor-5-cholestene;   W   128117

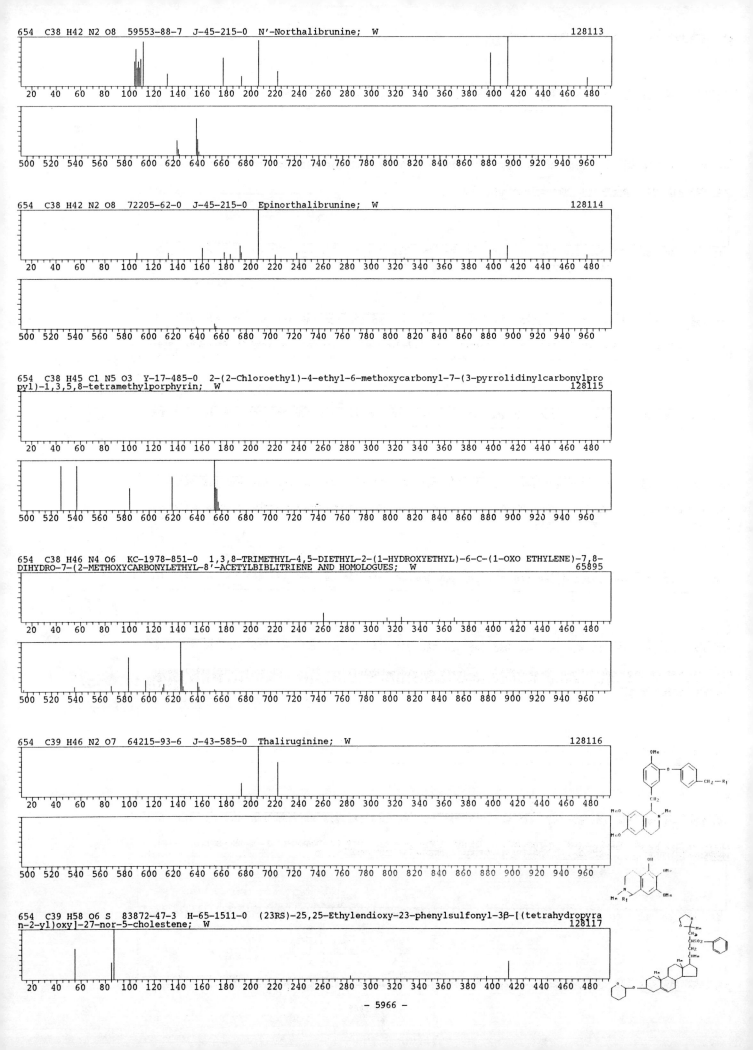

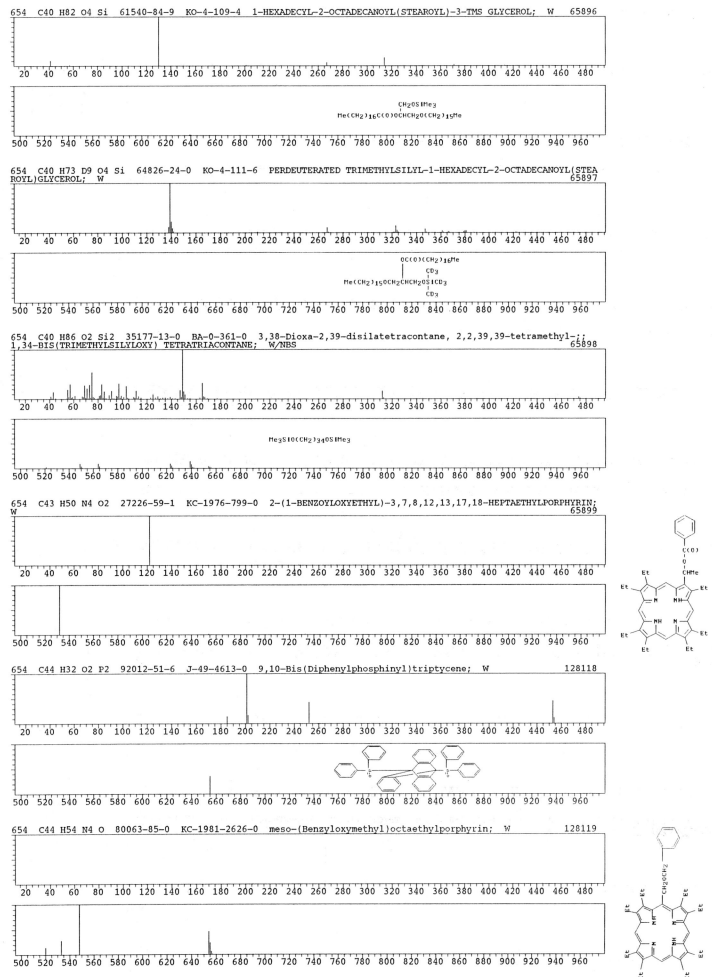

654   C40 H82 O4 Si   61540-84-9   KO-4-109-4   1-HEXADECYL-2-OCTADECANOYL(STEAROYL)-3-TMS GLYCEROL;   W   65896

CH2OSiMe3
Me(CH2)16C(O)OCHCH2O(CH2)15Me

654   C40 H73 D9 O4 Si   64826-24-0   KO-4-111-6   PERDEUTERATED TRIMETHYLSILYL-1-HEXADECYL-2-OCTADECANOYL(STEAROYL)GLYCEROL;   W   65897

OC(O)(CH2)16Me
CD3
Me(CH2)15OCH2CHCH2OSiCD3
CD3

654   C40 H86 O2 Si2   35177-13-0   BA-0-361-0   3,38-Dioxa-2,39-disilatetracontane, 2,2,39,39-tetramethyl-;;
1,34-BIS(TRIMETHYLSILYLOXY) TETRATRIACONTANE;   W/NBS   65898

Me3SiO(CH2)34OSiMe3

654   C43 H50 N4 O2   27226-59-1   KC-1976-799-0   2-(1-BENZOYLOXYETHYL)-3,7,8,12,13,17,18-HEPTAETHYLPORPHYRIN;
W   65899

654   C44 H32 O2 P2   92012-51-6   J-49-4613-0   9,10-Bis(Diphenylphosphinyl)triptycene;   W   128118

654   C44 H54 N4 O   80063-85-0   KC-1981-2626-0   meso-(Benzyloxymethyl)octaethylporphyrin;   W   128119

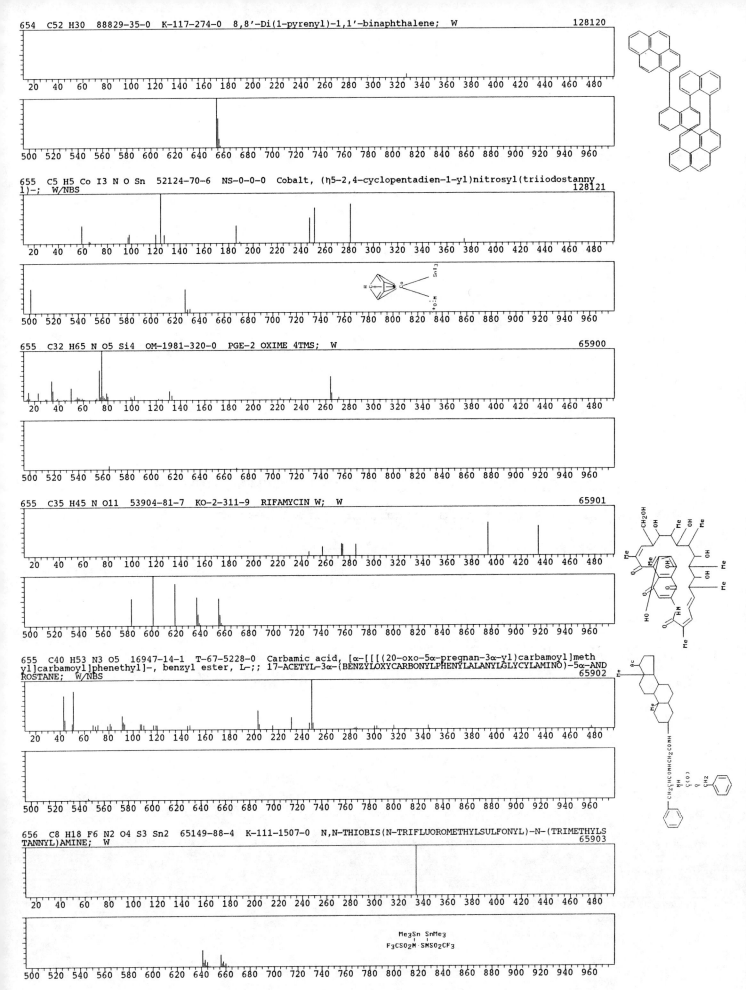

654   C52 H30   88829-35-0   K-117-274-0   8,8'-Di(1-pyrenyl)-1,1'-binaphthalene;   W   128120

655   C5 H5 Co I3 N O Sn   52124-70-6   NS-0-0-0   Cobalt, (η5-2,4-cyclopentadien-1-yl)nitrosyl(triiodostanny
l)-;   W/NBS   128121

655   C32 H65 N O5 Si4   OM-1981-320-0   PGE-2 OXIME 4TMS;   W   65900

655   C35 H45 N O11   53904-81-7   KO-2-311-9   RIFAMYCIN W;   W   65901

655   C40 H53 N3 O5   16947-14-1   T-67-5228-0   Carbamic acid, [α-[[[(20-oxo-5α-pregnan-3α-yl)carbamoyl]meth
yl]carbamoyl]phenethyl]-, benzyl ester, L-;;   17-ACETYL-3α-{BENZYLOXYCARBONYLPHENYLALANYLGLYCYLAMINO}-5α-AND
ROSTANE;   W/NBS   65902

656   C8 H18 F6 N2 O4 S3 Sn2   65149-88-4   K-111-1507-0   N,N-THIOBIS(N-TRIFLUOROMETHYLSULFONYL)-N-(TRIMETHYLS
TANNYL)AMINE;   W   65903

656 C10 H12 Fe2 I2 O6 P2 19599-83-8 AD-0-1596-0 Iron, hexacarbonylbis[.mu.-(dimethylphosphino)]diiodo di-;; .MU.-TETRAMETHYLDIPHOSPHINE-BIS(IODOTRICARBONYLIRON); W/NBS
65904

656 C16 H22 O2 Tl2 KF-3-241-991 DIMETHYLTHALLIUM PHENOXIDE DIMER; W
65905

656 C25 H22 Br2 O11 81410-21-1 KC-1982-975-0 6,8-Dibromo-3,3',4',5,7-penta-O-acetyl-(+)-catechin; W
128122

656 C29 H52 O7 S Si4 83228-91-5 KO-9-249-8 Per(trimethylsilyl)derivative of dihydromethylthionaphtholyl monoglucuronide; W
128123

656 C31 H54 N4 P2 Si4 57016-88-3 K-111-1819-0 1,3,2,4-Diazadiphosphetidine, 4-[bis(trimethylsilyl)amino]-2,2,4,4-tetrahydro-1-methyl-2,2,2-triphenyl-3-(trimethylsilyl)-4-[(trimethylsilyl)imino]-;; 4-(BIS(TRIMETHYLSILYL)AMINO)-1-METHYL-2,2,2-TRIPHENYL-3-(TRIMETHYLSILYL)-4-(TRIMETHYLSILYLIMINO)-1,3,2λ5,4λ5-DIAZADIPHOSPHETIDIN; W/NBS
65906

656 C33 H36 O9 Se 80115-40-8 H-64-1405-0 Limonol phenyl selenocarbonate; W
128124

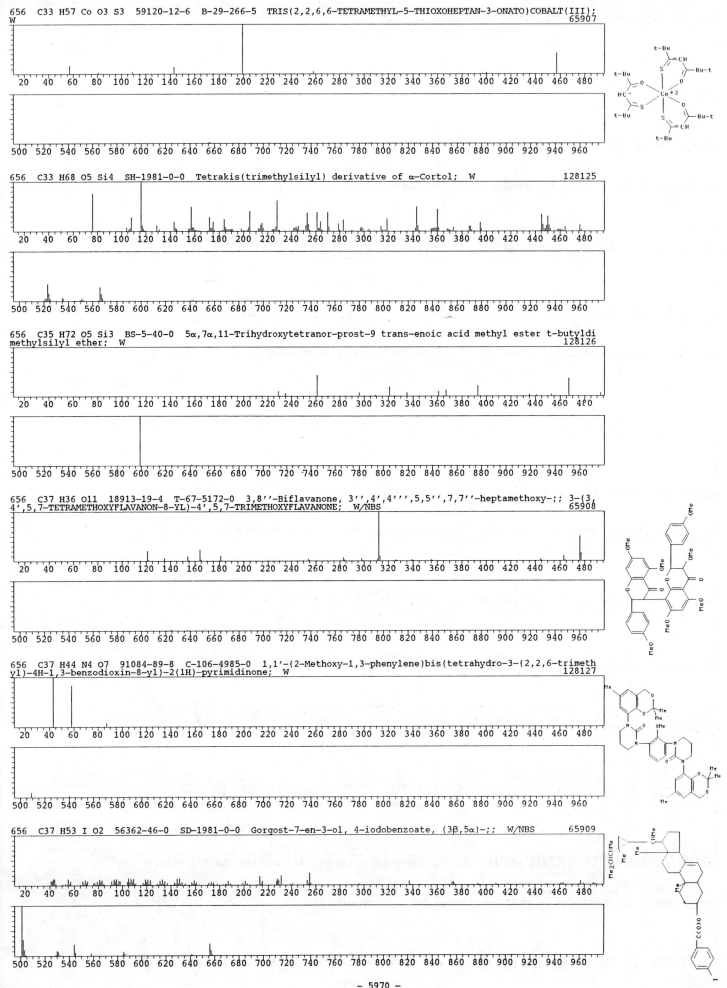

656 C33 H57 Co O3 S3 59120-12-6 B-29-266-5 TRIS(2,2,6,6-TETRAMETHYL-5-THIOXOHEPTAN-3-ONATO)COBALT(III);
W   65907

656 C33 H68 O5 Si4 SH-1981-0-0 Tetrakis(trimethylsilyl) derivative of α-Cortol; W   128125

656 C35 H72 O5 Si3 BS-5-40-0 5α,7α,11-Trihydroxytetranor-prost-9 trans-enoic acid methyl ester t-butyldi
methylsilyl ether; W   128126

656 C37 H36 O11 18913-19-4 T-67-5172-0 3,8''-Biflavanone, 3'',4',4''',5,5'',7,7''-heptamethoxy-;; 3-(3
',5,7-TETRAMETHOXYFLAVANON-8-YL)-4',5,7-TRIMETHOXYFLAVANONE; W/NBS   65908

656 C37 H44 N4 O7 91084-89-8 C-106-4985-0 1,1'-(2-Methoxy-1,3-phenylene)bis(tetrahydro-3-(2,2,6-trimeth
yl)-4H-1,3-benzodioxin-8-yl)-2(1H)-pyrimidinone; W   128127

656 C37 H53 I O2 56362-46-0 SD-1981-0-0 Gorgost-7-en-3-ol, 4-iodobenzoate, (3β,5α)-;; W/NBS   65909

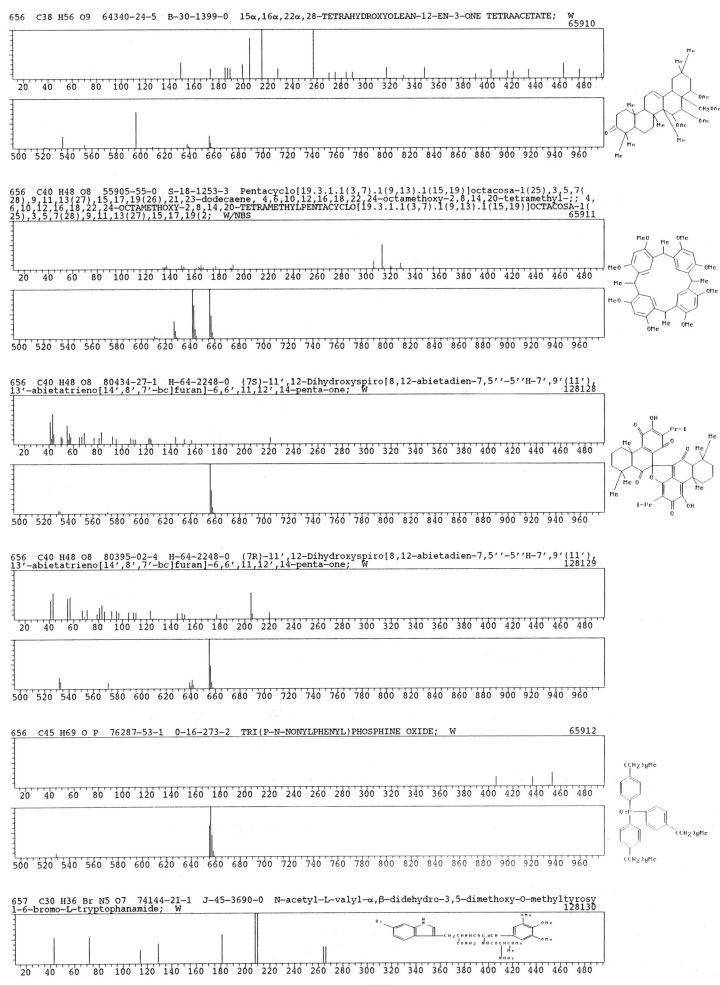

656   C38 H56 O9   64340-24-5   B-30-1399-0   15α,16α,22α,28-TETRAHYDROXYOLEAN-12-EN-3-ONE TETRAACETATE;   W
65910

656   C40 H48 O8   55905-55-0   S-18-1253-3   Pentacyclo[19.3.1.1(3,7).1(9,13).1(15,19)]octacosa-1(25),3,5,7(28),9,11,13(27),15,17,19(26),21,23-dodecaene, 4,6,10,12,16,18,22,24-octamethoxy-2,8,14,20-tetramethyl-;; 4,6,10,12,16,18,22,24-OCTAMETHOXY-2,8,14,20-TETRAMETHYLPENTACYCLO[19.3.1.1(3,7).1(9,13).1(15,19)]OCTACOSA-1(25),3,5,7(28),9,11,13(27),15,17,19(2;   W/NBS
65911

656   C40 H48 O8   80434-27-1   H-64-2248-0   (7S)-11',12-Dihydroxyspiro[8,12-abietadien-7,5''-5''H-7',9'(11')13'-abietatrieno[14',8',7'-bc]furan]-6,6',11,12',14-penta-one;   W
128128

656   C40 H48 O8   80395-02-4   H-64-2248-0   (7R)-11',12-Dihydroxyspiro[8,12-abietadien-7,5''-5''H-7',9'(11')13'-abietatrieno[14',8',7'-bc]furan]-6,6',11,12',14-penta-one;   W
128129

656   C45 H69 O P   76287-53-1   O-16-273-2   TRI(P-N-NONYLPHENYL)PHOSPHINE OXIDE;   W
65912

657   C30 H36 Br N5 O7   74144-21-1   J-45-3690-0   N-acetyl-L-valyl-α,β-didehydro-3,5-dimethoxy-O-methyltyrosyl-6-bromo-L-tryptophanamide;   W
128130

657   C33 H27 N3 O6 S3   56221-24-0   O-11-1222-1   1,3,5,-TRIS(1-NAPHTHYLSULPHONYL) HEXAHYDRO-1,3,5,-TRIAZINE;
W
65914

657   C33 H27 N3 O6 S3   56221-25-1   O-11-1222-1   1,3,5,-TRIS(2-NAPHTHYSULPHONYL) HEXAHYDRO-1,3,5,-TRIAZINE;
W
65915

657   C35 H35 Cl Cu N4 O3   96806-59-6   C-107-4951-0   Methyl 9,10-didehydro-9-chloro-10-formylmesopyropheopho
rbide; W
128131

657   C36 H51 N O10   CD-267-0-0   Veratroylzygadenine; W
128132

658   C14 C114   2142-32-7   O-5-887-4   PERCHLORO(4,4'-BITOLYL); W
65916

658   C22 H14 F12 N4 O6   77144-98-0   F-36-2510-0   1,5-Bis(1,3-bistrifluoroacetyl-4-imidazolin-2-yl)-2,4-di
methoxybenzene; W
128133

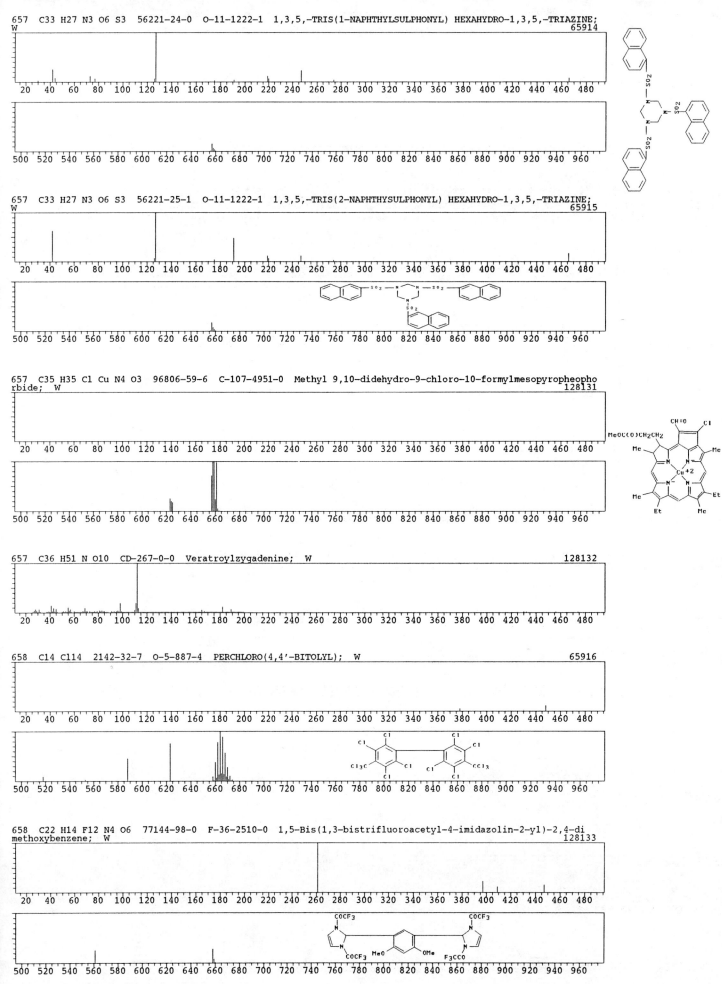

658   C26 H22 Mo2 S4   72186-27-7   C-101-5251-0   BIS(CYCLOPENTADIENYL MOLYBDENUM(III)-1-PHENYL 1,2-ETHENEDI
THIOLATE;   W   65917

658   C28 H38 N2 O16   28435-53-2   K-103-1606-1   4-AMINO-4-DESOXY-D-GLUCOSE-OCTAACETATE DIMER;   W   65918

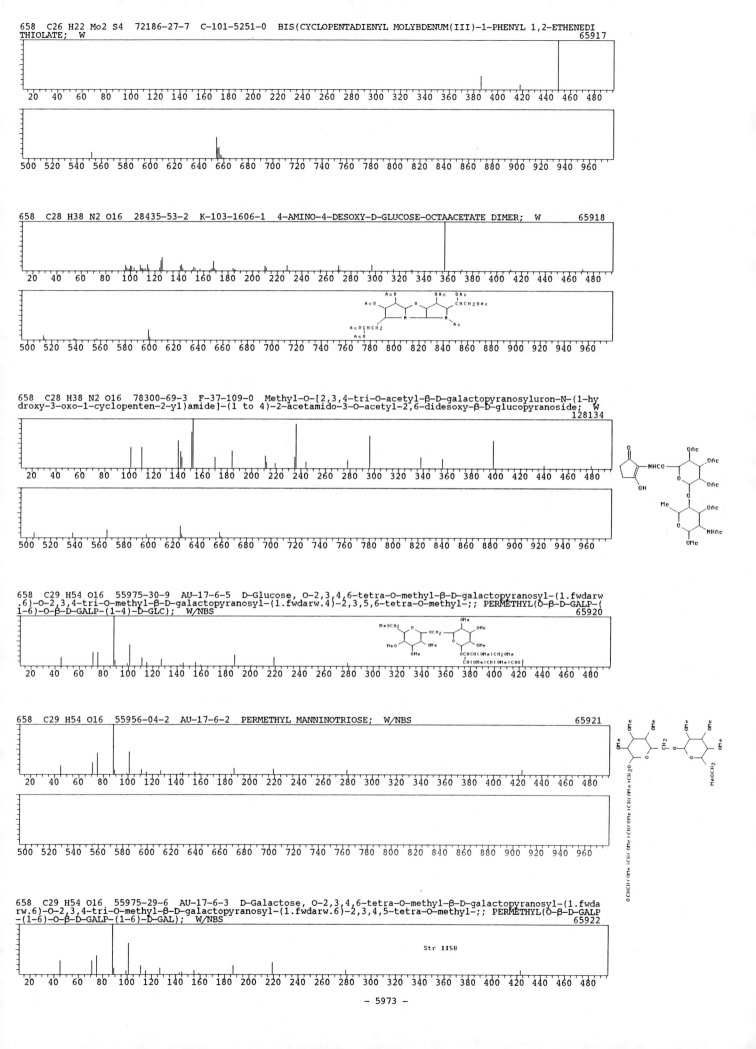

658   C28 H38 N2 O16   78300-69-3   F-37-109-0   Methyl-O-[2,3,4-tri-O-acetyl-β-D-galactopyranosyluron-N-(1-hy
droxy-3-oxo-1-cyclopenten-2-yl)amide]-(1 to 4)-2-acetamido-3-O-acetyl-2,6-didesoxy-β-D-glucopyranoside;   W
                                                                                                                 128134

658   C29 H54 O16   55975-30-9   AU-17-6-5   D-Glucose, O-2,3,4,6-tetra-O-methyl-β-D-galactopyranosyl-(1.fwdarw
.6)-O-2,3,4-tri-O-methyl-β-D-galactopyranosyl-(1.fwdarw.4)-2,3,5,6-tetra-O-methyl-;; PERMETHYL(O-β-D-GALP-(
1-6)-O-β-D-GALP-(1-4)-D-GLC);   W/NBS   65920

658   C29 H54 O16   55956-04-2   AU-17-6-2   PERMETHYL MANNINOTRIOSE;   W/NBS   65921

658   C29 H54 O16   55975-29-6   AU-17-6-3   D-Galactose, O-2,3,4,6-tetra-O-methyl-β-D-galactopyranosyl-(1.fwda
rw.6)-O-2,3,4-tri-O-methyl-β-D-galactopyranosyl-(1.fwdarw.6)-2,3,4,5-tetra-O-methyl-;; PERMETHYL(O-β-D-GALP
-(1-6)-O-β-D-GALP-(1-6)-D-GAL);   W/NBS   65922

Str 1158

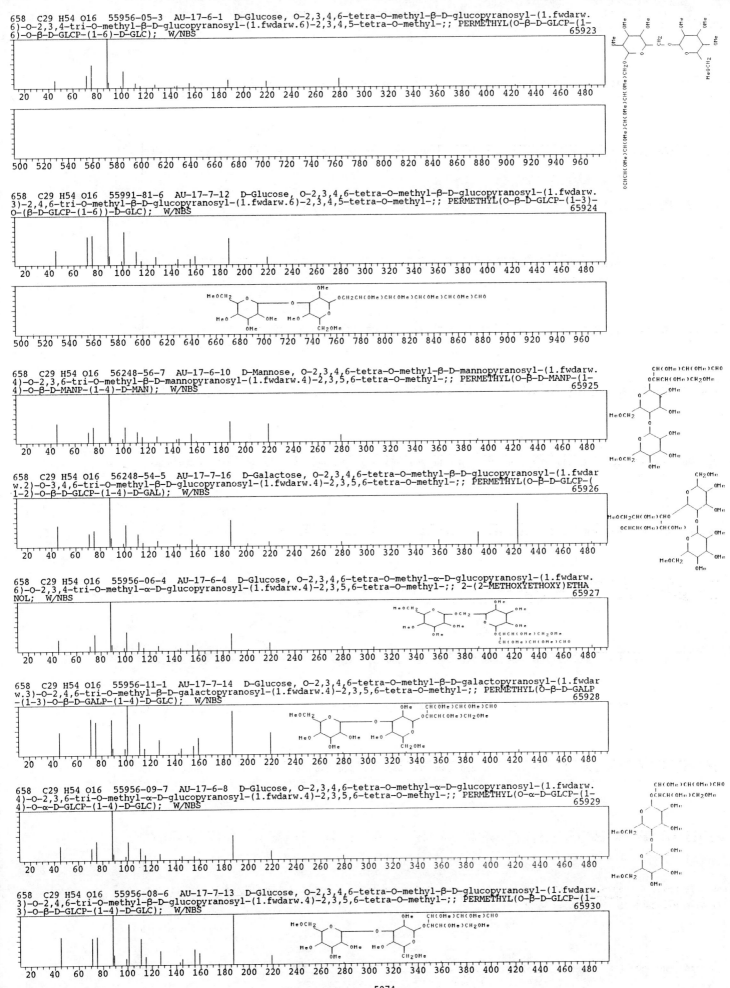

658 C29 H54 O16 55956-05-3 AU-17-6-1 D-Glucose, O-2,3,4,6-tetra-O-methyl-β-D-glucopyranosyl-(1.fwdarw.6)-O-2,3,4-tri-O-methyl-β-D-glucopyranosyl-(1.fwdarw.6)-2,3,4,5-tetra-O-methyl-;; PERMETHYL(O-β-D-GLCP-(1-6)-O-β-D-GLCP-(1-6)-D-GLC); W/NBS
65923

658 C29 H54 O16 55991-81-6 AU-17-7-12 D-Glucose, O-2,3,4,6-tetra-O-methyl-β-D-glucopyranosyl-(1.fwdarw.3)-2,4,6-tri-O-methyl-β-D-glucopyranosyl-(1.fwdarw.6)-2,3,4,5-tetra-O-methyl-;; PERMETHYL(O-β-D-GLCP-(1-3)-O-(β-D-GLCP-(1-6))-D-GLC); W/NBS
65924

658 C29 H54 O16 56248-56-7 AU-17-6-10 D-Mannose, O-2,3,4,6-tetra-O-methyl-β-D-mannopyranosyl-(1.fwdarw.4)-O-2,3,6-tri-O-methyl-β-D-mannopyranosyl-(1.fwdarw.4)-2,3,5,6-tetra-O-methyl-;; PERMETHYL(O-β-D-MANP-(1-4)-O-β-D-MANP-(1-4)-D-MAN); W/NBS
65925

658 C29 H54 O16 56248-54-5 AU-17-7-16 D-Galactose, O-2,3,4,6-tetra-O-methyl-β-D-glucopyranosyl-(1.fwdarw.2)-O-3,4,6-tri-O-methyl-β-D-glucopyranosyl-(1.fwdarw.4)-2,3,5,6-tetra-O-methyl-;; PERMETHYL(O-β-D-GLCP-(1-2)-O-β-D-GLCP-(1-4)-D-GAL); W/NBS
65926

658 C29 H54 O16 55956-06-4 AU-17-6-4 D-Glucose, O-2,3,4,6-tetra-O-methyl-α-D-glucopyranosyl-(1.fwdarw.6)-O-2,3,4-tri-O-methyl-α-D-glucopyranosyl-(1.fwdarw.4)-2,3,5,6-tetra-O-methyl-;; 2-(2-METHOXYETHOXY)ETHANOL; W/NBS
65927

658 C29 H54 O16 55956-11-1 AU-17-7-14 D-Glucose, O-2,3,4,6-tetra-O-methyl-β-D-galactopyranosyl-(1.fwdarw.3)-O-2,4,6-tri-O-methyl-β-D-galactopyranosyl-(1.fwdarw.4)-2,3,5,6-tetra-O-methyl-;; PERMETHYL(O-β-D-GALP-(1-3)-O-β-D-GALP-(1-4)-D-GLC); W/NBS
65928

658 C29 H54 O16 55956-09-7 AU-17-6-8 D-Glucose, O-2,3,4,6-tetra-O-methyl-α-D-glucopyranosyl-(1.fwdarw.4)-O-2,3,6-tri-O-methyl-α-D-glucopyranosyl-(1.fwdarw.4)-2,3,5,6-tetra-O-methyl-;; PERMETHYL(O-α-D-GLCP-(1-4)-O-α-D-GLCP-(1-4)-D-GLC); W/NBS
65929

658 C29 H54 O16 55956-08-6 AU-17-7-13 D-Glucose, O-2,3,4,6-tetra-O-methyl-β-D-glucopyranosyl-(1.fwdarw.3)-O-2,4,6-tri-O-methyl-β-D-glucopyranosyl-(1.fwdarw.4)-2,3,5,6-tetra-O-methyl-;; PERMETHYL(O-β-D-GLCP-(1-3)-O-β-D-GLCP-(1-4)-D-GLC); W/NBS
65930

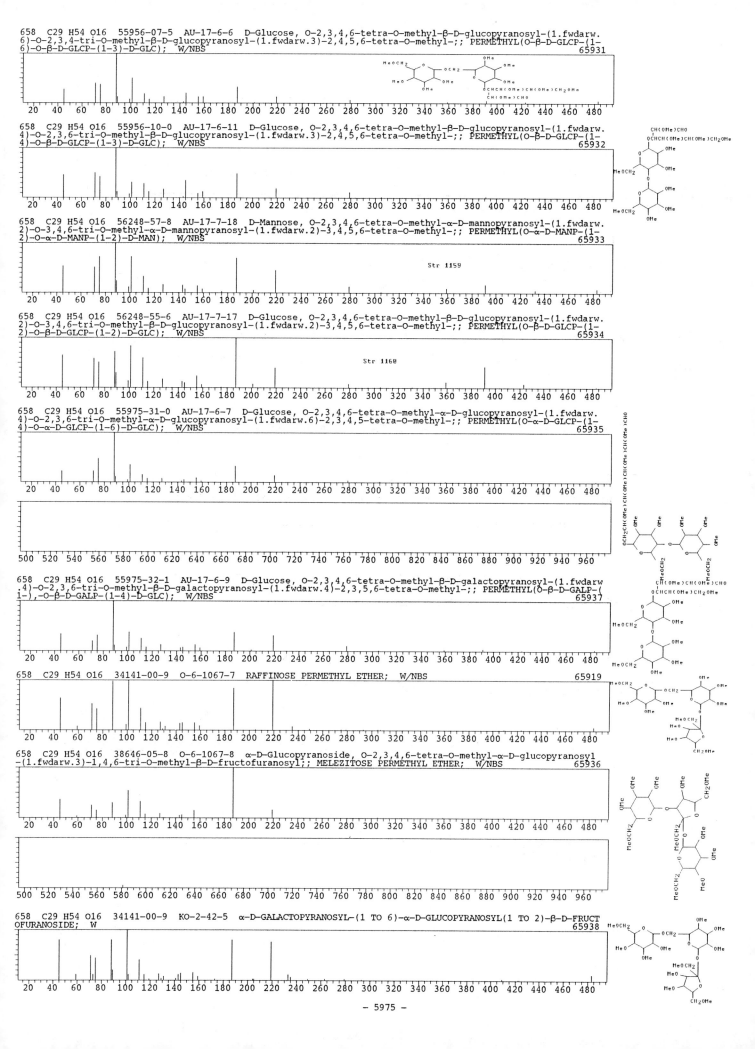

658   C29 H54 O16   55956-07-5   AU-17-6-6   D-Glucose, O-2,3,4,6-tetra-O-methyl-β-D-glucopyranosyl-(1.fwdarw.
6)-O-2,3,4-tri-O-methyl-β-D-glucopyranosyl-(1.fwdarw.3)-2,4,5,6-tetra-O-methyl-;; PERMETHYL(O-β-D-GLCP-(1-
6)-O-β-D-GLCP-(1-3)-D-GLC);   W/NBS   65931

658   C29 H54 O16   55956-10-0   AU-17-6-11   D-Glucose, O-2,3,4,6-tetra-O-methyl-β-D-glucopyranosyl-(1.fwdarw.
4)-O-2,3,6-tri-O-methyl-β-D-glucopyranosyl-(1.fwdarw.3)-2,4,5,6-tetra-O-methyl-;; PERMETHYL(O-β-D-GLCP-(1-
4)-O-β-D-GLCP-(1-3)-D-GLC);   W/NBS   65932

658   C29 H54 O16   56248-57-8   AU-17-7-18   D-Mannose, O-2,3,4,6-tetra-O-methyl-α-D-mannopyranosyl-(1.fwdarw.
2)-O-3,4,6-tri-O-methyl-α-D-mannopyranosyl-(1.fwdarw.2)-3,4,5,6-tetra-O-methyl-;; PERMETHYL(O-α-D-MANP-(1-
2)-O-α-D-MANP-(1-2)-D-MAN);   W/NBS   65933

Str 1159

658   C29 H54 O16   56248-55-6   AU-17-7-17   D-Glucose, O-2,3,4,6-tetra-O-methyl-β-D-glucopyranosyl-(1.fwdarw.
2)-O-3,4,6-tri-O-methyl-β-D-glucopyranosyl-(1.fwdarw.2)-3,4,5,6-tetra-O-methyl-;; PERMETHYL(O-β-D-GLCP-(1-
2)-O-β-D-GLCP-(1-2)-D-GLC);   W/NBS   65934

Str 1160

658   C29 H54 O16   55975-31-0   AU-17-6-7   D-Glucose, O-2,3,4,6-tetra-O-methyl-α-D-glucopyranosyl-(1.fwdarw.
4)-O-2,3,6-tri-O-methyl-α-D-glucopyranosyl-(1.fwdarw.6)-2,3,4,5-tetra-O-methyl-;; PERMETHYL(O-α-D-GLCP-(1-
4)-O-α-D-GLCP-(1-6)-D-GLC);   W/NBS   65935

658   C29 H54 O16   55975-32-1   AU-17-6-9   D-Glucose, O-2,3,4,6-tetra-O-methyl-β-D-galactopyranosyl-(1.fwdarw.
4)-O-2,3,6-tri-O-methyl-β-D-galactopyranosyl-(1.fwdarw.4)-2,3,5,6-tetra-O-methyl-;; PERMETHYL(O-β-D-GALP-(
1-),-O-β-D-GALP-(1-4)-D-GLC);   W/NBS   65937

658   C29 H54 O16   34141-00-9   O-6-1067-7   RAFFINOSE PERMETHYL ETHER;   W/NBS   65919

658   C29 H54 O16   38646-05-8   O-6-1067-8   α-D-Glucopyranoside, O-2,3,4,6-tetra-O-methyl-α-D-glucopyranosyl
-(1.fwdarw.3)-1,4,6-tri-O-methyl-β-D-fructofuranosyl;; MELEZITOSE PERMETHYL ETHER;   W/NBS   65936

658   C29 H54 O16   34141-00-9   KO-2-42-5   α-D-GALACTOPYRANOSYL-(1 TO 6)-α-D-GLUCOPYRANOSYL(1 TO 2)-β-D-FRUCT
OFURANOSIDE;   W   65938

- 5975 -

658 C29 H54 O16 38646-05-8 KO-2-42-5 α-D-GLUCOPYRANOSYL(1 TO 2)-β-D-FRUCTOFURANOSYL(3 TO 1)-α-D-GLUCOPY
RANOSIDE; W
65939

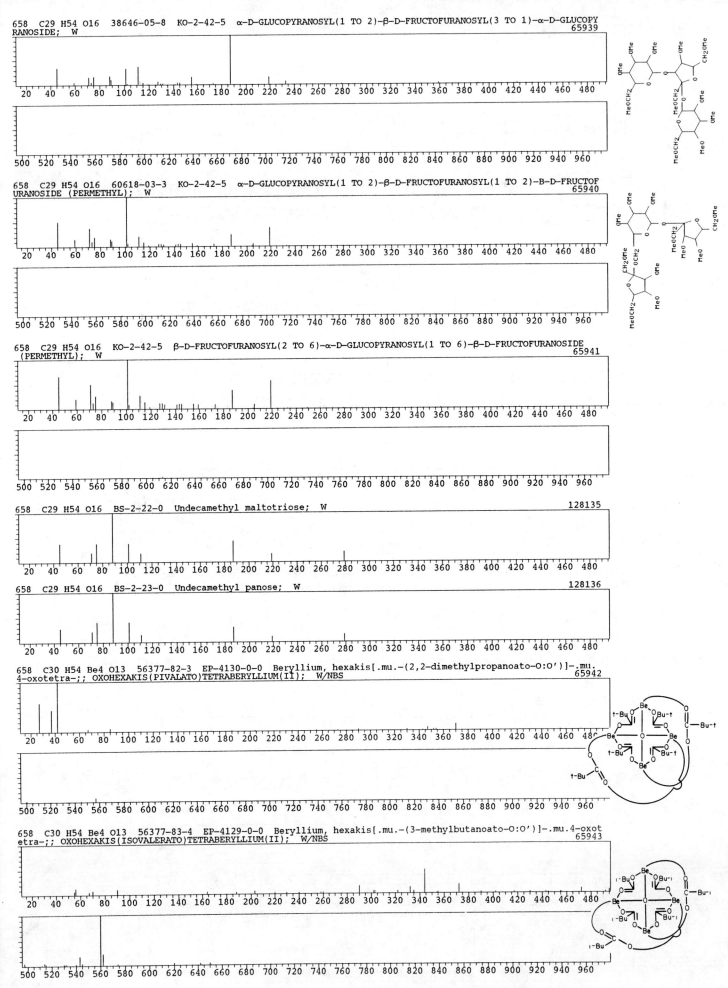

658 C29 H54 O16 60618-03-3 KO-2-42-5 α-D-GLUCOPYRANOSYL(1 TO 2)-β-D-FRUCTOFURANOSYL(1 TO 2)-B-D-FRUCTOF
URANOSIDE (PERMETHYL); W
65940

658 C29 H54 O16 KO-2-42-5 β-D-FRUCTOFURANOSYL(2 TO 6)-α-D-GLUCOPYRANOSYL(1 TO 6)-β-D-FRUCTOFURANOSIDE
(PERMETHYL); W
65941

658 C29 H54 O16 BS-2-22-0 Undecamethyl maltotriose; W
128135

658 C29 H54 O16 BS-2-23-0 Undecamethyl panose; W
128136

658 C30 H54 Be4 O13 56377-82-3 EP-4130-0-0 Beryllium, hexakis[.mu.-(2,2-dimethylpropanoato-O:O')]-.mu.
4-oxotetra-;; OXOHEXAKIS(PIVALATO)TETRABERYLLIUM(II); W/NBS
65942

658 C30 H54 Be4 O13 56377-83-4 EP-4129-0-0 Beryllium, hexakis[.mu.-(3-methylbutanoato-O:O')]-.mu.4-oxot
etra-;; OXOHEXAKIS(ISOVALERATO)TETRABERYLLIUM(II); W/NBS
65943

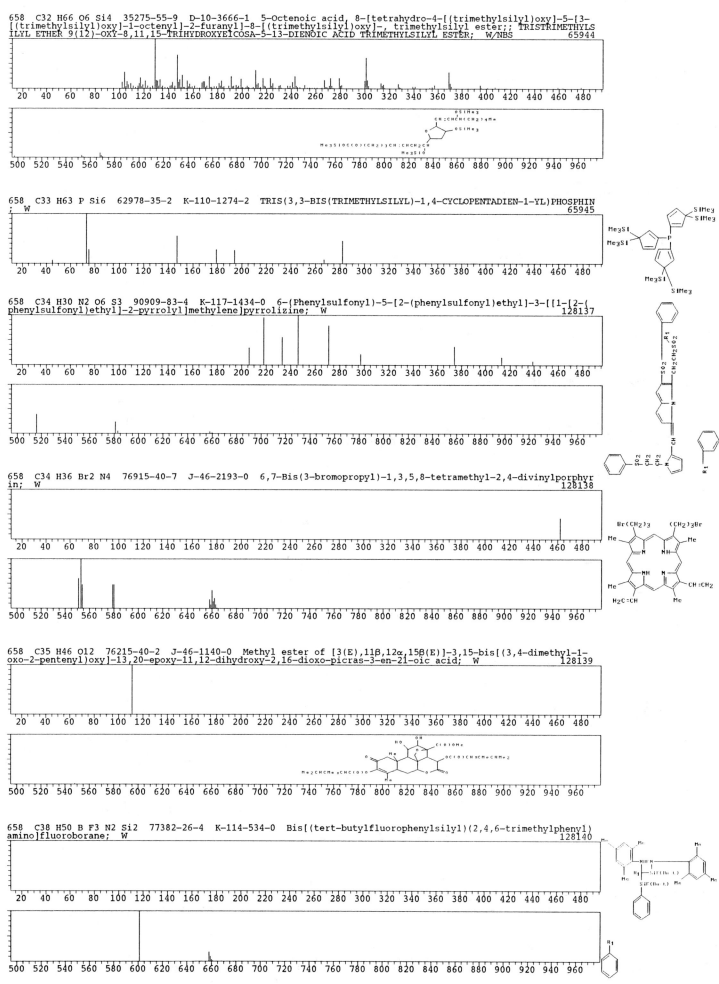

658 C32 H66 O6 Si4 35275-55-9 D-10-3666-1 5-Octenoic acid, 8-[tetrahydro-4-[(trimethylsilyl)oxy]-5-[3-[(trimethylsilyl)oxy]-1-octenyl]-2-furanyl]-8-[(trimethylsilyl)oxy]-, trimethylsilyl ester;; TRISTRIMETHYLSILYL ETHER 9(12)-OXY-8,11,15-TRIHYDROXYEICOSA-5-13-DIENOIC ACID TRIMETHYLSILYL ESTER; W/NBS   65944

658 C33 H63 P Si6 62978-35-2 K-110-1274-2 TRIS(3,3-BIS(TRIMETHYLSILYL)-1,4-CYCLOPENTADIEN-1-YL)PHOSPHIN; W   65945

658 C34 H30 N2 O6 S3 90909-83-4 K-117-1434-0 6-(Phenylsulfonyl)-5-[2-(phenylsulfonyl)ethyl]-3-[[1-[2-(phenylsulfonyl)ethyl]-2-pyrrolyl]methylene]pyrrolizine; W   128137

658 C34 H36 Br2 N4 76915-40-7 J-46-2193-0 6,7-Bis(3-bromopropyl)-1,3,5,8-tetramethyl-2,4-divinylporphyrin; W   128138

658 C35 H46 O12 76215-40-2 J-46-1140-0 Methyl ester of [3(E),11β,12α,15β(E)]-3,15-bis[(3,4-dimethyl-1-oxo-2-pentenyl)oxy]-13,20-epoxy-11,12-dihydroxy-2,16-dioxo-picras-3-en-21-oic acid; W   128139

658 C38 H50 B F3 N2 Si2 77382-26-4 K-114-534-0 Bis[(tert-butylfluorophenylsilyl)(2,4,6-trimethylphenyl)amino]fluoroborane; W   128140

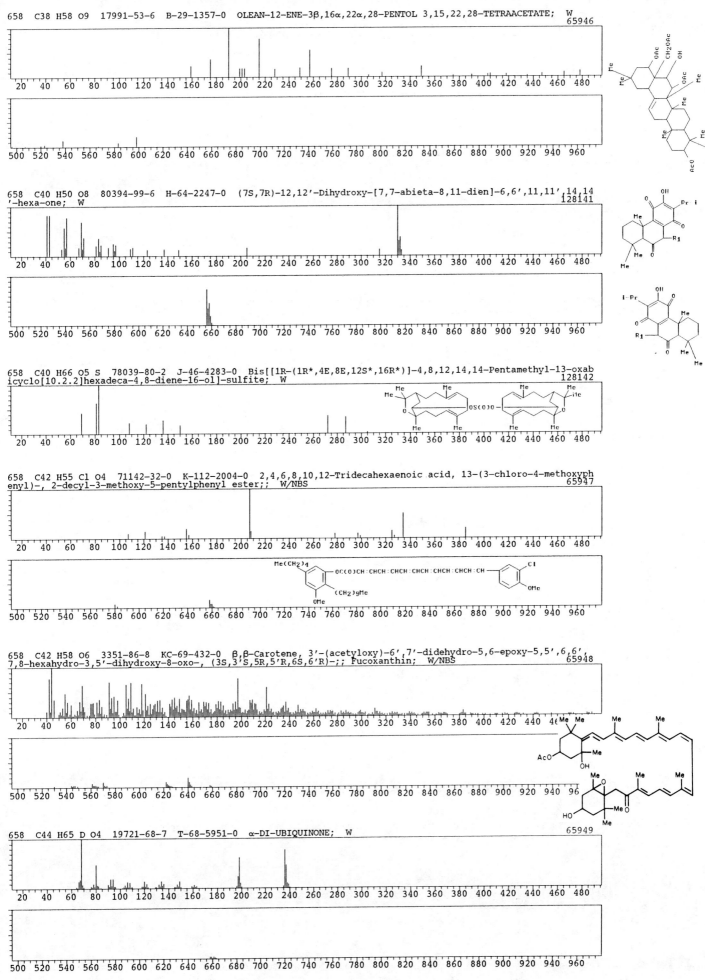

658   C38 H58 O9   17991-53-6   B-29-1357-0   OLEAN-12-ENE-3β,16α,22α,28-PENTOL 3,15,22,28-TETRAACETATE;   W
65946

658   C40 H50 O8   80394-99-6   H-64-2247-0   (7S,7R)-12,12'-Dihydroxy-[7,7-abieta-8,11-dien]-6,6',11,11',14,14
'-hexa-one;   W
128141

658   C40 H66 O5 S   78039-80-2   J-46-4283-0   Bis[[1R-(1R*,4E,8E,12S*,16R*)]-4,8,12,14,14-Pentamethyl-13-oxab
icyclo[10.2.2]hexadeca-4,8-diene-16-ol]-sulfite;   W
128142

658   C42 H55 Cl O4   71142-32-0   K-112-2004-0   2,4,6,8,10,12-Tridecahexaenoic acid, 13-(3-chloro-4-methoxyph
enyl)-, 2-decyl-3-methoxy-5-pentylphenyl ester;;   W/NBS
65947

658   C42 H58 O6   3351-86-8   KC-69-432-0   β,β-Carotene, 3'-(acetyloxy)-6',7'-didehydro-5,6-epoxy-5,5',6,6',
7,8-hexahydro-3,5'-dihydroxy-8-oxo-, (3S,3'S,5R,5'R,6S,6'R)-;; Fucoxanthin;   W/NBS
65948

658   C44 H65 D O4   19721-68-7   T-68-5951-0   α-DI-UBIQUINONE;   W
65949

658 C44 H66 O4 303-95-7 J-45-4103-0 2-(3,7,11,15,23,27-heptamethyl-2,6,10,14,18,22,26-octacosaheptaeny
l)-5,6-dimethoxy-3-methyl-2,5-cyclohexadiene-1,4-dione; W   128143

20 40 60 80 100 120 140 160 180 200 220 240 260 280 300 320 340 360 380 400 420 440 460 480

500 520 540 560 580 600 620 640 660 680 700 720 740 760 780 800 820 840 860 880 900 920 940 960

658 C50 H58 64354-78-5 F-33-383-0 20-(2,3,4-TRIMETHYLBENZAL)-RENIERAPURPURIN; W   65950

20 40 60 80 100 120 140 160 180 200 220 240 260 280 300 320 340 360 380 400 420 440 460 480

500 520 540 560 580 600 620 640 660 680 700 720 740 760 780 800 820 840 860 880 900 920 940 960

658 C52 H34 KC-1977-268-273 TRANS,TRANS,TRANS-1,2,2A,10B-TETRAHYDRO-1,2-BIS-(2-BENZO(C)PHENANTHRYL)CYCLO
BUTA(L)PHENANTHRENE; W   65951

20 40 60 80 100 120 140 160 180 200 220 240 260 280 300 320 340 360 380 400 420 440 460 480

500 520 540 560 580 600 620 640 660 680 700 720 740 760 780 800 820 840 860 880 900 920 940 960

658 C52 H34 KC-1977-268-273 TRANS,TRANS-0,0'-BIS-(2-2-BENZO(C)PHENANTHRYL)VINYL)BIPHENYL; W   65952

20 40 60 80 100 120 140 160 180 200 220 240 260 280 300 320 340 360 380 400 420 440 460 480

500 520 540 560 580 600 620 640 660 680 700 720 740 760 780 800 820 840 860 880 900 920 940 960

659 C22 H17 O8 P Re S K-114-3287-0 Diethylester of 5,5,5,5-Tetracarbonyl-3,3-diphenyl-4λ4-thia-3λ5-phosp
ha-5-rhena-1,3-cyclopentadien-1,2-dicarboxylic acid; W   128144

20 40 60 80 100 120 140 160 180 200 220 240 260 280 300 320 340 360 380 400 420 440 460 480

500 520 540 560 580 600 620 640 660 680 700 720 740 760 780 800 820 840 860 880 900 920 940 960

659 C33 H33 N5 O10 66981-29-1 B-31-1107-0 3-BENZYL-6-BENZAMIDO-7-(TETRA-O-ACETYL-β-D-GLUCOPYRANOSYL)PUR
INE; W   65953

20 40 60 80 100 120 140 160 180 200 220 240 260 280 300 320 340 360 380 400 420 440 460 480

500 520 540 560 580 600 620 640 660 680 700 720 740 760 780 800 820 840 860 880 900 920 940 960

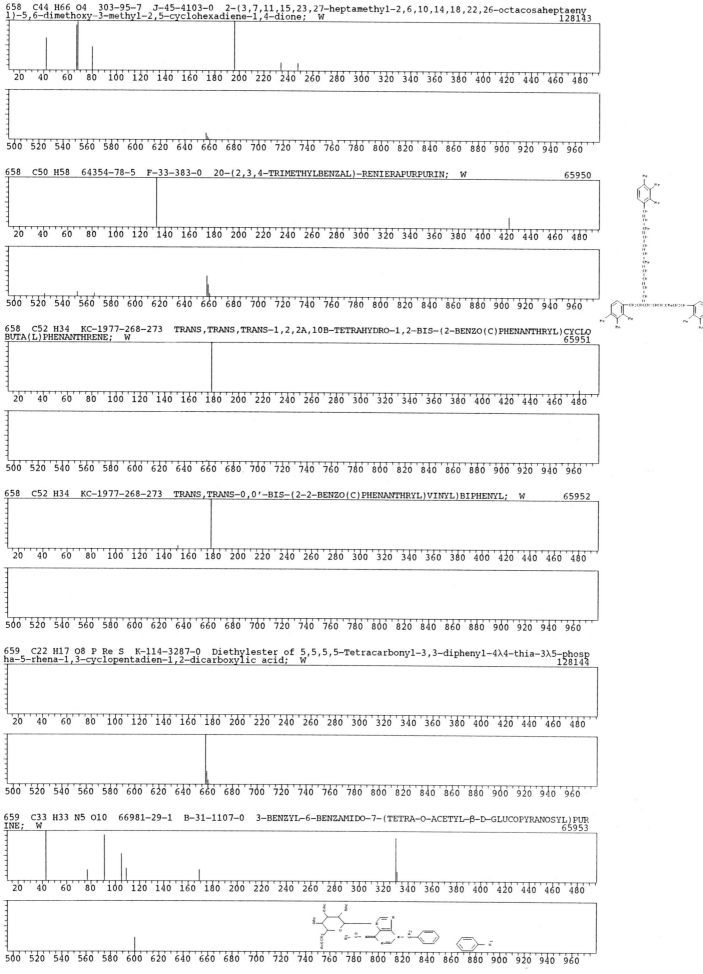

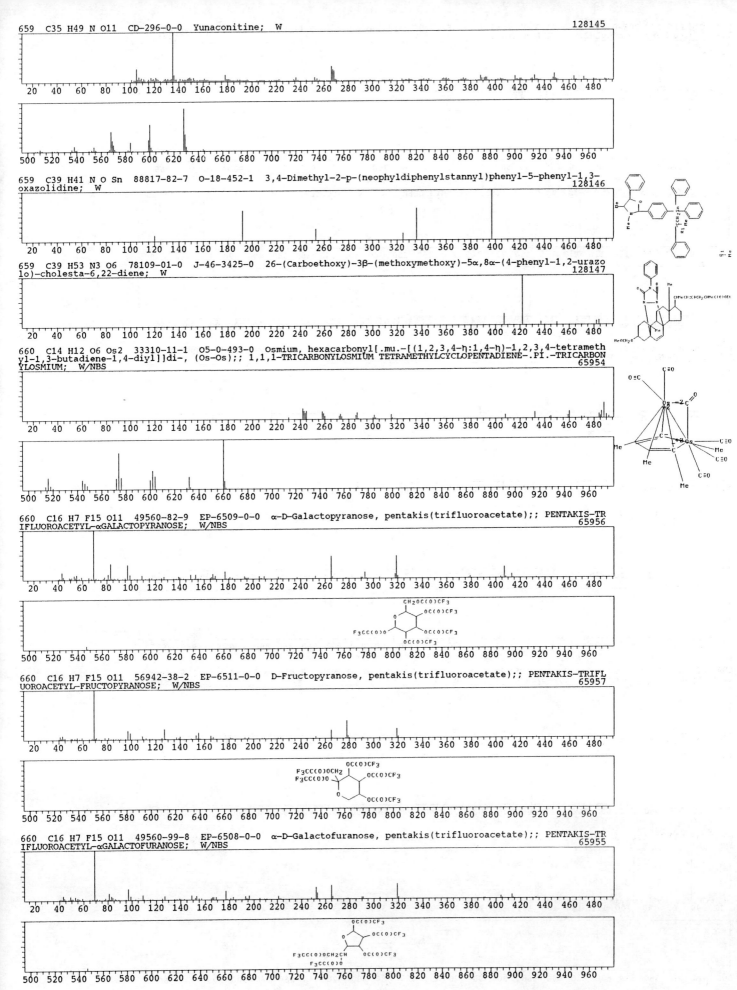

659   C35 H49 N O11   CD-296-0-0   Yunaconitine;   W   128145

659   C39 H41 N O Sn   88817-82-7   O-18-452-1   3,4-Dimethyl-2-p-(neophyldiphenylstannyl)phenyl-5-phenyl-1,3-oxazolidine;   W   128146

659   C39 H53 N3 O6   78109-01-0   J-46-3425-0   26-(Carboethoxy)-3β-(methoxymethoxy)-5α,8α-(4-phenyl-1,2-urazolo)-cholesta-6,22-diene;   W   128147

660   C14 H12 O6 Os2   33310-11-1   O5-0-493-0   Osmium, hexacarbonyl[.mu.-[(1,2,3,4-η:1,4-η)-1,2,3,4-tetramethyl-1,3-butadiene-1,4-diyl]]di-, (Os-Os);; 1,1,1-TRICARBONYLOSMIUM TETRAMETHYLCYCLOPENTADIENE-.PI.-TRICARBONYLOSMIUM;   W/NBS   65954

660   C16 H7 F15 O11   49560-82-9   EP-6509-0-0   α-D-Galactopyranose, pentakis(trifluoroacetate);; PENTAKIS-TRIFLUOROACETYL-αGALACTOPYRANOSE;   W/NBS   65956

660   C16 H7 F15 O11   56942-38-2   EP-6511-0-0   D-Fructopyranose, pentakis(trifluoroacetate);; PENTAKIS-TRIFLUOROACETYL-FRUCTOPYRANOSE;   W/NBS   65957

660   C16 H7 F15 O11   49560-99-8   EP-6508-0-0   α-D-Galactofuranose, pentakis(trifluoroacetate);; PENTAKIS-TRIFLUOROACETYL-αGALACTOFURANOSE;   W/NBS   65955

660   C27 H51 Fe2 O5 P3   55493-73-7   NS--27351-0-0   Iron, pentacarbonylbis[.mu.-(dimethylphosphino)](trihexy
lphosphine)di-, (Fe-Fe);;   NBS   65958

660   C30 H24 N6 O8 S2   IC-1301-0-0   1,4-Bis(2-nitro-4-anilinosulphonylanilino)benzene; W   128148

660   C31 H48 Mo O5 P2   56557-15-4   AG-94-60-0   (Dicyclohexylphosphine)(pentacarbonyl)molybdenum; W   128149

660   C32 H22 I2   59934-31-5   SB-30-370-0   4,20(21)-DI-IODO[2.2.2.2]PARACYCLOPHANETETRAENE;   W   65959

660   C32 H68 O6 Si4   35275-57-1   D-10-3667-3   2-Furanoctanoic acid, tetrahydro-η,4-bis[(trimethylsilyl)oxy]
-5-[3-[(trimethylsilyl)oxy]-1-octenyl]-, trimethylsilyl ester;; TRISTRIMETHYLSILYL ETHER 9(12)-OXY-8,11,15-
TRIHYDROXYEICOSA-13-ENOIC ACID TRIMETHYLSILYL ESTER; W/NBS   65960

660   C35 H48 O12   88972-34-3   KC-1983-2835-0   3'-Epiafroside 2',3',15-triacetate; W   128150

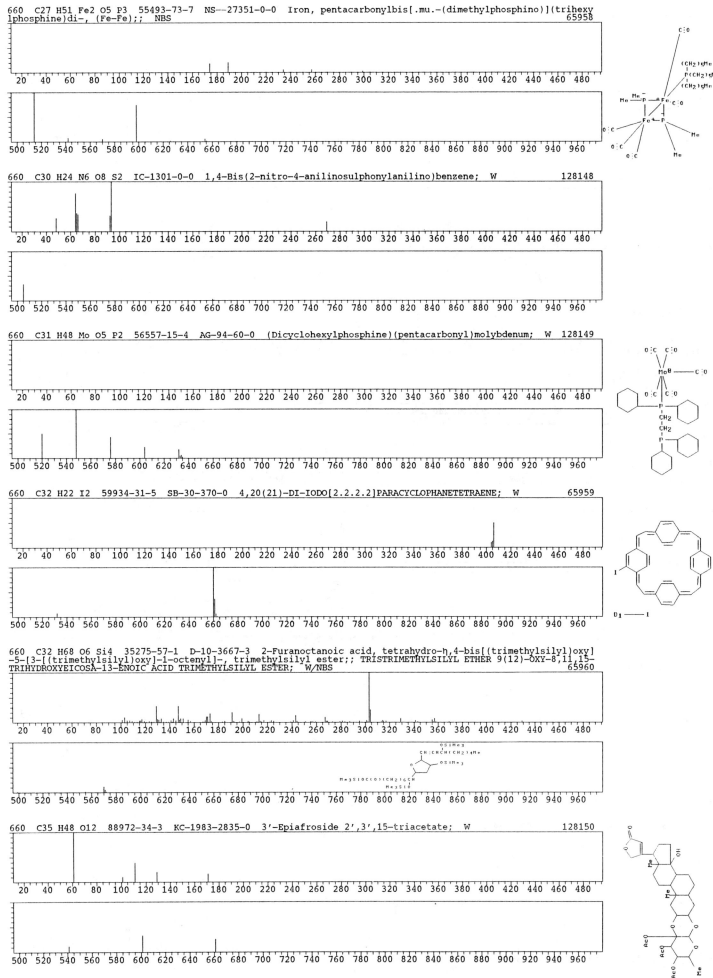

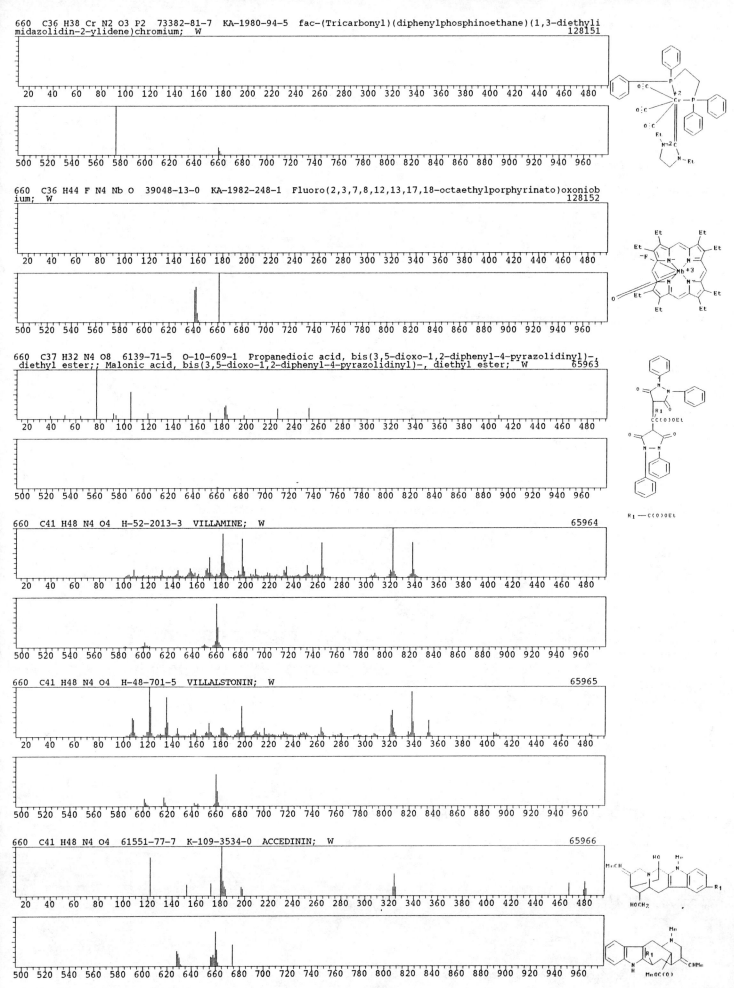

660   C36 H38 Cr N2 O3 P2   73382-81-7   KA-1980-94-5   fac-(Tricarbonyl)(diphenylphosphinoethane)(1,3-diethyli
midazolidin-2-ylidene)chromium;   W   128151

660   C36 H44 F N4 Nb O   39048-13-0   KA-1982-248-1   Fluoro(2,3,7,8,12,13,17,18-octaethylporphyrinato)oxoniob
ium;   W   128152

660   C37 H32 N4 O8   6139-71-5   O-10-609-1   Propanedioic acid, bis(3,5-dioxo-1,2-diphenyl-4-pyrazolidinyl)-,
diethyl ester;; Malonic acid, bis(3,5-dioxo-1,2-diphenyl-4-pyrazolidinyl)-, diethyl ester;   W   65963

660   C41 H48 N4 O4   H-52-2013-3   VILLAMINE;   W   65964

660   C41 H48 N4 O4   H-48-701-5   VILLALSTONIN;   W   65965

660   C41 H48 N4 O4   61551-77-7   K-109-3534-0   ACCEDININ;   W   65966

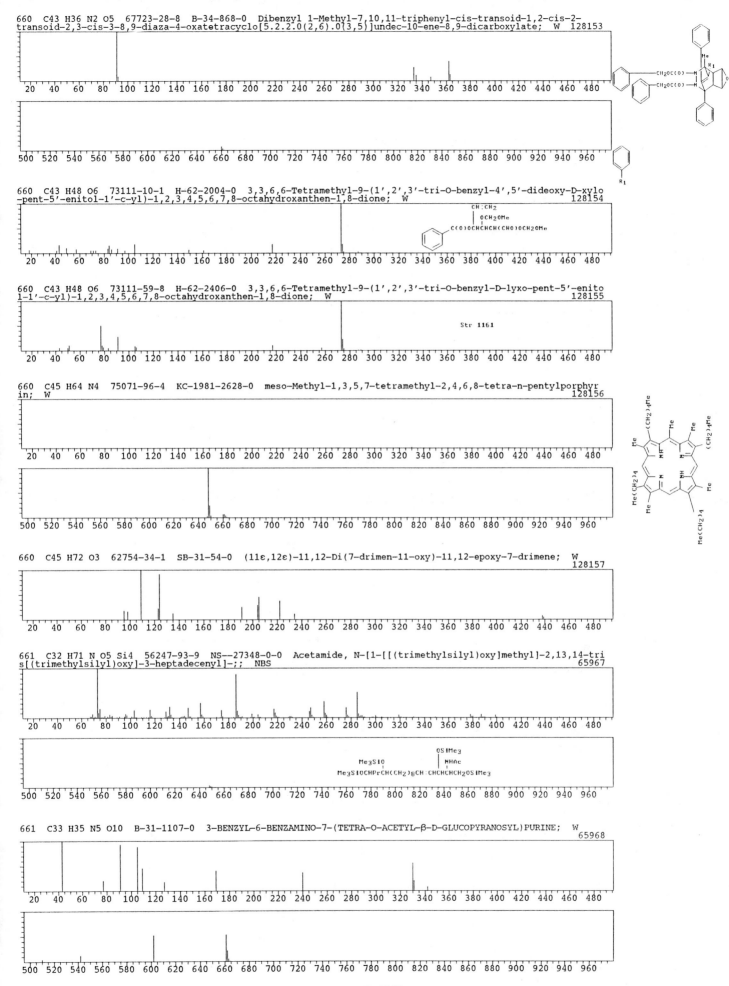

660   C43 H36 N2 O5   67723-28-8   B-34-868-0   Dibenzyl 1-Methyl-7,10,11-triphenyl-cis-transoid-1,2-cis-2-transoid-2,3-cis-3-8,9-diaza-4-oxatetracyclo[5.2.2.0(2,6).0(3,5)]undec-10-ene-8,9-dicarboxylate;   W 128153

660   C43 H48 O6   73111-10-1   H-62-2004-0   3,3,6,6-Tetramethyl-9-(1',2',3'-tri-O-benzyl-4',5'-dideoxy-D-xylo-pent-5'-enitol-1'-c-yl)-1,2,3,4,5,6,7,8-octahydroxanthen-1,8-dione;   W   128154

660   C43 H48 O6   73111-59-8   H-62-2406-0   3,3,6,6-Tetramethyl-9-(1',2',3'-tri-O-benzyl-D-lyxo-pent-5'-enitol-1'-c-yl)-1,2,3,4,5,6,7,8-octahydroxanthen-1,8-dione;   W   128155

Str 1161

660   C45 H64 N4   75071-96-4   KC-1981-2628-0   meso-Methyl-1,3,5,7-tetramethyl-2,4,6,8-tetra-n-pentylporphyrin; W   128156

660   C45 H72 O3   62754-34-1   SB-31-54-0   (11ε,12ε)-11,12-Di(7-drimen-11-oxy)-11,12-epoxy-7-drimene;   W   128157

661   C32 H71 N O5 Si4   56247-93-9   NS--27348-0-0   Acetamide, N-[1-[[(trimethylsilyl)oxy]methyl]-2,13,14-tris[(trimethylsilyl)oxy]-3-heptadecenyl]-;;   NBS   65967

661   C33 H35 N5 O10   B-31-1107-0   3-BENZYL-6-BENZAMINO-7-(TETRA-O-ACETYL-β-D-GLUCOPYRANOSYL)PURINE;   W   65968

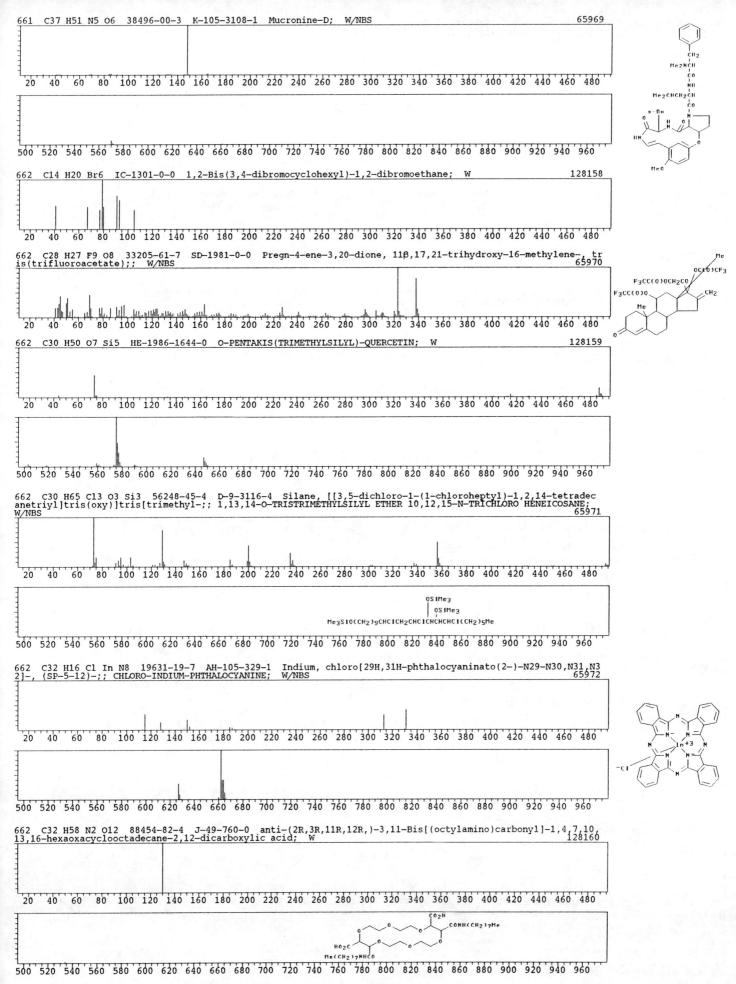

661 C37 H51 N5 O6 38496-00-3 K-105-3108-1 Mucronine-D; W/NBS   65969

662 C14 H20 Br6 IC-1301-0-0 1,2-Bis(3,4-dibromocyclohexyl)-1,2-dibromoethane; W   128158

662 C28 H27 F9 O8 33205-61-7 SD-1981-0-0 Pregn-4-ene-3,20-dione, 11β,17,21-trihydroxy-16-methylene-, tris(trifluoroacetate);; W/NBS   65970

662 C30 H50 O7 Si5 HE-1986-1644-0 O-PENTAKIS(TRIMETHYLSILYL)-QUERCETIN; W   128159

662 C30 H65 Cl3 O3 Si3 56248-45-4 D-9-3116-4 Silane, [[3,5-dichloro-1-(1-chloroheptyl)-1,2,14-tetradecanetriyl]tris(oxy)]tris[trimethyl-;; 1,13,14-O-TRISTRIMETHYLSILYL ETHER 10,12,15-N-TRICHLORO HENEICOSANE; W/NBS   65971

662 C32 H16 Cl In N8 19631-19-7 AH-105-329-1 Indium, chloro[29H,31H-phthalocyaninato(2-)-N29-N30,N31,N32]-, (SP-5-12)-;; CHLORO-INDIUM-PHTHALOCYANINE; W/NBS   65972

662 C32 H58 N2 O12 88454-82-4 J-49-760-0 anti-(2R,3R,11R,12R,)-3,11-Bis[(octylamino)carbonyl]-1,4,7,10,13,16-hexaoxacyclooctadecane-2,12-dicarboxylic acid; W   128160

- 5984 -

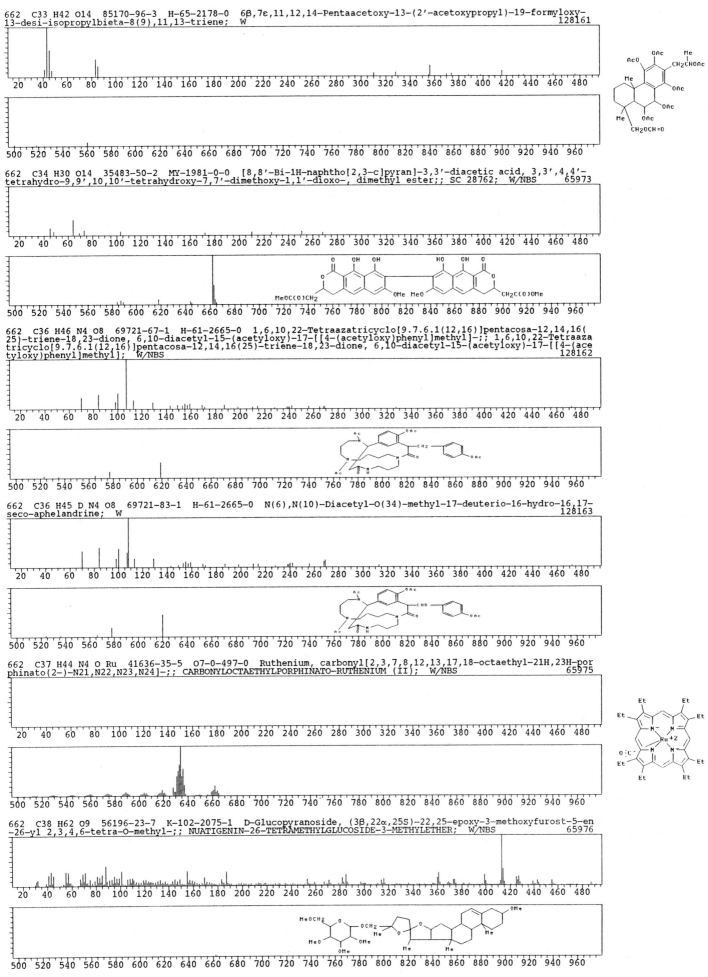

662 C33 H42 O14 85170-96-3 H-65-2178-0 6β,7ε,11,12,14-Pentaacetoxy-13-(2'-acetoxypropyl)-19-formyloxy-13-desi-isopropylbieta-8(9),11,13-triene; W   128161

662 C34 H30 O14 35483-50-2 MY-1981-0-0 [8,8'-Bi-1H-naphtho[2,3-c]pyran]-3,3'-diacetic acid, 3,3',4,4'-tetrahydro-9,9',10,10'-tetrahydroxy-7,7'-dimethoxy-1,1'-dioxo-, dimethyl ester;; SC 28762; W/NBS   65973

662 C36 H46 N4 O8 69721-67-1 H-61-2665-0 1,6,10,22-Tetraazatricyclo[9.7.6.1(12,16)]pentacosa-12,14,16(25)-triene-18,23-dione, 6,10-diacetyl-15-(acetyloxy)-17-[[4-(acetyloxy)phenyl]methyl]-;; 1,6,10,22-Tetraazatricyclo[9.7.6.1(12,16)]pentacosa-12,14,16(25)-triene-18,23-dione, 6,10-diacetyl-15-(acetyloxy)-17-[[4-(acetyloxy)phenyl]methyl]; W/NBS   128162

662 C36 H45 D N4 O8 69721-83-1 H-61-2665-0 N(6),N(10)-Diacetyl-O(34)-methyl-17-deuterio-16-hydro-16,17-seco-aphelandrine; W   128163

662 C37 H44 N4 O Ru 41636-35-5 O7-0-497-0 Ruthenium, carbonyl[2,3,7,8,12,13,17,18-octaethyl-21H,23H-porphinato(2-)-N21,N22,N23,N24]-;; CARBONYLOCTAETHYLPORPHINATO-RUTHENIUM (II); W/NBS   65975

662 C38 H62 O9 56196-23-7 K-102-2075-1 D-Glucopyranoside, (3β,22α,25S)-22,25-epoxy-3-methoxyfurost-5-en-26-yl 2,3,4,6-tetra-O-methyl-;; NUATIGENIN-26-TETRAMETHYLGLUCOSIDE-3-METHYLETHER; W/NBS   65976

662 C39 H42 N4 O6 83604-25-5 C-106-7161-0   34,38-Dioxo-33,35,36,37-tetramethoxy-31-methyl-3,7,23,27-tet
raazaheptacyclo(27.3.1.1.1.1.1.1)octaconta-1(33),8-(35),9,11,13(36),14,16,18(37),19,21,29,31-dodecaene; W
128164

662 C39 H54 O7 Si 80126-59-6 I-59-2735-0   Methyl 11R-acetoxy-8R-tert-butyldiphenylsilyloxy-15-oxo-9S,12S
-oxyeicosa-5Z,13E-dienoate; W
128165

662 C42 H30 O8 63910-27-0 KC-1977-1052-0   2,2'-[O-PHENYLENEDIOXYDI-(O-PHENYLENE)DIOXYDI-(O-PHENYLENE)-DI
OXY]DIPHENOL; W
65977

662 C42 H54 N4 O3 B-28-1820-0   16-DEMETHOXYCARBONYLDIHYDROVOACAMINE; W
65828

662 C42 H54 N4 O3 B-28-1820-0   16-DEMETHOXYCARBONYL-20'-EPIDIHYDROVOACAMINE; W
65829

662 C45 H58 O4 F-37-561-0   2-Dodecyl-3-hydroxy-5-(2-methylpropyl)phenylester of 17-(4-Hydroxyphenyl)-2,4,
6,8,10,12,14,16-heptadecaoctaenoic acid; W
128166

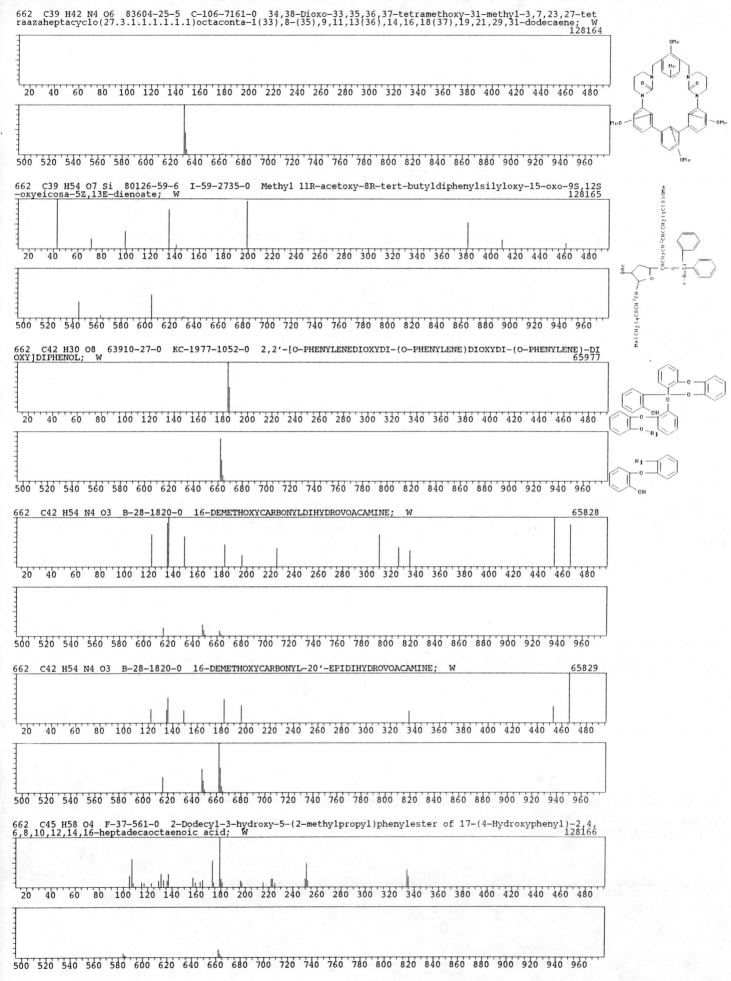

663　C21 H16 F15 N O6　KO-6-32-1　Tri-o-Pentafluoropropionyl derivative of Terbutaline;　W　　　128167

663　C27 H23 Br2 N O5 S2　75931-78-1　O-15-25-2　N,N-BIS(4'-P-BROMOPHENYLSULFONYL-2'-BUTYNYL)-4-ANISIDINE;
W　　65978

663　C32 H73 N O5 Si4　25307-60-2　D-8-1813-1　O-TETRAKISTRIMETHYLSILYL ETHER 14,15-DIHYDROXY N-ACETYL SPHI
NGOSINE;　W　　　　　　　　　　　　　　　　　　　　　　　　　　　　　　　　　　　　　65979

663　C32 H73 N O5 Si4　EP-1918-0-0　4,5-DIHYDROXY SPHINGANINE-TETRATMS;　W　　　65980

663　C34 H37 N3 O7 S2　36600-92-7　NS-0-0-0　1-Naphthalenesulfonic acid, 5-(dimethylamino)-, 4-[2-[[[5-(di
methylamino)-1-naphthalenyl]sulfonyl]methylamino]-1-hydroxyethyl]-2-methoxyphenyl ester;; BIS(1-DIMETHYLAMI
NONAPHTHALENE-5-SULPHONYL)METANEPHRINE;　W/NBS　　　　　　　　　　　　　　　128168

663　C39 H36 Br Cl N P　66966-64-1　J-43-4214-0　[(m-Chlorobenzyl)-(6-phenyl-7-methyl-2H-1,3,4,7-tetrahydro
azepin-2-ylidene)methyl]triphenylphosphonium bromide;　W　　　　　　　　　　128169

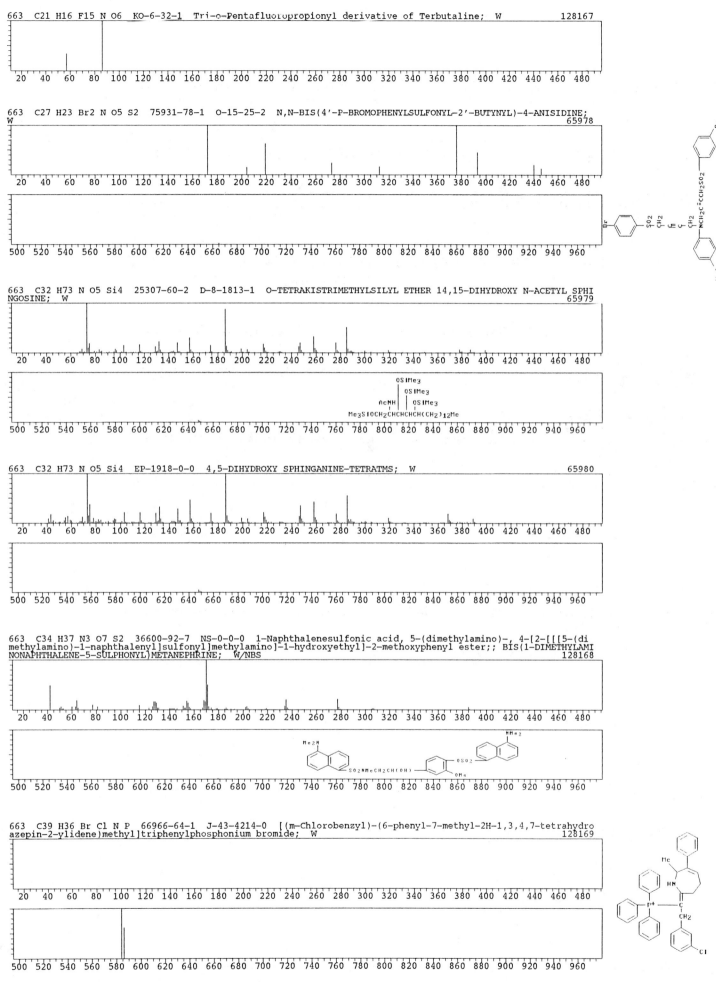

664 C10 H28 Hg2 Si4 28911-65-1 AG-120-177-1 2,2,4,4,5,5,8,8-Octamethyl-2,4,6,8-tetrasila-1,5-dimercurac
yclooctane; W
128170

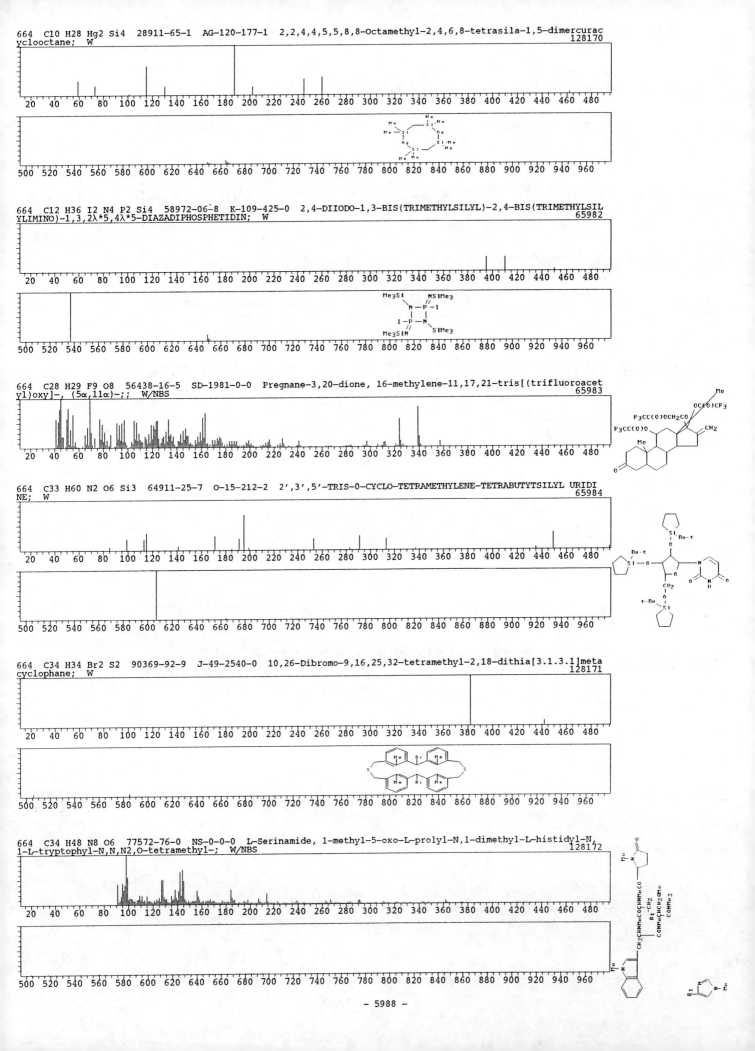

664 C12 H36 I2 N4 P2 Si4 58972-06-8 K-109-425-0 2,4-DIIODO-1,3-BIS(TRIMETHYLSILYL)-2,4-BIS(TRIMETHYLSIL
YLIMINO)-1,3,2λ*5,4λ*5-DIAZADIPHOSPHETIDIN; W
65982

664 C28 H29 F9 O8 56438-16-5 SD-1981-0-0 Pregnane-3,20-dione, 16-methylene-11,17,21-tris[(trifluoroacet
yl)oxy]-, (5α,11α)-;; W/NBS
65983

664 C33 H60 N2 O6 Si3 64911-25-7 O-15-212-2 2',3',5'-TRIS-O-CYCLO-TETRAMETHYLENE-TETRABUTYTSILYL URIDI
NE; W
65984

664 C34 H34 Br2 S2 90369-92-9 J-49-2540-0 10,26-Dibromo-9,16,25,32-tetramethyl-2,18-dithia[3.1.3.1]meta
cyclophane; W
128171

664 C34 H48 N8 O6 77572-76-0 NS-0-0-0 L-Serinamide, 1-methyl-5-oxo-L-prolyl-N,1-dimethyl-L-histidyl-N,
1-L-tryptophyl-N,N,N2,O-tetramethyl-; W/NBS
128172

- 5988 -

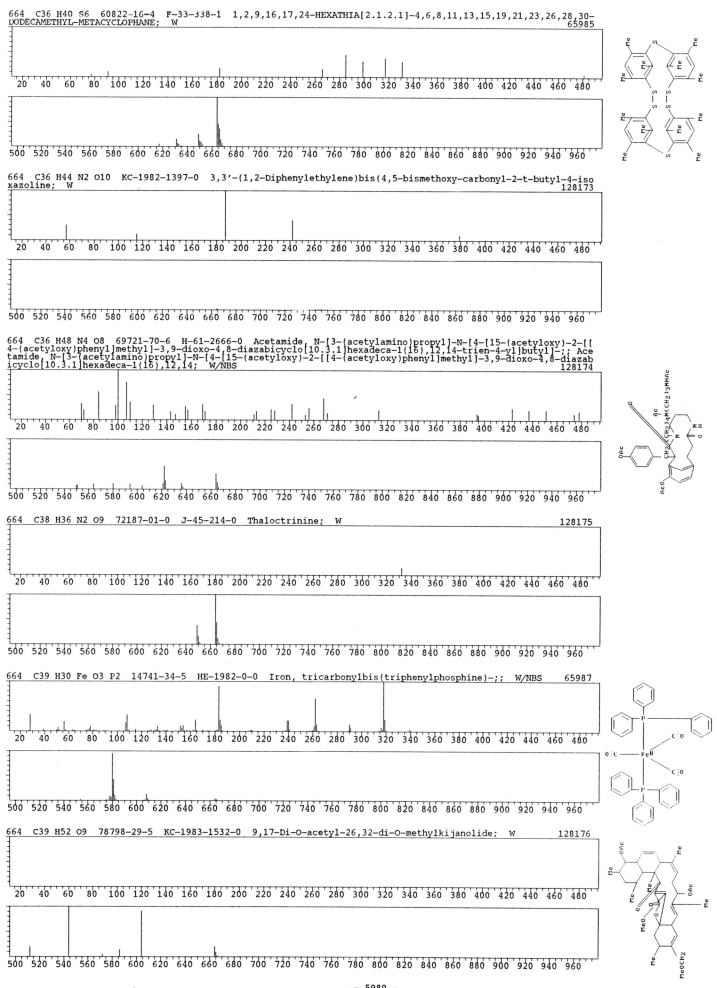

664   C36 H40 S6   60822-16-4   F-33-338-1   1,2,9,16,17,24-HEXATHIA[2.1.2.1]-4,6,8,11,13,15,19,21,23,26,28,30-
DODECAMETHYL-METACYCLOPHANE;   W   65985

664   C36 H44 N2 O10   KC-1982-1397-0   3,3'-(1,2-Diphenylethylene)bis(4,5-bismethoxy-carbonyl-2-t-butyl-4-iso
xazoline;   W   128173

664   C36 H48 N4 O8   69721-70-6   H-61-2666-0   Acetamide, N-[3-(acetylamino)propyl]-N-[4-[15-(acetyloxy)-2-[[
4-(acetyloxy)phenyl]methyl]-3,9-dioxo-4,8-diazabicyclo[10.3.1]hexadeca-1(16),12,14-trien-4-yl]butyl]-;; Ace
tamide, N-[3-(acetylamino)propyl]-N-[4-[15-(acetyloxy)-2-[[4-(acetyloxy)phenyl]methyl]-3,9-dioxo-4,8-diazab
icyclo[10.3.1]hexadeca-1(16),12,14;   W/NBS   128174

664   C38 H36 N2 O9   72187-01-0   J-45-214-0   Thaloctrinine;   W   128175

664   C39 H30 Fe O3 P2   14741-34-5   HE-1982-0-0   Iron, tricarbonylbis(triphenylphosphine)-;;   W/NBS   65987

664   C39 H52 O9   78798-29-5   KC-1983-1532-0   9,17-Di-O-acetyl-26,32-di-O-methylkijanolide;   W   128176

- 5989 -

664   C41 H64 O3 S Si   79409-80-6   J-46-5186-0   (3β,20S,22E)-24-.xi.-(Phenylthio)-3-[(tetrahydro-2H-pyran-2-yl)oxy]-24-.xi.-(trimethylsilyl)cholesta-5,22-dien-20-ol;   W   128177

664   C44 H56 O5   68420-67-7   H-61-1966-0   2,3-Didehydrofritschiellaxanthin-diacetate;   W   128178

664   C44 H72 O4   71142-43-3   NS--27334-0-0   Benzenepentadecanoic acid, 4-methoxy-, 2-decyl-3-methoxy-5-pentylphenyl ester;;   NBS   65989

664   C44 H72 O4   71142-41-1   K-112-2008-0   Benzenetridecanoic acid, 4-methoxy-, 3-methoxy-2-(9-methyldecyl)-5-(4-methylpentyl)phenyl ester;;   3-METHOXY-2-(9-METHYLDECYL)-5-(4-METHYLPENTYL)PHENYLESTER OF 13-(4-METHOXYPHENYL)TRIDECANCARBOXYLIC ACID;   W/NBS   65988

664   C45 H60 O4   K-112-2006-0   MIXTURE OF 2-(9-ME-DECYL)-5-PENTYL- AND 2-DECYL-5-(4-ME-PENTYL)-3-MEO-PHENYL 15-(4-MEO-PHENYL)-2,4,6,8,10,12,14-PENTADECAHEPTAENCARBOXYLATES;   W   65990

664   C50 H32 O2   75155-51-0   KC-1983-462-0   Dimer of 1,3-Diphenylbenz[f]indan-2-one;   W   128179

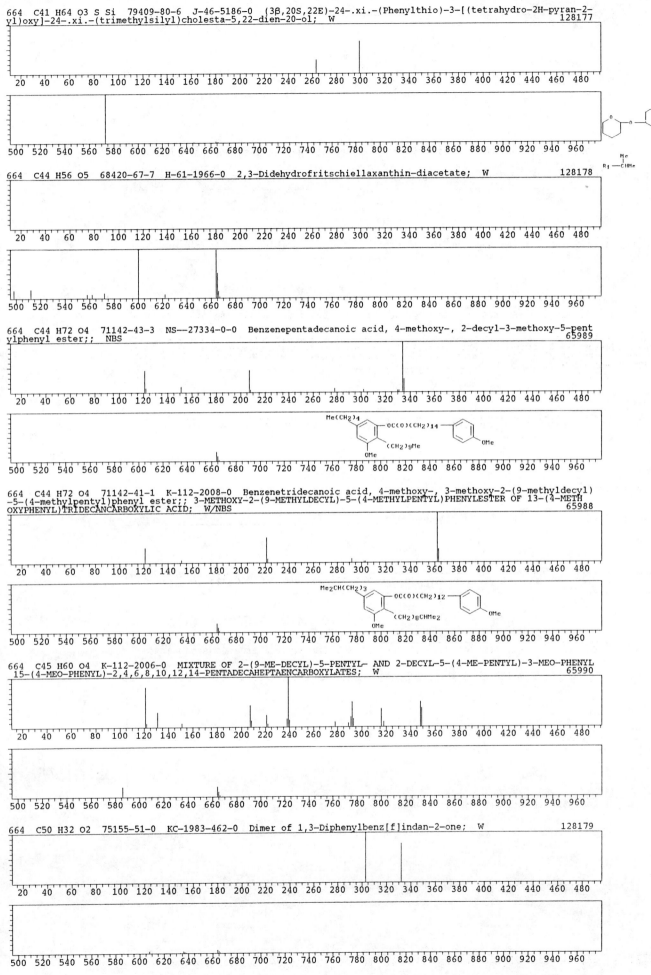

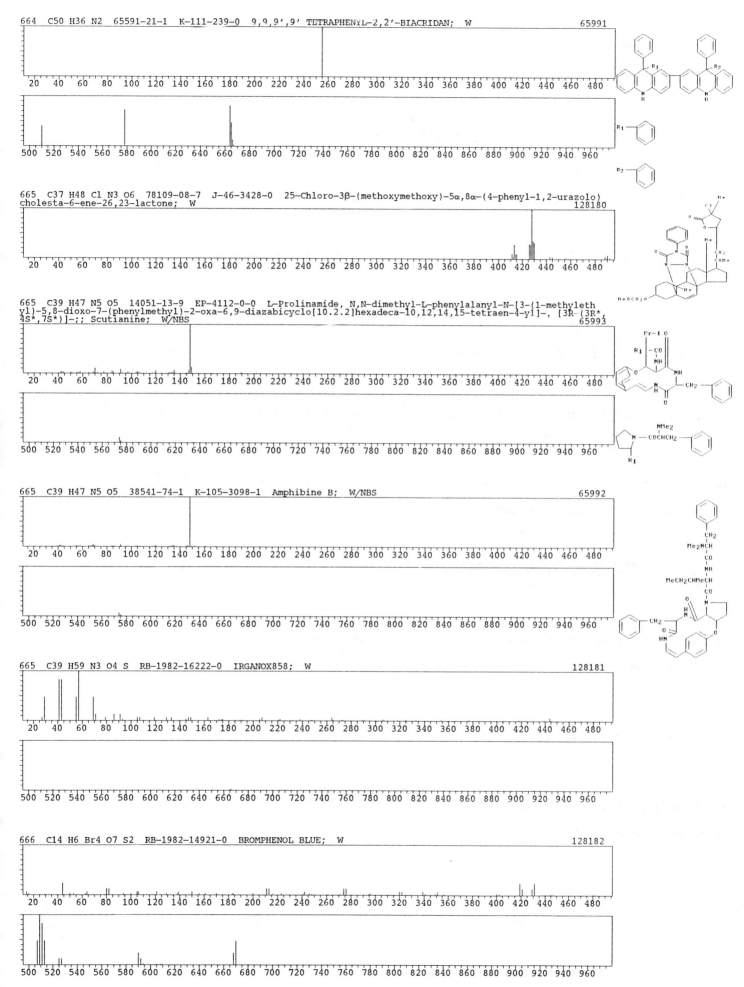

664   C50 H36 N2   65591-21-1   K-111-239-0   9,9,9',9' TETRAPHENYL-2,2'-BIACRIDAN;   W   65991

665   C37 H48 Cl N3 O6   78109-08-7   J-46-3428-0   25-Chloro-3β-(methoxymethoxy)-5α,8α-(4-phenyl-1,2-urazolo)
cholesta-6-ene-26,23-lactone;   W   128180

665   C39 H47 N5 O5   14051-13-9   EP-4112-0-0   L-Prolinamide, N,N-dimethyl-L-phenylalanyl-N-[3-(1-methyleth
yl)-5,8-dioxo-7-(phenylmethyl)-2-oxa-6,9-diazabicyclo[10.2.2]hexadeca-10,12,14,15-tetraen-4-yl]-, [3R-(3R*,
4S*,7S*)]-;; Scutianine;   W/NBS   65993

665   C39 H47 N5 O5   38541-74-1   K-105-3098-1   Amphibine B;   W/NBS   65992

665   C39 H59 N3 O4 S   RB-1982-16222-0   IRGANOX858;   W   128181

666   C14 H6 Br4 O7 S2   RB-1982-14921-0   BROMPHENOL BLUE;   W   128182

666   C14 H20 Br2 Mo2 N2 O2 S2   40671-84-9   KA-72-2590-1   Molybdenum, dibromobis(η5-2,4-cyclopentadien-1-yl)
bis[.mu.-(ethanethiolato)]dinitrosyldi-;;   BIS(BROMO(ETHANETHIOLATO)(NITROSYL)-.PI.-CYCLOPENTADIENYLMOLYBDEN
IUM) (STRUCTURE?);   W/NBS   65994

666   C15 H12 Br6   73255-12-6   C-102-1061-0   1,3,4,6,7,9-Hexabromotrindanes;   W   128183

666   C18 H54 O9 Si9   556-71-8   HE-1986-2363-0   OCTADECAMETHYLCYCLONONASILOXANE;   W   128184

666   C19 H10 Br4 O5 S   JS-1986-1-0   Bromphenol blue;   W   128185

666   C21 H24 F14 N2 O6   OD-1984-276-4   N-Heptafluorobutyryl-n-propyl derivative of diaminopimelic acid;   W
                                                                                                       128186

666   C22 H10 Br F15 N2   82633-63-4   AC-1982-94-0   4-Bromo-1,5-diphenyl-3-perfluoroheptyl-pyrazole;   W
                                                                                                       128187

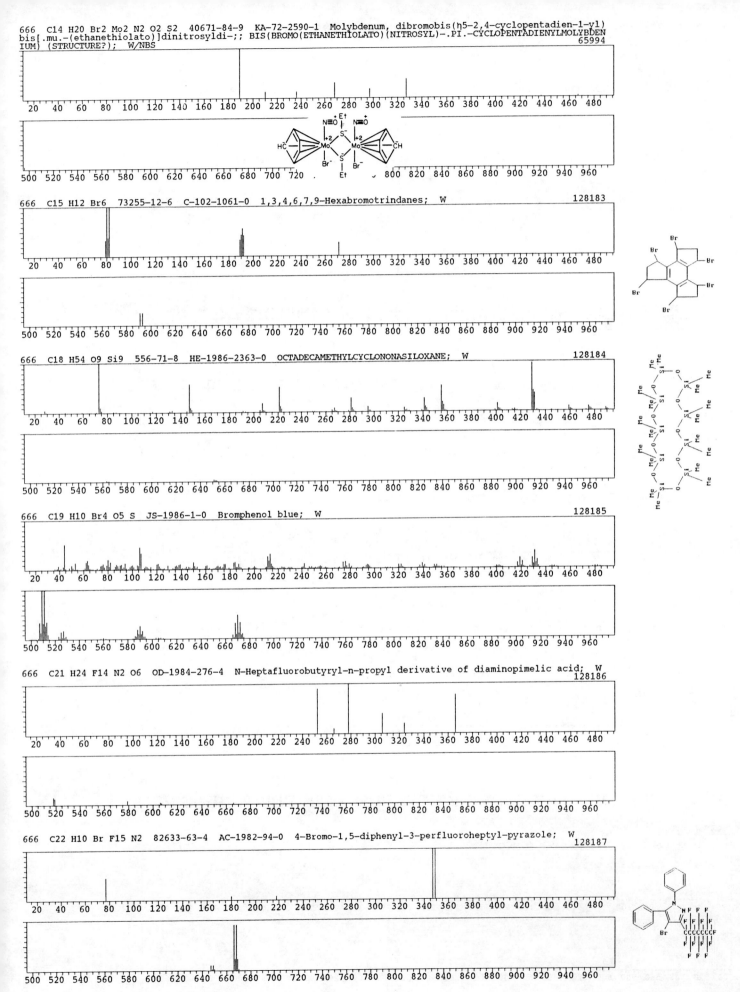

666   C36 H40 N4 Ni O3 S   96806-62-1   C-107-4953-0   Nickel(II) methyl δ-meso-((methylthio)methyl)mesopyropheophorbide a;   W

128188

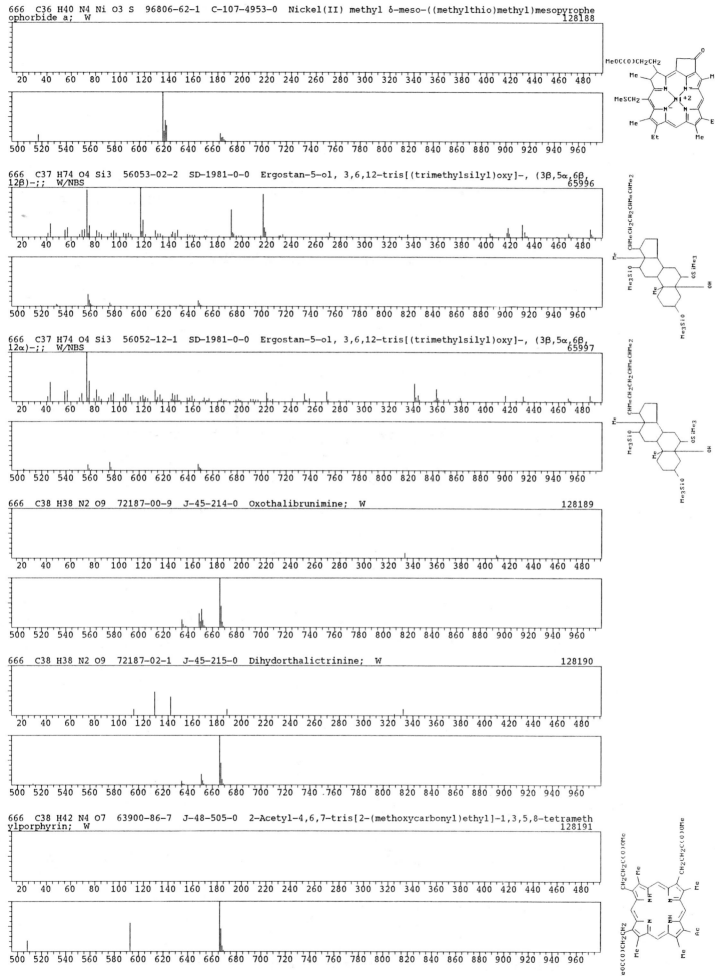

666   C37 H74 O4 Si3   56053-02-2   SD-1981-0-0   Ergostan-5-ol, 3,6,12-tris[(trimethylsilyl)oxy]-, (3β,5α,6β,12β)-;;   W/NBS

65996

666   C37 H74 O4 Si3   56052-12-1   SD-1981-0-0   Ergostan-5-ol, 3,6,12-tris[(trimethylsilyl)oxy]-, (3β,5α,6β,12α)-;;   W/NBS

65997

666   C38 H38 N2 O9   72187-00-9   J-45-214-0   Oxothalibrunimine;   W

128189

666   C38 H38 N2 O9   72187-02-1   J-45-215-0   Dihydorthalictrinine;   W

128190

666   C38 H42 N4 O7   63900-86-7   J-48-505-0   2-Acetyl-4,6,7-tris[2-(methoxycarbonyl)ethyl]-1,3,5,8-tetramethylporphyrin;   W

128191

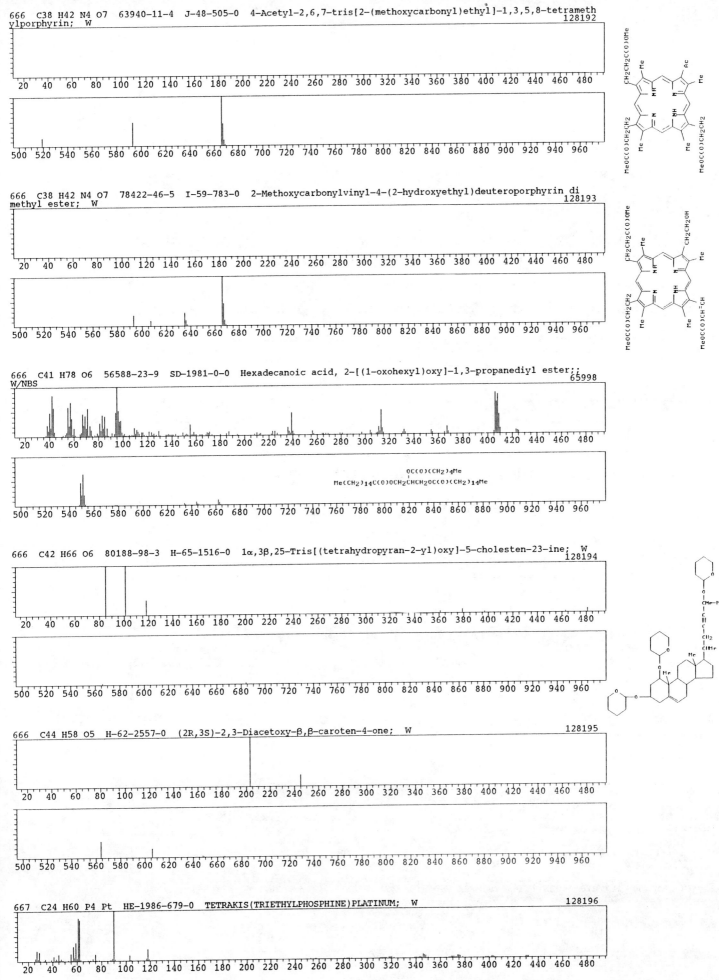

666   C38 H42 N4 O7   63940-11-4   J-48-505-0   4-Acetyl-2,6,7-tris[2-(methoxycarbonyl)ethyl]-1,3,5,8-tetrameth
ylporphyrin; W   128192

666   C38 H42 N4 O7   78422-46-5   I-59-783-0   2-Methoxycarbonylvinyl-4-(2-hydroxyethyl)deuteroporphyrin di
methyl ester; W   128193

666   C41 H78 O6   56588-23-9   SD-1981-0-0   Hexadecanoic acid, 2-[(1-oxohexyl)oxy]-1,3-propanediyl ester;;
W/NBS   65998

666   C42 H66 O6   80188-98-3   H-65-1516-0   1α,3β,25-Tris[(tetrahydropyran-2-yl)oxy]-5-cholesten-23-ine; W
                                                                                                             128194

666   C44 H58 O5   H-62-2557-0   (2R,3S)-2,3-Diacetoxy-β,β-caroten-4-one; W   128195

667   C24 H60 P4 Pt   HE-1986-679-0   TETRAKIS(TRIETHYLPHOSPHINE)PLATINUM; W   128196

667 C33 H57 N5 O9 30859-37-1 O-4-179-7 β-Alanine, N-[N-[N-[N-[1-(D-2-hydroxy-4-methylvaleryl)-L-prolyl]
-L-isoleucyl]-N-methyl-L-valyl]-N-methyl-L-alanyl]-, methyl ester, acetate (ester);; METHYL O-ACETYLDESTRUX
INATE B; W/NBS   65999

667 C37 H47 Fe N4 O2 S 82008-29-5 KA-1984-571-4 Methylsulfinato(2,3,7,8,12,13,17,18-octaethylprophyrina
to)iron; W   128197

667 C42 H25 N3 O6 128-89-2 NS-0-0-0 Benzamide, N-[4-[[5-(benzoylamino)-9,10-dihydro-9,10-dioxo-1-anthra
cenyl]amino]-9,10-dihydro-9,10-dioxo-1-anthracenyl]-; W/NBS   128198

Str 1162

667 C42 H25 N3 O6 129-28-2 NS--27326-0-0 Benzamide, N,N'-[iminobis(9,10-dihydro-9,10-dioxo-5,1-anthrace
nediyl)]bis-;; 5,5'-Dibenzamido-1,1'-dianthrimide; NBS   66000

667 C42 H69 N O5 71609-93-3 B-32-606-0 (25S)-22,26-PIYALOYLEPIMINO-5α-CHOLEST-22-ENE-3β,16α-DIOL DIPIVA
LATE; W   66001

667 C44 H28 Mn N4 31004-82-7 KO-3-74-3 Manganese, [5,10,15,20-tetraphenyl-21H,23H-porphinato(2-)-N21,N2
2,N23,N24]-, (SP-4-1)-;; (MNTPP)MN M5-TETRAPHENYLPORPHIN; W/NBS   66002

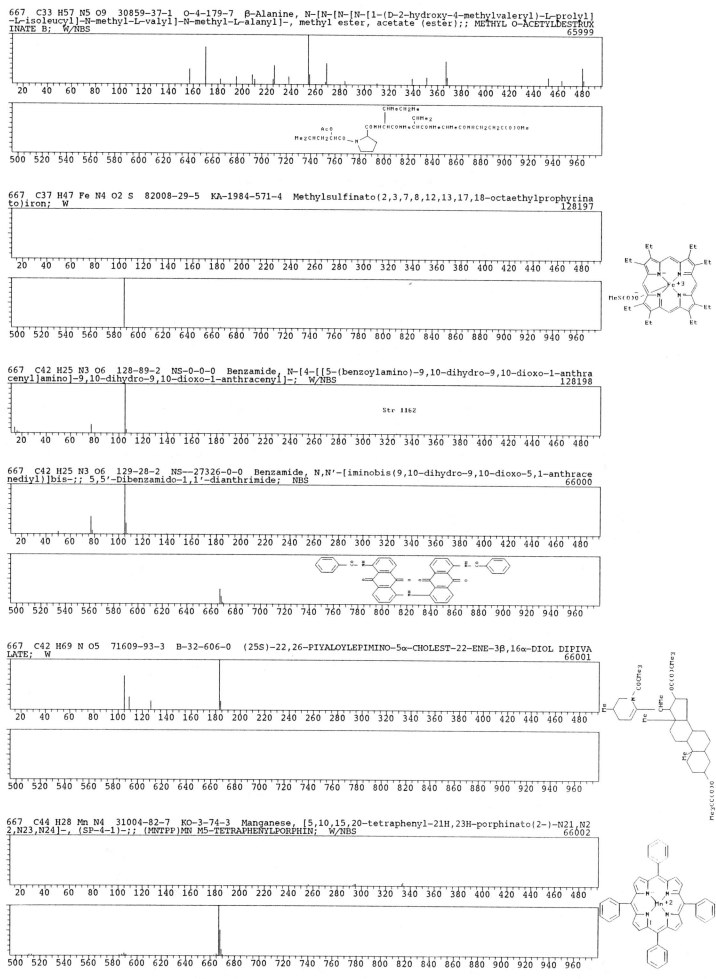

668　C8 Cl2 O8 Re2　15189-52-3　AD-0-1569-0　Rhenium, octacarbonyldi-.mu.-chlorodi-;; DI-.MU.-CHLORO-OCTACA
RBONYL-DIRHENIUM;　W/NBS
66003

668　C27 H48 Br Co O6 P2　HE-1986-2196-0　(2-BROMO-HAPTO-5-INDENYL)-BIS(TRIISOPROPYLPHOSPHITE)COBALT;　W
128199

668　C29 H47 B3 O15　HE-1986-786-0　MANNO-LYXO-GALACTO-TRISACCHARIDE;　W
128200

668　C30 H24 As Bi　62335-93-7　AG-122-354-1　(o-Diphenylarsinophenyl)diphenylbismuthine;　W
128201

668　C32 H28 O16　81757-70-2　B-35-654-0　2,2',4,6,6'-Pentaacetoxy-1'-(3,5-diacetoxyphenoxy)biphenyl ether;
W
128202

668　C33 H32 O15　CD-342-0-0　Puerarin acetate;　W
128203

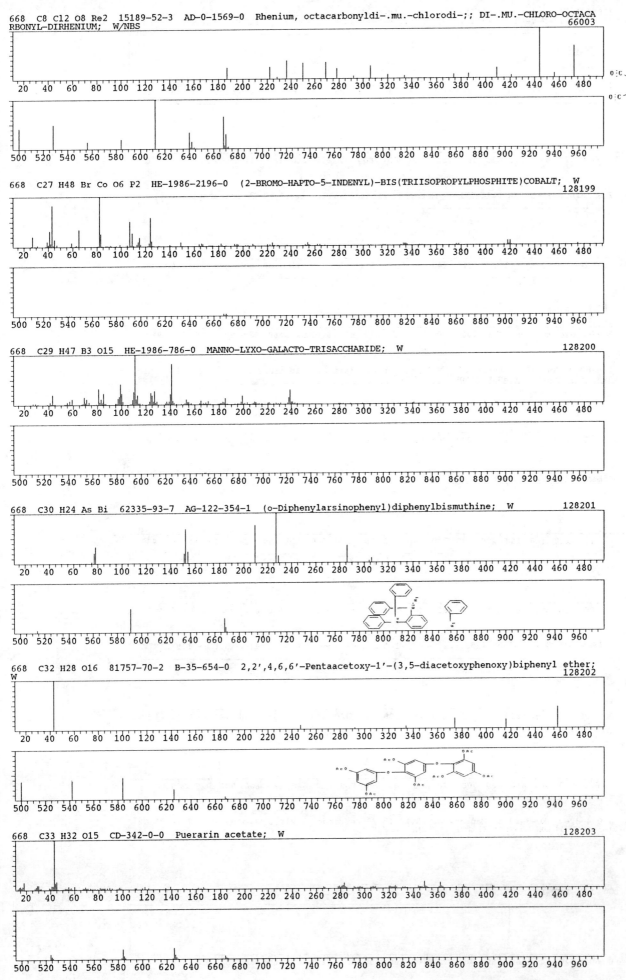

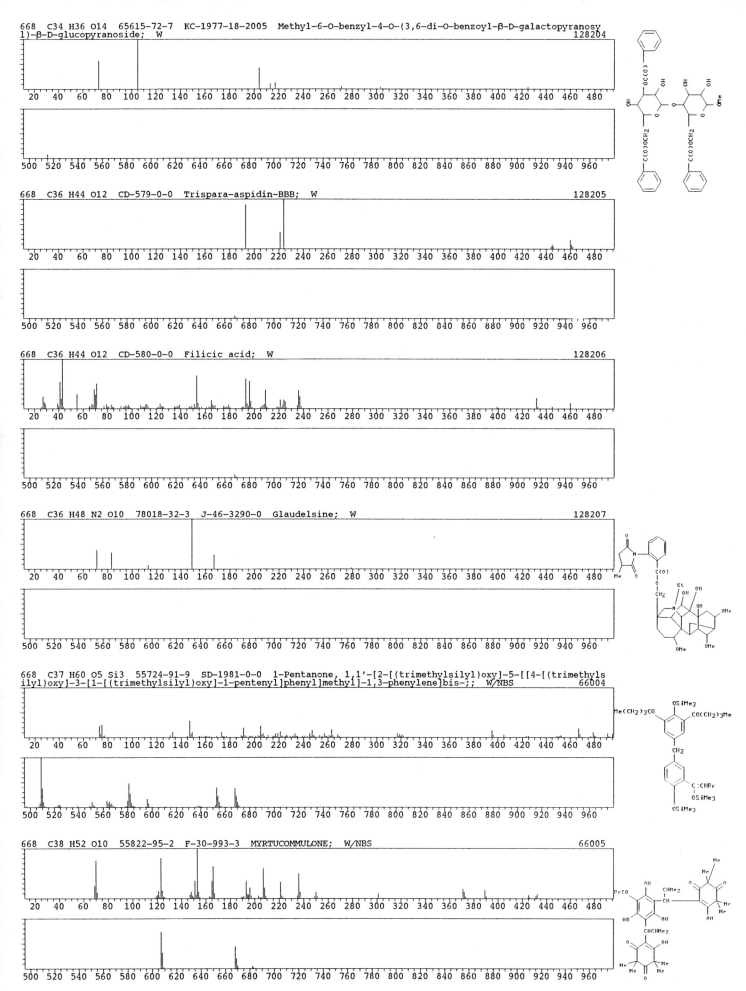

668 C34 H36 O14 65615-72-7 KC-1977-18-2005 Methyl-6-O-benzyl-4-O-(3,6-di-O-benzoyl-β-D-galactopyranosyl)-β-D-glucopyranoside; W   128204

668 C36 H44 O12 CD-579-0-0 Trispara-aspidin-BBB; W   128205

668 C36 H44 O12 CD-580-0-0 Filicic acid; W   128206

668 C36 H48 N2 O10 78018-32-3 J-46-3290-0 Glaudelsine; W   128207

668 C37 H60 O5 Si3 55724-91-9 SD-1981-0-0 1-Pentanone, 1,1'-[2-[(trimethylsilyl)oxy]-5-[[4-[(trimethylsilyl)oxy]-3-[1-[(trimethylsilyl)oxy]-1-pentenyl]phenyl]methyl]-1,3-phenylene]bis-;; W/NBS   66004

668 C38 H52 O10 55822-95-2 F-30-993-3 MYRTUCOMMULONE; W/NBS   66005

668   C38 H56 N2 O4 S2   60405-24-5   H-59-1565-12   N-PHENETHYL-1,16-HEXADECANDIAMIN-DI-P-TOLUOLSULFONIC ACID
AMIDE;   W   66006

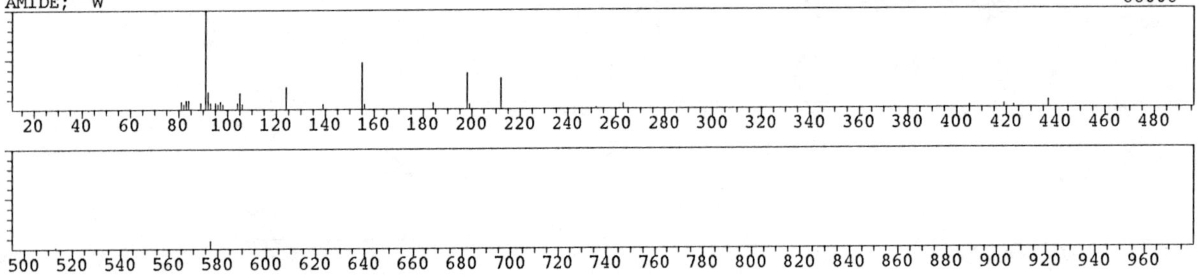

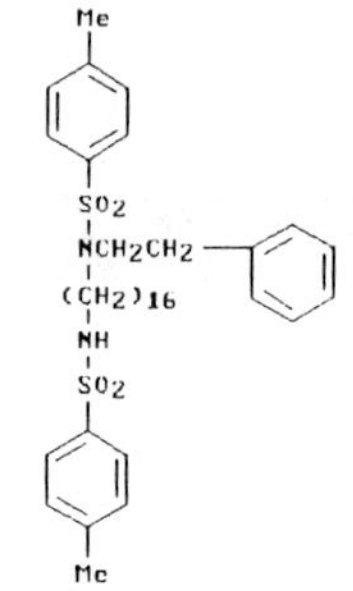

668   C39 H48 N4 O6   50865-93-5   AK-133-497-1   BILIRUBIN-DI-ISOPROPYL ESTER;   W/NBS   66007

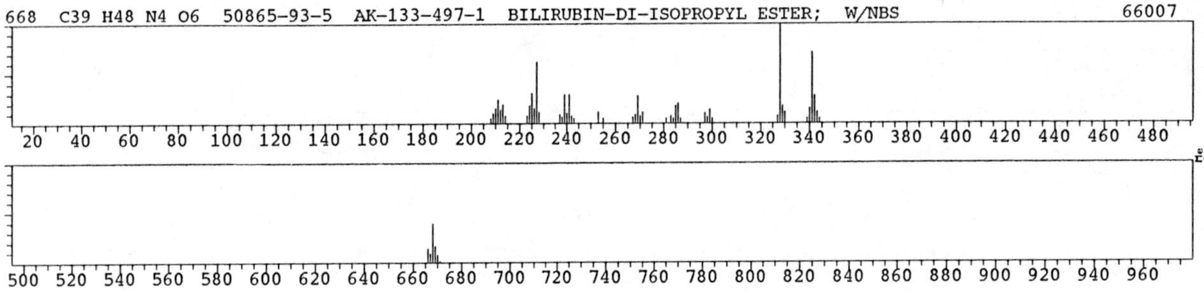

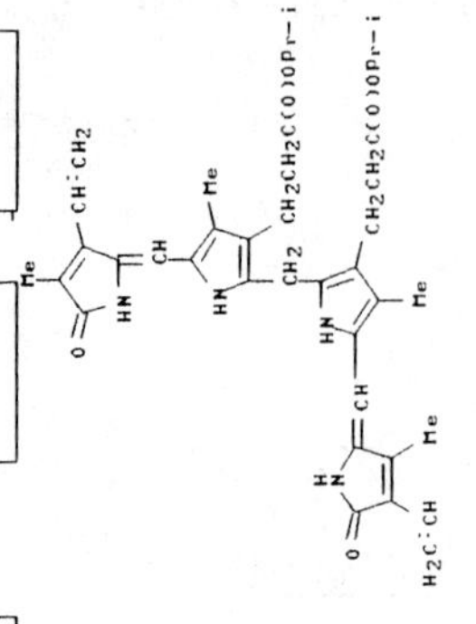

668   C40 H60 O8   F-32-374-0   METHYL 16,16'-BIS(ENT-13,16α-DIHYDROXY-17-NOR-KAURAN-19-OATE-16-YL);   W   66008

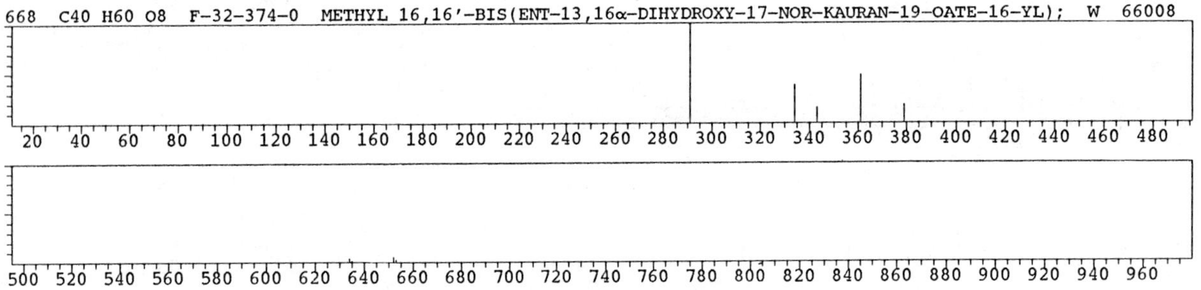

668   C40 H80 O5 Si   W-9-235-9   2-STEARO-1-PALMITIN TMS ETHER;   W   66009

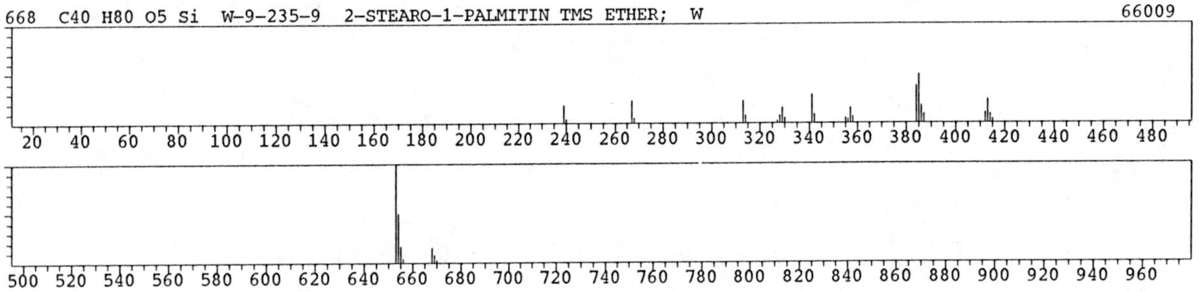

668   C40 H80 O5 Si   W-9-235-9   2-PALMITO-1-STEARIN TMS ETHER;   W   66010

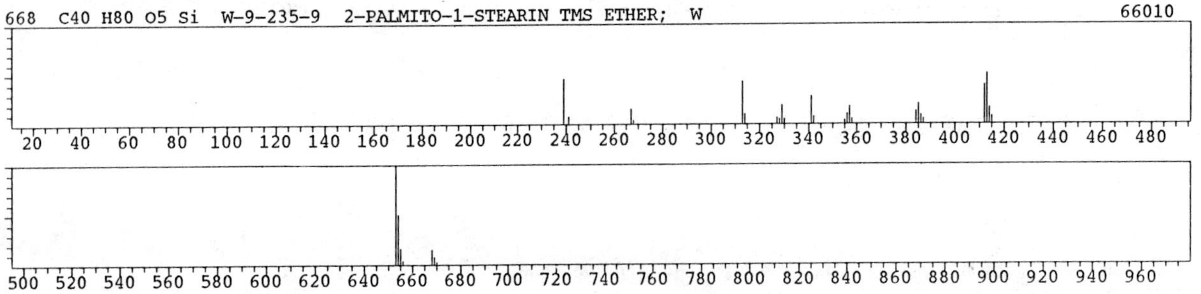

668   C42 H88 O3 Si   67604-83-5   KO-5-87-0   1,2-DIOCTADECYL-3-TMS-GLYCEROL;   W   66011

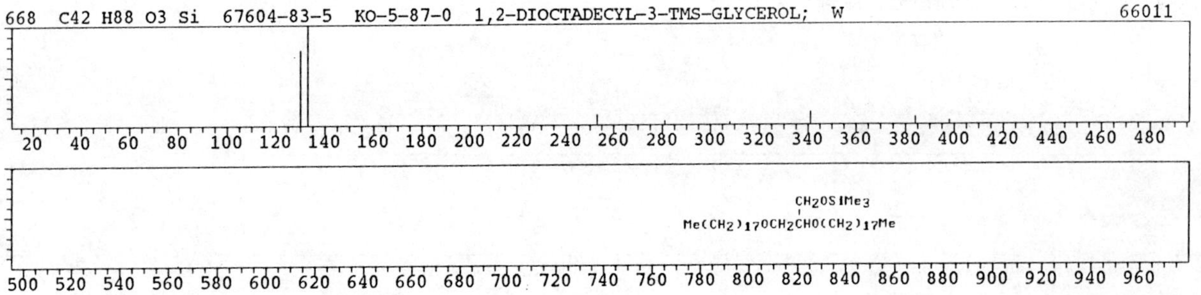

CH2OSiMe3
|
Me(CH2)17OCH2CHO(CH2)17Me

668   C43 H56 O4 S   88904-71-6   J-49-1088-0   (3β,22E)-25-(4-Methoxyphenyl)-25-(phenylthio)-3-[(tetrahydro-2H
-pyran-2-yl)oxy]26,27-dinorcholesta-5,22,24-trien-20-ol;   W   128208

668   C44 H28 Fe N4   16591-56-3   KO-3-74-3   (FETPP)FE MS-TETRAPHENYLPORPHIN;   W/NBS   66012

668   C44 H60 O5   61229-20-7   B-29-1814-0   Lupane-3β,16β,28-triol 3,16-dibenzoate;   W   66013

668   C45 H80 O3   58870-28-3   NS--27320-0-0   Cholest-5-ene-3,7-diol, 7-octadecanoate, (3β,7α)-;;   NBS   66014

668   C48 H28 O4   69699-04-3   F-34-1991-0   2,3,5,6-TETRAPHENYL-1,4-BIS(2,7-DIOXOBENZ(3,4)CYCLOPENTANYLIDENE)
-2,5-CYCLOHEXADIENE;   W   66015

668   C48 H36 N4   69395-37-5   H-61-2821-0   Hexabenzo[b,f,j,p,t,x]-1,4,15,18-tetraazacyclooctacosene;   W
                                                                                                          128209

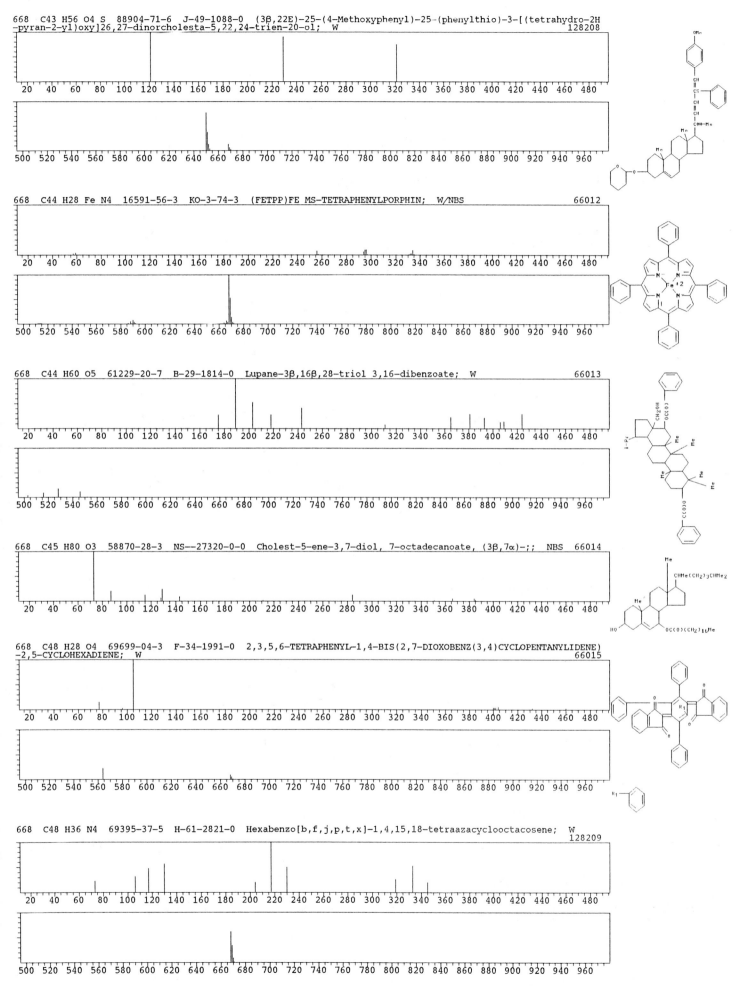

668   C50 H68   30627-86-2   SB-33-556-0   Tetraanhydrobacterioruberin;   W   128210

20   40   60   80   100 120 140 160 180 200 220 240 260 280 300 320 340 360 380 400 420 440 460 480

500 520 540 560 580 600 620 640 660 680 700 720 740 760 780 800 820 840 860 880 900 920 940 960

668   C50 H68   73365-74-9   SB-33-557-0   (2'S)-Δ-1-Tetraanhydrobacterioruberin;   W   128211

20   40   60   80   100 120 140 160 180 200 220 240 260 280 300 320 340 360 380 400 420 440 460 480

500 520 540 560 580 600 620 640 660 680 700 720 740 760 780 800 820 840 860 880 900 920 940 960

668   C17 13C36 O2   SE-4-0-0   1-13C-Heptadecanoate (53.6% 13C);   W   128212

20   40   60   80   100 120 140 160 180 200 220 240 260 280 300 320 340 360 380 400 420 440 460 480

669   C15 F25 N   30649-87-7   KC-73-1545-3   1-Azatetracyclo[2.2.0.0(2,6).0(3,5)]hexane, 2,3,4,5,6-pentakis(pe
ntafluoroethyl)-;;   PENTAKIS(PENTAFLUOROETHYL)-1-AZAPRISMANE;   W/NBS   66017

20   40   60   80   100 120 140 160 180 200 220 240 260 280 300 320 340 360 380 400 420 440 460 480

500 520 540 560 580 600 620 640 660 680 700 720 740 760 780 800 820 840 860 880 900 920 940 960

669   C15 F25 N   30567-91-0   KC-73-1545-2   1-Azabicyclo[2.2.0]hexa-2,5-diene, 2,3,4,5,6-pentakis(pentafluoro
ethyl)-;;   PENTAKIS(PENTAFLUOROETHYL)-1-AZABICYCLO(2,2,O)HEXA-2,5-DIENE;   W/NBS   66018

20   40   60   80   100 120 140 160 180 200 220 240 260 280 300 320 340 360 380 400 420 440 460 480

500 520 540 560 580 600 620 640 660 680 700 720 740 760 780 800 820 840 860 880 900 920 940 960

669   C15 F25 N   20017-53-2   KC-73-1545-1   Pyridine, pentakis(pentafluoroethyl)-;;   PENTAKIS (PENTAFLUOROETH
YL)PYRIDINE;   W/NBS   66019

20   40   60   80   100 120 140 160 180 200 220 240 260 280 300 320 340 360 380 400 420 440 460 480

500 520 540 560 580 600 620 640 660 680 700 720 740 760 780 800 820 840 860 880 900 920 940 960

669   C27 H21 Br2 N5 O6   64775-03-7   H-60-1992-0   9-(2-O-ACETYL-5-O-BENZOYL-3-DEOXY-3-C-DIBROMOMETHYLIDENE-β
-D-ERYTHRO-1-PENTOFURANOSYL)-6-BENZAMIDO-PURINE;   W   66020

20   40   60   80   100 120 140 160 180 200 220 240 260 280 300 320 340 360 380 400 420 440 460 480

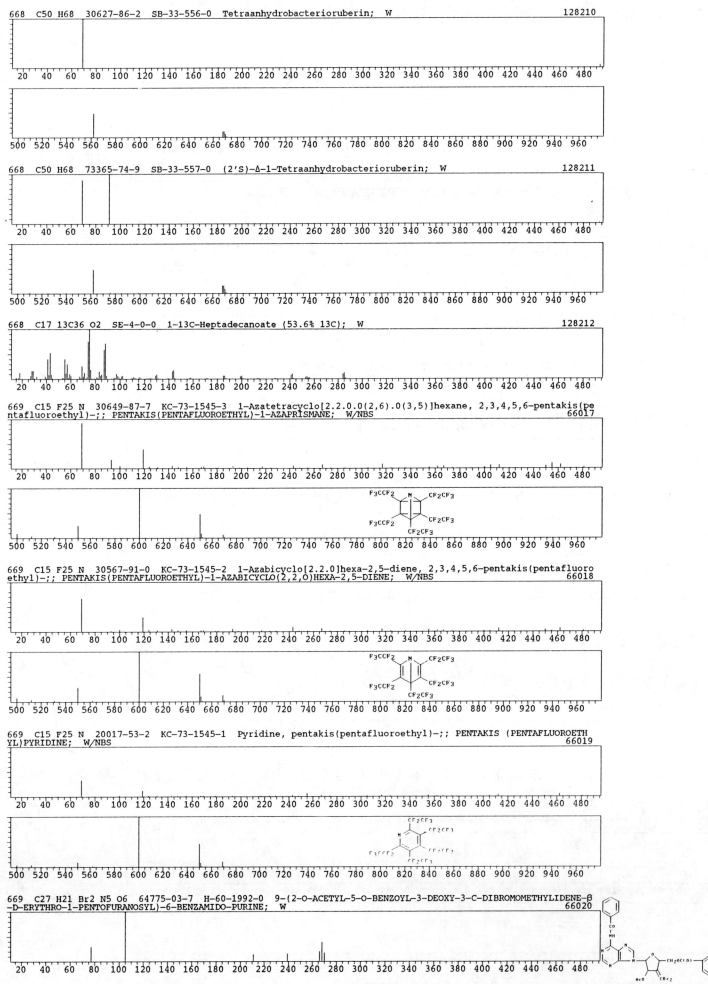

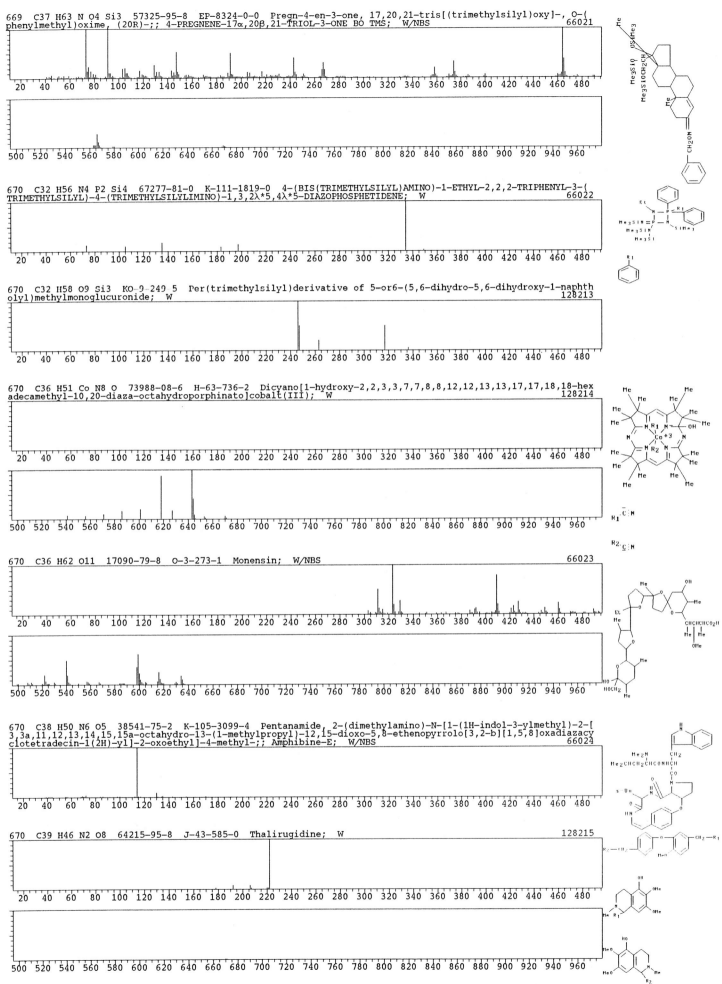

669　C37 H63 N O4 Si3　57325-95-8　EP-8324-0-0　Pregn-4-en-3-one, 17,20,21-tris[(trimethylsilyl)oxy]-, O-(phenylmethyl)oxime, (20R)-;; 4-PREGNENE-17α,20β,21-TRIOL-3-ONE BO TMS; W/NBS　66021

670　C32 H56 N4 P2 Si4　67277-81-0　K-111-1819-0　4-(BIS(TRIMETHYLSILYL)AMINO)-1-ETHYL-2,2,2-TRIPHENYL-3-(TRIMETHYLSILYL)-4-(TRIMETHYLSILYLIMINO)-1,3,2λ*5,4λ*5-DIAZOPHOSPHETIDENE; W　66022

670　C32 H58 O9 Si3　KO-9-249-5　Per(trimethylsilyl)derivative of 5-or6-(5,6-dihydro-5,6-dihydroxy-1-naphtholyl)methylmonoglucuronide; W　128213

670　C36 H51 Co N8 O　73988-08-6　H-63-736-2　Dicyano[1-hydroxy-2,2,3,3,7,7,8,8,12,12,13,13,17,17,18,18-hexadecamethyl-10,20-diaza-octahydroporphinato]cobalt(III); W　128214

670　C36 H62 O11　17090-79-8　O-3-273-1　Monensin; W/NBS　66023

670　C38 H50 N6 O5　38541-75-2　K-105-3099-4　Pentanamide, 2-(dimethylamino)-N-[1-(1H-indol-3-ylmethyl)-2-[3,3a,11,12,13,14,15,15a-octahydro-13-(1-methylpropyl)-12,15-dioxo-5,8-ethenopyrrolo[3,2-b][1,5,8]oxadiazacyclotetradecin-1(2H)-yl]-2-oxoethyl]-4-methyl-;; Amphibine-E; W/NBS　66024

670　C39 H46 N2 O8　64215-95-8　J-43-585-0　Thalirugidine; W　128215

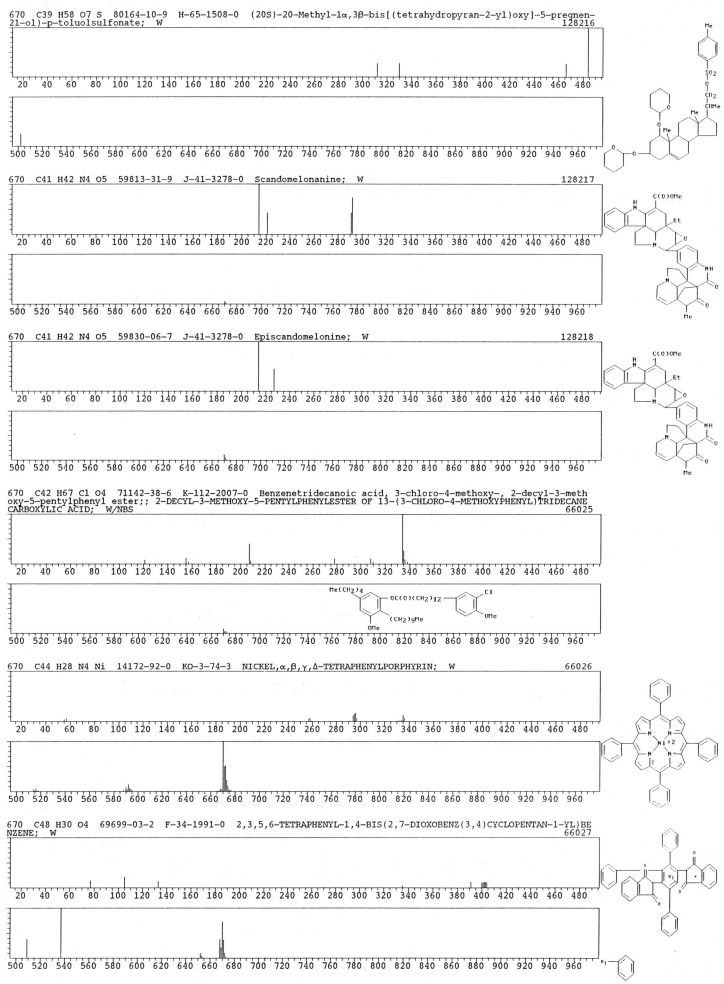

670   C39 H58 O7 S   80164-10-9   H-65-1508-0   (20S)-20-Methyl-1α,3β-bis[(tetrahydropyran-2-yl)oxy]-5-pregnen-21-ol)-p-toluolsulfonate;   W   128216

670   C41 H42 N4 O5   59813-31-9   J-41-3278-0   Scandomelonanine;   W   128217

670   C41 H42 N4 O5   59830-06-7   J-41-3278-0   Episcandomelonine;   W   128218

670   C42 H67 Cl O4   71142-38-6   K-112-2007-0   Benzenetridecanoic acid, 3-chloro-4-methoxy-, 2-decyl-3-meth
oxy-5-pentylphenyl ester;; 2-DECYL-3-METHOXY-5-PENTYLPHENYLESTER OF 13-(3-CHLORO-4-METHOXYPHENYL)TRIDECANE
CARBOXYLIC ACID;   W/NBS   66025

670   C44 H28 N4 Ni   14172-92-0   KO-3-74-3   NICKEL,α,β,γ,Δ-TETRAPHENYLPORPHYRIN;   W   66026

670   C48 H30 O4   69699-03-2   F-34-1991-0   2,3,5,6-TETRAPHENYL-1,4-BIS(2,7-DIOXOBENZ(3,4)CYCLOPENTAN-1-YL)BE
NZENE;   W   66027

671   C12 F27 N   311-89-7   HE-1982-0-0   Perfluorotributylamine;   W
66029

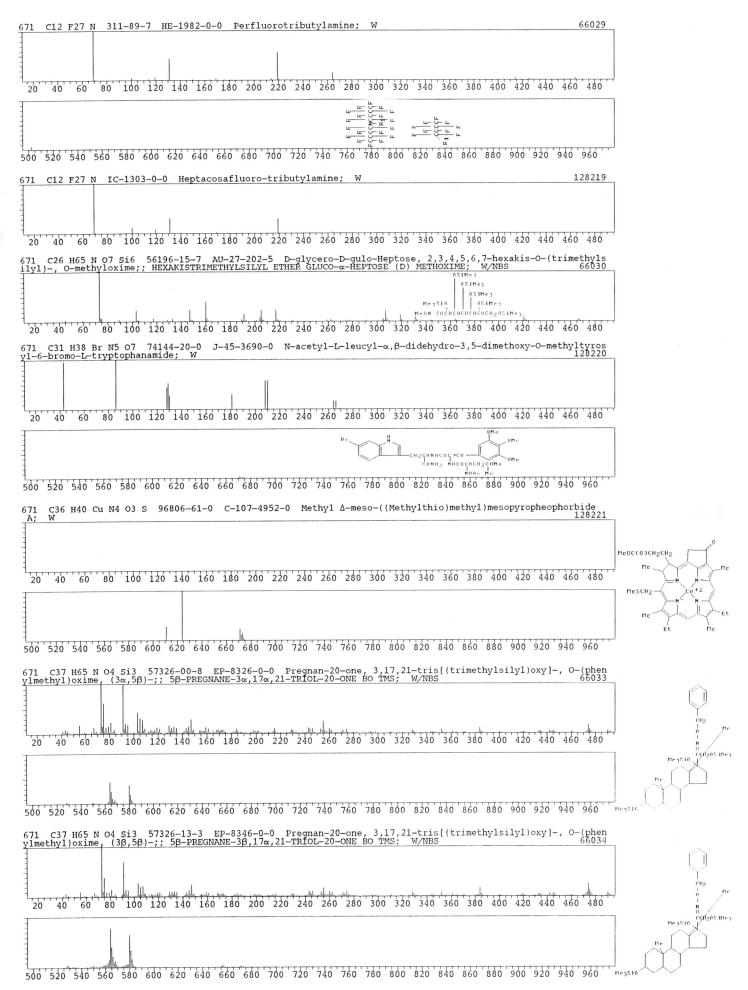

671   C12 F27 N   IC-1303-0-0   Heptacosafluoro-tributylamine;   W
128219

671   C26 H65 N O7 Si6   56196-15-7   AU-27-202-5   D-glycero-D-gulo-Heptose, 2,3,4,5,6,7-hexakis-O-(trimethyls
ilyl)-, O-methyloxime;;   HEXAKISTRIMETHYLSILYL ETHER GLUCO-α-HEPTOSE (D) METHOXIME;   W/NBS
66030

671   C31 H38 Br N5 O7   74144-20-0   J-45-3690-0   N-acetyl-L-leucyl-α,β-didehydro-3,5-dimethoxy-O-methyltyros
yl-6-bromo-L-tryptophanamide;   W
128220

671   C36 H40 Cu N4 O3 S   96806-61-0   C-107-4952-0   Methyl Δ-meso-((Methylthio)methyl)mesopyropheophorbide
A;   W
128221

671   C37 H65 N O4 Si3   57326-00-8   EP-8326-0-0   Pregnan-20-one, 3,17,21-tris[(trimethylsilyl)oxy]-, O-(phen
ylmethyl)oxime, (3α,5β)-;;   5β-PREGNANE-3α,17α,21-TRIOL-20-ONE BO TMS;   W/NBS
66033

671   C37 H65 N O4 Si3   57326-13-3   EP-8346-0-0   Pregnan-20-one, 3,17,21-tris[(trimethylsilyl)oxy]-, O-(phen
ylmethyl)oxime, (3β,5β)-;;   5β-PREGNANE-3β,17α,21-TRIOL-20-ONE BO TMS;   W/NBS
66034

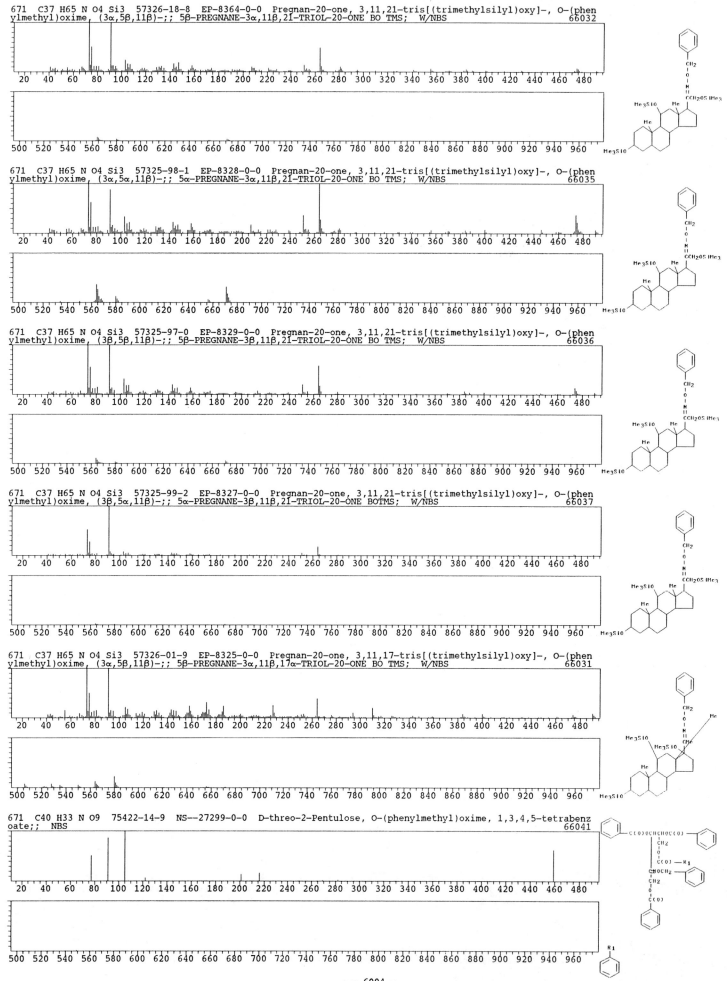

671 C37 H65 N O4 Si3 57326-18-8 EP-8364-0-0 Pregnan-20-one, 3,11,21-tris[(trimethylsilyl)oxy]-, O-(phenylmethyl)oxime, (3α,5β,11β)-;; 5β-PREGNANE-3α,11β,21-TRIOL-20-ONE BO TMS; W/NBS   66032

671 C37 H65 N O4 Si3 57325-98-1 EP-8328-0-0 Pregnan-20-one, 3,11,21-tris[(trimethylsilyl)oxy]-, O-(phenylmethyl)oxime, (3α,5α,11β)-;; 5α-PREGNANE-3α,11β,21-TRIOL-20-ONE BO TMS; W/NBS   66035

671 C37 H65 N O4 Si3 57325-97-0 EP-8329-0-0 Pregnan-20-one, 3,11,21-tris[(trimethylsilyl)oxy]-, O-(phenylmethyl)oxime, (3β,5β,11β)-;; 5β-PREGNANE-3β,11β,21-TRIOL-20-ONE BO TMS; W/NBS   66036

671 C37 H65 N O4 Si3 57325-99-2 EP-8327-0-0 Pregnan-20-one, 3,11,21-tris[(trimethylsilyl)oxy]-, O-(phenylmethyl)oxime, (3β,5α,11β)-;; 5α-PREGNANE-3β,11β,21-TRIOL-20-ONE BOTMS; W/NBS   66037

671 C37 H65 N O4 Si3 57326-01-9 EP-8325-0-0 Pregnan-20-one, 3,11,17-tris[(trimethylsilyl)oxy]-, O-(phenylmethyl)oxime, (3α,5β,11β)-;; 5β-PREGNANE-3α,11β,17α-TRIOL-20-ONE BO TMS; W/NBS   66031

671 C40 H33 N O9 75422-14-9 NS--27299-0-0 D-threo-2-Pentulose, O-(phenylmethyl)oxime, 1,3,4,5-tetrabenzoate;; NBS   66041

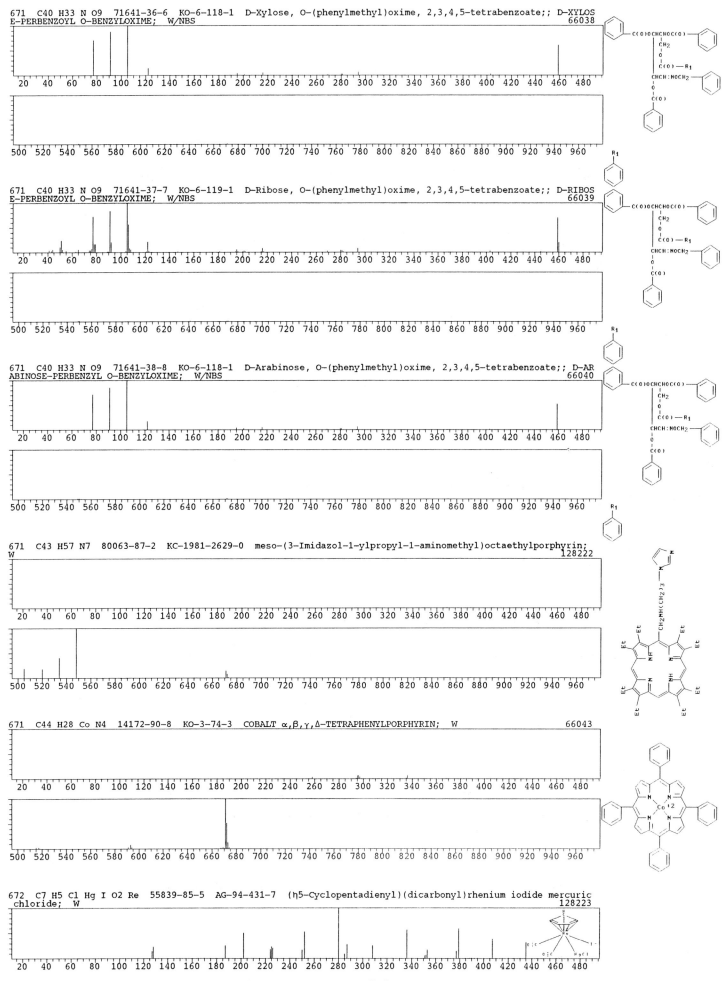

671　C40 H33 N O9　71641-36-6　KO-6-118-1　D-Xylose, O-(phenylmethyl)oxime, 2,3,4,5-tetrabenzoate;; D-XYLOS
E-PERBENZOYL O-BENZYLOXIME;　W/NBS　　　　　　　　　　　　　　　　　　　　　　　　　　　　66038

671　C40 H33 N O9　71641-37-7　KO-6-119-1　D-Ribose, O-(phenylmethyl)oxime, 2,3,4,5-tetrabenzoate;; D-RIBOS
E-PERBENZOYL O-BENZYLOXIME;　W/NBS　　　　　　　　　　　　　　　　　　　　　　　　　　　　66039

671　C40 H33 N O9　71641-38-8　KO-6-118-1　D-Arabinose, O-(phenylmethyl)oxime, 2,3,4,5-tetrabenzoate;; D-AR
ABINOSE-PERBENZYL O-BENZYLOXIME;　W/NBS　　　　　　　　　　　　　　　　　　　　　　　　66040

671　C43 H57 N7　80063-87-2　KC-1981-2629-0　meso-(3-Imidazol-1-ylpropyl-1-aminomethyl)octaethylporphyrin;
W　　128222

671　C44 H28 Co N4　14172-90-8　KO-3-74-3　COBALT α,β,γ,Δ-TETRAPHENYLPORPHYRIN;　W　　　　66043

672　C7 H5 Cl Hg I O2 Re　55839-85-5　AG-94-431-7　(η5-Cyclopentadienyl)(dicarbonyl)rhenium iodide mercuric
chloride;　W　　　　　　　　　　　　　　　　　　　　　　　　　　　　　　　　　　　　128223

672 C13 H36 Sn4 56177-41-4 O-10-19-2 Stannane, methanetetrayltetrakis[trimethyl-;; TETRAKIS(TRIMETHYLST
ANNANE)METHANE; W/NBS   66044

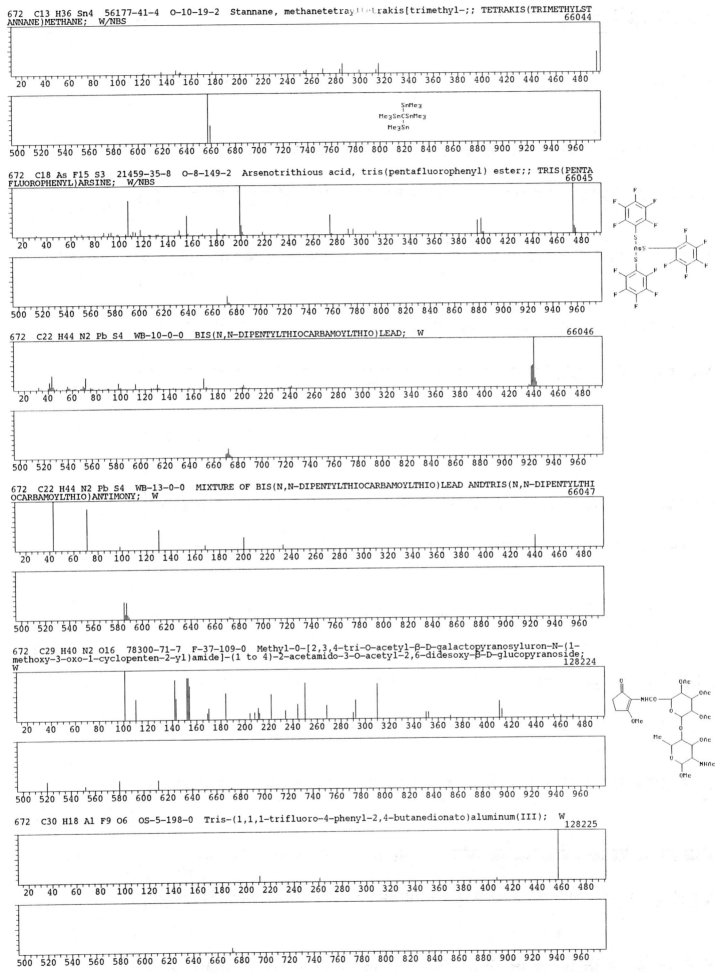

672 C18 As F15 S3 21459-35-8 O-8-149-2 Arsenotrithious acid, tris(pentafluorophenyl) ester;; TRIS(PENTA
FLUOROPHENYL)ARSINE; W/NBS   66045

672 C22 H44 N2 Pb S4 WB-10-0-0 BIS(N,N-DIPENTYLTHIOCARBAMOYLTHIO)LEAD; W   66046

672 C22 H44 N2 Pb S4 WB-13-0-0 MIXTURE OF BIS(N,N-DIPENTYLTHIOCARBAMOYLTHIO)LEAD ANDTRIS(N,N-DIPENTYLTHI
OCARBAMOYLTHIO)ANTIMONY; W   66047

672 C29 H40 N2 O16 78300-71-7 F-37-109-0 Methyl-O-[2,3,4-tri-O-acetyl-β-D-galactopyranosyluron-N-(1-
methoxy-3-oxo-1-cyclopenten-2-yl)amide]-(1 to 4)-2-acetamido-3-O-acetyl-2,6-didesoxy-β-D-glucopyranoside;
W   128224

672 C30 H18 Al F9 O6 OS-5-198-0 Tris-(1,1,1-trifluoro-4-phenyl-2,4-butanedionato)aluminum(III); W
                                                                                                128225

672 C32 H32 O16 35286-54-5 EP-2240-0-0 Vogelin pentaacetate; W/NBS   66048

672 C33 H68 O6 Si4 OM-1981-661-0 19-R-HYDROXY PGF2A ME ESTER 4TMS; W   66049

672 C34 H20 Cl4 N4 O3 IC-1304-0-0 2-Hydroxy-3-(naphth-1-ylaminocarbonyl)-1-[2,3,5,6-tetrachloro-4-(phen
ylaminocarbonyl)phenylazo]naphthalene; W   128226

672 C35 H52 N4 O5 S2 67171-90-8 H-61-1351-0 Benzenesulfonamide, N-(2-cyanoethyl)-4-methyl-N-[3-[[(4-
methylphenyl)sulfonyl][3-(2-oxoazacyclotridec-1-yl)propyl]amino]propyl]-;; Azacyclotridecane, benzenesulfon
amide deriv.; W/NBS   128227

672 C35 H72 O6 Si3 BS-5-36-0 5α,7α,16-Trihydroxy-11-oxotetranor-prostanoic acid methyl ester tert-butyl
dimethylsilyl ether; W   128228

672 C36 H48 O12 64379-05-1 C-99-6445-0 Δ6-TETRAHYDROCANNABINOL-C-4'-GLUCURONIDE METHYL ESTER TETRAACET
ATE; W   66051

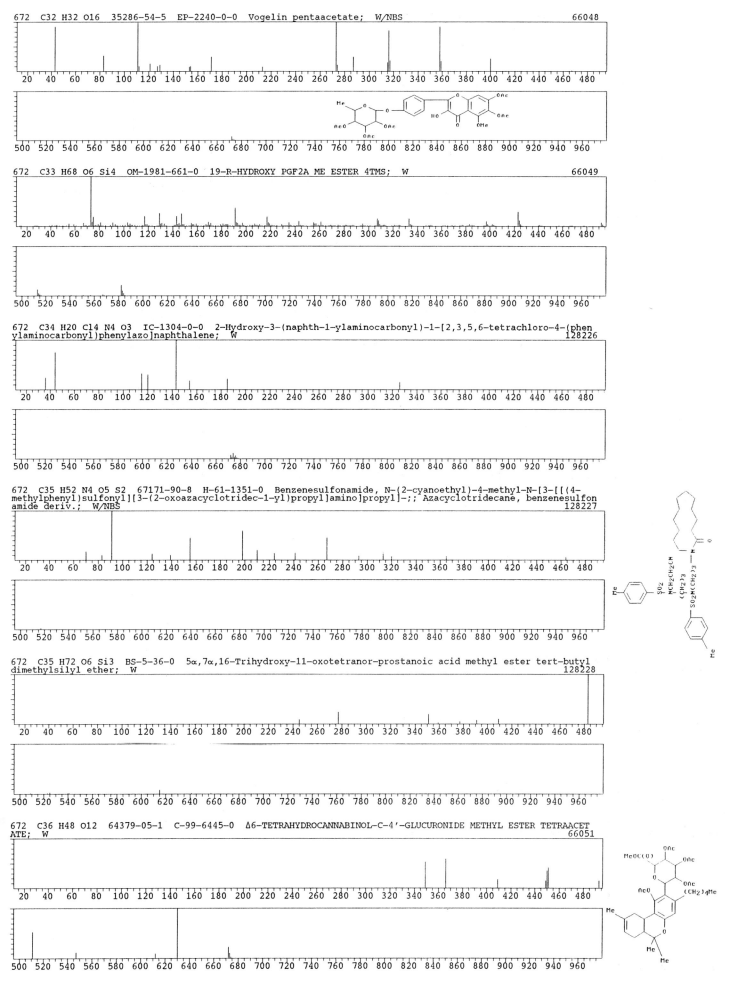

672 C36 H48 O12 64379-06-2 C-99-6446-0 Δ6-TETRAHYDROCANNABINOL-C-6'-GLUCURONIDE METHYL ESTER TETRAACET
ATE; W   66052

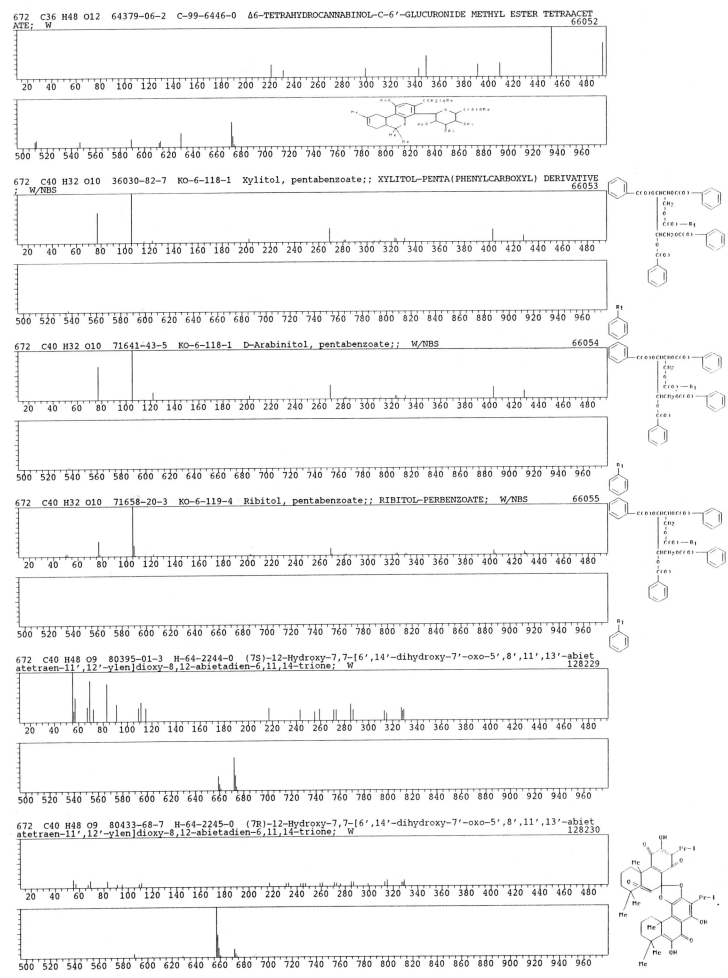

672 C40 H32 O10 36030-82-7 KO-6-118-1 Xylitol, pentabenzoate;; XYLITOL-PENTA(PHENYLCARBOXYL) DERIVATIVE
; W/NBS   66053

672 C40 H32 O10 71641-43-5 KO-6-118-1 D-Arabinitol, pentabenzoate;; W/NBS   66054

672 C40 H32 O10 71658-20-3 KO-6-119-4 Ribitol, pentabenzoate;; RIBITOL-PERBENZOATE; W/NBS   66055

672 C40 H48 O9 80395-01-3 H-64-2244-0 (7S)-12-Hydroxy-7,7-[6',14'-dihydroxy-7'-oxo-5',8',11',13'-abiet
atetraen-11',12'-ylen]dioxy-8,12-abietadien-6,11,14-trione; W   128229

672 C40 H48 O9 80433-68-7 H-64-2245-0 (7R)-12-Hydroxy-7,7-[6',14'-dihydroxy-7'-oxo-5',8',11',13'-abiet
atetraen-11',12'-ylen]dioxy-8,12-abietadien-6,11,14-trione; W   128230

672 C40 H48 O9 80395-00-2 H-64-2245-0 (7S)-12-Hydroxy-7,7-[14'-hydroxy-6',7'-dioxo-8',11',13'-abietatri
en-11',12'-ylen]dioxy-8,12-abietadien-6,11,14-trione; W   128231

672 C40 H48 O9 80433-67-6 H-64-2245-0 (7R)-12-Hydroxy-7,7-[14'-hydroxy-6',7'-dioxo-8',11',13'-abietatri
en-11',12'-ylen]dioxy-8,12-abietadien-6,11,14-trione; W   128232

672 C40 H64 O8 27536-56-7 KO-5-166-3 4Aα-PHORBOL DIDECANOATE; W   66056

672 C40 H64 O8 24928-17-4 NS--26258-0-0 Decanoic acid, 1,1a,1b,4,4a,5,7a,7b,8,9-decahydro-4a,7b-dihydr
oxy-3-(hydroxymethyl)-1,1,6,8-tetramethyl-5-oxo-9aH-cyclopropa[3,4]benz[1,2-e]azulene-9,9a-diyl ester, [1aR
-(1aα,1bβ,4aβ,7aα,7bα,8α,9β,9aα)]-; NBS   128233

672 C40 H64 O8 BS-5-7-0 4aα-Phorbol-9,9a-didecanoate; W   128234

672 C41 H44 N4 O5 77745-12-1 J-46-3501-0 Bis(methyl ester) of 2-(and 4-)[1-(2-hydroxy-5-methylphenyl)
ethyl]deuteroporphyrin; W   128235

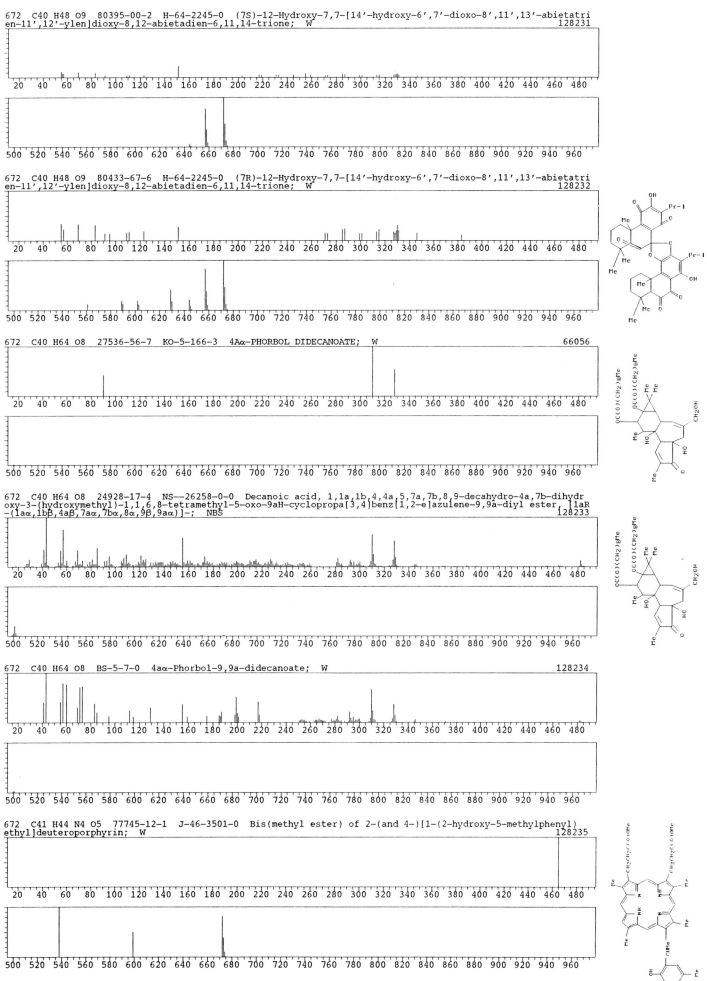

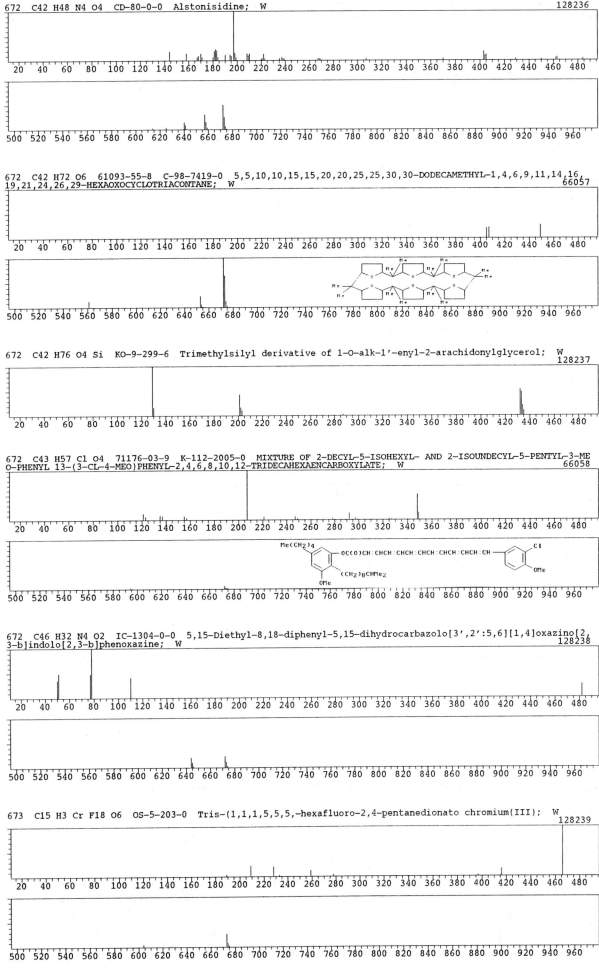

672   C42 H48 N4 O4   CD-80-0-0   Alstonisidine;   W   128236

672   C42 H72 O6   61093-55-8   C-98-7419-0   5,5,10,10,15,15,20,20,25,25,30,30-DODECAMETHYL-1,4,6,9,11,14,16,
19,21,24,26,29-HEXAOXOCYCLOTRIACONTANE;   W   66057

672   C42 H76 O4 Si   KO-9-299-6   Trimethylsilyl derivative of 1-O-alk-1'-enyl-2-arachidonylglycerol;   W
                                                                              128237

672   C43 H57 Cl O4   71176-03-9   K-112-2005-0   MIXTURE OF 2-DECYL-5-ISOHEXYL- AND 2-ISOUNDECYL-5-PENTYL-3-ME
O-PHENYL 13-(3-CL-4-MEO)PHENYL-2,4,6,8,10,12-TRIDECAHEXAENCARBOXYLATE;   W   66058

672   C46 H32 N4 O2   IC-1304-0-0   5,15-Diethyl-8,18-diphenyl-5,15-dihydrocarbazolo[3',2':5,6][1,4]oxazino[2,
3-b]indolo[2,3-b]phenoxazine;   W   128238

673   C15 H3 Cr F18 O6   OS-5-203-0   Tris-(1,1,1,5,5,5,-hexafluoro-2,4-pentanedionato chromium(III);   W
                                                                              128239

673   C26 H55 N5 O5 18O Si5   AV-50-118-4   PENTAKISTRIMETHYLSILYL-7-METHYL-8-OXOGUANOSINE(O8-18O);   W   66059

673   C26 H55 N5 O6 Si5   AV-50-118-3   PENTAKISTRIMETHYLSILYL-7-METHYLGUANOSINE;   W   66060

673   C26 H55 N5 O6 Si5   56273-02-0   NS--27291-0-0   Guanosine, 7,8-dihydro-7-methyl-7,8-oxo, pentakis(tri
methylsilyl) deriv.;;   NBS   66061

673   C34 H26 F6 N O5 P   65138-52-5   K-113-64-0   Dimethylester of 2,2,5,5-Tetraphenyl-7,7-bis(trifluormeth
yl)-6-oxa-1-aza-5λ(5)-phosphabicyclo(3.2.0)hept-3-ene-3,4-dicarboxylic acid;   W   128240

673   C37 H55 N O10   CD-268-0-0   Stenophylline;   W   128241

673   C39 H47 N O9   73111-23-6   H-62-2008-0   (1L)-(1,2,4/3,5)-1',2'-Anhydro-3,4,5-tri-O-benzyl-2-[N-hydroxy-
N-(2',3':5',6'-di-O-isopropylidene-α-D-mannofuranosyl)amino]-1-hydroxymethyl-2,4,5-cyclopentantriol;   W
128242

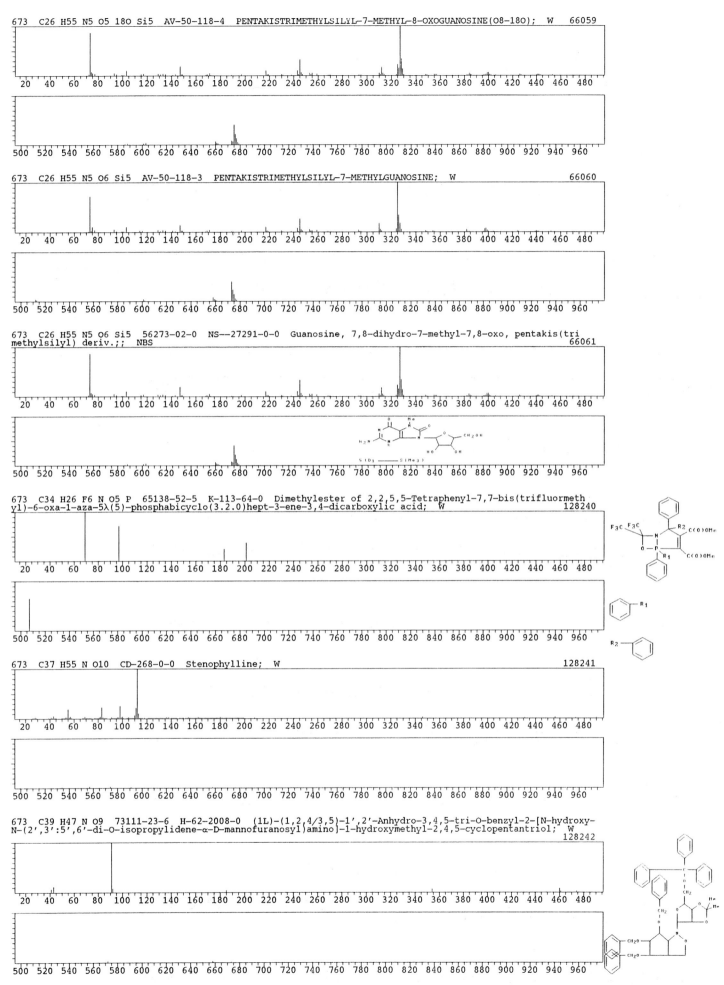

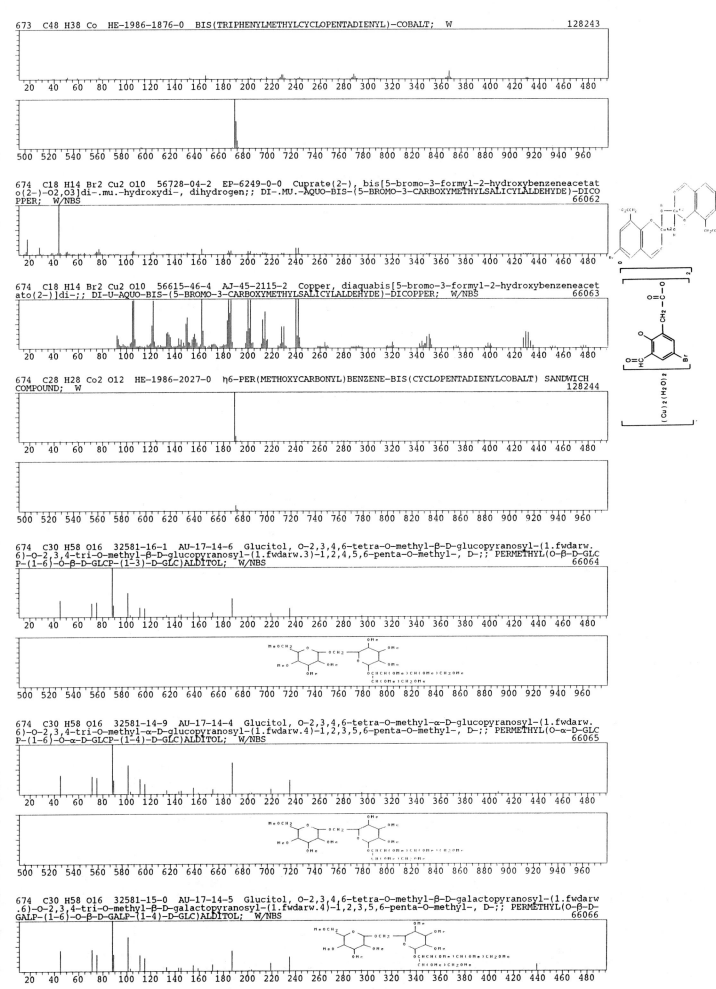

673   C48 H38 Co   HE-1986-1876-0   BIS(TRIPHENYLMETHYLCYCLOPENTADIENYL)-COBALT;   W   128243

674   C18 H14 Br2 Cu2 O10   56728-04-2   EP-6249-0-0   Cuprate(2-), bis[5-bromo-3-formyl-2-hydroxybenzeneacetat
o(2-)-O2,O3]di-.mu.-hydroxydi-, dihydrogen;; DI-.MU.-AQUO-BIS-(5-BROMO-3-CARBOXYMETHYLSALICYLALDEHYDE)-DICO
PPER;   W/NBS   66062

674   C18 H14 Br2 Cu2 O10   56615-46-4   AJ-45-2115-2   Copper, diaquabis[5-bromo-3-formyl-2-hydroxybenzeneacet
ato(2-)]di-;; DI-U-AQUO-BIS-(5-BROMO-3-CARBOXYMETHYLSALICYLALDEHYDE)-DICOPPER;   W/NBS   66063

674   C28 H28 Co2 O12   HE-1986-2027-0   η6-PER(METHOXYCARBONYL)BENZENE-BIS(CYCLOPENTADIENYLCOBALT) SANDWICH
COMPOUND;   W   128244

674   C30 H58 O16   32581-16-1   AU-17-14-6   Glucitol, O-2,3,4,6-tetra-O-methyl-β-D-glucopyranosyl-(1.fwdarw.
6)-O-2,3,4-tri-O-methyl-β-D-glucopyranosyl-(1.fwdarw.3)-1,2,4,5,6-penta-O-methyl-, D-;; PERMETHYL(O-β-D-GLC
P-(1-6)-O-β-D-GLCP-(1-3)-D-GLC)ALDITOL;   W/NBS   66064

674   C30 H58 O16   32581-14-9   AU-17-14-4   Glucitol, O-2,3,4,6-tetra-O-methyl-α-D-glucopyranosyl-(1.fwdarw.
6)-O-2,3,4-tri-O-methyl-α-D-glucopyranosyl-(1.fwdarw.4)-1,2,3,5,6-penta-O-methyl-, D-;; PERMETHYL(O-α-D-GLC
P-(1-6)-O-α-D-GLCP-(1-4)-D-GLC)ALDITOL;   W/NBS   66065

674   C30 H58 O16   32581-15-0   AU-17-14-5   Glucitol, O-2,3,4,6-tetra-O-methyl-β-D-galactopyranosyl-(1.fwdarw
.6)-O-2,3,4-tri-O-methyl-β-D-galactopyranosyl-(1.fwdarw.4)-1,2,3,5,6-penta-O-methyl-, D-;; PERMETHYL(O-β-D-
GALP-(1-6)-O-β-D-GALP-(1-4)-D-GLC)ALDITOL;   W/NBS   66066

674 C30 H58 O16 32581-20-7 AU-17-14-11 Glucitol, O-2,3,4,6-tetra-O-methyl-β-D-glucopyranosyl-(1.fwdarw.
4)-O-2,3,6-tri-O-methyl-β-D-glucopyranosyl-(1.fwdarw.3)-1,2,4,5,6-penta-O-methyl-, D-;; PERMETHYL(O-β-D-GLC
P-(1-4)-O-β-D-GLCP-(1-3)-D-GLC)ALDITOL; W/NBS   66067

674 C30 H58 O16 32581-19-4 AU-17-14-10 Mannitol, O-2,3,4,6-tetra-O-methyl-β-D-mannopyranosyl-(1.fwdarw.
4)-O-2,3,6-tri-O-methyl-β-D-mannopyranosyl-(1.fwdarw.3)-1,2,4,5,6-penta-O-methyl-, D-;; PERMETHYL(O-β-D-MAN
P-(1-4)-O-β-D-MANP-(1-4)-D-MAN)ALDITOL; W/NBS   66068

674 C30 H58 O16 32581-17-2 AU-17-14-8 D-Glucitol, O-2,3,4,6-tetra-O-methyl-α-D-glucopyranosyl-(1.fwdarw
.4)-O-2,3,6-tri-O-methyl-α-D-glucopyranosyl-(1.fwdarw.4)-1,2,3,5,6-penta-O-methyl-;; PERMETHYL(O-α-D-GLCP-(
1-4)-O-α-D-GLCP-(1-4)-D-GLC)ALDITOL; W/NBS   66069

Str 1163

674 C30 H58 O16 32581-18-3 AU-17-14-9 Glucitol, O-2,3,4,6-tetra-O-methyl-β-D-galactopyranosyl-(1.fwdarw
.4)-O-2,3,6-tri-O-methyl-β-D-galactopyranosyl-(1.fwdarw.4)-1,2,3,5,6-penta-O-methyl-, D-;; PERMETHYL(O-β-D-
GALP-(1-4)-O-β-D-GALP-(1-4)-D-GLC)ALDITOL; W/NBS   66070

674 C30 H58 O16 32831-67-7 AU-17-15-14 Glucitol, O-2,3,4,6-tetra-O-methyl-β-D-galactopyranosyl-(1.fwdar
w.3)-O-2,4,6-tri-O-methyl-β-D-galactopyranosyl-(1.fwdarw.4)-1,2,3,5,6-penta-O-methyl-, D-;; PERMETHYL(O-β-
D-GALP-(1-3)-O-β-D-GALP-(1-4)-D-GLC)ALDITOL; W/NBS   66071

674 C30 H58 O16 32581-22-9 AU-17-15-13 Glucitol, O-2,3,4,6-tetra-O-methyl-β-D-glucopyranosyl-(1.fwdarw.
3)-O-2,4,6-tri-O-methyl-β-D-glucopyranosyl-(1.fwdarw.4)-1,2,3,5,6-penta-O-methyl-, D-;; PERMETHYL(O-β-D-GLC
P-(1-3)-O-β-D-GLCP-(1-4)-D-GLC)ALDITOL; W/NBS   66072

674 C30 H58 O16 55658-19-0 AU-17-15-15 D-Glucitol, O-2,3,4,6-tetra-O-methyl-β-D-glucopyranosyl-(1.fwdar
w.3)-O-2,4,6-tri-O-methyl-β-D-glucopyranosyl-(1.fwdarw.3)-1,2,4,5,6-penta-O-methyl-;; PERMETHYL(O-β-D-GLCP
-(1-3)-O-β-D-GLCP-(1-3)-D-GLC)ALDITOL; W/NBS   66073

674 C30 H58 O16 32581-24-1 AU-17-15-17 Glucitol, O-2,3,4,6-tetra-O-methyl-β-D-glucopyranosyl-(1.fwdarw.
2)-3,4,6-tri-O-methyl-β-D-glucopyranosyl-(1.fwdarw.2)-1,3,4,5,6-penta-O-methyl-, D-;; PERMETHYL(O-β-D-GLCP
-(1-2)-O-β-D-GLCP-(1-2)-D-GLC)ALDITOL; W/NBS   66074

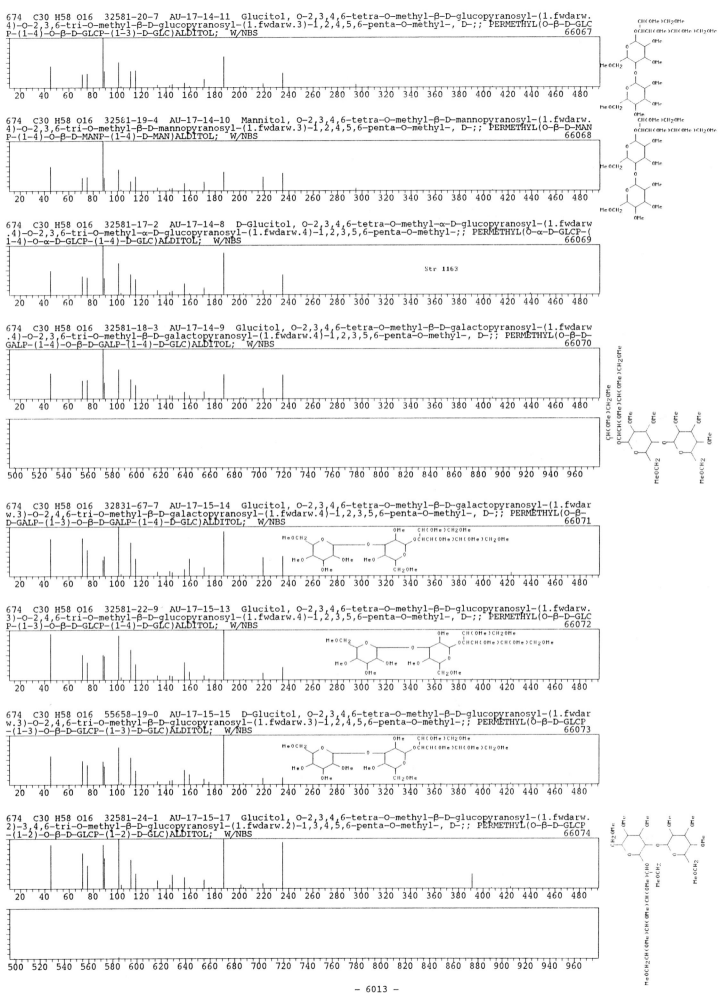

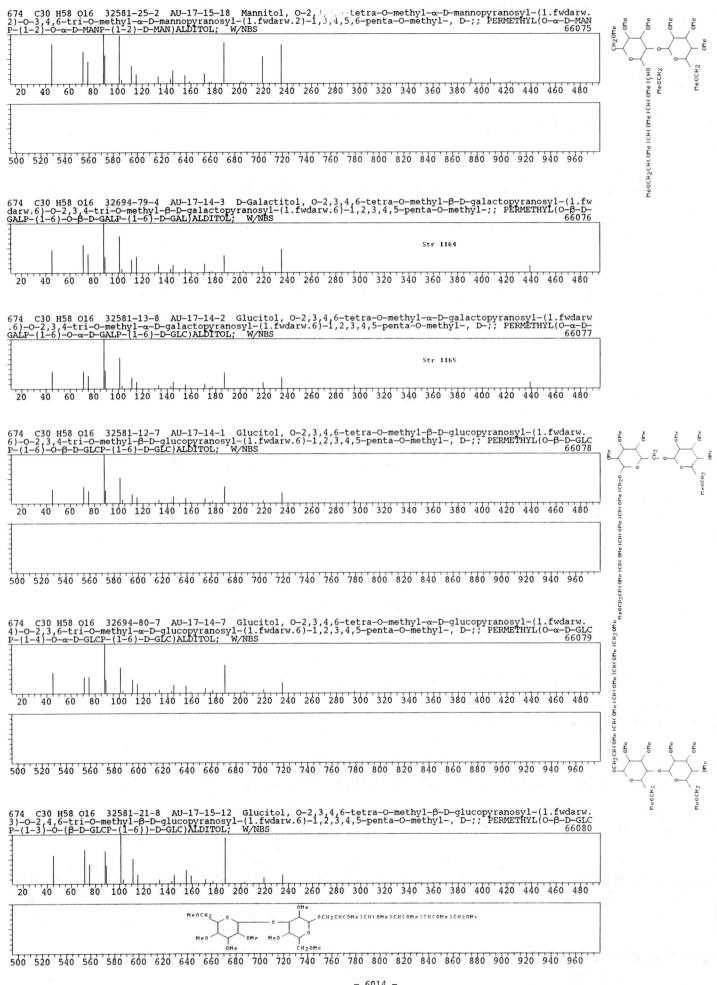

674   C30 H58 O16   32581-25-2   AU-17-15-18   Mannitol, O-2,3,4,6-tetra-O-methyl-α-D-mannopyranosyl-(1.fwdarw.
2)-O-3,4,6-tri-O-methyl-α-D-mannopyranosyl-(1.fwdarw.2)-1,3,4,5,6-penta-O-methyl-, D-;; PERMETHYL(O-α-D-MAN
P-(1-2)-O-α-D-MANP-(1-2)-D-MAN)ALDITOL; W/NBS   66075

674   C30 H58 O16   32694-79-4   AU-17-14-3   D-Galactitol, O-2,3,4,6-tetra-O-methyl-β-D-galactopyranosyl-(1.fw
darw.6)-O-2,3,4-tri-O-methyl-β-D-galactopyranosyl-(1.fwdarw.6)-1,2,3,4,5-penta-O-methyl-;; PERMETHYL(O-β-D-
GALP-(1-6)-O-β-D-GALP-(1-6)-D-GAL)ALDITOL; W/NBS   66076

Str 1164

674   C30 H58 O16   32581-13-8   AU-17-14-2   Glucitol, O-2,3,4,6-tetra-O-methyl-α-D-galactopyranosyl-(1.fwdarw
.6)-O-2,3,4-tri-O-methyl-α-D-galactopyranosyl-(1.fwdarw.6)-1,2,3,4,5-penta-O-methyl-, D-;; PERMETHYL(O-α-D-
GALP-(1-6)-O-α-D-GALP-(1-6)-D-GLC)ALDITOL; W/NBS   66077

Str 1165

674   C30 H58 O16   32581-12-7   AU-17-14-1   Glucitol, O-2,3,4,6-tetra-O-methyl-β-D-glucopyranosyl-(1.fwdarw.
6)-O-2,3,4-tri-O-methyl-β-D-glucopyranosyl-(1.fwdarw.6)-1,2,3,4,5-penta-O-methyl-, D-;; PERMETHYL(O-β-D-GLC
P-(1-6)-O-β-D-GLCP-(1-6)-D-GLC)ALDITOL; W/NBS   66078

674   C30 H58 O16   32694-80-7   AU-17-14-7   Glucitol, O-2,3,4,6-tetra-O-methyl-α-D-glucopyranosyl-(1.fwdarw.
4)-O-2,3,6-tri-O-methyl-α-D-glucopyranosyl-(1.fwdarw.6)-1,2,3,4,5-penta-O-methyl-, D-;; PERMETHYL(O-α-D-GLC
P-(1-4)-O-α-D-GLCP-(1-6)-D-GLC)ALDITOL; W/NBS   66079

674   C30 H58 O16   32581-21-8   AU-17-15-12   Glucitol, O-2,3,4,6-tetra-O-methyl-β-D-glucopyranosyl-(1.fwdarw.
3)-O-2,4,6-tri-O-methyl-β-D-glucopyranosyl-(1.fwdarw.6)-1,2,3,4,5-penta-O-methyl-, D-;; PERMETHYL(O-β-D-GLC
P-(1-3)-O-(β-D-GLCP-(1-6))-D-GLC)ALDITOL; W/NBS   66080

- 6014 -

674　C30 H58 O16　32581-23-0　AU-17-15-16　D-Galactitol, O-2,3,4,6-tetra-O-methyl-β-D-glucopyranosyl-(1.fwd
arw.2)-O-3,4,6-tri-O-methyl-β-D-glucopyranosyl-(1.fwdarw.4)-1,2,3,5,6-penta-O-methyl-;;　PERMETHYL(O-β-D-GLC
P-(1-2)-O-β-D-GLCP-(1-4)-D-GAL)ALDITOL;　W/NBS
66081

674　C30 H53 D16 F3 N4 O3 Si3　KO-7-177-4　Trimethylsilyl, trifluoroacetyl derivative of deutero-Leu-Gla-Al
a-Glu;　W
128245

674　C33 H70 O6 Si4　66447-98-1　KO-4-362-0　TRIMETHYLSILYL DERIVATINE OF 6-HYDROXY-PGF 1α METHYL ESTER;　W
66082

674　C35 H47 Cl N2 O9　78987-28-7　J-46-4402-0　Demethyltrewiasine;　W
128246

674　C36 H30 N6 O8　87880-46-4　KC-1983-2008-0　Dibenzyl 3-methyl-5-[4,5-methylenedioxy-2-(2-methyl-6,7-
methylenedioxyquinazolin-4-ylamino)phenyl]-Δ2-1,2,4-triazoline-1,4-dicarboxylate;　W
128247

674　C40 H42 N4 O6　77745-05-2　J-46-3500-0　Bis(methyl ester) of 2-(and 4-)[1-(2,4-dihydroxyphenyl)ethyl]
deuteroporphyrin;　W
128248

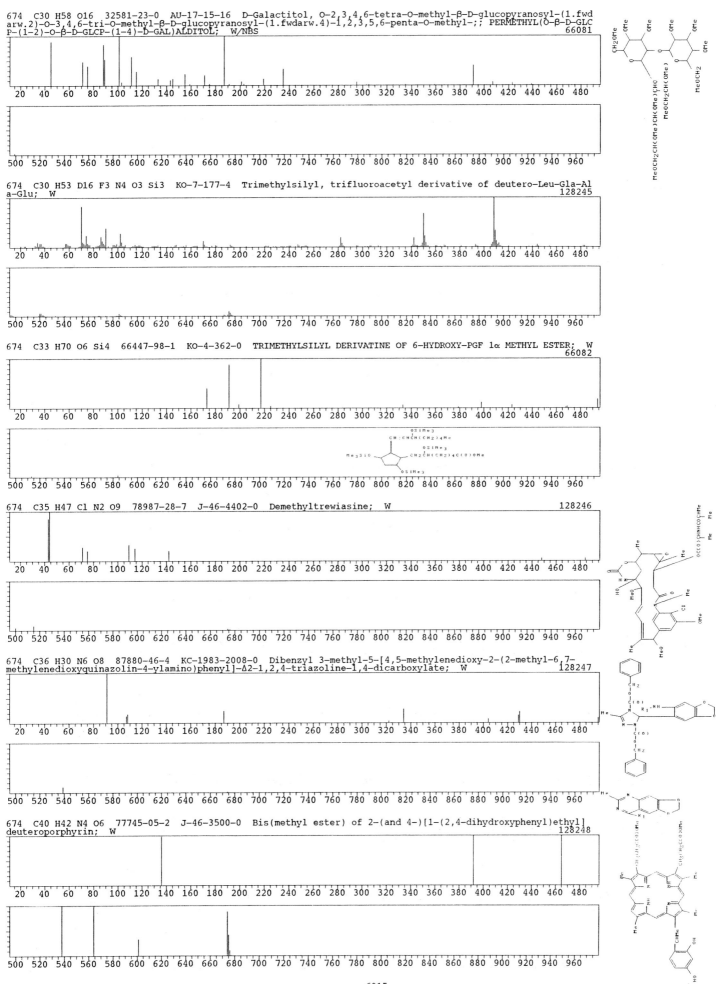

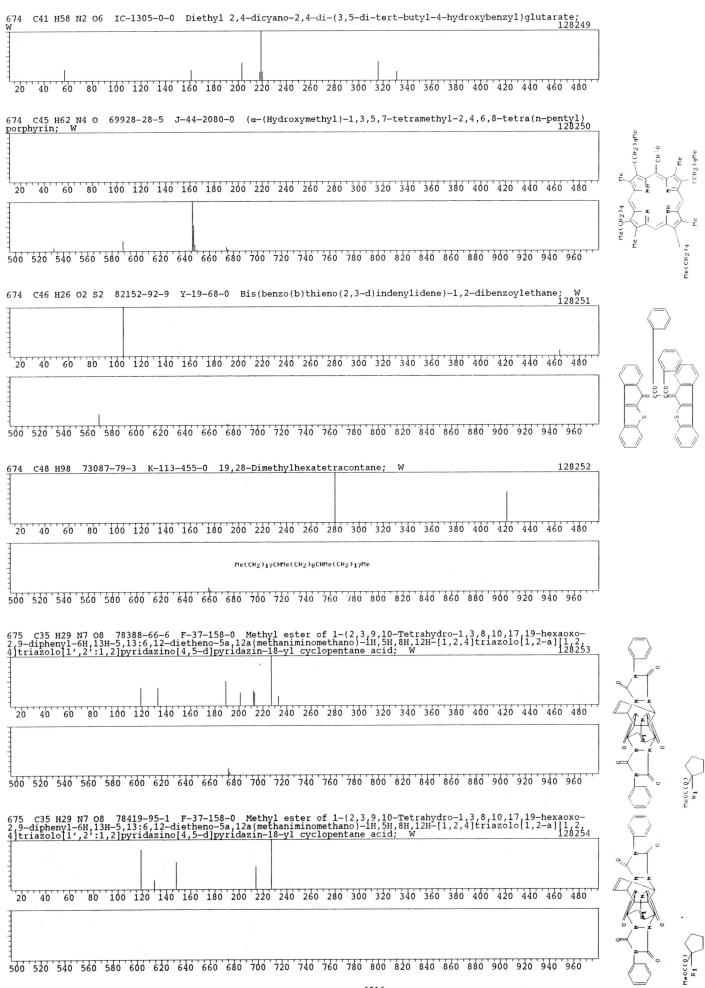

674   C41 H58 N2 O6   IC-1305-0-0   Diethyl 2,4-dicyano-2,4-di-(3,5-di-tert-butyl-4-hydroxybenzyl)glutarate; W
128249

674   C45 H62 N4 O   69928-28-5   J-44-2080-0   (α-(Hydroxymethyl)-1,3,5,7-tetramethyl-2,4,6,8-tetra(n-pentyl)porphyrin; W
128250

674   C46 H26 O2 S2   82152-92-9   Y-19-68-0   Bis(benzo(b)thieno(2,3-d)indenylidene)-1,2-dibenzoylethane; W
128251

674   C48 H98   73087-79-3   K-113-455-0   19,28-Dimethylhexatetracontane; W
128252

Me(CH2)17CHMe(CH2)8CHMe(CH2)17Me

675   C35 H29 N7 O8   78388-66-6   F-37-158-0   Methyl ester of 1-(2,3,9,10-Tetrahydro-1,3,8,10,17,19-hexaoxo-2,9-diphenyl-6H,13H-5,13:6,12-dietheno-5a,12a(methaniminomethano)-1H,5H,8H,12H-[1,2,4]triazolo[1,2-a][1,2,4]triazolo[1′,2′:1,2]pyridazino[4,5-d]pyridazin-18-yl cyclopentane acid; W
128253

675   C35 H29 N7 O8   78419-95-1   F-37-158-0   Methyl ester of 1-(2,3,9,10-Tetrahydro-1,3,8,10,17,19-hexaoxo-2,9-diphenyl-6H,13H-5,13:6,12-dietheno-5a,12a(methaniminomethano)-1H,5H,8H,12H-[1,2,4]triazolo[1,2-a][1,2,4]triazolo[1′,2′:1,2]pyridazino[4,5-d]pyridazin-18-yl cyclopentane acid; W
128254

675   C35 H29 N7 O8   78403-31-3   F-37-158-0   Methyl ester of 1-(octahydro-3,5,8,10,11,13-hexaoxo-4,9-diphen
yl-1,7:2,6-ethanediylidene-6a,10c-(methaniminomethano)-3H,6H,7H,8H-2b,4,5a,7a,9,10a-hexaazadicyclopent[d,i]
acenaphthylen-12-yl)cyclopentane carboxylic acid;   W   128255

675   C36 H57 N3 O5 S2   96624-97-4   H-67-2184-0   N-[4-tosyl-8-(tosylamino)-4-azaoctyl]-15-pentadecanlactam;
W   128256

675   C39 H41 N5 O6   78422-48-7   I-59-784-0   2-Methoxycarbonylvinyl-4-(2-bromoethyl)deuteroporphyrin dimeth
yl ester;   W   128257

675   C39 H46 Cl N O7   60281-94-9   0-16-404-1   PENITREM A ACETATE;   W   66083

HO2CMe

675   C39 H53 N3 O7   78109-02-1   J-46-3426-0   26-(Carboethoxy)-3β-(methoxymethoxy)-5α,8α-(4-phenyl-1,2-urazo
lo)-cholesta-6,22-diene-25-ol;   W   128258

675   C40 H57 N3 O6   87286-81-5   H-66-1088-0   N,N'-Diheptyl-N,N',5-trimethyl-5-(2,3-diphenyl-maleinimido)-3,
7-dioxanonandiamide;   W   128259

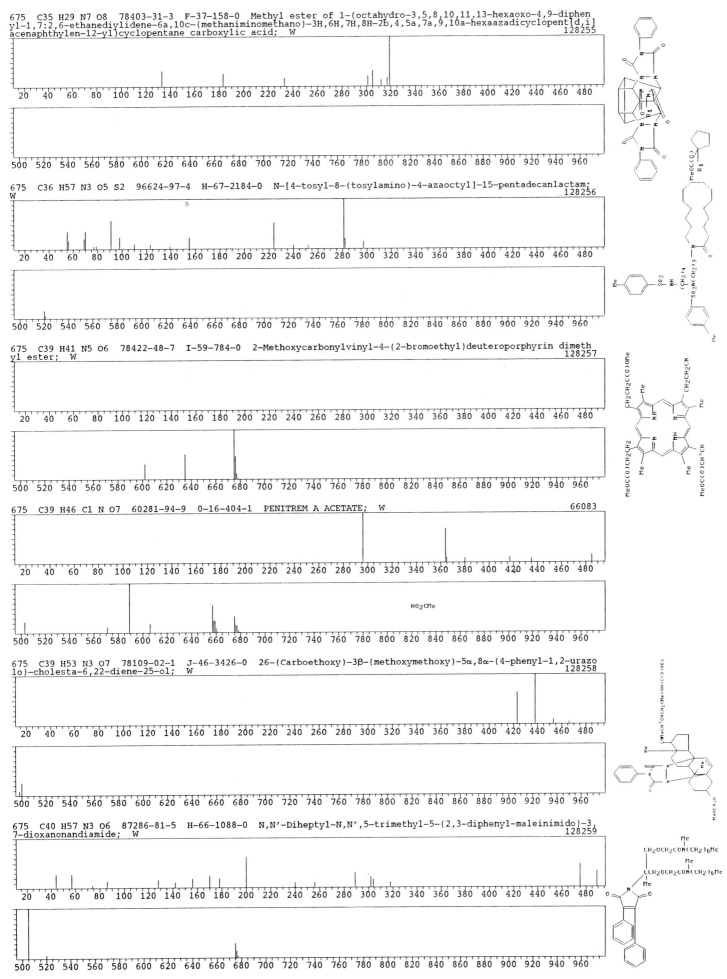

675　C43 H65 N O5　F-36-829-0　Severine palmitate;　W　　　　　　　　　128260

20 40 60 80 100 120 140 160 180 200 220 240 260 280 300 320 340 360 380 400 420 440 460 480

675　C44 H16 D12 Cu N4　50914-73-3　O-9-79-5　CUT(2,4,6-TRIDEUTERIO)PP;　W　　　66084

20 40 60 80 100 120 140 160 180 200 220 240 260 280 300 320 340 360 380 400 420 440 460 480

500 520 540 560 580 600 620 640 660 680 700 720 740 760 780 800 820 840 860 880 900 920 940 960

675　C44 H20 D8 Cu N4　51068-97-4　O-9-79-5　CUT(3,5-DIDEUTERIO)PP;　W　　　66085

20 40 60 80 100 120 140 160 180 200 220 240 260 280 300 320 340 360 380 400 420 440 460 480

500 520 540 560 580 600 620 640 660 680 700 720 740 760 780 800 820 840 860 880 900 920 940 960

675　C44 H28 Cu N4　14172-91-9　KO-3-74-3　Copper tetraphenylporphyrin;　W　　　66086

20 40 60 80 100 120 140 160 180 200 220 240 260 280 300 320 340 360 380 400 420 440 460 480

500 520 540 560 580 600 620 640 660 680 700 720 740 760 780 800 820 840 860 880 900 920 940 960

675　C44 H85 N O3　82280-28-2　H-65-61-0　cis-2-(N,N-Dioctadecylcarbamoyl)-1-cyclohexanecarboxylic acid;　W
128261

20 40 60 80 100 120 140 160 180 200 220 240 260 280 300 320 340 360 380 400 420 440 460 480

500 520 540 560 580 600 620 640 660 680 700 720 740 760 780 800 820 840 860 880 900 920 940 960

676　C29 H56 N2 O10 Si3　55556-79-1　BA-0-160-0　D-Glucopyranosiduronic acid, 3-(5-ethylhexahydro-1,3-di
methyl-2,4,6-trioxo-5-pyrimidinyl)-1-methylbutyl 2,3,4-tris-O-(trimethylsilyl)-, methyl ester;; 1,3-DIMETH
YL-5-ETHYL-5-(3-HYDROXY-1-METHYLBUTYL) BARBITURIC ACID O-TRIS(TRIMETHYLSILYL)GLUCURONIDE METHYL ESTER;
W/NBS　　　　　　　　　　　　　　　　　　　　　　　　　　　　　　　　66090

20 40 60 80 100 120 140 160 180 200 220 240 260 280 300 320 340 360 380 400 420 440 460 480

676　C29 H56 N2 O10 Si3　BL-0-53-0　HYDROXYAMOBARBITAL GLUCURONIDE METHYL TMS DERIVATIVE (ISOMER 2);　W
66088

20 40 60 80 100 120 140 160 180 200 220 240 260 280 300 320 340 360 380 400 420 440 460 480

500 520 540 560 580 600 620 640 660 680 700 720 740 760 780 800 820 840 860 880 900 920 940 960

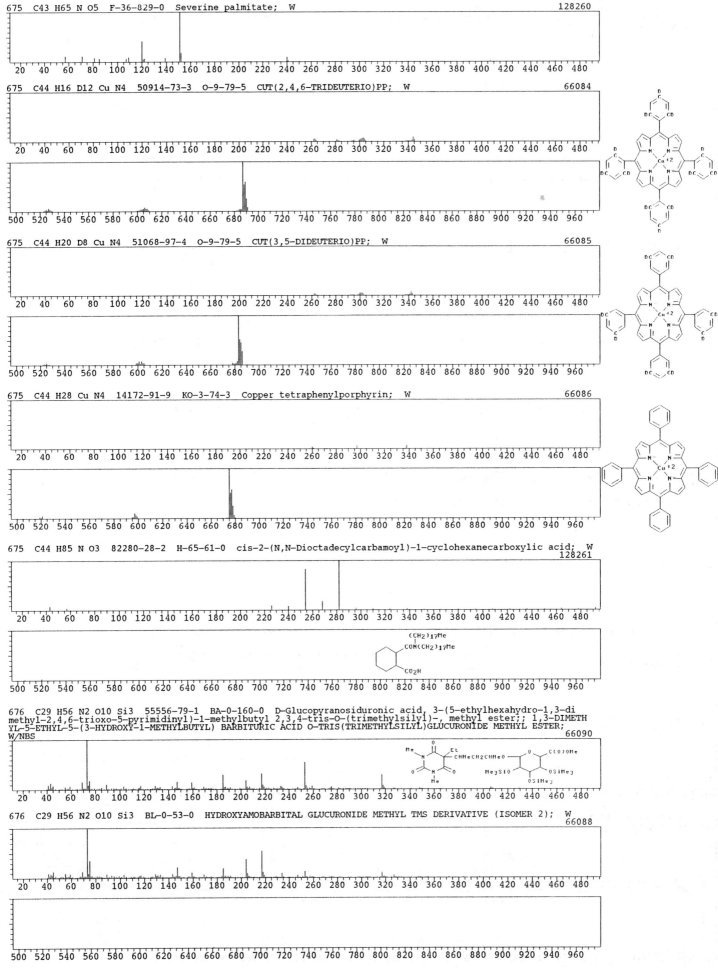

676 C29 H56 N2 O10 Si3 BL-0-52-0   HYDROXYAMOBARBITAL GLUCURONIDE METHYL TMS DERIVATIVE (ISOMER 1); W
66089

676 C29 H56 N2 O10 Si3 CR-41-0-0   HYDROXYPENTOBARBITAL GLUCURONIDE 1,3-DIMETHYLDERIV METHYL ESTER TMS ETH
ER; W
66091

676 C32 H44 N4 O12 70445-66-8 J-44-2696-0   2,5,8,11,14,17,23,26,29,32,35,38-dodecaoxa-43,44-diazatricycl
o[37.3.1.1(18,22)]tetratetraconta-1(43),18(44),19,21,39,41-hexaene-19,42-dicarbonitrile; W   128262

676 C35 H56 N4 O5 S2 75422-06-9 H-62-814-0   Benzenesulfonamide, N-(3-aminopropyl)-4-methyl-N-[3-[[(4-
methylphenyl)sulfonyl][3-(2-oxoazacyclotridec-1-yl)propyl]amino]propyl]-;; N-(11-Amino-4,8-ditosyl-4,8-diaz
aundecyl)-12-dodecanlactam; W/NBS   128263

676 C39 H76 O3 Si3 71899-28-0 O-14-162-4   Silane, [[(2α,3β,8α,9β,13α,14β,17α,18β,22α)-gammacerane-2,3,22
-triyl]tris(oxy)]tris[trimethyl-;; (8α,14β,18β)-21,21-DIMETHYL-2α,3β,22α-TRI(TRIMETHYLSILOXY)-29,30-DINORGA
MMACERANE; W/NBS   66093

676 C41 H48 N4 O5 75112-51-5 J-45-5202-0   2-Formyl-4-vinyldeuteroporphyrin di-tert-butyl ester; W
128264

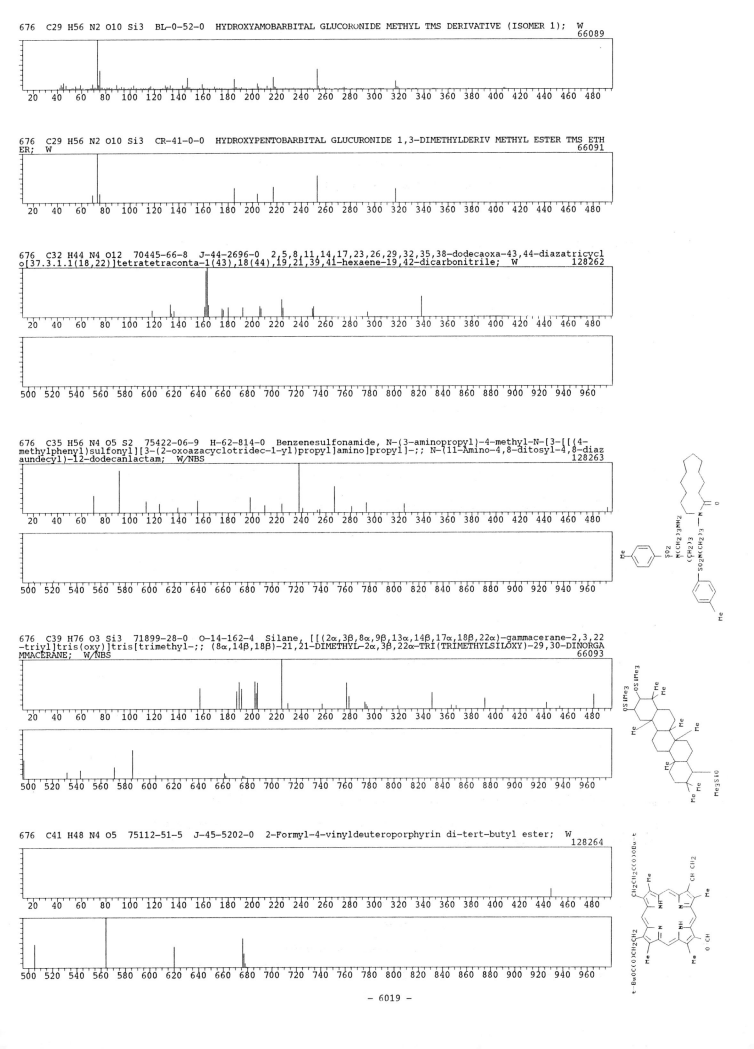

676 C41 H48 N4 O5   75125-19-8   J-45-5202-0   2-Vinyl-4-formyldeuteroporphyrin di-tert-butyl ester; W
128265

676 C44 H28 N4 Zn   14074-80-7   KO-3-74-3   ZINC α,β,γ,Δ-TETRAPHENYLPORPHIYRIN; W
66094

676 C44 H56 N2 O4   84389-95-7   KC-1982-2321-0   3,8-Dimethoxy-4,7-di-(4-methoxy-3,5-di-t-butylphenyl)-benzo
[c]cinnoline; W
128266

676 C44 H68 O5   19721-60-9   T-68-5952-0   p-Benzoquinone, 2-(3-hydroxy-3,7,11,15,19,23,27-heptamethyl-6,10,
14,18,22,26-octacosahexaenyl)-5,6-dimethoxy-3-methyl-;; γ-HYDROXY-β,γ-DIHYDRO-UBIQUINONE; W/NBS   66095

676 C45 H64 N4 O   69912-17-0   J-44-2080-0   (α-(Hydroxymethyl)-1,3,5,7-tetramethyl-2,4,6,8-tetra(n-pentyl)
porphyrin; W
128267

676 C46 H60 O4   69505-95-9   K-112-204-0   2,4,6,8,10,12,14,16-Heptadecaoctaenoic acid, 17-(4-methoxyphenyl)
-, 2-decyl-3-methoxy-5-pentylphenyl ester;; 2-DECYL-3-METHOXY-5-PENTYLPHENYLESTER OF 17-(4-METHOXYPHENYL)-
2,4,6,8,10,12,14,16-HEPTADECAOCTAENE CARBOXYLIC ACID; W/NBS
66096

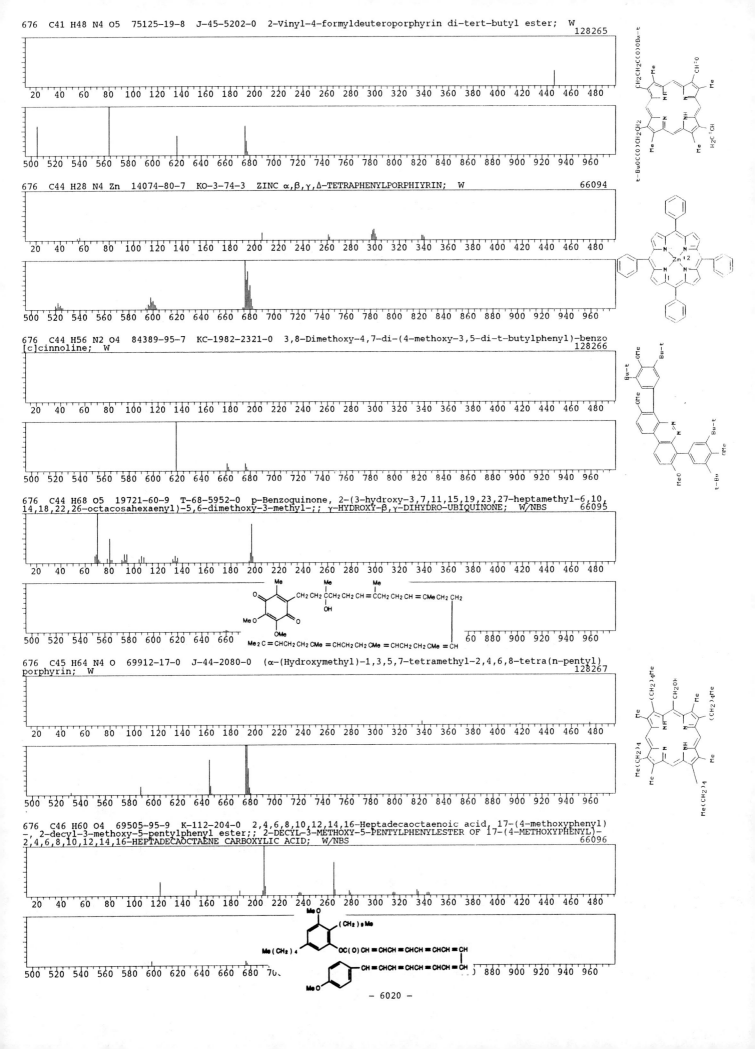

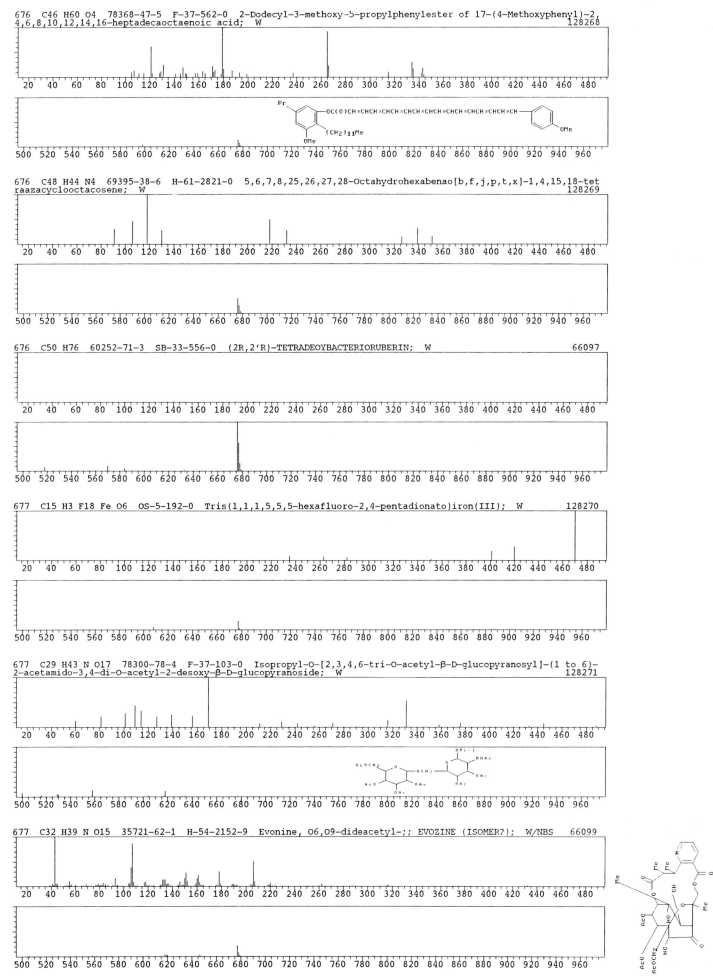

676 C46 H60 O4 78368-47-5 F-37-562-0 2-Dodecyl-3-methoxy-5-propylphenylester of 17-(4-Methoxyphenyl)-2,
4,6,8,10,12,14,16-heptadecaoctaenoic acid; W   128268

676 C48 H44 N4 69395-38-6 H-61-2821-0 5,6,7,8,25,26,27,28-Octahydrohexabenao[b,f,j,p,t,x]-1,4,15,18-tet
raazacyclooctacosene; W   128269

676 C50 H76 60252-71-3 SB-33-556-0 (2R,2'R)-TETRADEOYBACTERIORUBERIN; W   66097

677 C15 H3 F18 Fe O6 OS-5-192-0 Tris(1,1,1,5,5,5-hexafluoro-2,4-pentadionato)iron(III); W   128270

677 C29 H43 N O17 78300-78-4 F-37-103-0 Isopropyl-O-[2,3,4,6-tri-O-acetyl-β-D-glucopyranosyl]-(1 to 6)-
2-acetamido-3,4-di-O-acetyl-2-desoxy-β-D-glucopyranoside; W   128271

677 C32 H39 N O15 35721-62-1 H-54-2152-9 Evonine, O6,O9-dideacetyl-;; EVOZINE (ISOMER?); W/NBS 66099

677　C35 H31 N7 O8　78413-96-4　F-37-159-0　Methyl ester of 2-(octahydro-3,5,8,10,11,13-hexaoxo-4,9-diphen
yl-1,7:2,6-ethanediylidene-6a,10c-(methaniminomethano)-3H,6H,7H,8H-2b,4,5a,7a,9,10a-hexaazadicyclopent[d,i]
acenaphthylen-12-yl)-4-methylpentanoic acid;　W　　　　　　　　　　　　　　　　　　　128272

677　C35 H31 N7 O8　78478-16-7　F-37-159-0　Methyl ester of 2-(2,3,9,10-Tetrahydro-1,3,8,10,17,19-hexaoxo-
2,9-diphenyl-6H,13H-5,13:6,12-dietheno-5a,12a-(methaniminomethano)-1H,5H,8H,12H-[1,2,4]triazolo[1,2-a][1,2,
4]triazolo[1',2':1,2]pyridazino[4,5-d]pyridazin-18-yl)-4-methylpentanoic acid;　W　　　　　　　128273

677　C35 H31 N7 O8　78403-29-9　F-37-159-0　Methyl ester of 2-(2,3,9,10-Tetrahydro-1,3,8,10,17,19-hexaoxo-
2,9-diphenyl-6H,13H-5,13:6,12-dietheno-5a,12a-(methaniminomethano)-1H,5H,8H,12H-[1,2,4]triazolo[1,2-a][1,2,
4]triazolo[1',2':1,2]pyridazino[4,5-d]pyridazin-18-yl)-4-methylpentanoic acid;　W　　　　　　　128274

677　C36 H72 N O8 P　18194-24-6　EP-5252-0-0　L-Dimyrsitoyllecithin;　W/NBS　　　　　　　　　　66100

677　C37 H71 N O4 Si3　33745-60-7　H-54-1604-8　Pyrrolidine, 1-[3α,7α,12α-tris(trimethylsiloxy)-5β-cholan-
24-oyl]-;;　TRIS-O-MONOTRIMETHYLSILYL-CHOLIC ACID PYRROLIDIDE;　W/NBS　　　　　　　　　　　　66101

678　C14 Co4 F6 O10　12564-26-0　AD-0-3684-0　Cobalt, decacarbonyl(1,1,1,4,4,4-hexafluoro-2-butyne)tetra-;;
.MU.-(HEXAFLUOROBUT-2-YNE)-DECACARBONYLTETRACOBALT;　W/NBS　　　　　　　　　　　　　　　66102

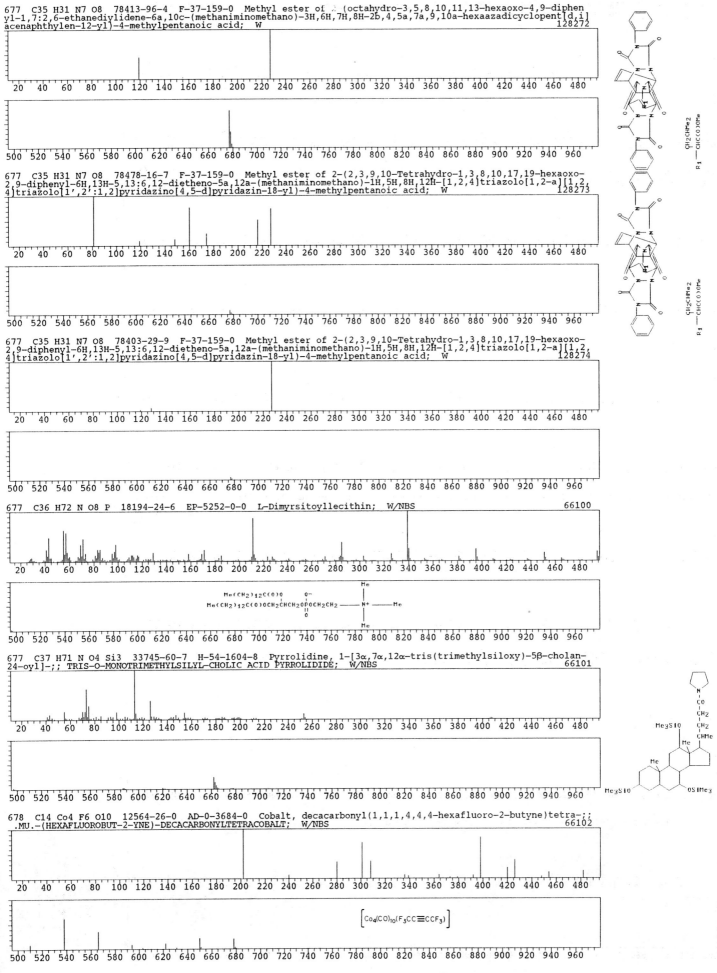

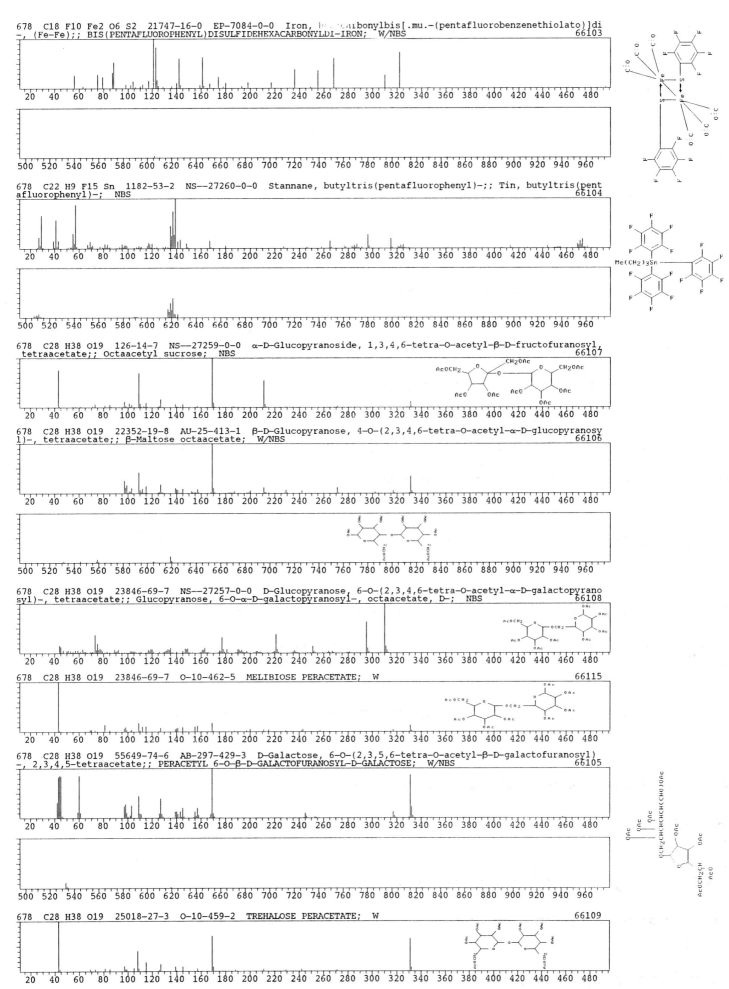

678 C18 F10 Fe2 O6 S2   21747-16-0   EP-7084-0-0   Iron, hexacarbonylbis[.mu.-(pentafluorobenzenethiolato)]di
-, (Fe-Fe);; BIS(PENTAFLUOROPHENYL)DISULFIDEHEXACARBONYLDI-IRON; W/NBS   66103

678 C22 H9 F15 Sn   1182-53-2   NS--27260-0-0   Stannane, butyltris(pentafluorophenyl)-;; Tin, butyltris(pent
afluorophenyl)-; NBS   66104

678 C28 H38 O19   126-14-7   NS--27259-0-0   α-D-Glucopyranoside, 1,3,4,6-tetra-O-acetyl-β-D-fructofuranosyl,
tetraacetate;; Octaacetyl sucrose; NBS   66107

678 C28 H38 O19   22352-19-8   AU-25-413-1   β-D-Glucopyranose, 4-O-(2,3,4,6-tetra-O-acetyl-α-D-glucopyranosy
l)-, tetraacetate;; β-Maltose octaacetate; W/NBS   66106

678 C28 H38 O19   23846-69-7   NS--27257-0-0   D-Glucopyranose, 6-O-(2,3,4,6-tetra-O-acetyl-α-D-galactopyrano
syl)-, tetraacetate;; Glucopyranose, 6-O-α-D-galactopyranosyl-, octaacetate, D-; NBS   66108

678 C28 H38 O19   23846-69-7   O-10-462-5   MELIBIOSE PERACETATE; W   66115

678 C28 H38 O19   55649-74-6   AB-297-429-3   D-Galactose, 6-O-(2,3,5,6-tetra-O-acetyl-β-D-galactofuranosyl)
-, 2,3,4,5-tetraacetate;; PERACETYL 6-O-β-D-GALACTOFURANOSYL-D-GALACTOSE; W/NBS   66105

678 C28 H38 O19   25018-27-3   O-10-459-2   TREHALOSE PERACETATE; W   66109

- 6023 -

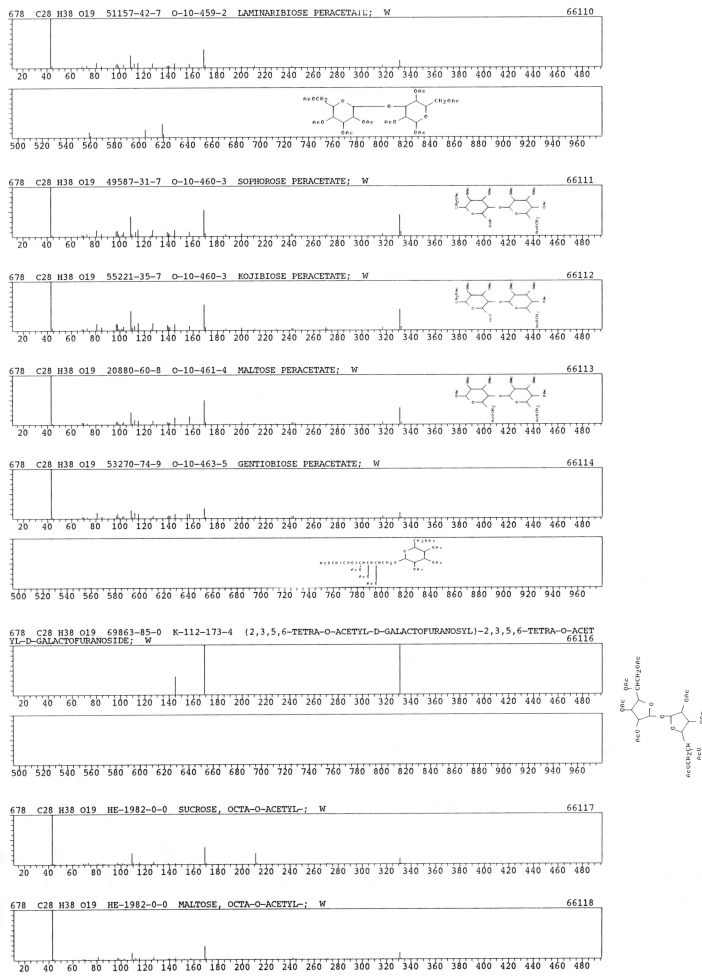

678 C28 H38 O19 51157-42-7 O-10-459-2 LAMINARIBIOSE PERACETATE; W 66110

678 C28 H38 O19 49587-31-7 O-10-460-3 SOPHOROSE PERACETATE; W 66111

678 C28 H38 O19 55221-35-7 O-10-460-3 KOJIBIOSE PERACETATE; W 66112

678 C28 H38 O19 20880-60-8 O-10-461-4 MALTOSE PERACETATE; W 66113

678 C28 H38 O19 53270-74-9 O-10-463-5 GENTIOBIOSE PERACETATE; W 66114

678 C28 H38 O19 69863-85-0 K-112-173-4 (2,3,5,6-TETRA-O-ACETYL-D-GALACTOFURANOSYL)-2,3,5,6-TETRA-O-ACETYL-D-GALACTOFURANOSIDE; W 66116

678 C28 H38 O19 HE-1982-0-0 SUCROSE, OCTA-O-ACETYL-; W 66117

678 C28 H38 O19 HE-1982-0-0 MALTOSE, OCTA-O-ACETYL-; W 66118

678　C30 H26 F12 O4　49842-32-2　NS--27255-0-0　Androsta␣␣␣␣iene-3,17-diol, 3-(heptafluorobutanoate) 17-(
pentafluorobenzoate), (17β)-;;　NBS　　　　　　　　　　　　　　　　　　　　　66119

678　C32 H54 O15　90634-01-8　KC-1984-426-0　1,3,6-Tri-O-pivaloylβ-D-fructofuranosyl 6-O-pivaloylα-D-glucop
yranoside;　W　　　　　　　　　　　　　　　　　　　　　　　　　　　　　128275

678　C32 H54 O15　90634-02-9　KC-1984-426-0　1,4,6-Tri-O-pivaloylβ-D-fructofuranosyl 6-O-pivaloylα-D-glucop
yranoside;　W　　　　　　　　　　　　　　　　　　　　　　　　　　　　　128276

678　C33 H42 O15　37294-05-6　NS-0-0-0　Azadirachtin, 3-O-deacetyl-;　W/NBS　　　　　　128277

678　C34 H58 N6 O8　BS-1-65-0　N-Acetyl-(N,0-permethyl)-prolyl-leucyl-valyl-alanyl-prolyl-alanine;　W
　　　　　　　　　　　　　　　　　　　　　　　　　　　　　　　　　128278

678　C34 H40 D18 N6 O8　BS-1-57-0　N-Trideuteroacetyl-(N,0-pertrideuteromethyl)-prolyl-leucyl-valyl-alanyl-
propyl-alanine;　W　　　　　　　　　　　　　　　　　　　　　　　　　128279

678　C35 H34 O14　20688-98-6　AD-0-1974-0　Salicin, 3',4',6'-triacetate 2'-benzoate α-salicylate acetate;;
SALICYLOYL TREMULOIDIN TETRAACETATE;　W/NBS　　　　　　　　　　　　66120

678　C35 H34 O14　20689-01-4　AD-0-1973-0　Salicin, 2',3',4'-triacetate 6'-benzoate α-salicylate acetate;;
SALICYLOYL POPULIN TETRAACETATE;　W/NBS　　　　　　　　　　　　　　66121

678　C40 H42 N2 O8　76947-04-1　J-46-2389-0　(R,R)-O,O-Diacetylbebeerine;　W　　　　　128280

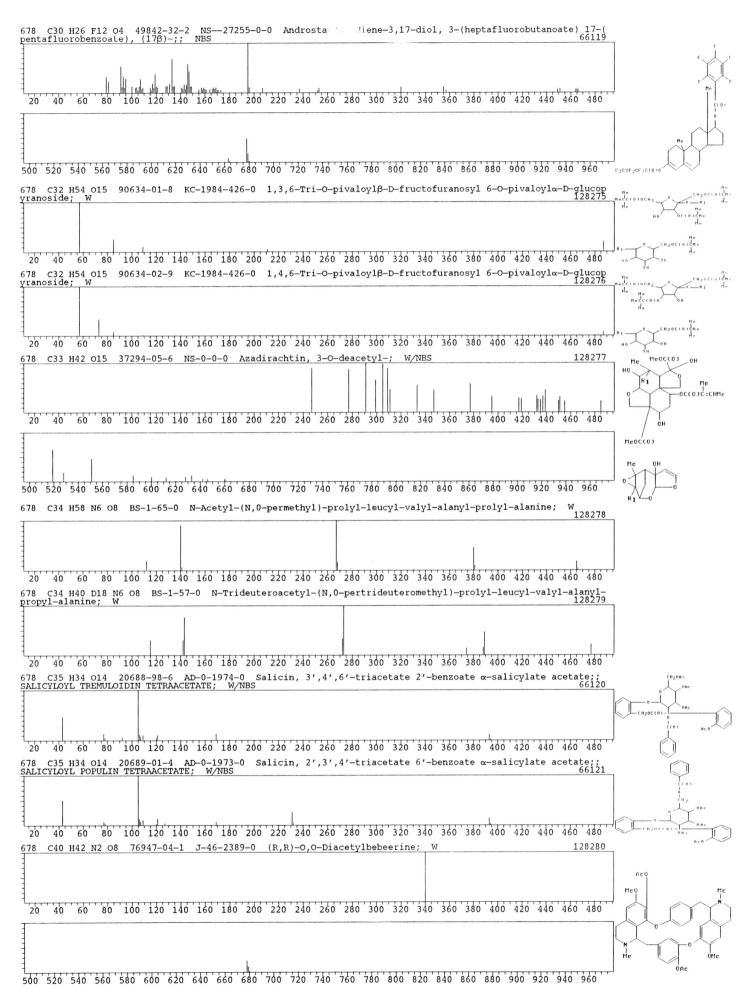

678 C42 H82 O4 Si 64861-36-5 KO-4-110-5 1-OCTADEC-1-ENYL-2-OCTADECENOYL(OLEOYL)-3-TMS GLYCEROL; W

66122

CH₂OSiMe₃
Me(CH₂)₇CH:CH(CH₂)₇C(O)OCHCH₂OCH:CH(CH₂)₁₅Me

20 40 60 80 100 120 140 160 180 200 220 240 260 280 300 320 340 360 380 400 420 440 460 480

678 C45 H58 O5 72251-68-4 C-103-3789-0 7,13,19,25-Tetra-tert-butyl-27,28,29,30-tetrahydroxy-2,3-bishomo
-3-oxacalix[4]arene; W

128281

20 40 60 80 100 120 140 160 180 200 220 240 260 280 300 320 340 360 380 400 420 440 460 480

500 520 540 560 580 600 620 640 660 680 700 720 740 760 780 800 820 840 860 880 900 920 940 960

678 C45 H74 O4 K-112-2009-0 MIXTURE OF 2-ISOUNDECYL-5-PENTYL- AND 2-DECYL-5-ISOHEXYL-3-MEO-PHENYLESTERS
OF 15-(4-MEO-PHENYL)PENTADECANCARBOXYLIC ACID; W

66123

20 40 60 80 100 120 140 160 180 200 220 240 260 280 300 320 340 360 380 400 420 440 460 480

500 520 540 560 580 600 620 640 660 680 700 720 740 760 780 800 820 840 860 880 900 920 940 960

678 C46 H62 O4 71394-73-5 K-112-2007-0 2,4,6,8,10,12,14-Pentadecaheptaenoic acid, 15-(4-methoxyphenyl)
-, 3-methoxy-2-(9-methyldecyl)-5-(4-methylpentyl)phenyl ester;; 3-METHOXY-2-(9-METHYLDECYL)-5-(4-METHYLPENT
YL)PHENYLESTER OF 15-(4-METHOXYPHENYL)-2,4,6,8,10,12,14-PENTADECAHEPTAENCARBOXYLIC ACID; W/NBS

66124

20 40 60 80 100 120 140 160 180 200 220 240 260 280 300 320 340 360 380 400 420 440 460 480

500 520 540 560 580 600 620 640 660 680 700 720 740 760 780 800 820 840 860

679 C38 H41 N5 O7 91084-76-3 C-106-4985-0 7,8,17,18-Tetrahydro-1,23-bis(hydroxymethyl)-35-methoxy-3,21-
dimethyl-16H,31H-5,9:15,19-dimetheno-10,14-metheno-26,30-nitrilo-6H,25H-dibenzo[b,s][1,21,4,8,14,18]dioxate
traazacyclooctacosine-34,36-dione; W

128282

20 40 60 80 100 120 140 160 180 200 220 240 260 280 300 320 340 360 380 400 420 440 460 480

500 520 540 560 580 600 620 640 660 680 700 720 740 760 780 800 820 840 860 880 900 920 940 960

679 C39 H44 Cu N4 O3 76703-04-3 KC-1982-584-0 α,β,γ-triformyloctaethylporphyrin; W

128283

20 40 60 80 100 120 140 160 180 200 220 240 260 280 300 320 340 360 380 400 420 440 460 480

500 520 540 560 580 600 620 640 660 680 700 720 740 760 780 800 820 840 860 880 900 920 940 960

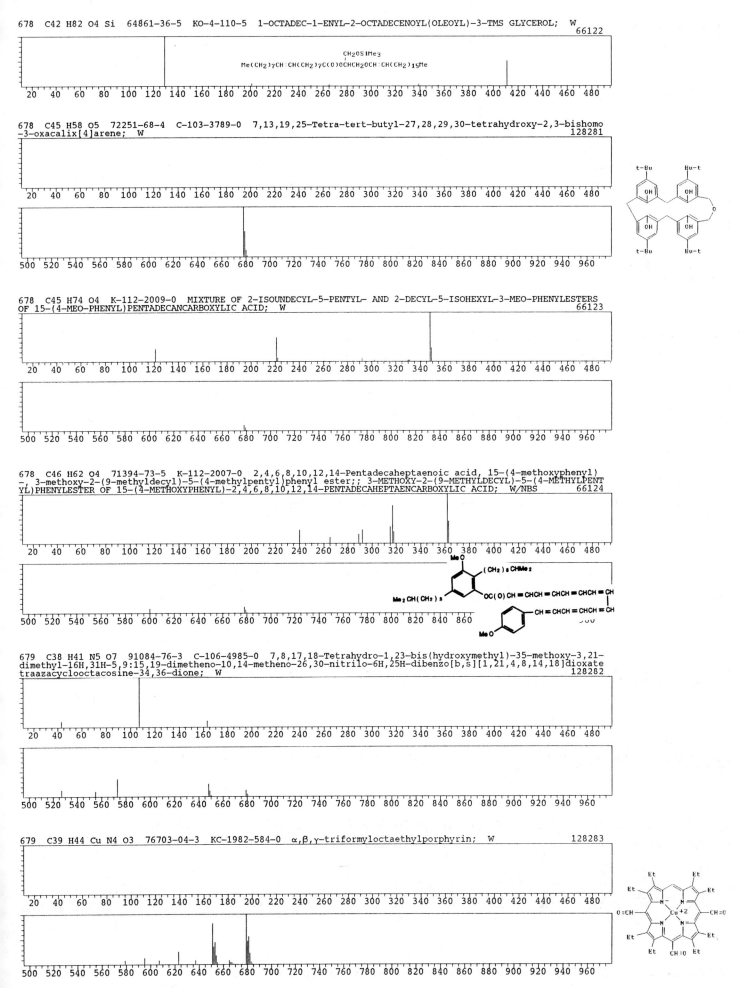

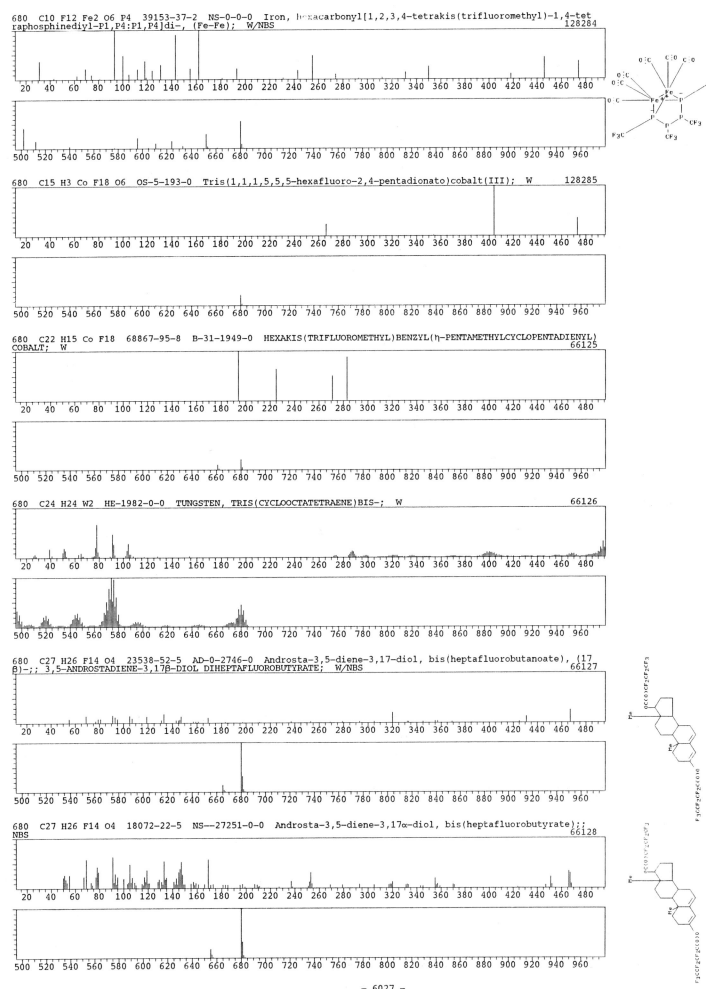

680   C10 F12 Fe2 O6 P4   39153-37-2   NS-0-0-0   Iron, hexacarbonyl[1,2,3,4-tetrakis(trifluoromethyl)-1,4-tet
raphosphinediyl-P1,P4:P1,P4]di-, (Fe-Fe); W/NBS   128284

680   C15 H3 Co F18 O6   OS-5-193-0   Tris(1,1,1,5,5,5-hexafluoro-2,4-pentadionato)cobalt(III); W   128285

680   C22 H15 Co F18   68867-95-8   B-31-1949-0   HEXAKIS(TRIFLUOROMETHYL)BENZYL(η-PENTAMETHYLCYCLOPENTADIENYL)
COBALT; W   66125

680   C24 H24 W2   HE-1982-0-0   TUNGSTEN, TRIS(CYCLOOCTATETRAENE)BIS-; W   66126

680   C27 H26 F14 O4   23538-52-5   AD-0-2746-0   Androsta-3,5-diene-3,17-diol, bis(heptafluorobutanoate), (17
β)-;; 3,5-ANDROSTADIENE-3,17β-DIOL DIHEPTAFLUOROBUTYRATE; W/NBS   66127

680   C27 H26 F14 O4   18072-22-5   NS--27251-0-0   Androsta-3,5-diene-3,17α-diol, bis(heptafluorobutyrate);;
NBS   66128

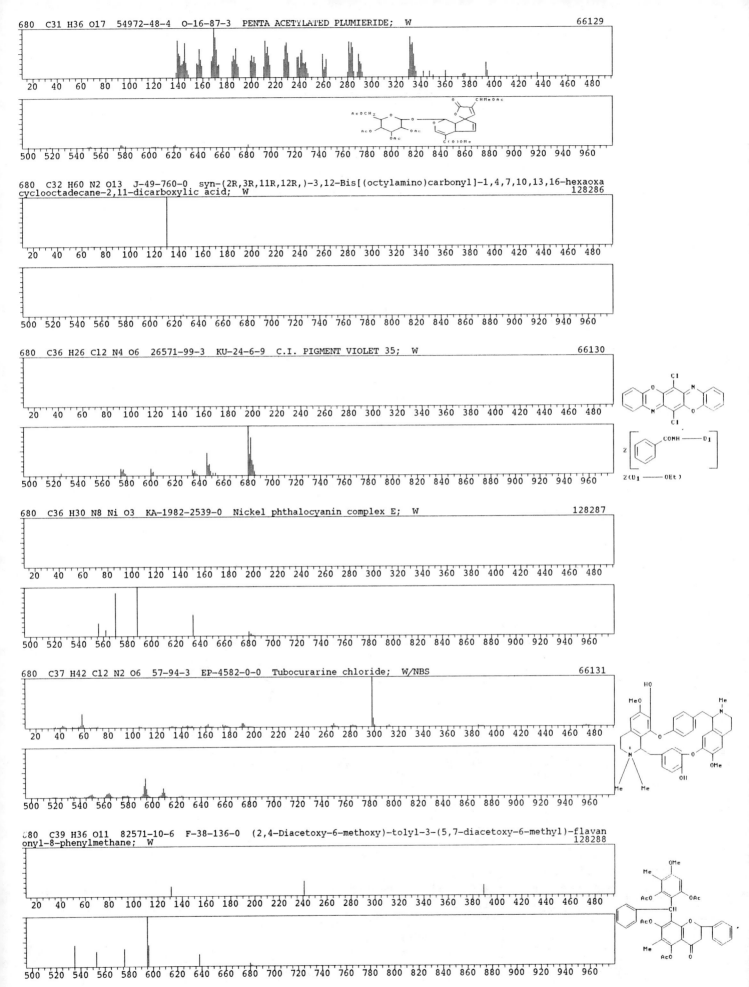

680   C31 H36 O17   54972-48-4   O-16-87-3   PENTA ACETYLATED PLUMIERIDE;   W   66129

680   C32 H60 N2 O13   J-49-760-0   syn-(2R,3R,11R,12R,)-3,12-Bis[(octylamino)carbonyl]-1,4,7,10,13,16-hexaoxa
cyclooctadecane-2,11-dicarboxylic acid;   W   128286

680   C36 H26 Cl2 N4 O6   26571-99-3   KU-24-6-9   C.I. PIGMENT VIOLET 35;   W   66130

680   C36 H30 N8 Ni O3   KA-1982-2539-0   Nickel phthalocyanin complex E;   W   128287

680   C37 H42 Cl2 N2 O6   57-94-3   EP-4582-0-0   Tubocurarine chloride;   W/NBS   66131

680   C39 H36 O11   82571-10-6   F-38-136-0   (2,4-Diacetoxy-6-methoxy)-tolyl-3-(5,7-diacetoxy-6-methyl)-flavan
onyl-8-phenylmethane;   W   128288

680   C39 H36 O11   60715-60-8   F-37-2605-0   2,4-Diacetoxy-6-methoxy-tolyl-3-(5,7-diacetoxy-6-methyl)-flavano
nyl-8-phenylmethane;   W   128289

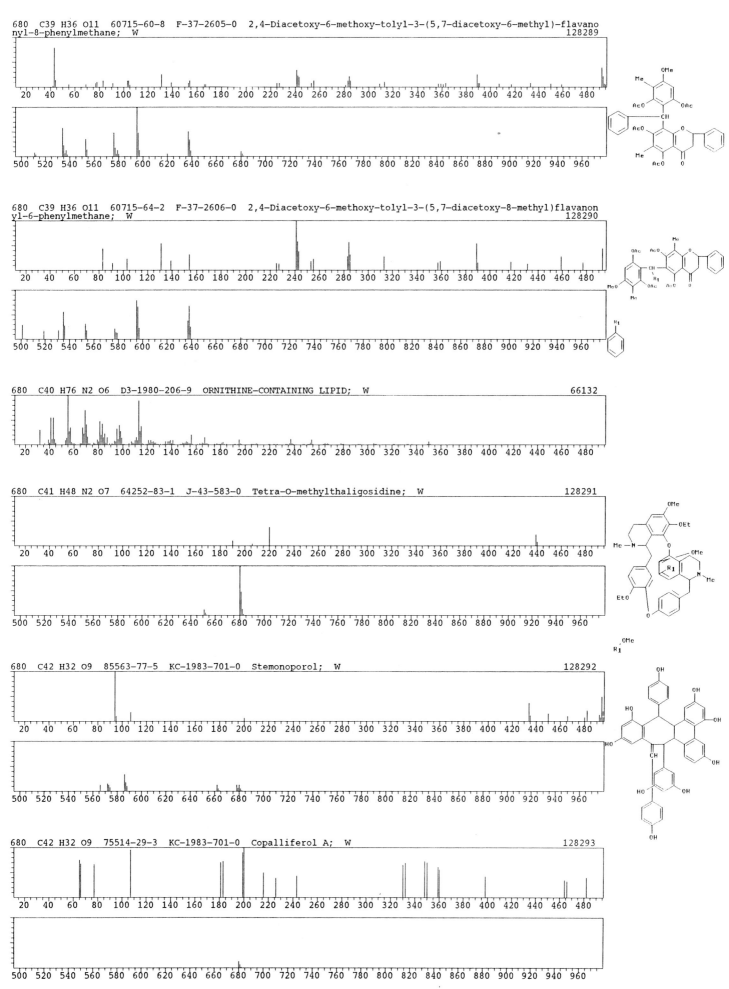

680   C39 H36 O11   60715-64-2   F-37-2606-0   2,4-Diacetoxy-6-methoxy-tolyl-3-(5,7-diacetoxy-8-methyl)flavanon
yl-6-phenylmethane;   W   128290

680   C40 H76 N2 O6   D3-1980-206-9   ORNITHINE-CONTAINING LIPID;   W   66132

680   C41 H48 N2 O7   64252-83-1   J-43-583-0   Tetra-O-methylthaligosidine;   W   128291

680   C42 H32 O9   85563-77-5   KC-1983-701-0   Stemonoporol;   W   128292

680   C42 H32 O9   75514-29-3   KC-1983-701-0   Copalliferol A;   W   128293

- 6029 -

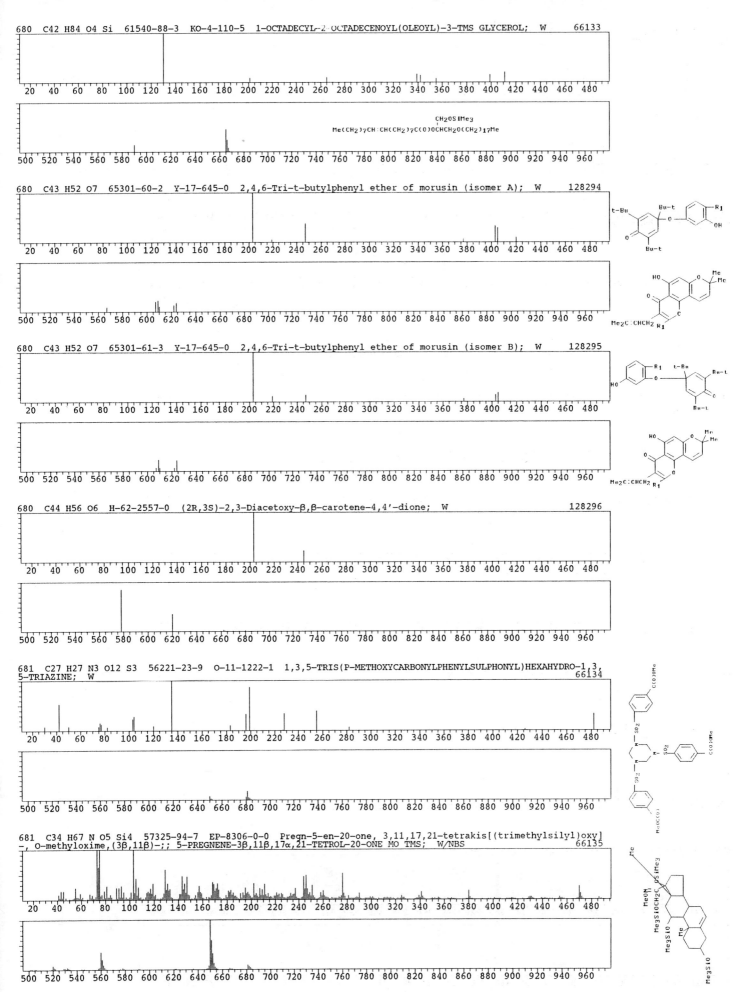

680   C42 H84 O4 Si   61540-88-3   KO-4-110-5   1-OCTADECYL-2-OCTADECENOYL(OLEOYL)-3-TMS GLYCEROL;   W   66133

CH2OSiMe3
Me(CH2)7CH:CH(CH2)7C(O)OCHCH2O(CH2)17Me

680   C43 H52 O7   65301-60-2   Y-17-645-0   2,4,6-Tri-t-butylphenyl ether of morusin (isomer A);   W   128294

680   C43 H52 O7   65301-61-3   Y-17-645-0   2,4,6-Tri-t-butylphenyl ether of morusin (isomer B);   W   128295

680   C44 H56 O6   H-62-2557-0   (2R,3S)-2,3-Diacetoxy-β,β-carotene-4,4'-dione;   W   128296

681   C27 H27 N3 O12 S3   56221-23-9   O-11-1222-1   1,3,5-TRIS(P-METHOXYCARBONYLPHENYLSULPHONYL)HEXAHYDRO-1,3,
5-TRIAZINE;   W   66134

681   C34 H67 N O5 Si4   57325-94-7   EP-8306-0-0   Pregn-5-en-20-one, 3,11,17,21-tetrakis[(trimethylsilyl)oxy]
-, O-methyloxime,(3β,11β)-;;   5-PREGNENE-3β,11β,17α,21-TETROL-20-ONE MO TMS;   W/NBS   66135

681 C34 H67 N O5 Si4 57325-93-6 EP-8307-0-0 Pregn-4-en-3-one, 11,17,20,21-tetrakis[(trimethylsilyl)oxy]
-, O-methyloxime,(11β,20S)-;; 4-PREGNENE-11β,17α,20α,21-TETROL-3-ONE MO TMS; W/NBS
66136

681 C34 H67 N O5 Si4 57325-92-5 EP-8308-0-0 Pregn-4-en-3-one, 11,17,20,21-tetrakis[(trimethylsilyl)oxy]
-, O-methyloxime,(11β,20R)-;; 4-PREGNENE-11β,17α,20β,21-TETROL-3-ONE MO TMS; W/NBS
66137

681 C34 H67 N O5 Si4 57325-92-5 SH-1981-0-0 Methyloxime, tetrakis(trimethylsilyl) derivative of 11β,17
α,20α,21-Tetrahydroxy-4-pregnen-3-one; W
128297

681 C36 H63 N3 O9 AT-64-958-31 CYCLOHEXADESIPEPTIDES; W
66138

681 C47 H71 N O2 33569-81-2 AK-123-441-5 Benzoic acid, 4-amino-3-(3,7,11,15,19,23,27,31-octamethyl-2,6,
10,14,18,22,26,30-dotriacontaoctaenyl)-, (all-E)-;; 3-Octaprenyl-4-aminobenzoic acid; W/NBS
66139

682 C14 H10 F6 Hg N2 O4 S3 82396-82-5 K-115-1465-0 N-(Phenylmercurio)-N,N'-bis(trifluoromethylsulfonyl)
benzolsulfinamidine; W
128298

682 C24 H30 N10 O14 H-60-2283-0 3,3,6,6-TETRAMETHYL PIPERAZINE-2,5-BIS(N,N-DIMETHYLIMINIUM)DIPICRATE; W
66140

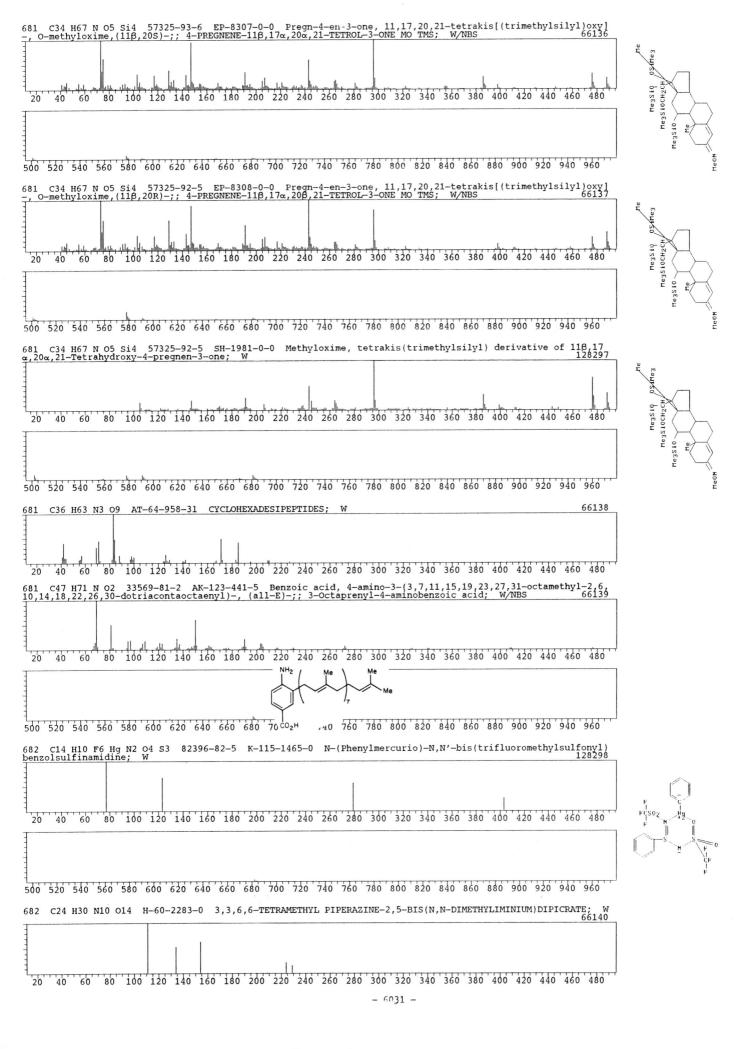

682 C25 H27 F9 N4 O8 56772-21-5 EP-6496-0-0 L-Phenylalanine, N-[N-[N2,N5-bis(trifluoroacetyl)-L-ornithy
l]-O-(trifluoroacetyl)-D-allothreonyl]-, methyl ester;; TRIS-TRIFLUOROACETYL-L-ORNITHYL-D-ALLO-THREONYL-L-
PHENYLALANINE METHYL ESTER; W/NBS   66141

682 C36 H45 D N2 O11 55256-20-7 AD-0-2587-0   3-DEUTERIO-ISORESERPINE METHACETATE; W   66142

682 C36 H46 N2 O11 55256-19-4 AD-0-2584-0 Yohimbanium, 11,17-dimethoxy-16-(methoxycarbonyl)-4-methyl-18
-[(3,4,5-trimethoxybenzoyl)oxy]-, (3β,16β,17α,18β,20α)-, acetate;; RESERPINE METHACETATE; W/NBS   66143

682 C38 H35 O10 P 81095-44-5 C-104-2501-0 3,4:8,9-Bis(4'-methoxybenzo)-5,7-bis(2',4'dimethoxyphenyl)-1-
phenyl-2,6,10,11-teteroxa-1-phospha(V)tricyclo[5.3.1.0]undecane; W   128299

682 C39 H46 N4 O3 Zn 63984-77-0 F-33-632-0 ZINC(II)-α-ACETOXY-β-FORMYLOCTAETHYLPORPHRIN; W   66144

682 C40 H46 N2 O8 66408-23-9 F-33-2922-0 THALILUTIDINE; W   66145

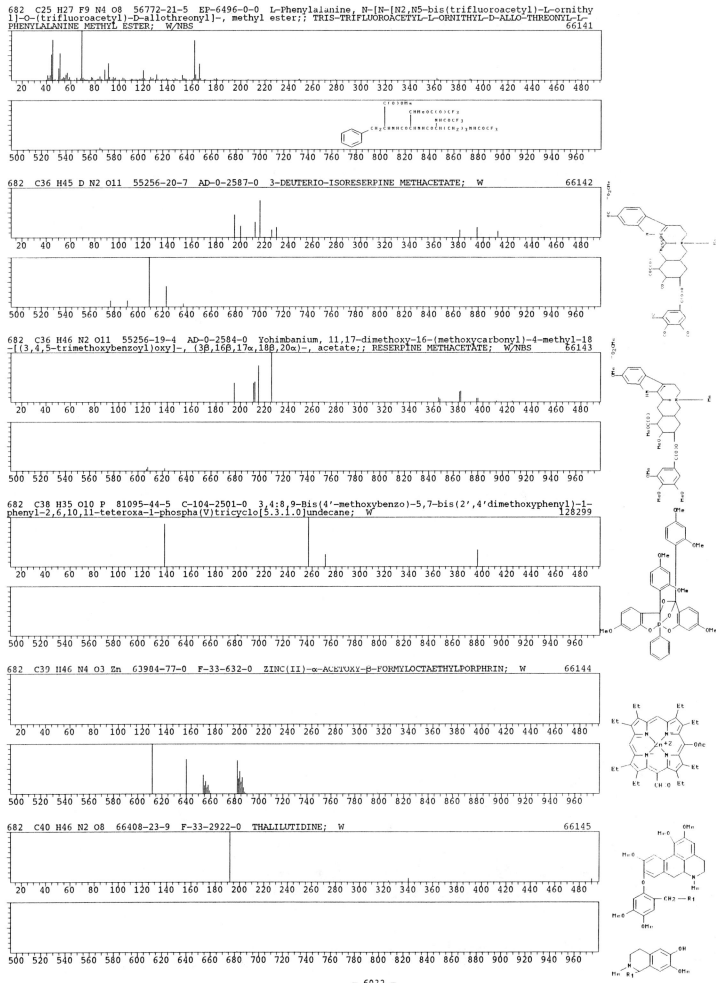

682　C40 H46 N2 O8　65853-12-5　F-33-2921-0　THALIREVOLINE;　W　66146

682　C41 H62 O8　60419-33-2　B-29-1362-0　Olean-12-ene-3β,16α,22α,28-tetrol 3,22,28-triacetate 16-angelate;
W　66147

682　C41 H62 O8　60419-34-3　B-29-1363-0　OLEAN-12-ENE-3β,16α,22α,28-TETROL 3,16,28-TRIACETATE 2I-ANGELATE;
W　66148

682　C41 H82 O5 Si　65615-86-3　E-50-559-3　1,2-DIPALMITOYL-RAC-GLYCEROL TERT-BUTYL-DIMETHYLSILYL;　W 66149

682　C41 H82 O5 Si　65615-87-4　E-50-559-3　1,3-DIPALMITOYL-RAC-GLYCEROL TER-BUTYL-DIMETHYLSILYL;　W　66150

682　C42 H86 O4 Si　64826-22-8　KO-4-110-5　1-OCTADECYL-2-OCTADECANOYL(STEAROYL)-3-TMS GLYCEROL;　W　66151

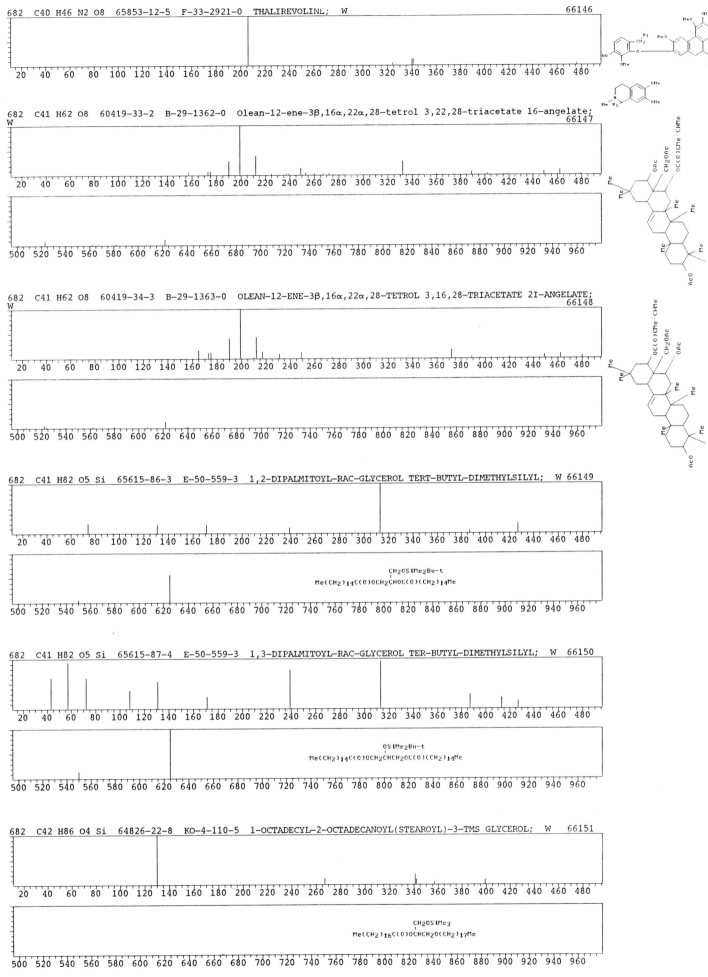

- 6033 -

682   C44 H55 Cl O4   75013-60-4   F-37-561-0   2-Dodec[...] [...]droxy-5-propylphenylester of 17-(3-Chloro-4-hydr
oxyphenyl-2,4,6,8,10,12,14,16-heptadecaoctaenoic acid;   W
128300

683   C19 Cl13   32390-14-0   O-5-887-6   Fluoren-9-yl, 1,2,3,4,5,6,7,8-octachloro-9-(pentachlorophenyl)-;; PER
CHLORO-9-PHENYLFLUORENYL RADICAL;   W/NBS
66152

683   C24 H54 N3 O8 P Si5   32653-16-0   BA-0-260-0   5'-Cytidylic acid, N-(trimethylsilyl)-2',3'-bis-O-(tri
methylsilyl)-,bis(trimethylsilyl) ester;;   5'-(O,O-BIS(TRIMETHYLSILYL)PHOSPHATYL)-N4,O2',O3'-TRIS(TRIMETHYLS
ILYL)CYTIDINE;   W/NBS
66153

683   C24 H54 N3 O8 P Si5   BS-5-125-0   Hexakis(trimethylsilyl) derivative of cytidylic acid 3'-monophosphate
;   W
128301

683   C24 H54 N3 O8 P Si5   BS-5-126-0   Hexakis(trimethylsilyl) derivative of cytidylic acid 5'-monophosphate
;   W
128302

683   C27 H61 N O9 Si5   57397-00-9   EP-831-0-0   D-glycero-D-galacto-2-Nonulosonic acid, 5-(acetylamino)-3,5-
dideoxy-4,6,7,8,9-pentakis-O-(trimethylsilyl)-, methyl ester;;   N-ACETYLNEURAMINIC ACID-METHYL ESTER-PENTAT
MS ETHER;   W
79517

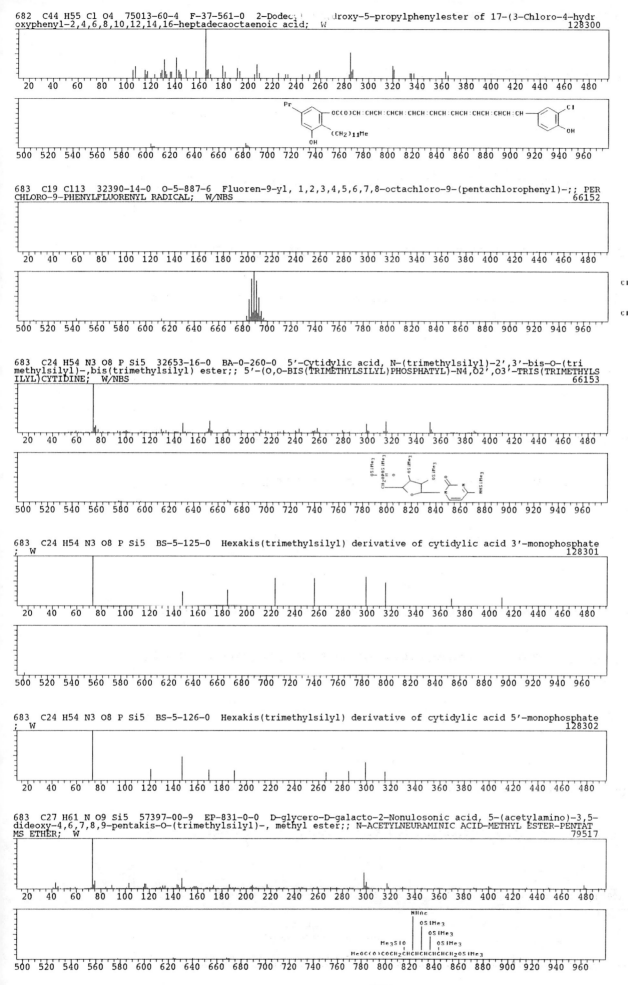

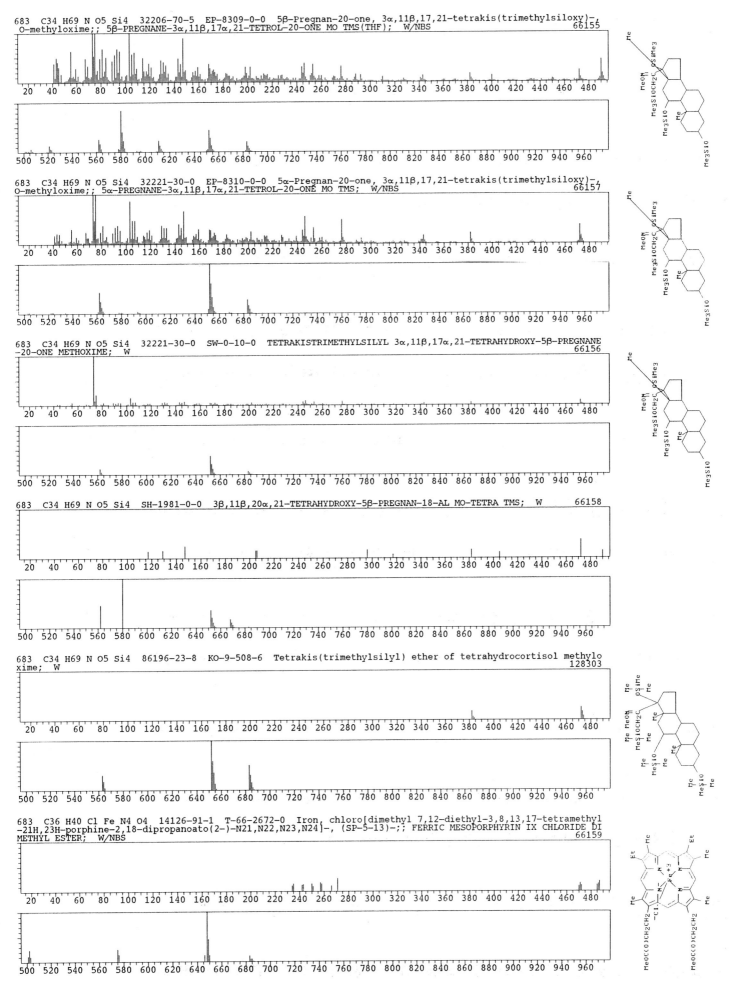

683   C34 H69 N O5 Si4   32206-70-5   EP-8309-0-0   5β-Pregnan-20-one, 3α,11β,17,21-tetrakis(trimethylsiloxy)-
O-methyloxime;; 5β-PREGNANE-3α,11β,17α,21-TETROL-20-ONE MO TMS(THF); W/NBS   66155

683   C34 H69 N O5 Si4   32221-30-0   EP-8310-0-0   5α-Pregnan-20-one, 3α,11β,17,21-tetrakis(trimethylsiloxy)-
O-methyloxime;; 5α-PREGNANE-3α,11β,17α,21-TETROL-20-ONE MO TMS; W/NBS   66157

683   C34 H69 N O5 Si4   32221-30-0   SW-0-10-0   TETRAKISTRIMETHYLSILYL 3α,11β,17α,21-TETRAHYDROXY-5β-PREGNANE
-20-ONE METHOXIME; W   66156

683   C34 H69 N O5 Si4   SH-1981-0-0   3β,11β,20α,21-TETRAHYDROXY-5β-PREGNAN-18-AL MO-TETRA TMS; W   66158

683   C34 H69 N O5 Si4   86196-23-8   KO-9-508-6   Tetrakis(trimethylsilyl) ether of tetrahydrocortisol methylo
xime; W   128303

683   C36 H40 Cl Fe N4 O4   14126-91-1   T-66-2672-0   Iron, chloro[dimethyl 7,12-diethyl-3,8,13,17-tetramethyl
-21H,23H-porphine-2,18-dipropanoato(2-)-N21,N22,N23,N24]-, (SP-5-13)-;; FERRIC MESOPORPHYRIN IX CHLORIDE DI
METHYL ESTER; W/NBS   66159

683 C37 H47 Fe N4 O3 S 82008-25-1 KA-1984-571-4 Methylsulfonato(2,3,7,8,12,13,17,18-octaethylprophyrina
to)iron; W   128304

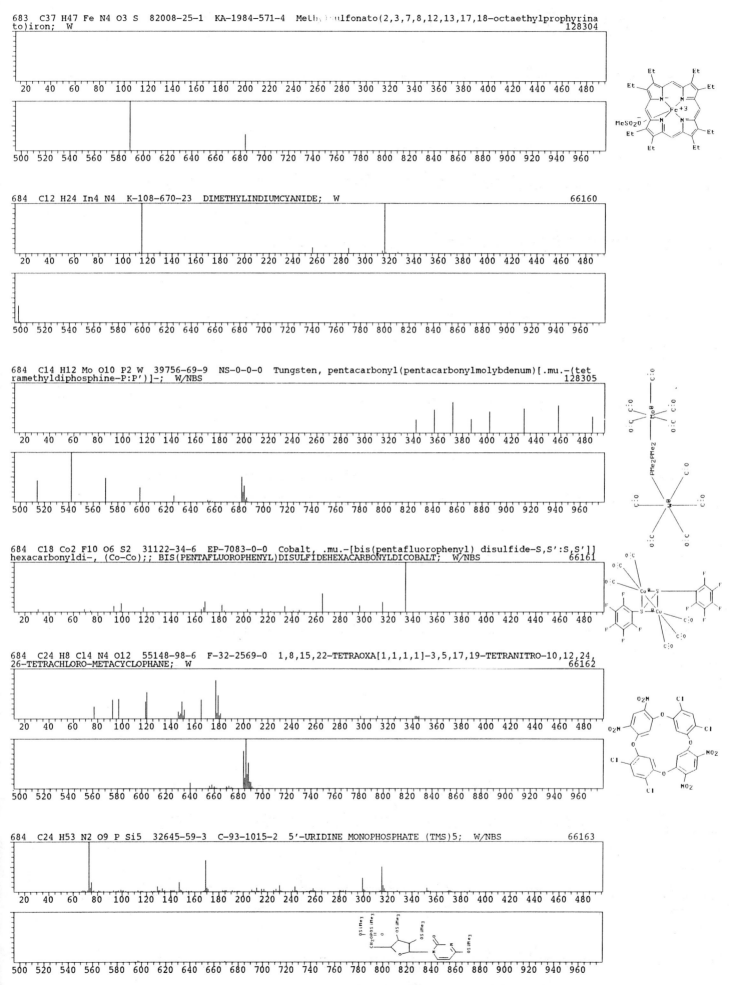

684 C12 H24 In4 N4 K-108-670-23 DIMETHYLINDIUMCYANIDE; W   66160

684 C14 H12 Mo O10 P2 W 39756-69-9 NS-0-0-0 Tungsten, pentacarbonyl(pentacarbonylmolybdenum)[.mu.-(tet
ramethyldiphosphine-P:P')]-; W/NBS   128305

684 C18 Co2 F10 O6 S2 31122-34-6 EP-7083-0-0 Cobalt, .mu.-[bis(pentafluorophenyl) disulfide-S,S':S,S']]
hexacarbonyldi-, (Co-Co);; BIS(PENTAFLUOROPHENYL)DISULFIDEHEXACARBONYLDICOBALT; W/NBS   66161

684 C24 H8 Cl4 N4 O12 55148-98-6 F-32-2569-0 1,8,15,22-TETRAOXA[1,1,1,1]-3,5,17,19-TETRANITRO-10,12,24,
26-TETRACHLORO-METACYCLOPHANE; W   66162

684 C24 H53 N2 O9 P Si5 32645-59-3 C-93-1015-2 5'-URIDINE MONOPHOSPHATE (TMS)5; W/NBS   66163

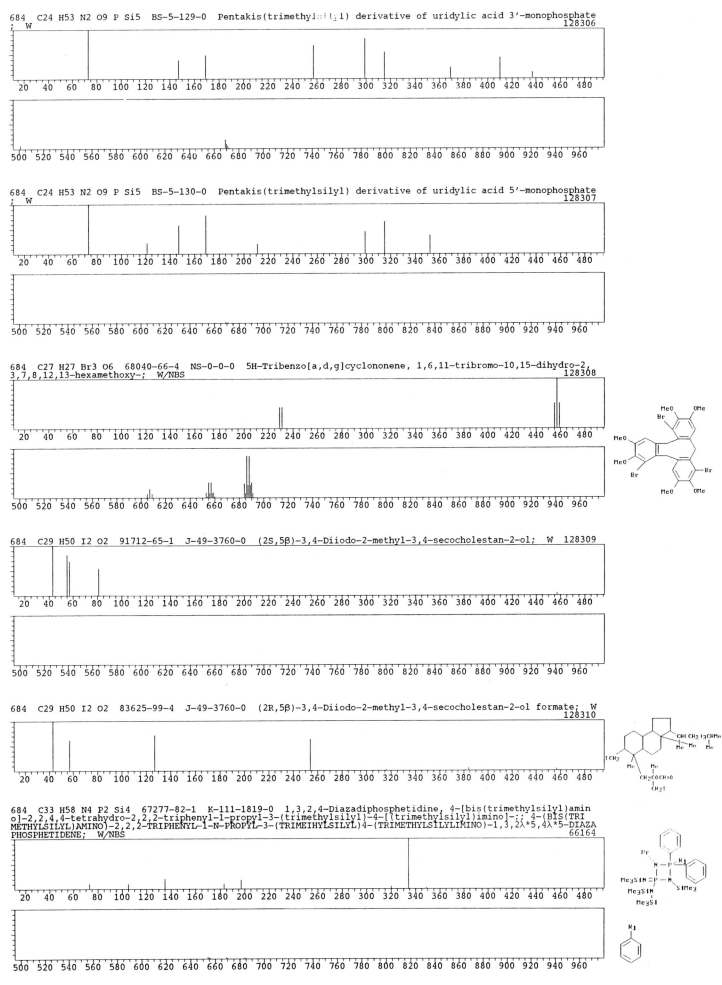

684 C24 H53 N2 O9 P Si5 BS-5-129-0 Pentakis(trimethylsilyl) derivative of uridylic acid 3'-monophosphate
; W   128306

684 C24 H53 N2 O9 P Si5 BS-5-130-0 Pentakis(trimethylsilyl) derivative of uridylic acid 5'-monophosphate
; W   128307

684 C27 H27 Br3 O6 68040-66-4 NS-0-0-0 5H-Tribenzo[a,d,g]cyclononene, 1,6,11-tribromo-10,15-dihydro-2,
3,7,8,12,13-hexamethoxy-; W/NBS   128308

684 C29 H50 I2 O2 91712-65-1 J-49-3760-0 (2S,5β)-3,4-Diiodo-2-methyl-3,4-secocholestan-2-ol; W 128309

684 C29 H50 I2 O2 83625-99-4 J-49-3760-0 (2R,5β)-3,4-Diiodo-2-methyl-3,4-secocholestan-2-ol formate; W
                                                                                       128310

684 C33 H58 N4 P2 Si4 67277-82-1 K-111-1819-0 1,3,2,4-Diazadiphosphetidine, 4-[bis(trimethylsilyl)amin
o]-2,2,4,4-tetrahydro-2,2,2-triphenyl-1-propyl-3-(trimethylsilyl)-4-[(trimethylsilyl)imino]-;; 4-(BIS(TRI
METHYLSILYL)AMINO)-2,2,2-TRIPHENYL-1-N-PROPYL-3-(TRIMEIHYLSILYL)4-(TRIMETHYLSILYLIMINO)-1,3,2λ*5,4λ*5-DIAZA
PHOSPHETIDENE; W/NBS   66164

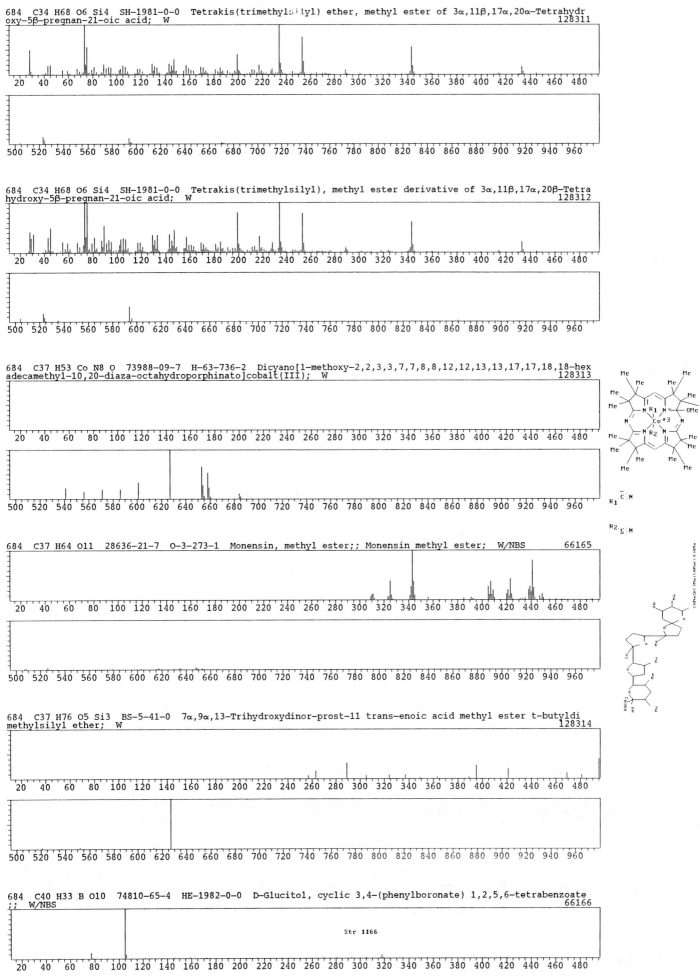

684　C34 H68 O6 Si4　SH-1981-0-0　Tetrakis(trimethylsilyl) ether, methyl ester of 3α,11β,17α,20α-Tetrahydroxy-5β-pregnan-21-oic acid;　W
128311

684　C34 H68 O6 Si4　SH-1981-0-0　Tetrakis(trimethylsilyl), methyl ester derivative of 3α,11β,17α,20β-Tetrahydroxy-5β-pregnan-21-oic acid;　W
128312

684　C37 H53 Co N8 O　73988-09-7　H-63-736-2　Dicyano[1-methoxy-2,2,3,3,7,7,8,8,12,12,13,13,17,17,18,18-hexadecamethyl-10,20-diaza-octahydroporphinato]cobalt(III);　W
128313

684　C37 H64 O11　28636-21-7　O-3-273-1　Monensin, methyl ester;; Monensin methyl ester;　W/NBS
66165

684　C37 H76 O5 Si3　BS-5-41-0　7α,9α,13-Trihydroxydinor-prost-11 trans-enoic acid methyl ester t-butyldimethylsilyl ether;　W
128314

684　C40 H33 B O10　74810-65-4　HE-1982-0-0　D-Glucitol, cyclic 3,4-(phenylboronate) 1,2,5,6-tetrabenzoate ;;　W/NBS
66166

Str 1166

684   C40 H44 O10   73636-23-4   J-45-3456-0   2,3-bis(2-ben   2,2-dicarbethoxyethyl)-6-methoxy-1-indenone;   W
128315

684   C41 H30 Cr O3 P2   82322-00-7   K-115-1387-0   Tricarbonyl[[diphenylphosphino]methyl](1,4a,9a,η6-fluoreny
lidene)diphenylphosphorane]chromium;   W
128316

684   C43 H69 Cl O4   K-112-2007-0   MIXTURE OF 2-ISOUNDECYL-5-PENTYL- AND 2-DECYL-5-ISOHEXYL-3-METHOXYPHENYL
13-(3-CL-4-MEO-PHENYL)TRIDECACARBOXYLATES;   W
66167

684   C44 H57 Cl O4   71142-36-4   K-112-2006-0   2,4,6,8,10,12,14-Pentadecaheptaenoic acid, 15-(3-chloro-4-
methoxyphenyl)-, 2-decyl-3-methoxy-5-pentylphenyl ester;; 2-DECYL-3-METHOXY-5-PENTYLPHENYL ESTER OF 15-(3-
CHLORO-4-METHOXYPHENYL)-2,4,6,8,10,12,14-PENTADECAHEXENE CARBOXYLIC ACID;   W/NBS
66168

684   C44 H60 O6   H-67-2153-0   Di-O-acetal of (8'R)-Luteoxanthin;   W
128317

684   C45 H30 N4 Ni   C-97-6180-0   EXO(1'-PHENYLCYCLOPROPANO(2',3':3,4))3,4-DIHYDRONICKELHOMOPORPHYRIN;   W
66169

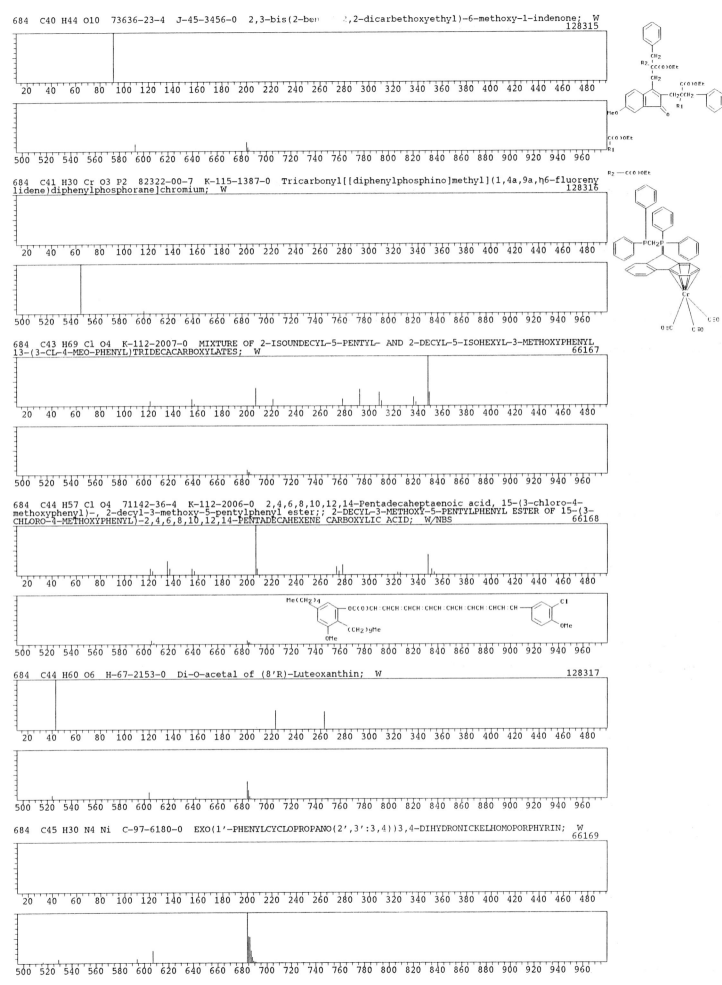

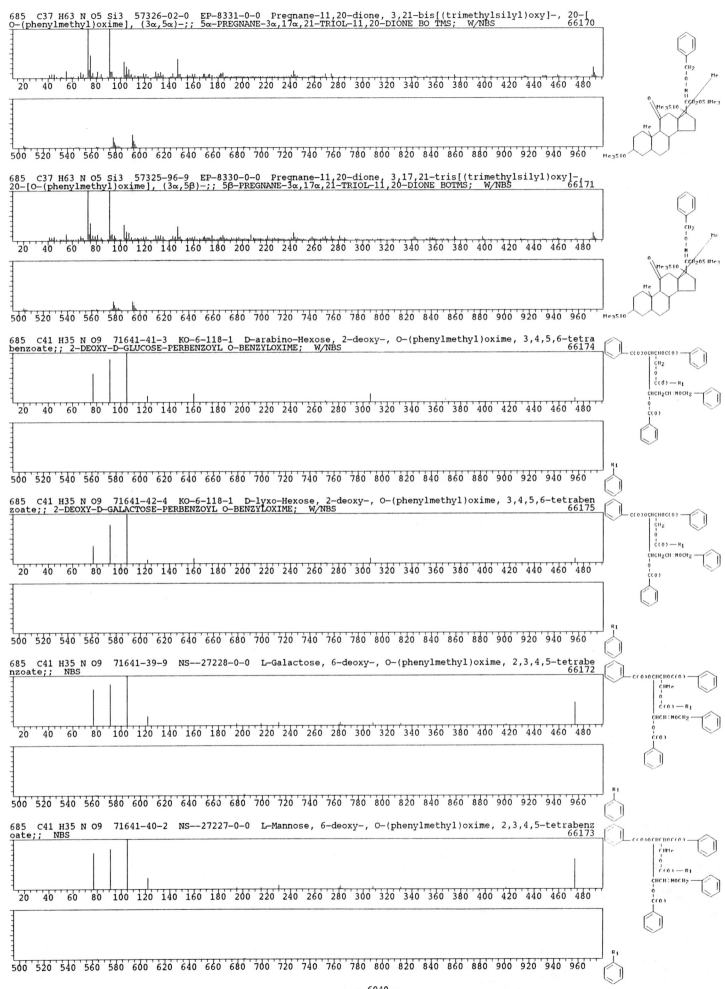

685　C37 H63 N O5 Si3　57326-02-0　EP-8331-0-0　Pregnane-11,20-dione, 3,21-bis[(trimethylsilyl)oxy]-, 20-[
O-(phenylmethyl)oxime], (3α,5α)-;; 5α-PREGNANE-3α,17α,21-TRIOL-11,20-DIONE BO TMS;　W/NBS　　　66170

685　C37 H63 N O5 Si3　57325-96-9　EP-8330-0-0　Pregnane-11,20-dione, 3,17,21-tris[(trimethylsilyl)oxy]-,
20-[O-(phenylmethyl)oxime], (3α,5β)-;; 5β-PREGNANE-3α,17α,21-TRIOL-11,20-DIONE BOTMS;　W/NBS　　66171

685　C41 H35 N O9　71641-41-3　KO-6-118-1　D-arabino-Hexose, 2-deoxy-, O-(phenylmethyl)oxime, 3,4,5,6-tetra
benzoate;; 2-DEOXY-D-GLUCOSE-PERBENZOYL O-BENZYLOXIME;　W/NBS　　　66174

685　C41 H35 N O9　71641-42-4　KO-6-118-1　D-lyxo-Hexose, 2-deoxy-, O-(phenylmethyl)oxime, 3,4,5,6-tetraben
zoate;; 2-DEOXY-D-GALACTOSE-PERBENZOYL O-BENZYLOXIME;　W/NBS　　　66175

685　C41 H35 N O9　71641-39-9　NS--27228-0-0　L-Galactose, 6-deoxy-, O-(phenylmethyl)oxime, 2,3,4,5-tetrabe
nzoate;;　NBS　　　66172

685　C41 H35 N O9　71641-40-2　NS--27227-0-0　L-Mannose, 6-deoxy-, O-(phenylmethyl)oxime, 2,3,4,5-tetrabenz
oate;;　NBS　　　66173

685   C41 H35 N O9   71641-39-9   KO-6-118-1   L-FUCOSE-PERBENZOYL O-BENZYLOXIME;   W   66176

685   C41 H35 N O9   71641-40-2   KO-6-118-1   L-THAMNOSE-PERBENZOYL O-BENZYLOXIME;   W   66177

685   C43 H59 N O6   C-104-5739-0   (1S,2R,3S,5R,6R,7S,8R,10R,11S,14R,15S,18R)-13,14-Dibenzyl-18-(2-trimethyls
ilyl)ethoxy)-6,7,10-trihydroxy-5,7,16,17-tetramethyl-12-oxo-13-aza-19-oxapentacyclo[9.7.0.1.0.0]nonadec-16-
ene-6,7-acetonide;   W   128318

686   C28 H26 Mo2 S4   72186-28-8   C-101-5251-0   BIS(CYCLOPENTADIENYL)MOLYBDENUM(III)-1-PHENYL-1-PROPEN-1,2-
DITHIOLATE;   W   66178

686   C29 H38 N2 O17   78300-67-1   F-37-109-0   O-[2,3,4-Tri-O-acetyl-β-D-galactopyranosyluron-N-(1-hydroxy-3-
oxo-1-cyclopenten-2-yl)amide]-(1 to 4)-2-acetamido-1,3-di-O-acetyl-2,6-didesoxy-α-D-glucopyranose;   W
                                                                                                                    128319

686   C30 H42 N2 O16   78300-68-2   F-37-109-0   2-Propyl-O-[2,3,4-tri-O-acetyl-β-D-galactopyranosyluron-N-(1-
hydroxy-3-oxo-1-cyclopenten-2-yl)amide]-(1 to 4)-2-acetamido-3-O-acetyl-2,6-didesoxy-β-D-glucopyranoside;
W   128320

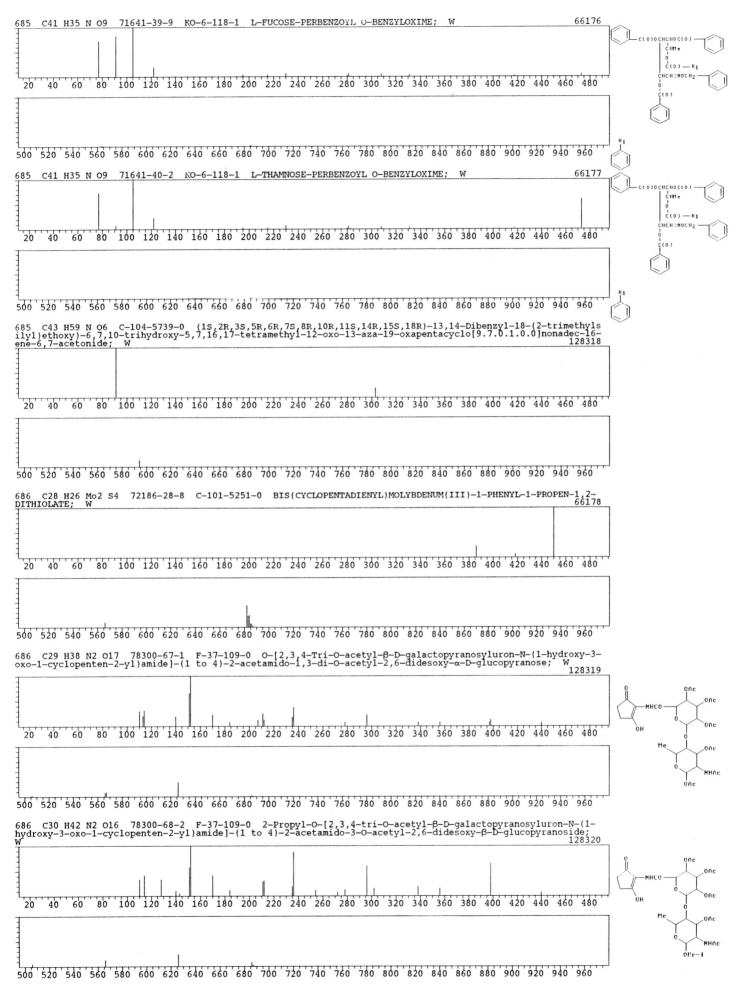

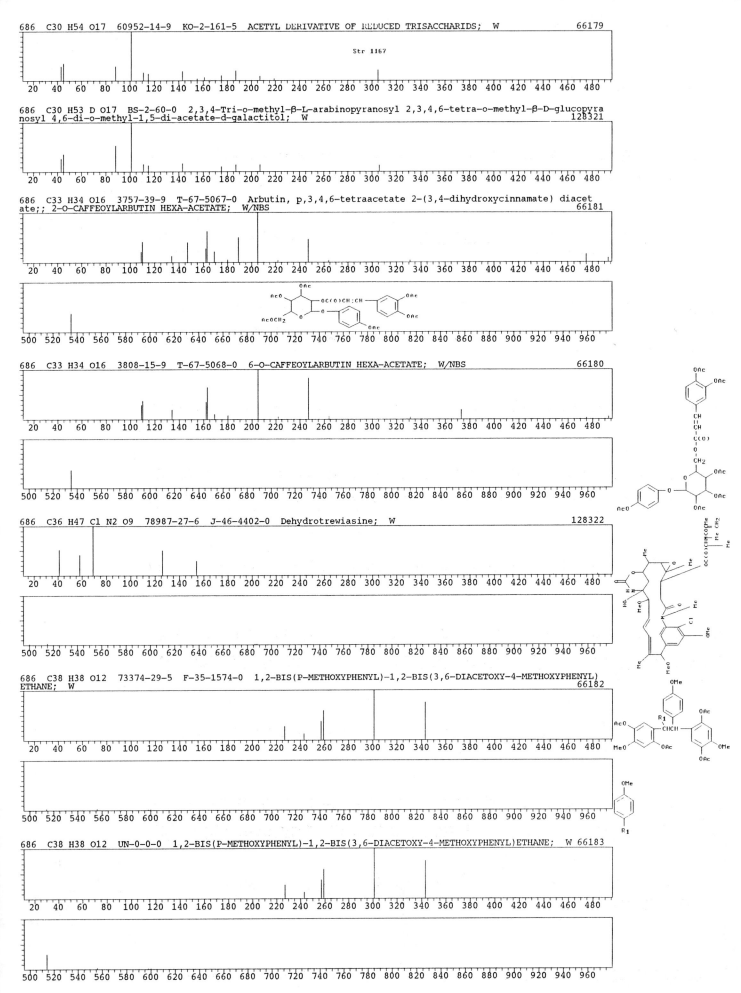

686   C30 H54 O17   60952-14-9   KO-2-161-5   ACETYL DERIVATIVE OF REDUCED TRISACCHARIDS;   W   66179

Str 1167

686   C30 H53 D O17   BS-2-60-0   2,3,4-Tri-o-methyl-β-L-arabinopyranosyl 2,3,4,6-tetra-o-methyl-β-D-glucopyra
nosyl 4,6-di-o-methyl-1,5-di-acetate-d-galactitol;   W   128321

686   C33 H34 O16   3757-39-9   T-67-5067-0   Arbutin, p,3,4,6-tetraacetate 2-(3,4-dihydroxycinnamate) diacet
ate;;   2-O-CAFFEOYLARBUTIN HEXA-ACETATE;   W/NBS   66181

686   C33 H34 O16   3808-15-9   T-67-5068-0   6-O-CAFFEOYLARBUTIN HEXA-ACETATE;   W/NBS   66180

686   C36 H47 Cl N2 O9   78987-27-6   J-46-4402-0   Dehydrotrewiasine;   W   128322

686   C38 H38 O12   73374-29-5   F-35-1574-0   1,2-BIS(P-METHOXYPHENYL)-1,2-BIS(3,6-DIACETOXY-4-METHOXYPHENYL)
ETHANE;   W   66182

686   C38 H38 O12   UN-0-0-0   1,2-BIS(P-METHOXYPHENYL)-1,2-BIS(3,6-DIACETOXY-4-METHOXYPHENYL)ETHANE;   W 66183

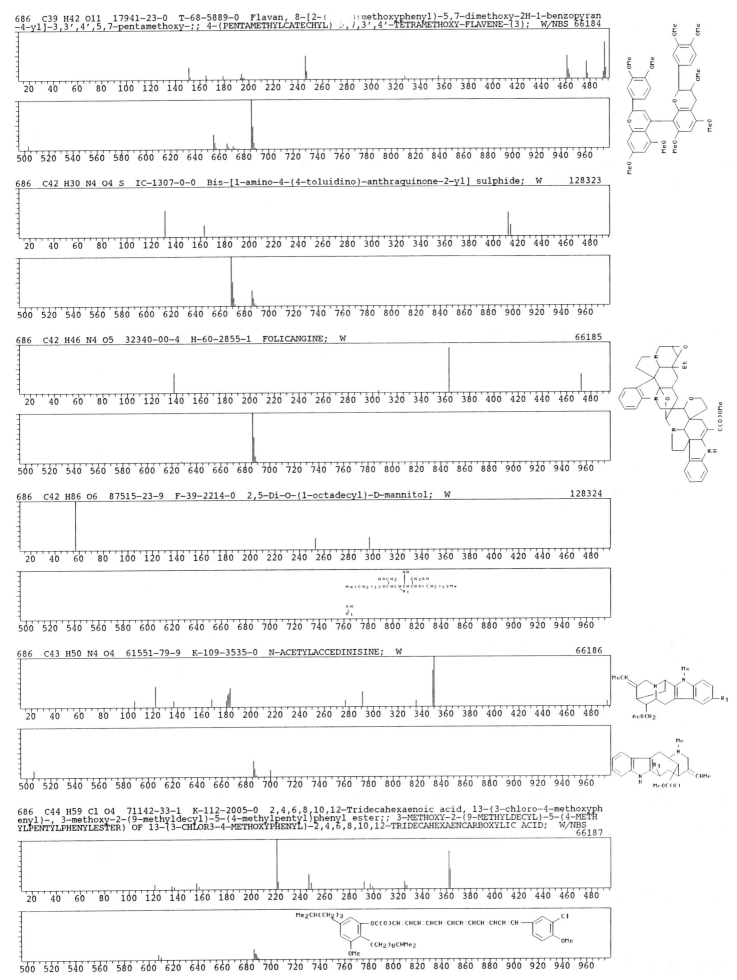

686 C39 H42 O11 17941-23-0 T-68-5889-0 Flavan, 8-[2-(imethoxyphenyl)-5,7-dimethoxy-2H-1-benzopyran
-4-yl]-3,3',4',5,7-pentamethoxy-;; 4-(PENTAMETHYLCATECHYL) 5,7,3',4'-TETRAMETHOXY-FLAVENE-(3); W/NBS 66184

686 C42 H30 N4 O4 S IC-1307-0-0 Bis-[1-amino-4-(4-toluidino)-anthraquinone-2-yl] sulphide; W   128323

686 C42 H46 N4 O5 32340-00-4 H-60-2855-1 FOLICANGINE; W   66185

686 C42 H86 O6 87515-23-9 F-39-2214-0 2,5-Di-O-(1-octadecyl)-D-mannitol; W   128324

686 C43 H50 N4 O4 61551-79-9 K-109-3535-0 N-ACETYLACCEDINISINE; W   66186

686 C44 H59 Cl O4 71142-33-1 K-112-2005-0 2,4,6,8,10,12-Tridecahexaenoic acid, 13-(3-chloro-4-methoxyph
enyl)-, 3-methoxy-2-(9-methyldecyl)-5-(4-methylpentyl)phenyl ester;; 3-METHOXY-2-(9-METHYLDECYL)-5-(4-METH
YLPENTYLPHENYLESTER) OF 13-(3-CHLOR3-4-METHOXYPHENYL)-2,4,6,8,10,12-TRIDECAHEXAENCARBOXYLIC ACID; W/NBS
                                                                                                        66187

687 C29 H57 N O8 Si5 55836-51-6 EP-1749-0-0 β-D-Glucopyranosiduronic acid, 4-[acetyl(trimethylsilyl)ami
nolphenyl 2,3,4-tris-O-(trimethylsilyl)-, trimethylsilyl ester;; 4-GLUCURONOSIDO-N-ACETYLANILINE-PENTATMS;
W/NBS   66188

687 C33 H69 N O6 Si4 66447-97-0 KO-4-360-2 TRIMETHYLSILYLOXIME DERIVATIVE OF 6-OXO-PGF1α; W   66189

688 C28 H25 O7 P W 96826-89-0 C-107-4705-0 (1-(Ethoxycarbonyl)-3,4-dimethyl-6,7-diphenyl-7-phosphabicyc
lo-[4.1.0]-3-heptene)pentacarbonyltungsten; W   128325

688 C30 H56 N2 O10 Si3 BA-0-171-0 5-ALLYL-1,3-DIMETHYL-5-(3-HYDROXY-1-METHYLBUTYL) BARBITURIC ACID O-TRI
S(TRIMETHYLSILYL)GLUCURONIDE METHYL ESTER; W   66190

688 C30 H56 N2 O10 Si3 CR-212-0-0 HYDROXYSECOBARBITAL GLUCURONIDE1,3-DI-METHYLDERIVATIVE METHYL ESTER
TMS ETHER; W   66191

688 C32 H48 N2 O9 Si3 BL-0-44-0 5-(4-HYDROXYPHENYL)-3-METHYL-5-PHENYLHYDANTOINGLUCURONIDE METHYL ESTER
TMS ETHER; W   66192

688 C34 H15 Co F10 O 31760-83-5 AD-0-5746-0 Cobalt, [2,4-bis(pentafluorophenyl)-3,5-diphenyl-2,4-cyclop
entadien-1-one]-.pi.-cyclopentadienyl-;; 2,4-BIS(PENTAFLUOROPHENYL)-3,5-DIPHENYLCYCLOPENTADIENEONE-.PI.-CYC
LOPENTADIENYLCOBALT; W/NBS   66193

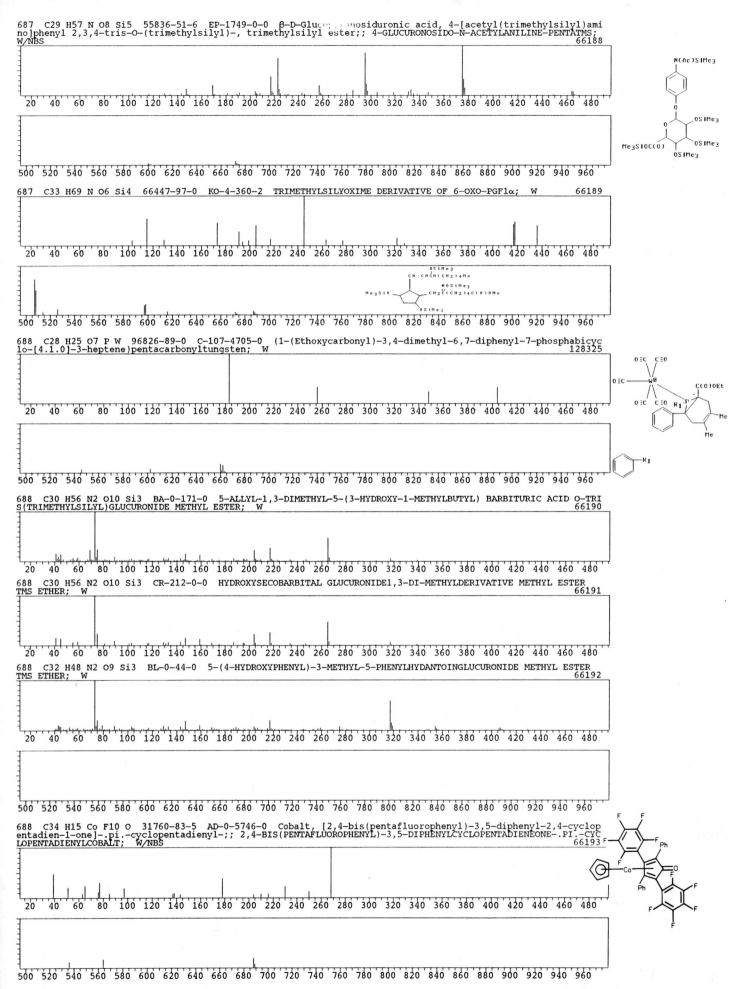

688   C35 H45 Cl N2 O10   79101-55-6   J-46-4403-0   Treflorine;   W   128326

20 40 60 80 100 120 140 160 180 200 220 240 260 280 300 320 340 360 380 400 420 440 460 480

500 520 540 560 580 600 620 640 660 680 700 720 740 760 780 800 820 840 860 880 900 920 940 960

688   C36 H49 Cl N2 O9   78987-26-5   J-46-4402-0   Trewiasine;   W   128327

20 40 60 80 100 120 140 160 180 200 220 240 260 280 300 320 340 360 380 400 420 440 460 480

500 520 540 560 580 600 620 640 660 680 700 720 740 760 780 800 820 840 860 880 900 920 940 960

688   C42 H48 N4 O5   32063-91-5   H-60-2855-1   VOAFOLIDINE;   W   66194

20 40 60 80 100 120 140 160 180 200 220 240 260 280 300 320 340 360 380 400 420 440 460 480

500 520 540 560 580 600 620 640 660 680 700 720 740 760 780 800 820 840 860 880 900 920 940 960

688   C42 H48 N4 O5   33055-38-8   H-60-2855-1   ISOVOAFOLIDINE;   W   66195

20 40 60 80 100 120 140 160 180 200 220 240 260 280 300 320 340 360 380 400 420 440 460 480

500 520 540 560 580 600 620 640 660 680 700 720 740 760 780 800 820 840 860 880 900 920 940 960

688   C42 H47 D N4 O5   H-60-2855-1   3-DEUTERIO-ISOVOAFOLIDINE;   W   66196

20 40 60 80 100 120 140 160 180 200 220 240 260 280 300 320 340 360 380 400 420 440 460 480

500 520 540 560 580 600 620 640 660 680 700 720 740 760 780 800 820 840 860 880 900 920 940 960

688   C44 H64 O6   92011-63-7   J-49-4302-0   [3α(Z),6α(Z)]-1,1'-[1,2,4,5-tetroxane-3,6-diylbis[[1,2-bis(1,1-di
methylethyl)-2,1-ethenediyl]-2,1-phenylene]]bis[2,2-dimethyl-1-propanone];   W   128328

Str 1168

20 40 60 80 100 120 140 160 180 200 220 240 260 280 300 320 340 360 380 400 420 440 460 480

688   C44 H64 O6   92077-26-4   J-49-4302-0   [1α(Z),4β(Z)]-1,1'-[1,2,4,5-tetroxane-3,6-diylbis[[1,2-bis(1,1-di
methylethyl)-2,1-ethenediyl]-2,1-phenylene]]bis[2,2-dimethyl-1-propanone];   W   128329

Str 1169

20 40 60 80 100 120 140 160 180 200 220 240 260 280 300 320 340 360 380 400 420 440 460 480

6045 -

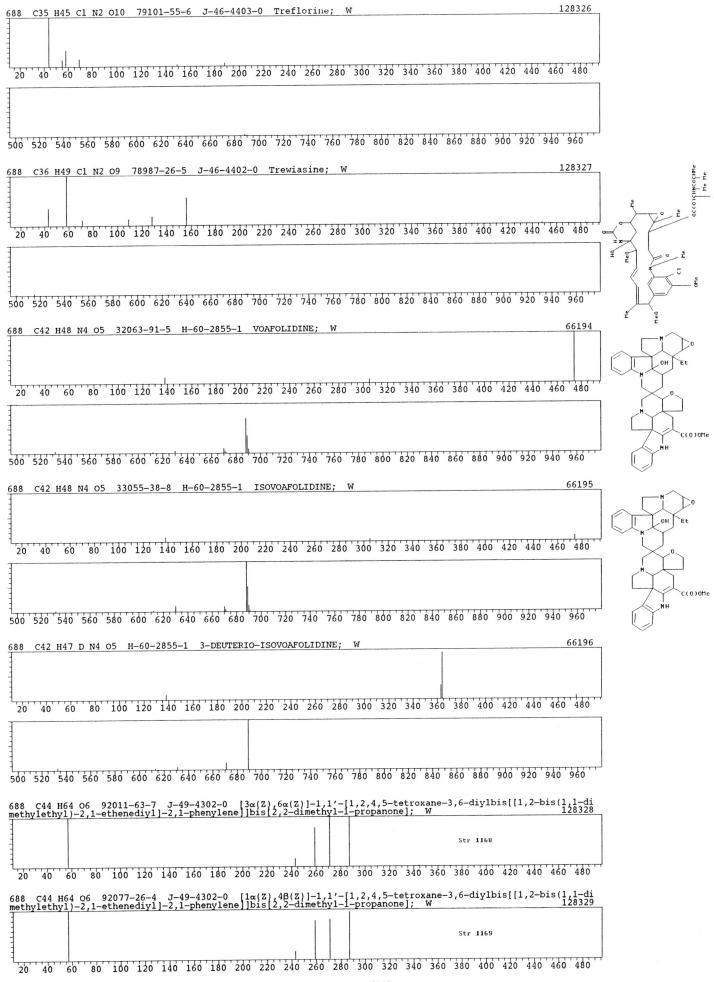

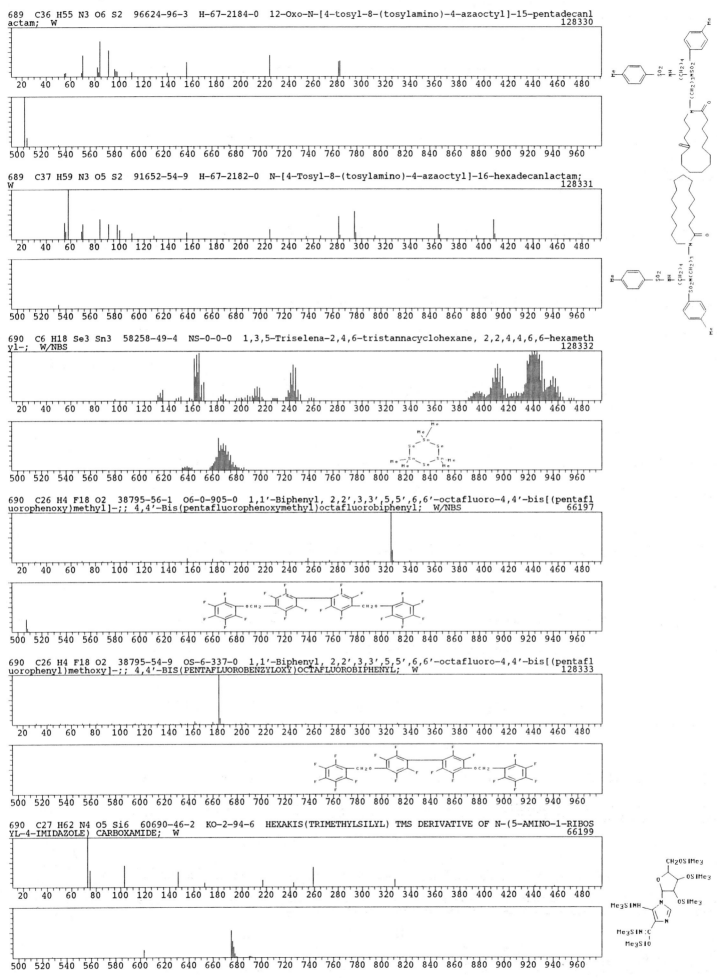

689 C36 H55 N3 O6 S2 96624-96-3 H-67-2184-0 12-Oxo-N-[4-tosyl-8-(tosylamino)-4-azaoctyl]-15-pentadecanlactam; W
128330

689 C37 H59 N3 O5 S2 91652-54-9 H-67-2182-0 N-[4-Tosyl-8-(tosylamino)-4-azaoctyl]-16-hexadecanlactam; W
128331

690 C6 H18 Se3 Sn3 58258-49-4 NS-0-0-0 1,3,5-Triselena-2,4,6-tristannacyclohexane, 2,2,4,4,6,6-hexamethyl-; W/NBS
128332

690 C26 H4 F18 O2 38795-56-1 O6-0-905-0 1,1'-Biphenyl, 2,2',3,3',5,5',6,6'-octafluoro-4,4'-bis[(pentafluorophenoxy)methyl]-;; 4,4'-Bis(pentafluorophenoxymethyl)octafluorobiphenyl; W/NBS
66197

690 C26 H4 F18 O2 38795-54-9 OS-6-337-0 1,1'-Biphenyl, 2,2',3,3',5,5',6,6'-octafluoro-4,4'-bis[(pentafluorophenyl)methoxy]-;; 4,4'-BIS(PENTAFLUOROBENZYLOXY)OCTAFLUOROBIPHENYL; W
128333

690 C27 H62 N4 O5 Si6 60690-46-2 KO-2-94-6 HEXAKIS(TRIMETHYLSILYL) TMS DERIVATIVE OF N-(5-AMINO-1-RIBOSYL-4-IMIDAZOLE) CARBOXAMIDE; W
66199

690 C30 H66 O6 Si6 HE-1986-848-0 HEXAKIS(TRIMETHYLSILYLOXIMETHYL)-BENZENE; W   128334

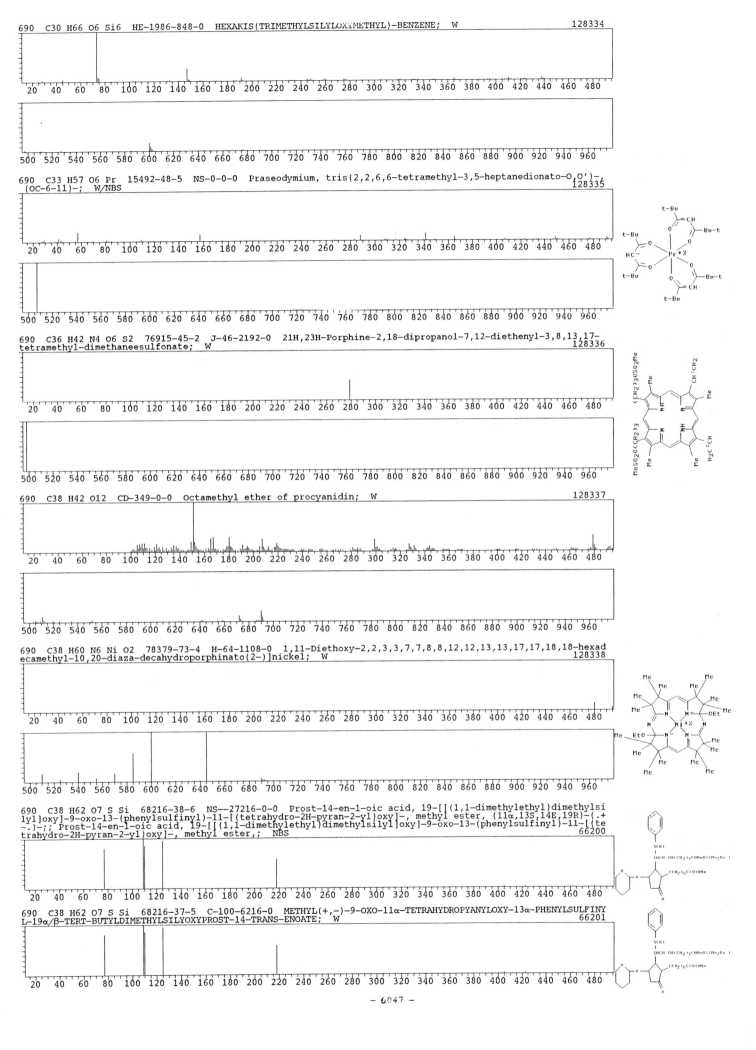

690 C33 H57 O6 Pr 15492-48-5 NS-0-0-0 Praseodymium, tris(2,2,6,6-tetramethyl-3,5-heptanedionato-O,O')-,
(OC-6-11)-; W/NBS   128335

690 C36 H42 N4 O6 S2 76915-45-2 J-46-2192-0 21H,23H-Porphine-2,18-dipropanol-7,12-diethenyl-3,8,13,17-
tetramethyl-dimethaneesulfonate; W   128336

690 C38 H42 O12 CD-349-0-0 Octamethyl ether of procyanidin; W   128337

690 C38 H60 N6 Ni O2 78379-73-4 H-64-1108-0 1,11-Diethoxy-2,2,3,3,7,7,8,8,12,12,13,13,17,17,18,18-hexad
ecamethyl-10,20-diaza-decahydroporphinato(2-)]nickel; W   128338

690 C38 H62 O7 S Si 68216-38-6 NS--27216-0-0 Prost-14-en-1-oic acid, 19-[[(1,1-dimethylethyl)dimethylsi
lyl]oxy]-9-oxo-13-(phenylsulfinyl)-11-[(tetrahydro-2H-pyran-2-yl)oxy]-, methyl ester, (11α,13S,14E,19R)-(.+
-.)-;; Prost-14-en-1-oic acid, 19-[[(1,1-dimethylethyl)dimethylsilyl]oxy]-9-oxo-13-(phenylsulfinyl)-11-[(te
trahydro-2H-pyran-2-yl)oxy]-, methyl ester,; NBS   66200

690 C38 H62 O7 S Si 68216-37-5 C-100-6216-0 METHYL(+,-)-9-OXO-11α-TETRAHYDROPYANYLOXY-13α-PHENYLSULFINY
L-19α/β-TERT-BUTYLDIMETHYLSILYOXYPROST-14-TRANS-ENOATE; W   66201

690 C41 H58 N2 O7 89759-11-5 H-67-247-0 O-Acetyl-O',O''-dimethylkopsirachin; W 128339

690 C41 H55 D3 N2 O7 89759-12-6 H-67-247-0 O-Trideuterioacetyl-O',O''-dimethylkopsirachin; W 128340

690 C42 H50 N4 O5 38899-91-1 EP-6421-0-0 Ajmalan-16-carboxylic acid, 19,20-didehydro-1-[(20α)-20,21-di hydro-21-hydroxy-21-methyl-18-noralstophyllan-19-yl]-17-hydroxy-, methyl ester, (2α,17S,19E)-;; MACROLINE: QUEBRACHIDINE ADDUCT; W/NBS 66202

690 C42 H50 N4 O5 61589-94-4 K-109-3533-0 N-DEMETHYLVOACAMIN; W 66203

690 C43 H46 O8 77495-83-1 NS-0-0-0 5H-Cyclopropa[3,4]benz[1,2-e]azulen-5-one, 9,9a-bis(acetyloxy)-1,1a, 1b,4,4a,7a,7b,8,9,9a-decahydro-4a,7b-dihydroxy-1,1,6,8-tetramethyl-3-[(triphenylmethoxy)methyl]-, [1aR-(1a α,1bβ,4aβ,7aα,7bα,8α,9β,9aα)]-; W/NBS 128341

690 C43 H66 O3 S Si 79409-81-7 J-46-5186-0 (3β,22E)-24-Cyclopentyl-24-(phenylthio)-24-(trimethylsilyl)- 3-[(tetrahydro-2H-pyran-2-yl)oxy]chola-5,22-dien-20-ol; W 128342

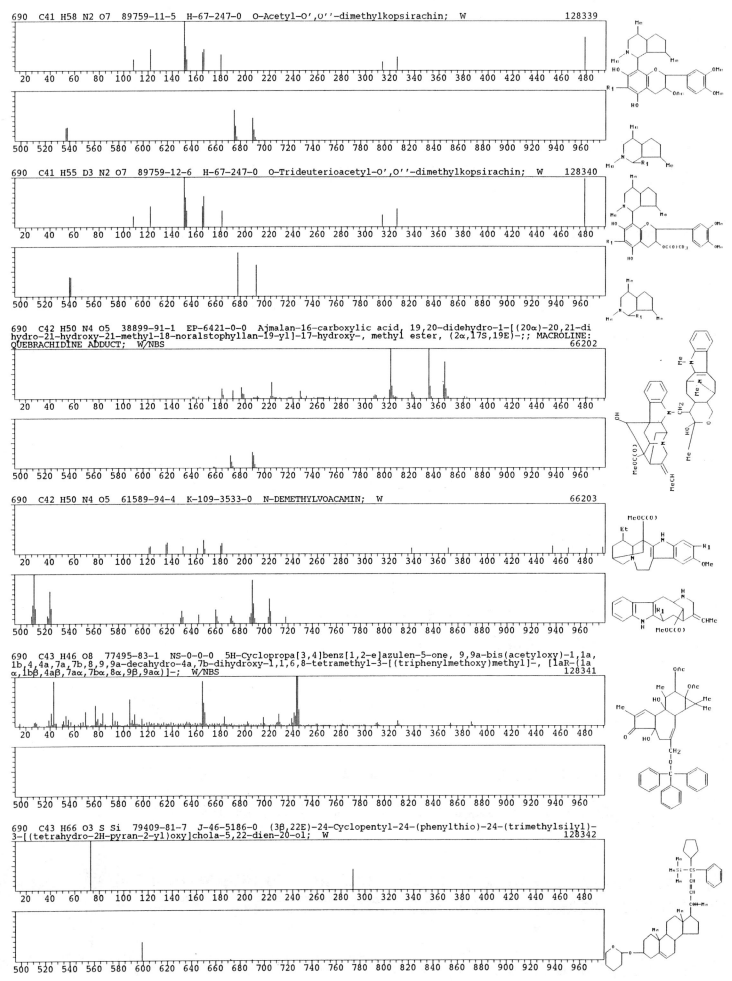

690   C44 H34 O8   63910-24-7   KC-1977-1052-0   2,2'-(O-PHENYLENEDIOXYDI-(O-PHENYLENE)DIOXYDI-(O-PHENYLENE)DIO
XY)DIPHENOL DIMETHYL ETHER;   W   66204

690   C44 H50 N8   69937-02-6   KC-1978-1678-0   QUADRIGEMINE-A;   W   66205

690   C44 H50 N8   69937-10-6   KC-1978-1679-0   QUADRIGEMINE-B;   W   66206

690   C47 H62 O4   69505-96-0   K-112-204-0   MIXTURE OF 2-DECYL-5-ISOHEXYL- AND 2-ISOUNDECYL-5-PENTYL-3-MEO-PH
ENYL17-(4-MEO-PHENYL)-2,4,6,8,10,12,14-HEPTADECANOCTAENCARBOXYLATES;   W   66207

690   C47 H62 O4   78368-46-4   F-37-562-0   2-Dodecyl-3-methoxy-5-(2-methylpropyl)phenylester of 17-(4-Methoxy
phenyl)-2,4,6,8,10,12,14,16-heptadecaoctaenoic acid;   W   128343

690   C50 H74 O   64364-74-5   F-33-383-0   (2R,2'R)-2,2'-BIS(3-METHYLBUTYL)-3,4,3',4'-TETRAHYDRO-.PSI.,.PSI.-
CAROTEN-20-AL;   W   66208

691   C Ag Hf3 S   811-68-7   NS-16761-0-0   Methanethiol, trifluoro-, silver(1+) salt;;   NBS   66209

F3CSH

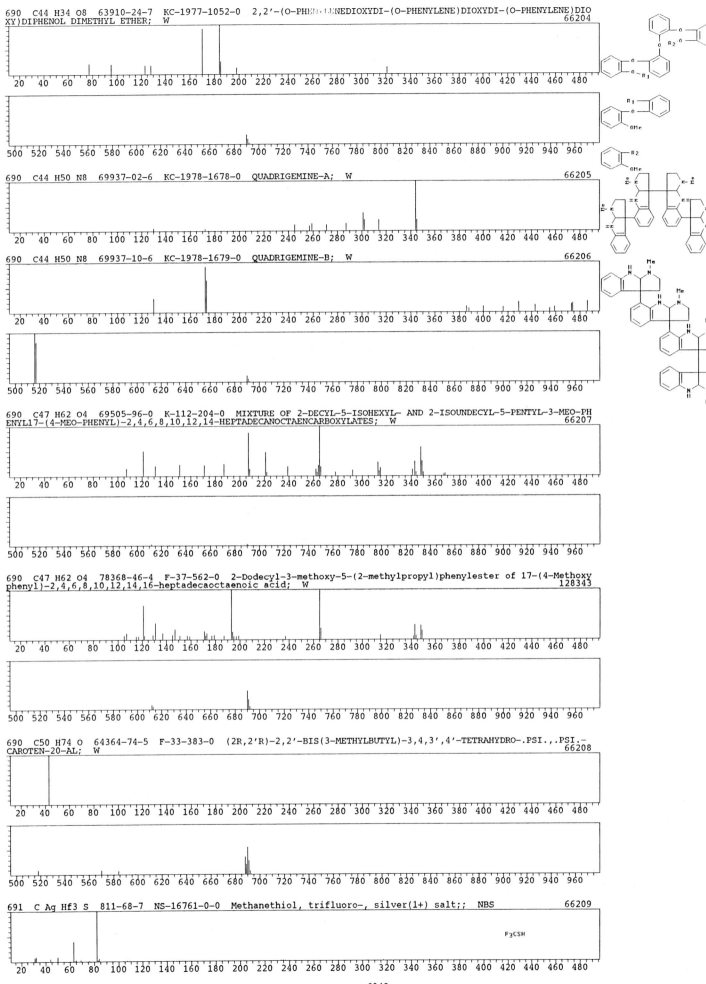

691 C17 H19 I2 N O W 35004-45-6 EP-7338-0-0 Tungsten, (α-tert-butylbenzylideniminato)carbonyl-.pi.-cyclopentadienyldiiodo-;; 2,2-DIMETHYL-1-PHENYLPROPYLIDENEAMINO-CARBONYL-CYCLOPENTADIENYL-TUNGSTEN DI-IODIDE; W/NBS   66210

691 C24 H62 N O8 P Si6 55530-82-0 BA-0-337-0 D-Glucose, 2-deoxy-3,4,5-tris-O-(trimethylsilyl)-2-[(trimethylsilyl)amino]-, 6-[bis(trimethylsilyl) phosphate];; O-PENTAKIS(TRIMETHYLSILYL)-2-TRIMETHYLSILYLAMINO-2-DEOXY-D-GLUCOSE-6-PHOSPHATE; W/NBS   66211

691 C27 H61 N3 O6 Si6 72360-96-4 Y-10-845-2 1H-Pyrazole-5-carboxamide, N,1-bis(trimethylsilyl)-4-[(trimethylsilyl)oxy]-3-[2,3,5-tris-O-(trimethylsilyl)-β-D-ribofuranosyl]-;; HEXAKISTRIMETHYLSILYL PYRAZOMYCIN; W/NBS   66212

691 C30 H30 F12 N3 P 70758-15-5 I-57-1016-0 TRIS(4-N,N-DIETHYLAMINO-2,3,5,6-TETRAFLUOROPHENYL)PHOSPHINE ; W   66213

691 C34 H46 Cl N3 O10 ZH-613-1-0 MAYTANSINE; W   66214

691 C37 H74 N O8 P 3026-45-7 EP-5247-0-0 Hexadecanoic acid, 1-[[[(2-aminoethoxy)hydroxyphosphinyl]oxy]methyl]-1,2-ethanediyl ester;; DIPALMITOYL PHOSPHATIDYL ETHANOLAMINE; W/NBS   66215

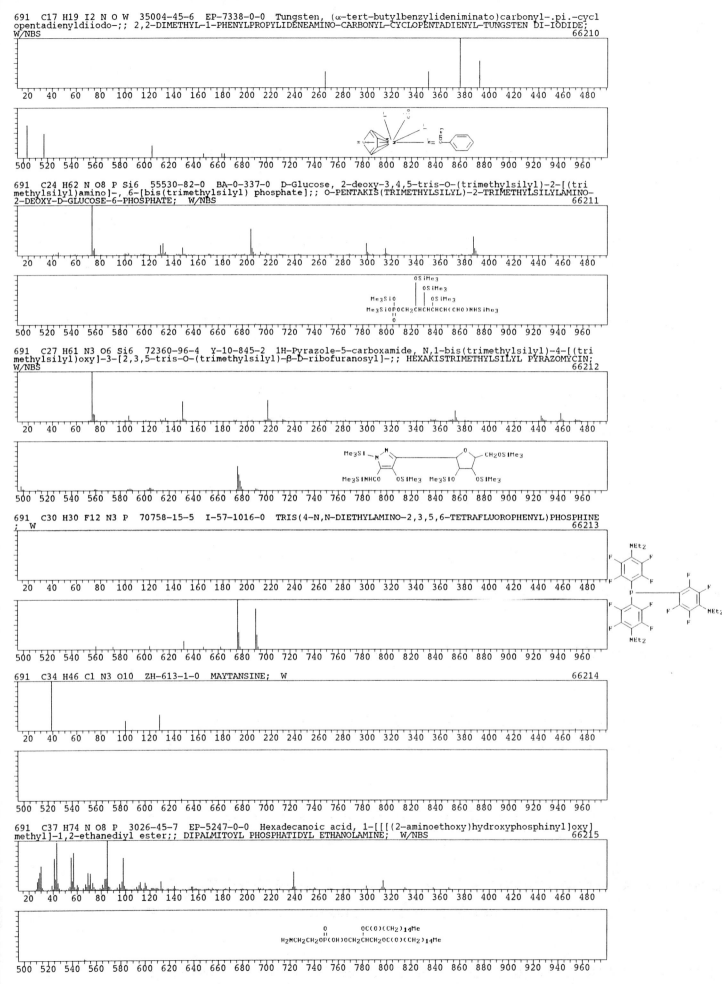

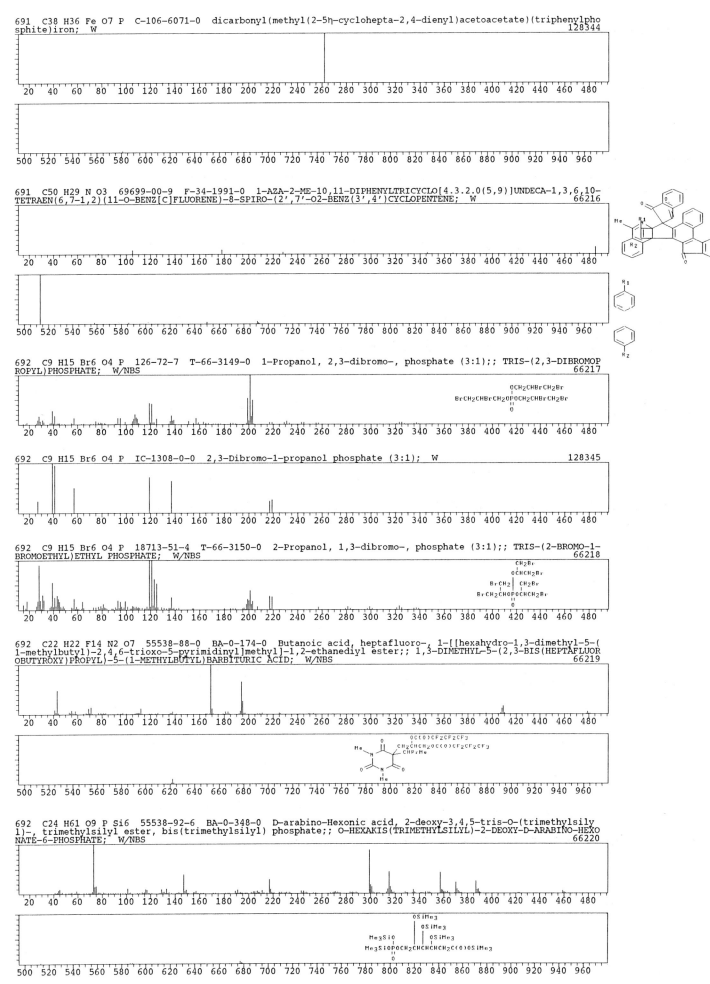

691 C38 H36 Fe O7 P C-106-6071-0 dicarbonyl(methyl(2-5η-cyclohepta-2,4-dienyl)acetoacetate)(triphenylpho
sphite)iron; W   128344

691 C50 H29 N O3 69699-00-9 F-34-1991-0 1-AZA-2-ME-10,11-DIPHENYLTRICYCLO[4.3.2.0(5,9)]UNDECA-1,3,6,10-
TETRAEN(6,7-1,2)(11-O-BENZ[C]FLUORENE)-8-SPIRO-(2',7'-O2-BENZ(3',4')CYCLOPENTENE; W   66216

692 C9 H15 Br6 O4 P 126-72-7 T-66-3149-0 1-Propanol, 2,3-dibromo-, phosphate (3:1);; TRIS-(2,3-DIBROMOP
ROPYL)PHOSPHATE; W/NBS   66217

OCH₂CHBrCH₂Br
BrCH₂CHBrCH₂OPOCH₂CHBrCH₂Br
O

692 C9 H15 Br6 O4 P IC-1308-0-0 2,3-Dibromo-1-propanol phosphate (3:1); W   128345

692 C9 H15 Br6 O4 P 18713-51-4 T-66-3150-0 2-Propanol, 1,3-dibromo-, phosphate (3:1);; TRIS-(2-BROMO-1-
BROMOETHYL)ETHYL PHOSPHATE; W/NBS   66218

CH₂Br
OCHCH₂Br
BrCH₂ | CH₂Br
BrCH₂CHOPOCHCH₂Br
O

692 C22 H22 F14 N2 O7 55538-88-0 BA-0-174-0 Butanoic acid, heptafluoro-, 1-[[hexahydro-1,3-dimethyl-5-(
1-methylbutyl)-2,4,6-trioxo-5-pyrimidinyl]methyl]-1,2-ethanediyl ester;; 1,3-DIMETHYL-5-(2,3-BIS(HEPTAFLUOR
OBUTYROXY)PROPYL)-5-(1-METHYLBUTYL)BARBITURIC ACID; W/NBS   66219

OC(O)CF₂CF₂CF₃
CH₂CHCH₂OC(O)CF₂CF₂CF₃
CHPrMe

692 C24 H61 O9 P Si6 55538-92-6 BA-0-348-0 D-arabino-Hexonic acid, 2-deoxy-3,4,5-tris-O-(trimethylsily
l)-, trimethylsilyl ester, bis(trimethylsilyl) phosphate;; O-HEXAKIS(TRIMETHYLSILYL)-2-DEOXY-D-ARABINO-HEXO
NATE-6-PHOSPHATE; W/NBS   66220

OSiMe₃
OSiMe₃
Me₃SiO   OSiMe₃
Me₃SiOPOCH₂CHCHCHCH₂C(O)OSiMe₃
O

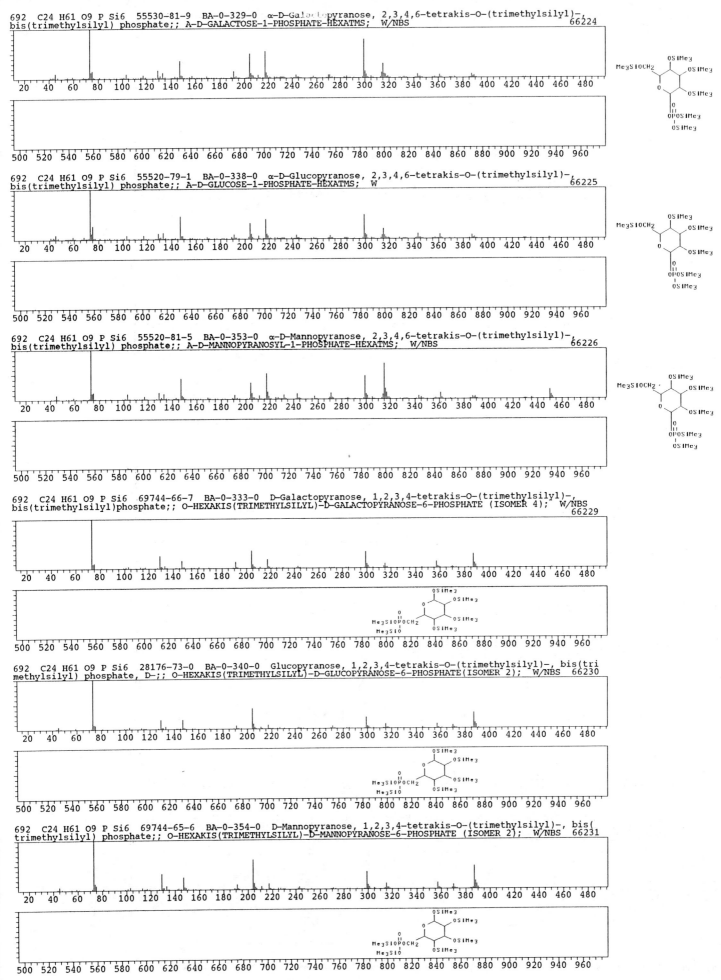

692 C24 H61 O9 P Si6 55530-81-9 BA-0-329-0 α-D-Galactopyranose, 2,3,4,6-tetrakis-O-(trimethylsilyl)-
bis(trimethylsilyl) phosphate;; A-D-GALACTOSE-1-PHOSPHATE-HEXATMS; W/NBS
66224

692 C24 H61 O9 P Si6 55520-79-1 BA-0-338-0 α-D-Glucopyranose, 2,3,4,6-tetrakis-O-(trimethylsilyl)-
bis(trimethylsilyl) phosphate;; A-D-GLUCOSE-1-PHOSPHATE-HEXATMS; W
66225

692 C24 H61 O9 P Si6 55520-81-5 BA-0-353-0 α-D-Mannopyranose, 2,3,4,6-tetrakis-O-(trimethylsilyl)-
bis(trimethylsilyl) phosphate;; A-D-MANNOPYRANOSYL-1-PHOSPHATE-HEXATMS; W/NBS
66226

692 C24 H61 O9 P Si6 69744-66-7 BA-0-333-0 D-Galactopyranose, 1,2,3,4-tetrakis-O-(trimethylsilyl)-,
bis(trimethylsilyl)phosphate;; O-HEXAKIS(TRIMETHYLSILYL)-D-GALACTOPYRANOSE-6-PHOSPHATE (ISOMER 4); W/NBS
66229

692 C24 H61 O9 P Si6 28176-73-0 BA-0-340-0 Glucopyranose, 1,2,3,4-tetrakis-O-(trimethylsilyl)-, bis(tri
methylsilyl) phosphate, D-;; O-HEXAKIS(TRIMETHYLSILYL)-D-GLUCOPYRANOSE-6-PHOSPHATE(ISOMER 2); W/NBS 66230

692 C24 H61 O9 P Si6 69744-65-6 BA-0-354-0 D-Mannopyranose, 1,2,3,4-tetrakis-O-(trimethylsilyl)-, bis(
trimethylsilyl) phosphate;; O-HEXAKIS(TRIMETHYLSILYL)-D-MANNOPYRANOSE-6-PHOSPHATE (ISOMER 2); W/NBS 66231

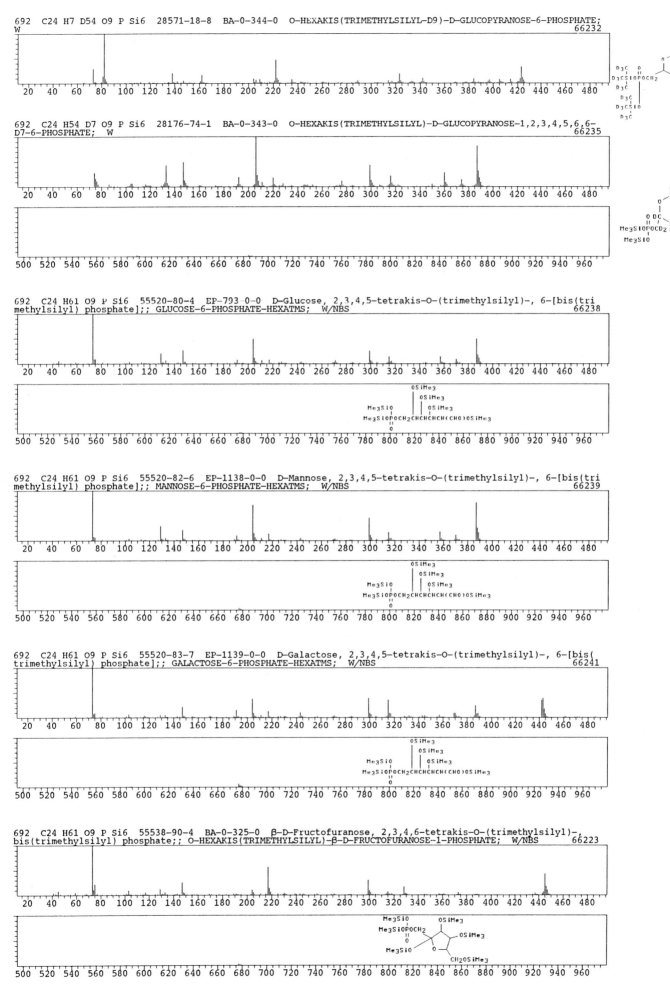

692   C24 H7 D54 O9 P Si6   28571-18-8   BA-0-344-0   O-HEXAKIS(TRIMETHYLSILYL-D9)-D-GLUCOPYRANOSE-6-PHOSPHATE;
W   66232

692   C24 H54 D7 O9 P Si6   28176-74-1   BA-0-343-0   O-HEXAKIS(TRIMETHYLSILYL)-D-GLUCOPYRANOSE-1,2,3,4,5,6,6-
D7-6-PHOSPHATE;   W   66235

692   C24 H61 O9 P Si6   55520-80-4   EP-793-0-0   D-Glucose, 2,3,4,5-tetrakis-O-(trimethylsilyl)-, 6-[bis(tri
methylsilyl) phosphate];; GLUCOSE-6-PHOSPHATE-HEXATMS;   W/NBS   66238

692   C24 H61 O9 P Si6   55520-82-6   EP-1138-0-0   D-Mannose, 2,3,4,5-tetrakis-O-(trimethylsilyl)-, 6-[bis(tri
methylsilyl) phosphate];; MANNOSE-6-PHOSPHATE-HEXATMS;   W/NBS   66239

692   C24 H61 O9 P Si6   55520-83-7   EP-1139-0-0   D-Galactose, 2,3,4,5-tetrakis-O-(trimethylsilyl)-, 6-[bis(
trimethylsilyl) phosphate];; GALACTOSE-6-PHOSPHATE-HEXATMS;   W/NBS   66241

692   C24 H61 O9 P Si6   55538-90-4   BA-0-325-0   β-D-Fructofuranose, 2,3,4,6-tetrakis-O-(trimethylsilyl)-,
bis(trimethylsilyl) phosphate;; O-HEXAKIS(TRIMETHYLSILYL)-β-D-FRUCTOFURANOSE-1-PHOSPHATE;   W/NBS   66223

692 C24 H61 O9 P Si6 28176-76-3 BA-0-326-0 Fructofuranose, 1,2,3,4-tetrakis-O-(trimethylsilyl)-, bis(tr
imethylsilyl) phosphate, β-D-;; O-HEXAKIS(TRIMETHYLSILYL)-β-D-FRUCTOFURANOSE-6-PHOSPHATE; W/NBS   66236

692 C24 H61 O9 P Si6 69688-48-8 BA-0-331-0 D-Galactofuranose, 1,2,3,5-tetrakis-O-(trimethylsilyl)-,
bis(trimethylsilyl)phosphate;; O-HEXAKIS(TRIMETHYLSILYL)-D-GALACTOFURANOSE-6-PHOSPHATE (ISOMER 2); W/NBS
66222

692 C24 H61 O9 P Si6 55530-83-1 EP-804-0-0 D-Fructose, 1,3,4,5-tetrakis-O-(trimethylsilyl)-, 6-[bis(tri
methylsilyl) phosphate];; D-FRUCTOSE-6-PHOSPHATE-HEXATMS; W/NBS   66234

692 C24 H61 O9 P Si6 EP-803-0-0 A-D-FRUCTOSE-1-PHOSPHATE-HEXATMS; W   66221

692 C24 H61 O9 P Si6 BA-0-332-0 O-HEXAKIS(TRIMETHYLSILYL)-D-GALACTOPYRANOSE-6-PHOSPHATE (ISOMER 3); W
66227

692 C24 H56 D5 O9 P Si6 28215-69-2 BA-0-342-0 O-HEXAKIS(TRIMETHYLSILYL)-D-GLUCOPYRANOSE-3,4,5,6,6-D5-6-
PHOSPHATE; W   66228

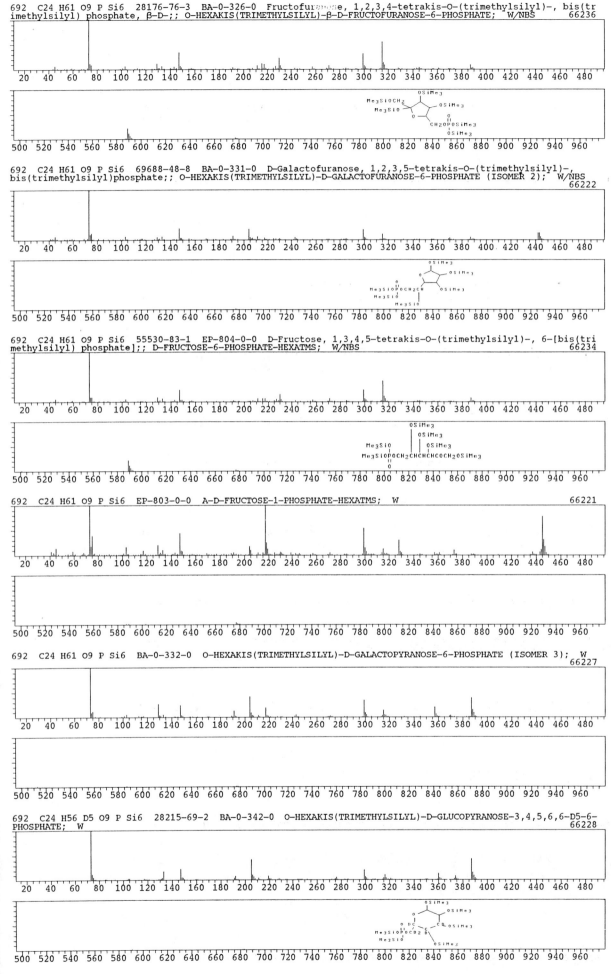

- 6054 -

692   C24 H59 D2 O9 P Si6   28215-68-1   BA-0-341-0   O-HEXAKIS(TRIMETHYLSILYL)-D-GLUCOPYRANOSE-6,6-D2-6-PHOSPHATE;   W
66233

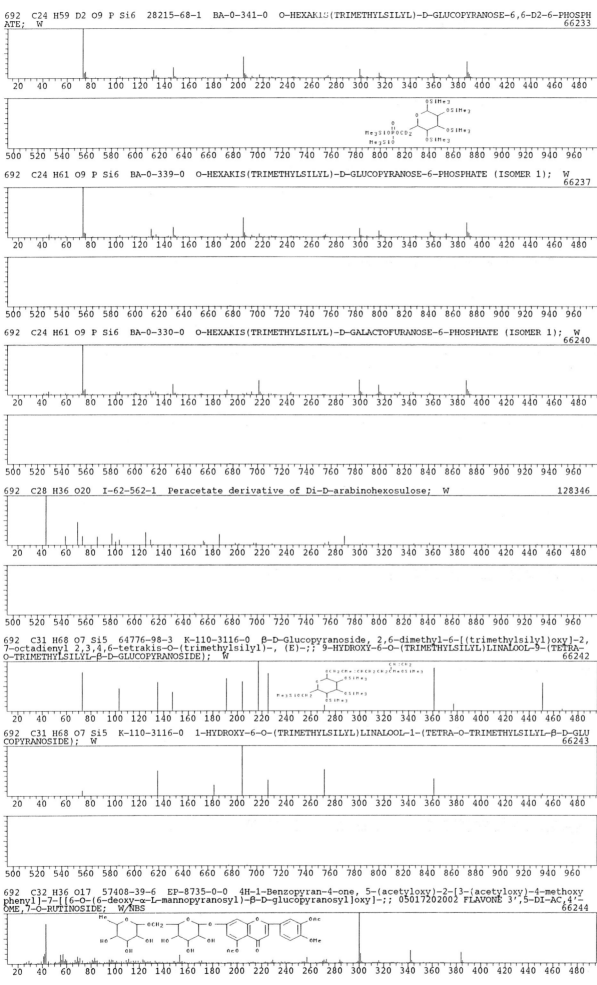

692   C24 H61 O9 P Si6   BA-0-339-0   O-HEXAKIS(TRIMETHYLSILYL)-D-GLUCOPYRANOSE-6-PHOSPHATE (ISOMER 1);   W
66237

692   C24 H61 O9 P Si6   BA-0-330-0   O-HEXAKIS(TRIMETHYLSILYL)-D-GALACTOFURANOSE-6-PHOSPHATE (ISOMER 1);   W
66240

692   C28 H36 O20   I-62-562-1   Peracetate derivative of Di-D-arabinohexosulose;   W
128346

692   C31 H68 O7 Si5   64776-98-3   K-110-3116-0   β-D-Glucopyranoside, 2,6-dimethyl-6-[(trimethylsilyl)oxy]-2,7-octadienyl 2,3,4,6-tetrakis-O-(trimethylsilyl)-, (E)-;; 9-HYDROXY-6-O-(TRIMETHYLSILYL)LINALOOL-9-(TETRA-O-TRIMETHYLSILYL-β-D-GLUCOPYRANOSIDE);   W
66242

692   C31 H68 O7 Si5   K-110-3116-0   1-HYDROXY-6-O-(TRIMETHYLSILYL)LINALOOL-1-(TETRA-O-TRIMETHYLSILYL-β-D-GLUCOPYRANOSIDE);   W
66243

692   C32 H36 O17   57408-39-6   EP-8735-0-0   4H-1-Benzopyran-4-one, 5-(acetyloxy)-2-[3-(acetyloxy)-4-methoxyphenyl]-7-[[6-O-(6-deoxy-α-L-mannopyranosyl)-β-D-glucopyranosyl]oxy]-;; 05017202002 FLAVONE 3',5-DI-AC,4'-OME,7-O-RUTINOSIDE;   W/NBS
66244

692 C37 H68 N6 O6 75422-11-6 NS--27194-0-0 Acetamide, N-[3-[acetyl[3-(acetylamino)propyl]amino]propyl]-N-[3-[acetyl[3-[acetyl[3-(2-oxoazacyclotridec-1-yl)propyl]amino]propyl]amino]propyl]-;; Acetamide, N-[3-[acetyl[3-(acetylamino)propyl]amino]propyl]-N-[3-[acetyl[3-[acetyl[3-(2-oxoazacyclotridec-1-yl)propyl]amino]propyl]amino]propyl]; NBS 66245

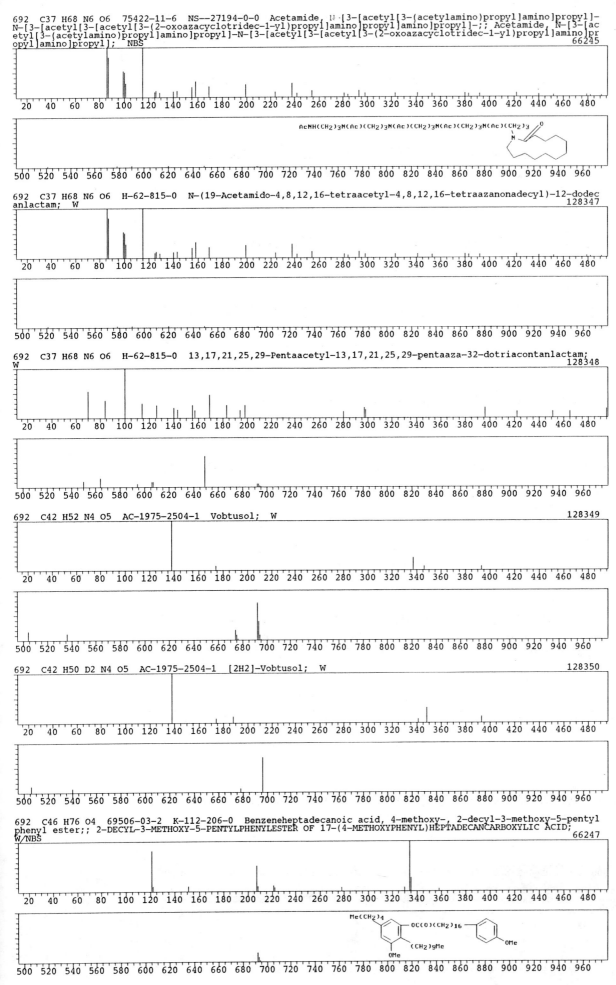

AcNH(CH2)3N(Ac)(CH2)3N(Ac)(CH2)3N(Ac)(CH2)3N(Ac)(CH2)3

692 C37 H68 N6 O6 H-62-815-0 N-(19-Acetamido-4,8,12,16-tetraacetyl-4,8,12,16-tetraazanonadecyl)-12-dodecanlactam; W 128347

692 C37 H68 N6 O6 H-62-815-0 13,17,21,25,29-Pentaacetyl-13,17,21,25,29-pentaaza-32-dotriacontanlactam; W 128348

692 C42 H52 N4 O5 AC-1975-2504-1 Vobtusol; W 128349

692 C42 H50 D2 N4 O5 AC-1975-2504-1 [2H2]-Vobtusol; W 128350

692 C46 H76 O4 69506-03-2 K-112-206-0 Benzeneheptadecanoic acid, 4-methoxy-, 2-decyl-3-methoxy-5-pentylphenyl ester;; 2-DECYL-3-METHOXY-5-PENTYLPHENYLESTER OF 17-(4-METHOXYPHENYL)HEPTADECANCARBOXYLIC ACID; W/NBS 66247

Me(CH2)4 OC(O)(CH2)16 OMe
(CH2)9Me
OMe

692 C46 H76 O4 63953-41-3 K-112-2009-0 Benzenepentadecanoic acid, 4-methoxy-, 3-methoxy-2-(9-methyldecyl)-5-(4-methylpentyl)phenyl ester;; 3-METHOXY-2-(9-METHYLDECYL)-5-(4-METHYLPENTYL)PHENYLESTER OF15-(4-METHOXYPHENYL)PENTADECANCARBOXYLIC ACID; W/NBS
66246

692 C50 H76 O 64354-75-2 F-33-382-0 (2R,2'R)-2,2'-BIS(3-METHYLBUTYL)-3,4,3',4'-TETRAHYDRO-.PSI.,.PSI.-CAROTEN--20-OL; W
66248

693 C34 H27 N7 O10 78413-95-3 F-37-158-0 Methyl ester of 3-(octahydro-3,5,8,10,11,13-hexaoxo-4,9-diphenyl-1,7:2,6-ethanediylidene-6a,10c-(methaniminomethano)-3H,6H,7H,8H-2b,4,5a,7a,9,10a-hexaazadicyclopent[d,i]acenaphthylen-12-yl)-3-methoxycarbonyl propanoic acid; W
128351

693 C34 H27 N7 O10 78420-48-1 F-37-158-0 Methyl ester of 3-(2,3,9,10-Tetrahydro-1,3,8,10,17,19-hexaoxo-2,9-diphenyl-6H,13H-5,13:6,12-dietheno-5a,12a-(methaniminomethano)-1H,5H,8H,12H-[1,2,4]triazolo[1,2-a][1,2,4]triazolo[1',2':1,2]pyridazino[4,5-d]pyridazin-18-yl)-3-methoxycarbonyl propanoic acid stereoisomer; W
128352

693 C34 H27 N7 O10 78388-63-3 F-37-158-0 Methyl ester of 3-(2,3,9,10-Tetrahydro-1,3,8,10,17,19-hexaoxo-2,9-diphenyl-6H,13H-5,13:6,12-dietheno-5a,12a-(methaniminomethano)-1H,5H,8H,12H-[1,2,4]triazolo[1,2-a][1,2,4]triazolo[1',2':1,2]pyridazino[4,5-d]pyridazin-18-yl)-3-methoxycarbonyl propanoic acid stereoisomer; W
128353

693 C36 H62 Na O11 22373-78-0 O-3-277-3 Monensin, monosodium salt;; Monensin sodium salt; W
66249

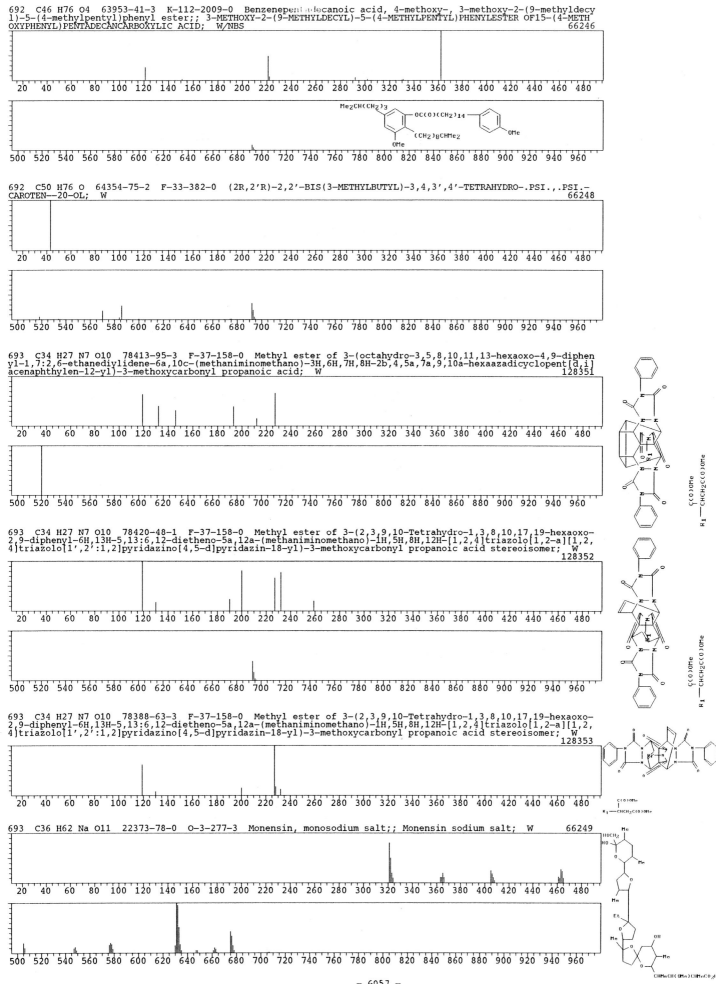

693 C39 H79 N O3 Si3 KO-6-32-1 Tri-o-tri-n-propylsilyl derivative of Terbutaline; W   128354

20 40 60 80 100 120 140 160 180 200 220 240 260 280 300 320 340 360 380 400 420 440 460 480

500 520 540 560 580 600 620 640 660 680 700 720 740 760 780 800 820 840 860 880 900 920 940 960

693 C43 H48 B F4 N3 K-116-1995-0 1,7,10-Tris(1,3,3-trimethyl-2-indolinyl)-[3.3.3]decamethinium-tetrafluo
roborate; W   128355

20 40 60 80 100 120 140 160 180 200 220 240 260 280 300 320 340 360 380 400 420 440 460 480

500 520 540 560 580 600 620 640 660 680 700 720 740 760 780 800 820 840 860 880 900 920 940 960

694 C26 H2 F20 5736-50-5 T-68-5213-0 Benzene, 1,1',1'',1'''-(1,2-ethanediylidene)tetrakis[2,3,4,5,6-pen
tafluoro-;; TETRA-PENTAFLUOROPHENYLETHANE; W/NBS   66250

20 40 60 80 100 120 140 160 180 200 220 240 260 280 300 320 340 360 380 400 420 440 460 480

500 520 540 560 580 600 620 640 660 680 700 720 740 760 780 800 820 840 860 880 900 920 940 960

694 C27 H21 I2 O4 P 84530-89-2 KC-1982-2297-0 3,7-Bis-(2-iodobenzyloxy)-5-methoxydibenzophosphole 5-oxi
de; W   128356

20 40 60 80 100 120 140 160 180 200 220 240 260 280 300 320 340 360 380 400 420 440 460 480

500 520 540 560 580 600 620 640 660 680 700 720 740 760 780 800 820 840 860 880 900 920 940 960

694 C28 H35 Br N6 O10 IC-1309-0-0 N,N-Bis(butyryloxyethyl)-2-ethoxy-4-acetamido-3-(6-bromo-2,4-dinitroph
enylazo)aniline; W   128357

20 40 60 80 100 120 140 160 180 200 220 240 260 280 300 320 340 360 380 400 420 440 460 480

500 520 540 560 580 600 620 640 660 680 700 720 740 760 780 800 820 840 860 880 900 920 940 960

694 C28 H42 N2 O18 56195-97-2 NS--27188-0-0 D-Glucose, 4-amino-4-deoxy-, 2,3,5,6-tetraacetate, dimer;;
NBS   66251

20 40 60 80 100 120 140 160 180 200 220 240 260 280 300 320 340 360 380 400 420 440 460 480

500 520 540 560 580 600 620 640 660 680 700 720 740 760 780 800 820 840 860 880 900 920 940 960

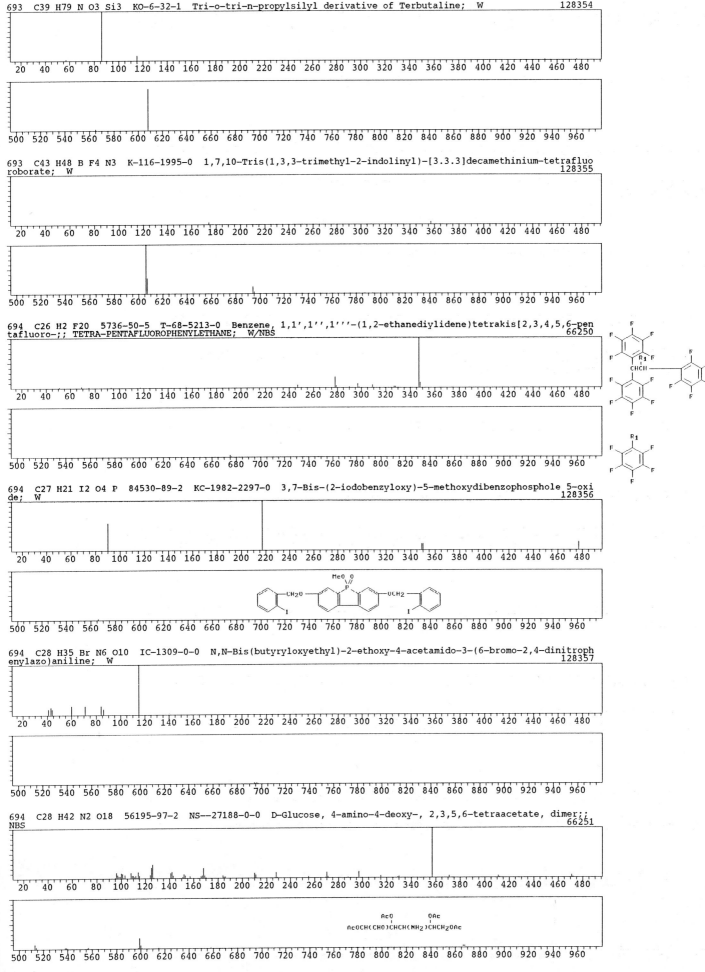

694 C34 H56 Co2 Si4 63537-28-0 C-99-4061-0 1,4-BIS-(CYCLOPENTADIENYL COBALT-2,3-DI-(TRIMETHYL SILYL)CYC
LOBUTADIENYL)ETHANE; W   66252

694 C37 H47 In N4 S 75845-78-2 KA-1980-2084-2 Methylthio[2,3,7,8,12,13,17,18-octaethylporphyrinato]indi
um; W   128358

694 C38 H74 O5 Si3 69519-40-0 KO-5-472-2 CHOLIC ACID ETHYL ESTER DIMETHYLETHYLSILYL ETHER; W   66253

694 C40 H46 N4 O7 69721-72-8 H-61-2669-0 4H-1,16-Etheno-5,15-(propaniminoethano)furo[3,4-1][1,5,10]tria
zacyclohexadecine-4,21-dione, 10,14-diacetyl-22-benzoyl-3,3a,6,7,8,9,10,11,12,13,14,15-dodecahydro-3-(4-
methoxyphenyl)-;; 4H-1,16-Etheno-5,15-(propaniminoethano)furo[3,4-1][1,5,10]triazacyclohexadecine-4,21-di
one, 10,14-diacetyl-22-benzoyl-3,3a,6,7,8,9,10,11,12,13,1; W/NBS   128359

694 C40 H46 N4 O7 73745-68-3 H-62-2722-0 N(6),N(10)-Diacetyl-N(23)-benzoyl-O(34)-methyl-orantin; W
                                                                                                128360

694 C42 H34 N2 S4 71981-66-3 B-32-1250-0 1,2,9,10-TETRABENZYLTHIO[2,2](2,6)CYRIDINOPHANE-1,9-DIENE; W
                                                                                                66255

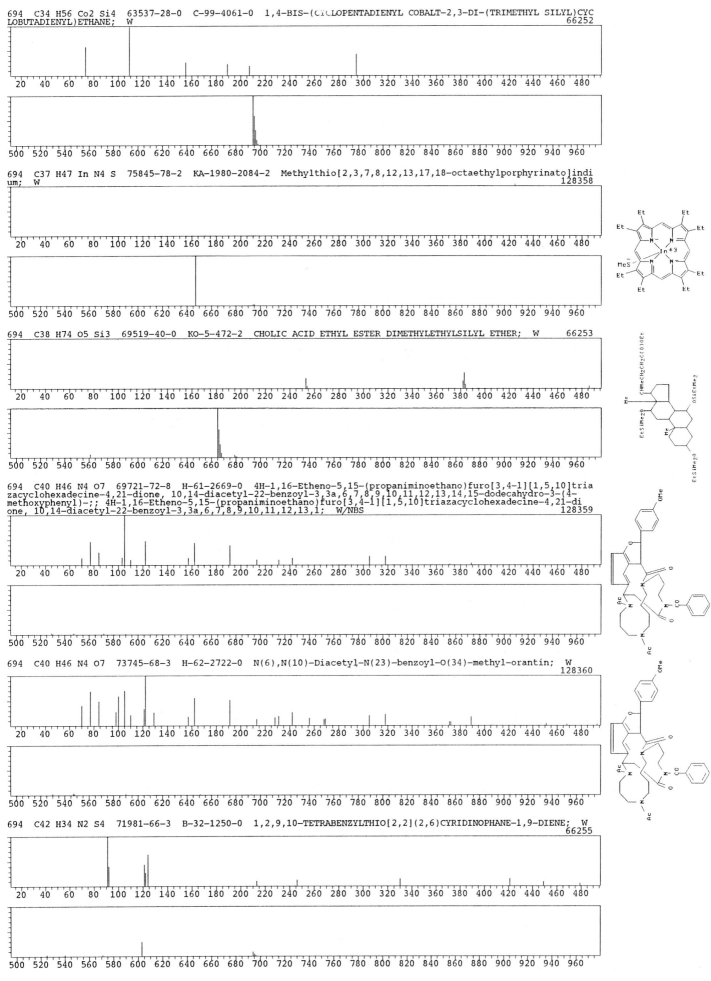

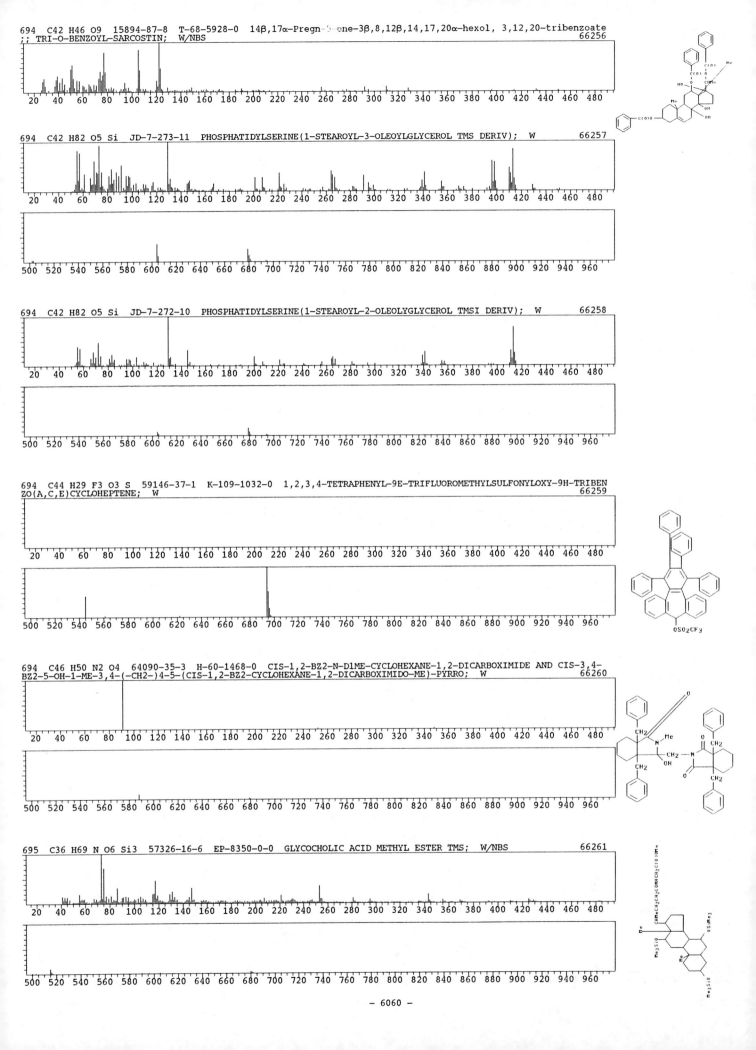

694   C42 H46 O9   15894-87-8   T-68-5928-0   14β,17α-Pregn-5-one-3β,8,12β,14,17,20α-hexol, 3,12,20-tribenzoate
;; TRI-O-BENZOYL-SARCOSTIN;   W/NBS   66256

694   C42 H82 O5 Si   JD-7-273-11   PHOSPHATIDYLSERINE(1-STEAROYL-3-OLEOYLGLYCEROL TMS DERIV);   W   66257

694   C42 H82 O5 Si   JD-7-272-10   PHOSPHATIDYLSERINE(1-STEAROYL-2-OLEOLYGLYCEROL TMSI DERIV);   W   66258

694   C44 H29 F3 O3 S   59146-37-1   K-109-1032-0   1,2,3,4-TETRAPHENYL-9E-TRIFLUOROMETHYLSULFONYLOXY-9H-TRIBEN
ZO(A,C,E)CYCLOHEPTENE;   W   66259

694   C46 H50 N2 O4   64090-35-3   H-60-1468-0   CIS-1,2-BZ2-N-D1ME-CYCLOHEXANE-1,2-DICARBOXIMIDE AND CIS-3,4-
BZ2-5-OH-1-ME-3,4-(-CH2-)4-5-(CIS-1,2-BZ2-CYCLOHEXANE-1,2-DICARBOXIMIDO-ME)-PYRRO;   W   66260

695   C36 H69 N O6 Si3   57326-16-6   EP-8350-0-0   GLYCOCHOLIC ACID METHYL ESTER TMS;   W/NBS   66261

696   C12 H F24 O4 P   80281-11-4   K-114-3622-0   4,4,4',4',5,5,5',5'-Octakis(trifluoromethyl)spiro[1,3,2λ5-di
oxaphospholan-2,2'-[1,3,2λ5]dioxaphospholan];   W   128361

696   C16 H10 Hg 96Mo2 O6   AD-0-2869-0   DI(.PI.-CYCLOPENTADIENYL-TRICARBONYL MOLYBDENUM)MERCURY;   W   66262

696   C16 H10 Hg Mo2 O6   12194-13-7   NS--27183-0-0   Molybdenum, hexacarbonylbis(η5-2,4-cyclopentadien-1-yl)(
mercury)di-, (2Hg-Mo);; Molybdenum, hexacarbonyldi-.pi.-cyclopentadienyl-.mu.-mercuriodi-;   NBS   66263

696   C18 H54 Hg Si8   61576-80-5   AG-120-174-1   Bis[tri(trimethyl)silylsilyl]mercury;   W   128362

696   C19 H3 F12 O2 Tl   B-31-1720-0   2,3,5,6-TETRAFLUOROBENZOATOBIS(2,3,5,6-TETRAFLUOROPHENYL)THALLIUM(III);
W   66264

696   C19 H3 F12 O2 Tl   B-31-1720-0   2,3,4,6-TETRAFLUOROBENZOATOBIS(2,3,4,6-TETRAFLUOROPHENYL)THALLIUM(III);
W   66265

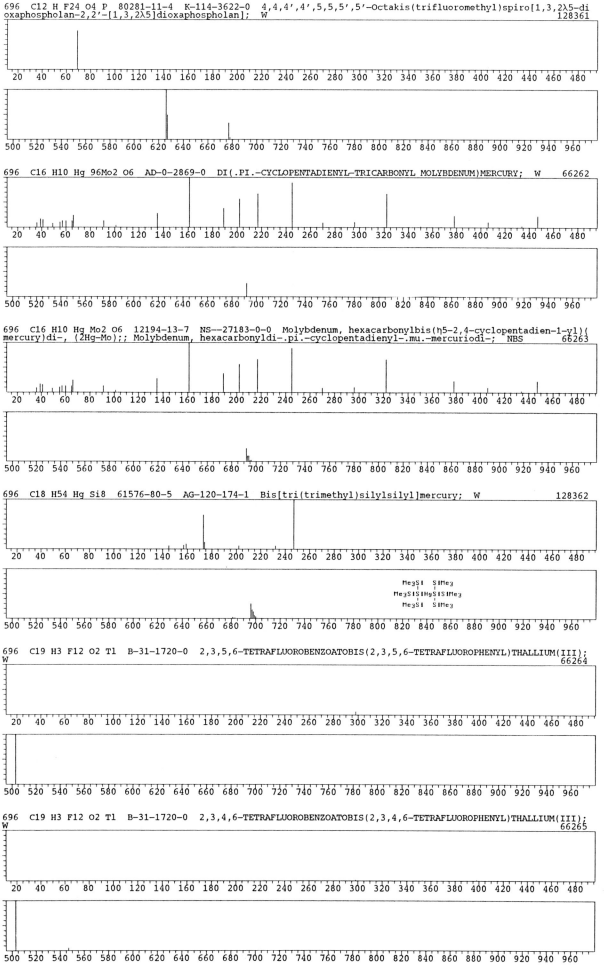

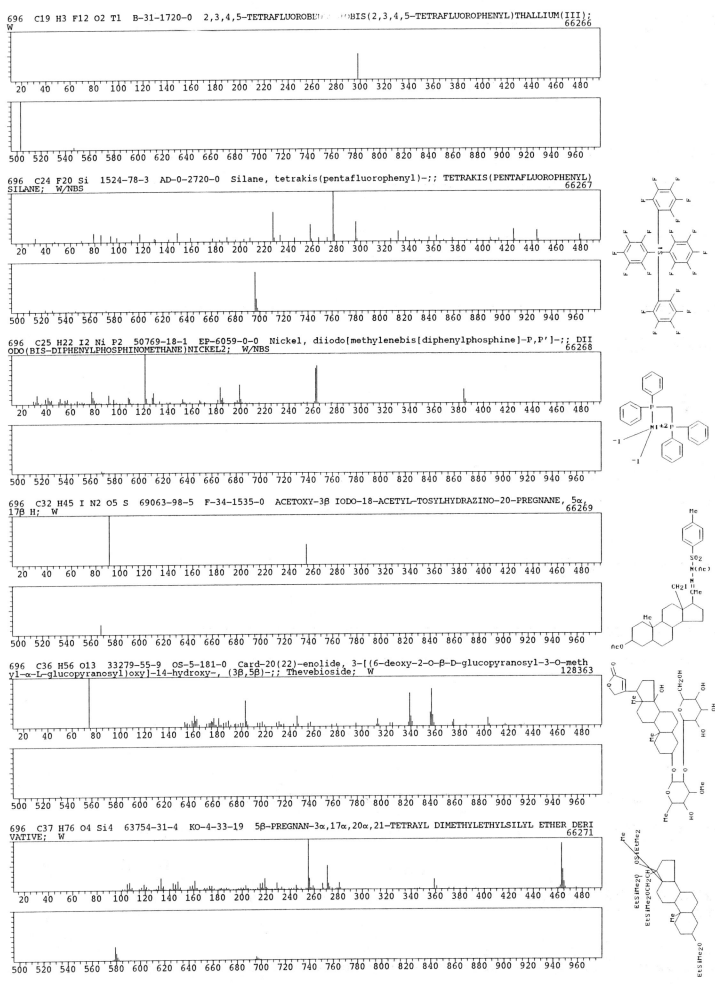

696   C19 H3 F12 O2 Tl   B-31-1720-0   2,3,4,5-TETRAFLUOROBE[...]OBIS(2,3,4,5-TETRAFLUOROPHENYL)THALLIUM(III);
W
66266

696   C24 F20 Si   1524-78-3   AD-0-2720-0   Silane, tetrakis(pentafluorophenyl)-;; TETRAKIS(PENTAFLUOROPHENYL)
SILANE; W/NBS
66267

696   C25 H22 I2 Ni P2   50769-18-1   EP-6059-0-0   Nickel, diiodo[methylenebis[diphenylphosphine]-P,P']-;; DII
ODO(BIS-DIPHENYLPHOSPHINOMETHANE)NICKEL2; W/NBS
66268

696   C32 H45 I N2 O5 S   69063-98-5   F-34-1535-0   ACETOXY-3β IODO-18-ACETYL-TOSYLHYDRAZINO-20-PREGNANE, 5α,
17β H; W
66269

696   C36 H56 O13   33279-55-9   OS-5-181-0   Card-20(22)-enolide, 3-[(6-deoxy-2-O-β-D-glucopyranosyl-3-O-meth
yl-α-L-glucopyranosyl)oxy]-14-hydroxy-, (3β,5β)-;; Thevebioside; W
128363

696   C37 H76 O4 Si4   63754-31-4   KO-4-33-19   5β-PREGNAN-3α,17α,20α,21-TETRAYL DIMETHYLETHYLSILYL ETHER DERI
VATIVE; W
66271

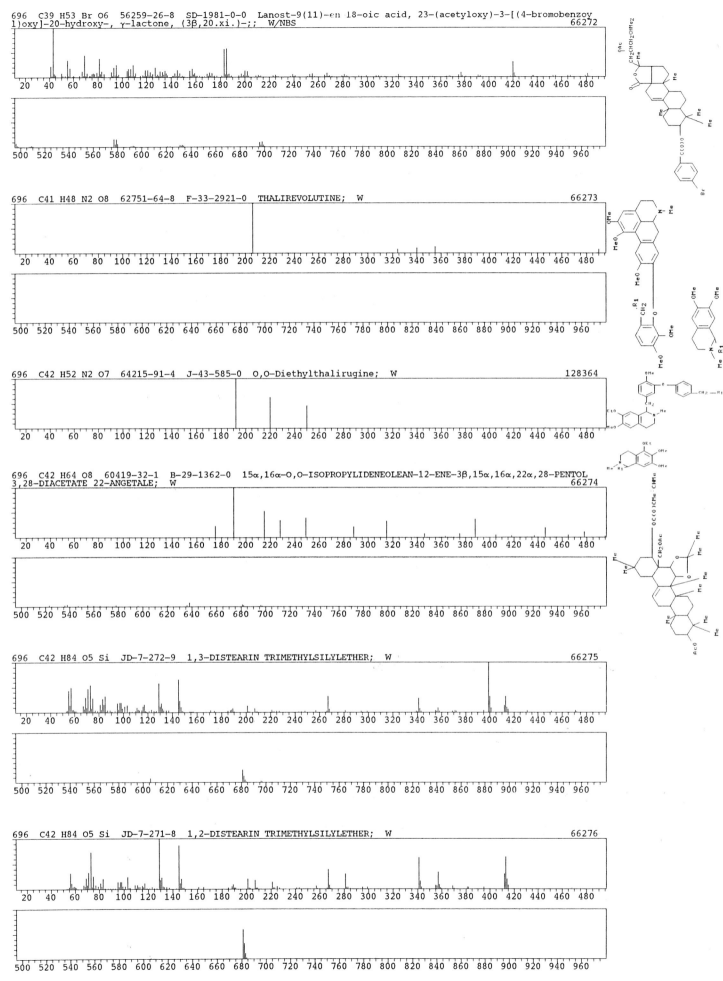

696   C39 H53 Br O6   56259-26-8   SD-1981-0-0   Lanost-9(11)-en 18-oic acid, 23-(acetyloxy)-3-[(4-bromobenzoyl)oxy]-20-hydroxy-, γ-lactone, (3β,20.xi.)-;;   W/NBS
66272

696   C41 H48 N2 O8   62751-64-8   F-33-2921-0   THALIREVOLUTINE;   W
66273

696   C42 H52 N2 O7   64215-91-4   J-43-585-0   O,O-Diethylthalirugine;   W
128364

696   C42 H64 O8   60419-32-1   B-29-1362-0   15α,16α-O,O-ISOPROPYLIDENEOLEAN-12-ENE-3β,15α,16α,22α,28-PENTOL 3,28-DIACETATE 22-ANGETALE;   W
66274

696   C42 H84 O5 Si   JD-7-272-9   1,3-DISTEARIN TRIMETHYLSILYLETHER;   W
66275

696   C42 H84 O5 Si   JD-7-271-8   1,2-DISTEARIN TRIMETHYLSILYLETHER;   W
66276

696　C42 H84 O5 Si　W-9-236-10　2-STEARO-1-STEARIN TMS ETHER;　W　66277

696　C43 H50 N4 Ni O　67168-58-5　KC-1978-369-0　NICKEL MESO-α-HYDROXYBENZYL-OCTACETHYLPORPHYRIN;　W　66278

696　C44 H60 N2 O5　78009-97-9　I-59-277-0　1-[2,3,5-Tri-O-benzyl-4-amino-4-cyano-D-arabinosyl]-9-ethylenedioxypentadec-1-ene (isomer A);　W　128365

696　C44 H60 N2 O5　78086-20-1　I-59-277-0　1-[2,3,5-Tri-O-benzyl-4-amino-4-cyano-D-arabinosyl]-9-ethylenedioxypentadec-1-ene (isomer B);　W　128366

696　C45 H33 Al O6　14405-36-8　T-68-5956-0　Aluminum, tris(1,3-diphenyl-1,3-propanedionato-O,O')-, (OC-6-11)-;; TRIS(1,3-DIPHENYL-1,3-PROPANEDIONATO)ALUMINUM(III);　W/NBS　66279

696　C45 H57 Cl O4　78380-00-4　F-37-561-0　2-Dodecyl-3-hydroxy-5-(2-methylpropyl)-phenylester of 17-(3-Chloro-4-hydroxyphenyl-2,4,6,8,10,12,14,16-heptadecaoctaenoic acid;　W　128367

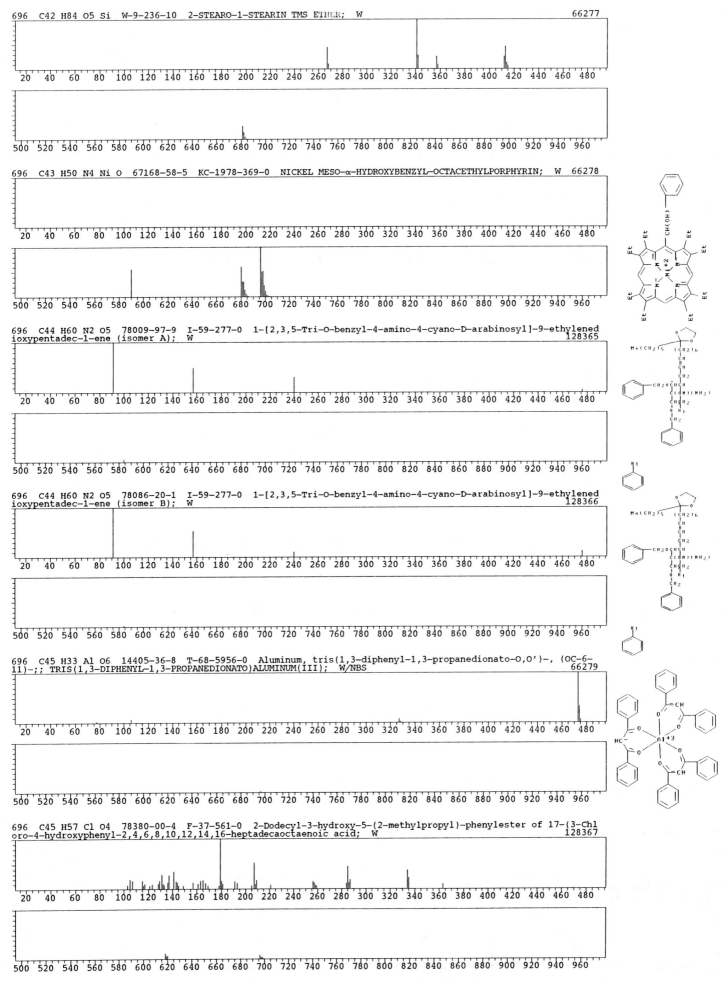

696　C51 H68 O　82526-86-1　F-38-381-0　26-Trityloxy-3-methylenecyclolaudanes;　W　　　　128368

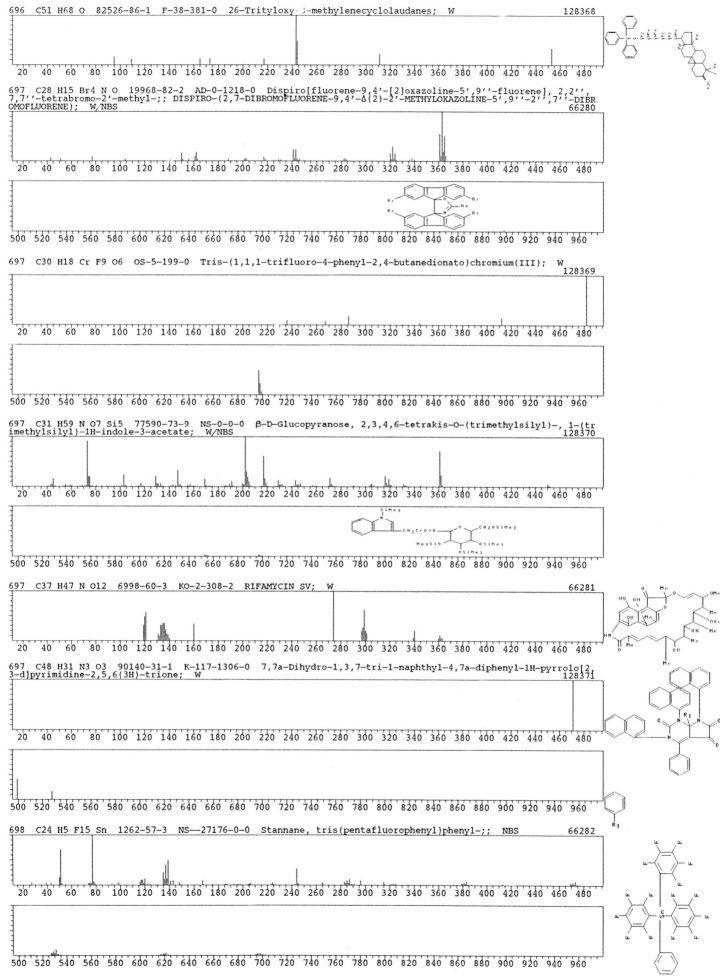

697　C28 H15 Br4 N O　19968-82-2　AD-0-1218-0　Dispiro[fluorene-9,4'-[2]oxazoline-5',9''-fluorene], 2,2'',
7,7''-tetrabromo-2'-methyl-;; DISPIRO-(2,7-DIBROMOFLUORENE-9,4'-Δ(2)-2'-METHYLOXAZOLINE-5',9''-2'',7''-DIBR
OMOFLUORENE);　W/NBS　　　　　　　　　　　　　　　　　　　　　　　　　　　　66280

697　C30 H18 Cr F9 O6　OS-5-199-0　Tris-(1,1,1-trifluoro-4-phenyl-2,4-butanedionato)chromium(III);　W　　128369

697　C31 H59 N O7 Si5　77590-73-9　NS-0-0-0　β-D-Glucopyranose, 2,3,4,6-tetrakis-O-(trimethylsilyl)-, 1-(tr
imethylsilyl)-1H-indole-3-acetate;　W/NBS　　　　　　　　　　　　　　　　　　128370

697　C37 H47 N O12　6998-60-3　KO-2-308-2　RIFAMYCIN SV;　W　　　　　　　　　　66281

697　C48 H31 N3 O3　90140-31-1　K-117-1306-0　7,7a-Dihydro-1,3,7-tri-1-naphthyl-4,7a-diphenyl-1H-pyrrolo[2,
3-d]pyrimidine-2,5,6(3H)-trione;　W　　　　　　　　　　　　　　　　　　　128371

698　C24 H5 F15 Sn　1262-57-3　NS--27176-0-0　Stannane, tris(pentafluorophenyl)phenyl-;;　NBS　　　66282

698   C26 H35 Br O17   14187-83-8   H-67-384-0   O-(2,3,4,6-Tetra-O-acetyl-β-D-glucopyranosyl)-(1-6)-2,3,4-tri-
O-acetyl-α-D-glucopyranosyl-bromide;   W
128372

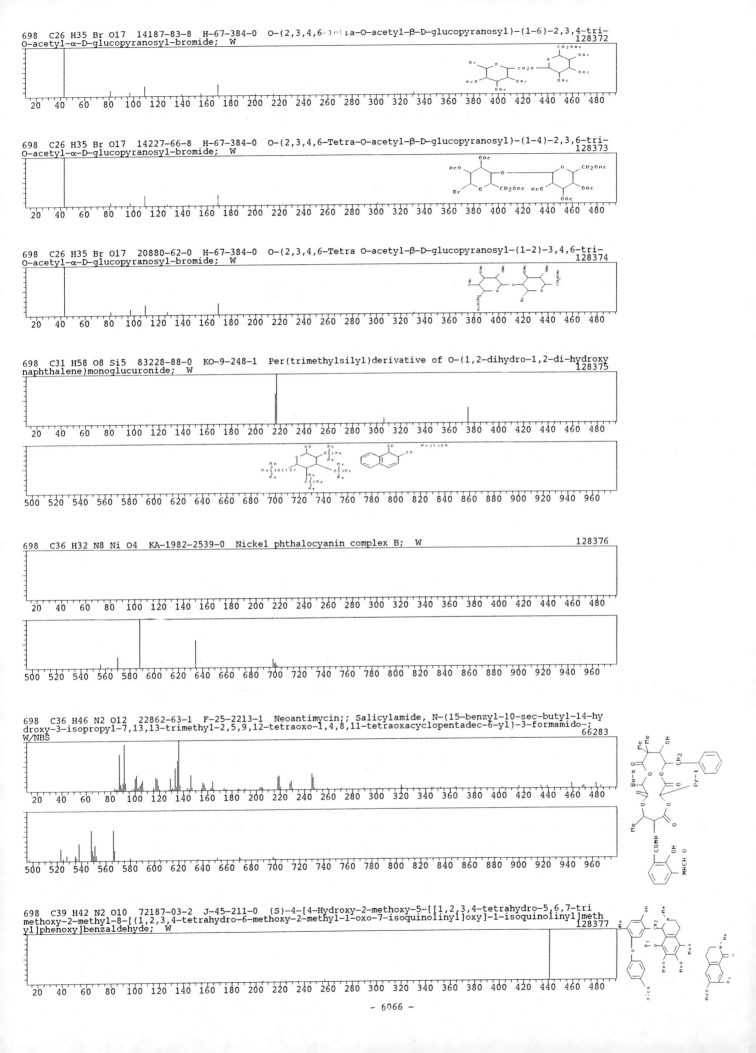

698   C26 H35 Br O17   14227-66-8   H-67-384-0   O-(2,3,4,6-Tetra-O-acetyl-β-D-glucopyranosyl)-(1-4)-2,3,6-tri-
O-acetyl-α-D-glucopyranosyl-bromide;   W
128373

698   C26 H35 Br O17   20880-62-0   H-67-384-0   O-(2,3,4,6-Tetra O-acetyl-β-D-glucopyranosyl-(1-2)-3,4,6-tri-
O-acetyl-α-D-glucopyranosyl-bromide;   W
128374

698   C31 H58 O8 Si5   83228-88-0   KO-9-248-1   Per(trimethylsilyl)derivative of O-(1,2-dihydro-1,2-di-hydroxy
naphthalene)monoglucuronide;   W
128375

698   C36 H32 N8 Ni O4   KA-1982-2539-0   Nickel phthalocyanin complex B;   W
128376

698   C36 H46 N2 O12   22862-63-1   F-25-2213-1   Neoantimycin;;   Salicylamide, N-(15-benzyl-10-sec-butyl-14-hy
droxy-3-isopropyl-7,13,13-trimethyl-2,5,9,12-tetraoxo-1,4,8,11-tetraoxacyclopentadec-6-yl)-3-formamido-;
W/NBS
66283

698   C39 H42 N2 O10   72187-03-2   J-45-211-0   (S)-4-[4-Hydroxy-2-methoxy-5-[[1,2,3,4-tetrahydro-5,6,7-tri
methoxy-2-methyl-8-[(1,2,3,4-tetrahydro-6-methoxy-2-methyl-1-oxo-7-isoquinolinyl)oxy]-1-isoquinolinyl]meth
yl]phenoxy]benzaldehyde;   W
128377

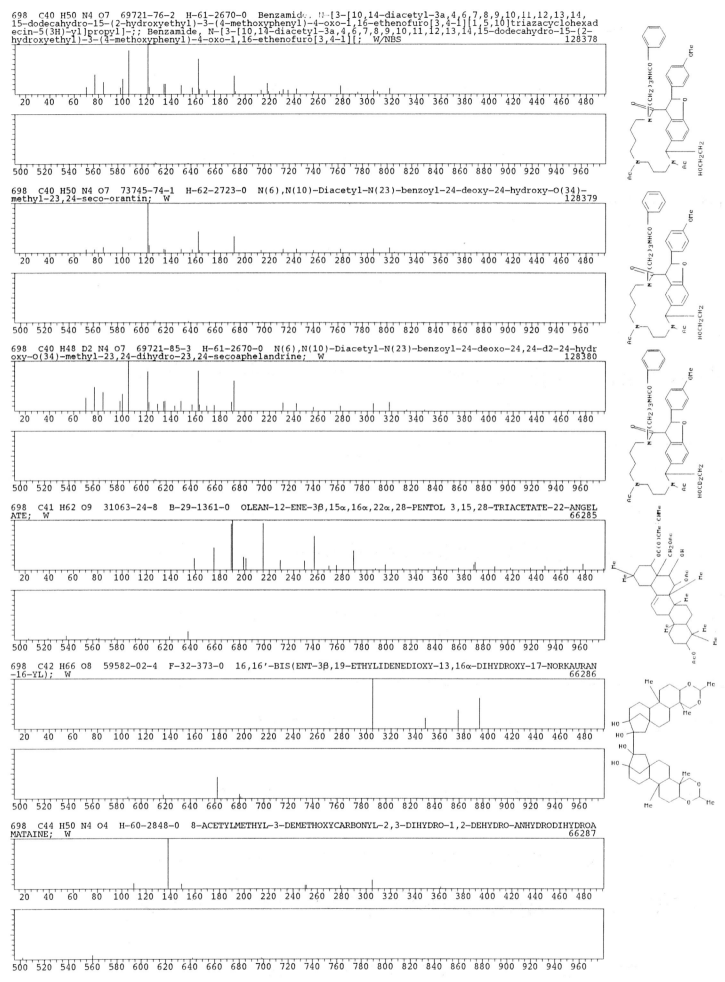

698   C40 H50 N4 O7   69721-76-2   H-61-2670-0   Benzamide, N-[3-[10,14-diacetyl-3a,4,6,7,8,9,10,11,12,13,14,
15-dodecahydro-15-(2-hydroxyethyl)-3-(4-methoxyphenyl)-4-oxo-1,16-ethenofuro[3,4-l][1,5,10]triazacyclohexad
ecin-5(3H)-yl]propyl]-;; Benzamide, N-[3-[10,14-diacetyl-3a,4,6,7,8,9,10,11,12,13,14,15-dodecahydro-15-(2-
hydroxyethyl)-3-(4-methoxyphenyl)-4-oxo-1,16-ethenofuro[3,4-l][;   W/NBS   128378

698   C40 H50 N4 O7   73745-74-1   H-62-2723-0   N(6),N(10)-Diacetyl-N(23)-benzoyl-24-deoxy-24-hydroxy-O(34)-
methyl-23,24-seco-orantin;   W   128379

698   C40 H48 D2 N4 O7   69721-85-3   H-61-2670-0   N(6),N(10)-Diacetyl-N(23)-benzoyl-24-deoxo-24,24-d2-24-hydr
oxy-O(34)-methyl-23,24-dihydro-23,24-secoaphelandrine;   W   128380

698   C41 H62 O9   31063-24-8   B-29-1361-0   OLEAN-12-ENE-3β,15α,16α,22α,28-PENTOL 3,15,28-TRIACETATE-22-ANGEL
ATE;   W   66285

698   C42 H66 O8   59582-02-4   F-32-373-0   16,16'-BIS(ENT-3β,19-ETHYLIDENEDIOXY-13,16α-DIHYDROXY-17-NORKAURAN
-16-YL);   W   66286

698   C44 H50 N4 O4   H-60-2848-0   8-ACETYLMETHYL-3-DEMETHOXYCARBONYL-2,3-DIHYDRO-1,2-DEHYDRO-ANHYDRODIHYDROA
MATAINE;   W   66287

698   C44 H45 D5 N4 O4   H-60-2848-0   8-(TRIDEUTERIOACETYLDIDEUTERIOMETHYL)-3-DEMETHOXYCARBONYL-2,3-DIHYDRO-
1,2-DEHYDRO-ANHYDRODIHYDROAMATAINE;   W
66288

698   C44 H71 Cl O4   71142-42-2   K-112-2009-0   2-DECYL-3-METHOXY-5-PENTYLPHENYLESTER OF 15-(3-CHLORO-4-METH
OXYPHENYL)PENTADECANCARBOXYLIC ACID;   W
66291

698   C44 H71 Cl O4   71142-39-7   K-112-2008-0   Benzenetridecanoic acid, 3-chloro-4-methoxy-, 3-methoxy-2-(9-
methyldecyl)-5-(4-methylpentyl)phenyl ester;;   3-METHOXY-2-(9-METHYLDECYL)-5-(4-METHYLPENTYL)PHENYLESTER OF
13-(3-CHLORO-4-METHOXYPHENYL)TRIDECANCARBOXYLIC ACID;   W/NBS
66289

698   C45 H59 Cl O4   71142-37-5   K-112-2006-0   MIXTURE OF 2-DECYL-5-ISOHEXYL- AND 2-ISOUNDECYL-5-PENTYL-3-ME
O-PHENYL 15-(3-CL-4-MEO)PHENYL-2,4,6,8,10,10,12,14-PENTADECAHEPTAENCARBOXYLATES;   W
66292

699   C27 H33 N O3 P3 Re   33990-59-9   KA-71-2365-16   Rhenium, dicarbonyltris(dimethylphenylphosphine)(isocya
nato)-, stereoisomer;;   ISOCYANATO(DICARBONYL)TRIS(DIMETHYLPHENYLPHOSPHINE)RHENIUM(I);   W/NBS
66293

699   C34 H41 N3 O13   5123-47-7   T-67-5112-0   Glutamine, N-(2-acetamido-2-deoxy-β-D-glucopyranosyl)-N2-carbo
xy-, dibenzyl ester, 3',4',6'-triacetate, L-;;   N-(α-BENZYL N-BENZYLOXYCARBONYL-L-α-GLUTAMYL)-2-DEOXY-2-ACET
AMIDO-3,4,6-TRI-O-ACETYL-β-D-GLUCOSYL-AMINE;   W/NBS
66294

699   C34 H53 N9 O7   L2-1975-82-4   O,N-PERMETHYLATED MODIFIED AC-SER-ARG-HIS-PRO;   W   66295

699   C39 H45 N3 O9   79087-34-6   C-103-5135-0   1,3,5-Tris(N-methyl-N-(2,3-dimethoxybenzoyl)aminomethyl)benzene;   W   128381

700   C30 H40 N2 O17   78300-70-6   F-37-109-0   O-[2,3,4-Tri-O-acetyl-β-D-galactopyranosyluron-N-(1-methoxy-3-oxo-1-cyclopenten-2-yl)amide]-(1 to 4)-2-acetamido-1,3-di-O-acetyl-2,6-didesoxy-α-D-glucopyranose;   W   128382

700   C31 H60 O8 Si5   83228-92-6   KO-9-250-11   Per(trimethylsilyl)derivative of tetrahydronapthyldiol monoglucuronide;   W   128383

700   C33 H32 O17   F-33-1409-0   2-(2-ACETOXY-1-(3',4',5'-TRIACETOXY-2-ACETOXYCARBONYLPYRAN-6-YLOXY)PHEN-4-YL-5,7-DIHYDROXY-3,6-DIMETHYLBENZO[B]PYRAN OR SPINATINPENTAACETATE;   W   66296

700   C34 H68 O14   C-104-4141-0   Peracetyl derivative of 1,3,7,9,11,15,17-heptahydroxy-2,6,8,10-tetramethylheptadec-4-ene;   W   128384

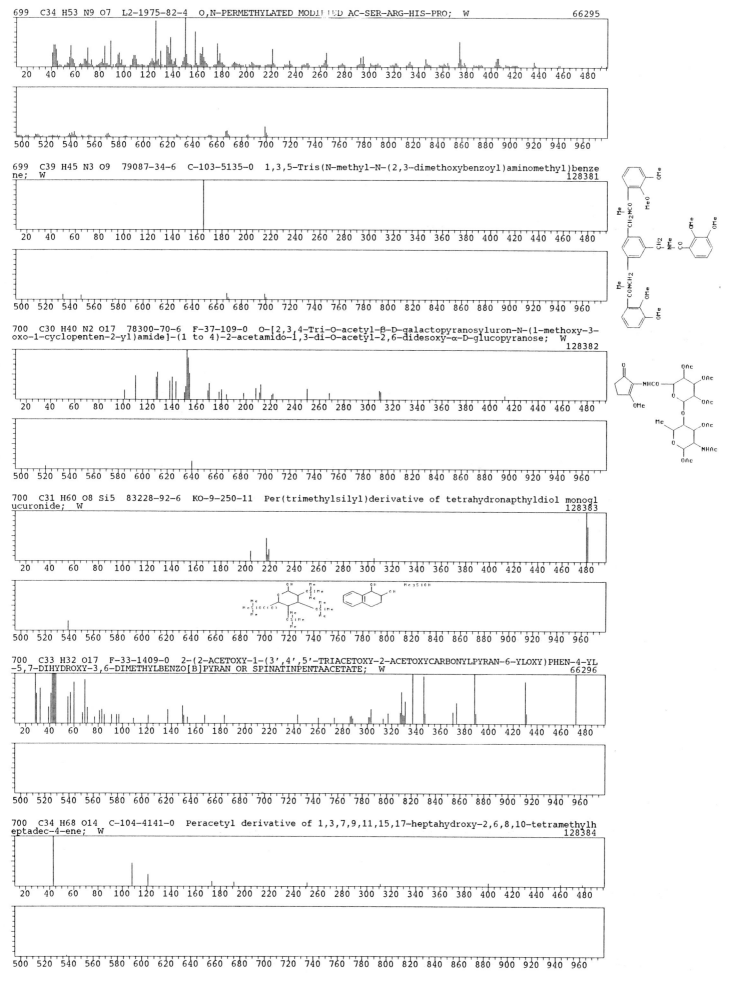

- 6069 -

700 C35 H72 O6 Si4 OM-1981-668-0 19-OH-16,16-DIME F2A ME 4TMS; W   66297

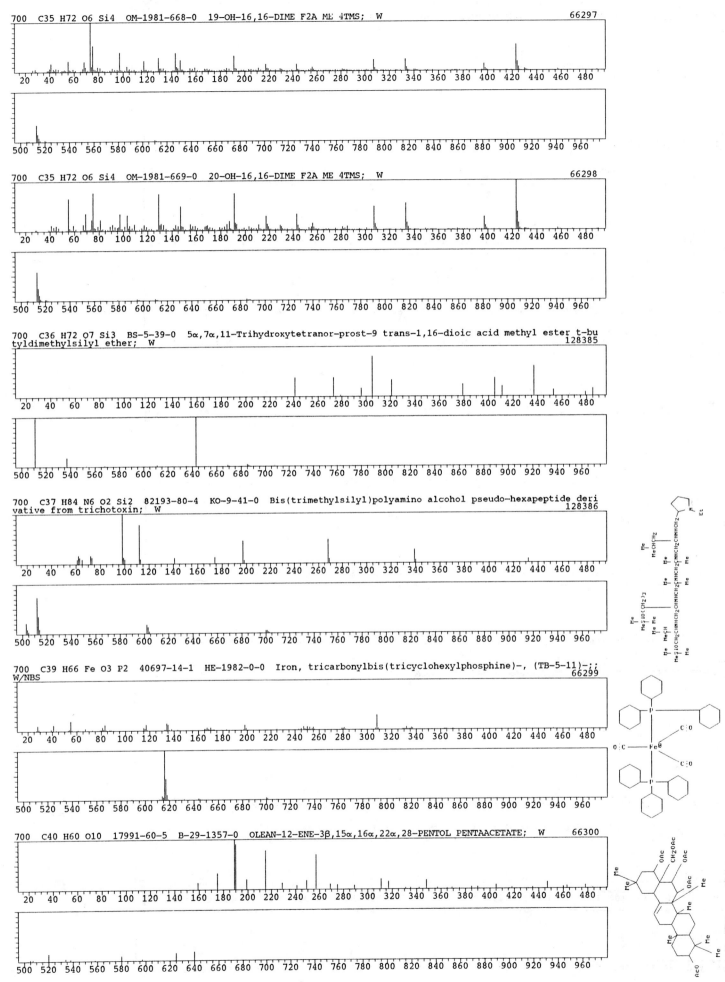

700 C35 H72 O6 Si4 OM-1981-669-0 20-OH-16,16-DIME F2A ME 4TMS; W   66298

700 C36 H72 O7 Si3 BS-5-39-0 5α,7α,11-Trihydroxytetranor-prost-9 trans-1,16-dioic acid methyl ester t-bu
tyldimethylsilyl ether; W   128385

700 C37 H84 N6 O2 Si2 82193-80-4 KO-9-41-0 Bis(trimethylsilyl)polyamino alcohol pseudo-hexapeptide deri
vative from trichotoxin; W   128386

700 C39 H66 Fe O3 P2 40697-14-1 HE-1982-0-0 Iron, tricarbonylbis(tricyclohexylphosphine)-, (TB-5-11)-;;
W/NBS   66299

700 C40 H60 O10 17991-60-5 B-29-1357-0 OLEAN-12-ENE-3β,15α,16α,22α,28-PENTOL PENTAACETATE; W   66300

700 C42 H40 N2 O8   68152-30-7   J-44-295-0   4,4'-Bis[N-3,4-dimethoxyphenylmethylene]-1-(6,7-dimethoxy)napht
hylamine;   W   128387

700 C43 H48 N4 O5   25662-91-3   T-66-2764-0   Vobtusine, 2',3'-didehydro-2'-deoxy-;; ANHYDROVOBTUSIN;   W/NBS
                                                                                                           66301

700 C43 H48 N4 O5   31222-09-0   H-60-2849-0   ANHYDRODIHYDROAMATAINE;   W   66302

700 C43 H47 D N4 O5   65967-12-6   H-60-2849-0   8-DEUTERIO-ANHYDRODIHYDROAMATAINE;   W   66303

700 C48 H36 N4 O2   69813-02-1   F-34-2298-2300   HOMOPORPHYRINE;   W   66304

700 C48 H36 N4 O2   C-97-6180-0   ENDO(CYCLOPROPANO (2',3' : 3,4)3,4-DIHYDRO NICKEL HOMOPERPHYRIN 1'-ETHYL
ESTER.;   W   66305

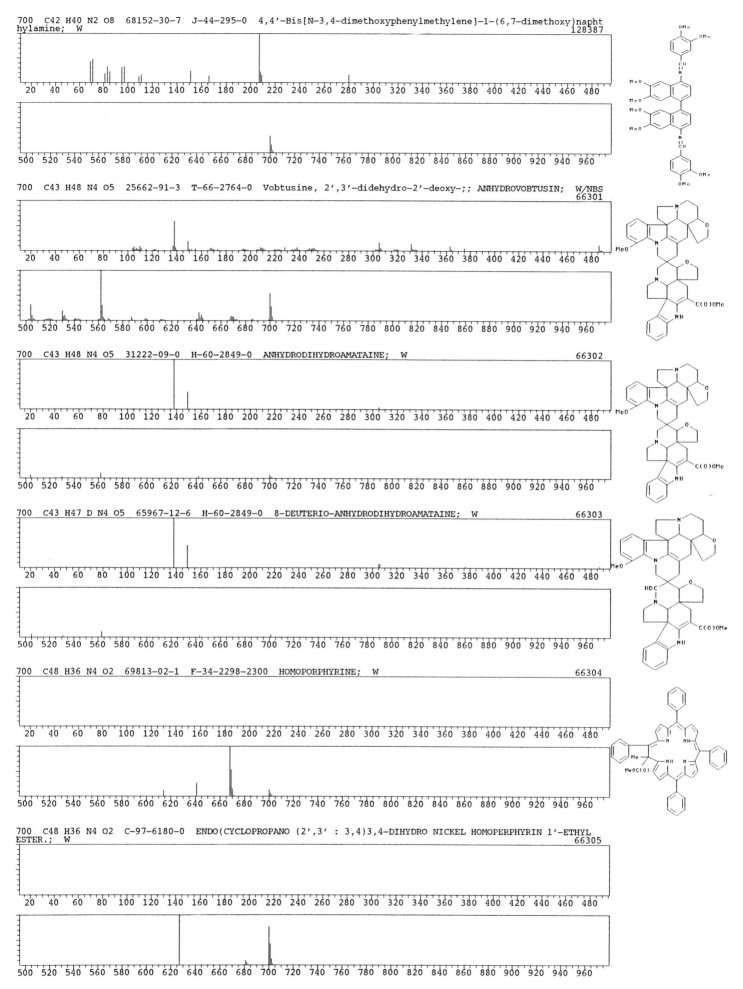

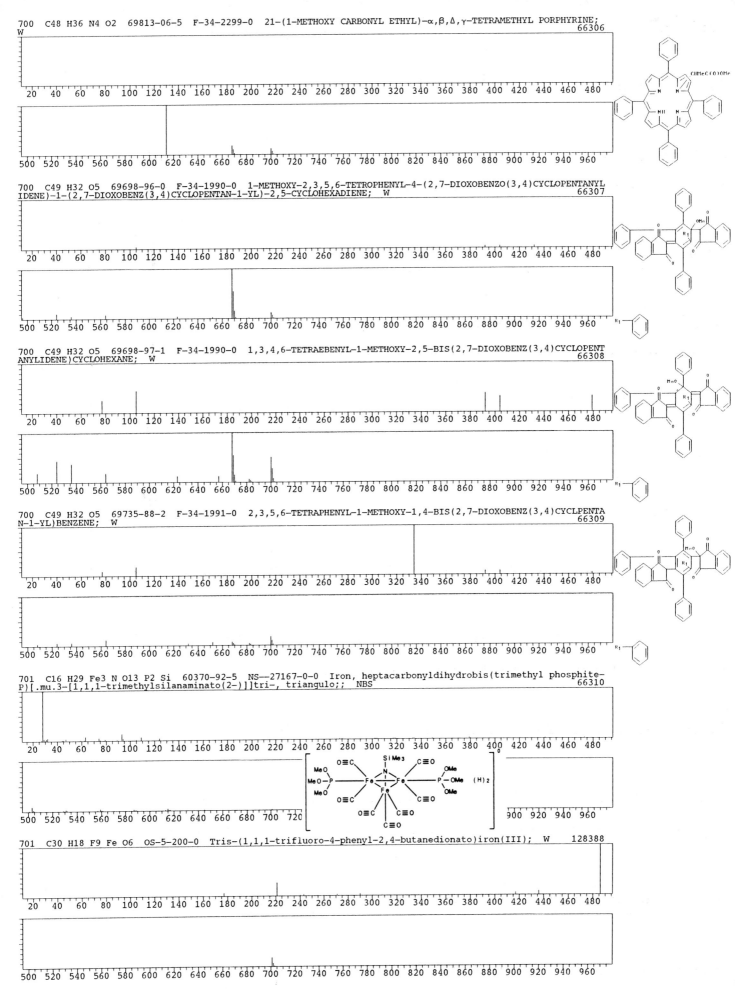

700   C48 H36 N4 O2   69813-06-5   F-34-2299-0   21-(1-METHOXY CARBONYL ETHYL)-α,β,Δ,γ-TETRAMETHYL PORPHYRINE;
W   66306

700   C49 H32 O5   69698-96-0   F-34-1990-0   1-METHOXY-2,3,5,6-TETROPHENYL-4-(2,7-DIOXOBENZO(3,4)CYCLOPENTANYL
IDENE)-1-(2,7-DIOXOBENZ(3,4)CYCLOPENTAN-1-YL)-2,5-CYCLOHEXADIENE;   W   66307

700   C49 H32 O5   69698-97-1   F-34-1990-0   1,3,4,6-TETRAEBENYL-1-METHOXY-2,5-BIS(2,7-DIOXOBENZ(3,4)CYCLOPENT
ANYLIDENE)CYCLOHEXANE;   W   66308

700   C49 H32 O5   69735-88-2   F-34-1991-0   2,3,5,6-TETRAPHENYL-1-METHOXY-1,4-BIS(2,7-DIOXOBENZ(3,4)CYCLPENTA
N-1-YL)BENZENE;   W   66309

701   C16 H29 Fe3 N O13 P2 Si   60370-92-5   NS--27167-0-0   Iron, heptacarbonyldihydrobis(trimethyl phosphite-
P)[.mu.3-[1,1,1-trimethylsilanaminato(2-)]]tri-, triangulo;;   NBS   66310

701   C30 H18 F9 Fe O6   OS-5-200-0   Tris-(1,1,1-trifluoro-4-phenyl-2,4-butanedionato)iron(III);   W   128388

702 C24 H49 F3 N6 Si8 62978-30-7 K-110-1283-0 (3-(DIFLUOROPHENYLSILYL)-2,2,4,4,6,6-HEXAMETHYLCYCLOTRISI
LAZAN-1-YL)FLUORO(2,2,4,4,6,6-HEXAMETHYLCYCLOTRISILAZAN-1-YL)PHENYLSILANE; W   66311

702 C28 H24 Br2 N4 O4 S2 90867-71-3 J-49-2971-0 2,2'-[1,2-ethanediylbis(thio)]bis[6-[2-(6-bromo-2-pyrid
inyl)-1,3-dioxolan-2-yl]pyridine; W   128389

Str 1170

702 C33 H57 Eu O6 15522-71-1 NS-0-0-0 Europium, tris(2,2,6,6-tetramethyl-3,5-heptanedionato-O,O')-, (OC
-6-11)-; W/NBS   128390

702 C36 H39 Br N4 O6 73228-69-0 F-35-1326-0 (SYN)12-AZA-3-BROMO-7,10-(2,5-DIOXO-1-PHENYL-1,3,4-TRIAZOLI
DIN-3,4-DIYL)-12-(4-METHYL-1-BORMYLOXYCARBONYLBUTYL)[4.4.3]PROPELLA-2,4,8-TRIENE; W   66312

702 C36 H39 Br N4 O6 73228-69-0 F-35-1326-0 (ANTI)-12-AZA-3-BROMO-7,10-(2,5-DIOXO-1-PHENYL-1,3,4-TRIAZO
LIDIN-3,4-DIYL)-12-(4-METHYL-1-BORNYLOXYCARBONYLBUTYL(4.4.3)PROPELLA-2,4,8-TRIENE; W   66313

702 C36 H39 Br N4 O6 78420-56-1 F-37-165-0 exobornyl ester of (5α,5aβ,9aβ,10α)-2-(7-bromo-2,3-dihydro-
1,3, 14,16-tetraoxo-2-phenyl-5H,10H-5,10-etheno-5a,9a-(methaniminomethano)-1H-[1,2,4]triazolo[1,2-b]phthali
zin-15-yl)-4-methylpentanoic acid; W   128391

702 C36 H39 Br N4 O6 78420-57-2 F-37-165-0 exobornyl ester of 2-(7-bromo-2,3-dihydro-1,3, 14,16-tetraox
o-2-phenyl-5H,10H-5,10-etheno-5a,9a-(methaniminomethano)-1H-[1,2,4]triazolo[1,2-b]phthalizin-15-yl)-4-meth
ylpentanoic acid stereoisomer; W   128392

702 C39 H60 N6 Ni O2 78362-51-3 H-64-1107-0 [1,11-Pentamethylendioxy-2,2,3,3,7,7,8,8,12,12,13,13,17,17,
18,18-hexadecamethyl-10,20-diaza-decahydroporphinato(2-)]nickel; W   128393

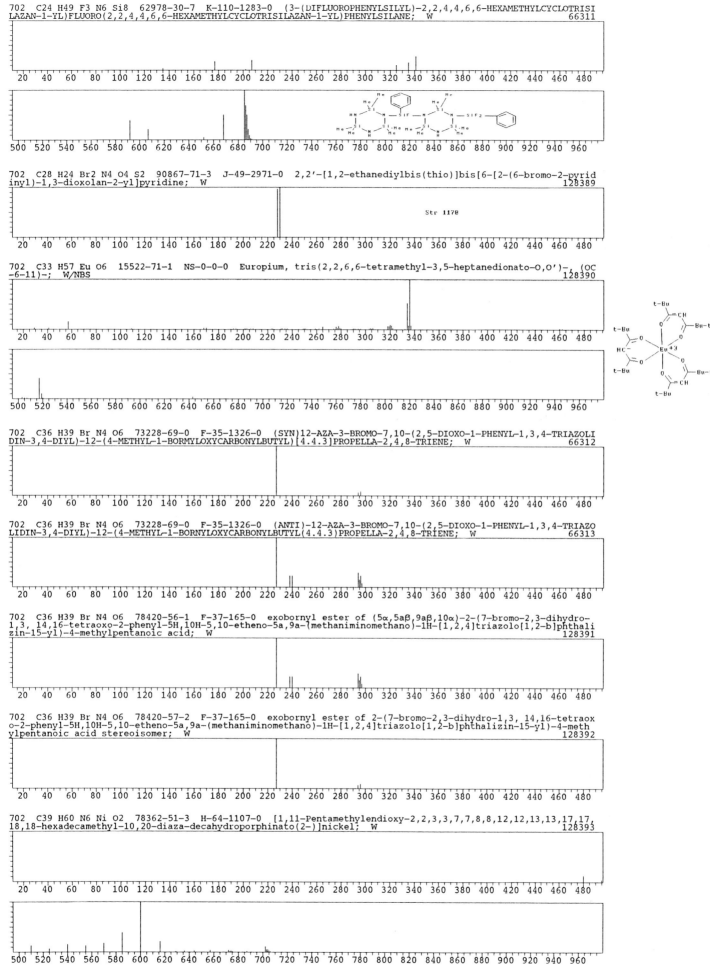

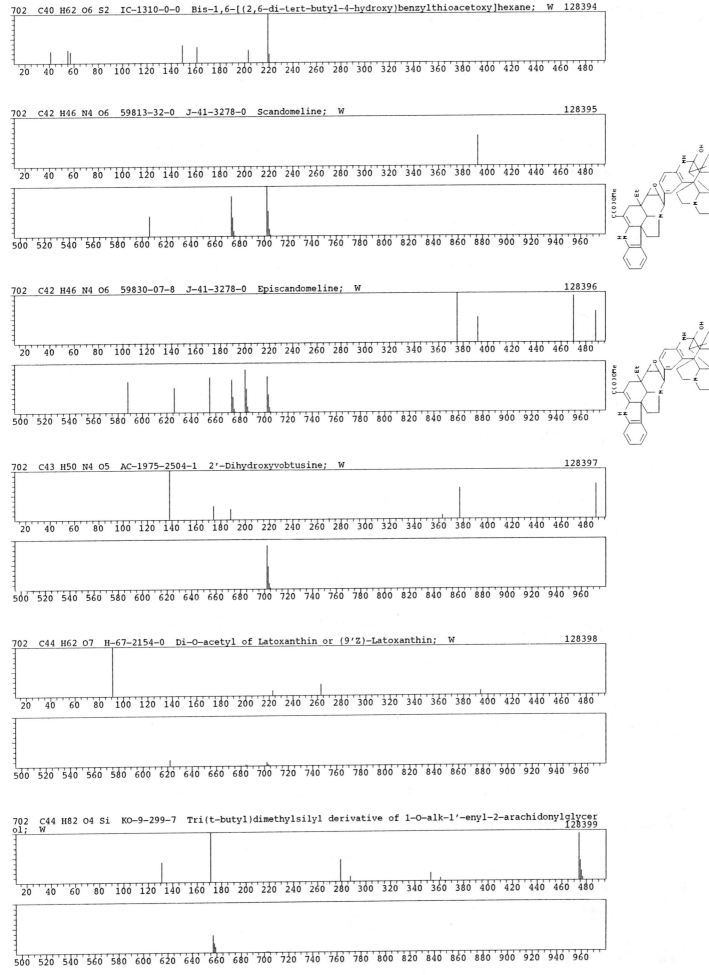

702   C40 H62 O6 S2   IC-1310-0-0   Bis-1,6-[(2,6-di-tert-butyl-4-hydroxy)benzylthioacetoxy]hexane;   W   128394

702   C42 H46 N4 O6   59813-32-0   J-41-3278-0   Scandomeline;   W   128395

702   C42 H46 N4 O6   59830-07-8   J-41-3278-0   Episcandomeline;   W   128396

702   C43 H50 N4 O5   AC-1975-2504-1   2'-Dihydroxyvobtusine;   W   128397

702   C44 H62 O7   H-67-2154-0   Di-O-acetyl of Latoxanthin or (9'Z)-Latoxanthin;   W   128398

702   C44 H82 O4 Si   KO-9-299-7   Tri(t-butyl)dimethylsilyl derivative of 1-O-alk-1'-enyl-2-arachidonylglycer
ol;   W   128399

702　C48 H30 O2 S2　82152-93-0　Y-19-68-0　Bis(benzo(b)thieno(2,3-d)indenylidene)-1,2-di-p-toluylithane;　W
128400

702　C50 H70 O2　29790-47-4　D-8-4654-1　2-DEMETHYLMENAGUINONE-8;　W/NBS
66314

702　C50 H102　55256-09-2　AA-0-2264-1　Triacontane, 11,20-didecyl-;;　W
66315

702　C50 H102　55256-07-0　AA-0-2265-1　Tetratriacontane, 17-hexadecyl-;;　17-N-HEXADECYLTETRATRIACONTANE;
W/NBS
66316

702　C50 H102　55256-08-1　AA-0-2266-1　Tritriacontane, 13-decyl-13-heptyl-;;　13-N-HEPTYL-13-N-DECYLTRITRIA
CONTANE;　W/NBS
66317

702　C51 H42 O3　61539-53-5　KC-1976-1886-0　1,2,4,5,7,8-HEXAPHENYL-TRI(CYCLOBUTA)[A.D.G]NONANE-3,6,9-TRIO
NE;　W
66318

Str 1171

703　C27 H77 N5 O4 Si6　GC-12-38-4　Pentakis(trimethylsilyl) derivative of 3'-hydroxysepiapterin;　W 128401

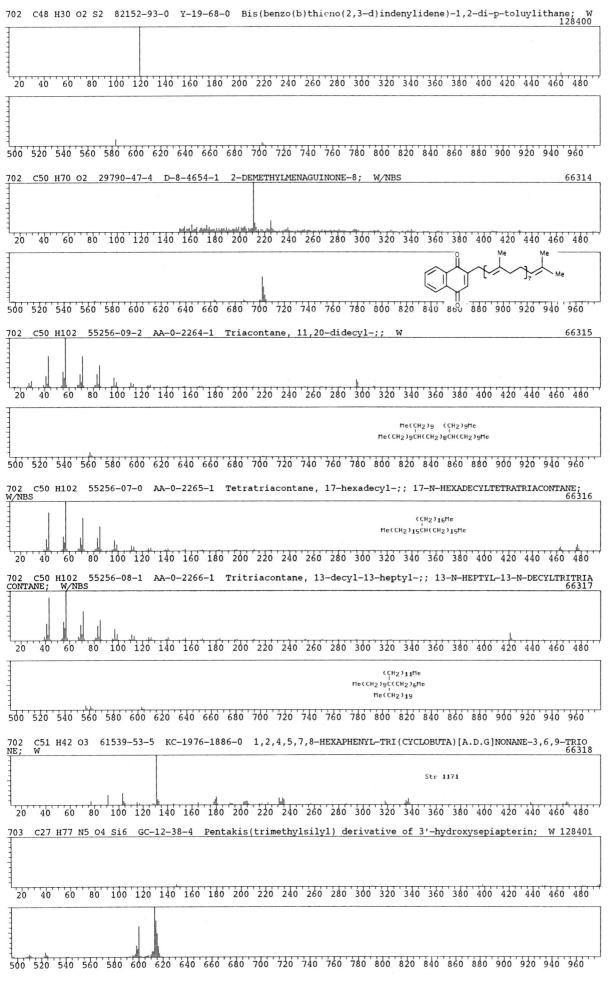

703   C37 H57 N3 O6 S2   91652-53-8   H-67-2182-0   13-Oxo-N-[4-tosyl-8-(tosylamino)-4-azaoctyl]-16-hexadecanlactam; W
128402

703   C42 H61 N3 O6   87286-82-6   H-66-1088-0   N,N'-Diheptyl-N,N',6-trimethyl-6-(2,3-diphenyl-maleinimido)-4,8-dioxaundecandiamide; W
128403

704   C28 H18 Cl10   8017-34-3   NS--27162-0-0   Benzene, 1-chloro-2-[2,2,2-trichloro-1-(4-chlorophenyl)ethyl]-, mixt. with 1,1'-(2,2,2-trichloroethylidene)bis[4-chlorobenzene];; Technical chlorophenothane; NBS 66319

704   C30 H18 Co F9 O6   OS-5-201-0   Tris-(1,1,1-trifluoro-4-phenyl-2,4-butanedionato)cobalt(III); W 128404

704   C32 H20 Mn2 O8 P2   AD-0-1588-0   DI-.mu.-DIPHENYL-PHOSPHINO-BIS(TETRACARBONYLMANGANESE); W   66320

704   C35 H45 Cl N2 O11   79101-56-7   J-46-4402-0   Trenudine; W
128405

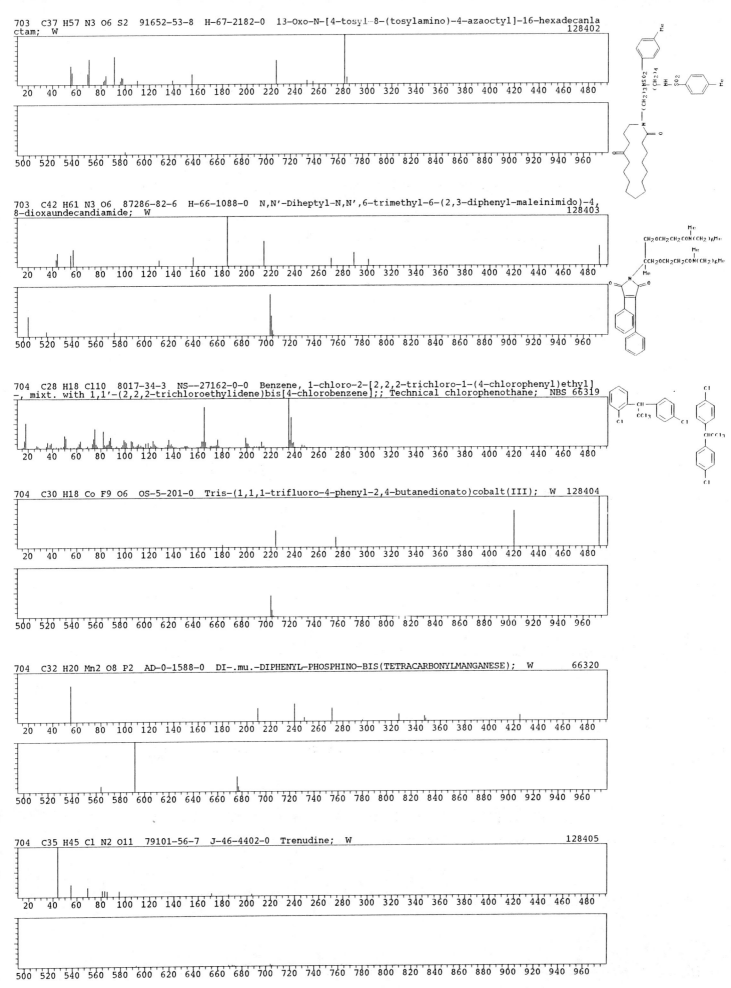

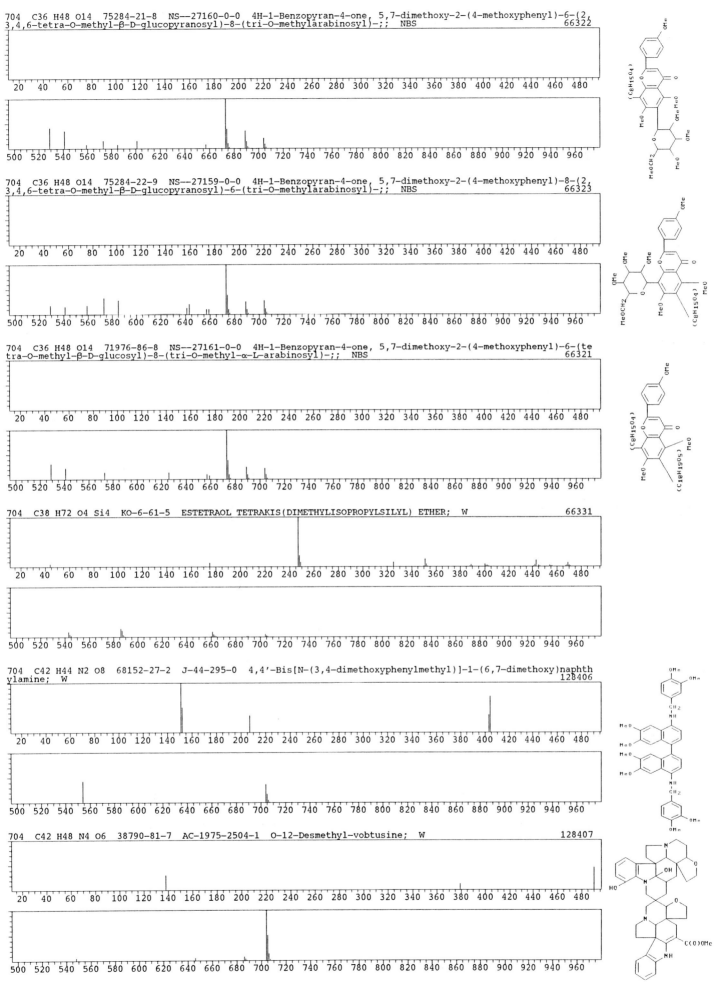

704   C36 H48 O14   75284-21-8   NS--27160-0-0   4H-1-Benzopyran-4-one, 5,7-dimethoxy-2-(4-methoxyphenyl)-6-(2,
3,4,6-tetra-O-methyl-β-D-glucopyranosyl)-8-(tri-O-methylarabinosyl)-;;   NBS   66322

704   C36 H48 O14   75284-22-9   NS--27159-0-0   4H-1-Benzopyran-4-one, 5,7-dimethoxy-2-(4-methoxyphenyl)-8-(2,
3,4,6-tetra-O-methyl-β-D-glucopyranosyl)-6-(tri-O-methylarabinosyl)-;;   NBS   66323

704   C36 H48 O14   71976-86-8   NS--27161-0-0   4H-1-Benzopyran-4-one, 5,7-dimethoxy-2-(4-methoxyphenyl)-6-(te
tra-O-methyl-β-D-glucosyl)-8-(tri-O-methyl-α-L-arabinosyl)-;;   NBS   66321

704   C38 H72 O4 Si4   KO-6-61-5   ESTETRAOL TETRAKIS(DIMETHYLISOPROPYLSILYL) ETHER;   W   66331

704   C42 H44 N2 O8   68152-27-2   J-44-295-0   4,4'-Bis[N-(3,4-dimethoxyphenylmethyl)]-1-(6,7-dimethoxy)naphth
ylamine;   W   128406

704   C42 H48 N4 O6   38790-81-7   AC-1975-2504-1   O-12-Desmethyl-vobtusine;   W   128407

704   C43 H52 N4 O5   3371-85-5   AC-63-1902-6   Ibogamine-18-carboxylic acid, 12-methoxy-13-[(3α)-17-methoxy-17-oxovobasan-3-yl]-, methyl ester;; Voacamine;   W/NBS
66324

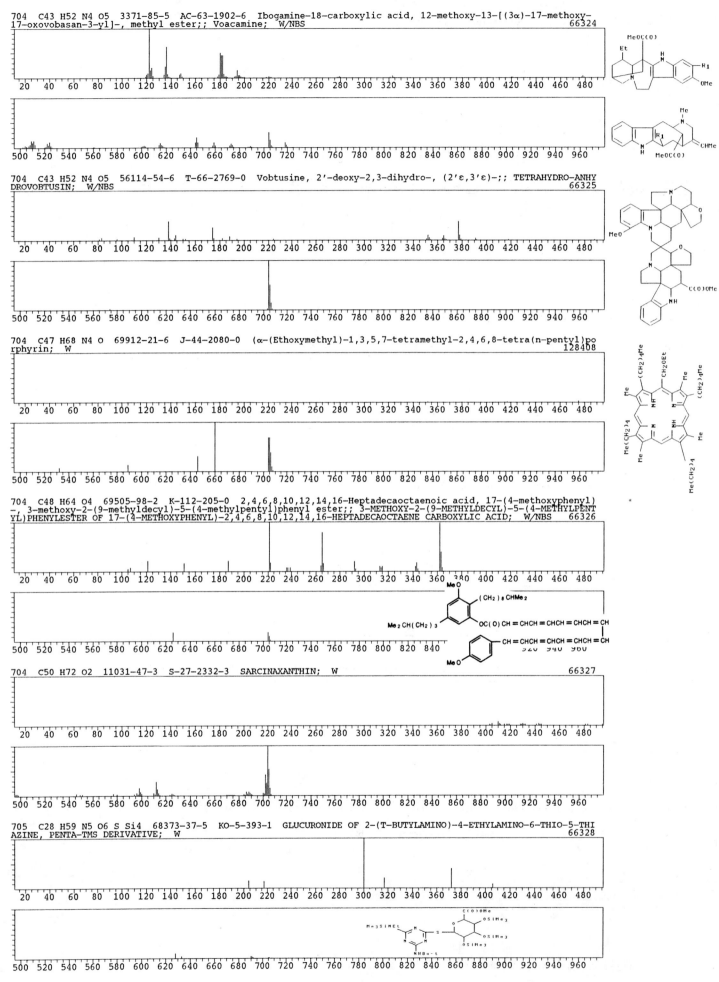

704   C43 H52 N4 O5   56114-54-6   T-66-2769-0   Vobtusine, 2'-deoxy-2,3-dihydro-, (2'ε,3'ε)-;; TETRAHYDRO-ANHY
DROVOBTUSIN;   W/NBS
66325

704   C47 H68 N4 O   69912-21-6   J-44-2080-0   (α-(Ethoxymethyl)-1,3,5,7-tetramethyl-2,4,6,8-tetra(n-pentyl)po
rphyrin;   W
128408

704   C48 H64 O4   69505-98-2   K-112-205-0   2,4,6,8,10,12,14,16-Heptadecaoctaenoic acid, 17-(4-methoxyphenyl)
-, 3-methoxy-2-(9-methyldecyl)-5-(4-methylpentyl)phenyl ester;; 3-METHOXY-2-(9-METHYLDECYL)-5-(4-METHYLPENT
YL)PHENYLESTER OF 17-(4-METHOXYPHENYL)-2,4,6,8,10,12,14,16-HEPTADECAOCTAENE CARBOXYLIC ACID;   W/NBS   66326

704   C50 H72 O2   11031-47-3   S-27-2332-3   SARCINAXANTHIN;   W
66327

705   C28 H59 N5 O6 S Si4   68373-37-5   KO-5-393-1   GLUCURONIDE OF 2-(T-BUTYLAMINO)-4-ETHYLAMINO-6-THIO-5-THI
AZINE, PENTA-TMS DERIVATIVE;   W
66328

705　C35 H48 Cl N3 O10　ZH-614-3-0　MAYTANPRINE;　W　　　　　　　　　66329

20 40 60 80 100 120 140 160 180 200 220 240 260 280 300 320 340 360 380 400 420 440 460 480

500 520 540 560 580 600 620 640 660 680 700 720 740 760 780 800 820 840 860 880 900 920 940 960

705　C37 H67 N O6 Si3　59403-26-8　KO-4-360-2　BENZYLOXIME DERIVATIVE OF 6-OXO-PGF1α;　W　　66330

20 40 60 80 100 120 140 160 180 200 220 240 260 280 300 320 340 360 380 400 420 440 460 480

500 520 540 560 580 600 620 640 660 680 700 720 740 760 780 800 820 840 860 880 900 920 940 960

705　C40 H52 Cl N O6 Si　81025-20-9　O-16-404-1　PENITREM A MONOTRIMETHYLSILYL ETHER;　W　　66332

20 40 60 80 100 120 140 160 180 200 220 240 260 280 300 320 340 360 380 400 420 440 460 480

Me3SiOH

500 520 540 560 580 600 620 640 660 680 700 720 740 760 780 800 820 840 860 880 900 920 940 960

705　C43 H67 N3 O5　IC-1311-0-0　1-(1-Hydroxynaphth-2-ylcarboxylaminohexyl)-4-(N-isobutyryl)octadecylamino-
2,5-dioxopyrrole;　W　　　　　　　　　　　　　　　　　　　　　　128409

20 40 60 80 100 120 140 160 180 200 220 240 260 280 300 320 340 360 380 400 420 440 460 480

500 520 540 560 580 600 620 640 660 680 700 720 740 760 780 800 820 840 860 880 900 920 940 960

706　C16 H36 Bi2 N2 S　91173-33-0　K-117-2004-0　Bis(di-tert-butylbismutino)sulfurimide;　W　　128410

20 40 60 80 100 120 140 160 180 200 220 240 260 280 300 320 340 360 380 400 420 440 460 480

t-Bu　　Bu-t
t-BuBiN=S=NBiBu-t

500 520 540 560 580 600 620 640 660 680 700 720 740 760 780 800 820 840 860 880 900 920 940 960

706　C18 Cl14　29629-84-3　EC-0-2-0　Terphenyl, tetradecachloro-;;　W/NBS　　　　　66333

20 40 60 80 100 120 140 160 180 200 220 240 260 280 300 320 340 360 380 400 420 440 460 480

500 520 540 560 580 600 620 640 660 680 700 720 740 760 780 800 820 840 860 880 900 920 940 960

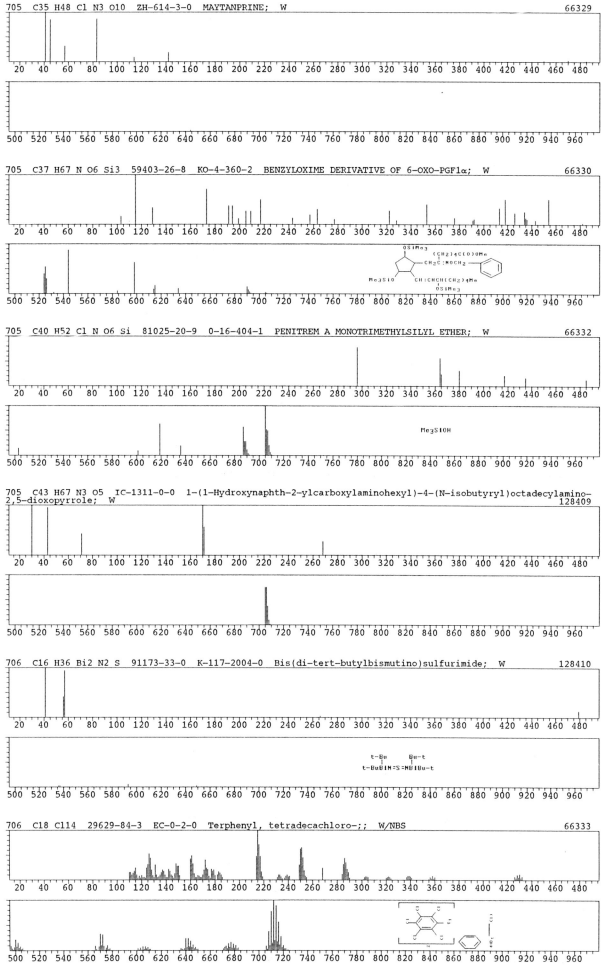

706   C24 H59 O10 P Si6   69745-86-4   BA-0-352-0   1-PHOSPHOGALACTURONIC ACID-HEXATMS;   W/NBS   66334

706   C24 H59 O10 P Si6   69727-23-7   BA-0-351-0   D-Glucopyranuronic acid, 2,3,4-tris-O-(trimethylsilyl)-, tr
imethylsilyl ester, bis(trimethylsilyl) phosphate;; O-HEXAKIS(TRIMETHYLSILYL)-D-GLUCOPYRANURONATE-1-PHOSPH
ATE;   W/NBS   66335

706   C31 H66 O8 Si5   83233-57-2   KO-9-251-12   Per(trimethylsilyl)derivative of dihydroxydecalin monoglucuro
nide; W   128411

706   C36 H58 Ni O4 P2 S   HE-1986-858-0   [ETHYLENEBIS(DICYCLOHEXYLPHOSPHINE)][TRANS-β-PHENYLSULFONYLMETHYLAC
RYLATE]NICKEL;   W   128412

706   C38 H55 Cl O10   83637-86-9   AN-91-328-0   Triacetate dimethyl ester of atroxigenine;   W   128413

706   C40 H42 N4 O8   92284-74-7   J-49-4605-0   6,7-Bis[2-(methoxycarbonyl)ethyl]-β,4-[β'-(methoxycarbonyl)eth
ylene]-2-[2-(methoxycarbonyl)vinyl]-1,3,5,8-tetramethylporphyrin;   W   128414

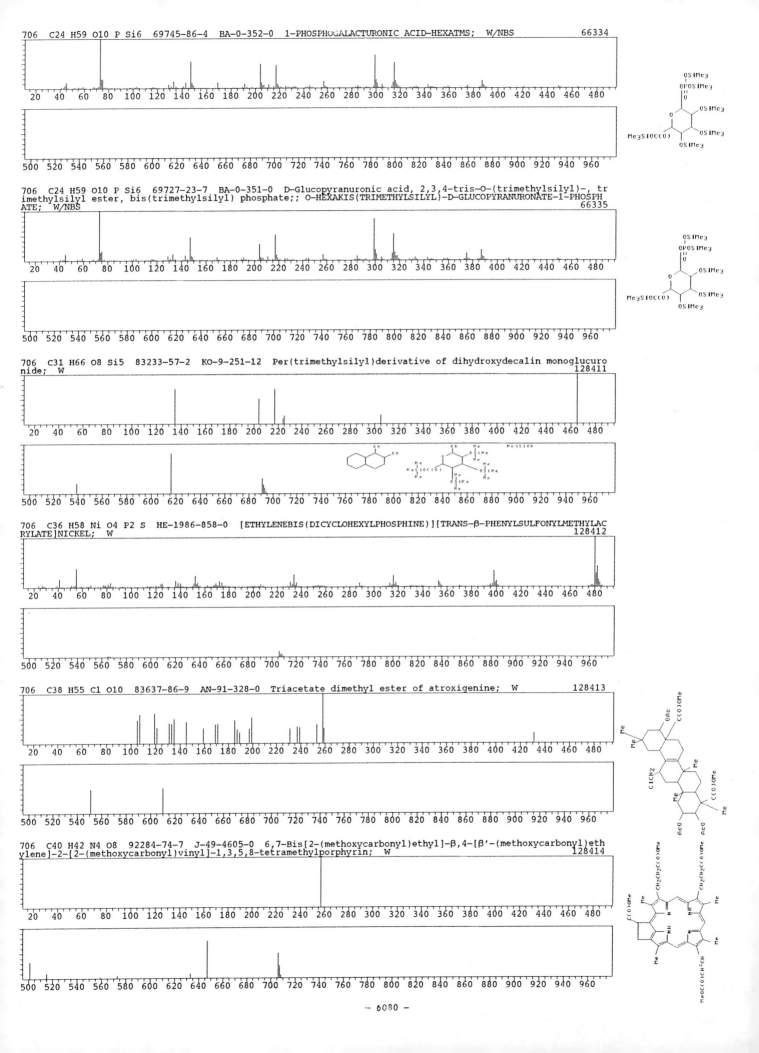

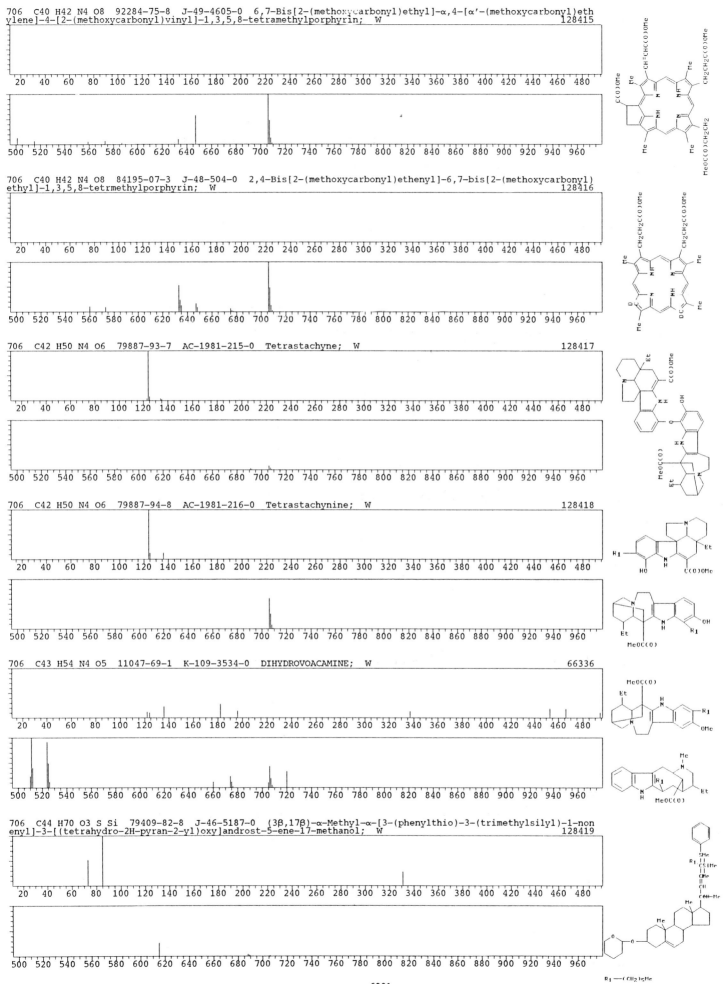

706 C40 H42 N4 O8 92284-75-8 J-49-4605-0 6,7-Bis[2-(methoxycarbonyl)ethyl]-α,4-[α'-(methoxycarbonyl)ethylene]-4-[2-(methoxycarbonyl)vinyl]-1,3,5,8-tetramethylporphyrin; W   128415

706 C40 H42 N4 O8 84195-07-3 J-48-504-0 2,4-Bis[2-(methoxycarbonyl)ethenyl]-6,7-bis[2-(methoxycarbonyl)ethyl]-1,3,5,8-tetrmethylporphyrin; W   128416

706 C42 H50 N4 O6 79887-93-7 AC-1981-215-0 Tetrastachyne; W   128417

706 C42 H50 N4 O6 79887-94-8 AC-1981-216-0 Tetrastachynine; W   128418

706 C43 H54 N4 O5 11047-69-1 K-109-3534-0 DIHYDROVOACAMINE; W   66336

706 C44 H70 O3 S Si 79409-82-8 J-46-5187-0 (3β,17β)-α-Methyl-α-[3-(phenylthio)-3-(trimethylsilyl)-1-nonenyl]-3-[(tetrahydro-2H-pyran-2-yl)oxy]androst-5-ene-17-methanol; W   128419

706   C47 H46 O6   H-62-2002-0   Methyl 2,3,4-tri-O-benzyl-6-O-trityl-β-D-glucopyranoside;   W   128420

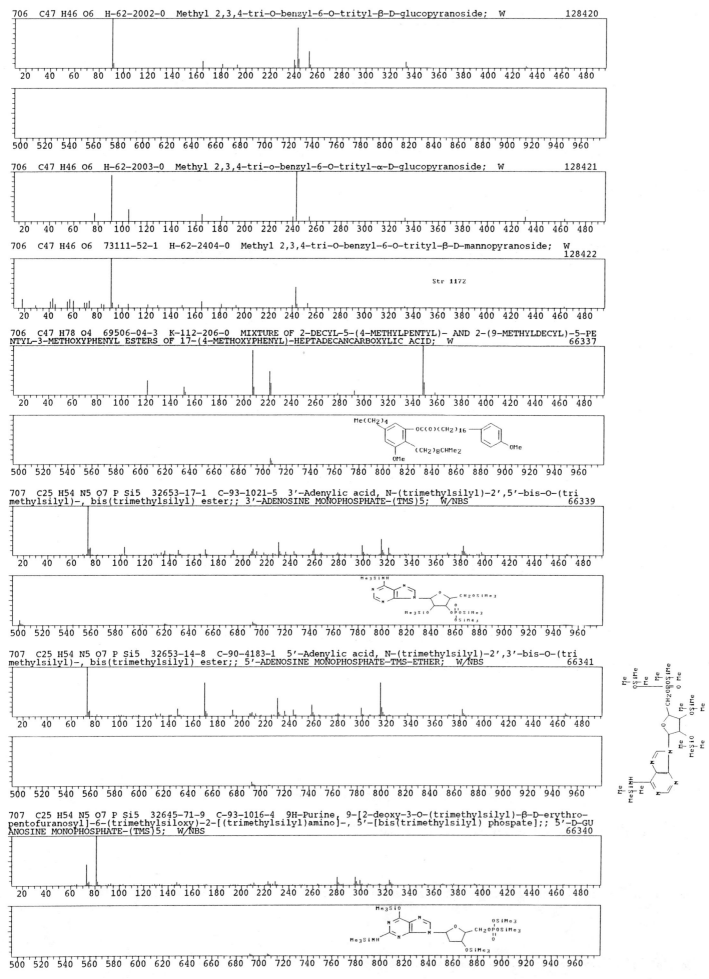

706   C47 H46 O6   H-62-2003-0   Methyl 2,3,4-tri-o-benzyl-6-O-trityl-α-D-glucopyranoside;   W   128421

706   C47 H46 O6   73111-52-1   H-62-2404-0   Methyl 2,3,4-tri-o-benzyl-6-O-trityl-β-D-mannopyranoside;   W
128422

Str 1172

706   C47 H78 O4   69506-04-3   K-112-206-0   MIXTURE OF 2-DECYL-5-(4-METHYLPENTYL)- AND 2-(9-METHYLDECYL)-5-PE
NTYL-3-METHOXYPHENYL ESTERS OF 17-(4-METHOXYPHENYL)-HEPTADECANCARBOXYLIC ACID;   W   66337

707   C25 H54 N5 O7 P Si5   32653-17-1   C-93-1021-5   3'-Adenylic acid, N-(trimethylsilyl)-2',5'-bis-O-(tri
methylsilyl)-, bis(trimethylsilyl) ester;; 3'-ADENOSINE MONOPHOSPHATE-(TMS)5;   W/NBS   66339

707   C25 H54 N5 O7 P Si5   32653-14-8   C-90-4183-1   5'-Adenylic acid, N-(trimethylsilyl)-2',3'-bis-O-(tri
methylsilyl)-, bis(trimethylsilyl) ester;; 5'-ADENOSINE MONOPHOSPHATE-TMS-ETHER;   W/NBS   66341

707   C25 H54 N5 O7 P Si5   32645-71-9   C-93-1016-4   9H-Purine, 9-[2-deoxy-3-O-(trimethylsilyl)-β-D-erythro-
pentofuranosyl]-6-(trimethylsilyloxy)-2-[(trimethylsilyl)amino]-, 5'-[bis(trimethylsilyl) phospate];; 5'-D-GU
ANOSINE MONOPHOSPHATE-(TMS)5;   W/NBS   66340

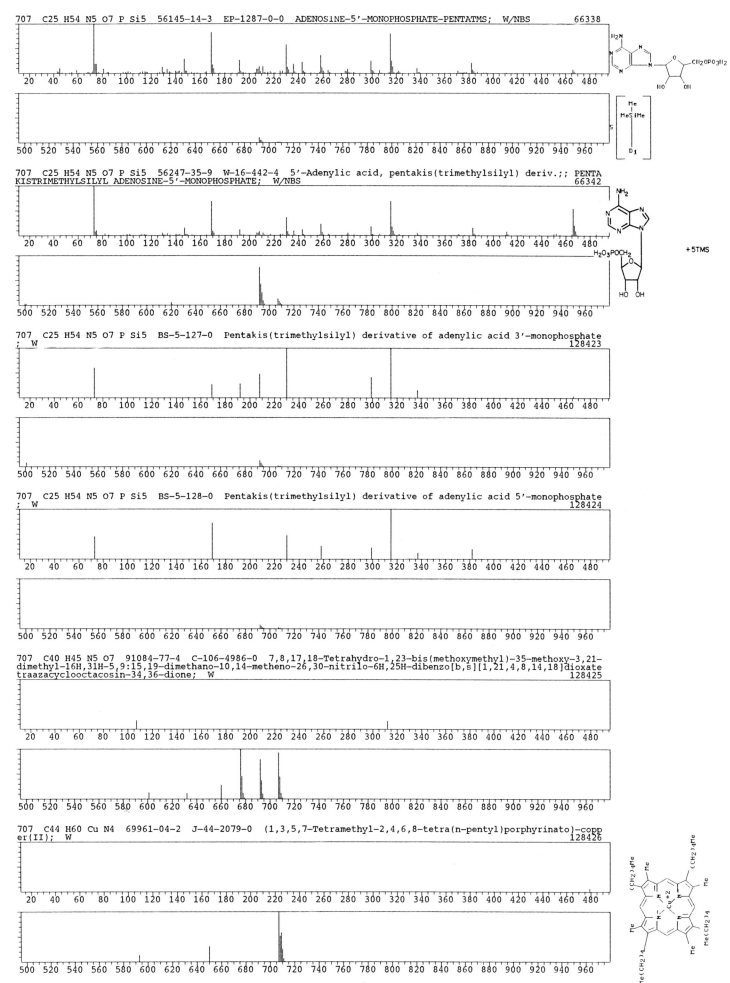

707   C25 H54 N5 O7 P Si5   56145-14-3   EP-1287-0-0   ADENOSINE-5'-MONOPHOSPHATE-PENTATMS;   W/NBS   66338

707   C25 H54 N5 O7 P Si5   56247-35-9   W-16-442-4   5'-Adenylic acid, pentakis(trimethylsilyl) deriv.;; PENTA
KISTRIMETHYLSILYL ADENOSINE-5'-MONOPHOSPHATE;   W/NBS   66342

+5TMS

707   C25 H54 N5 O7 P Si5   BS-5-127-0   Pentakis(trimethylsilyl) derivative of adenylic acid 3'-monophosphate
; W   128423

707   C25 H54 N5 O7 P Si5   BS-5-128-0   Pentakis(trimethylsilyl) derivative of adenylic acid 5'-monophosphate
; W   128424

707   C40 H45 N5 O7   91084-77-4   C-106-4986-0   7,8,17,18-Tetrahydro-1,23-bis(methoxymethyl)-35-methoxy-3,21-
dimethyl-16H,31H-5,9:15,19-dimethano-10,14-metheno-26,30-nitrilo-6H,25H-dibenzo[b,s][1,21,4,8,14,18]dioxate
traazacyclooctacosin-34,36-dione; W   128425

707   C44 H60 Cu N4   69961-04-2   J-44-2079-0   (1,3,5,7-Tetramethyl-2,4,6,8-tetra(n-pentyl)porphyrinato)-copp
er(II); W   128426

708   C6 H I5   608-96-8   B-30-1709-0   PENTAIODOBENZENE;   W   66343

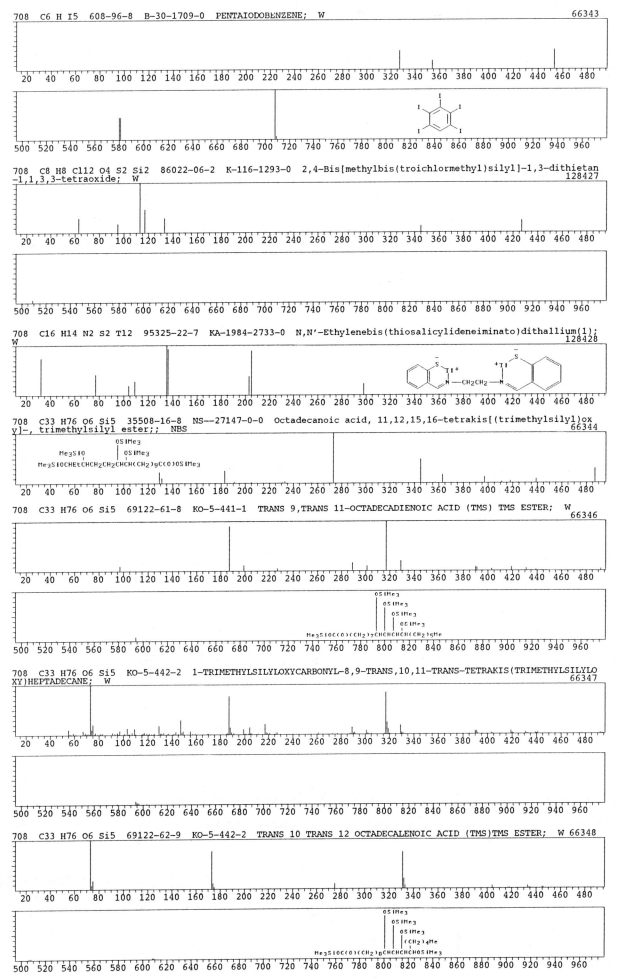

708   C8 H8 Cl12 O4 S2 Si2   86022-06-2   K-116-1293-0   2,4-Bis[methylbis(troichlormethyl)silyl]-1,3-dithietan
-1,1,3,3-tetraoxide;   W   128427

708   C16 H14 N2 S2 Tl2   95325-22-7   KA-1984-2733-0   N,N'-Ethylenebis(thiosalicylideneiminato)dithallium(1);
W   128428

708   C33 H76 O6 Si5   35508-16-8   NS--27147-0-0   Octadecanoic acid, 11,12,15,16-tetrakis[(trimethylsilyl)ox
y]-, trimethylsilyl ester;;   NBS   66344

708   C33 H76 O6 Si5   69122-61-8   KO-5-441-1   TRANS 9,TRANS 11-OCTADECADIENOIC ACID (TMS) TMS ESTER;   W
                                                                                                             66346

708   C33 H76 O6 Si5   KO-5-442-2   1-TRIMETHYLSILYLOXYCARBONYL-8,9-TRANS,10,11-TRANS-TETRAKIS(TRIMETHYLSILYLO
XY)HEPTADECANE;   W   66347

708   C33 H76 O6 Si5   69122-62-9   KO-5-442-2   TRANS 10 TRANS 12 OCTADECALENOIC ACID (TMS)TMS ESTER;   W 66348

708   C35 H58 Co2 Si4   63517-00-0   C-99-4061-1   1,5-BIS-(CYCLOPENTADIENYL COBALT-2,3-DI(TRIMETHYL SILYL)CYCL
OBUTADIENYL)ETHANE;   W   66349

708   C39 H46 Cl2 N2 O6   53935-72-1   F-30-308-4   Berbamanium, 6,6',7,12-tetramethoxy-2,2',?-trimethyl-, chlo
ride, monohydrochloride, (1β)-;; MONOMETHYLTETRANDRINIUMCHLORIDE;   W/NBS   66350

MeCl

708   C40 H44 N4 O8   31034-73-8   KC-1976-2500-0   4,6,7-TRIS-(2-METHOXYCARBONYLETHYL)-2-(2-METHOXYCARBONYLVIN
YL)-1,3,5,8-TETRAMETHYLPORPHIN;   W   66351

708   C40 H44 N4 O8   92284-76-9   J-49-4605-0   4,6,7-Tris[2-(methoxycarbonyl)ethyl]-α,2-[α'-(methoxycarbonyl)
ethylene]-1,3,5,8-tetramethylporphyrin;   W   128429

708   C47 H48 O6   77573-18-3   NS-0-0-0   1H-2,8a-Methanocyclopenta[a]cyclopropa[e]cyclodecen-11-one, 5-(benzo
yloxy)-1a,2,5,5a,6,9,10,10a-octahydro-5a-hydroxy-6-methoxy-1,1,7,9-tetramethyl-4-[(triphenylmethoxy)methyl]
-, [1aR-(1aα,2α,5β,5aβ,6β,8aα,9α,10aα)]-;   W/NBS   128430

Str 1173

708   C49 H40 O5   69699-05-4   F-34-1991-0   1-METHOXY-2,3,5,6-TETRAPHENYL-4-(2,7-DIHYDROXYBENZO(3,4)CYCLOPENT
ANYLIDENE)-1-(2,7-DIHYDROXYBENZ(3,4)CYCLOPENTAN-1-YL)-2,5-CYCLOHEXADIENE;   W   66352

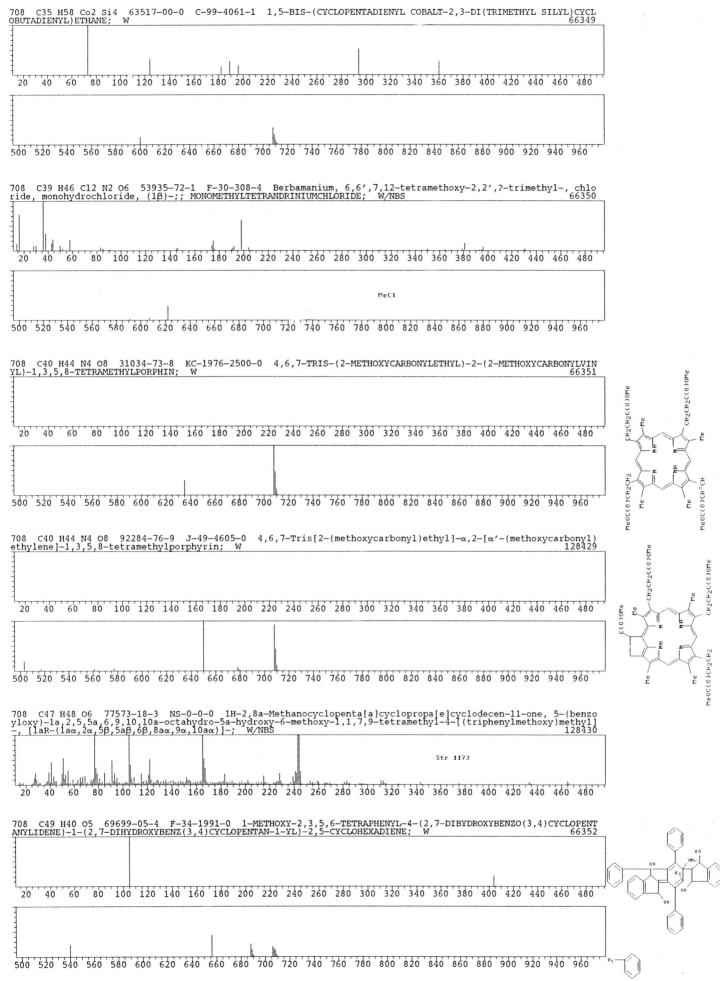

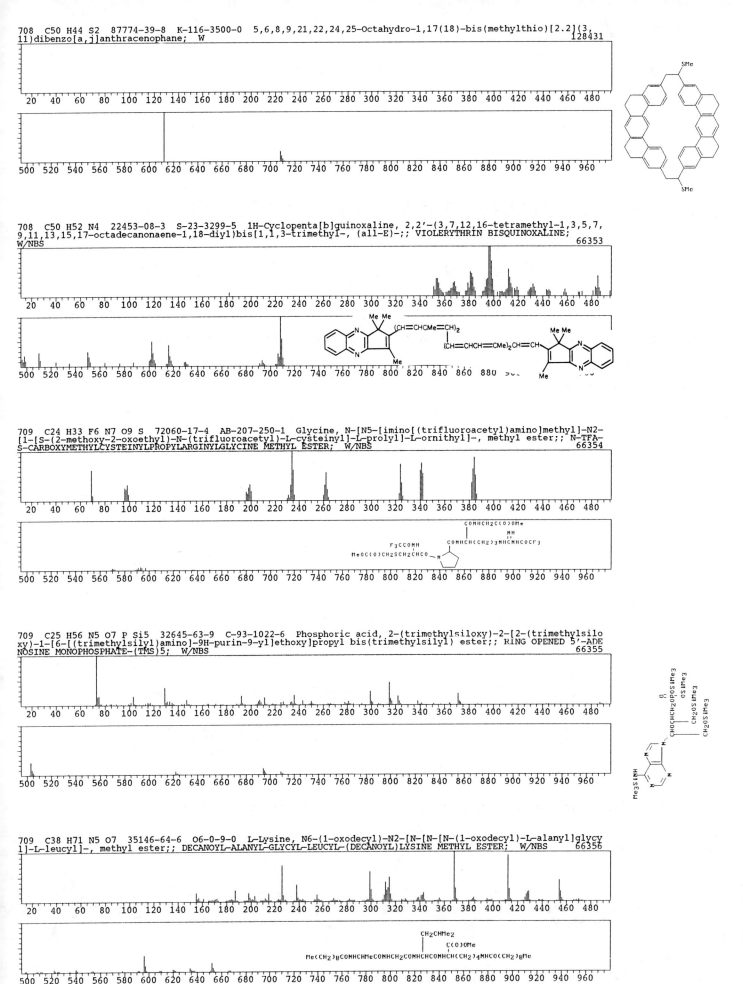

708　C50 H44 S2　87774-39-8　K-116-3500-0　5,6,8,9,21,22,24,25-Octahydro-1,17(18)-bis(methylthio)[2.2](3,11)dibenzo[a,j]anthracenophane;　W　128431

708　C50 H52 N4　22453-08-3　S-23-3299-5　1H-Cyclopenta[b]quinoxaline, 2,2'-(3,7,12,16-tetramethyl-1,3,5,7,9,11,13,15,17-octadecanonaene-1,18-diyl)bis[1,1,3-trimethyl-, (all-E)-;;　VIOLERYTHRIN BISQUINOXALINE;　W/NBS　66353

709　C24 H33 F6 N7 O9 S　72060-17-4　AB-207-250-1　Glycine, N-[N5-[imino[(trifluoroacetyl)amino]methyl]-N2-[1-[S-(2-methoxy-2-oxoethyl)-N-(trifluoroacetyl)-L-cysteinyl]-L-prolyl]-L-ornithyl]-, methyl ester;;　N-TFA-S-CARBOXYMETHYLCYSTEINYLPROPYLARGINYLGLYCINE METHYL ESTER;　W/NBS　66354

709　C25 H56 N5 O7 P Si5　32645-63-9　C-93-1022-6　Phosphoric acid, 2-(trimethylsiloxy)-2-[2-(trimethylsiloxy)-1-[6-[(trimethylsilyl)amino]-9H-purin-9-yl]ethoxy]propyl bis(trimethylsilyl) ester;;　RING OPENED 5'-ADENOSINE MONOPHOSPHATE-(TMS)5;　W/NBS　66355

709　C38 H71 N5 O7　35146-64-6　O6-0-9-0　L-Lysine, N6-(1-oxodecyl)-N2-[N-[N-[N-(1-oxodecyl)-L-alanyl]glycyl]-L-leucyl]-, methyl ester;;　DECANOYL-ALANYL-GLYCYL-LEUCYL-(DECANOYL)LYSINE METHYL ESTER;　W/NBS　66356

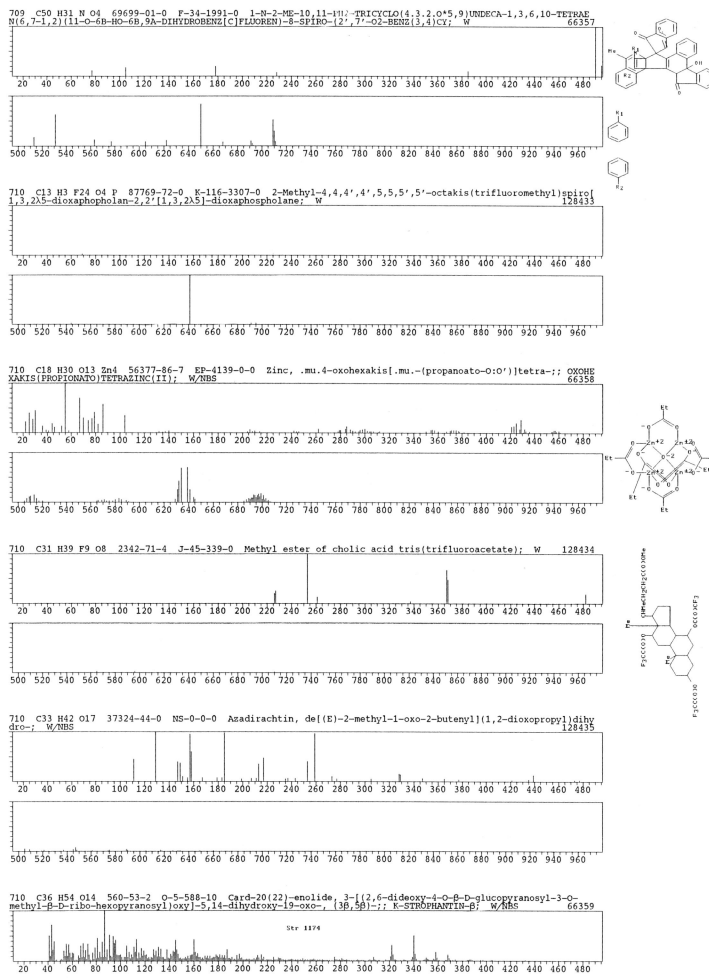

709 C50 H31 N O4 69699-01-0 F-34-1991-0 1-N-2-ME-10,11-PH2-TRICYCLO(4.3.2.O*5,9)UNDECA-1,3,6,10-TETRAE
N(6,7-1,2)(11-O-6B-HO-6B,9A-DIHYDROBENZ[C]FLUOREN)-8-SPIRO-(2',7'-O2-BENZ(3,4)CY; W   66357

710 C13 H3 F24 O4 P 87769-72-0 K-116-3307-0 2-Methyl-4,4,4',4',5,5,5',5'-octakis(trifluoromethyl)spiro[
1,3,2λ5-dioxaphopholan-2,2'[1,3,2λ5]-dioxaphospholane; W   128433

710 C18 H30 O13 Zn4 56377-86-7 EP-4139-0-0 Zinc, .mu.4-oxohexakis[.mu.-(propanoato-O:O')]tetra-;; OXOHE
XAKIS(PROPIONATO)TETRAZINC(II); W/NBS   66358

710 C31 H39 F9 O8 2342-71-4 J-45-339-0 Methyl ester of cholic acid tris(trifluoroacetate); W   128434

710 C33 H42 O17 37324-44-0 NS-0-0-0 Azadirachtin, de[(E)-2-methyl-1-oxo-2-butenyl](1,2-dioxopropyl)dihy
dro-; W/NBS   128435

710 C36 H54 O14 560-53-2 O-5-588-10 Card-20(22)-enolide, 3-[(2,6-dideoxy-4-O-β-D-glucopyranosyl-3-O-
methyl-β-D-ribo-hexopyranosyl)oxy]-5,14-dihydroxy-19-oxo-, (3β,5β)-;; K-STROPHANTIN-β; W/NBS   66359

Str 1174

- 6087 -

710   C37 H46 N2 O12   22862-49-3   F-25-2204-0   o-Anisamide, N-(15-benzyl-10-sec-butyl-3-isopropyl-7,13,13-tr
imethyl-2,5,9,12,14-pentaoxo-1,4,8,11-tetraoxacyclopentadec-6-yl)-3-formamido-;; OME-NEOOXOANTIMYCIN;
W/NBS   66360

710   C39 H78 O5 Si3   OM-1981-552-0   PGF-2A ME 3TBDMS; W   66361

710   C39 H78 O5 Si3   BS-5-43-0   Prostaglandin F2α methyl ester t-butyldimethylsilyl ether; W   128436

710   C40 H46 N4 O8   25767-20-8   NS--27138-0-0   Coproporphyrin I tetramethyl ester;   NBS   66362

710   C40 H46 N4 O8   BS-4-14-0   Coproporphyrin I permethylester; W   128437

710   C42 H63 O7 P   IC-1312-0-0   Tris-(3,5-di-tert-butyl-4-hydroxyphenyl) phosphate; W   128438

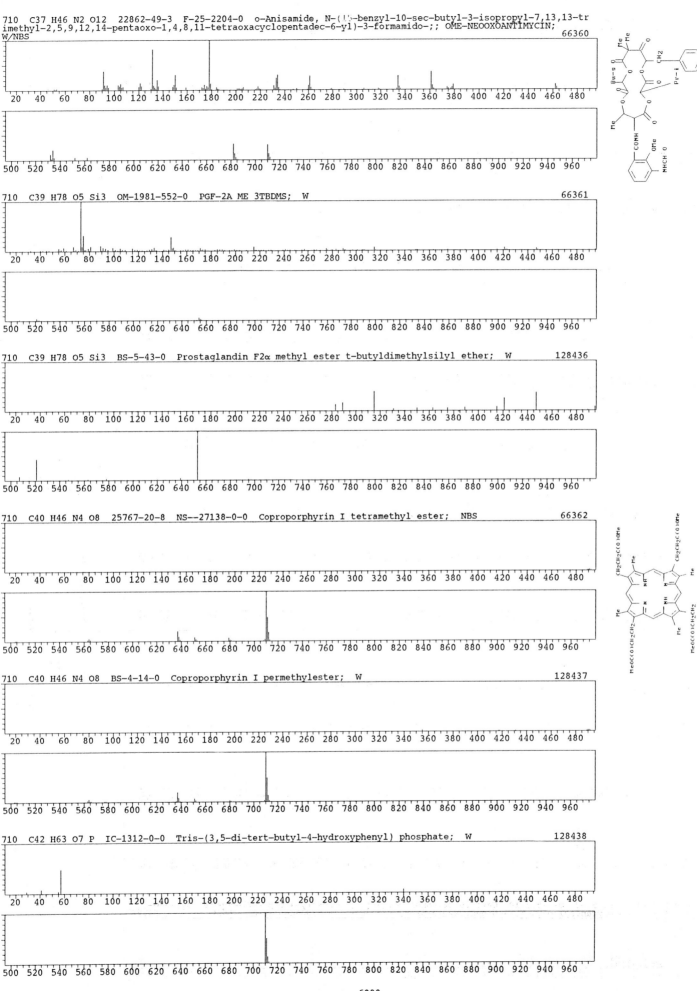

- 6088 -

710   C43 H86 O5 Si   D3-1980-178-9   T-BUTYLDIMETHYLSILYL ETHER OF 1-PALMITOYL-2-STEAROYL-RAC-GLYCEROL; W
66363

710   C43 H86 O5 Si   65615-85-2   E-50-559-3   1-STEAROYL-2-PALMITOYL-RAC-GLYCEROL TERT-BUTYL-DIMETHYLSILYL ET
HER; W
66364

CH2OSIMe2Bu-t
|
Me(CH2)16C(O)OCH2CHOC(O)(CH2)14Me

710   C43 H86 O5 Si   65615-84-1   E-50-559-3   1-PALMITOYL-2-STEAROYL-RAC-GLYCEROL-BUTYL-DIMETHYLSILYL; W
66365

CH2OSIMe2Bu-t
|
Me(CH2)14C(O)OCH2CHOC(O)(CH2)16Me

710   C44 H86 O4 S   3895-83-8   EP-2318-0-0   Butyric acid, 4,4'-thiodi-, dioctadecyl ester;; DI-STEARYL THIO-
DI-BUTYRATE; W/NBS
66366

Me(CH2)17OC(O)(CH2)3S(CH2)3C(O)O(CH2)17Me

710   C44 H86 O4 S   IC-1313-0-0   Distearyl thiodibutyrate; W
128439

710   C46 H59 Cl O4   69505-91-5   K-112-204-0   2,4,6,8,10,12,14,16-Heptadecaoctaenoic acid, 17-(3-chloro-4-
methoxyphenyl)-, 2-decyl-3-methoxy-5-pentylphenyl ester;; 2-DECYL-3-METHOXY-5-PENTYLPHENYLESTER OF 17-(3-CH
LORO-4-METHOXYPHENYL)-2,4,6,8,10,12,14,16-HEPTADECAOCTAENCARBOXYLIC ACID; W/NBS
66367

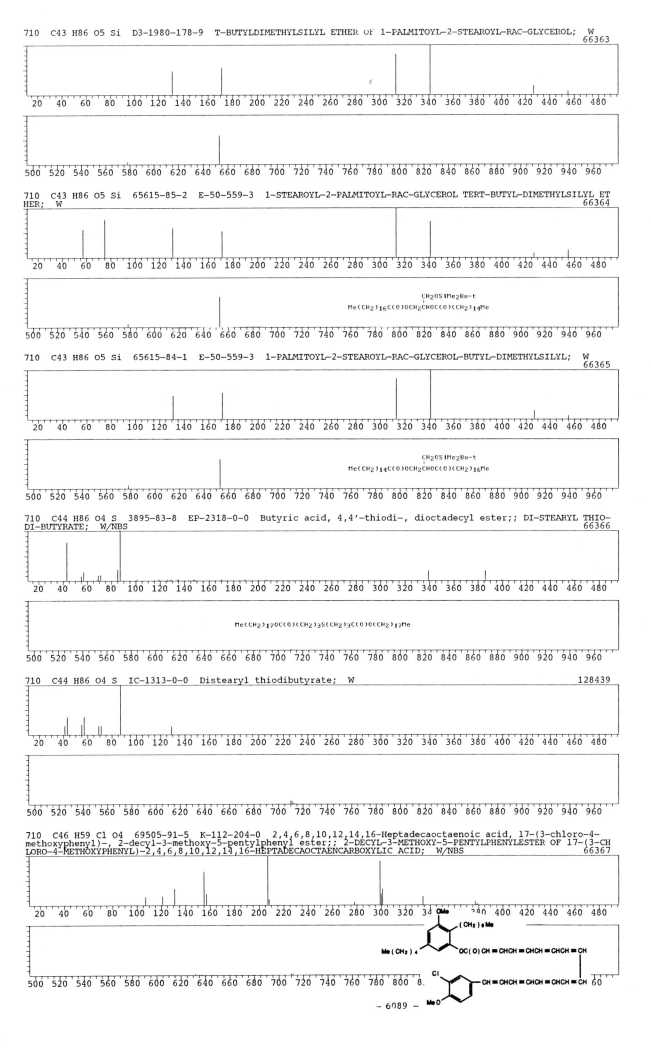

- 6089 -

710 C46 H59 Cl O4 78368-45-3 F-37-562-0 2-Dodecyl-3-methoxy-5-propylphenylester of 17-(3-Chloro-4-meth
oxyphenyl)-2,4,6,8,10,12,14,16-heptadecaoctaenoic acid; W   128440

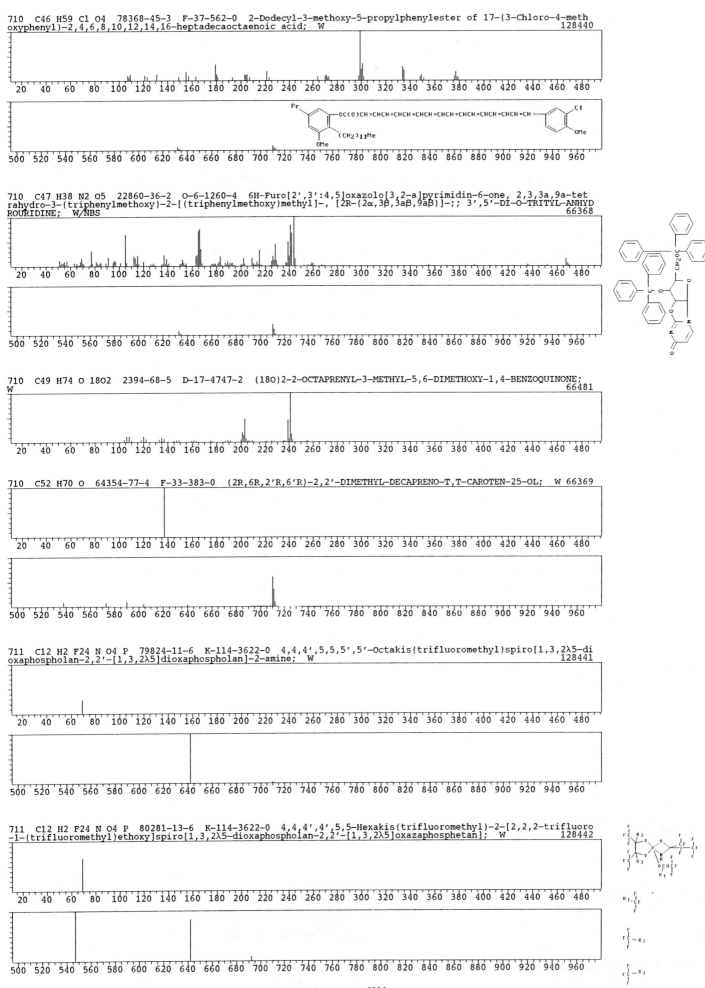

710 C47 H38 N2 O5 22860-36-2 O-6-1260-4 6H-Furo[2',3':4,5]oxazolo[3,2-a]pyrimidin-6-one, 2,3,3a,9a-tet
rahydro-3-(triphenylmethoxy)-2-[(triphenylmethoxy)methyl]-, [2R-(2α,3β,3aβ,9aβ)]-;; 3',5'-DI-O-TRITYL-ANHYD
ROURIDINE; W/NBS   66368

710 C49 H74 O 18O2 2394-68-5 D-17-4747-2 (18O)2-2-OCTAPRENYL-3-METHYL-5,6-DIMETHOXY-1,4-BENZOQUINONE;
W   66481

710 C52 H70 O 64354-77-4 F-33-383-0 (2R,6R,2'R,6'R)-2,2'-DIMETHYL-DECAPRENO-T,T-CAROTEN-25-OL; W 66369

711 C12 H2 F24 N O4 P 79824-11-6 K-114-3622-0 4,4,4',5,5,5'-Octakis(trifluoromethyl)spiro[1,3,2λ5-di
oxaphospholan-2,2'-[1,3,2λ5]dioxaphospholan]-2-amine; W   128441

711 C12 H2 F24 N O4 P 80281-13-6 K-114-3622-0 4,4,4',4',5,5-Hexakis(trifluoromethyl)-2-[2,2,2-trifluoro
-1-(trifluoromethyl)ethoxy]spiro[1,3,2λ5-dioxaphospholan-2,2'-[1,3,2λ5]oxazaphosphetan]; W   128442

- 6090 -

711   C14 H3 F18 N W   12086-82-7   AD-0-2802-0   Tungsten, (acetonitrile)tris(hexafluoro-2-butyne)-;; ACETONIT
RILO-TRIS(.PI.-HEXAFLUOROBUT-2-YNE)TUNGSTEN;   W/NBS   66370

711   C32 H41 N O17   35427-07-7   O-5-1158-1   D-Glucopyranosylamine, N-phenyl-6-O-(2,3,4,6-tetra-O-acetyl-α-
D-galactopyranosyl)-, 2,3,4-triacetate;; N-PHENYLMELIBIOSYLAMINE ACETATE;   W/NBS   66371

711   C32 H41 N O17   35573-83-2   O-5-1158-4   D-Glucopyranosylamine, N-phenyl-4-O-(2,3,4,6-tetra-O-acetyl-β-
D-galactopyranosyl)-, 2,3,6-triacetate;; N-PHENYLLACTOSYLAMINE ACETATE;   W/NBS   66372

711   C43 H57 N O6 Si   82770-12-5   C-104-5739-0   (1R,3S,5R,6R,7S,8R,11S,14R,15S,18R)-13,14-Dibenzyl-18-(2-(
trimethylsilyl)ethoxy)-6,7-dihydroxy-5,7,16,17-tetramethyl-2,10,12-reioxo-13-azatetracyclo[9.7.0.0]ocadec-
16-ene 6,7-acetonide;   W   128443

712   C12 Cl2 F12 O4 Rh2   B-31-669-0   MIXTURE OF DICARBONYL-DI-.MU.-CHLORO-BIS-3,4-DITRIFLUOROMETHYL-1-RHODI
NA-3-CYCLOPENTEN-2,5-DIONE AND TETRACARBONYL ISOMER;   W   66373

712   C29 H34 F14 O4   18072-44-1   AD-0-2747-0   5β-Pregnane-3α,20α-diol, bis(heptafluorobutyrate);; 5β-PREGN
ANE-3α,20α-DIOL DIHEPTAFLUOROBUTYRATE;   W/NBS   66374

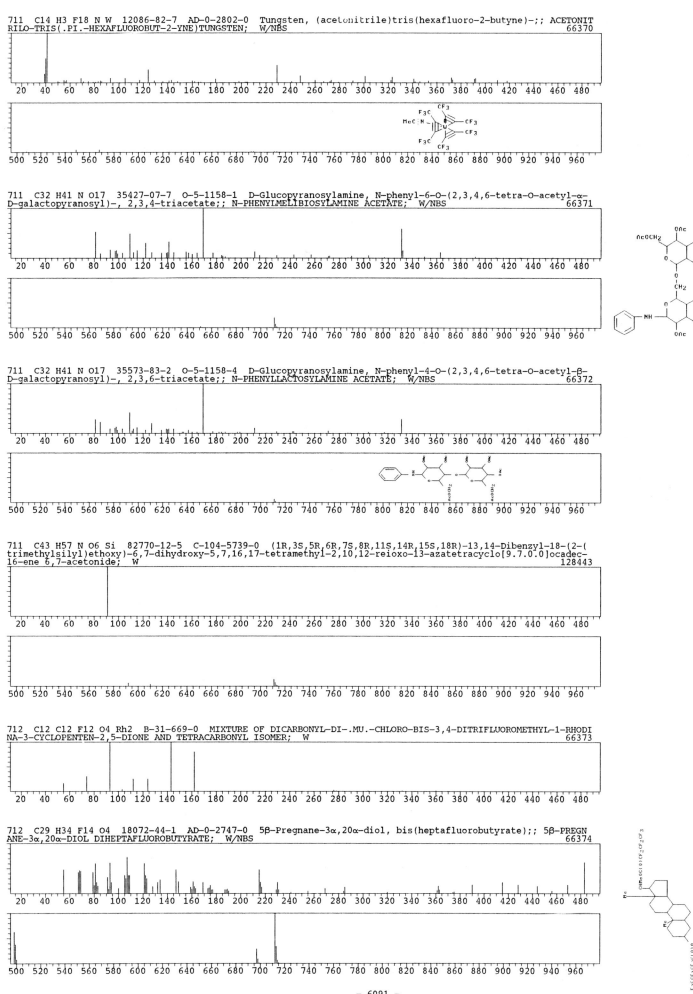

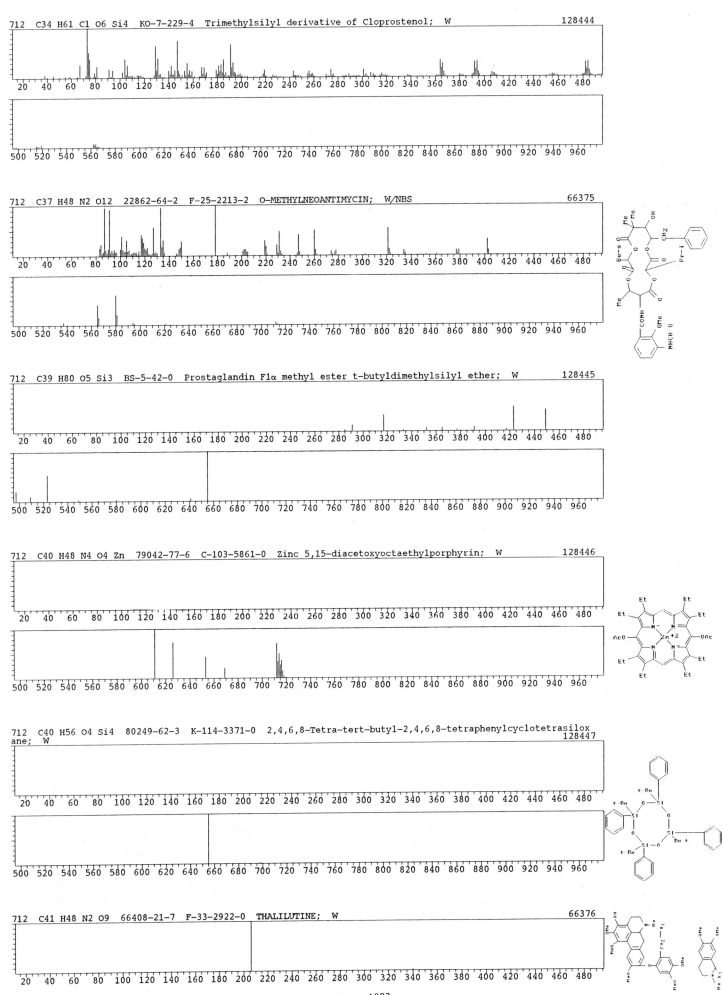

712　C34 H61 Cl O6 Si4　KO-7-229-4　Trimethylsilyl derivative of Cloprostenol;　W　　128444

712　C37 H48 N2 O12　22862-64-2　F-25-2213-2　O-METHYLNEOANTIMYCIN;　W/NBS　　66375

712　C39 H80 O5 Si3　BS-5-42-0　Prostaglandin F1α methyl ester t-butyldimethylsilyl ether;　W　　128445

712　C40 H48 N4 O4 Zn　79042-77-6　C-103-5861-0　Zinc 5,15-diacetoxyoctaethylporphyrin;　W　　128446

712　C40 H56 O4 Si4　80249-62-3　K-114-3371-0　2,4,6,8-Tetra-tert-butyl-2,4,6,8-tetraphenylcyclotetrasiloxane;　W　　128447

712　C41 H48 N2 O9　66408-21-7　F-33-2922-0　THALILUTINE;　W　　66376

712   C42 H30 Cr O4 P2   82307-59-3   K-115-1387-0   Tetracarbonyl[[(diphenylphosphino)methyl](9-fluorenyliden e)diphenylphosphorane]chromium;   W
128448

712   C43 H52 O9   65301-59-9   Y-17-645-0   Cyclic mono(2,4,6-tri-t-butylphenol) derivative of morusin;   W
128449

712   C44 H60 P4   57978-17-3   HE-1982-0-0   1,6,11,16-Tetraphosphacycloeicosane, 1,6,11,16-tetrakis(phenyl methyl)-;; 1,6,11,16-TETRAPHOSPHACYCLOEICOSAN, TETRA-P-BENZYL-;   W/NBS
66377

712   C45 H60 O7   H-60-2787-0   AGELAXANTHIN-C-DIACETATE;   W
66378

712   C45 H73 Cl O4   K-112-2009-0   MIX OF 2-DECYL-3-MEO-5-(4-ME PENTYL)PHESTER OF AND 3-MEO-2-(9-ME DECYL)- 5-PENTYL PHESTER OF 15-(3-CHLORO-4-MEOPHENYL)PENTADECANCARBOXYLIC ACID;   W
66379

712   C46 H61 Cl O4   71246-77-0   K-112-2006-0   2,4,6,8,10,12,14-Pentadecaheptaenoic acid, 15-(3-chloro-4- methoxyphenyl)-, 3-methoxy-2-(9-methyldecyl)-5-(4-methylpentyl)phenyl ester;; 3-METHOXY-2-(9-METHYLDECYL)- 5-(4-METHYLPENTYL)PHENLYESTER OF 15-(3-CHLORO-4-METHOXYPHENYL)-2,4,6,8,10,12,14-PENTADECAHEPTAENCARBOXYLIC ACID;   W/NBS
66380

712   C52 H72 O   64354-76-3   F-33-383-0   (2R,6R,2'R,6'R)-2,2'-DIMETHYLDECAPRENO-ε,ε-CAROTEN-25-OL;   W   66381

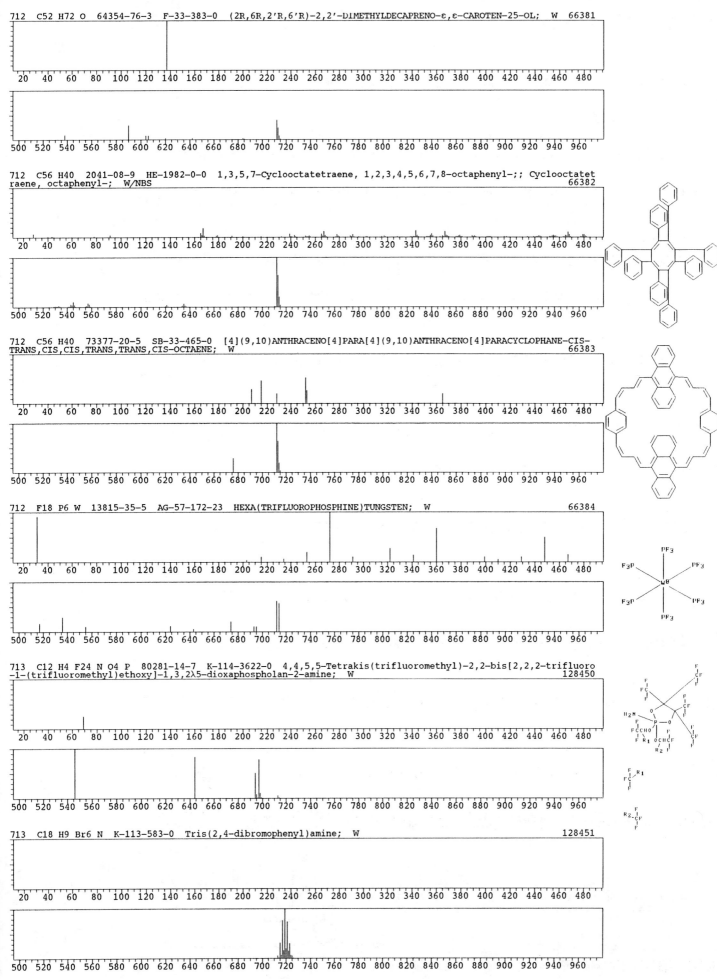

712   C56 H40   2041-08-9   HE-1982-0-0   1,3,5,7-Cyclooctatetraene, 1,2,3,4,5,6,7,8-octaphenyl-;; Cyclooctatet
raene, octaphenyl-;   W/NBS   66382

712   C56 H40   73377-20-5   SB-33-465-0   [4](9,10)ANTHRACENO[4]PARA[4](9,10)ANTHRACENO[4]PARACYCLOPHANE-CIS-
TRANS,CIS,CIS,TRANS,TRANS,CIS-OCTAENE;   W   66383

712   F18 P6 W   13815-35-5   AG-57-172-23   HEXA(TRIFLUOROPHOSPHINE)TUNGSTEN;   W   66384

713   C12 H4 F24 N O4 P   80281-14-7   K-114-3622-0   4,4,5,5-Tetrakis(trifluoromethyl)-2,2-bis[2,2,2-trifluoro
-1-(trifluoromethyl)ethoxy]-1,3,2λ5-dioxaphospholan-2-amine;   W   128450

713   C18 H9 Br6 N   K-113-583-0   Tris(2,4-dibromophenyl)amine;   W   128451

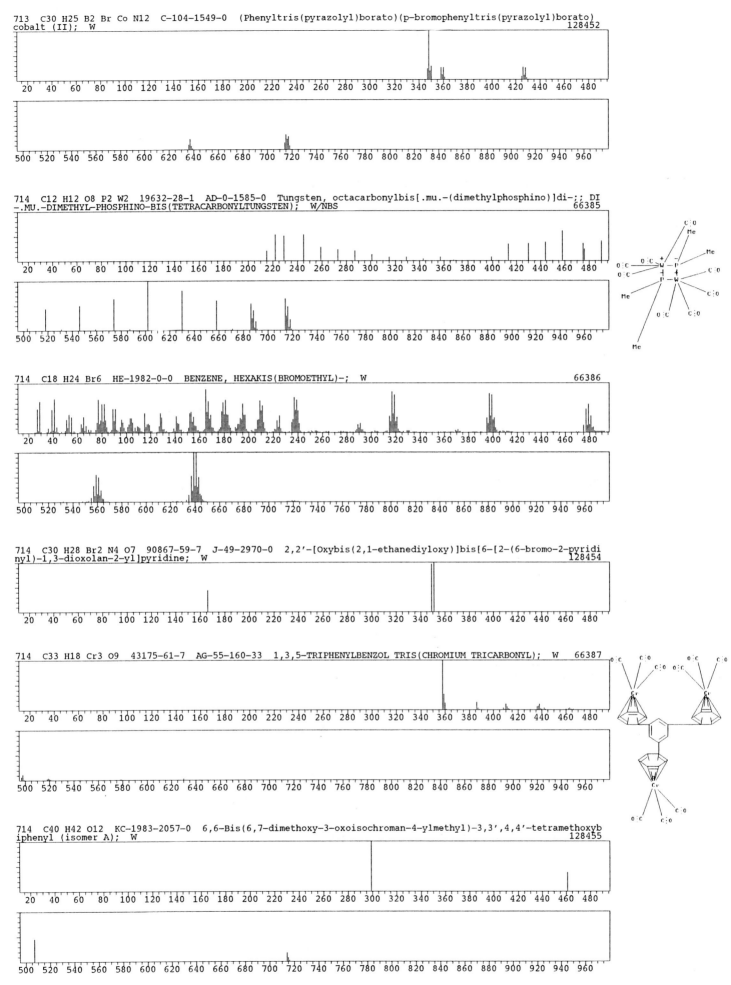

713 C30 H25 B2 Br Co N12 C-104-1549-0 (Phenyltris(pyrazolyl)borato)(p-bromophenyltris(pyrazolyl)borato) cobalt (II); W   128452

714 C12 H12 O8 P2 W2 19632-28-1 AD-0-1585-0 Tungsten, octacarbonylbis[.mu.-(dimethylphosphino)]di-;; DI -.MU.-DIMETHYL-PHOSPHINO-BIS(TETRACARBONYLTUNGSTEN); W/NBS   66385

714 C18 H24 Br6 HE-1982-0-0 BENZENE, HEXAKIS(BROMOETHYL)-; W   66386

714 C30 H28 Br2 N4 O7 90867-59-7 J-49-2970-0 2,2'-[Oxybis(2,1-ethanediyloxy)]bis[6-[2-(6-bromo-2-pyridi nyl)-1,3-dioxolan-2-yl]pyridine; W   128454

714 C33 H18 Cr3 O9 43175-61-7 AG-55-160-33 1,3,5-TRIPHENYLBENZOL TRIS(CHROMIUM TRICARBONYL); W   66387

714 C40 H42 O12 KC-1983-2057-0 6,6-Bis(6,7-dimethoxy-3-oxoisochroman-4-ylmethyl)-3,3',4,4'-tetramethoxyb iphenyl (isomer A); W   128455

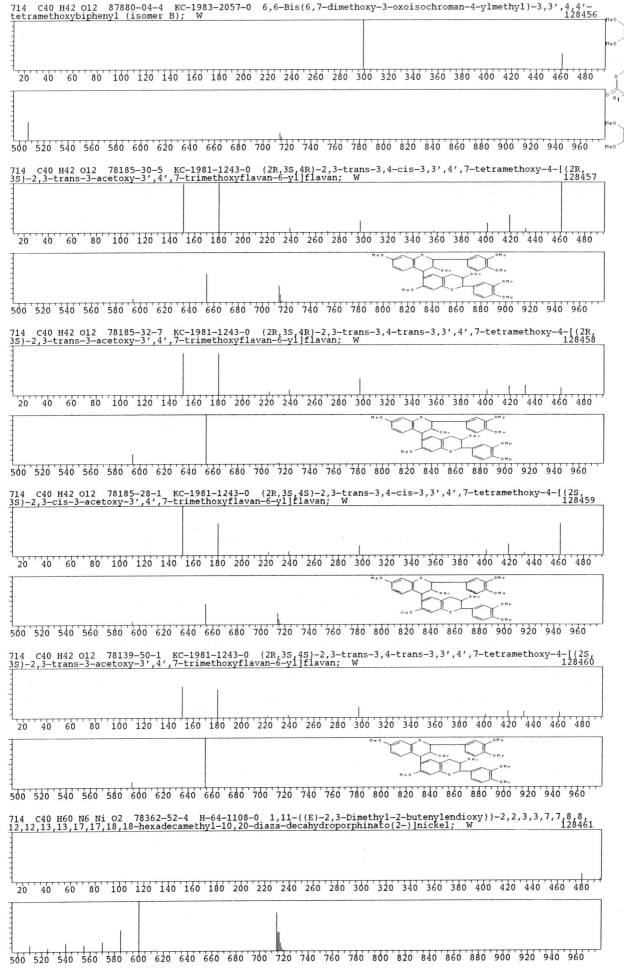

714 C40 H42 O12 87880-04-4 KC-1983-2057-0 6,6-Bis(6,7-dimethoxy-3-oxoisochroman-4-ylmethyl)-3,3',4,4'-tetramethoxybiphenyl (isomer B); W 128456

714 C40 H42 O12 78185-30-5 KC-1981-1243-0 (2R,3S,4R)-2,3-trans-3,4-cis-3,3',4',7-tetramethoxy-4-[(2R,3S)-2,3-trans-3-acetoxy-3',4',7-trimethoxyflavan-6-yl]flavan; W 128457

714 C40 H42 O12 78185-32-7 KC-1981-1243-0 (2R,3S,4R)-2,3-trans-3,4-trans-3,3',4',7-tetramethoxy-4-[(2R,3S)-2,3-trans-3-acetoxy-3',4',7-trimethoxyflavan-6-yl]flavan; W 128458

714 C40 H42 O12 78185-28-1 KC-1981-1243-0 (2R,3S,4S)-2,3-trans-3,4-cis-3,3',4',7-tetramethoxy-4-[(2S,3S)-2,3-cis-3-acetoxy-3',4',7-trimethoxyflavan-6-yl]flavan; W 128459

714 C40 H42 O12 78139-50-1 KC-1981-1243-0 (2R,3S,4S)-2,3-trans-3,4-trans-3,3',4',7-tetramethoxy-4-[(2S,3S)-2,3-trans-3-acetoxy-3',4',7-trimethoxyflavan-6-yl]flavan; W 128460

714 C40 H60 N6 Ni O2 78362-52-4 H-64-1108-0 1,11-((E)-2,3-Dimethyl-2-butenylendioxy))-2,2,3,3,7,7,8,8,12,12,13,13,17,17,18,18-hexadecamethyl-10,20-diaza-decahydroporphinato(2-)]nickel; W 128461

714   C42 H50 O10   80395-05-7   H-64-2249-0   (7S)-12-Acetoxy-7,7-[6',14'-dihydroxy-7'-oxo-5',8',11',13'-abiet
atetraen-11',12'-ylen]dioxy-8,12-abietadien-6,11,14-trione;   W   128462

714   C43 H46 N4 O6   J-46-3502-0   Bis(methyl ester) of 2-(and 4-)[1-(2-acetoxy-5-methylphenyl)ethyl]deuterop
orphyrin;   W   128463

714   C44 H46 N10   61327-89-7   B-29-1857-0   2,5-Di[N-phenyl-N-(phenylazoethyl)aminomethyl]-1,4-diphenylhexa
hydro-s-tetrazine;   W   128464

714   C44 H58 S4   C-104-7145-0   1,1,4,4-Tetrakis(p-tert-butylphenyl)thio)butane;   W   128453

714   C45 H33 O6 Sc   15133-54-7   T-68-5959-0   Scandium, tris(1,3-diphenyl-1,3-propanedionato)-;; TRIS(1,3-DI
PHENYL-1,3-PROPANEDIONATO)SCANDIUM(III);   W/NBS   66388

714   C48 H54 Si3   80584-59-4   C-104-1151-0   Tetrakis-1,1,2,2-(2,6-dimethylphenyl)disilane;   W   128465

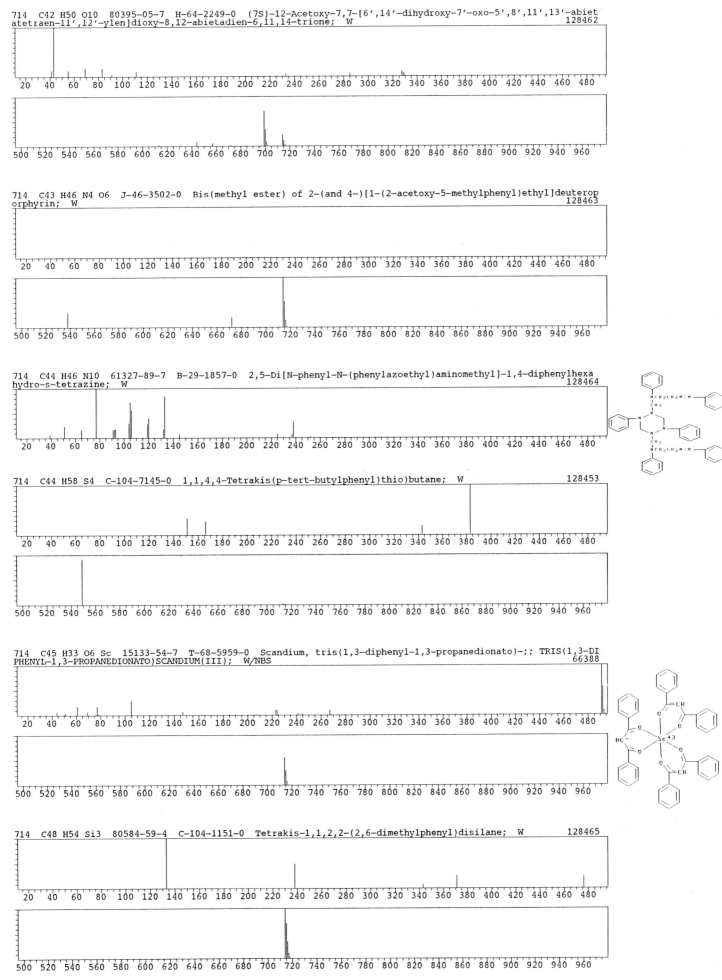

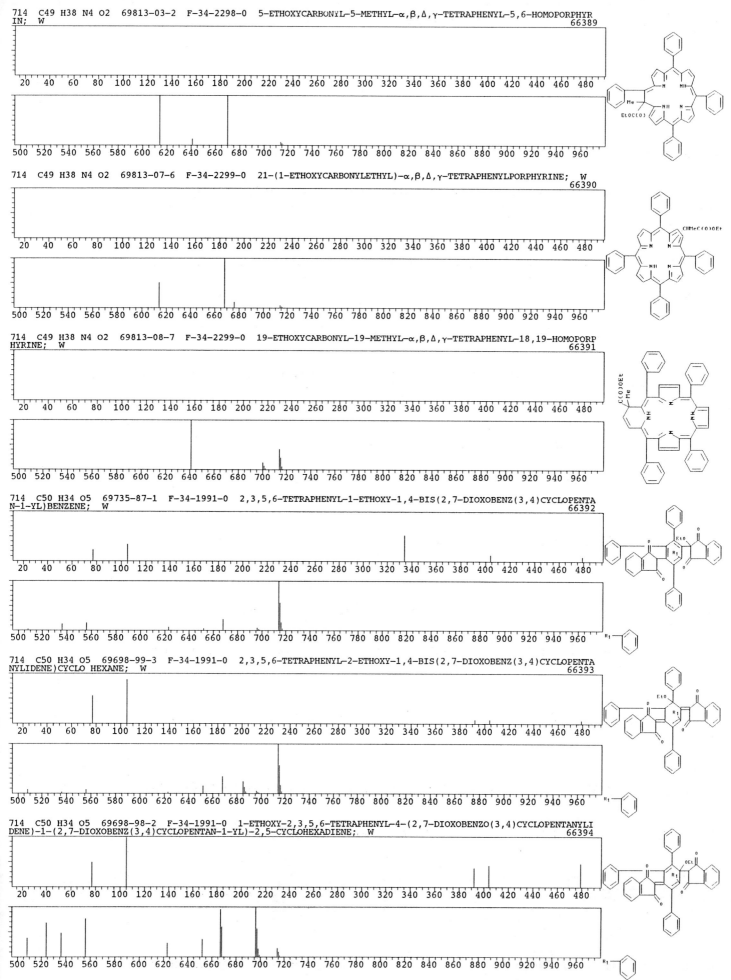

714   C49 H38 N4 O2   69813-03-2   F-34-2298-0   5-ETHOXYCARBONYL-5-METHYL-α,β,Δ,γ-TETRAPHENYL-5,6-HOMOPORPHYR
IN;   W
66389

714   C49 H38 N4 O2   69813-07-6   F-34-2299-0   21-(1-ETHOXYCARBONYLETHYL)-α,β,Δ,γ-TETRAPHENYLPORPHYRINE;   W
66390

714   C49 H38 N4 O2   69813-08-7   F-34-2299-0   19-ETHOXYCARBONYL-19-METHYL-α,β,Δ,γ-TETRAPHENYL-18,19-HOMOPORP
HYRINE;   W
66391

714   C50 H34 O5   69735-87-1   F-34-1991-0   2,3,5,6-TETRAPHENYL-1-ETHOXY-1,4-BIS(2,7-DIOXOBENZ(3,4)CYCLOPENTA
N-1-YL)BENZENE;   W
66392

714   C50 H34 O5   69698-99-3   F-34-1991-0   2,3,5,6-TETRAPHENYL-2-ETHOXY-1,4-BIS(2,7-DIOXOBENZ(3,4)CYCLOPENTA
NYLIDENE)CYCLO HEXANE;   W
66393

714   C50 H34 O5   69698-98-2   F-34-1991-0   1-ETHOXY-2,3,5,6-TETRAPHENYL-4-(2,7-DIOXOBENZO(3,4)CYCLOPENTANYLI
DENE)-1-(2,7-DIOXOBENZ(3,4)CYCLOPENTAN-1-YL)-2,5-CYCLOHEXADIENE;   W
66394

715 C28 H17 F12 N5 O4 77145-01-8 F-36-2510-0 3,6-Bis-(1,3-bis(trifluoroacetyl)-4-imidazolin-2-yl)9-eth
ylcarbazole; W   128466

715 C32 H61 N O16 BS-2-105-0 Permethylated alditol derivative of α-D-mannopyranoside-(1-3)-β-D-mannopyra
noside-(1-4)-2-acetamido-2-deoxy-D-glucose; W   128467

715 C44 H28 N4 Rh KO-3-75-3 RHODIUM α,β,γ,Δ-TETRAPHENYLPORPHYRIN; W   66395

716 C24 H64 Ni2 P8 HE-1986-951-0 TETRAKIS[BIS(1,2-DIMETHYLPHOSPHINOETHANE)]BIS-NICKEL; W   128468

716 C30 H32 Cl2 F6 N10 GC-13-10-3 Triazine N-methyl derivative of chlorohexidine; W   128469

716 C37 H60 O8 Si3 65929-51-3 KO-4-191-1 TRIS(TRIMETHYLSILYL) CANNABINOL GLUCURONIDE METHYL ESTER; W
                                                                                     66396

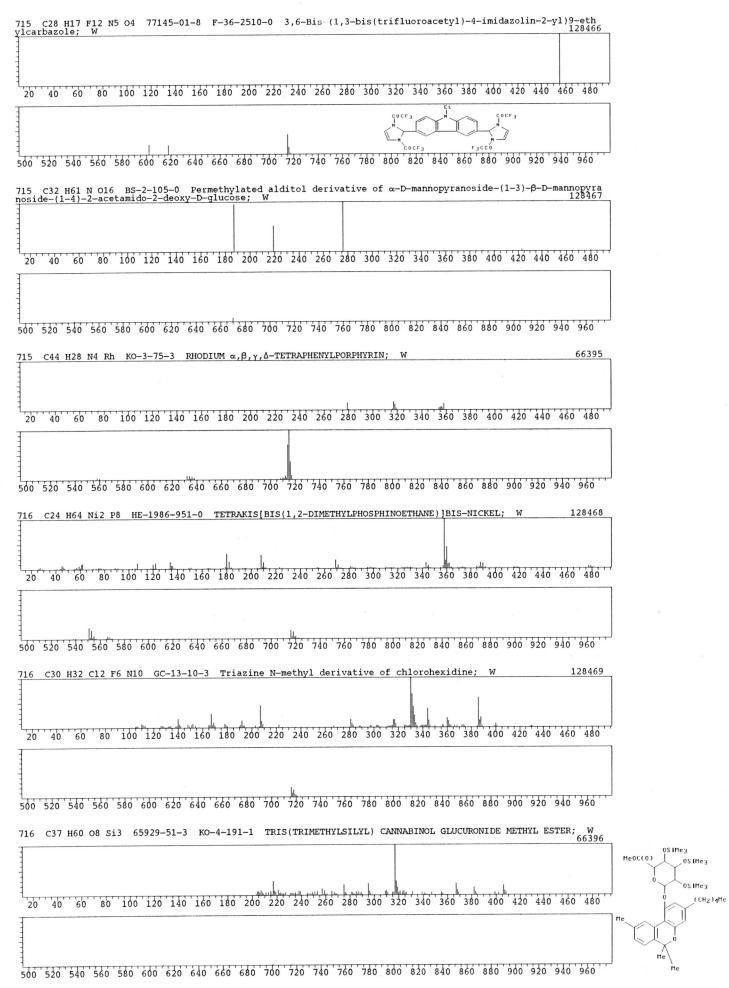

716   C37 H60 O8 Si3   KO-4-369-5   ME-TMS DERIVATIVE OF CBN GLUCURONIDE;   W   66397

716   C40 H44 O12   58912-59-7   B-29-207-0   Benzyl 3-methoxy-2',4,4'-tri-O-acetylisophysodate;   W   128470

716   C41 H64 O10   6907-84-2   KC-1976-1360-0   CHOLEST-5-EN-3β-YL TETRA-O-ACETYL-β-O-GLUCOPYRANOSIDE;   W
                                                                                                            66399

Str 1175

716   C43 H48 N4 O6   31148-60-4   H-60-2848-0   AMATAINE;   W   66400

716   C43 H48 N4 O6   50924-02-2   AC-1975-2504-1   3-Oxovobtusine;   W   128471

716   C43 H72 O4 S2   63693-27-6   KC-1977-6-656   4,7-DIACETOXY-6-(3,7,11,15,19,23-HEXAMETHYLTETRACOSYL)-5-
METHYLTHIO-4,7-DIHYDROBENZO(B)THIOPHENE;   W   66401

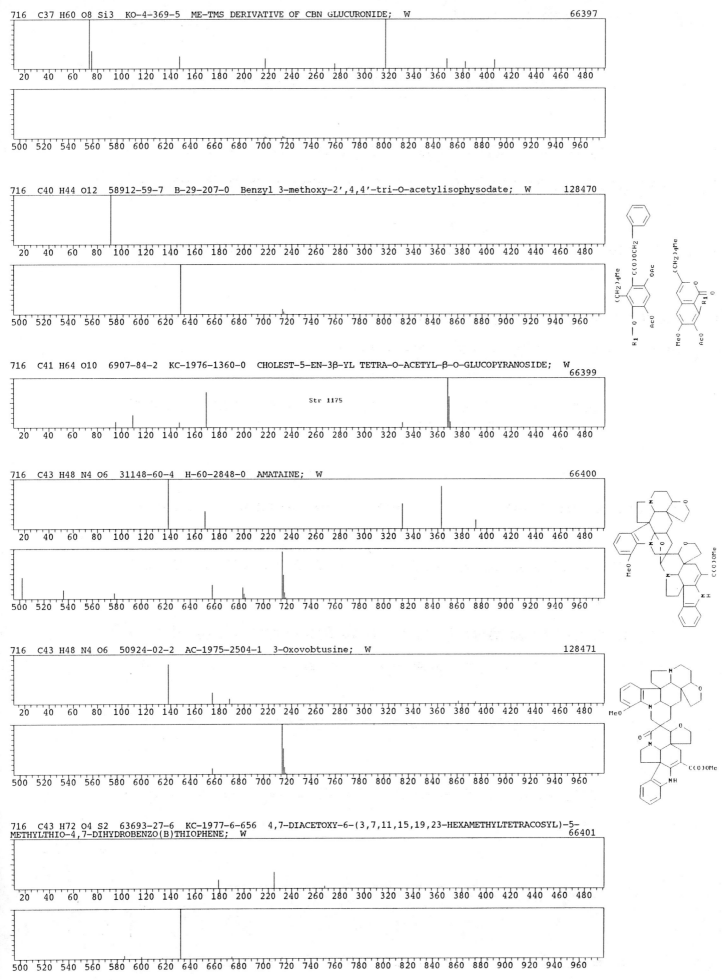

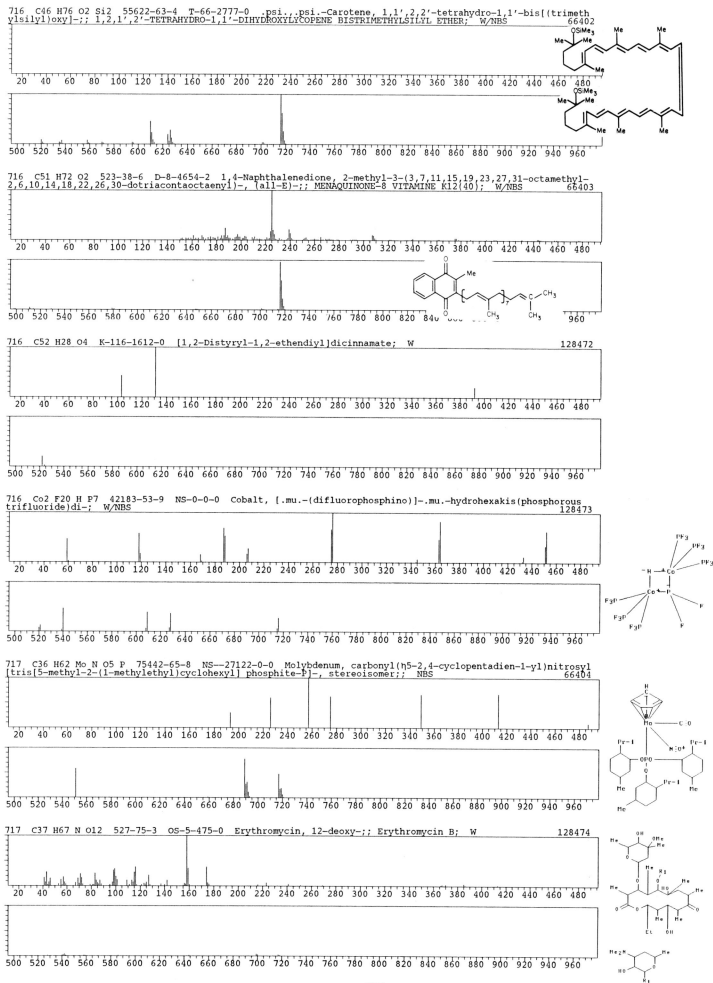

716 C46 H76 O2 Si2 55622-63-4 T-66-2777-0 .psi.,.psi.-Carotene, 1,1',2,2'-tetrahydro-1,1'-bis[(trimeth
ylsilyl)oxy]-;; 1,2,1',2'-TETRAHYDRO-1,1'-DIHYDROXYLYCOPENE BISTRIMETHYLSILYL ETHER; W/NBS
66402

716 C51 H72 O2 523-38-6 D-8-4654-2 1,4-Naphthalenedione, 2-methyl-3-(3,7,11,15,19,23,27,31-octamethyl-
2,6,10,14,18,22,26,30-dotriacontaoctaenyl)-, (all-E)-;; MENAQUINONE-8 VITAMINE K12(40); W/NBS
66403

716 C52 H28 O4 K-116-1612-0 [1,2-Distyryl-1,2-ethendiyl]dicinnamate; W
128472

716 Co2 F20 H P7 42183-53-9 NS-0-0-0 Cobalt, [.mu.-(difluorophosphino)]-.mu.-hydrohexakis(phosphorous
trifluoride)di-; W/NBS
128473

717 C36 H62 Mo N O5 P 75442-65-8 NS--27122-0-0 Molybdenum, carbonyl(η5-2,4-cyclopentadien-1-yl)nitrosyl
[tris[5-methyl-2-(1-methylethyl)cyclohexyl] phosphite-P]-, stereoisomer;; NBS
66404

717 C37 H67 N O12 527-75-3 OS-5-475-0 Erythromycin, 12-deoxy-;; Erythromycin B; W
128474

717   C38 H63 N5 O8   BS-7-153-4   N-Carbobenzoxyleucyl-leucyl-leucyl-leucyl-leucine;   W   128475

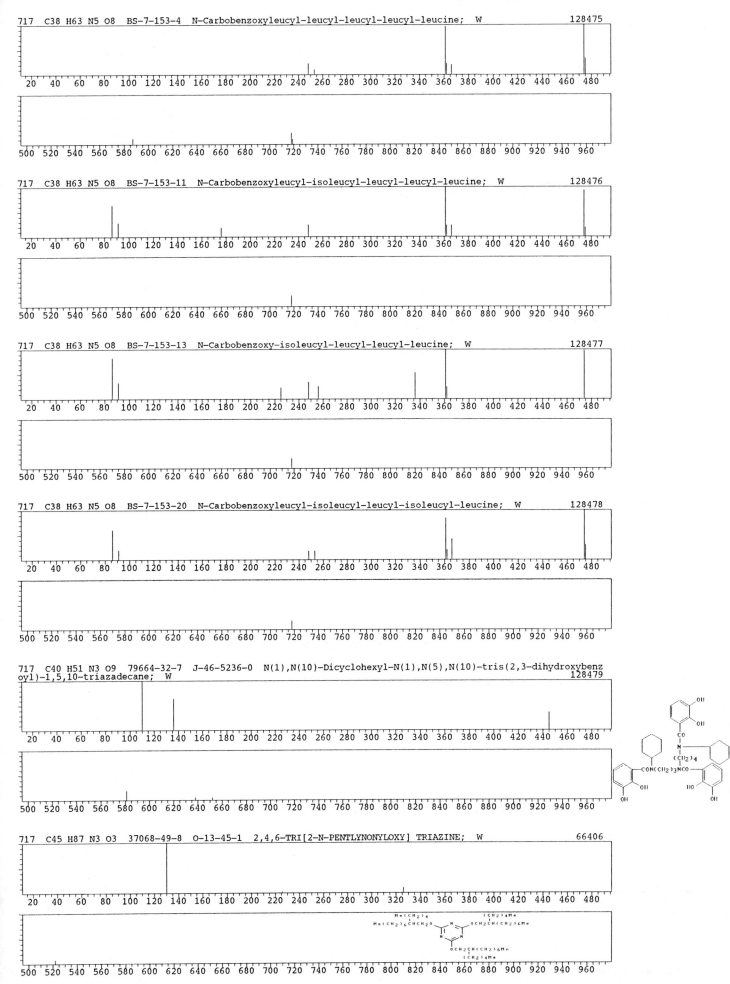

717   C38 H63 N5 O8   BS-7-153-11   N-Carbobenzoxyleucyl-isoleucyl-leucyl-leucyl-leucine;   W   128476

717   C38 H63 N5 O8   BS-7-153-13   N-Carbobenzoxy-isoleucyl-leucyl-leucyl-leucine;   W   128477

717   C38 H63 N5 O8   BS-7-153-20   N-Carbobenzoxyleucyl-isoleucyl-leucyl-isoleucyl-leucine;   W   128478

717   C40 H51 N3 O9   79664-32-7   J-46-5236-0   N(1),N(10)-Dicyclohexyl-N(1),N(5),N(10)-tris(2,3-dihydroxybenz
oyl)-1,5,10-triazadecane;   W   128479

717   C45 H87 N3 O3   37068-49-8   O-13-45-1   2,4,6-TRI[2-N-PENTLYNONYLOXY] TRIAZINE;   W   66406

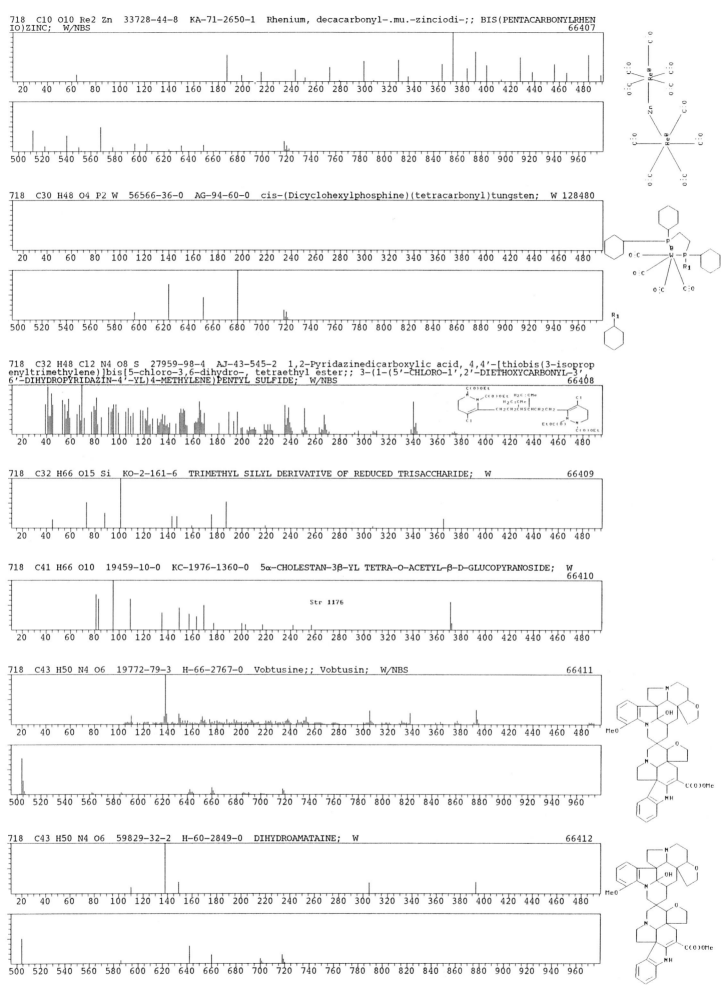

718   C10 O10 Re2 Zn   33728-44-8   KA-71-2650-1   Rhenium, decacarbonyl-.mu.-zinciodi-;; BIS(PENTACARBONYLRHEN
IO)ZINC; W/NBS   66407

718   C30 H48 O4 P2 W   56566-36-0   AG-94-60-0   cis-(Dicyclohexylphosphine)(tetracarbonyl)tungsten; W 128480

718   C32 H48 Cl2 N4 O8 S   27959-98-4   AJ-43-545-2   1,2-Pyridazinedicarboxylic acid, 4,4'-[thiobis(3-isoprop
enyltrimethylene)]bis[5-chloro-3,6-dihydro-, tetraethyl ester;; 3-(1-(5'-CHLORO-1',2'-DIETHOXYCARBONYL-3'
6'-DIHYDROPYRIDAZIN-4'-YL)4-METHYLENE)PENTYL SULFIDE; W/NBS   66408

718   C32 H66 O15 Si   KO-2-161-6   TRIMETHYL SILYL DERIVATIVE OF REDUCED TRISACCHARIDE;   W   66409

718   C41 H66 O10   19459-10-0   KC-1976-1360-0   5α-CHOLESTAN-3β-YL TETRA-O-ACETYL-β-D-GLUCOPYRANOSIDE; W
                                                                                                         66410
Str 1176

718   C43 H50 N4 O6   19772-79-3   H-66-2767-0   Vobtusine;; Vobtusin; W/NBS   66411

718   C43 H50 N4 O6   59829-32-2   H-60-2849-0   DIHYDROAMATAINE; W   66412

718 C43 H49 D N4 O6 65967-11-5 H-60-2849-0 8-DEUTERIO-DIHYDROAMATAINE; W   66415

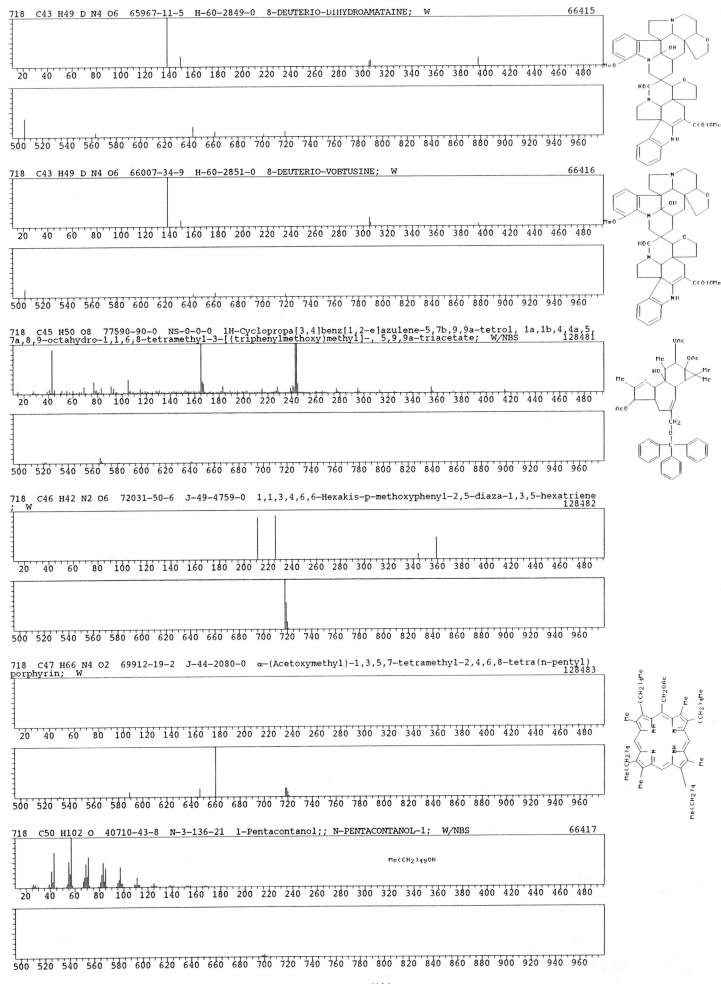

718 C43 H49 D N4 O6 66007-34-9 H-60-2851-0 8-DEUTERIO-VOBTUSINE; W   66416

718 C45 H50 O8 77590-90-0 NS-0-0-0 1H-Cyclopropa[3,4]benz[1,2-e]azulene-5,7b,9,9a-tetrol, 1a,1b,4,4a,5,
7a,8,9-octahydro-1,1,6,8-tetramethyl-3-[(triphenylmethoxy)methyl]-, 5,9,9a-triacetate; W/NBS   128481

718 C46 H42 N2 O6 72031-50-6 J-49-4759-0 1,1,3,4,6,6-Hexakis-p-methoxyphenyl-2,5-diaza-1,3,5-hexatriene
; W   128482

718 C47 H66 N4 O2 69912-19-2 J-44-2080-0 α-(Acetoxymethyl)-1,3,5,7-tetramethyl-2,4,6,8-tetra(n-pentyl)
porphyrin; W   128483

718 C50 H102 O 40710-43-8 N-3-136-21 1-Pentacontanol;; N-PENTACONTANOL-1; W/NBS   66417

Me(CH2)49OH

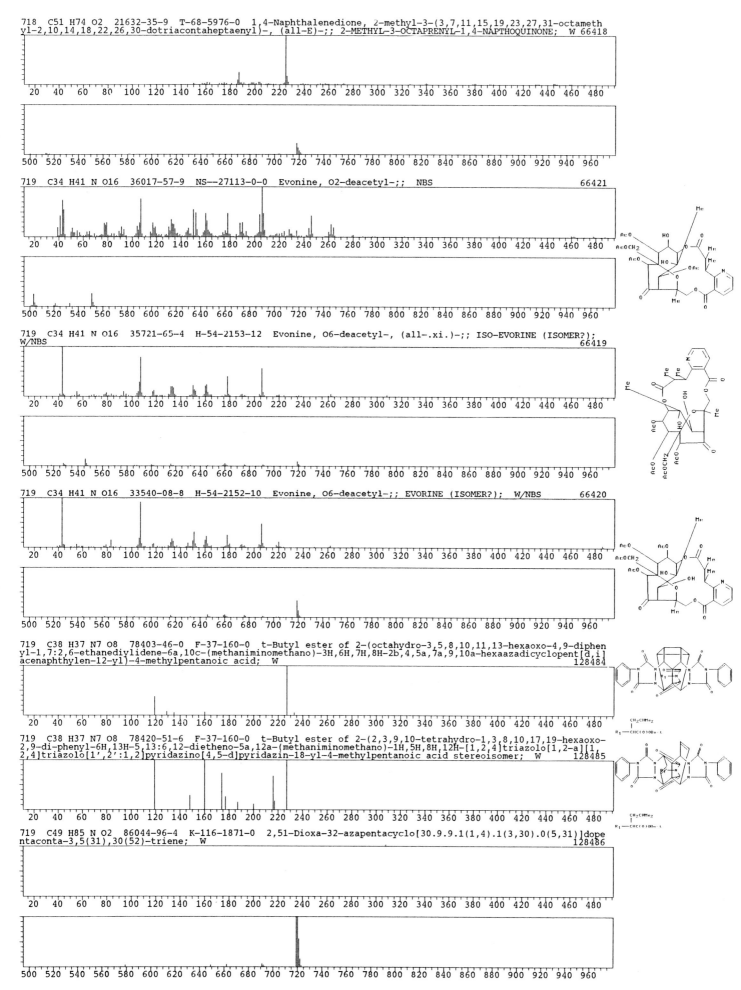

718   C51 H74 O2   21632-35-9   T-68-5976-0   1,4-Naphthalenedione, 2-methyl-3-(3,7,11,15,19,23,27,31-octameth
yl-2,10,14,18,22,26,30-dotriacontaheptaenyl)-, (all-E)-;; 2-METHYL-3-OCTAPRENYL-1,4-NAPTHOQUINONE;   W 66418

719   C34 H41 N O16   36017-57-9   NS--27113-0-0   Evonine, O2-deacetyl-;;   NBS   66421

719   C34 H41 N O16   35721-65-4   H-54-2153-12   Evonine, O6-deacetyl-, (all-.xi.)-;; ISO-EVORINE (ISOMER?);
W/NBS   66419

719   C34 H41 N O16   33540-08-8   H-54-2152-10   Evonine, O6-deacetyl-;; EVORINE (ISOMER?);   W/NBS   66420

719   C38 H37 N7 O8   78403-46-0   F-37-160-0   t-Butyl ester of 2-(octahydro-3,5,8,10,11,13-hexaoxo-4,9-diphen
yl-1,7:2,6-ethanediylidene-6a,10c-(methaniminomethano)-3H,6H,7H,8H-2b,4,5a,7a,9,10a-hexaazadicyclopent[d,i]
acenaphthylen-12-yl)-4-methylpentanoic acid;   W   128484

719   C38 H37 N7 O8   78420-51-6   F-37-160-0   t-Butyl ester of 2-(2,3,9,10-tetrahydro-1,3,8,10,17,19-hexaoxo-
2,9-di-phenyl-6H,13H-5,13:6,12-dietheno-5a,12a-(methaniminomethano)-1H,5H,8H,12H-[1,2,4]triazolo[1,2-a][1,
2,4]triazolo[1',2':1,2]pyridazino[4,5-d]pyridazin-18-yl-4-methylpentanoic acid stereoisomer;   W   128485

719   C49 H85 N O2   86044-96-4   K-116-1871-0   2,51-Dioxa-32-azapentacyclo[30.9.9.1(1,4).1(3,30).0(5,31)]dope
ntaconta-3,5(31),30(52)-triene;   W   128486

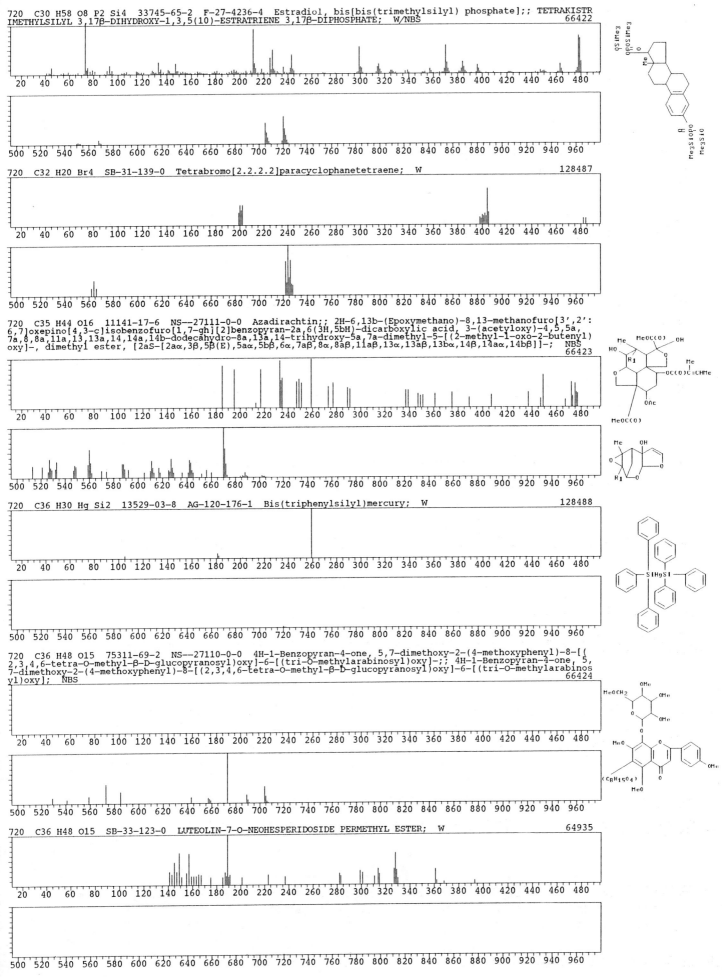

720 C30 H58 O8 P2 Si4 33745-65-2 F-27-4236-4 Estradiol, bis[bis(trimethylsilyl) phosphate];; TETRAKISTR
IMETHYLSILYL 3,17β-DIHYDROXY-1,3,5(10)-ESTRATRIENE 3,17β-DIPHOSPHATE; W/NBS   66422

720 C32 H20 Br4 SB-31-139-0 Tetrabromo[2.2.2.2]paracyclophanetetraene; W   128487

720 C35 H44 O16 11141-17-6 NS--27111-0-0 Azadirachtin;; 2H-6,13b-(Epoxymethano)-8,13-methanofuro[3',2':
6,7]oxepino[4,3-c]isobenzofuro[1,7-gh][2]benzopyran-2a,6(3H,5bH)-dicarboxylic acid, 3-(acetyloxy)-4,5,5a,
7a,8,8a,11a,13,13a,14,14a,14b-dodecahydro-8a,13a,14-trihydroxy-5a,7a-dimethyl-5-[(2-methyl-1-oxo-2-butenyl)
oxy]-, dimethyl ester, [2aS-[2aα,3β,5β(E),5aα,5bβ,6α,7aβ,8α,8aβ,11aβ,13α,13aβ,13bα,14β,14aα,14bβ]]-; NBS
                                                                                                 66423

720 C36 H30 Hg Si2 13529-03-8 AG-120-176-1 Bis(triphenylsilyl)mercury; W   128488

720 C36 H48 O15 75311-69-2 NS--27110-0-0 4H-1-Benzopyran-4-one, 5,7-dimethoxy-2-(4-methoxyphenyl)-8-[(
2,3,4,6-tetra-O-methyl-β-D-glucopyranosyl)oxy]-6-[(tri-O-methylarabinosyl)oxy]-;; 4H-1-Benzopyran-4-one, 5,
7-dimethoxy-2-(4-methoxyphenyl)-8-[(2,3,4,6-tetra-O-methyl-β-D-glucopyranosyl)oxy]-6-[(tri-O-methylarabinos
yl)oxy]; NBS   66424

720 C36 H48 O15 SB-33-123-0 LUTEOLIN-7-O-NEOHESPERIDOSIDE PERMETHYL ESTER; W   64935

720   C36 H56 N4 O7 S2   96624-95-2   H-67-2183-0   12-Nitro-N-[4-tosyl-8-(tosylamino)-4-azaoctyl]-15-pentadeca
nlactam;   W   128489

720   C37 H64 O8 Si3   65929-54-6   KO-4-193-6   TRIS(TRIMETHYLSILYL)Δ9-TETRAHYDROCANNABINOL GLUCURONIDE METHYL
ESTER;   W   66425

720   C37 H64 O8 Si3   65929-53-5   KO-4-193-8   TRIS(TRIMETHYLSILYL)Δ8-TETRAHYDROCANNABINOL GLUCURONIDE METHYL
ESTER;   W   66426

720   C38 H68 O7 Si3   56009-08-6   O-7-148-3   Card-20(22)-enolide, 3-[[2,6-dideoxy-3,4-bis-O-(trimethylsilyl)
-ribo-hexopyranosyl]oxy]-14-[(trimethylsilyl)oxy]-, (3β,5β)-;; TRISTRIMETHYLSILYL DIGITOXIGENIN MONODIGITOX
OSIDE;   W/NBS   66427

720   C38 H68 O7 Si3   57289-27-7   O7-0-141-0   Card-20(22)-enolide, 3-[[2,6-dideoxy-3,4-bis-O-(trimethylsily
l)-D-ribo-hexopyranosyl]oxy]-14-[(trimethylsilyl)oxy]-, (3β,5β)-;; DIGITOXIGENIN MONODIGITOXOSIDE TMS3;
W/NBS   66428

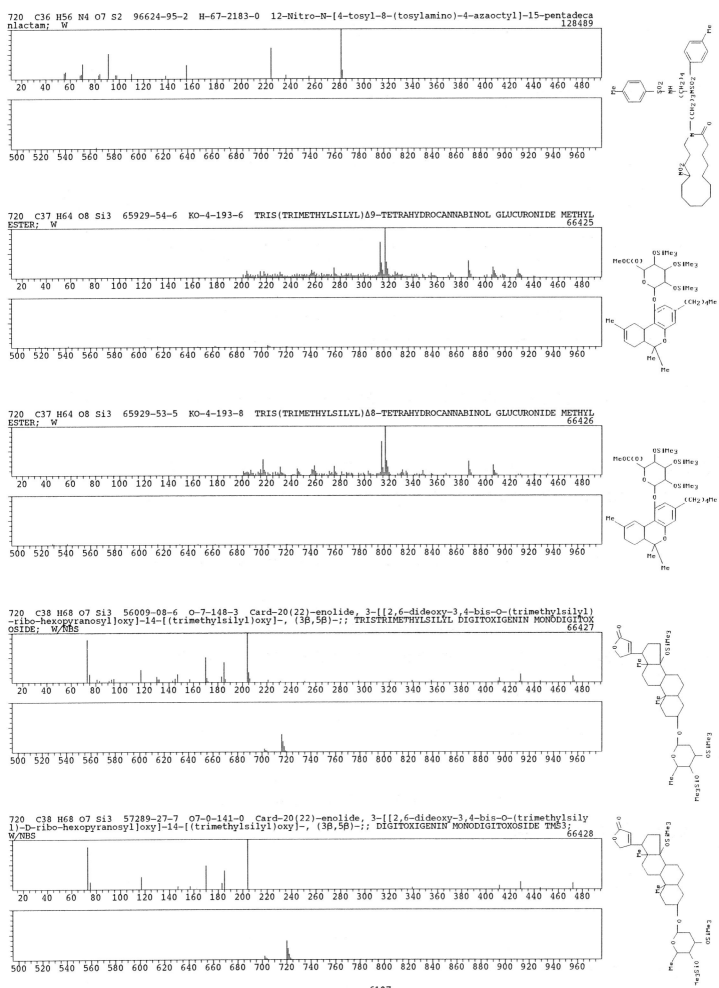

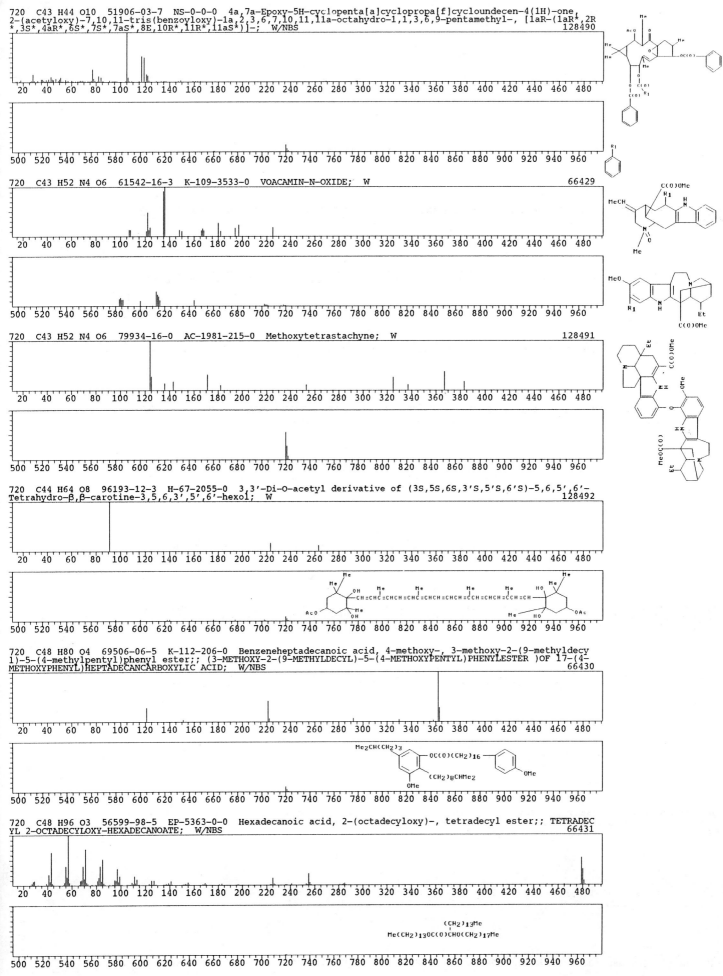

720   C43 H44 O10   51906-03-7   NS-0-0-0   4a,7a-Epoxy-5H-cyclopenta[a]cyclopropa[f]cycloundecen-4(1H)-one, 2-(acetyloxy)-7,10,11-tris(benzoyloxy)-1a,2,3,6,7,10,11,11a-octahydro-1,1,3,6,9-pentamethyl-, [1aR-(1aR*,2R *,3S*,4aR*,6S*,7S*,7aS*,8E,10R*,11R*,11aS*)]-;   W/NBS
128490

720   C43 H52 N4 O6   61542-16-3   K-109-3533-0   VOACAMIN-N-OXIDE;   W
66429

720   C43 H52 N4 O6   79934-16-0   AC-1981-215-0   Methoxytetrastachyne;   W
128491

720   C44 H64 O8   96193-12-3   H-67-2055-0   3,3'-Di-O-acetyl derivative of (3S,5S,6S,3'S,5'S,6'S)-5,6,5',6'-Tetrahydro-β,β-carotine-3,5,6,3',5',6'-hexol;   W
128492

720   C48 H80 O4   69506-06-5   K-112-206-0   Benzeneheptadecanoic acid, 4-methoxy-, 3-methoxy-2-(9-methyldecy l)-5-(4-methylpentyl)phenyl ester;; (3-METHOXY-2-(9-METHYLDECYL)-5-(4-METHOXYPENTYL)PHENYLESTER)OF 17-(4-METHOXYPHENYL)HEPTADECANCARBOXYLIC ACID;   W/NBS
66430

720   C48 H96 O3   56599-98-5   EP-5363-0-0   Hexadecanoic acid, 2-(octadecyloxy)-, tetradecyl ester;; TETRADEC YL 2-OCTADECYLOXY-HEXADECANOATE;   W/NBS
66431

720   C51 H94 N   K-116-1872-0   Dimethyl ester of 4-(1-azacycloeicosanyl)-3,5-tetracosamethyllenbrenzacatequin;   W
128493

721   C25 H64 N O9 P Si6   55530-74-0   BA-0-327-0   D-Fructose, 1,3,4,5-tetrakis-O-(trimethylsilyl)-, O-methyl oxime, 6-[bis(trimethylsilyl) phosphate];; O-HEXAKIS(TRIMETHYLSILYL)-D-FRUCTOSE-6-PHOSPHATE METHYLOXIME; W/NBS
66432

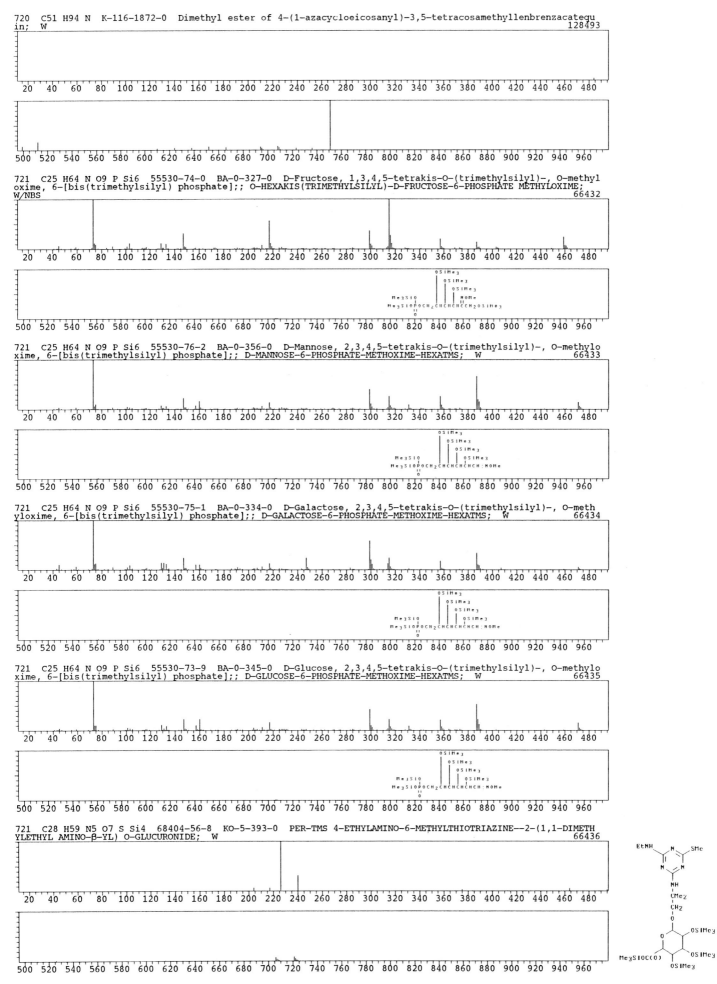

```
                                             OSiMe3
                                          OSiMe3
                                       OSiMe3
                                    HOMe
                        Me3SiO                ||
                        Me3SiOPOCH2CHCHCHCCH2OSiMe3
                              ||
                              O
```

721   C25 H64 N O9 P Si6   55530-76-2   BA-0-356-0   D-Mannose, 2,3,4,5-tetrakis-O-(trimethylsilyl)-, O-methyloxime, 6-[bis(trimethylsilyl) phosphate];; D-MANNOSE-6-PHOSPHATE-METHOXIME-HEXATMS;   W
66433

```
                                          OSiMe3
                                       OSiMe3
                                    OSiMe3
                        Me3SiO          OSiMe3
                        Me3SiOPOCH2CHCHCHCH:NOMe
                              ||
                              O
```

721   C25 H64 N O9 P Si6   55530-75-1   BA-0-334-0   D-Galactose, 2,3,4,5-tetrakis-O-(trimethylsilyl)-, O-methyloxime, 6-[bis(trimethylsilyl) phosphate];; D-GALACTOSE-6-PHOSPHATE-METHOXIME-HEXATMS;   W
66434

```
                                          OSiMe3
                                       OSiMe3
                                    OSiMe3
                        Me3SiO          OSiMe3
                        Me3SiOPOCH2CHCHCHCH:NOMe
                              ||
                              O
```

721   C25 H64 N O9 P Si6   55530-73-9   BA-0-345-0   D-Glucose, 2,3,4,5-tetrakis-O-(trimethylsilyl)-, O-methyloxime, 6-[bis(trimethylsilyl) phosphate];; D-GLUCOSE-6-PHOSPHATE-METHOXIME-HEXATMS;   W
66435

```
                                          OSiMe3
                                       OSiMe3
                                    OSiMe3
                        Me3SiO          OSiMe3
                        Me3SiOPOCH2CHCHCHCH:NOMe
                              ||
                              O
```

721   C28 H59 N5 O7 S Si4   68404-56-8   KO-5-393-0   PER-TMS 4-ETHYLAMINO-6-METHYLTHIOTRIAZINE--2-(1,1-DIMETHYLETHYL AMINO-β-YL) O-GLUCURONIDE;   W
66436

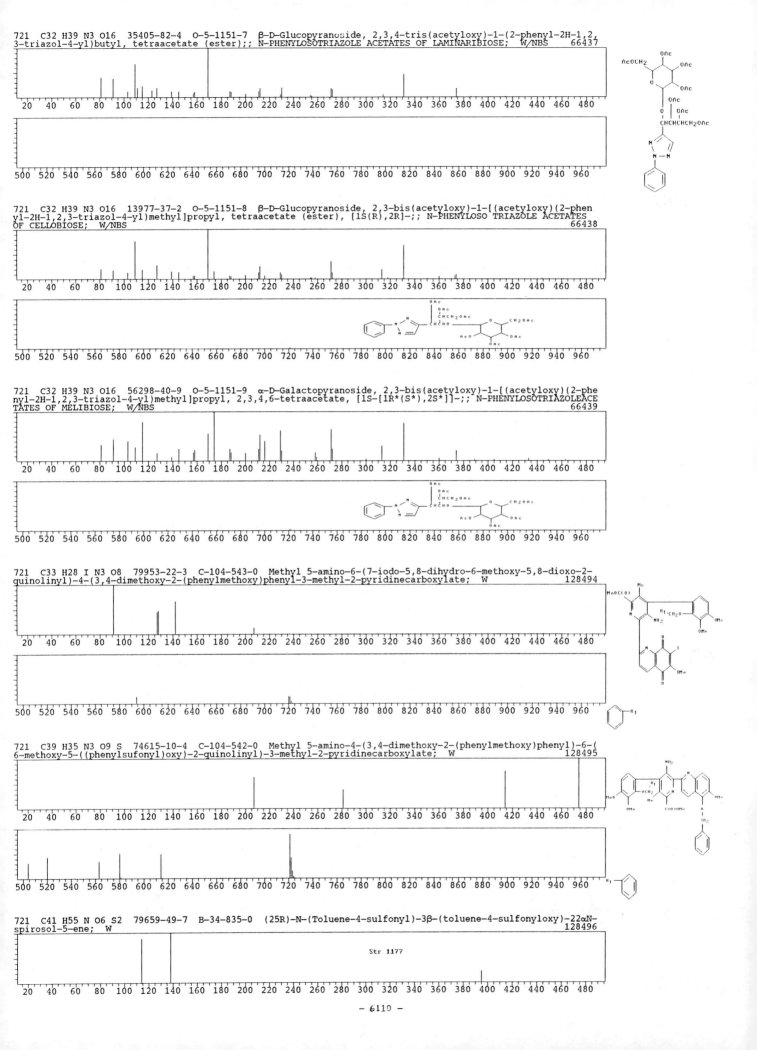

721 C32 H39 N3 O16 35405-82-4 O-5-1151-7 β-D-Glucopyranoside, 2,3,4-tris(acetyloxy)-1-(2-phenyl-2H-1,2,3-triazol-4-yl)butyl, tetraacetate (ester);; N-PHENYLOSOTRIAZOLE ACETATES OF LAMINARIBIOSE; W/NBS   66437

721 C32 H39 N3 O16 13977-37-2 O-5-1151-8 β-D-Glucopyranoside, 2,3-bis(acetyloxy)-1-[(acetyloxy)(2-phenyl-2H-1,2,3-triazol-4-yl)methyl]propyl, tetraacetate (ester), [1S(R),2R]-;; N-PHENYLOSO TRIAZOLE ACETATES OF CELLOBIOSE; W/NBS   66438

721 C32 H39 N3 O16 56298-40-9 O-5-1151-9 α-D-Galactopyranoside, 2,3-bis(acetyloxy)-1-[(acetyloxy)(2-phenyl-2H-1,2,3-triazol-4-yl)methyl]propyl, 2,3,4,6-tetraacetate, [1S-[1R*(S*),2S*]]-;; N-PHENYLOSOTRIAZOLEACETATES OF MELIBIOSE; W/NBS   66439

721 C33 H28 I N3 O8 79953-22-3 C-104-543-0 Methyl 5-amino-6-(7-iodo-5,8-dihydro-6-methoxy-5,8-dioxo-2-quinolinyl)-4-(3,4-dimethoxy-2-(phenylmethoxy)phenyl-3-methyl-2-pyridinecarboxylate; W   128494

721 C39 H35 N3 O9 S 74615-10-4 C-104-542-0 Methyl 5-amino-4-(3,4-dimethoxy-2-(phenylmethoxy)phenyl)-6-(6-methoxy-5-((phenylsufonyl)oxy)-2-quinolinyl)-3-methyl-2-pyridinecarboxylate; W   128495

721 C41 H55 N O6 S2 79659-49-7 B-34-835-0 (25R)-N-(Toluene-4-sulfonyl)-3β-(toluene-4-sulfonyloxy)-22αN-spirosol-5-ene; W   128496

Str 1177

- 6110 -

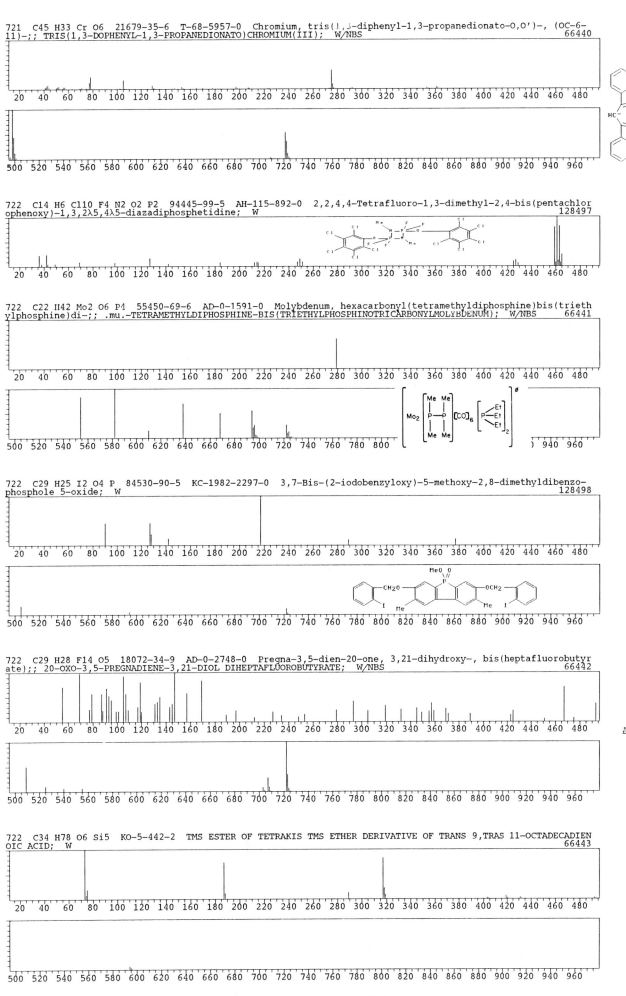

721 C45 H33 Cr O6 21679-35-6 T-68-5957-0 Chromium, tris(1,3-diphenyl-1,3-propanedionato-O,O')-, (OC-6-11)-;; TRIS(1,3-DOPHENYL-1,3-PROPANEDIONATO)CHROMIUM(III); W/NBS   66440

722 C14 H6 Cl10 F4 N2 O2 P2 94445-99-5 AH-115-892-0 2,2,4,4-Tetrafluoro-1,3-dimethyl-2,4-bis(pentachlorophenoxy)-1,3,2λ5,4λ5-diazadiphosphetidine; W   128497

722 C22 H42 Mo2 O6 P4 55450-69-6 AD-0-1591-0 Molybdenum, hexacarbonyl(tetramethyldiphosphine)bis(triethylphosphine)di-;; .mu.-TETRAMETHYLDIPHOSPHINE-BIS(TRIETHYLPHOSPHINOTRICARBONYLMOLYBDENUM); W/NBS   66441

722 C29 H25 I2 O4 P 84530-90-5 KC-1982-2297-0 3,7-Bis-(2-iodobenzyloxy)-5-methoxy-2,8-dimethyldibenzophosphole 5-oxide; W   128498

722 C29 H28 F14 O5 18072-34-9 AD-0-2748-0 Pregna-3,5-dien-20-one, 3,21-dihydroxy-, bis(heptafluorobutyrate);; 20-OXO-3,5-PREGNADIENE-3,21-DIOL DIHEPTAFLUOROBUTYRATE; W/NBS   66442

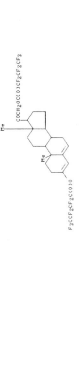

722 C34 H78 O6 Si5 KO-5-442-2 TMS ESTER OF TETRAKIS TMS ETHER DERIVATIVE OF TRANS 9,TRAS 11-OCTADECADIENOIC ACID; W   66443

722   C35 H46 O16   37324-45-1   NS-0-0-0   Azadirachtin, 22,23-dihydro-;   W/NBS   128499

722   C37 H54 O14   60492-16-2   KC-1976-1360-0   3β,17β-DIACETOXY-5α-ANDROSTAN-6α-YL TETRA-O-ACETYL-β-D-GLUCOP
YRANOSIDE; W   66444

722   C40 H68 O7 P2   38005-61-7   AE-248-2759-5   Prelycopersene pyrophosphate;   W/NBS   66445

CH:CMeCH₂CH₂CH:CMeCH₂CH₂CH:CMeCH₂CH₂CH:CMe₂
              |
              Me
HO[P(O)(OH)O]₂CH₂   CH₂CH₂CH:CMeCH₂CH₂CH:CMeCH₂CH₂CH:CMe₂

722   C43 H46 O10   77495-84-2   NS-0-0-0   4a,7a-Epoxy-5H-cyclopenta[a]cyclopropa[f]cycloundecene-2,4,7,10,11-
pentol, 1,1a,2,3,4,6,7,10,11,11a-decahydro-1,1,3,6,9-pentamethyl-, 2-acetate 7,10,11-tribenzoate;   W/NBS
          128500

722   C45 H86 O6   555-45-3   EP-5339-0-0   Tetradecanoic acid, 1,2,3-propanetriyl ester;; GLYCERYL TRITETRADEC
ANOATE;   W/NBS   66446

OC(O)(CH₂)₁₂Me
|
Me(CH₂)₁₂C(O)OCH₂CHCH₂OC(O)(CH₂)₁₂Me

722   C45 H85 D O6   55256-14-9   HO-0-1650-0   2-DEUTERIOGLYCERYL 1-HEXANOATE-2,3-DIOCTADECANOATE;   W   66447

OC(O)(CH₂)₁₆Me
|
Me(CH₂)₁₆C(O)OCH₂CDCH₂OC(O)(CH₂)₄Me

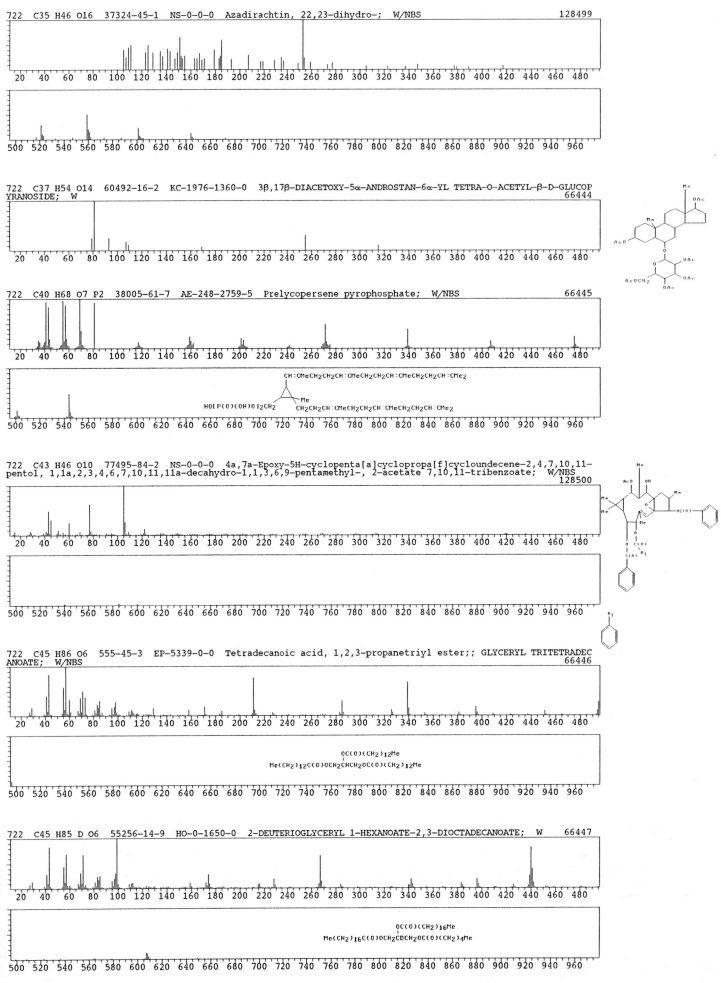

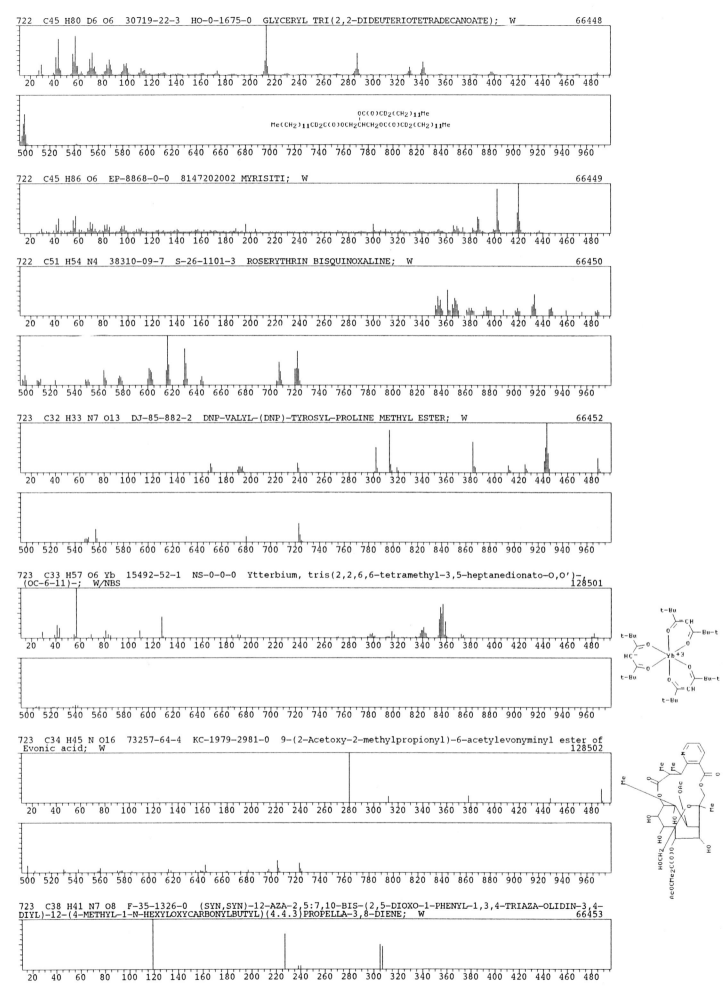

722   C45 H80 D6 O6   30719-22-3   HO-0-1675-0   GLYCERYL TRI(2,2-DIDEUTERIOTETRADECANOATE);   W   66448

OC(O)CD₂(CH₂)₁₁Me
|
Me(CH₂)₁₁CD₂C(O)OCH₂CHCH₂OC(O)CD₂(CH₂)₁₁Me

722   C45 H86 O6   EP-8868-0-0   8147202002 MYRISITI;   W   66449

722   C51 H54 N4   38310-09-7   S-26-1101-3   ROSERYTHRIN BISQUINOXALINE;   W   66450

723   C32 H33 N7 O13   DJ-85-882-2   DNP-VALYL-(DNP)-TYROSYL-PROLINE METHYL ESTER;   W   66452

723   C33 H57 O6 Yb   15492-52-1   NS-0-0-0   Ytterbium, tris(2,2,6,6-tetramethyl-3,5-heptanedionato-O,O')-,
(OC-6-11)-;   W/NBS   128501

723   C34 H45 N O16   73257-64-4   KC-1979-2981-0   9-(2-Acetoxy-2-methylpropionyl)-6-acetylevonyminyl ester of
Evonic acid;   W   128502

723   C38 H41 N7 O8   F-35-1326-0   (SYN,SYN)-12-AZA-2,5:7,10-BIS-(2,5-DIOXO-1-PHENYL-1,3,4-TRIAZA-OLIDIN-3,4-
DIYL)-12-(4-METHYL-1-N-HEXYLOXYCARBONYLBUTYL)(4.4.3)PROPELLA-3,8-DIENE;   W   66453

724 C22 H15 F18 Rh 68867-94-7 B-31-1950-0 (η-PENTAMETHYLCYCLOPENTADIENYL)(HEXAKIS(TRIFLUOROMETHYL)BENZYL)RHODIUM; W   66454

724 C30 H68 Co O9 P3 HE-1986-1310-0 HAPTO-3-ALLYL-TRIS(TRIISOPROPYLPHOSPHITE)-COBALT; W   128503

724 C33 H40 O18 KC-1977-13-1600 7-GLUCOSYLOXY-1,6-DIHYDROXY-3,5-DIMETHOXYXANTHONE,HEXAACETATE; W 66455

724 C38 H34 N8 Ni O4 KA-1982-2539-0 Nickel phthalocyanin complex F; W   128504

724 C39 H76 O6 Si3 56053-01-1 SD-1981-0-0 Ergostane-5,25-diol, 3,6,12-tris[(trimethylsilyl)oxy]-, 25-acetate, (3β,5α,6β,12β)-;; W/NBS   66456

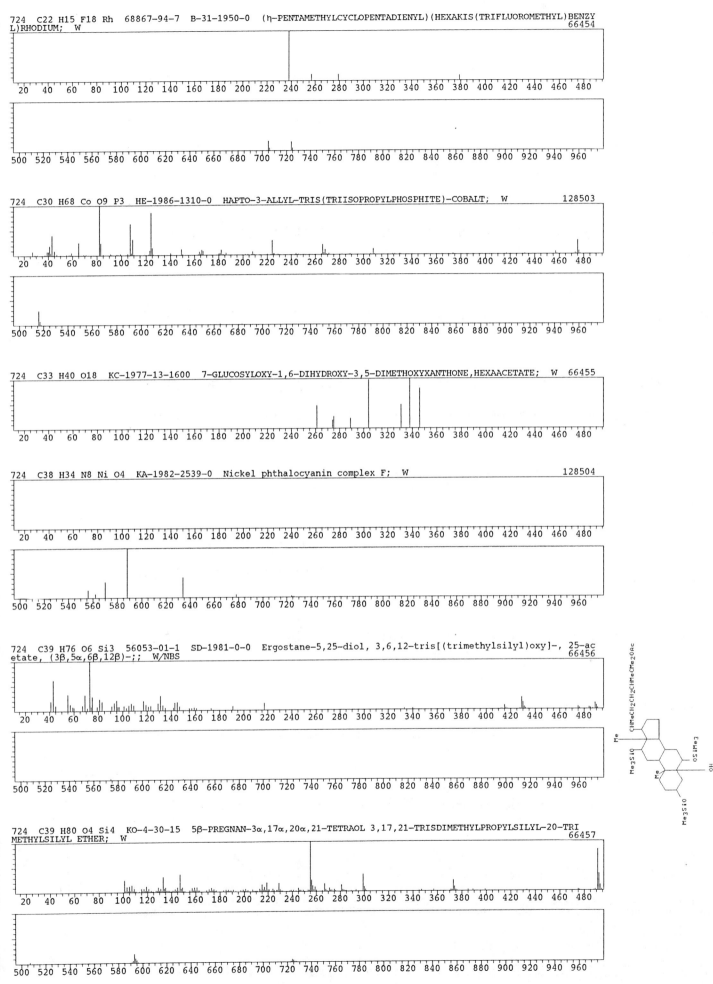

724 C39 H80 O4 Si4 KO-4-30-15 5β-PREGNAN-3α,17α,20α,21-TETRAOL 3,17,21-TRISDIMETHYLPROPYLSILYL-20-TRIMETHYLSILYL ETHER; W   66457

724   C39 H80 O4 Si4   D3-1980-1042-33   5β-CHOLESTANE-3α,7α,12α,25-TETROL TMS ETHER;   W   66458

724   C40 H68 O11   28380-24-7   W-17-461-16   Nigericin;; X 464;   W/NBS   66459

724   C42 H28 S6   C-104-7147-0   1,1,1,6,6,6-Hexakis(phenylthio)hexane;   W   128505

724   C46 H76 O6   55401-92-8   SD-1981-0-0   Lup-20(29)-en-21-ol, 3,28-bis[(tetrahydro-2H-pyran-2-yl)oxy]-, 3,
3-dimethylbutanoate, (3β,21β)-;;   W/NBS   66460

724   C47 H61 Cl O4   69505-92-6   K-112-204-0   MIXTURE OF 2-DECYL-5-ISOHEXYL- AND 2-ISOUNDECYL-5-PENTYL-3-MEO
-PHENYL17-(3-CL-4-MEO-PHENYL)-2,4,6,8,10,12,14,16-HEPTADECANOCTAENCARBOXYLATES;   W   66461

724   C47 H61 Cl O4   F-37-562-0   2-Dodecyl-3-methoxy-5-(2-methylpropyl)phenylester of 17-(3-Chloro-4-methoxy
phenyl)-2,4,6,8,10,12,14,16-heptadecaoctaenoic acid;   W   128506

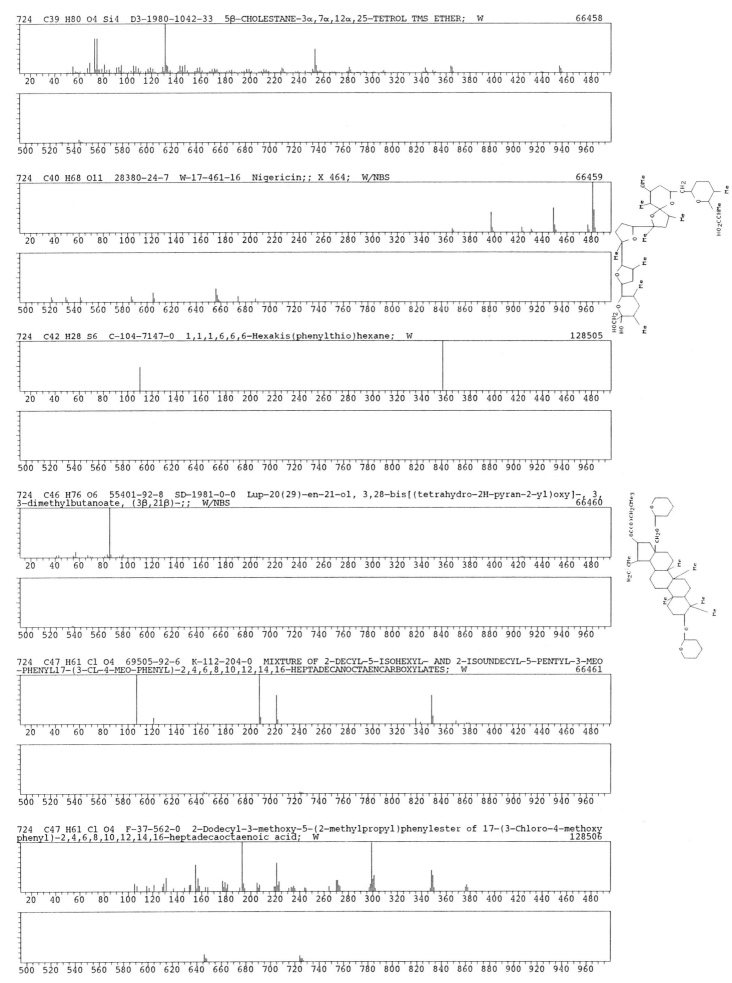

724 C48 H84 O4 B-28-170-0 OLEAN-12-ENE-3β,16β,28-TRIOL 3β-HEPTADECANOATE; W   66462

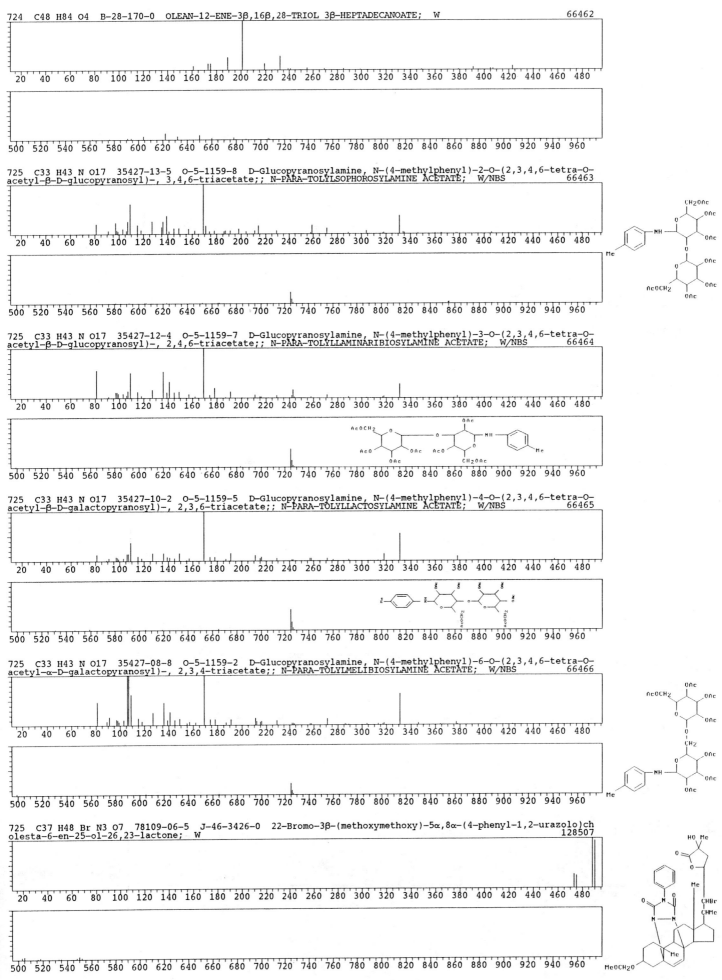

725 C33 H43 N O17 35427-13-5 O-5-1159-8 D-Glucopyranosylamine, N-(4-methylphenyl)-2-O-(2,3,4,6-tetra-O-
acetyl-β-D-glucopyranosyl)-, 3,4,6-triacetate;; N-PARA-TOLYLSOPHOROSYLAMINE ACETATE; W/NBS   66463

725 C33 H43 N O17 35427-12-4 O-5-1159-7 D-Glucopyranosylamine, N-(4-methylphenyl)-3-O-(2,3,4,6-tetra-O-
acetyl-β-D-glucopyranosyl)-, 2,4,6-triacetate;; N-PARA-TOLYLLAMINARIBIOSYLAMINE ACETATE; W/NBS   66464

725 C33 H43 N O17 35427-10-2 O-5-1159-5 D-Glucopyranosylamine, N-(4-methylphenyl)-4-O-(2,3,4,6-tetra-O-
acetyl-β-D-galactopyranosyl)-, 2,3,6-triacetate;; N-PARA-TOLYLLACTOSYLAMINE ACETATE; W/NBS   66465

725 C33 H43 N O17 35427-08-8 O-5-1159-2 D-Glucopyranosylamine, N-(4-methylphenyl)-6-O-(2,3,4,6-tetra-O-
acetyl-α-D-galactopyranosyl)-, 2,3,4-triacetate;; N-PARA-TOLYLMELIBIOSYLAMINE ACETATE; W/NBS   66466

725 C37 H48 Br N3 O7 78109-06-5 J-46-3426-0 22-Bromo-3β-(methoxymethoxy)-5α,8α-(4-phenyl-1,2-urazolo)ch
olesta-6-en-25-ol-26,23-lactone; W   128507

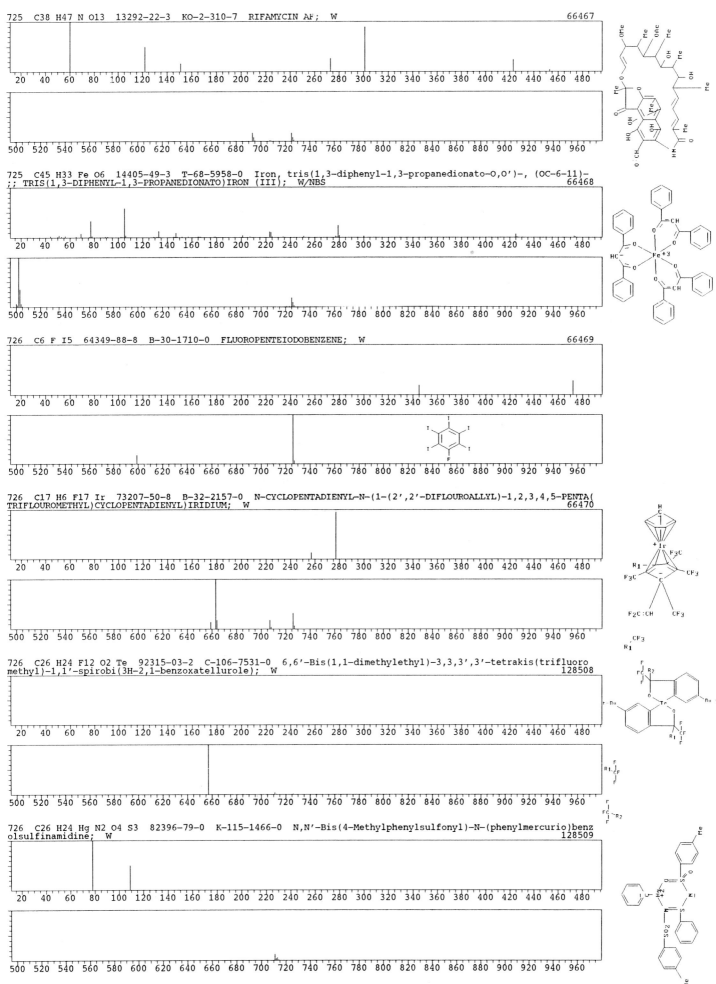

725 C38 H47 N O13 13292-22-3 KO-2-310-7 RIFAMYCIN AF; W   66467

725 C45 H33 Fe O6 14405-49-3 T-68-5958-0 Iron, tris(1,3-diphenyl-1,3-propanedionato-O,O')-, (OC-6-11)-
;; TRIS(1,3-DIPHENYL-1,3-PROPANEDIONATO)IRON (III); W/NBS   66468

726 C6 F I5 64349-88-8 B-30-1710-0 FLUOROPENTEIODOBENZENE; W   66469

726 C17 H6 F17 Ir 73207-50-8 B-32-2157-0 N-CYCLOPENTADIENYL-N-(1-(2',2'-DIFLOUROALLYL)-1,2,3,4,5-PENTA(
TRIFLOUROMETHYL)CYCLOPENTADIENYL)IRIDIUM; W   66470

726 C26 H24 F12 O2 Te 92315-03-2 C-106-7531-0 6,6'-Bis(1,1-dimethylethyl)-3,3,3',3'-tetrakis(trifluoro
methyl)-1,1'-spirobi(3H-2,1-benzoxatellurole); W   128508

726 C26 H24 Hg N2 O4 S3 82396-79-0 K-115-1466-0 N,N'-Bis(4-Methylphenylsulfonyl)-N-(phenylmercurio)benz
olsulfinamidine; W   128509

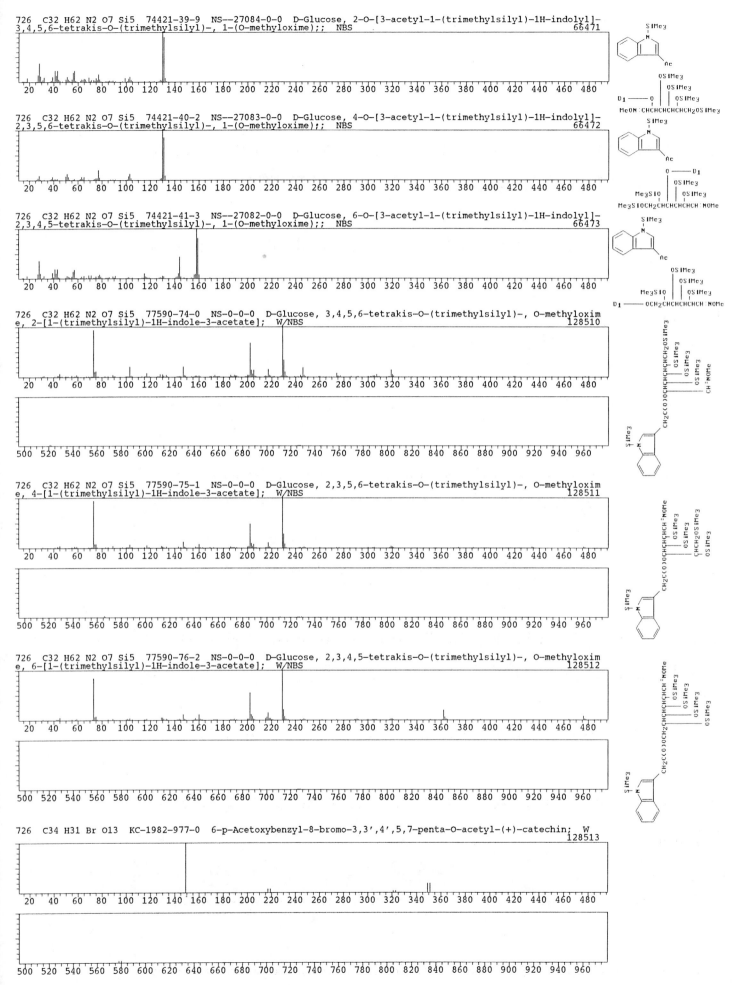

726   C32 H62 N2 O7 Si5   74421-39-9   NS--27084-0-0   D-Glucose, 2-O-[3-acetyl-1-(trimethylsilyl)-1H-indolyl]-
3,4,5,6-tetrakis-O-(trimethylsilyl)-, 1-(O-methyloxime);;   NBS   66471

726   C32 H62 N2 O7 Si5   74421-40-2   NS--27083-0-0   D-Glucose, 4-O-[3-acetyl-1-(trimethylsilyl)-1H-indolyl]-
2,3,5,6-tetrakis-O-(trimethylsilyl)-, 1-(O-methyloxime);;   NBS   66472

726   C32 H62 N2 O7 Si5   74421-41-3   NS--27082-0-0   D-Glucose, 6-O-[3-acetyl-1-(trimethylsilyl)-1H-indolyl]-
2,3,4,5-tetrakis-O-(trimethylsilyl)-, 1-(O-methyloxime);;   NBS   66473

726   C32 H62 N2 O7 Si5   77590-74-0   NS-0-0-0   D-Glucose, 3,4,5,6-tetrakis-O-(trimethylsilyl)-, O-methyloxim
e, 2-[1-(trimethylsilyl)-1H-indole-3-acetate];   W/NBS   128510

726   C32 H62 N2 O7 Si5   77590-75-1   NS-0-0-0   D-Glucose, 2,3,5,6-tetrakis-O-(trimethylsilyl)-, O-methyloxim
e, 4-[1-(trimethylsilyl)-1H-indole-3-acetate];   W/NBS   128511

726   C32 H62 N2 O7 Si5   77590-76-2   NS-0-0-0   D-Glucose, 2,3,4,5-tetrakis-O-(trimethylsilyl)-, O-methyloxim
e, 6-[1-(trimethylsilyl)-1H-indole-3-acetate];   W/NBS   128512

726   C34 H31 Br O13   KC-1982-977-0   6-p-Acetoxybenzyl-8-bromo-3,3',4',5,7-penta-O-acetyl-(+)-catechin;   W
                                                                                             128513

726   C37 H47 In N4 O2 S   70619-63-5   KA-1980-2084-9   Methylsulfinato[2,3,7,8,12,13,17,18-octaethylporphyrin
ato]indium;   W   128514

726   C41 H37 O2 P2 Rh   HE-1986-1374-0   ACETYLACETONATO-BIS(TRIPHENYLPHOSPHINE)-RHODIUM;   W   128515

726   C41 H50 N4 O8   69721-75-1   H-61-2669-0   1,16-Ethenofuro[3,4-1][1,5,10]triazacyclohexadecine-15-acetic
acid, 10,14-diacetyl-5-[3-(benzoylamino)propyl]-3,3a,4,5,6,7,8,9,10,11,12,13,14,15-tetradecahydro-3-(4-meth
oxyphenyl)-4-oxo-, methyl ester;; 1,16-Ethenofuro[3,4-1][1,5,10]triazacyclohexadecine-15-acetic acid, 10,14
-diacetyl-5-[3-(benzoylamino)propyl]-3,3a,4,5,6,7,8,9,10,11,12,13,14,15;   W/NBS   128516

726   C41 H50 N4 O8   73745-73-0   H-62-2723-0   N(6),N(10)-Diacetyl-N(23)-benzoyl-24-methoxy-O(34)-methyl-23,
24-seco-orantin;   W   128517

726   C41 H47 D3 N4 O8   69721-87-5   H-61-2670-0   N(6),N(10)-Diacetyl-N(23)-benzoyl-24-(methoxy-d3)-O(34)-
methyl-23,24-dihydro-23,24-secoaphelandrine;   W   128518

726   C43 H59 Br N4 O   80056-59-3   KC-1981-2627-0   meso-(6-Bromohexyloxymethyl)octaethylporphyrin;   W 128519

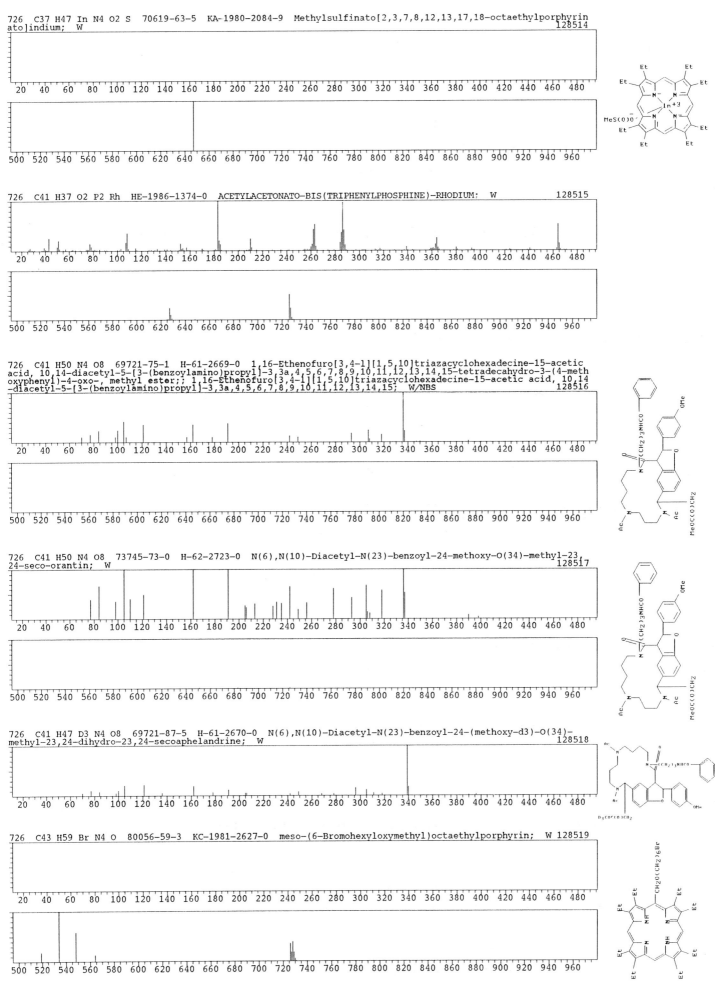

726   C44 H28 Cd N4   14977-07-2   KO-3-75-3   CADMIUM α,β,γ,Δ-TETRAPHENYL PORPHYRIN;   W   66475

726   C46 H62 O7   33474-10-1   O5-0-565-0   Siphonaxanthin triacetate;   W/NBS   66476

726   C46 H75 Cl O4   69505-99-3   K-112-205-0   Benzeneheptadecanoic acid, 3-chloro-4-methoxy-, 2-decyl-3-meth
oxy-5-pentylphenyl ester;; 2-DECYL-3-METHOXY-5-PENTYLPHENYLESTER OF 17-(3-CHLORO-4-METHOXYPHENYL)HEPTADECAN
CARBOXYLIC ACID;   W/NBS   66478

726   C46 H75 Cl O4   63953-40-2   K-112-2009-0   Benzenepentadecanoic acid, 3-chloro-4-methoxy-, 3-methoxy-2-(
9-methyldecyl)-5-(4-methylpentyl)phenyl ester;; 3-METHOXY-2-(9-METHYLDECYL)-5-(4-METHYLPENTYL)PHENYLESTEROF
15-(3-CHLORO-4-METHOXYPHENYL)PENTADECANCARBOXYLIC ACID;   W/NBS   66477

726   C49 H42 O6   55018-95-6   C-98-5385-0   2,2',4,4',6-PENTAKIS(BENZYLOXY)-6'-METHYL-BENZOPHENONE;   W   66479

726   C49 H74 O4   2394-68-5   J-45-4103-0   2-OCTAPRENYL-3-METHYL-5,6-DIMETHOXY-1,4-BENZOQUINONE;   W   128520

727   C15 H29 Br4 N3 O10   95763-24-9   C-107-2920-0   Psammaplysin-B acetamide diacetate;   W   127639

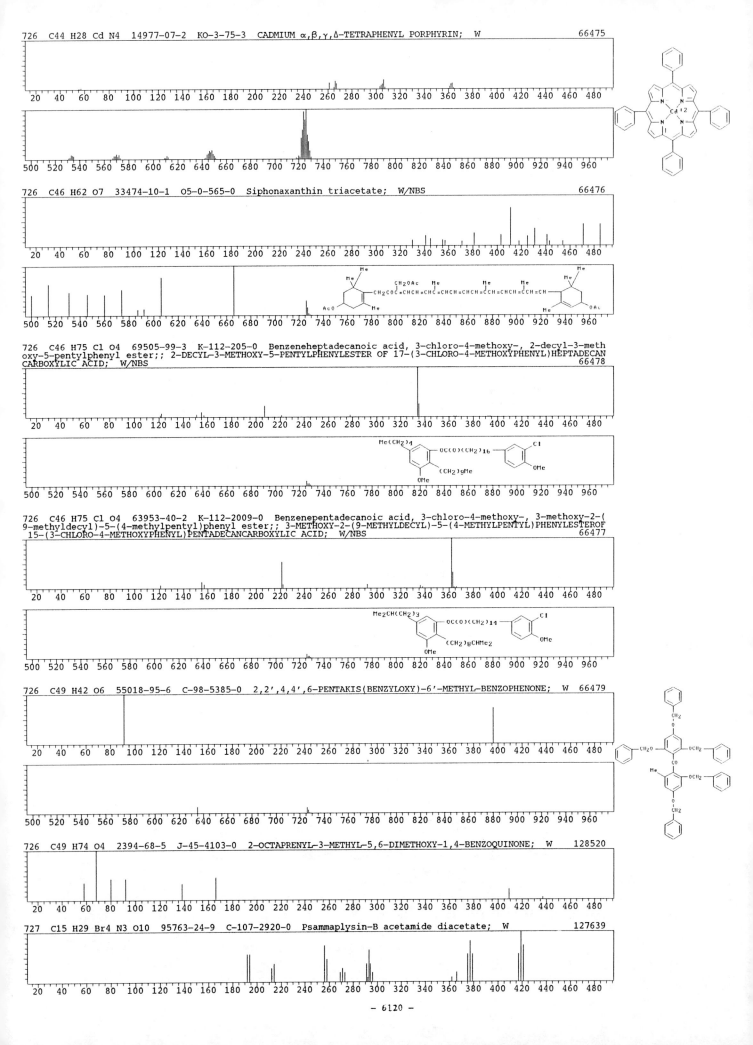

727 C25 H46 Cl5 N O5 Si4   KO-5-410-1   TETRAKIS(TRIMETHYLSILYL)PENTACHLORO ANILINE GLUCURONIDE; W   66482

727 C42 H33 N O11 88278-49-3 KC-1983-1617-0 2-Benzoylamino-2-benzoyloxymethyl-2-deoxy-3,5,6-tri-O-benzo
yl-D-mannono-1,4-lac-tone; W   128521

727 C49 H61 N O4 77588-22-8 K-114-957-0 10-[2-(4-Cyanobenzyloxy)-3-(octadecyloxy)propyloxy]-10-phenyl-
9(10H)-anthracenone; W   128522

728 C16 H8 O9 Os2 94472-14-7 H-67-1635-0 cis-.mu.-[(1R,2R,3S,4S)-(5,6-dimethylidene-7-oxabicyclo[2.2.1]
heptane-2,3-diyl)]bis(tetracarbonylosmium); W   128523

728 C33 H44 O18 89332-48-9 H-67-165-0 7-O-[(E)-β-(4''-(β-D-Glucopyranosyloxy)-3''-methoxyphenyl)propeny
l]-loganin; W   128524

728 C34 H20 O6 Ru2 33310-08-6 O5-0-493-0 Ruthenium, hexacarbonyl[.mu.-(1,2,3,4-tetraphenyl-1,3-butadien
ylene)]di-, (Ru-Ru);; 1,1,1-TRICARBONYLRUTHENIUM TETRAPHENYLCYCLOPENTADIENE .PI.-TRICARBONYLRUTHENIUM;
W/NBS   66483

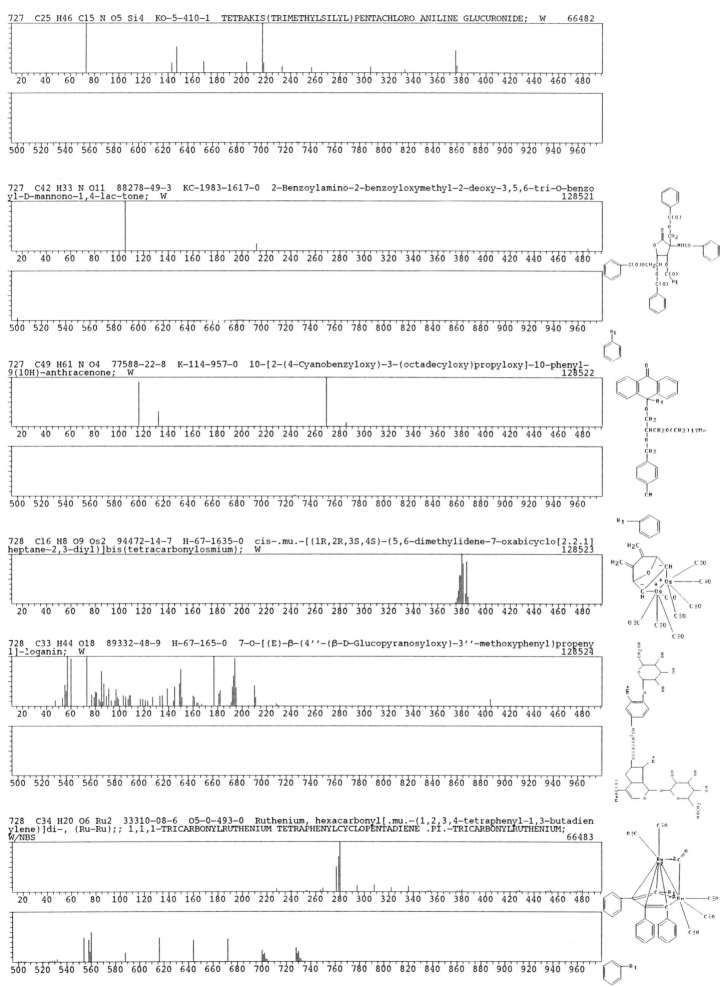

728 C36 H40 S8 60822-17-5 F-32-2569-2   1,2,9,10,17,18,25,26-OCTATHIO[2,2,2,2]-4,6,8,12,14,16,20,22,24,
28,30,32-DODECAMETHYLMETACYCLOPHANE; W   66484

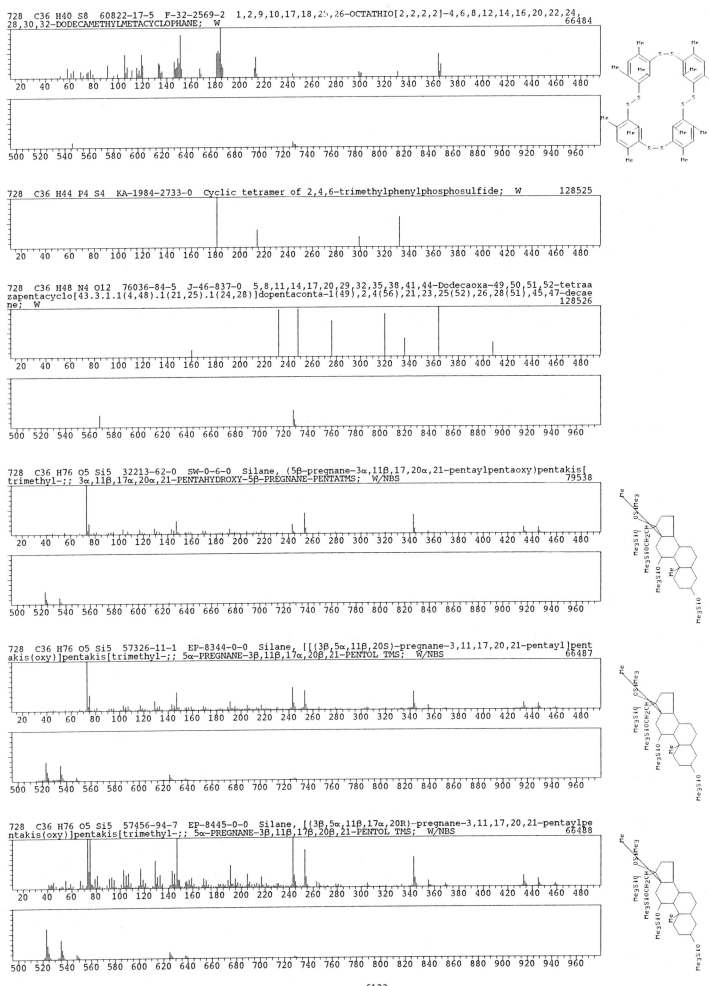

728 C36 H44 P4 S4 KA-1984-2733-0 Cyclic tetramer of 2,4,6-trimethylphenylphosphosulfide; W   128525

728 C36 H48 N4 O12 76036-84-5 J-46-837-0 5,8,11,14,17,20,29,32,35,38,41,44-Dodecaoxa-49,50,51,52-tetraa
zapentacyclo[43.3.1.1(4,48).1(21,25).1(24,28)]dopentaconta-1(49),2,4(56),21,23,25(52),26,28(51),45,47-decae
ne; W   128526

728 C36 H76 O5 Si5 32213-62-0 SW-0-6-0 Silane, (5β-pregnane-3α,11β,17,20α,21-pentaylpentaoxy)pentakis[
trimethyl-;; 3α,11β,17α,20α,21-PENTAHYDROXY-5β-PREGNANE-PENTATMS; W/NBS   79538

728 C36 H76 O5 Si5 57326-11-1 EP-8344-0-0 Silane, [[(3β,5α,11β,20S)-pregnane-3,11,17,20,21-pentayl]pent
akis(oxy)]pentakis[trimethyl-;; 5α-PREGNANE-3β,11β,17α,20β,21-PENTOL TMS; W/NBS   66487

728 C36 H76 O5 Si5 57456-94-7 EP-8445-0-0 Silane, [(3β,5α,11β,17α,20R)-pregnane-3,11,17,20,21-pentaylpe
ntakis(oxy)]pentakis[trimethyl-;; 5α-PREGNANE-3β,11β,17β,20β,21-PENTOL TMS; W/NBS   66488

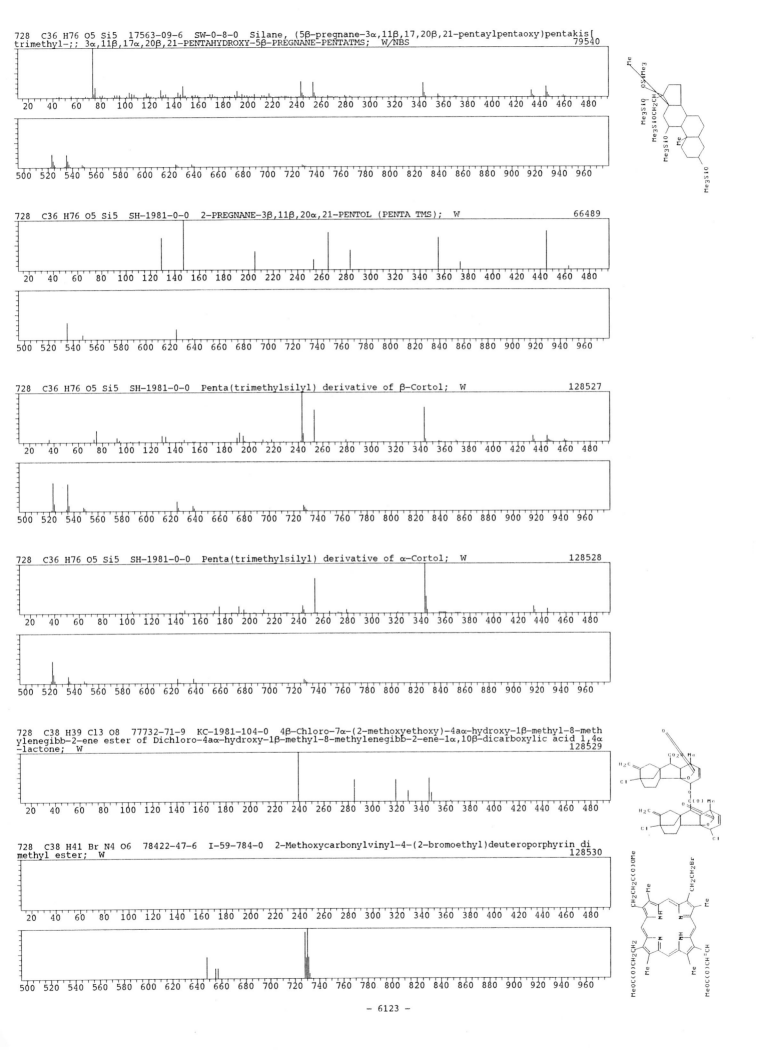

728 C36 H76 O5 Si5 17563-09-6 SW-0-8-0 Silane, (5β-pregnane-3α,11β,17,20β,21-pentaylpentaoxy)pentakis[
trimethyl-;; 3α,11β,17α,20β,21-PENTAHYDROXY-5β-PREGNANE-PENTATMS; W/NBS   79540

728 C36 H76 O5 Si5 SH-1981-0-0 2-PREGNANE-3β,11β,20α,21-PENTOL (PENTA TMS); W   66489

728 C36 H76 O5 Si5 SH-1981-0-0 Penta(trimethylsilyl) derivative of β-Cortol; W   128527

728 C36 H76 O5 Si5 SH-1981-0-0 Penta(trimethylsilyl) derivative of α-Cortol; W   128528

728 C38 H39 Cl3 O8 77732-71-9 KC-1981-104-0 4β-Chloro-7α-(2-methoxyethoxy)-4aα-hydroxy-1β-methyl-8-meth
ylenegibb-2-ene ester of Dichloro-4aα-hydroxy-1β-methyl-8-methylenegibb-2-ene-1α,10β-dicarboxylic acid 1,4α
-lactone; W   128529

728 C38 H41 Br N4 O6 78422-47-6 I-59-784-0 2-Methoxycarbonylvinyl-4-(2-bromoethyl)deuteroporphyrin di
methyl ester; W   128530

- 6123 -

728   C38 H60 N6 O8   BS-1-62-0   N-Acetyl-(N,0-permethyl)-alanyl-prolyl-leucyl-phenylalanyl-valyl-glycine;   W
128531

728   C38 H39 D21 N6 O8   BS-1-56-0   N-Trideuteroacetyl-(N,0-pertrideuteromethyl)alanyl-prolyl-leucyl-phenyla
lanyl-valyl-glycine;   W
128532

728   C40 H64 O8 Si2   86668-52-2   J-48-3262-0   6-(tert-Butyldimethylsiloxy)-3,4-(isopropylidenedioxy)-10,11
-(di-tert-butylsilylnedioxy)-12,12-(2',2'-dimethylpropane-1',3'-diyldioxy)-2-oxo-1,2,3-(R*),4(R*)),4(S*),5,
5a(S*),6(S*),11(R*),11a(R*),12,12a-(S*)-dodecahydronaphthacene;   W
128533

728   C41 H60 O11   56312-53-9   SD-1981-0-0   Cholesta-9(11),20(22)-dien-23-one, 6-(acetyloxy)-3-[(2,3,4-tri-
O-acetyl-deoxy-β-D-galactopyranosyl)oxy]-, (3β,5α,6α)-;;   W/NBS
66490

728   C43 H69 D15 O8   55256-18-3   AD-0-4874-0   PENTA-(TRIDEUTEROMETHYL)-6-O-CORYNOMYCOLYLGLUCOSIDE;   W 66491

728   C45 H61 O6 P   84251-45-6   KC-1982-2311-0   Methyl Bis-(4',6-dimethoxy-3',5'-di-t-butylbiphenyl-2-yl)-ph
osphinate;   W
128534

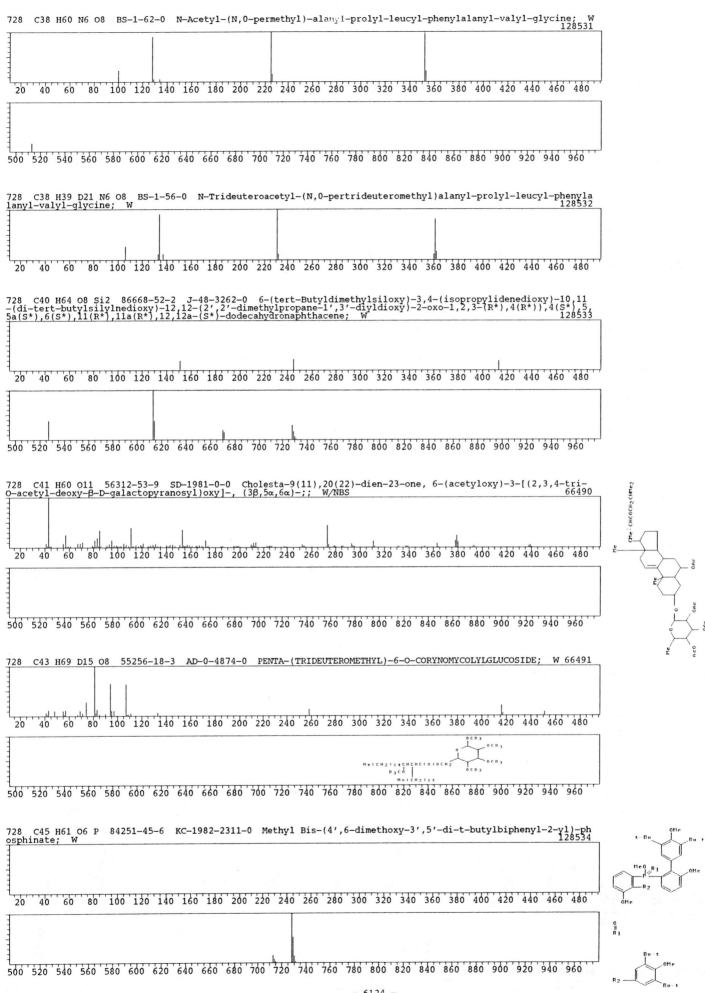

728 C45 H80 O5 Si 65572-94-3 E-50-558-2 1-PALMITOYL-2-EICOSAPENTAENOYL-SN-GLYCEROL, TERT-BUTYL-DIMETHYL SILYL DERIVATIVE; W
66492

CH2OSiMe2Bu-t
|
Me(CH2)14C(O)OCH2CHOC(O)(CH2)18Me

729 C12 H12 F18 N3 O6 P3 1065-05-0 KO-4-285-1 HEXAKIS(2,2,2-TRIFLUOROETHOXY)CYCLOTRIPHOSPHAZENE; W
66493

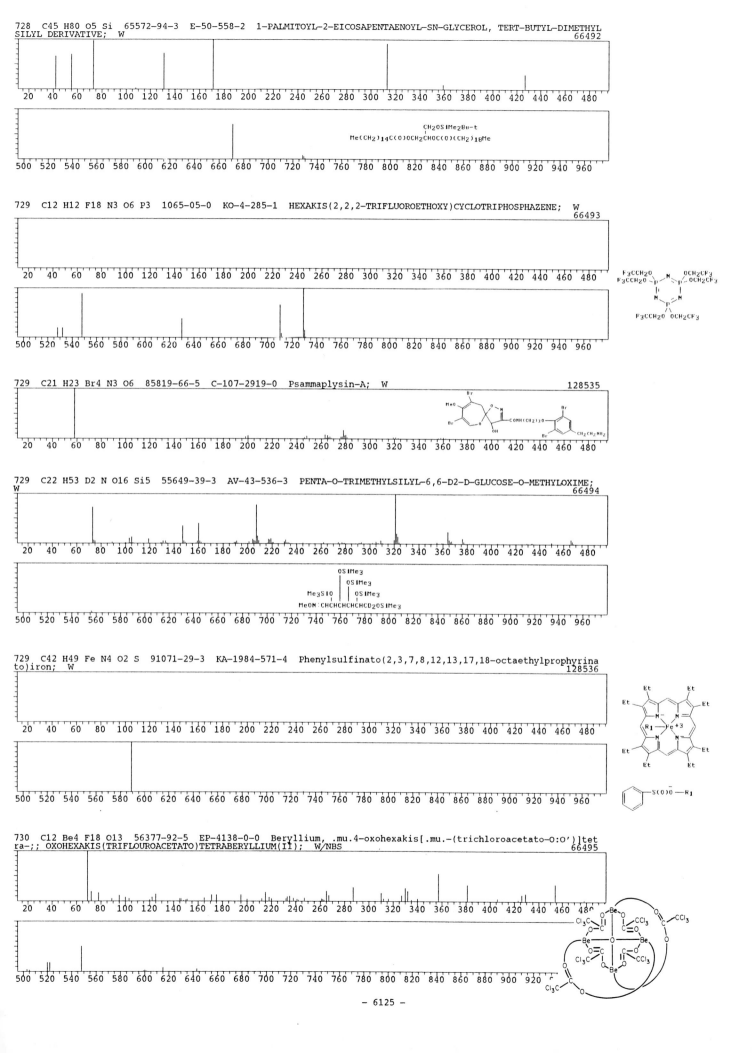

F3CCH2O   OCH2CF3
F3CCH2O—P   P—OCH2CF3
         N   N
           N   P
F3CCH2O OCH2CF3

729 C21 H23 Br4 N3 O6 85819-66-5 C-107-2919-0 Psammaplysin-A; W
128535

729 C22 H53 D2 N O16 Si5 55649-39-3 AV-43-536-3 PENTA-O-TRIMETHYLSILYL-6,6-D2-D-GLUCOSE-O-METHYLOXIME; W
66494

OSiMe3
OSiMe3
Me3SiO OSiMe3
MeON:CHCHCHCHCHCD2OSiMe3

729 C42 H49 Fe N4 O2 S 91071-29-3 KA-1984-571-4 Phenylsulfinato(2,3,7,8,12,13,17,18-octaethylprophyrinato)iron; W
128536

730 C12 Be4 F18 O13 56377-92-5 EP-4138-0-0 Beryllium, .mu.4-oxohexakis[.mu.-(trichloroacetato-O:O')]tetra-;; OXOHEXAKIS(TRIFLOUROACETATO)TETRABERYLLIUM(II); W/NBS
66495

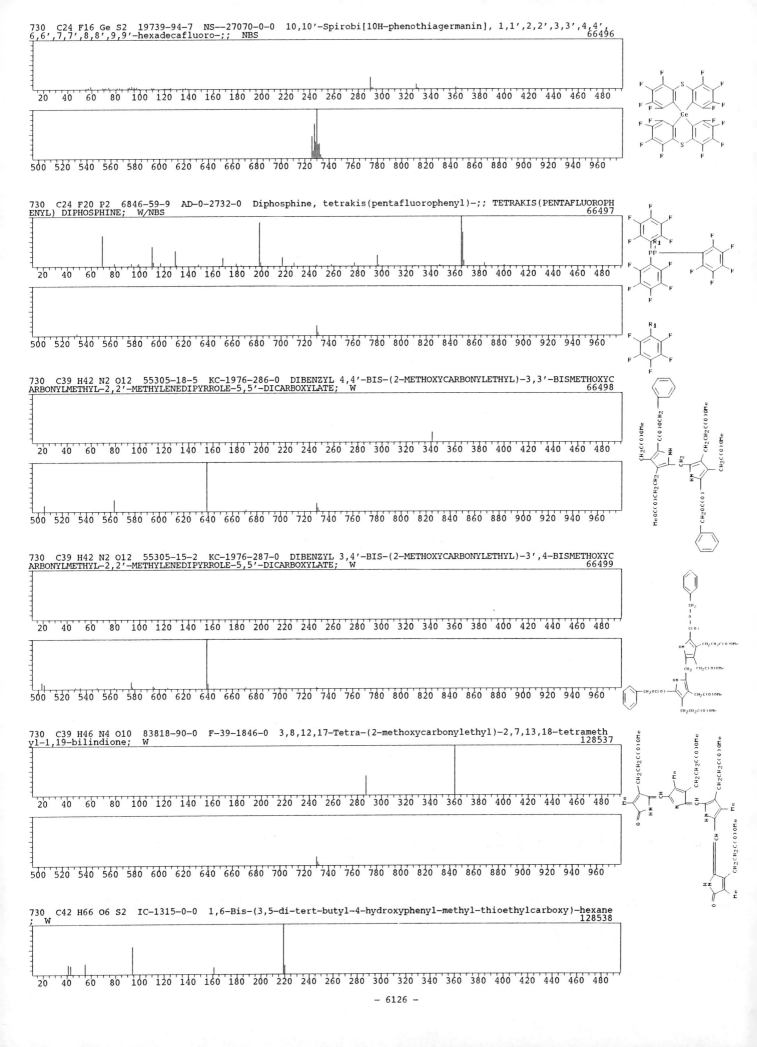

730   C24 F16 Ge S2   19739-94-7   NS--27070-0-0   10,10'-Spirobi[10H-phenothiagermanin], 1,1',2,2',3,3',4,4',
6,6',7,7',8,8',9,9'-hexadecafluoro-;;   NBS   66496

730   C24 F20 P2   6846-59-9   AD-0-2732-0   Diphosphine, tetrakis(pentafluorophenyl)-;;   TETRAKIS(PENTAFLUOROPH
ENYL) DIPHOSPHINE;   W/NBS   66497

730   C39 H42 N2 O12   55305-18-5   KC-1976-286-0   DIBENZYL 4,4'-BIS-(2-METHOXYCARBONYLETHYL)-3,3'-BISMETHOXYC
ARBONYLMETHYL-2,2'-METHYLENEDIPYRROLE-5,5'-DICARBOXYLATE;   W   66498

730   C39 H42 N2 O12   55305-15-2   KC-1976-287-0   DIBENZYL 3,4'-BIS-(2-METHOXYCARBONYLETHYL)-3',4-BISMETHOXYC
ARBONYLMETHYL-2,2'-METHYLENEDIPYRROLE-5,5'-DICARBOXYLATE;   W   66499

730   C39 H46 N4 O10   83818-90-0   F-39-1846-0   3,8,12,17-Tetra-(2-methoxycarbonylethyl)-2,7,13,18-tetrameth
yl-1,19-bilindione;   W   128537

730   C42 H66 O6 S2   IC-1315-0-0   1,6-Bis-(3,5-di-tert-butyl-4-hydroxyphenyl-methyl-thioethylcarboxy)-hexane
;   W   128538

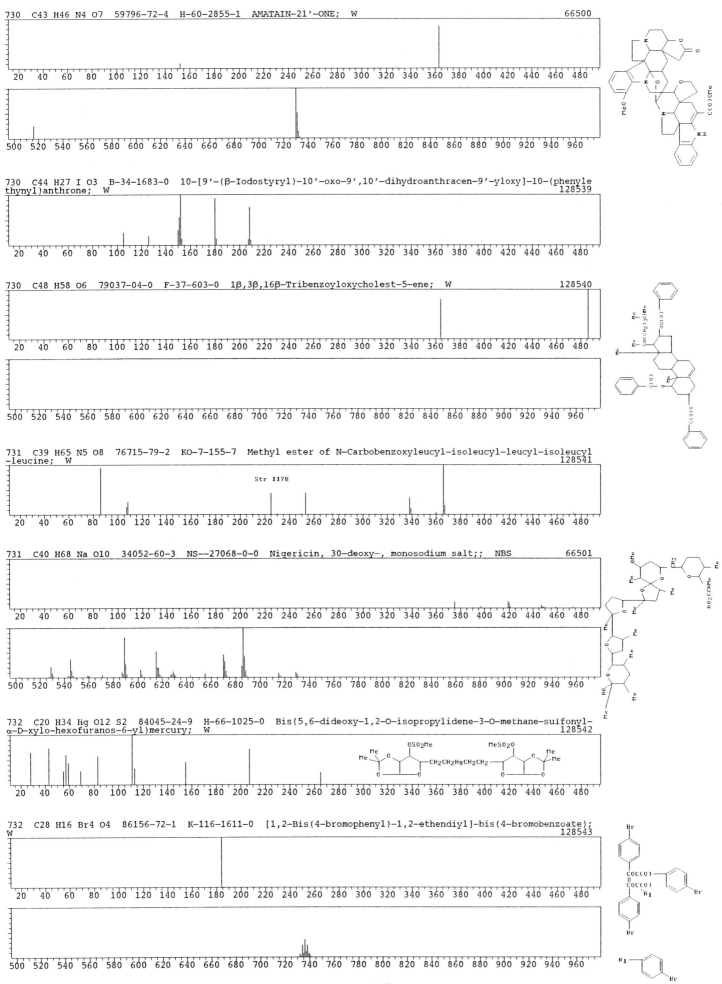

730 C43 H46 N4 O7 59796-72-4 H-60-2855-1 AMATAIN-21'-ONE; W   66500

730 C44 H27 I O3 B-34-1683-0 10-[9'-(β-Iodostyryl)-10'-oxo-9',10'-dihydroanthracen-9'-yloxy]-10-(phenyle thynyl)anthrone; W   128539

730 C48 H58 O6 79037-04-0 F-37-603-0 1β,3β,16β-Tribenzoyloxycholest-5-ene; W   128540

731 C39 H65 N5 O8 76715-79-2 KO-7-155-7 Methyl ester of N-Carbobenzoxyleucyl-isoleucyl-leucyl-isoleucyl -leucine; W   128541

Str 1178

731 C40 H68 Na O10 34052-60-3 NS--27068-0-0 Nigericin, 30-deoxy-, monosodium salt;; NBS   66501

732 C20 H34 Hg O12 S2 84045-24-9 H-66-1025-0 Bis(5,6-dideoxy-1,2-O-isopropylidene-3-O-methane-sulfonyl- α-D-xylo-hexofuranos-6-yl)mercury; W   128542

732 C28 H16 Br4 O4 86156-72-1 K-116-1611-0 [1,2-Bis(4-bromophenyl)-1,2-ethendiyl]-bis(4-bromobenzoate); W   128543

732 C34 H15 F10 O Rh 31760-85-7 AD-0-5750-0 Rhodium, [2,4-bis(pentafluorophenyl)-3,5-diphenyl-2,4-cyclo
pentadien-1-one]-.pi.-cyclopentadienyl;; 2,4-BIS(PENTAFLUOROPHENYL)-3,5-DIPHENYLCYCLOPENTADIENONE-.PI.-CYCL
OPENTADIENYLRHODIUM; W/NBS   66502

732 C40 H60 O12 BS-2-19-0 Yononin peracetate; W   128544

732 C41 H64 O11 60492-17-3 KC-1976-1360-0 6-OXO-5α-CHOLESTAN-3β-YL TETRA-O-ACETYL-β-D-GLUCOPYRANOSIDE;
W   66503

Str 1179

732 C43 H48 N4 O7 19772-81-7 H-60-2855-1 (14S)-VOBTUSIN-21'-ONE; W   66504

732 C43 H48 N4 O7 19772-81-7 H-60-2855-1 (14R)-VOBTUSIN-21'-ONE; W   66505

732 C43 H48 N4 O7 50924-04-4 AC-1975-2504-1 Vobtusine-lactone; W   128545

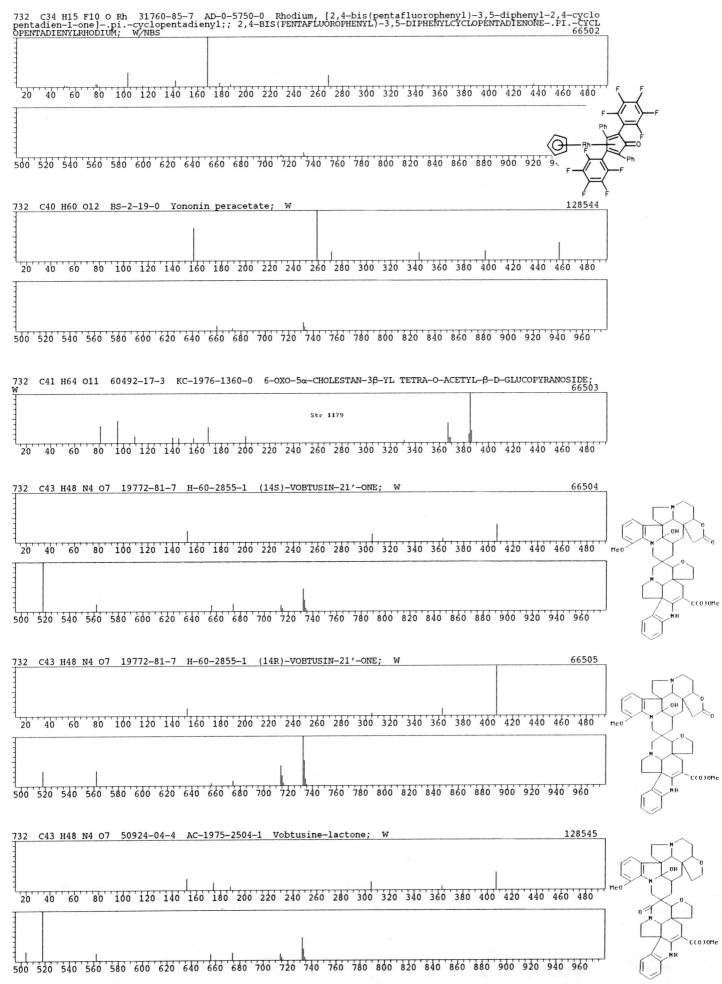

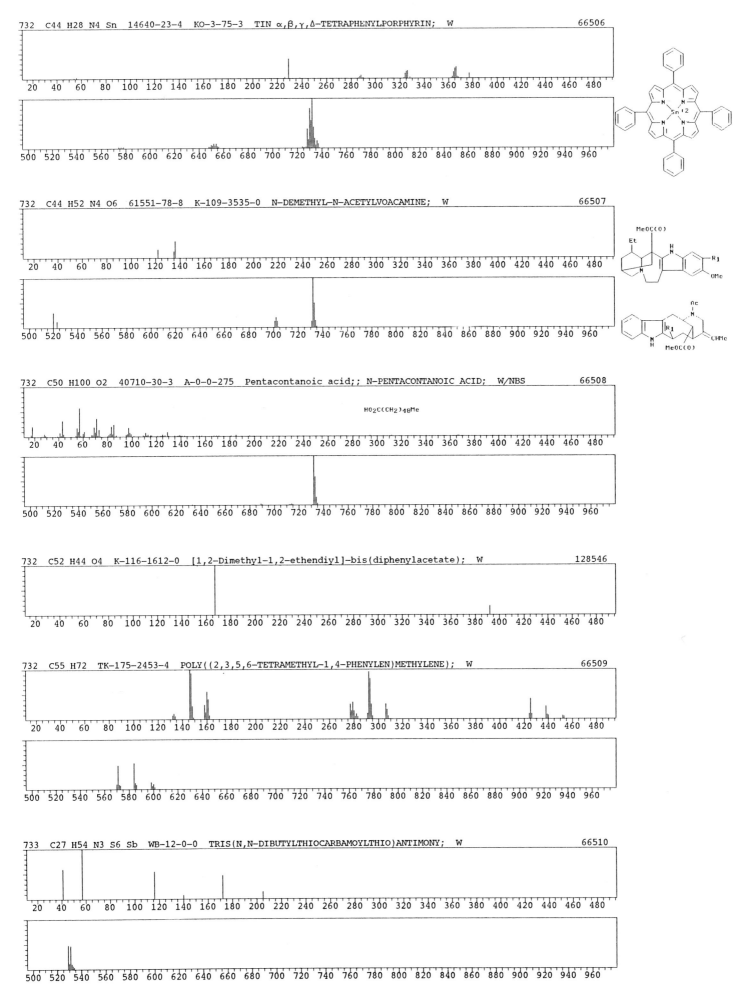

732   C44 H28 N4 Sn   14640-23-4   KO-3-75-3   TIN α,β,γ,Δ-TETRAPHENYLPORPHYRIN;   W   66506

732   C44 H52 N4 O6   61551-78-8   K-109-3535-0   N-DEMETHYL-N-ACETYLVOACAMINE;   W   66507

732   C50 H100 O2   40710-30-3   A-0-0-275   Pentacontanoic acid;; N-PENTACONTANOIC ACID;   W/NBS   66508

HO2C(CH2)48Me

732   C52 H44 O4   K-116-1612-0   [1,2-Dimethyl-1,2-ethendiyl]-bis(diphenylacetate);   W   128546

732   C55 H72   TK-175-2453-4   POLY((2,3,5,6-TETRAMETHYL-1,4-PHENYLEN)METHYLENE);   W   66509

733   C27 H54 N3 S6 Sb   WB-12-0-0   TRIS(N,N-DIBUTYLTHIOCARBAMOYLTHIO)ANTIMONY;   W   66510

733 C37 H67 N O13 114-07-8 EP-1297-0-0 Erythromycin;; Erythromycin A; W/NBS   66512

733 C37 H67 N O13 21395-58-4 O5-0-1229-0 Erythromycin, 12-deoxy-, N-oxide;; Erythromycin B N-oxide;
W/NBS   66511

733 C40 H80 N O8 P 2644-64-6 EP-4664-0-0 3,5,9-Trioxa-4-phosphapentacosan-1-aminium, 4-hydroxy-N,N,N-tr
imethyl-10-oxo-7-[(1-oxohexadecyl)oxy]-, hydroxide, inner salt, 4-oxide;; DIPALMITOYL LECITHIN; W/NBS
66513

733 C42 H43 N3 O9 79664-30-5 J-46-5236-0 N(1),N(10)-Dibenzyl-N(1),N(5),N(10)-tris(2,3-dihydroxybenzoyl)
-1,5,10-triazadecane; W   128547

733 C44 H28 Cl2 N4 V 83582-00-7 KA-1982-1453-2 Dichloro(5,10,15,20-tetraphenylporphyrinato)vanadium; W
128548

734 C21 H54 F6 N6 P2 Si6 59199-28-9 AG-104-25-0 2-Trimethylsilyl-4-bis(trifluormethyl)azomethan-1,3-bis
[bis(trimethylsilyl)amino]-1-trimethylsilylimino-2,4-1.lamda.5,3.lamda.3-diazadiphosphetidine; W 128549

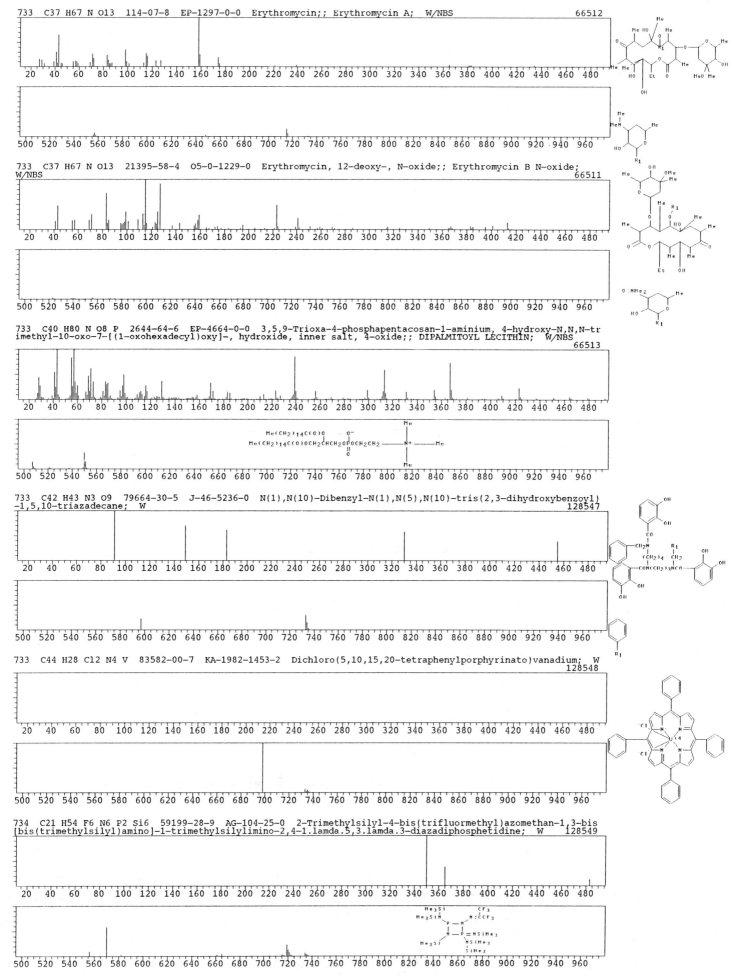

734 C22 H26 O8 Ru2 Si4 59369-25-4 H-59-612-11 1,1,3,3,5,5,7,7-OCTAMETHYL-2,2,2,2,6,6,6,6-OCTACARBONYL-
3,7-DIHYDRO-1H,5H-BENZO(1,2-D:4,5-D)BIS(2,13)RUTHENA(II)DISILOL; W   66514

734 C37 H50 O15 60767-82-0 NS-0-0-0 4H-1-Benzopyran-4-one, 5,7-dimethoxy-2-(4-methoxyphenyl)-6-[3,4,6-
tri-O-methyl-2-O-(2,3,4,6-tetra-O-methyl-β-D-glucopyranosyl)-β-D-glucopyranosyl]-; W/NBS   128550

734 C37 H58 N4 O7 S2 91653-21-3 H-67-2182-0 13-Nitro-N-[4-tosyl-8-(tosylamino)-4-azaoctyl]-16-hexadecan
lactam; W   128551

734 C40 H30 O14 77681-11-9 KC-1981-557-0 Tetra-acetate of 4''',7-Di-O-methyltaiwaniaflavone; W 128552

734 C43 H50 N4 O7 65967-13-7 H-60-2850-0 HYDRATOAMATAINE; W   66515

734 C43 H50 N4 O7 65967-15-9 H-60-2852-0 VOBTUSINE-N(B')-OXIDE; W   66516

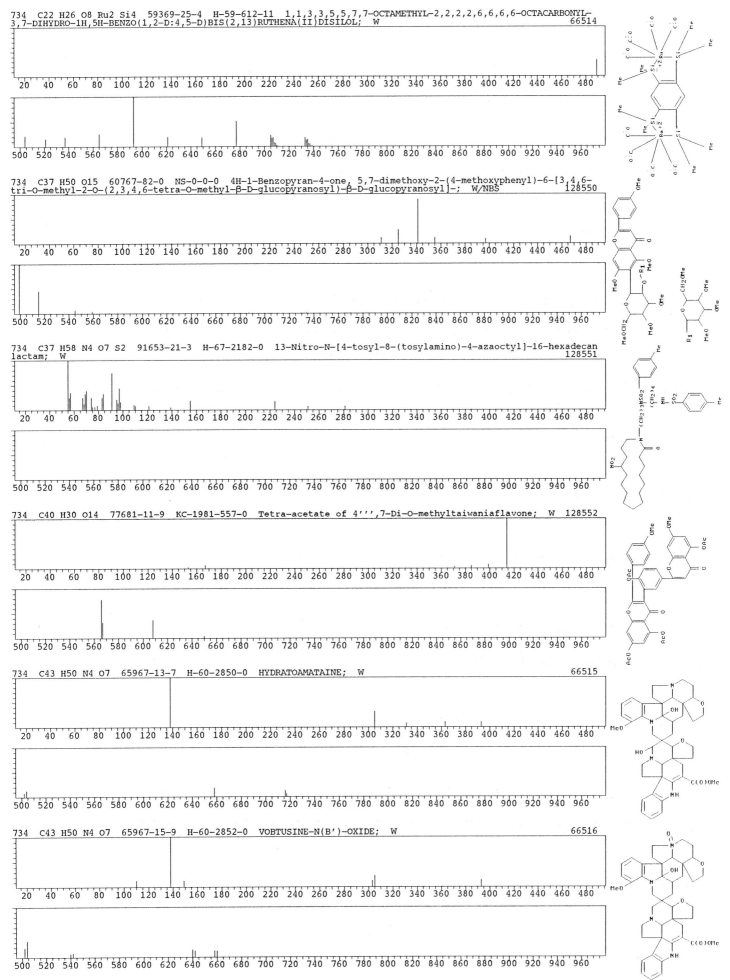

734 C43 H50 N4 O7 AC-1975-2504-1 2,16-Dihydro-vobtusine-lactame; W   128553

```
|    |    |    |    |    |    |    |    |    |    |    |    |    |    |    |    |    |    |    |    |    |    |    |
20   40   60   80  100  120  140  160  180  200  220  240  260  280  300  320  340  360  380  400  420  440  460  480
```

```
|    |    |    |    |    |    |    |    |    |    |    |    |    |    |    |    |    |    |    |    |    |    |    |
500  520  540  560  580  600  620  640  660  680  700  720  740  760  780  800  820  840  860  880  900  920  940  960
```

734 C44 H54 N4 O6 79921-96-3 AC-1981-216-0 O,O-Dimethyltetrastachynine; W   128554

```
|    |    |    |    |    |    |    |    |    |    |    |    |    |    |    |    |    |    |    |    |    |    |    |
20   40   60   80  100  120  140  160  180  200  220  240  260  280  300  320  340  360  380  400  420  440  460  480
```

```
|    |    |    |    |    |    |    |    |    |    |    |    |    |    |    |    |    |    |    |    |    |    |    |
500  520  540  560  580  600  620  640  660  680  700  720  740  760  780  800  820  840  860  880  900  920  940  960
```

734 C46 H94 O2 Si2 OD-1984-83-5 Bis-trimethylsilyl derivative of biphytanol; W   128555

```
|    |    |    |    |    |    |    |    |    |    |    |    |    |    |    |    |    |    |    |    |    |    |    |
20   40   60   80  100  120  140  160  180  200  220  240  260  280  300  320  340  360  380  400  420  440  460  480
```

```
|    |    |    |    |    |    |    |    |    |    |    |    |    |    |    |    |    |    |    |    |    |    |    |
500  520  540  560  580  600  620  640  660  680  700  720  740  760  780  800  820  840  860  880  900  920  940  960
```

734 C48 H38 N4 O4 22112-78-3 O-9-79-4 21H,23H-Porphine, 5,10,15,20-tetrakis(4-methoxyphenyl)-;; (T-(-P-
OME)PP) MS-TETRA(P-METHOXY)PHENYLPORPHIN; W/NBS   66517

```
|    |    |    |    |    |    |    |    |    |    |    |    |    |    |    |    |    |    |    |    |    |    |    |
20   40   60   80  100  120  140  160  180  200  220  240  260  280  300  320  340  360  380  400  420  440  460  480
```

```
|    |    |    |    |    |    |    |    |    |    |    |    |    |    |    |    |    |    |    |    |    |    |    |
500  520  540  560  580  600  620  640  660  680  700  720  740  760  780  800  820  840  860  880  900  920  940  960
```

734 C48 H94 O4 69688-49-9 EP-5364-0-0 Octadecanoic acid, 1-[(tetradecyloxy)carbonyl]pentadecyl ester;;
TETRADECYL 2-OCTADECANOYLOXY-HEXADECANOATE; W/NBS   66519

```
|    |    |    |    |    |    |    |    |    |    |    |    |    |    |    |    |    |    |    |    |    |    |    |
20   40   60   80  100  120  140  160  180  200  220  240  260  280  300  320  340  360  380  400  420  440  460  480
```

Me(CH2)16C(O)OCHC(O)O(CH2)13Me
                 |
            (CH2)13Me

```
|    |    |    |    |    |    |    |    |    |    |    |    |    |    |    |    |    |    |    |    |    |    |    |
500  520  540  560  580  600  620  640  660  680  700  720  740  760  780  800  820  840  860  880  900  920  940  960
```

734 C48 H94 O4 56599-97-4 EP-5356-0-0 Hexadecanoic acid, 1-tetradecyl-1,2-ethanediyl ester;; 1,2-DIHEXA
DECANOYLOXY-HEXADECANE; W/NBS   66518

```
|    |    |    |    |    |    |    |    |    |    |    |    |    |    |    |    |    |    |    |    |    |    |    |
20   40   60   80  100  120  140  160  180  200  220  240  260  280  300  320  340  360  380  400  420  440  460  480
```

Me(CH2)14C(O)OCH2CHOC(O)(CH2)14Me
                 |
            (CH2)13Me

```
|    |    |    |    |    |    |    |    |    |    |    |    |    |    |    |    |    |    |    |    |    |    |    |
500  520  540  560  580  600  620  640  660  680  700  720  740  760  780  800  820  840  860  880  900  920  940  960
```

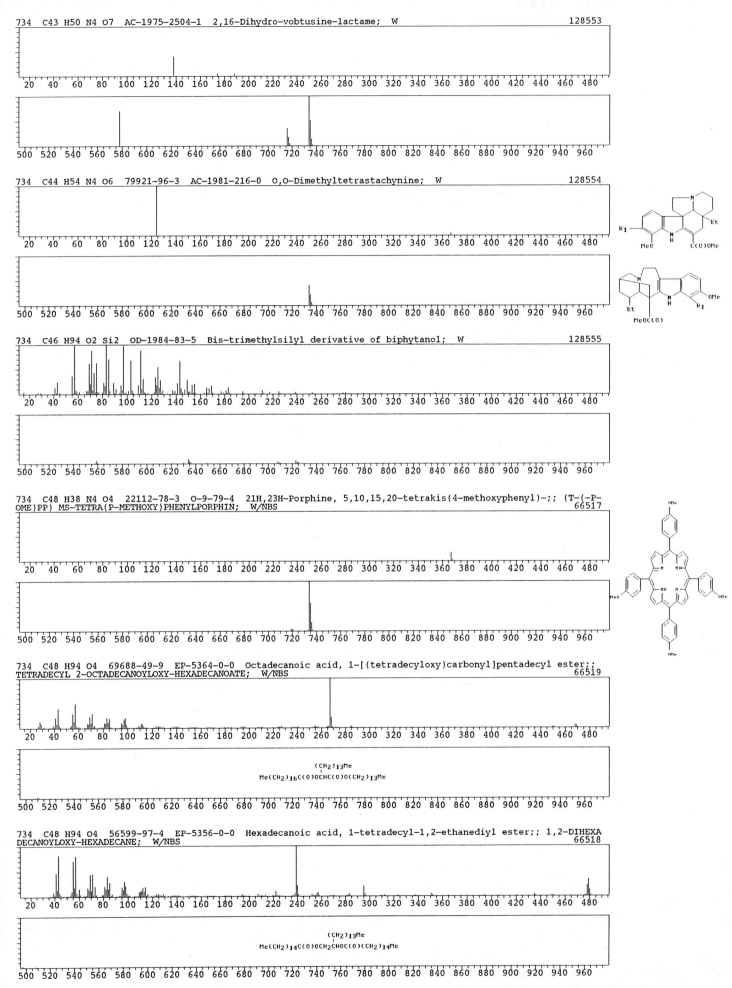

734 C51 H74 O3 F-33-382-0 (2R,2'R)-2,2'-BIS(3-METHYLBUTYL)-3,4,3',4'-TETRADIHYDRO-.PSI.,.PSI.-CAROTEN-2-
ACETATE; W   66520

734 C51 H74 O3 F-33-382-0 (2R,2'R)-2,2'-BIS(3-METHYL BUTYL)-3,4,3',4'-TETRADIHYDRO-.PSI.,.PSI.-CAROTEN-
20-ACETATE; W   66521

735 C26 H66 N O9 P Si6 55530-77-3 BA-0-347-0 D-Fructose, 1,3,4,5-tetrakis-O-(trimethylsilyl)-, O-ethylo
xime, 6-[bis(trimethylsilyl) phosphate];; O-HEXAKIS(TRIMETHYLSILYL)-D-FRUCTOSE-6-PHOSPHATE ETHYLOXIME;
W/NBS   66522

735 C28 H61 N5 O6 S Si5 68373-38-6 KO-5-393-1 GLUCONERIDE OF 2-AMINO-4-(T-BUTYLAMINO)-6-THIO-S-TRIAZINE
; W   66523

735 C28 H61 N5 O6 S Si5 BS-5-69-0 Pertrimethylsilyl derivative of the s-glucuronide of 2-amino-4-(tert-
butylamino)-6-thio-s-triazine; W   128556

735 C45 H60 Cu N4 O J-40-2080-0 (α-Formyl-1,3,5,7-tetramethyl-2,4,6,8-tetra(n-pentyl)porphyrinato)copper
-(II); W   128557

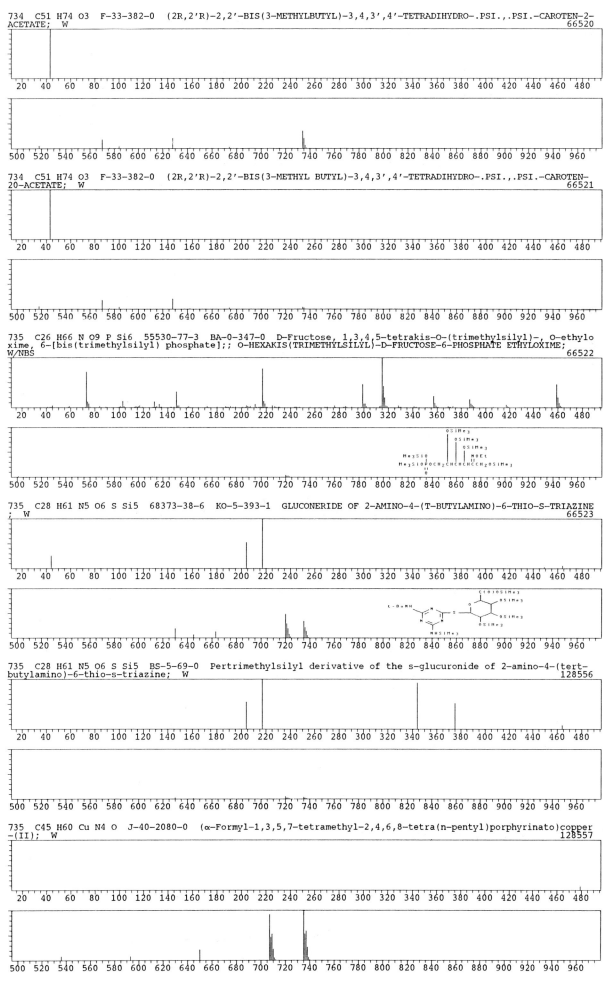

736 C31 H43 F3 N4 O11 S 69782-85-0 AB-207-251-3 L-Glutamic acid, N-[N-[N-[S-(2-methoxy-2-oxoethyl)-N-(trifluoroacetyl)-L-cysteinyl]-O-methyl-L-tyrosyl]-L-isoleucyl]-, dimethyl ester;; N-TFA-S-CARBOXYMETHYLCYSTEINYL-O-METHYLTYROSYLISOLEUCYL-GLUTAMIC-ACID METHYL ESTER; W/NBS
66524

736 C36 H32 O2 Sn2 E-37-27-1 TRIPHENYLTIN HYDROXIDE DIMER; W
66525

736 C40 H53 In N4 S 75845-79-3 KA-1980-2084-2 t-Butylthio[2,3,7,8,12,13,17,18-octaethylporphyrinato]indium; W
128558

736 C40 H64 O12 6833-84-7 KZ-221-353-7 Nonactin;; 4,13,22,31,37,38,39,40-Octaoxapentacyclo[32.2.1.1(7,10).1(16,19).1(25,28)]tetracontane-3,12,21,30-tetrone, 2,5,11,14,20,23,29,32-octamethyl-, [1R-(1R*,2R*,5R*,7R*,10S*,11S*,14S*,16S*,19R*,20R*,23R*,25R*,28S*,29S*,32S*,34S*)]-; W/NBS
66526

736 C52 H56 N4 28081-91-6 S-23-3299-6 ASTACENE BISPHENAZINE; W/NBS
66527

736 C52 H80 O2 10232-06-1 D-11-902-2 Phenol, 2-methoxy-6-(3,7,11,15,19,23,27,31,35-nonamethyl-2,6,10,14,18,22,26,30,34-hexatriacontanonaenyl)-;; 2-Nonaprenyl-6-methoxyphenol; W/NBS
66528

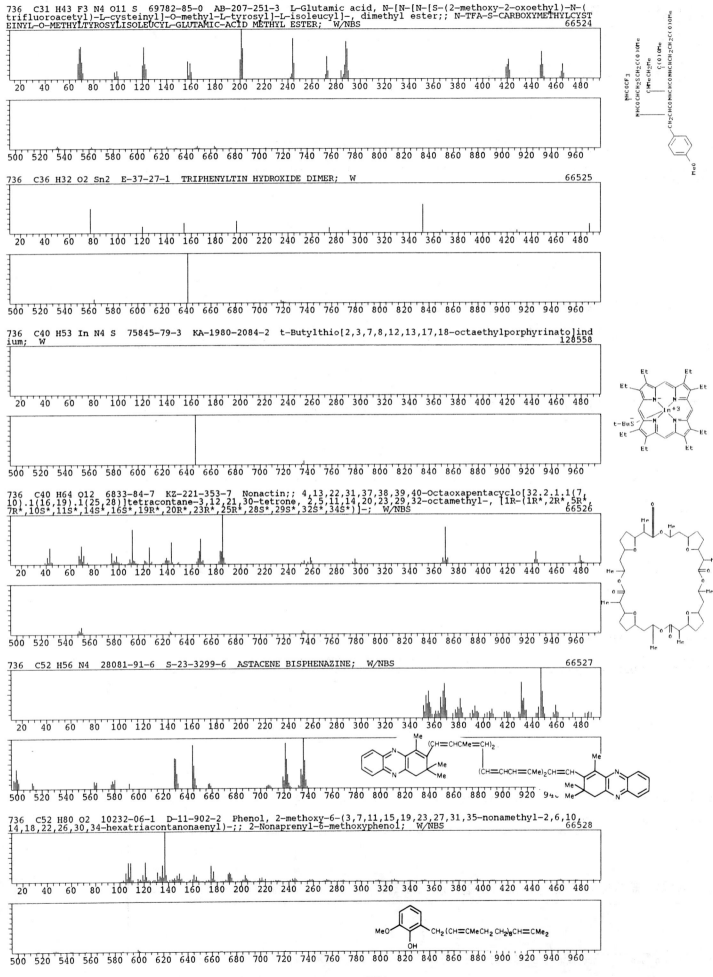

- 6134 -

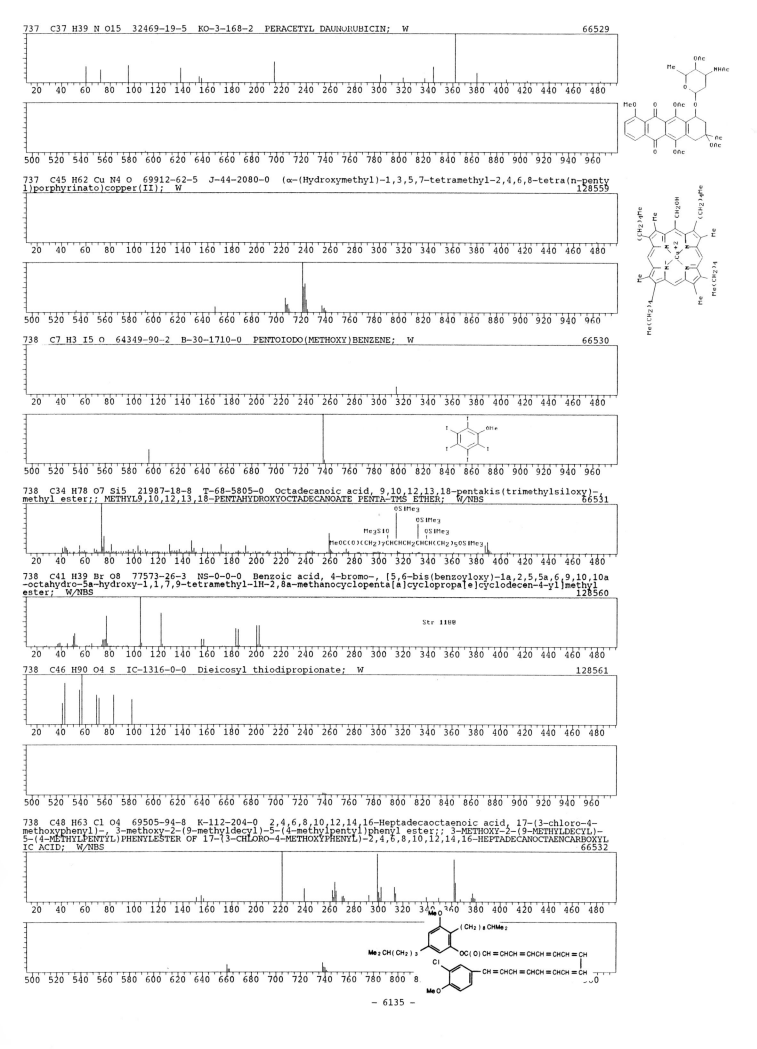

737   C37 H39 N O15   32469-19-5   KO-3-168-2   PERACETYL DAUNORUBICIN;   W   66529

737   C45 H62 Cu N4 O   69912-62-5   J-44-2080-0   (α-(Hydroxymethyl)-1,3,5,7-tetramethyl-2,4,6,8-tetra(n-penty
l)porphyrinato)copper(II);   W   128559

738   C7 H3 I5 O   64349-90-2   B-30-1710-0   PENTOIODO(METHOXY)BENZENE;   W   66530

738   C34 H78 O7 Si5   21987-18-8   T-68-5805-0   Octadecanoic acid, 9,10,12,13,18-pentakis(trimethylsiloxy)-,
methyl ester;; METHYL9,10,12,13,18-PENTAHYDROXYOCTADECANOATE PENTA-TMS ETHER;   W/NBS   66531

738   C41 H39 Br O8   77573-26-3   NS-0-0-0   Benzoic acid, 4-bromo-, [5,6-bis(benzoyloxy)-1a,2,5,5a,6,9,10,10a
-octahydro-5a-hydroxy-1,1,7,9-tetramethyl-1H-2,8a-methanocyclopenta[a]cyclopropa[e]cyclodecen-4-yl]methyl
ester;   W/NBS   128560

Str 1188

738   C46 H90 O4 S   IC-1316-0-0   Dieicosyl thiodipropionate;   W   128561

738   C48 H63 Cl O4   69505-94-8   K-112-204-0   2,4,6,8,10,12,14,16-Heptadecaoctaenoic acid, 17-(3-chloro-4-
methoxyphenyl)-, 3-methoxy-2-(9-methyldecyl)-5-(4-methylpentyl)phenyl ester;; 3-METHOXY-2-(9-METHYLDECYL)-
5-(4-METHYLPENTYL)PHENYLESTER OF 17-(3-CHLORO-4-METHOXYPHENYL)-2,4,6,8,10,12,14,16-HEPTADECANOCTAENCARBOXYL
IC ACID;   W/NBS   66532

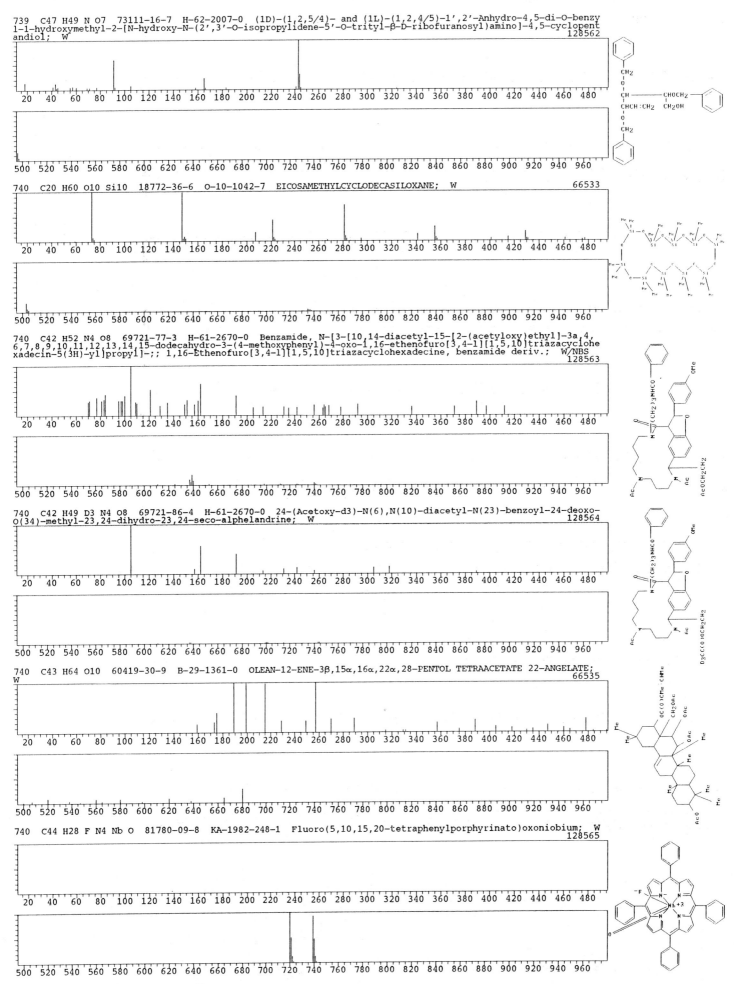

739 C47 H49 N O7 73111-16-7 H-62-2007-0 (1D)-(1,2,5/4)- and (1L)-(1,2,4/5)-1',2'-Anhydro-4,5-di-O-benzyl-1-hydroxymethyl-2-[N-hydroxy-N-(2',3'-O-isopropylidene-5'-O-trityl-β-D-ribofuranosyl)amino]-4,5-cyclopentandiol; W   128562

740 C20 H60 O10 Si10 18772-36-6 O-10-1042-7 EICOSAMETHYLCYCLODECASILOXANE; W   66533

740 C42 H52 N4 O8 69721-77-3 H-61-2670-0 Benzamide, N-[3-[10,14-diacetyl-15-[2-(acetyloxy)ethyl]-3a,4,6,7,8,9,10,11,12,13,14,15-dodecahydro-3-(4-methoxyphenyl)-4-oxo-1,16-ethenofuro[3,4-l][1,5,10]triazacyclohexadecin-5(3H)-yl]propyl]-;; 1,16-Ethenofuro[3,4-l][1,5,10]triazacyclohexadecine, benzamide deriv.; W/NBS   128563

740 C42 H49 D3 N4 O8 69721-86-4 H-61-2670-0 24-(Acetoxy-d3)-N(6),N(10)-diacetyl-N(23)-benzoyl-24-deoxo-O(34)-methyl-23,24-dihydro-23,24-seco-alphelandrine; W   128564

740 C43 H64 O10 60419-30-9 B-29-1361-0 OLEAN-12-ENE-3β,15α,16α,22α,28-PENTOL TETRAACETATE 22-ANGELATE; W   66535

740 C44 H28 F N4 Nb O 81780-09-8 KA-1982-248-1 Fluoro(5,10,15,20-tetraphenylporphyrinato)oxoniobium; W   128565

740   C44 H52 O10   80395-06-8   H-64-2250-0   (7S)-11',12-Diacetoxyspiro[8,12-abietadien-7,5''-5''H-7',9'(11
'),13'-abietatrieno[14',8',7'-bc]furan]-6,6',11,12',14-penta-one; W   128566

740   C44 H56 N2 O8   84251-57-0   KC-1982-2314-0   4,4'',4''',6'-Tetramethoxy-2',2''-dinitro-3,3''',5,5'''-tet
ra-t-butyl-m-quaterphenyl; W   128567

740   C47 H77 Cl O4   69506-00-9   K-112-205-0   MIXTURE OF 2-ISOUNDECYL-5-PENTYL- AND 2-DECYL-5-ISOHEXYL-3-MEO
-PHENYL17-(3-CL-4-MEO-PHENYL)HEPTADECANCARBOXYLATES; W   66536

740   C50 H76 O4   32719-43-0   SB-33-557-0   BACTERIORUBERIN; W   66537

741   C20 H8 F21 N O5   55538-89-1   AD-0-5206-0   Butanoic acid, heptafluoro-, 4-[2-[(2,2,3,3,4,4,4-heptafluor
o-1-oxobutyl)amino]ethyl]-1,2-phenylene ester;; 2-(3',4'-HEPTAFLUOROBUTYRYLPHENYL)-N-HEPTAFLUOROBUTYRYL-ETH
YLAMINE; W/NBS   66539

741   C20 H8 F21 N O5   55538-87-9   AD-0-5207-0   Butanoic acid, heptafluoro-, 4-[2-[(2,2,3,3,4,4,4-heptafluor
o-1-oxobutyl)amino]-1-(2,2,3,3,4,4,4-heptafluoro-1-oxobutoxy)ethyl]phenyl ester;; O,O,N,-TRIS-HEPTAFLUOROBU
TYRYL-OCTOPAMINE; W/NBS   66538

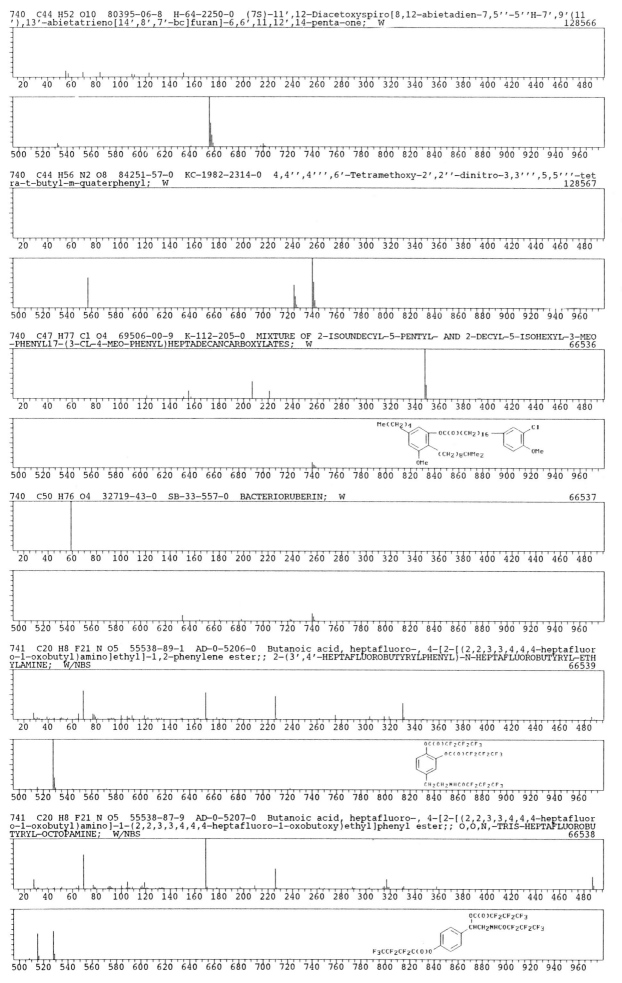

741   C29 H67 N O9 Si6   55538-91-5   AM-0-73-0   D-glycero-D-galacto-2-Nonulosonic acid, 5-(acetylamino)-3,5-dideoxy-4,6,7,8,9-pentakis-O-(trimethylsilyl)-, trimethylsilyl ester;; N-ACETYL NEURAMINIC ACID-HEXA-TMS; W/NBS
66540

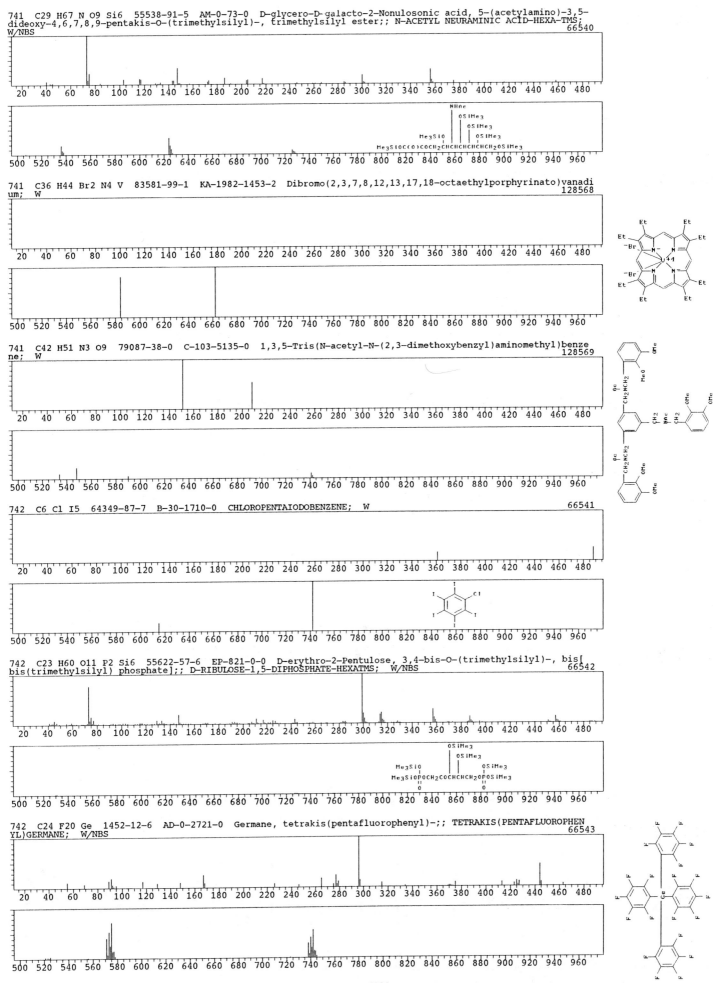

741   C36 H44 Br2 N4 V   83581-99-1   KA-1982-1453-2   Dibromo(2,3,7,8,12,13,17,18-octaethylporphyrinato)vanadium; W
128568

741   C42 H51 N3 O9   79087-38-0   C-103-5135-0   1,3,5-Tris(N-acetyl-N-(2,3-dimethoxybenzyl)aminomethyl)benzene; W
128569

742   C6 Cl I5   64349-87-7   B-30-1710-0   CHLOROPENTAIODOBENZENE; W
66541

742   C23 H60 O11 P2 Si6   55622-57-6   EP-821-0-0   D-erythro-2-Pentulose, 3,4-bis-O-(trimethylsilyl)-, bis[bis(trimethylsilyl) phosphate];; D-RIBULOSE-1,5-DIPHOSPHATE-HEXATMS; W/NBS
66542

742   C24 F20 Ge   1452-12-6   AD-0-2721-0   Germane, tetrakis(pentafluorophenyl)-;; TETRAKIS(PENTAFLUOROPHENYL)GERMANE; W/NBS
66543

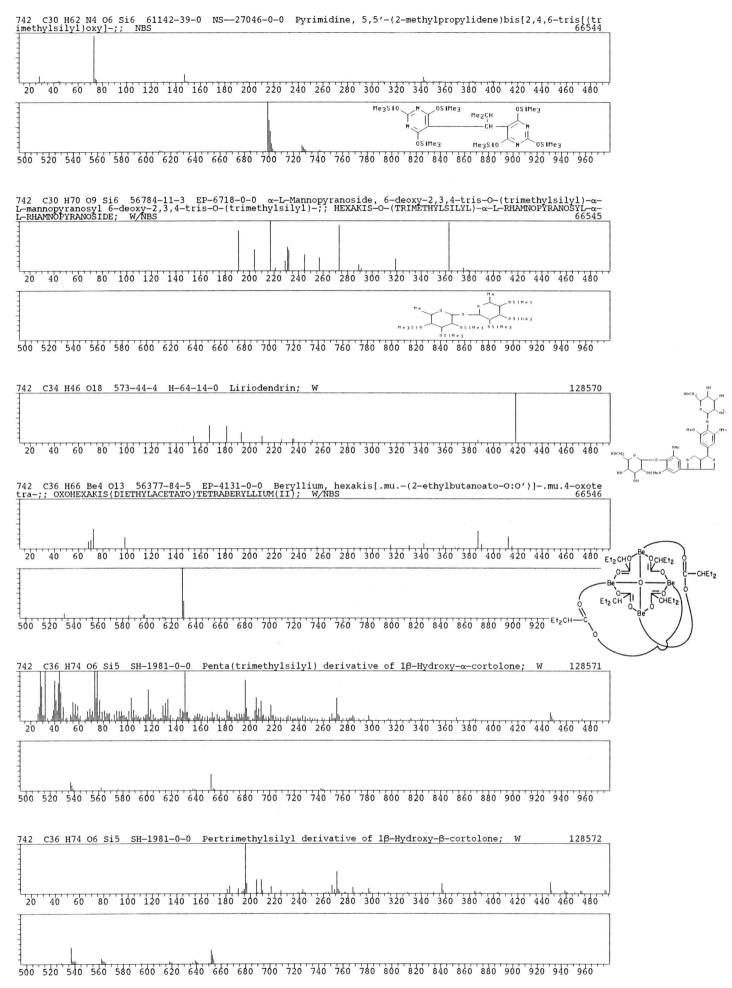

742   C30 H62 N4 O6 Si6   61142-39-0   NS--27046-0-0   Pyrimidine, 5,5'-(2-methylpropylidene)bis[2,4,6-tris[(trimethylsilyl)oxy]-;;   NBS
66544

742   C30 H70 O9 Si6   56784-11-3   EP-6718-0-0   α-L-Mannopyranoside, 6-deoxy-2,3,4-tris-O-(trimethylsilyl)-α-L-mannopyranosyl 6-deoxy-2,3,4-tris-O-(trimethylsilyl)-;; HEXAKIS-O-(TRIMETHYLSILYL)-α-L-RHAMNOPYRANOSYL-α-L-RHAMNOPYRANOSIDE; W/NBS
66545

742   C34 H46 O18   573-44-4   H-64-14-0   Liriodendrin; W
128570

742   C36 H66 Be4 O13   56377-84-5   EP-4131-0-0   Beryllium, hexakis[.mu.-(2-ethylbutanoato-O:O')]-.mu.4-oxotetra-;; OXOHEXAKIS(DIETHYLACETATO)TETRABERYLLIUM(II); W/NBS
66546

742   C36 H74 O6 Si5   SH-1981-0-0   Penta(trimethylsilyl) derivative of 1β-Hydroxy-α-cortolone; W   128571

742   C36 H74 O6 Si5   SH-1981-0-0   Pertrimethylsilyl derivative of 1β-Hydroxy-β-cortolone; W   128572

742   C36 H74 O6 Si5   SH-1981-0-0   Pertrimethylsilyl derivative of 6α-Hydroxy-β-cortolone;   W   128573

742   C37 H47 In N4 O3 S   KA-1980-2084-2   Methylsulphonato[2,3,7,8,12,13,17,18-octaethylporphyrinato]indium; W   128574

742   C42 H54 N4 O8   69721-74-0   H-61-2669-0   3,7,12-Triazabicyclo[13.3.1]nonadeca-1(19),15,17-triene-2-acet
ic acid, 3,7-diacetyl-12-[3-(benzoylamino)propyl]-16-methoxy-14-[(4-methoxyphenyl)methyl]-13-oxo-, methyl
ester;; 3,7,12-Triazabicyclo[13.3.1]nonadeca-1(19),15,17-triene-2-acetic acid, 3,7-diacetyl-12-[3-(benzoyla
mino)propyl]-16-methoxy-14-[(4-methoxyphenyl)];   W/NBS   128575

742   C46 H62 S4   C-104-7146-0   1,1,6,6-Tetrakis(p-tert-butylphenyl)thio)hexane;   W   128432

742   C46 H94 O6   87515-24-0   F-39-2214-0   2,5-Di-O-(2,3-dihydrophytyl)-D-mannitol;   W   128576

743   C29 H57 N5 O8 Si5   60690-43-9   KO-2-92-2   N*6-(2-SUCCINYL)ADENOSINE;   W   66548

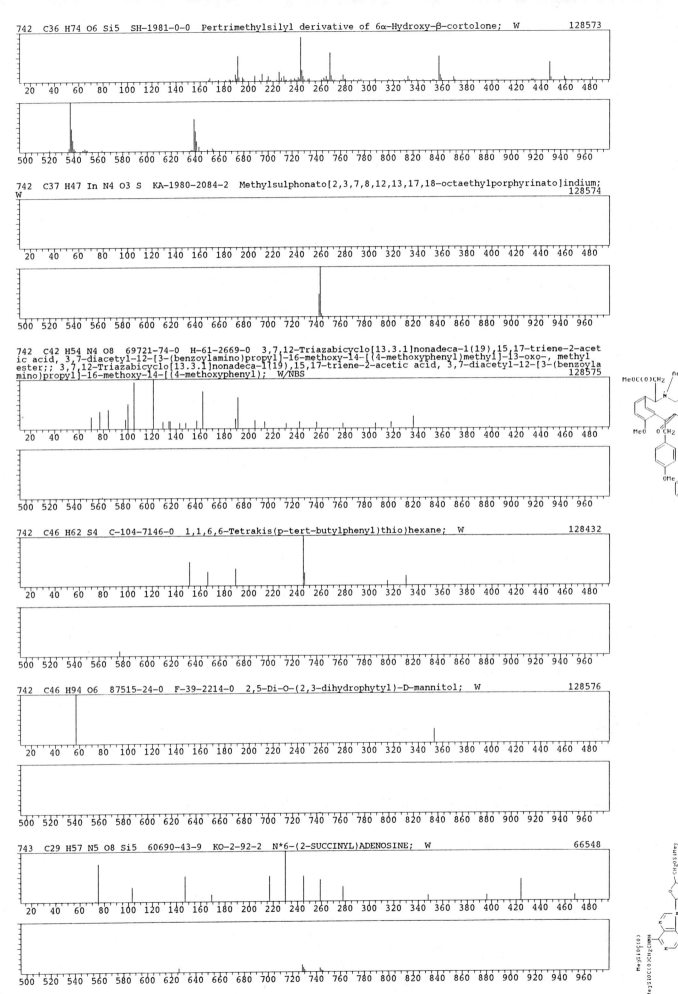

743 C36 H35 Mo2 O5 AG-134-287-3 (Dicarbonyl)(1-3-η-cycloheptatrienyl)(1-7-η-cycloheptatrienyl)-tris-.mu.-phenoxydimolybdenum; W   128577

743 C39 H65 N7 O7 42547-82-0 O7-0-817-0 L-Leucine, N-[N-[N-[N-[N-(2-pyridinylmethylene)-L-valyl]-L-isoleucyl]-L-alanyl]-L-leucyl]-L-leucyl]-, methyl ester;; 2-PYRIDYLMETHYLIDENE VALYLISOLEUCYLALANYLLEUCYLLEUCYLLEUCINE; W/NBS   66549

743 C41 H69 N5 O7 35146-60-2 O-6-14-3 L-Phenylalanine, N-[N-[N-[N2,N5-bis(1-oxodecyl)-L-ornithyl]-L-alanyl]-L-alanyl]-, methyl ester;; BISDECANOYLORNITHYLALANYLALANYLPHENYLALANINE METHYL ESTER; W/NBS   66550

744 C30 H25 Fe3 O8 P S 81877-19-2 K-115-1303-0 .mu.(3)-(tert-Butylthiolato)-octacarbonyl-.mu.-hydrido-(triphenylphosphine)triiron; W   128578

744 C41 H44 O13 69127-07-7 KC-1981-1241-0 (2R,3S,4R)-2,3-trans-3,4-cis-3-Acetoxy-4-[(2R,3S)-2,3-trans-3-acetoxy-3',4',5,7-tetramethoxyflavan-8-yl]-3',4',7-trimethoxyflavan; W   128579

744 C41 H44 O13 69127-08-8 KC-1981-1241-0 (2R,3S,4R)-2,3-trans-3,4-cis-3-Acetoxy-4-[(2R,3S)-2,3-trans-3-acetoxy-3',4',5,7-tetramethoxyflavan-8-yl]-3',4',7-trimethoxyflavan; W   128580

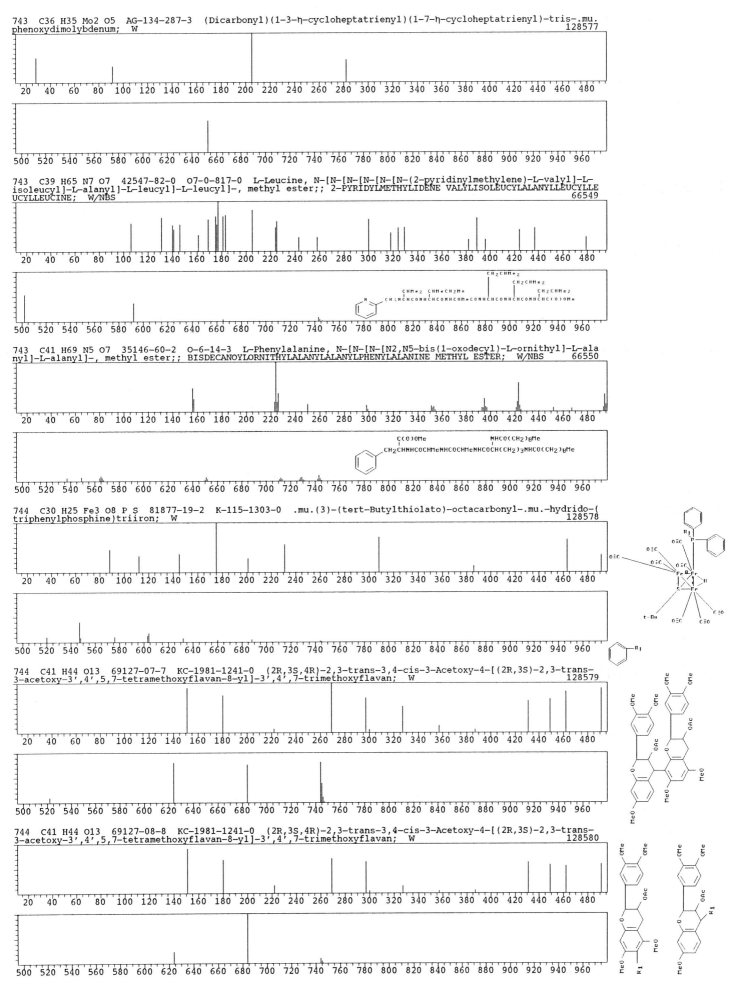

744   C41 H44 O13   69082-83-3   KC-1981-1241-0   (2R,3S,4S)-2,3-trans-3,4-trans-3-Acetoxy-4-[(2R,3S)-2,3-trans
-3-acetoxy-3',4',5,7-tetramethoxyflavan-8-yl]-3',4',7-trimethoxyflavan;   W   128581

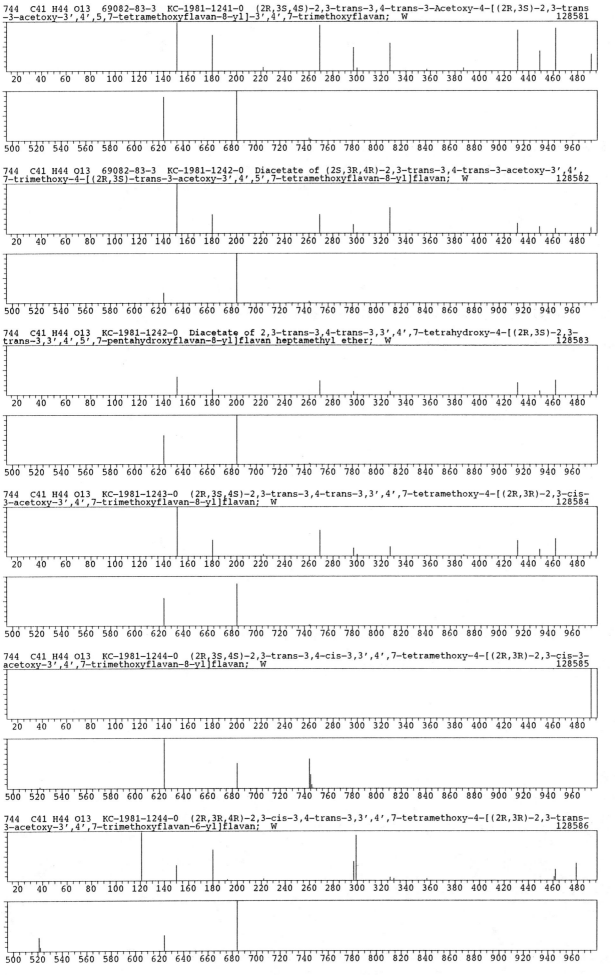

744   C41 H44 O13   69082-83-3   KC-1981-1242-0   Diacetate of (2S,3R,4R)-2,3-trans-3,4-trans-3-acetoxy-3',4'
7-trimethoxy-4-[(2R,3S)-trans-3-acetoxy-3',4',5',7-tetramethoxyflavan-8-yl]flavan;   W   128582

744   C41 H44 O13   KC-1981-1242-0   Diacetate of 2,3-trans-3,4-trans-3,3',4',7-tetrahydroxy-4-[(2R,3S)-2,3-
trans-3,3',4',5',7-pentahydroxyflavan-8-yl]flavan heptamethyl ether;   W   128583

744   C41 H44 O13   KC-1981-1243-0   (2R,3S,4S)-2,3-trans-3,4-trans-3,3',4',7-tetramethoxy-4-[(2R,3R)-2,3-cis-
3-acetoxy-3',4',7-trimethoxyflavan-8-yl]flavan;   W   128584

744   C41 H44 O13   KC-1981-1244-0   (2R,3S,4S)-2,3-trans-3,4-cis-3,3',4',7-tetramethoxy-4-[(2R,3R)-2,3-cis-3-
acetoxy-3',4',7-trimethoxyflavan-8-yl]flavan;   W   128585

744   C41 H44 O13   KC-1981-1244-0   (2R,3R,4R)-2,3-cis-3,4-trans-3,3',4',7-tetramethoxy-4-[(2R,3R)-2,3-trans-
3-acetoxy-3',4',7-trimethoxyflavan-6-yl]flavan;   W   128586

744 C41 H44 O13 KC-1981-1244-0 (2R,3R,4R)-2,3-cis-3,4-trans-3,3',4',7-tetramethoxy-4-[(2R,3R)-2,3-trans-3-acetoxy-3',4',7-trimethoxyflavan-8-yl]flavan; W   128587

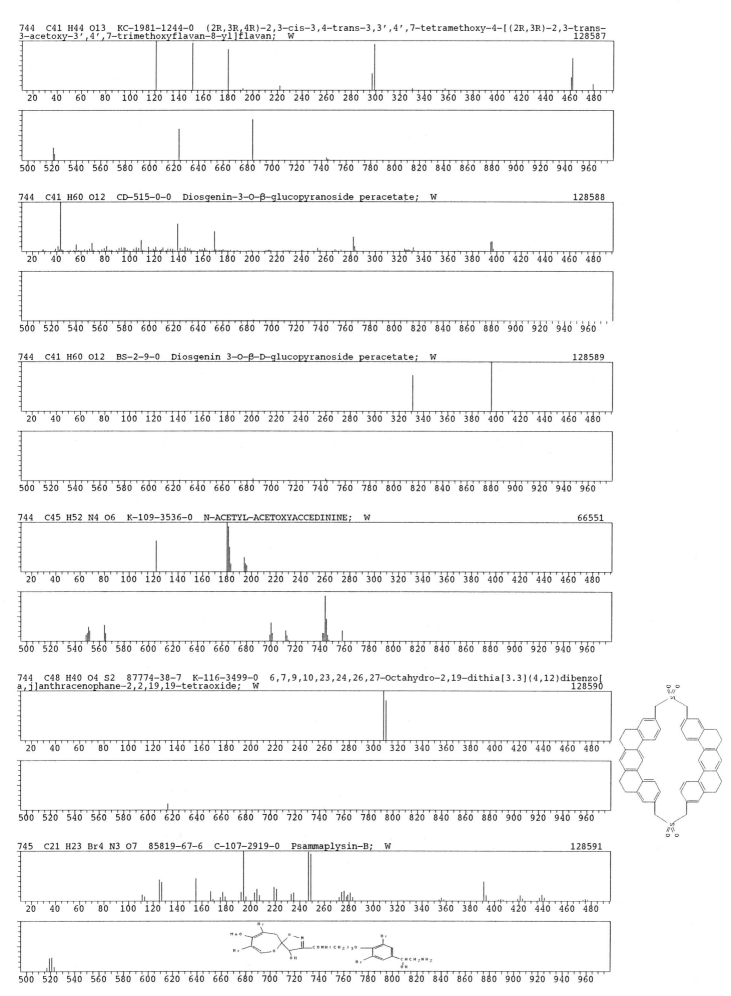

744 C41 H60 O12 CD-515-0-0 Diosgenin-3-O-β-glucopyranoside peracetate; W   128588

744 C41 H60 O12 BS-2-9-0 Diosgenin 3-O-β-D-glucopyranoside peracetate; W   128589

744 C45 H52 N4 O6 K-109-3536-0 N-ACETYL-ACETOXYACCEDININE; W   66551

744 C48 H40 O4 S2 87774-38-7 K-116-3499-0 6,7,9,10,23,24,26,27-Octahydro-2,19-dithia[3.3](4,12)dibenzo[a,j]anthracenophane-2,2,19,19-tetraoxide; W   128590

745 C21 H23 Br4 N3 O7 85819-67-6 C-107-2919-0 Psammaplysin-B; W   128591

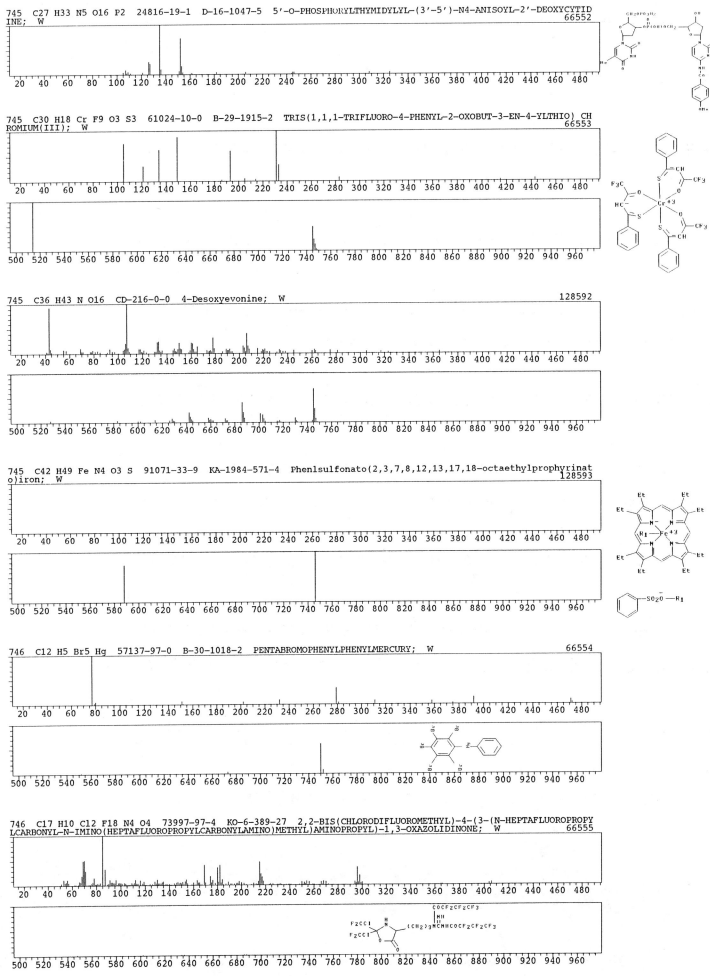

745 C27 H33 N5 O16 P2 24816-19-1 D-16-1047-5 5'-O-PHOSPHORYLTHYMIDYLYL-(3'-5')-N4-ANISOYL-2'-DEOXYCYTID
INE; W   66552

745 C30 H18 Cr F9 O3 S3 61024-10-0 B-29-1915-2 TRIS(1,1,1-TRIFLUORO-4-PHENYL-2-OXOBUT-3-EN-4-YLTHIO) CH
ROMIUM(III); W   66553

745 C36 H43 N O16 CD-216-0-0 4-Desoxyevonine; W   128592

745 C42 H49 Fe N4 O3 S 91071-33-9 KA-1984-571-4 Phenlsulfonato(2,3,7,8,12,13,17,18-octaethylprophyrinat
o)iron; W   128593

746 C12 H5 Br5 Hg 57137-97-0 B-30-1018-2 PENTABROMOPHENYLPHENYLMERCURY; W   66554

746 C17 H10 Cl2 F18 N4 O4 73997-97-4 KO-6-389-27 2,2-BIS(CHLORODIFLUOROMETHYL)-4-(3-(N-HEPTAFLUOROPROPY
LCARBONYL-N-IMINO(HEPTAFLUOROPROPYLCARBONYLAMINO)METHYL)AMINOPROPYL)-1,3-OXAZOLIDINONE; W 66555

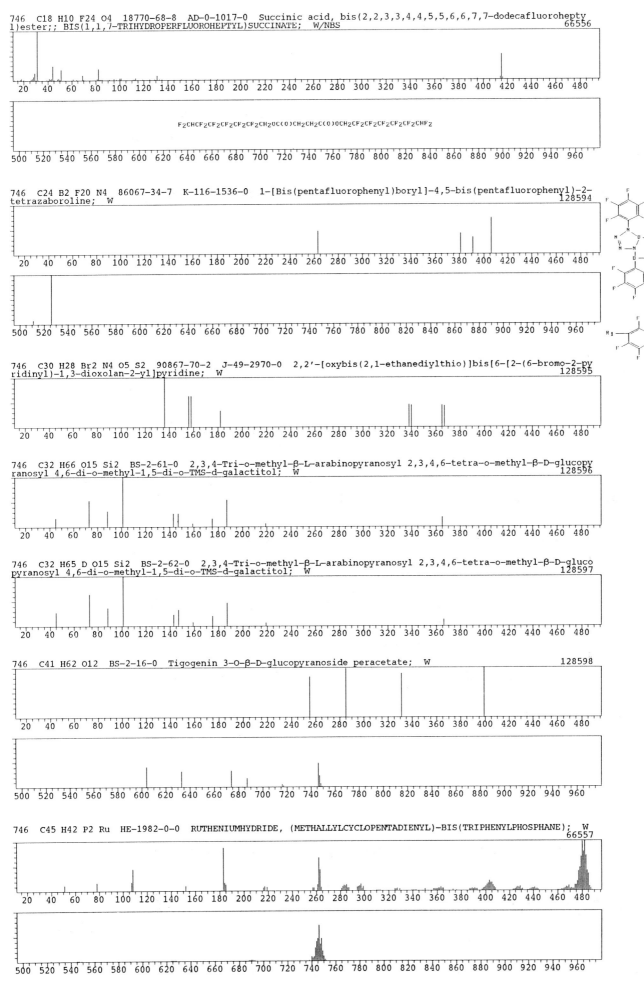

746   C18 H10 F24 O4   18770-68-8   AD-0-1017-0   Succinic acid, bis(2,2,3,3,4,4,5,5,6,6,7,7-dodecafluorohepty
l)ester;; BIS(1,1,7-TRIHYDROPERFLUOROHEPTYL)SUCCINATE; W/NBS   66556

F2CHCF2CF2CF2CF2CF2CH2OC(O)CH2CH2C(O)OCH2CF2CF2CF2CF2CF2CHF2

746   C24 B2 F20 N4   86067-34-7   K-116-1536-0   1-[Bis(pentafluorophenyl)boryl]-4,5-bis(pentafluorophenyl)-2-
tetrazaboroline; W   128594

746   C30 H28 Br2 N4 O5 S2   90867-70-2   J-49-2970-0   2,2'-[oxybis(2,1-ethanediylthio)]bis[6-[2-(6-bromo-2-py
ridinyl)-1,3-dioxolan-2-yl]pyridine; W   128595

746   C32 H66 O15 Si2   BS-2-61-0   2,3,4-Tri-o-methyl-β-L-arabinopyranosyl 2,3,4,6-tetra-o-methyl-β-D-glucopy
ranosyl 4,6-di-o-methyl-1,5-di-o-TMS-d-galactitol; W   128596

746   C32 H65 D O15 Si2   BS-2-62-0   2,3,4-Tri-o-methyl-β-L-arabinopyranosyl 2,3,4,6-tetra-o-methyl-β-D-gluco
pyranosyl 4,6-di-o-methyl-1,5-di-o-TMS-d-galactitol; W   128597

746   C41 H62 O12   BS-2-16-0   Tigogenin 3-O-β-D-glucopyranoside peracetate; W   128598

746   C45 H42 P2 Ru   HE-1982-0-0   RUTHENIUMHYDRIDE, (METHALLYLCYCLOPENTADIENYL)-BIS(TRIPHENYLPHOSPHANE); W
                                                                                           66557

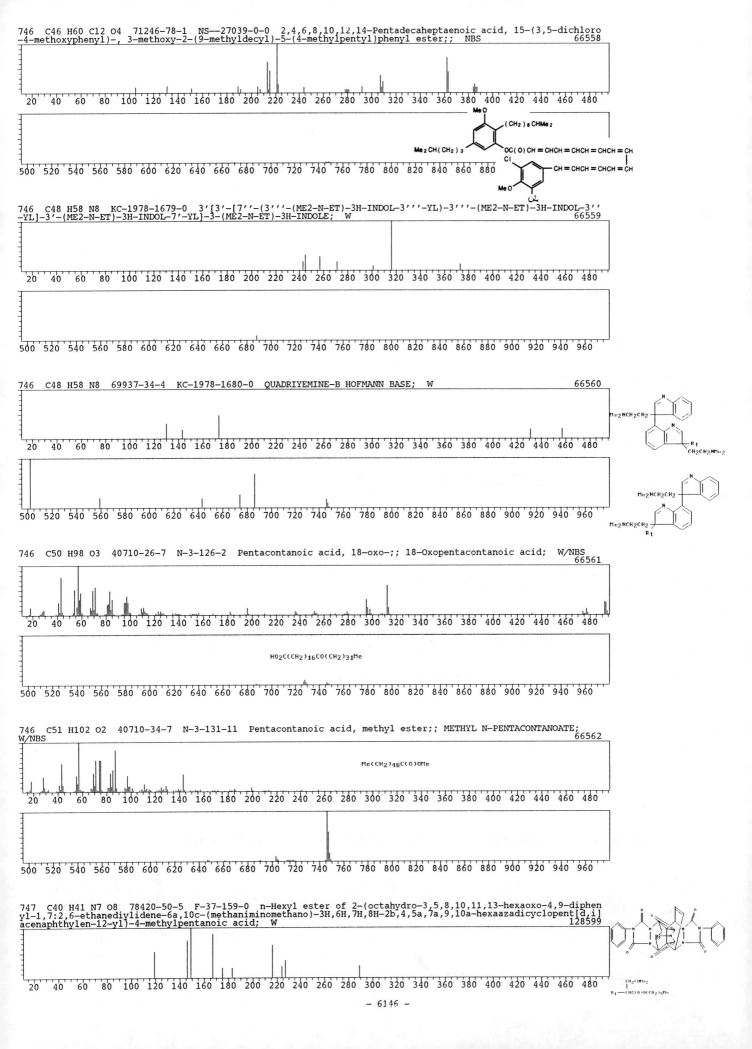

746 C46 H60 Cl2 O4 71246-78-1 NS--27039-0-0 2,4,6,8,10,12,14-Pentadecaheptaenoic acid, 15-(3,5-dichloro
-4-methoxyphenyl)-, 3-methoxy-2-(9-methyldecyl)-5-(4-methylpentyl)phenyl ester;; NBS
66558

746 C48 H58 N8 KC-1978-1679-0 3'{3'-[7''-(3'''-(ME2-N-ET)-3H-INDOL-3'''-YL)-3'''-(ME2-N-ET)-3H-INDOL-3''
-YL]-3'-(ME2-N-ET)-3H-INDOL-7'-YL}-3-(ME2-N-ET)-3H-INDOLE; W
66559

746 C48 H58 N8 69937-34-4 KC-1978-1680-0 QUADRIYEMINE-B HOFMANN BASE; W
66560

746 C50 H98 O3 40710-26-7 N-3-126-2 Pentacontanoic acid, 18-oxo-;; 18-Oxopentacontanoic acid; W/NBS
66561

HO2C(CH2)16CO(CH2)31Me

746 C51 H102 O2 40710-34-7 N-3-131-11 Pentacontanoic acid, methyl ester;; METHYL N-PENTACONTANOATE;
W/NBS
66562

Me(CH2)48C(O)OMe

747 C40 H41 N7 O8 78420-50-5 F-37-159-0 n-Hexyl ester of 2-(octahydro-3,5,8,10,11,13-hexaoxo-4,9-diphen
yl-1,7:2,6-ethanediylidene-6a,10c-(methaniminomethano)-3H,6H,7H,8H-2b,4,5a,7a,9,10a-hexaazadicyclopent[d,i]
acenaphthylen-12-yl)-4-methylpentanoic acid; W
128599

- 6146 -

747 C40 H68 Na O11 28643-80-3 O-3-282-4 Nigericin, monosodium salt;; Nigericin sodium salt; W/NBS
66563

747 C42 H49 N3 O4 S Si2 95694-59-0 C-107-3260-0 1,4-Dibenzyl-3-(2'-thiopyridyl)-6-(1''-(((tert-butyldi phenylsilyl)oxy)-methyl)-3'-(((tert-butyldimethylsilyl)oxy)propyl))-2,5-piperazinedione; W
128600

747 C44 H37 N5 O7 81352-48-9 Y-19-601-0 6,3'-N-O-Dibenzoyl-5'-O-monomethyoxytrityl-9β-D-xylofurannosyla denine; W
128601

747 C45 H31 Fe N4 O2 S 82008-30-8 KA-1984-571-4 Methylsulfinato(5,10,15,20-tetraphenylporphyrinato)iron ; W
128602

748 C12 H36 Ge6 P4 28133-43-9 C-97-6365-0 HEXA(DIMETHYLGERMA)TETRAPHOSPHIDE; W
66564

748 C24 H8 Cl4 N4 O8 S4 60822-10-8 F-32-2569-0 1,8,15,22-TETRATHIA[1,1,1,1]-3,5,17,19-TETRANITRO-10,12, 24,26-TETRACHLORO-METACYCLOPHANE; W
66565

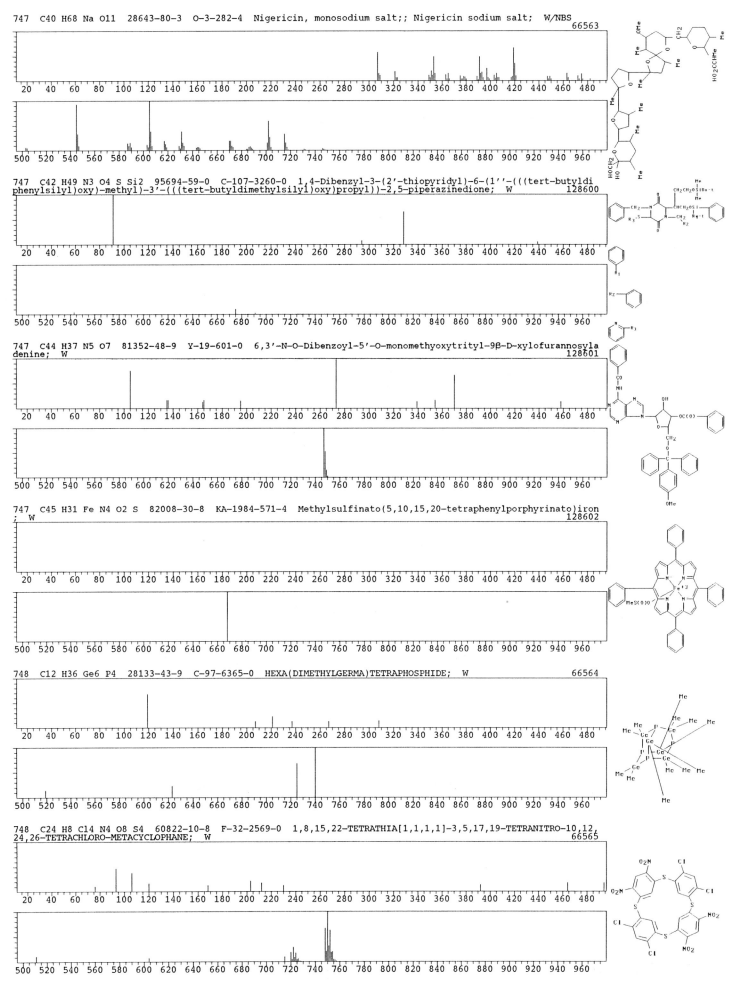

748 C32 H32 Cl4 Ti4 56732-57-1 AG-90-331-1 Clclooctatetraene titanium chloride; W   128603

748 C38 H52 O15 SB-32-96-0 APIGENIN 6,8-DI-C-β-D-GLUCOPYRANOSIDE PERMETHYL ETHER; W   66566

748 C38 H19 D33 O15 SB-32-96-0 APIGENIN 6,8-DI-C-β-D-GLUCOPYRANOSIDE PERDEUTERIOMETHYL ETHER; W 66567

748 C38 H52 O15 56254-40-1 NS-0-0-0 4H-1-Benzopyran-4-one, 5,7-dimethoxy-2-(4-methoxyphenyl)-6,8-bis(2,3,4,6-tetra-O-methyl-β-D-glucopyranosyl)-; W/NBS   128604

748 C39 H56 O14 42011-33-6 O-8-92-1 ACETYL ATRACTYLOSIDE; W   66568

748 C42 H68 O11 KO-3-273-2 SALINOMYCIN; W   66569

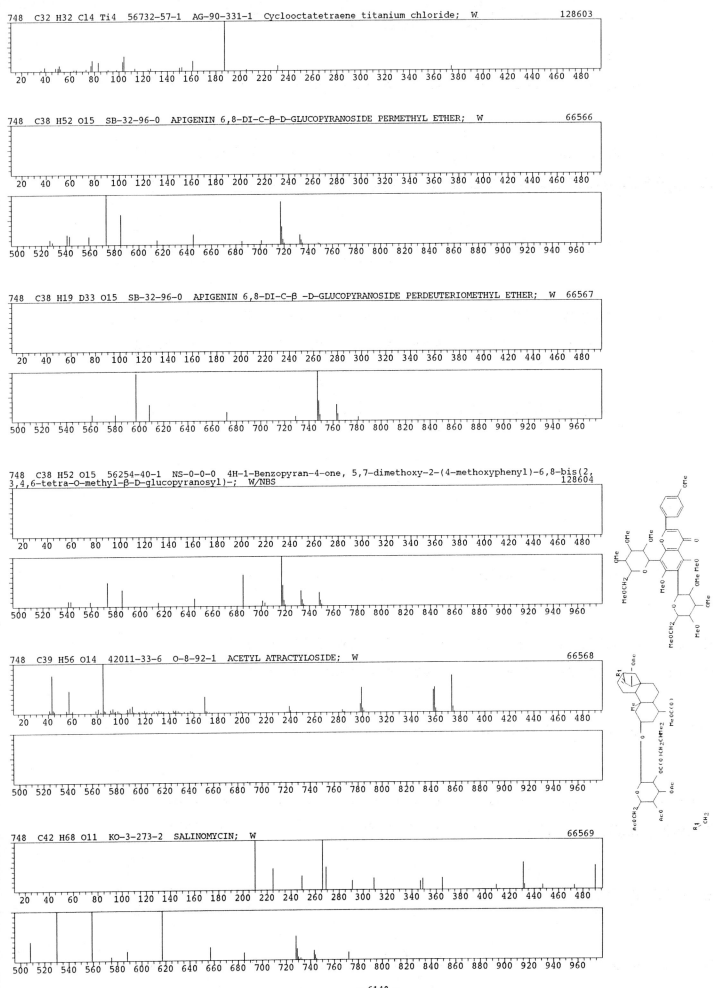

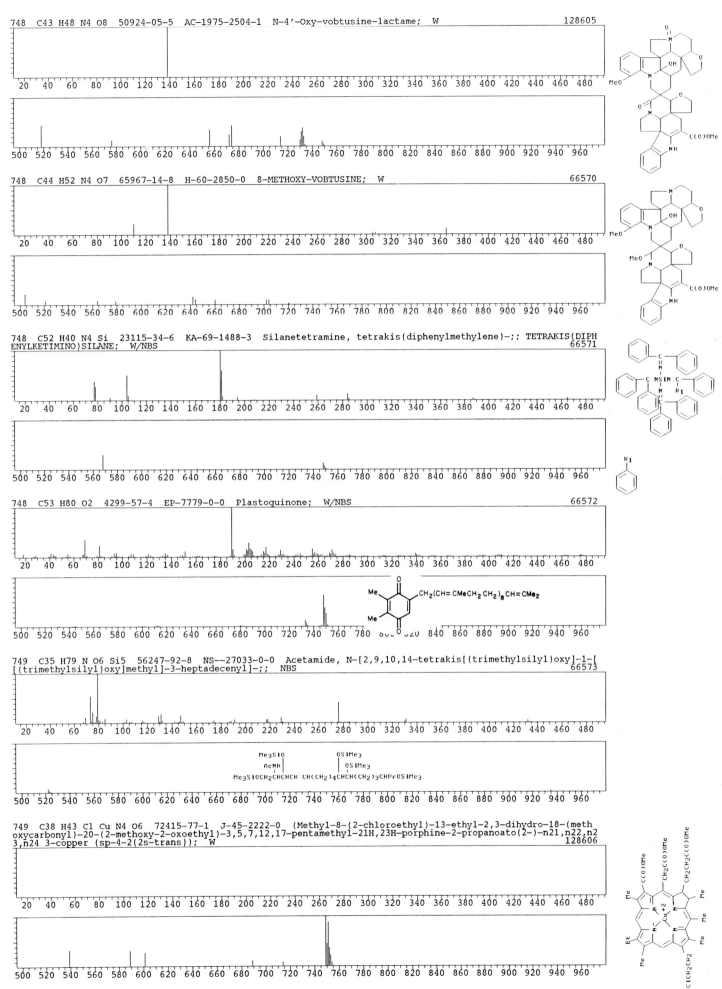

748　C43 H48 N4 O8　50924-05-5　AC-1975-2504-1　N-4'-Oxy-vobtusine-lactame;　W　128605

748　C44 H52 N4 O7　65967-14-8　H-60-2850-0　8-METHOXY-VOBTUSINE;　W　66570

748　C52 H40 N4 Si　23115-34-6　KA-69-1488-3　Silanetetramine, tetrakis(diphenylmethylene)-;; TETRAKIS(DIPHENYLKETIMINO)SILANE;　W/NBS　66571

748　C53 H80 O2　4299-57-4　EP-7779-0-0　Plastoquinone;　W/NBS　66572

749　C35 H79 N O6 Si5　56247-92-8　NS--27033-0-0　Acetamide, N-[2,9,10,14-tetrakis[(trimethylsilyl)oxy]-1-[[(trimethylsilyl)oxy]methyl]-3-heptadecenyl]-;;　NBS　66573

749　C38 H43 Cl Cu N4 O6　72415-77-1　J-45-2222-0　(Methyl-8-(2-chloroethyl)-13-ethyl-2,3-dihydro-18-(methoxycarbonyl)-20-(2-methoxy-2-oxoethyl)-3,5,7,12,17-pentamethyl-21H,23H-porphine-2-propanoato(2-)-n21,n22,n23,n24 3-copper (sp-4-2(2s-trans));　W　128606

749   C38 H44 B2 Co N12   80593-38-0   C-104-1549-0   Bis(4-n-butylphenyltris(pyrazolyl)borato)cobalt (II); W
128607

750   C19 F15 O2 Tl   B-31-1720-0   PENTAFLUOROBENZOATOBIS(PENTAFLUOROPHENYL)THALLIUM(III); W
66574

750   C22 H54 F6 N4 O2 P2 Si6   68222-40-2   K-111-3111-0   2,4-BIS(DI-TMS-AMINO)-6,7-BIS(TRIFLUOROMETHYL)-1,3-
BIS-TMS-5,8-DIOXA-1,3-DIAZA-2λ*3,4λ*5-DIPHOSPHASPIRO[3.4]OCT-6-ENE; W
66575

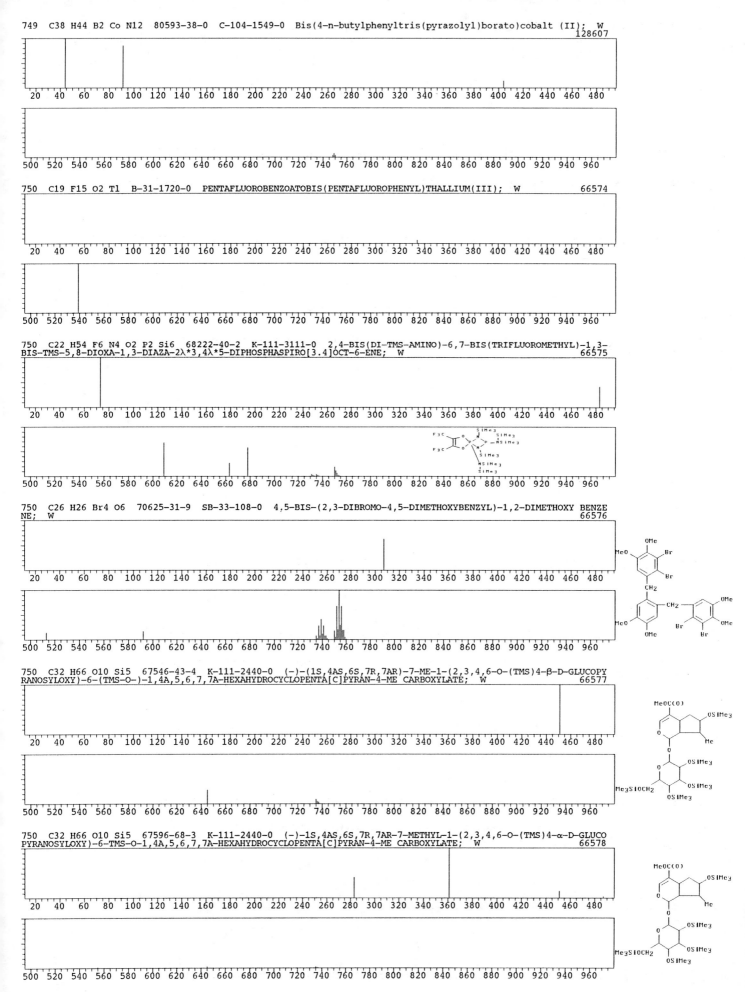

750   C26 H26 Br4 O6   70625-31-9   SB-33-108-0   4,5-BIS-(2,3-DIBROMO-4,5-DIMETHOXYBENZYL)-1,2-DIMETHOXY BENZE
NE; W
66576

750   C32 H66 O10 Si5   67546-43-4   K-111-2440-0   (-)-(1S,4AS,6S,7R,7AR)-7-ME-1-(2,3,4,6-O-(TMS)4-β-D-GLUCOPY
RANOSYLOXY)-6-(TMS-O-)-1,4A,5,6,7,7A-HEXAHYDROCYCLOPENTA[C]PYRAN-4-ME CARBOXYLATE; W
66577

750   C32 H66 O10 Si5   67596-68-3   K-111-2440-0   (-)-1S,4AS,6S,7R,7AR-7-METHYL-1-(2,3,4,6-O-(TMS)4-α-D-GLUCO
PYRANOSYLOXY)-6-TMS-O-1,4A,5,6,7,7A-HEXAHYDROCYCLOPENTA[C]PYRAN-4-ME CARBOXYLATE; W
66578

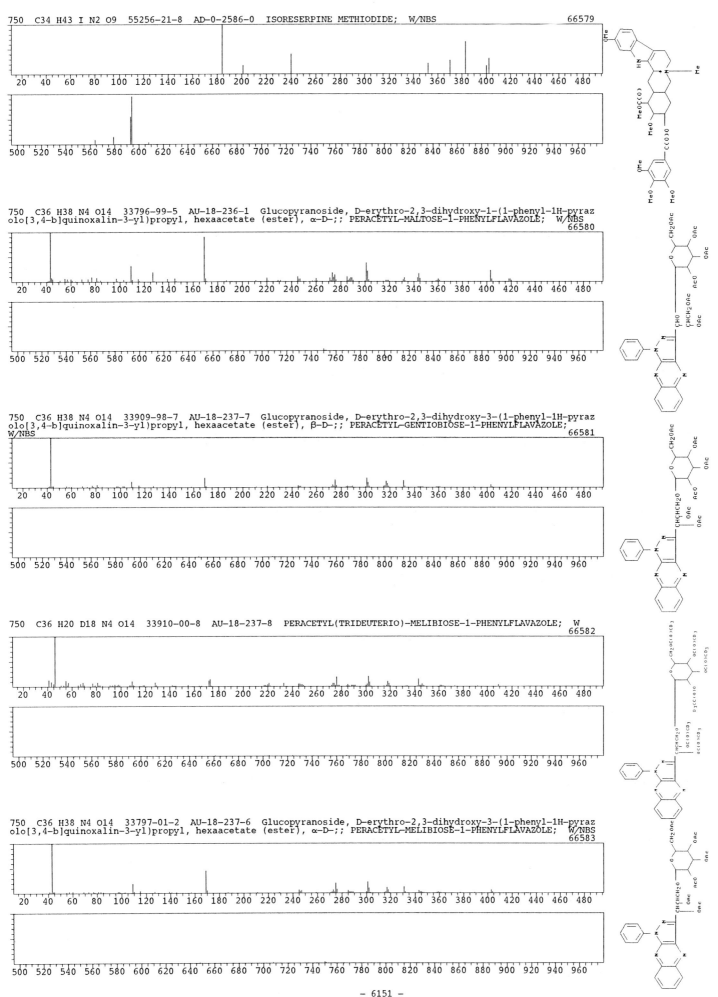

750　C34 H43 I N2 O9　55256-21-8　AD-0-2586-0　ISORESERPINE METHIODIDE;　W/NBS　66579

750　C36 H38 N4 O14　33796-99-5　AU-18-236-1　Glucopyranoside, D-erythro-2,3-dihydroxy-1-(1-phenyl-1H-pyraz
olo[3,4-b]quinoxalin-3-yl)propyl, hexaacetate (ester), α-D-;; PERACETYL-MALTOSE-1-PHENYLFLAVAZOLE;　W/NBS
66580

750　C36 H38 N4 O14　33909-98-7　AU-18-237-7　Glucopyranoside, D-erythro-2,3-dihydroxy-3-(1-phenyl-1H-pyraz
olo[3,4-b]quinoxalin-3-yl)propyl, hexaacetate (ester), β-D-;; PERACETYL-GENTIOBIOSE-1-PHENYLFLAVAZOLE;
W/NBS　66581

750　C36 H20 D18 N4 O14　33910-00-8　AU-18-237-8　PERACETYL(TRIDEUTERIO)-MELIBIOSE-1-PHENYLFLAVAZOLE;　W
66582

750　C36 H38 N4 O14　33797-01-2　AU-18-237-6　Glucopyranoside, D-erythro-2,3-dihydroxy-3-(1-phenyl-1H-pyraz
olo[3,4-b]quinoxalin-3-yl)propyl, hexaacetate (ester), α-D-;; PERACETYL-MELIBIOSE-1-PHENYLFLAVAZOLE;　W/NBS
66583

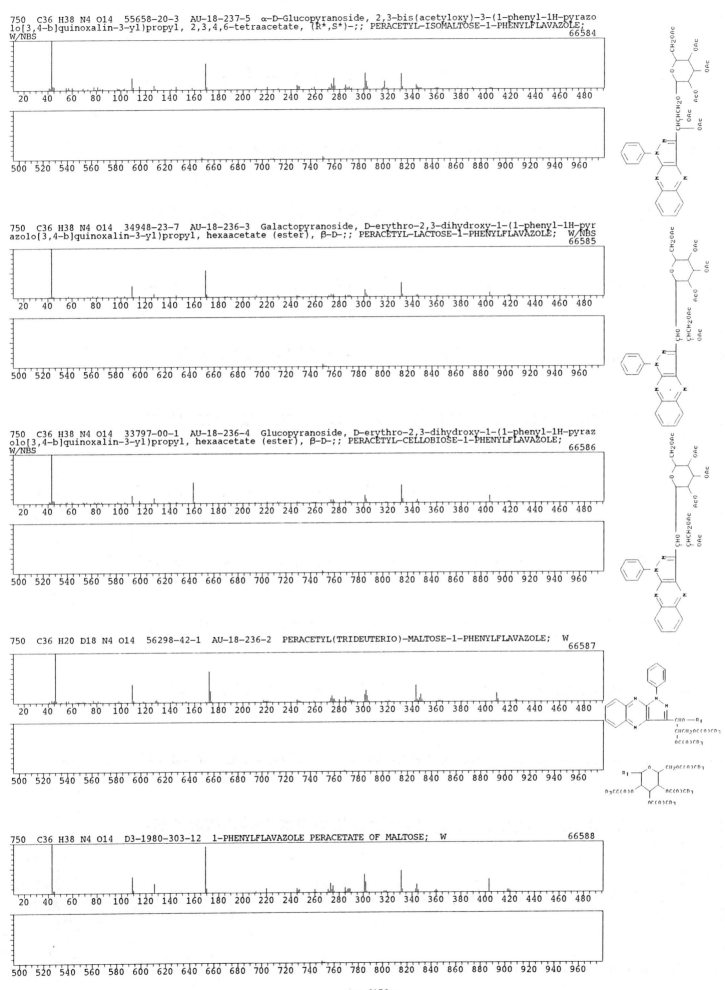

750   C36 H38 N4 O14   55658-20-3   AU-18-237-5   α–D–Glucopyranoside, 2,3–bis(acetyloxy)–3–(1–phenyl–1H–pyrazo
lo[3,4–b]quinoxalin–3–yl)propyl, 2,3,4,6–tetraacetate, (R*,S*)–;; PERACETYL–ISOMALTOSE–1–PHENYLFLAVAZOLE;
W/NBS   66584

750   C36 H38 N4 O14   34948-23-7   AU-18-236-3   Galactopyranoside, D–erythro–2,3–dihydroxy–1–(1–phenyl–1H–pyr
azolo[3,4–b]quinoxalin–3–yl)propyl, hexaacetate (ester), β–D–;; PERACETYL–LACTOSE–1–PHENYLFLAVAZOLE; W/NBS
                                                                                                          66585

750   C36 H38 N4 O14   33797-00-1   AU-18-236-4   Glucopyranoside, D–erythro–2,3–dihydroxy–1–(1–phenyl–1H–pyraz
olo[3,4–b]quinoxalin–3–yl)propyl, hexaacetate (ester), β–D–;; PERACETYL–CELLOBIOSE–1–PHENYLFLAVAZOLE;
W/NBS   66586

750   C36 H20 D18 N4 O14   56298-42-1   AU-18-236-2   PERACETYL(TRIDEUTERIO)–MALTOSE–1–PHENYLFLAVAZOLE;   W
                                                                                                          66587

750   C36 H38 N4 O14   D3-1980-303-12   1–PHENYLFLAVAZOLE PERACETATE OF MALTOSE;   W   66588

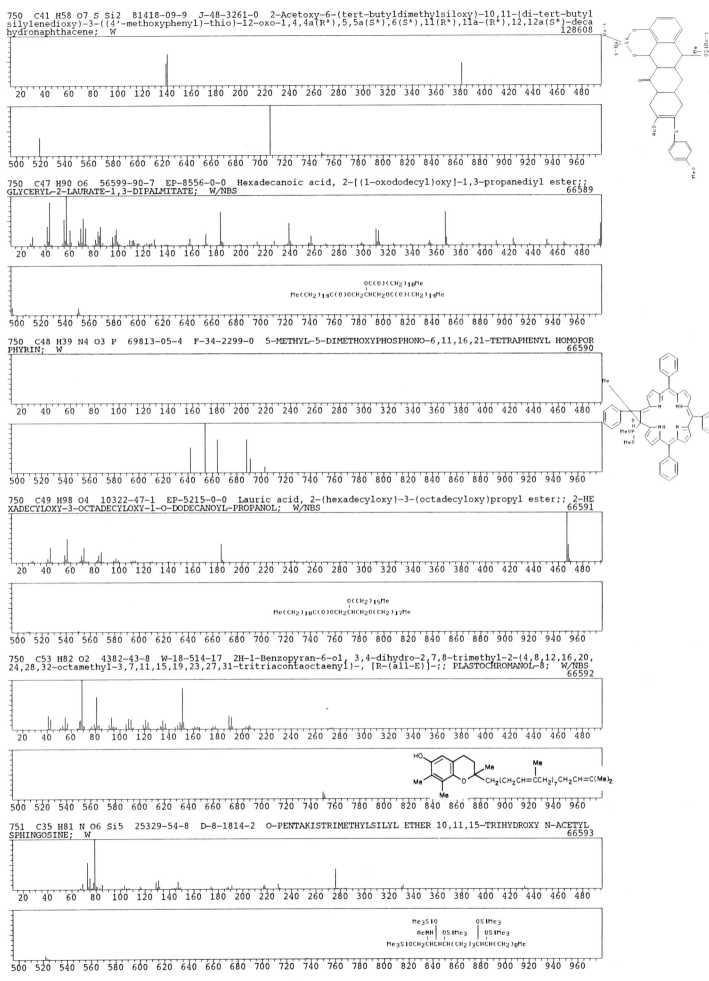

750   C41 H58 O7 S Si2   81418-09-9   J-48-3261-0   2-Acetoxy-6-(tert-butyldimethylsiloxy)-10,11-(di-tert-butyl silylenedioxy)-3-((4'-methoxyphenyl)-thio)-12-oxo-1,4,4a(R*),5,5a(S*),6(S*),11(R*),11a-(R*),12,12a(S*)-deca hydronaphthacene;   W
128608

750   C47 H90 O6   56599-90-7   EP-8556-0-0   Hexadecanoic acid, 2-[(1-oxododecyl)oxy]-1,3-propanediyl ester;; GLYCERYL-2-LAURATE-1,3-DIPALMITATE;   W/NBS
66589

750   C48 H39 N4 O3 P   69813-05-4   F-34-2299-0   5-METHYL-5-DIMETHOXYPHOSPHONO-6,11,16,21-TETRAPHENYL HOMOPOR PHYRIN;   W
66590

750   C49 H98 O4   10322-47-1   EP-5215-0-0   Lauric acid, 2-(hexadecyloxy)-3-(octadecyloxy)propyl ester;; 2-HE XADECYLOXY-3-OCTADECYLOXY-1-O-DODECANOYL-PROPANOL;   W/NBS
66591

750   C53 H82 O2   4382-43-8   W-18-514-17   2H-1-Benzopyran-6-ol, 3,4-dihydro-2,7,8-trimethyl-2-(4,8,12,16,20, 24,28,32-octamethyl-3,7,11,15,19,23,27,31-tritriacontaoctaenyl)-, [R-(all-E)]-;; PLASTOCHROMANOL-8;   W/NBS
66592

751   C35 H81 N O6 Si5   25329-54-8   D-8-1814-2   O-PENTAKISTRIMETHYLSILYL ETHER 10,11,15-TRIHYDROXY N-ACETYL SPHINGOSINE;   W
66593

752   C7 H I5 O2   64385-02-0   B-30-1711-0   PENTAIODOBENZOIC ACID;   W   66594

752   C8 H24 F10 N8 O2 P6 Si4   61510-02-9   K-109-3963-0   2,2,4,4,6,6,8,8-Octamethyl-3,7-bis(pentafluoro-1,3,
5,2λ5,4λ5,6λ5-triazatriphosphorin-2-yl)-1,5-dioxa-3,7-diaza-2,4,6,8-tetrasilacyclooctane;   W   128609

752   C36 H20 F8 Ge2   19638-32-5   NS--27022-0-0   Dibenzo[b,e][1,4]digermanin, 1,2,3,4,6,7,8,9-octafluoro-5,
10-dihydro-5,5,10,10-tetraphenyl-;;   NBS   66595

752   C41 H84 O4 Si4   63754-73-4   KO-4-33-19   5β-PREGNAN-3α,17α,20α,21-TETRAOL DIMETHYLPROPYLSILYL ETHER DER
IVATIVE;   W   66596

752   C45 H32 N6 O6   3812-22-4   O-10-608-1   [4,4':4',4''-Terpyrazolidine]-3,3',3'',5,5',5''-hexone, 1,1',1'
',2,2',2''-hexaphenyl-;;   Tripyraphene;   W/NBS   66597

752   C45 H40 N2 O9   CD-158-0-0   1,3-Bis-(11-Hydrochelerythrinyl)-Acetone;   W   128610

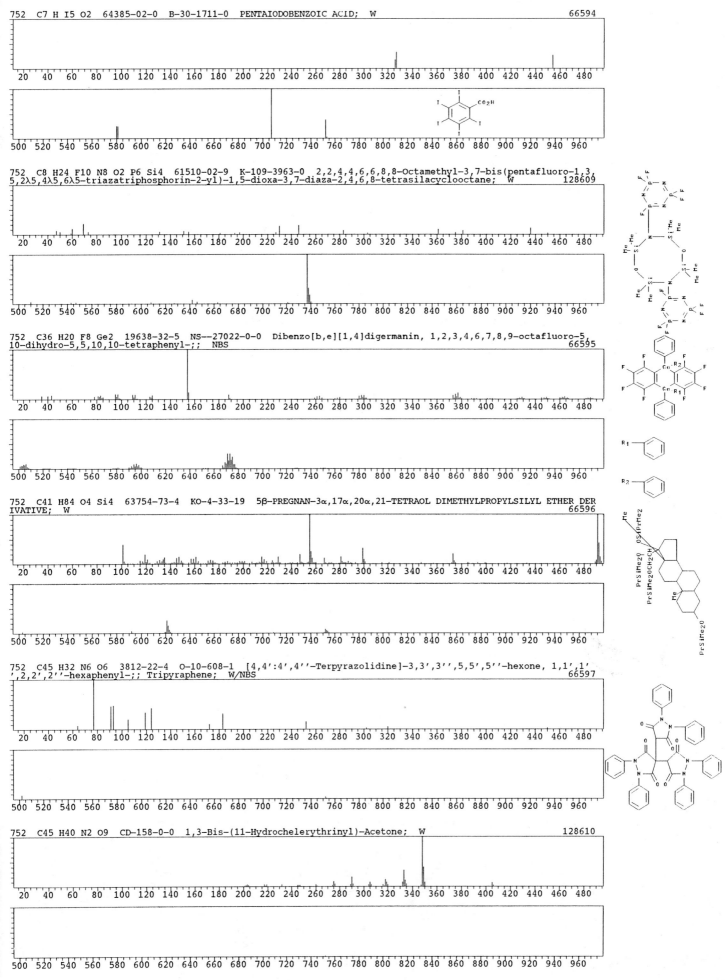

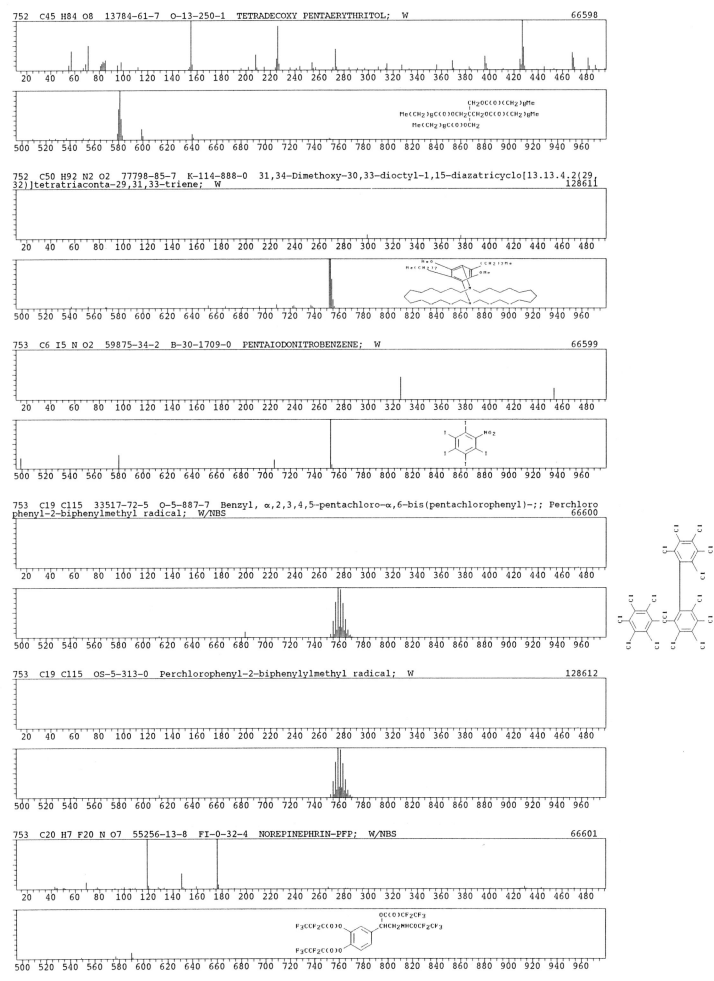

752   C45 H84 O8   13784-61-7   O-13-250-1   TETRADECOXY PENTAERYTHRITOL;   W   66598

CH2OC(O)(CH2)8Me
|
Me(CH2)8C(O)OCH2CCH2OC(O)(CH2)8Me
|
Me(CH2)8C(O)OCH2

752   C50 H92 N2 O2   77798-85-7   K-114-888-0   31,34-Dimethoxy-30,33-dioctyl-1,15-diazatricyclo[13.13.4.2(29,
32)]tetratriaconta-29,31,33-triene;   W   128611

753   C6 I5 N O2   59875-34-2   B-30-1709-0   PENTAIODONITROBENZENE;   W   66599

753   C19 Cl15   33517-72-5   O-5-887-7   Benzyl, α,2,3,4,5-pentachloro-α,6-bis(pentachlorophenyl)-;;   Perchloro
phenyl-2-biphenylmethyl radical;   W/NBS   66600

753   C19 Cl15   OS-5-313-0   Perchlorophenyl-2-biphenylylmethyl radical;   W   128612

753   C20 H7 F20 N O7   55256-13-8   FI-0-32-4   NOREPINEPHRIN-PFP;   W/NBS   66601

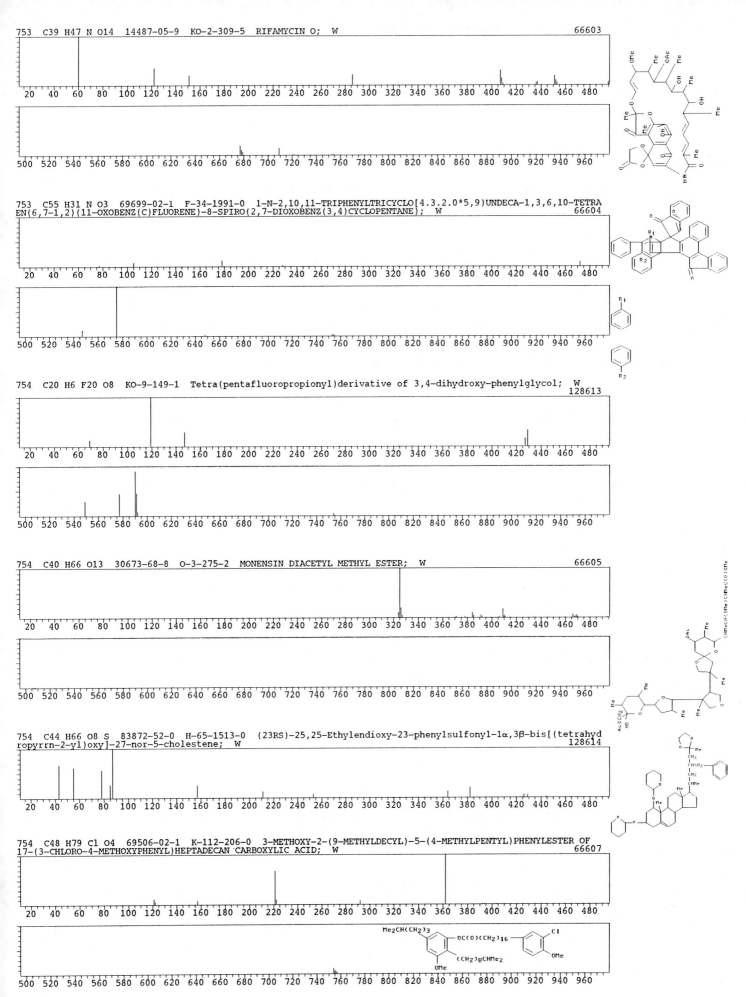

753   C39 H47 N O14   14487-05-9   KO-2-309-5   RIFAMYCIN O;   W   66603

753   C55 H31 N O3   69699-02-1   F-34-1991-0   1-N-2,10,11-TRIPHENYLTRICYCLO[4.3.2.0*5,9]UNDECA-1,3,6,10-TETRA
EN(6,7-1,2)(11-OXOBENZ(C)FLUORENE)-8-SPIRO(2,7-DIOXOBENZ(3,4)CYCLOPENTANE);   W   66604

754   C20 H6 F20 O8   KO-9-149-1   Tetra(pentafluoropropionyl)derivative of 3,4-dihydroxy-phenylglycol;   W
                                                                                                      128613

754   C40 H66 O13   30673-68-8   O-3-275-2   MONENSIN DIACETYL METHYL ESTER;   W   66605

754   C44 H66 O8 S   83872-52-0   H-65-1513-0   (23RS)-25,25-Ethylendioxy-23-phenylsulfonyl-1α,3β-bis[(tetrahyd
ropyrrn-2-yl)oxy]-27-nor-5-cholestene;   W   128614

754   C48 H79 Cl O4   69506-02-1   K-112-206-0   3-METHOXY-2-(9-METHYLDECYL)-5-(4-METHYLPENTYL)PHENYLESTER OF
17-(3-CHLORO-4-METHOXYPHENYL)HEPTADECAN CARBOXYLIC ACID;   W   66607

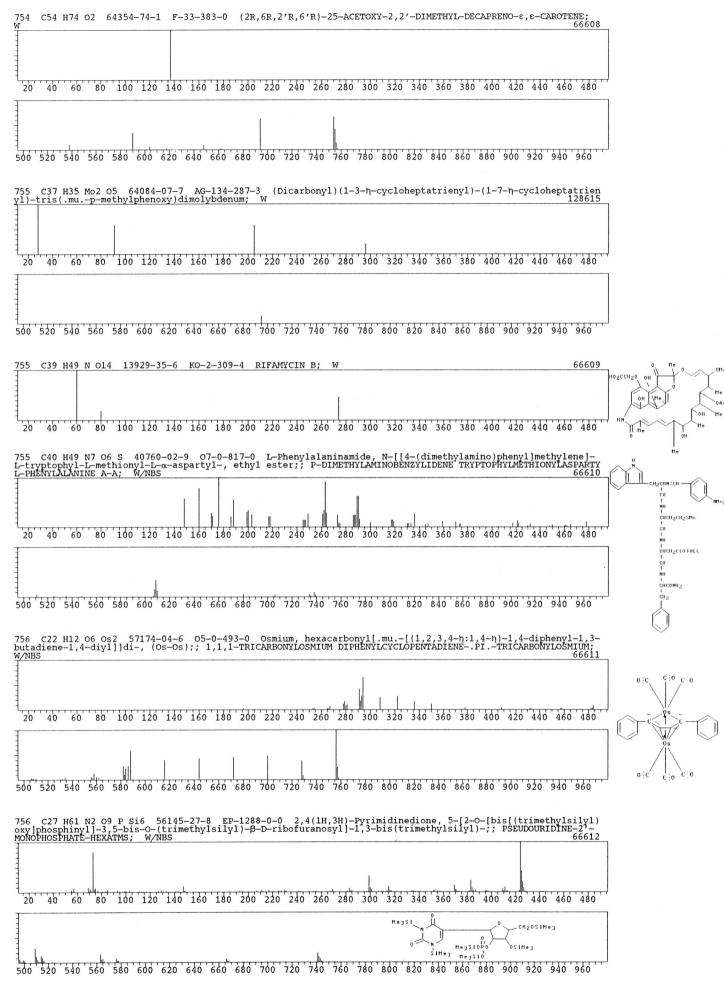

754   C54 H74 O2   64354-74-1   F-33-383-0   (2R,6R,2'R,6'R)-25-ACETOXY-2,2'-DIMETHYL-DECAPRENO-ε,ε-CAROTENE;
W   66608

755   C37 H35 Mo2 O5   64084-07-7   AG-134-287-3   (Dicarbonyl)(1-3-η-cycloheptatrienyl)-(1-7-η-cycloheptatrien
yl)-tris(.mu.-p-methylphenoxy)dimolybdenum;   W   128615

755   C39 H49 N O14   13929-35-6   KO-2-309-4   RIFAMYCIN B;   W   66609

755   C40 H49 N7 O6 S   40760-02-9   O7-0-817-0   L-Phenylalaninamide, N-[[4-(dimethylamino)phenyl]methylene]-
L-tryptophyl-L-methionyl-L-α-aspartyl-, ethyl ester;; P-DIMETHYLAMINOBENZYLIDENE TRYPTOPHYLMETHIONYLASPARTY
L-PHENYLALANINE A-A;   W/NBS   66610

756   C22 H12 O6 Os2   57174-04-6   O5-0-493-0   Osmium, hexacarbonyl[.mu.-[(1,2,3,4-η:1,4-η)-1,4-diphenyl-1,3-
butadiene-1,4-diyl]]di-, (Os-Os);; 1,1,1-TRICARBONYLOSMIUM DIPHENYLCYCLOPENTADIENE-.PI.-TRICARBONYLOSMIUM;
W/NBS   66611

756   C27 H61 N2 O9 P Si6   56145-27-8   EP-1288-0-0   2,4(1H,3H)-Pyrimidinedione, 5-[2-O-[bis[(trimethylsilyl)
oxy]phosphinyl]-3,5-bis-O-(trimethylsilyl)-β-D-ribofuranosyl]-1,3-bis(trimethylsilyl)-;; PSEUDOURIDINE-2'-
MONOPHOSPHATE-HEXATMS;   W/NBS   66612

756   C33 H40 O20   70404-46-5   SB-33-123-5   LUTEOLIN-7-O-NEOHESPERIDOSIDE-4'-O-β-D-GLUCOPYRANOSIDE;   W 66613

Str 1181

756   C36 H57 Cl N4 O7 S2   H-67-2183-0   2-[8-Tosyl-12-(tosylamino)-4,8-diazadodecyl]-2-nitrocyclododecanone;
W   128616

756   C42 H49 In N4 S   63221-27-2   KA-1980-2084-2   Phenylthio[2,3,7,8,12,13,17,18-octaethylporphyrinato]indi
um;   W   128617

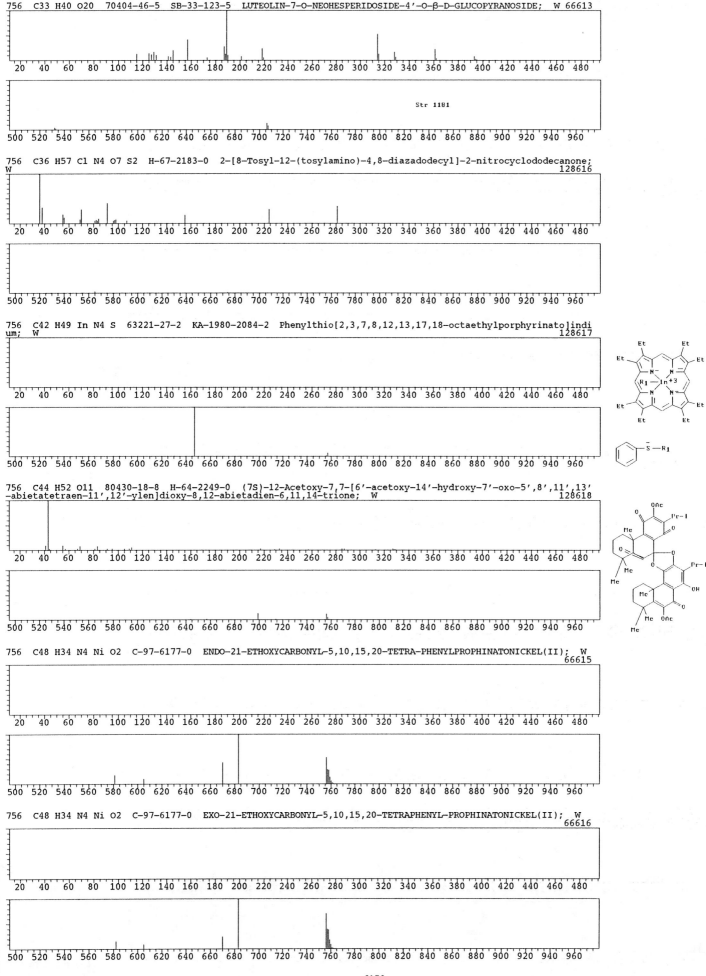

756   C44 H52 O11   80430-18-8   H-64-2249-0   (7S)-12-Acetoxy-7,7-[6'-acetoxy-14'-hydroxy-7'-oxo-5',8',11',13'
-abietatetraen-11',12'-ylen]dioxy-8,12-abietadien-6,11,14-trione;   W   128618

756   C48 H34 N4 Ni O2   C-97-6177-0   ENDO-21-ETHOXYCARBONYL-5,10,15,20-TETRA-PHENYLPROPHINATONICKEL(II);   W
                                                                                                      66615

756   C48 H34 N4 Ni O2   C-97-6177-0   EXO-21-ETHOXYCARBONYL-5,10,15,20-TETRAPHENYL-PROPHINATONICKEL(II);   W
                                                                                                      66616

756   C48 H34 N4 Ni O2   C-97-6180-0   21-ETHOXYCARBONYL-5,10,15 ENDO,20-TETRA-PHENYLPROPHINATONICKEL(II); W
66617

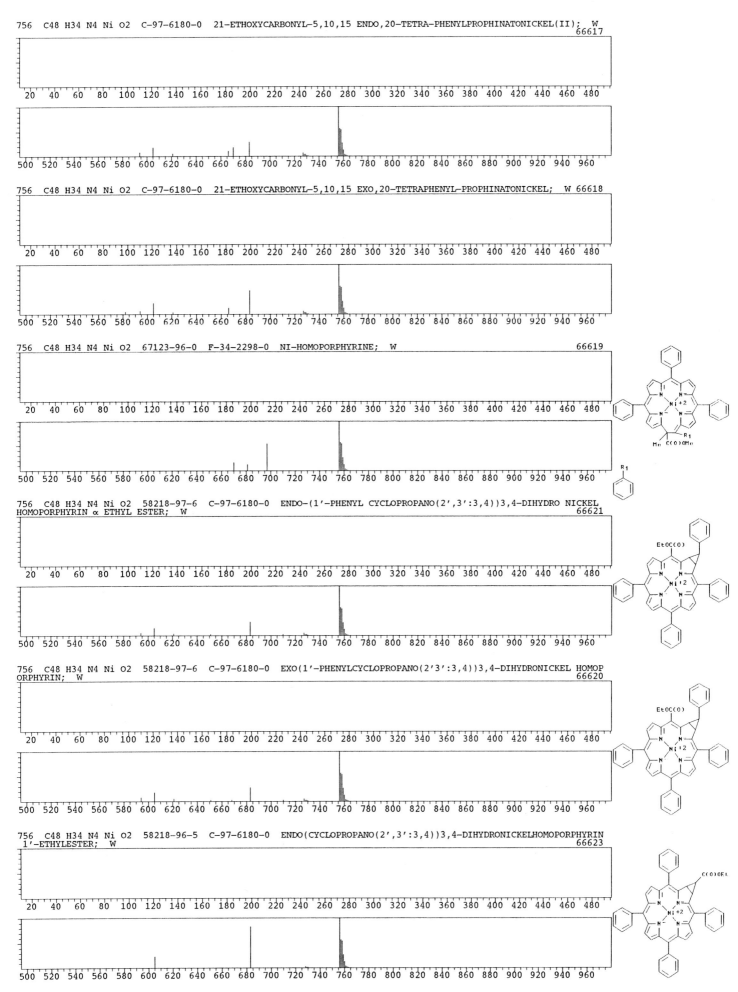

756   C48 H34 N4 Ni O2   C-97-6180-0   21-ETHOXYCARBONYL-5,10,15 EXO,20-TETRAPHENYL-PROPHINATONICKEL; W 66618

756   C48 H34 N4 Ni O2   67123-96-0   F-34-2298-0   NI-HOMOPORPHYRINE; W
66619

756   C48 H34 N4 Ni O2   58218-97-6   C-97-6180-0   ENDO-(1'-PHENYL CYCLOPROPANO(2',3':3,4))3,4-DIHYDRO NICKEL
HOMOPORPHYRIN α ETHYL ESTER; W
66621

756   C48 H34 N4 Ni O2   58218-97-6   C-97-6180-0   EXO(1'-PHENYLCYCLOPROPANO(2'3':3,4))3,4-DIHYDRONICKEL HOMOP
ORPHYRIN; W
66620

756   C48 H34 N4 Ni O2   58218-96-5   C-97-6180-0   ENDO(CYCLOPROPANO(2',3':3,4))3,4-DIHYDRONICKELHOMOPORPHYRIN
1'-ETHYLESTER; W
66623

756   C48 H34 N4 Ni O2   69811-92-3   F-34-2298-0   5-METHYL-5-METHOXYCARBONYL-α,β,Δ,γ-TETRAPHENYLNICKEL-5,6-HO
MOPORPHYRINE;   W   66624

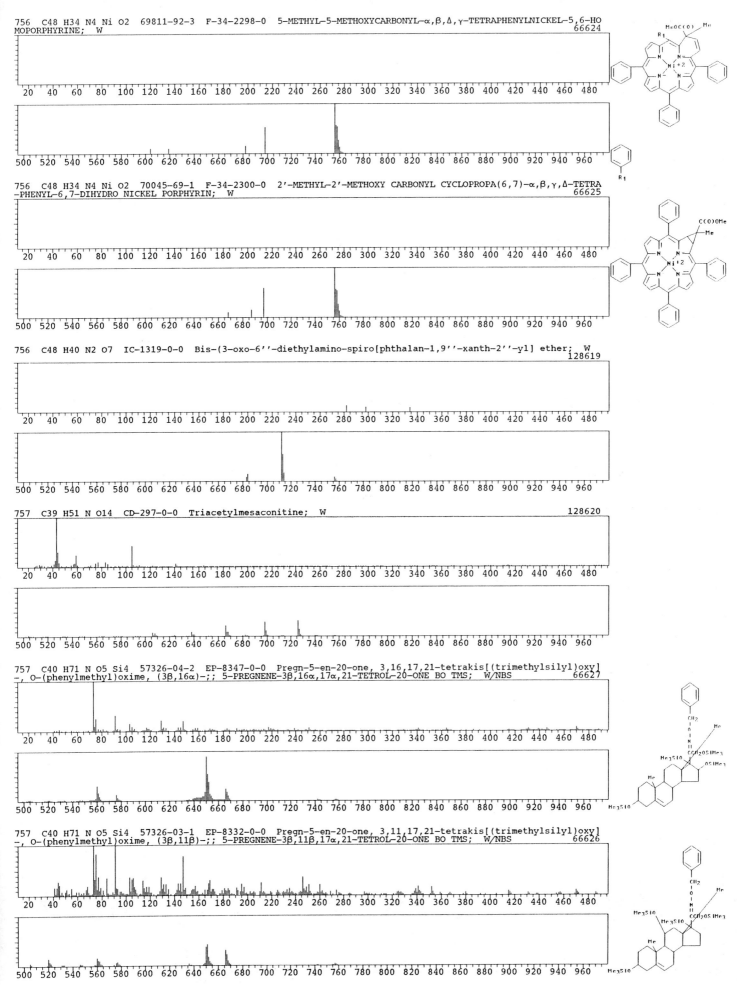

756   C48 H34 N4 Ni O2   70045-69-1   F-34-2300-0   2'-METHYL-2'-METHOXY CARBONYL CYCLOPROPA(6,7)-α,β,γ,Δ-TETRA
-PHENYL-6,7-DIHYDRO NICKEL PORPHYRIN;   W   66625

756   C48 H40 N2 O7   IC-1319-0-0   Bis-(3-oxo-6''-diethylamino-spiro[phthalan-1,9''-xanth-2''-yl] ether;   W
128619

757   C39 H51 N O14   CD-297-0-0   Triacetylmesaconitine;   W   128620

757   C40 H71 N O5 Si4   57326-04-2   EP-8347-0-0   Pregn-5-en-20-one, 3,16,17,21-tetrakis[(trimethylsilyl)oxy]
-, O-(phenylmethyl)oxime, (3β,16α)-;;   5-PREGNENE-3β,16α,17α,21-TETROL-20-ONE BO TMS;   W/NBS   66627

757   C40 H71 N O5 Si4   57326-03-1   EP-8332-0-0   Pregn-5-en-20-one, 3,11,17,21-tetrakis[(trimethylsilyl)oxy]
-, O-(phenylmethyl)oxime, (3β,11β)-;;   5-PREGNENE-3β,11β,17α,21-TETROL-20-ONE BO TMS;   W/NBS   66626

- 6160 -

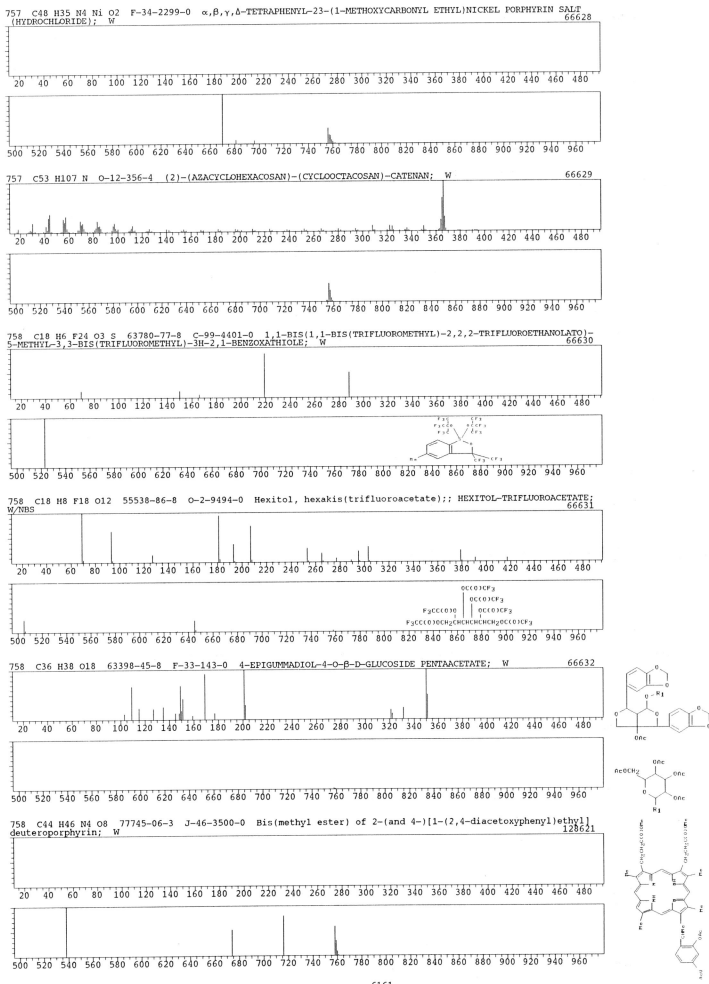

757   C48 H35 N4 Ni O2   F-34-2299-0   α,β,γ,Δ-TETRAPHENYL-23-(1-METHOXYCARBONYL ETHYL)NICKEL PORPHYRIN SALT
(HYDROCHLORIDE);   W   66628

757   C53 H107 N   O-12-356-4   (2)-(AZACYCLOHEXACOSAN)-(CYCLOOCTACOSAN)-CATENAN;   W   66629

758   C18 H6 F24 O3 S   63780-77-8   C-99-4401-0   1,1-BIS(1,1-BIS(TRIFLUOROMETHYL)-2,2,2-TRIFLUOROETHANOLATO)-
5-METHYL-3,3-BIS(TRIFLUOROMETHYL)-3H-2,1-BENZOXATHIOLE;   W   66630

758   C18 H8 F18 O12   55538-86-8   O-2-9494-0   Hexitol, hexakis(trifluoroacetate);;   HEXITOL-TRIFLUOROACETATE;
W/NBS   66631

758   C36 H38 O18   63398-45-8   F-33-143-0   4-EPIGUMMADIOL-4-O-β-D-GLUCOSIDE PENTAACETATE;   W   66632

758   C44 H46 N4 O8   77745-06-3   J-46-3500-0   Bis(methyl ester) of 2-(and 4-)[1-(2,4-diacetoxyphenyl)ethyl]
deuteroporphyrin;   W   128621

758 C45 H33 O6 Y 15688-50-3 T-68-5960-0 Yttrium, tris(1,3-diphenyl-1,3-propanedionato)-;; TRIS(1,3-DIPH
ENYL-1,3-PROPANEDIONATO)YTTRIUM(III); W/NBS   66633

758 C50 H38 N4 O4 69813-04-3 F-34-2299-0 5-METHOXYCARBONYL-5-METHOXYCARBONYLMETHYL-α,β,Δ,γ-TETRAPHENYL-
5,6-HOMOPORPHYRINE; W   66634

758 C52 H78 N4 69912-07-8 J-44-2079-0 2,4,6,8-Tetra(n-heptyl)-1,3,5,7-tetramethylporphyrin; W 128622

759 C40 H73 N O5 Si4 57326-05-3 EP-8334-0-0 Pregnan-20-one, 3,11,17,21-tetrakis[(trimethylsilyl)oxy]-,
O-(phenylmethyl)oxime, (3α,5β,11β)-;; 5β-PREGNANE-3α,11β,17α,21-TETROL-20-ONE BO TMS; W/NBS 66635

760 C37 H60 N8 O9 60008-50-6 KO-1-184-5 N,O-PERMETHYL-N-ACETYLGLUTAMINYLGLUTAMINYLGLYCYLGLYCYLTYROSINAM
IDE; W   66636

760 C39 H44 N4 O8 S2 H-62-2322-0 11-Benzyloxycarbonylamino-1-phthalimido-4,8-ditosyl-4,8-diazaundecane;
W   128623

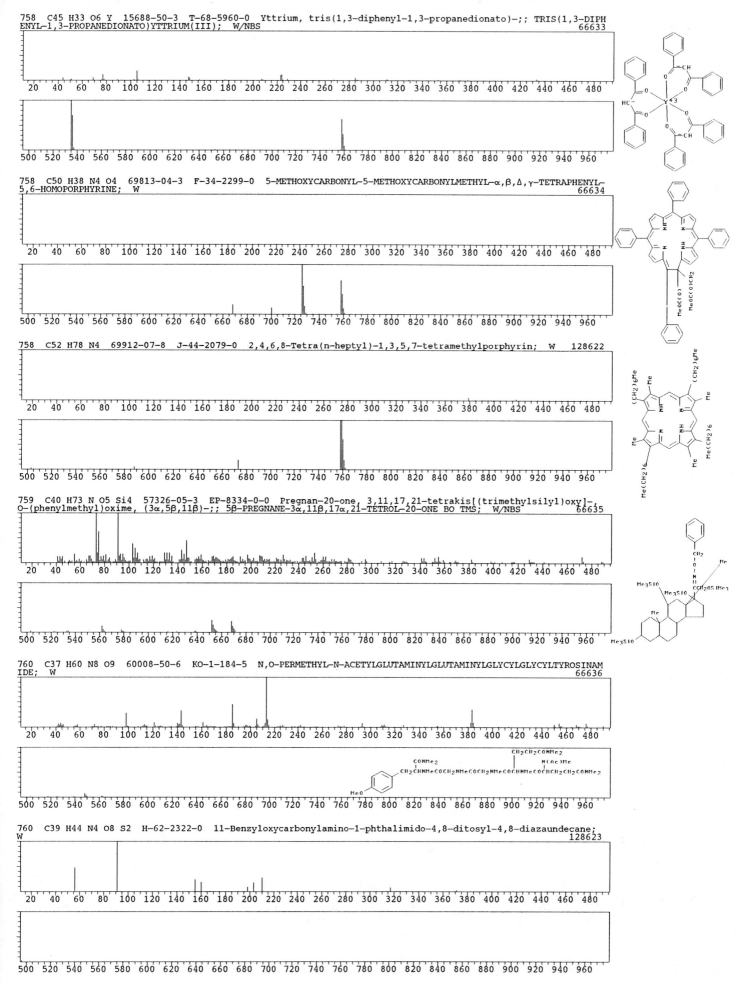

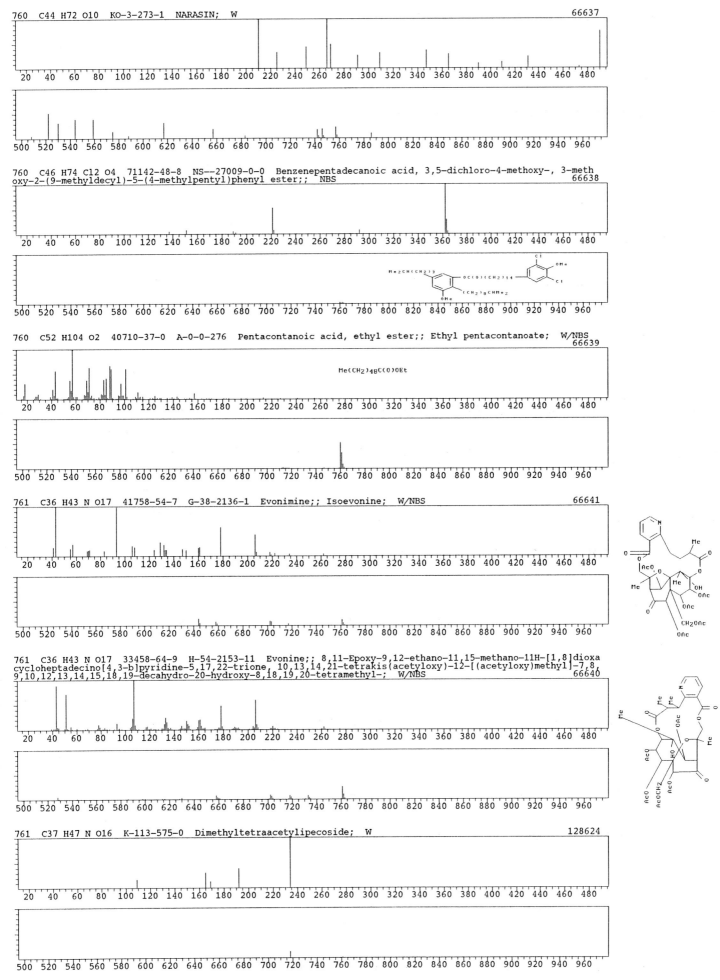

760   C44 H72 O10   KO-3-273-1   NARASIN;   W   66637

760   C46 H74 Cl2 O4   71142-48-8   NS--27009-0-0   Benzenepentadecanoic acid, 3,5-dichloro-4-methoxy-, 3-meth
oxy-2-(9-methyldecyl)-5-(4-methylpentyl)phenyl ester;;   NBS   66638

760   C52 H104 O2   40710-37-0   A-0-0-276   Pentacontanoic acid, ethyl ester;;   Ethyl pentacontanoate;   W/NBS
                                                                                                               66639

Me(CH2)48C(O)OEt

761   C36 H43 N O17   41758-54-7   G-38-2136-1   Evonimine;;   Isoevonine;   W/NBS   66641

761   C36 H43 N O17   33458-64-9   H-54-2153-11   Evonine;;   8,11-Epoxy-9,12-ethano-11,15-methano-11H-[1,8]dioxa
cycloheptadecino[4,3-b]pyridine-5,17,22-trione, 10,13,14,21-tetrakis(acetyloxy)-12-[(acetyloxy)methyl]-7,8,
9,10,12,13,14,15,18,19-decahydro-20-hydroxy-8,18,19,20-tetramethyl-;   W/NBS   66640

761   C37 H47 N O16   K-113-575-0   Dimethyltetraacetylipecoside;   W   128624

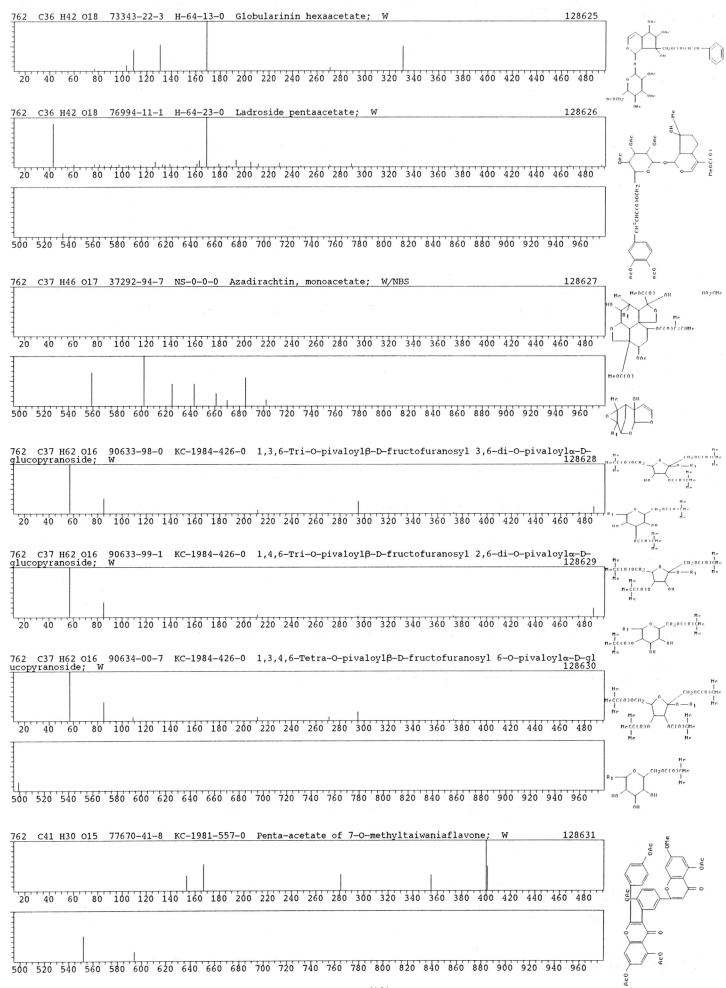

762   C36 H42 O18   73343-22-3   H-64-13-0   Globularinin hexaacetate;   W   128625

762   C36 H42 O18   76994-11-1   H-64-23-0   Ladroside pentaacetate;   W   128626

762   C37 H46 O17   37292-94-7   NS-0-0-0   Azadirachtin, monoacetate;   W/NBS   128627

762   C37 H62 O16   90633-98-0   KC-1984-426-0   1,3,6-Tri-O-pivaloylβ-D-fructofuranosyl 3,6-di-O-pivaloylα-D-
glucopyranoside;   W   128628

762   C37 H62 O16   90633-99-1   KC-1984-426-0   1,4,6-Tri-O-pivaloylβ-D-fructofuranosyl 2,6-di-O-pivaloylα-D-
glucopyranoside;   W   128629

762   C37 H62 O16   90634-00-7   KC-1984-426-0   1,3,4,6-Tetra-O-pivaloylβ-D-fructofuranosyl 6-O-pivaloylα-D-gl
ucopyranoside;   W   128630

762   C41 H30 O15   77670-41-8   KC-1981-557-0   Penta-acetate of 7-O-methyltaiwaniaflavone;   W   128631

763 C24 H12 Cr F9 O3 S6 61024-11-1 B-29-1915-2 TRIS(1,1,1-TRIFLUORO-4-(2'-THIENYL)-2-OXOBUT-3-EN-4-YL
THIO)CHROMIUM(III); W
66642

763 C30 H65 N5 O6 S Si5 68373-36-4 KO-5-393-1 GLUCURONIDE OF 2-(T-BUTYL AMINO)-4-ETHYAMINO)-6-THIO-S-TR
IAZINE; W
66643

763 C36 H37 N5 O14 35426-85-8 J-7-3527-7 3H-Pyrazol-3-one, 2,4-dihydro-4-[[5-hydroxy-1-phenyl-3-[1,2,3-
tris(acetyloxy)propyl]-1H-pyrazol-4-yl]imino]-2-phenyl-5-[1,2,3-tris(acetyloxy)propyl]-, stereoisomer;;
BIS(L-THREO-TRIACETOXYPROPYL)RUBIAZONIC ACID; W/NBS
66644

763 C37 H50 Cl N3 O12 82390-95-2 C-104-4934-0 Treflorine; W
128632

763 C38 H49 N7 O10 85640-90-0 H-65-2286-0 Virginiamycin S5; W
128633

763 C45 H31 Fe N4 O3 S 82008-26-2 KA-1984-571-4 Methylsulfonato(5,10,15,20-tetraphenylporphyrinato)iron
; W
128634

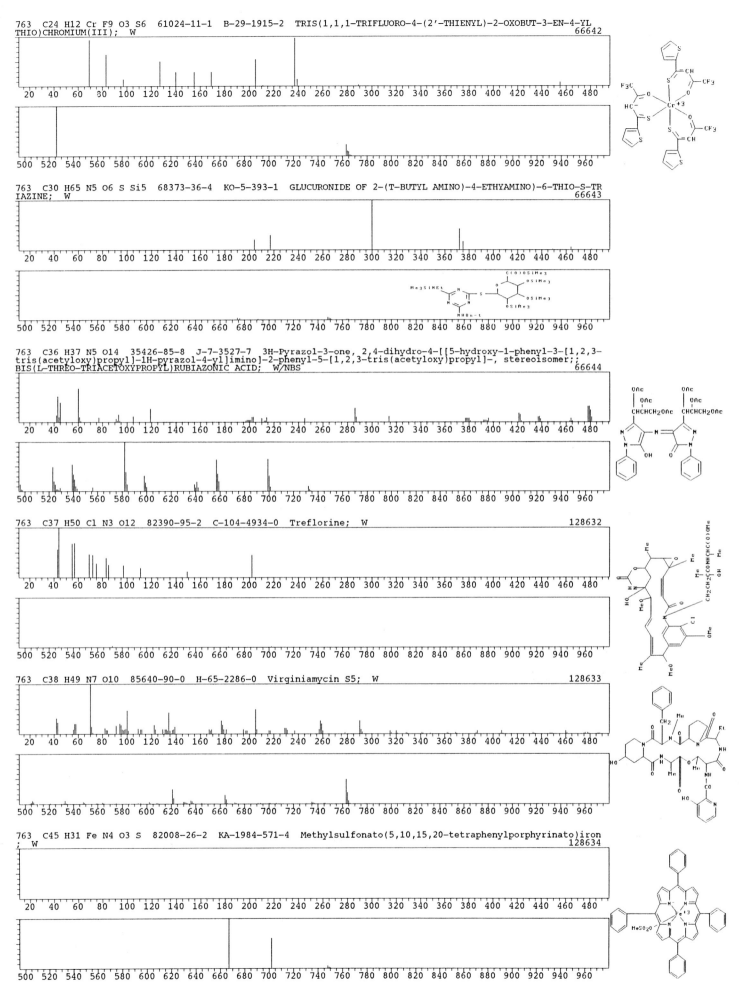

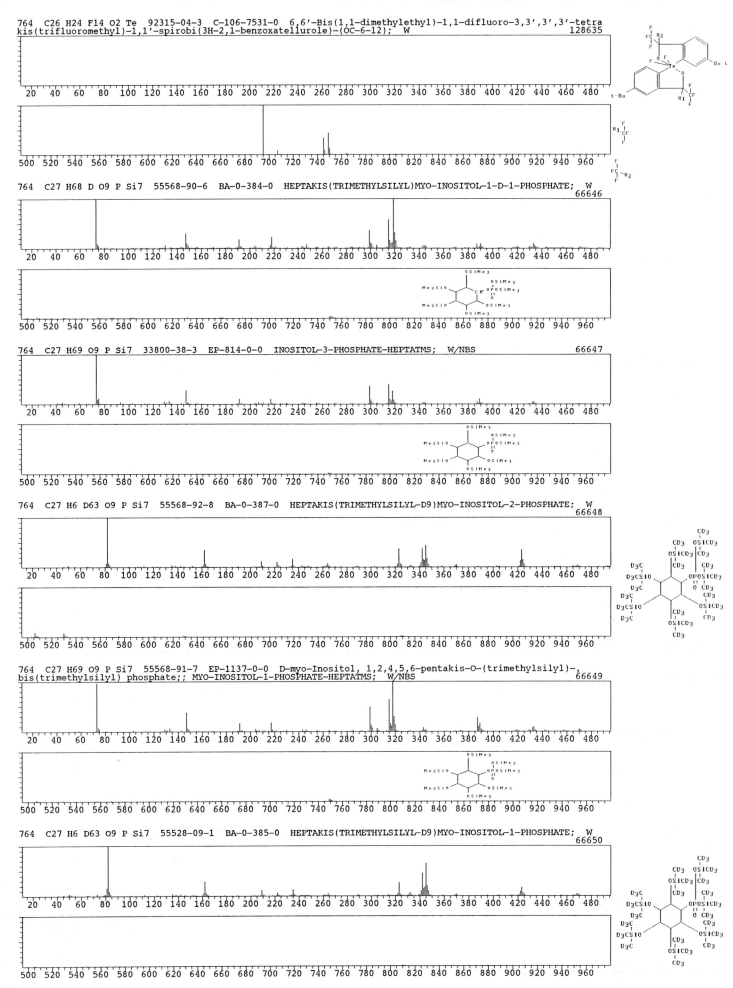

764 C26 H24 F14 O2 Te 92315-04-3 C-106-7531-0 6,6'-Bis(1,1-dimethylethyl)-1,1-difluoro-3,3',3',3'-tetra
kis(trifluoromethyl)-1,1'-spirobi(3H-2,1-benzoxatellurole)-(OC-6-12); W   128635

764 C27 H68 D O9 P Si7 55568-90-6 BA-0-384-0 HEPTAKIS(TRIMETHYLSILYL)MYO-INOSITOL-1-D-1-PHOSPHATE; W
66646

764 C27 H69 O9 P Si7 33800-38-3 EP-814-0-0 INOSITOL-3-PHOSPHATE-HEPTATMS; W/NBS   66647

764 C27 H6 D63 O9 P Si7 55568-92-8 BA-0-387-0 HEPTAKIS(TRIMETHYLSILYL-D9)MYO-INOSITOL-2-PHOSPHATE; W
66648

764 C27 H69 O9 P Si7 55568-91-7 EP-1137-0-0 D-myo-Inositol, 1,2,4,5,6-pentakis-O-(trimethylsilyl)-,
bis(trimethylsilyl) phosphate;; MYO-INOSITOL-1-PHOSPHATE-HEPTATMS; W/NBS   66649

764 C27 H6 D63 O9 P Si7 55528-09-1 BA-0-385-0 HEPTAKIS(TRIMETHYLSILYL-D9)MYO-INOSITOL-1-PHOSPHATE; W
66650

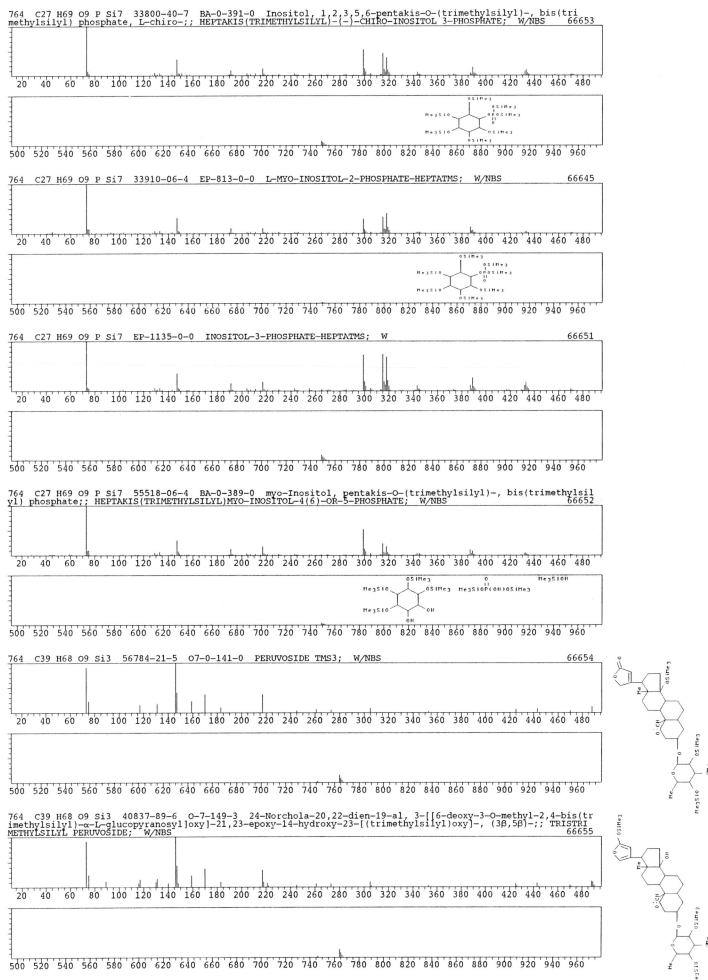

764 C27 H69 O9 P Si7 33800-40-7 BA-0-391-0 Inositol, 1,2,3,5,6-pentakis-O-(trimethylsilyl)-, bis(tri
methylsilyl) phosphate, L-chiro-;; HEPTAKIS(TRIMETHYLSILYL)-(-)-CHIRO-INOSITOL 3-PHOSPHATE; W/NBS   66653

764 C27 H69 O9 P Si7 33910-06-4 EP-813-0-0 L-MYO-INOSITOL-2-PHOSPHATE-HEPTATMS; W/NBS   66645

764 C27 H69 O9 P Si7 EP-1135-0-0 INOSITOL-3-PHOSPHATE-HEPTATMS; W   66651

764 C27 H69 O9 P Si7 55518-06-4 BA-0-389-0 myo-Inositol, pentakis-O-(trimethylsilyl)-, bis(trimethylsil
yl) phosphate;; HEPTAKIS(TRIMETHYLSILYL)MYO-INOSITOL-4(6)-OR-5-PHOSPHATE; W/NBS   66652

764 C39 H68 O9 Si3 56784-21-5 O7-0-141-0 PERUVOSIDE TMS3; W/NBS   66654

764 C39 H68 O9 Si3 40837-89-6 O-7-149-3 24-Norchola-20,22-dien-19-al, 3-[[6-deoxy-3-O-methyl-2,4-bis(tr
imethylsilyl)-α-L-glucopyranosyl]oxy]-21,23-epoxy-14-hydroxy-23-[(trimethylsilyl)oxy]-, (3β,5β)-;; TRISTRI
METHYLSILYL PERUVOSIDE; W/NBS   66655

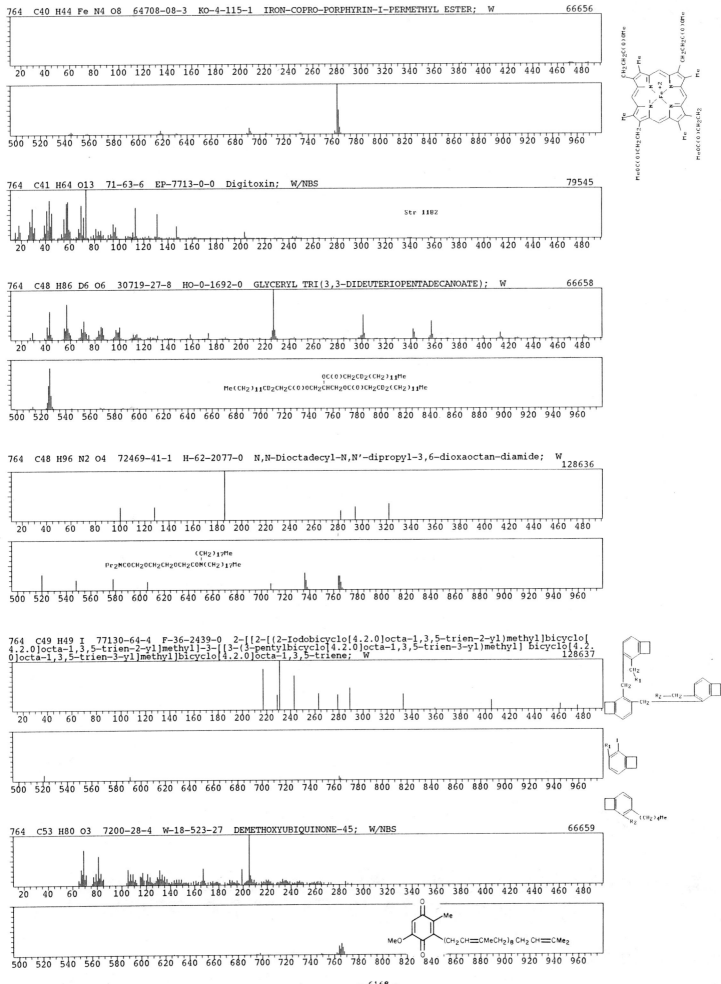

764   C40 H44 Fe N4 O8   64708-08-3   KO-4-115-1   IRON-COPRO-PORPHYRIN-I-PERMETHYL ESTER;   W   66656

764   C41 H64 O13   71-63-6   EP-7713-0-0   Digitoxin;   W/NBS   79545

Str 1182

764   C48 H86 D6 O6   30719-27-8   HO-0-1692-0   GLYCERYL TRI(3,3-DIDEUTERIOPENTADECANOATE);   W   66658

764   C48 H96 N2 O4   72469-41-1   H-62-2077-0   N,N-Dioctadecyl-N,N'-dipropyl-3,6-dioxaoctan-diamide;   W
                                                                                                  128636

764   C49 H49 I   77130-64-4   F-36-2439-0   2-[[2-[(2-Iodobicyclo[4.2.0]octa-1,3,5-trien-2-yl)methyl]bicyclo[
4.2.0]octa-1,3,5-trien-2-yl]methyl]-3-[[3-(3-pentylbicyclo[4.2.0]octa-1,3,5-trien-3-yl)methyl] bicyclo[4.2.
0]octa-1,3,5-trien-3-yl]methyl]bicyclo[4.2.0]octa-1,3,5-triene;   W   128637

764   C53 H80 O3   7200-28-4   W-18-523-27   DEMETHOXYUBIQUINONE-45;   W/NBS   66659

- 6168 -

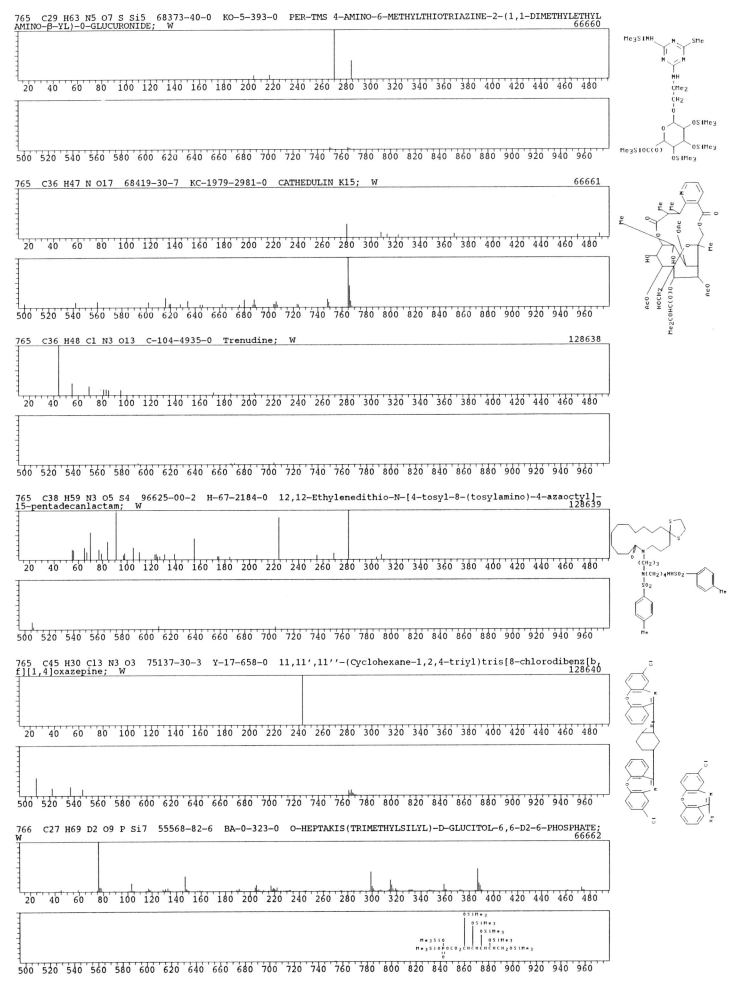

765   C29 H63 N5 O7 S Si5   68373-40-0   KO-5-393-0   PER-TMS 4-AMINO-6-METHYLTHIOTRIAZINE-2-(1,1-DIMETHYLETHYL
AMINO-β-YL)-0-GLUCURONIDE;   W   66660

765   C36 H47 N O17   68419-30-7   KC-1979-2981-0   CATHEDULIN K15;   W   66661

765   C36 H48 Cl N3 O13   C-104-4935-0   Trenudine;   W   128638

765   C38 H59 N3 O5 S4   96625-00-2   H-67-2184-0   12,12-Ethylenedithio-N-[4-tosyl-8-(tosylamino)-4-azaoctyl]-
15-pentadecanlactam;   W   128639

765   C45 H30 Cl3 N3 O3   75137-30-3   Y-17-658-0   11,11',11''-(Cyclohexane-1,2,4-triyl)tris[8-chlorodibenz[b,
f][1,4]oxazepine;   W   128640

766   C27 H69 D2 O9 P Si7   55568-82-6   BA-0-323-0   O-HEPTAKIS(TRIMETHYLSILYL)-D-GLUCITOL-6,6-D2-6-PHOSPHATE;
W   66662

766 C27 H64 D7 O9 P Si7 55568-83-7 BA-0-324-0 O-HEPTAKIS(TRIMETHYLSILYL)-D-GLUCITOL-1,2,3,4,5,6,6-D7-6-
PHOSPHATE; W
66663

766 C27 H71 O9 P Si7 33800-37-2 BA-0-319-0 D-MANNITOL-1-PHOSPHATE-HEPTATMS; W/NBS
66664

766 C27 H71 O9 P Si7 33800-35-0 EP-817-0-0 D-SORBITOL-6-PHOSPHATE-HEPTATMS; W/NBS
66665

766 C27 H71 O9 P Si7 55528-11-5 BA-0-321-0 D-Glucitol, 2,3,4,5,6-pentakis-O-(trimethylsilyl)-, bis(tri
methylsilyl) phosphate;; O-HEPTAKIS(TRIMETHYLSILYL)-D-GLUCITOL-1-PHOSPHATE; W/NBS
66666

766 C32 H46 O21 HE-1986-310-0 DECA-O-ACETYL-D-GLUCITOL-(1-4)-D-GLUCITOL; W
128641

766 C32 H46 O21 HE-1986-1371-0 DECA-O-ACETYL-D-GALACTITE-(1,4)-D-GLUCITOL; W
128642

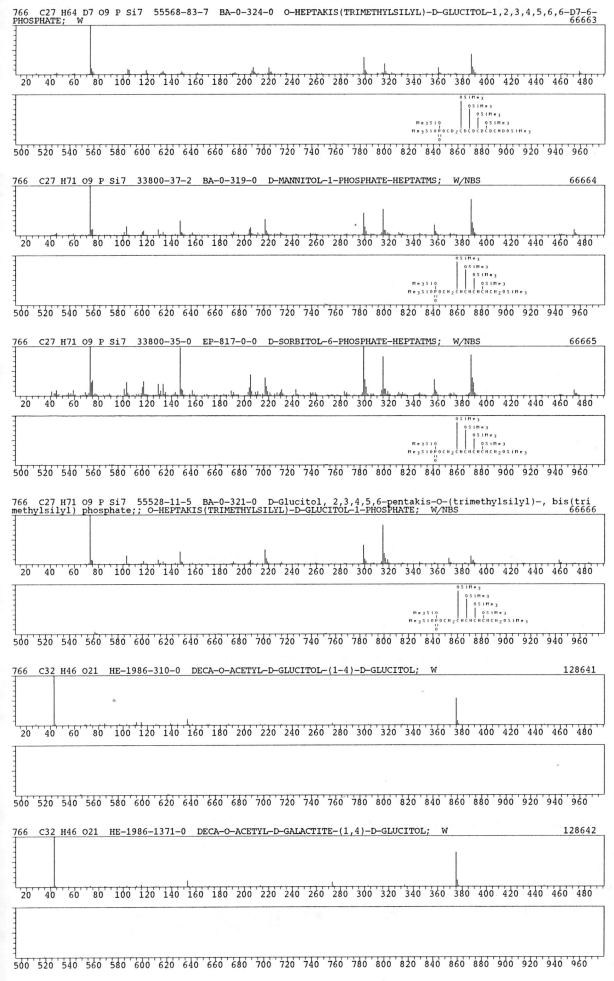

766   C40 H44 N4 Ni O8   64973-98-4   KO-4-115-1   NICKEL-COPRO-PORPHYRIN-I-PERMETHYL ESTER;   W   66667

766   C47 H50 N4 O6   84379-22-6   C-106-7163-0   34,38-Dioxo-35,37-dimethoxy-36-(phenylmethoy)-10,15,20,31-tet
ramethyl-3,7,23,27-tetraazaheptacyclo(27.3.1.1.1.1.1.1)octaconta-1(33),8(35),9,11,13(36),14,16,18(37),19,
21,29,31-dodecaen-33-ol;   W   128643

767   C21 H9 F20 N O7   74231-48-4   NS--26990-0-0   Propanoic acid, pentafluoro-, 4-[1-(2,2,3,3,3-pentafluoro-
1-oxopropoxy)-2-[(2,2,3,3,3-pentafluoro-1-oxopropyl)amino]propyl]-1,2-phenylene ester;;   NBS   66670

767   C21 H9 F20 N O7   59785-42-1   NS--26992-0-0   Propanoic acid, pentafluoro-, 4-[2-[methyl(2,2,3,3,3-penta
fluoro-1-oxopropyl)amino]-1-(2,2,3,3,3-pentafluoro-1-oxopropoxy)ethyl]-1,2-phenylene ester, (R)-;; Propano
ic acid, pentafluoro-, 4-[2-[methyl(2,2,3,3,3-pentafluoro-1-oxopropyl)amino]-1-(2,2,3,3,3-pentafluoro-1-oxo
propoxy)ethyl]-1,2-phenylene e;   NBS   66669

767   C21 H9 F20 N O7   1995-97-7   NS--25958-0-0   Propionic acid, pentafluoro-, triester with 2,2,3,3,3-penta
fluoro-N-methyl-N-(β,3,4-trihydroxyphenethyl)propionamide;   NBS   128644

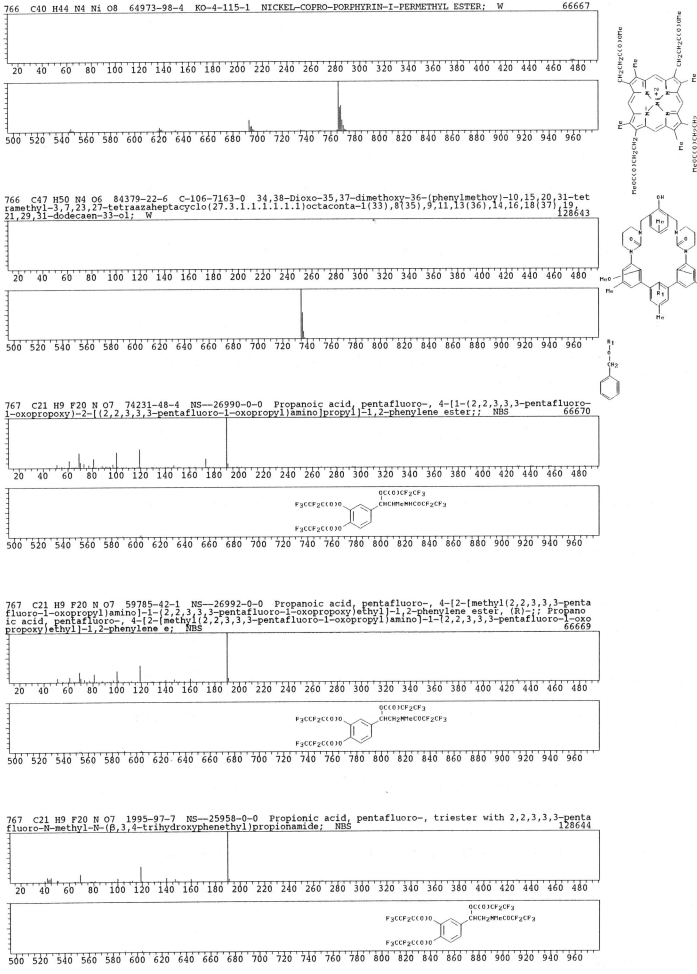

767   C21 H9 F20 N O7   72347-74-1   FI-0-32-3   Propanamide, N-[2-(3,4-dihydroxyphenyl)-2-hydroxy-1-methyleth yl]-2,2,3,3,3-pentafluoro-, tris(2,2,3,3,3-pentafluoro-1-oxopropyl) deriv.;; α-METHYLNOREPINEPHRIN-PFP; W/NBS
66668

767   C40 H44 Co N4 O8   64973-99-5   KO-4-115-1   COBALT-COPRO-POPHYNIN-PERMETHYL ESTER; W
66671

767   C40 H61 N7 O8   55822-85-0   O-5-25-2   DL-Alanine, N-[N-[1-[N-[N-[N-(1-oxopropyl)-DL-tryptophyl]-DL-leuc yl]-DL-valyl]-DL-prolyl]-DL-leucyl]-, methyl ester;; PROPIONYLTRYPTOPHYLLEUCYLVALYLPROLYLLEUCYLALANINE METH YL ESTER; W/NBS
66672

767   C40 H61 N7 O8   31944-47-5   NS--26989-0-0   Alanine, N-[N-[1-[N-[N-(N-propionyl-L-tryptophyl)-L-leucyl]- L-valyl]-L-prolyl]-L-leucyl]-, methyl ester, L-;; NBS
66673

768   C10 Cd O10 Re2   33728-45-9   KA-71-2650-2   Rhenium, .mu.-cadmiodecacarbonyldi-;; BIS(PENTACARBONYLRHENI O)CADMIUM; W/NBS
66674

768   C14 Cl2 F12 O6 Rh2   B-31-668-0   MIXTURE OF TETRACARBONYL-DI-.MU.-CHLORO-BIS-HEXAFLUOROBUT-2-EN-2,3-DIY LDIRHODINIUM AND DICARBONYL ISOMER; W
66675

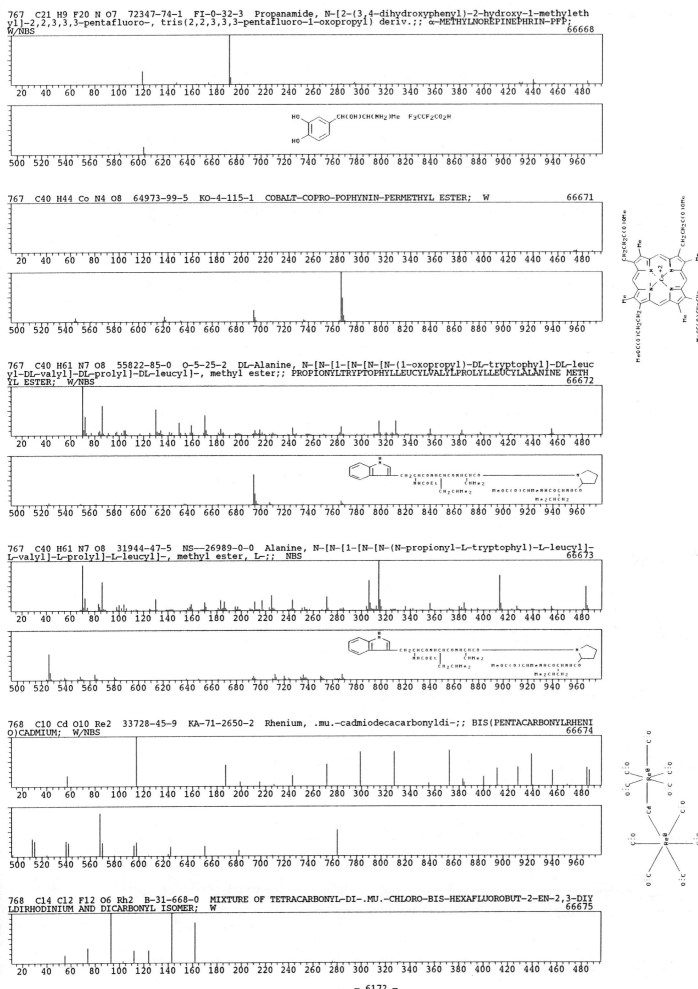

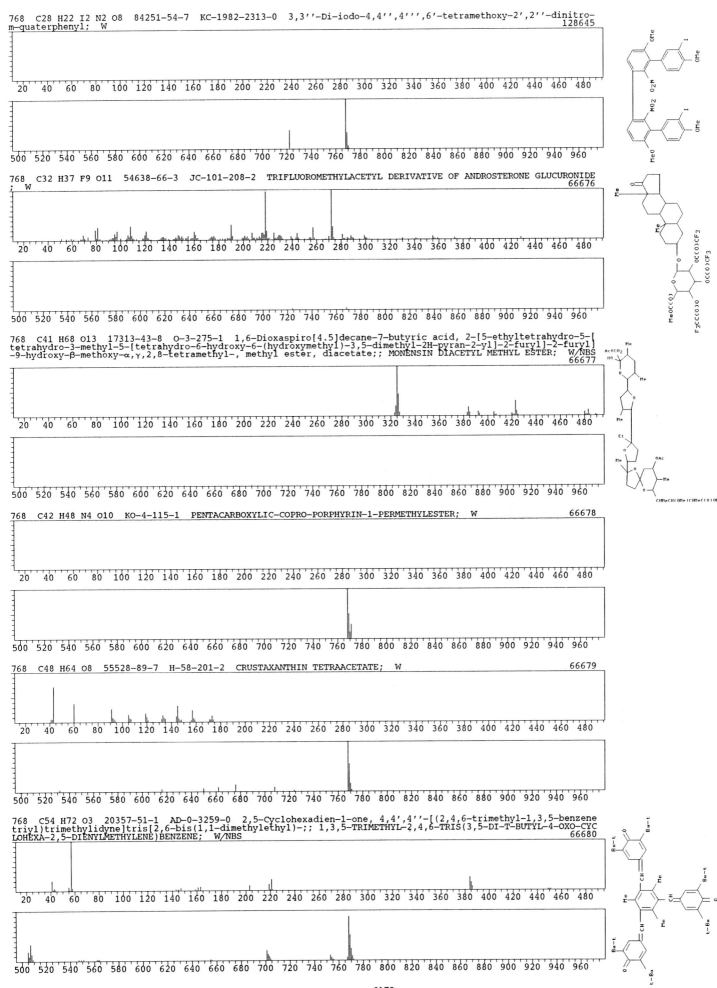

768   C28 H22 I2 N2 O8   84251-54-7   KC-1982-2313-0   3,3''-Di-iodo-4,4'',4''',6'-tetramethoxy-2',2''-dinitro-m-quaterphenyl;   W
128645

768   C32 H37 F9 O11   54638-66-3   JC-101-208-2   TRIFLUOROMETHYLACETYL DERIVATIVE OF ANDROSTERONE GLUCURONIDE ;   W
66676

768   C41 H68 O13   17313-43-8   O-3-275-1   1,6-Dioxaspiro[4.5]decane-7-butyric acid, 2-[5-ethyltetrahydro-5-[tetrahydro-3-methyl-5-[tetrahydro-6-hydroxy-6-(hydroxymethyl)-3,5-dimethyl-2H-pyran-2-yl]-2-furyl]-2-furyl]-9-hydroxy-β-methoxy-α,γ,2,8-tetramethyl-, methyl ester, diacetate;; MONENSIN DIACETYL METHYL ESTER;   W/NBS
66677

768   C42 H48 N4 O10   KO-4-115-1   PENTACARBOXYLIC-COPRO-PORPHYRIN-1-PERMETHYLESTER;   W
66678

768   C48 H64 O8   55528-89-7   H-58-201-2   CRUSTAXANTHIN TETRAACETATE;   W
66679

768   C54 H72 O3   20357-51-1   AD-0-3259-0   2,5-Cyclohexadien-1-one, 4,4',4''-[(2,4,6-trimethyl-1,3,5-benzenetriyl)trimethylidyne]tris[2,6-bis(1,1-dimethylethyl)-;; 1,3,5-TRIMETHYL-2,4,6-TRIS(3,5-DI-T-BUTYL-4-OXO-CYCLOHEXA-2,5-DIENYLMETHYLENE)BENZENE;   W/NBS
66680

769 C20 H60 Na7 O7 Si7 K-110-3395-0 NATRUMTRIMETHYLSILANOLATE; W   66681

769 C39 H35 N3 O12 S 74615-09-1 C-104-542-0 Methyl 5-amino-4-(3,4-dimethoxy-2-(phenylmethoxy)phenyl)-
6-(3-(3-methoxy-6-nitro-2-((phenylsulfonyl)oxy)phenyl)-1-oxo-2-propenyl)-3-methyl-2-pyridinecarboxylate; W
                                                                                   128646

770 C2 H5 F20 Fe2 P7 S 39733-63-6 EP-6186-0-0 Iron, [.mu.-(difluorophosphino)][.mu.-(ethanethiolato)]he
xakis(phosphorous trifluoride)di-, (Fe-Fe);; .mu.-ETHYLMERCAPTO-.mu.-DIFLUOROPHOSPHIDO-HEXAKIS(TRIFLUOROPHO
SPHINE)IRON DIHYDRIDE; W/NBS   66683

770 C14 H12 O10 P2 W2 18129-38-9 AD-0-1581-0 Tungsten, decacarbonyl[.mu.-(tetramethyldiphosphine-P:P')]
di-;; .mu.-TETRAMETHYLDIPHOSPHINE-BIS(PENTACARBONYLTUNGSTEN); W/NBS   66684

770 C34 H41 F3 N4 O11 S 69782-86-1 AB-207-251-2 L-Glutamic acid, N-[N-[N-[S-(2-methoxy-2-oxoethyl)-N-(
trifluoroacetyl)-L-cysteinyl]-O-methyl-L-tyrosyl]-L-phenylalanyl]-, dimethyl ester;; N-TFA-S-CARBOMETHOXYCY
STEINYL-O-METHYLTYROSYLPHENYLALANYL GLUTAMIC ACID METHYL ESTER; W/NBS   66685

770 C40 H66 O14 61773-55-5 K-109-3460-0 METHYL ESTER OF 15-O-METHYL-2-O-(3-O-(2,3,4,6-TETRA-O-METHYL-β-
D-GLUCOPYRANOSYL)-2,4,6-TRI-O-METHYL-β-D-GLUCOPYRANOSYL)ATRACTYLIGENIN; W   66686

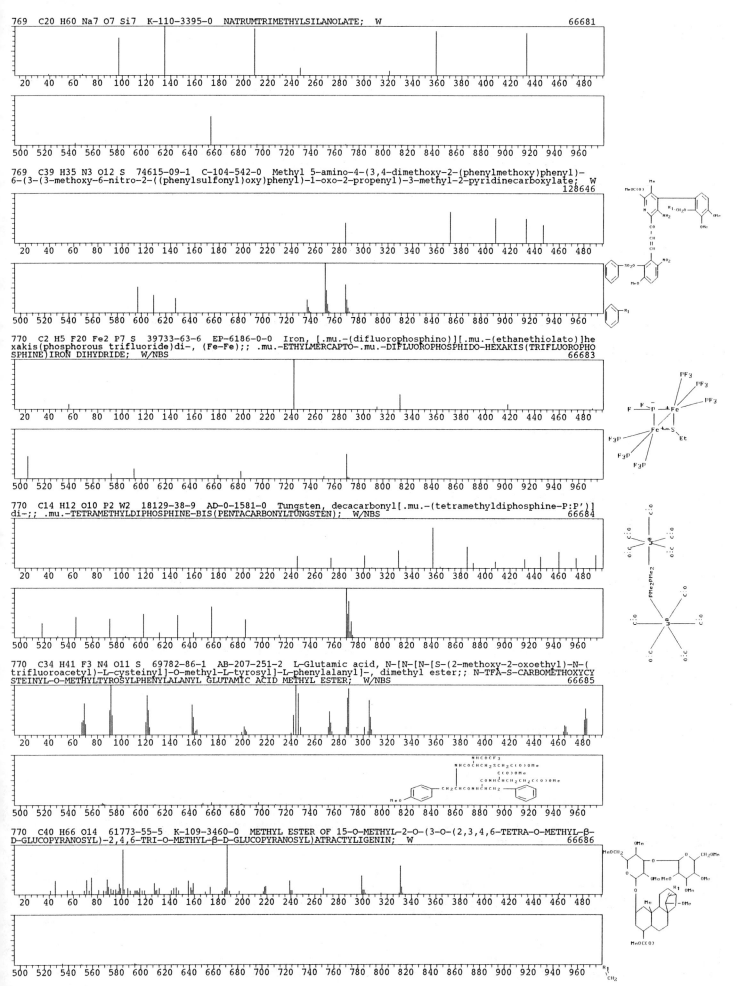

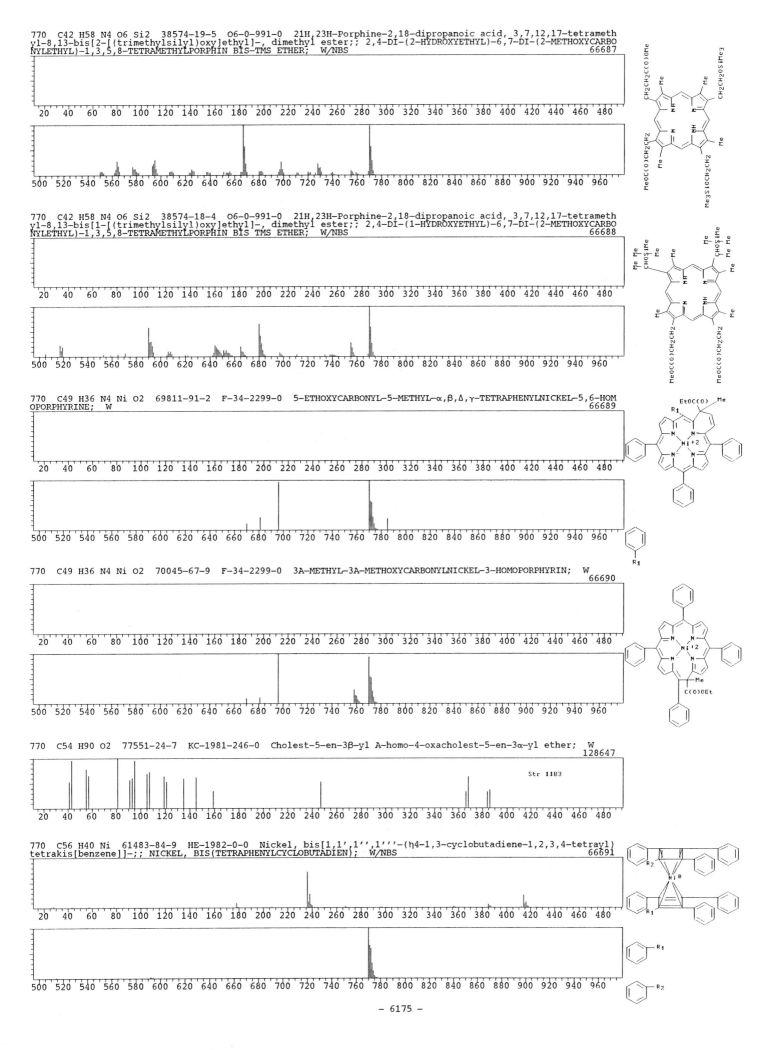

770 C42 H58 N4 O6 Si2 38574-19-5 O6-0-991-0 21H,23H-Porphine-2,18-dipropanoic acid, 3,7,12,17-tetrameth
yl-8,13-bis[2-[(trimethylsilyl)oxy]ethyl]-, dimethyl ester;; 2,4-DI-(2-HYDROXYETHYL)-6,7-DI-(2-METHOXYCARBO
NYLETHYL)-1,3,5,8-TETRAMETHYLPORPHIN BIS-TMS ETHER; W/NBS
66687

770 C42 H58 N4 O6 Si2 38574-18-4 O6-0-991-0 21H,23H-Porphine-2,18-dipropanoic acid, 3,7,12,17-tetrameth
yl-8,13-bis[1-[(trimethylsilyl)oxy]ethyl]-, dimethyl ester;; 2,4-DI-(1-HYDROXYETHYL)-6,7-DI-(2-METHOXYCARBO
NYLETHYL)-1,3,5,8-TETRAMETHYLPORPHIN BIS TMS ETHER; W/NBS
66688

770 C49 H36 N4 Ni O2 69811-91-2 F-34-2299-0 5-ETHOXYCARBONYL-5-METHYL-α,β,Δ,γ-TETRAPHENYLNICKEL-5,6-HOM
OPORPHYRINE; W
66689

770 C49 H36 N4 Ni O2 70045-67-9 F-34-2299-0 3A-METHYL-3A-METHOXYCARBONYLNICKEL-3-HOMOPORPHYRIN; W
66690

770 C54 H90 O2 77551-24-7 KC-1981-246-0 Cholest-5-en-3β-yl A-homo-4-oxacholest-5-en-3α-yl ether; W
128647

Str 1183

770 C56 H40 Ni 61483-84-9 HE-1982-0-0 Nickel, bis[1,1',1'',1'''-(η4-1,3-cyclobutadiene-1,2,3,4-tetrayl)
tetrakis[benzene]]-;; NICKEL, BIS(TETRAPHENYLCYCLOBUTADIEN); W/NBS
66691

771   C21 H10 F21 N O6   55255-98-6   AD-0-5205-0   Butanoic acid, heptafluoro-, 4-[1-(2,2,3,3,4,4,4-heptafluor
o-1-oxobutoxy)-2-[(2,2,3,3,4,4,4-heptafluoro-1-oxobutyl)amino]ethyl]-2-methoxyphenyl ester;; O,O,N-TRIS-HEP
TAFLUOROBUTYRYL-NORMETANEPHRINE;   W/NBS   66692

771   C40 H44 Cu N4 O8   54111-68-1   KO-4-115-1   Copper, [tetramethyl 3,8,13,18-tetramethyl-21H,23H-porphine-
2,7,12,17-tetrapropanoato(2-)-N21,N22,N23,N24]-, (SP-4-1)-;; 21H,23H-Porphine-2,7,12,17-tetrapropanoic aci
d, 3,8,13,18-tetramethyl-, tetramethylester, copper complex;   W   66694

772   C18 Cl15 P   16716-14-6   AD-0-2741-0   Phosphine, tris(pentachlorophenyl)-;; TRIS(PENTACHLOROPHENYL)PHOS
PHINE (IMPURITY ?);   W/NBS   66695

772   C36 H47 N4 O2 Tl   33339-93-4   O6-0-1401-0   Thallium, aquahydroxy[2,3,7,8,12,13,17,18-octaethyl-21H,23H
-porphinato(2-)-N21,N22,N23,N24]-, (OC-6-23)-;; Aquo octaethylporphinatothallium(III)hydroxide;   W/NBS
                                                                                               66696

772   C37 H40 O18   CD-335-0-0   Heptaacetate derivative of 2'-hydroxyfarrerol glucoside;   W   128648

772   C40 H44 N4 O8 Zn   19441-64-6   KO-4-115-1   ZINC-COPRO-PORPHYRIN-PERMETHYL ESTER;   W   66697

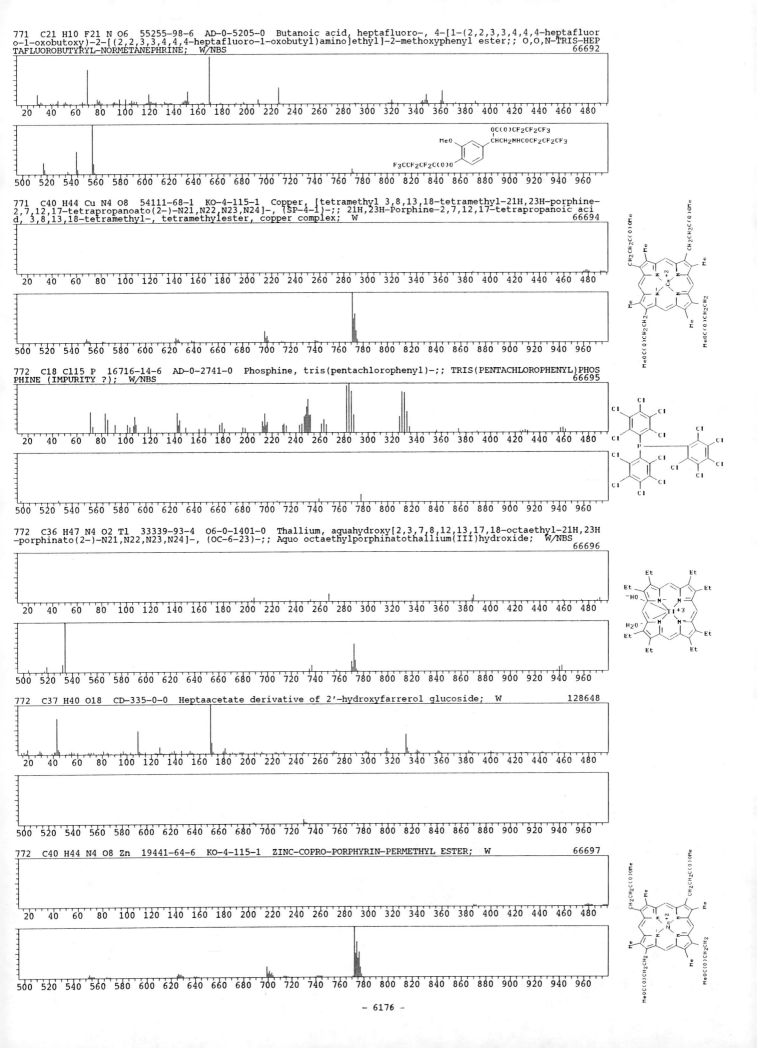

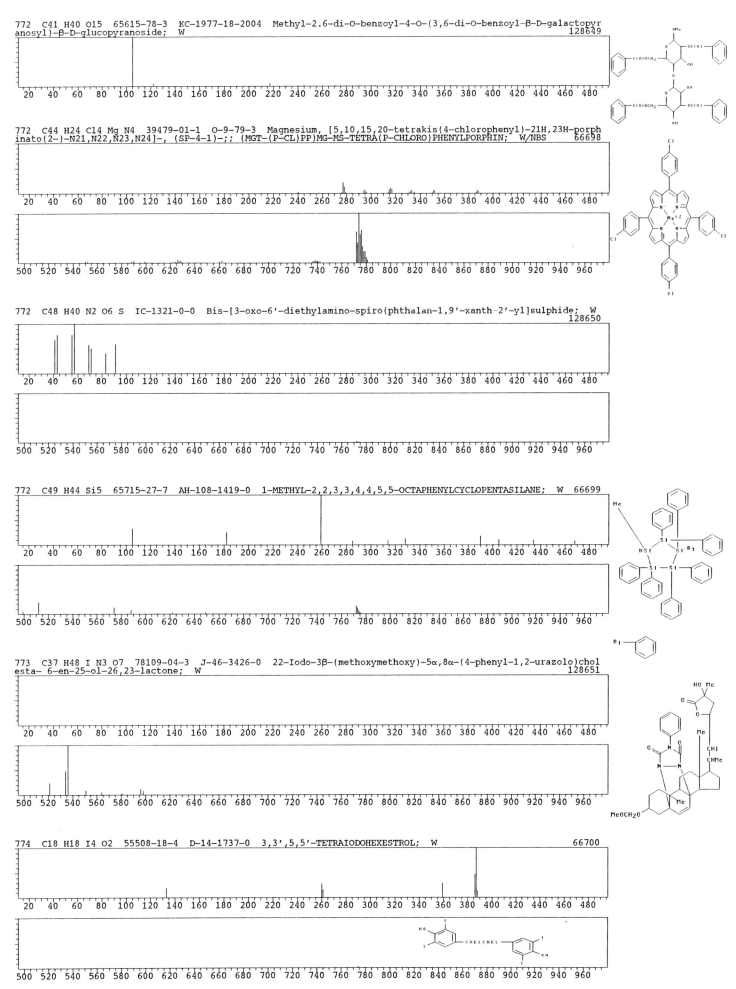

772   C41 H40 O15   65615-78-3   KC-1977-18-2004   Methyl-2.6-di-O-benzoyl-4-O-(3,6-di-O-benzoyl-β-D-galactopyr
anosyl)-β-D-glucopyranoside;   W   128649

772   C44 H24 Cl4 Mg N4   39479-01-1   O-9-79-3   Magnesium, [5,10,15,20-tetrakis(4-chlorophenyl)-21H,23H-porph
inato(2-)-N21,N22,N23,N24]-, (SP-4-1)-;; (MGT-(P-CL)PP)MG-MS-TETRA(P-CHLORO)PHENYLPORPHIN; W/NBS   66698

772   C48 H40 N2 O6 S   IC-1321-0-0   Bis-[3-oxo-6'-diethylamino-spiro(phthalan-1,9'-xanth-2'-yl]sulphide;   W
                                                                                             128650

772   C49 H44 Si5   65715-27-7   AH-108-1419-0   1-METHYL-2,2,3,3,4,4,5,5-OCTAPHENYLCYCLOPENTASILANE;   W   66699

773   C37 H48 I N3 O7   78109-04-3   J-46-3426-0   22-Iodo-3β-(methoxymethoxy)-5α,8α-(4-phenyl-1,2-urazolo)chol
esta- 6-en-25-ol-26,23-lactone;   W   128651

774   C18 H18 I4 O2   55508-18-4   D-14-1737-0   3,3',5,5'-TETRAIODOHEXESTROL;   W   66700

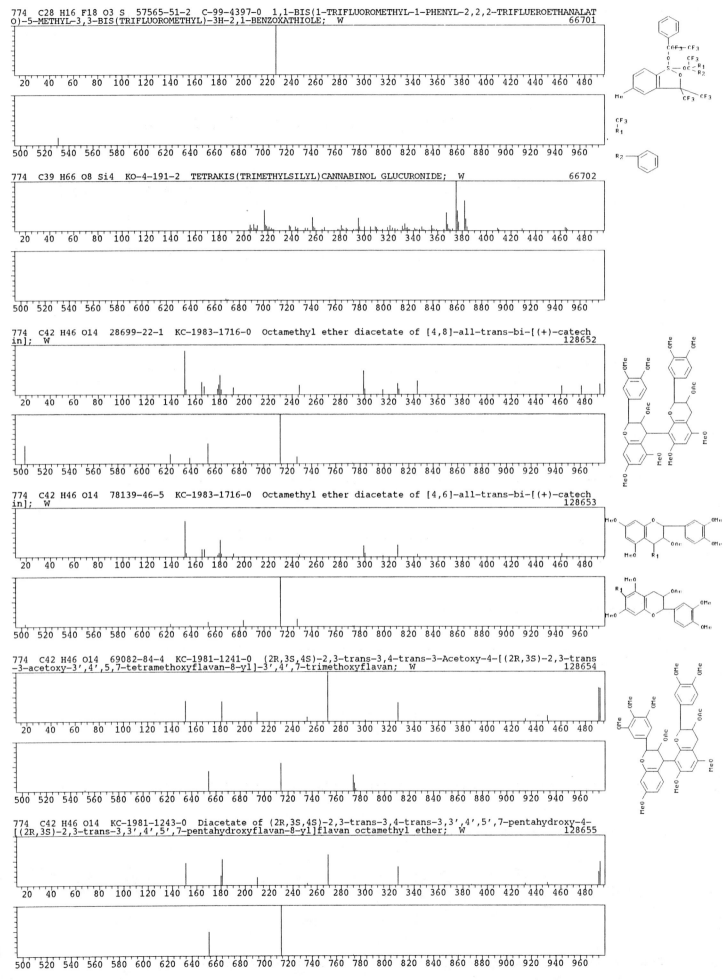

774 C28 H16 F18 O3 S 57565-51-2 C-99-4397-0 1,1-BIS(1-TRIFLUOROMETHYL-1-PHENYL-2,2,2-TRIFLUEROETHANALAT
O)-5-METHYL-3,3-BIS(TRIFLUOROMETHYL)-3H-2,1-BENZOXATHIOLE; W   66701

774 C39 H66 O8 Si4 KO-4-191-2 TETRAKIS(TRIMETHYLSILYL)CANNABINOL GLUCURONIDE; W   66702

774 C42 H46 O14 28699-22-1 KC-1983-1716-0 Octamethyl ether diacetate of [4,8]-all-trans-bi-[(+)-catech
in]; W   128652

774 C42 H46 O14 78139-46-5 KC-1983-1716-0 Octamethyl ether diacetate of [4,6]-all-trans-bi-[(+)-catech
in]; W   128653

774 C42 H46 O14 69082-84-4 KC-1981-1241-0 (2R,3S,4S)-2,3-trans-3,4-trans-3-Acetoxy-4-[(2R,3S)-2,3-trans
-3-acetoxy-3',4',5,7-tetramethoxyflavan-8-yl]-3',4',7-trimethoxyflavan; W   128654

774 C42 H46 O14 KC-1981-1243-0 Diacetate of (2R,3S,4S)-2,3-trans-3,4-trans-3,3',4',5',7-pentahydroxy-4-
[(2R,3S)-2,3-trans-3,3',4',5',7-pentahydroxyflavan-8-yl]flavan octamethyl ether; W   128655

774 C42 H46 O14 KC-1981-1244-0 (2R,3S,4R)-2,3-trans-3,4-trans-3,3',4',7-tetramethoxy-4-[(2R,3S)-2,3-
trans-3-acetoxy-3',4',7-trimethoxyflavan-6-yl]flavan; W   128656

774 C44 H72 N6 Ni O2 78379-74-5 H-64-1108-0 1,11-Dineopentyloxy-2,2,3,3,7,7,8,8,12,12,13,13,17,17,18,18
-hexadecamethyl-10,20-diaza-decahydroporphinato(2-)]nickel; W   128657

774 C45 H31 In N4 S 75845-80-6 KA-1980-2084-2 Methylthio[5,10,15,20-tetraphenylporphyrinato]indium; W
                                                                                                   128658

774 C45 H62 N2 O9 89759-10-4 H-67-246-0 O,O',O''-Triacetyl-O''',O''''-dimethylkopsirachin; W   128659

774 C50 H54 N4 O4 59942-58-4 KC-1976-799-0 BIS-(1-BENZOYLOXYETHYL)HEXAETHYLPORPHYRIN; W   66703

774 C50 H54 N4 O4 59942-59-5 KC-1976-800-0 MESO-BENZOYLOXY-(1-BENZOYLOXYETHYL)HEPTAETHYLPORPHYRIN; W
                                                                                                       66704

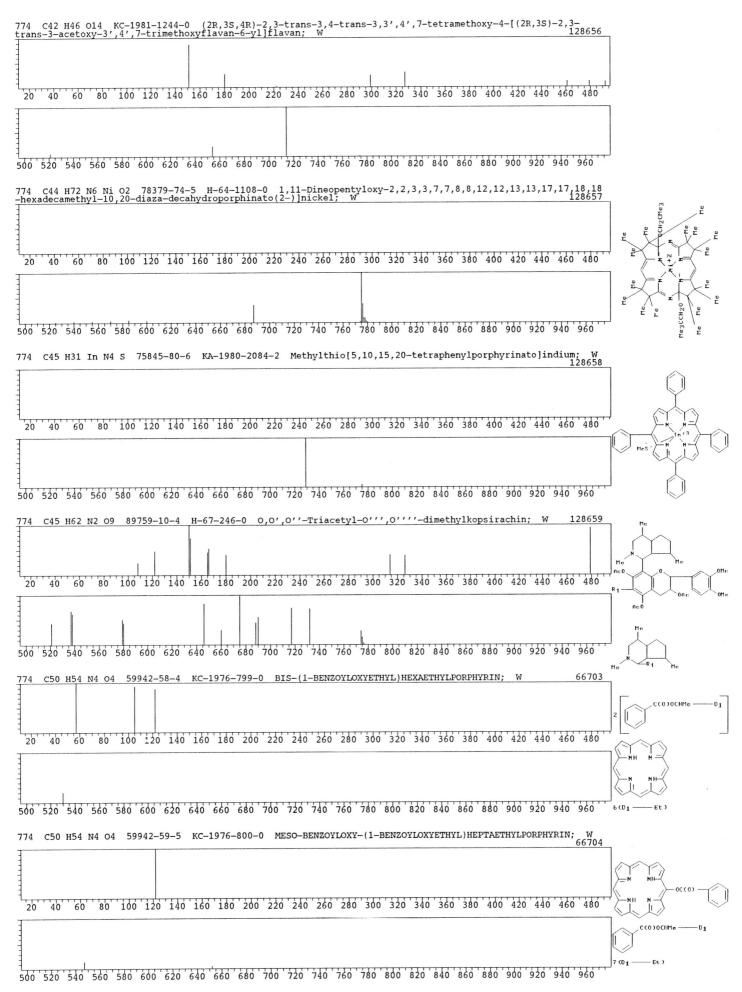

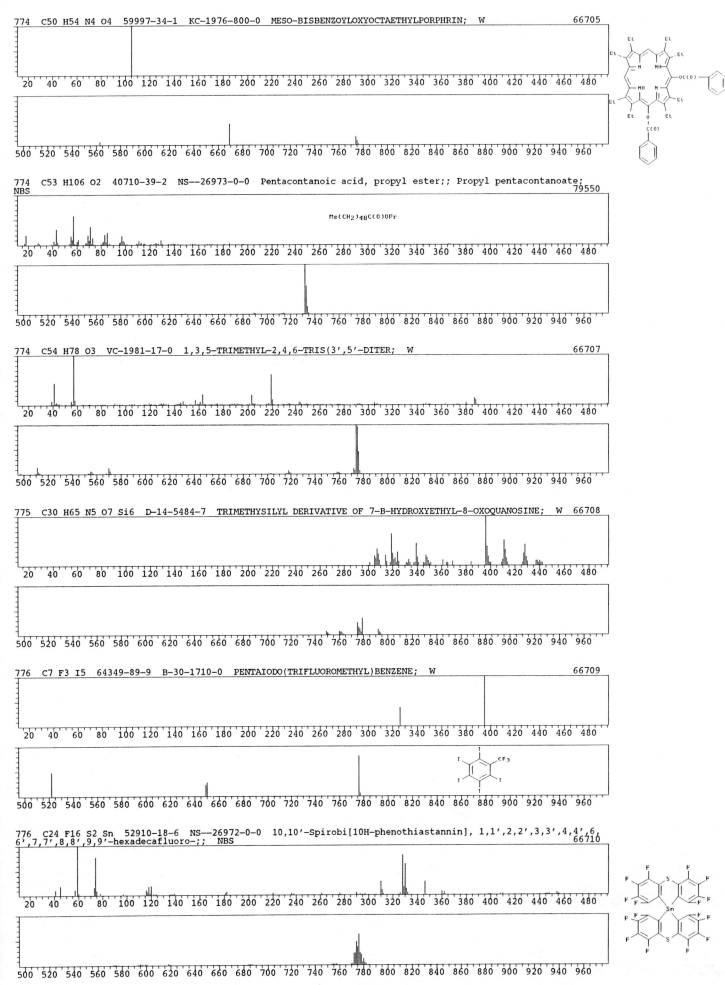

774   C50 H54 N4 O4   59997-34-1   KC-1976-800-0   MESO-BISBENZOYLOXYOCTAETHYLPORPHRIN;   W   66705

774   C53 H106 O2   40710-39-2   NS--26973-0-0   Pentacontanoic acid, propyl ester;; Propyl pentacontanoate;
NBS   79550

Me(CH2)48C(O)OPr

774   C54 H78 O3   VC-1981-17-0   1,3,5-TRIMETHYL-2,4,6-TRIS(3',5'-DITER;   W   66707

775   C30 H65 N5 O7 Si6   D-14-5484-7   TRIMETHYSILYL DERIVATIVE OF 7-B-HYDROXYETHYL-8-OXOQUANOSINE;   W   66708

776   C7 F3 I5   64349-89-9   B-30-1710-0   PENTAIODO(TRIFLUOROMETHYL)BENZENE;   W   66709

776   C24 F16 S2 Sn   52910-18-6   NS--26972-0-0   10,10'-Spirobi[10H-phenothiastannin], 1,1',2,2',3,3',4,4',6,
6',7,7',8,8',9,9'-hexadecafluoro-;;   NBS   66710

776   C32 H60 O10 S Si5   83228-79-9   KO-9-249-7   Per(trimethylsilyl)derivative of O-(dihydrodihydroxymethyls
ulfonylnaphthylene)monoglucuronide;   W   128660

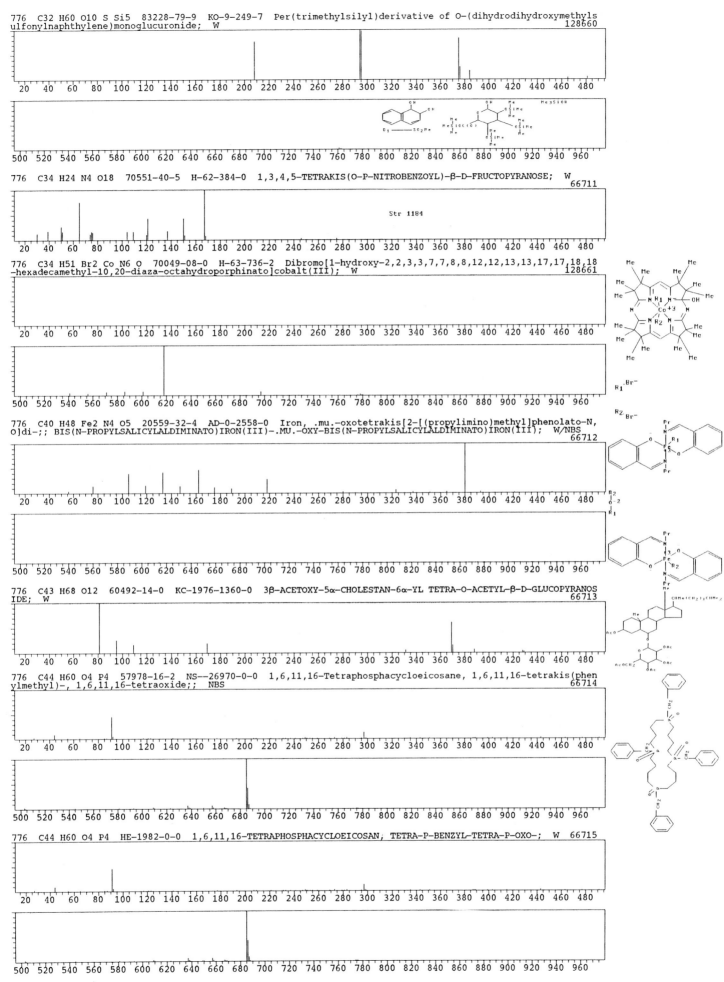

776   C34 H24 N4 O18   70551-40-5   H-62-384-0   1,3,4,5-TETRAKIS(O-P-NITROBENZOYL)-β-D-FRUCTOPYRANOSE;   W
66711

Str 1184

776   C34 H51 Br2 Co N6 O   70049-08-0   H-63-736-2   Dibromo[1-hydroxy-2,2,3,3,7,7,8,8,12,12,13,13,17,17,18,18
-hexadecamethyl-10,20-diaza-octahydroporphinato]cobalt(III);   W   128661

776   C40 H48 Fe2 N4 O5   20559-32-4   AD-0-2558-0   Iron, .mu.-oxotetrakis[2-[(propylimino)methyl]phenolato-N,
O]di-;;   BIS(N-PROPYLSALICYLALDIMINATO)IRON(III)-.MU.-OXY-BIS(N-PROPYLSALICYLALDIMINATO)IRON(III);   W/NBS
66712

776   C43 H68 O12   60492-14-0   KC-1976-1360-0   3β-ACETOXY-5α-CHOLESTAN-6α-YL TETRA-O-ACETYL-β-D-GLUCOPYRANOS
IDE;   W   66713

776   C44 H60 O4 P4   57978-16-2   NS--26970-0-0   1,6,11,16-Tetraphosphacycloeicosane, 1,6,11,16-tetrakis(phen
ylmethyl)-, 1,6,11,16-tetraoxide;;   NBS   66714

776   C44 H60 O4 P4   HE-1982-0-0   1,6,11,16-TETRAPHOSPHACYCLOEICOSAN, TETRA-P-BENZYL-TETRA-P-OXO-;   W   66715

776   C45 H52 N4 O8   72253-59-9   I-57-2576-0   3'-VINDOLINYL-5-NOR-CATHARANTHINE;   W   66716

776   C45 H52 N4 O8   72154-67-7   I-57-2576-0   6'-VINDOLINYL-5-NOR-CATHARANTHINE;   W   66717

776   C46 H34 Br2 O2   J-44-3693-0   Dimer of 9-(α-Bromoanisylidene)-10-methylene-9,10-dihydroanthracene;   W
                                                                                       128662

777   C37 H48 Cl N3 O13   82400-19-9   C-104-4934-0   N-Methyltrenudone;   W   128663

777   C43 H60 Cl N O6 Si2   81025-19-6   O-16-404-1   PENITREM A DITRIMETHYLSILYL ETHER;   W   66718

Me3SiOH

777   C44 H63 N3 O9   79664-34-9   J-46-5236-0   N(1),N(10)-Di-n-octyl-N(1),N(5),N(10)-tris(2,3-dihydroxybenzoyl)-1,5,10-triazadecane;   W
                                                                                      128664

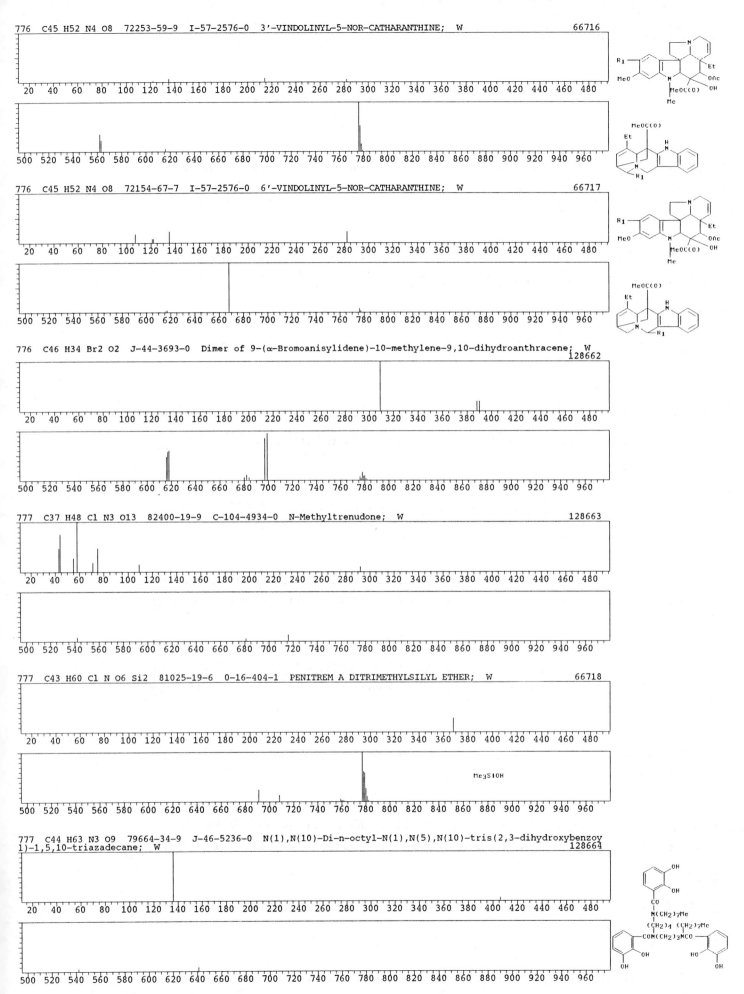

778   C24 H10 F8 Ge I2   69688-61-5   NS--26969-0-0   Germane, diphenylbis(2,3,4,5-tetrafluoro-6-iodophenyl)-;;
NBS   66719

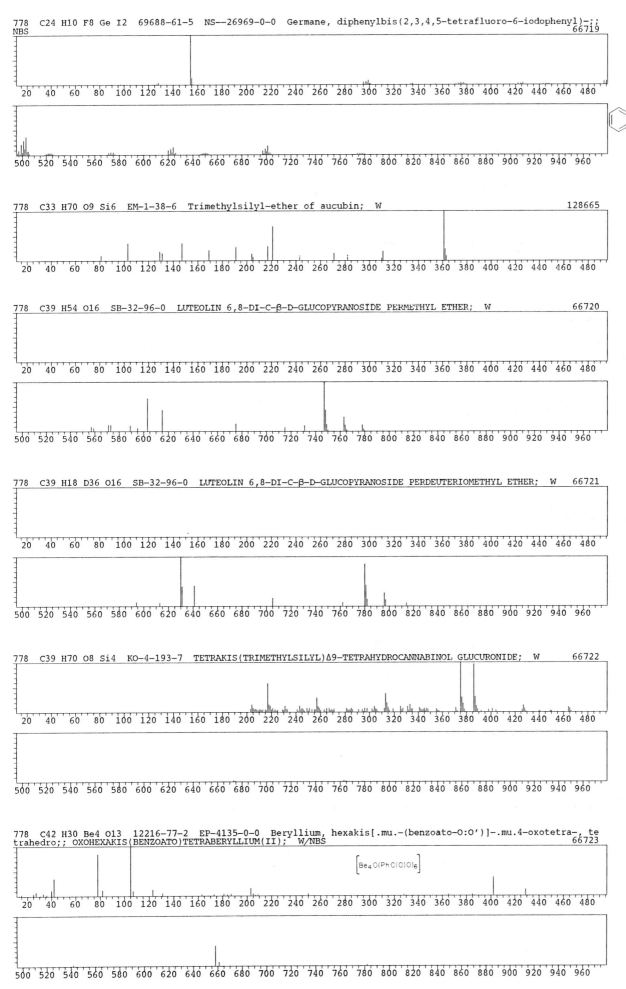

778   C33 H70 O9 Si6   EM-1-38-6   Trimethylsilyl-ether of aucubin;   W   128665

778   C39 H54 O16   SB-32-96-0   LUTEOLIN 6,8-DI-C-β-D-GLUCOPYRANOSIDE PERMETHYL ETHER;   W   66720

778   C39 H18 D36 O16   SB-32-96-0   LUTEOLIN 6,8-DI-C-β-D-GLUCOPYRANOSIDE PERDEUTERIOMETHYL ETHER;   W   66721

778   C39 H70 O8 Si4   KO-4-193-7   TETRAKIS(TRIMETHYLSILYL)Δ9-TETRAHYDROCANNABINOL GLUCURONIDE;   W   66722

778   C42 H30 Be4 O13   12216-77-2   EP-4135-0-0   Beryllium, hexakis[.mu.-(benzoato-O:O')]-.mu.4-oxotetra-, tetrahedro;;   OXOHEXAKIS(BENZOATO)TETRABERYLLIUM(II);   W/NBS   66723

[Be₄O(PhC(O)O)₆]

778   C44 H74 O11   BS-3-36-0   Methylnarasin;   W   128666

20 40 60 80 100 120 140 160 180 200 220 240 260 280 300 320 340 360 380 400 420 440 460 480

500 520 540 560 580 600 620 640 660 680 700 720 740 760 780 800 820 840 860 880 900 920 940 960

778   C49 H94 O6   56599-89-4   EP-5346-0-0   Hexadecanoic acid, 2-[(1-oxotetradecyl)oxy]-1,3-propanediyl ester
;; GLYCERYL 1,3-DIHEXADECANOATE-2-TETRADECANOATE;   W/NBS   66724

20 40 60 80 100 120 140 160 180 200 220 240 260 280 300 320 340 360 380 400 420 440 460 480

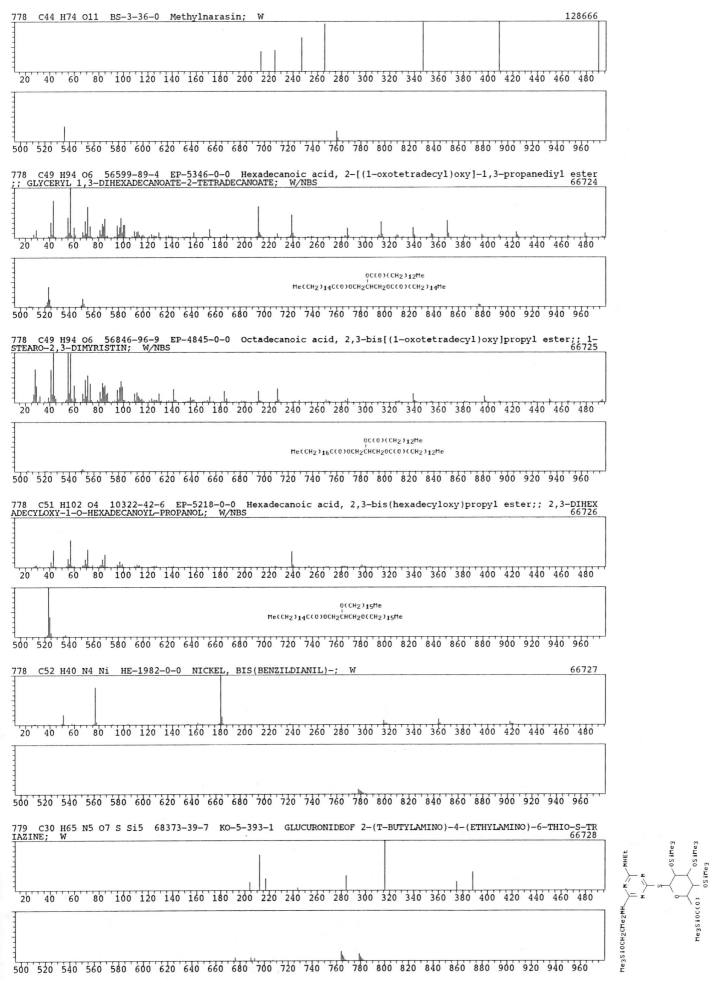

OC(O)(CH₂)₁₂Me
Me(CH₂)₁₄C(O)OCH₂CHCH₂OC(O)(CH₂)₁₄Me

500 520 540 560 580 600 620 640 660 680 700 720 740 760 780 800 820 840 860 880 900 920 940 960

778   C49 H94 O6   56846-96-9   EP-4845-0-0   Octadecanoic acid, 2,3-bis[(1-oxotetradecyl)oxy]propyl ester;; 1-
STEARO-2,3-DIMYRISTIN;   W/NBS   66725

20 40 60 80 100 120 140 160 180 200 220 240 260 280 300 320 340 360 380 400 420 440 460 480

OC(O)(CH₂)₁₂Me
Me(CH₂)₁₆C(O)OCH₂CHCH₂OC(O)(CH₂)₁₂Me

500 520 540 560 580 600 620 640 660 680 700 720 740 760 780 800 820 840 860 880 900 920 940 960

778   C51 H102 O4   10322-42-6   EP-5218-0-0   Hexadecanoic acid, 2,3-bis(hexadecyloxy)propyl ester;; 2,3-DIHEX
ADECYLOXY-1-O-HEXADECANOYL-PROPANOL;   W/NBS   66726

20 40 60 80 100 120 140 160 180 200 220 240 260 280 300 320 340 360 380 400 420 440 460 480

O(CH₂)₁₅Me
Me(CH₂)₁₄C(O)OCH₂CHCH₂O(CH₂)₁₅Me

500 520 540 560 580 600 620 640 660 680 700 720 740 760 780 800 820 840 860 880 900 920 940 960

778   C52 H40 N4 Ni   HE-1982-0-0   NICKEL, BIS(BENZILDIANIL)-;   W   66727

20 40 60 80 100 120 140 160 180 200 220 240 260 280 300 320 340 360 380 400 420 440 460 480

500 520 540 560 580 600 620 640 660 680 700 720 740 760 780 800 820 840 860 880 900 920 940 960

779   C30 H65 N5 O7 S Si5   68373-39-7   KO-5-393-1   GLUCURONIDEOF 2-(T-BUTYLAMINO)-4-(ETHYLAMINO)-6-THIO-S-TR
IAZINE;   W   66728

20 40 60 80 100 120 140 160 180 200 220 240 260 280 300 320 340 360 380 400 420 440 460 480

500 520 540 560 580 600 620 640 660 680 700 720 740 760 780 800 820 840 860 880 900 920 940 960

779 C30 H65 N5 O7 S Si5 BS-5-70-0 Pertrimethylsilyl derivative of the s-glucuronide of 2-(ethylamino)-4-[(2-hydroxy-1,1-dimethyl)amino]-6-thio-s-triazine; W   128667

779 C32 H10 Cl6 Cu N8 F-20-1360-6 COPPER 3:6 HEXACHLOROPHTHALOCYANINE; W   66729

779 C37 H50 Cl N3 O13 82390-96-3 C-104-4934-0 Methyl ester of decyclotrenudine; W   128668

779 C39 H61 N3 O5 S4 91652-55-0 H-67-2182-0 13,13-(Ethylenedithio)-N-[4-tosyl-8-(tosylamino)-4-azaoctyl]-16-hexadecanlactam; W   128669

779 C53 H81 N O3 CD-309-0-0 Rhodoquinone-9; W   128670

780 C10 H18 Bi2 Mo O4 P2 65060-45-9 K-110-3437-0 CIS-TETRACARBONYL(1,2-BIS(DIMETHYLPHOSPHINO)-1,2-DIMETHYLBISMUTIN-P,P')MOLYBDANUM; W   66730

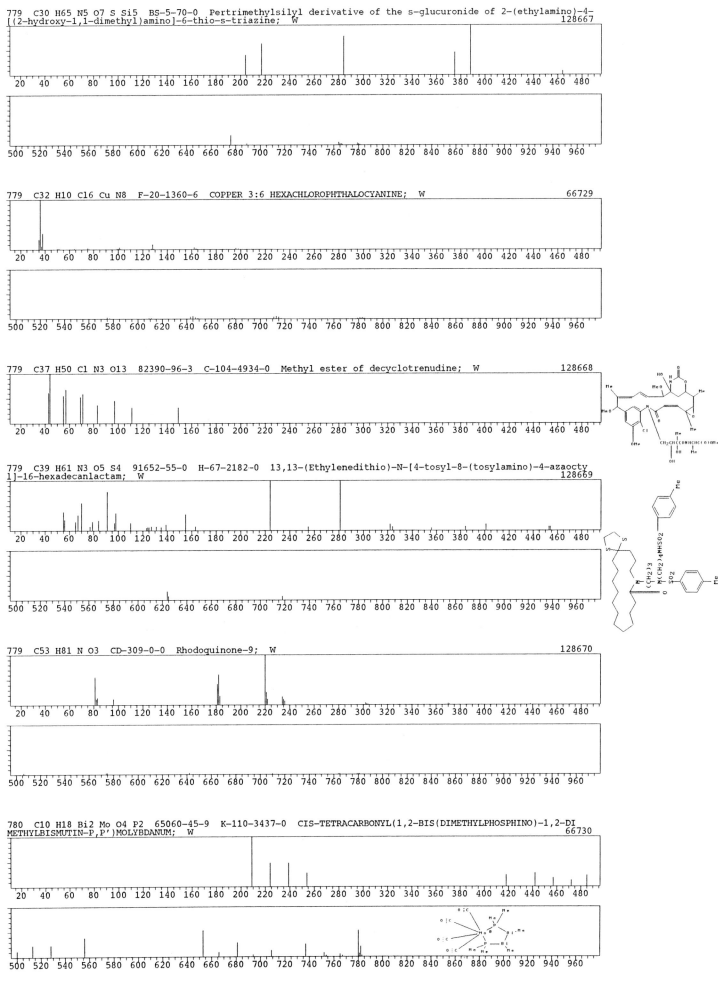

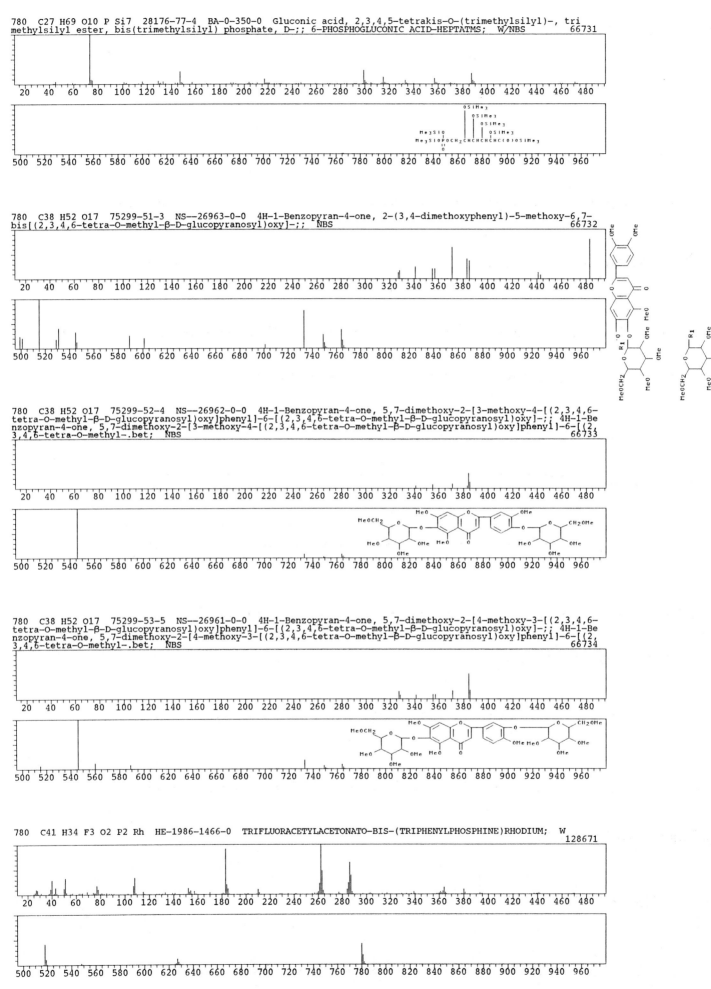

780 C27 H69 O10 P Si7 28176-77-4 BA-0-350-0 Gluconic acid, 2,3,4,5-tetrakis-O-(trimethylsilyl)-, tri
methylsilyl ester, bis(trimethylsilyl) phosphate, D-;; 6-PHOSPHOGLUCONIC ACID-HEPTATMS; W/NBS   66731

780 C38 H52 O17 75299-51-3 NS--26963-0-0 4H-1-Benzopyran-4-one, 2-(3,4-dimethoxyphenyl)-5-methoxy-6,7-
bis[(2,3,4,6-tetra-O-methyl-β-D-glucopyranosyl)oxy]-;; NBS   66732

780 C38 H52 O17 75299-52-4 NS--26962-0-0 4H-1-Benzopyran-4-one, 5,7-dimethoxy-2-[3-methoxy-4-[(2,3,4,6-
tetra-O-methyl-β-D-glucopyranosyl)oxy]phenyl]-6-[(2,3,4,6-tetra-O-methyl-β-D-glucopyranosyl)oxy]-;; 4H-1-Be
nzopyran-4-one, 5,7-dimethoxy-2-[3-methoxy-4-[(2,3,4,6-tetra-O-methyl-β-D-glucopyranosyl)oxy]phenyl]-6-[(2,
3,4,6-tetra-O-methyl-.bet; NBS   66733

780 C38 H52 O17 75299-53-5 NS--26961-0-0 4H-1-Benzopyran-4-one, 5,7-dimethoxy-2-[4-methoxy-3-[(2,3,4,6-
tetra-O-methyl-β-D-glucopyranosyl)oxy]phenyl]-6-[(2,3,4,6-tetra-O-methyl-β-D-glucopyranosyl)oxy]-;; 4H-1-Be
nzopyran-4-one, 5,7-dimethoxy-2-[4-methoxy-3-[(2,3,4,6-tetra-O-methyl-β-D-glucopyranosyl)oxy]phenyl]-6-[(2,
3,4,6-tetra-O-methyl-.bet; NBS   66734

780 C41 H34 F3 O2 P2 Rh HE-1986-1466-0 TRIFLUORACETYLACETONATO-BIS-(TRIPHENYLPHOSPHINE)RHODIUM; W
          128671

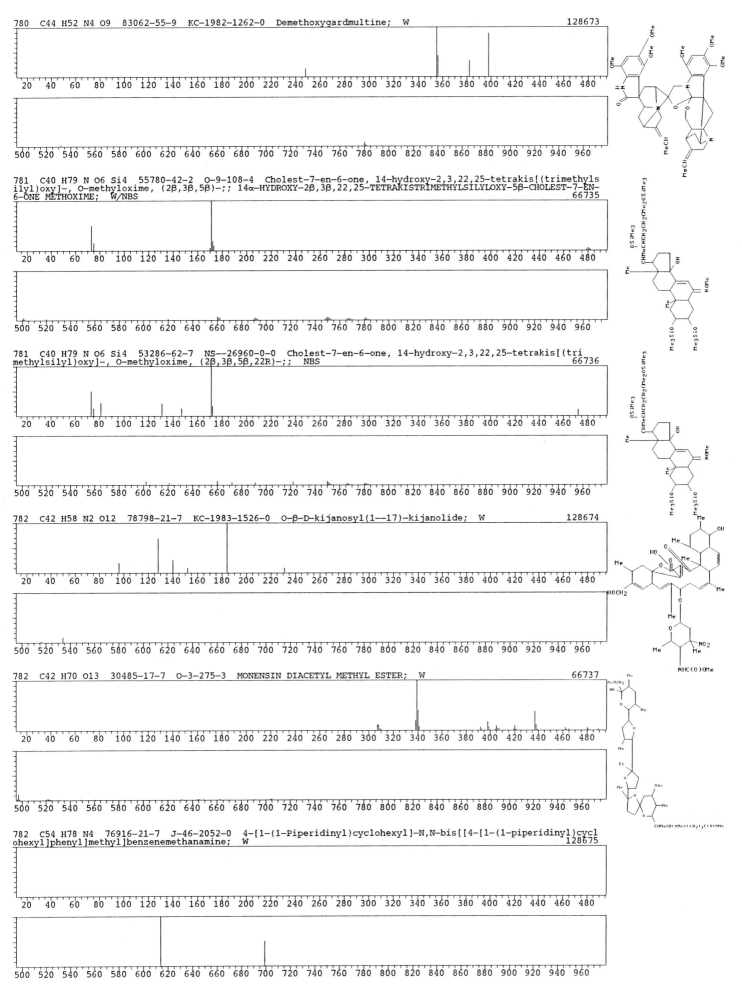

780   C44 H52 N4 O9   83062-55-9   KC-1982-1262-0   Demethoxygardmultine;   W   128673

781   C40 H79 N O6 Si4   55780-42-2   O-9-108-4   Cholest-7-en-6-one, 14-hydroxy-2,3,22,25-tetrakis[(trimethyls ilyl)oxy]-, O-methyloxime, (2β,3β,5β)-;; 14α-HYDROXY-2β,3β,22,25-TETRAKISTRIMETHYLSILYLOXY-5β-CHOLEST-7-EN- 6-ONE METHOXIME;   W/NBS   66735

781   C40 H79 N O6 Si4   53286-62-7   NS--26960-0-0   Cholest-7-en-6-one, 14-hydroxy-2,3,22,25-tetrakis[(tri methylsilyl)oxy]-, O-methyloxime, (2β,3β,5β,22R)-;;   NBS   66736

782   C42 H58 N2 O12   78798-21-7   KC-1983-1526-0   O-β-D-kijanosyl(1--17)-kijanolide;   W   128674

782   C42 H70 O13   30485-17-7   O-3-275-3   MONENSIN DIACETYL METHYL ESTER;   W   66737

782   C54 H78 N4   76916-21-7   J-46-2052-0   4-[1-(1-Piperidinyl)cyclohexyl]-N,N-bis[[4-[1-(1-piperidinyl)cycl ohexyl]phenyl]methyl]benzenemethanamine;   W   128675

784 C24 H5 Co4 F5 O10 12572-37-1 AD-0-3685-0 Cobalt, decacarbonyl[(pentafluorophenyl)phenylacetylene]te
tra-;; .MU.-(2,3,4,5,6-PENTAFLUORODIPHENYLACETYLENE)-DECACARBONYLTETRACOBALT; W/NBS   66738

784 C32 H40 B4 O12 P4 HE-1982-0-0 PENTACYCLO[9,5,1,1E3,9-1E5,15-1E7,13]EICOSAN, 3,7,11,15-TETRABORATA-TE
TRA-B-ETHYL-DODECAOXA-1,5,9,13-TETRAPHOSPHONIA-TETRA-P-PHENYL-; W   66739

784 C40 H53 In N4 O3 S KA-1980-2084-2 t-Butylsulfonato[2,3,7,8,12,13,17,18-octaethylporphyrinato]indium;
W   128676

784 C43 H69 Na O11 KO-3-276-6 SODIUM SALT OF A28086B; W   66740

785 C22 H12 F21 N O6 55255-99-7 AD-0-5204-0 Butanoic acid, heptafluoro-, 4-[1-(2,2,3,3,4,4,4-heptafluor
o-1-oxobutoxy)-2-[(2,2,3,3,4,4,4-heptafluoro-1-oxobutyl)methylamino]ethyl]-2-methoxyphenyl ester;; O,O,N-TR
IS-HEPTAFLUOROBUTYRYL-METANEPHRINE; W/NBS   66741

785 C26 H24 P2 Pt S6 80679-75-0 J-47-1492-0 Dimethylbis(tri-2-thienylphosphine-P)-(SP-4-2)-platinum; W
128677

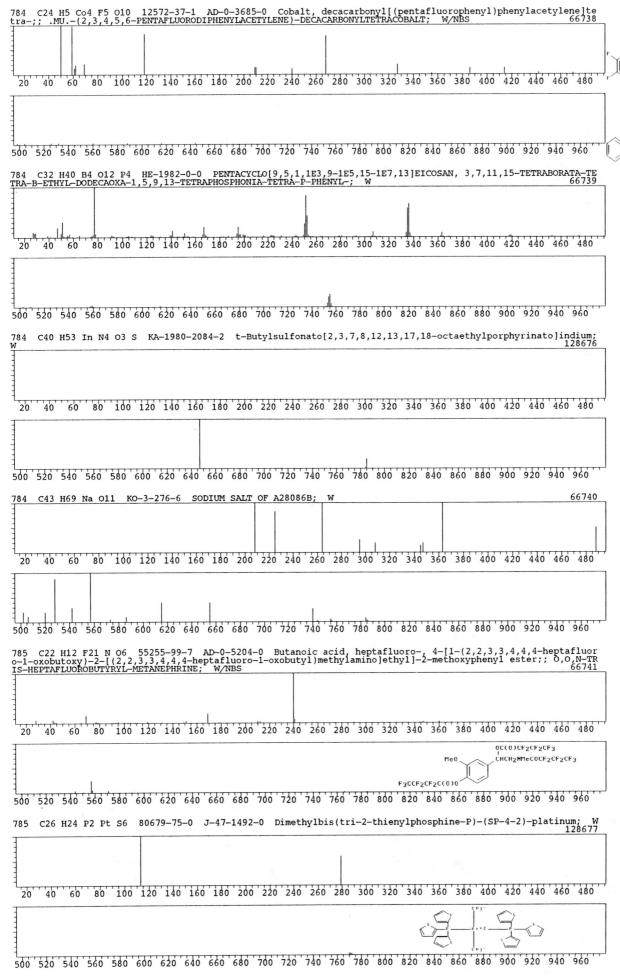

785   C44 H84 N O8 P   56648-95-4   EP-5266-0-0   Ethanaminium, 2-[[[2,3-bis[(1-oxo-9-octadecenyl)oxy]propoxy] hydroxyphosphinyl]oxy]-N,N,N-trimethyl-, hydroxide, inner salt, (R)-;; L-1,2-DIOLEOYLGLYCERYLPHOSPHORYLCHOL INE; W/NBS   66742

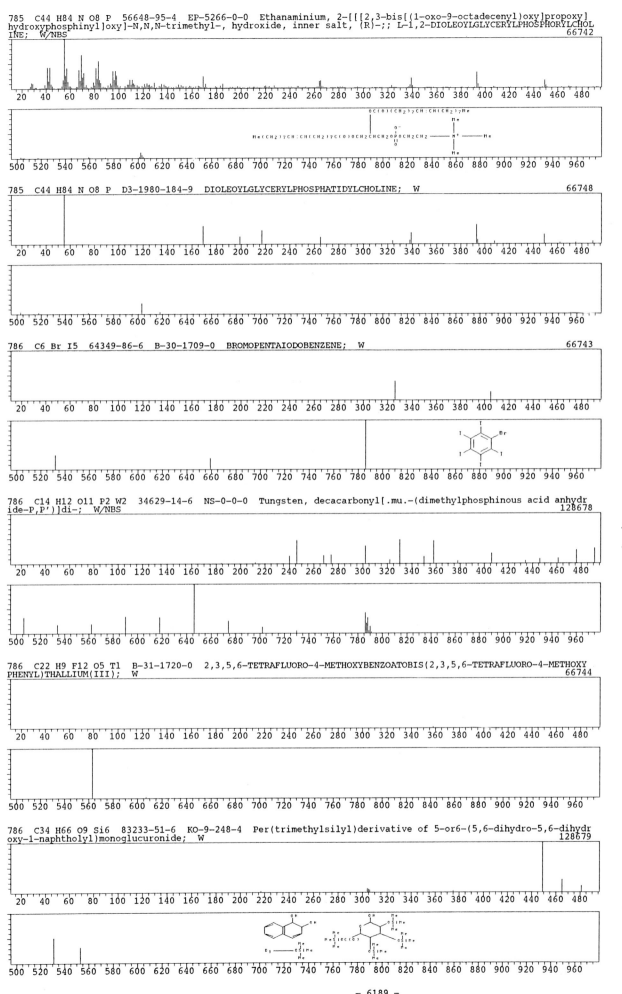

785   C44 H84 N O8 P   D3-1980-184-9   DIOLEOYLGLYCERYLPHOSPHATIDYLCHOLINE; W   66748

786   C6 Br I5   64349-86-6   B-30-1709-0   BROMOPENTAIODOBENZENE; W   66743

786   C14 H12 O11 P2 W2   34629-14-6   NS-0-0-0   Tungsten, decacarbonyl[.mu.-(dimethylphosphinous acid anhydr ide-P,P')]di-; W/NBS   128678

786   C22 H9 F12 O5 Tl   B-31-1720-0   2,3,5,6-TETRAFLUORO-4-METHOXYBENZOATOBIS(2,3,5,6-TETRAFLUORO-4-METHOXY PHENYL)THALLIUM(III); W   66744

786   C34 H66 O9 Si6   83233-51-6   KO-9-248-4   Per(trimethylsilyl)derivative of 5-or6-(5,6-dihydro-5,6-dihydr oxy-1-naphtholyl)monoglucuronide; W   128679

786   C34 H12 D54 O9 Si6   KO-9-249-6   Per(perdeuterotrimethylsilyl)derivative of 5-or6-(5,6-dihydro-5,6-dihy
droxy-1-naphtholyl)monoglucuronide;   W   128680

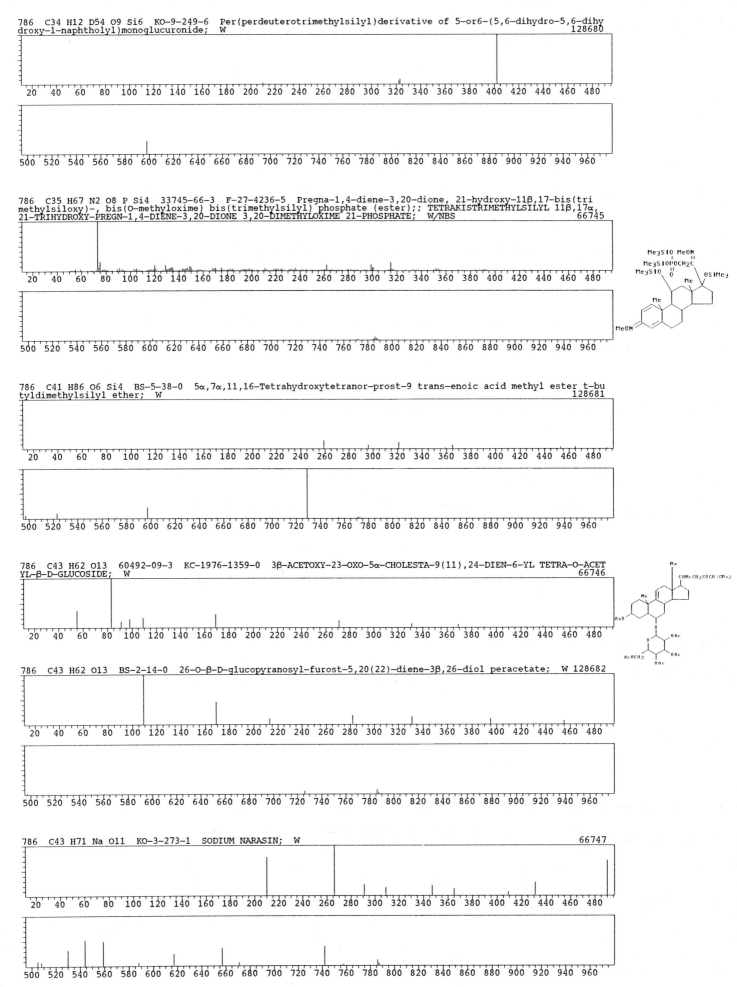

786   C35 H67 N2 O8 P Si4   33745-66-3   F-27-4236-5   Pregna-1,4-diene-3,20-dione, 21-hydroxy-11β,17-bis(tri
methylsiloxy)-, bis(O-methyloxime) bis(trimethylsilyl) phosphate (ester);; TETRAKISTRIMETHYLSILYL 11β,17α,
21-TRIHYDROXY-PREGN-1,4-DIENE-3,20-DIONE 3,20-DIMETHYLOXIME 21-PHOSPHATE;   W/NBS   66745

786   C41 H86 O6 Si4   BS-5-38-0   5α,7α,11,16-Tetrahydroxytetranor-prost-9 trans-enoic acid methyl ester t-bu
tyldimethylsilyl ether;   W   128681

786   C43 H62 O13   60492-09-3   KC-1976-1359-0   3β-ACETOXY-23-OXO-5α-CHOLESTA-9(11),24-DIEN-6-YL TETRA-O-ACET
YL-β-D-GLUCOSIDE;   W   66746

786   C43 H62 O13   BS-2-14-0   26-O-β-D-glucopyranosyl-furost-5,20(22)-diene-3β,26-diol peracetate;   W 128682

786   C43 H71 Na O11   KO-3-273-1   SODIUM NARASIN;   W   66747

786   C46 H74 O6 S2   IC-1322-0-0   1,10-Bis-(3,5-di-t-butyl-4-hydroxyphenylmethyl-thioethylcarboxy)-decane;
W
128683

786   C48 H66 N4 O2 Si2   62786-49-6   KC-1977-1-71   2-(1-(2,4-BIS-O-TRIMETHYLSILYLPHENYL)ETHYL)HEPTAETHYL POR
PHYRIN; W
66749

786   C53 H78 N4 O   69912-16-9   J-44-2080-0   α-Formyl-2,4,6,8-tetra(n-heptyl)-1,3,5,7-tetramethylporphyrin;
W
128684

788   C24 F20 Sn   1065-49-2   AD-0-2722-0   Stannane, tetrakis(pentafluorophenyl)-;; Tetrakis(pentafluorophen
yl)stannane; W/NBS
66750

788   C35 H69 N2 O8 P Si4   33745-67-4   F-27-4237-6   Pregn-4-ene-3,20-dione, 21-hydroxy-11β,17-bis(trimethyls
iloxy)-, bis(O-methyloxime), bis(trimethylsilyl) phosphate (ester);; TETRAKISTRIMETHYLSILYL 11β,17α,21-TRI
HYDROXY-PREGN-4-ENE-3,20-DIONE 3,20-DIMETHYLOXIME 21-PHOSPHATE; W/NBS
66751

788   C36 H48 Cu4 N4   AG-84-133-1   2-[(Dimethylamino)methyl]phenylcopper; W
128685

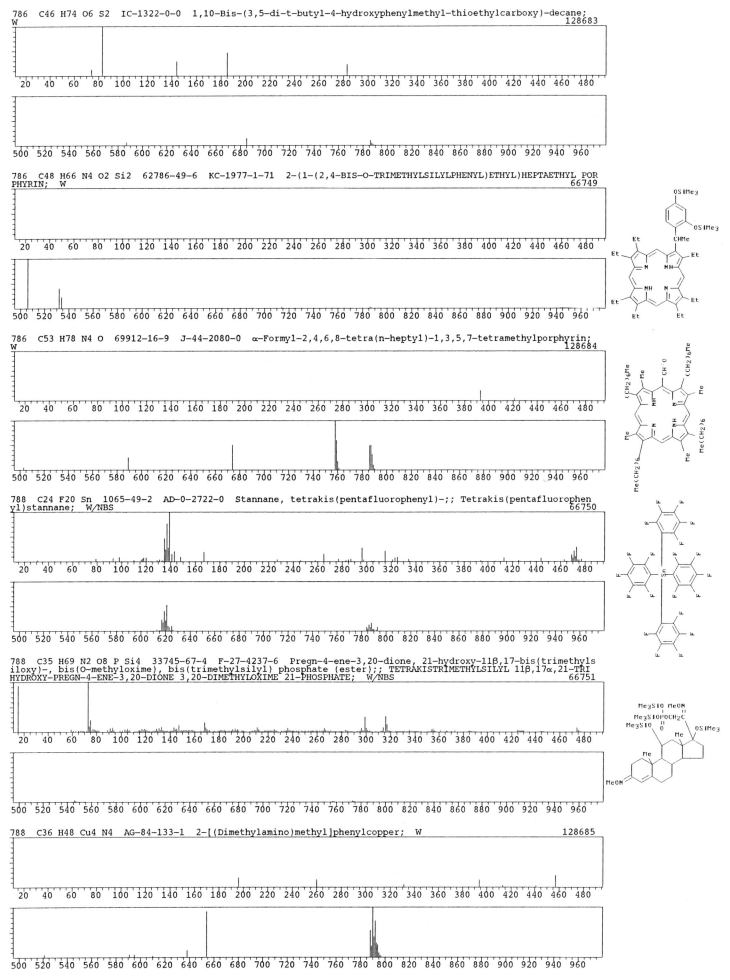

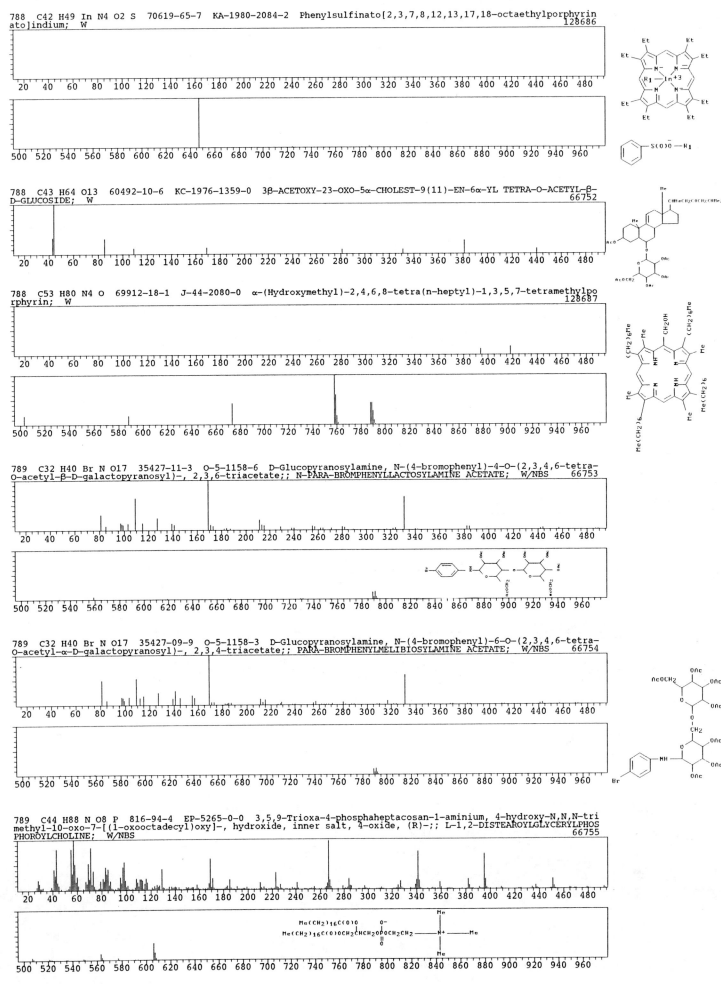

788   C42 H49 In N4 O2 S   70619-65-7   KA-1980-2084-2   Phenylsulfinato[2,3,7,8,12,13,17,18-octaethylporphyrin
ato]indium; W   128686

788   C43 H64 O13   60492-10-6   KC-1976-1359-0   3β-ACETOXY-23-OXO-5α-CHOLEST-9(11)-EN-6α-YL TETRA-O-ACETYL-β-
D-GLUCOSIDE; W   66752

788   C53 H80 N4 O   69912-18-1   J-44-2080-0   α-(Hydroxymethyl)-2,4,6,8-tetra(n-heptyl)-1,3,5,7-tetramethylpo
rphyrin; W   128687

789   C32 H40 Br N O17   35427-11-3   O-5-1158-6   D-Glucopyranosylamine, N-(4-bromophenyl)-4-O-(2,3,4,6-tetra-
O-acetyl-β-D-galactopyranosyl)-, 2,3,6-triacetate;; N-PARA-BROMPHENYLLACTOSYLAMINE ACETATE; W/NBS   66753

789   C32 H40 Br N O17   35427-09-9   O-5-1158-3   D-Glucopyranosylamine, N-(4-bromophenyl)-6-O-(2,3,4,6-tetra-
O-acetyl-α-D-galactopyranosyl)-, 2,3,4-triacetate;; PARA-BROMPHENYLMELIBIOSYLAMINE ACETATE; W/NBS   66754

789   C44 H88 N O8 P   816-94-4   EP-5265-0-0   3,5,9-Trioxa-4-phosphaheptacosan-1-aminium, 4-hydroxy-N,N,N-tri
methyl-10-oxo-7-[(1-oxooctadecyl)oxy]-, hydroxide, inner salt, 4-oxide, (R)-;; L-1,2-DISTEAROYLGLYCERYLPHOS
PHOROYLCHOLINE; W/NBS   66755

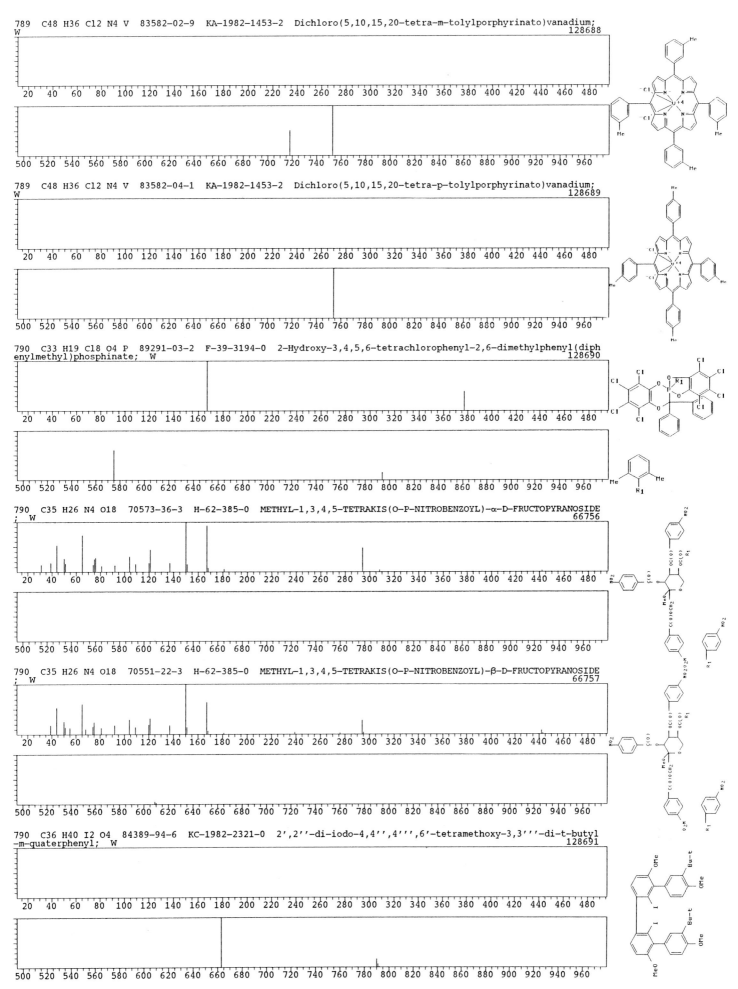

789   C48 H36 Cl2 N4 V   83582-02-9   KA-1982-1453-2   Dichloro(5,10,15,20-tetra-m-tolylporphyrinato)vanadium;
W
128688

789   C48 H36 Cl2 N4 V   83582-04-1   KA-1982-1453-2   Dichloro(5,10,15,20-tetra-p-tolylporphyrinato)vanadium;
W
128689

790   C33 H19 Cl8 O4 P   89291-03-2   F-39-3194-0   2-Hydroxy-3,4,5,6-tetrachlorophenyl-2,6-dimethylphenyl(diph
enylmethyl)phosphinate;   W
128690

790   C35 H26 N4 O18   70573-36-3   H-62-385-0   METHYL-1,3,4,5-TETRAKIS(O-P-NITROBENZOYL)-α-D-FRUCTOPYRANOSIDE
;   W
66756

790   C35 H26 N4 O18   70551-22-3   H-62-385-0   METHYL-1,3,4,5-TETRAKIS(O-P-NITROBENZOYL)-β-D-FRUCTOPYRANOSIDE
;   W
66757

790   C36 H40 I2 O4   84389-94-6   KC-1982-2321-0   2',2''-di-iodo-4,4'',4''',6'-tetramethoxy-3,3'''-di-t-butyl
-m-quaterphenyl;   W
128691

790   C37 H42 O19   70639-62-2   H-62-538-0   MINECOSIDE HEXAACETATE;   W   66758

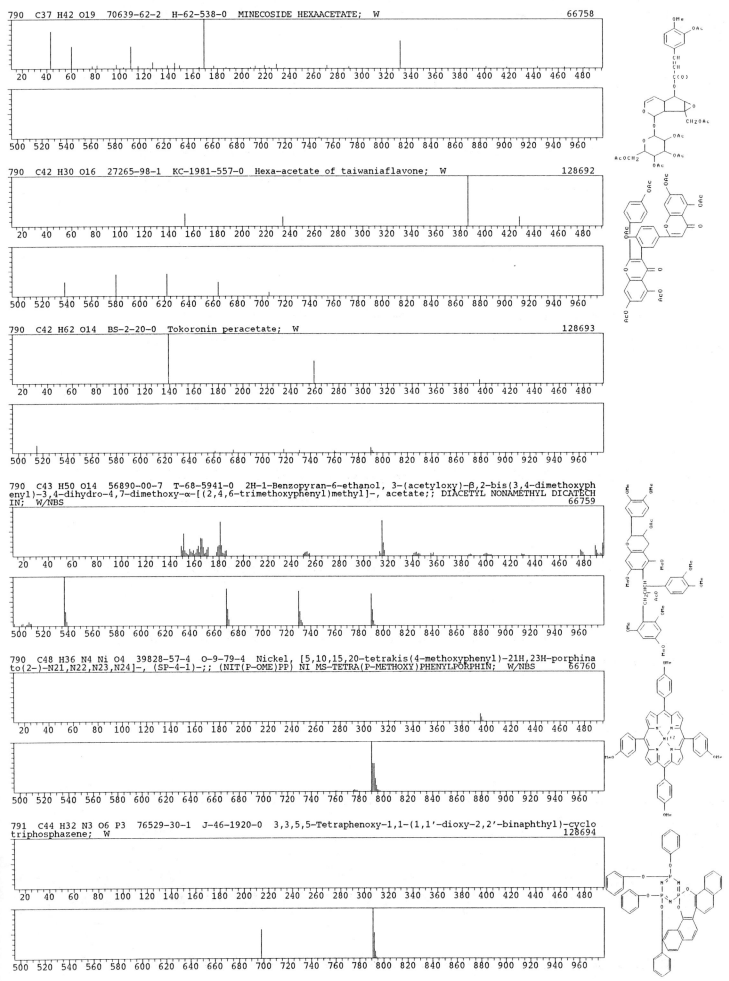

790   C42 H30 O16   27265-98-1   KC-1981-557-0   Hexa-acetate of taiwaniaflavone;   W   128692

790   C42 H62 O14   BS-2-20-0   Tokoronin peracetate;   W   128693

790   C43 H50 O14   56890-00-7   T-68-5941-0   2H-1-Benzopyran-6-ethanol, 3-(acetyloxy)-β,2-bis(3,4-dimethoxyph
enyl)-3,4-dihydro-4,7-dimethoxy-α-[(2,4,6-trimethoxyphenyl)methyl]-, acetate;; DIACETYL NONAMETHYL DICATECH
IN;   W/NBS   66759

790   C48 H36 N4 Ni O4   39828-57-4   O-9-79-4   Nickel, [5,10,15,20-tetrakis(4-methoxyphenyl)-21H,23H-porphina
to(2-)-N21,N22,N23,N24]-, (SP-4-1)-;;   (NIT(P-OME)PP) NI MS-TETRA(P-METHOXY)PHENYLPORPHIN;   W/NBS   66760

791   C44 H32 N3 O6 P3   76529-30-1   J-46-1920-0   3,3,5,5-Tetraphenoxy-1,1-(1,1'-dioxy-2,2'-binaphthyl)-cyclo
triphosphazene;   W   128694

791   C44 H32 N3 O6 P3   76529-31-2   J-46-1920-0   3,3,5,5-Tetraphenoxy-1,1-(2,2'-dioxy-1,1'-binaphthyl)-cyclo
triphosphazene;   W   128695

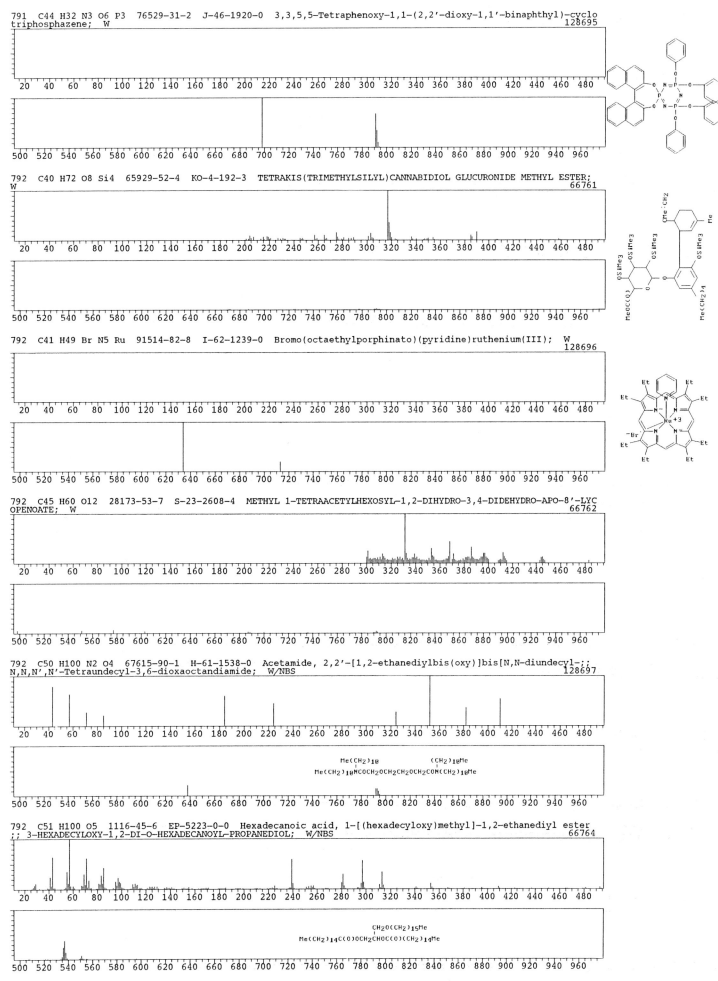

792   C40 H72 O8 Si4   65929-52-4   KO-4-192-3   TETRAKIS(TRIMETHYLSILYL)CANNABIDIOL GLUCURONIDE METHYL ESTER;
W   66761

792   C41 H49 Br N5 Ru   91514-82-8   I-62-1239-0   Bromo(octaethylporphinato)(pyridine)ruthenium(III);   W
                                                                                                           128696

792   C45 H60 O12   28173-53-7   S-23-2608-4   METHYL 1-TETRAACETYLHEXOSYL-1,2-DIHYDRO-3,4-DIDEHYDRO-APO-8'-LYC
OPENOATE;   W   66762

792   C50 H100 N2 O4   67615-90-1   H-61-1538-0   Acetamide, 2,2'-[1,2-ethanediylbis(oxy)]bis[N,N-diundecyl-;;
N,N,N',N'-Tetraundecyl-3,6-dioxaoctandiamide;   W/NBS   128697

792   C51 H100 O5   1116-45-6   EP-5223-0-0   Hexadecanoic acid, 1-[(hexadecyloxy)methyl]-1,2-ethanediyl ester
;; 3-HEXADECYLOXY-1,2-DI-O-HEXADECANOYL-PROPANEDIOL;   W/NBS   66764

792   C51 H100 O5   10322-32-4   EP-5214-0-0   Octadecanoic acid, 1-[(dodecyloxy)methyl]-1,2-ethanediyl ester;;
3-DODECYLOXY-1,2-DI-O-OCTADECANOYL-PROPANEDIOL;   W/NBS   66765

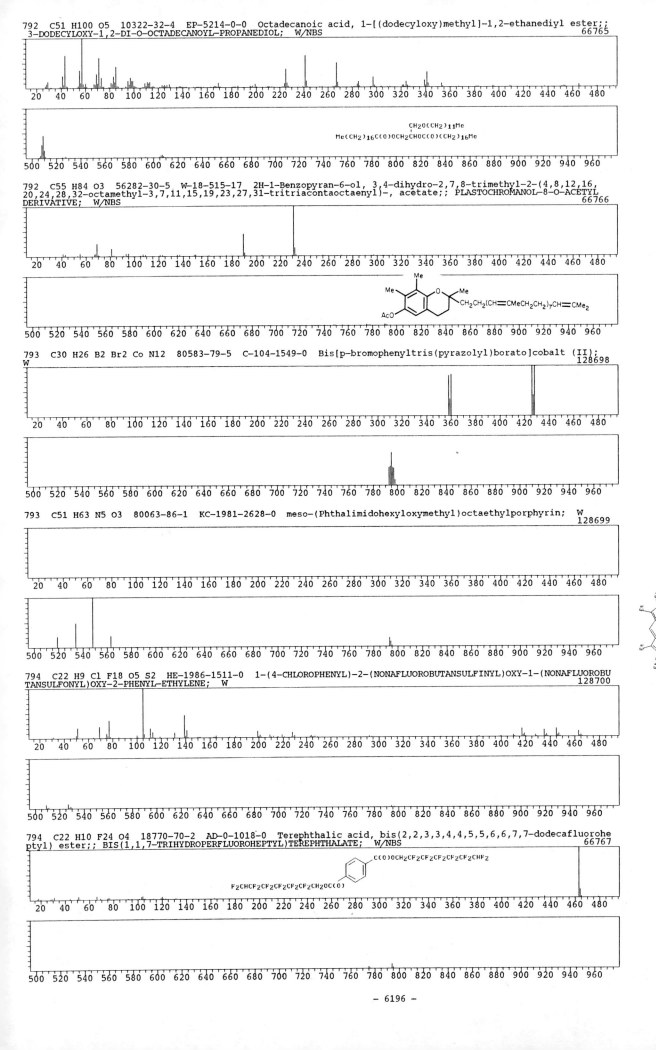

CH₂O(CH₂)₁₁Me
Me(CH₂)₁₆C(O)OCH₂CHOC(O)(CH₂)₁₆Me

792   C55 H84 O3   56282-30-5   W-18-515-17   2H-1-Benzopyran-6-ol, 3,4-dihydro-2,7,8-trimethyl-2-(4,8,12,16,
20,24,28,32-octamethyl-3,7,11,15,19,23,27,31-tritriacontaoctaenyl)-, acetate;; PLASTOCHROMANOL-8-O-ACETYL
DERIVATIVE;   W/NBS   66766

793   C30 H26 B2 Br2 Co N12   80583-79-5   C-104-1549-0   Bis[p-bromophenyltris(pyrazolyl)borato]cobalt (II);
W   128698

793   C51 H63 N5 O3   80063-86-1   KC-1981-2628-0   meso-(Phthalimidohexyloxymethyl)octaethylporphyrin;   W
                               128699

794   C22 H9 Cl F18 O5 S2   HE-1986-1511-0   1-(4-CHLOROPHENYL)-2-(NONAFLUOROBUTANSULFINYL)OXY-1-(NONAFLUOROBU
TANSULFONYL)OXY-2-PHENYL-ETHYLENE;   W   128700

794   C22 H10 F24 O4   18770-70-2   AD-0-1018-0   Terephthalic acid, bis(2,2,3,3,4,4,5,5,6,6,7,7-dodecafluorohe
ptyl) ester;; BIS(1,1,7-TRIHYDROPERFLUOROHEPTYL)TEREPHTHALATE;   W/NBS   66767

C(O)OCH₂CF₂CF₂CF₂CF₂CF₂CHF₂
F₂CHCF₂CF₂CF₂CF₂CF₂CH₂OC(O)

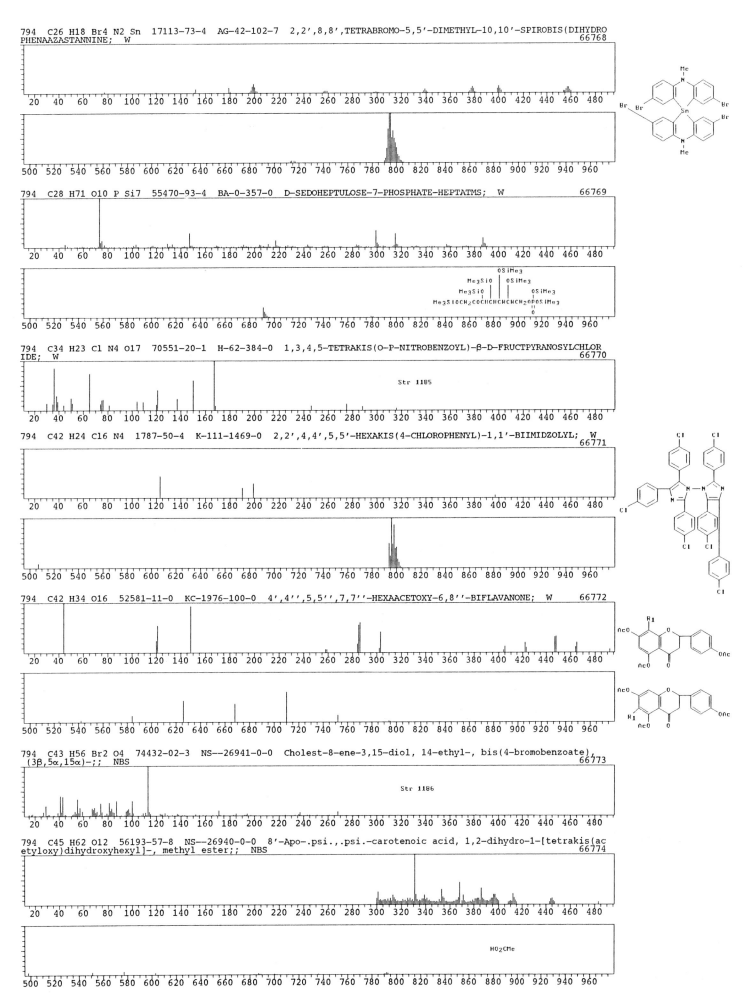

794 C26 H18 Br4 N2 Sn 17113-73-4 AG-42-102-7 2,2′,8,8′,TETRABROMO-5,5′-DIMETHYL-10,10′-SPIROBIS(DIHYDRO
PHENAAZASTANNINE; W   66768

794 C28 H71 O10 P Si7 55470-93-4 BA-0-357-0 D-SEDOHEPTULOSE-7-PHOSPHATE-HEPTATMS; W   66769

794 C34 H23 Cl N4 O17 70551-20-1 H-62-384-0 1,3,4,5-TETRAKIS(O-P-NITROBENZOYL)-β-D-FRUCTPYRANOSYLCHLOR
IDE; W   66770

Str 1185

794 C42 H24 Cl6 N4 1787-50-4 K-111-1469-0 2,2′,4,4′,5,5′-HEXAKIS(4-CHLOROPHENYL)-1,1′-BIIMIDZOLYL; W
66771

794 C42 H34 O16 52581-11-0 KC-1976-100-0 4′,4″,5,5″,7,7″-HEXAACETOXY-6,8″-BIFLAVANONE; W   66772

794 C43 H56 Br2 O4 74432-02-3 NS--26941-0-0 Cholest-8-ene-3,15-diol, 14-ethyl-, bis(4-bromobenzoate),
(3β,5α,15α)-;; NBS   66773

Str 1186

794 C45 H62 O12 56193-57-8 NS--26940-0-0 8′-Apo-.psi.,.psi.-carotenoic acid, 1,2-dihydro-1-[tetrakis(ac
etyloxy)dihydroxyhexyl]-, methyl ester;; NBS   66774

HO2CMe

794 C54 H82 O4 303-97-9 J-45-4103-0 2,3-Dimethoxy-5-methyl-6-(3,7,11,15,19,23,27,31,35-nonamethyl-2,6,
10,14,18,22,26,30,34-hexatriacontanonaenyl)-2,5-cyclohexadiene-1,4-dione; W   128701

795 C27 H37 I2 N2 O2 Sn O-18-455-1 p-(Iodomethyl)methylphenylstannyl acetophenone menthydrazone; W
                                                                                                  128702

795 C28 H62 N5 O8 P Si6 32653-15-9 BA-0-264-0 9H-Purine, 9-[2,3-bis-O-(trimethylsilyl)-β-D-ribofuranosy
l]-6-(trimethylsiloxy)-2-[(trimethylsilyl)amino]-, 5'-[bis(trimethylsilyl)phospate];; 5'-(O,O-BIS(TRIMETHYL
SILYL)PHOSPHATYL)-N2,O6,O2',O3'-TETRAKIS(TRIMETHYLSILYL)GUANOSINE; W/NBS   66775

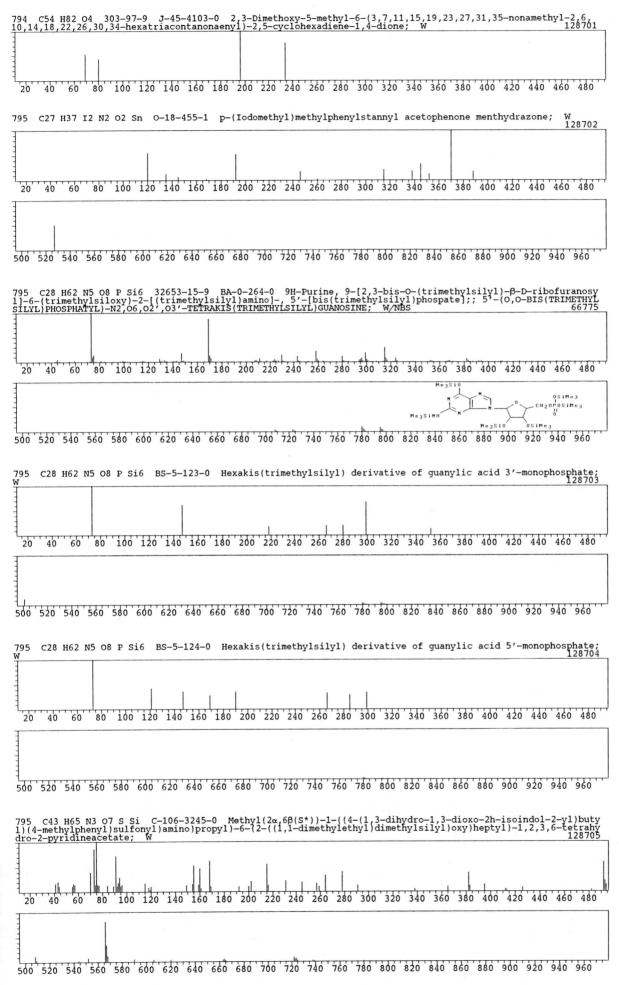

795 C28 H62 N5 O8 P Si6 BS-5-123-0 Hexakis(trimethylsilyl) derivative of guanylic acid 3'-monophosphate;
W   128703

795 C28 H62 N5 O8 P Si6 BS-5-124-0 Hexakis(trimethylsilyl) derivative of guanylic acid 5'-monophosphate;
W   128704

795 C43 H65 N3 O7 S Si C-106-3245-0 Methyl(2α,6β(S*))-1-((4-(1,3-dihydro-1,3-dioxo-2h-isoindol-2-yl)buty
l)(4-methylphenyl)sulfonyl)amino)propyl)-6-(2-((1,1-dimethylethyl)dimethylsilyl)oxy)heptyl)-1,2,3,6-tetrahy
dro-2-pyridineacetate; W   128705

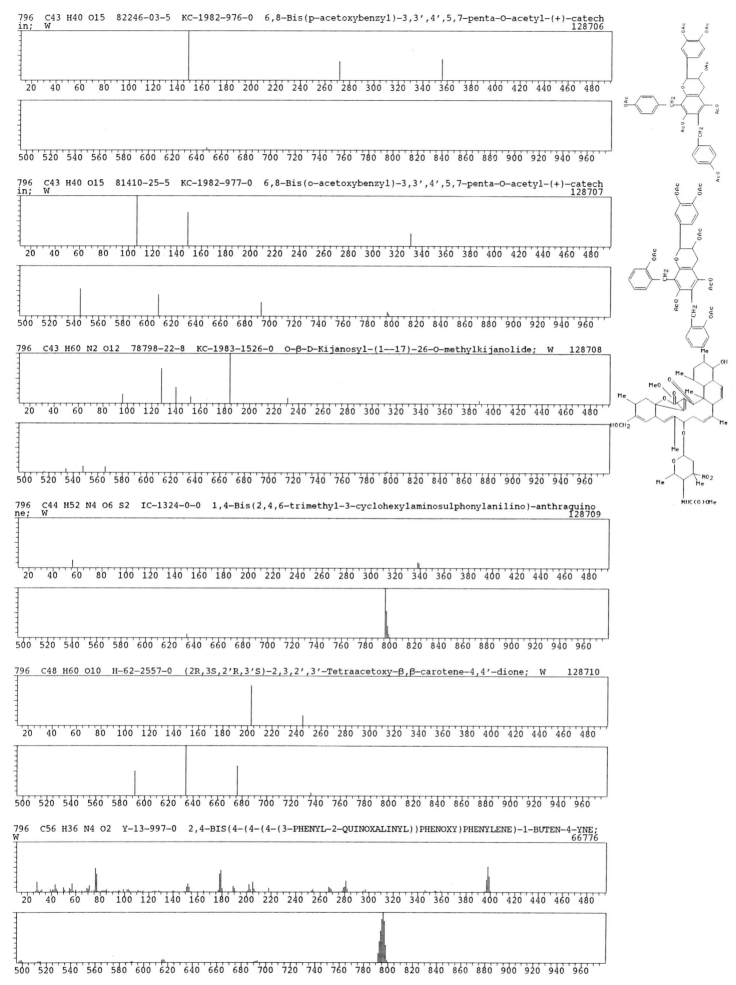

796   C43 H40 O15   82246-03-5   KC-1982-976-0   6,8-Bis(p-acetoxybenzyl)-3,3',4',5,7-penta-O-acetyl-(+)-catech
in; W   128706

796   C43 H40 O15   81410-25-5   KC-1982-977-0   6,8-Bis(o-acetoxybenzyl)-3,3',4',5,7-penta-O-acetyl-(+)-catech
in; W   128707

796   C43 H60 N2 O12   78798-22-8   KC-1983-1526-0   O-β-D-Kijanosyl-(1--17)-26-O-methylkijanolide; W   128708

796   C44 H52 N4 O6 S2   IC-1324-0-0   1,4-Bis(2,4,6-trimethyl-3-cyclohexylaminosulphonylanilino)-anthraquino
ne; W   128709

796   C48 H60 O10   H-62-2557-0   (2R,3S,2'R,3'S)-2,3,2',3'-Tetraacetoxy-β,β-carotene-4,4'-dione; W   128710

796   C56 H36 N4 O2   Y-13-997-0   2,4-BIS(4-(4-(4-(3-PHENYL-2-QUINOXALINYL))PHENOXY)PHENYLENE)-1-BUTEN-4-YNE;
W   66776

797 C30 H18 F9 O6 Sm 22681-42-1 B-29-1847-0 Samarium(II) chelate of 1,1,1-trifluoro-4-phenylbutane-2,4-dione; W   128711

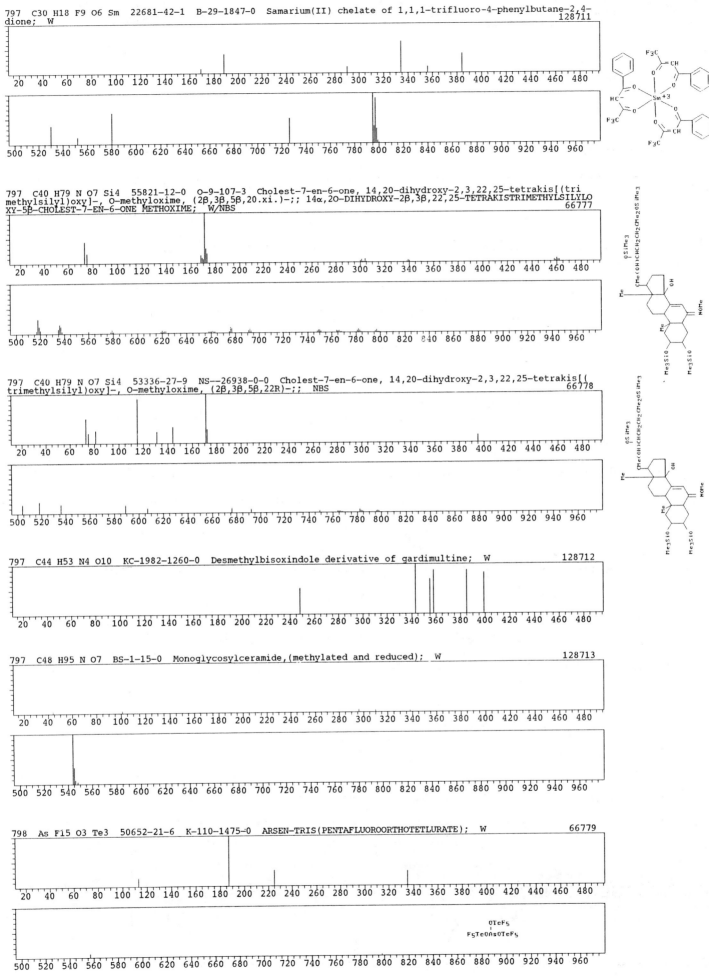

797 C40 H79 N O7 Si4 55821-12-0 O-9-107-3 Cholest-7-en-6-one, 14,20-dihydroxy-2,3,22,25-tetrakis[(tri methylsilyl)oxy]-, O-methyloxime, (2β,3β,5β,20.xi.)-;; 14α,2O-DIHYDROXY-2β,3β,22,25-TETRAKISTRIMETHYLSILYLO XY-5β-CHOLEST-7-EN-6-ONE METHOXIME; W/NBS   66777

797 C40 H79 N O7 Si4 53336-27-9 NS--26938-0-0 Cholest-7-en-6-one, 14,20-dihydroxy-2,3,22,25-tetrakis[(trimethylsilyl)oxy]-, O-methyloxime, (2β,3β,5β,22R)-;; NBS   66778

797 C44 H53 N4 O10 KC-1982-1260-0 Desmethylbisoxindole derivative of gardimultine; W   128712

797 C48 H95 N O7 BS-1-15-0 Monoglycosylceramide,(methylated and reduced); W   128713

798 As F15 O3 Te3 50652-21-6 K-110-1475-0 ARSEN-TRIS(PENTAFLUOROORTHOTETLURATE); W   66779

798   C30 H18 Eu F9 O6   14552-19-3   B-29-1847-4   Europium(II) chelate of 1,1,1-trifluoro-4-phenylbutane-2,4-dione; W   128714

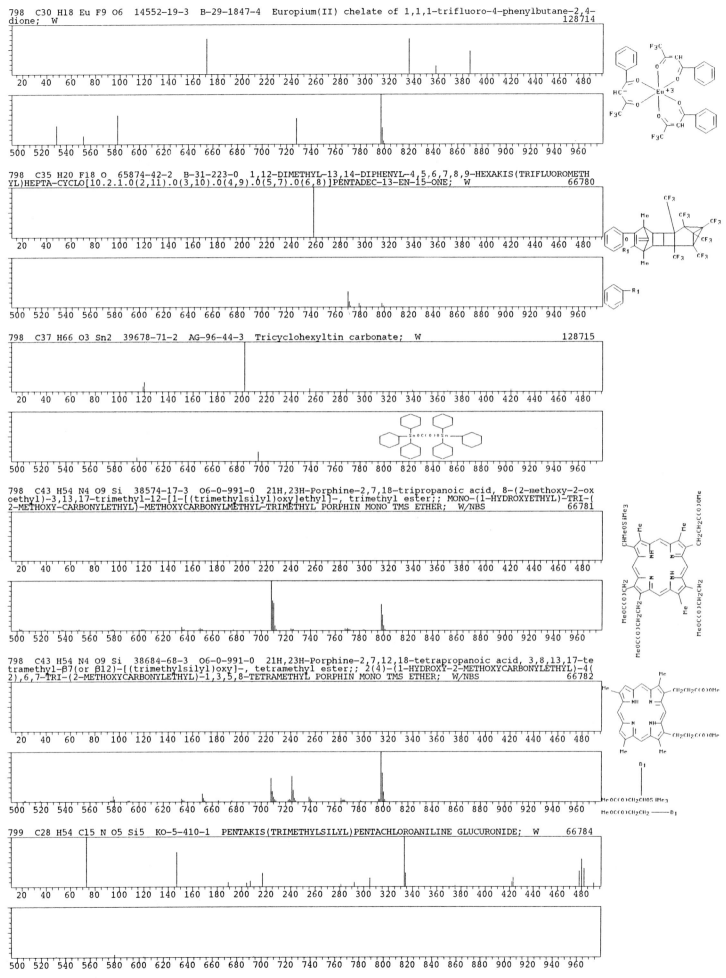

798   C35 H20 F18 O   65874-42-2   B-31-223-0   1,12-DIMETHYL-13,14-DIPHENYL-4,5,6,7,8,9-HEXAKIS(TRIFLUOROMETHYL)HEPTA-CYCLO[10.2.1.0(2,11).0(3,10).0(4,9).0(5,7).0(6,8)]PENTADEC-13-EN-15-ONE; W   66780

798   C37 H66 O3 Sn2   39678-71-2   AG-96-44-3   Tricyclohexyltin carbonate; W   128715

798   C43 H54 N4 O9 Si   38574-17-3   O6-0-991-0   21H,23H-Porphine-2,7,18-tripropanoic acid, 8-(2-methoxy-2-oxoethyl)-3,13,17-trimethyl-12-[1-[(trimethylsilyl)oxy]ethyl]-, trimethyl ester;; MONO-(1-HYDROXYETHYL)-TRI-(2-METHOXY-CARBONYLETHYL)-METHOXYCARBONYLMETHYL-TRIMETHYL PORPHIN MONO TMS ETHER; W/NBS   66781

798   C43 H54 N4 O9 Si   38684-68-3   O6-0-991-0   21H,23H-Porphine-2,7,12,18-tetrapropanoic acid, 3,8,13,17-tetramethyl-β7(or β12)-[(trimethylsilyl)oxy]-, tetramethyl ester;; 2(4)-(1-HYDROXY-2-METHOXYCARBONYLETHYL)-4(2),6,7-TRI-(2-METHOXYCARBONYLETHYL)-1,3,5,8-TETRAMETHYL PORPHIN MONO TMS ETHER; W/NBS   66782

799   C28 H54 Cl5 N O5 Si5   KO-5-410-1   PENTAKIS(TRIMETHYLSILYL)PENTACHLOROANILINE GLUCURONIDE; W   66784

799　C44 H45 N7 O8　78413-98-6　F-37-160-0　exobornyl ester of 2-(2,3,9,10-Tetrahydro-1,3,8,10,17,19-hexaox
o-2,9-diphenyl-6H,13H-5,13:6,12-dietheno-5a,12a-(methaniminomethano)-1H,5H,8H,12H-[1,2,4]triazolo[1,2-a][1,
2,4]triazolo[1',2':1,2]pyridazino[4,5-d]pyridazin-18-yl)-4-methylpentanoic acid;　W　　　　134183

799　C55 H109 N O　O-12-356-4　(2)-(N-ACETYL-AZACYCLOHEXACOSAN)-(CYCLOOCTACOSAN)-CATENAN;　W　　　66785

799　C55 H106 D3 N O　O-12-357-4　(2)-(N-TRIDEUTEROACETYL-AZACYCLOHEXACOSAN)-(CYCLOOCTACOSAN)-CATENAN;　W
　　　　66786

800　C20 H20 Mn4 O12 S4　22450-91-5　AD-0-1599-0　Manganese, dodecacarbonyltetrakis(.mu.3-ethanethiolato)te
tra-;; ETHANETHIOLATO-TRICARBONYLMANGANESE-TETRAMER;　W/NBS　　　　66787

800　C30 H54 Cl2 N2 O11 Si4　40619-05-4　AF-16-31-2　TETRAKISTRIMETHYLSILYL CHLORAMPHENICOL;　W/NBS　66788

800　C37 H36 O20　CD-329-0-0　Hyperin octaacetate;　W　　　　128717

800　C45 H64 N6 O7　93398-11-9　AH-115-1079-0　Amide of (3R)-(4Z,9Z,15Z)-17-Ethyl-1,19-dioxo-2,2,7,8,12,13,
18-heptamethyl-1,2,3,19,23,24-hexahydro-21H-bilin-3-N2-(tert-butylester of N6-tert-butoxycarbonyl-(S)-lysi
n)-acetic acid;　W　　　　128718

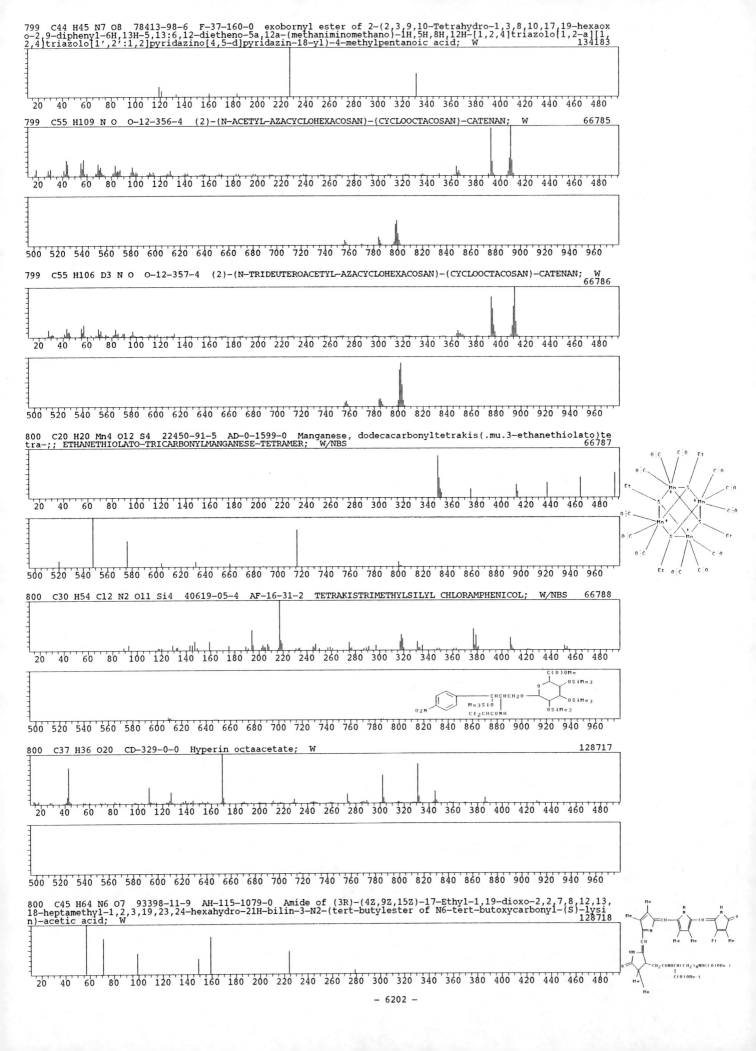

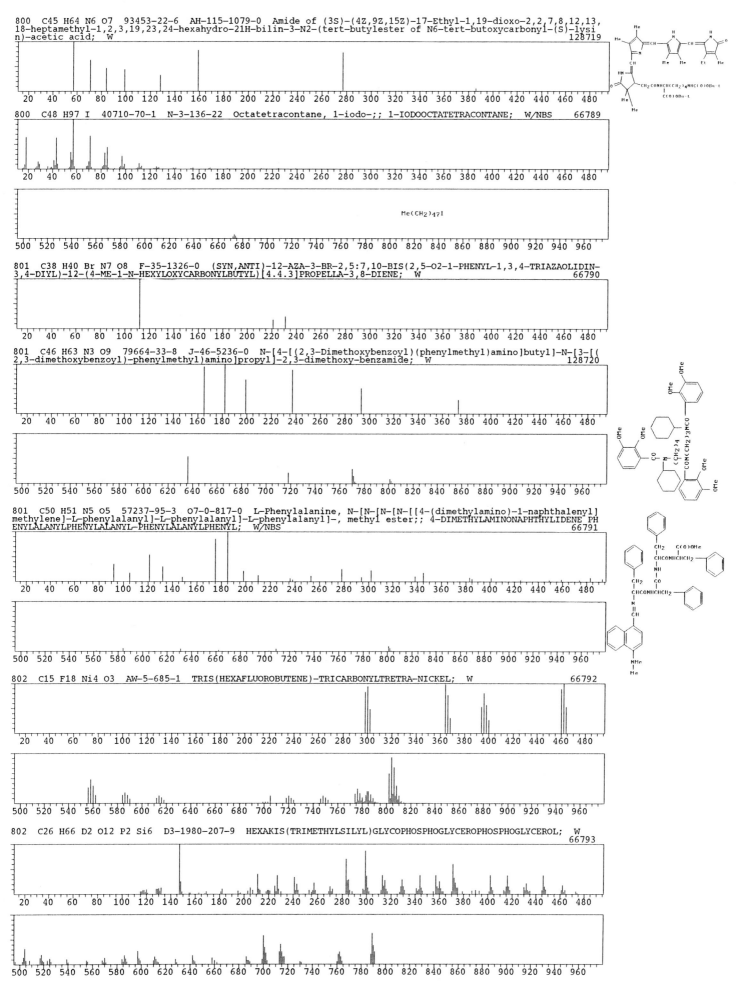

800 C45 H64 N6 O7 93453-22-6 AH-115-1079-0 Amide of (3S)-(4Z,9Z,15Z)-17-Ethyl-1,19-dioxo-2,2,7,8,12,13,
18-heptamethyl-1,2,3,19,23,24-hexahydro-21H-bilin-3-N2-(tert-butylester of N6-tert-butoxycarbonyl-(S)-lysi
n)-acetic acid; W   128719

800 C48 H97 I 40710-70-1 N-3-136-22 Octatetracontane, 1-iodo-;; 1-IODOOCTATETRACONTANE; W/NBS   66789

Me(CH2)47I

801 C38 H40 Br N7 O8 F-35-1326-0 (SYN,ANTI)-12-AZA-3-BR-2,5:7,10-BIS(2,5-O2-1-PHENYL-1,3,4-TRIAZAOLIDIN-
3,4-DIYL)-12-(4-ME-1-N-HEXYLOXYCARBONYLBUTYL)[4.4.3]PROPELLA-3,8-DIENE; W   66790

801 C46 H63 N3 O9 79664-33-8 J-46-5236-0 N-[4-[(2,3-Dimethoxybenzoyl)(phenylmethyl)amino]butyl]-N-[3-[(
2,3-dimethoxybenzoyl)-phenylmethyl)amino]propyl]-2,3-dimethoxy-benzamide; W   128720

801 C50 H51 N5 O5 57237-95-3 O7-0-817-0 L-Phenylalanine, N-[N-[N-[N-[[4-(dimethylamino)-1-naphthalenyl]
methylene]-L-phenylalanyl]-L-phenylalanyl]-L-phenylalanyl]-, methyl ester;; 4-DIMETHYLAMINONAPHTHYLIDENE PH
ENYLALANYLPHENYLALANYL-PHENYLALANYLPHENYL; W/NBS   66791

802 C15 F18 Ni4 O3 AW-5-685-1 TRIS(HEXAFLUOROBUTENE)-TRICARBONYLTRETRA-NICKEL; W   66792

802 C26 H66 D2 O12 P2 Si6 D3-1980-207-9 HEXAKIS(TRIMETHYLSILYL)GLYCOPHOSPHOGLYCEROPHOSPHOGLYCEROL; W
                                                                                                   66793

802   C30 H21 F18 O3 P   71401-74-6   C-101-4626-0   TRIS(2-(1-TRIFLUOROMETHYL-2,2,2-TRIFLUOROETHYL)PHENYL)PHOS
PHINE; W
66794

802   C43 H62 O14   68673-74-5   K-111-3305-0   NUATIGENIN-3-(2,3,4,6-TETRA-O-ACETYL-β-D-GLUCOPYRANOSID)-26-ACE
TATE; W
66795

802   C58 H42 O2 S   56728-01-9   NS--26930-0-0   Benzene, 1,1',1'',1''',1'''',1''''',1'''''',1'''''''-(sulfony
ldi-1,3-cyclopentadiene-5,1,2,3,4-pentayl)octakis-;;   NBS
66796

Str 1187

803   C43 H73 N5 O7 S   19729-26-1   T-68-5946-0   Glycine, N-[N-[N-[3-[(2-decanamidoethyl)thio]-N-decanoyl-L-
alanyl]-3-phenyl-L-alanyl]-L-leucyl]-, methyl ester;; DECANOYL(S-DECANOYLAMINOETHYL)CYSTEYLPHENYLLEUCYLGLYC
INE METHYL ESTER; W/NBS
66797

803   C49 H39 Fe N4 O2 S   82008-31-9   KA-1984-571-4   Methylsulfinato(5,10,15,20-tetra-m-tolylporphyrinato)ir
on; W
128721

803   C49 H39 Fe N4 O2 S   82008-32-0   KA-1984-571-4   Methylsulfinato(5,10,15,20-tetra-p-tolylporphyrinato)ir
on; W
128722

804   C38 H44 O19   73343-23-4   H-64-13-0   Globularinin heptaacetate; W
128723

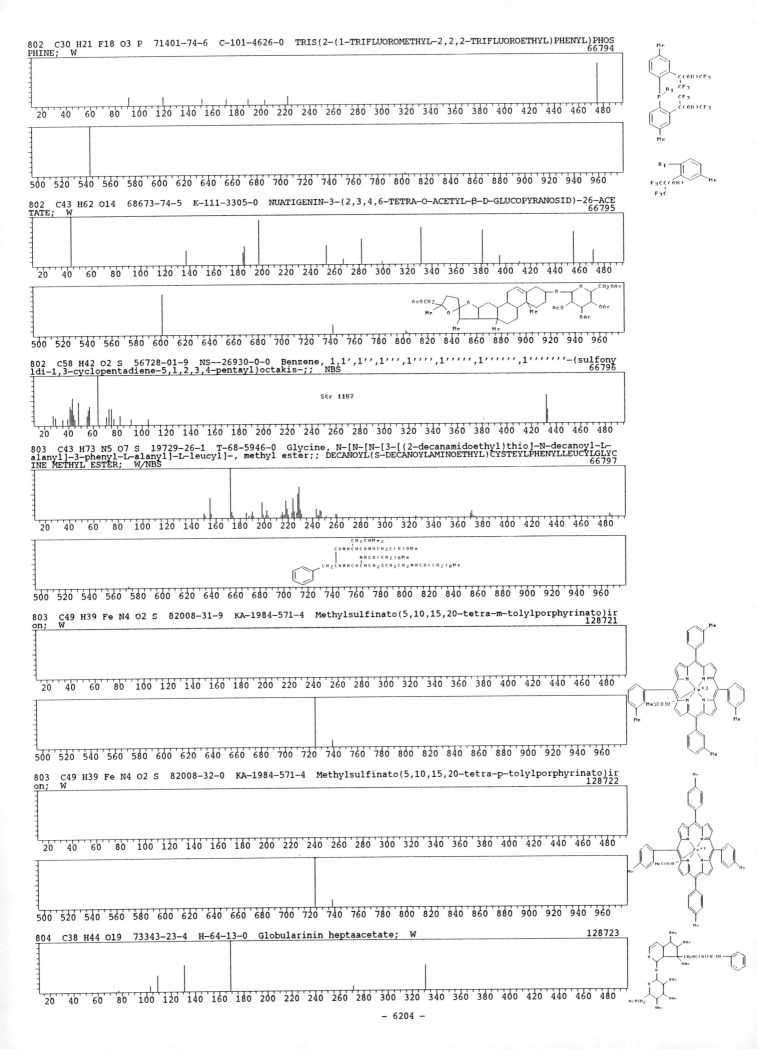

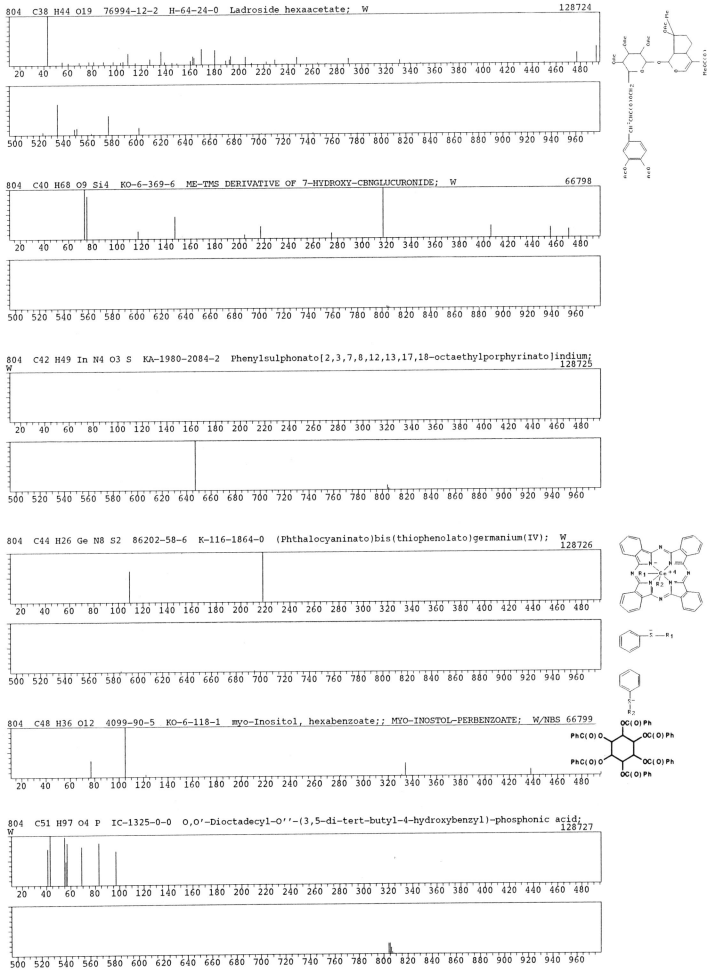

804   C38 H44 O19   76994-12-2   H-64-24-0   Ladroside hexaacetate;   W   128724

804   C40 H68 O9 Si4   KO-6-369-6   ME-TMS DERIVATIVE OF 7-HYDROXY-CBNGLUCURONIDE;   W   66798

804   C42 H49 In N4 O3 S   KA-1980-2084-2   Phenylsulphonato[2,3,7,8,12,13,17,18-octaethylporphyrinato]indium;
W   128725

804   C44 H26 Ge N8 S2   86202-58-6   K-116-1864-0   (Phthalocyaninato)bis(thiophenolato)germanium(IV);   W   128726

804   C48 H36 O12   4099-90-5   KO-6-118-1   myo-Inositol, hexabenzoate;;   MYO-INOSTOL-PERBENZOATE;   W/NBS 66799

804   C51 H97 O4 P   IC-1325-0-0   O,O'-Dioctadecyl-O''-(3,5-di-tert-butyl-4-hydroxybenzyl)-phosphonic acid;
W   128727

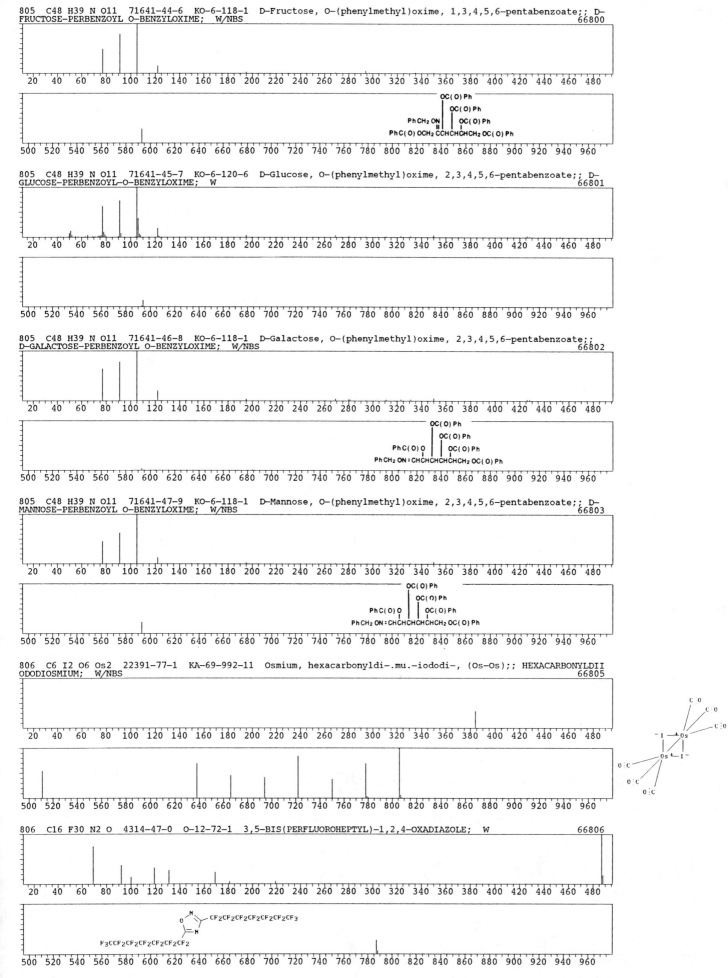

805 C48 H39 N O11 71641-44-6 KO-6-118-1 D-Fructose, O-(phenylmethyl)oxime, 1,3,4,5,6-pentabenzoate;; D-
FRUCTOSE-PERBENZOYL O-BENZYLOXIME; W/NBS
66800

805 C48 H39 N O11 71641-45-7 KO-6-120-6 D-Glucose, O-(phenylmethyl)oxime, 2,3,4,5,6-pentabenzoate;; D-
GLUCOSE-PERBENZOYL-O-BENZYLOXIME; W
66801

805 C48 H39 N O11 71641-46-8 KO-6-118-1 D-Galactose, O-(phenylmethyl)oxime, 2,3,4,5,6-pentabenzoate;;
D-GALACTOSE-PERBENZOYL O-BENZYLOXIME; W/NBS
66802

805 C48 H39 N O11 71641-47-9 KO-6-118-1 D-Mannose, O-(phenylmethyl)oxime, 2,3,4,5,6-pentabenzoate;; D-
MANNOSE-PERBENZOYL O-BENZYLOXIME; W/NBS
66803

806 C6 I2 O6 Os2 22391-77-1 KA-69-992-11 Osmium, hexacarbonyldi-.mu.-iododi-, (Os-Os);; HEXACARBONYLDII
ODODIOSMIUM; W/NBS
66805

806 C16 F30 N2 O 4314-47-0 O-12-72-1 3,5-BIS(PERFLUOROHEPTYL)-1,2,4-OXADIAZOLE; W
66806

- 6206 -

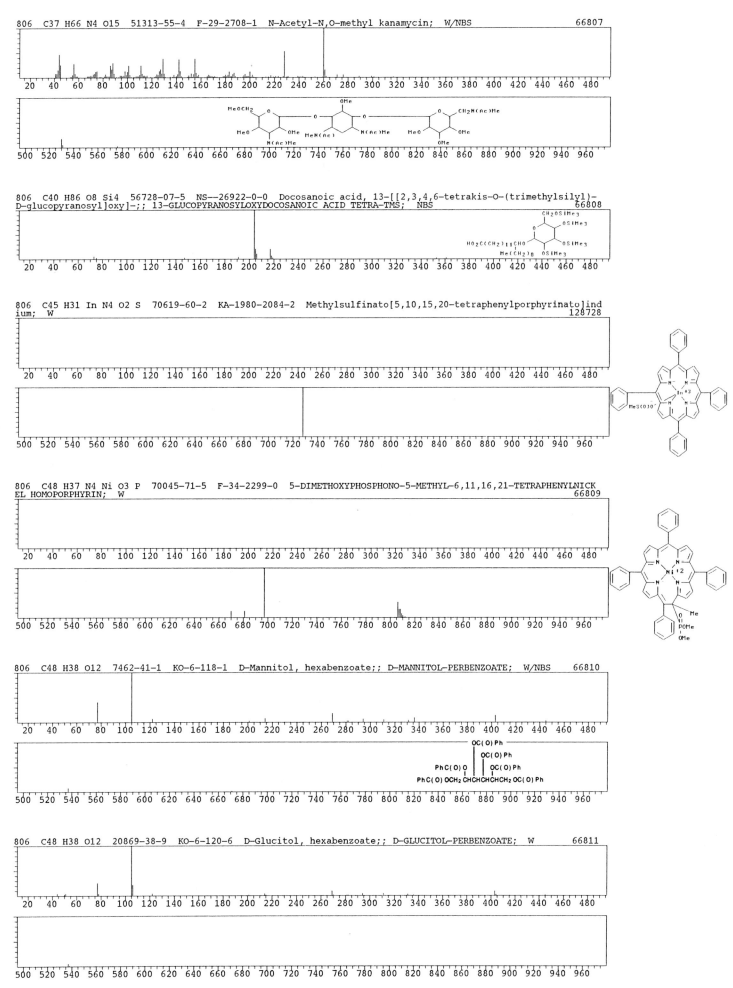

806   C37 H66 N4 O15   51313-55-4   F-29-2708-1   N-Acetyl-N,O-methyl kanamycin;   W/NBS   66807

806   C40 H86 O8 Si4   56728-07-5   NS--26922-0-0   Docosanoic acid, 13-[[2,3,4,6-tetrakis-O-(trimethylsilyl)-
D-glucopyranosyl]oxy]-;;   13-GLUCOPYRANOSYLOXYDOCOSANOIC ACID TETRA-TMS;   NBS   66808

806   C45 H31 In N4 O2 S   70619-60-2   KA-1980-2084-2   Methylsulfinato[5,10,15,20-tetraphenylporphyrinato]ind
ium;   W   128728

806   C48 H37 N4 Ni O3 P   70045-71-5   F-34-2299-0   5-DIMETHOXYPHOSPHONO-5-METHYL-6,11,16,21-TETRAPHENYLNICK
EL HOMOPORPHYRIN;   W   66809

806   C48 H38 O12   7462-41-1   KO-6-118-1   D-Mannitol, hexabenzoate;;   D-MANNITOL-PERBENZOATE;   W/NBS   66810

806   C48 H38 O12   20869-38-9   KO-6-120-6   D-Glucitol, hexabenzoate;;   D-GLUCITOL-PERBENZOATE;   W   66811

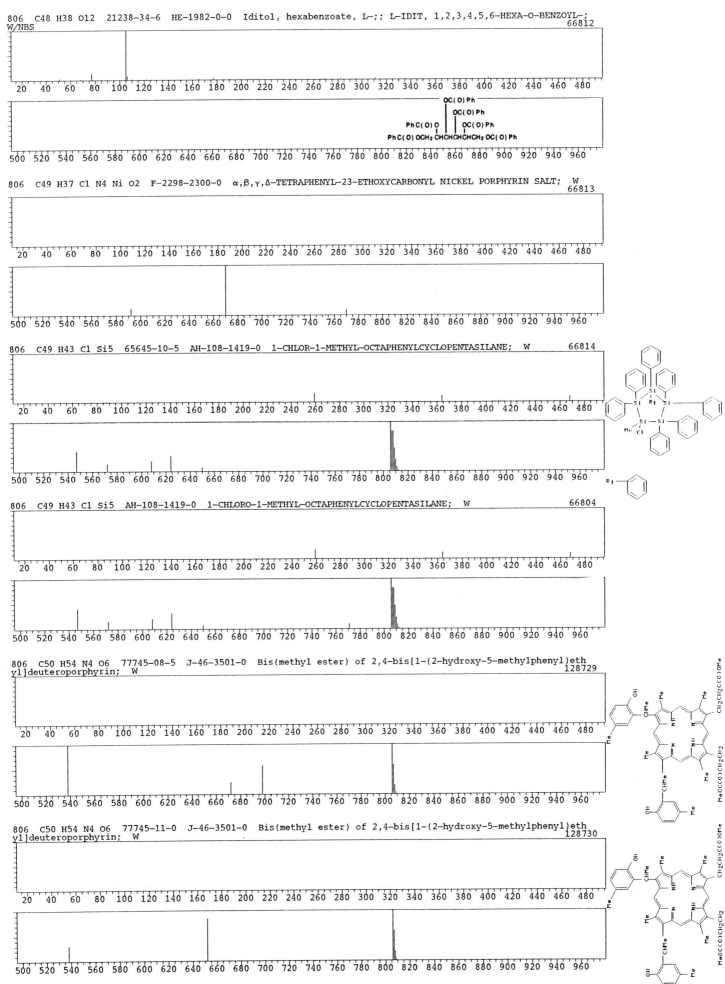

806   C48 H38 O12   21238-34-6   HE-1982-0-0   Iditol, hexabenzoate, L-;; L-IDIT, 1,2,3,4,5,6-HEXA-O-BENZOYL-;
W/NBS   66812

OC(O)Ph
OC(O)Ph
PhC(O)O   OC(O)Ph
PhC(O)OCH₂CHCHCHCHCH₂OC(O)Ph

806   C49 H37 Cl N4 Ni O2   F-2298-2300-0   α,β,γ,Δ-TETRAPHENYL-23-ETHOXYCARBONYL NICKEL PORPHYRIN SALT;   W
                                                                                                            66813

806   C49 H43 Cl Si5   65645-10-5   AH-108-1419-0   1-CHLOR-1-METHYL-OCTAPHENYLCYCLOPENTASILANE;   W   66814

806   C49 H43 Cl Si5   AH-108-1419-0   1-CHLORO-1-METHYL-OCTAPHENYLCYCLOPENTASILANE;   W   66804

806   C50 H54 N4 O6   77745-08-5   J-46-3501-0   Bis(methyl ester) of 2,4-bis[1-(2-hydroxy-5-methylphenyl)eth
yl]deuteroporphyrin;   W   128729

806   C50 H54 N4 O6   77745-11-0   J-46-3501-0   Bis(methyl ester) of 2,4-bis[1-(2-hydroxy-5-methylphenyl)eth
yl]deuteroporphyrin;   W   128730

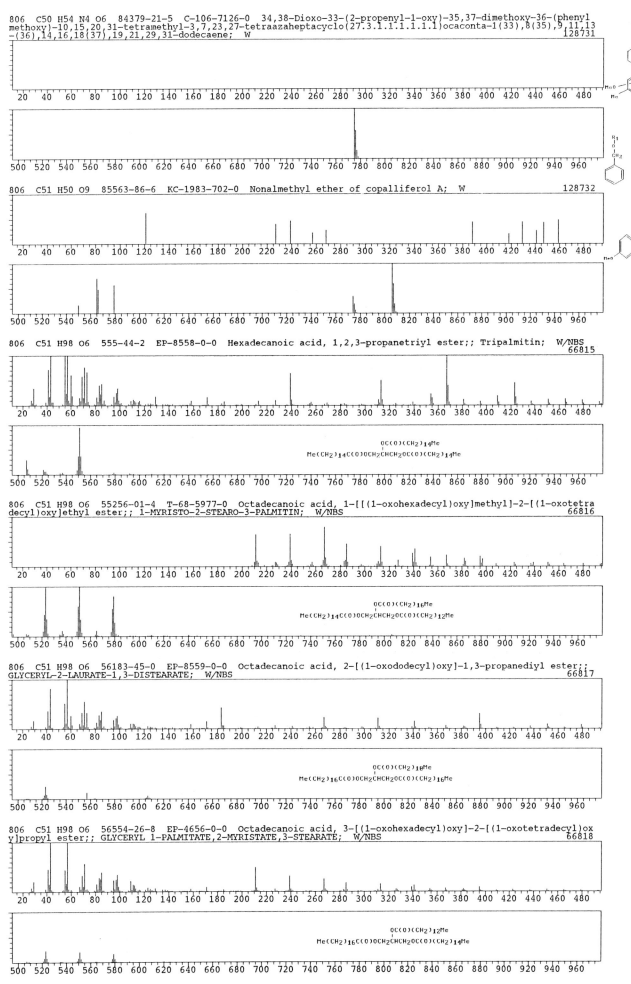

806   C50 H54 N4 O6   84379-21-5   C-106-7126-0   34,38-Dioxo-33-(2-propenyl-1-oxy)-35,37-dimethoxy-36-(phenyl methoxy)-10,15,20,31-tetramethyl-3,7,23,27-tetraazaheptacyclo(27.3.1.1.1.1.1.1)ocaconta-1(33),8(35),9,11,13 -(36),14,16,18(37),19,21,29,31-dodecaene;   W   128731

806   C51 H50 O9   85563-86-6   KC-1983-702-0   Nonalmethyl ether of copalliferol A;   W   128732

806   C51 H98 O6   555-44-2   EP-8558-0-0   Hexadecanoic acid, 1,2,3-propanetriyl ester;; Tripalmitin;   W/NBS
                                                                                                              66815

806   C51 H98 O6   55256-01-4   T-68-5977-0   Octadecanoic acid, 1-[[(1-oxohexadecyl)oxy]methyl]-2-[(1-oxotetra decyl)oxy]ethyl ester;; 1-MYRISTO-2-STEARO-3-PALMITIN;   W/NBS   66816

806   C51 H98 O6   56183-45-0   EP-8559-0-0   Octadecanoic acid, 2-[(1-oxododecyl)oxy]-1,3-propanediyl ester;;
GLYCERYL-2-LAURATE-1,3-DISTEARATE;   W/NBS   66817

806   C51 H98 O6   56554-26-8   EP-4656-0-0   Octadecanoic acid, 3-[(1-oxohexadecyl)oxy]-2-[(1-oxotetradecyl)ox y]propyl ester;; GLYCERYL 1-PALMITATE,2-MYRISTATE,3-STEARATE;   W/NBS   66818

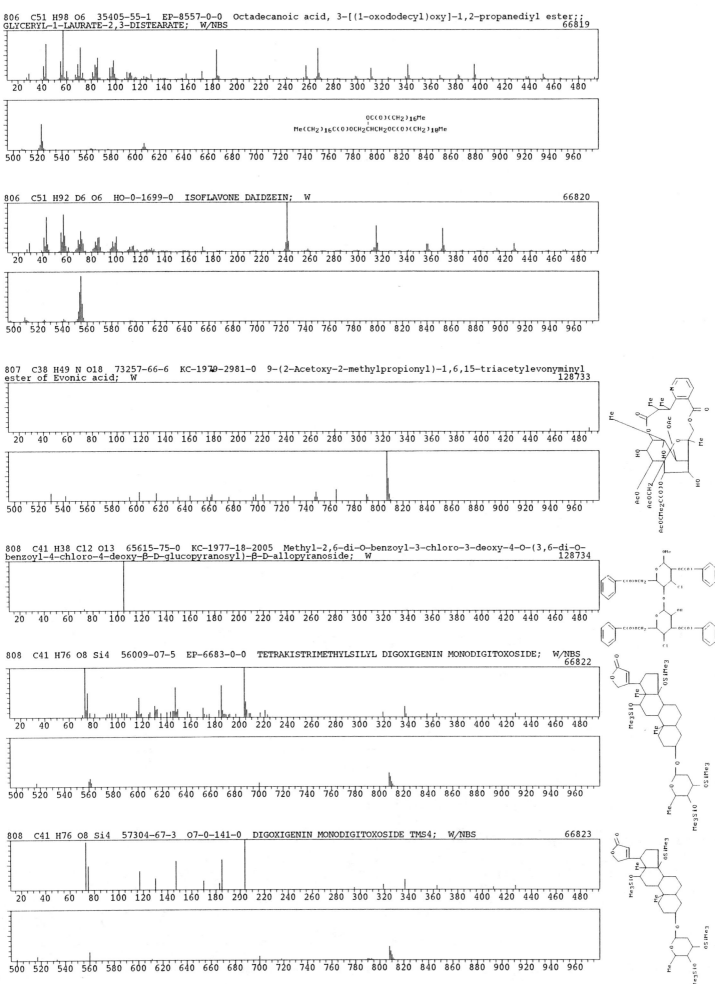

806   C51 H98 O6   35405-55-1   EP-8557-0-0   Octadecanoic acid, 3-[(1-oxododecyl)oxy]-1,2-propanediyl ester;;
GLYCERYL-1-LAURATE-2,3-DISTEARATE;   W/NBS   66819

OC(O)(CH2)16Me
Me(CH2)16C(O)OCH2CHCH2OC(O)(CH2)18Me

806   C51 H92 D6 O6   HO-0-1699-0   ISOFLAVONE DAIDZEIN;   W   66820

807   C38 H49 N O18   73257-66-6   KC-1979-2981-0   9-(2-Acetoxy-2-methylpropionyl)-1,6,15-triacetylevonyminyl
ester of Evonic acid;   W   128733

808   C41 H38 Cl2 O13   65615-75-0   KC-1977-18-2005   Methyl-2,6-di-O-benzoyl-3-chloro-3-deoxy-4-O-(3,6-di-O-
benzoyl-4-chloro-4-deoxy-β-D-glucopyranosyl)-β-D-allopyranoside;   W   128734

808   C41 H76 O8 Si4   56009-07-5   EP-6683-0-0   TETRAKISTRIMETHYLSILYL DIGOXIGENIN MONODIGITOXOSIDE;   W/NBS
                                                                                          66822

808   C41 H76 O8 Si4   57304-67-3   O7-0-141-0   DIGOXIGENIN MONODIGITOXOSIDE TMS4;   W/NBS   66823

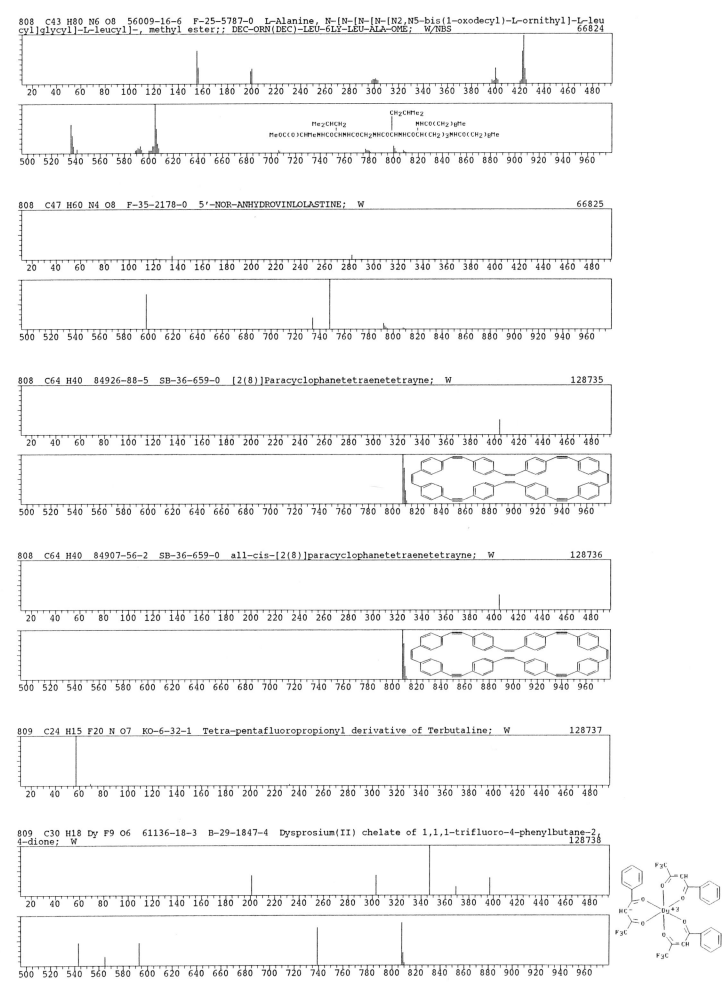

808   C43 H80 N6 O8   56009-16-6   F-25-5787-0   L-Alanine, N-[N-[N-[N-[N2,N5-bis(1-oxodecyl)-L-ornithyl]-L-leu
cyl]glycyl]-L-leucyl]-, methyl ester;; DEC-ORN(DEC)-LEU-GLY-LEU-ALA-OME;   W/NBS   66824

CH2CHMe2
Me2CHCH2   NHCO(CH2)8Me
MeOC(O)CHMeNHCOCHNHCOCH2NHCOCHNHCOCH(CH2)3NHCO(CH2)8Me

808   C47 H60 N4 O8   F-35-2178-0   5'-NOR-ANHYDROVINLOLASTINE;   W   66825

808   C64 H40   84926-88-5   SB-36-659-0   [2(8)]Paracyclophanetetraenetetrayne;   W   128735

808   C64 H40   84907-56-2   SB-36-659-0   all-cis-[2(8)]paracyclophanetetraenetetrayne;   W   128736

809   C24 H15 F20 N O7   KO-6-32-1   Tetra-pentafluoropropionyl derivative of Terbutaline;   W   128737

809   C30 H18 Dy F9 O6   61136-18-3   B-29-1847-4   Dysprosium(II) chelate of 1,1,1-trifluoro-4-phenylbutane-2,
4-dione;   W   128738

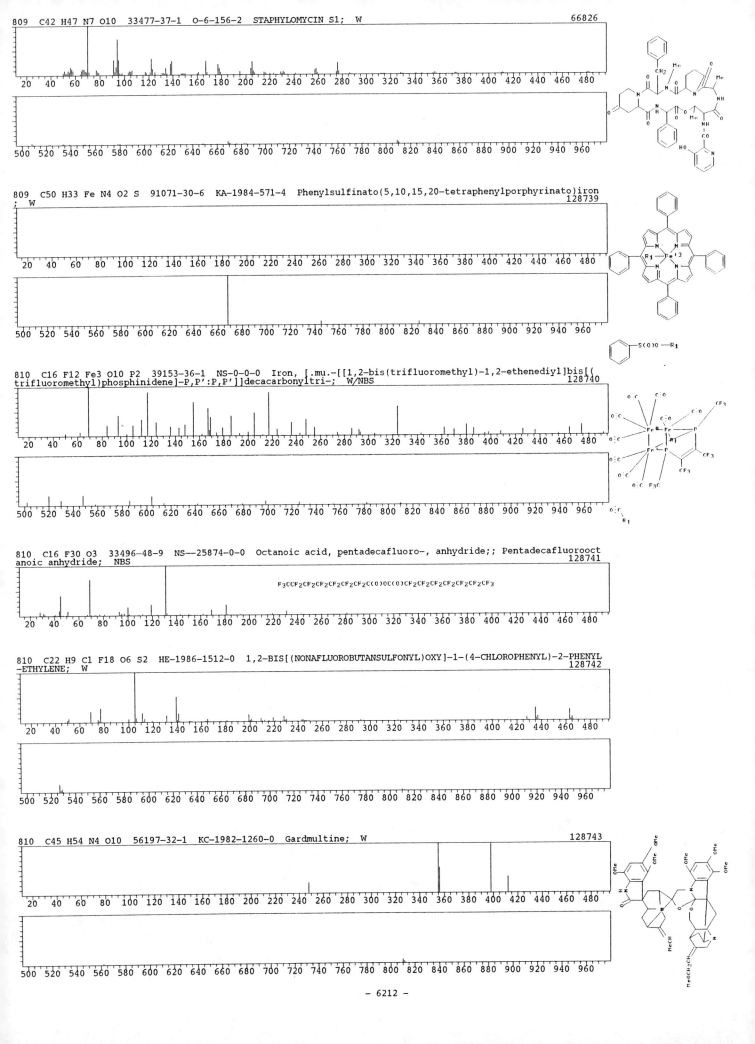

809 C42 H47 N7 O10 33477-37-1 O-6-156-2 STAPHYLOMYCIN S1; W
66826

809 C50 H33 Fe N4 O2 S 91071-30-6 KA-1984-571-4 Phenylsulfinato(5,10,15,20-tetraphenylporphyrinato)iron
; W
128739

810 C16 F12 Fe3 O10 P2 39153-36-1 NS-0-0-0 Iron, [.mu.-[[1,2-bis(trifluoromethyl)-1,2-ethenediyl]bis[(
trifluoromethyl)phosphinidene]-P,P':P,P']]decacarbonyltri-; W/NBS
128740

810 C16 F30 O3 33496-48-9 NS--25874-0-0 Octanoic acid, pentadecafluoro-, anhydride;; Pentadecafluorooct
anoic anhydride; NBS
128741

F3CCF2CF2CF2CF2CF2CF2C(O)OC(O)CF2CF2CF2CF2CF2CF2CF3

810 C22 H9 Cl F18 O6 S2 HE-1986-1512-0 1,2-BIS[(NONAFLUOROBUTANSULFONYL)OXY]-1-(4-CHLOROPHENYL)-2-PHENYL
-ETHYLENE; W
128742

810 C45 H54 N4 O10 56197-32-1 KC-1982-1260-0 Gardmultine; W
128743

811   C30 H18 Er F9 O6   61136-19-4   B-29-1847-4   Erbium(II) chelate of 1,1,1-trifluoro-4-phenylbutane-2,4-dione;   W
128744

811   C42 H49 N7 O10   33477-38-2   O-6-157-3   Staphylomycin S2;   W/NBS
66827

811   C44 H65 N O11 Si   74581-62-7   I-59-2734-0   2,2-Dimethyl-4-(2-nitro-1-acetoxy-4,4-(ethylenedioxy)nonyl)-6-(methoxy-1,3-dioxane;   W
128745

811   C48 H93 N O8   BS-1-14-0   Monoglycosylceramide,(methylated);   W
128746

812   Br8 N4 P4   14621-11-5   AD-0-1013-0   1,3,5,7,2,4,6,8-Tetrazatetraphosphocine, 2,2,4,4,6,6,8,8-octabromo-2,2,4,4,6,6,8,8-octahydro-;;   OCTABROMOTETRAPHOSPHONITRILE (MONOISOTOPIC);   W
66828

812   C32 H40 N6 O19   88261-49-8   Y-20-1309-0   1-(2,3,4,6-Tetra-O-acetyl-β-D-glucopyranosyl)-4-[1-(2,3,4,6-tetra-O-acetyl-β-D-glucopyranosyl)-1,2,4-triazol-5-yl]-1,2,4-triazolin-5-one;   W
128747

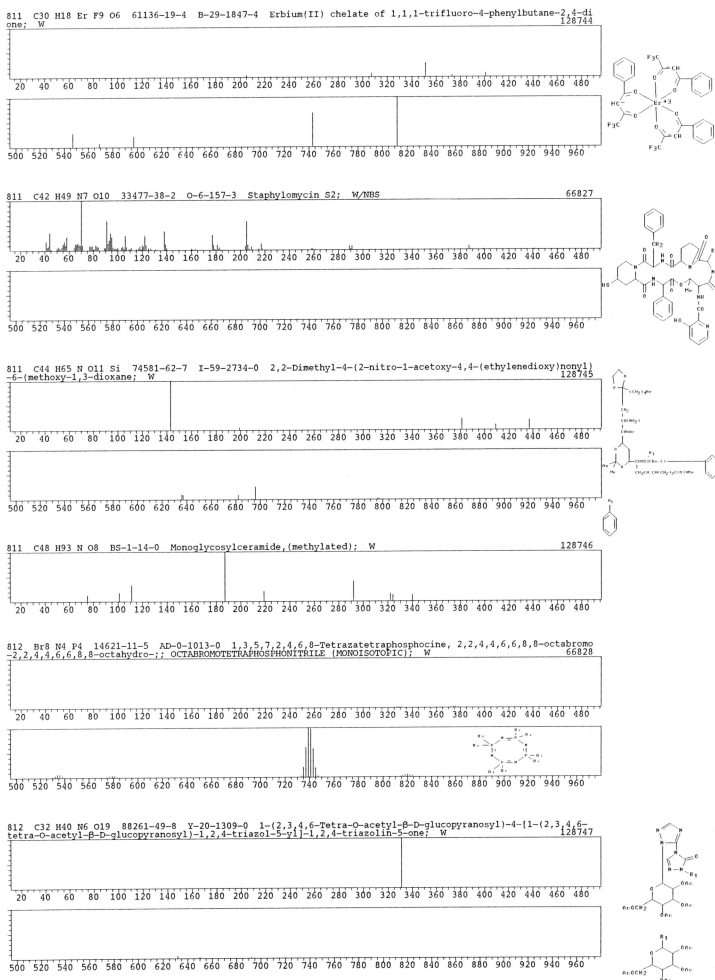

812   C52 H68 N4 O4   68464-56-2   KC-1978-850-0   FARNESYL 5-ETHYL-2-(1-HYDROXYETHYL)-5-DEMETHYL-Δ-METHYL-2-DE
VINYLPYROPHAEOPHORBIDE A AND HOMOLOGUES;   W   66829

Str 1188

813   C24 H16 F21 N O6   KO-6-32-1   Tri-o-heptafluorobutanoyl derivative of Terbutaline;   W   128748

813   C25 H27 Br4 N3 O8   95739-54-1   C-107-2919-0   Psammaplysin-A acetamide acetate;   W   128749

814   C12 H6 Be4 Cl12 O13   56377-93-6   EP-4137-0-0   Beryllium, hexakis[.mu.-(dichloroacetato-O:O')]-.mu.4-ox
otetra-;;   OXOHEXAKIS(DICHLOROACETATO)TETRABERYLLIUM(II);   W/NBS   66830

814   C42 H66 N6 O10   BS-7-153-18   N-Hydroxysuccinimide ester of N-Carbobenzoxyleucyl-leucyl-leucyl-leucyl-
leucine;   W   128750

814   C42 H66 N6 O10   BS-7-153-19   N-Hydroxysuccinimide ester of N-Carbobenzoleucyl-isoleucyly-leucyl-leucyl
-leucyl-leucine;   W   128751

814   C44 H34 Cl4 O7   72911-64-9   SB-33-481-0   2,4-DIBENZYLOXY-3,5-DICHLORO-6-METHYLBENZOIC ANHYRIDE;   W
                                                                                                     66831

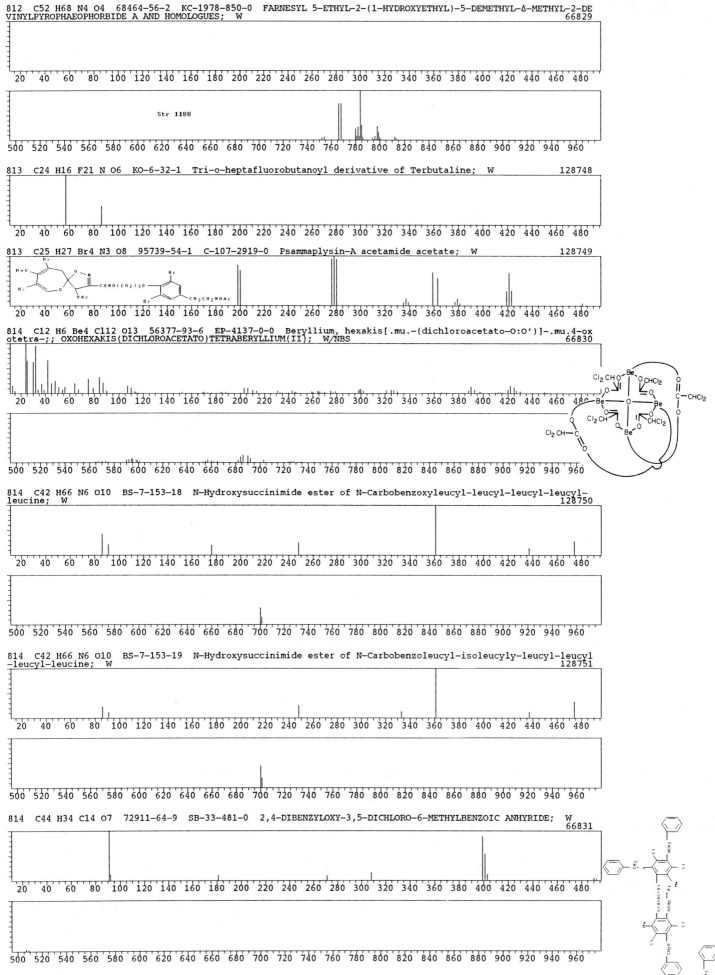

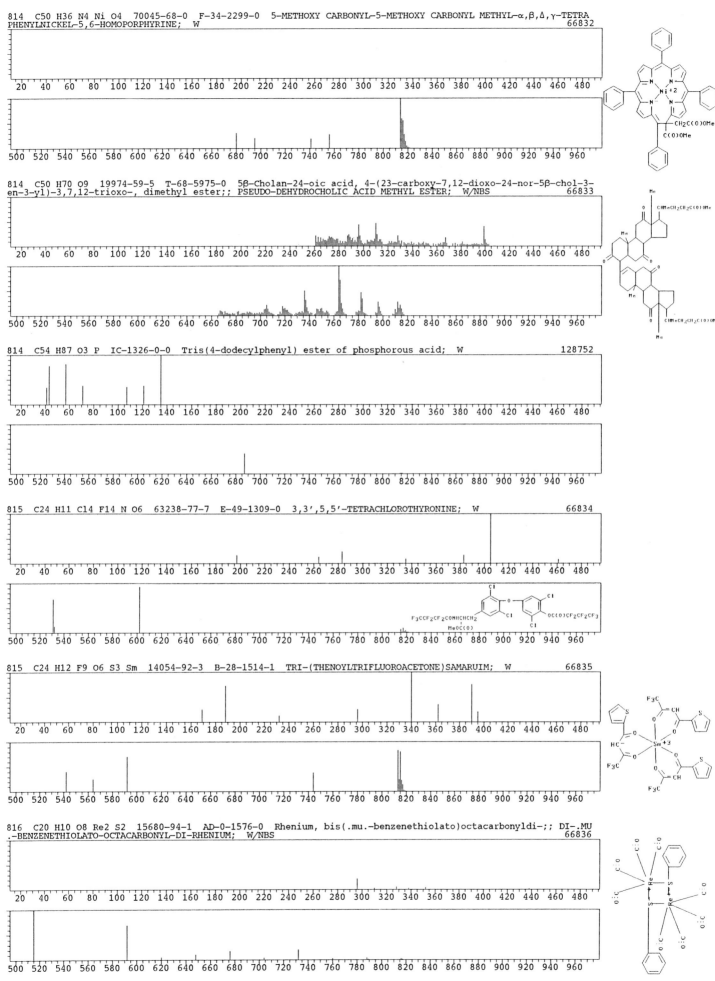

814   C50 H36 N4 Ni O4   70045-68-0   F-34-2299-0   5-METHOXY CARBONYL-5-METHOXY CARBONYL METHYL-α,β,Δ,γ-TETRA
PHENYLNICKEL-5,6-HOMOPORPHYRINE;   W   66832

814   C50 H70 O9   19974-59-5   T-68-5975-0   5β-Cholan-24-oic acid, 4-(23-carboxy-7,12-dioxo-24-nor-5β-chol-3-
en-3-yl)-3,7,12-trioxo-, dimethyl ester;; PSEUDO-DEHYDROCHOLIC ACID METHYL ESTER;   W/NBS   66833

814   C54 H87 O3 P   IC-1326-0-0   Tris(4-dodecylphenyl) ester of phosphorous acid;   W   128752

815   C24 H11 Cl4 F14 N O6   63238-77-7   E-49-1309-0   3,3',5,5'-TETRACHLOROTHYRONINE;   W   66834

815   C24 H12 F9 O6 S3 Sm   14054-92-3   B-28-1514-1   TRI-(THENOYLTRIFLUOROACETONE)SAMARUIM;   W   66835

816   C20 H10 O8 Re2 S2   15680-94-1   AD-0-1576-0   Rhenium, bis(.mu.-benzenethiolato)octacarbonyldi-;; DI-.MU
.-BENZENETHIOLATO-OCTACARBONYL-DI-RHENIUM;   W/NBS   66836

- 6215 -

816   C24 H12 Eu F9 O6 S3   14054-87-6   B-28-1514-1   TRI-(THENOYLTRIFLUOROACETONE)EUROOPIUM;   W   66837

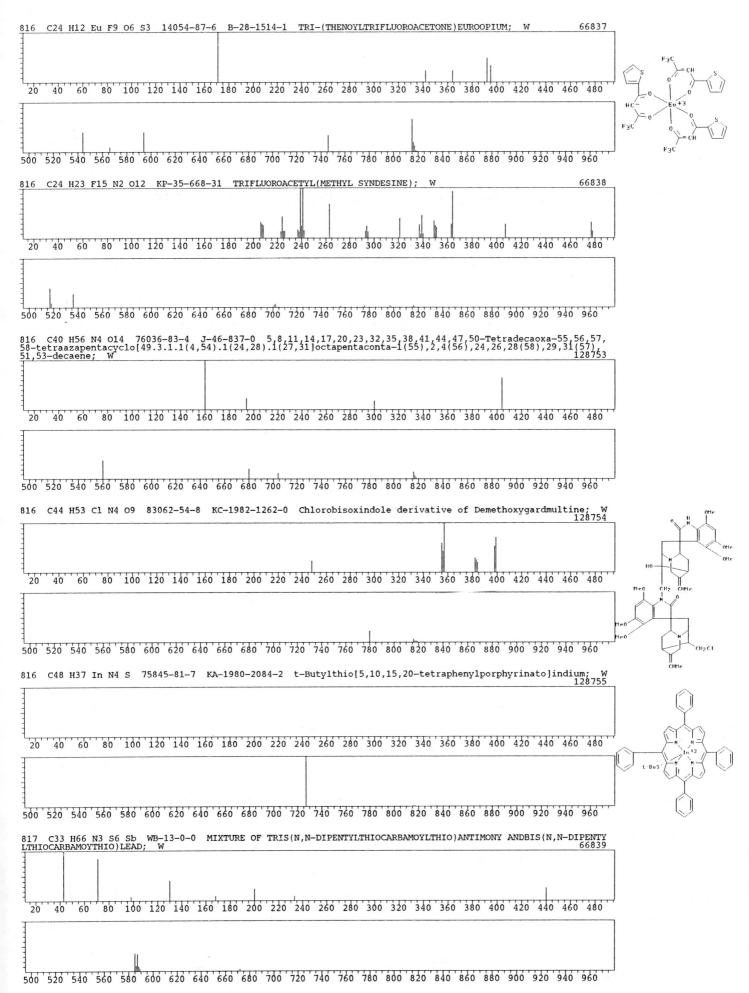

816   C24 H23 F15 N2 O12   KP-35-668-31   TRIFLUOROACETYL(METHYL SYNDESINE);   W   66838

816   C40 H56 N4 O14   76036-83-4   J-46-837-0   5,8,11,14,17,20,23,32,35,38,41,44,47,50-Tetradecaoxa-55,56,57,
58-tetraazapentacyclo[49.3.1.1(4,54).1(24,28).1(27,31)]octapentaconta-1(55),2,4(56),24,26,28(58),29,31(57),
51,53-decaene;   W   128753

816   C44 H53 Cl N4 O9   83062-54-8   KC-1982-1262-0   Chlorobisoxindole derivative of Demethoxygardmultine;   W
          128754

816   C48 H37 In N4 S   75845-81-7   KA-1980-2084-2   t-Butylthio[5,10,15,20-tetraphenylporphyrinato]indium;   W
          128755

817   C33 H66 N3 S6 Sb   WB-13-0-0   MIXTURE OF TRIS(N,N-DIPENTYLTHIOCARBAMOYLTHIO)ANTIMONY ANDBIS(N,N-DIPENTY
LTHIOCARBAMOYTHIO)LEAD;   W   66839

817 C48 H55 N3 O9 79664-31-6 J-46-5236-0 N-[4-[(2,3-Dimethoxybenzoyl)(phenylmethyl)amino]butyl]-N-[3-[(2,3-dimethoxybenzoyl)-phenylmethyl)amino]propyl]-2,3-dimethoxy-benzamide; W 128756

817 C50 H63 N O7 Si C-104-2739-0 (1S,2R,3S,5R,6R,7S,8R,10R,11S,14R,15S,18R)-10-(Benzoyloxy)-13,14-dibenzyl-18-(2-trimethylsilyl)ethoxy)-6,7-dihydroxy[9.7.0.0.0]nonadec-16-ene 6,7-acetonide; W 128757

817 C50 H63 N O7 Si 82770-08-9 C-104-5739-0 (1S,2R,3S,5R,6R,7S,8R,11R,14R,15S,18R)-2-(Benzoyloxy)-13,14-dibenzyl-18-(2-trimethylsilyl)ethoxy-6,7-dihydroxy-5,7,16,17-tetramethyl-10,12-dioxo-13-azatetracyclo[9.7.0.0]octadec-16-ene-6,7-acetonide; W 128758

818 C38 H42 O20 70748-48-0 H-62-539-0 MERMINOSIDHEPTAACETATE; W 66840

818 C40 H82 O16 56247-74-6 S-25-3782-1 Docosane, 1-(2,3-dimethoxypropoxy)-2,4,5,7,8,10,11,13,14,16,17,19,20-tridecamethoxy-;; 1-(2,3-DIMETHOXYPROPOXY)-2,4,5,7,8,10,11,13,14,16,17,19,20-TRIDECAMETHOXYDOCOSANE; W/NBS 66841

MeOCH₂CH(OMe)CH₂OCH₂CH(OMe) [CH₂CH(OMe)CH(OMe)]₆ CH₂CH₃

818 C46 H74 O12 B-30-1319-0 (21β,22α-DITIGLOYLBARRINGTOGENOL C-28-O-HYXOSIDE); W 66842

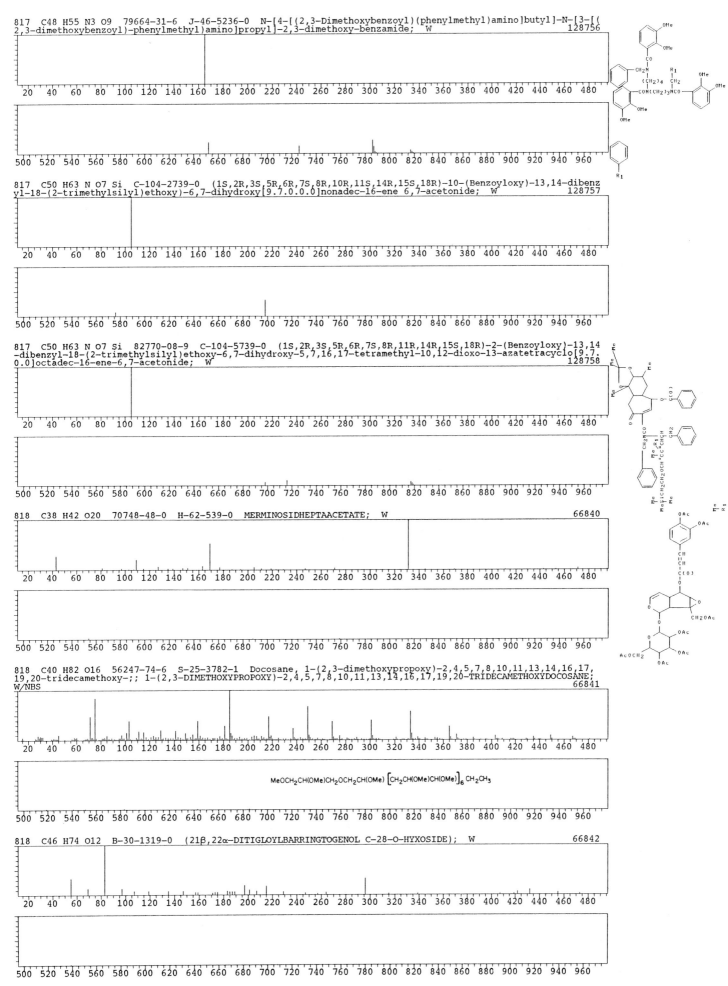

818   C50 H36 Co2 S2   64538-22-3   AG-135-233-0   Bis[(η5-cyclopentadienyl)(η4-trans-diphenyl-2-thienylcyclobu
tadienyl)]cobalt;   W   128759

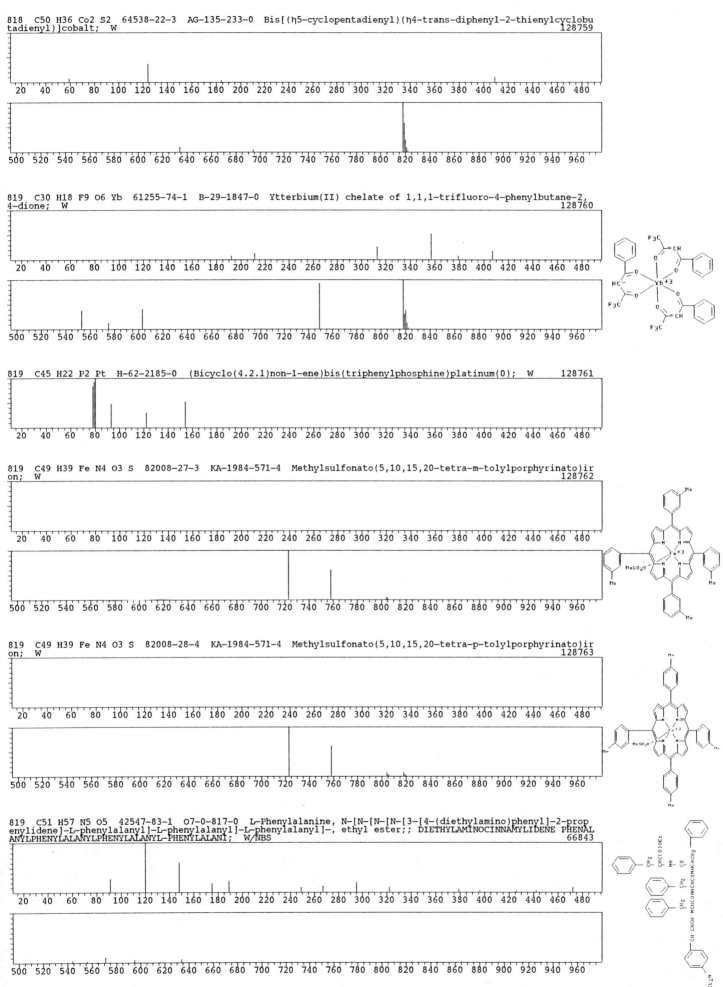

819   C30 H18 F9 O6 Yb   61255-74-1   B-29-1847-0   Ytterbium(II) chelate of 1,1,1-trifluoro-4-phenylbutane-2,
4-dione;   W   128760

819   C45 H22 P2 Pt   H-62-2185-0   (Bicyclo(4.2.1)non-1-ene)bis(triphenylphosphine)platinum(0);   W   128761

819   C49 H39 Fe N4 O3 S   82008-27-3   KA-1984-571-4   Methylsulfonato(5,10,15,20-tetra-m-tolylporphyrinato)ir
on;   W   128762

819   C49 H39 Fe N4 O3 S   82008-28-4   KA-1984-571-4   Methylsulfonato(5,10,15,20-tetra-p-tolylporphyrinato)ir
on;   W   128763

819   C51 H57 N5 O5   42547-83-1   O7-0-817-0   L-Phenylalanine, N-[N-[N-[N-[3-[4-(diethylamino)phenyl]-2-prop
enylidene]-L-phenylalanyl]-L-phenylalanyl]-L-phenylalanyl]-, ethyl ester;; DIETHYLAMINOCINNAMYLIDENE PHENAL
ANYLPHENYLALANYLPHENYLALANYL-PHENYLALANI;   W/NBS   66843

819   C52 H76 Cu N4   69912-59-0   J-44-2080-0   (2,4,6,8-Tetra(n-heptyl)-1,3,5,7-tetramethylporphyrinato)-copper(II);   W
128764

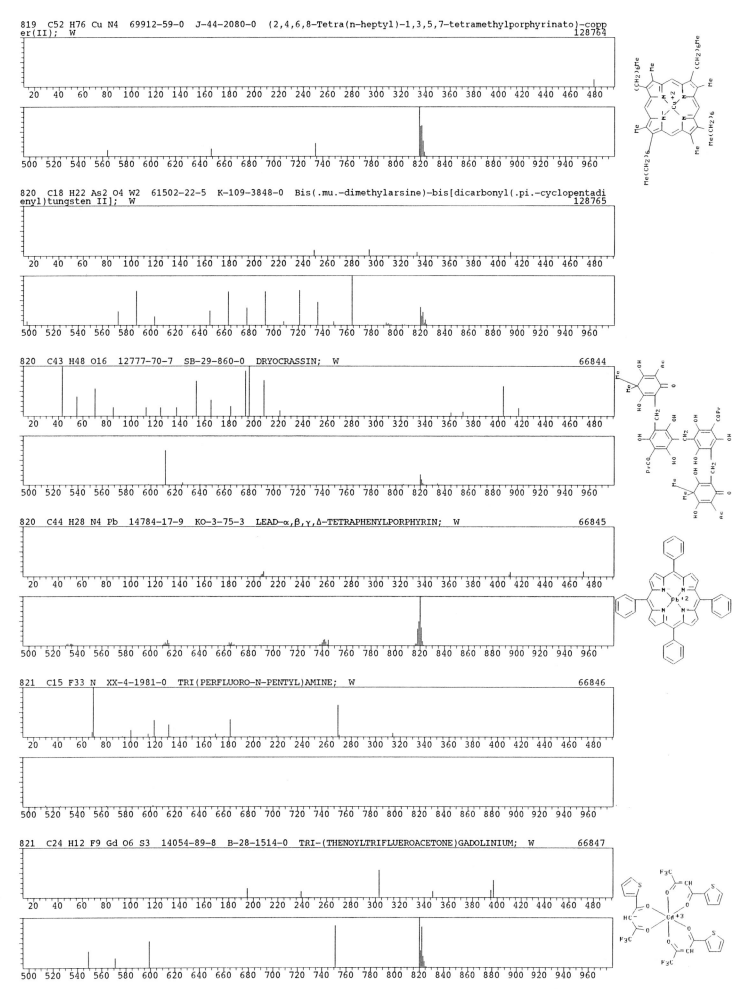

820   C18 H22 As2 O4 W2   61502-22-5   K-109-3848-0   Bis(.mu.-dimethylarsine)-bis[dicarbonyl(.pi.-cyclopentadienyl)tungsten II];   W
128765

820   C43 H48 O16   12777-70-7   SB-29-860-0   DRYOCRASSIN;   W
66844

820   C44 H28 N4 Pb   14784-17-9   KO-3-75-3   LEAD-α,β,γ,Δ-TETRAPHENYLPORPHYRIN;   W
66845

821   C15 F33 N   XX-4-1981-0   TRI(PERFLUORO-N-PENTYL)AMINE;   W
66846

821   C24 H12 F9 Gd O6 S3   14054-89-8   B-28-1514-0   TRI-(THENOYLTRIFLUEROACETONE)GADOLINIUM;   W
66847

- 6219 -

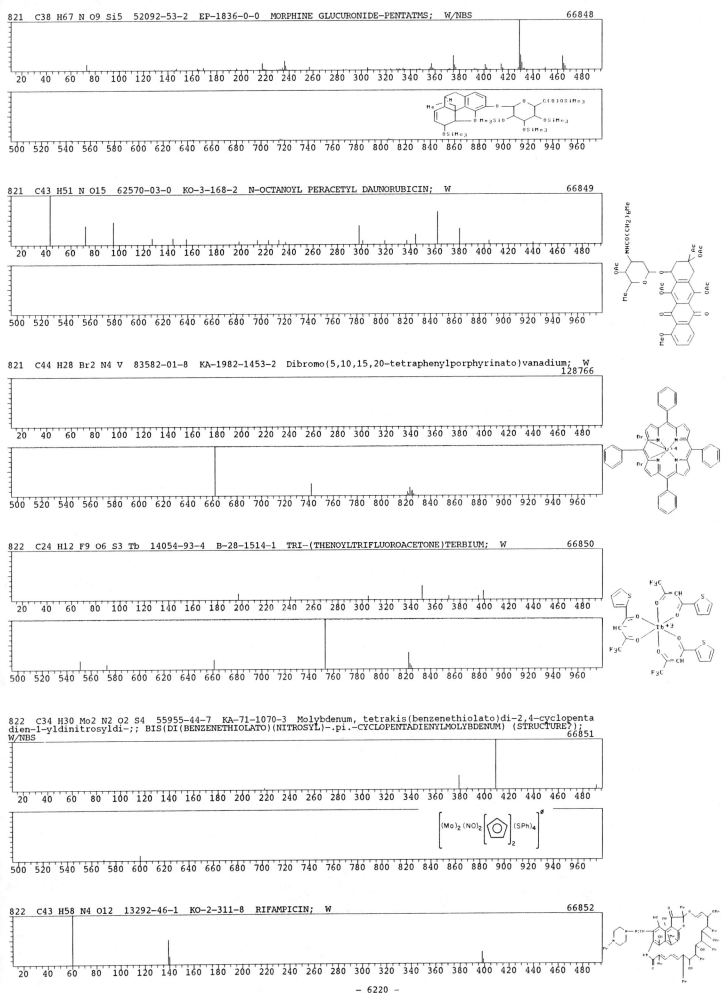

821 C38 H67 N O9 Si5 52092-53-2 EP-1836-0-0 MORPHINE GLUCURONIDE-PENTATMS; W/NBS   66848

821 C43 H51 N O15 62570-03-0 KO-3-168-2 N-OCTANOYL PERACETYL DAUNORUBICIN; W   66849

821 C44 H28 Br2 N4 V 83582-01-8 KA-1982-1453-2 Dibromo(5,10,15,20-tetraphenylporphyrinato)vanadium; W
128766

822 C24 H12 F9 O6 S3 Tb 14054-93-4 B-28-1514-1 TRI-(THENOYLTRIFLUOROACETONE)TERBIUM; W   66850

822 C34 H30 Mo2 N2 O2 S4 55955-44-7 KA-71-1070-3 Molybdenum, tetrakis(benzenethiolato)di-2,4-cyclopenta
dien-1-yldinitrosyldi-;; BIS(DI(BENZENETHIOLATO)(NITROSYL)-.pi.-CYCLOPENTADIENYLMOLYBDENUM) (STRUCTURE?);
W/NBS   66851

822 C43 H58 N4 O12 13292-46-1 KO-2-311-8 RIFAMPICIN; W   66852

822   C45 H31 In N4 O3 S   KA-1980-2084-2   Methylsulfonato[5,10,15,20-tetraphenylporphyrinato]indium;   W
128767

822   C45 H74 O13   BS-2-18-0   Pennogenin 3-O-α-L-rhamnopyranosyl-(1-2)-β-D-glucopyranoside methylate;   W
128768

822   C52 H110 O2 Si2   35177-38-9   BA-0-362-0   3,50-Dioxa-2,51-disiladopentacontane, 2,2,51,51-tetramethyl-
;; 1,46-BIS(TRIMETHYLSILYLOXY) HEXATETRACONTANE;   W/NBS
66853

Me3SIO(CH2)46OSIMe3

823   C29 H74 N O10 P Si7   55470-94-5   BA-0-358-0   D-altro-2-Heptulose, 1,3,4,5,6-pentakis-O-(trimethylsily
l)-, O-methyloxime, 7-[bis(trimethylsilyl) phosphate];; D-SEDOHEPTULOSE-7-PHOSPHATE-METHOXIME-HEPTATMS;   W
66854

823   C43 H49 N7 O10   23152-29-6   O-6-155-1   Virginiamycin S1;; Staphylomycin S;   W/NBS
66855

825   C40 H40 Br N7 O8   78403-53-9   F-37-165-0   n-Hexyl ester of 2-(7-bromo-octahydro-3,5,8,10,11,13-hexaoxo
-4,9-diphenyl-1,7:2,6-ethanediylidene-6a,10c-(methaniminomethano)-3H,6H,7H,8H-2b,4,5a,7a,9,10a-hexaazadicyc
lopent[d,i]acenaphthylen-12-yl)-4-methylpentanoic acid;   W
128769

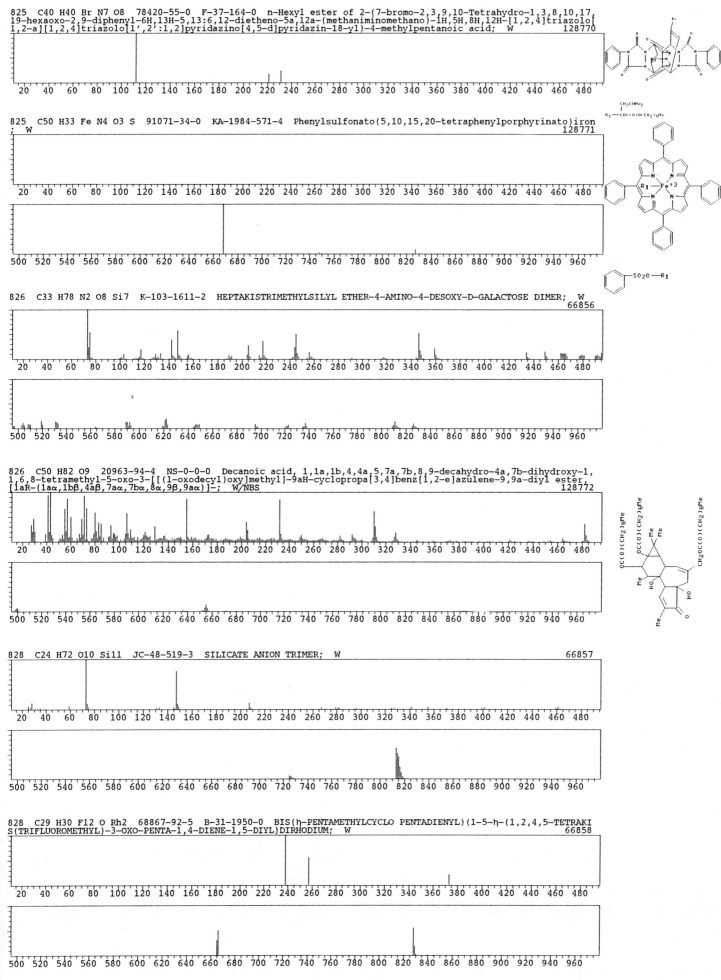

825   C40 H40 Br N7 O8   78420-55-0   F-37-164-0   n-Hexyl ester of 2-(7-bromo-2,3,9,10-Tetrahydro-1,3,8,10,17,19-hexaoxo-2,9-diphenyl-6H,13H-5,13:6,12-dietheno-5a,12a-(methaniminomethano)-1H,5H,8H,12H-[1,2,4]triazolo[1,2-a][1,2,4]triazolo[1',2':1,2]pyridazino[4,5-d]pyridazin-18-yl)-4-methylpentanoic acid;   W   128770

825   C50 H33 Fe N4 O3 S   91071-34-0   KA-1984-571-4   Phenylsulfonato(5,10,15,20-tetraphenylporphyrinato)iron ;   W   128771

826   C33 H78 N2 O8 Si7   K-103-1611-2   HEPTAKISTRIMETHYLSILYL ETHER-4-AMINO-4-DESOXY-D-GALACTOSE DIMER;   W   66856

826   C50 H82 O9   20963-94-4   NS-0-0-0   Decanoic acid, 1,1a,1b,4,4a,5,7a,7b,8,9-decahydro-4a,7b-dihydroxy-1,1,6,8-tetramethyl-5-oxo-3-[[(1-oxodecyl)oxy]methyl]-9aH-cyclopropa[3,4]benz[1,2-e]azulene-9,9a-diyl ester, [1aR-(1aα,1bβ,4aβ,7aα,7bα,8α,9β,9aα)]-;   W/NBS   128772

828   C24 H72 O10 Si11   JC-48-519-3   SILICATE ANION TRIMER;   W   66857

828   C29 H30 F12 O Rh2   68867-92-5   B-31-1950-0   BIS(η-PENTAMETHYLCYCLO PENTADIENYL)(1-5-η-(1,2,4,5-TETRAKIS(TRIFLUOROMETHYL)-3-OXO-PENTA-1,4-DIENE-1,5-DIYL)DIRHODIUM;   W   66858

828 C30 H14 F16 N4 O2 P2 70393-77-0 K-112-1369-0 3,8-DIMETHYL-4,5-BIS(PENTAFLUOROPHENYL)-1,6-BIS(3-TRIF
LUOROMETHYL)PHENY)1,3,6,8-TETRAAZA-4λ5,5λ3-DIPHOSPHASPIRO(3,4)OCTAN-2,7-DIONE; W   66859

20 40 60 80 100 120 140 160 180 200 220 240 260 280 300 320 340 360 380 400 420 440 460 480

500 520 540 560 580 600 620 640 660 680 700 720 740 760 780 800 820 840 860 880 900 920 940 960

828 C37 H48 O21 78077-29-9 H-65-1684-0 Octaacetylsinuatol; W   128773

20 40 60 80 100 120 140 160 180 200 220 240 260 280 300 320 340 360 380 400 420 440 460 480

828 C47 H72 O12 64285-77-4 B-30-1321-0 16α-ACETYL-21β,22α-DITIGLOYLBARRINGTOGENOL C-28-O-HYXOSIDE; W
                                                                                        66860

20 40 60 80 100 120 140 160 180 200 220 240 260 280 300 320 340 360 380 400 420 440 460 480

500 520 540 560 580 600 620 640 660 680 700 720 740 760 780 800 820 840 860 880 900 920 940 960

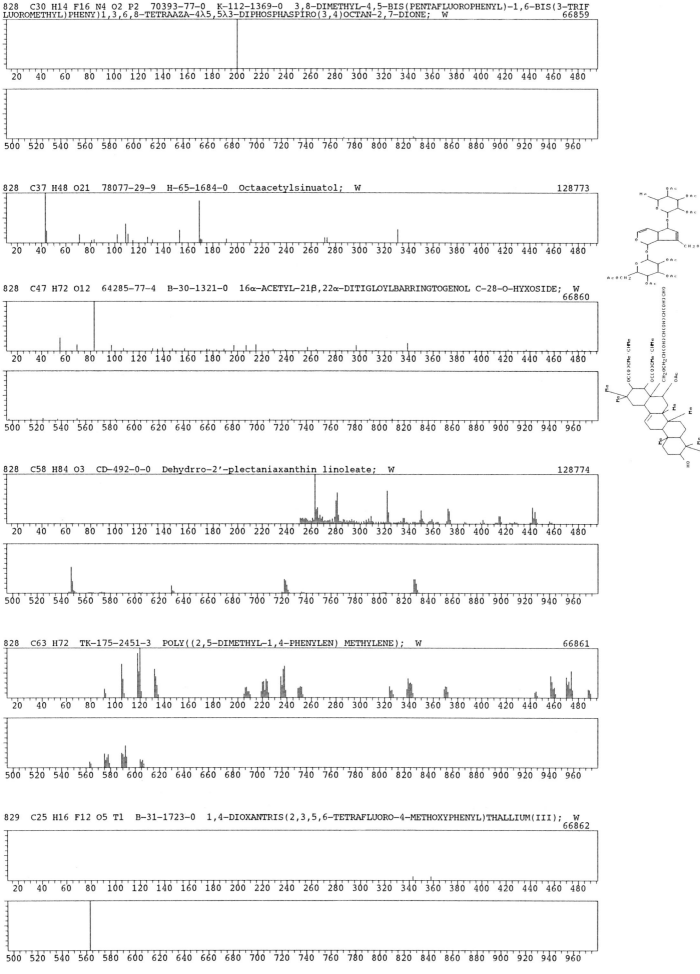

828 C58 H84 O3 CD-492-0-0 Dehydrro-2'-plectaniaxanthin linoleate; W   128774

20 40 60 80 100 120 140 160 180 200 220 240 260 280 300 320 340 360 380 400 420 440 460 480

500 520 540 560 580 600 620 640 660 680 700 720 740 760 780 800 820 840 860 880 900 920 940 960

828 C63 H72 TK-175-2451-3 POLY((2,5-DIMETHYL-1,4-PHENYLEN) METHYLENE); W   66861

20 40 60 80 100 120 140 160 180 200 220 240 260 280 300 320 340 360 380 400 420 440 460 480

500 520 540 560 580 600 620 640 660 680 700 720 740 760 780 800 820 840 860 880 900 920 940 960

829 C25 H16 F12 O5 Tl B-31-1723-0 1,4-DIOXANTRIS(2,3,5,6-TETRAFLUORO-4-METHOXYPHENYL)THALLIUM(III); W
                                                                                        66862

20 40 60 80 100 120 140 160 180 200 220 240 260 280 300 320 340 360 380 400 420 440 460 480

500 520 540 560 580 600 620 640 660 680 700 720 740 760 780 800 820 840 860 880 900 920 940 960

829   C26 H26 Br5 O6   SB-33-107-0   Hexamethyl ether of 4,5-bis(2,3-dibromo-4,5-dihydroxybenzyl)pyrocatechol;
W   128775

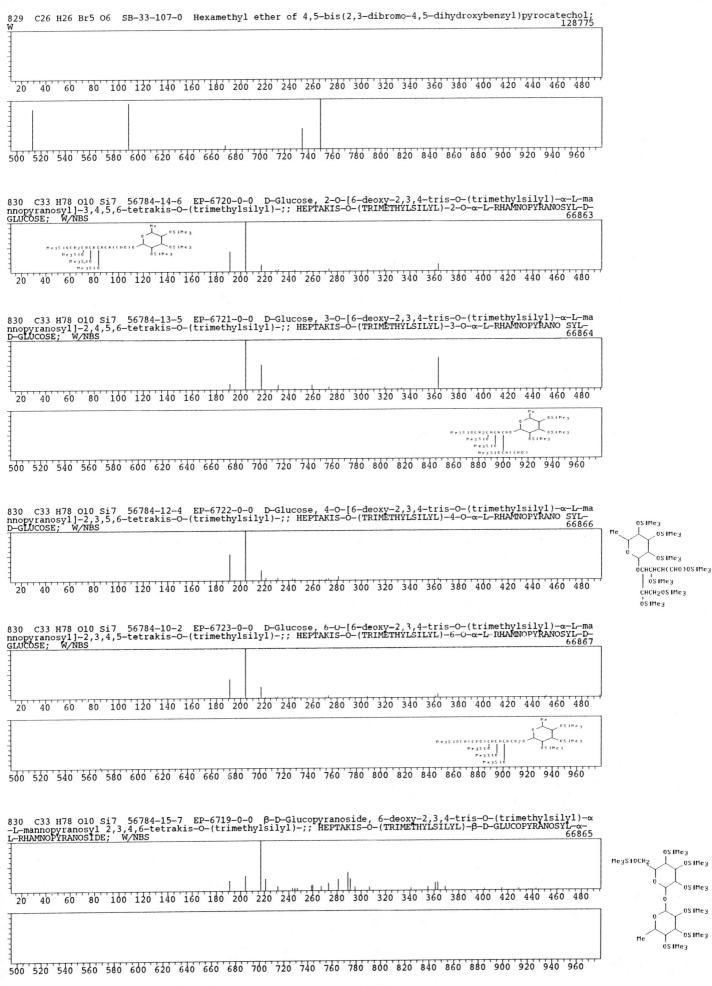

830   C33 H78 O10 Si7   56784-14-6   EP-6720-0-0   D-Glucose, 2-O-[6-deoxy-2,3,4-tris-O-(trimethylsilyl)-α-L-ma
nnopyranosyl]-3,4,5,6-tetrakis-O-(trimethylsilyl)-;; HEPTAKIS-O-(TRIMETHYLSILYL)-2-O-α-L-RHAMNOPYRANOSYL-D-
GLUCOSE;   W/NBS   66863

830   C33 H78 O10 Si7   56784-13-5   EP-6721-0-0   D-Glucose, 3-O-[6-deoxy-2,3,4-tris-O-(trimethylsilyl)-α-L-ma
nnopyranosyl]-2,4,5,6-tetrakis-O-(trimethylsilyl)-;; HEPTAKIS-O-(TRIMETHYLSILYL)-3-O-α-L-RHAMNOPYRANO SYL-
D-GLUCOSE;   W/NBS   66864

830   C33 H78 O10 Si7   56784-12-4   EP-6722-0-0   D-Glucose, 4-O-[6-deoxy-2,3,4-tris-O-(trimethylsilyl)-α-L-ma
nnopyranosyl]-2,3,5,6-tetrakis-O-(trimethylsilyl)-;; HEPTAKIS-O-(TRIMETHYLSILYL)-4-O-α-L-RHAMNOPYRANO SYL-
D-GLUCOSE;   W/NBS   66866

830   C33 H78 O10 Si7   56784-10-2   EP-6723-0-0   D-Glucose, 6-O-[6-deoxy-2,3,4-tris-O-(trimethylsilyl)-α-L-ma
nnopyranosyl]-2,3,4,5-tetrakis-O-(trimethylsilyl)-;; HEPTAKIS-O-(TRIMETHYLSILYL)-6-O-α-L-RHAMNOPYRANOSYL-D-
GLUCOSE;   W/NBS   66867

830   C33 H78 O10 Si7   56784-15-7   EP-6719-0-0   β-D-Glucopyranoside, 6-deoxy-2,3,4-tris-O-(trimethylsilyl)-α
-L-mannopyranosyl 2,3,4,6-tetrakis-O-(trimethylsilyl)-;; HEPTAKIS-O-(TRIMETHYLSILYL)-β-D-GLUCOPYRANOSYL-α-
L-RHAMNOPYRANOSIDE;   W/NBS   66865

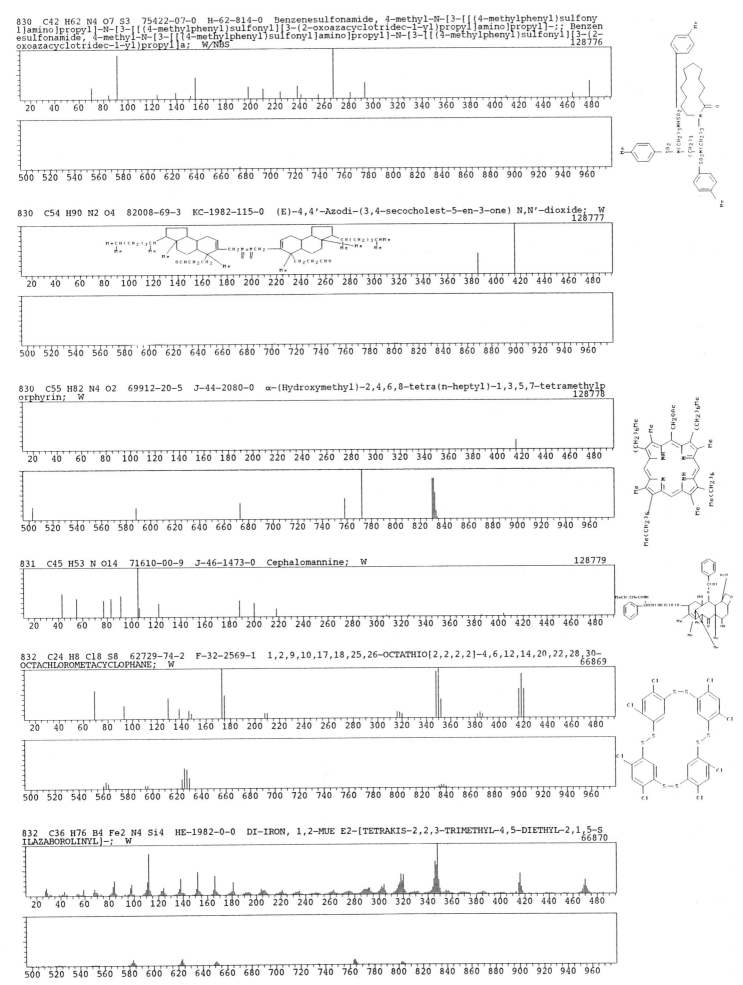

830   C42 H62 N4 O7 S3   75422-07-0   H-62-814-0   Benzenesulfonamide, 4-methyl-N-[3-[[(4-methylphenyl)sulfonyl]amino]propyl]-N-[3-[[(4-methylphenyl)sulfonyl][3-(2-oxoazacyclotridec-1-yl)propyl]amino]propyl]-;; Benzenesulfonamide, 4-methyl-N-[3-[[(4-methylphenyl)sulfonyl]amino]propyl]-N-[3-[[(4-methylphenyl)sulfonyl][3-(2-oxoazacyclotridec-1-yl)propyl]a;   W/NBS
128776

830   C54 H90 N2 O4   82008-69-3   KC-1982-115-0   (E)-4,4'-Azodi-(3,4-secocholest-5-en-3-one) N,N'-dioxide;   W
128777

830   C55 H82 N4 O2   69912-20-5   J-44-2080-0   α-(Hydroxymethyl)-2,4,6,8-tetra(n-heptyl)-1,3,5,7-tetramethylporphyrin;   W
128778

831   C45 H53 N O14   71610-00-9   J-46-1473-0   Cephalomannine;   W
128779

832   C24 H8 Cl8 S8   62729-74-2   F-32-2569-1   1,2,9,10,17,18,25,26-OCTATHIO[2,2,2,2]-4,6,12,14,20,22,28,30-OCTACHLOROMETACYCLOPHANE;   W
66869

832   C36 H76 B4 Fe2 N4 Si4   HE-1982-0-0   DI-IRON, 1,2-MUE E2-[TETRAKIS-2,2,3-TRIMETHYL-4,5-DIETHYL-2,1,5-SILAZABOROLINYL]-;   W
66870

- 6225 -

832　C46 H30 Ge N8 S2　86292-49-1　K-116-1864-0　Bis(4-methylthiophenolato)(phthalocyaninato)germanium(IV);
W
128780

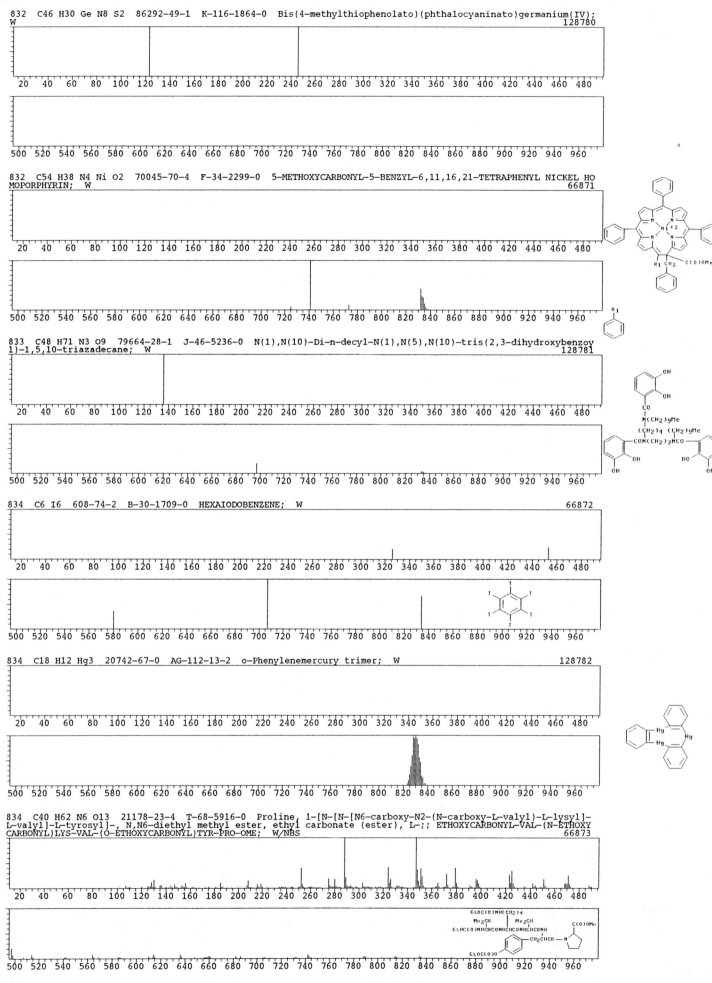

832　C54 H38 N4 Ni O2　70045-70-4　F-34-2299-0　5-METHOXYCARBONYL-5-BENZYL-6,11,16,21-TETRAPHENYL NICKEL HO
MOPORPHYRIN;　W
66871

833　C48 H71 N3 O9　79664-28-1　J-46-5236-0　N(1),N(10)-Di-n-decyl-N(1),N(5),N(10)-tris(2,3-dihydroxybenzoy
l)-1,5,10-triazadecane;　W
128781

834　C6 I6　608-74-2　B-30-1709-0　HEXAIODOBENZENE;　W
66872

834　C18 H12 Hg3　20742-67-0　AG-112-13-2　o-Phenylenemercury trimer;　W
128782

834　C40 H62 N6 O13　21178-23-4　T-68-5916-0　Proline, 1-[N-[N-[N6-carboxy-N2-(N-carboxy-L-valyl)-L-lysyl]-
L-valyl]-L-tyrosyl]-, N,N6-diethyl methyl ester, ethyl carbonate (ester), L-;; ETHOXYCARBONYL-VAL-(N-ETHOXY
CARBONYL)LYS-VAL-(O-ETHOXYCARBONYL)TYR-PRO-OME;　W/NBS
66873

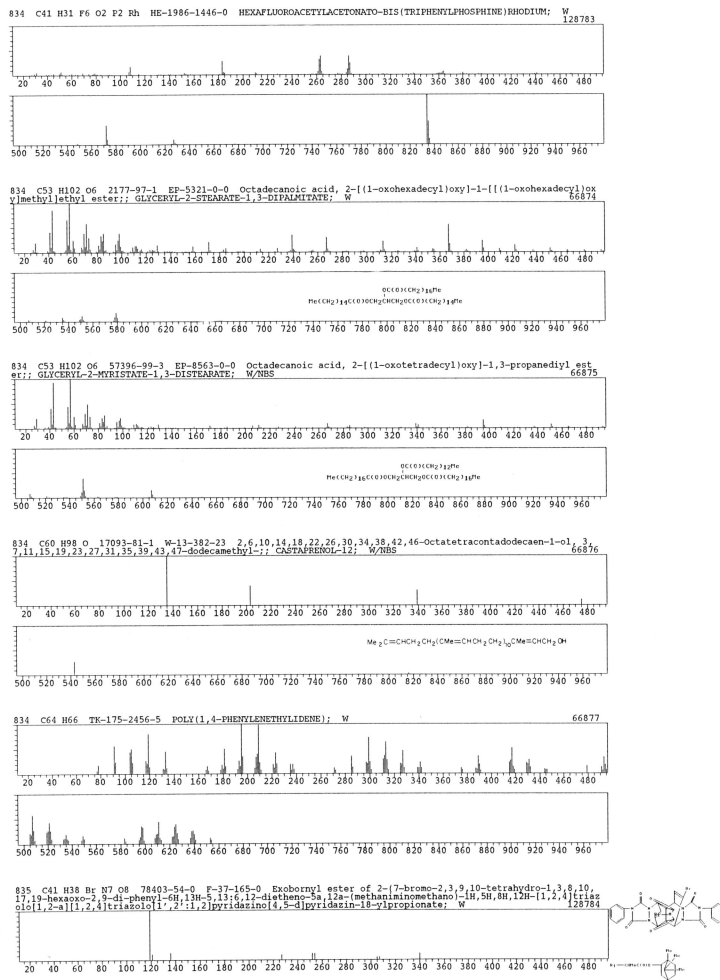

834 C41 H31 F6 O2 P2 Rh HE-1986-1446-0 HEXAFLUOROACETYLACETONATO-BIS(TRIPHENYLPHOSPHINE)RHODIUM; W
128783

834 C53 H102 O6 2177-97-1 EP-5321-0-0 Octadecanoic acid, 2-[(1-oxohexadecyl)oxy]-1-[[(1-oxohexadecyl)oxy]methyl]ethyl ester;; GLYCERYL-2-STEARATE-1,3-DIPALMITATE; W
66874

834 C53 H102 O6 57396-99-3 EP-8563-0-0 Octadecanoic acid, 2-[(1-oxotetradecyl)oxy]-1,3-propanediyl ester;; GLYCERYL-2-MYRISTATE-1,3-DISTEARATE; W/NBS
66875

834 C60 H98 O 17093-81-1 W-13-382-23 2,6,10,14,18,22,26,30,34,38,42,46-Octatetracontadodecaen-1-ol, 3,7,11,15,19,23,27,31,35,39,43,47-dodecamethyl-;; CASTAPRENOL-12; W/NBS
66876

834 C64 H66 TK-175-2456-5 POLY(1,4-PHENYLENETHYLIDENE); W
66877

835 C41 H38 Br N7 O8 78403-54-0 F-37-165-0 Exobornyl ester of 2-(7-bromo-2,3,9,10-tetrahydro-1,3,8,10,17,19-hexaoxo-2,9-di-phenyl-6H,13H-5,13:6,12-dietheno-5a,12a-(methaniminomethano)-1H,5H,8H,12H-[1,2,4]triazolo[1,2-a][1,2,4]triazolo[1',2':1,2]pyridazino[4,5-d]pyridazin-18-ylpropionate; W
128784

- 6227 -

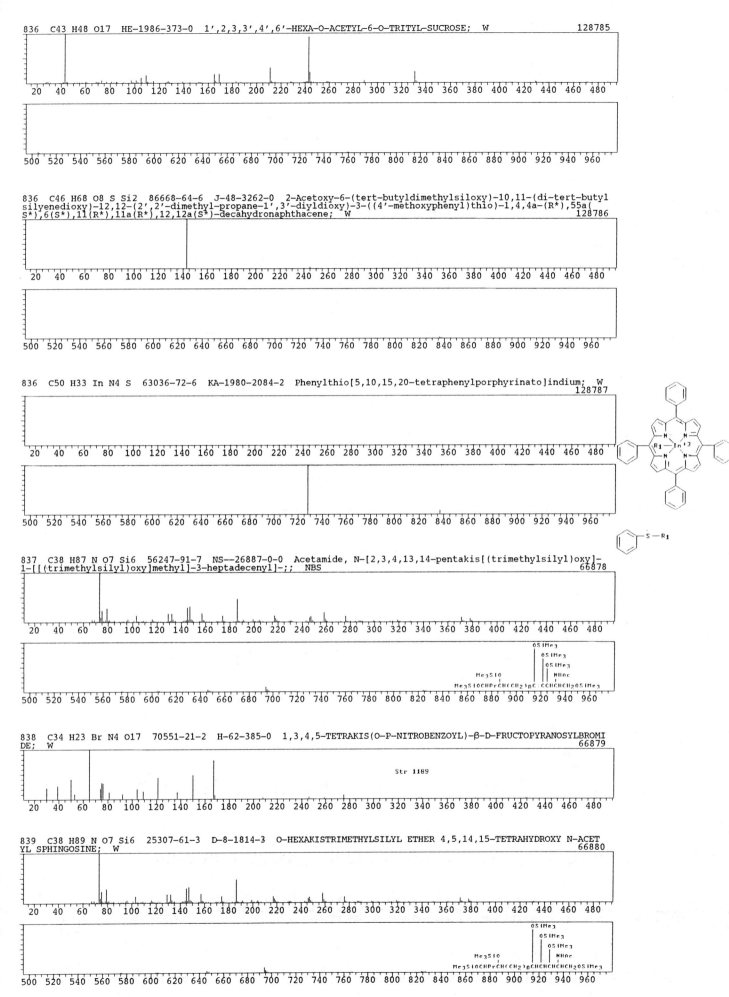

836 C43 H48 O17 HE-1986-373-0 1',2,3,3',4',6'-HEXA-O-ACETYL-6-O-TRITYL-SUCROSE; W   128785

836 C46 H68 O8 S Si2 86668-64-6 J-48-3262-0 2-Acetoxy-6-(tert-butyldimethylsiloxy)-10,11-(di-tert-butyl
silyenedioxy)-12,12-(2',2'-dimethyl-propane-1',3'-diyldioxy)-3-((4'-methoxyphenyl)thio)-1,4,4a-(R*),55a(
S*),6(S*),11(R*),11a(R*),12,12a(S*)-decahydronaphthacene; W   128786

836 C50 H33 In N4 S 63036-72-6 KA-1980-2084-2 Phenylthio[5,10,15,20-tetraphenylporphyrinato]indium; W
128787

837 C38 H87 N O7 Si6 56247-91-7 NS--26887-0-0 Acetamide, N-[2,3,4,13,14-pentakis[(trimethylsilyl)oxy]-
1-[[(trimethylsilyl)oxy]methyl]-3-heptadecenyl]-;; NBS   66878

838 C34 H23 Br N4 O17 70551-21-2 H-62-385-0 1,3,4,5-TETRAKIS(O-P-NITROBENZOYL)-β-D-FRUCTOPYRANOSYLBROMI
DE; W   66879

Str 1189

839 C38 H89 N O7 Si6 25307-61-3 D-8-1814-3 O-HEXAKISTRIMETHYLSILYL ETHER 4,5,14,15-TETRAHYDROXY N-ACET
YL SPHINGOSINE; W   66880

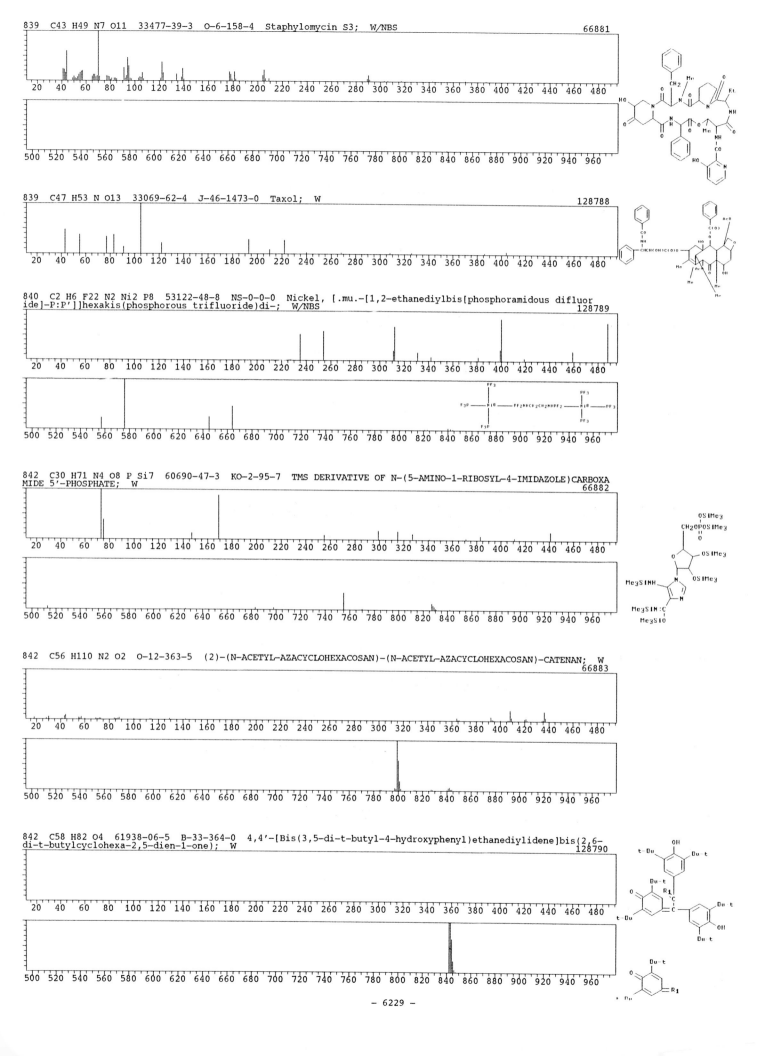

839 C43 H49 N7 O11 33477-39-3 O-6-158-4 Staphylomycin S3; W/NBS   66881

839 C47 H53 N O13 33069-62-4 J-46-1473-0 Taxol; W   128788

840 C2 H6 F22 N2 Ni2 P8 53122-48-8 NS-0-0-0 Nickel, [.mu.-[1,2-ethanediylbis[phosphoramidous difluor
ide]-P:P']]hexakis(phosphorous trifluoride)di-; W/NBS   128789

842 C30 H71 N4 O8 P Si7 60690-47-3 KO-2-95-7 TMS DERIVATIVE OF N-(5-AMINO-1-RIBOSYL-4-IMIDAZOLE)CARBOXA
MIDE 5'-PHOSPHATE; W   66882

842 C56 H110 N2 O2 O-12-363-5 (2)-(N-ACETYL-AZACYCLOHEXACOSAN)-(N-ACETYL-AZACYCLOHEXACOSAN)-CATENAN; W
                                                                                    66883

842 C58 H82 O4 61938-06-5 B-33-364-0 4,4'-[Bis(3,5-di-t-butyl-4-hydroxyphenyl)ethanediylidene]bis(2,6-
di-t-butylcyclohexa-2,5-dien-1-one); W   128790

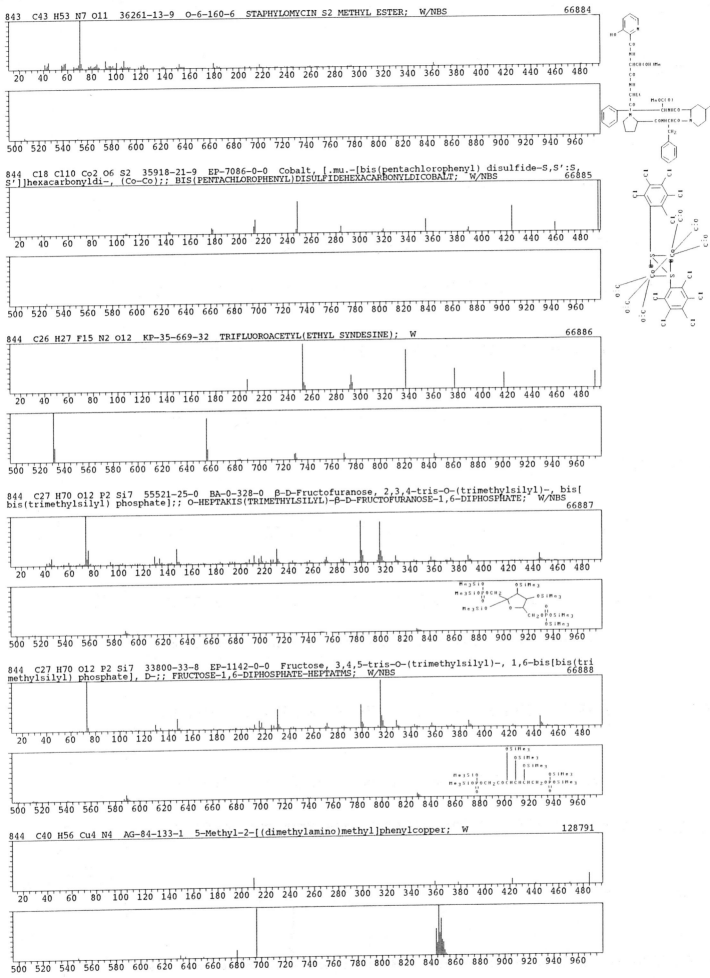

843 C43 H53 N7 O11 36261-13-9 O-6-160-6 STAPHYLOMYCIN S2 METHYL ESTER; W/NBS 66884

844 C18 Cl10 Co2 O6 S2 35918-21-9 EP-7086-0-0 Cobalt, [.mu.-[bis(pentachlorophenyl) disulfide-S,S':S,
S']]hexacarbonyldi-, (Co-Co);; BIS(PENTACHLOROPHENYL)DISULFIDEHEXACARBONYLDICOBALT; W/NBS 66885

844 C26 H27 F15 N2 O12 KP-35-669-32 TRIFLUOROACETYL(ETHYL SYNDESINE); W 66886

844 C27 H70 O12 P2 Si7 55521-25-0 BA-0-328-0 β-D-Fructofuranose, 2,3,4-tris-O-(trimethylsilyl)-, bis[
bis(trimethylsilyl) phosphate];; O-HEPTAKIS(TRIMETHYLSILYL)-β-D-FRUCTOFURANOSE-1,6-DIPHOSPHATE; W/NBS 66887

844 C27 H70 O12 P2 Si7 33800-33-8 EP-1142-0-0 Fructose, 3,4,5-tris-O-(trimethylsilyl)-, 1,6-bis[bis(tri
methylsilyl) phosphate], D-;; FRUCTOSE-1,6-DIPHOSPHATE-HEPTATMS; W/NBS 66888

844 C40 H56 Cu4 N4 AG-84-133-1 5-Methyl-2-[(dimethylamino)methyl]phenylcopper; W 128791

844   C49 H72 N2 O8 Si   95739-42-7   C-107-3262-0   8,10-Dibenzyl-8,10-diaza-1-(1'-O-(tert-butyldimethylsilyl)
-2',3'-O-isopropylidene)-5-(((tert-butyldimethylsilyl)oxy)methyl)-2-oxabicyclo[4.2.2]decane-7,9-dione;   W
                                                                                                      128792

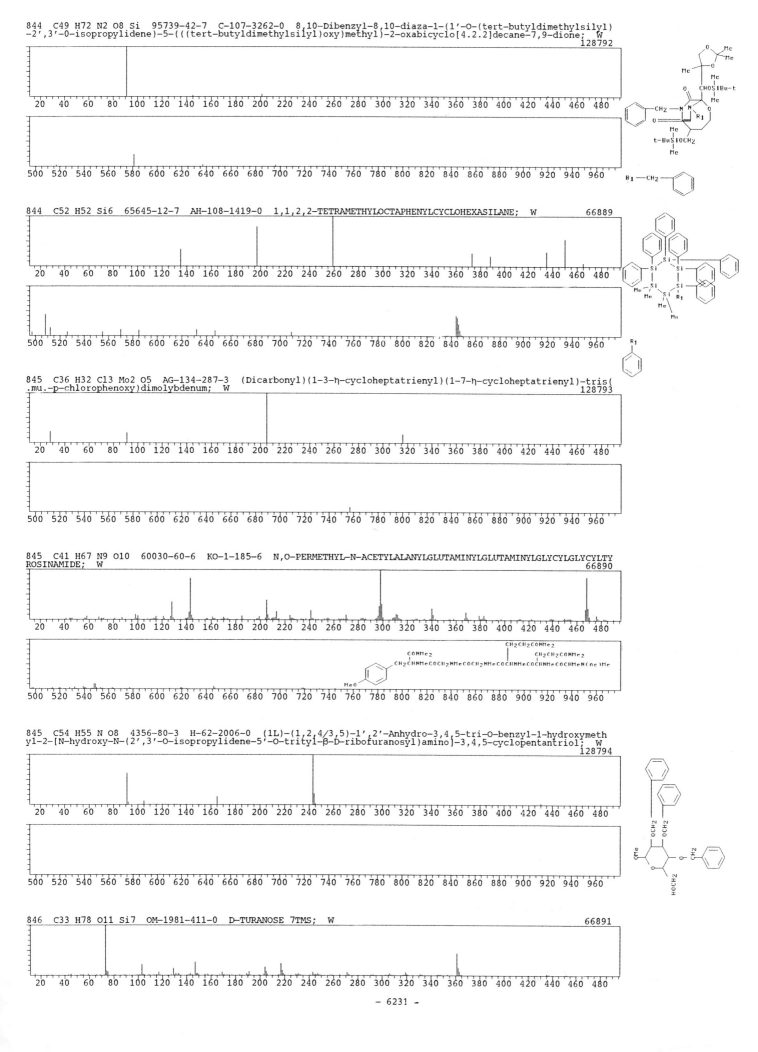

844   C52 H52 Si6   65645-12-7   AH-108-1419-0   1,1,2,2-TETRAMETHYLOCTAPHENYLCYCLOHEXASILANE;   W   66889

845   C36 H32 Cl3 Mo2 O5   AG-134-287-3   (Dicarbonyl)(1-3-η-cycloheptatrienyl)(1-7-η-cycloheptatrienyl)-tris(
.mu.-p-chlorophenoxy)dimolybdenum;   W   128793

845   C41 H67 N9 O10   60030-60-6   KO-1-185-6   N,O-PERMETHYL-N-ACETYLALANYLGLUTAMINYLGLUTAMINYLGLYCYLGLYCYLTY
ROSINAMIDE;   W   66890

845   C54 H55 N O8   4356-80-3   H-62-2006-0   (1L)-(1,2,4/3,5)-1',2'-Anhydro-3,4,5-tri-O-benzyl-1-hydroxymeth
yl-2-[N-hydroxy-N-(2',3'-O-isopropylidene-5'-O-trityl-β-D-ribofuranosyl)amino]-3,4,5-cyclopentantriol;   W
                                                                                                      128794

846   C33 H78 O11 Si7   OM-1981-411-0   D-TURANOSE 7TMS;   W   66891

- 6231 -

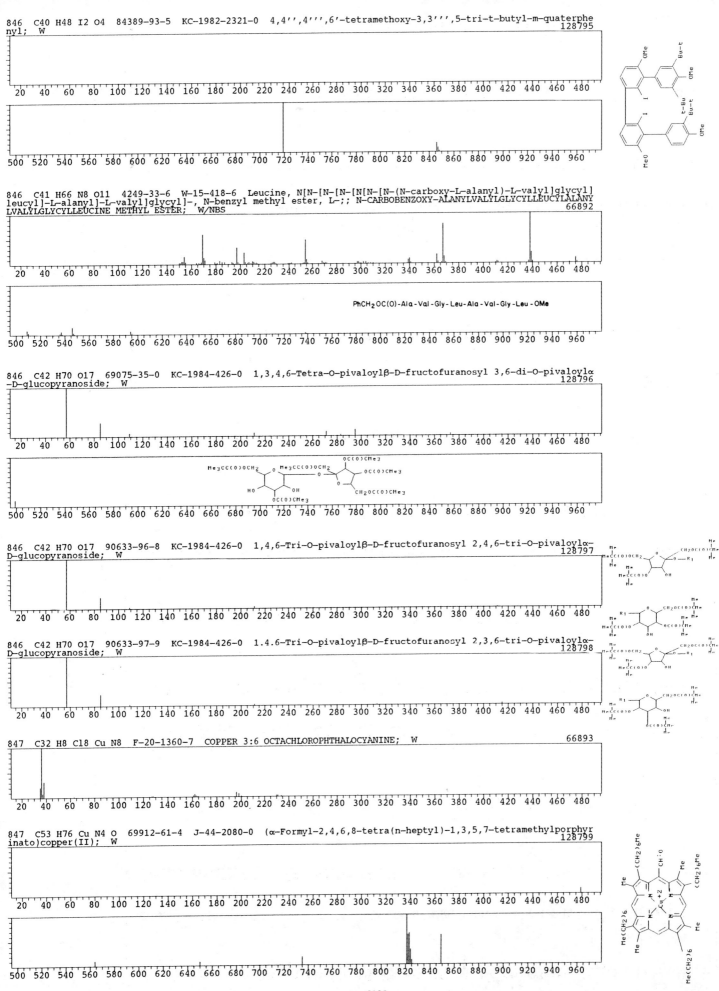

846 C40 H48 I2 O4 84389-93-5 KC-1982-2321-0 4,4''',4''',6'-tetramethoxy-3,3''',5-tri-t-butyl-m-quaterphenyl; W
128795

846 C41 H66 N8 O11 4249-33-6 W-15-418-6 Leucine, N[N-[N-[N-[N[N-[N-(N-carboxy-L-alanyl)-L-valyl]glycyl]leucyl]-L-alanyl]-L-valyl]glycyl]-, N-benzyl methyl ester, L-;; N-CARBOBENZOXY-ALANYLVALYLGLYCYLLEUCYLALANYLVALYLGLYCYLLEUCINE METHYL ESTER; W/NBS
66892

PhCH₂OC(O)-Ala-Val-Gly-Leu-Ala-Val-Gly-Leu-OMe

846 C42 H70 O17 69075-35-0 KC-1984-426-0 1,3,4,6-Tetra-O-pivaloylβ-D-fructofuranosyl 3,6-di-O-pivaloylα-D-glucopyranoside; W
128796

846 C42 H70 O17 90633-96-8 KC-1984-426-0 1,4,6-Tri-O-pivaloylβ-D-fructofuranosyl 2,4,6-tri-O-pivaloylα-D-glucopyranoside; W
128797

846 C42 H70 O17 90633-97-9 KC-1984-426-0 1.4.6-Tri-O-pivaloylβ-D-fructofuranosyl 2,3,6-tri-O-pivaloylα-D-glucopyranoside; W
128798

847 C32 H8 Cl8 Cu N8 F-20-1360-7 COPPER 3:6 OCTACHLOROPHTHALOCYANINE; W
66893

847 C53 H76 Cu N4 O 69912-61-4 J-44-2080-0 (α-Formyl-2,4,6,8-tetra(n-heptyl)-1,3,5,7-tetramethylporphyrinato)copper(II); W
128799

848   C28 H24 Mo2 O2 Te2   64065-71-0   AG-134-287-3   Dicarbonyldi(1-7-η-cycloheptatrienyl)-di-.mu.-phenyltell
urodimolybdenum; W
128800

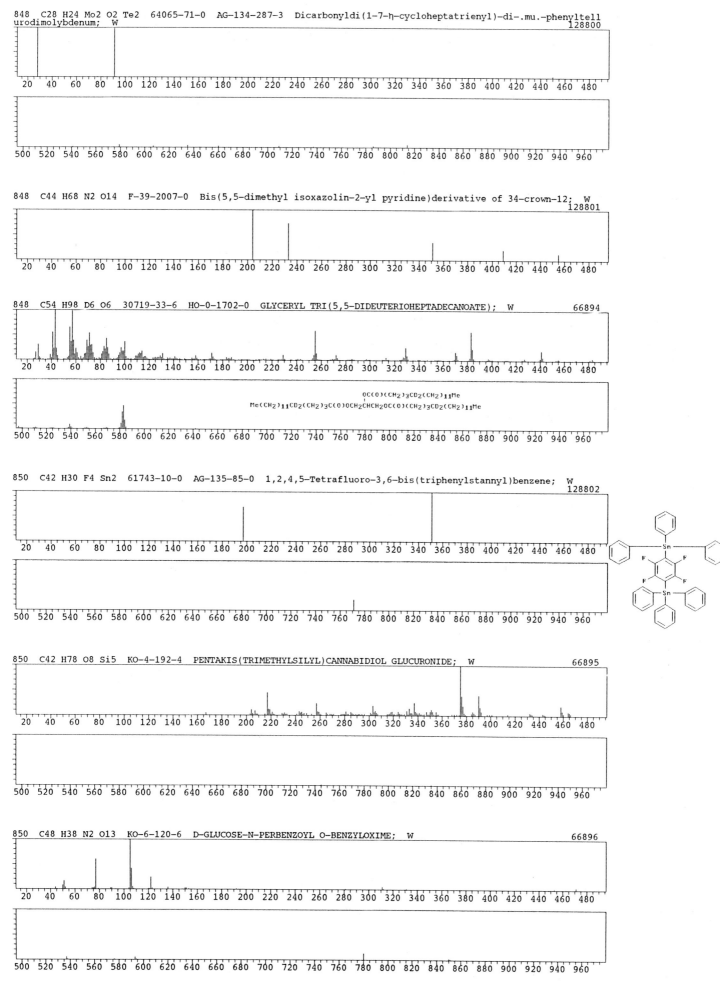

848   C44 H68 N2 O14   F-39-2007-0   Bis(5,5-dimethyl isoxazolin-2-yl pyridine)derivative of 34-crown-12; W
128801

848   C54 H98 D6 O6   30719-33-6   HO-0-1702-0   GLYCERYL TRI(5,5-DIDEUTERIOHEPTADECANOATE); W
66894

$$OC(O)(CH_2)_3CD_2(CH_2)_{11}Me$$
$$Me(CH_2)_{11}CD_2(CH_2)_3C(O)OCH_2CHCH_2OC(O)(CH_2)_3CD_2(CH_2)_{11}Me$$

850   C42 H30 F4 Sn2   61743-10-0   AG-135-85-0   1,2,4,5-Tetrafluoro-3,6-bis(triphenylstannyl)benzene; W
128802

850   C42 H78 O8 Si5   KO-4-192-4   PENTAKIS(TRIMETHYLSILYL)CANNABIDIOL GLUCURONIDE; W
66895

850   C48 H38 N2 O13   KO-6-120-6   D-GLUCOSE-N-PERBENZOYL O-BENZYLOXIME; W
66896

850 C52 H82 O9 75311-71-6 NS--26880-0-0 9,12-Octadecadienoic acid (Z,Z)-, [9a-(acetyloxy)-1a,1b,4,4a,5,
7a,7b,8,9,9a-decahydro-4a,7b-dihydroxy-1,1,6,8-tetramethyl-5-oxo-1H-cyclopropa[3,4]benz[1,2-e]azulen-3-yl]
methyl ester, [1aR-(1aα,1bβ,4aβ,7aα,7bα,8a,9β,9aα)]-;; 9,12-Octadecadienoic acid (Z,Z)-, [9a-(acetyloxy)-
1a,1b,4,4a,5,7a,7b,8,9,9a-decahydro-4a,7b-dihydroxy-1,1,6,8-tetramethyl-5-oxo-1H-cyclopropa[3,; NBS 66897

850 C55 H62 O8 79037-07-3 F-37-602-0 1β,3β,16β,26-Tetrabenzoyloxycholest-5-ene; W   128803

Str 1190

850 Co F20 H Ir P7 42230-58-0 NS-0-0-0 Iridium, .mu.-hydro[.mu.-(phosphonous difluoride-P]tris(phosph
orous trifluoride)[tris(phosphorous trifluoride)cobalt]-; W/NBS   128804

852 C8 I2 O8 Re2 15189-53-4 AD-0-1571-0 Rhenium, octacarbonyldi-.mu.-iododi-;; DI-.MU.-IODO-OCTACARBONY
L-DIRHENIUM; W/NBS   66898

852 C34 H34 I2 N2 O8 84251-55-8 KC-1982-2314-0 2,2'''-Di-iodo-4,4'',4''',6-tetramethyl-3,3''',5,5''',6,
6''''-hexamethyl-2',2''-dinitro-; W   128805

852 C38 H50 Cl2 O4 Pd2 34700-03-3 B-33-2763-0 Di-.mu.-Chloro-bis-[α-4-6-η-(3,17-dioxoandrostenyl)pallad
ium(II)]; W   128806

Str 1191

852 C41 H72 O18 78810-60-3 KC-1983-1533-0 Tetrasaccharide fragment of kijanimicin; W   128807

Str 1192

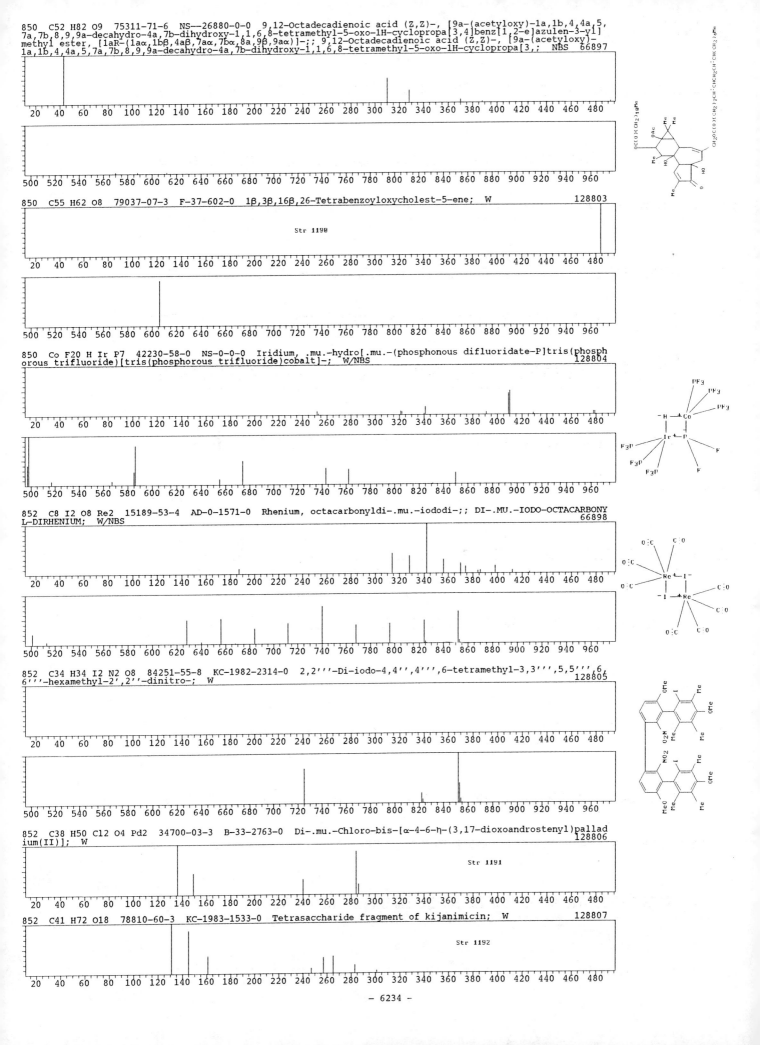

853   C55 H106 F3 N O   O-12-357-4   (2)-(N-TRIFLUORACETYL-AZACYCLOHEXACOSAN)-(CYCLOOCTACOSAN)-CATENAN;   W
66899

855   C44 H53 N7 O11   36261-12-8   O-6-159-5   STAPHYLOMYCIN S METHYL ESTER;   W/NBS
66900

856   C10 Hg O10 Re2   33728-46-0   KA-71-2650-3   Rhenium, decacarbonyl-.mu.-mercuriodi-;; BIS(PENTACARBONYLRH
ENIO)MERCURY;   W/NBS
66901

856   C30 H30 F12 O2 Rh2   68867-93-6   B-31-1951-0   BIS(η-PENTAMETHYLCYCLOPENTADIENYL)(1-6-η-(1,2,5,6-TETRAKI
S(TRIFLUOROMETHYL)-3,4-DIOXO-HEXA-1,5-DIENE-1,6-DIYL))DIRHODIUM;   W
66902

858   C42 H66 O18   OS-5-182-0   Cerberoside;   W
128808

858   C47 H70 O14   64285-81-0   B-30-1322-0   16-DEOXYBARRINGTOZENOL C-28-O-LYXOSIDE HEXAACETATE;   W
66903

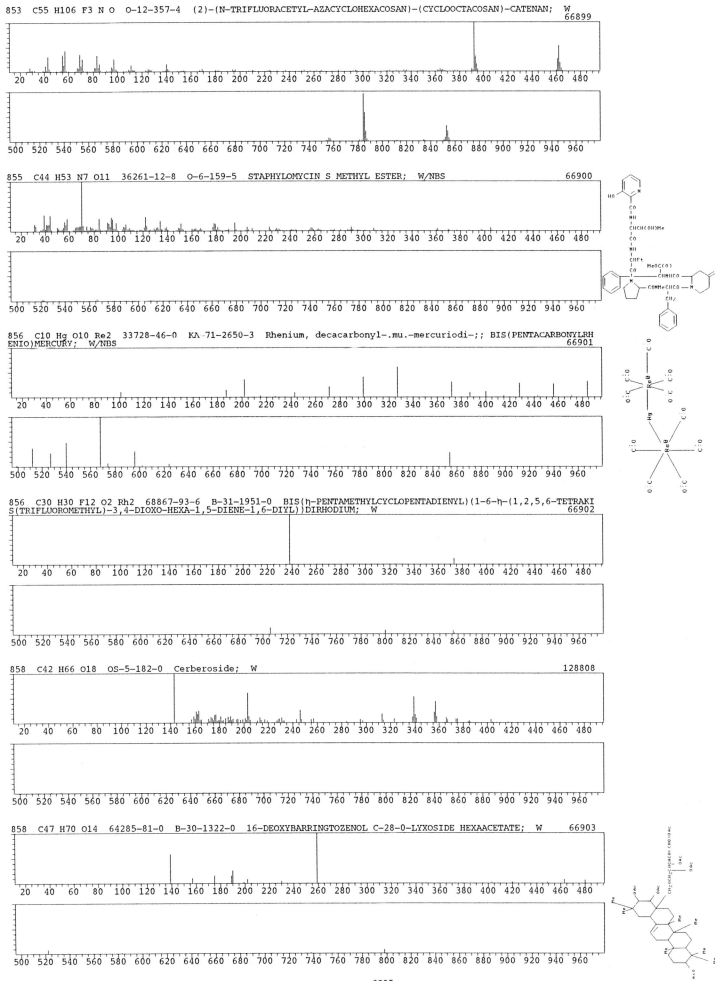

859 C34 H32 Co N12 O12 C-104-1549-0 Bis(4-hydroxycarbonytlphenyltris(pyrazolyl)borato)cobalt (II); W
128809

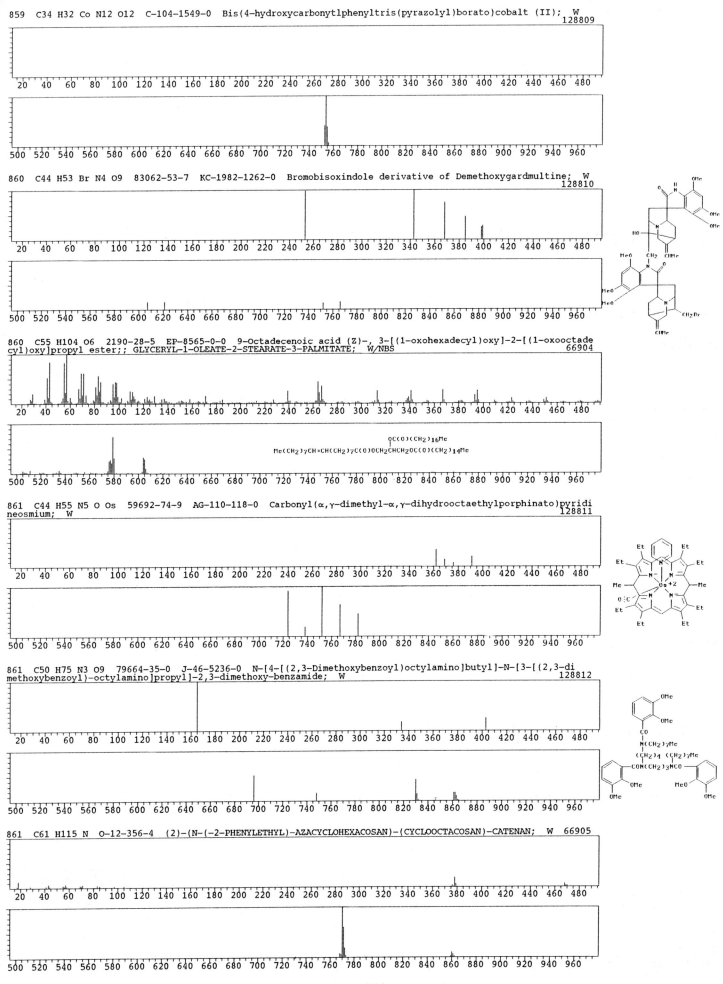

860 C44 H53 Br N4 O9 83062-53-7 KC-1982-1262-0 Bromobisoxindole derivative of Demethoxygardmultine; W
128810

860 C55 H104 O6 2190-28-5 EP-8565-0-0 9-Octadecenoic acid (Z)-, 3-[(1-oxohexadecyl)oxy]-2-[(1-oxooctade
cyl)oxy]propyl ester;; GLYCERYL-1-OLEATE-2-STEARATE-3-PALMITATE; W/NBS
66904

Me(CH₂)₇CH=CH(CH₂)₇C(O)OCH₂CHCH₂OC(O)(CH₂)₁₄Me
OC(O)(CH₂)₁₆Me

861 C44 H55 N5 O Os 59692-74-9 AG-110-118-0 Carbonyl(α,γ-dimethyl-α,γ-dihydrooctaethylporphinato)pyridi
neosmium; W
128811

861 C50 H75 N3 O9 79664-35-0 J-46-5236-0 N-[4-[(2,3-Dimethoxybenzoyl)octylamino]butyl]-N-[3-[(2,3-di
methoxybenzoyl)-octylamino]propyl]-2,3-dimethoxy-benzamide; W
128812

861 C61 H115 N O-12-356-4 (2)-(N-(-2-PHENYLETHYL)-AZACYCLOHEXACOSAN)-(CYCLOOCTACOSAN)-CATENAN; W 66905

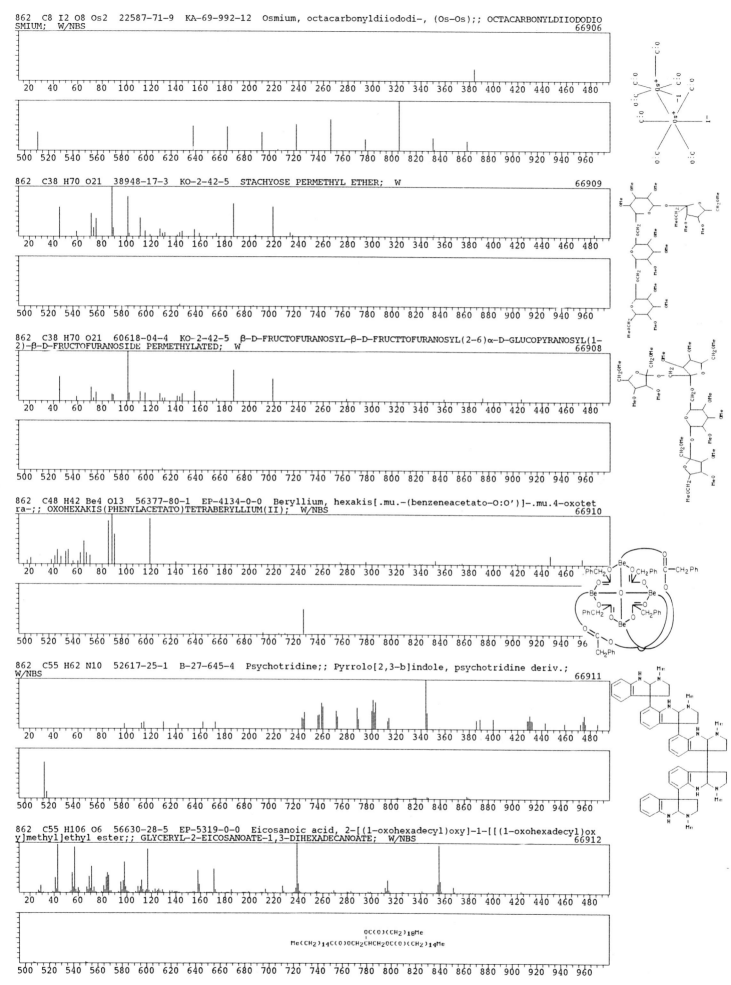

862   C8 I2 O8 Os2   22587-71-9   KA-69-992-12   Osmium, octacarbonyldiiododi-, (Os-Os);; OCTACARBONYLDIIODODIO
SMIUM; W/NBS   66906

862   C38 H70 O21   38948-17-3   KO-2-42-5   STACHYOSE PERMETHYL ETHER; W   66909

862   C38 H70 O21   60618-04-4   KO-2-42-5   β-D-FRUCTOFURANOSYL-β-D-FRUCTTOFURANOSYL(2-6)α-D-GLUCOPYRANOSYL(1-
2)-β-D-FRUCTOFURANOSIDE PERMETHYLATED; W   66908

862   C48 H42 Be4 O13   56377-80-1   EP-4134-0-0   Beryllium, hexakis[.mu.-(benzeneacetato-O:O')]-.mu.4-oxotet
ra-;; OXOHEXAKIS(PHENYLACETATO)TETRABERYLLIUM(II); W/NBS   66910

862   C55 H62 N10   52617-25-1   B-27-645-4   Psychotridine;; Pyrrolo[2,3-b]indole, psychotridine deriv.;
W/NBS   66911

862   C55 H106 O6   56630-28-5   EP-5319-0-0   Eicosanoic acid, 2-[(1-oxohexadecyl)oxy]-1-[[(1-oxohexadecyl)ox
y]methyl]ethyl ester;; GLYCERYL-2-EICOSANOATE-1,3-DIHEXADECANOATE; W/NBS   66912

862 C55 H106 O6 2190-24-1 EP-5320-0-0 Octadecanoic acid, 2-[(1-oxohexadecyl)oxy]-1,3-propanediyl ester
;; GLYCERYL-2-PALMITATE-1,3-DISTEARATE; W/NBS   66913

OC(O)(CH2)14Me
|
Me(CH2)16C(O)OCH2CHCH2OC(O)(CH2)16Me

862 C55 H106 O6 2177-99-3 EP-8566-0-0 Octadecanoic acid, 1-[[(1-oxohexadecyl)oxy]methyl]-1,2-ethanediyl
ester;; GLYCERYL-1-PALMITATE-2,3-DISTEARATE; W/NBS   79567

OC(O)(CH2)16Me
|
Me(CH2)16C(O)OCH2CHCH2OC(O)(CH2)14Me

862 C56 H94 O6 82534-70-1 F-39-2214-0 1,3(R): 4,6(R)-Di-O-benzylidene-2,5-di-O-(1-octadecyl)-D-mannitol
; W   128813

862 C59 H90 O4 303-98-0 T-67-5331-0 2,5-Cyclohexadiene-1,4-dione, 2-(3,7,11,15,19,23,27,31,35,39-deca
methyl-2,6,10,14,18,22,26,30,34,38-tetracontadecaenyl)-5,6-dimethoxy-3-methyl-, (all-E)-;; 2,3-DIMETHOXY-5-
METHYL-6-(3,7,11,15,19,23,27,31,35,39-DECAMETHYL-2,6,10,14,18,22,26,30,34,38-TETRACONTADECAENYL)-1,4-BENZOQ
UINONE; W/NBS   66915

864 C39 H41 N4 O7 Re 39452-93-2 C-97-3954-2 (MONOHYDROGENMESOPORPHRIN IX DIMETHYL ESTERATO)-TRICARBONYL
RHENIUM (I); W   66916

864 C41 H60 O16 Si2 37293-15-5 NS-0-0-0 Azadirachtin, O,O-bis(trimethylsilyl)-; W/NBS   128814

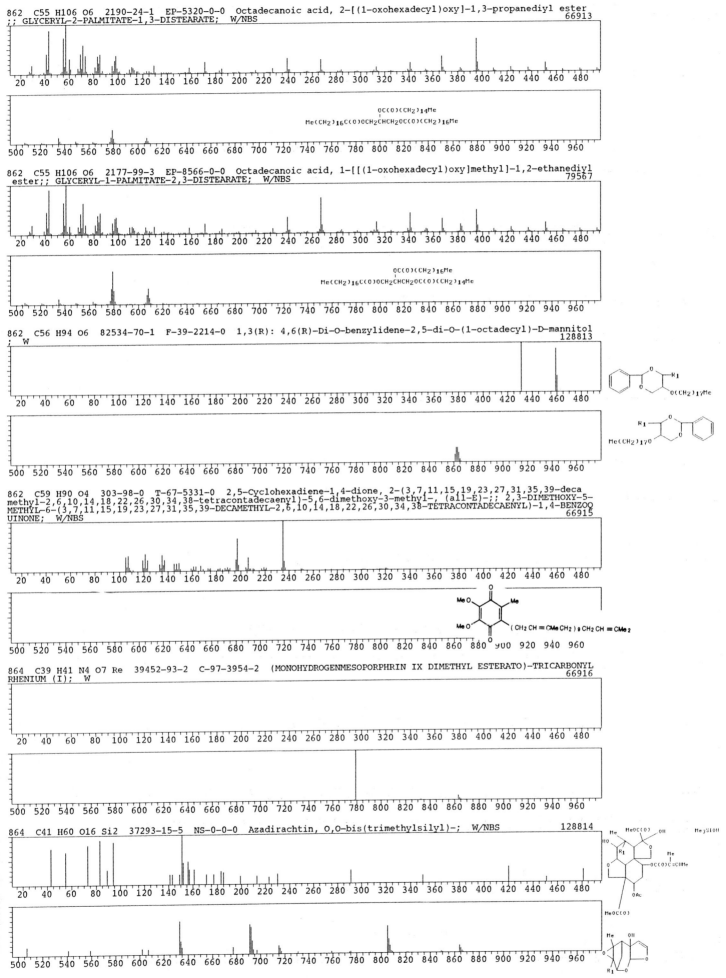

864   C43 H60 N8 O11   85640-89-7   H-65-2286-0   Viridogrisein II;   W   128815

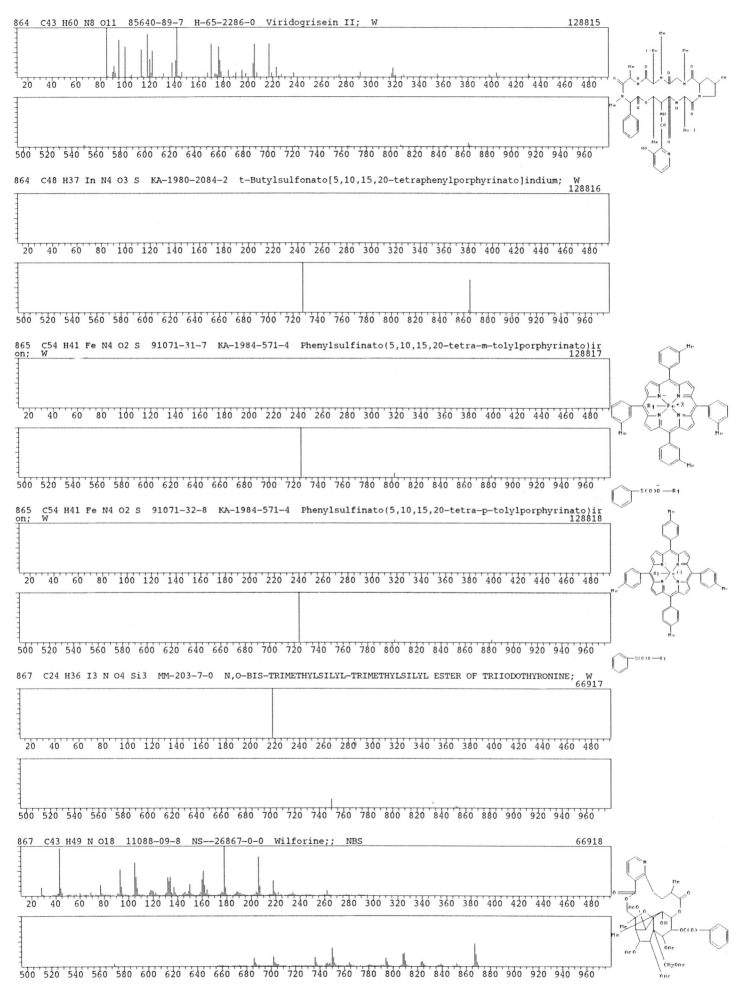

864   C48 H37 In N4 O3 S   KA-1980-2084-2   t-Butylsulfonato[5,10,15,20-tetraphenylporphyrinato]indium;   W
                                                                                                          128816

865   C54 H41 Fe N4 O2 S   91071-31-7   KA-1984-571-4   Phenylsulfinato(5,10,15,20-tetra-m-tolylporphyrinato)iron;   W
                                                                                                          128817

865   C54 H41 Fe N4 O2 S   91071-32-8   KA-1984-571-4   Phenylsulfinato(5,10,15,20-tetra-p-tolylporphyrinato)iron;   W
                                                                                                          128818

867   C24 H36 I3 N O4 Si3   MM-203-7-0   N,O-BIS-TRIMETHYLSILYL-TRIMETHYLSILYL ESTER OF TRIIODOTHYRONINE;   W
                                                                                                          66917

867   C43 H49 N O18   11088-09-8   NS--26867-0-0   Wilforine;;   NBS   66918

- 6239 -

867 C54 H81 N3 O6 IC-1330-0-0 N,N',N'',-Tris-(4-hydroxy-3,5-di-tert-butylphenylcarbonyl)triazine; W 128819

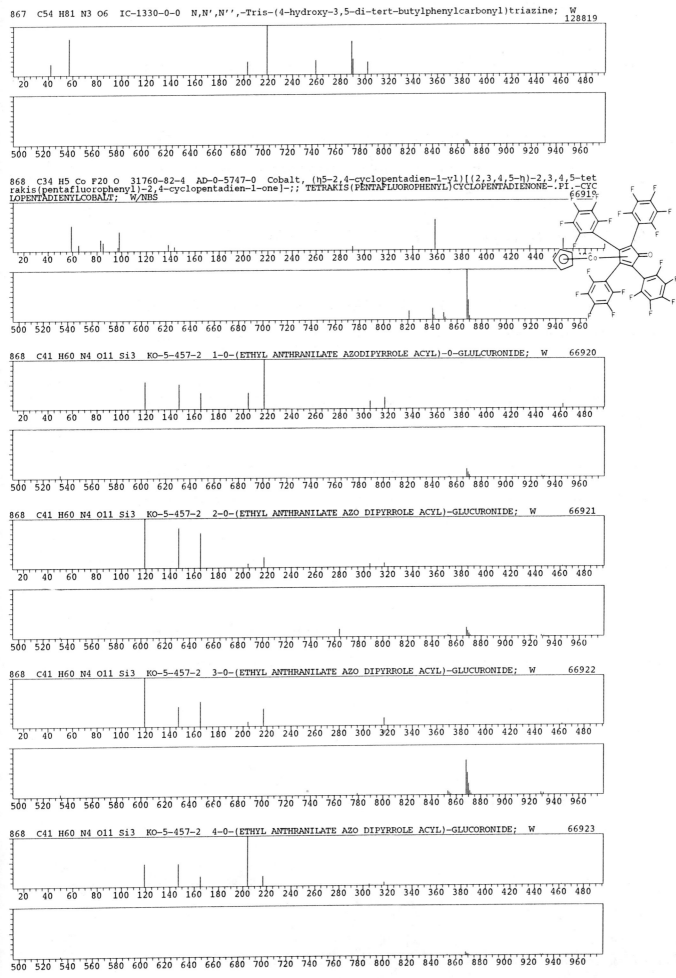

868 C34 H5 Co F20 O 31760-82-4 AD-0-5747-0 Cobalt, (η5-2,4-cyclopentadien-1-yl)[(2,3,4,5-η)-2,3,4,5-tetrakis(pentafluorophenyl)-2,4-cyclopentadien-1-one]-;; TETRAKIS(PENTAFLUOROPHENYL)CYCLOPENTADIENONE-.PI.-CYCLOPENTADIENYLCOBALT; W/NBS 6691°

868 C41 H60 N4 O11 Si3 KO-5-457-2 1-O-(ETHYL ANTHRANILATE AZODIPYRROLE ACYL)-0-GLULCURONIDE; W 66920

868 C41 H60 N4 O11 Si3 KO-5-457-2 2-O-(ETHYL ANTHRANILATE AZO DIPYRROLE ACYL)-GLUCURONIDE; W 66921

868 C41 H60 N4 O11 Si3 KO-5-457-2 3-O-(ETHYL ANTHRANILATE AZO DIPYRROLE ACYL)-GLUCURONIDE; W 66922

868 C41 H60 N4 O11 Si3 KO-5-457-2 4-O-(ETHYL ANTHRANILATE AZO DIPYRROLE ACYL)-GLUCORONIDE; W 66923

868   C46 H51 Br Cl P Sn   15652-38-7   PG-1982-1790-0   Decafentin;   W/NBS   66924

Str 1193

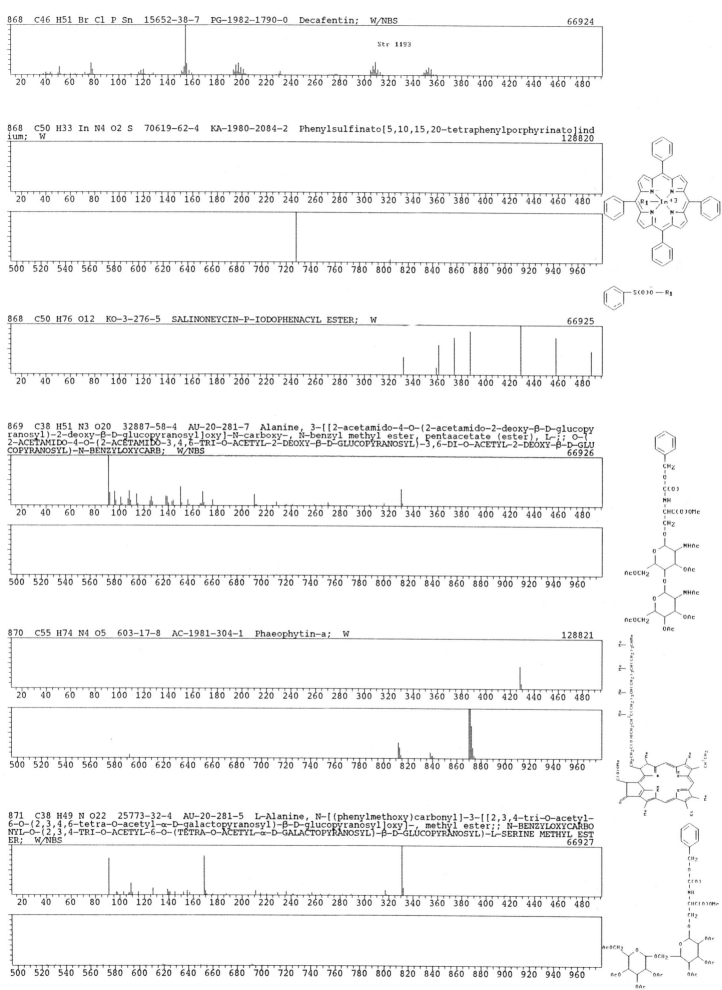

868   C50 H33 In N4 O2 S   70619-62-4   KA-1980-2084-2   Phenylsulfinato[5,10,15,20-tetraphenylporphyrinato]ind
ium;   W   128820

868   C50 H76 O12   KO-3-276-5   SALINONEYCIN-P-IODOPHENACYL ESTER;   W   66925

869   C38 H51 N3 O20   32887-58-4   AU-20-281-7   Alanine, 3-[[2-acetamido-4-O-(2-acetamido-2-deoxy-β-D-glucopy
ranosyl)-2-deoxy-β-D-glucopyranosyl]oxy]-N-carboxy-, N-benzyl methyl ester, pentaacetate (ester), L-;; O-(
2-ACETAMIDO-4-O-(2-ACETAMIDO-3,4,6-TRI-O-ACETYL-2-DEOXY-β-D-GLUCOPYRANOSYL)-3,6-DI-O-ACETYL-2-DEOXY-β-D-GLU
COPYRANOSYL)-N-BENZYLOXYCARB;   W/NBS   66926

870   C55 H74 N4 O5   603-17-8   AC-1981-304-1   Phaeophytin-a;   W   128821

871   C38 H49 N O22   25773-32-4   AU-20-281-5   L-Alanine, N-[(phenylmethoxy)carbonyl]-3-[[2,3,4-tri-O-acetyl-
6-O-(2,3,4,6-tetra-O-acetyl-α-D-galactopyranosyl)-β-D-glucopyranosyl]oxy]-, methyl ester;; N-BENZYLOXYCARBO
NYL-O-(2,3,4-TRI-O-ACETYL-6-O-(TETRA-O-ACETYL-α-D-GALACTOPYRANOSYL)-β-D-GLUCOPYRANOSYL)-β-D-GLUCOPYRANOSYL)-L-SERINE METHYL EST
ER;   W/NBS   66927

871 C38 H49 N O22 26034-15-1 AU-20-281-6 L-Serine, N-[(phenylmethoxy)carbonyl]-O-[2,3,6-tri-O-acetyl-4-
O-(2,3,4,6-tetra-O-acetyl-α-D-glucopyranosyl)-β-D-glucopyranosyl]-, methyl ester;; N-BENZYLOXYCARBONYL-O-(
2,3,6-TRI-O-ACETYL-4-O-(TETRA-O-ACETYL-α-D-GLUCOPYRANOSYL)-β-D-GLUCOPYRANOSYL)-L-SERINE METHYL ESTER;
W/NBS   66928

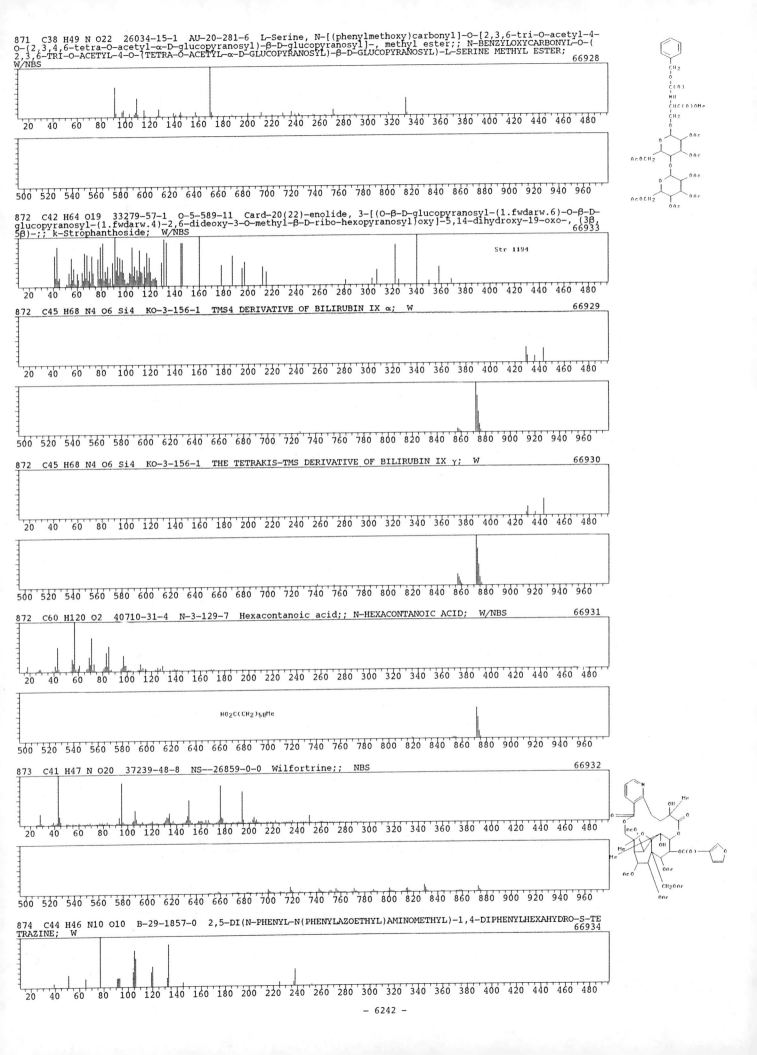

872 C42 H64 O19 33279-57-1 O-5-589-11 Card-20(22)-enolide, 3-[(O-β-D-glucopyranosyl-(1.fwdarw.6)-O-β-D-
glucopyranosyl-(1.fwdarw.4)-2,6-dideoxy-3-O-methyl-β-D-ribo-hexopyranosyl)oxy]-5,14-dihydroxy-19-oxo-, (3β,
5β)-;; k-Strophanthoside; W/NBS   66933

Str 1194

872 C45 H68 N4 O6 Si4 KO-3-156-1 TMS4 DERIVATIVE OF BILIRUBIN IX α; W   66929

872 C45 H68 N4 O6 Si4 KO-3-156-1 THE TETRAKIS-TMS DERIVATIVE OF BILIRUBIN IX γ; W   66930

872 C60 H120 O2 40710-31-4 N-3-129-7 Hexacontanoic acid;; N-HEXACONTANOIC ACID; W/NBS   66931

HO2C(CH2)58Me

873 C41 H47 N O20 37239-48-8 NS--26859-0-0 Wilfortrine;; NBS   66932

874 C44 H46 N10 O10 B-29-1857-0 2,5-DI(N-PHENYL-N(PHENYLAZOETHYL)AMINOMETHYL)-1,4-DIPHENYLHEXAHYDRO-S-TE
TRAZINE; W   66934

874   C47 H70 O15   64285-80-9   B-30-1320-0   BARRINGTOGENOL C-28-O-LYXOSIDE HEXAACETATE;   W   66935

Str 1195

874   C55 H58 N2 O8   73111-41-8   H-62-2014-0   N,N'-Methylendi-[(1L)-(1,2,4/3,)-1',2'-anhydro-3,4,5-tri-O-ben
zyl-2-hydroxyamino-11-hydroxymethyl-3,4,5-cyclopentantriol;   W   128822

876   C24 F20 Pb   1111-02-0   AD-0-2723-0   Plumbane, tetrakis(pentafluorophenyl)-;;   TETRAKIS(PENTAFLUOROPHEN
YL)PLUMBANE;   W/NBS   66936

876   C45 H72 N4 O6 Si4   KO-3-156-1   THE TETRAKIS-TMS DERIVATIVE OF BILIRUBIN MESO-IX Δ;   W   66937

876   C48 H44 O16   65615-77-2   KC-1977-18-2004   Methyl-2,6-di-O-benzoyl-4-O-(2,3,6-tri-O-benzoyl-β-D-galacto
pyranosyl)-β-D-glucopyranoside;   W   128823

876   C52 H52 O2 Si6   65645-13-8   AH-108-1420-0   1,2-DIMETHOXY-1,2-DIMETHYL-3,3,4,4,5,5,6,6-OCTAPHENYLCYCLOH
EXASILANE;   W   66938

877   C44 H44 Br N7 O8   73228-71-4   F-35-1327-0   CAGE PRODUCT 24B;   W   66939

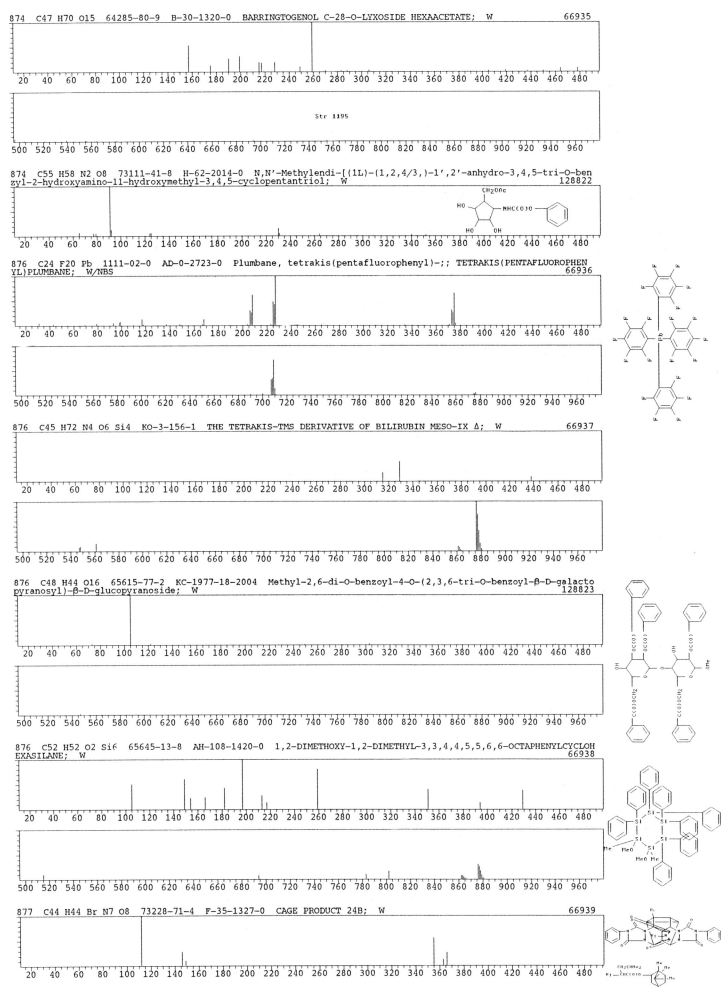

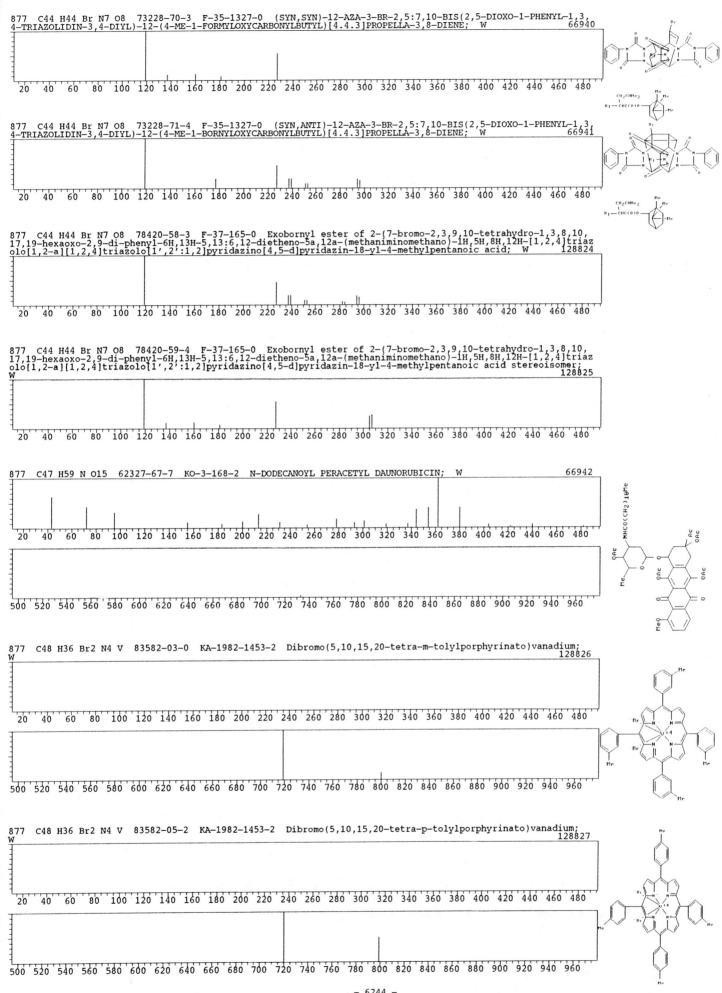

877   C44 H44 Br N7 O8   73228-70-3   F-35-1327-0   (SYN,SYN)-12-AZA-3-BR-2,5:7,10-BIS(2,5-DIOXO-1-PHENYL-1,3,
4-TRIAZOLIDIN-3,4-DIYL)-12-(4-ME-1-FORMYLOXYCARBONYLBUTYL)[4.4.3]PROPELLA-3,8-DIENE;   W   66940

877   C44 H44 Br N7 O8   73228-71-4   F-35-1327-0   (SYN,ANTI)-12-AZA-3-BR-2,5:7,10-BIS(2,5-DIOXO-1-PHENYL-1,3,
4-TRIAZOLIDIN-3,4-DIYL)-12-(4-ME-1-BORNYLOXYCARBONYLBUTYL)[4.4.3]PROPELLA-3,8-DIENE;   W   66941

877   C44 H44 Br N7 O8   78420-58-3   F-37-165-0   Exobornyl ester of 2-(7-bromo-2,3,9,10-tetrahydro-1,3,8,10,
17,19-hexaoxo-2,9-di-phenyl-6H,13H-5,13:6,12-dietheno-5a,12a-(methaniminomethano)-1H,5H,8H,12H-[1,2,4]triaz
olo[1,2-a][1,2,4]triazolo[1',2':1,2]pyridazino[4,5-d]pyridazin-18-yl-4-methylpentanoic acid;   W   128824

877   C44 H44 Br N7 O8   78420-59-4   F-37-165-0   Exobornyl ester of 2-(7-bromo-2,3,9,10-tetrahydro-1,3,8,10,
17,19-hexaoxo-2,9-di-phenyl-6H,13H-5,13:6,12-dietheno-5a,12a-(methaniminomethano)-1H,5H,8H,12H-[1,2,4]triaz
olo[1,2-a][1,2,4]triazolo[1',2':1,2]pyridazino[4,5-d]pyridazin-18-yl-4-methylpentanoic acid stereoisomer;
W   128825

877   C47 H59 N O15   62327-67-7   KO-3-168-2   N-DODECANOYL PERACETYL DAUNORUBICIN;   W   66942

877   C48 H36 Br2 N4 V   83582-03-0   KA-1982-1453-2   Dibromo(5,10,15,20-tetra-m-tolylporphyrinato)vanadium;
W   128826

877   C48 H36 Br2 N4 V   83582-05-2   KA-1982-1453-2   Dibromo(5,10,15,20-tetra-p-tolylporphyrinato)vanadium;
W   128827

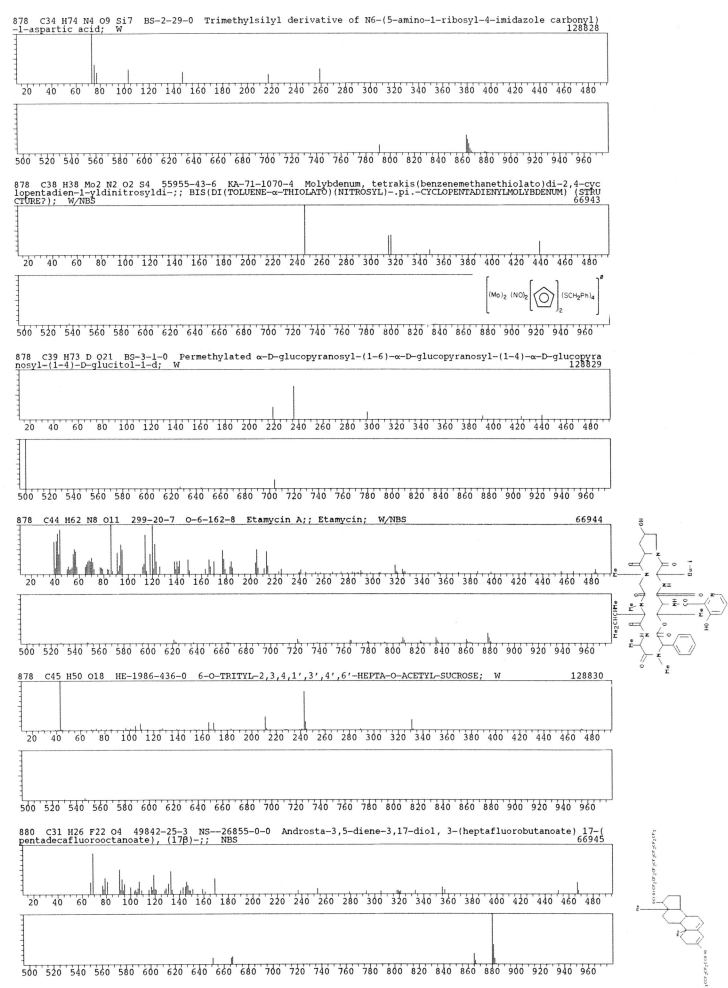

878   C34 H74 N4 O9 Si7   BS-2-29-0   Trimethylsilyl derivative of N6-(5-amino-1-ribosyl-4-imidazole carbonyl)
-l-aspartic acid;   W   128828

878   C38 H38 Mo2 N2 O2 S4   55955-43-6   KA-71-1070-4   Molybdenum, tetrakis(benzenemethanethiolato)di-2,4-cyc
lopentadien-1-yldinitrosyldi-;;   BIS(DI(TOLUENE-α-THIOLATO)(NITROSYL)-.pi.-CYCLOPENTADIENYLMOLYBDENUM) (STRU
CTURE?);   W/NBS   66943

$[(Mo)_2\ (NO)_2\ [\bigcirc]_2\ (SCH_2Ph)_4]^\delta$

878   C39 H73 D O21   BS-3-1-0   Permethylated α-D-glucopyranosyl-(1-6)-α-D-glucopyranosyl-(1-4)-α-D-glucopyra
nosyl-(1-4)-D-glucitol-1-d;   W   128829

878   C44 H62 N8 O11   299-20-7   O-6-162-8   Etamycin A;;   Etamycin;   W/NBS   66944

878   C45 H50 O18   HE-1986-436-0   6-O-TRITYL-2,3,4,1',3',4',6'-HEPTA-O-ACETYL-SUCROSE;   W   128830

880   C31 H26 F22 O4   49842-25-3   NS--26855-0-0   Androsta-3,5-diene-3,17-diol, 3-(heptafluorobutanoate) 17-(
pentadecafluorooctanoate), (17β)-;;   NBS   66945

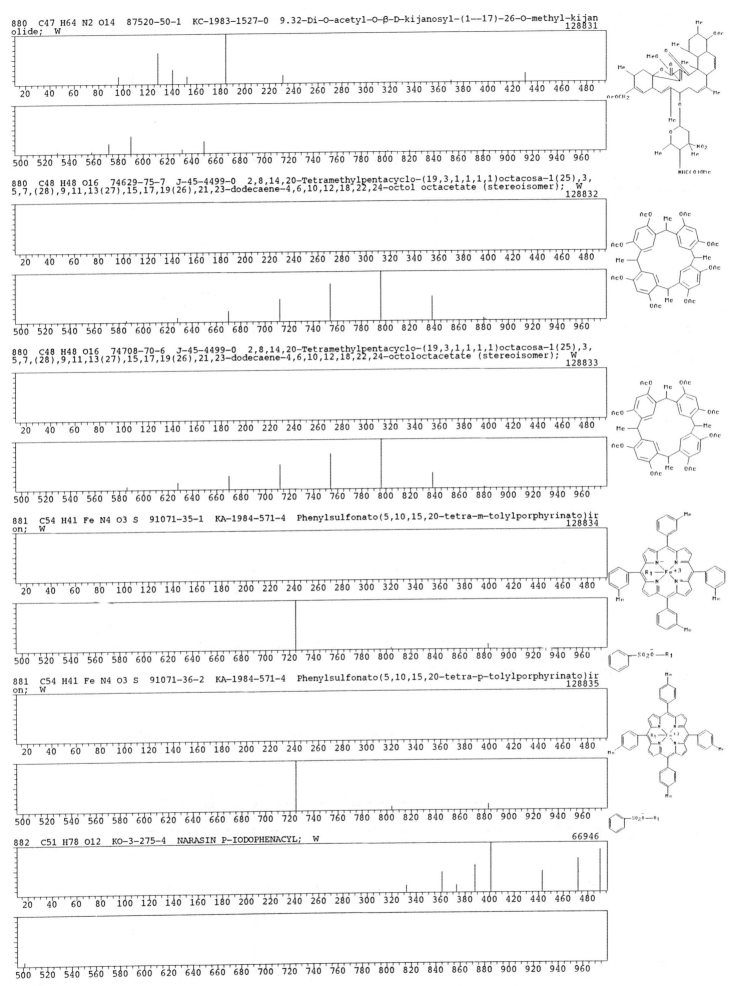

880   C47 H64 N2 O14   87520-50-1   KC-1983-1527-0   9.32-Di-O-acetyl-O-β-D-kijanosyl-(1--17)-26-O-methyl-kijan
olide; W
                                                                                    128831

880   C48 H48 O16   74629-75-7   J-45-4499-0   2,8,14,20-Tetramethylpentacyclo-(19,3,1,1,1,1)octacosa-1(25),3,
5,7,(28),9,11,13(27),15,17,19(26),21,23-dodecaene-4,6,10,12,18,22,24-octol octacetate (stereoisomer); W
                                                                                    128832

880   C48 H48 O16   74708-70-6   J-45-4499-0   2,8,14,20-Tetramethylpentacyclo-(19,3,1,1,1,1)octacosa-1(25),3,
5,7,(28),9,11,13(27),15,17,19(26),21,23-dodecaene-4,6,10,12,18,22,24-octoloctacetate (stereoisomer); W
                                                                                    128833

881   C54 H41 Fe N4 O3 S   91071-35-1   KA-1984-571-4   Phenylsulfonato(5,10,15,20-tetra-m-tolylporphyrinato)ir
on; W
                                                                                    128834

881   C54 H41 Fe N4 O3 S   91071-36-2   KA-1984-571-4   Phenylsulfonato(5,10,15,20-tetra-p-tolylporphyrinato)ir
on; W
                                                                                    128835

882   C51 H78 O12   KO-3-275-4   NARASIN P-IODOPHENACYL; W
                                                                                    66946

883   C43 H49 N O19   37239-51-3   NS--26854-0-0   Evonimine, 8-(acetyloxy)-O2-benzoyl-O2-deacetyl-8-deoxo-26-
hydroxy-, (8α)-;;   NBS
66947

884   C46 H52 N4 O14   14670-22-5   KC-1976-1013-0   HEPTA METHYL PHYRIAPORPHYRIN-III HEPTACARBOXYLIC ACID;   W
66948

884   C50 H33 In N4 O3 S   KA-1980-2084-2   Phenylsulfonato[5,10,15,20-tetraphenylporphyrinato]indium;   W
128836

884   C55 H72 N4 O6   3147-18-0   AC-1981-304-1   Phaeophytin-b;   W
128837

Str 1196

886   C31 H67 N4 O12 P Si6   BS-2-30-0   Trimethylsilyl derivative of N-(5-amino-1-ribosyl-4-imidazole carbony
l)-1-aspartic acid 5'-phosphate;   W
128838

886   C39 H50 O23   SB-33-123-3   APIGERUN-7-O-NEOHESPERIDOSIDE-4'-O-SOPHOROSIDE;   W
66949

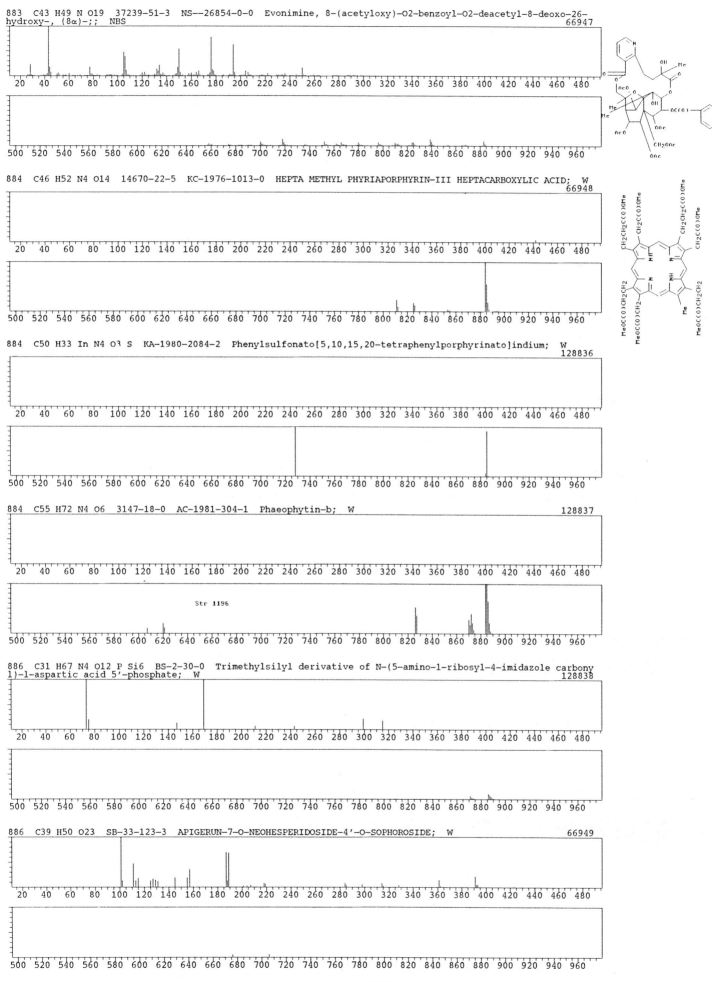

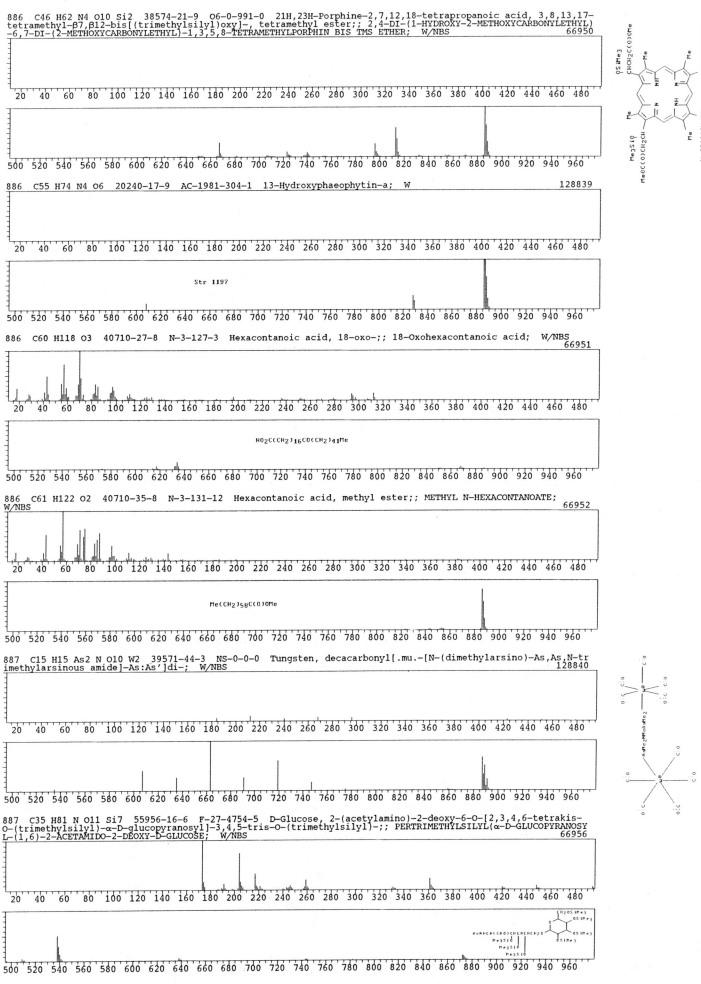

886   C46 H62 N4 O10 Si2   38574-21-9   O6-0-991-0   21H,23H-Porphine-2,7,12,18-tetrapropanoic acid, 3,8,13,17-
tetramethyl-β7,β12-bis[(trimethylsilyl)oxy]-, tetramethyl ester;; 2,4-DI-(1-HYDROXY-2-METHOXYCARBONYLETHYL)
-6,7-DI-(2-METHOXYCARBONYLETHYL)-1,3,5,8-TETRAMETHYLPORPHIN BIS TMS ETHER; W/NBS   66950

886   C55 H74 N4 O6   20240-17-9   AC-1981-304-1   13-Hydroxyphaeophytin-a; W   128839

Str 1197

886   C60 H118 O3   40710-27-8   N-3-127-3   Hexacontanoic acid, 18-oxo-;; 18-Oxohexacontanoic acid; W/NBS
                                                                                                        66951

HO2C(CH2)16CO(CH2)41Me

886   C61 H122 O2   40710-35-8   N-3-131-12   Hexacontanoic acid, methyl ester;; METHYL N-HEXACONTANOATE;
W/NBS   66952

Me(CH2)58C(O)OMe

887   C15 H15 As2 N O10 W2   39571-44-3   NS-0-0-0   Tungsten, decacarbonyl[.mu.-[N-(dimethylarsino)-As,As,N-tr
imethylarsinous amide]-As:As']di-; W/NBS   128840

887   C35 H81 N O11 Si7   55956-16-6   F-27-4754-5   D-Glucose, 2-(acetylamino)-2-deoxy-6-O-[2,3,4,6-tetrakis-
O-(trimethylsilyl)-α-D-glucopyranosyl]-3,4,5-tris-O-(trimethylsilyl)-;; PERTRIMETHYLSILYL(α-D-GLUCOPYRANOSY
L-(1,6)-2-ACETAMIDO-2-DEOXY-D-GLUCOSE; W/NBS   66956

887 C35 H81 N O11 Si7 55956-13-3 F-27-4253-3 D-Galactopyranose, 4-O-[2-(acetylamino)-2-deoxy-3,4,6-tris
-O-(trimethylsilyl)-β-D-galactopyranosyl]-1,2,3,6-tetrakis-O-(trimethylsilyl)-;; PERTRIMETHYLSILYL(β-D-2-AC
ETAMIDO-2-DEOXYGALACTOPYRANOSYL-(1,4)-D-GALACTOPYRANOSE); W/NBS   66953

887 C35 H81 N O11 Si7 55956-15-5 F-27-4752-2 D-Galactose, 3-O-[2-(acetylamino)-2-deoxy-3,4,6-tris-O-(tr
imethylsilyl)-β-D-galactopyranosyl]-2,4,5,6-tetrakis-O-(trimethylsilyl)-;; PERTRIMETHYLSILYL(β-D-2-ACETAMID
O-2-DEOXYGALACTOPYRANOSYL-(1,3)-D-GALACTOSE); W/NBS   66954

887 C35 H81 N O11 Si7 55956-14-4 F-27-4253-4 D-Galactose, 6-O-[2-(acetylamino)-2-deoxy-3,4,6-tris-O-(tr
imethylsilyl)-β-D-galactopyranosyl]-2,3,4,5-tetrakis-O-(trimethylsilyl)-;; PERTRIMETHYLSILYL(β-D-2-ACETAMID
O-2-DEOXYGALACTOPYRANOSYL-(1,6)-D-GALACTOSE); W/NBS   66955

887 C35 H81 N O11 Si7 55956-12-2 F-27-4752-1 D-Mannose, 2-O-[2-(acetylamino)-2-deoxy-3,4,6-tris-O-(tri
methylsilyl)-β-D-glucopyranosyl]-3,4,5,6-tetrakis-O-(trimethylsilyl)-;; PERTRIMETHYLSILYL(β-D-2-ACETAMIDO-
2-DEOXY-GLUCOPYRANOSYL-(1,2)-D-MANNOSE); W/NBS   66957

888 C24 H72 O12 Si12 JC-48-519-4 SILICATE ANION TETRAMER; W   66958

888 C24 H72 O12 Si12 18919-94-3 RB-1982-15790-0 TETRACOSAMETHYLCYCLODODECASILOXANE; W   128841
Str 1198

890 C45 H55 Br N4 O10 83048-32-2 KC-1982-1260-0 Bromobisoxindole derivative of gardimultine; W 128842
Str 1199

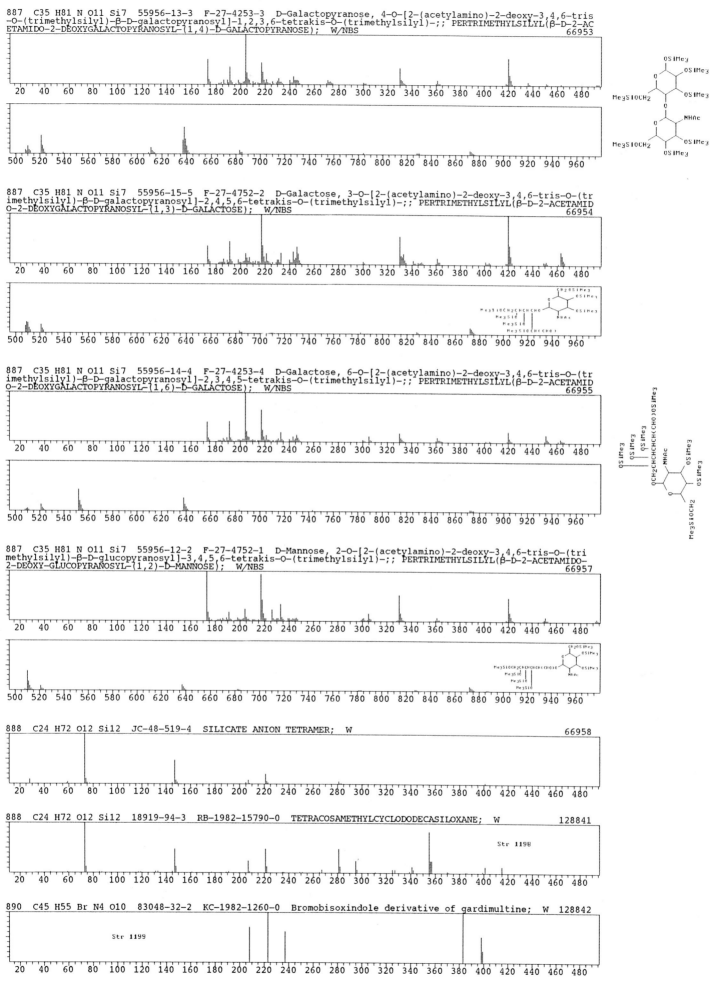

- 6249 -

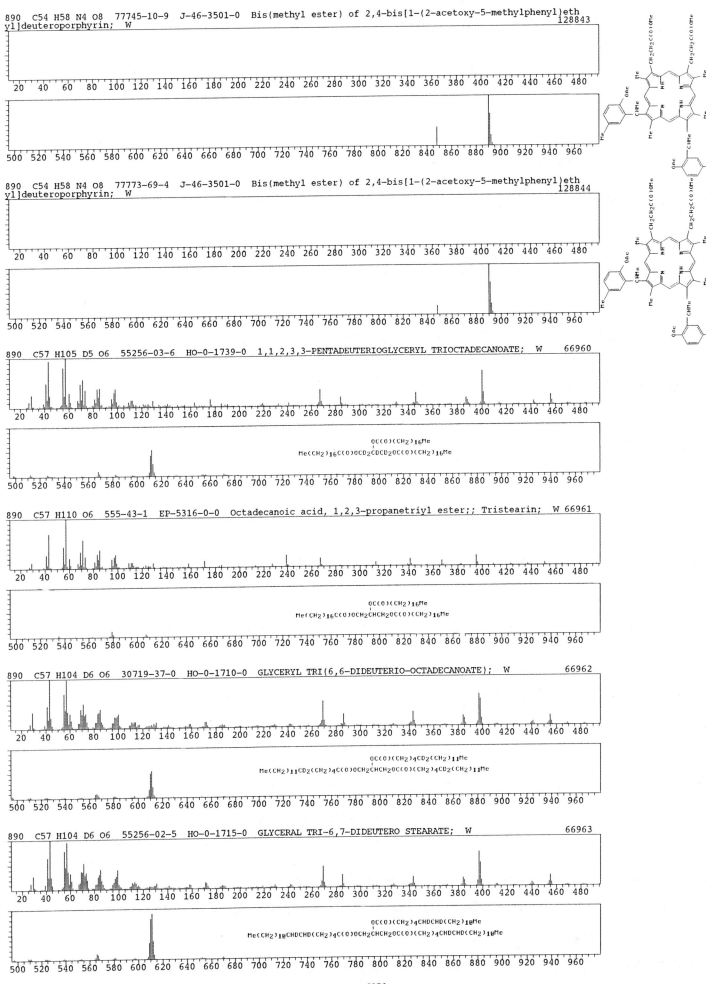

890 C54 H58 N4 O8 77745-10-9 J-46-3501-0 Bis(methyl ester) of 2,4-bis[1-(2-acetoxy-5-methylphenyl)ethyl]deuteroporphyrin; W
128843

890 C54 H58 N4 O8 77773-69-4 J-46-3501-0 Bis(methyl ester) of 2,4-bis[1-(2-acetoxy-5-methylphenyl)ethyl]deuteroporphyrin; W
128844

890 C57 H105 D5 O6 55256-03-6 HO-0-1739-0 1,1,2,3,3-PENTADEUTERIOGLYCERYL TRIOCTADECANOATE; W 66960

OC(O)(CH2)16Me
Me(CH2)16C(O)OCD2CDCD2OC(O)(CH2)16Me

890 C57 H110 O6 555-43-1 EP-5316-0-0 Octadecanoic acid, 1,2,3-propanetriyl ester;; Tristearin; W 66961

OC(O)(CH2)16Me
Me(CH2)16C(O)OCH2CHCH2OC(O)(CH2)16Me

890 C57 H104 D6 O6 30719-37-0 HO-0-1710-0 GLYCERYL TRI(6,6-DIDEUTERIO-OCTADECANOATE); W 66962

OC(O)(CH2)4CD2(CH2)11Me
Me(CH2)11CD2(CH2)4C(O)OCH2CHCH2OC(O)(CH2)4CD2(CH2)11Me

890 C57 H104 D6 O6 55256-02-5 HO-0-1715-0 GLYCERAL TRI-6,7-DIDEUTERO STEARATE; W 66963

OC(O)(CH2)4CHDCHD(CH2)10Me
Me(CH2)10CHDCHD(CH2)4C(O)OCH2CHCH2OC(O)(CH2)4CHDCHD(CH2)10Me

890   C57 H109 D O6   30719-38-1   HO-0-1638-0   2-DEUTERIOGLYCERYL TRIOCTADECANOATE;   W   66964

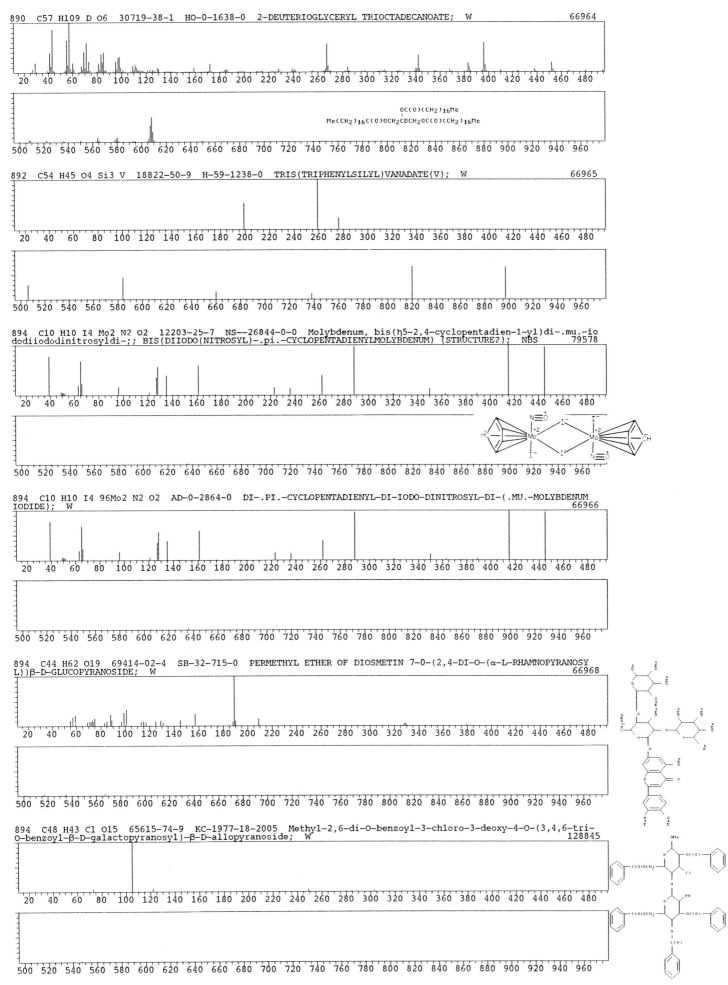

```
                                    OC(O)(CH2)16Me
                                    |
                      Me(CH2)16C(O)OCH2CDCH2OC(O)(CH2)16Me
```

892   C54 H45 O4 Si3 V   18822-50-9   H-59-1238-0   TRIS(TRIPHENYLSILYL)VANADATE(V);   W   66965

894   C10 H10 I4 Mo2 N2 O2   12203-25-7   NS--26844-0-0   Molybdenum, bis(η5-2,4-cyclopentadien-1-yl)di-.mu.-io
dodiiododinitrosyldi-;;   BIS(DIIODO(NITROSYL)-.pi.-CYCLOPENTADIENYLMOLYBDENUM) (STRUCTURE?);   NBS   79578

894   C10 H10 I4 96Mo2 N2 O2   AD-0-2864-0   DI-.PI.-CYCLOPENTADIENYL-DI-IODO-DINITROSYL-DI-(.MU.-MOLYBDENUM
IODIDE);   W   66966

894   C44 H62 O19   69414-02-4   SB-32-715-0   PERMETHYL ETHER OF DIOSMETIN 7-O-(2,4-DI-O-(α-L-RHAMNOPYRANOSY
L))β-D-GLUCOPYRANOSIDE;   W   66968

894   C48 H43 Cl O15   65615-74-9   KC-1977-18-2005   Methyl-2,6-di-O-benzoyl-3-chloro-3-deoxy-4-O-(3,4,6-tri-
O-benzoyl-β-D-galactopyranosyl)-β-D-allopyranoside;   W   128845

894 C57 H58 N4 O6 59997-36-3 KC-1976-800-0 5,10,15-TRISBENZOYLOXY-2,3,7,8,12,13,17,18-OCTAETHYLPORPHYR
IN; W   66969

20 40 60 80 100 120 140 160 180 200 220 240 260 280 300 320 340 360 380 400 420 440 460 480

500 520 540 560 580 600 620 640 660 680 700 720 740 760 780 800 820 840 860 880 900 920 940 960

894 C59 H94 N2 O4 72469-42-2 H-62-2077-0 N-Benzyl-N-phenyl-N',N'-dioctadecyl-1,2-phenylendioxydiacetami
de; W   128846

20 40 60 80 100 120 140 160 180 200 220 240 260 280 300 320 340 360 380 400 420 440 460 480

500 520 540 560 580 600 620 640 660 680 700 720 740 760 780 800 820 840 860 880 900 920 940 960

895 C32 H66 N5 O11 P Si6 60690-44-0 KO-2-92-3 N*6-(2-SUCCINYL)ADENOSINE 5'-PHOSPHATE; W   66970

20 40 60 80 100 120 140 160 180 200 220 240 260 280 300 320 340 360 380 400 420 440 460 480

500 520 540 560 580 600 620 640 660 680 700 720 740 760 780 800 820 840 860 880 900 920 940 960

895 C42 H24 Cr F9 O3 S3 61024-12-2 B-29-1915-2 TRI(1,1,1-TRIFLUORO-4-(2'-NAPHTHYL)-2-OXO-BUT-3-EN-4-YL
THIO)CHROMIUM(III); W   66971

20 40 60 80 100 120 140 160 180 200 220 240 260 280 300 320 340 360 380 400 420 440 460 480

500 520 540 560 580 600 620 640 660 680 700 720 740 760 780 800 820 840 860 880 900 920 940 960

898 C31 H17 F12 N2 O3 Tl 67840-76-0 B-31-1722-0 4,4'-BIPYRIDYLTRIS(2,3,5,6-TETRAFLUORO-4-METHOXYPHENYL)
THALLIUM(III); W   66972

20 40 60 80 100 120 140 160 180 200 220 240 260 280 300 320 340 360 380 400 420 440 460 480

500 520 540 560 580 600 620 640 660 680 700 720 740 760 780 800 820 840 860 880 900 920 940 960

898 C35 H74 N4 O13 Si5 56051-47-9 AE-245-4160-1 L-Asparagine, N-[2-(acetylamino)-4-O-[2-(acetylamino)-
2-deoxy-3,4,6-tris-O-(trimethylsilyl)-β-D-glucopyranosyl]-2-deoxy-3,6-bis-O-(trimethylsilyl)-β-D-glucopyran
osyl]-;; PER(TRIMETHYLSILYL)-1-N-(4-L-ASPARTYL)-2-ACETAMIDO-4-O-(2-ACETAMIDO-2-DEOXY-β-D-GLUCOPYRANOSYL)-2-
DEOXY-β-D-GLUCOPYRANOSYLAMINE; W/NBS   66973

20 40 60 80 100 120 140 160 180 200 220 240 260 280 300 320 340 360 380 400 420 440 460 480

500 520 540 560 580 600 620 640 660 680 700 720 740 760 780 800 820 840 860 880 900 920 940 960

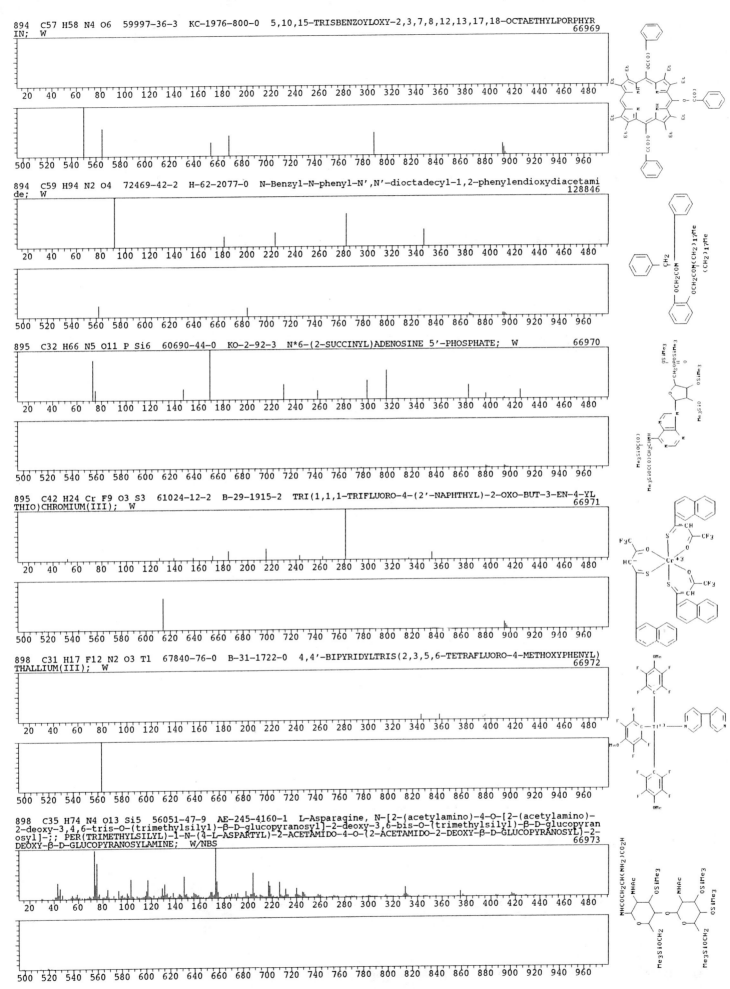

900 C12 H3 O12 Re3 12146-47-3 AD-0-1543-0 Rhenium, dodecacarbonyltrihydrotri-;; HYDRIDORHENIUM TETRACAR
BONYL TRIMER; W/NBS   66974

$H_3Re_3(CO)_{12}$

900 C46 H80 O15 Si 85026-78-4 C-105-2434-0 t-Butyldimethylsilyl, polymethyl ether derivative of tylosin
; W   128847

900 C55 H72 N4 O7 79684-24-5 AC-1981-304-1 13-Hydroxyphaeophytin-b; W   128848

901 C43 H51 N O20 73364-66-6 I-59-2939-0 Nona-acetate derivative of 4-[(1-deoxylcellobit-1-yl)-amino]be
nzophenone; W   128849

902 C44 H56 I2 O4 84389-92-4 KC-1982-2321-0 2',2''-Di-Iodo-4,4'',4''',6'tetramethoxy-3,5,3''',5'''-tet
ra-t-butyl-m-quaterphenyl; W   128850

902 C70 H62 TK-175-2445-1 POLY(1,4-PHENYLENMETHYLENE); W   66976

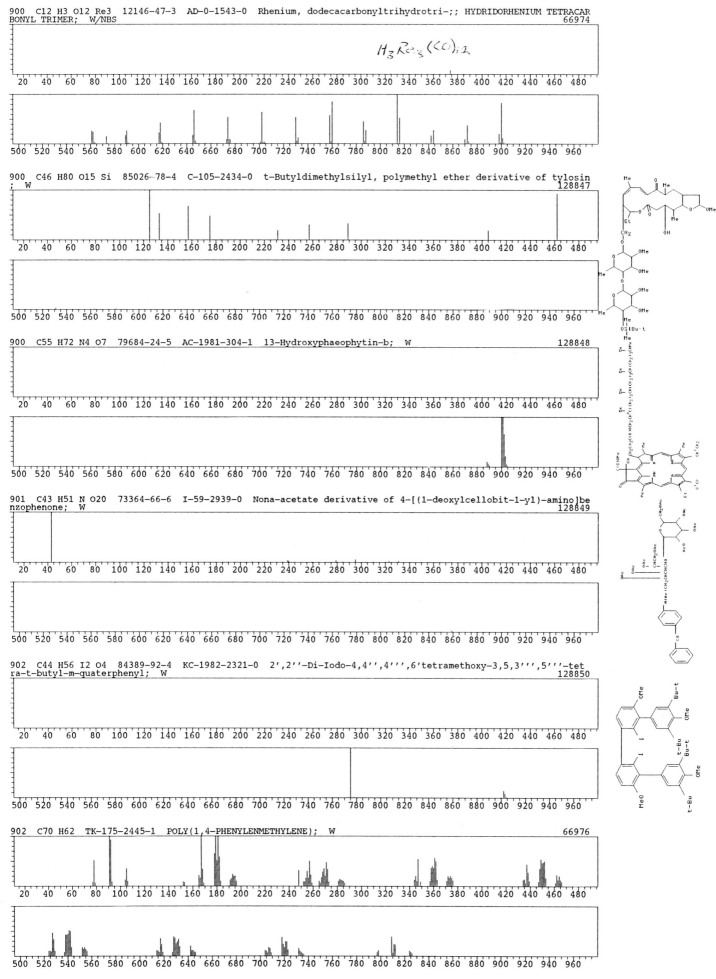

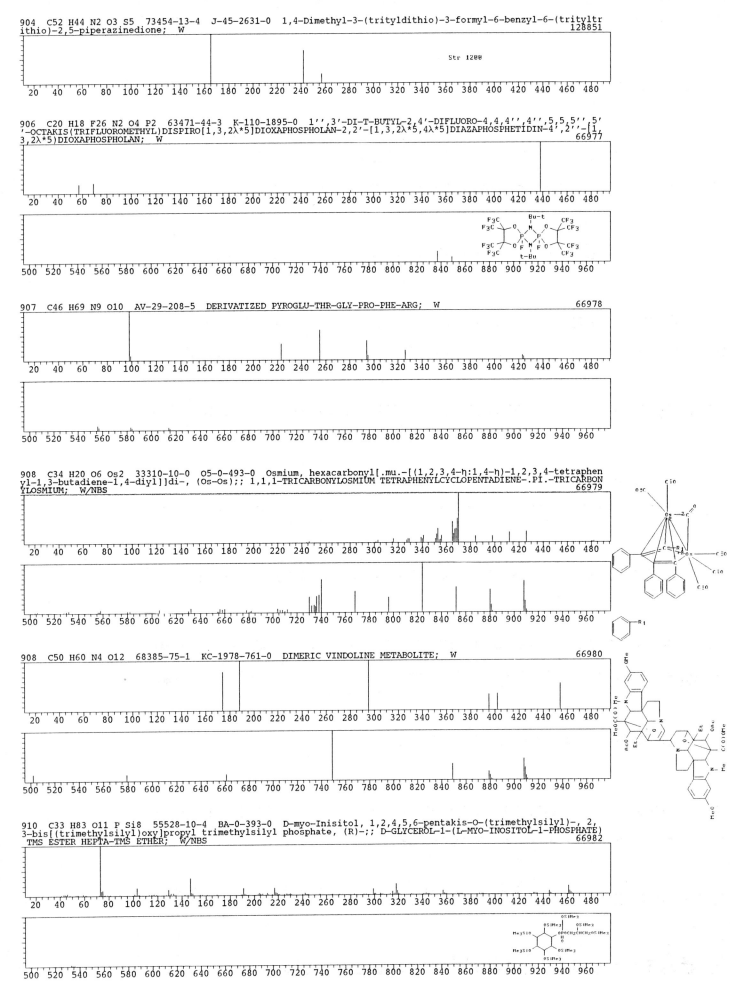

904   C52 H44 N2 O3 S5   73454-13-4   J-45-2631-0   1,4-Dimethyl-3-(trityldithio)-3-formyl-6-benzyl-6-(trityltr ithio)-2,5-piperazinedione;   W   128851

Str 1200

906   C20 H18 F26 N2 O4 P2   63471-44-3   K-110-1895-0   1'',3'-DI-T-BUTYL-2,4'-DIFLUORO-4,4,4'',4'',5,5,5'',5' '-OCTAKIS(TRIFLUOROMETHYL)DISPIRO[1,3,2λ*5]DIOXAPHOSPHOLAN-2,2'-[1,3,2λ*5,4λ*5]DIAZAPHOSPHETIDIN-4',2''-[1, 3,2λ*5)DIOXAPHOSPHOLAN;   W   66977

907   C46 H69 N9 O10   AV-29-208-5   DERIVATIZED PYROGLU-THR-GLY-PRO-PHE-ARG;   W   66978

908   C34 H20 O6 Os2   33310-10-0   O5-0-493-0   Osmium, hexacarbonyl[.mu.-[(1,2,3,4-η:1,4-η)-1,2,3,4-tetraphen yl-1,3-butadiene-1,4-diyl]]di-, (Os-Os);; 1,1,1-TRICARBONYLOSMIUM TETRAPHENYLCYCLOPENTADIENE-.PI.-TRICARBON YLOSMIUM; W/NBS   66979

908   C50 H60 N4 O12   68385-75-1   KC-1978-761-0   DIMERIC VINDOLINE METABOLITE;   W   66980

910   C33 H83 O11 P Si8   55528-10-4   BA-0-393-0   D-myo-Inisitol, 1,2,4,5,6-pentakis-O-(trimethylsilyl)-, 2, 3-bis[(trimethylsilyl)oxy]propyl trimethylsilyl phosphate, (R)-;; D-GLYCEROL-1-(L-MYO-INOSITOL-1-PHOSPHATE) TMS ESTER HEPTA-TMS ETHER;   W/NBS   66982

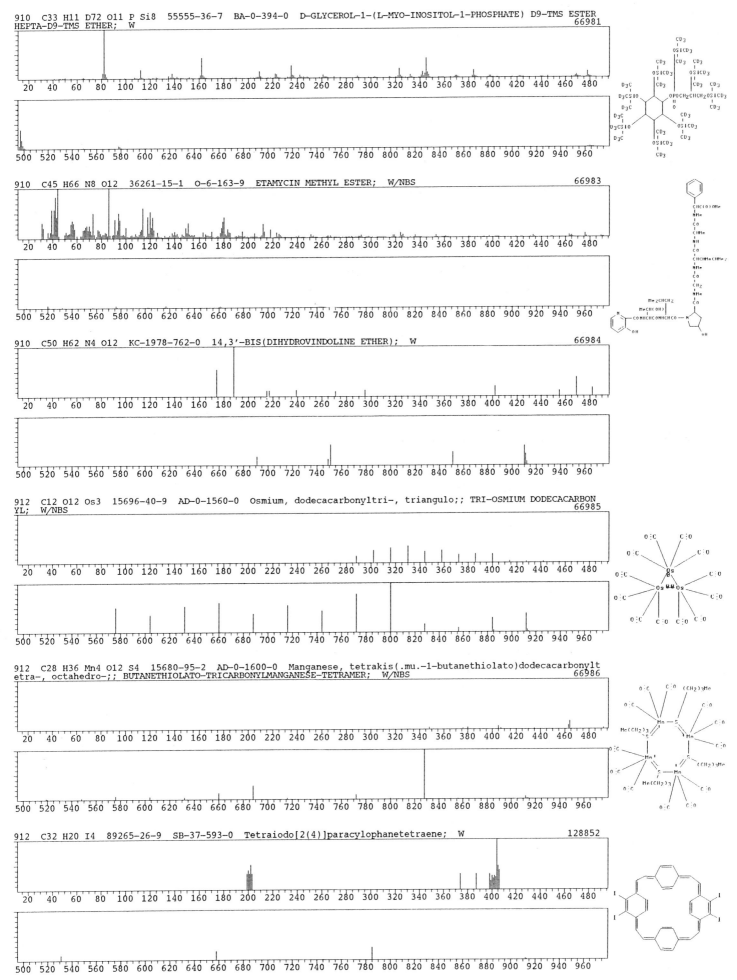

910 C33 H11 D72 O11 P Si8 55555-36-7 BA-0-394-0 D-GLYCEROL-1-(L-MYO-INOSITOL-1-PHOSPHATE) D9-TMS ESTER HEPTA-D9-TMS ETHER; W
66981

910 C45 H66 N8 O12 36261-15-1 O-6-163-9 ETAMYCIN METHYL ESTER; W/NBS
66983

910 C50 H62 N4 O12 KC-1978-762-0 14,3'-BIS(DIHYDROVINDOLINE ETHER); W
66984

912 C12 O12 Os3 15696-40-9 AD-0-1560-0 Osmium, dodecacarbonyltri-, triangulo;; TRI-OSMIUM DODECACARBONYL; W/NBS
66985

912 C28 H36 Mn4 O12 S4 15680-95-2 AD-0-1600-0 Manganese, tetrakis(.mu.-1-butanethiolato)dodecacarbonyltetra-, octahedro-;; BUTANETHIOLATO-TRICARBONYLMANGANESE-TETRAMER; W/NBS
66986

912 C32 H20 I4 89265-26-9 SB-37-593-0 Tetraiodo[2(4)]paracylophanetetraene; W
128852

912   C34 H5 F20 O Rh   31760-84-6   AD-0-5749-0   Rhodium, (η5-2,4-cyclopentadien-1-yl)[(2,3,4,5-η)-2,3,4,5-te
trakis(pentafluorophenyl)-2,4-cyclopentadien-1-one]-;;   TETRAKIS(PENTAFLUOROPHENYL)CYCLOPENTADIENONE-.PI.-CY
CLOPENTADIENYLRHODIUM;   W/NBS   66987

912   C48 H42 Cl2 O14   65615-73-8   KC-1977-18-2005   Methyl-2,6-di-O-benzoyl-3-chloro-3-deoxy-4-O-(2,3,6-tri-
O-benzoyl-4-chloro-4-deoxy-β-D-glucopyranosyl)-β-D-allopyranoside;   W   128853

912   C49 H80 N6 O10   56272-43-6   T-67-5296-0   L-Proline, 1-[O-(1-oxohexyl)-N-[N-[N6-(1-oxohexyl)-N2-[N-(1-
oxohexyl)-L-valyl]-L-lysyl]-L-valyl]-L-tyrosyl]-, methyl ester;;   CAPROVALYL-(O-CAPROYL)LYSYLVALYL-(O-CAPROY
L)TYROSYLPROLINE METHYL ESTER;   W/NBS   66988

914   C50 H22 O2 Te2   C-103-5199-0   Bis[2-(5-methyl-2,3-dihydrobenzofuranyl)methyl]ditelluride;   W   128854

914   C63 H126 O2   40710-40-5   N-3-134-18   Hexacontanoic acid, propyl ester;;   Propyl hexacontanoate;   W/NBS
                                                                                                       66989

915   C48 H71 Br N4 P Ru   87794-46-5   I-62-1239-0   Bromo(octaethylporphinato)(tri-n-butylphosphine)rutheniu
m(III);   W   128855

916   C24 H60 Ge6 P4   C-98-6365-0   HEXA(DIETHYLGERMA) TETRAPHOSPHIDE;   W   66990

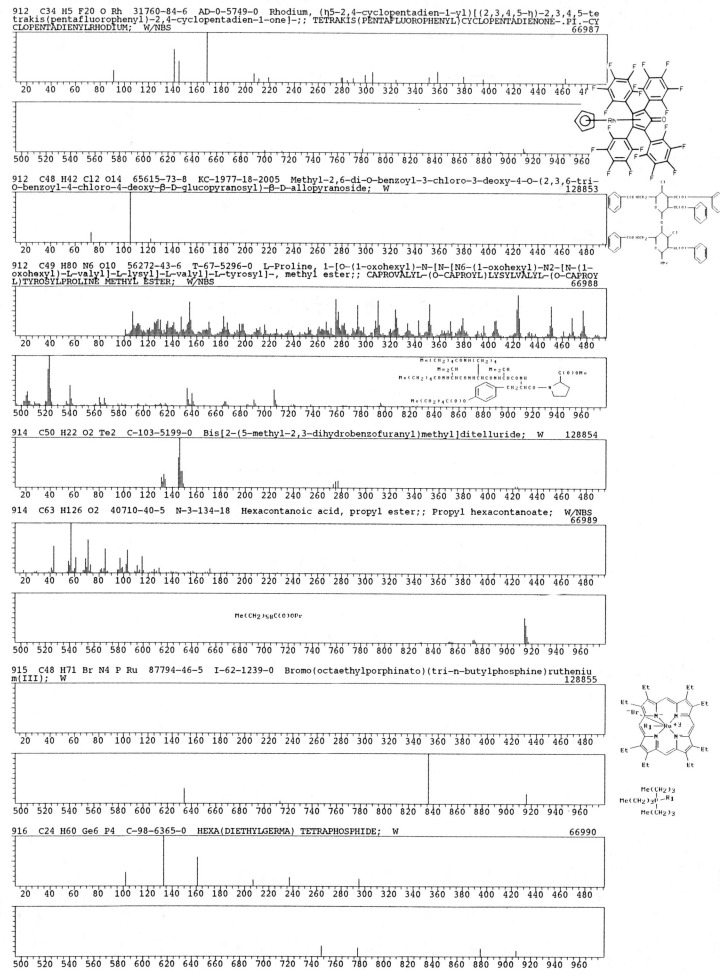

916 C30 H78 O12 P2 Si8 33981-93-0 BA-0-390-0 Inositol, 1,2,4,5-tetrakis-O-(trimethylsilyl)-, bis[bis(tr
imethylsilyl) phosphate], myo-;; OCTAKIS(TRIMETHYLSILYL)MYO-INOSITOL-1,4-DIPHOSPHATE; W/NBS   66991

916 C44 H52 O21 83529-70-8 H-65-1684-0 Heptaacetylnigroside I; W   128856

916 C50 H46 N6 P6 59902-48-6 K-109-1916-0 2,2'-DIMETHYL-4,4,4',4',6,6,6',6'-OCTAPHENYL-2,2'-BI(CYCLOTR
I(λ*5-PHOSPHAZEN)YL); W   66992

917 C54 H83 N3 O9 79664-29-2 J-46-5236-0 N-[4-[(2,3-Dimethoxybenzoyl)decylamino]butyl]-N-[3-[(2,3-di
methoxybenzoyl)-decylamino]propyl]-2,3-dimethoxy-benzamide; W   128857

918 C36 H86 O11 Si8 55191-01-0 F-27-4281-6 TMS(6-O-α-D-GLUCOPYRANOSYL-D-GLUCOSE); W/NBS   66997

918 C36 H86 O11 Si8 55191-02-1 F-27-4280-3 TMS-(3-O-β-D-GLUCOPYRANOSYL-D-GLUCOSE); W/NBS   66999

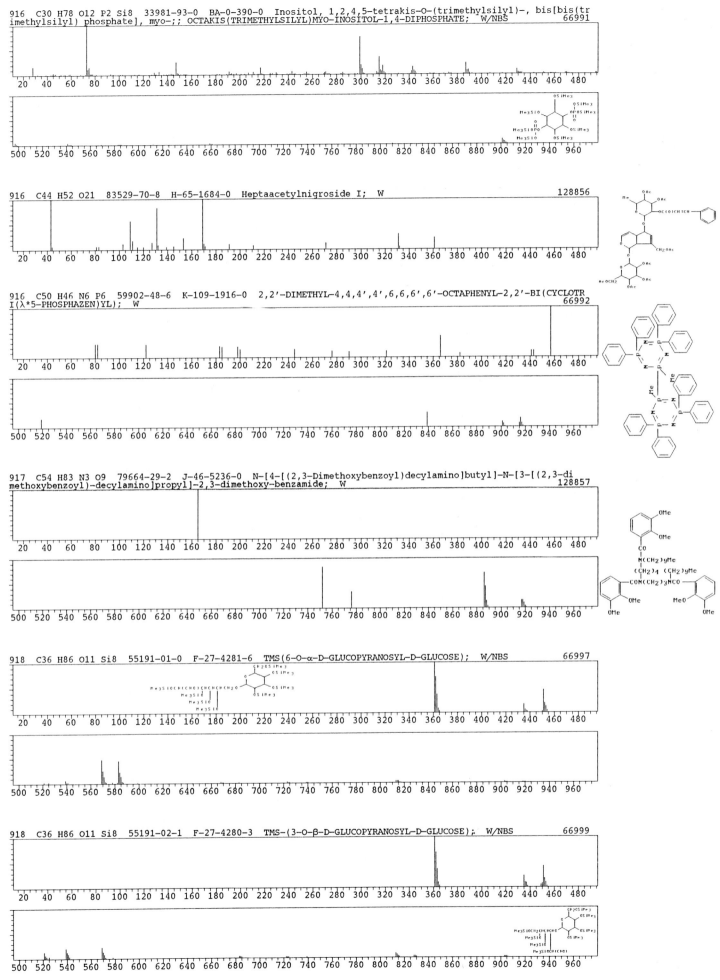

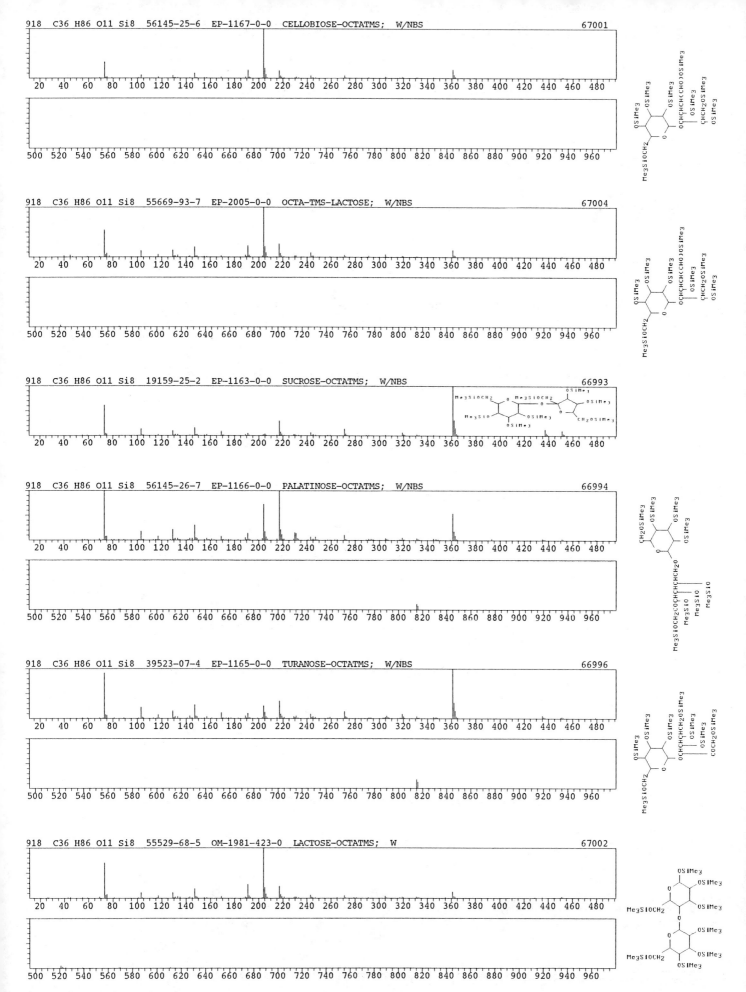

918   C36 H86 O11 Si8   56145-25-6   EP-1167-0-0   CELLOBIOSE-OCTATMS; W/NBS   67001

918   C36 H86 O11 Si8   55669-93-7   EP-2005-0-0   OCTA-TMS-LACTOSE; W/NBS   67004

918   C36 H86 O11 Si8   19159-25-2   EP-1163-0-0   SUCROSE-OCTATMS; W/NBS   66993

918   C36 H86 O11 Si8   56145-26-7   EP-1166-0-0   PALATINOSE-OCTATMS; W/NBS   66994

918   C36 H86 O11 Si8   39523-07-4   EP-1165-0-0   TURANOSE-OCTATMS; W/NBS   66996

918   C36 H86 O11 Si8   55529-68-5   OM-1981-423-0   LACTOSE-OCTATMS; W   67002

918   C36 H86 O11 Si8   17225-43-3   AM-0-178-0   LACTOSE OCTA-TMS;   W/NBS   67003

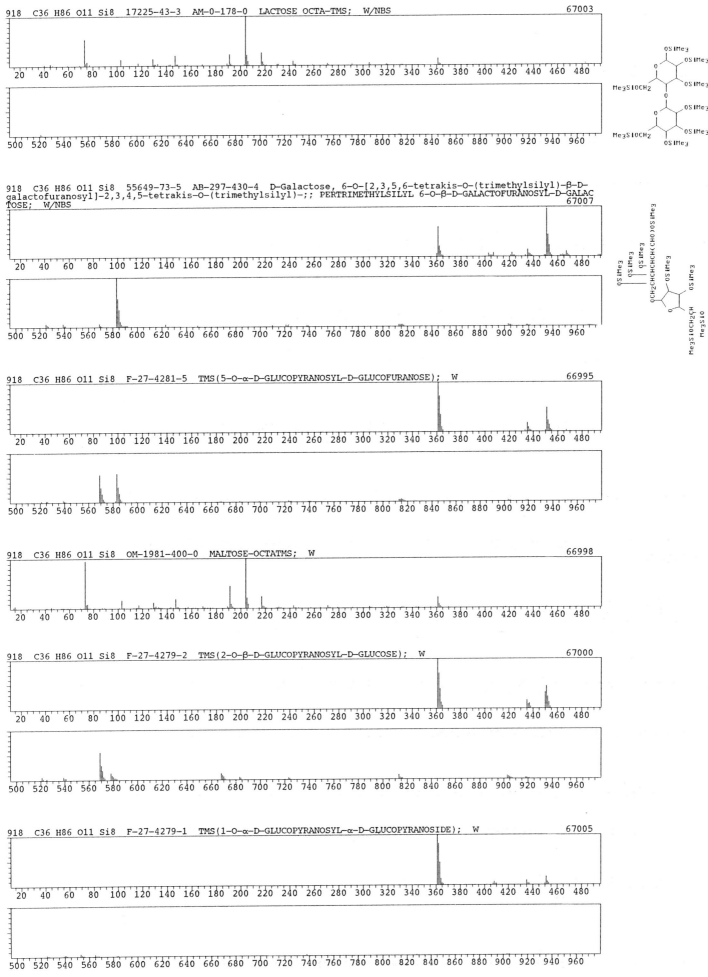

918   C36 H86 O11 Si8   55649-73-5   AB-297-430-4   D-Galactose, 6-O-[2,3,5,6-tetrakis-O-(trimethylsilyl)-β-D-galactofuranosyl]-2,3,4,5-tetrakis-O-(trimethylsilyl)-;; PERTRIMETHYLSILYL 6-O-β-D-GALACTOFURANOSYL-D-GALACTOSE; W/NBS   67007

918   C36 H86 O11 Si8   F-27-4281-5   TMS(5-O-α-D-GLUCOPYRANOSYL-D-GLUCOFURANOSE);   W   66995

918   C36 H86 O11 Si8   OM-1981-400-0   MALTOSE-OCTATMS;   W   66998

918   C36 H86 O11 Si8   F-27-4279-2   TMS(2-O-β-D-GLUCOPYRANOSYL-D-GLUCOSE);   W   67000

918   C36 H86 O11 Si8   F-27-4279-1   TMS(1-O-α-D-GLUCOPYRANOSYL-α-D-GLUCOPYRANOSIDE);   W   67005

918   C36 H86 O11 Si8   F-27-4280-4   β-CELLOBIOSE(TMS ETHER)TMS(4-O-β-D-GLUCOPYRANOSYL-D-GLUCOSE);   W   67006

918   C36 H86 O11 Si8   OM-1981-399-0   MELIBIOSE 8TMS;   W   67008

918   C60 H102 O6   82521-18-4   F-39-2214-0   1,3(R): 4,6(R)-Di-O-benzyliden-2,5-di-O-(2,3-dihydrophytyl)-D-mannitol;   W   128858

919   C60 H105 N O5   15462-29-0   T-67-5334-0   Azacyclohexacosan-14-one, 1-(28,29-dihydroxybicyclo[25.3.1]hentriaconta-1(31),27,29-trien-31-yl)-, diacetate (ester);; N,N-(13-OXOPENTACOSYLENE)-2,6-PENTACOSYLENE-3,4-DIACETOXYANILINE;   W/NBS   67009

920   C45 H69 Na O18   KC-1981-1860-0   Sepositoside A;   W   128859

920   C57 H108 O8   CB-0-266-3   1,2-DIHEPTADECA-3-HEPTANEOCTOATEGLYCERYLTRIESTER;   W   67010

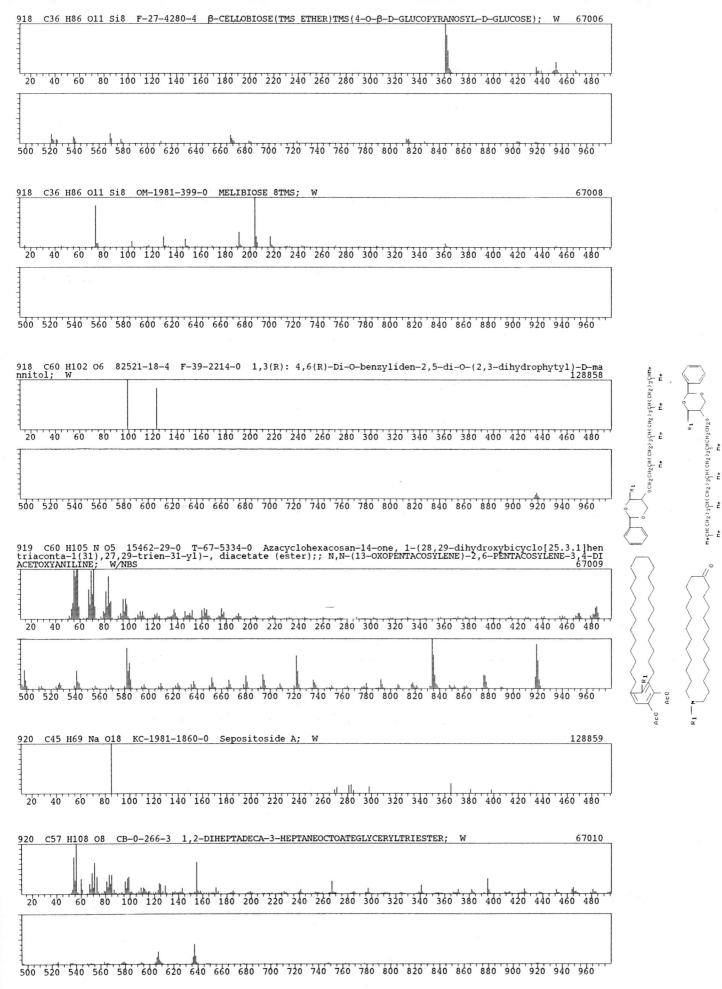

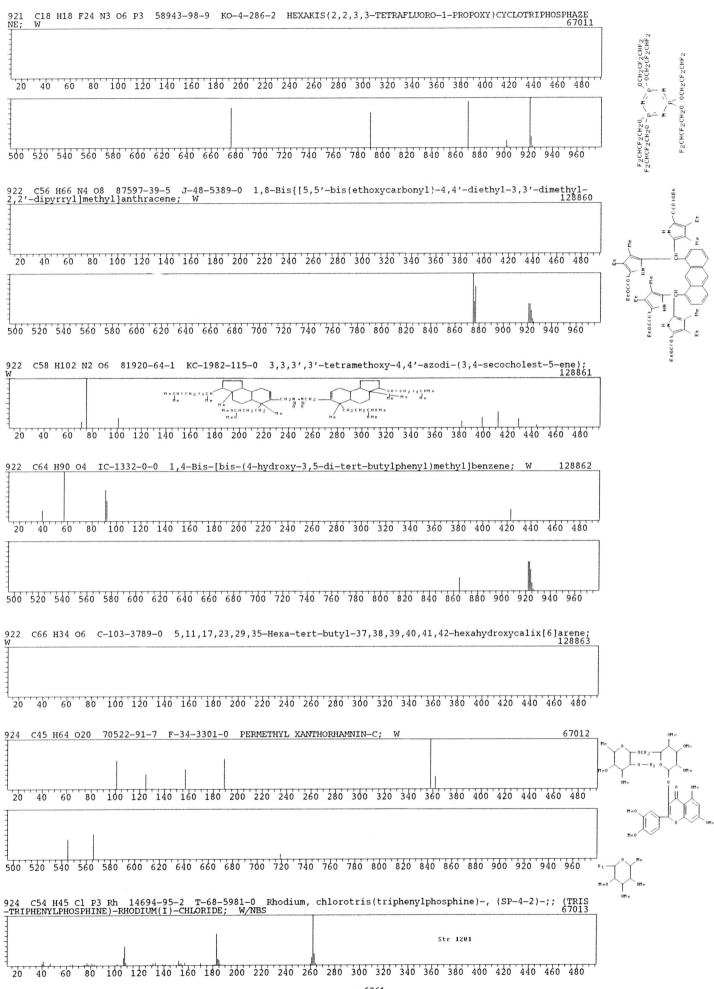

921   C18 H18 F24 N3 O6 P3   58943-98-9   KO-4-286-2   HEXAKIS(2,2,3,3-TETRAFLUORO-1-PROPOXY)CYCLOTRIPHOSPHAZE
NE;   W   67011

922   C56 H66 N4 O8   87597-39-5   J-48-5389-0   1,8-Bis{[5,5'-bis(ethoxycarbonyl)-4,4'-diethyl-3,3'-dimethyl-
2,2'-dipyrryl]methyl}anthracene;   W   128860

922   C58 H102 N2 O6   81920-64-1   KC-1982-115-0   3,3,3',3'-tetramethoxy-4,4'-azodi-(3,4-secocholest-5-ene);
W   128861

922   C64 H90 O4   IC-1332-0-0   1,4-Bis-[bis-(4-hydroxy-3,5-di-tert-butylphenyl)methyl]benzene;   W   128862

922   C66 H34 O6   C-103-3789-0   5,11,17,23,29,35-Hexa-tert-butyl-37,38,39,40,41,42-hexahydroxycalix[6]arene;
W   128863

924   C45 H64 O20   70522-91-7   F-34-3301-0   PERMETHYL XANTHORHAMNIN-C;   W   67012

924   C54 H45 Cl P3 Rh   14694-95-2   T-68-5981-0   Rhodium, chlorotris(triphenylphosphine)-, (SP-4-2)-;; (TRIS
-TRIPHENYLPHOSPHINE)-RHODIUM(I)-CHLORIDE;   W/NBS   67013

Str 1201

928 C36 H84 Al3 La O12 20538-49-2 KA-70-847-2 Lanthanum, tris[bis(2-propanolato)aluminum]hexakis[.mu.-(2-propanolato)]-;; LANTHANUM ALUMINUM ISOPROPOXIDE; W/NBS   67014

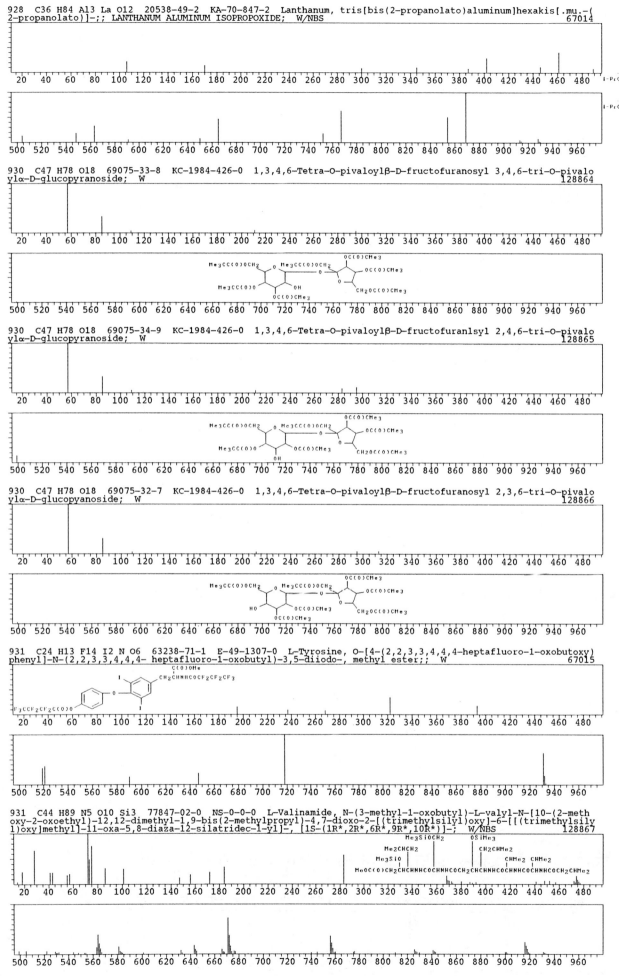

930 C47 H78 O18 69075-33-8 KC-1984-426-0 1,3,4,6-Tetra-O-pivaloylβ-D-fructofuranosyl 3,4,6-tri-O-pivaloylα-D-glucopyranoside; W   128864

930 C47 H78 O18 69075-34-9 KC-1984-426-0 1,3,4,6-Tetra-O-pivaloylβ-D-fructofuranlsyl 2,4,6-tri-O-pivaloylα-D-glucopyranoside; W   128865

930 C47 H78 O18 69075-32-7 KC-1984-426-0 1,3,4,6-Tetra-O-pivaloylβ-D-fructofuranosyl 2,3,6-tri-O-pivaloylα-D-glucopyanoside; W   128866

931 C24 H13 F14 I2 N O6 63238-71-1 E-49-1307-0 L-Tyrosine, O-[4-(2,2,3,3,4,4,4-heptafluoro-1-oxobutoxy)phenyl]-N-(2,2,3,3,4,4,4- heptafluoro-1-oxobutyl)-3,5-diiodo-, methyl ester;; W   67015

931 C44 H89 N5 O10 Si3 77847-02-0 NS-0-0-0 L-Valinamide, N-(3-methyl-1-oxobutyl)-L-valyl-N-[10-(2-methoxy-2-oxoethyl)-12,12-dimethyl-1,9-bis(2-methylpropyl)-4,7-dioxo-2-[(trimethylsilyl)oxy]-6-[[(trimethylsilyl)oxy]methyl]-11-oxa-5,8-diaza-12-silatridec-1-yl]-, [1S-(1R*,2R*,6R*,9R*,10R*)]-; W/NBS   128867

931 C54 H59 Cl N4 P Ru 91514-79-3 I-62-1239-0 Chloro(octaethylporphinato)(triphenylphosphine)ruthenium(III); W
128868

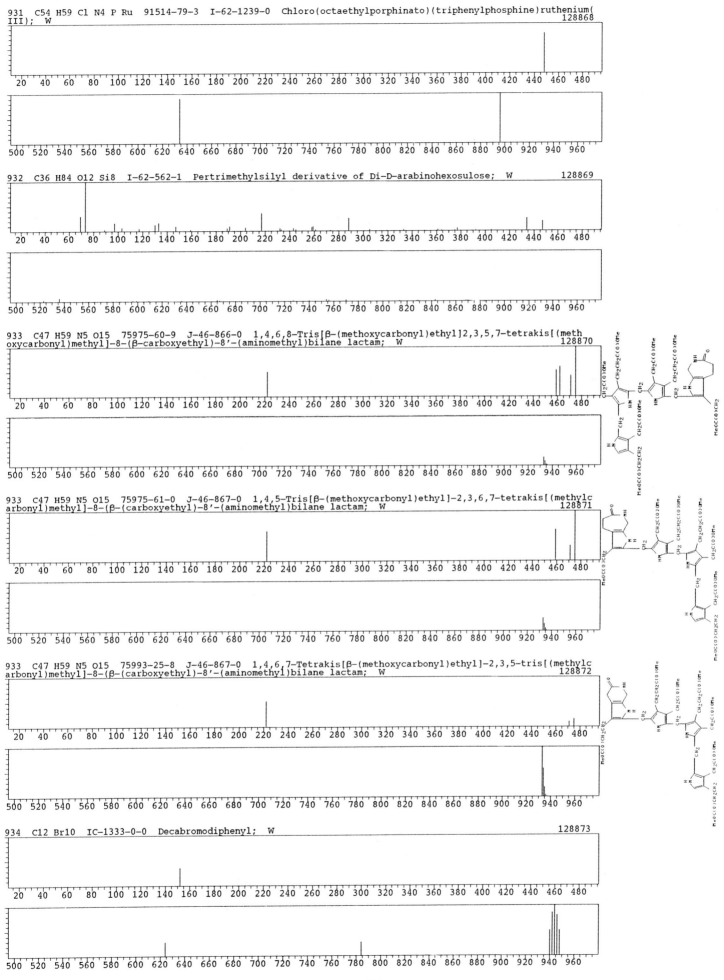

932 C36 H84 O12 Si8 I-62-562-1 Pertrimethylsilyl derivative of Di-D-arabinohexosulose; W
128869

933 C47 H59 N5 O15 75975-60-9 J-46-866-0 1,4,6,8-Tris[β-(methoxycarbonyl)ethyl]2,3,5,7-tetrakis[(methoxycarbonyl)methyl]-8-(β-carboxyethyl)-8′-(aminomethyl)bilane lactam; W
128870

933 C47 H59 N5 O15 75975-61-0 J-46-867-0 1,4,5-Tris[β-(methoxycarbonyl)ethyl]-2,3,6,7-tetrakis[(methylcarbonyl)methyl]-8-(β-(carboxyethyl)-8′-(aminomethyl)bilane lactam; W
128871

933 C47 H59 N5 O15 75993-25-8 J-46-867-0 1,4,6,7-Tetrakis[β-(methoxycarbonyl)ethyl]-2,3,5-tris[(methylcarbonyl)methyl]-8-(β-(carboxyethyl)-8′-(aminomethyl)bilane lactam; W
128872

934 C12 Br10 IC-1333-0-0 Decabromodiphenyl; W
128873

934   C49 H62 N2 O16   62327-69-9   KO-3-168-2   N-(N'-DODECANOYLGLYCYL)PERACETYL DANORUBICIN;   W   67016

935   C22 H29 I4 N O4 Si2   MM-203-6-0   N,O-BIS-TRIMETHYLSILYL-METHYL ESTER OF TETRAIODOTHYRONINE;   W   67017

936   C44 H68 O16 Si3   37300-11-1   NS-0-0-0   Azadirachtin, O,O,O-tris(trimethylsilyl)-;   W/NBS   128874

936   C47 H60 N4 O16   63341-07-1   J-46-867-0   1,4,6,8-Tetrakis[β-(methoxycarbonyl)ethyl]-2,3,5,7-tetrakis[(
metHoxycarbonyl)methyl]bilane;   W   128875

937   C31 H38 Cl2 N4 S6 Si Ta   KA-1984-1071-0   Tris(dimethyldithiocarbamato)trimethylsilylimidotantalum(V)-
dichloromethane;   W   128876

938   C18 H18 F26 N2 O4 P2 Si2   61091-55-2   K-110-1895-0   2,4'-F2-4,4,4'',4''',5,5,5'',5''-(F3 ME)8-1',3'-(TM
S)2-DISPIRO[1,3,2L5]DI-O-PHOSPHOLAN-2,2'-[1,3,2L5]DIAZAPHOSPHETIDIN-4',2''-[1,3,2L5]DI-O-PHOSPHOLA;   W   67018

938   C44 H30 F4 O4 Sn2   AG-135-84-0   Bis(triphenyltin) 2,3,4,5-tetrafluoroterephthalate hydrate;   W   128877

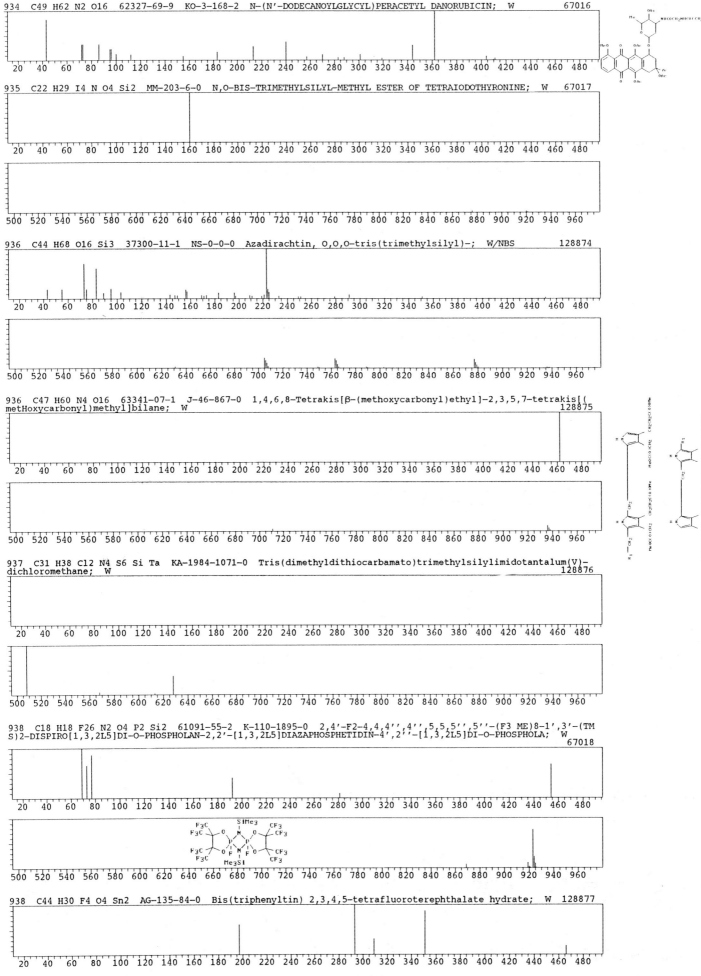

940 C31 H35 F15 N4 O12 59275-67-1 AB-310-133-3 5-Undecenedioic acid, 5-[[[6-methoxy-6-oxo-5-[(trifluoro
acetyl)amino]-2-[(trifluoroacetyl)oxy]hexyl](trifluoroacetyl)amino]methyl]-2,10-bis[(trifluoroacetyl)amino]
-, dimethyl ester;; HYDROXYMERODESMOSINE TFA METHYL ESTER; W/NBS   67019

```
                              COCF3      C(O)OMe
                                |          |
         F3CC(O)O            CH(CH2)3CHNHCOCF3
           |       |          |
       MeOC(O)    |           C(O)OMe
           |    |    |  |      |
       F3CCONHCHCH2CH2CHCH2NCH2CCH2CH2CHNHCOCF3
```

940 C58 H55 O6 P3 63051-15-0 K-110-1503-0 4,4-BIS(1,1-DIMETHOXY-2,6-DIPHENYL-λ5-PHOSPHORIN-4-YLMETHYL)-
1-METHOXY-1-OXO-2,6-DIPHENYL-1,4-DIHYDRO-λ5-PHOSPHORIN; W   67020

941 C60 H111 N O6 PG-1982-200-0 TRIETHANOL AMINE OLEATE; W   67021

942 C48 H54 N4 O16 BS-4-15-0 Uroporphyrin I permethylester; W   128878

942 C69 H98 O EP-8571-0-0 GLYCERYL TRIDOCASAHEXAENOATE; W   67022

943 C46 H57 N5 O13 Si2 KO-5-457-2 4-O-(ETHYL ANTHRANILATE AZO DIPYRROLE ACYL)-1-(ETHYL ANTHRANILATE)-N-
GLUCURONIDE; W   67023

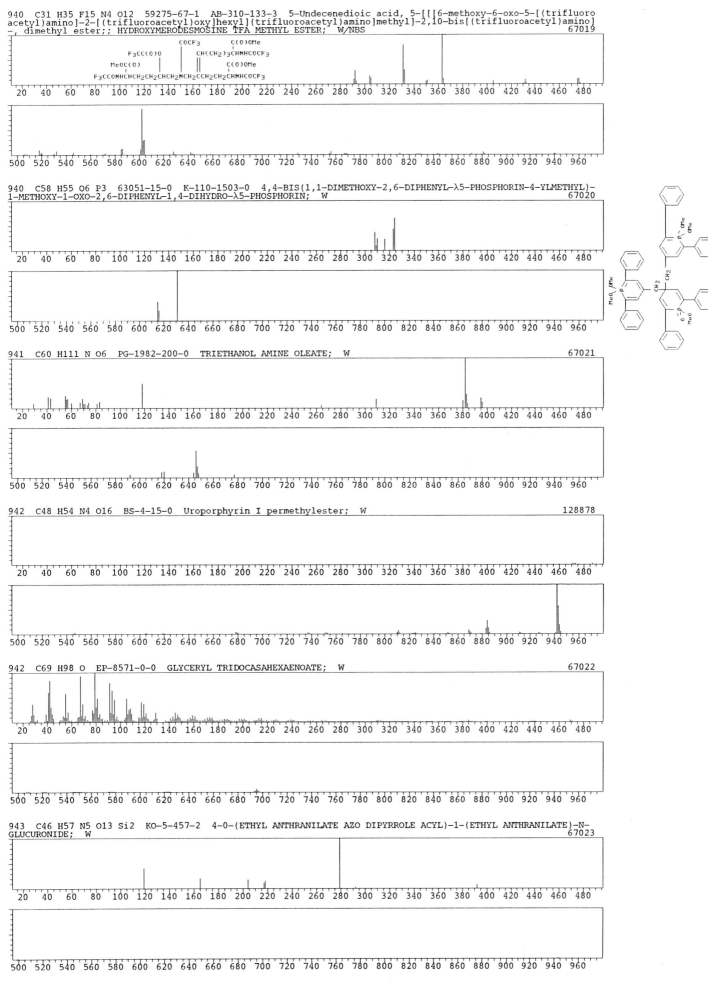

943 C46 H57 N5 O13 Si2 KO-5-457-2 3-O-(ETHYL ANTHRANILATE AZO DIPYRROLE ACYL)-1-(ETHYL ANTHRANILATE)-N-
GLUCURONIDE TMIS ETHER; W   67024

943 C46 H57 N5 O13 Si2 KO-5-457-0 2-O-(ETHYLANTHRANILATE AZO DIPYRROLE ACYL)-1-(ETHYLANTHRANILATE)-N-GLU
CURONIDE, DI-TMS ETHER; W   67025

943 C47 H61 N5 O12 Si2 BS-5-107-0 Ethylanthranilateazopigment (γ3) of bilirubin-1Xα 2-O-acylglucuronide
1-N-ethylanthranilate bis-trimethylsilyl ether methyl ester; W   128879

943 C47 H61 N5 O12 Si2 BS-5-108-0 Ethylanthranilateazopigment (β4) of bilirubin-1Xα 2-O-acylglucuronide
1-N-ethylanthranilate bis-trimethylsilyl ether methyl ester; W   128880

946 C48 H74 N4 O8 Si4 38574-20-8 O6-0-991-0 21H,23H-Porphine-2,18-dipropanoic acid, 7,12-bis[1,2-bis[(
trimethylsilyl)oxy]ethyl]-3,8,13,17-tetramethyl-, dimethyl ester;; 2,4-DI-(1,2-DIHYDROXYETHYL)-6,7-DI-(2-
METHOXYCARBONYLETHYL)-1,3,5,8-TETRAMETHYLPORPHIN TETRA TMS ETHER; W/NBS   67026

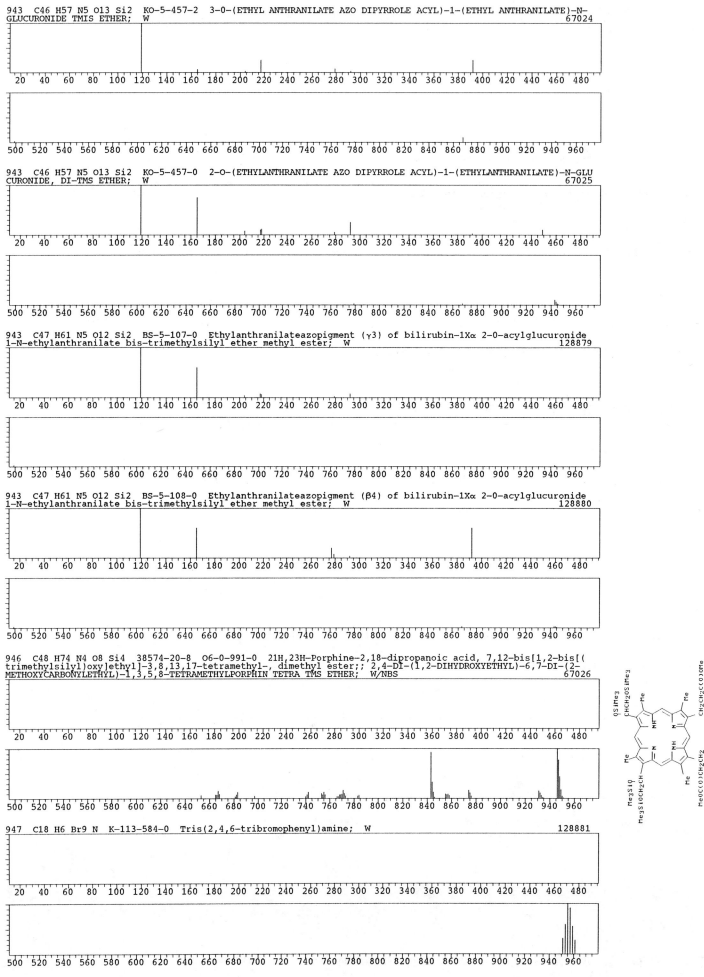

947 C18 H6 Br9 N K-113-584-0 Tris(2,4,6-tribromophenyl)amine; W   128881

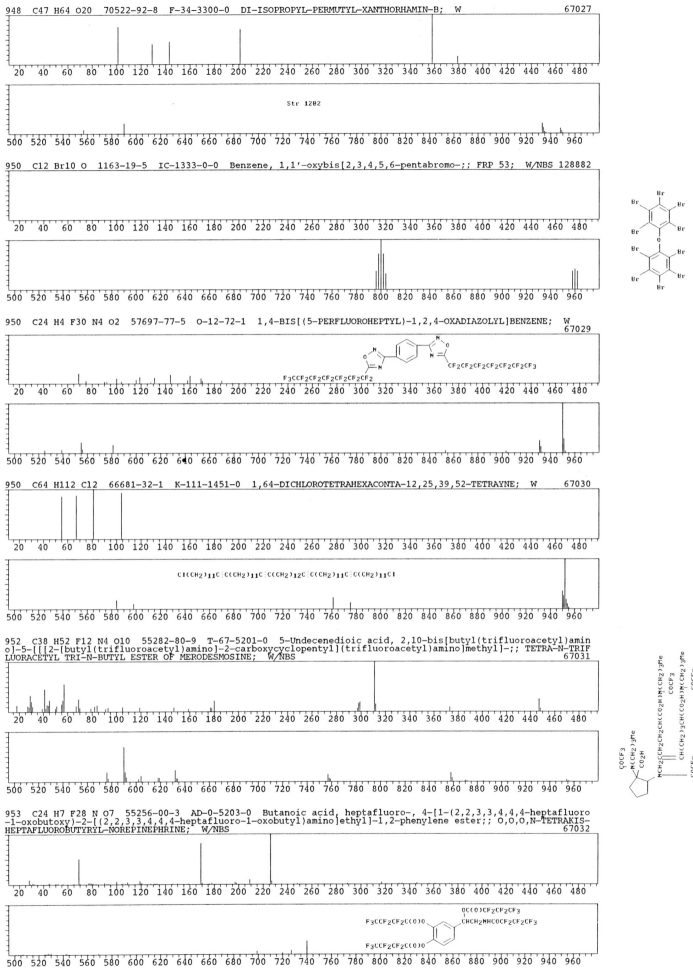

948 C47 H64 O20   70522-92-8   F-34-3300-0   DI-ISOPROPYL-PERMUTYL-XANTHORHAMIN-B;   W   67027

Str 1202

950 C12 Br10 O   1163-19-5   IC-1333-0-0   Benzene, 1,1'-oxybis[2,3,4,5,6-pentabromo-;; FRP 53;   W/NBS 128882

950 C24 H4 F30 N4 O2   57697-77-5   O-12-72-1   1,4-BIS[(5-PERFLUOROHEPTYL)-1,2,4-OXADIAZOLYL]BENZENE;   W   67029

950 C64 H112 Cl2   66681-32-1   K-111-1451-0   1,64-DICHLOROTETRAHEXACONTA-12,25,39,52-TETRAYNE;   W   67030

Cl(CH2)11C⋮C(CH2)11C⋮C(CH2)12C⋮C(CH2)11C⋮C(CH2)11Cl

952 C38 H52 F12 N4 O10   55282-80-9   T-67-5201-0   5-Undecenedioic acid, 2,10-bis[butyl(trifluoroacetyl)amino]-5-[[[2-[butyl(trifluoroacetyl)amino]-2-carboxycyclopentyl](trifluoroacetyl)amino]methyl]-;; TETRA-N-TRIFLUOROACETYL TRI-N-BUTYL ESTER OF MERODESMOSINE;   W/NBS   67031

953 C24 H7 F28 N O7   55256-00-3   AD-0-5203-0   Butanoic acid, heptafluoro-, 4-[1-(2,2,3,3,4,4,4-heptafluoro-1-oxobutoxy)-2-[(2,2,3,3,4,4,4-heptafluoro-1-oxobutyl)amino]ethyl]-1,2-phenylene ester;; O,O,O,N-TETRAKIS-HEPTAFLUOROBUTYRYL-NOREPINEPHRINE;   W/NBS   67032

954   C42 H30 Co4 N12 O   56802-37-0   B-28-1364-0   OXOHEXA(1H-PYRROLO(2,3-B)PYRIDINATO)TETRA COBALT(II); W
67033

954   C42 H52 I2 O9   72049-90-2   J-45-339-0   Methyl 3α-(Carbethoxyoxy)-7α,12α-bis(m-iodobenzoxy)-5β-cholano ate; W
128883

955   C40 H81 N5 O8 Si7   D-14-4202-5   2-TMS-AMINO-5-[(N-AC-(4,5-TMS-2-CYCLOPENTEN-1-YL)AMINOME]-4-(TMS-O)-7-Tβ-D-(TMS)3-RIBOFURANOSYLPYRROLL[2,3-D]PYRIMIDINE AC-(TMS); W
67034

955   C51 H69 N7 O11   36261-14-0   O-6-161-7   PERMETHYL STAPHYLOMYCIN S2 METHYL ESTER; W/NBS
67035

956   C33 H33 F21 O8   52058-18-1   JC-90-174-7   PGF2α-METHYL ESTER-9,11,15-TRI(HEPTAFLUOROBUTYRYL)DERIVATIVE; W
67036

958   C33 H35 F21 O8   52058-17-0   JC-90-174-6   PGF1α-METHYL ESTER-9,11,15-TRI(HEPTAFLUOROBUTYRYL)DERIVATIVE; W
67037

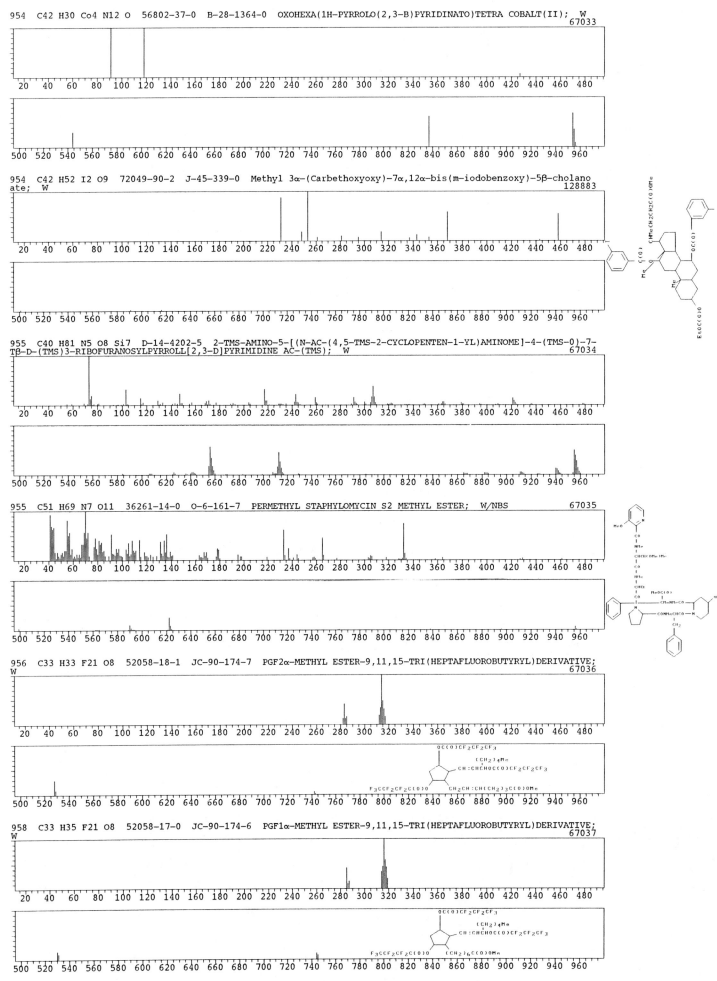

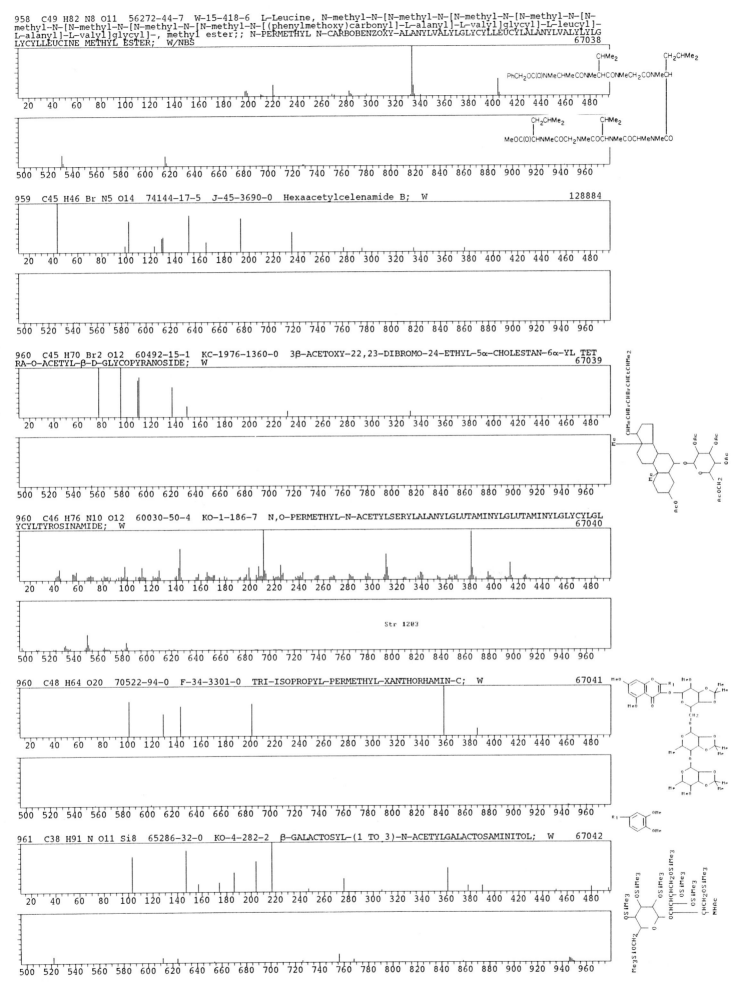

958 C49 H82 N8 O11 56272-44-7 W-15-418-6 L-Leucine, N-methyl-N-[N-methyl-N-[N-methyl-N-[N-methyl-N-[N-methyl-N-[N-methyl-N-[N-methyl-N-[N-methyl-N-[(phenylmethoxy)carbonyl]-L-alanyl]-L-valyl]glycyl]-L-leucyl]-L-alanyl]-L-valyl]glycyl]-, methyl ester;; N-PERMETHYL N-CARBOBENZOXY-ALANYLVALYLGLYCYLLEUCYLALANYLVALYLYLGLYCYLLEUCINE METHYL ESTER; W/NBS
67038

959 C45 H46 Br N5 O14 74144-17-5 J-45-3690-0 Hexaacetylcelenamide B; W
128884

960 C45 H70 Br2 O12 60492-15-1 KC-1976-1360-0 3β-ACETOXY-22,23-DIBROMO-24-ETHYL-5α-CHOLESTAN-6α-YL TETRA-O-ACETYL-β-D-GLYCOPYRANOSIDE; W
67039

960 C46 H76 N10 O12 60030-50-4 KO-1-186-7 N,O-PERMETHYL-N-ACETYLSERYLALANYLGLUTAMINYLGLUTAMINYLGLYCYLGLYCYLTYROSINAMIDE; W
67040

Str 1203

960 C48 H64 O20 70522-94-0 F-34-3301-0 TRI-ISOPROPYL-PERMETHYL-XANTHORHAMIN-C; W
67041

961 C38 H91 N O11 Si8 65286-32-0 KO-4-282-2 β-GALACTOSYL-(1 TO 3)-N-ACETYLGALACTOSAMINITOL; W
67042

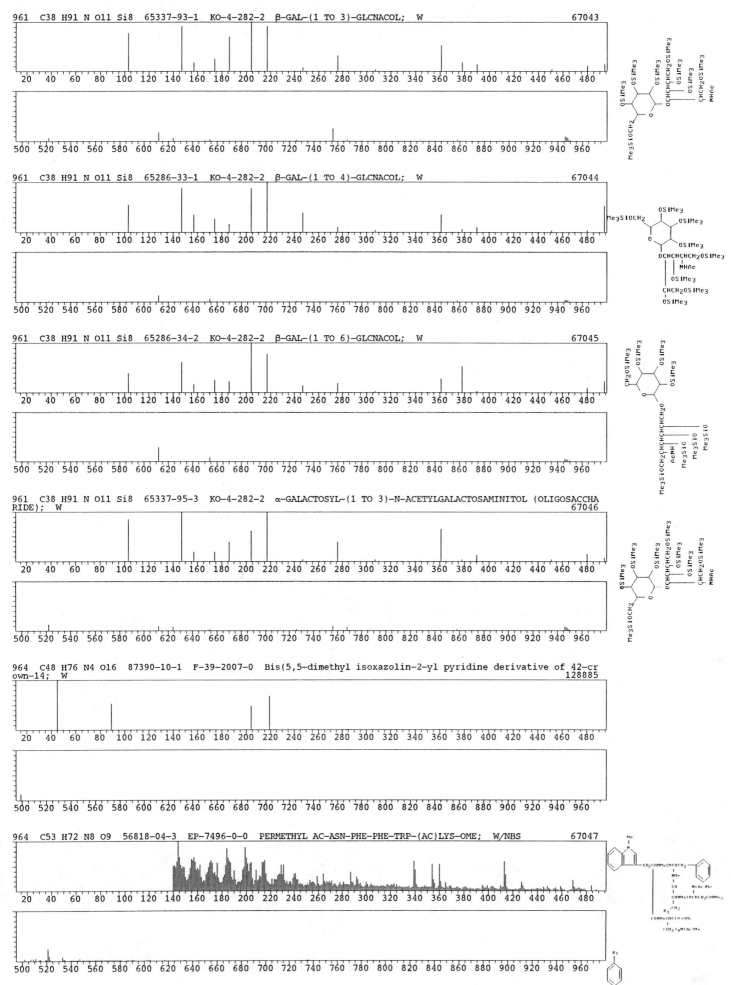

961   C38 H91 N O11 Si8   65337-93-1   KO-4-282-2   β-GAL-(1 TO 3)-GLCNACOL;   W   67043

961   C38 H91 N O11 Si8   65286-33-1   KO-4-282-2   β-GAL-(1 TO 4)-GLCNACOL;   W   67044

961   C38 H91 N O11 Si8   65286-34-2   KO-4-282-2   β-GAL-(1 TO 6)-GLCNACOL;   W   67045

961   C38 H91 N O11 Si8   65337-95-3   KO-4-282-2   α-GALACTOSYL-(1 TO 3)-N-ACETYLGALACTOSAMINITOL (OLIGOSACCHA
RIDE);   W   67046

964   C48 H76 N4 O16   87390-10-1   F-39-2007-0   Bis(5,5-dimethyl isoxazolin-2-yl pyridine derivative of 42-cr
own-14;   W   128885

964   C53 H72 N8 O9   56818-04-3   EP-7496-0-0   PERMETHYL AC-ASN-PHE-PHE-TRP-(AC)LYS-OME;   W/NBS   67047

966   C40 H54 O27   37698-65-0   H-67-384-0   O-(2,3,4,6,-Tetra-O-acetyl-β-D-glucopyranosyl)-(1-6)-O-(2,3,4-tri
-O-acetyl-β-D-glucopyranosyl)-(1-6)-1,2,3,4-tetra-O-acetyl-β-D-glucopyanose; W   128886

20 40 60 80 100 120 140 160 180 200 220 240 260 280 300 320 340 360 380 400 420 440 460 480

500 520 540 560 580 600 620 640 660 680 700 720 740 760 780 800 820 840 860 880 900 920 940 960

966   C40 H54 O27   55287-00-8   H-67-385-0   O-(2,3,4,6-Tetra-O-acetyl-β-D-glucopyranosyl)-(1-4)-O-(2,3,6-tri-
O-acetyl-β-D-glucopyranosyl)-(1-6)-1,2,3,4-tetra-O-acetyl-β-D-glucopyranose; W   128887

20 40 60 80 100 120 140 160 180 200 220 240 260 280 300 320 340 360 380 400 420 440 460 480

500 520 540 560 580 600 620 640 660 680 700 720 740 760 780 800 820 840 860 880 900 920 940 960

966   C40 H54 O27   89367-11-3   H-67-385-0   O-(2,3,4,6-Tetra-O-acetyl-β-D-glucopyranosyl)-(1-3)-O-(2,4,6-tri-
O-acetyl-β-D-glucopyranosyl)-(1-6)-1,2,3,4-tetra-O-acetyl-β-D-glucopyranose; W   128888

20 40 60 80 100 120 140 160 180 200 220 240 260 280 300 320 340 360 380 400 420 440 460 480

500 520 540 560 580 600 620 640 660 680 700 720 740 760 780 800 820 840 860 880 900 920 940 960

966   C40 H54 O27   89945-12-0   H-67-385-0   O-(2,3,4,6-Tetra-O-acetyl-β-D-glucopyranosyl-(1-2)-O-(3,4,6-tri-
O-acetyl-β-D-glucopyranosyl)-(1-6)-1,2,3,4-tetra-O-acetyl-β-Dgluco-pyranose; W   128889

20 40 60 80 100 120 140 160 180 200 220 240 260 280 300 320 340 360 380 400 420 440 460 480

500 520 540 560 580 600 620 640 660 680 700 720 740 760 780 800 820 840 860 880 900 920 940 960

966   C40 H54 O27   90332-61-9   H-67-391-0   O-(2,3,4,6-Tetra O-acetyl-β-D-glucopyranosyl)-(1-2)-O-[(2,3,4,6-
tetra-O-acetyl-β-D-glucopyranosyl)-(1-6)]-1,3,4-tri-O-acetyl-α-D-glucopyranose; W   128890

20 40 60 80 100 120 140 160 180 200 220 240 260 280 300 320 340 360 380 400 420 440 460 480

500 520 540 560 580 600 620 640 660 680 700 720 740 760 780 800 820 840 860 880 900 920 940 960

966   C52 H86 O16   BS-2-6-0   Diosgenin 3-O-α-L-rhamnopyranosyl-(1-2)-[α-L-arabinofuranosyl-(1-4)]-β-D-glucop
yranoside permethylate; W   128891

20 40 60 80 100 120 140 160 180 200 220 240 260 280 300 320 340 360 380 400 420 440 460 480

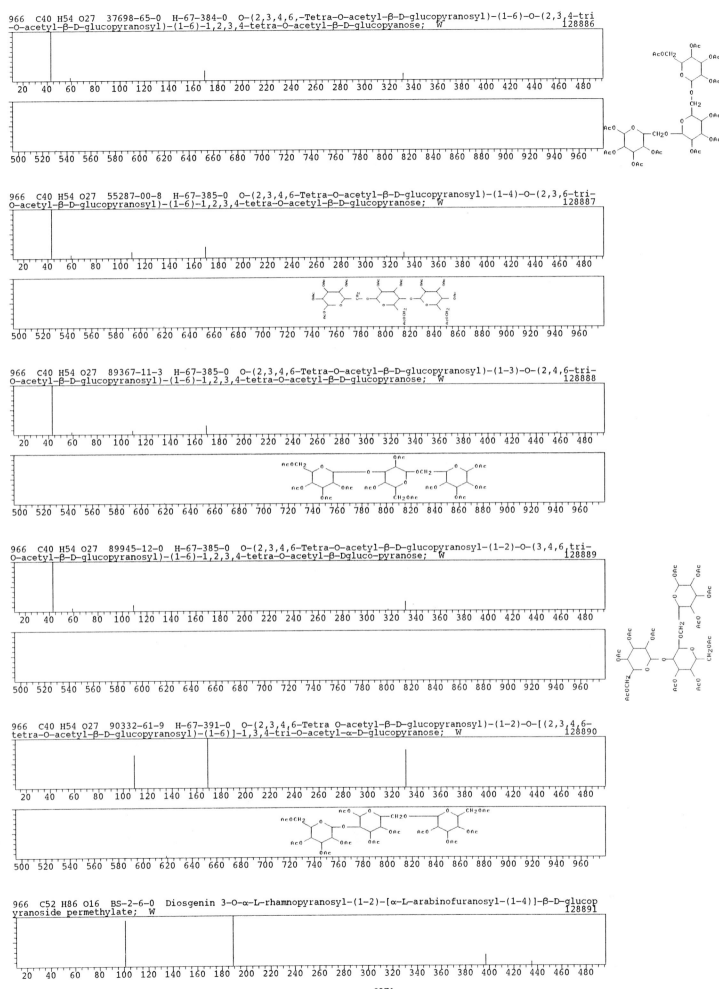

- 6271 -

967　C25 H9 F28 N O7　55282-79-6　AD-O-5202-0　Butanoic acid, heptafluoro-, 4-[1-(2,2,3,3,4,4,4-heptafluoro
-1-oxobutoxy)-2-[(2,2,3,3,4,4,4-heptafluoro-1-oxobutyl)methylamino]ethyl]-1,2-phenylene ester, (-)-;; O,O,
O,N-TETRAKIS-HEPTAFLUOROBUTYRYL-EPINEPHRINE;　W/NBS　　　　　　　　　　　　　　　　　　　67048

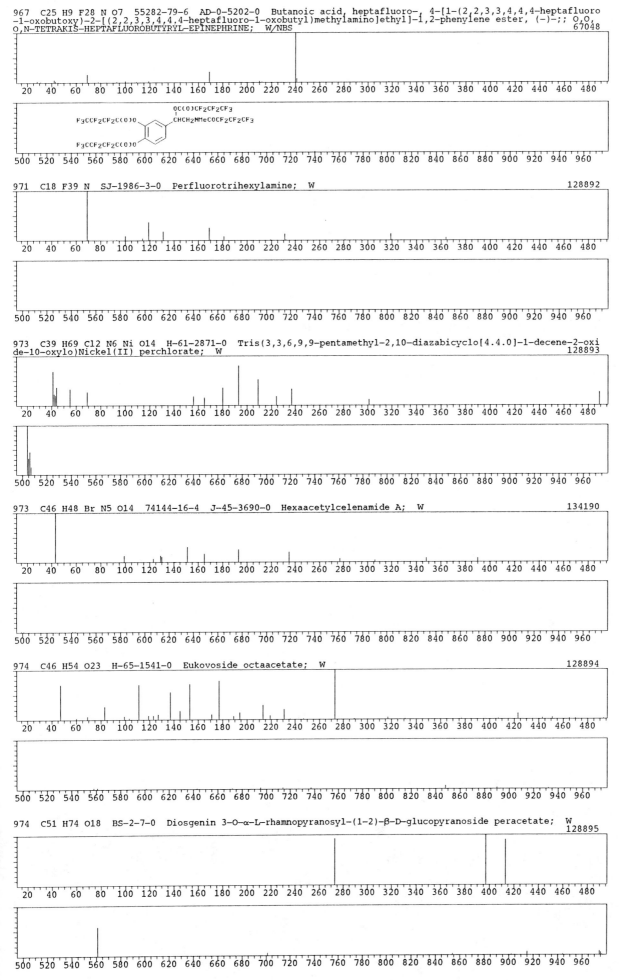

971　C18 F39 N　SJ-1986-3-0　Perfluorotrihexylamine;　W　　　　　　　　　　　　　　　　128892

973　C39 H69 Cl2 N6 Ni O14　H-61-2871-0　Tris(3,3,6,9,9-pentamethyl-2,10-diazabicyclo[4.4.0]-1-decene-2-oxi
de-10-oxylo)Nickel(II) perchlorate;　W　　　　　　　　　　　　　　　　　　　　128893

973　C46 H48 Br N5 O14　74144-16-4　J-45-3690-0　Hexaacetylcelenamide A;　W　　　　　134190

974　C46 H54 O23　H-65-1541-0　Eukovoside octaacetate;　W　　　　　　　　　　　　128894

974　C51 H74 O18　BS-2-7-0　Diosgenin 3-O-α-L-rhamnopyranosyl-(1-2)-β-D-glucopyranoside peracetate;　W
　　　　　　　　　　　　　　　　　　　　　　　　　　　　　　　　　　　　　128895

974   C51 H74 O18   BS-2-8-0   Diosgenin 3-O-α-L-rhamnopyranosyl-(1-4)-β-D-glucopyranoside peracetate;   W
128896

975   C54 H59 Br N4 P Ru   I-62-1238-0   Bromo(octaethylporphinato)(triphenylphosphine)ruthenium(III);   W
128897

976   C51 H76 O18   BS-2-15-0   Tigogenin 3-O-α-L-rhamnopyranosyl-(1-4)-β-D-glucopyranoside peracetate;   W
128898

978   C47 H40 Fe3 O7 P2 S   81877-20-5   K-115-1303-0   .mu.(3)-(tert-Butylthiolato)-heptacarbonyl-.mu.-hydrido
-bis(triphenylphosphine)-triiron;   W
128899

978   C51 H55 Br N4 O11   62786-85-0   KC-1976-2499-0   DIBENZYL 2-BROMO-4,6,7-TRIS-(2-METHOXYCARBONYLETHYL)-1,
3,5,8-TETRAMETHYL-B-OXOBILANE-1',8'-DICARBOXYLATE;   W
67049

Str 1204

978   C54 H44 Ir P3   38316-81-3   EP-7447-0-0   Iridium, [2-(diphenylphosphino)phenyl-C,P]bis(triphenylphosphi
ne)-, (SP-4-2)-;; O-DIPHENYLPHOSPHINYLPHENYL-BIS-(TRIPHENYLPHOSPHINE)IRIDIUM(I);   W/NBS
67050

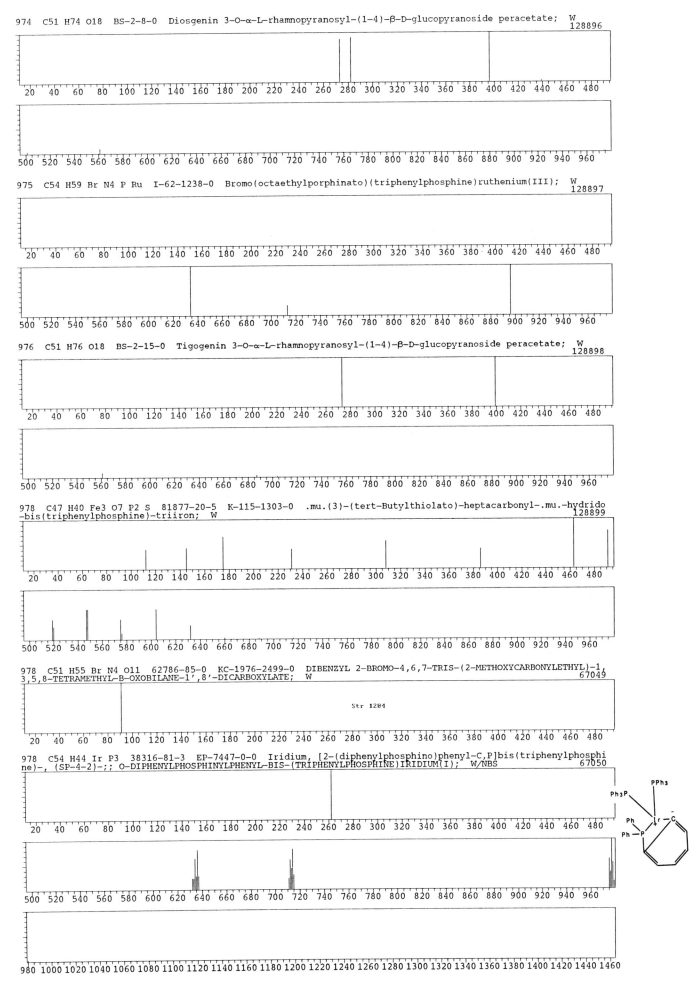

978   C56 H58 N4 O12   77773-68-3   J-46-3501-0   Bis(methyl ester) of 2,4-bis[1-(2,4-dihydroxyphenyl)ethyl]
deuteroporphyrin;   W   128900

978   C72 H82 O2   65549-96-4   K-111-270-0   2,4-BIS(4-TERT-BUTYLPHENYL)-6-(4-TERT-BUTYL-2-(2,4,6-TRIS(4-TERT-
BUTYLPHENYL)PHENOXY)PHENYL)PHENOL;   W   67051

978   C72 H82 O2   K-111-269-0   2,4,6-TRIS(4-TERT-BUTYLPHENYL)-4-[2,4,6-TRIS(4-TERT-BUTYLPHENYL)PHENOXY]-2,5-
CYCLOHEXADIEN-1-ONE;   W   67052

981   C57 H63 N3 O12   66885-69-6   C-100-2843-0   ETHYL 1-CN-3-S-BUTOXY-C(O)-2-(P-S-BUTYL-3'-(E)-PROPENOATE)PH
ENYL-4-(P-(3-(P-ET-2-CN-PROPENOATE)PHENYL-1-CN-3-S-BUTOXY-C(O)-CYCLOBUTANOIC ACID-;   W   67053

983   C32 H4 Cl12 Cu N8   F-20-1360-5   COPPER DODECACHLOROPHTHALOCYANINE;   W   67054

984   C50 H80 N8 O12   83712-17-8   J-49-240-0   Majusculamide C;   W   128903

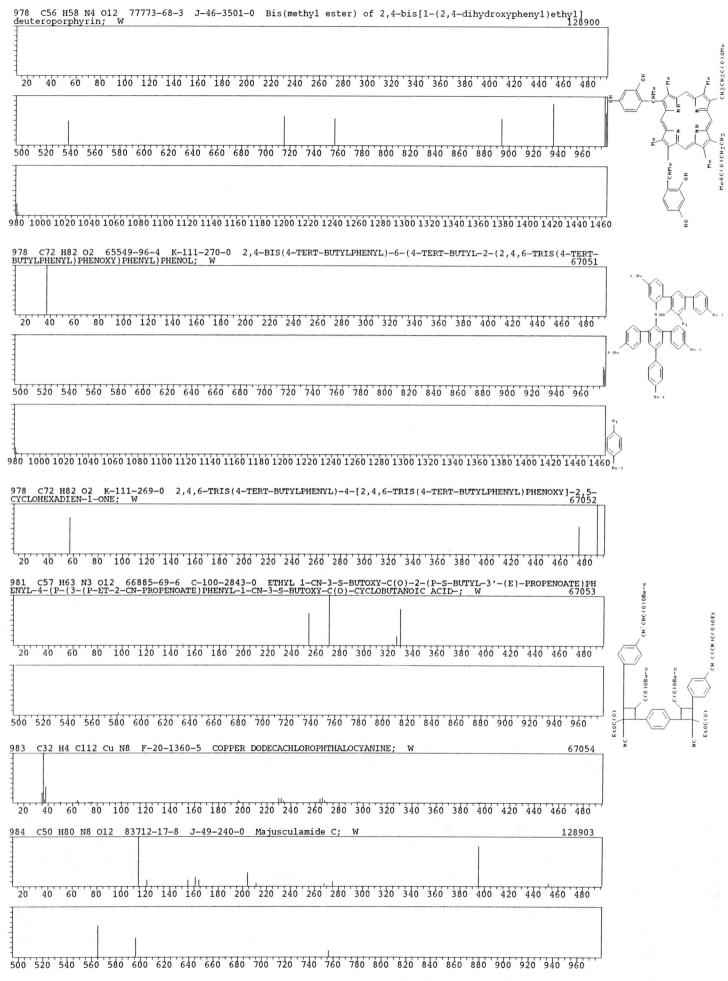

984 C56 H56 O16 88038-09-9 KC-1983-2034-0 6-(2,3-trans-3,4-trans-3-acetoxy-3',4',7-trimethoxyflavan-4-yl)derivative of [3,4':3',4]-O,O-linked-2,3-trans-3,4-cis:2',3'-trans-3',4'-cis-profisetinidin; W 128904

984 C56 H56 O16 88082-59-1 KC-1983-2034-0 6-(2,3-trans-3,4-cis-3-acetoxy-3',4',7-trimethoxyflavan-4-yl) derivative of [3,4':3',4]-O,O-linked-2,3-trans-3,4-cis:2',3'-trans-3',4'-cis-profisetinidin; W 128905

988 C18 F24 N2 O2 S8 90594-27-7 K-117-1898-0 1,1'-Oxalylbis[2,3,4,5-tetrakis(trifluoromethylthio)pyrrole]; W 128906

988 C48 H60 O22 72504-49-5 H-62-1951-0 Di-(2,3,4,6-tetra-O-acetyl-β-D-glucosyl) ester of Crocetin; W 128907

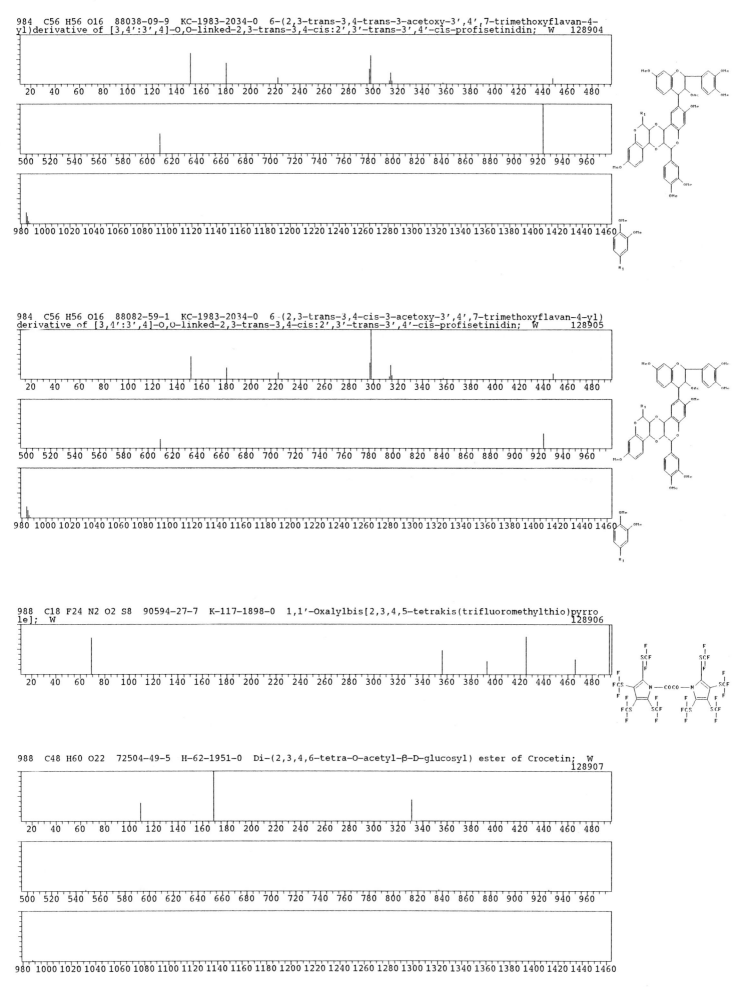

990   C51 H74 O19   BS-2-17-0   Pennogenin 3-O-α-L-rhamnopyranosyl-(1-2)-β-D-glucopyranoside acetate;   W
128908

991   C52 H93 N7 O11   56272-51-6   T-66-2782-0   L-Threonine, N-[N-[N-[1-[N-[N-[N-(3-hydroxy-1-oxoeicosyl)-L-threonyl]-L-valyl]-L-alanyl]-L-prolyl]-L-leucyl]-L-valyl]-, .psi.-lactone;; VAL6-PEPTIDOLIPINE (ISOLATED FROM NOCARDIA ASTEROIDES);   W/NBS
67055

992   C39 H96 O11 Si9   55649-72-4   AB-297-430-5   D-Galactitol, 6-O-[2,3,5,6-tetrakis-O-(trimethylsilyl)-β-D-galactofuranosyl]-1,2,3,4,5-pentakis-O-(trimethylsilyl)-;; PERTRIMETHYLSILYL 6-O-β-D-GALACTOFURANOSYL-D-GALACTOSE ALDITOL;   W/NBS
67056

992   C39 H96 O11 Si9   55823-10-4   AB-297-431-6   D-Galactitol, 1,2,3,4,6-pentakis-O-(trimethylsilyl)-, anhydride with 1,2,3,6-tetrakis-O-(trimethylsilyl)-β-D-galactofuranose;; TRIMETHYLSILYL-5-O-β-D-GALACTOFURANOSYL-D-GALACTITOL;   W/NBS
67057

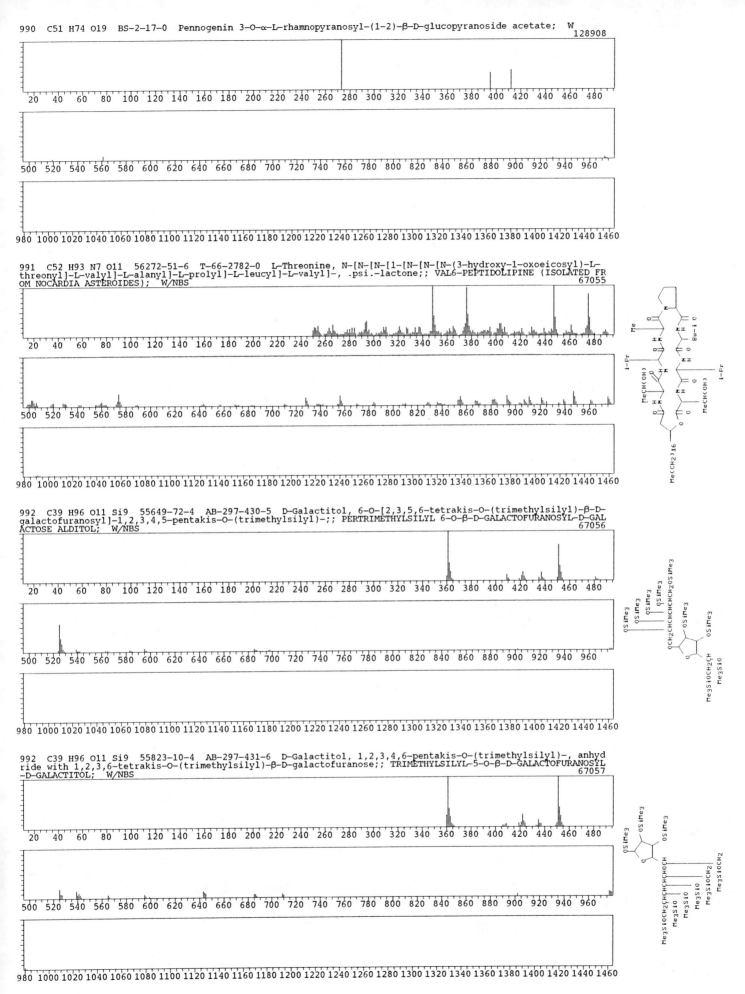

992   C56 H64 O16   74629-76-8   J-45-4499-0   2,8,14,20-Tetramethylpentacyclo-(19,3,1,1,1,1)octacosa-1(25),3,
5,7,(28),9,11,13(27),15,17,19(26),21,23-dodecaene-4,6,10,12,18,22,24-octol octapropanoate(stereoisomer);   W
128909

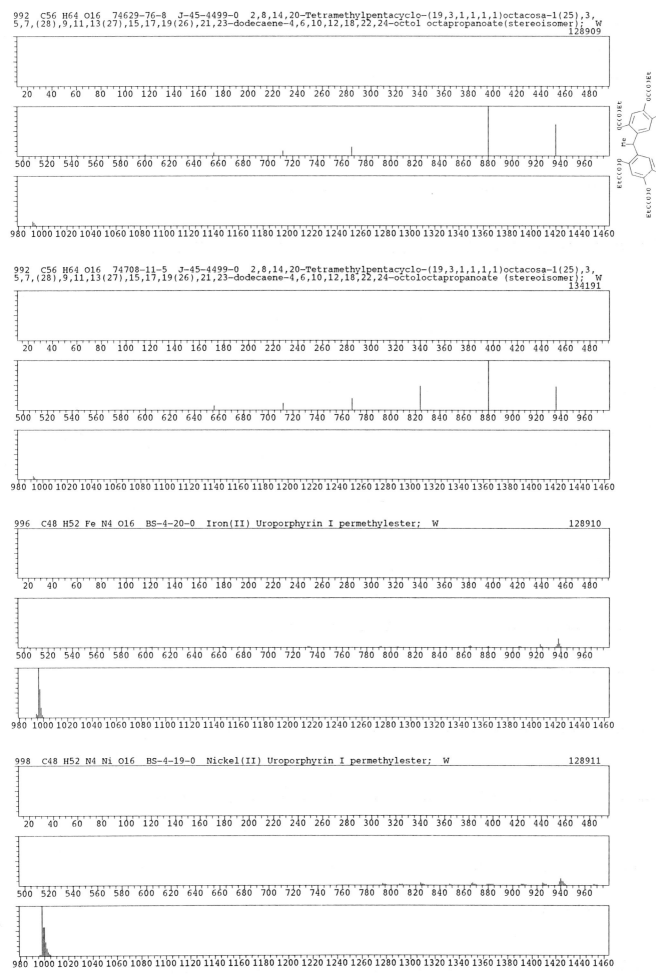

992   C56 H64 O16   74708-11-5   J-45-4499-0   2,8,14,20-Tetramethylpentacyclo-(19,3,1,1,1,1)octacosa-1(25),3,
5,7,(28),9,11,13(27),15,17,19(26),21,23-dodecaene-4,6,10,12,18,22,24-octoloctapropanoate (stereoisomer);   W
134191

996   C48 H52 Fe N4 O16   BS-4-20-0   Iron(II) Uroporphyrin I permethylester;   W   128910

998   C48 H52 N4 Ni O16   BS-4-19-0   Nickel(II) Uroporphyrin I permethylester;   W   128911

998　C55 H47 Cl O16　65615-76-1　KC-1977-18-2005　Methyl-2,6-di-O-benzoyl-3-chloro-deoxy-4-O-(2,3,4,6-tetra
-O-benzoyl-β-D-galactopyranosyl-β-D-allopyranoside;　W　　　　　　　　　128912

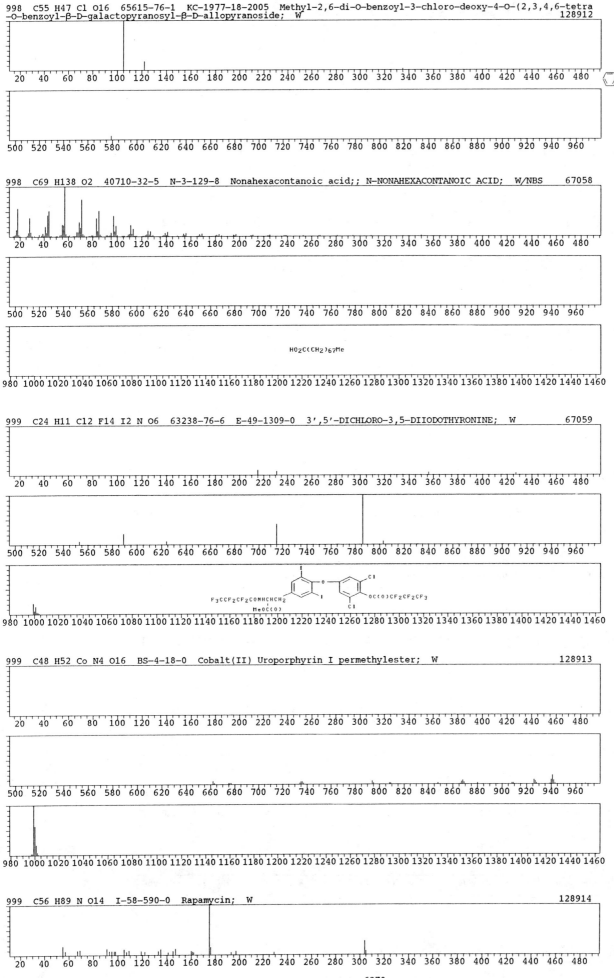

998　C69 H138 O2　40710-32-5　N-3-129-8　Nonahexacontanoic acid;; N-NONAHEXACONTANOIC ACID;　W/NBS　　67058

HO2C(CH2)67Me

999　C24 H11 Cl2 F14 I2 N O6　63238-76-6　E-49-1309-0　3',5'-DICHLORO-3,5-DIIODOTHYRONINE;　W　　67059

F3CCF2CF2CONHCHCH2　MeOC(O)　OC(O)CF2CF2CF3

999　C48 H52 Co N4 O16　BS-4-18-0　Cobalt(II) Uroporphyrin I permethylester;　W　　　　　128913

999　C56 H89 N O14　I-58-590-0　Rapamycin;　W　　　　　　　　　　　　128914

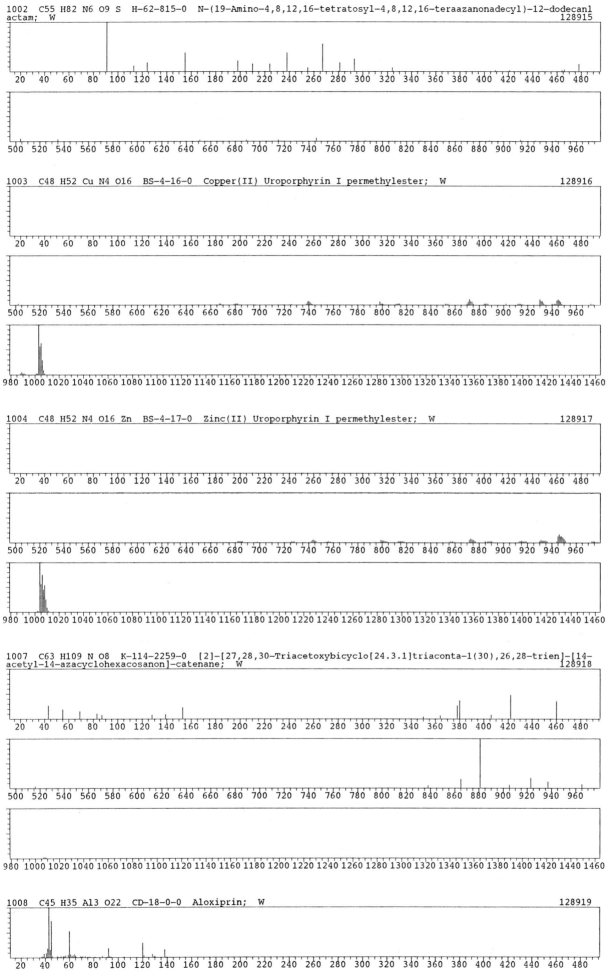

1002   C55 H82 N6 O9 S   H-62-815-0   N-(19-Amino-4,8,12,16-tetratosyl-4,8,12,16-teraazanonadecyl)-12-dodecanl
actam;   W   128915

1003   C48 H52 Cu N4 O16   BS-4-16-0   Copper(II) Uroporphyrin I permethylester;   W   128916

1004   C48 H52 N4 O16 Zn   BS-4-17-0   Zinc(II) Uroporphyrin I permethylester;   W   128917

1007   C63 H109 N O8   K-114-2259-0   [2]-[27,28,30-Triacetoxybicyclo[24.3.1]triaconta-1(30),26,28-trien]-[14-
acetyl-14-azacyclohexacosanon]-catenane;   W   128918

1008   C45 H35 Al3 O22   CD-18-0-0   Aloxiprin;   W   128919

1009　C28 H15 F28 N O7　KO-6-32-1　Tetraheptafluorobutanoyl derivative of Terbutaline;　W　128920

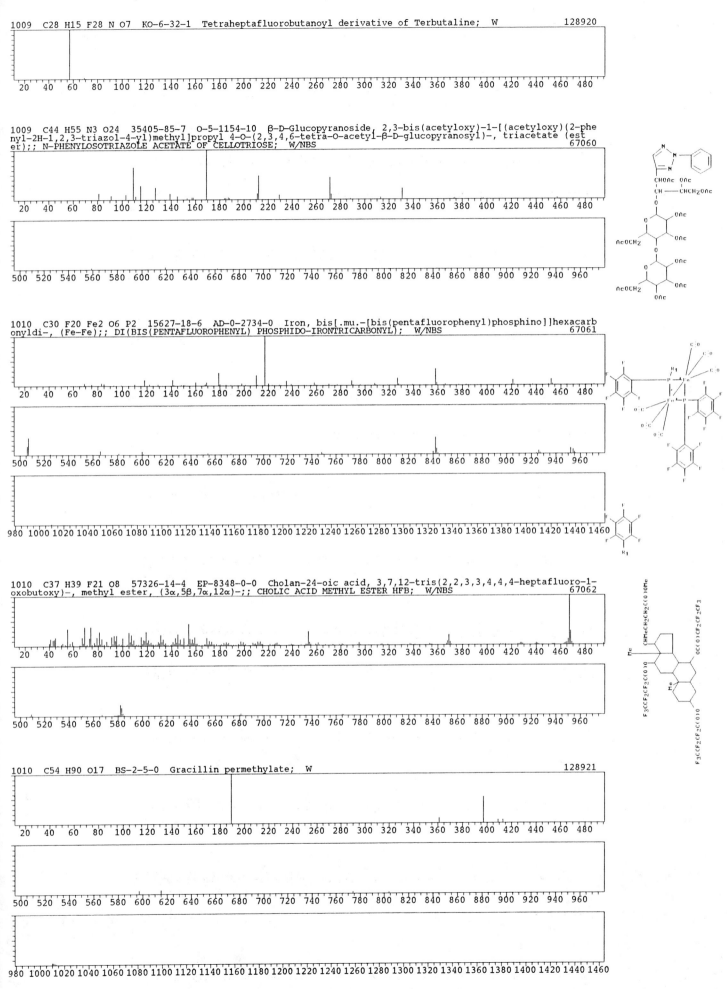

1009　C44 H55 N3 O24　35405-85-7　O-5-1154-10　β-D-Glucopyranoside, 2,3-bis(acetyloxy)-1-[(acetyloxy)(2-phenyl-2H-1,2,3-triazol-4-yl)methyl]propyl 4-O-(2,3,4,6-tetra-O-acetyl-β-D-glucopyranosyl)-, triacetate (ester);; N-PHENYLOSOTRIAZOLE ACETATE OF CELLOTRIOSE;　W/NBS　67060

1010　C30 F20 Fe2 O6 P2　15627-18-6　AD-0-2734-0　Iron, bis[.mu.-[bis(pentafluorophenyl)phosphino]]hexacarbonyldi-, (Fe-Fe);; DI(BIS(PENTAFLUOROPHENYL) PHOSPHIDO-IRONTRICARBONYL);　W/NBS　67061

1010　C37 H39 F21 O8　57326-14-4　EP-8348-0-0　Cholan-24-oic acid, 3,7,12-tris(2,2,3,3,4,4,4-heptafluoro-1-oxobutoxy)-, methyl ester, (3α,5β,7α,12α)-;; CHOLIC ACID METHYL ESTER HFB;　W/NBS　67062

1010　C54 H90 O17　BS-2-5-0　Gracillin permethylate;　W　128921

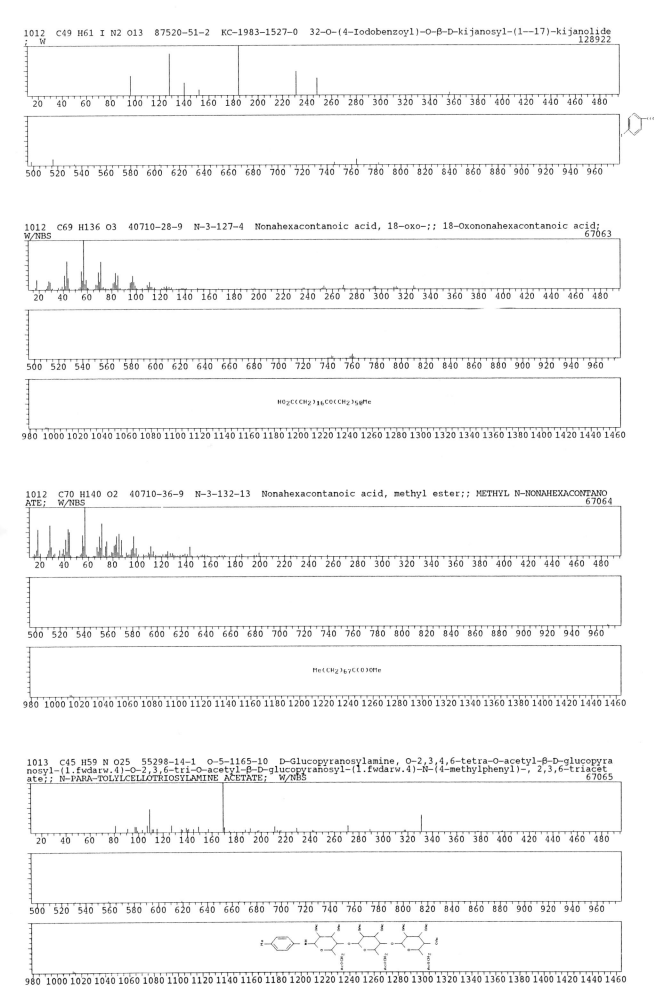

1012   C49 H61 I N2 O13   87520-51-2   KC-1983-1527-0   32-O-(4-Iodobenzoyl)-O-β-D-kijanosyl-(1--17)-kijanolide
; W
128922

1012   C69 H136 O3   40710-28-9   N-3-127-4   Nonahexacontanoic acid, 18-oxo-;; 18-Oxononahexacontanoic acid;
W/NBS
67063

HO2C(CH2)16CO(CH2)58Me

1012   C70 H140 O2   40710-36-9   N-3-132-13   Nonahexacontanoic acid, methyl ester;; METHYL N-NONAHEXACONTANO
ATE;   W/NBS
67064

Me(CH2)67C(O)OMe

1013   C45 H59 N O25   55298-14-1   O-5-1165-10   D-Glucopyranosylamine, O-2,3,4,6-tetra-O-acetyl-β-D-glucopyra
nosyl-(1.fwdarw.4)-O-2,3,6-tri-O-acetyl-β-D-glucopyranosyl-(1.fwdarw.4)-N-(4-methylphenyl)-, 2,3,6-triacet
ate;; N-PARA-TOLYLCELLOTRIOSYLAMINE ACETATE;   W/NBS
67065

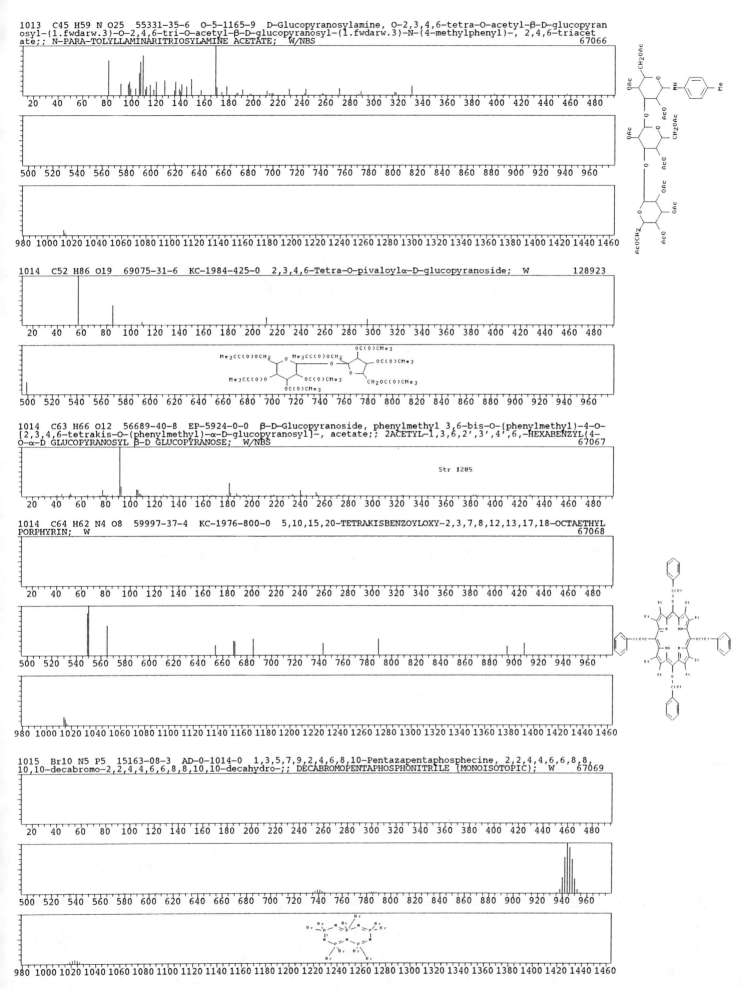

1013 C45 H59 N O25 55331-35-6 O-5-1165-9 D-Glucopyranosylamine, O-2,3,4,6-tetra-O-acetyl-β-D-glucopyran
osyl-(1.fwdarw.3)-O-2,4,6-tri-O-acetyl-β-D-glucopyranosyl-(1.fwdarw.3)-N-(4-methylphenyl)-, 2,4,6-triacet
ate;; N-PARA-TOLYLLAMINARITRIOSYLAMINE ACETATE; W/NBS   67066

1014 C52 H86 O19 69075-31-6 KC-1984-425-0 2,3,4,6-Tetra-O-pivaloylα-D-glucopyranoside; W   128923

1014 C63 H66 O12 56689-40-8 EP-5924-0-0 β-D-Glucopyranoside, phenylmethyl 3,6-bis-O-(phenylmethyl)-4-O-
[2,3,4,6-tetrakis-O-(phenylmethyl)-α-D-glucopyranosyl]-, acetate;; 2ACETYL-1,3,6,2',3',4',6,-HEXABENZYL(4-
O-α-D GLUCOPYRANOSYL β-D GLUCOPYRANOSE; W/NBS   67067

Str 1205

1014 C64 H62 N4 O8 59997-37-4 KC-1976-800-0 5,10,15,20-TETRAKISBENZOYLOXY-2,3,7,8,12,13,17,18-OCTAETHYL
PORPHYRIN; W   67068

1015 Br10 N5 P5 15163-08-3 AD-0-1014-0 1,3,5,7,9,2,4,6,8,10-Pentazapentaphosphecine, 2,2,4,4,6,6,8,8,
10,10-decabromo-2,2,4,4,6,6,8,8,10,10-decahydro-;; DECABROMOPENTAPHOSPHONITRILE (MONOISOTOPIC); W 67069

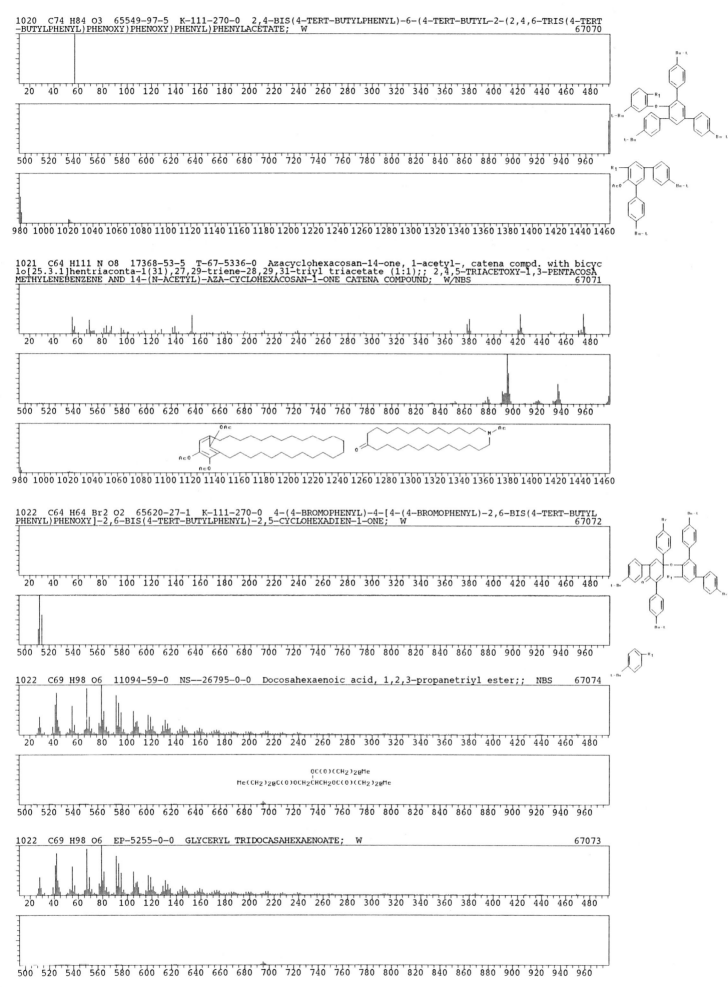

1020 C74 H84 O3 65549-97-5 K-111-270-0 2,4-BIS(4-TERT-BUTYLPHENYL)-6-(4-TERT-BUTYL-2-(2,4,6-TRIS(4-TERT
-BUTYLPHENYL)PHENOXY)PHENOXY)PHENYL)PHENYLACETATE; W   67070

1021 C64 H111 N O8 17368-53-5 T-67-5336-0 Azacyclohexacosan-14-one, 1-acetyl-, catena compd. with bicyc
lo[25.3.1]hentriaconta-1(31),27,29-triene-28,29,31-triyl triacetate (1:1);; 2,4,5-TRIACETOXY-1,3-PENTACOSA
METHYLENEBENZENE AND 14-(N-ACETYL)-AZA-CYCLOHEXACOSAN-1-ONE CATENA COMPOUND; W/NBS   67071

1022 C64 H64 Br2 O2 65620-27-1 K-111-270-0 4-(4-BROMOPHENYL)-4-[4-(4-BROMOPHENYL)-2,6-BIS(4-TERT-BUTYL
PHENYL)PHENOXY]-2,6-BIS(4-TERT-BUTYLPHENYL)-2,5-CYCLOHEXADIEN-1-ONE; W   67072

1022 C69 H98 O6 11094-59-0 NS--26795-0-0 Docosahexaenoic acid, 1,2,3-propanetriyl ester;; NBS   67074

1022 C69 H98 O6 EP-5255-0-0 GLYCERYL TRIDOCASAHEXAENOATE; W   67073

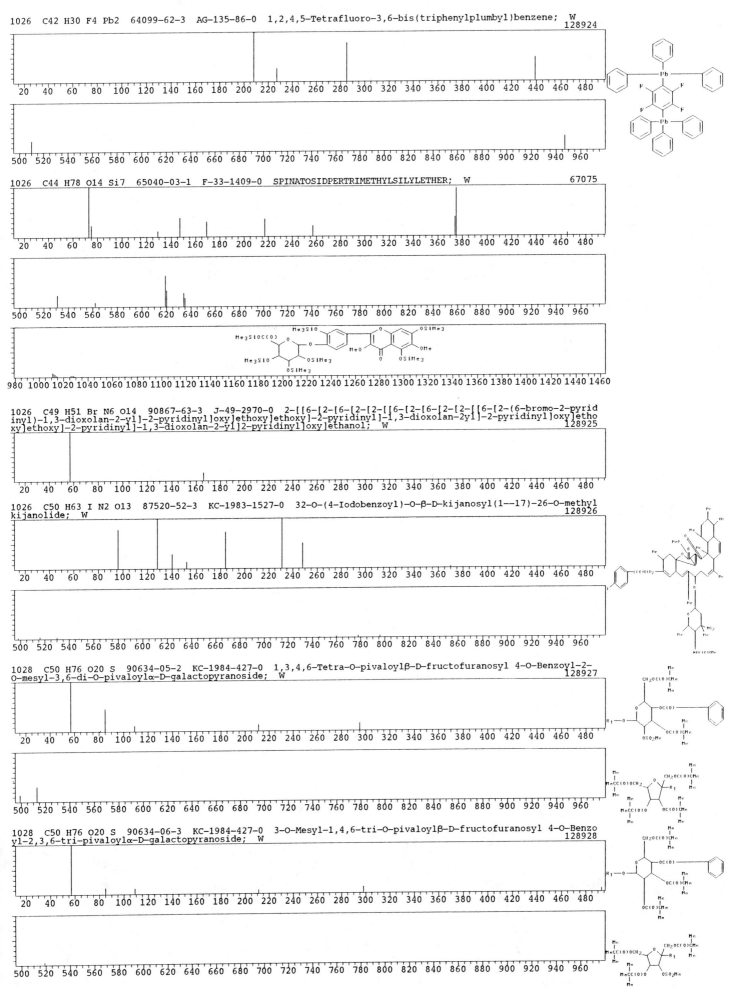

1026   C42 H30 F4 Pb2   64099-62-3   AG-135-86-0   1,2,4,5-Tetrafluoro-3,6-bis(triphenylplumbyl)benzene;   W
128924

1026   C44 H78 O14 Si7   65040-03-1   F-33-1409-0   SPINATOSIDPERTRIMETHYLSILYLETHER;   W
67075

1026   C49 H51 Br N6 O14   90867-63-3   J-49-2970-0   2-[[6-[2-[6-[2-[2-[[6-[2-[6-[2-[2-[[6-[2-(6-bromo-2-pyrid
inyl]-1,3-dioxolan-2-yl]-2-pyridinyl]oxy]ethoxy]ethoxy]-2-pyridinyl]-1,3-dioxolan-2yl]-2-pyridinyl]oxy]etho
xy]ethoxy]-2-pyridinyl]-1,3-dioxolan-2-yl]2-pyridinyl]oxy]ethanol;   W
128925

1026   C50 H63 I N2 O13   87520-52-3   KC-1983-1527-0   32-O-(4-Iodobenzoyl)-O-β-D-kijanosyl(1--17)-26-O-methyl
kijanolide;   W
128926

1028   C50 H76 O20 S   90634-05-2   KC-1984-427-0   1,3,4,6-Tetra-O-pivaloylβ-D-fructofuranosyl 4-O-Benzoyl-2-
O-mesyl-3,6-di-O-pivaloylα-D-galactopyranoside;   W
128927

1028   C50 H76 O20 S   90634-06-3   KC-1984-427-0   3-O-Mesyl-1,4,6-tri-O-pivaloylβ-D-fructofuranosyl 4-O-Benzo
yl-2,3,6-tri-pivaloylα-D-galactopyranoside;   W
128928

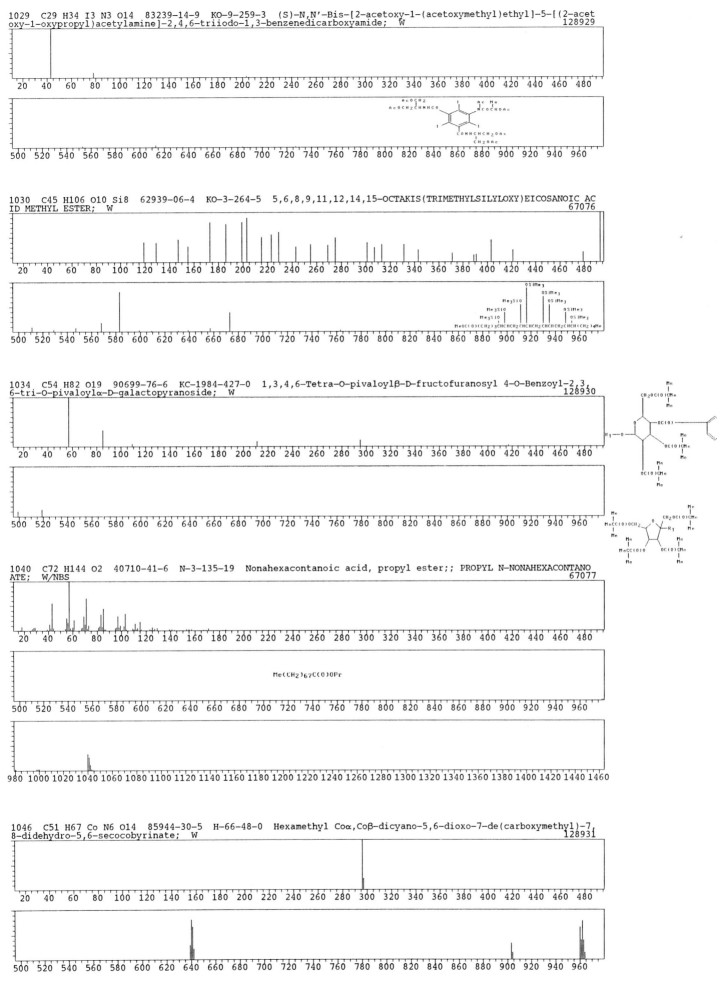

1029 C29 H34 I3 N3 O14 83239-14-9 KO-9-259-3 (S)-N,N'-Bis-[2-acetoxy-1-(acetoxymethyl)ethyl]-5-[(2-acet
oxy-1-oxypropyl)acetylamine]-2,4,6-triiodo-1,3-benzenedicarboxyamide; W   128929

1030 C45 H106 O10 Si8 62939-06-4 KO-3-264-5 5,6,8,9,11,12,14,15-OCTAKIS(TRIMETHYLSILYLOXY)EICOSANOIC AC
ID METHYL ESTER; W   67076

1034 C54 H82 O19 90699-76-6 KC-1984-427-0 1,3,4,6-Tetra-O-pivaloylβ-D-fructofuranosyl 4-O-Benzoyl-2,3,
6-tri-O-pivaloylα-D-galactopyranoside; W   128930

1040 C72 H144 O2 40710-41-6 N-3-135-19 Nonahexacontanoic acid, propyl ester;; PROPYL N-NONAHEXACONTANO
ATE; W/NBS   67077

1046 C51 H67 Co N6 O14 85944-30-5 H-66-48-0 Hexamethyl Coα,Coβ-dicyano-5,6-dioxo-7-de(carboxymethyl)-7,
8-didehydro-5,6-secocobyrinate; W   128931

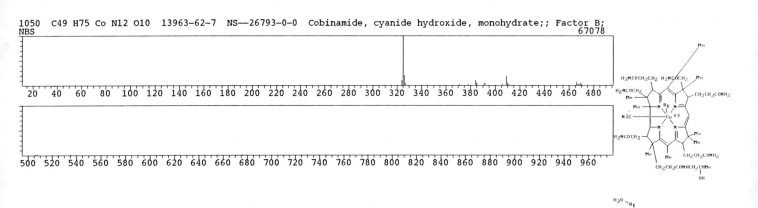

1050 C49 H75 Co N12 O10 13963-62-7 NS--26793-0-0 Cobinamide, cyanide hydroxide, monohydrate;; Factor B;
NBS 67078

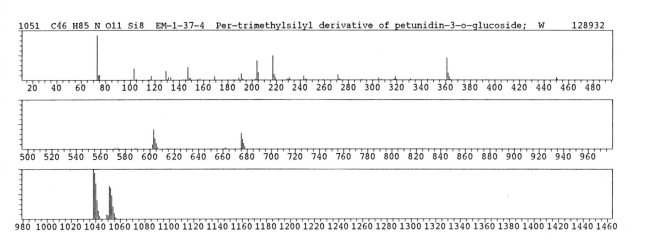

1051 C46 H85 N O11 Si8 EM-1-37-4 Per-trimethylsilyl derivative of petunidin-3-o-glucoside; W 128932

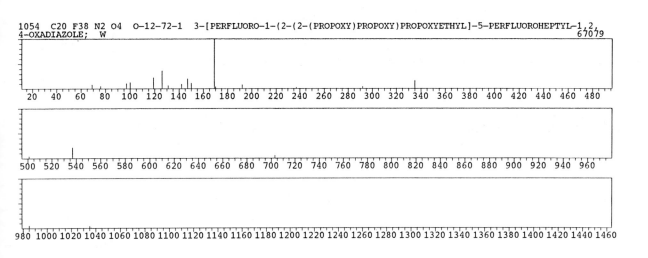

1054 C20 F38 N2 O4 O-12-72-1 3-[PERFLUORO-1-(2-(2-(PROPOXY)PROPOXY)PROPOXYETHYL]-5-PERFLUOROHEPTYL-1,2,
4-OXADIAZOLE; W 67079

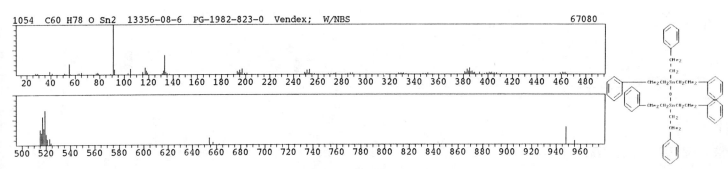

1054 C60 H78 O Sn2 13356-08-6 PG-1982-823-0 Vendex; W/NBS 67080

1057   C24 H12 F14 I3 N O6   63238-72-2   E-49-1307-0   N,O-BIS(HEPTAFLUOROBUTYRYL)-3,5,3'-TRIIODOTHYRONINE, METHYL ESTER;   W   67081

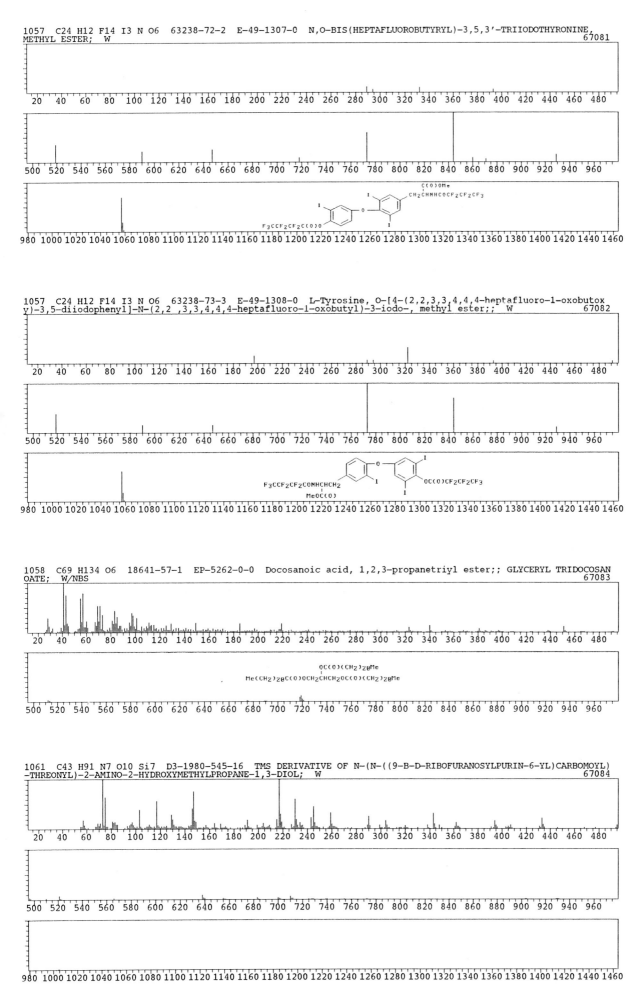

1057   C24 H12 F14 I3 N O6   63238-73-3   E-49-1308-0   L-Tyrosine, O-[4-(2,2,3,3,4,4,4-heptafluoro-1-oxobutoxy)-3,5-diiodophenyl]-N-(2,2,3,3,4,4,4-heptafluoro-1-oxobutyl)-3-iodo-, methyl ester;;   W   67082

1058   C69 H134 O6   18641-57-1   EP-5262-0-0   Docosanoic acid, 1,2,3-propanetriyl ester;; GLYCERYL TRIDOCOSANOATE;   W/NBS   67083

1061   C43 H91 N7 O10 Si7   D3-1980-545-16   TMS DERIVATIVE OF N-(N-((9-B-D-RIBOFURANOSYLPURIN-6-YL)CARBOMOYL)-THREONYL)-2-AMINO-2-HYDROXYMETHYLPROPANE-1,3-DIOL;   W   67084

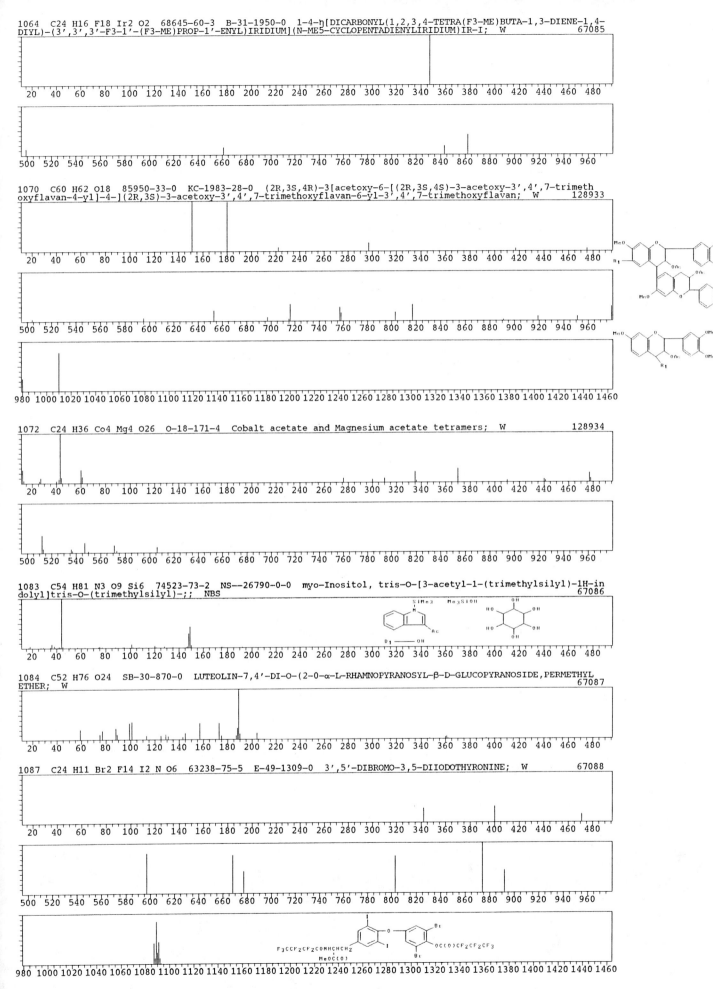

1064　C24 H16 F18 Ir2 O2　68645-60-3　B-31-1950-0　1-4-η[DICARBONYL(1,2,3,4-TETRA(F3-ME)BUTA-1,3-DIENE-1,4-DIYL)-(3',3',3'-F3-1'-(F3-ME)PROP-1'-ENYL)IRIDIUM](N-ME5-CYCLOPENTADIENYLIRIDIUM)IR-I;　W　67085

1070　C60 H62 O18　85950-33-0　KC-1983-28-0　(2R,3S,4R)-3[acetoxy-6-[(2R,3S,4S)-3-acetoxy-3',4',7-trimethoxyflavan-4-yl]-4-](2R,3S)-3-acetoxy-3',4',7-trimethoxyflavan-6-yl-3',4',7-trimethoxyflavan;　W　128933

1072　C24 H36 Co4 Mg4 O26　O-18-171-4　Cobalt acetate and Magnesium acetate tetramers;　W　128934

1083　C54 H81 N3 O9 Si6　74523-73-2　NS--26790-0-0　myo-Inositol, tris-O-[3-acetyl-1-(trimethylsilyl)-1H-indolyl]tris-O-(trimethylsilyl)-;;　NBS　67086

1084　C52 H76 O24　SB-30-870-0　LUTEOLIN-7,4'-DI-O-(2-0-α-L-RHAMNOPYRANOSYL-β-D-GLUCOPYRANOSIDE,PERMETHYL ETHER;　W　67087

1087　C24 H11 Br2 F14 I2 N O6　63238-75-5　E-49-1309-0　3',5'-DIBROMO-3,5-DIIODOTHYRONINE;　W　67088

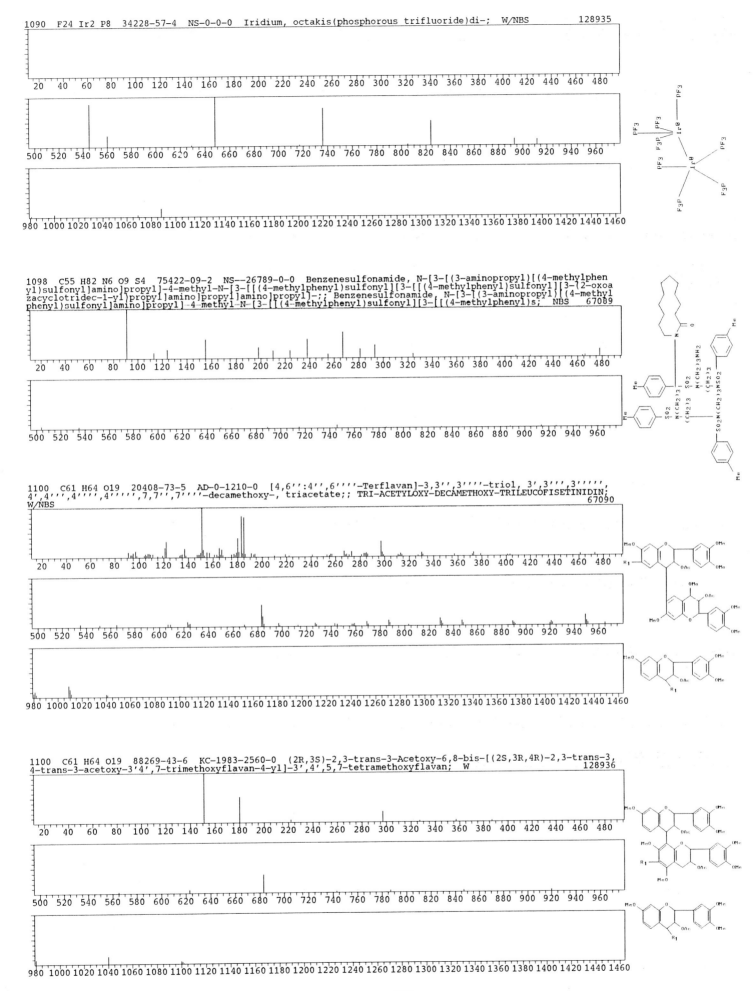

1090 F24 Ir2 P8 34228-57-4 NS-0-0-0 Iridium, octakis(phosphorous trifluoride)di-; W/NBS   128935

1098 C55 H82 N6 O9 S4 75422-09-2 NS--26789-0-0 Benzenesulfonamide, N-[3-[(3-aminopropyl)[(4-methylphen yl)sulfonyl]amino]propyl]-4-methyl-N-[3-[[(4-methylphenyl)sulfonyl][3-(2-oxoa zacyclotridec-1-yl)propyl]amino]propyl]amino]propyl]-;; Benzenesulfonamide, N-[3-[(3-aminopropyl)][(4-methyl phenyl)sulfonyl]amino]propyl]-4-methyl-N-[3-[[(4-methylphenyl)sulfonyl][3-[[(4-methylphenyl)s; NBS   67089

1100 C61 H64 O19 20408-73-5 AD-0-1210-0 [4,6'':4',6''''-Terflavan]-3,3'',3''''-triol, 3',3''',3''''', 4',4''',4''''',4''''',7,7'',7''''-decamethoxy-, triacetate;; TRI-ACETYLOXY-DECAMETHOXY-TRILEUCOFISETINIDIN; W/NBS   67090

1100 C61 H64 O19 88269-43-6 KC-1983-2560-0 (2R,3S)-2,3-trans-3-Acetoxy-6,8-bis-[(2S,3R,4R)-2,3-trans-3, 4-trans-3-acetoxy-3'4',7-trimethoxyflavan-4-yl]-3',4',5,7-tetramethoxyflavan; W   128936

- 6289 -

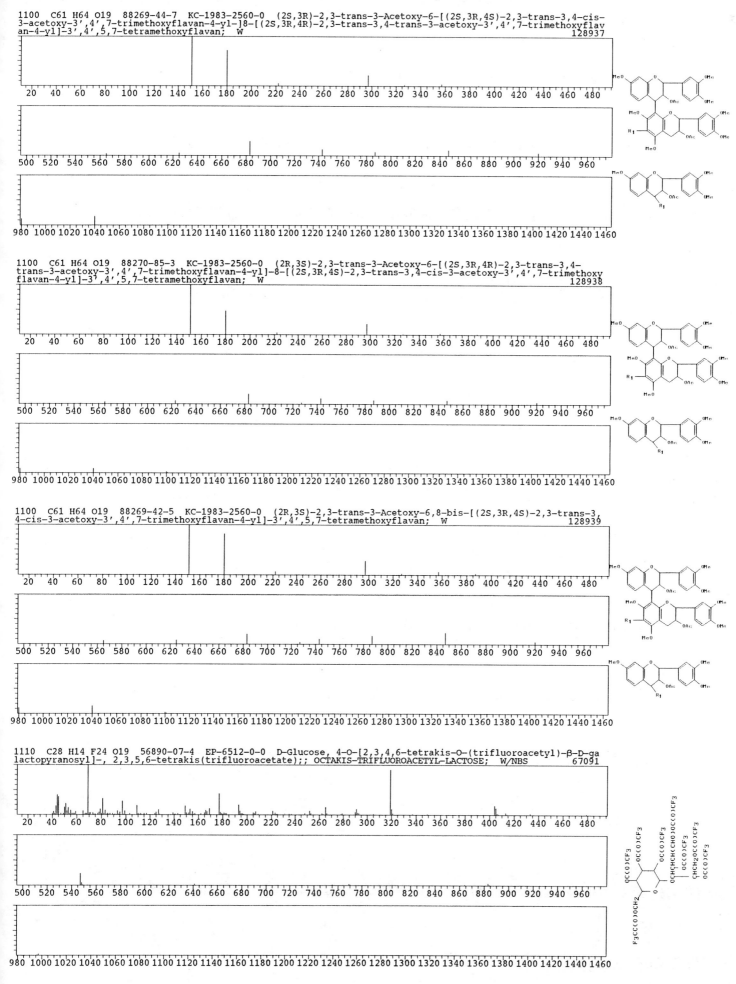

1100   C61 H64 O19   88269-44-7   KC-1983-2560-0   (2S,3R)-2,3-trans-3-Acetoxy-6-[(2S,3R,4S)-2,3-trans-3,4-cis-3-acetoxy-3',4',7-trimethoxyflavan-4-yl]-8-[(2S,3R,4R)-2,3-trans-3,4-trans-3-acetoxy-3',4',7-trimethoxyflavan-4-yl]-3',4',5,7-tetramethoxyflavan;   W   128937

1100   C61 H64 O19   88270-85-3   KC-1983-2560-0   (2R,3S)-2,3-trans-3-Acetoxy-6-[(2S,3R,4R)-2,3-trans-3,4-trans-3-acetoxy-3',4',7-trimethoxyflavan-4-yl]-8-[(2S,3R,4S)-2,3-trans-3,4-cis-3-acetoxy-3',4',7-trimethoxyflavan-4-yl]-3',4',5,7-tetramethoxyflavan;   W   128938

1100   C61 H64 O19   88269-42-5   KC-1983-2560-0   (2R,3S)-2,3-trans-3-Acetoxy-6,8-bis-[(2S,3R,4S)-2,3-trans-3,4-cis-3-acetoxy-3',4',7-trimethoxyflavan-4-yl]-3',4',5,7-tetramethoxyflavan;   W   128939

1110   C28 H14 F24 O19   56890-07-4   EP-6512-0-0   D-Glucose, 4-O-[2,3,4,6-tetrakis-O-(trifluoroacetyl)-β-D-galactopyranosyl]-, 2,3,5,6-tetrakis(trifluoroacetate);; OCTAKIS-TRIFLUOROACETYL-LACTOSE;   W/NBS   67091

1110   C30 H90 O15 Si15   23523-14-0   O-10-1043-9   TRICONTAMETHYLCYCLOPENTADECASILOXANE;   W   67092

1114   C26 H4 F34 N4 O6   O-12-72-1   1,4-BIS[5-(PERFLUORO-1-(2-(PROPOXY)PROPOXY)ETHYL)-1,2,4-OXADIAZOL-3-YL]
BENZENE;   W   67093

1114   C62 H66 O19   90634-07-4   KC-1984-427-0   3,4-Di-O-benzoyl-1,6-di-O-pivaloylβ-D-fructofuranosyl 2,3,4-
tri-O-benzoyl-6-O-pivaloylα-D-glucopyranoside;   W   128940

1124   C60 H44 O8 Th   12105-01-0   T-68-5990-0   Thorium, tetrakis(1,3-diphenyl-1,3-propanedionato-O,O')-;; TE
TRAKIS(1,3-DIPHENYL-1,3-PROPANEDIONATO)THORIUM(IV);   W/NBS   67095

1131   C18 F36 N3 O6 P3   80281-15-8   K-114-3621-0   2,2-Dichloro-4,4,5,5-tetrakis(trifluoromethyl)-1,3,2λ5-di
oxaphospholan-2-amine;   W   128941

1132   F22 O4 Te5   60788-79-6   K-110-1476-0   TRANS-TETRAKIS(PENTAFLUOROORTHOTELLURATE)DIFLUOROTELLURIUM;   W
                                                                                                              67096

1134   C42 H40 N4 O10 Re2   56535-01-4   C-97-3954-2   U-(MESOPORPHYRIN IX DIMETHYL ESTERATO)-BIS(TRICARBONYLRH
ENIUM (I));   W   67097

1136   C12 Br10 Hg   57137-96-9   B-30-1018-2   BIS(PENTABROMOPHENYL)MERCURY;   W   67098

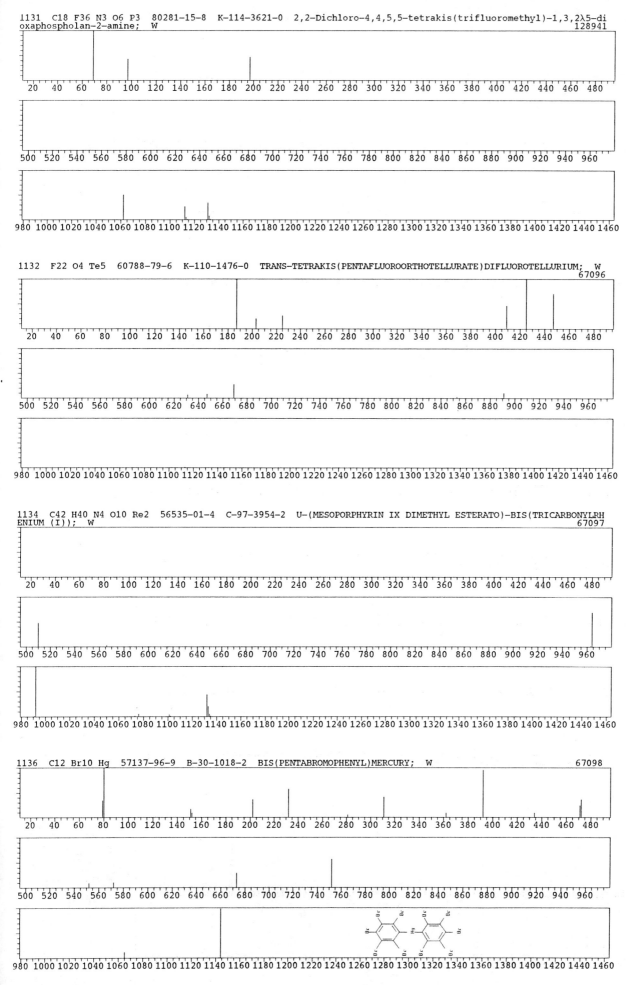

1136   C21 F40 N2 O6   O-12-72-1   5-[PERFLUORO-1-(2-(PROPOXY)PROPOXY)ETHYL]-3-[PERFLUORO-1-(2-(2-(PROPOXY)PRO
POXY)PROPOXYETHYL]-1,2,4-OXADIAZOLE;   W   67099

1138   C57 H90 O14 Si5   EP-6286-0-0   13-GLUCOPYRANOSYLGLUCOPYRANOSYLOXYDOCOSANOIC ACID DIACETATE PENTA-TMS;
W   67100

1140   C27 H15 O9 Re3 S3   17685-50-6   AD-0-1577-0   Rhenium, tris(.mu.-benzenethiolato)nonacarbonyltri-, tria
ngulo-;;   PHENYLTHIO RHENIUM-TRICARBONYL-TRIMER;   W/NBS   67101

1140   C60 H100 O20   61089-73-4   KO-2-157-1   PERMETHYLATED SAPONOSIDE;   W   67102

1146   C24 H54 N6 Se9   92074-83-4   K-117-2692-0   3,5,8,10,13,15-Hexa-tert-butyl-1,2,4,6,7,9,11,12,14-nonasel
ena-3,5,8,10,13,15-hexaazacyclopentadecane;   W   128942

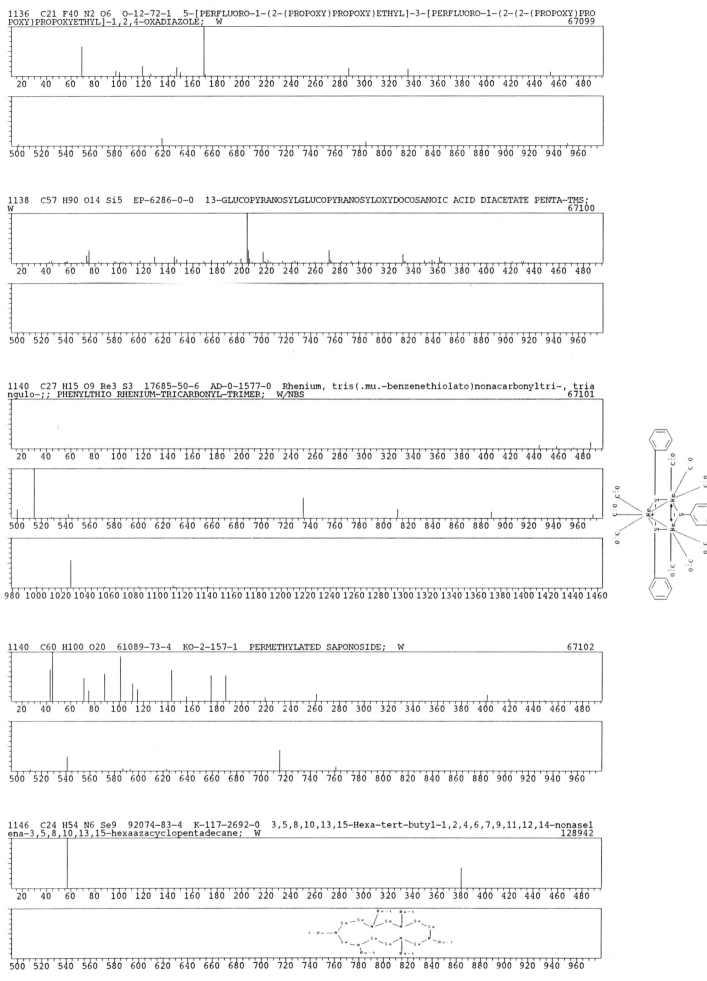

1154 C61 H102 O20 BS-2-2-0 Diosgenin 3-O-α-L-rhamnopyranosyl-(1-4)-α-L-rhamnopyranosyl-(1-4)-α-L-rhamnop
yranosyl-(1-2)-β-D-glycopyranoside permethylate; W   128943

1156 C47 H100 N4 O15 Si7 51246-90-3 F-29-2712-2 D-Streptamine, O-3-(acetylamino)-3-deoxy-2,4,6-tris-O-(
trimethylsilyl)-α-D-glucopyranosyl-(1.fwdarw.6)-O-[6-(acetylamino)-6-deoxy-2,3,4-tris-O-(trimethylsilyl)-α-
D-glucopyranosyl-(1.fwdarw.4)]-N,N'-diacetyl-2-deoxy-5-O-(trimethylsilyl)-;; N-ACETYL-O-HEPTAKISTRIMETHYLSI
LYL KANNAMYCIN A; W/NBS   67103

1158 C56 H94 N12 O14 KO-1-187-8 N,O-PERMETHYL-Nα-ACETYLLYSYLSERYLALANYLGLUTAMINYLGLUTAMINYLGLYCYLGLYCYLT
RYROSINAMIDE; W   67104

1160 C63 H68 O21 85820-28-6 KC-1983-21-0 [4,6:4,8]-all-trans-Bi-[(-)-robinetinidol]-(+)-catechin dodeca
methyl ether triacetate; W   128944

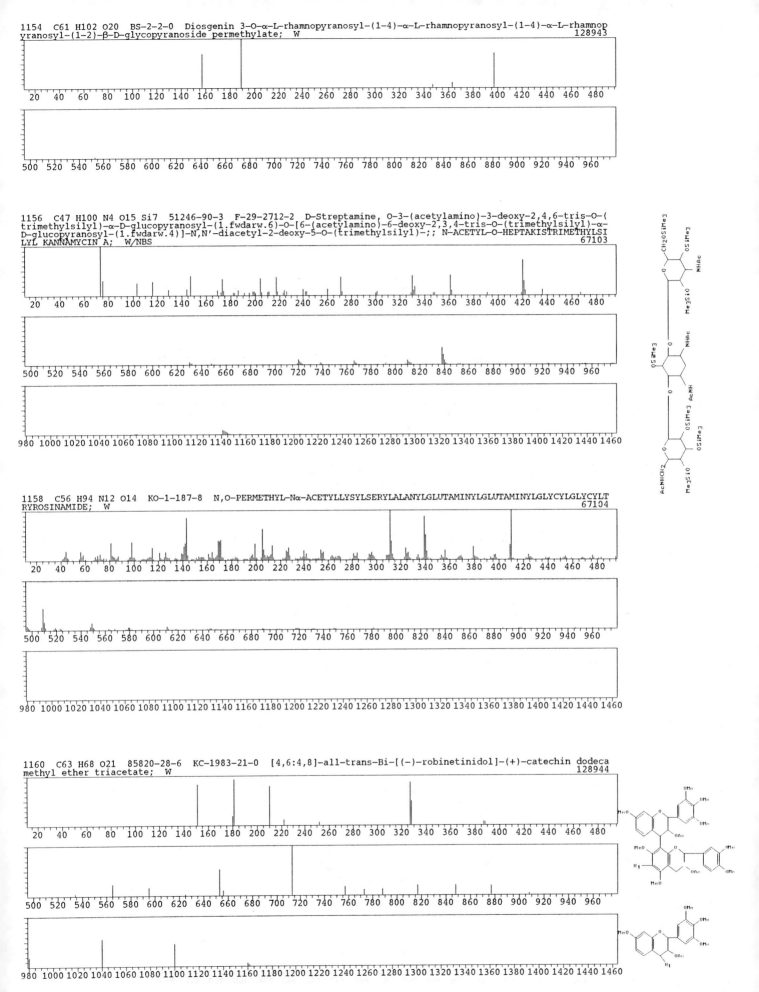

- 6294 -

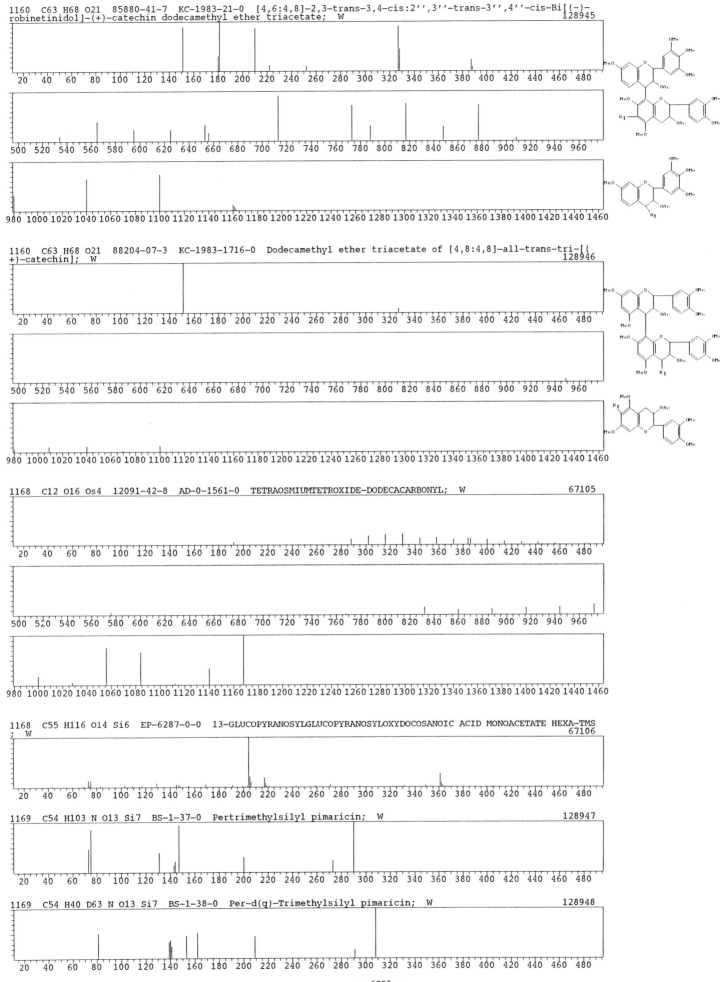

1160 C63 H68 O21 85880-41-7 KC-1983-21-0 [4,6:4,8]-2,3-trans-3,4-cis:2'',3''-trans-3'',4''-cis-Bi[(-)-robinetinidol]-(+)-catechin dodecamethyl ether triacetate; W   128945

1160 C63 H68 O21 88204-07-3 KC-1983-1716-0 Dodecamethyl ether triacetate of [(4,8:4,8]-all-trans-tri-[(+)-catechin]; W   128946

1168 C12 O16 Os4 12091-42-8 AD-0-1561-0 TETRAOSMIUMTETROXIDE-DODECACARBONYL; W   67105

1168 C55 H116 O14 Si6 EP-6287-0-0 13-GLUCOPYRANOSYLGLUCOPYRANOSYLOXYDOCOSANOIC ACID MONOACETATE HEXA-TMS ; W   67106

1169 C54 H103 N O13 Si7 BS-1-37-0 Pertrimethylsilyl pimaricin; W   128947

1169 C54 H40 D63 N O13 Si7 BS-1-38-0 Per-d(q)-Trimethylsilyl pimaricin; W   128948

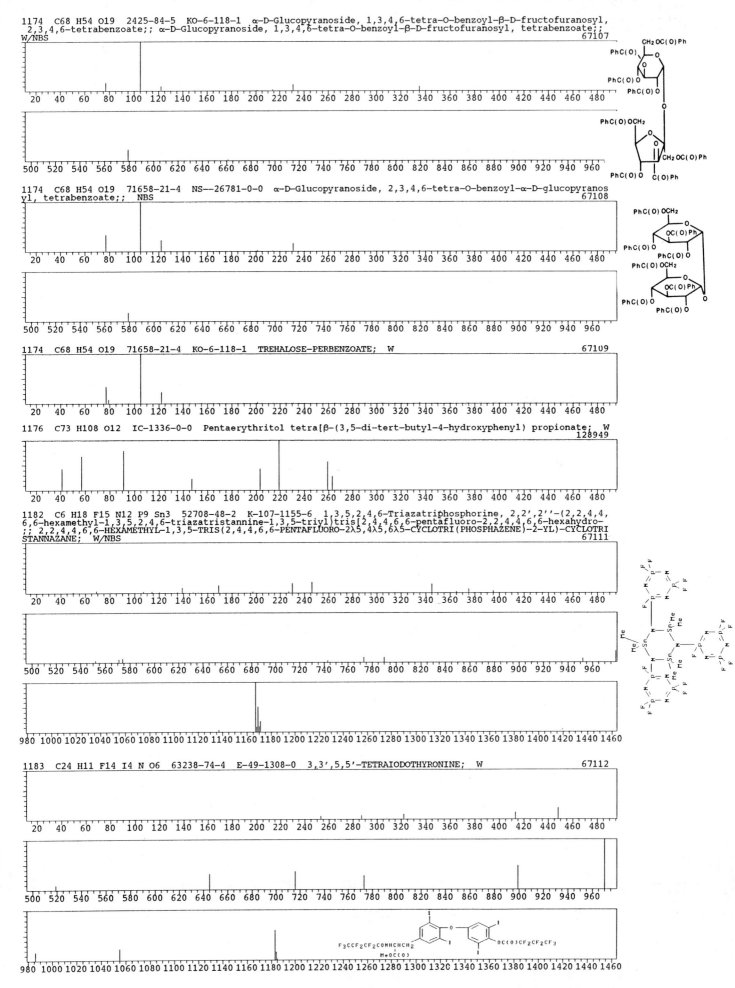

1174   C68 H54 O19   2425-84-5   KO-6-118-1   α-D-Glucopyranoside, 1,3,4,6-tetra-O-benzoyl-β-D-fructofuranosyl,
2,3,4,6-tetrabenzoate;; α-D-Glucopyranoside, 1,3,4,6-tetra-O-benzoyl-β-D-fructofuranosyl, tetrabenzoate;;
W/NBS   67107

1174   C68 H54 O19   71658-21-4   NS--26781-0-0   α-D-Glucopyranoside, 2,3,4,6-tetra-O-benzoyl-α-D-glucopyranos
yl, tetrabenzoate;;   NBS   67108

1174   C68 H54 O19   71658-21-4   KO-6-118-1   TREHALOSE-PERBENZOATE;   W   67109

1176   C73 H108 O12   IC-1336-0-0   Pentaerythritol tetra[β-(3,5-di-tert-butyl-4-hydroxyphenyl) propionate;   W
                                                                                                         128949

1182   C6 H18 F15 N12 P9 Sn3   52708-48-2   K-107-1155-6   1,3,5,2,4,6-Triazatriphosphorine, 2,2',2''-(2,2,4,4,
6,6-hexamethyl-1,3,5,2,4,6-triazatristannine-1,3,5-triyl)tris[2,4,4,6,6-pentafluoro-2,2,4,4,6,6-hexahydro-
;; 2,2,4,4,6,6-HEXAMETHYL-1,3,5-TRIS(2,4,4,6,6-PENTAFLUORO-2λ5,4λ5,6λ5-CYCLOTRI(PHOSPHAZENE)-2-YL)-CYCLOTRI
STANNAZANE;   W/NBS   67111

1183   C24 H11 F14 I4 N O6   63238-74-4   E-49-1308-0   3,3',5,5'-TETRAIODOTHYRONINE;   W   67112

- 6296 -

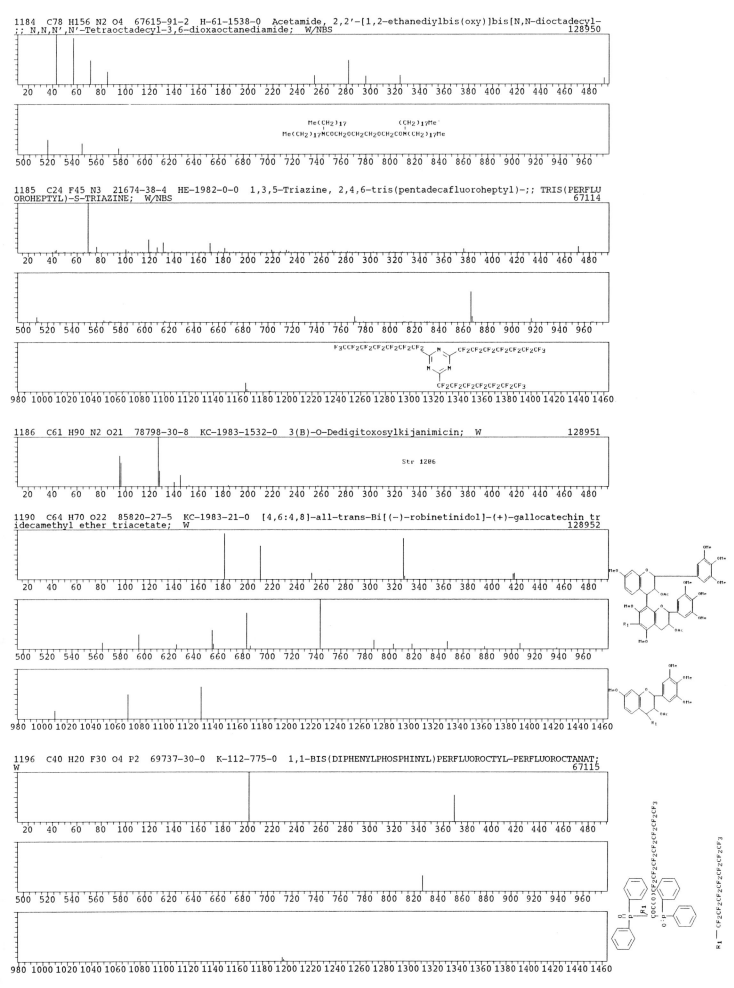

1184　C78 H156 N2 O4　67615-91-2　H-61-1538-0　Acetamide, 2,2'-[1,2-ethanediylbis(oxy)]bis[N,N-dioctadecyl-
;; N,N,N',N'-Tetraoctadecyl-3,6-dioxaoctanediamide;　W/NBS　　　　128950

Me(CH2)17　　　　　　(CH2)17Me
　　｜　　　　　　　　　｜
Me(CH2)17NCOCH2OCH2CH2OCH2CON(CH2)17Me

1185　C24 F45 N3　21674-38-4　HE-1982-0-0　1,3,5-Triazine, 2,4,6-tris(pentadecafluoroheptyl)-;; TRIS(PERFLU
OROHEPTYL)-S-TRIAZINE;　W/NBS　　　　67114

F3CCF2CF2CF2CF2CF2CF2　　CF2CF2CF2CF2CF2CF2CF3
CF2CF2CF2CF2CF2CF2CF3

1186　C61 H90 N2 O21　78798-30-8　KC-1983-1532-0　3(B)-O-Dedigitoxosylkijanimicin;　W　　　　128951

Str 1206

1190　C64 H70 O22　85820-27-5　KC-1983-21-0　[4,6:4,8]-all-trans-Bi[(-)-robinetinidol]-(+)-gallocatechin tr
idecamethyl ether triacetate;　W　　　　128952

1196　C40 H20 F30 O4 P2　69737-30-0　K-112-775-0　1,1-BIS(DIPHENYLPHOSPHINYL)PERFLUOROCTYL-PERFLUOROCTANAT;
W　　　　67115

R1 ── CF2CF2CF2CF2CF2CF2CF2CF3

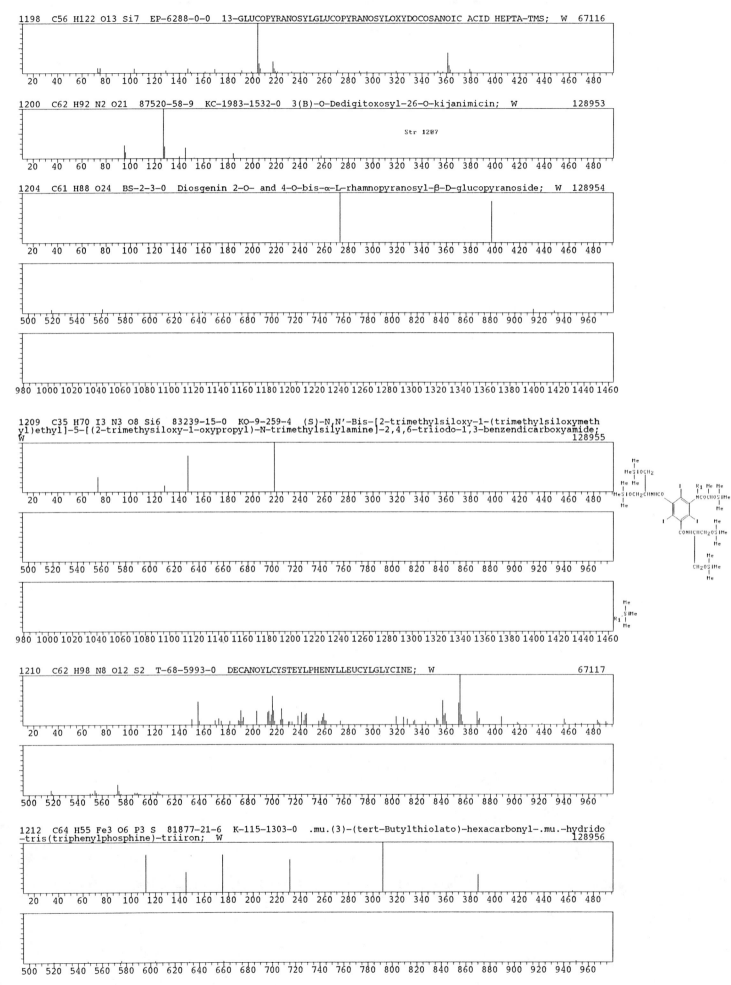

1198   C56 H122 O13 Si7   EP-6288-0-0   13-GLUCOPYRANOSYLGLUCOPYRANOSYLOXYDOCOSANOIC ACID HEPTA-TMS;   W   67116

1200   C62 H92 N2 O21   87520-58-9   KC-1983-1532-0   3(B)-O-Dedigitoxosyl-26-O-kijanimicin;   W   128953

Str 1207

1204   C61 H88 O24   BS-2-3-0   Diosgenin 2-O- and 4-O-bis-α-L-rhamnopyranosyl-β-D-glucopyranoside;   W   128954

1209   C35 H70 I3 N3 O8 Si6   83239-15-0   KO-9-259-4   (S)-N,N'-Bis-[2-trimethylsiloxy-1-(trimethylsiloxymethyl)ethyl]-5-[(2-trimethysiloxy-1-oxypropyl)-N-trimethylsilylamine]-2,4,6-triiodo-1,3-benzendicarboxyamide;   W   128955

1210   C62 H98 N8 O12 S2   T-68-5993-0   DECANOYLCYSTEYLPHENYLLEUCYLGLYCINE;   W   67117

1212   C64 H55 Fe3 O6 P3 S   81877-21-6   K-115-1303-0   .mu.(3)-(tert-Butylthiolato)-hexacarbonyl-.mu.-hydrido-tris(triphenylphosphine)-triiron;   W   128956

1218 Br12 N6 P6 17497-82-4 NS--26777-0-0 1,3,5,7,9,11-Hexaaza-2,4,6,8,10,12-hexaphosphacyclododeca-1,3,
5,7,9,11-hexaene, 2,2,4,4,6,6,8,8,10,10,12,12-dodecabromo-2,2,4,4,6,6,8,8,10,10,12,12-dodecahydro-;; DODECA
BROMOHEXAPHOSPHONITRILE (MONOISOTOPIC); NBS   79589

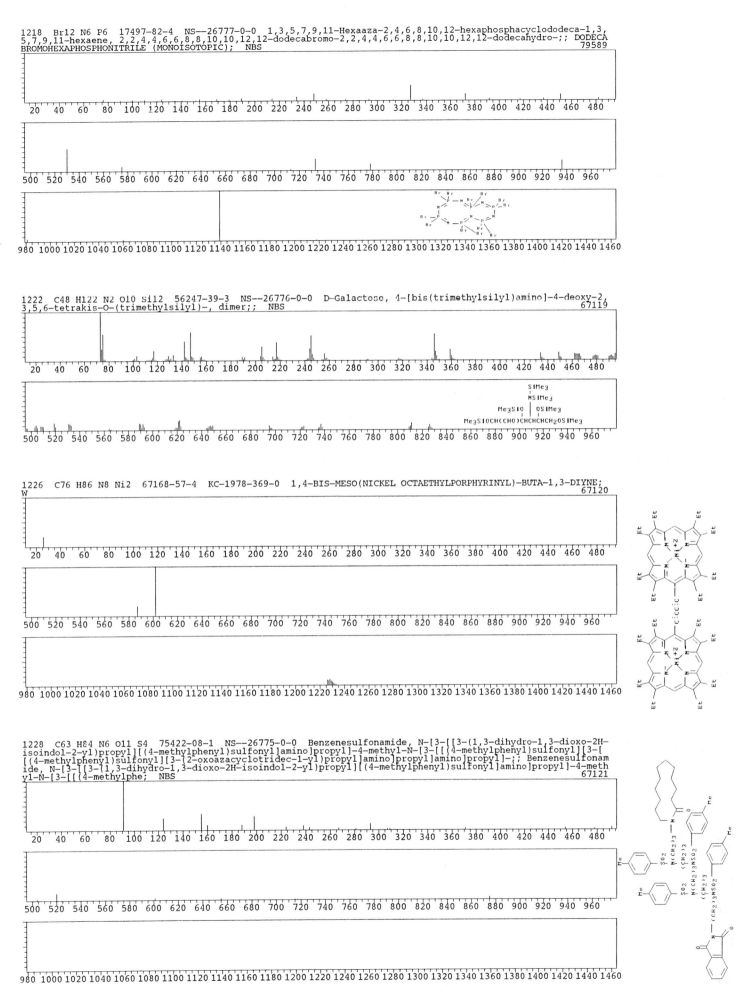

1222 C48 H122 N2 O10 Si12 56247-39-3 NS--26776-0-0 D-Galactose, 4-[bis(trimethylsilyl)amino]-4-deoxy-2,
3,5,6-tetrakis-O-(trimethylsilyl)-, dimer;; NBS   67119

1226 C76 H86 N8 Ni2 67168-57-4 KC-1978-369-0 1,4-BIS-MESO(NICKEL OCTAETHYLPORPHYRINYL)-BUTA-1,3-DIYNE;
W   67120

1228 C63 H84 N6 O11 S4 75422-08-1 NS--26775-0-0 Benzenesulfonamide, N-[3-[[3-(1,3-dihydro-1,3-dioxo-2H-
isoindol-2-yl)propyl][(4-methylphenyl)sulfonyl]amino]propyl]-4-methyl-N-[3-[[(4-methylphenyl)sulfonyl][3-[
(4-methylphenyl)sulfonyl][3-(2-oxoazacyclotridec-1-yl)propyl]amino]propyl]amino]propyl]-;; Benzenesulfonam
ide, N-[3-[[3-(1,3-dihydro-1,3-dioxo-2H-isoindol-2-yl)propyl][(4-methylphenyl)sulfonyl]amino]propyl]-4-meth
yl-N-[3-[[(4-methylphe; NBS   67121

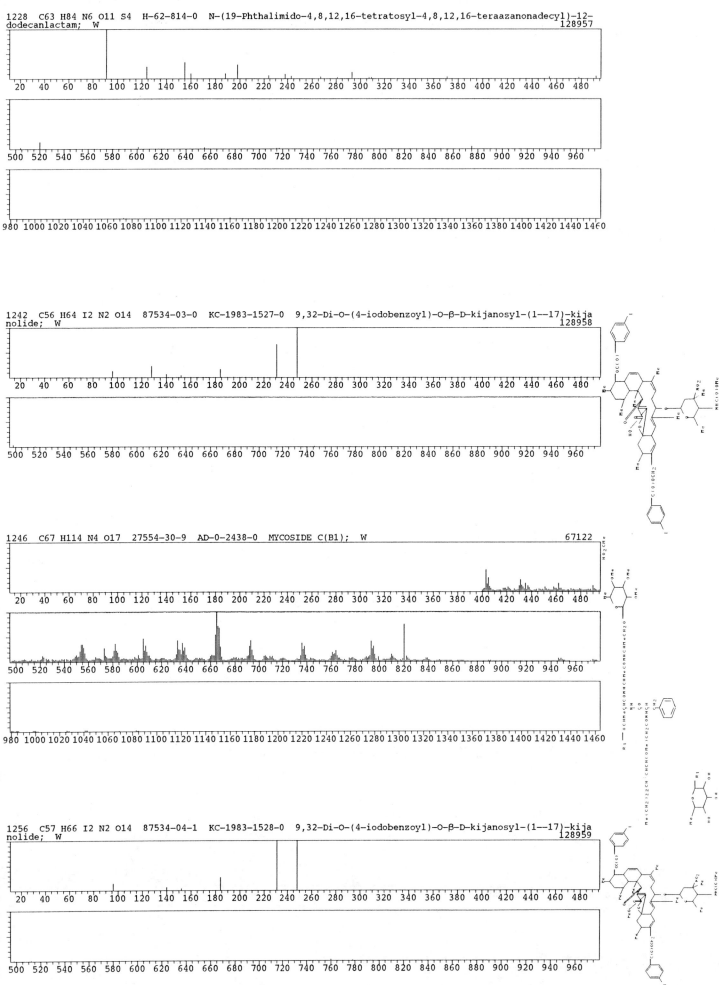

1228 C63 H84 N6 O11 S4 H-62-814-0 N-(19-Phthalimido-4,8,12,16-tetratosyl-4,8,12,16-teraazanonadecyl)-12-dodecanlactam; W   128957

1242 C56 H64 I2 N2 O14 87534-03-0 KC-1983-1527-0 9,32-Di-O-(4-iodobenzoyl)-O-β-D-kijanosyl-(1--17)-kijanolide; W   128958

1246 C67 H114 N4 O17 27554-30-9 AD-0-2438-0 MYCOSIDE C(B1); W   67122

1256 C57 H66 I2 N2 O14 87534-04-1 KC-1983-1528-0 9,32-Di-O-(4-iodobenzoyl)-O-β-D-kijanosyl-(1--17)-kijanolide; W   128959

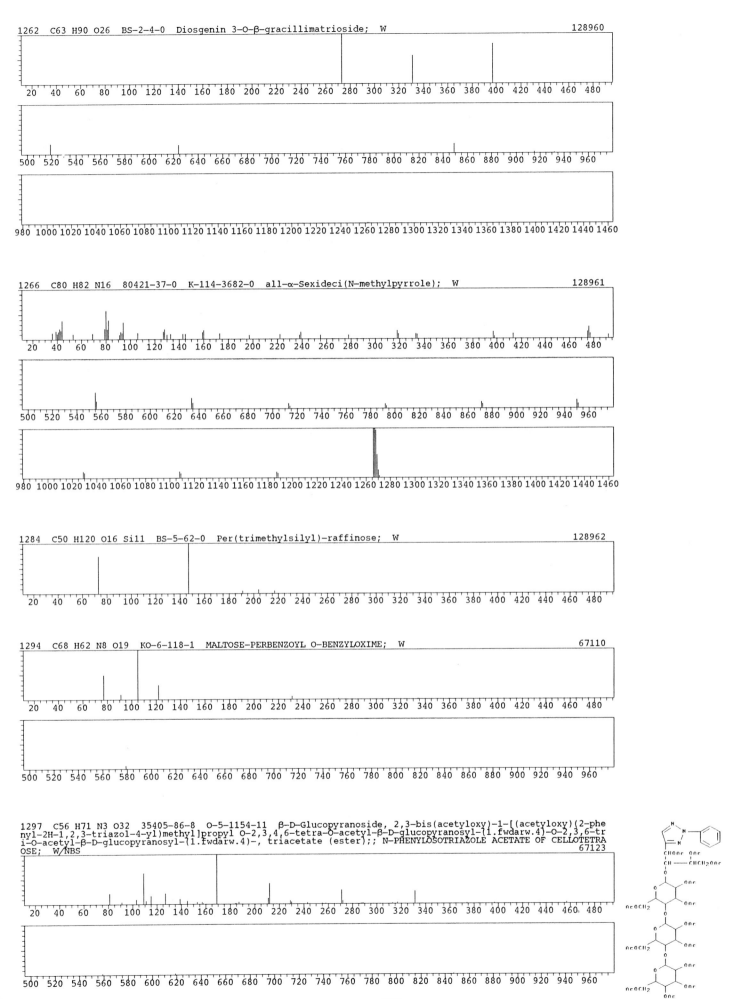

1262   C63 H90 O26   BS-2-4-0   Diosgenin 3-O-β-gracillimatrioside;   W   128960

1266   C80 H82 N16   80421-37-0   K-114-3682-0   all-α-Sexideci(N-methylpyrrole);   W   128961

1284   C50 H120 O16 Si11   BS-5-62-0   Per(trimethylsilyl)-raffinose;   W   128962

1294   C68 H62 N8 O19   KO-6-118-1   MALTOSE-PERBENZOYL O-BENZYLOXIME;   W   67110

1297   C56 H71 N3 O32   35405-86-8   O-5-1154-11   β-D-Glucopyranoside, 2,3-bis(acetyloxy)-1-[(acetyloxy)(2-phe
nyl-2H-1,2,3-triazol-4-yl)methyl]propyl O-2,3,4,6-tetra-O-acetyl-β-D-glucopyranosyl-(1.fwdarw.4)-O-2,3,6-tr
i-O-acetyl-β-D-glucopyranosyl-(1.fwdarw.4)-, triacetate (ester);; N-PHENYLOSOTRIAZOLE ACETATE OF CELLOTETRA
OSE;   W/NBS   67123

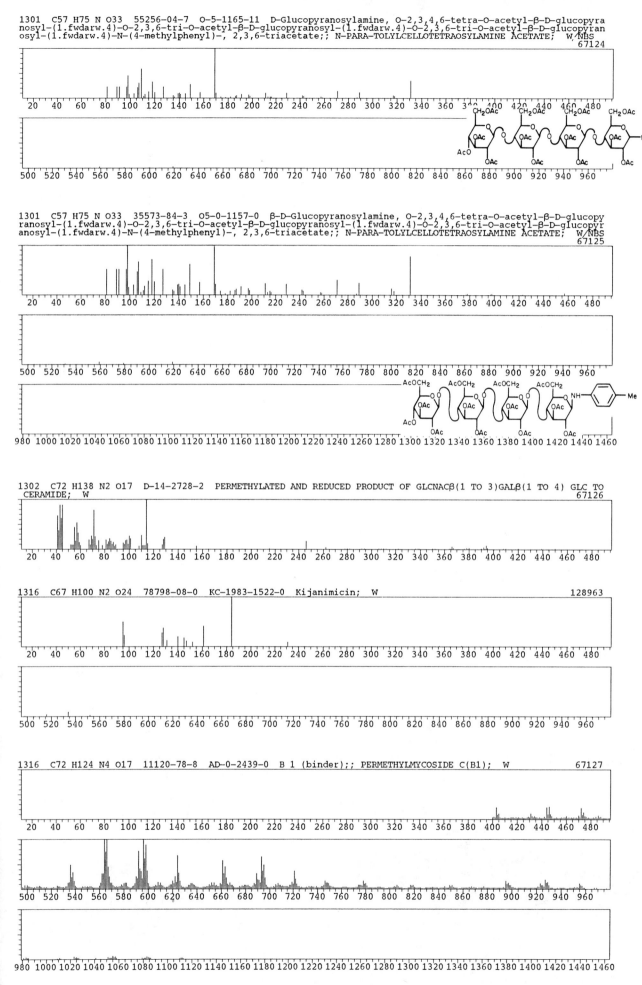

1301   C57 H75 N O33   55256-04-7   O-5-1165-11   D-Glucopyranosylamine, O-2,3,4,6-tetra-O-acetyl-β-D-glucopyra
nosyl-(1.fwdarw.4)-O-2,3,6-tri-O-acetyl-β-D-glucopyranosyl-(1.fwdarw.4)-O-2,3,6-tri-O-acetyl-β-D-glucopyran
osyl-(1.fwdarw.4)-N-(4-methylphenyl)-, 2,3,6-triacetate;; N-PARA-TOLYLCELLOTETRAOSYLAMINE ACETATE; W/NBS
67124

1301   C57 H75 N O33   35573-84-3   O5-0-1157-0   β-D-Glucopyranosylamine, O-2,3,4,6-tetra-O-acetyl-β-D-glucopy
ranosyl-(1.fwdarw.4)-O-2,3,6-tri-O-acetyl-β-D-glucopyranosyl-(1.fwdarw.4)-O-2,3,6-tri-O-acetyl-β-D-glucopyr
anosyl-(1.fwdarw.4)-N-(4-methylphenyl)-, 2,3,6-triacetate;; N-PARA-TOLYLCELLOTETRAOSYLAMINE ACETATE; W/NBS
67125

1302   C72 H138 N2 O17   D-14-2728-2   PERMETHYLATED AND REDUCED PRODUCT OF GLCNACβ(1 TO 3)GALβ(1 TO 4) GLC TO
CERAMIDE; W
67126

1316   C67 H100 N2 O24   78798-08-0   KC-1983-1522-0   Kijanimicin; W
128963

1316   C72 H124 N4 O17   11120-78-8   AD-0-2439-0   B 1 (binder);; PERMETHYLMYCOSIDE C(B1); W
67127

- 6302 -

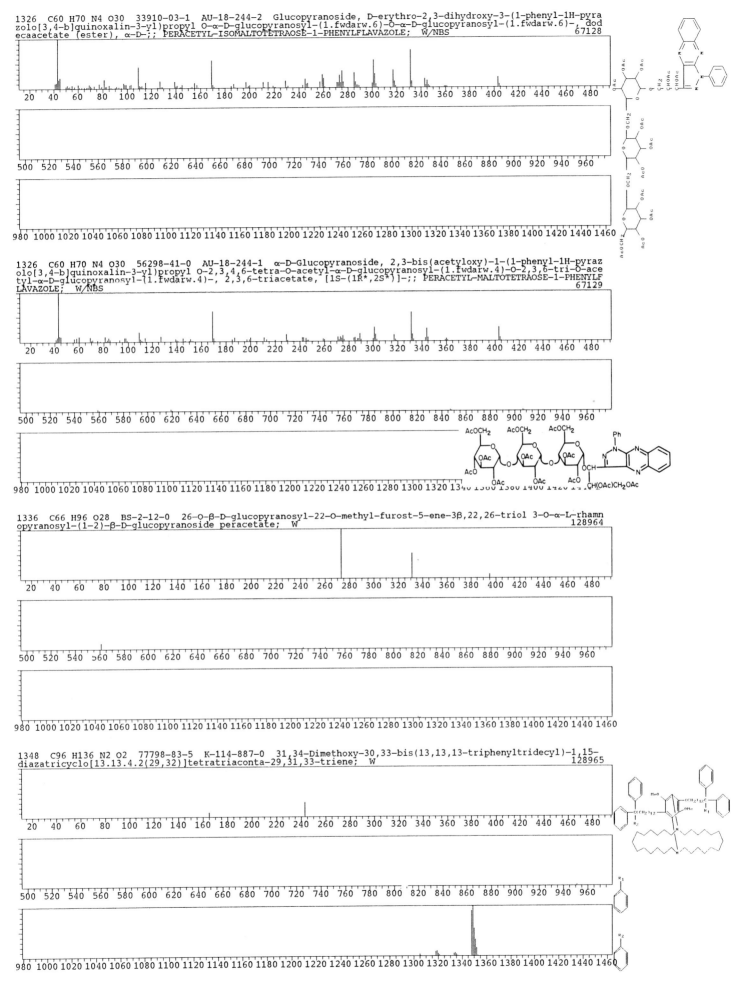

1326 C60 H70 N4 O30 33910-03-1 AU-18-244-2 Glucopyranoside, D-erythro-2,3-dihydroxy-3-(1-phenyl-1H-pyra
zolo[3,4-b]quinoxalin-3-yl)propyl O-α-D-glucopyranosyl-(1.fwdarw.6)-O-α-D-glucopyranosyl-(1.fwdarw.6)-, dod
ecaacetate (ester), α-D-;; PERACETYL-ISOMALTOTETRAOSE-1-PHENYLFLAVAZOLE; W/NBS   67128

1326 C60 H70 N4 O30 56298-41-0 AU-18-244-1 α-D-Glucopyranoside, 2,3-bis(acetyloxy)-1-(1-phenyl-1H-pyraz
olo[3,4-b]quinoxalin-3-yl)propyl O-2,3,4,6-tetra-O-acetyl-α-D-glucopyranosyl-(1.fwdarw.4)-O-2,3,6-tri-O-ace
tyl-α-D-glucopyranosyl-(1.fwdarw.4)-, 2,3,6-triacetate, [1S-(1R*,2S*)]-;; PERACETYL-MALTOTETRAOSE-1-PHENYLF
LAVAZOLE; W/NBS   67129

1336 C66 H96 O28 BS-2-12-0 26-O-β-D-glucopyranosyl-22-O-methyl-furost-5-ene-3β,22,26-triol 3-O-α-L-rhamn
opyranosyl-(1-2)-β-D-glucopyranoside peracetate; W   128964

1348 C96 H136 N2 O2 77798-83-5 K-114-887-0 31,34-Dimethoxy-30,33-bis(13,13,13-triphenyltridecyl)-1,15-
diazatricyclo[13.13.4.2(29,32)]tetratriaconta-29,31,33-triene; W   128965

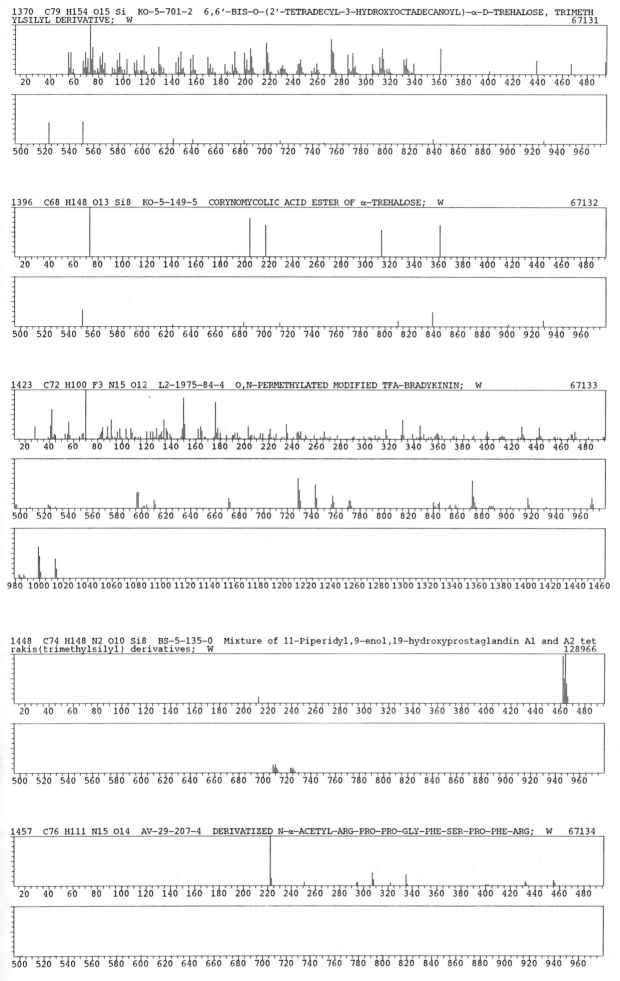

1370 C79 H154 O15 Si KO-5-701-2 6,6'-BIS-O-(2'-TETRADECYL-3-HYDROXYOCTADECANOYL)-α-D-TREHALOSE, TRIMETHYLSILYL DERIVATIVE; W   67131

1396 C68 H148 O13 Si8 KO-5-149-5 CORYNOMYCOLIC ACID ESTER OF α-TREHALOSE; W   67132

1423 C72 H100 F3 N15 O12 L2-1975-84-4 O,N-PERMETHYLATED MODIFIED TFA-BRADYKININ; W   67133

1448 C74 H148 N2 O10 Si8 BS-5-135-0 Mixture of 11-Piperidyl,9-enol,19-hydroxyprostaglandin A1 and A2 tetrakis(trimethylsilyl) derivatives; W   128966

1457 C76 H111 N15 O14 AV-29-207-4 DERIVATIZED N-α-ACETYL-ARG-PRO-PRO-GLY-PHE-SER-PRO-PHE-ARG; W 67134

1460 C96 H148 O10 R-33-83-1 ETHYL-2-O-MYCOLYL-3-O-METHYL-4-O-ACETYL-6-DEOXYPYRANOSIDE; W 67135

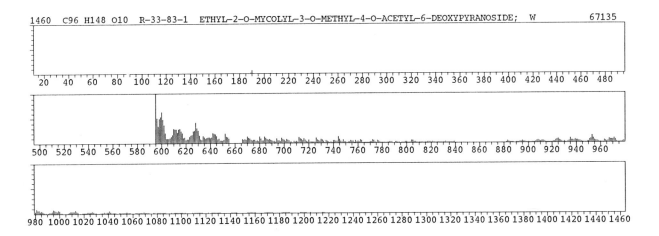

1474 C81 H158 N2 O18 Si BS-1-13-0 Hematoside,(methylated, reduced and trimethylsilylated); W 128967

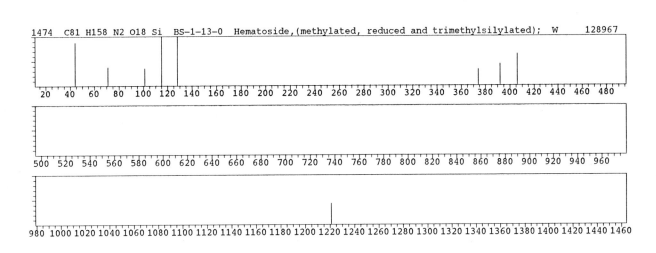

1485 C30 F57 N3 57101-59-4 KO-4-289-9 TRIS(PERFLUORONONYL)-S-TRIAZINE; W 67136

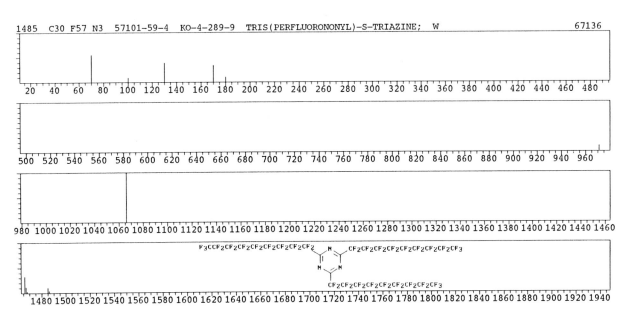

1493 C101 H139 N O8 K-106-231-5 (2)-(1,2,4-TRIACETOXY-3,5-BIS(13,13,13-TRIPHENYLTRIDECYL)BENZOYL)(1-ACET
YL-1-AZA-14-CYCLOHEXACOSANONE)-ROTAXANE; W
67137

1506 C81 H154 N2 O22 D-14-2728-2 PERMETHYLATED AND REDUCED PRODUCT OF GLOBOSIDE; W
67138

1520 C36 H20 O12 Re4 S4 23686-30-8 AD-0-1578-0 Rhenium, tetrakis[.mu.3-(benzenethiolato)]dodecacarbonyl
tetra-;; PHENYLTHIO RHENIUM-TRICARBONYL-TETRAMER; W/NBS
67139

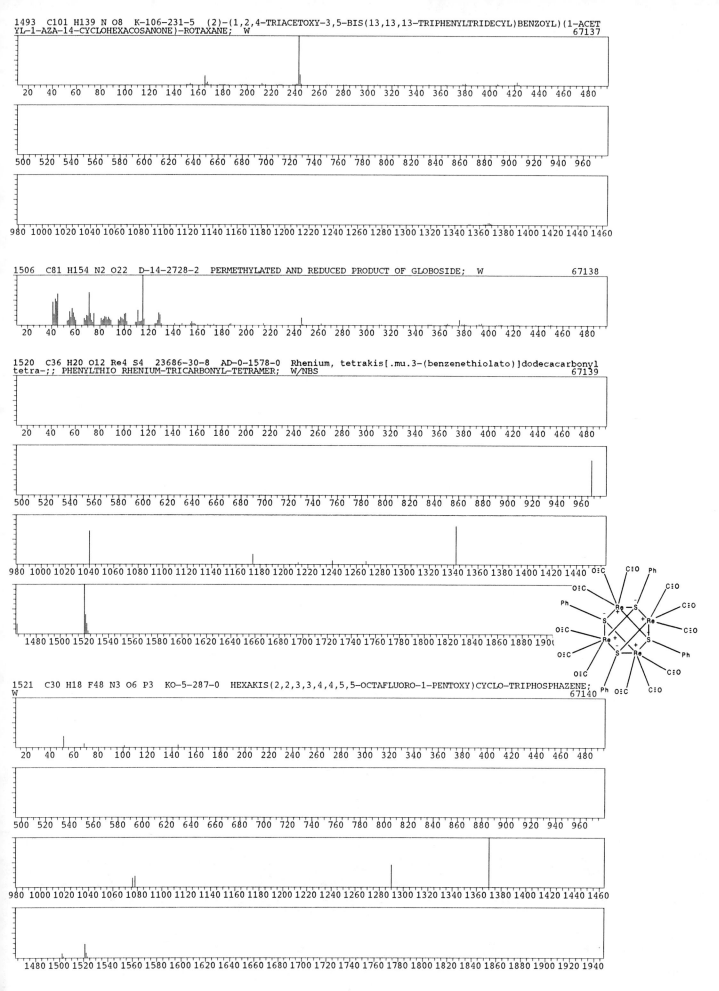

1521 C30 H18 F48 N3 O6 P3 KO-5-287-0 HEXAKIS(2,2,3,3,4,4,5,5-OCTAFLUORO-1-PENTOXY)CYCLO-TRIPHOSPHAZENE;
W
67140

1536   C74 H104 O34   BS-2-1-0   Diosgenin 3-o-β-lycotetraoside;   W

128968

| 20 | 40 | 60 | 80 | 100 | 120 | 140 | 160 | 180 | 200 | 220 | 240 | 260 | 280 | 300 | 320 | 340 | 360 | 380 | 400 | 420 | 440 | 460 | 480 |

| 500 | 520 | 540 | 560 | 580 | 600 | 620 | 640 | 660 | 680 | 700 | 720 | 740 | 760 | 780 | 800 | 820 | 840 | 860 | 880 | 900 | 920 | 940 | 960 |

1536   C86 H114 Cl2 Fe2 N8 O2 S2   80057-75-6   KC-1981-2628-0   Bis-[6-(meso-octaethylporphyrinylmethoxy)hexyl]disulphide iron;   W

128969

| 20 | 40 | 60 | 80 | 100 | 120 | 140 | 160 | 180 | 200 | 220 | 240 | 260 | 280 | 300 | 320 | 340 | 360 | 380 | 400 | 420 | 440 | 460 | 480 |

| 500 | 520 | 540 | 560 | 580 | 600 | 620 | 640 | 660 | 680 | 700 | 720 | 740 | 760 | 780 | 800 | 820 | 840 | 860 | 880 | 900 | 920 | 940 | 960 |

1566   C76 H110 O34   BS-2-10-0   26-O-β-D-glucopyranosyl-22-O-methyl-furost-5-ene-3β,22,26-triol 3-O-β-chacotrioside peracetate;   W

128970

| 20 | 40 | 60 | 80 | 100 | 120 | 140 | 160 | 180 | 200 | 220 | 240 | 260 | 280 | 300 | 320 | 340 | 360 | 380 | 400 | 420 | 440 | 460 | 480 |

| 500 | 520 | 540 | 560 | 580 | 600 | 620 | 640 | 660 | 680 | 700 | 720 | 740 | 760 | 780 | 800 | 820 | 840 | 860 | 880 | 900 | 920 | 940 | 960 |

| 980 | 1000 | 1020 | 1040 | 1060 | 1080 | 1100 | 1120 | 1140 | 1160 | 1180 | 1200 | 1220 | 1240 | 1260 | 1280 | 1300 | 1320 | 1340 | 1360 | 1380 | 1400 | 1420 | 1440 | 1460 |

| 1480 | 1500 | 1520 | 1540 | 1560 | 1580 | 1600 | 1620 | 1640 | 1660 | 1680 | 1700 | 1720 | 1740 | 1760 | 1780 | 1800 | 1820 | 1840 | 1860 | 1880 | 1900 | 1920 | 1940 |

1585   C68 H87 N3 O40   35405-87-9   O-5-1154-12   β-D-Glucopyranoside, 2,3-bis(acetyloxy)-1-[(acetyloxy)(2-phenyl-2H-1,2,3-triazol-4-yl)methyl]propyl O-2,3,4,6-tetra-O-acetyl-β-D-glucopyranosyl-(1.fwdarw.4)-O-2,3,6-tri-O-acetyl-β-D-glucopyranosyl-(1.fwdarw.4)-O-2,3,6-tri-O-acetyl-β-D-glucopyranosyl-(1.fwdarw.4)-, triacetate (ester);; N-PHENYLOSOTRIAZOLE ACETATE OF CELLOPENTAOSEE;   W/NBS

67141

| 20 | 40 | 60 | 80 | 100 | 120 | 140 | 160 | 180 | 200 | 220 | 240 | 260 | 280 | 300 | 320 | 340 | 360 | 380 | 400 | 420 | 440 | 460 | 480 |

| 500 | 520 | 540 | 560 | 580 | 600 | 620 | 640 | 660 | 680 | 700 | 720 | 740 | 760 | 780 | 800 | 820 | 840 | 860 | 880 | 900 | 920 | 940 | 960 |

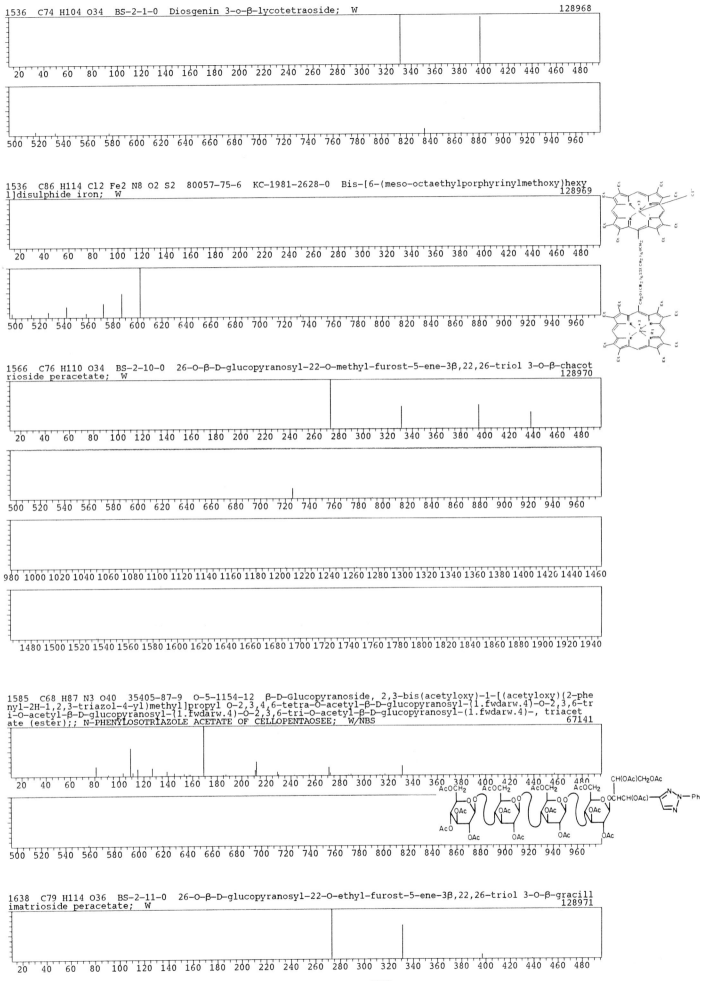

1638   C79 H114 O36   BS-2-11-0   26-O-β-D-glucopyranosyl-22-O-ethyl-furost-5-ene-3β,22,26-triol 3-O-β-gracillimatrioside peracetate;   W

128971

| 20 | 40 | 60 | 80 | 100 | 120 | 140 | 160 | 180 | 200 | 220 | 240 | 260 | 280 | 300 | 320 | 340 | 360 | 380 | 400 | 420 | 440 | 460 | 480 |

1674　C66 H154 O21 Si14　32831-69-9　AU-25-295-1　Glucopyranoside, O-1,3,4,6-tetrakis-O-(trimethylsilyl)-β-D-fructofuranosyl-(2.fwdarw.1)-O-3,4,6-tris-O-(trimethylsilyl)-β-D-fructofuranosyl-(2.fwdarw.1)-3,4,6-tris-O-(trimethylsilyl)-β-D-fructofuranosyl, α-D-;; PER-O-TRIMETHYLSILYL NYSTOSE;　W/NBS　　　　67142

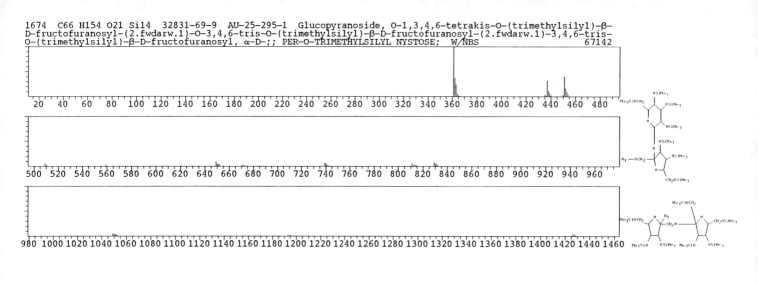

1679　C83 H61 N O17 Si12　BS-1-36-0　Per-d(g)-Trimethylsilyl amphotericin B;　W　　　　128972

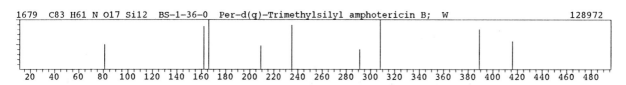

1737　C89 H163 N3 O29　BS-4-4-0　Sialyl-sialyl-dihexosyl ceramide permethylate;　W　　　　128973

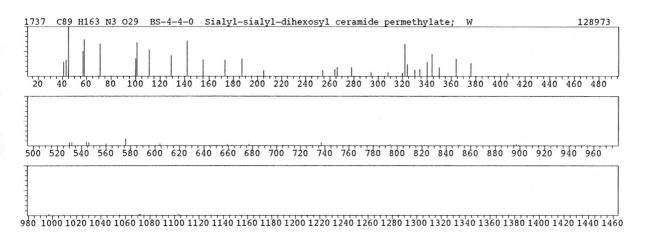

1760　C102 H146 Cl2 Fe2 N8 O2 S2　80057-76-7　KC-1981-2628-0　Bis-[6-(meso-1,3,5,7-tetramethyl-2,4,6,8-tetra-(n-pentyl)porphyrinylmethoxy)hexyl]disulphide iron;　W　　　　128974

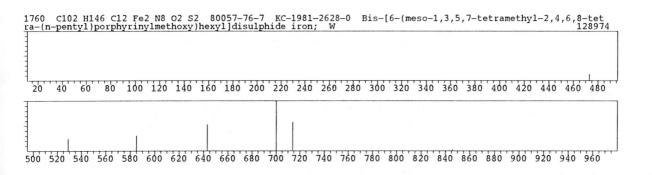

1780　C94 H180 N4 O26　D-14-272-2　PERMETHYLATED AND REDUCED PRODUCT OF DEGRADATION PRODUCT FROM H3-GLYCOLIPID BY L-L-FUCOSIDASE AND BY B-GALACTOSIDASE;　W　　　　67143

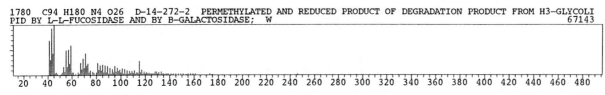

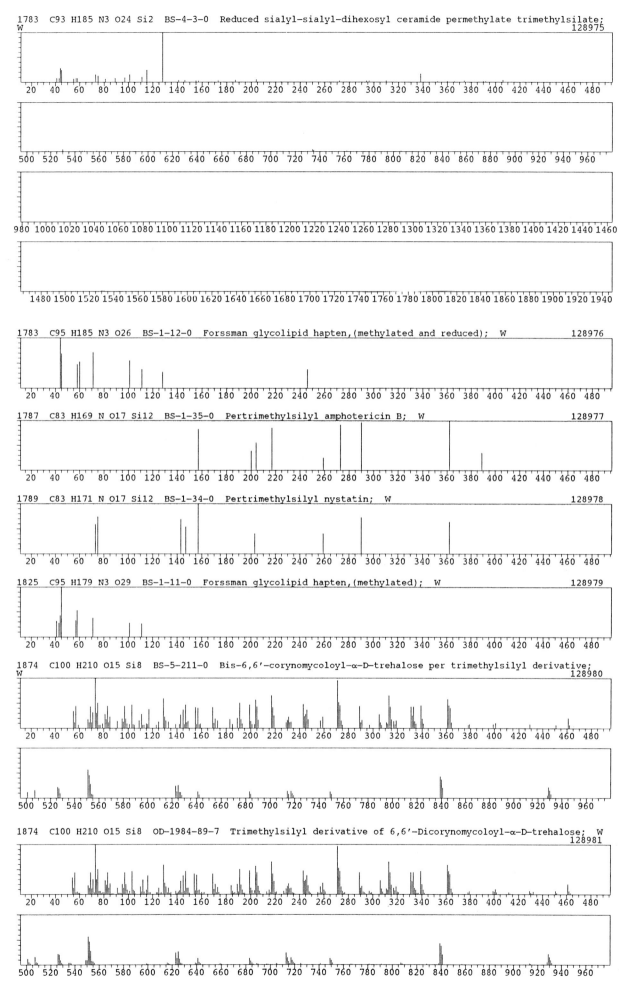

1783　C93 H185 N3 O24 Si2　BS-4-3-0　Reduced sialyl-sialyl-dihexosyl ceramide permethylate trimethylsilate;
W
128975

1783　C95 H185 N3 O26　BS-1-12-0　Forssman glycolipid hapten,(methylated and reduced);　W　128976

1787　C83 H169 N O17 Si12　BS-1-35-0　Pertrimethylsilyl amphotericin B;　W　128977

1789　C83 H171 N O17 Si12　BS-1-34-0　Pertrimethylsilyl nystatin;　W　128978

1825　C95 H179 N3 O29　BS-1-11-0　Forssman glycolipid hapten,(methylated);　W　128979

1874　C100 H210 O15 Si8　BS-5-211-0　Bis-6,6'-corynomycoloyl-α-D-trehalose per trimethylsilyl derivative;
W
128980

1874　C100 H210 O15 Si8　OD-1984-89-7　Trimethylsilyl derivative of 6,6'-Dicorynomycoloyl-α-D-trehalose;　W
128981

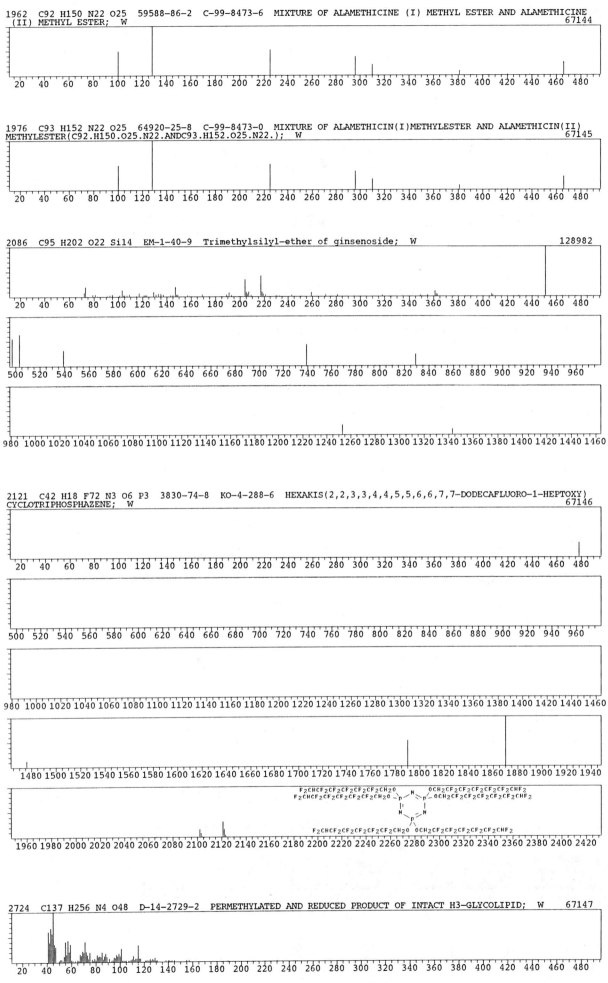

1962   C92 H150 N22 O25   59588-86-2   C-99-8473-6   MIXTURE OF ALAMETHICINE (I) METHYL ESTER AND ALAMETHICINE (II) METHYL ESTER;   W
67144

1976   C93 H152 N22 O25   64920-25-8   C-99-8473-0   MIXTURE OF ALAMETHICIN(I)METHYLESTER AND ALAMETHICIN(II)METHYLESTER(C92.H150.O25.N22.ANDC93.H152.O25.N22.);   W
67145

2086   C95 H202 O22 Si14   EM-1-40-9   Trimethylsilyl-ether of ginsenoside;   W
128982

2121   C42 H18 F72 N3 O6 P3   3830-74-8   KO-4-288-6   HEXAKIS(2,2,3,3,4,4,5,5,6,6,7,7-DODECAFLUORO-1-HEPTOXY)CYCLOTRIPHOSPHAZENE;   W
67146

2724   C137 H256 N4 O48   D-14-2729-2   PERMETHYLATED AND REDUCED PRODUCT OF INTACT H3-GLYCOLIPID;   W   67147

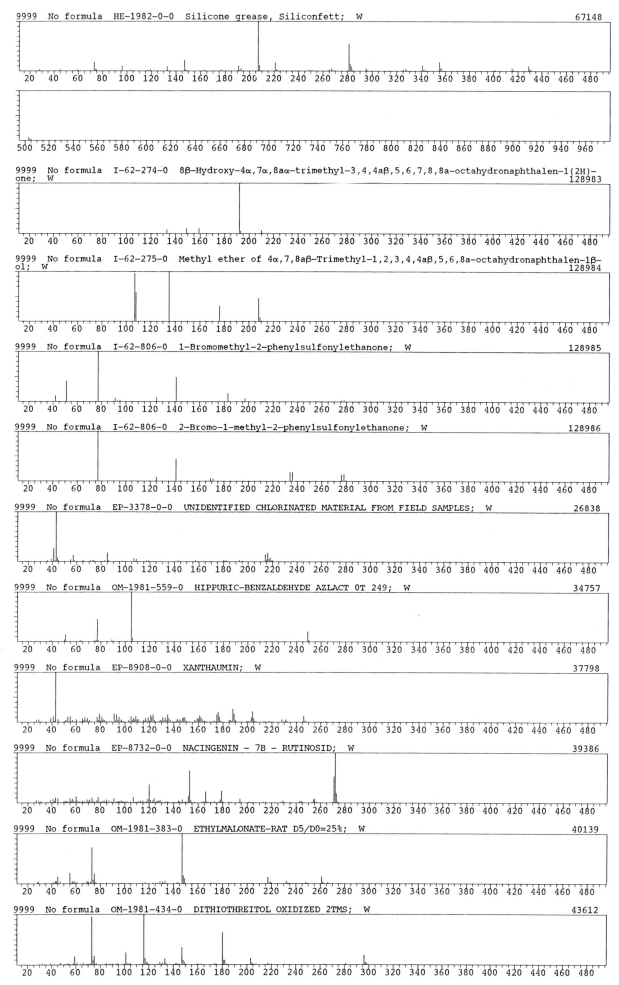

9999   No formula   HE-1982-0-0   Silicone grease, Siliconfett;   W   67148

9999   No formula   I-62-274-0   8β-Hydroxy-4α,7α,8aα-trimethyl-3,4,4aβ,5,6,7,8,8a-octahydronaphthalen-1(2H)-one;   W   128983

9999   No formula   I-62-275-0   Methyl ether of 4α,7,8aβ-Trimethyl-1,2,3,4,4aβ,5,6,8a-octahydronaphthalen-1β-ol;   W   128984

9999   No formula   I-62-806-0   1-Bromomethyl-2-phenylsulfonylethanone;   W   128985

9999   No formula   I-62-806-0   2-Bromo-1-methyl-2-phenylsulfonylethanone;   W   128986

9999   No formula   EP-3378-0-0   UNIDENTIFIED CHLORINATED MATERIAL FROM FIELD SAMPLES;   W   26838

9999   No formula   OM-1981-559-0   HIPPURIC-BENZALDEHYDE AZLACT OT 249;   W   34757

9999   No formula   EP-8908-0-0   XANTHAUMIN;   W   37798

9999   No formula   EP-8732-0-0   NACINGENIN - 7B - RUTINOSID;   W   39386

9999   No formula   OM-1981-383-0   ETHYLMALONATE-RAT D5/D0=25%;   W   40139

9999   No formula   OM-1981-434-0   DITHIOTHREITOL OXIDIZED 2TMS;   W   43612

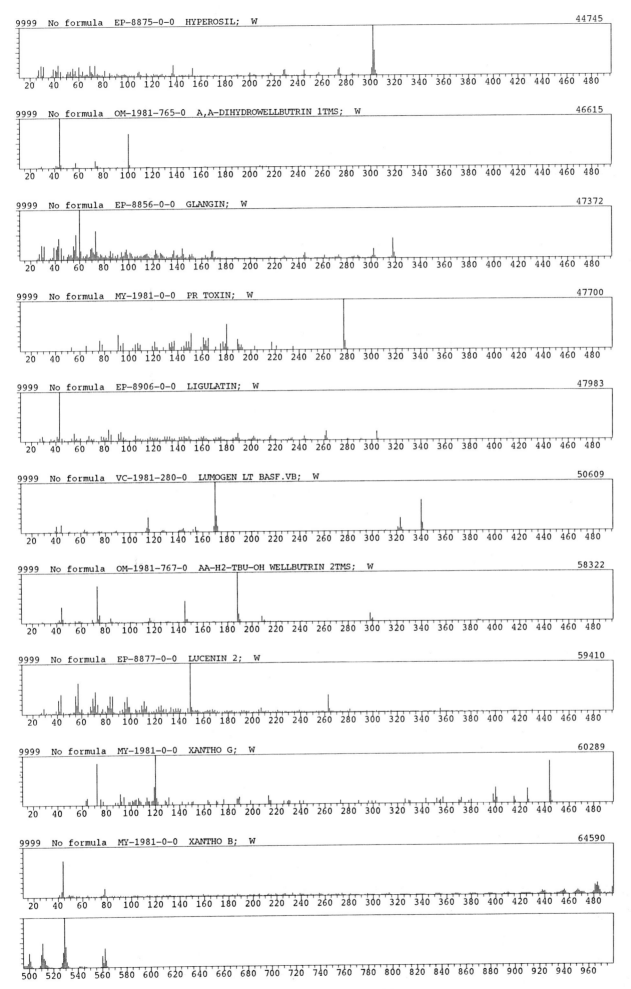

9999   No formula   EP-8875-0-0   HYPEROSIL;   W   44745

9999   No formula   OM-1981-765-0   A,A-DIHYDROWELLBUTRIN 1TMS;   W   46615

9999   No formula   EP-8856-0-0   GLANGIN;   W   47372

9999   No formula   MY-1981-0-0   PR TOXIN;   W   47700

9999   No formula   EP-8906-0-0   LIGULATIN;   W   47983

9999   No formula   VC-1981-280-0   LUMOGEN LT BASF.VB;   W   50609

9999   No formula   OM-1981-767-0   AA-H2-TBU-OH WELLBUTRIN 2TMS;   W   58322

9999   No formula   EP-8877-0-0   LUCENIN 2;   W   59410

9999   No formula   MY-1981-0-0   XANTHO G;   W   60289

9999   No formula   MY-1981-0-0   XANTHO B;   W   64590

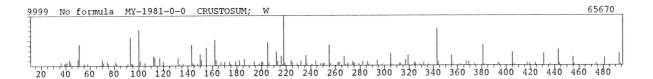

9999   No formula   MY-1981-0-0   CRUSTOSUM;   W   65670

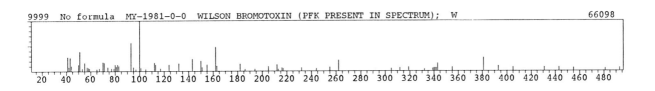

9999   No formula   MY-1981-0-0   WILSON BROMOTOXIN (PFK PRESENT IN SPECTRUM);   W   66098

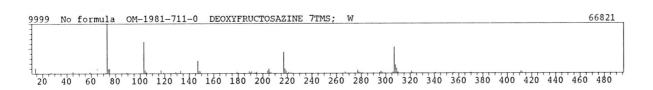

9999   No formula   OM-1981-711-0   DEOXYFRUCTOSAZINE 7TMS;   W   66821

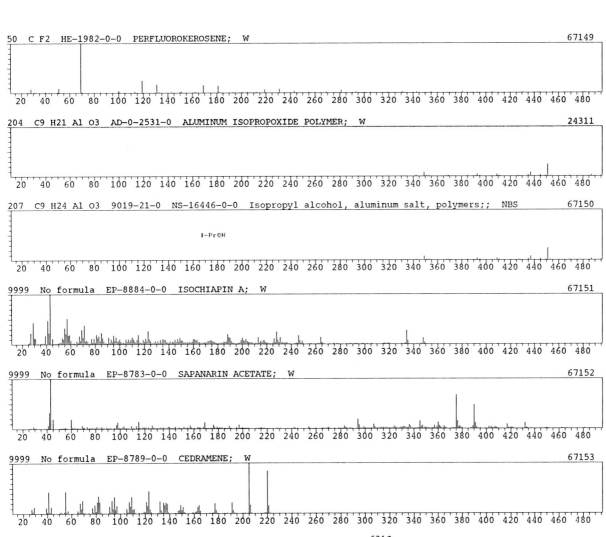

50   C F2   HE-1982-0-0   PERFLUOROKEROSENE;   W   67149

204   C9 H21 Al O3   AD-0-2531-0   ALUMINUM ISOPROPOXIDE POLYMER;   W   24311

207   C9 H24 Al O3   9019-21-0   NS-16446-0-0   Isopropyl alcohol, aluminum salt, polymers;;   NBS   67150

i-PrOH

9999   No formula   EP-8884-0-0   ISOCHIAPIN A;   W   67151

9999   No formula   EP-8783-0-0   SAPANARIN ACETATE;   W   67152

9999   No formula   EP-8789-0-0   CEDRAMENE;   W   67153

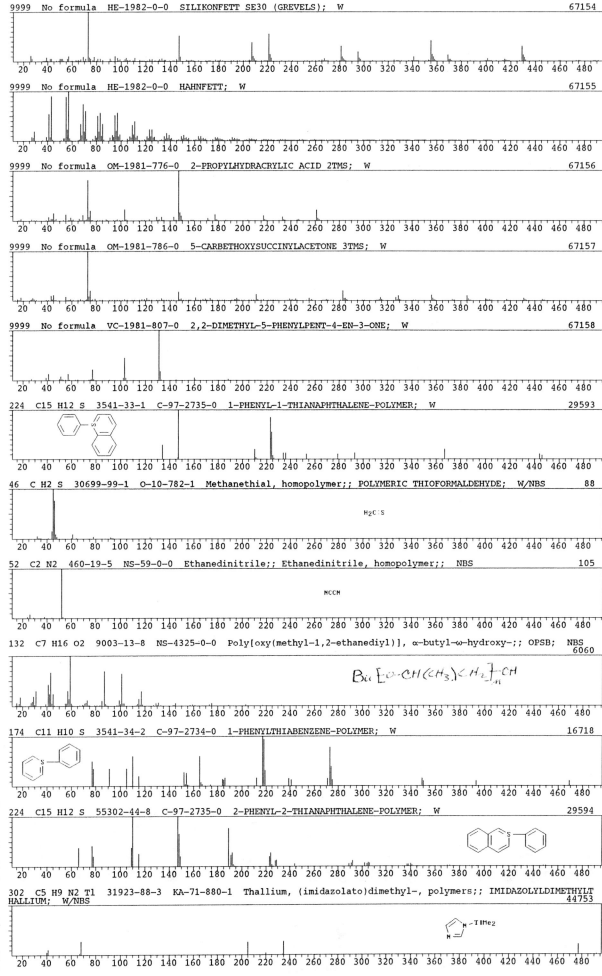

9999 No formula HE-1982-0-0   SILIKONFETT SE30 (GREVELS); W   67154

9999 No formula HE-1982-0-0   HAHNFETT; W   67155

9999 No formula OM-1981-776-0   2-PROPYLHYDRACRYLIC ACID 2TMS; W   67156

9999 No formula OM-1981-786-0   5-CARBETHOXYSUCCINYLACETONE 3TMS; W   67157

9999 No formula VC-1981-807-0   2,2-DIMETHYL-5-PHENYLPENT-4-EN-3-ONE; W   67158

224 C15 H12 S 3541-33-1 C-97-2735-0   1-PHENYL-1-THIANAPHTHALENE-POLYMER; W   29593

46 C H2 S 30699-99-1 O-10-782-1   Methanethial, homopolymer;; POLYMERIC THIOFORMALDEHYDE; W/NBS   88

H2C:S

52 C2 N2 460-19-5 NS-59-0-0   Ethanedinitrile;; Ethanedinitrile, homopolymer;; NBS   105

NCCN

132 C7 H16 O2 9003-13-8 NS-4325-0-0   Poly[oxy(methyl-1,2-ethanediyl)], α-butyl-ω-hydroxy-;; OPSB; NBS
                                                                                                6060

174 C11 H10 S 3541-34-2 C-97-2734-0   1-PHENYLTHIABENZENE-POLYMER; W   16718

224 C15 H12 S 55302-44-8 C-97-2735-0   2-PHENYL-2-THIANAPHTHALENE-POLYMER; W   29594

302 C5 H9 N2 Tl 31923-88-3 KA-71-880-1   Thallium, (imidazolato)dimethyl-, polymers;; IMIDAZOLYLDIMETHYLT
HALLIUM; W/NBS   44753

350 C25 H18 S 41959-22-2 C-97-2735-0 9,10-DIPHENYL-10-THIOXANTHYLIUM-POLYMER; W   52021

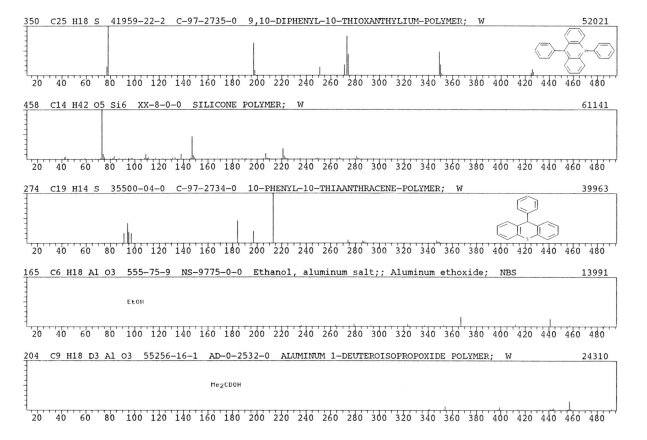

458 C14 H42 O5 Si6 XX-8-0-0 SILICONE POLYMER; W   61141

274 C19 H14 S 35500-04-0 C-97-2734-0 10-PHENYL-10-THIAANTHRACENE-POLYMER; W   39963

165 C6 H18 Al O3 555-75-9 NS-9775-0-0 Ethanol, aluminum salt;; Aluminum ethoxide; NBS   13991

EtOH

204 C9 H18 D3 Al O3 55256-16-1 AD-0-2532-0 ALUMINUM 1-DEUTEROISOPROPOXIDE POLYMER; W   24310

Me2CDOH

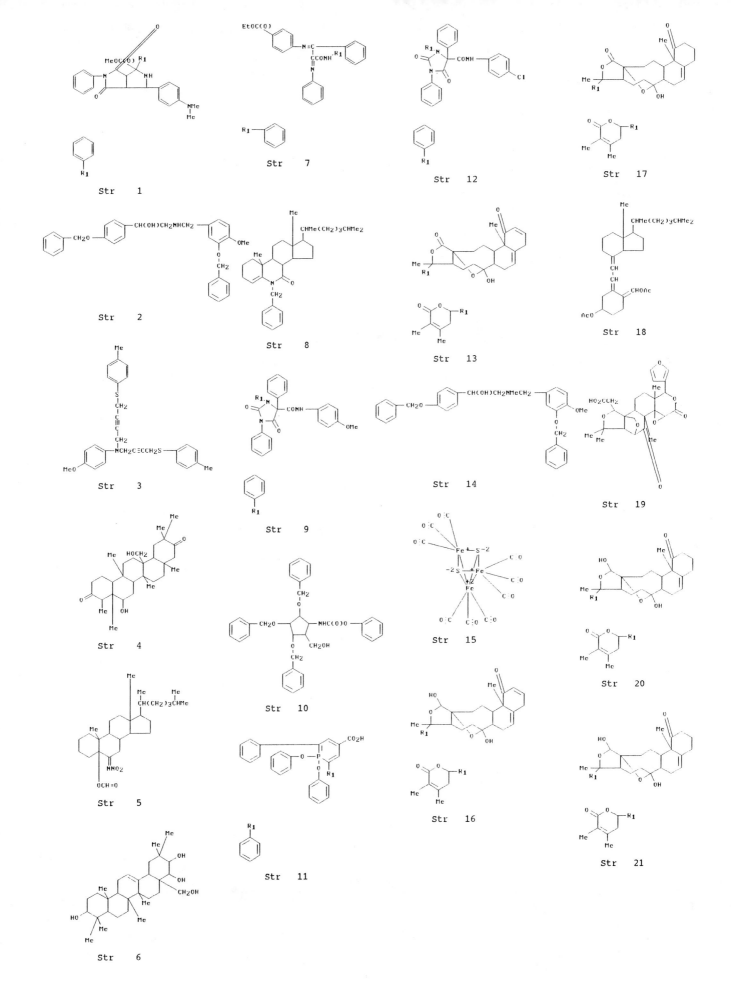

Str 1

Str 2

Str 3

Str 4

Str 5

Str 6

Str 7

Str 8

Str 9

Str 10

Str 11

Str 12

Str 13

Str 14

Str 15

Str 16

Str 17

Str 18

Str 19

Str 20

Str 21

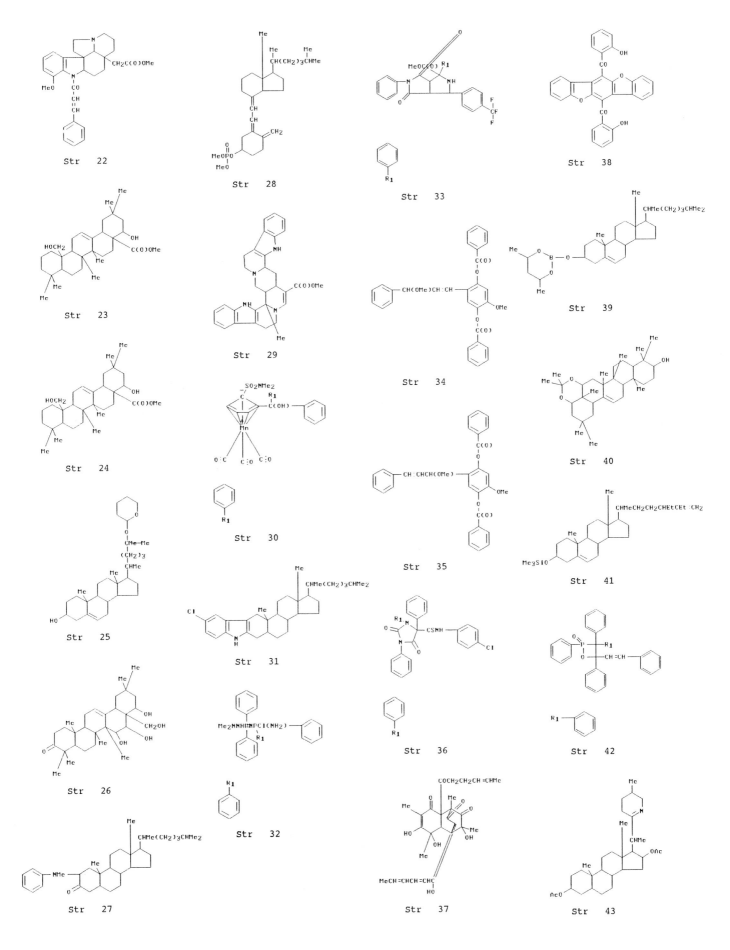

Str 22

Str 28

Str 33

Str 38

Str 23

Str 29

Str 34

Str 39

Str 24

Str 30

Str 35

Str 40

Str 25

Str 31

Str 36

Str 41

Str 26

Str 32

Str 37

Str 42

Str 27

Str 43

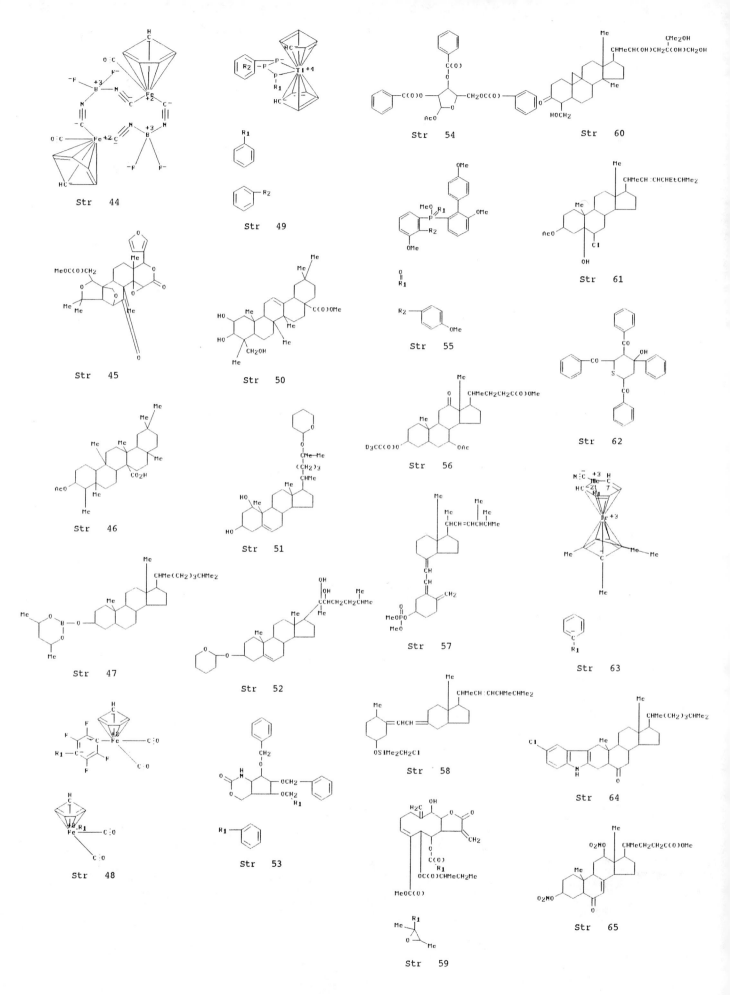

Str 44

Str 49

Str 54

Str 60

Str 45

Str 50

Str 55

Str 61

Str 46

Str 51

Str 56

Str 62

Str 47

Str 52

Str 57

Str 63

Str 48

Str 53

Str 58

Str 64

Str 59

Str 65

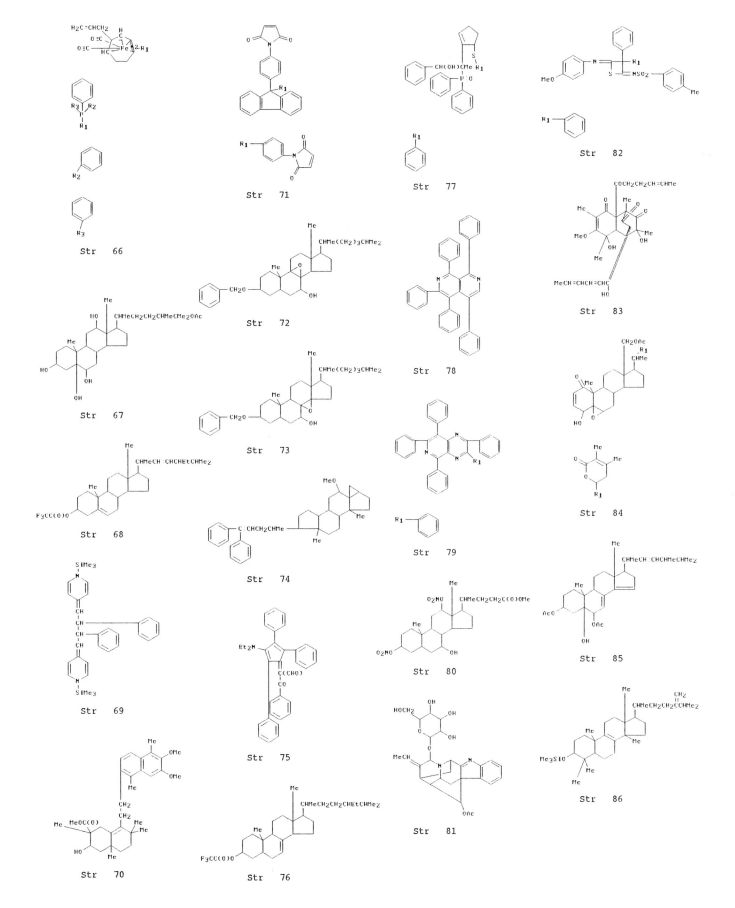

Str   66

Str   67

Str   68

Str   69

Str   70

Str   71

Str   72

Str   73

Str   74

Str   75

Str   76

Str   77

Str   78

Str   79

Str   80

Str   81

Str   82

Str   83

Str   84

Str   85

Str   86

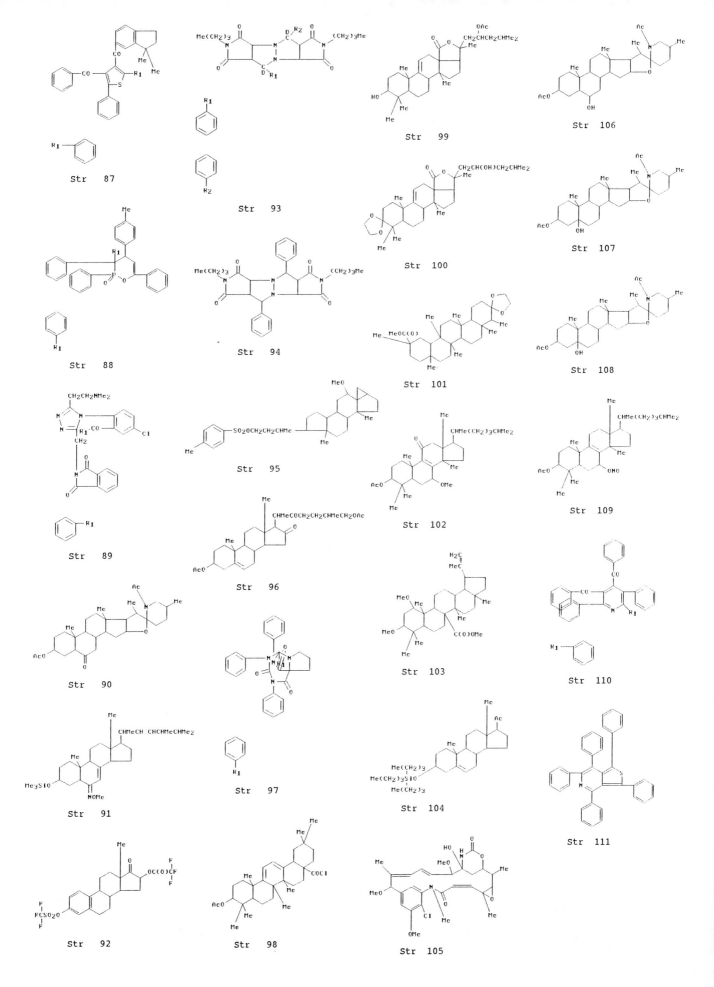

Str 87

Str 88

Str 89

Str 90

Str 91

Str 92

Str 93

Str 94

Str 95

Str 96

Str 97

Str 98

Str 99

Str 100

Str 101

Str 102

Str 103

Str 104

Str 105

Str 106

Str 107

Str 108

Str 109

Str 110

Str 111

VI 87-111

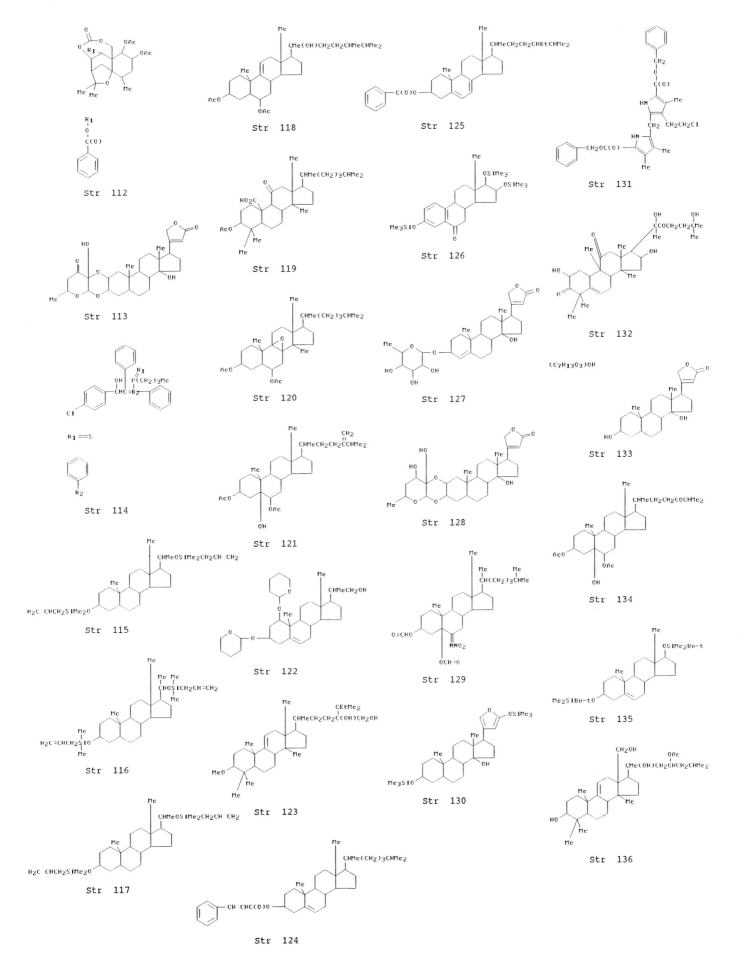

Str 112

Str 113

Str 114

R₁ ═ S

Str 115

Str 116

Str 117

Str 118

Str 119

Str 120

Str 121

Str 122

Str 123

Str 124

Str 125

Str 126

Str 127

Str 128

Str 129

Str 130

Str 131

Str 132

(C₇H₁₃O₃)OH

Str 133

Str 134

Str 135

Str 136

VI 112-136

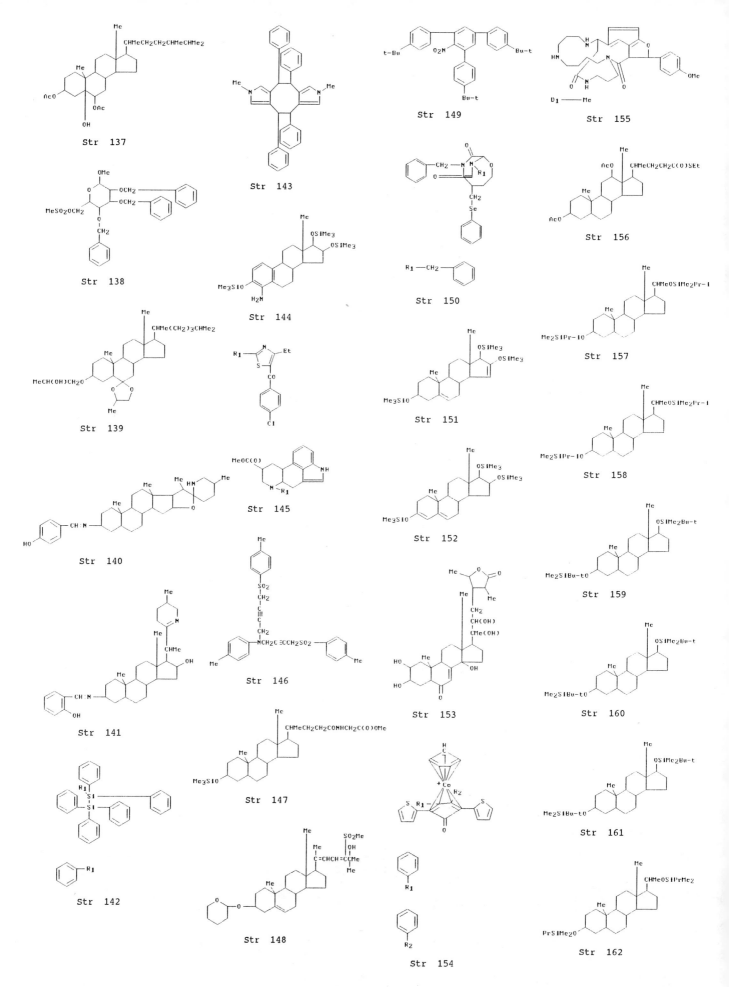

Str 137

Str 138

Str 139

Str 140

Str 141

Str 142

Str 143

Str 144

Str 145

Str 146

Str 147

Str 148

Str 149

Str 150

Str 151

Str 152

Str 153

Str 154

Str 155

Str 156

Str 157

Str 158

Str 159

Str 160

Str 161

Str 162

VI 137–162

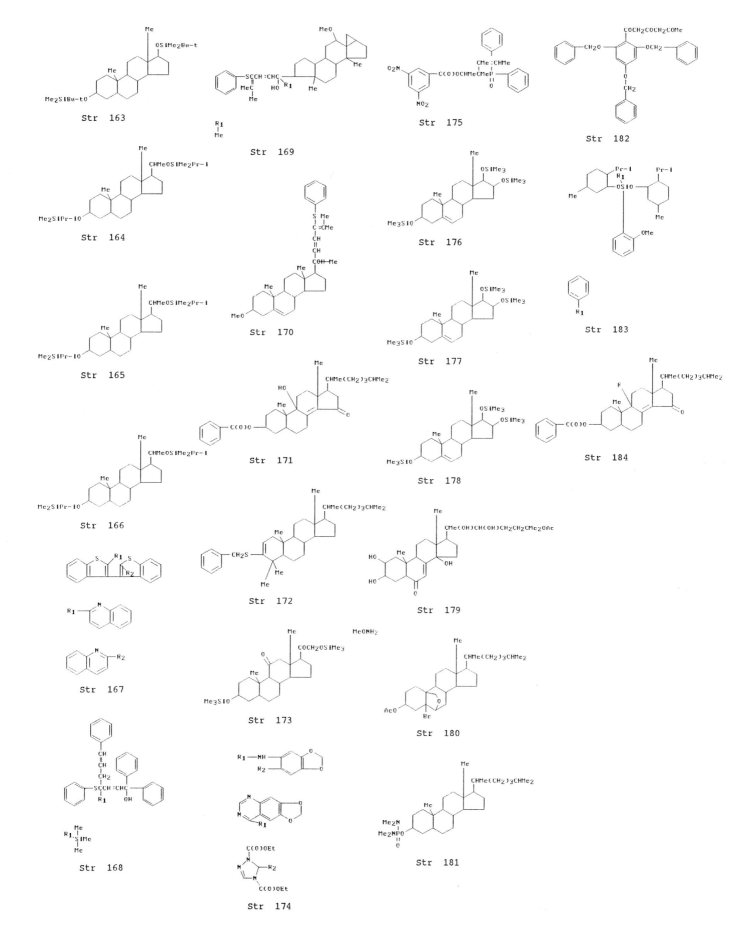

Str 163

Str 164

Str 165

Str 166

Str 167

Str 168

Str 169

Str 170

Str 171

Str 172

Str 173

Str 174

Str 175

Str 176

Str 177

Str 178

Str 179

Str 180

Str 181

Str 182

Str 183

Str 184

VI 163-184

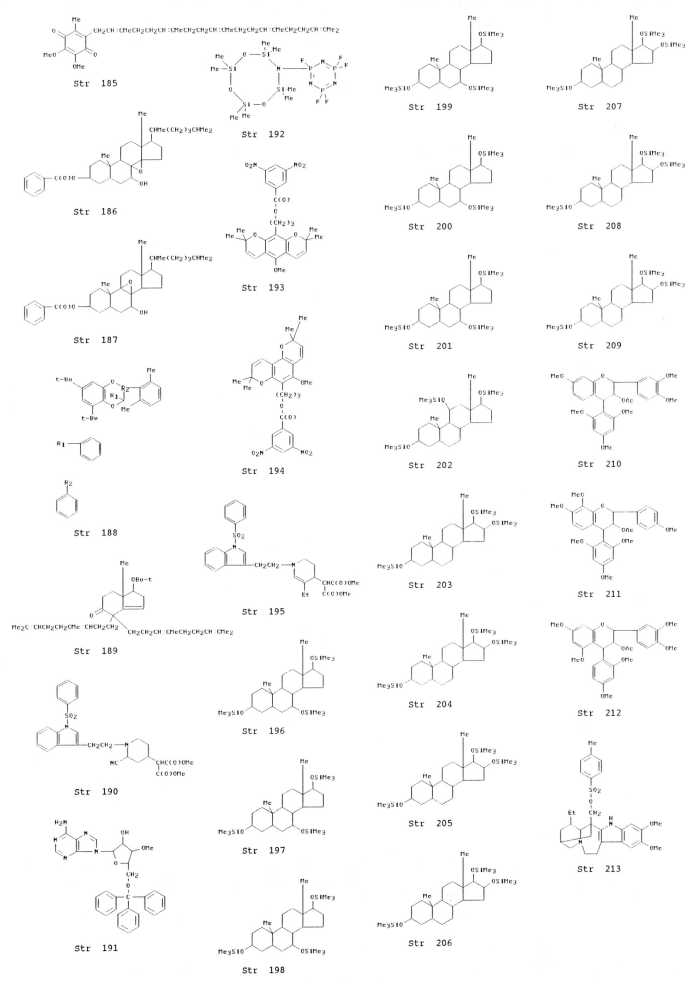

Str 185

Str 186

Str 187

Str 188

Str 189

Str 190

Str 191

Str 192

Str 193

Str 194

Str 195

Str 196

Str 197

Str 198

Str 199

Str 200

Str 201

Str 202

Str 203

Str 204

Str 205

Str 206

Str 207

Str 208

Str 209

Str 210

Str 211

Str 212

Str 213

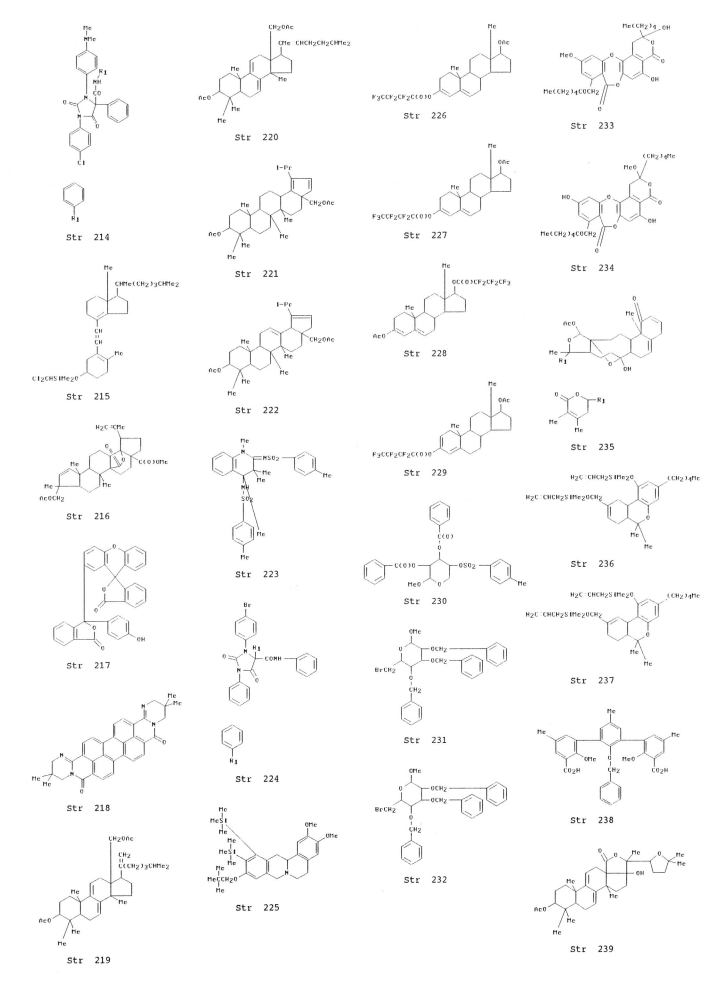

Str 214

Str 215

Str 216

Str 217

Str 218

Str 219

Str 220

Str 221

Str 222

Str 223

Str 224

Str 225

Str 226

Str 227

Str 228

Str 229

Str 230

Str 231

Str 232

Str 233

Str 234

Str 235

Str 236

Str 237

Str 238

Str 239

VI 214-239

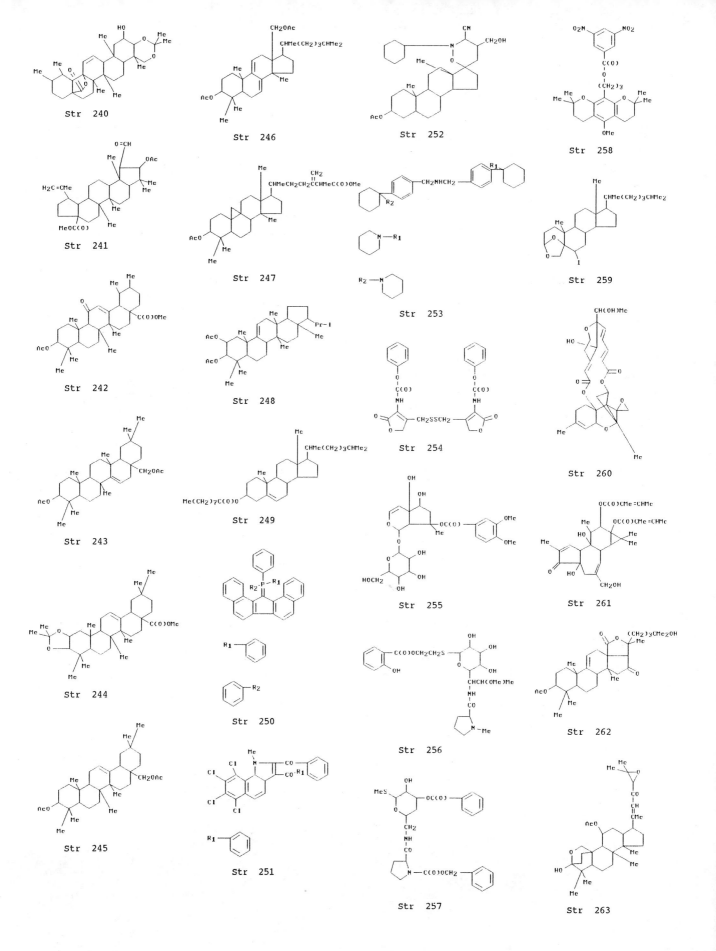

Str 240

Str 246

Str 252

Str 258

Str 241

Str 247

Str 253

Str 259

Str 242

Str 248

Str 254

Str 260

Str 243

Str 249

Str 255

Str 261

Str 244

Str 250

Str 256

Str 262

Str 245

Str 251

Str 257

Str 263

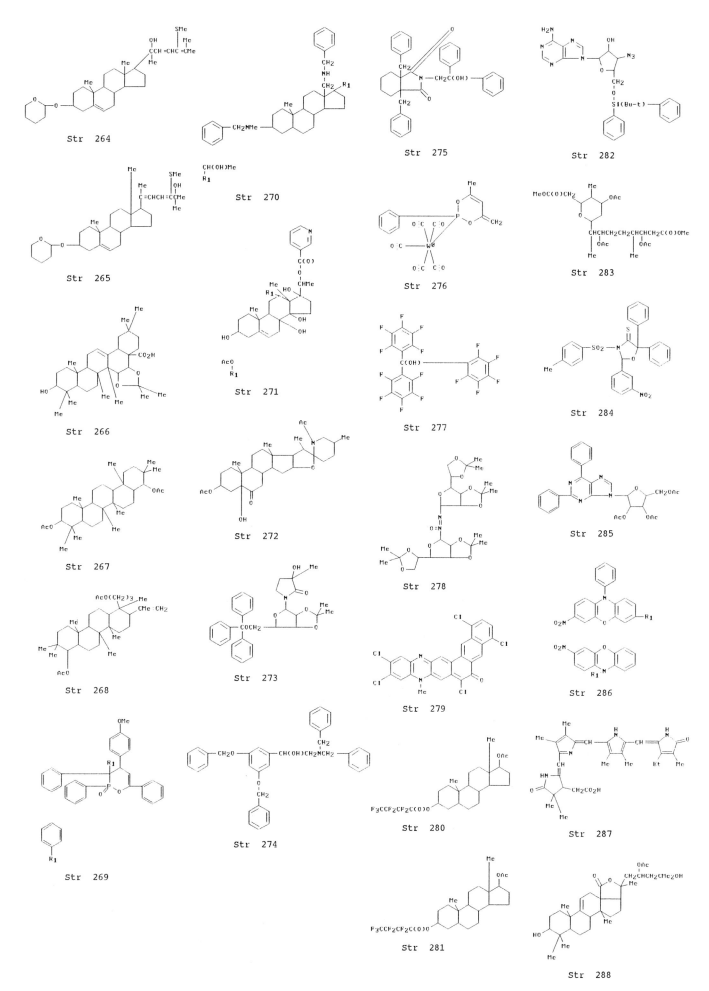

Str 264

Str 265

Str 266

Str 267

Str 268

Str 269

Str 270

Str 271

Str 272

Str 273

Str 274

Str 275

Str 276

Str 277

Str 278

Str 279

Str 280

Str 281

Str 282

Str 283

Str 284

Str 285

Str 286

Str 287

Str 288

VI 264-288

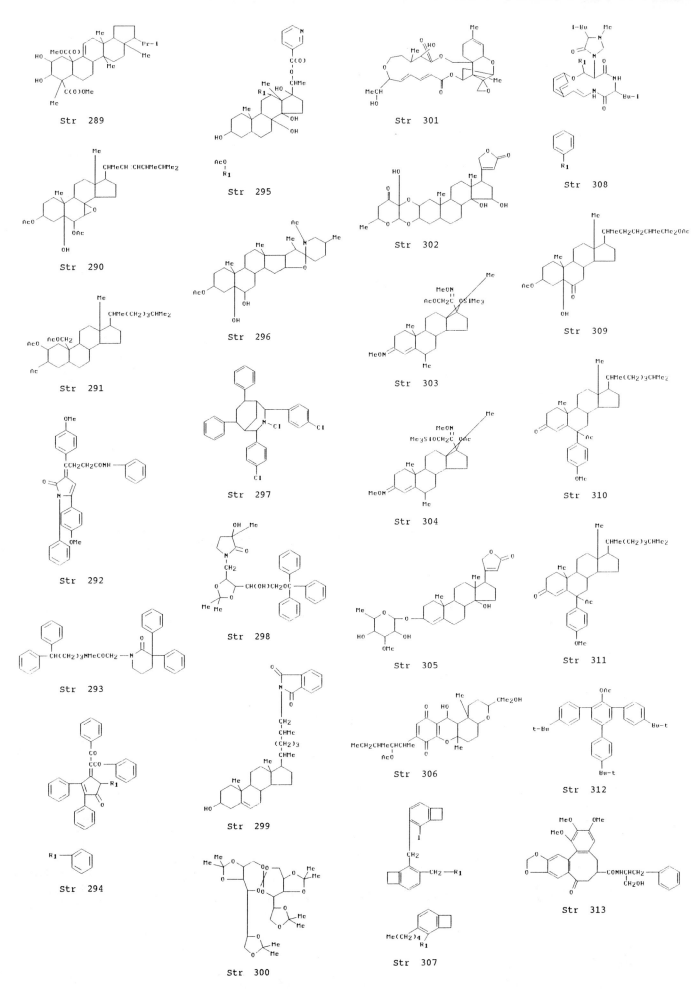

Str 289

Str 290

Str 291

Str 292

Str 293

Str 294

Str 295

Str 296

Str 297

Str 298

Str 299

Str 300

Str 301

Str 302

Str 303

Str 304

Str 305

Str 306

Str 307

Str 308

Str 309

Str 310

Str 311

Str 312

Str 313

VI 289-313

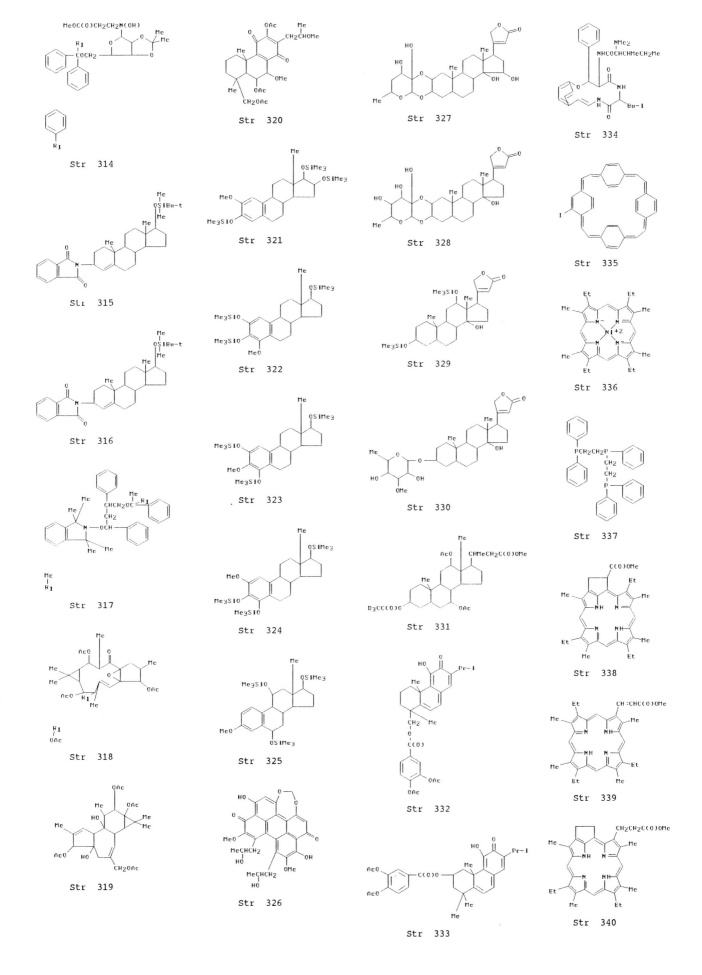

Str 314

Str 315

Str 316

Str 317

Str 318

Str 319

Str 320

Str 321

Str 322

Str 323

Str 324

Str 325

Str 326

Str 327

Str 328

Str 329

Str 330

Str 331

Str 332

Str 333

Str 334

Str 335

Str 336

Str 337

Str 338

Str 339

Str 340

VI 314-340

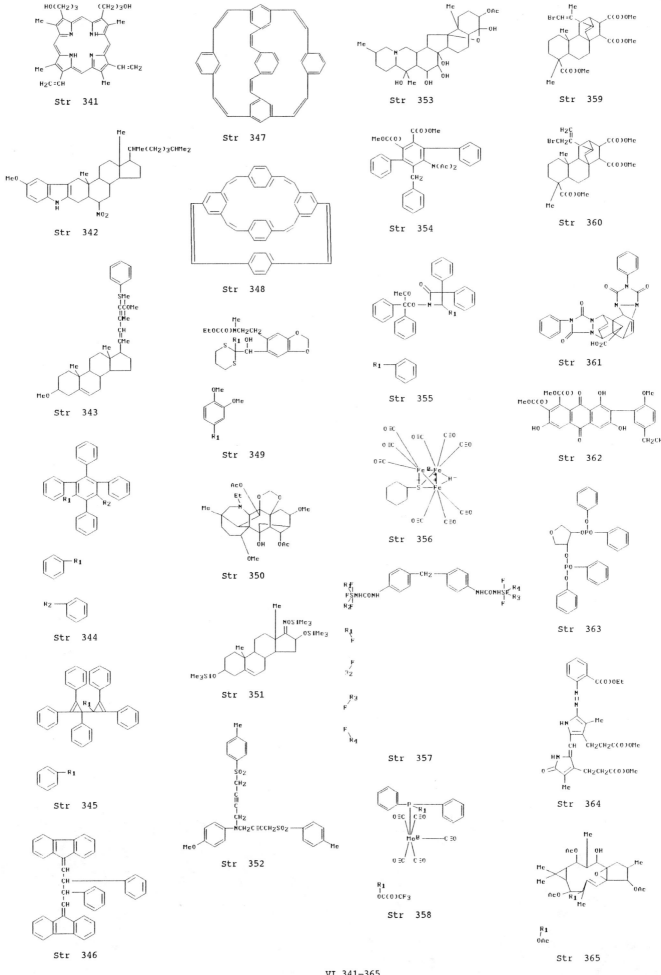

Str 341

Str 342

Str 343

Str 344

Str 345

Str 346

Str 347

Str 348

Str 349

Str 350

Str 351

Str 352

Str 353

Str 354

Str 355

Str 356

Str 357

Str 358

Str 359

Str 360

Str 361

Str 362

Str 363

Str 364

Str 365

VI 341-365

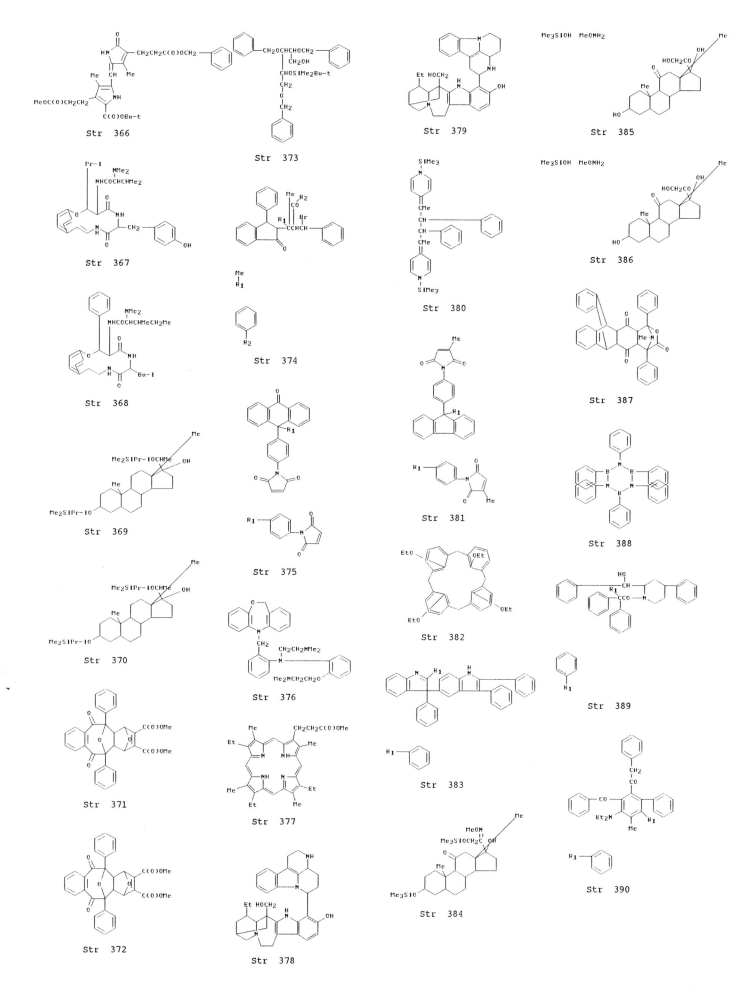

Str 366

Str 373

Str 367

Str 368

Str 369

Str 370

Str 371

Str 372

Str 374

Str 375

Str 376

Str 377

Str 378

Str 379

Str 380

Str 381

Str 382

Str 383

Str 384

Str 385

Str 386

Str 387

Str 388

Str 389

Str 390

VI 366-390

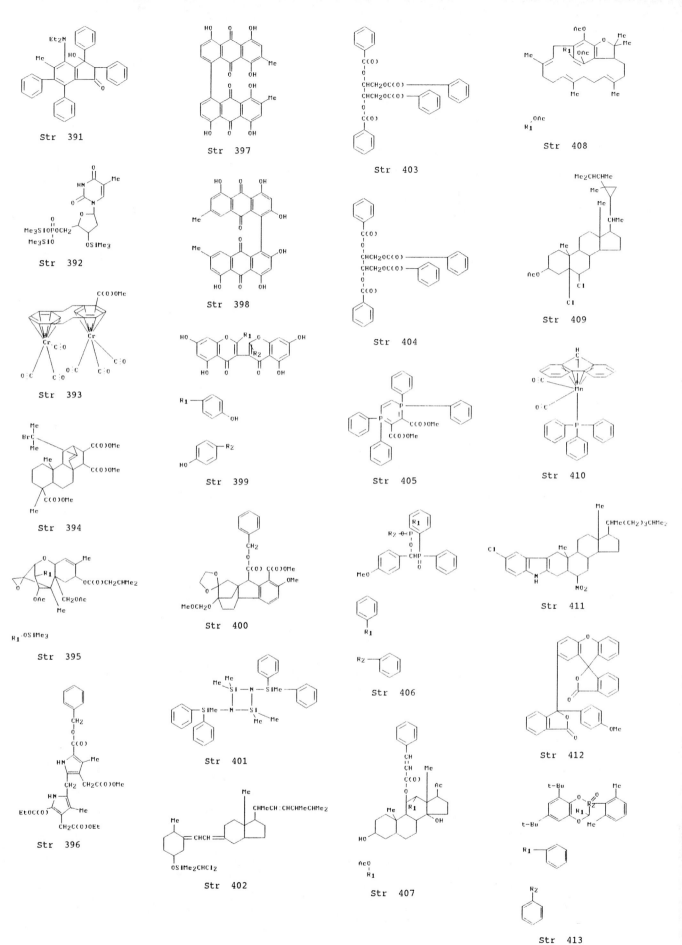

Str 391

Str 392

Str 393

Str 394

Str 395

Str 396

Str 397

Str 398

Str 399

Str 400

Str 401

Str 402

Str 403

Str 404

Str 405

Str 406

Str 407

Str 408

Str 409

Str 410

Str 411

Str 412

Str 413

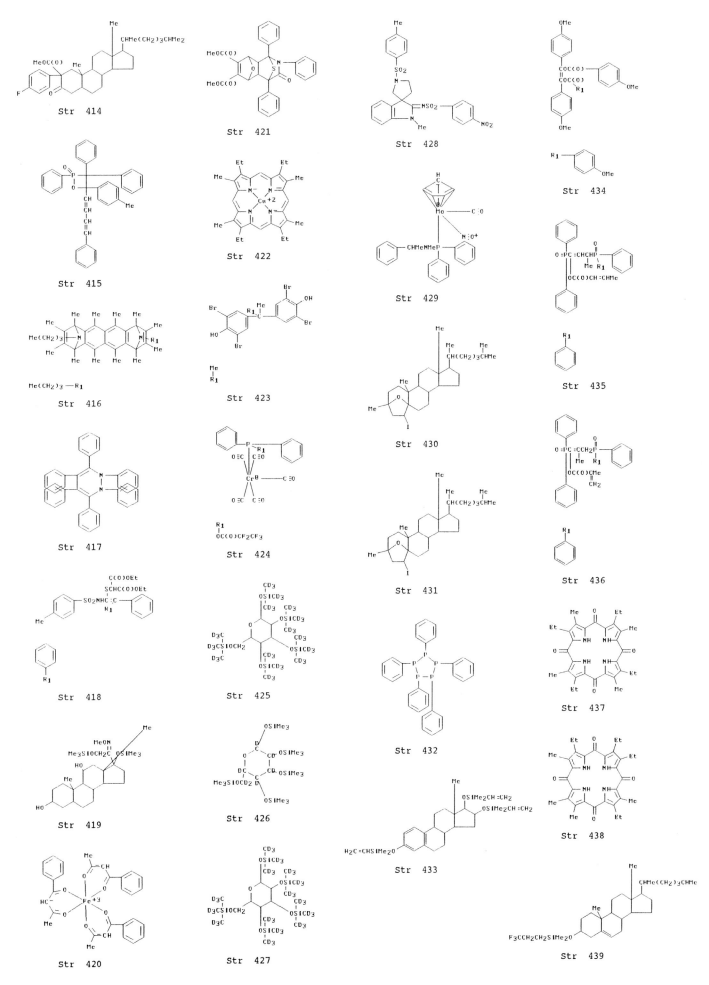

Str 414

Str 421

Str 428

Str 434

Str 415

Str 422

Str 429

Str 435

Str 416

Str 423

Str 430

Str 436

Str 417

Str 424

Str 431

Str 437

Str 418

Str 425

Str 432

Str 438

Str 419

Str 426

Str 433

Str 439

Str 420

Str 427

VI 414-439

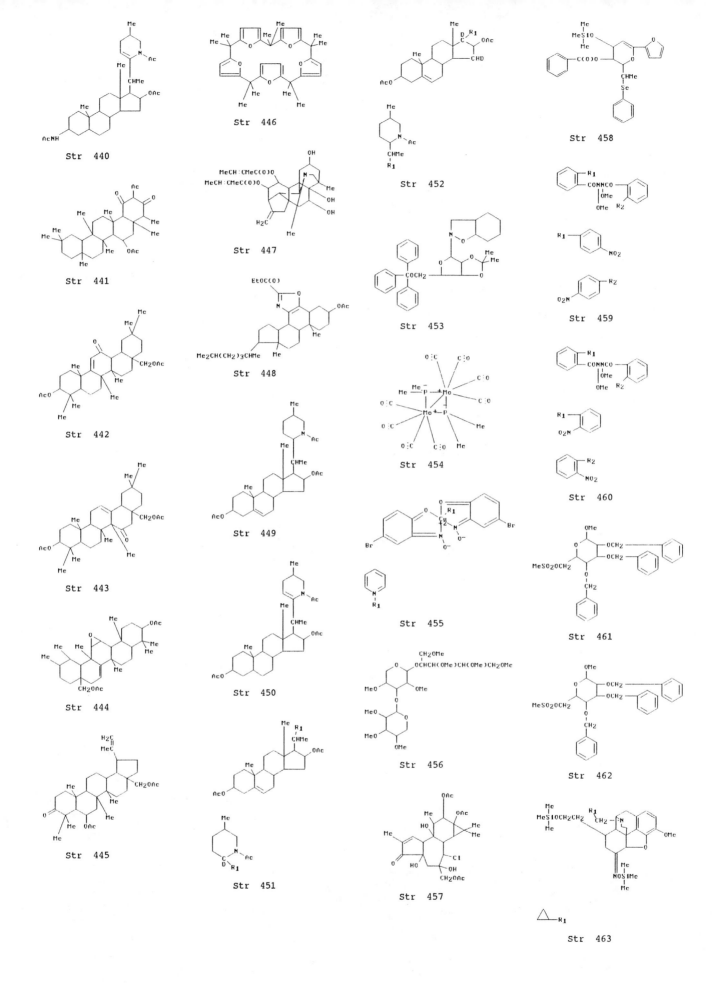

Str 440

Str 446

Str 452

Str 458

Str 441

Str 447

Str 453

Str 459

Str 442

Str 448

Str 454

Str 460

Str 443

Str 449

Str 455

Str 461

Str 444

Str 450

Str 456

Str 462

Str 445

Str 451

Str 457

Str 463

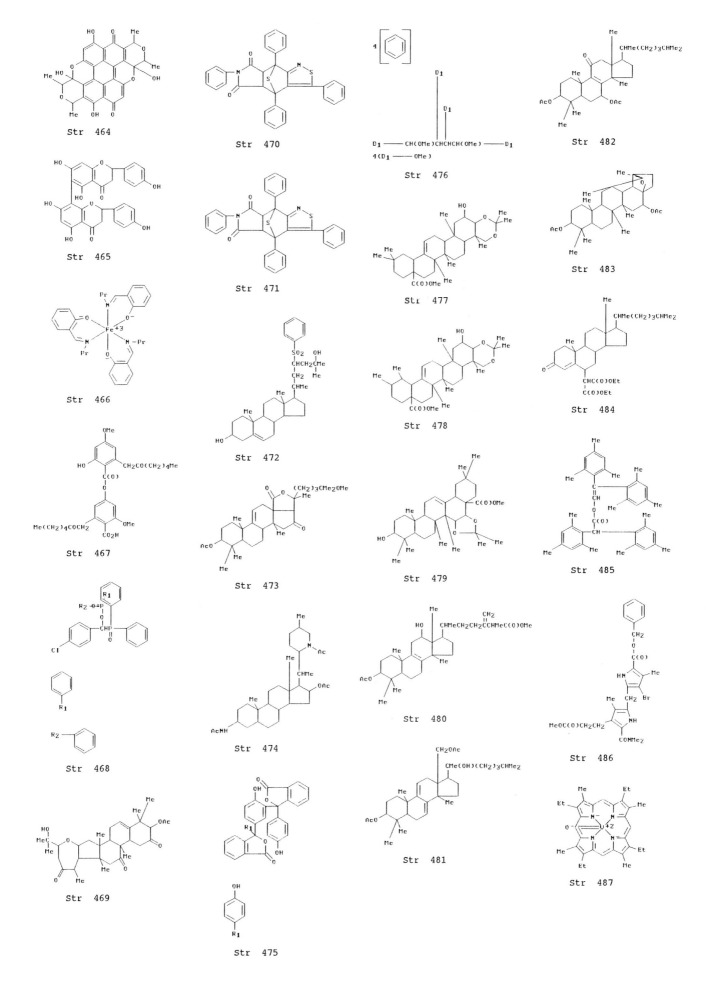

Str 464

Str 465

Str 466

Str 467

Str 468

Str 469

Str 470

Str 471

Str 472

Str 473

Str 474

Str 475

Str 476

Str 477

Str 478

Str 479

Str 480

Str 481

Str 482

Str 483

Str 484

Str 485

Str 486

Str 487

VI 464-487

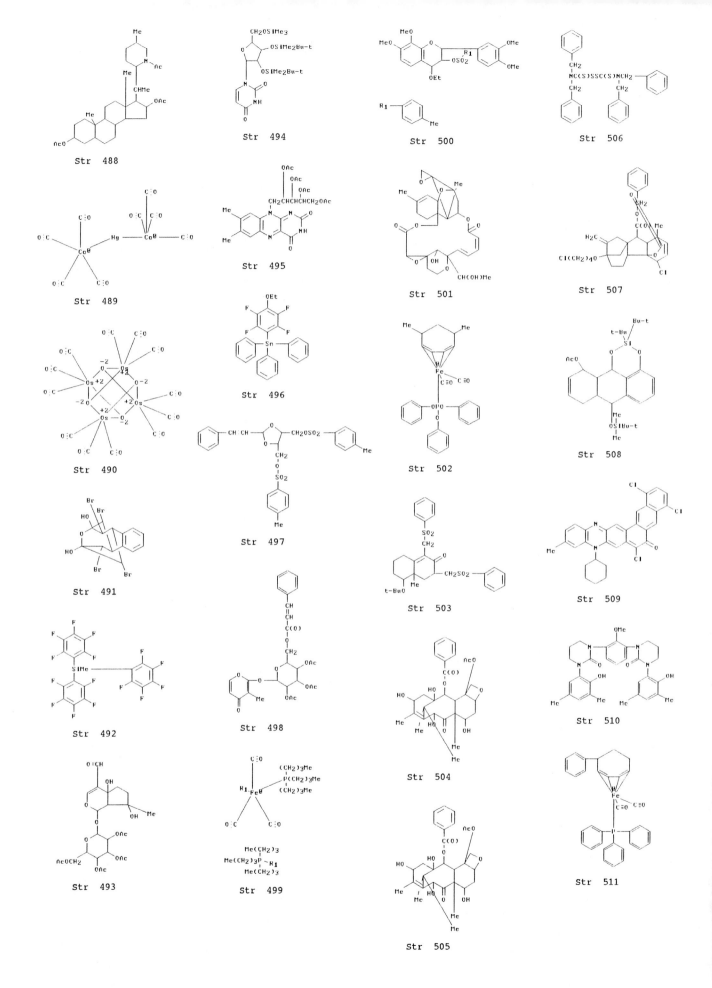

Str 488

Str 494

Str 500

Str 506

Str 489

Str 495

Str 501

Str 507

Str 490

Str 496

Str 502

Str 508

Str 491

Str 497

Str 503

Str 509

Str 492

Str 498

Str 504

Str 510

Str 493

Str 499

Str 505

Str 511

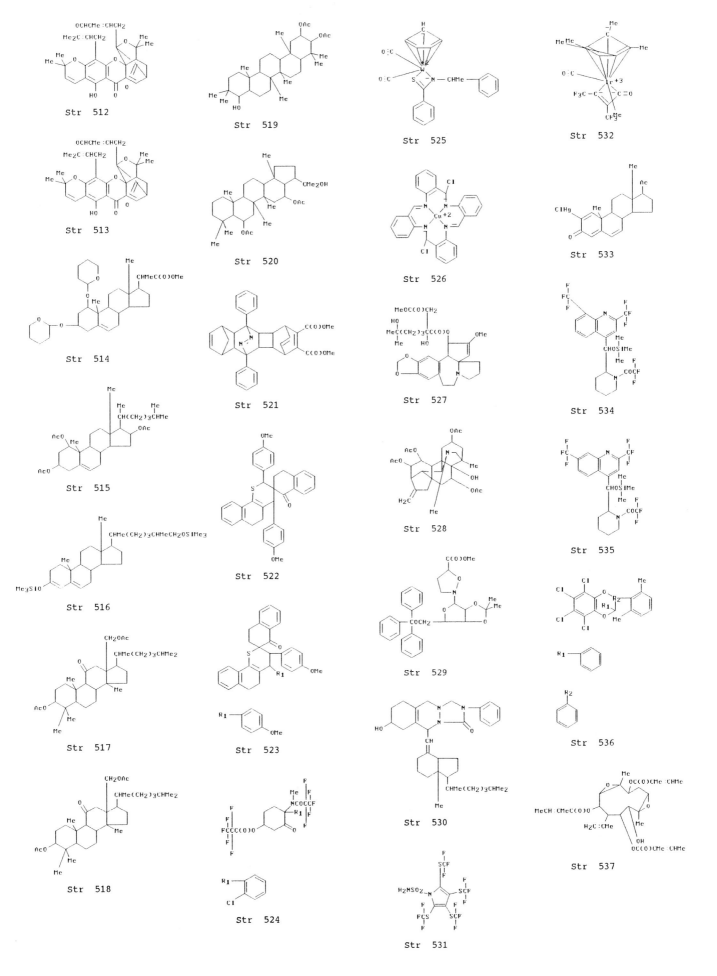

Str 512

Str 513

Str 514

Str 515

Str 516

Str 517

Str 518

Str 519

Str 520

Str 521

Str 522

Str 523

Str 524

Str 525

Str 526

Str 527

Str 528

Str 529

Str 530

Str 531

Str 532

Str 533

Str 534

Str 535

Str 536

Str 537

VI 512-537

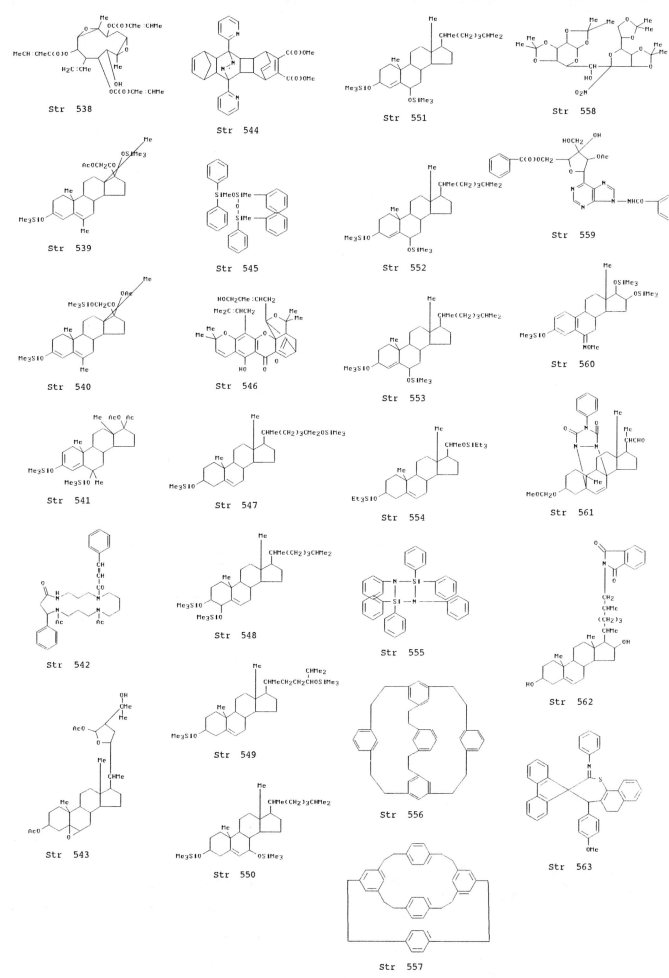

Str 538

Str 544

Str 551

Str 558

Str 539

Str 545

Str 552

Str 559

Str 540

Str 546

Str 553

Str 560

Str 541

Str 547

Str 554

Str 561

Str 542

Str 548

Str 555

Str 562

Str 543

Str 549

Str 556

Str 563

Str 550

Str 557

VI 538-563

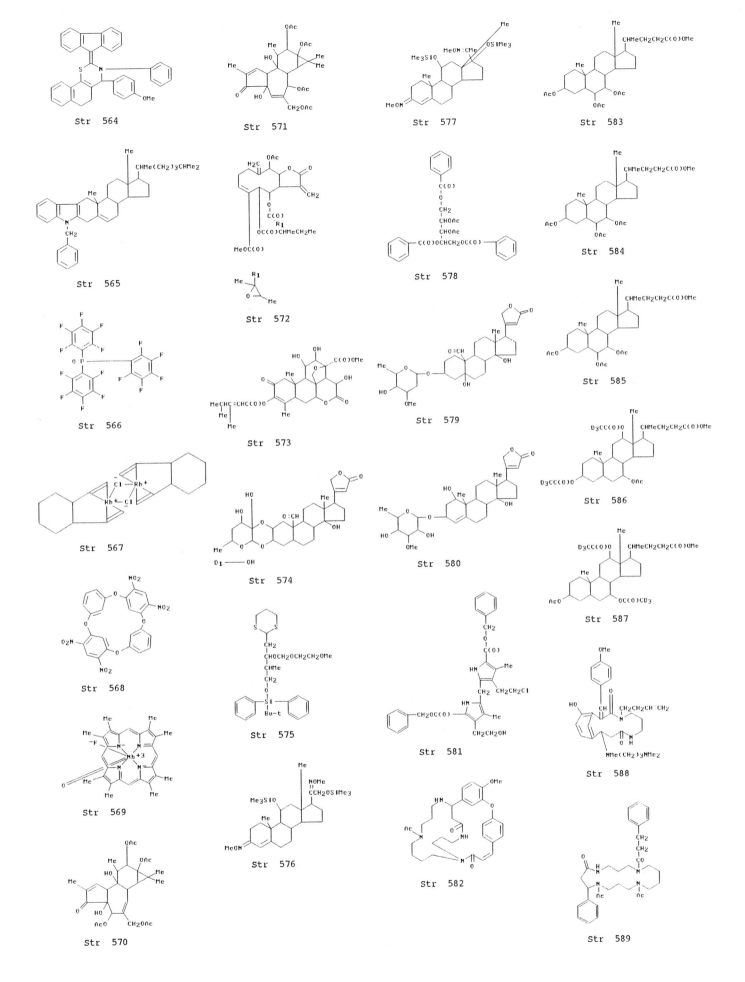

Str   564

Str   571

Str   577

Str   583

Str   565

Str   572

Str   578

Str   584

Str   566

Str   573

Str   579

Str   585

Str   567

Str   574

Str   580

Str   586

Str   568

Str   575

Str   581

Str   587

Str   569

Str   576

Str   582

Str   588

Str   570

Str   589

VI 564-589

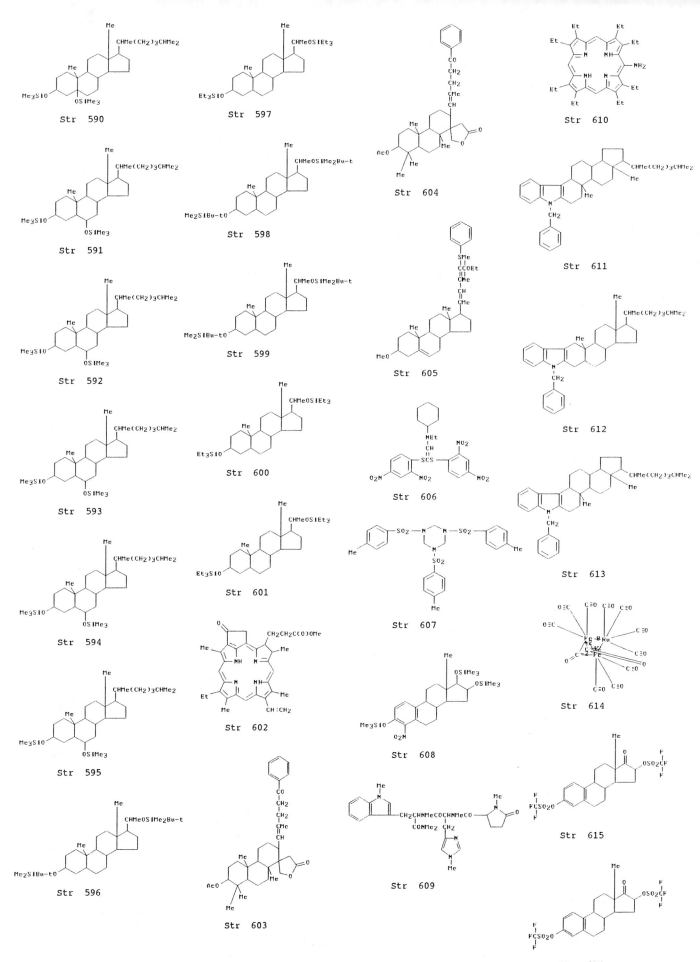

Str 590

Str 591

Str 592

Str 593

Str 594

Str 595

Str 596

Str 597

Str 598

Str 599

Str 600

Str 601

Str 602

Str 603

Str 604

Str 605

Str 606

Str 607

Str 608

Str 609

Str 610

Str 611

Str 612

Str 613

Str 614

Str 615

Str 616

VI 590-616

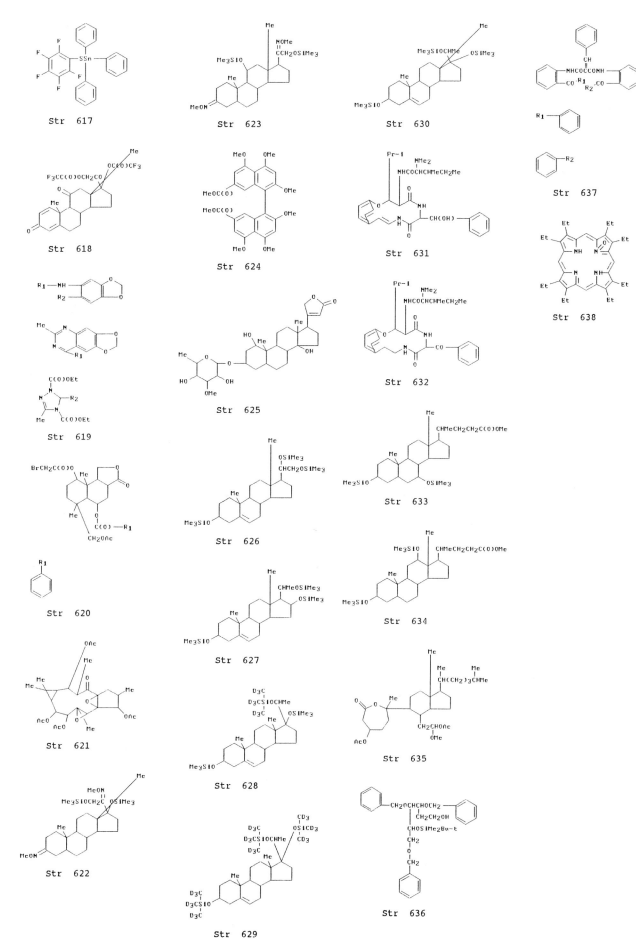

Str 617

Str 623

Str 630

Str 618

Str 624

Str 631

Str 637

Str 619

Str 625

Str 632

Str 638

Str 620

Str 626

Str 633

Str 621

Str 627

Str 634

Str 622

Str 628

Str 635

Str 629

Str 636

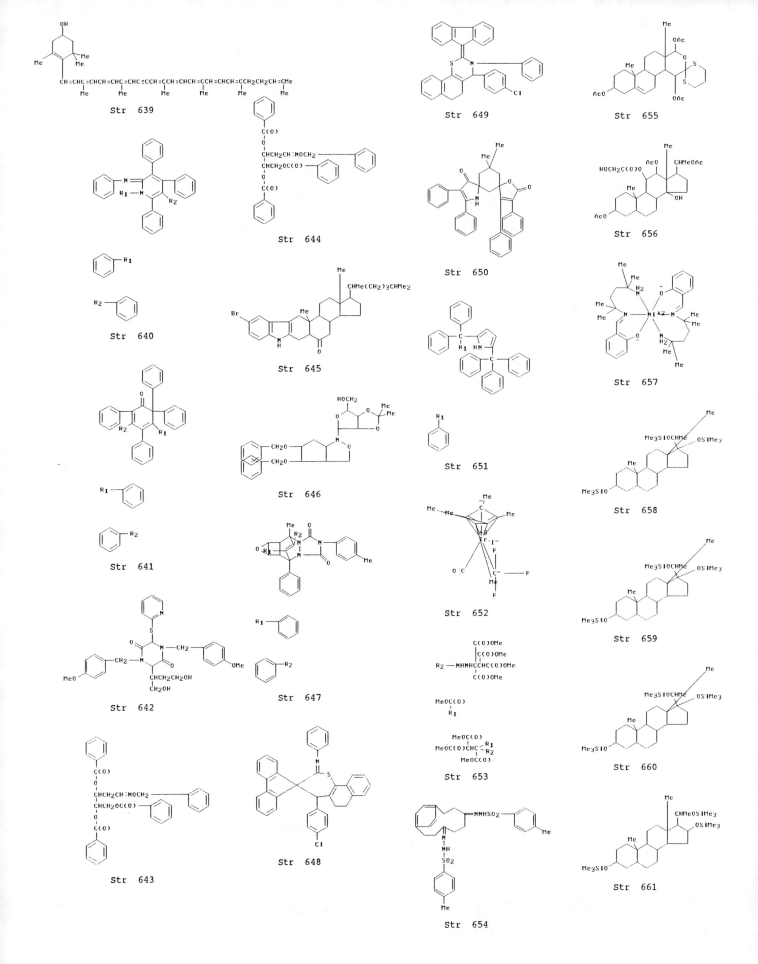

Str 639

Str 644

Str 640

Str 641

Str 642

Str 643

Str 649

Str 650

Str 645

Str 646

Str 647

Str 648

Str 651

Str 652

Str 653

Str 654

Str 655

Str 656

Str 657

Str 658

Str 659

Str 660

Str 661

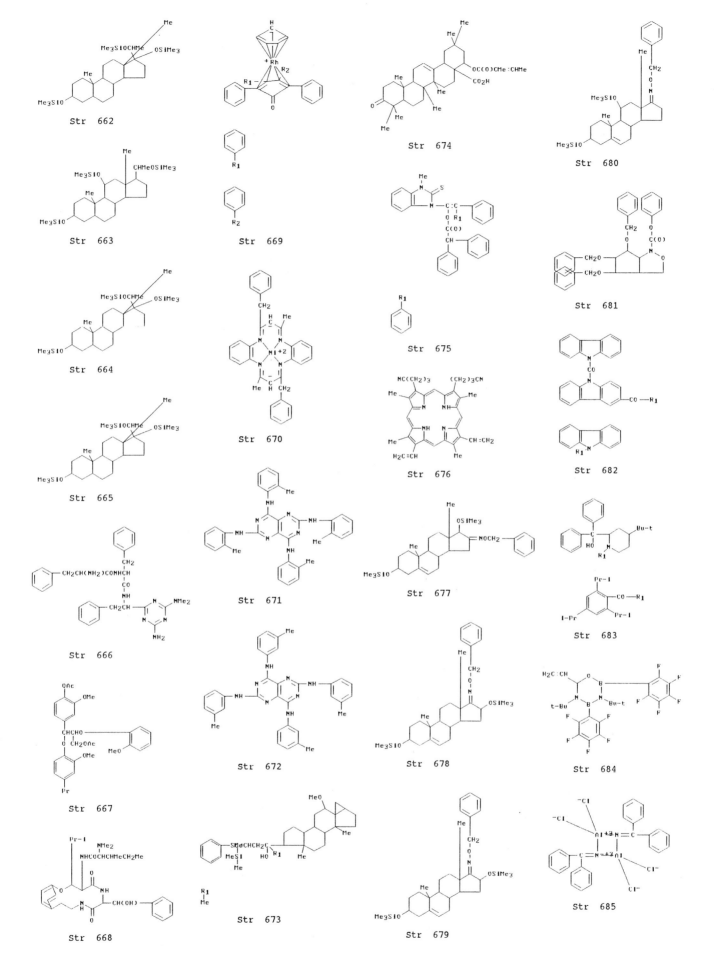

Str 662

Str 663

Str 664

Str 665

Str 666

Str 667

Str 668

Str 669

Str 670

Str 671

Str 672

Str 673

Str 674

Str 675

Str 676

Str 677

Str 678

Str 679

Str 680

Str 681

Str 682

Str 683

Str 684

Str 685

VI 662-685

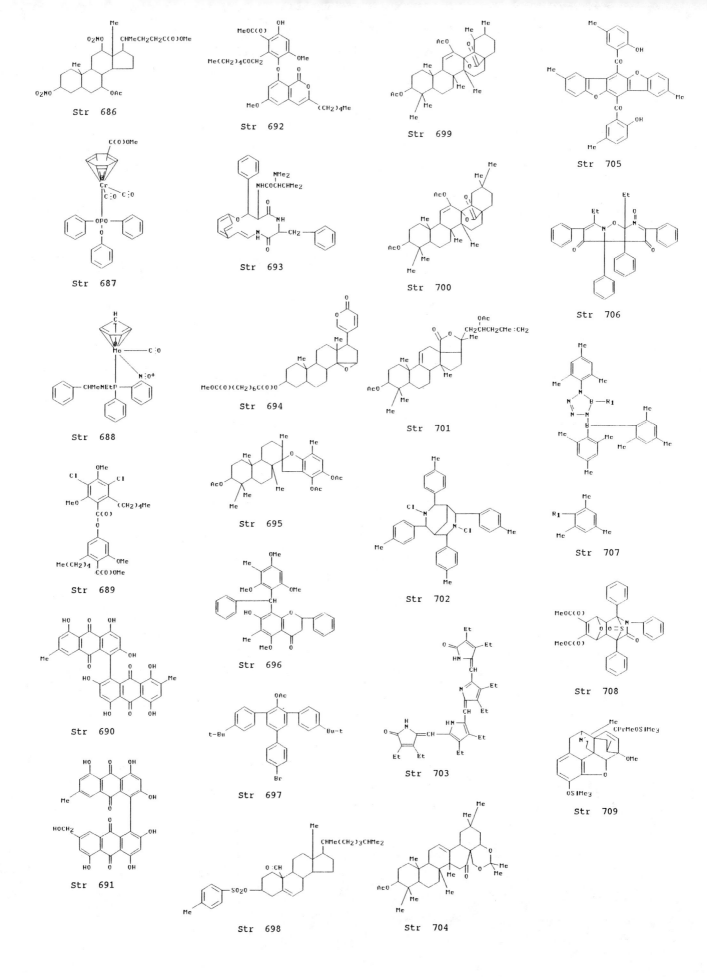

Str 686

Str 692

Str 699

Str 705

Str 687

Str 693

Str 700

Str 706

Str 688

Str 694

Str 701

Str 707

Str 689

Str 695

Str 702

Str 708

Str 690

Str 696

Str 703

Str 709

Str 691

Str 697

Str 698

Str 704

VI 686—709

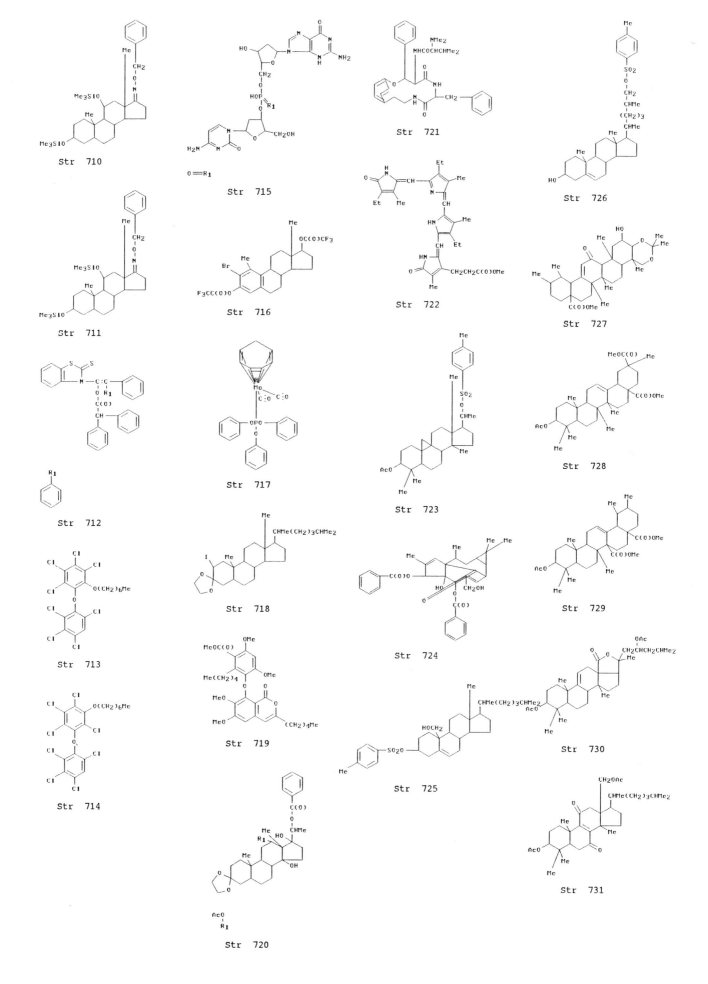

Str 710

Str 711

Str 712

Str 713

Str 714

Str 715

O=R₁

Str 716

Str 717

Str 718

Str 719

Str 720

Str 721

Str 722

Str 723

Str 724

Str 725

Str 726

Str 727

Str 728

Str 729

Str 730

Str 731

VI 710-731

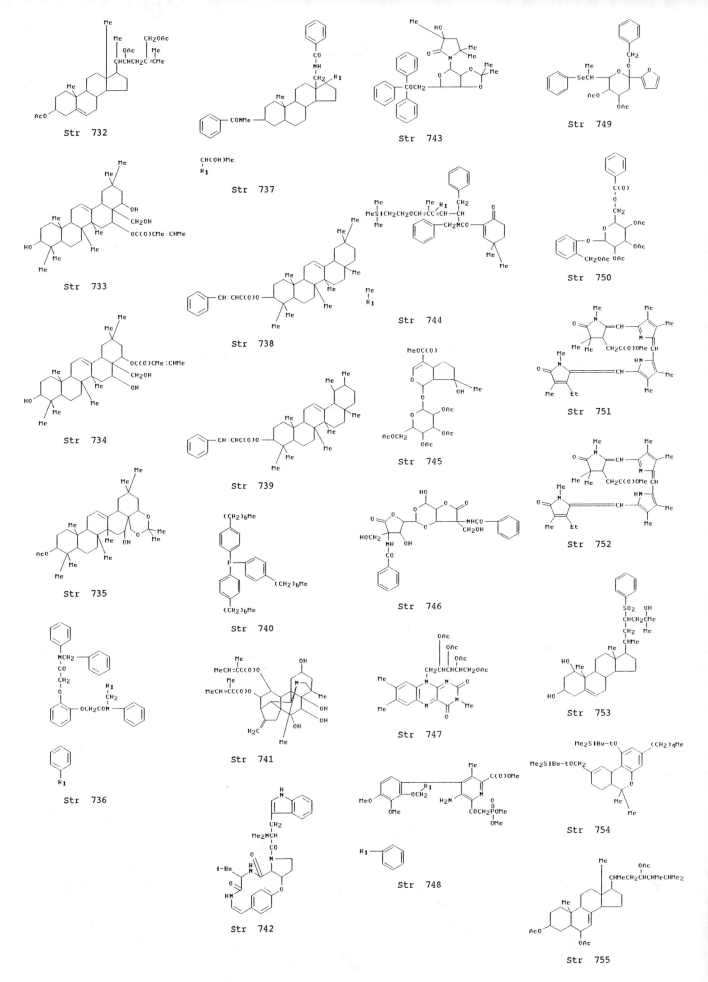

Str 732

Str 733

Str 734

Str 735

Str 736

Str 737

Str 738

Str 739

Str 740

Str 741

Str 742

Str 743

Str 744

Str 745

Str 746

Str 747

Str 748

Str 749

Str 750

Str 751

Str 752

Str 753

Str 754

Str 755

VI 732-755

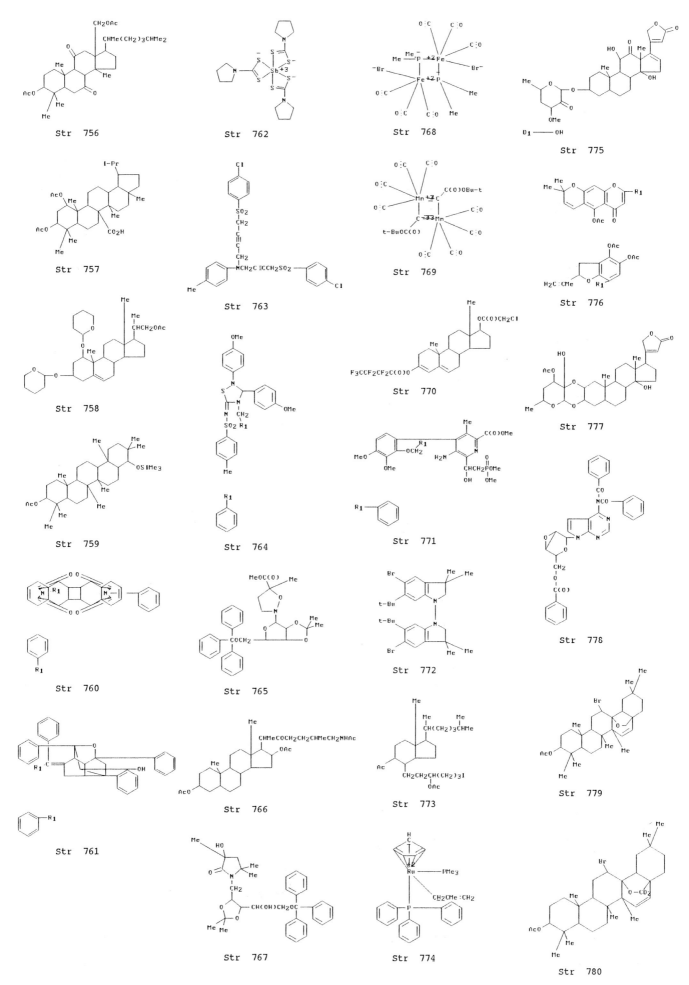

Str 756

Str 762

Str 768

Str 775

Str 757

Str 763

Str 769

Str 776

Str 758

Str 764

Str 771

Str 777

Str 759

Str 765

Str 772

Str 778

Str 760

Str 766

Str 773

Str 779

Str 761

Str 767

Str 774

Str 780

Str 770

VI 756-780

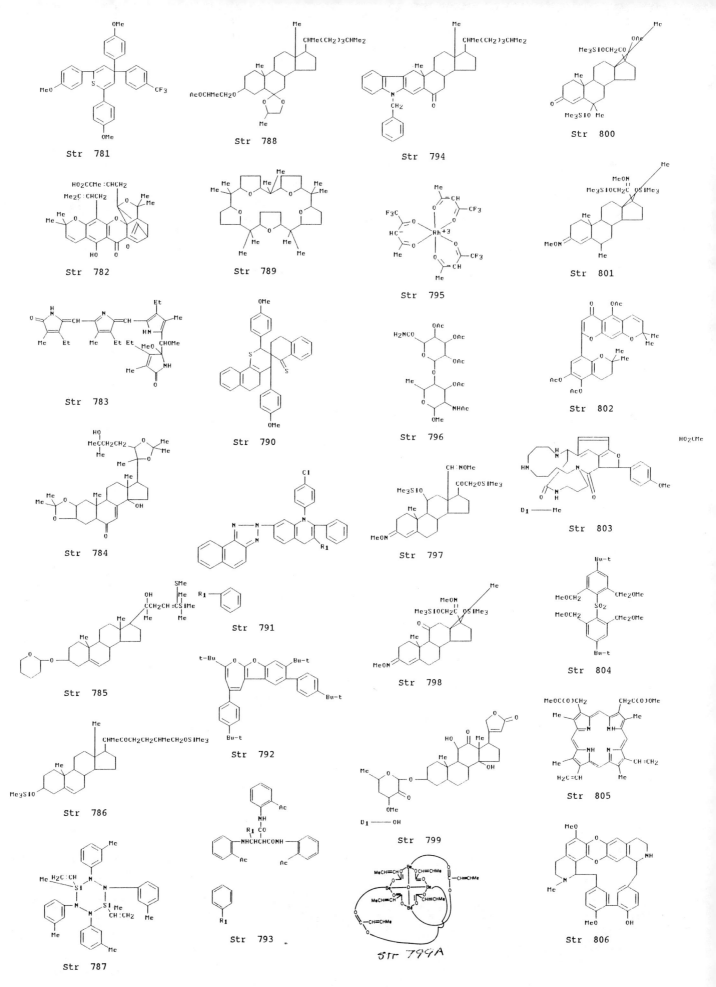

Str 781

Str 788

Str 794

Str 800

Str 782

Str 789

Str 795

Str 801

Str 783

Str 790

Str 796

Str 802

Str 784

Str 791

Str 797

Str 803

Str 785

Str 792

Str 798

Str 804

Str 786

Str 793

Str 799

Str 799A

Str 805

Str 787

Str 806

VI 781-806

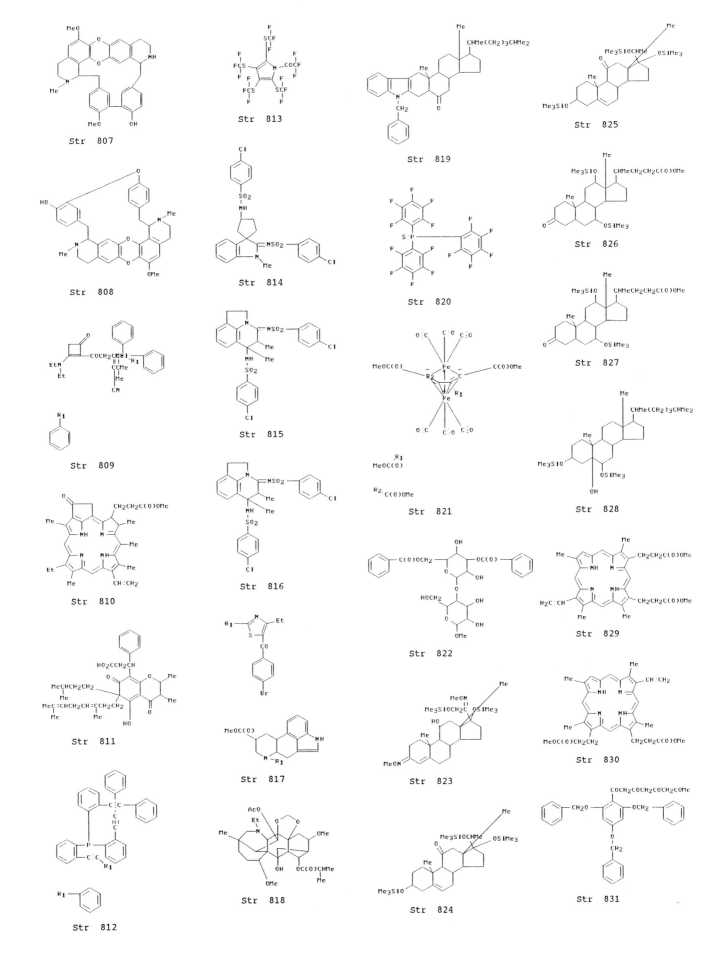

Str 807

Str 813

Str 819

Str 825

Str 808

Str 814

Str 820

Str 826

Str 809

Str 815

Str 821

Str 827

Str 810

Str 816

Str 822

Str 828

Str 811

Str 817

Str 823

Str 829

Str 812

Str 818

Str 824

Str 830

Str 831

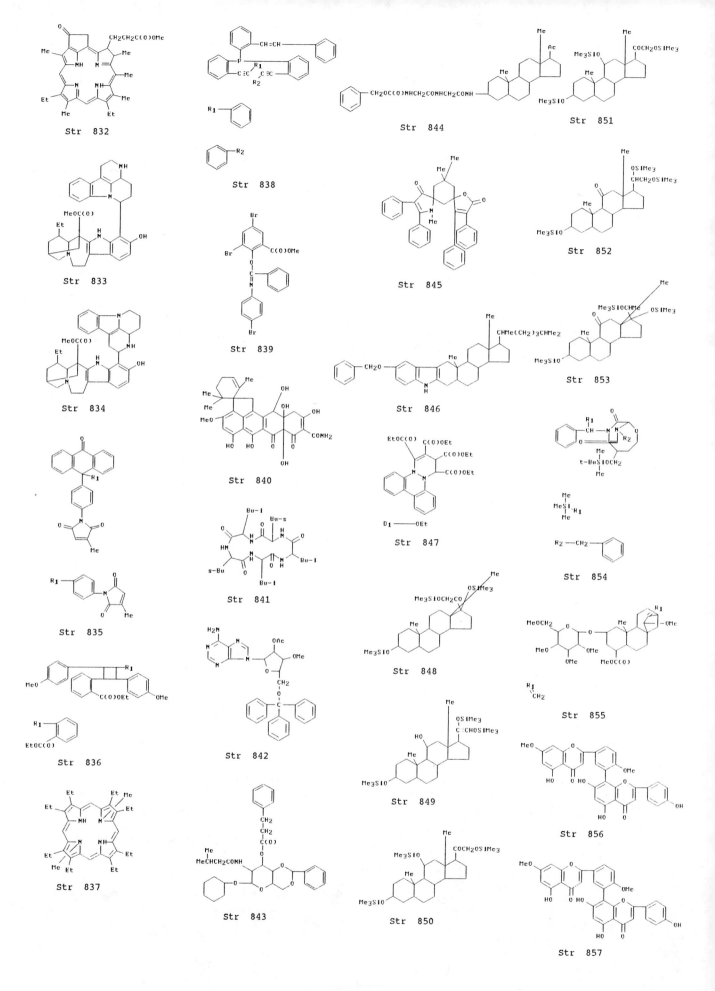

Str 832

Str 833

Str 834

Str 835

Str 836

Str 837

Str 838

Str 839

Str 840

Str 841

Str 842

Str 843

Str 844

Str 845

Str 846

Str 847

Str 848

Str 849

Str 850

Str 851

Str 852

Str 853

Str 854

Str 855

Str 856

Str 857

VI 832–857

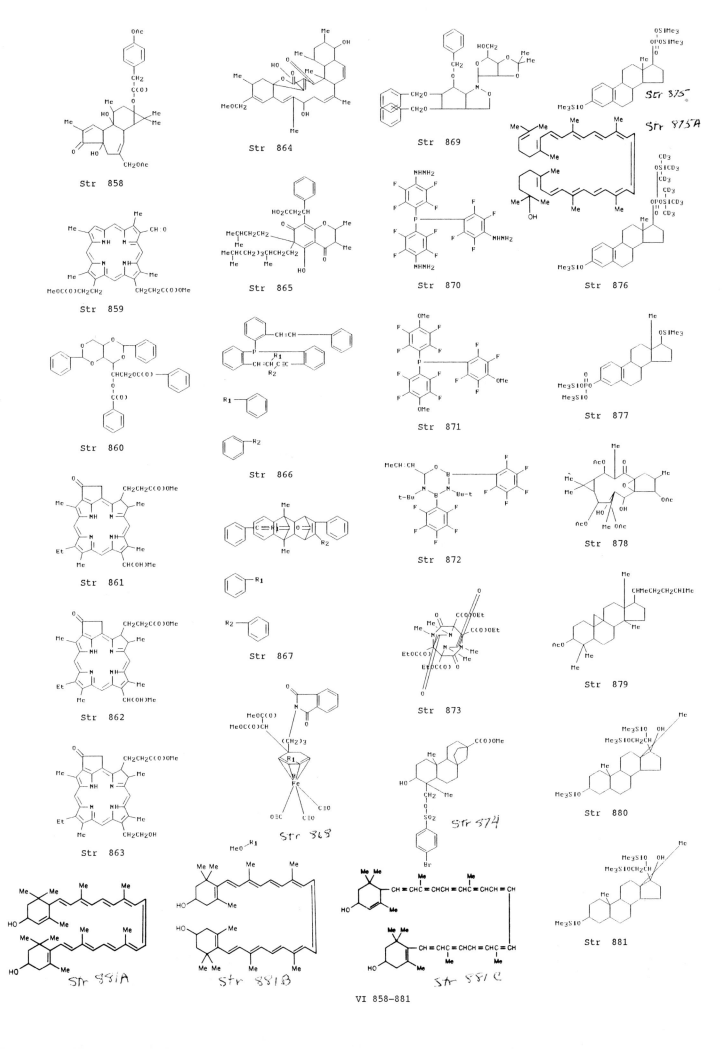

Str 858

Str 864

Str 869

Str 875

Str 875A

Str 859

Str 865

Str 870

Str 876

Str 860

Str 866

Str 871

Str 877

Str 861

Str 867

Str 872

Str 878

Str 862

Str 868

Str 873

Str 879

Str 863

Str 874

Str 880

Str 881

Str 881A

Str 881B

Str 881C

VI 858-881

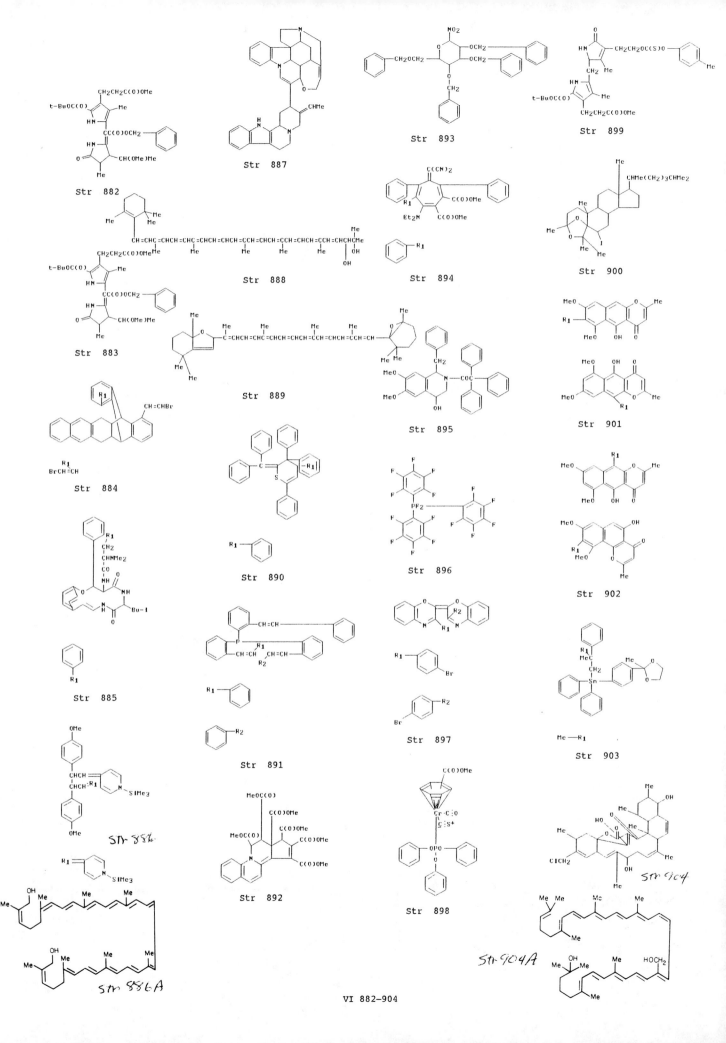

Str 882

Str 883

Str 884

Str 885

Str 886

Str 886A

Str 887

Str 888

Str 889

Str 890

Str 891

Str 892

Str 893

Str 894

Str 895

Str 896

Str 897

Str 898

Str 899

Str 900

Str 901

Str 902

Str 903

Str 904

Str 904A

VI 882-904

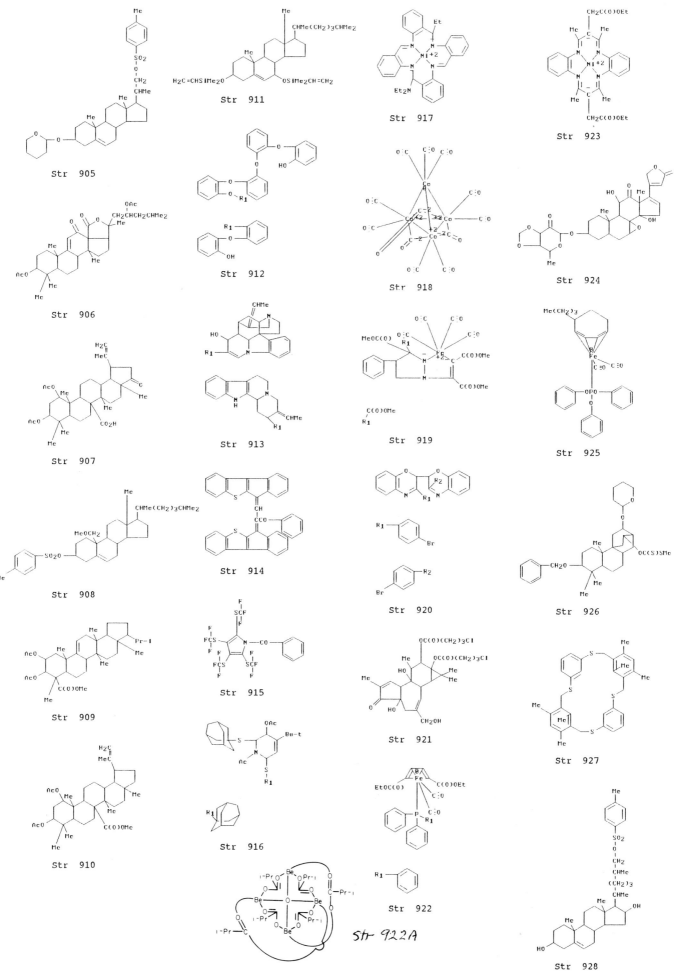

Str 905

Str 906

Str 907

Str 908

Str 909

Str 910

Str 911

Str 912

Str 913

Str 914

Str 915

Str 916

Str 917

Str 918

Str 919

Str 920

Str 921

Str 922

Str 922A

Str 923

Str 924

Str 925

Str 926

Str 927

Str 928

VI 905-928

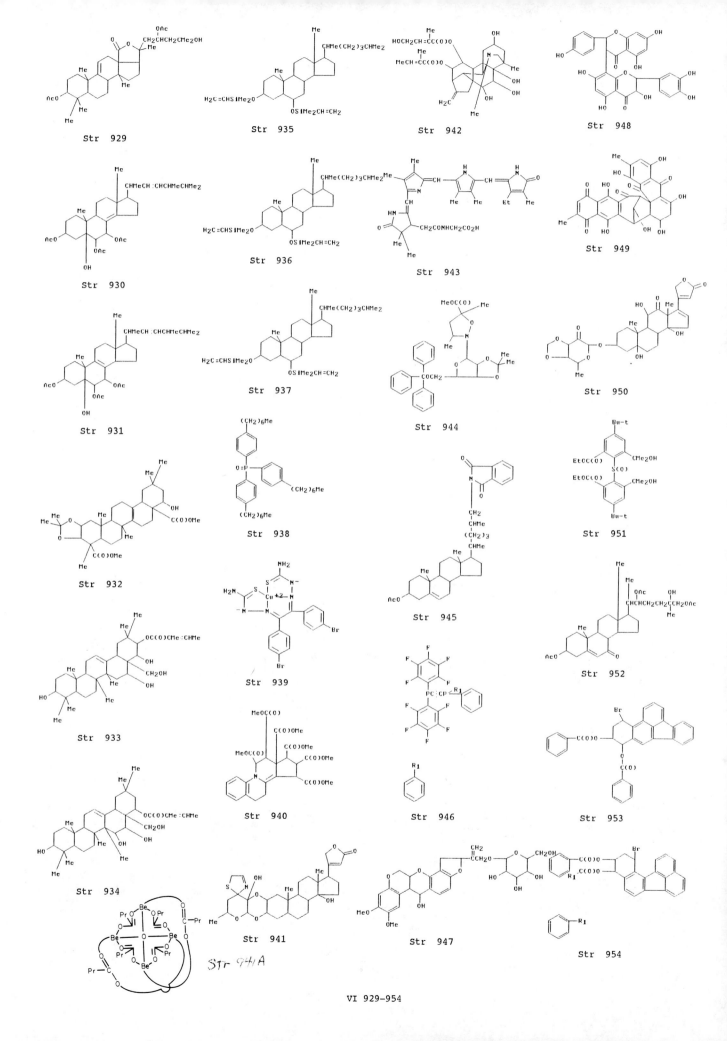

Str 929

Str 935

Str 942

Str 948

Str 930

Str 936

Str 943

Str 949

Str 931

Str 937

Str 944

Str 950

Str 932

Str 938

Str 945

Str 951

Str 933

Str 939

Str 946

Str 952

Str 934

Str 940

Str 941

Str 953

Str 941A

Str 947

Str 954

VI 929-954

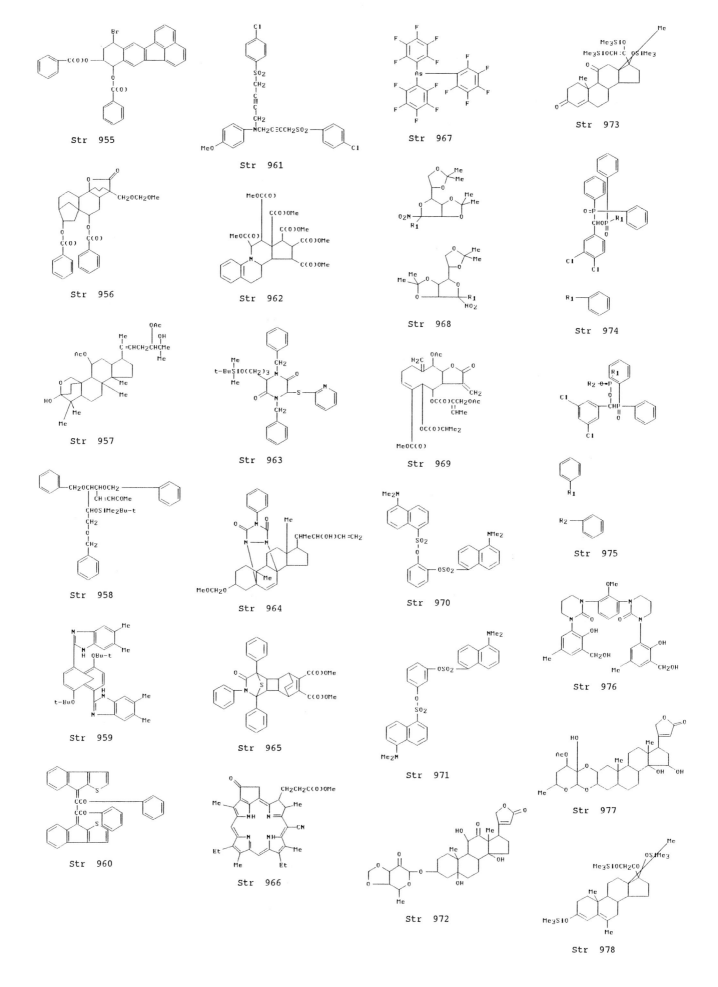

Str 955

Str 961

Str 967

Str 973

Str 956

Str 962

Str 968

Str 974

Str 957

Str 963

Str 969

Str 975

Str 958

Str 964

Str 970

Str 976

Str 959

Str 965

Str 971

Str 977

Str 960

Str 966

Str 972

Str 978

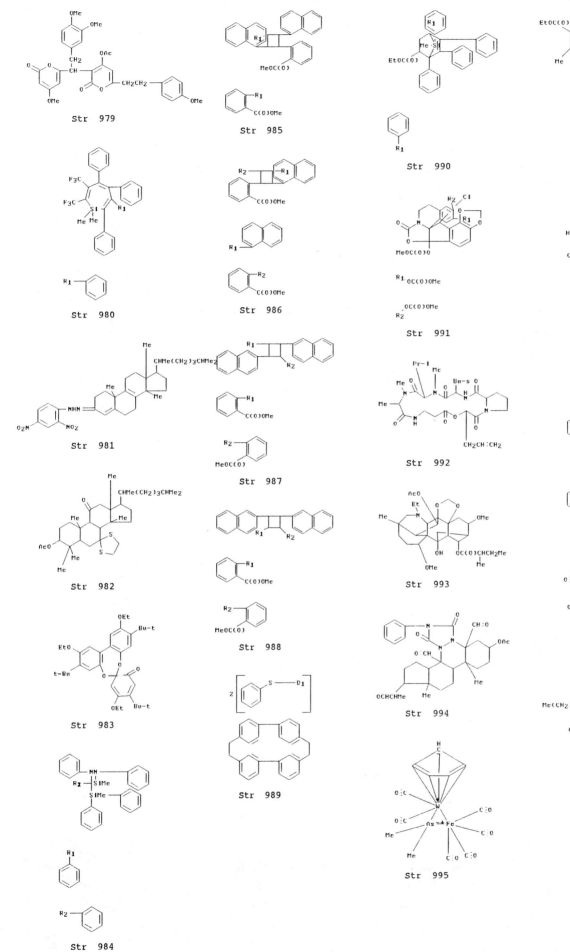

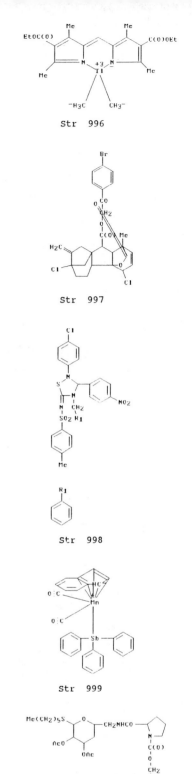

Str 979

Str 985

Str 990

Str 996

Str 980

Str 986

Str 997

Str 981

Str 991

Str 998

Str 982

Str 987

Str 992

Str 983

Str 988

Str 993

Str 999

Str 984

Str 989

Str 994

Str 995

Str 1000

VI 979-1000

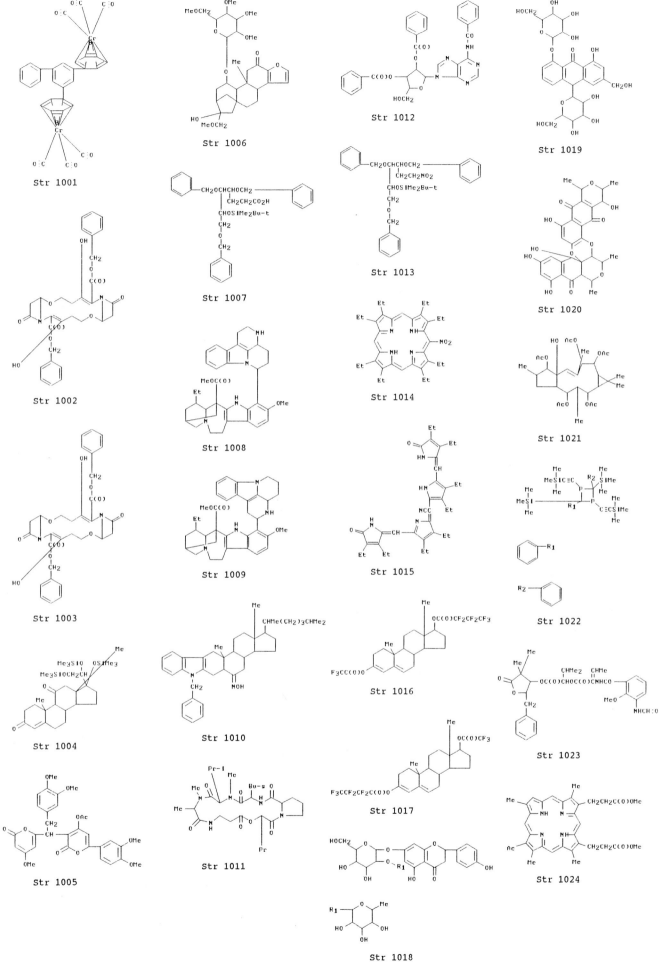

Str 1001

Str 1002

Str 1003

Str 1004

Str 1005

Str 1006

Str 1007

Str 1008

Str 1009

Str 1010

Str 1011

Str 1012

Str 1013

Str 1014

Str 1015

Str 1016

Str 1017

Str 1018

Str 1019

Str 1020

Str 1021

Str 1022

Str 1023

Str 1024

VI 1001-1024

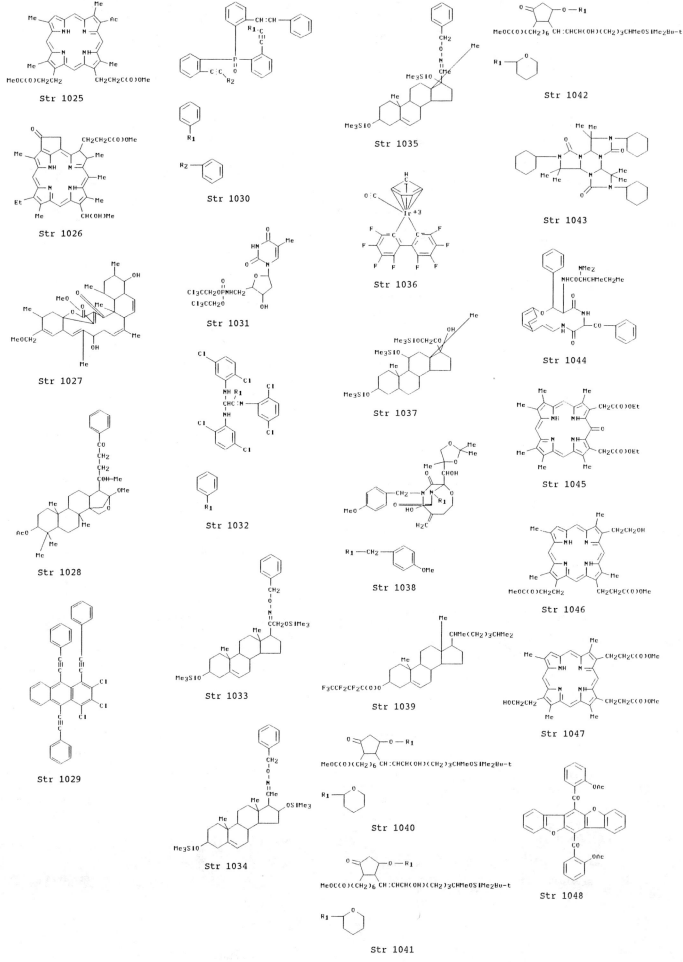

Str 1025

Str 1026

Str 1027

Str 1028

Str 1029

Str 1030

Str 1031

Str 1032

Str 1033

Str 1034

Str 1035

Str 1036

Str 1037

Str 1038

Str 1039

Str 1040

Str 1041

Str 1042

Str 1043

Str 1044

Str 1045

Str 1046

Str 1047

Str 1048

VI 1025-1048

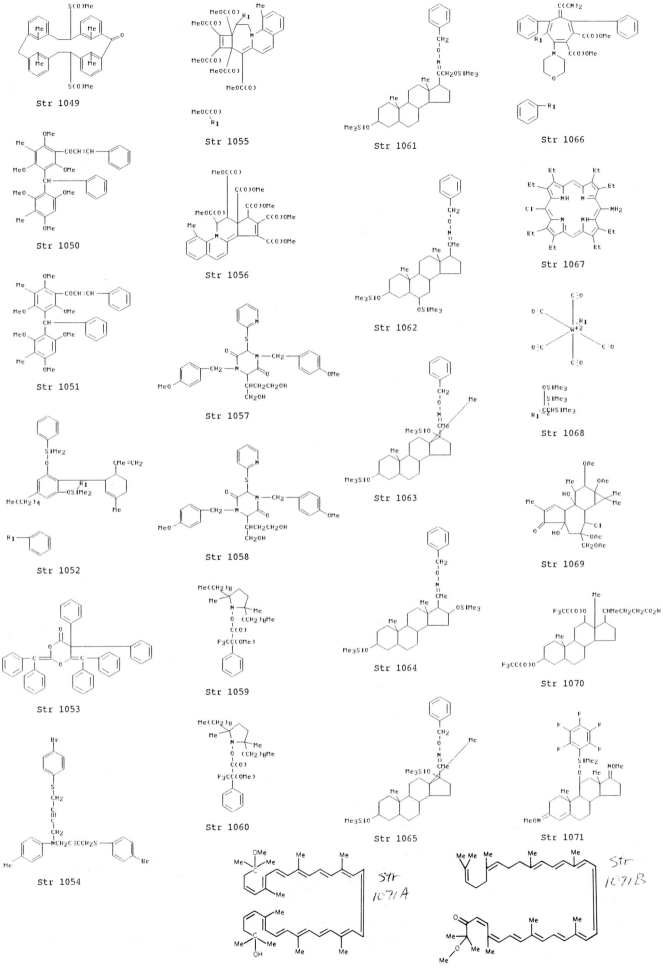

Str 1049

Str 1050

Str 1051

Str 1052

Str 1053

Str 1054

Str 1055

Str 1056

Str 1057

Str 1058

Str 1059

Str 1060

Str 1061

Str 1062

Str 1063

Str 1064

Str 1065

Str 1066

Str 1067

Str 1068

Str 1069

Str 1070

Str 1071

Str 1071A

Str 1071B

VI 1049–1071

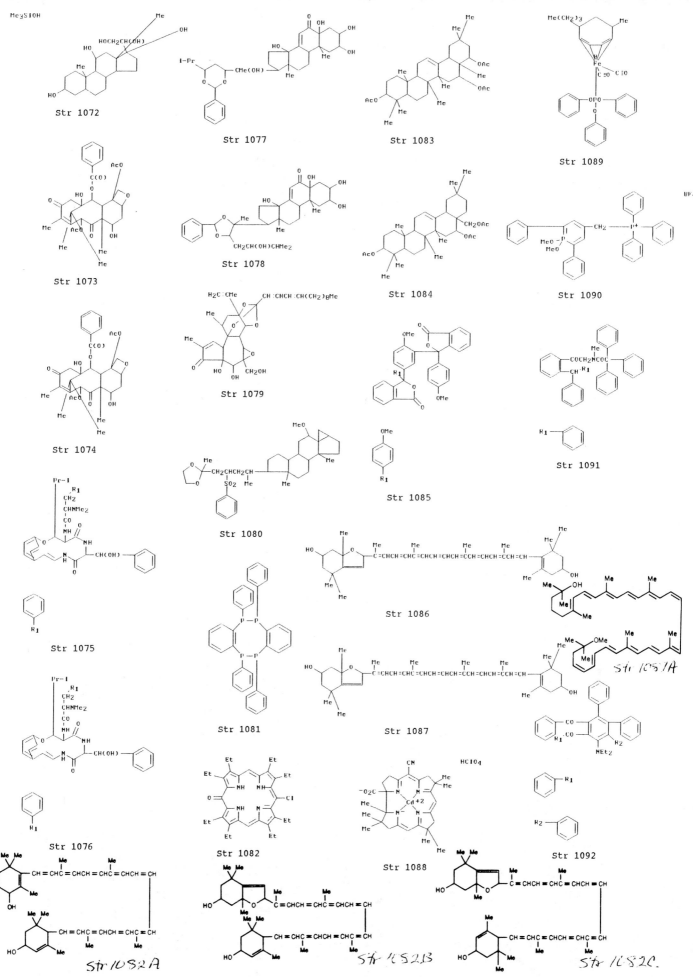

Str 1072

Str 1077

Str 1083

Str 1089

Str 1073

Str 1078

Str 1084

Str 1090

Str 1074

Str 1079

Str 1085

Str 1091

Str 1075

Str 1080

Str 1086

Str 1082A

Str 1076

Str 1081

Str 1087

Str 1082B

Str 1082

Str 1088

Str 1092

Str 1082C

VI 1072-1092

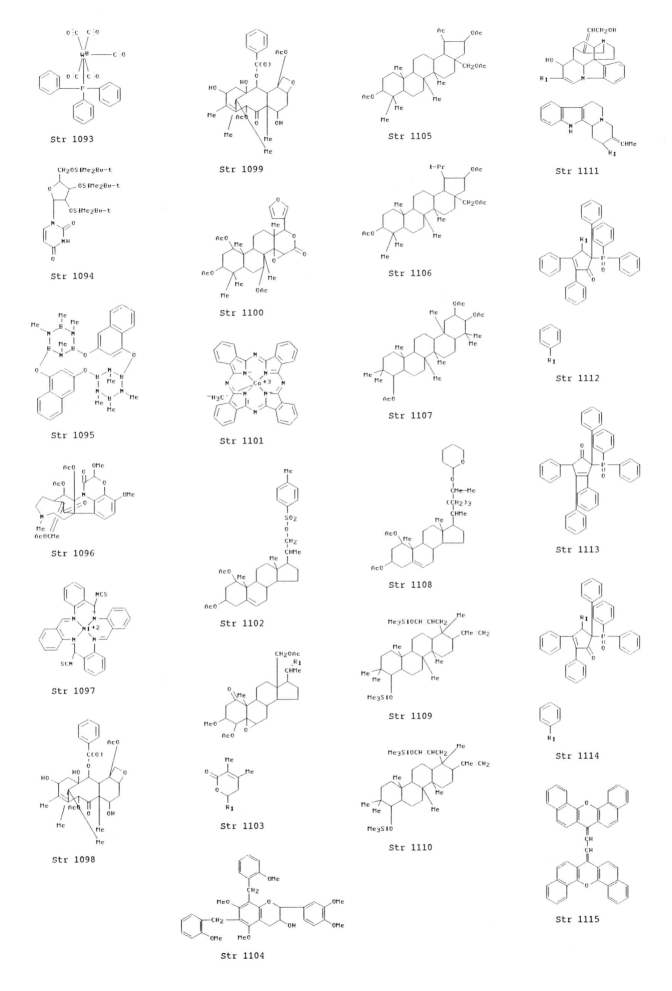

Str 1093

Str 1094

Str 1095

Str 1096

Str 1097

Str 1098

Str 1099

Str 1100

Str 1101

Str 1102

Str 1103

Str 1104

Str 1105

Str 1106

Str 1107

Str 1108

Str 1109

Str 1110

Str 1111

Str 1112

Str 1113

Str 1114

Str 1115

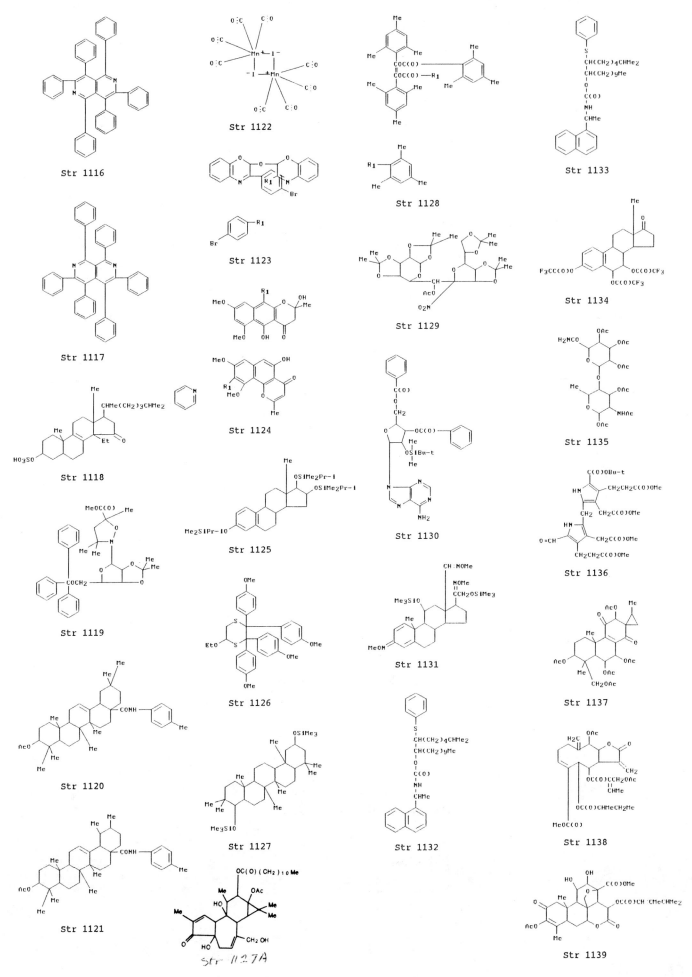

Str 1116

Str 1117

Str 1118

Str 1119

Str 1120

Str 1121

Str 1122

Str 1123

Str 1124

Str 1125

Str 1126

Str 1127

Str 1127A

Str 1128

Str 1129

Str 1130

Str 1131

Str 1132

Str 1133

Str 1134

Str 1135

Str 1136

Str 1137

Str 1138

Str 1139

VI 1116–1139

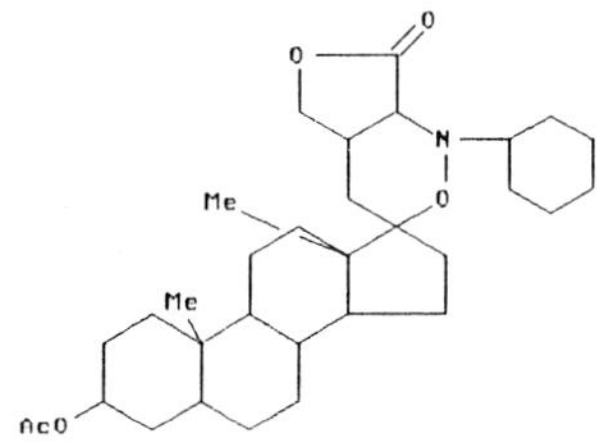

Str 1140

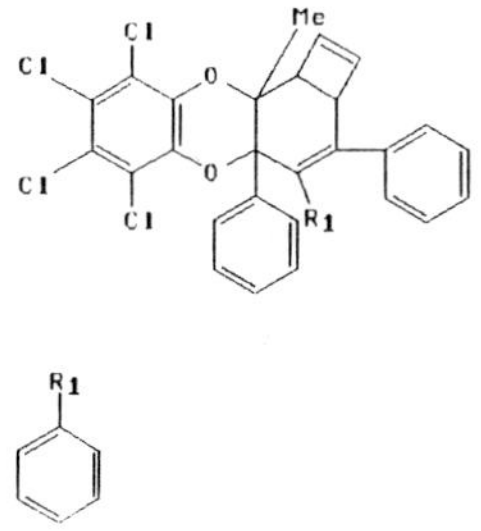

Str 1146

CH2O —(benzyl)

CH2C(O)OCHMe(CH2)5C(O)OCH2CH2SiMe3

CH2O —(benzyl)

Str 1141

Cl Cl Cl Cl
Me
R1
R1

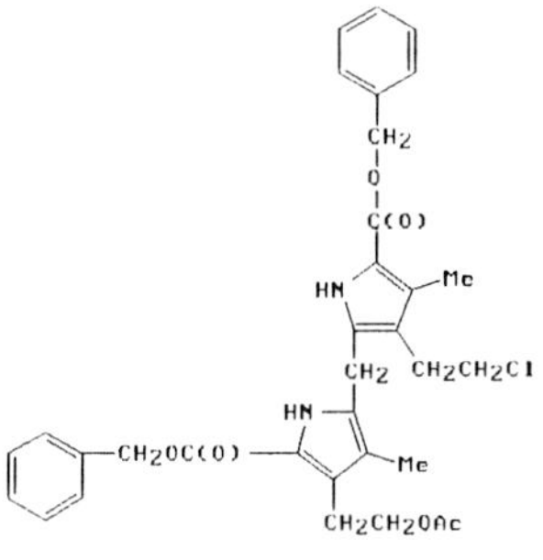

Str 1151

R1

Str 1147

R1—O
SO2
(phenyl)

O
OH
R1

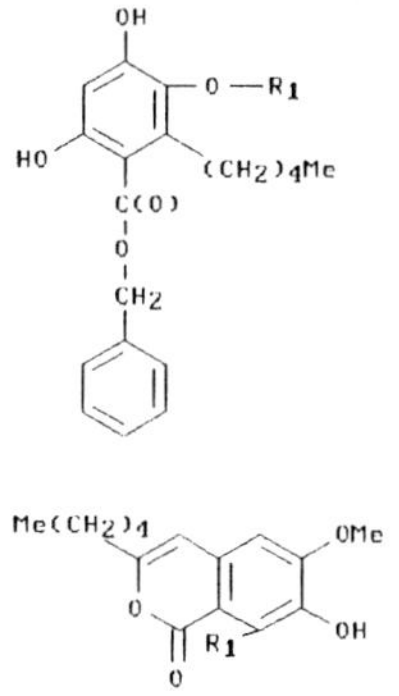

Str 1152

Me Me Me Me Me Me
C=C=CHCH=CHCH=CHCH=CHCH=C—CHCH=CHCOCH2
HO OH OH
Me Me Me

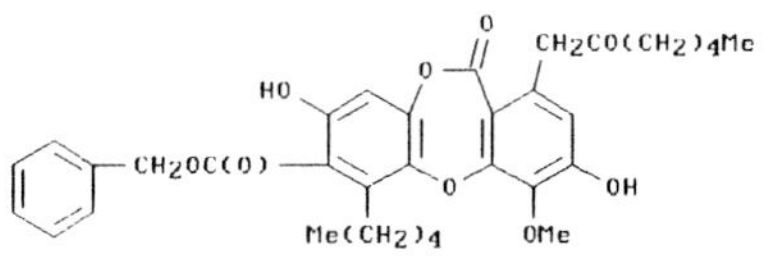

Str 1142

CH2
O(C=O)
HN
Me
CH2
CH2CH2Cl
HN
CH2OC(O)
Me
CH2CH2OAc

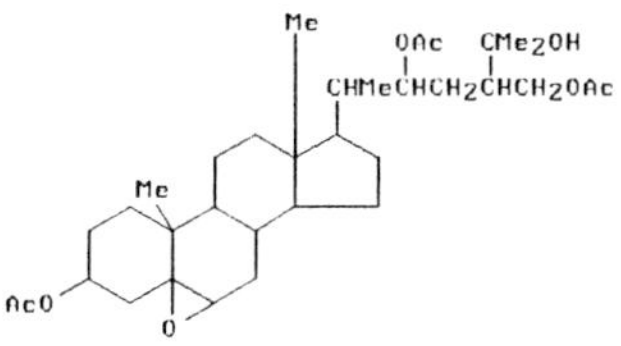

Str 1148

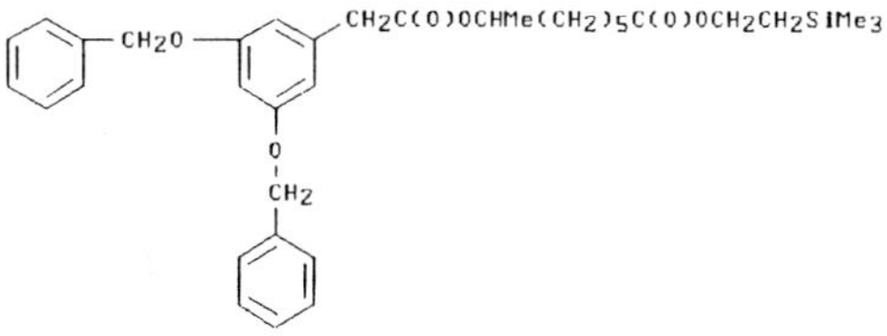

MeOC(O)CH2CH2   CH2CH2C(O)OMe
Me   Me
NH
N   N
NH
Me   Me
H2C=CH   CH=CH2
Me

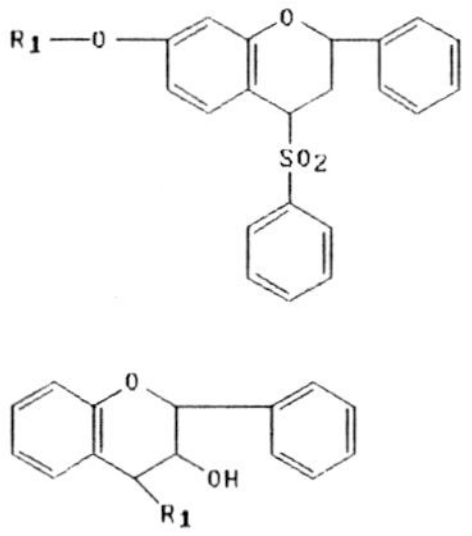

Str 1153

Me
OH
OH
CH2OCH2CH(OH)CH(OH)CH(OH)CHO
HO
OH
Me Me

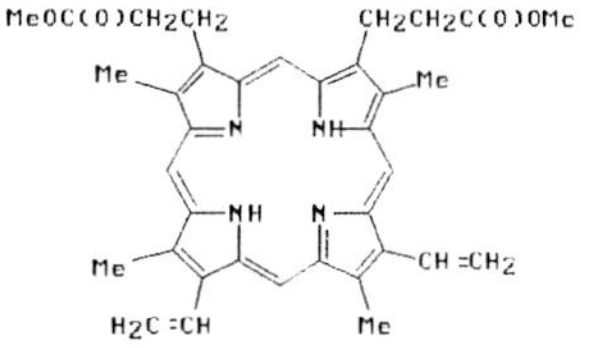

Str 1143

OH
O—R1
HO
(CH2)4Me
C(O)
O
CH2
(phenyl)

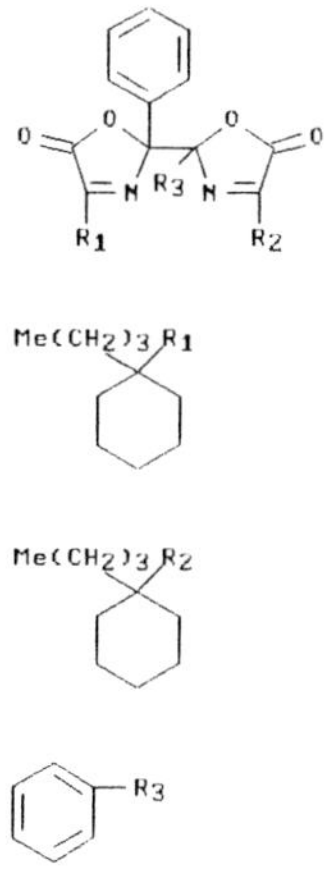

Me(CH2)4
OMe
OH
R1
O

Str 1149

O   O
(phenyl)
N R3
R1   R2

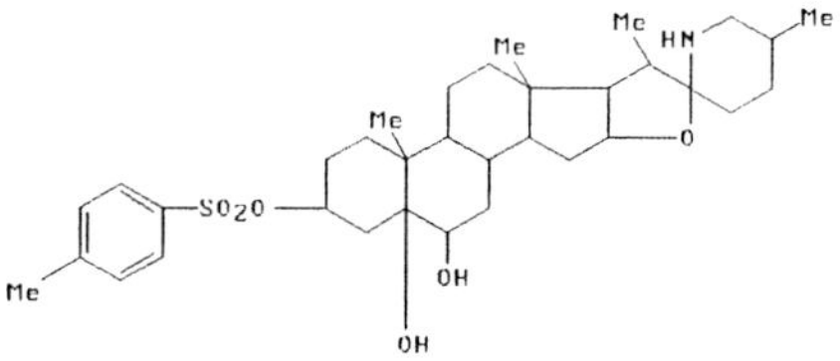

Me(CH2)3R1

Me(CH2)3R2

R3

Str 1144

HO
CH2CO(CH2)4Me
O
CH2OC(O)
Me(CH2)4
O
OH
OMe

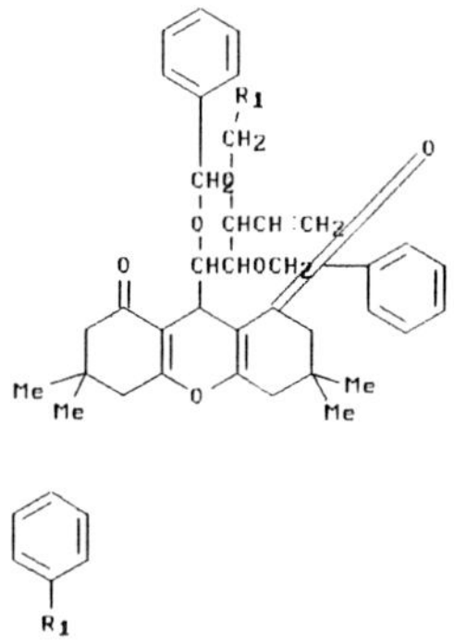

Str 1150

Me
SO2O
Me
Me Me HN Me
O
OH
OH

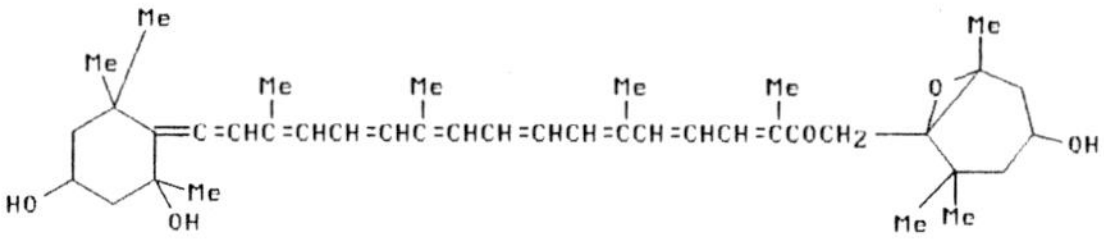

Str 1154

R2—CH2
O
CH2—R1
MeO
OH

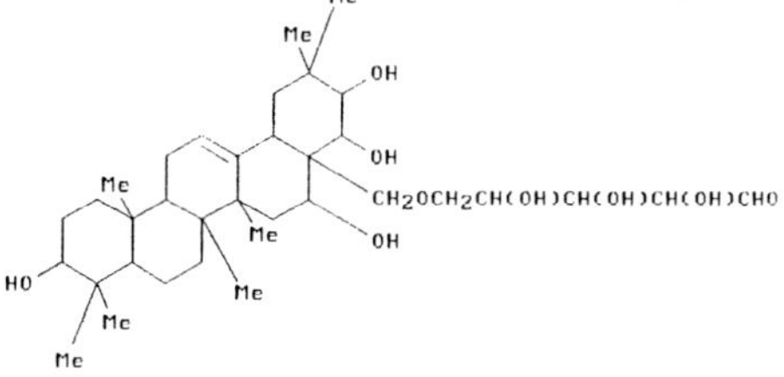

OMe
N
Me R1
OMe

MeO
HO
N
Me
R2

Str 1155

OMe
O
CH2—R1
HO
CH2
N
Me
MeO

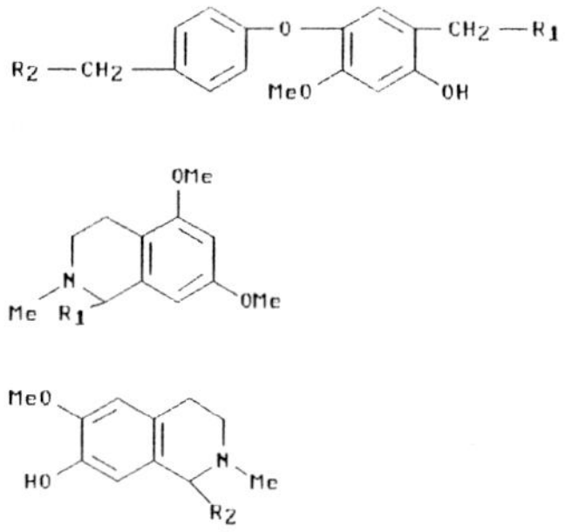

OH
OMe
N
Me R1
OMe

Str 1145

Me
ONc CMe2OH
CHMeCH2CH2CHCH2OAc
Me
OAc
O

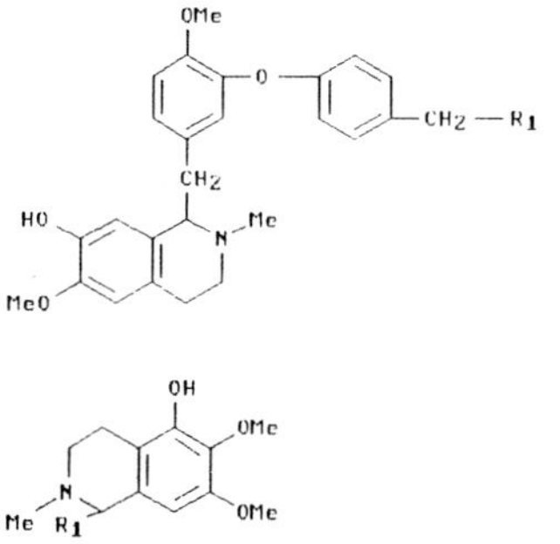

VI 1140–1155

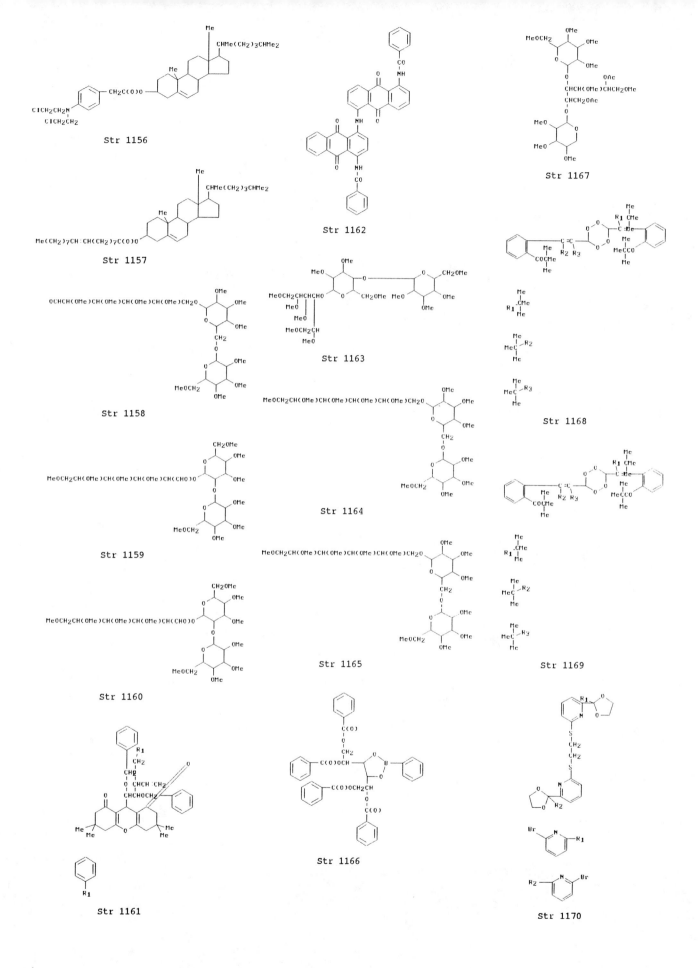

Str 1156

Str 1157

Str 1158

Str 1159

Str 1160

Str 1161

Str 1162

Str 1163

Str 1164

Str 1165

Str 1166

Str 1167

Str 1168

Str 1169

Str 1170

VI 1156–1170

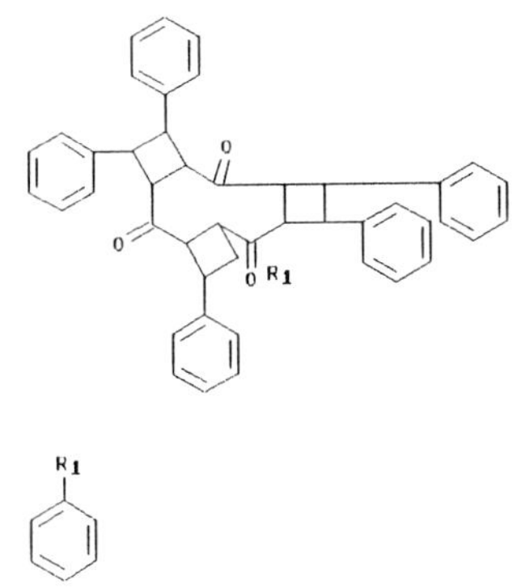

Str 1176

Str 1181

Str 1171

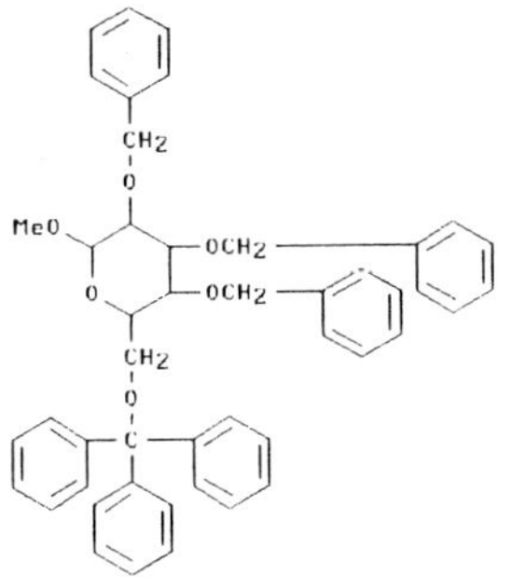

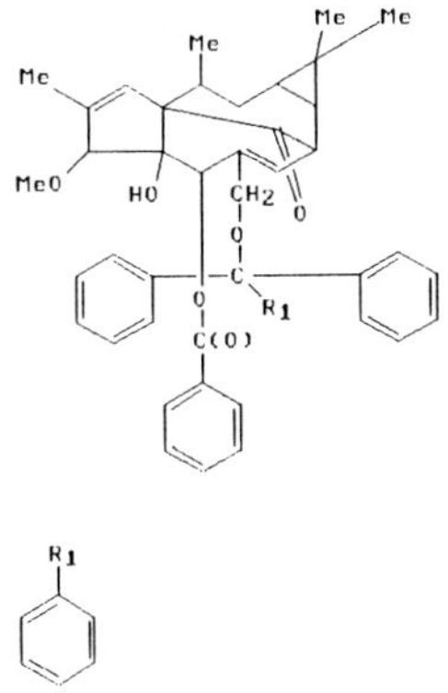

Str 1177

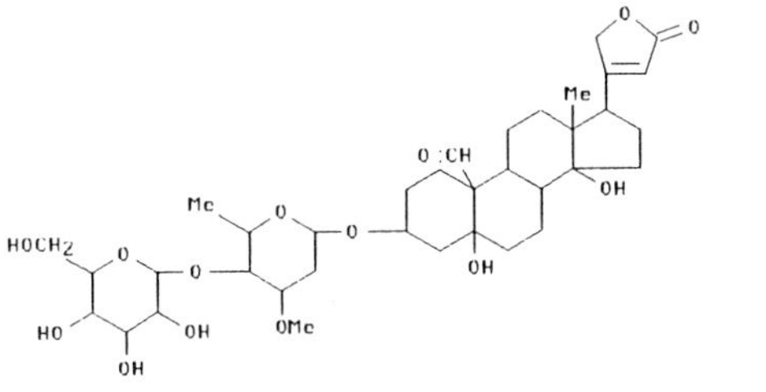

Str 1182

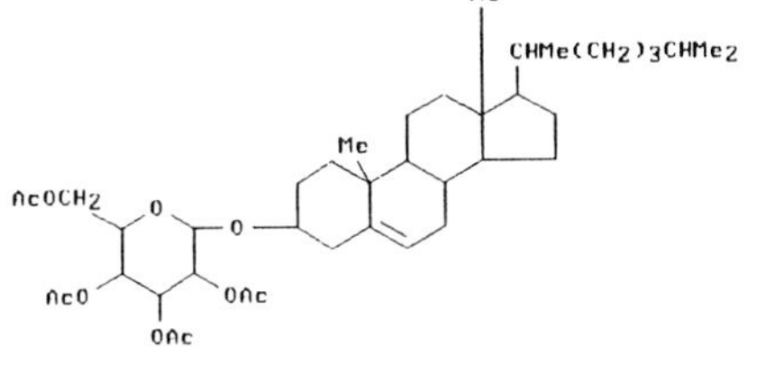

Str 1172

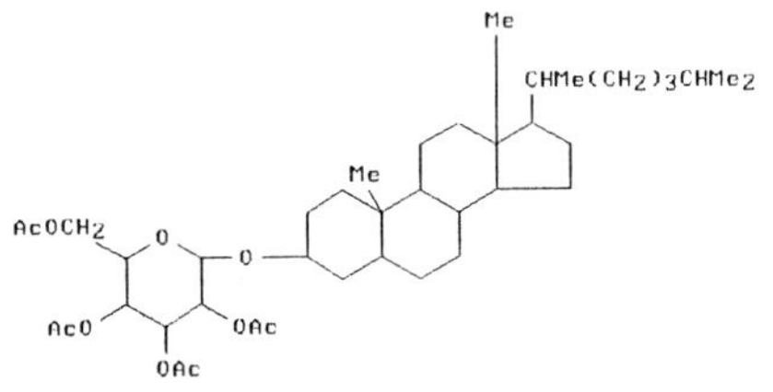

Str 1183

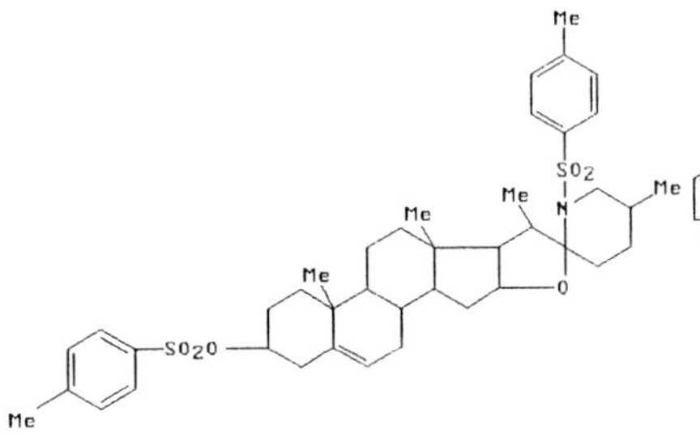

Str 1178

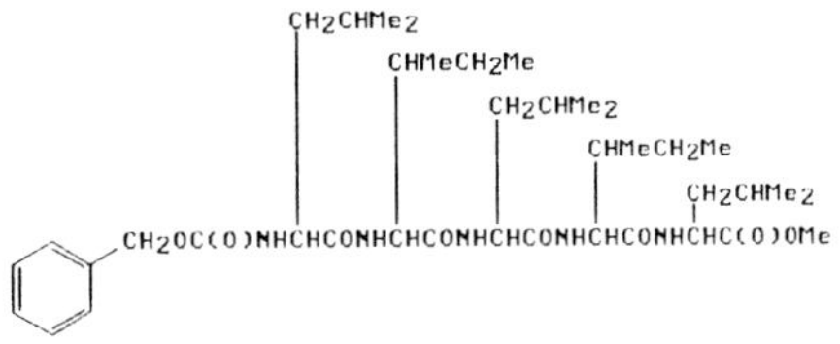

Str 1173

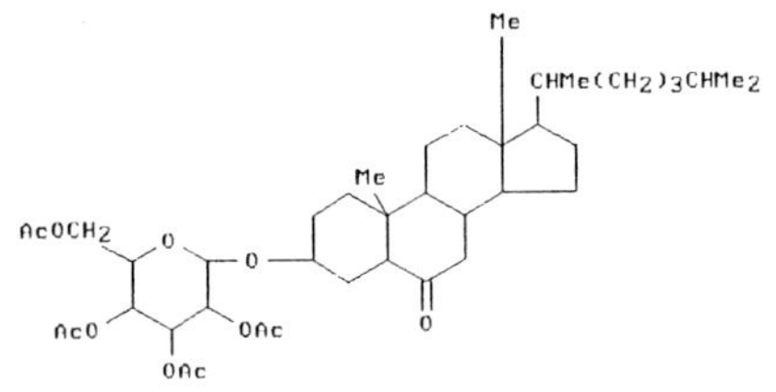

Str 1179

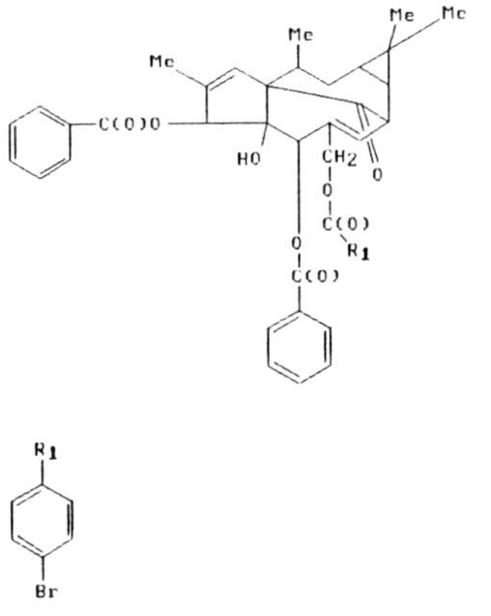

Str 1184

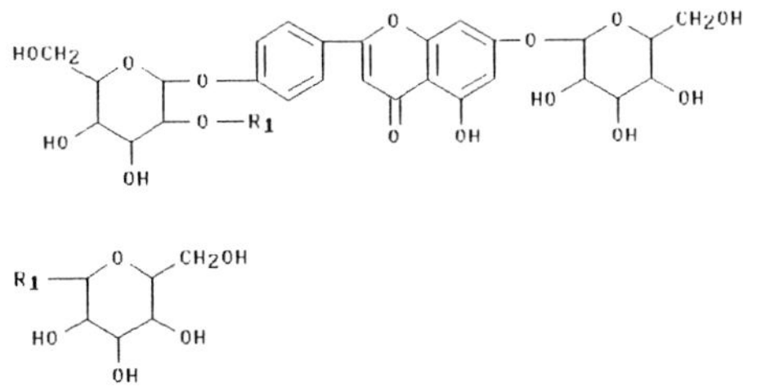

Str 1174

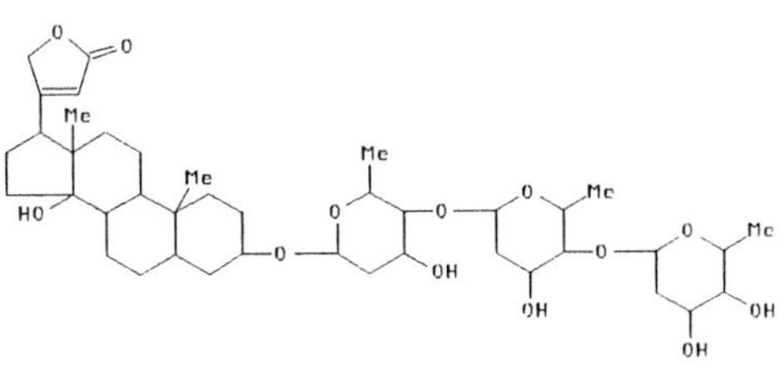

Str 1180

Str 1185

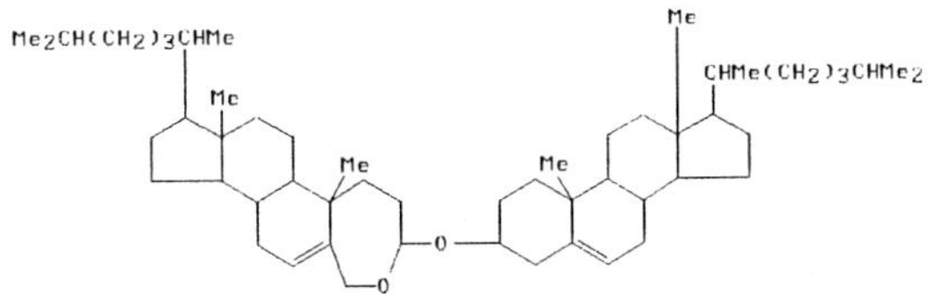

Str 1175

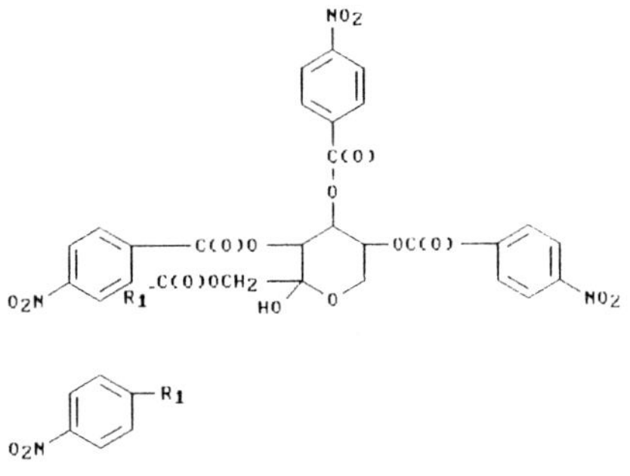

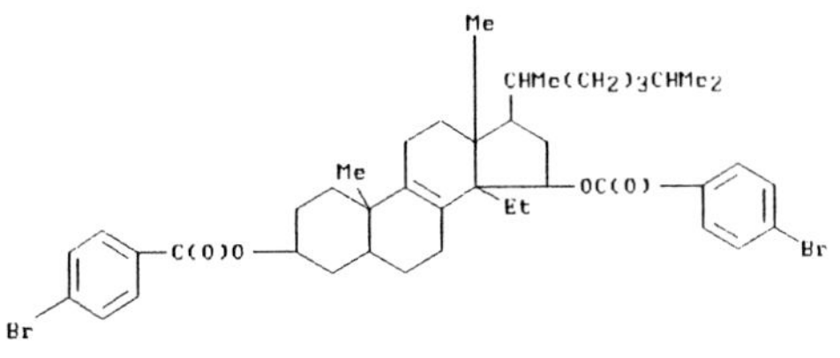

Str 1186

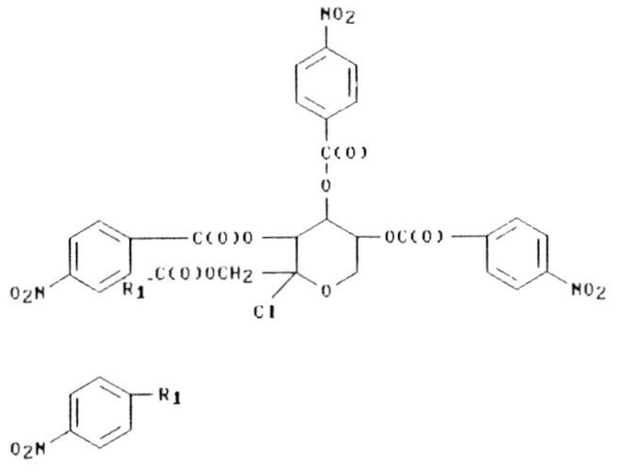

VI 1171–1186

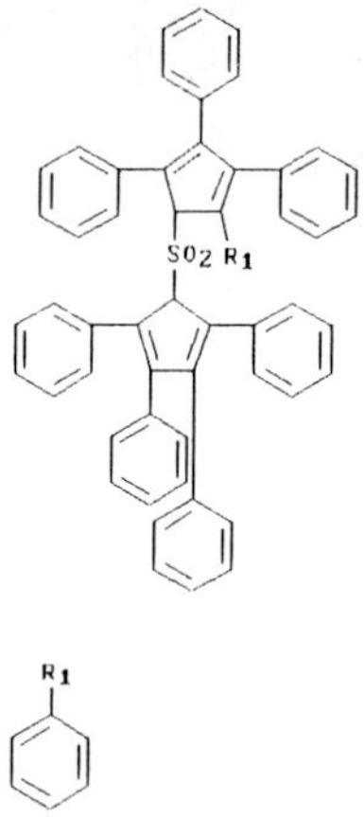

Str 1187

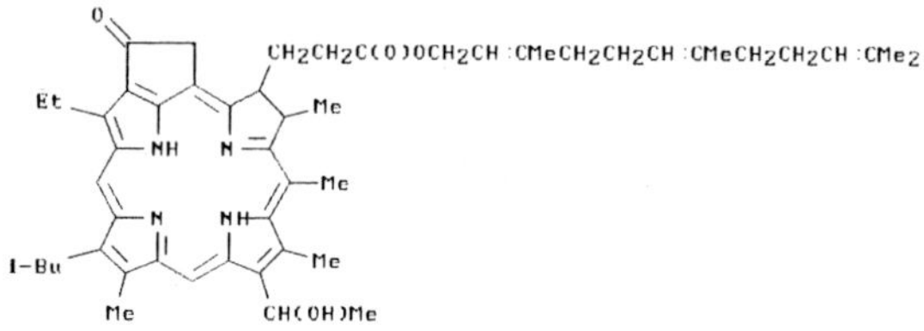

R₁

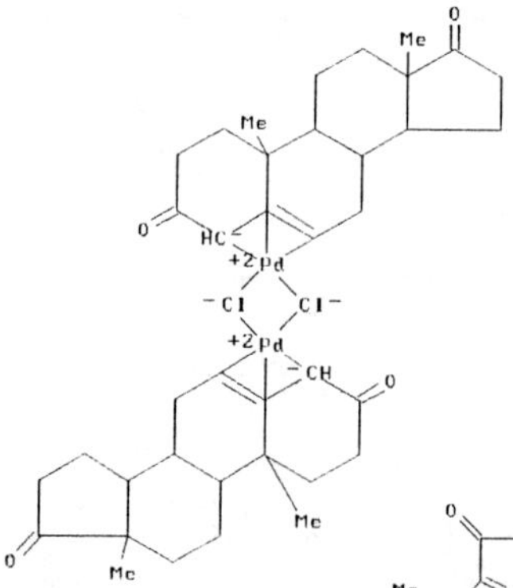

Str 1191

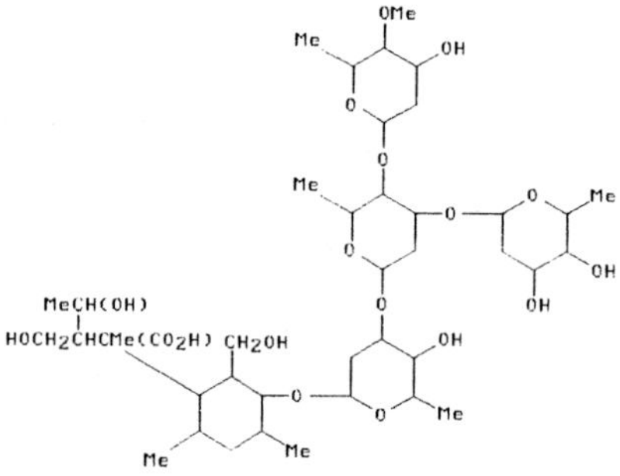

Str 1195

$CH_2OCH_2CHCHCH(CHO)OAc$

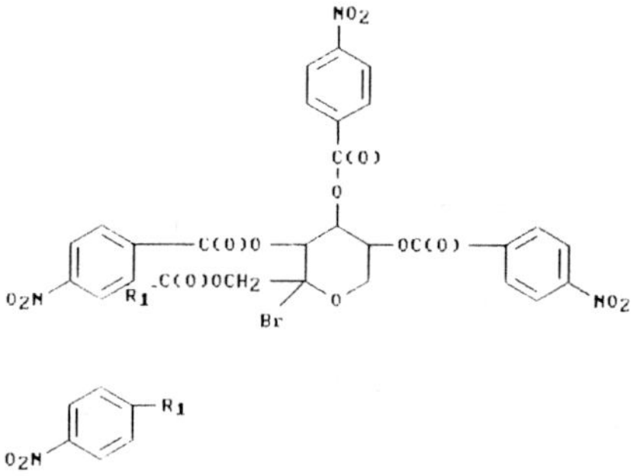

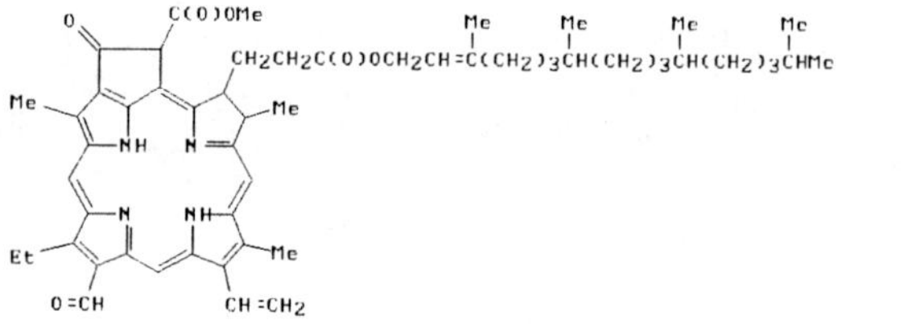

Str 1196

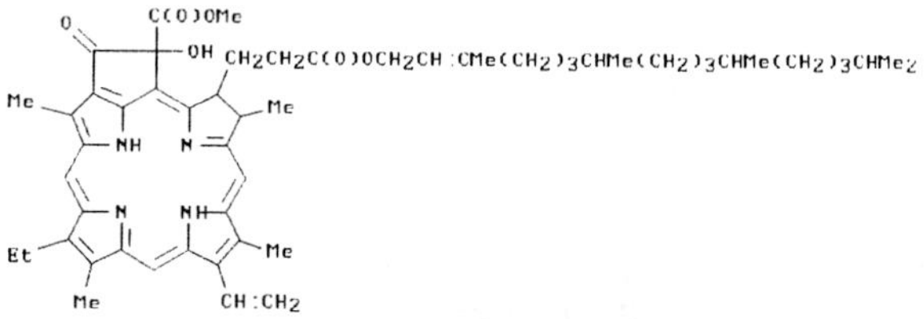

$C(O)OMe$

$CH_2CH_2C(O)OCH_2CH=C(CH_2)_3CH(CH_2)_3CH(CH_2)_3CHMe$

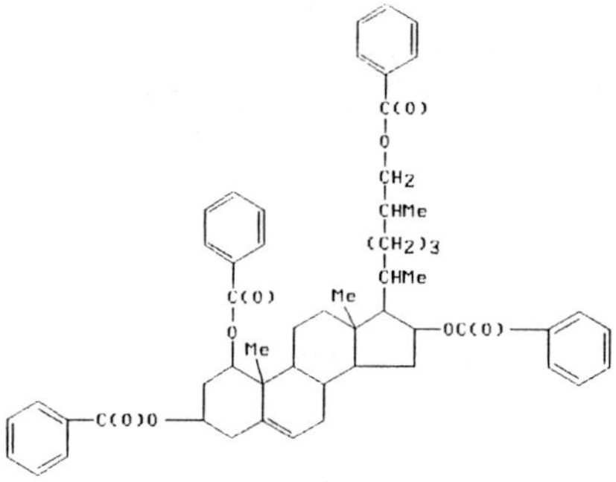

Str 1188

$CH_2CH_2C(O)OCH_2CH:CMeCH_2CH_2CH:CHCH_2CH_2CH:CMe_2$

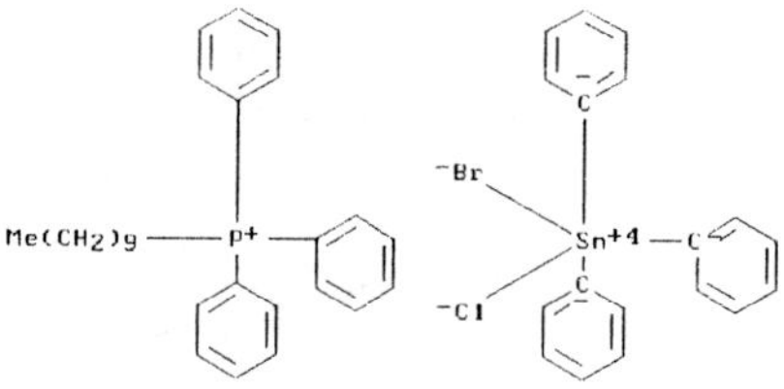

Str 1197

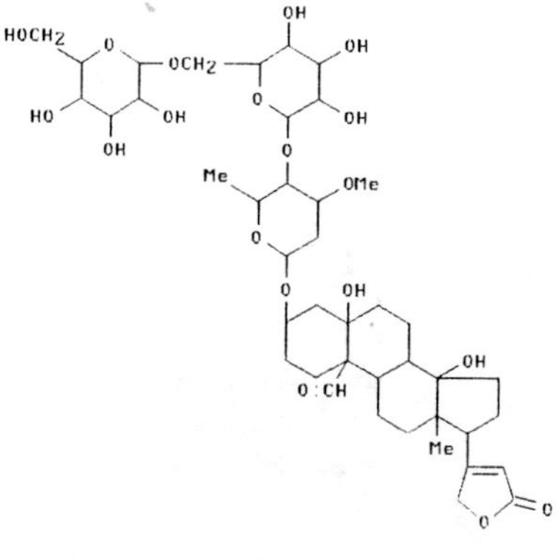

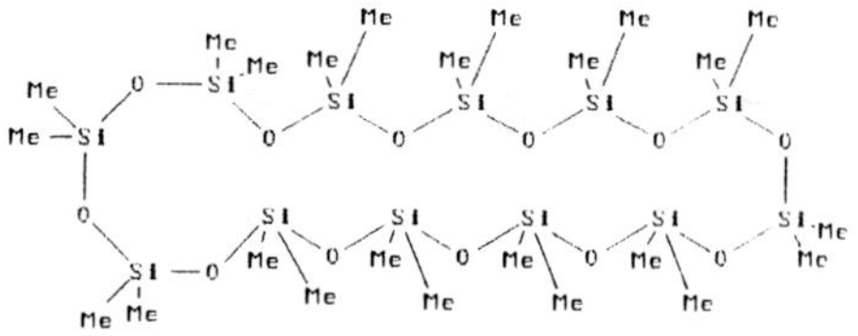

$C(O)OMe$

OH $CH_2CH_2C(O)OCH_2CH:CMe(CH_2)_3CHMe(CH_2)_3CHMe(CH_2)_3CHMe_2$

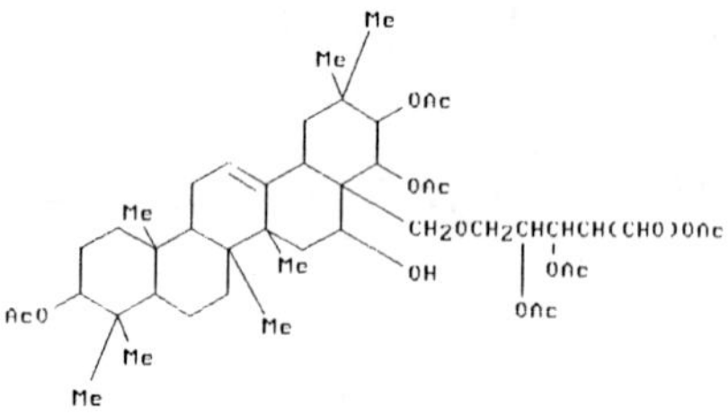

Str 1189

Str 1192

Str 1198

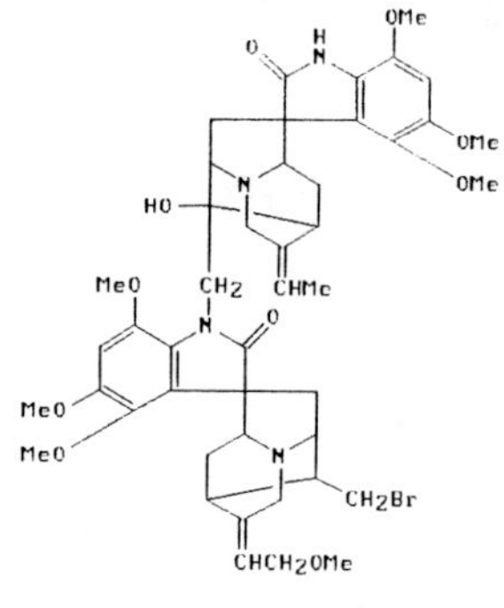

Str 1190

$Me(CH_2)_9 P^+$

^-Br

Sn^{+4}

^-Cl

Str 1193

Str 1199

R_1

Str 1194

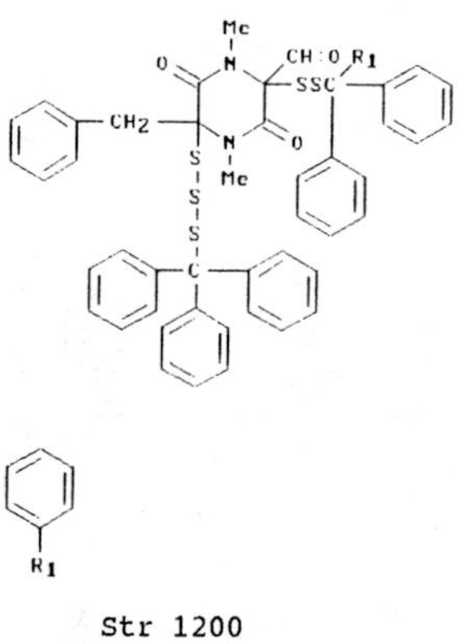

Str 1200

VI 1187–1200

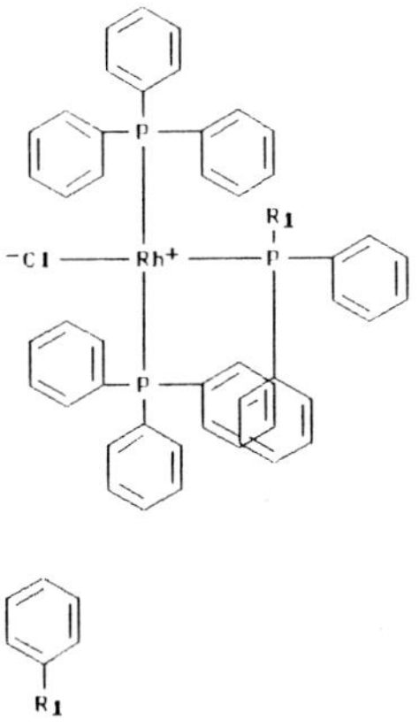

Str 1201

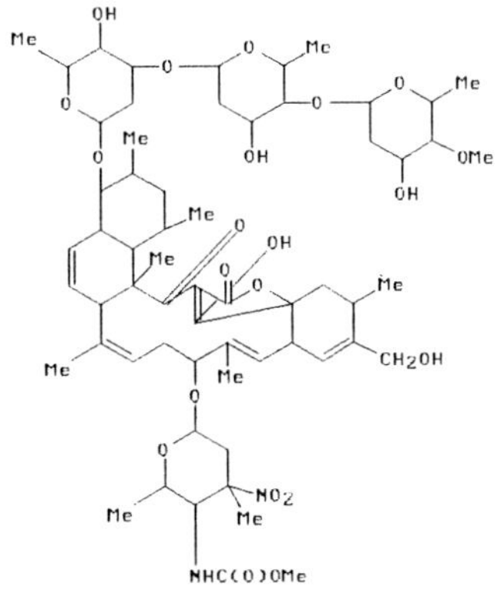

Str 1206

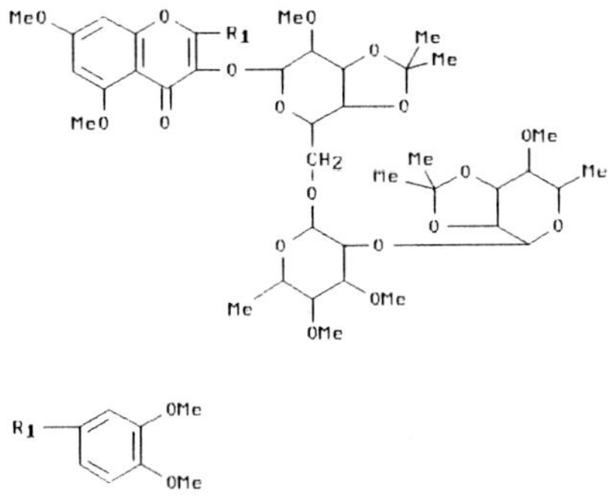

Str 1202

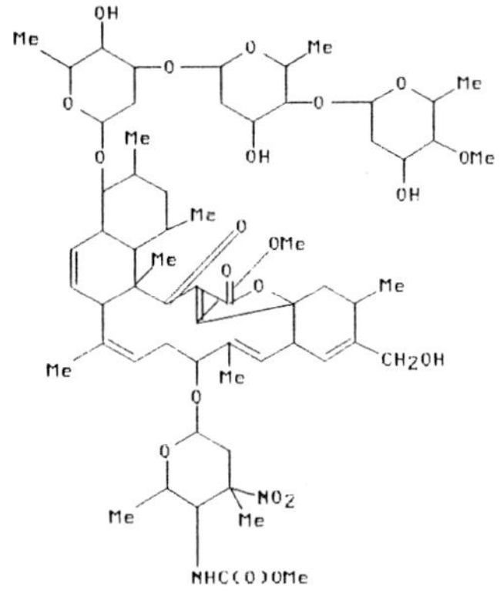

Str 1207

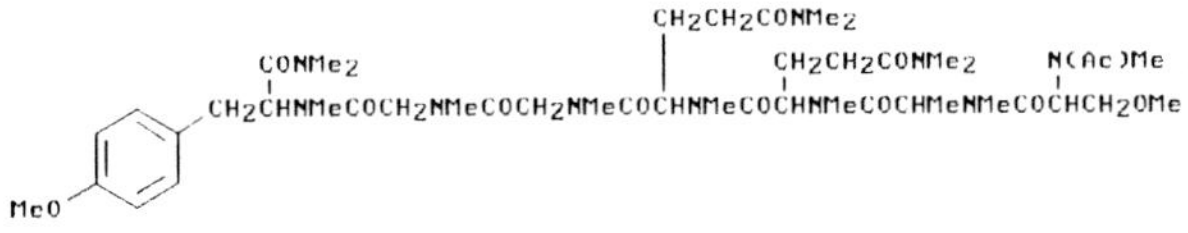

Str 1203

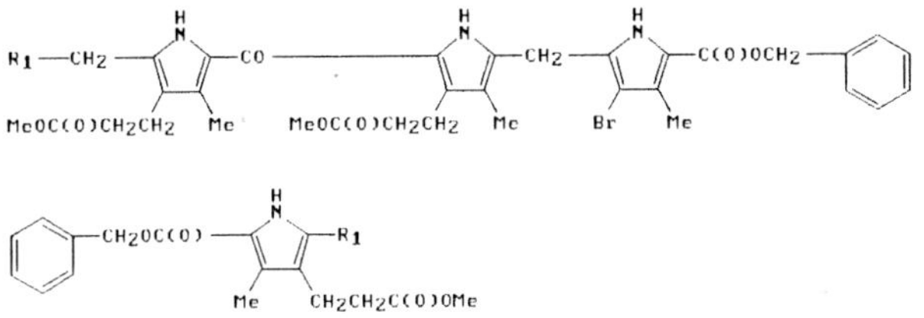

Str 1204

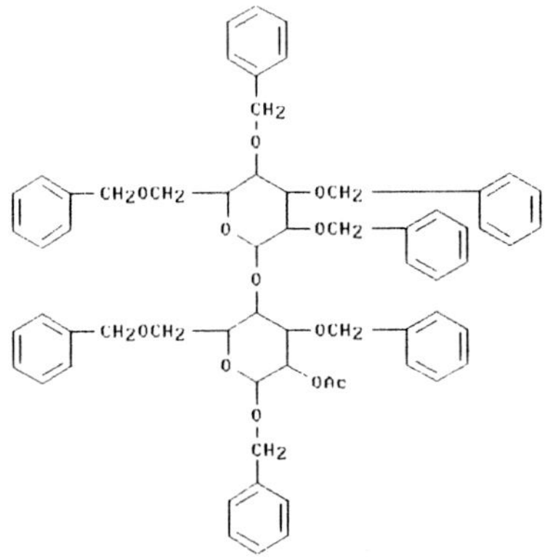

Str 1205

VI 1201–1207